Standard Atomic Weights of the Elements 1999 • Based on relative atomic mass of $^{12}C = 12$, where ^{12}C is a neutral atom in its nuclear and electronic ground state.†

Name	Symbol	Atomic Number	Atomic Weight	Name	Symbol	Atomic Number	Atomic Weight
Actinium*	Ac	89	(227)	Molybdenum	Mo	42	95.94(1)
Aluminium	Al	13	26.981538(2)	Neodymium	Nd	60	144.24(3)
Americium*	Am	95	(243)	Neon	Ne	10	20.1797(6)
Antimony	Sb	51	121.760(1)	Neptunium*	Np	93	(237)
Argon	Ar	18	39.948(1)	Nickel	Ni	28	58.6934(2)
Arsenic	As	33	74.92160(2)	Niobium	Nb	41	92.90638(2)
Astatine*	At	85	(210)	Nitrogen	N	7	14.0067(2)
Barium	Ba	56	137.327(7)	Nobelium*	No	102	(259)
Berkelium*	Bk	97	(247)	Osmium	Os	76	190.23(3)
Beryllium	Be	4	9.012182(3)	Oxygen	O	8	15.9994(3)
Bismuth	Bi	83	208.98038(2)	Palladium	Pd	46	106.42(1)
Bohrium	Bh	107	(264)	Phosphorus	P	15	30.973761(2)
Boron	B	5	10.811(7)	Platinum	Pt	78	195.078(2)
Bromine	Br	35	79.904(1)	Plutonium*	Pu	94	(244)
Cadmium	Cd	48	112.411(8)	Polonium*	Po	84	(209)
Cesium	Cs	55	132.90545(2)	Potassium	K	19	39.0983(1)
Calcium	Ca	20	40.078(4)	Praseodymium	Pr	59	140.90765(2)
Californium*	Cf	98	(251)	Promethium*	Pm	61	(145)
Carbon	C	6	12.0107(8)	Protactinium*	Pa	91	231.03588(2)
Cerium	Ce	58	140.116(1)	Radium*	Ra	88	(226)
Chlorine	Cl	17	35.453(2)	Radon*	Rn	86	(222)
Chromium	Cr	24	51.9961(6)	Rhenium	Re	75	186.207(1)
Cobalt	Co	27	58.933200(9)	Rhodium	Rh	45	102.90550(2)
Copper	Cu	29	63.546(3)	Rubidium	Rb	37	85.4678(3)
Curium*	Cm	96	(247)	Ruthenium	Ru	44	101.07(2)
Dubnium	Db	105	(262)	Rutherfordium	Rf	104	(261)
Dysprosium	Dy	66	162.50(3)	Samarium	Sm	62	150.36(3)
Einsteinium*	Es	99	(252)	Scandium	Sc	21	44.955910(8)
Erbium	Er	68	167.259(3)	Seaborgium	Sg	106	(266)
Europium	Eu	63	151.964(1)	Selenium	Se	34	78.96(3)
Fermium*	Fm	100	(257)	Silicon	Si	14	28.0855(3)
Fluorine	F	9	18.9984032(5)	Silver	Ag	47	107.8682(2)
Francium*	Fr	87	(223)	Sodium	Na	11	22.989770(2)
Gadolinium	Gd	64	157.25(3)	Strontium	Sr	38	87.62(1)
Gallium	Ga	31	69.723(1)	Sulfur	S	16	32.065(5)
Germanium	Ge	32	72.64(1)	Tantalum	Ta	73	180.9479(1)
Gold	Au	79	196.96655(2)	Technetium*	Tc	43	(98)
Hafnium	Hf	72	178.49(2)	Tellurium	Te	52	127.60(3)
Hassium	Hs	108	(277)	Terbium	Tb	65	158.92534(2)
Helium	He	2	4.002602(2)	Thallium	Tl	81	204.3833(2)
Holmium	Ho	67	164.93032(2)	Thorium*	Th	90	232.0381(1)
Hydrogen	H	1	1.00794(7)	Thulium	Tm	69	168.93421(2)
Indium	In	49	114.818(3)	Tin	Sn	50	118.710(7)
Iodine	I	53	126.90447(3)	Titanium	Ti	22	47.867(1)
Iridium	Ir	77	192.217(3)	Tungsten	W	74	183.84(1)
Iron	Fe	26	55.845(2)	Ununnilium	Uun	110	(281)
Krypton	Kr	36	83.80(1)	Unununium	Uuu	111	(272)
Lanthanum	La	57	138.9055(2)	Ununbium	Uub	112	(285)
Lawrencium*	Lr	103	(262)	Ununquadium	Uuq	114	(289)
Lead	Pb	82	207.2(1)	Uranium*	U	92	238.02
Lithium	Li	3	6.941(2)	Vanadium	V	23	891(3)
Lutetium	Lu	71	174.967(1)	Xenon	Xe	54	131.293(6)
Magnesium	Mg	12	24.3050(6)	Ytterbium	Yb	70	173.04(3)
Manganese	Mn	25	54.938049(9)	Yttrium	Y	39	88.90585(2)
Meitnerium	Mt	109	(268)	Zinc	Zn	30	65.39(2)
Mendelevium*	Md	101	(258)	Zirconium	Zr	40	91.224(2)
Mercury	Hg	80	200.59(2)				

†The atomic weights of many elements can vary depending on the origin and treatment of the sample. This is particularly true for Li; commercially available lithium-containing materials have Li atomic weights in the range of 6.96 and 6.99. The uncertainties in atomic weight values are given in parentheses following the last significant figure to which they are attributed.

*Elements with no stable nuclide; the value given in parentheses is the atomic mass number of the isotope of longest known half-life. However, three such elements (Th, Pa, and U) have a characteristic terrestial isotopic composition, and the atomic weight is tabulated for these.

Chemistry

&

Chemical Reactivity

Fifth Edition

John C. Kotz
SUNY Distinguished Teaching Professor
State University of New York
College at Oneonta

Paul M. Treichel, Jr.
Professor of Chemistry
University of Wisconsin–Madison

Patrick A. Harman
Art Development and Design

THOMSON

BROOKS/COLE

AUSTRALIA   CANADA   MEXICO   SINGAPORE
SPAIN   UNITED KINGDOM   UNITED STATES

THOMSON

★ ™

BROOKS/COLE

Executive Editor: Angus McDonald
Developmental Editor: Peter McGahey
Consulting Editor: Mary Castellion
Technology Project Manager: Ericka Yeoman
Assistant Editor: Alyssa White
Editorial Assistant: Jessica Howard
Marketing Manager: Jeff Ward
Marketing Assistant: Mona Weltmer
Advertising Project Manager: Stacey Purviance
Production Manager: Charlene Catlett Squibb
Project Editor: Bonnie Boehme
Print/Media Buyer: Kristine Waller

Permissions Editor: Robert Kauser
Art Director: Caroline McGowan
Art Development and Design: Patrick Harman
Photo Researcher: Kathrine Kotz
Copy Editor: Linda Davoli
Illustrator: Rolin Graphics
Cover Designer: Larry Didona
Cover Image: J. Pinkston and L. Stern/U.S. Geological Survey
Cover Printer: Phoenix
Compositor: Techbooks
Printer: Quebecor/World

Printed in the United States of America
1 2 3 4 5 6 7 05 04 03 02 01

For more information about our products, contact us at:

Thomson Learning Academic Resource Center
1-800-423-0563

For permission to use material from this text, contact us by:
Phone: 1-800-730-2214
Fax: 1-800-730-2215
Web: http://www.thomsonrights.com

Library of Congress Control Number: 2002102108
CHEMISTRY AND CHEMICAL REACTIVITY, Fifth Edition,
ISBN: 0-03-033604-X

Asia
Thomson Learning
60 Albert Street, #15-01
Albert Complex
Singapore 189969

Australia
Nelson Thomson Learning
102 Dodds Street
South Melbourne, Victoria 3205
Australia

Canada
Nelson Thomson Learning
1120 Birchmount Road
Toronto, Ontario M1K 5G4
Canada

Europe/Middle East/Africa
Thomson Learning
Berkshire House
168-173 High Holborn
London WC1 V7AA
United Kingdom

Latin America
Thomson Learning
Seneca, 53

This book is dedicated to the memory of John Vondeling.

John exhibited the traits of a life well-lived: wit, grace, beauty, and elegance. He left us with a treasury of witty stories, was a model for a graceful life, and produced a library of beautiful and elegant books that made a difference.

Brief Contents

Interactive Modules

The included *General Chemistry Interactive CD-ROM* contains hundreds of interactive simulations, tutors, and exercises. These modules are useful explorations and study tools that cover the main concepts for each chapter.

Chapter 1: Matter and Measurement

Chapter 2: Atoms and Elements

Chapter 3: Molecules, Ions, and Compounds

Chapter 19: Principles of Reactivity: Entropy and Free Energy

Chapter 20: Principles of Reactivity: Electron Transfer Reactions

Essays and Applications

Chemical Perspectives

Problem Solving Tips

A Closer Look

Contents

The Structure of Atoms and Molecules

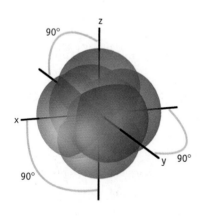

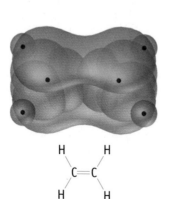

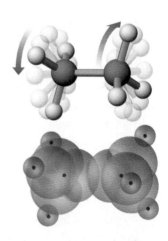

11   Carbon: More than Just Another Element   *418*

States of Matter

12   Gases and Their Properties   *470*

The Control of Chemical Reactions

18   Principles of Reactivity: Other Aspects of Aqueous Equilibria   *738*

Roses Are Red, Violets Are Blue, and Hydrangeas Are Red or Blue   *738*

19   Principles of Reactivity: Entropy and Free Energy   *788*

Perpetual Motion Machines   *788*

The Chemistry of the Elements

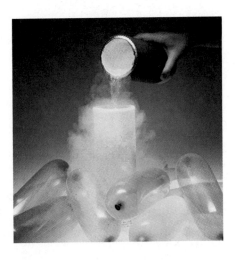

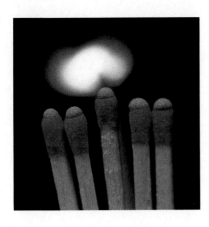

Appendices

Preface

This is the Fifth Edition of *Chemistry & Chemical Reactivity*. The principal goals of the book, beginning with the first edition almost 15 years ago, are to provide a broad overview of the principles of chemistry, the reactivity of the chemical elements and their compounds, and the applications of chemistry. We have organized this approach around the close relation between the observations chemists make in the laboratory and in nature of chemical and physical changes and the way these changes are viewed at the atomic and molecular level.

Another of our goals has been to convey a sense of chemistry as a field that not only has a lively history but also one that is currently dynamic, with important new developments on the horizon. Furthermore, we want to provide some insight into the chemical aspects of the world around us. What materials are important to our economy, what are some reactions in plants and animals and in our environment, and what role do chemists play in protecting the environment? By tackling the principles leading to answers to these questions, you can come to a better understanding of nature and to an appreciation for consumer products. Indeed, one of the objectives of this book is to provide the tools needed for you to function as a chemically literate citizen. Learning something of the chemical world is just as important as understanding some basic mathematics and biology, and as important as having an appreciation for history, music, and literature.

Computers and the World Wide Web are becoming an ever more powerful way to organize and convey information. The first edition of our *Interactive General Chemistry CD-ROM* has been used by thousands of students worldwide and is the most successful attempt to date to allow students to interact with chemistry. The new version of the CD-ROM, with many new features, is included with this edition of the book. In addition, the online homework system, OWL, which was first used by students at a number of universities during 2001–2002 and is keyed to *Chemistry & Chemical Reactivity,* is an optional part of the learning package.

The authors of this book became chemists because, simply put, it is exciting to discover new compounds and find new ways to apply chemical principles. We hope we have conveyed that sense of enjoyment in this book as well as our awe at what is known about chemistry and, just as important, what is not known!

AUDIENCE FOR *CHEMISTRY & CHEMICAL REACTIVITY, THE INTERACTIVE GENERAL CHEMISTRY CD-ROM,* AND OWL

The textbook, CD-ROM, and OWL are designed for introductory courses in chemistry for students interested in further study in science, whether that science is biology, chemistry, engineering, geology, physics, or related subjects. Our assumption is that students beginning this course have had some preparation in algebra and in general science. Although undeniably helpful, a previous exposure to chemistry is neither assumed nor required.

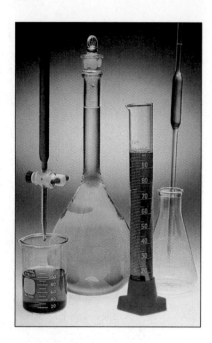

PHILOSOPHY AND APPROACH OF THE BOOK

Starting with the first edition of this book, this book has had two major, but not independent, goals. The first was to construct a book that students would enjoy reading and that would offer, at a reasonable level of rigor, chemistry and chemical principles in a format and organization typical of college and university courses today. Second, we wanted to convey the utility and importance of chemistry by introducing the properties of the elements, their compounds, and their reactions as early as possible and by focusing the discussion as much as possible on these subjects.

A glance at the introductory chemistry texts currently available shows that there is a generally common order of treatment of chemical principles used by educators. With a few minor variations we have followed that order as well. That is not to say that the chapters cannot be used in some other order. For example, the chapter on the behavior of gases (Chapter 12) is placed with chapters on liquids, solids, and solutions because it logically fits with these other topics. It can easily be read and understood, however, after covering only the first four or five chapters of the book.

The discussion of organic chemistry (Chapter 11) is often left to one of the final chapters in chemistry textbooks. We believe, however, that the importance of organic compounds in biochemistry and in the chemical industry means that we should present that material earlier in the sequence of chapters. Therefore, it follows the chapters on structure and bonding because organic chemistry illustrates the application of models of chemical bonding and molecular structure.

The order of topics in the text was also devised to introduce as early as possible the background required for the laboratory experiments usually done in General Chemistry courses. For this reason, chapters on chemical and physical properties, common reaction types, and stoichiometry begin the book. In addition, because an understanding of energy is so important in the study of chemistry, thermochemistry is introduced in Chapter 6.

The American Chemical Society has been urging educators to put "chemistry" back into introductory chemistry courses. As inorganic chemists, we agree wholeheartedly. Therefore, we have tried to describe the elements, their compounds, and their reactions as early and as often as possible in several ways. First, there are numerous color photographs of reactions occurring, of the elements and common compounds, and of common laboratory operations and industrial processes. Second, we have tried to bring material on the properties of elements and compounds as early as possible into the Exercises and Study Questions and to introduce new principles using realistic chemical situations. Finally, relevant highlights are given in Chapters 21 and 22 as a capstone to the principles described earlier.

Additionally, *Current Perspective* attempt to bring relevance and perspective to a study of chemistry. These include such topics as nanotechnology, using isotopes, what it means to be in the "limelight," the importance of sulfuric acid in the world economy, sunscreens, the newly recognized importance of the NO molecule, and LEDs.

Finally, *Closer Look* boxes describe ideas that form the background to material under discussion or provide another dimension of the subject. For example, in chapter 11 on Organic Chemistry the *Closer Look* boxes are devoted to a discussion of structural aspects of biochemically important molecules such as amino acids and proteins. In other chapters we delve into molecular modeling and mass spectrometry.

ORGANIZATION OF THE BOOK

Chemistry & Chemical Reactivity has two overarching themes: *Chemical Reactivity* and *Bonding and Molecular Structure*. The chapters on *Principles of Reactivity* introduce you to the factors that lead chemical reactions to be successful in producing products. Thus, under this topic you will study common types of reactions, the energy involved in reactions, and the factors that affect the speed of a reaction. One reason for the enormous advances in chemistry and molecular biology in the last several decades has been an understanding of molecular structure. Thus, sections of the book on *Principles of Bonding and Molecular Structure* lay the groundwork for understanding these developments. Particular attention is paid to an understanding of the structural aspects of such biologically important molecules as DNA.

The chapters of *Chemistry & Chemical Reactivity* are organized into five sections, each grouping with a common theme.

Part 1: The Basic Tools of Chemistry

Certain basic ideas and methods form the fabric of chemistry, and these are introduced in Part 1. Chapter 1 defines important terms and is a review of units and mathematical methods. Chapters 2 and 3 introduce basic ideas of atoms, molecules, and ions, and Chapter 2 introduces one of the most important organizational devices of chemistry, the periodic table. In Chapters 4 and 5 we begin to discuss the principles of chemical reactivity and to introduce the numerical methods used by chemists to extract quantitative information from chemical reactions. Chapter 6 is an introduction to the energy involved in chemical processes.

Part 2: The Structure of Atoms and Molecules

The goal of this section is to outline the current theories of the arrangement of electrons in atoms and some of the historical developments that led to these ideas (Chapters 7 and 8). With this background, you can understand why atoms and their ions have different chemical and physical properties. This discussion is tied closely to the arrangement of elements in the periodic table so that these properties can be recalled and predictions made. In Chapter 9 we discuss for the first time how the electrons of atoms in a molecule lead to chemical bonding and the properties of these bonds. In addition, we show how to derive the three-dimensional structure of simple molecules. Finally, Chapter 10 considers the major theories of chemical bonding in more detail.

This part of the book is completed with a discussion of organic chemistry (Chapter 11), primarily from a structural point of view. Organic chemistry is such an enormous area of chemistry that we cannot hope to cover it in detail in this book. Therefore, we have focused on compounds of particular importance, including synthetic polymers, and the structures of these materials.

In this section of the book you will find the molecular modeling software on the *General Chemistry Interactive CD-ROM* to be especially useful. Models of most of the molecules in the book were done using with programs from CAChe/Fujitsu Group, and we thank them for providing a version for us to use. Portions of this software, and hundreds of models, are included on the CD-ROM along with instructions for their use.

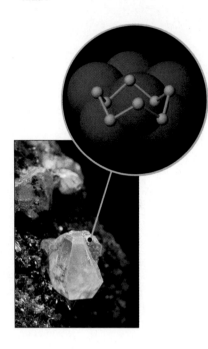

Part 3: States of Matter

The behavior of the three states of matter — gases, liquids, and solids — is described in that order in Chapters 12 and 13. The discussion of liquids and solids is tied to gases through the description of intermolecular forces, with particular attention given to liquid and solid water. Chapter 13 also considers the solid state, an area of chemistry currently undergoing a renaissance. In Chapter 14 we talk about the properties of solutions, intimate mixtures of gases, liquids, and solids.

Part 4: The Control of Chemical Reactions

This section is wholly concerned with the *Principles of Reactivity.* Chapter 15 examines the important question of the rates of chemical processes and the factors controlling these rates. With this in mind, we move to Chapters 16–18, a group of chapters that describes chemical reactions at equilibrium. After an introduction to equilibrium in Chapter 16, we highlight the reactions involving acids and bases in water (Chapters 17 and 18) and reactions leading to insoluble salts (Chapter 18). To tie together the discussion of chemical equilibria, we again explore thermodynamics in Chapter 19. As a final topic in this section we describe in Chapter 20 a major class of chemical reactions, those involving the transfer of electrons, and the use of these reactions in electrochemical cells.

Part 5: The Chemistry of the Elements and Their Compounds

Although the chemistry of the various elements has been described throughout the book to this point, Part 5 considers this topic in a more systematic way. Chapter 21 is devoted to the chemistry of the representative elements, whereas Chapter 22 is a discussion of the transition elements and their compounds. Finally, Chapter 23 is a brief discussion of nuclear chemistry.

NEW TO THE FIFTH EDITION

Colleagues and students often ask why yet another edition of the book has been prepared. We all understand, however, that even the most successful books can be improved. In addition, our experience in the classroom suggests that student interests change and that there are ever more effective ways to help our students learn chemistry. Therefore, the changes in this edition are focused on those two areas: topics of interest to the majority of our students and new ways to organize the information for effective learning.

Changes in Chapter Organization

Chapter Opening Essay

Each chapter in the book opens with an essay that describes a person, an event, or a development relevant to the subject of the chapter. The objective is to place the chapter's subject in the context of the history of science, of recent events or observations, or questions that students have about the world around them.

Chapter Goals

A list of 4–6 goals for each chapter is given at the beginning of the chapter. When material relevant to an objective begins within the chapter, the list appears again, and the objective being addressed is highlighted. Students tell us this is useful because it lets them know where they are in the development of the topics, where they have come from, and what topics are yet to come.

"Before You Begin"

This is a list of 2–4 items that you should know, or review, from previous chapters before you begin the new material.

Chapter Focus Pages

The Chapter Focus page illustrates important ideas to be covered in the chapter.

New Layout of Example Problems

All of the Example problems in the book are now divided into 3 or 4 sections. The question is presented as succinctly as possible in the PROBLEMS statement. Next, the STRATEGY section describes generally how the question is to be approached, what data are relevant, and where the student might look in the text for assistance. The third section outlines the SOLUTION to the problem. Finally, if appropriate, there is a COMMENT on the nature of the solution. This represents the way that we ask out students to approach questions.

New Organization of Study Questions

As in previous editions, end-of-chapter Study Questions are basically divided into Review Questions, questions organized according to the relevant section of the chapter, and a list of general questions with no indication of their type. This same organization is followed in this edition. However, more information is now given for questions organized according to chapter section. For questions on a given section or subsection of the chapter, we note which Example questions or Exercises are relevant. Also, we refer to a particular screen or screens of the GENERAL CHEMISTRY INTERACTIVE CD-ROM. Student feedback on this approach has been extremely positive.

History of Chemistry

We believe it is important for students to put developments in chemistry in their historical context. Therefore, we have included a number of marginal notes with vignettes of historic figures.

Early Coverage of pH

In a number of universities students take only the first semester of general chemistry. One topic previously covered later in the book, but that would be useful for students taking only the first term, is pH. Therefore, the definition of pH and examples of its use are now covered in Chapter 5 as part of the discussion of solution concentration.

Chemical Equilibria

In the four previous editions there were four chapters on chemical equilibria. This subject is now covered in three chapters by reducing slightly the coverage of insoluble compounds and incorporating the topic into a new chapter titled *Other*

Aspects of Aqueous Equilibria (Chapter 18). This new chapter covers buffer solutions, acid-base titrations, and compound solubility.

The Writing Program

A hallmark of the first four editions of *Chemistry & Chemical Reactivity* has been its readability. Nonetheless, the book has been almost completely rewritten with an eye to shortening the material without reducing content coverage or readability.

Revision to Art, Illustrations, and Book Design

For this edition, we have paid particular attention to the art and illustrations. Patrick Harman, the art and design director on the GENERAL CHEMISTRY INTERACTIVE CD-ROM, was brought into the book project. We believe one reason the earlier version of the CD-ROM was so successful is that it was well designed. Good design leads the user to know clearly how each part of the presentation functions. We wanted to bring this same clarity and ease of use to the book.

To learn how best to improve the illustration program, we spent considerable time talking with students about how they used the book, particularly in regard to the illustrations, look layout, and typography. Based on these discussions, we have paid special attention to the following items.

Molecular Models

We have continued the use of molecular models. However, many ball-and-stick models are now enclosed in a light "screened" version of the space-filling version of the molecule. This gives the student a clear view of the atom-to-atom connections in the molecule as well as a sense of how that molecule fills space. In addition, many of the models are now accompanied by a letter-and-line version of the structure for easier interpretation of the model. And finally, changes in the models used with photos better connect them with the subject of the photo.

Photos

Many photographs in previous editions of the book have been replaced by new photos. (There are roughly 300 new photos by Charles Winters.) One aspect of the revised photo program is to illustrate a sequence of events in a reaction or a chemical analysis. Another aspect of the photo program is that photos are sometimes reused at various places in the book to remind students of a previously covered subject and its relevance to a new topic.

Labeling Photos and Illustrations

Illustrations in science texts can be complex. Also, at times students may be uncertain on what to focus in an illustration. Therefore, we have worked to clarify our illustrations, to make them consistent throughout the book, and, in particular, make them illustrate the connection between the world we can see and the world of atoms, molecules, and ions.

Typography and Book Design

Again from our conversations with students we have made changes in typography and book design. In particular, marginal notes are now titled to highlight their relevance. Also, references to material before or after a topic, and to useful CD-ROM screens, are incorporated into the text.

SUPPORTING MATERIALS FOR STUDENTS

General Chemistry Interactive CD-ROM, version 3.0

This multimedia companion to *Chemistry & Chemical Reactivity* was originally designed by John Kotz and William Vining (University of Massachusetts). William Vining is the primary author of the newest version, and Patrick A. Harman, a design consultant, is now a co-author.

The CD-ROM is divided into chapters that closely follow the organization of *Chemistry & Chemical Reactivity*. Instead of consisting solely of text and illustrations, however, the CD-ROM presents ideas and concepts with which the user can interact. One can watch a reaction in process, change a variable in an experiment and experience the result, follow stepwise solutions to problems, explore the periodic table, and listen to tips and suggestions on problem solving and understanding concepts. The CD-ROM includes original graphics, over 100 video clips of chemical experiments, which are enhanced by sound and narration, and several hundred molecular models and animations.

In addition to the descriptive material developed for the earlier edition, the new version of the CD-ROM includes Simulations, Tutorials, and Exercises in which the user can interact with chemical information. For example, the reactants and their concentrations can be changed in a reaction and the effect on reaction rate can be visualized. The user is led through a series of questions on the CD-ROM that lead to fuller understanding of the experiment. So that a student knows when the CD-ROM supports book material, references (such as [CD-ROM, Screen 1.5]) are given at appropriate places in the book.

The CD-ROM also includes molecular modeling software from the CAChe/Fujitsu Group. This software can be used to view hundreds of models, rotate the models for a fuller understanding of their structures, and measure bond lengths and bond angles.

Additionally, the CD-ROM has a plotting tool, molar mass and molarity calculators, an extensive database of compounds with their thermodynamic properties, and copies of the important tables from the textbook.

Earlier versions of the CD-ROM, used by thousands of students worldwide, were sold separately. The newest version is included in each copy of the textbook.

The ISBN number for the CD-ROM as a stand-alone product is 0-03-035319-X

OWL — Online Web-based Learning System

Learning chemistry takes practice and that usually means completing homework assignments on various sections of the book. The Web-based OWL system presents the student with a series of questions on a given topic and the student responds by indicating a numerical answer or selecting from a menu of choices. The questions are generated from a database of information, so each student in a course is given a different question each time they access a instructional unit. Extensive feedback is available online for each question. Instructors can set up the system to deliver questions on certain dates, specify the number of tries, and specify the number of questions a student must answer successfully before the student is considered to have mastered the topic. Students find it is an excellent way to review for examinations, and studies at the University of Massachusetts-Amherst show a strong correlation between use of the OWL system and course performance.

Use the ISBN number 0-534-45634-0 to order the textbook and OWL.

Student Study Guide, Paul Hunter, Michigan State University, ISBN 0-03-035023-9
This *Guide* has been designed around the key objectives of the book. It provides
section summaries, review questions and answers, study questions and problems
with answers, crossword puzzles, and a sample test at the end of each of the five
parts of the book.

Student Solution Manual, Alton Banks, North Carolina State University, ISBN 0-
03-035016-6 This *Manual* contains detailed solutions to the even-numbered, end-
of-chapters Study Questions.

Student Lecture Outline, Ronald Ragsdale, University of Utah, ISBN 0-03-034977-X
The *Outline* is an aid for students to take better notes, help focus on key concepts,
and streamline reviewing for quizzes, tests, and exams.

WebTutor™ Advantage and Web Tutor™ Advantage Plus.
WebTutor offers real-time access to an array of study tools, including flash cards (with
audio), practice quizzes, online tutorials, and Web links. Professors can use it to
provide virtual office hours, post syllabi, set up threaded discussions, track student
progress with quizzing material, and more. *WebTutor Plus* adds an integrated elec-
tronic version of the complete textbook.
Advantage on WebCT: ISBN 0-534-40142-2
Advantage on Blackboard: ISBN 0-534-40143-0
Advantage Plus on WebCT: ISBN 0-534-40144-9
Advantage Plus on Blackboard: ISBN 0-534-40145-7

Student PowerPoint® Notebook: ISBN 0-03-035008-5
The *Notebook* provides black and white copies of the PowerPoint slides with note -
taking space.

Essential Math for Chemistry Students, David W. Ball, Cleveland State University,
ISBN 0-314-09604-3
This text is for students who lack confidence and/or competency in the essential
mathematics skills necessary to survive in general chemistry. Each chapter focuses
on a specific type of skill and has workedout examples to show how these skills
translate to chemical problem solving.

Supporting Materials for Instructors

Instructor's Resource Manual, Susan Young, Hartwick College, ISBN 0-03-34954-0
This *Manual* suggests alternative organizations of the course, classroom demonstrations, and contains workedout solutions to *all* end-of-chapter Study Questions.

PowerPoint® Slides, John Kotz, State University of New York, Oneonta.
Download from http://chemistry.brookscole.com
The author has presented his class lectures for several years using PowerPoint™.
Hundreds of slides that cover the entire year of introductory chemistry have been
created for lecture presentations. They use the cover the entire year of introductory chemistry have been created for lecture presentations. They use the full power
of PowerPoint® and incorporate videos, animations, and photos from the *General
Chemistry Interactive CD-ROM*. Professors can customize their lecture presentations by adding their own slides or by deleting or changing existing slides. The

PowerPoint® files are available on the Brooks/Cole Chemistry Resources Center at http://chemistry.brookscole.com. (The Microsoft® PowerPoint® Notebook is a printed version of the slides, with space for taking notes.)

Overhead Transparency Set, ISBN 0-03-034984-2
This is a collection of 150 full-color transparencies with large labels for viewing in large lecture halls.

TestBank, David Treichel, Nebraska Wesleyan University, ISBN is 0-03-035031-X
The *Test Bank* is comprised of more than 1250 questions, over two-thirds of which are newly written for this edition. Questions range in difficulty and variety. Numerical, open-ended, or conceptual problems are written in the multiple choice, fill-in-the-blank, or short-answer formats. Both single and multiple steps problems are presented for each chapter.

ExamView® Computerized Testing, cross platform, ISBN 0-03-034947-8.
The computerized version of the Test Bank allows instructors to create, deliver and customize tests and study guides (both print and online) with this assessment and tutorial system. *Quick Test Wizard* and an *Online Test Wizard* easily guide you through the step-by-step process of creating tests that print exactly as you see them on your computer screen."

MultiMedia Manager CD-ROM, cross-platform ISBN: 0-03-034962-1
The *Manager* is a digital library and presentation tool. Included are already-created text-specific presentations and a library of resources valuable to instructors such as text art, photos, and tables. Electronic files are easily exported into other software packages so you can create your own materials.

For the Laboratory: A selection of laboratory manuals and materials are available from Brooks/Cole.

Chemical Education Resources (CER) (catalog ISBN: 0-534-97709-X) allows you to customize a laboratory manual for your course from a wide range of more than 300 experiments. Contact your local Brooks/Cole representative or visit http://www.textchoice.com.

Chemical Principles in the Laboratory, seventh edition (ISBN: 0-03-031167-5) by Emil J. Slowinski, Wayne C. Wolsey, and William Masterton provides detailed directions and study assignments. The manual contains 42 experiments that have been thoroughly class-tested and selected with regard to safety and cost. An Instructor's Manual provides lists of equipment and chemical needs for each experiment.

Experiments in General Chemistry, third edition (ISBN: 0-534-36102-1) by Steven L. Murov, contains tested and clearly-presented experiments and lab exercises covering a wide range of topics. Safe laboratory practices are emphasized, and the use of hazardous substances is limited to the minimum amount required for reliable results.

Laboratory Experiments for General Chemistry, third edition (ISBN: 0-03-032906-X) by Toby F. Block, George M. McKelvy, and Harold R. Hunt, focuses on using non-hazardous materials to teach the experimental nature of general chemistry to students of various academic backgrounds and differing interests and abilities. Multiple week experiments illustrate that chemical systems do not work at an arbitrary schedule. An instructor's manual provides suggestions for combining experiments of shorter length and similar pedagogy as well as for using experiments for group or collaborative environments.

Laboratory Inquiry in Chemistry (ISBN: 0-534-37694-0) by Rich Bauer, Jim Birk, and Doug Sawyer, provides a unique set of guided-inquiry investigations that focus on constructing knowledge about the conceptual basis of laboratory techniques, instead of simply learning techniques. The experiments provide a realistic laboratory experience, typical of the practicing chemist.

Standard and Microscale Experiments in General Chemistry, third edition (ISBN: 0-03-021017-8) by Carl B. Bishop, Muriel B. Bishop, and Kenneth W. Whitten, contains a large number of experiments without sacrificing organizational clarity. All experiments have undergone massive testing and many microscale experiments allow more efficient and cost-effective means of conducting experiments.

ACKNOWLEDGEMENTS

Preparing the fifth edition of *Chemistry & Chemical Reactivity* took almost three years of continuous effort. However, as in our work on the first four editions, we have had the support and encouragement of our families and of some wonderful friends, colleagues, and students.

Saunders College Publishing, now Brooks/Cole

The first four editions of this book were published by Saunders College Publishing, a part of Harcourt Brace, and the fifth edition was begun under their auspices. About 12 months before publication of this edition, however, the company came under new ownership, the Brooks/Cole group of Thomson Learning. While the transition has been a trying time for all concerned, everyone involved worked tirelessly to ensure the successful publication of this book, and our new colleagues have been very supportive.

Another difficulty experienced by the author team on this book was the death of our long-time Editor, Publisher and friend, John Vondeling, in January 2001. Much of the credit for the success of our earlier editions goes to John. We had worked with him for many years and became fast friends. His support, confidence, loyalty, and good humor are not forgotten. John had an enormous impact on college textbook publishing in the sciences. Not only will we miss him personally, but higher education in the sciences is the poorer for his passing.

Our publisher for this edition is Angus McDonald. He has successfully guided the former Saunders group and our book and others through the editorial and production process, Angus has a long history in publishing, first in marketing and then in editorial. Like his predecessor John Vondeling, Angus will make his mark in college textbook publishing. We thank him for his guidance and encouragement over the past months and applaud his wisdom in helping us through a difficult period.

The Developmental Editors for this edition were Peter McGahey and Mary Castellion. Peter is relatively new in publishing but he has a bright future. He is blessed with energy, creativity, enthusiasm, intelligence, and good humor. Mary Castellion has been involved in the development and authorship of college chemistry books for some time and brought that wealth of experience to this project. Her guidance has had a significant influence on this book. Both Peter and Mary have been very pleasant colleagues, and they are trusted friends and confidants.

Our Managing Editor for much of the work on this edition was Bonnie Boehme. Her attention to detail and her energy and enthusiasm for the project will surely help make this edition as successful as previous editions. John Probst of TechBooks took over the project in its final months and guided it to a successful conclusion.

No book can be successful without proper marketing. Julie Conover and Jeff Ward came into the project just prior to publication and have been a delight to work with. They are knowledgeable about the market and have worked tirelessly to bring the book to everyone's attention.

Caroline McGowan was again in charge of the art program at Brooks/Cole. Her assistance was invaluable in helping us learn how to produce art for the book, in helping to ensure that we had chosen appropriate materials, and in designing the final product. All agree it is a beautiful book.

Our team at Brooks/Cole is completed with Charlene Squibb, Production Manager. Charlene has helped us through previous edition and has gently kept us on schedule. We certainly appreciate her patience and organizational skills.

Photography, Art, and Design

Most of the color photographs for this edition were again beautifully done by Charles D. Winters. He produced dozens of new images for this book, often under great deadline pressure and always with a creative eye. Charlie's work gets better and better with each edition. A great leap forward was made with this edition by moving entirely to digital photography. The effect on the quality of work has been dramatic. We have worked with Charlie for some years and have become close friends. We listen to his jokes, both new and old — and always forget them. When we finish the book, we look forward to a kayaking trip.

When this edition was being planned, we brought in Patrick Harman as a member of the team. Pat designed the first edition of the *General Chemistry Interactive CD-ROM,* and we believe its success is in no small way connected to his design skill. For this edition of the book Pat went over almost every figure, and almost every word, to bring a fresh perspective to ways to communicate chemistry. The result in seen in uses of different typography, in a new approach to molecular models, and in greatly improved illustrations throughout. His work, with Charles Hamper, on the illustrations of atomic and molecular orbitals is without parallel in introductory chemistry textbooks. Pat brings a dimension to the book that the authors simply could not have supplied. It is a vastly better product for his work. And Pat has opened our eyes to modern music.

Finally, we are very pleased to again work with Susan Young. Susan helped to produce some of the molecular models and worked with the authors and Charlie Winters on the photography. In addition, she read the manuscript, the galley and page proofs, and provided good advice. We simply could not have done this project without Susan's help, creativity, good humor, and energy.

Others

Publishing a book is a complicated process, and a large team of people is needed to carry out the task. At least one more member of our team deserves special thanks. Katie Kotz kept the work in Oneonta, NY, Santa Cruz, CA, and Madison, WI — photography, art and design, text preparation, and photo research — organized. Her organizational skills and her expertise in maintaining a large database to keep track of hundreds of photographs and art pieces has been crucial to our success. In addition, she has been the wonderful wife of one of the authors for 41 years.

REVIEWERS

We believe the success of any book is due in no small way to the quality of the reviewers of the manuscript. The reviewers of previous editions made important contributions that are still part of this book. Reviewers of the fifth edition continued that tradition. We wish to acknowledge with gratitude the efforts of all those listed below. Several should be noted in particular, however. Gary Riley and Steve Lnaders worked all of the problems in the book. Proof readers Susan Young, Larry Fishel, and Marie Nguyen were invaluable. With thousands of words and numbers to be checked, careful and competent proof readers are crucial.

David W. Ball, Cleveland State University
Roger Barth, West Chester University
John G. Berberian, Saint Joseph's University
Don A. Berkowitz, University of Maryland
Simon Bott, University of Houston
Wendy Clevenger Cory, University of Tennessee at Chattanooga
Richard Cornelius, Lebanon Valley College
James S. Falcone, West Chester University
Martin Fossett, Tabor Academy
Michelle Fossum, Laney College
Sandro Gambarotta, University of Ottawa
Robert Garber, California State University, Long Beach
Michael D. Hampton, University of Central Florida
Paul Hunter, Michigan State University
Michael E. Lipschutz, Purdue University
Shelley D. Minteer, Saint Louis University
Jessica N. Orvis, Georgia Southern University
David Spurgeon, University of Arizona
Stephen P. Tanner, University of West Florida
John Townsend, West Chester University
John A. Weyh, Western Washington University
Marcy Whitney, The University of Alabama
Sheila Woodgate, The University of Auckland

Reviewers of the Fourth Edition
Alton Banks, North Carolina State University
John DeKorte, Glendale Community College
L, Peter Gold, Pennsylvania State University
Paul Hunter, Michigan State University
Donald Kleinfelter, University of Tennessee-Knoxville

Michael E. Lipschutz, Purdue University
Mark E. Noble, University of Louisville
Lee G. Pedersen, University of North Carolina, Chapel Hill
Ronald O. Ragsdale, University of Utah
Wayne Tikkanen, California State University, Los Angeles
John B. Vincent, University of Alabama

About the Authors

John C. Kotz, a State University of New York Distinguished Teaching Professor at the College at Oneonta, was educated at Washington and Lee University and Cornell University. He held National Institutes of Health postdoctoral appointments at UMIST in England and at Indiana University.

He has coauthored three textbooks in several editions (*Inorganic Chemistry, Chemistry & Chemical Reactivity,* and *The Chemical World*) and the *General Chemistry Interactive CD-ROM.* He has also published on his research in inorganic chemistry and electrochemistry.

He was a Fulbright Lecturer and Research Scholar in Lisbon, Portugal in 1979 and a Visiting Professor there in 1992. In 1991–1992 he was a Visiting Professor at the Institute for Chemical Education (University of Wisconsin) and in 1999 was an Overseas Visitor at Auckland University in New Zealand. He was also recently an invited plenary lecturer at the annual meeting of the South African Chemical Society and at the biennial conference of secondary school chemistry teachers in New Zealand.

He has received several awards, among them a National Catalyst Award for Excellence in Teaching in 1992, the 1998 Estee Lecturership in Chemical Education at the University of South Dakota, the 1999 Visiting Scientist Award from the Western Connecticut Section of the American Chemical Society, and, in 2001, the first annual Distinguished Education Award from the Binghamton (NY) Section of the American Chemical Society. He may be contacted by e-mail: kotzjc@oneonta.edu

Paul M. Treichel, Jr., received his BS degree from the University of Wisconsin in 1958 and a Ph.D. from Harvard University in 1962. After a year of postdoctoral study in London, he assumed a faculty position at the University of Wisconsin-Madison, where he is currently Helfaer Professor of Chemistry. He served as department chair from 1986 through 1995. He has held visiting faculty positions in South Africa (1975) and in Japan (1995). Currently, he teaches courses in general chemistry, inorganic chemistry, and scientific ethics. Dr. Treichel's research in organometallic and metal cluster chemistry and in mass spectrometry, aided by 75 graduate and undergraduate students, has led to over 170 papers in scientific journals. He may be contacted by e-mail: treichel@chem.wisc.edu

Patrick Harman, Information and Media Design Consultant, studied communication design, film and animation as an undergraduate and graduate student at the University of Illinois. He joined the faculty of the University of Illinois at Chicago for 8 1/2 years as an adjunct lecturer and assistant professor, teaching a variety of communication design and motion graphics courses. In addition to teaching,

Patrick co-founded an educational animation studio to produce graphic design, animation, sound design and interface design for the Encyclopaedia Britannica Educational Corporation, Jacques Cousteau Society, Chicago Symphony Orchestra, Public Broadcasting System, and others. He relocated to California to design and co-produce the original Saunders Interactive General Chemistry CD-ROM, and spent six years directing the Information Design Group and developing college-level educational media products for Archipelago Productions, including distance-learning solutions for Chemistry, Physics, Biology and Economics.

Chemistry & Chemical Reactivity

Fifth Edition

A Preface to Students

The Double Helix

DNA is the substance in every plant and animal that carries the exact blueprint of that plant or animal. The structure of this cornerstone of life was uncovered by James D. Watson, Francis Crick, and Maurice Wilkins, who shared the 1962 Nobel Prize in medicine or physiology for their work. It was one of the most important scientific discoveries of the 20th century, and their story has been told by Watson in his book *The Double Helix*.

When Watson was a graduate student at Indiana University he had an interest in the gene and said he hoped that the puzzle of its biological role might be solved "without my learning any chemistry." Later, however, he and Crick found out just how valuable chemistry can be when they used their imaginations and a classic text on organic compounds to unravel the structure of DNA.

Solving important problems requires teamwork among scientists of many kinds, so Watson went to Cambridge University in England in 1951. There he met Crick, who, Watson said, talked louder and faster than anyone else. Crick shared Watson's belief in the fundamental importance of DNA, and they soon learned that Maurice Wilkins and Rosalind Franklin at King's College in London were using the technique called x-ray crystallography to learn more about DNA's structure. Watson and Crick believed that understanding the structure was crucial to understanding genetics, which is the chemical basis of heredity. To solve the structural problem, however, they needed experimental data of the type that could come from the experiments at King's College.

The King's College group was initially reluctant to share their data, and, what is more, did not seem to share Watson and Crick's sense of urgency. And there was an ethical dilemma: Could Watson and Crick work on a problem others had claimed as theirs? "The English sense of fair play would not allow Francis to move in on Maurice's problem," said Watson.

Watson and Crick recognized early on that the overall structure of DNA was a helix; that is, the atomic-level building blocks formed chains that twisted in space like the strands of a grapevine.

▲ **Helical objects.** The threads of a drill or screw twist along the axis of a helix, and some plants also climb by sending out tendrils that twist helically. The backbone of DNA twists in a helical fashion in a similar way. *(Charles D. Winters)*

▲ **James D. Watson and Francis Crick.** In a photo taken in 1953 Watson *(left)* and Crick *(right)* stand by their model of the DNA double helix. Together with Maurice Wilkins, Watson and Crick received the Nobel Prize in medicine and physiology in 1962. *(A. Barrington Brown/Science Source/Photo Researchers, Inc.)*

"I believe there remains general ignorance about how science is 'done'."

James Watson in the Preface to *The Double Helix*, 1968.

▲ **Rosalind Franklin of King's College, London.** She died in 1958 at the age of 37 and so did not share in the Nobel Prize because the prizes are never awarded posthumously. *(Courtesy American Society for Microbiology)*

They also knew what chemical elements it contained and roughly how they were grouped together. What they did not know was the detailed structure of the helix. By the spring of 1953, however, they had the answer. The atomic level building blocks of DNA form two chains twisted together in a double helix.

Watson and Crick approached the problem using a technique chemists now use frequently—model building. They built models of the pieces of the DNA chain and tried various chemically reasonable ways of fitting them together. Finally, they discovered that one arrangement was "too pretty not to be true." However, it was the experimental

evidence of Wilkins and Franklin that confirmed the "pretty structure." Models are useful to chemists, but experimental evidence is definitive.

The story of how Watson, Crick, Wilkins, and Franklin ultimately came to share information and insight is an interesting human drama and illustrates how scientific progress is often made. As Watson later said in *The Double Helix*, Rosalind Franklin eventually came to appreciate the fact that his and Wilkins's "past hooting about model building represented a serious approach to science, not the easy resort of slackers who wanted to avoid the hard work necessitated by an honest scientific career."

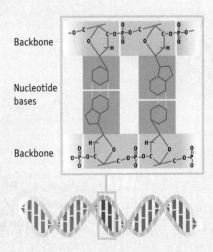

▶ **A DNA model.** This model of the DNA chains shows the basic arrangement of the groups of atoms in the chains. The backbone of each chain is assembled from oxygen (O) and phosphorus (P) atoms along with sugar molecules. Attached to each chain are a series of molecules called nucleotide bases, which are made up of atoms of carbon (C), nitrogen (N), hydrogen (H), and oxygen. Atoms and elements are defined and described in Chapters 2, 7, and 8. Molecules and their structures are a theme of this book, but details are especially found in Chapters 3, 9, and 10.

Chemistry is about change. It was once only about changing one natural substance into another—wood and oil burn, grape juice turns into wine, and cinnabar, a red mineral from the earth, changes into shiny quicksilver. It is still about change, but now chemists focus on the change of one pure substance, whether natural or synthetic, into another (Figure 1).

Although chemistry is endlessly fascinating—at least to chemists—why should you study it? Each person probably has a different answer, but many of you may be taking this chemistry course because someone else has decided it is an important part of preparing for a particular career. Chemistry is especially useful because it is central to our understanding of disciplines as diverse as biology, geology, materials science, medicine, physics, and many branches of engineering. In addition, chemistry plays a major role in our economy; chemistry and chemicals affect our daily lives in a wide variety of ways. Furthermore, a course in chemistry can help you see how a scientist thinks about the world and how to solve problems. The knowledge and skills developed in such a course will benefit you in many career paths and will help you become a better informed citizen in a world that is becoming technologically more complex—and more interesting. Therefore, to begin your study of chemistry, this Preface discusses some fundamental ideas used by scientists of all kinds.   ●

SCIENCE AND ITS METHODS

The story of the way the basic structure of DNA was uncovered illustrates the way science works.

Hypotheses, Laws, and Theories

As scientists we study questions of our own choosing or ones that someone else poses in the hope of finding an answer or of discovering some useful information. Although it may not seem so as you begin to study chemistry, it is easy to

(a)   (b)   (c)

Figure 1   Chemistry is the study of atoms and molecules, and their transformations. Here metallic aluminum and orange-brown, liquid bromine **(a)** combine with each other **(b)**, and are transformed into white, solid aluminum bromide **(c)**. Molecular models give us an idea of how the atoms are assembled in aluminum metal, in bromine, and in the aluminum bromide. (For a video of this reaction, see the *General Chemistry Interactive CD-ROM*, Version 3.0, Screen 4.2.) *(Charles D. Winters)*

conceive an idea to study. The goal should be to state clearly a problem that is worth studying, that is narrow enough in scope so that a useful conclusion may be reached, and for which there is some hope of finding a solution. James Watson and Francis Crick did just that in the 1950s when they set out to understand the structure of DNA, a story told in Watson's book *The Double Helix* and outlined on pages 2 and 3.

Having posed a reasonable question, scientists first look at the work that others have already done to provide some notion of the most productive direction to take. Watson and Crick did this by reading papers published by Linus Pauling, among others, and by talking with Maurice Wilkins and Rosalind Franklin, colleagues who were also studying DNA. They then proceeded to the next step, forming a *hypothesis,* a tentative explanation or prediction of experimental observations.

After formulating one or more hypotheses, scientists perform experiments that are designed to give results that may confirm some hypotheses and invalidate others. In chemistry this usually requires that both quantitative and qualitative information be collected. *Quantitative* information is numerical data, such as the temperature at which a chemical substance melts. *Qualitative* information, in contrast, consists of nonnumerical observations, such as the color of a substance or its physical appearance. Watson and Crick built physical models of parts of DNA to see how the atoms would fit together and if the arrangement matched the experimental data provided by Wilkins and Franklin.

Preliminary experiments usually lead to revision and extension of the original hypothesis. This creates the need for new experiments. After a number of experiments have been done and the results checked to ensure they are reproducible, a pattern of behavior or results may emerge. At this point it may be possible to summarize the observations in the form of a general rule. Finally, after numerous experiments by many scientists over an extended period, the original hypothesis may become a *law*—a concise verbal or mathematical statement of a relation that seems always to be the same under the same conditions.

We base much of what we do in science on laws because they help us predict what may occur under a new set of circumstances. For example, we know from experience that if we allow the chemical element sodium to come in contact with the water, a violent reaction occurs and several new substances are formed (Figure 2). But the result of an experiment might be different from what is expected based on a general rule. When that happens, chemists get excited because experiments that do not follow the known rules of chemistry are the most interesting. We know that understanding the exceptions almost invariably gives new insights.

Once enough reproducible experiments have been conducted, and experimental results have been generalized as a law or general rule, it may be possible to conceive a theory to explain the observation. A *theory* is a unifying principle that explains a body of facts and the laws based on them. It is capable of suggesting new hypotheses. Theories abound, not only in the sciences, but also in other disciplines like economics and sociology. Laws summarize the facts of nature and rarely change. Theories are inventions of the human mind. Theories can and do change as new facts are uncovered.

Goals of Science

The sciences, including chemistry, have two goals. The first of these is prediction and control. We do experiments and seek generalities because we want to be able to predict what may occur under a given set of circumstances. We also want to know how we might control the outcome of a chemical reaction or process.

Hypotheses, laws, and theories. Richard Feynman (1918–1988), one of the preeminent scientists of the 20th century, uncovered the problems that led to the explosion of the Space Shuttle Challenger in 1986. The steps leading to a solution are a classic example of the scientific method. See the *General Chemistry Interactive CD-ROM,* Version 3.0, Screen 0.2. *(Marilynn K. Yee/NYT pictures)*

Figure 2   The metallic element sodium reacts vigorously with water. For a video of the reactions of lithium, sodium, and potassium with water, see the *General Chemistry Interactive CD-ROM,* Version 3.0, Screen 8.15. *(Charles D. Winters)*

● **Atoms and Elements**
Atoms are the building blocks of all matter. Everything is made of atoms of different kinds assembled in distinctive ways. *Elements* are types of matter composed of just one kind of atom. See Section 1.1, Elements and Atoms.

Figure 3 **Discovery of Teflon.** In a photo taken of a reenactment of the actual event in 1938, Roy Plunkett *(right)* (1910–1994) and his assistants find a white solid coating the inside of a gas cylinder. The solid, now called Teflon, was discovered by accident. *(Hagley Museum and Library)*

Figure 4 **Teflon products.** Teflon was discovered by accident in 1938. It is now the basis of an enormous industry of industrial and consumer products. *(Charles D. Winters)*

The second goal is explanation and understanding. We know, for example, that certain elements will react vigorously with water (see Figure 2). But why should this be true? And why is this extreme reactivity unique to these elements? To explain and understand this, we turn to theories such as those developed in Chapters 9 and 10.

The Importance of Serendipity

People who work outside of science usually have the idea that science is an intensely logical field. They picture white-coated chemists moving logically from hypothesis to experiment and then to laws and theories without human emotion or foibles. This picture is a great simplification! Watson and Crick worked many months and made numerous errors before they understood DNA's structure.

Often, scientific results and understanding arise quite by accident, otherwise known as *serendipity*. Creativity and insight are needed to transform a fortunate accident into useful and exciting results. The discovery of the cancer drug cisplatin by Barnett Rosenberg in 1965 or of penicillin by Alexander Fleming (1881–1955) in 1928 is each a wonderful example of serendipity.

A material familiar to many of you—Teflon—was found by a combination of serendipity and curiosity. In 1938, Dr. Roy Plunkett was a young scientist working in a DuPont laboratory on the chemistry of fluorine-containing refrigerants (which we now know by their trademark name, Freon). For one experiment, Plunkett and his assistants opened the valve on a tank of tetrafluoroethylene gas. The tank supposedly held 1000 g of gas, but only 990 g came out. What happened to the other 10 g? Curiosity is the mark of a good scientist, so they sawed open the tank. A white, waxy substance coated the inside (Figure 3). Following his curiosity further, Plunkett tested the material and found it had remarkable properties. It was more inert than sand! Strong acids and bases did not affect it, nothing could dissolve it, and it was resistant to heat. Unlike sand, it was slippery.

Were it not so expensive, the remarkable properties of this new substance should have led to an immediate search for uses in consumer products. However, it found its first use in the World War II atomic bomb project as a sealer in the equipment used in the separation of uranium. The project was of such national importance, that the expense of the material was of no concern. It was not until the 1960s that Teflon began to show up in consumer items (Figure 4). One of its most important uses is now in medical products. Because it is one of the few substances the body does not reject, it can be used to line artificial hip and knee joints and for heart valves and many other body parts.

Dilemmas and Integrity in Science

You may think research in science is straightforward: Do an experiment, draw a conclusion. But, research is seldom that easy. Frustrations and disappointments are common enough, and results can be inconclusive. Complicated experiments often contain some level of uncertainty, and spurious or contradictory data can be collected. For example, suppose you do an experiment expecting to find a direct relation between two experimental quantities. You collect six data sets. When plotted, four of the sets lie on a straight line, but two others lie far away from the line. Should you ignore the last two points? Or should you do more experiments when you know the time they take will mean someone else could publish first and thus get the credit for a new scientific principle? Or should you consider that the two points not on the line might indicate that your original hypothesis is wrong, and that you will have to abandon a favorite idea you have worked on for a year? Scientists have a responsibility to remain objective in the face of these difficulties, but it is sometimes hard to do.

It is important to remember that scientists are subject to the same moral pressures and dilemmas as any other person. To help ensure integrity in science, some simple principles have emerged over time that guide scientific practice: Experimental results should be reproducible, conclusions should be reasonable and unbiased, credit should be given where it is due, and results should be reported in sufficient detail that they can be used or reproduced by others.

Moral and ethical issues frequently arise in science. In the search for DNA's structure, Watson questioned whether they should work in an area of research already "claimed" by a colleague. A more recent example is the story of silicone breast implants. Because silicone materials had been found to be medically useful, the Dow Corning Corporation began in the 1960s to market a silicone-based breast implant. By the 1990s so many law suits, claiming injury, had been brought against the company that it stopped manufacturing the items and, because of the size of the judgments against it, declared bankruptcy. Medical issues were clearly a factor of concern for the recipients of the implants. But was the incidence of medical problems greater for them than among women without implants? Were the implants responsible for these medical conditions? In 1998, a scientific panel appointed by the court said it saw no proven links between the implants and disease. Nonetheless, public opinion expressed through law suits had forced Dow Corning and others out of the business of manufacturing this particular product.

There are many, many moral and ethical issues for chemists. Chemistry has extended and improved lives for millions of people. But just as clearly, chemicals can cause harm, particularly when misused. It is incumbent on all of us to understand enough science to ask pertinent questions and to evaluate sources of information sufficiently to reach reasonable conclusions regarding our own health and safety and that of our communities.

A Final Word to Students

Why study chemistry? The reasons are clear. Whether you want to become a biologist, a geologist, an engineer, or a physician, or pursue any of dozens of other professions, chemistry will be at the core of your discipline. It will always be useful to you, sometimes when least expected.

In addition, you will be called on to make many decisions in your life for your own good or for that of those in your community—whether that be your neighborhood or the world. An understanding of the nature of science in general, and of chemistry in particular, can only serve to help in these decisions.

Because the authors of this book were students once—and still are—we know chemistry can be a challenging area of study. However, like anything worthwhile, it takes time and effort to reach genuine understanding. Be sure to give it time, and consult with your professors and your fellow students. We are certain you will find it as exciting, as useful, and as interesting as we do.

The chrysotile form of asbestos. *(Charles D. Winters)*

● **Risks and Benefits of Asbestos**
The asbestos family of minerals has important applications. Unfortunately, overuse or misuse can cause medical problems, particularly with one type of asbestos. The offending variety of asbestos is not mined in North America and so is not used in consumer and building products in North America. Nonetheless, asbestos removal is widely practiced in our homes and schools, even though the annual death rate from asbestos exposure in schools is about 0.005 per million people, as opposed to that resulting from participation in high school football, for which the death rate is 10 per million people. What is the responsibility of scientists in this situation?

● **ChemCases**
ChemCases is a Web-based series of stories about products such as Gatorade, Nutrasweet, silicones, and Olestra and industries such as nuclear power. The cases also explore the moral and ethical issues surrounding these products and industries. See http://www.chemcases.com.

1 Matter and Measurement

Out of Gas!

On July 23, 1983, Air Canada Flight 143 was flying at an altitude of 26,000 ft from Montreal to Edmonton. Warning buzzers sounded in the cockpit of the Boeing 767. One of the world's largest planes was now a glider — the plane had run out of fuel!

How did this modern jet airplane, having the latest technology, run out of fuel? A simple mistake had been made in calculating the amount of fuel required for the flight.

Like all Boeing 767s, this plane had a sophisticated fuel gauge, but it was not working properly. The plane was still allowed to fly, however, because there is an alternative method of determining the quantity of fuel in the tanks. Mechanics can use a stick, much like the oil dipstick in an automobile engine, to measure the fuel level in each of the three tanks. The mechanics in Montreal read the dipsticks, which were calibrated in centimeters, and translated those readings to a volume in liters. According to their calculation, the plane had a total of 7682 L of fuel.

Pilots always calculate fuel quantities in units of mass because they need to know the total mass of the plane before take-off. Air Canada pilots had always calculated the mass of fuel in pounds, but the new 767's fuel consumption was given in kilograms. The pilots knew that 22,300 kg of fuel was required for the trip. If 7682 L of fuel remained in the tanks, how much had to be added? This involved using the fuel's density to convert 7682 L to a mass in kilograms. The mass of fuel to be added could then be calculated, and that mass converted to a volume of fuel to be added.

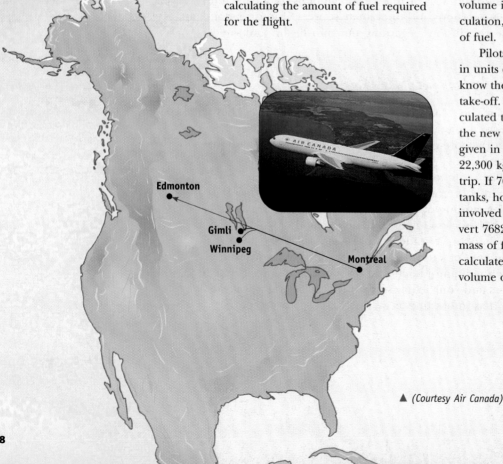

Edmonton

Gimli
Winnipeg

Montreal

▲ *(Courtesy Air Canada)*

▲ After running out of fuel, Air Canada Flight 143 glided 30 min before landing on an abandoned airstrip at Gimli, Manitoba, near Winnipeg. *(AP/Wide World Photos)*

The First Officer of the plane asked a mechanic for the conversion factor to do the volume-to-mass conversion, and the mechanic replied "1.77." Using that number, the First Officer and the mechanics calculated that 4917 L of fuel should be added. But later calculations showed that this is only about one fourth of the required amount of fuel (about 16,200 kg)! Why? Because no one thought about the *units* of the number 1.77. They realized later that 1.77 has units of pounds per liter and *not* kilograms per liter.

Out of fuel, the plane could not make it to Winnipeg, Manitoba, so controllers directed them to the town of Gimli and to the small airport there abandoned by the Royal Canadian Air Force. After gliding for almost 30 min, the plane approached the Gimli runway. The runway, however, had been converted to a race course for cars, and a race was underway. Furthermore, a steel barrier had been erected across the runway. Nonetheless, the pilot managed to touch down very near the end of the runway. The plane sped down the concrete strip, the nose wheel collapsed, several tires blew — and the plane skidded safely to a stop just before the barrier. The Gimli glider had made it! And somewhere an aircraft mechanic is paying more attention to units on numbers.

Thinking About Matter Chemistry is the study of the nature of matter and its interactions. Matter is ultimately composed of chemical elements and their compounds. Most matter can exist in one or more states: solid, liquid, and gas.

Although chemists make observations at the macroscopic level, we are concerned with the properties of matter at the microscopic and submicroscopic scale as well.

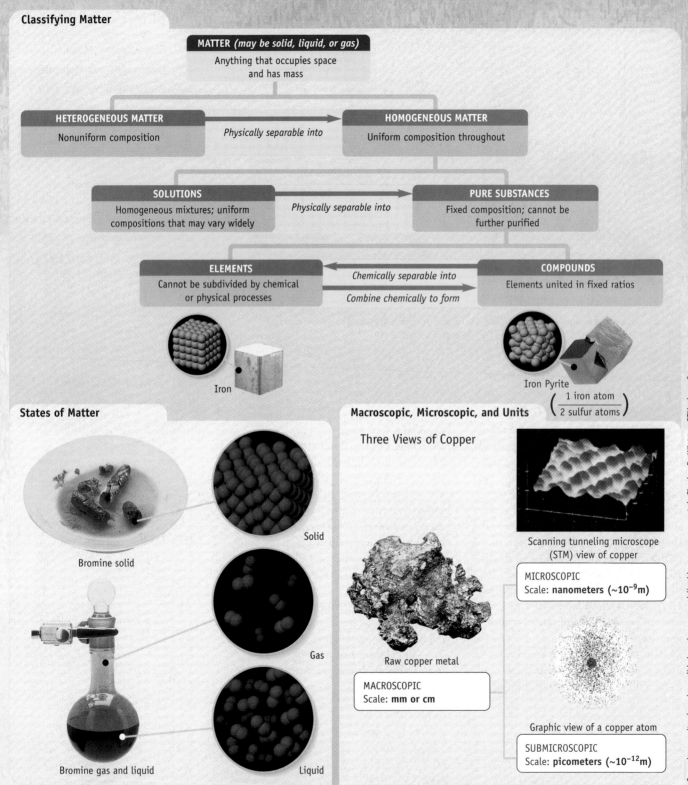

Classifying Matter

MATTER *(may be solid, liquid, or gas)*
Anything that occupies space and has mass

HETEROGENEOUS MATTER
Nonuniform composition

Physically separable into

HOMOGENEOUS MATTER
Uniform composition throughout

SOLUTIONS
Homogeneous mixtures; uniform compositions that may vary widely

Physically separable into

PURE SUBSTANCES
Fixed composition; cannot be further purified

ELEMENTS
Cannot be subdivided by chemical or physical processes

Chemically separable into
Combine chemically to form

COMPOUNDS
Elements united in fixed ratios

Iron

Iron Pyrite
$\left(\dfrac{1 \text{ iron atom}}{2 \text{ sulfur atoms}} \right)$

States of Matter

Bromine solid

Solid

Bromine gas and liquid

Gas

Liquid

Macroscopic, Microscopic, and Units

Three Views of Copper

Scanning tunneling microscope (STM) view of copper

MICROSCOPIC
Scale: **nanometers** $(\sim 10^{-9}\text{m})$

Raw copper metal

MACROSCOPIC
Scale: **mm or cm**

Graphic view of a copper atom

SUBMICROSCOPIC
Scale: **picometers** $(\sim 10^{-12}\text{m})$

Iron, iron pyrite, bromine solid, bromine gas and liquid, raw copper metal: Charles D. Winters. STM view of copper: Reproduced with permission from X. Xu, S. M. Vesecky, and D. W. Goodman, Science, Vol. 258, p. 788, 1992.

Imagine a tall glass filled with a clear liquid. Sunlight from a nearby window causes the liquid to sparkle, and the glass is cool to the touch. A drink of water would certainly taste good, but should you take a sip? If the glass were sitting in your kitchen you might say yes. But what if this scene occurred in a chemical laboratory? How would you know that the glass held pure water? Or, to pose a more "chemical" question: How would you *prove* this liquid is water?

We usually think of the water we drink as being pure, but this is not strictly true. In some instances material may be suspended in it or bubbles of gases such as oxygen may be visible to the eye. Some tap water has a slight color from dissolved iron. In fact, drinking water is almost always a mixture of substances, some dissolved and some not. As with any mixture, we could ask many questions. What are the components of the mixture — dust particles, bubbles of oxygen, dissolved sodium, calcium, or iron salts — and what are their relative amounts? How can these substances be separated from one another, and how are the properties of one substance changed when it is mixed with another?

This chapter is the beginning of our discussion of how chemists think about matter. We will explore some basic ideas about elements, atoms, compounds, and molecules, the focus of chemistry. Next we shall ask how chemists characterize these building blocks of matter and then begin to see how we can use numerical information.

Thinking about matter. Is this a glass of pure water? How can you prove it is pure? *(Charles D. Winters)*

Chapter Goals • Revisited

- **Recognize elements, atoms, compounds, and molecules.**
- Identify physical and chemical properties and changes.
- Apply the kinetic-molecular theory to the properties of matter.
- Use metric units and significant figures properly.

1.1   ELEMENTS AND ATOMS

Passing an electric current through pure water can decompose it to gaseous hydrogen and oxygen (Figure 1.1a). Substances like hydrogen and oxygen that are composed of only one type of atom are classified as **elements** [CD-ROM, Screen 1.5]. Currently 113 elements are known. Of these, only about 90 are found in nature; the remainder have been created by scientists. The name and symbol for each

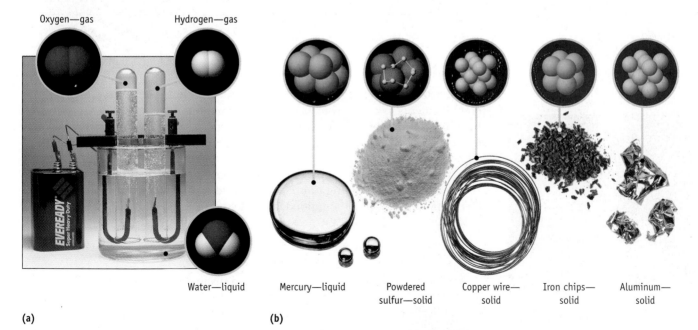

Figure 1.1   Elements. (a) Passing an electric current through water produces the elements hydrogen (*test tube on the right*) and oxygen (*on the left*). **(b)** Chemical elements can often be distinguished by their color and their state at room temperature. *(Charles D. Winters)*

• **Writing Element Symbols**

Be sure to notice that only the first letter of an element's symbol is capitalized. For example, cobalt is Co and not CO. The notation CO represents the chemical compound carbon monoxide.

• **Periodic Table**

See the periodic table on the *General Chemistry Interactive CD-ROM,* Version 3.0. It can be accessed from Screen 1.5 or from the Toolbox. See also Professor Frank DiSalvo describe the significance of the periodic table on Screen 1.6.

• **Exercise Answers**

In each chapter of the book you will find one or more Exercises at the end of a section or within a section. Their purpose is to help you to check your knowledge of the material in that section. Solutions to the Exercises are found in Appendix N.

element are listed in the tables at the front of the book. Carbon (C), sulfur (S), iron (Fe), copper (Cu), silver (Ag), tin (Sn), gold (Au), mercury (Hg), and lead (Pb) were known in relatively pure form to the early Greeks and Romans and to the alchemists of ancient China, the Arab world, and medieval Europe. However, many others — such as aluminum (Al), silicon (Si), iodine (I), and helium (He) — were not discovered until the 18th and 19th centuries. (Some of these elements are pictured in Figure 1.1b.) Finally, artificial elements, such as technetium (Tc), plutonium (Pu), and americium (Am), were made in the 20th century using the techniques of modern physics.

Many elements have names and symbols with Latin or Greek origins. Examples include helium (He), named from the Greek word for sun, *helios,* and lead, whose symbol, Pb, comes from the Latin word for heavy, *plumbum.* More recently discovered elements, though, have been named for their place of discovery or for a person or place of significance. Examples of these are americium (Am), californium (Cf), and curium (Cm).

The table inside the front cover of this book, in which the symbol and other information for all but the newest of the elements are enclosed in a box, is called the **periodic table** [🖰 CD-ROM, Toolbox]. We will describe this important tool of chemistry in more detail beginning in Chapter 2.

An **atom** is the smallest particle of an element that retains the characteristic properties of that element [🖰 CD-ROM, Screen 1.5]. Modern chemistry is based on an understanding and exploration of nature at the atomic level. We will have much more to say about atoms and atomic properties in Chapters 2, 7, and 8, in particular.

Exercise 1.1 Elements

Using the periodic table inside the front cover of this book,

(a) find the names of the elements having the symbols Na, Cl, and Cr.

(b) find the symbols for the elements zinc, nickel, and potassium.

1.2   COMPOUNDS AND MOLECULES

A pure substance like sugar, salt, or water, which is composed of two or more different elements, is referred to as a **chemical compound** [🖰 CD-ROM, Screen 1.6]. Even though only 113 elements are known, there appears to be no limit to the number of compounds that can be made from those elements. More than 20 million compounds are now known, with about a half million added to the list each year.

When elements become part of a compound, their original properties, such as color, hardness, and melting point, are replaced by the characteristic properties of the compound. Consider common table salt (sodium chloride), which is composed of two elements (Figure 1.2).

- Sodium is a shiny metal that interacts violently with water. It is composed of sodium atoms tightly packed together.

- Chlorine is a light yellow gas that has a distinctive, suffocating odor and is a powerful irritant to lungs and other tissues. The element is composed of Cl_2 units in which two chlorine atoms are tightly bound together.

- Sodium chloride or common salt, is a crystalline solid with properties completely unlike those of the two elements from which it is made (Figure 1.2).

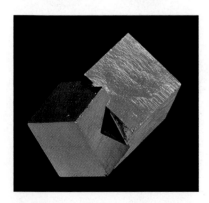

Iron pyrite is a chemical compound composed of iron and sulfur. It is often found in nature as perfect, golden cubes. *(Charles D. Winters)*

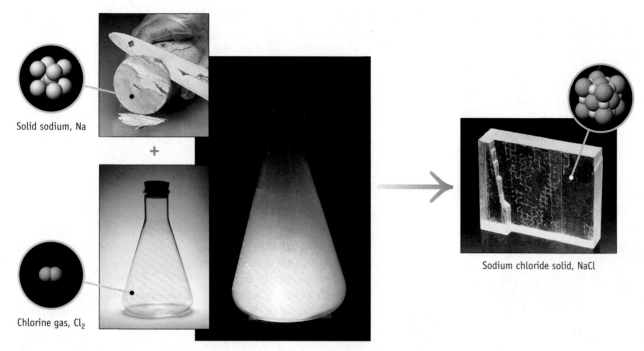

Solid sodium, Na

Chlorine gas, Cl₂

Sodium chloride solid, NaCl

Figure 1.2   Forming a chemical compound. Sodium chloride, table salt, can be made by combining sodium metal (Na) and yellow chlorine gas (Cl₂). The result is a crystalline solid. *(Charles D. Winters)*

Salt is composed of sodium and chlorine bound tightly together. (The meaning of chemical formulas such as NaCl is explored in Sections 3.3 and 3.4.)

It is important to make a careful distinction between a mixture of elements (see Section 1.5) and a chemical compound of two or more elements. Pure metallic iron and yellow, powdered sulfur (Figure 1.1b) can be mixed in varying proportions. In the chemical compound known as iron pyrite, however, this kind of variation cannot occur. Not only does iron pyrite exhibit properties peculiar to itself and different from those of either iron or sulfur, or a mixture of these elements, but it has a definite percentage composition by weight (46.55% Fe and 53.45% S, or 46.55 g of Fe and 53.45 g of S in 100.00 g of sample). Thus, two major differences exist between mixtures and pure compounds: Compounds have distinctly different characteristics from their parent elements and have a definite percentage composition (by mass) of their combining elements.

Some compounds — such as salt, NaCl — are composed of **ions,** which are electrically charged atoms or groups of atoms [➡ CHAPTER 3]. Other compounds — such as water and sugar — consist of **molecules,** the smallest, discrete units that retain the composition and chemical characteristics of the compound.

The composition of any compound can be represented by its **chemical formula.** In the formula for water, H_2O, for example, the symbol for hydrogen, H, is followed by a subscript "2" indicating that two atoms of hydrogen occur in a single water molecule. The symbol for oxygen appears without a subscript; this indicates that one oxygen atom occurs in the water molecule.

As you shall see throughout this book, molecules can be represented with models that depict their composition and structure. Figure 1.3 illustrates the names, formulas, and models of the structures of a few common molecules.

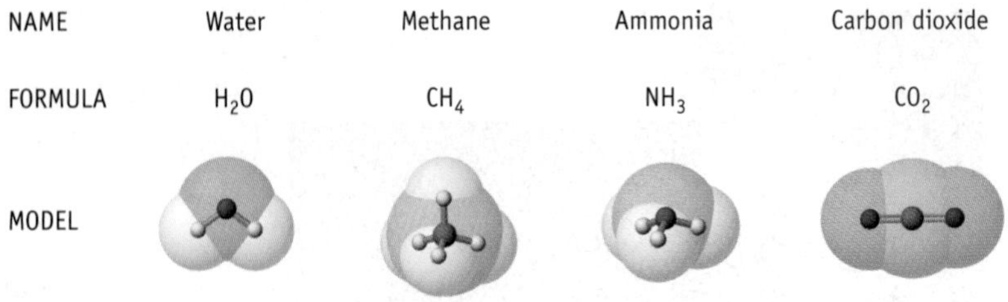

Figure 1.3 Names, formulas, and models of the structures of some common molecules. Models of molecules are widely used, and many appear throughout the book. In such models C atoms are gray, H atoms white, N atoms blue, and O atoms red. The color scheme commonly used in chemistry is given in the Preface.

NAME	Water	Methane	Ammonia	Carbon dioxide
FORMULA	H_2O	CH_4	NH_3	CO_2
MODEL				

Figure 1.4 Physical properties. An ice cube and a piece of lead can be easily differentiated by their physical properties (such as density, color, and melting point). *(Charles D. Winters)*

1.3 PHYSICAL PROPERTIES

You recognize your friends by their physical appearance: their height and weight and the color of their eyes and hair. The same is true of chemical substances. You can tell the difference between an ice cube and a cube of lead of the same size, not only because of their appearance (one is clear and colorless and the other is a shiny metal) but also because one is much heavier (lead) than the other (ice) (Figure 1.4). Properties such as these, which can be observed and measured without changing the composition of a substance, are called **physical properties** [CD-ROM, Screen 1.2]. The chemical elements in Figures 1.1 and 1.2, for example, clearly differ in color, appearance, and their state, that is, whether they are a solid, liquid, or gas. Physical properties allow us to classify and identify substances of the material world. Table 1.1 lists a few physical properties of matter that chemists commonly use.

Exercise 1.2 Physical Properties

Identify as many physical properties in Table 1.1 as you can for the following common substances: (a) iron, (b) water, (c) table salt (whose chemical name is sodium chloride), and (d) oxygen.

Table 1.1 • Some Physical Properties

Property	Using the Property to Distinguish Substances
Color	Is the substance colored or colorless? What is the color and what is its intensity?
State of matter	Is it a solid, liquid, or gas? If it is a solid, what is the shape of the particles?
Melting point	At what temperature does a solid melt?
Boiling point	At what temperature does a liquid boil?
Density	What is its density (mass per unit volume)?
Solubility	What mass of substance can dissolve in a given volume of water or other solvent?
Electric conductivity	Is it a conductor of electricity or an insulator?
Malleability	How easily can a solid be deformed?
Ductility	How easily can a solid be drawn into a wire?
Viscosity	How susceptible is a liquid to flow?

Density

Density, the ratio of the mass of an object to its volume, is a physical property useful for identifying substances [🖱 CD-ROM, Screen 1.8].

$$\text{Density} = \frac{\text{mass}}{\text{volume}} \qquad\qquad \textbf{(1.1)}$$

Your brain unconsciously uses the density of an object you want to pick up by estimating volume visually and preparing your muscles to lift the expected mass. For example, you can readily tell the difference between an ice cube and a cube of lead of identical size (see Figure 1.4). Lead has a high density, $11.35\ \text{g/cm}^3$ (11.35 grams per cubic centimeter), whereas the density of ice is slightly less than $0.917\ \text{g/cm}^3$. An ice cube with a volume of $16.0\ \text{cm}^3$ has a mass of 14.7 g, whereas a lead cube with the same volume has a mass of 180 g.

The density of a substance relates the mass and volume of substance. If any two of three quantities — mass, volume, and density — are known for a sample of matter, the third can be calculated. For example, the mass of an object is the product of its density and volume.

$$\text{Mass (g)} = \text{volume} \times \text{density} = \text{volume (cm}^3) \times \frac{\text{mass (g)}}{\text{volume (cm}^3)}$$

You can use this approach to find the mass of $24\ \text{cm}^3$ [or 24 mL (milliliters)] of mercury in the graduated cylinder in the photo. A handbook of information for chemistry lists the density of mercury as $13.534\ \text{g/cm}^3$ (at 20 °C).

$$\text{Mass (g)} = 24\ \text{cm}^3 \times \frac{13.534\ \text{g}}{1\ \text{cm}^3} = 320\ \text{g}$$

What is the mass of 24 mL of mercury? See text for the answer. *(Charles D. Winters)*

Example 1.1 **Using Density**

Problem • Ethylene glycol, $C_2H_6O_2$, is widely used in automobile antifreeze. It has a density of $1.11\ \text{g/cm}^3$ (or 1.11 g/mL). What is the mass, in grams, of 56 L (56,000 mL = 56,000 cm³) of ethylene glycol?

Strategy • You know the density and volume of the sample. Because density is the ratio of the mass of a sample to its volume, then mass = volume × density. Therefore, to find the sample's mass, multiply the volume by the density.

Solution •

$$56{,}000\ \text{cm}^3 \times \frac{1.11\ \text{g}}{1\ \text{cm}^3} = 62{,}000\ \text{g}$$

Comment • Notice that units of cm³ cancel in the calculation. See the tutorials on Screen 1.8 of the *General Chemistry Interactive CD-ROM*, Version 3.0.

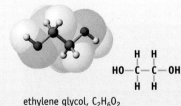

ethylene glycol, $C_2H_6O_2$
density = 1.11 g/cm³ (or 1.11 g/mL)

Exercise 1.3 **Density**

The density of dry air is $1.18 \times 10^{-3}\ \text{g/cm}^3$ (= 0.00118 g/cm³; see Appendix A on using scientific notation). What volume of air, in cubic centimeters, has a mass of 15.5 g?

Temperature

● **Heat Transfer and Temperature**
As is described in Chapter 6, heat energy transfers spontaneously only from a hotter object to a cooler one, that is, from an object at a higher temperature to one at a lower temperature.

● **Fahrenheit Scale**
Gabriel Fahrenheit (1686–1736), a German physicist, defined 0 °F as the freezing point of a solution with the maximum amount of salt in water (because this was the lowest temperature he could reproduce reliably), and he intended 100 °F to be the normal human body temperature (but this turned out to be 98.6 °F). Today, the reference points are 32 °F and 212 °F, the freezing and boiling points of pure water. For more details see J. Pellicer, M. A. Gilabert, and E. Lopez-Baeza: *Journal of Chemical Education.* Vol. 76, pp. 911–912, 1999.

History

Lord Kelvin

William Thomson (1824–1907), who was known as Lord Kelvin, was a professor of natural philosophy at the University in Glasgow, Scotland, from 1846 to 1899. He was best known for his studies on heat and work, from which came the concept of the absolute temperature scale. *(E. F. Smith Collection/Van Pelt Library/University of Pennsylvania)*   ●

Another useful physical property of pure elements and compounds is the temperature at which a solid melts (its melting point) or a liquid boils (its boiling point). **Temperature** is the property of matter that determines whether heat energy can be transferred from one body to another and the direction of that transfer [CD-ROM, Screen 1.10].

Three temperature measurement scales are commonly used: the Fahrenheit, Celsius, and Kelvin scales (Figure 1.5). The Celsius scale is generally used for measurements in the laboratory. When calculations incorporate temperature data, however, the Kelvin scale must be used.

The Celsius Temperature Scale

In the United States everyday temperatures are reported using the Fahrenheit scale, but the **Celsius scale** is used in most other countries and in scientific notation. The latter scale was first suggested by Anders Celsius (1701–1744), a Swedish astronomer, and based on the properties of water. The size of the Celsius degree is defined by assigning zero as the freezing point of pure water (0 °C) and 100 as its boiling point (100 °C).

The Kelvin Temperature Scale

Winter temperatures in many places can easily drop to negative values on the Celsius scale. In the laboratory, much colder temperatures can be achieved easily, and the temperatures are given by even larger negative numbers. There is a limit to how low the temperature can go, however, and many experiments have found that the limiting temperature is −273.15 °C (or −459.67 °F).

William Thomson, known as Lord Kelvin, first suggested a temperature scale that does not use negative numbers. The **Kelvin scale,** now adopted as the international standard for science, uses the same size unit as the Celsius scale, but it takes the lowest possible temperature as its zero, a point called **absolute zero.** Because kelvin units and Celsius degrees are the same size, the freezing point of water is reached 273.15 K or °C above the starting point; that is, 0 °C = 273.15 K. The boiling point of pure water is 373.15 K. Temperatures in Celsius degrees are

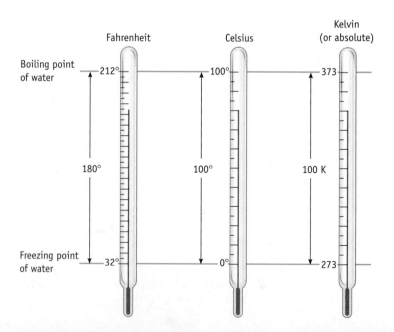

Figure 1.5 **A comparison of Fahrenheit, Celsius, and Kelvin temperature scales.** The reference, or starting point, for the Kelvin scale is absolute zero (0 K = −273.15 °C), which has been shown theoretically and experimentally to be the lowest possible temperature.

readily converted to kelvins, and vice versa, using the relation

$$T(\text{K}) = \frac{1 \text{ K}}{1 \text{ °C}}(T\text{ °C} + 273.15 \text{ °C}) \tag{1.2}$$

Thus, a common room temperature of 23.5 °C is

$$T(\text{K}) = \frac{1 \text{ K}}{1 \text{ °C}}(23.5 \text{ °C} + 273.15 \text{ °C}) = 296.7 \text{ °C}$$

Finally, notice that the degree symbol (°) is not used with Kelvin temperatures. The name of the unit on this scale is the kelvin (not capitalized), and such temperatures are designated with a capital K.

● **Temperature Conversions**
When converting 23.5 °C to kelvins, adding the two numbers gives 296.65. However, the rules of "significant figures" tell us that the sum or difference of two numbers can have no more decimal places than the number with the fewest decimal places. (See page 33.) Thus, we round the answer to 296.7 K.

Exercise 1.4 Temperature Changes

Liquid nitrogen boils at 77 K. What is this temperature in Celsius degrees?

Temperature Dependence of Physical Properties

The temperature of a sample of matter often affects the numerical values of its properties. Density is a particularly important example. Although the change in water density with temperature seems small, it affects our environment profoundly. For example, as the water in a lake cools, the density of the water increases, and the denser water sinks (Figure 1.6a). This continues until the water reaches 3.98 °C, the temperature at which it has its maximum density (0.999973 g/cm³). If the water temperature drops further, the density decreases slightly, and the colder water floats on top of water at 3.98 °C.

If water is cooled to 0 °C, solid ice forms. Ice is much less dense than water, so ice floats on water.

Because the density of liquids changes with temperature, it is necessary to report the temperature when you make accurate volume measurements. Laboratory glassware used to make such measurements always specifies the temperature at which they were calibrated (Figure 1.6b).

Temperature Dependence of Water Density

Temperature (°C)	Density of Water (g/cm³)
0 (ice)	0.917
0 (liquid water)	0.99984
2	0.99994
4	0.99997
10	0.99970
25	0.99707
100	0.95836

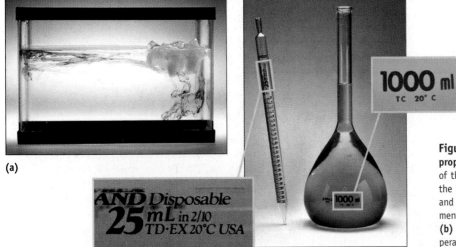

(a)

(b)

Figure 1.6 Temperature dependence of physical properties. (a) Ice cubes were placed in the right side of the tank and blue dye at the left. The water beneath the ice is cooler and denser than the surrounding water and sinks. The convection current created by this movement of water is traced by the dye movement.
(b) Laboratory glassware is calibrated for specific temperatures. It will contain the specified volume only if the temperature is as marked. *(Charles D. Winters)*

Exercise 1.5 **Density and Temperature**

(a) The density of air at 0 °C (and 1 atmosphere [atm] pressure) is 1.293×10^{-3} g/cm^3. What is the density of air at this temperature in grams per liter?

(b) The density of mercury at 0 °C is 13.595 g/cm^3, at 10 °C it is 13.570 g/cm^3 and at 20 °C it is 13.546 g/cm^3. Estimate the density of mercury at 30 °C.

Extensive and Intensive Properties

Extensive properties depend on the amount of a substance present. Thus, the mass and volume of each sample of an element in Figures 1.1 and 1.2 are extensive properties. In contrast, **intensive properties** are those that do not depend on the amount of substance. A sample of ice will melt at 0 °C, no matter whether you have an ice cube or an iceberg.

Density is an intensive property, which seems curious at first because it is the quotient of two extensive properties. On reflection, however, you recognize that the density of gold, for example (19.3 g/cm^3), does not depend on the size of the sample. Whether you have a flake of pure gold or a solid gold ring, both have the same ratio of mass to volume.

Chapter Goals • Revisited

- Recognize elements, atoms, compounds, and molecules.
- **Identify physical and chemical properties and changes.**
- Apply the kinetic-molecular theory to the properties of matter.
- Use metric units and significant figures properly.

1.4 PHYSICAL AND CHEMICAL CHANGE

Changes in physical properties are called **physical changes.** In a physical change the identity of a substance is preserved even though it may have changed its physical state or the gross size and shape of its pieces. An example of a physical change is the melting of a solid, and the temperature at which this occurs (the melting point) is often so characteristic that it can be used to identify the solid (Figure 1.7).

A physical property of hydrogen gas (H_2) is its low density, so a balloon filled with H_2 floats in air (Figure 1.8). Suppose a lighted candle is brought up to the balloon. When the heat causes the skin of the balloon to rupture, the hydrogen combines with the oxygen (O_2) in the air, and the heat of the candle sets off an explosion (Figure 1.8), producing water, H_2O [CD-ROM, Screen 1.11]. The explosion is an example of a **chemical change,** or **chemical reaction,** because one or more substances (the **reactants**) has been transformed into one or more different substances (the **products**).

The reaction of H_2 with O_2 is an example of a chemical property of hydrogen. A **chemical property** involves a change in the identity of a substance. Here the H

Problem-Solving Tip 1.1

Finding Data

All the information you may need to solve a problem in this book may not be presented in the problem. For example, we could have left out the value of the density in Example 1.1 and assumed you would (a) recognize that you needed density to convert a volume to a mass and (b) know where to find the information. The Appendices of this book contain a wealth of information, and even more is available on the *General Chemistry*

Interactive CD-ROM, Version 3.0. Various handbooks of information are available in most libraries, and among the best are the *Handbook of Chemistry and Physics* (CRC Press) and *Lange's Handbook of Chemistry* (McGraw-Hill). Perhaps the most accurate source of data is produced by the National Institutes for Standards and Technology (http://www.nist.gov). See also the World Wide Web site Webelements (www.webelements.com).

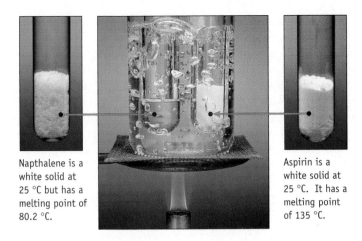

Figure 1.7   A physical property used to distinguish compounds. Aspirin and naphthalene are both white solids at 25 °C. You can tell them apart by, among other things, a difference in physical properties. At the temperature of boiling water, 100 °C, naphthalene is a liquid *(left)*, whereas aspirin is a solid *(right)*.   *(Charles D. Winters)*

Napthalene is a white solid at 25 °C but has a melting point of 80.2 °C.

Aspirin is a white solid at 25 °C. It has a melting point of 135 °C.

atoms of the gaseous H_2 molecules have become incorporated into H_2O. Similarly, a chemical change occurs when gasoline burns in air in an automobile engine or an old car rusts in the air. Burning of gasoline or rusting of iron are characteristic chemical properties of these substances.

At the particulate level a chemical change produces a new arrangement of atoms without a gain or loss in the number of atoms of each kind [➡ CHAPTER 3] [🖫 CD-ROM, Screen 1.12]. The particles (atoms, molecules, or ions) present after the reaction, however, are different from those present before the reaction. The reaction of hydrogen and oxygen molecules to form water molecules can be represented as

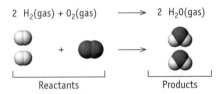

$$2\ H_2(gas) + O_2(gas) \longrightarrow 2\ H_2O(gas)$$

Reactants   Products

The representation of the change with chemical formulas is called a **chemical equation.** It shows that the substances on the left (reactants) produce, or "give," the substances on the right (products). This equation shows that if there are four atoms

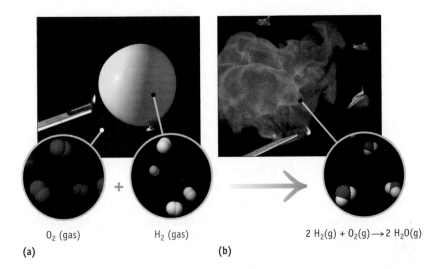

O_2 (gas) H_2 (gas) $2\ H_2(g) + O_2(g) \longrightarrow 2\ H_2O(g)$

(a)   **(b)**

Figure 1.8   The explosion of a hydrogen/oxygen mixture. (a) A balloon filled with molecules of hydrogen gas, H_2, and surrounded by molecules of oxygen, O_2, in the air. (The balloon floats in air because gaseous H_2 is less dense than air.) **(b)** When H_2 and O_2 mix, and the mixture is ignited, it explodes, producing water, H_2O. *(Photos [Charles D. Winters] are from a video on Screen 1.11 of the* General Chemistry Interactive CD-ROM, *Version 3.0.)*

A pot of water has been put on a camp fire. What chemical and physical changes are occurring here? *(Charles D. Winters)*

of H and two atoms of O before reaction, the same number must be present after reaction, albeit in a different arrangement.

In contrast to a chemical change, a physical change does not result in a new chemical substance. The substances (atoms, molecules, or ions) present before and after the change are the same, but their arrangement relative to one another (farther apart in a gas, closer together in a solid, for example) is different.

Finally, physical changes and chemical changes are often accompanied by a transfer of energy. The reaction of hydrogen and oxygen to give water (see Figure 1.8) transfers a tremendous amount of energy (in the form of heat and light) to its surroundings [➡ CHAPTER 6].

Exercise 1.6 **Chemical Reactions and Physical Changes**

When camping in the mountains, you boil a pot of water on a campfire. What physical and chemical changes take place in this process?

1.5 CLASSIFYING MATTER

A chemist looks at a glass of drinking water and sees a liquid. The liquid could be the chemical compound water. More likely, the liquid is a homogeneous mixture of water and dissolved substances, that is, a solution. Or, it is possible our water sample is a heterogeneous mixture, with solids suspended in the liquid. These represent some of the ways we can classify matter (see Chapter Focus).

States of Matter and the Kinetic-Molecular Theory

An easily observed property of matter is its **state,** that is, whether a substance is a solid, liquid, or gas (see Figures 1.1, 1.2, 1.7; Chapter Focus) [🖱 CD-ROM, Screen 1.3]. You recognize a solid because it has a rigid shape and a fixed volume that changes little as temperature and pressure change. Like solids, liquids have a fixed volume, but a liquid is fluid — it takes on the shape of its container and has no definite shape of its own. Gases are fluid also, but the volume of a gas is the volume of the container. The volume of a given amount of gas varies with temperature and pressure.

At low temperatures, virtually all matter is found in the solid state. As the temperature is raised, though, solids usually melt to form liquids. Eventually, if the temperature is raised high enough, liquids evaporate to form gases. Volume changes accompany changes in state. For a given mass of material, there is usually a small increase in volume on melting — water being a significant exception — and then a large increase in volume on evaporation.

The **kinetic-molecular theory** of matter helps us interpret the properties of solids, liquids, and gases (see Chapter Focus) [🖱 CD-ROM, Screen 1.7]. According to this theory, all matter consists of extremely tiny particles (atoms, molecules, or ions), which are in constant motion. In solids these particles are packed closely together, usually in a regular array. The particles vibrate back and forth about their average positions, but seldom does a particle in a solid squeeze past its immediate neighbors to come into contact with a new set of particles.

The atoms or molecules of liquids or gases are arranged randomly rather than in the regular patterns found in solids. Liquids and gases are fluid because the particles are not confined to specific locations and can move past one another. Under normal conditions, the particles in a gas are far apart.

A volume change with change of state. When water freezes, it is always accompanied by an increase in volume. (The density of ice is less than that of liquid water, so, for a given mass of water, the volume increases on freezing.) Here milk, which is mostly water, has frozen and the expansion in volume is so great that the milk pushed its way out of the bottle.

Gas molecules move extremely rapidly because they are not constrained by their neighbors. The molecules of a gas fly about, colliding with one another and with the container walls. This random motion allows gas molecules to fill their container, so the volume of the gas sample is the volume of the container.

Another important aspect of the kinetic-molecular theory is that the higher the temperature the faster the particles move. The particles' energy of motion (their **kinetic energy**) acts to overcome the forces of attraction between the particles. A solid melts to form a liquid when the temperature of the solid is raised to the point at which the particles vibrate fast enough and far enough to push one another out of the way and move out of their regularly spaced positions. As the temperature increases even more, the particles move even faster until finally they can escape the clutches of their comrades and enter the gaseous state. Increasing temperature corresponds to faster and faster motions of atoms and molecules, a general rule you will find useful in many future discussions [CD-ROM, Screens 1.10 and 6.10].

Matter at the Macroscopic and Particulate Levels

The characteristic properties of gases, liquids, and solids just described are observed by the unaided human senses. They are determined using samples of matter large enough to be seen, measured, and handled. Using such samples, we can also determine, for example, the color of a substance, whether it dissolves in water, or whether it conducts electricity or reacts with oxygen. Observations and manipulation generally take place in the **macroscopic** world of chemistry (see Chapter Focus)(Figure 1.9). It is the world of experiments and observations.

Now let us move to the level of atoms, molecules, and ions, a world of chemistry we cannot see. Take a macroscopic sample of material and divide it, again and again, past the point where the amount of sample can be seen by the naked eye, past the point where it can be seen using an optical microscope. Eventually you reach that level of individual particles that make up all of matter, a level that chemists refer to as the **submicroscopic** or **particulate** world of atoms and molecules (Figure 1.9) [CD-ROM, Screen 1.4].

Chemists are interested in the structure of matter at the particulate level. Atoms, molecules, and ions cannot be "seen" in the same way that one views the macroscopic world, but they are no less real to chemists. Chemists imagine what

Chapter Goals • Revisited

- Recognize elements, atoms, compounds, and molecules.
- Identify physical and chemical properties and changes.
- **Apply the kinetic-molecular theory to the properties of matter.**
- Use metric units and significant figures properly.

- **Solids, Liquids, and Gases**
An animation on Screen 1.7 of the *General Chemistry Interactive CD-ROM*, Version 3.0, shows the basic differences between solid, liquid, and gas in terms of the kinetic molecular theory.

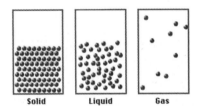

Solid   Liquid   Gas

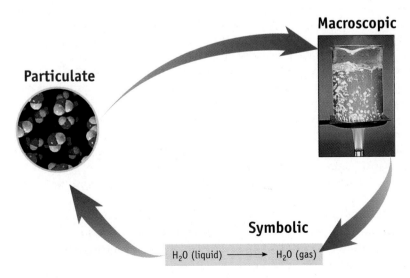

Particulate

Macroscopic

Symbolic

H_2O (liquid) ⟶ H_2O (gas)

Figure 1.9   Chemistry, a matrix of concepts. We observe chemical processes on the macroscopic scale and then write symbols to represent those observations. To understand those processes we try to view or imagine what has occurred at the particulate — atomic, ionic, and molecular — level. See also the Chapter Focus. *(Charles D. Winters)*

A Closer Look

Symbolism in Chemistry

So that a piece of music can be played by others, the composer expresses the music in symbols — musical notes — that are understood by all musicians. In the same way, chemists convey observations to one another using symbols as a shorthand. For example, we could indicate the evaporation of water by

$$H_2O(liquid) \longrightarrow H_2O(gas)$$

Here H and O are the symbols for the chemical elements hydrogen and oxygen, respec-

tively, and water molecules are composed of H and O atoms in a 2-to-1 ratio. The arrow connecting H_2O(liquid) and H_2O(gas) means that liquid water changes to water vapor.

Symbols are widely used for convenience. For example, the symbol mph stands for "miles per hour," or a rate of progress. Changing m to k in mph (to obtain kph, or "kilometers per hour") leads to a different value. A speed of 50 mph is about 82 kph.

The symbols you use in chemistry also have precise meanings and must be used carefully. Changing the H to K in H_2O (to get K_2O), for example, produces the formula for a *very* different substance.

atoms must look like and how they might fit together to form molecules. They create models to represent atoms and molecules (see Figures 1.3 and 1.9) and use these models to think about chemistry and to explain the observations they have made about the macroscopic world.

It has been said that chemists carry out experiments at the macroscopic level, but they think about chemistry at the particulate level. They then write down their observations as "symbols," the letters that signify the elements and compounds involved. This is a useful perspective that will help you as you study chemistry. Indeed, one of our goals is to help you make these connections (as in Figure 1.9).

Pure Substances

● **Thinking About Matter**
How can you decide if the glass contains pure water? What are the physical and chemical properties of pure water?

Let us think again about a glass of drinking water. How would you tell whether the water is pure (a single substance) or is a mixture of substances? Begin by making a few simple observations. Is solid material floating in the liquid? Does the liquid have an odor or unexpected taste or color?

Every pure substance has two features [CD-ROM, Screen 1.13]. First, it has a set of unique properties by which it can be recognized. Pure water, for example, is colorless, odorless, and certainly does not contain suspended solids. If you wanted to identify a substance conclusively as water, you would have to examine its properties carefully and compare them against the known properties of pure water. Melting point and boiling point serve the purpose well here. If you could show that the substance melts at 0 °C and boils at 100 °C at atmospheric pressure, you can be certain it is water. No other substance melts and boils at precisely these temperatures.

The second feature of a pure substance is that it cannot be separated into two or more different pure substances by any physical technique such as heating in a Bunsen flame (see Section 1.4 and Chapter Focus). If this were true, our sample would be classified as a mixture.

Mixtures: Homogeneous and Heterogeneous

A cup of noodle soup is obviously a mixture of solids and liquids (Figure 1.10). A mixture in which the uneven texture of the material can be detected is called a **heterogeneous mixture**. Heterogeneous mixtures may appear completely uniform but on closer examination are not. Blood, for example, may not look heterogeneous until you examine it under a microscope and red and white blood cells are revealed (Figure 1.10b). Milk appears smooth in texture to the unaided eye, but magnification would reveal fat and protein globules within the liquid. In a heterogeneous mixture the properties in one region are different from those in another region, and the properties differ from sample to sample.

A **homogeneous mixture** is completely uniform at the particulate level and consists of one or more substances in the same phase (Figure 1.10c). No amount of optical magnification reveals a homogeneous mixture to have different properties in one region than in another.

The composition of a homogeneous mixture is the same everywhere in the sample. Such mixtures are often called **solutions**, and common examples include air (mostly a mixture of nitrogen and oxygen gases), gasoline (a mixture of carbon- and hydrogen-containing compounds called hydrocarbons), or an unopened soft drink.

When a mixture is separated into its pure components, the components are said to be **purified** (Figure 1.11). Efforts at separation are often not complete in a single step, however, and repetition almost always will give an increasingly pure substance. For example, soil particles can be separated from water by filtration (see Figure 1.11a) [🖰 CD-ROM, Screen 1.14]. When the mixture is passed through a filter, many of the particles are removed. Repeated filtrations will lead to water in a higher and higher state of purity. This purification process uses a property of the mixture, its clarity, as one measure of the extent of purification. When a perfectly clear sample of water is obtained, all the soil particles are assumed to have been removed.

> ● **Compounds and Mixtures of Elements — a Difference?**
> Recall that there is a difference between a mixture of elements and a compound formed from those elements (page 13). A mixture, say of copper and sulfur, can have any composition. A compound formed by these elements, however, has one Cu atom for every S atom (CuS).

(a)   **(b)**   **(c)**

Figure 1.10   Mixtures. (a) A cup of noodle soup is a heterogeneous mixture. **(b)** A sample of blood may look homogeneous, but examination with an optical microscope shows it is in fact a heterogeneous mixture of liquids and suspended particles (blood cells). **(c)** A homogeneous solution, here of salt in water. (The model shows that salt consists of separate particles (ions) in water.) The particles cannot be seen with an optical microscope. *(a, c, Charles D. Winters; b, Dec Breger/Science Source/Photo Researchers, Inc.)*

(b)

(a)

Figure 1.11   Purifying water by filtration. (a) A laboratory setup. A beaker full of muddy water is passed through a paper filter, and the mud and dirt are removed. **(b)** A water treatment plant uses filtration to remove suspended particles from the water. *(a, Charles D. Winters; b, Littleton, Massachusetts, Spectacle Pond Iron and Manganese Treatment Facility)*

Homogeneous and heterogeneous mixtures. Which is homogeneous? See Exercise 1.7. *(Charles D. Winters)*

Exercise 1.7   **Mixtures and Pure Substances**

The photo in the margin shows mixtures in two beakers. Which one is homogeneous and which is heterogeneous? Which one is a solution?

1.6   UNITS OF MEASUREMENT

Doing chemistry requires observing chemical reactions and physical changes. Suppose you mix two solutions in the laboratory and see a golden yellow solid form, and, because the solid is denser than water, the solid drops to the bottom of the test tube. The color and appearance of the substances, if heat was involved, or if the process occurred quickly or slowly, are **qualitative** observations. No measurements and numbers were involved.

To understand a chemical reaction more completely, chemists usually make **quantitative** observations. These involve numerical information. For example, if two compounds react with each other, how much product forms? Is heat evolved in the process and if so how much?

In chemistry, quantitative measurements of time, mass, volume, and distance, among other things, are common. On page 26 you can read about one of the fastest growing areas of science, nanotechnology, which is the creation and study of matter on the nanometer scale. A nanometer (nm), equivalent to 1×10^{-9} m (meter), is a common dimension in chemistry and biology. For example, a typical molecule is only about 1 nm across and a bacterium is about 1000 nm in length.

The scientific community has chosen a modified version of the **metric system** as the standard system for recording and reporting measurements. [CD-ROM, Screens

---

Chapter Goals • Revisited

- Recognize elements, atoms, compounds, and molecules.
- Identify physical and chemical properties and changes.
- Apply the kinetic-molecular theory to the properties of matter.
- **Use metric units and significant figures properly.**

Table 1.2 • Some SI Base Units

Measured Property	Name of Unit	Abbreviation
Mass	kilogram	kg
Length	meter	m
Time	second	s
Temperature	kelvin	K
Amount of substance	mole	mol
Electric current	ampere	A

Qualitative and quantitative. A new substance is formed by mixing two known substancs. *Qualitative observations:* yellow, fluffy solid. *Quantitative observations:* mass of solid formed. *(Charles D. Winters)*

1.15 and 1.16]. This decimal system, used internationally in science, is called the Système International d'Unités (International System of Units), abbreviated SI.

All SI units are derived from base units, some of which are listed in Table 1.2. Larger and smaller quantities are expressed by using appropriate prefixes with the base unit (Table 1.3). For instance, highway distances are given in kilometers, in which 1 kilometer (km) is exactly 1000 m (1×10^3 m). In chemistry, length is often given in subdivisions of the meter: centimeters (cm) or millimeters (mm). The prefix "centi-" means 1/100, so 1 centimeter is 1/100 of a meter (1 cm = 1×10^{-2} m); 1 millimeter is 1/1000 of a meter (1 mm = 1×10^{-3} m). On the atomic scale, dimensions are often given in nanometers (1 nm = 1×10^{-9} m) or picometers (1 pm = 1×10^{-12} m).

1.7 USING NUMERICAL INFORMATION

As part of your course in chemistry your preparation of materials in the laboratory will require you to make some calculations [CD-ROM, Screen 1.17]. You will collect numerical data and use those data to calculate a result, or you will look for correlations among pieces of data. This section describes some common calculations and proper ways to handle quantitative information.

Length

Suppose you want to find the density of a rectangular piece of aluminum in units of grams per cubic centimeter (g/cm^3). Because density is the ratio of mass to

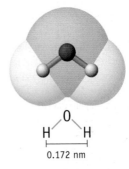

The distance between the two H atoms in a water molecule is 0.172 nm or 172 pm.

Table 1.3 • Selected Prefixes Used in the Metric System

Prefix	Abbreviation	Meaning	Example
mega-	M	10^6	1 megaton = 1×10^6 tons
kilo-	k	10^3	1 kilometer (km) = 1×10^3 m
deci-	d	10^{-1}	1 decimeter (dm) = 1×10^{-1} m
centi-	c	10^{-2}	1 centimeter (cm) = 1×10^{-2} m
milli-	m	10^{-3}	1 millimeter (mm) = 1×10^{-3} m
micro-	μ	10^{-6}	1 micrometer (μm) = 1×10^{-6} m
nano-	n	10^{-9}	1 nanometer (nm) = 1×10^{-9} m
pico-	p	10^{-12}	1 picometer (pm) = 1×10^{-12} m

● **Common Conversion Factors**

1 kg = 1000 g
1×10^9 nm = 1 m
1×10^{12} pm = 1 m
10 dm = 1 m
100 cm = 1 m
1000 m = km

Chemical Perspectives

It's a Nanoworld!

A nanometer is a billionth of a meter, a dimension in the realm of atoms and molecules — eight oxygen atoms in a row span a distance of about 1 nm. Nanotechnology is one of the hottest fields of science today because materials having those dimensions — nanomaterials — can have unique properties.

Professor Alex Zettl of University of California-Berkeley, holding a model of a carbon nanotube. *(Lawrence Berkeley Laboratory)*

Carbon nanotubes are excellent examples of nanomaterials. These are lattices of carbon atoms forming the walls of tubes having diameters of a few nanometers. Carbon nanotubes are at least 100 times stronger than steel, but only one-sixth as dense. In addition, they conduct heat and electricity far better than copper. So, carbon nanotubes could be used in tiny, physically strong, conducting devices. Recently carbon nanotubes have been filled with potassium atoms, making them even better electric conductors. And

even more recently molecular-sized bearings have been made by sliding one nanotube inside another.

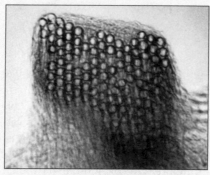

A bundle of carbon nanotubes. Each tube has a diameter of 1.4 nm, and the bundle is 10–20 nm thick. *(P. Nikolaev, Rice University, Center for Nanoscale Science and Technology)*

Nanomaterials are by no means new. For the last century tire companies have reinforced tires by adding nanosized particles called carbon black to rubber.

Atomic force microscopy (AFM) is becoming an important tool in chemistry and physics. A tiny probe, often a whisker of a carbon nanotube, moves over the surface of a substance and interacts with individual molecules. Mapping the surface of a silicon surface provided the image shown at the top of the next column.

A goal of nanotechnologists is to build machines or instruments at the nanometer level. As a demonstration, scientists at IBM have built a device with

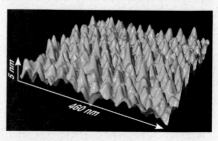

An AFM image of "nanobumps" on a silicon surface. The average spacing between nanobumps is 38 nm, or about 160 silicon atoms. The average nanobump width is 25 nm or 100 silicon atoms. *(Melissa A. Hines/Cornell University)*

eight cantilevers, each an AFM probe. Each probe is coated with a short chain of DNA. When these probes pass over the surface of a a DNA sample, interactions occur. Researchers hope to use this to identify minute differences in DNA samples, which can be important in detecting certain diseases or effecting treatments.

A "cantilever" device with 8 AFM probes. Each probe is 500 μm long. *(IBM Zurich Research Laboratory)*

---

volume, you need to measure the mass and determine the volume of the piece. The data in the margin were collected in the laboratory. To find the volume of the aluminum sample in cubic centimeters, you multiply its length by its width and its thickness. First, however, all the measurements must have the same units, meaning that the thickness must be converted to centimeters. Recognizing that there are 10 mm in 1 cm, the thickness is 0.31 cm.

$$3.1 \text{ mm} \times \frac{1 \text{ cm}}{10 \text{ mm}} = 0.31 \text{ cm}$$

With all the dimensions in the same unit, the volume and then the density can be calculated:

● Conversion Factors
Conversion factors for SI units are given in Appendix C and inside the back cover of the book.

$$\text{Length} \times \text{width} \times \text{thickness} = \text{volume}$$

$$6.45 \text{ cm} \times 2.50 \text{ cm} \times 0.31 \text{ cm} = 5.0 \text{ cm}^3$$

$$\text{Density} = \frac{13.56 \text{ g}}{5.0 \text{ cm}^3} = 2.7 \text{ g/cm}^3$$

● **Determining the Density of Aluminum**

Measurement	*Data Collected*
Mass of aluminum	13.56 g
Length	6.45 cm
Width	2.50 cm
Thickness	3.1 mm

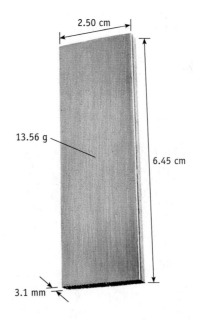

You will find that **dimensional analysis,** an approach explored further in Appendix A and used throughout the book, is useful in solving mathematical problems. This approach was used in a previous equation to change 3.1 mm to its equivalent in centimeters. We multiplied the number we wished to convert (3.1 mm) by a factor called a **conversion factor** (here 1 cm/10 mm) to produce the result with the desired unit (0.31 cm). Units are handled like numbers, and because the unit "mm" was in both the numerator and the denominator, dividing one by the other leaves a quotient of 1. The units are said to cancel out. Here this leaves the answer in centimeters, the desired unit.

A conversion factor expresses the equivalence of a measurement in two different units (1 cm ≡ 10 mm; 1 g ≡ 1000 mg; 12 eggs ≡ 1 dozen). Because the numerator and denominator describe the same quantity, the factor is equivalent to the number 1. Therefore, multiplication by this factor does not change the measured quantity, only its units. A conversion factor is always written so that it has the form "new units divided by units of original number."

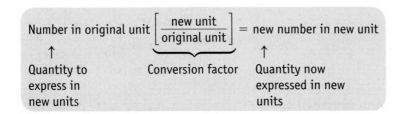

Number in original unit $\left[\dfrac{\text{new unit}}{\text{original unit}} \right]$ = new number in new unit

↑   ↑
Quantity to   Conversion factor   Quantity now
express in   expressed in new
new units   units

Example 1.2 Distances on the Molecular Scale

Problem • The distance between the O atom and an H atom in a water molecule is 98.5 pm. What is this distance in meters? In nanometers (nm)?

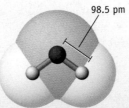

98.5 pm

Strategy • You can solve this problem by knowing the conversion factor between the units in the information you are given (picometers) and the units of the desired outcome (meters or nanometers). There is no direct conversion from nanometers to picometers given in Table 1.3, but relationships are listed between meters and picometers and between meters and nanometers (see

Table 1.3). Therefore, we first convert picometers to meters and then convert meters to nanometers.

$$\text{Picometers} \xrightarrow{\times \frac{m}{pm}} \text{Meters} \xrightarrow{\times \frac{nm}{m}} \text{Nanometers}$$

Solution • Using the appropriate conversion factors (1 pm = 1×10^{-12} m and 1 nm = 1×10^{-9} m), we have

$$98.5 \text{ pm} \cdot \frac{1 \times 10^{-12} \text{ m}}{1 \text{ pm}} = 9.85 \times 10^{-11} \text{ m}$$

$$9.85 \times 10^{-11} \text{ m} \cdot \frac{1 \text{ nm}}{1 \times 10^{-9} \text{ m}} = 9.85 \times 10^{-2} \text{ nm}$$

or 0.0985 nm

Comment • Notice how units cancel to leave an answer whose unit is that of the numerator of the conversion factor.

Exercise 1.8 Interconverting Units of Length

The pages of a typical textbook are 25.3 cm long and 21.6 cm wide. What is each length in meters? In millimeters? What is the area of a page in square centimeters? In square meters?

Exercise 1.9 **Using Density**

A platinum sheet is 2.50 cm square and has a mass of 1.656 g. The density of platinum is 21.45 g/cm³. What is the thickness of the platinum sheet, in millimeters?

● **Choosing Volume Units**
If you use cubic meters for volume measurements, a common laboratory beaker would have a volume of 0.0006 m³ and you would often work with volumes of chemicals in the range of 0.001 m³ or less in the laboratory.

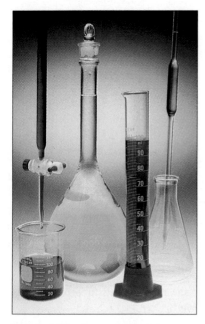

Figure 1.12   Some common laboratory glassware. Volumes are marked in units of milliliters (mL). Remember that 1 mL is equivalent to 1 cm³. *(Charles D. Winters)*

Volume

Chemists often handle chemicals in glassware such as beakers, flasks, pipets, graduated cylinders, and burets, which are marked in volume units (Figure 1.12). The SI unit of volume is the cubic meter (m³), which is too large for everyday laboratory use. Therefore, chemists usually use the **liter,** symbolized by **L.** A cube with sides equal to 10 cm (0.1 m) has a volume of 10 cm × 10 cm × 10 cm = 1000 cm³ (or 0.001 m³). This is defined as 1 liter.

$$1 \text{ liter (L)} = 1000 \text{ mL} = 1000 \text{ cm}^3$$

The liter is a convenient unit to use in the laboratory, as is the milliliter (mL). Because there are exactly 1000 mL and 1000 cm³ in a liter, this means that

$$1 \text{ cm}^3 = 0.001 \text{ L} = 1 \text{ milliliter (1 mL)}$$

The units *milliliter and cubic centimeter* (or "cc") *are interchangeable.* Therefore, a flask that contains exactly 125 mL has a volume of 125 cm³.

$$125 \text{ cm}^3 \cdot \frac{1 \text{ L}}{1000 \text{ cm}^3} = 0.125 \text{ L}$$

Although not widely used in the United States, the cubic decimeter (dm³) is a common unit in the rest of the world. A length of 10 cm is called a decimeter (dm) because it is 1/10 of a meter. Because a cube 10 cm on a side defines a volume of one liter, *a liter is equivalent to a cubic decimeter:* 1 L = 1 dm³. Products in Europe and other parts of the world are often sold by the cubic decimeter.

The *deciliter (dL),* which is exactly equivalent to 0.100 L or 100 mL, is widely used in medicine. For example, standards for amounts of environmental contaminants are often set as a certain mass per deciliter. The state of Massachusetts recommends that children with more than 10 micrograms (10 × 10⁻⁶ g) of lead per deciliter of blood be tested further for lead poisoning.

Example 1.3 **Units of Volume**

Problem ● A laboratory beaker has a volume of 0.6 L. What is its volume in cubic centimeters (cm³), milliliters (mL), and in deciliters (dL)?

Strategy ● The relation between liters and cubic centimeters is 1 L = 1000 cm³. Therefore, you should multiply 0.6 L by the conversion factor (1000 cm³/L). The units of L cancel to leave an answer with units of cm³. The deciliter is equivalent to 0.100 L or 100 mL.

Solution ●

$$0.6 \text{ L} \cdot \frac{1000 \text{ cm}^3}{1 \text{ L}} = 600 \text{ cm}^3$$

Because cubic centimeters and milliliters are equivalent, we can also say the volume of the beaker is 600 mL. In deciliters, the volume is

$$600 \text{ mL} \cdot \frac{1 \text{ dL}}{100 \text{ mL}} = 6 \text{ dL}$$

Exercise 1.10   Volume

(a) A standard wine bottle has a volume of 750 mL. How many liters does this represent? How many deciliters?

(b) One U.S. gallon is equivalent to 3.7865 L. How many liters are in a 2.0-qt carton of milk? (Remember there are 4 qt in a gallon.) How many cubic decimeters?

Mass

To determine the density of a piece of aluminum (pages 26–27), we needed to know its mass. The mass of a body is the fundamental measure of the quantity of matter in that body, and the SI unit of mass is the kilogram (kg). Smaller masses are expressed in grams (g) or milligrams (mg) (see Table 1.3).

● **Micrograms**
Very small masses are often given in micrograms. A microgram (μm) is 1/1000 of a milligram or a millionth of a gram.

$$1 \text{ kg} = 1000 \text{ g}$$
$$1 \text{ g} = 1000 \text{ mg}$$

Example 1.4   Mass in Kilograms, Grams, and Milligrams

Problem • A new U.S. quarter has a mass of 5.59 g. Express this mass in kilograms and milligrams.

Strategy • Here the relation between the unit of the desired answer and the unit of the information given is 1 kg = 1000 g and 1000 mg = 1 g. Multiply the mass in grams by a conversion factor that has the form "(units for answer/units of information given)."

Solution •

$$5.59 \text{ g} \cdot \frac{1 \text{ kg}}{1000 \text{ g}} = 5.59 \times 10^{-3} \text{ kg}$$

$$5.59 \text{ g} \cdot \frac{1000 \text{ mg}}{1 \text{ g}} = 5.59 \times 10^{3} \text{ mg}$$

Example 1.5   Density in Different Units

Problem • Oceanographers often express the density of sea water in units of kilograms per cubic meter. If the density of sea water is 1.025 g/cm^3 at 15 °C, what is its density in kilograms per cubic meter?

Strategy • To simplify this problem, break it into three steps. First change grams to kilograms and then convert cubic centimeters to cubic meters. Finally, the density is calculated by dividing the mass in kilograms by the volume in cubic meters.

Solution • First convert the mass in grams to kilograms.

$$1.025 \text{ g} \cdot \frac{1 \text{ kg}}{1000 \text{ g}} = 1.025 \times 10^{-3} \text{ kg}$$

No conversion factor is available in one of our tables to *directly* change units of cubic centimeters to cubic meters. You can find

one, however, by cubing (raising to the third power) the relation between the meter and the centimeter.

$$1 \text{ cm}^3 \left(\frac{1 \text{ m}}{100 \text{ cm}}\right)^3 = 1 \text{ cm}^3 \left(\frac{1 \text{ m}^3}{1 \times 10^6 \text{ cm}^3}\right)$$
$$= 1 \times 10^{-6} \text{ m}^3$$

Therefore, the density of sea water is

$$\text{Density} = \frac{1.025 \times 10^{-3} \text{ kg}}{1 \times 10^{-6} \text{ m}^3} = 1.025 \times 10^3 \text{ kg/m}^3$$

Exercise 1.11 Mass and Density

(a) The mass of aspirin in a standard tablet is 325 mg. What is this mass in grams? In kilograms?

(b) The density of gold is 19,320 kg/m^3. What is this density in g/cm^3?

(c) The density of platinum is 21.450 g/cm^3. What is the mass, in grams, of a piece of platinum cylinder 3.0 cm long with a diameter of 5.0 mm?

Making Measurements: Precision, Accuracy, and Experimental Error

The **precision** of a measurement indicates how well several determinations of the same quantity agree. This is illustrated by the results of throwing darts at a target (Figure 1.13). In Figure 1.13a the dart thrower was apparently not skillful (or threw the darts from a long distance away from the target), and the precision of their placement on the target is low. In Figure 1.13b the darts are all clustered together, indicating much better consistency on the part of the thrower, that is, greater precision.

Problem-Solving Tip 1.2

Using Scientific Notation

The number 0.001 or 1/1000 is written as 1×10^{-3}. This notation — called scientific, or exponential, notation — is used throughout the book and is explained further in Appendix A. Scientific notation makes it easier to handle very large or very small numbers.

Make sure you know how to use your calculator to solve problems with exponential numbers. When entering a number such as 1.23×10^{-4} into your calculator, you first enter 1.23 and then press a key marked EE or EXP (or something similar). This enters the "$\times 10$" portion of the notation for you. You then complete the entry by keying in the exponent of the number,

-4. (To change the exponent from $+4$ to -4, you need to press the "$+/-$" key.) A calculator display for the number 8×10^3 will appear something like the images shown below (depending on the model or manufacturer).

A common error made by students is to enter 1.23, then press the multiply key (x) and then key in 10 before finishing by pressing EE or EXP followed by -4. This gives you an entry that is 10 times too large. Try this! Experiment with your calculator so you are sure you are entering data correctly. See the tutorials on Screen 1.17 of the *General Chemistry Interactive CD-ROM*, Version 3.0.

(a)

(b)

(c)

(Charles D. Winters)

(a) Poor precision and poor accuracy

(b) Good precision and poor accuracy

(c) Good precision and good accuracy

Figure 1.13 Precision and accuracy. *(Charles D. Winters)*

Accuracy is the agreement of a measurement with the accepted value of the quantity. Figure 1.13c shows that our thrower was accurate as well as precise — all of the shots are close to the accepted position, namely the bull's eye.

Figure 1.13b shows it is possible to be precise without being accurate — the thrower has consistently missed the bull's eye, although all the darts are clustered precisely around one point on the target. This is analogous to an experiment with some flaw (either in design or in a measuring device) that causes all results to differ from the correct value by the same amount.

The precision of a measurement is often expressed by the **average deviation** (see Example 1.6). That is, we calculate the difference between each experimental result and the average result. These differences, each expressed as a positive quantity, are averaged, and the results of the experiment are reported as the average value plus or minus ($\pm$) the average deviation.

Who determines what is "accepted" so we know when a result is accurate? In the case of mass, for example, there is a standard kilogram mass kept by the national laboratories in several countries. In the United States this is the National Institute for Standards and Technology (NIST), and the balance in your laboratory was probably certified by its manufacturer against standard masses from NIST. In general, scientists establish as the standard of accuracy any result that has been determined reproducibly in many different laboratories.

If you are measuring a quantity in the laboratory, you may be required to report the error in the result, the difference between your result and the accepted value,

$$\text{Error} = \text{experimentally determined value} - \text{accepted value}$$

or the **percent error.**

$$\text{Percent error} = \frac{\text{error in measurement}}{\text{accepted value}} \times 100\% \qquad (1.3)$$

● **NIST**
The National Institute for Standards and Technology is the most important resource for the standards used in science. See http://www.nist.gov.

Example 1.6   Precision and Accuracy

Problem • Student A makes four measurements of the diameter of a coin using a precision tool called a micrometer. Student B measures the same coin using a simple plastic ruler. They report the following results:

Student A	Student B
28.246 mm	27.9 mm
28.244	28.0
28.246	27.8
28.248	28.1

Calculate the average value and the average deviation for each set of data.

Strategy • For each set of values we calculate the average of the results, the deviation of each result from the average, and the average deviation.

Solution • The average for each set of data is obtained by summing the four values and dividing by 4.

Student A	Deviation from Average	Student B	Deviation from Average
28.246 mm	0.000	27.9 mm	0.1
28.244	0.002	28.0	0.0
28.246	0.000	27.8	0.2
28.248	0.002	28.1	0.1
Average = 28.246	Average = 0.001	Average = 28.0	Average = 0.1

Student A would report the experimental results as 28.246 ± 0.001, whereas B would report 28.0 ± 0.1. Student A's results are more precise.

Example 1.7   Error, Precision, and Accuracy

Problem • Suppose the coin described in Example 1.6 has an "accepted" diameter of 28.054 mm. What are the experimental errors of students A and B? Which is more accurate?

Strategy • The error is the difference between the experimental and accepted values. The more accurate result is the one with the smaller error.

Solution •

Student A: Error = 28.246 mm − 28.054 mm = +0.192

Student B: Error = 28.0 mm − 28.1 mm = −0.1

Student B's results were more accurate.

Comment • Notice that Student A's results were more precise, but B's results were more accurate.

Exercise 1.12   Error, Precision, and Accuracy

Two students measured the freezing point of an unknown liquid. Student A used an ordinary laboratory thermometer calibrated in 0.1 °C units. Student B used a thermometer certified by NIST and calibrated in 0.01 °C units. Their results were as follows:

Student A: −0.3 °C; 0.2 °C; 0.0 °C; and −0.3 °C
Student B: 273.13 K; 273.17 K; 273.15 K; and 273.19 K

Calculate the average value and average deviation for each student. Knowing that the unknown liquid was water, calculate the error for each student. Which student has the more precise values? Which has the smaller error?

Significant Figures

In most experiments several kinds of measurements must be made, and some can be made more precisely than others. It is common sense that a result calculated from these data can be no more precise than the least precise piece of information that went into the calculation. This is where the rules for **significant figures** come in.

Consider again the calculation of the density of aluminum described on pages 26–27. The mass and dimensions were determined by standard techniques. All these numbers have two digits to the right of the decimal, but they have different numbers of significant figures.

Measurement	Data Collected	Significant Figures
Mass of aluminum	13.56 g	4
Length	6.45 cm	3
Width	2.50 cm	3
Thickness	3.1 mm	2

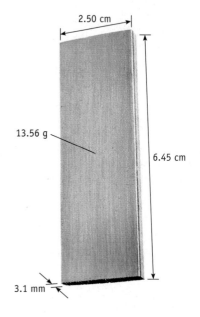

The quantity 0.31 cm has two significant figures. That is, the 3 in 0.31 is exactly right, but the 1 is not known exactly. Unless indicated otherwise, the standard convention used in science is that the final digit in a number is uncertain to the extent of ±1. In general, *in a number representing a scientific measurement, the last digit to the right is taken to be inexact, but all nonzero digits farther to the left are assumed to be exact.* This means the thickness of the aluminum piece could have been as small as 0.30 or as large as 0.32 cm.

When the data on the piece of aluminum are combined, the calculated density is 2.7 g/cm^3, a number with two significant figures. This follows from the previous statement that the calculated result can be no more precise than the least precise piece of information.

When doing calculations using measured quantities, we follow simple rules so that the results reflect the precision of all the measurements that go into the calculations. The rules used in this book are given in the *Closer Look* box titled *Guidelines for Using Significant Figures.*

A Closer Look

Guidelines for Using Significant Figures

Rule 1. To determine the number of significant figures in a measurement, read the number from left to right and count all digits, starting with the first digit that is not zero.

Example	Number of Significant Figures
1.23	3
0.00123 g	3; the zeros to the left of the 1 simply locate the decimal point. To avoid confusion, write numbers of this type in scientific notation; thus, $0.00123 = 1.23 \times 10^{-3}$.
2.0 and 0.020 g	Both have two significant digits. When a number is greater than 1, *all zeros to the right of the decimal point are significant.* For a number less than 1, only zeros to the right of the first significant digit are significant.
100 g	1; in numbers that do not contain a decimal point, "trailing" zeros may or may not be sig-

nificant. The practice followed in this book is to include a decimal point if the zeros are significant. Thus, 100. is used to represent three significant digits, whereas 100 has only one. To avoid confusion, an alternative method is to write numbers in scientific notation because all digits are significant when written in scientific notation. Thus, 1.00×10^2 has three significant digits, whereas 1×10^2 has only one.

100 cm/m	Infinite number of significant digits. This is a defined quantity. Defined quantities do not limit the number of significant figures in a calculated result.
$\pi = 3.1415926$	The value of π is known to a greater number of significant figures than you will ever use in a calculation.

(continued)

A Closer Look

Guidelines for Using Significant Figures *(continued)*

Rule 2. When adding or subtracting numbers, the number of decimal places in the answer is equal to the number of decimal places in the number with the fewest places.

0.12	two decimal places	two significant figures
+1.9	one decimal place	two significant figures
+10.925	three decimal places	five significant figures
12.945	three decimal places	five significant figures

The sum should be reported as 12.9, a number with one decimal place, because 1.9 has only one decimal place.

Rule 3. In multiplication or division, the number of significant figures in the answer should be the same as that in the quantity with the fewest significant figures.

$$\frac{0.01208}{0.0236} = 0.512 \text{ or, in scientific notation, } 5.12 \times 10^{-1}$$

Because 0.0236 has only three significant digits and 0.01208 has four, the answer should have three significant digits.

Rule 4. When a number is rounded off, the last digit to be retained is increased by one only if the following digit is 5 or greater.

Full Number	Number Rounded to Three Significant Digits
12.696	12.7
16.349	16.3
18.35	18.4
18.351	18.4

Standard laboratory balance. Such balances can determine the mass of an object to the nearest milligram. Thus, an object may have a mass of 13.456 g (13,456 mg, five significant figures), 0.123 g (123 mg, three significant figures), or 0.072 g (72 mg, two significant figures). *(Charles D. Winters)*

Example 1.8 Using Significant Figures

Problem • A 26-m tall statue of Buddha in Tibet is covered with 279 kg of gold. If the gold was applied to a thickness of 0.0015 mm, what surface area (in square meters) was covered?

Strategy • First, recognize that the density of gold is needed. The density of the metal (in kilograms per cubic meter) relates the mass of the gold to the volume of the covering. [Here d (gold) = 19,320 kg/m^3]. The volume of the gold is then the product of the area covered and the thickness (as explained on page 27).

Solution • Let us first decide on the number of significant figures in each number.

Number	Number of Significant Figures	Comments
279	3	—
0.0015	2	The first two 0's to the right of the decimal are not significant. Rule 1
19320	4	0 on the right is not significant. Number has no decimal point. Rule 1

Use the mass of gold and its density to calculate the volume covered.

$$\text{Volume of gold (m}^3) = 279 \text{ kg} \times \frac{1 \text{ m}^3}{19320 \text{ kg}}$$
$$= 1.44 \times 10^{-2} \text{ m}^3$$

Here the answer can have only three significant figures because 279 has only three, whereas 19,320 has four (rule 3).

The next step is to calculate the area of coverage from

$$\text{Volume (m}^3) = \text{Area (m}^2) \times \text{Thickness (m)}$$

To obtain area in units of square meters, the thickness, 0.0015 mm, must first be converted to units of meters.

$$\text{Thickness (m)} = 0.0015 \text{ mm} \times \frac{1 \text{ m}}{1000 \text{ mm}}$$
$$= 1.5 \times 10^{-6} \text{ m}$$

Finally, combine the volume and thickness and calculate the area of coverage.

(continued)

$$\text{Area (m}^2) = \frac{\text{Volume of gold (m}^3)}{\text{Thickness of covering (m)}}$$

$$= \frac{1.44 \times 10^{-2}\ \text{m}^3}{1.5 \times 10^{-6}\ \text{m}}$$

$$= 9.6 \quad \times \quad 10^3\ \text{m}^2\ \text{or } 9600\ \text{m}^2$$

Comment • A calculator would show 9.62733×10^3 as an answer, but the final answer can have only two significant figures as explained by rule 2. Also, recognize that the answer is a

reasonable one. The 26-m tall statue is large, so 9600 m², a sheet slightly less than the area of football or soccer field, is reasonable.

Finally, note that the more common unit of density in chemistry is g/cm³, although the SI unit is kg/m³. In other fields, tables of chemical data give density in kg/m³. If you are a geologist or an oceanographer, for example, you will often see densities in kg/m³. To convert from kg/m³ to g/cm³, divide by 1000.

Exercise 1.13   **Using Significant Figures**

(a) What is the sum and product of 10.26 and 0.063?

(b) What is the result of the following calculation?

$$x = \frac{(110.7 - 64)}{(0.056)(0.00216)}$$

One last word on significant figures and calculations. When working problems, you should do the calculation with all the digits allowed by your calculator and round off only at the end of the calculation. Rounding off in the middle can introduce errors.

1.8   PROBLEM SOLVING

Chemistry is a quantitative science. We make measurements, collect data, and then search for patterns in those data. We might calculate the quantity of a substance required for a chemical reaction or, knowing the density of a metal and its identity, calculate the size of its atoms. Some of these calculations can be complex, and our students find it is helpful to follow a definite plan of attack as illustrated in the previous examples and in examples throughout the book.

Step 1: Problem State the problem. Read it carefully.

Step 2: Strategy What key principles are involved? What information is known or not known? What information might be there just to place the question in the context of chemistry? Organize the information to see what is required and to discover the relationships among the data given. Try writing the information down in a table form. If it is numerical information, *be sure to include units.*

One of the greatest difficulties for a student in introductory chemistry is picturing what is being asked for. Try sketching a picture of the situation involved. For example, we sketched a picture of the piece of aluminum whose density we wanted to calculate, and put the dimensions on the drawing (pages 26–27).

Develop a plan. Have you done a problem of this type before? If not, perhaps the problem is really just a combination of several simpler ones you have seen before. Break it down into those simpler components. Try reasoning backward from the units of the answer. What data do you need to find an answer in those units?

● **Who Is Right — You or the Book?**
If your answer to a problem in this book does not quite agree with the answers in Appendix N or O, this may be the result of rounding the answer after each step and then using that rounded answer in the next step. This book follows these conventions:
(a) Final answers to numerical problems in this book result from retaining full calculator accuracy throughout the calculation and then rounding only at the end.
(b) In Example problems the answer to each step is given to the correct number of significant figures for that step, but the full calculator accuracy is carried to the next step. The number of significant figures in the final answer is dictated by the number of significant figures in the original data.

Step 3: Solution Execute the plan. Carefully write down each step of the problem, being sure to keep track of the units on numbers. (Do the units cancel to give you the answer in the desired units?) *Don't skip steps.* Don't do anything but the simplest steps in your head. Students often say they got a problem wrong because they "made a stupid mistake." Your instructor — and book authors — make them, too, and it is usually because they don't take the time to write down the steps of the problem clearly.

Step 4: Check Answer As a final check, ask yourself if the answer is reasonable.

Example 1.9 Problem Solving

Problem • A mineral oil has a density of 0.875 g/cm³. Suppose you spread 0.75 g of this oil over the surface of water in a large dish with a diameter of 21.6 cm. How thick is the oil layer? Express the thickness in centimeters.

Strategy • It is often useful to begin solving such problems by sketching a picture of the situation.

This helps recognize that the solution to the problem is to find the volume of the oil on the water. If we know the volume, then we can find the thickness because

Volume of oil layer = thickness of layer × area of oil layer

So, we need two things: (a) the volume of the oil layer and (b) the area of the layer.

Solution • First calculate the volume of oil. The mass of the oil layer is known, so combining the mass of oil with its density gives the volume of the oil used.

$$0.75 \text{ g} \cdot \frac{1 \text{ cm}^3}{0.875 \text{ g}} = 0.86 \text{ cm}^3$$

Next calculate the area of the oil layer. The oil is spread over a circular surface, whose area is given by

$$\text{Area} = \pi \,(\text{radius})^2$$

The radius of the oil layer is one-half its diameter (= 21.6 cm) or 10.8 cm, so

$$\text{Area of oil layer} = (3.142)(10.8 \text{ cm})^2 = 366 \text{ cm}^2$$

With the volume and area of the oil layer known, the thickness can be calculated.

$$\text{Thickness} = \frac{\text{Volume}}{\text{Area}} = \frac{0.86 \text{ cm}^3}{366 \text{ cm}^2} = 0.0023 \text{ cm}$$

Comment • In the volume calculation, the calculator shows 0.857143... . The quotient should have two significant figures because 0.75 has only two significant figures, so the result of this step is 0.86 cm³. In the area calculation, the calculator shows 366.435... . The answer to this step should have only three significant figures because 10.8 has three. When these are combined, however, the final result can have only two significant figures.

Exercise 1.14 Problem Solving

A particular paint has a density of 0.914 g/cm³. You need to a cover a wall that is 7.6 m long and 2.74 m high with a paint layer 0.13 mm thick. What volume of paint (in liters) is required? What is the mass (in grams) of the paint layer?

In Summary

When you have finished studying this chapter, you should ask if you have met the chapter goals. In particular, you should be able to

• Identify the name or symbol for an element, given its symbol or name (Section 1.1).

- Use the terms *atom, element, molecule,* and *compound* correctly (Sections 1.1 and 1.2).
- Identify physical properties of matter and give some examples (Section 1.3).
- Use density as a way to connect the volume and mass of a substance (Sections 1.3 and 1.7).
- Convert between temperatures on the Celsius and Kelvin scales (Section 1.3).
- Understand the difference between extensive and intensive properties and give examples (Section 1.3).
- Explain the difference between chemical and physical change (Section 1.4).
- Recognize the different states of matter (solids, liquids, and gases) and give their characteristics (Section 1.5).
- Understand the basic ideas of the kinetic-molecular theory (Section 1.5).
- Understand the difference between matter represented at the macroscopic level and at the particulate level (Section 1.5).
- Appreciate the difference between pure substances and mixtures and the difference between homogeneous and heterogeneous mixtures (Section 1.5).
- Recognize and know how to use the prefixes that modify metric units (Section 1.6).
- Use dimensional analysis to carry out unit conversions and other calculations (Section 1.7 and Appendix A).
- Know the difference between precision and accuracy and how to calculate percent error (Section 1.7).
- Understand the use of significant figures (Section 1.7).

Key Terms

Section 1.1
- elements
- periodic table
- atom

Section 1.2
- chemical compound
- ion
- molecule
- chemical formula

Section 1.3
- physical properties
- density
- temperature
- Celsius scale
- Kelvin scale
- absolute zero
- extensive properties
- intensive properties

Section 1.4
- physical change
- chemical change or chemical reaction
- reactant
- product
- chemical property
- chemical equation

Section 1.5
- state
- kinetic molecular theory
- kinetic energy
- macroscopic
- submicroscopic or particulate
- heterogeneous mixture
- homogeneous mixture
- solutions
- purified

Section 1.6
- qualitative
- quantitative
- metric system

Section 1.7
- dimensional analysis
- conversion factor
- liter
- precision
- accuracy
- average deviation
- percent error
- significant figures

Key Equations

Equation 1.1 (page 15)

Density is the quotient of the mass of an object and its volume. Units are usually g/cm^3.

$$\text{Density} = \frac{\text{mass}}{\text{volume}}$$

Equation 1.2 (page 17)

Conversion between the Kelvin and Celsius temperature scales.

$$T(K) = \frac{1\ K}{1\ °C}(T\ °C + 273.15\ °C)$$

Equation 1.3 (page 31)

The percent error of a measurement is the deviation of the measurement from the accepted value.

$$\text{Percent error} = \frac{\text{error in measurement}}{\text{accepted value}} \times 100\%$$

Study Questions

Questions with blue, bold-faced numbers have answers in Appendix O. See the tutorials on Screens 1.8, 1.10, 1.16, and 1.17 of the General Chemistry Interactive CD-ROM, *Version 3.0.*

Reviewing Important Concepts

1. The mineral fluorite contains the elements calcium and fluorine. What are the symbols of these elements? How would you describe the shape of the fluorite crystals in the photo? What can this tell us about the arrangement of the atoms inside the crystal?

The mineral fluorite, calcium fluoride. *(Charles D. Winters)*

2. What are the states of matter and how do they differ from one another?

3. Small chips of iron are mixed with sand (see the following photo). Is this a homogeneous or heterogeneous mixture? Suggest a way to separate the iron from the sand.

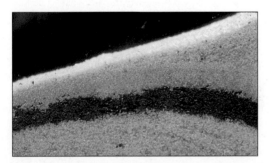

Chips of iron mixed with sand. *(Charles D. Winters)*

4. What is the difference between the terms *compound* and *molecule*? Use these words in a sentence.

5. In each case, decide if the underlined property is a physical or chemical property.

   (a) The normal color of elemental bromine is <u>orange</u>.

   (b) Iron <u>turns to rust</u> in the presence of air and water.

   (c) Hydrogen can <u>explode</u> when ignited in air (see Figure 1.8).

   (d) The <u>density</u> of titanium metal is 4.5 g/cm^3.

   (e) Tin metal <u>melts</u> at 505 K.

   (f) Chlorophyll, a plant pigment, is <u>green</u>.

6. In each case, decide if the change is a chemical or physical change.

   (a) A cup of household bleach changes the color of your favorite T-shirt from purple to pink.

   (b) Water vapor in your exhaled breath condenses in the air on a cold day.

   (c) Plants use carbon dioxide from the air to make sugar.

   (d) Butter melts when placed in the sun.

7. The following photo shows copper balls, immersed in water, floating on top of mercury. What are the liquids and solids in this photo? Which substance is most dense? Which is least dense?

Water, copper, and mercury. *(Charles D. Winters)*

8. Give the number of significant figures in each of the following numbers:
    (a) 6.2348   (c) 0.00823
    (b) 20,600   (d) 1.670×10^{-6}

9. The gemstone called aquamarine is composed of aluminum, silicon, and oxygen.

Aquamarine is the bluish crystal. It is surrounded by aluminum foil and crystalline silicon. *(Charles D. Winters)*

   (a) What are the symbols of the three elements that combine to make the gem aquamarine?
   (b) Based on the photo, describe some of the physical properties of the elements and the mineral. Are any the same? Are any properties different?

10. A piece of turquoise is a blue-green solid, has a density of 2.65 g/cm^3, a mass of 2.5 g, and a length of 4.6 cm. Which of these observations are qualitative and which are quantitative? Which of the observations are extensive and which are intensive?

11. In Figure 1.2 you see a piece of salt and a representation of its internal structure. Which is the macroscopic view and which is the particulate view? How are the macroscopic and particulate views related?

Practicing Skills

Elements and Atoms, Compounds and Molecules
(See Exercise 1.1)

12. Give the name of each of the following elements:
    (a) C   (c) Cl   (e) Mg
    (b) K   (d) P   (f) Ni

13. Give the name of each of the following elements:
    (a) Mn   (c) Na   (e) Xe
    (b) Cu   (d) Br   (f) Fe

14. Give the symbol for each of the following elements:
    (a) Barium   (d) Lead
    (b) Titanium   (e) Arsenic
    (c) Chromium   (f) Zinc

15. Give the symbol for each of the following elements:
    (a) Silver   (d) Tin
    (b) Aluminum   (e) Technetium
    (c) Plutonium   (f) Krypton

16. In each of the following pairs, decide which is an element and which is a compound.
    (a) Na and NaCl
    (b) Sugar and carbon
    (c) Gold and gold chloride

17. In each of the following pairs, decide which is an element and which is a compound.
    (a) $Pt(NH_3)_2Cl_2$ and Pt
    (b) Copper and copper(II) oxide
    (c) Silicon and sand

Physical and Chemical Properties
(See Exercises 1.2 and 1.6)

18. Which part of the description of a compound or element refers to its physical properties and which to its chemical properties?
    (a) The colorless liquid ethanol burns in air.
    (b) The shiny metal aluminum reacts readily with orange, liquid bromine.

19. Which part of the description of a compound or element refers to its physical properties and which to its chemical properties?
    (a) Calcium carbonate is a white solid with a density of 2.71 g/cm^3. It reacts readily with an acid to produce gaseous carbon dioxide.
    (b) Gray, powdered zinc metal reacts with purple iodine to give a white compound.

Using Density
(See Example 1.1 and CD-ROM Screen 1.8)

20. Ethylene glycol, $C_2H_6O_2$, is an ingredient of automobile antifreeze. Its density is 1.11 g/cm^3 at 20 °C. If you need exactly 500. mL of this liquid, what mass of the compound, in grams, is required?

21. A piece of silver metal has a mass of 2.365 g. If the density of silver is 10.5 g/cm^3, what is the volume of the silver?

22. A chemist needs 2.00 g of a liquid compound with a density of 0.718 g/cm^3. What volume of the compound is required?

23. The *cup* is a volume measure widely used by cooks in the United States. One cup is equivalent to 237 mL. If 1 cup of olive oil has a mass of 205 g, what is the density of the oil (in grams per cubic centimeter)?

24. A sample of unknown metal is placed in a graduated cylinder containing water (see figure on page 40). The mass of the sample is 37.5 g, and the water levels before and after adding the sample to the cylinder are as shown in the figure. Which metal in the following list is most likely the sample? (*d* is the density of the metal.)
    (a) Mg, $d = 1.74 \text{ g/cm}^3$ (d) Al, $d = 2.70 \text{ g/cm}^3$
    (b) Fe, $d = 7.87 \text{ g/cm}^3$ (e) Cu, $d = 8.96 \text{ g/cm}^3$
    (c) Ag, $d = 10.5 \text{ g/cm}^3$ (f) Pb, $d = 11.3 \text{ g/cm}$

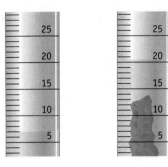

Graduated cylinders with unknown metal *(right).*

25. Iron pyrite is often called "fool's gold" because it looks like gold (see page 12). Suppose you have a solid that looks like gold, but you believe it to be fool's gold. The sample has a mass of 23.5 g. When the sample is lowered into the water in a graduated cylinder (see Study Question 24), the water level rises from 47.5 mL to 52.2 mL. Is the sample fool's gold ($d = 5.00$ g/cm^3) or "real" gold ($d = 19.3$ g/cm^3)?

Temperature
(See pages 16–17, Exercise 1.4, and CD-ROM Screen 1.10)

26. Many laboratories use 25 °C as a standard temperature. What is this temperature in kelvins?

27. The temperature on the surface of the sun is 5.5×10^3 °C. What is this temperature in kelvins?

28. Make the following temperature conversions:

°C	K
(a) 16	——
(b) ——	370
(c) 40	——

29. Make the following temperature conversions:

°C	K
(a) ——	77
(b) 63	——
(c) ——	1450

Units and Unit Conversions
(See Examples 1.2–1.5 and Exercises 1.8–1.11)

30. A marathon distance race covers a distance of 40.0 km. What is this distance in meters? In miles?

31. The average lead pencil, new and unused, is 19 cm long. What is its length in millimeters? In meters?

32. A standard U. S. postage stamp is 2.5 cm long and 2.1 cm wide. What is the area of the stamp in square centimeters? In square meters?

33. A compact disk has a diameter of 11.8 cm. What is the sur-face area of the disk in square centimeters? In square meters? [Area of a circle = (π) (radius)^2.]

34. A typical laboratory beaker has a volume of 250 mL. What is its volume in cubic centimeters? In liters? In cubic meters? In cubic decimeters?

35. Some soft drinks are sold in bottles with a volume of 1.5 L. What is this volume in milliliters? In cubic centimeters? In cubic decimeters?

36. A book has a mass of 2.52 kg. What is this mass in grams?

37. A new U. S. dime has a mass of 2.265 g. What is its mass in kilograms? In milligrams?

Accuracy, Precision, and Error
(See Example 1.7)

38. You and your lab partner are asked to determine the density of an aluminum bar. The mass is known accurately (to four significant figures). You use a simple metric ruler to measure its dimensions, and after calculating the volume, you determine the density given in the table (Method A). Your partner uses a precision micrometer to measure the dimensions and then calculates the density (Method B).

Method A (g/cm^3)	Method B (g/cm^3)
2.2	2.703
2.3	2.701
2.7	2.705
2.4	5.811

The accepted density of aluminum is 2.702 g/cm^3.

(a) Calculate the average density and average deviation for each method. Should all the experimental results be included in your calculations? If not, justify any omissions.

(b) Calculate the error for each method's average value.

(c) Which method's average value is more precise? Which method is more accurate?

39. The accepted value of the melting point of pure aspirin is 135 °C. Trying to verify that value, you obtain 134 °C, 136 °C, 133 °C, and 138 °C in four separate trials. Your partner finds 138 °C, 137 °C, 138 °C, and 138 °C.

(a) Calculate the average value and average deviation for you and your partner.

(b) Calculate the error for each set of data.

(c) Which of you is more precise? More accurate?

Significant Figures
(See Example 1.8)

40. Give the number of significant figures in each of the following numbers:

(a) 0.0123   (c) 1.6402

(b) 3.40×10^3 (d) 1.020

41. Give the number of significant figures in each of the following numbers:

    (a) 0.00546   (c) 2.300×10^{-4}

    (b) 1600   (d) 2.34×10^{9}

42. Carry out the following calculation, and report the answer in the correct number of significant figures.

$$(0.0546)(16.0000)\left(\frac{7.779}{55.85}\right)$$

43. Carry out the following calculation, and report the answer in the correct number of significant figures.

$$(1.68)\left(\frac{23.56 - 2.3}{1.248 \times 10^{3}}\right)$$

Problem Solving

(See Example 1.9 and CD-ROM Screen 1.17)

44. Diamond has a density of $3.513 \, \text{g/cm}^3$. The mass of diamonds is often measured in carats, 1 carat equaling 0.200 g. What is the volume (in cubic centimeters) of a 1.50-carat diamond?

45. The smallest repeating unit of a crystal of common salt is a cube (called a unit cell) with an edge length of 0.563 nm.

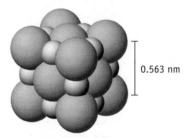

0.563 nm

sodium chloride, NaCl

(a) What is the volume of this cube in cubic nanometers? In cubic centimeters?

(b) The density of NaCl is $2.17 \, \text{g/cm}^3$. What is the mass of this smallest repeating unit (unit cell)?

(c) Each repeating unit is composed of four formula units of NaCl. What is the mass of one formula unit of NaCl?

46. An ancient gold coin is 2.2 cm in diameter and 3.0 mm thick. It is a cylinder for which volume = $(\pi)(\text{radius})^2(\text{thickness})$. If the density of gold is $19.3 \, \text{g/cm}^3$, what is the mass of the coin in grams?

47. Copper has a density of $8.94 \, \text{g/cm}^3$. An ingot of copper with a mass of 57 kg (125 lb) is drawn into wire with a diameter of 9.50 mm. What length of wire (in meters) can be produced? [Volume of wire = $(\pi)(\text{radius})^2(\text{length})$].

48. The Gimli Glider was a Boeing 767 that ran out of fuel. Read the description on page 8, and then verify that the ground crew should have had added 20,163 L of fuel (and not 4916 L). The crucial piece of information is the density of fuel. The crew used 1.77, but they did not recognize that the units were *pounds* per liter. To solve this problem you need first to find the fuel density in units of *kilograms* per liter (1 lb = 453.6 g).

49. When you heat popcorn, it pops because it loses water explosively. Assume a kernel of corn, with a mass of 0.125 g, has a mass of only 0.106 g after popping.

    (a) What percent of its mass did the kernel lose on popping?

    (b) Popcorn is sold by the pound in the U. S. Using 0.125 g as the average mass of a popcorn kernel, how many kernels are in a pound of popcorn? (1 lb = 453.6 g)

General Questions

These questions are not designated as to type or location in the chapter. They may combine several concepts. Questions 50–63 emphasize using numerical data whereas Questions 64–73 are more conceptual. More challenging questions are indicated by an underlined number (74–77).

50. The element gallium has a melting point of 29.8 °C. If you held a sample of gallium in your hand, should it melt? Explain briefly.

51. Neon, a gaseous element used in neon signs, has a melting point of −248.6 °C and a boiling point of −246.1 °C. Express these temperatures in kelvins.

52. Molecular distances are usually given in nanometers (1 nm = 1×10^{-9} m) or in picometers (1 pm = 1×10^{-12} m). However, the angstrom (Å) unit is sometimes used, where 1 Å = 1×10^{-10} m. (The angstrom unit is not an SI unit.) If the distance between the Pt atom and the N atom in the cancer chemotherapy drug cisplatin is 1.97 Å, what is this distance in nanometers? In picometers?

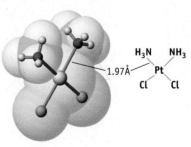

H_3N NH_3

1.97Å   Pt

Cl   Cl

cisplatin

53. The separation between carbon atoms in diamond is 0.154 nm. (a) What is their separation in meters? (b) What is the car-

bon atom separation in angstrom units (where $1 \text{ Å} = 1 \times 10^{-10}$ m)?

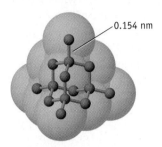

0.154 nm

A portion of the diamond structure

54. At 25 °C the density of water is 0.997 g/cm^3, whereas the density of ice at −10 °C is 0.917 g/cm^3.
    (a) If a soft drink can (volume = 250. mL) is filled completely with pure water and then frozen at −10 °C, what volume does the solid occupy?
    (b) Can the ice be contained within the can?

55. You can identify a metal by carefully determining its density (d). An unknown piece of metal, with a mass of 2.361 g, is 2.35 cm long, 1.34 cm wide, and 1.05 mm thick. Which of the following is the element?
    (a) Nickel, $d = 8.91$ g/cm^3
    (b) Titanium, $d = 4.50$ g/cm^3
    (c) Zinc, $d = 7.14$ g/cm^3
    (d) Tin, $d = 7.23$ g/cm^3

56. Which occupies a larger volume, 600 g of water (with a density of 0.995 g/cm^3) or 600 g of lead (with a density of 11.34 g/cm^3)?

57. You have a 100.0-mL graduated cylinder containing 50.0 mL of water. You drop a 154-g piece of brass (density = 8.56 g/cm^3) into the water. How high does the water rise in the graduated cylinder? See Study Question 24.

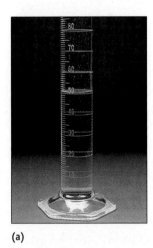

(a)

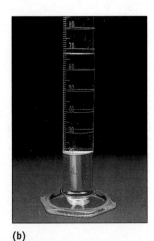

(b)

(a) A graduated cylinder with 50.0 mL of water. **(b)** A piece of brass is added to the cylinder. *(Charles D. Winters)*

58. The density of pure water is given at various temperatures.

T (°C)	d (g/cm^3)
4	0.99997
15	0.99913
25	0.99707
35	0.99406

Suppose your laboratory partner tells you the density of water at 20 °C is 0.99910 g/cm^3. Is this a reasonable number? Why or why not?

59. A red blood cell has a diameter of 7.5 μm (micrometer). What is this dimension in (a) meters, (b) nanometers, and (c) angstrom units (where $1 \text{ Å} = 1 \times 10^{-10}$ m)?

60. Carbon tetrachloride, CCl$_4$, a common liquid compound, has a density of 1.58 g/cm^3. If you place a piece of a plastic soda bottle ($d = 1.37$ g/cm^3) and a piece of aluminum ($d = 2.70$ g/cm^3) in liquid CCl$_4$, will the plastic and aluminum float or sink?

61. The platinum-containing cancer drug cisplatin contains 65.0% platinum. If you have 1.53 g of the compound, what mass of platinum (in grams) is contained in this sample?

62. The solder once used by plumbers to fasten copper pipes together consists of 67% lead and 33% tin. What is the mass of lead in a 250-g block of solder?

63. The anesthetic procaine hydrochloride is often used to deaden pain during dental surgery. The compound is packaged as a 10.% solution (by mass; $d = 1.0$ g/mL) in water. If your dentist injects 0.50 mL of the solution, what mass of procaine hydrochloride (in milligrams) is injected?

64. Make a drawing, based on the kinetic-molecular theory and the ideas about atoms and molecules presented in this chapter, of the arrangement of particles in each of the cases listed here. For each case draw ten particles of each substance. It is acceptable for your diagram to be two-dimensional. Represent each atom as a circle and distinguish each different kind of atom by shading.
    (a) A sample of solid iron (which consists of iron atoms)
    (b) A sample of *liquid* water (which consists of H$_2$O molecules)
    (c) A sample of pure water *vapor*
    (d) A homogeneous mixture of water vapor and helium gas (which consists of helium atoms)
    (e) A heterogeneous mixture consisting of liquid water and solid aluminum; show a region of the sample that includes both substances
    (f) A sample of brass (which is a homogeneous mixture of copper and zinc)

65. You are given a sample of a silvery metal. What information would you seek to prove that the metal is silver?

66. Suggest a way to determine if the colorless liquid in a beaker is water. If it is water, does it contain dissolved salt? How could you discover if there is salt dissolved in the water?

67. Describe an experimental method that can be used to determine the density of an irregularly shaped piece of metal.

68. Three liquids of different densities are mixed. Because they are not miscible (do not form a homogeneous solution with one another) they form discrete layers, one on top of the other. Sketch the result of mixing carbon tetrachloride (CCl_4, $d = 1.58 \text{ g/cm}^3$), mercury ($d = 13.546 \text{ g/cm}^3$), and water ($d = 1.00 \text{ g/cm}^3$).

69. Diabetes can alter the density of urine, and so urine density can be used as a diagnostic tool. Diabetics can excrete too much sugar or excrete too much water. What do you predict will happen to the density of urine under each of these conditions?

70. A copper-colored metal is found to conduct an electric current. Can you say with certainty that it is copper? Why or why not? Suggest additional information that could provide unequivocal confirmation that the metal is copper.

71. What experiment can you use to

    (a) Separate salt from water?

    (b) Separate iron filings from small pieces of lead?

    (c) Separate elemental sulfur from sugar?

72. The following photo shows the element potassium reacting with water to form the element hydrogen, a gas, and a solution of the compound potassium hydroxide.

Potassium reacting with water to produce hydrogen gas and potassium hydroxide. *(Charles D. Winters)*

    (a) What states of matter are involved in the reaction?

    (b) Is the observed change chemical or physical?

    (c) What are the reactants in this reaction and what are the products?

    (d) What qualitative observations can be made concerning this reaction?

73. Four balloons are each filled with a different gas of varying density:

    Helium, $d = 0.164 \text{ g/L}$

    Neon, $d = 0.825 \text{ g/L}$

    Argon, $d = 1.633 \text{ g/L}$

    Krypton, $d = 3.425 \text{ g/L}$

If the density of dry air is 1.12 g/L, which balloon or balloons float in air?

74. The aluminum in a package containing 75 ft^2 of kitchen foil weighs approximately 12 oz. Aluminum has a density of 2.70 g/cm^3. What is the approximate thickness of the aluminum foil in millimeters? (1 oz = 28.4 g)

75. The fluoridation of city water supplies has been practiced in the United States for several decades. This is done by continuously adding sodium fluoride to water as it comes from a reservoir. Assume you live in a medium-sized city of 150,000 people and that 660 L (170 gal) of water is consumed per person per day. What mass of sodium fluoride (in kilograms) must be added to the water supply each year (365 days) to have the required fluoride concentration of 1 ppm (part per million), that is, 1 kg of fluoride per million kilograms of water? (Sodium fluoride is 45.0% fluoride, and water has a density of 1.00 g/cm^3.)

76. About two centuries ago, Benjamin Franklin showed that 1 tsp of oil would cover about 0.5 acre of still water. If you know that $1.0 \times 10^4 \text{ m}^2 = 2.47$ acres, and that there are approximately 5 cm^3 in a teaspoon, what is the thickness of the layer of oil? How might this thickness be related to the sizes of molecules?

77. Automobile batteries are filled with sulfuric acid. What is the mass of the acid (in grams) in 500. mL of the battery acid solution if the density of the solution is 1.285 g/cm^3 and if the solution is 38.08% sulfuric acid by mass?

Using Electronic Resources

This question refers to the *General Chemistry Interactive CD-ROM*, Version 3.0.

78. See CD-ROM Screen 1.18: Chemical Puzzler. Many foods are fortified with vitamins and minerals. Some breakfast cereals have elemental iron added. Iron chips are used instead of iron compounds because the compounds can be converted by the oxygen in air to a form of iron that is not biochemically useful. Iron chips, on the other hand, are converted to iron compounds in the gut, and the iron can then be absorbed.

Outline a method by which you could remove the iron (as iron chips) from a box of cereal and determine the mass of iron in a given mass of cereal.

Some breakfast cereals contain iron in the form of elemental iron. *(Charles D. Winters)*

2 Atoms and Elements

Stardust

A wide array of different elements make up planet Earth and every living thing on it. What is science's view of the cosmic origin of these elements that we take for granted in our environment and in our lives?

The "big bang" theory is the generally accepted explanation for the origin of the universe [CD-ROM, Screen 2.2].

▲ **The supernova of 1987.** When a star becomes denser and denser, and hotter and hotter, it can become a red giant. The star is unstable and explodes as a supernova. One such spectacular event occurred in 1987. These explosions are the origin of the heavier elements, such as iron, nickel, and cobalt. *(Dr. Christopher Burrows, ESA/STSc1 and NASA)*

This theory holds that an unimaginably dense, grapefruit-sized sphere of matter exploded about 15 billion years ago, spewing the products of that explosion as a rapidly expanding cloud with a temperature in the range of 10^{30} K. Within a second, the universe was populated with the particles we explore in this chapter: protons, electrons, and neutrons. Within a few more seconds, the universe had cooled by millions and millions of degrees, and protons and neutrons began to combine to form helium. After only about 8 minutes scientists believe that the universe was about one quarter helium and about three quarters hydrogen. In fact, this is very close to the composition of the universe today, 15 billion years later. But humans, animals, and plants are built mainly from carbon, oxygen, nitrogen, sulfur, phosphorus, iron, and zinc, heavier elements that have only a trace abundance in the universe. Where do these heavier elements come from?

The cloud of hydrogen and helium cooled over a period of thousands of years and condensed into stars like our sun. Here hydrogen atoms fuse into more helium atoms and energy streams outward. Every second on the sun, 700 million tons of hydrogen are converted to 695 million tons of helium, and 3.9×10^{26} joules of energy are evolved.

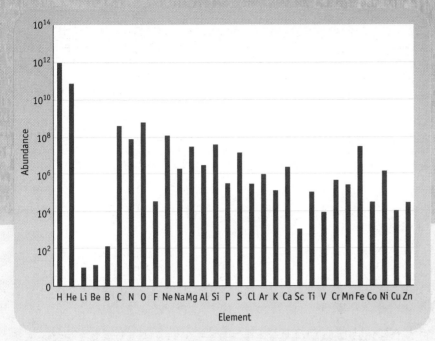

▲ **The abundance of the elements in the solar system from H to Zn.** Notice that Li, Be, and B are very low in abundance, C and O are relatively high, as is Fe. Beyond Fe, abundances steadily decline. (Abundances are plotted as "number of atoms per 10^{12} H atoms.")

Gradually, over millions of years, a hydrogen-burning star becomes denser and denser and hotter and hotter. The helium atoms initially formed now begin to fuse into heavier and heavier atoms, first carbon, then oxygen, and then neon, magnesium, silicon, phosphorus, and argon. The star becomes even denser and hotter. Hydrogen is forced to the outer reaches of the star, and the star becomes a red giant. Under certain circumstances, the star will explode, and earth-bound observers see it as a supernova. A supernova can be up to 10^8 times brighter than the original star. A single supernova is comparable in brightness to the whole of the galaxy in which it is formed! The supernova that appeared in 1987 gave astronomers an opportunity to study what happens in these element factories. It is here that the heavier atoms such as iron form. In fact, it is with iron that nature reaches its zenith of stability. To make heavier and heavier elements requires energy, rather than having energy as an outcome of element synthesis.

The elements spewing out of an exploding supernova move through space and gradually condense into planets, of which ours is just one.

The mechanism of element formation in stars is reasonably well understood, and much experimental evidence supports this process. However, the way in which these elements are then assembled out of stardust into living organisms on our planet — and perhaps other planets — is not yet understood at all.

Before You Begin

- Review names and uses of SI units (Section 1.6).
- Know how to solve numerical problems using dimensional analysis (Section 1.7).

The Structure of the Atom To chemists the atom is the fundamental building block of our universe. Our current understanding of atoms, their structure and their function, is built from many experiments, particularly in the 20th century. The experiments outlined here are three of the most important because they uncovered the nature of two of the particles of which atoms are composed: electrons and protons. Other experiments — chiefly those of Ernest Rutherford — are outlined later in the chapter.

Measuring the charge-to-mass ratio of the electron

J. J. Thomson
Experiment to measure the ratio of the electron's mass to its charge
1896–1897

1. A beam of electrons (cathode rays) passes through an electric field and a magnetic field.

2. The experiment is arranged so that the electric field causes the beam of electrons to be deflected in one direction. The magnetic field deflects the beam in the opposite direction.

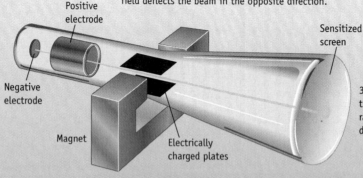

Positive electrode

Sensitized screen

Negative electrode

3. By balancing the effects of these fields, the charge-to-mass ratio of the electron can be determined.

Magnet

Electrically charged plates

Measuring the charge on the electron

Millikan Oil Drop Experiment to measure charge on the electron
1911–1913

1. A fine mist of oil drops is introduced into one chamber. The droplets fall one by one into the lower chamber.

2. The gas molecules in the bottom chamber are ionized (split into electrons and a positive fragment) by a beam of x-rays (not shown).

3. The electrons adhere to the oil drops, some droplets having one electron, some two, and so on.

4. These negatively charged droplets fall under the force of gravity.

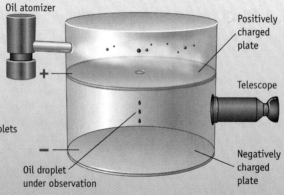

Oil atomizer

Positively charged plate

Telescope

Negatively charged plate

Oil droplet under observation

5. By carefully adjusting the voltage on the plates, the force of gravity on the droplet is exactly counterbalanced by the attraction of the negative droplet to the upper, positively charged plate. Analysis of these forces lead to a value for the charge on the electron.

Discovering the proton

Eugene Goldstein
Canal rays — Discovering the proton
1886

1. Electrons collide with gas molecules in this cathode-ray tube with a perforated cathode.

2. The molecules become positively charged, and these are attracted to the negatively charged, perforated cathode.

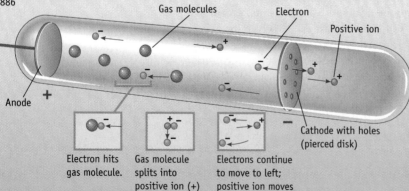

Gas molecules

Electron

Positive ion

Anode

Cathode with holes (pierced disk)

Electron hits gas molecule.

Gas molecule splits into positive ion (+) and electron (–).

Electrons continue to move to left; positive ion moves to right.

3. Some positive particles pass through the holes and form a beam, or "ray."

Like cathode rays, positive rays (or "canal rays") are deflected by electric and magnetic fields but much less so than cathode rays for a given value of the field because positive particles are much heavier than electrons.

The chemical elements are forged in stars. What are the similarities among these elements? What are their differences? What are their physical and chemical properties? How can we tell them apart? This chapter begins our exploration of the chemistry of the elements, the building blocks of the science of chemistry.   ●

2.1   PROTONS, ELECTRONS, AND NEUTRONS: DEVELOPMENT OF ATOMIC STRUCTURE

Around 1900 a series of experiments done by scientists such as Sir Joseph John Thomson (1856–1940) (see Chapter Focus) and Ernest Rutherford (1871–1937) in England established a model of the atom that is still the basis of modern atomic theory. Three **subatomic particles** make up all atoms: electrically positive protons, electrically neutral neutrons, and electrically negative electrons. The model places the more massive protons and neutrons in a very small nucleus, which means that the nucleus contains all the positive charge and almost all the mass of an atom. Electrons, with a much smaller mass than protons or neutrons, surround the nucleus and occupy most of the volume (Figure 2.1). Atoms have no net charge; the positive and negative charges within the atom balance. *The number of electrons outside the nucleus equals the number of protons in the nucleus* [🖱 CD-ROM, Screen 2.2].

What is the experimental basis of atomic structure? How did the work of Thomson, Rutherford, and others lead to this model?

Electricity

Electricity is involved in many of the experiments from which the theory of atomic structure was derived. The fact that objects can bear an electric charge was first observed by the ancient Egyptians, who noted that amber, when rubbed with wool or silk, attracted small objects. You can observe the same thing when you rub a balloon on your hair on a dry day — your hair is attracted to the balloon (Figure 2.2a). A bolt of lightning or the shock you get when touching a doorknob results when an electric charge moves from one place to another.

Two types of electric charge had been discovered by the time of Benjamin Franklin (1706–1790), the American statesman and inventor. He named them positive (+) and negative (−), because they appear as opposites and can neutralize each other. Experiments show that like charges repel each other and unlike charges attract each other. Franklin also concluded that charge is balanced: if a negative charge appears somewhere, a positive charge of the same size must appear somewhere else. The fact that a charge builds up when one substance is rubbed over another implies that the rubbing separates positive and negative charges (see Figure 2.2a). By the 19th century it was understood that positive and negative charges are somehow associated with matter — perhaps with atoms [🖱 CD-ROM, Screen 2.4].

Radioactivity

In 1896 the French physicist Henri Becquerel (1852–1908) discovered that a uranium ore emitted rays that could darken a photographic plate, even though the plate was covered by black paper to protect it from being exposed to light. In 1898, Marie Curie and coworkers isolated polonium and radium, which also emitted the same kind of rays, and in 1899 Madame Curie suggested that atoms of certain substances emit these unusual rays when they disintegrate. She named this phenomenon **radioactivity,** and substances that display this property are said to be *radioactive* [🖱 CD-ROM, Screen 2.5].

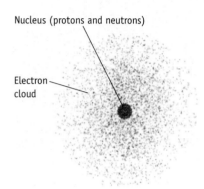

Nucleus (protons and neutrons)

Electron cloud

Figure 2.1   The structure of the atom. All atoms (except H) consist of one or more protons (positive electric charge) and at least as many neutrons (no electric charge) packed into an extremely small nucleus. Electrons (negative electric charge) are arranged in space as a "cloud" around the nucleus. In an electrically neutral atom the number of electrons equals the number of protons.

> *Chapter Goals* ● Revisited

- **Describe the structure of the atom and define atomic number and mass number.**
- Understand the nature of isotopes and calculate atomic weight from isotope abundances and exact atomic masses.
- Explain the concept of the mole and use molar mass in calculations.
- Know the terminology of the periodic table.

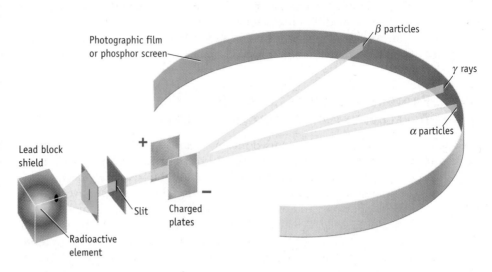

(a) If you brush a balloon against your hair, a static electric charge builds up on the surface of the balloon. Experiment shows that objects having opposite electric charges attract, whereas objects having the same electric charge repel.

(b) Alpha (α), beta (β), and gamma (γ) rays from a radioactive element are separated by passing them through electrically charged plates. Positively charged α particles are attracted to the negative plate, and negatively charged β particles are attracted to the positive plate. (Note that the heavier α particles are deflected less than the lighter β particles.) Gamma rays have no electric charge and pass undeflected between the charged plates.

Figure 2.2   Electricity and radioactivity. *(a, Charles D. Winters)*

Early experiments identified three kinds of radiation: alpha (α), beta (β), and gamma rays (γ). These behave differently when passed between electrically charged plates (Figure 2.2b). Alpha and β rays are deflected, but γ rays pass straight through. This implies that α and β rays are electrically charged particles, because charges are attracted or repelled by the charged plates. Even though an α particle was found to have an electric charge (+2) twice as large as that of a β particle (−1), α particles are deflected less, which implies that α particles must be heavier than β particles. Gamma rays have no detectable charge or mass; they behave like light rays.

Marie Curie's suggestion that atoms disintegrate contradicted ideas put forward in 1803 by John Dalton that atoms are indivisible. If atoms can break apart, there must be something smaller than an atom; that is, atoms must be composed of even smaller, subatomic particles.

Electrons

Further evidence that atoms are composed of smaller particles came from experiments with cathode ray tubes. These are glass tubes from which most of the air has been removed and that contain two metal electrodes. When a sufficiently high voltage is applied to the electrodes, a *cathode ray* flows from the negative electrode (cathode) to the positive electrode (anode). Experiments showed that cathode rays travel in straight lines, cause gases to glow, can heat metal objects red hot, can be deflected by a magnetic field, and are attracted toward positively charged plates. When cathode rays strike a fluorescent screen, light is given off in a series of tiny flashes. We can understand all these observations if a cathode ray is assumed to be a beam of the negatively charged particles we now know as **electrons** [CD-ROM, Screen 2.6].

You are already familiar with cathode rays. Television pictures and the images on a computer monitor are formed by the deflection of cathode rays by electrically charged plates inside the tube. Sir Joseph John Thomson used this principle to prove experimentally the existence of the electron and study its properties (see Chapter Focus). He applied electric and magnetic fields simultaneously to a beam of cathode rays. By balancing the effect of the electric field against that of the magnetic field and using basic laws of electricity and magnetism, he calculated the ratio of the charge to the mass for the particles in the beam. He was not able to determine either charge or mass independently. However, he found the same charge-to-mass ratio in experiments using 20 different metals as cathodes and several different gases. These results suggested that electrons are present in all kinds of matter and that they exist in atoms of all elements.

It remained for the American physicist Robert Andrews Millikan (1868–1953) to measure the charge on an electron and thereby enable scientists to calculate its mass [CD-ROM, Screen 2.7]. Tiny droplets of oil were sprayed into a chamber. As they settled slowly through the air, the droplets were exposed to x-rays, which caused them to acquire an electric charge. Millikan used a small telescope to observe individual droplets. If the electric charge on the plates above and below the droplets was adjusted, the electrostatic attractive force pulling a droplet upward could be balanced by the force of gravity pulling the droplet downward. From the equations describing these forces, Millikan calculated the charge on various droplets. Different droplets had different charges, but Millikan found that each was a whole-number multiple of the same smaller charge, 1.60×10^{-19} C (where C represents the coulomb, the SI unit of electric charge; Appendix D). Millikan assumed this to be the fundamental unit of charge, the charge on an electron. Because the charge-to-mass ratio of the electron was known, the mass of an electron could be calculated. The currently accepted value for the electron mass is 9.109382×10^{-28} g, and the electron charge is $-1.602176 \times 10^{-19}$ C. When talking about the properties of fundamental particles, we always express charge relative to the charge on the electron, which is given the value of -1.

Additional experiments showed that cathode rays had the same properties as the β particles emitted by radioactive elements. This provided further evidence that the electron is a fundamental particle of matter.

Protons

The first experimental evidence of a fundamental *positive* particle came from the study of *canal rays* (see Chapter Focus), which were observed in a special cathode-ray tube with a perforated cathode. When high voltage was applied to the tube, cathode rays were observed. On the other side of the perforated cathode, however, a different kind of ray was seen. Because these rays were deflected toward a negatively charged plate, they had to be composed of positively charged particles.

Each gas used in the tube gave a different charge-to-mass ratio for the positively charged particles (unlike cathode rays, which are the same no matter what gas is used). When hydrogen gas was used, the largest charge-to-mass ratio was obtained, suggesting that hydrogen provides positive particles with the smallest mass. These were considered to be the fundamental, positively charged particles of atomic structure and Ernest Rutherford later called them **protons** (from a Greek word meaning "the primary one") [CD-ROM, Screen 2.8].

History

John Dalton (1766–1844)

Dalton was the first to apply the notion of atoms to explain chemical phenomena. Among the postulates of his atomic theory were that all matter is made of atoms, all atoms of a given element are identical, and atoms are indivisible and indestructible. We know now the last two postulates are not correct [CD-ROM, Screen 2.3]. *(Oesper Collection in the History of Chemistry/University of Cincinnati)* •

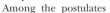

• **Experiments of Thomson and Millikan** The experiments of Thomson and Millikan that uncovered the nature of electrons and protons are illustrated in the Chapter Focus, [CD-ROM, Screens 2.6 and 2.7].

History

A readable account of the history of the discovery of atomic structure is given in the book by Richard Rhodes: *The Making of the Atomic Bomb*, New York, Simon and Schuster, 1986. See also B. M. Peake: *Journal of Chemical Education*, 1989, Vol. 66, p. 738, 1989. •

The experimentally determined mass of a proton is 1.672622×10^{-24} g. The relative charge on the proton, equal in size but opposite in sign to the charge on the electron, is +1.

Neutrons

Because atoms have no net electric charge, the number of positive protons must equal the number of negative electrons in an atom. Most atoms, however, have masses greater than would be predicted on the basis of only protons and electrons, and Rutherford suggested that atoms may contain relatively massive particles with no electric charge. In 1932, the British physicist James Chadwick (1891–1974), a student of Rutherford, succeeded in finding evidence for their existence. Chadwick found very penetrating radiation was released when particles from radioactive polonium struck a beryllium target. This radiation was directed at a paraffin wax target, and he observed protons coming from that target. He reasoned that only a heavy, noncharged particle emanating from the beryllium could have caused this effect. This particle, now known as the **neutron,** has no electric charge and has a mass of 1.674927×10^{-24} g, slightly greater than the mass of a proton.

The Nucleus of the Atom

J. J. Thomson had supposed that an atom was a uniform sphere of positively charged matter within which thousands of electrons circulated in coplanar rings. Thomson and his students thought the only question was the number of electrons circulating within this sphere. About 1910, Ernest Rutherford decided to test Thomson's model. Rutherford had discovered earlier that α rays (see Figure 2.2) consisted of positively charged particles having the same mass as helium atoms. He reasoned that, if Thomson's atomic model were correct, a beam of such massive particles would be deflected very little as it passed through the atoms in a thin sheet of gold foil. Rutherford's associates, Hans Geiger (1882–1945) and Ernst Marsden (1889–1970), set up the apparatus diagrammed in Figure 2.3 and observed what happened when α particles hit the foil. Most passed almost straight through, but a few were deflected at large angles, and some came almost straight back! Rutherford later described this unexpected result by saying, "It was about as credible as if you had fired a 15-inch [artillery] shell at a piece of paper and it came back and hit you" [CD-ROM, Screen 2.9].

The only way for Rutherford to account for their observations was to propose a new model of the atom, in which all of the positive charge and most of the mass of the atom is concentrated in a very small volume. Rutherford called this tiny core of the atom the **nucleus** (see Figure 2.1). The electrons occupy the rest of the space in the atom. From their results Rutherford, Geiger, and Marsden calculated that the nucleus of a gold atom had a positive charge in the range of 100 ± 20 and a radius of about 10^{-12} cm. The currently accepted values are +79 for the charge and about 10^{-13} cm for the radius.

Exercise 2.1 **Describing Atoms**

We know now that the radius of the nucleus is about 0.001 pm, and the radius of an atom is, say, 100 pm. If an atom were a macroscopic object with a diameter of 100 m, it would approximately fill a small football stadium. What would be the diameter of the nucleus of such an atom? Can you think of an object that is about that size?

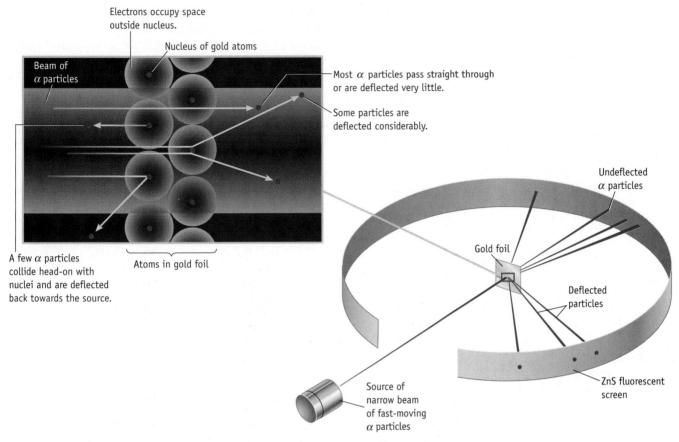

Figure 2.3   Rutherford's experiment to determine the structure of the atom. *(right)* A beam of positively charged α particles was directed at a thin gold foil. A luminescent screen coated with zinc sulfide (ZnS) was used to detect particles passing through. Most of the particles passed through the foil, but some were deflected from their path. A few were even deflected backward. *(left)* Interpretation of the experimental results. See *General Chemistry Interactive CD-ROM, Version 3.0, Screen 2.9.*

2.2   ATOMIC NUMBER AND ATOMIC MASS

Atomic Number

All atoms of the same element have the same number of protons in the nucleus. Hydrogen is the simplest element, with one nuclear proton. All helium atoms have two protons, whereas all lithium atoms have three and all beryllium atoms have four. The number of protons in the nucleus of an element is its **atomic number,** generally given the symbol **Z.**

Currently known elements are listed in the periodic table on the inside front cover of this book. The integer number at the top of the box for each element is its atomic number. A sodium atom, for example, has an atomic number of 11, so its nucleus contains 11 protons. A uranium atom has 92 nuclear protons and $Z = 92$.

● **The Periodic Table Entry for Copper**

29	← atomic number
Cu	← symbol

Relative Atomic Mass and the Atomic Mass Unit

What is the mass of an atom? Chemists in the 18th and 19th centuries recognized that careful experiments could give *relative* atomic masses. For example, the mass

Chapter Goals • Revisited

- **Describe the structure of the atom and define atomic number and mass number.**
- Understand the nature of isotopes and calculate atomic weight from isotope abundances and exact atomic masses.
- Explain the concept of the mole and use molar mass in calculations.
- Know the terminology of the periodic table.

of an oxygen atom was found to be 1.33 times the mass of a carbon atom, and a calcium atom is 2.5 times the mass of an oxygen atom.

Chemistry in the 21st century still uses a system of relative masses. After trying several standards, scientists settled on the current one: a carbon atom having six protons and six neutrons in the nucleus is assigned a mass value of exactly 12. An oxygen atom having eight protons and eight neutrons has 1.3333 times the mass of carbon, and so has a relative mass of 16.000. Masses of atoms of other elements have been assigned in a similar manner.

Masses of fundamental atomic particles are often expressed in **amu** or **atomic mass units.** *One atomic mass unit, 1 amu, is 1/12th of the mass of an atom of carbon with six protons and six neutrons.* Thus, such a carbon atom has a mass of 12.000 amu. The atomic mass unit can be related to other units of mass using a conversion factor; that is, 1 amu = 1.661×10^{-24} g.

Mass Number

Protons and neutrons have masses very close to 1 amu (Table 2.1). The electron, in contrast, has a mass only about 1/2000th of this value. Because proton and neutron masses are so close to 1 amu, the relative mass of an atom can be estimated if the number of neutrons and protons is known. The sum of the number of protons and neutrons for an atom is called its **mass number** and is given the symbol A.

$$A = \text{mass number} = \text{number of protons} + \text{number of neutrons}$$

For example, a sodium atom that has 11 protons and 12 neutrons in its nucleus, has $A = 23$. The most common atom of uranium has 92 protons and 146 neutrons, and $A = 238$. Using this information, we often symbolize atoms with the notation

$$\begin{matrix}\textbf{Mass number} \longrightarrow \\ \textbf{Atomic number} \longrightarrow\end{matrix} {}^{A}_{Z}\text{X} \longleftarrow \textbf{Element symbol}$$

(The subscript Z is optional because the element symbol tells us what the atomic number must be.) For example, the atoms described previously have the symbols ${}^{23}_{11}\text{Na}$ or ${}^{238}_{92}\text{U}$, or just ${}^{23}\text{Na}$ or ${}^{238}\text{U}$. In words, we say "sodium-23" or "uranium-238" [CD-ROM, Screen 2.11].

Example 2.1 Atomic Composition

Problem • What is the composition of an atom of phosphorus with 16 neutrons? What is its mass number? What is the symbol for such an atom? If the atom has an actual mass of 30.9738 amu, what is its mass in grams?

Strategy • All P atoms have the same number of protons, 15, which is given by the atomic number (see the periodic table on the inside front cover). The mass number is the sum of the number of protons and neutrons. The mass of the atom, in grams, can be obtained from the mass in amu (1 amu = 1.661×10^{-24}g) [CD-ROM, Screen 2.11].

Solution • A phosphorus atom has 15 protons, and, because it is electrically neutral, also has 15 electrons.

$$\begin{aligned}\text{Mass number} &= \text{number of protons} + \text{number of neutrons}\\ &= 15 + 16 = 31\end{aligned}$$

The atom's complete symbol is ${}^{31}_{15}\text{P}$.

$$\begin{aligned}\text{Mass of one } {}^{31}\text{P atom} &= (30.9738 \text{ amu})(1.661 \times 10^{-24} \text{ g/amu})\\ &= 5.145 \times 10^{-23} \text{ g}\end{aligned}$$

Table 2.1 • **Properties of Subatomic Particles***

| Particle | Mass | | Charge | Symbol |
	Grams	Relative Mass		
Electron	9.109382×10^{-28}	0.0005485799	-1	$_{-1}^{0}e$ or e^{-}
Proton	1.672622×10^{-24}	1.007276	$+1$	$_{1}^{1}p$ or p^{+}
Neutron	1.674927×10^{-24}	1.008665	0	$_{0}^{1}n$ or n^{0}

*These constants and others in the book are taken from the National Institute of Standards and Technology Web site at http://physics.nist.gov/cuu/Constants/index.html

Exercise 2.2 **Atomic Composition**

(a) What is the mass number of an iron atom with 30 neutrons?

(b) A nickel atom with 32 neutrons has a mass of 59.930788 amu. What is its mass in grams?

(c) How many protons, neutrons, and electrons are in a ^{64}Zn atom?

2.3   ISOTOPES

In only a few instances (for example, aluminum, fluorine, and phosphorus) do all atoms in a naturally occurring sample of a given element have the same mass. Most elements consist of atoms having several different mass numbers. For example, there are two kinds of boron atoms, one with a mass of about 10 amu (^{10}B) and the second with a mass of about 11 amu (^{11}B). Atoms of tin can have any of ten different masses. Atoms with the same atomic number but different mass numbers are called **isotopes** [CD-ROM, Screens 2.10 and 2.17].

All atoms of an element have the same number of protons, five in the case of boron. To have different masses, isotopes must have different numbers of neutrons. The nucleus of a ^{10}B atom ($Z = 5$) contains five protons and five neutrons, whereas the nucleus of a ^{11}B atom contains five protons and six neutrons.

Scientists often refer to a particular isotope by giving its mass number (for example, uranium-238, ^{238}U), but the isotopes of hydrogen are so important that they have special names and symbols. Hydrogen atoms all have one proton. When that is the only nuclear particle, the isotope is called *protium*, or more usually "hydrogen." The isotope of hydrogen with one neutron, $_{1}^{2}H$, is called *deuterium*, or "heavy hydrogen" (symbol = D). The nucleus of radioactive hydrogen, $_{1}^{3}H$, or *tritium* (symbol = T), contains one proton and two neutrons.

The substitution of one isotope of an element for another of that element in a compound sometimes has an interesting effect (Figure 2.4). This is especially true when deuterium is substituted for hydrogen because the mass of deuterium is double that of hydrogen.

Isotope Abundance

A sample of water from a stream or lake will consist almost entirely of H_2O in which the H atoms are the ^{1}H isotope. A few molecules, however, will have deuterium (^{2}H) substituted for ^{1}H. We can predict this because we know that ^{1}H atoms make

• Isotopes
Isotopes are atoms having the same atomic number but different mass numbers.

Boron–10, ^{10}B
5 protons and
5 neutrons

Boron–11, ^{11}B
5 protons and
6 neutrons

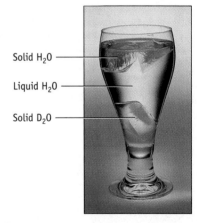

Solid H_2O ———

Liquid H_2O ———

Solid D_2O ———

Figure 2.4   Ice made from "heavy water." Water containing ordinary hydrogen ($_{1}^{1}H$, protium) forms a solid that is less dense ($d = 0.917$ g/cm³ at 0 °C) than liquid H_2O ($d = 0.997$ g/cm³ at 25 °C) and so floats in the liquid. (Water is unique in this regard. The solid phase of virtually all other substances sinks in the liquid phase of that substance.) Similarly, "heavy ice" (D_2O, deuterium oxide) floats in "heavy water." D_2O-ice is denser than H_2O, however, so cubes made of D_2O sink in the liquid phase of H_2O. (*Charles D. Winters*)

Chemical Perspectives

Tritium — A Powerful Tool in Science

Tritium is a radioactive isotope of hydrogen (^{3}H or T) that occurs to the extent of about 1 atom for every 10^{18} hydrogen atoms. The tritium is produced naturally in the upper atmosphere by cosmic radiation. It combines with oxygen to give radioactive water, which reaches the earth in rain. The rain soaks into the ground, where the tritium gradually decays, losing half of its radioactivity in 12.43 years.

Because it is known precisely how rapidly the tritium decays, it is used as a geological "clock" to discover, for example, the age of groundwater. The tritium content of groundwater newly replenished by rain is almost the same as that of rain. As the water sinks into the ground, however, the tritium decays, so "older" water has less tritium. Geochemists tell us that groundwater spreads into the ground relatively slowly. Therefore, by the time it reaches a deep aquifer, the tritium content will likely be considerably diminished. If a deep aquifer is sampled for tritium, and almost none is detected, a geochemist knows the water in the aquifer is at least 100 years old. On the other hand, if tritium is detected in a deep aquifer, then it is clear that it readily replenished by a source of water on the surface.

A geologist in Illinois has said that "Characterization of the groundwater system is essential because it is a principal source of drinking water and the most likely conveyer of contaminants from wastes stored beneath the ground."

After a water sample has been collected, testing for tritium is complicated and laborious. After first removing impurities, the water is then electrolyzed. That is, electricity is passed through the sample, converting the water to elemental hydrogen and oxygen. Normal water, H_2O, is more readily converted than is T_2O, so the sample eventually becomes more concentrated in tritium. The process normally takes about a week to complete, but the tritium content is increased about 20 times, and measurements are then more readily made.

Isotopes of many elements have been used to study biological, agricultural, and industrial processes; in archaeology, fisheries, ocean, and atmospheric research; and in geological and environmental studies. Learn more from sites on the World Wide Web such as the USGS, the United States Geological Survey (www.usgs.gov).

A volcano. Radioactive tritium can be used to discover the source of water vapor vented by the volcano. *(Stephen and Donna O'Meara/Cra-072/Photo Researchers, Inc.)*

---

up 99.895% of the H atoms on the earth. That is, the **percent abundance** of ^{1}H atom, the percentage of atoms of that kind in a sample, is 99.895% [CD-ROM, Screen 2.12].

$$\text{Percent abundance} = \frac{\text{number of atoms of a given isotope}}{\text{total number of atoms of all isotopes of that element}} \times 100\% \quad \textbf{(2.1)}$$

The remainder of naturally occurring hydrogen is deuterium, whose abundance is only 0.015% of the total hydrogen atoms.

Consider the two isotopes of boron. The boron-10 isotope has an abundance of 19.91%; that of boron-11 is 80.09%. This means that if you could count out 10,000 boron atoms from an "average" natural sample, 1991 of them would be boron-10 atoms and 8009 of them would be boron-11 atoms.

Example 2.2   **Isotopes**

Problem • Silver has two isotopes, one with 60 neutrons (percent abundance = 51.839%) and the other with 62 neutrons. What are the mass numbers and symbols of these isotopes? What is the percent abundance of the isotope with 62 neutrons?

Strategy • Recall that the mass number is the sum of the number of proton and neutrons. The symbol is written as ^A_ZX, where X is the one or two-letter element symbol. The percent abundances of all isotopes must add up to 100% [CD-ROM, Screen 2.10].

Solution • Silver has an atomic number of 47, so it has 47 protons in the nucleus. The two isotopes, therefore, have mass numbers of 107 and 109.

Isotope 1, with 47 protons and 60 neutrons

$$A = 47 \text{ protons} + 60 \text{ neutrons} = 107$$

Isotope 2, with 47 protons and 62 neutrons

$$A = 47 \text{ protons} + 62 \text{ neutrons} = 109$$

The first isotope has a symbol $^{107}_{47}Ag$ and the second is $^{109}_{47}Ag$.
Silver-107 has a percent abundance of 51.839%. Therefore, the percent abundance of silver-109 is

$$\% \text{ abundance of } ^{109}Ag = 100.000\% - 51.839\%$$
$$= 48.161\%$$

Exercise 2.3   **Isotopes**

(a) Argon has three isotopes with 18, 20, and 22 neutrons, respectively. What are the mass numbers and symbols of these three isotopes?

(b) Gallium has two isotopes: ^{69}Ga and ^{71}Ga. How many protons and neutrons are in the nuclei of each of these isotopes? If the abundance of ^{69}Ga is 60.1%, what is the abundance of ^{71}Ga?

Determining Exact Atomic Mass and Isotope Abundance

The mass number of an isotope is an approximation of the **exact atomic mass** of the isotope, the experimentally determined mass of the atom. For example, the exact atomic mass of a boron atom with 6 neutrons, ^{11}B, is 11.0093 amu, and the exact atomic mass of an iron atom with 32 neutrons, ^{58}Fe, is 57.9333 amu.

The exact masses of isotopes and their percent abundance are determined experimentally using a mass spectrometer (Figure 2.5). These instruments, crucial in modern chemistry, resemble the canal ray apparatus in the Chapter Focus.

There are two important points regarding exact atomic masses:

• The exact mass of any atom is not an integral number (except for ^{12}C, which is exactly 12 by definition). For example, the exact mass of a gold atom with 118 neutrons, ^{197}Au, is 196.966543 amu, slightly less than the mass number.

• The exact mass of an atom is always slightly less than the sum of the masses of the protons, neutrons, and electrons in the atom. The difference, sometimes called the "mass defect," is related to the energy binding the particles of the nucleus together. See *A Closer Look: Atomic Mass and the Mass Defect* and Chapter 23 [CD-ROM, Screen 2.18].

• **Exact Masses of Some Isotopes**

Atom	Exact Mass (amu)
^4He	4.0092603
^{13}C	13.003355
^{16}O	15.994915
^{58}Ni	57.935346
^{60}Ni	59.930788
^{79}Br	78.918336
^{81}Br	80.916289
^{197}Au	196.966543
^{238}U	238.050784

2.4   ATOMIC WEIGHT

Because every sample of boron has some atoms with a mass of 10.0129 amu and others with a mass of 11.0093 amu, the average atomic mass must be somewhere between these values. The **atomic mass** or **atomic weight** is the average mass of a

• **Atomic Weight and Units**
Values of atomic weight are relative to the mass of the carbon-12 isotope and so are unitless numbers.

A sample is introduced as a vapor into the ionization chamber. There it is bombarded with high-energy electrons that strip electrons from the atoms or molecules of the sample.

The resulting positive particles are accelerated by a series of negatively charged accelerator plates into an analyzing chamber. This chamber is in a magnetic field, which is perpendicular to the direction of the beam of charged particles.

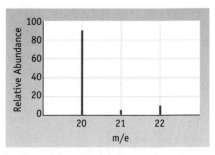

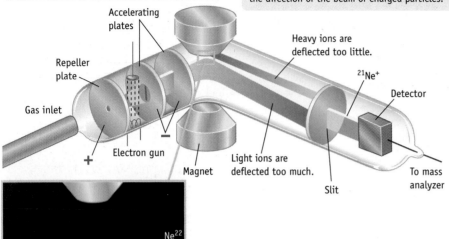

Accelerating plates

Repeller plate

Gas inlet

Electron gun

Magnet

Heavy ions are deflected too little.

^{21}Ne$^+$

Detector

Light ions are deflected too much.

Slit

To mass analyzer

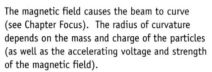

Ne^{22}
Ne^{21}
Ne^{20}

The magnetic field causes the beam to curve (see Chapter Focus). The radius of curvature depends on the mass and charge of the particles (as well as the accelerating voltage and strength of the magnetic field).

A *mass spectrum* is a plot of the relative abundance of the charged particles versus the ratio of mass/charge. Here particles of ^{21}Ne$^+$ are focused on the detector, whereas beams of ions of ^{20}Ne$^+$ and ^{22}Ne$^+$ (of lighter or heavier mass) experience greater and lesser curvature, respectively, and so fail to be detected. By changing the magnetic field, a beam of charged particles of different mass can be focused on the detector, and a spectrum of masses is observed.

Figure 2.5   Mass spectrometer.

A Closer Look

Atomic Mass and the Mass Defect

You might expect that the mass of a deuterium nucleus, ^{2}H, would be the sum of the masses of its constituent particles, a proton and a neutron.

$^1_1$p (1.007276 amu) $+$ $^1_0$n (1.008665 amu)
$\longrightarrow$ $^2_1$H (2.01355 amu)

However, the mass of ^{2}H is *less* than the sum of its constituents!

Change in mass

= mass of product − total mass of reactants

= 2.01355 amu − 2.015941 amu

= −0.00239 amu

The theory is that this "missing mass" is converted to energy, the *binding energy*. This energy can be calculated from Einstein's equation that relates the change in mass (Δm) to the change in energy (ΔE).

$$\Delta E = (\Delta m)c^2$$

where c is the velocity of light. For deuterium, the energy is enormous, 2.1×10^8 kJ/mol of deuterium nuclei [CD-ROM, Screen 2.18].

The mass of an atom is *always* smaller than the sum of the subatomic particles from which it is assembled. For example, the fusion of two H atom nuclei and two neutrons to form a helium nucleus

$$2\,^1_1H + 2\,^1_0n \longrightarrow\,^4_2He$$

results in a mass loss of 0.03038 amu, much larger than for assembling a deuterium nucleus. In fact, this enormous mass loss is the ultimate energy source for life on our planet — the energy provided by our sun. (See the story, Stardust, on page 44.)

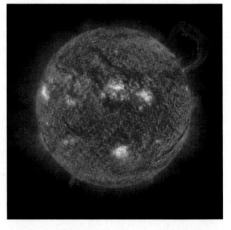

The energy provided by the sun comes from the fusion of hydrogen atoms to form helium. *(NASA)*

representative sample of atoms. For boron, the atomic mass or atomic weight is 10.81, as shown by the following calculation, in which the mass of each isotope is multiplied by its percent abundance divided by 100:

$$\text{Atomic weight} = \left(\frac{19.91}{100}\right)(10.0129) + \left(\frac{80.09}{100}\right)(11.0093)$$
$$= 10.81$$

In general, the atomic weight of an element can be calculated using the equation

$$\text{Atomic weight} = \left(\frac{\text{\% abundance isotope 1}}{100}\right)(\text{atomic weight of isotope 1})$$
$$+ \left(\frac{\text{\% abundance isotope 2}}{100}\right)(\text{atomic weight of isotope 2}) + \cdots \quad \textbf{(2.2)}$$

where (% abundance/100) is the **fractional abundance.** This equation gives an average weighted for the abundance of each isotope for the element. As illustrated by the data in Table 2.2, the atomic weight of an element always reflects the mass of the more abundant isotope or isotopes [CD-ROM, Screen 2.12].

The atomic weight of each stable element has been determined by experiment, and these numbers appear in the periodic table in the front of the book. In the periodic table, each element's box contains the atomic number, the element symbol, and the atomic weight. For unstable (radioactive) elements, the mass number of the most stable isotope is given in parentheses.

(*Chapter Goals* • Revisited)

- Describe the structure of the atom and define atomic number and mass number.
- **Understand the nature of isotopes and calculate atomic weight from isotope abundances and exact atomic masses.**
- Explain the concept of the mole and use molar mass in calculations.
- Know the terminology of the periodic table.

- **Atomic Mass or Weight?**
Chemists usually refer to the "atomic weight" of an element rather than the "atomic mass." Although the quantity is more properly called a "mass" than a "weight," "atomic weight" is commonly used.

- **The Periodic Table Entry for Copper**

29	← atomic number
Cu	← symbol
63.546	← atomic weight

Table 2.2 • Isotope Abundance and Atomic Weight

Element	Symbol	Atomic Weight	Mass Number	Isotopic Mass (amu)	Natural Abundance (%)
Hydrogen	H	1.00794	1	1.0078	99.985
	D*		2	2.0141	0.015
	T†		3	3.0161	0
Boron	B	10.811	10	10.0129	19.91
			11	11.0093	80.09
Neon	Ne	20.1797	20	19.9924	90.48
			21	20.9938	0.27
			22	21.9914	9.25
Magnesium	Mg	24.305	24	23.9850	78.99
			25	24.9858	10.00
			26	25.9826	11.01

*D = deuterium; †T = tritium, radioactive

Problem-Solving Tip 2.1

Atomic Weight from Isotope Data

The atomic weight is a "weighted" average of isotope masses. A good analogy to this is the calculation of your grade. Let us say you have five quizzes in chemistry; three of them have a grade of 9 (out of 10), another is a 5, and another is a 6. You calculate the quiz average by

$$\text{Average grade} = \frac{9 + 9 + 9 + 5 + 6}{5}$$

$$= \frac{(3)(9) + 5 + 6}{5} = \frac{3}{5}(9) + \frac{1}{5}(5) + \frac{1}{5}(6)$$

The last equation tells us that three fifths of your grades were 9's and one fifth each of your grades were a 5 and a 6. Another way to write this is

$$\text{Average grade} = (0.6)(9) + (0.2)(5) + (0.2)(6)$$

$$= 7.6$$

Your average grade reflects the fact that 60% of your quiz grades were 9's. That is, your quiz average was "weighted" more heavily by the three most abundant 9 grades and less heavily by the less abundant grades of 5 and 6. In the same way, the atomic weight of an element reflects the fact that the average is weighted by the more abundant isotope or isotopes.

Example 2.3 Calculating Atomic Weight from Isotope Abundance

Problem • Bromine (used to make silver bromide, the important component of photographic film) has two naturally occurring isotopes, one with a mass of 78.918336 amu and an abundance of 50.69%. The other isotope, of mass 80.916289 amu, has an abundance of 49.31%. Calculate the atomic weight of bromine.

Strategy • The atomic weight of any element is the weighted average of the masses of all the isotopes in a representative sample. To calculate the atomic mass, use Equation 2.2 [CD-ROM, Screen 2.12].

Solution •

Average atomic weight of bromine

$$= (0.5069)(78.918336) + (0.4931)(80.916289)$$

$$= 79.90$$

Exercise 2.4 Calculating Atomic Weight

Verify that the atomic weight of chlorine is 35.45, given the following information:

^{35}Cl, mass = 34.96885 amu; percent abundance = 75.77%
^{37}Cl, mass = 36.96590 amu; percent abundance = 24.23%

Chapter Goals • Revisited

- Describe the structure of the atom and define atomic number and mass number.
- Understand the nature of isotopes and calculate atomic weight from isotope abundances and exact atomic masses.
- **Explain the concept of the mole and use molar mass in calculations.**
- Know the terminology of the periodic table.

2.5 ATOMS AND THE MOLE

One of the most exciting parts of chemistry is the discovery of some new substance. Chemistry is also a quantitative science, however. When two chemicals react with each other, we want to know how many atoms of each are used so that a formula of the product can be established. To do this, we need some method of counting atoms, no matter how small they are. That is, we must discover a way of connecting the macroscopic world, the world we can see, with the particulate world of atoms, molecules, and ions. The solution to this problem is to define a convenient unit of matter that contains a known number of particles. The chemical counting unit is the **mole** [CD-ROM, Screens 2.15 and 2.16].

The word "mole" was introduced about 1896 by Friedrich Wilhelm Ostwald (1853–1932), who derived the term from the Latin word *moles*, meaning a "heap" or "pile." The mole, whose symbol is mol, is the SI base unit for measuring an *amount of substance* (see Table 1.2) and is defined as follows:

> A **mole** is the amount of substance that contains as many elementary entities (atoms, molecules, or other particles) as there are atoms in exactly 12 g of the carbon-12 isotope.

The key to understanding the concept of the mole is that *one mole always contains the same number of particles, no matter what the substance*. One mole of sodium contains the same number of atoms as one mole of iron. But how many particles? Many, many experiments over the years have established that number as

$$1 \text{ mole} = 6.02214199 \times 10^{23} \text{ particles}$$

This value is known as **Avogadro's number** in honor of Amedeo Avogadro, an Italian lawyer and physicist who conceived the basic idea (but never determined the number).

History

Amedeo Avogadro (1776–1856)
Amedeo Avogadro, conte di Quaregna, an Italian nobleman, was a lawyer. In about 1800, he turned to science and was the first professor of mathematical physics in Italy. *(E. F. Smith Collection/VanPelt Library/ University of Pennsylvania)*   ●

The Molar Mass

The mass in grams of one mole of atoms of any element ($6.02214199 \times 10^{23}$ atoms of that element) is the **molar mass** of that element. Molar mass is conventionally abbreviated with a capital italicized M and is expressed in units of grams per mole (g/mol). For elements *molar mass is an amount in grams numerically equal to the atomic mass in atomic mass units*. Using sodium and lead as examples,

$$
\begin{aligned}
\text{Molar mass of sodium (Na)} &= \text{mass of exactly 1 mol of Na atoms} \\
&= 22.99 \text{ g/mol} \\
&= \text{mass of } 6.022 \times 10^{23} \text{ Na atoms} \\
\text{Molar mass of lead (Pb)} &= \text{mass of exactly 1 mol of Pb atoms} \\
&= 207.2 \text{ g/mol} \\
&= \text{mass of } 6.022 \times 10^{23} \text{ Pb atoms}
\end{aligned}
$$

The relative physical sizes of 1-mol quantities of some common elements are shown in Figure 2.6. Although each of these "piles of atoms" has a different volume and different mass, each contains 6.022×10^{23} atoms.

The mole concept is the cornerstone of quantitative chemistry. It is essential to be able to convert from moles to mass and from mass to moles. Dimensional analysis, which is described in Section 1.7 and in Appendix A, shows that this can be done in the following way:

● **How Big Is Avogadro's Number?**
Avogadro's number of unpopped popcorn kernels spread over the continental United States would cover the country to a depth of about 9 miles. One mole of pennies divided equally among every man, woman, and child in the United States would allow each person to pay off the national debt ($5.7 trillion or 5.7×10^{12} dollars) and still have 15 trillion dollars left over!

● **Is Avogadro's Number Special?**
No. It is fixed by the definition of the mole as exactly 12 g of carbon-12. If one mole of carbon were defined to have some other mass, then Avogadro's number would have a different value.

MASS ⟷ MOLES CONVERSION

Moles to Mass	*Mass to Moles*
$\text{Moles} \cdot \dfrac{\text{grams}}{1 \text{ mol}} = \text{grams}$	$\text{Grams} \cdot \dfrac{1 \text{ mol}}{\text{grams}} = \text{moles}$
↑	↑
molar mass	1/molar mass

Figure 2.6   One-mole quantities of common elements. *(left to right)* Sulfur powder, 32.066 g; magnesium chips, 24.305 g; tin, 118.710 g; silicon, 28.0855 g. *(above)* Copper beads, 63.546 g. *(Charles D. Winters)*

For example, what mass, in grams, is represented by 0.35 mol of aluminum? Using the molar mass of aluminum (27.0 g/mol), you find that 0.35 mol of Al has a mass of 9.5 g.

$$0.35 \text{ mol Al} \cdot \frac{27.0 \text{ g Al}}{1 \text{ mol Al}} = 9.5 \text{ g Al}$$

● **Amount and Quantity**
The terms "amount" and "quantity" are used in a specific sense by chemists. The *amount* of a substance is the number of moles of that substance. *Quantity* refers to the mass of the substance. See W. G. Davies and J. W. Moore: *Journal of Chemical Education,* Volume 57, page 303, 1980. See also physics.nist.gov/PUBS/SP811 on the Internet.

Look at the periodic table in the front of the book and notice that some atomic masses are known to more significant figures and decimal places than others. When using molar masses in a calculation, the convention followed in this book is to use one more significant figure in the molar mass than in any of the other data. For example, if you weigh out 16.5 g of carbon, you use 12.01 g/mol for the molar mass of C to find the amount of carbon present.

$$16.5 \text{ g C} \cdot \frac{1 \text{ mol C}}{12.01 \text{ g C}} = 1.37 \text{ mol C}$$
$$\uparrow$$

Note that four significant figures are used in the molar mass, but there are three in the sample mass.

Using one more significant figure means the accuracy of the molar mass is greater than the other numbers and does not limit the accuracy of the result.

Example 2.4   **Mass, Moles, and Atoms**

Problem • Consider two elements in the same vertical column of the periodic table: lead and tin.

(a) What mass of lead, in grams, is equivalent to 2.50 mol of lead (Pb, atomic number = 82)?

(b) What amount of tin, in moles, is represented by 36.5 g of tin (Sn, atomic number = 50)? How many atoms of tin are in the sample?

(continued)

Strategy • The molar masses of lead (207.2 g/mol) and of tin (118.7 g/mol) are required and can be found in the periodic table in the front of the book. Avogadro's number is needed to convert the amount of each element to number of atoms [CD-ROM, Screen 2.16].

Solution •

(a) Convert the amount of lead in moles to mass in grams.

$$2.50 \text{ mol Pb} \cdot \frac{207.2 \text{ g Pb}}{1 \text{ mol Pb}} = 518 \text{ g Pb}$$

(b) First convert the mass of tin to the amount in moles.

$$36.5 \text{ g Sn} \cdot \frac{1 \text{ mol Sn}}{118.7 \text{ g Sn}} = 0.307 \text{ mol Sn}$$

A 150-mL beaker containing 2.50 mol or 518 g of lead. *(Charles D. Winters)*

Finally, use the relation between atoms and moles (Avogadro's number) to find the number of atoms in the sample.

$$0.307 \text{ mol Sn} \cdot \frac{6.022 \times 10^{23} \text{ atoms Sn}}{1 \text{ mol Sn}} = 1.85 \times 10^{23} \text{ atoms Sn}$$

A sample of tin having a mass of 36.5 g of the metal (and 1.85×10^{23} atoms). *(Charles D. Winters)*

Example 2.5   Mole Calculation

Problem • The graduated cylinder in the photo contains 32.0 cm³ of mercury. If the density of mercury at 25 °C is 13.534 g/cm³, what amount of mercury, in moles, is in the cylinder?

Strategy • Volume and moles of mercury are not directly connected. Therefore, you must use the density of mercury to first find the mass of the metal and then calculate the amount, in moles, from the mass.

$$\text{Volume, cm}^3 \xrightarrow[\text{use density}]{\times \frac{\text{g}}{\text{cm}^3}} \text{Mass, g} \xrightarrow[\text{use molar mass}]{\times \frac{\text{mol}}{\text{g}}} \text{Moles}$$

Solution • Combining the volume and density gives the mass of the mercury.

$$32.0 \text{ cm}^3 \cdot \frac{13.534 \text{ g}}{1 \text{ cm}^3} = 433 \text{ g}$$

Finally, the number of moles of mercury can be calculated.

$d = \frac{m}{v}$

$$433 \text{ g} \cdot \frac{1 \text{ mol Hg}}{200.6 \text{ g Hg}} = 2.16 \text{ mol Hg}$$

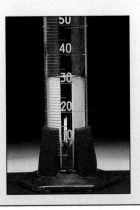

A graduated cylinder containing 32.0 cm³ of mercury. This is equivalent to 433 g or 2.16 mol of mercury. *(Charles D. Winters)*

Example 2.6   **Mass of an Atom**

Problem • What is the average mass of an atom of platinum (Pt)?

Strategy • The mass of 1 mol of platinum is 195.08 g. Each mole contains Avogadro's number of atoms.

Solution • Divide the mass of a mole by Avogadro's number.

$$\frac{195.08 \text{ g Pt}}{1 \text{ mol Pt}} \cdot \frac{1 \text{ mol Pt}}{6.02214 \times 10^{23} \text{ atoms}} = \frac{3.2394 \times 10^{-22} \text{ g}}{1 \text{ atom Pt}}$$

Exercise 2.5   **Mass/Mole Conversions**

(a) What is the mass, in grams, of 1.5 mol of silicon?

(b) What amount (in moles) of sulfur is represented by 454 g? How many atoms?

(c) What is the mass of one sulfur atom?

Exercise 2.6   **Atoms**

The density of gold, Au, is 19.32 g/cm^3. What is the volume (in cubic centimeters) of a piece of gold that contains 2.6×10^{24} atoms? If the piece of metal is a square with a thickness of 0.10 cm, what is the length, in centimeters, of one side of the piece?

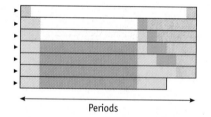

Periods

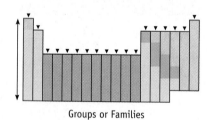

Groups or Families

● **Two Ways to Designate Groups**
One way to designate periodic table groups is to number them 1 through 18 from left to right. This method is used generally outside of the United States. The system predominant in the United States labels main group elements as Groups 1A–8A and transition elements as Groups 1B–8B. References to elements in this book shall use the A/B system.

2.6   THE PERIODIC TABLE

The periodic table of elements in the front of the book is one of the most useful tools in chemistry [CD-ROM, Screens 2.13 and 2.14]. Not only does it contain a wealth of information, but it can be used to organize many of the ideas of chemistry. It is important that you be familiar with its main features and terminology.

Features of the Periodic Table

The main organizational features of the periodic table are the following:

- Elements are arranged so that those with similar chemical and physical properties lie in vertical columns called **groups** or **families.** The table commonly used in the United States has groups numbered 1 through 8, with each number followed by a letter: A or B. The A groups are often called the **main group elements** and the B groups are the **transition elements.**

- The horizontal rows of the table are called **periods**, and they are numbered beginning with 1 for the period containing only H and He. For example, sodium, Na, is in Group 1A and is the first element in the third period. Mercury, Hg, is in Group 2B and in the sixth period (or sixth row).

The periodic table can be divided into several regions according to the properties of the elements. On the table on the inside front cover of this book, elements that behave as *metals* are indicated in purple, those that are *nonmetals* are indicated in yellow, and elements called *metalloids* are in green. Elements gradually become less metallic as one moves from left to right across a period, and the metalloids lie along the metal–nonmetal boundary. Some elements are shown in Figure 2.7.

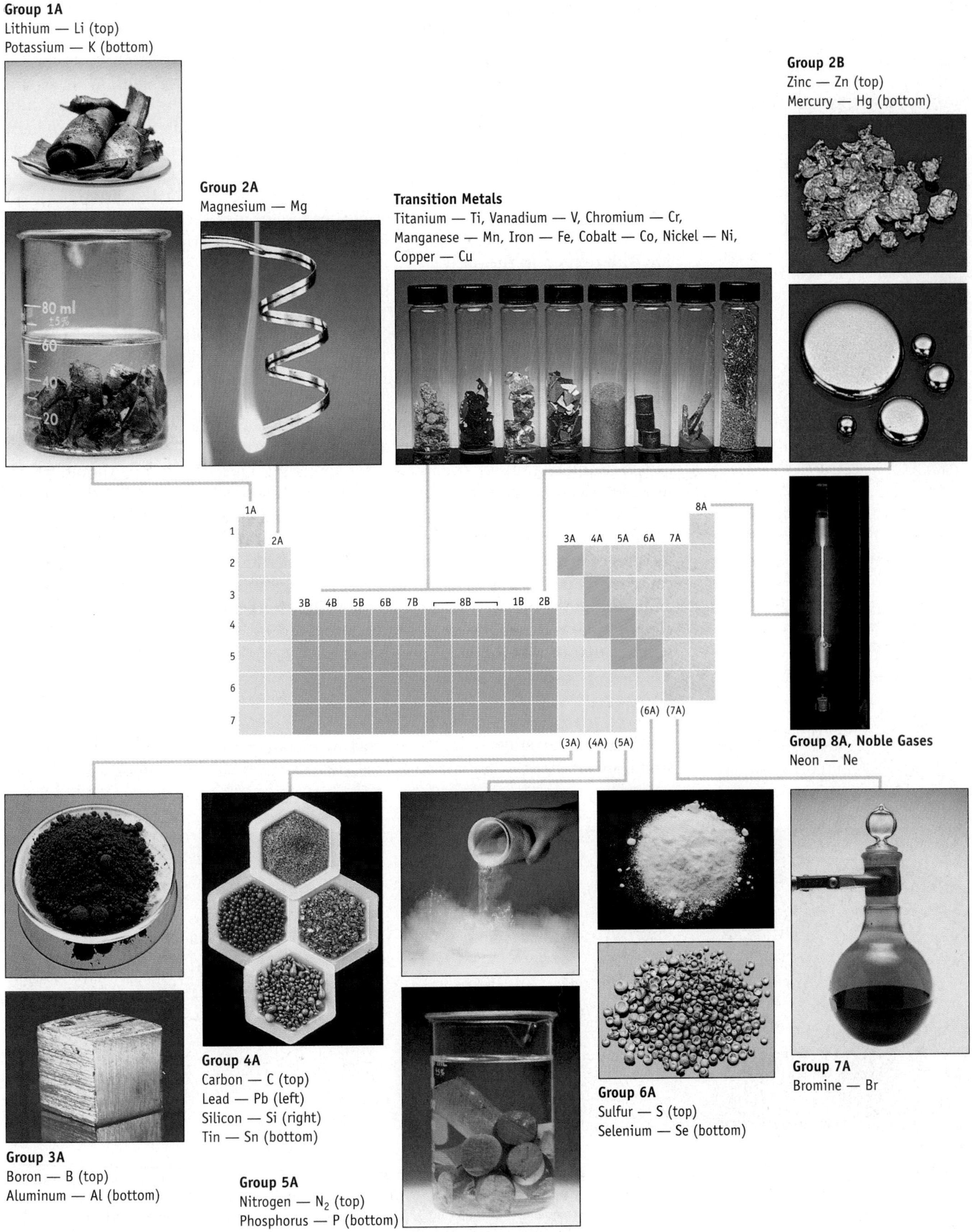
Group 1A
Lithium — Li (top)
Potassium — K (bottom)
80 ml
60
40
20
Group 2A
Magnesium — Mg
Transition Metals
Titanium — Ti, Vanadium — V, Chromium — Cr,
Manganese — Mn, Iron — Fe, Cobalt — Co, Nickel — Ni,
Copper — Cu
Group 2B
Zinc — Zn (top)
Mercury — Hg (bottom)
1A
2A
3B 4B 5B 6B 7B 8B 1B 2B
3A 4A 5A 6A 7A
8A
1
2
3
4
5
6
7
(3A) (4A) (5A)
(6A) (7A)
Group 8A, Noble Gases
Neon — Ne
Group 3A
Boron — B (top)
Aluminum — Al (bottom)
Group 4A
Carbon — C (top)
Lead — Pb (left)
Silicon — Si (right)
Tin — Sn (bottom)
Group 5A
Nitrogen — N_2 (top)
Phosphorus — P (bottom)
Group 6A
Sulfur — S (top)
Selenium — Se (bottom)
Group 7A
Bromine — Br

Figure 2.7 **Some of the 113 known elements.** *(Charles D. Winters)*

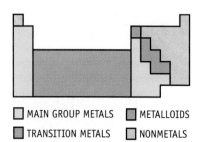

☐ MAIN GROUP METALS   ☐ METALLOIDS
☐ TRANSITION METALS   ☐ NONMETALS

The elements can be classified as metals, nonmetals, and metalloids. They occupy well-defined regions of the periodic table.

(**Chapter Goals ● Revisited**)

- Describe the structure of the atom and define atomic number and mass number.

- Understand the nature of isotopes and calculate atomic weight from isotope abundances and exact atomic masses.

- Explain the concept of the mole and use molar mass in calculations.

- **Know the terminology of the periodic table.**

You are probably familiar with many properties of **metals** from everyday experience (Figure 2.8a). Metals are solids (except for mercury), conduct electricity, are usually ductile (can be drawn into wires) and malleable (can be rolled into sheets), and can form alloys (solutions of one or more metals in another metal). Iron (Fe) and aluminum (Al) are used in automobile parts because of their ductility, malleability, and low cost relative to other metals. Copper (Cu) is used in electric wiring because it conducts electricity better than most metals. Chromium (Cr) is plated onto automobile parts, not only because its metallic luster makes cars look better but also because chrome-plating protects the underlying metal from reacting with oxygen in the air.

Nonmetals have a wide variety of properties. Some are solids; bromine is a liquid (Figure 2.8b), and a few, like nitrogen and oxygen, are gases at room temperature. With the exception of carbon in the form of graphite, nonmetals do not conduct electricity, which is one of the main features that distinguishes them from metals. All nonmetals lie to the right of a diagonal line that stretches from B to Te in the periodic table.

Some of the elements along the diagonal line from B to Te have properties that make them difficult to classify as a metal or nonmetal. Chemists have come to call them metalloids or, sometimes, semimetals (Figure 2.8c). You should know, however, that chemists often disagree, not only about what a metalloid is but also what elements fit this category. We shall define a **metalloid** as an element that has some of the physical characteristics of a metal but some of the chemical characteristics of a nonmetal, and we shall include only B, Si, Ge, As, Sb, and Te in the category. This may not be a completely satisfying distinction, but it does reflect the ambiguity in the behavior of these elements. Antimony (Sb), for example, conducts electricity as well as many elements that are truly metals. Its chemistry, however, resembles that of the nonmetal phosphorus.

Developing the Periodic Table

Although the arrangement of elements in the periodic table can now be understood on the basis of atomic structure [➡ CHAPTER 8], the table was originally de-

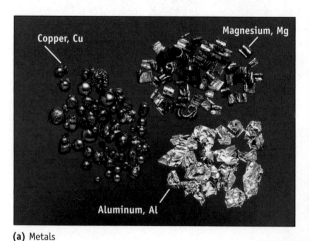

(a) Metals

(b) Nonmetals

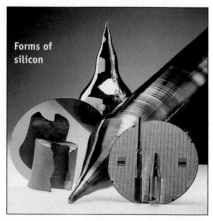

(c) Metalloids

Figure 2.8   Representative elements. (a) Magnesium, aluminum, and copper are metals. All can be drawn into wires and conduct electricity. **(b)** Only 15 or so elements can be classified as nonmetals. Here are orange liquid bromine and solid purple iodine. **(c)** Only six elements are generally classified as metalloids or semimetals. This photograph is of solid silicon in various forms including a wafer on which are printed electronic circuits. *(Charles D. Winters)*

veloped from many, many experimental observations of the chemical and physical properties of elements and is the result of the ideas of a number of chemists in the 18th and 19th centuries.

In 1869, at the University of St. Petersburg in Russia, Dmitri Ivanovitch Mendeleev was pondering the properties of the elements as he wrote a textbook on chemistry. On studying the chemical and physical properties of the elements, he realized that, if the elements were arranged in order of increasing atomic mass, elements with similar properties appeared in a regular pattern. That is, he saw a **periodicity,** or periodic repetition, of the properties of elements. Mendeleev organized the known elements into a table by lining them up in a horizontal row in order of increasing atomic mass. Every time he came to an element with properties similar to one already in the row, he started a new row. For example, the elements Li, Be, B, C, N, O, and F were in a row. When he came to Na, its properties closely resembled those of Li, so he started a new row. The columns, then, contained elements such as Li, Na, and K with similar properties [➡ CHAPTER 8].

The most important feature of Mendeleev's table — and a mark of his genius and daring — was that he left an empty space in a column when an element was not known but should exist and have properties similar to the element above it in his table. He deduced that these spaces would be filled by as-yet undiscovered elements. For example, a space was left between Si (silicon) and Sn (tin) in what is now Group 4A. Based on the progression of properties in this group, Mendeleev was able to predict the properties of this missing element. The discovery of germanium in 1886 verified one of Mendeleev's predictions.

A problem with Mendeleev's table was that the elements were ordered by increasing mass. A glance at a modern table shows that, on this basis, Ni and Co, among others, should be reversed. Mendeleev assumed the atomic masses known at that time were inaccurate, not a bad assumption based on analytical methods then in use. In fact his order was correct and what was wrong was his assumption that element properties were a function of their mass.

In 1913 H. G. J. Moseley (1887–1915), a young English scientist working with Ernest Rutherford, corrected Mendeleev's assumption. Moseley was doing experiments in which he bombarded many different metals with electrons in a cathode ray tube (see Chapter Focus) and examined the x-rays emitted in the process. In seeking some order in his data, he realized that the wavelength of the x-rays emitted by a given element were related in a precise manner to the *atomic number* of the element. Indeed, chemists quickly recognized that, if the elements were organized in a table by increasing atomic number, the defects in the Mendeleev table were corrected. The **law of chemical periodicity** is now stated as "the properties of the elements are periodic functions of atomic number."

2.7   AN OVERVIEW OF THE ELEMENTS, THEIR CHEMISTRY, AND THE PERIODIC TABLE

The vertical columns, or groups, of the periodic table contain elements having similar chemical and physical properties, and several groups of elements have distinctive names that are useful to know [💿 CD-ROM, Screens 2.16 and 2.17].

Group 1A, the Alkali Metals: Li, Na, K, Rb, Cs, Fr

Elements in the leftmost column, Group 1A, are known as the **alkali metals.** All are metals and are solids at room temperature. The metals of Group 1A are all reactive. For example, they react with water to produce hydrogen and alkaline solutions

● **Periodic Table and the Scientific Method**
The development of the periodic table illustrates the way chemistry has developed: experimental observations led to empirical correlations of properties and then to the prediction of results to be tested by further experiments. Once those predictions were tested, a theory could then be developed.

● **Placing H in the Periodic Table**
Where to place H? It is clearly not an alkali metal and so does not belong with the alkali metals. However, in its reactions it forms a 1+ ion just like the alkali metals. For this reason H is often placed in Group 1A.

Figure 2.9   Alkali metals. (a) Cutting a bar of sodium with a knife is about like cutting a stick of cold butter. **(b)** When an alkali metal such as potassium is treated with water, a vigorous reaction occurs, giving an alkaline solution and hydrogen gas, which burns in air. *(Charles D. Winters)*

(a)   **(b)**

(Figure 2.9). Because of their reactivity, these metals are only found in nature combined in compounds (such as NaCl) (◀ SECTION 1.2), never as the free element.

 ### Group 2A, the Alkaline Earth Metals: Be, Mg, Ca, Sr, Ba, Ra

● **Alkali and Alkaline**
The word "alkali" comes from the Arabic language; ancient Arabian chemists discovered that ashes of certain plants, which they called *al-qali,* gave water solutions that felt slippery and burned the skin. These ashes contain compounds of Group 1A elements that produce alkaline (basic) solutions.

Elements in Group 2A are called **alkaline earth elements.** These metallic elements occur naturally only in compounds (Figure 2.10). Except for beryllium (Be), these elements also react with water to produce alkaline solutions, and most of their oxides (such as lime, CaO) form alkaline solutions. Magnesium (Mg) and calcium (Ca) are the seventh and fifth most abundant elements in the earth's crust, respectively (Table 2.3). Calcium is especially well known. It is one of the important elements in teeth and bones, and it occurs in vast limestone deposits. Calcium carbonate ($CaCO_3$) is the chief constituent of limestone and of corals, sea shells, marble, and chalk (see Figure 2.10b). Radium (Ra), the heaviest alkaline earth element, is radioactive and is used to treat some cancers by radiation.

Figure 2.10   The alkaline earth metals. (a) When heated in air, magnesium burns to give magnesium oxide. The white sparks you see in burning fireworks are burning magnesium. **(b)** Some common calcium-containing substances: calcite (the clear crystal); a seashell; limestone; and an over-the-counter remedy for eliminating excess stomach acid. *(a, James Cowlin/ Image Enterprises, Phoenix, AZ; b, Charles D. Winters)*

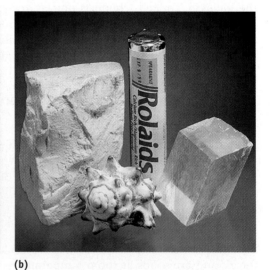

(a)   **(b)**

Group 3A: B, Al, Ga, In, Tl

Group 3A contains one element of great importance, aluminum (Figure 2.11). This element and three others (Ga, In, Tl) are metals, whereas boron (B) is a metalloid. Aluminum (Al) is the most abundant metal in the earth's crust. It is exceeded in abundance only by the nonmetals oxygen and silicon. These three elements are found combined in clays and other common minerals.

Boron occurs in the mineral borax, which is mined in Death Valley, California, for use as a cleaning agent, antiseptic, and flux for metal work. In the late 19th century the mineral was hauled out of the valley in wagons drawn by 20 mules, hence the name of a once-popular washing powder Twenty Mule Team Borax.

Group 4A: C, Si, Ge, Sn, Pb

Thus far all the elements we have described, except boron, have been metals. Beginning with Group 4A, however, the groups contain more and more nonmetals. In Group 4A there is a nonmetal, carbon (C), two metalloids, silicon (Si) and germanium (Ge), and two metals, tin (Sn) and lead (Pb). Because of the change from nonmetallic to metallic behavior, more variation occurs in the properties of the elements of this group than in most others.

Carbon is the basis for the great variety of chemical compounds that make up living things. On earth it is found in the atmosphere as CO_2, in carbonates like limestone (see Figure 2.10b), and in coal, petroleum, and natural gas — the fossil fuels.

One of the most interesting aspects of the chemistry of the nonmetals is that a particular element can often exist in several different and distinct forms, called **allotropes,** each having its own properties. Carbon has at least three allotropes, the best known of which are graphite and diamond (Figure 2.12). The flat sheets of carbon atoms in graphite (Figure 2.12a) cling only weakly to one another. One layer can slip easily over another, which explains why graphite is soft, is a good lubricant, and is used in pencil lead. (Pencil "lead" is not the element lead but a composite of clay and graphite that leaves a trail of graphite on the page as you write.)

In diamond each carbon atom is connected to four others at the corners of a tetrahedron, and this extends throughout the solid (see Figure 2.12b). This

Table 2.3 • The 10 Most Abundant Elements in the Earth's Crust

Rank	Element	Abundance (ppm)*
1	Oxygen	474,000
2	Silicon	277,000
3	Aluminum	82,000
4	Iron	41,000
5	Calcium	41,000
6	Sodium	23,000
7	Magnesium	23,000
8	Potassium	21,000
9	Titanium	5,600
10	Hydrogen	1,520

*ppm = g per 1000 kg

(a)

(b)

Figure 2.11   Group 3A elements.
(a) Boron is mined as borax, a natural compound used in soap. The borax was mined in Death Valley, California, at the end of the 19th century and was hauled from the mines in wagons drawn by teams of 20 mules. Boron is also a component of borosilicate glass, the glass used for laboratory glassware. **(b)** A ruby is composed of aluminum oxide with a trace of chromium oxide. Aluminum is abundant in the earth's crust. It is found in all clays and is present in many minerals and gems.
(a, Gregory P. Logiodice; b, Charles D. Winters)

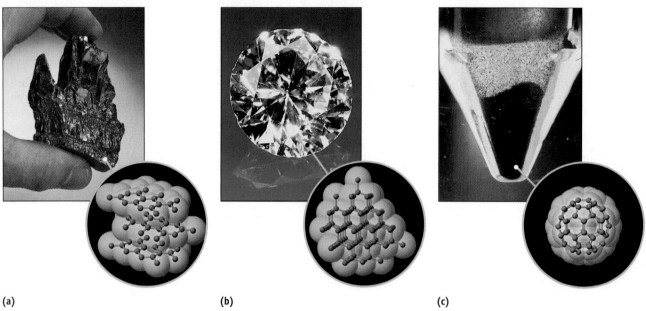

(a)   (b)   (c)

Figure 2.12   The allotropes of carbon. (a) Graphite consists of layers of carbon atoms. Each carbon atom is linked to three others to form a sheet of six-member, hexagonal rings. **(b)** In diamond the carbon atoms are also arranged in six-member rings, but the rings are not flat because each C atom is connected tetrahedrally to four other C atoms. **(c)** Buckyballs. A member of the family called buckminsterfullerenes, C_{60}, is an allotrope of carbon. Sixty carbon atoms are arranged in a spherical cage that resembles a hollow soccer ball. Notice that each six-member ring is part of three other six-member rings and three five-member rings. Chemists call this molecule a "buckyball." *(Charles D. Winters)*

Figure 2.13   Compounds containing silicon. Ordinary clay, sand, and many gemstones are based on compounds of silicon and oxygen. Here amethyst and quartz lie in a bed of sand. All are silicon dioxide, SiO_2. *(Charles D. Winters)*

structure causes diamonds to be extremely hard, denser than graphite ($d = 3.51$ g/cm^3 for diamond and $d = 2.22$ g/cm^3 for graphite), and chemically less reactive than graphite. Because diamonds are not only hard but are excellent conductors of heat, they are used on the tips of metal- and rock-cutting tools.

In the late 1980s another form of carbon was identified as a component of black soot, the stuff that collects when carbon-containing materials are burned in a deficiency of oxygen. This substance is made up of molecules with 60 carbon atoms arranged as a spherical "cage" (Figure 2.12c). You may recognize that the surface is made up of five- and six-member rings and resembles a soccer ball. The shape also reminded its discoverers of an architectural dome invented several decades ago by the American philosopher and engineer, R. Buckminster Fuller. The official name of the allotrope is therefore buckminsterfullerene, and chemists often call these molecules "buckyballs."

Silicon is the basis of many minerals such as clay, quartz, and beautiful gemstones like amethyst (Figure 2.13). Tin and lead have been known for centuries because they are easily obtained from their ores. Tin alloyed with copper makes bronze, which was used for hundreds of years in utensils and weapons. Lead has been used in water pipes and paint, even though the element is toxic to humans.

Group 5A: N, P, As, Sb, Bi

Nitrogen, in the form of N_2 (Figures 2.7 and 2.14), makes up about three-fourths of earth's atmosphere. It is also incorporated in biochemically important substances

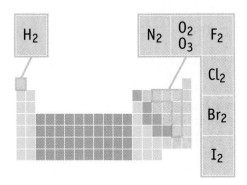

Figure 2.14   Elements that exist as diatomic or polyatomic molecules. Seven of the elements in the periodic table exist as diatomic, or two-atom, molecules. Ozone, an allotrope of oxygen, has three O atoms in the molecule.

such as chlorophyll, proteins, and DNA. Therefore, scientists have long sought ways of "fixing" atmospheric nitrogen (forming compounds from the element). Nature accomplishes this easily in plants, but severe methods (high temperatures, for example) must be used in the laboratory and in industry to cause N_2 to react with other elements (such as H_2 to make ammonia, NH_3, which is widely used as a fertilizer).

Phosphorus is also essential to life as an important constituent in bones and teeth. The element glows in the dark if it is in the air, and its name, based on Greek words meaning "light-bearing," reflects this. This element also has several allotropes, the most important being white and red phosphorus (see Figure 2.7).

Both forms of phosphorus are used commercially. White phosphorus ignites spontaneously in air and so is normally stored under water. However, when it does react with air, it forms P_4O_{10}, which is then turned into phosphoric acid (H_3PO_4) for use in food products such as soft drinks. Red phosphorus also reacts with oxygen in the air and so is used in the striking strips on match books.

Group 6A: O, S, Se, Te, Po

Group 6A begins with oxygen, which constitutes about 20% of earth's atmosphere and which combines readily with most other elements. Most of the energy that powers life on earth is derived from reactions in which oxygen combines with other substances.

Sulfur has been known in elemental form since ancient times as brimstone or "burning stone" (Figure 2.15). Sulfur, selenium, and tellurium are referred to collectively as *chalcogens* (from the Greek word, *khalkos*, for copper) because copper ores contain them. Their compounds can be foul-smelling and poisonous; nevertheless sulfur and selenium are essential components of the human diet. By far the most important compound of sulfur is sulfuric acid (H_2SO_4), the compound the chemical industry manufactures in the largest amount.

As in Group 5A, the second- and third-period elements have different structures. Like nitrogen, oxygen is also a diatomic molecule (see Figure 2.14). Unlike nitrogen, however, oxygen has an allotrope, the well-known compound ozone, O_3. Sulfur, which can be found in nature as a yellow solid, has many allotropes. The most common allotrope consists of eight-member, crown-shaped rings of sulfur atoms (see Figure 2.15).

Polonium, a radioactive element, was isolated in 1898 by Marie and Pierre Curie, who separated it from tons of a uranium-containing ore and named it for Madame Curie's native country, Poland.

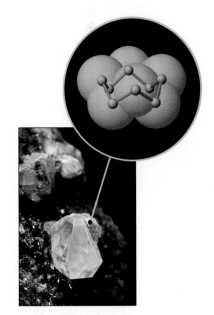

Figure 2.15   Sulfur. The most common allotrope of sulfur consists of eight-member, crown-shaped rings. *(Charles D. Winters)*

Group 7A, Halogens: F, Cl, Br, I, At

At the far right of the periodic table are two groups composed entirely of nonmetals. The Group 7A elements — fluorine, chlorine, bromine, and iodine — all exist as diatomic molecules (see Figure 2.14). All combine violently with alkali metals to form salts such as table salt, NaCl (see Figure 1.2). The name for this group, the **halogens,** comes from the Greek words *hals,* meaning "salt," and *genes,* for "forming." The halogens react with many other metals to form compounds, and they also combine with most nonmetals. They are among the most reactive of elements.

Group 8A, Noble Gases: He, Ne, Ar, Kr, Xe, Rn

The Group 8A elements — helium, neon, argon, krypton, xenon, and radon — are the least reactive elements (Figure 2.16). All are gases, and none is abundant on earth or in earth's atmosphere. Because of this, they were not discovered until the end of the 19th century. Helium, the second most abundant element in the universe after hydrogen, was detected in the sun in 1868 by analysis of the solar spectrum. (The name of the element comes from the Greek word for the sun, *helios.*) It was not found on earth until 1895, however. Until 1962, when a compound of xenon was first prepared, it was believed that none of these elements would combine chemically with any other element. This led to the name **noble gases** for this group, a term meant to denote their general lack of reactivity. For the same reason they are sometimes called the *inert gases* or, because of their low abundance, the *rare gases.*

The Transition Elements

Stretching between Groups 2A and 3A is a series of elements called the **transition elements.** These fill the B-groups in the fourth through the seventh periods in the center of the periodic table. All are metals (see Figure 2.7), and 13 of them are in the top 30 elements in terms of abundance in the earth's crust. Some, like iron (Fe), are abundant in nature (Table 2.4). Most occur naturally in combination with other elements, but a few — silver (Ag), gold (Au), and platinum (Pt) — are much less reactive and so can be found in nature as pure elements.

Virtually all of the transition elements have commercial uses. They are used as structural materials (iron, titanium, chromium, copper); in paints (titanium,

Table 2.4 • Abundance of the 10 Most Abundant Transition Elements in the Earth's Crust

Rank	Element	Abundance (ppm)*
4	Iron	41,000
9	Titanium	5,600
12	Manganese	950
18	Zirconium	190
19	Vanadium	160
21	Chromium	100
23	Nickel	80
24	Zinc	75
25	Cerium	68
26	Copper	50

*ppm = g per 1000 kg

Figure 2.16   Uses of the noble gases. Here you see a kit for detecting the presence of radioactive radon gas in the home. Neon is used in advertising signs, and helium-filled balloons are popular. *(Charles D. Winters)*

chromium); in the catalytic converters in automobile exhaust systems (platinum and rhodium); in coins (copper, nickel, zinc); and in batteries (manganese, nickel, cadmium, mercury).

A number of the transition elements play important biological roles as well. For example, iron, the fourth most abundant element in the earth's crust and most abundant transition element (see Table 2.3), is the central element in the chemistry of hemoglobin, the oxygen-carrying component of blood.

Two rows at the bottom of the table accommodate the **lanthanides** (the series of elements between the elements lanthanum [$Z = 57$] and hafnium [$Z = 72$]) and the **actinides** (the series of elements between actinium [$Z = 89$] and rutherfordium [$Z = 104$]). Some lanthanide compounds are used in color television picture tubes, uranium ($Z = 92$) is the fuel for atomic power plants, and americium ($Z = 95$) is used in smoke detectors.

Exercise 2.7 The Periodic Table

How many elements are in the third period of the periodic table? Give the name and symbol of each. Tell whether each element in the period is a metal, metalloid, or nonmetal.

2.8   ESSENTIAL ELEMENTS

As our knowledge of biochemistry — the chemistry of living systems — increases, we learn more and more about essential elements. These are so important to life that a deficiency in any one will result in death, severe developmental abnormalities, or chronic aliments. No other element can take the place of an essential element.

Of the known elements, only 11 are major elements; they are predominant in all biological systems and are present in approximately the same relative amounts (Table 2.5). In humans these 11 constitute 99.9% of the total number of atoms present, but 4 of these elements — C, H, N, and O — constitute 99% of the total. These are the elements found in the basic structure of all biochemical

Foods rich in essential elements. *(Charles D. Winters)*

Table 2.5 • Relative Amounts Of Essential Elements in the Human Body

Element	Percent by Mass
Oxygen	65
Carbon	18
Hydrogen	10
Nitrogen	3
Calcium	1.5
Phosphorus	1.2
Potassium, sulfur, chlorine	0.2
Sodium	0.1
Magnesium	0.05
Iron, cobalt, copper, zinc, iodine	<0.05
Selenium, fluorine	<0.01

molecules. The large atom percent of oxygen and hydrogen (25.4% and 62.8%, respectively) reflect the high water content of all living systems.

The other 7 elements of the group of 11 elements constitute only about 0.9% of the total atoms in the body. These are sodium, potassium, calcium, magnesium, phosphorus, sulfur, and chlorine. These generally occur in the form of ions such as Na^+, K^+, Mg^{2+}, Ca^{2+}, Cl^-, and HPO_4^{2-}.

The 11 major elements represent six of the groups of the periodic table, and all are "light" elements; they have atomic numbers less than 18. About 17 other elements are required in trace amounts by most but not all biological systems. Some may be required by plants, some by animals, or by only certain plants or animals. With a few exceptions, these elements are generally "heavier" elements, elements having an atomic number greater than 18. They are about evenly divided between metals and nonmetals (or metalloids).

• **Sources of Some Biologically Important Elements**

Element	Source	mg/100 g
Iron	Brewer's yeast	17.3
	Eggs	2.3
Zinc	Brazil nuts	4.2
	Chicken	2.6
Copper	Oysters	13.7
	Brazil nuts	2.3
Calcium	Swiss cheese	925
	Whole milk	118
	Broccoli	103
Selenium	Butter	0.15
	Cider vinegar	0.09

Major Elements 99.9% of all atoms (99.5% by mass)	Trace Elements 0.1% of all atoms (0.5% by mass)
C, H, N, O Na, Ca, P, S, Cl K, Mg	V, Cr, Mo, Mn, Fe, Co, Ni, Cu, Zn B, Si, Se, F, Br, I, As, Sn

Although many of the metals are required only in trace amounts, they are often an integral part of specific biological molecules — such as hemoglobin (Fe), myoglobin (Fe), and vitamin B_{12} (Co) — and activate or regulate their function.

Much of the 3 or 4 g of iron in the body is found in hemoglobin, the substance responsible for carrying oxygen to the cells of the body. Iron deficiency is marked by fatigue, infections, and mouth inflammation. The average person also contains about 2 g of zinc. A deficiency of this element will be noticed by loss of appetite, failure to grow, and skin changes.

Your body has about 75 mg of copper, about a third of which is found in the muscles and the remainder in other tissues. It is involved in many biological functions, so a deficiency shows up in a variety of ways: anemia, degeneration of the nervous system, impaired immunity, and defects in hair color and structure.

Vitamins are essential biological organic nutrients, and the essential elements are their inorganic counterparts. Like the vitamins, essential elements must be present in the diet. It is therefore not surprising that the relative quantities of essential elements in the human body reflect the relative abundances of the known elements in the earth's oceans and, to a lesser extent, in the crust [← PAGE 67].

In Summary

When you have finished studying this chapter, you should ask if you have met the chapter goals. In particular, you should be able to

- Explain the historical development of the atomic theory and identify some of the scientists who made important contributions (Section 2.1).
- Describe electrons, protons, and neutrons, and the general structure of the atom (Section 2.1).
- Understand the relative mass scale and the atomic mass unit (Section 2.2).
- Define isotope and give the mass number and number of neutrons for a specific isotope (Sections 2.2 and 2.3).

- Calculate the atomic weight (atomic mass) of an element from isotopic abundances and masses (Section 2.4).
- Understand that the molar mass of an element is the mass in grams of Avogadro's number of atoms of an element (Section 2.5).
- Know how to use the molar mass of an element in calculations (Section 2.5).
- Identify the periodic table location of groups, periods, metals, metalloids, nonmetals, alkali metals, alkaline earth metals, halogens, noble gases, and the transition elements (Section 2.6).

Key Terms

Section 2.1
subatomic particles
radioactivity
electrons
protons
neutrons
nucleus

Section 2.2
atomic number
atomic mass unit (amu)
mass number

Section 2.3
isotopes
percent abundance
exact atomic mass

Section 2.4
atomic mass, atomic weight
fractional abundance

Section 2.5
mole (mol)
Avogadro's number
molar mass

Section 2.6
groups, families
main group elements
transition elements
periods
metals
nonmetals

metalloids
periodicity
law of chemical periodicity

Section 2.7
alkali metals
alkaline earth elements
allotropes
halogens
noble gases
transition elements
lanthanides
actinides

Key Equations

Equation 2.1 (page 54)
Calculate the percent abundance of an isotope.

$$\text{Percent abundance} = \frac{\text{number of atoms of a given isotope}}{\text{total number of atoms of all isotopes of that element}} \times 100\%$$

Equation 2.2 (page 57)
Calculate the atomic weight from isotope abundances and the exact atomic mass of each isotope of an element.

$$\text{Atomic weight} = \left(\frac{\%\ \text{abundance isotope 1}}{100}\right)(\text{atomic weight of isotope 1}) + \left(\frac{\%\ \text{abundance isotope 2}}{100}\right)(\text{atomic weight of isotope 2}) + \cdots$$

Study Questions

Questions with blue, bold-faced numbers have answers in Appendix O.

Reviewing Important Concepts

1. What are the three fundamental particles from which atoms are built? What are their electric charges? Which of these particles constitute the nucleus of an atom? Which is the least massive particle of the three?

2. Define atomic mass unit (amu).

3. What determines the atomic number of an atom? The mass number of an atom?

4. What did the discovery of radioactivity reveal about the structure of atoms?

5. What is the relationship between the work of J. J. Thomson and that of Robert Millikan? How was Ernest Rutherford's work related to J. J. Thomson's?

6. If the nucleus of an atom were the size of a medium-sized orange (let us say with a diameter of about 6 cm), what would be the diameter of the atom?

7. The volcanic eruption of Mt. St. Helens in the state of Washington in 1980 produced a considerable quantity of a radioactive element in the gaseous state. The element has atomic number 86. What are the symbol and name of this element?

8. Titanium and thallium have symbols that are easily confused with each other. Give the symbol, atomic number, atomic weight, and group and period number of each. Are they metals, metalloids, or nonmetals?

9. Lithium has two stable isotopes: ^{6}Li and ^{7}Li. One of them has an abundance of 92.5%, and the other has an abundance of 7.5%. Knowing that the atomic mass of lithium is 6.941, which is the more abundant isotope?

10. In each case, decide which represents more mass:

    (a) 0.5 mol of Na or 0.5 mol of Si

    (b) 9.0 g of Na or 0.50 mol of Na

    (c) 10 atoms of Fe or 10 atoms of K

11. The famous Italian writer (and chemist) Primo Levi said that zinc "is not an element which says much to the imagination, it is gray and its salts are colorless, it is not toxic, nor does it produce striking chromatic reactions; in short, it is a boring metal." From this, and from reading this chapter, make a list of the properties of zinc. For example, include in your list the position of the element in the periodic table, and tell how many electrons and protons an atom of zinc has. What is its atomic number and atomic mass? Can you think of any uses of the element? Check a dictionary, a reference book such as *The Handbook of Chemistry and Physics*, or the World Wide Web.

12. What is the difference between a group or family and a period in the periodic table?

13. What was incorrect about Mendeleev's original concept of the periodic table? What is the modern "law of chemical periodicity," and how is this related to Mendeleev's ideas?

14. Name and give symbols for (a) three elements that are metals, (b) four elements that are nonmetals, (c) and two elements that are metalloids. In each case locate the element in the periodic table by giving the group and period in which the element is found.

15. Name and give symbols for three transition metals in the fourth period. Look up each of your choices on the World Wide Web and make a list of the uses you find.

16. Name three transition elements, a halogen, a noble gas, and an alkali metal.

17. Name two halogens. Look up each of your choices on the World Wide Web, and make a list of their properties and uses.

18. Name an element first discovered by Madame Curie. Give its name, symbol, and atomic number. Use a dictionary or the World Wide Web to find the origin of the name of the element.

19. Name several nonmetallic elements that have allotropes. Describe how the allotropes differ for one of those elements.

Practicing Skills

Composition of Atoms

(See Example 2.1, Exercise 2.2, and CD-ROM Screen 2.11)

20. Give the mass number of each of the following atoms: (a) magnesium with 15 neutrons, (b) titanium with 26 neutrons, and (c) zinc with 32 neutrons.

21. Give the mass number of (a) a nickel atom with 31 neutrons, (b) a plutonium atom with 150 neutrons, and (c) a tungsten atom with 110 neutrons.

22. Give the complete symbol (^A_ZX) for each of the following atoms: (a) potassium with 20 neutrons, (b) krypton with 48 neutrons, and (c) cobalt with 33 neutrons.

23. Give the complete symbol (^A_ZX) for each of the following atoms: (a) fluorine with 10 neutrons, (b) chromium with 28 neutrons, and (c) xenon with 78 neutrons.

24. How many electrons, protons, and neutrons are there in an atom of (a) magnesium-24, ^{24}Mg; (b) tin-119, ^{119}Sn; and (c) thorium-232, ^{232}Th?

25. How many electrons, protons, and neutrons are there in an atom of (a) carbon-13, ^{13}C; (b) copper-63, ^{63}Cu; and (c) bismuth-205, ^{205}Bi?

Isotopes

(See Example 2.2)

26. The synthetic radioactive element technetium is used in many medical studies. Give the number of electrons, protons, and neutrons in an atom of technetium-99.

27. Radioactive americium-241 is used in household smoke detectors and in bone mineral analysis. Give the number of electrons, protons, and neutrons in an atom of americium-241.

28. Cobalt has three radioactive isotopes used in medical studies. Atoms of these isotopes have 30, 31, and 33 neutrons, respectively. Give the symbol for each of these isotopes.

29. Which of the following are isotopes of element X, the atomic number for which is 9: $^{19}_9$X, $^{20}_9$X, $^9_{18}$X, and $^{21}_9$X?

Isotope Abundance and Atomic Mass

(See Exercises 2.3 and 2.4 and CD-ROM Screen 2.12)

30. Thallium has two stable isotopes, ^{203}Tl and ^{205}Tl. Which is the more abundant of the two?

31. Strontium has four stable isotopes. Strontium-84 has a very low natural abundance, but ^{86}Sr, ^{87}Sr, and ^{88}Sr are all reasonably abundant. Which of these more abundant isotopes predominates?

32. Verify that the atomic mass of Li is 6.94, given the following information:

    ^{6}Li, mass = 6.015121 amu; percent abundance = 7.50%

    ^{7}Li, mass = 7.016003 amu; percent abundance = 92.50%

33. Verify that the atomic mass of Mg is 24.31, given the following information:

^{24}Mg, mass = 23.985 amu; percent abundance = 78.99%

^{25}Mg, mass = 24.986 amu; percent abundance = 10.00%

^{26}Mg, mass = 25.983 amu; percent abundance = 11.01%

34. Gallium has two naturally occurring isotopes, ^{69}Ga and ^{71}Ga, with masses of 68.9257 amu and 70.9249 amu, respectively. Calculate the percent abundances of these isotopes of gallium.

35. Antimony has two stable isotopes, ^{121}Sb and ^{123}Sb, with masses of 120.9038 amu and 122.9042 amu, respectively. Calculate the percent abundances of these isotopes of antimony.

Atoms and the Mole
(See Examples 2.5–2.6 and CD-ROM Screens 2.15 and 2.16)

36. Calculate the mass, in grams, of

    (a) 2.5 mol of aluminum   (c) 0.015 mol of calcium

    (b) 1.25×10^{-3} mol of iron   (d) 653 mol of neon

37. Calculate the mass, in grams, of

    (a) 4.24 mol of gold   (c) 0.063 mol of platinum

    (b) 15.6 mol of He   (d) 3.63×10^{-4} mol of Pu

38. Calculate the amount (moles) represented by each of the following:

    (a) 127.08 g of Cu   (c) 5.0 mg of americium

    (b) 0.012 g of lithium   (d) 6.75 g of Al

39. Calculate the amount (moles) represented by each of the following:

    (a) 16.0 g of Na   (c) 0.0034 g of platinum

    (b) 0.876 g of tin   (d) 0.983 g of Xe

40. What is the average mass of one copper atom?

41. What is the average mass of one atom of titanium?

The Periodic Table
(See Section 2.6, Exercise 2.7, and CD-ROM Screens 2.13 and 2.14)

42. How many elements occur in Group 5A of the periodic table? Give the name and symbol of each of these elements. Tell whether each is a metal, nonmetal, or metalloid.

43. How many elements occur in the fourth period of the periodic table? Give the name and symbol of each. Tell if each is a metal, nonmetal, or metalloid.

44. How many periods of the periodic table have 8 elements, how many have 18 elements, and how many have 32 elements?

45. How many elements occur in the seventh period? What is the name given to the majority of these elements and what well-known property characterizes them?

General Questions

These questions are not designated as to type or location in the chapter. They may combine several concepts. More challenging questions are indicated by an underlined number.

46. Fill in the blanks in the table (one column per element).

Symbol	^{58}Ni	^{33}S	_____	_____
Number of protons	_____	_____	10	_____
Number of neutrons	_____	_____	10	30
Number of electrons	_____	_____	_____	25
Name of element	_____	_____	_____	_____

47. Fill in the blanks in the table (one column per element).

Symbol	^{65}Cu	^{86}Kr	_____	_____
Number of protons	_____	_____	78	_____
Number of neutrons	_____	_____	117	46
Number of electrons	_____	_____	_____	35
Name of element	_____	_____	_____	_____

48. Potassium has three naturally occurring isotopes (^{39}K, ^{40}K, and ^{41}K), but ^{40}K has a very low natural abundance. Which of the other two is the more abundant? Briefly explain your answer.

49. *Crossword Puzzle:* In the 2×2 box shown here, each answer must be correct four ways: horizontally, vertically, diagonally, and by itself. Instead of words, use symbols of elements. When the puzzle is complete, the four spaces will contain the overlapping symbols of ten elements. There is only one correct solution.

1	2
3	4

Horizontal

1–2: Two-letter symbol for a metal used in ancient times.

3–4: Two-letter symbol for a metal that burns in air and is found in Group 5A.

Vertical

1–3: Two-letter symbol for a metalloid.

2–4: Two-letter symbol for a metal used in U.S. coins.

Single squares: all one-letter symbols

1. A colorful nonmetal.

2. Colorless gaseous nonmetal.

3. An element that makes fireworks green.

4. An element that has medicinal uses.

Diagonal

1–4: Two-letter symbol for an element used in electronics.

2–3: Two-letter symbol for a metal used with Zr to make wires for superconducting magnets.

This puzzle first appeared in *Chemical & Engineering News,* December 14, 1987 (p. 86) (submitted by S. J. Cyvin) and in *Chem Matters,* October, 1988.

50. The chart shown in the story on Stardust (page 45) is a plot of the logarithm of the relative abundance of elements 1 through 30 in the solar system. (The abundances are given on a scale that gives hydrogen a relative abundance of 1×10^{12} [the logarithm of which is 12].)

    (a) What is the most abundant metal?

    (b) What is the most abundant nonmetal?

    (c) What is the most abundant metalloid?

    (d) Which of the transition elements is most abundant?

    (e) What halogens are included on this plot and which is the most abundant?

51. Again consider the plot of relative element abundances on page 45. Is there a relationship between abundance and atomic number? Is there any difference between the relative abundance of an element of even atomic number and those of odd atomic number?

52. Which of the following is impossible?

    (a) Silver foil that is 1.2×10^{-4} m thick.

    (b) A sample of potassium that contains 1.784×10^{24} atoms

    (c) A gold coin of mass 1.23×10^{-3} kg

    (d) 3.43×10^{-27} mol of S_8

53. The photo here depicts what happens when magnesium and calcium are placed in water. Based on their relative reactivities, what might you expect to see when barium, another Group 2A element, is placed in water? Give the period in which each element (Mg, Ca, and Ba) is found. What correlation do you think you might find between the reactivity of these elements and their position in the periodic table?

Magnesium *(left)* and calcium *(right)* in water. *(Charles D. Winters)*

54. The plot shows the variation in density with atomic number for the first 36 elements. Use this plot to answer the following questions:

    (a) What three elements in this series have the highest density. What is their approximate density? Are these elements metals or nonmetals?

    (b) Which element in the second period has the largest density? Which element in the third period has the largest density? What do these two elements have in common?

(c) Some elements have densities so low that they do not show up on the plot. What elements are these? What property do they have in common?

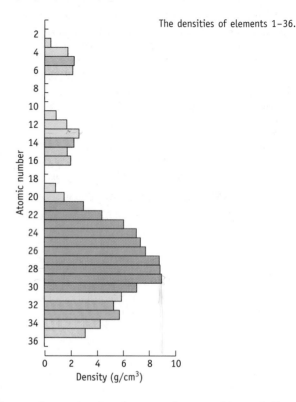

The densities of elements 1–36.

55. Draw a picture showing the approximate positions of all protons, electrons, and neutrons in an atom of helium-4. Make certain that your diagram indicates both the number and position of each type of particle.

56. Reviewing the periodic table:

    (a) Name an element in Group 2A.

    (b) Name an element in the third period.

    (c) What element is in the second period in Group 4A?

    (d) What element is in the third period in Group 6A?

    (e) What halogen is in the fifth period?

    (f) What alkaline earth element is in the third period?

    (g) What noble gas element is in the fourth period?

    (h) Name the nonmetal in Group 6A and the third period.

    (i) Name a metalloid in the fourth period.

57. Reviewing the periodic table.

    (a) Name an element in Group 2B.

    (b) Name an element in the fifth period.

    (c) What element is in the sixth period in Group 4A?

    (d) What element is in the fourth period in Group 6A?

    (e) What alkali metal is in the third period?

    (f) What noble gas element is in the fifth period?

    (g) Name the element in Group 6A and the fourth period. Is this a metal, nonmetal, or metalloid?

    (h) Name a metalloid in Group 5A.

58. If you have 0.00789 g of the gaseous element krypton how many moles does this represent? How many atoms?

59. The recommended daily allowance (RDA) of iron in your diet is 15 mg. How many moles is this? How many atoms?

60. In an experiment, you need 0.125 mol of sodium metal. Sodium can be cut easily with a knife (see Figure 2.9), so if you cut out a block of sodium, what should the volume of the block be in cubic centimeters? If you cut a perfect cube, what is the length of the edge of the cube? (The density of sodium is 0.971 g/cm^3.)

61. Dilithium is the fuel for the *Starship Enterprise*. Because its density is quite low, however, you need a large space to store a large mass. To estimate the volume required, we shall use the element lithium. If you need 256 mol for an interplanetary trip, what must the volume of a piece of lithium be? If the piece of lithium is a cube, what is the dimension of an edge of the cube? (The density for the element lithium is 0.534 g/cm^3 at 20°C.)

62. An object is coated with a layer of chromium, 0.015 cm thick. The object has a surface area of 15.3 cm^2. How many atoms of chromium are used in the coating? (Density of chromium = 7.19 g/cm^3.)

63. A cylindrical piece of sodium is 12.00 cm long and has a diameter of 4.5 cm. The density of sodium is 0.971 g/cm^3. How many atoms does the piece of sodium contain? (The volume of a cylinder is $V = \pi \cdot r^2 \cdot$ length.)

64. When a sample of phosphorus burns in air, the compound P_4O_{10} forms. One experiment showed that 0.744 g of phosphorus formed 1.704 g of P_4O_{10}. Use this information to determine the ratio of the atomic masses of phosphorus and oxygen (mass P/mass O). If the atomic mass of oxygen is assumed to be 16.000 amu, calculate the atomic mass of phosphorus.

65. Consider an atom of ^{64}Zn.
    (a) Calculate the density of the nucleus in grams per cubic centimeter, knowing that the nuclear radius is 4.8×10^{-6} nm and the mass of the ^{64}Zn atom is 1.06×10^{-22} g. Recall that the volume of a sphere is $(\frac{4}{3})\pi r^3$.
    (b) Calculate the density of the space occupied by the electrons in the zinc atom, given that the atomic radius is 0.125 nm and the electron mass is 9.11×10^{-28} g.
    (c) Having calculated the previous densities, what statement can you make about the relative densities of the parts of the atom?

66. Most standard analytical balances can measure accurately to the nearest 0.0001 g. Assume you have weighed out a 2.0000-g sample of carbon. How many atoms are contained in this sample? Assuming the indicated accuracy of the measurement, what is the largest number of atoms that can be present in the sample?

67. To estimate the radius of a lead atom:
    (a) You are given a cube of lead that is 1.000 cm on each side. The density of lead is 11.35 g/cm^3. How many atoms of lead are contained in the sample?

    (b) Atoms are spherical; therefore, the lead atoms in this sample cannot fill all the available space. As an approximation, assume that 60% of the space of the cube is filled with spherical lead atoms. Calculate the volume of one lead atom from this information. From the calculated volume (V), and the formula $V = \frac{4}{3}(\pi r^3)$, estimate the radius (r) of a lead atom.

68. The value of Avogadro's number is set by the fact that we use units of grams for molar mass. Suppose we use ton-mol instead of g-mol (where 1 ton = 1 metric ton = 1000 kg). You have 1.0 ton-mol of aluminum; how many atoms of aluminum are in this amount of the metal?

Using Electronic Resources

These questions refer to the *General Chemistry Interactive CD-ROM*, Version 3.0.

69. See CD-ROM Screen 2.20: Chemical Puzzler. A jar contains a number of jelly beans. To find out how many are in the jar you could dump them out and count them. But how could you estimate their number without counting each one? (Chemists need to do just this kind of "bean counting" when we work with atoms and molecules. They are too small to count one by one, so we have worked out other methods to "count atoms.")

How many jelly beans are in the jar?
(Charles D. Winters)

70. See CD-ROM Screen 2.17: Nuclear Stability. Examine the simulation regarding nuclear stability. How many stable isotopes does Ca have? What is the neutron/proton ratio for one of these isotopes? As the atomic number increases, what happens to the ratio of neutrons to protons? Is an isotope with the composition $^{108}_{54}Xe$ stable?

71. See CD-ROM Screen 2.18: Nuclear Stability, Binding Energy.
    (a) Use the simulation on this screen to examine the relative binding energy. (Here you can find the relative binding energy, that is, the energy per nuclear particle.) As the atomic number increases to 20, does the relative binding energy increase, decrease, or stay constant? What element or elements have the maximum binding energy? How is it related to their abundance in the earth's crust?
    (b) ^4He is the predominant helium isotope in nature. Which has the larger relative binding energy, ^3He or ^4He?
    (c) In the Tutorial section, calculate the relative binding energy for ^2H and for ^{16}O. Which is larger?

3 Molecules, Ions, and Their Compounds

▲ Robert F. Curl *(left)*, Richard E. Smalley *(middle)*, and Harold W. Kroto *(right)*. *(© The Nobel Foundation)*

Buckyball, A Startling New Molecule

In the 1970s astronomers detected strange, new carbon-containing molecules in outer space. Harold Kroto, a chemist at Sussex University in England, was studying similar molecules and realized he could use his expertise to work with astronomers to search for other, related molecules in space.

Early in the 1980s Kroto met Robert Curl, a chemist at Rice University in Texas. Kroto was very interested in talking with Curl, because Curl and Richard Smalley, also a chemist at Rice University, had built an instrument to study clusters of atoms under extreme conditions, just the kind of conditions that might exist in the vicinity of a star. So, in 1985 Kroto went to Rice to use the instrument. Kroto and several graduate students spent a week looking at the result of vaporizing graphite with a laser beam. Much to their amazement, their instrument detected a molecule that contained 60 carbon atoms as the major product.

Being chemists, the first thing Kroto, Curl, and Smalley asked themselves is what this C_{60} molecule might look like. And what was magic about 60 atoms? The answer came when they realized that the molecule was a "closed" cage based on rings of five and six carbon atoms. It looked just like a soccer ball! Although they toyed with the idea of naming it "soccerene," they finally gave it the name *buckminsterfullerene* after R. Buckminster Fuller, an engineer who invented the geodesic dome (like the dome at Epcot Center in Florida). Chemists quickly shortened the name to "buckyball."

Richard Smalley has said that "To a chemist, [the discovery of buckyballs] is

▲ A model of the C_{60} "buckyball" and a soccer ball. Both have 60 vertices or points and have a surface composed of five- and six-member rings. The dome at the Epcot Center is a design from R. Buckminster Fuller, whose name was given to this family of carbon cages. *(Soccer ball, Charles D. Winters; Epcot, Stanley Schoenberger/Grant Heilman)*

> "*The discovery of buckminsterfullerene ... ranks with some of the greatest moments in chemistry.*"

H. Aldersey-Williams: *The Most Beautiful Molecule*, New York, John Wiley & Sons, 1995, p. 2.

▲ A computer-generated model of HIV-protease, the enzyme that replicates the HIV virus. (Each sphere represents an atom.) The purple structure in the center is a C_{60} molecule in the active site of the enzyme. The presence of the C_{60} molecule prevents the enzyme from functioning and blocks the replication of the virus. This experiment was done to test a concept in research on HIV, not to produce a viable drug. (*Dr. Simon Friedman, UCSF [Science News, August 7, 1993, p. 87]*)

like Christmas." The discovery prompted vigorous research in the field, and soon other closed-cage molecules of carbon were discovered. Dozens of uses for the molecules have been proposed, ranging from microscopic ball bearings to new plastics and new lubricants. One idea was to attach groups to C_{60} to make it soluble in water. Indeed, this has been done, and biochemists have found the C_{60} derivative is effective in inhibiting the reproduction of the HIV virus. The experiment has not led to a drug in the treatment of AIDS, but it did prove an important concept in this research.

Finally, the circle has been closed on Kroto's idea that interesting new molecules can be found in outer space.

Recently buckyballs were found in a 1.8 billion-year-old meteorite in Canada. This was itself astonishing because it was thought buckyballs may have only been a laboratory curiosity. Now we know they are found in natural systems. What is more, the meteorite buckyballs contained helium atoms trapped inside! The ratio of the ^{3}He and ^{4}He isotopes for the trapped atoms was different from that in our solar system or on earth. Clearly, the He atoms trapped in the C_{60} cage came from outside our solar system and so, too, did the buckyballs.

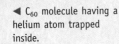

◀ C_{60} molecule having a helium atom trapped inside.

To learn more about buckyballs see R. F. Curl and R. E. Smalley: "Fullerenes," *Scientific American*, October 1991, p. 54; H. Aldersey-Williams: *The Most Beautiful Molecule*, John Wiley & Sons, New York, 1995; and "Race to Catch a Buckyball," a *Nova* presentation on the Public Broadcasting System.

Before You Begin

- Know how to calculate and use molar amounts (Section 2.5).

Modeling Molecules Much of this book is focused on understanding the structures and properties of molecules. A key discovery in the 20th century was the structure of DNA, deoxyribonucleic acid, the fundamental genetic material.

As outlined on pages 2–3, the discovery was made largely by the four chemists pictured here. It rested on experimental work by Franklin and Wilkins and modeling by Watson and Crick. As we progress through the book, we shall return to the subject of the structure of DNA.

Model Building and the Structure of DNA

◄ A model of DNA for classroom use.

◄ Francis Crick (right) and James Watson (left) examine a model of DNA in 1953. Their work, combined with that of Franklin and Wilkins, led to the discovery of the double helix nature of the DNA chain of atoms. Notice the drawing of a helix on the wall nearest Crick.

Rosalind Franklin, 1920–1958.

Maurice Wilkins, born in 1916 in New Zealand.

Structure of DNA: Sugar, Phosphate, and Bases

3.4 nm

DNA is a very large molecule that consists of two chains of atoms (P, C, and O) that twist together. The P, C, and O atoms are parts of phosphate ions (P) and sugar molecules (S). The chains are joined by four different molecules (adenine, thymine, guanine, and cytosine) belonging to a general class of molecules called bases.

▲ A sample of DNA.

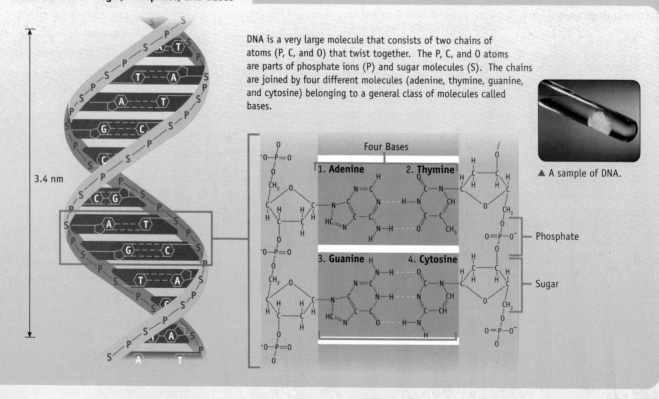

Four Bases

1. **Adenine**   2. **Thymine**

3. **Guanine**   4. **Cytosine**

Phosphate

Sugar

Model of DNA: Courtesy Indigo® Instruments (indigo.com). Watson and Crick: A. Barrington Brown/Science Source/Photo Researchers, Inc. Rosalind Franklin: American Society for Microbiology. Maurice Wilkins: © The Nobel Foundation. DNA sample: © BSIP/Emakoff/Science Source/Photo Researchers, Inc.

On June 26 of 2000, President Clinton called it "the most important, most wondrous map ever produced by humankind." That map was of the human genome. Chromosomes, which are present in the nuclei of almost all living cells, consist of giant molecules of DNA or *deoxyribonucleic acid.* Understanding the human genome, which is the complete structure of the DNA in every one of our 23 chromosomes, is widely expected to revolutionize the practice of medicine.

To understand the basic structure of DNA or why buckyballs can be cages of atoms and why they can just fit into the cavity of an important biological molecule means we need to begin to explore many kinds of chemical compounds and their properties.

● **Human Genome — An Inspiration**
In 2000, two research groups simultaneously announced completion of a working draft of the human genome. One group was an international consortium of nonprofit laboratories. J. Craig Venter, then head of Celera Laboratories, the other group that sequenced the human genome, said, "The complexities and wonder of how inanimate chemicals that are our genetic code give rise to the imponderables of the human spirit should keep poets and philosophers inspired for the milleniums."

3.1   MOLECULES, COMPOUNDS, AND FORMULAS

A *molecule* is the smallest identifiable unit into which a pure substance like sugar or water can be divided and still retain the composition and chemical properties of the substance. Such substances are composed of identical molecules consisting of atoms of two or more elements bound firmly together [◀ SECTION 1.4]. For example, atoms of the element aluminum combine with molecules of the element bromine, Br_2, to produce the compound aluminum bromide, Al_2Br_6 (Figure 3.1).

$$2\ Al(s) + 3\ Br_2(\ell) \longrightarrow Al_2Br_6(s)$$

aluminum + bromine $\longrightarrow$ aluminum bromide

To describe this chemical change (or chemical reaction) on paper, the composition of each element and compound is represented by a symbol or formula. Here one molecule of Al_2Br_6 is composed of two Al atoms and six Br atoms.

How do compounds differ from elements? The characteristics of the constituent elements are lost when a compound is produced from its elements. Solid, metallic aluminum and red-orange liquid bromine, for example, produce a white solid product (see Figure 3.1).

Chapter Goals ● Revisited

● **Interpret, predict, and write formulas for ionic and molecular compounds.**

● Name compounds.

● Calculate and use molar mass.

● Calculate percent composition for a compound and derive formulas from experimental data.

(a)

(b)

(c)

Figure 3.1   Reaction of the elements aluminum and bromine. (a) Solid aluminum and liquid bromine in beaker. **(b)** When the aluminum is added to the bromine, a vigorous chemical reaction produces white, solid aluminum bromide, Al_2Br_6 **(c)**. *(Photos, Charles D. Winters)*

NAME	MOLECULAR FORMULA	CONDENSED FORMULA	STRUCTURAL FORMULA	MOLECULAR MODEL
Ethanol	C_2H_6O	CH_3CH_2OH		
Dimethyl ether	C_2H_6O	CH_3OCH_3		

Figure 3.2   Four approaches to showing molecular formulas. Here the two molecules have the same molecular formula. However, once written as condensed or structural formulas and pictured with a molecular model, it is clear that these molecules are different.

● **Standard Colors for Atoms in Molecular Models**
The colors listed here are used in this book and are generally used by chemists. The colors of some common atoms are:

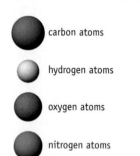

carbon atoms

hydrogen atoms

oxygen atoms

nitrogen atoms

chlorine atoms

Formulas

Water has the simple formula H_2O. However, for more complicated molecules, there is often more than one way to write the formula. For example, the formula of one of the bases of DNA, thymine, can be represented as $C_5H_6N_2O_2$. This **molecular formula** describes the composition of thymine molecules—five carbon atoms, six hydrogen atoms, and two atoms each of nitrogen and oxygen occur per molecule—but it gives us no structural information. Structural information is important, however, because that helps us understand how one molecule can interact with others, the essence of chemistry.

To provide some structural information, it is useful to write formulas in which certain atoms are grouped together. For example, C_2H_6O is the molecular formula of an alcohol called ethanol. Its **condensed formula,** CH_3CH_2OH, tells us more about how the atoms are grouped and identifies important parts of the molecule (Figure 3.2). This formula implies the molecule consists of three "groups," a CH_3 group, a CH_2 group, and an OH group. Writing the formula as CH_3CH_2OH also shows the compound is not dimethyl ether, CH_3OCH_3, a compound having the same molecular formula but with distinctly different properties.

You will often see simple **structural formulas** that present an even higher level of structural detail, showing how all the atoms are attached within a molecule. Several examples are given in Figure 3.2. The lines between atoms represent the **chemical bonds** that hold atoms together in molecules [➡ CHAPTERS 9 AND 10].

Example 3.1   Molecular Formulas

Problem • The acrylonitrile molecule is the building block for acrylic plastics (such as Orlon and Acrilan). Its structural formula is shown here. What is the molecular formula for acrylonitrile?

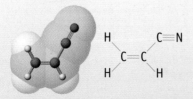

Strategy • Count the number of atoms of each type [💿 CD-ROM, Screen 3.4].

Solution • Acrylonitrile has three C atoms, three H atoms, and one N atom. Therefore, its molecular formula is C_3H_3N.

Comment • It is conventional when writing molecular formulas of organic compounds (compounds with C, H, and other elements) to write C first, then H, and finally other elements in alphabetical order.

Exercise 3.1 Molecular Formulas

The styrene molecule is the building block of polystyrene, a material familiar in the form of drinking cups and building insulation. What is the molecular formula of styrene?

Molecular model   Structural formula

3.2   MOLECULAR MODELS

Molecular structures are often beautiful in the same sense that art is beautiful. For example, there is something intrinsically attractive about the spherical cage of C_{60} or the pattern created as water molecules are assembled in ice (Figure 3.3).

More important, however, is the fact that the physical and chemical properties of a molecular compound are often closely related to its structure. For example, two well-known features of ice are easily related to its structure. The first is the shape of ice crystals. The six-fold symmetry of macroscopic ice crystals also appears at the particulate level, in the form of six-sided rings of hydrogen and oxygen atoms. The second is water's unique property of being less dense when solid than it is when liquid. The lower density of ice, which has enormous consequences for earth's climate, results from the fact that molecules of water are not packed together tightly in ice.

Because molecules are three-dimensional, it is often difficult to represent their shapes on paper. Certain conventions have been developed, however, that help represent three-dimensional structures on two-dimensional surfaces. Simple perspective drawings are often used (Figure 3.4) [🖱 CD-ROM, Screen 3.4].

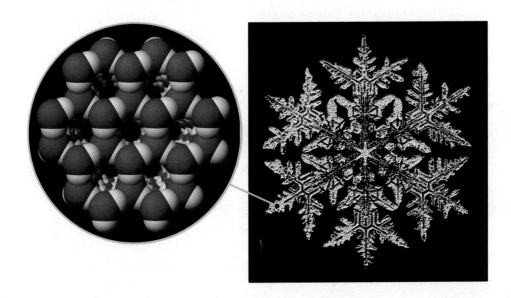

Figure 3.3   Ice. Snowflakes are six-sided structures, reflecting the underlying structure of ice. Ice consists of six-sided rings formed by water molecules, in which each side of a ring consists of two O atoms and an H atom. *(Mehau Kulyk/Science Photo Library/Photo Researchers, Inc.; model by S. M. Young)*

A Closer Look

Computer Resources for Molecular Modeling

With the availability of relatively low-cost, high-powered computers, the use of molecular-modeling programs has become common. Although the computer screen is two-dimensional, the perspective drawings obtained from molecular-modeling programs are usually quite accurate. In addition, most programs offer an option to rotate the molecule on the computer screen to allow the viewer to see the structure from any desired angle. Both ball-and-stick and space-filling representations can be portrayed. Most of the drawings in this book were prepared with the molecular modeling software from CAChe/Fujitsu. The *General Chemistry Interactive CD-ROM* that accompanies this book contains a portion of this software that you can use to visualize models of a large number of substances. The site on the World Wide Web for this textbook (http://www.brookscole.com/chemistry) contains a link to RasMol and Chime, molecular visualization software. Models of many of the compounds mentioned in this book are on the CD-ROM or are available at the site on the World Wide Web. You can visualize these molecules using the software on the CD-ROM, or, if you download RasMol or Chime and configure your browser properly, you can download files from the Internet that allow you to visualize these models on your own computer.

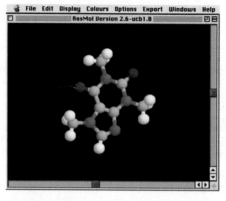 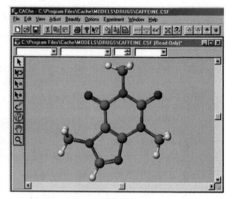

A model of caffeine as viewed with Rasmol *(left)* and the CAChe/Fujitsu software *(right)*.

Wood or plastic models are also a useful way of representing molecular structure. These models can be held in the hand and rotated to view all parts of the molecule.

There are several kinds of molecular models. One is called the **ball-and-stick model.** Spheres, usually in different colors, represent the atoms, and the sticks represent the bonds holding them together. These models make it easy to see how atoms are attached to one another.

Molecules can also be represented using **space-filling models.** These models are more realistic because they are a better representation of relative sizes of atoms and molecules. A disadvantage of pictures of space-filling models is that atoms can often be hidden from view.

Simple perspective drawing

Plastic model

Ball-and-stick model

Space-filling model

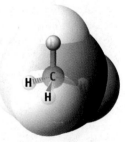

All visualizing techniques represent the same molecule.

Figure 3.4   Ways of depicting the methane molecule. *(Photo, Charles D. Winters)*

Example 3.2 Using Molecular Models

Problem • A model of uracil, an important biological molecule, is given here. Write its molecular formula.

A molecular model of uracil.

Strategy • The standard color codes used for the atoms are: carbon atoms = gray; hydrogen atoms = white; nitrogen atoms = blue; and oxygen atoms = red. See Preface.

Solution • Uracil has four C atoms, four H atoms, two N atoms, and two O atoms, giving a formula of $C_4H_4N_2O_2$.

Comment • Notice that uracil has almost the same structure as thymine (page 80), lacking only a CH_3 group.

Exercise 3.2 Formulas of Molecules

Cysteine, whose molecular model is illustrated below, is an important amino acid and a constituent of many living things. What is its molecular formula? See Example 3.2 and the Preface for the color coding of the model.

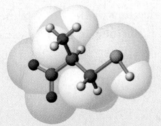

A molecular model of the amino acid cysteine.

3.3 IONIC COMPOUNDS: FORMULAS, NAMES, AND PROPERTIES

The compounds you have seen so far in this chapter are molecular compounds, that is, compounds that consist, at the particulate level, of discrete molecules. **Ionic compounds** are another major class of compounds. These consist of *ions,* atoms or groups of atoms that bear a positive or negative electric charge [🖱 CD-ROM, Screen 3.5]. Many familiar compounds are composed of ions (Figure 3.5). Table salt, or sodium chloride (NaCl), and lime (CaO) are just two. To recognize ionic compounds, and to be able to write formulas for these compounds, it is important to know the formulas and charges of common ions. You also need to know the names of ions and how to name the compounds they form.

Ions

Atoms of many elements can lose or gain electrons in the course of a chemical reaction [⬅ FIGURE 1.2] [🖱 CD-ROM, Screen 3.5]. So that you can predict the outcome of chemical reactions [➡ SECTION 5.6], you need to know if an element will likely gain or lose electrons and if so how many.

Salt, sodium chloride, NaCl
contains Na⁺ cations, Cl⁻ anions

Sodium fluoride, NaF
contains Na⁺ cations, F⁻ anions

Copper sulfate, CuSO₄
contains Cu²⁺ cations, SO₄²⁻ anions

Alum, KAl(SO₄)₂ • 12 H₂O
contains K⁺ cations, Al³⁺ cations,
SO₄²⁻ anions, and H₂O molecules

Figure 3.5   Some common ionic compounds. *(Charles D. Winters)*

Cations

If an atom loses an electron (which is transferred to an atom of another element in the course of a reaction), the atom now has one more positive proton than it has negative electrons. The result is a positively charged ion called a **cation** (Figure 3.6). (The name is pronounced "cat'-ion.") Because it has an excess of one positive charge, we write the cation's symbol as, for example, Li⁺.

$$\text{Li atom} \longrightarrow 1\ e^- + \text{Li}^+ \text{ cation}$$

(3 protons and 3 electrons)   (3 protons and 2 electrons)

Anions

Conversely, if an atom gains one or more electrons, there is now one or more negatively charged electrons than nuclear protons. The result is an **anion** (see Figure 3.6). (The name is pronounced "ann'-ion.")

$$\text{O atom} + 2\ e^- \longrightarrow \text{O}^{2-} \text{ anion}$$

(8 protons and 8 electrons)   (8 protons and 10 electrons)

Here the O atom has gained two electrons, giving it an excess of electrons, so we write the anion's symbol O²⁻.

How do you know if an atom is likely to form a cation or an anion? It depends on whether the element is a metal or nonmetal.

- Metals generally lose electrons in the course of their reactions to form cations.
- Nonmetals frequently gain one or more electrons to form anions in the course of their reactions.

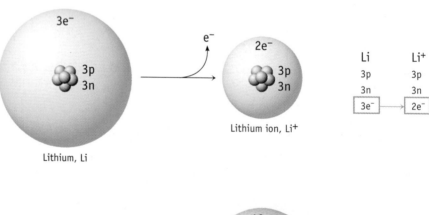

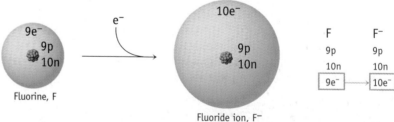

Figure 3.6   Ions. A lithium atom is electrically neutral because the number of positive charges (three protons) and negative charges (three electrons) is the same. When it loses one electron, it has one more positive charge than negative charge, so it has a net charge of +1. We symbolize the resulting lithium cation as Li⁺. A fluorine atom is also electrically neutral, having nine protons and nine electrons. A characteristic of nonmetals is to form anions. A fluorine atom can acquire an electron to form an F⁻ anion. This anion has one more electron than it has protons, so it has a net charge of −1.

Monatomic Ions

Monatomic ions are single atoms that have lost or gained electrons. How can you predict the number of electrons gained or lost? Typical charges on such ions are indicated in Figure 3.7. *Metals of Groups 1A–3A form positive ions having a charge equal to the group number of the metal.*

Group	Metal Atom	Electron Change		Resulting Metal Cation
1A	Na (11 protons, 11 electrons)	-1	$\longrightarrow$	Na^+ (11 protons, 10 electrons)
2A	Ca (20 protons, 20 electrons)	-2	$\longrightarrow$	Ca^{2+} (20 protons, 18 electrons)
3A	Al (13 protons, 13 electrons)	-3	$\longrightarrow$	Al^{3+} (13 protons, 10 electrons)

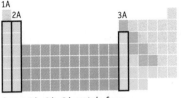

Group 1A, 2A, 3A metals form M^{n+} cations where n = group number.

Transition metals (B group elements) also form cations. Unlike A group metals, though, no easily predictable pattern of behavior occurs for transition metal cations. In addition, many of these metals form several different ions. An iron-containing compound, for example, may contain either Fe^{2+} or Fe^{3+} ions. Indeed, +2 and +3 ions are typical of many transition metals (see Figure 3.7).

Group	Metal Atom	Electron Change		Resulting Metal Cation
7B	Mn (25 protons, 25 electrons)	-2	$\longrightarrow$	Mn^{2+} (25 protons, 23 electrons)
8B	Fe (26 protons, 26 electrons)	-2	$\longrightarrow$	Fe^{2+} (26 protons, 24 electrons)
8B	Fe (26 protons, 26 electrons)	-3	$\longrightarrow$	Fe^{3+} (26 protons, 23 electrons)

Nonmetals often form ions having a negative charge equal to 8 minus the group number of the element. For example, nitrogen is in Group 5A and so forms an ion having a charge of -3 because a nitrogen atom can gain three electrons.

Group	Nonmetal Atom	Electron Change		Resulting Nonmetal Anion
5A	N (7 protons, 7 electrons)	$+3\,(=8-5)$	$\longrightarrow$	N^{3-} (7 protons, 10 electrons)
6A	S (16 protons, 16 electrons)	$+2\,(=8-6)$	$\longrightarrow$	S^{2-} (16 protons, 18 electrons)
7A	Br (35 protons, 35 electrons)	$+1\,(=8-7)$	$\longrightarrow$	Br^- (35 protons, 36 electrons)

Galena, lead sulfide, PbS contains Pb^{2+} cations, S^{2-} anions

Fluorite, calcium fluoride, CaF_2 contains Ca^{2+} cations, F^- anions

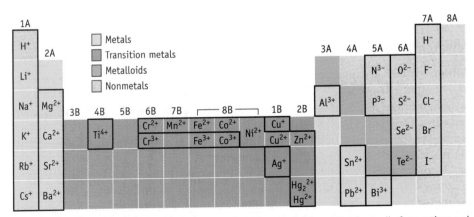

Figure 3.7   Charges on some common monatomic cations and anions. Metals usually form cations and nonmetals usually form anions.

Ruby, aluminum oxide, Al_2O_3 contains Al^{3+} cations, O^{2-} anions, and trace of Cr^{3+} ions

Common ionic compounds with monatomic cations and anions. *(Charles D. Winters)*

Notice that hydrogen appears at two locations in Figure 3.7; the H atom can either lose or gain an electron, depending on the other atoms it encounters.

Electron lost: H (1 proton, 1 electron) $\longrightarrow$ H$^+$ (1 proton, 0 electrons) + e$^-$

Electron gained: H (1 proton, 1 electron) + e$^-$ $\longrightarrow$ H$^-$ (1 proton, 2 electrons)

Finally, *the noble gases lose or gain electrons only in special cases,* reflecting their chemical inertness.

Ion Charges and the Periodic Table

As illustrated in Figure 3.7, the metals of Groups 1A, 2A, and 3A form ions having +1, +2, and +3 charges; that is, their atoms lose one, two, or three electrons, respectively. *For cations formed from A group elements, the number of electrons remaining on the ion is the same as the number of electrons in an atom of a noble gas.* For example, Mg^{2+} has ten electrons, the same number as in an atom of the noble gas neon (atomic number 10), which precedes magnesium in the periodic table.

An atom of a nonmetal at the right in the periodic table would have to lose a great many electrons to achieve the same number as a noble gas atom of lower atomic number. (For instance, Cl, whose atomic number is 17, would have to lose seven electrons to have the same number of electrons as Ne.) If a nonmetal atom were to gain just a few electrons, however, it would have the same number as a noble gas atom of higher atomic number. For example, an oxygen atom has eight electrons. By gaining two electrons per atom it forms O^{2-}, which has ten electrons, the same number as neon. This illustrates the observation that the negative charge on an anion, and so the number of electrons gained, is given by 8 minus the group number. As illustrated in Figure 3.7, *ions having the same number of electrons as a noble gas atom are especially favored in chemical compounds.*

Calcite, CaCO$_3$

CO$_3$$^{2-}$

Apatite, Ca$_5$(PO$_4$)$_3$F

PO$_4$$^{3-}$

Exercise 3.3 Predicting Ion Charges

Predict formulas for monatomic ions formed from (a) K, (b) Se, (c) Ba, and (d) Cs. In each case indicate the number of electrons gained or lost by an atom of the element in forming the anion or cation, respectively. For each ion, indicate the noble gas atom having the same total number of electrons.

Polyatomic Ions

Polyatomic ions are made up of two or more atoms, and the collection has an electric charge [⌨ CD-ROM, Screen 3.6] (Table 3.1 and Figure 3.8). For example, carbonate ion, CO$_3$$^{2-}$, is a common polyatomic anion consisting of one C atom and three O atoms. The ion has two units of negative charge because there are two more electrons (a total of 24) in the ion than there are protons (a total of 22) in the nuclei of one C atom and three O atoms.

A common polyatomic cation is NH$_4$$^+$, the ammonium ion. In this case, four H atoms surround an N atom, and the ion has a +1 electric charge. This ion has eight electrons, but there are nine positively charged protons in the nuclei of the N and H atoms (five and one each, respectively).

Celestite, SrSO$_4$

SO$_4$$^{2-}$

Figure 3.8 Common ionic compounds based on polyatomic ions. *(Photos, Charles D. Winters)*

Table 3.1 • Formulas and Names of Some Common Polyatomic Ions

Formula	Name	Formula	Name
CATION: Positive Ion			
NH_4^+	ammonium ion		

Formula	Name	Formula	Name
ANIONS: Negative Ions			
Based on a Group 4A element		***Based on a Group 7A element***	
CN^-	cyanide ion	ClO^-	hypochlorite ion
$CH_3CO_2^-$	acetate ion	ClO_2^-	chlorite ion
CO_3^{2-}	carbonate ion	ClO_3^-	chlorate ion
HCO_3^-	hydrogen carbonate ion (or bicarbonate ion)	ClO_4^-	perchlorate ion
Based on a Group 5A element		***Based on a transition metal***	
NO_2^-	nitrite ion	CrO_4^{2-}	chromate ion
NO_3^-	nitrate ion	$Cr_2O_7^{2-}$	dichromate ion
PO_4^{3-}	phosphate ion	MnO_4^-	permanganate ion
HPO_4^{2-}	hydrogen phosphate ion		
$H_2PO_4^-$	dihydrogen phosphate ion		
Based on a Group 6A element			
OH^-	hydroxide ion		
SO_3^{2-}	sulfite ion		
SO_4^{2-}	sulfate ion		
HSO_4^-	hydrogen sulfate ion (or bisulfate ion)		

Formulas of Ionic Compounds

Ionic compounds are composed of ions. For an ionic compound to be electrically neutral — to have no net charge — the numbers of positive and negative ions must be such that the positive and negative charges balance. In sodium chloride the sodium ion has a $+1$ charge (Na^+) and the chloride ion has a -1 charge (Cl^-). These ions must be present in a one-to-one ratio and the formula is NaCl.

The beautiful gem ruby is largely the compound formed from aluminum ions (Al^{3+}) and oxide ions (O^{2-}) (page 87). Here the ions have positive and negative charges that are of different absolute size. To have a compound with the same number of positive and negative charges, two Al^{3+} ions [total charge $= 6+ = 2 \times (3+)$] must combine with three O^{2-} ions [total charge $= 6- = 3 \times (2-)$] to give a formula of Al_2O_3.

Calcium metal is a member of Group 2A, and it forms a cation having a $+2$ charge. It can combine with a variety of anions to form ionic compounds such as those in the table shown here.

• **Balancing Ion Charges**
Aluminum is a metal in Group 3A and so loses three electrons to form the Al^{3+} cation. Oxygen is a nonmetal in Group 6A and so gains two electrons to form an O^{2-} anion. Notice that the charge on the cation is the subscript on the anion (and vice versa).

$$2\,Al^{3+} + 3\,O^{2-} \longrightarrow Al_2O_3$$

This often works well, but be careful. The subscripts in $Ti^{4+} + O^{2-}$ are reduced to the simplest ratio (1 Ti to 2 O and not 2 Ti to 4 O)

$$Ti^{4+} + 2\,O^{2-} \longrightarrow TiO_2$$

Compound	Ion Combination	Overall Charge on Compound
$CaCl_2$	$Ca^{2+} + 2\,Cl^-$	$(2+) + 2 \times (1-) = 0$
$CaCO_3$	$Ca^{2+} + CO_3^{2-}$	$(2+) + (2-) = 0$
$Ca_3(PO_4)_2$	$3\,Ca^{2+} + 2\,PO_4^{3-}$	$3 \times (2+) + 2 \times (3-) = 0$

In writing formulas, it is conventional that the symbol of the cation is given first, followed by the anion symbol. Also notice the use of parentheses when there is more than one polyatomic ion.

Example 3.3   Formulas of Ionic Compounds

Problem • For each of the following ionic compounds, write the symbols for the ions present and give the number of each: (a) $MgBr_2$, (b) Li_2CO_3, and (c) $Fe_2(SO_4)_3$.

Strategy • When looking at the formula of an ionic compound you must first recognize the cation and anion involved [CD-ROM, Screens 3.5, 3.6, and 3.8].

Solution •

(a) $MgBr_2$ is composed of one Mg^{2+} ion and two Br^- ions. When a halogen such as bromine is combined only with a metal, you can assume the halogen is an anion with a charge of -1. Magnesium is a metal in Group 2A and always has a charge of $+2$ in its compounds.

(b) Li_2CO_3 is composed of two lithium ions, Li^+, and one carbonate ion, CO_3^{2-}. Li is a Group 1A element and always has a $+1$ charge in its compounds. Because the two $+1$ charges balance the negative charge of the carbonate ion, the latter must be -2.

(c) $Fe_2(SO_4)_3$ contains two Fe^{3+} ions and three sulfate ions, SO_4^{2-}. The way to rationalize this is to recall that sulfate has a -2 charge. Because three sulfate ions are present (with a total charge of -6), the two iron cations must have a total charge of $+6$. This is possible only if each iron cation has a charge of $+3$.

Example 3.4   Formulas of Ionic Compounds

Problem • Write formulas for ionic compounds composed of an aluminum cation and each of the following anions: (a) fluoride ion, (b) sulfide ion, and (c) nitrate ion.

Strategy • First decide on the charge on the Al cation then write the formulas of each of the anions. Combine the Al cation with each type of anion in the correct ratio to form an electrically neutral compound [CD-ROM, Screens 3.5, 3.6, and 3.8].

Solution • An aluminum cation is predicted to have a charge of $+3$ because Al is a metal in Group 3A.

(a) Fluorine is a Group 7A element. The charge of the fluoride ion is predicted to be -1 (from $8 - 7 = 1$). Therefore, we

need three F^- ions to combine with one Al^{3+}. The formula of the compound is AlF_3.

(b) Sulfur is a nonmetal in Group 6A and so forms a -2 anion. Thus, we need to combine two Al^{3+} ions [total charge is $+6 = 2 \times (3+)$] with three S^{2-} ions [total charge is $-6 = 3 \times (-2)$]. The compound has the formula Al_2S_3.

(c) The nitrate ion has the formula NO_3^- (see Table 3.1). The answer here is therefore similar to the AlF_3 case, and the compound has the formula $Al(NO_3)_3$. Here we place parentheses around NO_3 to show that three polyatomic NO_3^- ions are involved.

Exercise 3.4   Formulas of Ionic Compounds

(a) Give the number and identity of the constituent ions in each of the following ionic compounds: (a) NaF, (b) $Cu(NO_3)_2$, and (c) $NaCH_3CO_2$.

(b) Iron, a transition metal, forms ions having at least two different charges. Write the formulas of the compounds formed between two different iron cations and chloride ions.

(c) Write the formulas of all of the neutral ionic compounds that can be formed by combining the cations Na^+ and Ba^{2+} with the anions S^{2-} and PO_4^{3-}.

Names of Ions

Naming Positive Ions (Cations)

With a few exceptions (such as NH_4^+), the positive ions described in this text are metal ions. Positive ions are named by the following rules:

Problem-Solving Tip 3.1

Formulas for Ions and Ionic Compounds

Writing formulas for ionic compounds takes practice, and it requires that you know the formulas and charges of the most common ions. The charges on monatomic ions are often evident from the position of the element in the periodic table, but you simply have to remember the formulas and charges of polyatomic ions, especially the most common ones such as nitrate, sulfate, carbonate, phosphate, and acetate.

If you cannot remember the formula of a polyatomic ion, or if you encounter an ion you have not seen before, you may be able to figure out its formula from the formula and the name of one of its compounds. For example, suppose you are told that $NaCHO_2$ is sodium formate. You know that the sodium ion is Na^+, so the formate ion must be the remaining portion of the compound and it must have a charge of -1 to balance the $+1$ charge on the sodium ion. Thus, the formate ion must be CHO_2^-.

Finally, when writing the formulas of ions, you must include the charge on the ion (except in an ionic compound formula). Writing Na when you mean sodium ion is incorrect. There is a vast difference in the properties of the element sodium (Na) and those of its ion (Na^+).

1. For a monatomic positive ion, that is, a metal cation, the name is that of the metal plus the word "ion." For example, we have already referred to Al^{3+} as the aluminum ion.

2. Some cases occur, especially in the transition series, in which a metal can form more than one type of positive ion. The charge of the ion is commonly indicated by a Roman numeral in parentheses immediately following the ion's name. For example, Co^{2+} is the cobalt(II) ion, and Co^{3+} is the cobalt(III) ion.

Finally, you will encounter the ammonium cation, NH_4^+, many times in this book and in the laboratory. Do not confuse the ammonium ion with the uncharged ammonia molecule, NH_3.

● **"-ous" and "-ic" Endings**
An older naming system for metal ions uses the ending *-ous* for the ion of lower charge and *-ic* for the ion of higher charge. For example, there are cobaltous (Co^{2+}) and cobaltic (Co^{3+}) ions, and ferrous (Fe^{2+}) and ferric (Fe^{3+}) ions. We do not use this system in this book, but some chemical manufacturers continue to use it.

Naming Negative Ions (Anions)

Two types of negative ions must be considered: those having only one atom (*monatomic*) and those having several atoms (*polyatomic*).

1. A monatomic negative ion is named by adding *-ide* to the stem of the name of the nonmetal element from which the ion is derived (Figure 3.9). The anions of the Group 7A elements, the halogens, are known as the fluoride, chloride, bromide, and iodide ions and as a group are called **halide ions.**

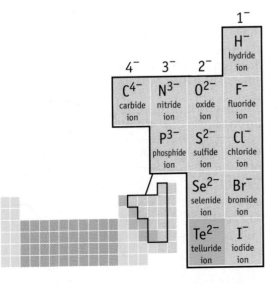

Figure 3.9 **Names and charges of some common monatomic anions.**

2. Polyatomic negative ions are common, especially those containing oxygen (called **oxoanions**). The names of some of the most common oxoanions are given in Table 3.1. Although most of these names must simply be learned, some guidelines can help. For example, consider the following pairs of ions:

NO_3^- is the nitrate ion, whereas NO_2^- is the nitrite ion

SO_4^{2-} is the sulfate ion, whereas SO_3^{2-} is the sulfite ion

The oxoanion having the *greater number of oxygen atoms* is given the suffix *-ate,* and the oxoanion having the *smaller number of oxygen atoms* has the suffix *-ite.* For a series of oxoanions having more than two members, the ion with the largest number of oxygen atoms has the prefix *per-* and the suffix *-ate.* The ion having the smallest number of oxygen atoms has the prefix *hypo-* and the suffix *-ite.* The oxoanions containing chlorine are good examples.

• **Naming Oxoanions**

increasing oxygen content →

per . . . ate

. . . ate

. . . ite

hypo . . . ite

ClO_4^-	*perchlorate* ion
ClO_3^-	*chlorate* ion
ClO_2^-	*chlorite* ion
ClO^-	*hypochlorite* ion

Oxoanions that contain hydrogen are named by adding the word "hydrogen" before the name of the oxoanion. If two hydrogens are in the compound, we say dihydrogen. Many of these hydrogen-containing oxoanions have common names that are often used. For example, the hydrogen carbonate ion, HCO_3^-, is often called the bicarbonate ion.

Ion	Systematic Name	Common Name
HPO_4^{2-}	hydrogen phosphate ion	
$H_2PO_4^-$	dihydrogen phosphate ion	
HCO_3^-	hydrogen carbonate ion	bicarbonate ion
HSO_4^-	hydrogen sulfate ion	bisulfate ion
HSO_3^-	hydrogen sulfite ion	bisulfite ion

Names of Ionic Compounds

The name of an ionic compound is built from the names of the positive and negative ions in the compound [🖱 CD-ROM, Screen 3.11]. The name of the positive ion is given first, followed by the name of the negative ion. Examples of ionic compound names are shown in the table, and others are shown in Figure 3.8.

Chapter Goals • Revisited

• Interpret, predict, and write formulas for ionic and molecular compounds.

• **Name compounds.**

• Calculate and use molar mass.

• Calculate percent composition for a compound and derive formulas from experimental data.

Ionic Compound	Ions Involved	Name
$CaBr_2$	Ca^{2+} and 2 Br^-	calcium bromide
$NaHSO_4$	Na^+ and HSO_4^-	sodium hydrogen sulfate
$(NH_4)_2CO_3$	2 NH_4^+ and CO_3^{2-}	ammonium carbonate
$Mg(OH)_2$	Mg^{2+} and 2 OH^-	magnesium hydroxide
$TiCl_2$	Ti^{2+} and 2 Cl^-	titanium(II) chloride
Co_2O_3	2 Co^{3+} and 3 O^{2-}	cobalt(III) oxide

Exercise 3.5 **Names of Ionic Compounds**

1. Give the formula for each of the following ionic compounds. Use Table 3.1 and Figure 3.9.

   (a) ammonium nitrate   **(d)** vanadium(III) oxide
   (b) cobalt(II) sulfate   **(e)** barium acetate
   (c) nickel(II) cyanide   **(f)** calcium hypochlorite

2. Name the following ionic compounds:

   (a) $MgBr_2$ **(d)** $KMnO_4$
   (b) Li_2CO_3 **(e)** $(NH_4)_2S$
   (c) $KHSO_3$ **(f)** $CuCl$ and $CuCl_2$

Properties of Ionic Compounds

What is the "glue" that causes ions of opposite electric charge to be held together, to form an orderly arrangement of ions in an ionic compound? As described in Section 2.1, when a substance having a negative electric charge is brought near a substance having a positive electric charge, a force of attraction occurs between them (Figure 3.10). Similarly, there is a force of repulsion when two substances, both positive or both negative, are brought together. These forces are called **electrostatic forces**, and the force of attraction or repulsion between ions is given by **Coulomb's law** (Equation 3.1) [⊕ CD-ROM, Screen 3.7].

$$\text{Force of attraction} = k\,\frac{(n^+\text{e})(n^-\text{e})}{d^2} \qquad\qquad (3.1)$$

charge on + and − ions — charge on electron
proportionality constant — distance between ions

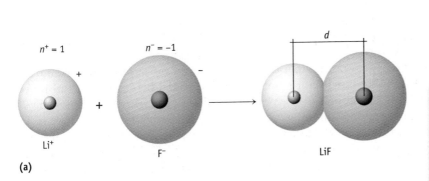

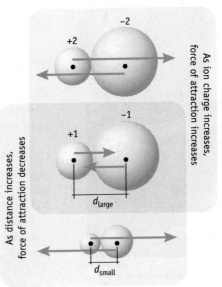

(a)

(b)

Figure 3.10   Coulomb's law and electrostatic forces. (a) Ions such as Li^+ and F^- are held together by an electrostatic force. Here a lithium ion is attracted to a fluoride ion. **(b)** Forces of attraction between ions of opposite charge increase with increasing ion charge and with decreasing distance. (As in the Simulation on CD-ROM Screen 3.7, the length of an arrow reflects the relative attractive force one ion exerts on another.)

Problem-Solving Tip 3.2

Is a Compound Ionic?

Students often ask how to know if a compound is ionic. No method works all of the time, but here are some useful guidelines.

1. Most metal-containing compounds are ionic. So, if a metal atom appears in the formula of a compound, a good first guess is that it is ionic. (There are interesting exceptions, but few come up in introductory chemistry.) It is helpful in this regard to recall trends in metallic behavior: all elements to the left of a diagonal line running from boron to tellurium in the periodic table are metallic.

2. If there is no metal in the formula, it is likely that the compound is not ionic. The exceptions here are compounds composed of polyatomic ions based on nonmetals (e.g., NH_4Cl or NH_4NO_3).

3. Learn to recognize the formulas of polyatomic ions (see Table 3.1) because they combine to form ionic compounds. Chemists write the formula of ammonium nitrate as NH_4NO_3 (and not as $N_2H_4O_3$) to alert others to the fact that it is an ionic compound composed of the common polyatomic ions NH_4^+ and NO_3^-.

As an example of these guidelines, you can be sure that Mg^{2+} with Br^- and K^+ with S^{2-} are combinations that are suitable for ionic compound formation. On the other hand, the compound CCl_4, formed from two nonmetals, C and Cl, is not ionic.

where, for example, n^+ is 3 for Al^{3+} and n^- is 2 for O^{2-}. Based on Coulomb's law, the force of attraction between oppositely charged ions increases

- As the ion charges (n^+ and n^-) increase. Thus, the attraction between ions having charges of +2 and −2 is greater than between ions having +1 and −1 charges (see Figure 3.10).

- As the distance between the ions becomes smaller (Figure 3.10; ➔ CHAPTER 8).

Unlike molecular compounds, ionic compounds do not consist of simple pairs or small groups of positive and negative ions as implied by their formulas. Rather, in the solid state the ions are generally arranged in an extended three-dimensional network called a **crystal lattice** [⊕ CD-ROM, Screen 3.9]. No discrete molecules exist. The structure of sodium chloride (Figure 3.11) is an excellent example.

The simplest ratio of cations to anions in an ionic compound is represented by its formula. For example, NaCl represents the ratio of sodium ions to chloride ions in the lattice, namely 1:1. In $CaCl_2$ the ratio of ions is one Ca^{2+} cation to two Cl^- anions.

Ionic compounds have characteristic properties that can be understood in terms of the charges of the ions and their arrangement in the lattice. Because each ion is surrounded by oppositely charged nearest neighbors, it is held tightly in its allotted location. At room temperature each ion can move just a bit around its average position, but much more energy must be added before an ion can move fast enough and far enough to escape the attraction of neighboring ions. Only if enough energy is added will the lattice structure collapse and the substance melt. Greater coulombic forces of attraction mean that ever more energy — higher and higher temperatures — are required to cause melting. Thus, Al_2O_3, a solid composed of Al^{3+} and O^{2-} ions, melts at a much higher temperature (2072 °C) than NaCl (801 °C), a solid composed of Na^+ and Cl^- ions.

Most ionic compounds are "hard" solids. That is, the solids are not pliable or soft. The reason for this is again the lattice of ions. The nearest neighbor of a cation in a lattice is an anion, and the force of attraction between them makes the lattice rigid. However, a blow with a hammer can cause the lattice to cleave cleanly along a sharp boundary. The hammer blow displaces layers of ions just enough to cause ions of like charge to become nearest neighbors [⊕ CD-ROM, Screen 3.9]. The repulsion between like charges forces the lattice apart (Figure 3.12).

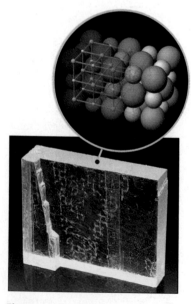

Figure 3.11   Sodium chloride. A crystal of sodium chloride consists of an extended lattice of sodium ions and chloride ions. *(Photo, Charles D. Winters)*

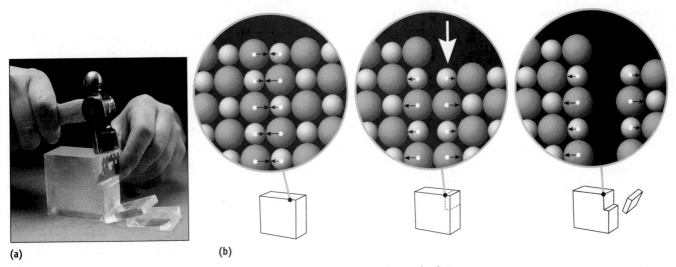

(a)   **(b)**

Figure 3.12   Ionic solids. (a) An ionic solid is normally rigid owing to the forces of attraction between oppositely charged ions. When struck sharply, however, the crystal can cleave cleanly. **(b)** When a crystal is struck, layers of ions move slightly, and ions of like charge become nearest neighbors. Repulsions between ions of similar charge cause the crystal to cleave. *(a, Charles D. Winters)*

Exercise 3.6   **Coulomb's Law**

Explain why the melting point of MgO, 2830 °C, is much higher than the melting point of NaCl (801 °C).

3.4   MOLECULAR COMPOUNDS: FORMULAS, NAMES, AND PROPERTIES

Many familiar compounds are not ionic, they are molecular: the water you drink, the sugar in your coffee or tea, or the aspirin you take for a headache. Some of these compounds have complicated formulas that you cannot, at this stage, predict or even decide if they are correct. However, you will often encounter many simpler compounds, and you should know how to name them and, in many cases, you should know their formulas.

Let us look first at molecules formed from combinations of two nonmetals. These "two-element" or **binary compounds** of nonmetals can be named in a systematic way [CD-ROM, Screens 3.12 and 3.13].

Hydrogen forms binary compounds with all the nonmetals except the noble gases. For compounds of oxygen, sulfur, and the halogens, the H atom is generally written first in the formula and is named first. The other nonmetal is named as if it were a negative ion.

Compound	Name
HF	hydrogen fluoride
HCl	hydrogen chloride
H_2S	dihydrogen sulfide

Virtually all binary molecular compounds of nonmetals are a combination of elements from Groups 4A–7A with one another or with hydrogen. The formula is generally written

Chapter Goals • Revisited

- Interpret, predict, and write formulas for ionic and molecular compounds.
- **Name compounds.**
- Calculate and use molar mass.
- Calculate percent composition for a compound and derive formulas from experimental data.

● Formulas of Binary Nonmetal Compounds

Simple hydrocarbons (compounds of C and H) such as methane and ethane have formulas written with H following C, and the formulas of ammonia and hydrazine have H following N. Water and the hydrogen halides, however, have the H atom preceding O or the halogen atom. Tradition is the only explanation for such oddities in chemistry.

● Hydrocarbons

Compounds such as methane, ethane, propane, and butane belong to a class of hydrocarbons called alkanes (see Chapter 11 and the *General Chemistry Interactive CD-ROM,* Version 3.0, Screen 3.13).

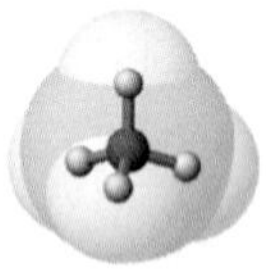

methane, CH_4 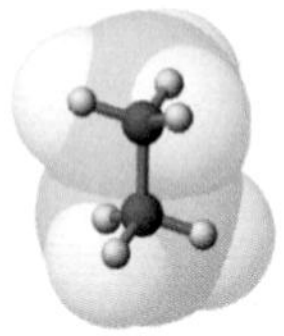   ethane, C_2H_6

propane, C_3H_8

butane, C_4H_{10}

by putting the elements in order of increasing group number. When naming the compound, the number of atoms of a given type in the compound is designated with a prefix, such as *di-, tri-, tetra-, penta-,* and so on.

Compound	Systematic Name
NF_3	nitrogen trifluoride
NO	nitrogen monoxide
NO_2	nitrogen dioxide
N_2O	dinitrogen monoxide
N_2O_4	dinitrogen tetraoxide
PCl_3	phosphorus trichloride
PCl_5	phosphorus pentachloride
SF_6	sulfur hexafluoride
S_2F_{10}	disulfur decafluoride

Finally, many of the binary compounds of nonmetals were discovered years ago and have names so common that they continue to be used.

Compound	Common Name
CH_4	methane
C_2H_6	ethane
C_3H_8	propane
C_4H_{10}	butane
NH_3	ammonia
N_2H_4	hydrazine
PH_3	phosphine
NO	nitric oxide
N_2O	nitrous oxide ("laughing gas")
H_2O	water

Exercise 3.7 **Naming Compounds of the Nonmetals**

1. Give the formula for each of the following binary, nonmetal compounds:
   (a) carbon dioxide   (d) boron trifluoride
   (b) phosphorus triiodide   (e) dioxygen difluoride
   (c) sulfur dichloride   (f) xenon trioxide
2. Name the following binary, nonmetal compounds:
   (a) N_2F_4 (c) SF_4 (e) P_4O_{10}
   (b) HBr   (d) BCl_3 (f) ClF_3

Ionic compounds are generally solids, whereas molecular compounds can range from gases to liquids to solids at ordinary temperatures (Figure 3.13). As size and molecular complexity increase, compounds generally exist as solids. Later in the book we shall explore some of the underlying causes of these general observations [➡ CHAPTER 13].

Water
H₂O

Nitrogen dioxide
NO₂

Sugar
C₁₂H₂₂O₁₁

(a)

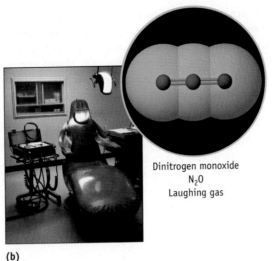

Dinitrogen monoxide
N₂O
Laughing gas

(b)

Figure 3.13   Molecular compounds.
(a) Nitrogen dioxide, NO_2, is a brown gas at room temperature, whereas water is a liquid. Sugar, $C_{12}H_{22}O_{11}$, is a solid. **(b)** Laughing gas is a common name for dinitrogen monoxide, N_2O, a simple compound often used as an anesthetic. *(Photos, Charles D. Winters)*

3.5   FORMULAS, COMPOUNDS, AND THE MOLE

The formula of a compound tells you the type of atoms or ions in the compound and the relative number of each. For example, one molecule of methane, CH_4, is made up of one atom of C and four atoms of H. But suppose you have Avogadro's number of C atoms (6.022×10^{23}) (◄ SECTION 2.5) combined with the proper number of H atoms. The compound's formula informs us that four times as many H atoms are required $(4 \times 6.022 \times 10^{23}$ H atoms) to give Avogadro's number of CH_4 molecules. What masses of atoms are combined, and what is the mass of this many CH_4 molecules?

C	+	4 H	⟶	CH₄
6.022×10^{23} C atoms		$4 \times 6.022 \times 10^{23}$ H atoms		6.022×10^{23} CH₄ molecules
= 1.000 mol of C		= 4.000 mol of H atoms		= 1.000 mol of CH₄ molecules
= 12.01 g of C atoms		= 4.032 g of H atoms		= 16.04 g of CH₄ molecules

Because we know the number of moles of C and H atoms, we know the masses of carbon and hydrogen that combine to form CH_4. It follows from the law of the conservation of mass that the mass of CH_4 is the sum of these masses (Section 2.1). That is, 1 mol of CH_4 has a mass equivalent to the mass of 1 mol of C atoms (12.01 g) plus 4 mol of H atoms (4.032 g). This is the *molar mass*, M, of CH_4 (◄ SECTION 2.5) [⊙ CD-ROM, Screen 3.14].

Molar and Molecular Masses

Element or Compound	Molar Mass (g/mol)	Average Mass of One Molecule* (g/molecule)
O₂	32.00	5.314×10^{-23}
P₄	123.9	2.057×10^{-22}
NH₃	17.03	2.828×10^{-23}
H₂O	18.02	2.992×10^{-23}
CH₂Cl₂	84.93	1.410×10^{-22}

*See text, page 99, for the calculation of the mass of one molecule.

Chapter Goals • Revisited

- Interpret, predict, and write formulas for ionic and molecular compounds.
- Name compounds.
- **Calculate and use molar mass.**
- Calculate percent composition for a compound and derive formulas from experimental data.

Ionic compounds such as NaCl do not exist as individual molecules. Thus, no molecular formula can be given. Rather, one can only write the simplest formula that shows the relative number of each kind of atom in a "formula unit" of the compound. Nonetheless, we talk about the molar mass of such compounds and calculate it from the simplest formula. To differentiate substances like NaCl that do not contain molecules, however, chemists sometimes refer to their **formula weight** instead of their molecular weight.

Figure 3.14 illustrates 1-mol quantities of several common compounds. To find the molar mass of any compound you need only add up the atomic masses for each element in one formula unit. As an example, let us find the molar mass of aspirin, $C_9H_8O_4$. In 1 mol of aspirin there are 9 mol of carbon atoms, 8 mol of hydrogen atoms, and 4 mol of oxygen atoms, which add up to 180.2 g/mol of aspirin

Aspirin has the molecular formula $C_9H_8O_4$ and a molar mass of 180.2 g/mol. Aspirin is the common name of the compound, but a more chemically correct name is acetylsalicylic acid.

$$\text{Mass of C in 1 mol } C_9H_8O_4 = 9 \text{ mol C} \cdot \frac{12.01 \text{ g C}}{1 \text{ mol C}} = 108.1 \text{ g C}$$

$$\text{Mass of H in 1 mol } C_9H_8O_4 = 8 \text{ mol H} \cdot \frac{1.008 \text{ g H}}{1 \text{ mol H}} = 8.064 \text{ g H}$$

$$\text{Mass of O in 1 mol } C_9H_8O_4 = 4 \text{ mol O} \cdot \frac{16.00 \text{ g O}}{1 \text{ mol O}} = 64.00 \text{ g O}$$

$$\text{Total mass of 1 mol of } C_9H_8O_4 = \text{molar mass of } C_9H_8O_4 = 180.2 \text{ g}$$

As was the case with elements [⬅ SECTION 2.5], it is important to be able to convert the mass of a compound to the equivalent number of moles (or moles to mass). For example, if you take 325 mg (0.325 g) of aspirin in one tablet, what amount of the compound have you ingested? Using the molar mass just calculated (180.2 g/mol), there is 0.00180 mol of aspirin per tablet.

$$0.325 \text{ g aspirin} \cdot \frac{1 \text{ mol aspirin}}{180.2 \text{ g aspirin}} = 0.00180 \text{ mol aspirin}$$

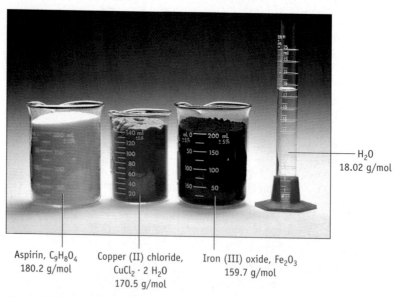

Aspirin, $C_9H_8O_4$
180.2 g/mol

Copper (II) chloride,
$CuCl_2 \cdot 2\ H_2O$
170.5 g/mol

Iron (III) oxide, Fe_2O_3
159.7 g/mol

H_2O
18.02 g/mol

Figure 3.14   One-mole quantities of some compounds. *(Charles D. Winters)*

The molar mass of a compound is the mass in grams of Avogadro's number of molecules (or of Avogadro's number of formula units of an ionic compound). With this knowledge, it is possible to determine the number of molecules in any sample from its mass or even to determine the average mass of one molecule. For example, the number of aspirin molecules in one tablet is

$$0.00180 \ \text{mol aspirin} \cdot \frac{6.022 \times 10^{23} \ \text{molecules}}{1 \ \text{mol aspirin}} = 1.09 \times 10^{21} \ \text{aspirin molecules}$$

and the mass of one molecule is

$$\left(\frac{180.2 \ \text{g aspirin}}{1 \ \text{mol aspirin}} \right) \left(\frac{1 \ \text{mol aspirin}}{6.022 \times 10^{23} \ \text{molecules}} \right) = 2.99 \times 10^{-22} \ \text{g/molecule}$$

Example 3.5 Molar Mass and Moles

Problem • You have 16.5 g of the common compound oxalic acid, $H_2C_2O_4$.

(a) What amount is represented by 16.5 g of oxalic acid?

(b) How many molecules of oxalic acid are in 16.5 g?

(c) How many atoms of carbon are in 16.5 g of oxalic acid?

(d) What is the mass of one molecule of oxalic acid?

Strategy • The first step in any problem involving the conversion of mass and moles is to find the molar mass of the compound in question. Then you can perform the other calculations as outlined by the scheme shown here (to find the number of molecules from the amount of substance and then to find the number of atoms of a particular kind):

[CD-ROM, Screens 3.14 and 3.15]

Solution •

(a) *Moles represented by 16.5 g*
Let us first calculate the molar mass of oxalic acid

$$2 \ \text{mol C per mol } H_2C_2O_4 \cdot \frac{12.01 \ \text{g C}}{1 \ \text{mol C}} = 24.02 \ \text{g C}$$

$$2 \ \text{mol H per mol } H_2C_2O_4 \cdot \frac{1.008 \ \text{g H}}{1 \ \text{mol H}} = 2.016 \ \text{g H}$$

$$4 \ \text{mol O per mol } H_2C_2O_4 \cdot \frac{16.00 \ \text{g O}}{1 \ \text{mol}} = 64.00 \ \text{g O}$$

$$\text{Molar mass of } H_2C_2O_4 = 90.04 \ \text{g}$$

Now calculate the amount in moles. The molar mass expressed in units of grams per mole is the conversion factor in all mass-to-mole conversions.

$$16.5 \ \text{g } H_2C_2O_4 \cdot \frac{1 \ \text{mol}}{90.04 \ \text{g } H_2C_2O_4} = 0.183 \ \text{mol } H_2C_2O_4$$

(b) *Number of molecules*
To find the number of oxalic acid molecules in 0.183 mol of $H_2C_2O_4$ we use Avogadro's number.

$$0.183 \ \text{mol} \cdot \frac{6.022 \times 10^{23} \ \text{molecules}}{1 \ \text{mol}} = 1.10 \times 10^{23} \ \text{molecules}$$

(c) *Number of C atoms*
Because each molecule contains two carbon atoms, the number of carbon atoms in 16.5 g (= 0.183 mol) of the acid is

$$1.10 \times 10^{23} \ \text{molecules} \cdot \frac{2 \ \text{C atoms}}{1 \ \text{molecule}} = 2.20 \times 10^{23} \ \text{C atoms}$$

(d) *Mass of one molecule*
The units of the desired answer are grams per molecule, which indicates that you should multiply the starting unit of molar mass (grams per mole) by (1/Avogadro's number) (units are mole/molecule), so that the unit "mol" cancels.

$$\left(\frac{90.04 \ \text{g}}{1 \ \text{mol}} \right) \left(\frac{1 \ \text{mol}}{6.022 \times 10^{23} \ \text{molecules}} \right) = 1.495 \times 10^{-22} \ \text{g/molecule}$$

Exercise 3.8 **Molar Mass and Moles-to-Mass Conversions**

(a) Calculate the molar mass of citric acid, $C_6H_8O_7$, and of $MgCO_3$.
(b) If you have 454 g of citric acid, what amount does this represent?
(c) To have 0.125 mol of $MgCO_3$, what mass must you have?

3.6 DETERMINING COMPOUND FORMULAS

Given a sample of an unknown compound, how can its formula be determined? The answer lies in chemical analysis, a major branch of chemistry that deals with the determination of formulas and structures.

Percent Composition

● **Molecular Composition**
This can be expressed as a percent (mass of an element in a 100-g sample) or as the mass of an element in a sample of any size. For example, NH_3 is 82.27% N. Therefore, it has 82.27 g of N in 100.0 g of compound or 41.14 g in 50.0 g.

The **law of constant composition** states that any sample of a pure compound always consists of the same elements combined in the same proportion by mass. Molecular composition can be expressed in at least three ways:

1. In terms of the number of atoms of each type per molecule or per formula unit; that is, by giving the formula of the compound.
2. In terms of the mass of each element per mole of compound.
3. By the mass of each element in the compound relative to the total mass of the compound. That is, as a **mass percent.**

Suppose you have 1.000 mol of NH_3, or 17.03 g. This mass of NH_3 is composed of 14.01 g of N (1.000 mol) and 3.024 g of H (3.000 mol). If you compare the mass of N, for example, to the total mass of compound, 82.27% of the total mass is N (and 17.76% is H).

$$\text{Mass percent N in } NH_3 = \frac{\text{mass of N in 1 mol } NH_3}{\text{mass of 1 mol of } NH_3} \times 100\%$$

$$= \frac{14.007 \text{ g N}}{17.031 \text{ g } NH_3} \times 100\%$$

$$= 82.24\% \text{ (or 82.24 g N in 100.0 g } NH_3)$$

$$\text{Mass percent H in } NH_3 = \frac{\text{mass of H in 1 mol } NH_3}{\text{mass of 1 mol of } NH_3} \times 100\%$$

$$= \frac{3(1.008 \text{ g H})}{17.03 \text{ g } NH_3} \times 100\%$$

$$= 17.76\% \text{ (or 17.76 g H in 100.0 g } NH_3)$$

82.27% of NH_3 mass
is **nitrogen.**

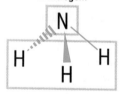

17.76% of NH_3 mass
is **hydrogen.**

These values represent the mass percent of each element, or **percent composition** by mass and tell you that in a 100.0-g sample there are 82.24 g of N and 17.76 g of H [🖱 CD-ROM, Screens 3.14 and 3.16].

Example 3.6 **Using Percent Composition**

Problem • What mass of nitrogen is contained in 454 g of NH_3?   **Solution** •

Strategy • Percent composition has units of (g N/100 g NH_3). Use this as the conversion factor to relate mass of N to the mass of NH_3 [🖱 CD-ROM, Screen 3.16].

$$454 \text{ g } NH_3 \cdot \frac{82.24 \text{ g N}}{100.0 \text{ g } NH_3} = 373 \text{ g N}$$

Exercise 3.9 **Expressing Composition as Percent Composition**

Express the composition of each of the following compounds in terms of the mass of each element in 1.00 mol of compound and the mass percent of each element:

(a) NaCl, sodium chloride

(b) C_8H_{18}, octane

(c) $(NH_4)_2CO_3$, ammonium carbonate

Exercise 3.10 **Using Percent Composition**

What is the mass of carbon in 454 g of octane, C_8H_{18}?

Empirical and Molecular Formulas from Percent Composition

Now let us consider the *reverse* of the procedure just described: using relative mass or percent composition data to find a molecular formula [⊕ CD-ROM, Screens 3.17 and 3.18]. Suppose you know the identity of the elements in a sample and have determined the mass of each element in a given mass of compound by chemical analysis [➡ SECTION 4.6]. You can then calculate the relative amount (moles) of each element and from this the relative number of atoms of each element in the compound. For example, for a compound composed of atoms of A and B, the steps from percent composition to a formula are

Convert weight percent to mass	**Convert mass to moles**	**Find mole ratio**	**Ratio gives formula**
$\%$ A $\longrightarrow$ g A	$\longrightarrow x$ mol A	$\dfrac{x \text{ mol A}}{y \text{ mol B}}$	A_xB_y
$\%$ B $\longrightarrow$ g B	$\longrightarrow y$ mol B		

Let us derive the formula for hydrazine, a close relative of ammonia and a compound used to remove metal ions from polluted water.

Convert weight percent to mass. The mass percentages in a sample of hydrazine are 87.42% N and 12.58% H. This means that in a 100.00-g sample of hydrazine, there are 87.42 g of N and 12.58 g of H.

Convert the mass of each element to moles. The amount of each element in the 100.00-g sample is

$$87.42 \text{ g N} \cdot \frac{1 \text{ mol N}}{14.007 \text{ g N}} = 6.241 \text{ mol N}$$

$$12.58 \text{ g H} \cdot \frac{1 \text{ mol H}}{1.0079 \text{ g H}} = 12.48 \text{ mol H}$$

Find the mole ratio of elements. Next, use the amount (moles) of each element in the 100.00-g of sample to find the amount of one element relative to the other. For hydrazine, this ratio is 2 mol of H to 1 mol of N,

$$\frac{12.48 \text{ mol H}}{6.241 \text{ mol N}} = \frac{2.000 \text{ mol H}}{1.000 \text{ mol N}} \longrightarrow NH_2$$

Chapter Goals • Revisited

• Interpret, predict, and write formulas for ionic and molecular compounds.

• Name compounds.

• Calculate and use molar mass.

• **Calculate percent composition for a compound and derive formulas from experimental data.**

• **Deriving Formulas from Percent Composition**
When finding the ratio of moles of one element relative to another, always divide the larger number by the smaller one.

Isooctane, C_8H_{18}, is the standard against which the octane rating of gasoline is determined. Octane numbers are assigned by comparing the burning performance of gasoline with the burning performance of mixtures of isooctane and heptane. Gasoline with an octane rating of 90 matches the burning characteristics of a mixture of 10% heptane and 90% isooctane. (*Photo, Charles D. Winters*)

showing that there are 2 mol of H atoms for every mole of N atoms in hydrazine. Thus, in one molecule two atoms of H occur for every atom of N, that is, the simplest atom ratio is represented by the formula NH_2.

Percent composition data allow us to calculate the atom ratios in a compound. A *molecular formula*, however, must convey *two* pieces of information: (1) the relative number of atoms of each element in a molecule (the atom ratios) and (2) the total number of atoms in the molecule. For hydrazine there are twice as many H atoms as N atoms. This means the molecular formula could be NH_2. Recognize, however, that percent composition data give only the *simplest possible ratio of atoms* in a molecule, that is, the **empirical formula.** The empirical formula of hydrazine may be NH_2, but its true *molecular formula* could also be N_2H_4, N_3H_6, N_4H_8, or any other formula having a 1-to-2 ratio of N to H.

To determine the molecular formula from the empirical formula, the molar mass must be obtained from experiment. For example, experiments show that the molar mass of hydrazine is 32.0 g/mol, twice the formula mass of NH_2, which is 16.0 g/mol. This must mean that the molecular formula of hydrazine is two times the empirical formula of NH_2, that is, N_2H_4.

As another example of the usefulness of percent composition data, let us say that you collected the following information in the laboratory for isooctane, the compound used as the standard for determining the octane rating of a fuel: % carbon = 84.12; % hydrogen = 15.88; molar mass = 114.2 g/mol. These data can be used to calculate the empirical and molecular formulas for the compound. The data inform us that 84.12 g of C and 15.88 g of H occur in a 100.0-g sample. From this, we find the amount (moles) of each element in this sample.

$$84.12 \text{ g C} \cdot \frac{1 \text{ mol C}}{12.011 \text{ g C}} = 7.004 \text{ mol C}$$

$$15.88 \text{ g H} \cdot \frac{1 \text{ mol H}}{1.0079 \text{ g H}} = 15.76 \text{ mol H}$$

This means that, in any sample of isooctane, the ratio of moles of H to C is

$$\text{Mole ratio} = \frac{15.76 \text{ mol H}}{7.004 \text{ mol C}} = \frac{2.250 \text{ mol H}}{1.000 \text{ mol C}}$$

Now the task is to turn this decimal fraction into a whole-number ratio of H to C. To do this, recognize that 2.25 is the same as $2\frac{1}{4}$ or $\frac{9}{4}$. Therefore, the ratio of C to H is

$$\text{Mole ratio} = \frac{2.25 \text{ mol H}}{1.00 \text{ mol C}} = \frac{2\frac{1}{4} \text{ mol H}}{1 \text{ mol C}} = \frac{9/4 \text{ mol H}}{1 \text{ mol C}} = \frac{9 \text{ mol H}}{4 \text{ mol C}}$$

and you know now that nine H atoms occur for every four C atoms in isooctane. Thus, the simplest, or *empirical, formula* is C_4H_9. If C_4H_9 were the molecular formula, the molar mass would be 57.12 g/mol. Your experiments gave the actual molar mass as 114.2 g/mol, however, twice the value for the empirical formula.

$$\frac{114.2 \text{ g/mol of isooctane}}{57.12 \text{ g/mol of } C_4H_9} = 2.00 \text{ mol } C_4H_9 \text{ per mol of isooctane}$$

The molecular formula is therefore $(C_4H_9)_2$, or C_8H_{18}.

> ### Problem-Solving Tip 3.3
>
> ### Finding Empirical and Molecular Formulas
>
> - The experimental data available to find a formula may be in the form of percent composition or the masses of elements combined in some mass of compound. No matter what the starting point, the first step is always to convert masses of elements to moles.
> - Be sure to use at least three significant figures when calculating empirical formulas. Fewer significant figures often give a misleading result.
>
> - When finding atom ratios, it is better to divide the larger number of moles by the smaller one.
> - Empirical and molecular formulas often differ for molecular compounds. In contrast, the formula of an ionic compound is generally the same as its empirical formula.
> - To determine the molecular formula of a compound after calculating the empirical formula, the molar mass must be obtained by some experimental method.

Example 3.7   Calculating a Formula from Percent Composition

Problem • Eugenol is the major component in oil of cloves. It has a molar mass of 164.2 g/mol and is 73.14% C and 7.37% H; the remainder is oxygen. What are the empirical and molecular formulas of eugenol?

Strategy • To derive a formula we need to know the mass percent of each element, or the mass of each element in a given sample mass. Here we have the former. Because the mass percent of all elements must add up to 100.00%, we find the mass percent of O from the difference between 100.00% and the mass percent of C and H. Next, we assume the mass percent of each element is equivalent to its mass in grams, and convert each mass to moles. Finally, the ratio of moles gives the empirical formula. The mass of a mole of compound having the calculated empirical formula is compared with the actual, experimental molar mass to find the true molecular formula [CD-ROM, Screens 3.17 and 3.18].

Eugenol, $C_{10}H_{12}O_2$, is an important component of oil of cloves. (Photo, Charles D. Winters)

Solution • The mass of O in a 100.0 g sample is

$$100.0 \text{ g} = 73.14 \text{ g C} + 7.37 \text{ g H} + \text{mass of O}$$

$$\text{Mass of O} = 19.49 \text{ g}$$

The amount of each element is:

$$73.14 \text{ g C} \cdot \frac{1 \text{ mol C}}{12.011 \text{ g C}} = 6.089 \text{ mol C}$$

$$7.37 \text{ g H} \cdot \frac{1 \text{ mol H}}{1.008 \text{ g H}} = 7.31 \text{ mol H}$$

$$19.49 \text{ g O} \cdot \frac{1 \text{ mol O}}{15.999 \text{ g O}} = 1.218 \text{ mol O}$$

To find the mole ratio, the best approach is to base the ratios on the smallest number of moles present, in this case oxygen.

$$\frac{\text{Mol C}}{\text{mol O}} = \frac{6.089 \text{ mol C}}{1.218 \text{ mol O}} = \frac{4.999 \text{ mol C}}{1.000 \text{ mol O}} = \frac{5 \text{ mol C}}{1 \text{ mol O}}$$

$$\frac{\text{Mol H}}{\text{mol O}} = \frac{7.31 \text{ mol H}}{1.218 \text{ mol O}} = \frac{6.00 \text{ mol H}}{1.000 \text{ mol O}} = \frac{6 \text{ mol C}}{1 \text{ mol O}}$$

Now we know there are 5 mol of C and 6 mol of H per mol of O. Thus, the empirical formula is C_5H_6O.

The experimentally determined molar mass of eugenol is 164.2 g/mol. This is twice the mass for C_5H_6O (82.10 g/mol).

$$\frac{164.2 \text{ g/mol of eugenol}}{82.10 \text{ g of } C_5H_6O} = 2.000 \text{ mol } C_5H_6O \text{ per mol of eugenol}$$

The molecular formula is $C_{10}H_{12}O_2$.

Exercise 3.11   Empirical and Molecular Formulas

(a) What is the empirical formula of naphthalene, $C_{10}H_8$?

(b) The empirical formula of acetic acid is CH_2O. If its molar mass is 60.05 g/mol, what is the molecular formula of acetic acid?

Exercise 3.12 · Calculating a Formula from Percent Composition

Isoprene is a liquid compound that can be polymerized to form natural rubber. It is composed of 88.17% carbon and 11.83% hydrogen. Its molar mass is 68.11 g/mol. What are its empirical and molecular formulas?

Exercise 3.13 · Calculating a Formula from Percent Composition

Camphor is the compound that creates the familiar smell of "mothballs." It is composed of 78.90% carbon and 10.59% hydrogen. The remainder is oxygen. What are its empirical and molecular formulas?

Determining a Formula from Mass Data

The composition of a compound in terms of mass percent gives us the mass of each element in a 100.0-g sample. In the laboratory we often collect information on the composition of compounds slightly differently. We can

1. *Combine known masses of elements to give a sample of the compound of known mass.* Element masses can be converted to moles, and the ratio of moles gives the combining ratio of atoms, that is, the molecular formula. This approach is described in Example 3.8.

2. *Decompose a known mass of a unknown compound into "pieces" of known composition.* If the masses of the pieces can be determined, the ratio of moles of the pieces gives the formula. An example is a decomposition such as

$$Ni(CO)_4(\ell) \longrightarrow Ni(s) + 4\ CO(g)$$

The masses of Ni and CO can be converted to moles, whose 1-to-4 ratio would reveal the formula of the compound. We shall describe this approach in Chapter 4 [➡ SECTION 4.6].

Example 3.8 · Determining a Formula from Mass Data

Problem • Tin metal (Sn) and purple iodine (I_2) combine to form orange, solid tin iodide with an unknown formula.

$$Sn\ metal + solid\ I_2 \longrightarrow solid\ Sn_xI_y$$

Weighed quantities of Sn and I_2 are combined, where the quantity of Sn is more than is needed to react with *all* of the iodine. After Sn_xI_y has been formed, it is isolated by filtration and weighed. The mass of excess tin is also determined. The following data were collected:

Mass of tin (Sn) in original mixture	1.056 g
Mass of iodine (I_2) in original mixture	1.947 g
Mass of tin (Sn) recovered after reaction	0.601 g

What is the empirical formula of this compound of tin and iodine?

Strategy • The first step is to find the masses of Sn and I that are combined in Sn_xI_y. The masses are then converted to moles, and the ratio of moles reveals the compound's empirical formula.

Solution • First, let us find the mass of tin that combined with iodine.

Mass of Sn in original mixture	1.056 g
Mass of Sn recovered	−0.601 g
Mass of Sn combined	0.455 g

Now convert the mass of tin to the amount of tin.

$$0.455\ g\text{-}Sn \cdot \frac{1\ mol\ Sn}{118.7\ g\text{-}Sn} = 0.00383\ mol\ Sn$$

(continued)

(a) Weighed samples of tin and iodine.

(b) The tin and iodine are heated in a solvent.

(c) The hot reaction mixture is filtered to recover unreacted tin.

(d) When the solvent cools, solid, orange tin iodide forms and is isolated.

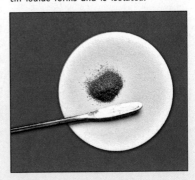

If tin and iodine can be combined quantitatively, the formula of the compound can be determined. *(Charles D. Winters)*

This amount of tin combined with 1.947 g of I_2. Because we want to know the amount of I that combined with this amount of Sn, we calculate the amount of I from the mass of I_2.

$$1.947 \text{ g } I_2 \cdot \frac{1 \text{ mol } I_2}{253.81 \text{ g } I_2} \cdot \frac{2 \text{ mol } I}{1 \text{ mol } I_2} = 0.01534 \text{ mol } I$$

Finally, we find the ratio of moles.

$$\frac{\text{Mol I}}{\text{Mol Sn}} = \frac{0.01534 \text{ mol I}}{0.00383 \text{ mol Sn}} = \frac{4.01 \text{ mol I}}{1.00 \text{ mol Sn}} = \frac{4 \text{ mol I}}{1 \text{ mol Sn}}$$

There are four times as many moles of I as moles of Sn in the sample. Therefore, there are four times as many atoms of I as atoms of Sn per formula unit. The empirical formula is SnI_4.

Example 3.9 · Formula of a Compound from Combining Masses

Problem • A gallium oxide, Ga_xO_y, forms when gallium is combined with oxygen. Suppose you allow 1.25 g of gallium (Ga) to react with oxygen and obtain 1.68 g of Ga_xO_y. What is the formula of the product?

Strategy • Calculate the mass of oxygen in 1.68 g of product (which you already know contains 1.25 g of Ga). Next calculate the amount of Ga and O (in moles) and find their ratio.

Solution • The masses of Ga and O combined in 1.68 g of product are

$$1.68 \text{ g product} - 1.25 \text{ g Ga} = 0.43 \text{ g O}$$

Next, calculate the amount of each reactant.

$$1.25 \text{ g Ga} \cdot \frac{1 \text{ mol Ga}}{69.72 \text{ g Ga}} = 0.0179 \text{ mol Ga}$$

$$0.43 \text{ g O} \cdot \frac{1 \text{ mol O}}{16.0 \text{ g O}} = 0.027 \text{ mol O}$$

Find the ratio of moles of O to moles of Ga,

$$\text{Mole ratio} = \frac{0.027 \text{ mol O}}{0.0179 \text{ mol Ga}} = \frac{1.5 \text{ mol O}}{1.0 \text{ mol Ga}}$$

and you see it is 1.5 mol O/1.0 mol Ga or 3 mol O to 2 mol Ga. Thus, the product is gallium oxide, Ga_2O_3.

Exercise 3.14 · Determining a Formula from Combining Masses

Analysis shows that 0.586 g of potassium metal combines with 0.480 g of O_2 gas to give a white solid having a formula of K_xO_y. What is the empirical formula of the compound?

Determining a Formula by Mass Spectrometry

There are many methods by which chemists determine the formulas of compounds. One way to make a new compound is to add a known molecule to a known compound by methods designed to give us a particular product. For example, we know that if we react an organic compound in a class of compounds called alkenes (here ethylene, C_2H_4) with water under the right conditions, the product will be an alcohol (here ethanol, CH_3CH_2OH).

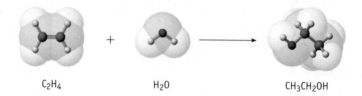

C_2H_4 H_2O CH_3CH_2OH

Our problem is usually to confirm the composition of a reaction product, and a number of instrumental methods can be used to do so. One of them is mass spectrometry. We introduced this technique in Chapter 2 where it was used to show the existence of isotopes and to measure their relative abundance [← FIGURE 2.5]. If a compound can be turned into a vapor, the vapor can be passed through an electron beam in a mass spectrometer where high energy electrons remove an electron from the compound, turning it into a positive ion. This ion usually breaks apart or fragments into smaller pieces. Here the cation created from ethanol ($CH_3CH_2OH^+$) expels a piece of itself (an H atom) to give another cation ($CH_3CH_2O^+$), which further fragments. A mass spectrometer detects and records the masses of the different fragments (Figure 3.15). Analysis of the spectrum gives an accurate molar mass for the original compound.

Figure 3.15   Mass spectrum of ethanol, CH_3CH_2OH. A prominent peak in the spectrum is the "parent" ion ($CH_3CH_2OH^+$) at mass 46. (The parent ion is the heaviest ion observed.) The mass designated by the peak for the parent ion confirms the formula of the molecule. Other peaks are for "fragment" ions. This pattern of lines can provide further, unambiguous evidence of the formula of the compound. (The horizontal axis is the mass-to-charge ratio of a given ion. Because almost all observed ions have a charge of +1, the value observed is the mass of the ion.) (See *A Closer Look: Mass Spectrometry, Molar Mass, and Isotopes*.)

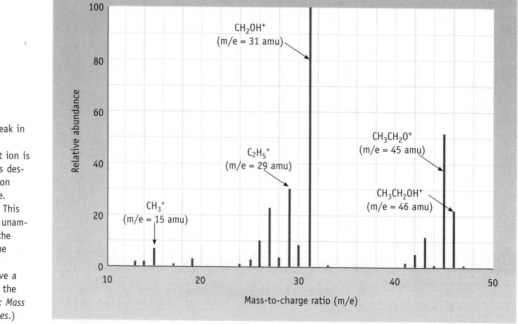

A Closer Look

Mass Spectrometry, Molar Mass, and Isotopes

Bromobenzene, C_6H_5Br, has a molecular weight of 157.010. Why then are there two prominent lines at 156 and 158 in the mass spectrum of the compound? The answer shows us the influence of isotopes on molecular weight.

Bromine has two naturally occurring isotopes: ^{79}Br and ^{81}Br. They are 50.7% and 49.3% abundant, respectively. What is the mass of C_6H_5Br based on each isotope? If we use the most abundant isotopes of C and H (^{12}C and ^{1}H), the mass of the compound having only the ^{79}Br isotope, $C_6H_5{}^{79}Br$, is 156. The mass of the compound containing only the ^{81}Br isotope, $C_6H_5{}^{81}Br$, is 158. These are the two lines having the highest mass-to-charge ratio in the spectrum.

The calculated molecular weight of bromobenzene is 157.010, a value calculated from the atomic weights of the elements. These atomic weights reflect the abundances of all isotopes. In contrast, the mass spectrum has a line for each possible combination of isotopes. This is the reason there are also small lines at a mass-to-charge ratio of 157 and 159. These arise from various combinations of ^{1}H, ^{12}C, ^{13}C,

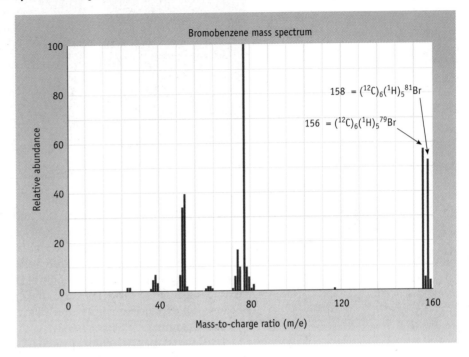

Bromobenzene mass spectrum

$158 = (^{12}C)_6(^1H)_5{}^{81}Br$

$156 = (^{12}C)_6(^1H)_5{}^{79}Br$

Relative abundance

Mass-to-charge ratio (m/e)

^{79}Br, and ^{81}Br atoms. In fact, careful analysis of such patterns can unambiguously identify a molecule.

3.7 HYDRATED COMPOUNDS

If ionic compounds are prepared in water solution and then isolated as solids, the crystals often have molecules of water trapped in the lattice. Compounds in which molecules of water are associated with the ions of the compound are called **hydrated compounds.** The beautiful blue copper(II) compound in Figure 3.14, for example, has a formula that is conventionally written as $CuCl_2 \cdot 2\,H_2O$. The dot between $CuCl_2$ and $2\,H_2O$ indicates that 2 mol of water is associated with every mole of $CuCl_2$; it is equivalent to writing the formula as $CuCl_2(H_2O)_2$. The name of the compound, copper(II) chloride dihydrate, reflects the presence of 2 mol of water per mole of $CuCl_2$. The molar mass of $CuCl_2 \cdot 2\,H_2O$ is 134.5 g/mol (for $CuCl_2$) plus 36.0 g/mol (for $2\,H_2O$) for a total mass of 170.5 g/mol.

Hydrated compounds are common. The walls of your home may be covered with wallboard, or "plaster board" (Figure 3.16). These sheets contain hydrated calcium sulfate, or gypsum ($CaSO_4 \cdot 2\,H_2O$), as well as unhydrated $CaSO_4$, sandwiched between paper. Gypsum is a mineral that can be mined. Now, however, it is more commonly a byproduct in the manufacture of hydrofluoric acid and phosphoric acid.

Figure 3.16 Gypsum wallboard. Gypsum is hydrated calcium sulfate, $CaSO_4 \cdot 2\,H_2O$. *(Charles D. Winters)*

Figure 3.17 Dehydrating hydrated cobalt(II) chloride, CoCl$_2$ · 6 H$_2$O. When water of hydration is lost, deep blue CoCl$_2$ is formed. *(Charles D. Winters)*

If gypsum is heated between 120 and 180 °C, the water is partly driven off to give CaSO$_4$ · $\frac{1}{2}$ H$_2$O, a compound commonly called "plaster of Paris." This compound is an effective casting material because, when added to water, it forms a thick slurry that can be poured into a mold or spread out over a part of the body. As it takes on more water, the material increases in volume and forms a hard, inflexible solid. These properties also make plaster of Paris a useful material to artists, because the expanding compound fills a mold completely and makes a high-quality reproduction.

Hydrated cobalt(II) chloride is the deep red solid in Figure 3.17. When heated it turns purple and then deep blue as it loses water to form **anhydrous** CoCl$_2$; "anhydrous" means a substance without water. On exposure to moist air, anhydrous CoCl$_2$ takes up water and is converted back into the red hydrated compound. It is this property that allows crystals of the blue compound to be used as a humidity indicator. You may have seen them in a small bag packed with a piece of electronic equipment. The compound also makes a good "invisible ink." A solution of cobalt(II) chloride in water is red, but if you write on paper with the solution it cannot be seen. When the paper is warmed, however, the cobalt compound dehydrates to give the deep blue anhydrous compound, and the writing becomes visible.

There is no simple way to predict how much water will be present in a hydrated compound, so it must be determined experimentally. Such an experiment may involve heating the hydrated material so that all the water is released from the solid and evaporated. Only the anhydrous compound is left. The formula of hydrated copper(II) sulfate, commonly known as "blue vitriol," is determined in this manner in Example 3.10.

Example 3.10 **Determining the Formula of a Hydrated Compound**

Problem • You want to know the value of x in blue, hydrated copper(II) sulfate, CuSO$_4$ · x H$_2$O, that is, the number of water molecules for each unit of CuSO$_4$. In the laboratory you weigh out 1.023 g of the solid. After heating the solid thoroughly in a porcelain crucible (see figure), 0.654 g of nearly white, anhydrous copper(II) sulfate, CuSO$_4$, remains.

$$1.023 \text{ g CuSO}_4 \cdot x \text{ H}_2\text{O} + \text{heat} \longrightarrow 0.654 \text{ g CuSO}_4 + ? \text{ g H}_2\text{O}$$

Strategy • To find x we need to know the amount of H$_2$O per mole CuSO$_4$. Therefore, first find the mass of water lost by the sample from the difference between the mass of hydrated compound and the anhydrous form. Finally, find the ratio of amount of water lost (moles) to the amount of anhydrous CuSO$_4$.

(continued)

Solution • Find the mass of water.

Mass of hydrated compound	1.023 g
Mass of anhydrous compound, $CuSO_4$	−0.654 g
Mass of water	0.369 g

Next convert the masses of $CuSO_4$ and H_2O to moles.

$$0.369 \text{ g } H_2O \cdot \frac{1 \text{ mol } H_2O}{18.02 \text{ g } H_2O} = 0.0205 \text{ mol } H_2O$$

$$0.654 \text{ g } CuSO_4 \cdot \frac{1 \text{ mol } CuSO_4}{159.6 \text{ g } CuSO_4} = 0.00410 \text{ mol } CuSO_4$$

The value of x is determined from the mole ratio.

$$\frac{0.0205 \text{ mol } H_2O}{0.00410 \text{ mol } CuSO_4} = \frac{5.00 \text{ mol } H_2O}{1.00 \text{ mol } CuSO_4}$$

The water-to-$CuSO_4$ ratio is 5-to-1, so the formula of the hydrated compound is $CuSO_4 \cdot 5 H_2O$. Its name is copper(II) sulfate pentahydrate.

White $CuSO_4$

Blue $CuSO_4 \cdot 5 H_2O$

The formula of a hydrated compound can be determined by heating a weighed sample enough to cause the compound to release its water of hydration. Knowing the mass of the hydrated compound before heating, and the mass of the anhydrous compound after heating, gives the mass of water in the original sample. *(Charles D. Winters)*

Exercise 3.15 **Determining the Formula of a Hydrated Compound**

Hydrated nickel(II) chloride is a beautiful green, crystalline solid. When heated strongly, the compound is dehydrated. If 0.235 g of $NiCl_2 \cdot x H_2O$ gives 0.128 g of $NiCl_2$ on heating, what is the value of x?

In Summary

When you have finished studying this chapter, you should ask if you have met the chapter goals. In particular, you should be able to

- Interpret the meaning of molecular formulas, condensed formulas, and structural formulas (Section 3.1).
- Recognize that metal atoms commonly lose one or more electrons to form positive ions (cations), and nonmetal atoms often gain electrons to form negative ions (anions) (see Figures 3.6 and 3.7).
- Recognize that the charge on a metal cation (other than the transition metals) is equal to the number of the group in which the element is found in the periodic table (M^{n+}, n = Group number) (Section 3.3). Transition metal cations are often +2 or +3, but other charges are observed.
- Recognize that the negative charge on a single-atom or monatomic anion, X^{n-}, is given by $n = 8 -$ Group number (Section 3.3).
- Give the names or formulas of polyatomic ions, knowing their formulas or names, respectively (Table 3.1 and Section 3.3).
- Write the formulas for ionic compounds by combining ions in the proper ratio to give no overall charge (Section 3.4).
- Understand the importance of Coulomb's law (Equation 3.1). Electrostatic forces are responsible for the attraction or repulsion of charged species (ions). The

magnitude of the force is given by Coulomb's law, which states that the force of attraction between oppositely charged species increases with electric charge and with decreasing distance between the species (Section 3.3).

- Name ionic compounds and simple binary compounds of the nonmetals (Section 3.4).
- Understand that the molar mass of a molecular compound is the mass in grams of Avogadro's number of molecules and that the molar mass of an ionic compound is the mass in grams of Avogadro's number of the combination of ions shown in the formula (Section 3.5).
- Calculate the molar mass of a compound from its formula and a table of atomic weights (Section 3.5).
- Calculate the amount (moles) of a compound represented by a given mass, and vice versa (Section 3.5).
- Express the composition of a compound in terms of percent composition (Section 3.6).
- Use percent composition to determine the empirical formula of a compound (Section 3.6).
- Understand how mass spectrometry can be used to find a molar mass (Section 3.6).
- Use experimental data to find the number of water molecules in a hydrated compound (Section 3.7).

Key Terms

Section 3.1
   molecular formula
   condensed formula
   structural formula
   chemical bonds

Section 3.2
   ball-and-stick model
   space-filling model

Section 3.3
   ionic compound
   cation

   anion
   monatomic ion
   polyatomic ions
   halide ion
   oxoanions
   electrostatic forces
   Coulomb's law
   crystal lattice

Section 3.4
   binary compound

Section 3.5
   formula weight

Section 3.6
   law of constant composition
   mass (or weight) percent
   percent composition
   empirical formula

Section 3.7
   hydrated compounds
   anhydrous

Key Equations

Equation 3.1 (page 93)
Coulomb's law describes the dependence of the force of attraction between ions of opposite charge (or the force of repulsion between ions of like charge) on ion properties and the distance between ions.

charge on + and − ions   charge on electron

$$\text{Force of attraction} = k\,\frac{(n^{+}e)(n^{-}e)}{d^{2}}$$

proportionality constant   distance between ions

Study Questions

Questions with blue, bold-faced numbers have answers in Appendix O.

Reviewing Important Concepts

1. A model of the cancer chemotherapy agent cisplatin is given here. How many N atoms are in one molecule? How many H atoms occur in one molecule? How many H atoms are in 1 mol of the compound? What is its molar mass?

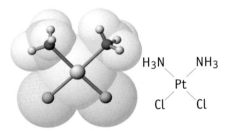

A molecular model and structural formula for cisplatin.

2. The molecule illustrated here is methanol. Using Figure 3.4 as a guide, decide which atoms are in the plane of the paper, which lie above the plane, and which lie below. Sketch a ball-and-stick model. If available to you, go to the textbook site on the World Wide Web (or the *General Chemistry Interactive CD-ROM*) and find the model of methanol.

A molecular model of methanol.

3. How many electrons are in a strontium atom (Sr)? Does an atom of Sr gain or lose electrons when forming an ion? How many electrons are gained or lost by the atom? When Sr forms an ion, the ion has the same number of electrons as which one of the noble gases?

4. The compound $(NH_4)_2SO_4$ consists of two different polyatomic ions. What are the names and electric charges of these ions? What is the molar mass of this compound?

5. The structure of adenine is shown here.
   (a) Show that 40 g of adenine represents less mass than 3.0×10^{23} molecules of the compound.
   (b) What structural feature of the molecule is common to the other bases (see Chapter Focus)?

A molecular model of adenine.

6. Knowing that the formula of sodium borate is Na_3BO_3, give the formula and charge of the borate ion. Is the borate ion a cation or an anion?

7. Which has the larger mass, 0.5 mol of $BaCl_2$ or 0.5 mol of $SiCl_4$?

8. What is the difference between an empirical and a molecular formula? Use the compound ethane, C_2H_6, to illustrate your answer.

9. Which compound has the larger weight percentage of oxygen, water or methanol (CH_3OH)?

Practicing Skills

Molecular Formulas
(See Example 3.1 and Exercises 3.1 and 3.2)

10. Write the molecular formula of each of the following compounds:
    (a) An organic compound, heptanol, has 7 carbon atoms, 16 hydrogen atoms, and 1 oxygen atom.
    (b) Vitamin C, ascorbic acid, has six carbon atoms, eight hydrogen atoms, and six oxygen atoms per molecule.
    (c) A molecule of aspartame, an artificial sweetener, has 14 carbon atoms, 18 hydrogen atoms, 2 nitrogen atoms, and 5 oxygen atoms.

11. Write the molecular formula for each of the following compounds:
    (a) Isoprene, the building block of rubber, has five carbon atoms and eight hydrogen atoms per molecule.
    (b) A molecule of BHA (butylated hydroxanisole, used as a preservative in foods) has 11 carbon atoms, 16 hydrogen atoms, and 2 oxygen atoms per molecule.
    (c) A molecule of camphor has 10 carbon atoms, 16 hydrogen atoms, and 1 oxygen atom per molecule.

12. Give the total number of atoms of each element in one formula unit of each of the following compounds:
    (a) CaC_2O_4 (c) $Co(NH_3)_5(NO_2)Cl_2$
    (b) C_6H_5CHO (d) $K_4Fe(CN)_6$

13. Give the total number of atoms of each element in each of the following molecules.
    (a) $Mn_2(CO)_{10}$
    (b) $HO_2CCH_2CH_2CO_2H$, succinic acid
    (c) $C_6H_2CH_3(NO_2)_3$, TNT, an explosive

Molecular Models
(See Example 3.2)

14. A ball-and-stick model of sulfuric acid is illustrated here. What is the formula for sulfuric acid? Describe the structure of the molecule. Is it flat? That is, are all the atoms in the plane of the paper? (Color code: sulfur atoms are

yellow; oxygen atoms are red; and hydrogen atoms are white.)

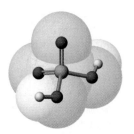

A molecular model of sulfuric acid.

15. A ball-and-stick model of toluene is illustrated here. What is its molecular formula? Describe the structure of the molecule. Is it flat or is only a portion of it flat? (Color code: carbon atoms are gray and hydrogen atoms are white.)

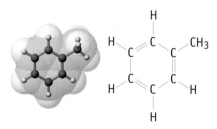

A molecular model and structural formula for toluene.

Ions and Ion Charges
(See Exercise 3.3, Figure 3.7, Table 3.1, and CD-ROM Screens 3.5 and 3.6)

16. What charges are most commonly observed for monatomic ions of the following elements?
    (a) magnesium   (c) nickel
    (b) zinc   (d) gallium

17. What charges are most commonly observed for monatomic ions of the following elements?
    (a) selenium   (c) iron
    (b) fluorine   (d) nitrogen

18. Give the symbol, including the correct charge, for each of the following ions:
    (a) barium ion   (e) sulfide ion
    (b) titanium(IV) ion   (f) perchlorate ion
    (c) phosphate ion   (g) cobalt(II) ion
    (d) hydrogen carbonate ion (h) sulfate ion

19. Give the symbol, including the correct charge, for each of the following ions:
    (a) permanganate ion   (d) ammonium ion
    (b) nitrite ion   (e) phosphate ion
    (c) dihydrogen phosphate ion (f) sulfite ion

20. When potassium becomes a monatomic ion, how many electrons does it lose or gain? What noble gas atom has the same number of electrons as a potassium ion?

21. When oxygen and sulfur become monatomic ions, how many electrons do they lose or gain? What noble gas atom has the same number of electrons as an oxygen ion? What noble gas atom has the same number of electrons as a sulfur ion? How do O and S resemble each other in their behavior?

Ionic Compounds
(See Examples 3.3 and 3.4 and CD-ROM Screen 3.8)

22. Predict the charges of the ions in an ionic compound containing the elements barium and bromine. Write the formula for the compound.

23. What are the charges of the ions in an ionic compound containing cobalt(III) and fluoride ions? Write the formula for the compound.

24. For each of the following compounds, give the formula, charge, and the number of each ion that makes up the compound:
    (a) K_2S (d) $(NH_4)_3PO_4$
    (b) $CoSO_4$ (e) $Ca(ClO)_2$
    (c) $KMnO_4$

25. For each of the following compounds, give the formula, charge, and the number of each ion that makes up the compound:
    (a) $Mg(CH_3CO_2)_2$ (d) KH_2PO_4
    (b) $Ti(SO_4)_2$ (e) $CuCO_3$
    (c) $Al(OH)_3$

26. Cobalt is a transition metal and forms ions with at least two different charges. Write the formulas for the two different cobalt oxides.

27. Platinum is a transition element and forms Pt^{2+} and Pt^{4+} ions. Write the formulas for the compounds of each of these ions with (a) chloride ions and (b) sulfide ions.

28. Which of the following are correct formulas for ionic compounds? For those that are not, give the correct formula.
    (a) $AlCl_2$ (b) KF_2 (c) Ga_2O_3 (d) MgS

29. Which of the following are correct formulas for ionic compounds? For those that are not, give the correct formula.
    (a) Ca_2O (b) $SrBr_2$ (c) Fe_2O_5 (d) Li_2O

30. Write formulas for all the compounds that can be made by combining the cations Mg^{2+} and Al^{3+} with the anions O^{2-} and PO_4^{3-}.

31. Write formulas for all the compounds that can be made by combining the cations NH_4^+ and Ni^{2+} with the anions CO_3^{2-} and SO_4^{2-}.

Naming Ionic Compounds
(See Exercise 3.5 and CD-ROM Screen 3.8)

32. Name each of the following ionic compounds:
    (a) K_2S (c) $(NH_4)_3PO_4$
    (b) $CoSO_4$ (d) $Ca(ClO)_2$

33. Name each of the following ionic compounds:
    (a) $Ca(CH_3CO_2)_2$ (c) $Al(OH)_3$
    (b) $Ni_3(PO_4)_2$ (d) KH_2PO_4

34. Give the formula for each of the following ionic compounds:
    (a) ammonium carbonate
    (b) calcium iodide
    (c) copper(II) bromide
    (d) aluminum phosphate
    (e) silver(I) acetate

35. Give the formula for each of the following ionic compounds:
    (a) calcium hydrogen carbonate
    (b) potassium permanganate
    (c) magnesium perchlorate
    (d) potassium hydrogen phosphate
    (e) sodium sulfite

36. Write the formulas for all the ionic compounds that can be made by combining the cations Na^+ and Ba^{2+} with the anions CO_3^{2-} and I^-. Name each compound formed.

37. Write the formulas for all the ionic compounds that can be made by combining the cations Mg^{2+} and Fe^{3+} with the anions PO_4^{3-} and NO_3^-. Name each compound formed.

Coulomb's Law
(See Equation 3.1, Figure 3.10, and CD-ROM Screen 3.7)

38. Sodium ion, Na^+, forms ionic compounds with fluoride, F^-, and iodide, I^-. You know that the radii of these ions are: $Na^+ = 116$ pm; $F^- = 119$ pm; and $I^- = 206$ pm. In which ionic compound, NaF or NaI, are the forces of attraction between cation and anion stronger? Explain your answer.

39. Consider the two ionic compounds NaCl and CaO. In which compound are the cation–anion attractive forces stronger? Explain your answer.

Naming Binary, Nonmetal Compounds
(See Exercise 3.6 and CD-ROM Screen 3.12)

40. Give the name for each of the following binary, nonionic compounds:
    (a) NF_3 (b) HI (c) BI_3 (d) PF_5

41. Give the name for each of the following binary, nonionic compounds:
    (a) N_2O_5 (b) P_4S_3 (c) OF_2 (d) XeF_4

42. Give the formula for each of the following nonmetal compounds:
    (a) sulfur dichloride
    (b) dinitrogen pentaoxide
    (c) silicon tetrachloride
    (d) diboron trioxide (commonly called boric oxide)

43. Give the formula for each of the following nonmetal compounds:
    (a) bromine trifluoride
    (b) xenon difluoride
    (c) hydrazine
    (d) diphosphorus tetrafluoride
    (e) butane

Molecules, Compounds, and the Mole
(See Example 3.5 and CD-ROM Screens 3.14 and 3.15)

44. Calculate the molar mass of each of the following compounds:
    (a) Fe_2O_3, iron(III) oxide
    (b) BCl_3, boron trichloride
    (c) $C_6H_8O_6$, ascorbic acid (vitamin C)

45. Calculate the molar mass of each of the following compounds:
    (a) $Fe(C_6H_{11}O_7)_2$, iron(II) gluconate, a dietary supplement
    (b) $CH_3CH_2CH_2CH_2SH$, butanethiol, has a skunk-like odor
    (c) $C_{20}H_{24}N_2O_2$, quinine, used as an antimalarial drug

46. Calculate the molar mass of each hydrated compound. Note that the water of hydration is included in the molar mass. (See Section 3.7.)
    (a) $Ni(NO_3)_2 \cdot 6\,H_2O$ (b) $CuSO_4 \cdot 5\,H_2O$

47. Calculate the molar mass of each hydrated compound. Note that the water of hydration is included in the molar mass. (See Section 3.7.)
    (a) $H_2C_2O_4 \cdot 2\,H_2O$ (b) $MgSO_4 \cdot 7\,H_2O$, epsom salts

48. What amount (moles) is represented by 1.00 g of each of the following compounds?
    (a) C_3H_7OH, propanol, rubbing alcohol
    (b) $C_{11}H_{16}O_2$, an antioxidant in foods, also known as BHA (butylated hydroxyanisole)
    (c) $C_9H_8O_4$, aspirin

49. Assume you have 0.250 g of each of the following compounds. What amount of each is present?
    (a) $C_{14}H_{10}O_4$, benzoyl peroxide, used in acne medications
    (b) $Pt(NH_3)_2Cl_2$, cisplatin, a cancer chemotherapy agent

50. Acetonitrile, CH_3CN, was found in the coma (the gaseous head) of Comet Hale-Bopp in 1997. What amount of acetonitrile is represented by 2.50 kg?

51. Acetone, $(CH_3)_2CO$, is an important industrial solvent. If 1260 million kg of this organic compound is produced annually, what amount (moles) is produced?

52. Sulfur trioxide, SO_3, is made industrially in enormous quantities by combining oxygen and sulfur dioxide, SO_2. What amount of SO_3 is represented by 1.00 kg of sulfur trioxide? How many molecules? How many sulfur atoms? How many oxygen atoms?

53. An Alka-Seltzer tablet contains 324 mg of aspirin ($C_9H_8O_4$), 1904 mg of $NaHCO_3$, and 1000. mg of citric acid ($C_6H_8O_7$). (The last two compounds react with each other to provide the "fizz," bubbles of CO_2, when the tablet is put into water.)
    (a) Calculate the amount (moles) of each substance in the tablet.
    (b) If you take one tablet, how many molecules of aspirin are you consuming?

Percent Composition
(See Exercise 3.8 and CD-ROM Screen 3.16)

54. Calculate the mass percent of each element in the following compounds.
    (a) PbS, lead(II) sulfide, galena

(b) C_3H_8, propane

(c) $C_{10}H_{14}O$, carvone, found in caraway seed oil

55. Calculate the mass percent of each element in the following compounds:

    (a) $C_8H_{10}N_2O_2$, caffeine   (c) $CoCl_2 \cdot 6 H_2O$

    (b) $C_{10}H_{20}O$, menthol

56. What mass of lead is present in 10.0 g of PbS?

57. What mass of iron is present in 25.0 g of Fe_2O_3?

58. If you wish to obtain 10.0 g of copper metal from copper(II) sulfide, CuS, what mass of CuS must you use?

59. The mineral ilmenite, $FeTiO_3$, is a source of titanium. What mass of ilmenite, in grams, is required if you wish to obtain 750 g of titanium?

Empirical and Molecular Formulas

(See Examples 3.6–3.8 and CD-ROM Screens 3.17–3.19)

60. Succinic acid occurs in fungi and lichens. Its empirical formula is $C_2H_3O_2$ and its molar mass is 118.1 g/mol. What is its molecular formula?

61. An organic compound has the empirical formula C_2H_4NO. If its molar mass is 116.1 g/mol, what is the molecular formula of the compound?

62. Give the empirical or molecular formula for each of the following:

Empirical Formula	Molar Mass (g/mol)	Molecular Formula
(a) CH	26.0	_____
(b) CHO	116.1	_____
(c) _____	_____	C_8H_{16}

63. Give the molecular formula for each of the following:

Empirical Formula	Molar Mass (g/mol)	Molecular Formula
(a) $C_2H_3O_3$	150.1	_____
(b) C_3H_8	44.1	_____
(c) B_5H_7	122.2	_____

64. Acetylene is a colorless gas used as a fuel in welding torches, among other things. It is 92.26% C and 7.74% H. Its molar mass is 26.02 g/mol. Calculate the empirical and molecular formulas of acetylene.

65. A large family of boron-hydrogen compounds has the general formula B_xH_y. One member of this family contains 88.5% B; the remainder is hydrogen. Which of the following is its empirical formula: BH_2, BH_3, B_2H_5, B_5H_7, or B_5H_{11}?

66. Cumene is a hydrocarbon, a compound composed only of C and H. It is 89.94% carbon, and the molar mass is 120.2 g/mol. What are the empirical and molecular formulas of cumene?

67. Nitrogen and oxygen form a series of oxides with the general formula N_xO_y. One of them, a blue solid, contains 36.84% N. What is the empirical formula of this oxide?

68. Mandelic acid is an organic acid composed of carbon (63.15%), hydrogen (5.30%), and oxygen (31.55%). Its molar mass is 152.14 g/mol. Determine the empirical and molecular formulas of the acid.

69. Nicotine, a poisonous compound found in tobacco leaves, is 74.0% C, 8.65% H, and 17.35% N. Its molar mass is 162 g/mol. What are the empirical and molecular formulas of nicotine?

70. If Epsom salt, $MgSO_4 \cdot x H_2O$, is heated to 250 °C, all the water of hydration is lost. On heating a 1.687-g sample of the hydrate, 0.824 g of $MgSO_4$ remains. How many molecules of water occur per formula unit of $MgSO_4$?

71. The alum used in cooking is potassium aluminum sulfate hydrate, $KAl(SO_4)_2 \cdot x H_2O$. To find the value of x, you can heat a sample of the compound to drive off all of the water and leave only $KAl(SO_4)_2$. Assume you heat 4.74 g of the hydrated compound and that the sample loses 2.16 g of water. What is the value of x?

72. A new compound containing xenon and fluorine was formed by shining sunlight on a mixture of Xe (0.526 g) and F_2 gas. If you isolate 0.678 g of the new compound, what is its empirical formula?

73. Elemental sulfur (1.256 g) is combined with fluorine, F_2, to give a compound with the formula SF_x, a very stable, colorless gas. If you have isolated 5.722 g of SF_x, what is the value of x?

74. Zinc metal (2.50 g) combines with 9.70 g of iodine to produce zinc iodide, Zn_xI_y. What is the formula of this ionic compound?

75. You combine 1.25 g of germanium, Ge, with excess chlorine, Cl_2. The mass of product, Ge_xCl_y, is 3.69 g. What is the formula of the product, Ge_xCl_y?

General Questions

More challenging questions are marked with an underlined number.

76. A drop of water has a volume of about 0.05 mL. How many molecules of water are in a drop of water? (Assume water has a density of 1.00 g/cm^3.)

77. Capsaicin, the compound that gives the hot taste to chili peppers, has the formula $C_{18}H_{27}NO_3$.

(a) Calculate the molar mass.

(b) If you eat 55 mg of capsaicin, what amount (moles) have you consumed?

(c) Calculate the mass percent of each element in the compound.

(d) What mass of carbon (mg) is in 55 mg of capsaicin?

78. Calculate the mass percent of each element in the blue solid $Cu(NH_3)_4SO_4 \cdot H_2O$. What is the mass in grams of copper and of water in 10.5 g of the compound?

79. Elemental phosphorus is made by heating calcium phosphate with carbon and sand in an electric furnace. What quantity of calcium phosphate, in kilograms, must be used to produce 15.0 kg of phosphorus?

80. Chromium is obtained by heating chromium(III) oxide with carbon. If you want to produce 850 kg of chromium metal, what quantity of Cr_2O_3 (in kilograms) is required?

81. Write the molecular formula and calculate the molar mass for each of the molecules shown here. Which has the larger percentage of carbon? Of oxygen?

(a) Ethylene glycol (used in antifreeze)

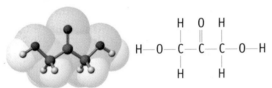

A molecular model and structural formula for ethylene glycol.

(b) Dihydroxyacetone (used in artificial tanning lotions)

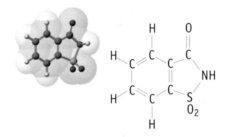

A molecular model and structural formula for dihydroxyacetone.

82. Malic acid, an organic acid found in apples, contains C, H, and O in the following ratios: $C_1H_{1.50}O_{1.25}$. What is the empirical formula of malic acid?

83. Your doctor has diagnosed you as being anemic, that is, as having too little iron in your blood. At the drugstore you find two iron-containing dietary supplements, one with iron(II) sulfate, $FeSO_4$, and the other with iron(II) gluconate, $Fe(C_6H_{11}O_7)_2$. If you take 100. mg of each compound, which delivers more atoms of iron?

84. Spinach is high in iron (2 mg/90-g serving). It is also a source of the oxalate ion, $C_2O_4^{2-}$, however, which combines with iron ions to form iron oxalate, $Fe_x(C_2O_4)_y$, a substance that prevents your body from absorbing the iron. Analysis of a 0.109-g sample of iron oxalate shows that it contains 38.82% iron. What is the empirical formula of the compound?

85. A sample of a compound composed of iron and carbon monoxide, $Fe_x(CO)_y$, is 30.70% iron. What is the empirical formula for the compound?

86. *Ma huang*, an extract from the ephedra species of plants, contains ephedrine. The Chinese have used it for over 5000 years to treat asthma. More recently the substance has been used in diet pills that can be purchased over the counter in herbal medicine shops. However, very serious concerns about these pills have been raised because of reports of serious heart problems associated with their use.

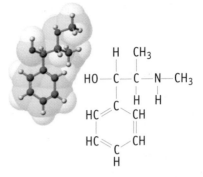

A molecular model and structural formula for ephedrine.

(a) Write the molecular formula for ephedrine and calculate its molar mass.

(b) What is the weight percent of carbon in ephedrine?

(c) Calculate the amount of ephedrine in a 0.125-g sample.

(d) How many molecules of ephedrine are in 0.125 g? How many C atoms?

87. Saccharin is over 300 times sweeter than sugar. It was first made in 1897, a time when it was common practice for chemists to record the taste of any new substances they synthesized.

A molecular model and structural formula for saccharin.

(a) Write the molecular formula for the compound.

(b) If you ingest 125 mg of saccharin, what amount (moles) of saccharin have you ingested?

(c) What mass of sulfur is contained in 125 mg of saccharin?

88. Which of the following pairs of elements are likely to form ionic compounds when allowed to react? Write appropriate formulas for the ionic compound you expect to form, and give the name of each.

(a) chlorine and bromine   (e) sodium and argon

(b) nitrogen and bromine   (f) sulfur and bromine

(c) lithium and sulfur   (g) calcium and fluorine

(d) indium and oxygen

89. Name each of the following compounds, and tell which ones are best described as ionic:

(a) ClF_3 (d) $Ca(NO_3)_2$ (g) KI (i) OF_2

(b) NCl_3 (e) XeF_4 (h) Al_2S_3 (j) K_3PO_4

(c) $SrSO_4$ (f) PCl_3

90. Write the formula for each of the following compounds, and tell which ones are best described as ionic:
    (a) sodium hypochlorite
    (b) boron triiodide
    (c) aluminum perchlorate
    (d) calcium acetate
    (e) potassium permanganate
    (f) ammonium sulfite
    (g) potassium dihydrogen phosphate
    (h) disulfur dichloride
    (i) chlorine trifluoride
    (j) phosphorus trifluoride

91. Complete the table by placing symbols, formulas, and names in the blanks.

Cation	Anion	Name	Formula
		ammonium bromide	
Ba^{2+}			BaS
	Cl^-	iron(II) chloride	
	F^-		PbF_2
Al^{3+}	CO_3^{2-}		
		iron(III) oxide	
			$LiClO_4$
		aluminum phosphate	
	Br^-	lithium bromide	
			$Ba(NO_3)_2$
Al^{3+}		aluminum oxide	
		iron(III) carbonate	

92. Azulene, a beautiful blue hydrocarbon, has 93.71% C and a molar mass of 128.16 g/mol. What are the empirical and molecular formulas of azulene?

93. Fluorocarbonyl hypofluorite is composed of 14.6% C, 39.0% O, and 46.3% F. If the molar mass of the compound is 82 g/mol, determine the empirical and molecular formulas of the compound.

94. The action of bacteria on meat and fish produces a compound called cadaverine. As its name and origin imply, it stinks! (It is also present in bad breath and adds to the odor of urine.) It is 58.77% C, 13.81% H, and 27.40% N. Its molar mass is 102.2 g/mol. Determine the molecular formula of cadaverine.

95. Cacodyl, a compound containing arsenic, was reported in 1842 by the German chemist Robert Wilhelm Bunsen. It has an almost intolerable garlic-like odor. Its molar mass is 210 g/mol, and it is 22.88% C, 5.76% H, and 71.36% As. Determine its empirical and molecular formulas.

96. A major oil company has used a gasoline additive called MMT to boost the octane rating of its gasoline. What is the empirical formula of MMT if it is 49.5% C, 3.2% H, 22.0% O, and 25.2% Mn?

97. Transition metals can combine with carbon monoxide (CO) to form compounds such as $Fe(CO)_5$ (Study Question 3.85). Assume that you combine 0.125 g of nickel with CO and isolate 0.364 g of $Ni(CO)_x$. What is the value of x?

98. Direct reaction of iodine (I_2) and chlorine (Cl_2) produces an iodine chloride, I_xCl_y, a bright yellow solid. If you completely used up 0.678 g of iodine, and produced 1.246 g of I_xCl_y, what is the empirical formula of the compound? A later experiment showed the molar mass of I_xCl_y was 467 g/mol. What is the molecular formula of the compound?

99. In a reaction 2.04 g of vanadium combines with 1.93 g of sulfur to give a pure compound. What is the empirical formula of the product?

100. Iron pyrite, often called "fool's gold," has the formula FeS_2. If you could convert 15.8 kg of iron pyrite to iron metal, what mass of the metal do you obtain?

101. Stibnite, Sb_2S_3, is a dark gray mineral from which antimony metal is obtained. If you have 1.00 kg of an ore that contains 10.6% antimony, what mass of Sb_2S_3 (in grams) is in the ore?

102. Which of the following statements about 57.1 g of octane, C_8H_{18}, is (are) not true.
     (a) 57.1 g is 0.500 mol of octane.
     (b) The compound has 84.1% C by weight.
     (c) The empirical formula of the compound is C_4H_9.
     (d) 57.1 g of octane contains 28.0 g of hydrogen atoms.

103. The formula of barium molybdate is $BaMoO_4$. What is the formula of sodium molybdate?
     (a) Na_4MoO (c) Na_2MoO_3 (e) Na_4MoO_4
     (b) $NaMoO$ (d) Na_2MoO_4

104. Pepto-Bismol, which helps provide soothing relief for an upset stomach, contains 300. mg of bismuth subsalicylate, $C_{21}H_{15}Bi_3O_{12}$, per tablet. If you take two tablets for your stomach distress, what amount of the "active ingredient" are you taking? What mass of Bi are you consuming in two tablets?

105. A metal M forms a compound with the formula MCl_4. If the compound is 74.75% chlorine, what is the identity of M?

106. The mass of 2.50 mol of a compound with the formula ECl_4, in which E is a nonmetallic element, is 385 g. What is the molar mass of ECl_4? What is the identity of E?

107. The weight percent of oxygen in an oxide that has the formula MO_2 is 15.2%. What is the molar mass of this compound? What element or elements are possible for M?

108. An ionic compound can dissolve in water because the cations and anions are attracted to water molecules. The drawing here shows how a cation and a water molecule, which has a negatively charged O atom and positively charged H atoms, can interact. Which of the following cations should be most strongly attracted to water: Na^+, Mg^{2+}, or Al^{3+}? Explain briefly.

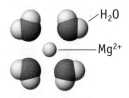

H₂O

Mg²⁺

Water molecules interacting with a metal cation such as Mg^{2+}.

109. When analyzed, an unknown compound gave these experimental results: C, 54.0%; H, 6.00%; and O, 40.0%. Four different students used these values to calculate the empirical formulas shown here. Which answer is correct? Why did some students not get the correct answer?

    (a) $C_4H_5O_2$ (c) $C_7H_{10}O_4$

    (b) $C_5H_7O_3$ (d) $C_9H_{12}O_5$

110. Two general chemistry students working together in the lab weigh out 0.832 g of $CaCl_2 \cdot 2\ H_2O$ into a crucible. After heating the sample for a short time and allowing the crucible to cool, these students determine that the sample has a mass of 0.739 g. They then do a quick calculation. On the basis of this calculation, what should they do next?

    (a) Congratulate themselves on a job well done.

    (b) Assume the bottle of $CaCl_2 \cdot 2\ H_2O$ was mislabeled; it actually contained something different.

    (c) Heat the crucible again, and then reweigh it.

111. The elements A and Z combine to produce two different compounds: A_2Z_3 and AZ_2. If 0.15 mol of A_2Z_3 has a mass of 15.9 g and 0.15 mol of AZ_2 has a mass of 9.3 g, what are the atomic masses of A and Z?

112. A piece of nickel foil, 0.550 mm thick and 1.25 cm square, is allowed to react with fluorine, F_2, to give a nickel fluoride. (a) What amount of nickel foil is used? (The density of nickel is 8.908 g/cm³.) (b) If you isolate 1.261 g of the nickel fluoride, what is its formula? (c) What is its name?

113. Uranium is used as a fuel, primarily in the form of uranium(IV) oxide, in nuclear power plants. This question considers some uranium chemistry.

    (a) A small sample of uranium metal (0.169 g) is heated to between 800 and 900 °C in air to give 0.199 g of a dark green oxide, U_xO_y. What amount of uranium metal was used? What is the empirical formula of the oxide, U_xO_y? What is the name of the oxide? What amount of U_xO_y must have been obtained?

    (b) The naturally occurring isotopes of uranium are ²³⁴U, ²³⁵U, and ²³⁸U. Which is the most abundant?

    (c) If the hydrated compound $UO_2(NO_3)_2 \cdot z\ H_2O$ is heated gently, the water of hydration is lost. If you have

0.865 g of the hydrated compound and obtain 0.679 g of $UO_2(NO_3)_2$ on heating, how many molecules of water of hydration are in each formula unit of the original compound? (The oxide U_xO_y is obtained if the hydrate is heated to temperatures over 800 °C in the air.)

Using Electronic Resources

These questions refer to the *General Chemistry Interactive CD-ROM,* Version 3.0.

114. See CD-ROM Screen 3.20: Chemical Puzzler. Cobalt(II) chloride hexahydrate dissolves readily in water to give a red solution. If we use this solution as an "ink," we can write a secret message on paper. The writing is not visible when the water evaporates from the paper. When the paper is heated, however, the message can be read. Explain the chemistry behind this observation.

A solution of $CoCl_2 \cdot 6\ H_2O$.

Using the secret ink to write on paper.

Heating the paper reveals the writing.

115. See CD-ROM Screen 3.4: Representing Compounds (see the sidebar on isomers). Isomers are compounds with the same molecular formula but with different atom-to-atom linkages or that fill space in different ways. Examine the *Closer Look* sidebar and the Exercise. What types of isomers are possible? Give an example of each.

116. See CD-ROM Screen 3.7: Coulomb's Law. The "simulation" section of this screen helps you explore Coulomb's law. You can change the charges on the ions and the distance between them. If the ions experience an attractive force, arrows point from one ion to the other. Repulsion is indicated by arrows pointing in opposite directions. Change the ion charges (from ±1 to ±2 to ±3). How does this affect the attractive force? How close can the ions approach before significant repulsive forces set in? How does this distance vary with ion charge?

4 Chemical Equations and Stoichiometry

Chapter Goals

- Balance equations for simple chemical reactions.
- Perform stoichiometry calculations using balanced chemical equations.
- Understand the meaning of a limiting reactant.
- Calculate the theoretical and percent yield of a chemical reaction.
- Use stoichiometry to analyze a mixture of compounds or to determine the formula of a compound.

▲ The oceans may have played a role in the origin of life on our planet. *(John C. Kotz)*

Black Smokers and the Origin of Life

The statement by Francis Crick on the facing page does not mean that chemists and biologists have not tried to find the conditions under which life might have begun. Charles Darwin thought life might have begun when simple molecules combined to produce ones of greater and greater complexity. Darwin's idea lives on in experiments such as those done by Stanley Miller in 1953. Attempting to recreate what was thought to be the atmosphere of the primeval earth, he filled a flask with the gases methane, ammonia, and hydrogen and added a bit of water. A discharge of electricity acted like lightning in the mixture. The inside of the flask was soon covered with a reddish slime, a mixture found to contain amino acids, the building blocks of proteins. Chemists thought they would soon know in more detail how living organisms began their development — but it was not to be. As Miller said recently, "The problem of the origin of life has turned out to be much more difficult than I, and most other people, envisioned."

Other theories have been advanced to account for the origin of life. The most recent relates to the discovery of highly active geological sites on the ocean floor. Could life have originated in such exotic environments? The evidence is tenuous. As in Miller's experiments, the evidence leading to this suggestion was the creation of complex carbon-based molecules from simple ones.

In 1977, scientists were exploring the junction of two of the tectonic plates that form the floor of the Pacific Ocean. There they found thermal springs gushing a hot, black soup of minerals. Water seeping into cracks in the thin surface is superheated to between 300 and 400 °C by the magma of the earth's core. This superhot water dissolves minerals in the crust and provides conditions for the conversion of sulfates in seawater to hydrogen sulfide, H_2S. When this hot water, now laden with dissolved minerals and rich in sulfides, gushes through the surface, it cools, and metal sulfides, such as those of copper, manganese, iron, zinc, and nickel, precipitate.

Many metal sulfides are black, and the plume of material coming from the sea bottom looks like black smoke; thus, the vents have been called "black smokers." The solid sulfides settle around the edges of the vent, and eventually form a "chimney" of precipitated minerals.

Scientists were amazed to find the vents were surrounded by primitive animals living in the hot, sulfide-rich environment. Because smokers are under hundreds of meters of water and sunlight does not penetrate to these depths, the animals have developed a way to live without energy from sun-

> "*The origin of life appears almost a miracle, so many are the conditions which would have had to be satisfied to get it going.*"
>
> Francis Crick, quoted by John Horgan, "In the Beginning," *Scientific American*, February, 1991, pages 116–125.

▲ A "black smoker" in the east Pacific Rise. *(National Oceanic and Atmospheric Administration/Department of Commerce)*

▲ Blacker smoker chimney and shrimp on the Mid-Atlantic Ridge. *(National Oceanic and Atmospheric Administration/ Department of Commerce)*

light. It is currently believed they derive the energy needed to make the organic compounds on which they depend from the reaction of oxygen with hydrogen sulfide, H_2S.

$$H_2S(aq) + 2\ O_2(aq) \longrightarrow H_2SO_4(aq) + energy$$

The hypothesis that life might have originated in this inhospitable location developed out of laboratory experiments by a German lawyer and scientist, G. Wächtershäuser and a colleague, Claudia Huber. They found that metal sulfides such as iron sulfide promote reactions that convert simple carbon-containing molecules to more complex molecules. If this could happen in the laboratory, perhaps similar chemistry might also occur in the exotic environment of black smokers.

Before You Begin

- Review names of common compounds and ions (Chapter 3).
- Know how to convert mass to moles and moles to mass (Chapters 2 and 3).

Chemical Reactions Chemistry is the study of change. Here we observe chemical change, the conversion of elements or compounds into different elements or compounds, often with the evolution of energy. The broad objective of this chapter is the study of the equations used to represent chemical change and the use of these chemical equations in quantitative studies.

Solid white phosphorus + gaseous chlorine ⟶ liquid phosphorus trichloride

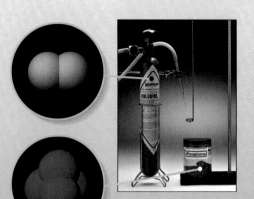

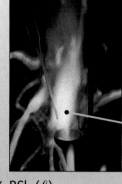
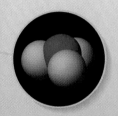

$$P_4(s) + 6\ Cl_2(g) \longrightarrow 4\ PCl_3(\ell)$$

Reactants   Products

Solid aluminum metal + liquid bromine ⟶ white solid aluminum bromide

$$2\ Al(s) + 3\ Br_2(\ell) \longrightarrow Al_2Br_6(s)$$

Reactants   Products

Hot iron + gaseous chlorine ⟶ solid iron (III) chloride

$$2\ Fe(s) + 3\ Cl_2(g) \longrightarrow 2\ FeCl_3(s)$$

Reactants   Products

Photos: Charles D. Winters

When you think about chemistry you think of chemical reactions. The image of a medieval alchemist mixing chemicals hoping to turn lead into gold lingers in the imagination. But of course there is much more. Just reading this sentence involves an untold number of chemical reactions in your body. Indeed, every activity of living things depends on carefully regulated chemical reactions.   Our objective in this chapter is to introduce the quantitative study of chemical reactions. Quantitative studies are needed to determine, for example, how much bromine is required for complete reaction with a given quantity of aluminum and what mass of aluminum bromide can be obtained (Chapter Focus). This part of chemistry is fundamental to much of what chemists, chemical engineers, biochemists, molecular biologists, geochemists, and many others do.   ●

4.1   CHEMICAL EQUATIONS

The first Chapter Focus reaction shows a stream of chlorine gas (Cl_2) directed onto solid phosphorus (P_4). The mixture bursts into flame, and the chemical reaction produces liquid phosphorus trichloride, PCl_3. We can depict this reaction using the balanced **chemical equation** [CD-ROM, Screens 4.2–4.4]

$$\underbrace{P_4(s) + 6\ Cl_2(g)}_{\text{Reactants}} \longrightarrow \underbrace{4\ PCl_3(\ell)}_{\text{Products}}$$

In a balanced equation, the formulas for the **reactants** (the substances combined in the reaction) are written to the left of the arrow and the formulas of the **products** (the substances produced) are written to the right of the arrow. The physical states of reactants and products can also be indicated. The symbol (s) indicates a solid, (g) a gas, and (ℓ) a liquid. A substance dissolved in water, that is, an *aqueous solution* of a substance, would be indicated by (aq). The relative amounts of the reactants and products are shown by numbers, the *coefficients*, before the formulas. Equations like the previous one do not, however, show the conditions of the experiment or if any energy (in the form of heat or light) is involved.

In the 18th century, the great French scientist Antoine Lavoisier introduced the **law of conservation of matter.** That is, he showed that matter can neither be created nor destroyed. This means that if you use 10 g of reactants, and if the reaction completely converts reactants to products, you must end up with 10 g of products. This also means that if 1000 atoms of a particular element are contained in the reactants, then those 1000 atoms must appear in the products in some fashion. When applied to the reaction of phosphorus and chlorine, the conservation of matter means that 1 molecule of phosphorus (with 4 phosphorus atoms) and 6 diatomic molecules of Cl_2 (with 12 atoms of Cl) are required to produce 4 molecules of PCl_3. Because each PCl_3 molecule contains 1 P atom and 3 Cl atoms, the 4 PCl_3 molecules are needed to account for 4 P atoms and 12 Cl atoms in the product.

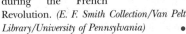

$$\underset{\text{4 P atoms}}{\underbrace{P_4(s)}} + \underset{\text{12 Cl atoms}}{\overbrace{6\ Cl_2(g)}} \times \underset{\text{4 P atoms}}{\underbrace{4\ PCl_3(\ell)}}$$

6 × 2 = 12 Cl atoms   4 × 3 = 12 Cl atoms

The numbers in front of each formula in a *balanced* chemical equation are required by the principle of the conservation of matter. Review the equation for the

reaction of phosphorus and chlorine, and then consider the balanced equation for the reaction of aluminum and bromine (see Chapter Focus).

$$2 \ Al(s) + 3 \ Br_2(\ell) \longrightarrow Al_2Br_6(s)$$

stoichiometric coefficients

The number in front of each chemical formula can be read as a number of atoms or molecules (two atoms of Al and three molecules of Br_2 form one molecule of Al_2Br_6). It can refer equally well to amounts of reactants and products: 2 mol of solid aluminum combine with 3 mol of liquid bromine to produce 1 mol of solid Al_2Br_6. The relationship between the amounts of chemical reactants and products is called **stoichiometry** (pronounced "stoy-key-AHM-uh-tree"), and the coefficients in a balanced equation are the **stoichiometric coefficients.**

Balanced chemical equations are fundamentally important in depicting the outcome of chemical reactions and in the quantitative understanding of chemistry.

Exercise 4.1   Chemical Reactions

The reaction of iron with chlorine is shown on page 120. The equation for the reaction is

$$2 \ Fe(s) + 3 \ Cl_2(g) \longrightarrow 2 \ FeCl_3(s)$$

(a) Name the reactants and products in this reaction and give their states.

(b) What are the stoichiometric coefficients in this equation?

(c) If you were to use 8000 atoms of Fe, how many molecules of Cl_2 are required to consume the iron completely?

4.2   BALANCING CHEMICAL EQUATIONS

A chemical equation must be balanced before useful quantitative information can be obtained about the reaction [CD-ROM, Screen 4.4]. Balancing an equation ensures that the same number of atoms of each element appear on both sides of the equation. Many chemical equations can be balanced by trial and error, although some will involve more trial than others.

One general class of chemical reactions is the reaction of metals or nonmetals with oxygen to give oxides of the general formula M_xO_y. For example, iron can react with oxygen to give iron(III) oxide (Figure 4.1a),

$$4 \ Fe(s) + 3 \ O_2(g) \longrightarrow 2 \ Fe_2O_3(s)$$

magnesium gives magnesium oxide (Figure 4.1b),

$$2 \ Mg(s) + O_2(g) \longrightarrow 2 \ MgO(s)$$

and phosphorus, P_4, reacts vigorously with oxygen to give tetraphosphorus decaoxide, P_4O_{10} (Figure 4.1c):

$$P_4(s) + 5 \ O_2(g) \longrightarrow P_4O_{10}(s)$$

These equations are balanced as we have written them because the same number of metal or phosphorus atoms and oxygen atoms occurs on each side of the equation.

The **combustion,** or burning, of a fuel in oxygen is accompanied by the evolution of heat. You are familiar with combustion reactions such as the burning of octane, C_8H_{18}, a component of gasoline, in an automobile engine:

Chapter Goals • Revisited

• **Balance equations for simple chemical reactions.**

• Perform stoichiometry calculations using balanced chemical equations.

• Understand the meaning of a limiting reactant.

• Calculate the theoretical and percent yield of a chemical reaction.

• Use stoichiometry to analyze a mixture of compounds or to determine the formula of a compound.

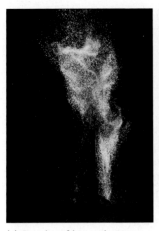

(a) Reaction of iron and oxygen to give iron(III) oxide, Fe_2O_3.

(b) Reaction of magnesium and oxygen to give magnesium oxide, MgO.

(c) Reaction of phosphorus and oxygen to give tetraphosphorus decaoxide, P_4O_{10}.

Figure 4.1 Reactions of metals and a nonmetal with oxygen. *(Charles D. Winters)*

$$2\ C_8H_{18}(\ell) + 25\ O_2(g) \longrightarrow 16\ CO_2(g) + 18\ H_2O(\ell)$$

In all combustion reactions involving oxygen, some or all the elements in the reactants end up as oxides, compounds containing oxygen. For hydrocarbons (compounds containing only C and H) the products of complete combustion are always carbon dioxide and water.

As an example of equation balancing, let us write the balanced equation for the complete combustion of propane, C_3H_8.

Step 1. *Write correct formulas for the reactants and products.*

$$C_3H_8(g) + O_2(g) \xrightarrow{\text{unbalanced equation}} CO_2(g) + H_2O(\ell)$$

Here propane and oxygen are the reactants, and carbon dioxide and water are the products.

Step 2. *Balance the C atoms.* In combustion reactions it is usually best to balance the carbon atoms first and leave the oxygen atoms until the end (because the oxygen atoms are often found in more than one product). In this case three carbon atoms are in the reactants, so three must occur in the products. Three CO_2 molecules are therefore required on the right side:

$$C_3H_8(g) + O_2(g) \xrightarrow{\text{unbalanced equation}} 3\ CO_2(g) + H_2O(\ell)$$

Step 3. *Balance the H atoms.* Eight H atoms are in the reactants. Each molecule of water has two hydrogen atoms, so four molecules of water account for the required eight hydrogen atoms on the right side:

$$C_3H_8(g) + O_2(g) \xrightarrow{\text{unbalanced equation}} 3\ CO_2(g) + 4\ H_2O(\ell)$$

Step 4. *Balance the number of O atoms.* Ten oxygen atoms are on the right side ($3 \times 2 = 6$ in CO_2 plus $4 \times 1 = 4$ in water). Therefore, five O_2 molecules are needed to supply the required ten oxygen atoms:

$$C_3H_8(g) + 5\ O_2(g) \xrightarrow{\text{balanced equation}} 3\ CO_2(g) + 4\ H_2O(\ell)$$

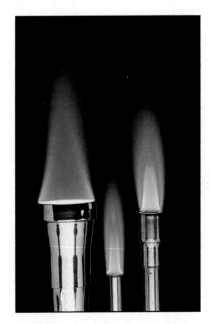

A combustion reaction. Here propane, C_3H_8, burns to give CO_2 and H_2O. These simple oxides are always the products of the combustion of a hydrocarbon. *(Charles D. Winters)*

Step 5. *Verify that the number of atoms of each element is balanced.* The equation shows three carbon atoms, eight hydrogen atoms, and ten oxygen atoms on each side.

When balancing chemical equations, there are two important things to remember.

- Formulas for reactants and products must be correct or the equation is meaningless.
- Subscripts in the formulas of reactants and products cannot be changed to balance equations. These formulas identify the particular substances, and changing the subscripts changes the identity of the substance. For example, you cannot change CO_2 to CO to balance an equation; carbon monoxide, CO, and carbon dioxide, CO_2, are different compounds.

Example 4.1   Balancing an Equation for a Combustion Reaction

Problem • Write the balanced equation for the combustion of ammonia ($NH_3 + O_2$) to give N_2 and H_2O.

Strategy • First write the unbalanced equation. Next, balance the N atoms, then the H atoms, and finally balance the O atoms.

Solution •

Step 1. *Write correct formulas for reactants and products.* The unbalanced equation for the combustion is

$$NH_3(g) + O_2(g) \xrightarrow{\text{unbalanced equation}} N_2(g) + H_2O(\ell)$$

Step 2. *Balance the N atoms.* Two N atoms on the right require two NH_3 molecules on the left.

$$2\,NH_3(g) + O_2(g) \xrightarrow{\text{unbalanced equation}} N_2(g) + H_2O(\ell)$$

Step 3. *Balance the H atoms.* There are six H atoms on the left, so three molecules of H_2O, each having two H atoms, are required on the right:

$$2\,NH_3(g) + O_2(g) \xrightarrow{\text{unbalanced equation}} N_2(g) + 3\,H_2O(\ell)$$

Step 4. *Balance the O atoms.* As the equation stands after Step 3, there are two O atoms on the left side and three on the right. That is, there is an even number of O atoms on the left and an odd number on the right. Because there cannot be an odd number of O atoms on the left (O atoms are paired in O_2 molecules), multiply each coefficient on both sides of the equation by 2 so that an even number of oxygen atoms (six) now occurs on the right side:

$$4\,NH_3(g) + \underline{}\,O_2(g) \xrightarrow{\text{unbalanced equation}} 2\,N_2(g) + 6\,H_2O(\ell)$$

Now the oxygen atoms can be balanced by having three O_2 molecules on the left side of the equation:

$$4\,NH_3(g) + 3\,O_2(g) \xrightarrow{\text{balanced equation}} 2\,N_2(g) + 6\,H_2O(\ell)$$

Step 5. *Verify the result.* Four N atoms, 12 H atoms, and 6 O atoms occur on each side of the equation.

Comment • An alternate way to write this equation is by using a fractional coefficient.

$$2\,NH_3(g) + \tfrac{3}{2}\,O_2(g) \longrightarrow N_2(g) + 3\,H_2O(\ell)$$

This equation is correctly balanced and will be useful under some circumstances. In general, however, we balance equations with whole-number coefficients.

Exercise 4.2   Balancing the Equation for a Combustion Reaction

(a) Butane gas, C_4H_{10}, can burn completely in air (use $O_2(g)$ as the reactant) to give carbon dioxide gas and water vapor. Write a balanced equation for this combustion reaction.

(b) Write a balanced chemical equation for the complete combustion of liquid tetraethyllead, $Pb(C_2H_5)_4$ (which was used until the 1970s as a gasoline additive). The products of combustion are $PbO(s)$, $H_2O(\ell)$, and $CO_2(g)$.

4.3   MASS RELATIONSHIPS IN CHEMICAL REACTIONS: STOICHIOMETRY

A balanced chemical equation shows the quantitative relationship between reactants and products in a chemical reaction. Let us apply this concept to the reaction of phosphorus and chlorine (see page 120). Suppose you use 1.00 mol of phosphorus (P_4, 124 g/mol) in this reaction. The balanced equation shows that 6.00 mol (= 425 g) of Cl_2 must be used and that 4.00 mol (= 549 g) of PCl_3 can be produced [CD-ROM, Screen 4.5].

	$P_4(s)$	+	$6\ Cl_2(g)$	$\longrightarrow$	$4\ PCl_3(\ell)$
Initial amount (mol)	1.00 mol (124 g)		6.00 mol (425 g)		0 mol (0 g)
Change in amount (mol)	−1.00 mol		−6.00 mol		+4.00 mol
After complete reaction (mol)	0 mol (0 g)		0 mol (0 g)		4.00 mol [549 g = 124 g + 425 g]

It is important to remember that the balanced equation shows mole relationships, not mass relationships, as illustrated by the preceding "amounts table."

The balanced equation for the reaction of phosphorus and chlorine applies no matter how much P_4 is used. If 0.0100 mol of P_4 (1.24 g) is used, then 0.0600 mol of Cl_2 (4.25 g) is required, and 0.0400 mol of PCl_3 (5.49 g) can form.

Following this line of reasoning, decide (a) what mass of Cl_2 is required to react completely with 1.45 g of phosphorus and (b) what mass of PCl_3 is produced?

(a) Mass of Cl_2 Required

Step 1. *Write the balanced equation* (using correct formulas for reactants and products). This is always the first step when dealing with chemical reactions.

$$P_4(s) + 6\ Cl_2(g) \longrightarrow 4\ PCl_3(\ell)$$

Step 2. *Calculate moles from masses.* From the mass of P_4, calculate the amount of P_4 available. Remember that the balanced equation shows mole relationships, not mass relationships.

$$1.45\ \text{g}\ P_4 \cdot \frac{1\ \text{mol}\ P_4}{123.9\ \text{g}} = 0.0117\ \text{mol}\ P_4$$

Step 3. *Use a stoichiometric factor.* The amount available of one reactant (P_4) is related to the amount of the other reactant (Cl_2) required by the balanced equation.

$$0.0117\ \text{mol}\ P_4 \cdot \frac{6\ \text{mol}\ Cl_2\ \text{required}}{1\ \text{mol}\ P_4\ \text{available}} = 0.0702\ \text{mol}\ Cl_2\ \text{required}$$

↑ stoichiometric factor (from balanced equation)

To perform this calculation the amount of phosphorus available has been multiplied by a **stoichiometric factor,** a *mole ratio based on the coefficients of reactants or products in the balanced equation.* This is the reason you must balance chemical equations before proceeding with calculations. Here the calculation shows that 0.0702 mol of Cl_2 is required to react with all the available phosphorus.

Chapter Goals • Revisited

- Balance equations for simple chemical reactions.
- **Perform stoichiometry calculations using balanced chemical equations.**
- Understand the meaning of a limiting reactant.
- Calculate the theoretical and percent yield of a chemical reaction.
- Use stoichiometry to analyze a mixture of compounds or to determine the formula of a compound.

● **Amounts Tables**
The "amounts table" describes the reactants and products in the $P_4 + Cl_2$ reaction. Amounts tables are not only useful here but will also be used extensively when you study chemical equilibria in Chapters 16–18.

● **Mass Balance**
Mass is always conserved in chemical reactions. The total mass before reaction is always the same as that after reaction. This does not mean, however, that the total amount of reactants (mol) is the same as that of the products. Atoms are rearranged into different molecules, ions, or ionic compounds in the course of a reaction. In the $P_4 + Cl_2$ reaction, 7 mol of reactants gives 4 mol of product.

● **Stoichiometric Factor**
A stoichiometric factor can relate the moles of a reactant to moles of a product or vice versa. Also note that the stoichiometric factor is a conversion factor (see page 27).

Problem-Solving Tip 4.1

Stoichiometry Calculations

You are asked to determine what mass of product can be formed from a given mass of reactant. Keep in mind that it is not possible to calculate the mass of product in a single step. Instead, you must follow a route such as that illustrated here for the reaction of a reactant A to give the product B according to an equation such as $x\,A \longrightarrow y\,B$. Here the mass of reactant A is converted to moles of A. Then, using the stoichiometric factor, you find moles of B. Finally, the mass of B is obtained by multiplying moles of B by its molar mass.

When solving a chemical stoichiometry problem, remember that you will always use a stoichiometric factor at some point.

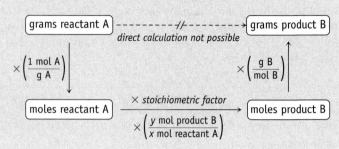

Step 4. *Calculate mass from moles.* Convert moles of Cl_2 (calculated in Step 3) to mass of Cl_2 required.

$$0.0702 \ \text{mol} \ Cl_2 \cdot \frac{70.91 \text{ g } Cl_2}{1 \ \text{mol} \ Cl_2} = 4.98 \text{ g } Cl_2$$

Because the objective of this example was to find the mass of Cl_2 required, the problem is solved.

(b) Mass of PCl_3 Produced from P_4 and Cl_2

What mass of PCl_3 can be produced in the reaction of 1.45 g of phosphorus with 4.98 g of Cl_2? Because matter is conserved, the answer can be obtained in this case by adding the masses of P_4 and Cl_2 used (giving 1.45 g + 4.98 g = 6.43 g of PCl_3 produced). Alternatively, Steps 3 and 4 can be repeated, but with the appropriate stoichiometric factor and molar mass.

Step 3. *Use a stoichiometric factor.* Convert amount of available P_4 to the amount of PCl_3 that can be produced.

$$0.0117 \ \text{mol} \ P_4 \cdot \frac{4 \text{ mol } PCl_3 \text{ produced}}{1 \ \text{mol} \ P_4 \text{ available}} = 0.0468 \text{ mol } PCl_3 \text{ produced}$$

↑ stoichiometric factor (from balanced equation)

Step 4. *Calculate mass from moles.* Convert amount of PCl_3 produced to a mass in grams.

$$0.0468 \ \text{mol} \ PCl_3 \cdot \frac{137.3 \text{ g } PCl_3}{1 \ \text{mol} \ PCl_3} = 6.43 \text{ g } PCl_3$$

• **Amount and Quantity**
When doing stoichiometry problems, recall from Chapter 2 that the terms "amount" and "quantity" are used in a specific sense by chemists. The *amount* of a substance is the number of moles of that substance. *Quantity* refers to the mass of the substance.

Example 4.2   Mass Relations in Chemical Reactions

Problem • Glucose reacts with oxygen to give CO_2 and H_2O.

$$C_6H_{12}O_6(s) + 6\ O_2(g) \longrightarrow 6\ CO_2(g) + 6\ H_2O(\ell)$$

What mass of oxygen (in grams) is required for complete reaction of 25.0 g of glucose? What masses of carbon dioxide and water (in grams) are formed?

Strategy • After referring to the balanced equation, you can perform the stoichiometric calculations. *Problem-Solving Tip 4.1*

suggests that you proceed in the following way to find the mass of O_2:

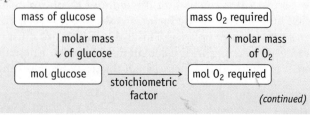

(continued)

First find the amount of glucose available, then relate this to the amount of O_2 required using the stoichiometric factor. Finally, find the mass of O_2 required from the moles of O_2. Then, follow the same procedure to find the masses of carbon dioxide and water.

Solution •
Step 1. *Write a balanced equation.*

$$C_6H_{12}O_6(s) + 6\ O_2(g) \longrightarrow 6\ CO_2(g) + 6\ H_2O(\ell)$$

Step 2. *Convert the mass of glucose to moles.*

$$25.0\ \text{g glucose} \cdot \frac{1\ \text{mol}}{180.2\ \text{g}} = 0.139\ \text{mol glucose}$$

Step 3. *Use the stoichiometric factor.* Here we calculate the amount of O_2 required.

$$0.139\ \text{mol glucose} \cdot \frac{6\ \text{mol}\ O_2}{1\ \text{mol glucose}} = 0.832\ \text{mol}\ O_2$$

Step 4. *Calculate mass from moles.* Convert the required amount of O_2 to a mass in grams.

$$0.832\ \text{mol}\ O_2 \cdot \frac{32.00\ \text{g}\ O_2}{1\ \text{mol}\ O_2} = 26.6\ \text{g}\ O_2$$

Repeat Steps 3 and 4 to find the mass of CO_2 produced in the combustion. First, relate the amount (moles) of glucose available to the amount of CO_2 produced by using a stoichiometric factor.

$$0.139\ \text{mol glucose} \cdot \frac{6\ \text{mol}\ CO_2}{1\ \text{mol glucose}} = 0.832\ \text{mol}\ CO_2$$

Then convert the amount of CO_2 produced to the mass in grams.

$$0.832\ \text{mol}\ CO_2 \cdot \frac{44.01\ \text{g}\ CO_2}{1\ \text{mol}\ CO_2} = 36.6\ \text{g}\ CO_2$$

Now, how can you find the mass of H_2O produced? You could go through Steps 3 and 4 again. However, recognize that the total mass of reactants

$$25.0\ \text{g}\ C_6H_{12}O_6 + 26.6\ \text{g}\ O_2 = 51.6\ \text{g of reactants}$$

must be the same as the total mass of products. The mass of water that can be produced is therefore

Total mass of products = 51.6 g
= 36.6 g CO_2 produced + ? g H_2O
Mass of H_2O produced = 15.0 g

Exercise 4.3   **Mass Relations in Chemical Reactions**

What mass of oxygen, O_2, is required to completely combust 454 g of propane, C_3H_8? What masses of CO_2 and H_2O are produced?

$$C_3H_8(g) + 5\ O_2(g) \longrightarrow 3\ CO_2(g) + 4\ H_2O(\ell)$$

4.4   REACTIONS IN WHICH ONE REACTANT IS PRESENT IN LIMITED SUPPLY

When chemists carry out reactions, the goal is usually to produce the largest possible quantity of a compound from a given quantity of starting material. As you go through this book, and learn about the factors affecting chemical reactions, you will see that it is often desirable to use an excess of one reactant over that required by stoichiometry. This is usually done to ensure that one of the reactants in the reaction is consumed completely, even though some of another reactant remains unused.

Consider the preparation of cisplatin, $Pt(NH_3)_2Cl_2$, a compound used to treat certain cancers.

$$\underset{\text{ammonia}}{(NH_4)_2PtCl_4(s) + 2\ NH_3(aq)} \longrightarrow 2\ NH_4Cl(aq) + \underset{\text{cisplatin}}{Pt(NH_3)_2Cl_2(s)}$$

This reaction is usually done by combining the more expensive chemical $(NH_4)_2PtCl_4$ (roughly $100 per gram) with a much greater amount of the less expensive chemical NH_3 (only pennies per gram) than is called for by the balanced equation. On completion of the reaction, all the $(NH_4)_2PtCl_4$ will have been converted to product, although some NH_3 will remain. How much $Pt(NH_3)_2Cl_2$ is formed? It depends on the amount of $(NH_4)_2PtCl_4$ present at the start, not on the

---

Chapter Goals • Revisited

- Balance equations for simple chemical reactions.
- Perform stoichiometry calculations using balanced chemical equations.
- **Understand the meaning of a limiting reactant.**
- Calculate the theoretical and percent yield of a chemical reaction.
- Use stoichiometry to analyze a mixture of compounds or to determine the formula of a compound.

amount of NH_3, because more NH_3 is present than is required by stoichiometry. A compound such as $(NH_4)_2PtCl_4$ in this example is called the **limiting reactant** because its amount determines, or limits, the amount of product formed [CD-ROM, Screens 4.6 and 4.7].

Another important reaction is the oxidation of carbon monoxide to carbon dioxide. The balanced equation for the reaction is

$$2\ CO(g) + O_2(g) \longrightarrow 2\ CO_2(g)$$

Suppose you have a mixture of four CO molecules and three O_2 molecules.

Reactants: 4 CO and 3 O_2 Products: 4 CO_2 and 1 O_2

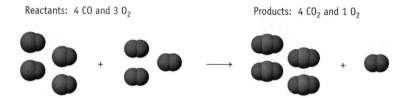

The four CO molecules require only two O_2 molecules (and produce four CO_2 molecules). This must mean that one O_2 molecule remains after the reaction is complete. Thus, CO is the limiting reactant, and O_2 is in excess.

Another example of a reaction that involves a limiting reactant is the oxidation of ammonia over a wire gauze made of platinum (Figure 4.2a).

$$4\ NH_3(g) + 5\ O_2(g) \longrightarrow 4\ NO(g) + 6\ H_2O(g)$$

Suppose that exactly 750 g NH_3 is mixed with 750 g of O_2. Are these reactants mixed in the correct stoichiometric ratio or is one of them in short supply? That is, will one of them limit the quantity of NO that can be produced? If so, how much NO can be formed if the reaction goes to completion? And how much of the excess reactant is left over when the maximum amount of NO has been formed?

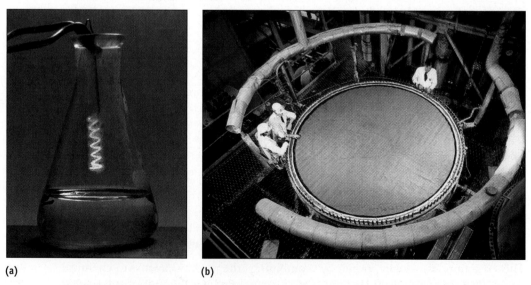

(a)   (b)

Figure 4.2   Oxidation of ammonia. (a) Burning ammonia on the surface of a platinum wire produces so much heat that the wire glows bright red. **(b)** Billions of kilograms of HNO_3 is made annually starting with the oxidation of ammonia over a wire gauze made of platinum. *(a, Charles D. Winters; b, Johnson Matthey)*

Because the quantities of both starting materials are given, the first step in answering these questions involves finding the amount of each.

$$750. \text{ g NH}_3 \cdot \frac{1 \text{ mol NH}_3}{17.03 \text{ g}} = 44.0 \text{ mol NH}_3 \text{ available}$$

$$750. \text{ g O}_2 \cdot \frac{1 \text{ mol O}_2}{32.00 \text{ g}} = 23.4 \text{ mol O}_2 \text{ available}$$

Are these reactants present in the correct stoichiometric ratio as given by the balanced equation?

$$\text{Stoichiometric ratio of reactants required by balanced equation} = \frac{5 \text{ mol O}_2}{4 \text{ mol NH}_3}$$

$$= \frac{1.25 \text{ mol O}_2}{1 \text{ mol NH}_3}$$

$$\text{Ratio of reactants actually available} = \frac{23.4 \text{ mol O}_2}{44.0 \text{ mol NH}_3} = \frac{0.532 \text{ mol O}_2}{1 \text{ mol NH}_3}$$

Dividing moles of O_2 available by moles of NH_3 available shows that ratio of available reactants is much smaller than the $(5 \text{ mol O}_2/4 \text{ mol NH}_3)$ ratio required by the balanced equation. This means that insufficient O_2 is available to react with all of the NH_3, and *oxygen, O_2, is the limiting reactant.*

Now that O_2 has been found to be the limiting reactant, we can calculate the mass of product, NO, expected based on the amount of O_2 available.

$$23.4 \text{ mol O}_2 \cdot \frac{4 \text{ mol NO}}{5 \text{ mol O}_2} \cdot \frac{30.01 \text{ g NO}}{1 \text{ mol NO}} = 562 \text{ g NO}$$

Ammonia is the "excess reactant" because more than enough NH_3 is available to react with 23.4 mol of O_2. Let us calculate the amount of NH_3 remaining after all the O_2 has been used. To do this, we first need to know the amount of NH_3 required to consume all the limiting reactant, O_2.

$$23.4 \text{ mol O}_2 \text{ available} \cdot \frac{4 \text{ mol NH}_3 \text{ required}}{5 \text{ mol O}_2 \text{ available}} = 18.8 \text{ mol NH}_3 \text{ required}$$

Because 44.0 mol of NH_3 is available, the amount of excess NH_3 can be calculated,

$$\text{Excess NH}_3 = 44.0 \text{ mol NH}_3 \text{ available} - 18.8 \text{ mol NH}_3 \text{ required}$$

$$= 25.2 \text{ mol NH}_3 \text{ remain}$$

and then converted to a mass.

$$25.2 \text{ mol NH}_3 \cdot \frac{17.03 \text{ g NH}_3}{1 \text{ mol NH}_3} = 429 \text{ g NH}_3 \text{ in excess of that required}$$

Finally, because 429 g of NH_3 is left over, this means that 321 g of NH_3 has been consumed $(= 750. \text{ g} - 429 \text{ g})$.

You may find it helpful in limiting reactant problems to summarize your results in an amounts table.

● **Simulation of Reactions with a Limiting Reactant**
Screen 4.8 of the *General Chemistry Interactive CD-ROM,* Version 3.0, has a very useful simulation of reactions involving a limiting reactant.

● **Follow Up on This Example**
See Study Question 4.52 for a calculation of the quantity of water formed in the NH_3/O_2 reaction.

• Conservation of Matter
The total quantity of matter present before reaction (1500. g) is the same as the total quantity of matter produced in the reaction *plus* the quantity of NH_3 remaining. That is, 562 g of NO (18.8 mol) and 506 g of H_2O (28.1 mol) are produced. Because 429 g of NH_3 (25.2 mol) remains, the total quantity of matter after reaction (562 g + 506 g + 429 g = 1500. g) is the same as the quantity of matter before reaction.

Reaction	$4\ NH_3(g)$	$+\ 5\ O_2(g)$	$\longrightarrow$	$4\ NO(g)$	$+\ 6\ H_2O(g)$
Initial amount (mol)	44.0	23.4		0	0
Change in amount (mol)	$-\frac{4}{5}(23.4)$	-23.4		$+\frac{4}{5}(23.4)$	$+\frac{6}{5}(23.4)$
	$= -18.8$			$= +18.8$	$= +28.1$
After complete reaction (mol)	25.2	0		18.8	28.1

All of the limiting reactant, O_2, has been consumed. Of the original 44.0 mol of NH_3, 18.8 mol has been consumed to leave 25.2 mol. Because the amount of NO produced is equal to the amount of NH_3 consumed, 18.8 mol of NO is produced. In addition, 28.1 mol of H_2O has been produced.

Example 4.3 A Reaction with a Limiting Reactant

Problem • Methanol, CH_3OH, which is used as a fuel, can be made by the reaction of carbon monoxide and hydrogen.

$$CO(g) + 2\ H_2(g) \longrightarrow CH_3OH(\ell)$$
$$\text{methanol}$$

Suppose 356 g of CO is mixed with 65.0 g of H_2. (a) Which is the limiting reactant? (b) What mass of methanol can be produced? (c) What mass of the excess reactant remains after the limiting reactant has been consumed?

Strategy • A limiting reactant problem usually has two steps:

1. After calculating the amount of each reactant, compare the ratio of reactant amounts to the required stoichiometric ratio, here 2 mol H_2/1 mol CO.

   • If [mol H_2 available/mol CO available] > 2/1, then CO is the limiting reactant.
   • If [mol H_2 available/mol CO available] < 2/1, then H_2 is the limiting reactant.

2. Use the amount of limiting reactant to find the amount of product.

Solution •
(a) *What is the limiting reactant?* The amount of each reactant is

$$\text{Amount of CO} = 356\ g\ CO \cdot \frac{1\ mol\ CO}{28.01\ g}$$
$$= 12.7\ mol\ CO$$

$$\text{Amount of } H_2 = 65.0\ g\ H_2 \cdot \frac{1\ mol\ H_2}{2.016\ g\ H_2}$$
$$= 32.2\ mol\ H_2$$

Are these reactants present in an exact stoichiometric ratio?

$$\frac{\text{mol } H_2 \text{ available}}{\text{mol CO available}} = \frac{32.2\ mol\ H_2}{12.7\ mol\ CO} = \frac{2.54\ mol\ H_2}{1.00\ mol\ CO}$$

The required mole ratio is 2 mol of H_2 to 1 mol of CO. Here we see that more hydrogen is available than is required to consume all the CO. It follows that not enough CO is present to use up all of the hydrogen. *CO is the limiting reactant.*

(b) *What is the maximum mass of CH_3OH that can be formed?* This calculation is based on the amount of limiting reactant.

$$12.7\ mol\ CO \cdot \frac{1\ mol\ CH_3OH\ formed}{1\ mol\ CO\ available} \cdot \frac{32.04\ g\ CH_3OH}{1\ mol\ CH_3OH} =$$
$$407\ g\ CH_3OH$$

(c) *What amount of H_2 remains when all the CO has been converted to product?* First, we must find the amount of H_2 required to react with all the CO.

$$12.7\ mol\ CO \cdot \frac{2\ mol\ H_2}{1\ mol\ CO} = 25.4\ mol\ H_2\ required$$

Because 32.2 mol of H_2 is available, but only 25.4 mol is required by the limiting reactant, 32.2 mol − 25.4 mol = 6.8 mol of H_2 is in excess. This is equivalent to 14 g of H_2.

$$6.8\ mol\ H_2 \cdot \frac{2.02\ g\ H_2}{1\ mol\ H_2} = 14\ g\ H_2\ remain$$

Comment • The amounts table for this reaction is

	$CO(g)$	$+\ 2\ H_2(g)$	$\longrightarrow$	$CH_3OH(\ell)$
Initial amount (mol)	12.7	32.2		0
Change (mol)	−12.7	−2(12.7)		+12.7
After complete reaction (mol)	0	6.8		12.7

The mass of product formed plus the mass of H_2 remaining after reaction (407 g CH_3OH + 14 g H_2 remaining = 421 g) is equal to the mass of reactants present before reaction (356 g CO + 65.0 g H_2 = 421 g).

A car that uses methanol as a fuel. In this van methanol is converted to hydrogen, which is then combined with oxygen in a fuel cell. The fuel cell generates electric energy to run the car. (See Chapter 20.) *(Courtesy of Ballard Power Systems)*

Example 4.4   A Reaction with a Limited Reactant

Problem • The thermite reaction produces iron metal and aluminum oxide from a mixture of powdered aluminum metal and iron(III) oxide.

$$Fe_2O_3(s) + 2\ Al(s) \longrightarrow 2\ Fe(s) + Al_2O_3(s)$$

A mixture of 50.0 g each of Fe_2O_3 and Al is used. (a) Which is the limiting reactant? (b) What mass of iron metal can be produced?

Strategy • As in Example 4.3, we first calculate the amount of each reactant and then compare their ratio with the required ratio of 2 mol Al/1 mol Fe_2O_3. If the ratio is less than 2/1, Al is the limiting reactant. If it is greater than 2/1, then Fe_2O_3 is the limiting reactant. Calculate the mass of iron from the amount of limiting reactant.

Solution •
(a) *What is the limiting reactant?* First calculate the amount of each reactant

$$50.0\ g\ Al \cdot \frac{1\ mol\ Al}{26.98\ g} = 1.85\ mol\ Al$$

$$50.0\ g\ Fe_2O_3 \cdot \frac{1\ mol\ Fe_2O_3}{159.7\ g} = 0.313\ mol\ Fe_2O_3$$

and then calculate the ratio of moles available.

$$\frac{mol\ Al\ available}{mol\ Fe_2O_3\ available} = \frac{1.85\ mol\ Al}{0.313\ mol\ Fe_2O_3} = \frac{5.91\ mol\ Al}{1.00\ mol\ Fe_2O_3}$$

The ratio (mol Al/mol Fe_2O_3) is much greater than 2/1, so the limiting reactant is Fe_2O_3.

(b) *What mass of iron is formed?* Using the amount of limiting reactant, calculate the quantity of iron produced.

Thermite reaction. Here iron(III) oxide reacts with aluminum metal to produce aluminum oxide and iron metal. The reaction produces so much heat that the iron melts and spews out of the reaction vessel. *(Charles D. Winters)*

$$0.313\ mol\ Fe_2O_3 \cdot \frac{2\ mol\ Fe}{1\ mol\ Fe_2O_3} \cdot \frac{55.85\ g\ Fe}{1\ mol\ Fe} = 35.0\ g\ Fe$$

Comment • The amounts table for this reaction is

	$Fe_2O_3(s)$ +	2 $Al(s)$ $\longrightarrow$	2 $Fe(s)$ +	$Al_2O_3(s)$
Initial amount (mol)	0.313	1.85	0	0
Change in amount (mol)	−0.313	2(−0.313)	2(+0.313)	+0.313
After complete reaction (mol)	0	1.22	0.626	0.313

The total mass of products formed (35.0 g Fe + 31.9 g Al_2O_3) plus aluminum remaining (33.1 g) is 100.0 g, equal to the mass of reactants before reaction.

Exercise 4.4   A Reaction with a Limiting Reactant

Titanium tetrachloride, $TiCl_4$, is an important industrial chemical. It is used, for example, to make TiO_2, the material used as a white pigment in paper and paints. Titanium tetrachloride can be made by combining titanium-containing ore (which is often impure TiO_2) with carbon and chlorine.

$$TiO_2(s) + 2\ Cl_2(g) + C(s) \longrightarrow TiCl_4(\ell) + CO_2(g)$$

Using 125 g each of Cl_2 and C, but plenty of TiO_2-containing ore, which is the limiting reactant in this reaction? What mass of $TiCl_4$, in grams, can be produced?

Exercise 4.5   A Reaction with a Limiting Reactant

Pure silicon, required for computer chips and solar cells, is made by the reaction

$$SiCl_4(\ell) + 2\ Mg(s) \longrightarrow Si(s) + 2\ MgCl_2(s)$$

If you begin with 225 g each of $SiCl_4$ and Mg, which is the limiting reactant in this reaction? What quantity of Si, in grams, can be produced?

(a)

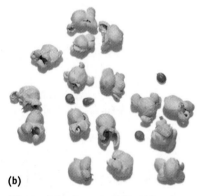

(b)

Figure 4.3   Percent yield. Although not a chemical reaction, popping corn is a good analogy to the difference between a theoretical yield and an actual yield. Here we began with 20 popcorn kernels and found that only 16 of them popped. The theoretical yield is 20, but the percent yield was only $(16/20) \times 100\%$, or 80%. *(Charles D. Winters)*

4.5   PERCENT YIELD

The maximum quantity of product that can be obtained from a chemical reaction is the **theoretical yield.** Frequently, the **actual yield** of a compound — the quantity of material that could be obtained in the laboratory or a chemical plant — is less than the theoretical yield. Some loss often occurs during the isolation and purification of products. In addition, some reactions do not go completely to products, and reactions are sometimes complicated by giving more than one set of products. For these reasons, the actual yield, the amount of product obtained, is likely to be less than the theoretical yield (Figure 4.3).

To provide information to other chemists who might want to carry out a reaction, it is customary to report a **percent yield,** which is defined as [🖱 CD-ROM, Screen 4.9]

$$\text{Percent yield} = \frac{\text{actual yield}}{\text{theoretical yield}} \times 100\% \qquad (4.1)$$

Suppose you made aspirin in the laboratory by the following reaction:

$$C_7H_6O_3(s) \quad + \quad C_4H_6O_3(\ell) \quad \longrightarrow \quad C_9H_8O_4(s) \quad + \quad CH_3CO_2H(\ell)$$

salicylic acid   acetic anhydride   aspirin   acetic acid

and that you began with 14.4 g of salicylic acid and an excess of acetic anhydride. That is, salicylic acid is the limiting reactant. If you obtain 6.26 g of aspirin, what is the percent yield of this product? The first step is to find the amount of the limiting reactant, salicylic acid ($C_7H_6O_3$).

Problem-Solving Tip 4.2

More on Reactions with a Limiting Reactant

Here is another method of solving limiting reactant problems that some students find works well: Calculate the mass of product expected based on each reactant. The limiting reactant is the reactant that gives the smallest quantity of product. For example, refer to the $NH_3 + O_2$ reaction on page 128. To confirm that O_2 is the limiting reactant, calculate the quantity of NO that can be formed starting with (a) 44.0 mol of NH_3 and unlimited O_2 or (b) with 23.4 mol of O_2 and unlimited NH_3.

1. Quantity of NO produced from 44.0 mol of NH_3 and unlimited O_2

$$44.0 \text{ mol } NH_3 \cdot \frac{4 \text{ mol NO}}{4 \text{ mol } NH_3} \cdot \frac{30.01 \text{ g NO}}{1 \text{ mol NO}} = 1320 \text{ g NO}$$

2. Quantity of NO produced from 23.4 mol O_2 and unlimited NH_3

$$23.4 \text{ mol } O_2 \cdot \frac{4 \text{ mol NO}}{5 \text{ mol } O_2} \cdot \frac{30.01 \text{ g NO}}{1 \text{ mol NO}} = 562 \text{ g NO}$$

Comparing the quantities of NO produced shows that the available O_2 is capable of producing less NO (562 g) than the available NH_3 (1320 g). This confirms the conclusion that O_2 is the limiting reactant.

As a final note, you will find this approach easier to use when there are more than two reactants, each present initially in some designated quantity.

$$14.4 \text{ g } \cancel{C_7H_6O_3} \cdot \frac{1 \text{ mol } C_7H_6O_3}{138.1 \text{ g } \cancel{C_7H_6O_3}} = 0.104 \text{ mol } C_7H_6O_3$$

Next, use the stoichiometric factor from the balanced equation to find the amount of aspirin expected based on the limiting reactant, $C_7H_6O_3$.

$$0.104 \text{ mol } \cancel{C_7H_6O_3} \cdot \frac{1 \text{ mol aspirin}}{1 \text{ mol } \cancel{C_7H_6O_3}} = 0.104 \text{ mol aspirin}$$

The maximum amount of aspirin that can be produced — the theoretical yield — is 0.104 mol. Because the quantity you measure in the laboratory is the mass of the product, it is customary to express the theoretical yield as a mass in grams.

$$0.104 \text{ mol } \cancel{\text{aspirin}} \cdot \frac{180.2 \text{ g aspirin}}{1 \text{ mol } \cancel{\text{aspirin}}} = 18.8 \text{ g aspirin}$$

Finally, with the actual yield known to be only 6.26 g, the percent yield of aspirin can be calculated.

$$\text{Percent yield} = \frac{6.26 \text{ g aspirin obtained}}{18.8 \text{ g aspirin expected}} \times 100\% = 33.3\% \text{ yield}$$

Exercise 4.6   Percent Yield

Methanol, CH_3OH, can be burned in oxygen to provide energy, or it can be decomposed to form hydrogen gas, which can then be used as a fuel.

$$CH_3OH(\ell) \longrightarrow 2 \text{ H}_2(g) + CO(g)$$

If 125 g of methanol is decomposed, what is the theoretical yield of hydrogen? If only 13.6 g of hydrogen is obtained, what is the percent yield of the gas?

4.6   CHEMICAL EQUATIONS AND CHEMICAL ANALYSIS

Analytical chemists use a variety of approaches to identify substances as well as to measure the quantities of components of mixtures. Analytical chemistry is often done now using instrumental methods (Figure 4.4), but classical chemical reactions and stoichiometry play a central role.

Quantitative Analysis of a Mixture

Quantitative chemical analyses generally depend on one or the other of two basic ideas:

1. A substance, present in unknown amount, can be allowed to react with a known quantity of another substance. If the stoichiometric ratio for their reaction is known, the unknown amount can be determined.

2. A material of unknown composition can be converted to one or more substances of known composition. Those substances can be identified, their amounts determined, and these amounts related to the amount of the original, unknown substance [CD-ROM, Screen 4.10].

Chapter Goals • Revisited

- Balance equations for simple chemical reactions.
- Perform stoichiometry calculations using balanced chemical equations.
- Understand the meaning of a limiting reactant.
- Calculate the theoretical and percent yield of a chemical reaction.
- **Use stoichiometry to analyze a mixture of compounds or to determine the formula of a compound.**

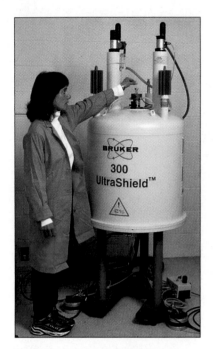

Figure 4.4 A modern analytical instrument. This nuclear magnetic resonance (NMR) spectrometer is closely related to a magnetic resonance imaging (MRI) instrument found in a hospital. The NMR is used to analyze compounds and to decipher their structures. *(Charles D. Winters)*

An example of the first type of analysis is the analysis of a sample of vinegar containing an unknown amount of acetic acid, the ingredient that makes vinegar acidic. The acid reacts readily and completely with sodium hydroxide.

$$CH_3CO_2H(aq) + NaOH(aq) \longrightarrow NaCH_3CO_2(aq) + H_2O(\ell)$$
acetic acid

If the exact quantity of sodium hydroxide used in the reaction can be measured, the quantity of acetic acid present is also known. This type of analysis is the subject of a major portion of Chapter 5 [➡ SECTION 5.10].

The second type of analysis is exemplified by the analysis of a sample of a compound found in consumer products. Suppose you have tablets that contain the artificial sweetener saccharin, $C_7H_5NO_3S$. The problem is to find the mass of saccharin in a packet.

A general analytical approach is to convert one of the substances in the mixture to a water-insoluble compound. The insoluble material is collected on a filter and weighed. To analyze for saccharin, we convert the sulfur in the sample to sulfate ion, SO_4^{2-}, which dissolves in water.

$$C_7H_5NO_3S + \text{compound to convert S to sulfate} \longrightarrow SO_4^{2-}(aq) + \text{other products}$$

We then treat this solution with barium chloride to form the water-insoluble compound barium sulfate, which is collected on a filter and weighed (Figure 4.5).

$$Na_2SO_4(aq) + BaCl_2(aq) \longrightarrow BaSO_4(s) + 2\ NaCl(aq)$$

We can then find the amount of saccharin because this is directly related to the amount of $BaSO_4$.

$$1\ \text{mol}\ C_7H_5NO_3S \longrightarrow 1\ \text{mol}\ S \longrightarrow 1\ \text{mol}\ SO_4^{2-} \longrightarrow 1\ \text{mol}\ BaSO_4$$

This approach to the analysis of saccharin is one of many examples of the use of stoichiometry in chemical analysis. All rely on converting some substance of unknown composition to products of known composition. The mass of these products is then related back to the substance being analyzed. Examples 4.5 and 4.6 further illustrate this method.

Saccharin, $C_7H_5NO_3S$, is 500 times sweeter than sugar. It is widely used even though it has a bitter, metallic aftertaste. It is 17.5% S by weight.

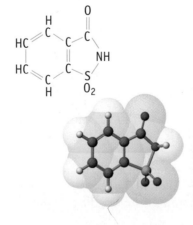

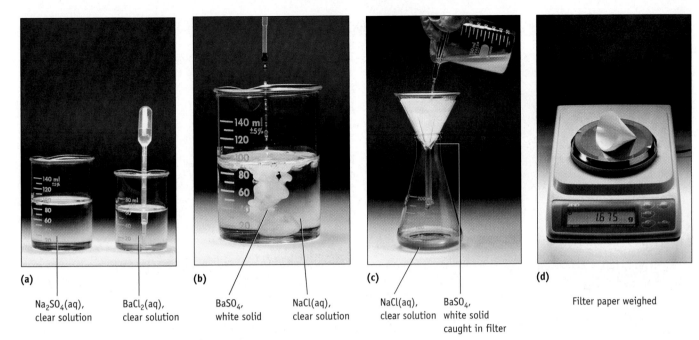

(a)
Na$_2$SO$_4$(aq),
clear solution

BaCl$_2$(aq),
clear solution

(b)
BaSO$_4$,
white solid

NaCl(aq),
clear solution

(c)
NaCl(aq),
clear solution

BaSO$_4$,
white solid
caught in filter

(d)
Filter paper weighed

Figure 4.5 **Analysis for the sulfur content of a sample.** The sulfur in a sample is converted to sulfate ion (SO$_4^{2-}$), which is dissolved in water as sodium sulfate **(a)**. Adding a solution of barium chloride to the sodium sulfate solution leads to a precipitate of BaSO$_4$ (and NaCl in solution) **(b)**. Solid barium sulfate is collected on a filter **(c)** and weighed **(d)**. The amount of BaSO$_4$ obtained can be related to the amount of sulfur in the sample. *(Charles D. Winters)*

Example 4.5 **Analysis of a Lead-Containing Mineral**

Problem • Cerrusite, a lead-containing mineral, is impure lead carbonate, PbCO$_3$. To analyze a sample of the mineral for its content of PbCO$_3$ a sample is first treated with nitric acid to dissolve the lead carbonate.

$$PbCO_3(s) + 2\ HNO_3(aq) \longrightarrow Pb(NO_3)_2(aq) + H_2O(\ell) + CO_2(g)$$

On adding sulfuric acid, lead sulfate precipitates.

$$Pb(NO_3)_2(aq) + H_2SO_4(aq) \longrightarrow PbSO_4(s) + 2\ HNO_3(aq)$$

The pure solid lead sulfate is isolated and weighed. Suppose a 0.583-g sample of mineral produced 0.628 g of PbSO$_4$. What is the mass percent of PbCO$_3$ in the mineral sample?

Strategy • The key is to recognize that 1 mol of PbCO$_3$ will yield 1 mol of PbSO$_4$. Thus, the mass of PbSO$_4$ will lead to the mass of PbCO$_3$ and then to the mass percent.

Solution • Let us first calculate the amount of PbSO$_4$.

$$0.628\ \text{g PbSO}_4 \cdot \frac{1\ \text{mol PbSO}_4}{303.3\ \text{g PbSO}_4} = 0.00207\ \text{mol PbSO}_4$$

From stoichiometry, we can relate the amount of PbSO$_4$ to the amount of PbCO$_3$.

$$0.00207\ \text{mol PbSO}_4 \cdot \frac{1\ \text{mol PbCO}_3}{1\ \text{mol PbSO}_4} = 0.00207\ \text{mol PbCO}_3$$

The mass of PbCO$_3$ is

$$0.00207\ \text{mol PbCO}_3 \cdot \frac{267.2\ \text{g PbCO}_3}{1\ \text{mol PbCO}_3} = 0.553\ \text{g PbCO}_3$$

Finally, the mass percent of PbCO$_3$ in the mineral sample is

$$\text{Mass percent of PbCO}_3 = \frac{0.553\ \text{g PbCO}_3}{0.583\ \text{g sample}} \times 100\%$$

$$= 94.9\%$$

Example 4.6 **Mineral Analysis**

Problem • Nickel(II) sulfide, NiS, occurs naturally as the relatively rare mineral millerite. One of its occurrences is in meteorites.

Millerite is a relatively rare mineral. It occurs as bunches of tiny, bronze-colored needles. *(Charles D. Winters)*

To analyze a sample containing millerite for the quantity of NiS, the sample is digested in nitric acid to release the nickel.

$$NiS(s) + 4\ HNO_3(aq) \longrightarrow$$
$$Ni(NO_3)_2(aq) + S(s) + 2\ NO_2(g) + 2\ H_2O(\ell)$$

The aqueous solution of $Ni(NO_3)_2$ is then treated with the organic compound dimethylglyoxime ($C_4H_8N_2O_2$, DMG) to give the red solid $Ni(C_4H_7N_2O_2)_2$.

$$Ni(NO_3)_2(aq) + 2\ C_4H_8N_2O_2(aq) \longrightarrow$$
$$Ni(C_4H_7N_2O_2)_2(s) + 2\ HNO_3(aq)$$

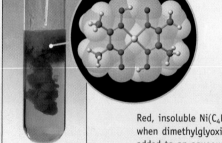

Red, insoluble $Ni(C_4H_7N_2O_2)_2$ precipitates when dimethylglyoxime ($C_4H_8N_2O_2$) is added to an aqueous solution of nickel(II) ions. *(Charles D. Winters)*

Suppose a 0.468-g sample containing millerite produces 0.206 g of red, solid $Ni(C_4H_7N_2O_2)_2$. What is the mass percent of NiS in the sample?

Strategy • The balanced equations show the following "road map."

$$1\ mol\ NiS \longrightarrow 1\ mol\ Ni(NO_3)_2 \longrightarrow 1\ mol\ Ni(C_4H_7N_2O_2)_2$$

Thus, if we know the mass of $Ni(C_4H_7N_2O_2)_2$, we can calculate its amount and then the amount of NiS. The amount of NiS allows us to calculate the mass of NiS.

Solution • The molar mass of $Ni(C_4H_7N_2O_2)_2$ is 288.9 g/mol. Thus, the amount of the red solid is

$$0.206\ g\ Ni(C_4H_7N_2O_2)_2 \cdot \frac{1\ mol}{288.9\ g\ Ni(C_4H_7N_2O_2)_2} =$$
$$7.13 \times 10^{-4}\ mol\ Ni(C_4H_7N_2O_2)_2$$

Because 1 mol of $Ni(C_4H_7N_2O_2)_2$ is ultimately produced by 1 mol of NiS, the amount of NiS in the sample must have been 7.13×10^{-4} mol.

With the amount of NiS known, we calculate the mass of NiS.

$$\left(7.13 \times 10^{-4}\ mol\ NiS \right)\left(\frac{90.76\ g}{1\ mol} \right) = 0.0647\ g\ NiS$$

Finally, the mass percent of NiS in the 0.468-g sample is

$$Mass\ percent\ NiS = \frac{0.0647\ g\ NiS}{0.468\ g\ sample} \times 100\%$$
$$= 13.8\%\ NiS$$

Exercise 4.7 **Analysis of a Mixture**

One method for determining the purity of a sample of titanium(IV) oxide, TiO_2, an important industrial chemical, is to combine the sample with bromine trifluoride.

$$3\ TiO_2(s) + 4\ BrF_3(\ell) \longrightarrow 3\ TiF_4(s) + 2\ Br_2(\ell) + 3\ O_2(g)$$

Oxygen gas is evolved quantitatively. The gas can be captured readily, and its mass can be determined. Suppose 2.367 g of a TiO_2-containing sample evolves 0.143 g of O_2. What is the mass percent of TiO_2 in the sample?

Determining the Formula of a Compound by Combustion

The empirical formula of a compound can be determined if the percent composition of the compound is known [← SECTION 3.6]. But where do the percent composition data come from? Various methods, including mass spectrometry [← PAGE 106], are used. Chemical methods are based on the following procedure, which relies on the second major approach to analysis:

- The unknown but pure compound is decomposed into known products.
- The reaction products are isolated in pure form and the amount of each is determined.
- The amount of each product is related to the amount of each element in the original compound to give the empirical formula.

One method that works well for compounds that burn in oxygen is *analysis by combustion*. Each element in the compound combines with oxygen to produce the appropriate oxide.

Consider an analysis of the hydrocarbon, methane, as an example of combustion analysis. A balanced equation for the combustion of methane shows that every mole of carbon in the original compound is converted to a mole of CO_2. Every mole of hydrogen in the original compound gives *half* a mole of H_2O. (Here the 4 mol of H atoms in CH_4 gives 2 mol of H_2O.)

$$CH_4(g) \;+\; 2\,O_2(g) \longrightarrow CO_2(g) \;+\; 2\,H_2O(\ell)$$

The gaseous carbon dioxide and water are separated and their masses determined as illustrated in Figure 4.6. These masses can then be converted to the moles of C and H in CO_2 and H_2O, respectively, and the ratio of the moles of C and H in a sample of the original compound can be found. This ratio gives the empirical formula:

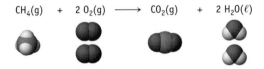

A key concept in using this procedure is mass balance. That is, for every mole of H_2O observed from combustion, there must have been *two* moles of H atoms in the unknown carbon–hydrogen (or carbon–hydrogen–oxygen) compound. Similarly, for every mole of CO_2 observed, there must have been one mole of carbon in the unknown compound.

Figure 4.6   Combustion analysis of a hydrocarbon. If a compound containing C and H is burned in oxygen, CO_2 and H_2O are formed, and the mass of each can be determined. The H_2O is absorbed by magnesium perchlorate, and the CO_2 is absorbed by finely divided NaOH supported on asbestos. The mass of each absorbent before and after combustion gives the masses of CO_2 and H_2O. Only a few milligrams of a combustible compound are needed for analysis.

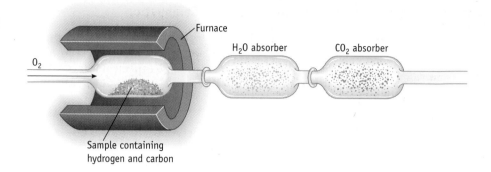

Example 4.7   Using Combustion Analysis to Determine the Formula of a Hydrocarbon

Problem • When 1.125 g of a liquid hydrocarbon, C_xH_y, was burned in an apparatus like that in Figure 4.6, 3.447 g of CO_2 and 1.647 g of H_2O were produced. The molar mass of the compound was found to be 86.2 g/mol in a separate experiment. Determine the empirical and molecular formulas for the unknown hydrocarbon, C_xH_y.

Strategy • As outlined in the diagram on page 137, we first calculate the amounts of CO_2 and H_2O. These are then converted to amount of C and H. The ratio (moles C/moles H) or (moles H/moles C) gives the empirical formula of the compound.

Solution • The amount of CO_2 and H_2O isolated from the combustion is:

$$3.447 \text{ g } CO_2 \cdot \frac{1 \text{ mol } CO_2}{44.010 \text{ g } CO_2} = 0.07832 \text{ mol } CO_2$$

$$1.647 \text{ g } H_2O \cdot \frac{1 \text{ mol } H_2O}{18.015 \text{ g } H_2O} = 0.09142 \text{ mol } H_2O$$

For every mole of CO_2 isolated, a mole of C must have existed in the compound C_xH_y.

$$0.07832 \text{ mol } CO_2 \cdot \frac{1 \text{ mol C in } C_xH_y}{1 \text{ mol } CO_2} = 0.07832 \text{ mol C}$$

and for every mole of H_2O isolated 2 mol of H must have existed in C_xH_y.

$$0.09142 \text{ mol } H_2O \cdot \frac{2 \text{ mol H in } C_xH_y}{1 \text{ mol } H_2O} = 0.1828 \text{ mol H in } C_xH_y$$

The original 1.125-g sample of compound therefore contained 0.07832 mol of C and 0.1828 mol of H. To determine the empirical formula of C_xH_y we find the ratio of moles of H to moles of C [← SECTION 3.7].

$$\frac{0.1828 \text{ mol H}}{0.07832 \text{ mol C}} = \frac{2.335 \text{ mol H}}{1.000 \text{ mol C}} = \frac{2\frac{1}{3} \text{ mol H}}{1 \text{ mol C}}$$

$$= \frac{7/3 \text{ mol H}}{1 \text{ mol C}} = \frac{7 \text{ mol H}}{3 \text{ mol C}}$$

The *empirical formula* of the hydrocarbon is therefore C_3H_7.

Comparing the experimental molar mass with the molar mass calculated for the empirical formula,

$$\frac{\text{Experimental molar mass}}{\text{Molar mass of } C_3H_7} = \frac{86.2 \text{ g / mol}}{43.1 \text{ g / mol}} = \frac{2}{1}$$

we find that the molecular formula is twice the empirical formula. That is, the *molecular formula* is $(C_3H_7)_2$, or C_6H_{14}.

Comment • The determination of the molecular formula does not end the problem for a chemist. In this case, the formula C_6H_{14} is appropriate for several distinctly different compounds. Two of the five compounds having this formula are shown here.

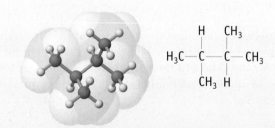

2,3-dimethylbutane, boiling point = 58.0°C

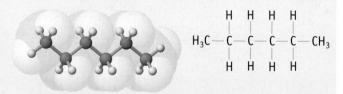

hexane, boiling point = 68.7°C

To decide finally the identity of the unknown compound, more laboratory experiments will have to be done.

Exercise 4.8 Determining the Empirical and Molecular Formulas for a Hydrocarbon

A 0.523-g sample of the unknown compound C_xH_y is burned in air to give 1.612 g of CO_2 and 0.7425 g of H_2O. A separate experiment gave a molar mass for C_xH_y of 114 g/mol. Determine the empirical and molecular formulas for the hydrocarbon.

In Summary

When you have finished studying this chapter, you should ask if you have met the chapter goals. In particular, you should be able to

- Interpret information conveyed by a balanced chemical equation (Section 4.1).
- Balance simple chemical equations (Section 4.2).
- Understand the principle of the conservation of matter, the basis of chemical stoichiometry (Section 4.3).
- Calculate the mass of one reactant or product from the mass of another reactant or product by using the balanced chemical equation (Section 4.3).
- Understand the impact of a limiting reactant on the outcome of a chemical reaction (Section 4.4).
- Determine which of two reactants is the limiting reactant (Section 4.4).
- Determine the quantity of a product based on the limiting reactant (Section 4.4).
- Explain the differences among actual yield, theoretical yield, and percent yield, and calculate percent yield (Section 4.5).
- Use stoichiometry principles to analyze a mixture or to find the empirical formula of an unknown compound (Section 4.6).

Key Terms

Section 4.1
chemical equation
reactants
products
aqueous solution
law of the conservation of matter
stoichiometry
stoichiometric coefficients

Section 4.2
combustion

Section 4.3
stoichiometric factor

Section 4.4
limiting reactant

Section 4.5
theoretical yield
actual yield
percent yield

Key Equations

Equation 4.1 (page 132)
Calculating percent yield.

$$\text{Percent yield} = \frac{\text{actual yield (g)}}{\text{theoretical yield (g)}} \times 100\%$$

Study Questions

Questions with blue, bold-faced numbers have answers in Appendix O.

Reviewing Important Concepts

1. What information is provided by a balanced chemical equation?

2. Write a balanced chemical equation for the production of ammonia $NH_3(g)$ from $N_2(g)$ and $H_2(g)$.

3. In the reaction of aluminum and bromine,

$$2\ Al(s) + 3\ Br_2(\ell) \longrightarrow Al_2Br_6(s)$$

(see Chapter Focus), how many molecules of Br_2 do you need for complete reaction if you use 2000 atoms of Al? How many molecules of Al_2Br_6 do you obtain from the reaction?

4. Given that you have 3 mol of N_2 and wish to calculate the amount of NH_3 produced in the reaction of N_2 and H_2 (see Study Question 2), what stoichiometric factor would you use in the calculation?

5. Explain how to calculate the theoretical yield for the reaction of zinc and iodine

$$Zn(s) + I_2(s) \longrightarrow ZnI_2(s)$$

knowing the mass of zinc (and using an excess of iodine). What experimental information do you need to calculate the percent yield?

6. Iron(III) oxide can be converted to iron by treatment with carbon monoxide.

$$Fe_2O_3(s) + 3\ CO(g) \longrightarrow 2\ Fe(s) + 3\ CO_2(g)$$

If 25 mol of Fe_2O_3 is combined with 65 mol of CO, which is the limiting reactant? What amount of iron, in moles, is expected based on 25 mol of Fe_2O_3? Based on 65 mol of CO?

7. Mixing aqueous solutions of silver nitrate and sodium chloride results in the formation of insoluble silver chloride.

$$AgNO_3(aq) + NaCl(aq) \longrightarrow AgCl(s) + NaNO_3(aq)$$

Describe how you would use this reaction to analyze a sample of a mineral for its NaCl content.

Practicing Skills

Balancing Equations
(See Example 4.1 and CD-ROM Screen 4.4)

8. Balance the following equations:
   (a) $Cr(s) + O_2(g) \longrightarrow Cr_2O_3(s)$
   (b) $Cu_2S(s) + O_2(g) \longrightarrow Cu(s) + SO_2(g)$
   (c) $C_6H_5CH_3(\ell) + O_2(g) \longrightarrow H_2O(\ell) + CO_2(g)$

9. Balance the following equations:
   (a) $Cr(s) + Cl_2(g) \longrightarrow CrCl_3(s)$
   (b) $SiO_2(s) + C(s) \longrightarrow Si(s) + CO(g)$
   (c) $Fe(s) + H_2O(g) \longrightarrow Fe_3O_4(s) + H_2(g)$

10. Balance the following equations and name each reactant and product:
    (a) $Fe_2O_3(s) + Mg(s) \longrightarrow MgO(s) + Fe(s)$
    (b) $AlCl_3(s) + H_2O(\ell) \longrightarrow Al(OH)_3(s) + HCl(aq)$
    (c) $NaNO_3(s) + H_2SO_4(\ell) \longrightarrow Na_2SO_4(s) + HNO_3(\ell)$
    (d) $NiCO_3(s) + HNO_3(aq) \longrightarrow$
    $$Ni(NO_3)_2(aq) + CO_2(g) + H_2O(\ell)$$

11. Balance the following equations and name each reactant and product:
    (a) $SF_4(g) + H_2O(\ell) \longrightarrow SO_2(g) + HF(\ell)$
    (b) $NH_3(aq) + O_2(g) \longrightarrow NO(g) + H_2O(\ell)$
    (c) $BF_3(g) + H_2O(\ell) \longrightarrow HF(aq) + H_3BO_3(\ell)$

Mass Relationships in Chemical Reactions: Basic Stoichiometry
(See Example 4.2 and CD-ROM Screens 4.5 and 4.6)

12. Aluminum reacts with oxygen to give aluminum oxide.

$$4\ Al(s) + 3\ O_2(g) \longrightarrow 2\ Al_2O_3(s)$$

What amount of O_2 (in moles) is needed for complete reaction with 6.0 mol of Al? What mass of Al_2O_3, in grams, can be produced?

13. What mass of HCl, in grams, is required to react with 0.750 g of $Al(OH)_3$? What mass of water, in grams, is produced?

$$Al(OH)_3(s) + 3\ HCl(aq) \longrightarrow AlCl_3(aq) + 3\ H_2O(\ell)$$

14. Like many metals, aluminum reacts with a halogen to give a metal halide (see Chapter Focus, page 120).

$$2\ Al(s) + 3\ Br_2(\ell) \longrightarrow Al_2Br_6(s)$$

What mass of Br_2, in grams, is required for complete reaction with 2.56 g of Al? What mass of white, solid Al_2Br_6 is expected?

15. The balanced equation for a reaction in the process of reducing iron ore to the metal is

$$Fe_2O_3(s) + 3\ CO(g) \longrightarrow 2\ Fe(s) + 3\ CO_2(g)$$

(a) What is the maximum mass of iron, in grams, that can be obtained from 454 g (1.00 lb) of iron(III) oxide?
(b) What mass of CO is required to react with 454 g of Fe_2O_3?

16. Iron metal reacts with oxygen to give iron(III) oxide, Fe_2O_3.
    (a) Write a balanced equation for the reaction.
    (b) If an ordinary iron nail (assumed to be pure iron) has a mass of 2.68 g, what mass (in grams) of Fe_2O_3 is produced if the nail is converted completely to the oxide?
    (c) What mass of O_2 (in grams) is required for the reaction?

17. Methane, CH_4, burns in oxygen.
    (a) What are the products of the reaction?

(b) Write the balanced equation for the reaction.

(c) What mass of O_2, in grams, is required for complete combustion of 25.5 g of methane?

(d) What is the total mass of products expected from the combustion of 25.5 g of methane?

18. Sulfur dioxide, a pollutant produced by burning coal and oil in power plants, can be removed by reaction with calcium carbonate.

$$2\ SO_2(g) + 2\ CaCO_3(s) + O_2(g) \longrightarrow$$
$$2\ CaSO_4(s) + 2\ CO_2(g)$$

(a) What mass of $CaCO_3$ is required to remove 155 g of SO_2?

(b) What mass of $CaSO_4$ is formed when 155 g of SO_2 is consumed completely?

19. The formation of water-insoluble silver chloride is useful in the analysis of chloride-containing substances. Consider the following *unbalanced* equation.

$$BaCl_2(aq) + AgNO_3(aq) \longrightarrow AgCl(s) + Ba(NO_3)_2(aq)$$

(a) Write the balanced equation.

(b) What mass $AgNO_3$, in grams, is required for complete reaction with 0.156 g of $BaCl_2$? What mass of $AgCl$ is produced?

Limiting Reactants
(See Examples 4.3 and 4.4 and CD-ROM Screens 4.7 and 4.8)

20. The compound SF_6 is made by burning sulfur in an atmosphere of fluorine. The balanced equation is

$$S_8(s) + 24\ F_2(g) \longrightarrow 8\ SF_6(g)$$

If you begin with 1.6 mol of sulfur, S_8, and 35 mol of F_2, which is the limiting reagent?

21. Disulfur dichloride, S_2Cl_2, is used to vulcanize rubber. It can be made by treating molten sulfur with gaseous chlorine:

$$S_8(\ell) + 4\ Cl_2(g) \longrightarrow 4\ S_2Cl_2(\ell)$$

Starting with a mixture of 32.0 g of sulfur and 71.0 g of Cl_2, which is the limiting reactant?

22. The reaction of methane and water is one way to prepare hydrogen for use as a fuel:

$$CH_4(g) + H_2O(g) \longrightarrow CO(g) + 3\ H_2(g)$$

If you begin with 995 g of CH_4 and 2510 g of water,

(a) Which reactant is the limiting reactant?

(b) What is the maximum mass of H_2 that can be prepared?

(c) What mass of the excess reactant remains when the reaction is completed?

23. Aluminum chloride, $AlCl_3$, is made by treating scrap aluminum with chlorine.

$$2\ Al(s) + 3\ Cl_2(g) \longrightarrow 2\ AlCl_3(s)$$

If you begin with 2.70 g of Al and 4.05 g of Cl_2,

(a) Which reactant is limiting?

(b) What mass of $AlCl_3$ can be produced?

(c) What mass of the excess reactant remains when the reaction is completed?

24. Ammonia gas can be prepared by the reaction of a metal oxide such as calcium oxide with ammonium chloride.

$$CaO(s) + 2\ NH_4Cl(s) \longrightarrow 2\ NH_3(g) + H_2O(g) + CaCl_2(s)$$

If 112 g of CaO and 224 g of NH_4Cl are mixed,

(a) What mass of NH_3 can be produced?

(b) What mass of the excess reactant remains after the ammonia has been formed?

25. Aspirin ($C_9H_8O_4$) is produced by the reaction of salicylic acid ($C_7H_6O_3$) and acetic anhydride ($C_4H_6O_3$) (page 132).

$$C_7H_6O_3(s) + C_4H_6O_3(\ell) \longrightarrow C_9H_8O_4(s) + CH_3CO_2H(\ell)$$

If you mix 100. g of each of the reactants, what is the maximum mass of aspirin that can be obtained?

Percent Yield
(See Exercise 4.6 and CD-ROM Screen 4.9)

26. In Example 4.3 you found that a mixture of CO and H_2 produced 407 g CH_3OH.

$$CO(g) + 2\ H_2(g) \longrightarrow CH_3OH(\ell)$$

If only 332 g of CH_3OH is actually produced, what is the percent yield of the compound?

27. Ammonia gas can be prepared by the reaction in Study Question 24. If 112 g of CaO and 224 g of NH_4Cl are mixed, the theoretical yield of NH_3 is 68.0 g. If only 16.3 g of NH_3 is actually obtained, what is its percent yield?

28. The deep blue compound $Cu(NH_3)_4SO_4$ is made by the reaction of copper(II) sulfate and ammonia.

$$CuSO_4(aq) + 4\ NH_3(aq) \longrightarrow Cu(NH_3)_4SO_4(aq)$$

(a) If you use 10.0 g of $CuSO_4$ and excess NH_3, what is the theoretical yield of $Cu(NH_3)_4SO_4$?

(b) If you obtain 12.6 g of $Cu(NH_3)_4SO_4$, what is the percent yield?

29. A reaction studied by Wächtershäuser and Huber (see "Black Smokers and the Origins of Life," page 119) is

$$2\ CH_3SH + CO \longrightarrow CH_3COSCH_3 + H_2S$$

If you begin with 10.0 g of CH_3SH, and excess CO,

(a) What is the theoretical yield of CH_3COSCH_3?

(b) If 8.65 g of CH_3COSCH_3 is isolated, what is its percent yield?

Analysis of Mixtures

(See Example 4.5 and CD-ROM Screen 4.10)

30. A mixture of $CuSO_4$ and $CuSO_4 \cdot 5\,H_2O$ has a mass of 1.245 g, but, after heating to drive off all the water, the mass is only 0.832 g. What is the mass percent of $CuSO_4 \cdot 5\,H_2O$ in the mixture? (See page 109.)

31. A 2.634-g sample containing $CuCl_2 \cdot 2\,H_2O$ and other materials, was heated. The sample mass after heating to drive off the water attached to $CuCl_2$ was 2.125 g. What was the mass percent of $CuCl_2 \cdot 2\,H_2O$ in the original sample?

32. A sample of limestone and other soil materials is heated, and the limestone decomposes to give calcium oxide and carbon dioxide.

$$CaCO_3(s) \longrightarrow CaO(s) + CO_2(g)$$

A 1.506-g sample of limestone-containing material gives 0.558 g of CO_2, in addition to CaO, after being heated at a high temperature. What is the mass percent of $CaCO_3$ in the original sample?

33. At higher temperatures $NaHCO_3$ is converted quantitatively to Na_2CO_3.

$$2\,NaHCO_3(s) \longrightarrow Na_2CO_3(s) + CO_2(g) + H_2O(g)$$

Heating a 0.7184-g sample of impure $NaHCO_3$ gives 0.4724 g of Na_2CO_3. What was the mass percent of $NaHCO_3$ in the original 0.7184-g sample?

34. A pesticide contains thallium(I) sulfate, Tl_2SO_4. Dissolving a 10.20-g sample of impure pesticide in water, and adding sodium iodide, precipitates 0.1964 g of thallium(I) iodide, TlI.

$$Tl_2SO_4(aq) + 2\,NaI(aq) \longrightarrow 2\,TlI(s) + Na_2SO_4(aq)$$

What is the mass percent Tl_2SO_4 in the original 10.20-g sample?

35. The aluminum in a 0.764-g sample of an unknown material was precipitated as aluminum hydroxide $Al(OH)_3$, which was then converted to Al_2O_3 by heating strongly. If 0.127 g of Al_2O_3 is obtained from the 0.764-g sample, what is the mass percent of aluminum in the sample?

Determination of Empirical and Molecular Formulas

(See Examples 4.6 and 4.7 and CD-ROM Screen 4.11)

36. Styrene, the building block of polystyrene, is a hydrocarbon, a compound consisting only of C and H. If 0.438 g of styrene is burned in oxygen and produces 1.481 g of CO_2 and 0.303 g of H_2O, what is the empirical formula of styrene?

37. Mesitylene is a liquid hydrocarbon. Burning 0.115 g of the compound in oxygen gives 0.379 g of CO_2 and 0.1035 g of H_2O. What is the empirical formula of mesitylene?

38. Cyclopentane is a simple hydrocarbon. If 0.0956 g of the compound is burned in oxygen, 0.300 g of CO_2 and 0.123 g of H_2O are isolated.
    (a) What is the empirical formula of cyclopentane?
    (b) If a separate experiment gave 70.1 g/mol as the molar mass of the compound, what is its molecular formula?

39. Azulene is a beautiful blue hydrocarbon, C_xH_y. If 0.106 g of the compound is burned in oxygen, 0.364 g of CO_2 and 0.0596 g of H_2O are isolated.
    (a) What is the empirical formula of azulene?
    (b) If a separate experiment gave 128.2 g/mol as the molar mass of the compound, what is its molecular formula?

40. Allowing potassium to react carefully with oxygen gives an oxide, K_xO_y. If 0.233 g of K produces 0.328 g of the oxide, what is the empirical formula for K_xO_y?

41. Sulfur reacts with chlorine to produce several different compounds. In one case you find that 0.125 g of elemental sulfur produces 0.263 g of a compound S_xCl_y. What is the empirical formula of S_xCl_y? If the molar mass of S_xCl_y was found to be 135.0 g/mol in another experiment, what is the molecular formula of the compound?

42. Nickel forms a compound with carbon monoxide, $Ni_x(CO)_y$. To determine its formula, you carefully heat a 0.0973-g sample in air to convert the nickel to NiO (0.0426 g) and the CO to 0.100 g of CO_2 (according to the following *unbalanced* equation). What is the empirical formula of $Ni_x(CO)_y$?

$$Ni_x(CO)_y(s) + \text{some } O_2(g) \longrightarrow x\,NiO(s) + y\,CO_2(g)$$

43. To find the formula of a compound composed of iron and carbon monoxide, $Fe_x(CO)_y$, the compound is burned in pure oxygen, a reaction that proceeds according to the following *unbalanced* equation.

$$Fe_x(CO)_y(s) + O_2(g) \longrightarrow Fe_2O_3(s) + CO_2(g)$$

If you burn 1.959 g of $Fe_x(CO)_y$ and obtain 0.799 g of Fe_2O_3 and 2.200 g of CO_2, what is the empirical formula of $Fe_x(CO)_y$?

General Questions

More challenging questions are marked with an underlined number.

44. Balance the following equations:
    (a) The synthesis of urea, a common fertilizer

$$CO_2(g) + NH_3(g) \longrightarrow NH_2CONH_2(s) + H_2O(\ell)$$

    (b) Reactions used to make uranium(VI) fluoride for the enrichment of natural uranium

$$UO_2(s) + HF(aq) \longrightarrow UF_4(s) + H_2O(\ell)$$
$$UF_4(s) + F_2(g) \longrightarrow UF_6(s)$$

    (c) The reaction to make titanium(IV) chloride, which is then converted to titanium metal

$$TiO_2(s) + Cl_2(g) + C(s) \longrightarrow TiCl_4(\ell) + CO(g)$$
$$TiCl_4(\ell) + Mg(s) \longrightarrow Ti(s) + MgCl_2(s)$$

45. Balance the following equations:
    (a) Reaction to produce "superphosphate" fertilizer.

$$Ca_3(PO_4)_2(s) + H_2SO_4(aq) \longrightarrow$$
$$Ca(H_2PO_4)_2(aq) + CaSO_4(s)$$

(b) Reaction to produce diborane, B_2H_6

$$NaBH_4(s) + H_2SO_4(aq) \longrightarrow$$
$$B_2H_6(g) + H_2(g) + Na_2SO_4(aq)$$

(c) Reaction to produce tungsten metal from tungsten(VI) oxide

$$WO_3(s) + H_2(g) \longrightarrow W(s) + H_2O(\ell)$$

(d) Decomposition of ammonium dichromate

$$(NH_4)_2Cr_2O_7(s) \longrightarrow N_2(g) + H_2O(\ell) + Cr_2O_3(s)$$

46. Suppose 16.04 g of methane, CH_4, is burned in oxygen.

(a) What are the products of the reaction?

(b) What is the balanced equation for the reaction?

(c) What mass of O_2, in grams, is required for complete combustion of methane?

(d) What is the total mass of products expected from 16.04 g of methane?

47. If 10.0 g of carbon is combined with an exact, stoichiometric amount of oxygen (26.6 g) to produce carbon dioxide, what is the theoretical yield of CO_2 (in grams)?

48. The metabolic disorder diabetes causes a buildup of acetone (CH_3COCH_3) in the blood of untreated victims. Acetone, a volatile compound, is exhaled, giving the breath of untreated diabetics a distinctive odor. The acetone is produced by a breakdown of fats in a series of reactions. The equation for the last step is

$$CH_3COCH_2CO_2H \longrightarrow CH_3COCH_3 + CO_2$$

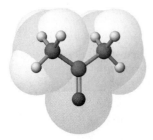

acetone, CH_3COCH_3

What mass of acetone can be produced from 125 mg of acetoacetic acid ($CH_3COCH_2CO_2H$)?

49. Your body deals with excess nitrogen by excreting it in the form of urea, NH_2CONH_2. The reaction producing it is the combination of arginine ($C_6H_{14}N_4O_2$) with water to give urea and ornithine ($C_5H_{12}N_2O_2$).

$$\underset{\text{arginine}}{C_6H_{14}N_4O_2} + H_2O \longrightarrow \underset{\text{urea}}{NH_2CONH_2} + \underset{\text{ornithine}}{C_5H_{12}N_2O_2}$$

If you excrete 95 mg of urea, what mass of arginine must have been used? What mass of ornithine must have been produced?

50. In the Chapter Focus (page 120), you saw the reaction of iron metal and chlorine gas to give iron(III) chloride.

(a) Write the balanced chemical equation for the reaction.

(b) Beginning with 10.0 g of iron, what mass of Cl_2, in grams, is required for complete reaction? What mass of $FeCl_3$ can be produced?

(c) If only 18.5 g of $FeCl_3$ is obtained, what is the percent yield?

51. Some metal halides react with water to produce the metal oxide and the appropriate hydrogen halide (see photo). For example,

$$TiCl_4(\ell) + 2 H_2O(\ell) \longrightarrow TiO_2(s) + 4 HCl(g)$$

Titanium tetrachloride, $TiCl_4$, is a liquid at room temperature. When exposed to moist air, it forms a dense fog of titanium(IV) oxide. *(Charles D. Winters)*

(a) Name the four compounds involved in this reaction.

(b) If you begin with 14.0 mL of $TiCl_4$ ($d = 1.73$ g/mL), what mass of water, in grams, is required for complete reaction?

(c) What mass of each product is expected?

52. The reaction of 750. g each of NH_3 and O_2 was found to produce 562 g of NO (see page 129).

$$4 NH_3(g) + 5 O_2(g) \longrightarrow 4 NO(g) + 6 H_2O(g)$$

(a) What mass of water is also produced by this reaction?

(b) What quantity of O_2 is required to consume 750. g of NH_3?

53. Saccharin, an artificial sweetener, has the formula $C_7H_5NO_3S$ (page 134). Suppose you have a sample of a saccharin-containing sweetener with a mass of 0.2140 g. After decomposition to free the sulfur and convert it to the SO_4^{2-} ion, the sulfate ion is trapped as water-insoluble $BaSO_4$ (see Figure 4.5). The quantity of $BaSO_4$ obtained is 0.2070 g. What mass percent of saccharin is in the sample of sweetener?

54. Boron forms an extensive series of compounds with hydrogen, all with the general formula B_xH_y.

$$B_xH_y(s) + \text{excess } O_2(g) \longrightarrow B_2O_3(s) + H_2O(g)$$

If 0.148 g of B_xH_y gives 0.422 g of B_2O_3 when burned in excess O_2, what is the empirical formula of B_xH_y?

55. Silicon and hydrogen form a series of compounds with the general formula Si_xH_y. To find the formula of one of them, a 6.22-g sample of the compound is burned in oxygen. On doing so, all of the Si is converted to 11.64 g of SiO_2 and all of the H to 6.980 g of H_2O. What is the empirical formula of the silicon compound?

56. Menthol, from oil of mint, has a characteristic odor. The compound contains only C, H, and O. If 95.6 mg of menthol burns completely in O_2 and gives 269 mg of CO_2 and 110 mg of H_2O, what is the empirical formula of menthol? (*Hint:* The C in menthol is converted to CO_2 and the H is converted to H_2O. The O atoms are found both in CO_2 and H_2O. To find the mass of O in the 95.6-mg sample, use the masses of CO_2 and H_2O to find the masses of C and H in the 95.6-mg sample. Whatever of the 95.6-mg sample is not C and H is the mass of O.)

57. Quinone, a chemical used in the dye industry and in photography, is an organic compound containing only C, H, and O. What is the empirical formula of the compound if 0.105 g of the compound gives 0.257 g of CO_2 and 0.0350 g of H_2O when burned completely in oxygen? (See *Hint* in Study Question 56.)

58. In an experiment 1.056 g of a metal carbonate, containing an unknown metal M, is heated to give the metal oxide and 0.376 g CO_2.

$$MCO_3(s) + \text{heat} \longrightarrow MO(s) + CO_2(g)$$

What is the identity of the metal M?
(a) M = Ni   (c) M = Zn
(b) M = Cu   (d) M = Ba

59. An unknown metal reacts with oxygen to give the metal oxide, MO_2. Identify the metal based on this information:

Mass of metal = 0.356 g

Mass of sample after converting metal completely to oxide = 0.452 g

60. Titanium(IV) oxide, TiO_2, is heated in hydrogen gas to give water and a new titanium oxide, Ti_xO_y. If 1.598 g of TiO_2 produces 1.438 g of Ti_xO_y, what is the formula of the new oxide?

61. The elements silver, molybdenum, and sulfur combine to form Ag_2MoS_4. What is the maximum mass of Ag_2MoS_4 that can be obtained if 8.63 g of silver, 3.36 g of molybdenum, and 4.81 g of sulfur are combined?

62. Thioridazine, $C_{21}H_{26}N_2S_2$, is a pharmaceutical used in the regulation of dopamine, a neurotransmitter in the brain. A chemist can analyze a sample for the thioridazine content by decomposing to convert the sulfur in the compound to sulfate ion. This is then "trapped" as water-insoluble barium sulfate (see Figure 4.5).

$$SO_4{}^{2-}(aq, \text{from thioridazine}) + BaCl_2(aq) \longrightarrow$$

$$BaSO_4(s) + 2\,Cl^-(aq)$$

Suppose a 12-tablet sample of the drug yields 0.301 g of $BaSO_4$. What is the thioridazine content, in milligrams, of each tablet?

63. An herbicide contains 2,4-D (2,4-dichlorophenoxyacetic acid), $C_8H_6Cl_2O_3$. A 1.236-g sample of the herbicide is decomposed to liberate the chlorine as Cl^- ion. This was precipitated as AgCl, with a mass of 0.1840 g. What is the mass percent of 2,4-D in the sample?

64. A weighed sample of iron (Fe) is added to liquid bromine (Br_2) and allowed to react completely. The reaction produces a single product, which can be isolated and weighed. The experiment is repeated a number of times with different masses of iron but with the same mass of bromine. (See the following graph.)

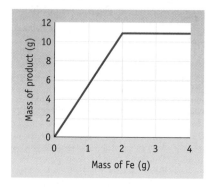

(a) What mass of Br_2 is used when the reaction consumes 2.0 g of Fe?
(b) What is the mole ratio of Br_2 to Fe in the reaction?
(c) What is the empirical formula of the product?
(d) Write the balanced chemical equation for the reaction of iron and bromine.
(e) What is the name of the reaction product?
(f) Which statement or statements best describe the experiments summarized by the graph?
   (i) When 1.00 g of Fe is added to the Br_2, Fe is the limiting reagent.
   (ii) When 3.50 g of Fe is added to the Br_2, there is an excess of Br_2.
   (iii) When 2.50 g of Fe is added to the Br_2, both reactants are used up completely.
   (iv) When 2.00 g of Fe is added to the Br_2, 10.0 g of product is formed. The percent yield must therefore be 20.0%.

65. Sodium hydrogen carbon, $NaHCO_3$, can be decomposed quantitatively by heating.

$$2\,NaHCO_3(s) \longrightarrow Na_2CO_3(s) + CO_2(g) + H_2O(g)$$

A 0.784-g sample of impure $NaHCO_3$ yields a solid residue (consisting of Na_2CO_3 and other solids) with a mass of 0.4724 g. What is the mass percent of $NaHCO_3$ in the sample?

Using Electronic Resources

This question refers to the *General Chemistry Interactive CD-ROM*, Version 3.0.

66. See CD-ROM Screen 4.8: Limiting Reactants. Let us explore a reaction with a limiting reactant. Here zinc metal is added to a flask containing aqueous HCl and H_2 gas is a product.

$$Zn(s) + 2\,HCl(aq) \longrightarrow ZnCl_2(aq) + H_2(g)$$

The three flasks each contain 0.100 mol of HCl. Zinc is added to each flask in the following quantities.

Flask 1: 7.00 g Zn

Flask 2: 3.27 g Zn

Flask 3: 1.31 g Zn

When the reactants are combined, the H_2 inflates the balloon attached to the flask. The results are shown in a photograph taken from the CD-ROM and are as follows:

Flask 1: Balloon inflates completely but some Zn remains when inflation ceases.

Flask 2: Balloon inflates completely. No Zn remains.

Flask 3: Balloon does not inflate completely. No Zn remains.

Explain these results completely. Perform calculations that support your explanation.

5

Reactions in Aqueous Solution

Chapter Goals

- Understand the nature of ionic substances dissolved in water.
- Recognize and write equations for the common types of reactions in aqueous solution.
- Recognize acids and bases and oxidizing and reducing agents.
- Define and use molarity in solution stoichiometry.

Why Is the Sea Salty?

Saltiness is one of the basic taste sensations, and a taste of sea water quickly reveals its nature. How did the oceans become salty? What, in addition to salt, is dissolved in sea water?

Sea water contains enormous amounts of dissolved salts. Ions of virtually every element are present as well as dozens of polyatomic ions. What is their origin? And why is chloride ion the most abundant ion?

The carbonate ion and its close relative HCO_3^- can come from the interaction of atmospheric CO_2 with water.

▲ A sample of albite, a sodium-containing mineral and one of many mineral sources for the sodium ions in the oceans of the Earth. *(José Manuel Sanches Calvete/CORBIS)*

mixture of $CaCO_3$ and $MgCO_3$) from terrestrial rocks such as those in Arizona's Grand Canyon and Italy's Dolomite Mountains.

Sodium ions arrive in the oceans by a similar reaction with sodium-bearing minerals such as albite, $NaAlSi_3O_8$.

▲ The Pacific Ocean Taveuni Island, the Fiji Islands. *(John C. Kotz)*

$$CO_2(g) + H_2O(\ell) \longrightarrow H_2CO_3(aq)$$

$$H_2CO_3(aq) \longrightarrow H^+(aq) + HCO_3^-(aq)$$

This is the reason rain is normally acidic. The slightly acidic water causes substances such as limestone or corals to dissolve, thus producing calcium ions and more bicarbonate ions.

$$CaCO_3(s) + CO_2(g) + H_2O(\ell)$$
$$\longrightarrow Ca^{2+}(aq) + 2\ HCO_3^-(aq)$$

Magnesium ions come from a similar reaction with the mineral dolomite (a

▲ The Dead Sea in the Middle East has the highest salt content of any body of water. *(Paul Stephan-Vierow/Photo Researchers, Inc.)*

▲ Volcanoes are the chief source of chloride ion in Earth's oceans. *(API/Explorer/Photo Researchers, Inc.)*

emitted from these volcanoes is very soluble in water and is quickly dissolved.

$$HCl(g) \longrightarrow H^+(aq) + Cl^-(aq)$$

The chloride ions from the dissolved HCl gas and the sodium ions from weathered rocks are the source of the salt in the sea.

(Byproducts are common clay, which is $Al_2Si_2O_5(OH)_4$, and sand, SiO_2.)

$$2\ NaAlSi_3O_8(s) + 2\ CO_2(g) + 3\ H_2O(\ell) \longrightarrow$$
$$Al_2Si_2O_5(OH)_4(s) + 2\ Na^+(aq) +$$
$$2\ HCO_3^-(aq) + 4\ SiO_2(s)$$

Rain water falling on land masses extracts ions that are carried to the oceans in rivers.

The average chloride content of rocks in the earth's crust is only 0.01%, so only a minute proportion of the chloride ion in the oceans can come from the weathering of rocks and minerals. What then is the origin of the chloride ion in sea water? The answer is volcanoes. Hydrogen chloride gas, HCl, is an important constituent of volcanic gases. Early in the earth's history, the planet was much hotter, and volcanoes were much more widespread. The HCl gas

Ion	Total Amount in the Oceans (tons)
Cl^-	2.6×10^{16}
Na^+	1.4×10^{16}
Mg^{2+}	1.7×10^{15}
SO_4^{2-}	1.2×10^{15}
Ca^{2+}	5.5×10^{14}
K^+	5.0×10^{14}
Br^-	8.9×10^{13}
HCO_3^-, CO_3^{2-}	3.7×10^{13}

Before You Begin

- Review names of common ions (Section 3.3 and Table 3.1).
- Know how to do mass-to-moles and moles-to-mass calculations.

Aqueous Solutions Many substances dissolve in water. Soluble ionic compounds (and strong acids and bases) dissociate into their constituent cations and anions, whereas soluble molecular compounds remain intact as molecules. Understanding the nature of substances in aqueous solution, and the reactions they undergo, is fundamental to understanding much of the chemistry of the natural world [CD-ROM, Screen 5.3].

Water, cations, and anions

A water molecule is electrically positive on one side (the H atoms) and electrically negative on the other (the O atom). These charges enable water to interact with negative and positive ions in aqueous solution.

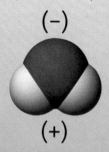

(−)

(+)

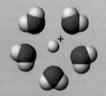

Water surrounding a cation

Water surrounding an anion

Copper chloride dissolved in water

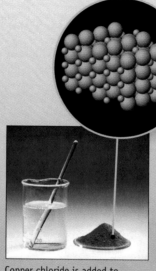

Copper chloride is added to water. Interactions between water and the Cu^{2+} and Cl^- ions allow the solid to dissolve.

The ions are now sheathed in water molecules.

Electrolytes

Strong Electrolyte	Weak Electrolyte	Nonelectrolyte

$CuCl_2$

Cu^{2+}

Cl^-

A strong electrolyte conducts electricity. $CuCl_2$ is completely dissociated into Cu^{2+} and Cl^- ions.

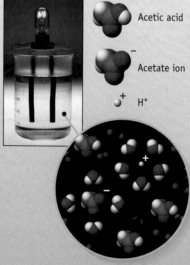

Acetic acid

Acetate ion

H^+

A weak electrolyte conducts electricity poorly because so few ions are present in solution.

Ethanol

A nonelectrolyte does not conduct electricity because no ions are present in solution.

Photos: Charles D. Winters

The human body is two-thirds water. Water is essential because it is involved in every function of the body. It assists in transporting nutrients and waste products in and out of cells and is necessary for all digestive, absorption, circulatory, and excretory functions. Therefore, we turn now to the study of *aqueous* systems, chemical systems in which water plays a major role.

5.1   PROPERTIES OF COMPOUNDS IN AQUEOUS SOLUTION

A **solution** is a homogeneous mixture of two or more substances, in which one is generally considered the **solvent,** the medium in which another substance — the **solute** — is dissolved [← SECTION 1.5]. Chemical reactions in plants and animals occur largely among substances dissolved in water, that is, in **aqueous solutions.** Many of the reactions you will see in the laboratory are also done in aqueous solution. Therefore, to understand such reactions, it is important first to understand something about the behavior of compounds in water. The focus here is on compounds that produce ions in aqueous solution [⏻ CD-ROM, Screen 5.2].

Ions in Aqueous Solution: Electrolytes

The water you drink every day and the oceans of the world contain small concentrations of many ions, most of which result from dissolving solid materials present in the environment (Table 5.1) [⏻ CD-ROM, Screen 4.5].

Dissolving an ionic solid requires separating each ion from the oppositely charged ions that surround it in the solid state [Chapter Focus]. Water is especially good at dissolving ionic compounds because each water molecule has a positively charged end and a negatively charged end. Therefore, a water molecule can attract a positive ion to its negative end, or it can attract a negative ion to its positive end. When an ionic compound dissolves in water, each negative ion becomes surrounded by water molecules with their positive ends pointing toward the ion, and each positive ion becomes surrounded by the negative ends of several water molecules.

Chapter Goals • Revisited

- **Understand the nature of ionic substances dissolved in water.**
- Recognize and write equations for the common types of reactions in aqueous solution.
- Recognize acids and bases and oxidizing and reducing agents.
- Define and use molarity in solution stoichiometry.

Table 5.1 • **Some Cations and Anions in Living Cells and Their Environment***

Element	Dissolved Species	Sea Water	*Valonia*[†]	Red-Blood Cells	Blood Plasma
Chlorine	Cl^-	550	50	50	100
Sodium	Na^+	460	80	11	160
Magnesium	Mg^{2+}	52	50	2.5	2
Calcium	Ca^{2+}	10	1.5	10^{-4}	2
Potassium	K^+	10	400	92	10
Carbon	HCO_3^-, CO_3^{2-}	30	<10	<10	30
Phosphorus	HPO_4^{2-}	<1	5	3	<3

*Data taken from J. J .R. Fraústo da Silva and R. J. P. Williams: *The Biological Chemistry of the Elements*, Oxford, England, Clarendon Press, 1991. Concentrations are given in millimoles per liter. (A millimole is 1/1000 of a mole.)

[†]*Valonia* are single-celled algae that live in sea water.

Positive H atoms attracted to negative ion

Negative O atom attracted to positive ion

Anion

Cation

Water can interact with positive cations and negative anions in aqueous solution.

The water-encased ions produced by dissolving an ionic compound are free to move about in solution. Under normal conditions, the movement of ions is random, and the cations and anions from a dissolved ionic compound are dispersed uniformly throughout the solution. If two **electrodes** (conductors of electricity such as copper wire) are placed in the solution and connected to a battery, however, cations migrate through the solution to the negative electrode and anions move to the positive electrode. If a light bulb is inserted into the circuit, the bulb lights, showing that ions are available to conduct charge in the solution just as electrons conduct charge in the wire part of the circuit. Compounds whose aqueous solutions conduct electricity are called **electrolytes,** and *all ionic compounds that are soluble in water are electrolytes* [CD-ROM, Screen 5.3].

Types of Electrolytes

Electrolytes can be classified as strong or weak. When sodium chloride and many other ionic compounds dissolve in water, the ions separate or dissociate. For every mole of NaCl that dissolves, 1 mol of Na^+ and 1 mol of Cl^- ions enter the solution (see Chapter Focus).

$$NaCl(s) \longrightarrow Na^+(aq) + Cl^-(aq)$$

100% Dissociation ≡ strong electrolyte

Because the solute has dissociated completely into ions, the solution can be a good conductor of electricity. Substances whose solutions are good electrical conductors are **strong electrolytes** (see Chapter Focus).

Other substances dissociate only partially in solution and so are poor conductors of electricity; they are known as **weak electrolytes** (see Chapter Focus). For example, when acetic acid — an important ingredient in vinegar — dissolves in water, only a few molecules in every 100 molecules of acetic acid are ionized to form acetate ions and hydrogen.

- **Ion Pairing in Electrolytes**
The idea that salts such as NaCl dissociate completely in solution is a simplification. In fact, a measurable concentration of species such as NaCl(aq), species called "ion pairs," also occurs in solutions.

- **Double Arrows,** ⇌
The double arrows in the equation for the ionization of acetic acid, and in many others, indicate the reactant produces product, but also that the product ions recombine to produce the original reactant. This is the subject of chemical equilibrium [➡ CHAPTERS 16–18.]

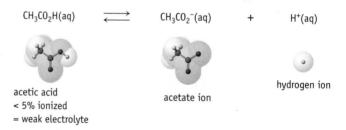

$CH_3CO_2H(aq)$ ⇌ $CH_3CO_2^-(aq)$ + $H^+(aq)$

acetic acid
< 5% ionized
= weak electrolyte

acetate ion

hydrogen ion

Many other substances dissolve in water but do not ionize. These are called **nonelectrolytes** because their solutions do not conduct electricity (see Chapter Focus). Examples of nonelectrolytes include sugar (sucrose, $C_{12}H_{22}O_{11}$), ethanol (CH_3CH_2OH), and antifreeze (ethylene glycol, $HOCH_2CH_2OH$).

Exercise 5.1 Electrolytes

Epsom salt, $MgSO_4 \cdot 7 H_2O$, is sold in drugstores and used, as a solution in water, for various medical purposes. Methanol, CH_3OH, is dissolved in gasoline in the winter in colder climates to prevent the formation of ice in automobile fuel lines. Which of these compounds is an electrolyte and which is a nonelectrolyte?

Solubility of Ionic Compounds in Water

Not all ionic compounds dissolve completely in water. Many dissolve only to a small extent, and still others are essentially insoluble. Fortunately, we can make some general statements about which types of ionic compounds are water-soluble [CD-ROM, Screen 5.4].

Figure 5.1 lists broad guidelines that help predict whether a particular ionic compound will be soluble in water. For example, sodium nitrate, $NaNO_3$, contains both an alkali metal cation, Na^+, and the nitrate anion, NO_3^-. According to Figure 5.1, the presence of either of these ions ensures that the compound is soluble in water. *Ionic compounds that dissolve in water are electrolytes,* which means that an aqueous solution of $AgNO_3$ (Figure 5.1a) also consists of the separated ions $Ag^+(aq)$

● **Solubility Guidelines**
Observations such as those shown in the photos in Figure 5.1 were used to create the solubility guidelines in that figure. Remember that an ionic compound will be moderately soluble in water if it contains at least one of the ions listed in the Soluble Compounds column of Figure 5.1.

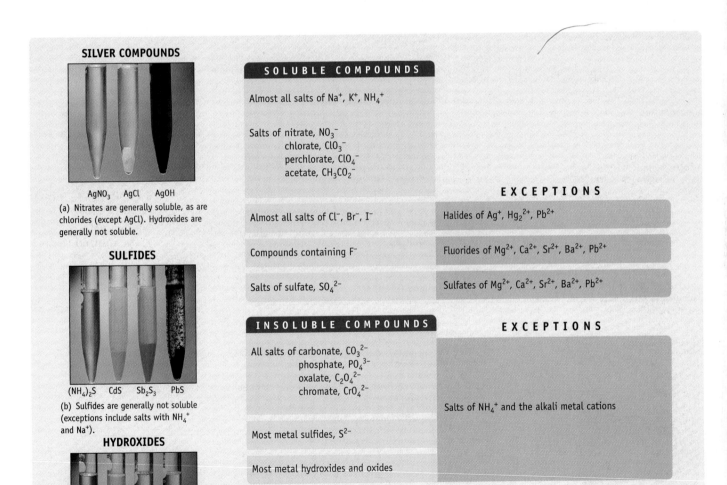

SILVER COMPOUNDS

$AgNO_3$ AgCl AgOH

(a) Nitrates are generally soluble, as are chlorides (except AgCl). Hydroxides are generally not soluble.

SULFIDES

$(NH_4)_2S$ CdS Sb_2S_3 PbS

(b) Sulfides are generally not soluble (exceptions include salts with NH_4^+ and Na^+).

HYDROXIDES

NaOH $Ca(OH)_2$ $Fe(OH)_3$ $Ni(OH)_2$

(c) Hydroxides are generally not soluble except when the cation is a Group 1A metal.

SOLUBLE COMPOUNDS

Almost all salts of Na^+, K^+, NH_4^+

Salts of nitrate, NO_3^-
        chlorate, ClO_3^-
        perchlorate, ClO_4^-
        acetate, $CH_3CO_2^-$

Almost all salts of Cl^-, Br^-, I^-

Compounds containing F^-

Salts of sulfate, SO_4^{2-}

EXCEPTIONS

Halides of Ag^+, Hg_2^{2+}, Pb^{2+}

Fluorides of Mg^{2+}, Ca^{2+}, Sr^{2+}, Ba^{2+}, Pb^{2+}

Sulfates of Mg^{2+}, Ca^{2+}, Sr^{2+}, Ba^{2+}, Pb^{2+}

INSOLUBLE COMPOUNDS

All salts of carbonate, CO_3^{2-}
        phosphate, PO_4^{3-}
        oxalate, $C_2O_4^{2-}$
        chromate, CrO_4^{2-}

Most metal sulfides, S^{2-}

Most metal hydroxides and oxides

EXCEPTIONS

Salts of NH_4^+ and the alkali metal cations

Figure 5.1 Guidelines to predict the solubility of ionic compounds. If a compound contains one of the ions in the column to the left in the top chart, the compound is predicted to be at least moderately soluble in water. There are a few exceptions, and these are noted at the right. Most ionic compounds formed by the anions listed in the bottom part of the chart are poorly soluble (with exceptions such as compounds with NH_4^+ and the alkali metal cations) [CD-ROM, Screen 5.4]. *(Photos, Charles D. Winters)*

and NO_3^-(aq). By contrast, calcium hydroxide is poorly soluble in water (Figure 5.1c). For example, if a spoonful of solid $Ca(OH)_2$ is added to 100 mL of water, only 0.17 g, or 0.0023 mol, will dissolve at 10 °C.

0.0023 mol $Ca(OH)_2$ dissolves in 100 mL water at 10 °C $\longrightarrow$
$$0.0023 \text{ mol Ca}^{2+}(aq) + (2 \times 0.0023) \text{ mol OH}^-(aq)$$

When solid $Ca(OH)_2$ is placed in water, nearly all of it remains as the solid, and a heterogeneous mixture is the result. This is generally true of all metal hydroxides (Figure 5.2) except those containing an alkali metal cation (see Figure 5.1c).

● **Dissolving Halides**
When an ionic compound with halide ions dissolves in water, the halide ions are released into aqueous solution. Thus, $BaCl_2$ produces one Ba^{2+} and two Cl^- ions for each Ba^{2+} ion (and not Cl_2 or Cl_2^{2-} ions).

Example 5.1 Solubility Guidelines

Problem • Predict whether each of the following ionic compounds is likely to be water-soluble. If a compound is soluble, list the ions present in solution.

(a) KCl   **(b)** $MgCO_3$ **(c)** Fe_2O_3 **(d)** $Cu(NO_3)_2$

Strategy • You must first recognize the cation and anion involved and then decide the probable water solubility based on the guidelines outlined in Figure 5.1.

Solution •
(a) KCl is composed of K^+ and Cl^- ions. The presence of *either* of these ions means that KCl is likely to be soluble in water. The solution consists of K^+ and Cl^- ions.

$$\text{KCl (s)} \longrightarrow K^+(aq) + Cl^-(aq)$$

(The solubility of KCl is about 35 g in 100 mL of water at 20 °C.)
(b) Magnesium carbonate is composed of Mg^{2+} and CO_3^{2-} ions. Mg^{2+} is in the alkaline earth group (Group 2A), a group that often forms water-insoluble compounds. The carbonate ion usually gives insoluble compounds (see Figure 5.1), unless combined with an ion like Na^+ or NH_4^+. $MgCO_3$ is therefore predicted to be insoluble in water. (The actual solubility of $MgCO_3$ is less than 0.2 g/100 mL of water.)

(c) Iron(III) oxide is composed of Fe^{3+} and O^{2-} ions. Again, Figure 5.1 suggests that oxides are soluble only when O^{2-} is combined with an alkali metal ion; Fe^{3+} is a transition metal ion, so Fe_2O_3 is insoluble.

(d) Copper(II) nitrate is composed of Cu^{2+} and NO_3^- ions. Almost all nitrates are soluble in water, so $Cu(NO_3)_2$ is water-soluble and produces copper(II) cations and nitrate anions in water.

$$\text{Cu(NO}_3)_2(s) \longrightarrow Cu^{2+}(aq) + 2\,NO_3^-(aq)$$

Comment • Notice that the $Cu(NO_3)_2$ gives one Cu^{2+} ion and *two* NO_3^- ions on dissolving in water.

Exercise 5.2 Solubility of Ionic Compounds

Predict whether each of the following ionic compounds is likely to be soluble in water. If soluble, write the formulas of the ions present in aqueous solution.

(a) $LiNO_3$ **(b)** $CaCl_2$ **(c)** CuO   **(d)** $NaCH_3CO_2$

Figure 5.2   Milk of magnesia, $Mg(OH)_2$. This common antacid is a good example of a metal hydroxide that is nearly insoluble in water. *(Charles D. Winters)*

5.2   PRECIPITATION REACTIONS

A **precipitation reaction** produces an insoluble product, a **precipitate** [🖱 CD-ROM, Screen 5.7]. The reactants in such reactions are generally water-soluble ionic compounds. When these are dissolved in water, therefore, they dissociate to give the appropriate cations and anions. If one of the cations can form an insoluble compound with one of the anions in the solution, a precipitation reaction occurs. For example, mixing solutions of silver nitrate and potassium chloride, leads to insoluble silver chloride and soluble potassium nitrate (Figure 5.3).

(a)

(b) The silver (Ag^+, gray) and chloride (Cl^-, green) ions are widely separated in $AgNO_3(aq)$.

(c) After Cl^- is added, Ag^+ and Cl^- ions approach and form ion pairs.

(d) As more and more Ag^+ and Cl^- ions come together, a precipitate of solid AgCl forms.

Figure 5.3   Precipitation of silver chloride. (a) Adding aqueous silver nitrate to potassium chloride produces white, insoluble silver chloride, AgCl. In **(b)** through **(d)** you see a model of the process. *(Photo, Charles D. Winters, model from animation by Roy Tasker, University Western Sydney, Australia)*

$$AgNO_3(aq) + KCl(aq) \longrightarrow AgCl(s) + KNO_3(aq)$$

Reactants	Products
$Ag^+(aq) + NO_3^-(aq)$	Insoluble AgCl
$K^+(aq) + Cl^-(aq)$	$K^+(aq) + NO_3^-(aq)$

Many precipitation reactions are possible because many combinations of positive and negative ions give insoluble substances (see Figures 5.3 and 5.4). For example, lead(II) chromate precipitates when a water-soluble lead(II) compound is combined with a water-soluble chromate compound (Figure 5.4a).

$$Pb(NO_3)_2(aq) + K_2CrO_4(aq) \longrightarrow PbCrO_4(s) + 2\ KNO_3(aq)$$

Reactants	Products
$Pb^{2+}(aq) + 2\ NO_3^-(aq)$	Insoluble $PbCrO_4$
$2\ K^+(aq) + CrO_4^{2-}(aq)$	$2\ K^+(aq) + 2\ NO_3^-(aq)$

● **Exchange Reactions**
Notice that, when two ionic compounds in aqueous solution react to form a solid precipitate, they do so by exchanging ions. For example, silver(I) ions exchange nitrate for chloride ions, and potassium ions exchange chloride ions for nitrate ions.

$$Ag^+ + NO_3^-$$
$$K^+ + Cl^-$$

(a) $Pb(NO_3)_2$ and K_2CrO_4 produce yellow, insoluble $PbCrO_4$ and soluble KNO_3.

(b) $Pb(NO_3)_2$ and $(NH_4)_2S$ produce black, insoluble PbS and soluble NH_4NO_3.

(c) $FeCl_3$ and NaOH produce red, insoluble $Fe(OH)_3$ and soluble NaCl.

(d) $AgNO_3$ and K_2CrO_4 produce red, insoluble Ag_2CrO_4 and soluble KNO_3. See Example 5.2.

Figure 5.4   Precipitation reactions. In **(a)** solutions of lead(II) nitrate and potassium chromate are mixed to give yellow, insoluble lead(II) chromate and soluble potassium nitrate. Lead(II) chromate is commonly known as chrome yellow and has been used as a pigment in paint. You may know it as "school bus yellow." *(Charles D. Winters)* [CD-ROM, Screen 5.7]

Chapter Goals • Revisited

- Understand the nature of ionic substances dissolved in water.
- **Recognize and write equations for the common types of reactions in aqueous solution.**
- Recognize acids and bases and oxidizing and reducing agents.
- Use correct units and stoichiometry principles for substances and reactions in solution.

Almost all metal sulfides are insoluble in water (Figure 5.4b). If a soluble metal compound in nature comes in contact with a source of sulfide ions (say from a volcano, a natural gas pocket in the earth, or a "black smoker" in the ocean, page 119), the metal sulfide precipitates.

$$Pb(NO_3)_2(aq) + (NH_4)_2S(aq) \longrightarrow PbS(s) + 2\ NH_4NO_3(aq)$$

Reactants	Products
$Pb^{2+}(aq) + 2\ NO_3^-(aq)$	Insoluble PbS
$2\ NH_4^+(aq) + S^{2-}(aq)$	$2\ NH_4^+(aq) + 2\ NO_3^-(aq)$

In fact, this is how many sulfur-containing minerals such as iron purite (FeS_2, see page 12) are believed to have been formed in nature.

Finally, with the exception of the alkali metal cations (and Ba^{2+}), all metal cations form insoluble hydroxides. Thus, water-soluble iron(III) chloride and sodium hydroxide react to give insoluble iron(III) hydroxide (Figures 5.1c and 5.4c).

$$FeCl_3(aq) + 3\ NaOH(aq) \longrightarrow Fe(OH)_3(s) + 3\ NaCl(aq)$$

Reactants	Products
$Fe^{3+}(aq) + 3\ Cl^-(aq)$	Insoluble $Fe(OH)_3$
$3\ Na^+(aq) + 3\ OH^-(aq)$	$3\ Na^+(aq) + 3\ Cl^-(aq)$

Example 5.2 Writing the Equation for a Precipitation Reaction

Problem • Is an insoluble product formed when aqueous solutions of potassium chromate and silver nitrate are mixed? If so, write the balanced equation.

Strategy • First decide what ions are formed from the reactants. Then determine if a cation from one reactant will combine with an anion from the other reactant to form an insoluble compound (according to Figure 5.1).

Solution • Both reactants are water-soluble, and the ions Ag^+, NO_3^-, K^+, and CrO_4^{2-} are released into solution when the compounds are dissolved.

$$AgNO_3(s) \longrightarrow Ag^+(aq) + NO_3^-(aq)$$
$$K_2CrO_4(s) \longrightarrow 2\ K^+(aq) + CrO_4^{2-}(aq)$$

Here Ag^+ could combine with CrO_4^{2-}, and K^+ could combine with NO_3^-. The former combination, Ag_2CrO_4, is an insoluble compound, whereas KNO_3 is soluble in water. Thus, the balanced equation for the reaction of silver nitrate and potassium chromate is (Figure 5.4d)

$$2\ AgNO_3(aq) + K_2CrO_4(aq) \longrightarrow Ag_2CrO_4(s) + 2\ KNO_3(aq)$$

Exercise 5.3 Precipitation Reactions

In each of the following cases, does a precipitation reaction occur when solutions of the two water-soluble reactants are mixed?

(a) Sodium carbonate is mixed with copper(II) chloride.

(b) Potassium carbonate is mixed with sodium nitrate.

(c) Nickel(II) chloride is mixed with potassium hydroxide.

Write a balanced chemical equation for the precipitation reactions that occur.

Net Ionic Equations

A solution of silver nitrate contains Ag^+ and NO_3^- ions. A solution of potassium chloride contains K^+ and Cl^- ions. When these solutions are mixed (Figure 5.3), insoluble AgCl precipitates, and the ions K^+ and NO_3^- remain in solution.

$$\underbrace{Ag^+(aq) + NO_3^-(aq) + K^+(aq) + Cl^-(aq)}_{\text{before reaction}} \longrightarrow AgCl(s) + \underbrace{K^+(aq) + NO_3^-(aq)}_{\text{after reaction}}$$

The K^+ and NO_3^- ions are present in solution before and after reaction and so appear on both the reactant and product sides of the balanced chemical equation. Such ions are often called **spectator ions** because they do not participate in the net reaction; they only "look on" from the sidelines. Little chemical information is lost if the equation is written without them, and so we can simplify the equation to

$$Ag^+(aq) + Cl^-(aq) \longrightarrow AgCl(s)$$

The balanced equation that results from leaving out the spectator ions is called a **net ionic equation** [CD-ROM, Screen 5.5]. *Only the aqueous ions and nonelectrolytes* (which can be insoluble compounds, soluble molecular compounds such as sugar, or gases) *that participate in a chemical reaction need to be included in the net ionic equation.*

Leaving out the spectator ions does not imply that K^+ and NO_3^- ions are totally unimportant in the $AgNO_3$ + KCl reaction. Indeed, Ag^+ and Cl^- ions cannot exist alone in solution; a negative ion of some kind must be present to balance the positive ion charge of Ag^+, for example. Any anion will do, however, as long as it forms a water-soluble compound with Ag^+. Thus, we could have used ClO_4^- and combined $AgClO_4$ and KCl. The net ionic equation would have been the same.

Finally, notice that there must always be a *charge balance* as well as a mass balance in a balanced chemical equation. Thus, in the Ag^+ + Cl^- net ionic equation, the cation and anion charges on the left add together to give a net charge of 0, the same as the 0 charge on AgCl(s) on the right.

● **Net Ionic Equations**
See Problem-Solving Tip 5.1, page 167.

Example 5.3 | Writing and Balancing Net Ionic Equations

Problem • Write a balanced, net ionic equation for the reaction of aqueous solutions of $BaCl_2$ and Na_2SO_4 to give $BaSO_4$ and NaCl.

Strategy • First write a balanced equation for the overall reaction. Next, decide what compounds are soluble in water and determine the ions these compounds produce in solution. Finally, eliminate ions that appear on both the reactant and product sides of the equation.

Solution •
Step 1. Write the balanced equation

$$BaCl_2 + Na_2SO_4 \longrightarrow BaSO_4 + 2\,NaCl$$

Step 2. Decide on the solubility of each compound using Figure 5.1. Compounds containing sodium ions are always water-soluble, and those containing chloride ions are almost always soluble. Sulfate salts are also usually soluble, one important exception being $BaSO_4$. We can therefore write

$$BaCl_2(aq) + Na_2SO_4(aq) \longrightarrow BaSO_4(s) + 2\,NaCl(aq)$$

Precipitation reaction. The reaction of barium chloride and sodium sulfate produces insoluble barium sulfate and water-soluble sodium chloride. *(Charles D. Winters)*

(Example continues on p. 156)

Step 3. Identify the ions in solution. All soluble ionic compounds dissociate to form ions in aqueous solution. (All are electrolytes.)

$$BaCl_2(s) \longrightarrow Ba^{2+}(aq) + 2\,Cl^-(aq)$$
$$Na_2SO_4(s) \longrightarrow 2\,Na^+(aq) + SO_4{}^{2-}(aq)$$
$$NaCl(s) \longrightarrow Na^+(aq) + Cl^-(aq)$$

This results in the following ionic equation:

$$Ba^{2+}(aq) + 2\,Cl^-(aq) + 2\,Na^+(aq) + SO_4{}^{2-}(aq)$$
$$\longrightarrow BaSO_4(s) + 2\,Na^+(aq) + 2\,Cl^-(aq)$$

Step 4. Identify and eliminate the spectator ions (Na^+ and Cl^-) to give the net ionic equation.

$$Ba^{2+}(aq) + SO_4{}^{2-}(aq) \longrightarrow BaSO_4(s)$$

Notice that the sum of ion charges is the same on both sides of the equation. On the left, $+2$ and -2 give zero; on the right the charge on $BaSO_4$ is also zero.

Comment • The steps followed in this Example represent a general approach to writing net ionic equations. See *Problem-Solving Tip* 5.1 on page 167 that describes the application of this approach to other kinds of chemical reactions.

Exercise 5.4   **Net Ionic Equations**

Write balanced net ionic equations for each of the following reactions:

(a) $AlCl_3 + Na_3PO_4 \longrightarrow AlPO_4 + NaCl$ (not balanced)

(b) Iron(III) chloride is mixed with potassium hydroxide to give iron(III) hydroxide and potassium chloride. See Figure 5.4c.

(c) Solutions of lead(II) nitrate and potassium chloride are mixed to give lead(II) chloride and potassium nitrate.

Chapter Goals • Revisited

- Understand the nature of ionic substances dissolved in water.
- Recognize and write equations for the common types of reactions in aqueous solution.
- **Recognize acids and bases and oxidizing and reducing agents.**
- Use correct units and stoichiometry principles for substances and reactions in solution.

5.3   ACIDS AND BASES

Acids and bases, two important classes of compounds, have some properties in common. Solutions of acids or bases, for example, can change the colors of vegetable pigments (Figure 5.5). You may have seen acids change the color of litmus, a dye derived from certain lichens, from blue to red. If an acid has made blue litmus paper turn red, adding a base reverses the effect, making the litmus blue again. Thus, acids and bases seem to be opposites. A base can neutralize the effect of an acid, and an acid can neutralize the effect of a base.

Acids

Acids have characteristic properties. They produce bubbles of CO_2 gas when added to limestone and react with many metals, producing hydrogen gas (H_2) at the same time (see Figure 5.5b and c). Although tasting substances is *never* done in a chemistry laboratory, you have also probably experienced the sour taste of acids such as citric acid, commonly found in fruits and added to candies and soft drinks.

The properties of acids can be interpreted in terms of a common feature of all acid molecules [CD-ROM, Screen 5.8].

An acid is a substance that, when dissolved in water, increases the concentration of hydrogen ions, $H^+(aq)$, in the water.

One of the most common acids is hydrochloric acid, which ionizes in water to form a hydrogen ion, $H^+(aq)$, and a chloride ion, $Cl^-(aq)$.

(a) The juice of a red cabbage is normally blue-purple. On adding acid, the juice becomes more red. Adding base produces a yellow color.

(b) A piece of coral (mostly $CaCO_3$) dissolves in acid to give CO_2 gas.

(c) Zinc reacts with hydrochloric acid to produce zinc chloride and hydrogen gas.

Figure 5.5   Some properties of acids and bases. (a) The colors of natural dyes, such as the juice from a red cabbage, are affected by acids and bases. **(b)** Acids react readily with limestone ($CaCO_3$) and other metal carbonates to produce gaseous CO_2 and a salt. **(c)** Acids react with many metals to produce hydrogen gas and a metal salt [🖱 CD-ROM, Screens 5.8–5.10]. *(Charles D. Winters)*

$$HCl(aq) \longrightarrow H^+(aq) + Cl^-(aq)$$

hydrochloric acid
strong electrolyte
= 100% ionized

Because it is completely converted to ions in aqueous solution, HCl is a **strong acid** (and a strong electrolyte). See Table 5.2 for other common acids.

Many acids, such as sulfuric acid, can provide more than 1 mol of H^+ per mole of acid.

Strong Acid: $\qquad H_2SO_4(aq) \longrightarrow \quad H^+(aq) \ + HSO_4^-(aq)$

$\qquad\qquad\qquad$ sulfuric acid $\qquad\qquad$ hydrogen ion $\quad$ hydrogen
$\qquad\qquad\qquad$ 100% ionized $\qquad\qquad\qquad\qquad\qquad$ sulfate ion

Weak Acid: $\qquad HSO_4^-(aq) \rightleftharpoons \quad H^+(aq) \ + SO_4^{2-}(aq)$

$\qquad\qquad\qquad$ hydrogen sulfate ion $\quad$ hydrogen ion $\quad$ sulfate ion
$\qquad\qquad\qquad$ <100% ionized

The first ionization reaction is essentially complete, so sulfuric acid is considered a strong acid (and, thus, a strong electrolyte). The hydrogen sulfate ion (HSO_4^-), like acetic acid, is only partially ionized in aqueous solution. Both the hydrogen sulfate ion and acetic acid are therefore classified as **weak acids.**

Bases

The hydroxide ion is a characteristic of bases so we can immediately recognize metal hydroxides as bases from their formulas. Although most metal hydroxides are insoluble (Figure 5.2), a few dissolve in water, which leads to an increase in the concentration of OH^- ions in solution [🖱 CD-ROM, Screen 5.9].

● **Weak Acids**
How do you know whether an acid or base is strong or weak? The common strong acids and bases are those listed in Table 5.2. There are many additional weak acids and bases, however, and many are natural substances. Oxalic and acetic acid are among them. Notice that these both contain CO_2H groups. (The H of this group is lost as H^+.) This structural feature is characteristic of organic acids. (See Chapter 11.)

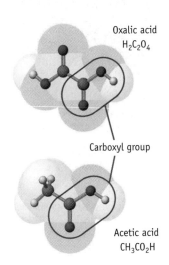

Oxalic acid
$H_2C_2O_4$

Carboxyl group

Acetic acid
CH_3CO_2H

Table 5.2 • Common Acids and Bases

Strong Acids (Strong Electrolytes)		Strong Bases (Strong Electrolytes)	
HCl	Hydrochloric acid	LiOH	Lithium hydroxide
HBr	Hydrobromic acid	NaOH	Sodium hydroxide
HI	Hydroiodic acid	KOH	Potassium hydroxide
HNO_3	Nitric acid		
$HClO_4$	Perchloric acid		
H_2SO_4	Sulfuric acid		

Weak Acids (Weak Electrolytes)*		Weak Base (Weak Electrolyte)	
H_3PO_4	Phosphoric acid	NH_3	Ammonia
H_2CO_3	Carbonic acid		
CH_3CO_2H	Acetic acid		
$H_2C_2O_4$	Oxalic acid		
$C_4H_6O_6$	Tartaric acid		
$C_6H_8O_7$	Citric acid		
$C_9H_8O_4$	Aspirin		

*These are representative of hundreds of weak acids.

A base is a substance that, when dissolved in water increases the concentration of hydroxide ion, OH^-, in the water.

These compounds are strong electrolytes and are called **strong bases.** Sodium and potassium hydroxide are common examples of strong bases.

$$NaOH(s) \longrightarrow Na^+(aq) + OH^-(aq)$$

sodium hydroxide, soluble   hydroxide ion
base, strong electrolyte
= 100% dissociated

Ammonia, another common base, does not have an OH^- ion as part of its formula. Instead it produces the OH^- ions on reacting with water. Only a small concentration of ions is present, so ammonia is a weak electrolyte and a weak base.

$$NH_3(aq) + H_2O(\ell) \rightleftharpoons NH_4^+(aq) + OH^-(aq)$$

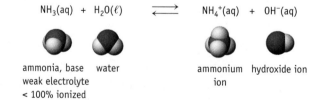

ammonia, base   water   ammonium   hydroxide ion
weak electrolyte   ion
< 100% ionized

Only a small concentration of ions is present, so ammonia is a **weak base** (and a weak electrolyte) (Figure 5.6).

Figure 5.6   Ammonia, NH_3, a weak electrolyte and weak base. Ammonia is a weak base because it interacts with water to produce a very small number of NH_4^+ and OH^- ions per mole of NH_3. Thus, the label on a bottle of commercial ammonia is misleading. It is best considered an aqueous solution of NH_3 with a small amount of NH_4^+ and OH^- ions. *(Charles D. Winters)*

Exercise 5.5   Acids and Bases

(a) What ions are produced when nitric acid dissolves in water?

(b) Barium hydroxide is moderately soluble in water. What ions are produced when it dissolves in water?

Sulfuric Acid — A Gauge of Business Conditions

Sulfuric acid is a colorless, syrupy liquid with a density of 1.84 g/mL and a boiling point of 270 °C. It has several desirable properties that have led to its widespread use: it is generally less expensive to produce than other acids, is a strong acid, can be handled in steel containers, reacts readily with many organic compounds to produce useful products, and reacts readily with lime (CaO), the least expensive and most readily available base.

The first step in the industrial preparation of sulfuric acid is combustion of sulfur in air to give sulfur dioxide.

$$S_8(s) + 8\ O_2(g) \longrightarrow 8\ SO_2(g)$$

This gas is then combined with more oxygen, in the presence of a catalyst, to give sulfur trioxide,

$$2\ SO_2(g) + O_2(g) \longrightarrow 2\ SO_3(g)$$

which can give sulfuric acid when absorbed in water.

$$SO_3(g) + H_2O(\ell) \longrightarrow H_2SO_4(aq)$$

For some years sulfuric acid has been the chemical produced in the largest quantity in the United States and in many other industrialized countries. About 40–50 billion kg (40–50 million metric tons) is made annually in the United States. Some economists have said the quantity

of sulfuric acid produced is a measure of a nation's industrial strength or general business conditions.

Currently, over two-thirds of the production is used in the phosphate fertilizer industry, which makes "superphosphate" fertilizer by treating phosphate rock with sulfuric acid.

$$Ca_{10}F_2(PO_4)_6(s) + 7\ H_2SO_4(aq) + 3\ H_2O(\ell) \longrightarrow$$

$$3\ Ca[H_2PO_4]_2 \cdot H_2O(s) + 7\ CaSO_4(s) + 2\ HF(g)$$

Sulfur is found in pure form in underground deposits along the coast of the United States in the Gulf of Mexico. It is recovered by pumping superheated steam into the sulfur beds to melt the sulfur. The molten sulfur is brought to the surface by means of compressed air. *(Farrel Grehan/Photo Researchers, Inc.)*

The remainder is used to make pigments, explosives, alcohol, pulp and paper, detergents, and as a component for storage batteries.

A sulfuric acid plant. *(Fleck Chemical, United Kingdom)*

Some products that depend on sulfuric acid for their manufacture. *(Charles D. Winters)*

Oxides of Nonmetals and Metals

Each acid shown in Table 5.2 has one or more H atoms in the molecular formula that can be released to form H^+ ions in water. There are, however, less obvious compounds that form acidic solutions [CD-ROM, Screen 5.9]. These are oxides of nonmetals, such as carbon dioxide and sulfur trioxide, which have no H atoms but which react with water to produce H^+ ions. Carbon dioxide, for example, dissolves in water to a small extent, and some of the dissolved molecules react with water to form the weak acid, carbonic acid. This acid then ionizes to a small extent to form the hydrogen ion, H^+, and the hydrogen carbonate (bicarbonate) ion, HCO_3^-,

$$CO_2(g) \quad + \quad H_2O(\ell) \quad \rightleftharpoons \quad H_2CO_3(aq)$$

A Closer Look

H⁺ Ions in Water

The H⁺ ion is a hydrogen atom that has lost its electron. Only the nucleus, a proton, remains. Because a proton is only about 1/100,000 as large as the average atom or ion, water molecules can approach closely, and the proton and the water molecules are strongly attracted. The H⁺ ion in water is better represented by the combination of H⁺ and H_2O, H_3O^+, an ion called the *hydronium ion*. Experiments also show, however, that other forms of the ion exist in water, one example being $[H_3O(H_2O)_3]^+$.

For simplicity we shall use H⁺(aq) in this text to indicate the presence of hydronium and similar ions. When discussing the functions of acids in detail, however, we shall use H_3O^+ [➡ CHAPTERS 17 and 18].

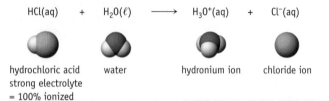

$$HCl(aq) \quad + \quad H_2O(\ell) \longrightarrow \quad H_3O^+(aq) \quad + \quad Cl^-(aq)$$

hydrochloric acid   water   hydronium ion   chloride ion
strong electrolyte
= 100% ionized

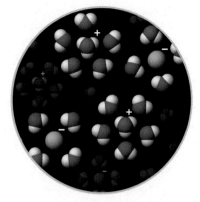

When an acid ionizes in water, it produces a hydronium ion, H_3O^+, which is surrounded by water molecules.

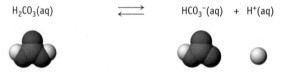

$$H_2CO_3(aq) \quad \rightleftharpoons \quad HCO_3^-(aq) \quad + \quad H^+(aq)$$

Like the HSO_4^- ion, the HCO_3^- ion is also an acid and ionizes to produce the carbonate ion, CO_3^{2-}.

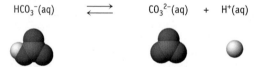

$$HCO_3^-(aq) \quad \rightleftharpoons \quad CO_3^{2-}(aq) \quad + \quad H^+(aq)$$

These reactions are important in our environment and in humans. Carbon dioxide is normally found in small amounts in the atmosphere, so rainwater is always slightly acidic. In the human body, carbon dioxide is dissolved in body fluids where the HCO_3^- and CO_3^{2-} ions perform an important "buffering" action [➡ CHAPTER 18].

Oxides like CO_2 that can react with water to produce H⁺ ions are known as **acidic oxides.** Other acidic oxides include those of sulfur and nitrogen, which can be present in significant amounts in polluted air and can lead ultimately to acids and other pollutants. For example, sulfur dioxide, SO_2, from human and natural sources can react with oxygen to give sulfur trioxide, SO_3, which then forms sulfuric acid with water.

$$2\ SO_2(g) + O_2(g) \longrightarrow 2\ SO_3(g)$$

$$SO_3(g) + H_2O(\ell) \longrightarrow H_2SO_4(aq)$$

Nitrogen dioxide, NO_2, reacts with water to give nitric and nitrous acids.

$$2\ NO_2(g) + H_2O(\ell) \longrightarrow HNO_3(aq) + HNO_2(aq)$$

nitric acid   nitrous acid

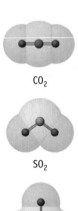

CO_2

SO_2

SO_3

NO_2

Some common nonmetal oxides that form acids in water.

These reactions are the origin of acid rain from the burning of fossil fuels such as coal and gasoline in the United States, Canada, and other industrialized countries. The gaseous oxides mix with water and other chemicals in the troposphere, and the rain that falls is more acidic than if it contained only dissolved CO_2. When the rain falls on areas that cannot easily tolerate this greater than normal acidity, such as the northeastern parts of the United States and the eastern provinces of Canada, serious environmental problems can occur.

Oxides of metals can give basic solutions if they dissolve appreciably in water. Perhaps the best known example is calcium oxide, CaO, often called *lime,* or *quicklime.* This metal oxide reacts with water to give calcium hydroxide, commonly called *slaked lime.* The latter compound, although only slightly soluble in water (0.17 g/100 g H_2O at 10 °C), is widely used in industry as a base because it is inexpensive.

$$CaO(s) + H_2O(\ell) \longrightarrow Ca(OH)_2(s)$$
$$\text{lime} \qquad\qquad\qquad\qquad \text{slaked lime}$$

Oxides like CaO that react with water to produce OH^- ions are known as **basic oxides.** Almost 20 billion kg of lime is produced annually in the United States for use in the metals and construction industries, in sewage and pollution control, in water treatment, and in agriculture.

Chemical Perspectives

Limelight and Metal Oxides

In the 1820s, Lt. Thomas Drummond (1797–1840) of the Royal Engineers was involved in a survey of Great Britain. During the winters he attended the famous public chemistry lectures and demonstrations by the great chemist Michael Faraday at the Royal Institution in London. There he apparently heard about the bright light that is emitted when a piece of lime, CaO, is heated to a high temperature. It occurred to him that it could be used to make distant surveying stations visible, especially at night. Soon he developed an apparatus in which a ball of lime was heated by an alcohol flame in a stream of oxygen gas. It was reported at the time that the light from a "ball of lime not larger than a boy's marble" could be seen at a distance of 70 miles! Such lights were soon adapted to lighthouses and became known as Drummond lights.

Many inventions are soon adapted to warfare and that was the case with limelights. They were used to illuminate targets in the battle of Charleston, SC, during the United States Civil War in the 1860s. The public came to know about

CaO dissolves only slightly in water, but when it does it gives a basic solution. A dye, phenolphthalein, which turns red in the presence of a base, has been added. *(Charles D. Winters)*

limelight, though, when it moved into theaters. Gaslights were used in the early 1800s to illuminate the stage, but they were clearly not adequate. Soon after Drummond's invention, though, actors trod the boards "in the limelight."

Metal oxides such as CaO and ThO_2 [thorium(IV) oxide] emit a brilliant white light when heated to incandescence. *(Charles D. Winters)*

Exercise 5.6 Acidic and Basic Oxides

For each of the following, indicate whether you expect an acidic or basic solution when the compound dissolves in water. Remember that compounds based on elements in the same group usually behave similarly.

(a) SeO_2 (b) BaO (c) P_4O_{10}

Chapter Goals • Revisited

• Understand the nature of ionic substances dissolved in water.

• **Recognize and write equations for the common types of reactions in aqueous solution.**

• Recognize acids and bases and oxidizing and reducing agents.

• Define and use molarity in solution stoichiometry.

5.4   REACTIONS OF ACIDS AND BASES

In general, acids react with strong bases in aqueous solution to produce a salt and water [⊕ CD-ROM, Screen 5.10]. For example (Figure 5.7),

$$HCl(aq) \quad + \quad NaOH(aq) \quad \longrightarrow \quad H_2O(\ell) + \quad NaCl(aq)$$
hydrochloric acid   sodium hydroxide   water   sodium chloride

The word "salt" has come into the language of chemistry to describe any ionic compound whose cation comes from a base (here Na^+ from NaOH) and whose anion comes from an acid (here Cl^- from HCl). Reaction of any of the acids listed in Table 5.1 with any of the strong bases listed there produces a salt and water.

Hydrochloric acid and sodium hydroxide are strong electrolytes in water (see Table 5.1), so the complete ionic equation for the reaction of HCl(aq) and NaOH(aq) should be written as

$$\underbrace{H^+(aq) + Cl^-(aq)}_{\text{from HCl(aq)}} + \underbrace{Na^+(aq) + OH^-(aq)}_{\text{from NaOH(aq)}} \longrightarrow \underset{\text{water}}{H_2O(\ell)} + \underbrace{Na^+(aq) + Cl^-(aq)}_{\text{from salt}}$$

Because Na^+ and Cl^- ions appear on both sides of the equation, the *net ionic equation* is just the combination of the ions H^+ and OH^- to give water.

$$H^+(aq) + OH^-(aq) \longrightarrow H_2O(\ell)$$

The net ionic equation for the reaction between any strong acid and any strong base.

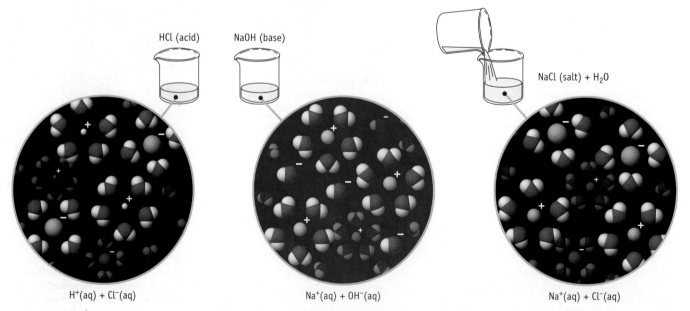

HCl (acid)   NaOH (base)   NaCl (salt) + H_2O

$H^+(aq) + Cl^-(aq)$ $Na^+(aq) + OH^-(aq)$ $Na^+(aq) + Cl^-(aq)$

Figure 5.7   An acid–base reaction: HCl and NaOH. The acid and base consist of ions in solution. On mixing the H^+ and OH^- ions combine to produce H_2O. The ions Na^+ and Cl^- remain in solution [⊕ CD-ROM, Screen 5.10].

Reactions between *strong acids* and *strong bases* are called **neutralization reactions** because, on completion of the reaction, the solution is neutral; it is neither acidic nor basic. The other ions (the cation of the base and the anion of the acid) remain unchanged. If the water is evaporated, however, the cation and anion form a solid salt. In the preceding example, NaCl can be obtained, whereas nitric acid, HNO_3, and NaOH give the salt sodium nitrate, $NaNO_3$.

$$HNO_3(aq) + NaOH(aq) \longrightarrow H_2O(\ell) + NaNO_3(aq)$$

Calcium oxide (lime) is inexpensive and is used in waste and pollution control. Indeed, one of its major uses is in "scrubbing" sulfur oxides from the exhaust gases of power plants fueled by coal and oil. The oxides of sulfur dissolve in water to produce acids (page 160), and these acids can react with a base. Lime produces the base calcium hydroxide when added to water. A water suspension of lime is sprayed into the exhaust stack of the power plant, where it reacts with sulfur-containing acids. One such reaction is

$$Ca(OH)_2(s) + H_2SO_4(aq) \longrightarrow CaSO_4 \cdot 2\ H_2O(s)$$

The compound $CaSO_4 \cdot 2\ H_2O$, hydrated calcium sulfate, is found in the earth as the mineral gypsum. Assuming the gypsum from a coal-burning power plant is not contaminated with compounds that contain "heavy" metals (e.g., Hg) or other pollutants, it is environmentally acceptable to put it into the earth.

Acetic acid, CH_3CO_2H, is an important compound. You know it as the substance that gives the taste and odor to vinegar. Fermentation of carbohydrates such as sugar produces ethanol (CH_3CH_2OH), and the action of bacteria on this alcohol results in acetic acid. Even a trace of the acid will ruin the taste of wine. This is the source of the name vinegar, which comes from the French phrase *vin egar* meaning "sour wine." In addition to its use in food products such as salad dressings, mayonnaise, pickles, and hair-coloring products, acetic acid is used in the preparation of cellulose acetate, a commonly used synthetic fiber.

Acetic acid is a weak acid. Only a few acetic acid molecules are ionized to form H^+ and $CH_3CO_2^-$ ions in water (page 150).

$$CH_3CO_2H(aq) \rightleftharpoons H^+(aq) + CH_3CO_2^-(aq)$$

Nonetheless, like all acids, acetic acid will react with metal carbonates such as calcium carbonate. This carbonate is a common residue from hard water in home heating systems and cooking utensils, so washing with vinegar is a good way to clean the system or utensils because the insoluble calcium carbonate is turned into water-soluble calcium acetate.

$$2\ CH_3CO_2H(aq) + CaCO_3(s) \longrightarrow Ca(CH_3CO_2)_2(aq) + H_2O(\ell) + CO_2(g)$$

What is the net ionic equation for this reaction? Acetic acid is a weak acid and so produces only a trace of ions in solution. Calcium carbonate is insoluble in water. Therefore, the two reactants are simply $CH_3CO_2H(aq)$ and $CaCO_3(s)$. The product, calcium acetate, is water-soluble and forms calcium and acetate ions.

$$2\ CH_3CO_2H(aq) + CaCO_3(s) \longrightarrow Ca^{2+}(aq) + 2\ CH_3CO_2^-(aq) + H_2O(\ell) + CO_2(g)$$

There are no spectator ions in this reaction.

Dissolving limestone (calcium carbonate, $CaCO_3$) in vinegar. This is an illustration of the use of vinegar as a household cleaning agent. It can be used, for example, to clean the filter in an electric coffee maker. *(Charles D. Winters)*

Example 5.4 | **Net Equation for an Acid–Base Reaction**

Problem • Ammonia, NH_3, is one of the most important chemicals in industrial economies. Not only is it used directly as a fertilizer but it is the raw material for the manufacture of nitric acid. As a base, it reacts with acids such as hydrochloric acid. Write a balanced, net ionic equation for this reaction.

Strategy • First, write a complete balanced equation for the reaction. Second, indicate whether each reactant or product is a solid, liquid, gas, or soluble in water (aq). Third, write each water-soluble species or any strong acids and bases as the ions they produce in water. *Insoluble solids and weak acids and bases are not written as ions.* Finally, eliminate any spectator ions to give the net ionic equation.

Solution • The complete balanced equation is

$$NH_3(aq) + \underset{\text{hydrochloric acid}}{HCl(aq)} \longrightarrow \underset{\text{ammonium chloride}}{NH_4Cl(aq)}$$
$$\underset{\text{ammonia}}{}$$

Notice that the reaction produces a salt, NH_4Cl, but not water. An H^+ ion from the acid transfers directly to the ammonia to give the ammonium ion. To write the net ionic equation start with the fact that hydrochloric acid is a strong acid and produces H^+ and Cl^- ions.

$$HCl(aq) \longrightarrow H^+(aq) + Cl^-(aq)$$

Ammonia, a weak base, produces NH_4^+ and OH^- ions in water (page 158).

$$NH_3(aq) + H_2O(\ell) \rightleftharpoons NH_4^+(aq) + OH^-(aq)$$

The OH^- ions produced by NH_3 react with the H^+ ions produced by HCl to form water.

$$OH^-(aq) + H^+(aq) + Cl^-(aq) \longrightarrow H_2O(\ell) + Cl^-(aq)$$

Combining this equation with the equation for ammonia ionization gives the balanced, net ionic equation for the reaction of the weak base ammonia with a strong acid such as HCl(aq).

$$NH_3(aq) + H^+(aq) \longrightarrow NH_4^+(aq)$$

Comment • The net ionic equation shows that the important aspect of the reaction between the weak base ammonia and the strong acid HCl is the transfer of an H^+ ion from the acid to the NH_3. This implies that any strong acid could be used here (HBr, HNO_3, $HClO_4$, H_2SO_4) and the net ionic equation would be the same.

Exercise 5.7 | **Acid–Base Reactions**

Write the balanced, overall equation and the net ionic equation for the reaction of magnesium hydroxide with hydrochloric acid.

Figure 5.8 Biscuits rise because of a gas-forming reaction. The acid and sodium bicarbonate in baking powder produce carbon dioxide gas. The acid used in many baking powders is $CaHPO_4$, but $NaAl(SO_4)_2$ is also common. (The aluminum-containing compound forms an acidic solution when placed in water. See Chapter 17.) *(Charles D. Winters)*

5.5 GAS-FORMING REACTIONS

Have you ever made biscuits or muffins? As you bake the dough, it rises in the oven (Figure 5.8). But what makes it rise? A gas-forming reaction occurs between an acid of some type and baking soda, sodium hydrogen carbonate (bicarbonate of soda, $NaHCO_3$). One possible acid used is tartaric acid, a weak acid found in many foods. The net ionic equation for a typical reaction would be

$$\underset{\text{tartaric acid}}{C_4H_6O_6(aq)} + \underset{\text{hydrogen carbonate ion}}{HCO_3^-(aq)} \longrightarrow \underset{\text{tartrate ion}}{C_4H_5O_6^-(aq)} + H_2O(\ell) + CO_2(g)$$

In dry baking powder, the acid and $NaHCO_3$ are kept apart by using starch as a filler. When mixed into the moist batter, however, the acid and the hydrogen carbonate ion react to produce CO_2, and the dough rises.

Several types of chemical reactions lead to gas formation (Table 5.3 [CD-ROM, Screen 5.11]), but the most common leads to CO_2 formation. All metal carbonates (and bicarbonates) react with acids to produce carbonic acid, H_2CO_3, which in turn decomposes to carbon dioxide and water (see Figure 5.5).

$$CaCO_3(s) + 2\ HCl(aq) \longrightarrow CaCl_2(aq) + H_2CO_3(aq)$$
$$H_2CO_3(aq) \rightleftharpoons H_2O(\ell) + CO_2(g)$$

Table 5.3 • Gas-Forming Reactions

Metal carbonate or bicarbonate + acid ⟶ metal salt + CO₂(g) + H₂O(ℓ)

$Na_2CO_3(aq) + 2\ HCl(aq) \longrightarrow 2\ NaCl(aq) + CO_2(g) + H_2O(\ell)$

Metal sulfide + acid ⟶ metal salt + H₂S(g)

$Na_2S(aq) + 2\ HCl(aq) \longrightarrow 2\ NaCl(aq) + H_2S(g)$

Metal sulfite + acid ⟶ metal salt + SO₂(g) + H₂O(ℓ)

$Na_2SO_3(aq) + 2\ HCl(aq) \longrightarrow 2\ NaCl(aq) + SO_2(g) + H_2O(\ell)$

Ammonium salt + strong base ⟶ metal salt + NH₃(g) + H₂O(ℓ)

$NH_4Cl(aq) + NaOH(aq) \longrightarrow NaCl(aq) + NH_3(g) + H_2O(\ell)$

• **Gas-Forming Reactions**

All metal carbonates such as $CaCO_3$ react with acids to produce a salt and CO_2 gas. See Figure 5.5 and [🖰 CD-ROM Screen 5.11].

The carbonic acid produced in such reactions is unstable and is rapidly converted to water and CO_2 gas. If the reaction is done in an open beaker, most of the gas bubbles out of the solution.

Example 5.5   Gas-Forming Reactions

Problem • Write a balanced equation for the reaction that occurs when nickel(II) carbonate is treated with sulfuric acid.

Strategy • First identify the reactants and write their formulas (here $NiCO_3$ and H_2SO_4). Next, recognize this as a typical gas-forming reaction (see Table 5.2) between a metal carbonate (or metal hydrogen carbonate) and an acid. According to Table 5.2,

the products are water and CO_2, as well as a metal salt. The anion of the metal salt is the anion from the acid (SO_4^{2-}), and the cation is from the metal carbonate (Ni^{2+}).

Solution • The complete, balanced equation is

$$NiCO_3(s) + H_2SO_4(aq) \longrightarrow NiSO_4(aq) + H_2O(\ell) + CO_2(g)$$

Exercise 5.8   A Gas-Forming Reaction

(a) Barium carbonate, $BaCO_3$, is widely used in the brick, ceramic, glass, and chemical manufacturing industries. Write a balanced equation that shows what happens when barium carbonate is treated with nitric acid. Give the name of each of the reaction products.

(b) Write a balanced equation for the reaction of ammonium sulfate with sodium hydroxide.

5.6   CLASSIFYING REACTIONS IN AQUEOUS SOLUTION

One goal of this chapter is to explore the most common types of reactions that can occur in aqueous solution [🖰 CD-ROM, Screen 5.6]. This helps you decide, for example, that a gas-forming reaction occurs when an Alka-Seltzer tablet (containing citric acid and $NaHCO_3$) is dropped into water.

$$\underset{\text{citric acid}}{C_6H_8O_7(aq)} + \underset{\text{hydrogen carbonate ion}}{HCO_3^{-}(aq)} \longrightarrow \underset{\text{dihydrogen citrate ion}}{C_6H_7O_7^{-}(aq)} + H_2O(\ell) + CO_2(g)$$

Reactions in aqueous solution are important not only because they provide a way to make useful products, but also because these kinds of reactions occur on the earth and in plants and animals. Therefore, it is useful to look for common

An Alka-Seltzer tablet contains citric acid and sodium hydrogen carbonate ($NaHCO_3$), the reactants in a gas-forming reaction. *(Charles D. Winters)*

reaction patterns to see what their "driving forces" might be. How can you know when you mix two chemicals in water that they will combine to produce one or more new compounds?

To organize what you have learned about reactions in aqueous solution, recognize that a common theme ties together all the reactions we have described thus far: in each case *the ions of the reactants changed partners.*

$$A^+B^- \; + \; C^+D^- \longrightarrow \; A^+D^- \; + \; C^+B^-$$

Thus, we refer to them as **exchange reactions,** which gives us a good way to predict the products of precipitation, acid–base, and gas-forming reactions.

Precipitation Reactions   (see Figure 5.4): Ions combine in solution to form an insoluble reaction product.

Overall Equation

$$Pb(NO_3)_2(aq) + 2\ KI(aq) \longrightarrow PbI_2(s) + 2\ KNO_3(aq)$$

Net Ionic Equation

$$Pb^{2+}(aq) + 2\ I^-(aq) \longrightarrow PbI_2(s)$$

Acid–Base Reactions   (see Figure 5.7): The cation of the base and the anion of the acid form a salt. For strong acids and bases, water is also a product.

Overall Equation

$$HNO_3(aq) + KOH(aq) \longrightarrow H_2O(\ell) + KNO_3(aq)$$

Net Ionic Equation for the Reaction of a Strong Acid and Strong Base

$$H^+(aq) + OH^-(aq) \longrightarrow H_2O(\ell)$$

Chapter Goals • Revisited

• Understand the nature of ionic substances dissolved in water.

• **Recognize and write equations for the common types of reactions in aqueous solution.**

• Recognize acids and bases and oxidizing and reducing agents.

• Define and use molarity in solution stoichiometry.

Gas-Forming Reactions   (see Figure 5.8): The most common examples involve metal carbonates and acids (but others exist; see Table 5.3). One product is carbonic acid, H_2CO_3, most of which decomposes to H_2O and CO_2. Carbon dioxide is the gas in the bubbles you see during this reaction.

Overall Equation

$$CuCO_3(s) + 2\ HNO_3(aq) \longrightarrow Cu(NO_3)_2(aq) + H_2CO_3(aq)$$
$$H_2CO_3(aq) \rightleftharpoons CO_2(g) + H_2O(\ell)$$

Net Ionic Equation

$$CuCO_3(s) + 2\ H^+(aq) \longrightarrow Cu^{2+}(aq) + CO_2(g) + H_2O(\ell)$$

Product- and Reactant-Favored Reactions

The driving force for a precipitation, acid–base, or gas-forming reaction is, in each case, the formation of a product that removes ions from solution: a solid precipitate, a water molecule, or a gas molecule. These, and all other reactions that occur from left to right as the chemical equation is written, and in which reactants are largely converted to products, are said to be **product-favored.**

Problem-Solving Tip 5.1

Writing Net Ionic Equations

Net ionic equations are commonly written for chemical reactions in aqueous solution because they describe the actual chemical species involved in a reaction. To write net ionic equations we must know what compounds exist as ions in solution.

1. Strong acids, soluble strong bases, and soluble salts exist as ions in solution. Examples include the acids HCl and HNO_3, a base such as NaOH, and salts such as NaCl and $MgCl_2$ (see Chapter Focus 5.6 and Table 5.1).

2. All other species should be represented by their complete formulas. Weak acids exist in solution primarily as molecules. Even though their structures are made up of ions, insoluble salts such as $CaCO_3(s)$ or insoluble bases such as $Mg(OH)_2(s)$ should not be written in ionic form.

   The best way to approach writing net ionic equations is to follow precisely a set of steps.

1. Write a complete, balanced equation. Indicate the state of each substance (aq, s, ℓ, g).

2. Next rewrite the whole equation, writing all strong acids, strong soluble bases, and soluble salts as ions. (Look carefully at species labeled with an "(aq)" suffix. Select from those all strong acids, soluble salts, and soluble bases, and rewrite them as ions.)

3. Finally, it is likely that some of the ions remain unchanged in the reaction (the ions that appear in the equation both as reactants or products). These "spectator ions" are not part of the chemistry that is going on. You can cancel them from each side of the equation.

   Here are three general net ionic equations it is helpful to remember:

   • The net ionic equation for the reaction between any strong acid and any soluble strong base is $H^+(aq) + OH^-(aq) \longrightarrow H_2O(\ell)$.
   • The net ionic equation for the reaction of any weak acid HX (such as HCN, HF, HOCl, CH_3CO_2H) and a soluble strong base is $HX + OH^-(aq) \longrightarrow H_2O(\ell) + X^-(aq)$.
   • The net ionic equation for the reaction of ammonia with any weak acid HX is $NH_3(aq) + HX(aq) \longrightarrow NH_4^+(aq) + X^-(aq)$ and with a strong acid it is $NH_3(aq) + H^+(aq) \longrightarrow NH_4^+(aq)$.

   Like molecular equations, net ionic equations must be balanced. The same number of atoms appears on each side of the arrow. But there is an additional requirement: The sum of the ion charges on the two sides must also be equal.

The opposite of a product-favored reaction is one that is **reactant-favored.** Such reactions lead to the conversion of little if any of the reactants to products. An example would be the formation of hydrochloric acid and sodium hydroxide in a solution of sodium chloride in water.

Reactant-favored: $NaCl(aq) + H_2O(\ell) \longrightarrow NaOH(aq) + HCl(aq)$

This reaction, which does *not* occur to any measurable extent, is of course the opposite of an acid–base reaction.

The title of this book is *Chemistry & Chemical Reactivity*. One aspect of chemical reactivity, and a goal of this book, is to be able to predict whether a chemical reaction is product- or reactant-favored. Thus far you have learned that certain common reactions are generally product-favored, and we shall return to this discussion many more times in this book.

Summary of Reaction Driving Forces

Three common "driving forces" responsible for reactions in aqueous solution were outlined earlier. A fourth, to be discussed in the next section (see Section 5.7), is the transfer of electrons from one substance to another rather than the exchange of partners. Such reactions are called oxidation–reduction processes.

Reaction Type	Driving Force
Precipitation	Formation of an insoluble compound (Section 5.2)
Acid–base; neutralization	Formation of a salt and water; proton transfer (Section 5.3)
Gas-forming	Evolution of a water-insoluble gas such as CO_2 (Section 5.5)
Oxidation–reduction	Electron transfer (Section 5.7)

These four types of reactions are usually easy to recognize, but keep in mind that a reaction may have more than one driving force. For example, barium hydroxide reacts readily with sulfuric acid to give barium sulfate and water, a reaction that is both a precipitation and an acid–base reaction.

$$Ba(OH)_2(aq) + H_2SO_4(aq) \longrightarrow BaSO_4(s) + 2\ H_2O(\ell)$$

Exercise 5.9 **Classifying Reactions**

Classify each of the following reactions as a precipitation, acid–base reaction, or gas-forming reaction. Predict the products of the reaction, and then balance the completed equation. Write the net ionic equation for each.

(a) $CuCO_3(s) + H_2SO_4(aq) \longrightarrow$

(b) $Ba(OH)_2(s) + HNO_3(aq) \longrightarrow$

(c) $CuCl_2(aq) + (NH_4)_2S(aq) \longrightarrow$

5.7   OXIDATION–REDUCTION REACTIONS

The terms "oxidation" and "reduction" come from reactions that have been known for centuries. Ancient civilizations learned how to change metal oxides and sulfides to the metal, that is, how to "reduce" ore to the metal. A modern example is the reduction of iron(III) oxide with carbon monoxide to give iron metal (Figure 5.9).

Fe$_2$O$_3$ loses oxygen and is reduced

$$Fe_2O_3(s) + 3\ CO(g) \longrightarrow 2\ Fe(s) + 3\ CO_2(g)$$

CO is the reducing agent. It gains oxygen and is oxidized.

In this reaction carbon monoxide is the agent that brings about the reduction of iron ore to iron metal, so carbon monoxide is called the **reducing agent.**

When Fe_2O_3 is reduced by carbon monoxide, oxygen is removed from the iron ore and added to the carbon monoxide, which is "oxidized" by the addition of oxygen to give carbon dioxide. *Any process in which oxygen is added to another substance is an oxidation.* In the reaction of oxygen with magnesium, for example (see Figure 5.9), oxygen is the **oxidizing agent** because it is the agent responsible for the oxidation.

Mg combines with oxygen and is oxidized.

$$2\ Mg(s) + O_2(g) \longrightarrow 2\ MgO(s)$$

O$_2$ is the oxidizing agent

Chapter Goals • Revisited

- Understand the nature of ionic substances dissolved in water.
- Recognize and write equations for the common types of reactions in aqueous solution.
- **Recognize acids and bases and oxidizing and reducing agents.**
- Define and use molarity in solution stoichiometry.

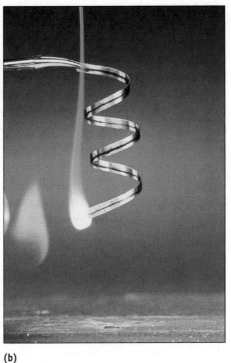

(a)   (b)

Figure 5.9   Oxidation–Reduction.
(a) Iron ore, which is largely Fe_2O_3, is reduced to metallic iron with carbon or carbon monoxide in a blast furnace, a process
done on a massive scale. **(b)** The oxidation
of magnesium converts metallic magnesium
to magnesium oxide. *(a, Jan Halaska/Photo
Researchers, Inc.,b, Charles D. Winters)*

The experimental observations outlined here point to several important
conclusions:

- If one substance is oxidized, another substance in the same reaction must
  be reduced. For this reason, such reactions are often called oxidation–
  reduction reactions, or **redox reactions** for short.
- The reducing agent is itself oxidized, and the oxidizing agent is reduced.
- Oxidation is the opposite of reduction. For example, the removal of oxygen
  is reduction and the addition of oxygen is oxidation.

● **Redox Reactions**
See the Chapter Focus page in Chapter 4
(page 120) for other redox reactions.

Redox Reactions and Electron Transfer

Not all redox reactions involve oxygen, but *all oxidation and reduction reactions involve transfer of electrons between substances.* When a substance accepts electrons, it is
said to be **reduced** because there is a reduction in the charge on an atom of the
substance [CD-ROM, Screen 5.12]. In the net ionic equation shown here, positively
charged Ag^+ is reduced to uncharged $Ag(s)$ on accepting electrons from copper metal.

Ag^+ accepts electrons from Cu and is
reduced to Ag. Ag^+ is the oxidizing agent.

$$Ag^+(aq) + e^- \longrightarrow Ag(s)$$

$$2\,Ag^+(aq) + Cu(s) \longrightarrow 2\,Ag(s) + Cu^{2+}(aq)$$

Cu donates electrons to Ag^+ and is oxidized to Cu^{2+}.
Cu is the reducing agent.

$$Cu(s) \longrightarrow Cu^{2+}(aq) + 2\,e^-$$

Because copper metal supplies the electrons and causes the Ag^+ ion to be reduced,
Cu is the reducing agent (Figure 5.10).

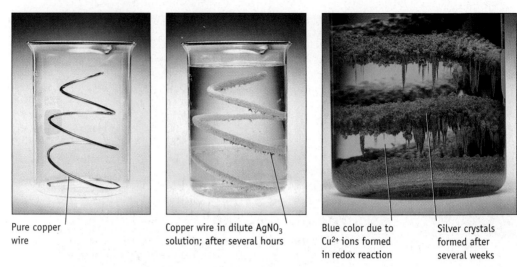

Pure copper wire

Copper wire in dilute $AgNO_3$ solution; after several hours

Blue color due to Cu^{2+} ions formed in redox reaction

Silver crystals formed after several weeks

Figure 5.10   The oxidation of copper metal by silver ions. A clean piece of copper wire is placed in a solution of silver nitrate, $AgNO_3$. With time, the copper reduces Ag^+ ions to silver metal crystals, and the copper metal is oxidized to copper ions, Cu^{2+}. The blue color of the solution is due to the presence of aqueous copper(II) ions [🖱 CD-ROM, Screen 5.12]. *(Charles D. Winters)*

When a substance *loses electrons*, the positive charge on an atom of the substance increases. The substance is said to have been **oxidized.** In our example, copper metal releases electrons on going to Cu^{2+}, so the copper has been oxidized. For this to happen, something must be available to take the electrons offered by the copper. In this case, Ag^+ is the electron acceptor, and its charge is reduced to zero in silver metal. Therefore, Ag^+ is the "agent" that causes Cu metal to be oxidized, so Ag^+ is the oxidizing agent. In every oxidation–reduction reaction, one reactant is reduced (and is therefore the oxidizing agent) and one reactant is oxidized (and is therefore the reducing agent). In summary,

• **Balancing Equations for Redox Reactions**
The notion that a redox reaction can be divided into an oxidizing portion and a reducing portion will lead us to a method of balancing more complex equations for redox reactions described in Chapter 20.

> The redox reaction $X + Y \longrightarrow X^{n+} + Y^{n-}$ can be divided into two parts:
> (a) $X \longrightarrow X^{n+} + n\,e^-$: X is **oxidized,** losing n electrons (to Y) to form X^{n+}. X is the **reducing agent** in the process.
> (b) $Y + n\,e^- \longrightarrow Y^{n-}$: Y is **reduced,** gaining n electrons (supplied by X) to form Y^{n-}. Y is the **oxidizing agent** in the process.

In the reaction of magnesium and oxygen (see Figure 5.9), oxygen is reduced because it gains electrons (four electrons per molecule) on going to the oxide ion. Thus, O_2 is the oxidizing agent.

Mg releases 2 e^- per atom. Mg is oxidized to Mg^{2+} and is the reducing agent.

$$2\ Mg(s) + O_2(g) \longrightarrow 2\ MgO(s)$$

O_2 gains 4 e^- per molecule to form O^{2-}. O_2 is reduced and is the oxidizing agent.

In the same reaction, magnesium is the reducing agent because it releases two electrons per atom on being oxidized to the Mg^{2+} ion. All redox reactions can be analyzed in a similar manner.

Oxidation Numbers

How can you tell an oxidation–reduction reaction when you see one? How can you tell which substance has gained or lost electrons and so decide which one is the oxidizing or reducing agent? Sometimes it is obvious. For example, if an uncombined element becomes part of a compound (Mg becomes part of MgO, for example), it's definitely a redox reaction. If it's not obvious, then the answer is to *look for a change in the oxidation number of an element in the course of the reaction.* The **oxidation number** of an atom in a molecule or ion is defined as the charge an atom has, or appears to have, as determined by some guidelines for assigning oxidation numbers [⟲ CD-ROM, Screen 5.13]. The guidelines for assigning oxidation numbers are

1. **Each atom in a pure element has an oxidation number of zero.** The oxidation number of Cu in metallic copper, as well as for each atom in I_2 or S_8, is 0.

2. **For monatomic ions, the oxidation number is equal to the charge on the ion.** Elements of Periodic Groups 1A–3A form monatomic ions with a positive charge and an oxidation number equal to the group number. Magnesium forms Mg^{2+}, and its oxidation number is therefore $+2$ [⟸ SECTION 3.3].

3. **Fluorine always has an oxidation number of -1 in compounds.**

4. **Cl, Br, and I always have oxidation numbers of -1 in compounds, except when combined with oxygen and fluorine.** This means that Cl has an oxidation number of -1 in NaCl (in which Na is $+1$, as predicted by the fact that it is an element of Group 1A). In the ion ClO^-, however, the Cl atom has an oxidation number of $+1$ (and O has an oxidation number of -2; see Guideline 5).

5. **The oxidation number of H is $+1$ and of O is -2 in most compounds.** Although this statement applies to many, many compounds, a few important exceptions occur.

   • When H forms a binary compound with a metal, the metal forms a positive ion and H becomes a hydride ion, H^-. Thus, in CaH_2 the oxidation number of Ca is $+2$ (equal to the group number) and that of H is -1.

   • Oxygen can have an oxidation number of -1 in a class of compounds called peroxides. For example, in H_2O_2, hydrogen peroxide, H is assigned its usual oxidation number of $+1$, and so O is -1.

● **Why Use Oxidation Numbers?**
The reason for learning about oxidation numbers at this point is to be able to identify which reactions are oxidation–reduction processes and to know which is the oxidizing agent and which is the reducing agent in a reaction. We return to a more detailed discussion of redox reactions in Chapter 20.

● **Writing Charges on Ions**
Conventionally, charges on ions are written as (number, sign) (Cu^{2+}), whereas oxidation numbers are written as (sign, number). For example, the oxidation number of the Cu^{2+} ion is $+2$.

A Closer Look

Are Oxidation Numbers "Real"?

Do oxidation numbers reflect the actual electric charge on an atom in a molecule or ion? With the exception of monatomic ions such as Cl^- or Na^+, the answer is no.

Oxidation numbers assume that the atoms in a molecule are positive or negative ions, which is not true. For example, in H_2O, the H atoms are not H^+ ions and the O atoms are not O^{2-} ions. This is not to say, however, that atoms in molecules do not bear an electric charge of any kind. In water, for example, calculations indicate the O atom has a charge of about -0.4 (or

Charge on O atom = –0.4

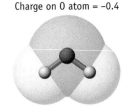

Charge on each H atom = +0.2

40% of the electron charge) and the H atoms are each about $+0.2$.

Why then use oxidation numbers? Oxidation numbers provide a way of dividing up the electrons in a molecule or polyatomic ion. In a redox reaction that distribution of electrons changes. We use this as a way to decide if a redox reaction has occurred, to distinguish the oxidizing and reducing agents, and, as you shall see in Chapter 20, as a way to balance equations for redox reactions.

6. The algebraic sum of the oxidation numbers in a neutral compound must be zero; in a polyatomic ion, the sum must be equal to the ion charge. Examples of this rule are the previously mentioned compounds and others found in Example 5.6.

Example 5.6 Determining Oxidation Numbers

Problem • Determine the oxidation number of the indicated element in each of the following compounds or ions:

(a) Aluminum in aluminum oxide, Al_2O_3

(b) Phosphorus in phosphoric acid, H_3PO_4

(c) Sulfur in the sulfate ion, SO_4^{2-}

(d) Each Cr atom in the dichromate ion, $Cr_2O_7^{2-}$

Strategy • Follow the guidelines in the text, paying particular attention to Guidelines 5 and 6.

Solution •

(a) Al_2O_3 is a neutral compound. Assuming that O has its usual oxidation number of −2, the oxidation number of Al must be +3, in agreement with its position in the periodic table.

$$\text{Net charge on } Al_2O_3 = 0$$
$$= \text{sum of oxidation number of Al atoms}$$
$$+ \text{ sum of oxidation numbers of O atoms}$$
$$= 2(+3) + 3(-2)$$

(b) H_3PO_4 has an overall charge of 0. If the oxygen atoms each have an oxidation number of −2 and that of the H atoms is +1, then the oxidation number of P must be +5.

$$\text{Net charge on } H_3PO_4 = 0$$
$$= \text{sum of oxidation numbers for}$$
$$\text{H atoms} + \text{oxidation number of}$$
$$P + \text{sum of oxidation numbers}$$
$$\text{for O atoms}$$
$$= 3(+1) + (+5) + 4(-2)$$

(c) The sulfate ion, SO_4^{2-}, has an overall charge of −2. Because this compound is not a peroxide, O is assigned an oxidation number of −2, which means that S has an oxidation number of +6

$$\text{Net charge on } SO_4^{2-} = -2$$
$$= \text{oxidation number of S} + \text{sum of}$$
$$\text{oxidation numbers for O atoms}$$
$$= (+6) + 4(-2)$$

(d) The net charge on the, $Cr_2O_7^{2-}$ ion is −2. Assigning each O atom an oxidation number of −2, means that each Cr atom must have an oxidation number of +6

$$\text{Net charge on } Cr_2O_7^{2-} = -2$$
$$= \text{sum of oxidation numbers for Cr atoms}$$
$$+ \text{ sum of oxidation numbers}$$
$$\text{for O atoms}$$
$$= 2(+6) + 7(-2)$$

Exercise 5.10 Determining Oxidation Numbers

Assign an oxidation number to the underlined atom in each ion or molecule.

(a) $\underline{Fe}_2O_3$ **(b)** $H_2\underline{S}O_4$ **(c)** $\underline{C}O_3^{2-}$ **(d)** $\underline{N}O_2^+$

Recognizing Oxidation–Reduction Reactions

• **Sodium/Chlorine Reaction**
Sodium metal reduces chlorine gas. The product is sodium chloride. See Figure 1.2, page 13.

Having learned some guidelines for determining oxidation numbers, you can tell which reactions can be classified as oxidation–reduction and which must be of some other type [CD-ROM, Screen 5.14]. In many cases, however, it will be obvious that the reaction is an oxidation–reduction because it involves an uncombined element or a well-known oxidizing or reducing agent (Table 5.4).

Like oxygen, O_2, the halogens (F_2, Cl_2, Br_2, and I_2) are always oxidizing agents in their reactions with metals and nonmetals. An example is the reaction of chlorine with sodium metal [◀ FIGURE 1.2].

Table 5.4 • Common Oxidizing and Reducing Agents

Oxidizing Agent	Reaction Product	Reducing Agent	Reaction Product
O_2, oxygen	O^{2-}, oxide ion or O combined in H_2O	H_2, hydrogen	$H^+(aq)$, hydrogen ion or H combined in H_2O or other molecule
Halogen, F_2, Cl_2, Br_2, or I_2	Halide ion, F^-, Cl^-, Br^-, or I^-	M, metals such as Na, K, Fe, and Al	M^{n+}, metal ions such as Na^+, K^+, Fe^{2+} or Fe^{3+}, and Al^{3+}
HNO_3, nitric acid	Nitrogen oxides* such as NO and NO_2	C, carbon (used to reduce metal oxides)	CO and CO_2
$Cr_2O_7{}^{2-}$, dichromate ion	Cr^{3+}, chromium(III) ion (in acid solution)		
$MnO_4{}^-$, permanganate ion	Mn^{2+}, manganese(II) ion (in acid solution)		

*NO is produced with dilute HNO_3, whereas NO_2 is a product of concentrated acid.

Na releases 1 e^- per atom.
Oxidation number increases.
Na is oxidized to Na^+ and is the reducing agent.

$$2\ Na(s)\ +\ Cl_2(g)\ \longrightarrow\ 2\ NaCl(s)$$

Cl_2 gains 2 e^- per molecule
Oxidation number decreases by 1 per Cl
Cl_2 is reduced to Cl^- and is the oxidizing agent

Chlorine ends up as Cl^-, having acquired two electrons (from two Na atoms) per Cl_2 molecule. Thus, the oxidation number of each Cl atom has decreased from 0 to −1. This means Cl_2 has been reduced and so is the oxidizing agent.

Chlorine gas is widely used as an oxidizing agent to treat water and sewage. For example, it can remove hydrogen sulfide, H_2S, from drinking water by oxidizing the sulfide to insoluble, elemental sulfur. (Hydrogen sulfide has a characteristic "rotten egg" odor; it comes from the decay of organic matter or underground mineral deposits.)

$$8\ Cl_2(g)\ +\ 8\ H_2S(aq)\ \longrightarrow\ S_8(s)\ +\ 16\ HCl(aq)$$

Figure 5.11 illustrates the chemistry of another excellent oxidizing agent, nitric acid, HNO_3. Here the acid oxidizes copper metal to give copper(II) nitrate, and the nitrate ion from HNO_3 is reduced to the brown gas NO_2. The net ionic equation for the reaction is

Oxidation number of Cu changes from 0 to +2. Cu is oxidized to Cu^{2+} and is the reducing agent.

$$Cu(s) + 2\ NO_3{}^-(aq) + 4\ H^+(aq) \longrightarrow Cu^{2+}(aq) + 2\ NO_2(g) + 2\ H_2O(\ell)$$

N in $NO_3{}^-$ changes from +5 to +4 in NO_2. $NO_3{}^-$ is reduced to NO_2 and is the oxidizing agent.

NO_2 gas

Copper metal oxidized to green $Cu(NO_3)_2$

Figure 5.11 The reaction of copper with nitric acid. Copper (a reducing agent) reacts vigorously with concentrated nitric acid, an oxidizing agent, to give the brown gas NO_2 and a deep green solution of copper(II) nitrate. Notice that nitric acid, a common acid, can be both an oxidizing agent and an acid. (*Charles D. Winters*)

Figure 5.12 **Thermite reaction.** Here Fe_2O_3 is reduced by aluminum metal to produce iron metal and aluminum oxide. *(Charles D. Winters)*

Nitrogen has been reduced from +5 (in the NO_3^- ion) to +4 (in NO_2); therefore, the nitrate ion in acid solution is an oxidizing agent. Copper metal, like all metals, is the reducing agent; here each metal atom has given up two electrons to produce the Cu^{2+} ion.

In the reactions of sodium with chlorine and copper with nitric acid, the metals are oxidized. This is typical of many metals, which are generally good reducing agents. Indeed, the alkali and alkaline earth metals are especially good reducing agents. Another example of this is the reaction of potassium with water. Here potassium reduces the hydrogen in water to H_2 gas (page 66).

$$2\ K(s) + 2\ H_2O(\ell) \longrightarrow 2\ KOH(aq) + H_2(g)$$

    reducing   oxidizing
    agent   agent

Aluminum metal is yet another good reducing agent and is capable of reducing iron(III) oxide to iron metal in a reaction called the *thermite reaction.*

$$Fe_2O_3(s) + 2\ Al(s) \longrightarrow 2\ Fe(\ell) + Al_2O_3(s)$$

    oxidizing   reducing
    agent   agent

Such a large quantity of heat is evolved in the reaction that the iron is produced in the molten state (Figure 5.12).

Table 5.5, as well as Table 5.4, may help you organize your thinking as you look for oxidation–reduction reactions and use their terminology.

Example 5.7   **Oxidation–Reduction Reaction**

Problem • For the reaction of iron(II) ion with permanganate ion in aqueous acid,

$5\ Fe^{2+}(aq) + MnO_4^-(aq) + 8\ H^+(aq) \longrightarrow$
$\qquad\qquad 5\ Fe^{3+}(aq) + Mn^{2+}(aq) + 4\ H_2O(\ell)$

decide which atoms are undergoing a change in oxidation number and identify the oxidizing and reducing agents.

Strategy • Define the oxidation number of the atoms in each ion or molecule involved in the reaction. Decide which atoms have increased in oxidation number (oxidation) and which have decreased in oxidation number (reduction).

Solution • The Mn oxidation number in MnO_4^- is +7, and it decreases to +2 in the product, the Mn^{2+} ion. Thus, the MnO_4^- ion has been reduced and is the oxidizing agent (see Table 5.4).

$5\ Fe^{2+}(aq) + MnO_4^-(aq) + 8\ H^+(aq) \longrightarrow$
  +2   +7, −2
$\qquad\qquad 5\ Fe^{3+}(aq) + Mn^{2+}(aq) + 4\ H_2O(\ell)$
$\qquad\qquad$ +3   +2

The oxidation number of iron has increased from +2 to +3, so the Fe^{2+} ion has lost electrons upon being oxidized to Fe^{3+}. This means the Fe^{2+} ion is the reducing agent.

Comment • One of the reactants in a redox reaction is often a simple substance (here Fe^{2+}), and it is usually evident whether

its oxidation number has increased or decreased. Once this species has been established as having been reduced (or oxidized), you know that a second species in the reaction must be oxidized (or reduced). Another tip is to recognize common oxidizing and reducing agents (Table 5.4).

  — $KMnO_4(aq)$ oxidizing agent

  — $Fe^{2+}(aq)$ reducing agent

The reaction of iron(II) ion and permanganate ion. The reaction of purple permanganate ion (MnO_4^-, the oxidizing agent) with the iron(II) ion (Fe^{2+}, the reducing agent) in acidified aqueous solution gives the nearly colorless manganese(II) ion (Mn^{2+}) and the iron(III) ion (Fe^{3+}). *(Charles D. Winters)*

Table 5.5 • Recognizing Oxidation–Reduction Reactions

	Oxidation	Reduction
In terms of oxidation number	Increase in oxidation number of an atom	Decrease in oxidation number of an atom
In terms of electrons	Loss of electrons by an atom	Gain of electrons by an atom
In terms of oxygen	Gain of one or more O atoms	Loss of one or more O atoms

Example 5.8 Types of Reactions

Problem • Classify each of the following reactions as precipitation, acid–base, gas-forming, or oxidation–reduction.

(a) $2 HNO_3(aq) + Ca(OH)_2(s) \longrightarrow Ca(NO_3)_2(aq) + 2 H_2O(\ell)$

(b) $SO_4^{2-}(aq) + 2 CH_2O(aq) + 2 H^+(aq) \longrightarrow$
$$H_2S(aq) + 2 CO_2(g) + 2 H_2O(\ell)$$

Strategy • A good strategy is first to see if a reaction is one of the three types of exchange reactions. An acid–base reaction is usually easy to distinguish. Next, check the oxidation numbers of each element. If they change, then it is a redox reaction. If there is no change, then it is a simple precipitation or gas-forming process.

Solution • Reaction (a) involves a common acid (nitric acid, HNO_3) and a common base [calcium hydroxide, $Ca(OH)_2$], and it produces a salt, calcium nitrate, and water. Thus, it is an acid–base

reaction. Reaction (b) is a redox reaction because the oxidation numbers of S and C change.

$$\underset{+6,\ -2 \qquad\quad 0,\ +1,\ -2 \qquad +1}{SO_4^{2-}(aq) + 2 CH_2O(aq) + 2 H^+(aq) \longrightarrow}$$
$$\underset{+1,\ -2 \qquad +4,\ -2 \qquad +1,\ -2}{H_2S(aq) + 2 CO_2(g) + 2 H_2O(\ell)}$$

The oxidation number of S changes from +6 to −2, and that of C changes from 0 to +4. Therefore, sulfate, SO_4^{2-}, has been reduced (and is the oxidizing agent), and CH_2O has been oxidized (and is the reducing agent).

No changes occur in the oxidation numbers of the elements in reaction (a).

$$\underset{+1,\ +5,\ -2 \quad +2,\ -2,\ +1 \qquad\quad +2,\ +5,\ -2 \qquad +1,\ -2}{HNO_3(aq) + Ca(OH)_2(s) \longrightarrow Ca(NO_3)_2(aq) + 2 H_2O(\ell)}$$

Exercise 5.11 Oxidation–Reduction and Other Reactions

Decide which of the following reactions are oxidation–reduction reactions. In each case explain your choice and identify the oxidizing and reducing agents.

(a) $NaOH(aq) + HNO_3(aq) \longrightarrow NaNO_3(aq) + H_2O(\ell)$

(b) $Cu(s) + Cl_2(g) \longrightarrow CuCl_2(s)$

(c) $Na_2CO_3(aq) + 2 HClO_4(aq) \longrightarrow CO_2(g) + H_2O(\ell) + 2 NaClO_4(aq)$

(d) $2 S_2O_3^{2-}(aq) + I_2(aq) \longrightarrow S_4O_6^{2-}(aq) + 2 I^-(aq)$

Exercise 5.12 Oxidation–Reduction Reactions

The following reaction occurs in a device for testing the breath for the presence of ethanol. Identify the oxidizing and reducing agents and the substance oxidized and the substance reduced (Figure 5.13).

$$\underset{\text{ethanol} \qquad \text{dichromate ion;} \qquad\qquad\qquad \text{acetic acid} \qquad \text{chromium(III)}}{3 C_2H_5OH(aq) + 2 Cr_2O_7^{2-}(aq) + 16 H^+(aq) \longrightarrow 3 CH_3CO_2H(aq) + 4 Cr^{3+}(aq) + 11 H_2O(\ell)}$$
$$\underset{\text{orange-red} \qquad\qquad\qquad\qquad\qquad\qquad\qquad\qquad \text{ion; green}}{}$$

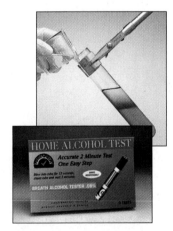

Figure 5.13 The redox reaction of ethanol and dichromate ion is the basis of the test used in a Breathalyzer. When ethanol, an alcohol, is poured into a solution of orange-red dichromate ion, it reduces the dichromate ion to green chromium(III) ion. The inset photo is a breath tester that can be purchased in grocery or drug stores. *(Charles D. Winters)*

5.8 MEASURING CONCENTRATIONS OF COMPOUNDS IN SOLUTION

Most kinds of chemical studies require quantitative measurements, and this includes experiments involving aqueous solutions. To accomplish this, we continue to use balanced equations and moles, but we measure volumes of solution rather than masses of solids, liquids, or gases. Solution concentration expressed as molarity relates the volume of solution in liters or milliliters to the amount of substance in moles.

Solution Concentration: Molarity

The concept of concentration is useful in many contexts. For example, about 4,900,000 people live in Wisconsin, and the state has a land area of roughly 56,000 square miles; therefore, the average concentration of people is about (4.9×10^6 people/5.6×10^4 square miles) or 88 people/mile^2. In chemistry the amount of solute dissolved in a given volume of solution, the concentration of the solution, can be found in the same way. A useful unit of solution concentration, c, is **molarity**, which is defined as *amount of solute per liter of solution* [⚙ CD-ROM, Screen 5.15].

$$\text{Concentration } (c_{\text{molarity}}) = \frac{\text{amount of solute (mol)}}{\text{volume of solution (L)}} \qquad (5.1)$$

For example, if 58.4 g, or 1.00 mol, of NaCl is dissolved in enough water to give a total solution volume of 1.00 L, the concentration, c, is 1.00 mol/L, or 1.00 molar. This is often abbreviated as 1.00 M, where the capital M stands for "moles per liter." Another common notation is to place the formula of the compound in square brackets; this implies that the concentration of the solute in moles of compound per liter of solution is being specified.

$$c_{\text{molarity}} = 1.00\ \text{M} = [\text{NaCl}]$$

It is important to notice that molarity refers to the amount of solute per liter of *solution* and not per liter of *solvent*. If one liter of water is added to one mole of a solid compound, the final volume probably will not be exactly one liter, and the final concentration will not be exactly one molar (Figure 5.14). Therefore, when making solutions of a given molarity, it is almost always the case that we dissolve the solute in a volume of solvent smaller than the desired volume of solution, then make up the final volume with more solvent.

Potassium permanganate, $KMnO_4$, which was used at one time as a germicide in the treatment of burns, is a shiny, purple-black solid that dissolves readily in water to give deep purple solutions. Suppose 0.435 g of $KMnO_4$ has been dissolved in enough water to give 250. mL of solution (Figure 5.15). What is the molar concentration of $KMnO_4$? The first step is to convert the mass of material to an amount of solute.

$$0.435\ \text{g } KMnO_4 \cdot \frac{1\ \text{mol } KMnO_4}{158.0\ \text{g } KMnO_4} = 0.00275\ \text{mol } KMnO_4$$

Now that the amount of substance is known, this can be combined with the volume of solution — which must be in liters — to give the molarity. Because 250. mL is equivalent to 0.250 L,

$$\text{Concentration of } KMnO_4 = \frac{0.00275\ \text{mol } KMnO_4}{0.250\ \text{L}} = 0.0110\ \text{M}$$

$$[KMnO_4] = 0.0110\ \text{M}$$

Chapter Goals • Revisited

• Understand the nature of ionic substances dissolved in water.

• Recognize and write equations for the common types of reactions in aqueous solution.

• Recognize acids and bases and oxidizing and reducing agents.

• **Define and use molarity in solution stoichiometry.**

• **Molar and Molarity**
Chemists use "molar" as an adjective to describe a solution. We use "molarity" as a noun. For example, we refer to a 0.1 molar solution or say the solution has a molarity of 0.1 mole per liter.

• **Volumetric Flask**
A special flask with a line marked on its neck (see Figures 5.14 and 5.15). If the flask is filled with a solution to this line (at a given temperature), it contains precisely the volume of solution specified.

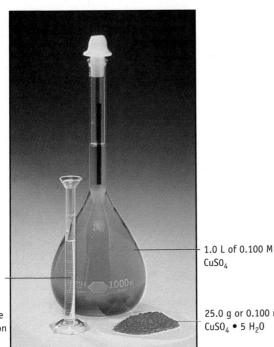

Figure 5.14   Volume of solution versus volume of solvent.
To make a 0.100 M solution of $CuSO_4$, 25.0 g or 0.100 mol of $CuSO_4 \cdot 5 H_2O$ (the blue crystalline solid) was placed in a 1.00-L volumetric flask.

For this photo we measured out exactly 1.00 L of water and this was slowly added to the volumetric flask containing $CuSO_4 \cdot 5 H_2O$. When enough water had been added so that the solution volume was exactly 1.00 L, approximately 8 mL of water (the quantity in the small graduated cylinder) was left over.

This emphasizes that molar concentrations are defined as moles of solute per liter of solution and not per liter of water or other solvent. *(Charles D. Winters)*

Volume of water remaining when 1.0 L of water was used to make 1.0 L of a solution

1.0 L of 0.100 M $CuSO_4$

25.0 g or 0.100 mol of $CuSO_4 \bullet 5 H_2O$

The $KMnO_4$ concentration is 0.0110 molar, or 0.0110 M. This is useful information, but it is often equally useful to know the concentration of each type of ion in a solution. Like all soluble ionic compounds, $KMnO_4$ dissociates completely into its ions, K^+ and MnO_4^-, when dissolved in water.

$$KMnO_4(aq) \longrightarrow K^+(aq) + MnO_4^-(aq)$$
100% dissociation

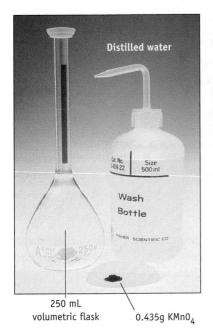

250 mL volumetric flask   0.435g $KMnO_4$

The $KMnO_4$ is first dissolved in a small amount of water.

Distilled water is added to fill the flask with solution just to the mark on the flask.

A mark on the neck of a volumetric flask indicates a volume of exactly 250 mL at 25°C.

Figure 5.15   Making a solution. A 0.0110 M solution of $KMnO_4$ is made by adding enough water to 0.435 g of $KMnO_4$ to make 0.250 L of solution [CD-ROM, Screen 5.16]. *(Charles D. Winters)*

One mole of $KMnO_4$ provides 1 mol of K^+ ions and 1 mol of MnO_4^- ions. Accordingly, 0.0110 M $KMnO_4$ gives a concentration of K^+ in the solution of 0.0110 M; similarly, the concentration of MnO_4^- is also 0.0110 M.

Another example of ion concentrations is provided by the dissociation of an ionic compound such as $CuCl_2$ (see Chapter Focus).

$$CuCl_2(aq) \xrightarrow{\hspace{1cm}} Cu^{2+}(aq) + 2\ Cl^-(aq)$$
100% dissociation

If 0.1 mol of $CuCl_2$ is dissolved in enough water to make 1 L of solution, the concentration of the copper(II) ion is $[Cu^{2+}] = 0.1$ M. Because the compound dissociates to provide 2 mol of Cl^- ions for each mole of $CuCl_2$, however, the concentration of chloride ion, $[Cl^-]$, is 0.2 M.

Example 5.9   Concentration

Problem • If 25.3 g of sodium carbonate, Na_2CO_3, is dissolved in enough water to make 250. mL of solution, what is the molar concentration of Na_2CO_3? What are the concentrations of the Na^+ and CO_3^{2-} ions?

Strategy • The molar concentration of Na_2CO_3 is defined as the amount of Na_2CO_3 per liter of solution. We know the volume of solution (0.250 L). We need the amount of Na_2CO_3. To find the concentrations of the individual ions, recognize that the salt dissociates completely.

$$Na_2CO_3(s) \longrightarrow 2\ Na^+(aq) + CO_3^{2-}(aq)$$

Thus, a 1 M solution of Na_2CO_3 is really a 2 M solution of Na^+ ions and 1 M solution of CO_3^{2-} ions.

Solution • Let us first find the amount of Na_2CO_3.

$$25.3\ \text{g Na}_2\text{CO}_3 \cdot \frac{1\ \text{mol Na}_2\text{CO}_3}{106.0\ \text{g Na}_2\text{CO}_3} = 0.239\ \text{mol Na}_2\text{CO}_3$$

and then the molar concentration of Na_2CO_3.

$$\text{Concentration} = \frac{0.239\ \text{mol Na}_2\text{CO}_3}{0.250\ \text{L}} = 0.955\ \text{M}$$

$$[Na_2CO_3] = 0.955\ \text{M}$$

The ion concentrations follow from the concentration of Na_2CO_3 and the knowledge that each mole of Na_2CO_3 produces 2 mol of Na^+ ions and 1 mol of CO_3^{2-} ions.

$$[Na^+] = (0.955\ \text{M Na}_2\text{CO}_3)\left(\frac{2\ \text{mol Na}^+}{1\ \text{mol Na}_2\text{CO}_3}\right) = 1.91\ \text{M}$$

$$[CO_3^{2-}] = (0.955\ \text{M Na}_2\text{CO}_3)\left(\frac{1\ \text{mol CO}_3^{2-}}{1\ \text{mol Na}_2\text{CO}_3}\right) = 0.955\ \text{M}$$

Exercise 5.13   Concentration

Sodium bicarbonate, $NaHCO_3$, is used in baking powder formulations and in the manufacture of plastics and ceramics, among other things. If 26.3 g of the compound is dissolved in enough water to make exactly 200 mL of solution, what is the molar concentration of $NaHCO_3$? What are the concentrations of the ions in solution?

Preparing Solutions of Known Concentration

A task chemists often must perform is preparing a given volume of solution of known concentration. There are two commonly used ways to do this [CD-ROM, Screen 5.16].

Combining a Weighed Solute with the Solvent

Suppose you wish to prepare 2.00 L of a 1.50 M solution of Na_2CO_3. You have some solid Na_2CO_3 and distilled water. You also have a 2.00-L volumetric flask (see Figures 5.14 and 5.15). To make the solution, you must weigh the necessary quantity of

Na_2CO_3 as accurately as possible, carefully place all the solid in the volumetric flask, and then add water to dissolve the solid. After the solid has dissolved completely, more water is added to bring the solution volume to 2.00 L. The solution then has the desired concentration and the volume specified.

But what mass of Na_2CO_3 is required to make 2.00 L of 1.50 M Na_2CO_3? First, calculate the amount of substance required,

$$2.00 \; \cancel{L} \cdot \frac{1.50 \; mol \; Na_2CO_3}{1.00 \; \cancel{L} \; solution} = 3.00 \; mol \; Na_2CO_3 \; required$$

and then the mass in grams.

$$3.00 \; \cancel{mol \; Na_2CO_3} \cdot \frac{106.0 \; g \; Na_2CO_3}{1 \; \cancel{mol \; Na_2CO_3}} = 318 \; g \; Na_2CO_3$$

Thus, to prepare the desired solution, you should dissolve 318 g of Na_2CO_3 in enough water to make 2.00 L of solution.

Exercise 5.14   **Preparing Solutions of Known Concentration**

An experiment in your laboratory requires 250. mL of a 0.0200 M solution of $AgNO_3$. You are given solid $AgNO_3$, distilled water, and a 250.-mL volumetric flask. Describe how to make up the required solution.

Diluting a More Concentrated Solution

Another method of making a solution of a given concentration is to *begin with a concentrated solution and add water until the desired, lower concentration is reached* (Figure 5.16). Many of the solutions prepared for your laboratory course are probably made

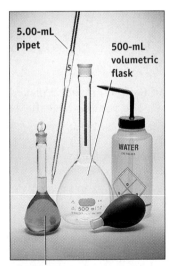

0.100 M $K_2Cr_2O_7$

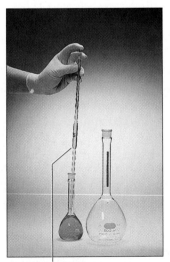

Use a 5.00-mL pipet to withdraw 5.00 mL of 0.100 M $K_2Cr_2O_7$ solution.

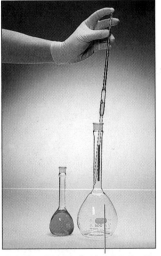

Add the 5.00-mL sample of 0.100 M $K_2Cr_2O_7$ solution to a 500-mL volumetric flask.

Fill the flask to the mark with distilled water to give 0.00100 M $K_2Cr_2O_7$ solution.

Figure 5.16   Making a solution by dilution. Here 5.00 mL of a $K_2Cr_2O_7$ solution is diluted to 500.0 mL. The result is dilution by a factor of 100, from 0.100 M to 0.00100 M [CD-ROM, Screen 5.16]. *(Charles D. Winters)*

Problem-Solving Tip 5.2

Preparing a Solution by Dilution

The preparation of the $K_2Cr_2O_7$ solution (below and in Figure 5.16) suggest a way to do the calculations for dilutions. The central idea is that the amount of solute in the final, dilute solution has to be equal to the amount of solute taken from the more concentrated solution. If c is the concentration (molarity) and V is the volume (and the subscripts d and c identify the dilute and concentrated solutions, respectively), the amount of solute in either solution (in the case of the $K_2Cr_2O_7$ example in the text) can be calculated as follows:

Amount of $K_2Cr_2O_7$ in the final dilute solution $= c_d V_d$
$$= 0.00050 \text{ mol}$$

Amount of $K_2Cr_2O_7$ taken from the more concentrated solution
$$= c_c V_c = 0.00050 \text{ mol}$$

Because both cV products are equal to the same amount of solute, we can use the following equation:

$$c_c \times V_c = c_d \times V_d$$

Amount of reagent in concentrated solution
$$= \text{Amount of reagent in dilute solution}$$

This equation is valid for all cases in which a more concentrated solution is used to make a more dilute one. It can be used to find, for example, the molarity of the dilute solution, c_d, from values of c_c, V_c, and V_d.

● **Diluting Concentrated H_2SO_4**
The direction that one can prepare a solution by adding water to a more concentrated solution is correct except for sulfuric acid solutions. When mixing water and sulfuric acid, the resulting solution becomes quite warm. If water is added to concentrated sulfuric acid, so much heat is evolved that the solution may boil over or splash and burn someone nearby. To avoid this, chemists always add concentrated sulfuric acid to water when making a dilute acid solution.

by this dilution method. It is more efficient to store a small volume of a concentrated solution and then, when needed, add water to make it into a much larger volume of a dilute solution.

Suppose you need 500. mL of 0.0010 M potassium dichromate, $K_2Cr_2O_7$, for use in chemical analysis. You have some 0.100 M $K_2Cr_2O_7$ solution available. To make the required 0.0010 M solution, place a measured volume of the more concentrated $K_2Cr_2O_7$ solution in a flask and then add water until the $K_2Cr_2O_7$ is contained in a larger volume of water, that is, until it is less concentrated (or more dilute) (see Figure 5.16).

What volume of a 0.100 M $K_2Cr_2O_7$ solution must be diluted to make the 0.0010 M solution? If the volume and concentration of a solution are known, the amount of solute is also known. Therefore, the amount of $K_2Cr_2O_7$ that must be in the final dilute solution is

$$\text{Amount of } K_2Cr_2O_7 \text{ in dilute solution} = (0.500 \text{ L})\left(\frac{0.0010 \text{ mol}}{L}\right)$$
$$= 0.00050 \text{ mol } K_2Cr_2O_7$$

A more concentrated solution containing this amount of $K_2Cr_2O_7$ must be placed in a 500.-mL flask and then diluted to the final volume. The volume of 0.100 M $K_2Cr_2O_7$ that must be transferred and diluted is 5.0 mL.

$$0.00050 \text{ mol } K_2Cr_2O_7 \cdot \frac{1.00 \text{ L}}{0.100 \text{ mol } K_2Cr_2O_7} = 0.0050 \text{ L, or } 5.0 \text{ mL}$$

Thus, to prepare 500. mL of 0.0010 M $K_2Cr_2O_7$, place 5.0 mL of 0.100 M $K_2Cr_2O_7$ in a 500.-mL flask and add water until a volume of 500. mL is reached (see Figure 5.16).

Example 5.10   Preparing a Solution by Dilution

Problem ● You need a 2.36×10^{-3} M solution of iron(III) ion. A lab procedure suggests this can be done by placing 1.00 mL of 0.236 M iron(III) nitrate in a volumetric flask and diluting to exactly 100.0 mL. Show that this method will work.

Strategy ● First calculate the amount of iron(III) ion in the 1.00-mL sample. The concentration of the ion in the final, dilute solution is equal to this amount of iron(III) divided by the new volume.

Solution • The amount of iron(III) ion in the 1.00-mL sample is

$$c_M \times V = \left(\frac{0.236 \text{ mol Fe}^{3+}}{\text{L}}\right)(1.00 \times 10^{-3} \text{ L})$$

$$= 2.36 \times 10^{-4} \text{ mol Fe}^{3+}$$

This amount of iron(III) ion is distributed in 100.0 mL, so the final concentration is

$$[\text{Fe}^{3+}] = \frac{2.36 \times 10^{-4} \text{ mol Fe}^{3+}}{0.100 \text{ L}} = 2.36 \times 10^{-3} \text{ M}$$

Comment • The original solution was diluted by a factor of 100, so the final concentration is 100 times smaller than the original concentration.

Exercise 5.15   Preparing a Solution by Dilution

In one of your laboratory experiments you are given a solution of $CuSO_4$ that has a concentration of 0.15 M. If you mix 6.0 mL of this solution with enough water to have a total volume of 10.0 mL, what is the concentration of $CuSO_4$ in this new solution?

Exercise 5.16   Preparing a Solution by Dilution

An experiment calls for you to use 250. mL of 1.00 M NaOH, but you are given a large bottle of 2.00 M NaOH. Describe how to make the 1.00 M NaOH in the desired volume.

5.9   pH: A CONCENTRATION SCALE FOR ACIDS AND BASES

Acids produce hydrogen ions, H^+, in aqueous solution (see Section 5.3). Naturally occurring acids and acidic solutions share the characteristic that the hydrogen ion concentration of the solution increases when the acid dissolves. Vinegar, which contains the weak acid acetic acid, has a hydrogen ion concentration of only 1.6×10^{-3} M and "pure" rain water has $[H^+] = 2.5 \times 10^{-6}$ M. These extremely small values can be expressed using scientific notation, but this is awkward. A more convenient way to express such numbers is the logarithmic pH scale.

The **pH** of a solution is the negative of the logarithm of the hydrogen ion concentration [⊙ CD-ROM, Screen 5.17].

$$pH = -\log[H^+] \tag{5.2}$$

Taking vinegar, pure water, blood, and ammonia as examples,

$$\text{pH of vinegar} = -\log(1.6 \times 10^{-3} \text{ M}) = -(-2.80) = 2.80$$
$$\text{pH of pure water (at 25 °C)} = -\log(1.0 \times 10^{-7} \text{ M}) = -(7.00) = 7.00$$
$$\text{pH of blood} = -\log(4.0 \times 10^{-8} \text{ M}) = -(7.40) = 7.40$$
$$\text{pH of ammonia} = -\log(1.0 \times 10^{-11} \text{ M}) = -(11.00) = 11.00$$

you see that acids have a relatively low pH, whereas ammonia, a common base, has a *very* low hydrogen ion concentration and a high pH. Blood, which your common sense tells you is likely to be neither acidic nor basic, has a pH near 7. Indeed, for aqueous solutions at 25 °C, we can say that acids will have pH values less than 7, bases will have values greater than 7, and a pH of 7 represents a neutral solution (Figure 5.17).

● **Logarithms**
Numbers less than 1 have negative logs. Therefore, defining pH as $-\log[H^+]$, produces a positive number. See Appendix A for more discussion of logs.

● **pH of Pure Water**
Highly purified water has a pH of exactly 7 at 25 °C. This is the "dividing line" between acidic substances (pH < 7) and basic substances (pH > 7).

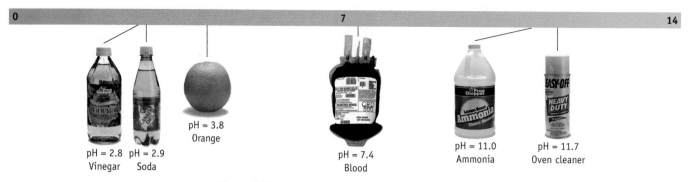

Figure 5.17   pH values of some common substances. Here the "bar" is colored red at one end and blue at the other. These are the colors of litmus paper, commonly used in the laboratory to decide if a solution is acidic (litmus is red) or basic (litmus is blue). *(Charles D. Winters)*

● **Logs and Your Calculator**
All scientific calculators have a key marked log. To find an antilog, use the key marked 10^x or the inverse log. When you enter the value of x for 10^x, make sure it has a negative sign.

● **pH-Indicating Dyes**
Many natural substances change color in solution as pH changes. See the extract of red cabbage in Figure 5.5. Tea changes color when acidic lemon juice is added.

Suppose you know the pH of a solution. To find the hydrogen ion concentration you take the antilog of the pH. That is,

$$[H^+] = 10^{-pH} \qquad\qquad (5.3)$$

For example, the pH of a diet soda is 3.12, and the hydrogen ion concentration of the solution is

$$[H^+] = 10^{-3.12} = 7.6 \times 10^{-4}\ M$$

The approximate pH of a solution may be determined using any of a variety of dyes or indicators. The litmus you use in the laboratory is a dye extracted from a variety of lichen, but many other dyes are available (Figure 5.18a). Accurate pH measurements are made with instruments such as that shown in Figure 5.18b. Here a pH electrode is immersed in the solution to be tested, and the pH is read from the instrument.

Figure 5.18   Determining pH.
(a) Some household products. Each solution contains a few drops of a dye called a pH indicator (in this case a "universal indicator"). A color of yellow or red indicates a pH less than 7. A green to purple color indicates a pH greater than 7.
(b) The pH of a soda is measured with a modern pH meter. Soft drinks are often quite acidic owing to the dissolved CO_2 and other ingredients. *(Charles D. Winters)*

(a)

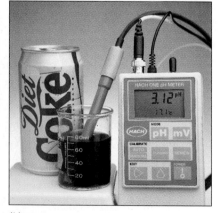

(b)

Example 5.11   pH of Solutions

Problem •
(a) Lemon juice has $[H^+] = 0.0032$ M. What is its pH?

(b) Sea water has a pH of 8.30. What is the hydrogen ion concentration of this solution?

(c) A solution of nitric acid has $[HNO_3] = 0.0056$ M. What is the pH of this solution?

Strategy • Use Equation 5.2 to calculate pH from the H^+ concentration. Use Equation 5.3 to find $[H^+]$ from the pH.

Solution •
(a) *Lemon juice:*

$$pH = -\log [H^+] = -\log (3.2 \times 10^{-3})$$
$$= -(-2.49) = 2.49$$

(b) *Sea water:*

$$[H^+] = 10^{-pH} = 10^{-8.30} = 5.0 \times 10^{-9} \text{ M}$$

(c) *Nitric acid:* Nitric acid is a common strong acid (page 158); it is completely ionized in aqueous solution. Therefore,

$$\text{because } [HNO_3] = 0.0056 \text{ M}$$
$$\text{then } [H^+] = [NO_3^-] = 0.0056 \text{ M}$$
$$pH = -\log [H_3O^+] = -\log (0.0056 \text{ M}) = 2.25$$

Comment • Be sure to understand the relationship of logarithms and significant figures (Appendix A). The number to the left of the decimal point in a logarithm is called the *characteristic*, and the number to the right is the *mantissa*. The mantissa has as many significant figures as the number whose log was found. For example, the logarithm of 3.2×10^{-3} (two significant figures) is 2.49 (two numbers to the right of the decimal point).

Exercise 5.17   pH of Solutions

(a) What is the pH of a solution of HCl, where $[HCl] = 2.6 \times 10^{-2}$ M?

(b) What is the hydrogen ion concentration in saturated calcium hydroxide with a pH of 12.45?

5.10   STOICHIOMETRY OF REACTIONS IN AQUEOUS SOLUTION

General Solution Stoichiometry

One type of exchange reaction involves a metal carbonate and an aqueous acid. The products are a salt and CO_2 gas (Figure 5.19) [🖲 CD-ROM, Screen 5.18].

$$CaCO_3(s) + 2 HCl(aq) \longrightarrow CaCl_2(aq) + H_2O(\ell) + CO_2(g)$$

Suppose we want to know what mass of $CaCO_3$ is required to react completely with 25 mL of 0.750 M HCl. This problem can be solved in the same way as all the stoichiometry problems you have seen so far, except that the amount of one reactant must be determined from the solution volume and concentration. The first step is to find the amount of HCl,

$$0.025 \text{ L} \cdot \frac{0.750 \text{ mol HCl}}{1 \text{ L HCl}} = 0.019 \text{ mol HCl}$$

and then to relate this to the amount of $CaCO_3$ required.

$$0.019 \text{ mol HCl} \cdot \frac{1 \text{ mol CaCO}_3}{2 \text{ mol HCl}} = 0.0094 \text{ mol CaCO}_3$$

Finally, the amount of $CaCO_3$ is converted to a mass in grams.

$$0.0094 \text{ mol CaCO}_3 \cdot \frac{100. \text{ g CaCO}_3}{1 \text{ mol CaCO}_3} = 0.94 \text{ g CaCO}_3$$

Figure 5.19   A commercial remedy for excess stomach acid. The tablet contains calcium carbonate, which reacts with hydrochloric acid, the acid present in the digestive system. The most obvious product is CO_2 gas. *(Charles D. Winters)*

Problem-Solving Tip 5.3

Stoichiometry Calculations Involving Solutions

In Problem-Solving Tip 4.1 you learned about a general approach to stoichiometry problems. We can now modify that scheme for a reaction such as $x A + y B \longrightarrow$ products occurring in solution.

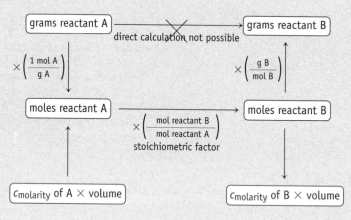

Chemists do such calculations many times in the course of their work in research and product development. If you follow the general scheme outlined in *Problem-Solving Tip 5.3* and pay attention to the units on the numbers, you can successfully carry out any kind of stoichiometry calculations involving concentrations.

Example 5.12 Stoichiometry of a Reaction in Solution

Problem • Metallic zinc reacts with aqueous HCl and other acids (see Figure 5.5, page 157).

$$Zn(s) + 2\,HCl(aq) \longrightarrow ZnCl_2(aq) + H_2(g)$$

What volume (mL) of 2.50 M HCl is required to convert 11.8 g of Zn completely to products?

Strategy • Here we have a mass of a known substance (Zn), so first calculate the amount of zinc. Next, use a stoichiometric factor (= 2 mol HCl/1 mol Zn) to relate amount of HCl to amount of Zn. Finally, calculate the volume of HCl from the amount of HCl and its concentration.

Solution • Begin by calculating the amount of Zn.

$$11.8\ \text{g Zn} \cdot \frac{1\ \text{mol Zn}}{65.39\ \text{g Zn}} = 0.180\ \text{mol Zn}$$

Use the stoichiometric factor to calculate the amount of HCl required.

$$0.180\ \text{mol Zn} \cdot \frac{2\ \text{mol HCl}}{1\ \text{mol Zn}} = 0.360\ \text{mol HCl}$$

Convert the amount of HCl required to a volume.

$$0.360\ \text{mol HCl} \cdot \frac{1.00\ \text{L solution}}{2.50\ \text{mol HCl}} = 0.144\ \text{L HCl}$$

The answer is requested in units of milliliters, so we find that 144 mL of 2.50 M HCl is required to convert 11.8 g of Zn completely to products.

Exercise 5.18 Solution Stoichiometry

If you combine 75.0 mL of 0.350 M HCl and an excess of Na_2CO_3, what mass of CO_2 (in grams) should be produced?

$$Na_2CO_3(s) + 2\,HCl(aq) \longrightarrow 2\,NaCl(aq) + H_2O(\ell) + CO_2(g)$$

Titration: A Method of Chemical Analysis

Chemists are often concerned with **quantitative chemical analysis**, the determination of the quantity of a given constituent of a mixture. You were first introduced to methods of analysis in Chapter 4 [◀ SECTION 4.6]. One method of analysis is that:

> A substance, present in unknown quantity, can be allowed to react with a known quantity of another substance. If the stoichiometric ratio for their reaction is known, the unknown quantity can be determined.

This is the essence of analysis by titration [💾 CD-ROM, Screen 5.19].

Oxalic acid, $H_2C_2O_4$, is a naturally occurring acid. Suppose you are asked to analyze a sample of oxalic acid to ascertain its purity. Because the compound is an acid, it reacts with a base such as sodium hydroxide (see Section 5.4)

$$H_2C_2O_4(aq) + 2\,NaOH(aq) \longrightarrow Na_2C_2O_4(aq) + 2\,H_2O(\ell)$$

You can use this reaction to determine the quantity of oxalic acid present in a given mass of sample if the following conditions are met:

- You can determine when the amount of sodium hydroxide added is just enough to react with all the oxalic acid present in solution.
- You know the concentration of the sodium hydroxide solution and volume that has been added at exactly the point of complete reaction.

These conditions are fulfilled in a **titration,** a procedure illustrated in Figure 5.20. The solution containing oxalic acid is placed in a flask along with an **acid–base indicator,** a dye that changes color when the reaction used for analysis is complete. In this case, the dye is colorless in acid solution but pink in basic solution. Aqueous sodium hydroxide of accurately known concentration is placed in a buret. As the sodium hydroxide in the buret is added slowly to the acid solution in the flask, the acid reacts with the base according to the net ionic equation

$$H_2C_2O_4(aq) + 2\,OH^-(aq) \longrightarrow C_2O_4^{2-}(aq) + 2\,H_2O(\ell)$$

As long as some acid is present in solution, all the base supplied from the buret is consumed, and the indicator remains colorless. At some point, however, the amount of OH^- added exactly equals the amount of H^+ that can be supplied by the acid. This is called the **equivalence point.** As soon as the slightest excess of base has been added beyond the equivalence point, the solution becomes basic, and the indicator changes color (Figure 5.20).

When the equivalence point has been reached in a titration, the volume of base added can be determined by reading the calibrated buret. From this volume and the concentration of the base, the amount of base used can be found:

Amount of base added (mol) = concentration of base (mol/L) × volume of base (L)

Then, using the stoichiometric factor from the balanced equation, the amount of base added is related to the amount of acid present in the original sample.

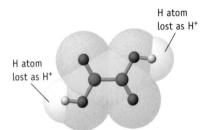

Oxalic acid $H_2C_2O_4$

Oxalate anion $C_2O_4^{2-}$

Oxalic acid can supply two H^+ ions in a reaction with a base. Hence, 1 mol of the acid requires 2 mol of NaOH for complete reaction.

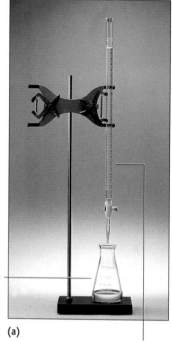

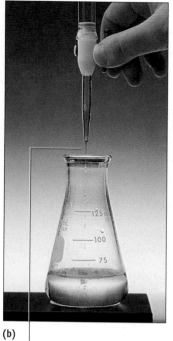

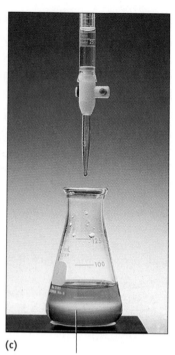

Flask containing aqueous solution of sample being analyzed

(a)

50-mL buret containing aqueous NaOH of accurately known concentration

(b)

A solution of NaOH is added slowly to the sample being analyzed. The sample is mixed.

(c)

When the amount of NaOH added from the buret exactly equals the amount of H⁺ supplied by the acid being analyzed, the dye (indicator) changes color.

Figure 5.20   Titration of an acid in aqueous solution with a base. (a) A buret, a volumetric measuring device calibrated in divisions of 0.1 mL, is filled with an aqueous solution of a base of known concentration. (b) Base is added slowly from the buret to the solution. (c) A change in the color of an indicator signals the equivalence point. (The indicator used here is phenolphthalein [CD-ROM, Screen 5.14].) *(Charles D. Winters)*

Example 5.13   Acid–Base Titration

Problem • A 1.034-g sample of impure oxalic acid is dissolved in water and an acid–base indicator added. The sample requires 34.47 mL of 0.485 M NaOH to reach the equivalence point. What is the mass of oxalic acid and what is its mass percent in the sample?

Strategy • The balanced equation for the reaction of NaOH and $H_2C_2O_4$ is given in the preceding text. The concentration of NaOH, and the volume used in the titration, are known and provide the amount of NaOH. Use a stoichiometric factor to relate the amount of NaOH to the amount of $H_2C_2O_4$. Finally, convert the amount of $H_2C_2O_4$ to a mass and then calculate the mass percent in the sample. See *Problem-Solving Tip 5.4*.

Solution • The amount of NaOH is given by

$$c_{NaOH} \times V_{NaOH} = \frac{0.485 \text{ mol NaOH}}{L} \cdot 0.03447 \text{ L}$$

$$= 0.0167 \text{ mol NaOH}$$

The balanced equation for the reaction shows that 1 mol of oxalic acid requires 2 mol of sodium hydroxide. This is the required stoichiometric factor to obtain the amount of oxalic acid present.

$$0.0167 \text{ mol NaOH} \cdot \frac{1 \text{ mol } H_2C_2O_4}{2 \text{ mol NaOH}} = 0.00836 \text{ mol } H_2C_2O_4$$

The mass of oxalic acid is found from the amount of the acid.

$$0.00836 \text{ mol } H_2C_2O_4 \cdot \frac{90.04 \text{ g } H_2C_2O_4}{1 \text{ mol } H_2C_2O_4} = 0.753 \text{ g } H_2C_2O_4$$

This mass of oxalic acid represents 72.8% of the total sample mass.

$$\frac{0.753 \text{ g } H_2C_2O_4}{1.034 \text{ g sample}} \times 100\% = 72.8\% \ H_2C_2O_4$$

Exercise 5.19 Acid–Base Titration

A 25.0-mL sample of vinegar requires 28.33 mL of a 0.953 M solution of NaOH for titration to the equivalence point. What mass (in grams) of acetic acid is in the vinegar sample, and what is the concentration of acetic acid in the vinegar?

$$CH_3CO_2H(aq) + NaOH(aq) \longrightarrow NaCH_3CO_2(aq) + H_2O(\ell)$$
acetic acid   sodium acetate

In Example 5.13 the concentration of the base used in the titration was given. In actual practice this usually has to be found by a prior measurement. The procedure by which the concentration of an analytical reagent is determined accurately is called **standardization,** and there are two general approaches.

One approach is to weigh accurately a sample of a pure, solid acid or base (known as a **primary standard**) and then titrate this sample with a solution of the base or acid to be standardized (Example 5.14). Another approach to standardizing a solution is to titrate it with another solution that is already standardized (Exercise 5.20). This is often done with standard solutions purchased from chemical supply companies.

Example 5.14 Standardizing an Acid by Titration

Problem • A sample of sodium carbonate (Na_2CO_3, 0.263 g) requires 28.35 mL of aqueous HCl for titration to the equivalence point. What is the molar concentration of the HCl?

Strategy • The balanced equation for the reaction is written first.

$$Na_2CO_3(aq) + 2\,HCl(aq) \longrightarrow 2\,NaCl(aq) + H_2O(\ell) + CO_2(g)$$

The amount of Na_2CO_3 can be calculated from its mass and then, using the stoichiometric factor, the amount of HCl in 28.35 mL can be calculated. The quotient of amount of HCl and its volume gives its concentration.

Solution • Convert the mass of Na_2CO_3 used as the standard to the amount of the base.

$$0.263 \text{ g } Na_2CO_3 \cdot \frac{1 \text{ mol } Na_2CO_3}{106.0 \text{ g}} = 0.00248 \text{ mol } Na_2CO_3$$

Use the stoichiometric factor to calculate amount of HCl in 28.35 mL.

$$0.00248 \text{ mol } Na_2CO_3 \cdot \frac{2 \text{ mol HCl required}}{1 \text{ mol } Na_2CO_3 \text{ available}} = 0.00496 \text{ mol HCl}$$

The 28.35-mL (or 0.02835-L) sample of aqueous HCl contains 0.00496 mol of HCl, and so the concentration of the HCl solution is 0.175 M.

$$[HCl] = \frac{0.00496 \text{ mol HCl}}{0.02835 \text{ L}} = 0.175 \text{ M}$$

Comment • Solutions of acids such as HCl can be standardized by titrating a base such as Na_2CO_3, a solid that can be obtained in pure form, that can be weighed accurately, and that reacts completely with an acid.

Exercise 5.20 Standardization of a Base

Hydrochloric acid, HCl, can be purchased from chemical supply houses with a concentration of 0.100 M, and such a solution can be used to standardize the solution of a base. If titrating 25.00 mL of a sodium hydroxide solution to the equivalence point requires 29.67 mL of 0.100 M HCl, what is the concentration of the base?

Oxidation–reduction reactions (see Section 5.7) also lend themselves well to chemical analysis by titration. Many of these reactions go rapidly to completion in aqueous solution, and methods exist for finding their equivalence points.

Example 5.15 Using an Oxidation–Reduction Reaction in a Titration

Problem • We wish to analyze an iron ore for its iron content. The iron in the sample can be converted quantitatively to the iron(II) ion, Fe^{2+}, in aqueous solution, and this solution can then be titrated with aqueous potassium permanganate, $KMnO_4$. The balanced net ionic equation for the reaction occurring in the course of this titration is

$$MnO_4^-(aq) + 5\ Fe^{2+}(aq) + 8\ H^+(aq) \longrightarrow$$

purple   colorless

$$Mn^{2+}(aq) + 5\ Fe^{3+}(aq) + 4\ H_2O(\ell)$$

colorless   pale yellow

A 1.026-g sample of iron-containing ore requires 24.35 mL of 0.0195 M $KMnO_4$ to reach the equivalence point. What is the mass percent of iron in the ore?

Strategy • Because the volume and molar concentration of the $KMnO_4$ solution are known, the amount of $KMnO_4$ used in the titration can be calculated. Using the stoichiometric factor, the amount of $KMnO_4$ is related to the amount of iron(II) ion. The amount of iron(II) is converted to its mass and then to a mass percent in the sample.

Solution • First, calculate the amount of $KMnO_4$.

$$c_{KMnO_4} \times V_{KMnO_4} = \frac{0.0195\ mol\ KMnO_4}{L} \cdot 0.02435\ L$$

$$= 0.000475\ mol\ KMnO_4$$

Use the stoichiometric factor to calculate the amount of iron(II) ion.

$$0.000475\ mol\ KMnO_4 \cdot \frac{5\ mol\ Fe^{2+}}{1\ mol\ KMnO_4} = 0.00237\ mol\ Fe^{2+}$$

The mass of iron in the ore can now be calculated,

Here purple, aqueous $KMnO_4$ is added to a solution containing Fe^{2+}. As $KMnO_4$ drops into the solution, colorless Mn^{2+} and pale yellow Fe^{3+} form. Here an area of the solution containing unreacted $KMnO_4$ is seen. As the solution is mixed, this color disappears until the equivalence point is reached. (*Charles D. Winters*)

$$0.00237\ mol\ Fe^{2+} \cdot \frac{55.85\ g\ Fe^{2+}}{1\ mol\ Fe^{2+}} = 0.133\ g\ Fe^{2+}$$

and finally the mass percent can be derived.

$$\frac{0.133\ g\ Fe^{2+}}{1.026\ g\ sample} \times 100\% = 12.9\%\ iron$$

Comment • This is a useful analytical reaction, because it is easy to detect when all the iron(II) has reacted. The MnO_4^- ion is deep purple, but when it reacts with Fe^{2+} the color disappears because the reaction product, the Mn^{2+} ion, is colorless. Thus, as $KMnO_4$ is added from a buret, the purple color disappears as the solutions mix. When all the Fe^{2+} has been converted to Fe^{3+}, any additional $KMnO_4$ will give the solution a permanent purple color. Therefore, $KMnO_4$ solution is added from the buret until the initially colorless, Fe^{2+}-containing solution just turns a faint purple color, the signal that the equivalence point has been reached.

Exercise 5.21 Using an Oxidation–Reduction Reaction in a Titration

Vitamin C, ascorbic acid ($C_6H_8O_6$), is a reducing agent. One way to determine the ascorbic acid content of a sample is to mix the acid with an excess of iodine,

$$C_6H_8O_6(aq) + I_2(aq) \longrightarrow C_6H_6O_6(aq) + 2\ H^+(aq) + 2\ I^-(aq)$$

and then titrate the iodine that did *not* react with the ascorbic acid with sodium thiosulfate. The balanced, net ionic equation for the reaction occurring in the course of this titration is

$$I_2(aq) + 2\ S_2O_3{}^{2-}(aq) \longrightarrow 2\ I^-(aq) + S_4O_6{}^{2-}(aq)$$

Suppose 50.00 mL of 0.0520 M I_2 was added to the sample containing ascorbic acid. After the ascorbic acid/I_2 reaction was complete, the I_2 not used in the reaction required 20.30 mL of 0.196 M $Na_2S_2O_3$ for titration to the equivalence point. Calculate the mass of ascorbic acid in the unknown sample.

In Summary

When you have finished studying this chapter, you should ask if you have met the chapter goals. In particular, you should be able to

- Explain the difference between electrolytes and nonelectrolytes (Section 5.1 and Chapter Focus).
- Predict the solubility of ionic compounds in water (Section 5.1 and Figure 5.1).
- Recognize what ions are formed when an ionic compound or acid or base dissolves in water (Sections 5.1–5.3).
- Predict the products of precipitation reactions (Section 5.2), the formation of an insoluble reaction product by the exchange of anions between the cations of the reactants.
- Write the net ionic equation for a given reaction (Sections 5.2 and 5.6).
- Recognize common acids and bases and understand their behavior in aqueous solution (Section 5.3 and Table 5.2).
- Predict the products of acid–base reactions involving common acids and strong bases (Section 5.4).
- Understand that the net ionic equation for the reaction of a strong acid with a strong base is $H^+(aq) + OH^-(aq) \longrightarrow H_2O(\ell)$ (Section 5.4).
- Predict the products of gas-forming reactions (Section 5.5), the most common of which are those between a metal carbonate and an acid.

$$NiCO_3(s) \;+\; 2\,HNO_3(aq) \;\longrightarrow\; Ni(NO_3)_2(aq) \;+\; CO_2(g) \;+\; H_2O(\ell)$$

- Use the ideas developed in Sections 5.2–5.7 as an aid in recognizing four of the common types of reactions that occur in aqueous solution and write balanced equations for such reactions (Section 5.6).

Reaction Type	Driving Force
Precipitation	Formation of an insoluble compound
Acid–strong base	Formation of a salt and water
Gas-forming	Evolution of a gas such as CO_2
Oxidation–reduction	Transfer of electrons

Note that the first three of these reaction types involve the exchange of anions between the cations involved,

$$A^+B^- \;+\; C^+D^- \;\longrightarrow\; A^+D^- \;+\; C^+B^-$$

and so are called exchange reactions. The fourth (redox reactions) involves the transfer of electrons.

- Recognize oxidation numbers of elements in a compound and understand that these numbers represent the charge an atom has, or appears to have, when the electrons of the compound are counted according to a set of guidelines (Section 5.7).
- Recognize oxidation–reduction reactions (often called redox reactions) (Section 5.7 and Tables 5.4 and 5.5).
- Calculate the concentration of a solute in a solution in units of moles per liter (mol/L) (molarity), and use concentrations in calculations (Section 5.8).
- Describe how to prepare a solution of a given molarity from the solute and a solvent or by dilution from a more concentrated solution (Section 5.8).

- Calculate the pH of a solution containing a strong acid or base and know what this means in terms of the relative amount of hydrogen ion in the solution. Calculate the hydrogen ion concentration of a solution from the pH (Section 5.9).
- Solve stoichiometry problems using solution volume and concentrations (Section 5.10).
- Explain how a titration is carried out, explain the procedure of standardization, and calculate concentrations or amounts of reactants from titration data (Section 5.10).

Key Terms

Section 5.1
aqueous solution
solvent
solute
electrodes
electrolytes
strong electrolytes
weak electrolytes
nonelectrolytes

Section 5.2
precipitation reaction
precipitate
spectator ions
net ionic equation

Section 5.3
strong acid

weak acid
strong base
weak base
acidic oxides
basic oxides

Section 5.4
neutralization reaction

Section 5.6
exchange reactions
product-favored
reactant-favored

Section 5.7
reducing agent
oxidizing agent
redox reactions
reduced

oxidized
oxidation number

Section 5.8
molarity

Section 5.9
pH

Section 5.10
acid–base indicator
quantitative chemical analysis
titration
equivalence point
standardization
primary standard

Key Equations

Equation 5.1 (page 176)
Definition of molarity, a measure of the concentration of a solute in a solution.

$$\text{Concentration } (c_{molarity}) = \frac{\text{amount of solute (mol)}}{\text{volume of solution (L)}}$$

A useful form of this equation is

$$\text{Amount of solute (moles)} = c_{molarity} \times \text{volume of solution (L)}$$

Related to this equation is the "shortcut" used when diluting a concentrated solution to obtain a more dilute solution. The product of the concentration and volume of a more concentrated solution (c) must be the same as that for the diluted solution (d).

$$c_c \times V_c = c_d \times V_d$$

If any three of these parameters is known (say c_c, V_c, and c_d), the fourth may be calculated (say V_d).

Equation 5.2 (page 181)
The pH of a solution is the negative logarithm of the hydrogen ion concentration.

$$pH = -\log [H^+]$$

Equation 5.3 (page 182)
The equation for calculating the hydrogen ion concentration of a solution from the pH of the solution.

$$[H^+] = 10^{-pH}$$

Study Questions

Questions with blue, bold-faced numbers have answers in Appendix O.

Reviewing Important Concepts

1. What is the difference between a solvent and a solute?

2. Find one example in the chapter of each of the following reaction types: acid–base, precipitation, gas-forming, and oxidation–reduction reactions. Describe how each reaction exemplifies its class. Name the reactants and products of each reaction.

3. What is an electrolyte? How can you differentiate experimentally between a weak and a strong electrolyte? Give an example of each.

4. Name two acids that are strong electrolytes and one that is a weak electrolyte. Name two bases that are strong electrolytes and one that is a weak electrolyte.

5. Which of the following copper(II) salts are soluble in water and which are insoluble: $Cu(NO_3)_2$, $CuCO_3$, $Cu_3(PO_4)_2$, and $CuCl_2$?

6. What anions combine with Al^{3+} ion to produce water-soluble compounds?.

7. Name the spectator ion or ions in the reaction of nitric acid and magnesium hydroxide and write the net ionic equation:

$$2\ H^+(aq) + 2\ NO_3^-(aq) + Mg(OH)_2(s) \longrightarrow$$
$$2\ H_2O(\ell) + Mg^{2+}(aq) + 2\ NO_3{}^-(aq)$$

What type of exchange reaction is this?

8. Name the water-insoluble product in each reaction:
   (a) $CuCl_2(aq) + H_2S(aq) \longrightarrow CuS + 2\ HCl$
   (b) $CaCl_2(aq) + K_2CO_3(aq) \longrightarrow 2\ KCl + CaCO_3$
   (c) $AgNO_3(aq) + NaI(aq) \longrightarrow AgI + NaNO_3$

9. Bromine is obtained from sea water by the following reaction:

$$Cl_2(g) + 2\ NaBr(aq) \longrightarrow 2\ NaCl(aq) + Br_2(\ell)$$

What has been oxidized? What has been reduced? Name the oxidizing and reducing agents.

10. Oxidation–reduction reactions:
    (a) Explain the difference between oxidation and reduction. Give an example of each.
    (b) Explain the difference between an oxidizing agent and a reducing agent. Give an example of each.

11. Identify each of the following substances as an oxidizing or reducing agent: HNO_3, Na, Cl_2, O_2, and $KMnO_4$.

12. Find an example of a gas-forming reaction in the chapter. Write a balanced equation for the reaction and name the reactants and products.

13. Which contains the greater mass of solute: 1 L of 0.1 M $NaCl$ or 1 L of 0.06 M Na_2CO_3?

14. Which solution contains the larger concentration of ions: 0.20 M $BaCl_2$ or 0.25 M $NaCl$?

15. Describe each of the following as product- or reactant-favored.
    (a) $BaBr_2(aq) + 2\ H_2O(\ell) \longrightarrow Ba(OH)_2(aq) + 2\ HBr(aq)$
    (b) $3\ NaOH(aq) + FeCl_3(aq) \longrightarrow$
    $$3\ NaCl(aq) + Fe(OH)_3(s)$$

16. You have a bottle of solid Na_2CO_3 and a 500.0-mL volumetric flask. Explain how to make a 0.20 M solution of sodium carbonate.

17. You have 0.500 mol of KCl, some distilled water, and a 250.-mL volumetric flask. Describe how to make a 0.500 M solution of KCl.

18. Which has the larger concentration of hydrogen ion, 0.015 M HCl or a hydrochloric acid solution with a pH of 1.2?

19. What is the equivalence point in a titration? What is the function of an indicator in a titration?

Practicing Skills

Solubility of Compounds
(See Example 5.1 and CD-ROM Screen 5.4)

20. Which compound or compounds in each of the following groups is (are) expected to be soluble in water?
    (a) CuO, $CuCl_2$, and $FeCO_3$
    (b) AgI, Ag_3PO_4, and $AgNO_3$
    (c) K_2CO_3, KI, and $KMnO_4$

21. Which compound or compounds in each of the following groups is (are) expected to be soluble in water?
    (a) $BaSO_4$, $Ba(NO_3)_2$, and $BaCO_3$
    (b) Na_2SO_4, $NaClO_4$, $NaCH_3CO_2$
    (c) $AgBr$, KBr, Al_2Br_6

22. Each of the following compounds is water-soluble. What ions are produced in water?
    (a) KOH (c) $LiNO_3$
    (b) K_2SO_4 (d) $(NH_4)_2SO_4$

23. Each of the following compounds is water-soluble. What ions are produced in water?
    (a) KI (c) K_2HPO_4
    (b) $Mg(CH_3CO_2)_2$ (d) $NaCN$

24. Decide whether each of the following is water-soluble. If soluble, tell what ions are produced.
    (a) Na_2CO_3 (c) NiS
    (b) $CuSO_4$ (d) $BaBr_2$

25. Decide whether each of the following is water-soluble. If soluble, tell what ions are produced.
    (a) $NiCl_2$ (c) $Pb(NO_3)_2$
    (b) $Cr(NO_3)_3$ (d) $BaSO_4$

Precipitation Reactions

(See Example 5.2 and CD-ROM Screens 5.6 and 5.7)

26. Balance the equation for the following precipitation reaction, and then write the net ionic equation. Indicate the state of each species (s, ℓ, aq, or g).

$$CdCl_2 + NaOH \longrightarrow Cd(OH)_2 + NaCl$$

27. Balance the equation for the following precipitation reaction, and then write the net ionic equation. Indicate the state of each species (s, ℓ, aq, or g).

$$Ni(NO_3)_2 + Na_2CO_3 \longrightarrow NiCO_3 + NaNO_3$$

28. Predict the products of each precipitation reaction, and then balance the completed equation.
    (a) $NiCl_2(aq) + (NH_4)_2S(aq) \longrightarrow$?
    (b) $Mn(NO_3)_2(aq) + Na_3PO_4(aq) \longrightarrow$?

29. Predict the products of each precipitation reaction, and then balance the completed equation.
    (a) $Pb(NO_3)_2(aq) + KBr(aq) \longrightarrow$?
    (b) $Ca(NO_3)_2(aq) + KF(aq) \longrightarrow$?
    (c) $Ca(NO_3)_2(aq) + Na_2C_2O_4(aq) \longrightarrow$?

Writing Net Ionic Equations

(See Example 5.3 and CD-ROM Screen 5.5)

30. Balance each of the following equations, and then write the net ionic equation:
    (a) $(NH_4)_2CO_3(aq) + Cu(NO_3)_2(aq) \longrightarrow$
    $$CuCO_3(s) + NH_4NO_3(aq)$$
    (b) $Pb(OH)_2(s) + HCl(aq) \longrightarrow PbCl_2(s) + H_2O(\ell)$
    (c) $BaCO_3(s) + HCl(aq) \longrightarrow$
    $$BaCl_2(aq) + H_2O(\ell) + CO_2(g)$$

31. Balance each of the following equations, and then write the net ionic equation:
    (a) $Zn(s) + HCl(aq) \longrightarrow H_2(g) + ZnCl_2(aq)$
    (b) $Mg(OH)_2(s) + HCl(aq) \longrightarrow MgCl_2(aq) + H_2O(\ell)$
    (c) $HNO_3(aq) + CaCO_3(s) \longrightarrow$
    $$Ca(NO_3)_2(aq) + H_2O(\ell) + CO_2(g)$$

32. Balance each of the following equations, and then write the net ionic equation. Show states for all reactants and products (s, ℓ, g, aq).
    (a) $AgNO_3 + KI \longrightarrow AgI + KNO_3$
    (b) $Ba(OH)_2 + HNO_3 \longrightarrow Ba(NO_3)_2 + H_2O$
    (c) $Na_3PO_4 + Ni(NO_3)_2 \longrightarrow Ni_3(PO_4)_2 + NaNO_3$

33. Write balanced equations for the following reactions, and then write net ionic equations. Show states for all reactants and products (s, ℓ, g, aq).
    (a) The reaction of sodium hydroxide and iron(II) chloride to give iron(II) hydroxide and sodium chloride.

(b) The reaction of barium chloride with sodium carbonate to give barium carbonate and sodium chloride.

Acids and Bases

(See Exercises 5.5 and 5.6 and CD-ROM Screens 5.8 and 5.9)

34. Write a balanced equation for the ionization of nitric acid in water.

35. Write a balanced equation for the ionization of perchloric acid in water.

36. Oxalic acid, $H_2C_2O_4$, which is found in certain plants, can provide two hydrogen ions in water. Write balanced equations (like those for sulfuric acid on page 157) to show how oxalic acid can supply one and then a second H^+ ion.

37. Phosphoric acid can supply one, two, or three H^+ ions in aqueous solution. Write balanced equations (like those for sulfuric acid on page 157) to show this successive loss of hydrogen ions.

38. Write a balanced equation for reaction of the basic oxide, magnesium oxide, with water.

39. Write a balanced equation for the reaction of sulfur trioxide with water.

Reactions of Acids and Bases

(See Example 5.4, Exercise 5.7, and CD-ROM Screens 5.6 and 5.10)

40. Complete and balance the following acid–base reactions. Name the reactants and products.
    (a) $CH_3CO_2H(aq) + Mg(OH)_2(s) \longrightarrow$
    (b) $HClO_4(aq) + NH_3(aq) \longrightarrow$

41. Complete and balance the following acid–base reactions. Name the reactants and products.
    (a) $H_3PO_4(aq) + KOH(aq) \longrightarrow$
    (b) $H_2C_2O_4(aq) + Ca(OH)_2(s) \longrightarrow$

42. Write a balanced equation for the reaction of barium hydroxide with nitric acid to give barium nitrate.

43. Write a balanced equation for the reaction of aluminum hydroxide with sulfuric acid.

Gas-Forming Reactions

(See Example 5.5 and CD-ROM Screens 5.6 and 5.11)

44. Siderite is a mineral consisting largely of iron(II) carbonate. Write a balanced equation for the reaction of the mineral with nitric acid, and name each reactant and product.

45. The beautiful red mineral rhodochrosite is manganese(II) carbonate. Write a balanced equation for the reaction of the mineral with hydrochloric acid. Name each reactant and product.

Rhodochrosite, a mineral consisting largely of $MnCO_3$.
(Charles D. Winters)

Types of Reactions in Aqueous Solution
(See Exercise 5.8 and CD-ROM Screen 5.6)

46. Balance the following equations and then classify each reaction as a precipitation reaction, an acid–base reaction, or a gas-forming reaction.
    (a) $Ba(OH)_2(s) + HCl(aq) \longrightarrow BaCl_2(aq) + H_2O(\ell)$
    (b) $HNO_3(aq) + CoCO_3(s) \longrightarrow$
    $\qquad Co(NO_3)_2(aq) + H_2O(\ell) + CO_2(g)$
    (c) $Na_3PO_4(aq) + Cu(NO_3)_2(aq) \longrightarrow$
    $\qquad Cu_3(PO_4)_2(s) + NaNO_3(aq)$

47. Balance the following equations and then classify each reaction as a precipitation reaction, an acid–base reaction, or a gas-forming reaction.
    (a) $K_2CO_3(aq) + Cu(NO_3)_2(aq) \longrightarrow$
    $\qquad CuCO_3(s) + KNO_3(aq)$
    (b) $Pb(NO_3)_2(aq) + HCl(aq) \longrightarrow PbCl_2(s) + HNO_3(aq)$
    (c) $MgCO_3(s) + HCl(aq) \longrightarrow$
    $\qquad MgCl_2(aq) + H_2O(\ell) + CO_2(g)$

48. Balance the following equations and then classify each reaction as a precipitation reaction, an acid–base reaction, or a gas-forming reaction. Show states for the products (s, ℓ, g, aq). Write the net ionic equation.
    (a) $MnCl_2(aq) + Na_2S(aq) \longrightarrow MnS + NaCl$
    (b) $K_2CO_3(aq) + ZnCl_2(aq) \longrightarrow ZnCO_3 + KCl$

49. Balance the following equations and then classify each reaction as a precipitation reaction, an acid–base reaction, or a gas-forming reaction. Write the net ionic equation.
    (a) $Fe(OH)_3(s) + HNO_3(aq) \longrightarrow Fe(NO_3)_3 + H_2O$
    (b) $FeCO_3(s) + HNO_3(aq) \longrightarrow Fe(NO_3)_2 + CO_2 + H_2O$

Product- or Reactant-Favored Reactions

50. What feature of each the following reactions causes it to be product-favored?
    (a) $CuCl_2(aq) + H_2S(aq) \longrightarrow CuS(s) + 2 HCl(aq)$
    (b) $H_3PO_4(aq) + 3 KOH(aq) \longrightarrow 3 H_2O(\ell) + K_3PO_4(aq)$

51. Which of the following reactions is predicted to be product-favored?
    (a) $Zn(s) + 2 HCl(aq) \longrightarrow H_2(g) + ZnCl_2(aq)$
    (b) $MgCl_2(aq) + 2 H_2O(\ell) \longrightarrow Mg(OH)_2(s) + 2 HCl(aq)$

Oxidation Numbers
(See Example 5.7 and CD-ROM Screen 5.13)

52. Determine the oxidation number of each element in the following ions or compounds:
    (a) BrO_3^- (d) CaH_2
    (b) $C_2O_4^{2-}$ (e) H_4SiO_4
    (c) F^- (f) HSO_4^-

53. Determine the oxidation number of each element in the following ions or compounds:
    (a) PF_6^- (d) N_2O_5
    (b) $H_2AsO_4^-$ (e) $POCl_3$
    (c) UO^{2+} (f) XeO_4^{2-}

Oxidation–Reduction Reactions
(See Examples 5.8 and 5.9 and CD-ROM Screens 5.12–5.14)

54. Which of the following reactions are oxidation–reduction reactions? Explain your answer in each case. Classify the remaining reaction.
    (a) $Zn(s) + 2 NO_3^-(aq) + 4 H^+(aq) \longrightarrow$
    $\qquad Zn^{2+}(aq) + 2 NO_2(g) + 2 H_2O(\ell)$
    (b) $Zn(OH)_2(s) + H_2SO_4(aq) \longrightarrow ZnSO_4(aq) + 2 H_2O(\ell)$
    (c) $Ca(s) + 2 H_2O(\ell) \longrightarrow Ca(OH)_2(s) + H_2(g)$

55. Which of the following reactions are oxidation–reduction reactions? Explain your answer briefly. Classify the remaining reaction.
    (a) $CdCl_2(aq) + Na_2S(aq) \longrightarrow CdS(s) + 2 NaCl(aq)$
    (b) $2 Ca(s) + O_2(g) \longrightarrow 2 CaO(s)$
    (c) $4 Fe(OH)_2(s) + 2 H_2O(\ell) + O_2(g) \longrightarrow 4 Fe(OH)_3(aq)$

56. In each of the following reactions, decide which reactant is oxidized and which is reduced. Identify the oxidizing agent and reducing agent.
    (a) $C_2H_4(g) + 3 O_2(g) \longrightarrow 2 CO_2(g) + 2 H_2O(g)$
    (b) $Si(s) + 2 Cl_2(g) \longrightarrow SiCl_4(\ell)$

57. In each of the following reactions, decide which reactant is oxidized and which is reduced. Identify the oxidizing agent and reducing agent.
    (a) $Cr_2O_7^{2-}(aq) + 3 Sn^{2+}(aq) + 14 H^+(aq) \longrightarrow$
    $\qquad 2 Cr^{3+}(aq) + 3 Sn^{4+}(aq) + 7 H_2O(\ell)$
    (b) $FeS(s) + 3 NO_3^-(aq) + 4 H^+(aq) \longrightarrow$
    $\qquad 3 NO(g) + SO_4^{2-}(aq) + Fe^{3+}(aq) + 2 H_2O(\ell)$

Solution Concentration
(See Example 5.10 and CD-ROM Screen 5.15)

58. If 6.73 g of Na_2CO_3 is dissolved in enough water to make 250. mL of solution, what is the molar concentration of the sodium carbonate? What are the molar concentrations of the Na^+ and CO_3^{2-} ions?

59. Some potassium dichromate ($K_2Cr_2O_7$), 2.335 g, is dissolved in enough water to make exactly 500. mL of solution. What is the molar concentration of the potassium dichromate? What are the molar concentrations of the K^+ and $Cr_2O_7^{2-}$ ions?

60. What is the mass, in grams, of solute in 250. mL of a 0.0125 M solution of $KMnO_4$?

61. What is the mass, in grams, of solute in 125 mL of a 1.023×10^{-3} M solution of Na_3PO_4? What are the molar concentrations of the Na^+ and PO_4^{3-} ions?

62. What volume of 0.123 M NaOH, in milliliters, contains 25.0 g of NaOH?

63. What volume of 2.06 M $KMnO_4$, in liters, contains 322 g of solute?

64. For each solution, identify the ions that exist in aqueous solution, and specify the concentration of each.
    (a) 0.25 M $(NH_4)_2SO_4$
    (b) 0.123 M Na_2CO_3
    (c) 0.056 M HNO_3

65. For each solution, identify the ions that exist in aqueous solution, and specify the concentration of each.
    (a) 0.12 M $BaCl_2$
    (b) 0.0125 M $CuSO_4$
    (c) 0.500 M $K_2Cr_2O_7$

Preparing Solutions
(See Exercise 5.14, Example 5.10, and CD-ROM Screen 5.16)

66. An experiment in your laboratory requires exactly 500. mL of a 0.0200 M solution of Na_2CO_3. You are given solid Na_2CO_3, distilled water, and a 500.-mL volumetric flask. Describe how to make up the required solution.

67. What mass of oxalic acid, $H_2C_2O_4$, is required to make 250. mL of a solution that has a concentration of 0.15 M in the acid?

68. If you dilute 25.0 mL of 1.50 M hydrochloric acid to 500. mL, what is the molar concentration of the dilute acid?

69. If 4.00 mL of 0.0250 M $CuSO_4$ is diluted to 10.0 mL with pure water, what is the molar concentration of copper(II) sulfate in the dilute solution?

70. Which of the following methods would you use to prepare 1.00 L of 0.125 M H_2SO_4?
    (a) Dilute 20.8 mL of 6.00 M H_2SO_4 to a volume of 1.00 L.
    (b) Add 950. mL of water to 50.0 mL of 3.00 M H_2SO_4.

71. Which of the following methods would you use to prepare 300. mL of 0.500 M $K_2Cr_2O_7$?
    (a) Add 30.0 mL of 1.50 M $K_2Cr_2O_7$ to 270. mL of water.
    (b) Dilute 250. mL of 0.600 M $K_2Cr_2O_7$ to a volume of 300. mL.

pH
(See Example 5.11 and CD-ROM Screen 5.17)

72. A table wine has a pH of 3.40. What is the hydrogen ion concentration of the wine? Is it acidic or basic?

73. A saturated solution of milk of magnesia, $Mg(OH)_2$, has a pH of 10.5. What is the hydrogen ion concentration of the solution? Is the solution acidic or basic?

74. What is the hydrogen ion concentration of a 0.0013 M solution of HNO_3? What is its pH?

75. What is the hydrogen ion concentration of a 1.2×10^{-4} M solution of $HClO_4$? What is its pH?

76. Make the following conversions. In each case, tell whether the solution is acidic or basic.

	pH	$[H^+]$
(a)	1.00	_____
(b)	10.50	_____
(c)	_____	1.3×10^{-5} M
(d)	_____	2.3×10^{-8} M

77. Make the following conversions. In each case, tell whether the solution is acidic or basic.

	pH	$[H^+]$
(a)	_____	6.7×10^{-10} M
(b)	_____	2.2×10^{-6} M
(c)	5.25	_____
(d)	_____	2.5×10^{-2} M

Stoichiometry of Reactions in Solution
(See Example 5.13 and CD-ROM Screen 5.18)

78. What volume of 0.109 M HNO_3, in milliliters, is required to react completely with 2.50 g of $Ba(OH)_2$?

$$2\,HNO_3(aq) + Ba(OH)_2(s) \longrightarrow 2\,H_2O(\ell) + Ba(NO_3)_2(aq)$$

79. What mass of Na_2CO_3, in grams, is required for complete reaction with 50.0 mL of 0.125 M HNO_3?

$$Na_2CO_3(aq) + 2\,HNO_3(aq) \longrightarrow 2\,NaNO_3(aq) + CO_2(g) + H_2O(\ell)$$

80. When an electric current is passed through an aqueous solution of NaCl, $H_2(g)$, $Cl_2(g)$, and NaOH, all valuable industrial chemicals are produced.

$$2\,NaCl(aq) + 2\,H_2O(\ell) \longrightarrow H_2(g) + Cl_2(g) + 2\,NaOH(aq)$$

What mass of NaOH can be produced from 15.0 L of 0.35 M NaCl? What mass of chlorine can be obtained?

81. Hydrazine, N_2H_4, a base like ammonia, can react with an acid such as sulfuric acid.

$$2\,N_2H_4(aq) + H_2SO_4(aq) \longrightarrow 2\,N_2H_5^+(aq) + SO_4^{2-}(aq)$$

What mass of hydrazine reacts with 250. mL of 0.146 M H_2SO_4?

82. In the photographic developing process, silver bromide is dissolved by adding sodium thiosulfate:

$$AgBr(s) + 2\,Na_2S_2O_3(aq) \longrightarrow$$
$$Na_3Ag(S_2O_3)_2(aq) + NaBr(aq)$$

If you want to dissolve 0.225 g of AgBr, what volume of 0.0138 M $Na_2S_2O_3$, in milliliters, should be used?

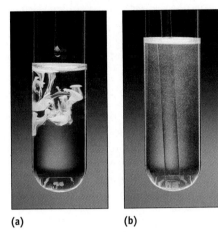

(a)   **(b)**

Silver chemistry. (a) A precipitate of AgBr formed by adding $AgNO_3$ (aq) to KBr(aq). **(b)** On adding $Na_2S_2O_3$(aq), sodium thiosulfate, the solid AgBr dissolves. *(Charles D. Winters)*

83. You can dissolve an aluminum soft drink can in an aqueous base such as potassium hydroxide.

$$2\,Al(s) + 2\,KOH(aq) + 6\,H_2O(\ell) \longrightarrow$$
$$2\,KAl(OH)_4(aq) + 3\,H_2(g)$$

If you place 2.05 g of aluminum in a beaker with 185 mL of 1.35 M KOH, will any aluminum remain? What mass of $KAl(OH)_4$ is produced?

84. What volume of 0.750 M $Pb(NO_3)_2$, in milliliters, is required to react completely with 1.00 L of 2.25 M NaCl solution? The balanced equation is

$$Pb(NO_3)_2(aq) + 2\,NaCl(aq) \longrightarrow PbCl_2(s) + 2\,NaNO_3(aq)$$

85. What volume of 0.125 M oxalic acid, $H_2C_2O_4$, is required to react with 35.2 mL of 0.546 M NaOH?

$$H_2C_2O_4(aq) + 2\,NaOH(aq) \longrightarrow$$
$$Na_2C_2O_4(aq) + 2\,H_2O(aq)$$

Titrations
(See Examples 5.14–5.16 and CD-ROM Screen 5.19)

86. What volume of 0.812 M HCl, in milliliters, is required to titrate 1.45 g of NaOH to the equivalence point?

$$NaOH(aq) + HCl(aq) \longrightarrow H_2O(\ell) + NaCl(aq)$$

87. What volume of 0.955 M HCl, in milliliters, is needed to titrate 2.152 g of Na_2CO_3 to the equivalence point?

$$Na_2CO_3(aq) + 2\,HCl(aq) \longrightarrow$$
$$H_2O(\ell) + CO_2(g) + 2\,NaCl(aq)$$

88. If 38.55 mL of HCl is used to titrate 2.150 g of Na_2CO_3 according to the following equation, what is the molarity of the HCl?

$$Na_2CO_3(aq) + 2\,HCl(aq) \longrightarrow$$
$$2\,NaCl(aq) + CO_2(g) + H_2O(\ell)$$

89. Potassium hydrogen phthalate, $KHC_8H_4O_4$, is used to standardize solutions of bases. The acidic anion reacts with strong bases according to the following net ionic equation:

$$HC_8H_4O_4^-(aq) + OH^-(aq) \longrightarrow C_8H_4O_4^{2-}(aq) + H_2O(\ell)$$

If a 0.902-g sample of potassium hydrogen phthalate is dissolved in water and titrated to the equivalence point with 26.45 mL of NaOH, what is the molar concentration of the NaOH?

90. You have 0.954 g of an unknown acid, H_2A, which reacts with NaOH according to the balanced equation

$$H_2A(aq) + 2\,NaOH(aq) \longrightarrow Na_2A(aq) + 2\,H_2O(\ell)$$

If 36.04 mL of 0.509 M NaOH is required to titrate the acid to the equivalence point, what is the molar mass of the acid?

91. An unknown solid acid is either citric acid or tartaric acid. To determine which acid you have, you titrate a sample of the solid with NaOH. The appropriate reactions are:

Citric acid:

$$H_3C_6H_5O_7(aq) + 3\,NaOH(aq) \longrightarrow$$
$$3\,H_2O(\ell) + Na_3C_6H_5O_7(aq)$$

Tartaric acid:

$$H_2C_4H_4O_6(aq) + 2\,NaOH(aq) \longrightarrow$$
$$2\,H_2O(\ell) + Na_2C_4H_4O_6(aq)$$

A 0.956-g sample requires 29.1 mL of 0.513 M NaOH for titration to the equivalence point. What is the unknown acid?

92. To analyze an iron-containing compound, you convert all the iron to Fe^{2+} in aqueous solution and then titrate the solution with standardized $KMnO_4$. The balanced, net ionic equation is:

$$MnO_4^-(aq) + 5\,Fe^{2+}(aq) + 8\,H^+(aq) \longrightarrow$$
$$Mn^{2+}(aq) + 5\,Fe^{3+}(aq) + 4\,H_2O(\ell)$$

A 0.598-g sample of the iron-containing compound requires 22.25 mL of 0.0123 M $KMnO_4$ for titration to the equivalence point. What is the mass percent of iron in the sample?

93. Vitamin C is the simple compound $C_6H_8O_6$. Besides being an acid, it is also a reducing agent. One method for determining the amount of vitamin C in a sample is therefore to titrate it with a solution of bromine, Br_2, an oxidizing agent.

$$C_6H_8O_6(aq) + Br_2(aq) \longrightarrow 2\,HBr(aq) + C_6H_6O_6(aq)$$

A 1.00-g "chewable" vitamin C tablet requires 27.85 mL of 0.102 M Br_2 for titration to the equivalence point. What is the mass of vitamin C in the tablet?

General Questions

More challenging questions are indicated by an underlined number.

94. What volume of 0.054 M H_2SO_4 is required to react completely with 1.56 g of KOH?

95. The mineral dolomite contains magnesium carbonate. Write the net ionic equation for the reaction of magnesium carbonate and hydrochloric acid and name the spectator ions. What type of reaction is this?

$$MgCO_3(s) + 2\,HCl(aq) \longrightarrow$$
$$MgCl_2(g) + CO_2(aq) + H_2O(\ell)$$

96. Mg metal reacts readily with HNO_3 according to the following *unbalanced* equation:

$$Mg(s) + HNO_3(aq) \longrightarrow$$
$$Mg(NO_3)_2(aq) + NO_2(g) + H_2O(\ell)$$

(a) Balance the equation for the reaction.

(b) Name each compound.

(c) Write the net ionic equation for the reaction.

(d) What are the oxidizing and reducing agents?

97. Ammonium sulfide, $(NH_4)_2S$, reacts with $Hg(NO_3)_2$ to produce HgS and NH_4NO_3.

(a) Write the overall balanced equation for the reaction. Indicate the state (s or aq) for each compound.

(b) Name each compound.

(c) What type of reaction is this?

98. What species (atoms, molecules, or ions) are present in an aqueous solution of each of the following compounds?

(a) NH_3 (c) NaOH

(b) CH_3CO_2H (d) HBr

99. Suppose an Alka-Seltzer tablet contains exactly 100 mg of citric acid, $H_3C_6H_5O_7$, plus some sodium bicarbonate. If a reaction occurs according to the following *(unbalanced)* equation, what mass of sodium bicarbonate must the tablet also contain?

$$H_3C_6H_5O_7(aq) + NaHCO_3(aq) \longrightarrow$$
$$H_2O(\ell) + CO_2(g) + Na_3C_6H_5O_7(aq)$$

100. Sodium bicarbonate and acetic acid react according to the equation

$$NaHCO_3(aq) + CH_3CO_2H(aq) \longrightarrow$$
$$NaCH_3CO_2(aq) + CO_2(g) + H_2O(\ell)$$

Suppose you add 15.0 g of $NaHCO_3$ to 125 mL of 0.15 M acetic acid. What is the limiting reactant?.

101. A noncarbonated soft drink contains an unknown amount of citric acid, $H_3C_6H_5O_7$. If 100. mL of the soft drink requires 33.51 mL of 0.0102 M NaOH to neutralize the citric acid completely, what mass of citric acid does the soft drink contain per 100. mL? The reaction of citric acid and NaOH is

$$H_3C_6H_5O_7(aq) + 3\,NaOH(aq) \longrightarrow$$
$$Na_3C_6H_5O_7(aq) + 3\,H_2O(\ell)$$

102. You have a 4.554-g sample that is a mixture of oxalic acid, $H_2C_2O_4$, and another solid that does not react with sodium hydroxide. If 29.58 mL of 0.550 M NaOH is required to titrate the oxalic acid in the sample to the equivalence point, what is the weight percent of oxalic acid in the mixture?

103. Sodium thiosulfate, $Na_2S_2O_3$, is used as a "fixer" in black-and-white photography. Suppose you have a bottle of sodium thiosulfate and want to determine its purity. The thiosulfate ion can be oxidized with I_2 according to the balanced equation

$$I_2(aq) + 2\,S_2O_3{}^{2-}(aq) \longrightarrow 2\,I^-(aq) + S_4O_6{}^{2-}(aq)$$

If you use 40.21 mL of 0.246 M I_2 in a titration, what is the weight percent of $Na_2S_2O_3$ in a 3.232-g sample of impure material?

104. Solubility

(a) Name two water-soluble compounds containing the Cu^{2+} ion. Name two water-insoluble compounds based on the Cu^{2+} ion.

(b) Name two water-soluble compounds containing the Ba^{2+} ion. Name two water-insoluble compounds based on the Ba^{2+} ion.

105. A reaction occurs when aqueous lead(II) nitrate is mixed with an aqueous solution of potassium hydroxide.

(a) Write an overall, balanced equation for the reaction.

(b) Name each reactant and product.

(c) Write the net ionic equation.

106. Balance these reactions and then classify each one as a precipitation, an acid–base reaction, or a gas-forming reaction. Show states for the products (s, ℓ, g, aq), and write the net ionic equation.

(a) $K_2CO_3(aq) + HClO_4(aq) \longrightarrow KClO_4 + CO_2 + H_2O$

(b) $FeCl_2(aq) + (NH_4)_2S(aq) \longrightarrow FeS + NH_4Cl$

(c) $Fe(NO_3)_2(aq) + Na_2CO_3(aq) \longrightarrow FeCO_3 + NaNO_3$

107. A reaction occurs when aqueous copper(II) nitrate is mixed with an aqueous solution of sodium carbonate.

(a) Write an overall, balanced equation for the reaction.

(b) Name each reactant and product.

(c) Write the net ionic equation.

108. pH Measurements

(a) What is the pH of a 0.105 M HCl solution?

(b) What is the hydrogen ion concentration in a solution with a pH of 2.56? Is the solution acidic or basic?

(c) A solution has a pH of 9.67. What is the hydrogen ion concentration in the solution? Is the solution acidic or basic?

109. A 10.0-mL sample of 2.56 M HCl is diluted with water to 250. mL.

(a) What is the concentration of HCl in the diluted solution?

(b) What is the pH of the dilute solution?

110. One-half liter (500. mL) of 2.50 M HCl is mixed with 250. mL of 3.75 M HCl. Assuming the total solution volume after mixing is 750. mL, what is the concentration of hydrochloric acid in the resulting solution? What is its pH?

111. A solution of hydrochloric acid has a volume of 125 mL and a pH of 2.56. What mass of $NaHCO_3$ must be added to completely consume the HCl?

112. A solution of hydrochloric acid has a volume of 250. mL and a pH of 1.92. Exactly 250. mL of 0.0105 M NaOH is added. What is the pH of the resulting solution?

113. Suppose you dilute 25.0 mL of a 0.110 M solution of Na_2CO_3 to exactly 100.0 mL. You then take exactly 10.0 mL of this diluted solution and add it to a 250-mL volumetric flask. After filling the volumetric flask to the mark with distilled water (indicating the volume of the new solution is exactly 250 mL), what is the concentration of the diluted Na_2CO_3 solution?

114. The following reaction can be used to prepare iodine in the laboratory. (See photos.)

$$2\ NaI(s) + 2\ H_2SO_4(aq) + MnO_2(s) \longrightarrow$$
$$Na_2SO_4(aq) + MnSO_4(aq) + I_2(s) + 2\ H_2O(\ell)$$

   (a) Determine the oxidation number of each atom in the equation.

   (b) What is the oxidizing agent and what has been oxidized? What is the reducing agent and what has been reduced?

   (c) What quantity of iodine can be obtained if 20.0 g of NaI is mixed with 10.0 g of MnO_2 (and a stoichiometric excess of sulfuric acid)?

Preparation of iodine. A mixture of sodium iodide and manganese(IV) oxide was placed in a flask in a hood *(left)*. On adding concentrated sulfuric acid *(right)*, brown I_2 was evolved. *(Charles D. Winters)*

115. You place 2.56 g of $CaCO_3$ in a beaker containing 250. mL of 0.125 M HCl (see Figure 5.19). When the reaction below has ceased, does any calcium carbonate remain? Explain your reasoning. What mass of $CaCl_2$ can be produced?

$$CaCO_3(s) + 2\ HCl(aq) \longrightarrow$$
$$CaCl_2(aq) + CO_2(g) + H_2O(\ell)$$

116. The types of reactions described in this chapter can be used to prepare compounds. For example, insoluble barium chromate can be made by a precipitation reaction.

$$BaCl_2(aq) + K_2CrO_4(aq) \longrightarrow BaCrO_4(s) + 2\ KCl(aq)$$

The product, $BaCrO_4$, can be separated from dissolved products or reactants by filtering the product mixture (see Figure 4.5). The insoluble $BaCrO_4$ is trapped on the filter paper, and aqueous KCl passes through the paper. Suggest a precipitation reaction and a gas-forming reaction by which barium sulfate can be made.

117. You want to prepare barium chloride, $BaCl_2$, using an exchange reaction of some type. To do this, you have the following reagents from which to select the reactants: $BaSO_4$, $BaBr_2$, $BaCO_3$, $Ba(OH)_2$, HCl, H_2SO_4, $AgNO_3$, and HNO_3.

   Write a complete, balanced equation to show how to prepare barium chloride using some combination of the preceding reagents. *(Note: There are several possibilities.)*

118. Describe how to prepare zinc chloride by (a) an acid–base reaction, (b) a gas-forming reaction, and (c) an oxidation–reduction reaction. The available starting materials are $ZnCO_3$, HCl, Cl_2, HNO_3, $Zn(OH)_2$, NaCl, $Zn(NO_3)_2$, and Zn.

119. A compound has been isolated that can have either of two possible formulas: (a) $K[Fe(C_2O_4)_2(H_2O)_2]$ or (b) $K_3[Fe(C_2O_4)_3]$. To find which is correct, you dissolve a weighed sample of the compound in acid and then titrate the oxalate ion $(C_2O_4^{2-})$ with potassium permanganate, $KMnO_4$ (the source of the MnO_4^- ion). The balanced, net ionic equation for the titration is

$$5\ C_2O_4^{2-}(aq) + 2\ MnO_4^-(aq) + 16\ H^+(aq) \longrightarrow$$
$$2\ Mn^{2+}(aq) + 10\ CO_2(g) + 8\ H_2O(\ell)$$

Titration of 1.356 g of the compound requires 34.50 mL of 0.108 M $KMnO_4$. Which is the correct formula of the iron-containing compound: (a) or (b)?

120. Two students titrate different samples of the same solution of HCl using 0.100 M NaOH solution and phenolphthalein indicator (see Figure 5.20). The first student pipets 20.0 mL of the HCl solution into a flask, adds 20 mL of distilled water and a few drops of phenolphthalein solution, and titrates until a lasting pink color appears. The second student pipets 20.0 mL of the HCl solution into a flask, adds 60 mL of distilled water and a few drops of phenolphthalein solution, and titrates to the first lasting pink color. Each student correctly calculates the molarity of a HCl solution. The second student's result will be

   (a) Four times less than the first student's

   (b) Four times more than the first student's

   (c) Two times less than the first student's

   (d) Two times more than the first student's

   (e) The same as the first student's

121. You need to know the volume of water in a small swimming pool, but, owing to the pool's irregular shape, it is not simple to determine its dimensions and calculate the volume.

To solve the problem you stir in a solution of a dye (1.0 g of methylene blue, $C_{16}H_{18}ClN_3S$, in 50.0 mL of water). After the dye has mixed with the water in the pool, you take a sample of the water. Using an instrument such as a spectrophotometer, you determine that the concentration of the dye in the pool is 4.1×10^{-8} M. What is the volume of water in the pool?

122. Gold can be dissolved from gold-bearing rock by treating the rock with sodium cyanide in the presence of oxygen.

$$4 Au(s) + 8 NaCN(aq) + O_2(g) + 2 H_2O(\ell) \longrightarrow$$
$$4 NaAu(CN)_2(aq) + 4 NaOH(aq)$$

(a) Name the oxidizing and reducing agents in this reaction. What has been oxidized and what has been reduced?

(b) If you have exactly one metric ton (1 metric ton = 1000 kg) of gold-bearing rock, what volume of 0.075 M NaCN, in liters, do you need to extract the gold if the rock is 0.019% gold?

123. You mix 25.0 mL of 0.234 M $FeCl_3$ with 42.5 mL of 0.453 M NaOH.

(a) Which reactant is in excess?

(b) What mass, in grams, of $Fe(OH)_3$ precipitates?

(c) What is the molar concentration of the excess reactant remaining in solution after the maximum mass of $Fe(OH)_3$ has been precipitated?

124. You wish to determine the weight percent of copper in a copper-containing alloy. After dissolving a sample of the alloy in acid, an excess of KI is added, and the Cu^{2+} and I^- ions undergo the reaction

$$2 Cu^{2+}(aq) + 5 I^-(aq) \longrightarrow 2 CuI(s) + I_3^-(aq)$$

The liberated I_3^- is titrated with sodium thiosulfate according to the equation

$$I_3^-(aq) + 2 S_2O_3^{2-}(aq) \longrightarrow S_4O_6^{2-}(aq) + 3 I^-(aq)$$

If 26.32 mL of 0.101 M $Na_2S_2O_3$ is required for titration to the equivalence point, what is the weight percent of Cu in 0.251 g of the alloy?

125. Chromium(III) ion forms many compounds with ammonia. To find the formula of one of these compounds, you titrate the NH_3 in the compound with standardized acid.

$$Cr(NH_3)_xCl_3(aq) + x HCl(aq) \longrightarrow$$
$$x NH_4^+(aq) + Cr^{3+}(aq) + (x + 3) Cl^-(aq)$$

Assume that 24.26 mL of 1.500 M HCl is used to titrate 1.580 g of $Co(NH_3)_xCl_3$. What is the value of x?

126. The cancer chemotherapy drug cisplatin, $Pt(NH_3)_2Cl_2$, can be made by reacting $(NH_4)_2PtCl_4$ with ammonia in aqueous solution. Besides cisplatin, the other product is NH_4Cl.

(a) Write a balanced equation for this reaction.

(b) To obtain 12.50 g of cisplatin, what mass of $(NH_4)_2PtCl_4$ is required? What volume of 0.125 M NH_3 is required?

(c) Cisplatin can react with the organic compound pyridine, C_5H_5N, to form a new compound.

$$Pt(NH_3)_2Cl_2(aq) + x C_5H_5N(aq) \longrightarrow$$
$$Pt(NH_3)_2Cl_2(C_5H_5N)_x(s)$$

Suppose you treat 0.150 g of cisplatin with what you believe is an excess of liquid pyridine (1.50 mL; $d = 0.979$ g/mL). When the reaction is complete, you can find out how much pyridine was not used by titrating the solution with standardized HCl. If 37.0 mL of 0.475 M HCl is required to titrate the excess pyridine,

$$C_5H_5N(aq) + HCl(aq) \longrightarrow C_5H_5NH^+(aq) + Cl^-(aq)$$

what is the formula of the unknown compound $Pt(NH_3)_2Cl_2(C_5H_5N)_x$?

Using Electronic Resources

These questions refer to the *General Chemistry Interactive CD-ROM*, Version 3.0.

127. See CD-ROM Screen 4.11, Chemical Puzzler: On this screen you can explore the reaction of baking soda ($NaHCO_3$) with the acetic acid in vinegar.

You place exactly 200 mL of vinegar in the beaker and add baking soda. The reaction occurring is

$$CH_3CO_2H(aq) + NaHCO_3(aq) \longrightarrow$$
$$NaCH_3CO_2(aq) + CO_2(g) + H_2O(\ell)$$

(Bottom) Vinegar, which contains acetic acid, and baking soda. *(Top)* Baking soda is added to vinegar. *(Charles D. Winters)*

How many spoonfuls of baking soda are required to consume the acetic acid in the 200-mL sample? (Assume there is 50.0 g of acetic acid per liter of vinegar and that a spoonful of baking soda has a mass of 3.8 g.) Are 3 spoonfuls sufficient? Are 4 spoonfuls enough?

128. See CD-ROM Screen 4.8, Limiting Reactants: You previously explored the reaction of zinc metal and hydrochloric acid. We added smaller and smaller amounts of zinc to three flasks, each containing exactly 100 mL of 0.10 M HCl.

    Flask 1: 7.00 g Zn

    Flask 2: 3.27 g Zn

    Flask 3: 1.31 g Zn

    The zinc and acid combined to produce hydrogen gas.

    $$Zn(s) + 2\,HCl(aq) \longrightarrow ZnCl_2(aq) + H_2(g)$$

    The same amount of H_2 gas was generated in Flasks 1 and 2, but a smaller amount was generated in Flask 3. The zinc was completely consumed in Flask 2 and 3, but some remained in Flask 1. Explain these observations.

129. See the Exercise on CD-ROM Screen 5.14, Recognizing Oxidation–Reduction Reactions.

    (a) Write the balanced equation for the reaction of iron metal and chlorine gas.

    (b) What is the oxidizing agent? The reducing agent?

    (c) Name the product.

    (d) Why was the iron heated before plunging it into the flask of Cl_2 gas?

130. See the Exercise on CD-ROM Screen 5.18, Stoichiometry of Reactions in Solution. The video shows the reaction of Fe^{2+} with MnO_4^- in aqueous solution.

    (a) What is the balanced equation for the reaction that occured?

    (b) What is the oxidizing agent and what is the reducing agent?

    (c) Equal volumes of the Fe^{2+}-containing solution and MnO_4^--containing solution were mixed. The amount of Fe^{2+} was just sufficient to consume all of the MnO_4^- ion. Which ion (Fe^{2+} or MnO_4^-) was initially present in larger concentration?

6 Principles of Reactivity: Energy and Chemical Reactions

Abba's Refrigerator

You know that if you put a pot of water on a kitchen stove or a campfire, or put it in the sun, the water will evaporate. You must supply energy in some form because evaporation requires the input of energy. This well-known principle was applied in a novel way by a young African teacher, Mohammed Bah Abba in Nigeria, to improve the life of his people.

Life is hard in northern Nigerian communities. In this rural semidesert area most people eke out a living by subsistence farming. Lacking modern refrigeration, food spoilage is a major problem. Using a simple thermodynamic principle, Abba developed a refrigerator that cost about 30 cents to make and does not use electricity.

Abba's refrigerator is made of two earthen pots, one inside the other, separated by a layer of sand. The pots are covered

◀ Mohammed Bah Abba's family were pot-makers. As a boy he was fascinated by earthenware objects and that they would absorb water but remain structurally intact. Abba earned a college degree in business, and while still in his 20's, became an instructor in a college of business in Jigwa, Nigeria, and a consultant to the United Nations Development Program. That brought him back into close contact with rural communities in northern Nigeria and made him aware of the hardships of the families there. Making and distributing the pot-in-pot device for safe food storage closed this interesting circle. (*The Rolex Awards for Enterprise/ Tomas Bertelsen/Scientific American, Nov. 2000, p. 26*)

with damp cloth and placed in a well-ventilated area. Water seeps through the pot's outer wall and rapidly evaporates in the dry desert air. The water remaining in the pot and its contents drop in temperature. Food in the inner pot can stay cool for days and not spoil.

In the 1990s, at his own expense, Abba made and distributed almost 10,000 pots in the villages of northern

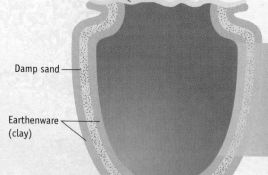

Damp cloth

Damp sand

Earthenware (clay)

Water evaporates from pot walls and damp sand.

◀ Water seeps through the outer pot from the damp sand layer separating the pots or from food stored in the inner pot. As the water evaporates from the surface of the outer pot, the food is cooled.

> "**A**bba's refrigerator is a lesson in simplicity: a porous clay pot, nestled inside a second porous clay pot, with moist sand separating the two vessels."

▲ "Swamp coolers" are often used as inexpensive air-conditioners in hot and dry climates. A trickle of water washes over a bed of straw or other porous material. As air is drawn over the moist material, the air is cooled as the water takes energy from the air to evaporate. *(James Cowlin/Image Enterprises, Phoenix, AZ)*

Nigeria. He estimates that about 75% of the families in this area are now using his refrigerator. The impact of this simple device has implications not only for the health of his people but for their economy and their social structure. Prior to the development of the pot-in-pot device for food storage, it was necessary to sell produce immediately on harvesting. The young girls in the family who sold food on the street daily could now be re-leased from this chore to attend school and improve their lives.

Every two years, the Rolex Company, the Swiss maker of timepieces, gives a series of awards for enterprise. For his pot-in-pot refrigerator, Abba was one of the five recipients of a Rolex Award in 2000.

▶ Evaporatative cooling. The same principle that cools Abba's refrigerator cools you down if you wear a strip of damp cloth, a "neck cooler," around your neck on a hot day. *(Charles D. Winters)*

Before You Begin

- Know how to balance chemical equations (Section 4.2).
- Review product-favored and reactant-favored reactions (Section 5.6).
- Know how to use Kelvin and Celsius temperature scales (Section 1.6).
- Review states of matter and changes of state (Section 1.5).

Energy changes in physical and chemical processes

Energy is invariably transferred during chemical and physical changes, most commonly in the form of heat. Changes of heat content, and the transfer of heat between objects, are major themes of thermodynamics, the science of heat and work.

Energy Transfer in Physical Processes

CHANGE • Water vapor condenses to liquid.

CAUSE • Heat is transferred from water vapor to the atmosphere.

CHANGE • Solid water changes to liquid water.

CAUSE • Heat is transferred to ice.

CHANGE • Cold pack absorbs heat.

CAUSE • Dissolving solid in water absorbs heat energy.

CHANGE • Hot pack evolves heat.

CAUSE • Crystallization produces heat energy.

Energy Transfer in a Chemical Reaction

STORED ENERGY

Chemical reactions can absorb or evolve energy, often in the form of heat.

ENERGY TRANSFER

If the Gummi Bear is placed in molten potassium chlorate ($KClO_3$) the sugar is oxidized to CO_2 and H_2O . . .

Chemical energy is stored in a Gummi Bear, which is primarily sugar.

RELEASED ENERGY

. . . and the energy change occurring in the chemical change is observed as heat.

Energy Transfer in Nature

Hurricanes and other forms of violent weather involve the storage and release of energy. The average hurricane releases heat energy equivalent to the annual U.S. production of electricity.

Water vapor condenses to liquid water, forming clouds and transferring energy to the surrounding atmosphere.

STORED ENERGY

Sunlight pumps energy into ocean water, converting liquid water to water vapor.

ENERGY TRANSFER

The energy stored in the atmosphere is converted to mechanical energy through wind and waves.

ENERGY TRANSFER

RELEASED ENERGY

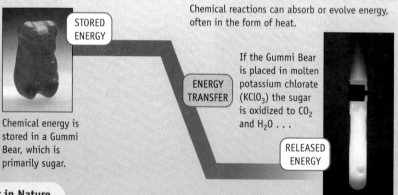

Energy transfer accompanies both chemical and physical changes (Chapter Focus). Our bodies are cooled when we perspire — the evaporation of water in sweat, a physical change, draws energy from our body and causes us to feel cooler. When water vapor condenses heat is given off, a process that has a significant impact on the weather. The sun's energy is stored as chemical energy by the formation of carbohydrates and oxygen from carbon dioxide and water in the process of photosynthesis, a chemical change.

$$6\ CO_2(g) + 6\ H_2O(\ell) + energy \longrightarrow C_6H_{12}O_6(s) + 6\ O_2(g)$$

This chemical energy can be released in a chemical reaction of carbohydrate and oxygen, whether in living tissue, in the laboratory (see Chapter Focus), or in a forest fire [CD-ROM, Screen 6.2].

$$C_6H_{12}O_6(s) + 6\ O_2(g) \longrightarrow 6\ CO_2(g) + 6\ H_2O(\ell) + energy$$

When studying chemistry, it is important to know something about energy. The most common form of energy we see in chemical processes is heat. Changes of heat content, and the transfer of heat between objects, is a major theme of **thermodynamics,** the science of heat and work — and the subject of this chapter and a later one [CHAPTER 19]. The principles of thermodynamics apply to energy use in your home, to ways of conserving energy, to recycling of materials, and to problems of current and future energy use in our economy. And, as told in the introductory story to this chapter, to refrigeration. ●

6.1 ENERGY: SOME BASIC PRINCIPLES

Energy is defined as the capacity to do work. You do work against the force of gravity when carrying yourself and hiking equipment up a mountain. You can do this work because you have the energy to do so, the energy having been provided by the food you have eaten. Food energy is chemical energy — energy stored in chemical compounds and released when the compounds undergo the chemical reactions of metabolism in your body.

Energy can be classified as kinetic or potential [CD-ROM, Screens 6.4 and 6.5]. Kinetic energy, as noted in the discussion of kinetic-molecular theory [SECTION 1.5], is energy associated with motion, such as:

- *Thermal energy* of atoms, molecules, or ions in motion at the submicroscopic level. All matter has thermal energy.
- *Mechanical energy* of a macroscopic object like a moving tennis ball or automobile.
- *Electric energy* of electrons moving through a conductor.
- *Sound,* which corresponds to compression and expansion of the spaces between molecules.

Potential energy, energy that results from an object's position (Figure 6.1), includes:

- *Chemical potential energy* resulting from attractions among electrons and atomic nuclei in molecules. Rearranging electrons and nuclei in a chemical reaction changes the potential energy.
- *Gravitational energy,* such as that possessed by a ball held above the floor and by water at the top of a waterfall.
- *Electrostatic energy* such as that of positive and negative ions a small distance apart.

Chapter Goals ● Revisited

- **Assess heat transfer associated with changes in temperature and changes of state.**
- Apply the first law of thermodynamics.
- Define and understand the state functions enthalpy and internal energy.
- Calculate the energy changes occurring in chemical reactions and learn how these changes are measured.
- Recognize how chemistry contributes to understanding societal problems involving energy.

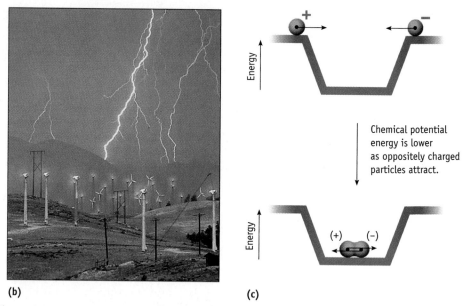

Figure 6.1   Energy and its conversion. (a) Water at the top of a waterfall represents stored, or potential, energy. As water falls, its potential energy is converted to mechanical energy. **(b)** Lightning converts electrostatic energy into radiant and thermal energy and wind produces electrical energy. **(c)** Chemical potential energy. *(a, James Cowlin/Image Enterprises, Phoenix AZ; b, Keith Kent/Science Photo Library/Photo*

Potential energy is stored energy and it can be converted into kinetic energy. For example, as water falls over a waterfall, its potential energy is converted into kinetic energy. Similarly, kinetic energy can be converted to potential energy: the kinetic energy of falling water can turn a turbine to produce electricity, which, in turn, can be used to convert water to H_2 and O_2 [⬅ FIGURE 1.1, PAGE 11]. The H_2 gas represents stored chemical potential energy because it can be burned to produce heat and light [⬅ FIGURE 1.8, PAGE 19] or used in a fuel cell (as in the Space Shuttle) to produce electric energy.

Conservation of Energy

Standing on a diving board, you have considerable potential energy because of your position above the water. Once you jump off the board, some of that potential energy is converted into kinetic energy (Figure 6.2). During the dive, the force of gravity accelerates your body to move faster and faster. Your kinetic energy increases and your potential energy decreases. At the moment you hit the water, your velocity is abruptly reduced, and much of your kinetic energy is converted to mechanical energy; the water splashes as your body moves it aside by doing work on it. Eventually you float on the surface, and the water becomes still again. If you could see them, however, you would find that the water molecules are moving a little faster in the vicinity of your dive; that is, the temperature of the water is slightly higher.

This series of energy conversions illustrates the **law of conservation of energy,** otherwise known as the **first law of thermodynamics.** These terms are synonymous; both state that energy can neither be created nor destroyed. That is to say, *the total energy of the universe is constant.* These statements summarize the results of a great many experiments in which heat, work, and other forms of energy transfer have been measured and the total energy content found to be the same before and after an event.

Figure 6.2   The law of energy conservation. The diver's potential energy is converted to kinetic energy and then to thermal and mechanical energy, illustrating the law of conservation of energy [🖱 CD-ROM, Screen 6.4].

The law of energy conservation is the reason we should be careful not to say, for example, that the energy of oil to heat your home is used up when it is burned. What has been consumed is an energy resource. That is, the oil's capacity to transfer heat energy to its surroundings when it is burned is lost. The chemical energy present in oil has been converted to an equal quantity of energy, now in the form of heat for your home and the thermal energy of the gases going up the chimney.

Exercise 6.1 **Energy**

A battery stores chemical potential energy. Into what types of energy can this potential energy be converted? [CD-ROM, Screen 6.5]

Temperature and Heat

Temperature changes can be measured with a mercury thermometer (Figure 6.3). When the thermometer is placed in hot water, heat is transferred from the water to the thermometer. The increased energy causes the mercury atoms to move about more rapidly and the space between atoms increases slightly. You observe this as an expansion in the volume of the mercury, and the column of mercury rises higher in the thermometer tube.

Three important aspects of thermal energy and temperature should be understood:

- Heat is not the same as temperature.
- The more thermal energy a substance has, the greater the motion of its atoms and molecules.
- The total thermal energy in an object is the sum of the individual energies of all the atoms, molecules, or ions in that object.

● **Volume and Temperature**
Virtually all substances, whether solid, liquid, or gas, expand when heated. A measure of this physical property is the coefficient of thermal expansion of the substance.

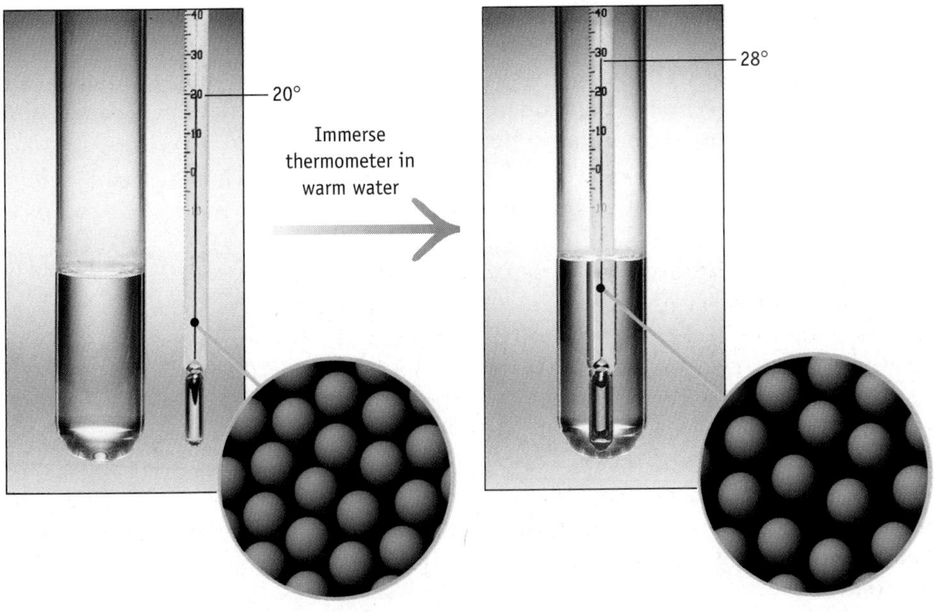

20°

Immerse thermometer in warm water

28°

Figure 6.3 **Measuring temperature.** The volume of liquid mercury in a thermometer increases slightly when immersed in warm water. The volume increase causes the mercury to rise in the thermometer, which is calibrated to give the temperature. *(Charles D. Winters)*

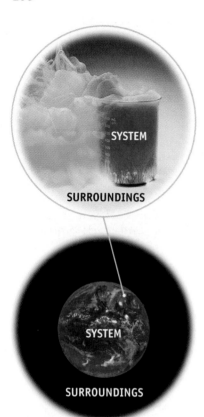

For a given substance, its thermal energy depends not only on temperature but also on the amount of substance. Thus, a cup of hot coffee may contain less thermal energy than a bathtub full of warm water, even though the coffee is at a higher temperature.

Systems and Surroundings

In thermodynamics, the terms "system" and "surroundings" have precise and important scientific meanings. A **system** is defined as the object, or collection of objects, being studied (Figure 6.4). The **surroundings** include everything outside the system that can exchange energy with the system. A system may be contained within an actual physical boundary, such as a flask or the cell wall in a cell in your body or, alternatively, the boundary may be purely imaginary. In discussions that follow, we will need to identify systems precisely. If we are studying the heat evolved in a chemical reaction, for example, the system might be defined as a reaction vessel and its contents. The surroundings would be the air in the room and anything else in contact with the vessel. At the atomic level, the system could be a single atom or molecule and the surroundings would be the atoms or molecules in its vicinity.

This concept of a system and its surroundings applies to nonchemical situations as well. If we want to study the energy balance on this planet, we might choose to define the earth as the system and outer space as the surroundings. On a cosmic level, the solar system might be defined as the system being studied and the rest of the galaxy would be the surroundings [CD-ROM, Screen 6.6].

Directionality of Heat Transfer: Thermal Equilibrium

Heat transfer occurs when two objects at different temperatures are brought into contact. In Figure 6.5, for example, the beaker of water and the piece of metal being heated in a Bunsen burner flame have different temperatures. When the hot metal is plunged into the cold water, heat is transferred from the metal to the water. Eventually, the two objects reach the same temperature. We say that the system has reached **thermal equilibrium.** The distinguishing feature of thermal equilibrium is that, on the macroscopic scale, no further temperature change occurs and the temperature throughout the entire system (metal plus water) is the same [CD-ROM, Screen 6.11].

The experiment with the hot metal bar and the beaker of water may seem like a rather simple experiment. Embedded in the experiment, however, are some principles that will be very important in our further discussion.

Figure 6.4 Systems and their surroundings. Earth can be considered a thermodynamic system, with the rest of the universe as its surroundings. A chemical reaction occurring in a laboratory is also a system, with the laboratory its surroundings. *(NASA, Charles D. Winters)*

Figure 6.5   Energy transfer. Heat is transferred from the hotter metal bar to the cooler water. Eventually the water and metal reach the same temperature and are said to be in thermal equilibrium [CD-ROM, Screen 6.11]. *(Charles D. Winters)*

A Closer Look

Why Doesn't the Heat in a Room Cause Your Cup of Coffee to Boil?

If a cup of coffee or tea is hotter than its surroundings, heat is transferred to the surroundings until the hot coffee cools off and the surroundings warm up a bit. It is interesting and useful to think about why the opposite process doesn't occur. Why doesn't the heat in a room cause a cup of cold coffee to boil? The law of energy conservation would not be violated if the coffee got hotter and hotter and the surroundings in the room got cooler and cooler. However, we know from experience that this will never happen. The directionality in heat

transfer — heat energy always transfers from a hotter object to a cooler one, never the reverse — corresponds to a spreading out of energy over the greatest possible number of atoms, ions, or molecules. Energy transfers from a relatively small number of molecules in a hot cup of coffee to a large number of atoms and molecules surrounding the cup.

Similarly, the large number of particles in the surrounding environment will heat a glass of ice water by transferring some of their energy to the glass, ice, and water

molecules. As in the previous example, the end result is to spread thermal energy more evenly over the maximum number of particles. The opposite process, concentrating energy in only a few particles at the expense of many, or concentrating energy over a large number of particles at the expense of a few, is never observed.

The directionality of energy transfer, which plays an important role in thermodynamics, will be discussed further in Chapter 19.

- Heat transfer always occurs from a hotter object to a cooler object. The directionality of heat transfer is an important principle of thermodynamics.

- Transfer of heat continues until both objects are at the same temperature (thermal equilibrium). (See "A Closer Look: Why Doesn't the Heat in a Room Cause Your Cup of Coffee to Boil?").

- The quantity of heat lost by a hotter object and the quantity of heat gained by a cooler object when they are in contact are numerically equal. (This is required by the law of conservation of energy.)

Two terms are useful when describing the directionality of heat transfer: exothermic and endothermic (Figure 6.6) [CD-ROM, Screen 6.6].

- In an **exothermic** process heat is transferred from a system to the surroundings.

- An **endothermic** process is the opposite of an exothermic process: heat is transferred from surroundings to the system.

● **Thermal Equilibrium**
Although no change is evident at the macroscopic level when thermal equilibrium is reached, on the molecular level transfer of energy between individual molecules will continue to occur. This feature — no change visible on a macroscopic level, but processes still occurring at the particulate level — is a general feature of equilibria that we will encounter again.

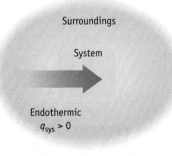

Surroundings

System

Endothermic
$q_{sys} > 0$

Endothermic: energy transferred from surroundings to system

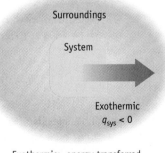

Surroundings

System

Exothermic
$q_{sys} < 0$

Exothermic: energy transferred from system to surroundings

Figure 6.6   Exothermic and endothermic processes. The symbol q represents heat transferred, and the subscript *sys* refers to the system [CD-ROM, Screen 6.6]. *(Photos, Charles D. Winters)*

Figure 6.7   Energy content of foods. In countries that use standardized SI units, food energy is measured in joules (J). The diet soda in this can (from Australia) is said to have an energy content of only 1 J. *(Charles D. Winters)*

● **Kinetic Energy**
Kinetic energy is calculated by the equation $KE = \frac{1}{2}mv^2$. The kinetic energy of a 2.0-kg mass (m) moving at a velocity of 1.0 m/s (v) is $\frac{1}{2}(2.0 \text{ kg})(1.0 \text{ m/s})^2 = 1.0 \text{ kg} \cdot \text{m}^2/\text{s}^2 = 1.0 \text{ J}$.

Figure 6.8   Food label with dietary information. The U.S. Food and Drug Administration (FDA) mandates that nutritional data, including energy contents, be included on almost all packaged food labels. *(Charles D. Winters)*

Energy Units

When expressing energy quantities, most chemists (and much of the world outside the United States) use the **joule** (J), the SI unit of thermal energy (Figure 6.7) [CD-ROM, Screen 6.7]. The joule is preferred in scientific study because it is related directly to the units used for mechanical energy: 1 J equals $1 \text{ kg} \cdot \text{m}^2/\text{s}^2$. However, the joule can be inconveniently small as a unit for use in chemistry and so the **kilojoule** (kJ), equivalent to 1000 J, is often used.

To give you some feeling for joules, suppose you drop a six-pack of soft drink cans, each full of liquid, on your foot. Although you probably will not take time to calculate the kinetic energy at the moment of impact, it is between 4 J and 10 J.

An older unit for measuring heat is the **calorie** (cal). This is defined as the heat required to raise the temperature of 1.00 g of pure liquid water from 14.5 °C to 15.5 °C. A **kilocalorie** (kcal) is equivalent to 1000 cal. The conversion factor relating joules and calories is

$$1 \text{ joule (J)} = 4.184 \text{ calorie (cal)}$$

The **dietary Calorie** (with a capital C) is often used in the United States to represent the energy content of foods (Figure 6.8). We encounter this unit when reading the nutritional information on a food label. The dietary Calorie (Cal) is equivalent to the kilocalorie or 1000 cal. Thus, a breakfast cereal that gives you 100.0 Cal of nutritional energy per serving provides 100.0 kcal, or 418.4 kJ.

Exercise 6.2   Energy Units

(a) In an old textbook you read that the oxidation of 1.00 g of hydrogen to form liquid water produces 3800 cal. What is this energy in units of joules?

(b) The label on a cereal box indicates that one serving (with skim milk) provides 250 Cal. What is this energy in kilojoules (kJ)?

6.2   SPECIFIC HEAT CAPACITY AND HEAT TRANSFER

The quantity of heat transferred to or from an object when its temperature changes depends on three things:

- The quantity of material.
- The size of the temperature change.
- The identity of the material gaining or losing heat.

The **specific heat capacity** (C) is related to these three parameters. The specific heat capacity, often called simply specific heat, is the quantity of heat required to raise the temperature of 1 gram of a substance by one kelvin. It has units of joules per gram per kelvin (J/g·K) [CD-ROM, Screens 6.8 and 6.9].

The quantity of heat gained or lost when a given mass of a substance is warmed or cooled is calculated using Equation 6.1.

$$q = C \cdot m \cdot \Delta T \qquad (6.1)$$

Specific heat capacity (J/g · K)

Change in temperature (K)

Heat transferred (J)

Mass of substance (g)

Here, q is the quantity of heat transferred to or from a given mass (m) of substance, C is the specific heat capacity, and ΔT is the change in temperature. The capital Greek letter delta, Δ, means "change in." The change in temperature, ΔT, is calculated as the final temperature minus the initial temperature.

$$\Delta T = T_{\text{final}} - T_{\text{initial}} \qquad (6.2)$$

As outlined in "A Closer Look: Sign Conventions," calculating a change in temperature as in Equation 6.2 will give a result with an algebraic sign that indicates the direction of heat transfer. For example, we can use the specific heat capacity of copper, 0.385 J/g · K, to calculate the heat gained by a 10.0-g sample of copper if its temperature is raised from 298 K (25 °C) to 598 K (325 °C).

$$q = \left(0.385 \, \frac{J}{g \cdot K}\right)(10.0 \text{ g})(598 \text{ K} - 298 \text{ K}) = +1160 \text{ J}$$

T_{final}
Final temp.

T_{initial}
Initial temp.

• **Change in Temperature, ΔT**

Sign of ΔT	Meaning
positive	$T_{\text{final}} > T_{\text{initial}}$, so T has increased, and q will be positive. Heat has been transferred to the object under study.
negative	$T_{\text{final}} < T_{\text{initial}}$, so T has decreased, and q will be negative. Heat has been transferred out of the object under study.

A Closer Look

Sign Conventions

Whenever you take the difference between two quantities in chemistry, you should always subtract the initial quantity from the final quantity. A consequence of this convention is that the algebraic sign of the result indicates an increase ($+$) or a decrease ($-$) in the quantity being studied. This is an important point, as you will see not only in this but also in other chapters of this book.

Thus far, we have described temperature changes and the direction of heat transfer. The following table summarizes the conventions used.

When discussing the quantity of heat, we use an *unsigned number*. If we want to indicate the *direction of transfer* in a process, however, we attach a sign, either negative (heat transferred from the substance) or positive (heat transferred into the substance), to q. The sign of q "signals" the direction of heat transfer. Heat, a quantity of energy, cannot be negative but the heat content of an object can increase or decrease, depending on the direction of heat transfer.

An analogy might make this point clearer. Consider your bank account. Assume

you have $260 in your account ($A_{\text{initial}}$) and after a withdrawal you have $200 ($A_{\text{final}}$). The cash flow is thus

$$\begin{aligned} \text{Cash flow} &= A_{\text{final}} - A_{\text{initial}} \\ &= \$200 - \$260 \\ &= -\$60 \end{aligned}$$

The negative sign on the $60 indicates that a withdrawal has been made; the cash itself is not a negative quantity.

ΔT of System	Sign of ΔT	Sign of q	Direction of Heat Transfer
increase	$+$	$+$	Heat transferred from surroundings to system (an endothermic process)
decrease	$-$	$-$	Heat transferred from system to surroundings (an exothermic process)

Notice that the answer has a positive sign. This indicates that the heat content of the sample of copper has *increased* by 1160 J because heat has transferred *to* the copper (the system) *from* the surroundings.

Specific heat capacities of some metals, compounds, and common substances are listed in Table 6.1. Notice that water has one of the highest values, 4.184 J/g·K. In contrast, the specific heats of metals are one-fifth of this value or smaller. For example the specific heat capacity of iron is 0.45 J/g·K; to raise the temperature of a gram of water by 1 K requires about nine times as much heat as is required to cause a 1 K change in temperature for a gram of iron. For comparison, notice that the specific heat capacities of common building materials are about the same as most metals.

The high specific heat capacity of water has major significance. A great deal of energy must be absorbed by a large body of water to raise its temperature just a degree or so. Thus, large bodies of water have a profound influence on our weather. In spring, lakes tend to warm up more slowly than the air. In autumn, the heat given off by a large lake moderates the drop in air temperature.

The greater the specific heat and the larger the mass, the more thermal energy a substance can store. This has numerous implications. For example, you might wrap some bread in aluminum foil and heat it in an oven. You can remove the foil with your fingers after taking the bread from the oven, even though the bread is very hot. A small quantity of aluminum foil is used and the metal has a low specific heat capacity so, when you touch the hot foil, only a small quantity of heat will be transferred to your fingers (which have a higher mass and higher specific heat capacity). This is also the reason a chain of fast-food restaurants warns you that the filling of an apple pie can be much warmer than the paper wrapper or the pie crust (Figure 6.9).

● **Molar Heat Capacity**
Heat capacities can be expressed on a per mole basis. The molar heat capacity is the heat required to raise the temperature of one mole of a substance by one kelvin. The molar heat capacity of metals at room temperature is always near 25 kJ/mol·K.

Figure 6.9   A practical example of specific heat capacity. The filling of the apple pie has a higher specific heat (and higher mass) than the pie crust and wrapper. Notice the warning on the wrapper. *(Charles D. Winters)*

Table 6.1 ● **Specific Heat Capacity Values for Some Elements, Compounds, and Common Solids**

Substance	Name	Specific Heat Capacity (J/g·K)
Elements		
Al	aluminum	0.897
C	graphite	0.685
Fe	iron	0.449
Cu	copper	0.385
Au	gold	0.129
Compounds		
$NH_3(\ell)$	ammonia	4.70
$H_2O(\ell)$	water (liquid)	4.184
$C_2H_5OH(\ell)$	ethanol	2.44
$HOCH_2CH_2OH(\ell)$	ethylene glycol (antifreeze)	2.39
$H_2O(s)$	water (ice)	2.06
Common Solids		
wood		1.8
cement		0.9
glass		0.8
granite		0.8

Example 6.1   **Specific Heat Capacity**

Problem • Determine the quantity of heat that must be added to raise the temperature of a cup of coffee (250 mL) from 20.5 °C (293.7 K) to 95.6 °C (368.8 K). Assume that water and coffee have the same density (1.00 g/mL) and same specific heat capacity.

Strategy • Use Equation 6.1. For this calculation, you will need the specific heat capacity for H_2O from Table 6.1 (4.184 J/g·K), the mass of the coffee (calculated from its density and volume), and the change in temperature ($T_{final} - T_{initial}$).

Solution •

Mass of coffee $= (250 \text{ mL})(1.00 \text{ g/mL}) = 250$ g

$$\Delta T = T_{final} - T_{initial} = 368.8 \text{ K} - 293.7 \text{ K}$$
$$= 75.1 \text{ K}$$
$$q = C \times m \times \Delta T$$
$$= (4.184 \text{ J/g} \cdot \text{K})(250 \text{ g})(75.1 \text{ K})$$
$$= 79,000 \text{ J (or 79 kJ)}$$

Comment • Notice that the answer is a positive number, indicating that heat has transferred into the coffee from the surroundings. The heat content of the coffee has increased.

Exercise 6.3   **Specific Heat Capacity**

In an experiment it was determined that 59.8 J was required to change the temperature of 25.0 g of ethylene glycol (a compound used as antifreeze in automobile engines) by 1.00 K. Calculate the specific heat capacity of ethylene glycol from these data.

Quantitative Aspects of Heat Transfer

Like melting point, boiling point, and density, the specific heat capacity is a characteristic property of a pure substance. The specific heat capacity of a substance can be determined experimentally by accurately measuring temperature changes that occur when heat is transferred from the substance to a known quantity of water (whose specific heat capacity is known) [CD-ROM, Screens 6.10 and 6.11].

Suppose a 55.0-g piece of metal is heated in boiling water to 99.8 °C and then dropped into water in an insulated beaker (in an experiment similar to that in Figure 6.5). Assume the beaker contains 225 g of water and its initial temperature (before the metal was dropped in) was 21.0 °C. The final temperature of the metal and water is 23.1 °C. What is the specific heat capacity of the metal? Here are the most important aspects of this experiment:

- The water and the metal bar end up at the same temperature (T_{final} is the same for both).
- We assume no heat is transferred to warm the glass beaker and that there is no heat transfer to the surroundings. (For a very accurate measurement, we would want to include these factors.)
- The heat transferred from the metal to the water, q_{metal}, has a negative value because the temperature of the metal dropped as heat was transferred out of it to the water. Conversely, q_{water} has a positive value because its temperature increased as heat was transferred into the water from the metal.
- The values of q_{water} and q_{metal} are numerically equal but of opposite sign, that is, $q_{water} = -q_{metal}$. Expressed another way, $q_{water} + q_{metal} = 0$. To paraphrase this equation: *The sum of thermal energy changes in this system is zero.*

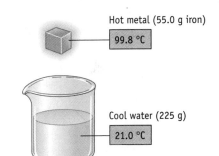

Hot metal (55.0 g iron)

99.8 °C

Cool water (225 g)

21.0 °C

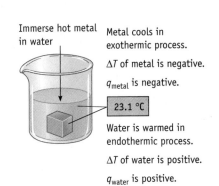

Immerse hot metal in water

Metal cools in exothermic process.

ΔT of metal is negative.

q_{metal} is negative.

23.1 °C

Water is warmed in endothermic process.

ΔT of water is positive.

q_{water} is positive.

When heat transfers from a hot metal to cool water, the heat transferred from the metal, q_{metal}, has a negative value. The heat transferred to the water, q_{water}, is positive.

Problems involving heat transfer can be approached by assuming that the sum of the heat content changes within a given system is zero (Equation 6.3).

$$q_1 + q_2 + q_3 + \cdots = 0 \qquad (6.3)$$

The quantities q_1, q_2, and so on represent the changes in thermal energy for the individual parts of the system. For this specific problem, there are two heat content changes, q_{water} and q_{metal}, and

$$q_{water} + q_{metal} = 0$$

● **Heat Transfer**
Remember that $T_{initial}$ for the metal and $T_{initial}$ for the water in this problem have different values.

Each of these quantities is related to the specific heat capacities, mass, and change of temperature of the water and metal, as defined by Equation 6.1. Thus

$$[C_{water} \times m_{water} \times (T_{final} - T_{initial, water})] +$$
$$[C_{metal} \times m_{metal} \times (T_{final} - T_{initial, metal})] = 0$$

The specific heat capacity of the metal is the unknown in this problem. Using the specific heat capacity of water from Table 6.1 and converting Celsius temperatures to kelvins gives

$$[(4.184 \text{ J/g} \cdot \text{K})(225 \text{ g})(296.3 \text{ K} - 294.2 \text{ K})] +$$
$$[(C_{metal})(55.0 \text{ g})(296.3 \text{ K} - 373.0 \text{ K})] = 0$$
$$C_{metal} = 0.47 \text{ J/g} \cdot \text{K}$$

Example 6.2   Using Specific Heat Capacity

Problem • A 88.5-g piece of iron whose temperature is 78.8 °C (352.0 K) is placed in a beaker containing 244 g of water at 18.8 °C (292.0 K). When thermal equilibrium is reached, what is the final temperature? (Assume no heat is lost to warm the beaker and no heat is lost to the surroundings.)

Strategy • Two changes occur in the system: iron gives up heat and water gains heat. The sum of the heat quantities of these changes must equal zero. Each quantity of heat is related to the specific heat capacity, mass, and temperature change of the substance using Equation 6.1 $[q = C \times m \times (T_{final} - T_{initial})]$. The final temperature is unknown. The specific heat capacities of iron and water are given in Table 6.1. Remember that temperatures must be used in kelvins.

Solution •

$$q_{metal} + q_{water} = 0$$
$$[C_{water} \times m_{water} \times (T_{final} - T_{initial, water})] +$$
$$[C_{metal} \times m_{Fe} \times (T_{final} - T_{initial, Fe})] = 0$$
$$[(4.184 \text{ J/g} \cdot \text{K})(244 \text{ g})(T_{final} - 292.0 \text{ K})] +$$
$$[(0.449 \text{ J/g} \cdot \text{K})(88.5 \text{ g})(T_{final} - 352.0 \text{ K})] = 0$$
$$T_{final} = 294.3 \text{ K} (21.1 \text{ °C})$$

Comment • The low specific heat capacity of iron and the small quantity of iron result in the temperature of iron being reduced by about 50 °C whereas the temperature of water has been raised by only a few degrees.

Exercise 6.4   Using Specific Heat Capacity

A 15.5-g piece of chromium, heated to 100.0 °C, is dropped into 55.5 g of water at 16.5 °C. The final temperature of the metal and the water is 18.9 °C. What is the specific heat capacity of chromium? (Assume no heat is lost to the container or to the surrounding air.)

Exercise 6.5   Heat Transfer Between Substances

A piece of iron (400. g) is heated in a flame and then dropped into a beaker containing 1000. g of water. The original temperature of the water was 20.0 °C, and the final temperature of the water and iron is 32.8 °C after thermal equilibrium has been attained. What was the original temperature of the hot iron bar? (Assume no heat is lost to the beaker or to the surrounding air.)

6.3   ENERGY AND CHANGES OF STATE

When a solid melts, its atoms, molecules, or ions move about vigorously enough to break free of the constraints imposed by their neighbors in the solid. When a liquid boils, particles move much farther apart from one another. A change between solid and liquid or between liquid and gas is called a **change of state** [🖰 CD-ROM, Screen 6.12]. In both cases, energy must be furnished to overcome attractive forces among the particles. The heat required to convert a substance from a solid at its melting point to a liquid is called the **heat of fusion.** The heat required to convert liquid at its boiling point to gas is called the **heat of vaporization.** Heats of fusion and vaporization for many pure substances are provided along with other physical properties in handbooks of chemical data.

For water, the heat of fusion at 0 °C is 333 J/g and the heat of vaporization at 100 °C is 2256 J/g. These values are used when calculating the quantity of heat required or evolved when water boils or freezes. For example, the heat required to convert 20.0 g of water from the liquid to gaseous state at 100 °C is

$$(2256 \text{ J/g})(20.0 \text{ g}) = 45{,}100 \text{ J (or } 45.1 \text{ kJ)}$$

If the same quantity of liquid water at 0 °C freezes to ice, the quantity of heat evolved is

$$(333 \text{ J/g})(20.0 \text{ g}) = 6660 \text{ J (or } 6.66 \text{ kJ)}$$

● **Heats of Fusion and Vaporization for H_2O**
Heat of fusion = 333 J/g
          = 6.00 kJ/mol
Heat of vaporization = 2256 J/g
          = 40.65 kJ/mol

Problem-Solving Tip 6.1

Units for *T* and Specific Heat Capacity

(a) *Calculating ΔT.* Notice that specific heat values are given in units of joules per gram per kelvin (J/g·K). In virtually all calculations in chemistry, temperature is expressed in kelvins. In calculating ΔT, however, we could use Celsius temperatures because a kelvin and a Celsius degree are the same size, so that the difference between two temperatures is the same on both scales. For example, the difference between the boiling and freezing points of water is

$$\Delta T \text{, Celsius} = 100 \text{ °C} - 0 \text{ °C} = 100 \text{ °C}$$
$$\Delta T \text{, Kelvin} = 373 \text{ K} - 273 \text{ K} = 100 \text{ K}$$

(b) *Units of Specific Heat Capacity.*   Specific heat capacities are given in this book in units of joules per gram per kelvin (J/g·K). Often, however, specific heat capacity values found in handbooks (such as the *CRC Handbook of Chemistry and Physics*) or the NIST Webbook (http://webbook.nist.gov) will have units of J/mol·K, that is, they are molar heat capacities. For example, liquid water has a specific heat capacity of 4.184 J/g·K or 75.40 J/mol·K. The values are related as follows:

$$\left(4.184 \frac{\text{J}}{\text{g·K}}\right)\left(18.02 \frac{\text{g}}{\text{mol}}\right) = 75.40 \text{ J/mol·K}$$

Figure 6.10   Heat transfer and the temperature change for water. This graph shows the quantity of heat absorbed and the consequent temperature change as 500. g of water warms from −50 to 200 °C.

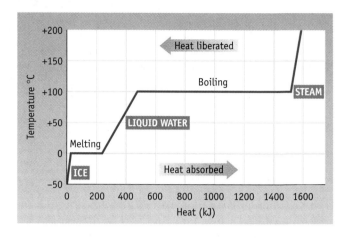

Figure 6.10 illustrates the quantity of heat absorbed and the consequent temperature change as 500. g of water is warmed from −50 to 200 °C. First, the temperature of the ice increases as heat is added. On reaching 0 °C, however, the temperature remains constant as sufficient heat is absorbed to melt the ice to liquid water. When all the ice has melted, the liquid absorbs heat and is warmed to 100°C, the boiling point of water. The temperature is again constant as heat is absorbed to convert the liquid completely to vapor. Any further heat added raises the temperature of the water vapor. The heat absorbed at each step is calculated in Example 6.3.

It is important to notice that *temperature is constant throughout a change of state.* During a change of state, the added energy is used to break down the forces holding one molecule to another, not to increase the temperature of the substance (Figure 6.11).

● **Temperature and Changes of State**
Be sure to notice that the temperature of water does not change during melting and vaporization.

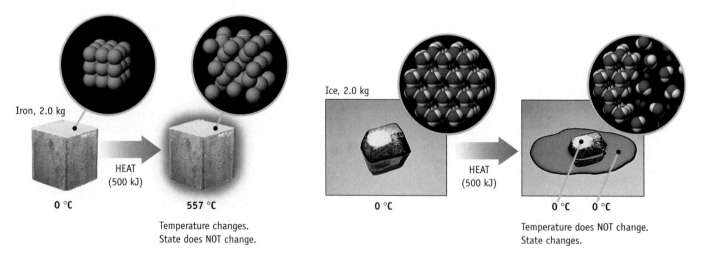

Iron, 2.0 kg

HEAT
(500 kJ)

0 °C   557 °C

Temperature changes.
State does NOT change.

Ice, 2.0 kg

HEAT
(500 kJ)

0 °C   0 °C   0 °C

Temperature does NOT change.
State changes.

Figure 6.11   Heat and changes of state. Adding 500 kJ of heat to 2.0 kg of iron at 0 °C will cause the iron's temperature to increase to 557 °C (and the metal expands slightly). In contrast, adding 500 kJ of heat to 2.0 kg of ice will cause 1.5 kg of ice to melt to water at 0 °C (and 0.5 kg of ice will remain). No temperature change occurs [CD-ROM, Screen 6.12]. *(Photos: Charles D. Winters)*

Example 6.3 **Energy and Changes of State**

Problem • Calculate the quantity of heat involved in each step shown in Figure 6.10 and the total quantity of heat required to convert 500. g of ice at −50.0 °C to steam at 200 °C. The heat of fusion of water is 333 J/g and the heat of vaporization is 2256 J/g. Specific heat capacities are given in Table 6.1.

Strategy • The problem is broken down into a series of steps: (1) warm the ice from −50 to 0 °C; (2) melt the ice at 0 °C; (3) raise the temperature of the liquid water from 0 to 100 °C; (4) evaporate the water at 100 °C; (5) raise the temperature of the steam from 100 to 200 °C. Use Equation 6.1 to calculate the heats associated with temperature changes. Use the heat of fusion and the heat of vaporization for heats associated with changes of state. The total heat required is the sum of the heats of the individual steps.

Solution •
Step 1

q_1 (to warm ice from −50 to 0 °C)
$$= (2.06 \text{ J/g} \cdot \text{K})(500. \text{ g})(273.2 \text{ K} − 223.2 \text{ K})$$
$$= 5.15 \times 10^4 \text{ J}$$

Step 2

q_2 (to melt ice at 0 °C) $= (333 \text{ J/g}) (500. \text{ g})$
$$= 1.67 \times 10^5 \text{ J}$$

Step 3

q_3 (to raise temperature of water from 0 to 100 °C)
$$= (4.184 \text{ J/g} \cdot \text{K})(500. \text{ g})(373.2 \text{ K} − 273.2 \text{ K})$$
$$= 2.09 \times 10^5 \text{ J}$$

Step 4

q_4 (to evaporate water at 100 °C)
$$= (2256 \text{ J/g})(500. \text{ g}) = 1.13 \times 10^6 \text{ J}$$

Step 5

q_5 (to raise temperature of steam from 100 to 200 °C)
$$= (2.03 \text{ J/g} \cdot \text{K})(500. \text{ g})(473.2 \text{ K} − 373.2 \text{ K})$$
$$= 1.02 \times 10^5 \text{ J}$$

The total thermal energy required is the sum of the thermal energy required in each step.

$$q_{\text{total}} = q_1 + q_2 + q_3 + q_4 + q_5$$
$$= 1.66 \times 10^6 \text{ J (or 1660 kJ)}$$

Comment • The conversion of liquid water to steam is the largest increment of energy added by a considerable margin. (You may have noticed that it does not take much time to heat water to boiling on a stove, but to boil off the water takes a much greater time.)

Example 6.4 **Change of State**

Problem • What is the minimum amount of ice at 0 °C that must be added to the contents of a can of diet cola (340. mL) to cool it from 20.5 to 0 °C? Assume that the specific heat capacity and density of diet cola are the same as for water, and that no heat is gained or lost to the surroundings.

Strategy • Two energy quantities, the heat change in cooling the soda and the heat change in melting the ice, are needed. The first is calculated using the specific heat capacity and Equation 6.1 ($q_{\text{cola}} = C_{\text{cola}} \times m \times \Delta T$); the second uses the heat of fusion of water [$q_{\text{ice}} = $ (heat of fusion)(mass of ice)]. The law of conservation of energy requires that the sum of these two quantities of energy be zero (Equation 6.3).

Solution • The mass of cola is

$$(340. \text{ mL})(1.00 \text{ g/mL}) = 340. \text{ g}$$

and its temperature changes from 293.7 K to 273.2 K. The heat of

fusion of water is 333 J/g, and the mass of ice is the unknown.

$$q_{\text{cola}} + q_{\text{ice}} = 0$$
$$C_{\text{cola}} \times m \times (T_{\text{final}} − T_{\text{initial}}) + q_{\text{ice}} = 0$$
$$[(4.184 \text{ J/g} \cdot \text{K})(340.\text{g})(273.2 − 293.7)]$$
$$+ [(333 \text{ J/g})(m_{\text{ice}})] = 0$$
$$m_{\text{ice}} = 87.6 \text{ g}$$

Comment • This quantity of ice is just sufficient to cool the cola to 0 °C. If more than 87.6 g of ice is added then, when thermal equilibrium is reached, the temperature will be 0 °C and some ice will remain (see Exercise 6.7). If less than 87.6 g of ice is added, the final temperature will be greater than 0 °C. In this case, all the ice will melt and the liquid water formed by melting the ice will absorb additional heat to warm up to the final temperature (An example is given in Study Question 72, page 245.)

History

Work, Heat, and Benjamin Thompson (1753–1814)

Born in New Hampshire, Thompson fled to London before the American revolution because of his sympathy with the royalists. He later moved to Bavaria. Thompson was a government official, an inventor (he invented an efficient fireplace, the first double boiler and pressure cooker, and a candle that became the international standard for light level (known as candle power), a nutritional expert (stressing the benefits of the potato), and a scientist. He was the first to describe the relationship between work and heat and to show heat was not a substance. And he was a well-known scoundrel. See G. I. Brown: *Count Rumford, The Extraordinary Life of a Scientific Genius*, Trowbridge, England, Sutton Publishing, 1999. •

Figure 6.12   Work and heat. A classic experiment that showed the relationship between work and heat was performed by Benjamin Thompson using the apparatus shown here. Thompson measured the rise in temperature of water (in the vessel mostly hidden at the back of the apparatus) that resulted from the energy expended to turn the crank. *(Richard Howard)*

Exercise 6.6   Changes of State

How much heat must be absorbed to warm 25.0 g of liquid methanol, CH_3OH, from 25.0 °C to its boiling point (64.6 °C) and then to evaporate the methanol completely at that temperature? The specific heat of liquid methanol is 2.53 J/g·K. The heat of vaporization of methanol is 2.00×10^3 J/g.

Exercise 6.7   Changes of State

To make a glass of iced tea, you pour 250 mL of tea, whose temperature is 18.2 °C, into a glass containing 5 ice cubes. Each cube has a mass of about 15 g. What quantity of ice will melt, and how much ice will remain to float at the surface in this beverage? Iced tea has a density of 1.0 g/cm³ and a specific heat of 4.2 J/g·K. Assume that no heat is lost in cooling the glass or to the surroundings.

6.4   THE FIRST LAW OF THERMODYNAMICS

To this point, we have only considered energy in the form of heat. Now, we need to broaden the discussion. Recall the definition given on page 203: Thermodynamics is the study of heat and work. If a system does work on its surroundings, energy must be expended and the energy content of the system will decrease. Conversely, if work is done by the surroundings on a system, the energy content of the system increases. As with heat gained or lost, work done by a system or on a system will change the system's energy content (Figure 6.12). Therefore, we next want to introduce work into the equations for energy transfer.

An example of a system doing work on its surroundings is illustrated by the experiment shown in Figure 6.13. A small quantity of dry ice is sealed inside a plastic bag, and a weight (a book in Figure 6.13) is placed on top of the bag. Dry ice is solid carbon dioxide, $CO_2(s)$. It has the interesting property of changing directly from solid to gas at −78 °C when it absorbs heat from its surroundings, in a process called **sublimation:**

$$CO_2(s, -78\ °C) \xrightarrow[+\ heat]{} CO_2(gas, -78\ °C)$$

As the experiment proceeds, the gaseous CO_2 expands within the plastic bag, lifting the book. Work is done whenever something is moved against an opposing force. To lift the book against the force of gravity requires that work be done. The system (the CO_2 inside the bag) is expending energy to do this work [CD-ROM, Screen 6.13].

Even if the book had not been on top of the plastic bag, work would have been done by the expanding gas. This is because a gas must push back the atmosphere when it expands. Instead of raising a book, the expanding gas moves a part of the atmosphere.

Now let us recast this example as an experiment in thermodynamics. First, we must precisely identify system and surroundings. The system is the CO_2, a solid initially, and later a mixture of solid and gas. The surroundings consist of the objects that exchange energy with the system, that is, those objects in contact with the CO_2. This includes the plastic bag, the book, the tabletop, and the sur-

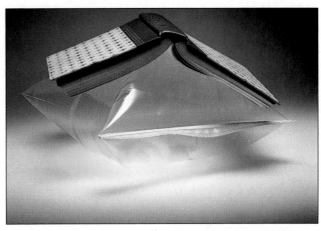

(a) Pieces of dry ice ($CO_2(s)$, $-78°C$) are placed in a plastic bag. The dry ice will sublime (change directly from a solid to a gas) upon the input of heat.

(b) The CO_2 gas formed in the sublimation of solid CO_2 inflates the bag and lifts a book placed on top.

Figure 6.13   Energy changes in the sublimation of dry ice. Heat is absorbed by $CO_2(s)$ when it sublimes, and the system (the contents of the bag) does work on its surroundings by lifting the book against the force of gravity [CD-ROM, Screen 6.13]. *(Charles D. Winters)*

rounding air. Thermodynamics focuses on the energy transfer that is occurring in the experiment. Sublimation of CO_2 requires heat, and this is transferred to the CO_2 from the surroundings. At the same time, the system is doing work on the surroundings by lifting the book. An energy balance for the system will include both quantities, that is, the change in energy content for the system (ΔE) will equal the sum of heat transferred (q) to or from the system and the work done by the system or to the system (w). We can express this explicitly as an equation:

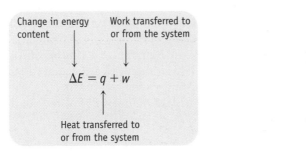

$$\Delta E = q + w \qquad (6.4)$$

Change in energy content

Work transferred to or from the system

Heat transferred to or from the system

Chapter Goals • Revisited

- Assess heat transfer associated with changes in temperature and changes of state.
- **Apply the first law of thermodynamics.**
- Define and understand the state functions enthalpy and internal energy.
- Calculate the energy changes occurring in chemical reactions and learn how these changes are measured.
- Recognize how chemistry contributes to understanding societal problems involving energy.

Equation 6.4 is a mathematical statement of the *first law of thermodynamics:* The energy change for a system is the sum of heat transferred between the system and its surroundings and the work done on the system by the surroundings or on the surroundings by the system. You will notice that this law is a version of the general principle of conservation of energy (see page 204) [CD-ROM, Screen 6.13].

The quantity E in Equation 6.4 has a formal name and a precise meaning in thermodynamics: **internal energy.** The internal energy in a chemical system is the sum of the potential and kinetic energies of the atoms, molecules, or ions in the system. Potential energy is the energy associated with the attractive and repulsive forces between all the nuclei and electrons in the system. It includes the energy

A Closer Look

P–V Work

Work is done when an object of some mass is moved against an external resisting force. We know this well from common experience, such as blowing up a balloon or a bicycle tire.

To evaluate the work done when a gas is compressed, we can use, for example, a cylinder with a movable piston as shown in the figure. The drawing on the left shows the initial position of the piston, and the one on the right the final position.

Work is defined as force (F) times distance (d) ($w = -F \times d$). The distance the piston moves is labeled d on the drawing.

Pressure is defined as force divided by the area over which the pressure is exerted; that is, $P = F/A$. The force is being applied to a piston with an area A. Substituting P times A for F in the equation for work gives $w = -(P \times A)d$. However, the product of A

times d equals the change of volume, ΔV, within the cylinder. Substituting, our equation for work becomes $w = -P\Delta V$ at constant pressure.

In this example, work is done on the gas inside the cylinder. The volume change ΔV ($V_{final} - V_{initial}$) is negative, so $P\Delta V$ has a positive value. As a result, the internal energy of the gas increases.

$$w = -F \cdot d$$
$$\underset{P\,=\,F/A}{}$$
$$w = -(P \cdot A) \cdot d$$
$$\underset{A \cdot d\,=\,\Delta V}{}$$
$$w = -P\Delta V$$

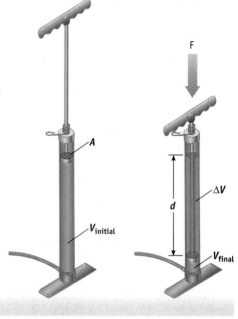

associated with bonds in molecules, forces between ions, and the forces between molecules in the liquid and solid state. Kinetic energy is the energy of motion of the atoms, ions, and molecules in the system. A value of internal energy is extremely hard to determine, but, fortunately, this is not necessary. As the equation indicates, we are evaluating the *change* of internal energy, ΔE, and this is a measurable quantity. In fact, the equation tells us how *to determine ΔE: measure the heat transferred and the work done to or by the system.*

The work represented in Equation 6.4 can be of a specific type, called **P–V (pressure–volume) work.** This is the work associated with a change in volume (ΔV) that occurs against a resisting external pressure (P). For a system in which the external pressure is constant, the value of P–V work can be calculated by Equation 6.5:

● **WORK**
Electrical work is another type of work commonly encountered in chemistry.

Work done on or by the system   Change in volume

$$w = -(P \cdot \Delta V)$$ **(6.5)**

Work is done at constant pressure

The origin of this relationship is explained in "A Closer Look: P–V Work."

The sign convention for Equation 6.4 is important. How the internal energy is affected by heat and work is summarized in the following table.

Sign Conventions for q and w of the System

Change	Sign convention	Effect on E_{system}
Heat transferred to system from surroundings	$q > 0 (+)$	E increases
Heat transferred from system to surroundings	$q < 0 (-)$	E decreases
Work done on system by the surroundings	$w > 0 (+)$	E increases
Work done by the system on surroundings	$w < 0 (-)$	E decreases

Enthalpy

Chemists and biochemists are interested in the energy changes that accompany chemical reactions. Most experiments in a chemical laboratory are carried out in vessels open to the atmosphere. Similarly, chemical processes that occur in living systems are open to the atmosphere. Therefore, experiments are most often run at the constant pressure of the atmosphere, an important consideration when applying the first law of thermodynamics to heat measurements.

Because heat at constant pressure is so frequently the focus of attention, it is useful to have a specific measure of heat transfer under these conditions. The heat content of a substance at constant pressure is called its enthalpy and is given the symbol H. In experiments at constant pressure, the **enthalpy change, ΔH,** is the difference between the final and initial enthalpy content. With enthalpy, as with internal energy, attention is focused on changes (that is, ΔH) rather than on the value of H itself. The value of H is not measured. Rather it is the enthalpy change, ΔH, that is measured in chemical and physical processes [CD-ROM, Screen 6.14].

Similar sign conventions apply to both ΔE and ΔH.

- Negative values of ΔE and ΔH specify that energy is transferred from the system to the surroundings.
- Positive values of ΔE and ΔH refer to energy transferred from the surroundings to the system.

Changes in internal energy and enthalpy are mathematically related by the general equation $\Delta E = \Delta H + w$, showing that ΔE and ΔH differ by the quantity of energy transferred to or from a system as work. Taking work to be $-P\Delta V$, we observe that in many processes—such as the melting of ice—ΔV is small and hence the amount of work is small. Under these circumstances, ΔE and ΔH are of similar magnitude. The amount of work can be significant, however, in processes in which the volume change is large. This usually occurs when gases are formed or consumed, as in the evaporation or condensation of water, the sublimation of CO_2 (see Figure 6.13), and chemical reactions in which gas volumes change. Under these conditions, ΔE and ΔH are significantly different.

State Functions

Both internal energy and enthalpy have a significant characteristic. Changes in these quantities that accompany chemical or physical changes do not depend on which path is chosen in going from the initial state to the final state. No matter how you go from reactants to products in a reaction, for example, the value of ΔH or ΔE for the reaction is always the same. A quantity that has this characteristic property is called a **state function.**

Chapter Goals • Revisited

- Assess heat transfer associated with changes in temperature and changes of state.
- Apply the first law of thermodynamics.
- **Define and understand the state functions enthalpy and internal energy.**
- Calculate the energy changes occurring in chemical reactions and learn how these changes are measured.
- Recognize how chemistry contributes to understanding societal problems involving energy.

● **Internal Energy and Enthalpy**
The heat transferred at constant pressure is often symbolized by q_p and is equivalent to ΔH. The heat transferred at constant volume, symbolized by q_v, is equivalent to ΔE. q_p and q_v differ by the amount of work, w, done on or by the system.

Figure 6.14   State functions. There are many ways to climb Mount Ranier, but the change in altitude from the base of the mountain to its summit is the same. The change in altitude is a state function. The distance traveled to reach the summit is not. *(James Kay Photographs)*

Many commonly measured quantities, such as the pressure of a gas, the volume of a gas or liquid, the temperature of a substance, and the size of your bank account, are state functions. For example, you could have arrived at a current bank balance of $25 by having deposited $25, or you could have deposited $100 and then withdrawn $75.

The volume of a balloon is also a state function. You can blow up a balloon to a large volume and then let some air out to arrive at the desired volume. Alternatively, you can blow up the balloon in stages, adding tiny amounts of air at each stage. The final volume does not depend on how you got there. For bank balances and balloons, an infinite number of paths can lead to the final state, but the final value depends only on the size of the bank balance or the balloon, and not on the path taken from the initial to the final state.

Not all quantities are state functions. Distance traveled is not a state function (Figure 6.14). The travel distance from Oneonta, NY, to Madison, WI, depends on the route taken. Neither is the elapsed time of travel between these two locations a state function. In contrast, the altitude above sea level is; in going from Oneonta (altitude, 331 m above sea level) to Madison (298 m above sea level) there is an altitude change of 33 m, regardless of the route followed. Interestingly, neither heat nor work individually are state functions although their sum, the change in internal energy, ΔE, is. The value of ΔE is fixed by $E_{initial}$ and E_{final}, but a transition between the initial and final state can be accomplished by different routes having different values of q and w. Enthalpy is also a state function. The enthalpy change occurring when 1.0 g of water is heated from 20 to 50 °C, or when 1.0 g of water is evaporated at 100 °C, is independent of the way in which the process is carried out.

6.5   ENTHALPY CHANGES FOR CHEMICAL REACTIONS

● **Fractional Stoichiometric Coefficients** When writing balanced equations to define thermodynamic quantities chemists often use fractional stoichiometric coefficients. For example, when we wish to define ΔH for the decomposition or formation of 1 mol of H_2O, the coefficient for O_2 must be $\frac{1}{2}$.

An enthalpy change accompanies all chemical reactions [⊕ CD-ROM, Screen 6.15]. For example, for the decomposition of 1 mol of water vapor to its elements, 1 mol of H_2 and $\frac{1}{2}$ mol of O_2, the enthalpy change, $\Delta H = +241.8$ kJ.

$$H_2O(g) \longrightarrow H_2(g) + \tfrac{1}{2} O_2(g) \qquad \Delta H = +241.8 \text{ kJ}$$

That is, the decomposition is an *endothermic* process that requires that 241.8 kJ be transferred to the system, $H_2O(g)$, from the surroundings.

Now consider the opposite reaction, the combination of hydrogen and oxygen to form water. The quantity of heat energy evolved in this reaction is the same as is required for the decomposition reaction, except that the sign of ΔH is reversed. The *exothermic* formation of 1 mol of water vapor from H_2 and $\frac{1}{2}$ mol of O_2 transfers 241.8 kJ to the surroundings (Figure 6.15).

$$H_2(g) + \tfrac{1}{2} O_2(g) \longrightarrow H_2O(g) \qquad \Delta H = -241.8 \text{ kJ}$$

The quantity of heat transferred during a chemical change depends on the amounts of reactants used or products formed. Thus, the formation of 2 mol of water vapor from the elements produces twice as much heat as the formation of 1 mol of water.

$$2 H_2(g) + O_2(g) \longrightarrow 2 H_2O(g) \qquad \Delta H = -483.6 \text{ kJ} \ (= 2 \times -241.8 \text{ kJ})$$

It is important to identify the states of reactants and products in a reaction because the magnitude of ΔH also depends on whether they are solids, liquids,

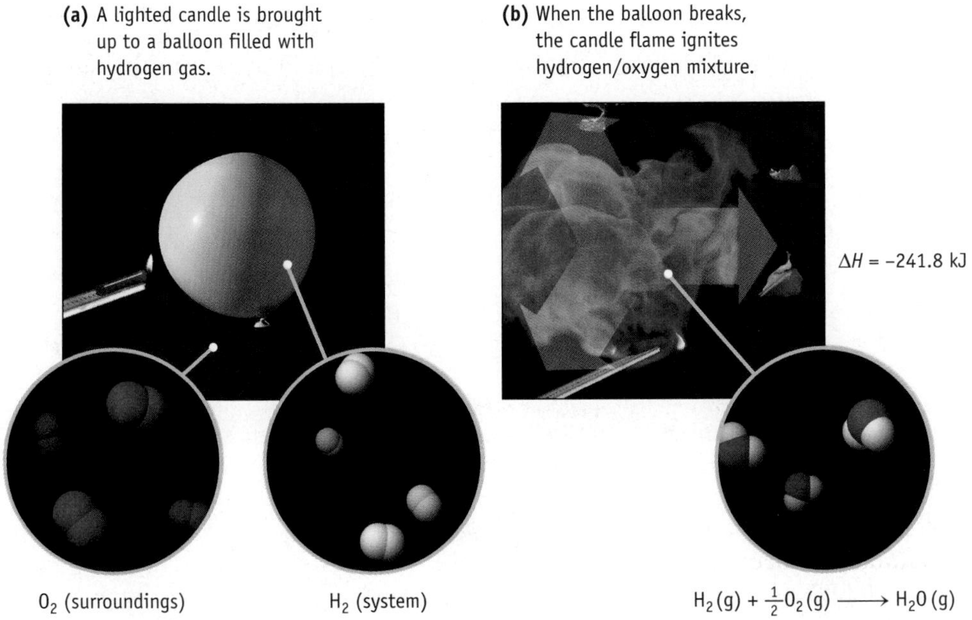

(a) A lighted candle is brought up to a balloon filled with hydrogen gas.

(b) When the balloon breaks, the candle flame ignites hydrogen/oxygen mixture.

Figure 6.15   The exothermic combustion of hydrogen in air. Energy transfer occurs in the form of heat, light, and work [CD-ROM, Screen 6.15]. *(Charles D. Winters)*

$\Delta H = -241.8$ kJ

O_2 (surroundings) H_2 (system) $H_2(g) + \frac{1}{2}O_2(g) \longrightarrow H_2O(g)$

or gases. Formation of 1 mol of liquid water from the elements is accompanied by the evolution of 285.8 kJ.

$$H_2(g) + \tfrac{1}{2} O_2(g) \longrightarrow H_2O(\ell) \qquad \Delta H = -285.8 \text{ kJ}$$

The additional energy relative to the formation of water vapor arises from the energy released when 1 mol of water vapor condenses to 1 mol of liquid water.

These examples illustrate several features of the enthalpy changes for chemical reactions.

- Enthalpy changes are specific to the reactants and products and their amounts. Both the identity of reactants and products and their states (s, ℓ, g) are important.
- ΔH has a negative value if heat is evolved (an exothermic reaction). It has a positive value if heat is required (an endothermic reaction).
- Values of ΔH are numerically the same, but opposite in sign, for chemical reactions that are the reverse of each other.
- The enthalpy change depends on the molar amounts of reactants and products. The formation of 2 mol of $H_2O(g)$ from the elements results in an enthalpy change that is twice as large as the enthalpy change in forming 1 mol of $H_2O(g)$.

Enthalpies of reactions are usually provided in one of two ways. They may be expressed as energy per mole of a reactant or per mole of a product. Alternatively, the enthalpy change may be given along with a balanced chemical equation, as was done in the preceding equation for the reaction of hydrogen with oxygen. In this case the value of ΔH is given for the equation as it is written. Whichever way the enthalpy change is presented, the value can be used to calculate the quantity of heat transferred by any given mass of a reactant or product. Suppose, for example, you want to know the quantity of heat evolved if 454 g of propane, C_3H_8, is burned, given the equation for the exothermic combustion and the enthalpy change for the reaction.

Chapter Goals • Revisited

- Assess heat transfer associated with changes in temperature and changes of state.
- Apply the first law of thermodynamics.
- Define and understand the state functions enthalpy and internal energy.
- **Calculate the energy changes occurring in chemical reactions and learn how these changes are measured.**
- Recognize how chemistry contributes to understanding societal problems involving energy.

$$C_3H_8(g) + 5\ O_2(g) \longrightarrow 3\ CO_2(g) + 4\ H_2O(\ell)\qquad \Delta H = -2220\ kJ$$

Two steps are needed. First, find the amount of propane present in the sample:

$$454\ g\ C_3H_8 \left(\frac{1\ mol\ C_3H_8}{44.10\ g\ C_3H_8}\right) = 10.3\ mol\ C_3H_8$$

Second, multiply the quantity of heat transferred per mole of propane by the amount of propane.

$$\Delta H = 10.3\ mol\ C_3H_8 \left(\frac{-2220\ kJ}{1\ mol\ C_3H_8}\right) = -22,900\ kJ$$

Example 6.5 ● Enthalpy Calculation

Problem • Sucrose (sugar, $C_{12}H_{22}O_{11}$) is oxidized to CO_2 and H_2O. The enthalpy change for the reaction can be measured in the laboratory.

$$C_{12}H_{22}O_{11}(s) + 12\ O_2(g) \longrightarrow 12\ CO_2(g) + 11\ H_2O(\ell)$$
$$\Delta H = -5645\ kJ$$

What is the enthalpy change for the oxidation of 5.0 g (1 tea-spoonful) of sugar?

Strategy • We will first determine the amount of sucrose in 5.0 g, then use this with the value given for the enthalpy change for the oxidation of 1 mol of sucrose.

Solution •

$$5.0\ g\ sucrose \cdot \frac{1\ mol\ sucrose}{342.3\ g\ sucrose} = 1.5 \times 10^{-2}\ mol\ sucrose$$

$$q = 1.5 \times 10^{-2}\ mol\ sucrose \left(\frac{-5645\ kJ}{1\ mol\ sucrose}\right)$$
$$= -82\ kJ$$

Comment • Persons concerned about their diets might be interested to note that a (level) teaspoon of sugar supplies about 2 Cal (dietary Calories, the conversion is 4.184 kJ = 1 Cal). As diets go, a single spoonful of sugar doesn't have a large caloric content. But, will you use a level teaspoonful? And, will you stop with just one?

Gummi Bears are mostly sugar, and you can see in the Chapter Focus (page 202) that their oxidation is clearly highly exothermic.

Exercise 6.8 ● Enthalpy Calculation

(a) What quantity of heat energy is required to decompose 12.6 g of liquid water to the elements?
(b) The combustion of ethane, C_2H_6, has an enthalpy change of -2857.3 kJ for the reaction as written below. Calculate the value of ΔH when 15.0 g of C_2H_6 is burned.

$$2\ C_2H_6(g) + 7\ O_2(g) \longrightarrow 4\ CO_2(g) + 6\ H_2O(g)\qquad \Delta H = -2857.3\ kJ$$

6.6 CALORIMETRY

The heat transferred in a chemical or physical process is determined by an experimental technique called **calorimetry** [◉ CD-ROM, Screen 6.16]. The apparatus used in this kind of experiment is a calorimeter, of which there are two basic types. A **constant pressure calorimeter** allows measurement of heats evolved or required under constant pressure conditions. A **constant volume calorimeter** is one in which the volume cannot change. The two types of calorimetry highlight the differences between enthalpy and internal energy. Heat transferred at constant pressure is, by definition, ΔH, whereas the heat transferred at constant volume is ΔE.

Constant Pressure Calorimetry: Measuring ΔH

For reactions occurring in solution, heat changes at constant pressure are often measured in a general chemistry laboratory using a "coffee-cup calorimeter" (Figure 6.16). This inexpensive device consists of two nested styrofoam coffee cups with a loose fitting lid and a temperature-measuring device such as a thermometer as shown in the figure. The cup contains a solution of the reactants. The mass and specific heat capacity of the solution, and the amount of reactants, must be known. If heat is evolved in the process being studied, the temperature of the solution rises. If heat is required, it is furnished by the solution, and a decrease in temperature will be seen. In each case the change in temperature is measured. From mass, specific heat capacity, and temperature change, the heat change for the contents of the calorimeter can be calculated.

In the terminology of thermodynamics, the contents of the coffee-cup calorimeter are the system, and the cup and the immediate environment around the apparatus are the surroundings. Two heat changes take place within the system. One is the change that occurs as the chemical (potential) energy stored in the reactants is released as heat during the reaction. We label this heat quantity q_{rxn} (where "rxn" is an abbreviation for "reaction"). The other is the heat gained or lost by the solution ($q_{solution}$). Assuming no heat transfer between the system and the surroundings, the sum of the heat changes within the system is zero.

$$q_{rxn} + q_{solution} = 0$$

The change in heat content of the solution ($q_{solution}$) can be calculated from its heat capacity, mass, and change in temperature. The quantity of heat evolved or required for the reaction (q_{rxn}) is the unknown in the equation. Because the reaction is carried out at constant pressure, the heat being measured is an enthalpy change, ΔH.

The accuracy of the result in a calorimeter experiment depends on the accuracy of the measured quantities (temperature, mass, specific heat capacity). In addition, it depends on how closely the assumption of no heat transfer between system and surroundings is followed. A coffee cup calorimeter is an unsophisticated apparatus and the results obtained are not highly accurate, largely because the assumption of no heat transfer is not exactly met. In research laboratories, scientists utilize calorimeters that more effectively limit the heat transfer between system and surroundings, and they may also estimate and correct for any such minimal heat transfer that does occur.

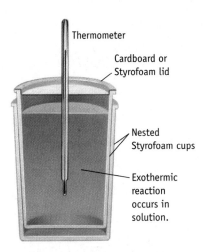

Figure 6.16   A coffee-cup calorimeter. A chemical reaction produces a change in temperature of the solution in the calorimeter. The styrofoam container is fairly effective in preventing heat transfer between the solution and its surroundings.

Example 6.6 Using a Coffee-Cup Calorimeter

Problem • Suppose you place 0.500 g of magnesium chips in a coffee-cup calorimeter and then add 100.0 mL of 1.00 M HCl. The reaction that occurs is

$$Mg(s) + 2\ HCl(aq) \longrightarrow H_2(g) + MgCl_2(aq)$$

The temperature of the solution increases from 22.2 (295.4 K) to 44.8 °C (318.0 K). What is the enthalpy change for the reaction per mole of Mg? (Assume that the specific heat capacity of the solution is 4.20 J/g·K and the density of the HCl solution is 1.00 g/mL.)

Strategy • Two changes in heat content take place within the system, the heat evolved in the reaction (q_{rxn}) and the heat gained by the solution to increase its temperature ($q_{solution}$). The problem solution has three steps. The first is to calculate $q_{solution}$ from values of the mass, specific heat capacity, and ΔT using Equation 6.1. Second, calculate q_{rxn}, assuming there is no energy transfer between system and surroundings (so the the sum of heat changes in the system, $q_{rxn} + q_{solution} = 0$). Third, use the value of q_{rxn} and the amount of Mg to calculate the enthalpy change per mole.

(Example continues on p. 224)

Solution •

Step 1: *Calculate $q_{solution}$.* The mass of the solution is the mass of the 100.0 mL of HCl plus the mass of magnesium or 100.5 g

$$q_{solution} = (100.5\ g)(4.20\ J/g \cdot K)(318.0\ K - 295.4\ K)$$
$$= 9.54 \times 10^3\ J$$

Step 2: *Calculate q_{rxn}.*

$$q_{rxn} + q_{solution} = 0$$
$$q_{rxn} + 9.54 \times 10^3\ J = 0$$
$$q_{rxn} = -9.54 \times 10^3\ J$$

Step 3: *Calculate the value of ΔH per mole.* The quantity of heat found in Step 2 is produced by the reaction of 0.500 g of Mg. Therefore, the heat produced by the reaction of 1.00 mol of Mg is

$$\Delta H = \left(\frac{-9.54 \times 10^3\ J}{0.500\ g\ Mg} \right)\left(\frac{24.31\ g\ Mg}{1\ mol\ Mg} \right)$$
$$= -4.64 \times 10^5\ J/mol\ Mg$$

Comment • The calculation will give the correct sign of q_{rxn} and ΔH. The negative sign indicates that this is an exothermic reaction.

Exercise 6.9   Using a Coffee-Cup Calorimeter

Assume you mix 200. mL of 0.400 M HCl with 200. mL of 0.400 M NaOH in a coffee-cup calorimeter. The temperature of the solutions before mixing was 25.10 °C; after mixing and allowing reaction to occur, the temperature is 27.78 °C. What is the molar enthalpy of neutralization of the acid? (Assume that the densities of all solutions are 1.00 g/mL and their specific heat capacities are 4.20 J/g·K.)

Constant Volume Calorimetry: Measuring ΔE

Constant volume calorimetry is commonly used to evaluate heats of combustion and the caloric value of foods [CD-ROM, Screen 6.16]. A weighed sample of a combustible solid or liquid is placed inside a "bomb," often a cylinder about the size of a large fruit juice can with thick steel walls and ends (Figure 6.17). The bomb is placed in a water-filled container with well-insulated walls. After filling the bomb with pure oxygen, the sample is ignited, usually by an electric spark. The heat generated by the combustion reaction warms the bomb and the water around it. The bomb, its contents, and the water are defined as the system. Assessment of heat transfer within the system shows that

$$q_{rxn} + q_{bomb} + q_{water} = 0$$

Because the volume does not change in a constant volume calorimeter, energy transfer as work cannot occur. Therefore, the heat measured at constant volume (q_v) is the change in internal energy, ΔE.

Example 6.7   Constant Volume Calorimetry

Problem • Octane, C_8H_{18}, a primary constituent of gasoline, burns:

$$C_8H_{18}(\ell) + 25/2\ O_2(g) \longrightarrow 8\ CO_2(g) + 9\ H_2O(\ell)$$

A 1.00-g sample of octane is burned in a constant volume calorimeter (similar to that shown in Figure 6.17). The calorimeter contains 1.20 kg of water. The temperature of the water and the bomb rises from 25.00 °C (298.15 K) to 33.20 °C (306.35 K). The heat capacity of the bomb, C_{bomb}, is 837 J/K. (a) What is the heat of combustion per gram of octane? (b) What is the heat of combustion per mole of octane?

Strategy • (a) The sum of all heat changes in the system will be zero, that is, $q_{rxn} + q_{bomb} + q_{water} = 0$. The first term, q_{rxn}, is the unknown. The second and third terms in the equation can be calculated from the data given: q_{bomb} is calculated from the bomb's heat capacity and ΔT, and q_{water} is determined from the specific heat capacity, mass, and ΔT for water. (b) The value of q_{rxn} calculated in part (a), is the heat evolved in the combustion of 1.00 g of octane. Use this and the molar mass of octane (114.2 g/mol) to calculate the heat evolved per mole of octane.

Solution •

(a) $q_{water} = (C_{water})(m_{water})(\Delta T)$

$= (4.184 \text{ J/g} \cdot \text{K})(1.20 \times 10^3 \text{ g})(306.35 \text{ K} - 298.15 \text{ K})$

$= +41.2 \times 10^3 \text{ J}$

$q_{bomb} = (C_{bomb})(\Delta T) = 837 \text{ J/K} (306.35 \text{ K} - 298.15 \text{ K})$

$= +6.86 \times 10^3 \text{ J}$

$q_{rxn} + q_{water} + q_{bomb} = 0$

$q_{rxn} + (41.2 \times 10^3 \text{ J}) + (6.86 \times 10^3 \text{ J}) = 0$

$q_{rxn} = -48.1 \times 10^3 \text{ J}$

(or -48.1 kJ)

Heat of combustion per gram $= -48.1$ kJ

(b) Heat of combustion per mol $= (-48.1 \text{ kJ/g})(114.2 \text{ g/mol})$

$= -5.49 \times 10^3 \text{ kJ/mol}$

Comment • Because the volume does not change, no energy transfer in the form of work occurs. The change of internal energy, ΔE, for the combustion of $C_8H_{18}(\ell)$ is -5.49×10^3 kJ/mol. Also note that C_{bomb} has no mass units. It is the heat required to warm the whole object by 1 kelvin.

Exercise 6.10   Constant Volume Calorimetry

A 1.00-g sample of ordinary table sugar (sucrose, $C_{12}H_{22}O_{11}$) is burned in a bomb calorimeter. The temperature of 1.50×10^3 g of water in the calorimeter rises from 25.00 °C to 27.32 °C. The heat capacity of the bomb is 837 J/K and the specific heat capacity of the water is 4.20 J/g·K. Calculate (a) the heat evolved per gram of sucrose and (b) the heat evolved per mole of sucrose.

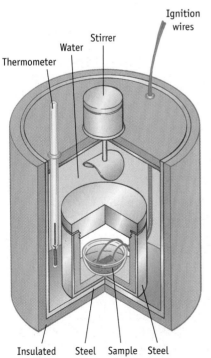
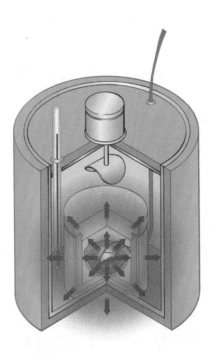

Figure 6.17   Constant volume calorimeter. A combustible sample is burned in pure oxygen in a sealed metal container, or "bomb." The heat generated warms the bomb and the water surrounding it. By measuring the increase in temperature the heat evolved in the reaction can be determined [CD-ROM, Screen 6.16].

Thermometer   Water   Stirrer   Ignition wires

Insulated outside container   Steel container   Sample dish   Steel bomb

6.7   HESS'S LAW

Measuring a heat of reaction using a calorimeter is impossible for many chemical reactions. Consider, for example, the oxidation of carbon to carbon monoxide.

$$C(s) \ + \ \tfrac{1}{2}O_2(g) \ \longrightarrow \ CO(g)$$

Some CO_2 will always form in reactions of carbon and oxygen, even if there is a deficiency of oxygen. The reaction of CO and O_2 is very favorable; thus, as soon as CO is formed, it will react with O_2 to form CO_2. Therefore, using calorimetry to measure the heat evolved in the formation of CO is impossible.

Fortunately, the heat evolved in the reaction forming CO(g) from C(s) and $O_2(g)$ can be calculated from heats measured for other reactions. The calculation is based on **Hess's law,** which states that *if a reaction is the sum of two or more other reactions, ΔH for the overall process is the sum of the ΔH values of those reactions* [CD-ROM, Screen 6.17].

The oxidation of C(s) to $CO_2(g)$ can be viewed as occurring in two steps, first the oxidation of C(s) to CO(g) (Equation 1), and then the oxidation of CO(g) to $CO_2(g)$ (Equation 2). Adding these two equations gives the equation for the oxidation of C(s) to $CO_2(g)$ (Equation 3).

Equation 1:	$C(s) + \tfrac{1}{2}O_2(g) \longrightarrow CO(g)$	$\Delta H_1 = ?$
Equation 2:	$CO(g) + \tfrac{1}{2}O_2(g) \longrightarrow CO_2(g)$	$\Delta H_2 = -283.0$ kJ
Equation 3:	$C(s) + O_2(g) \longrightarrow CO_2(g)$	$\Delta H_3 = -393.5$ kJ

Hess's law tells us that the enthalpy change for overall reaction (ΔH_3) will equal the sum of the enthalpy changes for Equations 1 and 2 ($\Delta H_1 + \Delta H_2$). Both ΔH_2 and ΔH_3 can be measured. These values are used to determine the enthalpy change for reaction 1.

$$\Delta H_3 \ = \ \Delta H_1 \ + \ \Delta H_2$$
$$-393.5 \text{ kJ} = \Delta H_1 + (-283.0 \text{ kJ})$$
$$\Delta H_1 = -110.5 \text{ kJ}$$

Hess's law applies to physical processes, too. The enthalpy change for the reaction of $H_2(g)$ and $O_2(g)$ to form 1 mol of liquid H_2O is different from the enthalpy change to form 1 mol of H_2O vapor (page 221). The difference is the heat of vaporization of water, ΔH_2.

Equation 1:	$H_2(g) + \tfrac{1}{2}O_2(g) \longrightarrow H_2O(\ell)$	$\Delta H_1 = -285.8$ kJ
Equation 2:	$H_2O(\ell) \longrightarrow H_2O(g)$	$\Delta H_2 = ?$
Equation 3:	$H_2(g) + \tfrac{1}{2}O_2(g) \longrightarrow H_2O(g)$	$\Delta H_3 = -241.8$ kJ

The relationship $\Delta H_3 = \Delta H_1 + \Delta H_2$ makes it possible to calculate the value of ΔH_2, the heat of vaporization of water (44.0 kJ, with all substances at 25 °C).

Energy Level Diagrams

When using Hess's law, it is often helpful to represent enthalpy data schematically in an **energy level diagram.** This is a drawing in which various substances, for example the reactants and products in a chemical reaction, are placed on an arbitrary (potential) energy scale. The energy of each substance is identified by a horizontal line, and numerical differences in energy between them are shown by the vertical arrows. If they are drawn to scale, the spacing between the substances shows the differences in their energies. Such a diagram provides an easy-to-read perspective on the magnitude and direction of energy changes and shows how energy of the substances are related [CD-ROM, Screen 6.17].

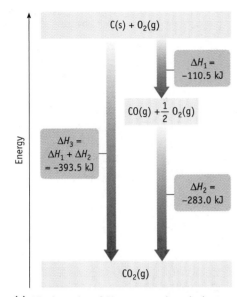

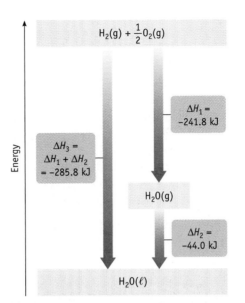

Figure 6.18   Energy level diagrams.
(a) Relating enthalpy changes in the formation of $CO_2(g)$. **(b)** Relating enthalpy changes in the formation of $H_2O(\ell)$. Enthalpy changes associated with changes between energy levels are given alongside the vertical arrows.

(a) The formation of CO_2 can occur in a single step or in a succession of steps. ΔH for the overall process is –393.5 kJ, no matter which path is followed.

(b) The formation of $H_2O(\ell)$ can occur in a single step or in a succession of steps. ΔH for the overall process is –285.8 kJ, no matter which path is followed.

Energy level diagrams that summarize the two examples of Hess's law above are given in Figure 6.18. In Figure 6.18a, the elements, C(s) and O_2(g), are at the highest potential energy. Converting carbon and oxygen to CO_2 lowers the potential energy by 393.5 kJ. This can occur either in a single step, shown on the left, or in two steps, as shown on the right. Similarly, in Figure 6.18b, the elements are at the highest potential energy. The product, liquid or gaseous water, has a lower potential energy, with the difference between the two being the heat of vaporization.

Example 6.8   Using Hess's Law

Problem • Suppose you want to know the enthalpy change for the formation of methane, CH_4, from solid carbon (as graphite) and hydrogen gas:

$$C(s) \ + \ 2\,H_2(g) \ \longrightarrow \ CH_4(g) \qquad \Delta H = ?$$

The enthalpy change for this reaction cannot be measured in the laboratory because the reaction is very slow. We can, however, measure enthalpy changes for the combustion of carbon, hydrogen, and methane.

Equation 1: $C(s) + O_2(g) \longrightarrow CO_2(g)$ $\Delta H_1 = -393.5$ kJ

Equation 2: $H_2(g) + \frac{1}{2} O_2(g) \longrightarrow H_2O(\ell)$ $\Delta H_2 = -285.8$ kJ

Equation 3: $CH_4(g) + 2\,O_2(g) \longrightarrow CO_2(g) + 2\,H_2O(\ell)$

$\Delta H_3 = -890.3$ kJ

Use these energies to obtain ΔH for the formation of methane from its elements.

Strategy • The three equations (1, 2, and 3), as they are written, cannot be added together to obtain the equation for the formation of CH_4 from its elements. Methane, CH_4, is a product in a reaction whose enthalpy is sought, but it is a reactant in Equation 3. Water appears in two of these equations although it is not a component of the reaction forming CH_4 from carbon and hydrogen. To use Hess's law to solve this problem, we will have to manipulate the equations and adjust the heats accordingly. Recall, from Section 6.5, that writing an equation in the reverse direction changes the sign of ΔH and that doubling the amount of reactants and products doubles the value of ΔH. Adjustments to Equations 2 and 3 will produce new equations that, along with Equation 1, can be combined to give the desired net reaction.

Solution • To make CH_4 a product in the overall reaction, we reverse Equation 3 while changing the sign of ΔH. (If a reaction is exothermic in one direction, its reverse must be endothermic):

(Example continues on p. 228)

Equation 3': $CO_2(g) + 2 H_2O(\ell) \longrightarrow CH_4(g) + 2 O_2(g)$
$$\Delta H_3' = -\Delta H_3 = +890.3 \text{ kJ}$$

Next, we see that 2 mol of $H_2(g)$ is on the reactant side in our desired equation. Equation 2 is written for only 1 mol of $H_2(g)$ as a reactant, however. We therefore multiply the stoichiometric coefficients in Equation 2 by 2 and multiply the value of ΔH by 2.

Equation 2 × 2: $2 H_2(g) + O_2(g) \longrightarrow 2 H_2O(\ell)$
$$2 \Delta H_2 = 2(-285.8 \text{ kJ}) = -571.6 \text{ kJ}$$

With these modifications, we rewrite the three equations. When added together, $O_2(g)$, $H_2O(\ell)$, and $CO_2(g)$ all cancel to give the equation for the formation of methane from its elements.

Equation 1: $C(s) + O_2(g) \longrightarrow CO_2(g)$
$$\Delta H_1 = -393.5 \text{ kJ}$$

Equation 2 × 2: $2 H_2(g) + O_2(g) \longrightarrow 2 H_2O(\ell)$
$$2 \Delta H_2 = 2(-285.8 \text{ kJ}) = -571.6 \text{ kJ}$$

Equation 3: $CO_2(g) + 2 H_2O(\ell) \longrightarrow CH_4(g) + 2 O_2(g)$
$$\Delta H_3' = -\Delta H_3 = +890.3 \text{ kJ}$$

Net Equation: $C(s) + 2 H_2(g) \longrightarrow CH_4(g)$
$$\Delta H_{net} = -74.8 \text{ kJ}$$

Comment • You can construct an energy level diagram summarizing the energies of this process.

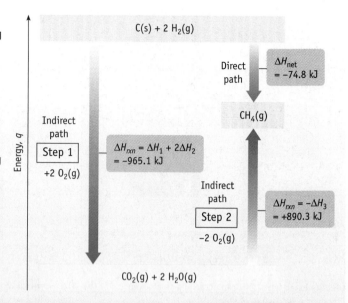

This diagram shows that one can get to the product, $CH_4(g)$, from $C(g)$ and $H_2(g)$ by a direct path or by two-step indirect path. In the indirect path, Step 1 is the exothermic formation of CO_2 and $H_2O(g)$, and Step 2 is the endothermic formation of $CH_4(g)$ and O_2 from CO_2 and H_2O. The value of ΔH for the latter step is known because it is the reverse of the combustion of CH_4.

Exercise 6.11   **Using Hess's Law**

What is the enthalpy change for the formation of ethane, C_2H_6, from elemental carbon and hydrogen?

$$2 C(s) + 3 H_2(g) \longrightarrow C_2H_6(g) \qquad \Delta H = ?$$

Use the equations (and enthalpies) for the combustion of carbon and hydrogen in Example 6.8 along with the experimentally determined ΔH value for the combustion of ethane given here to solve this exercise.

$$C_2H_6(g) + \tfrac{7}{2} O_2(g) \longrightarrow 2 CO_2(g) + 3 H_2O(\ell) \qquad \Delta H = -1559.7 \text{ kJ}$$

Exercise 6.12   **Using Hess's Law**

Graphite and diamond are two allotropes of carbon. The enthalpy change for the process

$$C(\text{graphite}) \longrightarrow C(\text{diamond})$$

cannot be measured directly, but it can be evaluated using Hess's law. (a) Determine this enthalpy change, using experimentally measured heats of combustion of graphite (-393.5 kJ/mol) and diamond (-395.4 kJ/mol). (b) Draw an energy level diagram for this system.

Problem-Solving Tip 6.2

Using Hess's Law

How did we know how to adjust the three equations in Example 6.8? Here is a general strategy to follow for this type of problem.

Step 1. Inspect the equation whose ΔH you wish to calculate, identifying the reactants and products, and locate those substances in the equations available to be added. In Example 6.8 the reactants, C(s) and $H_2(g)$, are reactants in Equations 1 and 2, and the product, $CH_4(g)$, is a reactant in Equation 3. Equation 3 was reversed to get CH_4 on the product side where it is located in the target equation.

Step 2. Get the correct amount of reactants and products on each side. In Example 6.8 only one adjustment was needed. There was 1 mol of H_2 on the left in Equation 2. We needed

2 mol of H_2 in the overall equation, which required doubling the quantities in Equation 2.

Step 3. Make sure other reactants and products in the equations will cancel when the equations are added. In Example 6.8 the equations at this point showed that equal amounts of O_2 and H_2O appeared on the left- and right-hand sides in the three equations, so they canceled when the equations were added together.

Each manipulation requires adjustment of the energy quantities. Summing the equations and the adjusted enthalpies gives the overall equation and its enthalpy change.

Exercise 6.13   **Using Hess's Law**

Use Hess's law to calculate the enthalpy change for the formation of $CS_2(\ell)$ from C(s) and S(s) from the following enthalpy values.

$$C(s) + O_2(g) \longrightarrow CO_2(g) \qquad \Delta H = -393.5 \text{ kJ}$$
$$S(s) + O_2(g) \longrightarrow SO_2(g) \qquad \Delta H = -296.8 \text{ kJ}$$
$$CS_2(g) + 3\,O_2(g) \longrightarrow CO_2(g) + 2\,SO_2(g) \qquad \Delta H = -1103.9 \text{ kJ}$$
$$C(s) + 2\,S(s) \longrightarrow CS_2(g) \qquad \Delta H = ?$$

6.8   STANDARD ENTHALPIES OF FORMATION

Calorimetry and the application of Hess's law have made available a great many ΔH values for chemical reactions. Often, these values are assembled into tables, to make it easy to retrieve and use the data (see Table 6.3 or Appendix L). A very useful table contains **standard molar enthalpies of formation, ΔH_f°.** The standard molar enthalpy of formation is the enthalpy change for the formation of 1 mol of a compound directly from its component elements in their standard states. The **standard state** of an element or a compound is defined as the most stable form of the substance in the physical state that exists at a pressure of 1 bar and at a specified temperature. Most tables report standard molar enthalpies of formation at 25 °C (298 K) [CD-ROM, Screen 6.18].

Several examples of standard molar enthalpies of formation will be helpful to illustrate the meaning of these definitions.

ΔH_f° for $CO_2(g)$:   At 25 °C and 1 bar, the standard state of carbon is solid graphite, the most stable form of this element. The most stable form of oxygen is $O_2(g)$. The standard enthalpy of formation of $CO_2(g)$ is defined as the enthalpy change that occurs in the formation of 1 mol of $CO_2(g)$ from 1 mol of C(s, graphite) and 1 mol of $O_2(g)$; that is, it is the enthalpy change for the process

$$C(s) + O_2(g) \longrightarrow CO_2(g) \qquad \Delta H_f^\circ = -393.5 \text{ kJ}$$

● **ΔH_f° Values**
Enthalpy of formation values are found in this book in Table 6.2 or Appendix L. A useful source of values is the National Institute for Standards and Technology (webbook. nist.gov).

● **ΔH Under Standard Conditions**
The superscript ° indicating standard conditions is applied to other types of thermodynamic data such as the heat of fusion and vaporization (ΔH_{fus}° and ΔH_{vap}°) and the heat of a reaction (ΔH_{rxn}°).

● **Pressure and Standard Conditions**
The bar is the unit of pressure for thermo-dynamic quantities. One bar is approximately one atmosphere. (1 atm = 1.013 bar; see Appendix B.)

ΔH_f° for NaCl(s): At 25 °C and 1 bar, Na is a solid and Cl_2 is a gas. The standard enthalpy of formation of NaCl(s) is defined as the enthalpy change that occurs if 1 mol of NaCl(s) is formed from 1 mol Na(s) and $\frac{1}{2}$ mol $Cl_2(g)$.

$$Na(s) + \tfrac{1}{2} Cl_2(g) \longrightarrow NaCl(s) \qquad \Delta H_f^\circ = -411.1 \text{ kJ}$$

ΔH_f° for $C_2H_5OH(\ell)$: At 25 °C and 1 bar, the standard states of the elements are C(s, graphite), $H_2(g)$ and $O_2(g)$. The standard enthalpy of formation of $C_2H_5OH(\ell)$ is defined as the enthalpy change that occurs if 1 mol of $C_2H_5OH(\ell)$ is formed from 2 mol C(s), 3 mol $H_2(g)$, and $\frac{1}{2}$ mol $O_2(g)$.

$$2 \, C(s) + 3 \, H_2(g) + \tfrac{1}{2} O_2(g) \longrightarrow C_2H_5OH(\ell) \qquad \Delta H_f^\circ = -277.0 \text{ kJ}$$

Notice that the reaction defining the heat of formation need not be (and most often is not) a reaction that you are likely to carry out in laboratory. Ethanol, for example, is not made by a reaction of the elements.

Table 6.2 (and Appendix L) list values of ΔH_f°, obtained from the National Institute for Standards and Technology (NIST), for some common substances. These values are for the formation of one mole of the compound in its standard state from its elements in their standard states. A survey of these values leads to some important observations.

- The standard enthalpy of formation for an element in its standard state is zero.
- Values for compounds in solution refer to the enthalpy change for the formation of a 1 M solution of the compound from the elements making up the compound plus the enthalpy change occurring when the substance dissolves in water.
- Most ΔH_f° values are negative, indicating that formation of most compounds from the elements is exothermic. Heat evolution generally indicates that forming compounds from their elements (under standard conditions) is product-favored (see Section 6.9, page 233).
- Values of ΔH_f° can be used to compare thermal stabilities of related compounds. Consider the values of ΔH_f° for the hydrogen halides in Table 6.3. Hydrogen fluoride is the most stable of these compounds, whereas HI the least stable.

Exercise 6.14 Standard States

What are the standard states of the following elements or compounds (at 25 °C): bromine, mercury, sodium sulfate, ethanol?

Exercise 6.15 Standard Heats of Formation

Write equations for the reactions that define the standard enthalpy of formation of $FeCl_3(s)$ and sucrose (sugar, $C_{12}H_{22}O_{11}$). What are the standard states of the reactants in each equation?

Table 6.2 • Selected Standard Molar Enthalpies of Formation at 298 K

Substance	Name	Standard Molar Enthalpy of Formation (kJ/mol)
C(graphite)	graphite	0
C(diamond)	diamond	+1.8
$CH_4(g)$	methane	−74.87
$C_2H_6(g)$	ethane	−83.85
$C_3H_8(g)$	propane	−104.7
$C_4H_{10}(g)$	butane	−127.1
$C_2H_4(g)$	ethene (ethylene)	+52.47
$CH_3OH(\ell)$	methanol	−238.4
$C_2H_5OH(\ell)$	ethanol	−277.0
$C_{12}H_{22}O_{11}(s)$	sucrose	−2,221.2
$CO(g)$	carbon monoxide	−110.53
$CO_2(g)$	carbon dioxide	−393.51
$CaCO_3(s)$*	calcium carbonate	−1207.6
$CaO(s)$	calcium oxide	−635.0
$H_2(g)$	hydrogen	0
$HCl(g)$	hydrogen chloride	−92.31
$HCl(aq)$*	hydrochloric acid (1 M)	−167.2
$H_2O(\ell)$	liquid water	−285.83
$H_2O(g)$	water vapor	−241.83
$N_2(g)$	nitrogen	0
$NH_3(g)$	ammonia	−45.90
$NH_4Cl(s)$	ammonium chloride	−314.55
$NO(g)$	nitric oxide	+90.29
$NO_2(g)$	nitrogen dioxide	+33.10
$NaCl(s)$	sodium chloride	−411.12
$NaCl(aq)$*	sodium chloride (1 M)	−407.3
$NaOH(s)$	sodium hydroxide	−425.93
$NaOH(aq)$*	sodium hydroxide (1 M)	−470.1
$S_8(s)$	sulfur	0
$SO_2(g)$	sulfur dioxide	−296.84
$SO_3(g)$	sulfur trioxide	−395.77

Source: Data from the NIST Webbook (http://webbook.nist.gov). Values marked with an * are from *Lange's Handbook of Chemistry*, J. Dean (editor): 14th ed., New York, McGraw-Hill, 1992.

Table 6.3 • Standard Molar Enthalpies of Formation of the Hydrogen Halides (kJ/mol, at 298 K)

Compound	ΔH_f° (kJ/mol)
HF(g)	−273.3
HCl(g)	−92.3
HBr(g)	−36.3
HI(g)	+26.5

Enthalpy Change for a Reaction

The enthalpy change for a reaction under standard conditions can be calculated using Equation 6.6 if the standard molar enthalpies of formation are known for *all* reactants and products [CD-ROM, Screen 6.18].

$$\Delta H^\circ_{rxn} = \sum \left[\Delta H_f^\circ \text{ (products)} \right] - \sum \left[\Delta H_f^\circ \text{ (reactants)} \right] \qquad \text{(6.6)}$$

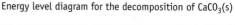

Hess's Law and Equation 6.6

Why is Equation 6.6 a convenient way to apply Hess's law when the enthalpies of formation of *all* the reactants and products are known? Let us look again at the decomposition of calcium carbonate,

$$CaCO_3(s) \longrightarrow CaO(s) + CO_2(g) \qquad \Delta H^\circ_{rxn} = ?$$

and think about an alternative route from the reactant to the products. Imagine the reaction as occurring by first breaking up $CaCO_3$ into its elements (Step 1 in the table below), and then recombining the elements to produce CO_2 and CaO (Steps 2 and 3). Summing the equations for Steps 1, 2, and 3 gives the equation for the net reaction. Most importantly, ΔH°_{rxn} for the net reaction is the sum of the enthalpy changes for each step. For the net reaction,

$$\Delta H^\circ_{net} = \Delta H^\circ_1 + \Delta H^\circ_2 + \Delta H^\circ_3$$
$$\Delta H^\circ_{net} = -\Delta H^\circ_f[CaCO_3(s)] + \Delta H^\circ_f[CaO(s)] + \Delta H^\circ_f[CO_2(g)]$$

or

$$\Delta H^\circ_{net} = \Delta H^\circ_f[CaO(s)] + \Delta H^\circ_f[CO_2(g)] - \Delta H^\circ_f[CaCO_3(s)]$$

This is exactly the result given by applying Equation 6.6. The enthalpy change for the reaction is indeed the sum of the enthalpies of formation of the products minus that of the reactant.

Energy level diagram for the decomposition of $CaCO_3(s)$

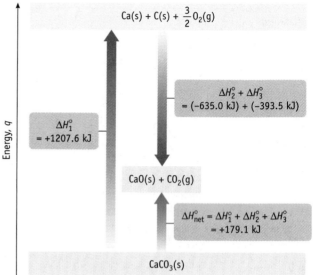

	Reaction	ΔH°_{rxn}
Step 1.	$CaCO_3(s) \longrightarrow Ca(s) + C(s) + \frac{3}{2}O_2(g)$	$\Delta H^\circ_1 = -\Delta H^\circ_f[CaCO_3(s)] = -(-1207.6 \text{ kJ})$
Step 2.	$C(s) + O_2(g) \longrightarrow CO_2(g)$	$\Delta H^\circ_2 = \Delta H^\circ_f[CO_2(g)] = -393.5 \text{ kJ}$
Step 3.	$Ca(s) + \frac{1}{2}O_2(g) \longrightarrow CaO(s)$	$\Delta H^\circ_3 = \Delta H^\circ_f[CaO(s)] = -635.0 \text{ kJ}$
Net	$CaCO_3(s) \longrightarrow CaO(s) + CO_2(g)$	$\Delta H^\circ_{rxn} = +179.1 \text{ kJ}$

• Δ = Final − Initial
Equation 6.6 is another example of the principle that a change (Δ) is always calculated by subtracting the initial state (the reactants here) from the final state (the products in this case).

In this equation, the symbol Σ (the Greek capital letter sigma) means "take the sum." To find ΔH°_{rxn}, add up the molar enthalpies of formation of the products and subtract from this the sum of the molar enthalpies of formation of the reactants. This equation is a logical consequence of the definition of ΔH°_f (See "A Closer Look: Hess's Law and Equation 6.6").

Suppose you want to know how much heat is required to decompose 1 mol of calcium carbonate (limestone) to calcium oxide (lime) and carbon dioxide under standard conditions:

$$CaCO_3(s) \longrightarrow CaO(s) + CO_2(g) \qquad \Delta H^\circ_{rxn} = ?$$

To do this, you find the following enthalpies of formation in a table such as Table 6.2 or Appendix L:

Compound	ΔH°_f (kJ/mol)
$CaCO_3(s)$	−1207.6
$CaO(s)$	−635.0
$CO_2(g)$	−393.5

and then use Equation 6.6 to find the standard enthalpy change for the reaction, ΔH°_{rxn}.

$$\Delta H^\circ_{rxn} = \Delta H^\circ_f\,[CaO(s)] + \Delta H^\circ_f\,[CO_2(g)] - \Delta H^\circ_f\,[CaCO_3(s)]$$
$$= [1\ mol\,(-635.0\ kJ/mol) + 1\ mol\,(-393.5\ kJ/mol)] - [1\ mol\,(-1207.6\ kJ/mol)]$$
$$= +179.1\ kJ$$

The decomposition of limestone to lime and CO_2 is endothermic. That is, energy (+179.1 kJ/mol of $CaCO_3$) must be supplied to decompose 1 mol of $CaCO_3(s)$ to $CaO(s)$ and $CO_2(g)$.

Example 6.9   Using Enthalpies of Formation

Problem • Nitroglycerin is a powerful explosive that forms four different gases when detonated:

$$2\ C_3H_5(NO_3)_3(\ell) \longrightarrow 3\ N_2(g) + \tfrac{1}{2}\ O_2(g) + 6\ CO_2(g) + 5\ H_2O(g)$$

Calculate the enthalpy change when 10.0 g of nitroglycerin is detonated. (The enthalpy of formation of nitroglycerin, ΔH°_f, is −364 kJ/mol. Use Table 6.2 or Appendix L to find other ΔH°_f values that are needed.)

Strategy • Use values of ΔH°_f for the reactants and products in Equation 6.6 to calculate the enthalpy change produced by the detonation of 2 mol of nitroglycerin (ΔH°_{rxn}). From Table 6.2, $\Delta H^\circ_f[CO_2(g)] = -393.5\ kJ/mol$, $\Delta H^\circ_f[H_2O(g)] = -241.8\ kJ/mol$, and $\Delta H^\circ_f = 0$ for $N_2(g)$ and $O_2(g)$. Determine the amount represented by 10.0 g of nitroglycerin, then use this value with ΔH°_{rxn} to obtain the answer.

Solution • Using Equation 6.6, we find the enthalpy change for the explosion of 2 mol of nitroglycerin is

$$\Delta H^\circ_{rxn} = (6\ mol)(\Delta H^\circ_f\,[CO_2(g)]) + (5\ mol)(\Delta H^\circ_f\,[H_2O(g)])$$
$$- (2\ mol)\,(\Delta H^\circ_f\,[C_3H_5(NO_3)_3(\ell)])$$

$$= 6\ mol\,(-393.5\ kJ/mol) + 5\ mol\,(-241.8\ kJ/mol)$$
$$- 2\ mol\,(-364\ kJ/mol)$$
$$= -2842\ kJ\ \text{for 2 mol nitroglycerin}$$

The problem asks for the enthalpy change using 10.0 g of nitroglycerin. We next need to determine the amount of nitroglycerin in 10.0 g.

$$10.0\ g\ \text{nitroglycerin}\left(\frac{1\ mol\ \text{nitroglycerin}}{227.1\ g\ \text{nitroglycerin}}\right) =$$
$$0.0440\ mol\ \text{nitroglycerin}$$

The enthalpy change for the detonation of 0.0440 mol is

$$\Delta H^\circ_{rxn} = 0.0440\ mol\ \text{nitroglycerin}\left(\frac{-2842\ kJ}{2\ mol\ \text{nitroglycerin}}\right)$$
$$= -62.6\ kJ$$

Comment • The large exothermic value of ΔH°_{rxn} is in accord with the fact that this reaction is highly product-favored.

Exercise 6.16   Using Enthalpies of Formation

Calculate the standard enthalpy of combustion for benzene, C_6H_6.

$$C_6H_6(\ell) + 7\tfrac{1}{2}\ O_2(g) \longrightarrow 6\ CO_2(g) + 3\ H_2O(\ell) \qquad \Delta H^\circ_{rxn} = \ ?$$

$\Delta H^\circ_f\,[C_6H_6(\ell)] = +48.95\ kJ/mol$. Other values needed can be found in Table 6.2 and Appendix L.

6.9   PRODUCT- OR REACTANT-FAVORED REACTIONS AND THERMOCHEMISTRY

Reactions that occur from left to right as the chemical equation is written, and in which reactants are largely converted to products, are said to be *product-favored* (see page 166). One aspect of chemical reactivity, and a goal of this book, is to be able to predict whether a chemical reaction is product- or reactant-favored. In Chapter

Figure 6.19 The product-favored oxidation of iron. Iron powder, sprayed into a Bunsen burner flame, is rapidly oxidized. The reaction is exothermic and product-favored. *(Charles D. Winters)*

• **Reactant- or Product-Favored**
In most cases exothermic reactions are product-favored and endothermic reactions are reactant-favored.

5 you learned that certain common reactions occurring in aqueous solution — precipitation, acid–base, and gas-forming — and reactions such as combustions are generally product-favored. Our discussion of the energy involved in chemical reactions allows us to begin to understand more about predicting which reactions may be product-favored [🖰 CD-ROM, Screen 6.2].

The oxidation reactions of hydrogen and carbon (see Figures 6.15 and 6.18), Gummi Bears (see Chapter Focus), and iron (Figure 6.19)

$$4\,Fe(s) + 3\,O_2(g) \longrightarrow 2\,Fe_2O_3(s)$$
$$\Delta H^\circ_{rxn} = 2\,\Delta H^\circ_f[Fe_2O_3(s)] = 2(-824.2\ kJ) = -1648.4\ kJ$$

are exothermic. All have negative values for ΔH°_{rxn} and transfer energy to their surroundings. They are also all product-favored reactions.

Conversely, the reactant-favored decomposition of calcium carbonate is endothermic. Heat is required for the reaction to occur, and ΔH°_{rxn} is positive.

$$CaCO_3(s) \longrightarrow CaO(s) + CO_2(g) \qquad \Delta H^\circ_{rxn} = +179.1\ kJ$$

Are all exothermic reactions product-favored and all endothermic reactions reactant-favored? From these examples, we might formulate this as a hypothesis that can be tested by experiment and by examining many other examples. We would find that *in most cases product-favored reactions have negative values of ΔH°_{rxn} and reactant-favored reactions have positive values of ΔH°_{rxn}*. But this is not *always* true; there are exceptions, and we shall return to the issue in Chapter 19.

Exercise 6.17 **Product- or Reactant-Favored**

Calculate ΔH°_{rxn} for each of the following reactions and decide if the reaction may be product- or reactant-favored.

(a) $2\,HBr(g) \longrightarrow H_2(g) + Br_2(g)$

(b) $C(diamond) \longrightarrow C(graphite)$

6.10 ENERGY RESOURCES

Thermochemistry is a topic that has renewed importance in the 21st century because of concerns about energy resources for the developed and developing world. Some useful and interesting information can put the problem in perspective:

• Global demand for energy has tripled in the past 50 years and may triple again in the next 50. The majority of the demand has been in industrialized countries, and 90% has been met with fossil fuels.

• With only 4.6% of the world's population, the United States consumes about 25% of the energy used in the world. This is equivalent to the consumption of 7 gallons of oil or 70 pounds of coal per person per day.

• About 85% of the energy used in the United States comes from fossil fuels: coal, natural gas, and oil.

• About 8% of the energy consumed in the United States is generated by nuclear power plants.

• The smallest fraction of energy consumed in the United States (7%) comes from renewable energy resources, mostly hydroelectric power and biomass (wood). Smaller amounts still are based on solar, wind, and geothermal power.

• **Nuclear Power**
At the end of the 1990s the United States had 104 nuclear power plants in operation.

Fossil Fuels: What Is New?

Methane, CH_4, is the lightest carbon-containing molecule. This hydrocarbon is a colorless, odorless, flammable gas under ordinary conditions. You probably know it as the major constituent of natural gas (70–95% methane). The gas is found deep under the earth's surface, especially in Texas and Louisiana. You might also know methane as *swamp gas,* or *marsh gas,* because it is formed by bacteria working on organic matter in an anaerobic environment (one in which the oxygen concentration is low). These conditions occur in sedimentary layers in coastal waters and in marshes. The gas can escape if the sediment layer is thin, and you see it as bubbles in a marsh. It may be trapped under thicker sediment layers, however. Over time it can be squeezed into porous rocks such as sandstone or limestone, and here it waits to be tapped by drilling into the rock layer.

Chemical Perspectives

Thermodynamics and Consumer Products

Does your elbow hurt from too much tennis? Are your hands or feet cold after a day of skiing? Pull out a cold pack or a hot pack.

How do these things work? Cold packs generally contain the salt, ammonium nitrate, and water in two separate packages. Ammonium nitrate, which is readily soluble in water, dissolves when the inner bag is broken. The dissolving process is endothermic, and the heat required is transferred from the surroundings — your elbow — to the system — the solution [CD-ROM, Screen 6.6].

$$NH_4NO_3(s) \longrightarrow NH_4^+(aq) + NO_3^-(aq)$$
$$\Delta H^\circ_{rxn} = +25.69 \text{ kJ/mol}$$

Hot packs come in several varieties, one of which uses sodium acetate. When the hydrated salt crystallizes from water, heat is evolved.

$$Na^+(aq) + CH_3CO_2^-(aq) + 3 H_2O(\ell)$$
$$\longrightarrow NaCH_3CO_2 \cdot 3 H_2O(s)$$
$$\Delta H^\circ_{rxn} = -37.86 \text{ kJ/mol}$$

Another variety of hot pack uses an oxidation–reduction reaction.

$$4 Fe(s) + 3 O_2(g) \longrightarrow 2 Fe_2O_3(s)$$
$$\Delta H^\circ_{rxn} = -1648 \text{ kJ}$$

This very exothermic reaction allows the hand warmer to maintain a temperature of 57–69 °C for several hours if oxygen flow is somewhat restricted, as by keeping the device in a glove or pocket.

(Charles D. Winters) ▶

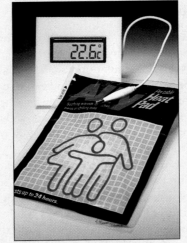

Cold packs absorb heat as ammonium nitrate dissolves in water.

A hot pack relies on the heat evolved by the crystallization of sodium acetate.

A hand warmer uses the oxidation of iron as the source of heat.

Figure 6.20 Methane hydrate. (a) This interesting substance is found in huge deposits hundreds of feet down on the floor of the ocean. When a sample is brought to the surface, the methane oozes out of the solid, and the gas readily burns. **(b)** The structure of the solid hydrate consists of methane molecules trapped in a lattice of water molecules. Each point of the lattice shown here is an O atom of an H_2O molecule. The edges are O—H—O hydrogen bonds. Such structures are often called "clathrates" and are also found with substances other than methane. See E. Suess, G. Bohrmann, J. Greinert, and E. Lausch: *Scientific American*, November 1999, pp. 76–83. See also the cover of this book. *(a, John Pinkston and Laura Stern/U.S. Geological Survey/Science News, 11-9-96)*

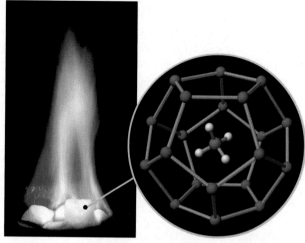

(a) Methane hydrate burns as methane gas escapes from the solid hydrate.

(b) Methane hydrate consists of a lattice of water molecules with methane molecules trapped in the cavity.

Chapter Goals • Revisited

- Assess heat transfer associated with changes in temperature and changes of state.
- Apply the first law of thermodynamics.
- Define and understand the state functions enthalpy and internal energy.
- Calculate the energy changes occurring in chemical reactions and learn how these changes are measured.
- **Recognize how chemistry contributes to understanding societal problems involving energy.**

When natural gas pipelines were laid across states and across the continent, pipeline operators soon found that, unless water was carefully kept out of the line, chunks of *methane hydrate* would form and clog the pipes. In solid methane hydrate the hydrocarbon is trapped in cavities in the molecular structure of ice (Figure 6.20). When methane hydrate melts, the volume of gas released (at normal pressure and temperature) is about 165 times larger than the volume of the hydrate. So, methane hydrate is an efficient method for storing methane gas.

If methane hydrate can form in a pipeline, is it found in nature? In May 1970, oceanographers, drilling into the sea bed off the coast of South Carolina, pulled up samples of the hydrate that fizzed and oozed out of the drill casing. Since then, methane hydrate has been found in many parts of the oceans as well as under permafrost in the Arctic. Indeed, it is estimated that 1.5×10^{13} tons of methane hydrate is buried under the sea floor around the world. In fact, the energy content of this gas may surpass that of the known fossil fuel reserves by as much as a factor of 2!

Solar Energy

The sun is the origin of all current major energy sources except nuclear energy. Chemical potential energy in fossil fuels was collected on earth over many millions of years. Biomass derives from solar energy collected more recently on a shorter time scale, and water and wind power come indirectly from the sun's energy. Significantly, though, we are making very inefficient direct use of the sun's energy. Photosynthesis by plants uses only about 1–3% of the incident sunlight. This process, in which solar energy is collected and stored by plants as chemical potential energy, is not efficient.

Considering the importance of solar energy to our lives, perhaps we should be giving more attention to this resource. It has potential that has yet to be effectively exploited. Every year the earth's surface receives about ten times as much energy from sunlight as is contained in all the known reserves of coal, oil, natural gas, and uranium combined. This is equivalent to about 15,000 times the world's annual consumption by humans.

Example 6.10 Using Energy Units

Problem • The average solar energy received by a horizontal surface in Madison, Wisconsin, in the summer is 2.3×10^7 J/m² · day. If a home in Madison has a horizontal roof that measures 10.0 m by 25.0 m, what quantity of energy, measured in kilojoules per day, strikes the roof?

Strategy • Units will guide us to the answer. We know the quantity of energy received per square meter. Multiplying the energy per square meter (J/m² · day) by the area in square meters (m²) will give the energy in joules (J/day); this can be converted to kilojoules per day using the conversion factor relating J and kJ.

Solution •

$$\text{Roof area} = (10.0 \text{ m})(25.0 \text{ m}) = 2.50 \times 10^2 \text{ m}^2$$

$$\text{Energy received} = (2.50 \times 10^2 \text{ m}^2)$$
$$(2.3 \times 10^7 \text{ J/m}^2 \cdot \text{day})(1 \text{ kJ}/10^3 \text{ J})$$
$$= 5.8 \times 10^6 \text{ kJ/day}$$

Comment • It is also interesting to calculate the energy received in kilowatt-hours, the energy unit used by your local electric utility. The conversion factor needed is 1 kwh = 3.61×10^3 kJ.

$$\text{Energy received} = (5.8 \times 10^6 \text{ kJ/day})(1 \text{ kwh}/3.61 \times 10^3 \text{ kJ})$$
$$= 1.6 \times 10^3 \text{ kwh/day}$$

We find that the roof receives about 1600 kwh of energy in a single day. This is equivalent to the energy used to keep several 100-W light bulbs lighted 24 hours a day for a year.

Exercise 6.18 Energy Resources

A lake has a surface area of 2.6×10^6 m² (about 1 square mile) and an average depth of 12 m. What quantity of heat (in kilojoules) is given off if the temperature of the lake water decreases by 1.0 °C? Assume the density of the water is 1.0 g/cm³.

How might the sun's energy be more efficiently exploited? A number of problems need to be addressed: collection, storage, and transmission of energy come first to mind, and in all of these chemistry (and chemists) have a role to play in developing the necessary technology. The bottom line will ultimately be cost, relative to the cost of other energy resources.

One way to collect solar energy is to use photovoltaic cells, devices capable of transforming solar energy directly into electric energy. These devices are now used in spacecraft and in pocket calculators, and they have been tested for large-scale commercial use (Figure 6.21). Research has produced photovoltaic cells that use 20–30% of the light. High efficiency is necessary to offset the high cost of making the devices, which are usually based on highly purified silicon.

(a) House with photovoltaic panels on the roof.

(b) A solar-powered Honda. From the competition of university students in Australia.

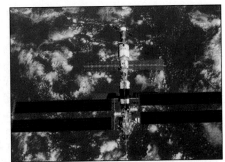

(c) Artist's rendition of the completed Space Station with the solar panels in place.

Figure 6.21 **Solar energy.** Three uses of photovoltaic panels. You may also have one on your electronic calculator. *(a, Brian Parker/Tom Stack & Associates; b, Greg Vaughn/Tom Stack & Associates; c, NASA)*

Figure 6.22 Artificial photosynthesis.
Research has produced a prototype device that can generate electricity or produce a chemical fuel. Hydrogen gas is generated when light shines on a light-sensitive device immersed in water. *(Nathan Lewis/California Institute of Technology)*

The Hydrogen Economy

A problem with electrical energy is the need for an economical and convenient way to store large amounts of it. This is one of the major obstacles to the construction and utilization of electric cars. It is also a problem for power companies that are required to face both low- and high-peak demands. A desirable situation for power companies would be to store power during low-use periods and have it available during high-use periods.

We know how to store large amounts of energy as chemical energy — as fuels. One option to consider uses sunlight to produce a storable chemical fuel. For example, sunlight might be used to decompose water into its elements (Figure 6.22). The hydrogen produced could then be stored and later burned in a power plant. Or it could be used in an automobile.

This brings us to a concept talked about, off and on, called the **hydrogen economy,** which is a technology powered by energy from the oxidation of H_2. If hydrogen could be obtained from water at low cost, there would be advantages to its use as a fuel. Hydrogen could be stored and transported to places where it can be used. It could serve as a fuel on a large scale (in a power plant), or on a small scale (as in an automobile). Combustion of hydrogen provides far more energy per gram than any other common fuel (Table 6.4), and the product of oxidation, water, is nonpolluting.

Currently the best known use of hydrogen as a fuel is in the Space Shuttle (Figure 6.23a). The tank strapped to the shuttle contains 1.46×10^6 L of hydrogen and 5.43×10^5 L of oxygen. The Space Shuttle also uses a device called a fuel cell [➡ CHAPTER 20] that produces electricity directly from the chemical reaction between hydrogen and oxygen.

Someday hydrogen-powered automobiles may be available. The BMW company has built prototype vehicles that run on liquid hydrogen, which is stored in a pressurized tank (Figure 6.23b). The Ford Motor Company is also working on such a car. Of course, significant problems still need to be solved, such as a distribution network for hydrogen, a safe way of storing hydrogen in an automobile, and a method of making hydrogen that is competitive in cost with gasoline.

But a hydrogen-fueled economy is possible in another way. In 2001, Iceland announced that the country would become a "carbon-free economy." They plan to rely on hydrogen-powered electric fuel cells to run vehicles and fishing boats.

(a) The main engine of the Space Shuttle burns hydrogen. The flame is colorless, so the flames you see are from the solid-fuel booster rockets.

(b) A prototype of a hydrogen-powered BMW.

Figure 6.23 Hydrogen-powered vehicles. *(a, NASA; b, Martin Bond/Science Photo Library/Photo Researchers, Inc.)*

Figure 6.24 Iceland, a "carbon-free," hydrogen-based economy. A geothermal field in Iceland. The country plans to use such renewable resources to produce hydrogen from water and then use the hydrogen to produce electricity in fuel cells. (*©2002 Corbis*)

Table 6.4 • Energy Released by Combustion of Some Common Fuels

Substance	Energy Released (kJ/g)
hydrogen	142
gasoline	48
crude petroleum	43
coal	29
paper	20
dried biomass	16
air-dried wood	15

Source: Data taken in part from J. Harte: *Consider a Spherical Cow,* University Science Books, Mill Valley, CA, 1988, page 242.

Iceland is fortunate in that two-thirds of its energy already comes from renewable sources — hydroelectric and geothermal energy (Figure 6.24). The country has decided to use the electricity produced by geothermal heat or hydropower to separate water into hydrogen and oxygen (see Figure 1.1, page 11). The hydrogen will then be used in fuel cells or combined with CO_2 to make methanol, CH_3OH, a liquid fuel.

As the world's population expands, and the standard of living increases around the world, more and more complex problems arise for chemists and other scientists to solve: producing inexpensive and effective medicines, finding ways to produce the materials used in modern economies in inexpensive and nonpolluting ways, and providing energy to developing countries while avoiding the pollution problems endemic in some emerging economies.

In Summary

When you have finished studying this chapter, you should ask if you have met the chapter goals. In particular, you should be able to

- Describe various forms of energy and the nature of heat and thermal energy transfer (Section 6.1).
- Use the most common energy unit, the joule, and convert other energy units to joules (Section 6.1).
- Recognize and use the language of thermodynamics: the system and its surroundings; exothermic and endothermic reactions (Section 6.1).
- Use specific heat capacity in calculations of heat transfer and temperature changes (Section 6.2).
- Understand the sign conventions in thermodynamics (Section 6.2).
- Use heat of fusion and heat of vaporization to find the quantity of thermal energy involved in changes of state (Section 6.3).
- Understand the basis of the first law of thermodynamics (Section 6.4).
- Recognize state functions whose value is determined only by the state of the system and not by the pathway taken by the process (Section 6.4).

- Recognize that when a process is carried out under constant pressure conditions, the heat transferred is the enthalpy change, ΔH (Section 6.5).
- Describe how to measure the quantity of heat energy transferred in a reaction by using calorimetry (Section 6.6).
- Apply Hess's law to find the enthalpy change for a reaction (Sections 6.7).
- Know how to draw and interpret energy level diagrams (Section 6.7).
- Use standard molar enthalpy of formation, ΔH_f°, to calculate the enthalpy change for a reaction, ΔH_{rxn}° (Section 6.8).
- Discuss uses of energy in our economy and problems and opportunities for developing new energy resources (Section 6.10).

Key Terms

thermodynamics

Section 6.1

energy
potential energy
law of conservation of energy
first law of thermodynamics
system
surroundings
thermal equilibrium
exothermic
endothermic
joule
kilojoule
calorie (with a small c)
kilocalorie
dietary Calorie (with a capital C)

Section 6.2

specific heat capacity

Section 6.3

change of state
heat of fusion
heat of vaporization

Section 6.4

sublimation
internal energy
P–V (pressure–volume) work
enthalpy change (ΔH)
state function

Section 6.6

calorimetry
constant pressure calorimeter
constant volume calorimeter

Section 6.7

Hess's law
energy level diagrams

Section 6.8

standard molar enthalpy of
formation
standard state

Section 6.10

hydrogen economy

Key Equations

Equation 6.1 (page 209)

The heat transferred when the temperature of a substance changes (q). Calculated from the specific heat capacity (C), mass (m), and change in temperature (ΔT).

$$q(J) = C(J/g \cdot K) \times m(g) \times \Delta T(K)$$

Equation 6.2 (page 209)

Temperature changes are always calculated as final temperature minus initial temperature.

$$\Delta T = T_{final} - T_{initial}$$

Equation 6.3 (page 212)

If no heat is transferred between a system and its surroundings, the sum of heat changes within the system equals zero.

$$q_1 + q_2 + q_3 + \cdots = 0$$

Equation 6.4 (page 217)

The first law of thermodynamics: The change of internal energy (ΔE) in a system is the sum of heat transferred (q) and work done (w).

$$\Delta E = q + w$$

Equation 6.5 (page 218)

Work (w) at constant pressure is the product of pressure (P) and change in volume (ΔV).

$$w = -P \times \Delta V$$

Equation 6.6 (page 231)

This equation is used to calculate the standard enthalpy change of a reaction (ΔH°_{rxn}) when the heats of formation of all of the reactants and products are known.

$$\Delta H^\circ_{rxn} = \sum [\Delta H^\circ_f (\text{products})] - \sum [\Delta H^\circ_f (\text{reactants})]$$

Study Questions

Questions with blue, bold-faced numbers have answers in Appendix O. See the descriptions, tutorials, and simulations in Chapter 6 of the General Chemistry Interactive CD-ROM, *Version 3.0.*

Reviewing Important Concepts

1. The following terms are used extensively in thermodynamics. Define each and give an example.
   (a) Exothermic and endothermic
   (b) System and surroundings
   (c) Specific heat capacity
   (d) State function
   (e) Standard state
   (f) Enthalpy change, ΔH
   (g) Standard enthalpy of formation

2. For each of the following, tell whether the process is exothermic or endothermic.
   (a) $H_2O(\ell) \longrightarrow H_2O(s)$
   (b) $2\,H_2(g) + O_2(g) \longrightarrow 2\,H_2O(g)$
   (c) $H_2O(\ell, 25\,°C) \longrightarrow H_2O(\ell, 15\,°C)$
   (d) $H_2O(\ell) \longrightarrow H_2O(g)$

3. For each of the following, define a system and its surroundings and give the direction of heat transfer between system and surroundings.
   (a) Methane is burning in a gas furnace in your home.
   (b) Water drops, sitting on your skin after a dip in a swimming pool, evaporate.
   (c) Water, at 25 °C, is placed in the freezing compartment of a refrigerator where it cools and eventually solidifies.
   (d) Aluminum and $Fe_2O_3(s)$ are mixed in a flask sitting on a laboratory bench. A reaction occurs, and a large quantity of heat is evolved.

4. Which of the following are state functions?
   (a) The volume of a balloon
   (b) The time it takes to drive from your home to your college or university
   (c) The temperature of the water in a coffee cup
   (d) The potential energy of a ball held in your hand

5. Define the first law of thermodynamics using a mathematical equation and explain the meaning of each term in the equation.

6. The first law of thermodynamics is often described as another way of stating the law of conservation of energy. Discuss whether this is an accurate portrayal.

7. Many people have tried to make a perpetual motion machine, but none have succeeded although some have claimed success. Use the law of conservation of energy to explain why such a device is impossible.

8. What does the term "standard state" mean? What are the standard states of the following substances at 298 K: H_2O, NaCl, Hg, CH_4?

9. Use Appendix L to find the standard enthalpies of formation of oxygen atoms, oxygen molecules (O_2), and ozone (O_3). What is the standard state of oxygen? Is the formation of oxygen atoms from O_2 exothermic? What is the enthalpy change for the formation of 1 mol of O_3 from O_2?

10. How do product- and reactant-favored reactions differ? What is the relation of these terms to the enthalpy change for a reaction?

11. Without doing calculations, decide if each of the following is product- or reactant-favored.
    (a) The combustion of natural gas
    (b) The decomposition of sugar, $C_{12}H_{22}O_{11}$, to carbon and water

Practicing Skills

Energy Units
(See Exercise 6.2 and CD-ROM Screen 6.7)

12. You are on a diet that calls for eating no more than 1200 Cal/day. How many joules would this be?

13. A 2-in. piece of chocolate cake with frosting provides 1670 kJ of energy. What is this in dietary calories (Cal)?

Specific Heat Capacity
(See Examples 6.1 and 6.2 and CD-ROM Screens 6.8–6.11)

14. The molar heat capacity of mercury is 28.1 J/mol·K. What is the specific heat capacity of this metal in J/g·K?

15. The specific heat capacity of benzene is 1.74 J/g·K. What is its molar heat capacity (J/mol·K)?

16. The specific heat capacity of copper is 0.385 J/g·K. What quantity of heat is required to heat 168 g of copper from −12.2 to +25.6 °C?

17. What quantity of heat is required to raise the temperature of 50.00 mL of water from 25.52 to 28.75 °C? The density of water at this temperature is 0.997 g/mL.

18. The initial temperature of a 344-g sample of iron is 18.2 °C. If the sample absorbs 2.25 kJ of heat, what is its final temperature?

19. After absorbing 1.850 kJ of heat, the temperature of a 0.500 kg block of copper is 37 °C. What was its initial temperature?

20. A 45.5-g sample of copper at 99.8 °C is dropped into a beaker containing 152 g of water at 18.5 °C. When thermal equilibrium is reached, what is the final temperature?

21. A 182-g sample of gold at some temperature is added to 22.1 g of water. The initial water temperature is 25.0 °C, the final temperature is 27.5 °C. If the specific heat capacity of gold is 0.128 J/g·K, what was the initial temperature of the gold?

22. One beaker contains 156 g of water at 22 °C, and a second beaker contains 85.2 g of water at 95 °C. The water in the two beakers is mixed. What is the final water temperature?

23. When 108 g of water at a temperature of 22.5 °C is mixed with 65.1 g of water at an unknown temperature, the final temperature of the resulting mixture is 47.9 °C. What was the initial temperature of the second sample of water?

24. A 13.8-g piece of zinc was heated to 98.8 °C in boiling water and then dropped into a beaker containing 45.0 g of water at 25.0 °C. When the water and metal come to thermal equilibrium, the temperature is 27.1 °C. What is the specific heat capacity of zinc?

25. A 237-g piece of molybdenum, initially at 100.0 °C, is dropped into 244 g of water at 10.0 °C. When the system comes to thermal equilibrium, the temperature is 15.3 °C. What is the specific heat capacity of molybdenum?

Changes of State

(See Examples 6.3 and 6.4 and CD-ROM Screen 6.12)

26. What quantity of heat is evolved when 1.0 L of water at 0 °C solidifies to ice? The heat of fusion of water is 333 J/g.

27. The heat energy required to melt 1.00 g of ice at 0 °C is 333 J. If one ice cube has a mass of 62.0 g, and a tray contains 16 ice cubes, what quantity of energy is required to melt a tray of ice cubes to form liquid water at 0 °C?

28. What quantity of heat is required to vaporize 125 g of benzene, C_6H_6, at its boiling point, 80.1 °C? The heat of vaporization of benzene is 30.8 kJ/mol.

29. Chloromethane, CH_3Cl, is used as a topical anesthetic. What quantity of heat must be absorbed to convert 92.5 g of liquid to a vapor at its boiling point, −24.09 °C? The heat of vaporization of CH_3Cl is 21.40 kJ/mol.

30. The freezing point of mercury is −38.8 °C. What quantity of heat energy (in joules) is released to the surroundings if 1.00 mL of mercury is cooled from 23.0 to −38.8 °C and then frozen to a solid? (The density of liquid mercury is 13.6 g/cm³. Its specific heat capacity is 0.140 J/g·K and its heat of fusion is 11.4 J/g.)

31. What quantity of heat energy (in joules) is required to raise the temperature of 454 g of tin from room temperature (25.0 °C) to its melting point, 231.9 °C, and then melt the tin at that temperature? The specific heat capacity of tin is 0.227 J/g·K, and the heat of fusion of this metal is 59.2 J/g.

32. Ethanol, C_2H_5OH, boils at 78.29 °C. What quantity of heat energy (in joules) is required to raise the temperature of 1.00 kg of ethanol from 20.0 °C to the boiling point and then change the liquid to vapor at that temperature? (The specific heat capacity of liquid ethanol is 2.44 J/g·K, and its enthalpy of vaporization is 855 J/g.)

33. A 25.0-mL sample of benzene at 19.9 °C was cooled to its melting point (5.5 °C) and then frozen. How much heat was given off in this process? The density of benzene is 0.80 g/mL, its specific heat capacity is 1.74 J/g·K, and its heat of fusion is 127 J/g.

Enthalpy

(See Example 6.5 and CD-ROM Screens 6.14 and 6.15)

34. Nitrogen monoxide, a gas recently found to be involved in a wide range of biological processes, reacts with oxygen to give brown NO_2 gas.

$$2\ NO(g) + O_2(g) \longrightarrow 2\ NO_2(g) \qquad \Delta H°_{rxn} = -114.1\ kJ$$

Is the reaction endothermic or exothermic? If 1.25 g of NO is converted completely to NO_2, what quantity of heat is absorbed or evolved?

35. Calcium carbide, CaC_2, is manufactured by the reaction of CaO with carbon at a high temperature. (Calcium carbide is then used to make acetylene.)

$$CaO(s) + 3\ C(s) \longrightarrow CaC_2(s) + CO(g)$$
$$\Delta H°_{rxn} = -464.8\ kJ$$

Is the reaction endothermic or exothermic? If 10.0 g of CaO is allowed to react with an excess of carbon, what quantity of heat is absorbed or evolved by the reaction?

36. Isooctane (2,2,4-trimethylpentane), one of many hydrocarbons that make up gasoline, burns in air to give water and carbon dioxide.

$$2\ C_8H_{18}(\ell) + 25\ O_2(g) \longrightarrow 16\ CO_2(g) + 18\ H_2O(\ell)$$
$$\Delta H°_{rxn} = -10{,}922\ kJ$$

If you burn 1.00 L of isooctane (density = 0.69 g/mL), what quantity of heat is evolved?

37. Acetic acid, CH_3CO_2H, is made industrially by the reaction of methanol and carbon monoxide.

$$CH_3OH(\ell) + CO(g) \longrightarrow CH_3CO_2H(\ell)$$
$$\Delta H°_{rxn} = -355.9\ kJ$$

If you produce 1.00 L of acetic acid (density = 1.044 g/mL) by this reaction, what quantity of heat is evolved?

Calorimetry

(See Examples 6.6 and 6.7 and CD-ROM Screens 6.11 and 6.16)

38. Assume you mix 100.0 mL of 0.200 M CsOH with 50.0 mL of 0.400 M HCl in a coffee-cup calorimeter. The following reaction occurs.

$$CsOH(aq) + HCl(aq) \longrightarrow CsCl(aq) + H_2O(\ell)$$

The temperature of both solutions before mixing was 22.50 °C, and it rises to 24.28 °C after the acid–base reaction takes place. What is the enthalpy of the reaction per mole of CsOH? Assume the densities of the solutions are all 1.00 g/mL and the specific heat capacities of the solutions are 4.2 J/g·K.

39. You mix 125 mL of 0.250 M CsOH with 50.0 mL of 0.625 M HF in a coffee-cup calorimeter, and the temperature of both solutions rises from 21.50 °C before mixing to 24.40 °C after the reaction.

$$CsOH(aq) + HF(aq) \longrightarrow CsF(aq) + H_2O(\ell)$$

What is the enthalpy of reaction per mole of CsOH? Assume the densities of the solutions are all 1.00 g/mL and the specific heats of the solutions are 4.2 J/g·K.

40. A piece of titanium metal with a mass of 20.8 g is heated in boiling water to 99.5 °C and then dropped into a coffee-cup calorimeter containing 75.0 g of water at 21.7 °C. When thermal equilibrium is reached the final temperature is 24.3 °C. Calculate the specific heat capacity of titanium.

41. A piece of chromium metal with a mass of 24.26 g is heated in boiling water to 98.3 °C and then dropped into a coffee-cup calorimeter containing 82.3 g of water at 23.3 °C. When thermal equilibrium is reached the final temperature is 25.6 °C. Calculate the specific heat capacity of chromium.

42. Adding 5.44 g $NH_4NO_3(s)$ to 150.0 g of water in a coffee-cup calorimeter (with stirring to dissolve the salt) resulted in a decrease in temperature from 18.6 to 16.2 °C. Calculate the enthalpy change for dissolving $NH_4NO_3(s)$ in water, in kilojoules per mole. Assume that the solution (whose mass is 155.4 g) has a specific heat capacity of 4.2 J/g·K. (Cold packs take advantage of the fact that dissolving ammonium nitrate in water is an endothermic process. See page 235.)

43. Care must be taken when dissolving H_2SO_4 in water because the process is highly exothermic. To measure the enthalpy change, 5.2 g $H_2SO_4(\ell)$ was added to 135 g of water in a coffee-cup calorimeter and stirred to ensure mixing. This resulted in an increase in temperature from 20.2 to 28.8 °C. Calculate the enthalpy change for the process $H_2SO_4(\ell) \longrightarrow H_2SO_4(aq)$, in kilojoules per mole.

44. Sulfur (2.56 g) is burned in a bomb calorimeter with excess $O_2(g)$. The temperature increases from 21.25 to 26.72 °C. The bomb has a heat capacity of 923 J/K, and the calorimeter contains 815 g of water. Calculate the heat evolved, per mole of SO_2 formed, for the reaction

$$S_8(s) + 8\ O_2(g) \longrightarrow 8\ SO_2(g)$$

Sulfur burns in oxygen with a bright blue flame to give $SO_2(g)$. *(Charles D. Winters)*

45. You can find the amount of heat evolved in the combustion of carbon by carrying out the reaction in a combustion calorimeter. Suppose you burn 0.300 g of C(graphite) in an excess of $O_2(g)$ to give $CO_2(g)$.

$$C(graphite) + O_2(g) \longrightarrow CO_2(g)$$

The temperature of the calorimeter, which contains 775 g of water, increases from 25.00 to 27.38 °C. The heat capacity of the bomb is 893 J/K. What quantity of heat is evolved per mole of carbon?

46. Suppose you burn 1.500 g of benzoic acid, $C_6H_5CO_2H$, in a constant volume calorimeter and find that the temperature increases from 22.50 to 31.69 °C. The calorimeter contains 775 g of water, and the bomb has a heat capacity of 893 J/K. What quantity of heat is evolved in the combustion, per mole of benzoic acid?

Benzoic acid, $C_6H_5CO_2H$, occurs naturally in many berries.

47. A 0.692-g sample of glucose, $C_6H_{12}O_6$, is burned in a constant volume calorimeter. The temperature rises from 21.70 to 25.22 °C. The calorimeter contains 575 g of water and the bomb has a heat capacity of 650 J/K. What quantity of heat is evolved per mole of glucose?

48. An "ice calorimeter" can be used to determine the specific heat capacity of a metal. A piece of hot metal is dropped into a weighed quantity of ice. The quantity of heat transferred from the metal to the ice can be determined from the amount of ice melted. Suppose you heat a 50.0-g piece of silver to 99.8 °C and then drop it onto ice. When the

metal's temperature has dropped to 0.0 °C, you find that 3.54 g of ice has melted. What is the specific heat capacity of silver?

49. A 9.36-g piece of platinum is heated to 98.6 °C in a boiling water bath and then dropped onto ice. (See Study Question 48.) When the metal's temperature has dropped to 0.0 °C, you find that 0.37 g of ice has melted. What is the specific heat capacity of platinum?

Hess's Law
(See Example 6.7 and CD-ROM Screen 6.17)

50. The enthalpies of the following reactions can be measured.

$$CH_4(g) + 2\,O_2(g) \longrightarrow CO_2(g) + 2\,H_2O(g)$$
$$\Delta H° = -802.4\ kJ$$

$$CH_3OH(g) + \tfrac{3}{2}\,O_2(g) \longrightarrow CO_2(g) + 2\,H_2O(g)$$
$$\Delta H° = -676\ kJ$$

(a) Use these values and Hess's law to determine the enthalpy change for the reaction

$$CH_4(g) + \tfrac{1}{2}\,O_2(g) \longrightarrow CH_3OH(g)$$

(b) Draw an energy level diagram that shows the relationship between the energy quantities involved in this problem.

51. The enthalpies of the following reactions can be measured.

$$C_2H_4(g) + 3\,O_2(g) \longrightarrow 2\,CO_2(g) + 2\,H_2O(\ell)$$
$$\Delta H° = -1411.1\ kJ$$

$$C_2H_5OH(\ell) + 3\,O_2(g) \longrightarrow 2\,CO_2(g) + 3\,H_2O\,(\ell)$$
$$\Delta H° = -1367.5\ kJ$$

(a) Use these values and Hess's law to determine the enthalpy change for the reaction

$$C_2H_4(g) + H_2O(\ell) \longrightarrow C_2H_5OH(\ell)$$

(b) Draw an energy level diagram that shows the relationship between the energy quantities involved in this problem.

52. Enthalpy changes for the following reactions can be determined experimentally.

$$N_2(g) + 3\,H_2(g) \longrightarrow 2\,NH_3(g) \quad \Delta H° = -91.8\ kJ$$

$$4\,NH_3(g) + 5\,O_2(g) \longrightarrow 4\,NO(g) + 6\,H_2O(g)$$
$$\Delta H° = -906.2\ kJ$$

$$H_2(g) + \tfrac{1}{2}\,O_2(g) \longrightarrow H_2O(g) \quad \Delta H° = -241.8\ kJ$$

Use these values to determine the enthalpy change for the formation of NO(g) from the elements (an enthalpy that cannot be measured directly because the reaction is reactant-favored).

$$\tfrac{1}{2}\,N_2(g) + \tfrac{1}{2}\,O_2(g) \longrightarrow NO(g) \quad \Delta H° = ?$$

53. You wish to know the enthalpy change for the formation of liquid PCl_3 from the elements.

$$P_4(s) + 6\,Cl_2(g) \longrightarrow 4\,PCl_3(\ell) \quad \Delta H° = ?$$

The enthalpy change for the formation of PCl_5 from the elements can be determined experimentally, as can the enthalpy change for the reaction of $PCl_3(\ell)$ with more chlorine to give $PCl_5(s)$:

$$P_4(s) + 10\,Cl_2(g) \longrightarrow 4\,PCl_5(s) \quad \Delta H° = -1774.0\ kJ$$

$$PCl_3(\ell) + Cl_2(g) \longrightarrow PCl_5(s) \quad \Delta H° = -123.8\ kJ$$

Use these data to calculate the enthalpy change for the formation of 1.00 mol of $PCl_3(\ell)$ from phosphorus and chlorine.

Standard Enthalpies of Formation
(See Example 6.8 and CD-ROM Screen 6.18)

54. The standard heat of formation of a compound is defined as the enthalpy change for the formation of the compound from its elements in their standard states. Write a balanced chemical equation for the reaction for the formation of $CH_3OH(\ell)$ and find the value for $\Delta H_f°$ for $CH_3OH(\ell)$ in Appendix L.

55. The standard heat of formation of $CaCO_3(s)$ is defined as the enthalpy change for the formation of $CaCO_3(s)$ from its elements in their standard states. Write a balanced chemical equation for this reaction and find the value for $\Delta H_f°$ for $CaCO_3(s)$ in Appendix L.

56. (a) Write a balanced chemical equation for the formation of 1 mol of $Cr_2O_3(s)$ from Cr and O_2 in their standard states and find the value for $\Delta H_f°$ for $Cr_2O_3(s)$ in Appendix L.

(b) What is the standard enthalpy change if 2.4 g of chromium is oxidized to $Cr_2O_3(s)$?

57. (a) Write a balanced chemical equation for the formation of 1 mol of MgO(s) from the elements in their standard states and find the value for $\Delta H_f°$ for MgO(s) in Appendix L.

(b) What is the standard enthalpy change for the reaction of 2.5 mol of Mg with oxygen?

58. Use standard heats of formation in Appendix L to calculate standard enthalpy changes for the following:
(a) 1.0 g of white phosphorus burns, forming $P_4O_{10}(s)$
(b) 0.20 mol of NO(g) decomposes to $N_2(g)$ and $O_2(g)$
(c) 2.40 g of NaCl(s) is formed from Na(s) and excess $Cl_2(g)$
(d) 250 g of iron is oxidized with oxygen to $Fe_2O_3(s)$

59. Use standard heats of formation in Appendix L to calculate standard enthalpy changes for the following:
(a) 0.054 g of sulfur burns, forming $SO_2(g)$
(b) 0.20 mol of HgO(s) decomposes to Hg(ℓ) and $O_2(g)$
(c) 2.40 g of $NH_3(g)$ is formed from $N_2(g)$ and excess $H_2(g)$
(d) 1.05×10^{-2} mol of carbon is oxidized to $CO_2(g)$

60. The first step in the production of nitric acid from ammonia involves the oxidation of NH_3.

$$4\,NH_3(g) + 5\,O_2(g) \longrightarrow 4\,NO(g) + 6\,H_2O(g)$$

(a) Use standard enthalpies of formation to calculate the standard enthalpy change for this reaction.

(b) What quantity of heat is evolved or absorbed in the oxidation of 10.0 g of NH_3?

61. The Romans used calcium oxide (CaO) to produce a very strong mortar in stone structures. The CaO was mixed with water to give $Ca(OH)_2$, which reacted slowly with CO_2 in the air to give $CaCO_3$.

$$Ca(OH)_2(s) + CO_2(g) \longrightarrow CaCO_3(s) + H_2O(g)$$

(a) Calculate the standard enthalpy change for this reaction.

(b) What quantity of heat is evolved or absorbed if 1.00 kg of $Ca(OH)_2$ reacts with a stoichiometric amount of CO_2?

62. The standard enthalpy of formation of solid barium oxide, BaO, is -553.5 kJ/mol, and the enthalpy of formation of barium peroxide, BaO_2, is -634.3 kJ/mol.

(a) Calculate the standard enthalpy change for the following reaction. Is the reaction exothermic or endothermic?

$$BaO_2(s) \longrightarrow BaO(s) + \tfrac{1}{2} O_2(g)$$

(b) Draw an energy level diagram that shows the relationship between the enthalpy of this reaction and the heats of formation of BaO(s) and BaO_2(s).

63. An important step in the production of sulfuric acid is the oxidation of SO_2 to SO_3.

$$SO_2(g) + \tfrac{1}{2} O_2(g) \longrightarrow SO_3(g)$$

Formation of SO_3 from the air pollutant SO_2 is also a key step in the formation of acid rain.

(a) Use standard enthalpies of formation to calculate the enthalpy change for the reaction. Is the reaction exothermic or endothermic?

(b) Draw an energy level diagram that shows the relationship between the enthalpies of S(s) and $\tfrac{3}{2} O_2(g)$, $SO_2(s) + \tfrac{1}{2} O_2(g)$, and $SO_3(g)$.

64. The enthalpy change for the oxidation of naphthalene, $C_{10}H_8$, is measured by calorimetry.

$$C_{10}H_8(s) + 12 O_2(g) \longrightarrow 10 CO_2(g) + 4 H_2O(\ell)$$
$$\Delta H^\circ_{rxn} = -5156.1 \text{ kJ}$$

Use this value, along with standard heats of formation of $CO_2(g)$ and $H_2O(\ell)$ to calculate the enthalpy of formation of naphthalene (in kilojoules per mole).

65. The enthalpy change for the oxidation of styrene, C_8H_8, is measured by calorimetry.

$$C_8H_8(\ell) + 10 O_2(g) \longrightarrow 8 CO_2(g) + 4 H_2O(\ell)$$
$$\Delta H^\circ_{rxn} = -4395.0 \text{ kJ}$$

Use this value, along with standard heats of formation of $CO_2(g)$ and $H_2O(\ell)$ to calculate the enthalpy of formation of styrene (in kilojoules per mole).

Product- and Reactant-Favored Reactions
(See Exercise 6.17)

66. Use your "chemical sense" and decide if each of the following reactions is product- or reactant-favored. Calculate ΔH°_{rxn} in each case, and draw an energy level diagram like those in Figure 6.18.

(a) The reaction of aluminum and chlorine to produce $AlCl_3$(s)

(b) The decomposition of mercury(II) oxide to produce liquid mercury and oxygen gas

67. Use your "chemical sense" and decide if each of the following reactions is product- or reactant-favored. Calculate ΔH°_{rxn} in each case, and draw an energy level diagram like those in Figure 6.18.

(a) The formation of ozone, O_3, and from oxygen molecules and O atoms

(b) Dissolving solid NaOH in water to give aqueous NaOH [NaOH(aq)]

(c) The decomposition of $MgCO_3$(s) to give MgO(s) and CO_2(g)

General Questions on Thermochemistry

These questions are not designated as to type or location in the chapter. They may combine several concepts. More challenging questions are indicated by an underlined number.

68. A piece of lead with a mass of 27.3 g was heated to 98.90 °C and then dropped into 15.0 g of water at 22.50 °C. The final temperature is 26.32 °C. Calculate the specific heat capacity of lead from these data.

69. Which gives up more heat on cooling from 50 to 10 °C, 50.0 g of water or 100. g of ethanol (specific heat capacity of ethanol $(\ell) = 2.46$ J/g·K)?

70. A 192-g piece of copper is heated to 100.0 °C in a boiling water bath and then dropped into a beaker containing 751 g of water (density = 1.00 g/cm³) at 4.0 °C. What is the final temperature of the copper and water after thermal equilibrium is reached? (The specific heat capacity of solid copper is 0.385 J/g·K.)

71. You determine that 187 J of heat is required to raise the temperature of 93.45 g of silver from 18.5 to 27.0 °C. What is the specific heat capacity of solid silver?

72. Calculate the quantity of heat required to convert 60.1 g of H_2O(s) at 0.0 °C to H_2O(g) at 100.0 °C. The heat of fusion of ice at 0 °C is 333 J/g; the heat of vaporization of liquid water at 100 °C is 2260 J/g.

73. You add 100.0 g of water at 60.0 °C to 100.0 g of ice at 0.00 °C. Some of the ice melts and cools the water to 0.00 °C. When the ice and water mixture has come to a uniform temperature of 0 °C, how much ice has melted?

74. Three 45-g ice cubes at 0 °C are dropped into 5.00×10^2 mL of tea to make ice tea. The tea was initially at 20.0 °C; when thermal equilibrium was reached the final temperature was 0 °C. How much of the ice melted and how much remained floating in the beverage? Assume the specific heat capacity of tea is the same as that of pure water.

75. Suppose that only two 45-g ice cubes were added to your glass containing 5.00×10^2 mL of tea. When thermal equilibrium is reached, all of the ice will have melted and the temperature of the mixture will be somewhere between 20.0 and 0 °C. Calculate the final temperature of the beverage. (Be sure to note that the 90 g of water formed when the ice melts must be warmed from 0 °C to the final temperature.)

76. You take a diet cola from the refrigerator, and pour 240 mL into a glass. The temperature of the beverage is 10.5 °C. You then add one ice cube (45 g). Which of the following describes the system when thermal equilibrium is reached? (a) The temperature is 0 °C and some ice remains; (b) The temperature is 0 °C and no ice remains; or (c) The temperature is above 0 °C and no ice remains. Determine the final temperature and the amount of ice remaining, if any.

77. A commercial product called "Instant Car Kooler" contains 10% by weight ethanol, C_2H_5OH, and 90% by weight water. You spray the "Kooler" inside an overheated car. It works because thermal energy of the air in the car will be used to evaporate some of the alcohol and water. If the air inside an average size car must lose 3.6 kJ of heat to drop the air temperature from 55 to 25 °C, what mass of the ethanol–water mixture must evaporate to absorb this heat? (The enthalpy of vaporization for ethanol is 850 J/g and for water it is 2260 J/g.)

A commercial product uses the heat of vaporization of water and ethanol as a way to cool the interior of a car on a hot day. *(Charles D. Winters)*

78. Insoluble AgCl(s) precipitates when solutions of $AgNO_3$(aq) and NaCl(aq) are mixed.

$$AgNO_3(aq) + NaCl(aq) \longrightarrow AgCl(s) + NaNO_3(aq)$$
$$\Delta H^\circ_{rxn} = ?$$

To measure the heat evolved in this reaction, 250. mL of 0.18 M $AgNO_3$(aq) and 125 mL of 0.36 M NaCl(aq) are mixed in a coffee-cup calorimeter. The temperature of the mixture rises from 21.15 to 22.90 °C. Calculate the enthalpy change for the precipitation of AgCl(s), in kilojoules per mole. (Assume the density of the solution is 1.0 g/mL and its specific heat capacity is 4.2 J/g·K.)

79. Insoluble $PbBr_2$(s) precipitates when solutions of $Pb(NO_3)_2$(aq) and NaBr(aq) are mixed.

$$Pb(NO_3)_2(aq) + 2\,NaBr(aq) \longrightarrow$$
$$PbBr_2(s) + 2\,NaNO_3(aq) \qquad \Delta H^\circ_{rxn} = ?$$

To measure the heat evolved, 200. mL of 0.75 M $Pb(NO_3)_2$(aq) and 200. mL of 1.5 M NaBr(aq) are mixed in a coffee-cup calorimeter. The temperature of the mixture rises 2.44 °C. Calculate the enthalpy change for the precipitation of $PbBr_2$(s), in kilojoules per mole. Assume the density of the solution is 1.0 g/mL and its specific heat capacity is 4.2 J/g·K.

80. The heat evolved in the decomposition of 7.647 g of ammonium nitrate can be measured in a bomb calorimeter. The reaction that occurs is

$$NH_4NO_3(s) \longrightarrow N_2O(g) + 2\,H_2O(g)$$

The temperature of the calorimeter, which contains 415 g of water, increases from 18.90 to 20.72 °C. The heat capacity of the bomb is 155 J/K. What quantity of heat is evolved in this reaction, in kilojoules?

81. A bomb calorimetric experiment was run to determine the heat of combustion of ethanol (a common fuel additive). The reaction is

$$C_2H_5OH(\ell) + 3\,O_2(g) \longrightarrow 2\,CO_2(g) + 3\,H_2O(\ell)$$

The bomb had a heat capacity of 550 J/K, and the calorimeter contained 650 g of water. Burning 4.20 g of ethanol, $C_2H_5OH(\ell)$ resulted in a rise in temperature from 18.5 to 22.3 °C. Calculate the heat of combustion of ethanol, in kilojoules per mole.

82. The standard molar enthalpy of formation of diborane, B_2H_6(g), cannot be determined directly because the compound cannot be prepared by the reaction of boron and hydrogen. It can be calculated from other enthalpy changes, however. The following enthalpies can be measured.

$$4\,B(s) + 3\,O_2(g) \longrightarrow 2\,B_2O_3(s) \qquad \Delta H^\circ_{rxn} = -2543.8\ kJ$$
$$H_2(g) + \tfrac{1}{2}\,O_2(g) \longrightarrow H_2O(g) \qquad \Delta H^\circ_{rxn} = -241.8\ kJ$$
$$B_2H_6(g) + 3\,O_2(g) \longrightarrow B_2O_3(s) + 3\,H_2O(g)$$
$$\Delta H^\circ_{rxn} = -2032.9\ kJ$$

(a) Modify these equations to give a new set of equations, which, when added together, give the equation for the formation of B_2H_6(g) from B(s) and H_2(g) in their standard states. Assign enthalpy changes to each reaction.

(b) Calculate ΔH°_f for B_2H_6(g).

(c) Draw an energy level diagram that shows how the various enthalpies in this problem are related.

(d) Is the formation of $B_2H_6(g)$ from its elements product- or reactant-favored?

83. The standard molar enthalpy of formation of $CS_2(g)$ cannot be determined directly because the compound cannot be prepared by the reaction of carbon and sulfur. It can be calculated from other enthalpy changes, however. The following enthalpies can be measured.

$$C(s) + O_2(g) \longrightarrow CO_2(g) \qquad \Delta H^\circ_{rxn} = -393.5 \text{ kJ}$$

$$S(s) + O_2(g) \longrightarrow SO_2(g) \qquad \Delta H^\circ_{rxn} = -296.8 \text{ kJ}$$

$$CS_2(g) + 3\,O_2(g) \longrightarrow CO_2(g) + 2\,SO_2(g)$$
$$\Delta H^\circ_{rxn} = -1103.9 \text{ kJ}$$

(a) Modify these equations to give a new set of equations, which, when added together, give the equation for the formation of $CS_2(g)$ from $C(s)$ and $S(s)$ in their standard states. Assign enthalpy changes to each reaction.

(b) Calculate ΔH°_f for $CS_2(g)$.

(c) Draw an energy level diagram that shows how the various enthalpies in this problem are related.

(d) Is the formation of $CS_2(g)$ from its elements product- or reactant-favored?

84. The meals-ready-to-eat (MREs) in the military can be heated on a flameless heater. The source of energy in the heater is

$$Mg(s) + 2\,H_2O(\ell) \longrightarrow Mg(OH)_2(s) + H_2(g)$$

Calculate the enthalpy change under standard conditions (in joules) for this reaction. What quantity of magnesium is needed to supply the heat required to warm 250 mL of water ($d = 1.00$ g/mL) from 25 to 85 °C? (See W. Jensen: *Journal of Chemical Education*, Vol. 77, pages 713–717, 2000.)

85. Hydrazine, $N_2H_4(\ell)$, is an efficient oxygen scavenger. It is sometimes added to steam boilers to remove traces of oxygen that can cause corrosion in these systems. Combustion of hydrazine gives the following information

$$N_2H_4(\ell) + O_2(g) \longrightarrow N_2(g) + 2\,H_2O(g)$$
$$\Delta H^\circ_{rxn} = -534.3 \text{ kJ}$$

(a) Is the reaction product- or reactant-favored?

(b) Use the value for ΔH°_{rxn} with the enthalpy of formation of $H_2O(g)$ to calculate the molar enthalpy of formation of $N_2H_4(\ell)$.

86. When heated to a high temperature, coke (mainly carbon, obtained by heating coal in the absence of air) and steam produces a mixture called coal gas, which can be used as a fuel or as a chemical feedstock for other reactions. The equation for the production of coal gas is

$$C(s) + H_2O(g) \longrightarrow CO(g) + H_2(g)$$

(a) Use standard heats of formation to determine the enthalpy change for this reaction.

(b) Is the reaction product- or reactant-favored?

(c) What quantity of heat is involved if 1.0 metric ton (1000.0 kg) of carbon is converted to coal gas?

87. Camping stoves are fueled by propane (C_3H_8), butane [$C_4H_{10}(g)$, $\Delta H^\circ_f = -127.1$ kJ/mol)], gasoline, or ethanol (C_2H_5OH). Calculate the heat of combustion per gram of each of these fuels. (Assume that gasoline is represented by isooctane, $C_8H_{18}(\ell)$, with $\Delta H^\circ_f = -259.2$ kJ/mol.) Do you notice any great differences among these fuels? Are these differences related to their composition?

88. Methanol, CH_3OH, a compound that can be made relatively inexpensively from coal, is a promising substitute for gasoline. The alcohol has a smaller energy content than gasoline, but, with its higher octane rating, it burns more efficiently than gasoline in combustion engines. (It has the added advantage of contributing to a lesser degree to some air pollutants.) Compare the heat of combustion per gram of CH_3OH and C_8H_{18} (isooctane), the latter being representative of the compounds in gasoline. ($\Delta H^\circ_f = -259.2$ kJ/mol for isooctane.)

89. Hydrazine and 1,1-dimethylhydrazine both react spontaneously with O_2 and can be used as rocket fuels.

$$N_2H_4(\ell) + O_2(g) \longrightarrow N_2(g) + 2\,H_2O(g)$$
hydrazine

$$N_2H_2(CH_3)_2(\ell) + 4\,O_2(g) \longrightarrow$$
1,1-dimethylhydrazine $\qquad 2\,CO_2(g) + 4\,H_2O(g) + N_2(g)$

The molar enthalpy of formation of $N_2H_4(\ell)$ is $+50.6$ kJ/mol, and that of $N_2H_2(CH_3)_2(\ell)$ is $+48.9$ kJ/mol. Use these values, with other ΔH°_f values, to decide whether the reaction of hydrazine or dimethylhydrazine with oxygen gives more heat per gram.

A control rocket in the Space Shuttle uses hydrazine as the fuel. *(NASA)*

90. Calculate the molar heat capacity, in joules per mole per kelvin, for the metals in Table 6.1. What observation can you make about these values — specifically, are they widely different or very similar? Using this information, estimate the specific heat capacity for silver. Compare this estimate with the correct value for silver, 0.236 J/g·K.

91. Prepare a graph of molar heat capacities for metals versus their atomic weight. Use the data in Table 6.1 and the values in the following table. Does any relation exist between specific heat capacity and atomic weight? Use this relation to predict the specific heat capacity of platinum. (The

specific heat capacity for platinum is given in the literature as $0.133\,J/g \cdot K$.) How good is the agreement between the predicted and actual values?

Metal	Specific heat capacity (J/g·K)
chromium	0.450
lead	0.127
silver	0.236
tin	0.227
titanium	0.522

92. Suppose you are attending summer school and are living in a very old dormitory. The day is oppressively hot. There is no air conditioner, and you can't open the windows of your room because they are stuck shut from layers of paint. There is a refrigerator in the room, however. In a stroke of genius you open the door of the refrigerator, and cool air cascades out. The relief does not last long, though. Soon the refrigerator motor and condenser begin to run, and not long thereafter the room is hotter than it was before. Why did the room warm up?

93. You want to determine the value for the enthalpy of formation of $CaSO_4(s)$.

$$Ca(s) + \tfrac{1}{8} S_8(s) + 2\,O_2(g) \longrightarrow CaSO_4(s)$$

The reaction cannot be done directly. You know, however, that both calcium and sulfur react with oxygen to produce oxides in reactions that can be studied calorimetrically. You also know that the basic oxide CaO reacts with the acidic oxide SO_3 to produce $CaSO_4(s)$, with $\Delta H^{\circ}_{rxn} = -502.7\,kJ$. Outline a method for determining ΔH°_f for $CaSO_4(s)$, and identify the information that must be collected by experiment. Using information in Table 6.2 confirm that ΔH°_f for $CaSO_4(s) = -1434.5\,kJ/mol$.

94. You drink 350 mL of diet soda that is at a temperature of 5 °C.

(a) How much energy will your body expend to raise the temperature of this liquid to body temperature (37 °C)? Assume that the density and specific heat capacity of diet soda is the same as water.

(b) Compare the value in part (a) with the caloric content of beverage. (The label says that it has a caloric content of 1 Cal.) What is the net energy change in your body resulting from drinking this beverage.

95. Carry out a comparison similar to that in Study Question 94 with a nondiet beverage whose label indicates a caloric content of 240 Cal.

96. Suppose you want to heat the air in your house with natural gas (CH_4). Assume your house has $275\,m^2$ (about $2800\,ft^2$) of floor area and that the ceilings are 2.50 m from the floors. The air in the house has a molar heat capacity of $29.1\,J/mol \cdot K$. (The number of moles of air in the house can be found by assuming that the average molar mass of air is 28.9 g/mol and that the density of air at these temperatures is about 1.22 g/L.) What mass of methane do you have to burn to heat the air from 15.0 to 22.0 °C?

97. (a) Calculate the enthalpy change, ΔH°, for the formation of 1.00 mol of strontium carbonate (the material that gives the red color in fireworks) from its elements.

$$Sr(s) + C(graphite) + \tfrac{3}{2} O_2(g) \longrightarrow SrCO_3(s)$$

The experimental information available is:

$$Sr(s) + \tfrac{1}{2} O_2(g) \longrightarrow SrO(s) \qquad \Delta H^{\circ}_f = -592\,kJ$$
$$SrO(s) + CO_2(g) \longrightarrow SrCO_3(s) \qquad \Delta H^{\circ}_f = -234\,kJ$$
$$C(graphite) + O_2(g) \longrightarrow CO_2(g) \qquad \Delta H^{\circ}_f = -394\,kJ$$

(b) Draw an energy level diagram relating the energy quantities in this problem.

98. Isomers are molecules with the same elemental composition but a different atomic arrangement. Three ways of arranging the atoms in a compound with the formula C_4H_8 are shown on the models below. The enthalpy of combustion of each isomer, determined using a calorimeter, is:

Compound	$\Delta H_{combustion}$ (kJ/mol)
cis-2 butene	−2687.5
trans-2-butene	−2684.2
1-butene	−2696.7

Draw an energy level diagram relating the energy content of the three isomers to the energy content of the combustion products, $CO_2(g)$ and $H_2O(g)$.

$\Delta H_{combustion}$ = −2687.5 kJ/mol

cis-2-butene

$\Delta H_{combustion}$ = −2684.2 kJ/mol

trans-2-butene

$\Delta H_{combustion}$ = −2696.7 kJ/mol

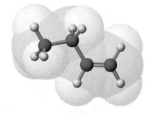

1-butene

99. (a) Use the $\Delta H_{combustion}$ data in Study Question 98, along with enthalpies of formation of $CO_2(g)$ and $H_2O(g)$ from Appendix L, to calculate the enthalpy of formation for each of the isomers.

    (b) Draw an energy level diagram that relates the heats of formation of the three isomers to the energy of the elements in their standard states.

    (c) What is the enthalpy change for the conversion of *cis*-2-butene to *trans*-2-butene?

100. Chloroform, $CHCl_3$, is formed from methane and chlorine in the following reaction.

$$CH_4(g) + 3\,Cl_2(g) \longrightarrow 3\,HCl(g) + CHCl_3(g)$$

Calculate ΔH°_{rxn}, the enthalpy change for this reaction, using the enthalpy of formation of $CHCl_3(g)$, $\Delta H^\circ_f = -103.1$ kJ/mol and the enthalpies of the following reactions.

$$CH_4(g) + 2\,O_2(g) \longrightarrow 2\,H_2O(\ell) + CO_2(g)$$
$$\Delta H^\circ_{rxn} = -890.3 \text{ kJ}$$
$$2\,HCl(g) \longrightarrow H_2(g) + Cl_2(g) \quad \Delta H^\circ_{rxn} = +184.6 \text{ kJ}$$
$$C(graphite) + O_2(g) \longrightarrow CO_2(g) \quad \Delta H^\circ = -393.5 \text{ kJ}$$
$$H_2(g) + \tfrac{1}{2}O_2(g) \longrightarrow H_2O(\ell) \quad \Delta H^\circ_f = -285.8 \text{ kJ}$$

101. Suppose that an inch of rain falls over a square mile of ground so that 6.6×10^{10} g of water has fallen (a density of 1.0 g/cm^3 is assumed). The heat of vaporization of water at 25 °C is 44.0 kJ/mol. Calculate the quantity of heat transferred to the surroundings from the condensation of water vapor in forming this quantity of liquid water. (The huge number tells you how much energy is "stored" in water vapor and why we think of storms as such great forces of nature. It is interesting to compare this result with the energy given off (4.2×10^6 kJ) when a ton of dynamite explodes.)

102. A parking lot in Los Angeles, California, receives an average of 2.6×10^7 J/m^2 of solar energy per day in the summer. If the parking lot is 325 m long and 50.0 m wide, what is the total quantity of energy striking the area per day?

103. Your home loses heat in the winter through doors and windows and any poorly insulated walls. A sliding glass door (6 ft $\times$ 6$\tfrac{1}{2}$ ft with $\tfrac{1}{2}$ in. of insulating glass) allows 1.0×10^6 J/h to pass through the glass if the inside temperature is 22 °C (72 °F) and the outside temperature is 0 °C (32 °F). What quantity of heat, expressed in kilojoules, is lost per day? If 1 kwh of energy is equal to 3.60×10^6 J, how many kilowatt-hours of energy are lost per day through the door?

Using Electronic Resources

These questions refer to the General Chemistry Interactive CD-ROM, *Version 3.0.*

104. *See CD-ROM Screens 6.1 and 6.19: Chemical Puzzler.* Peanuts and peanut oil are organic materials and burn in air. How many burning peanuts does it take to provide the energy to boil a cup of water (250 mL of water)? To solve this we assume that each peanut, with an average mass of 0.73 g, is 49% peanut oil and 21% starch; the remainder is non-combustible. We further assume that peanut oil is palmitic acid, $C_{16}H_{32}O_2$, with an enthalpy of formation of -124.8 kJ/mol. Starch is a long chain of $C_6H_{10}O_5$ units, each unit having an enthalpy of formation of -734 kJ.

How many burning peanuts are required to provide the heat to boil 250 mL of water? *(Charles D. Winters)*

105. *See CD-ROM Screen 6.3: Control of Chemical Reactions.* What is the difference between thermodynamics and kinetics? Is the reaction of $Si(s)$ with $O_2(g)$ to give sand (SiO_2) thermodynamically possible?

106. *See CD-ROM Screen 6.5: Forms of Energy.* Radiant energy may be converted into what other forms of energy?

107. *See CD-ROM Screen 6.11: Heat Transfer Between Substances.*

    (a) Explain what happens in terms of molecular motions when a hotter object comes in contact with a cooler one.

    (b) What does it mean when two objects have come to thermal equilibrium?

108. *See CD-ROM Screen 6.11: Heat Transfer Between Substances.* Use the Simulation section of this screen to do the following experiment: Add 10.0 g of Al at 80°C to 10.0 g of water at 20°C. What is the final temperature when equilibrium is achieved? Use this value to estimate the specific heat capacity of aluminum.

109. *See CD-ROM Screen 6.17: Hess's Law.* Use the Simulation section of this screen to find the value of ΔH°_{rxn} for

$$SnBr_2(s) + TiCl_4(\ell) \longrightarrow SnCl_4(\ell) + TiBr_2(s)$$

7

Atomic Structure

Chapter Goals

- Describe the properties of electromagnetic radiation.
- Understand the origin of light from excited atoms and how this is related to atomic structure.
- Describe the experimental evidence for the wave–particle duality.
- Describe the basic ideas of quantum mechanics.
- Define the three quantum numbers (n, ℓ, and m_ℓ) and their relation to atomic structure.

Colors in the Sky

Black powder, the predecessor of gunpowder, was discovered well before 1000 AD, most likely in China. It was not until the Middle Ages, however, that black powder was known in the western world. In 1252 Roger Bacon in England described the preparation of black powder from "saltpetre [potassium nitrate], young willow, and sulfur," and its use by the military and for fireworks spread to the European continent. By the time of the American Revolution, fireworks formulations and manufacturing methods had been worked out that are still in use today.

Typical fireworks have several important chemical components. First, there must be an oxidizer, and today this is usually potassium perchlorate ($KClO_4$), potassium chlorate ($KClO_3$), or potassium nitrate (KNO_3). Potassium salts are used instead of sodium salts because the latter have two important drawbacks. They are hygroscopic — they absorb water from the air — and so do not remain

▲ A fireworks display by the famous Grucci Company of New York. (Fireworks by Grucci)

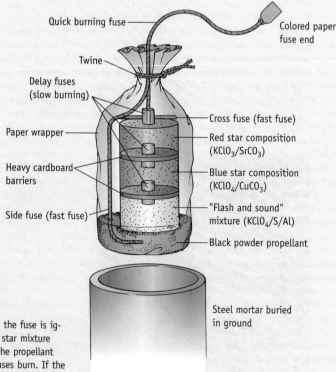

Quick burning fuse

Colored paper fuse end

Twine

Delay fuses (slow burning)

Paper wrapper

Heavy cardboard barriers

Side fuse (fast fuse)

Cross fuse (fast fuse)

Red star composition ($KClO_3/SrCO_3$)

Blue star composition ($KClO_4/CuCO_3$)

"Flash and sound" mixture ($KClO_4/S/Al$)

Black powder propellant

Steel mortar buried in ground

▶ The design of an aerial rocket for a fireworks display. When the fuse is ignited, it burns quickly to the delay fuse at the top of the red star mixture and burns on to the black powder propellant at the bottom. The propellant ignites, sending the shell into the air. Meanwhile, the delay fuses burn. If the timing is correct, the shell bursts high in the sky into a red star. This is followed by a blue burst and then a flash and sound.

> *"**O**f saltpetre take six parts, five of young willow, and five of sulfur and so you will make thunder and lightning."*
>
> **Roger Bacon, 1252**

▲ A collection of fireworks. *(Charles D. Winters)*

dry on storage. Second, when heated, sodium salts give off an intense, yellow light that is so bright it can mask other colors.

The parts of any fireworks display we remember best are the vivid colors and brilliant flashes. White light can be produced by oxidizing magnesium or aluminum metal at high temperatures, and the flashes you see at rock concerts or similar events typically result from the ignition of $Mg/KClO_4$ mixtures.

Yellow light is easiest to produce because sodium salts give an intense light with a wavelength of 589 nm. Fireworks mixtures usually contain sodium in the form of such nonhygroscopic compounds as cryolite, Na_3AlF_6. Strontium salts are most often used to produce a red light, and green is produced by barium salts such as barium nitrate, $Ba(NO_3)_2$.

The next time you see a fireworks display, watch for the ones that are blue. Blue has always been the most difficult color to produce. Recently, however, fireworks designers have learned that the best way to get a really good "blue" is to decompose copper(I) chloride at low temperatures. To achieve this effect, CuCl is mixed with $KClO_4$, copper powder, and the organic chlorine-containing compound hexachloroethane, C_2Cl_6.

Why are chemists — and many others — interested in fireworks? Because their colors arise from energetically excited molecules and atoms. The way that atoms can produce colored light provides insight into the structure of the atom, the subject of this chapter.

◀ Fireworks, some with the difficult-to-produce blue color. *(Fireworks by Grucci)*

Before You Begin

- Review metric units of measurements (Chapter 1).
- Review the structure of the atom (Section 2.1).

Atomic Structure The structure of the atom was largely revealed in studies by physicists early in the 20th century. In this chapter you will learn more about our current understanding of the arrangement of electrons in atoms and the evidence that led to this understanding. Some of the evidence for our modern view of atomic structure came from studies of the energy emitted by excited atoms.

The Colors of Excited Atoms

White, crystalline salts of sodium (NaCl) (left) and strontium (SrCl₂) (center), and white powdered boric acid [B(OH)₃)] (right).

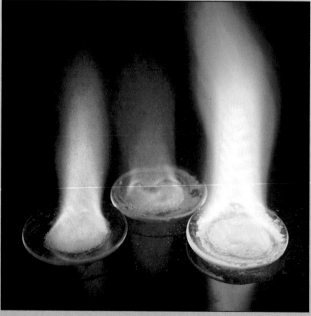

The solids pictured above were soaked in methanol (CH₃OH) and the methanol was set ablaze. The compounds are entrained in the burning liquid, and energy from the combustion excites the atoms. The colors you observe are characteristic of sodium (left), strontium (center), and boron (right)[🖱 CD-ROM, Screen 7.1].

The Colors of Excited Atoms

Electricity off. Colorless gases.

Electricity on. Excited electrons.

Gases such as neon are colorless. However, if electricity is passed through the gas, the atoms are excited, and the gas glows.

The Structure of the Atom

The modern view of the atom is of an exceedingly small but massive nucleus consisting of protons and neutrons at the core of the atom with electrons arranged in space about the nucleus. This chapter addresses the questions of how the electrons are arranged and how atoms can absorb and re-emit energy.

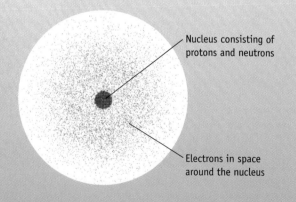

Nucleus consisting of protons and neutrons

Electrons in space around the nucleus

Photos: Charles D. Winters

Chemical elements that exhibit similar properties are found in the same column of the periodic table. But why should this be so? The discovery of the electron, proton, and neutron [← SECTION 2.1] prompted scientists to look for relationships between atomic structure and chemical behavior. As early as 1902 Gilbert N. Lewis (1875–1946) suggested the idea that electrons in atoms might be arranged in shells, starting close to the nucleus and building outward. Lewis explained the similarity of chemical properties for elements in a given group by assuming that all the elements of that group have the same number of electrons in the outer shell.

Lewis's model of the atom raises a number of questions. Where are the electrons located? Do they have different energies? What experimental evidence supports this model? These questions inspired many of the experimental and theoretical studies that began around 1900 and continue to this day. This chapter and the next outline the current theories of electronic structure.   •

7.1 ELECTROMAGNETIC RADIATION

You are no doubt familiar with water waves, and you may also know that some properties of radiation such as visible light can be explained by wave motion. Our understanding of light as waves came from the experiments of physicists in the 19th century, among them a Scot, James Clerk Maxwell (1831–1879). In 1864, he developed an elegant mathematical theory to describe all forms of radiation in terms of oscillating, or wave-like, electric and magnetic fields (Figure 7.1). Hence, radiation, such as light, microwaves, television and radio signals, and x-rays, is collectively called **electromagnetic radiation** [⊙ CD-ROM, Screen 7.3].

Wave Properties

The distance between successive crests or high points of a wave (or between successive troughs or low points) is the **wavelength** of a wave. This distance can be given in meters, nanometers, or whatever unit is convenient. The symbol for wavelength is the Greek letter λ (lambda).

Chapter Goals • Revisited

- **Describe the properties of electromagnetic radiation.**
- Understand the origin of light from excited atoms and how this is related to atomic structure.
- Describe the experimental evidence for the wave–particle duality.
- Describe the basic ideas of quantum mechanics.
- Define the three quantum numbers (n, ℓ, and m_ℓ) and their relation to atomic structure.

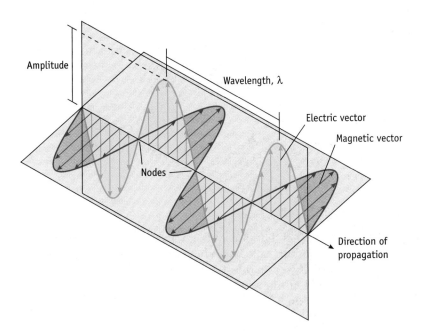

Figure 7.1   Electromagnetic radiation. In the 1860s James Clerk Maxwell developed the currently accepted theory that all forms of radiation are propagated through space as vibrating electric and magnetic fields at right angles to one another. Each of the fields is described by a sine wave (the mathematical function describing the wave). Such oscillating fields emanate from vibrating charges in the source [⊙ CD-ROM, Screen 7.3].

● **Hertz**
The existence of electromagnetic radiation was proved experimentally by Heinrich Hertz (1857–1894). In his honor scientists use the "hertz" as the unit of frequency, or number of cycles per second.

Waves are also characterized by their **frequency,** symbolized by the Greek letter ν (nu). For waves passing some point in space, the frequency is equal to the number of complete waves passing the point in a fixed amount of time. Thus, we usually refer to the frequency as the number of events per second (such as the number of waves that pass per second)(see Figure 7.1). The unit for frequency is often written as s^{-1} (standing for 1 per second, $1/s$) and is now called the **hertz,** Hz.

If you enjoy water sports, you are familiar with the height of waves. In more scientific terms, the maximum height of a wave is its **amplitude.** In Figure 7.1, notice that the wave has zero amplitude at certain intervals along the wave. Points of zero amplitude, called **nodes,** occur at intervals of $\lambda/2$.

Finally, the speed of a moving wave is an important factor. As an analogy, consider cars in a traffic jam traveling bumper to bumper. If each car is 5 m long, and if a car passes you every 4 s (that is, the frequency is 1 per 4 seconds, or $\frac{1}{4} s^{-1}$), then the traffic is moving at the speed of $(5 \text{ m}) \times (\frac{1}{4} s^{-1})$, or $1.25 \text{ m} \cdot s^{-1}$. The speed for any periodic motion, including a wave, is the product of the wavelength and the frequency of the wave:

$$\text{Speed } (m \cdot s^{-1}) = \text{wavelength } (m) \times \text{frequency } (s^{-1})$$

This equation also applies to electromagnetic radiation, where the speed of light, c, is the product of the wavelength and frequency of a light wave.

● **Speed of Light**
The speed of light passing through a substance (air, glass, water, and so on) depends on the chemical constitution of the substance and the wavelength of the light. This is the basis for using a glass prism to disperse light and is the explanation for rainbows. The speed of sound also depends on the material through which it passes.

Speed of light $(m \cdot s^{-1})$

$$c = \lambda \times \nu \tag{7.1}$$

Wavelength (m)   Frequency (s^{-1})

The speed of visible light and all other forms of electromagnetic radiation in a vacuum is a constant, $c\ (= 2.99792458 \times 10^8 \text{ m} \cdot s^{-1}$; approximately $186,000 \text{ miles} \cdot s^{-1})$. Given this value, and knowing the wavelength of a light wave, you can calculate the frequency, and vice versa. For example, what is the frequency of orange light, which has a wavelength of 625 nm? Because the speed of light is expressed in meters per second, the wavelength in nanometers must be changed to meters before substituting into Equation 7.1:

● **Speed of Light and Significant Figures**
The speed of light is known to nine significant figures. For most calculations, however, we shall use four at most.

$$625 \text{ nm} \cdot \frac{1 \times 10^{-9} \text{ m}}{1 \text{ nm}} = 6.25 \times 10^{-7} \text{ m}$$

$$\nu = \frac{c}{\lambda} = \frac{2.998 \times 10^8 \text{ m} \cdot s^{-1}}{6.25 \times 10^{-7} \text{ m}} = 4.80 \times 10^{14} \text{ s}^{-1}$$

Standing Waves

The wave motion described so far is that of traveling waves such as sound or water waves. Another type of wave motion, called **standing,** or stationary, **waves,** is relevant to modern atomic theory. If you tie down a string at both ends, as you would the string of a guitar, and pluck it, the string vibrates as a standing wave (Figure 7.2). Several important points about standing waves are relevant to our discussion of electrons in atoms:

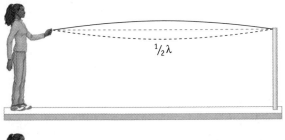

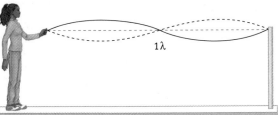

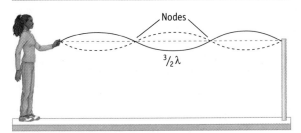

Figure 7.2   Standing waves. In the first wave, the end-to-end distance is $(1/2)\lambda$, in the second wave it is λ, and in the third wave it is $(3/2)\lambda$.

- A standing wave is characterized by having two or more points of no movement; that is, the wave amplitude is zero at the nodes. As with traveling waves, the distance between consecutive nodes is always $\lambda/2$.

- In the first of the vibrations illustrated in Figure 7.2, the distance between the ends of the string is half a wavelength, or $\lambda/2$. In the second vibration the string length equals one complete wavelength, or $2(\lambda/2)$. In the third, the string length is $3(\lambda/2)$, or $(3/2)\lambda$. Could the distance between the ends of a standing wave vibration ever be $(3/4)\lambda$? For standing waves, only certain wavelengths are possible. Because the ends of a standing wave must be nodes, the only allowed vibrations are those in which the distance from one end, or "boundary," to the other is $n(\lambda/2)$, where n is an integer $(1, 2, 3, \ldots)$.

• **Standing Waves**
Only certain wavelengths are allowed for standing waves. This is an example of *quantization,* a concept we turn to in the sections that follow.

Exercise 7.1   **Standing Waves**

The line shown here is 10 cm long.

|←——————————————— 10 cm ———————————————→|

Using this line,

(a) Draw a standing wave with one node between the ends. What is the wavelength of this wave?

(b) Draw a standing wave with three nodes between the ends. What is its wavelength?

(c) If the wavelength of the standing wave is 2.5 cm, how many waves fit within the boundaries? How many nodes are there?

Visible Spectrum of Light

We are bathed constantly in electromagnetic radiation, including the radiation you can see, visible light. As you know, visible light consists of a spectrum of colors, ranging from red at the long-wavelength end of the spectrum to violet at the short-wavelength end (Figure 7.3). Visible light is, however, only a small portion of the total electromagnetic spectrum. Ultraviolet (UV) radiation, the radiation that can lead to sunburn, has wavelengths shorter than those of visible light; x-rays and γ rays, the latter emitted in the process of radioactive disintegration of some atoms, have even shorter wavelengths. At longer wavelengths than visible light, we first encounter infrared radiation (IR), the type that is sensed as heat. Longer still is the wavelength of the radiation used in microwave ovens and in television and radio transmissions [🖰 CD-ROM, Screen 7.4].

● **ROY G BIV**
You can remember the colors of visible light, in order of decreasing wavelength, by the famous mnemonic phrase ROY G BIV, which stands for red, orange, yellow, green, blue, indigo, and violet.

Example 7.1   Wavelength–Frequency Conversions

Problem • The frequency of the radiation used in all microwave ovens sold in the United States is 2.45 GHz. (The unit GHz stands for gigahertz; 1 GHz is a billion cycles per second, or $10^9\ s^{-1}$.) What is the wavelength (in meters) of this radiation? How much longer or shorter is this than the wavelength of orange light (625 nm)?

Strategy • The wavelength of microwave radiation in meters can be calculated directly from Equation 7.1. Convert 625 nm to a wavelength in meters so that units are compatible.

Solution •

$$\lambda = \frac{c}{\nu} = \frac{2.998 \times 10^8\ m \cdot s^{-1}}{2.45 \times 10^9\ s^{-1}} = 0.122\ m$$

Orange light has a wavelength, in meters, of

$$625\ nm \cdot \frac{1 \times 10^{-9}\ m}{1\ nm} = 6.25 \times 10^{-7}\ m$$

A comparison of the wavelengths of microwave radiation and orange light shows that the microwaves have wavelengths almost 200,000 times longer than the wavelength of orange light.

$$\frac{\lambda\ (microwaves)}{\lambda\ (orange\ light)} = \frac{0.122\ m}{6.25 \times 10^{-7}\ m} = 195,000$$

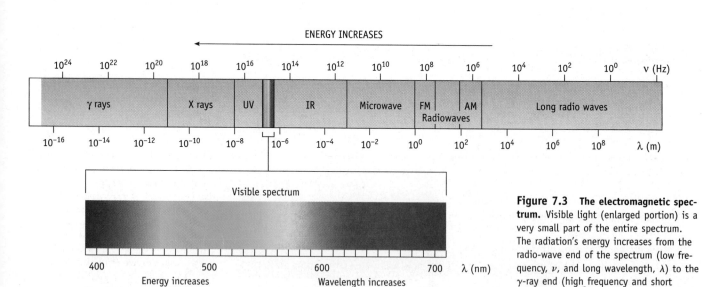

Figure 7.3 The electromagnetic spectrum. Visible light (enlarged portion) is a very small part of the entire spectrum. The radiation's energy increases from the radio-wave end of the spectrum (low frequency, ν, and long wavelength, λ) to the γ-ray end (high frequency and short wavelength) [🖰 CD-ROM, Screen 7.4].

Exercise 7.2 **Radiation, Wavelength, and Frequency**

(a) Which color in the visible spectrum has the highest frequency? Which has the lowest frequency?

(b) Is the frequency of the radiation used in a microwave oven higher or lower than that from your favorite FM radio station (91.7 MHz) (where MHz (megahertz) = 10^6 s^{-1}) (see Figure 7.3)?

(c) Is the wavelength of x-rays longer or shorter than that of ultraviolet light?

7.2 PLANCK, EINSTEIN, ENERGY, AND PHOTONS

Planck's Equation

If you heat a piece of metal, it emits electromagnetic radiation, with wavelengths that depend on temperature. At first its color is a dull red. At higher temperatures, the red color brightens, and at still higher temperatures the redness turns to a brilliant white light. For example, the heating element of a toaster becomes "red hot," and the filament of an incandescent light bulb glows "white hot" (Figure 7.4a).

Your eyes detect the radiation from a piece of heated metal that occurs in the visible region of the electromagnetic spectrum. These are not the only wavelengths of the light emitted by the metal, though. Radiation is also emitted with wavelengths both shorter (in the ultraviolet) and longer (in the infrared, Figure 7.4b) than those of visible light. That is, a spectrum of electromagnetic radiation is emitted (Figure 7.5), with some wavelengths more intense than others. As the metal is heated, the maximum in the curve of light intensity versus wavelength is shifted more and more to the ultraviolet region, so the color of the glowing object shifts from red to yellow and finally to white hot.

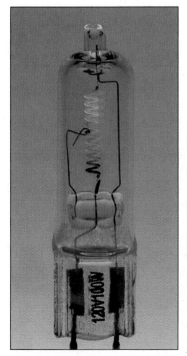

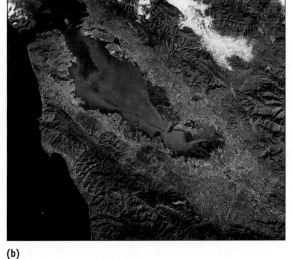

Figure 7.4 Infrared radiation. IR radiation has longer wavelengths than light of the visible region of the spectrum.
(a) The filament of an incandescent light bulb emitting radiation at the long wavelength, or red, end of the visible spectrum.
(b) A photo of the San Francisco Bay area taken from a satellite. The film responds to wavelengths in the infrared region. *(a, Charles D. Winters; b, Earth Satellite Corp./Photo Researchers, Inc.)*

(a)   **(b)**

Figure 7.5 The spectrum of the radiation given off by a heated body. When an object such as the filament in a light bulb is heated, it emits radiation. The emitted radiation covers a spectrum of wavelengths. For a given temperature, some of the radiation is emitted at long wavelengths and some at short wavelengths. Most, however, is emitted at some intermediate wavelength, the maximum in the curve. As the temperature of the object increases, the maximum moves from the red to the violet end of the spectrum. At still higher temperatures intense light is emitted at all wavelengths in the visible region, and the maximum in the curve is in the ultraviolet region. The object is described as "white hot." (Stars are often referred to as "red giants" or "white dwarfs," a reference to their temperatures and relative sizes.)

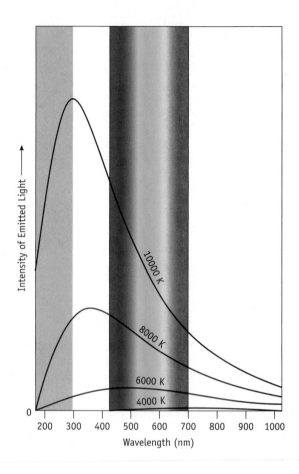

History

Max Planck (1858–1947)

His solution to the ultraviolet catastrophe was announced two weeks before Christmas in 1900, and he was awarded the Nobel Prize in physics in 1918 for this work. Einstein later said it was a longing to find harmony and order in nature, a "hunger in his soul," that spurred Planck on. *(E. F. Smith Collection/Van Pelt Library/ University of Pennsylvania)* ●

At the end of the 19th century, scientists were trying to explain the relationship between the intensity and wavelength for radiation given off by heated objects. All attempts were unsuccessful, however. Theories available at the time predicted that the intensity of radiation should increase continuously with decreasing wavelength. This perplexing situation became known as the "ultraviolet catastrophe" because predictions failed in the ultraviolet region. Classical physics did not provide a satisfactory explanation, and a new way to look at matter and energy was needed.

In 1900, the German physicist, Max Planck, offered an explanation. Following classical theory, he assumed that vibrating atoms in a heated object give rise to the emitted electromagnetic radiation. He also introduced an important new assumption, however, that these vibrations are **quantized.** In Planck's model, quantization means that only certain vibrations, with specific frequencies, are allowed.

Planck introduced an important equation, now called Planck's equation, that states that the energy of a vibrating system is proportional to the frequency of vibration. The proportionality constant h is called **Planck's constant,** in his honor. It has the value $6.62606876 \times 10^{-34}$ J · s [⊕ CD-ROM, Screen 7.5].

$$\underset{\text{Frequency (s}^{-1})}{\overset{\overset{\text{Energy (J)}\quad\text{Planck's constant (J·s)}}{\downarrow\qquad\qquad\downarrow}}{E = h\nu}}$$

(7.2)

Now, assume as Planck did that there must be a *distribution* of vibrations of atoms in an object — some atoms are vibrating at a high frequency, some at a low frequency, but most have some intermediate frequency. The few atoms with high-frequency vibrations are responsible for some of the light, as are those few with low-frequency vibrations. Most of the light must come, however, from the majority of the atoms that have intermediate vibrational frequencies. That is, the emitted light should have its maximum intensity at the wavelength of these intermediate frequencies. The intensity should not become greater and greater on approaching the ultraviolet region. The ultraviolet catastrophe was solved.

Einstein and the Photoelectric Effect

As almost always occurs, the explanation of a fundamental phenomenon — such as the spectrum of light from a hot object — leads to another fundamental discovery. A few years after Planck's work, Albert Einstein incorporated Planck's ideas into an explanation of the photoelectric effect.

The **photoelectric effect** occurs when light strikes the surface of a metal and electrons are ejected (Figure 7.6). Here an electric potential is applied to the cell.

History

Albert Einstein (1879–1955)
Einstein received the Nobel Prize in physics in 1921 for "services to theoretical physics, especially for the discovery of the law of the photoelectric effect." *(U.S. Library of Congress/Mark Marten/Photo Researchers, Inc.)*

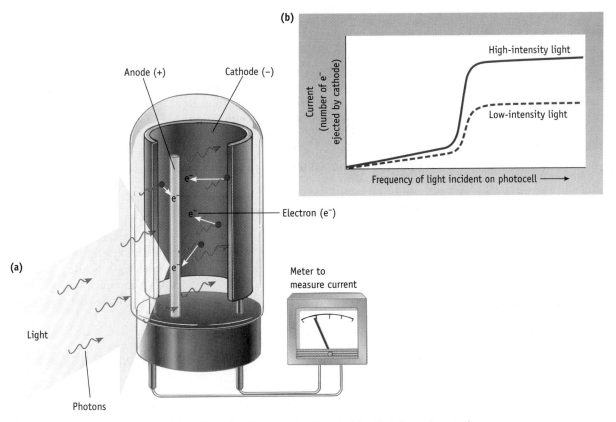

Figure 7.6   Photoelectric effect. (a) A photocell operates by the photoelectric effect. The main part of the cell is a light-sensitive cathode. This is a material, usually a metal, that ejects electrons if struck by photons of light of sufficient energy. The ejected electrons move to the anode and a current flows in the cell. Such a device can be used as a switch in electric circuits. **(b)** No current is observed until the critical frequency is reached. If higher intensity light is used, the only effect is to cause more electrons to be released from the surface; the onset of current is observed at the same frequency as with lower intensity light. When light of higher frequency than the minimum is used, the excess energy of the photon allows the electron to escape the atom with greater velocity.

When light strikes the cathode of the cell, electrons are ejected from the cathode surface and move to a positively charged anode. A stream of electrons — a current — flows through the cell. Thus, the cell can act as a light-activated switch in an electric circuit. Such cells are used in automatic door openers in stores and elevators.

Experiments with photoelectric cells show that electrons are ejected from the surface *only* if the frequency of the light is high enough. If lower frequency light is used, no effect is observed, regardless of the light's intensity (brightness). If the frequency is above the minimum, however, increasing the light intensity causes a higher current to flow because more and more electrons are ejected.

Einstein decided the experimental observations could be explained by combining Planck's equation ($E = h\nu$) with a new idea: that light could be described not only as having wave-like properties but also as having particle-like properties. Einstein assumed these massless "particles," now called **photons,** are packets of energy. The energy of each photon is proportional to the frequency of the radiation, as given by Planck's relation.

Einstein's proposal helps us understand the photoelectric effect. It is reasonable to suppose that a high-energy particle would have to bump into an atom to cause the atom to lose an electron. It is also reasonable to suppose that an electron can be torn away from the atom only if some minimum amount of energy is used. If electromagnetic radiation can be thought of as a stream of photons, as Einstein said, then the greater the intensity of light, the more photons there are to strike a surface per unit of time. It follows that the atoms of a metal surface do not lose electrons when the metal is bombarded by millions of photons if no individual photon has enough energy to remove an electron from an atom. Once the critical minimum energy (that is, minimum light frequency) is exceeded, the energy content of each photon is sufficient to displace an electron from a metal atom. The greater the number of photons with this energy that strike the surface, the greater the number of electrons dislodged. Thus, the connection is made between light intensity and the number of electrons ejected.

Energy and Chemistry: Using Planck's Equation

Compact disc players use lasers that emit red light with a wavelength of 685 nm. What is the energy of one photon of this light? What is the energy of a mole of photons of red light? To answer this, first convert the wavelength to the frequency of the radiation and then use the frequency to calculate the energy per photon. Finally, the energy of a mole of photons is obtained by multiplying the energy per photon by Avogadro's number:

$$\boxed{\lambda,\ nm} \xrightarrow[\times\ \frac{10^{-9}\ m}{nm}]{} \boxed{\lambda,\ m} \xrightarrow[\nu = c/\lambda]{} \boxed{\nu,\ s^{-1}} \xrightarrow[E = h\nu]{} \boxed{E,\ J/photon} \xrightarrow[\times\ \text{Avogadro's number}]{} \boxed{E,\ J/mol}$$

$$685\ nm\ (10^{-9}\ m/nm) = 6.85 \times 10^{-7}\ m$$

$$\nu = \frac{c}{\lambda} = \frac{2.998 \times 10^8\ m \cdot s^{-1}}{6.85 \times 10^{-7}\ m} = 4.38 \times 10^{14}\ s^{-1}$$

$$E = h\nu$$

$$= (6.626 \times 10^{-34}\ J \cdot s/photon)(4.38 \times 10^{14}\ s^{-1}) = 2.90 \times 10^{-19}\ J/photon$$

$$E = (2.90 \times 10^{-19}\ J/photon)(6.022 \times 10^{23}\ photons/mol) = 1.75 \times 10^5\ J/mol$$

Chemical Perspectives

UV Radiation, Skin Damage, and Sunscreens

Most of us are well aware of the effects of exposure to the sun. A sunburn results, and over a long period, permanent skin damage can occur. Most of this problem is the result of the damage to organic molecules caused by ultraviolet (UV) radiation from the sun.

UV radiation is often divided into three categories: UVA (315–400 nm), UVB (290–315 nm), and UVC (100–290 nm). UVC radiation is dangerous, but most of it is absorbed by earth's ozone layer. UVB light is responsible for your sunburn. Tanning occurs when the light strikes your skin and activates the melanocytes in the skin so that they produce melanin. UVA light also produces skin damage such as the alteration of connective tissue in the dermis.

Using the methods in the text, we calculate that the energy of a mole of photons in the ultraviolet region (at 300 nm) is about 400 kJ/mol.

Sun screens block very energetic ultraviolet radiation and help to prevent skin damage. *(Bachmann/Photo Researchers, Inc.)*

For $\lambda = 300$ nm, $\nu = 9.99 \times 10^{14}$ s^{-1}

$$E = h\nu$$
$$= (6.626 \times 10^{-34} \text{ J} \cdot \text{s/photon})$$
$$\times (9.99 \times 10^{14} \text{ s}^{-1})$$
$$= 6.62 \times 10^{-19} \text{ J/photon}$$
$$E = 3.99 \times 10^{5} \text{ J/mol}$$

This energy is more than twice the energy of photons of light in the visible region (say red light). Indeed, the energy of UV light is in the range of the energies necessary to break the chemical bonds in proteins.

Various manufacturers have developed mixtures of compounds that protect skin from UVA and UVB radiation. These sunscreens are given "sun protection factor" (SPF) labels that indicate how long the user can stay in the sun without burning. Sunscreens produced by Coppertone, for example, contain the organic compounds 2-ethylhexyl-*p*-methoxycinnamate and oxybenzene. These molecules absorb UV radiation, preventing it from reaching your skin.

The energy of a mole of photons of red light is equivalent to 175 kJ. A mole of photons of blue light ($\lambda = 400$ nm) has an energy of about 300 kJ. These energies are in a range that can affect the bonds between atoms in molecules. It should not be surprising therefore that light can cause chemical reactions to occur. For example, you may have seen cases in which sunlight causes paint or dye to fade or cloth to decompose, changes that result from chemical changes in dye and cloth molecules.

The previous calculation shows that, as the frequency of radiation increases, the energy of the radiation also increases (see Figure 7.3). Similarly, the energy increases as the wavelength of radiation decreases:

As frequency (ν) increases, energy (E) increases

$$E = h\nu = \frac{hc}{\lambda}$$

As wavelength (λ) decreases, energy (E) increases

Therefore, photons of ultraviolet radiation — with wavelengths shorter than those of visible light — have higher energy than visible light. Because visible light has enough energy to affect bonds between atoms, it is obvious that ultraviolet light does as well. That is the reason ultraviolet radiation can cause a sunburn. In contrast, photons of infrared radiation — with wavelengths longer than those of visible light — have lower energy than visible light. They are generally not energetic enough to cause chemical reactions, but they can affect the vibrations of molecules. We sense infrared radiation as heat, such as the heat given off by a glowing burner on an electric stove.

Exercise 7.3   Photon Energies

Compare the energy of a mole of photons of blue light (4.00×10^2 nm) with the energy of a mole of photons of microwave radiation having a frequency of 2.45 GHz (1 GHz = 10^9 s^{-1}). Which has the greater energy? By what factor is one greater than the other?

7.3   ATOMIC LINE SPECTRA AND NIELS BOHR

Atomic Line Spectra

The final piece of experimental information that played a major role in developing the modern view of atomic structure is the observation of the light emitted by atoms after they absorb energy. The spectrum of white light, such as that from an incandescent light bulb or from the sun, produces the rainbow display of colors shown in Figure 7.7. Such a spectrum, consisting of light of all wavelengths, is called a **continuous spectrum.**

If a high voltage is applied to atoms of an element in the gas phase at low pressure, the atoms absorb energy and are said to be "excited." The excited atoms emit light. This light is different, however, from the continuous spectrum of wavelengths. Excited atoms in the gas phase emit only certain wavelengths of light. We know this because when this light is passed through a prism, only a few colored lines are

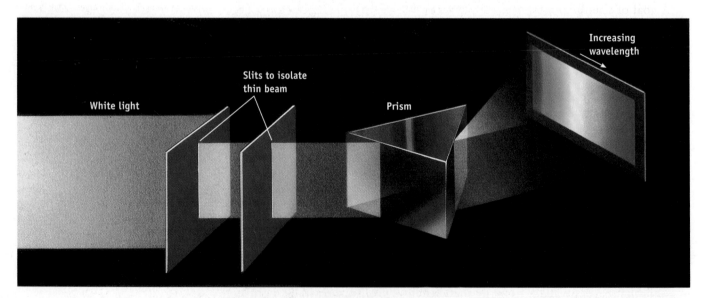

Figure 7.7   A spectrum of white light, produced by refraction in a prism. The light is observed with a spectroscope or spectrometer by passing the light through a narrow slit to isolate a thin beam, or line, of light. The beam is then passed through a device (a prism or, in modern instruments, a diffraction grating) that separates the light into its component wavelengths. See the spectrum of visible light in Figure 7.3.

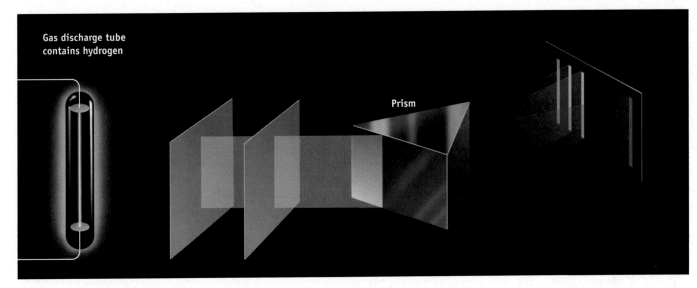

Figure 7.8   The line emission spectrum of hydrogen. The emitted light is passed through a series of slits to create a narrow beam of light, which is then separated into its component wavelengths by a prism. A photographic plate or photocell can be used to detect the separate wavelengths as individual lines. Hence, the name "line spectrum" for the light emitted by a glowing gas [🖱 CD-ROM, Screen 7.6].

seen. This is called a **line emission spectrum** (Figure 7.8). A familiar example of this is the light from a neon advertising sign, in which excited neon atoms emit orange-red light. [🖱 CD-ROM Screen 7.6]

Line emission spectra of hydrogen, mercury, and neon are shown in Figure 7.9. Every element has a unique line spectrum. Indeed, the characteristic lines in the emission spectrum of an element can be used in chemical analysis, both to identify the element and to determine how much of it is present.

A goal of scientists in the late 19th century was to explain why gaseous atoms emit light of only certain frequencies and to find a mathematical relationship among the observed frequencies. (It is always useful if experimental data can be related by a mathematical equation because a regular pattern of information implies a logical explanation for the observations.) The first steps in this direction

● **Neon Signs**
Gas discharge signs are often referred to as neon signs even though many do not contain that gas. Colors are emitted by excited atoms of noble gases: neon, reddish orange; argon, blue; helium, yellowish white. A helium-argon mixture emits an orange light, and a neon-argon mixture gives a dark lavender color.

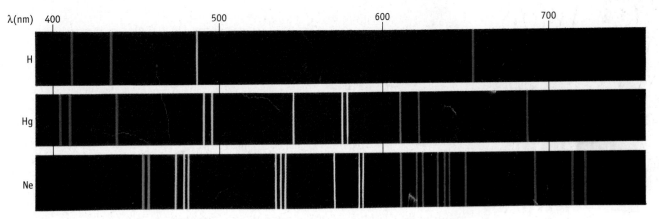

Figure 7.9   Line emission spectra of hydrogen, mercury, and neon. Excited gaseous atoms produce characteristic spectra that can be used to identify the element as well as to determine how much of each element is present in a sample.

came from Johann Balmer (1825–1898) and later Johannes Rydberg (1854–1919). They developed an equation — now called the **Rydberg equation** — from which it was possible to calculate the wavelength of the red, green, and blue lines in the visible emission spectrum of hydrogen atoms (see Figure 7.9).

$$\frac{1}{\lambda} = R\left(\frac{1}{2^2} - \frac{1}{n^2}\right), \text{ when } n > 2 \tag{7.3}$$

In this equation n is an integer, and R, now called the **Rydberg constant,** has the value $1.0974 \times 10^7 \text{ m}^{-1}$. If $n = 3$, the wavelength of the red line in the hydrogen spectrum is obtained (6.563×10^{-7} m, or 656.3 nm). If $n = 4$, the wavelength for the green line is obtained, and $n = 5$ gives the wavelength of the blue line. This group of visible lines in the spectrum of hydrogen atoms (and others for which $n = 6, 7, 8$, and so on) is now called the **Balmer series.**

Bohr Model of the Hydrogen Atom

Niels Bohr, a Danish physicist, provided the first connection between the spectra of excited atoms and the quantum ideas of Planck and Einstein. From Rutherford's work [← SECTION 2.1], it was known that electrons are arranged in space outside the atom's nucleus. For Bohr the simplest model of a hydrogen atom was one in which the electron moved in a circular orbit around the nucleus. In proposing this, however, he had to contradict the laws of classical physics. According to the theories at the time, a charged electron moving in the positive electric field of the nucleus should lose energy. Eventually the electron would crash into the nucleus, much in the same way a satellite in earth orbit eventually crashes into the earth as the satellite loses energy by rubbing up against the earth's atmosphere. But electrons don't behave this way; if it were so, matter would eventually be destroyed [CD-ROM, Screen 7.7].

To solve the contradiction with the laws of classical physics, Bohr introduced the condition that an electron orbiting the nucleus could occupy only *certain* orbits or energy levels in which it is stable. That is, the energy of the electron in the atom is quantized. By combining this *quantization postulate* with the laws of motion from classical physics, Bohr showed that the potential energy possessed by the single electron in the nth orbit or energy level of the H atom is given by the equation

Planck's constant

Rydberg constant   Speed of light

$$\text{Potential energy of electron in the } n\text{th level} = E_n = -\frac{Rhc}{n^2} \tag{7.4}$$

Principal quantum number

which gives the energy in units of joules per atom. Each allowed orbit was assigned a value of n, a unitless integer having values of 1, 2, 3, and so on. This integer is now known as the **principal quantum number** for the electron.

Equation 7.4 has several important features. First, the potential energy of the electron has a negative value, This follows from Coulomb's law [← SECTION 3.3]. The energy of attraction between oppositely charged bodies (a negative electron and the positive nuclear proton) has a negative value, and that value becomes more negative as the bodies move closer together.

Next, note the relation between potential energy and the value of n. An outcome of Bohr's model is that the radius of the circular orbits increases as n increases (Figure 7.10). An electron in the $n = 1$ orbit is closest to the nucleus and thus has the lowest or most negative energy. The electron of the hydrogen atom is normally in this energy level. An atom with its electrons in the lowest possible energy levels is said to be in its **ground state.** When the electron of a hydrogen atom occupies an orbit with n greater than 1, the electron is farther from the nucleus, the value of its energy is less negative, and it is said to be in an **excited state.** The energies of the ground state and an excited state are calculated in Example 7.2.

Example 7.2   Energies of the Ground and Excited States of the H Atom

Problem • Calculate the energies of the $n = 1$ and $n = 2$ states of the hydrogen atom in joules per atom and in kilojoules per mole.

Strategy • Here we use Equation 7.4 with the following constants: $R = 1.097 \times 10^7 \text{ m}^{-1}$, $h = 6.626 \times 10^{-34} \text{ J} \cdot \text{s}$, and $c = 2.998 \times 10^8 \text{ m/s}$.

Solution • When $n = 1$, the energy of an electron in a single H atom is

$$E_1 = -\frac{Rhc}{n^2} = -\frac{Rhc}{1^2} = -Rhc$$

$$= -(1.097 \times 10^7 \text{ m}^{-1})(6.626 \times 10^{-34} \text{ J} \cdot \text{s})(2.998 \times 10^8 \text{ m/s})$$

$$= -2.179 \times 10^{-18} \text{ J/atom}$$

In units of kilojoules per mole, we have

$$E_1 = \frac{-2.179 \times 10^{-18} \text{ J}}{\text{atom}} \cdot \frac{6.022 \times 10^{23} \text{ atoms}}{\text{mol}} \cdot \frac{1 \text{ kJ}}{1000 \text{ J}}$$

$$= -1312 \text{ kJ/mol}$$

When $n = 2$, the energy is

$$E_2 = -\frac{Rhc}{2^2} = -\frac{E_1}{4} = -\frac{2.179 \times 10^{-18} \text{ J/atom}}{4}$$

$$= -5.448 \times 10^{-19} \text{ J/atom}$$

Finally, because $E_2 = E_1/4$, we calculate E_2 to be -328.0 kJ/mol.

Comment • Notice that the calculated energies are negative for an electron in $n = 1$ or in $n = 2$, with E_1 more negative than E_2.

Exercise 7.4   Electron Energies

Calculate the energy of the $n = 3$ state of the H atom in (a) joules per atom and (b) kilojoules per mole.

You can think of the energy levels in the Bohr model as the rungs of a ladder climbing out of the basement of an "atomic building" (where the energy of the H atom is -2.18×10^{-18} J/atom; see Example 7.2) to the ground level (where the energy is 0) (see Figure 7.10). Each step represents a quantized energy level; as you climb the ladder, you can stop on any rung but not between them. Unlike the rungs of a real ladder, however, Bohr's energy levels get closer and closer together as n increases.

The Bohr Theory and the Spectra of Excited Atoms

A major assumption of Bohr's theory was that an electron in an atom would remain in its lowest energy level unless disturbed. Energy is absorbed or evolved if the electron changes from one energy level to another, and it is this idea that allowed Bohr to explain the spectra of excited gases.

When the H atom electron has $n = 1$, and so is in its ground state, the energy has a large negative value. As we climb the ladder (see Figure 7.10) to the $n = 2$

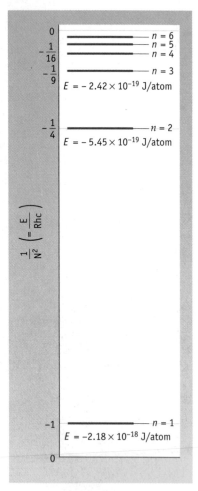

Figure 7.10   Energy levels for the H atom in the Bohr model. The energies of the electron in the hydrogen atom depend on the value of the principal quantum number n ($E_n = -Rhc/n^2$). The larger the value of n the larger the Bohr radius and the less negative the value of the energy. Energies are given in joules per atom (J/atom). Notice that the difference between successive energy levels becomes smaller as n becomes larger.

level, the electron is less strongly attracted to the nucleus, and the energy of an $n = 2$ electron is less negative. Therefore, to move an electron in the $n = 1$ state to the $n = 2$ state, the atom must absorb energy, just as energy must be expended in climbing a ladder. The electron must be excited (Figure 7.11).

Using Bohr's equation we can calculate the energy required to carry the H atom from the ground state ($n = 1$) to its first excited state ($n = 2$). As you learned in Chapter 6, the difference in energy between two states is always

$$\Delta E = E_{final} - E_{initial}$$

Because E_{final} has $n = 2$, and $E_{initial}$ has $n = 1$, ΔE is

$$\Delta E = E_2 - E_1 = \left(-\frac{Rhc}{2^2}\right) - \left(-\frac{Rhc}{1^2}\right)$$
$$= \left(\tfrac{3}{4}\right)Rhc = 0.75\,Rhc = 0.75(1312\text{ kJ})$$
$$= 984\text{ kJ/mol of H atoms}$$

where we used the energy of the $n = 1$ state, as calculated in Example 7.2, for the value of Rhc. The amount of energy that must be absorbed by the atom so that an electron can move from the first to the second energy state is $0.75Rhc$, no more and no less. If $0.7Rhc$ or $0.8Rhc$ is provided, no transition between states is possible. *Energy levels in the H atom are quantized*, with the consequence that only certain amounts of energy may be absorbed or emitted.

Moving an electron from a state of low n to one of higher n requires that energy is absorbed, and the sign of the value of ΔE is positive. The opposite process, an electron falling from a level of higher n to a lower one, therefore emits energy (see Figure 7.11). For example, for a transition from $n = 2$ to $n = 1$,

$$\Delta E = E_{final} - E_{initial}$$
$$= E_1 - E_2 = \left(-\frac{Rhc}{1^2}\right) - \left(-\frac{Rhc}{2^2}\right)$$
$$= -\left(\tfrac{3}{4}\right)Rhc$$

The negative sign indicates energy is evolved; that is, 984 kJ must be *emitted* per mole of H atoms.

Depending on how much energy is added to a collection of H atoms, some atoms have their electrons excited from the $n = 1$ to the $n = 2$ or 3 or higher states. After absorbing energy, these electrons naturally move back down to lower levels (either directly or in a series of steps to $n = 1$) and release the energy the atom originally absorbed. In doing so they *emit energy*, and the energy is observed as light. *This is the source of the lines observed in the emission spectrum of H atoms*, and the same basic explanation holds for the spectra of other elements and for the colors of fireworks.

Figure 7.11   Absorption of energy by the atom as the electron moves to an excited state. Energy is absorbed when an electron moves from the $n = 1$ state to the $n = 2$ state ($\Delta E > 0$). When the electron returns to the $n = 1$ state from $n = 1$, energy is emitted ($\Delta E < 0$).

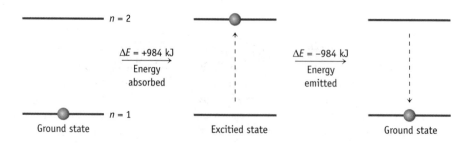

For hydrogen, the series of emission lines having energies in the ultraviolet region (called the **Lyman series,** Figure 7.12) arises from electrons moving from states with n greater than 1 to that with $n = 1$. The series of lines that have energies in the visible region — the Balmer series — arises from electrons moving from states with $n = 3$ or greater to the lower state with $n = 2$.

In summary, we now recognize that *the origin of atomic spectra is the movement of electrons between quantized energy states.* If an electron is excited from a lower energy state to a higher one, energy is absorbed. On the other hand, if an electron moves from a higher energy state to a lower one, energy is emitted. If the energy is emitted as electromagnetic radiation, an emission line is observed. The energy of a given emission line in excited hydrogen atoms is

$$\Delta E = E_{final} - E_{initial} = -Rhc\left(\frac{1}{n_{final}^2} - \frac{1}{n_{initial}^2}\right) \qquad (7.5)$$

where Rhc is 1312 kJ/mol.

Bohr was able to use his model of the atom to calculate the wavelengths of the lines in the hydrogen spectrum. He had tied the unseen (the interior of the atom) to the seen (the observable lines in the hydrogen spectrum) — a fantastic achievement! In addition, he introduced the concept of energy quantization in describing atomic structure, a concept that remains an important part of modern science.

History

When Bohr's paper describing his ideas was published in 1913, Einstein declared it to be "one of the great discoveries." •

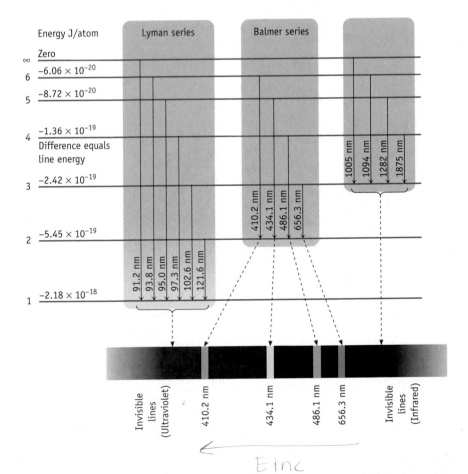

Figure 7.12   Some of the electronic transitions that can occur in an excited H atom. The Lyman series of lines in the ultraviolet region result from transitions to the $n = 1$ level. Transitions from levels with values of n greater than 2 to $n = 2$ occur in the visible region (Balmer series; see Figure 7.8). Lines in the infrared region result from transitions from levels with n greater than 3 or 4 to the $n = 3$ or 4 levels. (Only the series ending at $n = 3$ is illustrated [CD-ROM, Screen 7.7].)

As mentioned previously, agreement between theory and experiment is taken as evidence that the theoretical model is valid. It soon became apparent, however, that a flaw existed in Bohr's theory. It explained only the spectrum of H atoms and of other systems having one electron (such as He$^+$). Furthermore, *the idea that the electron moves about the nucleus with a path of fixed radius, like that of the planets about the sun, is no longer the accepted model for the atom.*

Example 7.3 Energies of Emission Lines for Excited Atoms

Problem • Calculate the wavelength of the green line in the visible spectrum of excited H atoms using the Bohr theory.

Strategy • First locate the green line in Figure 7.12 and determine the quantum states involved. That is, decide on $n_{initial}$ and n_{final}. Use Equation 7.5 to calculate the difference in energy, ΔE, between these states. Finally, express ΔE in terms of a wavelength.

Solution • The green line is the second most energetic line in the visible spectrum of hydrogen and arises from electrons moving from $n = 4$ to $n = 2$. Using Equation 7.5 where $n_{final} = 2$ and $n_{initial} = 4$, we have

$$\Delta E = E_{final} - E_{initial} = \left(-\frac{Rhc}{2^2}\right) - \left(-\frac{Rhc}{4^2}\right)$$

$$= -Rhc\left(\frac{1}{4} - \frac{1}{16}\right) = -Rhc(0.1875)$$

In the preceding text, we found that Rhc is 1312 kJ/mol, so the $n = 4$ to $n = 2$ transition involves an energy change of

$$\Delta E = -(1312 \text{ kJ/mol})(0.1875) = -246.0 \text{ kJ/mol}$$

The wavelength can now be calculated. First the photon energy, E_{photon}, is expressed as J/photon.

$$E_{photon} = \frac{\left(246.0\,\frac{\text{kJ}}{\text{mol}}\right)\left(1 \times 10^3\,\frac{\text{J}}{\text{kJ}}\right)}{\left(6.022 \times 10^{23}\,\frac{\text{photons}}{\text{mol}}\right)} = 4.085 \times 10^{-19}\,\frac{\text{J}}{\text{photon}}$$

Now apply Planck's equation.

$$\lambda = \frac{hc}{E_{photon}} = \frac{\left(6.626 \times 10^{-34}\,\frac{\text{J} \cdot \text{s}}{\text{photon}}\right)\left(2.998 \times 10^8\,\frac{\text{m}}{\text{s}}\right)}{4.085 \times 10^{-19}\,\frac{\text{J}}{\text{photon}}}$$

$$= 4.863 \times 10^{-7}\,\text{m}$$
$$= 4.863 \times 10^{-7}\,\text{m}\,(1 \times 10^9\,\text{nm/m})$$
$$= 486.3\,\text{nm}$$

The experimental value is 486.1 nm (see Figure 7.12). This represents excellent agreement between experiment and theory.

Exercise 7.5 Energy of an Atomic Spectral Line

The Lyman series of spectral lines for the H atom occurs in the ultraviolet region. They arise from transitions from higher levels down to $n = 1$. Calculate the frequency and wavelength of the least energetic line in this series.

A Closer Look

Experimental Evidence for Bohr's Theory

Niels Bohr's model of the hydrogen atom was powerful because it could reproduce experimentally observed line spectra. But there was additional experimental confirmation.

If the electron in the hydrogen atom is moved from the ground state, where $n = 1$, to the energy level where $n =$ infinity, the electron is considered to have been removed from the atom. That is, the atom has been ionized.

$$\text{H}(g) \longrightarrow \text{H}^+(g) + e^-$$

We can calculate the energy for this process from Equation 7.5 where $n_{final} = \infty$ and $n_{initial} = 1$.

$$\Delta E = -Rhc\left(\frac{1}{n_{final}^2} - \frac{1}{n_{initial}^2}\right)$$

$$\Delta E = -Rhc\left(\frac{1}{\infty^2} - \frac{1}{1^2}\right) = +Rhc$$

Because $Rhc = 1312$ kJ/mol, the energy to move an electron from $n = 1$ to $n = \infty$ is 1312 kJ/mol of H atoms. We now call this the *ionization energy* of the atom

(➡ SECTION 8.6) and can measure it in the laboratory. The experimental value is 1312 kJ/mol, in exact agreement with the result calculated from Bohr's theory!

H$^+$(g) —— $n = \infty$ $E = 0$ kJ/mol

$\Delta E = +Rhc = +1312$ kJ/mol

H(g) —— $n = 1$ $E = -1312$ kJ/mol

7.4   WAVE PROPERTIES OF THE ELECTRON

Einstein used the photoelectric effect to demonstrate that light, usually thought of as having wave properties, can also have the properties of particles, albeit without mass (see page 259). This fact was pondered by Louis Victor de Broglie. If light can be considered as having both wave and particle properties, would matter behave similarly? That is, could a tiny object such as an electron, normally considered a particle, also exhibit wave properties in some circumstances? In 1925, de Broglie proposed that a free electron of mass m moving with a velocity v should have an associated wavelength given by the equation [CD-ROM, Screen 7.8].

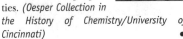

History

Louis Victor de Broglie (1892–1987)
De Broglie recognized the possibility that atomic-sized particles can have wave properties. *(Oesper Collection in the History of Chemistry/University of Cincinnati)*

$$\lambda = \frac{h}{mv} \qquad (7.6)$$

This idea was revolutionary because it linked the particle properties of the electron (m and v) with a wave property (λ). Experimental proof was soon produced. In 1927, C. J. Davisson (1881–1958) and L. H. Germer (1896–1971), working at the Bell Telephone Laboratories in New Jersey, found that a beam of electrons was diffracted like light waves by the atoms of a thin sheet of metal foil (Figure 7.13) and that de Broglie's relation was followed quantitatively. Because diffraction is an effect best explained based on the wave properties of light, it follows that *electrons also can be described as having wave properties under some circumstances.*

De Broglie's equation suggests that any moving particle has an associated wavelength. If λ is to be large enough to measure, however, the product of m and v must be very small because h is so small. For example, a 114-g baseball traveling at 110 mph has a large mv product (5.6 kg·m/s) and therefore the incredibly small wavelength of 1.2×10^{-34} m! This tiny value cannot be measured with any instrument now available. This means we will never assign wave properties to a baseball or any other massive object. It is only possible to observe wave-like properties for particles of extremely small mass, such as protons, electrons, and neutrons.

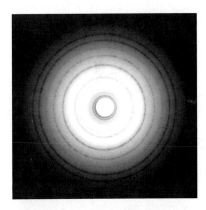

Figure 7.13   Electron diffraction pattern obtained for magnesium oxide (MgO). *(R. K. Bohn, Department of Chemistry, University of Connecticut)*

Example 7.4 **Using De Broglie's Equation**

Problem • Calculate the wavelength associated with an electron of mass $m = 9.109 \times 10^{-28}$ g that travels at 40.0% of the speed of light.

Strategy • First, consider the units involved. Wavelength is calculated from h/mv, where h is Planck's constant expressed in units of joule seconds (J · s). As discussed in Chapter 6, 1 J = 1 kg · m² · s⁻². Therefore, the mass must be in kilograms and speed in meters per second.

Solution •

Electron mass = 9.109×10^{-31} kg

Electron speed (40.0% of light speed)
$$= (0.400)(2.998 \times 10^8 \text{ m/s}) = 1.20 \times 10^8 \text{ m/s}$$

Substituting these values into de Broglie's equation, we have

$$\lambda = \frac{h}{mv}$$
$$= \frac{(6.626 \times 10^{-34})(\text{kg} \cdot \text{m}^2 \cdot \text{s}^{-2})(\text{s})}{(9.109 \times 10^{-31}\text{ kg})(1.20 \times 10^8\text{ m/s})}$$
$$= 6.06 \times 10^{-12}\text{ m}$$

In nanometers, the wavelength is

$$\lambda = (6.06 \times 10^{-12}\text{ m})(1.00 \times 10^9\text{ nm/m}) = 6.06 \times 10^{-3}\text{ nm}$$

Comment • The calculated wavelength is about 1/12 of the diameter of the H atom.

Exercise 7.6 **De Broglie's Equation**

Calculate the wavelength associated with a neutron with a mass of 1.675×10^{-24} g and a kinetic energy of 6.21×10^{-21} J. (Recall that the kinetic energy of a moving particle is $E = \frac{1}{2}(mv^2)$.)

7.5 QUANTUM MECHANICAL VIEW OF THE ATOM

In Copenhagen, Denmark, after World War I, Niels Bohr assembled a group of physicists who set out to derive a comprehensive theory for the behavior of electrons in atoms from the viewpoint of the electron as a particle. Erwin Schrödinger, an Austrian, independently worked toward the same goal, but he used de Broglie's hypothesis that an electron in an atom could be described by equations for wave motion. Although both Bohr and Schrödinger were successful in predicting some aspects of electron behavior, Schrödinger's approach gave correct results for some properties for which Bohr's failed. For this reason theoreticians today primarily use Schrödinger's concept. The general theoretical approach to understanding atomic behavior, developed by Bohr, Schrödinger, and their associates, has come to be called **quantum mechanics** or **wave mechanics.**

Uncertainty Principle

De Broglie's suggestion that an electron can be described as having wave properties was confirmed by experiment [SECTION 7.4]. J. J. Thomson's experiments showed the particle-like nature of the electron (see page 48). But how can an electron be both a particle and a wave? Although no single experiment can be done to show the electron behaves *simultaneously* as a wave *and* a particle, the electron indeed has the properties of both.

What does this wave–particle duality have to do with electrons in atoms? Werner Heisenberg and Max Born (1882–1970) provided the answer. Heisenberg concluded, in what is now known as the **uncertainty principle,** that it is impossible to fix both the position of an electron in an atom and its energy with any degree of certainty if the electron is described as a wave. Attempting to determine accurately either the location or the energy leaves the other uncertain. (Contrast this with the world around you. For objects larger than those on the atomic level — say an automobile — you can determine, with considerable accuracy, both their energy and location at a given time.)

Based on Heisenberg's idea, Max Born proposed that the results of quantum mechanics should be interpreted as follows: If we choose to know the energy of an electron in an atom with only a small uncertainty, then we must accept a correspondingly large uncertainty in its position in the space about the atom's nucleus. In practical terms, this means we can only assess the likelihood, or probability, of finding an electron with a given energy within a given region of space. In the next section you will see that the result of this viewpoint is the definition of the regions around an atom's nucleus in which there is the highest probability of finding a given electron [CD-ROM, Screen 7.9].

Schrödinger's Model of the Hydrogen Atom and Wave Functions

Schrödinger's model of the hydrogen atom is based on the premise that the electron can be described as a wave and not as a tiny particle. Unlike Bohr's model, Schrödinger's approach resulted in mathematical equations that are complex and difficult to solve except in simple cases. We need not be concerned here with the mathematics, but the solutions to the equation — called **wave functions** and symbolized by the Greek letter ψ(psi) — are chemically important. Understanding the implications of these wave functions is essential to understanding the modern view of the atom [CD-ROM, Screen 7.10]. The following important points can be made concerning wave functions:

1. The behavior of the electron in the atom is best described as a standing wave. In a vibrating string only certain vibrations or standing waves (see Figure 7.2) can be observed. Similarly, *only certain wave functions are allowed for the electron in the atom.*

2. Each wave function ψ is associated with an allowed energy value, E_n, for the electron.

3. Taken together, points 1 and 2 say that *the energy of the electron is quantized,* that is, that the electron can have only certain values of energy. This concept of energy quantization enters Schrödinger's theory naturally with the basic assumption of an electron as a standing wave. This is in contrast with Bohr's theory in which quantization was imposed as a postulate at the start.

4. The square of the wave function (ψ^2) is related to the probability of finding the electron within a given region of space. Scientists refer to this as the **electron density** in a given region.

5. Schrödinger's theory defines the energy of the electron precisely. The uncertainty principle, however, tells us there is a large uncertainty in the electron's position. Thus, we can only describe the *probability* of the electron being at a certain point in space when in a given energy state. The region of space in which an electron of a given energy is most probably located is called its **orbital.**

6. To solve Schrödinger's equation for an electron in three-dimensional space, three integer numbers — the **quantum numbers n, ℓ, and m_ℓ** — are an integral part of the mathematical solution. These quantum numbers may have only certain combinations of values, as outlined in the following sections.

Quantum numbers are used to define the energy states and orbitals available to the electron. They are analogous to an atomic "zip code," giving us the relative energy and approximate location of the electron. Let us first describe the quantum numbers and the information they provide. Then, we turn to the connection between quantum numbers and the shapes of atomic orbitals.

Quantum Numbers

Before looking into the meanings of the three quantum numbers n, ℓ, and m_ℓ [🖱 CD-ROM, Screens 7.11 and 7.12], it is important to say that

- The quantum numbers are all integers (that is, whole numbers), but their values cannot be selected randomly.
- The three quantum numbers (and their values) are not parameters that scientists dreamed up. Instead, when the behavior of the electron in the hydrogen atom is described mathematically as a wave, the quantum numbers are a natural consequence.

n, the Principal Quantum Number = 1, 2, 3, . . .

The principal quantum number n can have any integer value from 1 to infinity. The value of n is the primary factor in determining the energy of an electron. Indeed, for the hydrogen atom (with its single electron) the energy of the electron varies *only* with the value of n and is given by the same equation derived by Bohr for the H atom: $E_n = -Rhc/n^2$.

The value of n is also a measure of the size of an orbital: the greater the value of n, the larger the electron's orbital.

Each electron is labeled according to its value of n. In atoms having more than one electron, two or more electrons may have the same n value. These electrons are then said to be in the same **electron shell.**

● **Wave Functions and Energy**
In Bohr's theory, the electron energy for the H atom is given by $E_n = -Rhc/n^2$. This same result came from Schrödinger's electron wave model.

x-, y-, z-coordinates

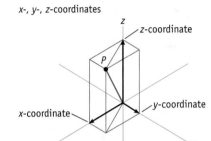

Longitude, Latitude, and Distance

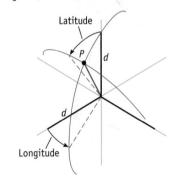

Defining the position of an electron relative to the nucleus requires three parameters. The position P of any object relative to an origin can be specified by giving its x, y, and z coordinates. Alternatively, its latitude, longitude, and distance from the origin can be specified.

● **Electron Energy and Quantum Numbers**
The electron energy in the H atom depends *only* on the value of n. In atoms with more than one electron, the energy depends on *both* n and ℓ.

ℓ, the Angular Momentum Quantum Number = 0, 1, 2, 3, . . . , $n - 1$

The electrons of a given shell can be grouped into **subshells,** each of which is characterized by a different value of the quantum number ℓ and by a characteristic shape. *Each value of ℓ corresponds to a different type of orbital with a different shape.*

The value of n limits the number of subshells possible for the nth shell because ℓ can be no larger than $n - 1$. Thus, for $n = 1$, ℓ must equal 0 and only 0. Because ℓ has only one value when $n = 1$, only one subshell is possible for an electron assigned to $n = 1$. When $n = 2$, ℓ can be either 0 or 1. Because two values of ℓ are now possible, there are two subshells in the $n = 2$ electron shell.

The values of ℓ are usually coded by letters according to the following scheme:

Value of ℓ	Corresponding Subshell Label
0	s
1	p
2	d
3	f

For example, an ℓ = 1 subshell is called a "*p* subshell," and an orbital found in that subshell is called a "*p* orbital." Conversely, an electron assigned to a *p* subshell has an ℓ value of 1.

m_ℓ, the Magnetic Quantum Number = 0, ±1, ±2, ±3, . . . , ±ℓ

The magnetic quantum number, m_ℓ, is related to the orientation of the orbitals within a subshell. *Orbitals in a given subshell differ only in their orientation in space, not in their energy.*

The value of ℓ limits the integer values assigned to m_ℓ: m_ℓ can range from +ℓ to −ℓ with 0 included. For example, when ℓ = 2, m_ℓ has five values: −2, −1, 0, +1, and +2. The number of values of m_ℓ for a given subshell (= 2ℓ + 1) specifies the number of orientations that exist for the orbitals of that subshell and thus the number of orbitals in the subshell.

Useful Information from Quantum Numbers

The three quantum numbers introduced thus far, as we have noted, are a kind of zip code for electrons. For example, suppose you live in an apartment building. You could specify your location as being on a particular floor (n), in a particular apartment on that floor (ℓ), and in a particular room in the apartment (m_ℓ). That is, n describes the shell to which an electron is assigned in an atom, ℓ describes the subshell within that shell, and m_ℓ specifies the number of orbitals within that subshell.

Allowed values of the three quantum numbers are summarized in Table 7.1. Before describing the composition of the first four electron shells ($n = 1, 2, 3$, and 4), let us summarize some useful points:

● Electrons in atoms are assigned to orbitals, which are grouped into subshells. Depending on the value of n, one or more subshells constitute an electron shell.

● Electron subshells are labeled by first giving the value of n and then the value of ℓ in the form of its letter code. For $n = 1$ and ℓ = 0, for example, the label is 1s.

Table 7.1 • **Summary of the Quantum Numbers, Their Interrelationships, and the Orbital Information Conveyed**

Principal Quantum Number	Angular Momentum Quantum Number	Magnetic Quantum Number	Number and Type of Orbitals in the Subshell
Symbol = n Values = 1, 2, 3, ... n = number of subshells	Symbol = ℓ Values = $0 \ldots n - 1$	Symbol = m_ℓ Values = $-\ell \ldots 0 \ldots +\ell$	Number of orbitals in shell = n^2 and number of orbitals in subshell = $2\ell + 1$
1	0	0	one 1s orbital (one orbital of one type in the $n = 1$ shell)
2	0 1	0 +1, 0, −1	one 2s orbital three 2p orbitals (four orbitals of two types in the $n = 2$ shell)
3	0 1 2	0 +1, 0, −1 +2, +1, 0, −1, −2	one 3s orbital three 3p orbitals five 3d orbitals (nine orbitals of three types in the $n = 3$ shell)
4	0 1 2 3	0 +1, 0, −1 +2, +1, 0, −1, −2 +3, +2, +1, 0, −1, −2, −3	one 4s orbital three 4p orbitals five 4d orbitals seven 4f orbitals (16 orbitals of four types in the $n = 4$ shell)

If you describe sets of quantum numbers, starting with a given value of n and then deciding the values of ℓ and then m_ℓ that follow (see Table 7.1), you would discover the following:

- n = the number of subshells in a shell
- $2\ell + 1$ = the number of orbitals in a subshell = the number of values of m_ℓ
- n^2 = the number of orbitals in a shell

First Electron Shell, $n = 1$

When $n = 1$ the value of ℓ can only be 0, and so m_ℓ must also have a value of 0. This means that in the electron shell closest to the nucleus, only one subshell exists, and that subshell consists of only a single orbital, the 1s orbital.

For the Second Shell, $n = 2$

When $n = 2$, ℓ can have two values (0 and 1), so two subshells or two types of orbitals occur in the second shell. One of these is the 2s subshell ($n = 2$ and $\ell = 0$), and the other is the 2p subshell ($n = 2$ and $\ell = 1$). Because the values of m_ℓ can be −1, 0, and +1 when $\ell = 1$, three p orbitals exist. Because all three have $\ell = 1$, they all have the same shape, but the different m_ℓ values tell us they differ in their orientation in space. Thus, when $\ell = 1$, p orbitals are indicated, and three of them always occur.

For the Third Shell, $n = 3$

When $n = 3$, three subshells, or orbital types, are possible for an electron because ℓ has the values 0, 1, and 2. Because you see ℓ values of 0 and 1 again, you know that two of the subshells within the $n = 3$ shell are 3s (one orbital) and 3p (three

• **Subshells and Orbitals**

Subshell	Number of Orbitals in Subshell
s	1
p	3
d	5
f	7

orbitals). The third subshell is *d*, indicated by $\ell = 2$. Because m_ℓ has five values $(-2, -1, 0, +1,$ and $+2)$ when $\ell = 2$, five *d* orbitals (no more and no less) occur in the $\ell = 2$ subshell.

For the Fourth Shell, *n* = 4, and Beyond

Table 7.1 shows that there are four subshells in the $n = 4$ shell. Besides *s*, *p*, and *d* subshells, we now encounter the *f* electron subshell, that is, orbitals for which $\ell = 3$. Seven such orbitals exist because there are seven values of m_ℓ when $\ell = 3 (-3, -2, -1, 0, +1, +2,$ and $+3)$.

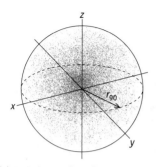

(a) Dot picture of an electron in a 1s orbital. Each dot represents the position of the electron at a different instant in time. Note that the dots cluster closest to the nucleus. r_{90} is the radius of a sphere within which the electron is found 90% of the time.

Exercise 7.7 **Using Quantum Numbers**

Complete the following statements:

(a) When $n = 2$, the values of ℓ can be _____ and _____.

(b) When $\ell = 1$, the values of m_ℓ can be _____, _____, and _____, and the subshell has the letter label _____.

(c) When $\ell = 2$, the subshell is called a _____ subshell.

(d) When a subshell is labeled *s*, the value of ℓ is _____ and m_ℓ has the value _____.

(e) When a subshell is labeled *p*, _____ orbitals occur within the subshell.

(f) When a subshell is labeled *f*, there are _____ values of m_ℓ, and _____ orbitals occur within the subshell.

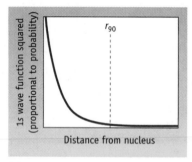

(b) A plot of the probability density as a function of distance for a one-electron atom with a 1s electron wave.

7.6 SHAPES OF ATOMIC ORBITALS

The chemistry of an element and of its compounds is determined by the electrons of the element's atoms, particularly the electrons with the highest value of *n*, which are often called *valence electrons* (➡ SECTION 9.1). The types of orbitals to which these electrons are assigned are also important, so we turn now to the question of orbital shape and orientation [🔘 CD-ROM, Screen 7.13].

s Orbitals

When an electron has $\ell = 0$, we often say the electron is assigned to, or "occupies," an *s* orbital. But what does this mean? What is an *s* orbital? What does it look like? To answer these questions, we begin with the wave function for an electron with $n = 1$ and $\ell = 0$, that is, with a 1s orbital. If we assume the electron is a tiny particle and not a wave, and if we could photograph the 1s electron at 1-second intervals for a few thousand seconds, the composite picture would resemble the drawing in Figure 7.14a. This resembles a cloud of dots, so chemists refer to such representations of electron orbitals as **electron cloud pictures.**

The fact that the density of dots is greater close to the nucleus (the electron cloud is denser close to the nucleus) indicates that the electron is most often found near the nucleus (or, conversely, it is less likely to be found farther away). Putting this in the language of quantum mechanics, we say the electron density is greater closer to the nucleus or that the greatest probability of finding the electron is in a tiny volume of space around the nucleus. Conversely, the electron density falls off on moving away from the nucleus; it is less probable that the electron is farther away. The thinning of the electron cloud at increasing distance, shown by the de-

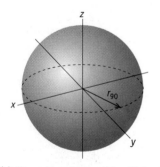

(c) The surface of the sphere within which the electron is found 90% of the time for a 1s orbital. This surface is often called a "boundary surface." (A 90% surface was chosen arbitrarily. If the choice was the surface within which the electron is found 50% of the time, the sphere would be considerably smaller.)

Figure 7.14 **Different views of a 1s** (*n* = 1 and ℓ = 0) orbital.

creasing density of dots in Figure 7.14a, is illustrated in a different way in Figure 7.14b. Here we plotted the square of the wave function for the electron in a 1s orbital as a function of the distance of the electron from the nucleus. The units of ψ^2 at each point are 1/volume, so the vertical axis of this plot represent the probability of finding the electron in each cubic nanometer, for example, at a given distance from the nucleus. For this reason, ψ^2 is called the **probability density.** For the 1s orbital, ψ^2 is very high for points immediately around the nucleus, but it drops off rapidly as the distance from the nucleus increases. Notice that the probability approaches but never quite reaches zero, even at very large distances.

For the 1s orbital, Figure 7.14a shows the electron is most likely found within a sphere with the nucleus at the center. No matter in which direction you proceed from the nucleus, the probability of finding an electron is the same at the same distance from the nucleus (Figure 7.14b). *The 1s orbital is spherical in shape.*

The visual image of Figure 7.14a is that of a cloud whose density is small at large distances from the center; there is no sharp boundary beyond which the electron is never found. The s and other orbitals, however, are often depicted as having a sharp boundary surface (Figure 7.14c), largely because it is easier to draw such pictures. To arrive at the diagram in Figure 7.14c we drew a sphere about the nucleus in such a way that the chance of finding the electron somewhere inside is 90%.

Many misconceptions exist about pictures such as Figure 7.14c. First, the surface within which the electron is "contained" is not impenetrable. Second, when we talk about electron clouds and electron distributions, we are referring to the probability of finding the electron at a given location. And third, the probability of finding the electron is not the same throughout the volume enclosed by the surface. For example, the electron in the H atom 1s orbital has a greater probability of being close to the nucleus than further away.

An important feature of s orbitals (1s, 2s, 3s, and so on) is that they are all spherical in shape. There are important differences between s orbitals of different n, however; *the size of s orbitals and their energy increases as n increases* (Figure 7.15). Thus, the 1s orbital is more compact than the 2s orbital, which is more compact than the 3s orbital.

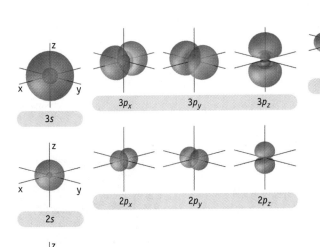

$3s$

$3p_x$ $3p_y$ $3p_z$

$3d_{z^2}$ $3d_{xz}$ $3d_{yz}$ $3d_{xy}$ $3d_{x^2-y^2}$

$2s$

$2p_x$ $2p_y$ $2p_z$

$1s$

Figure 7.15 Atomic orbitals. Boundary surface diagrams for electron densities of 1s, 2s, 2p, 3s, 3p, and 3d orbitals for a hydrogen atom. For the p orbitals, the subscript letter on the orbital notation (x, y, z) indicates the cartesian axis along which the orbital lies. The plane passing through the nucleus (perpendicular to this axis) is called a nodal surface ($\ell = 1$). The d orbitals all have two nodal surfaces ($\ell = 2$)[CD-ROM, Screen 7.13].

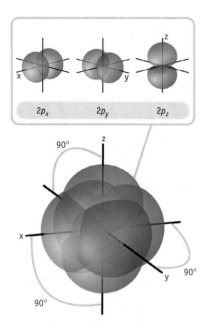

Atomic *p* orbitals. The three *p* orbitals in a given shell of an atom are oriented at 90° to each other. *(Orbital Models: Patrick A. Harman and Charles F. Hamper)*

● **Standing Waves and Nodal Surfaces**
Recall that standing waves have nodes (see Figure 7.2). Similarly, the electron waves in an atom have nodes.

● **Nodal Surfaces**
Nodal surfaces occur for all *p*, *d*, and *f* orbitals. These surfaces are usually flat and so are referred to as nodal planes. In some cases (for example, d_{z^2}), however, the "plane" is not flat and so is better referred to as a "surface."

p Orbitals

Atomic orbitals for which $\ell = 1$ are *p* orbitals and all have the same basic shape. *All p orbitals have one imaginary plane that slices through the nucleus and that divides the region of electron density in half* (Figures 7.15 and 7.16). This imaginary plane is called a **nodal surface,** a surface on which there is zero probability of finding the electron. The electron can never be found on the nodal surface; the regions of electron density lie on either side of the nucleus. A plot of electron probability (ψ^2) versus distance would start at zero at the nucleus, rise to a maximum, and then drop off at still greater distances.

If you enclose 90% of the electron density in a *p* orbital within a surface, the view in Figure 7.16 is appropriate. The electron cloud has a shape that resembles a weight lifter's dumbbell, and so chemists often describe *p* orbitals as having dumbbell shapes.

According to Table 7.1, when $\ell = 1$, then m_ℓ can only be $-1, 0$, or $+1$. That is, three orientations are possible for $\ell = 1$ or *p* orbitals. There are three mutually perpendicular directions in space (*x*, *y*, and *z*), and the *p* orbitals are commonly visualized as lying along those directions (with the nodal surface perpendicular to the axis). The orbitals are labeled according to the axis along which they lie (p_x, p_y, or p_z) (see Figure 7.16).

d Orbitals

The value of ℓ is equal to the number of nodal surfaces that slice through the nucleus. Thus, *s* orbitals, for which $\ell = 0$, have no nodal surfaces, and *p* orbitals, for which $\ell = 1$, have one planar nodal surface. It follows that the five *d* orbitals, for which $\ell = 2$, have two nodal surfaces, which results in four regions of electron density. The d_{xy} orbital, for example, lies in the *xy*-plane and the two nodal surfaces are the *xz*- and *yz*-planes (Figure 7.16). Two other orbitals, d_{xz} and d_{yz}, lie in planes defined by the *xz*- and *yz*-axes, respectively, and also have two, mutually perpendicular nodal surfaces (see Figure 7.15).

Of the two remaining *d* orbitals, the $d_{x^2-y^2}$ orbital is easier to visualize. Like the d_{xy} orbital, the $d_{x^2-y^2}$ orbital results from two vertical planes slicing the electron density into quarters. Now, however, the planes bisect the *x*- and *y*-axes, so the regions of electron density lie along the *x*- and *y*-axes.

The final *d* orbital, d_{z^2}, has two main regions of electron density along the *z*-axis, but a "donut" of electron density also occurs in the *xy*-plane. This orbital has two nodal surfaces, but the surfaces are not flat. Think of an ice cream cone sitting with its tip at the nucleus. One of the electron clouds along the *z*-axis sits inside the cone. If you have another cone pointing in the opposite direction from the

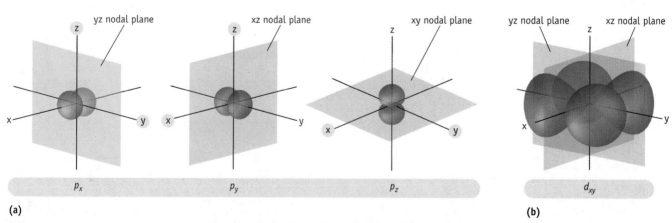

(a) p_x p_y p_z d_{xy} **(b)**

Figure 7.16 **Nodal surfaces in *p* and *d* orbitals.** **(a)** The three *p* orbitals each have one nodal surface ($\ell = 1$). **(b)** The d_{xy} orbital. All five *d* orbitals have two nodal surfaces ($\ell = 2$). Here the nodal surfaces are the *xz*- and *yz*-planes, so the regions of electron density lie in the *xy*-plane and between the *x*- and *y*-axes. *(Orbital Models: Patrick A. Harman and Charles F. Hamper)*

first cone, again with its tip at the nucleus, another region of electron density fits inside this second cone. The region outside both cones defines the remaining, donut-shaped region of electron density.

f Orbitals

The seven *f* orbitals all have $\ell = 3$. The three nodal surfaces cause the electron density to lie in eight regions of space. These orbitals are less easily visualized, but one *f* orbital is illustrated in Figure 7.17.

ℓ and Nodal Surfaces

Orbital	ℓ	Number of Nodal Surfaces
s	0	0
p	1	1
d	2	2
f	3	3

Exercise 7.8   Orbital Shapes

(a) What are the *n* and ℓ values for each of the following orbitals: 6*s*, 4*p*, 5*d*, and 4*f*?

(b) How many nodal planes exist for a 4*p* orbital? For a 6*d* orbital?

7.7   ATOMIC ORBITALS AND CHEMISTRY

We close with some questions to ponder. When an element is part of a molecule, are the orbitals the same as in a free atom? Do they have the same shapes? What does the shape of orbitals have to do with the chemistry of an element? These are questions we take up in the rest of the book, but a few answers are in order here.

Schrödinger's wave equation can be solved exactly for the hydrogen atom but not for heavier atoms or their ions. Nonetheless, chemists make the assumption that orbitals in other atoms are hydrogen-like, even when those atoms are part of a molecule. At least this is the best way chemists have found so far to interpret experimental information on how molecules react.

Figure 7.17   One of the seven possible *f* orbitals. Notice the presence of three nodal planes as required by an orbital with $\ell = 3$. *(Charles D. Winters)*

A Closer Look

Electron Density Plots

Theoreticians use several ways of visualizing atomic orbitals. In Figure 7.14, we chose to show the variation in the square of the wave function (ψ^2) versus distance from the nucleus. Because ψ^2 has units of 1/volume, this gives the probability of finding the electron in some given volume of space.

Another way to present information on electron density is a plot of $4\pi r^2 \psi^2$ versus the distance from the nucleus *r*. Such a plot for a 1*s* electron of H atom looks like the figure shown here. This product will have units of 1/distance, which we can see by assessing the units for each term [for example, $(cm^2) \times (1/cm^3) = 1/cm$]. The plot, called a *radial distribution plot*, reflects the probability of finding the electron at a given distance *r* from the nucleus. The maximum in the curve is the most probable distance of the electron from the nucleus. Interestingly, this value for the 1*s* orbital for H (0.0529 nm) corresponds to the first radius of the Bohr atom. The graph also shows that there is a finite probability that the electron can be found both closer to and farther from the nucleus.

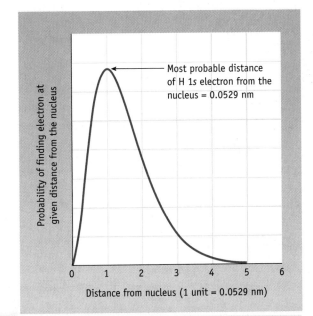

Plot of $4\pi r^2 \psi^2$ versus distance from the nucleus for the H 1*s* electron. ▶

Chemistry is the study of molecules and their transformations. By thinking about the orbitals of the atoms in molecules, and by making the simple assumption that they resemble those of the hydrogen atom, we can understand much of the chemistry of even complex systems such as those in plants and animals.

In Summary

When you have finished studying this chapter, you should ask if you have met the chapter goals. In particular, you should be able to

- Use the terms wavelength, frequency, amplitude, and node (Section 7.1).

- Use Equation 7.1 ($c = \lambda \times \nu$), the relationship between the wavelength (λ) and frequency (ν) of electromagnetic radiation and the speed of light (c).

- Recognize the relative wavelength (or frequency) of the various types of electromagnetic radiation (Figure 7.3).

- Understand that the energy of a photon, a massless particle of radiation, is proportional to its frequency (Planck's equation, Equation 7.2). This is an extension of Planck's idea that energy at the atomic level is quantized (Section 7.2).

- Describe the Bohr model of the atom, how it can account for the emission line spectra of excited hydrogen atoms, and the limitations of the model (Section 7.3).

- Understand that, in the Bohr model of the H atom, the electron can occupy only certain energy levels, each with an energy proportional to $1/n^2$ ($E = -Rhc/n^2$), where n is the principal quantum number (Equation 7.4) (Section 7.3). If an electron moves from one energy state to another, the amount of energy absorbed or emitted in the process is equal to the difference in energy between the two states (Equation 7.5) (Section 7.3).

- Understand that in the modern view of the atom, electrons are described by the physics of waves (Section 7.4). The wavelength of an electron or any subatomic particle is given by de Broglie's equation (Equation 7.6).

- Recognize the significance of quantum mechanics in describing the modern view of atomic structure (Section 7.5).

- Understand that an orbital for an electron in an atom corresponds to the allowed energy of that electron.

- Understand that the position of the electron is not known with certainty; only the probability of the electron being at a given point of space can be calculated. This is the interpretation of the quantum mechanical model and embodies the postulate called the uncertainty principle.

- Describe the allowed energy states of the electron in an atom using three quantum numbers n, ℓ, and m_ℓ (Section 7.5).

- Describe the shapes of the orbitals (Section 7.6).

Key Terms

Section 7.1
electromagnetic radiation
wavelength
frequency
hertz
amplitude
node
standing wave

Section 7.2
quantized
Planck's constant
photoelectric effect
photons

Section 7.3
continuous spectrum
line emission spectrum
Rydberg equation
Rydberg constant
Balmer series
principal quantum number
ground state
excited state
Lyman series

Section 7.5
quantum mechanics
wave mechanics

uncertainty principle
wave functions
electron density
orbital
quantum number (n, ℓ, m_ℓ)
electron shell
subshell

Section 7.6
electron cloud pictures
probability density
nodal surface

Key Equations

Equation 7.1 (page 254)
This equation states that the product of the wavelength (λ) and frequency (ν) of electromagnetic radiation is equal to the speed of light (c, 2.998×10^8 m $\cdot$ s^{-1}).

$$c = \lambda\nu$$

Equation 7.2 (page **258**)
Planck's equation states that the energy (E) of a photon, a massless particle of radiation, is proportional to its frequency (ν).

$$E = h\nu$$

where h is Planck's constant (6.626×10^{-34} J $\cdot$ s).

Equation 7.4 (page 264)
In a hydrogen atom, the potential energy of the electron in the nth quantum level, E_n, is proportional to $1/n^2$.

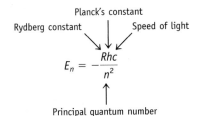

where n is an integer equal to or greater than 1 and $Rhc = 2.179 \times 10^{-18}$ J/atom or 1312 kJ/mol.

Equation 7.5 (page 267)
The change in energy for an electron moving between two quantum levels (n_{final} and n_{initial}) in the H atom. (Note that Equation 7.3 is derived from Equation 7.5 if n_{final} is 2.)

$$\Delta E = E_{\text{final}} - E_{\text{initial}} = -Rhc\left(\frac{1}{n^2_{\text{final}}} - \frac{1}{n^2_{\text{initial}}}\right)$$

Equation 7.6 (page 269)
De Broglie's equation relating the wavelength of the electron (λ) to its mass (m) and speed (v). h is Planck's constant.

$$\lambda = \frac{h}{mv}$$

Study Questions

Questions with blue, bold-faced numbers have answers in Appendix O. See Example Problems, Tutorials, and Simulations in Chapter 7 on the General Chemistry Interactive CD-ROM, *Version 3.0.*

Reviewing Important Concepts

1. Our modern view of atomic structure was developed through many experiments. Name at least three of these experiments, their outcomes, and the person most associated with that experiment. (You may have to review Chapter 2.)

2. Give the equation for each of the following important mathematical relations in this chapter:
   (a) The relationship among wavelength, frequency, and speed of radiation
   (b) The relation between energy and frequency of radiation
   (c) The energy of an electron in a given energy state of the H atom

3. State Planck's equation in words and as a mathematical equation.

4. Name the colors of visible light beginning with that of highest energy.

5. Draw a picture of a standing wave, and use this to define the terms wavelength, amplitude, and node.

6. Which of the following are applicable when explaining the photoelectric effect? Correct any statements that are wrong.
   (a) Light is electromagnetic radiation.
   (b) The intensity of a light beam is related to its frequency.
   (c) Light can be thought of as consisting of massless particles whose energy is given by Planck's equation, $E = h\nu$.

7. What is a photon? Explain how the photoelectric effect implies the existence of photons.

8. What are two major assumptions of Bohr's theory of atomic structure?

9. In what region of the electromagnetic spectrum is the Lyman series of lines found? The Balmer series?

10. Light is given off by a sodium- or mercury-containing streetlight when the atoms are excited in some way. The light you see arises for which of the following reasons:
    (a) Electrons moving from a given energy level to one of higher n
    (b) Electrons being removed from the atom, thereby creating a metal cation
    (c) Electrons moving from a given level to one of lower n

11. What is incorrect about the Bohr model of the atom?

12. What is Heisenberg's uncertainty principle? Explain how it applies to our modern view of atomic structure.

13. How do we interpret the physical meaning of the square of the wave function? What are the units of ψ^2?

14. What are the three quantum numbers used to describe an orbital? What property of an orbital is described by each quantum number? Specify the rules that govern the values of each quantum number.

15. Give the number of nodal surfaces for each orbital type: s, p, d, and f.

16. What is the maximum number of s orbitals found in a given electron shell? The maximum number of p orbitals? Of d orbitals? Of f orbitals?

17. Match the values of ℓ shown in the table with orbital type (s, p, d, or f).

ℓ Value	Orbital Type
3	————
0	————
1	————
2	————

18. Sketch a picture of the 90% boundary surface of an s orbital and the p_x orbital. Be sure the latter drawing shows why the p orbital is labeled p_x and not p_y, for example.

19. Complete the following table.

Orbital Type	Number of Orbitals in a Given Subshell	Number of Nodal Surfaces
s	————	————
p	————	————
d	————	————
f	————	————

Practicing Skills

Electromagnetic Radiation

(See Example 7.1, Exercise 7.1, Figure 7.3 and CD-ROM Screen 7.3)

20. Answer the following questions based on Figure 7.3:
    (a) Which type of radiation involves less energy, x-rays or microwaves?
    (b) Which radiation has the higher frequency, radar or red light?
    (c) Which radiation has the longer wavelength, ultraviolet or infrared light?

21. Consider the colors of the visible spectrum.
    (a) What colors of light involve less energy than green light?
    (b) Which color of light has photons of greater energy, yellow or blue?
    (c) Which color of light has the greater frequency, blue or green?

22. Traffic signals are often now made of LEDs (light-emitting diodes). Amber and green ones are pictured here.

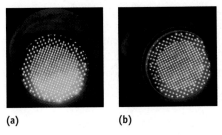

(a)   **(b)**

(Mike Condren/UW/MRSEC)

(a) The light from an amber signal has a wavelength of 595 nm, and that from a green signal is 500 nm. Which has the higher frequency?

(b) Calculate the frequency of amber light.

23. Suppose you are standing 225 m from a radio transmitter. What is your distance from the transmitter in terms of the number of wavelengths if

    (a) The station is broadcasting at 1150 kHz (on the AM radio band) (1 kHZ = 1×10^3 Hz, or 1000 cycles per second.)

    (b) The station is broadcasting at 98.1 MHz (on the FM radio band) (1 MHz = 10^6 Hz, or 1 million cycles per second)

Electromagnetic Spectrum and Planck's Equation

(See page 260, Exercise 7.3 and CD-ROM Screens 7.4 and 7.5)

24. Green light has a wavelength of 5.0×10^2 nm. (See Study Question 22.) What is the energy in joules of one photon of green light? What is the energy in joules of 1.0 mol of photons of green light?

25. Violet light has a wavelength of about 410 nm. What is its frequency? Calculate the energy of one photon of violet light. What is the energy of 1.0 mol of violet photons? Compare the energy of photons of violet light with those of red light. Which is more energetic and by what factor?

26. The most prominent line in the line spectrum of aluminum is found at 396.15 nm. What is the frequency of this line? What is the energy of one photon with this wavelength? Of 1.00 mol of these photons?

27. The most prominent line in the spectrum of magnesium is 285.2 nm. Others are found at 383.8 and 518.4 nm. In what region of the electromagnetic spectrum are these lines found? Which is the most energetic line? What is the energy of 1 mol of photons of the most energetic line?

28. Place the following types of radiation in order of increasing energy per photon:

    (a) Yellow light from a sodium lamp

    (b) X-rays from an instrument in a dentist's office

    (c) Microwaves in a microwave oven

    (d) Your favorite FM music station at 91.7 MHz

29. Place the following types of radiation in order of increasing energy per photon:

    (a) Radar signals

    (b) Radiation within a microwave oven

    (c) Gamma rays from a nuclear reaction

    (d) Red light from a neon sign

    (e) Ultraviolet radiation from a sun lamp

Photoelectric Effect

(See page 259 and Exercise 7.3)

30. An energy of 2.0×10^2 kJ/mol is required to cause a cesium atom on a metal surface to lose an electron. Calculate the longest possible wavelength of light that can ionize a cesium atom. What is the region of the electromagnetic spectrum in which this radiation is found?

31. You are an engineer designing a switch that works by the photoelectric effect. The metal you wish to use in your device requires 6.7×10^{-19} J/atom to remove an electron. Will the switch work if the light falling on the metal has a wavelength of 540 nm or greater? Why or why not?

Atomic Spectra and the Bohr atom

(See Examples 7.2 and 7.3, Figures 7.9 and 7.12, and CD-ROM Screens 7.6 and 7.7)

32. The most prominent line in the spectrum of mercury is found at 253.652 nm. Other lines are located at 365.015 nm, 404.656 nm, 435.833 nm, and 1013.975 nm.

    (a) Which of these lines represents the most energetic light?

    (b) What is the frequency of the most prominent line? What is the energy of one photon with this wavelength?

    (c) Are any of these lines found in the spectrum of mercury shown in Figure 7.9? What color or colors are these lines?

33. The most prominent line in the spectrum of neon is found at 865.438 nm. Other lines are located at 837.761 nm, 878.062 nm, 878.375 nm, and 1885.387 nm.

    (a) In what region of the electromagnetic spectrum are these lines found?

    (b) Are any of these lines found in the spectrum of neon shown in Figure 7.9? What color or colors are these lines?

    (c) Which of these lines represents the most energetic light?

    (d) What is the frequency of the most prominent line? What is the energy of one photon with this wavelength?

34. A line in the Balmer series of emission lines of excited H atoms has a wavelength of 410.2 nm (see Figure 7.12). What color is the light emitted in this transition? What quantum levels are involved in this emission line? What are the values of $n_{initial}$ and n_{final}?

35. What are the wavelength and frequency of the radiation involved in the least energetic emission line in the Lyman spectrum? What quantum levels are involved in this emission line? What are the values of $n_{initial}$ and n_{final}?

36. The energy emitted when an electron moves from a higher energy state to a lower one in any atom can be observed as electromagnetic radiation.

    (a) Which involves the emission of less energy in the H atom, an electron moving from $n = 4$ to $n = 2$ or an electron moving from $n = 3$ to $n = 2$?

(b) Which involves the emission of the greater energy in the H atom, an electron changing from $n = 4$ to $n = 1$ or an electron changing from $n = 5$ to $n = 2$? Explain fully.

37. If energy is absorbed by a hydrogen atom in its ground state, the atom is excited to a higher energy state. For example, the excitation of an electron from the level with $n = 1$ to the level with $n = 3$ requires radiation with a wavelength of 102.6 nm. Which of the following transitions would require radiation of *longer wavelength* than this?

    (a) $n_1 = 2$ to $n_2 = 4$ (c) $n = 1$ to $n = 5$
    (b) $n = 1$ to $n = 4$ (d) $n = 3$ to $n = 5$

38. Consider only transitions involving the $n = 1$ through $n = 4$ energy levels for the hydrogen atom (where the line spacing is very approximate).

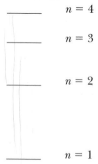

    _____ $n = 4$

    _____ $n = 3$

    _____ $n = 2$

    _____ $n = 1$

    (a) How many emission lines are possible, considering only the four quantum levels?

    (b) Photons of the lowest energy are emitted in a transition from the level with $n =$ _____ to a level with $n =$ _____.

    (c) The emission line having the shortest wavelength corresponds to a transition from the level with $n =$ _____ to the level with $n =$ _____.

39. Consider only transitions involving the $n = 1$ through $n = 5$ energy levels for the H atom (where the line spacing is very approximate).

    _____ $n = 5$
    _____ $n = 4$

    _____ $n = 3$

    _____ $n = 2$

    _____ $n = 1$

    (a) How many emission lines are possible, considering only the five quantum levels?

    (b) Photons of the highest frequency are emitted in a transition from the level with $n =$ _____ to a level with $n =$ _____.

    (c) The emission line having the longest wavelength corresponds to a transition from the level with $n =$ _____ to the level with $n =$ _____.

40. Calculate the wavelength and frequency of light emitted when an electron changes from $n = 3$ to $n = 1$ in the H atom. In what region of the spectrum is this radiation found?

41. Calculate the wavelength and frequency of light emitted when an electron changes from $n = 4$ to $n = 3$ in the H atom. In what region of the spectrum is this radiation found?

De Broglie and Matter Waves
(See Example 7.4 and CD-ROM Screen 7.8)

42. An electron moves with a velocity of 2.5×10^8 cm/s. What is its wavelength?

43. A beam of electrons ($m = 9.11 \times 10^{-31}$ kg/electron) has an average speed of 1.3×10^8 m $\cdot$ s^{-1}. What is the wavelength of electrons having this average speed?

44. Calculate the wavelength (in nanometers) associated with a 1.0×10^2-g golf ball moving at 30. m $\cdot$ s^{-1} (about 67 mph). How fast must the ball travel to have a wavelength of 5.6×10^{-3} nm?

45. A rifle bullet (mass = 1.50 g) has a velocity of 7.00×10^2 mph. What is the wavelength associated with this bullet?

Quantum Mechanics
(See Sections 7.5 and 7.6 and CD-ROM Screens 7.9–7.14)

46. Quantum numbers:

    (a) When $n = 4$, what are the possible values of ℓ?
    (b) When ℓ is 2, what are the possible values of m_ℓ?
    (c) For a 4s orbital, what are the possible values of n, ℓ, and m_ℓ?
    (d) For a 4f orbital, what are the possible values of n, ℓ, and m_ℓ?

47. Quantum numbers:

    (a) When $n = 4$, $\ell = 2$, and $m_\ell = -1$, to what orbital type does this refer? (Give the orbital label, such as 1s.)
    (b) How many orbitals occur in the $n = 5$ electron shell? How many subshells? What are the letter labels of the subshells?
    (c) How many orbitals occur in an f subshell? What are the values of m_ℓ?

48. A possible excited state of the H atom has the electron in a 4p orbital. List all possible sets of quantum numbers n, ℓ, and m_ℓ for this electron.

49. A possible excited state for the H atom has an electron in a 5d orbital. List all possible sets of quantum numbers n, ℓ, and m_ℓ for this electron.

50. How many subshells occur in the electron shell with the principal quantum number $n = 4$?

51. How many subshells occur in the electron shell with the principal quantum number $n = 5$?

52. Explain briefly why each of the following is not a possible set of quantum numbers for an electron in an atom.

    (a) $n = 2$, $\ell = 2$, $m_\ell = 0$
    (b) $n = 3$, $\ell = 0$, $m_\ell = -2$
    (c) $n = 6$, $\ell = 0$, $m_\ell = 1$

53. Which of the following represent valid sets of quantum numbers? For a set that is invalid, explain briefly why it is not correct.
    (a) $n = 3, \ell = 3, m_\ell = 0$
    (b) $n = 2, \ell = 1, m_\ell = 0$
    (c) $n = 6, \ell = 5, m_\ell = -1$
    (d) $n = 4, \ell = 3, m_\ell = -4$

54. What is the maximum number of orbitals that can be identified by each of the following sets of quantum numbers? When "none" is the correct answer, explain your reasoning.
    (a) $n = 3, \ell = 0, m_\ell = +1$
    (b) $n = 5, \ell = 1$
    (c) $n = 7, \ell = 5$
    (d) $n = 4, \ell = 2, m_\ell = -2$

55. What is the maximum number of orbitals that can be identified by each of the following sets of quantum numbers? When "none" is the correct answer, explain your reasoning.
    (a) $n = 4, \ell = 3$
    (b) $n = 2, \ell = 2$
    (c) $n = 5$
    (d) $n = 3, \ell = 1, m_\ell = -1$

56. State which of the following orbitals cannot exist according to the quantum theory: $2s$, $2d$, $3p$, $3f$, $4f$, and $5s$. Briefly explain your answers.

57. State which of the following are incorrect designations for orbitals according to the quantum theory and briefly explain your choices: $3p$, $4s$, $2f$, and $1p$.

58. Write a complete set of quantum numbers (n, ℓ, and m_ℓ) that quantum theory allows for each of the following orbitals: (a) $2p$, (b) $3d$, and (c) $4f$.

59. Write a complete set of quantum numbers (n, ℓ, and m_ℓ) for each of the following orbitals: (a) $5f$, (b) $4d$, and (c) $2s$.

60. A particular orbital has $n = 4$ and $\ell = 2$. This orbital must be: (a) $3p$, (b) $4p$, (c) $5d$, or (d) $4d$.

61. A given orbital has a magnetic quantum number of $m_\ell = -1$. This could not be a (an)
    (a) f orbital
    (b) d orbital
    (c) p orbital
    (d) s orbital

62. How many nodal surfaces are associated with each of the following orbitals: (a) $2s$, (b) $5d$, and (c) $5f$?

63. How many nodal surfaces are associated with each of the following atomic orbitals: (a) $4f$, (b) $2p$, and (c) $6s$?

General Questions on Atomic Structure

These questions are not designated as to type or location in the chapter. They may combine several concepts. More challenging questions are indicated by an underlined number.

64. An electron moves from the $n = 5$ to the $n = 1$ quantum level and emits a photon with an energy of 2.093×10^{-18} J. How much energy must the atom absorb to move an electron from $n = 1$ to $n = 5$?

65. Excited H atoms have many emission lines. One series of lines, called the Pfund series, occurs in the infrared region. It results when an electron changes from higher levels to a level with $n = 5$. Calculate the wavelength and frequency of the lowest energy line of this series.

66. An advertising sign gives off red light and green light.
    (a) Which light has the higher energy photons?
    (b) One of the colors has a wavelength of 680 nm and the other has a wavelength of 500 nm. Identify which color has which wavelength.
    (c) Which light has the higher frequency?

67. Radiation in the ultraviolet region of the electromagnetic spectrum is quite energetic. It is this radiation that causes dyes to fade and your skin to burn. If you are bombarded with 1.00 mol of photons with a wavelength of 375 nm, what amount of energy (in kilojoules per mole of photons) are you being subjected to?

68. A cell phone sends signals at about 850 MHz (1 MHz $= 1 \times 10^6$ Hz, or cycles per second).
    (a) What is the wavelength of this radiation?
    (b) What is the energy of 1.0 mol of photons with a frequency of 850 MHz?

    (c) Compare the energy in part (b) with the energy of a mole of photons of blue light (420 nm).
    (d) Comment on the difference in energy between that of 850 MHz radiation and blue light.

69. Assume your eyes receive a signal consisting of blue light, $\lambda = 470$ nm. The energy of the signal is 2.50×10^{-14} J. How many photons reach your eyes?

70. If sufficient energy is absorbed by an atom, an electron can be lost by the atom and a positive ion formed. The amount of energy required is called the ionization energy. In the H atom, the ionization energy is that required to change the electron from $n = 1$ to $n =$ infinity. (See "A Closer Look: Experimental Evidence for Bohr's Theory", page 268.) Calculate the ionization energy for He$^+$ ion. Is the ionization energy of He$^+$ more or less than that of H? (Bohr's theory applies to He$^+$ because it, like the H atom, has a single electron. The electron energy, however is now given by $E = -Z^2 Rhc / n^2$, where Z is the atomic number of helium.)

71. What is the shortest wavelength photon an excited H atom can emit? Explain briefly.

72. Hydrogen atoms absorb energy so that electrons are excited to the $n = 7$ energy level. Electrons then undergo these transitions, among others: (a) $n = 7$ to $n = 1$; (b) $n = 7$ to $n = 6$; and (c) $n = 2$ to $n = 1$. Which transition produces a photon with (i) the smallest energy; (ii) the highest frequency; (iii) the shortest wavelength?

73. Rank the following orbitals in the H atom in order of increasing energy: 3s, 2s, 2p, 4s, 3p, 1s, and 3d.

74. How many orbitals correspond to each of the following designations?

    (a) 3p   (d) 6d   (g) $n = 5$

    (b) 4p   (e) 5d   (h) 7s

    (c) $4p_x$ (f) 5f

75. Bohr pictured the electrons of the atom as being located in definite orbits about the nucleus, just as the planets orbit the sun. Criticize this model.

76. In what way does Bohr's model of the atom violate the uncertainty principle?

77. What does the wave–particle duality mean? What are its implications in our modern view of atomic structure?

78. Which of these are observable with accuracy?

    (a) Position of electron in H atom

    (b) Frequency of radiation emitted by H atoms

    (c) Path of electron in H atom

    (d) Wave motion of electrons

    (e) Diffraction patterns produced by electrons

    (f) Diffraction patterns produced by light

    (g) Energy required to remove electrons from H atoms

    (h) An atom

    (i) A molecule

    (j) A water wave

79. In principle, which of the following can be determined?

    (a) The energy of an electron in the H atom with high precision and accuracy

    (b) The position of a high-speed electron with high precision and accuracy

    (c) At the same time, both the position and energy of a high-speed electron with high precision and accuracy.

80. Suppose you live in a different universe where a different set of quantum numbers is required to describe the atoms of that universe. These quantum numbers have the following rules:

    N, principal   1, 2, 3, . . . , ∞

    L, orbital $= N$

    M, magnetic $-1, 0, +1$

    How many orbitals are there altogether in the first three electron shells?

81. Assume an electron is assigned to the 1s orbital in the H atom. Is the electron density zero at a distance of 0.40 nm from the nucleus? (See *A Closer Look: Electron Density Plots,* page 277)

82. Cobalt-60 is a radioactive isotope used in medicine for the treatment of certain cancers. It produces β particles and γ rays, the latter having energies of 1.173 and 1.332 MeV. (1 MeV = 1 million electron volts and 1 eV ≡ 9.6485×10^4 J/mol.) What are the wavelength and frequency of a γ-ray photon with an energy of 1.173 MeV?

83. Exposure to high doses of microwaves can cause damage. Estimate how many photons, with $\lambda = 12$ cm, must be absorbed to raise the temperature of your eye by 3.0 °C. Assume the mass of an eye is 11 g and its specific heat capacity is 4.0 J/g·K.

84. When the *Sojourner* spacecraft landed on Mars in 1997, the planet was approximately 7.8×10^7 km from the earth. How long did it take for the television picture signal to reach earth from Mars?

85. The most prominent line in the emission spectrum of chromium is found at 425.4 nm. Other lines in the chromium spectrum are found at 357.9 nm, 359.3 nm, 360.5 nm, 427.5 nm, 429.0 nm, and 520.8 nm.

    (a) Which of these lines represents the most energetic light?

    (b) What color is light of 425.4 nm, the most prominent line in the spectrum?

86. Answer the following questions as a summary quiz on the chapter.

    (a) The quantum number n describes the _____ of an atomic orbital.

    (b) The shape of an atomic orbital is given by the quantum number _____.

    (c) A photon of green light has _____ (less or more) energy than a photon of orange light.

    (d) The maximum number of orbitals that may be associated with the set of quantum numbers $n = 4$ and $\ell = 3$ is _____.

    (e) The maximum number of orbitals that may be associated with the quantum number set $n = 3$, $\ell = 2$, and $m_\ell = -2$ is _____.

    (f) Label each of the following orbital pictures with the appropriate letter.

    (g) When $n = 5$, the possible values of ℓ are _____.

    (h) The number of orbitals in the $n = 4$ shell is _____.

87. Answer the following questions as a review of this chapter:

    (a) The quantum number n describes the _____ of an atomic orbital and the quantum number ℓ describes its _____.

    (b) When $n = 3$, the possible values of ℓ are _____.

    (c) What type of orbital corresponds to $\ell = 3$? _____

    (d) For a 4d orbital, the value of n is _____, the value of ℓ is _____, and a possible value of m_ℓ is _____.

    (e) Each drawing represents a type of atomic orbital. Give the letter designation for the orbital, its value of ℓ, and specify the number of nodal surfaces.

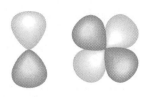

letter   = _____   _____

ℓ value   = _____   _____

nodal surfaces = _____   _____

(f) An atomic orbital with three nodal surfaces is _____.

(g) Which of the following orbitals *cannot* exist according to modern quantum theory: $2s$, $3p$, $2d$, $3f$, $5p$, $6p$?

(h) Which of the following is *not* a valid set of quantum numbers?

n	ℓ	m_ℓ
3	2	1
2	1	2
4	3	0

(i) What is the maximum number of orbitals that can be associated with each of the following sets of quantum numbers? (One possible answer is "none.")

   (i) $n = 2$ and $\ell = 1$

   (ii) $n = 3$

   (iii) $n = 3$ and $\ell = 3$

   (iv) $n = 2$, $\ell = 1$, and $m_\ell = 0$

88. Technetium is not found naturally on earth; it must be synthesized in the laboratory. Nonetheless, because it is radioactive it has valuable medical uses. For example, the element in the form of sodium pertechnetate ($NaTcO_4$) is used in imaging studies of the brain, thyroid, and salivary glands and in renal blood flow studies, among other things.

(a) In what group and period of the periodic table is the element found?

(b) The valence electrons of technetium are found in the $5s$ and $4d$ subshells. What is a set of quantum numbers (n, ℓ, and m_ℓ) for one of the electrons of the $5s$ subshell?

(c) Technetium emits a γ ray with an energy of 0.141 MeV. (1 MeV = 1 million electron volts, where 1 eV/particle $\equiv$ 9.6485×10^4 J/mol.) What are the wavelength and frequency of a γ-ray photon with an energy of 0.141 MeV?

(d) To make $NaTcO_4$, the metal is dissolved in nitric acid

$$7\,HNO_3(aq) + Tc(s) \longrightarrow$$
$$HTcO_4(aq) + 7\,NO_2(g) + 3\,H_2O(\ell)$$

and the product, $HTcO_4$, is treated with NaOH to make $NaTcO_4$.

   (i) Write a balanced equation for the reaction of $HTcO_4$ with NaOH.

   (ii) If you begin with 4.5 mg of Tc metal, how much $NaTcO_4$ can be made? What mass of NaOH (in grams) is required to convert all of the $HTcO_4$ into $NaTcO_4$?

Using Electronic Resources

These questions refer to the General Chemistry Interactive CD-ROM, *Version 3.0.*

89. See CD-ROM Screen 7.1: Chemical Puzzler. This screen shows that light of different colors can come from a "neon" sign or from certain salts when placed in a burning organic liquid. ("Neon" signs are glass tubes filled with neon, argon, and other gases, and the gases are excited by an electric current. They are very similar in this regard to common fluorescent lights, although the color of the latter comes from the phosphor that coats the inside of the tube.) What do these two sources of colored light have in common? How is the light generated in each case?

90. See CD-ROM Screen 7.7: Bohr's Model of the Hydrogen Atom. A large pickle is attached to two electrodes and these are attached to a 110-V power supply. As the voltage is increased across the pickle, it begins to glow with a yellow color (see Step Problem). Knowing that pickles are made by soaking the vegetable in a concentrated salt solution, describe why the pickle might emit light when electrical energy is added.

The "electric pickle."
(Charles D. Winters)

8 Atomic Electron Configurations and Chemical Periodicity

Chapter Goals

- Understand the role magnetism plays in determining and revealing atomic structure.
- Understand effective nuclear charge and its role in determining atomic properties.
- Write the electron configuration for any element or monatomic ion.
- Understand the fundamental physical properties of the elements and their periodic trends.

Everything in Its Place

The periodic table of elements has "put everything in its place" according to Oliver Sacks. Sacks is well known in medicine as a neurologist, but he is also a writer of books such as *The Man Who Mistook His Wife for a Hat* and *Awakenings*. Less well known is that he has had a love affair with chemistry since he was a boy growing up in London during World

▲ Oliver Sacks was born in London in 1933, the son of two physicians. He now lives in New York City where he is a practicing neurologist. He is a member of the American Academy of Arts and Letters and is the author of several books, including *The Man Who Mistook His Wife for a Hat*, and *Awakenings*. His most recent book, *Uncle Tungsten* (Alfred Knopf, New York, 2001), describes his life-long fascination with chemistry. (*Courtesy Chemical and Engineering News*)

War II. On a trip to the London Science Museum, he saw a wall-sized periodic table that displayed samples of many of the 92 chemical elements known at that time. Sacks said that "Seeing the table, with its actual samples of the elements, was one of the formative experiences of my boyhood and showed me, with the force of revelation, the beauty of science. The periodic table seemed so economical and simple: everything, the whole 92-ishness, reduced to two axes, and yet along each axis an ordered progression of different properties."

Dmitri Mendeleev, one of two people responsible for the periodic table, was born in Tobolsk in western Siberia on February 8, 1834. He was the youngest of 14 or 17 children, the number is not certain. His father was incapacitated shortly after Dmitri's birth, so, to support the large family, his mother took over a glass manufacturing business begun by her father.

Catastrophe struck the family in 1848 and 1849, however, when Mendeleev's father died and the glass factory burned. Young Mendeleev's mother was determined to ensure that he be schooled properly, so they journeyed 1300 miles to Moscow so he could enroll in the university. Once in Moscow they found students from Siberia were not permitted at the university, so they went another 400 miles to St. Petersburg. There his mother was able to secure a place for him at the Central Pedagogical Institute. She died shortly thereafter.

Mendeleev was an extraordinary student of science and published original work before he was 20, even though he

> "**M**y kitchen is papered with periodic tables of every size and sort . . . and on the kitchen table, a very favorite periodic table made of wood that I can spin like a prayer wheel."
>
> (Oliver Sacks: *New York Times Magazine*, April 18, 1999, pp. 126–130.)

was afflicted with tuberculosis and had to do much of his writing in bed. He took the gold medal as the top student at the Institute in 1855 and shortly thereafter was sent to the Crimea as a teacher. The climate in the Crimea was hospitable, much suited to recovering from his illness. However, one reason his school in St. Petersburg sent him there, far from St. Petersburg, was because he had a terrible temper and was less than beloved by his former teachers and colleagues.

Within a few years Mendeleev returned to St. Petersburg as a lecturer at the University, but soon thereafter went to study and do research in France and Germany. In Germany he worked briefly with Robert Bunsen, the inventor of a burner used for spectroscopic studies. There also Mendeleev's temper got the better of him, and he was forced to retreat to a small room where he worked in isolation.

A defining moment for Mendeleev came in 1860 at a conference in Karlsruhe, Germany, where chemists from all over Europe came to settle on a system for determining atomic weights. This system, once in place, was crucial to Mendeleev's discovery a few years later of the periodic law and his publication of the first periodic table.

In 1861, Mendeleev returned to St. Petersburg and joined the faculty of the Technical Institute. His love of chemistry, as well as his intense blue eyes and flowing beard and hair, made him a popular teacher. He also realized that the teaching of chemistry in Russia was in a sorry state, so, in only 60 days, he wrote a 500-page textbook of organic chemistry!

▲ Dmitri Mendeleev seated at his desk. Every picture of him shows his long hair and beard. He was in the habit of having it cut only once a year. For more on the story of Mendeleev and the periodic table see *Mendeleyev's Dream* by P. Strathern, St. Martin's Press, New York, 2001. *(Novosti/Science Photo Library/ Photo Researchers, Inc.)*

At the age of 32 Mendeleev was appointed professor of general chemistry at the University of St. Petersburg. By 1869 he completed the first volume of a new textbook, *The Principles of Chemistry*, a book that was translated into all the major languages of the world. As he began the second volume, he was searching for an organizing principle. To look for patterns in chemical and physical behavior of the elements, he wrote lists of those properties on small cards, one for each element. After four days of pondering the problem for hours on end, he was so exhausted he fell asleep at his desk. In his words "I saw in a dream a table where all the elements fell into place as required. Awakening, I immediately wrote it down on a piece of paper." This was the beginning of the periodic table chemists use today.

Periodic Trends Dmitri Mendeleev developed the first periodic table by observing that properties of elements were repeated every so often as the atomic weight increased. Elements with similar properties were arranged in vertical columns or groups in a periodic table.

Examples of Periodicity with Group 1A and Group 7A Elements

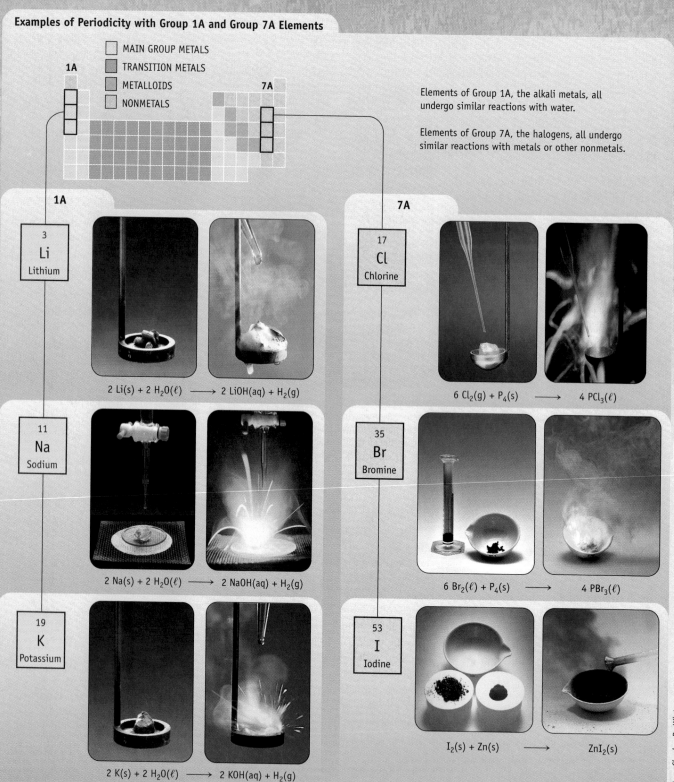

- ☐ MAIN GROUP METALS
- ☐ TRANSITION METALS
- ☐ METALLOIDS
- ☐ NONMETALS

1A **7A**

Elements of Group 1A, the alkali metals, all undergo similar reactions with water.

Elements of Group 7A, the halogens, all undergo similar reactions with metals or other nonmetals.

1A

3 Li Lithium

$2\ Li(s) + 2\ H_2O(\ell) \longrightarrow 2\ LiOH(aq) + H_2(g)$

11 Na Sodium

$2\ Na(s) + 2\ H_2O(\ell) \longrightarrow 2\ NaOH(aq) + H_2(g)$

19 K Potassium

$2\ K(s) + 2\ H_2O(\ell) \longrightarrow 2\ KOH(aq) + H_2(g)$

7A

17 Cl Chlorine

$6\ Cl_2(g) + P_4(s) \longrightarrow 4\ PCl_3(\ell)$

35 Br Bromine

$6\ Br_2(\ell) + P_4(s) \longrightarrow 4\ PBr_3(\ell)$

53 I Iodine

$I_2(s) + Zn(s) \longrightarrow ZnI_2(s)$

Photos: Charles D. Winters

The wave mechanical model of the atom accurately describes atoms or ions that have a single electron, such as H and He$^+$. It is obvious, however, that a truly useful model must be applicable to atoms with more than one electron, that is, to all the other known elements. Because the chemical properties of atoms depend on their electronic structure — the number and arrangement of electrons in the atom — one goal of this chapter is to describe the modern model for the electronic structure of elements other than hydrogen.

Another goal of this chapter is to explore some of the physical properties of elements, among them the ease with which atoms lose or gain electrons to form ions and the sizes of atoms and ions. These properties are directly related to the arrangement of electrons in atoms and thus to the chemistry of the elements and their compounds.

8.1 ELECTRON SPIN

Around 1920 it was demonstrated experimentally that the electron behaves as though it has a spin, just as the earth has a spin. To understand this property and its relation to atomic structure requires understanding some aspects of the general phenomenon of magnetism. You will see that electron spin must be represented by a fourth quantum number, the **electron spin magnetic quantum number, m_s** [CD-ROM, Screen 8.2]. That is, the complete description of an electron in an atom requires *four* quantum numbers: n, ℓ, m_ℓ, and m_s.

Magnetism

In 1600, William Gilbert (1544–1603) concluded that the earth is a large spherical magnet giving rise to a magnetic field that surrounds the planet (Figure 8.1). The needle of a compass, itself a small magnet, lines up with earth's magnetic field, one end of the needle pointing approximately to the earth's geographic North Pole. We say the end of the compass needle pointing north is the magnet's "magnetic north pole" or simply its "north pole" (N). The other end of the needle is its

Chapter Goals • Revisited

- **Understand the role magnetism plays in determining and revealing atomic structure.**

- Understand effective nuclear charge and its role in determining atomic properties.

- Write the electron configuration for any element or monatomic ion.

- Understand the fundamental physical properties of the elements and their periodic trends.

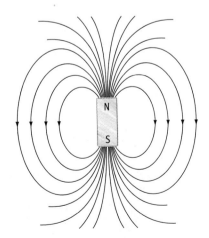

Figure 8.1 The magnetic fields of the earth and of a bar magnet. The lines of magnetic force around the earth come from one pole, arbitrarily called the "north magnetic pole" (N) and loop toward the "south magnetic pole" (S). (The geographic North Pole of the earth, named before the introduction of the term "magnetic pole," is actually the magnetic south pole.)

Figure 8.2   Observing and measuring paramagnetism. (a) A magnetic balance is used to measure the magnetic properties of a sample. The sample is first weighed with the electromagnet turned off. The magnet is then turned on and the sample reweighed. If the substance is paramagnetic, the sample is drawn into the magnetic field and the apparent weight increases. **(b)** Liquid oxygen (boiling point 90.2 K) clings to the poles of a strong magnet. Elemental oxygen is paramagnetic because it has unpaired electrons. (See Chapter 10 and Figure 10.20.) *(Charles D. Winters)*

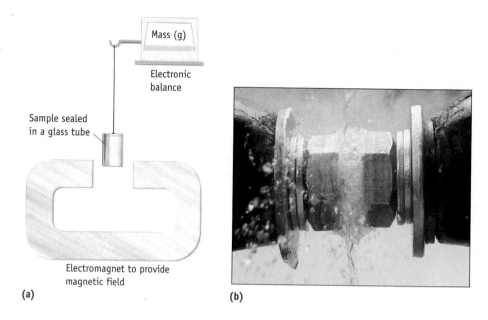

(a)   (b)

"south pole" (S). Because opposite poles (N–S) attract, this means that earth's geographic North Pole is its magnetic south pole.

Paramagnetism and Unpaired Electrons

Most substances are slightly repelled by a strong magnet. They are said to be **diamagnetic.** In contrast, some metals and compounds are attracted to a magnetic field. Such substances are generally called **paramagnetic,** and the magnitude of the effect can be determined with an apparatus such as that illustrated in Figure 8.2a.

The magnetism of most paramagnetic materials is so weak that you can only observe the effect in the presence of a strong magnetic field. For example, the oxygen we breathe is paramagnetic; it sticks to the poles of a strong magnet (Figure 8.2b) [CD-ROM, Screen 8.3].

Paramagnetism arises from electron spins. An electron in an atom has the magnetic properties expected for a spinning, charged particle (Figure 8.3). What is important here is the relation of that property to the arrangement of electrons in atoms. Experiments have shown that, if an atom with a single unpaired electron is placed in a magnetic field, only two orientations are possible for the electron spin: aligned with the field or opposed to the field. One orientation is associated with an electron spin quantum number value of $m_s = +\frac{1}{2}$ and the other with an m_s value of $-\frac{1}{2}$. *Electron spin is quantized.*

When one electron is assigned to an orbital in an atom, the electron's spin orientation can take either value of m_s. We observe experimentally that hydrogen atoms, each of which has a single electron, are paramagnetic; when an external magnetic field is applied, the electron magnets align with the field — like the needle of a compass — and experiences an attractive force. In helium, two electrons are assigned to the same 1s orbital, and we can confirm by experiment that helium is diamagnetic. To account for this observation, we assume that *the two electrons assigned to the same orbital have opposite spin orientations.* We say their *spins are paired,* which means that the magnetic field of one electron is canceled out by the magnetic field of the second of opposite spin.

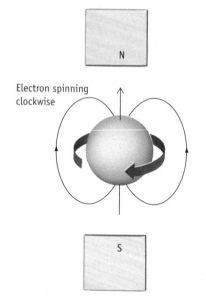

Electron spinning clockwise

Figure 8.3   Electron spin and magnetism. The electron, with its spin and negative electric charge, acts as a "micromagnet." Relative to a magnetic field only two spin directions are possible: clockwise or counterclockwise. The north pole of the spinning electron can therefore be either aligned with an external magnetic field or opposed to that field.

A Closer Look

Paramagnetism and Ferromagnetism

Magnetic materials are relatively common and many are important in our economy. Magnets are also found in stereo speakers and in telephone handsets, and magnetic oxides are used in recording tapes and computer disks. Other examples are the magnets in the little objects you stick on refrigerator doors.

Many common consumer products contain magnets. The stripped, round cylinder in the foreground is a cow magnet. It is "fed" to a cow and in the cow's stomach it attracts magnetic debris that the cow may eat and prevents the debris from traveling any further in the cow's digestive system and possibly injuring the animal. *(Charles D. Winters)*

The magnetic materials we use are **ferromagnetic**. The magnetic effect of fer-romagnetic materials is much larger than for paramagnetic ones. Ferromagnetism occurs when the spins of unpaired electrons in a cluster of atoms (called a *domain*) in the solid align themselves in the same direction. Only the metals of the iron, cobalt, and nickel subgroups, as well as a few other metals such as neodymium, exhibit this property. They are also unique in that, once the domains are aligned in a magnetic field, the metal is *permanently* magnetized.

Many alloys exhibit greater ferromagnetism than do the pure metals themselves. One example is alnico (an alloy of aluminum, nickel, and cobalt), and another is an alloy of neodymium, iron, and boron.

Audio and video tapes are plastics coated with crystals of ferromagnetic Fe_2O_3, CrO_2, or another metal oxide. The recording head uses an electromagnet to create a varying magnetic field based on signals from a microphone. This magnetizes the tape as it passes through the head, the strength and direction of magnetization varying with the frequency of the sound to be recorded. When the tape is played back, the magnetic field of the moving tape induces a current, which is amplified and sent to the speakers.

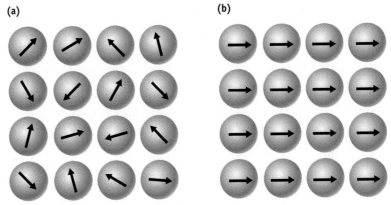

(a) Paramagnetism: the centers (atoms or ions) with magnetic moments are not aligned unless the substance is in a magnetic field. (b) Ferromagnetism: the spins of unpaired electrons in a cluster of atoms or ions align in the same direction.

In summary, *paramagnetism is the attraction to a magnetic field of substances in which the constituent ions or atoms contain unpaired electrons.* Substances in which all electrons are paired with partners of opposite spin are diamagnetic. This explanation opens the way to understanding the arrangement of electrons in atoms with more than one electron.

8.2   PAULI EXCLUSION PRINCIPLE

To make the quantum theory consistent with experiment, the Austrian physicist Wolfgang Pauli (1900–1958) stated in 1925 his **exclusion principle** [CD-ROM, Screen 8.4]:

> **Pauli Exclusion Principle**
> No two electrons in an atom can have the same set of
> four quantum numbers (n, ℓ, m_ℓ, and m_s).
>
> ⇓ which leads to
>
> No atomic orbital can contain
> more than two electrons.

The $1s$ orbital of the H atom has the set of quantum numbers $n = 1$, $\ell = 0$, and $m_\ell = 0$. *No other set is possible.* If an electron is in this orbital, the electron spin direction must also be specified. Let us represent an orbital by a box and the electron by an arrow (↑ or ↓). We represent the H atom as follows, in which the "electron spin arrow" is shown arbitrarily as pointing upward:

Electron in $1s$ orbital: ↑   Quantum number set

                   $1s$ $n = 1$, $\ell = 0$, $m_\ell = 0$, $m_s = +\frac{1}{2}$

If only one electron is in a given orbital, the electron spin arrow may point either up or down. Thus, an equally valid description would be

Electron in $1s$ orbital: ↓   Quantum number set

                   $1s$ $n = 1$, $\ell = 0$, $m_\ell = 0$, $m_s = -\frac{1}{2}$

● **Orbitals Are Not Boxes**
Orbitals are not literally things or boxes in which electrons are placed. Orbitals are electron waves. Thus, it is not conceptually correct to talk about electrons *in* orbitals or *occupying* orbitals, although it is commonly done for the sake of simplicity.

● **Spin Quantum Number and ↑**
We arbitrarily use $m_s = +\frac{1}{2}$ for an arrow pointing up (↑), and $m_s = -\frac{1}{2}$ for an arrow pointing down (↓). However, the opposite convention is also perfectly correct.

Table 8.1 • Number of Electrons Accommodated in Electron Shells and Subshells With $n = 1$ to 6

Electron Shell (n)	Subshells Available	Orbitals Available ($2\ell + 1$)	Number of Electrons Possible in Subshell [$2(2\ell + 1)$]	Maximum Electrons Possible for nth Shell ($2n^2$)
1	s	1	2	2
2	s	1	2	8
	p	3	6	
3	s	1	2	18
	p	3	6	
	d	5	10	
4	s	1	2	32
	p	3	6	
	d	5	10	
	f	7	14	
5	s	1	2	50
	p	3	6	
	d	5	10	
	f	7	14	
	$g*$	9	18	
6	s	1	2	72
	p	3	6	
	d	5	10	
	$f*$	7	14	
	$g*$	9	18	
	$h*$	11	22	

*These orbitals are not used in the ground state of any known element.

The preceding **orbital box diagrams** are equally appropriate for the H atom in its ground state: one electron is in the $1s$ orbital. For a helium atom, which has two electrons, both electrons are assigned to the $1s$ orbital. From the Pauli principle, you know that each electron must have a different set of quantum numbers, so the orbital box picture now is:

1s

Two electrons in 1s orbital: ↑↓ ←——— This electron has $n = 1$, $\ell = 0$, $m_\ell = 0$, $m_s = -\frac{1}{2}$

——— This electron has $n = 1$, $\ell = 0$, $m_\ell = 0$, $m_s = +\frac{1}{2}$

Each of the two electrons in the $1s$ orbital of an He atom has a different set of the four quantum numbers. The first three numbers of a set describe this as a $1s$ orbital. There are only two choices for the fourth number, $m_s = +\frac{1}{2}$ or $-\frac{1}{2}$. Thus, *the $1s$ orbital, and any other atomic orbital, can be occupied by no more than two electrons, and these two electrons must have opposite spin directions.* The consequence is that the helium atom is diamagnetic, as experimentally observed.

Our understanding of orbitals [← TABLE 7.1, PAGE 273], and the knowledge that an orbital can accommodate no more than two electrons, tells us the maximum number of electrons that can occupy each electron shell or subshell. As just demonstrated, only two electrons can be assigned to an s orbital. Because each of the three orbitals in a p subshell can hold two electrons, that subshell can hold a maximum of six electrons. The five orbitals of a d subshell can accommodate a total of ten electrons. Recall that there are always n subshells in the nth shell, and that there are n^2 orbitals in that shell [← TABLE 7.1, PAGE 273]. Thus, the maximum number of electrons in any shell is $2n^2$. The relationships among the quantum numbers and the numbers of electrons are shown in Table 8.1.

8.3 ATOMIC SUBSHELL ENERGIES AND ELECTRON ASSIGNMENTS

Our goal is to understand and predict the distribution of electrons in atoms with many electrons. The basic principle involved is the *aufbau,* or "building up," principle in which electrons are assigned to shells (defined by the quantum number n) of higher and higher energy [💿 CD-ROM, Screen 8.5]. Within a given shell, electrons are assigned to subshells (defined by the quantum number ℓ) of successively higher energy. Electrons are assigned in such a way that the total energy of the atom is as low as possible. Now the relevant question is the order of energy of shells and subshells.

Order of Subshell Energies and Electron Assignments

Quantum theory and the Bohr model of the atom state that the energy of the H atom, with a single electron, depends only on the value of n ($E = -Rhc/n^2$, Equation 7.4). For heavier atoms, however, the situation is more complex. The experimentally determined order of subshell energies in Figure 8.4 shows that the *subshell energies of multielectron atoms depend on both n and ℓ.* The subshells with $n = 3$, for example, have different energies; for a given atom they are in the order $3s < 3p < 3d$.

The subshell energy order in Figure 8.4 and the actual electron arrangements of the elements lead to two general rules that help us predict these arrangements:

- Electrons are assigned to subshells in order of increasing "$n + \ell$" value.
- For two subshells with the same value of "$n + \ell$" electrons are assigned first to the subshell of lower n.

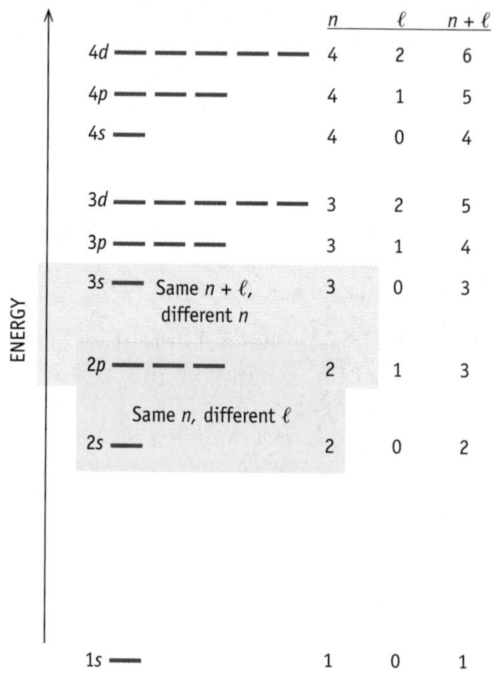

Figure 8.4 Experimentally determined order of subshell energies. In a multielectron atom, energies of electron shells increase with increasing n, and, within a shell, subshell energies increase with increasing ℓ. (The energy axis not to scale.) The energy gaps between subshells of a given shell become smaller as n increases. Note that the order of orbital energies does correspond to the order of orbital occupation for the heavier elements. For the filling order see Figure 8.5.

	n	ℓ	$n + \ell$
4d	4	2	6
4p	4	1	5
4s	4	0	4
3d	3	2	5
3p	3	1	4
3s Same $n + \ell$, different n	3	0	3
2p	2	1	3
Same n, different ℓ			
2s	2	0	2
1s	1	0	1

ENERGY

The following are examples of these rules.

- Electrons are assigned to the $2s$ subshell $(n + \ell = 2 + 0 = 2)$ before the $2p$ subshell $(n + \ell = 2 + 1 = 3)$. That is, the energy of the $2s$ subshell is less than the energy of the $2p$ subshell.
- Electrons are assigned in the order $3s (n + \ell = 3 + 0 = 3)$ before $3p (n + \ell = 3 + 1 = 4)$ before $3d (n + \ell = 3 + 2 = 5)$. That is, the subshell energies increase in the order $3s < 3p < 3d$.
- Electrons fill the $4s$ subshell $(n + \ell = 4)$ before filling the $3d$ subshell $(n + \ell = 5)$.

These filling orders, summarized in Figure 8.5, have been amply verified by experiment.

Chapter Goals • Revisited

- Understand the role magnetism plays in determining and revealing atomic structure.
- **Understand effective nuclear charge and its role in determining atomic properties.**
- Write the electron configuration for any element or monatomic ion.
- Understand the fundamental physical properties of the elements and their periodic trends.

Exercise 8.1 Order of Subshell Assignments

Using the "$n + \ell$," rules you can generally predict the order of subshell assignments (the electron filling order) for a multielectron atom. To which of the following subshells should an electron be assigned first?

(a) 4s or 4p   **(b)** 5d or 6s   **(c)** 4f or 5s

Effective Nuclear Charge, Z*

The order in which electrons are assigned to subshells in an atom, and many atomic properties, can be rationalized by the concept of **effective nuclear charge (Z*)** [CD-ROM, Screen 8.6]. This is the nuclear charge experienced by a particular

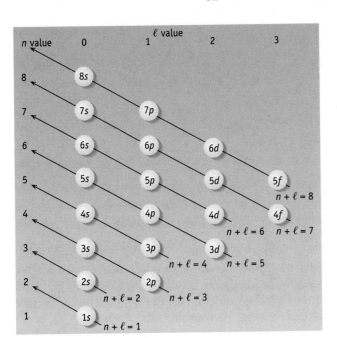

Figure 8.5   Subshell filling order. Subshells in atoms are filled in order of increasing $n + \ell$. When two subshells have the same $n + \ell$ value, the subshell of lower n is filled first. To use the diagram, begin at 1s and follow the arrows of increasing $n + \ell$. (Thus, the order of filling is $1s \Rightarrow 2s \Rightarrow 2p \Rightarrow 3s \Rightarrow 3p \Rightarrow 4s \Rightarrow 3d$ and so on.)

electron in a multielectron atom, as modified by the presence of the other electrons.

In the hydrogen atom, with only one electron, the $2s$ and $2p$ subshells have the same energy. However, for lithium, an atom with three electrons, the presence of the $1s$ electrons alters the energy of the $2s$ and $2p$ subshells. Why should this be true? This question can be answered in part by referring to Figure 8.6.

In Figure 8.6 we have plotted the square of the wave function for a $2s$ electron. The probability of finding the electron (vertical axis) changes as one moves away from the nucleus (horizontal axis). Lightly shaded on this figure is the region

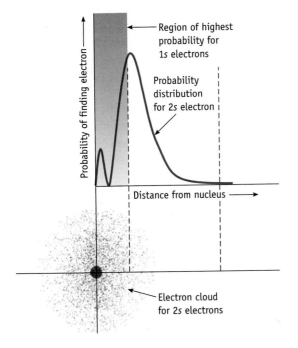

Figure 8.6   Effective nuclear charge, Z* for Li. The two $1s$ electrons of lithium approximately occupy the shaded region, but this region is penetrated by the $2s$ electron (whose approximate probability distribution curve is shown here). When the $2s$ electron is at some distance from the nucleus, it experiences a charge of $+1$ because the $+3$ charge of the lithium nucleus is screened by the two $1s$ electrons. As the $2s$ electron penetrates the $1s$ region, however, the $2s$ electron experiences a larger and larger charge, to a maximum of $+3$. On average, the $2s$ electron experiences a charge, called the effective nuclear charge ($Z^* = 1.28$) that is much smaller than $+3$ but greater than $+1$.

● **More About Z***

For a more complete discussion of effective nuclear charge, see *Essential Trends in Inorganic Chemistry,* by D. M. P. Mingos, Oxford University Press, 1998.

Table 8.2 • Effective Nuclear Charges, Z*, for the 2s and 2p Electrons of Second Period Elements

Atom	$Z^*(2s)$	$Z^*(2p)$
Li	1.28	
B	2.58	2.42
C	3.22	3.14
N	3.85	3.83
O	4.49	4.45
F	5.13	5.10

● **Z* for s and p Subshells**

Z* is greater for s electrons than for p electrons in the same shell. This difference becomes larger as n becomes larger. For example, compare Z* for the Group 4A elements.

Atom	Z^* (ns)	Z^* (np)	Value of n
C	3.22	3.14	2
Si	4.90	4.29	3
Ge	8.04	6.78	4

● **Electron Configurations**

See the interactive tool "Electron Configuration" in the Periodic Table Tools menu, accessible on any screen of the *General Chemistry Interactive CD-ROM.*

occupied by the 1s electrons of lithium. Observe that the 2s electron wave occurs partly within the region occupied by 1s electrons. Chemists say that the 2s electron density region *penetrates* the 1s electron density region. This alters the energy of the 2s orbital relative to what it would be in the H atom, where there are no other electrons. As more and more electrons are added to an atom, the outermost electrons will penetrate the region occupied by the inner electrons, but the penetration is different for ns, np, and nd orbitals, and their energies are altered by differing amounts.

Lithium has three protons in the nucleus. Suppose the two 1s electrons have been added to the atom, and the third electron (a 2s electron) approaches (see Figure 8.6). This third electron would initially experience a +1 charge because there are two electrons (total charge = −2) between the 2s electron and the +3 charge on the nucleus. Chemists say the 1s electrons *screen* the effect of the nuclear charge from the 2s electron.

The screening of the nuclear charge varies with the distance of the 2s electron from the nucleus, however. If a 2s electron were to penetrate the 1s electron region, it would experience a higher and higher positive charge, finally seeing a charge of +3 if it comes very close to the nucleus. Figure 8.6 shows, however, that a 2s electron has some probability of being both inside and outside the region occupied by the 1s electrons. Thus, on average, a 2s electron experiences a positive charge greater than +1 but smaller than +3. Because of the penetration of the inner electron region by outer electrons and the screening of the nuclear charge by the inner electrons, the outer electrons experience an *average* nuclear charge, the *effective nuclear charge, Z*.

Values of Z* for s and p electrons for most second period elements are listed in Table 8.2. In each case Z* is greater for s electrons than for p electrons and this explains why, in general, s electrons have a lower energy than p electrons in the same quantum shell (see Figure 8.4).

Another observation regarding Z* for the second period elements in Table 8.2 is that the value of Z* increases across the period. This effect will become important in understanding the change in properties of elements on proceeding across a period.

Extending these arguments to other subshells, it is observed that the relative penetration of subshells is $s > p > d > f$, and so the effective nuclear charge experienced by orbitals is in the order $ns > np > nd > nf$. A consequence of the differences in orbital penetration and electron shielding is that subshells within an electron shell are filled in the order ns before np before nd before nf.

What emerges from this analysis is the order of shell and subshell energies depicted in Figure 8.4 and the filling order in Figure 8.5. With this understanding, we turn to the periodic table and its use as a guide to electron arrangements in atoms.

8.4 ATOMIC ELECTRON CONFIGURATIONS

The arrangement of electrons in the elements up to 109 — the **electron configurations** of the elements — is given in Table 8.3. These are the ground state electron configurations, in which electrons are found in the shells, subshells, and orbitals that result in the lowest energy for the atom. In general, the guiding principle in assigning electrons to available orbitals is to do so in order of increasing $n + \ell$ (see Figures 8.4 and 8.5). Our emphasis, however, will be to connect the configurations of the elements with their position in the periodic table because this will allow us ultimately to organize a large number of chemical facts [CD-ROM, Screen 8.7].

Table 8.3 • Electron Configurations of Atoms in the Ground State

Z	Element	Configuration	Z	Element	Configuration	Z	Element	Configuration
1	H	$1s^1$	37	Rb	$[Kr]5s^1$	74	W	$[Xe]4f^{14}5d^46s^2$
2	He	$1s^2$	38	Sr	$[Kr]5s^2$	75	Re	$[Xe]4f^{14}5d^56s^2$
3	Li	$[He]2s^1$	39	Y	$[Kr]4d^15s^2$	76	Os	$[Xe]4f^{14}5d^66s^2$
4	Be	$[He]2s^2$	40	Zr	$[Kr]4d^25s^2$	77	Ir	$[Xe]4f^{14}5d^76s^2$
5	B	$[He]2s^22p^1$	41	Nb	$[Kr]4d^45s^1$	78	Pt	$[Xe]4f^{14}5d^96s^1$
6	C	$[He]2s^22p^2$	42	Mo	$[Kr]4d^55s^1$	79	Au	$[Xe]4f^{14}5d^{10}6s^1$
7	N	$[He]2s^22p^3$	43	Tc	$[Kr]4d^55s^2$	80	Hg	$[Xe]4f^{14}5d^{10}6s^2$
8	O	$[He]2s^22p^4$	44	Ru	$[Kr]4d^75s^1$	81	Tl	$[Xe]4f^{14}5d^{10}6s^26p^1$
9	F	$[He]2s^22p^5$	45	Rh	$[Kr]4d^85s^1$	82	Pb	$[Xe]4f^{14}5d^{10}6s^26p^2$
10	Ne	$[He]2s^22p^6$	46	Pd	$[Kr]4d^{10}$	83	Bi	$[Xe]4f^{14}5d^{10}6s^26p^3$
11	Na	$[Ne]3s^1$	47	Ag	$[Kr]4d^{10}5s^1$	84	Po	$[Xe]4f^{14}5d^{10}6s^26p^4$
12	Mg	$[Ne]3s^2$	48	Cd	$[Kr]4d^{10}5s^2$	85	At	$[Xe]4f^{14}5d^{10}6s^26p^5$
13	Al	$[Ne]3s^23p^1$	49	In	$[Kr]4d^{10}5s^25p^1$	86	Rn	$[Xe]4f^{14}5d^{10}6s^26p^6$
14	Si	$[Ne]3s^23p^2$	50	Sn	$[Kr]4d^{10}5s^25p^2$	87	Fr	$[Rn]7s^1$
15	P	$[Ne]3s^23p^3$	51	Sb	$[Kr]4d^{10}5s^25p^3$	88	Ra	$[Rn]7s^2$
16	S	$[Ne]3s^23p^4$	52	Te	$[Kr]4d^{10}5s^25p^4$	89	Ac	$[Rn]6d^17s^2$
17	Cl	$[Ne]3s^23p^5$	53	I	$[Kr]4d^{10}5s^25p^5$	90	Th	$[Rn]6d^27s^2$
18	Ar	$[Ne]3s^23p^6$	54	Xe	$[Kr]4d^{10}5s^25p^6$	91	Pa	$[Rn]5f^26d^17s^2$
19	K	$[Ar]4s^1$	55	Cs	$[Xe]6s^1$	92	U	$[Rn]5f^36d^17s^2$
20	Ca	$[Ar]4s^2$	56	Ba	$[Xe]6s^2$	93	Np	$[Rn]5f^46d^17s^2$
21	Sc	$[Ar]3d^14s^2$	57	La	$[Xe]5d^16s^2$	94	Pu	$[Rn]5f^67s^2$
22	Ti	$[Ar]3d^24s^2$	58	Ce	$[Xe]4f^15d^16s^2$	95	Am	$[Rn]5f^77s^2$
23	V	$[Ar]3d^34s^2$	59	Pr	$[Xe]4f^36s^2$	96	Cm	$[Rn]5f^76d^17s^2$
24	Cr	$[Ar]3d^54s^1$	60	Nd	$[Xe]4f^46s^2$	97	Bk	$[Rn]5f^97s^2$
25	Mn	$[Ar]3d^54s^2$	61	Pm	$[Xe]4f^56s^2$	98	Cf	$[Rn]5f^{10}7s^2$
26	Fe	$[Ar]3d^64s^2$	62	Sm	$[Xe]4f^66s^2$	99	Es	$[Rn]5f^{11}7s^2$
27	Co	$[Ar]3d^74s^2$	63	Eu	$[Xe]4f^76s^2$	100	Fm	$[Rn]5f^{12}7s^2$
28	Ni	$[Ar]3d^84s^2$	64	Gd	$[Xe]4f^75d^16s^2$	101	Md	$[Rn]5f^{13}7s^2$
29	Cu	$[Ar]3d^{10}4s^1$	65	Tb	$[Xe]4f^96s^2$	102	No	$[Rn]5f^{14}7s^2$
30	Zn	$[Ar]3d^{10}4s^2$	66	Dy	$[Xe]4f^{10}6s^2$	103	Lr	$[Rn]5f^{14}6d^17s^2$
31	Ga	$[Ar]3d^{10}4s^24p^1$	67	Ho	$[Xe]4f^{11}6s^2$	104	Rf	$[Rn]5f^{14}6d^27s^2$
32	Ge	$[Ar]3d^{10}4s^24p^2$	68	Er	$[Xe]4f^{12}6s^2$	105	Db	$[Rn]5f^{14}6d^37s^2$
33	As	$[Ar]3d^{10}4s^24p^3$	69	Tm	$[Xe]4f^{13}6s^2$	106	Sg	$[Rn]5f^{14}6d^47s^2$
34	Se	$[Ar]3d^{10}4s^24p^4$	70	Yb	$[Xe]4f^{14}6s^2$	107	Bh	$[Rn]5f^{14}6d^57s^2$
35	Br	$[Ar]3d^{10}4s^24p^5$	71	Lu	$[Xe]4f^{14}5d^16s^2$	108	Hs	$[Rn]5f^{14}6d^67s^2$
36	Kr	$[Ar]3d^{10}4s^24p^6$	72	Hf	$[Xe]4f^{14}5d^26s^2$	109	Mt	$[Rn]5f^{14}6d^77s^2$
			73	Ta	$[Xe]4f^{14}5d^36s^2$			

Electron Configurations of the Main Group Elements

Hydrogen, the first element in the periodic table, has one electron in a $1s$ orbital. One way to depict its electron configuration is with the orbital box diagram used earlier, but an alternative and more frequently used method is the *spdf* **notation.** Using the latter method, the electron configuration of H is $1s^1$, or "one ess one."

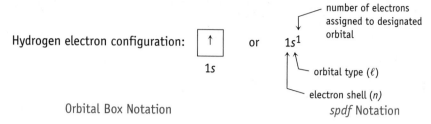

Hydrogen electron configuration: ↑   or $1s^1$

1s

Orbital Box Notation   *spdf* Notation

number of electrons
assigned to designated
orbital

orbital type (ℓ)

electron shell (n)

Let us look at electron configurations for the elements of the second period using both notation methods.

Lithium (Li) and Other Group 1A Elements

Lithium, with three electrons, is the first element in the second period of the periodic table. The first two electrons are in the $1s$ subshell, and the third electron must be in the $n = 2$ shell. According to the energy level diagram in Figure 8.4, that electron must be in the $2s$ subshell. The *spdf* notation, $1s^2 2s^1$, is read "one ess two, two ess one."

Lithium: *spdf* notation $1s^2 2s^1$

Box notation   ⇅   ↑   ☐☐☐

1s   2s   2p

Electron configurations are often written in abbreviated form by combining the **noble gas notation** with the *spdf*, or orbital box, notation. The arrangement preceding the $2s$ electron is that of the noble gas helium so, instead of writing out $1s^2 2s^1$, the completed electron shell is represented by placing the symbol of the corresponding noble gas in brackets. Thus, lithium's configuration would be written as $[\text{He}]2s^1$.

The electrons included in the noble gas notation are often referred to as the **core electrons** of the atom. Not only is it a time-saving way to write electron configurations, but the noble gas notation also conveys the idea that the core electrons can generally be ignored when considering the chemistry of an element. The electrons beyond the core electrons, the $2s^1$ electron in the case of lithium, are the **valence electrons,** the electrons that determine the chemical properties of an element.

The position of lithium in the periodic table tells you its configuration immediately. All the elements of Group 1A have one electron assigned to an s orbital of the nth shell, for which n is the number of the period in which the element is found (Figure 8.7). For example, potassium is the first element in the $n = 4$ row (the fourth period), so potassium has the electron configuration of the element preceding it in the table (Ar) plus an electron assigned to the $4s$ orbital: $[\text{Ar}]4s^1$.

Beryllium (Be) and Other Elements of Group 2A

Beryllium, in Group 2A, has two electrons in the $1s$ orbital plus two additional electrons.

Beryllium: *spdf* notation $1s^2 2s^2$ or $[\text{He}]2s^2$

Box notation   ⇅   ⇅   ☐☐☐

1s   2s   2p

Chapter Goals • Revisited

- Understand the role magnetism plays in determining and revealing atomic structure.
- Understand effective nuclear charge and its role in determining atomic properties.
- **Write the electron configuration for any element or monatomic ion.**
- Understand the fundamental physical properties of the elements and their periodic trends.

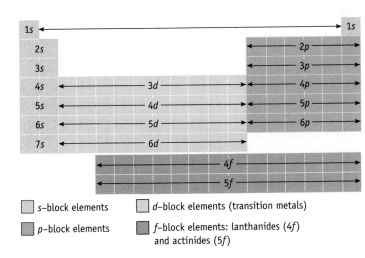

Figure 8.7   Electron configurations and the periodic table. The outermost electrons of an element are assigned to the indicated orbitals. See Table 8.2

All elements of Group 2A have electron configurations of [*electrons of preceding noble gas*] ns^2, where n is the period in which the element is found in the periodic table. Because all the elements of Group 1A have the valence electron configuration ns^1, and those in Group 2A have ns^2, these elements are called *s*-**block elements.**

Boron (B) and Other Elements of Group 3A

At boron (Group 3A) we come to the first element in the block of elements on the right side of the periodic table. Because $1s$ and $2s$ orbitals are filled in a boron atom, the fifth electron must be assigned to a $2p$ orbital.

Boron: *spdf* notation $1s^2 2s^2 2p^1$ or $[He]2s^2 2p^1$

Box notation

In fact, all the elements from Group 3A through Group 8A are adding electrons to p orbitals, so these elements are sometimes called the *p*-**block elements.** All have the general valence electron configuration $ns^2 np^x$, where x varies from one to six (and is equal to the group number minus 2).

Carbon (C) and Other Elements of Group 4A

Carbon (Group 4A) is the second element in the p block, so a second electron is assigned to the $2p$ orbitals. For carbon to be in its lowest energy or ground state, this electron must be assigned to either of the remaining p orbitals, and it must have the same spin direction as the first p electron.

Carbon: *spdf* notation $1s^2 2s^2 2p^2$ or $[He]2s^2 2p^2$

Box notation

In general, when electrons are assigned to p, d, or f orbitals, each successive electron is assigned to a different orbital of the subshell, and each electron has the same spin as the previous one; this pattern continues until the subshell is half full.

Additional electrons must be assigned to half-filled orbitals. This procedure follows **Hund's rule,** which states that the most stable arrangement of electrons is that with the maximum number of unpaired electrons, all with the same spin direction. This arrangement makes the total energy of an atom as low as possible.

Notice that carbon is the second element in the p block of elements, so there must be two p electrons (besides the two $2s$ electrons already present in the $n = 2$ shell). Because carbon is a second-period element, these are $2p$ electrons. Thus, you can immediately write the carbon electron configuration by referring to the periodic table: starting at H and moving from left to right across the successive periods, you write $1s^2$ to reach the end of period 1, and then $2s^2$ and finally $2p^2$ to bring the electron count to six. Carbon is in Group 4A of the periodic table, and it has four electrons in the $n = 2$ shell.

Nitrogen (N), Oxygen (O), and Elements of Groups 5A and 6A

Nitrogen (Group 5A) has three electrons, all with the same spin, in three different $2p$ orbitals.

Nitrogen: *spdf* notation $1s^2 2s^2 2p^3$ or $[He]2s^2 2p^3$

Box notation $\boxed{\uparrow\downarrow}$ $\boxed{\uparrow\downarrow}$ $\boxed{\uparrow\,|\,\uparrow\,|\,\uparrow}$
            $1s$ $2s$ $2p$

Oxygen (Group 6A) has yet another $2p$ electron. Two of the six electrons in oxygen's outer shell are assigned to the $2s$ orbital, and, as it is the fourth element in the p block, the other four electrons are assigned to $2p$ orbitals.

Oxygen: *spdf* notation $1s^2 2s^2 2p^4$ or $[He]2s^2 2p^4$

Box notation $\boxed{\uparrow\downarrow}$ $\boxed{\uparrow\downarrow}$ $\boxed{\uparrow\downarrow\,|\,\uparrow\,|\,\uparrow}$
            $1s$ $2s$ $2p$

This means the fourth $2p$ electron must pair up with one already present. It makes no difference to which orbital this electron is assigned (the $2p$ orbitals all have the same energy), but it must have a spin opposite to the other electron already assigned to that orbital so that each electron has a different set of quantum numbers (the Pauli exclusion principle).

Fluorine (F), Neon (Ne), and Elements of Groups 7A and 8A

Fluorine (Group 7A) has seven electrons in the $n = 2$ shell. Two of these electrons occupy the $2s$ subshell, and the remaining five electrons occupy the $2p$ subshell. All halogen atoms have a similar configuration, $ns^2 np^5$, where n is again the period in which the element is located.

Fluorine: *spdf* notation $1s^2 2s^2 2p^5$ or $[He]2s^2 2p^5$

Box notation $\boxed{\uparrow\downarrow}$ $\boxed{\uparrow\downarrow}$ $\boxed{\uparrow\downarrow\,|\,\uparrow\downarrow\,|\,\uparrow}$
            $1s$ $2s$ $2p$

Like all the other elements in Group 8A, neon is a noble gas. All Group 8A elements (except helium) have eight electrons in the shell of highest n value, so all have the valence electron configuration $ns^2 np^6$, where n is the period in which the element is found. That is, all the noble gases have filled ns and np subshells. As you will see, the nearly complete chemical inertness of the noble gases correlates with this electron configuration.

Neon: *spdf* notation $1s^2 2s^2 2p^6$ or $[He]2s^2 2p^6$

Box notation   ⊞ ⊞ ⊞⊞⊞
1s   2s   2p

Elements of Period 3 and Beyond

The next element after neon is sodium, and with it a new period is begun (recall Figure 8.4). Because sodium is the first element with $n = 3$, the added electron must be assigned to the 3s orbital. (Remember that all elements in Group 1A have the ns^1 configuration.) Thus, the complete electron configuration of sodium is that of neon (the preceding element) plus one 3s electron.

Sodium: *spdf* notation $1s^2 2s^2 2p^6 3s^1$ or $[Ne]3s^1$

Box notation   ⊞ ⊞ ⊞⊞⊞ ⊡
1s   2s   2p   3s

Moving across the third period, we come to silicon. This element is in Group 4A and so has four electrons beyond the neon core. Because it is the second element in the *p* block, it has two electrons in 3p orbitals. Thus, its electron configuration is

Silicon: *spdf* notation $1s^2 2s^2 2p^6 3s^2 3p^2$ or $[Ne]3s^2 3p^2$

Box notation   ⊞ ⊞ ⊞⊞⊞ ⊞ ⊡⊡
1s   2s   2p   3s   3p

From silicon to the end of the third period, electrons are added to the 3p orbitals in the same manner as for the elements in the second period. Finally, at argon the 3p subshell is completed with six electrons.

Example 8.1   Electron Configurations

Problem • Give the electron configuration of sulfur, using the *spdf*, noble gas, and orbital box notations.

Strategy • Sulfur, atomic number 16, is the sixth element in the third period ($n = 3$), and it is in the *p* block. The last six electrons assigned to the atom, therefore, have the configuration $3s^2 3p^4$. These are preceded by the completed shells $n = 1$ and $n = 2$, the electron arrangement for Ne.

Solution • The electron configuration of sulfur is

Complete *spdf* notation: $1s^2 2s^2 2p^6 3s^2 3p^4$

spdf with noble gas notation: $[Ne]3s^2 3p^4$

Orbital box notation: [Ne] ⊞ ⊞⊡⊡
3s   3p

Example 8.2   Electron Configurations and Quantum Numbers

Problem • Write the electron configuration for Al using the noble gas notation, and give a set of quantum numbers for each of the electrons with $n = 3$ (the valence electrons).

Strategy • Aluminum is the third element in the third period. It therefore has three electrons with $n = 3$, and because Al is in

the *p* block of elements, two of the electrons are assigned to 3s and the remaining electron is assigned to 3p.

Solution • The element is preceded by the noble gas neon, so the electron configuration is $[Ne]3s^2 3p^1$. Using box notation, the configuration is

(Example continues on p. 302)

Aluminum configuration: [Ne] ⇅   ↑ ☐ ☐

3s   3p

The possible sets of quantum numbers for the two 3s electrons are given in the table. For the single 3p electron, one of six possible sets would be $n = 3$, $\ell = 1$, $m_\ell = +1$, and $m_s = +\frac{1}{2}$.

	n	ℓ	m_ℓ	m_s
For ↑	3	0	0	$+\frac{1}{2}$
For ↓	3	0	0	$-\frac{1}{2}$

Exercise 8.2   *spdf* **Notation, Orbital Box Diagrams, and Quantum Numbers**

(a) What element has the configuration $1s^2 2s^2 2p^6 3s^2 3p^5$?

(b) Using the *spdf* notation and a box diagram, show the electron configuration of phosphorus.

(c) Write one possible set of quantum numbers for the valence electrons of calcium.

Electron Configurations of the Transition Elements

- **Writing Electron Configurations**
Although it does not necessarily reflect the filling order, we follow the convention of writing the orbitals in order of increasing n when writing electron configurations. For a given n, the subshells are listed in order of increasing ℓ (*s, p, d, f*).

The elements of the fourth through the seventh periods must use *d* or *f* subshells, in addition to *s* and *p* subshells, to accommodate electrons (see Figure 8.7 and Table 8.4) [CD-ROM, Screen 8.7]. Elements whose atoms are filling *d* subshells are described as *transition elements*. Those for which *f* subshells are filling are sometimes called the *inner transition elements* or, more usually, the **lanthanides** (filling 4*f* orbitals) and **actinides** (filling 5*f* orbitals).

In a given period, the transition elements are always preceded by two *s*-block elements. Accordingly, scandium, the first transition element, has the configuration $[Ar]3d^1 4s^2$, and titanium follows with $[Ar]3d^2 4s^2$ (Table 8.4).

The expected configuration of the chromium atom is $[Ar]3d^4 4s^2$. The actual configuration, however, has one electron assigned to each of the five available 3*d* orbitals and the 4*s* orbital: $[Ar]3d^5 4s^1$. This is explained by assuming that the 4*s* and the 3*d* orbitals have approximately the same energy at this point, thus giving rise to six orbitals of nearly the same energy. Each of the six valence electrons of chromium is assigned to a separate orbital. This electron configuration illustrates the fact that minor differences occasionally occur between the predicted and actual configurations. These differences have little or no effect on the chemistry of the elements, however.

Following chromium, atoms of manganese, iron, cobalt, and nickel have configurations that are expected from the order of orbital energies in Figure 8.4. The Group 1B element copper, however, has a single electron in the 4*s* orbital, in accord with its group number, and the remaining ten electrons beyond the argon core are assigned to the 3*d* orbitals. Zinc ends the first transition series. This Group 2B element has two electrons assigned to 4*s*, and the 3*d* orbitals are completely filled with ten electrons.

Lanthanides and Actinides

The fifth period ($n = 5$) follows the pattern of the fourth period with minor variations. The sixth period, however, includes the lanthanide series beginning with lanthanum, La. As the first element in the *d* block, lanthanum has the configuration $[Xe]5d^1 6s^2$. The next element, cerium (Ce), is set out in a separate row at the bottom of the periodic table, and it is with the elements in this row (from Ce through Lu) that electrons are first assigned to *f* orbitals. This means the configuration of cerium is $[Xe]4f^1 5d^1 6s^2$. Moving across the lanthanide series, the pattern

Table 8.4 • Orbital Box Diagrams for the Elements Ca Through Zn

		3d	4s
Ca	$[\text{Ar}]4s^2$	☐ ☐ ☐ ☐ ☐	↑↓
Sc	$[\text{Ar}]3d^1 4s^2$	↑ ☐ ☐ ☐ ☐	↑↓
Ti	$[\text{Ar}]3d^2 4s^2$	↑ ↑ ☐ ☐ ☐	↑↓
V	$[\text{Ar}]3d^3 4s^2$	↑ ↑ ↑ ☐ ☐	↑↓
Cr*	$[\text{Ar}]3d^5 4s^1$	↑ ↑ ↑ ↑ ↑	↑
Mn	$[\text{Ar}]3d^5 4s^2$	↑ ↑ ↑ ↑ ↑	↑↓
Fe	$[\text{Ar}]3d^6 4s^2$	↑↓ ↑ ↑ ↑ ↑	↑↓
Co	$[\text{Ar}]3d^7 4s^2$	↑↓ ↑↓ ↑ ↑ ↑	↑↓
Ni	$[\text{Ar}]3d^8 4s^2$	↑↓ ↑↓ ↑↓ ↑ ↑	↑↓
Cu*	$[\text{Ar}]3d^{10} 4s^1$	↑↓ ↑↓ ↑↓ ↑↓ ↑↓	↑
Zn	$[\text{Ar}]3d^{10} 4s^2$	↑↓ ↑↓ ↑↓ ↑↓ ↑↓	↑↓

*These configurations do not follow the "$n + \ell$" rule.

continues with some variation, with 14 electrons being assigned to the seven $5f$ orbitals in lutetium, Lu ($[\text{Xe}]4f^{14}5d^1 6s^2$) (see Table 8.3).

The seventh period also includes an extended series of elements utilizing f orbitals, the actinides, which begins with actinium (Ac: $[\text{Rn}]6d^1 7s^2$). The next element is thorium (Th), followed by protactinium (Pa) and uranium (U). The electron configuration of uranium is $[\text{Rn}]5f^3 6d^1 7s^2$. This is the third element in the actinide series and so has three $5f$ electrons.

When you have completed this section, you should be able to depict the electron configuration of any element in the s and p blocks using the periodic table as a guide. Regarding the prediction of electron configurations for atoms of elements in the d and f blocks, you have seen that minor differences can occur between actual (see Table 8.3) and predicted configurations. You are reminded that these small "anomalies" have little effect on the chemical behavior of the elements.

Example 8.3 Electron Configurations of the Transition Elements

Problem • Using the *spdf* and noble gas notations, give electron configurations for (a) technetium (Tc) and (b) osmium (Os).

Strategy • Base your answer on the positions of the elements in the periodic table. That is, for each element, find the preceding noble gas and then note the number of s, p, d, and f electrons that lead from the noble gas to the element.

Solution •
(a) *Technetium, Tc:* The noble gas that precedes Tc is krypton, Kr, at the end of the $n = 4$ row. After the 36 electrons of Kr are

assigned, 7 electrons remain. According to the periodic table, two of these seven electrons are in the $5s$ orbital, and the remaining five are in $4d$ orbitals. Therefore, the technetium configuration is $[\text{Kr}]4d^5 5s^2$.

(b) *Osmium, Os:* This is a sixth-period element and the 22nd element following the noble gas xenon. Of the 22 electrons to be added after the Xe core, 2 are assigned to the $6s$ orbital and 14 to $4f$ orbitals. The remaining six are assigned to $5d$ orbitals. Thus, the osmium configuration is $[\text{Xe}]4f^{14}5d^6 6s^2$.

Using the periodic table and without looking at Table 8.3, write electron configurations for the following elements: (a) P, (b) Zn, (c) Zr, (d) In, (e) Pb, and (f) U. Use the *spdf* and noble gas notations. When you have finished, check your answers with Table 8.2 or CD-ROM, Screen 8.7.

8.5   ELECTRON CONFIGURATIONS OF IONS

Much of the chemistry of the elements is that of their ions. To form a cation from a neutral atom, one or more of the valence electrons are removed; that is, electrons are removed from the electron shell of highest n [CD-ROM, Screen 8.8]. If there are several subshells within the nth shell, the electron or electrons of maximum ℓ are removed. Thus, a sodium ion is formed by removing the $3s^1$ electron from the Na atom,

$$\text{Na: } [1s^2 2s^2 2p^6 3s^1] \longrightarrow \text{Na}^+: [1s^2 2s^2 2p^6] + e^-$$

and P^{3+} would be formed by removing three $3p$ electron from a phosphorus atom:

$$\text{P: } [1s^2 2s^2 2p^6 3s^2 3p^3] \longrightarrow \text{P}^{3+}: [1s^2 2s^2 2p^6 3s^2] + 3e^-$$

The same general rule applies to transition metal atoms. This means the titanium(II) cation has the configuration $[Ar]3d^2$, for example,

$$\text{Ti: } [Ar]3d^2 4s^2 \longrightarrow \text{Ti}^{2+}: [Ar]3d^2 + 2e^-$$

and the iron(II) and iron(III) cations have the configurations $[Ar]3d^6$ and $[Ar]3d^5$, respectively.

$$\text{Fe: } [Ar]3d^6 4s^2 \longrightarrow \text{Fe}^{2+}: [Ar]3d^6 + 2e^-$$
$$\text{Fe}^{2+}: [Ar]3d^6 \longrightarrow \text{Fe}^{3+}: [Ar]3d^5 + e^-$$

All common transition metal cations have electron configurations of the general type [noble gas core]$(n-1)d^x$. It is very important to remember this because the chemical and physical properties of transition metal cations are determined by the presence of electrons in d orbitals [CD-ROM, Screen 8.8].

Atoms and ions with unpaired electrons are paramagnetic, that is, they are capable of being attracted to a magnetic field (see Section 8.1). Paramagnetism is important here because it provides experimental evidence that transition metal ions with charges of $2+$ or greater have no ns electrons. For example, the Fe^{2+} ion is paramagnetic to the extent of four unpaired electrons, and the Fe^{3+} ion has five unpaired electrons. If three $3d$ electrons had been removed instead to form Fe^{3+}, the ion would still be paramagnetic but only to the extent of three unpaired electrons.

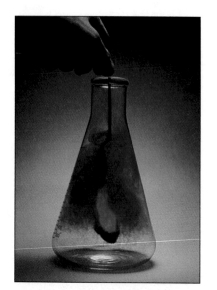

When iron reacts with chlorine (Cl_2) to produce $FeCl_3$, each iron atom loses three electrons to give a paramagnetic Fe^{3+} ion with the configuration $[Ar]3d^5$. *(Charles D. Winters)*

Example 8.4   **Configurations of Transition Metal Ions**

Problem • Give the electron configurations for copper, Cu, and for its $+1$ and $+2$ ions. Are either of these ions paramagnetic? How many unpaired electrons does each have?

Strategy • Observe the configuration of copper in Table 8.4.

Recall that s and then d electrons are removed when forming an ion from a transition metal atom.

Solution • Copper has only one electron in the $4s$ orbital and ten electrons in $3d$ orbitals:

Cu: [Ar]$3d^{10}4s^1$ ⊞⊞⊞⊞⊞ ⊞
                                    3d   4s

When copper is oxidized to Cu⁺, the 4s electron is lost.

Cu⁺: [Ar]$3d^{10}$ ⊞⊞⊞⊞⊞ ☐
                                 3d   4s

The copper(II) ion is formed from copper(I) by removal of one of the 3d electrons.

Cu²⁺: [Ar]$3d^9$ ⊞⊞⊞⊞⊞ ☐
                                3d   4s

Copper(II) ions (Cu²⁺) have one unpaired electron and so should be paramagnetic. In contrast, Cu⁺ has no unpaired electrons, so the ion and its compounds are diamagnetic.

Exercise 8.4 Metal Ion Configurations

Depict electron configurations for V²⁺, V³⁺, and Co³⁺. Use orbital box diagrams and the noble gas notation. Are any of the ions paramagnetic? If so, give the number of unpaired electrons.

8.6 ATOMIC PROPERTIES AND PERIODIC TRENDS

Once electron configurations were understood, chemists realized that *similarities in properties of the elements are the result of similar valence shell electron configurations* [🖱 CD-ROM, Screens 8.9–8.13]. An objective of this section is to describe how atomic electron configurations are related to some of the physical and chemical properties of the elements and why those properties change in a reasonably predictable manner when moving down groups and across periods. This background should make the periodic table an even more useful tool in your study of chemistry. With an understanding of electron configurations and their relation to properties, you should be able to organize and predict chemical and physical properties of the elements and their compounds. We will concentrate on physical properties in this section and then look briefly at chemical behavior in Section 8.7.

Atomic Size

An electron orbital has no sharp boundary beyond which the electron never strays [⬅ FIGURE 7.14, PAGE 274]. How then can we define the size of an atom? There are actually several ways, and they can give slightly different results.

One of the simplest and most useful ways to define atomic size is as the distance between atoms in a sample of the element. Let us take a diatomic molecule such as Cl₂ (Figure 8.8). The radius of a Cl atom is assumed to be one half the experimentally determined distance between the centers of the two atoms. This distance is 198 pm, so the radius of one Cl atom is 99 pm. Similarly, the C—C distance in diamond is 154 pm, so a radius of 77 pm can be assigned to carbon. To test these estimates, we can add them together to estimate the distance between Cl and C in CCl₄. The predicted distance of 176 pm agrees with the experimentally measured C—Cl distance of 176 pm.

This approach to determining atomic radii will only apply if molecular compounds of the element exist. For metals the atomic radius can be estimated from measurements of the atom-to-atom distance in a crystal of the element.

A reasonable set of atomic radii has been assembled (Figure 8.9), and some interesting periodic trends are seen immediately. *For the main group elements, atomic*

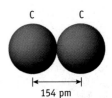
C   C

154 pm

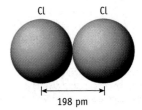
Cl   Cl

198 pm

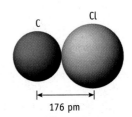
C   Cl

176 pm

Figure 8.8 Atomic radius. The sum of the atomic radii of C and Cl provides a good estimate of the C—Cl distance in a molecule having such a bond.

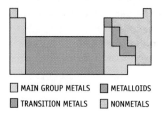

Figure 8.9 **Atomic radii in picometers for main group elements.** 1 pm $= 1 \times 10^{-12}$ m $= 1 \times 10^{-3}$ nm. *(Data taken from J. Emsley: The Elements, Clarendon Press, Oxford, 1998, 3rd ed)*

☐ MAIN GROUP METALS   ☐ METALLOIDS
☐ TRANSITION METALS   ☐ NONMETALS

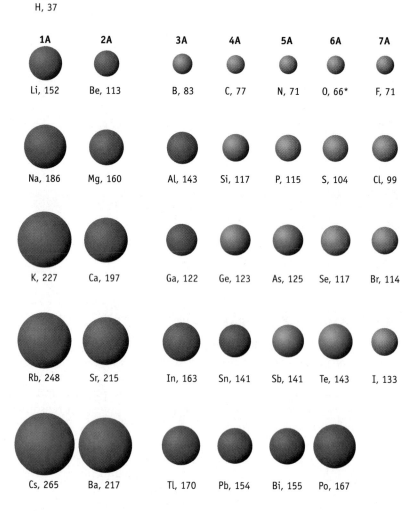

1A

H, 37

1A	2A	3A	4A	5A	6A	7A
Li, 152	Be, 113	B, 83	C, 77	N, 71	O, 66*	F, 71
Na, 186	Mg, 160	Al, 143	Si, 117	P, 115	S, 104	Cl, 99
K, 227	Ca, 197	Ga, 122	Ge, 123	As, 125	Se, 117	Br, 114
Rb, 248	Sr, 215	In, 163	Sn, 141	Sb, 141	Te, 143	I, 133
Cs, 265	Ba, 217	Tl, 170	Pb, 154	Bi, 155	Po, 167	

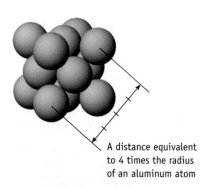

A distance equivalent to 4 times the radius of an aluminum atom

Measuring atomic radius. Pictured here is a tiny piece of an aluminum crystal. Each sphere represents an aluminum atom. Measuring the distance shown, for example, allows a scientist to estimate the radius of an aluminum atom.

● **Trends in Atomic Radii**
General trends in atomic radii of *s*- and *p*-block elements with position in the periodic table.

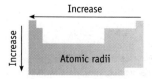

radii increase going down a group in the periodic table and decrease going across a period. These trends reflect two important effects [🖰 CD-ROM, Screen 8.10]:

- The size of an atom is determined by the outermost electrons. In going from the top to the bottom of a group in the periodic table, the outermost electrons are assigned to orbitals with higher and higher values of the principal quantum number, *n*. The underlying electrons require some space, so the electrons of the outer shell must be farther from the nucleus.

- For main group elements of a given period, the principal quantum number, *n*, of the valence electron orbitals is the same. In going from one element to the next across a period a proton is added to each nucleus and an electron to each outer shell. In each step, the effective nuclear charge, *Z** (see page 296) increases slightly because the effect of each additional proton is more important than the effect of an additional electron. The result is that attraction between the nucleus and electrons increases, and atomic radius decreases.

The periodic trend in the atomic radii of transition metal atoms (Figure 8.10) is somewhat different from that for main group elements. Going from left to right across a given period, the radii initially decrease across the first few elements. The sizes of the elements in the middle of a transition series then change very little until a small increase in size occurs at the end of the series. The size of the atom is determined largely by electrons in the outermost shell, that is, by the electrons of

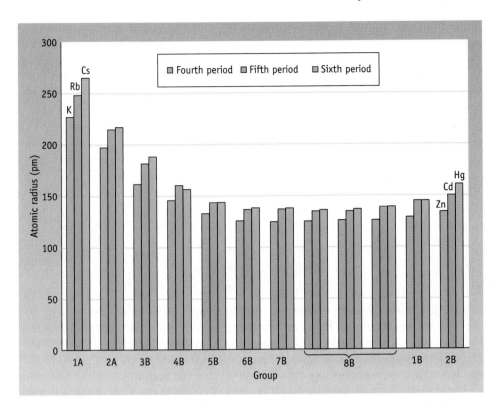

Figure 8.10   Trends in atomic radii for transition elements. Atomic radii of the Group 1A and 2A metals and the transition metals of the fourth, fifth, and sixth periods.

the *ns* subshell. In the first transition series, for example, the outer shell contains the 4*s* electrons, but electrons are being added to the 3*d* orbitals across the series. The increased nuclear charge on the atoms as one moves from left to right should cause the radius to decrease. This effect, however, is mostly cancelled out by increased electron–electron repulsion among the electrons. On reaching Groups 1B and 2B elements at the end of the series, the size increases slightly because the *d* subshell is filled, and electron–electron repulsions cause the size to increase.

Exercise 8.5   **Periodic Trends in Atomic Radii**

Place the three elements Al, C, and Si in order of increasing atomic radius.

Exercise 8.6   **Estimating Atom–Atom Distances**

(a) Using Figure 8.9, estimate the H—O and H—S distances in H_2O and H_2S, respectively.

(b) If the interatomic distance in Br_2 is 228 pm, what is the radius of Br? Using this value, and that for Cl (99 pm), estimate the distance between atoms in BrCl.

Ionization Energy

Ionization energy is the energy required to remove an electron from an atom in the gas phase.

$$\text{Atom in ground state(g)} \longrightarrow \text{Atom}^+(g) + e^-$$

$$\Delta E \equiv \text{ionization energy, } IE$$

- **Atomic Radii**

Caution: There are numerous tabulations of atomic and covalent radii, and the values quoted may differ. The variation comes about because several methods are used to determine the radii of atoms, and the different methods can give slightly different values.

Chapter Goals • Revisited

- Understand the role magnetism plays in determining and revealing atomic structure.

- Understand effective nuclear charge and its role in determining atomic properties.

- Write the electron configuration for any element or monatomic ion.

- **Understand the fundamental physical properties of the elements and their periodic trends.**

To separate an electron from an atom, energy must be supplied to overcome the attraction of the nuclear charge. Because energy must be supplied, the sign of the ionization energy is always positive.

Each atom (except the hydrogen atom) can have a series of ionization energies, because more than one electron can always be removed [◀ PAGE 87]. For example, the first three ionization energies of magnesium are

$$\text{First ionization energy, } IE_1 = 738 \text{ kJ/mol}$$

$$\underset{1s^22s^22p^63s^2}{\text{Mg}(g)} \longrightarrow \underset{1s^22s^22p^63s^1}{\text{Mg}^+(g)} + e^-$$

$$\text{Second ionization energy, } IE_2 = 1451 \text{ kJ/mol}$$

$$\underset{1s^22s^22p^63s^1}{\text{Mg}^+(g)} \longrightarrow \underset{1s^22s^22p^63s^0}{\text{Mg}^{2+}(g)} + e^-$$

$$\text{Third ionization energy, } IE_3 = 7733 \text{ kJ/mol}$$

$$\underset{1s^22s^22p^6}{\text{Mg}^{2+}(g)} \longrightarrow \underset{1s^22s^22p^5}{\text{Mg}^{3+}(g)} + e^-$$

Notice that removing each subsequent electron requires more energy because the electron is being removed from an increasingly positive ion. Most importantly, notice the large increase in ionization energy for removing the third electron to give Mg^{3+}. *This large increase is one piece of experimental evidence for the electron shell structure of atoms.* The first two ionization steps are for the removal of electrons from the outermost, or valence, shell of electrons. The third electron, however, must come from the $2p$ subshell. This subshell is significantly lower in energy than the $3s$ subshell (see page 294), and considerably more energy is required to remove the $n = 2$ electron than the $n = 3$ electrons.

As another example, consider the first two ionization energies for lithium.

$$\text{First ionization energy, } IE_1 = 513.3 \text{ kJ/mol}$$

$$\underset{1s^22s^1}{\text{Li}(g)} \longrightarrow \underset{1s^22s^°}{\text{Li}^+(g)} + e^-$$

$$\text{Second ionization energy, } IE_2 = 7298 \text{ kJ/mol}$$

$$\underset{1s^2}{\text{Li}^+(g)} \longrightarrow \underset{1s^1}{\text{Li}^{2+}(g)} + e^-$$

● Valence and Core Electrons
Removal of core electrons requires much more energy than removal of a valence electron. Loss of core electrons does not occur in ordinary chemical reactions.

Here the large increase in ionization energy occurs for the removal of the second electron, which in this case is also the first electron to come from a much lower energy, inner subshell.

For main group (*s*- and *p*-block) elements, *first ionization energies generally increase across a period and decrease down a group* (Figure 8.11 and Appendix F). The trend *across a period* is rationalized by the increase in effective nuclear charge, Z^*, with increasing atomic number. Not only does this mean that the atomic radius decreases, but the energy required to remove an electron increases. The general decrease in ionization energy *down a group* occurs because the electron removed is farther and farther from the nucleus, thus reducing the nucleus–electron attractive force [💿 CD-ROM, Screens 8.9 and 8.12].

● Trends in Ionization Energy
General trends in first ionization energies of *s*- and *p*-block elements with position in the periodic table.

A closer look at ionization energies reveals that the trend across a given period is not smooth, particularly in the second period. Variations are seen on going from *s*-block to *p*-block elements, from beryllium to boron, for example. This occurs because the $2p$ electrons are slightly higher in energy than the $2s$ electrons (see page 294), and so the ionization for boron is lower than for beryllium.

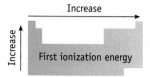

Increase

First ionization energy

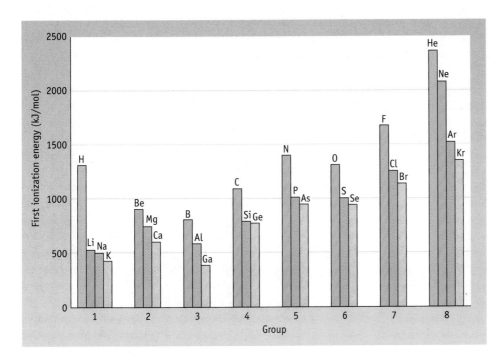

Figure 8.11   First ionization energies of the main group elements of the first four periods. (For data on all the elements see Appendix F.)

Moving from boron to carbon and then to nitrogen, the effective nuclear charge increases (see Table 8.2), which again means an increase in ionization energy. Another dip to lower ionization energy occurs on passing from Group 5A to Group 6A. This is especially noticeable in the second period (N and O). No change occurs in either n or ℓ, but electron–electron repulsions increase for the following reason. In Groups 3A–5A, electrons are assigned to separate p orbitals (p_x, p_y, and p_z). Beginning in Group 6A, however, two electrons are assigned to the same p orbital. The fourth p electron shares an orbital with another electron and thus experiences greater repulsion than it would if it had been assigned to an orbital of its own:

$$\text{O (oxygen atom)} \xrightarrow{\ +1314 \text{ kJ/mol}\ } \text{O}^+ \text{ (oxygen cation)} + e^-$$

$$[\text{Ne}]\,\boxed{\uparrow\downarrow}\quad \boxed{\uparrow\downarrow\ \uparrow\ \uparrow} \qquad\qquad [\text{Ne}]\,\boxed{\uparrow\downarrow}\quad \boxed{\uparrow\ \uparrow\ \uparrow}$$
$$\quad\ \ 2s \qquad\quad 2p \qquad\qquad\qquad\ \ \ 2s \qquad\quad 2p$$

The greater repulsion experienced by the fourth $2p$ electron makes it easier to remove, and each of the remaining p electrons has an orbital of its own. The normal trend resumes on going from oxygen to fluorine to neon, however, as the increase in Z* overcomes the effect of pairing electrons in the $2p$ subshell.

Electron Affinity

Some atoms have an affinity, or "liking," for electrons and can acquire one or more electrons to form a negative ion. The **electron affinity,** *EA,* of an atom is defined as the energy of a process in which an electron is acquired by the atom in the gas phase (Figure 8.12 and Appendix F) [CD-ROM, Screens 8.9 and 8.13].

$$\text{A(g)} + e^- \longrightarrow \text{A}^-(\text{g}) \qquad \Delta E \equiv \text{electron affinity, } EA$$

• **Factors Controlling Trends in Ionization Energies**
The ionization energy of an atom is always a balance between electron–nuclear attraction (which depends on Z*) and electron-electron repulsion.

Figure 8.12 Electron affinity. The larger the affinity (*EA*) of an atom for an electron, the more negative the value. For numerical data, see Appendix F. *(Data were taken from H. Hotop and W. C. Lineberger: "Binding energies of atomic negative ions,"* Journal of Physical Chemistry, *Reference Data, Vol. 14, p. 731, 1985.)*

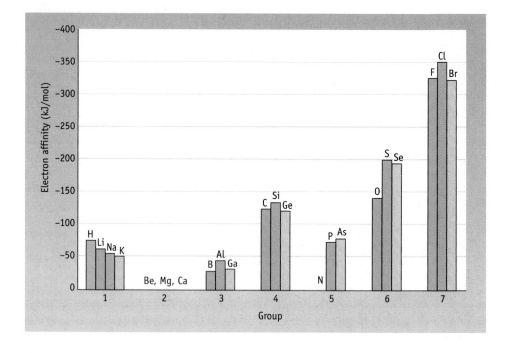

● **Electron Affinity and Sign Conventions** Electron affinity has been the subject of confusion in introductory chemistry for some years. For a useful discussion see J. C. Wheeler: "Electron affinities of the alkaline earth metals and the sign convention for electron affinity." *Journal of Chemical Education*, Vol. 74, pp. 123–127, 1997.

● **EA Values of Zero** The value of *EA* for Be is not measurable because the Be⁻ anion does not exist. Most tables assign a value of 0 to the *EA* for this element and similar cases (in particular the Group 2A elements and N).

The greater the affinity an atom has for an electron, the more negative the value of *EA* will be. For example, the electron affinity of fluorine is −328 kJ/mol, a large value indicating an exothermic, product-favored reaction to form a stable anion, F⁻. Boron has a much lower affinity for an electron, however, as indicated by a much less negative *EA* value of −26.7 kJ/mol.

Electron affinity and ionization energy represent the energy involved in the gain or loss of an electron by an atom, respectively. It is therefore not surprising that periodic trends in electron affinity are related to those for ionization energy. The effective nuclear charge of atoms increases across a period (see Table 8.2), not only making it more difficult to ionize the atom but also increasing the attraction of the atom for an additional electron. Thus, *an element with a high ionization energy generally has a high affinity for an electron.* As seen in Figure 8.12, the values of *EA* generally become more negative on moving across a period as the affinity for electrons increases.

One result of increasing Z* across a period is that nonmetals generally have much more negative values of *EA* than metals, reflecting the greater affinity of nonmetals for electrons. This agrees with chemical experience, which tells us that metals generally do not form negative ions and that nonmetals have an increasing tendency to form anions as we proceed across a period.

The trend to more negative electron affinity values across a period is not smooth. For example, beryllium has no affinity for an electron. A beryllium anion, Be⁻, is not stable because the added electron must be assigned to a higher energy subshell (2*s*) than the valence electrons (2*s*) (see page 294). Nitrogen atoms also have no affinity for electrons. Here an electron pair must be formed when an N atom acquires an electron. Significant electron–electron repulsions occur in an N⁻ ion, making the ion much less stable. The increase in Z* on going from carbon to nitrogen cannot overcome the effect of these electron–electron repulsions.

The noble gases are not included in our discussion of electron affinity. They have no affinity for electrons because any additional electron must be added to the next higher quantum shell. The higher Z* of the noble gases is insufficient to overcome this effect.

On descending a group of the periodic table the affinity for an electron generally declines. Electrons are added farther and farther from the nucleus, so the attractive force between the nucleus and electrons decreases. Figure 8.12 shows that this is the case for Cl and Br or P and As, for example. However, the affinity of the fluorine atom for an electron is lower than that of chlorine (*EA* for F is less negative than *EA* for Cl); the same phenomenon is observed in Groups 3A through 6A as well. One explanation is that significant electron–electron repulsions occur in the F⁻ ion. Adding an electron to the seven already present in the $n = 2$ shell of the small F atom leads to considerable repulsion between electrons. Chlorine has a larger atomic volume than fluorine, so adding an electron does not result in such significant electron–electron repulsions in the Cl⁻ anion.

No atom has a negative electron affinity for a *second* electron. Attaching a second electron to an ion that already has a negative charge leads to severe repulsions. The attachment of a second electron is *always* an endothermic process. So how can you account for ions such as O^{2-}, which is present in so many naturally occurring substances (such as CaO)? The answer is that doubly charged anions can sometimes be stabilized in crystalline environments by electrostatic attraction to neighboring positive ions [➡ CHAPTERS 9 AND 13].

● **Trends in *EA***
General trends in electron affinities of A-group elements. Exceptions occur at Groups 2A and 5A.

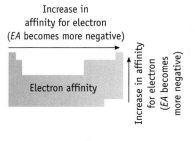

Example 8.5   **Periodic Trends**

Problem • Compare the three elements C, O, and Si.

(a) Place them in order of increasing atomic radius.

(b) Which has the largest ionization energy?

(c) Which has the more negative electron affinity, O or C?

Strategy • Review the trends in atomic properties in Figures 8.9, 8.11, and 8.12.

Solution •
(a) *Atomic size.* Atomic radius declines on moving across a period, so oxygen must have a smaller radius than carbon. However, radius increases down a periodic group. Because C and Si are in the same group (Group 4A), Si must be larger than C. In order of increasing size, the trend is therefore O < C < Si.

(b) *Ionization energy (IE).* Ionization energy generally increases across a period and decreases down a group; a large decrease in *IE* occurs from the second- to the third-period elements. Thus, the trend in ionization energies should be Si < C < O.

(c) *Electron affinity (EA).* Electron affinity values generally become more negative across a period and less negative down a group. Therefore, the *EA* for O should be more negative than the *EA* for C. That is, O (*EA* = −141.0 kJ/mol) has a greater affinity for an electron than does C (*EA* = −121.9 kJ/mol).

Comment • *EA* for third period elements is generally slightly more negative than *EA* for second period elements. This occurs because of electron–electron repulsions; such repulsions are larger in the small C⁻ ion than in the larger Si⁻ ion, for example.

Exercise 8.7   **Periodic Trends**

Compare the three elements B, Al, and C.

(a) Place the three elements in order of increasing atomic radius.

(b) Rank the elements in order of increasing ionization energy. (Try to do this without looking at Figure 8.13; then compare your estimates with the graph.)

(c) Which element is expected to have the most negative electron affinity value?

Figure 8.13 **Relative sizes of some common ions.** Radii are given in picometers (1 pm = 1 × 10^{-12} m). *(Data taken from J. Emsley: The Elements, Clarendon Press, Oxford, 1998, 3rd ed.)*

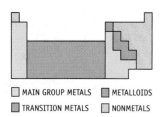

☐ MAIN GROUP METALS ☐ METALLOIDS
☐ TRANSITION METALS ☐ NONMETALS

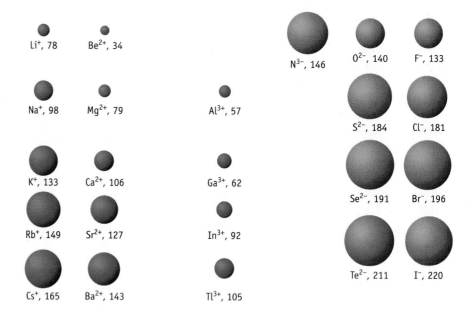

Li$^+$, 78   Be^{2+}, 34   N^{3-}, 146   O^{2-}, 140   F$^-$, 133

Na$^+$, 98   Mg^{2+}, 79   Al^{3+}, 57   S^{2-}, 184   Cl$^-$, 181

K$^+$, 133   Ca^{2+}, 106   Ga^{3+}, 62   Se^{2-}, 191   Br$^-$, 196

Rb$^+$, 149   Sr^{2+}, 127   In^{3+}, 92   Te^{2-}, 211   I$^-$, 220

Cs$^+$, 165   Ba^{2+}, 143   Tl^{3+}, 105

Ion Sizes

Having considered the energies involved in forming positive and negative ions, let us now look at the periodic trends in ion radii [🖰 CD-ROM, Screens 8.9 and 8.11].

Figure 8.13 shows clearly that periodic trends in the sizes of a few common ions in the same group are the same as those for neutral atoms: positive or negative ions increase in size when descending the group. Pause for a moment, however, and compare ionic radii in Figure 8.13 with atomic radii in Figure 8.9. When an electron is removed from an atom to form a cation, the size shrinks considerably; *the radius of a cation is always smaller than that of the atom from which it is derived.* For example, the radius of Li is 152 pm, whereas that for Li$^+$ is only 78 pm. This occurs because when an electron is removed from a Li atom the attractive force of three protons is now exerted on only two electrons, and the remaining electrons are drawn toward the nucleus. The decrease in ion size is especially great when the last electron of a particular shell is removed, as is the case for Li. The loss of the 2s electron from Li leaves Li$^+$ with no electrons in the n = 2 shell.

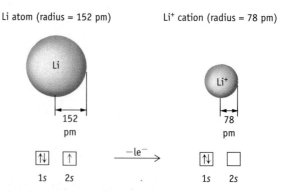

Li atom (radius = 152 pm)   Li$^+$ cation (radius = 78 pm)

Li   Li$^+$

152   78
pm   pm

⇅ ↑ $\xrightarrow{-1e^-}$ ⇅ ☐
1s 2s 1s 2s

The shrinkage will also be great when two or more electrons are removed, as for Al^{3+} in which it is over 50%:

Al atom (radius = 143 pm)   Al^{3+} cation (radius = 57 pm)

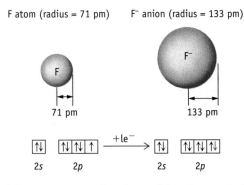

You can also see by comparing Figures 8.9 and 8.13 that *anions are always larger than the atoms from which they are derived.* Here the argument is the opposite of that used to explain positive ion radii. The F atom, for example, has nine protons and nine electrons. On forming the anion, the nuclear charge is still +9, but now ten electrons are in the anion. The F$^-$ ion is much larger than the F atom because of increased electron–electron repulsions.

F atom (radius = 71 pm)   F$^-$ anion (radius = 133 pm)

Finally, it is useful to compare the sizes of isoelectronic ions across the periodic table. **Isoelectronic ions** have the same number of electrons, and one such series of commonly occurring ions is O^{2-}, F$^-$, Na$^+$, and Mg^{2+}:

Ion	O^{2-}	F$^-$	Na$^+$	Mg^{2+}
Number of electrons	10	10	10	10
Number of nuclear protons	8	9	11	12
Ionic radius (pm)	140	133	98	79

All these ions have a total of ten electrons. The O^{2-} ion, however, has only 8 protons in its nucleus to attract these electrons, whereas F$^-$ has 9, Na$^+$ has 11, and Mg^{2+} has 12. As the number of protons increases in a series of isoelectronic ions, the balance in electron–proton attraction and electron–electron repulsion shifts in favor of attraction, and the radius decreases. As you can see in Figure 8.13, this is generally true for all isoelectronic series of ions.

Exercise 8.8   **Ion Sizes**

What is the trend in sizes of the ions N^{3-}, O^{2-}, and F$^-$? Briefly explain the reason for this trend.

8.7   PERIODIC TRENDS AND CHEMICAL PROPERTIES

Atomic and ionic radii, ionization energies, and electron affinities are properties associated with atoms and their ions. It is reasonable to expect that knowledge of these properties will be useful as we explore the chemistry of the elements

[⏻ CD-ROM, Screen 8.15]. Let us consider just one example here, the formation of ionic compounds.

As related in Section 2.6, the periodic table was created by grouping together elements having similar chemical properties. Alkali metals, for example, characteristically form compounds in which the metal is in the form of a +1 ion, such as Li^+, Na^+ or K^+. Thus, the reaction between sodium and chlorine gives the ionic compound, NaCl (composed of Na^+ and Cl^- ions) [← FIGURE 1.2, PAGE 13], and potassium and water react to form an aqueous solution of KOH (a solution containing the hydrated ions $K^+(aq)$ and $OH^-(aq)$) (page 288).

$$2\ Na(s) + Cl_2(g) \longrightarrow 2\ NaCl(s)$$
$$2\ K(s) + 2\ H_2O(\ell) \longrightarrow 2\ K^+(aq) + 2\ OH^-(aq) + H_2(g)$$

Both of these observations agree with the fact that alkali metals have electron configurations of the type [noble gas core]ns^1 and have low ionization energies.

You might have wondered, however, why sodium doesn't form $NaCl_2$ when it reacts with chlorine, or why potassium gives KOH and not a compound with a formula such as $K(OH)_2$ on reacting with water. Reflecting on the nature of ionization energies gives us an answer. Reactions generally provide the most stable product, and formation of a Na^{2+} or K^{2+} ion is clearly a very unfavorable process. Although the first ionization energy of the Group 1A elements is low, removing a second electron from the metal requires a great deal of energy because this electron must come from the atom's core electrons. Indeed, removal of core electrons from any atom is exceedingly unfavorable. This is the underlying reason that *main group metals generally form cations with an electron configuration equivalent to that of the nearest noble gas.*

Why isn't Na_2Cl another possible product from the sodium and chlorine reaction? This formula implies that the compound is formed from Na^+ and Cl^{2-} ions. Chlorine atoms have a relatively high electron affinity, but only for the addition of one electron. Adding two electrons per atom means the second must enter the next higher quantum shell, a shell of much higher energy. An anion such as Cl^{2-} is simply not stable. This example leads to the general statement that *nonmetals generally acquire enough electrons to form an anion with the electron configuration of the next, higher noble gas.*

We can use similar logic to rationalize results of other reactions. Ionization energies increase on going from left to right across a period. We have seen that Group 1A and 2A elements form ionic compounds, an observation directly related to the low ionization energies for these elements. Ionization energies for elements toward the middle and right side of a period, however, are sufficiently large that cation formation is not favorable. Thus, we generally do not expect to encounter ionic compounds containing carbon; instead we find carbon *sharing* electrons with other elements in compounds like CO_2 and CCl_4. At the right side of the second period, oxygen and fluorine much prefer taking on electrons than giving them up; these elements have high ionization energies and relatively large electron affinities. Thus, oxygen and fluorine form anions and not cations when they react.

Finally, what does it mean when chemists speak of the tendency of atoms "to achieve a noble gas configuration" or to "complete an electron octet." All noble gases have the valence electron configuration ns^2np^6 (plus any $(n-1)d^{10}$ and $(n-2)f^{14}$ electrons, as appropriate), and so have a total of eight ns and np valence electrons. The connection between the noble gas configuration and chemical reactivity should now be apparent.

- The noble gases themselves have high ionization energies and low electron affinities. They are unreactive.

- If an anion has the noble gas configuration, an additional electron would have to be added to the next higher quantum shell. This is not energetically favorable.
- Sufficient energy is available in ordinary chemical reactions to remove electrons to form cations with a noble gas configuration. For a cation to lose electrons beyond this, however, means that the additional electrons must come from an inner electron shell. The required energy is greater than is available under ordinary conditions.

Exercise 8.9 | Energies and Compound Formation

Give a plausible explanation for the observation that magnesium and chlorine react to form $MgCl_2$ and not $MgCl_3$.

In Summary

When you have finished studying this chapter, you should ask if you have met the chapter goals. In particular, you should be able to

- Classify substances as paramagnetic (attracted to a magnetic field; characterized by unpaired electron spins) or diamagnetic (repelled by a magnetic field; all electrons paired) (Section 8.1).
- Recognize that each electron in an atom has a different set of the four quantum numbers, n, ℓ, m_ℓ, and m_s, where m_s, the spin quantum number, has values of $+\frac{1}{2}$ or $-\frac{1}{2}$ (Section 8.2).
- Understand that the Pauli exclusion principle leads to the conclusion that no atomic orbital can be assigned more than two electrons and that the two electrons in an orbital must have opposite spins (different values of m_s) (Section 8.2).
- Understand effective nuclear charge, Z^*, and how it can be used to explain why different subshells in the same shell have different energies. Also, understand the role of Z^* in determining the properties of atoms (Sections 8.3 and 8.6).
- Use the periodic table as a guide to depict electron configurations of the elements and monatomic ions using an orbital box notation or an *spdf* notation. (In both cases, configurations can be abbreviated with the noble gas notation) (Sections 8.3 and 8.4).
- Recognize that electrons are assigned to the subshells of an atom in order of increasing subshell energy. In the H atom the subshell energies increase with increasing n, but, in a many-electron atom, the energies depend on both n and ℓ (see Figure 8.4).
- Apply the Pauli exclusion principle and Hund's rule when assigning electrons to atomic orbitals (Sections 8.2 and 8.4).
- Predict how properties of atoms — size, ionization energy (*IE*), and electron affinity (*EA*) — change on moving down a group or across a period of the periodic table (Section 8.6). The general periodic trends for these properties are
  (a) Atomic size decreases across a period and increases down a group.
  (b) *IE* increases across a period and decreases down a group.
  (c) The affinity for an electron increases generally across a period (the value of *EA* becomes more negative) and decreases down a group.

- Recognize the role that ionization energy and electron affinity play in the chemistry of the elements (Section 8.7).

Key Terms

Section 8.1
electron spin magnetic quantum
   number, m_s
diamagnetic
paramagnetic
ferromagnetic

Section 8.2
Pauli exclusion principle
orbital box diagram

Section 8.3
effective nuclear charge, Z*

Section 8.4
electron configurations
spdf notation
noble gas notation
core electrons
valence electrons
s-block elements

p-block elements
Hund's rule
lanthanides
actinides

Section 8.6
electron affinity
ionization energy
isoelectronic

Study Questions

Questions with blue, bold-faced numbers have answers in Appendix O. See also the Example Problems, Tutorials, and Simulations in Chapter 8 of the General Chemistry Interactive *CD-ROM, Version 3.0.*

Reviewing Important Concepts

1. List the four quantum numbers, specify their allowed values, and tell what property of the electron they describe.

2. What is the Pauli exclusion principle?

3. Using lithium as an example, show the two methods of depicting electron configurations (orbital box diagram and *spdf* notation).

4. What is Hund's rule? Give an example of its application.

5. What is the noble gas notation? Write an electron configuration using this notation.

6. Name an element of Group 3A. What does the group designation tell you about the electron configuration of the element?

7. Name an element of Group 7B. What does the group designation tell you about the electron configuration of the element?

8. What element is located in the fourth period in Group 4A? What does the element's location tell you about its electron configuration?

9. Describe the trends in atomic size, ionization energy, and electron affinity when proceeding across a period and down a group.

Practicing Skills

Writing Electron Configurations of Atoms
(See Examples 8.1–8.3, Tables 8.1, 8.3, and 8.4 and CD-ROM Screen 8.7)

10. Write the electron configurations for P and Cl, using both the *spdf* notation and orbital box diagrams. Describe the re-

lation of the atom's electron configuration to its position in the periodic table.

11. Write the electron configurations for Mg and Ar, using both the *spdf* notation and orbital box diagrams. Describe the relation of the atom's electron configuration to its position in the periodic table.

12. Using the *spdf* notation, write the electron configurations for atoms of chromium and iron, two of the major components of stainless steel.

13. Using the *spdf* notation, give the electron configuration of vanadium, V, an element found in some brown and red algae and some toadstools.

14. Depict the electron configuration for each of the following atoms using the *spdf* and noble gas notations.

    (a) Arsenic, As. A deficiency of As can impair growth in animals even though larger amounts are poisonous.

    (b) Krypton, Kr. It ranks seventh in abundance of the gases in the earth's atmosphere.

15. Using the *spdf* and noble gas notations, write electron configurations for atoms of the following elements and then check your answers with Table 8.2:

    (a) Strontium, Sr. This element is named for a town in Scotland.

    (b) Zirconium, Zr. The metal is exceptionally resistant to corrosion and so has important industrial applications. Moon rocks show a surprisingly high zirconium content compared with rocks on earth.

    (c) Rhodium, Rh. This metal is used in jewelry and in catalysts in industry.

    (d) Tin, Sn. The metal was used in the ancient world. Alloys of tin (solder, bronze, and pewter) are important.

16. Use the noble gas and *spdf* notations to predict electron configurations for the following metals of the third transition series:

    (a) Tantalum, Ta. The metal and its alloys resist corrosion and are often used in surgical and dental tools.

    (b) Platinum, Pt. This metal was used by pre-Columbian Indians in jewelry. It is used now in jewelry and for anticancer drugs and industrial catalysts.

17. The lanthanides, once called rare earths, are really only "medium rare." Using the noble gas and *spdf* notations, predict reasonable electron configurations for the following elements:

    (a) Samarium, Sm. This lanthanide is used in magnetic materials.

    (b) Ytterbium, Yb. This element was named for the village of Ytterby in Sweden where a mineral source of the element was found.

18. The actinide americium, Am, is a radioactive element that has found use in home smoke detectors. Depict its electron configuration using the noble gas and *spdf* notations.

19. Predict reasonable electron configurations for the following elements of the actinide series of elements. Use the noble gas and *spdf* notations.

    (a) Plutonium, Pu. The element is best known as a byproduct of nuclear power plant operation.

    (b) Curium, Cm. This actinide was named for Madame Curie (page 48).

Electron Configurations of Atoms and Ions and Magnetic Behavior
(See Example 8.4 and CD-ROM Screens 8.3, 8.7, and 8.8)

20. Using orbital box diagrams, depict an electron configuration for each of the following ions: (a) Mg^{2+}, (b) K^+, (c) Cl^-, and (d) O^{2-}.

21. Using orbital box diagrams, depict an electron configuration for each of the following ions: (a) Na^+, (b) Al^{3+}, (c) Ge^{2+}, and (d) F^-.

22. Using orbital box diagrams and the noble gas notation, depict the electron configurations of (a) V, (b) V^{2+}, and (c) V^{5+}. Are any of the ions paramagnetic?

23. Using orbital box diagrams and the noble gas notation, depict the electron configurations of (a) Ti, (b) Ti^{2+}, and (c) Ti^{4+}. Are any of the ions paramagnetic?

24. Manganese is found as MnO_2 in deep ocean deposits.

    (a) Depict the electron configuration of this element using the noble gas notation and an orbital box diagram.

    (b) Using an orbital box diagram, show the electrons beyond those of the preceding noble gas for the +2 ion.

    (c) Is the +2 ion paramagnetic?

    (d) How many unpaired electrons does the Mn^{2+} ion have?

25. Copper plays an important biochemical role, particularly in electron transfer. It exists as both +1 and +2 ions. Using orbital box diagrams and the noble gas notation, show elec-tron configurations of these ions. Are either of these ions paramagnetic?

Quantum Numbers and Electron Configurations
(See Example 8.2 and CD-ROM Screens 7.12, 8.4, and 8.7)

26. Explain briefly why each of the following is not a possible set of quantum numbers for an electron in an atom. In each case, change the incorrect value (or values) to make the set valid.

    (a) $n = 4$, $\ell = 2$, $m_\ell = 0$, $m_s = 0$

    (b) $n = 3$, $\ell = 1$, $m_\ell = -3$, $m_s = -\frac{1}{2}$

    (c) $n = 3$, $\ell = 3$, $m_\ell = -1$, $m_s = +\frac{1}{2}$

27. Explain briefly why each of the following is not a possible set of quantum numbers for an electron in an atom. In each case, change the incorrect value (or values) to make the set valid.

    (a) $n = 2$, $\ell = 2$, $m_\ell = 0$, $m_s = +\frac{1}{2}$

    (b) $n = 2$, $\ell = 1$, $m_\ell = -1$, $m_s = 0$

    (c) $n = 3$, $\ell = 1$, $m_\ell = +2$, $m_s = +\frac{1}{2}$

28. What is the maximum number of electrons that can be identified with each of the following sets of quantum numbers? In one case, the answer is "none." Explain why this is true.

    (a) $n = 4$, $\ell = 3$

    (b) $n = 6$, $\ell = 1$, $m_\ell = -1$

    (c) $n = 3$, $\ell = 3$, $m_\ell = -3$

29. What is the maximum number of electrons that can be identified with each of the following sets of quantum numbers? In some cases, the answer may be "none." In such cases, explain why "none" is the correct answer.

    (a) $n = 3$

    (b) $n = 3$ and $\ell = 2$

    (c) $n = 4$, $\ell = 1$, $m_\ell = -1$, and $m_s = -\frac{1}{2}$

    (d) $n = 5$, $\ell = 0$, $m_\ell = +1$

30. Depict the electron configuration for magnesium using the orbital box and noble gas notations. Give a complete set of four quantum numbers for each of the electrons beyond those of the preceding noble gas.

31. Depict the electron configuration for phosphorus using the orbital box and noble gas notations. Give one possible set of four quantum numbers for each of the electrons beyond those of the preceding noble gas.

32. Using an orbital box diagram and noble gas notation, show the electron configuration of gallium, Ga. Give a set of quantum numbers for the highest energy electron.

33. Using an orbital box diagram and the noble gas notation, show the electron configuration of titanium. Give one possible set of four quantum numbers for each of the electrons beyond those of the preceding noble gas.

Periodic Properties
(See Section 8.6, Example 8.5, and CD-ROM Screens 8.9– 8.13)

34. Arrange the following elements in order of increasing size: Al, B, C, K, and Na. (Try doing it without looking at Figure

8.9, then check yourself by looking up the necessary atomic radii.)

35. Arrange the following elements in order of increasing size: Ca, Rb, P, Ge, and Sr. (Try doing it without looking at Figure 8.9, then check yourself by looking up the necessary atomic radii.)

36. Select the atom or ion in each pair that has the larger radius.
    (a) Cl or Cl⁻
    (b) Al or O
    (c) In or I

37. Select the atom or ion in each pair that has the larger radius.
    (a) Cs or Rb
    (b) O^{2-} or O
    (c) Br or As

38. Which of the following groups of elements is arranged correctly in order of increasing ionization energy?
    (a) C < Si < Li < Ne   (c) Li < Si < C < Ne
    (b) Ne < Si < C < Li   (d) Ne < C < Si < Li

39. Arrange the following atoms in order of increasing ionization energy: Li, K, C, and N.

40. Compare the elements Na, Mg, O, and P.
    (a) Which has the largest atomic radius?
    (b) Which has the most negative electron affinity?

(c) Place the elements in order of increasing ionization energy.

41. Compare the elements B, Al, C, and Si.
    (a) Which has the most metallic character?
    (b) Which has the largest atomic radius?
    (c) Which has the most negative electron affinity?
    (d) Place the three elements B, Al, and C in order of increasing first ionization energy.

42. Periodic trends. Explain each answer briefly.
    (a) Place the following elements in order of increasing ionization energy: F, O, and S.
    (b) Which has the largest ionization energy: O, S, or Se?
    (c) Which has the most negative electron affinity: Se, Cl, or Br?
    (d) Which has the largest radius: O^{2-}, F⁻, or F?

43. Periodic trends. Explain each answer briefly.
    (a) Rank the following in order of increasing atomic radius: O, S, and F.
    (b) Which has the largest ionization energy: P, Si, S, or Se?
    (c) Place the following in order of increasing radius: O^{2-}, N^{3-}, or F⁻.
    (d) Place the following in order of increasing ionization energy: Cs, Sr, Ba.

General Questions on Electron Configurations and Periodic Trends

These questions are not designated as to type or location in the chapter. They may combine several concepts. More challenging questions are indicated by an underlined number.

44. The name rutherfordium, Rf, has been given to element 104 to honor the physicist Ernest Rutherford (page 50). Depict its electron configuration using the *spdf* and noble gas notations.

45. Using an orbital box diagram and the noble gas notation, show the electron configuration of uranium and of the uranium(IV) ion. Is either of these paramagnetic?

46. The rare earth elements, or lanthanides, commonly exist as +3 ions. Using an orbital box diagram and the noble gas notation, show the electron configurations of the following elements and ions:
    (a) Ce and Ce^{3+} (cerium)
    (b) Ho and Ho^{3+} (holmium)

47. A neutral atom has two electrons with n = 1, eight electrons with n = 2, eight electrons with n = 3, and two electrons with n = 4. Assuming this element is in its ground state, supply the following information:
    (a) Atomic number
    (b) Total number of s electrons
    (c) Total number of p electrons
    (d) Total number of d electrons
    (e) Is the element a metal, metalloid, or nonmetal?

48. Element 109, now named meitnerium (named in honor of the Austrian–Swedish physicist, Lise Meitner [1878–1968]), was produced in August, 1982, by a team at Germany's Institute for Heavy Ion Research. Depict its electron configuration using the *spdf* and noble gas notations. Name another element found in the same group as meitnerium.

49. Which of the following is *not* an allowable set of quantum numbers? Explain your answer briefly.

	n	ℓ	m_ℓ	m_s
(a)	2	0	0	$-\frac{1}{2}$
(b)	1	1	0	$+\frac{1}{2}$
(c)	2	1	-1	$-\frac{1}{2}$
(d)	4	3	$+2$	$-\frac{1}{2}$

50. A possible excited state for the H atom has an electron in a 4p orbital. List all possible sets of quantum numbers (n, ℓ, m_ℓ, and m_s) for this electron.

51. How many *complete* electron shells are
    (a) In element 71?
    (b) In copper?

52. Name the element corresponding to each of the following characteristics:

    (a) The element with the electron configuration $1s^2 2s^2 2p^6 3s^2 3p^3$

    (b) The alkaline earth element with the smallest atomic radius

    (c) The element with the largest ionization energy in Group 5A

    (d) The element whose +2 ion has the configuration $[Kr]4d^5$

    (e) The element with the most negative electron affinity in Group 7A

    (f) The element whose electron configuration is $[Ar]3d^{10}4s^2$

53. Arrange the following atoms in order of increasing ionization energy: Si, K, As, and Ca.

54. Rank the following in order of increasing ionization energy: Cl, Ca^{2+}, and Cl^-. Briefly explain your answer.

55. Answer the following questions about the elements A and B, which have the electron configurations shown.

    $$A = [Kr]5s^1 \qquad B = [Ar]3d^{10}4s^2 4p^4$$

    (a) Is element A a metal, nonmetal, or metalloid?

    (b) Which element has the greater ionization energy?

    (c) Which element has the more negative electron affinity?

    (d) Which element has the larger atomic radius?

56. Answer the following questions about the elements with the electron configurations shown here:

    $$A = [Ar]4s^2 \qquad B = [Ar]3d^{10}4s^2 4p^5$$

    (a) Is element A a metal, metalloid, or nonmetal?

    (b) Is element B a metal, metalloid, or nonmetal?

    (c) Which element is expected to have the larger ionization energy?

    (d) Which element has the smaller atomic radius?

57. Which of the following ions are unlikely to be found in a chemical compound: Cs^+, In^{4+}, Fe^{6+}, Te^{2-}, Sn^{5+}, and I^-? Explain briefly.

58. Place the following elements and ions in order of decreasing size: K^+, Cl^-, S^{2-}, and Ca^{2+}.

59. Answer each of the following questions:

    (a) Of the elements S, Se, and Cl, which has the largest atomic radius?

    (b) Which has the larger radius, Br or Br^-?

    (c) Which should have the largest difference between the first and second ionization energy: Si, Na, P, or Mg?

    (d) Which has the largest ionization energy: N, P, or As?

    (e) Which of the following has the largest radius: O^{2-}, N^{3-}, or F^-?

60. The following are isoelectronic species: Cl^-, K^+, and Ca^{2+}. Rank them in order of increasing (a) size, (b) ionization energy, and (c) electron affinity.

61. Compare the elements Na, B, Al, and C with regard to the following properties:

    (a) Which has the largest atomic radius?

    (b) Which has the largest electron affinity?

    (c) Place the elements in order of increasing ionization energy.

62. Two elements in the second transition series (Y through Cd) have four unpaired electrons in their +3 ions. What elements fit this description?

63. Explain how the ionization energy of atoms changes and why the change occurs when proceeding down a group of the periodic table.

64. The configuration for an element is given here.

    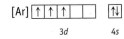

    (a) What is the identity of the element with this configuration?

    (b) Is a sample of the element paramagnetic or diamagnetic?

    (c) How many unpaired electrons does a +3 ion of this element have?

65. The configuration of an element is given here.

    [Ar] ↑ ↑ ↑ | □ □ | ↑↓
          3d   4s

    (a) What is the identity of the element?

    (b) In what group and period is the element found?

    (c) Is the element a nonmetal, a main group element, a transition element, a lanthanide element, or an actinide element?

    (d) Is the element diamagnetic or paramagnetic? If paramagnetic, how many unpaired electrons are there?

    (e) Write a complete set of quantum numbers (n, ℓ, m_ℓ, m_s) for each of the valence electrons.

    (f) What is the configuration of the +2 ion formed from this element? Is the ion diamagnetic or paramagnetic?

66. Trends in size:

    (a) Explain why the sizes of atoms change when proceeding across a period of the periodic table.

    (b) Explain why the sizes of transition metal atoms change, but not greatly, across a period.

67. Write electron configurations to show the first two ionization processes for potassium. Explain why the second ionization energy is much greater than the first.

68. Which of the following elements has the greatest difference between the first and second ionization energy: C, Li, N, and Be? Explain your answer.

69. Why is the radius of Li^+ so much smaller than the radius of Li? Why is the radius of F^- so much larger than the radius of F?

70. Which ions in the following list are not likely to be found in chemical compounds: K^{2+}, Cs^+, Al^{4+}, F^{2-}, and Se^{2-}? Explain briefly.

71. What arguments would you use to convince another student in general chemistry that MgO consists of the ions Mg^{2+} and O^{2-} and not the ions Mg^+ and O^-? What experiments could be done to provide some evidence that the correct formulation of magnesium oxide is $Mg^{2+}O^{2-}$?

72. Explain why the first ionization energy of Ca is greater than that of K, whereas the second ionization energy of Ca is lower than the second ionization energy of K.

73. In general, as you move across a periodic table, the affinity of the elements for an electron increases. One exception to this trend, however, is the large decrease in the electron affinity when going from Group 4A elements to those in Group 5A. Explain this decrease.

74. Explain why the reaction of calcium and fluorine does *not* form CaF_3.

75. Explain why lithium does not reduce argon to form LiAr, a compound having Li^+ and Ar^- ions.

76. Using your knowledge of the trends in element sizes on going across the periodic table, explain briefly why the density of the elements increases from K through V.

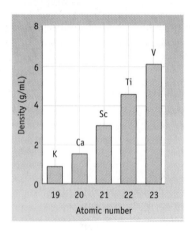

77. The reaction of cobalt metal with HCl gives $CoCl_2$, whereas the reaction with nitric acid gives $Co(NO_3)_3$. Using the magnetic behavior of these compounds, describe how to tell that neither $CoCl_3$ nor $Co(NO_3)_2$ is a reaction product.

78. The effective nuclear charge increases across the periodic table (see Table 8.2). The ionization energies of the first four second-period elements are in the order Li < Be > B < C. Explain this observation.

79. The ionization energies for the removal of the first electron in Si, P, S, and Cl are as listed in the table. Briefly rationalize this trend.

First Ionization Energy	
Element	**(kJ/mol)**
Si	780
P	1060
S	1005
Cl	1255

80. Suppose that elements on planet Suco have the same periodicity as elements on earth. Also suppose the planet Suco has a periodic table of elements. Shown here is a part of Suco's table and an incomplete table of atomic radii. Decide on the most likely radius for each element.

Element	Atomic Radius (pm)
_____	90
_____	120
_____	140
_____	180

81. Thionyl chloride, $SOCl_2$, is an important chlorinating and oxidizing agent in organic chemistry. It is prepared industrially by oxygen atom transfer from SO_3 to SCl_2.

$$SO_3(g) + SCl_2(g) \longrightarrow SO_2(g) + SOCl_2(g)$$

(a) Give the electron configuration for an atom of sulfur using the orbital box notation. Do not use the noble gas notation.

(b) Using the configuration given in part (a), write a set of quantum numbers for the highest energy electron in a sulfur atom.

(c) What element involved in this reaction (O, S, Cl) should have the smallest ionization energy? The smallest radius?

(d) Which should be smaller, the sulfide ion, S^{2-}, or a sulfur atom, S?

Using Electronic Resources

These questions refer to the General Chemistry Interactive CD-ROM, *Version 3.0.*

82. See CD-ROM Screen 8.16: Chemical Puzzler. Sodium metal reacts readily with chlorine gas to give sodium chloride.

$$Na(s) + \tfrac{1}{2} Cl_2(g) \longrightarrow NaCl(s)$$

(Charles D. Winters)

(a) What is the reducing agent in this reaction? What property of the element contributes to this ability as a reducing agent?

(b) What is the oxidizing agent in this reaction? What property of the element contributes to this ability as a reducing agent?

(c) Why does the reaction produce NaCl and not a compound such as Na_2Cl or $NaCl_2$?

83. See CD-ROM Screen 8.12: Ionization Energy. The ionization energy of oxygen is less than that of nitrogen because the addition of an electron to an already half-filled $2p$ orbital leads to greater electron–electron repulsions. In view of this observation, why is the ionization energy of beryllium higher than that of boron?

84. See CD-ROM Screen 8.6, Simulation: Effective Nuclear Charge.

(a) Is the effective nuclear charge Z^* ever the same as the nuclear charge?

(b) Describe the trend in effective nuclear charge on proceeding across the periodic table.

(c) How does the effective nuclear charge change on going from Ne to Na? What effect does this have on the chemical and physical properties of Na relative to Ne?

85. See CD-ROM Screen 8.9, Simulation: Orbital Energies.

(a) What happens to the energy of the $2s$ atomic orbital on moving across the second period elements?

(b) What happens to the highest occupied atomic orbital on moving down a group of elements?

86. See CD-ROM Screen 8.12, Simulation: Ionization Energy.

(a) Use the orbital energy diagrams on this screen to explain why the ionization energy of Mg is greater than that of Al.

(b) What happens to the $3p$ orbital energies on proceeding from Al to Ar? What effect does this have on the ionization energy of these elements?

87. If a C atom is attached, or "bonded," to a Cl atom, the calculated distance between the atoms is the sum of their radii. Calculate the expected distance between the pairs of atoms in the following table. Then use model molecules in the Molecular Models (CAChe) folder on the *General Chemistry Interactive CD-ROM*, Version 3.0, to examine the appropriate distance in the designated molecules. Is there reasonably good agreement between the calculated and measured distances?

Molecule	Atom Distance	Calculated (pm)	Measured (pm)
BF_3	B—F	_____	_____
PF_3	P—F	_____	_____
CH_4	C—H	_____	_____
H_3COH	C—O	_____	_____

(Note that BF_3 and PF_3 are molecular compounds in the Inorganic folder. CH_4, an alkane, and CH_3OH, methanol (an alcohol), are in the Organic folder. Directions for using the CAChe software are in the Documents folder. Finally, note that the distances given on these models are in Ångstrom units, where 1 Å = 100 pm.)

9 Bonding and Molecular Structure: Fundamental Concepts

Sugar in Space

Life is based on simple molecules like water and ammonia, slightly more complex ones like sugars, and very complex ones like DNA and hemoglobin. Where do they come from? How are they formed? What do they look like? Are their properties connected to how they look, that is, to their structures?

Where they come from is a topic eagerly studied by astronomers. Since the 1960s some space scientists have surmised that comets bring water, ammonia, and even more complex molecules of all kinds to earth from outer space. Every day an average of about 30 tons of organic material arrives on earth from space.

▲ The 12-meter radio telescope at the National Radio Astronomy Observatory. Used to search for molecules in deep space. *(National Radio Astronomy Observatory)*

▲ **The Eagle Nebula.** These pillar-like structures are vast columns of gas and dust, within which new stars have recently formed, and it is here that many molecules are created. The tallest of the pillars (at left) is about one light year in length from base to tip. The Eagle Nebula is a star-forming region 7000 light years away in the constellation Serpens. *(J. Hester and P. Scowan, of Arizona State University, and NASA)*

Over 120 molecules have been identified by radio astronomers in the far reaches of our galaxy. These range from hydrogen molecules to other simple molecules such as CO, H_2O, NH_3, and HCl.

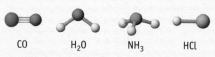

| CO | H_2O | NH_3 | HCl |

▲ Some molecules from deep space.

More recently more complex molecules have been observed, including a simple sugar, glycolaldehyde, $C_2H_4O_2$, discovered in 2001 in a cloud of gas and dust

"The discovery of this sugar molecule in a cloud from which new stars are forming means it is increasingly likely that the chemical precursors to life are formed in such clouds long before planets develop around the stars."

Jan M. Hollis, NASA Goddard Space Flight Center in Greenbelt, MD

Chapter Outline

about 26,000 light-years from earth. The compound is a member of the carbohydrate family, all of which have the general formula $C_a(H_2O)_b$. The molecule has two C atoms in its "backbone." One C atom is attached to an H atom and an O atom. The other C atom has two H atoms and one OH group attached. The structure and shape of the molecule are quite predictable, and such predictions are one objective of this chapter. We can explain, for example, what determines the angles between chemical bonds and why there are different kinds ofcarbon–oxygen bonds. We would also like to know how structure influences chemical and physical properties, so that we might predict how a molecule will interact with other molecules.

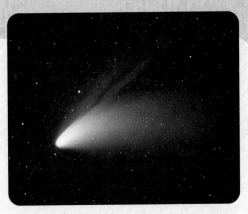

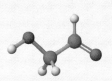

$HOCH_2CHO$

▲ Glycolaldehyde.

How are some of these complex molecules formed in deep space? Temperatures in deep space hover near absolute zero, and astronomers believe that simple molecules such as water, CO, CO_2, and CH_3OH (methanol) freeze onto the surface of minute pieces of interstellar dust. These dust particles are subjected to intense radiation from nearby stars and the molecules fragment (much as you saw in the mass spectrum in Figure 3.15). The fragments rearrange and combine, forming larger molecules such as glycolaldehyde.

Other compounds found in space are hydrocarbons (compounds composed only of C and H) such as anthracene. Anthracene is a member of a large class of compounds called polycyclic aromatic hydrocarbons. You may be aware of them because they are carcinogenic pollutants on earth, and you produce minute quantities when you cook a hamburger on a charcoal grill.

Why are compounds like anthracene flat? What relation do they have to other carbon-based compounds? These are just a few of the subjects we begin to explore in this and subsequent chapters.

▲ Comets deliver many complex molecules to earth. This is the Hale–Bopp comet in 1997. *(©1997, Fred Espenak, www.mreclipse.com)*

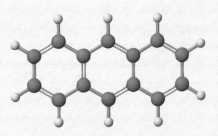

▲ Anthracene, a polycyclic aromatic hydrocarbon.

Before You Begin

- Know names of common compounds and ions (Chapter 3)
- Review Coulomb's law (page 93)
- Understand energy changes in chemical reactions (Chapter 6)

A major part of modern chemistry is the study of the relation between the structure of molecules and their function. This chapter begins the study of bonds that allow atoms to be assembled into molecules. Here we also begin a closer look at molecules, their structures, and their properties.

Insect Defense

Many insects have developed marvelous chemical defenses, and among them is the whip scorpion, often called a vinegarone. Its nickname comes from the fact that it sprays its enemies with a fine mist that is 85% acetic acid. This is the same acid found in common vinegar, but vinegar contains only about 3-5% acetic acid. So the scorpion uses a "supervinegar" as a part of its defense arsenal.

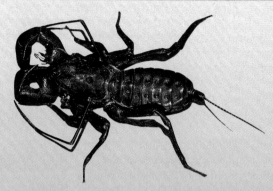

▲ The whip scorpion has two weapons in its arsenal. The defensive spray from the beetle consists not only of supervinegar but also of *n*-octanoic acid.

Structure

Acetic acid, CH_3CO_2H

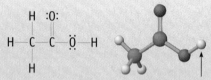

The acid loses this H atom as H^+ and functions as an acid.

In water, acetic acid is a weak acid. Less than 5% of the molecules ionize by loss of the H attached to the O atom.

n-Octanoic acid, $CH_3(CH_2)_6CO_2H$

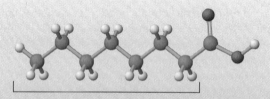

This portion of the molecule is nonpolar. This, combined with the $-CO_2H$ polar end of the molecule, allows it to penetrate the exoskeleton of insects and other enemies.

Function

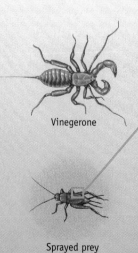

Vinegerone

Sprayed prey

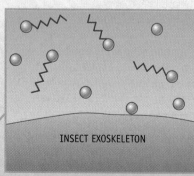

INSECT EXOSKELETON

The structure of the acetic acid does not allow it to interact with the waxy exoskeleton of another insect.

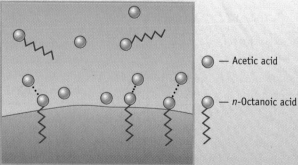

⬤ — Acetic acid

⬤ — *n*-Octanoic acid

However, the long carbon chain of *n*-octanoic acid, which is nonpolar, does interact with the enemy's exoskeleton and the acid is absorbed. The absorbed octanoic acid interacts with the acetic acid, and the two acids together penetrate the body of the prey.

Scientists have long known that the key to interpreting the properties of a chemical substance is first to recognize and understand its structure and bonding. **Structure** refers to the way atoms are arranged in space, and **bonding** defines the forces that hold adjacent atoms together. The introduction to this book told the story of how the basic structure of DNA was uncovered. But this structure raises many interesting questions, such as why it has a helical shape. The answer is related to the geometry of chemical bonds around each of its atoms. The specifics of that geometry will become more evident as you learn more about the topics of structure and bonding.

The goal of this and the next two chapters is to explain how atoms are arranged in chemical compounds and what holds them together. At the same time, we want to begin to show you how to relate the structure and bonding in a molecule to its chemical and physical properties.

Our discussion of structure and bonding begins with small molecules and ions and then progresses to larger molecules. You will see that from compound to compound, atoms of the same element participate in bonding and structure in the same manner. This consistency allows us to develop a group of principles that apply to many different chemical compounds, including those with complex structures such as DNA.   •

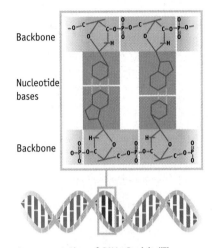

A representation of DNA. Revisit "The Double Helix" on page 3.

9.1   VALENCE ELECTRONS

The electrons in an atom can be divided into two groups: *valence electrons* and *core electrons*. Valence electrons are those in the outermost shell of an atom; they determine the chemical properties of the atom because chemical reactions result in the loss, gain, or rearrangement of valence electrons (◄ PAGE 298). The remaining electrons, those in inner shells, are the core electrons; they are not involved in chemical behavior [CD-ROM, Screen 9.4].

For main group elements (elements of the A groups in the periodic table), the valence electrons are the *s* and *p* electrons in the outermost shell (Table 9.1). All electrons in inner shells are core electrons. In addition, any electrons in *filled d* subshells are core electrons. Here is a useful guideline: for *main group elements the number of valence electrons is equal to the group number.* The fact that all elements in a periodic group have the same number of valence electrons accounts for the similarity of chemical properties among members of the group.

Table 9.1 • Core and Valence Electrons for Several Common Elements

Element	Periodic Group	Core Electrons	Valence Electrons	Total Configuration
Main Group Elements				
Na	1A	$1s^2 2s^2 2p^6 = [\text{Ne}]$	$3s^1$	$[\text{Ne}]3s^1$
Si	4A	$1s^2 2s^2 2p^6 = [\text{Ne}]$	$3s^2 3p^2$	$[\text{Ne}]3s^2 3p^2$
As	5A	$1s^2 2s^2 2p^6 3s^2 3p^6 3d^{10} = [\text{Ar}]3d^{10}$	$4s^2 4p^3$	$[\text{Ar}]3d^{10}4s^2 4p^3$
Transition Elements				
Ti	4B	$1s^2 2s^2 2p^6 3s^2 3p^6 = [\text{Ar}]$	$3d^2 4s^2$	$[\text{Ar}]3d^2 4s^2$
Co	8B	$[\text{Ar}]$	$3d^7 4s^2$	$[\text{Ar}]3d^7 4s^2$
Mo	6B	$[\text{Kr}]$	$4d^5 5s^1$	$[\text{Kr}]4d^5 5s^1$

• **H Atoms and Electron Octets**
Hydrogen cannot be surrounded by an octet of electrons. An atom of H, which has only a 1s valence electron orbital, can accommodate only one pair of electrons.

Valence electrons for transition elements include the electrons in the ns and $(n-1)d$ orbitals (see Table 9.1). The remaining electrons are core electrons. As with main group elements, the valence electrons for transition metals determine the chemical properties of these elements.

Lewis Symbols for Atoms

G. N. Lewis introduced a useful way to represent electrons in the valence shell of an atom. The element's symbol represents the atomic nucleus together with the core electrons. Up to four valence electrons, represented by dots, are placed one at a time around the symbol; then, if any valence electrons remain, they are placed next to ones already there. Chemists now refer to these pictures as **Lewis electron dot symbols.** Lewis symbols for main group elements of the second and third periods are shown in Table 9.2.

Arranging the valence electrons around an atom in four groups suggests that the valence shell of a main group element can accommodate a maximum of four pairs of electrons. Because this represents eight electrons in all, this is referred to as an *octet* of electrons. Indeed, information about many compounds of the main group elements can be organized based on an octet of electrons around each atom.

An octet of electrons surrounding an atom is regarded as a stable configuration. The noble gases, with the exception of helium, have eight valence electrons, and demonstrate a notable lack of reactivity. Helium, neon, and argon do not undergo any chemical reactions, and the other noble gases have very limited chemical reactivity. Because chemical reactions involve changes in the valence electron shell, the limited reactivity of the noble gases is taken as evidence of the stability of their noble gas $(ns^2 np^6)$ electron configuration. Hydrogen, which in its compounds has two electrons in its valence shell, obeys the spirit of this rule by matching the electron configuration of He.

Example 9.1 **Valence Electrons**

Problem • Give the number of valence electrons for Ca and Se. Draw the Lewis electron dot symbol for each element.

Strategy • Locate the elements in the periodic table and note that, for main group elements, the number of valence electrons equals the group number.

Solution • Calcium, in Group 2A, has two valence electrons, and selenium, in Group 6A, has six. Dots representing electrons are placed around the element symbol one at a time until there are four electrons. Subsequent electrons are paired with those already present:

$$\cdot Ca \cdot \qquad \cdot \overset{\cdots}{Se} \cdot$$

calcium   selenium

Exercise 9.1 **Electrons**

Give the number of valence electrons for Ba, As, and Br. Draw the Lewis dot symbol for each of these elements.

9.2 CHEMICAL BOND FORMATION

When a chemical reaction occurs between two atoms, their valence electrons are reorganized so that a net attractive force — a chemical bond — occurs between atoms (← CHAPTER 3). Bonds come in two general types, ionic and covalent, and their formation can be depicted using Lewis symbols [🖱 CD-ROM, Screen 9.5].

Table 9.2 • Lewis Dot Symbols for Main Group Atoms

1A ns^1	2A ns^2	3A ns^2np^1	4A ns^2np^2	5A ns^2np^3	6A ns^2np^4	7A ns^2np^5	8A ns^2np^6
Li·	·Be·	·Ḃ·	·Ċ·	·N̈·	:Ö·	:F̈·	:N̈e:
Na·	·Mg·	·Al̇·	·Si̇·	·P̈·	:S̈·	:Cl̈·	:Är:

An **ionic bond** forms when *one or more valence electrons is transferred from one atom to another,* creating positive and negative ions. When sodium and chlorine react (Figure 9.1a), an electron is transferred from a sodium atom to a chlorine atom to form Na^+ and Cl^- [CD-ROM, Screens 9.2 and 9.3].

$$Na· + ·\ddot{\underset{..}{Cl}}: \longrightarrow \left[Na· \overset{\frown}{} ·\ddot{\underset{..}{Cl}}: \right] \longrightarrow \left[Na^+ \quad :\ddot{\underset{..}{Cl}}:^- \right]$$

| Metal atom | Nonmetal atom | Electron transfer from reducing agent to oxidizing agent. | Ionic compound. Ions have noble gas electron configurations. |

The "bond" is the attractive force between the positive and negative ions.

Covalent bonding, in contrast, *involves sharing of valence electrons between atoms.* Two chlorine atoms, for example, share a pair of electrons, one electron from each atom, to form a covalent bond.

$$:\ddot{\underset{..}{Cl}}· + ·\ddot{\underset{..}{Cl}}: \longrightarrow :\ddot{\underset{..}{Cl}}:\ddot{\underset{..}{Cl}}:$$

It is useful to reflect on the differences in the Lewis electron dot structure representations for ionic and covalent bonding. In both processes, unpaired electrons in the reactants are paired up. Both processes give products in which each atom is surrounded by eight electrons (an octet). The position of the electron pair between the two bonded atoms differs significantly, however. In a chlorine molecule

Chapter Goals • Revisited

- **Understand the difference between ionic and covalent bonds.**
- Draw Lewis electron dot structures for small molecules and ions.
- Use electronegativity to predict the polarity of bonds and molecules.
- Understand the properties of covalent bonds and their influence on molecular properties.
- Use VSEPR to predict the shapes of simple molecules and ions and to understand the structures of more complex molecules.

(a) The reaction of elemental sodium and chlorine to give sodium chloride [CD-ROM, Screen 9.2]: $Na(s) + 1/2 \, Cl_2(g) \rightarrow NaCl(s)$

(b) The reaction of elemental calcium and oxygen to give calcium oxide. $Ca(s) + 1/2 \, O_2(g) \rightarrow CaO(s)$

Figure 9.1 Formation of ionic compounds. Notice that both reactions are quite exothermic, as reflected by the very negative molar enthalpies of formation for the reaction products.

$\Delta H_f^o[NaCl(s)] = -411.12 \text{ kJ/mol}$

$\Delta H_f^o[CaO(s)] = -635.09 \text{ kJ/mol}$

(Charles D. Winters)

(Cl_2), the electron pair is shared equally by the two atoms. In contrast, the electron pair in sodium chloride has become part of the valence shell of chlorine.

As bonding is described in greater detail, you will discover that the two types of bonding — complete electron transfer and the equal sharing of electrons — are extreme cases. In real compounds electrons are generally shared unequally. Not surprisingly, this results in a gradation of properties for the compounds as well.

9.3 BONDING IN IONIC COMPOUNDS

Metallic sodium reacts vigorously with gaseous chlorine to give sodium chloride [✇ CD-ROM, Screen 9.2], and calcium metal and oxygen react to give calcium oxide (Figure 9.1b). In each case, the product is an ionic compound: NaCl contains Na^+ and Cl^- ions, whereas CaO is composed of Ca^{2+} and O^{2-} ions.

$$Na(s) + \tfrac{1}{2}Cl_2(g) \longrightarrow NaCl(s) \qquad \Delta H° = -411.12 \text{ kJ/mol}$$
$$Ca(s) + \tfrac{1}{2}O_2(g) \longrightarrow CaO(s) \qquad \Delta H° = -635.09 \text{ kJ/mol}$$

These exothermic reactions, which are examples of the general chemical behavior for elements in these periodic groups, can be understood based on the atomic properties described in Chapter 8. The alkali and alkaline earth metals have low ionization energies. Thus, relatively little energy is required to remove the *ns* valence electrons from these elements to form cations with a noble gas configuration. In contrast, elements immediately preceding Group 8A (the halogens and the Group 6A elements) have high affinities for electrons. These elements typically form anions by adding electrons, in most instances giving an ion with an electron configuration equivalent to that of the next noble gas.

The tendency to achieve a noble gas configuration by gain or loss of electrons is important in the chemistry of main group elements. An even more important feature is the favorable energetics of compound formation. Generally,

- In all product-favored chemical reactions (◀ SECTION 6.9), the products have a lower chemical potential energy than the reactants.
- The structure of a compound, either ionic or covalent, is the one having the lowest potential energy, that is, the greatest thermodynamic stability.

Ion Attraction and Energy

To understand bonding in ionic compounds, it is useful to think about the energy involved in their formation. We shall first analyze the energy change as a cation and anion form an ionic bond and then apply that to the formation of a mole of solid, ionic compound.

Let us begin by asking what the energy change is for the formation of the ion pair $[Na^+, Cl^-]$ in the gas phase starting with sodium and chlorine atoms, also in the gas phase. The overall energy of this reaction can be thought of as the sum of three individual steps: (1) the ionization of sodium atoms to form Na^+ ions (the energy of this process is the *ionization energy* of the element), (2) the addition of electrons to chlorine atoms to form the Cl^- ions (the energy here is the *electron affinity* of the element), and (3) the formation of $[Na^+, Cl^-](g)$ ion pairs from $Na^+(g)$ and $Cl^-(g)$.

1. Formation of $Na^+(g)$ and an electron

$$Na(g) \longrightarrow Na^+(g) + e^- \qquad \Delta E_{ion} = \text{ionization energy of Na} = +496 \text{ kJ/mol}$$

● **Valence Electron Configurations and Ionic Compound Formation**
For the formation of NaCl:
Na changes from $1s^2 2s^2 2p^6 3s^1$ to Na^+ with a $1s^2 2s^2 2p^6$ configuration.
Cl changes from $[Ne]3s^2 3p^5$ to Cl^- with a $[Ne]3s^2 3p^6$ configuration.

2. Formation of $Cl^-(g)$ from a Cl atom and an electron

$$Cl(g) + e^- \longrightarrow Cl^-(g) \qquad \Delta E_{EA} = \text{electron affinity of Cl} = -349 \text{ kJ/mol}$$

3. Formation of the ion pair

$$Na^+(g) + Cl^-(g) \longrightarrow [Na^+, Cl^-](g) \qquad \Delta E_{\text{ion pair}} = -498 \text{ kJ/mol}$$

Examination of the energies for these three steps reveals that the formation of the two ions individually is *endo*thermic ($\Delta E_{ion} + \Delta E_{EA} = +153.0$ kJ). The formation of these ions would not be favorable were it not for the fact that ion pair formation is a very *exo*thermic process ($\Delta E_{\text{ion pair}}$ has a large, negative value) (Figures 9.2 and 9.3). Thus, the overall change in energy is

$$Na(g) + Cl(g) \longrightarrow [Na^+, Cl^-](g)$$
$$\Delta E_{net} = +496 \text{ kJ} + (-349 \text{ kJ}) + (-498 \text{ kJ}) = -351 \text{ kJ}$$

The energy associated with the formation of $[Na^+, Cl^-](g)$ (Step 3 above) cannot be measured directly, but it can be calculated from an equation derived from Coulomb's law in which ΔE represents the energy of attraction between pairs of ions (← EQUATION 3.1, PAGE 93).

$$\Delta E_{\text{ion pair}} = C\left(\frac{(n^+e)(n^-e)}{d}\right)$$

The symbol C is a constant, d represents the distance between the ion centers, n is the number of positive (n^+) and negative (n^-) charges on an ion, and e is the charge on an electron. Because the charges are opposite in sign, the energy value is negative. Inspecting this equation, you find that the energy of attraction between ions of opposite charge depends on two factors.

- *The magnitude of the ion charges.* The higher the ion charges, the greater the attraction, so ΔE has a larger negative value. For example, the attraction between Ca^{2+} and O^{2-} ions will be about four times larger $[(2+) \times (2-)]$ than the attraction between Na^+ and Cl^- ions, and the energy will be more negative by a factor of about 4.

- *The distance between the ions.* This is an inverse relationship because, as the distance between ions becomes greater (d becomes larger), the attractive force between the ions declines and the energy is less negative. The distance is determined by the sizes of the ions (← page 93 and FIGURE 8.14).

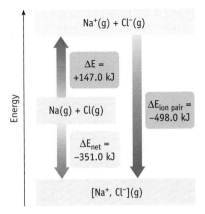

Figure 9.2 Energy level diagram for formation of Na^+, Cl^- ion pair.

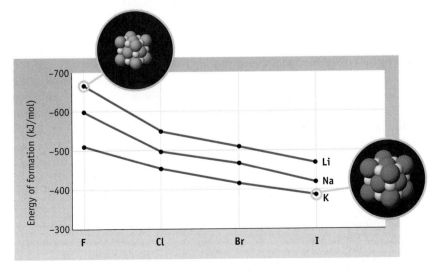

Figure 9.3 Energy of ion pair formation.
$\Delta E_{\text{ion pair}}$ is illustrated for the formation of the alkali metal halides, MX(g), from the ions $M^+(g) + X^-(g)$.

Figure 9.4   Models of the sodium chloride crystal lattice. The model represents only a small portion of the lattice. Ideally, it extends infinitely in all directions. Sodium ions are silver and chloride ions are yellow. The sticks in the model are there to help identify the locations of the atoms. *(Charles D. Winters)*

The effect of ion size is apparent in a plot of $\Delta E_{\text{ion pair}}$ for the alkali metal halides (Figure 9.3). The variations are regular and predictable based on trends in the sizes of ions. For example, values of $\Delta E_{\text{ion pair}}$ for the chlorides become progressively more negative going from KCl to NaCl to LiCl because the alkali metal ion radii decrease in the order $K^+ > Na^+ > Li^+$ (and so d decreases in this order). The smaller the positive ion, the more negative the value of $\Delta E_{\text{ion pair}}$. Similarly, the value of $\Delta E_{\text{ion pair}}$ for the halides of a given alkali metal ion becomes more negative with a decrease in size of the halide ion.

Lattice Energy

Table 9.3 • Lattice Energies of Some Ionic Compounds

Compound	$\Delta E_{\text{lattice}}$ (kJ/mol)
LiF	−1037
LiCl	−852
LiBr	−815
LiI	−761
NaF	−926
NaCl	−786
NaBr	−752
NaI	−702
KF	−821
KCl	−717
KBr	−689
KI	−649

Source: D. Cubicciotti: "Lattice energies of the alkali halides and electron affinities of the halogens." *Journal of Chemical Physics,* Vol. 31, p. 1646, 1959. Reprinted with permission.

Ionic compounds exist as solids under normal conditions. Their structures contain positive and negative ions arranged in a three-dimensional lattice (➡ CHAPTER 13). Models of a small segment of a NaCl lattice are pictured in Figure 9.4 (and Figure 3.11). In crystalline NaCl, each Na^+ cation is surrounded by six Cl^- anions, and six Na^+ ions are nearest neighbors to each Cl^-.

For ionic compounds, a quantity called the lattice energy is a measure of bonding energy. The **lattice energy,** $\Delta E_{\text{lattice}}$, is defined as the energy of formation of one mole of a solid crystalline ionic compound when ions in the gas phase combine (see *A Closer Look: Using a Born–Haber Cycle to Calculate Lattice Energies and Table 9.3*) [🖰 CD-ROM, Screen 9.3].

$$Na^+(g) + Cl^-(g) \longrightarrow NaCl(s) \qquad \Delta E_{\text{lattice}} = -786 \text{ kJ/mol}$$

The lattice energy for an ionic compound results from the attraction between the cations and anions *in a crystal.* Lattice energy cannot be measured directly because the reaction describing it cannot be carried out in a laboratory. It is possible to *calculate* lattice energies, however. The mathematical approach takes into account *all* possible interactions between cations and anions in the crystal, the largest being the attractive forces between neighboring ions of opposite charges. Attractive forces between ions of opposite charge that are farther apart, as well as the forces of repulsion between ions of the same charge in the lattice, are also included in the calculation.

What is important to us here is the dependence of the strength of ionic bonding in solid compounds on ion sizes and ion charges, as measured by $\Delta E_{\text{lattice}}$. Values of $\Delta E_{\text{lattice}}$ are closely related to those for $\Delta E_{\text{ion pair}}$, and both vary predictably

A Closer Look

Using a Born–Haber Cycle to Calculate Lattice Energies

As mentioned in the preceding discussion, values of $\Delta E_{lattice}$ can be calculated based on the structure of an ionic lattice and the charges and sizes of the ions. A value can also be obtained using measurable thermochemical quantities. This latter approach, described here, uses a Born–Haber cycle, so named for Max Born (1882–1970), and Fritz Haber (1868–1934). The calculation uses enthalpy values, and the value obtained is $\Delta H_{lattice}$. The difference between ΔE and ΔH is generally not significant and can be corrected for, if desired.

The calculation is an application of Hess's law because the energies involved in one pathway from reactants to products (ΔH_f°) is the sum of the energies involved in another pathway (Steps 1–3). The procedure is illustrated for solid sodium chloride. The overall process of lattice formation is first broken into a series of steps. Steps 1 and 2 involve the enthalpy of formation of the $Na^+(g)$ and $Cl^-(g)$ ions, values that can be measured experimentally. Step 3 is the enthalpy of formation of the solid ionic lattice. ΔH_f° is the standard molar enthalpy of formation of NaCl(s), obtained by calorimetry. The enthalpy values for each step are

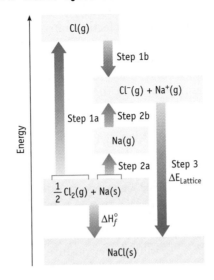

related by the equation:

$$\Delta H_f^\circ = \Delta H_{Step\ 1a} + \Delta H_{Step\ 1b}$$
$$+ \Delta H_{Step\ 2a} + \Delta H_{Step\ 2b} + \Delta H_{Step\ 3}$$

The value for Step 3 can be calculated using this equation because all other values are known (Appendix L).

Step 1a. Enthalpy of formation of Cl(g) = +121.68 kJ/mol

Step 1b. ΔH for $Cl(g) + e^- \rightarrow Cl^-(g)$ = −349 kJ/mol

Step 2a. Enthalpy of formation of Na(g) = +107.3 kJ/mol

Step 2b. ΔH for $Na(g) \rightarrow Na^+(g) + e^-$ = +496 kJ/mol

ΔH_f° Standard heat of formation of NaCl(s) = −411.12 kJ/mol

This means the enthalpy of formation of NaCl(s) from the ions in the gas phase is given in Step 3.

Step 3. Formation of NaCl(s) from the ions in the gas phase = $\Delta E_{lattice}$ = −786 kJ/mol

In this calculation, Steps 1a, 2a, and 2b are endothermic. Energy is required to vaporize and ionize sodium atoms, and to separate the Cl_2 molecule into two chlorine atoms. Steps 1b and 3 are both exothermic; energy is evolved when a chlorine atom acquires an electron and when 1 mol of $Na^+(g)$ and 1 mol of $Cl^-(g)$ ions come together to form the solid crystalline lattice.

with the charge on the ions. The value of $\Delta E_{lattice}$ for MgO (-4050 kJ/mol), for example, is about four times more negative than the value for NaF (-926 kJ/mol) because the charges on the Mg^{2+} and O^{2-} ions are each twice as large as those on Na^+ and F^- ions. The effect of ion size on lattice energy is also predictable: A lattice built from smaller ions generally leads to a more negative value for the lattice energy. Data in Table 9.3 illustrate this point. For the alkali metal halides, for example, the lattice energy for a lithium compound is generally more negative than that for a potassium compound. As we shall see, these energy values are closely related to the temperatures required to melt ionic compounds and to their solubilities in water (➧ SECTION 13.9, SECTION 14.2).

Exercise 9.2 Using Lattice Energies

Calculate the molar enthalpy of formation, ΔH_f°, of solid sodium iodide using the approach outlined in *A Closer Look: Using a Born–Haber Cycle to Calculate Lattice Energies*. The required data can be found in Appendices F and L and in Table 9.3.

Why Compounds Such as NaCl₂ and NaNe Don't Exist

Let us revisit the question of why a compound such as $NaCl_2$, in which sodium is present as the Na^{2+} ion, is unlikely (SECTION 8.7). The formation of Na^{2+} (with the configuration $1s^2 2s^2 2p^5$) would require the loss of *two electrons* from sodium. Because the second electron must be removed from the $n = 2$ shell, formation of Na^{2+} requires a substantial amount of energy. That is, the total energy for Step 3 in the Born–Haber cycle in *A Closer Look* would be approximately the sum of the first and second ionization energies for Na (496 kJ + 4562 kJ). Making the reasonable assumption that the lattice energy for $NaCl_2$ is at least double that of NaCl (the increase coming because the cation charge has doubled, and the size of Na^{2+} is less than that of Na^+), we would estimate a very positive value for ΔH_f° of $NaCl_2$ (about +3000 kJ/mol). A positive value of ΔH_f° means that formation of $NaCl_2$ from Na(s) and Cl_2(g) is unfavorable.

Because sodium is a good reducing agent, why doesn't it reduce neon to Ne^- and form NaNe? Again, we can think about this question in terms of the Born–Haber cycle for NaCl (see "A Closer Look"). Put Ne in place of Cl_2. Because neon exists in the form of atoms, $\Delta E_{Step\ 1a}$ is not required. More important to the outcome is the fact that neon's affinity for an electron should be extremely low, and $\Delta E_{Step\ 1b}$ will be very positive. (This is because the additional electron has to be placed in the next higher electron shell, and the $n = 3$ shell is much higher in energy than the $n = 2$ shell.)

$$Ne\ (1s^2 2s^2 2p^6) + e^- \longrightarrow Ne^-\ (1s^2 2s^2 2p^6 3s^1) \qquad \Delta E \gg 0$$

The lattice energy of NaNe is not expected to be exothermic enough to overcome this and other endothermic steps, so an overall positive enthalpy change is again expected. The formation of NaNe is energetically unfavorable.

9.4 COVALENT BONDING AND LEWIS STRUCTURES

The remainder of this chapter is concerned with covalent bonding, in which electron pairs are shared between bonded atoms. Examples of compounds having covalent bonds include gases in our atmosphere (O_2, N_2, H_2O, and CO_2), common fuels (CH_4), and most of the compounds in your body. Covalent bonding also serves to hold the atoms together in common ions such as CO_3^{2-}, CN^-, NH_4^+, NO_3^-, and PO_4^{3-}. We will develop the basic principles of structure and bonding using as examples molecules and ions made up of only a few atoms, but the same principles apply to larger molecules from aspirin to proteins and DNA with thousands of atoms.

The molecules and ions just mentioned are composed entirely of *nonmetal* atoms. A point that needs special emphasis is that, in molecules or ions made up *only* of nonmetal atoms, the atoms are attached by covalent bonds. Conversely, the presence of a metal in a formula is a signal that the compound is likely to be ionic.

Lewis Electron Dot Structures

A covalent bond results when one or more electron pairs are shared between two atoms. Thus, the electron pair bond between the two atoms of a H_2 molecule is represented by a pair of dots or, alternatively, a line.

Aspirin

Zingerone

This chapter discusses bonding and structures from A to Z, from aspirin to zingerone (the latter a compound isolated from ginger root).

Electron pair bond

H:H   H—H

These representations are called Lewis electron dot structures or just **Lewis structures** [CD-ROM, Screens 9.6 and 9.7].

Simple Lewis structures can be drawn starting with Lewis dot symbols for atoms and arranging the valence electrons to form bonds. To create the Lewis structure for F_2, for example, we start with the Lewis dot symbol for a fluorine atom. Fluorine, an element in Group 7A, has seven valence electrons. The Lewis symbol shows that an F atom has a single unpaired electron along with three electron pairs. In F_2, the single electrons, one on each F atom, pair up in the covalent bond.

$$:\ddot{F}\cdot + \cdot\ddot{F}: \longrightarrow :\ddot{F}:\ddot{F}: \quad \text{or} \quad :\ddot{F}-\ddot{F}:$$

Lone pair of electrons

Shared or bonding electron pair

In the Lewis dot structure for F_2 the pair of electrons in the F—F bond is the bonding pair, or **bond pair.** The other six pairs reside on single atoms and are called **lone pairs.** Because they are not involved in bonding, they are also called **nonbonding electrons.**

Carbon dioxide, CO_2, and dinitrogen, N_2, are important examples of molecules in which two atoms are multiply bonded, that is, they share more than one electron pair.

$$\ddot{O}=C=\ddot{O} \qquad :N\equiv N:$$

In carbon dioxide, the carbon atom shares two pairs of electrons with each oxygen and so is linked to each O atom by a **double bond.** The valence shell of each oxygen atom in CO_2 has two bonding pairs and two lone pairs. In dinitrogen, the two nitrogen atoms share three pairs of electrons, so they are linked by a **triple bond.** In addition, each N atom has a single lone pair.

Octet Rule

An important observation can be made about the molecules you have seen so far: *Each atom (except H) has a share in four pairs of electrons, so each has achieved a noble gas configuration. Each atom is surrounded by an octet of eight electrons.* (Hydrogen typically forms a bond to only one other atom, resulting in two electrons in its valence shell.) *The tendency of molecules and polyatomic ions to have structures in which eight electrons surround each atom* is known as the **octet rule.** As an example, a triple bond is necessary in dinitrogen in order to have an octet around each nitrogen atom. The carbon atom and both oxygen atoms in CO_2 achieve the octet configuration by forming double bonds.

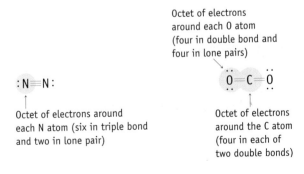

Octet of electrons around each O atom (four in double bond and four in lone pairs)

:N≡N:

Octet of electrons around each N atom (six in triple bond and two in lone pair)

Octet of electrons around the C atom (four in each of two double bonds)

Chapter Goals • Revisited

- Understand the difference between ionic and covalent bonds.
- **Draw Lewis electron dot structures for small molecules and ions.**
- Use electronegativity to predict the polarity of bonds and molecules.
- Understand the properties of covalent bonds and their influence on molecular properties.
- Use VSEPR to predict the shapes of simple molecules and ions and to understand the structures of more complex molecules.

• **Importance of Lone Pairs**
Lone pairs can be important in a structure. Being in the same valence electron shell as the bonding electrons, they can influence molecular shape. See Section 9.9.

• **Exceptions to the Octet Rule**
Although the octet rule is widely applicable, there are exceptions. Fortunately, many will be obvious, such as when an element has more than four bonds or when an odd number of electrons occur.

There is a systematic approach to constructing Lewis structures of molecules and ions [🖱 CD-ROM, Screen 9.7]. Let us take formaldehyde, CH_2O, as an example.

● **Choosing the Central Atom**

1. The relative electronegativities of atoms can also be used to choose the central atom. Electronegativity is discussed in Section 9.7.

2. For simple compounds, the first atom in a formula is often the central atom (e.g., SO_2, NH_4^+, NO_3^-). This is not always a reliable predictor, however. Notable exceptions include water (H_2O) and most common acids (HNO_3, H_2SO_4), in which the acidic hydrogen is usually written first but where N or S is the central atom.

1. *Decide on the central atom.* The central atom is *usually* the one with the lowest electron affinity. In CH_2O the central atom is C, for example. You will come to recognize that certain elements often appear as the center atom, among them C, N, P, and S. Halogens are often terminal atoms forming a single bond to one other atom, but they can be the central atom when combined with O in oxoacids (such as $HClO_4$). Oxygen is the central atom in water, but in conjunction with carbon, nitrogen, phosphorus, and the halogens it is usually a terminal atom. Hydrogen is a terminal atom because it typically bonds to only one other atom.

2. *Determine the total number of valence electrons in the molecule or ion.* In a neutral molecule this number will be the sum of the valence electrons for each atom. For an anion, *add* a number of electrons equal to the negative charge; for a cation, *subtract* the number of electrons equal to the positive charge. The number of valence electron pairs will be half the total number of valence electrons. For CH_2O,

$$\text{Valence electrons} = 12 \text{ electrons (or 6 electron pairs)}$$
$$= 4 \text{ for C} + (2 \times 1 \text{ for two H atoms}) + 6 \text{ for O}$$

3. *Place one pair of electrons between each pair of bonded atoms to form a single bond.*

Here three electron pairs are used to make three single bonds (which are represented by single lines). Three pairs of electrons remain to be used.

4. *Use any remaining pairs as lone pairs around each terminal atom (except H) so that each atom is surrounded by eight electrons.* If, after this is done, there are electrons left over, assign them to the central atom. If the central atom is an element in the third or higher period, it can have more than eight electrons.

Here all six pairs have been assigned, but the C atom has a share in only three pairs.

5. *If the central atom has fewer than eight electrons at this point, move one or more of the lone pairs on the terminal atoms into a position intermediate between the center and the terminal atom to form multiple bonds.*

As a general rule double or triple bonds are formed when both atoms are from the following list: C, N, O, or S. That is, bonds such as C=C, C=N, C=O, and S=O will be encountered frequently.

Example 9.2   Drawing Lewis Structures

Problem • Draw Lewis structures for ammonia (NH_3), the hypochlorite ion (ClO^-), and the nitronium ion (NO_2^+).

Strategy • Follow the five steps outlined for CH_2O on page 334.

Solution for NH_3 •

1. *Decide on the central atom.* Hydrogen atoms are always terminal atoms, so nitrogen must be the central atom in the molecule.

2. *Count the number of valence electrons.* The total is eight [or 4 valence pairs].

   Valence electrons = 5 (for N) + 3 (1 for each H)

3. *Form single covalent bonds between each pair of atoms.* This uses three of the four pairs available.

$$H-N-H$$
$$\quad\ \ |$$
$$\quad\ \ H$$

4. *Place the remaining pair of electrons on the central atom.*

$$H-\overset{..}{N}-H$$
$$\quad\ \ |$$
$$\quad\ \ H$$

Each H atom has a share in one pair of electrons as required, and the central N atom has achieved an octet configuration with four electron pairs. No further steps are required; this is the correct Lewis structure.

Solution for ClO^- Ion •

1. With two atoms, there is no "central" atom.

2. Valence electrons = 14 (or 7 valence pairs)

   = 7 (for Cl) + 6 (for O)
   +1 (for the negative charge on the ion)

3. One electron pair is used in the Cl—O bond: Cl—O.

4. Distribute the six remaining electron pairs around the terminal atoms.

$$\left[:\overset{..}{\underset{..}{Cl}}-\overset{..}{\underset{..}{O}}:\right]^-$$

5. Because no electrons remain to be assigned and both atoms have an octet, this is the correct Lewis structure.

Solution for NO_2^+ Ion •

1. Nitrogen is the center atom because its electron affinity is lower than that of oxygen.

2. Valence electrons = 16 (or 8 valence pairs)

   = 5 (for N) + 12 (six for each O)
   − 1 (for the positive charge)

3. Two electron pairs form the single bonds from the nitrogen to each oxygen:

$$O-N-O$$

4. Distribute the remaining six pairs of electrons on the terminal O atoms:

$$\left[:\overset{..}{\underset{..}{O}}-N-\overset{..}{\underset{..}{O}}:\right]^+$$

5. The central nitrogen atom is two electron pairs short of an octet. Thus, a lone pair of electrons on each oxygen atom is converted to a bonding electron pair to give two N=O double bonds. Each atom in the ion now has four electron pairs. Nitrogen has four bonding pairs, and each oxygen atom has two lone pairs and shares two bond pairs.

Move lone pairs to create double bonds and satisfy the octet for N.

$$\left[:\overset{..}{\underset{..}{O}}-N-\overset{..}{\underset{..}{O}}:\right]^+ \longrightarrow \left[\overset{..}{\underset{..}{O}}=N=\overset{..}{\underset{..}{O}}\right]^+$$

Exercise 9.3   Drawing Lewis Structures

Draw Lewis structures for NH_4^+, CO, NO^+, and SO_4^{2-}.

Predicting Lewis Structures

Lewis structures are useful in gaining a perspective on the structure and chemistry of a molecule or ion. The guidelines for drawing Lewis structures are helpful, but chemists also rely on patterns of bonding in related molecules.

Hydrogen Compounds

Some common compounds and ions formed from second-period nonmetal elements and hydrogen are shown in Table 9.4. Their Lewis structures illustrate the fact that the Lewis symbol for an element is a useful guide in determining

Table 9.4 • **Common Hydrogen-Containing Compounds and Ions of the Second-Period Elements**

Group 4A	Group 5A	Group 6A	Group 7A
CH_4 methane H—C—H (with H above and below)	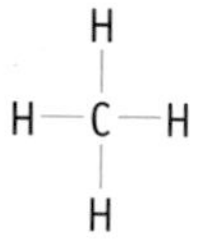 NH_3 ammonia H—N—H (with H below, lone pair above) 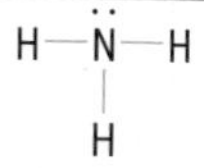	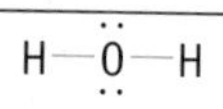 H_2O water H—O—H (lone pairs above and below)	HF hydrogen fluoride H—F: (lone pairs) 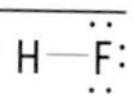
C_2H_6 ethane H—C—C—H (with H's)	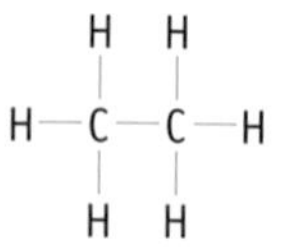 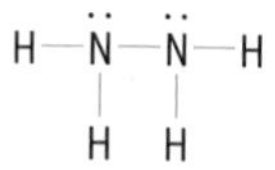 N_2H_4 hydrazine H—N—N—H (with H's and lone pairs)	H_2O_2 hydrogen peroxide H—O—O—H (lone pairs)	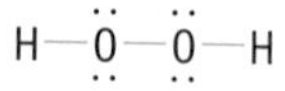
C_2H_4 ethylene H—C=C—H (with H's)	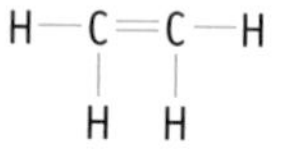 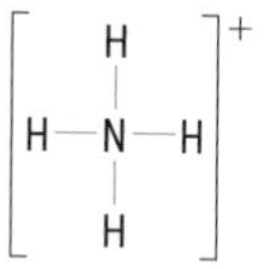 NH_4^+ ammonium ion [H—N—H with H above and below]$^+$	H_3O^+ hydronium ion [H—O—H with H below]$^+$	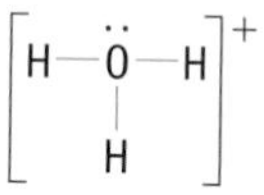
C_2H_2 acetylene H—C≡C—H	 NH_2^- amide ion [H—N—H with lone pairs]$^-$	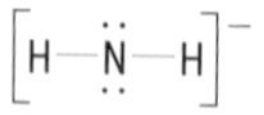 OH^- hydroxide ion [:O—H]$^-$	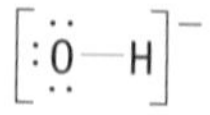

the number of bonds formed by the element. For example, if there is no charge, nitrogen has five valence electrons. Two electrons occur as a lone pair; the other three occur as unpaired electrons. To reach an octet, it is necessary to pair each of the unpaired electrons with an electron from another atom. Thus, N is predicted to form three bonds in uncharged molecules, and this is indeed the case. Similarly, carbon is expected to form four bonds, oxygen two, and fluorine one.

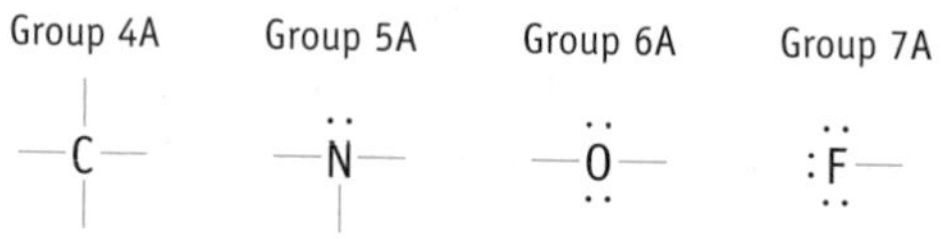

Group 4A	Group 5A	Group 6A	Group 7A
—C—	—N— (lone pair)	—O— (lone pairs)	:F— (lone pairs)

Hydrocarbons are compounds formed from carbon and hydrogen, and the first two members of the series called the *alkanes* are CH_4 and C_2H_6 (see Table 9.4). What is the Lewis structure of the third member of the series, propane, C_3H_8? We can rely on the idea that the atoms in this species each bond in a predictable way. Carbon is expected to form four bonds, and hydrogen can bond to only one other atom. The only arrangement of atoms that meets these criteria has three atoms of carbon linked together by carbon–carbon single bonds. The remaining positions around the carbon atoms are filled in with hydrogens — three hydrogen atoms on the end carbons and two on the middle carbon:

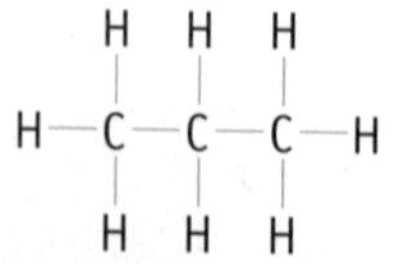

H—C—C—C—H (with H's above and below each carbon)

propane, C_3H_8

Example 9.3 Predicting Lewis Structures

Problem • Draw Lewis electron dot structures for CCl_4 and NF_3.

Strategy • One way to answer this is to recognize that the formulas are similar to CH_4 and NH_3, except that H atoms have been replaced by halogen atoms.

Solution • Recall that carbon is expected to form four bonds and nitrogen three bonds to give an octet of electrons. In addition, halogen atoms have seven valence electrons, so both Cl and F can attain an octet by forming one covalent bond, just as hydrogen does.

carbon tetrachloride   nitrogen trifluoride

As a check, count the number of valence electrons for each molecule and verify that all are present.

CCl_4: Valence electrons = 4 for C + (4 × 7) (for Cl) = 32 electrons

The structure shows 8 electrons in single bonds and 24 electrons as lone pair electrons, for a total of 32 electrons. The structure is correct.

NF_3: Valence electrons = 5 for N + (3 × 7) (for F) = 26 electrons

The structure shows 6 electrons in single bonds and 20 electrons as lone pair electrons, for a total of 26 electrons. The structure is correct.

Exercise 9.4 Predicting Lewis Structures

Predict Lewis structures for methanol, CH_3OH and hydroxylamine, NH_2OH. (*Hint:* The formulas of these compounds are written to guide you in choosing the correct arrangement of atoms.)

Oxo Acids and Their Anions

Lewis structures of common acids and their anions are illustrated in Table 9.5. In the absence of water these acids are covalently bonded molecular compounds (a conclusion that we should draw because all elements in the formulas are nonmetals). Nitric acid, for example, is a colorless liquid with a boiling point of 83 °C. In aqueous solution, however, HNO_3, H_2SO_4, and $HClO_4$ are ionized to give a hydrogen ion and the appropriate anions. A Lewis structure for the nitrate ion, for example, can be created using the guidelines on page 334, and the result is a structure with two N—O single bonds and one N=O double bond. To form nitric acid, a hydrogen ion is attached to one of the O atoms that has a single bond to the central N.

nitrate ion   nitric acid

A characteristic property of acids in aqueous solution is their ability to donate a hydrogen ion (H^+). The NO_3^- anion is formed when the acid, HNO_3, loses a hydrogen ion. The H^+ ion separates from the acid by breaking the H—O bond, the electrons of the bond staying with the O atom. As a result, HNO_3 and NO_3^- have the same number of electrons, 24, and their structures are closely related.

Exercise 9.5 Lewis Structures of Acids and their Anions

Draw a Lewis structure for the anion $H_2PO_4^-$, derived from phosphoric acid.

Table 9.5 • Lewis Structures of Common Oxoacids and Their Anions

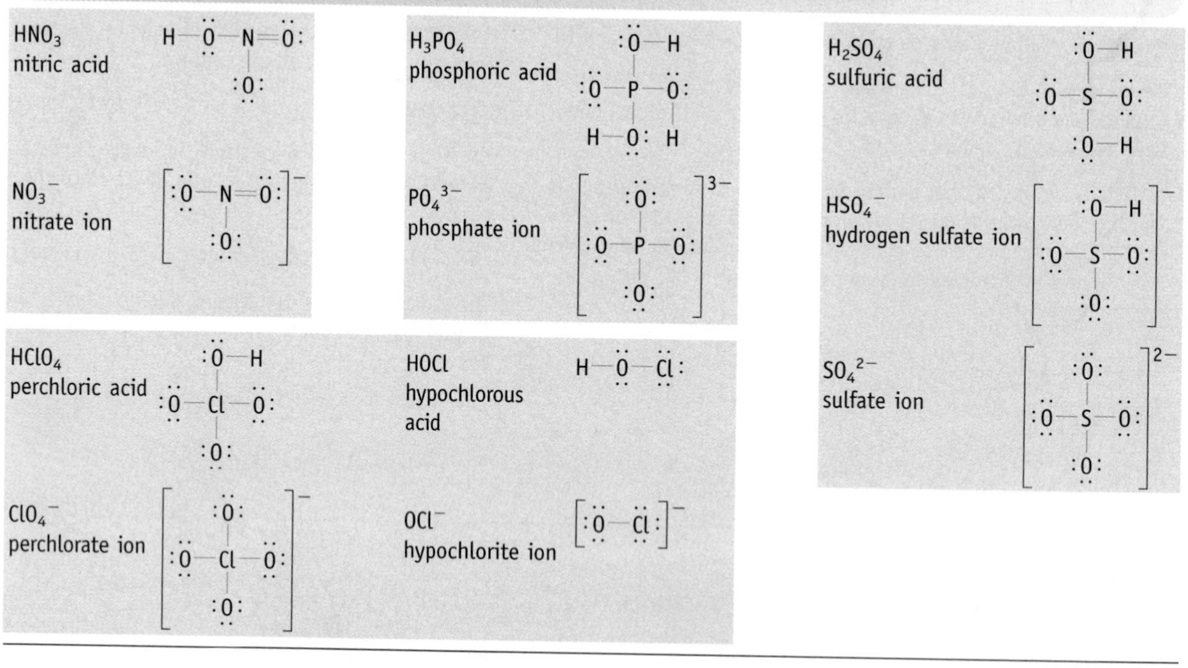

HNO₃ nitric acid	H—O—N=O with :O: below	H₃PO₄ phosphoric acid	structure
NO₃⁻ nitrate ion	structure	PO₄³⁻ phosphate ion	structure

(Lewis structures shown for: HNO₃ nitric acid, NO₃⁻ nitrate ion, HClO₄ perchloric acid, ClO₄⁻ perchlorate ion, H₃PO₄ phosphoric acid, PO₄³⁻ phosphate ion, HOCl hypochlorous acid, OCl⁻ hypochlorite ion, H₂SO₄ sulfuric acid, HSO₄⁻ hydrogen sulfate ion, SO₄²⁻ sulfate ion)

Isoelectronic Species

In what way are NO^+, N_2, CO, and CN^- similar? Most importantly, they each have two atoms and the same total number of valence electrons, ten, which leads to the same Lewis structure for each molecule or ion. The two atoms in each are linked with a triple bond. With three bonding pairs and one lone pair, each atom thus has an octet of electrons.

$$[:N\equiv O:]^+ \qquad :N\equiv N: \qquad :C\equiv O: \qquad [:C\equiv N:]^-$$

Molecules and ions having the *same number of valence electrons and the same Lewis structures* are said to be **isoelectronic** (Table 9.6). You will find it helpful to think in terms of isoelectronic molecules and ions because this is another way to see relationships among common chemical substances.

Isoelectronic species have similarities and important differences in their chemical properties. For example, both carbon monoxide, CO, and cyanide ion, CN^-, are very toxic because they can bind to the iron of hemoglobin in blood and block the uptake of oxygen. They are different, though, in their acid–base chemistry. In aqueous solution cyanide ion readily adds H^+ to form hydrogen cyanide, whereas CO does not do so. The isoelectronic species Cl_2 and ClO^- provide a similar example. Attachment of H^+ to ClO^- forms hypochlorous acid, HClO. In contrast, Cl_2 does not add a proton.

● **Isoelectronic and Isostructural**
The term *isostructural* is often used in conjunction with isoelectronic species. Species that are isostructural have the same structure. For example, the PO_4^{3-}, SO_4^{2-}, and ClO_4^- ions in Table 9.6 all have four oxygens bonded to the central atom. In addition they are isoelectronic in that all have 32 valence electrons.

Exercise 9.6 Identifying Isoelectronic Species

(a) Is the acetylide ion, C_2^{2-}, isoelectronic with N_2?

(b) Identify a common molecular (uncharged) species that is isoelectronic with nitrite ion, NO_2^-. Identify a common ion that is isoelectronic with HF.

Table 9.6 • Some Common Isoelectric Molecules and Ions

Formulas	Representative Lewis Structure	Formulas	Representative Lewis Structure
BH_4^-, CH_4, NH_4^+	$\left[\begin{array}{c} H \\ H-N-H \\ H \end{array}\right]^+$	CO_3^{2-}, NO_3^-	$\left[\ :\overset{..}{\underset{..}{O}}-N=\overset{..}{\underset{..}{O}}:\ \right]^-$ $\overset{..}{\underset{..}{O}}:$
NH_3, H_3O^+	$H-\overset{..}{N}-H$ H	PO_4^{3-}, SO_4^{2-}, ClO_4^-	$\left[\begin{array}{c} :\overset{..}{O}: \\ :\overset{..}{O}-P-\overset{..}{O}: \\ :\overset{..}{O}: \end{array}\right]^{3-}$
CO_2, OCN^-, SCN^-, N_2O, NO_2^+, COS, CS_2	$\overset{..}{O}=C=\overset{..}{O}$		

9.5   RESONANCE

Ozone, O_3, an unstable, blue, diamagnetic gas with a characteristic pungent odor, protects the earth and its inhabitants from intense ultraviolet radiation from the sun. An important feature of its structure is that the two oxygen–oxygen bonds are the same length, suggesting that the two oxygen–oxygen bonds are equivalent. That is, equal O—O bond lengths imply an equal number of bond pairs in each O—O bond. Using the guidelines for drawing Lewis structures, however, you might come to a different conclusion. There are two possible ways of writing the Lewis structure for the molecule:

Alternative ways of creating the Lewis structure of ozone

Double bond on the right: $:\overset{..}{O}-\overset{..}{O}-\overset{..}{O}: \longrightarrow :\overset{..}{O}-\overset{..}{O}=\overset{..}{O}:$

Double bond on the left: $:\overset{..}{O}-\overset{..}{O}-\overset{..}{O}: \longrightarrow :\overset{..}{O}=\overset{..}{O}-\overset{..}{O}:$

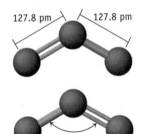

127.8 pm   127.8 pm

116.8°

Ozone, O_3, is a bent molecule with oxygen–oxygen bonds of the same length.

These structures are equivalent in that each has a double bond on one side of the central oxygen atom and a single bond on the other side. If either were the actual structure of ozone, one bond should be shorter (O=O) than the other (O—O). The actual structure of ozone shows this is not the case. The inescapable conclusion is that neither of these Lewis structures correctly represents the bonding in ozone.

Linus Pauling proposed the theory of **resonance** to reconcile the problem. *Resonance structures represent bonding in a molecule or ion when a single Lewis structure fails to describe accurately the actual electronic structure.* The alternative structures shown for ozone are called **resonance structures.** They have identical patterns of bonding and equal energy. The actual structure of the ozone molecule is a *composite,* or **resonance hybrid,** of the equivalent resonance structures [CD-ROM, Screen 9.8]. This is a reasonable conclusion because we see that the O—O bonds both have a length of 127.8 pm, intermediate between the average length of an O=O double bond (121 pm) and an O—O single bond (132 pm).

Benzene is a classic example in which resonance is used to represent a structure. The benzene molecule is a six-member ring of carbon atoms with six equivalent carbon–carbon bonds (and a hydrogen atom attached to each carbon atom).

Problem-Solving Tip 9.1

Useful Ideas to Consider When Drawing Lewis Electron Dot Structures

- The octet rule is a useful guideline when drawing Lewis structures.

- Carbon forms four bonds (four single bonds; two single bonds and one double bond; or one single bond and one triple bond). In uncharged species, nitrogen forms three bonds and oxygen two. Hydrogen typically forms only one bond to another atom.

- When multiple bonds are formed, both of the atoms involved are usually one of the following: C, N, O, and S. Oxygen has the ability to form multiple bonds with a variety of ele-

ments. Carbon forms many compounds having multiple bonds to another carbon or to N or O.

- Nonmetals may form single, double, and triple bonds but never quadruple bonds.

- Always account for single bonds and lone pairs before forming multiple bonds.

- Be alert for the possibility that the molecule or ion you are working on is isoelectronic with a species you have seen before.

The carbon–carbon bonds are 144 pm long, intermediate between the average length of a $C=C$ double bond (134 pm) and a $C—C$ single bond (154 pm).

Resonance structures of benzene, C_6H_6 Abbreviated representation of resonance structures

Two resonance structures can be written for the molecule that differ only in double bond placement. A hybrid of these two structures, however, gives a molecule with six equivalent carbon–carbon bonds.

Let us apply the concepts of resonance to describe bonding in the carbonate ion, CO_3^{2-}, an anion with 24 valence electrons (12 pairs).

Three equivalent structures can be drawn for this ion, differing only in the location of the $C=O$ double bond. This fits the classical situation for resonance, so it is appropriate to conclude that no single structure correctly describes this ion. Instead, the actual structure is the average of the three structures, in good agreement with experimental results. In the CO_3^{2-} ion, all three carbon–oxygen bond distances are 129 pm, intermediate between $C—O$ single bond (143 pm) and $C=O$ double bond (122 pm) distances.

In aqueous solution, a hydrogen ion can be attached to the carbonate ion to give the hydrogen carbonate, or bicarbonate, ion. This ion can be described as a resonance hybrid of two Lewis structures.

● **Depicting Resonance Structures** The use of an arrow (⟷) as a symbol to link resonance structures and the name "resonance" are somewhat unfortunate. An arrow seems to imply that a change is occurring, and the term resonance has the connotation of vibrating or alternating back and forth between different forms. Neither view is correct. Resonance is simply a way of representing a structure. Electron pairs are not actually moving from one place to another.

A Closer Look

Resonance Structures, Lewis Structures, and Molecular Models

When drawing structures of molecules or ions that have resonance structures, or when illustrating their structures with computer-based molecular models, we generally only show one resonance structure. Thus, a model of benzene, C_6H_6, would have alternating double bonds. This is one of the two possible resonance structures. A model of nitrate ion would have one double bond and two single bonds, as in each one of the three possible resonance structures.

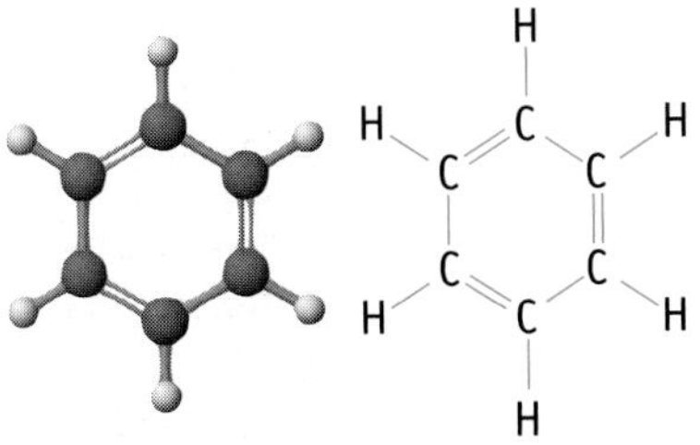

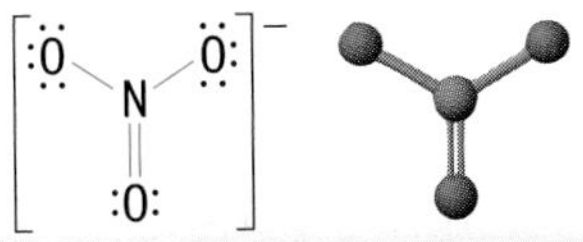

Finally, notice that in each of the examples of resonance structures, the structures have been linked by a double-headed arrow ($\longleftrightarrow$). This convention is followed throughout chemistry.

Example 9.4 • Drawing Resonance Structures

Problem • Draw resonance structures for the nitrite ion, NO_2^-. Are the N—O bonds single, double, or intermediate in value?

Strategy • Draw the dot structure in the usual manner. If multiple bonds are required, resonance structures may exist. This will be the case if the octet of an atom can be completed by using an electron pair from more than one terminal atom to form a multiple bond. Bonds to the central atom cannot then be "pure" single or double bonds but rather are somewhere between the two.

Solution • Nitrogen is the center atom in the nitrite ion, which has a total of 18 valence electrons (nine pairs).

Valence electrons = 5 (for the N atom) + 12 (6 for each O atom)
                            + 1 (for negative charge)

After forming N—O single bonds, and distributing lone pairs on the terminal O atoms, a pair remains, which is placed on the central N atom.

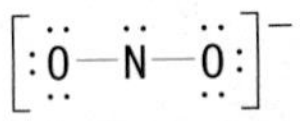

To complete the octet of electrons about the N atom, form an N=O double bond.

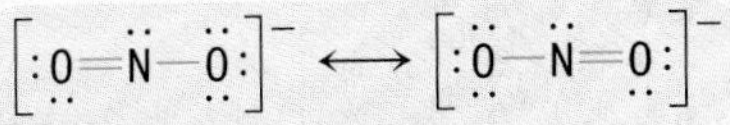

Because there are two ways to do this, two equivalent structures can be drawn, and the actual structure must be a resonance hybrid of these two structures. The nitrogen–oxygen bonds are neither single nor double bonds but have an intermediate value.

Exercise 9.7 • Drawing Resonance Structures

Draw resonance structures for the nitrate ion, NO_3^-. Sketch a plausible Lewis dot structure for nitric acid, HNO_3.

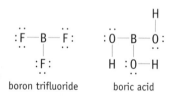

Problem-Solving Tip 9.2

Resonance Structures

- Resonance is a means of representing the bonding when a single Lewis structure fails to give an accurate picture.
- The atoms must have the same arrangement in space in all resonance structures. Moving the atoms around or attaching them in a different fashion creates a different compound.
- Resonance structures differ only in the assignment of electron-pair positions, never atom positions.
- Resonance structures differ in the number of bond pairs between a given pair of atoms.
- Even though the formal process of converting one resonance structure to another seems to move electrons about, resonance is not meant to indicate the motion of electrons.
- The actual structure of a molecule is a composite or hybrid of the resonance structures.
- There will always be at least one multiple bond (double or triple) in each resonance structure.

9.6   EXCEPTIONS TO THE OCTET RULE

Although the vast majority of molecular compounds and ions obey the octet rule, there are exceptions. These include molecules and ions that have fewer than four pairs of electrons on a central atom, those that have more than four pairs, and those that have an odd number of electrons [CD-ROM, Screens 9.9 and 9.10].

Compounds in Which an Atom Has Fewer Than Eight Valence Electrons

Boron, a nonmetal in Group 3A, has three valence electrons and so is expected to form three covalent bonds with other nonmetallic elements. This results in a valence shell for boron in its compounds with only six electrons, two short of an octet. Many boron compounds of this type are known, including such common compounds as boric acid ($B(OH)_3$), borax ($Na_2B_4O_5(OH)_4 \cdot 8 H_2O$) (Figure 9.5), and the boron trihalides (BF_3, BCl_3, BBr_3, and BI_3).

boron trifluoride   boric acid

Boron compounds such as BF_3 that are two electrons short of an octet can be quite reactive. The boron atom can accommodate a fourth electron pair, but only when that pair is provided by another atom. In general, molecules or ions with lone pairs can fulfill this role. Ammonia, for example, reacts with BF_3 to form $H_3N \rightarrow BF_3$. The bond between the B and N atoms in this compound uses an electron pair that originated on the N atom. The reaction of an F^- ion with BF_3 to form BF_4^- is another example.

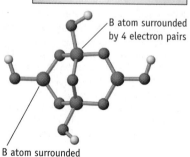

Figure 9.5   Borax. This common mineral, which is used in soaps, contains an interesting anion, $B_4O_5(OH)_4{}^{2-}$. This boron–oxygen ion has two B atoms surrounded by four electron pairs, and two B atoms surrounded by only three pairs. *(Charles D. Winters)*

B atom surrounded by 4 electron pairs

B atom surrounded by 3 electron pairs

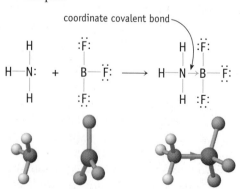

coordinate covalent bond

Table 9.7 • Lewis Structures in Which the Central Atom Exceeds an Octet

Group 4A	Group 5A	Group 6A	Group 7A	Group 8
SiF_5^-	PF_5	SF_4	ClF_3	XeF_2
SiF_6^{2-}	PF_6^-	SF_6	BrF_5	XeF_4

If the bonding pair of electrons originates on one of the bonded atoms, the bond is called a **coordinate covalent bond.** In Lewis structures, a coordinate covalent bond can be designated by an arrow that points away from the atom donating the electron pair.

Compounds in Which an Atom Has More Than Eight Valence Electrons

Elements in the third or higher periods often form compounds and ions in which the central element is surrounded by more than four valence electron pairs (Table 9.7). With most compounds and ions in this category, the central atom is bonded to small atoms such as fluorine, chlorine, or oxygen.

It is often obvious from the formula of a compound that an octet around an atom has been exceeded. As an example, consider sulfur hexafluoride, SF_6, a gas formed by the reaction of elemental sulfur and excess fluorine. Sulfur is the central atom in this compound, and fluorine typically bonds to only one other atom with a single electron pair bond (as in HF and CF_4). Six S—F bonds are required in SF_6, meaning there will be six electron pairs in the valence shell of the sulfur atom.

More than four groups bonded to a central atom is a reliable signal that there are more than eight electrons around a central atom. But be careful — the central atom octet can also be exceeded with four or fewer atoms bonded to the central atom when lone pairs are present. Consider three examples from Table 9.7: the central atoms in SF_4, ClF_3, and XeF_2 have five electron pairs in their valence shell.

A useful observation is that *only elements of the third and higher periods in the periodic table form compounds and ions in which an octet is exceeded.* Second-period elements (B, C, N, O, F) are restricted to a maximum of eight electrons in their compounds. For example, nitrogen forms compounds and ions such as NH_3, NH_4^+, and NF_3, but NF_5 is unknown. Phosphorus, the third-period element just below nitrogen in the periodic table, forms many compounds similar to nitrogen (PH_3, PH_4^+, PF_3), but it also readily accommodates five or six valence electron pairs in compounds such as PF_5 or in ions such as PF_6^-. Arsenic, antimony, and bismuth, the elements below phosphorus in Group 5A, resemble phosphorus in their behavior.

The usual explanation for the contrasting behavior of second- and third-period elements centers on the number of orbitals in the valence shell of an atom. Second-period elements have four valence orbitals (one $2s$ and three $2p$ orbitals). Two elec-

● **Xenon Compounds**

Compounds of xenon are among the more interesting entries in Table 9.7. When the authors of this book were undergraduates we were taught that the noble gases did not form chemical compounds. Then several noble gas compounds were discovered in the early 1960s. One of the more intriguing compounds is XeF_2, in part because of the simplicity of its synthesis. Xenon difluoride can be made by placing a flask containing xenon gas and fluorine gas in the sunlight. After several weeks, crystals of colorless XeF_2 are found in the flask.

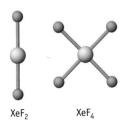

XeF_2 XeF_4

trons per orbital result in a total of eight electrons being accommodated around an atom. For elements in the third and higher periods, the d orbitals in the outer shell are traditionally included among valence orbitals for the elements. Thus, for phosphorus, the $3d$ orbitals are included with the $3s$ and $3p$ orbitals as valence orbitals. The extra orbitals provide the element with an opportunity to accommodate up to 12 electrons (or even more).

Example 9.5 **Lewis Structures in Which the Central Atom Has More Than Eight Electrons**

Problem • Sketch the Lewis structure of the $[ClF_4]^-$ ion.

Strategy • Use the guidelines on page 334.

Solution •

1. The Cl atom is the central atom.

2. This ion has 36 valence electrons [= 7 for Cl + 4 × (7 for F) + 1 for ion charge] or 18 pairs.

3. Draw the ion with four single covalent Cl—F bonds.

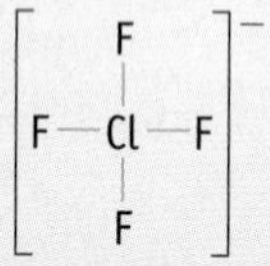

4. Place lone pairs on the terminal atoms. Because two electron pairs remain after placing lone pairs on the four F atoms, and because we know that Cl can accommodate more than four pairs, these two pairs are placed on the central Cl atom.

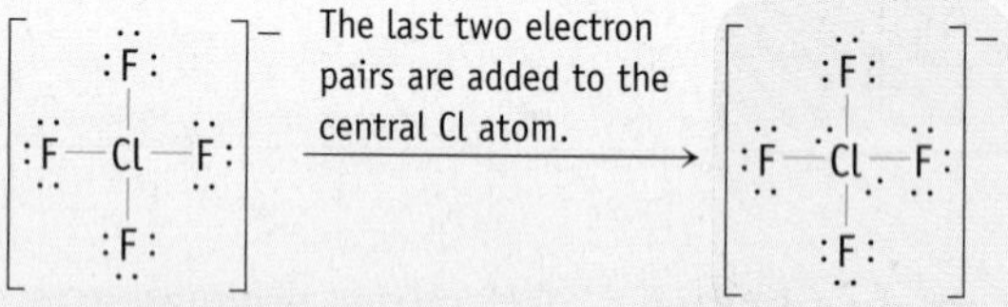

The last two electron pairs are added to the central Cl atom.

Exercise 9.8 **Lewis Structures in Which the Central Atom Has More Than Eight Electrons**

Sketch the Lewis structures for $[ClF_2]^+$ and $[ClF_2]^-$. How many lone pairs and bond pairs surround the Cl atom in each ion?

Molecules with an Odd Number of Electrons

Two nitrogen oxides — NO, with 11 valence electrons, and NO_2, with 17 — are among a small group of stable molecules with an odd number of electrons. An odd number of electrons makes it impossible to draw structures obeying the octet rule for these molecules; at least one electron must be unpaired.

Even though NO_2 does not obey the octet rule, an electron dot structure can be written that approximates the bonding in the molecule. This Lewis structure places the unpaired electron on nitrogen. Two resonance structures show that the nitrogen–oxygen bonds are equivalent, as observed experimentally.

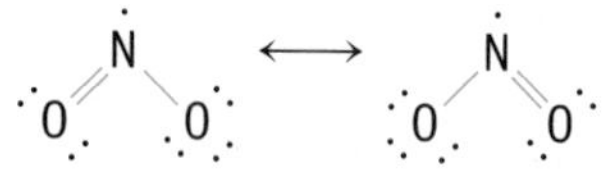

Experimental evidence for NO indicates that the bonding between N and O is intermediate between a double and a triple bond. It is impossible to write a Lewis structure for NO that is in accord with the properties of this substance, so a different theory is needed to understand bonding in this molecule. We shall return to compounds of this type when molecular orbital theory is introduced in Chapter 10.

The two nitrogen oxides, NO and NO_2, are members of a class of chemical substances called free radicals. **Free radicals** are chemical species — both atoms and molecules — with an unpaired electron. How do these unpaired electrons af-

Chemical Perspectives

NO: A Small Molecule with a Big Biological Role

Small molecules such as H_2, O_2, H_2O, CO, and CO_2 are among the most important molecules commercially, environmentally, and biologically. Imagine the surprise of chemists and biologists when it was discovered a few years ago that nitrogen monoxide (nitric oxide, NO), which was widely considered toxic, also has an important biological role.

Nitric oxide is a colorless, paramagnetic gas that is moderately soluble in water. In the laboratory, it can be synthesized by the reduction of nitrite ion with iodide ion:

$$KNO_2(aq) + KI(aq) + H_2SO_4(aq) \longrightarrow$$
$$NO(g) + K_2SO_4(aq) + H_2O(\ell) + \tfrac{1}{2} I_2(aq)$$

The formation of NO from the elements is an unfavorable, energetically uphill reaction ($\Delta H_f^\circ = 90.2$ kJ/mol). Nevertheless, small quantities of this compound form from nitrogen and oxygen at high temperatures. For example, conditions in an internal combustion engine are favorable for this to happen.

Nitric oxide reacts rapidly with O_2 to form the reddish brown gas NO_2.

$$2 \text{ NO}(g) + O_2(g) \longrightarrow 2 \text{ NO}_2(g)$$
colorless gas   brown gas

The result is that compounds such as NO_2 and HNO_3 arising from reactions of NO with O_2 and H_2O are among the air pollutants produced by automobiles.

A few years ago chemists learned that NO is synthesized in a biological process by animals as diverse as barnacles, fruit flies, horseshoe crabs, chickens, trout, and humans. Even more recently chemists have found that NO is important in an astonishing range of physiological processes in humans and other animals. These include a role in neurotransmission, blood clotting, and blood pressure control as well as in the immune system's ability to kill tumor cells.

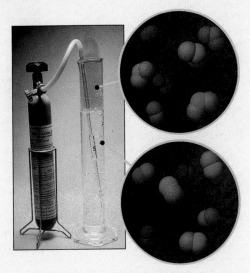

The colorless gas NO is bubbled into water from a high-pressure tank. When the gas emerges into the air, the NO reacts rapidly with O_2 to give brown NO_2 gas. *(Charles D. Winters)*

---

fect reactivity? Free atoms such as H and Cl are free radicals and are very reactive, readily combining with other atoms to give molecules such as H_2, Cl_2, and HCl. We therefore expect free radical molecules to be more reactive than molecules with paired electrons, and most are.

Free radicals are involved in many reactions in the environment. For example, small amounts NO are released from vehicle exhausts. The NO rapidly forms NO_2, which decomposes in the presence of sunlight and oxygen to give more NO as well as ozone, O_3, which is an air pollutant at ground level that affects the respiratory system.

$$NO_2(g) + O_2(g) \longrightarrow NO(g) + O_3(g)$$

The two nitrogen oxides, NO and NO_2, are unique in that they can be isolated and neither has the extreme reactivity of most free radicals. When cooled, however, two NO_2 molecules join or "dimerize" to form colorless N_2O_4; the unpaired electrons combine to form an N — N bond in N_2O_4. Even though this bond is weak, the reaction is easily observed in the laboratory (Figure 9.6).

9.7 CHARGE DISTRIBUTION IN COVALENT BONDS AND MOLECULES

Lewis structures generally provide a fairly good picture of bonding in a covalently bonded molecule or ion. It is possible to fine tune this picture, however, to get a more precise description of the distribution of the electrons. This more precise

Figure 9.6   Free radical chemistry. When cooled, the brown gas NO_2, a free radical, forms colorless N_2O_4, a molecule with an N — N single bond. The coupling of two free radicals is a common type of chemical reactivity. Because two identical free radicals come together, the product is called a dimer, and the process is called a dimerization. *(Charles D. Winters)*

A flask of NO_2 gas in warm water
(Charles D. Winters)

NO_2 free radicals couple to form N_2O_4 molecules.
$\longrightarrow$

A flask of NO_2 gas in ice water
(Charles D. Winters)

description will, in turn, provide further insight into the chemical and physical properties of covalent molecules.

Closer analysis of covalently bonded molecules reveals that the valence electrons are not distributed among the atoms as evenly as Lewis structures might suggest. Some atoms may have a slight negative charge and others a slight positive charge. This occurs because the electron pair or pairs in a given bond may be drawn more strongly toward one atom than the other. The way the electrons are distributed in the molecule is called its *charge distribution.*

Charge distribution can determine the properties of a molecule. Consider a diatomic (two-atom) molecule in which one atom is partially positive and the other is partially negative. In the solid state, for example, the molecules would be expected to line up with the positive end of one molecule near the negative end of another. The intermolecular, or "between molecule," force of attraction would be enhanced by the attraction of opposite charges, and properties of the substance that are related to intermolecular forces such as boiling point would be affected (➡ SECTIONS 13.2 AND 13.5).

The positive or negative charges in a molecule or ion will influence, among other things, the site at which reactions occur. For example, does a positive H^+ ion attach itself to the O or the Cl of OCl^-? Is the product HOCl or HClO? It is reasonable to expect H^+ to attach to the more negatively charged atom. We can perdict this by evaluating atom formal charges in molecules and ions.

Formal Charges on Atoms

The **formal charge** for an atom in a molecule or ion is the charge calculated for that atom based on the Lewis structure of the molecule or ion using Equation 9.1 [🖱 CD-ROM, Screen 9.15].

● **Calculating Formal Charge**
We shall apply formal charge calculations only to atoms of main group elements.

Formal charge of an atom in a molecule or ion =
group number − [number LPE + ½ (number BE)]

(9.1)

In this equation

- The group number gives the number of valence electrons brought to the molecule or ion by the atom.

- Number of LPE = number of lone pair electrons on an atom.
- Number of BE = number of bonding electrons around an atom.

The term in square brackets is the number of electrons assigned by the Lewis structure to an atom in a molecule or ion. The difference between this term and the group number is the formal charge. An atom will be positive if it "contributes" more electrons to bonding than it "gets back" when in a molecule or ion. The atom's formal charge will be negative if the reverse is true.

Equation 9.1 is based on two important assumptions. First, lone pairs are assumed to belong to the element on which they reside in the Lewis structure. Second, bond pairs are assumed to be divided equally between the bonded atoms. (The factor of $\frac{1}{2}$ divides the bonding electrons equally between the atoms linked by the bond.)

The sum of the formal charges on the atoms in a molecule or ion always equals the net charge on the molecule or ion. Consider the hydroxide ion. Oxygen is in Group 6A and so has six valence electrons. In the hydroxide ion, however, oxygen can lay claim to seven electrons (six lone pair electrons and one bonding electron), and so the atom has a formal charge of -1. The O atom has "formally" gained an electron as part of the hydroxide ion.

$$\text{Formal charge} = -1 = 6 - [6 + \tfrac{1}{2}(2)]$$

$$\left[:\!\ddot{O}\!-\!H \right]^{-} \qquad \textit{Sum of formal charges} = -1$$

$$\text{Formal charge} = 0 = 1 - [0 + \tfrac{1}{2}(2)]$$

Assume a covalent bond, so bonding electrons are divided equally between O and H

The formal charge on the hydrogen atom in OH^{-} is zero. (We have -1 for oxygen and 0 for hydrogen, which equals the net charge of -1 for the ion.) An important conclusion we can draw from this is that, if an H^{+} ion approaches an OH^{-} ion, it attaches itself to the O atom. This of course leads to water, as is indeed observed.

Formal charges can be calculated for more complicated species such as the nitrate ion. Using one of the resonance structures for the ion, we find that the central N atom has a formal charge of $+1$, and the singly bonded O atoms are both -1. The doubly bonded O atom has no charge. The net charge for the ion is thus -1.

$$\text{Formal charge} = 0 = 6 - [4 + \tfrac{1}{2}(4)]$$

$$\left[\begin{array}{c} :\!\ddot{O}: \\ \parallel \\ :\!\ddot{O}\!-\!N\!-\!\ddot{O}: \end{array} \right]^{-} \qquad \textit{Sum of formal charges} = -1$$

$$\text{Formal charge} = +1 = 5 - [0 + \tfrac{1}{2}(8)]$$
$$\text{Formal charge} = -1 = 6 - [6 + \tfrac{1}{2}(2)]$$

Is this a reasonable representation of the charge distribution for the nitrate ion? The answer is no. The problem is that the actual structure of the nitrate ion is a resonance hybrid of three equivalent resonance structures. Because the three oxygen atoms in NO_3^- are equivalent, the charge on one oxygen atom should not be different from the other two. This can be resolved, however, if the formal charges on the oxygens are averaged to give a formal charge of $-(\frac{2}{3})$. Summing the charges on the three oxygen atoms and the $+1$ charge on the nitrogen atom then gives -1, the charge on the ion.

In the resonance structures for O_3, CO_3^{2-}, and NO_3^- for example, all the possible resonance structures are equally likely; they are "equivalent" structures. The molecule or ion therefore has a symmetrical distribution of electrons over all the atoms involved — that is, its electronic structure consists of an equal "mixture," or "hybrid," of the resonance structures.

Example 9.6 Calculating Formal Charges

Problem • Calculate formal charges for the atoms in (a) NH_4^+ and (b) in one resonance structure of CO_3^{2-}.

Strategy • The first step is always to write the Lewis structure for the molecule or ion. Only then can you calculate the formal charges.

Solution •
(a) *Formal charge in the NH_4^+ ion*

Formal charge = 0
$= 1 - [0 + \frac{1}{2}(2)]$

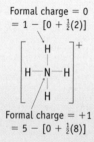

Formal charge = +1
$= 5 - [0 + \frac{1}{2}(8)]$

(b) *Formal charges for the CO_3^{2-} ion*

Formal charge = 0
$= 6 - [4 + \frac{1}{2}(4)]$

Formal charge = 0
$= 4 - [0 + \frac{1}{2}(8)]$

Formal charge = −1
$= 6 - [6 + \frac{1}{2}(2)]$

In each case notice that the sum of the atom formal charges is the charge on the ion. In the carbonate ion, which has three resonance structures, the average charge on the O atoms is $-(\frac{2}{3})$.

Exercise 9.9 Calculating Formal Charges

Calculate formal charges on each atom in (a) CN^- and (b) SO_3.

Bond Polarity and Electronegativity

The models used to represent covalent and ionic bonding are the extreme situations in bonding. Pure covalent bonding, in which atoms share an electron pair equally, occurs *only* when two identical atoms are bonded. When two dissimilar atoms form a covalent bond, the electron pair will be unequally shared. The result is a **polar covalent bond,** a bond in which the two atoms have residual or partial charges (Figure 9.7) [CD-ROM, Screen 9.13].

Why are bonds polar? Because not all atoms hold onto their valence electrons with the same force. Recall from the discussion of atom properties that different elements have different values of ionization energy and electron affinity (Section 8.6). These differences in behavior for free atoms carry over to atoms in molecules.

If a bond pair is not equally shared between atoms, the bonding electrons are nearer one of the atoms. The atom toward which the pair is displaced has a larger share of the electron pair and thus acquires a partial negative charge. At the same time, the atom at the other end of the bond is depleted in electrons and acquires a partial positive charge. The bond between the two atoms has a positive end and a negative end; that is, it has negative and positive poles and is a **polar covalent bond.** The term **dipolar** (having two poles) is also used. If there is no net displacement of the bond electron pair, the bond is **nonpolar covalent.**

In ionic compounds, displacement of the bonding pair to one of the two atoms is essentially complete, and + and − symbols are written alongside the atom sym-

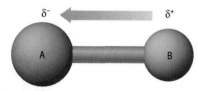

Figure 9.7   A polar covalent bond. Element A has a larger share of the bonding electrons and element B has the smaller share. The result is that A has a partial negative charge ($-\delta$), and B has a partial positive charge ($+\delta$).

A Closer Look

Formal Charge and Oxidation Number

In Chapter 5 you learned to calculate the oxidation number of an atom as a way to tell if a reaction is an oxidation–reduction. What is the difference between an oxidation number and an atom's formal charge?

To answer this question, look again at the hydroxide ion, OH^-. The formal charges (page 341) are

$$\text{Formal charge} = -1 = 6 - [6 + \tfrac{1}{2}(2)]$$

$$\left[\ddot{\underset{\displaystyle \cdot\cdot}{O}} - H\right]^- \quad \textit{Sum of formal charges} = -1$$

$$\text{Formal charge} = 0 = 1 - [0 + \tfrac{1}{2}(2)]$$

Recall that these formal charges are calculated assuming the $O-H$ bond electrons are shared equally; the $O-H$ is covalent.

In contrast, in Chapter 5 (page 171), you learned that O has an oxidation number of -2 and H has a number of $+1$. Oxidation numbers are determined by assuming that the bond between a pair of atoms is ionic, not covalent. For OH^- this means the pair of electrons between O and H is located fully on O. Thus, the O atom now has eight valence electrons instead of six and a charge of -2. The H atom now has no valence electrons and a charge of $+1$.

Oxidation number = -2

$$\left[\ddot{\underset{\displaystyle \cdot\cdot}{O}}\colon \; H\right]^-$$

Sum of oxidation numbers = -1

Assume an ionic bond   Oxidation number = $+1$

Formal charges and oxidation numbers are calculated using different assumptions. Both are useful, but for different purposes. Oxidation numbers allow us to follow changes in redox reactions. Formal charges more nearly resemble atom charges in molecules and ions.

---

bols in the Lewis structures. For a polar covalent bond the polarity is indicated by writing the symbols $+\delta$ and $-\delta$ alongside the atom symbols, where δ (the Greek letter delta) stands for a *partial* charge. Hydrogen fluoride, water, and ammonia are three simple molecules having polar, covalent bonds (Figure 9.8).

With so many atoms to use in covalent bond formation, it is not surprising that bonds between atoms can fall anywhere in a continuum from pure ionic to pure covalent.

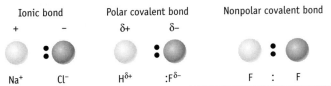

Ionic bond   Polar covalent bond   Nonpolar covalent bond

Electron pair shared more and more equally.
Bond becoming less ionic and more covalent.

There is no sharp dividing line between, for example, an ionic bond and a covalent bond.

In the 1930s, Linus Pauling proposed a parameter called atom electronegativity that allows us to decide if a bond is polar, which atom of the bond is negative and which is positive, and if one bond is more polar than another. The **electronegativity**, χ, of an atom is defined as a measure of *the ability of an atom in a molecule to attract electrons to itself.*

Values of electronegativity are given in Figure 9.9. Several features and periodic trends are apparent. The element with the largest electronegativity is fluorine; it is assigned a value of $\chi = 4.0$. The element with the smallest value is the alkali metal francium. Electronegativities generally increase from left to right across a period and decrease down a group. This is the opposite of the trend observed for metallic character. Metals typically have low values of electronegativity, ranging from slightly less than 1 to about 2. Electronegativity values for the metalloids are around 2, whereas nonmetals have values greater than 2. No values are given for He, Ne, and Ar because these elements are not known to form chemical compounds.

There will be a large *difference* in electronegativities if atoms from the left- and right-hand sides of the periodic table form a chemical compound. For cesium

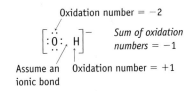

Figure 9.8 Three simple molecules with polar covalent bonds. In each case F, O, and N are more electronegative than H. See Figure 9.9.

● **Electronegativity —**
And Bond Ionic Character
The greater the difference in electronegativity of two elements, the greater the ionic character of the bond between them.

And Electron Affinity
Electronegativity and electron affinity may seem very similar, but they are not. *Electronegativity is a parameter that applies only to atoms in molecules.* As described in Section 8.6, *electron affinity is a measurable energy quantity that refers to isolated atoms.*

And Lewis Structures
A useful criterion for determining the central atom in a molecule is that it is the atom of lowest electronegativity.

													H 2.1						

1A	2A							8B			1B	2B	3A	4A	5A	6A	7A
Li 1.0	Be 1.5												B 2.0	C 2.5	N 3.0	O 3.5	F 4.0
Na 0.9	Mg 1.2	3B	4B	5B	6B	7B					1B	2B	Al 1.5	Si 1.8	P 2.1	S 2.5	Cl 3.0
K 0.8	Ca 1.0	Sc 1.3	Ti 1.5	V 1.6	Cr 1.6	Mn 1.5	Fe 1.8	Co 1.8	Ni 1.8	Cu 1.9	Zn 1.6	Ga 1.6	Ge 1.8	As 2.0	Se 2.4	Br 2.8
Rb 0.8	Sr 1.0	Y 1.2	Zr 1.4	Nb 1.6	Mo 1.8	Tc 1.9	Ru 2.2	Rh 2.2	Pd 2.2	Ag 1.9	Cd 1.7	In 1.7	Sn 1.8	Sb 1.9	Te 2.1	I 2.5
Cs 0.7	Ba 0.9	La 1.1	Hf 1.3	Ta 1.5	W 1.7	Re 1.9	Os 2.2	Ir 2.2	Pt 2.2	Au 2.4	Hg 1.9	Tl 1.8	Pb 1.8	Bi 1.9	Po 2.0	At 2.2

☐ <1.0   ☐ 1.5–1.9   ☐ 2.5–2.9
☐ 1.0–1.4   ☐ 2.0–2.4   ☐ 3.0–4.0

Figure 9.9   Electronegativity values for the elements according to Pauling. Trends for electronegativities are the opposite of the trends defining metallic character. Nonmetals have high values of electronegativity, the metalloids have intermediate values, and the metals have low values.

(Chapter Goals • Revisited)

● Understand the difference between ionic and covalent bonds.

● Draw Lewis electron dot structures for small molecules and ions.

● **Use electronegativity to predict the polarity of bonds and molecules.**

● Understand the properties of covalent bonds and their influence on molecular properties.

● Use VSEPR to predict the shapes of simple molecules and ions and to understand the structures of more complex molecules.

---

fluoride, for example, the difference in electronegativity values, $\Delta\chi$, is 3.3 $[= 4.0$ (for F) $- 0.7$ (for Cs)$]$. The bond is ionic, Cs is positive (Cs^+) and F negative (F^-). In contrast, the electronegativity difference between H and Cl in HCl is only 0.9 $[= 3.0$ (for Cl) $- 2.1$ (for H)$]$. We conclude that bonding in HCl must be more covalent than ionic, as expected for a compound formed from two nonmetals. The H—Cl bond is polar, however, with hydrogen being the positive end of the molecule and chlorine the negative end ($H^{\delta+}$ —$Cl^{\delta-}$).

Predicting trends in bond polarity in groups of related compounds is possible using values of electronegativity. Among the hydrogen halides, for example, the trend in polarity is HF ($\Delta\chi = 1.9$) > HCl ($\Delta\chi = 0.9$) > HBr ($\Delta\chi = 0.7$) > HI ($\Delta\chi = 0.4$).

Example 9.7   Estimating Bond Polarities

Problem • For each of the following bond pairs, decide which is the more polar and indicate the negative and positive poles.

(a) B—F and B—Cl
(b) Si—O and P—P
(c) C=O and C=S

Strategy • Locate the elements in the periodic table. Recall that electronegativity generally increases across a period and up a group.

Solution •

(a) B and F lie relatively far apart in the periodic table. B is a metalloid and F is a nonmetal. Here χ for B = 2.0 and χ for F = 4.0. Similarly, B and Cl are are relatively far apart in the periodic table, but Cl is below F in the periodic table (χ for Cl = 3.0) and is therefore less electronegative than F.

The difference in electronegativity for B—F is 2.0, for B—Cl 1.0. Both bonds are expected to be polar, with B positive and the halide atom negative, but the B—F bond will be more polar than the B—Cl bond.

(b) Because the bond is between two atoms of the same kind, the P—P bond is nonpolar. Silicon is in Group 4A and the third period, whereas O is in Group 6A and the second period. Consequently, O has a greater electronegativity (3.5) than Si (1.8), so the Si—O bond is highly polar ($\Delta\chi = 1.7$), with O the more negative atom.

(c) Oxygen lies above sulfur in the periodic table, so oxygen is more electronegative than S. This means the C—O bond is more polar than the C—S bond. For the C—O bond, O is the more negative atom. The value of $\Delta\chi$ (1.0) for CO indicates a moderately polar bond.

Exercise 9.10 Bond Polarity

For each of the following pairs of bonds, decide which is the more polar. For each polar bond, indicate the positive and negative poles. First make your prediction from the relative atom positions in the periodic table; then check your prediction by calculating $\Delta\chi$.

(a) H—F and H—I   **(b)** B—C and B—F   **(c)** C—Si and C—S

Combining Formal Charge and Bond Polarity

Using formal charge calculations alone to locate the site of a charge in an ion can sometimes lead to results that seem incorrect. The ion BF_4^- illustrates this point. Boron has a formal charge of -1 in this ion, whereas the formal charge calculated for the fluorine atoms is 0. This is not logical: fluorine is the more electronegative atom so the negative charge should reside on F and not on B.

The way to resolve this dilemma is to consider electronegativity in conjunction with formal charge. Based on the electronegativity difference between fluorine and boron ($\Delta\chi = 2.0$) the B—F bonds are expected to be polar, with fluorine being the negative end of the bond, $B^{\delta+}—F^{\delta-}$. So, for BF_4^-, predictions based on electronegativity and formal charge work in opposite directions. The formal charge calculation places the negative charge on boron, but the electronegativity difference says that the charge on boron ends up being distributed onto the fluorine atoms. In effect the charge is "spread out" over the molecule.

Linus Pauling proposed an important idea that applies to the problem with BF_4^- and to all other molecules: the **electroneutrality principle.** This declares that the electrons in a molecule are distributed in such a way that the charges on the atoms are as close to zero as possible. Furthermore, when a negative charge occurs, it should be placed on the most electronegative atom. Similarly, a positive charge should be on the least electronegative atom. For BF_4^-, this means the negative charge should not be located on the boron atom alone but spread out over the more electronegative F atoms.

Considering the concepts of electronegativity and formal charge together can also help to decide which of several resonance structures is the more important. For example, when drawing the Lewis structure for CO_2, structure A is the logical one to draw. But what is wrong with structure B, in which each atom also has an octet of electrons?

History

Linus Pauling (1901–1994)

Born in Portland, Oregon, Pauling earned a B.Sc. degree in chemical engineering from Oregon State College in 1922 and completed his Ph.D. in chemistry at the California Institute of Technology in 1925. In chemistry he is best known for his work on chemical bonding. For more on Linus Pauling, see page 378. *(Oesper Collection in the History of Chemistry/University of Cincinnati)*

Formal charges	0 0 0	+1 0 −1
Resonance structures	Ö=C=Ö	:Ö=C—Ö:
	A	B

For structure A, each atom has a formal charge of 0, a favorable situation. In B, however, one oxygen atom has a formal charge of $+1$ and the other has -1, contrary to the principle of electroneutrality. In addition, B places a positive charge on a very electronegative atom. Thus, we can conclude that B is a less satisfactory structure than A.

Now use what you have learned with CO_2 to decide which of the three possible resonance structures for the OCN^- ion is the most reasonable. Formal charges for each atom are given above the element's symbol.

● **Formal Charges in OCN⁻**

Example of formal charge calculation: For resonance form *C* for OCN⁻, we have

$$O = 6 - [2 + (\tfrac{1}{2})(6)] = +1$$
$$C = 4 - [0 + (\tfrac{1}{2})(8)] = 0$$
$$N = 5 - [6 + (\tfrac{1}{2})(2)] = -2$$

Sum of formal charges $= -1$
                    $=$ charge on the ion

$$
\underset{A}{\left[\overset{-1}{\underset{..}{\overset{..}{:O}}}\!-\!\overset{0}{C}\!\equiv\!\overset{0}{\underset{..}{N}}:\right]^{-}}
\longleftrightarrow
\underset{B}{\left[\overset{0}{\underset{..}{\overset{..}{:O}}}\!=\!\overset{0}{C}\!=\!\overset{-1}{\underset{..}{N}}:\right]^{-}}
\longleftrightarrow
\underset{C}{\left[\overset{+1}{\overset{..}{O}}\!=\!\overset{0}{C}\!-\!\overset{-2}{\underset{..}{\overset{..}{N}}}:\right]^{-}}
$$

Structure C will not contribute significantly to the overall electronic structure of the ion. It has a -2 formal charge on the N atom and a $+1$ formal charge on the O atom. Structure A is more significant than structure B because the negative charge in A is placed on the most electronegative atom. We predict, therefore, that the carbon–nitrogen bond will resemble a triple bond.

The result for OCN⁻ also allows us to predict that protonation of the ion will lead to HOCN and not HNCO. That is, an H⁺ ion will add to the more negative oxygen atom.

Example 9.8 **Calculating Formal Charges**

Problem • Boron-containing compounds often have a boron atom with only three bonds (and no lone pairs). Why not form a double bond with a terminal atom to complete the boron octet? To answer this, consider the resonance structures of BF₃ and calculate the atoms' formal charges. Are the bonds polar in BF₃? If so, which is the more negative atom?

Strategy • Calculate the formal charges on each atom in the resonance structures. The preferred structure will have atoms with low formal charges. Negative formal charges should be on the most electronegative atoms.

Solution • The two possible structures for BF₃ are illustrated here with the calculated formal charges on the B and F atoms.

Formal charge $= 0$
$\qquad = 7 - [6 + \tfrac{1}{2}(2)]$

Formal charge $= +1$
$\qquad = 7 - [4 + \tfrac{1}{2}(4)]$

Formal charge $= 0$
$\qquad = 3 - [0 + \tfrac{1}{2}(6)]$

Formal charge $= -1$
$\qquad = 3 - [0 + \tfrac{1}{2}(8)]$

The structure on the left is preferred because all atoms have a zero formal charge, and the very electronegative F atom does not have a charge of $+1$.

F $(\chi = 4.0)$ is more electronegative than B $(\chi = 2.0)$ so the B—F bond is polar, the F atom being partially negative and the B atom partially positive.

Exercise 9.11 **Formal Charge, Bond Polarity, and Electronegativity**

Consider all possible resonance structures for SO₂. What are the formal charges on each atom in each resonance structure? What are the bond polarities? Do they agree with the formal charges?

Chapter Goals • Revisited

● Understand the difference between ionic and covalent bonds.

● Draw Lewis electron dot structures for small molecules and ions.

● Use electronegativity to predict the polarity of bonds and molecules.

● **Understand the properties of covalent bonds and their influence on molecular properties.**

● Use VSEPR to predict the shapes of simple molecules and ions and to understand the structures of more complex molecules.

9.8 **BOND PROPERTIES**

Bond Order

The **order of a bond** is the number of bonding electron pairs shared by two atoms in a molecule (Figure 9.10). You will encounter bond orders of 1, 2, and 3, as well as fractional bond orders.

Where there is only a single covalent bond between a pair of atoms, the bond order is 1. Examples are the bonds in molecules such as H₂, NH₃, and CH₄. The bond order is 2 when two electron pairs are shared between atoms, such as the C=O bonds in CO₂ and the C=C bond in ethylene, H₂C=CH₂. The bond order is 3 when two atoms are connected by three bonds. Examples include the carbon–oxygen bond in carbon monoxide, CO and the nitrogen–nitrogen bond in N_2.

Fractional bond orders occur in molecules and ions having resonance structures. For example, what is the bond order for each oxygen–oxygen bond in O_3? Each resonance structure of O_3 has one O—O single bond and one O=O double bond, for a total of three shared bonding pairs accounting for two oxygen–oxygen links.

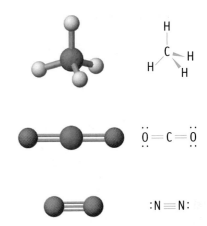

Figure 9.10   **Bond order.** The four C—H bonds in methane each have a bond order of 1. The two C=O bonds of CO_2 each have a bond order of 2, whereas the nitrogen–nitrogen bond in N_2 has an order of 3.

Bond order = 1

Bond order = 2

Average bond order for the resonance hybrid = $\frac{3}{2}$, or 1.5

One resonance structure

If we define the bond order between any bonded pair of atoms X and Y as

$$\text{Bond order} = \frac{\text{number of shared pairs linking X and Y}}{\text{number of X—Y links in the molecule or ion}} \qquad (9.2)$$

then the order is seen to be $\frac{3}{2}$, or 1.5, for ozone.

Bond Length

Bond length is the distance between the nuclei of two bonded atoms. Single bond lengths are largely determined by the sizes of the atoms (see Section 8.6). For a given pair of elements, the order of the bond determines the value of the distance [CD-ROM, Screen 9.11].

Table 9.8 lists average bond lengths for a number of common chemical bonds. It is important to recognize that these are *approximate* values. Neighboring parts of a molecule can affect the length of a particular bond. For example, Table 9.8 specifies that the average C—H bond has a length of 110 pm. In methane, CH_4, the actual C—H bond length is 109.4 pm, whereas it is only 105.9 pm in acetylene, H—C≡C—H. Variations as great as 10% from the approximate values listed in Table 9.8 are possible.

Because atom sizes vary in a regular fashion with the position of the element in the periodic table (FIGURE 8.9), predictions of trends in bond length can be made quickly. For example, the H—X distance in the hydrogen halides increases in the order predicted by the relative sizes of the halogens: H—F < H—Cl < H—Br < H—I. Likewise, bonds between carbon and another element in a given period decrease going from left to right, in a predictable fashion; for example, C—C > C—N > C—O > C—F. Trends involving multiple bonds are similar. A C=O bond is shorter than a C=S bond, and a C=N bond is shorter than a C=C bond.

The effect of bond order on bond length is evident when bonds between the same two atoms are compared. For example, the bonds become shorter as the bond order increases in the series C—O, C=O, and C≡O:

Bond	C—O	C=O	C≡O
Bond Order	1	2	3
Bond Length (pm)	143	122	113

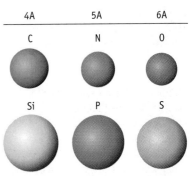

4A	5A	6A
C	N	O
Si	P	S

Relative sizes of some atoms of Groups 4A, 5A, and 6A.

Table 9.8 • Some Approximate Single and Multiple Bond Lengths*

Single Bond Lengths

Group

	1A	4A	5A	6A	7A	4A	5A	6A	7A	7A	7A
	H	C	N	O	F	Si	P	S	Cl	Br	I
H	74	110	98	94	92	145	138	132	127	142	161
C		154	147	143	141	194	187	181	176	191	210
N			140	136	134	187	180	174	169	184	203
O				132	130	183	176	170	165	180	199
F					128	181	174	168	163	178	197
Si						234	227	221	216	231	250
P							220	214	209	224	243
S								208	203	218	237
Cl									200	213	232
Br										228	247
I											266

Multiple Bond Lengths

$C=C$	134	$C\equiv C$	121
$C=N$	127	$C\equiv N$	115
$C=O$	122	$C\equiv O$	113
$N=O$	115	$N\equiv O$	108

*In picometers (pm); 1 pm $= 10^{-12}$ m.

Adding a second bond to a $C-O$ single bond to make a $C=O$ double bond shortens the bond by 21 pm. A further 9-pm reduction in bond length occurs from $C=O$ to $C\equiv O$. Double bonds are shorter than single bonds between the same set of atoms, and triple bonds between those same atoms are shorter still.

The carbonate ion, CO_3^{2-}, has three equivalent resonance structures. It has a CO bond order of 1.33 (or $\frac{4}{3}$) because four electron pairs link the central carbon with the three oxygen atoms. The CO bond distance (129 pm) is intermediate between a $C-O$ single bond (143 pm) and a $C=O$ double bond (122 pm).

Exercise 9.12   Bond Order and Bond Length

(a) Give the bond order of each of the following bonds and arrange them in order of decreasing bond distance: $C=N$, $C\equiv N$, and $C-N$.

(b) Draw resonance structures for NO_2^-. What is the NO bond order in this ion? Consult Table 9.8 for $N-O$ and $N=O$ bond lengths. Compare these with the NO bond length in NO_2^- (124 pm). Account for any differences you observe.

Bond Energy

The **bond dissociation energy,** symbolized by D, is the enthalpy change for breaking a bond in a molecule with the reactants and products in the gas phase under standard conditions [💾 CD-ROM, Screen 9.12].

$$\text{Molecule (g)} \underset{\text{Energy released} = -D}{\overset{\text{Energy supplied} = D}{\rightleftharpoons}} \text{Molecular fragments (g)}$$

Suppose you wish to break the carbon–carbon bonds in ethane (H_3C—CH_3), ethylene (H_2C=CH_2), and acetylene (HC≡CH), for which the bond orders are 1, 2, and 3, respectively. For the same reason that the ethane C—C bond is the longest of the series and the acetylene C≡C bond is the shortest, bond breaking requires the least energy for ethane and the most for acetylene.

$$H_3C—CH_3(g) \longrightarrow H_3C(g) + CH_3(g) \qquad \Delta H = D = +346 \text{ kJ}$$
$$H_2C=CH_2(g) \longrightarrow H_2C(g) + CH_2(g) \qquad \Delta H = D = +610 \text{ kJ}$$
$$HC≡CH(g) \longrightarrow HC(g) + CH(g) \qquad \Delta H = D = +835 \text{ kJ}$$

Because D represents the energy transferred to the molecule from its surroundings, D has a positive value; that is, *the process of breaking bonds in a molecule is always endothermic.*

The energy supplied to break carbon–carbon bonds must be the same as the energy released when the same bonds form. *The formation of bonds from atoms or radicals in the gas phase is always exothermic.* This means, for example, that ΔH for the formation of H_3C—CH_3 from two $CH_3(g)$ radicals is -346 kJ/mol.

$$H_3C \cdot (g) + \cdot CH_3(g) \longrightarrow H_3C—CH_3(g) \qquad \Delta H = -D = -346 \text{ kJ}$$

Generally, the bond energy for a given type of bond (C—C bonds, for example) varies somewhat depending on the compound, just as bond lengths vary from one molecule to another. They are sufficiently similar, however, that it is possible to create a table of *average bond energies* (Table 9.9). The values in such tables may be used to *estimate* enthalpies of reactions, as described below.

In reactions between molecules, bonds in reactants are broken and new bonds are made as products form. If the total energy released when new bonds form exceeds the energy required to break the original bonds, the overall reaction is exothermic. If the opposite is true, then the overall reaction is endothermic. Let us see how this works by using bond energies to estimate the enthalpy change for the hydrogenation of propene to propane:

propene → propane

The first step is to examine the reactants and product to see which bonds are broken and which are formed. In this case, the C=C bond in propene and the H—H bond in hydrogen are broken. A C—C bond and two C—H bonds in propane are formed.

Bonds broken: 1 mol of C=C bonds and 1 mol of H—H bonds

• Bond Energy and Electronegativity

Pauling derived electronegativity values from a consideration of bond energies. He recognized that the energy of a bond between two different atoms is often greater than what would be expected if the bond electrons are shared equally. He postulated that the "extra energy" arises from the fact that the atoms do not share electrons equally. One atom is slightly positive and the other slightly negative. This means that a small coulombic force of attraction occurs in addition to the force of attraction arising from the sharing of electrons. This coulombic force enhances the overall force of attraction.

• Hydrogenation Reactions

Adding hydrogen to a double (or triple) bond is called a hydrogenation reaction. It is commonly done to convert vegetable oils, whose molecules contain C=C double bonds, to solid fats.

Table 9.9 • **Some Average Single- and Multiple-Bond Energies (kJ/mol)**

Single Bonds

	H	C	N	O	F	Si	P	S	Cl	Br	I
H	436	413	391	463	565	328	322	347	432	366	299
C		346	305	358	485	—	—	272	339	285	213
N			163	201	283	—	—	—	192	—	—
O				146	—	452	335	—	218	201	201
F					155	565	490	284	253	249	278
Si						222	—	293	381	310	234
P							201	—	326	—	184
S								226	255	—	—
Cl									242	216	208
Br										193	175
I											151

Multiple Bonds

$N＝N$	418	$C＝C$	610
$N≡N$	945	$C≡C$	835
$C＝N$	615	$C＝O$	745
$C≡N$	887	$C≡O$	1046
$O＝O$ (in O_2)	498		

Source: I. Klotz and R. M. Rosenberg: *Chemical Thermodynamics,* 5th ed., p. 55. New York, John Wiley, 1994, and J. E. Huheey, E. A. Keiter, and R. L. Keiter, *Inorganic Chemistry,* 4th ed., New York, Harper Collins, 1993, Table E.1.

Energy required = 610 kJ for C＝C bonds + 436 kJ for H—H bonds = 1046 kJ

Bonds formed: 1 mol of C—C bonds and 2 mol of C—H bonds

$$
\begin{array}{ccc}
\text{H} & \text{H} & \text{H} \\
| & | & | \\
\text{H—C—C—C—H(g)} \\
| & | & | \\
\text{H} & \text{H} & \text{H}
\end{array}
$$

Energy evolved = 346 kJ for C—C bonds + 2 mol × 413 kJ/mol for C—H bonds
= 1172 kJ

By combining the energy required to break bonds and the energy evolved in making bonds, we can find the value of ΔH°_{rxn} and see that the reaction is exothermic.

$$\Delta H^\circ_{rxn} = 1046 \text{ kJ} - 1172 \text{ kJ} = -126 \text{ kJ}$$

This value for ΔH°_{rxn} compares favorably with the value calculated from enthalpies of formation for propene and propane (-123.8 kJ).

As we have just done for the propene–hydrogen reaction, the enthalpy change for any reaction can be estimated using the equation

$$\Delta H^\circ_{rxn} = \sum D(\text{bonds broken}) - \sum D(\text{bonds formed}) \qquad (9.3)$$

To use this equation, first identify all the bonds in the reactants that are broken and add up their bond energies. Then, identify all the new bonds formed in the products and add up their bond energies. The difference between the energy required to break bonds $[= \Sigma\, D\,(\text{bonds broken})]$ and the energy evolved when bonds are made $[= \Sigma\, D\,(\text{bonds formed})]$ gives the estimated enthalpy change for the reaction. Bond energy calculations can give acceptable results in many cases.

Example 9.9 Using Bond Energies

Problem • Acetone, a common industrial solvent, can be converted to isopropanol, rubbing alcohol, by hydrogenation. Calculate the enthalpy change for the following reaction using bond energies.

$$H_3C - \overset{\overset{O}{\|}}{C} - CH_3(g) + H-H(g) \longrightarrow H_3C - \overset{\overset{H}{\underset{O}{|}}}{\underset{|}{C}} - CH_3(g)$$

acetone   isopropanol

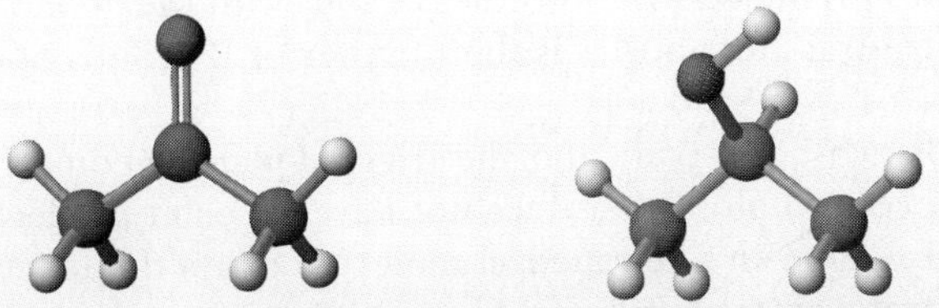

Strategy • Examine the reactants and products to determine which bonds are broken and which are formed. Add up the energies required to break bonds in the reactants and the energy evolved to form bonds in the product. The difference in these energies is an estimate of the enthalpy change of the reaction (Equation 9.3).

Solution •
Bonds broken: 1 mol of C=O bonds and 1 mol of H—H bonds

$$H_3C - \overset{\overset{O}{\|}}{C} - CH_3(g) + H-H(g)$$

$\Sigma\, D(\text{bonds broken}) = 745$ kJ for C=O bonds
$+\ 436$ kJ for H—H bonds
$= 1181$ kJ

Bonds formed: 1 mol of C—H bonds, 1 mol of C—O bonds, and 1 mol of O—H bonds

$$H_3C - \overset{\overset{H}{\underset{O}{|}}}{\underset{|}{C}} - CH_3(g)$$

$\Sigma\, D(\text{bonds formed}) = 413$ kJ for C—H + 358 kJ for C—O
$+\ 463$ kJ for O—H
$= 1234$ kJ

$\Delta H^\circ_{rxn} = \Sigma\, D(\text{bonds broken}) - \Sigma\, D(\text{bonds made})$
$= 1181$ kJ $-\ 1234$ kJ
$= -53$ kJ

The overall reaction is predicted to be exothermic by 53 kJ per mol of product formed. This is in good agreement with the value calculated from ΔH°_f values $(= -55.8$ kJ$)$.

Exercise 9.13 Using Bond Energies

Using the bond energies in Table 9.9, estimate the heat of combustion of gaseous methane, CH_4. That is, estimate ΔH°_{rxn} for the reaction of methane with O_2 to give water vapor and carbon dioxide gas.

9.9 MOLECULAR SHAPES

An important reason for drawing Lewis dot structures is to be able to predict the three-dimensional geometry of molecules and ions. Because the physical and chemical properties of compounds are tied to their structures, the importance of this subject cannot be overstated [CD-ROM, Screens 9.16–9.18].

The **valence shell electron-pair repulsion (VSEPR)** model provides a reliable method for predicting the shapes of covalent molecules and polyatomic ions. The VSEPR model is based on the idea that *bond and lone electron pairs in the valence shell of an element repel each other and seek to be as far apart as possible.* The positions assumed by the valence electrons of an atom thus define the angles between bonds to surrounding atoms. VSEPR is remarkably successful in predicting structures of molecules and ions of main group elements. However, it is less effective (and seldom used) to predict structures of compounds containing transition metals.

To get a sense of how valence shell electron pairs repel and determine structure, blow up several balloons to a similar size. Imagine that each balloon represents an electron cloud. A repulsive force prevents other balloons from occupying the same space. When two, three, four, five, or six balloons are tied together at a central point (representing the nucleus and core electrons of a central atom), the balloons naturally form the shapes shown in Figure 9.11. These geometric arrangements minimize interactions between the balloons.

Central Atoms Surrounded Only by Single-Bond Pairs

The simplest application of VSEPR theory is to molecules and ions in which all the electron pairs around the central atom are involved in single covalent bonds. Figure 9.12 illustrates the geometries predicted for molecules or ions with the general formulas AB_n, where A is the central atom and n is the number of B groups bonded to it.

The **linear** geometry for two bond pairs and the **trigonal-planar** geometry for three bond pairs apply to a central atom that does not have an octet of electrons (see Section 9.4). The central atom in a **tetrahedral** molecule obeys the octet rule with four bond pairs. The central atoms in **trigonal-bipyramidal** and **octahedral** molecules have five and six bonding pairs, respectively, and are expected only when the central atom is an element in Period 3 or higher of the periodic table (➡ PAGE 361).

Example 9.10   **Predicting Molecular Shapes**

Problem • Predict the shape of silicon tetrachloride, $SiCl_4$.

Strategy • The first step in predicting the shape of a molecule or ion is to draw its Lewis structure. The Lewis structure does not need to be drawn in any particular way. Its purpose is to describe the number of bonds around an atom and whether any lone pairs occur. The number of bond and lone pairs of electrons around the central atom determines the molecular shape (see Figure 9.12).

Solution • The Lewis structure of $SiCl_4$ has four electron pairs, all of them bond pairs, around the central Si atom. Therefore, a tetrahedral structure is predicted for the $SiCl_4$ molecule, with Cl—Si—Cl bond angles of 109.5°. This agrees with the actual structure for $SiCl_4$.

Lewis structure   Molecular geometry

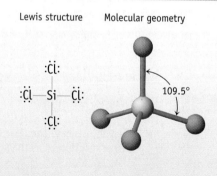

Exercise 9.14   **Predicting Molecular Shapes**

What is the shape of the dichloromethane (CH_2Cl_2) molecule? Predict the Cl—C—Cl bond angle.

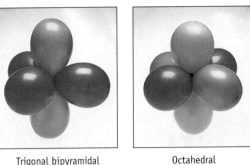

| Linear | Trigonal planar | Tetrahedral | Trigonal bipyramidal | Octahedral |

Figure 9.11   **Balloon models of electron pair geometries for two to six electron pairs.** If two to six balloons of similar size and shape are tied together, they will naturally assume the arrangements shown. These pictures illustrate the predictions of VSEPR. *(Charles D. Winters)*

Central Atoms with Single-Bond Pairs and Lone Pairs

To see how *lone pairs* affect the geometry of the molecule or polyatomic ion, return to the balloon models in Figure 9.11. Recall that the balloons represented *all* the electron pairs in the valence shell. The balloon model therefore predicts the "electron-pair geometry" rather than the "molecular geometry." The **electron-pair geometry** is the geometry taken up by *all* the valence electron pairs around a central atom, whereas the **molecular geometry** describes the arrangement in space of the central atom and the atoms directly attached to it. It is important to recognize that *lone pairs of electrons on the central atom occupy spatial positions even though their location is not included in the verbal description of the shape of the molecule or ion.*

Let us use the VSEPR model to predict the molecular geometry and bond angles in the NH_3 molecule, which has a lone pair on the central atom. First, draw the Lewis structure and count the total number of electron pairs around the central nitrogen atom. Four pairs of electrons are present in the nitrogen valence shell, so the *electron-pair geometry* is predicted to be tetrahedral. We have drawn a tetrahedron with nitrogen as the central atom and the three bond pairs represented by lines. The lone pair is included here to indicate its spatial position in the tetrahedron. The *molecular geometry* is described as a *trigonal-pyramid*. The nitrogen atom

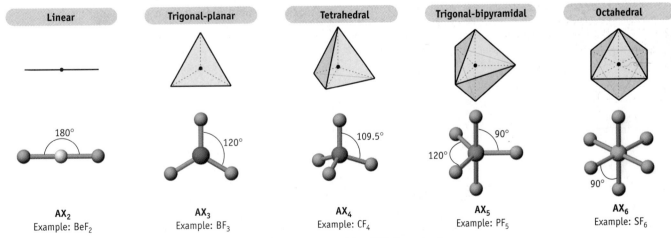

Figure 9.12   Various geometries predicted by VSEPR. Geometries predicted by VSEPR for molecules that contain only single covalent bonds around the central atom.

is at the apex of the pyramid, and the three hydrogen atoms form the trigonal base.

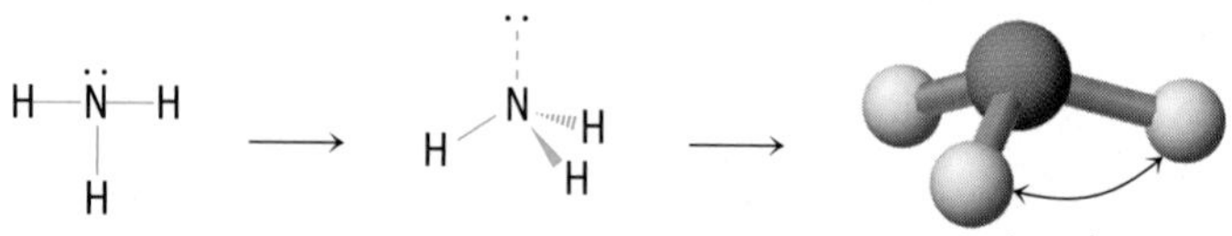

Lewis structure

Electron-pair geometry, tetrahedral

Actual H–N–H angle = 107.5°

Molecular geometry, trigonal pyramidal

Effect of Lone Pairs on Bond Angles

Because the electron-pair geometry in NH_3 is tetrahedral, we would expect the H—N—H bond angle to be 109.5°. The experimentally determined bond angles in NH_3, however, are 107.5°, and the H—O—H angle in water is smaller still (104.5°). These angles are close to the tetrahedral angle but not exactly that value. This highlights the fact that VSEPR is not a precise model; it can only predict the approximate geometry. Small variations in geometry (e.g., bond angles a few degrees different from predicted) are quite common and often arise because there is a difference between the spatial requirements of lone pairs and bond pairs. Lone pairs of electrons seem to occupy a larger volume than bonding pairs, and the increased volume of lone pairs causes bond pairs to squeeze closer together. In general, the relative strengths of repulsions are in the order

Lone pair–lone pair < lone pair–bond pair < bond pair–bond pair

The different spatial requirements of lone pairs and bond pairs are important and are included as part of the VSEPR model. For example, the VSEPR model can be used to predict variations in the bond angles in the series of molecules CH_4, NH_3, and H_2O. The bond angles decrease in the series CH_4, NH_3, and H_2O as the number of lone pairs on the central atom increases (Figure 9.13).

Example 9.11 — **Finding the Shapes of Molecules**

Problem • What are the shapes of the ions H_3O^+ and ClF_2^+?

Strategy • Draw the Lewis structure for each ion. Count the number of lone and bond pairs around the central atom. Use Figure 9.12 to decide on the electron-pair geometry. Finally, the location of the atoms in the ion — which are determined by the bond and lone pairs — gives the geometry of the ion.

Solution •

(a) The Lewis structure of the hydronium ion, H_3O^+, shows that the oxygen atom is surrounded by four electron pairs, so the electron-pair geometry is tetrahedral.

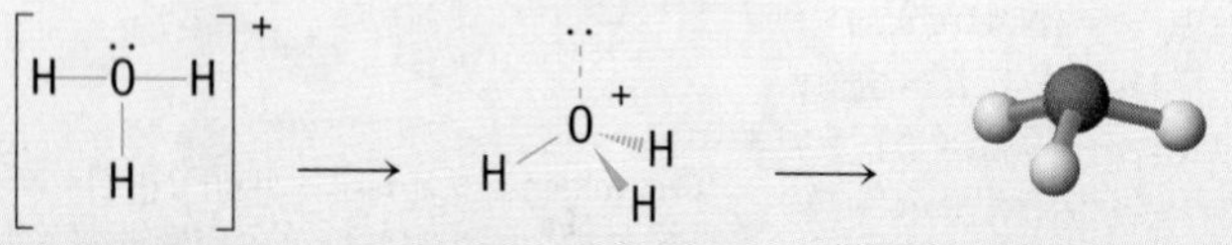

Because three of the four pairs are used to bond terminal atoms, the central O atom and the three H atoms form a trigonal-pyramidal molecular shape like that of NH_3.

(b) Chlorine is the central atom in ClF_2^+. It is surrounded by four electron pairs, so the electron-pair geometry around chlorine is tetrahedral. Because only two of the four pairs are bonding pairs, the ion has a bent geometry.

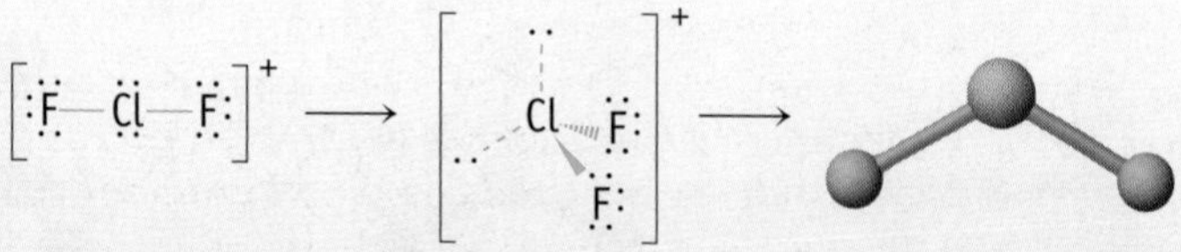

Lewis structure

Electron-pair geometry, tetrahedral

Molecular geometry, bent or angular

Lewis structure

Electron-pair geometry, tetrahedral

Molecular geometry, trigonal pyramid

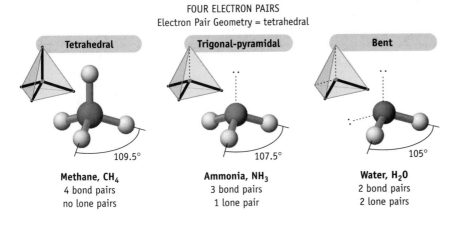

FOUR ELECTRON PAIRS
Electron Pair Geometry = tetrahedral

Tetrahedral

Trigonal-pyramidal

Bent

Methane, CH_4
4 bond pairs
no lone pairs

109.5°

Ammonia, NH_3
3 bond pairs
1 lone pair

107.5°

Water, H_2O
2 bond pairs
2 lone pairs

105°

Figure 9.13 The molecular geometries of methane, ammonia, and water. All have four electron pairs around the central atom, so all have a tetrahedral electron-pair geometry. **(a)** Methane has four bond pairs and so has a tetrahedral molecular shape. **(b)** Ammonia has three bond pairs and one lone pair, so it has a trigonal-pyramidal molecular shape. **(c)** Water has two bond pairs and two lone pairs, so it has a bent, or angular, molecular shape. The decrease in bond angles in the series can be explained by the fact that the lone pairs have a larger spatial requirement than the bond pairs.

Exercise 9.15 VSEPR and Molecular Shape

Give the electron-pair geometry and molecular shape for BF_3 and BF_4^-. What is the effect on the molecular geometry of adding an F^- ion to BF_3 to give BF_4^-?

Central Atoms with More Than Four Valence Electron Pairs

The situation becomes more complicated if the central atom has five or six electron pairs, some of which are lone pairs. A trigonal-bipyramidal structure (see Figure 9.12) has two sets of positions that are not equivalent. The positions in the trigonal plane lie in the equator of an imaginary sphere around the central atom and are called the *equatorial* positions. The north and south poles in this representation are called the *axial* positions. Each equatorial atom has two neighboring groups (the axial atoms) at 90°, and each axial atom has three groups (the equatorial atoms) at 90°. The result is that the lone pairs, which require more space than bonding pairs, prefer to occupy equatorial positions rather than axial positions.

The entries in the top line of Figure 9.14 show species having a total of five valence electron pairs, with zero, one, two, and three lone pairs. In SF_4, with one lone pair, the molecule assumes a seesaw shape with the lone pair in one of the equatorial positions. The ClF_3 molecule has three bond pairs and two lone pairs. The two lone pairs in ClF_3 are in equatorial positions; two bond pairs are axial and the third is in the equatorial plane, so the molecular geometry is T-shaped. The third molecule shown is XeF_2. Here, all three equatorial positions are occupied by lone pairs so the molecular geometry is linear.

The geometry assumed by six electron pairs is octahedral (see Figure 9.14), and all the angles at adjacent positions are 90°. Unlike the trigonal bipyramid, the octahedron has no distinct axial and equatorial positions; all positions are the same. Therefore, if the molecule has one lone pair, as in BrF_5, it makes no difference which position it occupies. The lone pair is often drawn in the top or bottom position to make it easier to visualize the molecular geometry, which in this case is square-pyramidal. If two pairs of the electrons in an octahedral arrangement are lone pairs, they seek to be as far apart as possible. The result is a square-planar molecule, as illustrated by XeF_4.

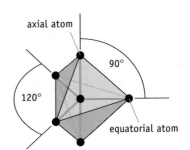

axial atom

90°

120°

equatorial atom

The trigonal bipyramid showing the axial and equatorial atoms. The angles between atoms equatorial positions is 120°. The angles between equatorial and axial atoms are 90°.

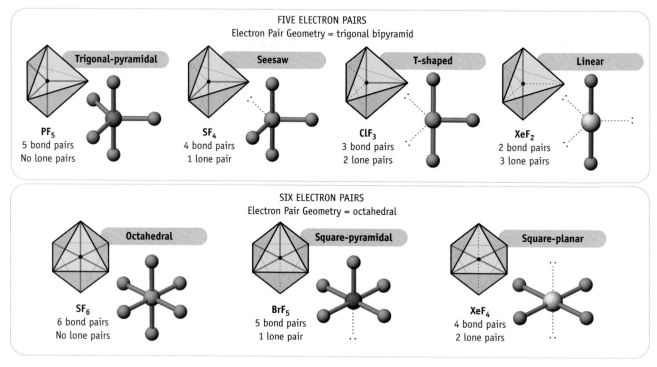

Figure 9.14   Electron-pair geometries and molecular shapes for molecules and ions with five (top) or six (bottom) electron pairs around the central atom.

Example 9.12   Predicting Molecular Shape

Problem • What is the shape of the ICl_4^- ion?

Strategy • Draw the Lewis structure and then decide on the electron-pair geometry. The position of the atoms gives the geometry of the ion. See Example 9.11.

Solution • A Lewis structure for the ICl_4^- ion shows that the central iodine atom has six electron pairs in its valence shell. Two of these are lone pairs. Placing the lone pairs on opposite sides leaves the four chlorine atoms in a square-planar geometry.

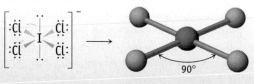

Electron-pair geometry, octahedral

Molecular geometry, square planar

Exercise 9.16   Predicting Molecular Shape

Draw the Lewis structure for ICl_2^- and then decide on the geometry of the ion.

Multiple Bonds and Molecular Geometry

Double and triple bonds involve more electron pairs than single bonds, but this does not affect the overall molecular shape. Electron pairs involved in a multiple bond are all shared between the same two nuclei and therefore occupy the same region of space. Because they must remain in that region, two electron pairs in a double bond (or three in a triple bond) behave like a single balloon (see Figure 9.11), rather than two or three balloons. All electron pairs in a multiple bond contribute to molecular geometry the same as a single bond does. For example, the carbon atom in CO_2 has no lone pairs and participates in two double bonds. Each double bond counts as one for the purpose of predicting geometry, so the structure of CO_2 is linear.

Lewis structure, electron   Molecular structure, linear
pair geometry = linear

When resonance structures are possible, the geometry can be predicted from any of the Lewis resonance structures or from the resonance hybrid structure. For example, the geometry of the CO_3^{2-} ion is predicted to be trigonal-planar because the carbon atom has three sets of bonds and no lone pairs.

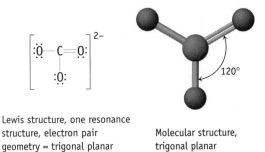

Lewis structure, one resonance
structure, electron pair   Molecular structure,
geometry = trigonal planar   trigonal planar

The NO_2^- ion also has a trigonal-planar electron-pair geometry. Because there is a lone pair on the central nitrogen atom and two bonds in the other two positions, the geometry of the ion is angular, or bent.

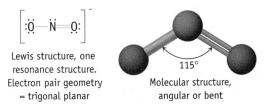

Lewis structure, one
resonance structure.
Electron pair geometry   Molecular structure,
= trigonal planar   angular or bent

The techniques just outlined can be used to find the geometries of much more complicated molecules. Consider, for example, cysteine, one of the natural amino acids.

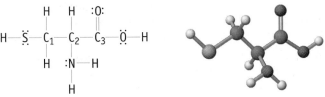

Cysteine, $HSCH_2CH(NH_2)CO_2H$

Four pairs of electrons occur around the S, N, C-1, and C-2 atoms, so the electron-pair geometry around each is tetrahedral. Thus, the H—S—C and H—N—H angles are predicted to be approximately 109°. The O atom in the grouping C—O—H also is surrounded by four pairs, and so this angle is likewise approximately 109°. Finally, the angle made by O—C-3—O is 120° because the electron-pair geometry around C-3 is planar and trigonal.

Example 9.13 Finding the Shapes of Molecules and Ions

Problem • What are the shapes of the nitrate ion, NO_3^-, and $XeOF_4$?

Strategy • Draw the Lewis structure and then decide on the electron-pair geometry. The position of the atoms gives the molecular geometry of the ion. Follow the procedure used in Examples 9.11 and 9.12.

Solution •

(a) The NO_3^- ion and CO_3^{2-} ion are isoelectronic. Thus, like the carbonate ion, the electron-pair geometry and molecular shape of NO_3^- are trigonal-planar.

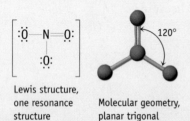

Lewis structure, one resonance structure

Molecular geometry, planar trigonal

(b) The $XeOF_4$ molecule has a Lewis structure with a total of six electron pairs about the central Xe atom, one of which is a lone pair. It has a square-pyramidal molecular structure. (Two structures are possible based on the position occupied by the oxygen, but there is no way to predict which is correct. The actual structure is the one shown, with the oxygen in the apex of the square pyramid.)

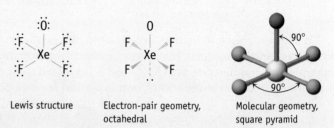

Lewis structure

Electron-pair geometry, octahedral

Molecular geometry, square pyramid

Exercise 9.17 Determining Molecular Shapes

Use Lewis structures and the VSEPR model to determine the electron-pair and molecular geometries for

(a) the phosphate ion, PO_4^{3-}; (b) the sulfite ion, SO_3^{2-}, and (c) IF_5.

9.10 MOLECULAR POLARITY

The term "polar" was used in Section 9.7 to describe a bond in which one atom had a partial positive charge and the other a partial negative charge. Because most molecules have polar bonds, molecules as a whole can also be polar. In a polar molecule, electron density accumulates toward one side of the molecule, giving that side a negative charge, $-\delta$, and leaving the other side with a positive charge of equal value, $+\delta$ (Figure 9.15) [CD-ROM, Screen 9.19].

Figure 9.15   Polar molecules in an electric field. (a) A representation of a polar molecule. To indicate the direction of molecular polarity, an arrow is drawn with the head pointing to the negative side and a plus sign placed at the positive end. **(b)** When placed between charged plates in an electric field, polar molecules experience a force that tends to align them with the field. The negative end of the molecules is drawn to the positive plate, and vice versa. This property affects the electrical capacitance of the plates (their ability to hold a charge) and provides a way to measure experimentally the magnitude of the dipole.

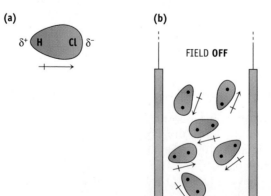

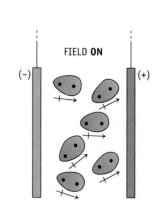

Before describing the factors that determine whether a molecule is polar, let us look at the experimental measurement of the polarity of a molecule. When placed in an electric field, polar molecules experience a force that tends to align them with the field (see Figure 9.15). When the electric field is created by a pair of oppositely charged plates, the positive end of each molecule is attracted to the negative plate, and the negative end is attracted to the positive plate. The extent to which the molecules line up with the field depends on their **dipole moment, μ,** which is defined as the product of the magnitude of the partial charges ($+\delta$ and $-\delta$) and the distance by which they are separated. The SI unit of the dipole moment is the coulomb-meter, but dipole moments have traditionally been given using a derived unit called the debye (D; 1 D $= 3.34 \times 10^{-30}$ C $\cdot$ m). Experimental values of some dipole moments are listed in Table 9.10.

The force of attraction between the negative end of one polar molecule and the positive end of another (called a *dipole–dipole force* and discussed in Section 13.2) affects the properties of polar compounds. Intermolecular forces (forces between molecules) influence the temperatures at which a liquid freezes or boils, for example. These forces will also help determine whether a liquid dissolves certain gases or solids or whether it mixes with other liquids, and whether it adheres to glass or other solids.

To predict if a molecule is polar, we need to consider if the molecule has polar bonds and how these bonds are positioned relative to one another. Diatomic molecules composed of two atoms with different electronegativities are always polar (see Table 9.10); there is one bond and the molecule has a positive and a negative end. But what happens with a molecule with three or more atoms, in which there are two or more polar bonds? Let us look at a series of molecules with stoichiometry AB_2, AB_3, and AB_4, evaluating how the choice of substituent or "terminal" groups (B) and molecular geometry influence the molecular polarity.

Consider first a linear triatomic molecule such as carbon dioxide, CO_2 (Figure 9.16). Here each C=O bond is polar, with the oxygen atom the negative end of the bond dipole. The terminal atoms are at the same distance from the C atom, they both have the same $-\delta$ charge, and they are symmetrically arranged around the central C atom. Therefore, CO_2 has no molecular dipole, even though each

Table 9.10 • **Dipole Moments of Selected Molecules**

Molecule (AB)	Moment (μ, D)	Geometry	Molecule (AB_2)	Moment (μ, D)	Geometry
HF	1.78	linear	H_2O	1.85	bent
HCl	1.07	linear	H_2S	0.95	bent
HBr	0.79	linear	SO_2	1.62	bent
HI	0.38	linear	CO_2	0	linear
H_2	0	linear			

Molecule (AB_3)	Moment (μ, D)	Geometry	Molecule (AB_4)	Moment (μ, D)	Geometry
NH_3	1.47	trigonal-pyramidal	CH_4	0	tetrahedral
NF_3	0.23	trigonal-pyramidal	CH_3Cl	1.92	tetrahedral
BF_3	0	trigonal-planar	CH_2Cl_2	1.60	tetrahedral
			$CHCl_3$	1.04	tetrahedral
			CCl_4	0	tetrahedral

Figure 9.16   **Polarity of triatomic molecules, *AB₂*.**

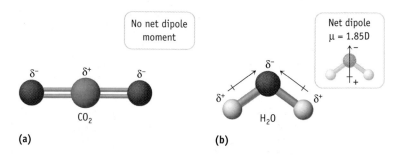

(a)

(b)

Chapter Goals • Revisited

- Understand the difference between ionic and covalent bonds.
- Draw Lewis electron dot structures for small molecules and ions.
- **Use electronegativity to predict the polarity of bonds and molecules.**
- Understand the properties of covalent bonds and their influence on molecular properties.
- Use VSEPR to predict the shapes of simple molecules and ions and to understand the structures of more complex molecules.

bond is polar. This is analogous to a tug-of-war in which the people at opposite ends of the rope are pulling with equal force.

In contrast, water is a bent triatomic molecule. Because O has a larger electronegativity ($\chi = 3.5$) than H ($\chi = 2.1$), each of the O—H bonds is polar, with the H atoms having the same $\delta+$ charge and oxygen having a negative charge ($\delta-$) (see Figure 9.16). Electron density accumulates on the O side of the molecule, making the molecule electrically "lopsided" and therefore polar ($\mu = 1.85$ D).

In trigonal-planar BF_3, the B—F bonds are highly polar because F is much more electronegative than B (χ of B = 2.0 and χ of F = 4.0) (Figure 9.17). The molecule is nonpolar, however, because the three terminal F atoms have the same $\delta-$ charge, are the same distance from the boron atom, and are arranged symmetrically and in the same plane as the central boron atom. In contrast, the trigonal-planar molecule phosgene is polar (Cl_2CO, $\mu = 1.17$ D) (see Figure 9.17). Here the angles are all about 120°, so the O and Cl atoms are symmetrically arranged around the C atom. The electronegativities of the three atoms in the molecule differ, however: $\chi(O) > \chi(Cl) > \chi(C)$. There is therefore a net displacement of electron density away from the center of the molecule, mostly toward the O atom.

Ammonia, like BF_3, has AB_3 stoichiometry and polar bonds. In contrast to BF_3, however, NH_3 is a trigonal-pyramidal molecule. The slightly positive H atoms are located in the base of the pyramid, and the slightly negative N atom is on the apex of the pyramid. As a consequence, NH_3 is polar (see Figure 9.17). Indeed, trigonal-pyramidal molecules are generally polar.

Molecules like carbon tetrachloride, CCl_4, and methane, CH_4, are nonpolar owing to their symmetrical, tetrahedral structures. The four atoms bonded to C have the same partial charge and are the same distance from the C atom. Tetrahedral molecules with both Cl and H atoms ($CHCl_3$, CH_2Cl_2, and CH_3Cl) are polar, however. The electronegativity for H atoms (2.1) is less than that of Cl atoms

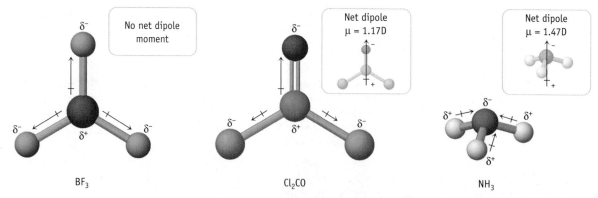

Figure 9.17   **Polar and nonpolar molecules of the type *AB₃*.**

Cooking with Microwaves

Microwave ovens are common appliances in homes, dorm rooms, and offices. They work because water is polar.

Microwaves are generated in a magnetron, a device invented during World War II for antiaircraft radar. The microwaves (frequency = $2.45 \times 10^{10}\ s^{-1}$) bounce off the metal walls of the oven and strike the food from many angles. They pass through glass or plastic dishes with no effect. Because electromagnetic radiation consists of oscillating electric (and magnetic) fields (see Figure 7.1), however, microwaves can affect mobile,

charged particles such as dissolved ions or the polar water molecules commonly found in food. As each wave crest approaches a water molecule, the molecule turns to align with the wave and continues turning over or rotating as the trough of the wave passes. The friction from the rotating water molecules heats the surrounding food.

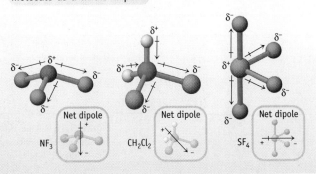

(Photo: Charles D. Winters) ▶

(3.0), and the carbon–hydrogen distance is different from the carbon – chlorine distances. Because Cl is more electronegative than H, the Cl atoms are on the more negative side of the molecule, and the positive end of the molecular dipole is toward the H atom.

To summarize this discussion of molecular polarity, look again at Figure 9.12 (page 359). These are sketches of molecules of the type AB_n, where A is the central atom and B is a terminal atom. You can predict that a molecule AB_n will *not* be polar, regardless of whether the A—B bonds are polar, if

- All the terminal atoms (or groups), B, are identical, and
- All the B atoms (or groups) are arranged symmetrically around the central atom, A, in the geometries shown.

On the other hand, if one of the B atoms (or groups) is different in the structures in Figure 9.12, or if one of the B positions is occupied by a lone pair, the molecule will generally be polar.

Example 9.14   Molecular Polarity

Problem • Are nitrogen trifluoride (NF_3), dichloromethane (CH_2Cl_2), and sulfur tetrafluoride (SF_4) polar or nonpolar? If polar, indicate the negative and positive sides of the molecule.

Strategy • You cannot decide if a molecule is polar without determining its structure. Therefore, start with the Lewis structure, decide on the electron-pair geometry, and then decide on the molecular geometry. If the molecular geometry is one of the highly symmetrical geometries in Figure 9.12, the molecule is *not* polar. If it does not fit one of these categories, it will be polar.

Solution •
(a) NF_3 has the same pyramidal structure as NH_3. Because F is more electronegative than N, each bond is polar, the more

negative end being the F atom. This means that the NF_3 molecule as a whole is polar.

Net dipole

NF_3

Net dipole

CH_2Cl_2

Net dipole

SF_4

(Example continues on p. 368)

(b) In CH_2Cl_2 the electronegativities are in the order Cl (3.0) > C (2.5) > H (2.1). This means the bonds are polar, H$(+\delta)$ — C$(-\delta)$ and C$(+\delta)$ — Cl$(-\delta)$, with a net displacement of electron density away from the H atoms and toward the Cl atoms. Although the electron-pair geometry around the C atom is tetrahedral, the polar bonds cannot be totally symmetrical in their arrangement. The CH_2Cl_2 molecule must be polar, with the negative end toward the two Cl atoms and the positive end toward the two H atoms.

(c) Sulfur tetrafluoride, SF_4, has an electron-pair geometry of a trigonal bipyramid (see Figure 9.14). Because the lone pair occupies one of the positions, the S — F bonds are not arranged symmetrically. Furthermore, the S — F bonds are highly polar, the bond dipole having F as the negative end (χ for S is 2.5 and χ for F is 4.0). SF_4 is therefore a polar molecule. The axial S — F bond dipoles cancel each other because they point in opposite directions. The equatorial S — F bond dipoles, however, both point to one side of the molecule.

Exercise 9.18 | **Molecular Polarity**

For each of the following molecules, decide whether the molecule is polar and which side is positive and which negative: $BFCl_2$, NH_2Cl, and SCl_2.

9.11   THE DNA STORY — REVISITED

This book opened with the story of the discovery of the structure of DNA, the molecule that determines our hereditary traits. The tools are now in place to say more about the structure of this important molecule and why it looks the way it does.

As described in Figure 9.18, each strand of the double-stranded DNA molecule consists of three units: a phosphate, a deoxyribose molecule (a sugar molecule with a five-member ring), and a nitrogen-containing base. (The bases in DNA can be one of four molecules: adenine, guanine, cytosine, or thymine. In Figure 9.18 the base is adenine.) Two units of the backbone (without the adenine on the deoxyribose ring) are illustrated in Figure 9.19.

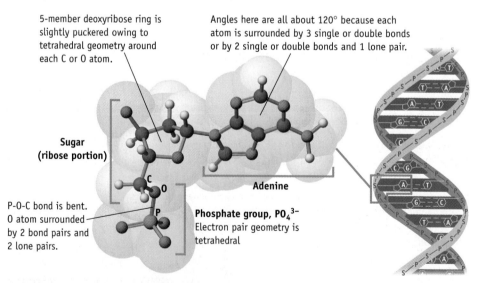

5-member deoxyribose ring is slightly puckered owing to tetrahedral geometry around each C or O atom.

Angles here are all about 120° because each atom is surrounded by 3 single or double bonds or by 2 single or double bonds and 1 lone pair.

Sugar (ribose portion)

P-O-C bond is bent. O atom surrounded by 2 bond pairs and 2 lone pairs.

Adenine

Phosphate group, PO_4^{3-}
Electron pair geometry is tetrahedral

Figure 9.18 A portion of the DNA molecule. It consists of a phosphate portion, a deoxyribose portion (a sugar molecule with a five-member ring), and a nitrogen-containing base (here adenine) attached to the ribose ring.

The important point here is that the repeating unit in the backbone of DNA consists of the atoms O — P — O — C — C — C. Each atom has a tetrahedral electron-pair geometry. Therefore, the chain cannot be linear. In fact, inspection reveals that the chain twists as one moves along the the backbone. It is this twisting that gives DNA its helical shape.

Why does DNA have two strands with the O — P — O — C — C — C backbone on the outside and the nitrogen-containing bases on the inside? This arises from the polarity of the bonds in the base molecules attached to the backbone. For example, the H atoms attached to N in the adenine molecule are very positively charged, and this leads to a special form of bonding, hydrogen bonding, to the base molecule in the neighboring chain. More about this in Chapter 13 when we again revisit the DNA structure.

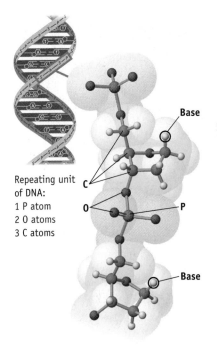

Repeating unit
of DNA:
1 P atom
2 O atoms
3 C atoms

Figure 9.19 **The backbone of the DNA molecule.**

In Summary

When you have finished studying this chapter, ask if you have met the chapter goals. In particular, you should be able to

- Describe the basic forms of chemical bonding, ionic and covalent, and the differences between them (Section 9.1).

- Predict from the formula whether a compound has ionic or covalent bonding, based on whether a metal is part of the formula (Section 9.1).

- Write Lewis symbols for atoms (Section 9.1).

- Describe the basic ideas of ionic bonding and how such bonds are affected by the sizes and charges of the ions (Section 9.3).

- Understand lattice energy and know how lattice energies are calculated (Born–Haber cycle); recognize trends in lattice energy and how melting points of ionic compounds are correlated with lattice energy (Section 9.3).

- Draw Lewis structures for molecular compounds and ions (Section 9.4).

- Understand and apply the octet rule; recognize exceptions to the octet rule (Sections 9.4 and 9.6).

- Write resonance structures, understand what resonance means, and how and when to use this method of representing bonding (Section 9.5).

- Calculate formal charges for atoms in a molecule based on the Lewis structure (Section 9.7).

- Define electronegativity and understand how it is used to describe the unequal sharing of electrons between atoms in a bond (Section 9.7).

- Combine formal charge and electronegativity to gain a perspective on the charge distribution in covalent molecules and ions (Section 9.7).

- Define and predict trends in bond order, bond length, and bond dissociation energy (Section 9.8).

- Use bond dissociation energies, D, in calculations (Section 9.8).

- Predict the shape or geometry of molecules and ions of main group elements using VSEPR theory (Section 9.9). Table 9.11 shows a summary of the relationships among valence electron pairs and electron-pair, molecular geometry, and molecular polarity.

- Understand why some molecules are polar and others are nonpolar (Section 9.10). See Tables 9.10 and 9.11.

- Predict the polarity of a molecule (Section 9.10).

Table 9.11 • Summary of Molecular Shapes and Molecular Polarity

Valence Electron Pairs	Electron-Pair Geometry	Number of Bond Pairs	Number of Lone Pairs	Molecular Geometry	Molecular Dipole?*	Examples
2	linear	2	0	linear	no	$BeCl_2$
3	trigonal-planar	3	0	trigonal-planar	no	BF_3, BCl_3
		2	1	bent (angular)	yes	SO_2
4	tetrahedral	4	0	tetrahedral	no	CH_4, BF_4^-
		3	1	trigonal-pyramidal	yes	NH_3, PF_3
		2	2	bent (angular)	yes	H_2O, SCl_2
5	trigonal-bipyramid	5	0	trigonal-bipyramidal	no	PF_5
		4	1	seesaw	yes	SF_4
		3	2	T-shaped	yes	ClF_3
		2	3	linear	no	XeF_2
6	octahedral	6	0	octahedral	no	SF_6
		5	1	square-pyramidal	yes	ClF_5
		4	2	square-planar	no	XeF_4

*For molecules of AB_n, where the B atoms are identical.

Key Terms

Introduction
structure
bonding

Section 9.1
Lewis electron dot symbols

Section 9.2
chemical bond
ionic bond
covalent bond

Section 9.3
lattice energy

Section 9.4
Lewis structure (electron dot structure)
bond pair
lone pair
nonbonding electrons
double bond

triple bond
octet rule
isoelectronic

Section 9.5
resonance
resonance structures
resonance hybrid

Section 9.6
coordinate covalent bond
free radical

Section 9.7
formal charge
polar covalent bond
dipolar
nonpolar covalent bond
electronegativity
electroneutrality principle

Section 9.8
bond order
bond length
bond dissociation energy (D)

Section 9.9
valence shell electron pair repulsion theory (VSEPR)
linear
trigonal-planar
tetrahedral
trigonal-bipyramidal
octahedral
electron-pair geometry
molecular geometry

Section 9.10
dipole moment

Key Equations

Equation 9.1 (page 346)
Calculate the formal charge on an atom in a molecule or ion

Formal charge of an atom in a molecule or ion = group number − [number LPE + $\frac{1}{2}$ (number BE)]

where the group number gives the number of valence electrons brought to the molecule or ion by the atom, LPE represents the number of lone pair electrons on an atom, and BE is the number of bonding electrons around an atom.

Equation 9.2 (page 353)

Calculate bond order

$$\text{Bond order} = \frac{\text{number of shared pairs linking X and Y}}{\text{number of X — Y links in the molecule or ion}}$$

Equation 9.3 (page 356)

Calculate the enthalpy change for a reaction using bond dissociation energies (*D*)

$$\Delta H^\circ_{\text{rxn}} = \sum D(\text{bonds broken}) - \sum D(\text{bonds formed})$$

where *D* is the bond dissociation energy.

Study Questions

Questions with blue, bold-faced numbers have answers in Appendix O. More challenging questions are indicated by an underlined number.

Reviewing Important Concepts

1. Specify the number of valence electrons for Li, Ti, Zn, Si, and Cl.

2. Write Lewis symbols for K, Mg, S, and Ar.

3. Describe the formation of KF from K and F atoms using Lewis symbols. Is bonding in KF ionic or covalent?

4. Predict whether the following compounds are ionic or covalent: KI, MgS, CS_2, P_4O_{10}.

5. Define lattice energy. Which should have the more negative lattice energy, LiF or CsF? Explain.

6. Which of the following compounds is not likely to exist: $CaCl_2$ or $CaCl_4$? Explain.

7. Refer to Table 9.4 to answer the following questions:
   (a) How many lone pairs and how many bond pairs exist in the amide ion (NH_2^-) and in hydrogen fluoride?
   (b) How many single bonds and how many double bonds exist in N_2H_4 and in C_2H_4?

8. In boron compounds the B atom often is not surrounded by four valence electron pairs. Illustrate this with BCl_3. Show how the molecule can achieve an octet configuration by forming a coordinate covalent bond with ammonia (NH_3).

9. Which of the following compounds or ions do *not* have an octet of electrons surrounding the central atom: BF_4^-, SiF_4, SeF_4, BrF_4^-, XeF_4?

10. In which of the following does the central atom obey the octet rule: NO_2, SF_4, NH_3, SO_3, O_2^-? Are any of them odd-electron molecules or ions?

11. Why is a single Lewis structure for benzene not an accurate representation of the structure of benzene?

12. Give the bond order of each bond in acetylene, $H-C\equiv C-H$, and phosgene, Cl_2CO.

13. Consider the following resonance structures for the formate ion, HCO_2^-. What is the $C-O$ bond order in the ion?

14. Determine the $N-O$ bond order in the nitrate ion, NO_3^-.

15. Consider a series of molecules in which carbon is bonded by single bonds to atoms of second-period elements: $C-O$, $C-F$, $C-N$, $C-C$, and $C-B$. Place these bonds in order of increasing bond length.

16. Define bond dissociation energy. Does the enthalpy change for a bond-breaking reaction [e.g., $C-H(g) \longrightarrow C(g) + H(g)$] always have a positive sign, always have a negative sign, or does the sign vary? Explain briefly.

17. What is the relationship among bond order, bond length, and bond energy for a series of related bonds, say carbon–nitrogen bonds?

18. To estimate the enthalpy change for the reaction

$$O_2(g) + 2\,H_2(g) \longrightarrow 2\,H_2O(g)$$

what bond energies do you need? Outline the calculation, being careful to show correct algebraic signs.

19. Define and give an example of a polar covalent bond. Give an example of a nonpolar bond.

20. How are electronegativity and electron affinity different?

21. Describe the trends in electronegativity in the periodic table.

22. What is the principle of electroneutrality? Use this rule to exclude a possible resonance structure of CO_2.

23. What is the VSEPR theory? What is the physical basis of the theory?

24. What is the difference between the electron-pair geometry and the molecular geometry of a molecule? Use the water molecule as an example in your discussion.

25. What is the molecular geometry for each of the following:

Estimate the $H-X-H$ bond angle for each.

26. A molecule has four electron pairs around a central atom. Explain how the molecule can have a pyramidal structure.

How can the molecule have a bent structure? What bond angles are predicted in each case?

27. Does SO_2 have a dipole moment? If so, what is the direction of the net dipole in SO_2?

Practicing Skills

Valence Electrons and the Octet Rule
(See Example 9.1 and CD-ROM Screen 9.4)

28. Give the periodic group number and number of valence electrons for each of the following atoms:
    (a) O   (d) Mg
    (b) B   (e) F
    (c) Na   (f) S

29. Give the periodic group number and number of valence electrons for each of the following atoms:
    (a) C   (d) Si
    (b) Cl   (e) Se
    (c) Ne   (f) Al

30. For elements in Groups 3A–7A of the periodic table, give the number of bonds an element is expected to form if it obeys the octet rule.

31. Which of the following elements are capable of forming compounds in which the indicated atom has more than four valence electron pairs?
    (a) C   (d) F   (g) Se
    (b) P   (e) Cl   (h) Sn
    (c) O   (f) B

Ionic Compounds
(See Section 9.3 and CD-ROM Screen 9.3)

32. Which compound has the most negative energy of ion pair formation? Which has the least negative value?
    (a) NaCl   (b) MgS   (c) MgF_2

33. Which of the following ionic compounds are not likely to exist: $MgCl$, $ScCl_3$, BaF_3, $CsKr$, Na_2O? Explain your choices.

34. Place the following compounds in order of increasing lattice energy (from least negative to most negative): LiI, LiF, CaO, RbI.

35. Calculate the molar enthalpy of formation, ΔH_f°, of solid lithium fluoride using the approach outlined in "A Closer Look: Using a Born–Haber Cycle to Calculate Lattice Energies." The required data can be found in Appendices F and L and Table 9.3. In addition, $\Delta H_f^\circ [Li(g)] = 159.37$ kJ/mol. (See also Exercise 9.2.)

36. To melt an ionic solid, energy must be supplied to disrupt the forces between ions so the regular array of ions collapses, loses order, and becomes a liquid. If the distance between the anion and cation in a crystalline solid decreases (but ion charges remain the same), should the melting point decrease or increase? Explain.

37. Which compound in each of the following pairs should require the higher temperature to melt? (See Study Question 36.)

(a) NaCl or RbCl
(b) BaO or MgO
(c) NaCl or MgS

Lewis Electron Dot Structures
(See Examples 9.2–9.5 and CD-ROM Screens 9.6–9.10)

38. Draw Lewis structures for the following molecules or ions:
    (a) NF_3 (b) HOBr   (c) ClO_3^- (d) SO_3^{2-}

39. Draw Lewis structures for the following molecules or ions:
    (a) CS_2 (b) NO_2^- (c) BF_4^- (d) Cl_2SO

40. Draw Lewis structures for the following molecules:
    (a) Chlorodifluoromethane, a $CHClF_2$ (C is the central atom)
    (b) Acetic acid, CH_3CO_2H. Its basic structure is pictured.

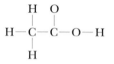

    (c) Acetonitrile, CH_3CN (the framework is $H_3C\!-\!C\!-\!N$)
    (d) Tetrafluoroethylene, $F_2C\!=\!CF_2$, the molecule from which Teflon is made.

41. Draw Lewis structures for each of the following molecules:
    (a) Methanol, CH_3OH (C is the central atom)
    (b) Vinyl chloride, $H_2C\!=\!CHCl$, the molecule from which PVC plastics are made. Its structure resembles that of C_2F_4 in Study Question 40.
    (c) Acrylonitrile, $H_2C\!=\!CHCN$, the molecule from which materials such as Orlon are made

42. Show all possible resonance structures for each of the following molecules or ions:
    (a) SO_2 (b) NO_2^- (c) SCN^-

43. Show all possible resonance structures for each of the following molecules or ions:
    (a) Nitrate ion, NO_3^-
    (b) Nitric acid, HNO_3
    (c) Nitrous oxide (laughing gas), N_2O

44. Draw Lewis structures for each of the following molecules or ions:
    (a) BrF_3 (b) I_3^- (c) XeO_2F_2 (d) XeF_3^+

45. Draw Lewis structures for each of the following molecules or ions:
    (a) BrF_5 (b) IF_3 (c) IBr_2^- (d) BrF_2^+

Formal Charge
(See Example 9.6 and CD-ROM Screen 9.15)

46. Determine the formal charge on each atom in each of the following molecules or ions:
    (a) N_2H_4 (b) PO_4^{3-} (c) BH_4^- (d) NH_2OH

47. Determine the formal charge on each atom in each of the following molecules or ions:
    (a) SCO   (c) O_3
    (b) HCO_2^- (formate ion)   (d) HCO_2H (formic acid)

48. Determine the formal charge on each atom in the following molecules and ions:
    (a) NO_2^+ (b) NO_2^- (c) NF_3 (d) HNO_3

49. Determine the formal charge on each atom in the following molecules and ions:
    (a) SO_2 (b) $SOCl_2$ (c) SO_2Cl_2 (d) FSO_3^-

Bond Polarity and Electronegativity
(See Example 9.7 and CD-ROM Screen 9.13)

50. For each pair of bonds, indicate the more polar bond and use an arrow to show the direction of polarity in each bond.
    (a) C—O and C—N   (c) B—O and B—S
    (b) P—Br and P—Cl   (d) B—F and B—I

51. For each of the following bonds, tell which atom is the more negatively charged.
    (a) C—N (b) C—H (c) C—Br (d) S—O

52. Acrolein, C_3H_4O, is the starting material for certain plastics.

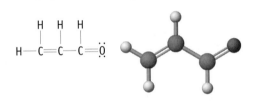

    (a) Which bonds in the molecule are polar and which are nonpolar?
    (b) Which is the most polar bond in the molecule? Which is the more negative atom of this bond?

53. Urea, $(NH_2)_2CO$, is used in plastics and fertilizers. It is also the primary nitrogen-containing substance excreted by humans.
    (a) Which bonds in the molecule are polar and which are nonpolar?
    (b) Which is the most polar bond in the molecule? Which atom is the negative end of the bond dipole?

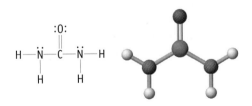

Bond Polarity and Formal Charge
(See Example 9.8 and CD-ROM Screens 9.13 and 9.15)

54. Considering both formal charges and bond polarities, predict on which atom or atoms the negative charge resides in the following anions:
    (a) BF_4^- (b) BH_4^- (c) OH^- (d) $CH_3CO_2^-$

55. Considering both formal charge and bond polarities, predict on which atom or atoms the positive charge resides in the following cations:
    (a) H_3O^+ (b) NH_4^+ (c) NO_2^+ (d) NF_4^+

56. Three resonance structures are possible for dinitrogen oxide, N_2O.

(a) Draw the three resonance structures.
(b) Calculate the formal charge on each atom in each resonance structure.
(c) Based on formal charges and electronegativity, predict which resonance structure is the most reasonable.

57. Compare the electron dot structures of the carbonate (CO_3^{2-}) and borate (BO_3^{3-}) ions.
    (a) Are these ions isoelectronic?
    (b) How many resonance structures does each ion have?
    (c) What are the formal atom charges in each ion?
    (d) If an H^+ ion attaches to CO_3^{2-} to form the bicarbonate ion, HCO_3^-, does it attach to an O atom or to the C atom?

58. Two resonance structures are possible for NO_2^-. Draw them and then find the formal charge on each atom in each resonance structure. If an H^+ ion is attached to NO_2^- (to form the acid HNO_2), does it attach to O or N?

59. Draw the resonance structures for the formate ion, HCO_2^- and find the formal charge on each atom. If an H^+ ion is attached to HCO_2^- (to form formic acid), does it attach to C or O?

Bond Order and Bond Length
(See Exercise 9.12 and CD-ROM Screen 9.11)

60. Specify the number of bonds for each of the following molecules or ions. Give the bond order for each bond.
    (a) H_2CO (c) NO_2^+
    (b) SO_3^{2-} (d) $NOCl$

61. Specify the number of bonds for each of the following molecules or ions. Give the bond order for each bond.
    (a) CN^- (c) SO_3
    (b) CH_3CN (d) $CH_3CH=CH_2$

62. In each pair of bonds, predict which is shorter.
    (a) B—Cl or Ga—Cl   (c) P—S or P—O
    (b) Sn—O or C—O   (d) C=O or C≡N

63. In each pair of bonds, predict which is shorter.
    (a) Si—N or Si—O
    (b) Si—O or C—O
    (c) C—F or C—Br
    (d) The C—N bond or the C≡N bond in $H_2NCH_2C≡N$

64. Consider the nitrogen–oxygen bond lengths in NO_2^+, NO_2^-, and NO_3^-. In which ion is the bond predicted to be longest? Which is predicted to be the shortest? Explain briefly.

65. Compare the carbon–oxygen bond lengths in the formate ion (HCO_2^-), in methanol (CH_3OH), and in the carbonate ion (CO_3^{2-}). In which species is the carbon–oxygen bond predicted to be longest? In which is it predicted to be shortest? Explain briefly.

Bond Energy
(See Table 9.9, Example 9.9, and CD-ROM Screen 9.12)

66. Consider the carbon–oxygen bond in formaldehyde (CH_2O) and carbon monoxide (CO). In which molecule is the CO bond shorter? In which molecule is it stronger?

67. Compare the nitrogen–nitrogen bond in hydrazine, H_2NNH_2, with that in "laughing gas," N_2O. In which molecule is the nitrogen–nitrogen bond shorter? In which is the bond stronger?

68. Hydrogenation reactions, the addition of H_2 to a molecule, are widely used in industry to transform one compound into another. For example, 1-butene (C_4H_8) is converted to butane (C_4H_{10}) by addition of H_2.

$$\underset{\substack{| \quad |\\ H \quad H}}{H-\overset{\substack{H \quad H \quad H \quad H\\ | \quad | \quad | \quad |}}{C-C-C=C}-H(g)} + H_2(g) \longrightarrow$$

$$\underset{\substack{| \quad | \quad | \quad |\\ H \quad H \quad H \quad H}}{H-\overset{\substack{H \quad H \quad H \quad H\\ | \quad | \quad | \quad |}}{C-C-C-C}-H(g)}$$

Use the bond energies of Table 9.9 to estimate the enthalpy change for this hydrogenation reaction.

69. Phosgene, Cl_2CO, is a highly toxic gas that was used as a weapon in World War I. Using the bond energies of Table 9.9, estimate the enthalpy change for the reaction of carbon monoxide and chlorine to produce phosgene. (*Hint:* First draw the electron dot structures of the reactants and products so you know the types of bonds involved.)

$$CO(g) + Cl_2(g) \longrightarrow Cl_2CO(g)$$

70. The compound oxygen difluoride is quite reactive, giving oxygen and HF when treated with water:

$$OF_2(g) + H_2O(g) \longrightarrow O_2(g) + 2\,HF(g) \qquad \Delta H^\circ_{rxn} = -318 \text{ kJ}$$

Calculate the bond dissociation energy of the $O-F$ bond in OF_2.

71. Oxygen atoms can combine with ozone to form O_2:

$$O_3(g) + O(g) \longrightarrow 2\,O_2(g) \qquad \Delta H^\circ_{rxn} = -394 \text{ kJ}$$

Using ΔH°_{rxn} and the bond energy data in Table 9.9, estimate the bond energy for the oxygen–oxygen bond in ozone, O_3. How does your estimate compare with the energies of an $O-O$ single bond and an $O=O$ double bond? Does the oxygen–oxygen bond energy in ozone correlate with its bond order?

Molecular Geometry

[See Examples 9.10–9.13 and CD Screens 9.16–9.18.
Note that many of these molecular structures are available on the General Chemistry Interactive CD-ROM, *Version 3.0 (in the MODELS directory.)]*

72. Draw the Lewis structure for each of the following molecules or ions. Describe the electron-pair geometry and the molecular geometry.
    (a) NH_2Cl
    (b) Cl_2O (O is the central atom)
    (c) SCN^-
    (d) HOF

73. Draw the Lewis structure for each of the following molecules or ions. Describe the electron-pair geometry and the molecular geometry.
    (a) ClF_2^+ (c) PO_4^{3-}
    (b) $SnCl_3^-$ (d) CS_2

74. The following molecules or ions all have two oxygen atoms attached to a central atom. Draw the Lewis structure for each one and then describe the electron-pair geometry and the molecular geometry. Comment on similarities and differences in the series.
    (a) CO_2 (b) NO_2^- (c) O_3 (d) ClO_2^-

75. The following molecules or ions all have three oxygen atoms attached to a central atom. Draw the Lewis structure for each one and then describe the electron-pair geometry and the molecular geometry. Comment on similarities and differences in the series.
    (a) CO_3^{2-} (c) SO_3^{2-}
    (b) NO_3^- (d) ClO_3^-

76. Draw the Lewis structure of each of the following molecules or ions. Describe the electron-pair geometry and the molecular geometry.
    (a) ClF_2^- (c) ClF_4^-
    (b) ClF_3 (d) ClF_5

77. Draw the Lewis structure of each of the following molecules or ions. Describe the electron-pair geometry and the molecular geometry.
    (a) SiF_6^{2-} (b) PF_5 (c) SF_4 (d) XeF_4

78. Give approximate values for the indicated bond angles.
    (a) $O-S-O$ in SO_2
    (b) $F-B-F$ angle in BF_3
    (c) $Cl-C-Cl$ angle in Cl_2CO
    (d) $H-C-H$ (angle 1) and $C-C\equiv N$ (angle 2) in acetonitrile

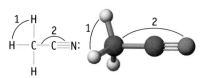

79. Give approximate values for the indicated bond angles.
    (a) $Cl-S-Cl$ in SCl_2
    (b) $N-N-O$ in N_2O
    (c) Bond angles in vinyl alcohol (a component of polymers and another molecule found in space).

80. Phenylalanine is one of the natural amino acids and is a "breakdown" product of aspartame. Estimate the values of the indicated angles in the amino acid. Explain why the $-CH_2-CH(NH_2)-CO_2H$ chain is not linear.

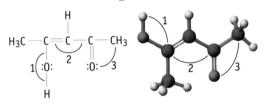

81. Acetylacetone has the structure shown here. Estimate the values of the indicated angles.

Molecular Polarity
(See Example 9.14 and CD-ROM Screen 9.19)

82. Consider the following molecules:

   (a) H_2O (c) CO_2 (e) CCl_4
   (b) NH_3 (d) ClF

   (i) In which compound are the bonds most polar?
   (ii) Which compounds in the list are *not* polar?
   (iii) Which atom in ClF is more negatively charged?

83. Consider the following molecules:

   (a) CH_4 (c) BF_3
   (b) NCl_3 (d) CS_2

   (i) Which compound has bonds with the greatest degree of polarity?
   (ii) Which compounds in the list are not polar?

84. Which of the following molecules is (are) polar? For each polar molecule indicate the direction of polarity, that is, which is the negative and which is the positive end of the molecule.

   (a) $BeCl_2$ (c) CH_3Cl
   (b) HBF_2 (d) SO_3

85. Which of the following molecules is (are) not polar? Which molecule has bonds with the greatest degree of polarity?

   (a) CO   (d) PCl_3
   (b) BCl_3 (e) GeH_4
   (c) CF_4

General Questions on Bonding and Molecular Structure

These questions are not designated as to type or location in the chapter. They may combine several concepts. More challenging questions are indicated by an underlined number.

86. Draw Lewis structures (and resonance structures where appropriate) for the following molecules and ions. What similarities and differences are there in this series?

   (a) CO_2
   (b) N_3^-
   (c) OCN^-

87. What are the orders of the N—O bonds in NO_2^- and NO_2^+? The nitrogen–oxygen bond length in one of these ions is 110 pm and in the other 124 pm. Which bond length corresponds to which ion? Explain briefly.

88. Which has the greater O—N—O bond angle, NO_2^- or NO_2^+? Explain briefly.

89. Compare the F—Cl—F angles in ClF_2^+ and ClF_2^-. Using Lewis structures determine the approximate bond angle in each ion. Decide which ion has the greater bond angle and explain your reasoning.

90. Draw an electron dot structure for the cyanide ion, CN^-. In aqueous solution this ion interacts with H^+ to form an acid. Should the acid formula be written as HCN or CNH?

91. Draw the electron dot structure for the sulfite ion, SO_3^{2-}. In aqueous solution the ion interacts with H^+. Does H^+ attach itself to the S atom or the O atom of SO_3^{2-}?

92. In principle, dinitrogen monoxide, N_2O, can decompose to nitrogen and oxygen gas:

$$2 N_2O(g) \longrightarrow 2 N_2(g) + O_2(g)$$

Use bond energies to estimate the enthalpy change for this reaction. (Assume N_2O has the structure: $:N{\equiv}N{-}\ddot{O}:$.)

93. The equation for the combustion of gaseous methanol is:

$$2 CH_3OH(g) + 3 O_2(g) \longrightarrow 2 CO_2(g) + 4 H_2O(g)$$

   (a) Using the bond energies in Table 9.9, estimate the enthalpy change for this reaction. What is the heat of combustion of 1 mol of gaseous methanol?
   (b) Compare your answer in part (a) with a calculation of ΔH°_{rxn} using thermochemical data and the methods of Chapter 6 (see Equation 6.6).

94. Examine the trends in lattice energy in Table 9.3. The value of the lattice energy becomes somewhat more negative on going from NaI to NaBr to NaCl, and all are in the range of -700 to -800 kJ/mol. Suggest a reason for the observation that the lattice energy of NaF ($\Delta E_{lattice} = -926$ kJ/mol) is much more negative than those of the other sodium halides.

95. The cyanate ion, NCO^-, has the least electronegative atom, C, in the center. The very unstable fulminate ion, CNO^-, has the same formula, but the N atom is in the center.

   (a) Draw the three possible resonance structures of CNO^-.
   (b) On the basis of formal charges, decide on the resonance structure with the most reasonable distribution of charge.
   (c) Mercury fulminate is so unstable it is used in blasting caps. Can you offer an explanation for this instability? (*Hint:* Are the formal charges in any resonance structure reasonable in view of the relative electronegativities of the atoms?)

96. Acrylonitrile, C_3H_3N, is the building block of the synthetic fiber Orlon.

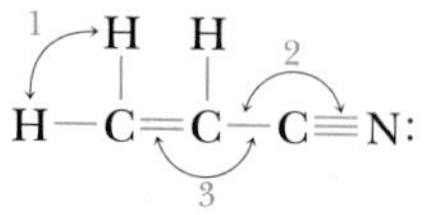

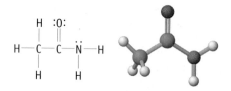

(a) Give the approximate values of angles 1, 2, and 3.

(b) Which is the shorter carbon–carbon bond?

(c) Which is the stronger carbon–carbon bond?

(d) Which is the most polar bond?

97. Vanillin is the flavoring agent in vanilla extract and in vanilla ice cream. Its structure is shown here:

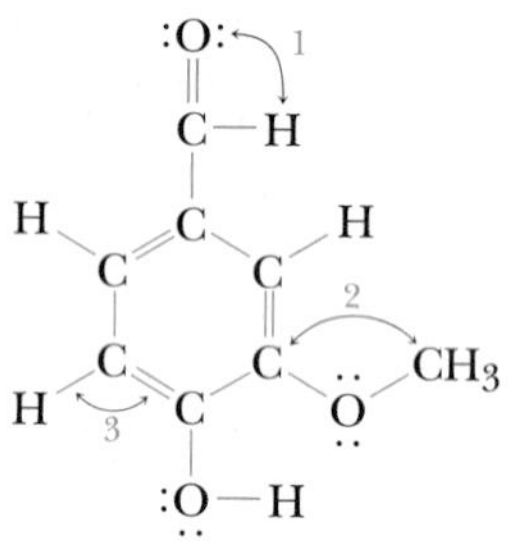

(a) Give values for the three bond angles indicated.

(b) Indicate the shortest carbon–oxygen bond in the molecule.

(c) Indicate the most polar bond in the molecule.

98. Given that the spatial requirement of a lone pair is much greater than that of a bond pair, explain why

(a) XeF_2 is a linear molecule.

(b) ClF_3 is a T-shaped molecule.

99. The formula for nitryl chloride is $ClNO_2$. Draw the Lewis structure for the molecule, including all resonance structures. Describe the electron-pair and molecular geometries, and give values for all bond angles.

100. Hydroxyproline is a less common amino acid.

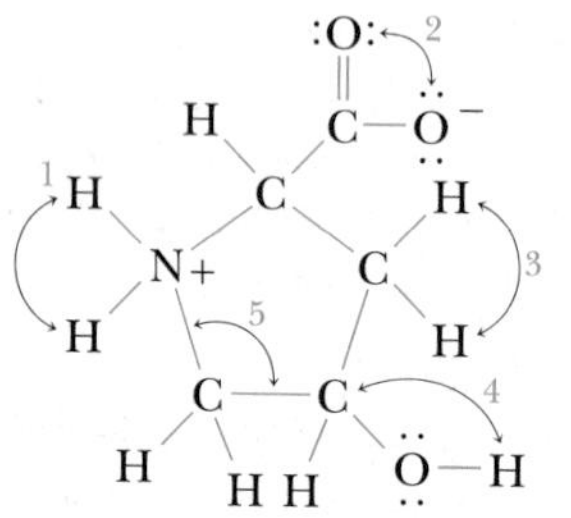

(a) Give approximate values for the indicated bond angles.

(b) Which are the most polar bonds in the molecule?

101. The effect of bond order is evident when bonds between the same two atoms are compared. For example, the bonds become shorter as the bond order increases in the series $C-O$, $C=O$, and $C\equiv O$.

Bond	$C-O$	$C=O$	$C\equiv O$
Bond length (pm)	143	122	113

Adding a second bond to the single bond in $C-O$ shortens the bond by only 21 pm on going to $C=O$. The third bond results in a further 9-pm reduction in bond length from $C=O$ to $C\equiv O$. Suggest why the bond shortening is much less than expected based on the fact that the bond order has doubled and then tripled.

102. Use the bond energies in Table 9.9 to calculate the enthalpy change for the decomposition of urea, $(NH_2)_2CO$ (Study Question 53) to hydrazine, H_2N-NH_2, and carbon monoxide. (Assume all compounds are in the gas phase.)

103. Amides are an important class of organic molecules. They are usually drawn as sketched here, but another resonance structure is possible.

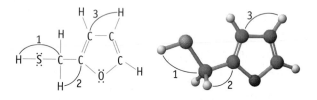

(a) Draw that structure, and then suggest why it is usually not pictured.

(b) Suggest a reason for the fact that the $H-N-H$ angle is close to 120°.

104. The molecule shown here, 2-furylmethanethiol, is responsible for the odor of coffee:

(a) What are the formal charges on the S and O atoms?

(b) Give approximate values of angles 1, 2, and 3.

(c) Which are the shorter carbon–carbon bonds in the molecule?

(d) Which bond in this molecule is the most polar?

(e) Is the molecule polar or nonpolar?

(f) The molecular model makes it clear that the four C atoms of the ring are all in a plane. Is the O atom in that same plane (making the five-member ring planar), or is the O atom bent above or below the plane?

105. Dihydroxyacetone is a component of quick-tanning lotions. (It reacts with the amino acids in the upper layer of skin and colors them brown in a reaction similar to that occurring when food is browned as it cooks.)

(a) Supposing you can make this compound by treating acetone with oxygen, use bond energies to estimate the enthalpy change for the following reaction (which is assumed to occur in the gas phase). Is the reaction exothermic or endothermic?

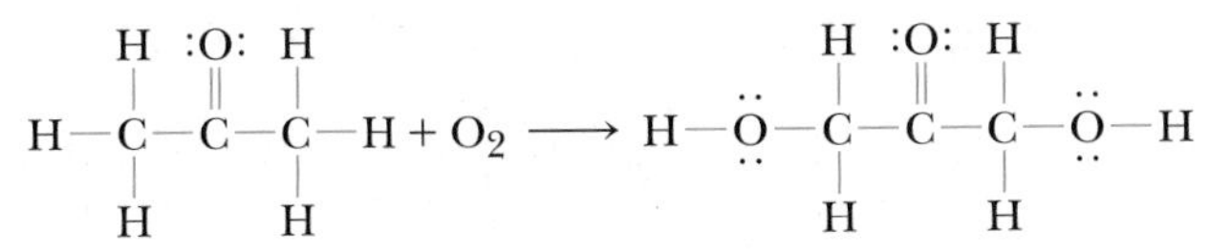

acetone → dihydroxyacetone

(b) Is acetone polar?

(c) Positive H atoms can sometimes be removed (as H^+) from molecules with strong bases (which is in part what happens in the tanning reaction). Which H atoms are the most positive in dihydroxyacetone?

106. Nitric acid, HNO_3, has three resonance structures. One of them, however, contributes much less to the resonance hybrid than the other two. Sketch the three resonance structures and assign a formal charge to each atom. Which one of your structures is the least important?

107. Molecules in space

(a) Glycolaldehyde was featured in the story "Sugar in Space" at the beginning of the chapter. Indicate the unique bond angles in this molecule.

(b) One molecule found in the 1995 Hale–Bopp comet is HC_3N. Suggest a structure for this molecule. (*Hint:* It is based on a chain of atoms.)

108. Acrolein (Study Question 52) is used to make plastics. Suppose the compound can be prepared by inserting a carbon monoxide molecule into the C—H bond of ethylene.

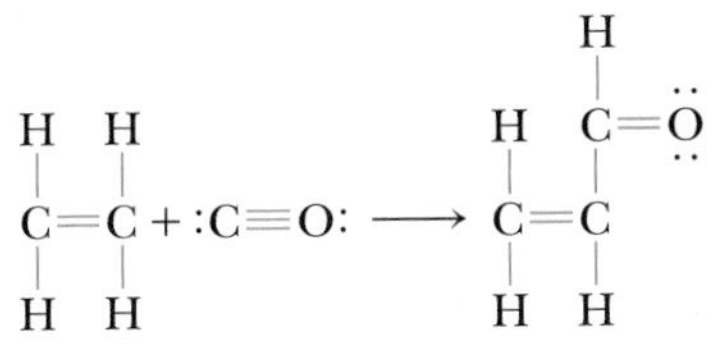

ethylene + :C≡O: → acrolein

(a) Which is the stronger carbon–carbon bond in acrolein?

(b) Which is the longer carbon–carbon bond in acrolein?

(c) Is ethylene or acrolein polar?

(d) Is the reaction of CO with C_2H_4 to give acrolein endothermic or exothermic?

109. 1,2–Dichloroethylene can be synthesized by adding Cl_2 to the carbon–carbon triple bond of acetylene.

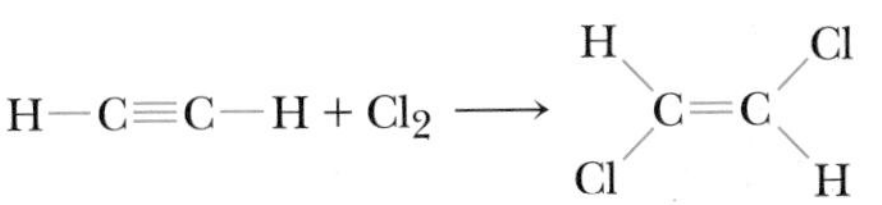

$$H—C≡C—H + Cl_2 \longrightarrow$$

Using bond energies, estimate the enthalpy change for this reaction in the gas phase.

110. The molecule shown here is epinephrine, a compound used as a bronchodilator (a medication to open the airways) and antiglaucoma agent.

(a) Give a value for each of the indicated bond angles.

(b) What are the most polar bonds in the molecule?

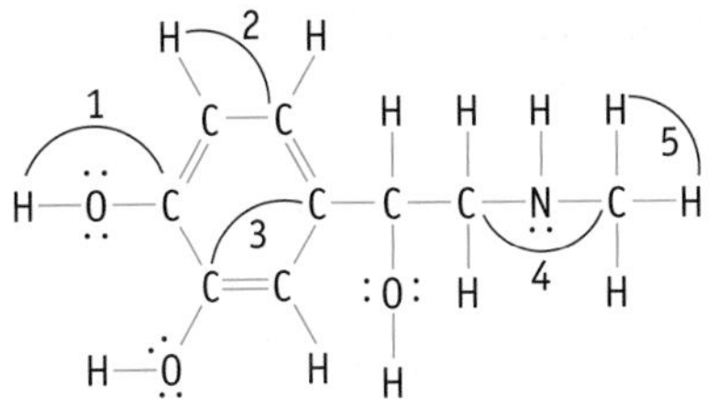

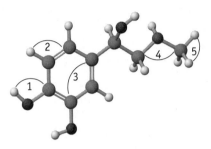

Using Electronic Resources

These questions refer to the General Chemistry Interactive CD-ROM, *Version 3.0.*

111. Locate the molecules in the table shown here in the *Molecular Models* folder on the *General Chemistry Interactive CD-ROM*, Version 3.0. Measure the carbon–carbon bond length in each and complete the table.

Formula	Measured Bond Distance (Å)	Bond Order
ethane, C_2H_6	_____	_____
butane, C_4H_{10}	_____	_____
ethylene, C_2H_4	_____	_____
acetylene, C_2H_2	_____	_____
benzene, C_6H_6	_____	_____

What relationship between bond order and carbon–carbon bond length do you observe?

112. Locate the following molecules in the *Molecular Models* folder. In each case, measure unique bond angles and bond lengths and use them to label a sketch of the molecule.

(a) Tylenol (*Drugs* folder)

(b) ClF_3 (*Inorganic* folder)

(c) ethylene glycol (*Organic Alcohols* folder)

113. See CD-ROM Screen 9.17: Molecular Polarity. Use the *Molecular Polarity* tool on this screen to explore the polarity of molecules.

(a) Is BF_3 a polar molecule? Does the molecular polarity change as F is replaced by H on BF_3? What happens when two F atoms are replaced by H?

(b) Is $BeCl_2$ a polar molecule? Does the polarity change when Cl is replaced by Br?

10

Bonding and Molecular Structure: Orbital Hybridization and Molecular Orbitals

Chapter Goals

- Understand the differences between valence bond and molecular orbital theory.
- Identify the hybridization of an atom in a molecule or ion.
- Understand the differences between bonding and antibonding molecular orbitals.
- Write the molecular orbital configuration for simple diatomic molecules.
- Understand the differences among a metal, an insulator, and a semiconductor.

Linus Pauling: A Life of Chemical Thought

Linus Pauling received the Nobel Prize for chemistry in 1954, an honorary high school diploma in 1962, and the Nobel Peace Prize in 1962. That is an extraordinary sequence but then he was an extraordinary man. He was born on February 28, 1901 in Portland, Oregon. His father, an itinerant pharmaceutical salesman, died when he was nine, and he soon became partial provider for his mother and two younger sisters. Though an excellent high school student, he refused to wait around in order to complete a civics requirement and did not graduate. It was only the first of many civil disobediencies.

Against the wishes of his mother in 1917 he enrolled as a chemical engineering major at Oregon Agricultural College. His interests soon turned to chemistry and, with a year off to help support his family, he graduated in 1922. He decided to embark on graduate work at the California Institute of Technology, then a fledgling institution, not today's research powerhouse. His research involved the use of the relatively new technique of x-ray crystallography to determine the atomic level structure of crystals. Experiment alone, however, was not sufficient; he also needed to master the relevant theory. Consequently he followed his Ph.D. studies with a tour of European centers of the emerging discipline of quantum mechanics. He was superbly, perhaps uniquely, equipped for his life's work.

Returning to Cal Tech, Pauling began an intensive program of structural determination, using x-ray crystallography for solids and electron-diffraction for vapors. He determined and analyzed interatomic distances and angles. He made semiempirical quantum mechanical calculations. Prediction became possible. All this was superbly summarized in his 1939 book *The Nature of the Chemical Bond, and the Structure of Molecules and Crystals.* It was to prove the most influential chemistry text of the 20th century.

Having largely solved the structural chemistry of inorganic and simple organic substances, Pauling now turned his attention to biochemical materials. He was to become, in Francis Crick's words, "one of the founders of molecular biology." Pauling, together with coworker Robert Corey, systematically tackled the basic structural chemistry of proteins.

▲ Linus Pauling 1901–1994 *(Thomas Hollyman/ Photo Researchers, Inc.)*

"So you must always be skeptical—always think for yourself."

Linus Pauling, on the occasion of accepting the Nobel Prize for chemistry in 1954.

▲ Linus and Ava Helen Pauling. Dr. Pauling received the Nobel Peace Prize in 1962, but his wife also played a major role in the effort to bring about a nuclear test-ban treaty. *(From the Ava and Linus Pauling Papers, Special Collections, Oregon State University.)*

On his 50th birthday he communicated his landmark paper on the α helix to the *Proceedings of the National Academy of Sciences*. It was his work on proteins, together with that on the nature of the chemical bond, that was cited in the award of the 1954 Nobel Prize for chemistry.

Pauling's scientific career was not yet half over. There were to be disappointments (his failure to solve the structure of DNA and the nonacceptance of his spheron model of nuclear stability) but also successes (a diagnosis of sickle cell anemia as a "molecular disease" and the introduction of the molecular evolutionary clock). But from about 1950 science was to be a part, at times even a minor part, of his active life.

Born into a relatively conservative family Pauling now became, under the urgent prompting of his wife, Ava Helen,

an active political propagandist and agitator. In particular he played a major role in bringing about the nuclear test-ban treaty of 1962. For this he received the 1962 Nobel Prize for peace. Although his chemistry prize was universally praised his peace prize was widely denounced in the conservative press. Only much later have his contributions been given their due.

The last years of Linus Pauling's life were mainly spent in the advocacy of what he called "ortho-molecular medicine" — the optimization of levels of various minerals and vitamins in the human body. The most familiar was the prescription of megadoses of vitamin C to treat various ailments, in particular the common cold. The nutritional establishment was outraged. The RDA value was vastly less than the 1–10 g a day recommended by Pauling who argued that the amount of vitamin C necessary to prevent scurvy was not necessarily sufficient to contribute maximally to bodily health. Although the jury is still out on some of Pauling's more extreme claims, the medical establishment seems to have developed a fondness for antioxidants such as vitamin C.

Active and optimistic to the end, Linus Pauling died at his ranch on the Big Sur coast of California on August 19, 1994.

Essay by Derek Davenport, Professor Emeritus of Chemistry, Purdue University

Bonding and Molecular Structure In this chapter we explore the details of chemical bonding and learn how atomic orbitals are involved in bond formation. One aspect of this is the formation and properties of double bonds. You shall see that rotation around a carbon-carbon double bond requires significant energy input. It is just such rotation around a double bond that is involved in the biochemistry of vision.

Rhodopsin

Rhodopsin

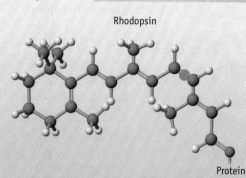

Meta-Rhodopsin

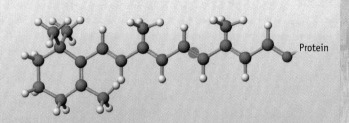

Protein

Protein

The yellow-orange compound β-carotene breaks down in your liver to give vitamin A, also called retinol. This is oxidized to *cis*-retinal, which binds to the protein opsin in your eyes to give rhodopsin.

Rhodopsin absorbs light in the visible region, and this triggers rotation around a carbon-carbon double bond (marked in red) to give meta-rhodopsin.

Light Absorption

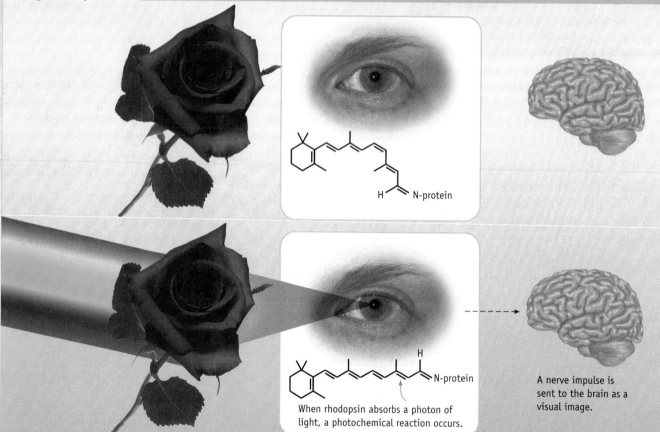

H
N-protein

When rhodopsin absorbs a photon of light, a photochemical reaction occurs.

A nerve impulse is sent to the brain as a visual image.

When rhodopsin absorbs light, it is transformed into meta-rhodopsin by a light-driven change in molecular structure. This triggers a nerve impulse which the brain decodes as a visual image. The meta-rhodopsin then decays to *trans*-retinal and the opsin protein. *Trans*-retinal is converted back to *cis*-retinal by a biochemical catalyst, an enzyme, and the retinal is ready to again form rhodopsin.

Rose: Charles D. Winters

Just how are molecules held together? How can two distinctly different molecules have the same formula, such as C_4H_8? Why is oxygen paramagnetic and how is this connected with bonding in the molecule? These are just a few of some fundamental and interesting questions that are raised in this chapter and that require us to take a more advanced look at bonding.   •

10.1   ORBITALS AND BONDING THEORIES

Orbitals, both atomic and molecular, are the focus of this chapter. The quantum mechanical model for the atom, the most successful way to explain the properties of atoms that scientists have yet devised, describes electrons in atoms as waves. An atomic orbital has a specific energy related to electrostatic forces: an attractive force on an electron in that orbital due to the positively charged atomic nucleus, and a repulsive force on the electron due to the other electrons in the atom. If the energy of the orbital is known accurately, its position is known less well (the Heisenberg uncertainty principle), so we think of orbitals as regions in space in which there is a high probability of finding the electron (← FIGURES 7.14 AND 7.15).

From Chapter 9 you know that valence electrons are responsible for the bonds in molecules, and from Chapters 7 and 8 you know that the location of the valence electrons in atoms is described by an orbital model. Therefore, it seems reasonable that an orbital model could be adopted to describe electrons in molecules.

Two common approaches to rationalizing chemical bonding based on orbitals are the **valence bond (VB) theory** and the **molecular orbital (MO) theory** [CD-ROM, Screen 10.2]. The former was developed largely by Linus Pauling (page 378) and the latter by another American chemist, Robert S. Mulliken. The valence bond approach is closely tied to Lewis's idea of bonding electron pairs between atoms and lone pairs of electrons localized on a particular atom. In contrast, Mulliken's approach was to derive molecular orbitals that are "spread out" or *delocalized,* over the molecule. The atomic orbitals of the atoms in the molecule combine to form a set of orbitals that are the property of the molecule, and the electrons of the molecule are distributed within these orbitals.

Why are two theories used? Isn't one more correct than the other? Actually, both give good descriptions of the bonding in molecules and polyatomic ions, but they are used for different purposes. Valence bond theory is generally the method of choice to provide a qualitative, visual picture of molecular structure and bonding. This theory is particularly useful for molecules made up of many atoms. In contrast, molecular orbital theory is used when a more quantitative picture of bonding is needed. Furthermore, valence bond theory provides a good description of bonding for molecules in their ground, or lowest, energy state. On the other hand, MO theory is essential if we want to describe molecules in higher energy excited states. Among other things, this is important in explaining the colors of compounds. Finally, for a few molecules such as NO and O_2, molecular orbital theory is the only way to describe their bonding accurately.

Chapter Goals • Revisited

- **Understand the differences between valence bond and molecular orbital theory.**
- Identify the hybridization of an atom in a molecule or ion.
- Understand the differences between bonding and antibonding molecular orbitals.
- Write the molecular orbital configuration for simple diatomic molecules.
- Understand the differences among a metal, an insulator, and a semiconductor.

History

Robert Mulliken (1896–1986)

Mulliken received the 1966 Nobel Prize in chemistry for the development of molecular orbital theory. He put forward many concepts and terms now widely used in chemistry, among them orbital, molecular orbital, electron affinity, and electron donor and acceptor. (© *The Nobel Foundation)*   •

10.2   VALENCE BOND THEORY

Orbital Overlap Model of Bonding

What happens if two atoms at an infinite distance apart are brought together to form a bond? This process is often illustrated with H_2 because, with two electrons and two nuclei, this is the simplest molecular compound (Figure 10.1). Initially, when two hydrogen atoms are widely separated, they do not interact. If the atoms

Figure 10.1 Potential energy change during H—H bond formation from isolated hydrogen atoms. The lowest energy is reached at an H—H separation of 74 pm, where overlap of 1*s* orbitals occurs. At greater distances the overlap is less, and the bond is weaker. At H—H distances less than 74 pm, repulsions between the nuclei and between the electrons of the two atoms increase rapidly, and the potential energy curve rises steeply. Thus, an H_2 molecule is expected to be less stable when the distance between the atoms is very small [CD-ROM, Screen 10.3].

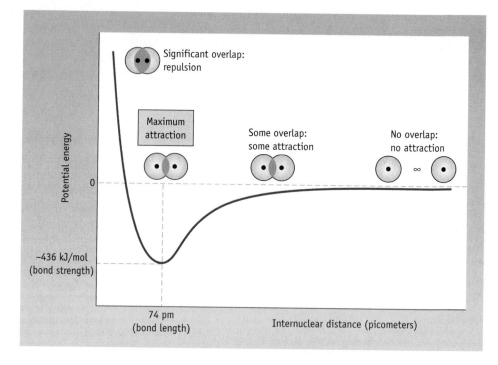

• Bonds in Valence Bond Theory
In the language of valence bond theory, a pair of electrons of opposite spin located between a pair of atoms constitutes a bond.

move closer together, however, the electron on one atom begins to experience an attraction to the positive charge of the nucleus of the other atom. Because of the attractive forces, the electron clouds on the atoms distort as the electron of one atom is drawn toward the nucleus of the second atom, and the potential energy of the system is lowered. Calculations show that when the distance between the H atoms is 74 pm, the potential energy reaches a minimum and the H_2 molecule is most stable. Significantly, 74 pm corresponds to the experimentally measured bond distance in the H_2 molecule [CD-ROM, Screen 10.3].

Individual hydrogen atoms each have a single electron. In H_2 the two electrons pair up to form the bond. There is a net stabilization, representing the extent to which the energies of the two electrons are lowered from their value in the free atoms. The net stabilization (the extent by which the potential energy is lowered) can be calculated, and the calculated value approximates the experimentally determined bond energy [◄ SECTION 9.8]. Agreement between theory and experiment on both bond distance and energy is evidence that this theoretical approach has merit.

Bond formation is depicted in Figures 10.1 and 10.2 as occurring when the electron clouds on the two atoms interpenetrate, or overlap. This **orbital overlap** increases the probability of finding the bonding electrons in the region of space between the two nuclei. *The idea that bonds are formed by overlap of atomic orbitals is the basis for valence bond theory.*

When the single covalent bond is formed in H_2, the electron cloud of each atom is distorted in a way that gives the electrons a higher probability of being in the region between the two hydrogen atoms. This makes sense because this distortion results in the electrons being situated so they can be attracted equally to the two positively charged nuclei. Placing the electrons between the nuclei also matches the Lewis electron dot model.

The covalent bond that arises from the overlap of two orbitals, one from each of two atoms as in H_2, is called a **sigma (σ) bond.** *The electron density of a σ bond is greatest along the axis of the bond.*

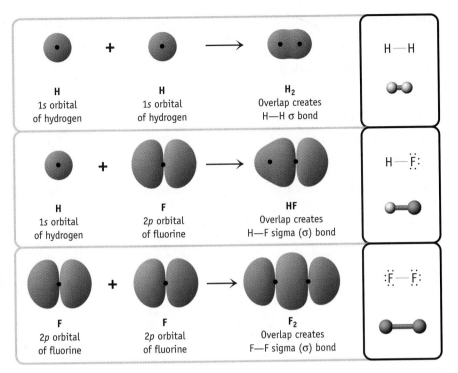

Figure 10.2 **Covalent bond formation in H_2, HF, and F_2.**

(a) Overlap of hydrogen 1s orbitals to form the H—H sigma (σ) bond.

(b) Overlap of hydrogen 1s and fluorine 2p orbitals to form the σ bond in HF.

(c) Overlap of 2p orbitals on two fluorine atoms forming the σ bond in F_2.

In summary, the main points of the valence bond approach to bonding are

- Orbitals overlap to form a bond between two atoms (see Figure 10.1).
- Two electrons, of opposite spin, can be accommodated in the overlapping orbitals. Usually one electron is supplied by each of the two bonded atoms.
- Because of orbital overlap, the bonding electrons have a higher probability of being found within a region of space influenced by both nuclei. Both electrons are simultaneously attracted to both nuclei.

What happens for elements beyond hydrogen? In the Lewis structure of HF, for example, a bonding electron pair is placed between H and F, and three lone pairs of electrons are depicted as localized on the F atom (Figure 10.2b). To use an orbital approach, look at the valence shell electrons and orbitals for each atom that will overlap. The hydrogen atom will use its 1s orbital in bond formation. The electron configuration of fluorine is $1s^2 2s^2 2p^5$, and the unpaired electron for this atom is assigned to one of the 2p orbitals. A σ bond results from overlap of the hydrogen 1s and the fluorine 2p orbital.

Formation of the H—F bond is similar to formation of an H—H bond. A hydrogen atom approaches a fluorine atom along the axis containing the 2p orbital with a single electron (see Figure 10.2b). The orbitals (1s on H and 2p on F) distort as each atomic nucleus influences the electron and orbital of the other atom. Still closer together, the 1s and 2p orbitals overlap, and the two electrons pair up to give a σ bond. There is an optimum distance (92 pm) at which the energy is lowest, and this corresponds to the bond distance in HF. The net stabilization achieved in this process is the energy for the H—F bond.

The remaining electrons on the fluorine atom (two electrons in the 2s orbital and four electrons in the other two 2p orbitals) are not involved in bonding. They are nonbonding electrons, the lone pairs associated with this element in the Lewis structure.

Extension of this model gives a description of bonding in F_2. The $2p$ orbitals on the two atoms overlap, and the single electron from each atom is paired in the resulting σ bond (Figure 10.2c). The $2s$ and the $2p$ electrons not involved in the bond are the lone pairs on each atom.

Hybridization of Atomic Orbitals

The simple picture using orbital overlap to describe bonding in H_2, HF, and F_2 works well, but we run into difficulty when molecules with more atoms are considered. For example, a Lewis dot structure of methane, CH_4, shows four C—H covalent bonds. VSEPR theory predicts, and experiments confirm, that the electron-pair geometry of the C atom in CH_4 is tetrahedral, with an angle of 109.5° between the bond pairs. The hydrogens are identical in this structure. This means that four equivalent bonding electron pairs occur around the C atom. An orbital picture of the bonds should convey both the geometry and the fact that all C—H bonds are the same.

● **Valence Shell Electron Configuration of Carbon in Its Ground State**

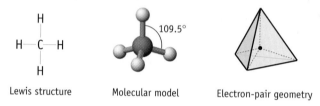

Lewis structure   Molecular model   Electron-pair geometry

If we apply the orbital overlap model used for H_2 and F_2 without modification to describe the bonding in CH_4, a problem arises. The three orbitals for the $2p$ valence electrons of carbon are at right angles, 90° (Figure 10.3) and do not match the tetrahedral angle of 109.5°. The spherical $2s$ orbital could bond in any direction. Furthermore, a carbon atom in its ground state ($1s^2 2s^2 2p^2$) has only two unpaired electrons, not the four that are needed to allow formation of four bonds.

To describe the bonding in methane and other molecules, Linus Pauling proposed the theory of **orbital hybridization** (Figure 10.4). He suggested that a new set of orbitals, called **hybrid orbitals,** could be created by mixing the s, p, and (sometimes) d atomic orbitals on an atom. Two important principles govern the outcome. First, the number of hybrid orbitals is always the same as the number of atomic orbitals that are mixed to create the hybrid orbital set. Second, the hybrid orbitals are more directed from the central atom toward the terminal atoms than are the unhybridized atomic orbitals, thus leading to better orbital overlap and a stronger bond between the central and terminal atoms.

The sets of hybrid orbitals that arise from mixing s, p, and d atomic orbitals are illustrated in Figure 10.5, and the following features are important:

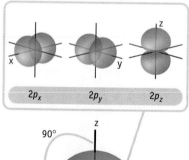

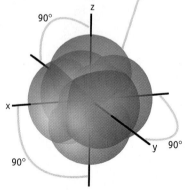

Figure 10.3 **The 2p orbitals on an atom.** The $2p_x$, $2p_y$, and $2p_z$ orbitals lie long the x-, y-, and z-axes, 90° to each other. *(Orbital Models: Patrick A. Harman and Charles F. Hamper.)*

- The number of hybrid orbitals required by an atom in a molecule or ion is determined by the electron-pair geometry around that atom. A hybrid orbital is required for each electron pair on the central atom (whether it is a bond pair or lone pair).

- If the valence shell s orbital on the central atom in a molecule or ion is mixed with a valence shell p orbital on that same atom, two hybrid orbitals are created. They are separated by 180°. The set of two orbitals is labeled sp.

- If an s orbital is combined with two p orbitals, all in the same valence shell, three hybrid orbitals are created. They are separated by 120°, and the set of three orbitals is labeled sp^2.

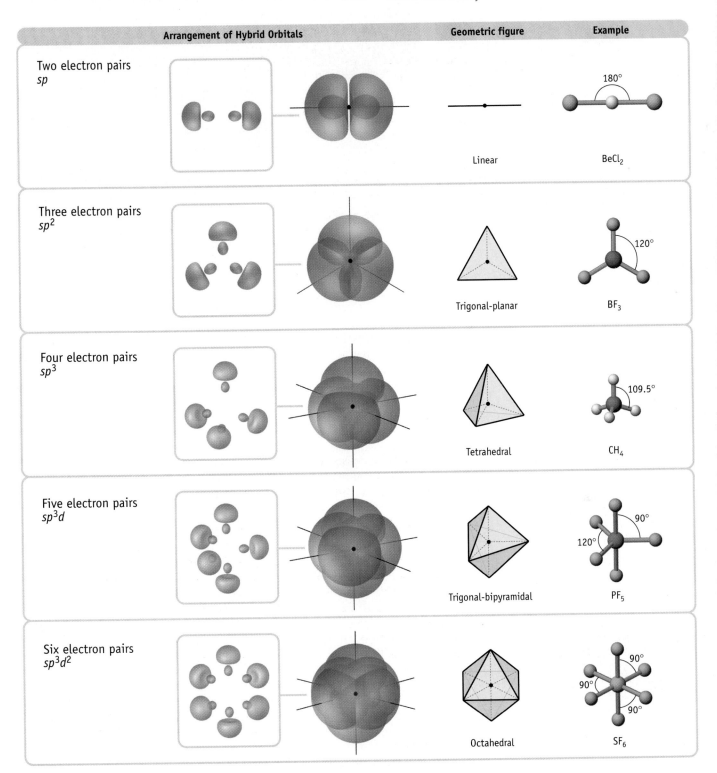

Arrangement of Hybrid Orbitals	Geometric figure	Example
Two electron pairs *sp*	Linear	180° BeCl₂
Three electron pairs *sp²*	Trigonal-planar	120° BF₃
Four electron pairs *sp³*	Tetrahedral	109.5° CH₄
Five electron pairs *sp³d*	Trigonal-bipyramidal	90° 120° PF₅
Six electron pairs *sp³d²*	Octahedral	90° 90° 90° SF₆

Figure 10.5   Hybrid orbitals for two through six electron pairs. The geometry of the hybrid orbital sets for two to six valence shell electron pairs is given in the right column. In forming a hybrid orbital set, the *s* orbital is always used, plus as many *p* orbitals (and *d* orbitals) as required to give the necessary number of σ-bonding and lone pair orbitals [CD-ROM, Screen 10.6]. *(Orbital Models: Patrick A. Harman and Charles F. Hamper)*

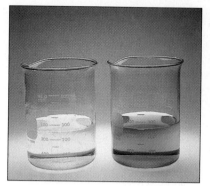  

Figure 10.4   Hybridization. Atomic orbitals can mix, or hybridize, to form hybrid orbitals. When two atomic orbitals on an atom combine, two new orbitals are produced on that atom, the new orbitals having a different direction in space than the original orbitals. An analogy is mixing two different colors (*left*) to produce a third color, which is a "hybrid" of the original colors (*center*). After mixing there are still two beakers (*right*), each containing the same volume of solution as before, but the color is a "hybrid" color [CD-ROM, Screens 10.4 and 10.6]. *(Charles. D. Winters)*

Chapter Goals • Revisited

• Understand the differences between valence bond and molecular orbital theory.

• **Identify the hybridization of an atom in a molecule or ion.**

• Understand the differences between bonding and antibonding molecular orbitals.

• Write the molecular orbital configuration for simple diatomic molecules.

• Understand the differences among a metal, an insulator, and a semiconductor.

• **Hybrid Orbitals and Atomic Orbitals**
Be sure to notice that *four* atomic orbitals produce *four* hybrid orbitals. The number of hybrid orbitals produced is always the same as the number of atomic orbitals used.

• When the *s* orbital in a valence shell is combined with three *p* orbitals, the result is four hybrid orbitals, each labeled sp^3. The hybrid orbitals are separated by $109.5°$, the tetrahedral angle.

• If one or two *d* orbitals is added to the sp^3 set, then two other hybrid orbital sets are created. These are used by the central atom of a molecule or ion with a trigonal-bipyramidal (sp^3d) or octahedral (sp^3d^2) electron pair geometry.

Let us examine a case of each type of hybridization in simple molecules, returning first to the case of methane. Keep in mind, however, that these principles apply to atoms in even the most complex molecules, such as DNA.

Valence Bond Theory for Methane, CH_4

In methane, four orbitals directed to the corners of a tetrahedron are needed to match the electron-pair geometry on the central carbon atom. By mixing the four valence shell orbitals, the $2s$ and all three of the $2p$ orbitals on carbon, a new set of four hybrid orbitals is created that has tetrahedral geometry (Figures 10.5 and 10.6). Each of the four hybrid orbitals is labeled sp^3 to indicate the atomic orbital combination (an *s* orbital and three *p* orbitals) from which they are derived. The four sp^3 orbitals have identical shapes, and the angle between them is $109.5°$, the tetrahedral angle. Because the orbitals have the same energy, one electron can be assigned to each according to Hund's rule (◄ SECTION 8.4).

The four valence electrons of the C atom in CH_4 are placed singly in the sp^3 orbitals. Each C—H bond is then formed by overlap of one of the carbon sp^3 hybrid orbitals with the $1s$ orbital of a hydrogen atom; one electron from the C atom is paired with an electron from an H atom.

Valence Bond Theory for Ammonia, NH_3

The Lewis structure for ammonia shows four electron pairs in the valence shell of nitrogen: three bond pairs and a lone pair (Figure 10.7). VSEPR theory predicts a tetrahedral electron-pair geometry and a trigonal-pyramidal molecular geometry. Structure evidence is a close match to prediction; the H—N—H bond angles are $107.5°$ in this molecule.

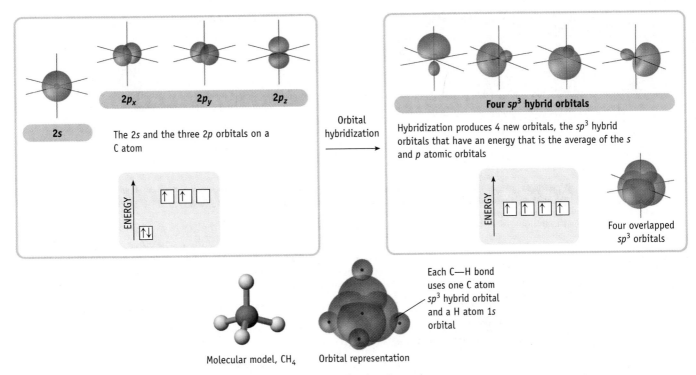

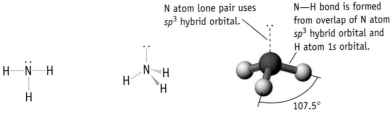

Molecular model, CH₄   Orbital representation

Figure 10.6   Bonding in the methane (CH₄) molecule. See the *General Chemistry Interactive*
CD-ROM, Version 3.0, Screen 10.5. *(Orbital Models: Patrick A. Harman and Charles F. Hamper)*

Based on the electron-pair geometry of NH_3, we predict sp^3 hybridization to accommodate the four electron pairs on the N atom. The lone pair is assigned to one of the hybrid orbitals, and each of the other three hybrid orbitals is occupied by a single electron. Overlap of each of the singly occupied, sp^3 hybrid orbitals with a $1s$ orbital for hydrogen, and pairing of the electrons, creates the N—H bonds about 109° apart.

Valence Bond Theory for Water, H_2O

The oxygen atom of water has two bonding pairs and two lone pairs in its valence shell, and the H—O—H angle is 104.5° (Figure 10.8). Four sp^3 hybrid orbitals are created from the $2s$ and $2p$ atomic orbitals of oxygen. Two of these sp^3 orbitals are occupied by unpaired electrons and are used to form O—H bonds. Lone pairs occupy the other two hybrid orbitals.

● **Hybridization and Geometry**
Hybridization reconciles the electron-pair geometry with the orbital overlap criterion of bonding. A statement such as "the atom is tetrahedral because it is sp^3-hybridized" is backward. That the electron-pair geometry around the atom is tetrahedral is a fact. Hybridization is one way to rationalize that fact.

N atom lone pair uses
sp^3 hybrid orbital.

N—H bond is formed
from overlap of N atom
sp^3 hybrid orbital and
H atom $1s$ orbital.

107.5°

Lewis structure   Electron-pair geometry   Molecular model

**Figure 10.7   Bonding in ammonia,
NH₃.**

Figure 10.8 **The bonding in the water molecule, H₂O.**

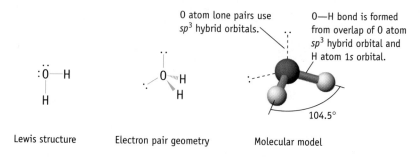

O atom lone pairs use sp^3 hybrid orbitals.

O—H bond is formed from overlap of O atom sp^3 hybrid orbital and H atom 1s orbital.

104.5°

Lewis structure   Electron pair geometry   Molecular model

Example 10.1 Valence Bond Description of Bonding in Ethane

Problem • Describe the bonding in ethane, C₂H₆, using valence bond theory.

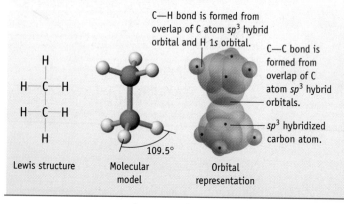

C—H bond is formed from overlap of C atom sp^3 hybrid orbital and H 1s orbital.

C—C bond is formed from overlap of C atom sp^3 hybrid orbitals.

sp^3 hybridized carbon atom.

109.5°

Lewis structure   Molecular model   Orbital representation

Strategy • First, draw the Lewis structure and predict the electron-pair geometry at both carbon atoms. Next, assign a hybridization to these atoms. Finally, describe covalent bonds that arise based on orbital overlap, and place electron pairs in their proper locations.

Solution • Each carbon atom has an octet configuration, sharing electron pairs with three hydrogen atoms and with the other carbon atom. The electron pairs around carbon have tetrahedral geometry, so carbon is assigned sp^3 hybridization. The C—C bond is formed by overlap of sp^3 orbitals on each C atom, and each of the C—H bonds is formed by overlap of an sp^3 orbital on carbon with a hydrogen 1s orbital.

◄ *(Orbital Models: Patrick A. Harman and Charles F. Hamper)*

Example 10.2 Valence Bond Description of Bonding in Methanol

Problem • Describe the bonding in the methanol molecule, CH₃OH, using valence bond theory.

Strategy • First construct the Lewis structure for the molecule. The electron-pair geometry around each atom determines the hybrid orbital set used by that atom.

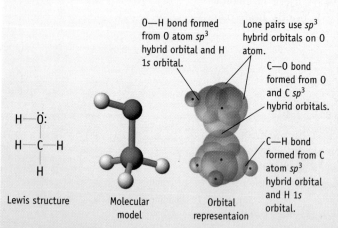

O—H bond formed from O atom sp^3 hybrid orbital and H 1s orbital.

Lone pairs use sp^3 hybrid orbitals on O atom.

C—O bond formed from O and C sp^3 hybrid orbitals.

C—H bond formed from C atom sp^3 hybrid orbital and H 1s orbital.

Lewis structure   Molecular model   Orbital representaion

Solution • The Lewis structure of CH₃OH implies that the electron-pair geometry around both the C and O atoms is tetrahedral. This geometry is confirmed by experiment. Thus, we may assign sp^3 hybridization to each atom, and the C—O bond is formed by overlap of sp^3 orbitals on these atoms. Each C—H bond is formed by overlap of a carbon sp^3 orbital with a hydrogen 1s orbital, and the O—H bond is formed by overlap of an oxygen sp^3 orbital with the hydrogen 1s orbital. Two lone pairs on oxygen occupy the remaining sp^3 orbitals.

Comment • Notice that one end of the CH₃OH molecule (the CH₃, or methyl group) is just like the CH₃ groups in the ethane molecule (Example 10.1), and the OH group resembles the OH group in water. It is helpful to recognize pieces of molecules and their bonding descriptions.

This example also shows how to predict the structure and bonding in a complicated molecule by looking at each atom separately. This important principle is essential when dealing with molecules made up of many atoms.

◄ *(Orbital Models: Patrick A. Harman and Charles F. Hamper)*

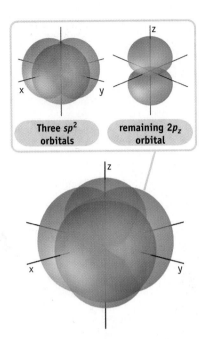

Figure 10.9 Unhybridized *p* atomic orbital on an atom with *sp²* hybridization. *(Orbital Models: Patrick A. Harman and Charles F. Hamper)*

Exercise 10.1 Valence Bond Description of Bonding

Use valence bond theory to describe the bonding in the hydronium ion, H_3O^+, and methylamine, CH_3NH_2.

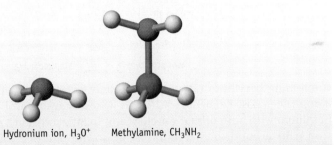

Hydronium ion, H_3O^+ Methylamine, CH_3NH_2

Hybrid Orbitals for Molecules and Ions with Trigonal-Planar Electron-Pair Geometries

Trigonal-planar geometries are commonly encountered in molecules and ions. For example, BF_3 and other boron halides are trigonal-planar, as are a number of other species, such as NO_3^- and CO_3^{2-}. The carbon atoms in ethylene, $H_2C{=}CH_2$, are also trigonal-planar, and the electron-pair geometry of O_3 and NO_2^- is trigonal-planar.

A trigonal-planar electron-pair geometry requires a central atom with three hybrid orbitals in a plane, 120° apart. Three hybrid orbitals mean three atomic orbitals must be combined, and the combination of an *s* orbital with two *p* orbitals is appropriate (see Figure 10.5). If p_x and p_y orbitals are used in hybrid orbital formation, the three hybrid *sp²* orbitals will lie in the *xy*-plane. The p_z orbital not used to form these hybrid orbitals is perpendicular to the plane containing the three *sp²* orbitals (Figure 10.9).

Boron trifluoride has a trigonal-planar electron-pair and molecular geometry. Each boron–fluorine bond in this compound results from overlap of an *sp²* orbital on boron with a *p* orbital on fluorine. Notice that the p_z orbital, which is not used to form the *sp²* hybrid orbitals, is not occupied by electrons.

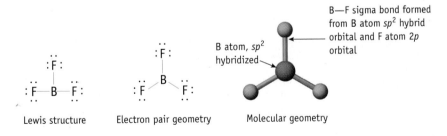

Lewis structure   Electron pair geometry   Molecular geometry

B—F sigma bond formed from B atom *sp²* hybrid orbital and F atom 2*p* orbital

B atom, *sp²* hybridized

Hybrid Orbitals for Molecules and Ions with Linear Electron-Pair Geometries

For molecules in which the central atom has a linear electron-pair geometry two hybrid orbitals, 180° apart, are required. One *s* and one *p* orbital can be hybridized to form two *sp* hybrid orbitals (Figure 10.10). If the p_y orbital is used, then the *sp* orbitals are oriented along the *y*-axis. The p_x and p_z orbitals are perpendicular to this axis.

Beryllium dichloride, $BeCl_2$, is a solid under ordinary conditions. When it is heated to over 520 °C, however, it vaporizes to give $BeCl_2$ vapor. In the gas phase,

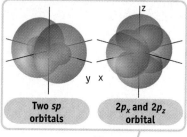

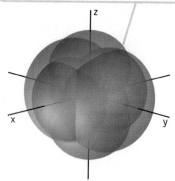

Figure 10.10   Unhybridized *p* atomic orbitals on an atom with *sp* hybridization. Because only one *p* orbital is incorporated in the hybrid orbital, two *p* orbitals remain. These are perpendicular to each other and to the axis along which the two *sp* hybrid orbitals lie. *(Orbital Models: Patrick A. Harman and Charles F. Hamper)*

$BeCl_2$ is a linear molecule, so *sp* hybridization is appropriate for the beryllium atom.

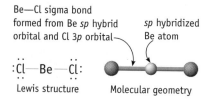

Combining beryllium's 2*s* and 2*p*$_y$ orbitals gives the two *sp* hybrid orbitals that lie along the *y*-axis. Each Be — Cl bond arises by overlap of an *sp* hybrid orbital on beryllium with a 3*p* orbital on chlorine. In this molecule, there are only two electron pairs around the beryllium atom, so the *p*$_x$ and *p*$_z$ orbitals are not occupied (see Figure 10.10).

Hybrid Orbitals Involving *s*, *p*, and *d* Atomic Orbitals

A basic assumption of Pauling's valence bond theory is that *the number of hybrid orbitals equals the number of valence orbitals used in their creation*. This means the maximum number of hybrid orbitals that can be created from the *s* and *p* orbitals for an atom is four.

How then should we deal with compounds like PF_5 or SF_6, compounds with more than four electron pairs in their valence shell? To describe five or six electron pairs requires the central atom to have five or six hybrid orbitals, which must be created from five or six atomic orbitals. This is possible if additional atomic orbitals from the *d* subshell are used in hybrid orbital formation. The *d* orbitals are considered to be valence shell orbitals for main group elements of the third and higher periods.

To accommodate six electron pairs in the valence shell of an element, six sp^3d^2 hybrid orbitals can be created from the *s*, three *p*, and two *d* orbitals. The six sp^3d^2 hybrid orbitals are directed to the corners of an octahedron (see Figure 10.5). Thus, they are oriented to accommodate the electron pairs for a compound that has an octahedral electron-pair geometry. Five coordination and trigonal-bipyramidal geometry are matched to sp^3d hybridization. One *s*, three *p*, and one *d* orbital combine to produce five sp^3d hybrid orbitals.

Example 10.3   Hybridization Involving *d* Orbitals

Problem • Describe the bonding in PF_5 using valence bond theory.

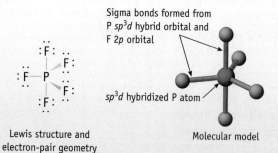

Lewis structure and electron-pair geometry

Sigma bonds formed from P sp^3d hybrid orbital and F 2*p* orbital

sp^3d hybridized P atom

Molecular model

Strategy • The first step is to establish the electron-pair and molecular geometry of PF_5. The electron-pair geometry around the P atom gives the number of hybrid orbitals required. If five hybrid orbitals are required, the combination of atomic orbitals is sp^3d.

Solution • Here the P atom is surrounded by five electron pairs, so PF_5 has a trigonal-bipyramidal electron-pair and molecular geometry. The hybridization scheme is therefore sp^3d.

Example 10.4   Recognizing Hybridization

Problem • Identify the hybridization of the central atom in the following compounds and ions:

(a) SF_3^+ (c) SF_4
(b) SO_4^{2-} (d) I_3^-

Strategy • The hybrid orbitals used by a central atom are determined by the electron-pair geometry (see Figure 10.5). Thus, to answer this question first write the Lewis structure and then predict the electron-pair geometry.

Solution • Following the procedures in Chapter 9 the Lewis structures for SF_3^+, and SO_4^{2-} can be written as follows:

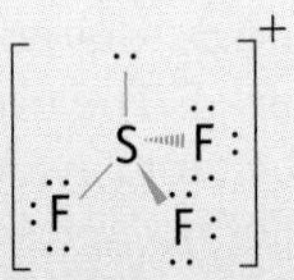

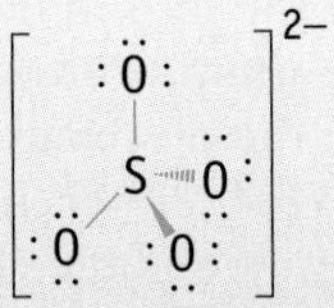

Four electron pairs surround the center atom in each of these ions, and the electron-pair geometry for these atoms is tetrahedral. Thus, sp^3 hybridization for the central atom is used to describe the bonding.

For SF_4 and I_3^-, five pairs of electrons are in the valence shell of the center atom. For these, sp^3d hybridization is appropriate for the central S or I atom.

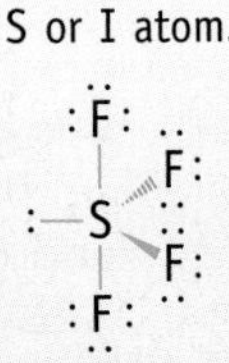

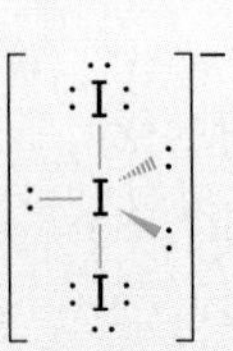

Exercise 10.2   Hybridization Involving d Orbitals

Describe the bonding in XeF_4 using hybrid orbitals. Remember to consider first the Lewis structure, then the electron-pair geometry (based on VSEPR theory), and then the molecular shape.

Exercise 10.3   Recognizing Hybridization

Identify the hybridization of the central atom in the following compounds and ions:

(a) BH_4^- (d) ClF_3
(b) SF_5^- (e) BCl_3
(c) OSF_4 (f) XeO_6^{4-}

Multiple Bonds

According to valence bond theory, bond formation requires that two orbitals on adjacent atoms overlap. Many molecules have two or three bonds between pairs of atoms. Therefore, according to valence bond theory, a double bond requires *two sets of overlapping orbitals* and *two electron pairs*. For a triple bond, *three* sets of atomic orbitals are required, each set accommodating a pair of electrons.

Double Bonds

Consider ethylene, $H_2C = CH_2$, one of the more common molecules with a double bond. The molecular structure of ethylene places all six atoms in a plane, with $H - C - H$ and $H - C - C$ angles of approximately 120°. Each carbon atom has trigonal-planar geometry, so sp^2 hybridization is assumed for these atoms. Thus,

● **Multiple Bonds**

$C = C$

Double bond requires two sets of overlapping orbitals and two pairs of electrons.

$C \equiv C$

Triple bond requires three sets of overlapping orbitals and three pairs of electrons.

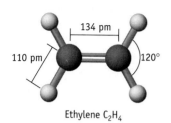

Ethylene C_2H_4

the model of bonding in ethylene starts with each carbon atom having three sp^2 hybrid orbitals in the molecular plane and an unhybridized p orbital perpendicular to that plane (see Figure 10.9). Because each carbon atom is involved in four bonds, a single unpaired electron is placed in each of these orbitals.

☐↑   Unhybridized p orbital. Used for π bonding in C_2H_4.

☐↑☐↑☐↑   Three sp^2 hybrid orbitals. Used for C—H and C—C σ bonding in C_2H_4.

Now we can visualize the C—H bonds, which arise from overlap of sp^2 orbitals on carbon with hydrogen $1s$ orbitals. After accounting for the C—H bonds, one sp^2 orbital on each carbon atom remains. These orbitals point toward each other and overlap to form one of the bonds linking the carbon atoms (Figure 10.11b). This leaves only one other orbital on each carbon, an unhybridized p orbital (see Figure 10.9), to be used to create the second bond between carbon atoms in C_2H_4. If they are aligned correctly, the unhybridized p orbitals on the two carbons can overlap, allowing the electrons in these orbitals to be paired. The overlap does not occur directly along the C—C axis, however. Instead, the atom arrangement compels these orbitals to overlap sideways, and the electron pair occupies an orbital with electron density above and below the plane containing the six atoms (Figure 10.11c).

This description results in two types of bonds in C_2H_4. One type is the C—H and C—C bonds that arise from the overlap of atomic orbitals so that the bonding electrons that lie along the bond axis form σ bonds. The other is the bond formed by sideways overlap of p atomic orbitals, called a **pi (π) bond.** In a π bond, the overlap region is above and below the internuclear axis, and the electron density of the π bond is above and below the bond axis (Figures 10.11a and 11.c).

Be sure to notice that a π bond can form *only* if (a) unhybridized p orbitals occur on adjacent atoms and (b) the p orbitals are perpendicular to the plane of the molecule and parallel to one another. This happens only if the sp^2 orbitals of both carbon atoms are in the same plane. Thus, *the π bond requires that the molecule be planar* (or have a planar region if it is a large molecule).

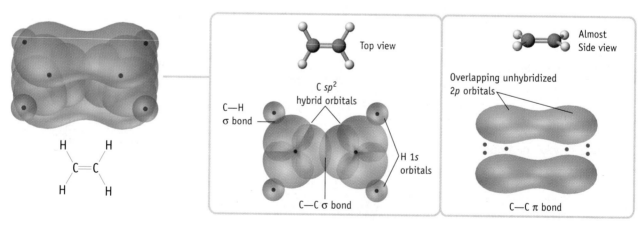

(a) Lewis structure and bonding of ethylene, C_2H_4.

(b) The C—H σ bonds are formed by overlap of C atom sp^2 hybrid orbitals with H atom $1s$ orbitals. The σ bond between C atoms arises from overlap of sp^2 orbitals.

(c) The carbon-carbon π bond is formed by overlap of an unhybridized $2p$ orbital on each atom. Note the lack of electron density along the C—C bond axis.

Figure 10.11   The valence bond model of bonding in ethylene, C_2H_4. Each C atom is assumed to be sp^2-hybridized. See the *General Chemistry Interactive CD-ROM*, Version 3.0, Screen 10.7, for an animation of these orbitals. *(Orbital Models: Patrick A. Harman and Charles F. Hamper)*

Double bonds between carbon and oxygen, sulfur, or nitrogen are quite common. To illustrate this, consider formaldehyde, CH_2O, in which a carbon–oxygen π bond occurs (Figure 10.12, page 394). A trigonal-planar electron-pair geometry indicates sp^2 hybridization for the C atom. The σ bonds from carbon to the O atom and the two H atoms form by overlap of sp^2 hybrid orbitals with half-filled orbitals from the oxygen and two hydrogen atoms. An unhybridized p orbital on carbon is oriented perpendicular to the molecular plane (just as for the carbon atoms of C_2H_4). This p orbital is available for π bonding, this time with an oxygen orbital.

What orbitals on oxygen are used in this model? The approach in Figure 10.12 assumes sp^2 hybridization for oxygen.* This uses one O atom sp^2 orbital in σ bond formation, leaving two sp^2 orbitals to accommodate lone pairs. The remaining p orbital on the O atom participates in the π bond.

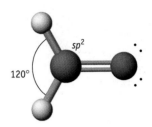

Formaldehyde, CH_2O

Example 10.5   **Bonding in Acetic Acid**

Problem • Using valence bond theory, describe the bonding in acetic acid, CH_3CO_2H, the important ingredient in vinegar.

Strategy • Write a Lewis electron dot structure and determine the geometry around each atom using VSEPR. Use this geometry to decide on the hybrid orbitals used in σ bonding. If an unhybridized p orbital is available, then $C{=}O$ π bonding can occur.

Lewis dot structure   Molecular model

Solution • The carbon atom of the CH_3 group has tetrahedral electron-pair geometry, which means that it is sp^3-hybridized. Three sp^3 orbitals are used to form the C—H bonds. The fourth sp^3 orbital is used to bond to the adjacent carbon atom. This carbon atom has a trigonal-planar electron-pair geometry; it must be sp^2-hybridized. The C—C bond is formed using one of these orbitals, and the other two sp^2 orbitals are used to form the σ bonds to the two oxygens. The oxygen of the O—H group has four electron pairs; it must be tetrahedral and sp^3-hybridized. Thus, this O atom uses two sp^3 orbitals to bond to the adjacent carbon and the hydrogen, and two sp^3 orbitals accommodate the two lone pairs.

Finally, the carbon–oxygen double bond can be described exactly as in the CH_2O molecule (see Figure 10.12). Both the C and O atoms are assumed to be sp^2-hybridized, and the unhybridized p orbital remaining on each atom is used to form the carbon–oxygen π bond.

Exercise 10.4   **Bonding in Acetone**

Use valence bond theory to describe the bonding in acetone, CH_3COCH_3.

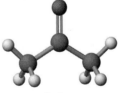

Acetone

Triple Bonds

Acetylene, $H{-}C{\equiv}C{-}H$, is an example of a molecule with a triple bond. VSEPR allows us to predict that the four atoms lie in a straight line with H—C—C angles of 180°. This implies that the carbon atom is sp-hybridized (see Figure 10.10).

---

*A second equally valid approach is to use unhybridized orbitals on oxygen in bonding. If unhybridized oxygen is assumed, the two p orbitals that are oriented at right angles and contain a single electron are used to create σ and π bonds. The argument favoring hybridization for oxygen is that this would add consistency to the valence bond approach; because hybridization is required for some atoms it makes sense to use it for all. The objection is that hybridization was introduced only to explain molecular geometry. The O atom is bonded to only one other atom, so there is no geometry to explain; hybridization does not add anything to the explanation, and it could be regarded as an additional complication.

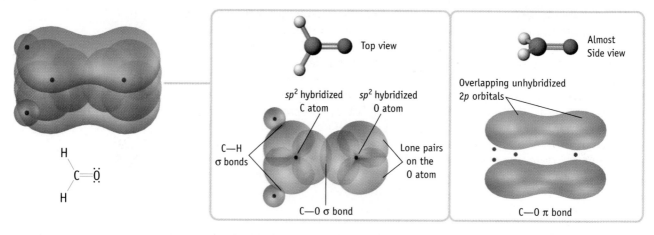

(a) Lewis structure and bonding of formaldehyde, CH₂O.

(b) The C—H σ bonds are formed by overlap of C atom sp^2 hybrid orbitals with H atom $1s$ orbitals. The σ bond between C and O atoms arises from overlap of sp^2 orbitals.

(c) The C—O π bond comes from the side-by-side overlap of p orbitals on the two atoms.

Figure 10.12   Valence bond description of bonding in formaldehyde, CH₂O. *(Orbital Models: Patrick A. Harman and Charles F. Hamper)*

For each carbon atom, there are two sp orbitals: one directed toward hydrogen and used to create the C—H σ bond, and the second directed toward the other carbon and used to create a σ bond between the two carbon atoms. Two unhybridized p orbitals remain on each carbon, and they are oriented so that it is possible to form *two* π bonds in HC≡CH (Figure 10.13).

⊞ Two unhybridized p orbitals. Used for π bonding in C₂H₂.

⊞ Two sp hybrid orbitals. Used for C—H and C—C σ bonding in C₂H₂.

These π bonds are perpendicular to the molecular axis and perpendicular to each other. Three electrons on each carbon atom are paired to form the triple bond consisting of a σ bond and two π bonds (see Figure 10.13).

Now that we have examined two cases of multiple bonds, let us summarize several important points:

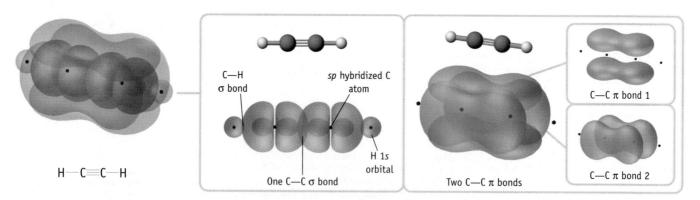

Figure 10.13   Bonding in acetylene. *(Orbital Models: Patrick A. Harman and Charles F. Hamper)*

- Pi bonds do not occur without the bonded atoms also being joined by a σ bond. Thus, a double bond always consists of a σ bond and a π bond. Similarly, a triple bond always consists of a σ bond and *two* π bonds.
- A π bond may form only if unhybridized p orbitals remain on the bonded atoms.
- If a Lewis structure shows multiple bonds, the atoms involved must therefore be either sp^2- or sp-hybridized. Only in this manner will unhybridized p orbitals be available to form a π bond.

Exercise 10.5 **Triple Bonds Between Atoms**

Describe the bonding in a nitrogen molecule, N_2.

Exercise 10.6 **Bonding and Hybridization**

Estimate values for the H—C—H, H—C—C, and C—C—N angles in acetonitrile, CH_3CN. Indicate the hybridization of both carbon atoms and the nitrogen atom, and analyze the bonding using valence bond theory.

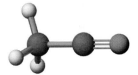

Acetonitrile, CH_3CN

Cis–Trans Isomerism: A Consequence of π Bonding

Ethylene, C_2H_4, is a planar molecule. This geometry allows the unhybridized p orbitals on the two carbon atoms to line up and form a π bond (see Figure 10.11). What would happen if one end of the ethylene molecule were twisted relative to the other end (Figure 10.14)? This action would distort the molecule away from planarity, and the p orbitals would rotate out of alignment. Rotation would diminish the extent of overlap of these orbitals, and if a twist of 90° were achieved, the two p orbitals would no longer overlap; the π bond would be broken. However, so much energy is required to break this bond (about 260 kJ/mol) that rotation around a C=C bond is not expected to occur at room temperature.

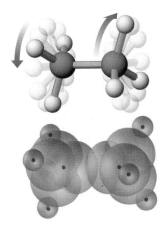

(a) Free rotation can occur around the axis of a single (σ) bond.

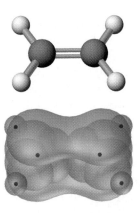

(b) In contrast, rotation is severely restricted around double bonds because doing so would break the π bond, a process generally requiring a great deal of energy.

Figure 10.14   Rotation around bonds. *(Orbital Models: Patrick A. Harman and Charles F. Hamper)*

● *Cis* and *Trans* Isomers
Compounds having the same formula, but different structures, are isomers. *Trans* isomers have distinguishing groups on opposite sides of a double bond. *Cis* isomers have these groups on the same side of the double bond.

A consequence of restricted rotation is that isomers occur for many compounds containing a C=C bond. **Isomers** are compounds that have the same formula but different structures. In this case, the two isomeric compounds differ with respect to the orientation of the groups attached to the carbons of the double bond. Two of the compounds with the formula $C_2H_2Cl_2$, *cis-* and *trans*-1,2-dichloroethylene, are isomers. Their structures resemble ethylene, except that two hydrogen atoms have been replaced by chlorine atoms. Because a large amount of energy is required to break the π bond, the *cis* compound cannot rearrange to the *trans* compound under ordinary conditions. Each compound can be obtained separately, and each has its own identity. *cis*-1,2-Dichloroethylene boils at 60.3 °C, whereas *trans*-1,2-dichloroethylene boils at 47.5 °C.

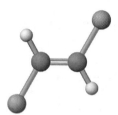

cis-1,2-dichloroethylene   trans-1,2-dichloroethylene

Although *cis* and *trans* isomers do not interconvert at ordinary temperatures, they may do so at higher temperatures. According to the kinetic theory of matter (◀ SECTION 12.6), molecules in the gas and liquid phases in particular move rapidly and often collide with one another. Molecules also constantly flex or vibrate along or around the σ bonds holding them together. If the temperature is sufficiently high, the molecular motions can become sufficiently energetic that rotation around the C=C bond can occur. It may also occur under other special conditions, such as when the molecule absorbs light energy. Indeed, this specific situation is found to occur in the physiological process that allows us to see (see Chapter Focus, page 380).

Benzene: A Special Case of π Bonding

Benzene, C_6H_6, is the simplest member of a large group of substances known as *aromatic* compounds, a historical reference to their odor. They occupy a pivotal place in the history and practice of chemistry.

To 19th-century chemists benzene was a perplexing substance with an unknown structure. Based on its chemical reactions, however, August Kekulé (1829–1896) suggested that the molecule has a planar, symmetrical ring structure. We know now he was correct. The ring is flat, and all the carbon–carbon bonds are the same length, 139 pm, a distance intermediate between the average single bond (154 pm) and double bond (134 pm) lengths. If we assume the molecule has two resonance structures with alternating double bonds, this would rationalize the observed structure (◀ SECTION 9.5). The C—C bond order in C_6H_6 (1.5) is the average of a single and a double bond.

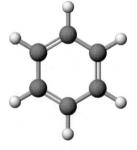

Benzene, C_6H_6

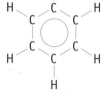

resonance structures   resonance hybrid

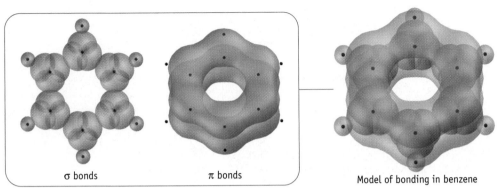

σ bonds   π bonds

σ and π bonding in benzene

Model of bonding in benzene

Figure 10.15 Bonding in benzene, C_6H_6. (left) The C atoms of the ring are bonded to each other through σ bonds using sp^2 C atom hybrid orbitals. The C—H bonds also use C atom sp^2 hybrid orbitals. The π framework of the molecule arises from overlap of C atom *p* orbitals not used in hybrid orbital formation. As these orbitals are perpendicular to the ring, π electron density is above and below the plane of the ring. (right) A composite of σ and π bonding in benzene. (*Orbital Models: Patrick A. Harman and Charles F. Hamper*)

Understanding the bonding in benzene is important because benzene is the basis of an enormous number of chemical compounds (Figure 10.15). We assume that the trigonal-planar carbon atoms are sp^2-hybridized. Each C—H bond is formed by overlap of an sp^2 orbital of a carbon atom with a 1*s* orbital of hydrogen, and the C—C σ bonds arise by overlap of sp^2 orbitals on adjacent carbon atoms. After accounting for the σ bonding, an unhybridized *p* orbital remains on each C atom, and each is occupied by a single electron. These six orbitals and six electrons form three π bonds. Because all carbon–carbon bond lengths are the same, each *p* orbital overlaps equally well with the *p* orbitals of both adjacent carbons, and the π interaction is unbroken around the six-member ring.

The orbital picture of benzene underscores an important point. The basis of valence bond theory, that a bond is described as a pair of electrons between two atoms, does not work well for the π electrons in benzene, nor does it work whenever resonance is needed to describe a structure. However, molecular orbital theory does give us a better view, and that is the subject of the next section.

10.3 MOLECULAR ORBITAL THEORY

Molecular orbital (MO) theory is an alternative way to view orbitals in molecules. In contrast to the localized bond and lone pair electrons of valence bond theory, MO theory assumes that pure atomic orbitals of the atoms in the molecule combine to produce orbitals that are spread out, or delocalized, over several atoms or even over an entire molecule. These new orbitals are called **molecular orbitals.**

One reason for learning about the MO concept is that it correctly predicts the electronic structures of molecules such as O_2 that do not follow the electron-pairing assumptions of the Lewis approach. The rules of Chapter 9 would guide you to draw the electron dot structure of O_2 with all the electrons paired, which fails to explain its paramagnetism (Figure 10.16). The molecular orbital approach can account for this property, but valence bond theory cannot. To see how MO theory can be used to describe the bonding in O_2 and other diatomic molecules, we shall first describe four principles used to develop the theory.

● **A Failure of the Valence Bond Theory** Lewis electron dot structures fail to describe the bonding correctly in one highly visible case, the O_2 molecule. The O_2 molecule is paramagnetic, which requires the presence of unpaired electrons. The Lewis structure, however, would have all electrons paired. The molecular orbital approach shows that the molecule has two unpaired electrons.

● **Diatomic Molecules** Molecules such as H_2, Li_2, and N_2, in which two identical atoms are bonded, are called homonuclear diatomic molecules

Principles of Molecular Orbital Theory

In MO theory we begin with a given arrangement of atoms in the molecule at the known bond distances. We then determine the *sets* of molecular orbitals. One way to do this is to combine available valence orbitals on all the constituent atoms. These molecular orbitals more or less encompass all the atoms of the molecule,

Figure 10.16   Liquid oxygen. Oxygen gas condenses to a liquid at −183 °C (*left*). Notice that liquid oxygen is very pale blue (*middle*). Oxygen in the liquid state is paramagnetic and clings to the poles of a magnet (*right*). *(Charles D. Winters)*

and the valence electrons for all the atoms in the molecule are assigned to the molecular orbitals. Just as with orbitals in atoms, electrons are assigned according to the Pauli principle and Hund's rule (◀ SECTIONS 8.2 AND 8.4).

The ***first principle of molecular orbital theory*** is that *the total number of molecular orbitals is always equal to the total number of atomic orbitals contributed by the atoms that have combined.* To illustrate this orbital conservation principle, let us consider the H_2 molecule.

Molecular Orbitals for H_2

Molecular orbital theory specifies that when the 1*s* orbitals of two hydrogen atoms overlap, *two* molecular orbitals result from the combination of the overlapping orbitals. In the molecular orbital resulting from *addition* of the atomic orbitals, the 1*s* regions of electron density add together, leading to an increased probability that electrons will reside in the bond region between the two nuclei (Figure 10.17). This is called a **bonding molecular orbital** and is the same as that described as a

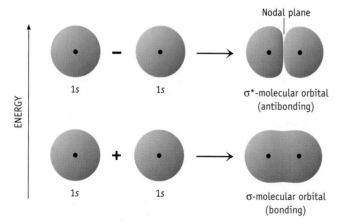
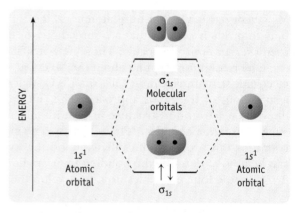

Figure 10.17   Molecular orbitals. (a) Bonding and antibonding σ molecular orbitals are formed from two 1*s* atomic orbitals on adjacent atoms. Notice the presence of a node in the antibonding orbital. (The node is a plane on which there is zero probability of finding an electron.) **(b)** A molecular orbital diagram for H_2. The two electrons are placed in the bonding σ_{1s} orbital, the molecular orbital lower in energy [🖱 CD-ROM, Screen 10.9].

chemical bond by valence bond theory. It is also a σ orbital because the region of electron probability lies directly along the bond axis. This molecular orbital is labeled σ_{1s}, the subscript $1s$ indicating that $1s$ atomic orbitals were used to create the molecular orbital.

The other molecular orbital is constructed by *subtracting* one atomic orbital from the other (see Figure 10.17). When this happens, the probability of finding an electron between the nuclei in the molecular orbital is reduced, and the probability of finding the electron in other regions is higher. Without significant electron density between them, the nuclei repel each other. This type of orbital is called an **antibonding molecular orbital.** Because it is also a σ orbital, it is labeled σ_{1s}^*. The asterisk signifies that it is antibonding. *Antibonding orbitals have no counterpart in valence bond theory.*

A ***second principle of molecular orbital theory*** is that *the bonding molecular orbital is lower in energy than the parent orbitals, and the antibonding orbital is higher in energy* (see Figure 10.17). This means that the energy of a group of atoms is lower than the energy of the separated atoms when electrons are assigned to bonding molecular orbitals. Chemists say the system is "stabilized" by chemical bond formation. Conversely, the system is "destabilized" when electrons are assigned to antibonding orbitals because the energy of the system is higher than that of the atoms themselves.

A ***third principle of molecular orbital theory*** is that the *electrons of the molecule are assigned to orbitals of successively higher energy* according to the Pauli exclusion principle and Hund's rule. This is analogous to the procedure for building up electronic structures of atoms. Thus, electrons occupy the lowest energy orbitals available, and when two electrons are assigned to an orbital, their spins are paired. Because the energy of the electrons in the bonding orbital of H_2 is lower than that of either parent $1s$ electron (Figure 10.17), the H_2 molecule is stable. We write the electron configuration of H_2 as $(\sigma_{1s})^2$.

What would happen if we try to combine two helium atoms to form dihelium, He_2 (Figure 10.18)? Both He atoms have a $1s$ valence orbital that can be added and subtracted to produce the same kind of molecular orbitals as in H_2. Unlike H_2, however, four electrons need to be assigned to these orbitals (see Figure 10.18). The pair of electrons in the σ_{1s} orbital stabilizes He_2. The two electrons in σ_{1s}^*, however, destabilize the He_2 molecule. The energy decrease from the electrons in the σ_{1s}-bonding molecular orbital is offset by the energy increase due to the electrons in the σ_{1s}^*-antibonding molecular orbital. Thus, molecular orbital theory predicts that He_2 has no net stability; two He atoms have no tendency to combine. This confirms what we already know, that elemental helium exists in the form of single atoms and not as a diatomic molecule.

Bond Order

Bond order was defined in Chapter 9 as the net number of bonding electron pairs linking a pair of atoms. This same concept can be applied directly to molecular orbital theory, but now bond order is defined as

$$\text{Bond order} = \frac{1}{2}\,(\text{number of electrons in bonding MOs} - \text{number of electrons in antibonding MOs})$$

(10.1)

In the H_2 molecule, there are two electrons in a bonding orbital and none in an antibonding orbital, so H_2 has a bond order of 1. In contrast, in the He_2 the

• **Orbitals and Electron Waves**
Orbitals are characterized as electron waves; therefore, a way to view molecular orbital formation is to assume that two electron waves, one from each atom, interfere with each other. The interference can be constructive, giving a bonding MO, or destructive, giving an antibonding MO.

Chapter Goals • **Revisited**

- Understand the differences between valence bond and molecular orbital theory.
- Identify the hybridization of an atom in a molecule or ion.
- **Understand the differences between bonding and antibonding molecular orbitals.**
- Write the molecular orbital configuration for simple diatomic molecules.
- Understand the differences among a metal, an insulator, and a semiconductor.

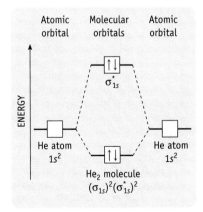

Figure 10.18 A molecular orbital energy level diagram for the dihelium molecule, He_2. This diagram provides a rationalization for the nonexistence of the molecule. In He_2 both the bonding (σ_{1s}) and antibonding orbitals (σ_{1s}^*) would be fully occupied.

stabilizing effect of the σ_{1s} pair is canceled by the destabilizing effect of the σ_{1s}^* pair, so the bond order is 0.

Fractional bond orders are possible. Consider the ion He_2^+. Its molecular orbital electron configuration is $(\sigma_{1s})^2(\sigma_{1s}^*)^1$. In this ion, two electrons are in a bonding molecular orbital, but only one is in an antibonding orbital. MO theory predicts that He_2^+ should have a bond order of 0.5, that is, a weak bond should exist between helium atoms in such a species. Interestingly, this ion has been identified in the gas phase using special experimental techniques.

Example 10.6 Molecular Orbitals and Bond Order

Problem • Write the electron configuration of the H_2^- ion in molecular orbital terms. What is the bond order of the ion?

Strategy • Count the number of valence electrons in the ion and then place those electrons in the MO diagram for the H_2 molecule. Find the bond order using Equation 10.1.

Solution • This ion has three electrons (one each from the H atoms plus one for the negative charge). Therefore, its electronic configuration is $(\sigma_{1s})^2(\sigma_{1s}^*)^1$, identical with the configuration for He_2^+. This means H_2^- also has a net bond order of 0.5. The H_2^- ion is thus predicted to exist under special circumstances.

Exercise 10.7 Molecular Orbitals and Bond Order

What is the electron configuration of the H_2^+ ion? Compare the bond order of this ion with He_2^+ and H_2^-. Do you expect H_2^+ to exist?

Molecular Orbitals of Li$_2$ and Be$_2$

A *fourth principle of molecular orbital theory* is that *atomic orbitals combine to form molecular orbitals most effectively when the atomic orbitals are of similar energy*. This principle becomes important when we move past He_2 to Li_2, dilithium, and heavier molecules still.

A lithium atom has electrons in two orbitals of the s type ($1s$ and $2s$), so a $1s \pm 2s$ combination is theoretically possible. Because the $1s$ and $2s$ orbitals are quite different in energy, however, this interaction can be disregarded. Thus, the molecular orbitals come only from $1s \pm 1s$ and $2s \pm 2s$ combinations (Figure 10.19). This means the molecular orbital electron configuration of dilithium, Li_2, is

$$Li_2 \text{ MO configuration:} \quad (\sigma_{1s})^2(\sigma_{1s}^*)^2(\sigma_{2s})^2$$

The bonding effect of the σ_{1s} electrons is canceled by the antibonding effect of the σ_{1s}^* electrons, so these pairs make no net contribution to bonding in Li_2. Bonding in Li_2 is due to the electron pair assigned to the σ_{2s} orbital, and the bond order is 1.

The fact that the σ_{1s} and σ_{1s}^* electron pairs of Li_2 make no net contribution to bonding is exactly what you observed in drawing electron dot structures in Chapter 9: *Core electrons are ignored.* In molecular orbital terms, core electrons are assigned to bonding and antibonding molecular orbitals that offset one another.

A diberyllium molecule, Be_2, is not expected to exist. Its electron configuration is

$$Be_2 \text{ MO configuration:} \quad [\text{core electrons}](\sigma_{2s})^2(\sigma_{2s}^*)^2$$

The effects of σ_{2s} and σ_{2s}^* electrons cancel, and there is no net bonding. The bond order is 0, so the molecule does not exist.

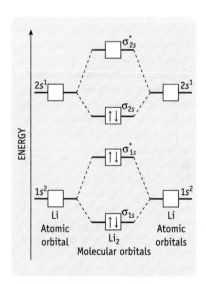

Figure 10.19 Energy level diagram for the combination of two Li atoms with 1s and 2s atomic orbitals. Notice that the molecular orbitals are created by combining orbitals of similar energies. The electron configuration is shown for Li$_2$.

Example 10.7 Molecular Orbitals in Diatomic Molecules

Problem • Be_2 does not exist. But what about the Be_2^+ ion? Describe its electron configuration in molecular orbital terms and give the net bond order. Do you expect the ion to exist?

Strategy • Count the number of electrons in the ion and place them in the MO diagram in Figure 10.19. Write the electron configuration and calculate the bond order from Equation 10.1.

Solution • The Be_2^+ ion has only seven electrons (in contrast to eight for Be_2), of which four are core electrons. (The core electrons are assigned to σ_{1s} and σ_{1s}^* molecular orbitals.) The remaining three electrons are assigned to the σ_{2s} and σ_{2s}^* molecular orbitals, so the MO electron configuration is [core electrons]$(\sigma_{2s})^2(\sigma_{2s}^*)^1$. This means the net bond order is 0.5, and so Be_2^+ is predicted to exist under special circumstances.

Exercise 10.8 Molecular Orbitals in Diatomic Molecules

Could the anion Li_2^- exist? What is the ion's bond order?

Molecular Orbitals from Atomic *p* Orbitals

With the principles of molecular orbital theory in place, we are ready to account for bonding in such important **homonuclear diatomic molecules** as N_2, O_2, and F_2. First, however, we need to see what types of molecular orbitals form when elements have both *s* and *p* valence orbitals. Three types of interactions are possible for two atoms that both have *s* and *p* orbitals. Sigma-bonding and antibonding molecular orbitals are formed by *s* orbitals interacting as in Figure 10.19. Similarly, it is possible for a *p* orbital on one atom to interact with a *p* orbital on the other atom in a head-to-head fashion to produce a σ-bonding and σ^*-antibonding pair of molecular orbitals (Figure 10.20). And finally, each atom has two *p* orbitals in planes perpendicular to the σ bond connecting the two atoms. These *p* orbitals can interact sideways to give π-bonding and π-antibonding molecular orbitals (Figure 10.21, page 402). The two *p* orbitals on each atom produce *two* π-bonding molecular orbitals (π_p) and *two* π-antibonding molecular orbitals (π_p^*).

Electron Configurations for Homonuclear Molecules for Boron Through Fluorine

Orbital interactions in a second-period, homonuclear, diatomic molecule lead to the energy level diagram in Figure 10.22 (page 402).

> ### Chapter Goals • Revisited
>
> - Understand the differences between valence bond and molecular orbital theory.
> - Identify the hybridization of an atom in a molecule or ion.
> - Understand the differences between bonding and antibonding molecular orbitals.
> - **Write the molecular orbital configuration for simple diatomic molecules.**
> - Understand the differences among a metal, an insulator, and a semiconductor.

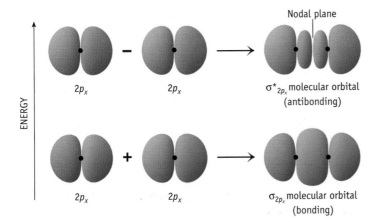

Figure 10.20 Sigma molecular orbitals from *p* atomic orbitals. Sigma-bonding (σ_{2p}) and antibonding (σ_{2p}^*) molecular orbitals arise from overlap of 2p orbitals. Each orbital can accommodate two electrons. The *p* orbitals in electron shells of higher *n* give molecular orbitals of the same basic shape.

Figure 10.21   Formation of π molecular orbitals. Sideways overlap of atomic $2p$ orbitals that lie in the same direction in space gives rise to π-bonding (π_{2p}) and π-antibonding (π_{2p}^*) molecular orbitals. The p orbitals in shells of higher n give molecular orbitals of the same basic shape.

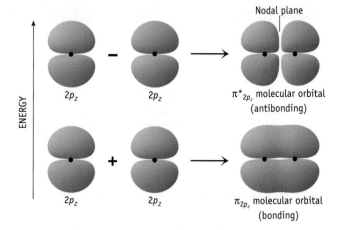

Electron assignments can be made using this diagram, and the results for the diatomic molecules B_2 through F_2 are tabulated in Table 10.1, which has two noteworthy features.

First, notice the correlation between the electron configurations and the bond orders, bond lengths, and bond energies at the bottom of Table 10.1. As the bond order between a pair of atoms increases, the energy required to break the bond increases, and the bond distance decreases. Dinitrogen, N_2, with a bond order of 3, has the largest bond energy and shortest bond distance.

Second, notice the configuration for dioxygen, O_2. Dioxygen has 12 valence electrons (6 from each atom), so it has the molecular orbital configuration

$$O_2 \text{ MO configuration:} \quad [\text{core electrons}](\sigma_{2s})^2(\sigma_{2s}^*)^2(\pi_{2p})^4(\sigma_{2p})^2(\pi_{2p}^*)^2$$

This configuration leads to a bond order of 2 in agreement with experiment, and it specifies two unpaired electrons (in π_{2p}^* molecular orbitals). Thus, molecular orbital theory succeeds where valence bond theory fails. MO theory explains both the observed bond order and the paramagnetic behavior of O_2.

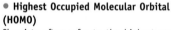

● **Highest Occupied Molecular Orbital (HOMO)**
Chemists often refer to the highest energy MO that contains electrons as the HOMO. For O_2 this is the π_{2p}^* orbital. Chemists also use the term LUMO for the lowest unoccupied molecular orbital. For O_2 this would be σ_{2p}^*.

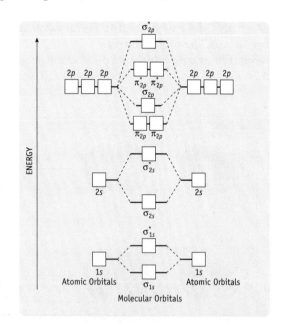

Figure 10.22   Molecular orbital energy level diagram for homonuclear diatomic molecules of second-period elements. See A Closer Look: Molecular Orbitals for Compounds Formed from p-Block Elements, page 404 [CD-ROM, Screen 10.11].

Table 10.1 • **Molecular Orbital Occupations and Physical Data for Homonuclear Diatomic Molecules of Second-Period Elements**

	B_2	C_2	N_2	O_2	F_2
σ^*_{2p}	☐	☐	☐	☐	☐
π^*_{2p}	☐☐	☐☐	☐☐	↑ ↑	↑↓ ↑↓
σ_{2p}	☐	☐	↑↓	↑↓	↑↓
π_{2p}	↑ ↑	↑↓ ↑↓	↑↓ ↑↓	↑↓ ↑↓	↑↓ ↑↓
σ^*_{2s}	↑↓	↑↓	↑↓	↑↓	↑↓
σ_{2s}	↑↓	↑↓	↑↓	↑↓	↑↓
Bond order	One	Two	Three	Two	One
Bond-dissociation energy (kJ/mol)	290	620	945	498	155
Bond distance (pm)	159	131	110	121	143
Observed magnetic behavior (paramagnetic or diamagnetic)	Para	Dia	Dia	Para	Dia

Example 10.8   Electron Configuration for a Homonuclear Diatomic Ion

Problem • When potassium reacts with O_2, potassium superoxide, KO_2, is one of the products. This is an ionic compound, and the anion is the superoxide ion, O_2^-. Write the molecular orbital electron configuration for the ion. Predict its bond order and magnetic behavior.

Strategy • Use the energy level diagram of Figure 10.22 to generate the configuration of this ion.

Solution • The MO configuration for O_2^- is

O_2^- *MO configuration:*
$$[\text{core electrons}]\ (\sigma_{2s})^2(\sigma^*_{2s})^2(\pi_{2p})^4(\sigma_{2p})^2(\pi^*_{2p})^3$$

The ion is predicted to be paramagnetic to the extent of one unpaired electron, a prediction confirmed by experiment. The bond order is 1.5, because there are eight bonding electrons and five antibonding electrons. The bond order for O_2^- is lower than O_2, so we predict the O—O bond length in O_2^- should be longer than the oxygen–oxygen bond length in O_2. The superoxide ion in fact has an O—O bond length of 134 pm, whereas the bond length in O_2 is 121 pm.

Comment • You should quickly spot the fact that the superoxide ion (O_2^-), contains an odd number of electrons. This is a another diatomic species (in addition to NO and O_2) for which it is not possible to write a Lewis structure that accurately represents the bonding.

Exercise 10.9   Molecular Electron Configurations

The cations O_2^+ and N_2^+ are important components of earth's upper atmosphere. Write the electron configuration of O_2^+. Predict its bond order and magnetic behavior.

Electron Configurations for Heteronuclear Diatomic Molecules

The molecules NO, CO, and ClF are examples of simple diatomic molecules formed from two elements of different kinds. These compounds are called **heteronuclear diatomic molecules.** MO descriptions for heteronuclear diatomic molecules generally resemble those for homonuclear diatomic molecules. As a consequence, an energy level diagram like Figure 10.22 can be used to judge the bond order and magnetic behavior for heteronuclear diatomics.

Let us do this for nitrogen monoxide, NO. Nitrogen monoxide has 11 molecular valence electrons. If these are assigned to the MOs for a homonuclear diatomic molecule, the molecular electron configuration is

NO MO configuration: $[\text{core electrons}](\sigma_{2s})^2(\sigma_{2s}^*)^2(\pi_{2p})^4(\sigma_{2p})^2(\pi_{2p}^*)^1$

The net bond order is 2.5, in accord with bond length information. The single, unpaired electron is assigned to the π_{2p}^* molecular orbital. The molecule is paramagnetic, as predicted for a molecule with an odd number of electrons.

Resonance and Molecular Orbital Theory

Ozone, O_3, is a simple triatomic molecule with equal oxygen–oxygen bond lengths. Equal X — O bond lengths are also observed in other molecules and ions, such as SO_2, NO_2^-, and HCO_2^-. Valence bond theory introduced resonance to rationalize the equivalent bonding to the oxygen atoms in these structures. MO theory provides another view of this problem.

O_3 SO_2 NO_2^- HCO_2^-

A Closer Look

Molecular Orbitals for Compounds Formed from *p*-Block Elements

Several features of the molecular orbital energy level diagram in Figure 10.22 can be described in more detail.

- The bonding and antibonding σ orbitals from 2s interactions are lower in energy than the σ and π MOs from 2p interactions. The reason is that 2s orbitals have a lower energy than 2p orbitals in the separated atoms.

- The energy separation of the bonding and antibonding orbitals is greater for σ_{2p} than for π_{2p}. This happens because p orbitals overlap to a greater extent when they are oriented head to head (to give σ_{2p} MOs) than when they are side by side

(to give π_{2p} MOs). The greater the orbital overlap, the greater the stabilization of the bonding MO and the greater the destabilization of the antibonding MO.

Figure 10.22 shows an energy ordering of molecular orbitals that you might not have expected, but there are reasons for this. A more sophisticated approach takes into account the "mixing" of s and p atomic orbitals, which have similar energies. This causes the σ_{2s} and σ_{2s}^* molecular orbitals to be lower in energy than otherwise expected, and the σ_{2p} and σ_{2p}^* orbitals to be higher in energy. This is the reason the en-

ergy lowering and raising for the σ_{2s} and σ_{2s}^* orbitals (and for the σ_{2p} and σ_{2p}^* orbitals) in Figure 10.22 is not symmetrical with respect to the 2s and 2p atomic orbital energies.

Another refinement concerning Figure 10.22 is that, because s and p orbital mixing is important for B_2, C_2, and N_2, the figure applies strictly only to these molecules. For O_2 and F_2, σ_{2p} is lower in energy than π_{2p}. Nonetheless, Figure 10.22 gives the correct bond order and magnetic behavior for these two molecules.

To visualize the bonding in ozone, begin by assuming that all three O atoms are sp^2-hybridized. The central atom uses its sp^2 hybrid orbitals to form two σ bonds and to accommodate a lone pair. The terminal atoms use their sp^2 hybrid orbitals to form one σ bond and to accommodate two lone pairs. In all, the lone pairs and bonding pairs in the σ framework of O_3 account for seven of the nine valence electron pairs in O_3.

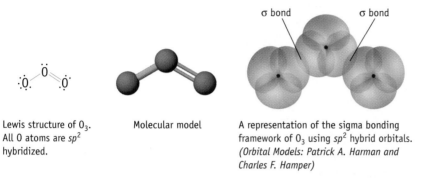

Lewis structure of O_3.
All O atoms are sp^2
hybridized.

Molecular model

A representation of the sigma bonding framework of O_3 using sp^2 hybrid orbitals. *(Orbital Models: Patrick A. Harman and Charles F. Hamper)*

The π bond in ozone arises from the two remaining pairs (Figure 10.23). Because we have assumed each oxygen atom in O_3 is sp^2-hybridized, an unhybridized p orbital perpendicular to the O_3 plane remains on each of the three oxygen atoms. The orbitals are in the correct orientation to form π bonds. *A principle of MO theory is that the number of molecular orbitals must equal the number of atomic orbitals.* Thus, the three $2p$ atomic orbitals must be combined in a way that forms three molecular orbitals.

One π_p MO for ozone is a bonding orbital because the three p orbitals are "in phase" across the molecule. Another π_p MO is antibonding because the central atomic orbital is "out of phase" with the terminal atom p orbitals. The third π_p MO is nonbonding because the middle p orbital does not participate in the MO. The bonding π_p MO is filled by a pair of electrons, which is delocalized, or "spread

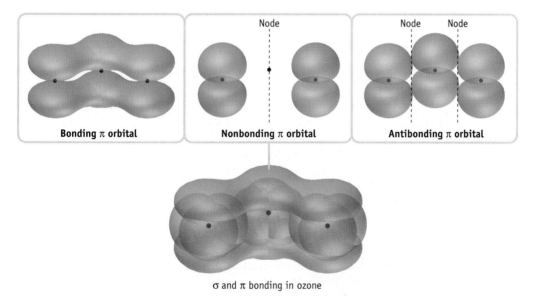

Figure 10.23   Pi-bonding in ozone, O_3. Each O atom in O_3 is sp^2-hybridized. The three $2p$ orbitals, one on each atom, are used to create the three π molecular orbitals. Two pairs of electrons are assigned to the orbitals: one pair in the bonding orbital and one pair in the nonbonding orbital. The π bond order is 0.5, one bonding pair spread across two bonds. *(Orbital Models: Patrick A. Harman and Charles F. Hamper)*

Figure 10.24 Molecular orbital energy level diagram for benzene. Because benzene has six unhybridized *p* orbitals, six π molecular orbitals can be formed. Three are bonding and three are antibonding. The three bonding molecular orbitals accommodate the six π electrons. (See the benzene π molecular orbitals in the molecular-modeling directory on the *General Chemistry Interactive CD-ROM, Version 3.0.*) *(Orbital Models: Patrick A. Harman and Charles F. Hamper)*

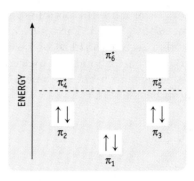

over," the molecule, just as the resonance hybrid implies. The nonbonding orbital is also occupied, but the electrons in this orbital are concentrated near the two terminal oxygens. As the name implies, electrons in this molecular orbital neither help nor hinder the bonding in the molecule. The π bond order of O_3 is 0.5 because one bond pair is spread over two O — O linkages. Because the σ bond order is 1.0 and the π bond order is 0.5, the net oxygen–oxygen bond order is 1.5, the same value given by valence bond theory.

The observation that two of the π molecular orbitals for ozone extend over three atoms illustrates an important point regarding molecular orbital theory: Orbitals can extend beyond two atoms. In valence bond theory, all representations for bonding were based on being able to localize pairs of electrons in bonds between two atoms. To further illustrate the MO approach, look again at benzene (Figure 10.24). On page 397 the point was made that the π electrons in this molecule were spread out over all six carbon atoms. We can now see how the same case can be made with MO theory. Six *p* orbitals contribute to the π system. Based on the premise that there must be the same number of molecular orbitals as atomic orbitals, benzene must have six π molecular orbitals. An energy level diagram for benzene shows that the six *p* electrons reside in the three lowest energy (bonding) molecular orbitals.

10.4 METALS AND SEMICONDUCTORS

Metals are among the oldest materials used by humankind. It was not until the last century, however, that new types of materials — semiconductors — were developed. Because metals and semiconductors are so important to modern economies, it is useful to understand more of their properties. So, let us extend the ideas of molecular orbital theory to learn why metals are good conductors of electricity and why semiconductors have special properties.

Conductors, Insulators, and Band Theory

Metal crystals can be viewed as "supermolecules" held together by delocalized bonds formed from the atomic orbitals of all the atoms in a crystal. Even the tiniest piece of metal contains a very large number of atoms, and an even larger number of orbitals is available to form molecular orbitals. If four lithium atoms, for example, are in a group, and each Li atom contributes one 2*s* and three 2*p* orbitals to metallic bonding, 16 molecular orbitals are possible. Four hundred Li atoms taken together would lead to 1600 molecular orbitals, and a single crystal containing a mole of lithium atoms would have $4 \times (6.022 \times 10^{23})$ molecular orbitals.

In a metal, the molecular orbitals spread out over many atoms and blend into a *band* of molecular orbitals, the energies of which are closely spaced within a range

Chapter Goals • Revisited

- Understand the differences between valence bond and molecular orbital theory.
- Identify the hybridization of an atom in a molecule or ion.
- Understand the differences between bonding and antibonding molecular orbitals.
- Write the molecular orbital configuration for simple diatomic molecules.
- **Understand the differences among a metal, an insulator, and a semiconductor.**

of energies (Figure 10.25). The band is composed of as many orbitals as there are contributing atomic orbitals, and each molecular orbital can accommodate two electrons of opposite spin. The idea that the molecular orbitals of the band of energy levels are spread out, or *delocalized,* over all the atoms in a piece of metal accounts for the bonding in metallic solids. This theory of metallic bonding is called **band theory.**

In a metal, the band of energy levels is only partially filled; there are not enough electrons to fill all the orbitals (see Figure 10.25). Electrons fill the lowest energy molecular orbitals, but the lowest energy for a system (with all electrons in orbitals with the lowest possible energy) is reached only at 0 K. The highest filled level at 0 K is called the **Fermi level** (Figure 10.26). A small input of energy (for example, raising the temperature above 0 K), can cause electrons to move from the filled portion of the band to the unfilled portion. For each electron promoted, two singly occupied levels result: a negative electron above the Fermi level and a positive "hole" — from the absence of an electron — below the Fermi level. When an electric field is applied to the metal, electrons move toward the positive side, and "holes" move to the negative side. Positive holes "move" because an electron from an adjacent level can move into the hole, thus creating a fresh "hole." *The electrical conductivity of metals arises from the movement of electrons in singly occupied states close to the Fermi level in the presence of an applied electric field.*

Because the band of unfilled energy levels in a metal is essentially continuous, that is, the energy gaps between orbitals are extremely small, a metal can absorb energy of nearly any wavelength. When light causes an electron in a metal to move to a higher energy state, the now-excited system can immediately emit a photon of the same energy, and the electron returns to the original energy level. It is this rapid and efficient reemission of light that makes polished metal surfaces reflective and appear lustrous.

A metal is characterized by a band structure in which the highest occupied band, called the **valence band,** is only partially filled. In contrast an electrical **insulator** has a completely filled valence band, and the next available empty levels are at much higher energy (Figure 10.26, page 408). The consequence of this is that the promotion of an electron to a higher energy level is not likely, and the solid does not conduct electricity.

Diamonds are electrical insulators. Each carbon atom in a diamond is surrounded by four other carbon atoms at the corners of a tetrahedron (Figure 10.27). In the valence bond model of bonding, each atom is viewed as being sp^3-hybridized, forming four localized carbon–carbon bonds with other carbon atoms. An alternative view, however, is that the orbitals of each carbon atom form molecular orbitals that are delocalized over the solid, representing bonding extending over the whole solid. That is, a band model can also be applied when discussing this solid. In this picture the levels are split into two bands: a *filled* valence band of bonding molecular orbitals and a higher energy *unfilled* band of antibonding orbitals called the **conduction band** (for a reason that will become apparent in the following discussion). (Notice the resemblance of this picture to the benzene molecular orbital diagram in Figure 10.24.) The energy difference separating the two bands is called the **band gap** (see Figure 10.26, page 408).

Semiconductors

The Group 4A elements silicon and germanium have structures similar to diamond (see Figure 10.27). In contrast to diamond, however, silicon and germanium are semiconductors. **Semiconductors** are materials that are able to conduct small quantities of current. They are much poorer conductors than metals, but they are not

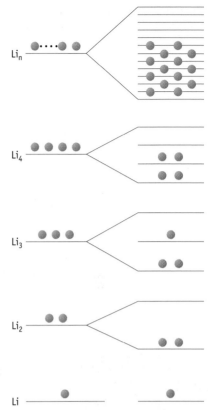

Figure 10.25 Bands of molecular orbitals in a metal crystal. Here Li atoms, each having one valence orbital (say 2s), are combined. As more and more atoms with the same valence orbitals are added, the number of molecular orbitals grows until the orbitals are so close in energy that they merge into a band. A band of molecular orbitals is formed [CD-ROM, Screen 10.14].

Silicon. Elemental silicon is a shiny material. At the rear is a thin wafer onto which circuits have been etched. The gold-colored object and the long object are both finished circuits fashioned from silicon. *(Charles D. Winters)*

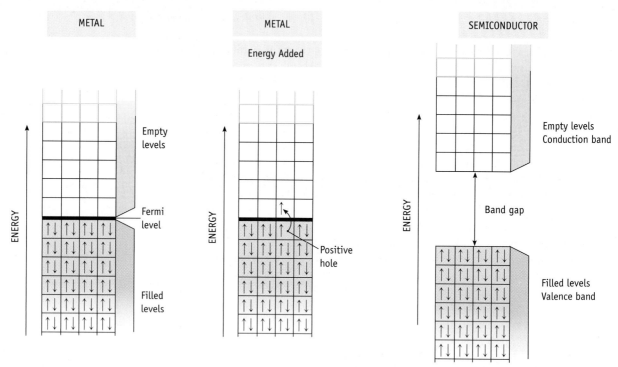

Figure 10.26   **Band theory applied to metals and semiconductors.** (*Metals, left*) The highest filled level at 0 K is referred to as the Fermi level. The molecular orbitals are delocalized over the metal, a fact accounting for the bonding in metals. (*Semiconductors and insulators, right*) In contrast to metals, the band of filled levels (the valence band) is separated from the band of empty levels (the conduction band) by a band gap [🖱 CD-ROM, Screen 10.16].

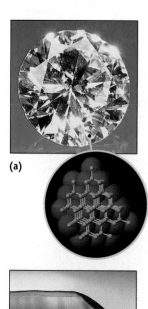

(a)

(b)

Figure 10.27   **Diamond and silicon.** Both elements have a "diamond-like" structure with tetrahedrally bonded C or Si atoms. Diamond **(a)**, pure carbon, is an electrical insulator. Pure silicon **(b)** is a semiconductor. (*Charles D. Winters*)

insulators. Why are these three structurally similar substances — diamond, silicon, and germanium — different in their conductivity?

The answer lies in the size of the band gap. If an electron is promoted from the valence band to the conduction band, singly occupied states in the valence and conduction bands are formed. As with metals, such states allow for electrical conductivity. The band gap in diamond is large, however, on the order of 500 kJ/mol, reflecting the strong bonds between carbon atoms. Semiconductors have band gaps in the range of 50 to 300 kJ/mol. The band gap narrows as one descends Group 4A, and the band gaps of silicon and germanium fall in this range. At least a few electrons can be promoted into the conduction band with the input of only modest amounts of energy, and electric conduction can occur.

Pure silicon and germanium belong to a class of materials called **intrinsic semiconductors.** The usual view of such materials is that the promotion of an electron from the valence band to the conduction band creates a positive hole in the valence band (Figure 10.28). As in a metal, the semiconductor conducts electricity because the electrons in the conduction band migrate in one direction and the positive holes in the valence band migrate in the opposite direction.

In intrinsic semiconductors, the number of electrons in the conduction band is governed by the temperature and the magnitude of the band gap. The smaller the band gap, the smaller the energy required to promote a significant number of electrons. As the temperature increases, a larger number of electrons can be promoted into the conduction band.

In contrast to intrinsic semiconductors are materials known as **extrinsic semiconductors.** The conductivity of these materials is controlled by adding small numbers of atoms of different kinds of impurities called **dopants.** For example, suppose a few silicon atoms in the silicon lattice are replaced by aluminum atoms

Molecular Orbitals, Semiconductors, and LEDs

The lights you see on the dashboard of your car and in its the rear warning lights, in traffic lights, and children's sneakers are LEDs, or "light-emitting diodes."

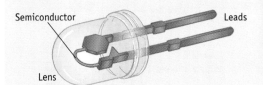

Design for an LED.

They are semiconducting devices made by combining elements such as gallium, phosphorus, arsenic, and aluminum. When attached to a low voltage source, they emit light with a wavelength that depends on their composition.

Composition	Color Emitted
GaAs	Red
AlGaInP, GaP	Red to yellow
InGaN	Green to blue

An LED itself is very small and simple in construction. A low voltage, say 6–12 V, is applied. The semiconductor emits light with a brightness now rivaling standard in-

A red traffic light illuminated by LEDs. *(Charles D. Winters)*

candescent lights, and the light is focused with a tiny plastic lens.

An LED contains an *n*-type and a *p*-type semiconductor. When an electric field is applied, electrons are supplied by the negative electrode and are attracted to the positive electrode. The positive electrode supplies holes, which are attracted to the negative electrode. Occasionally electrons in the valence band will move across the band gap and combine with a hole, and energy is evolved as light. In fact, the energy of the light is roughly equal to the band gap energy, so by adjusting the band gap, the color can be altered. (Note that the mechanism of light emission by an LED is

similar to that described for excited atoms in Chapter 7.)

Not only can the color of LEDs be altered, but they are efficient and long-lived devices. It is estimated that widespread replacement of incandescent and fluorescent lighting with LEDs could decrease electricity consumption significantly.

See S. M. Condren, et al.: *Journal of Chemical Education*, Volume 78, No. 8, pp. 1033–1040, 2001.

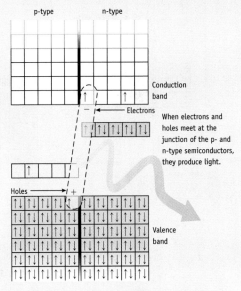

Mechanism for the emission of light from an LED constructed from *n*- and *p*-type semiconductors.

(or atoms of some other Group 3A element). Aluminum has only three valence electrons, whereas silicon has four. Four Si—Al bonds are created per aluminum atom in the lattice, but these bonds must be deficient in electrons. According to band theory, the Si—Al bonds form a discrete band at an energy level higher than the valence band. This level is referred to as an **acceptor level** because it can accept electrons. The gap between the valence band and the acceptor level is usually quite small, so electrons can be promoted readily to the acceptor level. The positive holes created in the valence band are able to move about under the influence of an electric potential. Because positive holes are created in an aluminum-doped semiconductor, this is called a ***p*-type semiconductor** (Figure 10.28 right).

Suppose phosphorus atoms are incorporated into the silicon lattice instead of aluminum atoms. The material is still a semiconductor, but it now has *extra* electrons because phosphorus has one more valence electron than silicon. Semiconductors doped in this manner have a discrete, *filled* donor level that is just below the conduction band. Electrons are promoted readily to the conduction band, and electrons in the conduction band carry the charge. Such a material is a negative charge carrier and is called an ***n*-type semiconductor** (Figure 10.28 right).

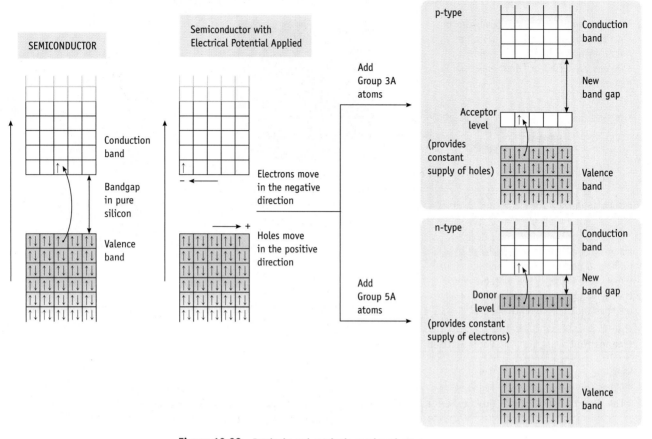

Figure 10.28 Intrinsic and extrinsic semiconductors.

In Summary

When you have finished studying this chapter, you should ask if you have met the chapter goals. In particular, you should be able to

- Describe the main features of valence bond theory and molecular orbital theory, the two commonly used theories for covalent bonding (Section 10.1).

- Recognize that the premise for valence bond theory is that bonding results from the overlap of atomic orbitals. By virtue of the overlap of orbitals, electrons are concentrated (or localized) between two atoms (Section 10.2).

- Distinguish how sigma (σ) and pi (π) bonds arise. For σ bonding, orbitals overlap in a head-to-head fashion, concentrating electrons along the bond axis. Sideways overlap of p atomic orbitals results in π bond formation, with electrons above and below the molecular plane (Section 10.2).

- Use the concept of hybridization to rationalize molecular structure (Section 10.2).

Hybrid Orbitals	Atomic Orbitals Used	Number of Hybrid Orbitals	Electron-Pair Geometry
sp	$s + p$	2	Linear
sp^2	$s + p + p$	3	Trigonal-planar
sp^3	$s + p + p + p$	4	Tetrahedral
sp^3d	$s + p + p + p + d$	5	Trigonal-bipyramidal
sp^3d^2	$s + p + p + p + d + d$	6	Octahedral

- Understand molecular orbital theory (Section 10.3), in which atomic orbitals are combined to form bonding orbitals, nonbonding orbitals, or antibonding orbitals that are delocalized over several atoms. In this description, the electrons of the molecule or ion are assigned to the orbitals beginning with the one at lowest energy, according to the Pauli exclusion principle and Hund's rule.
- Use molecular orbital theory to explain the properties of O_2 and other diatomic molecules (Section 10.3).
- Appreciate band theory and how it applies to solids, especially metals (Section 10.4).
- Understand the difference among an electric conductor, a semiconductor, and an insulator (Section 10.4).
- Recognize how dopants such as Group 3A and Group 5A elements affect semiconductor properties of Group 4A materials (Section 10.4).

Key Terms

Section 10.1
- valence bond (VB) theory
- molecular orbital (MO) theory

Section 10.2
- orbital overlap
- sigma (σ) bond
- orbital hybridization
- hybrid orbitals
- pi (π) bond
- isomers

Section 10.3
- molecular orbital
- bonding molecular orbital
- antibonding molecular orbital
- homonuclear diatomic molecules
- heteronuclear diatomic molecules

Section 10.4
- band theory
- Fermi level
- valence band

- insulator
- conduction band
- band gap
- semiconductor
- intrinsic semiconductor
- extrinsic semiconductor
- dopant
- acceptor level
- *p*-type semiconductor
- *n*-type semiconductor

Key Equations

Equation 10.1 (page 399)

Calculating the order of a bond from the molecular orbital (MO) electron configuration

$$\text{Bond order} = \frac{1}{2}(\text{number of electrons in bonding MOs} - \text{number of electrons in antibonding MOs})$$

Study Questions

Questions with blue, bold-faced numbers have answers in Appendix O and in the Student Solution Manual.
Models of many of the molecules and ions in the following Study Questions are available on the General Chemistry Interactive CD-ROM, *Version 3.0.*

Reviewing Important Concepts

1. What is the difference between a sigma (σ) and a pi (π) bond?

2. What is the maximum number of hybrid orbitals a carbon atom may form? What is the minimum number? Explain briefly.

3. What are the approximate angles between regions of electron density in sp, sp^2, and sp^3 hybrid orbital sets?

4. If an atom is sp-hybridized, how many pure p orbitals remain on the atom? How many π bonds can the atom form?

5. For each of the following electron-pair geometries, tell what hybrid orbital set is used: tetrahedral, linear, trigonal-planar, octahedral, and trigonal-bipyramidal.

6. Consider the three fluorides, BF_4^-, SiF_4, and SF_4.

    (a) Identify a molecule that is isoelectronic with BF_4^-.

    (b) Are SiF_4 and SF_4 isoelectronic?

    (c) What is the hybridization of the central atom in each of these species?

7. Give an example of a molecule with a central atom that has more than four valence electron pairs. Tell what hybrid orbitals are used by the atom.

8. What is the maximum number of hybrid orbitals a third-period element, say sulfur, can form? Explain briefly.

9. What is one important difference between molecular orbital theory and valence bond theory?

10. Describe four principles of molecular orbital theory.

11. Sketch a picture of the bonding and antibonding molecular orbitals of H_2 and describe how they differ.

12. What is the connection between bond order, bond length, and bond energy? Use ethane (C_2H_6), ethylene (C_2H_4), and acetylene (C_2H_2) as examples.

13. When is it desirable to use MO theory rather than valence bond theory?

14. What is meant by the terms "localized" and "delocalized" as they pertain to the two bonding theories?

15. How do valence bond theory and molecular orbital theory explain the bond order of 1.5 for ozone?

16. How do metals and insulators differ in terms of the band gap?

17. What is the difference between an intrinsic and an extrinsic semiconductor?

Practicing Skills

Valence Bond Theory

(See Examples 10.1–10.5 and CD Screens 10.3–10.8)

18. Draw the Lewis structure for NF_3. What are its electron-pair and molecular geometries? What orbitals on N and F overlap to form bonds between these elements?

19. Draw the Lewis structure for the ClF_2^+ ion. What are its electron-pair and molecular geometries? What orbitals on Cl and F overlap to form bonds between these elements?

20. Draw the Lewis structure for chloroform, $CHCl_3$. What are its electron-pair and molecular geometries? What orbitals on C, H, and Cl overlap to form bonds involving these elements?

21. Draw the Lewis structure for AlF_4^-. What are its electron-pair and molecular geometries? What orbitals on Al and F overlap to form bonds between these elements?

22. What hybrid orbital set is used by the underlined atom in each of the following molecules or ions?
    (a) $\underline{B}Br_3$ (b) $\underline{C}O_2$ (c) $\underline{C}H_2Cl_2$ (d) $\underline{C}O_3^{2-}$

23. What hybrid orbital set is used by the underlined atom in each of the following molecules or ions?
    (a) $\underline{C}Se_2$ (b) $\underline{S}O_2$ (c) $\underline{C}H_2O$ (d) $\underline{N}H_4^+$

24. Describe the hybrid orbital set used by each of the indicated atoms in the following molecules:
    (a) The carbon atoms and the oxygen atom in dimethyl ether, $H_3C-O-CH_3$
    (b) Each carbon atom in propene

$$H_3C-\underset{\underset{H}{|}}{C}=CH_2$$

    (c) The two carbon atoms and the nitrogen atom in the amino acid glycine.

$$H-\underset{\underset{H}{|}}{N}-\underset{\underset{H}{|}}{C}-\overset{\overset{:O:}{\|}}{C}-\overset{..}{\underset{..}{O}}-H$$

25. Give the hybrid orbital set used by each of the underlined atoms in the following molecules
    (a)
$$H-\underset{\underset{H}{|}}{N}-\overset{\overset{:O:}{\|}}{C}-\underset{\underset{..}{|}}{N}-H$$
    (b)
$$H_3\underline{C}-\underset{\underset{H}{|}}{C}=\underset{\underset{H}{|}}{C}-\underline{C}=\overset{..}{\underset{..}{O}}$$
    (c)
$$H-\underline{C}=\underset{\underset{H}{|}}{C}-\underline{C}\equiv N:$$

26. Draw the Lewis structure and then specify the electron-pair and molecular geometries for each of the following molecules or ions. Identify the hybridization of the central atom.
    (a) SiF_6^{2-} (b) SeF_4 (c) ICl_2^- (d) XeF_4

27. Draw the Lewis structure and then specify the electron-pair and molecular geometries for each of the following molecules or ions. Identify the hybridization of the central atom.
    (a) $XeOF_4$ (b) BrF_5
    (c) OSF_4 (d) Central Br in Br_3^-

28. Draw Lewis structures of the acid HPO_2F_2 and its anion $PO_2F_2^-$. What is the number of valence electrons in each species? What is the molecular geometry and hybridization for the phosphorus atom in each species? (H is bonded to the O atom in the acid.)

29. Draw Lewis structures of HSO_3F and SO_3F^-. What is the number of valence electrons in each species? What is the molecular geometry and hybridization for the sulfur atom in each species? (H is bonded to the O atom in the acid.)

30. What is the hybridization of the carbon atom in phosgene, Cl_2CO? Give a complete description of the σ and π bonding in this molecule.

31. What is the hybridization of the sulfur atom in sulfuryl fluoride, SO_2F_2?

32. The arrangement of groups attached to the C atoms involved in a $C=C$ double bond leads to *cis* and *trans* isomers. For each of the following compounds, draw the other isomer.

    (a)
$$\underset{H}{\overset{H_3C}{}}C=C\underset{CH_3}{\overset{H}{}}$$
    (b)
$$\underset{H}{\overset{Cl}{}}C=C\underset{H}{\overset{CH_3}{}}$$

33. For each of the following compounds decide if *cis* and *trans* isomers are possible. If isomerism is possible, draw the other isomer.

    (a)
$$\underset{H}{\overset{H_3C}{}}C=C\underset{CH_2CH_3}{\overset{H}{}}$$
    (b)
$$\underset{H}{\overset{H}{}}C=C\underset{H}{\overset{CH_3}{}}$$
    (c)
$$\underset{H}{\overset{Cl}{}}C=C\underset{H}{\overset{CH_2OH}{}}$$

Molecular Orbital Theory

(See Examples 10.7–10.9, CD Screens 10.9–10.13)

34. The hydrogen molecular ion, H_2^+, can be detected spectroscopically. Write the electron configuration of the ion in molecular orbital terms. What is the bond order of the ion? Is the hydrogen–hydrogen bond stronger or weaker in H_2^+ than in H_2?

35. Give the electron configurations for the ions Li_2^+ and Li_2^- in molecular orbital terms. Compare the Li—Li bond order in these ions with the bond order in Li_2.

36. Calcium carbide, CaC_2, contains the acetylide ion, C_2^{2-}. Sketch the molecular orbital energy level diagram for the ion. How many net σ and π bonds does the ion have? What

is the carbon–carbon bond order? How has the bond order changed on adding electrons to C_2 to obtain C_2^{2-}? Is the C_2^{2-} ion paramagnetic?

37. Oxygen, O_2, can acquire one or two electrons to give O_2^- (superoxide ion) or O_2^{2-} (peroxide ion). Write the electron configuration for the ions in molecular orbital terms, and then compare them with the O_2 molecule on the following basis: (a) magnetic character, (b) net number of σ and π bonds, (c) bond order, and (d) oxygen–oxygen bond length.

38. Assume the energy level diagram for homonuclear diatomic molecules (see Figure 10.22) can be applied to heteronuclear diatomics such as CO.
    (a) Write the electron configuration for carbon monoxide, CO.
    (b) What is the highest energy, occupied molecular orbital? (Chemists call this the HOMO.)
    (c) Is the molecule diamagnetic or paramagnetic?
    (d) What is the net number of σ and π bonds? What is the bond order in CO?

39. The nitrosyl ion, NO^+, has an interesting chemistry.
    (a) Is NO^+ diamagnetic or paramagnetic? If paramagnetic, how many unpaired electrons does it have?
    (b) Assume the molecular orbital diagram for a homonuclear diatomic molecule (see Figure 10.22) applies to NO^+. What is the highest energy molecular orbital occupied by electrons?
    (c) What is the nitrogen–oxygen bond order?
    (d) Is the N—O bond in NO^+ stronger or weaker than the bond in NO?

Metals and Semiconductors
(CD Screens 10.14–10.16)

40. Considering only valence atomic orbitals, how many molecular orbitals are formed by 100 Mg atoms? How many of these orbitals are populated by pairs of electrons?

41. A mole of lithium atoms will form a band having $4 \times (6.022 \times 10^{23})$ molecular orbitals (if only valence atomic orbitals are considered). What fraction of these orbitals are filled up to the Fermi level?

General Questions on Valence Bond and Molecular Orbital Theory

These questions are not designated as to type or location in the chapter. They may combine several concepts. More challenging questions are indicated by an underlined number.

42. Describe the O—S—O angle and the hybrid orbital set used by sulfur in each of the following molecules or ions:
    (a) SO_2 (b) SO_3 (c) SO_3^{2-} (d) SO_4^{2-}
    Do all have the same value of the O—S—O angle? Does the S atom in all these species use the same hybrid orbitals?

43. Sketch the Lewis structures of ClF_2^+ and ClF_2^-. What are the electron-pair and molecular geometries of each ion? Do they both have the same F—Cl—F angle? What hybrid orbital set is used by Cl in each ion?

44. Sketch the resonance structures for the nitrite ion, NO_2^-. Give the electron-pair and molecular geometry of the ion. From these geometries, decide on the O—N—O bond angle, the average NO bond order, and the N atom hybridization.

45. Sketch the resonance structures for the nitrate ion, NO_3^-. Is the hybridization of the N atom the same or different in each structure? Describe the orbitals involved in bond formation by the central N atom.

46. Sketch the resonance structures for the N_2O molecule. Is the hybridization of the N atom the same or different in each structure? Describe the orbitals involved in bond formation by the central N atom.

47. Compare the structure and bonding in CO_2 and CO_3^{2-} with regard to the O—C—O bond angles, the CO bond order, and the C atom hybridization.

48. Acrolein, a component of photochemical smog, has a pungent odor and irritates eyes and mucous membranes.

$$\begin{array}{ccccc} & H & H & :O: & \\ & _A| & |_B & ||_C & \\ H\!-\!\!&C&=\!C&\;\;C&\!\!-\!H \\ & _1 & & _2 & \end{array}$$

    (a) What are the hybridizations of carbon atoms 1 and 2?
    (b) What are the approximate values of angles A, B, and C?
    (c) Is *cis–trans* isomerism possible here?

49. The following organic compound is a member of a class known as oximes:

$$\begin{array}{ccc} H & & \ddot{\,}\!\!:\!\ddot{O}\!-\!H \\ | & & | \\ H\!-\!C&-\!C&=\!N: \\ | & | & \\ H & H & \end{array}$$

    (a) What are the hybridizations of the two C atoms and of the N atom?
    (b) What is the approximate C—N—O angle?

50. Numerous molecules are detected in deep space (page 322). Three of them are illustrated here.

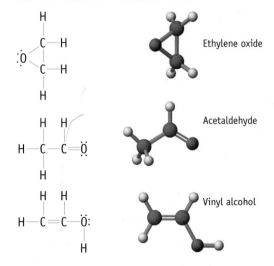

Ethylene oxide

Acetaldehyde

Vinyl alcohol

    (a) Comment on the similarities or differences in the formulas of these compounds. Are they isomers?

(b) Indicate the hybridization of each C atom in each molecule.

(c) Indicate the value of the H—C—H angles in the three molecules.

(d) Are any of these molecules polar?

(e) Which molecule should have the strongest carbon–carbon bond? The strongest carbon–oxygen bond?

51. The compound sketched here is acetylsalicylic acid, better known by its common name, aspirin:

(a) What are the approximate values of the angles marked *A*, *B*, *C*, and *D*?

(b) What hybrid orbitals are used by carbon atoms 1, 2, and 3?

52. Lactic acid is a natural compound found in sour milk.

(a) How many π bonds occur in lactic acid? How many σ bonds?

(b) Identify the hybridization of each atom 1 through 3.

(c) Which CO bond is the shortest in the molecule? Which CO bond is the strongest?

(d) What are the approximate values of the bond angles *A* through *C*?

53. Phosphoserine is a less common amino acid.

(a) Describe the hybridizations of atoms 1 through 5.

(b) What are the approximate values of the bond angles *A* through *D*?

(c) What are the most polar bonds in the molecule?

54. Boron trifluoride, BF_3, can accept a pair of electrons from another molecule to form a coordinate covalent bond, as in the following reaction with ammonia:

(a) What is the geometry about the boron atom in BF_3? In $H_3N \rightarrow BF_3$?

(b) What is the hybridization of the boron atom in the two compounds?

(c) Does the boron atom hybridization change on formation of the coordinate covalent bond?

55. The sulfamate ion, $H_2N \rightarrow SO_3^-$, can be thought of as having been formed from the amide ion, NH_2^-, and sulfur trioxide, SO_3.

(a) Sketch a structure for the sulfamate ion. Include estimates of bond angles.

(b) What changes in hybridization do you expect for N and S in the course of the reaction $NH_2^- + SO_3 \longrightarrow H_2N \rightarrow SO_3^-$?

56. Antimony pentafluoride reacts with HF according to the equation

$$2\ HF + SbF_5 \longrightarrow [H_2F]^+[SbF_6]^-$$

(a) What is the hybridization of the Sb atom in the reactant and product?

(b) Draw a Lewis structure for H_2F^+. What is the geometry of H_2F^+? What is the hybridization of F in H_2F^+?

57. Iodine and oxygen form a complex series of ions, among them IO_4^- and IO_5^{3-}. Draw Lewis structures for these ions and specify their electron-pair geometries and the shapes of the ions. What is the hybridization of the I atom in these ions?

58. Cinnamaldehyde occurs naturally in cinnamon oil.

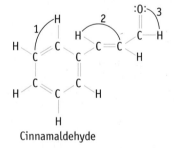

Cinnamaldehyde

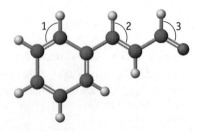

(a) What is the most polar bond in the molecule?

(b) How many σ and how many π bonds are there?

(c) Is *cis–trans* isomerism possible? If so, draw the isomers of the molecule.

(d) Give the hybridization of the C atoms in the molecule.

(e) What are the values of the indicated bond angles 1, 2, and 3?

59. Xenon forms well-characterized compounds. Two xenon–oxygen compounds are XeO_3 and XeO_4. Draw Lewis structures of each of these compounds and give their electron-pair and molecular geometries. What are the hybrid orbital sets used by xenon in these two oxides?

60. The simple valence bond picture of O_2 does not agree with the molecular orbital view. Compare these two theories with regard to the peroxide ion, O_2^{2-}.

(a) Draw an electron dot structure for O_2^{2-}. What is the bond order of the ion?

(b) Write the molecular orbital electron configuration for O_2^{2-}. What is the bond order based on this approach?

(c) Do the two theories of bonding lead to the same magnetic character and bond order for O_2^{2-}?

61. Nitrogen, N_2, can ionize to form N_2^+ or add an electron to give N_2^-. Using molecular orbital theory, compare these species with regard to (a) their magnetic character, (b) net number of π bonds, (c) bond order, (d) bond length, and (e) bond strength.

62. Which of the homonuclear, diatomic molecules of the second-period elements (from Li_2 to Ne_2) are paramagnetic? Which ones have a bond order of 1? Which ones have a bond order of 2? What diatomic molecule has the highest bond order?

63. Which of the following molecules or ions should be paramagnetic? What is the HOMO in each one? Assume the molecular orbital diagram in Figure 10.22 applies to all of them.

(a) NO   (c) OF^- (e) O_2^{2-}

(b) Ne_2^+ (d) CN

64. The CN molecule has been found in interstellar space. Assuming the electronic structure of the molecule can be described using the molecular orbital energy level diagram in Figure 10.22, answer the following questions.

(a) What is the HOMO to which an electron or electrons is (are) assigned?

(b) What is the bond order of the molecule?

(c) How many net σ bonds are there? How many net π bonds?

(d) Is the molecule paramagnetic or diamagnetic?

65. Amphetamine is a stimulant. Replacing one H atom on the NH_2, or amino, group with CH_3 gives methamphetamine, a particularly dangerous drug commonly known as "speed."

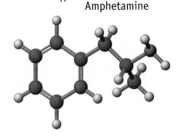

Amphetamine

(a) What are the hybrid orbitals used by the C atoms of the C_6 ring, by the C atoms of the side chain, and by the N atom?

(b) Give approximate values for the bond angles A, B, and C.

(c) How many σ bonds and π bonds are in the molecule?

(d) Is the molecule polar or nonpolar?

(e) Amphetamine reacts readily with a proton (H^+) in aqueous solution. Where does this proton attach to the molecule?

66. Menthol is used in soaps, perfumes, and foods. It is present in the common herb mint, and it can be prepared from turpentine.

(a) What are the hybridizations used by the C atoms in the molecule?

(b) What is the approximate C—O—H bond angle?

(c) Is the molecule polar or nonpolar?

(d) Is the six-member carbon ring planar or nonplanar? Explain why or why not.

Menthol

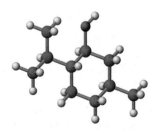

67. The elements of the second period from boron to oxygen form compounds of the type X_nE—EX_n, where X can be H

or a halogen. Sketch possible molecular structures for B_2F_4, C_2H_4, N_2H_4, and O_2H_2. Give the hybridizations of E in each molecule and specify approximate X—E—E bond angles.

68. What is the primary difference in the band structures of conductors and insulators?

69. Solid silicon has an electronic structure much like that of carbon in the diamond form, with a filled valence band and an empty conduction band. However, silicon is a semiconductor and diamond is an insulator. Explain the difference. (See *CD Screen 10.16*)

70. Solid germanium has a filled valence band and an empty conduction band, but the band gap is much smaller than in diamond or silicon. How do you expect germanium to act as a conductor in relation to diamond and to a metal such as lithium? (See *CD Screen 10.16*)

71. Why is band theory not used (or not necessary) to explain the bonding in N_2?

72. When two amino acids react with each other, they form a linkage called an amide group, or a peptide link. (If more linkages are added, a protein, or polypeptide is formed.)

(a) What are the hybridizations of the C and N atoms in the peptide linkage?

(b) Is the structure illustrated the only resonance structure possible for the peptide linkage? If another resonance structure is possible, compare it with the one shown. Decide which is the more important structure.

(c) The computer-generated structure shown here, which contains a peptide linkage, shows that the linkage is flat. This is an important feature of proteins. Speculate on why the CO—NH linkage is planar.

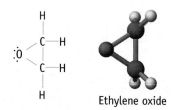

73. The compound whose structure is shown here is acetylacetone. It exists in two forms: the enol form and the keto form.

enol form   *keto* form

The molecule reacts with OH^- to form an anion $[CH_3COCHCOCH_3]^-$ (often abbreviated acac$^-$ for acetylacetonate ion). One of the most interesting aspects of this anion is that one or more of them can react with a transition metal cation to give very stable, highly colored compounds.

(a) Are the keto and enol forms of acetylacetone resonance forms? Explain your answer.

(b) What is the hybridization of each atom (except H) in the enol form? What changes in hybridization occur when this is transformed into the keto form?

(c) What is the electron-pair geometry and molecular geometry around each C atom in the keto and enol forms? What changes in geometry occur when the keto form changes to the enol form?

(d) Draw three possible resonance structures for the acac$^-$ ion.

(e) Is *cis–trans* isomerism possible in either the enol or the keto form?

74. Ethylene oxide has a three-member ring of two C atoms and an O atom.

Ethylene oxide

(a) What are the expected bond angles in the ring?

(b) What is the hybridization of each atom in the ring?

(c) Comment on the relation between the bond angles expected based on hybridization and the bond angles expected for a three-member ring.

Using Electronic Resources

These questions refer to the General Chemistry Interactive CD-ROM, *Version 3.0.*

75. See CD-ROM Screen 10.3: Valence Bond Theory.

(a) This screen describes attractive and repulsive forces that occur when two atoms approach each other. What must be true about the relative strengths of those attractive and repulsive forces if a covalent bond is to form? (The following figure is from Screen 10.3 and shows the change in potential energy as a function of the H—H distance.)

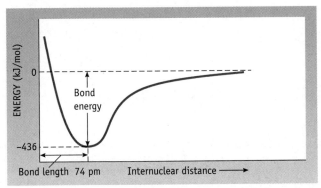

(b) When two atoms are widely separated, the energy of the system is defined as zero. As the atoms approach each other, the energy drops, reaches a minimum, and then increases as they approach still more closely. Explain these observations.

(c) It is stated that for a bond to form, orbitals on adjacent atoms must overlap, and each pair of overlapping orbitals will contain two electrons. Explain why neon does not form a diatomic molecule, Ne_2, whereas fluorine forms F_2.

76. See CD-ROM Screen 10.6: Determining Hybrid Orbitals. Examine the *Hybrid Orbitals* tool on this screen. Use this tool to systematically combine atomic orbitals to form hybrid atomic orbitals.

(a) What is the relationship between the number of hybrid orbitals produced and the number of atomic orbitals used to create them?

(b) Do hybrid atomic orbitals form between different p orbitals without involving s orbitals?

(c) What is the relationship between the energy of hybrid atomic orbitals and the atomic orbitals from which they are formed?

(d) Compare the shapes of the hybrid orbitals formed from an s orbital and a p_x orbital with the hybrid atomic orbitals formed from an s orbital and a p_z orbital.

(e) Compare the shape of the hybrid orbitals formed from s, p_x, and p_y orbitals with the hybrid atomic orbitals formed from s, p_x, and p_z orbitals.

77. See CD-ROM Screen 10.7: Multiple Bonding. After examining the bonding in ethylene on the main screen go to the "A Closer Look" auxiliary screen.

(a) Explain why the allene molecule is not flat. That is, explain why the CH_2 groups at opposite ends do not lie in the same plane.

(b) Based on the theory of orbital hybridization, explain why benzene is a planar, symmetrical molecule.

(c) What are the hybrid orbitals used by the three C atoms of allyl alcohol?

$$\overset{\text{H}}{\underset{\text{H}}{\overset{|}{\underset{|}{\text{C}}}}} = \overset{\text{H}}{\underset{\text{H}}{\overset{|}{\underset{|}{\text{C}}}}} - \overset{\text{H}}{\underset{}{\overset{|}{\text{C}}}} - \overset{..}{\underset{..}{\text{O}}} - \text{H}$$

78. See CD-ROM Screen 10.8: Molecular Fluxionality.

(a) Observe the animations of the rotations of *trans*-2-butene and butane about their carbon–carbon bonds.

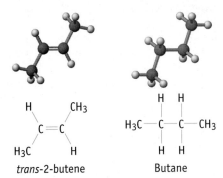

As one end of *trans*-2-butene rotates relative to the other, the energy increases greatly (from 27 kJ/mol to 233 kJ/mol) and then drops to 30 kJ/mol when the rotation has produced *cis*-2-butene. In contrast, the rotation of the butane molecule requires much less energy for rotation (only about 60 kJ/mol). When butane has reached the halfway point in its rotation, the energy has reached a maximum. Why does *trans*-2-butene require so much more energy to rotate about the central carbon–carbon bond than does butane?

(b) The structure of propene, C_3H_6, is pictured here. Which carbon–hydrogen group (CH_3 or CH_2) can rotate freely with respect to the rest of the molecule?

$$\text{H}_3\text{C} - \overset{\text{H}}{\overset{|}{\text{C}}} = \text{CH}_2$$

(c) Can the two CH_2 fragments of allene (see *Screen 10.7SB*) rotate with respect to each other? Briefly explain why or why not.

79. See CD-ROM Screen 10.14: Metallic Bonding: Band Theory. The example shown in the animation is that of solid lithium, which forms from the overlap of 2s orbitals on each of many Li atoms. In this case, the band would be half full of electrons (the upper half of the band would contain vacant orbitals). Considering the electron configuration of lithium, explain this.

80. See CD-ROM Screen 10.16: Semiconductors.

(a) Describe how semiconductors differ from conductors and from insulators.

(b) The animation on this screen shows how atoms with unfilled orbitals can be inserted into pure silicon to form synthetic semiconductors. In this case, vacant orbitals now exist to which electrons in the valence band can jump. What types of atoms can be used for this process, those from periodic Group 3A or those from Group 5A?

11 Carbon: More Than Just Another Element

Chapter Goals

- Classify organic compounds based on formula and structure.
- Name and draw structures of common organic compounds.
- Relate properties to molecular structure.
- Identify common polymers and compounds of biological significance.

▲ William Henry Perkin (1838–1907). See the book on Perkin's life, *Mauve*, by Simon Garfield (W. W. Norton, New York, 2001). *(Science & Society Picture Library/Science Museum/London)*

A Colorful Beginning

Among the roots of modern organic chemistry was the discovery, in 1856, of the compound mauveine (or mauve) by William Henry Perkin (1838–1907). This discovery dates from before the first periodic table was created; before the discovery of electrons, protons, and neutrons; before chemists knew anything about bonding; and even before the tetrahedral geometry of carbon was recognized. It led to a flourishing dye industry in the latter part of the 19th century, one of the first chemical industries to gain major importance. By 1900, more than a thousand synthetic dyes were known and in use.

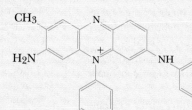

Mauveine cation.

Before the discovery of mauve, almost all dyes were from natural sources. Because the dye used for the color purple, Tyrian purple, was the rarest and most expensive, it became the exclusive color of royalty. Tyrian purple was the origin of both fame and fortune for the ancient empire of Tyre because the dye was obtained only from a small mollusk found in the Mediterranean Sea in that region. Over 9000 mollusks were needed to obtain a single gram of dye!

◄ Original, stoppered bottle of mauveine prepared by Perkin. *(Science & Society Picture Library/Science Museum/London)*

The discovery of mauve by Perkin is an interesting tale of serendipity. At the age of 13, Perkin enrolled at the City of London School. His father paid an extra fee for him to attend a lunchtime chemistry course and set up a lab at home for him to do experiments. Hooked on chemistry, Perkin attended the public lectures that Michael Faraday gave on Saturdays at the Royal Institution. At 15, Perkin enrolled in the Royal College of Science in London to study chemistry under the famous chemist August Wilhelm von Hofmann. Perkin completed his studies at age 17 (the field of chemistry being a lot smaller than it is today) and took a position at the college as Hofmann's assistant, rather a great honor.

Perkin's first chemistry project was to synthesize quinine ($C_{20}H_{24}N_2O_2$), an antimalarial drug. The route he pro-

The color purple was once associated with royalty because the dyes were rare and expensive. William Henry Perkin changed everything.

◀ This silk dress was dyed with Perkin's original sample of mauve in 1862, at the dawning of the synthetic dye industry. *(Science & Society Picture Library/Science Museum/London)*

posed involved oxidizing anilinium sulfate [$(C_6H_5NH_3)_2SO_4$]. From the reaction, he obtained a black solid that dissolved in a water–ethanol mixture to give a purple solution.

Using a cloth to mop up a spill on the lab bench, he noticed that the substance stained the cloth a beautiful purple color. Furthermore, the color didn't wash out, an essential feature of a useful dye. Later it was learned that the anilinium sulfate used in the original reaction was impure and that the impurity was essential in the synthesis. Had Perkin used a pure sample as his starting reagent, the discovery of mauve would not have happened, at least not in this way.

A study in 1994 on samples of mauve preserved in museums determined that Perkin's mauve was a mixture of primarily two compounds, which have closely related structures, along with traces of several others.

At the age of 18, Perkin quit his assistantship to exploit this new discovery. It was not an easy decision because it incurred the great displeasure of his mentor, Professor Hofmann. With financial help from his family Perkin set up a factory outside of London. Although the road to success was not smooth, Perkin persevered and by the age of 36 he was a very wealthy man. He then retired from the dye business and devoted the rest of his life to chemical research on various topics including the synthesis of fragrances. He also studied optical activity, the ability of certain compounds to rotate polarized light. During his life he received numerous honors for his research. One honor, however, came many years after his death. In 1972, when the Chemical Society of London renamed its research journals after famous society members, it chose Perkin's name for the journals in which organic chemists publish their research.

◀ Many flowers have colors best described as mauve. At the left is a hydrangea and at the right a day lily. *(Left, Terry Donnelly/Dembinsky Photo Associates; right, Michael P. Gadomski/Photo Researchers, Inc.)*

Before You Begin

- Review writing Lewis structures and predicting molecular structures (Section 9.4).
- Recall how to draw structures of molecules (Section 9.9).
- Review covalent bonding: valence bond and molecular orbital theory (Chapter 10).

Organic Chemistry Many things in our world have pleasant odors, whereas others are not as good — we say they stink! Most of these odors are from carbon-based compounds that chemists classify as organic, as opposed to compounds such as H_2SO_4 that are inorganic. Organic compounds can be subdivided by the type of functional groups incorporated in the molecule. As you read this chapter see what functional groups you can identify in the compounds responsible for odors, some good and some bad. [See D.R. Kimbrough, *Chem Matters*, December 2001, pages 8–11]

Carvone

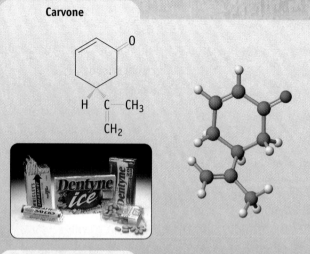

Isomers of Limonene

Oranges   Lemons

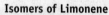

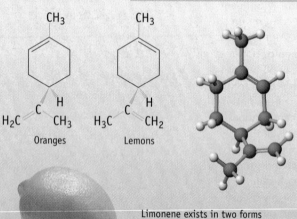

Limonene exists in two forms that differ very little but which have clearly different odors.

Butanedione

$$H_3C - \overset{O}{\underset{}{C}} - \overset{O}{\underset{}{C}} - CH_3$$

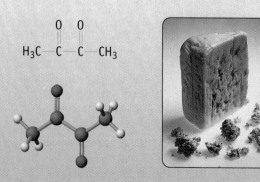

Cinnamaldehyde

$$\overset{O}{CH = CH - CH}$$

Allyl propyl disulfide

$$H_2C = CHCH_2 - S - S - CH_2CH_2CH_3$$

Trimethylamine

$$H_3C - \overset{N}{\underset{CH_3}{\,}} - CH_3$$

3-Methylbutanethiol

$$H_3C - \overset{CH_3}{\underset{}{CH}} - CH_2 - CH_2 - SH$$

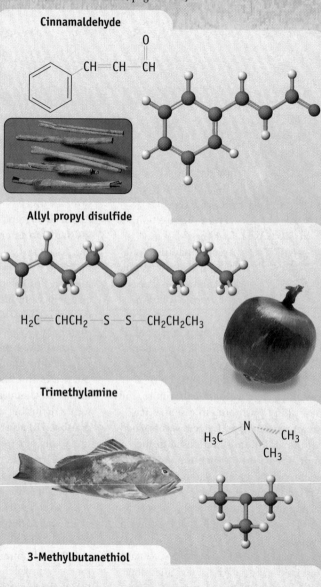

Photo/skunk: W. Perry Conway/CORBIS
Other photos: Charles D. Winters

The vast majority of the 20 million chemical compounds currently known are organic, that is, they are compounds built on a carbon framework. Organic compounds vary greatly in size and complexity, from the simplest hydrocarbon, methane, to molecules made up of many thousands of atoms. A glance at the Chapter Focus shows that the range of possible materials is huge.

11.1   WHY CARBON?

We begin this discussion with a question: What features of carbon lead to both the abundance and complexity of organic compounds? Answers fall into two categories: structural diversity and stability.

Structural Diversity

With four electrons in its outer shell, carbon will form four bonds to reach an octet configuration. In contrast, the elements boron and nitrogen form three bonds in molecular compounds, oxygen two bonds, and hydrogen and the halogens one. With a larger number of bonds comes the opportunity to create more complex structures. This will become increasingly evident as we evaluate the types of organic compounds.

A carbon atom can reach an octet of electrons in various ways (Figure 11.1). These include:

- *Forming four single bonds.* A carbon atom can bond to four other atoms, and these can be either atoms of other elements (often H, N, O) or other carbon atoms.
- *Forming a double bond and two single bonds.* The carbon atoms in ethylene, $H_2C{=}CH_2$, are linked in this way.
- *Forming two double bonds* as in the compound CO_2.

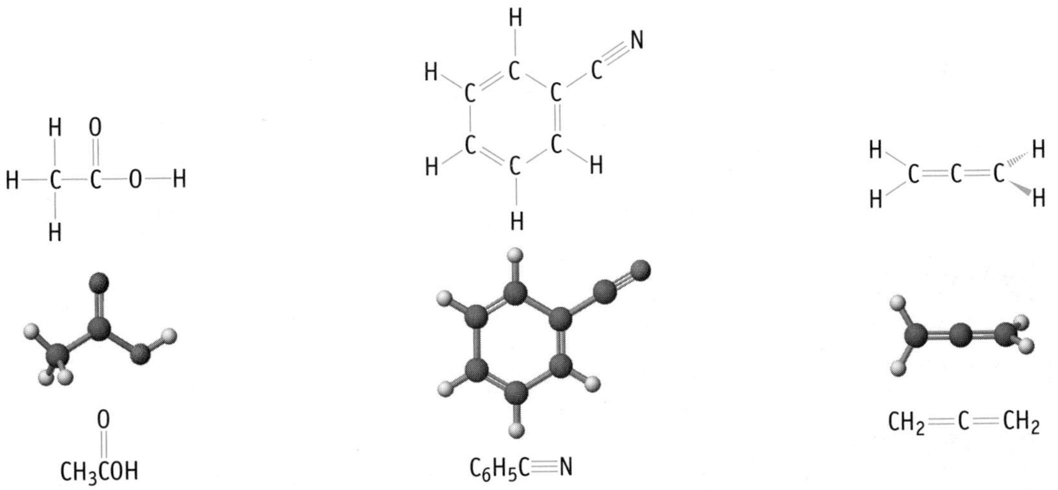

● **The Special Nature of Carbon**
More than any other element, carbon atoms bond to other carbon atoms. You have already seen this in the allotropic forms of carbon: diamond, graphite, buckyballs (page 78), and nanotubes (page 26). There seems to be no limit to the number of carbon atoms that can be assembled into molecules.

(a) Acetic acid. One carbon atom in this compound is attached to 4 other atoms by single bonds and has tetrahedral geometry. The second carbon atom, connected by a double bond to one oxygen, and by single bonds to the other oxygen and to carbon, has trigonal planar geometry.

(b) Benzonitrile. Six trigonal planar carbon atoms make up the benzene ring. The seventh C atom, bonded by a single bond to carbon and a triple bond to nitrogen, has a linear geometry.

(c) Carbon is linked by double bonds to two other carbon atoms in C_3H_4, a linear molecule commonly called allene.

Figure 11.1   **Ways that carbon atoms bond.**

A Closer Look

Writing Formulas and Drawing Structures

You learned in Chapter 3 that there are various ways of presenting structures (page 84), and it is appropriate to return to that point as we look at organic compounds. Consider methane and ethane, for example. We can represent these molecules in many ways:

1. *Molecular formula:* CH_4 or C_2H_6. This type of formula gives information only on molecular composition.

2. *Condensed formula:* For ethane this would be written CH_3CH_3. This method of writing the formula gives some information on the way atoms are connected.

3. *Structural formula:* You will recognize this as the Lewis structure. An elaboration on the condensed formula in (2), this representation defines more clearly

how each atom is connected, but it fails to describe the shapes of the molecules.

Methane, CH_4 Ethane, C_2H_6

4. *Perspective drawings:* These are used to convey three-dimensional character when drawing structures. Bonds extending out of the plane of the paper are drawn with solid wedges, and bonds behind the plane of the paper are represented as dashed wedges (page 84). Using these guidelines, the struc-

tures of methane and ethane could be drawn as follows:

5. *Computer-drawn ball-and-stick and space-filling models.*

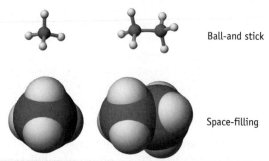

Ball-and stick

Space-filling

• *Forming a triple bond and a single bond,* an arrangement seen in acetylene, $HC \equiv CH$.

Recognize, with each of these arrangements, the various possible geometries around carbon: tetrahedral, trigonal-planar, and linear. Carbon's tetrahedral geometry is of special significance because it leads to three-dimensional chains and rings of carbon atoms as in propane and cyclopentane. The ability to form multiple bonds leads to whole families of compounds based on structures such as ethylene, acetylene, and benzene:

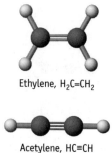

Ethylene, $H_2C=CH_2$

Acetylene, $HC \equiv CH$

propane, C_3H_8 cyclopentane, C_5H_{10} benzene, C_6H_6

Isomers

A hallmark of carbon chemistry is the remarkable array of isomers that can exist. **Isomers** are compounds that have identical composition but different structures. There are two broad categories of isomers: structural isomers and stereoisomers.

Structural isomers are compounds with the same elemental composition but the atoms are linked together in different ways. Ethanol and dimethyl ether (page 82) are structural isomers as are 1-butene and 2-methylpropene.

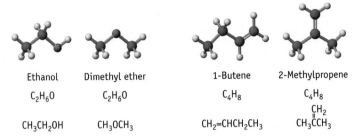

Ethanol	Dimethyl ether	1-Butene	2-Methylpropene
C_2H_6O	C_2H_6O	C_4H_8	C_4H_8
CH_3CH_2OH	CH_3OCH_3	$CH_2=CHCH_2CH_3$	$\overset{CH_2}{\underset{\parallel}{CH_3CCH_3}}$

Stereoisomers are compounds with the same formula and in which there is a similar attachment of atoms. However, the atoms have different orientations in space. There are two types of stereoisomers: geometric isomers and optical isomers.

Cis- and *trans-*2-butene are **geometric isomers.** Geometric isomerism in these compounds occurs as a result of the $C=C$ double bond. Recall that the carbon atom and its attached groups cannot rotate around a double bond (page 395). Thus, the geometry around the $C=C$ double bond is fixed in space. If two groups occur on adjacent carbon atoms and are on the same side of the double bond, a *cis* isomer is produced. Groups on opposite sides of a double bond produce a *trans* isomer.

Cis-2-butene, C_4H_8 *Trans*-2-butene, C_4H_8

Optical isomerism is a second type of stereoisomerism. Optical isomers are molecules that are nonsuperimposable mirror images (Figure 11.2). Molecules (and other objects) that have nonsuperimposable mirror images are termed **chiral.** Pairs of nonsuperimposable molecules are **enantiomers.**

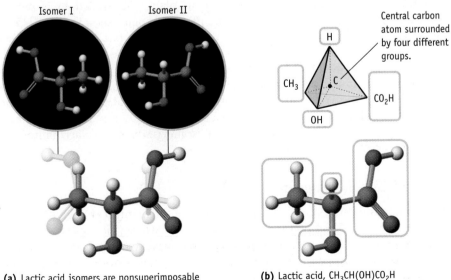

Isomer I   Isomer II

Central carbon atom surrounded by four different groups.

(a) Lactic acid isomers are nonsuperimposable

(b) Lactic acid, $CH_3CH(OH)CO_2H$

Figure 11.2   Optical isomers.
(a) Optical isomerism occurs if a molecule and its mirror image cannot be superimposed. The situation is seen if four different groups are attached to carbon.
(b) Lactic acid, a chiral molecule. Four different groups (H, OH, CH_3 and CO_2H) are attached to the central carbon atom.

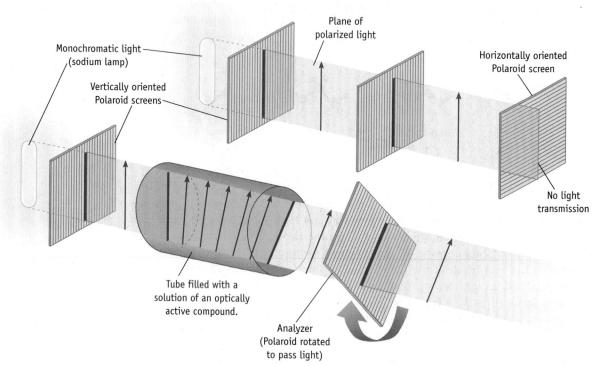

Figure 11.3   Rotation of plane-polarized light by an optical isomer. (*Top*) Monochromatic light (light of only one wavelength) is produced by a sodium lamp. After it passes through a polarizing filter, the light is vibrating in only one direction — it is polarized. Polarized light will pass through a second polarizing filter if this filter is parallel to the first filter, but not if the second filter is perpendicular. (*Bottom*) A solution of an optical isomer placed between the first and second polarizing filters causes rotation of the plane of polarized light. The angle of rotation can be determined by rotating the second filter until maximum light transmission occurs. The magnitude and direction of rotation are unique physical properties of the optical isomer being tested.

Pure samples of enantiomers have the same physical properties, such as melting point, boiling point, density, and solubility in common solvents. They differ in one significant way, however: When a beam of plane-polarized light is passed through a solution of a pure enantiomer, the plane of polarization is rotated. The two enantiomers rotate polarized light to an equal extent, but in opposite directions (Figure 11.3). The term "optical isomerism" is used because this effect involves light (see *A Closer Look: Optical Isomers*).

The most common examples of chiral compounds are those in which four different atoms (or groups of atoms) are attached to a tetrahedral carbon atom. Lactic acid, found in milk and a product of normal human metabolism, is an example of a chiral compound (see Figure 11.2). Optical isomerism is particularly important in the amino acids and other biological molecules.

Stability of Carbon Compounds

Carbon compounds are notable for their resistance to chemical change. Were this not so, the number of compounds of carbon would certainly be far less.

Strong bonds are needed for molecules to survive in their environment. Molecular collisions in gases, liquids and solutions often provide enough energy

● **Lactic Acid, a Chiral Compound**
Lactic acid is produced from milk when milk is fermented to make cheese. It is also found in other sour foods such as sauerkraut and is a preservative in pickled foods such as onions and olives. In our bodies it is produced by muscle activity and normal metabolism.

Lactic acid, $CH_3CH(OH)CO_2H$

A Closer Look

Optical Isomers and Chirality

Everyone has accidentally put a left shoe on a right foot, or a left-handed glove on a right hand. It doesn't work very well. Even though our two hands and two feet appear generally similar, there is a very important distinction between them. Left hands and feet are mirror images of right hands and feet, and, most importantly, these mirror images cannot be superimposed. We describe them as chiral.

Many common objects have this property. Some seashells are chiral. Wood screws and machine bolts are also chiral, distinguished by left-handed or right-handed threads.

Certain molecules have the same characteristic as gloves and hands: A given structure and its mirror image — its enan-

tiomer — cannot be superimposed. There are various ways to visualize that two enantiomers are different. Imagine a tetrahedral carbon atom attached to four other atoms or groups, all different. For simplicity the atoms bonded to carbon are shown as 1, 2, 3 and 4 in the drawing. Sight down one of the bonds to carbon (say the bond from atom 1 to C) in one enantiomer, the other three atoms (2, 3, and 4) will appear in a clock-

wise order. In the second enantiomer, 2, 3, and 4 will appear in counterclockwise order. This is illustrated here for the amino acid alanine.

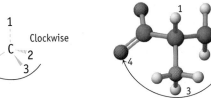

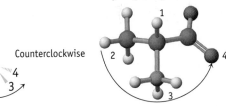

Clockwise arrangement of CH_3, NH_3^+, CO_2^-

Counterclockwise arrangement of CH_3, NH_3^+, CO_2^-

Right- and left-handed seashells are mirror images.

◀ The helical chain of DNA is like the threads of a screw. It twists to the left or it twists to the right. Here it twists to the right. If you curl your right hand around the chain, with your thumb extended, your fingers will show the direction of the twist and your thumb will point along the chain. *(Charles D. Winters)*

to break some chemical bonds, and bonds can be broken if the energy associated with photons of visible and ultraviolet light exceeds the bond energy. Carbon–carbon bonds are relatively strong, however, and so are bonds between carbon and most other atoms. The average C — C bond energy is 346 kJ/mol, the C — H bond energy is 413 kJ/mol, and carbon–carbon double and triple bond energies are even higher (◀ SECTION 9.0). Contrast these values with bond energies for the Si — H bond (318 kJ/mol) and the Si — Si bond (222 kJ/mol). The consequence of high bond energies for bonds to carbon is that, for the most part, organic compounds do not degrade under normal conditions.

Oxidation of most organic compounds is strongly product-favored, but most organic compounds survive contact with O_2. The reason is that these reactions are slow. Most organic compounds burn only if their combustion is initiated by heat or by a spark. The consequence is that oxidative degradation is also not a barrier to the existence of organic compounds.

11.2   HYDROCARBONS

Hydrocarbons, compounds made up of carbon and hydrogen only, are classified into several subgroups: alkanes, cycloalkanes, alkenes, alkynes, and aromatic compounds (Table 11.1) [CD-ROM, Screen 11.3]. We begin with compounds that have carbon atoms with four single bonds: the alkanes and cycloalkanes.

Alkanes

Alkanes have the general formula C_nH_{2n+2}, with n having integer values (Table 11.2). Formulas of specific compounds can be generated from the general formula, and the first four are CH_4 (methane), C_2H_6 (ethane), C_3H_8 (propane), and C_4H_{10} (butane) (Figure 11.4). Methane has four hydrogen atoms arranged tetrahedrally around a single carbon atom. Replacing a hydrogen atom in methane by a $—CH_3$ group gives ethane. If an H atom of ethane is replaced by yet another $—CH_3$ group, propane results. Butane is derived from propane by replacing an H atom of one of the chain-ending carbon atoms with a $—CH_3$ group.

Each C atom is attached to four other atoms, either C or H, and so alkanes are often called **saturated compounds.**

$$CH_3CH_2CH_2CH_3 \qquad CH_3\overset{\displaystyle CH_3}{\underset{|}{C}}HCH_3$$

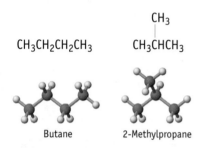

Butane   2-Methylpropane

Structural isomers of butane, C_4H_{10}.

Structural Isomers

The formulas for alkanes do not hint at the structural diversity of the alkanes. With all the alkanes after propane structural isomers are possible. These compounds have the same formula, but the atoms are connected in a different order. There are, for example, two structural isomers for C_4H_{10}, and three for C_5H_{12}. As the number of carbon atoms in an alkane increases, the number of possible structural isomers rapidly increases; there are 5 isomers possible for C_6H_{14}, 9 isomers for C_7H_{16}, 18 for C_8H_{18}, 75 for $C_{10}H_{22}$, and 366,319 structural isomers for $C_{20}H_{42}$.

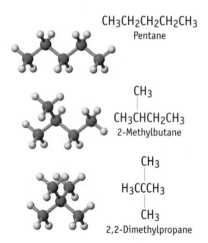

$$CH_3CH_2CH_2CH_2CH_3$$
Pentane

$$CH_3\overset{\displaystyle CH_3}{\underset{|}{C}}HCH_2CH_3$$
2-Methylbutane

$$H_3C\overset{\displaystyle CH_3}{\underset{|}{\underset{|}{C}}}\underset{\displaystyle CH_3}{CH_3}$$
2,2-Dimethylpropane

Structural isomers of pentane, C_5H_{12}.

Table 11.1 • Some Types of Hydrocarbons

Type of Hydrocarbon	Characteristic Feature	General Formula	Example
alkanes	C — C single bonds All C atoms surrounded by four single bonds	C_nH_{2n+2}	CH_4, methane C_2H_6, ethane
cycloalkanes	C — C single bonds All C atoms surrounded by four single bonds	C_nH_{2n}	C_6H_{12}, cyclohexane
alkenes	C = C double bond	C_nH_{2n}	$H_2C = CH_2$, ethylene
alkynes	C ≡ C triple bond	C_nH_{2n-2}	$HC ≡ CH$, acetylene
aromatics	Rings of C atoms with π bonding extending over several C atoms	—	C_6H_6, benzene

Table 11.2 • Selected Hydrocarbons of the Alkane Family, C_nH_{2n+2}*

Name	Molecular Formula	State at Room Temperature
methane	CH_4	
ethane	C_2H_6	gas
propane	C_3H_8	
butane	C_4H_{10}	
pentane	C_5H_{12} (pent- = 5)	
hexane	C_6H_{14} (hex- = 6)	
heptane	C_7H_{16} (hept- = 7)	liquid
octane	C_8H_{18} (oct- = 8)	
nonane	C_9H_{20} (non- = 9)	
decane	$C_{10}H_{22}$ (dec- = 10)	
octadecane	$C_{18}H_{38}$ (octadec- = 18)	solid
eicosane	$C_{20}H_{42}$ (eicos- = 20)	

*This table lists only selected alkanes. Liquid compounds with 11–16 carbon atoms are also known, and there are many solid alkanes with more than 18 C atoms.

Chapter Goals • Revisited

- Classify organic compounds based on formula and structure.
- Name and draw structures of common organic compounds.
- Relate properties to molecular structure.
- Identify common polymers and compounds of biological significance.

To recognize the isomers corresponding to a given formula, bear in mind the following points:

- Each alkane is built up with a framework of tetrahedral carbon atoms, and each carbon must have four single bonds.
- An effective approach is to create a framework of carbon atoms and then fill the remaining positions around carbon with H atoms so that each C atom has four bonds.
- Free rotation occurs around carbon–carbon single bonds. Therefore, when atoms are assembled to form the skeleton of an alkane, the emphasis is on how carbon atoms are attached to one another and not on how they might lie relative to each other in the plane of the paper.

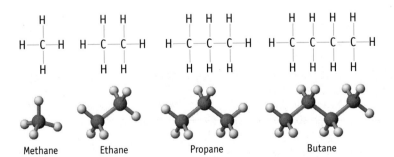

Methane   Ethane   Propane   Butane

Figure 11.4 Alkanes. The lowest molecular weight alkanes, all gases under normal conditions, are methane, ethane, propane, and butane.

Example 11.1 Drawing Structural Isomers of Alkanes

Problem • Draw structures of the five isomers of C_6H_{14}.

Strategy • Focus first on the different frameworks that can be built from six carbon atoms. Having created a carbon framework, fill hydrogen atoms into the structure so that each carbon has four bonds.

Solution •

Step 1. Placing six carbon atoms in a chain gives the framework for the first isomer. Now fill in hydrogen atoms: three on the carbons on the ends of the chain, two on each of the carbons in the middle. You have created the first isomer, hexane.

carbon framework of hexane   hexane

Step 2. Next draw a chain of five carbon atoms, then add the sixth carbon atom to one of the carbons in the middle of the chain. (Adding it to a carbon at the end of the chain gives a six-carbon chain, the same carbon framework drawn in step 1.) Two different carbon frameworks can be built from the five-carbon chain, depending on whether the sixth carbon is linked to the 2 or 3 position. For each of these frameworks, fill in the hydrogens.

carbon framework
of methylpentane isomers   2-methylpentane

3-methylpentane

Step 3. Draw a chain of four carbon atoms. Add in the two remaining carbons, again being careful not to extend the chain length. Two different structures are possible, one with the remaining carbon atoms in the 2 and 3 positions and another with both extra carbon atoms attached at the 2 position. Fill in the 14 hydrogens. You have now drawn the fourth and fifth isomers.

carbon atom frameworks
for dimethylbutane isomers   2,3-dimethylbutane

2,2-dimethylbutane

Comment • Should we look for structures in which the longest chain is three carbon atoms? Try it, but you will see that it is not possible to add the three remaining carbons to a three-carbon chain without creating one of the carbon chains already drawn in a previous step. Thus, we have completed the analysis, with five isomers of this compound being identified.

Names of each compound are given. See the text that follows this example and Appendix E for guidelines on nomenclature.

One possible isomer of an alkane with the formula C_7H_{16}.

Exercise 11.1 Drawing Structural Isomers of Alkanes

(a) Draw the nine isomers having the formula C_7H_{16}. [*Hint:* There is one structure with a seven-carbon chain, two with six-carbon chains, five in which the longest chain has five carbons (one is illustrated in the margin), and one with a four-carbon chain.]

(b) Identify the isomers of C_7H_{16} that are chiral.

Problem-Solving Tip 11.1

Drawing Structural Formulas

An error our students sometimes make is to suggest that the three carbon skeletons drawn here are different. They are in fact the same. All are five-carbon chains with one C atom in the 2 position.

Remember that Lewis structures do not indicate the geometry of molecules.

```
        C                    C                              C
        |                    |                              |
 C——C——C——C——C        C——C——C——C           C——C——C——C——C
 1   2   3   4   5      |2   3   4   5        5   4   3   2   1
                        C
                        1
```

Naming Alkanes

With so many possible isomers for a given alkane, chemists need a systematic way of naming them. The rules for naming alkanes and their derivatives are:

- The names of alkanes end in "-ane."
- Names of alkanes with chains of one to ten carbon atoms are given in Table 11.2. After the first four compounds, the names derive from Latin numbers — pentane, hexane, heptane, octane, nonane, decane — and this regular naming continues for higher alkanes.
- When naming a specific alkane, the root of the name corresponds to the longest carbon chain in the compound. One isomer of C_5H_{12} has a three-carbon chain with two — CH_3 groups on the second C atom of the chain. Thus, its name is based on propane.

$$
\begin{array}{c}
CH_3 \\
| \\
H_3C — C — CH_3 \\
| \\
CH_3
\end{array}
$$

2,2-dimethylpropane

- Substituent groups on a hydrocarbon chain are identified by a name and the position of substitution; this information precedes the root of the name. The position of the substituent is indicated by a number that refers to the carbon atom to which it is attached. (Numbering of the carbon atoms in a chain should begin at the end of the carbon chain that allows the substituent groups to have the lowest numbers.) Both — CH_3 groups in 2,2-dimethylpropane are located at the 2 position.
- Names of hydrocarbon substituents, called **alkyl groups,** are derived from the name of the hydrocarbon. For example, the group — CH_3, derived by taking a hydrogen from methane, is called the methyl group; the C_2H_5 group is the ethyl group.
- If two or more of the same substituent groups occur, the prefixes di-, tri-, and tetra- are added. When different substituent groups are present, they are generally listed in alphabetical order.

● **Naming Guidelines**
For more details on naming organic compounds see Appendix E.

● **Systematic and Common Names**
Many organic compounds are known by common names. For example, 2,2-dimethylpropane is also called neopentane. However, the IUPAC (International Union of Pure and Applied Chemistry) has formulated rules for systematic names, and these are generally used in this book.

Example 11.2   Naming Alkanes

Problem • Give the systematic name for

$$
\begin{array}{c}
CH_3 \qquad\quad C_2H_5 \\
| \qquad\qquad\quad | \\
CH_3CHCH_2CH_2CHCH_2CH_3
\end{array}
$$

Strategy • Identify the longest carbon chain and base the name of the compound on that alkane. Identify the substituent groups on the chain and their location.

(Example continues on page 430)

Solution • Here the longest chain has seven C atoms, so the root of the name is *heptane*. There is a methyl group on the second C atom and an ethyl group on the fifth C atom. Therefore, giving the substituents in alphabetic order, the systematic name is 5-ethyl-2-methylheptane.

Comment • The name here involves one more naming rule. If there are two or more different substituents, number the chain from the end that gives the lower number to the substituent encountered first. Thus, CH_3 is at C atom 2 and C_2H_5 is at C atom 5.

Exercise 11.2 **Naming Alkanes**

Name the nine isomers of C_7H_{16} in Exercise 11.1.

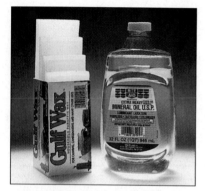

Figure 11.5 Paraffin wax and mineral oil. These common consumer products are mixtures of alkanes. *(Charles D. Winters)*

Properties of Alkanes

Although methane, ethane, propane, and butane are gases at normal temperature and pressure, the higher molecular weight alkanes are liquids or solids (see Table 11.2). An increase in melting point and boiling point with molecular weight is a general phenomenon that reflects the increased forces of attraction between molecules (➡ SECTION 13.2).

You already know about alkanes in a nonscientific context because several are common fuels. Gasoline, kerosene, fuel oils, and lubricating oils are mixtures of various alkanes. White mineral oil is also a mixture of alkanes, as is paraffin wax (Figure 11.5).

Pure alkanes are colorless. (The colors seen in gasoline and other petroleum products are due to additives.) The gases and liquids have noticeable but not unpleasant odors, and all of these substances are insoluble in water. Water-insolubility is typical of compounds that have low polarity. Low polarity is expected for alkanes because the electronegativities of carbon ($\chi = 2.5$) and hydrogen ($\chi = 2.1$) are not greatly different(⬅ SECTION 9.7).

All alkanes burn readily in air to give CO_2 and H_2O in very exothermic reactions. This is of course the reason they are widely used as fuels.

$$CH_4(g) + 2\ O_2(g) \longrightarrow CO_2(g) + 2\ H_2O(\ell) \qquad \Delta H^\circ_{rxn} = -890.3\ kJ$$

Other than combustion reactions, alkanes exhibit relatively low chemical reactivity. One reaction that does occur, however, is the replacement of the hydrogen atoms of an alkane by chlorine atoms on reaction with Cl_2. This is formally an oxidation because Cl_2, like O_2, is a strong oxidizing agent. These reactions, which can be initiated by ultraviolet radiation, are free radical reactions; highly reactive Cl atoms are formed from Cl_2 under UV radiation. Reaction of methane with Cl_2 under these conditions proceeds in a series of steps, eventually yielding CCl_4, commonly known as carbon tetrachloride. (HCl is the other product of these reactions.)

CH_4	$\xrightarrow[\text{UV}]{Cl_2}$	CH_3Cl	$\xrightarrow[\text{UV}]{Cl_2}$	CH_2Cl_2	$\xrightarrow[\text{UV}]{Cl_2}$	$CHCl_3$	$\xrightarrow[\text{UV}]{Cl_2}$	CCl_4
Systematic name		chloromethane		dichloromethane		trichloromethane		tetrachloromethane
Common name		methyl chloride		methylene chloride		chloroform		carbon tetrachloride

The last three compounds (methylene chloride, chloroform, and carbon tetrachloride) are used as solvents, although less frequently now because of their toxicity. Carbon tetrachloride was also once widely used as a dry cleaning fluid and, because it does not burn, in fire extinguishers.

Cycloalkanes, C_nH_{2n}

Cycloalkanes are constructed with tetrahedral carbon atoms joined together to form a ring. For example, cyclopentane, C_5H_{10}, consists of a ring of five carbon atoms. Each carbon atom is bonded to two adjacent carbon atoms and to two hydrogen atoms. Notice that the five carbon atoms fall very nearly in a plane. This is because the internal angles of a pentagon, 110°, very closely match the tetrahedral angle of 109.5°. The very small distortion from planarity allows hydrogen atoms on adjacent carbon atoms to be a little farther apart.

Cyclohexane, with a ring with six CH_2 groups, is well known. Its six carbon atoms do not lie in a plane. If the carbon atoms were in the form of a regular hexagon with all carbon atoms in one plane, the $C—C—C$ bond angles would be 120°. To have tetrahedral bond angles of 109.5° around each C atom, the ring has to pucker. The six-carbon ring is flexible, however, and exists in two interconverting forms (see "*A Closer Look: Flexible Molecules*").

Interestingly, cyclobutane and cyclopropane are also known, although the bond angles in these species are much less than 109.5°. These compounds are examples of **strained hydrocarbons,** so named because an unfavorable geometry is imposed around carbon. One of the features of strained hydrocarbons is that the $C—C$ bonds are weaker and the molecules readily undergo ring-opening reactions that relieve the bond angle strain.

Alkenes and Alkynes

The abundance and diversity of alkanes are repeated with **alkenes,** hydrocarbons with one or more $C=C$ double bonds. The presence of the double bond adds two features missing in alkanes: the possibility of geometric isomerism and increased reactivity [CD-ROM, Screen 11.3].

Cyclopentane, C_5H_{10}

Cyclohexane, C_6H_{12}

The structures of cyclopentane, C_5H_{12}, and cyclohexane, C_6H_{12}. Notice that the five-carbon ring is nearly planar, whereas the demands of the tetrahedral geometry of carbon mean the six-carbon ring is puckered.

Cyclopropane, C_3H_6

Cyclobutane, C_4H_8

A Closer Look

Flexible Molecules

Most organic molecules are flexible; they can twist and bend in various ways. Few molecules better illustrate this than cyclohexane. Two structures are possible, the so-called "chair" and "boat" forms. These forms can interconvert by partial rotation of several bonds. The more stable structure is the chair form because this allows the hydrogen atoms to be as far apart as possible. A side view of this form of cyclohexane reveals two sets of hydrogen atoms in this molecule. Six hydrogen atoms, called the *equatorial* hydrogens, lie in a plane around the carbon ring. The other six hydrogens are positioned above and below the plane and are called *axial* hydrogens. Flexing the ring (a rotation around the $C—C$ single bonds) moves the hydrogen atoms between axial and equatorial environments.

The general formula for alkenes is C_nH_{2n}. The first two members of the series of alkenes, and their systematic names, are ethene, C_2H_4 (also called ethylene), and propene (propylene), C_3H_6. Only a single structure can be drawn for these compounds. As with alkanes, the occurrence of isomers begins with the four-carbon species. The four alkene isomers with the formula C_4H_8, have distinct chemical and physical properties (Table 11.3).

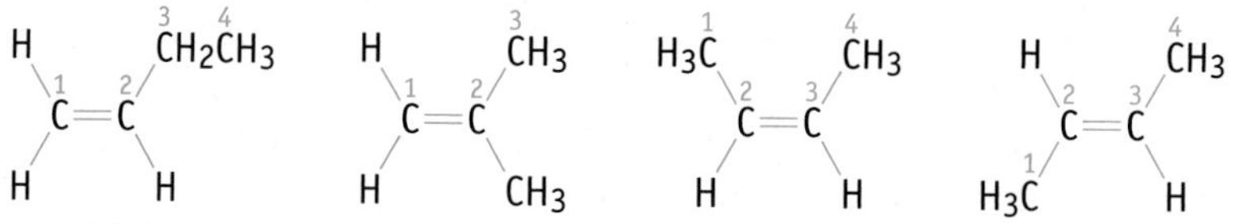

C_2H_4
Systematic name:
Ethene
Common name:
Ethylene

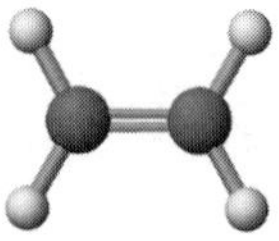

C_3H_6
Systematic name:
Propene
Common name:
Propylene

1-butene   2-methylpropene   *cis*-2-butene   *trans*-2-butene

Names of alkenes all end in "–ene." As with alkanes, the root name for alkene derivatives is that of the longest carbon chain. In addition, the position of the double bond is indicated with a number (and, when appropriate, the prefix *cis* or *trans*). Three of the C_4H_8 isomers have four-carbon chains and thus are butenes. One has a three-carbon chain and is a propene. Notice that the carbon chain is numbered from the end that gives the double bond the lowest number. In the first isomer at the left, the double bond is between C-1 and C-2, so the name is 1-butene and not 3-butene.

Table 11.3 • Properties of Butene Isomers

Name	Boiling Point	Melting Point	Dipole Moment (D)	ΔH_f° (gas) (kJ/mol)
1-butene	−6.26 °C	−185.4 °C	—	−20.5
2-methylpropene	−6.95 °C	−140.4 °C	0.503	−37.5
cis-2-butene	3.71 °C	−138.9 °C	0.253	−29.7
trans-2-butene	0.88 °C	−105.5 °C	0	−33.0

Example 11.3   Determining Isomers of Alkenes from a Formula

Problem • Draw structures for the six possible alkene isomers with the formula C_5H_{10}. Give the systematic name of each.

Strategy • A procedure that involved drawing the carbon skeleton and then adding hydrogen atoms served well when drawing structures of alkanes (Example 11.1), and a similar approach can be used here. It will be necessary to put one double bond into the framework and to be alert for *cis–trans* isomerism.

Solution •
1. A five-carbon chain with one double bond can be constructed in two ways. One gives rise to *cis–trans* isomers.

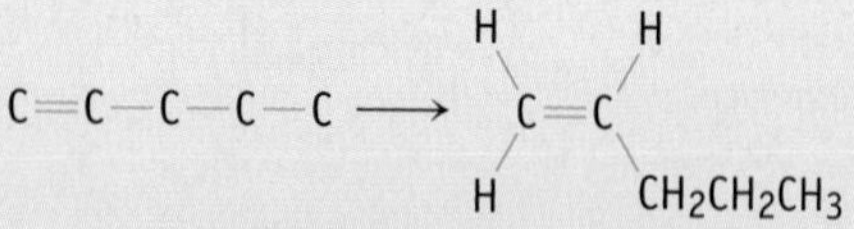

C=C—C—C—C ⟶   1-pentene

C—C=C—C—C ⟶   *cis*-2-pentene   /   *trans*-2-pentene

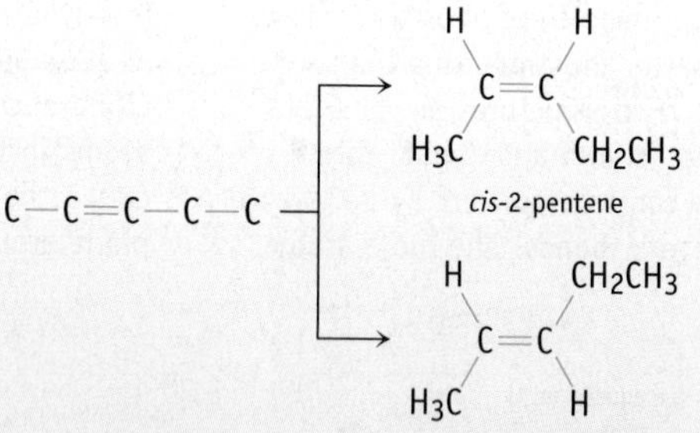

2. Now draw the possible four-carbon chains containing a double bond. Add the fifth carbon atom to either the 2 or 3 position. When all three possible combinations are found, fill in the hydrogen atoms. This results in three more structures:

C=C—C—C ⟶ (2-methyl-1-butene structure)

1 2 3 4

2-methyl-1-butene

C—C=C—C ⟶ (2-methyl-2-butene structure)

1 2 3 4

2-methyl-2-butene

C=C—C—C ⟶ (3-methyl-1-butene structure)

1 2 3 4

3-methyl-1-butene

Exercise 11.3 **Determining Structural Isomers of Alkenes from a Formula**

There are 17 possible alkene isomers with the formula C_6H_{12}. Draw structures of the five isomers in which the longest chain has six carbon atoms and give the name of each. Which of these isomers is chiral? (There are also eight isomers in which the longest chain has five carbon atoms, and four isomers in which the longest chain has four carbon atoms. How many can you find?)

Cyclohexene, C_6H_{10}

$H_2C=CHCH=CH_2$

1,3-Butadiene, C_4H_6

More than one double bond is present in some hydrocarbons. Butadiene, for example, has two double bonds and is known as a *diene*. Many natural products have numerous double bonds (Figure 11.6). There are also cyclic hydrocarbons with double bonds such as cyclohexene.

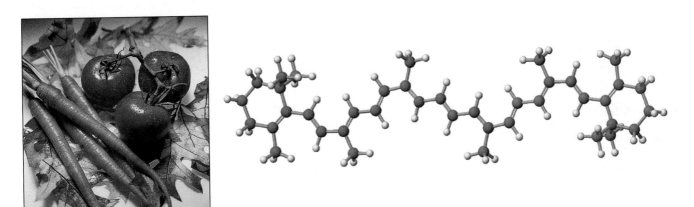

Figure 11.6   Carotene, a naturally occurring compound with eleven C=C bonds. The π electrons can be excited by visible light in the blue-violet region of the spectrum, and so carotene appears orange-yellow to the observer. Carotene or carotene-like molecules are partnered with chlorophyll in nature in the role of assisting in the harvesting of sunlight. Green leaves have a high concentration of carotene. In autumn, green chlorophyll molecules are destroyed and the yellows and reds of carotene and related molecules are seen. The red color of tomatoes comes from a molecule very closely related to carotene. As a tomato ripens, its chlorophyll disintegrates and the green color is replaced by the red of the carotene-like molecule. *(Charles D. Winters)*

Figure 11.7   An oxy-acetylene torch. The reaction of ethyne (acetylene) with oxygen produces a very high temperature. Oxyacetylene torches, used in welding, take advantage of this fact. *(Charles D. Winters)*

Table 11.4 • Some Simple Alkynes

Structure	Common Name	Systematic Name	BP (°C)
$HC{\equiv}CH$	acetylene	ethyne	−75
$CH_3C{\equiv}CH$	methylacetylene	propyne	−23
$CH_3CH_2C{\equiv}CH$	ethylacetylene	1-butyne	9
$CH_3C{\equiv}CCH_3$	dimethylacetylene	2-butyne	27

Alkynes, compounds with a carbon–carbon triple bond, have the general formula (C_nH_{2n-2}). Table 11.4 lists alkynes that have four or fewer carbon atoms. The first member of this family is ethyne (acetylene), a gas used as a fuel in metal-cutting torches (Figure 11.7).

Properties of Alkenes and Alkynes

Like alkanes, alkenes and alkynes are colorless. Low-molecular-weight compounds are gases, whereas compounds with higher molecular weights are liquids or solids. Alkanes, alkenes, and alkynes are all oxidized by O_2 to give CO_2 and H_2O.

In contrast to alkanes, alkenes and alkynes have an elaborate chemistry. We gain an insight into their chemical behavior from the fact that they are called **unsaturated compounds.** Carbon atoms are capable of being bonded to a maximum of four other atoms and they do so in alkanes and cycloalkanes. In alkenes, however, the carbon atoms linked by a double bond are bonded to only three atoms, and in alkynes the number is two. It is possible to increase the number of atoms bonded to carbon to the maximum of four by **addition reactions** in which molecules with the general formula X—Y (such as hydrogen, halogens, hydrogen halides, and water) add across the carbon–carbon double bond. The result is a compound with four atoms bonded to each carbon.

$$\underset{H}{\overset{H}{C}}{=}\underset{H}{\overset{H}{C}} + X{-}Y \longrightarrow H{-}\underset{H}{\overset{X}{C}}{-}\underset{H}{\overset{Y}{C}}{-}H$$

X—Y = H_2, Cl_2, Br_2; H—Cl, H—Br, H—OH, HO—Cl

The fat in bacon is partially unsaturated. Like other unsaturated compounds, bacon reacts with Br_2. Here you see the color of Br_2 vapor fade when a strip of bacon is introduced. *(Charles D. Winters)*

The products of addition reactions are substituted alkanes. For example, the addition of bromine to ethylene forms 1,2-dibromoethane

$$\underset{H}{\overset{H}{C}}{=}\underset{H}{\overset{H}{C}} + Br_2 \longrightarrow H{-}\underset{H}{\overset{Br}{C}}{-}\underset{H}{\overset{Br}{C}}{-}H$$

1,2-dibromoethane

and the addition of chlorine to acetylene gives 1,1,2,2-tetrachloroethane.

$$HC{\equiv}CH + Cl_2 \longrightarrow Cl{-}\underset{H}{\overset{Cl}{C}}{-}\underset{H}{\overset{Cl}{C}}{-}Cl$$

1,1,2,2-tetrachloroethane

If the reagent added to a double bond is hydrogen ($X—Y = H_2$), the reaction is called **hydrogenation,** and the product is an alkane. Hydrogenation is usually a very slow reaction, but it can be speeded up in the presence of a catalyst, often a specially prepared form of a metal, such as platinum, palladium, or rhodium. You may have heard the term "hydrogenation" because certain foods contain "partially hydrogenated" ingredients. One brand of crackers has a label that says "Made with 100% pure vegetable shortening . . . (partially hydrogenated soybean oil with hydrogenated cottonseed oil)." One reason for hydrogenating an oil is to make it less susceptible to spoilage; another is to convert it from a liquid to a solid.

● **Catalysts**
A substance that causes a reaction to occur at a faster rate is called a catalyst. We will describe catalysts in more detail in the chapter on kinetics, Chapter 15.

Example 11.4 **Reaction of an Alkene**

Problem • Draw the structure of the compound obtained from the reaction of Br_2 with propene, and name the compound.

Strategy • Bromine will add across the C=C double bond. The name will include the name of the carbon chain and indicate the positions of the Br atoms.

Solution •

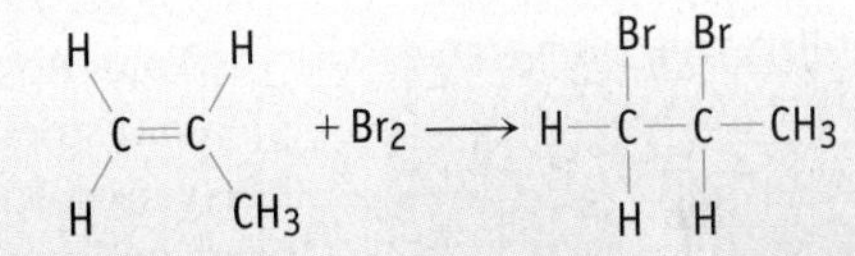

propene   1,2-dibromopropane

Exercise 11.4 **Reactions of Alkenes**

(a) Draw the structure of the compound obtained from the reaction of HBr with ethylene, and name the compound.

(b) Draw the structure for the compound that comes from the reaction of Br_2 with *cis*-2-butene, and name this compound.

Aromatic Compounds

Benzene, C_6H_6, is a key molecule in chemistry. It is the simplest **aromatic compound,** a class of compounds so named because they have significant, and usually not unpleasant, odors. Other members of this class include toluene and naphthalene. A source of many aromatic compounds is coal and the volatile substances that distill when coal is heated to a high temperature in the absence of air (Table 11.5).

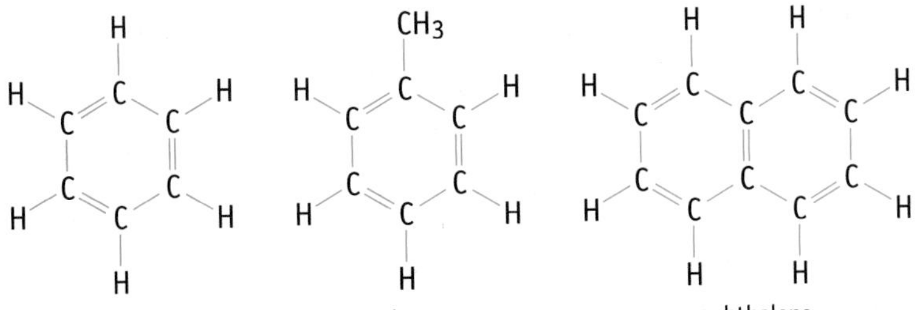

benzene   toluene   naphthalene

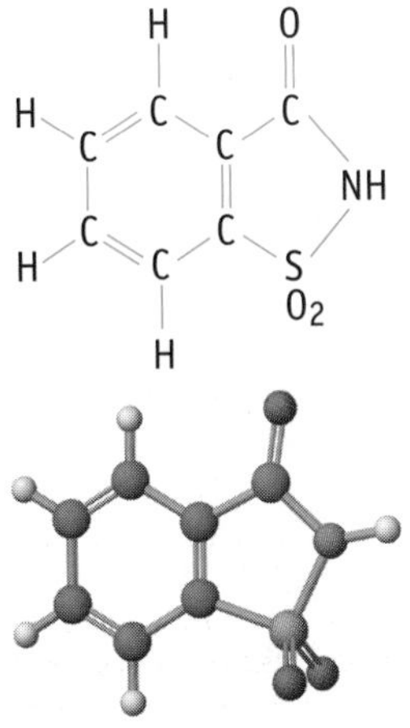

Saccharin ($C_7H_5NO_3S$), the artificial sweetener, is a benzene derivative.

Benzene occupies a pivotal place in the history and practice of chemistry. Michael Faraday discovered this compound in 1825 as a byproduct of illuminating gas, itself produced by heating coal. Today benzene is an important industrial chemical, usually in the top 25 chemicals produced annually in the United States. It is used as a solvent and is also the starting point for making thousands of different compounds by replacing the H atoms of the ring.

Table 11.5 • Some Aromatic Compounds from Coal Tar

Common Name	Formula	Boiling Point (°C)	Melting Point (°C)
benzene	C_6H_6	80	+6
toluene	$C_6H_5CH_3$	111	−95
o-xylene	$1,2\text{-}C_6H_4(CH_3)_2$	144	−27
m-xylene	$1,3\text{-}C_6H_4(CH_3)_2$	139	−54
p-xylene	$1,4\text{-}C_6H_4(CH_3)_2$	138	+13
naphthalene	$C_{10}H_8$	218	+80

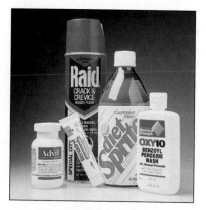

Some products containing compounds based on benzene. Examples include sodium benzoate in soft drinks, ibuprofen in Advil, and benzoyl peroxide in Oxy-10. *(Charles D. Winters)*

Toluene was originally obtained from Tolu balsam, which is derived from the pleasant-smelling gum of a South American tree, *Toluifera balsamum*. This balsam has been used in cough syrups and perfumes. Naphthalene is one of the ingredients used in "moth balls," although 1,4-dichlorobenzene is now more commonly used. Aspartame and another artificial sweetener, saccharin, are also benzene derivatives.

The Structure of Benzene

The formula of benzene suggested to 19th-century chemists that this compound should be unsaturated, but, if viewed this way, its chemistry was perplexing. Whereas an alkene readily adds Br_2, for example, benzene does not do so under similar conditions. Instead, benzene reacts with Br_2 at higher temperatures in the presence of other chemicals such as $FeBr_3$ to give a product in which a Br atom has been *substituted* for an H atom.

Substitution Reaction:

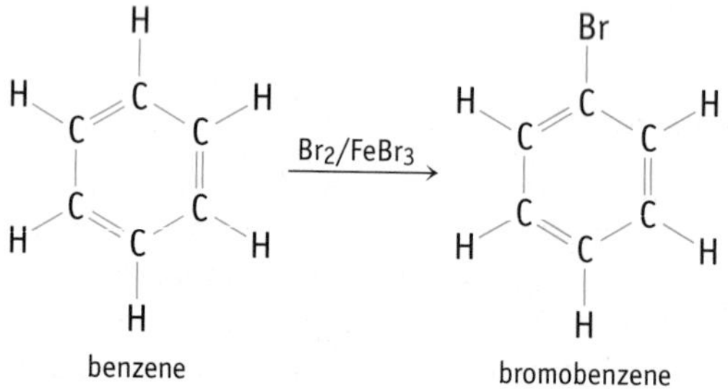

$$\text{benzene} \xrightarrow{Br_2/FeBr_3} \text{bromobenzene}$$

The benzene structural question was solved by August Kekulé (1829–1896), and we now recognize that benzene's lack of reactivity relates to its structure and bonding. Benzene has equivalent carbon–carbon bonds, 139 pm in length, intermediate between a C—C single bond (154 pm) and a C=C double bond (134 pm). This is the result of π bonding which extends over the six carbon atoms in the ring. The π bonds are formed by overlap of the *p* orbitals on the six carbon atoms (page 396), and two resonance are possible.

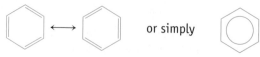

or simply

Representations of benzene, C_6H_6

Benzene Derivatives

Chlorobenzene, benzoic acid, aniline, styrene, and phenol are common examples of benzene derivatives.

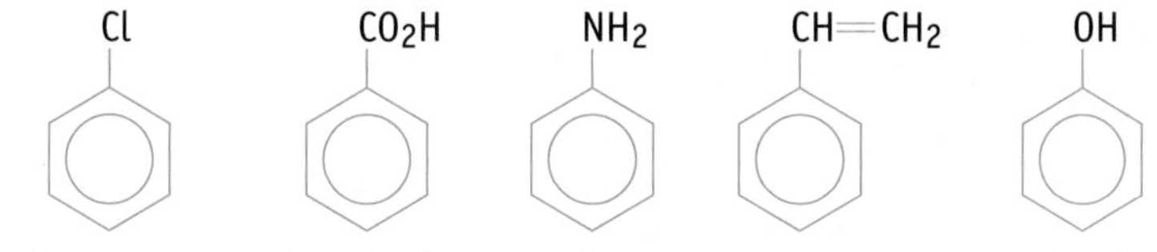

chlorobenzene   benzoic acid   aniline   styrene   phenol

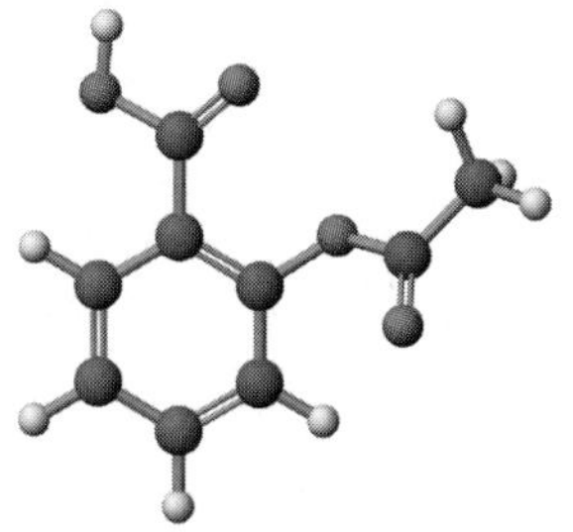

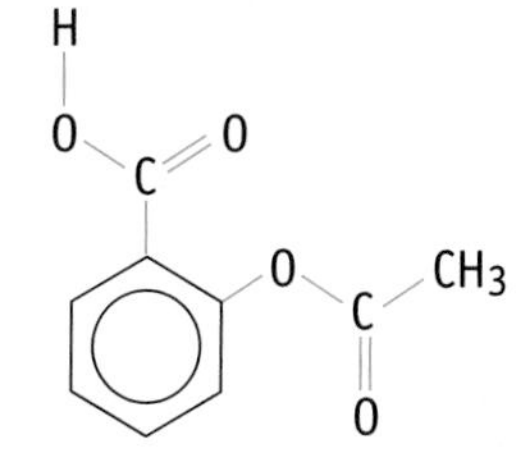

If more than one H atom of benzene is replaced, isomers can arise. Thus, the systematic nomenclature for benzene derivatives involves naming substituent groups and identifying their position on the ring by numbering the six carbon atoms (➡ APPENDIX E). Some common names, which are based on an older naming scheme, are also regularly used. This scheme identified isomers of disubstituted benzenes with the prefixes *ortho* (*o-*, substituent groups on adjacent carbons in the benzene ring), *meta* (*m-*,), and *para* (*p-*, substituent groups on carbons on opposite sides of the ring).

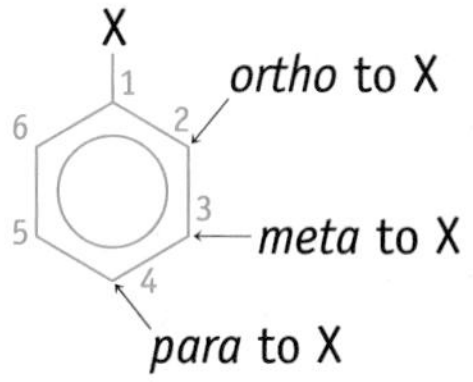

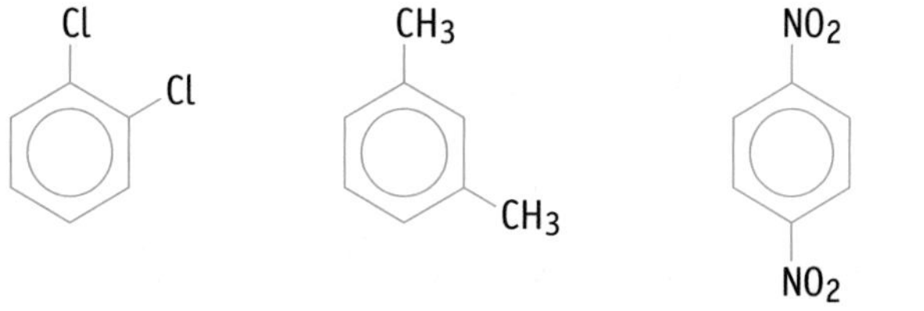

Aspirin, a commonly used analgesic. It is based on benzoic acid with an acetate group, —$OC(O)CH_3$, in the *ortho* position.

| Systematic name: | 1,2-dichlorobenzene | 1,3-dimethylbenzene | 1,4-dinitrobenzene |
| Common name: | *o*-dichlorobenzene | *m*-xylene | *p*-dinitrobenzene |

Example 11.5   Isomers of Substituted Benzenes

Problem • How many isomers are possible for trisubstituted benzenes? Draw and name the isomers of $C_6H_3Cl_3$.

Strategy • Begin by drawing the structure of C_6H_5Cl. Place a second Cl atom on the ring in the *ortho, meta,* and *para* positions. Add the third Cl in one of the positions that remain.

Solution • The three isomers of $C_6H_3Cl_3$ are shown here. They are named as derivatives of benzene by specifying the number of substituent groups by the prefix "tri-", the name of the substituent, and the positions of the three groups around the six-member ring.

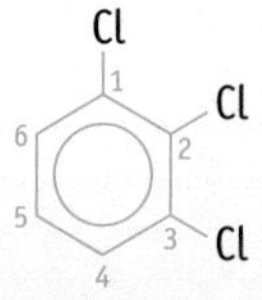

1,2,3-trichlorobenzene

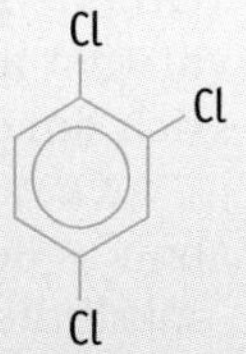

1,2,4-trichlorobenzene

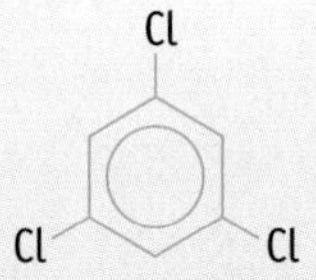

1,3,5-trichlorobenzene

Exercise 11.5   Isomers of Substituted Benzenes

Aniline, $C_6H_5NH_2$, is the common name for aminobenzene. Draw a structure for *p*-diaminobenzene, a compound also used in dye manufacture. What is the systematic name for *p*-diaminobenzene?

Properties of Aromatic Compounds

Benzene is a colorless liquid, and simple substituted benzenes are liquids or solids under normal conditions. The properties of aromatic compounds are typical of hydrocarbons generally: They are insoluble in water, soluble in nonpolar solvents, and oxidized by O_2 to form CO_2 and H_2O.

One of the most important properties of benzene and other aromatic compounds is an unusual stability that is associated with the unique π bonding in this molecule. Because the π bonding in benzene is usually described using resonance structures, the extra stability is termed **resonance stabilization.** The extent of resonance stabilization in benzene is evaluated by comparing the energy evolved in the hydrogenation of benzene to form cyclohexane

with the energy evolved in hydrogenation of three isolated double bonds.

$$3\ H_2C{=}CH_2(g) + 3\ H_2(g) \longrightarrow 3\ C_2H_6(g) \qquad \Delta H° = -410.8\ kJ$$

Hydrogenation of benzene is about 200 kJ less exothermic than the hydrogenation of 3 moles of ethylene. The difference is attributed to the added stability of benzene associated with its π bonding.

Although aromatic compounds are unsaturated hydrocarbons, they do not undergo the addition reactions typical of alkenes and alkynes. Instead, substitution reactions occur, in which one or more hydrogen atoms can be replaced by other groups. Such reactions require a second reagent, such as H_2SO_4, $AlCl_3$, or $FeCl_3$.

Nitration: $\quad C_6H_6(\ell) + HNO_3(\ell) \xrightarrow{H_2SO_4} C_6H_5NO_2(\ell) + H_2O(\ell)$

Alkylation: $\quad C_6H_6(\ell) + CH_3Cl(\ell) \xrightarrow{AlCl_3} C_6H_5CH_3(\ell) + HCl(g)$

Halogenation: $\quad C_6H_6(\ell) + Br_2(\ell) \xrightarrow{FeBr_3} C_6H_5Br(\ell) + HBr(g)$

11.3 ALCOHOLS, ETHERS, AND AMINES

Other types of organic compounds arise if elements other than carbon and hydrogen are introduced. Two elements in particular, oxygen and nitrogen, add a rich dimension to carbon chemistry [CD-ROM, Screens 11.5 and 11.6].

Organic chemistry organizes compounds containing elements other than carbon as derivatives of hydrocarbons. Formulas (and structures) are represented by substituting one or more hydrogens in a hydrocarbon molecule by a **functional group,** an atom or group of atoms attached to a carbon atom in the hydrocarbon. Formulas of hydrocarbon derivatives are then written as R—X, in which R is a hydrocarbon lacking a hydrogen atom, and X is the functional group that has replaced the hydrogen in the structures. The chemical and physical properties of the hydrocarbon derivatives are a blend of the properties associated with hydrocarbons and the group that has been substituted for hydrogen.

Table 11.6 identifies common functional groups and the families of organic compounds resulting from their attachment to an alkane.

Table 11.6 • Common Functional Groups and Derivatives of Alkanes

Functional Group*	General Formula*	Class of Compound	Examples
F, Cl, Br, I	RF, RCl, RBr, RI	haloalkane	CH_3CH_2Cl, chloroethane
OH	ROH	alcohol	CH_3CH_2OH, ethanol
OR'	ROR'	ether	$(CH_3CH_2)_2O$, diethyl ether
NH_2†	RNH_2	(primary) amine	$CH_3CH_2NH_2$, ethylamine
$\overset{O}{\overset{\|}{-CH}}$	RCHO	aldehyde	CH_3CHO, ethanal (acetaldehyde)
$\overset{O}{\overset{\|}{-C}}-R'$	RCOR'	ketone	CH_3COCH_3, propanone (acetone)
$\overset{O}{\overset{\|}{-C}}-OH$	RCO_2H	carboxylic acid	CH_3CO_2H, ethanoic acid (acetic acid)
$\overset{O}{\overset{\|}{-C}}-OR'$	RCO_2R'	ester	$CH_3CO_2CH_3$, methyl acetate
$\overset{O}{\overset{\|}{-C}}-NH_2$	$RCONH_2$	amide	CH_3CONH_2, acetamide

*R and R' can be the same or different hydrocarbon groups.
†Secondary amines (R_2NH) and tertiary amines (R_3N) are also possible, see discussion in the text.

Alcohols and Ethers

If one of the hydrogen atoms of an alkane is replaced by a hydroxyl (—OH) group, the result is an **alcohol,** ROH. Methanol, CH_3OH, and ethanol, CH_3CH_2OH, are the most important of the alcohols, but other alcohols are also commercially important (Table 11.7). Notice that several have more than one —OH group.

More than 5×10^8 kg of methanol is produced in the United States annually. Most is used to make formaldehyde (CH_2O) and acetic acid (CH_3CO_2H), both im-

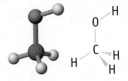

Methanol, CH_3OH, is used as the fuel in cars of the type that race in Indianapolis. *(David Young/Tom Stack & Associates)*

Methanol, CH_3OH, is the simplest alcohol.

Table 11.7 • Some Important Alcohols

Condensed Formula	BP (°C)	Common Name	Systematic Name	Use
CH_3OH	65.0	methyl alcohol	methanol	fuel, gasoline additive, making formaldehyde
CH_3CH_2OH	78.5	ethyl alcohol	ethanol	beverages, gasoline additive, solvent
$CH_3CH_2CH_2OH$	97.4	propyl alcohol	1-propanol	industrial solvent
$CH_3CH(OH)CH_3$	82.4	isopropyl alcohol	2-propanol	rubbing alcohol
$HOCH_2CH_2OH$	198	ethylene glycol	1,2-ethanediol	antifreeze
$HOCH_2CH(OH)CH_2OH$	290	glycerol (glycerin)	1,2,3-propanetriol	moisturizer in consumer products

portant chemicals in their own right. Methanol is also used as a solvent, as a de-icer in gasoline, and as a fuel in high-powered racing cars. Methanol is often called "wood alcohol" because it was originally produced by heating wood in the absence of air. It is found in low concentration in new wine, where it contributes to the odor, or "bouquet." Like ethanol, methanol causes intoxication, but methanol differs in being more poisonous, largely because the human body converts it to formic acid (HCO_2H) and formaldehyde (CH_2O). These compounds attack the cells of the retina in the eye, leading to permanent blindness.

Ethanol is the "alcohol" of alcoholic beverages, in which it is formed by the anaerobic (without air) fermentation of sugar. For many years, industrial alcohol, which is used as a solvent and as a starting material for the synthesis of other compounds, was made by fermentation. In the last several decades, however, it has become cheaper to make ethanol from petroleum byproducts, specifically by the addition of water to ethylene.

A Closer Look

Petroleum Chemistry

Much of current chemical technology relies on petroleum. Burning fuels derived from petroleum provides by far the largest amount of energy in the industrial world (Section 6.11). Petroleum and natural gas are also the chemical raw materials used in the manufacture of plastics, rubber, pharmaceuticals and a vast array of other compounds.

Petroleum as it is pumped out of the ground is a complex mixture whose composition varies greatly with source. The primary components of petroleum are alkanes, but, to varying degrees, nitrogen and sulfur-containing compounds are also present. Aromatic compounds are also present, but alkenes and alkynes are *not*.

An early step in the petroleum refining process is distillation (Chapter 14), a

A modern petrochemical plant. *(Thomas Kitchin/Tom Stack & Associates)*

process in which the crude mixture is separated into a series of fractions based on boiling point: first a gaseous fraction (mostly alkanes with one to four carbon atoms; this fraction is often burned off), and then gasoline, kerosene, and fuel oils. After distillation, considerable material, in the form of a semisolid, tar-like residue, is likely to remain.

The petrochemical industry seeks to maximize amounts of the higher valued fractions of petroleum and to make specific compounds for which there is a particular need. This means carrying out chemistry on the raw materials on a huge scale. One of the processes to which petroleum is subjected is known as *cracking*. At very high temperatures, this bond breaking, or "cracking," can occur, and longer chain hydrocarbons fragment to smaller molecular units. (These reactions are carried out in the presence of a wide array of catalysts, materials that speed up reactions and direct them toward specific products.) Among the important products of cracking are ethylene and other alkenes, the important raw materials for the formation of materials such as polyethylene. Cracking also produces gaseous hydrogen, another widely used raw material in the chemical industry.

Other important reactions involving petroleum are run at elevated temperature and in the presence of specific catalysts. Such

reactions include isomerization reactions in which the carbon skeleton of an alkane rearranges to a new isomeric species, and reformation processes in which smaller molecules combine to form new molecules. Each process is directed to a specific goal, such as increasing the proportion of branched-chain hydrocarbons in gasoline to obtain higher octane ratings. A great amount of chemical research has gone into developing and understanding these highly specialized processes.

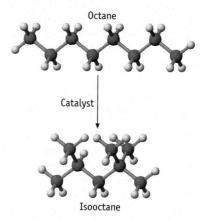

Octane

Catalyst

Isooctane

An important process in producing gasoline is the isomerization of octane to isooctane, 2,2,4–trimethylpentane, a branched hydrocarbon.

$$\underset{\text{ethylene}}{\overset{\displaystyle H}{\underset{\displaystyle H}{}}C=C\overset{\displaystyle H}{\underset{\displaystyle H}{}}\ (g) + H_2O(g) \xrightarrow{\text{catalyst}} \underset{\text{ethanol}}{H-\overset{\displaystyle H}{\underset{\displaystyle H}{C}}-\overset{\displaystyle H}{\underset{\displaystyle H}{C}}-OH(\ell)}$$

Beginning with three-carbon alcohols, structural isomers are possible. For example, 1-propanol and 2-propanol (see Table 11.7) are distinctly different compounds.

Ethylene glycol and glycerol are common alcohols having two or more —OH groups. Ethylene glycol is used as antifreeze in automobiles. Glycerol's most common use is as a softener in soaps and lotions. It is also a raw material for the preparation of nitroglycerin (Figure 11.8).

$H-\overset{H}{\underset{OH}{C}}-\overset{H}{\underset{OH}{C}}-H$	$H-\overset{H}{\underset{OH}{C}}-\overset{H}{\underset{OH}{C}}-\overset{H}{\underset{OH}{C}}-H$
Systematic name: 1,2-ethanediol	1,2,3-propanetriol
Common name: ethylene glycol	glycerol or glycerin

2-Propanol is the alcohol found in common rubbing alcohol. *(Charles D. Winters)*

Example 11.6   Structural Isomers of Alcohols

Problem • How many different alcohols are possible with one —OH group on a five-carbon unbranched chain? Draw structures and name each alcohol.

Strategy • Pentane, C_5H_{12}, has a five-carbon chain, and an —OH group is successively placed on one of the carbon atoms. Alcohols are named as derivatives of the alkane (pentane) by replacing the "-e" at the end with "-ol" and indicating the position of the —OH group by a numerical prefix (Appendix E).

Solution • Three different alcohols are possible, depending on whether the —OH group is placed on the first, second, or third carbon atom in the chain. (The fourth and fifth positions are identical to the second and first positions in the chain, respectively.)

$$HO-\overset{H}{\underset{H}{\overset{|}{C}}}^{1}-\overset{H}{\underset{H}{\overset{|}{C}}}^{2}-\overset{H}{\underset{H}{\overset{|}{C}}}^{3}-\overset{H}{\underset{H}{\overset{|}{C}}}^{4}-\overset{H}{\underset{H}{\overset{|}{C}}}^{5}-H$$
1-pentanol

$$H-\overset{H}{\underset{H}{C}}-\overset{OH}{\underset{H}{C}}-\overset{H}{\underset{H}{C}}-\overset{H}{\underset{H}{C}}-\overset{H}{\underset{H}{C}}-H$$
2-pentanol

$$H-\overset{H}{\underset{H}{C}}-\overset{H}{\underset{H}{C}}-\overset{OH}{\underset{H}{C}}-\overset{H}{\underset{H}{C}}-\overset{H}{\underset{H}{C}}-H$$
3-pentanol

Comment • Additional structural isomers with the formula $C_5H_{11}OH$ are possible in which the longest carbon chain has three C atoms (one isomer) or four C atoms (four isomers).

Exercise 11.6   Structures of Alcohols

Draw the structure of 1-butanol and any alcohols that are structural isomers of the compound.

Properties of Alcohols

Methane, CH_4, is a gas (boiling point, −161 °C) with low solubility in water. Methanol, CH_3OH, on the other hand, is a liquid that is *miscible* with water in all proportions. The boiling point of methanol, 65 °C, is 226 °C higher than the boil-

Figure 11.8 Nitroglycerin. (a) Concentrated nitric acid and glycerin react to form an oily, highly unstable compound, nitroglycerin, $C_3H_5(ONO_2)_3$. Nitroglycerin is more stable if absorbed onto an inert solid, a combination called dynamite. **(b)** The fortune of Alfred Nobel (1833–1896), built on the manufacture of dynamite, now funds the Nobel Prizes. *(a, Charles D. Winters; b, The Nobel Foundation)*

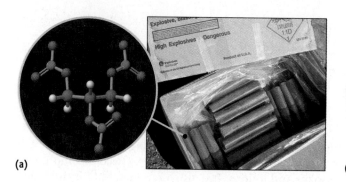

(a)   (b)

● **Alcohol–Water Mixtures**
Methanol and ethanol are soluble in all proportions in water. The term "miscible" is used to describe this property.

● **Hydrogen Bonding**
The intermolecular forces of attraction of compounds with hydrogen attached to a highly electronegative atom, like O, N, or F, are so exceptional that this circumstance is accorded a special name: hydrogen bonding. We will discuss hydrogen bonding further in Section 13.3.

⎛ **Chapter Goals ● Revisited** ⎞

● Classify organic compounds based on formula and structure.
● Name and draw structures of common organic compounds.
● **Relate properties to molecular structure.**
● Identify common polymers and compounds of biological significance.

ing point of methane. What a difference the addition of a single atom into the structure can make in the properties of simple molecules!

Alcohols are related to water with one of the H atoms of H_2O replaced by an organic group. If a methyl group is substituted for one of the hydrogens of water, methanol results. Ethanol has a $-C_2H_5$ (ethyl) group, and propanol a $-C_3H_7$ (propyl) group in place of one of the hydrogens of water. This perspective will also help in understanding the properties of alcohols.

The two parts of methanol, the $-CH_3$ group and the $-OH$ group, contribute to its properties. Methanol will burn, a property associated with hydrocarbons. On the other hand, its boiling point is more like that of water. The temperature at which a substance boils is related to the forces of attraction between molecules, the *intermolecular forces:* the stronger the attractive, intermolecular forces in a sample, the higher the boiling point [➡ SECTION 13.5]. These forces are particularly strong in water, a result of the polarity of the O—H bond in this molecule (⬅ SECTION 9.8). Methanol is also a polar molecule, and it is the polar $-OH$ group that leads to methanol's high boiling point. In contrast, methane is nonpolar and its low boiling point reflects low intermolecular forces.

It is also possible to explain the differences in the solubility of methane and methanol in water. The solubility of methanol is conferred by the polar $-OH$ portion of the molecule. Methane, which is nonpolar, has low water-solubility.

As the size of the alkyl group in an alcohol increases, the boiling point rises, a general trend in families of similar compounds (see Table 11.7). The solubility in water in this series decreases. Methanol and ethanol are completely miscible with water, whereas 1-propanol is moderately water-soluble, and 1-butanol is less soluble than 1-propanol. With an increase in the size of the hydrocarbon group, the organic group (the nonpolar part of the molecule) has become a larger fraction of the molecule, and properties associated with nonpolarity begin to dominate. Space-filling models show that in methanol, the polar and nonpolar parts of the molecule are approximately similar in size, but in 1-butanol the $-OH$ group is less than 20% of the molecule. The molecule is less like water and more "organic."

Nonpolar
hydrocarbon Polar
portion   portion

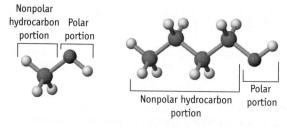

Nonpolar hydrocarbon   Polar
portion   portion

Attaching more than one $-OH$ group to a hydrocarbon framework has an effect that is opposite to the effect produced by increased hydrocarbon size. Two $-OH$ groups on a three-carbon framework, as in propylene glycol ($CH_3CHOHCH_2OH$), leads to complete miscibility with water, in contrast to the limited water solubility of 1-propanol and 2-propanol (Figure 11.9).

Ethers have the general formula ROR'. The best known ether is diethyl ether, $CH_3CH_2OCH_2CH_3$. Lacking an —OH group, the properties of ethers sharply contrast with alcohols. Diethyl ether, for example, has a lower boiling point (34.5 °C) than ethanol (CH_3CH_2OH, 78.3 °C) and is only slightly soluble in water.

Amines

It is often convenient to think about water and ammonia as similar molecules: They are the simplest hydrogen compounds of adjacent second-period elements and both are polar. They also exhibit some similar chemistry, such as protonation (giving H_3O^+ and NH_4^+) and deprotonation (giving OH^- and NH_2^-).

The comparison of water and ammonia can be extended to alcohols and amines. Alcohols have formulas related to water in which one hydrogen in H_2O is replaced with an organic group (R—OH). In organic **amines,** one or more hydrogen atoms of NH_3 are replaced with an organic group. Amine structures are similar to ammonia's structure; that is, the geometry about the N atom is trigonal-pyramidal.

Amines are categorized by the number of organic substituents as primary (one organic group), secondary (two organic groups), or tertiary (three organic groups). As examples, consider the three amines with methyl groups: CH_3NH_2, $(CH_3)_2NH$, and $(CH_3)_3N$.

Propylene glycol, CH₃CHOHCH₂OH.
Antifreeze consists of about 95% ethylene glycol. Cats and dogs are attracted by the smell and taste of the compound, but it is toxic. Only a few milliliters can be fatal to a small animal or child. In the first stage of poisoning, an animal may appear drunk, but within 12–36 hours the kidneys stop functioning and the animal slips into a coma. To avoid accidental poisoning of domestic and wild animals, you can use less toxic propylene glycol antifreeze. *(Charles D. Winters)*

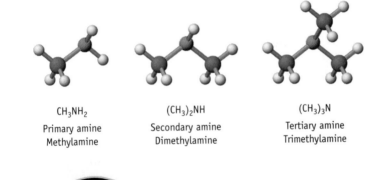

CH_3NH_2	$(CH_3)_2NH$	$(CH_3)_3N$
Primary amine	Secondary amine	Tertiary amine
Methylamine	Dimethylamine	Trimethylamine

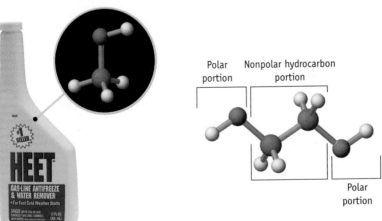

Polar portion   Nonpolar hydrocarbon portion

Polar portion

Methanol is often added to automobile gasoline tanks in the winter to prevent fuel lines from freezing. The methanol is soluble in water and lowers the water freezing point.

Ethylene glycol is used in automobile radiators. It is soluble in water and lowers the freezing point and raises the boiling point of the water in the cooling system. (See Section 14.4)

Ethylene glycol, a major component of automobile antifreeze, is completely miscible with water. The colorless glycol is dyed green for visibility.

Figure 11.9 Properties and uses of methanol and ethylene glycol. *(a, c, Charles D. Winters)*

Properties of Amines

Amines usually have offensive odors. You know what the odor is if you have ever smelled decaying fish. Two appropriately named amines, putrescine and cadaverine, add to the odor of urine, rotten meat, and bad breath.

$$H_2NCH_2CH_2CH_2CH_2NH_2 \qquad\qquad H_2NCH_2CH_2CH_2CH_2CH_2NH_2$$

putrescine   cadaverine

1,4-butanediamine   1,5-pentanediamine

The smallest amines are water-soluble, but most amines are not. All amines are bases, however, and react with acids to give salts, many of which are water-soluble. As with ammonia, the reactions involve adding H^+ to the lone pair of electrons on the N atom. This is illustrated by the reaction of aniline (aminobenzene) with H_2SO_4 to give anilinium sulfate.

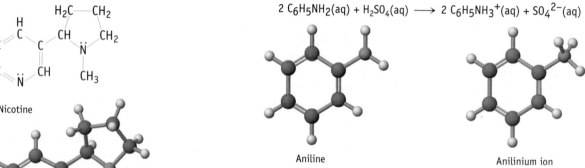

$$2\ C_6H_5NH_2(aq) + H_2SO_4(aq) \longrightarrow 2\ C_6H_5NH_3^+(aq) + SO_4^{2-}(aq)$$

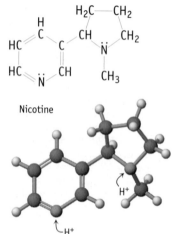

Nicotine

Two nitrogen atoms in the nicotine molecule can be protonated. It is in this form that nicotine is normally found. The protons can be removed, however, by treating it with a base. This "free-base" form is much more poisonous and addictive. See J. F. Pankow: *Environmental Science & Technology,* August 1997.

Aniline   Anilinium ion

Recall that Perkin started with this salt in his serendipitous discovery of the dye mauve (page 418).

The protonation of an amine, and the removal of the proton by treating with a base, has practical and physiological importance. Nicotine in cigarettes is normally found in the protonated form. (This water-soluble form is often used in insecticides.) Adding a base such as ammonia removes the H^+ ion to leave nicotine in its "free-base" form.

$$NicH_2^{2+}(aq) + 2\ NH_3(aq) \longrightarrow Nic(aq) + 2\ NH_4^+(aq)$$

In this form, nicotine (Nic) is much more readily absorbed by the skin and mucous membranes so the compound is a much more potent poison.

11.4   COMPOUNDS WITH A CARBONYL GROUP

Formaldehyde, acetic acid, and acetone are among the organic compounds referred to in previous examples. These compounds have a common structural feature: each contains a trigonal-planar carbon atom doubly bonded to an oxygen. The $C{=}O$ group is called the **carbonyl group,** and all of these compounds are members of a large class of compounds called **carbonyl compounds** [CD-ROM, Screen 11.5].

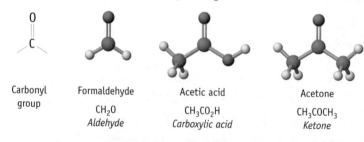

Carbonyl group

Formaldehyde

CH_2O
Aldehyde

Acetic acid

CH_3CO_2H
Carboxylic acid

Acetone

CH_3COCH_3
Ketone

In this section, we will examine five groups of carbonyl compounds (see Table 11.6):

- *Aldehydes* (RCHO) have an organic group (—R) and an H atom attached to a carbonyl group.
- *Ketones* (RCOR′) have two —R groups attached to the carbonyl carbon; they may be the same groups, as in acetone, or different groups.
- *Carboxylic acids* (RCO$_2$H) have an —R group and an —OH group attached to the carbonyl carbon.
- *Esters* (RCO$_2$R′), derivatives of acids, have —R and —OR′ groups attached to the carbonyl carbon.
- *Amides* (RCONR$_2$′, RCONHR′, and RCONH$_2$), derivatives of acids, have an —R group and an amino group (—NH$_2$, —NHR, —NR$_2$) bonded to the carbonyl carbon.

Aldehydes, ketones, and carboxylic acids are oxidation products of alcohols and, indeed, are commonly made by this route. The product obtained on oxidation of an alcohol depends on the alcohol's structure, which is classified according to the number of carbon atoms bonded to the C atom bearing the —OH group. *Primary alcohols* have one carbon and two hydrogen atoms attached, whereas *secondary alcohols* have two carbons and one hydrogen atom attached. *Tertiary alcohols* have three carbon atoms attached to the C atom bearing the —OH group.

The oxidation of a primary alcohol occurs in two steps. It is first oxidized to an aldehyde and then in a second step to a carboxylic acid:

$$
\underset{\substack{\text{primary} \\ \text{alcohol}}}{\text{R}-\text{CH}_2-\text{OH}} \xrightarrow[\text{agent}]{\text{oxidizing}} \underset{\text{aldehyde}}{\text{R}-\overset{\overset{\text{O}}{\|}}{\text{C}}-\text{H}} \xrightarrow[\text{agent}]{\text{oxidizing}} \underset{\text{carboxylic acid}}{\text{R}-\overset{\overset{\text{O}}{\|}}{\text{C}}-\text{OH}}
$$

The air oxidation of ethanol in wine produces wine vinegar. Because acids have a sour taste, the word "vinegar" means "sour wine" (from the French *vin aigre*). A device to test one's breath for alcohol relies on a similar oxidation of ethanol (Figures 5.13 and 11.10).

$$
\underset{\text{ethanol}}{\text{H}-\overset{\overset{\text{H}}{|}}{\underset{\underset{\text{H}}{|}}{\text{C}}}-\overset{\overset{\text{H}}{|}}{\underset{\underset{\text{H}}{|}}{\text{C}}}-\text{OH}(\ell)} \xrightarrow{+ \text{ oxidizing agent}} \underset{\text{acetic acid}}{\text{H}-\overset{\overset{\text{H}}{|}}{\underset{\underset{\text{H}}{|}}{\text{C}}}-\overset{\overset{\text{O}}{\|}}{\text{C}}-\text{OH}(\ell)}
$$

In contrast, the oxidation of a secondary alcohol produces a ketone:

$$
\underset{\text{secondary alcohol}}{\text{R}-\overset{\overset{\text{OH}}{|}}{\underset{\underset{\text{H}}{|}}{\text{C}}}-\text{R}'} \xrightarrow[\text{agent}]{\text{oxidizing}} \underset{\text{ketone}}{\text{R}-\overset{\overset{\text{O}}{\|}}{\text{C}}-\text{R}'}
$$

(—R and —R′ are organic groups. They may be the same or different.)

The oxidizing agents used for these reactions are common reagents such as KMnO$_4$ and K$_2$Cr$_2$O$_7$ (Table 5.4).

Finally, tertiary alcohols do not react with the usual oxidizing agents.

$$
(\text{CH}_3)_3\text{COH} \xrightarrow{\text{oxidizing agent}} \text{no reaction}
$$

Figure 11.10   Alcohol tester. This device for testing breath for the presence of ethanol relies on the oxidation of the alcohol. If present, ethanol is oxidized by potassium dichromate, K$_2$Cr$_2$O$_7$, to acetaldehyde, and then to acetic acid. The yellow-orange dichromate ion is reduced to green Cr^{3+}(aq), the color change indicating that ethanol was present. *(Charles D. Winters)*

Aldehydes and Ketones

Aldehydes and **ketones,** both classes of compounds having a carbonyl group, have pleasant odors and are often used as the basis of fragrances. Benzaldehyde is responsible for the odor of almonds and cherries, cinnamaldehyde is found in the bark of the cinnamon tree, and the ketone *p*-hydroxyphenyl-2-butanone is responsible for the odor of ripe raspberries (a favorite of the authors of this book). Table 11.8 lists several simple aldehydes and ketones.

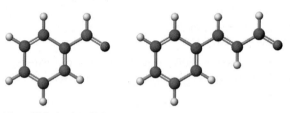

Benzaldehyde, C_6H_5CHO *trans*-Cinnamaldehyde, $C_6H_5CH=CHCHO$

Aldehydes and ketones are the oxidation products of primary and secondary alcohols, respectively. The reverse reactions, reduction of aldehydes to primary alcohols, and ketones to secondary alcohols, are also known. Commonly used reagents for such reductions are $NaBH_4$ or $LiAlH_4$, although H_2 is used on an industrial scale.

The odors of almonds and cinnamon are due to aldehydes, but the odor of fresh raspberries comes from a ketone. *(Charles D. Winters)*

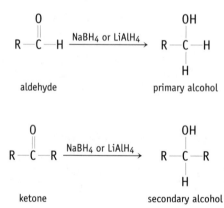

Exercise 11.7 Aldehydes and Ketones

(a) Draw the structural formula for 2-pentanone. Draw structures for a ketone and two aldehydes that are isomers of 2-pentanone, and name each of these compounds.

(b) What is the product of the reduction of 2-pentanone with $NaBH_4$?

Exercise 11.8 Aldehydes and Ketones

Draw structures and name the aldehyde or ketone that is formed on oxidation of the following alcohols: **(a)** 1-butanol, **(b)** 2-butanol, **(c)** 2-methyl-1-propanol. Are these three compounds structural isomers?

Table 11.8 • Simple Aldehydes and Ketones

Structure	Common Name	Systematic Name	BP (°C)
$\overset{O}{\overset{\|}{HCH}}$	formaldehyde	methanal	−21
$\overset{O}{\overset{\|}{CH_3CH}}$	acetaldehyde	ethanal	21
$\overset{O}{\overset{\|}{CH_3CCH_3}}$	acetone	propanone	56
$\overset{O}{\overset{\|}{CH_3CCH_2CH_3}}$	methyl ethyl ketone	butanone	80
$\overset{O}{\overset{\|}{CH_3CH_2CCH_2CH_3}}$	diethyl ketone	3-pentanone	102

Carboxylic Acids

Acetic acid is the most common and most important **carboxylic acid.** For years acetic acid was made by oxidizing ethanol produced by fermentation. Now, however, acetic acid is generally made by combining carbon monoxide and methanol in the presence of a catalyst:

$$\underset{\text{methanol}}{CH_3OH(\ell)} + CO(g) \xrightarrow{\text{catalyst}} \underset{\text{acetic acid}}{CH_3CO_2H(\ell)}$$

About a billion kilograms of acetic acid is produced annually in the United States for use in plastics, synthetic fibers, and fungicides.

Many organic acids are found naturally (Table 11.9). Acids are recognizable by their sour taste (Figure 11.11), and you will encounter them in common foods: citric acid in fruits, acetic acid in vinegar, and tartaric acid in grapes are just three examples.

Carboxylic acids often have common names derived from the source of the acid (see Table 11.9). Because formic acid is found in ants, its name comes from the Latin word for ant *(formica)*. Butyric acid gives rancid butter its unpleasant odor, and the name is related to the Latin word for butter *(butyrum)*. The systematic names of acids (Table 11.10) are formed by dropping the "-e" on the name of the corresponding alkane and adding "-oic" (and the word "acid").

Because of the substantial electronegativity of oxygen, we expect the two O atoms of the carboxylic acid group to be slightly negatively charged, and the H atom of the — OH group to be positively charged. This distribution of charges has several important implications:

- The polar acetic acid molecule dissolves readily in water, which you already know because vinegar is an aqueous solution of acetic acid. (Acids with larger organic groups are less soluble, however.)

Figure 11.11 Acetic acid in bread. Acetic acid is produced in bread that has been leavened with yeast. Additionally a group of bacteria, *Lactobacillus sanfrancisco,* contribute to the flavor of sourdough bread. These bacteria metabolize the sugar maltose, excreting acetic acid and lactic acid, $CH_3CH(OH)CO_2H$, thereby giving the bread its unique sour taste. *(Charles D. Winters)*

Formic acid, HCO_2H, is the sting in ant bites. *(Ted Nelson/Dembinsky Photo Associates)*

Table 11.9 • Some Naturally Occurring Carboxylic Acids

Name	Structure	Natural Source
benzoic acid	⬡—CO_2H	berries
citric acid	$HO_2C-CH_2-\overset{\overset{\displaystyle OH}{\|}}{\underset{\underset{\displaystyle CO_2H}{\|}}{C}}-CH_2-CO_2H$	citrus fruits
lactic acid	$H_3C-\underset{\underset{\displaystyle OH}{\|}}{CH}-CO_2H$	sour milk
malic acid	$HO_2C-CH_2-\underset{\underset{\displaystyle OH}{\|}}{CH}-CO_2H$	apples
oleic acid	$CH_3(CH_2)_7-CH=CH-(CH_2)_7-CO_2H$	vegetable oils
oxalic acid	HO_2C-CO_2H	rhubarb, spinach, cabbage, tomatoes
stearic acid	$CH_3(CH_2)_{16}-CO_2H$	animal fats
tartaric acid	$HO_2C-\underset{\underset{\displaystyle OH}{\|}}{CH}-\underset{\underset{\displaystyle OH}{\|}}{CH}-CO_2H$	grape juice, wine

---

Chemical Perspectives

Aspirin Is Over 100 Years Old

Aspirin is one of the most successful non-prescription drugs ever made. Americans swallow more than 50 million aspirin tablets a day, mostly for its pain-relieving (analgesic) effects. Aspirin wards off heart disease and thrombosis (blood clots), and it has even been suggested as a possible treatment for certain cancers and for senile dementia.

Hippocrates (c. 460–370 BC), the famous physician from ancient Greece, recommended an infusion of willow bark to ease the pain of childbirth. It was not until the 19th century that an Italian chemist, Raffaele Piria, isolated salicylic acid, the active compound, from the bark. Soon thereafter, it was found that the acid could be extracted from a wild flower *Spiraea ulmaria*. It is from the name of this plant that the name "aspirin" (a + spiraea) is derived.

Hippocrates's willow bark extract, salicylic acid, is an analgesic, but it is also very irritating to the stomach lining. It was therefore an important advance when in 1897 Felix Hoffmann and Henrich Dreser of Bayer Chemicals in Germany found that a derivative of salicylic acid, acetylsalicylic acid, was also a useful drug and had fewer side effects. This is the compound we now call "aspirin."

Acetylsalicylic acid slowly reverts to salicylic acid and acetic acid in the presence of moisture; therefore, if you smell the characteristic odor of acetic acid in an old bottle of aspirin tablets, they are too old and should be discarded.

Aspirin is a component of various over-the-counter medicines, such as Anacin, Ecotrin, Excedrin, and Alka-Seltzer. The latter is a combination of aspirin with citric acid and sodium bicarbonate. Sodium bicarbonate is a base and reacts with the acid to produce a sodium salt, a form of aspirin that is water-soluble and quicker acting.

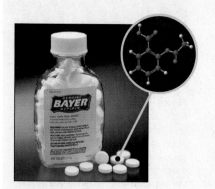

Acetylsalicylic acid, aspirin. *(Charles D. Winters)*

Glucose and Sugars

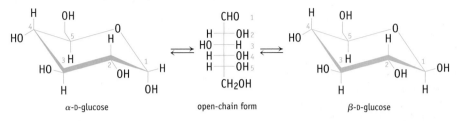

Having described alcohols and carbonyl compounds, we can now pause to look at glucose, the most common, naturally occurring carbohydrate.

As their name implies, formulas of carbohydrates can be written as though they are a combination of carbon and water, $C_x(H_2O)_y$. Thus, the formula of glucose, $C_6H_{12}O_6$, is equivalent to $C_6(H_2O)_6$. Glucose is a sugar, or, more accurately, a monosaccharide.

Carbohydrates are polyhydroxy aldehydes or ketones. Glucose is an interesting molecule in that it exists in three different isomeric forms. Two of the isomers contain six-member rings, and one is a chain structure. In solution, the three forms rapidly interconvert.

Be sure to notice that glucose is a chiral molecule. In the chain structure, four of the carbon atoms are bonded to four differ-

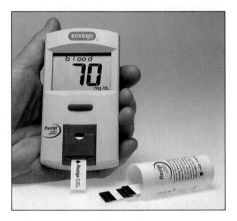

Laboratory test for glucose. *(Charles D. Winters)*

ent groups. In nature, glucose occurs in just one of its enantiomeric forms; thus, a solution of glucose rotates polarized light.

Knowing glucose's structure allows us to predict some of its properties. With five polar —OH groups in the molecule it is not surprising that glucose is soluble in water.

The aldehyde group is susceptible to chemical oxidation to form a carboxylic acid. Detection of glucose (in urine or blood) takes advantage of this fact; diagnostic tests for glucose use oxidation as part of the detection method.

Glucose is in a class of sugar molecules called hexoses, molecules having six carbon atoms. Ribose, the sugar in the backbone of the DNA molecule is a pentose, a molecule with five carbon atoms.

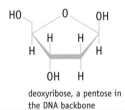

deoxyribose, a pentose in the DNA backbone

Glucose and other monosaccharides are the building blocks for larger carbohydrates. Sucrose, a disaccharide, is formed from a molecule of glucose and one of fructose, another monosaccharide. Starch, a polysaccharide, is composed of many dozens of monosaccharide units.

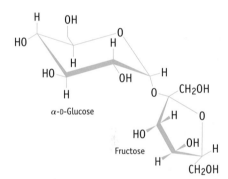

The structure of sucrose, which is formed from the sugars α-D-glucose and fructose. An ether linkage is formed by loss of H_2O from two —OH groups.

- The hydrogen of the —OH group is the acidic hydrogen. The interaction with water produces the hydronium ion and a carboxylate ion. As noted in Chapter 5, acetic acid is a weak acid in water, as are most organic acids.

Esters

Carboxylic acids (RCO_2H) react with alcohols ($R'OH$) in the presence of a strong acid to form **esters** (RCO_2R') in an **esterification** reaction.

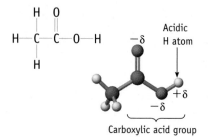

The H atom of the carboxylic acid group
($-CO_2H$) is the acidic proton of acetic
acid and of all organic carboxylic acids.

Table 11.10 • Some Simple Carboxylic Acids

Structure	Common Name	Systematic Name	BP (°C)
$\overset{O}{\underset{\|}{\|}}$ HCOH	formic acid	methanoic acid	101
$\overset{O}{\underset{\|}{\|}}$ CH₃COH	acetic acid	ethanoic acid	118
$\overset{O}{\underset{\|}{\|}}$ CH₃CH₂COH	propionic acid	propanoic acid	141
$\overset{O}{\underset{\|}{\|}}$ CH₃(CH₂)₂COH	butyric acid	butanoic acid	163
$\overset{O}{\underset{\|}{\|}}$ CH₃(CH₂)₃COH	valeric acid	pentanoic acid	187

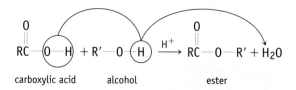

Table 11.11 lists a few common esters and the acid and alcohol from which they are formed. The two-part name of an ester is given by (1) the name of the alkyl group from the alcohol and (2) the name of the carboxylate group derived from the acid name by replacing "-ic" with "-ate". For example, ethanol (commonly called ethyl alcohol) and acetic acid combine to give the ester ethyl acetate.

An important reaction of esters is their **hydrolysis** (literally, a reaction with water), a reaction that is the reverse of the formation of the ester. The reaction, generally done in the presence of a base such as NaOH, produces the alcohol and a salt of the carboxylic acid

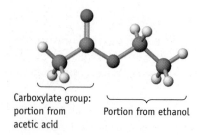

Carboxylate group:
portion from
acetic acid   Portion from ethanol

Ethyl acetate, an ester
CH₃CO₂CH₂CH₃

Table 11.11 • Some Acids, Alcohols, and Their Esters

Acid	Alcohol	Ester	Odor of Ester
CH_3CO_2H acetic acid	$CH_3\overset{CH_3}{\underset{|}{CH}}CH_2CH_2OH$ 3-methyl-1-butanol	$CH_3\overset{O}{\overset{\|}{C}}OCH_2CH_2\overset{CH_3}{\underset{|}{CH}}CH_3$ 3-methylbutyl acetate	banana
$CH_3CH_2CH_2CO_2H$ butanoic acid	$CH_3CH_2CH_2CH_2OH$ 1-butanol	$CH_3CH_2CH_2\overset{O}{\overset{\|}{C}}OCH_2CH_2CH_2CH_3$ butyl butanoate	pineapple
$CH_3CH_2CH_2CO_2H$ butanoic acid	⬡—CH_2OH benzyl alcohol	$CH_3CH_2CH_2\overset{O}{\overset{\|}{C}}OCH_2$—⬡ benzyl butanoate	rose

The carboxylic acid can be recovered if the sodium salt is treated with a strong acid such as HCl:

$$CH_3\overset{O}{\overset{\|}{C}}O^-Na^+(aq) + HCl(aq) \longrightarrow CH_3\overset{O}{\overset{\|}{C}}OH(aq) + NaCl(aq)$$

sodium acetate   acetic acid

Unlike the acids from which they are derived, esters often have pleasant odors (see Table 11.11). Typical examples are methyl salicylate, or "oil of wintergreen," and benzyl acetate. Methyl salicylate is derived from salicylic acid, the parent compound of aspirin.

$$\underset{\underset{OH}{|}}{⬡}\!\!-\!\overset{O}{\overset{\|}{C}}OH + CH_3OH \longrightarrow \underset{\underset{OH}{|}}{⬡}\!\!-\!\overset{O}{\overset{\|}{C}}OCH_3 + H_2O$$

salicylic acid   methanol   methyl salicylate, oil of wintergreen

Benzyl acetate, the active component of "oil of jasmine," is formed from benzyl alcohol ($C_6H_5CH_2OH$) and acetic acid. The chemicals are inexpensive, so synthetic jasmine is a common fragrance in less expensive perfumes and toiletries.

$$CH_3\overset{O}{\overset{\|}{C}}OH + ⬡\!\!-\!CH_2OH \longrightarrow CH_3\overset{O}{\overset{\|}{C}}OCH_2\!\!-\!⬡ + H_2O$$

acetic acid   benzyl alcohol   benzyl acetate oil of jasmine

Esters. Many fruits such as bananas and strawberries as well as consumer products (here perfume and oil of wintergreen) contain esters. *(Charles D. Winters)*

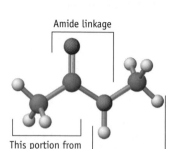

Amide linkage

This portion from acetic acid

This portion from methylamine

An amide *N*-methylacetamide, CH$_3$CONHCH$_3$.

● **Naming Amides**

In *N*-methylacetamide, the *N*-methyl portion of the name derives from the amine portion of the molecule, the *N* indicating the methyl group is attached to the nitrogen atom. The "-aceta" portion of the name indicates the acid on which the amide is based.

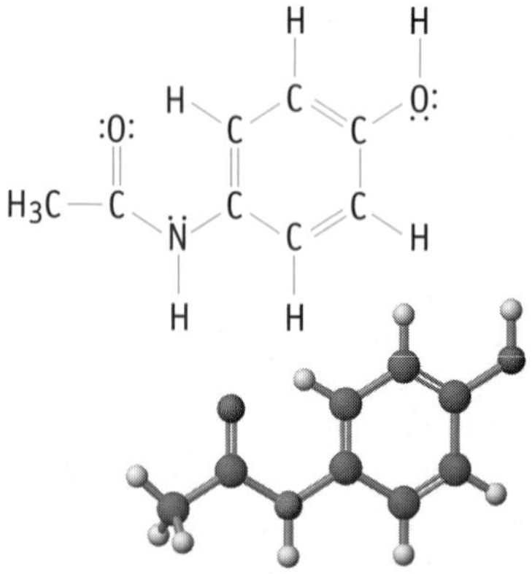

Acetaminophen is an amide. It is the analgesic used in over-the-counter painkillers such as Tylenol.

Exercise 11.9   Esters

Draw the structure and name the ester formed from each of the following reactions:

(a) propanoic acid and methanol

(b) butanoic acid and 1-butanol

(c) hexanoic acid and ethanol

Exercise 11.10   Esters

Draw the structure and name the acid and alcohol from which the following esters are derived:

(a) propyl acetate

(b) 3-methylpentyl benzoate

(c) ethyl salicylate

Amides

An acid and an alcohol react by loss of water to form an ester. In a similar manner, another class of organic compounds — **amides** — form when an acid reacts with an amine, again with loss of water.

$$R-\overset{\overset{\displaystyle O}{\|}}{C}-OR' \xrightleftharpoons[-H_2O]{+ \text{ alcohol, } R'OH} R-\overset{\overset{\displaystyle O}{\|}}{C}-OH \xrightarrow[-H_2O]{+ \text{ amine, } NHR'_2} R-\overset{\overset{\displaystyle O}{\|}}{C}-NR'_2$$

ester   acid   amide

Amides have an organic group and an amino group ($-NH_2$, $-NHR'$, or $-NR'R''$) attached to the carbonyl group.

The amide grouping is particularly important in some synthetic polymers (Section 11.5) and in many naturally occurring compounds, especially proteins (page 461). The compound *N*-acetyl-*p*-aminophenol, an analgesic known by the generic name acetaminophen and sold under the names Tylenol, Datril, and Momentum, among others, is another amide. Use of this compound as an analgesic was apparently discovered by accident when a common organic compound called acetanilide (like acetaminophen but without the $-OH$ group) was mistakenly put into a prescription for a patient. Acetanilide acts as an analgesic, but it can be toxic. An $-OH$ group *para* to the amide group makes the compound nontoxic, an interesting example of the relation between molecular structure and chemical function.

Example 11.7   Functional Group Chemistry

Problem ●

(a) Name the product of the reaction between ethylene and HCl.

(b) Draw the structure of the product of the reaction between propanoic acid and 1-propanol. What is the systematic name of the reaction product and what functional group does it contain?

(c) What is the result of reacting 2-butanol with an oxidizing agent? Give the name and draw the structure of the reaction product.

Strategy ● Ethylene is an alkene (page 431), propanoic acid is a carboxylic acid (page 447), and 2-butanol is an alcohol (page 439). Consult the discussions regarding their chemistry.

Solution •

(a) HCl will add to the double bond of ethylene to produce chloroethane.

$$H_2C=CH_2 + HCl \longrightarrow H-\underset{\underset{H}{|}}{\overset{\overset{H}{|}}{C}}-\underset{\underset{H}{|}}{\overset{\overset{H}{|}}{C}}-Cl$$

ethylene   chloroethane

(b) Carboxylic acids such as propanoic acid react with alcohols, such as 1-propanol, to give esters.

$$CH_3CH_2\overset{O}{\overset{\|}{C}}OH + CH_3CH_2CH_2OH \longrightarrow CH_3CH_2\overset{O}{\overset{\|}{C}}OCH_2CH_2CH_3 + H_2O$$

propanoic acid   1-propanol   propyl propanoate, an ester

(c) 2-Butanol is a secondary alcohol. Such alcohols are oxidized to ketones.

$$CH_3\overset{OH}{\overset{|}{C}}HCH_2CH_3 \xrightarrow{\text{oxidizing agent}} CH_3\overset{O}{\overset{\|}{C}}CH_2CH_3$$

2-butanol   butanone, a ketone

Exercise 11.11 Functional Groups

(a) Name each of the following compounds and its functional group.

1. $CH_3CH_2CH_2OH$ 2. $CH_3\overset{O}{\overset{\|}{C}}OH$ 3. $CH_3CH_2NH_2$

(b) Name the reaction product between compounds 1 and 2.

(c) What is the name and structure of the product from the oxidation of 1?

(d) What compound could result from combining compounds 2 and 3?

(e) What is the result of adding an acid (say HCl) to compound 3?

11.5 POLYMERS

We now turn to the very large molecules known as polymers. These can be either synthetic materials or naturally occurring molecules such as proteins or nucleic acids. Although these materials have widely varying compositions, their structures and properties are understandable based on the principles developed for small molecules.

Classifying Polymers

The word "polymer" means "many parts" (from the Greek, *poly* and *meros*). **Polymers** are giant molecules made by chemically joining many small molecules called **monomers.** Polymer molecular weights range from thousands to millions.

Extensive use of synthetic polymers is a fairly recent development. A few synthetic polymers (Bakelite, rayon, and celluloid) were made early in the 20th century, but most of the products you are familiar with originated in the last 50 years. By 1976, synthetic polymers outstripped steel as the most widely used material in the United States. The average production of synthetic polymers in the United States is approximately 150 kg per person annually.

The polymer industry classifies polymers in several different ways. One is their response to heating. **Thermoplastics** (like polyethylene) soften and flow when they are heated and harden when they are cooled. **Thermosetting plastics** (such as Formica) are initially soft but set to a solid when heated and cannot be resoftened. Another classification depends on the end use of the polymer. These alternative classes are plastics, fibers, elastomers, coatings, and adhesives.

Chapter Goals • Revisited

- Classify organic compounds based on formula and structure.
- Name and draw structures of common organic compounds.
- Relate properties to molecular structure.
- **Identify common polymers and compounds of biological significance.**

Fats and Oils

Fats and oils are among the many compounds found in plants and animal tissues. In the body, these substances serve several functions, a primary one being the storage of molecules that can be metabolized for energy.

Fats (solids) and oils (liquids) are triesters formed from glycerol (1,2,3-propanetriol) and three carboxylic acids that can be the same or different.

$$
\begin{array}{c}
\quad\quad\quad O \\
\quad\quad\quad \| \\
H_2C - O - CR \\
\quad\quad\quad O \\
\quad\quad\quad \| \\
HC - O - CR \\
\quad\quad\quad O \\
\quad\quad\quad \| \\
H_2C - O - CR
\end{array}
$$

The carboxylic acids in fats and oils, known as *fatty acids*, have a lengthy carbon chain (R), usually 12–18 carbon atoms long. The hydrocarbon chains can be saturated or may include one or more double bonds. The latter are referred to as monounsaturated or polyunsaturated, depending on the number of double bonds. Saturated compounds are

Common Fatty Acids

Name	Number of Carbon Atoms	Formula
Saturated Acids		
lauric	12	$CH_3(CH_2)_{10}CO_2H$
myristic	14	$CH_3(CH_2)_{12}CO_2H$
palmitic	16	$CH_3(CH_2)_{14}CO_2H$
stearic	18	$CH_3(CH_2)_{16}CO_2H$
Unsaturated Acid		
oleic	18	$CH_3(CH_2)_7CH = CH(CH_2)_7CO_2H$

more common in animal products, whereas unsaturated fats and oils are more common in plants.

In general, fats containing saturated fatty acids are solids and those with unsaturated fatty acids are liquids at room temperature. The difference in melting point relates to the molecular structure. With only single bonds linking carbon atoms, the hydrocarbon group is flexible, allowing the molecules to pack more closely together. The double bonds in unsaturated fats introduce kinks that make the hydrocarbon group less flexible and the molecules pack less tightly together.

Food companies often hydrogenate vegetable oils to reduce unsaturation. A chemical reason is that double bonds are reactive and unsaturated compounds are more susceptible to oxidation, which results

in unpleasant odors and fragrances. There are also esthetic reasons. Food processors often want solid fats to improve quality and appearance. If liquid vegetable oils are used in a cake icing, for example, the icing may slide off the cake.

Like other esters, fats and oils can undergo hydrolysis. This process is catalyzed by enzymes in the body. In industry, hydrolysis is carried out using aqueous NaOH or KOH to produce a mixture of glycerol and the alkali metal salts of the fatty acids. This reaction is called *saponification*, a term meaning "soap making."

Glyceryl stearate, a fat
$R = -(CH_2)_{16}CH_3$

$$
\begin{array}{c}
\quad\quad\quad O \\
\quad\quad\quad \| \\
H_2C - O - CR \\
\quad\quad\quad O \\
\quad\quad\quad \| \\
HC - O - CR + 3\ NaOH \\
\quad\quad\quad O \\
\quad\quad\quad \| \\
H_2C - O - CR \\
\downarrow
\end{array}
$$

$$
\begin{array}{c}
H_2C - O - H \quad\quad\quad O \\
\quad\quad\quad\quad\quad\quad\quad\quad \| \\
HC - O - H + 3\ RC - O^- Na^+ \\
H_2C - O - H
\end{array}
$$

glycerol   sodium stearate, a soap

Simple soaps are sodium salts of fatty acids. The anion in these compounds has an ionic end (the carboxylate group) and a nonpolar end (the large hydrocarbon tail). The ionic end allows these molecules to interact with water, the nonpolar end enables them to mix with oily and greasy substances.

About 94% of the fatty acids in olive oil are monounsaturated. The major fatty acid in olive oil is oleic acid. (Charles D. Winters)

Polar bears feed primarily on seal blubber and build up a huge fat reserve during winter. During summer, they maintain normal activity but eat nothing, relying entirely on body fat for sustenance. A polar bear will burn about 1 to 1.5 kg of fat per day. (Dan Guravich/CORBIS)

(a)

(b)

(c)

Figure 11.12   Common consumer products. (a) Packaging materials from high-density polyethylene; **(b)** objects made from polystyrene **(c)** and from polyvinylchloride. Recycling information is provided on most plastics (usually molded into the bottom of the bottle). High-density polyethylene is designated with a "2" inside a triangular symbol and the letters "HDPE." PVC is designated with a "3" inside a triangular symbol with the letter "V" below. *(Charles D. Winters)*

A more chemical approach to polymer classification is based on their method of synthesis. **Addition polymers** are made by directly adding monomer units together. **Condensation polymers** are made by combining monomer units and splitting out a small molecule, often water.

Addition Polymers

Polyethylene, polystyrene, and polyvinylchloride (PVC) are common addition polymers (Figure 11.12) [⏚ CD-ROM, Screen 11.9]. These are built by "adding together" simple alkenes such as ethylene ($H_2C=CH_2$), styrene ($C_6H_5CH=CH_2$), and vinyl chloride ($H_2C=CHCl$). These and other addition polymers (Table 11.12), all derived from alkenes, have widely varying properties and uses.

Polyethylene and Other Polyolefins

Polyethylene is by far the production leader among addition polymers. Ethylene (C_2H_4), the monomer from which polyethylene is made, is a product of petroleum refining and among the top five chemicals produced in the United States. When ethylene is heated to between 70 and 100 °C at a pressure of 1 to 20 atm in the presence of a catalyst, polymers with molecular weights up to several million are formed. The reaction can be expressed as a balanced chemical equation:

$$n\ H_2C=CH_2 \longrightarrow \left(\begin{matrix} H & H \\ | & | \\ C & C \\ | & | \\ H & H \end{matrix}\right)_n$$

ethylene   polyethylene

The abbreviated formula of the reaction product, $-(CH_2CH_2-)_n$, shows that polyethylene is a chain of carbon atoms, each bearing two hydrogens. The chain length for polyethylene can be very long. A polymer with a molecular weight of one million would contain almost 36,000 ethylene molecules linked together.

Polyethylene film is produced by extruding the molten plastic through a ring-like gap and inflating the film like a balloon. *(Christopher Springmann/corbisstockmarket. com)*

Table 11.12 • Ethylene Derivatives That Undergo Addition Polymerization

Formula	Monomer Common Name	Polymer Name (Trade Names)	Uses	U.S. Polymer Production (Metric tons/year)*
$H_2C=CH_2$	ethylene	polyethylene (polythene)	squeeze bottles, bags, films, toys and molded objects, electric insulation	7 million
$H_2C=CHCH_3$	propylene	polypropylene (Vectra, Herculon)	bottles, films, indoor–outdoor carpets	1.2 million
$H_2C=CHCl$	vinyl chloride	polyvinyl chloride (PVC)	floor tile, raincoats, pipe	1.6 million
$H_2C=CHCN$	acrylonitrile	polyacrylonitrile (Orlon, Acrilan)	rugs, fabrics	0.5 million
$H_2C=CH(C_6H_5)$	styrene	polystyrene (Styrofoam, Styron)	food and drink coolers, building material insulation	0.9 million
$H_2C=CH\,O-C(=O)-CH_3$	vinyl acetate	polyvinyl acetate (PVA)	latex paint, adhesives, textile coatings	200,000
$H_2C=C(CH_3)\,C(=O)-O-CH_3$	methyl methacrylate	polymethyl methacrylate (Plexiglas, Lucite)	high-quality transparent objects, latex paints, contact lenses	200,000
$F_2C=CF_2$	tetrafluoroethylene	polytetrafluoroethylene (Teflon)	gaskets, insulation, bearings, pan coatings	6,000

*One metric ton = 1000 kg

Polyethylenes formed under various pressures and catalytic conditions have different properties, a result of different molecular structures. For example, when chromium oxide is used as a catalyst, the product is almost exclusively a linear chain (Figure 11.13a). If ethylene is heated to 230 °C at high pressure, however, irregular branching occurs. Still other conditions lead to cross-linked polyethylene, in which different chains are linked together (Figure 11.13b and c).

The high-molecular-weight chains of linear polyethylene pack closely together and result in a material with a density of 0.97 g/mL. This material, referred to as high-density polyethylene (HDPE), is hard and tough, which makes it suitable for things like milk bottles. If the polyethylene chain contains branches, however, the chains cannot pack as closely together, and a lower density material (0.92 g/mL) known as low-density polyethylene (LDPE) results. This material is softer and more flexible than HDPE and is used in such things as sandwich bags. Linking up the

• Teflon
See page 6 for the story of the discovery of Teflon.

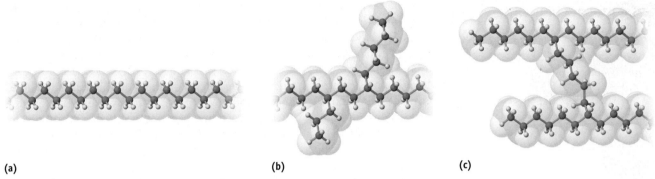

(a)   **(b)**   **(c)**

Figure 11.13   Polyethylene. (a) The linear form, high-density polyethylene (HDPE). **(b)** Branched chains make low-density polyethylene (LDPE). **(c)** Cross-linked polyethylene (CLPE).

polymer chains in cross-linked polyethylene (CLPE) causes the material to be even more rigid and inflexible. Plastic bottle caps are often made of CLPE.

Polymers formed from substituted ethylenes ($H_2C{=}CHX$) have a range of properties and uses (see Table 11.12). Often the properties are predictable based on structure. Polymers without polar substituent groups, like polystyrene, often dissolve in organic solvents, a property useful for some types of fabrication (Figure 11.14).

Polymers based on substituted ethylenes, $H_2C{=}CHX$

$$\left(\!CH_2CH\!\right)_n \qquad \left(\!CH_2CH\!\right)_n \qquad \left(\!CH_2CH\!\right)_n$$
$$\quad OH \qquad\qquad OCCH_3 \qquad\qquad\quad$$
$$\qquad\qquad\qquad\qquad O$$

polyvinyl alcohol   polyvinyl acetate   polystyrene

Polyvinyl alcohol is a polymer with little affinity for nonpolar solvents but an affinity for water, which is not surprising based on the large number of polar — OH groups (Figure 11.15). Vinyl alcohol itself is not a stable compound (it isomerizes to CH_3CHO), so polyvinyl alcohol cannot be made from this compound. Instead, it is made by hydrolyzing the ester groups in polyvinyl acetate.

Solubility in water or organic solvents can be a liability. The many uses of poly-tetrafluoroethylene [Teflon, $\left(\!CF_2CF_2\!\right)_n$] are a consequence of the fact that it does not interact with water or organic solvents (PAGE 486).

Polystyrene, with $n = 5700$, is a clear, hard, colorless solid that can be molded easily at 250 °C (Figure 11.12b). You are probably more familiar with the very light, foam-like material (Styrofoam) used widely for food and beverage containers and for home insulation. Styrofoam is produced by a process called "expansion molding." Polystyrene beads containing 4–7% of a low-boiling liquid like pentane are placed in a mold and heated with steam or hot air. Heat causes the solvent to vaporize and create a foam in the molten polymer that expands to fill the shape of the mold.

Natural and Synthetic Rubber

Natural rubber was first introduced in Europe in 1740, but it remained a curiosity until 1823 when Charles Mackintosh invented a way of using it to waterproof cotton cloth. The mackintosh, as rain coats are still sometimes called, became popu-

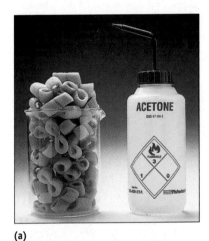

(a)

(b)

Figure 11.14   Polystyrene. (a) The polymer is a clear, hard, colorless solid, but it may be more familiar as a light, foam-like material called Styrofoam. **(b)** Styrofoam has no polar groups and thus dissolves well in organic solvents such as acetone. *(Charles D. Winters)*

Figure 11.15   Slime. When boric acid, B(OH)₃, is added to an aqueous suspension of polyvinyl alcohol, (CH₂CHOH)$_n$, the mixture becomes very viscous. This is because boric acid reacts with the —OH groups on the polymer chain, causing cross-linking to occur. (The model shows an idealized structure of a portion of the polymer.) *(Photo, Charles D. Winters)*

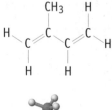

Isoprene, 2-methyl-1,3-butadiene.

The covering on a golf ball consists of gutta-percha rubber. In this hard material, produced from the sap of a certain kind of rubber tree, all the double bonds in poly-isoprene have *trans* configurations. *(Charles D. Winters)*

lar despite major problems: natural rubber is notably weak and is soft and tacky when warm but brittle at low temperatures. In 1839, after five years of work on natural rubber, the American inventor Charles Goodyear (1800–1860) discovered that heating gum rubber with sulfur produces a material that is elastic, water-repellent, resilient, and no longer sticky.

Rubber is a naturally occurring polymer, the monomers of which are molecules of 2-methyl-1,3-butadiene, commonly called *isoprene.* In natural rubber, isoprene monomers are linked together through carbon atoms 1 and 4, that is, through the end carbon atoms of the four-carbon chain (Figure 11.16). This leaves a double bond between carbon atoms 2 and 3. In natural rubber, all these double bonds have a *cis* configuration.

In vulcanized rubber, the material that Goodyear discovered, the polymer chains of natural rubber are cross-linked by short chains of sulfur atoms. Cross-linking helps to align the polymer chains so the material does not undergo a permanent change when stretched, and it springs back when the stress is removed. Substances that behave this way are called **elastomers.**

With a knowledge of the composition and structure of natural rubber, chemists began searching for ways to make synthetic rubber. When they first tried to make synthetic rubber by linking isoprene monomers together, however, what they made was sticky and useless. The problem was that synthetic procedures gave a mixture of *cis* and *trans* polyisoprene. In 1955, however, chemists at the Goodyear and Firestone companies discovered special catalysts to prepare the all -*cis* polymer. This synthetic material, structurally identical to natural rubber, is now manufactured cheaply, and over 8.0×10^8 kg of synthetic polyisoprene is produced annually in the United States.

Other kinds of polymers add to the repertoire of elastomeric materials now available. Polybutadiene, for example, is currently used in the production of tires, hoses, and belts. Some elastomers, called **copolymers,** are formed by polymerization of two (or more) different monomers. A copolymer of styrene and butadiene, made with a 1-to-3 ratio of these raw materials, is the most important synthetic rubber now made; more than one billion kilograms of styrene-butadiene rubber (SBR) is produced each year in the United States for making tires.

1,3-butadiene   styrene

sytrene-butadiene rubber (SBR)

And a little is left over each year to make bubble gum. The stretchiness of bubble gum once came from natural rubber, but SBR is now used to help you blow bubbles.

Condensation Polymers

A chemical reaction in which two molecules react by splitting out, or eliminating, a small molecule is called a **condensation reaction** [⊕ CD-ROM, Screen 11.10]. A reaction of an alcohol with a carboxylic acid to give an ester is an example. A polymer can be formed in a condensation reaction if two different reactant molecules, each containing *two* functional groups, are used. This is the route used to form polyesters and polyamides, two important types of condensation polymers.

Polyesters

Terephthalic acid has two carboxylic acid groups, and ethylene glycol has two alcohol groups. When mixed, the acid and alcohol functional groups at both ends of these molecules can react to form ester linkages, splitting out water. The result is a polymer called polyethylene terephthalate (PET) (Figure 11.17). The multiple ester linkages in the product make this substance a **polyester.**

Figure 11.16   Natural rubber. The sap that comes from the rubber tree is a natural polymer of isoprene. All the linkages in the carbon chain are *cis*. *(Photo, Kevin Schafer/Tom Stack & Associates)*

$$n \text{ HOC} \longrightarrow \text{COH} + n \text{ HOCH}_2\text{CH}_2\text{OH} \longrightarrow \left(\text{C} \longrightarrow \text{COCH}_2\text{CH}_2\text{O} \right)_n + n \text{ H}_2\text{O}$$

terephthalic acid   ethylene glycol   polyethylene terephthalate (PET), a polyester

Polyester textile fibers made from PET are marketed as Dacron and Terylene. The inert, nontoxic, noninflammatory, and non–blood-clotting properties of Dacron polymers make Dacron tubing an excellent substitute for human blood vessels in heart bypass operations, and Dacron sheets are sometimes used as temporary skin for burn victims. A polyester film, Mylar, has unusual strength and can be rolled into sheets one-thirtieth the thickness of a human hair. Magnetically coated Mylar films are used to make audio and video tapes (see Figure 11.17).

Polyamides

In 1928, the Du Pont Company embarked on a basic research program headed by Dr. Wallace Carothers (1896–1937). Carothers, who left a faculty position at Harvard to go to Du Pont, was interested in high-molecular-weight compounds, such as rubbers, proteins, and resins. In 1935, his research yielded nylon-6,6 (Figure

The elasticity of bubble gum comes from a copolymer of styrene and butadiene, SBR rubber. *(Richard Hutchings/Photo Researchers, Inc.)*

Figure 11.17   Polyesters. Polyethylene terephthalate is used to make clothing and soda bottles. Film made of Mylar, also a polyester, is used to make recording tape as well as balloons. Because the film has very tiny pores, it can be used for helium-filled balloons; the atoms of gaseous helium move through the pores very slowly. (The two students are wearing jackets made from recycled PET soda bottles.) *(Charles D. Winters)*

Figure 11.18   Nylon-6,6.
Hexamethylenediamine is dissolved in water (*bottom layer*), and adipoyl chloride (a derivative of adipic acid) is dissolved in hexane (*top layer*). The two compounds react at the interface between the layers to form nylon, which is being wound onto a stirring rod. *(Charles D. Winters)*

11.18), a **polyamide** prepared from adipoyl chloride, a derivative of adipic acid, a diacid, and hexamethylenediamine, a diamine:

$$n\ \text{ClC(CH}_2)_4\text{CCl} + n\ \text{H}_2\text{N(CH}_2)_6\text{NH}_2 \longrightarrow \left(\text{C(CH}_2)_4\text{C} - \text{N(CH}_2)_6\text{N} \right)_n + n\ \text{HCl}$$

adipoyl chloride   hexamethylenediamine   amide link in nylon-66, a polyamide

Nylon can be extruded easily into fibers that are stronger than natural fibers and chemically more inert. The discovery of nylon jolted the American textile industry at a critical time. Natural fibers were not meeting 20th-century needs. Silk was expensive and not durable, wool was scratchy, linen crushed easily, and cotton did not have a high-fashion image. Perhaps the most identifiable use for the new fiber was in nylon stockings. The first public sale of nylon hosiery took place on October 24, 1939, in Wilmington, Delaware (the site of Du Pont's main office.) This use of nylon in commercial products ended shortly thereafter, however, with the start of World War II. All nylon was diverted to making parachutes and other military gear. It was not until about 1952 that nylon reappeared in the consumer market place.

Figure 11.19 illustrates why nylon makes such a good fiber. To have good tensile strength (the ability to resist tearing) the polymer chains should be able to attract one another, though not so strongly that the plastic cannot be initially extended to form fibers. Ordinary covalent bonds between the chains (cross-linking) would be too strong. Instead, cross-linking occurs by a somewhat weaker intermolecular force (called *hydrogen bonding*) (➜ SECTION 13.3) between the hydrogens of N—H groups on one chain and carbonyl oxygens on another chain. The polarities of the $\text{N}(\delta^-)-\text{H}(\delta^+)$ group and the $\text{C}(\delta^+)=\text{O}(\delta^-)$ group lead to attractive forces between the polymer chains of the desired magnitude.

Example 11.8   Condensation Polymers

Problem • What is the repeating unit of the condensation polymer obtained by combining $\text{HO}_2\text{CCH}_2\text{CH}_2\text{CO}_2\text{H}$ (succinic acid) and $\text{H}_2\text{NCH}_2\text{CH}_2\text{NH}_2$ (1,2-ethylenediamine)?

Strategy • Recognize that the polymer will link the two parts of the polymer through the amide linkage. The smallest repeating unit of the chain will contain two parts: one from the diacid and the other from the diamine.

Solution • The repeating unit is of this polyamide is

amide linkage

$$\left(\text{CCH}_2\text{CH}_2\text{C} - \text{NCH}_2\text{CH}_2\text{N} \right)_n$$

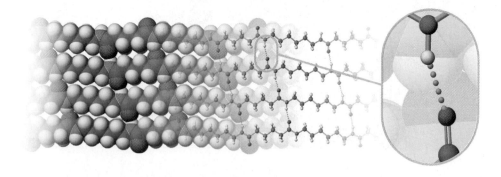

Figure 11.19   Hydrogen bonding between polyamide chains. Carbonyl oxygen atoms with a partial negative charge on one chain interact with an amine hydrogen with a partial positive charge on a neighboring chain. (Hydrogen bonding is described in more detail in Section 13.3.)

A Closer Look

Proteins — Natural Condensation Polymers

There are about 100,000 different proteins in your body; and about 50% of the dry weight of your body consists of proteins. These versatile compounds fulfill a number of functions. Some proteins are structural (muscle, collagen), others serve as catalysts (enzymes) and regulators (hormones), and still others assist in storage and transport of materials such as oxygen (hemoglobin).

Proteins are polymers. The building blocks for proteins are α-*amino acids*, the symbol α indicating that the amino group is attached to the carbon adjacent to the carboxyl group.

Formation of a dipeptide. Two amino acids condense to form a peptide bond. Proteins are polypeptides, polymers consisting of many amino acid units linked through peptide bonds.

In aqueous solution, amino acids exist in two isomeric forms, $H_2NCHRCO_2H$ and $H_3N^+CHRCO_2^-$, in equilibrium. The second form, called the *zwittererionic form*, is typically the more abundant of the two, and we have used this form here in the drawings of amino acid structures. The 20 amino acids that make up proteins differ only in the identity of the R group. All naturally occurring amino acids except glycine, where R = H, have a chiral α-carbon atom and, interestingly, all occur in nature in a single enantiomeric form. Amino acids are further classified by whether the R group is nonpolar or polar and whether the polar R group is acidic, neutral, or basic. The wide varia-

tion in characteristics of the R groups is important in determining the structure and function of proteins.

Proteins are formed from amino acids by condensation reactions. These reactions result in the elimination of water and the formation of an amide linkage between the carboxylic acid carbon on one amino acid and the amino group of another. The amide

linkage in proteins is often called a *peptide bond,* and the polymer (the protein) is a **polypeptide.** Naturally occurring proteins typically have molar masses of 5000 or greater and consist of one or more polymer chains. The protein insulin, for example, has two peptide chains, each with 51 amino acid units that are connected by a few bonds between R groups. Human hemoglobin has four polypeptide chains: two identical ones with 141 amino acids, and two other identical chains with 146 amino acids.

Because molecular structure is the key to understanding the properties of compounds, a great deal of effort has gone into structural determination. The starting point in protein structures is to determine its *primary structure*, the amino acid sequence. Equally important, however, is the protein's three-dimensional structure. This includes how the polymer chains coil or line up to form sheets (*secondary structure*), how protein chains fold (*tertiary structure*), and how two or more folded subunits interact (*quaternary structure*).

These five α-amino acids are representative of the 20 α-amino acids from which proteins are made.

Exercise 11.12   **Condensation Polymers**

Draw the structure of the repeating unit in the condensation polymer obtained from the reaction of propylene glycol with maleic acid:

$$
\begin{array}{ccc}
& \text{H}_3\text{C} \quad \text{H} & \text{O} \quad \text{H} \quad \text{H} \quad \text{O} \\
\text{HO}-\text{C}-\text{C}-\text{OH} & + & \text{HO}-\text{C}-\text{C}=\text{C}-\text{C}-\text{OH} \\
& \text{H} \quad \text{H} &
\end{array}
$$

propylene glycol   maleic acid

(A closely related material is combined with glass fibers [fiberglass] to make hulls for small boats and automobile panels and parts.) Is this a polyamide or a polyester?

New Polymer Materials

Few plastics are used today without some modification. Body panels for the Saturn and Corvette automobiles, for example, are made of **reinforced plastics,** which contain fibers in a polymer matrix. These materials are referred to as **composites.** The strongest geometry for a solid is a wire or fiber, and the use of a polymer matrix prevents the fiber from buckling or bending. As a result, reinforced plastics are stronger than steel. In addition, the composites have a low density, from 1.5 to 2.25 g/cm^3, compared with 2.7 g/cm^3 for aluminum, 7.9 g/cm^3 for steel, and 2.5 g/cm^3 for concrete. Wood is the only common structural material with a lower density. The low density, high strength, and resistance to corrosion of composites favor use of these materials.

Glass fibers currently account for more than 90% of the fibrous material used to reinforce plastics. Glass is inexpensive and glass fibers possess high strength, low density, good chemical resistance, and good insulating properties. In principle, any polymer can be used as the matrix material. Polyesters are the number one polymer matrix at the present time, and glass-reinforced polymer composites are commonly used for structural applications, such as boat hulls, airplanes, and automobile body panels.

Other fibers and polymers have been used, and the trend is toward increased utilization of composites in automobiles and aircraft. One of the best examples are the *graphite composites* used in the construction of aircraft parts and in a number of sporting goods such as golf club shafts, tennis racquets, fishing rods, and skis. Carbon or graphite fibers are made from polyacrylonitrile, PAN (see Table 11.12). The PAN fibers are first oxidized around 200 °C and then carbonized at temperatures higher than 1000 °C. This process makes long fibers that are about 90% carbon. Further treatment at temperatures as high as 3000 °C yields fibers that are about 99% carbon. Carbon fiber composites can replace metals because the composites are strong and can be conductive. They also provide strength without the problem of metal fatigue and do not corrode. These composites will find much wider use when the price decreases.

Composite materials made from graphite fibers and synthetic polymers have found many uses, such as aircraft parts and golf clubs. Here you are looking up at the large sail of a racing yacht made from carbon and aramid fibers. (Aramid fibers are like nylon, except that 1,4–disubstituted benzene groups and not C_6 chains provide the carbon backbone of the polymer.) *(Doyle Sailmakers, Inc.)*

In Summary

When you have finished studying this chapter, ask if you have met the chapter goals. In particular you should be able to

• Understand the factors that contribute to the large numbers of organic compounds and the wide array of structures (Section 11.1).

- Recognize and draw structures of structural isomers and stereoisomers for carbon compounds, including geometric isomers and optical isomers (Section 11.1).
- Draw structural formulas and name simple hydrocarbons, including alkanes, alkenes, alkynes, and aromatic compounds (Section 11.2).
- Identify possible isomers that have a given formula (Section 11.2).
- Summarize some reactions for the various types of hydrocarbons (Section 11.2).
- Describe the physical and chemical properties of hydrocarbons (Section 11.2).
- Name and draw structures of alcohols and amines (Section 11.3).
- Name and draw structures of carbonyl compounds: aldehydes, ketones, acids, esters, and amides (Section 11.4).
- Know the structures and properties of several natural products including carbohydrates, fats and oils, and proteins pages 448, 454, and 461.
- Write equations for the formation of addition polymers and condensation polymers, and describe the structures (Section 11.5).
- Relate properties of polymers to their structures (Section 11.5).

Key Terms

Section 11.1
isomers
structural isomers
stereoisomers
geometric isomers
optical isomerism
chiral
enantiomers

Section 11.2
hydrocarbons
alkane
saturated compound
alkyl groups
cycloalkane
strained hydrocarbon
alkene
alkyne
unsaturated compound

addition reaction
hydrogenation
aromatic compound
ortho, meta, and *para*
resonance stabilization

Section 11.3
functional group
alcohol
ether
amine

Section 11.4
carbonyl group
carbonyl compound
aldehyde
ketone
carboxylic acid
ester
esterification

hydrolysis
amide

Section 11.5
polymer
monomer
thermoplastic
thermosetting plastics
addition polymer
condensation polymer
elastomers
copolymer
condensation reaction
polyester
polyamide
reinforced plastic
composite

Study Questions

Questions with blue, bold-faced numbers have answers in Appendix O. Naming organic compounds is described in Appendix E. Structures of many of the compounds used in these questions are found on the General Chemistry Interactive CD-ROM, *Version 3, in the* Models *folder.*

Reviewing Important Concepts

1. Carbon atoms appear in organic compounds in several different ways with single, double, and triple bonds combining to give an octet configuration. Describe the various ways that carbon can bond to reach an octet, and give the name and draw the structure of a compound that illustrates that mode of bonding.

2. There is a high barrier to rotation around a carbon–carbon double bond, whereas the barrier to rotation around a carbon–carbon single bond is very small. Use the orbital overlap model of bonding (Chapter 10) to explain why there is restricted rotation around a double bond.

3. Three different compounds with the formula $C_2H_2Cl_2$ are known.

(a) Two of these compounds are geometric isomers. Draw their structures.

(b) The third compound is a structural isomer of the other two. Draw its structure.

4. Draw the structure of 2-butanol. Identify the chiral carbon atom in this compound. Draw the mirror image of the structure you first drew. Are the two molecules superimposable?

5. Draw Lewis structures and name the three isomers with the formula C_6H_{14}. Are any of these isomers chiral?

6. Draw structures and name the four alkenes that have the formula C_4H_8.

7. Write equations for the reactions of *cis*-2-butene with the following reagents, representing the reactants and products using structural formulas.

(a) H_2O

(b) $H_2(g)$ (in the presence of a catalyst)

(c) HBr

(d) Cl_2

8. Draw the structure and name the product formed if the following alcohols are oxidized. Assume an excess of the oxidizing agent is used. If the alcohol is not expected to react with a chemical oxidizing agent, write NR (no reaction).

(a) $CH_3CH_2CH_2CH_2OH$

(b) 2-Butanol

(c) 2-Methyl-2-propanol

(d) 2-Methyl-1-propanol

9. Write equations for the following reactions, representing the reactants and products using structural formulas.

(a) The reaction of acetic acid and sodium hydroxide

(b) The reaction of methylamine with HCl.

10. Write equations for the following reactions, representing the reactants and products using structural formulas.

(a) The formation of ethyl acetate from acetic acid and ethanol

(b) The hydrolysis of glyceryl tristearate (the triester of glycerol with stearic acid, a fatty acid)

11. Write an equation for the formation of the following polymers.

(a) Polystyrene, from styrene ($C_6H_5CH = CH_2$)

(b) PET (polyethylene terephthalate), from ethylene glycol and terephthalic acid

12. Write equations for the following reactions, representing the reactants and products using structural formulas.

(a) The hydrolysis of the amide, $C_6H_5CONHCH_3$ to form benzoic acid and methylamine.

(b) The hydrolysis of nylon-66,

$$-[CO(CH_2)_4CONH(CH_2)_6NH]_x$$

a polyamide, to give a carboxylic acid and an amine.

13. Identify all of the functional groups in the following compounds:

(a) Aspirin, whose structure is shown on page 449

(b) Alanine, whose structure is shown on page 461

(c) Acetaminophen, whose structure is shown on page 452

14. What important properties do the following characteristics impart on polymer?

(a) Cross-linking in polyethylene

(b) The OH groups in polyvinyl alcohol

(c) Hydrogen bonding in a peptide

15. Draw the structure of cysteine, an α-amino acid where $R = -CH_2SH$. Designate the chiral carbon atom.

Practicing Skills

Alkanes and Cycloalkanes
(See Examples 11.1 and 11.2, CD-ROM Screen 11.3)

16. What is the name of the straight (unbranched) chain alkane with the formula C_7H_{16}?

17. What is the molecular formula for an alkane with 12 carbon atoms?

18. Which of the following is an alkane? Which could be a cycloalkane?

(a) C_2H_4 (c) $C_{14}H_{30}$

(b) C_5H_{10} (d) C_7H_8

19. Isooctane, 2,2,4-trimethylpentane, is one of the possible structural isomers with the formula C_8H_{18}. Draw the structure of this isomer, and draw and name structures of two other isomers of C_8H_{18} in which the longest carbon chain is five atoms.

20. Give the systematic name for the following alkane:

$$CH_3$$
$$|$$
$$CH_3CHCHCH_3$$
$$|$$
$$CH_3$$

21. Give the systematic name for the following alkane. Draw a structural isomer of the compound and give its name.

$$CH_3$$
$$|$$
$$CH_3CHCH_2CH_2CHCH_3$$
$$|$$
$$CH_2CH_3$$

22. Draw the structure of each of the following compounds:

(a) 2,3-Dimethylhexane

(b) 2,3-Dimethyloctane

(c) 3-Ethylheptane

(d) 2-Methyl-3-ethylhexane

23. Draw a structure for cycloheptane. Is the seven-member ring planar? Explain your answer.

24. Draw Lewis structures and name all possible compounds that have a seven-carbon chain with one methyl substituent group. Which of these isomers has a chiral carbon center?

25. Draw a structure for cyclohexane in a perspective that allows you to distinguish between axial and equatorial hydrogens.

26. There are two ethylheptanes (compounds with a seven carbon chain and one ethyl substituent). Draw the structure, and name these compounds. Is either isomer chiral?

27. Among the 18 structural isomers with the formula C_8H_{18} are two with a five-carbon chain having one ethyl and one methyl substituent group. Draw structures and name these two isomers.

28. List several typical physical properties of C_4H_{10}. Predict the following physical properties of dodecane, $C_{12}H_{26}$: color, state (s, ℓ, g), solubility in water, solubility in a nonpolar solvent.

29. Write balanced equations for the following reactions of alkanes.

    (a) The reaction of methane with excess chlorine

    (b) Complete combustion of cyclohexane, C_6H_{12}, with excess oxygen

Alkenes and Alkynes
(See Examples 11.3 and 11.4, CD-ROM Screens 11.3 and 11.4)

30. Draw structures for the *cis* and *trans* isomers of 4-methyl-2-hexene.

31. What structural requirement is necessary for an alkene to have *cis* and *trans* isomers? Can *cis* and *trans* isomers exist for an alkane? For an alkyne?

32. A hydrocarbon with the formula C_5H_{10} can be either an alkene or a cycloalkane.

    (a) Draw a structure for each of the isomers possible for C_5H_{10}, assuming it is an alkene. Six isomers are possible. Give the systematic name of each isomer you have drawn.

    (b) Draw a structure for a cycloalkane having the formula C_5H_{10}.

33. Five alkenes have the formula C_7H_{14} with a seven-carbon atom chain. Draw their structures and name them.

34. Draw the structure and give the systematic name for the products of the following reactions:

    (a) $CH_3CH=CH_2 + Br_2 \longrightarrow$

    (b) $CH_3CH_2CH=CHCH_3 + H_2 \longrightarrow$

35. Draw the structure and give the systematic name for the products of the following reactions:

    (a)
    $$\begin{array}{c} H_3C \\ C=C \\ H_3C \end{array} \begin{array}{c} CH_2CH_3 \\ \\ H \end{array} + H_2 \longrightarrow$$

    (b) $CH_3C\equiv CCH_2CH_3 + 2\ Br_2 \longrightarrow$

36. The compound $CH_3CH_2CH_2CH_2Br$ is a product of addition of HBr to an alkene. Identify the alkene.

37. The compound 2,3-dibromo-2-methylhexane is formed by addition of Br_2 to an alkene. Identify the alkene, and write an equation for this reaction.

38. Draw and name alkenes that have the formula C_3H_5Cl (derivatives of propene in which a chlorine atom replaces one hydrogen).

39. Elemental analysis of a colorless liquid has given its formula as C_5H_{10}. You recognize that this could be either a cy-

cloalkane or an alkene. A chemical test to determine the class to which this compound belongs involves adding bromine. Explain how this would allow you to distinguish between the two classes.

Aromatic Compounds
(See Example 11.5 and Exercise 11.4, CD-ROM Screen 11.3)

40. Draw structural formulas for the following compounds:

    (a) *m*-Dichlorobenzene (alternatively called 1,3-dichlorobenzene)

    (b) *p*-Bromotoluene (alternatively called 1-bromo-4-methylbenzene)

41. Give the systematic name for each of the following compounds:

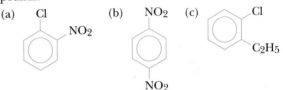

42. Write an equation for the preparation of ethylbenzene from benzene and an appropriate compound containing an ethyl group.

43. Write an equation for the preparation of hexylbenzene from benzene and other appropriate reagents.

44. A single compound is formed by alkylation of *p*-xylene (1,4-dimethylbenzene). Write the equation for the reaction of this compound with CH_3Cl and $AlCl_3$. What is the structure and name of the product?

45. Nitration of toluene gives a mixture of two products: one with the nitro group ($-NO_2$) in the *ortho* position and one with the nitro group in a *para* position. Draw structures of the two products.

Alcohols, Ethers, and Amines
(See Example 11.6 and CD-ROM Screen 11.5)

46. Give the systematic name for each of the following alcohols, and tell if each is a primary, secondary, or tertiary alcohol:

    (a) $CH_3CH_2CH_2OH$

    (b) $CH_3CH_2CH_2CH_2OH$

    (c)
    $$\begin{array}{c} CH_3 \\ | \\ H_3C-C-OH \\ | \\ CH_3 \end{array}$$
    (d)
    $$\begin{array}{c} CH_3 \\ | \\ H_3C-C-CH_2CH_3 \\ | \\ OH \end{array}$$

47. Draw structural formulas for the following alcohols, and tell if each is primary, secondary, or tertiary:

    (a) 1-Butanol

    (b) 2-Butanol

    (c) 3,3-Dimethyl-2-butanol

    (d) 3,3-Dimethyl-1-butanol

48. Give formulas and structures of the following amines:

    (a) Ethylamine

    (b) Dipropylamine

(c) Butyldimethylamine

(d) Triethylamine

49. Name the following amines
   (a) $CH_3CH_2CH_2NH_2$
   (b) $(CH_3)_3N$
   (c) $(CH_3)(C_2H_5)NH$
   (d) $C_6H_{13}NH_2$

50. Draw structural formulas for all the alcohols with the formula $C_4H_{10}O$. Give the systematic name of each.

51. Draw structural formulas for all primary amines with the formula $C_4H_9NH_2$.

52. Complete the following equations:
   (a) $C_6H_5NH_2(aq) + HCl(aq) \longrightarrow$
   (b) $(CH_3)_3N(aq) + H_2SO_4(aq) \longrightarrow$

53. Aldehydes and carboxylic acids are formed by oxidation of primary alcohols and ketones are formed when secondary alcohols are oxidized. Give the name and formula for the alcohol that when oxidized gives the following compounds.
   (a) $CH_3CH_2CH_2CHO$
   (b) 2-hexanone

Compounds with a Carbonyl Group
(See Exercises 11.7–11.10 and CD-ROM Screen 11.5)

54. Draw structural formulas for (a) 2-pentanone, (b) hexanal, and (c) pentanoic acid.

55. Identify the class of each the following compounds, and give the systematic name for each:
   (a)

   $CH_3\overset{\overset{\displaystyle O}{\|}}{C}CH_3$

   (b)

   $CH_3CH_2CH_2\overset{\overset{\displaystyle O}{\|}}{C}H$

   (c)

   $CH_3\overset{\overset{\displaystyle O}{\|}}{C}CH_2CH_2CH_3$

56. Identify the class of each the following compounds, and give the systematic name for each:
   (a)

   $CH_3CH_2\overset{\overset{\displaystyle CH_3}{|}}{C}HCH_2CO_2H$

   (b)

   $CH_3CH_2\overset{\overset{\displaystyle O}{\|}}{C}OCH_3$

   (c)

   $CH_3\overset{\overset{\displaystyle O}{\|}}{C}OCH_2CH_2CH_2CH_3$

   (d)

   $Br-\langle\bigcirc\rangle-\overset{\overset{\displaystyle O}{\|}}{C}OH$

57. Draw structural formulas for the following acids and esters:
   (a) 2-Methylhexanoic acid
   (b) Pentyl butanoate (which has the odor of apricots)
   (c) Octyl acetate (which has the odor of oranges)

58. Give the structural formula and systematic name for the product, if any, from each of the following reactions:
   (a) Pentanal and $KMnO_4$
   (b) Pentanal and $LiAlH_4$
   (c) 2-Octanone and $LiAlH_4$
   (d) 2-Octanone and $KMnO_4$

59. Describe how to prepare 2-pentanol beginning with the appropriate ketone.

60. Describe how to prepare propyl propanoate beginning with 1-propanol as the only carbon-containing reagent.

61. Give the name and structure of the product of the reaction of benzoic acid and 2-propanol.

62. Draw structural formulas and give the names for the products of the following reaction:

$$CH_3\overset{\overset{\displaystyle O}{\|}}{C}OCH_2CH_2CH_2CH_3 + NaOH$$

63. Draw structural formulas and give the names for the products of the following reaction:

$$\bigcirc-\overset{\overset{\displaystyle O}{\|}}{C}-O-\overset{\overset{\displaystyle CH_3}{|}}{\underset{\underset{\displaystyle CH_3}{|}}{C}}H + NaOH$$

64. The Lewis structure of phenylalanine, one of the 20 amino acids that make up proteins, is drawn here (without lone pairs of electrons). The carbon atoms are numbered for the purpose of this question.
   (a) What is the geometry of C-3?
   (b) What is the $O-C-O$ bond angle?
   (c) Is this molecule chiral? If so, which carbon atom is chiral?
   (d) Which hydrogen atom in this compound is acidic?

65. The Lewis structure of vitamin C, whose chemical name is ascorbic acid, is drawn below (without lone pairs of electrons).

   (a) What is the approximate value for the $O-C-O$ bond angle in the five-member ring?

(b) There are four —OH groups in this structure. Estimate the C—O—H bond angles for these groups. Will all four have the same value (more or less), or should there be significant differences in these bond angles?

(c) Is the molecule chiral? How many chiral carbon atoms can be identified in this structure?

(d) Identify the shortest carbon-carbon bond in this molecule.

(e) What are the functional groups of the molecule?

Functional Groups
(See Example 11.7 and CD-ROM Screen 11.5)

66. Identify the functional groups in the following molecules.

(a) $CH_3CH_2CH_2OH$

(b)
$$H_3CCNHCH_3$$
(with O double-bonded above C)

(c)
$$CH_3CH_2COH$$
(with O double-bonded above C)

(d)
$$CH_3CH_2COCH_3$$
(with O double-bonded above C)

67. Consider the following molecules:

1. OH
    |
    $CH_3CH_2CHCH_3$

2. O
    ‖
    CH_3CH_2COH

3. $H_2C{=}CHCH_2OH$

(a) What is the result of treating compound 1 with $NaBH_4$? What is the functional group in the product? Name the product.

(b) Draw the structure of the reaction product from combining compounds 1 and 2. What is the functional group in the product?

(c) What compound results from adding H_2 to compound 3? Name the reaction product.

(d) What compound results from adding NaOH to compound 2?

Polymers
(See Example 11.5 and Exercise 11.10, CD-ROM Screens 11.9 and 11.10)

68. Polyvinyl acetate is the binder in water-based paints.

(a) Write an equation for its formation from vinyl acetate.

(b) Show a portion of this polymer with three monomer units.

(c) Describe how to make polyvinyl alcohol from polyvinyl acetate.

69. Neoprene (polychloroprene, a kind of rubber) is a polymer formed from the chlorinated butadiene $H_2C{=}CHCCl{=}CH_2$.

(a) Write an equation showing the formation of polychloroprene from the monomer.

(b) Show a portion of this polymer with three monomer units.

70. Saran is a copolymer of 1,1-dichloroethylene and chloroethylene (vinyl chloride). Draw a possible structure for this polymer.

71. The structure of methyl methacrylate is given in Table 11.12. Draw the structure of a polymethyl methacrylate (PMMA) polymer that has four monomer units. (PMMA has excellent optical properties and is used to make hard contact lenses.)

72. Draw the structure for one of the tripeptides that can form from a mixture of glycine, histidine, and alanine.

73. Draw the structures of the four dipeptides that can form when serine and alanine are mixed.

General Questions on Organic Chemistry

These questions are not designated as to type or location in the chapter. They may combine several concepts. More challenging questions are indicated by an underlined number.

74. Draw the structure of each of the following compounds:

(a) 2,2-Dimethylpentane

(b) 3,3-Diethylpentane

(c) 2-Methyl-3-ethylpentane

(d) 4-Ethylhexane

75. Structural isomers

(a) Draw all of the isomers possible for C_3H_8O. Give the systematic name of each, and tell into what class of compound it fits.

(b) Draw the structural formula for an aldehyde and a ketone with the molecular formula C_4H_8O. Give the systematic name of each.

76. Draw structural formulas for possible isomers of the dichlorinated propane, $C_3H_6Cl_2$. Name each compound.

77. Draw structural formulas for possible isomers with the formula C_3H_6ClBr, and name each isomer.

78. Give structural formulas and systematic names for the three structural isomers of trimethylbenzene, $C_6H_3(CH_3)_3$.

79. Give structural formulas and systematic names for possible isomers of dichlorobenzene, $C_6H_4Cl_2$.

80. Voodoo lilies depend on carrion beetles for pollination. Carrion beetles are attracted to dead animals, and because dead and putrefying animals give off the horrible smelling amine cadaverine, the lily likewise releases cadaverine (and the closely related compound putrescine) (page 444). A biological catalyst, an enzyme, converts the naturally occurring amino acid lysine to cadaverine.

lysine

What group of atoms must be replaced in lysine to make cadaverine? (Lysine is essential to human nutrition but is not synthesized in the human body.)

81. Benzoic acid occurs in many berries. When humans eat berries, benzoic acid is converted to hippuric acid in the body by reaction with the amino acid glycine. Draw the structure of hippuric acid, knowing it is an amide formed by reaction of the carboxylic acid group of benzoic acid and the amino group of glycine. Why is hippuric acid referred to as an acid?

82. Kevlar is a polyamide made from *p*-phenylenediamine and terephthalic acid. (It is used to make bullet-proof vests, among other things.) Draw the repeating unit of the Kevlar polymer.

p-phenylenediamine   terephthalic acid

83. A well-known company selling outdoor clothing has recently introduced jackets made of recycled polyethylene terephthalate (PET), the principal material in many soft drink bottles. Another company makes PET fibers by treating recycled bottles with methanol to give the diester dimethyl terephthalate and ethylene glycol and then repolymerizes these compounds to give new PET. Write a chemical equation to show how the reaction of PET with methanol can give dimethyl terephthalate and ethylene glycol.

84. Draw the structure of glyceryl trilaurate. When this triester is saponified (page 454), what are the products?

85. Write a chemical equation describing the reaction between glycerol and stearic acid to give glyceryl tristearate.

86. You have a liquid that is either cyclohexene or benzene. When dark red bromine is added to the sample, it immediately decolorizes. What is the identity of the liquid? Write an equation for the chemical reaction that has occurred.

87. Hydrolysis of an unknown ester of butyric acid, $CH_3CH_2CH_2CO_2R$, produces an alcohol A and butanoic acid. Oxidation of this alcohol forms an acid B that is a structural isomer of butanoic acid. Give the names and structures for alcohol A and acid B.

88. You are asked to identify an unknown colorless, liquid carbonyl compound. Analysis has determined that the formula for this unknown is C_3H_6O. Only two compounds match this formula.

    (a) Draw structures for the two possible compounds.

    (b) To decide which of the two structures is correct, you react the compound with an oxidizing agent and isolate from that reaction a compound that is found to give an acidic solution in water. Use this result to identify the structure of the unknown.

    (c) Name the acid formed by oxidation of the unknown.

89. An unknown colorless liquid has the formula C_3H_8O. Draw structures of three compounds that have this formula.

90. Addition of water to alkene X gives an alcohol Y. Oxidation of Y produces 3,3-dimethyl-2-pentanone. Identify X and Y, and write equations for the two reactions.

91. An unknown ester has the formula $C_4H_8O_2$. Hydrolysis gave methanol as one product. Identify the ester and write an equation for the hydrolysis reaction.

92. Identify the reaction products and write an equation for the following reactions of $H_2C=CHCH_2OH$.

    (a) H_2 (hydrogenation, in the presence of a catalyst)

    (b) Oxidation (excess oxidizing agent)

    (c) Addition polymerization

    (d) Ester formation, using acetic acid

93. Recently, the commercialization of a new polyester was reported in the news media. It was prepared from lactic acid ($CH_3CH(OH)CO_2H$, whose structure is shown on page 424.) Write a balanced chemical equation for the formation of this polymer.

94. One of the resonance structures for pyridine is illustrated here. Draw another resonance structure for the molecule. Comment on the similarity between this compound and benzene.

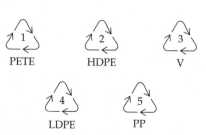

pyridine

95. Write balanced equations for the combustion of ethane and ethanol.

    (a) Calculate the heat of combustion of each compound. Which has the more negative enthalpy change for combustion per gram?

    (b) If ethanol is assumed to be partially oxidized ethane, what effect does this have on the heat of combustion?

96. Describe a simple chemical test to tell the difference between $CH_3CH_2CH_2CH=CH_2$ and its isomer cyclopentane.

97. Describe a simple chemical test to tell the difference between 2-propanol and its isomer methyl ethyl ether.

98. Plastics make up about 20% of the volume of landfills. There is, therefore, considerable interest in reusing or recycling these materials. To identify common plastics, a set of universal symbols is now used, five of which are illustrated here. They symbolize low- and high-density polyethylene, polyvinyl chloride, polypropylene, and polyethylene terephthalate.

(a) Tell which symbol belongs to which type of plastic.

(b) Find an item in the grocery or drug store made from each of these plastics.

(c) Properties of several plastics are listed in the table. Based on this information, describe how to separate samples of these plastics from one another.

Plastic	Density (g/cm^3)	Melting Point (°C)
Polypropylene	0.92	170
High-density polyethylene	0.97	135
Polyethylene terephthalate	1.34–1.39	245

99. Maleic acid is prepared by the catalytic oxidation of benzene. It is a dicarboxylic acid, that is, it has two carboxylic acid groups.

(a) Combustion of 0.125 g of the acid gives 0.190 g of CO_2 and 0.0388 g of H_2O. Calculate the empirical formula of the acid.

(b) A 0.261-g sample of the acid requires 34.60 mL of 0.130 M NaOH for complete titration (so that the H^+ ions from both carboxylic acid groups are neutralized). What is the molecular formula of the acid?

(c) Draw a Lewis structure for the acid.

(d) Describe the hybridization used by the C atoms.

(e) What are the bond angles around each C atom?

Using Electronic Resources

These questions refer to the General Chemistry Interactive CD-Rom, *Version 3.0.*

100. See CD-ROM Screen 11.2: Carbon–Carbon Bonds. Benzene, C_6H_6, is a planar molecule (Screen 10.7SB). Here we see that another six-carbon cyclic molecule, cyclohexane (C_6H_{12}), is not planar.

(a) Contrast the carbon atom hybridization in these two molecules.

(b) Why is π electron delocalization possible in benzene?

(c) Why is cyclohexane not planar?

101. See CD-ROM Screen 11.4: Hydrocarbons and Addition Reactions.

(a) Why are addition reactions termed "addition?"

(b) Alkenes can be hydrogenated to form alkanes. Alkynes can be hydrogenated as well. What class or classes of hydrocarbons can be formed from the hydrogenation of an alkyne?

(c) Alkanes do not undergo addition reactions. Why not?

102. See CD-ROM Screen 11.5: Functional Groups. Describe the hybrid orbitals used by the indicated atoms:

(a) The O atom in an alcohol

(b) The $C=O$ carbon in an aldehyde

(c) The $C=O$ carbon in a carboxylic acid

(d) The $C-O-C$ oxygen atom in an ester

(e) The N atom in an amine

103. See CD-ROM Screen 11.6: Reactions of Alcohols.

Addition and substitution reactions are described in this chapter. Another type of reaction of organic compounds, elimination, is described on the *General Chemistry Interactive CD-ROM,* Version 3.0.

(a) What is the difference between a substitution reaction and an elimination reaction?

(b) Compare the elimination reaction shown on this screen with the hydrogenation reaction shown on Screen 11.4. In what ways are they similar or dissimilar?

104. See CD-ROM Screen 11.7: Fats and Oils.

(a) What type of reaction is used to make a fat or oil from glycerol and a fatty acid: addition, substitution, or elimination?

(b) What is the primary structural difference between fats and oils? What types of functional groups do each contain?

(c) What structural feature of oil molecules prevents them from coiling up on themselves as fat molecules do?

105. See Screen 11.8: Amino Acids and Proteins.

(a) What two functional groups do amino acids contain?

(b) The functional group in a peptide, $-NHCO-$, is planar. Use hybrid orbitals and the concept of resonance to explain why this is the case.

(c) What type of reaction is the condensation reaction shown in the animation on this screen: addition, substitution, or elimination?

(d) The peptide shown in the animation is composed of three amino acids. Could this "tripeptide" be further reacted with one or more additional amino acid molecules?

(e) Draw the tripeptide that can be formed by linking together three molecules of the amino acid alanine.

106. See CD-ROM Screen 11.9: Addition Polymerization.

(a) What is the primary structural feature of the molecules used to form addition polymers?

(b) Consider the animation of a polymerization reaction shown on this screen. The polymer made here has a chain of 14 carbon atoms. Could the chain have been shorter than this or longer than this? Explain briefly.

(c) What controls the length of the polymer chains formed?

(d) Can the addition polymerization reaction be classified as one of the reaction types studied earlier: addition, substitution, or elimination?

107. See CD-ROM Screen 11.10: Condensation Polymerization.

(a) What is the primary structural feature necessary for a molecule to be useful in a condensation polymerization reaction?

(b) Describe the appearance of the nylon being made in this video.

(c) What does the designation "6,6" mean in nylon-6,6?

12

Gases and Their Properties

Chapter Goals

- Understand the basis of the gas laws and how to use those laws: Boyle's, Charles's, Avogadro's, Dalton's.
- Use the ideal gas law.
- Apply the gas laws to stoichiometric calculations.
- Understand kinetic-molecular theory applied to gases, especially the distribution of molecular speeds (energies).
- Recognize why gases do not behave as ideal gases.

◀ **Modern hot-air balloons.** A hot-air balloon rises because the heated air in the balloon has a lower density than the atmosphere. (Greg Gawlowski/Dembinsky Photo Associates)

Up, Up and Away!

The first flight of a manned hot-air balloon took place in Paris, France, on November 21, 1783. The balloon was designed by Joseph and Étienne Montgolfier and piloted by Pilatre de Rozier and the Marquis d'Arlandes. In June 1783 the brothers first tested a balloon without passengers that consisted of a spherical linen sack lined with paper. It was 36 ft in diameter, weighed about 500 lb, and was inflated with the hot air from a fire fueled by bundles of straw. It "ascended to a considerable height," and stayed aloft for about 10 min. The manned flight several months later was the successful culmination of their experiments.

As described in this chapter, a hot-air balloon can ascend because heating affects the density of the air in the balloon. When the air inside a hot-air balloon is heated (usually with a propane heater in a modern hot-air balloon), the gas expands. Rather than increasing the volume of the balloon, however, air is forced from the inside of the balloon. The result is that less air remains inside. The lower mass of air in the same volume means that the gas inside the

balloon has a lower density than the surrounding atmosphere, and the balloon ascends.

Historians have speculated that the Montgolfier brothers got their idea for a hot-air balloon after reading about the experiments on gases made by Joseph Black (1728–1799) of Scotland. Indeed, the 18th century was a time of great discoveries in chemistry. Experiments on gases by people such as Black, Henry Cavendish (1731–1810), Joseph Priestley (1733–1804), and Antoine Lavoisier (1743–1794) gave birth to modern chemistry. Among those chemists working on gases was Jacques Charles (1746–1823). In August 1783, Charles exploited his recent studies on hydrogen gas by inflating a balloon with this gas. Because hydrogen would escape easily from a paper bag, Charles made a silk bag coated with rubber. Inflating the bag took several days and required nearly 500 lb of sulfuric acid and 1000 lb of iron to produce the H_2 gas. The balloon stayed aloft for almost 45 min and traveled about 15 miles. When it landed in a village, however, the people were so terrified they tore it to shreds. Several months later Charles and a passenger flew a new hydrogen-filled balloon some distance across the French countryside and ascended to the then-incredible altitude of 2 miles.

The typical modern hot-air balloon is about 18 m tall (60 ft) and has a volume of about 2250 m^3 heated with a propane burner. These balloons can carry enough propane fuel to fly for about 2 h.

> "*The invention of the balloon appears ... to be a discovery of great importance.*"
>
> Benjamin Franklin, about 1784

▲ Jacques Charles filling a balloon with hydrogen (1783). (*Photo Researchers, Inc.*)

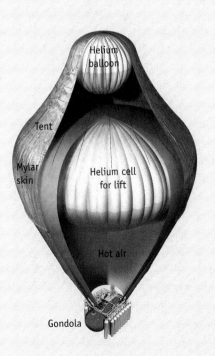

Gondola

Balloons that remain aloft for a long time will usually need to use a lighter-than-air gas such as helium or hydrogen to produce lift. This was first tried in 1785 by Montgolfier's first pilot, de Rozier. He tried to fly from Paris to London in a balloon containing a hydrogen-filled cell and an air-filled cell heated by a flame. He didn't make it. Nonetheless, the latest balloons designed for long-distance flight are called Rozier balloons. These have one or more cells filled with nonflammable helium as well as a cell in which the air can be heated. The helium provides much of the lift for the balloon, and the cell containing the hot air allows for adjustments to be made in the amount of lift: to move to a different altitude or to compensate for the cooling of the helium at night (and the consequent contraction in the volume of the helium). When the air is colder, the pilot heats the air in an un-sealed cell with a propane burner, which in turn transfers heat to upper helium cells. Such a balloon was used in the first successful circumnavigation of the globe by a balloon in March 1999.

◀ The Rozier design balloon set a ballooning distance record by flying around the globe in 1999. The upper helium cell "stakes out the tent" around the larger helium cell, which helps to insulate the latter. Lift is provided by both the helium cells and heated air. The gondola below the balloon is insulated and provided life-support for the crew. The balloon flies at an altitude of 6000–15,000 m where the outside temperature can be about −60 °C. (*Don Foley*)

Before You Begin

• Review chemical stoichiometry in Chapters 4 and 5.

Gases The air bag is a safety device now common in most automobiles. In the event of an accident, the air bag is rapidly inflated with nitrogen gas generated by a chemical reaction.

Automobile Airbags

Automobile crashes do occur, and protection of the driver and occupants is paramount.

Most automobiles are equipped with airbags for both driver and passenger. Some cars now have side airbags as well as bags for legs. The airbag is completely inflated in about 0.050 second. This is important because the typical automobile collision lasts about 0.125 second. *(Courtesy: Saab Cars USA, Inc.)*

Airbag Chemistry

The electrical signal from the sensor in the airbag unit triggers an explosion of sodium azide in the airbag unit in the steering wheel or on the passenger side of the car

$$2\ NaN_3(s) \rightarrow 2\ Na(s) + 3\ N_2(g)$$

The sodium from the detonation is very reactive and would produce NaOH or Na_2O_2 in the air. As both of these can be skin irritants, some iron(III) oxide is added to the mixture to produce a harmless byproduct.

Airbag Construction

When the car decelerates in a collision, an electrical contact is made in the sensor unit. The propellant (green solid) detonates, releasing nitrogen gas. The folded, nylon bag explodes out of the plastic housing, arresting the forward movement of driver and passenger. *(Courtesy: Autoliv ASP)*

Driver side airbags inflate with 35-70 L of N_2, whereas passenger airbags hold 60-160 L. *(Charles D. Winters)*

The bag deflates within 0.2 seconds, the gas escaping through holes in the bottom of the bag. *(Charles D. Winters)*

Aside from understanding automobile air bags (described in the Chapter Focus), there are at least three reasons for studying gases. First, some common elements and compounds exist in the gaseous state under normal conditions of pressure and temperature (Table 12.1). Furthermore, many common liquids can be vaporized, and the properties of these vapors are important. Second, our gaseous atmosphere provides one means of transferring energy and material throughout the globe, and it is the source of life-giving chemicals.

The third reason for studying gases is compelling, however. Of the three states of matter, gas behavior is reasonably simple when viewed at the molecular level and, as a result, gases are well understood. It is possible to describe the properties of gases qualitatively in terms of the behavior of the molecules that make up the gas. Even more impressive, it is possible to describe the properties of gases quantitatively using simple mathematical models. One objective of scientists is to develop models of natural phenomena, and a study of gas behavior will introduce you to this approach.

Table 12.1 • **Some Common Gaseous Elements and Compounds (at 1 Atm Pressure and 25 °C)**

He	CO
Ne	CO_2
Ar	CH_4
Kr	C_2H_4
Xe	C_3H_8
H_2	HF
O_2	HCl
O_3	HBr
N_2	HI
F_2	NO
Cl_2	N_2O
SO_2	H_2S

12.1 PROPERTIES OF GASES

To describe gases, chemists have learned that only four quantities are needed: the gas pressure (P), its volume (V) and temperature (T, kelvins), and the amount of gas (n, mol). Let us examine the first of these: gas pressure and its units [⊕ CD-ROM, Screen 12.2].

Gas Pressure

You are familiar with pressure. Weather people tell us that the pressure of the atmosphere is rising when nice weather approaches and that it is falling when a storm approaches. They also often speak of rising or falling barometric pressure because a barometer is the device we use to measure atmospheric pressure.

A barometer can be made by filling a tube with a liquid, often mercury, and inverting the tube in a dish containing the same liquid (Figure 12.1). If the air has been removed completely from the vertical tube, the liquid assumes a level in the tube such that the pressure exerted by the mass of the column of liquid in the tube is balanced by the pressure of the atmosphere pressing down on the surface of the liquid in the dish.

At sea level, the mercury in a mercury-filled barometer will rise about 760 mm above the surface of the mercury in the dish. Thus, pressure is often reported in units of **millimeters of mercury (mm Hg),** a unit sometimes called the *torr* in honor of Evangelista Torricelli (1608–1647), who invented the mercury barometer in 1643. Pressures are also reported as **standard atmospheres (atm),** a unit defined as follows:

1 standard atmosphere = 1 atm = 760 mm Hg (exactly)

The atmosphere is the unit used in this book.

The SI unit of pressure is the **pascal (Pa),** named for the French mathematician and philosopher Blaise Pascal (1623–1662). Pressure is defined as the force per unit area, and the pascal is expressed in these terms.

1 pascal (Pa) = 1 newton/meter^2

(The newton is the SI unit of force.) Because the pascal is a very small unit compared with ordinary pressures, the kilopascal (kPa) is more often used. Finally,

• **Gas or Vapor?**
A *gas* is a substance that is normally in the gaseous state at ordinary pressures and temperatures. A *vapor* is the gaseous form of a substance that is normally a liquid or solid at ordinary pressures and temperatures. Thus, we often speak of helium gas and water vapor.

• **Pressure Units**
In English units, pressure is often expressed as pounds per square inch (psi), where 1 atm = 14.7 lb/in.^{2}

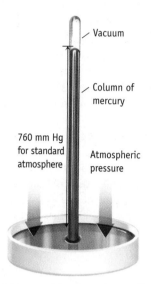

Vacuum

Column of mercury

760 mm Hg for standard atmosphere

Atmospheric pressure

The pressure of the atmosphere on the surface of the mercury in the dish is balanced by the downward pressure exerted by the column of mercury.

A simple hand pump is able to raise water from about 10 m underground. This is the maximum height of a column of water that can be supported by the atmosphere.

Figure 12.1   Using pressure — a barometer and a water pump.
(b, Charles D. Winters)

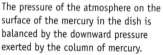

A Closer Look

Measuring Gas Pressure

Pressure is defined as the force exerted on an object divided by the area over which the force is exerted:

$$\text{Pressure} = \frac{\text{force}}{\text{area}}$$

This book, for example, weighs more than 4 lb and has an area of 82 in.², so it exerts a pressure of about 0.05 lb/in.² when it lies flat on a surface. (In metric units, the pressure is about 3 g/cm².)

Consider the pressure that the column of mercury exerts on the mercury in the dish in a barometer shown in Figure 12.1a. This pressure exactly balances the pressure of the atmosphere. Thus, the atmospheric pressure (or the pressure of any other gas) can be measured by measuring the height of the column of mercury (or any other liquid) the gas can support.

Mercury is the liquid of choice for barometers because of its high density. A barometer filled with water would be over 10 m in height. [The water column is about 13.6 times as high as a column of mercury because mercury's density (13.53 g/cm³) is about 13.6 times that of water (density = 0.997 g/cm³, at 25 °C).]

You may have used a tire gauge to check the pressure in your car or bike tires. This gauge indicates the pressure in pounds per square inch (psi), where 1 atm = 14.7 psi. Some newer gauges give the pressure in kilopascals as well. The reading on the scale refers to the pressure *in excess of atmospheric pressure*. (A flat tire is not a vacuum; it contains air at atmospheric pressure.) For example, if the gauge reads 35 psi (2.4 atm), the pressure in the tire is actually about 50 psi or 3.4 atm.

A tire gauge. The "lb" scale measures pressure in pounds per square inch (psi).

atmospheric pressures are sometimes reported in the unit called the **bar,** where 1 bar = 100,000 Pa. The thermodynamic data in Chapter 6 and in Appendix L are given for gas pressures of 1 bar. To summarize, the units used in science for pressure are

$$1 \text{ atm} = 760 \text{ mm Hg (exactly)} = 101.325 \text{ kilopascals (kPa)} = 1.01325 \text{ bar}$$

or

$$1 \text{ bar} = 1 \times 10^5 \text{ Pa (exactly)} = 1 \times 10^2 \text{ kPa} = 0.9869 \text{ atm}$$

Example 12.1   Pressure Unit Conversions

Problem • Convert a pressure of 635 mm Hg into its corresponding value in units of atmospheres (atm), bars, and kilopascals (kPa).

Strategy • Use the relationships between millimeters of mercury, bars, pascals, and atmospheres in the preceding text.

Solution • The relation between millimeters of mercury and atmospheres is 1 atm = 760 mm Hg. Notice that the given pressure is less than 760 mm Hg, that is, less than 1 atm:

$$635 \text{ mm Hg} \cdot \frac{1 \text{ atm}}{760 \text{ mm Hg}} = 0.836 \text{ atm}$$

The relationship between atmospheres and bars is 1 atm = 1.013 bar. We have

$$0.836 \text{ atm} \cdot \frac{1.013 \text{ bar}}{1 \text{ atm}} = 0.847 \text{ bar}$$

The factor relating units of millimeters of mercury and kilopascals is 101.325 kPa = 760 mm Hg. Therefore,

$$635 \text{ mm Hg} \cdot \frac{101.325 \text{ kPa}}{760 \text{ mm Hg}} = 84.7 \text{ kPa}$$

Exercise 12.1   Pressure Unit Conversions

Rank the following pressures in decreasing order of magnitude (largest first, smallest last): 75 kPa, 250 mm Hg, 0.83 bar, and 0.63 atm.

12.2   GAS LAWS: EXPERIMENTAL BASIS

Experimentation in the 17th and 18th centuries led to three gas laws that provide the basis for understanding gas behavior [CD-ROM, Screen 12.3].

Compressibility of Gases: Boyle's Law

When you pump up the tires of your bicycle, the pump squeezes the air into a smaller volume (Figure 12.2). This property of a gas is called its **compressibility.** While studying the compressibility of gases, Robert Boyle observed that the volume of a fixed amount of gas at a given temperature is inversely proportional to the pressure exerted by the gas. All gases behave in this manner, and we now refer to this relationship as **Boyle's law.**

Boyle's law can be demonstrated in many ways (Figures 12.3 and 12.4). In Figure 12.3 a hypodermic syringe is filled with air and sealed. When pressure is applied to the movable plunger of the syringe, the air inside is compressed. As the pressure (*P*) increases on the syringe, the gas volume in the syringe (*V*) decreases. If the pressure of the gas in the syringe were plotted as a function of $1/V$, a straight

History

Robert Boyle (1627–1691)

Born in Ireland as the 14th and last child of the first Earl of Cork, Boyle first published his studies of gases in 1660, and his book, *The Sceptical Chymist,* was published in 1680. Boyle was the first to try to define elements in modern terms, although he had medieval views about what elements were. He was also a physiologist and was the first to show that the healthy human body has a constant temperature. *(Oesper Collection in the History of Chemistry, University of Cincinnati)*

Figure 12.2   A bicycle pump — Boyle's law in action. A bicycle pump works by compressing air into a smaller volume. You experience Boyle's law because you can feel the increasing pressure of the gas as you press down on the plunger. *(Charles D. Winters)*

line would result. This type of plot demonstrates that the pressure and volume of the gas are inversely proportional; that is, they change in opposite directions. Mathematically, we can write this as (where the symbol $\propto$ means "proportional to"):

$$V \propto \frac{1}{P} \quad \text{when } n \text{ and } T \text{ are constant}$$

For a given amount of gas (n) at a fixed temperature (T), the gas volume decreases if the pressure increases. Conversely, if the pressure is lowered, then the gas volume increases.

Boyle's experimentally determined relationship can be put into a useful mathematical form as follows. When two quantities are proportional to each other, they can be equated if a *proportionality constant* is introduced. Thus,

$$P = C_B\left(\frac{1}{V}\right) \quad \text{or} \quad PV = C_B \quad \text{when } n \text{ and } T \text{ are constant}$$

This form of Boyle's law expresses the fact that *the product of the pressure and volume of a gas sample is a constant at a given temperature*, where the constant C_B is deter-

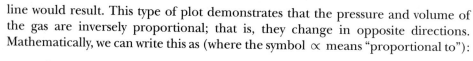

Increasing P

Figure 12.3   An experiment to demonstrate Boyle's law. A syringe filled with air was sealed and then placed in a flask. A bicycle pump was used to increase the air pressure inside the bottle. As the increased pressure pressed down on the syringe barrel, the air inside the syringe was compressed. *(Charles D. Winters)*

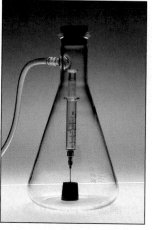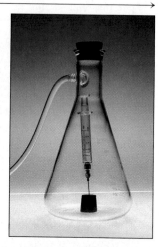

Figure 12.4   Demonstrating Boyle's law. **(a)** Marshmallows were placed in a laboratory flask. **(b)** The flask was evacuated, and the lowered pressure allowed the air in the marshmallows to expand. **(c)** When the pressure in the flask was increased, however, the increased pressure crushes the marshmallows. *(Charles D. Winters)*

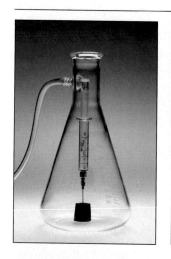

(a)   **(b)**   **(c)**

mined by the amount of gas (in moles) and its temperature (in kelvins). It follows that, if the pressure–volume product is known for one set of conditions (P_1 and V_1), then it is known for another set of conditions (P_2 and V_2). Under either set of conditions, the PV product is equal to C_B, so

$$P_1V_1 = P_2V_2 \quad \text{at constant } n \text{ and } T \qquad \textbf{(12.1)}$$

This form of Boyle's law is useful when we want to know, for example, what happens to the volume of a given amount of gas when the pressure changes at a constant temperature.

Example 12.2 Boyle's Law

Problem • A sample of gaseous nitrogen in a 65.0-L automobile air bag has a pressure of 745 mm Hg. If this sample is transferred to a 25.0-L bag with the same temperature as before, what is the pressure of the gas in the new bag?

Strategy • Here we use Boyle's law, Equation 12.1. The original pressure and volume are known (P_1 and V_1) as well as the new volume (V_2).

Solution • It is often useful to make a table of the information provided.

Original Conditions	Final Conditions
P_1 = 745 mm Hg	P_2 = ?
V_1 = 65.0 L	V_2 = 25.0 L

You know that $P_1V_1 = P_2V_2$. Therefore,

$$P_2 = \frac{P_1V_1}{V_2} = \frac{(745 \text{ mm Hg})(65.0 \text{ L})}{25.0 \text{ L}} = 1940 \text{ mm Hg}$$

Comment • The essence of Boyle's law is that P and V change *in opposite directions*. Because the volume has decreased, you know the new pressure (P_2) must be greater than the original pressure (P_1); thus, P_1 must be multiplied by a volume factor that has a value *greater than 1* to reflect the fact that P_2 must be greater than P_1.

$$P_2 = P_1 \cdot \frac{65.0 \text{ L}}{25.0 \text{ L}} = 1940 \text{ mm Hg}$$

Exercise 12.2 Boyle's Law

A sample of CO_2 has a pressure of 55 mm Hg in a volume of 125 mL. The sample is compressed so that the new pressure of the gas is 78 mm Hg. What is the new volume of the gas? (The temperature does not change in this process.)

Effect of Temperature on Gas Volume: Charles's Law

In 1787, the French scientist Jacques Charles discovered that the volume of a fixed quantity of gas at constant pressure decreases with decreasing temperature (Figure 12.5).

Figure 12.6 illustrates how the volumes of two different gas samples change with temperature (at a constant pressure). When the plots of volume versus temperature are extended to lower temperatures, they all extrapolate to zero volume at the same temperature, −273.15 °C. (Of course gases will not actually reach zero volume; they liquefy well above that temperature.) This temperature is significant, however. William Thomson (1824–1907), also known as Lord Kelvin, proposed a temperature scale — now known as the Kelvin scale — for which the zero point is −273.15 °C [◀, PAGE 16].

History

Jacques Alexandre César Charles (1746–1823)

Charles is known for his work with gases and for being the first to fly in a hydrogen-filled balloon. See the chapter-opening article on page 470. *(The Bettmann Archive/Corbis)* •

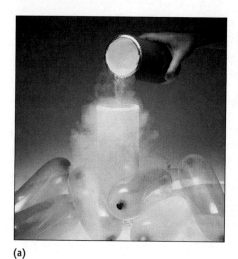

(a)   (b)   (c)

Figure 12.5   A dramatic illustration of Charles's law. (a) Air-filled balloons are placed in liquid nitrogen (77 K). The volume of the gas in the balloons is dramatically reduced at this temperature. **(b)** After all of the balloons have been placed in the liquid nitrogen, **(c)** they are removed; as they warm to room temperature they reinflate to their original volume. *(Charles D. Winters)*

● **Boyle's & Charles's Laws**
Neither Boyle's nor Charles's law depends on the identity of the gas being studied. These laws describe the behavior of any gaseous substance, regardless of its identity.

When Kelvin temperatures are used with volume measurements, the volume–temperature relationship is $V = C_c T$, where C_c is a proportionality constant (which depends on the amount of gas and its pressure). Written another way,

$$\frac{V}{T} = C_c \quad \text{at constant } n \text{ and } P$$

we have **Charles's law,** which states that if a given quantity of gas is held at a constant pressure, its volume is directly proportional to the Kelvin temperature. The quotient V/T always has the same value (C_c) for a given sample of gas at a specified pressure. Therefore, if we know the volume and temperature of a given quantity of gas (V_1 and T_1), we can find the volume, V_2, at some other temperature, T_2, using the equation

$$\frac{V_1}{T_1} = \frac{V_2}{T_2} \quad \text{at constant } n \text{ and } P \qquad (12.2)$$

Figure 12.6   Charles's law. The solid lines represent the volumes of samples of hydrogen and oxygen at different temperatures. The volumes decrease as the temperature is lowered (at constant pressure). These lines, if extended, intersect the temperature axis at approximately −273 °C.

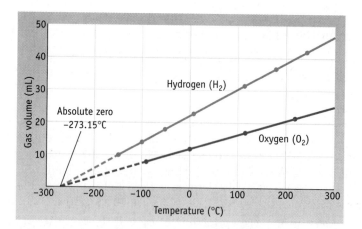

Calculations using Charles's law are illustrated by the following example and exercise. Be sure to notice that the temperature *T must always be expressed in kelvins.*

Example 12.3 **Charles's Law**

Problem • Suppose you have a sample of CO_2 in a gas-tight syringe (as in Figure 12.3). The gas volume is 25.0 mL at room temperature (20.0 °C). What is the final volume of the gas if you hold the syringe in your hand to raise its temperature to 37 °C?

Strategy • This question asks for the outcome of heating a given amount of gas (at a constant pressure). Therefore, Charles's law applies. Because we know the original V and T and want to calculate a new volume at a new, but known, temperature, use Equation 12.2.

Solution • Organize the information in a table. Notice that the temperature *must* be converted to kelvins.

Original Conditions	Final Conditions
$V_1 = 25.0$ mL	$V_2 = ?$
$T_1 = 20.0 + 273.2 = 293.2$ K	$T_2 = 37 + 273 = 310.$ K

Substitute the known quantities into Equation 12.2 and solve for V_2:

$$V_2 = T_2\left(\frac{V_1}{T_1}\right) = (310.\text{ K})\left(\frac{25.0\text{ mL}}{293.2\text{ K}}\right) = 26.4\text{ mL}$$

Comment • As expected the volume of the gas increased with a temperature increase. As in Example 12.2, this again suggests a logical approach to the problem. The new volume (V_2) must equal the original volume (V_1) multiplied by a temperature fraction that is greater than 1 to reflect the effect of the temperature increase. That is,

$$V_2 = V_1\left(\frac{310.\text{ K}}{293.2\text{ K}}\right) = 26.4\text{ mL}$$

Exercise 12.3 **Charles's Law**

A balloon is inflated with helium to a volume of 45 L at room temperature (25 °C). If the balloon is cooled to −10. °C, what is the new volume of the balloon? Assume that the pressure does not change.

Combining Boyle's and Charles's Laws: General Gas Law

For a given amount of gas, its volume is inversely proportional to its pressure at constant temperature (Boyle's law) and directly proportional to the Kelvin temperature at constant pressure (Charles's law). But what if we need to know what happens to the gas when two of the three parameters (P, V, and T) change? For example, what would happen to the pressure of a sample of nitrogen in an automobile air bag if the same amount of gas is placed in a smaller bag and heated to a higher temperature? You can deal with this situation by combining the two equations that express Boyle's and Charles's laws.

$$\frac{P_1V_1}{T_1} = \frac{P_2V_2}{T_2} \quad \text{for a given amount of gas, } n \qquad \textbf{(12.3)}$$

This equation is sometimes called the **general gas law** or **combined gas law.** It applies specifically to situations in which the *amount of gas does not change.*

Chapter Goals • Revisited

● **Understand the basis of the gas laws and how to use those laws: Boyle's, Charles's, Avogadro's, Dalton's.**

● Use the ideal gas law.

● Apply the gas laws to stoichiometric calculations.

● Understand kinetic-molecular theory applied to gases, especially the distribution of molecular speeds (energies).

● Recognize why gases do not behave as ideal gases.

Example 12.4 General Gas Law

Problem • Helium-filled balloons are used to carry scientific instruments high into the atmosphere. Suppose a balloon is launched when the temperature is 22.5 °C and the barometric pressure is 754 mm Hg. If the balloon's volume is 4.19×10^3 L (and no helium escapes from the balloon), what will the volume be at a height of 20 miles, where the pressure is 76.0 mm Hg and the temperature is −33.0 °C?

Strategy • Here we know the initial volume, temperature, and pressure of the gas. We want to know the volume of the same amount of gas at a new pressure and temperature. It is most convenient to use Equation 12.3, the general gas law.

Solution • Begin by setting out the information given in a table.

Initial Conditions	Final Conditions
$V_1 = 4.19 \times 10^3$ L	$V_2 = ?$ L
$P_1 = 754$ mm Hg	$P_2 = 76.0$ mm Hg
$T_1 = 22.5$ °C (295.7 K)	$T_2 = -33.0$ °C (240.2 K)

We can rearrange the general gas law to calculate the new volume V_2:

$$V_2 = \left(\frac{T_2}{P_2}\right)\left(\frac{P_1 V_1}{T_1}\right) = V_1 \left(\frac{P_1}{P_2}\right)\left(\frac{T_2}{T_1}\right)$$

$$= 4.19 \times 10^3 \text{ L}\left(\frac{754 \text{ mm Hg}}{76.0 \text{ mm Hg}}\right)\left(\frac{240.2 \text{ K}}{295.7 \text{ K}}\right)$$

$$= 3.38 \times 10^4 \text{ L}$$

Comment • The pressure decreased by almost a factor of 10, which should lead to about a tenfold volume increase. This increase is partly offset by a drop in temperature that leads to a volume decrease. On balance, the volume increases because the pressure has dropped so substantially.

Notice that the solution was to multiply the original volume (V_1) by a pressure factor (larger than 1 because the volume increases with a lower pressure) and a temperature factor (smaller than 1 because volume decreases with a decrease in temperature).

Exercise 12.4 The General Gas Law

You have a 22-L cylinder of helium at a pressure of 150 atm and at 31 °C. How many balloons can you fill, each with a volume of 5.0 L, on a day when the atmospheric pressure is 755 mm Hg and the temperature is 22 °C?

The general gas law leads to other useful predictions of gas behavior. For example, if a given amount of gas is held in a closed container, the pressure of the gas will increase with increasing temperature.

$$\frac{P_1}{T_1} = \frac{P_2}{T_2} \quad \text{when } V_1 = V_2 \text{ and so } P_2 = P_1\left(\frac{T_2}{T_1}\right)$$

That is, when T_2 is greater than T_1, P_2 will be greater than P_1. In fact, this is the reason tire manufacturers recommend checking tire pressures when the tires are cold. After driving for some distance, friction warms a tire and increases the internal pressure. Thus, filling a warm tire to the recommended pressure may lead to a dangerously underinflated tire.

Avogadro's Hypothesis

In 1811, Amedeo Avogadro used work on gases by the chemist (and early experimenter with hot-air balloons) Joseph Gay-Lussac (1778–1850) to propose that *equal volumes of gases under the same conditions of temperature and pressure have equal numbers of molecules*. This came to be known as **Avogadro's hypothesis.** Stated another way, the volume of a gas, at a given temperature and pressure, is directly proportional to the amount (mol) of gas.

$$V \propto n \text{ at constant } T \text{ and } P$$

A weather balloon is filled with helium. As it ascends into the troposphere, does the volume increase or decrease? *(NASA/Science Source/Photo Researchers, Inc.)*

Example 12.5   Avogadro's Hypothesis

Problem • Ammonia can be made directly from the elements:

$$N_2(g) + 3 H_2(g) \longrightarrow 2 NH_3(g)$$

If you begin with 15.0 L of $H_2(g)$, what volume of $N_2(g)$ is required for complete reaction (both gases being at the same T and P)? What is the theoretical yield of NH_3, in liters, under the same conditions?

Strategy • From Avogadro's hypothesis we know that gas volume is proportional to the amount of gas (at constant temperature and pressure). Therefore, we can substitute gas volumes for moles in this stoichiometry problem.

Solution • Calculate the volumes of N_2 required and NH_3 produced (in liters) by multiplying the volume of H_2 available by a stoichiometric factor (also in units of liters) obtained from the balanced chemical equation:

$$V(N_2 \text{ required}) = 15.0 \text{ L } H_2 \text{ available} \left(\frac{1 \text{ L } N_2}{3 \text{ L } H_2} \right)$$

$$= 5.00 \text{ L } N_2$$

$$V(NH_3 \text{ produced}) = 15.0 \text{ L } H_2 \text{ available} \left(\frac{2 \text{ L } NH_3}{3 \text{ L } H_2} \right)$$

$$= 10.0 \text{ L } NH_3$$

Exercise 12.5   Avogadro's Hypothesis

Methane burns in oxygen to give CO_2 and H_2O, according to the equation

$$CH_4(g) + 2 O_2(g) \longrightarrow CO_2(g) + 2 H_2O(g)$$

If 22.4 L of gaseous CH_4 is burned, what volume of O_2 is required for complete combustion? What volumes of CO_2 and H_2O are produced? Assume all gases have the same temperature and pressure.

12.3   IDEAL GAS LAW

Four interrelated quantities can be used to describe a gas: pressure, volume, temperature, and amount (moles). We know from experiments that three gas laws can be used to describe the relationship of these properties.

Boyle's Law	Charles's Law	Avogadro's Hypothesis
$V \propto (1/P)$	$V \propto T$	$V \propto n$
(constant T, n)	(constant P, n)	(constant T, P)

If all three laws are combined, the result is

$$V \propto \frac{nT}{P}$$

This can be made into a mathematical equation by introducing a proportionality constant, now labeled **R.** This constant, called the **gas constant,** is a *universal constant,* a number that you can use to interrelate the properties of any gas:

$$V = R \left(\frac{nT}{P} \right)$$

or

$$PV = nRT \qquad\qquad \textbf{(12.4)}$$

Chapter Goals • Revisited

- Understand the basis of the gas laws and how to use those laws: Boyle's, Charles's, Avogadro's, Dalton's.
- **Use the ideal gas law.**
- Apply the gas laws to stoichiometric calculations.
- Understand kinetic-molecular theory applied to gases, especially the distribution of molecular speeds (energies).
- Recognize why gases do not behave as ideal gases.

The equation $PV = nRT$ is called the **ideal gas law.** It describes the behavior of a so-called "ideal" gas. As you will learn in Section 12.8, there is no such thing as an ideal gas. However, real gases at pressures around 1 atm or less and temperatures around room temperature usually behave close enough to ideality that $PV = nRT$ adequately describes their behavior [⊕ CD-ROM, Screen 12.4].

To use the equation $PV = nRT$, we need a value for R. This is readily determined experimentally. By carefully measuring P, V, n, and T for a sample of gas we can calculate the value of R from these values using the ideal gas law equation. For example, under conditions of **standard temperature and pressure (STP)** (a gas temperature of 0 °C or 273.15 K and a pressure of 1 atm) 1 mol of gas occupies 22.414 L, a quantity called the **standard molar volume.** Substituting these values into the ideal gas law gives a value for R:

$$R = \frac{PV}{nT} = \frac{(1.0000 \text{ atm})(22.414 \text{ L})}{(1.0000 \text{ mol})(273.15 \text{ K})} = 0.082057 \text{ L} \cdot \text{atm/K} \cdot \text{mol}$$

With a value for R, we can now use the ideal gas law in calculations.

● **STP — What Is It?**
A gas is at STP, or standard temperature and pressure, when its temperature is 0 °C or 273.15 K and its pressure is 1 atm. Under these conditions, exactly 1 mol of a gas occupies 22.414 L.

Example 12.6 | Ideal Gas Law

Problem • The nitrogen gas in an automobile air bag, with a volume of 65 L, exerts a pressure of 829 mm Hg at 25 °C. What amount of N_2 gas (in moles) is in the air bag?

Strategy • You are given P, V, and T and want to calculate the amount of gas (n). Use the ideal gas law, Equation 12.4.

Solution • First list the information provided.

$\quad V = 65 \text{ L} \qquad P = 829 \text{ mm Hg} \qquad T = 25 \text{ °C} \qquad n = ?$

To use the ideal gas law with R having units of (L · atm/K · mol), the pressure must be expressed in atmospheres and the temperature in kelvins. Therefore,

$$P = 829 \text{ mm Hg} \cdot \frac{1 \text{ atm}}{760 \text{ mm Hg}} = 1.09 \text{ atm}$$

$$T = 25 + 273 = 298 \text{ K}$$

Now substitute the values of P, V, T, and R into the ideal gas law and solve for the amount of gas, n,

$$n = \frac{PV}{RT} = \frac{(1.09 \text{ atm})(65 \text{ L})}{(0.082057 \text{ L} \cdot \text{atm/K} \cdot \text{mol})(298 \text{ K})}$$

$$= 2.9 \text{ mol } N_2$$

Notice that units of atmospheres, liters, and kelvins cancel to leave the answer in units of moles.

Exercise 12.6 | Ideal Gas Law

The balloon used by Jacques Charles in his historic flight in 1783 was filled with about 1300 mol of H_2. If the temperature of the gas was 23 °C, and its pressure was 750 mm Hg, what was the volume of the balloon?

Density of Gases

The density of a gas at a given temperature and pressure (Figure 12.7) is a useful quantity. Let us see how density is related to the ideal gas law. Because the amount (n, mol) of any compound is given by its mass (m) divided by its molar mass (M), we can substitute m/M for n in the ideal gas equation.

$$PV = \left(\frac{m}{M}\right)RT$$

Figure 12.7   Gas density. (a) The two balloons are filled with nearly equal quantities of gas at the same temperature and pressure. The pink balloon contains low-density helium ($d = 0.178$ g/L), and the yellow balloon contains air, a higher density gas ($d = 1.2$ g/L). **(b)** A hot-air balloon rises because the heated air has a lower density. *(a, John C. Kotz; b, Greg Gawlowski/Dembinsky Photo Associates)*

(a)   **(b)**

Density (d) is defined as mass divided by volume (m/V); therefore, if we rearrange this equation so that m/V appears on one side,

$$d = \frac{m}{V} = \frac{PM}{RT} \tag{12.5}$$

we find the density of a gas is related to the pressure and temperature of the gas and to its molar mass [CD-ROM, Screen 12.5]. This fact is useful because a measurement of the density of a gas at a given pressure and temperature can be used to calculate its molar mass.

Example 12.7   **Density and Molar Mass**

Problem • Calculate the density of CO_2 at STP. Is it more or less dense than air at STP?

Strategy • Refer to Equation 12.5, the equation relating gas density and molar mass. Here we know the molar mass (44.0 g/mol), the pressure ($P = 1.00$ atm), the temperature ($T = 273$ K), and the gas constant (R). Only the density (d) is unknown.

Solution • The known values are substituted into Equation 12.5, which is then solved for the density (d):

$$d = \frac{PM}{RT} = \frac{(1.00 \text{ atm})(44.0 \text{ g/mol})}{(0.082057 \text{ L} \cdot \text{atm/K} \cdot \text{mol})(273 \text{ K})}$$

$$= 1.96 \text{ g/L}$$

The density of CO_2 is considerably greater than that of dry air at STP (1.2 g/L).

Exercise 12.7   **Gas Density Calculation**

Calculate the density of dry air at 15.0 °C and 1.00 atm if its molar mass (average) is 28.96 g/mol.

Gas density has practical implications. Among other things it is important to a hot-air balloonist (see Figure 12.7). As explained in "Up, Up, and Away" (page 470), a hot-air balloon rises because the gas inside the balloon has a lower density than the surrounding air.

Another way to explain why a hot-air balloon can rise starts with the equation for gas density, $d = PM/RT$, which states that gas density is inversely proportional to temperature. Raising the temperature of a gas results in a decrease in density. (This assumes the pressure remains the same in a balloon. This must be the case because gas can escape from the balloon, keeping the pressure inside equal to the pressure outside the balloon.)

From the equation $d = PM/RT$ we also see that the density of a gas is directly proportional to its molar mass. Dry air, which has an average molar mass of about 29 g/mol, has a density of about 1.2 g/L at 1 atm and 25 °C. Gases or vapors with molar masses greater than 29 g/mol have densities larger than 1.2 g/L under these same conditions (1 atm and 25 °C). Gases such as CO_2, SO_2, and gasoline vapor settle along the ground if released into the atmosphere (Figure 12.8). Conversely, gases such as H_2, He, CO, CH_4 (methane), and NH_3 rise if released into the atmosphere.

The significance of gas density was tragically revealed in several recent events. One such instance occurred in the African country of Cameroon in 1984 when Lake Nyos expelled a huge bubble of CO_2 into the atmosphere. Because CO_2 is denser than air, the CO_2 cloud hugged the ground, killing almost 1700 people living in a nearby village (Figure 12.8b and page 556).

Calculating Molar Mass from *P*, *V*, and *T* Data

When a new compound is isolated in the laboratory, one of the first things to be done is to determine its molar mass. If the compound is in the gas phase, a classical method of determining the molar mass is to measure the pressure and volume exerted by a given mass of the gas in a given volume and at a given temperature [CD-ROM, Screen 12.6].

● **Molar Mass of Air**
The value for the average molar mass for dry air (29 g/mol) falls between the values for the primary components of the atmosphere, N_2 (28 g/mol) and O_2 (32 g/mol). The value is nearer 28 than 32 because nitrogen is the more abundant gas.

Figure 12.8   Carbon dioxide gas is denser than air. (a) Because carbon dioxide from fire extinguishers is denser than air, it settles on top of a fire and smothers it. (When CO_2 gas is released from the tank, it expands and cools significantly. The white cloud is solid CO_2 and condensed moisture from the air.) **(b)** Lake Nyos in Cameroon (west Africa) was the site of a large natural disaster. A huge bubble of CO_2, produced by natural volcanic activity, escaped from the lake and asphyxiated nearly 1700 people and many animals. *(a, Charles D. Winters; b, courtesy of George Kling)*

(a)

(b)

Example 12.8 | **Calculating the Molar Mass of a Gas from *P*, *V*, and *T* Data**

Problem • You are trying to determine, by experiment, the empirical formula of a gaseous compound that might be used to replace chlorofluorocarbons in air conditioners. Your results give a formula of CHF_2. Now you need the molar mass of the compound to find the molecular formula. You do another experiment and find that a 0.100-g sample of the compound exerts a pressure of 70.5 mm Hg in a 256-mL container at 22.3 °C. What is the molar mass of the compound? What is its molecular formula?

Strategy • Here you know the mass of a gas in a given volume (*V*), so you can calculate its density, *d*. Then, knowing the gas pressure and temperature, you can use Equation 12.5 to calculate the molar mass.

Solution • Begin by organizing the data:

$$m = \text{mass of gas} = 0.100 \text{ g}$$
$$P = 70.5 \text{ mm Hg, or } 0.0928 \text{ atm}$$
$$V = 256 \text{ mL, or } 0.256 \text{ L}$$
$$T = 22.3 \text{ °C, or } 295.5 \text{ K}$$

The density of the gas is the mass of the gas divided by the volume:

$$d = \frac{0.100 \text{ g}}{0.256 \text{ L}} = 0.391 \text{ g/L}$$

Use this value of density along with the values of pressure and temperature in the equation $d = PM/RT$ and solve for the molar mass (*M*).

$$M = \frac{dRT}{P}$$

$$= \frac{(0.391 \text{ g/L})(0.082057 \text{ L} \cdot \text{atm/K} \cdot \text{mol})(295.5 \text{ K})}{0.0928 \text{ atm}}$$

$$= 102 \text{ g/mol}$$

With this result, you can compare the experimentally determined molar mass with the mass of a mole of gas having the empirical formula CHF_2.

$$\frac{\text{Experimental molar mass}}{\text{Mass of 1 mol of } CHF_2} = \frac{102 \text{ g/mol}}{51.0 \text{ g/formula unit}}$$

$$= 2 \text{ formula units of } CHF_2 \text{ per mol}$$

Therefore, the formula of the compound is $C_2H_2F_4$.

Comment • Alternatively, you can use the ideal gas law. Here you know the *P* and *T* of a gas in a given volume (*V*), so you can calculate the amount of gas (*n* mol).

$$n = \frac{PV}{RT} = \frac{(0.0928 \text{ atm})(0.256 \text{ L})}{(0.082057 \text{ L} \cdot \text{atm/K} \cdot \text{mol})(295.5 \text{ K})}$$

$$= 9.80 \times 10^{-4} \text{ mol}$$

You now know that 0.100 g of gas is equivalent to 9.80×10^{-4} mol. Therefore,

$$\text{Molar mass} = \frac{0.100 \text{ g}}{9.80 \times 10^{-4} \text{ mol}} = 102 \text{ g/mol}$$

Exercise 12.8 | **Molar Mass from *P*, *V*, and *T* Data**

A 0.105-g sample of a gaseous compound exerts a pressure of 561 mm Hg in a volume of 125 mL at 23.0 °C. What is its molar mass?

Chapter Goals • Revisited

12.4   GAS LAWS AND CHEMICAL REACTIONS

Many industrially important reactions involve gases. Two examples are the combination of nitrogen and hydrogen to produce ammonia and the electrolysis of aqueous NaCl to produce NaOH and the gases hydrogen and chlorine.

$$N_2(g) + 3 \text{ H}_2(g) \longrightarrow 2 \text{ NH}_3(g)$$

$$2 \text{ NaCl(aq)} + 2 \text{ H}_2O \xrightarrow{\text{electrolysis}} 2 \text{ NaOH(aq)} + \text{H}_2(g) + \text{Cl}_2(g)$$

It is important to deal with such reactions quantitatively. The examples that follow explore stoichiometry calculations involving gases. The scheme in Figure 12.9 connects these calculations for gas reactions with calculations done in Chapters 4 and 5 [CD-ROM, Screen 12.7].

- Understand the basis of the gas laws and how to use those laws: Boyle's, Charles's, Avogadro's, Dalton's.
- Use the ideal gas law.
- **Apply the gas laws to stoichiometric calculations.**
- Understand kinetic-molecular theory applied to gases, especially the distribution of molecular speeds (energies).
- Recognize why gases do not behave as ideal gases.

Figure 12.9 A scheme for performing stoichiometry calculations. Here A and B may be either reactants or products. The amount of A (mol) can be calculated from its mass in grams, from *P*, *V*, and *T* data by using the ideal gas law, or from the concentration and volume of a solution. Once the amount of B is determined, this value can be converted to a mass or solution concentration or volume, or to a volume of gas at a given pressure and temperature.

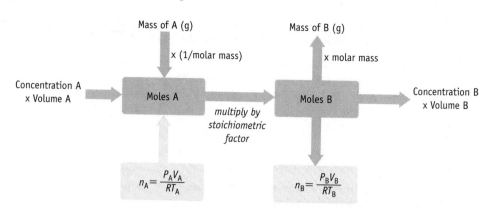

$$n_A = \frac{P_A V_A}{RT_A}$$

$$n_B = \frac{P_B V_B}{RT_B}$$

Example 12.9 Gas Laws and Stoichiometry

Problem • You are asked to design an air bag for a car. You know that the bag should be filled with gas with a pressure higher than atmospheric pressure, say 829 mm Hg, at a temperature of 22.0 °C. The bag has a volume of 45.5 L. What quantity of sodium azide, NaN_3, should be used to generate the required quantity of gas? The gas-producing reaction is

$$2\ NaN_3(s) \longrightarrow 2\ Na(s) + 3\ N_2(g)$$

Strategy • The general logic to be used here follows one of the pathways in Figure 12.9 (lower left to upper right).

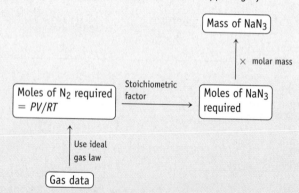

Solution • The first step is to find the amount (mol) of gas required so that this can be related to the quantity of sodium azide required:

$$P = 829\ \text{mm Hg} \ (1\ \text{atm}/760\ \text{mm Hg}) = 1.09\ \text{atm}$$

$$V = 45.5\ \text{L}$$

$$T = 22.0\ °\text{C, or } 295.2\ \text{K}$$

$$n = N_2 \text{ required (mol)} = \frac{PV}{RT}$$

$$= \frac{(1.09\ \text{atm})(45.5\ \text{L})}{(0.082057\ \text{L} \cdot \text{atm}/\text{K} \cdot \text{mol})(295.2\ \text{K})}$$

$$= 2.05\ \text{mol } N_2$$

Now that the required amount of nitrogen has been calculated, we can calculate the quantity of sodium azide that will produce 2.05 mol of N_2 gas.

$$\text{Mass of } NaN_3 = 2.05\ \text{mol } N_2 \cdot \frac{2\ \text{mol } NaN_3}{3\ \text{mol } N_2} \cdot \frac{65.01\ \text{g}}{1\ \text{mol } NaN_3}$$

$$= 88.8\ \text{g } NaN_3$$

Example 12.10 Gas Laws and Stoichiometry

Problem • You wish to prepare some deuterium gas, D_2, for use in an experiment. One way to do this is to react heavy water, D_2O, with an active metal such as lithium.

$$2\ Li(s) + 2\ D_2O(\ell) \longrightarrow 2\ LiOD(aq) + D_2(g)$$

Suppose you place 0.125 g of Li metal in 15.0 mL of D_2O ($d = 1.11$ g/mL). What amount of D_2 (in moles) can be prepared? If dry D_2 gas is captured in a 1450-mL flask at 22.0 °C, what is the pressure of the gas in millimeters of mercury? (Deuterium has an atomic weight of 2.0147 g/mol.)

Lithium metal (in the spoon) reacts with drops of water, H_2O, to produce LiOH and hydrogen gas, H_2. If heavy water, D_2O is used, deuterium gas, D_2, can be produced. *(Charles D. Winters)*

Strategy • You are combining two reactants with no guarantee that they are in the correct stoichiometric ratio. This reaction must therefore be approached as a *limiting reactant problem*. You have to find the amount of each reactant and then see if one of them is present in a limited amount. Once the limiting reactant is known, the amount of D_2 produced and its pressure under the conditions given can be calculated.

Solution •

Step 1. *Calculate the amount (mol) of Li and of D_2O.*

$$0.125 \text{ g Li} \cdot \frac{1 \text{ mol Li}}{6.941 \text{ g Li}} = 0.0180 \text{ mol Li}$$

$$15.0 \text{ mL } D_2O \cdot \frac{1.11 \text{ g } D_2O}{1 \text{ mL } D_2O} \cdot \frac{1 \text{ mol } D_2O}{20.03 \text{ g } D_2O} = 0.831 \text{ mol } D_2O$$

Step 2. *Decide which reactant is the limiting reactant.*

$$\text{Ratio of moles of reactants available} = \frac{0.831 \text{ mol } D_2O}{0.0180 \text{ mol Li}}$$

$$= \frac{46.2 \text{ mol } D_2O}{1 \text{ mol Li}}$$

The balanced equation shows that the ratio should be 1 mol of D_2O to 1 mol of Li. From the calculated values, we can see that D_2O is in large excess and Li is the limiting reactant. Therefore, further calculations are based on the amount of Li available.

Step 3. *Use the limiting reactant to calculate the amount of D_2 produced.*

$$0.0180 \text{ mol Li} \cdot \frac{1 \text{ mol } D_2 \text{ produced}}{2 \text{ mol Li}} = 0.00900 \text{ mol } D_2$$

Step 4. *Calculate the pressure of D_2.*

$P = ?$

$V = 1450 \text{ mL, or } 1.45 \text{ L}$

$T = 22.0 \text{ °C, or } 295.2 \text{ K}$

$n = 0.00900 \text{ mol } D_2$

$$P = \frac{nRT}{V}$$

$$= \frac{(0.00900 \text{ mol})(0.082057 \text{ L} \cdot \text{atm/K} \cdot \text{mol})(295.2 \text{ K})}{1.45 \text{ L}}$$

$$= 0.150 \text{ atm}$$

Exercise 12.9 **Gas Laws and Stoichiometry**

Gaseous ammonia is synthesized by the reaction

$$N_2(g) + 3 H_2(g) \xrightarrow[\text{500 °C}]{\text{iron catalyst}} 2 NH_3(g)$$

Assume that 355 L of H_2 gas at 25.0 °C and 542 mm Hg is combined with excess N_2 gas. What amount (mol) of NH_3 gas can be produced? If this amount of NH_3 gas is stored in a 125-L tank at 25.0 °C, what is the pressure of the gas?

Exercise 12.10 **Gas Laws and Stoichiometry**

Oxygen reacts with aqueous hydrazine to produce water and gaseous nitrogen according to the balanced equation

$$N_2H_4(aq) + O_2(g) \longrightarrow 2 H_2O(\ell) + N_2(g)$$

If a solution contains 180 g of N_2H_4, what is the maximum volume of O_2 that will react with the hydrazine if the oxygen is measured at a pressure of 750 mm Hg and a temperature of 21 °C?

Hc
when he was ⌐.
and shortly after began
scientific investigations, chiefly regarding
weather. In October 1803 he read the paper introducing his "Chemical Atomic
Theory" to the Literary and Philosophical
Society in Manchester. Thus, Dalton is
credited with first introducing the basic
ideas of an atomic theory of matter.
*(Oesper Collection in the History of
Chemistry/University of Cin-cinnati)*   •

12.5   GAS MIXTURES AND PARTIAL PRESSURES

The air you breathe is a mixture of nitrogen, oxygen, argon, carbon dioxide,
water vapor, and small amounts of other gases (Table 12.2). Each of these gases
exerts a pressure, and atmospheric pressure is the sum of the pressures exerted by
each gas. The pressure of each gas in the mixture is called its **partial pressure**
[🖰 CD-ROM, Screen 12.8].

John Dalton was the first to deduce that the pressure of a mixture of gases is
the sum of the partial pressures of the different gases in the mixture. This
observation is now known as **Dalton's law of partial pressures** (Figure 12.10).
Mathematically, we can write Dalton's law of partial pressures as

$$P_{total} = P_1 + P_2 + P_3 + \cdots \tag{12.6}$$

where P_1, P_2, and P_3 are the pressures of the different gases in a mixture and P_{total}
is the total pressure.

The pressure of a gas in a given volume and at a specific temperature depends
only on the amount of gas. There is no reference in the ideal gas law to the identity of the gas. It is also true that each gas behaves independently of all others in
a mixture of ideal gases. In a gas mixture, therefore, we can consider the behavior
of each gas separately. As an example let us take a mixture of three ideal gases, labeled A, B, and C. There are n_A moles of A, n_B moles of B, and n_C moles of C.
Assume that the mixture ($n_{total} = n_A + n_B + n_C$) is contained in a given volume
(V) at a given temperature (T). We can calculate the pressure exerted by each gas
from the ideal gas law equation:

$$P_A V = n_A RT \qquad P_B V = n_B RT \qquad P_C V = n_C RT$$

According to Dalton's law, the total pressure exerted by the mixture is the sum of
the pressures exerted by each component:

$$P_{total} = P_A + P_B + P_C = n_A\left(\frac{RT}{V}\right) + n_B\left(\frac{RT}{V}\right) + n_C\left(\frac{RT}{V}\right)$$

$$= (n_A + n_B + n_C)\left(\frac{RT}{V}\right)$$

$$P_{total} = n_{total}\left(\frac{RT}{V}\right) \tag{12.7}$$

For mixtures of gases, it is convenient to introduce a quantity called the **mole
fraction, X,** which is defined as the number of moles of a particular substance in

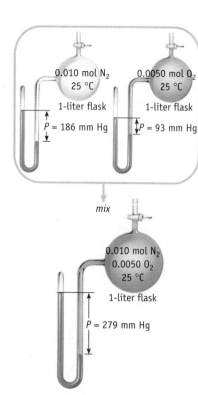

Figure 12.10   **Dalton's law.** In a 1.0-L
flask at 25 °C, 0.010 mol of N_2 exerts a
pressure of 186 mm Hg, and 0.0050 mol of
O_2 in a 1.0-L flask at 25 °C exerts a pressure of 93 mm Hg (*left and middle*). The
N_2 and O_2 samples are mixed in a 1.0-L
flask at 25 °C (*right*). The total pressure,
279 mm Hg, is the sum of the pressures
that each gas alone exerts in the flask.

Table 12.2 • Components of Atmospheric Dry Air

Constituent	Molar Mass*	Mole Percent	Partial Pressure at STP (atm)
N_2	28.01	78.08	0.7808
O_2	32.00	20.95	0.2095
CO_2	44.01	0.031	0.00031
Ar	39.95	0.934	0.00934

*The average molar mass of dry air = 28.96 g/mol.

Chemical Perspectives

Oxygen in Our Atmosphere — Where Did It Come From?

Earth's atmosphere is 21% O_2 gas, and life on our planet depends on close adherence to this level. How did the earth first have an atmosphere with O_2 gas?

Most scientists agree that oxygen was first present on Earth as water and the water was converted to O_2 gas by photosynthesizing bacteria (cyanobacteria or "blue-green algae"). This process is thought to have begun about 2.7 billion years ago. However, O_2 did not become abundant until about 2.3 billion years ago. Why did it take 400 million years for oxygen to begin to build up?

This is a complex problem, and there are certainly no definitive answers. However, in 2001 it was suggested that the process began with the combination of CO_2 and water to produce organic

matter (where CH_2O represents organic matter) and oxygen.

$$CO_2 + H_2O \longrightarrow CH_2O + O_2$$

The O_2 gas produced in the reaction would have been consumed rapidly by combining with sulfur or sulfur-containing compounds produced by volcanoes or by oxidation of earth's crust. The organic matter would have been broken down eventually in an anerobic ("without air") decomposition process driven by light to give the simple organic molecule methane, CH_4 (a process called "methanogenesis") and CO_2. (This reaction still occurs in landfills and swamps, and the product is widely known as "swamp gas.")

$$2\,CH_2O \longrightarrow CH_4 + CO_2$$

If the methane, a good greenhouse gas, eventually built up in the atmosphere, the planet would have become warmer. This is consistent with geological evidence.

The latest theory is that the methane was decomposed by solar radiation. The key idea in this theory is that the H atoms produced in this photochemical reaction escaped to space,

$$CH_4 + \text{solar radiation} \longrightarrow$$
$$C + 4\,H\,\text{(to space)}$$
$$C + O_2 \longrightarrow CO_2$$

and the hydrogen was lost permanently from the earth.

If the four preceding chemical equations are summed (after multiplying the CO_2/H_2O equation by 2), the result is the

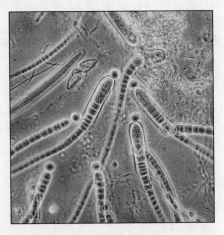

Cyanobacteria appeared on earth before chlorophyll and so were the basic system involved in photosynthesis. They are thought to be the source of O_2 on the early earth. *(Ken W. Davis/Tom Stack & Associates)*

net generation of oxygen in the atmosphere.

$$2\,H_2O + \text{"the biosphere"} +$$
$$\text{solar radiation} \longrightarrow O_2 + 4\,H\,\text{(to space)}$$

A final mystery: Why is the atmosphere 21% O_2? Why not 15% or 35%? If it were lower than 21%, life could not have occurred or would be very different. If it were higher than 21%, the atmosphere would be too oxidizing. Scientists are still searching for answers to this interesting question.

For more information see D. C. Catling, K. J. Zahle, and C. P. McKay: *Science,* Vol. 293, pp. 839–843, 2001.

Earth, with its oxygen atmosphere, as seen from space. *(NASA)*

a mixture divided by the total number of moles of all substances present. Mathematically, the mole fraction of a substance A in a mixture with B and C is expressed as

$$X_A = \frac{n_A}{n_A + n_B + n_C} = \frac{n_A}{n_{total}}$$

Now we can combine this equation (written as $n_{total} = n_A / X_A$) with the equations for P_A and P_{total} and derive the very useful equation

$$P_A = X_A(P_{total}) \tag{12.8}$$

that tells us that the partial pressure of a gas in a mixture of gases is the product of its mole fraction and the total pressure of the mixture. In other words, the mole fraction of O_2 in air is 0.78, so, at STP, its partial pressure is 0.78 atm or 590 mm Hg.

Example 12.11   Partial Pressures of Gases

Problem • Halothane, $C_2HBrClF_3$, is a nonflammable, nonexplosive, and nonirritating gas that is commonly used as an inhalation anesthetic.

Suppose you mix 15.0 g of halothane vapor with 23.5 g of oxygen gas. If the total pressure of the mixture is 855 mm Hg, what is the partial pressure of each gas?

Strategy • One way to solve this problem is to recognize that the partial pressure of a gas is given by the total pressure of the mixture multiplied by the mole fraction of the gas.

Solution • Let us first calculate the mole fractions of halothane and of O_2.

Step 1. *Calculate mole fractions:*

$$\text{Moles } C_2HBrClF_3 = 15.0 \text{ g}\left(\frac{1 \text{ mol}}{197.4 \text{ g}}\right) = 0.0760 \text{ mol}$$

$$\text{Moles } O_2 = 23.5 \text{ g}\left(\frac{1 \text{ mol}}{32.00 \text{ g}}\right) = 0.734 \text{ mol}$$

$$\text{Mole fraction } C_2HBrClF_3 = \frac{0.0760 \text{ mol } C_2HBrClF_3}{0.810 \text{ total moles}}$$

Because the sum of the mole fraction of halothane and of O_2 must equal 1.0000, this means that the mole fraction of oxygen is 0.9062.

$$X_{halothane} + X_{oxygen} = 1.0000$$

$$0.0938 + X_{oxygen} = 1.0000$$

$$X_{oxygen} = 0.9062$$

Step 2. *Calculate partial pressures:*

$$\text{Partial pressure of halothane} = P_{halothane}$$

$$= (X_{halothane})(P_{total})$$

$$P_{halothane} = 0.0938(P_{total}) = 0.0938(855 \text{ mm Hg})$$

$$P_{halothane} = 80.2 \text{ mm Hg}$$

The total pressure of the mixture is the sum of the partial pressures of the gases in the mixture.

$$P_{halothane} + P_{oxygen} = 855 \text{ mm Hg}$$

and so

$$P_{oxygen} = 855 \text{ mm Hg} - P_{halothane}$$

$$P_{oxygen} = 855 \text{ mm Hg} - 80.2 \text{ mm Hg} = 775 \text{ mm Hg}$$

cise 12.11   Partial Pressures

lothane–oxygen mixture described in Example 12.11 is placed in a 5.00-L tank at
. What is the total pressure (in millimeters of mercury) of the gas mixture in the
What are the partial pressures (in millimeters of mercury) of the gases?

KINETIC-MOLECULAR THEORY OF GASES

ve have discussed the macroscopic properties of gases. Now we turn to the
molecular theory [← SECTION 1.5], for a description of the behavior of mat-
he molecular or atomic level [⊕ CD-ROM, Screens 12.9 and 12.10]. Hundreds of
nental observations have led to the following postulates regarding the be-
of gases at the molecular and atomic level.

- Use the ideal gas law.
- Apply the gas laws to stoichiometric calculations.
- **Understand kinetic-molecular theory applied to gases, especially the distribution of molecular speeds (energies).**
- Recognize why gases do not behave as ideal gases.

- Gases consist of particles (molecules or atoms) whose separation is much greater than the size of the particles themselves (Figure 12.11).
- The particles of a gas are in continual, random, and rapid motion. As they move, they collide with one another and with the walls of their container.
- The average kinetic energy of gas particles is proportional to the gas temperature. *All gases, regardless of their molecular mass, have the same average kinetic energy at the same temperature.*

Let us discuss the behavior of gases from this point of view.

Molecular Speed and Kinetic Energy

If your friend is wearing a perfume or if a pizza smells good, how do you know it? In scientific terms, we know that molecules of perfume or the odor-causing molecules of food enter the gas phase and drift through space until they reach the cells of your body that react to odors. The same thing happens in the laboratory when containers of aqueous ammonia (NH_3) and hydrochloric acid (HCl) sit side by side (Figure 12.12). Molecules of the two compounds enter the gas phase and drift along until they encounter one another, at which time they react and form a cloud of tiny particles of solid ammonium chloride (NH_4Cl).

If you change the temperature of the environment of the containers in Figure 12.12 and measure the time needed for the cloud of ammonium chloride to form, you will find the time will be longer at lower temperatures. This change occurs because the speed at which molecules move depends on the temperature. Let us expand on this idea.

The kinetic energy of a single molecule of mass m in a gas sample is given by the equation [CD-ROM, Screen 12.10]

$$KE = \frac{1}{2}(\text{mass})(\text{speed})^2 = \frac{1}{2}mu^2$$

where u is the speed of that molecule. We can calculate the kinetic energy of a single gas molecule from this equation but not that of a collection of molecules because *not all the molecules in a gas sample move at the same speed*. Instead, there is a distribution of speeds, as shown for oxygen in Figure 12.13 [CD-ROM, Screen 12.11].

Two important observations can be made about Figure 12.13. First, some molecules have high speeds (and thus high kinetic energy) and others have low speeds (and low kinetic energy). The most probable speed corresponds to the maximum in the distribution curve. For oxygen gas at 25 °C, for example, the maximum is at a speed of about 400 m/s, but most of the molecules have speeds within the range 200–700 m/s.

The second observation from Figure 12.13 is that, as the temperature is increased, the most probable speed goes up, and the number of molecules traveling very fast goes up a great deal.

Experiments show that the average kinetic energy, $\overline{KE}$, of a sample of gas molecules depends only on the Kelvin temperature, a conclusion expressed by the following equation

$$\overline{KE} \propto T$$

(The horizontal bar over the symbol KE indicates an average value.) If we can calculate the kinetic energy of an individual molecule knowing its speed, we can do the same for a collection of molecules if we know their average speed, $\overline{u}$. If some molecules in a sample have a speed of u_1, some have a different speed u_2, and so on, the *average* speed, $\overline{u}$, of the molecules is therefore

$$\overline{u} = \frac{n_1 u_1 + n_2 u_2 + \cdots}{N}$$

where n_1 is the number of molecules with the speed u_1, n_2 is the number of molecules with speed u_2, and so on. N is the total number of molecules ($n_1 + n_2 + \cdots$). This means the *average* kinetic energy of the molecules in a gas sample, $\overline{KE}$, is related to $\overline{u^2}$, the *average of the squares of their speeds* (called the "mean square speed"). Therefore, the *average* kinetic energy for the gas sample is

$$\overline{KE} = \frac{1}{2}m\overline{u^2}$$

Figure 12.11 A molecular view of gases and liquids. The fact that a large volume of N_2 gas can be condensed to a small volume of liquid indicates that the distance between molecules in the gas phase is very large compared with the intermolecular distances in liquids. *(Charles D. Winters)*

Figure 12.12 The movement of gas molecules. Open dishes of aqueous ammonia and hydrochloric acid were placed side by side. When molecules of NH_3 and HCl escape from solution to the atmosphere and encounter one another, solid ammonium chloride, NH_4Cl, is formed. *(Charles D. Winters)*

Figure 12.13 **The distribution of molecular speeds.** A graph of the number of molecules with a given speed versus that speed shows the distribution of molecular speeds. The red curve shows the effect of increased temperature. Notice, however, that even though the curve for the higher temperature is "flatter" and broader than the one at a lower temperature, the areas under the curves are the same because the number of molecules in the sample is fixed.

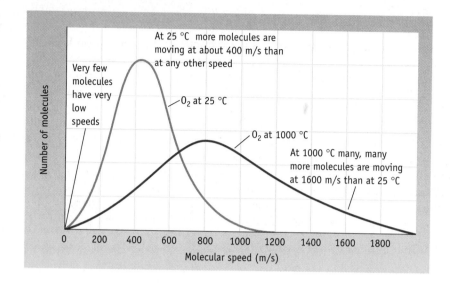

Now, because $\overline{KE}$ is proportional to T, $\frac{1}{2}\,m\overline{u}^{\,2}$ must also be proportional to T ($\frac{1}{2}\,m\overline{u}^{\,2} \propto T$). This relation among mass, average speed, and temperature is expressed in Equation 12.9. Here the square root of the mean square speed ($\sqrt{\overline{u^2}}$, called the **root-mean-square,** or **rms speed**), the temperature (T, in kelvins), and the molar mass (M) are related.

$$\sqrt{\overline{u^2}} = \sqrt{\frac{3RT}{M}}$$

(12.9)

● **Maxwell–Boltzmann Curves**
Curves of number of molecules versus speed or energy, such as those in Figure 12.13, are often called **Maxwell–Boltzmann distribution curves.** They are named after James Clerk Maxwell (1831–1879) and Ludwig Boltzmann (1844–1906). The distribution of speeds (or kinetic energies) of molecules illustrated by Figure 12.13 is an important concept that is often used when explaining chemical phenomena.

In this equation, sometimes called *Maxwell's equation* after James Clerk Maxwell (Section 7.1), R is the gas constant expressed in SI units; that is, $R = 8.314472$ J/K · mol. The speeds of gas molecules are indeed related directly to the temperature (see Figure 12.13).

All gases have the same average kinetic energy at the same temperature. However, if you compare a sample of one gas with another, say compare O_2 and N_2, this does not mean the molecules have the same average speed (Figure 12.14). Instead, Maxwell's equation shows that the smaller the molar mass of the gas the greater the rms speed.

Example 12.12 **Molecular Speed**

Problem • Calculate the rms speed of oxygen molecules at 25 °C.

Strategy • We must use Equation 12.9 with M in units of kilograms per mole. The reason for this is that R is in units of joules per kelvins per mole (J/K · mol) and 1 J = 1 kg · m²/s².

Solution • The molar mass of O_2 is 32.0×10^{-3} kg/mol.

$$\sqrt{\overline{u^2}} = \sqrt{\frac{3(8.3145\ \text{J/K} \cdot \text{mol})(298\ \text{K})}{32.0 \times 10^{-3}\ \text{kg/mol}}}$$

$$= \sqrt{2.32 \times 10^5\ \text{J/kg}}$$

To obtain the answer in meters per second, we use the relation 1 J = 1 kg · m²/s². This means we have

$$\sqrt{\overline{u^2}} = \sqrt{2.32 \times 10^5\ \text{kg} \cdot \text{m}^2/\text{kg} \cdot \text{s}^2}$$

$$= \sqrt{2.32 \times 10^5\ \text{m}^2/\text{s}^2}$$

$$= 482\ \text{m/s}$$

This speed is equivalent to about 1100 mph!

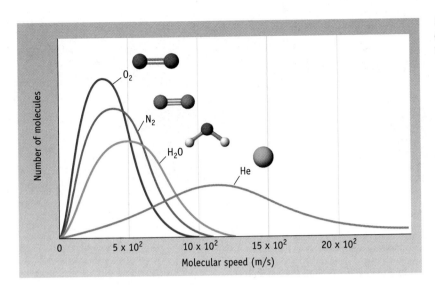

Figure 12.14 The effect of molecular mass on the distribution of speeds. At a given temperature, molecules with higher masses have lower speeds.

Exercise 12.12 **Molecular Speeds**

Calculate the rms speeds of helium atoms and N_2 molecules at 25 °C.

Kinetic-Molecular Theory and the Gas Laws

The gas laws, which come from experiment, can be explained by the kinetic-molecular theory. The starting place is to describe how pressure arises from collisions of gas molecules with the walls of the container holding the gas (Figure 12.15). The force developed by these collisions depends on the number of collisions and the average force per collision. When the temperature of a gas is increased, the average force of a collision on the walls of its container increases because the average kinetic energy of the molecules increases. Also, because the speed increases with temperature, more collisions occur per second. Thus, the collective force per square centimeter is greater, and the pressure increases. Mathematically, this means P is proportional to T when n and V are constant.

Increasing the number of molecules of a gas at a fixed temperature and volume does not change the average collision force, but it does increase the number of collisions occurring per second. Thus, the pressure increases, and we can say that P is proportional to n when V and T are constant.

If the pressure is not allowed to increase when either the number of molecules of gas or the temperature is increased, the volume of the container (and the area over which the collisions can take place) must increase. This is expressed by stating that V is proportional to nT when P is constant, a statement that is a *combination of Avogadro's hypothesis and Charles's law.*

Finally, if the temperature is constant, the average impact force of molecules of a given mass with the container walls must be constant. If n is kept constant while the volume of the container is made smaller, however, the number of collisions with the container walls per second must increase. This means the pressure increases, and so P is proportional to $1/V$ when n and T are constant, as stated by *Boyle's law.*

● **R in other Units**
The gas constant R can be expressed in different units. If P is measured in SI units of pascals (kg/m · s²), and V is measured in cubic meters (m³), then R has the value (at STP) of

$$8.3145 \text{ kg} \cdot \text{m}^2/\text{s}^2 \cdot \text{mol} \cdot \text{K}$$

Because 1 kg · m²/s² is 1 joule, the gas constant can be expressed as

$$R = 8.3145 \text{ J/mol} \cdot \text{K}.$$

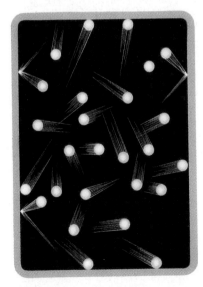

Figure 12.15 Gas pressure.
According to the kinetic-molecular theory, gas pressure is caused by gas molecules bombarding the container walls.

Figure 12.16 Diffusion. Brown NO_2 gas diffuses out of the flask in which it was generated (*left*) and into the attached tube within a few minutes (*right*). (The NO_2 is being made by the reaction of copper with nitric acid.) *(Photos, Charles D. Winters)*

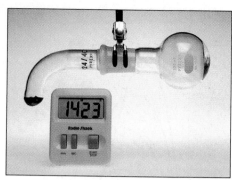

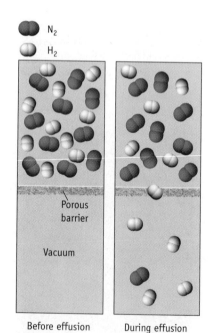

N_2

H_2

Porous barrier

Vacuum

Before effusion   During effusion

Figure 12.17 Effusion. Gas molecules effuse through the pores of a porous barrier. Lighter molecules (H_2) with higher average speeds strike the barrier more often and pass more rapidly through it than heavier, slower molecules (N_2) at the same temperature.

12.7 DIFFUSION AND EFFUSION

When a pizza is brought into a room, the volatile aroma-causing molecules vaporize into the atmosphere where they mix with the oxygen, nitrogen, carbon dioxide, water vapor, and other gases present. Even if there were no movement of the air in the room caused by fans or people moving about, the odor would eventually reach everywhere in the room. This mixing of molecules of two or more gases due to their molecular motions is called **diffusion;** it results from the random molecular motion of all gases. Given time, the molecules of one component in a gas mixture will thoroughly and completely mix with all other components of the mixture (Figure 12.16) [CD-ROM, Screen 12.12].

Closely related to diffusion is **effusion,** which is the movement of gas through a tiny opening in a container into another container where the pressure is very low (Figure 12.17). Thomas Graham (1805–1869), a Scottish chemist, studied the effusion of gases and found that the rate of effusion of a gas was inversely proportional to the square root of its molar mass. Based on Graham's experimental results, the rates of effusion of two gases can be compared:

$$\frac{\text{Rate of effusion of gas 1}}{\text{Rate of effusion of gas 2}} = \sqrt{\frac{\text{molar mass of gas 2}}{\text{molar mass of gas 1}}} \qquad (12.10)$$

Equation 12.10 is known as **Graham's law.** It is readily derived from Maxwell's equation by recognizing that the rate of effusion depends on the speed of the molecules. The ratio of the rms speeds is the same as the ratio of the effusion rates:

$$\frac{\text{Rate of effusion of gas 1}}{\text{Rate of effusion of gas 2}} = \frac{\text{rms for gas 1}}{\text{rms for gas 2}} = \frac{\sqrt{3RT/(M \text{ of gas 1})}}{\sqrt{3RT/(M \text{ of gas 2})}}$$

Canceling out like terms gives Equation 12.10.

Figure 12.17 illustrates the relative rates of effusion of H_2 and N_2 molecules, and Graham's law allows a quantitative comparison of their relative rates:

$$\frac{\text{Rate of effusion of } H_2}{\text{Rate of effusion of } N_2} = \sqrt{\frac{M \text{ of } N_2}{M \text{ of } H_2}} = \sqrt{\frac{28.0 \text{ g/mol}}{2.02 \text{ g/mol}}} = \frac{3.72}{1}$$

This calculation tells you that H_2 molecules will effuse through the barrier 3.72 times faster than N_2 molecules.

Example 12.13   Graham's Law of Effusion

Problem • Tetrafluoroethylene, C_2F_4, effuses through a barrier at a rate of 4.6×10^{-6} mol/h. An unknown gas, consisting only of boron and hydrogen, effuses at the rate of 5.8×10^{-6} mol/h under the same conditions. What is the molar mass of the unknown gas?

Strategy • From Graham's law we know that a light molecule will effuse more rapidly than a heavier one. Because the unknown gas effuses more rapidly than C_2F_4 ($M = 100.0$ g/mol), the unknown must have a molar mass less than 100 g/mol. Substitute the experimental data into Graham's law (Equation 12.10).

Solution •

$$\frac{\text{Rate of effusion of unknown}}{\text{Rate of effusion of } C_2F_4} = \sqrt{\frac{M \text{ of } C_2F_4}{M \text{ of unknown}}}$$

$$\frac{5.8 \times 10^{-6} \text{ mol/h}}{4.6 \times 10^{-6} \text{ mol/h}} = \sqrt{\frac{100.0 \text{ g/mol}}{M \text{ of unknown}}}$$

To solve for the unknown molar mass, square both sides of the equation and rearrange to find M for the unknown.

$$1.6 = \frac{100.0 \text{ g/mol}}{M \text{ of unknown}}$$

$$M = 63 \text{ g/mol}$$

Comment • A boron–hydrogen compound corresponding to this molar mass is B_5H_9, called pentaborane.

Exercise 12.13   Graham's Law

A sample of pure methane, CH_4, is found to effuse through a porous barrier in 1.50 min. Under the same conditions, an equal number of molecules of an unknown gas effuses through the barrier in 4.73 min. What is the molar mass of the unknown gas?

12.8   SOME APPLICATIONS OF THE GAS LAWS AND KINETIC-MOLECULAR THEORY

Hot-Air Balloons

A balloon will rise if the mass of heated air in the balloon is less than an equivalent mass of air at the temperature surrounding the balloon. Assuming the simple plastic balloon in Figure 12.18 has a volume of 140 L, the mass of air in the balloon is about 170 g at STP (assuming a density of air of 1.2 g/L). Suppose the balloon has a total mass of 40 g. If the balloon is to rise, a minimum of 40 g of air must be displaced from the bag by heating it. That is, the mass of air in the bag must be only about 130 g. Using the ideal gas law, you can show that a bag with a volume of 140 L will be filled with 130 g of air to 1 atm pressure at a temperature of about 110 °C.

Deep Sea Diving

Diving with a self-contained underwater breathing apparatus (SCUBA) is exciting. If you want to dive much beyond about 60 ft (18 m) or so, however, you need to take special precautions.

When you breathe air from a SCUBA tank (Figure 12.19), the pressure of the gas in your lungs is equal to the pressure exerted on your body. At the surface, air has an oxygen concentration of 21%, so the partial pressure of O_2 is about 0.21 atm. If you are at a depth of about 33 ft the water pressure is 2 atm. This means the oxygen partial pressure is double the surface partial pressure, or about 0.4 atm.

Figure 12.18   A hot-air balloon made from thin plastic bag. The bag was filled with air heated by a propane burner. *(Charles D. Winters)*

Figure 12.19   SCUBA diving. Ordinary recreational dives can be made with compressed air to depths of about 60 ft or so. With a gas mixture called Nitrox (which has up to 36% O$_2$) one can stay at such depths for a longer period. To go even deeper, however, divers must breath special gas mixtures such as Trimix. This is a breathing mixture consisting of oxygen, helium, and nitrogen. *(NOAA)*

Similarly, the partial pressure of N$_2$, which is about 0.8 atm at the surface, doubles to about 1.6 atm at a depth of 33 ft. What is the problem?

Nitrogen narcosis, also called "rapture of the deep," results from the toxic effect of high N$_2$ pressure on nerve conduction. Its effect is comparable to drinking alcohol on an empty stomach or taking laughing gas (nitrous oxide, N$_2$O) at the dentist; it makes you slightly giddy. In severe cases, it can impair a diver's judgment and even cause a diver to take the regulator out of his or her mouth and hand it to a fish! Some people can go as deep as 130 ft with no problem, but others experience nitrogen narcosis at 80 ft.

Another problem with breathing air at depths beyond 100 ft or so is oxygen toxicity. Our bodies are regulated for a partial pressure of O$_2$ of 0.21 atm. At a depth of 130 ft, the partial pressure of O$_2$ is comparable to breathing 100% oxygen at sea level. These higher partial pressures can harm the lungs and cause central nervous system damage. Oxygen toxicity is the reason very deep dives are done not with compressed air but with gas mixtures with a much lower percentage of O$_2$, say about 10%.

Because of the risk of nitrogen narcosis, divers going beyond about 130 ft, such as those who work for offshore oil drilling companies, use a mixture of oxygen and helium. This solves the nitrogen narcosis problem, but it introduces another. If the diver has a voice link to the surface, the diver's speech sounds like Donald Duck! Speech is altered because the velocity of sound in helium is different from that in air, and the density of gas at a depth of several hundred feet is much higher than at the surface. A search of the Internet will turn up several papers from the military and from commercial diving companies on "helium speech descrambling" devices.

12.9   NONIDEAL BEHAVIOR: REAL GASES

If you are working with a gas at approximately room temperature and a pressure of 1 atm or less, the ideal gas law is remarkably successful in relating the quantity of gas and its pressure, volume, and temperature. At higher pressures or lower temperatures, however, deviations from the ideal gas law occur. The origin of these deviations is explained by the breakdown of the assumptions used when describing ideal gases.

At standard temperature and pressure (STP), the volume occupied by a single molecule is *very* small relative to its share of the total gas volume. A helium atom with a radius of 31 pm (see Figure 8.10) has approximately the same space to move about as a pea has inside a basketball. Now suppose the pressure is increased significantly, to 1000 atm. The volume available to each molecule is a sphere with a radius of only about 200 pm, which means the situation is now like that of a pea inside a sphere a bit larger than a Ping-Pong ball.

The kinetic-molecular theory and the ideal gas law are concerned with the volume available to the molecules to move about, not the volume of the molecules themselves. It is clear that the volume occupied by gas molecules is not negligible at higher pressures (and this violates the first postulate of the kinetic-molecular theory; see page 490). For example, suppose you have a flask marked with a volume of 500 mL. This does not mean the space available to molecules is 500 mL. Rather, the available volume is less than 500 mL, especially at high gas pressures, because the molecules themselves occupy some of the volume.

Another assumption of the kinetic-molecular theory is that collisions between molecules are elastic, that is, that the atoms or molecules of the gas never stick to one another by some type of force. This is clearly not true as well. All gases can be liquefied — although some gases require a very low temperature — and the only way this can happen is if there are forces between the molecules (see Figure 12.11). When a molecule is about to strike the wall of its container, other molecules in its vicinity exert a slight pull on the molecule and pull it away from the wall (Figure 12.20). The effect of the intermolecular forces is that molecules strike the wall with less force than in the absence of intermolecular attractive forces. Thus, because collisions between molecules in a real gas and the wall are softer, the observed gas pressure is less than that predicted by the ideal gas law. This effect can be particularly pronounced when the temperature is close to the condensation temperature of the gas.

The Dutch physicist Johannes van der Waals (1837–1923) studied the breakdown of the ideal gas law equation and developed an equation to correct for the errors arising from nonideality. This equation is known as the **van der Waals equation:**

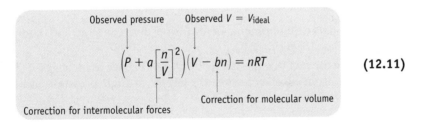

Observed pressure   Observed $V = V_{ideal}$

$$\left(P + a\left[\frac{n}{V}\right]^2\right)(V - bn) = nRT \qquad \textbf{(12.11)}$$

Correction for intermolecular forces   Correction for molecular volume

where a and b are experimentally determined constants. Although Equation 12.11 might seem complicated at first glance, the terms in parentheses are those of the ideal gas law, each corrected for the effects discussed previously. The pressure correction term, $a(n/V)^2$, corrects for intermolecular forces. Because the observed gas pressure is lower than the pressure calculated using the equation $PV = nRT$ owing to intermolecular forces, $a(n/V)^2$ is added to the observed pressure. The constant a, which is determined experimentally, typically has values in the range 0.01 to 10 atm(L/mol)^2. The bn term corrects the observed volume, $V_{observed}$, to a smaller value, the volume actually available to the gas molecules. Here n is the number of moles of gas, and b is an experimental quantity that corrects for the molecular volume. Typical values of b range from 0.01 to 0.1 L/mol, roughly increasing with increasing molecular size.

Chapter Goals • Revisited

- Understand the basis of the gas laws and how to use those laws: Boyle's, Charles's, Avogadro's, Dalton's.
- Use the ideal gas law.
- Apply the gas laws to stoichiometric calculations.
- Understand kinetic-molecular theory applied to gases, especially the distribution of molecular speeds (energies).
- **Recognize why gases do not behave as ideal gases.**

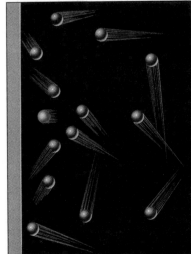

Figure 12.20   Effect of intermolecular forces. A gas molecule strikes the container wall with less force because of the attractive forces between it and its neighbors.

As an example of the importance of these corrections, consider a sample of 8.00 mol of chlorine gas, Cl_2, in a 4.00-L tank at 27.0 °C. (Cl_2 has $a = 6.49$ atm · L^2/mol^2 and $b = 0.0562$ L/mol.) The ideal gas law would lead you to expect a pressure of 49.2 atm. A better estimate of the pressure, obtained from the van der Waals equation, is 29.5 atm, about 20 atm less than the ideal pressure!

Exercise 12.14   **Using the van der Waals Equation**

Using both the ideal gas law and the van der Waals equation, calculate the pressure expected for 10.0 mol of helium gas in a 1.00-L container at 25 °C. For helium $a = 0.034$ (atm · L^2/mol^2) and $b = 0.0237$ L/mol.

In Summary

When you have finished studying this chapter, you should ask if you have met the chapter goals. In particular, you should be able to

- Describe how pressure measurements are made (Section 12.1).
- Use the units of pressure, especially atmospheres (atm) and millimeters of mercury (mm Hg) (Section 12.1).
- Understand the basis of the gas laws and how to use these laws (Section 12.2).
- Understand the ideal gas law and how to use this equation (Section 12.3).
- Calculate the molar mass of a compound from a knowledge of the pressure of a known quantity of a gas in a given volume at a known temperature (Section 12.3).
- Apply the gas laws to a study of the stoichiometry of reactions (Section 12.4).
- Use Dalton's law (Section 12.5).
- Apply the kinetic-molecular theory of gas behavior at the molecular level (Section 12.6).
- Understand the phenomena of diffusion and effusion and know how to use Graham's law (Section 12.7).
- Appreciate the fact that gases usually do not behave as ideal gases (Section 12.9). Deviations from ideal behavior are largest at high pressure and low temperature.

Key Terms

Section 12.1
  millimeters of mercury (mm Hg)
  standard atmosphere (atm)
  pascal
  bar

Section 12.2
  compressibility
  Boyle's law
  Charles's law
  general gas law, combined gas law
  Avogadro's hypothesis

Section 12.3
  gas constant, R
  ideal gas law
  standard temperature and pressure (STP)
  standard molar volume

Section 12.5
  partial pressure
  Dalton's law of partial pressures
  mole fraction, X

Section 12.6
  root mean square speed (rms speed)

Section 12.7
  diffusion
  effusion
  Graham's law

Section 12.9
  the van der Waals equation

Key Equations

Equation 12.1 (page 477)
Boyle's law (where P is the pressure and V is the volume)

$$P_1V_1 = P_2V_2$$

Equation 12.2 (page 478)
Charles's law (where T is the Kelvin temperature)

$$\frac{V_1}{T_1} = \frac{V_2}{T_2}$$

Equation 12.3 (page 479)
General gas law (combined gas law) for a fixed amount of gas

$$\frac{P_1V_1}{T_1} = \frac{P_2V_2}{T_2}$$

Equation 12.4 (page 481)
Ideal gas law (where n is the amount of gas (moles) and R is the universal gas constant, $0.082057 \, L \cdot atm/K \cdot mol$)

$$PV = nRT$$

Equation 12.5 (page 483)
Density of gases (where d is the gas density)

$$d = \frac{m}{V} = \frac{PM}{RT}$$

Equation 12.6 (page 488)
Dalton's law of partial pressures. The total pressure of a gas mixture is the sum of the partial pressures of the component gases (P_n).

$$P_{total} = P_1 + P_2 + P_3 + \cdots$$

Equation 12.7 (page 488)
The total pressure of a gas mixture is equal to the total number of moles of gases multiplied by (RT/V).

$$P_{total} = n_{total}\left(\frac{RT}{V}\right)$$

Equation 12.8 (page 489)
The partial pressure of a gas (A) in a mixture is the product of its mole fraction (X_A) and the total pressure of the mixture.

$$P_A = X_A(P_{total})$$

Equation 12.9 (page 492)
Maxwell's equation, which relates the rms speed ($\sqrt{\overline{u^2}}$) to the molar mass of a gas (M) and its temperature (T)

$$\sqrt{\overline{u^2}} = \sqrt{\frac{3RT}{M}}$$

Equation 12.10 (page 492)
Graham's law. The rate of effusion of a gas — the quantity of material moving from one place to another in a given amount of time — is inversely proportional to the square root of its molar mass.

$$\frac{\text{Rate of effusion of gas 1}}{\text{Rate of effusion of gas 2}} = \sqrt{\frac{\text{molar mass of gas 2}}{\text{molar mass of gas 1}}}$$

Equation 12.11 (page 497)

The van der Waals's equation. Relates pressure, volume, temperature, and amount of gas for a nonideal gas.

Observed pressure   Observed $V = V_{ideal}$

$$\left(P + a\left[\frac{n}{V}\right]^2\right)(V - bn) = nRT$$

Correction for intermolecular forces   Correction for molecular volume

Study Questions

Questions with blue, bold-faced numbers have answers in Appendix O. See Example Problems, Tutorials, and Simulations in Chapter 12 on the General Chemistry Interactive CD-ROM, *Version 3.0.*

Reviewing Important Concepts

1. Name the three gas laws that interrelate *P*, *V*, and *T*. Explain the relationships in words and in equations.

2. What conditions are represented by STP? What is the volume of a mole of ideal gas under these conditions?

3. Show how to calculate the molar mass of a gas from *P*, *V*, and *T* measurements and other information.

4. State Avogadro's hypothesis. Relate your discussion to the formation of water vapor from hydrogen and oxygen gas. For example, if 3 mol of H_2 is used at STP, what volume of O_2 gas (in L) is required at STP and what volume of H_2O vapor (in L) is produced at STP?

5. State Dalton's law. If the air you breathe is 78% N_2 and 21% O_2 (on a mole basis), what is the mole fraction of O_2? What is the partial pressure of O_2 when the atmospheric pressure is 748 mm Hg?

6. What are the basic assumptions of the kinetic-molecular theory? Explain Boyle's law on the basis of kinetic-molecular theory.

7. What assumptions are made in describing an ideal gas? Under what circumstances is the ideal gas law least accurate?

8. In the van der Waals equation, what properties of a real gas are accounted for by the constants *a* and *b*?

9. State Graham's law in words and in equation form. If an unknown gas effuses four times slower than H_2 gas at 25 °C, what is the molar mass of the unknown gas? What common gas has this molar mass?

Practicing Skills

Pressure

(See Example 12.1 and CD-ROM Screen 12.2)

10. The pressure of a gas is 440 mm Hg. Express this pressure in units of (a) atmospheres, (b) bars, and (c) kilopascals.

11. The average barometric pressure at an altitude of 10 km is 210 mm Hg. Express this pressure in atmospheres, bars, and kilopascals.

12. Which represents the higher pressure in each of the following pairs:
    (a) 534 mm Hg or 0.754 bar
    (b) 534 mm Hg or 650 kPa
    (c) 1.34 bar or 934 kPa

13. Put the following in order of increasing pressure: 363 mm Hg, 363 kPa, 0.256 atm, and 0.523 bar.

Boyle's Law and Charles's Law

(See Examples 12.2 and 12.3 and CD-ROM Screen 12.3)

14. A sample of nitrogen gas has a pressure of 67.5 mm Hg in a 500.-mL flask. What is the pressure of this gas sample when it is transferred to a 125-mL flask at the same temperature?

15. A sample of CO_2 gas has a pressure of 56.5 mm Hg in a 125-mL flask. The sample is transferred to a new flask where it has a pressure of 62.3 mm Hg at the same temperature. What is the volume of the new flask?

16. You have 3.5 L of NO at a temperature of 22.0 °C. What volume would the NO occupy at 37 °C? (Assume the pressure is constant.)

17. A 5.0-mL sample of CO_2 gas is enclosed in a gas-tight syringe (see Figure 12.3) at 22 °C. If the syringe is immersed in an ice bath (0 °C), what is the new gas volume, assuming that the pressure is held constant?

The General Gas Law

(See Example 12.4)

18. A sample of H_2 gas has a pressure of 380 mm Hg at 25 °C. What is the pressure of this gas sample in a 2.8-L flask at 0 °C?

19. You have a sample of CO_2 in a flask (A) with a volume of 25.0 mL. At 20.5 °C, the pressure of the gas is 436.5 mm Hg. To find the volume of another flask (B), you move the

CO_2 to that flask and find that its pressure is now 94.3 mm Hg at 24.5 °C. What is the volume of flask B?

20. You have a sample of gas in a flask with a volume of 250 mL. At 25.5 °C the pressure of the gas is 360 mm Hg. If you decrease the temperature to −5.0 °C, what is the gas pressure at the lower temperature?

21. A sample of gas occupies 135 mL at 22.5 °C; the pressure is 165 mm Hg. What is the pressure of the gas sample when it is placed in a 252-mL flask at a temperature of 0.0 °C?

22. One of the cylinders of an automobile engine has a volume of 400. cm^3. The engine takes in air at a pressure of 1.00 atm and a temperature of 15 °C and compresses the air to a volume of 50.0 cm^3 at 77 °C. What is the final pressure of the gas in the cylinder? (The ratio of before and after volumes, in this case 400:50 or 8:1, is called the compression ratio.)

23. A helium-filled balloon of the type used in long-distance flying contains 420,000 ft^3 (1.2×10^7 L) of helium. Let us say you fill the balloon with helium on the ground where the pressure is 737 mm Hg and the temperature is 16.0 °C. When the balloon ascends to a height of 2 miles where the pressure is only 600. mm Hg and the temperature is −33 °C, what volume is occupied by the helium gas? Assume the pressure inside the balloon matches the external pressure. Comment on the result.

Avogadro's Hypothesis
(See Example 12.5 and CD-ROM Screen 12.3)

24. Nitrogen monoxide reacts with oxygen to give nitrogen dioxide.

$$2 NO(g) + O_2(g) \longrightarrow 2 NO_2(g)$$

(a) If you mix NO and O_2 in the correct stoichiometric ratio, and NO has a volume of 150 mL, what volume of O_2 is required (at the same temperature and pressure)?

(b) After reaction is complete between 150 mL of NO and the stoichiometric volume of O_2, what is the volume of NO_2?

25. Ethane, C_2H_6, burns in air according to the equation

$$2 C_2H_6(g) + 7 O_2(g) \longrightarrow 4 CO_2(g) + 6 H_2O(g)$$

What volume of O_2 (in L) is required for complete reaction with 5.2 L of C_2H_6? What volume of H_2O vapor (in L) is produced? Assume all gases are measured at the same temperature and pressure.

Ideal Gas Law
(See Example 12.6 and CD-ROM Screen 12.4)

26. A 1.25-g sample of CO_2 is contained in a 750.-mL flask at 22.5 °C. What is the pressure of the gas?

27. A balloon holds 30.0 kg of helium. What is the volume of the balloon if the final pressure is 1.20 atm and the temperature is 22 °C?

28. A flask is first evacuated so that it contains no gas at all. Then, 2.2 g of CO_2 is introduced into the flask. On warming to 22 °C, the gas exerts a pressure of 318 mm Hg. What is the volume of the flask?

29. A steel cylinder holds 1.50 g of ethanol, C_2H_5QH. What is the pressure of the ethanol vapor if the cylinder has a volume of 251 cm^3 and the temperature is 250 °C? (Assume all the ethanol is in the vapor phase at this temperature.)

30. A balloon for long-distance flying contains 1.2×10^7 L of helium. If the helium pressure is 737 mm Hg at 25 °C, what mass of helium (in grams) does the balloon contain? (See Study Question 23 and page 471.)

31. What mass of helium, in grams, is required to fill a 5.0-L balloon to a pressure of 1.1 atm at 25 °C?

Gas Density
(See Example 12.7 and 12.8 and CD-ROM Screen 12.5)

32. Forty miles above the earth's surface the temperature is 250 K and the pressure is only 0.20 mm Hg. What is the density of air (in grams per liter) at this altitude? (Assume the molar mass of air is 28.96 g/mol.)

33. Diethyl ether, $(C_2H_5)_2O$, vaporizes easily at room temperature. If the vapor exerts a pressure of 233 mm Hg in a flask at 25 °C, what is the density of the vapor?

34. A gaseous organofluorine compound has a density of 0.355 g/L at 17 °C and 189 mm Hg. What is the molar mass of the compound?

35. Chloroform is a common liquid used in the laboratory. It vaporizes readily. If the pressure of chloroform vapor in a flask is 195 mm Hg at 25.0 °C, and the density of the vapor is 1.25 g/L, what is the molar mass of chloroform?

Ideal Gas Laws and Determining Molar Mass
(See Example 12.8 and CD-ROM Screen 12.6)

36. A 1.007-g sample of an unknown gas exerts a pressure of 715 mm Hg in a 452-mL container at 23 °C. What is the molar mass of the gas?

37. A 0.0125-g sample of a gas with an empirical formula of CHF_2 is placed in a 165-mL flask. It has a pressure of 13.7 mm Hg at 22.5 °C. What is the molecular formula of the compound?

38. A new boron hydride, B_xH_y, has been isolated. To find its molar mass, you measure the pressure of the gas in a known volume at a known temperature. The experimental data collected are:

Mass of gas = 12.5 mg	Temperature = 25 °C
Pressure of gas = 24.8 mm Hg	Volume of flask = 125 mL

Which formula corresponds to the calculated molar mass?

(a) B_2H_6 (d) B_6H_{10}

(b) B_4H_{10} (e) $B_{10}H_{14}$

(c) B_5H_9

39. Acetaldehyde is a common organic compound that vaporizes readily. Determine the molar mass of acetaldehyde from the following data:

Sample mass = 0.107 g	Volume of gas = 125 mL
Temperature = 0.0 °C	Pressure = 331 mm Hg

Gas Laws and Stoichiometry
(See Examples 12.9 and 12.10 and CD-ROM Screen 12.7)

40. Iron reacts with hydrochloric acid to produce iron(II) chloride and hydrogen gas:

$$Fe(s) + 2\ HCl(aq) \longrightarrow FeCl_2(aq) + H_2(g)$$

The H_2 gas from the reaction of 2.2 g of iron with excess acid is collected in a 10.0-L flask at 25 °C. What is the pressure of the H_2 gas in this flask?

41. Silane, SiH_4, reacts with O_2 to give silicon dioxide and water:

$$SiH_4(g) + 2\ O_2(g) \longrightarrow SiO_2(s) + 2\ H_2O(\ell)$$

A 5.20-L sample of SiH_4 gas at 356 mm Hg pressure and 25 °C is allowed to react with O_2 gas. What volume of O_2 gas, in liters, is required for complete reaction if the oxygen has a pressure of 425 mm Hg at 25 °C?

42. Sodium azide (NaN_3), the explosive compound in automobile air bags, decomposes according to the equation

$$2\ NaN_3(s) \longrightarrow 2\ Na(s) + 3\ N_2(g)$$

What mass of sodium azide is required to provide the nitrogen needed to inflate a 75.0-L bag to a pressure of 1.3 atm at 25 °C?

43. The hydrocarbon octane (C_8H_{18}) burns to give CO_2 and water vapor:

$$2\ C_8H_{18}(g) + 25\ O_2(g) \longrightarrow 16\ CO_2(g) + 18\ H_2O(g)$$

If a 0.095-g sample of octane burns completely in O_2, what will be the pressure of water vapor in a 4.75-L flask at 30.0 °C? If the O_2 gas needed for complete combustion was contained in a 4.75-L flask at 22 °C, what would its pressure be?

44. Hydrazine (N_2H_4) reacts with O_2 according to the equation

$$N_2H_4(g) + O_2(g) \longrightarrow N_2(g) + 2\ H_2O(\ell)$$

Assume the O_2 needed for the reaction is in a 450-L tank at 23 °C. What must the oxygen pressure be in the tank to have enough oxygen to consume 1.00 kg of hydrazine completely?

45. A self-contained breathing apparatus uses canisters containing potassium superoxide. The superoxide consumes the CO_2 exhaled by a person and replaces it with oxygen.

$$4\ KO_2(s) + 2\ CO_2(g) \longrightarrow 2\ K_2CO_3(s) + 3\ O_2(g)$$

What mass of KO_2, in grams, is required to react with 8.90 L of CO_2 at 22.0 °C and 767 mm Hg?

Gas Mixtures and Dalton's Law
(See Example 12.11 and CD-ROM Screen 12.8)

46. What is the total pressure in atmospheres of a gas mixture that contains 1.0 g of H_2 and 8.0 g of Ar in a 3.0-L container at 27 °C? What are the partial pressures of the two gases?

47. A cylinder of compressed gas is labeled "Composition (mole %): 4.5% H_2S, 3.0% CO_2, balance N_2." The pressure gauge attached to the cylinder reads 46 atm. Calculate the partial pressure of each gas, in atmospheres, in the cylinder.

48. A halothane–oxygen mixture ($C_2HBrClF_3 + O_2$) can be used as an anesthetic. A tank containing such a mixture has the following partial pressures: P(halothane) = 170 mm Hg and $P(O_2)$ = 570 mm Hg.
    (a) What is the ratio of the number of moles of halothane to the number of moles of O_2?
    (b) If the tank contains 160 g of O_2, what mass of $C_2HBrClF_3$ is present?

49. A collapsed balloon is filled with He to a volume of 12.5 L at a pressure of 1.00 atm. Oxygen (O_2) is then added so that the final volume of the balloon is 26 L with a total pressure of 1.00 atm. The temperature, constant throughout, is equal to 21.5 °C.
    (a) What mass of He does the balloon contain?
    (b) What is the final partial pressure of He in the balloon?
    (c) What is the partial pressure of O_2 in the balloon?
    (d) What is the mole fraction of each gas?

Kinetic-Molecular Theory
(See Section 12.6, Example 12.12, and CD-ROM Screens 12.9–12.12)

50. You have two flasks of equal volume. Flask A contains H_2 at 0 °C and 1 atm pressure. Flask B contains CO_2 gas at 25 °C and 2 atm pressure. Compare these two gases with respect to each of the following:
    (a) Average kinetic energy per molecule
    (b) Average molecular velocity
    (c) Number of molecules
    (d) Mass of gas

51. Equal masses of gaseous N_2 and Ar are placed in separate flasks of equal volume at the same temperature. Tell whether each of the following statements is true or false. Briefly explain your answer in each case.
    (a) More molecules of N_2 are present than atoms of Ar.
    (b) The pressure is greater in the Ar flask.
    (c) Ar atoms have a greater average speed than the N_2 molecules.
    (d) The molecules of N_2 collide more frequently with the walls of the flask than do the atoms of Ar.

52. If the average speed of an oxygen molecule is 4.28×10^4 cm/s at 25 °C, what is the average speed of a CO_2 molecule at the same temperature?

53. Calculate the rms speed for CO molecules at 25 °C. What is the ratio of this speed to that of Ar atoms at the same temperature?

54. Place the following gases in order of increasing average molecular speed at 25 °C: Ar, CH_4, N_2, and CH_2F_2.

55. The reaction of SO_2 with Cl_2 gives dichlorine oxide, which is used to bleach wood pulp and to treat wastewater:

$$SO_2(g) + 2\ Cl_2(g) \longrightarrow OSCl_2(g) + Cl_2O(g)$$

All the compounds involved in the reaction are gases. List them in order of increasing average speed.

Diffusion and Effusion
(See Example 12.13 and CD-ROM Screen 12.12)

56. In each of the following pairs of gases, tell which will effuse faster:
    (a) CO_2 or F_2
    (b) O_2 or N_2
    (c) C_2H_4 or C_2H_6
    (d) two CFCs: $CFCl_3$ or $C_2Cl_2F_4$

57. Argon gas is ten times denser than helium gas at the same temperature and pressure. Which gas is predicted to effuse faster? How much faster?

58. A gas whose molar mass you wish to know effuses through an opening at a rate one third as fast as that of helium gas. What is the molar mass of the unknown gas?

59. A sample of uranium fluoride is found to effuse at the rate of 17.7 mg/h. Under comparable conditions, gaseous I_2 effuses at the rate of 15.0 mg/h. What is the molar mass of the uranium fluoride? (*Hint:* Rates must be used in units of moles per time.)

Nonideal Gases
(See Section 12.9)

60. In the text it is stated that the pressure of 8.00 mol of Cl_2 in a 4.00-L tank at 27.0 °C should be 29.5 atm if calculated using the van der Waals equation. Verify this result and compare it with the pressure predicted by the ideal gas law.

61. You want to store 165 g of CO_2 gas in a 12.5-L tank at room temperature (25 °C). Calculate the pressure the gas would have using (a) the ideal gas law and (b) the van der Waals equation. (For CO_2, $a = 3.59$ atm · L^2/mol^2) and $b = 0.0427$ L/mol)

General Questions

These questions are not designated as to type or location in the chapter. They may combine several concepts. More challenging questions are indicated by an underlined number.

62. Complete the following table:

	atm	mm Hg	kPa	bar
Standard atmosphere	___	___	___	___
Partial pressure of N_2 in the atmosphere	___	593	___	___
Tank of compressed H_2	___	___	___	133
Atmospheric pressure at the top of Mt. Everest	___	___	33.7	___

63. You want to fill a cylindrical tank with CO_2 gas at 865 mm Hg and 25 °C. The tank is 20.0 m long with a 10.0-cm radius. What mass of CO_2 (in grams) is required?

64. On combustion 1.0 L of a gaseous compound of hydrogen, carbon, and nitrogen gives 2.82 g of H_2O as well as 2.0 L of CO_2 and 0.50 L of N_2 at STP. What is the empirical formula of the compound?

65. To what temperature, in degrees Celsius, must a 25.5-mL sample of oxygen at 90 °C be cooled for its volume to shrink to 21.5 mL? Assume the pressure and mass of the gas are constant.

66. You have a sample of helium gas at −33 °C, and you want to increase the average speed of helium atoms by 10.0%. To what temperature should the gas be heated to accomplish this?

67. If 12.0 g of O_2 is required to inflate a balloon to a certain size at 27 °C, what mass of O_2 is required to inflate it to the same size (and pressure) at 5.0 °C?

68. You have two gas-filled balloons, one containing He and the other H_2. The H_2 balloon is twice the size of the He balloon. The pressure of gas in the H_2 balloon is 1 atm, and that in the He balloon is 2 atm. The H_2 balloon is outside in the snow (−5 °C), and the He balloon is inside a warm building (23 °C).
    (a) Which balloon contains the greater number of molecules?
    (b) Which balloon contains the greater mass of gas?

69. A bicycle tire has an internal volume of 1.52 L and contains 0.406 mol of air. The tire will burst if its internal pressure reaches 7.25 atm. To what temperature, in degrees Celsius, does the air in the tire need to be heated to cause a blowout?

70. The temperature of the atmosphere on Mars can be as high as 27 °C at the equator at noon, and the atmospheric pressure is about 8 mm Hg. If a spacecraft could collect 10. m^3 of this atmosphere, compress it to a small volume, and send it back to Earth, how many moles of Martian gas would the sample contain?

71. If you place 2.25 g of solid silicon in a 6.56-L flask that contains CH_3Cl with a pressure of 585 mm Hg at 25 °C, what mass of $(CH_3)_2SiCl_2(g)$, dimethyldichlorosilane, can be formed?

$$Si(s) + 2 CH_3Cl(g) \longrightarrow (CH_3)_2SiCl_2(g)$$

What pressure of $(CH_3)_2SiCl_2(g)$ would you expect in this same flask at 95 °C on completion of the reaction? (Dimethyldichlorosilane is one starting material used to make silicones, polymeric substances used as lubricants, antistick agents, and in water-proofing caulk.)

72. $Ni(CO)_4$ can be made by reacting finely divided nickel with gaseous CO. If you have CO in a 1.50-L flask at a pressure of 418 mm Hg at 25.0 °C, along with 0.450 g of Ni powder, what is the theoretical yield of $Ni(CO)_4$?

73. The gas B_2H_6 burns in air to give H_2O and B_2O_3.

$$B_2H_6(g) + 3 O_2(g) \longrightarrow B_2O_3(s) + 3 H_2O(g)$$

    (a) Three gases are involved in this reaction. Place them in order of increasing molecular speed. (Assume all are at the same temperature.)

(b) A 3.26-L flask contains B_2H_6 with a pressure of 256 mm Hg at 25 °C. Suppose O_2 gas is added to the flask until B_2H_6 and O_2 are in the correct stoichiometric ratio for the combustion reaction. At this point, what is the the partial pressure of O_2?

74. Analysis of a gaseous chlorofluorocarbon, CCl_xF_y, shows it contains 11.79% C and 69.57% Cl. In another experiment you find that 0.107 g of the compound fills a 458-mL flask at 25 °C with a pressure of 21.3 mm Hg. What is the molecular formula of the compound?

75. Five compounds make up the family of sulfur–fluorine compounds with the general formula S_xF_y. One of these compounds is 25.23% S. If you place 0.0955 g of the compound in a 89-mL flask at 45 °C, the pressure of the gas is 83.8 mm Hg. What is the molecular formula of S_xF_y?

76. A miniature volcano can be made in the laboratory with ammonium dichromate. When ignited it decomposes in a fiery display.

$$(NH_4)_2Cr_2O_7(s) \longrightarrow N_2(g) + 4 H_2O(g) + Cr_2O_3(s)$$

If 0.95 g of ammonium dichromate is used, and if the gases from this reaction are trapped in a 15.0-L flask at 23 °C, what is the total pressure of the gas in the flask? What are the partial pressures of N_2 and H_2O?

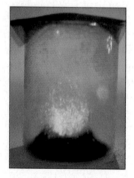

Ammonium dichromate, $(NH_4)_2Cr_2O_7$, decomposes on heating to give nitrogen gas, water vapor, and the green solid, chromium(III) oxide. *(Charles D. Winters)*

77. Which of the following gas samples contains the largest number of gas molecules? Which contains the smallest number? Which represents the largest mass of gas?

(a) 1.0 L of H_2 at STP

(b) 1.0 L of Ar at STP

(c) 1.0 L of H_2 at 27 °C and 760 mm Hg

(d) 1.0 L of He at 0 °C and 900 mm Hg

78. You are given a solid mixture of $NaNO_2$ and NaCl and are asked to analyze it for the amount of $NaNO_2$ present. To do so you allow it to react with sulfamic acid, HSO_3NH_2, in water according to the equation

$$NaNO_2(aq) + HSO_3NH_2(aq) \longrightarrow$$
$$NaHSO_4(aq) + H_2O(\ell) + N_2(g)$$

What is the weight percentage of $NaNO_2$ in 1.232 g of the solid mixture if reaction with sulfamic acid produces 295 mL of N_2 gas with a pressure of 713 mm Hg at 21.0 °C?

79. The density of air 20 km above the earth's surface is 92 g/m^3. The pressure of the atmosphere is 42 mm Hg and the temperature is −63 °C.

(a) What is the average molar mass of the atmosphere at this altitude?

(b) If the atmosphere at this altitude is only O_2 and N_2, what is the mole fraction of each gas?

80. A 3.0-L bulb containing He at 145 mm Hg is connected by a valve to a 2.0-L bulb containing Ar at 355 mm Hg. (See the figure.) Calculate the partial pressure of each gas and the total pressure after the valve between the flasks is opened.

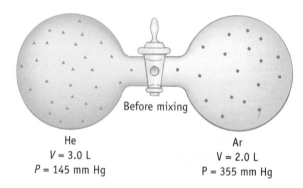

Before mixing

He
V = 3.0 L
P = 145 mm Hg

Ar
V = 2.0 L
P = 355 mm Hg

Valve open

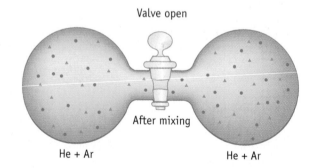

After mixing

He + Ar   He + Ar

81. Phosphine gas, PH_3, is toxic when it reaches a concentration of 7×10^{-5} mg/L. To what pressure does this correspond at 25 °C?

82. A xenon fluoride can be prepared by heating a mixture of Xe and F_2 gases to a high temperature in a pressure-proof container. Assume that xenon gas was added to a 0.25-L container until its pressure was 0.12 atm at 0.0 °C. Fluorine gas was then added until the total pressure was 0.72 atm at 0.0 °C. After the reaction was complete, the xenon was consumed completely and the pressure of the F_2 remaining in the container was 0.36 atm at 0.0 °C. What is the empirical formula of the xenon fluoride?

83. Chlorine dioxide, ClO_2, reacts with fluorine to give a new gas that contains Cl, O, and F. In an experiment you find that 0.150 g of this new gas has a pressure of 17.2 mm Hg in a 1850-mL flask at 21 °C. What is the empirical formula of the unknown gas?

84. A balloon is filled with helium gas to a gauge pressure of 22 mm Hg at 25 °C. The volume of the gas is 305 mL, and the barometric pressure is 755 mm Hg. What amount of helium is in the balloon? (Remember that gauge pressure = total pressure − barometric pressure. See page 474.)

85. Acetylene can be made by allowing calcium carbide to react with water:

$$CaC_2(s) + 2 H_2O(\ell) \longrightarrow C_2H_2(g) + Ca(OH)_2(s)$$

Suppose you react 2.65 g of CaC_2 with excess water. If you collect the acetylene and find that the gas has a volume of 795 mL at 25.2 °C with a pressure of 735.2 mm Hg, what is the percent yield of acetylene?

86. If you have a sample of water in a closed container some of the water will evaporate until the pressure of the water vapor, at 25 °C, is 23.8 mm Hg. How many molecules of water per cubic centimeter exist in the vapor phase?

87. You are given 1.56 g of a mixture of $KClO_3$ and KCl. When heated, the $KClO_3$ decomposes to KCl and O_2,

$$2 KClO_3(s) \longrightarrow 2 KCl(s) + 3 O_2(g)$$

and 327 mL of O_2 with a pressure of 735 mm Hg is collected at 19 °C. What is the weight percentage of $KClO_3$ in the sample?

88. A study of climbers who reached the summit of Mt. Everest without supplemental oxygen showed that the partial pressures of O_2 and CO_2 in their lungs were 35 mm Hg and 7.5 mm Hg, respectively. The barometric pressure at the summit was 253 mm Hg. Assume the lung gases are saturated with moisture at a body temperature of 37 °C (which means the partial pressure of water vapor in the lungs is $P(H_2O) = 47.1$ mm Hg). If you assume the lung gases are only O_2, N_2, CO_2, and H_2O, what is the partial pressure of N_2?

89. Nitrogen monoxide reacts with oxygen to give nitrogen dioxide.

$$2 NO(g) + O_2(g) \longrightarrow 2 NO_2(g)$$

(a) Place the three gases in order of increasing velocity at 298 K.

(b) If you mix NO and O_2 in the correct stoichiometric ratio, and NO has a partial pressure of 150 mm Hg, what is the partial pressure of O_2?

(c) After reaction between NO and O_2 is complete, what is the pressure of NO_2 if the NO originally had a pressure of 150 mm Hg and O_2 was added in the correct stoichiometric amount?

90. A 1.0-L flask contains 10.0 g each of O_2 and CO_2 at 25 °C.

(a) Which gas has the greater partial pressure, O_2 or CO_2, or are they the same?

(b) Which molecules have the greater average speed, or are they the same?

(c) Which molecules have the greater average kinetic energy, or are they the same?

91. If equal masses of O_2 and N_2 are placed in separate containers of equal volume at the same temperature, which of the following statements is true? If false, tell why it is false.

(a) The pressure in the N_2-containing flask is greater than that in the flask containing O_2.

(b) More molecules are in the flask containing O_2 than in the one containing N_2.

92. You have two pressure-proof steel cylinders of equal volume, one containing 1.0 kg of CO and the other 1.0 kg of acetylene, C_2H_2.

(a) In which cylinder is the pressure greater at 25 °C?

(b) Which cylinder contains the greater number of molecules?

93. Two flasks, each with a volume of 1.00 L, contain O_2 gas with a pressure of 380 mm Hg. Flask A is at 25 °C, and flask B is at 0 °C. Which flask contains the greater number of O_2 molecules?

94. State whether each of the following samples of matter is a gas. If there is not enough information for you to decide, write "insufficient information."

(a) A material is in a steel tank at 100 atm pressure. When the tank is opened to the atmosphere, the material suddenly expands, increasing its volume by 10%.

(b) A 1.0-mL sample of material weighs 8.2 g.

(c) A material is transparent and pale green in color.

(d) One cubic meter of material contains as many molecules as 1 m^3 of air at the same temperature and pressure.

95. Does the effect of intermolecular attraction on the properties of a gas become more significant or less significant if (a) the gas is compressed to a smaller volume at constant temperature; (b) more gas is forced into the same volume at constant temperature; (c) the temperature of the gas is raised at constant pressure?

96. Each of the four tires of a car is filled with a different gas. Each tire has the same volume and each is filled to the same pressure, 3.0 atm, at 25 °C. One tire contains 116 g of air, another tire has 80.7 g of neon, another tire has 16.0 g of helium, and the fourth tire has 160. g of an unknown gas.

(a) Do all four tires contain the same number of gas molecules? If not, which one has the greatest number of molecules?

(b) How many times heavier is a molecule of the unknown gas than an atom of helium?

(c) In which tire do the molecules have the largest kinetic energy? The greatest average speed?

97. You have a 550-mL tank of gas with a pressure of 1.56 atm at 24 °C. You thought the gas was pure carbon monoxide gas, CO, but you later found it was contaminated by small

quantities of gaseous CO_2 and O_2. Analysis shows that the tank pressure is 1.34 atm (at 24 °C) if the CO_2 is removed. Another experiment shows that 0.0870 g of O_2 can be removed chemically. What are the masses of CO and CO_2 in the tank, and what are the partial pressures of each of the three gases at 25 °C?

98. Methane is burned in a laboratory Bunsen burner to give CO_2 and water vapor. Methane gas is supplied to the burner at the rate of 5.0 L/min (at a temperature of 28 °C and a pressure of 773 mm Hg). At what rate must oxygen be supplied to the burner (at a pressure of 742 mm Hg and a temperature of 26 °C)?

99. Iron forms a series of compounds of the type $Fe_x(CO)_y$. In air they are oxidized to Fe_2O_3 and CO_2 gas. After heating a 0.142-g sample of $Fe_x(CO)_y$ in air, you isolate the CO_2 in a 1.50-L flask at 25 °C. The pressure of the gas is 44.9 mm Hg. What is the formula of $Fe_x(CO)_y$?

100. Group 2A metal carbonates are decomposed to the metal oxide and CO_2 on heating:

$$MCO_3(s) \longrightarrow MO(s) + CO_2(g)$$

You heat 0.158 g of a white, solid carbonate of a Group 2A metal (M) and find that the evolved CO_2 has a pressure of 69.8 mm Hg in a 285-mL flask at 25 °C. Identify M.

101. Silane, SiH_4, reacts with O_2 to give silicon dioxide and water vapor:

$$SiH_4(g) + 2\,O_2(g) \longrightarrow SiO_2(s) + 2\,H_2O(g)$$

If you mix SiH_4 with O_2 in the correct stoichiometric ratio, and if the total pressure of the mixture is 120 mm Hg, what are the partial pressures of SiH_4 and O_2? When the reactants have been completely consumed, what is the total pressure in the flask?

102. Chlorine trifluoride, ClF_3, is a valuable reagent because it can be used to convert metal oxides to metal fluorides:

$$6\,NiO(s) + 4\,ClF_3(g) \longrightarrow 6\,NiF_2(s) + 2\,Cl_2(g) + 3\,O_2(g)$$

(a) What mass of NiO will react with ClF_3 gas if the gas has a pressure of 250 mm Hg at 20 °C in a 2.5-L flask?

(b) If the ClF_3 described in part (a) is completely consumed, what are the partial pressures of Cl_2 and of O_2 in the 2.5-L flask at 20 °C (in millimeters of mercury)? What is the total pressure in the flask?

103. One way to synthesize diborane, B_2H_6, is the reaction

$$2\,NaBH_4(s) + 2\,H_3PO_4(aq) \longrightarrow$$
$$B_2H_6(g) + 2\,NaH_2PO_4(aq) + 2\,H_2(g)$$

(a) If you have 0.136 g of $NaBH_4$ and excess H_3PO_4, and collect the B_2H_6 in a 2.75 L flask at 25 °C, what is the pressure of the B_2H_6 in the flask?

(b) A byproduct of the reaction is H_2 gas. If both B_2H_6 and H_2 gas come from this reaction, what is the *total* pressure in the 2.75 L flask (after reaction of 0.136 g of $NaBH_4$ with excess H_3PO_4) at 25 °C?

104. The sodium azide required for automobile air bags is made by the reaction of sodium metal with dinitrogen oxide in liquid ammonia:

$$3\,N_2O(g) + 4\,Na(s) + NH_3(\ell) \longrightarrow$$
$$NaN_3(s) + 3\,NaOH(s) + 2\,N_2(g)$$

(a) You have 65.0 g of sodium and a 35.0-L flask containing N_2O gas with a pressure of 2.12 atm at 23 °C. What is the theoretical yield (in grams) of NaN_3?

(b) Draw a Lewis structure for the azide ion. Include all possible resonance structures. Which resonance structure is most likely?

(c) What is the shape of the azide ion?

105. Chlorine gas (Cl_2) is used as a disinfectant in municipal water supplies, although chlorine dioxide (ClO_2) and ozone are becoming more widely used. ClO_2 is better than Cl_2 because it leads to fewer chlorinated byproducts, which are themselves pollutants.

(a) How many valence electrons are in ClO_2?

(b) The chlorite ion, ClO_2^-, is obtained by reducing ClO_2. Draw a possible electron dot structure for the ion ClO_2^-. (Cl is the central atom.)

(c) What is the hybridization of the central Cl atom in ClO_2^-? What is the shape of the ion?

(d) Which species do you suppose has the larger bond angle, O_3 or ClO_2^-? Explain briefly.

(e) Chlorine dioxide, ClO_2, a yellow-green gas, can be made by the reaction of chlorine with sodium chlorite:

$$2\,NaClO_2(s) + Cl_2(g) \longrightarrow 2\,NaCl(s) + 2\,ClO_2(g)$$

Assume you react 15.6 g of $NaClO_2$ with chlorine gas, which has a pressure of 1050 mm Hg in a 1.45-L flask at 22 °C. What mass of ClO_2 can be produced?

Using Electronic Resources

These questions refer to the General Chemistry Interactive CD-ROM, *Version 3.0.*

106. See CD-ROM Screen 12.3: Gas Laws.

(a) In the animation of Charles's law, it is stated that if we extrapolate to a temperature of absolute zero a gas has no volume. Why can this experiment not be performed?

(b) Consider the animation of Avogadro's law. If we instead combined 12 H_2 molecules and 4 N_2 molecules to produce NH_3, what would be the final result? What would the relative volume be?

107. See CD-ROM Screen 12.4: The Ideal Gas Law. The "tool" on this screen allows you to explore the gas laws.

(a) If you did not already know the gas laws, describe how you could use this "tool" to discover them.

(b) Set the external pressure at 200 mm Hg and the amount of CO_2 gas at 180 mg. Change the temperature of the gas. What do you observe? What gas law does this illustrate?

(c) Verify the volume given on the screen for mass = 180 mg, $P = 200$ mm Hg, and temperature = 100°C.

108. See CD-ROM Screen 12.9: The Kinetic-Molecular Theory of Gases. Answer the questions after reading the screen and observing the animation.

   (a) What are all the pictures on this screen (of fish, coffee, garlic, blue cheese, and an orange) intended to make you think about? What does this have to do with the kinetic theory of gases?

   (b) If the absolute temperature of a gas doubles, by how much does the average speed of the gaseous molecules increase?

   (c) Do the principles of the kinetic-molecular theory include the shape, size, or chemical properties of gas molecules?

109. See CD-ROM Screen 12.10: Gas Laws and the Kinetic-Molecular Theory.

   (a) What is the nature of pressure on the molecular level assumed to be on this screen?

   (b) This screen shows animations describing the following relationships on the molecular scale: P versus n, P versus T, and P versus V. Sketch out a molecular scale animation for the relationship between n and V.

110. See CD-ROM Screen 12.11: Distribution of Molecular Speeds. Examine the Maxwell–Boltzmann distribution tool on this screen.

   (a) What trends do you observe about gas speeds and the molar mass of the gas?

   (b) What trends do you see for changes in temperature?

111. See CD-ROM Screen 12.12: Application of the Kinetic-Molecular Theory.

   (a) The presentation on this screen shows that NH_3 molecules diffuse through air more quickly than do HCl molecules. Why does this happen?

   (b) What compound is formed when the two gases, NH_3 and HCl, meet?

13 Intermolecular Forces, Liquids, and Solids

▲ Taking an official fingerprint at the local police station. *(Charles D. Winters)*

The Mystery of the Disappearing Fingerprints

The events of September 11, 2001, are etched in everyone's memory. The possibility of domestic terrorism, however, began almost two years before when a man was apprehended in late December 1999 at the Canadian border with bomb materials and a map of the Los Angeles International Airport. Although he claimed innocence, his fingerprints were on the bomb materials and he was convicted of attempting to bomb the airport.

Each of us has a unique fingerprint pattern, as first described by John Purkinji in 1823. Not long after, the English in India began using fingerprints on contracts because they believed it made the contract appear more binding. It was not until late in the 19th century, however, that fingerprinting was used as an identifier. Sir Francis Galton, a British anthropologist and cousin of Charles Darwin, established that a person's fingerprints do not change over the course of a lifetime and that no prints are exactly the same.

Fingerprinting has since become an accepted tool in forensic science. In 1993 in Knoxville, Tennessee, detective Art Bohanan thought he could use it to solve the case of the kidnapping of a young girl. The girl had been taken from her home and driven away in a green car. The girl soon managed to escape from her attacker and was able to describe the car to the police. After four days the police found the car and arrested its owner. But had the girl been in that car? Art Bohanan inspected the car for prints and even used the latest technique, fuming with superglue. No prints were found.

The abductor of the girl was eventually convicted on other evidence, but

▲ Dusting for fingerprints on a glass coffee mug. *(Charles D. Winters)*

"The finger-mark system of identification has been in use in India and China, to some extent, for twenty-five years."

Mark Twain, writing in 1897 in his book *Pudd'nhead Wilson* about the use of the newly discovered technique of fingerprinting.

▲ Close-up of a fingerprint. *(Charles D. Winters)*

Bohanan wondered why he had never found her prints in the car. He decided to test the permanence of children's fingerprints compared with those of adults. To his amazement, he found that children's prints disappeared in a few hours, whereas an adult's prints can last for days. Bohanan said, "It sounded like the compounds in children's fingerprints might simply be evaporating faster than adult's."

The residue deposited by fingerprints is 99% water. The other 1% contains oils, fatty acids, esters, salts, urea, and amino acids. Scientists at Oak Ridge National Laboratory in Tennessee studied the compounds in the prints of 50 child and adult volunteers, identifying the compounds by such techniques as mass spectrometry [← PAGE 107]. What they found clarified the mystery of the disappearing fingerprints.

Children's fingerprints contain more small fatty acids. Although one end of the molecule is polar (the carboxylic acid group, page 439), the long carbon chain is not. Due to the relatively low molecular weight and low polarity of these acids, their intermolecular forces are weak, and the molecules are volatile. Children's fingerprints simply evaporate.

In contrast, adult fingerprints contain esters of long-chain fatty acids with long-chain alcohols. These are waxes. The glossy surface of an apple results from a natural wax coating. Carnauba wax, from a Brazilian palm tree, is used in furniture polish, and lanolin is a component of wool wax.

$$CH_3(CH_2)_{30}-\overset{\displaystyle O}{\overset{\|}{C}}-O-(CH_2)_{33}CH_3$$
carnauba wax

The wax in an adult's fingerprints is deposited on the fingers when touching the face. Before puberty, children do not produce these waxy compounds. Adults, however, have such compounds on their skin and readily transfer them to the environment.

Before You Begin

- Review ion–ion attraction in ionic compounds (page 93).
- Know how to use electronegativity to determine the polarity of covalent bonds (Section 9.7).
- Be able to determine the polarity of molecules (Section 9.10).

Liquids and Solids This chapter is concerned with two of the phases or states of matter: liquids and solids. Here you will learn more about water and its truly unique properties. As solid ice, water is characterized by a highly regular structure. But another form of solid water is exemplified by a substance such as the methane-water clathrate pictured on the cover of this book.

Forms of water

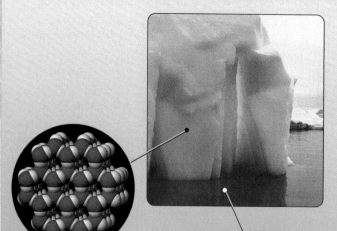

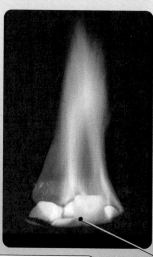

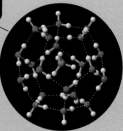

Solid ice (an iceberg in Antarctica) and liquid water

The methane-water clathrate. Methane (CH_4) is trapped in a cage of water molecules.

Common solids

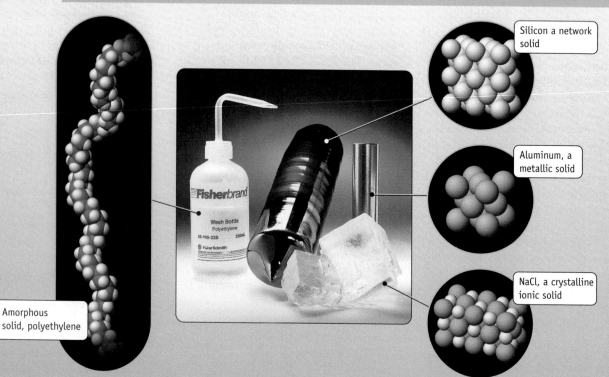

Silicon a network solid

Aluminum, a metallic solid

NaCl, a crystalline ionic solid

Amorphous solid, polyethylene

The vast majority of the known chemical elements are solids at 25 °C and 1 atm of pressure. Only 11 elements occur as gases under these conditions (H_2, N_2, O_2, F_2, Cl_2, and the six noble gases), and only two elements occur as liquids (Hg and Br_2), although two others (Cs and Ga) melt at only slightly higher temperatures. Many common compounds are gases (such as CO_2 and CH_4), and liquids (H_2O), but as is the case with the elements, the largest number of compounds are also solids.

The primary objective in this chapter is to provide an understanding of the liquid and solid states by looking at the particulate level: the level of atoms, molecules, and ions. You will find this chapter useful because it explains, among other things, why your body is cooled when you sweat, how bodies of water can influence local climate, why one form of pure carbon (diamond) is hard and another (graphite) is slippery, and why many solid compounds form beautiful crystals. ●

13.1   STATES OF MATTER AND THE KINETIC-MOLECULAR THEORY

The kinetic-molecular theory of gases [← SECTION 12.6] assumes that gas molecules or atoms are widely separated and that these particles can be considered independent of one another [⊕ CD-ROM, Screen 13.2]. Consequently, we can relate the properties of gases under most conditions by a simple mathematical equation, $PV = nRT$, the ideal gas law equation [← EQUATION 12.4]. Liquids and solids present a more complicated picture, however. In these states of matter, the particles are close together because forces of attraction occur between them, forces that we can mostly ignore when considering gases. When these features are introduced it is not possible to create a simple "ideal liquid equation" or "ideal solid equation" to describe these states of matter.

How different are the states of matter at the particulate level? We can get a sense of this by comparing volumes occupied by equal numbers of molecules of a material in different states. Figure 13.1 shows a flask containing about 300 mL of liquid nitrogen. If all this liquid were allowed to evaporate, the gaseous nitrogen would fill a large balloon (>200 L) to a pressure of 1 atm at room temperature. A large amount of space exists between molecules in a gas, whereas in liquids the molecules are close together, in fact touching one another.

The increase in volume when converting liquids to gases is strikingly large. In contrast, no dramatic change in volume occurs when a solid is converted to a liquid. Figure 13.2 shows the same amount of liquid and solid benzene side by side, and you see they are not appreciably different in volume. This means that the atoms in the liquid are packed together about as tightly as the atoms in the solid phase.

We know that gases are readily compressed, a process that forces the gas molecules closer together. The air–fuel mixture in your car's engine, for example, is routinely compressed by a factor of about 10 before it is ignited. In contrast, the molecules, ions, or atoms in liquid or solid phases strongly resist forces that would push them closer together. This is reflected in the lack of compressibility of liquids and solids. For example, the volume of liquid water changes only by 0.005% per atmosphere of pressure applied to it.

In the gaseous state, atoms or molecules are relatively far apart because forces between particles are not strong enough to pull them together and overcome their kinetic energy. In liquids and solids, much stronger forces pull the particles together and limit their motion. In solid ionic compounds, the positively and negatively charged ions are held together by electrostatic attraction [← PAGE 93]. In molecular

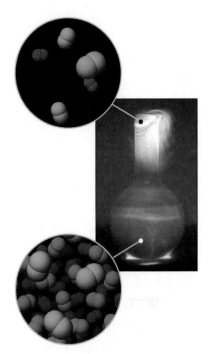

Figure 13.1   Liquid nitrogen. When this 300-mL sample of liquid nitrogen evaporates, it will produce over 200 L of gas at 25 °C and 1.0 atm. As a liquid, the molecules of N_2 are close together, but in the gas phase, they are far apart. *(Charles D. Winters)*

Liquid benzene   Solid benzene

Figure 13.2   Liquid and solid benzene. The same volume of liquid benzene, C_6H_6, is placed in two test tubes, and one tube (*right*) is cooled, freezing the liquid. The solid and liquid states have almost the same volume, showing that the molecules are packed together almost as tightly in the liquid state as they are in the solid. *(Charles D. Winters)*

solids and liquids, the forces *between* molecules, called **intermolecular forces,** are based on various electrostatic attractions that are weaker than the forces between oppositely charged ions. For comparison, the attractive forces between ions in ionic compounds are usually in the range of 700 to 1100 kJ/mol, and most covalent bond energies are in the range of 100 to 400 kJ/mol [← TABLE 9.9]. As a rough guideline, intermolecular forces are generally 15% (or less) of the values of covalent bond energies.

13.2   INTERMOLECULAR FORCES

Intermolecular forces influence chemistry in many ways [CD-ROM, Screen 13.3]. They are

- Directly related to properties such as melting point, boiling point, and the energy needed to overcome forces of attraction between particles in changes of state.
- Important in determining the solubility of gases, liquids, and solids in various solvents.
- Crucial in determining the structures of biologically important molecules such as DNA and proteins.

You will encounter examples of these features in this and subsequent chapters.

Bonding in ionic compounds depends on the electrostatic forces of attraction between oppositely charged ions. Similarly, the intermolecular forces attracting one molecule to another are electrostatic. Recall that molecules can have polar bonds owing to the differences in electronegativity of the bonded atoms [← SECTION 9.7]. Depending on the orientation of these polar bonds, an entire molecule can be polar, with the molecule having a negatively charged region and a positively charged region. Interactions between the polar molecules can have a profound effect on molecular properties and are the subject of this section.

Interactions Between Molecules with Permanent Dipoles

When a polar molecule encounters another polar molecule, of the same or different kind, the two molecules interact. The positive end of one molecule is attracted to the negative end of the other (Figure 13.3). Many molecules have dipoles and their interactions occur by **dipole–dipole attraction** [CD-ROM, Screen 13.4].

Dipole–dipole attractions influence the evaporation of a liquid or the condensation of a gas (Figure 13.4). In both processes an energy change occurs. Evaporation requires the addition of heat, specifically the enthalpy of vaporization (ΔH_{vap}) [← SECTION 6.3] (see also Section 13.5). The value for the enthalpy of vaporization has a positive sign, that is, evaporation is an endothermic process. The enthalpy change for the condensation process — the reverse of evaporation — has a negative value because heat is transferred out of the system on condensation.

The greater the forces of attraction between molecules in a liquid, the greater the energy that must be supplied to separate them. Thus, we expect polar compounds to have higher values of the enthalpy of vaporization than nonpolar compounds with similar molar masses. Comparisons between a few polar and nonpolar molecules that illustrate this are given in Table 13.1. For example, notice that $\Delta H^{\circ}_{\text{vap}}$ for polar CO is greater than for nonpolar N_2.

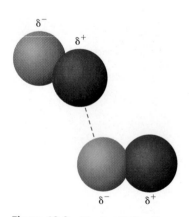

Figure 13.3   Dipole–dipole attractions. Two BrCl molecules are attracted to each other by dipole–dipole forces. Chlorine is more electronegative than bromine, so the covalent bond between these atoms is polar. The negative end of one BrCl molecule is attracted to the positive end of a second molecule.

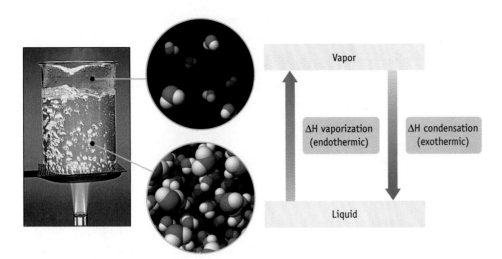

Figure 13.4 **Evaporation at the molecular level.** Energy must be supplied to separate molecules in the liquid state against intermolecular forces of attraction. *(Photo, Charles D. Winters)*

The boiling point of a liquid is also related to intermolecular forces of attraction. As the temperature of a substance is raised, its molecules gain kinetic energy. Eventually, when the boiling point is reached, the molecules have sufficient kinetic energy to escape the forces of attraction of their neighbors. The higher the forces of attraction, the higher the boiling point. In Table 13.1 you see that the boiling point for polar ICl is greater than that for nonpolar Br_2, for example.

Intermolecular forces also influence solubility. A qualitative observation on solubility is that "like dissolves like." This means *polar molecules are likely to dissolve in a polar solvent, and nonpolar molecules are likely to dissolve in a nonpolar solvent* (Figure 13.5) [➡ CHAPTER 14]. The converse is also true; that is, it is unlikely that polar molecules will dissolve in nonpolar solvents or that nonpolar molecules will dissolve in polar solvents.

For example, water and ethanol (C_2H_5OH) can be mixed in any ratio to give a homogeneous mixture. In contrast, water does not dissolve in gasoline to an appreciable extent. The difference in these two situations is that ethanol and water are polar molecules, whereas the hydrocarbon molecules in gasoline (for example, octane, C_8H_{18}) are nonpolar. Water–ethanol interactions are strong enough that the energy expended in pushing water molecules apart to make room for ethanol molecules is made up for by the energy of attraction between the two kinds of polar molecules. In contrast, water–hydrocarbon attractions are weak. The hydrocarbon molecules cannot disrupt the stronger water–water attractions.

Chapter Goals • Revisited

- **Describe intermolecular forces and their effects and the importance of hydrogen bonding.**
- Understand the properties of liquids.
- Understand cubic unit cells.
- Relate unit cells for ionic compounds to compound formulas.
- Describe the properties of solids.
- Understand the nature of phase diagrams.

● **Dissolving Substances**
Other factors besides energy are also important in the solution process. See Section 14.2.

Table 13.1 ● **Molar Masses and Boiling Points of Nonpolar and Polar Substances**

	Nonpolar				Polar		
	M (g/mol)	**BP** (°C)	**ΔH_{vap}** (kJ/mol)		**M** (g/mol)	**BP** (°C)	**ΔH_{vap}** (kJ/mol)
N_2	28	−196	5.57	CO	28	−192	6.04
SiH_4	32	−112	12.10	PH_3	34	−88	14.06
GeH_4	77	−90	14.06	AsH_3	78	−62	16.69
Br_2	160	59	29.96	ICl	162	97	—

(a) Ethylene glycol (HOCH$_2$CH$_2$OH), a polar compound used as antifreeze in automobiles, dissolves in water.

(b) Nonpolar motor oil (a hydrocarbon) dissolves in nonpolar solvents such as gasoline or CCl$_4$. It will not dissolve in a polar solvent such as water, however. Commercial spot removers use nonpolar solvents to dissolve oil and grease from fabrics.

Figure 13.5 "Like dissolves like." *(Photos, Charles D. Winters)*

Interactions Between Nonpolar Molecules

• **Dissolving O$_2$ in Water**
Oxygen dissolves in water to the extent of about 10 ppm (or about 0.001% by weight). This is important because microorganisms use oxygen dissolved in water to convert organic substances to simpler compounds. The quantity of oxygen required to oxidize a given quantity of organic material is called the biological oxygen demand (BOD). Highly polluted water often has a high concentration of organic matter and so has a high BOD.

Many important molecules such as O$_2$, N$_2$, and the halogens are not polar. Why, then, does O$_2$ dissolve in polar water? Perhaps even more difficult to imagine is how the N$_2$ of the atmosphere can be liquefied. There must be some intermolecular forces between O$_2$ and water and between N$_2$ molecules, but what is their nature [⚙ CD-ROM, Screen 13.7]?

Dipole/Induced Dipole Forces

Polar molecules such as water can *induce*, or create, a dipole in molecules that do not have a permanent dipole. To see how this can occur, picture a polar water molecule approaching a nonpolar molecule such as O$_2$ (Figure 13.6). The electron cloud of an isolated (gaseous) O$_2$ molecule is symmetrically distributed around the two oxygen atoms. As the negative end of the polar H$_2$O molecule approaches, however, the O$_2$ electron cloud distorts. In this process, the O$_2$ molecule itself becomes polar, that is, a dipole is induced, or created, in the otherwise nonpolar O$_2$ molecule. The result is that H$_2$O and O$_2$ molecules are now attracted to one another, although only weakly. Oxygen can dissolve in water because a force of attraction exists between a permanent dipole and the induced dipole. Chemists refer to such interactions as **dipole/induced dipole interactions.**

Figure 13.6 Dipole/induced dipole interaction. A polar molecule such as water can induce a dipole in nonpolar O$_2$.

The dipole of water induces a dipole in O$_2$ by distorting the O$_2$ electron cloud.

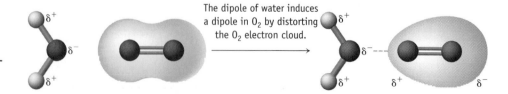

Table 13.2 • The Solubility of Some Gases in Water*		
Gas	Molar Mass (g/mol)	Solubility at 20 °C (g gas/100 g water)†
H_2	2.01	0.000160
N_2	28.0	0.000190
O_2	32.0	0.000434

*Data taken from J. A. Dean: *Lange's Handbook of Chemistry,* 15th ed., pp. 5.3–5.8. New York, McGraw-Hill, 1999.

†Measured under the following conditions: Pressure of gas + pressure of water vapor = 760 mm Hg.

The process of inducing a dipole is called **polarization,** and the degree to which the electron cloud of an atom or a molecule can be distorted to induce a dipole depends on the **polarizability** of the atom or molecule. This property is difficult to measure experimentally. It makes sense, however, that the electron cloud of an atom or molecule with a large, electron cloud, such as I_2, can be polarized, or distorted, more readily than the electron cloud in a much smaller atom or molecule, such as He or H_2, in which the valence electrons are close to the nucleus and more tightly held. In general, for an analogous series of compounds, say the halogens or alkanes (such as CH_4, C_2H_6, C_3H_8, and so on), *the higher the molar mass, the larger the electron cloud and the greater the polarizability of the molecule.*

The table of the solubilities of common gases in water (Table 13.2) illustrates the effect of interactions between a dipole and an induced dipole. Here you see a trend to higher solubility with increasing mass. As the molar mass of the gas increases, the polarizability of the electron cloud increases. Because a dipole is more readily induced as the polarizability increases, the strength of dipole/induced dipole interactions generally increases with mass. And finally, because the solubility of substances such as O_2 depends on the strength of the dipole/induced dipole interaction, the solubility of nonpolar substances in polar solvents generally increases with their mass.

London Dispersion Forces: Induced Dipole/Induced Dipole Forces

Iodine, I_2, is a solid and not a gas at STP, proving that nonpolar molecules must also experience intermolecular forces. An estimate of these forces is provided by the enthalpy of vaporization of the substance at its boiling point. The data in the following table suggest that the forces in this case can range from very weak forces (N_2, O_2, and CH_4 have low enthalpies of vaporization and very low boiling points) to more substantial forces (I_2 and benzene).

	ΔH_{vap} (kJ/mol)	Compound Boiling Point (°C)
N_2	5.57	−196
O_2	6.82	−183
I_2	41.95	185
CH_4 (methane)	8.2	−161.5
C_6H_6 (benzene)	30.7	80.1

Nonpolar iodine dissolves in the polar solvent ethanol, CH_3CH_2OH. The force of attraction between the two molecules is a dipole/induced dipole force. *(Charles D. Winters)*

The molecules of Br_2 (*left*) and I_2 (*right*) are both nonpolar. However, they are a liquid or solid, respectively, implying that the forces between the molecules are sufficient to cause them to be in a condensed phase. Such forces are known as London dispersion forces or induced dipole/induced dipole forces. *(Charles D. Winters)*

Figure 13.7   Induced dipole interactions. Momentary attraction and repulsion between nuclei and electrons creates induced dipoles and leads to a net stabilization due to attractive forces.

Two nonpolar atoms or molecules (Time-averaged shape is spherical)

Momentary attractions and repulsions between nuclei and electrons in neighboring molecules lead to induced dipoles.

Correlation of the electron motions between the two atoms or molecules (which are now dipolar) leads to a lower energy and stabilizes the system.

To understand how two nonpolar molecules can attract each other, remember that the electrons in atoms or molecules are in a state of constant motion. On average, their electron clouds are spherical (Figure 13.7). When two nonpolar atoms or molecules approach each other, however, attractions or repulsions between their electrons and nuclei can lead to distortions in their electron clouds. That is, dipoles can be induced momentarily in neighboring atoms or molecules, and these induced dipoles lead to intermolecular attraction. Thus, the intermolecular force of attraction in liquids and solids composed of nonpolar molecules is an **induced dipole/induced dipole force.** Chemists also call such forces **London dispersion forces.**

Example 13.1   Intermolecular Forces

Problem • Suppose you have solid iodine, I_2, and the liquids water and carbon tetrachloride (CCl_4). What intermolecular forces exist between each possible pair of compounds? Describe what you might see when these compounds are mixed.

Strategy • First decide if each substance is polar or nonpolar. Next, use the "like dissolves like" guideline to decide if iodine will dissolve in water or CCl_4 and if CCl_4 will dissolve in water.

Solution • Iodine, I_2, is nonpolar, and, being a molecule based on an element with large atoms, it has an extensive electron cloud. Thus, the molecule is easily polarized. Iodine could interact with water, a polar molecule, by induced dipole/dipole forces (Table 13.3).

Carbon tetrachloride, a tetrahedral molecule, is not polar (◀ SECTION 9.10). It can interact with iodine only by dispersion forces. Water and CCl_4 could interact by dipole/induced dipole forces, but the interaction is expected to be weak.

The photos here show the result of mixing these three compounds. Iodine does dissolve to a small extent in water to give a brown solution. When this brown solution is added to a test tube

containing CCl_4, the layers do not mix. (Polar water does not dissolve in nonpolar CCl_4.) When the test tube is shaken, however, nonpolar I_2 dissolves preferentially in nonpolar CCl_4, as evidenced by the disappearance of the color of I_2 in the water layer (*top right*) and the appearance of the purple I_2 color in the CCl_4 layer (*bottom right*).

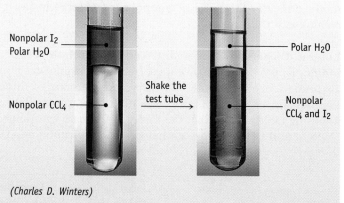

Nonpolar I_2
Polar H_2O

Nonpolar CCl_4

Shake the test tube

Polar H_2O

Nonpolar CCl_4 and I_2

(Charles D. Winters)

Exercise 13.1   Intermolecular Forces

You mix the liquids water, CCl_4, and hexane ($CH_3CH_2CH_2CH_2CH_2CH_3$). For each pair of compounds, what type of intermolecular forces can exist between the compounds? If you mix these three liquids, describe what observations you might make.

13.3   HYDROGEN BONDING

Hydrogen fluoride and many compounds with O—H and N—H bonds have exceptional properties. We can see this by examining the boiling points for hydrogen compounds of elements in Groups 4A through 7A (Figure 13.8). Generally, the boiling points of related compounds increase with molar mass because of increasing dispersion forces. This trend is seen in the boiling points of the hydrogen compounds of Group 4A elements ($CH_4 < SiH_4 < GeH_4 < SnH_4$). The same effect is also operating for the heavier molecules of the hydrogen compounds of elements of Groups 5A, 6A, and 7A. The boiling points of NH_3, H_2O, and HF, however, are greatly out of line with what might be expected based on molar mass alone. If we were to use the line from H_2Te to H_2Se to H_2S and extrapolate to the expected boiling point of water, we would predict that water would boil around $-90\ °C$. The boiling point of water is almost 200 °C higher than this value! Similarly, the boiling points of NH_3 and HF are much higher than would be expected based on molar mass. Why do the properties of water, ammonia, and hydrogen fluoride differ so from their expected values?

Because the temperature at which a substance boils depends on the attractive forces between molecules, the boiling points of H_2O, HF, and NH_3 clearly indicate strong intermolecular attractions. The unusually high boiling points in these compounds are due to hydrogen bonding. A **hydrogen bond** is an attraction between the hydrogen atom of an X—H bond and Y, where X and Y are atoms of highly electronegative elements and Y has a lone pair of electrons [CD-ROM, Screen 13.5]. Hydrogen bonds are an extreme form of dipole–dipole interaction in which one atom involved is always H and the other atom is most often O, N, or F. Such bonds can have profound effects on the properties of compounds.

A bond dipole arises as a result of a difference in electronegativity between bonded atoms [◀ SECTION 9.7]. The electronegativities of N (3.0), O (3.5), and F (4.0) are among the highest of all the elements, whereas the electronegativity of hydrogen is much lower (2.1). The large difference in electronegativity means that

Chapter Goals • Revisited

- **Describe intermolecular forces and their effects and the importance of hydrogen bonding.**
- Understand the properties of liquids.
- Understand cubic unit cells.
- Relate unit cells for ionic compounds to compound formulas.
- Describe the properties of solids.
- Understand the nature of phase diagrams.

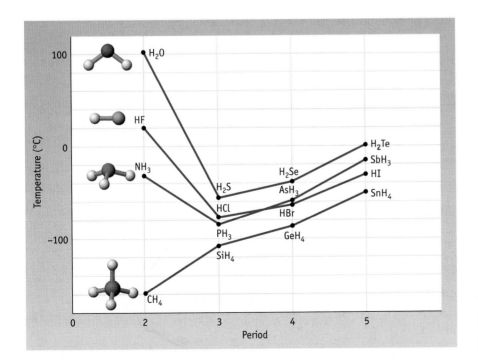

Figure 13.8   The boiling points of some simple hydrogen compounds. The effect of hydrogen bonding is apparent in the unusually high boiling points of H_2O, HF, and NH_3.

N—H, O—H, and F—H bonds are very polar. In bonds between H and N, O, or F, the more electronegative element takes on a partial negative charge and the hydrogen atom acquires a partial positive charge.

In hydrogen bonding, there is an unusually strong attraction between an electronegative atom with a lone pair of electrons (an N, O, or F atom in another molecule or even in the same molecule) and the hydrogen atom of the N—H, O—H, or F—H bond. A hydrogen bond can be represented as

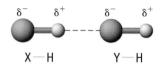

The hydrogen atom becomes a bridge between the two electronegative atoms X and Y, and the dotted line represents the hydrogen bond. The most pronounced effects of hydrogen bonding occur where X and Y are N, O, or F. Energies associated with most hydrogen bonds involving these elements are in the range of 20 to 30 kJ/mol.

Types of Hydrogen Bonds [X—H ⋯ :Y]		
N—H ⋯ :N—	O—H ⋯ :N—	F—H ⋯ :N—
N—H ⋯ :O—	O—H ⋯ :O—	F—H ⋯ :O—
N—H ⋯ :F—	O—H ⋯ :F—	F—H ⋯ :F—

Hydrogen bonding has important implications for any property of a compound that is influenced by intermolecular forces of attraction. For example, it is important in determining structures of molecular solids, one example of which is acetic acid. In the solid state, two molecules of CH_3CO_2H are joined to one another by hydrogen bonding (Figure 13.9).

Hydrogen bonding is also an important factor in the structure of some synthetic polymers including nylon, in which the N—H unit of the amide interacts with a carbonyl oxygen on an adjacent polymer chain [← FIGURE 11.19]. Hydrogen bonding in Kevlar, also a polyamide, gives this material the exceptional strength-to-weight ratio needed for its use in making canoes, ski equipment, and bullet-proof vests.

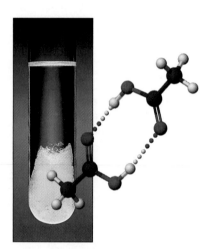

Figure 13.9 **Hydrogen bonding.** Acetic acid molecules can interact through hydrogen bonds. This photo shows partly solid glacial acetic acid. The solid is denser than the liquid, a property shared by virtually all substances, the notable exception being water. *(Charles D. Winters)*

Example 13.2   The Effect of Hydrogen Bonding

Problem • Ethanol, CH_3CH_2OH, and dimethyl ether, CH_3OCH_3, have the same formula but a different arrangement of atoms (they are *isomers*). Predict which of these compounds has the higher boiling point.

Ethanol, CH_3CH_2OH Dimethyl ether, CH_3OCH_3

Strategy • Inspect the structure of each molecule to decide if each is polar and, if polar, if hydrogen bonding is possible.

Solution • Although these two compounds have identical masses they have different structures. Ethanol possesses an O—H

group, and hydrogen bonding between ethanol molecules makes an important contribution to its intermolecular forces.

$$CH_3CH_2—\overset{..}{\underset{\underset{H}{|}}{O}}:\cdots H—\overset{..}{\underset{\underset{CH_2CH_3}{|}}{O}}:$$

hydrogen bonding in ethanol, CH_3CH_2OH

In contrast, dimethyl ether, although a polar molecule, presents no opportunity for hydrogen bonding because there is no O—H bond. We can predict, therefore, that intermolecular forces will be larger in ethanol than in dimethyl ether and that ethanol will have the higher boiling point. Indeed, ethanol boils at 78.3 °C, whereas dimethyl ether has a boiling point of −24.8 °C, more than 100 °C lower. Dimethyl ether is a gas, whereas ethanol is a liquid under standard conditions.

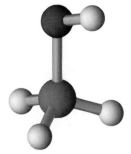

Exercise 13.2   Hydrogen Bonding

Using structural formulas, describe the hydrogen bonding between molecules of methanol, CH_3OH. What physical property or properties of methanol are likely to be affected by hydrogen bonding?

Methanol, CH_3OH

Hydrogen Bonding and the Unusual Properties of Water

One of the most striking differences between our planet and others in the solar system is the existence of liquid water on earth. Three fourths of the globe is covered by oceans, the polar regions are vast ice fields, and even soil and rocks hold large amounts of water. Although we tend to take water for granted, almost no other substance behaves in a similar manner. Water's unique features reflect the ability of H_2O molecules to cling tenaciously to one another by hydrogen bonding [CD-ROM, Screen 13.6].

One reason for ice's unusual structure, and water's unusual properties, is that each hydrogen atom of a water molecule can form a hydrogen bond to a lone pair of electrons on the oxygen atom of an adjacent water molecule. In addition, because the oxygen atom in water has two lone pairs of electrons, it can form two more hydrogen bonds with hydrogen atoms from adjacent molecules (Figure 13.10a). The result is a tetrahedral arrangement for the hydrogen atoms around the oxygen, involving two covalently bonded hydrogen atoms and two hydrogen-bonded hydrogen atoms.

To achieve the regular arrangement of water molecules linked by hydrogen bonding, ice has an open-cage structure with lots of empty space (see Figure 13.10). The result is that ice is about 10% less dense than liquid water, explaining why ice floats on liquid water. We can also see in this structure that the oxygen atoms are

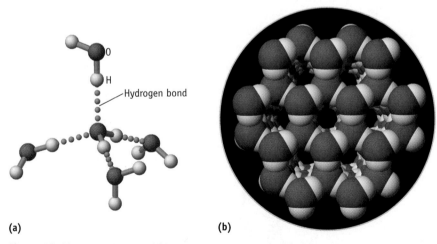

(a)   (b)

Figure 13.10   The structure of ice.
(a) The oxygen atom of a water molecule attaches to four other water molecules. Notice that the four groups that surround an oxygen atom are arranged tetrahedrally. Each oxygen atom is covalently bound to two hydrogen atoms and hydrogen-bonded to hydrogen atoms from two other molecules. The hydrogen bonds are longer than the covalent bonds.
(b) In ice, the structural unit shown in part (a) is repeated in the crystalline lattice. This computer-generated structure shows a small portion of the extensive lattice. Notice that six-member, hexagonal rings are formed. The corners of each hexagon are O atoms, and each side is composed of a normal O — H bond and a longer hydrogen bond. (The structure is on the *General Chemistry Interactive CD-ROM*, Version 3.0 in the Models/Solids folder; *b, S. M. Young*)

A Closer Look

Hydrogen Bonding in DNA

In 1953, when James Watson and Francis Crick finally felt confident they understood the structure of DNA, Crick proclaimed to the patrons of a pub in Cambridge, England, "We have discovered the secret of life." And indeed understanding DNA has turned out to be the key that has unlocked much of molecular biochemistry.

As we know now, DNA is a double helix of chains of phosphates linked to sugar molecules (pages 80 and 368). Bound to each of the sugar molecules is one of four possible base molecules: thymine, guanine, cytosine, and adenine.

The feature of DNA that Watson and Crick unraveled was that the base molecules on one chain interact through hydrogen bonding with the base molecules on the other chain and that only certain pairs of bases can interact in this manner. Owing to the demands of hydrogen bonding, adenine (A) can *only* interact with thymine (T), and cytosine (C) can *only* interact with guanine (G).

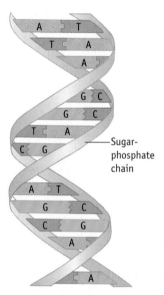

Pairing between bases on two chains leads to the *double* helix of DNA.

In humans, there are about 3 billion base pairs in all of our DNA.

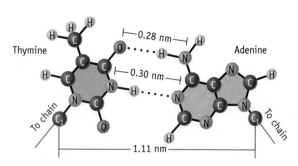

Base-pairing through hydrogen bonds.

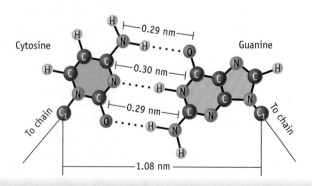

arranged at the corners of puckered, six-sided rings, or hexagons. Snowflakes are always based on six-sided figures [← SEE PAGE 83], a reflection of the internal molecular structure of ice.

When ice melts at 0 °C, a relatively large increase in density occurs, a result of the breakdown of the regular structure imposed on the solid state by hydrogen bonding (Figure 13.11). As the temperature of liquid water is raised from 0 to 4 °C, another surprising thing occurs: the density of water *increases*. The general rule followed by almost every other substance we know of is that volume increases and the density decreases as the temperature is raised. Once again, hydrogen bonding is the reason for water's seemingly odd behavior. At a temperature just above the melting point, some of the water molecules continue to cluster in ice-like arrangements, which require extra empty space. As the temperature is raised from 0 to 4 °C, the final vestiges of the ice structure disappear and the volume contracts further, giving rise to the increase in density. Water's density reaches a maximum at about 4 °C, and from this point the density declines with increasing temperature in the normal fashion.

Because of the way that water's density changes as the temperature approaches the freezing point, lakes do not freeze solidly from the bottom up in the winter. When lake water cools with the approach of winter, its density increases, the cooler water sinks, and the warmer water rises. This means the water and dissolved material turn over until all the water reaches 4 °C, the maximum density. (This process carries oxygen-rich water to the lake bottom to restore the oxygen used during the summer, and it brings nutrients to the top layers of the lake.) As the water cools further, it stays on the top of the lake, because water cooler than 4 °C is less dense than water at 4 °C. With further heat loss, ice can then begin to form on the surface, floating there and protecting the underlying water and aquatic life from further heat loss.

Extensive hydrogen bonding is also the origin of the extraordinarily high heat capacity of water. Although liquid water does not have the regular structure of ice, hydrogen bonding still occurs. With a rise in temperature, the extent of hydrogen bonding diminishes. Disrupting hydrogen bonds requires heat. The high heat capacity of water is, in large part, why oceans and lakes have such an enormous effect on weather. In autumn, when the temperature of the air is lower than the temperature of the ocean or lake, the ocean or lake gives up heat to the atmosphere, moderating the drop in air temperature. Furthermore, so much heat is given off for each degree drop in temperature (because so many hydrogen bonds form) that the decline in water temperature is gradual. For this reason the temperature of the ocean or a large lake is generally higher than the average air temperature until late in the autumn.

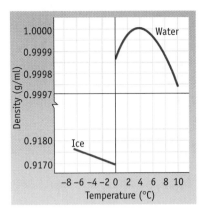

Figure 13.11 The temperature dependence of the densities of ice and pure water. In contrast with pure water, the density of sea water increases with falling temperature right down to the freezing point. This is a crucial distinction between fresh water and sea water, and it has an effect on oceanic circulation and on the formation of sea ice.

● **Methane hydrate**
Hydrogen bonding is responsible for the cage of water molecules enclosing methane in methane hydrate. See the model on the back cover of the book and on page 510.

13.4 SUMMARY OF INTERMOLECULAR FORCES

Intermolecular forces act among molecules that are polar or those in which polarity can be induced (Table 13.3). London dispersion forces are the only forces between nonpolar molecules. However, it is important to emphasize that dispersion forces are found in all molecules, both nonpolar and polar. In general, the strength of intermolecular forces is in the order

dipole–dipole (including H-bonding) > dipole/induced dipole >
induced dipole/induced dipole

Table 13.3 • Summary of Intermolecular Forces

Type of Interaction	Factors Responsible for Interaction	Example
Dipole–dipole	Dipole moment (depends on atom electronegativities and molecular structure)	H_2O, HCl
Hydrogen bonding, $X-H\cdots:Y$	Very polar $X-H$ bond (where X = F, N, O) and atom Y with lone pair of electrons An extreme form of dipole–dipole interaction	$H_2O\cdots H_2O$
Dipole/induced dipole	Dipole moment of polar molecule and polarizability of nonpolar molecule	$H_2O\cdots I_2$
Induced dipole/induced dipole (London dispersion forces)	Polarizability	$I_2\cdots I_2$

Example 13.3 Intermolecular Forces

Problem • Decide what type of intermolecular force is involved in each of these examples and place them in order of increasing strength of interaction: **(a)** liquid methane, CH_4; **(b)** a mixture of water and methanol, H_2O and CH_3OH; **(c)** a solution of bromine in water.

Strategy • For each molecule we consider its structure and then decide if it is polar. If polar, consider the possibility of hydrogen bonding.

Solution •

(a) Methane, CH_4, is a simple covalent molecule. After drawing the Lewis structure we can conclude that it must be a tetrahedral molecule and that it cannot be polar. The only way methane molecules can interact with one another is through induced dipole forces.

(b) Both water and methanol are covalently bonded molecules, both are polar, and both have an $O-H$ bond. They therefore interact through the special dipole–dipole force called hydrogen bonding.

(c) Bromine, Br_2, is a nonpolar molecule so it can only interact with another molecule by induced dipole forces. Water is a polar molecule; therefore, induced dipole/dipole forces are involved.

In order of increasing strength, the likely order of interactions is

liquid CH_4 < H_2O and Br_2 < H_2O and CH_3OH

Exercise 13.3 Intermolecular Forces

Decide what type of intermolecular force is involved in (a) liquid O_2; (b) liquid CH_3OH; (c) O_2 dissolved in H_2O. Place the interactions in order of increasing strength.

13.5 PROPERTIES OF LIQUIDS

Of the three states of matter, liquids are the most difficult to describe precisely. The molecules in a gas under normal conditions are far apart and may be considered more or less independent of one another. The particles that make up solids — atoms, molecules, or ions — are close together and are in an orderly arrangement. Thus, the structures of solids can be described readily. The particles of a liquid interact with their neighbors, like the particles in a solid, but unlike solids there is little order in their arrangement.

In spite of a lack of precision in describing liquids, we can still consider the behavior of liquids at the molecular level. In the following sections we will look further at the process of vaporization, at the vapor pressure of liquids, at their boiling points and critical properties, and at the behavior that results in their surface tension, capillary action, and viscosity.

Vaporization

Vaporization or **evaporation** is the process in which a substance in the liquid state becomes a gas [⊙ CD-ROM, Screen 13.8]. In this process, molecules escape from the liquid surface and enter the gaseous state.

To understand evaporation, we have to look at molecular energies. Molecules in a liquid have a range of energies (Figure 13.12) that closely resembles the distribution of energies for molecules of a gas [← FIGURE 12.13]. As with gases, the average energy for molecules in a liquid depends only on temperature: the higher the temperature, the higher the average energy and the greater the relative number of molecules with high kinetic energy. In a sample of a liquid, at least a few molecules have very high energy, and some of the molecules in the liquid state are going to have more kinetic energy than the potential energy of the intermolecular attractive forces holding liquid molecules to one another. If high-energy molecules find themselves at the surface of the liquid, and if they are moving in the right direction, they can break free of their neighbors and enter the gas phase (Figure 13.13).

Vaporization is an endothermic process because energy must be expended to separate the molecules from the intermolecular forces of attraction holding them together. The heat energy required to vaporize a sample is often given as the **standard molar enthalpy of vaporization, ΔH°_{vap}** (in units of kilojoules per mole; see Tables 13.1 and 13.4) (Figure 13.4).

$$\text{Liquid} \xrightarrow[\substack{\text{heat energy absorbed} \\ \text{by liquid}}]{\text{vaporization}} \text{Vapor} \qquad \Delta H^{\circ}_{vap} = \text{molar heat of vaporization}$$

A molecule in the gas phase will eventually transfer some of its kinetic energy by colliding with slower gaseous molecules and solid objects. If this molecule comes in contact with the surface of the liquid again, it can reenter the liquid phase in the process called **condensation.**

$$\text{Vapor} \xrightarrow[\substack{\text{heat energy released} \\ \text{by vapor}}]{\text{condensation}} \text{Liquid}$$

Chapter Goals • Revisited

- Describe intermolecular forces and their effects and the importance of hydrogen bonding.
- **Understand the properties of liquids.**
- Understand cubic unit cells.
- Relate unit cells for ionic compounds to compound formulas.
- Describe the properties of solids.
- Understand the nature of phase diagrams.

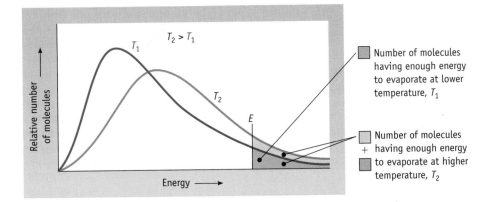

Figure 13.12 The distribution of energy among molecules in a liquid sample. T_2 is a higher temperature than T_1, and more molecules have an energy greater than the value marked E in the diagram at this temperature.

Figure 13.13   Evaporation. Some molecules at the surface of a liquid have enough energy to escape the attractions of their neighbors and enter the gaseous state. At the same time, some molecules in the gaseous state can reenter the liquid.

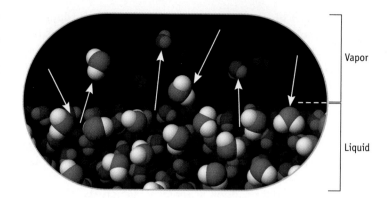

Vapor

Liquid

Condensation is the opposite of vaporization. Condensation is always exothermic; when it occurs, energy is transferred to the surroundings. The enthalpy change for condensation is equal but opposite in sign to the enthalpy of vaporization. For example, the enthalpy change for the vaporization of 1 mol of water at 100 °C is +40.7 kJ. On condensing 1 mol of water vapor to liquid water at 100 °C, the enthalpy change is −40.7 kJ.

In the discussion of intermolecular forces for polar and nonpolar molecules we pointed out the direct relationship between the ΔH°_{vap} values for various substances and the temperatures at which they boil (page 513) (see Tables 13.1 and 13.4 and page 515). Both properties reflect the attractive forces between particles in the liquid. The boiling points of nonpolar liquids (such as, hydrocarbons, atmospheric gases, and the halogens) increase with increasing atomic or molecular mass, a reflection of increased intermolecular dispersion forces. The alkanes in the list show this trend clearly. Similarly, the boiling points and enthalpies of vaporization of the heavier hydrogen halides (HX, where X = Cl, Br, and I) increase with increasing molecular mass. For these molecules, hydrogen bonding is not as important as it is in HF, so dispersion forces and ordinary dipole–dipole forces account for their intermolecular attractions. Because dispersion forces become increasingly important with increasing mass, the boiling points are in the order HCl < HBr < HI. Also notice in Table 13.4 the very high heats of vaporization of water and hydrogen fluoride that result from extensive hydrogen bonding.

Example 13.4   Enthalpy of Vaporization

Problem • You put 1.00 L of water (about 4 cups) in a pan at 100 °C, and the water slowly evaporates. How much heat must have been supplied to vaporize the water?

Strategy • Three pieces of information are needed to solve this problem:

1. ΔH°_{vap} for water = +40.7 kJ/mol at 100 °C

2. The density of water at 100 °C = 0.958 g/mL. (This is needed because ΔH°_{vap} has units of kilojoules per mole, so you first must find the mass of water and then the amount.)

3. Molar mass of water = 18.02 g/mol

Solution • Given the density of water, a volume of 1.00 L (or

1.00×10^3 cm³) is equivalent to 958 g, and this mass is in turn equivalent to 53.2 mol of water.

$$1.00 \text{ L} \left(\frac{1000 \text{ ml}}{1 \text{ L}} \right) \left(\frac{0.958 \text{ g}}{1 \text{ mL}} \right) \left(\frac{1 \text{ mol H}_2\text{O}}{18.02 \text{ g}} \right) = 53.2 \text{ mol H}_2\text{O}$$

Therefore, the amount of energy required is

$$53.2 \text{ mol H}_2\text{O} \left(\frac{40.7 \text{ kJ}}{\text{mol}} \right) = 2.16 \times 10^3 \text{ kJ}$$

2160 kJ is equivalent to about one quarter of the energy in your daily food intake.

Table 13.4 • Molar Enthalpy of Vaporization and Boiling Points for Common Substances*

Compound	Molar Mass (g/mol)	$\Delta H°_{vap}$ (kJ/mol)†	Boiling Point (°C) (Vapor pressure = 760 mm Hg)
Polar Compounds			
HF	20.0	25.2	19.7
HCl	36.5	16.2	−84.8
HBr	80.9	19.3	−66.4
HI	127.9	19.8	−35.6
NH_3	17.0	23.3	−33.3
H_2O	18.0	40.7	100.0
SO_2	64.1	24.9	−10.0
Nonpolar Compounds			
CH_4 (methane)	16.0	8.2	−161.5
C_2H_6 (ethane)	30.1	14.7	−88.6
C_3H_8 (propane)	44.1	19.0	−42.1
C_4H_{10} (butane)	58.1	22.4	−0.5
Monatomic Elements			
He	4.0	0.08	−268.9
Ne	20.2	1.7	−246.1
Ar	39.9	6.4	−185.9
Xe	131.3	12.6	−108.0
Diatomic Elements			
H_2	2.0	0.90	−252.9
N_2	28.0	5.6	−195.8
O_2	32.0	6.8	−183.0
F_2	38.0	6.6	−188.1
Cl_2	70.9	20.4	−34.0
Br_2	159.8	30.0	58.8

*Data taken from D. R. Lide: *Basic Laboratory and Industrial Chemicals,* Boca Raton, FL, CRC Press, 1993.

†$\Delta H°_{vap}$ is measured at the normal boiling point of the liquid.

Exercise 13.4 Enthalpy of Vaporization

The molar enthalpy of vaporization of methanol, CH_3OH, is 35.2 kJ/mol at 64.6 °C. How much energy is required to evaporate 1.00 kg of this alcohol?

Water is exceptional among the liquids listed in Table 13.4 in that an enormous amount of heat is required to convert liquid water to water vapor. This fact is important to your environment and your own physical well-being. When you exercise vigorously, your body responds by sweating to rid itself of the excess heat. Heat from your body is consumed in the process of evaporation, and your body is cooled.

Heats of vaporization and condensation of water also play an important role in weather (Figure 13.14). For example, if enough water condenses from the air to fall as an inch of rain on an acre of ground, the heat released is over 2.0×10^8 kJ! This is equivalent to about 50 tons of exploded dynamite or about the same as the energy released by a small bomb.

Vapor Pressure

If you put some water in an open beaker, it will eventually evaporate completely. If you put water in a sealed flask (Figure 13.15), however, the liquid will evaporate only until the rate of vaporization equals the rate of condensation. At this point, no further change will be observed in the system. The situation is an example of what chemists call a **dynamic equilibrium.** A system at equilibrium is represented using a set of double arrows connecting the reactant and product or the two phases of the substance.

$$\text{Liquid} \rightleftharpoons \text{Vapor}$$

The double arrows indicate that two processes occur at the same time and at the same rate. Molecules continuously move from the liquid to the vapor phase and from the vapor back to the liquid phase. Even though changes occur on the molecular level, no change can be detected on the macroscopic level. Furthermore, because the rate at which molecules move from liquid to vapor is the same as the rate at which they move from vapor to liquid, there is no net change in the masses of the two phases. In contrast to a closed flask, water in an open beaker does not reach an equilibrium with gas phase water molecules; air movement and gas diffusion remove the water vapor from the vicinity of the liquid surface so water molecules are not able to return to the liquid.

When a liquid–vapor equilibrium has been established, the pressure exerted by the vapor is called the **equilibrium vapor pressure** (often called just *vapor pressure*). The equilibrium vapor pressure of any substance is a measure of the tendency of its molecules to escape from the liquid phase and enter the vapor phase at a given temperature. This tendency is referred to qualitatively as the **volatility** of

Figure 13.14 Rainstorms release an enormous quantity of energy. When water vapor condenses, energy is evolved to the surroundings. The enthalpy of condensation of water is large, so a large quantity of heat is released in a rainstorm. *(Mark A. Schneider/Dembinsky Photo Associates)*

● **Equilibrium**
Equilibrium is a concept used throughout chemistry and one we shall return to often. In a system at equilibrium, equal but opposite changes are occurring simultaneously at the same rate. This situation is signaled by connecting the two states or the reactants and products by a set of double arrows ($\rightleftharpoons$). See Chapters 16–18 in particular.

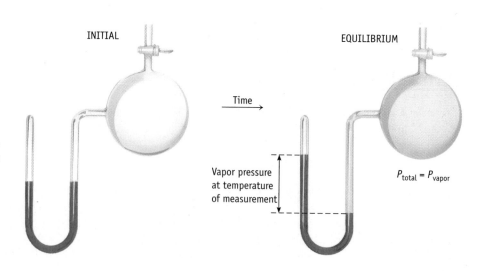

Figure 13.15 Vapor pressure. A volatile liquid is placed in an evacuated flask (*left*). At the beginning, no molecules of the liquid are in the vapor phase. After a short time, however, some of the molecules have evaporated, and the gaseous molecules exert a vapor pressure. The pressure of the vapor measured when the liquid and vapor are in equilibrium is called the equilibrium vapor pressure (*right*).

the compound. The higher the equilibrium vapor pressure at a given temperature, the more volatile the compound [CD-ROM, Screen 13.9].

As described previously (see Figure 13.12), the average energy of molecules in the liquid phase is a function of temperature. At a higher temperature, more molecules have sufficient energy to escape the surface of the liquid. The equilibrium vapor pressure must therefore also increase with temperature (Figure 13.16, page 528). All points along the vapor-pressure-versus-temperature curves in Figure 13.16 represent conditions of pressure and temperature at which liquid and vapor are in equilibrium. For example, at 60 °C the vapor pressure of water is 149 mm Hg [➡ APPENDIX G]. If water is placed in an evacuated flask that is maintained at 60 °C, liquid will evaporate until the pressure exerted by the vapor is 149 mm Hg (assuming there is enough water in the flask so some liquid remains when equilibrium is reached).

● **Equilibrium Vapor Pressure**
At the conditions of T and P given by any point on a curve in Figure 13.16, pure liquid and its vapor are in dynamic equilibrium. If T and P define a point not on the curve, only a single phase will be present.

Example 13.5   **Vapor Pressure**

Problem • You place 2.00 L of water in your dormitory room, which has a volume of 4.25×10^4 L. You seal the room and wait for the water to evaporate. Will all of the water evaporate at 25 °C? (At 25 °C the density of water is 0.997 g/mL, and its vapor pressure is 23.8 mm Hg.)

Strategy • One approach to solving this problem is to calculate the quantity of water that must evaporate in order to exert a pressure of 23.8 mm Hg in a volume of 4.25×10^4 L at 25 °C. Because water vapor is like any other gas, we use the ideal gas law for the calculation.

Solution • Calculate the amount and then the mass and volume of water that fulfills the following conditions:

$$P = 23.8 \text{ mm Hg}$$
$$V = 4.25 \times 10^4 \text{ L}$$

$T = 25$ °C (or 298 K)

$$P = 23.8 \text{ mm Hg} \cdot \frac{1 \text{ atm}}{760 \text{ mm Hg}} = 0.0313 \text{ atm}$$

$$n = \frac{PV}{RT} = \frac{(0.0313 \text{ atm})(4.25 \times 10^4 \text{ L})}{\left(0.082057 \dfrac{\text{L} \cdot \text{atm}}{\text{K} \cdot \text{mol}}\right)(298 \text{ K})} = 54.4 \text{ mol H}_2\text{O}$$

$$54.4 \text{ mol H}_2\text{O} \cdot \frac{18.02 \text{ g}}{1 \text{ mol H}_2\text{O}} = 980. \text{ g H}_2\text{O}$$

$$980. \text{ g H}_2\text{O} \cdot \frac{1 \text{ mL}}{0.997 \text{ g}} = 983 \text{ mL}$$

This calculation shows that only about half of the available water needs to evaporate in order to achieve an equilibrium water vapor pressure of 23.8 mm Hg at 25 °C in the dorm room.

Exercise 13.5   **Vapor Pressure Curves**

Look at the vapor pressure curve for ethanol in Figure 13.16 (page 529).

(a) What is the approximate vapor pressure of ethanol at 40 °C?

(b) Are liquid and vapor in equilibrium when the temperature is 60 °C and the pressure is 600 mm Hg? If not, does liquid evaporate to form more vapor, or does vapor condense to form liquid?

Exercise 13.6   **Vapor Pressure**

If 0.50 g of pure water is sealed in an evacuated 5.0-L flask, and the whole assembly is heated to 60 °C, will the pressure be equal to or less than the equilibrium vapor pressure of water at this temperature? What if you use 2.0 g of water? Under either set of conditions is any liquid water left in the flask, or does all of the water evaporate?

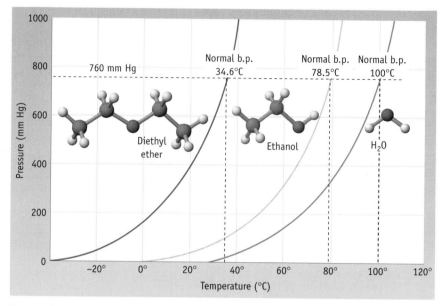

Figure 13.16   **Vapor pressure curves for diethyl ether (C₂H₅OC₂H₅), ethanol (C₂H₅OH), and water.** Each curve represents conditions of *T* and *P* at which the two phases, liquid and vapor, are in equilibrium. These compounds exist as liquids for temperatures and pressures to the left of the curve and as gases under conditions to the right of the curve.

Boiling Point

If you have a beaker of water open to the atmosphere, the mass of the atmosphere is pressing down on the surface. As heat is added, more and more water evaporates, pushing the molecules of the atmosphere aside. If enough heat is added, a temperature is eventually reached at which the vapor pressure of the liquid equals the atmospheric pressure. At this point, bubbles of vapor begin to form in the liquid, and the liquid boils (Figure 13.17).

The boiling point of a liquid is the temperature at which its vapor pressure is equal to the external pressure, and, if the external pressure is 1 atm, the temperature is designated the **normal boiling point** [🔘 CD-ROM, Screen 13.10]. Normal boiling points of some liquids are included in Table 13.4. Be sure to notice the relationships among normal boiling point, enthalpy of vaporization, and intermolecular forces.

The normal boiling point of water is 100 °C, and in a great many places in the United States, water boils at or near this temperature. If you live at higher altitudes, however, such as in Salt Lake City, Utah, where the barometric pressure is about 650 mm Hg, water will boil at a noticeably lower temperature. The curve for the equilibrium vapor pressure of water in Figure 13.16 shows that a pressure of 650 mm Hg corresponds to a boiling temperature of about 95 °C. Cooks know that food has to be cooked a little longer in Salt Lake City or Denver to achieve the same result as in New York City at sea level.

Critical Temperature and Pressure

The vapor pressure of a liquid will continue to increase above the normal boiling point. On first thought it might seem that vapor pressure–temperature curves (such as shown in Figure 13.16) should continue upward without limit, but this is not so.

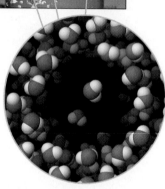

Figure 13.17   **Vapor pressure and boiling.** When the vapor pressure of the liquid equals the atmospheric pressure, bubbles of vapor begin to form within the body of liquid, and the liquid boils. *(Charles D. Winters)*

● **Cooking Under Pressure**
To shorten cooking time, a pressure cooker can be used. This is a sealed pot that allows water vapor to build up to pressures slightly greater than the external or atmospheric pressure. The boiling point of the water is higher than 100 °C, and foods cook faster.

A Closer Look

Vapor Pressure, Enthalpy of Vaporization, and the Clausius–Clapeyron Equation

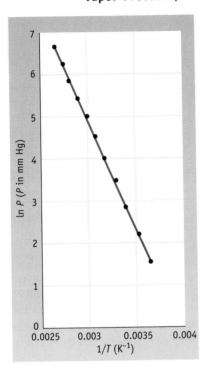

Plotting the vapor pressure for a liquid at a series of temperatures results in a curved line. However, the German physicist Rudolf Clausius (1822–1888) and the French scientist Émile Clapeyron (1799–1864) showed that, for a pure liquid, there is linear relationship between the reciprocal of the Kelvin temperature (T) and the logarithm of vapor pressure ($\ln P$).

$$\ln P = -(\Delta H_{vap}/RT) + C \quad \textbf{(13.1)}$$

Here ΔH_{vap} is the enthalpy of vaporization of the liquid, R is the ideal gas constant, and C is a constant characteristic of the compound in question. This equation, now called the **Clausius–Clapeyron equation,** provides a method of obtaining values for ΔH_{vap}. The equilibrium vapor pressure of a liquid can be measured at several different temperatures, and these pressures are plotted versus $1/T$. The result is a straight line with a slope of $-\Delta H_{vap}/R$.

As an alternative to plotting $\ln P$ versus $1/T$, we can write the following equation that allows us to calculate ΔH_{vap} knowing the vapor pressure of a liquid at two different temperatures.

$$\ln\frac{P_2}{P_1} = \frac{\Delta H_{vap}}{R}\left[\frac{1}{T_1} - \frac{1}{T_2}\right]$$

For example, ethylene glycol ($HOCH_2CH_2OH$), commonly used as an antifreeze in automobiles, has a vapor pressure of 14.9 mm Hg (P_1) at 373 K (T_1) and a vapor pressure of 49.1 mm Hg (P_2) at 398 K (T_2).

$$\ln\left(\frac{49.1 \text{ mm Hg}}{14.9 \text{ mm Hg}}\right) =$$

$$\frac{\Delta H_{vap}}{0.0083145 \text{ kJ/K} \cdot \text{mol}}\left[\frac{1}{373 \text{ K}} - \frac{1}{398 \text{ K}}\right]$$

$$1.192 = \frac{\Delta H_{vap}}{0.0083145 \text{ kJ/K} \cdot \text{mol}}(0.000168)\frac{1}{K}$$

$$\Delta H_{vap} = 59.0 \text{ kJ/mol}$$

As an exercise, you might try to calculate the enthalpy of vaporization of diethyl ether [$(C_2H_5)_2O$] (see Figure 13.16). This compound has vapor pressures of 57.0 mm Hg and 534 mm Hg at −22.8 °C and 25.0 °C, respectively. (Answer = 29.0 kJ/mol)

Instead, when a specific temperature and pressure are reached, the interface between the liquid and vapor disappears. This point is called the **critical point.** The temperature at which this occurs is the **critical temperature, T_c,** and the corresponding pressure is the **critical pressure, P_c** (Figure 13.18). The substance that exists under these conditions is called a **supercritical fluid.** It is like a gas under such high pressure in that its density resembles a liquid's, but its viscosity (ability to flow) remains close to that of a gas.

For most substances the critical point is at a very high temperature and pressure (see Table 13.5). Water, for instance, has a critical temperature of 374 °C and a critical pressure of 217.7 atm. Consider what the particulate level might look like under these conditions. At this high pressure, water molecules have been forced almost as close together as they are in the liquid state. The high temperature, however, means that each of these molecules has enough kinetic energy to exceed the forces holding molecules together. So, the critical fluid has a tightly packed molecular arrangement like a liquid, but the intermolecular forces of attraction that characterize the liquid state are less than the kinetic energy of the particles.

Supercritical fluids can have unexpected properties, such as the ability to dissolve normally insoluble materials. Supercritical CO_2 is especially useful. Carbon

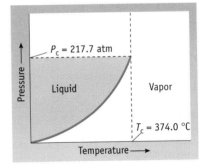

Figure 13.18 The vapor pressure curve for water. The curve representing equilibrium conditions for liquid and gaseous water ends at the critical point; above that temperature and pressure water becomes a supercritical fluid. The portion of the curve shown in Figure 13.16 is such a small part of this graph that it cannot be identified.

Table 13.5 • **Critical Temperatures and Pressures for Common Compounds***

Compound	T_c (°C)	P_c (atm)
CH_4 (methane)	−82.6	45.4
C_2H_6 (ethane)	32.3	49.1
C_3H_8 (propane)	96.7	41.9
C_4H_{10} (butane)	152.0	37.3
CCl_2F_2 (CFC-12)	111.8	40.9
NH_3	132.4	112.0
H_2O	374.0	217.7
CO_2	30.99	72.8
SO_2	157.7	77.8

*Data taken from D. R. Lide: *Basic Laboratory and Industrial Chemicals,* CRC Press, Boca Raton, FL, 1993.

Supercritical fluid extractions. A photograph of an industrial facility for extractions with supercritical CO_2. *(Thar Technologies, Inc., Pittsburgh, PA.)*

dioxide is readily available, essentially nontoxic, nonflammable, and inexpensive. It is relatively easy to reach the supercritical state (its critical temperature of 30.99 °C and pressure of 72.8 atm are easily achieved). The material is also easy to handle. The reason CO_2 is so useful is that it does not dissolve water or polar compounds such as sugar, but it does dissolve nonpolar oils, which constitute many of the flavoring or odor-causing compounds in foods. As a result, food companies now use supercritical CO_2 to extract caffeine from coffee, for example.

To decaffeinate coffee, the beans are treated with steam to bring the caffeine to the surface. The beans are then immersed in supercritical CO_2, which selectively dissolves the caffeine but leaves intact the compounds that give flavor to coffee. (Decaffeinated coffee contains less than 3% of the original caffeine.) The solution of caffeine in supercritical CO_2 is poured off, and the CO_2 is recovered by evaporation and reused.

Supercritical CO_2 has properties that make it attractive as a solvent so, not surprisingly, other uses are being sought for this substance. For example, more than 10 billion kilograms of organic and halogenated solvents are used worldwide every year as cleaning agents. These can have deleterious effects on the environment, so it is hoped that many can be replaced by supercritical CO_2.

Surface Tension, Capillary Action, and Viscosity

Molecules at the surface of a liquid behave differently from those in the interior. Molecules in the interior interact with molecules all around them (Figure 13.19) [CD-ROM, Screen 13.11]. In contrast, surface molecules are affected only by those below the surface layer. This phenomenon leads to a net inward force of attraction on the surface molecules, contracting the surface and making the liquid behave as though it had a skin. The toughness of the skin of a liquid is measured by its **surface tension** — the energy required to break through the surface or to disrupt a liquid drop and spread the material out as a film. It is surface tension that causes water drops to be spheres and not little cubes, for example (Figure 13.20a), because the sphere has a smaller surface area than any other shape of the same volume.

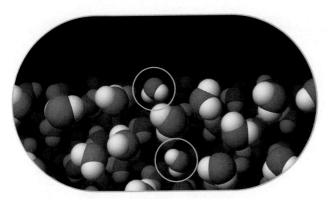

Water molecules on the surface are not completely surrounded by other water molecules.

Water molecules under the surface are completely surrounded by other water molecules.

Figure 13.19 Intermolecular forces in a liquid. There is a difference in the forces acting on a molecule at the surface of a liquid and those acting on a molecule in the interior of a liquid.

Capillary action is closely related to surface tension. When a small-diameter glass tube is placed in water, the water rises in the tube, just as water rises in a piece of paper in water (Figure 13.20b). Because polar Si—O bonds occur on the surface of glass, polar water molecules are attracted by **adhesive forces** between the two different substances. These forces are strong enough that they can compete with the **cohesive forces** between the water molecules themselves. Thus, some water molecules can adhere to the walls and others are attracted to these and build a "bridge" back into the liquid. The surface tension of the water (from cohesive forces) is great enough to pull the liquid up the tube, so the water level rises in the tube. The rise will continue until the attractive forces — adhesion between water and glass, cohesion between water molecules — are balanced by the force of gravity on the water column. It is these forces that lead to the characteristic concave, or downward-curving, *meniscus* that you see for water in a drinking glass or in a laboratory test tube (Figure 13.20c).

In some liquids cohesive forces (high surface tension) are much greater than adhesive forces with glass. Mercury is one example. Mercury does not climb the walls of a glass capillary, and when it is in a tube mercury will form a convex or upward-curving meniscus (Figure 13.20c).

Finally, one additional important property of liquids is their viscosity, the resistance of liquids to flow. When you turn over a glassful of water, it empties quickly. In contrast, it takes much more time to empty a glassful of honey or rubber cement. Although intermolecular forces play a significant role in determining viscosity, other factors are also clearly at work. For example, olive oil consists of molecules with long chains of carbon atoms (see Chapter 11), and it is about 70 times more viscous than ethanol, a small molecule with only two carbons and one oxygen. The long-chain molecules of natural oils are floppy and become entangled with one another; the longer the chain the greater the tangling and the greater the viscosity. Also, longer chains have greater intermolecular forces because there are more atoms to attract one another, each atom contributing to the total force.

Exercise 13.7 **Viscosity**

Glycerol ($HOCH_2CHOHCH_2OH$) is used in cosmetics. Do you expect its viscosity to be larger or smaller than the viscosity of ethanol, CH_3CH_2OH? Why or why not?

Glycerol

Figure 13.20 Adhesive and cohesive forces in liquids. *(a, S. R. Nagel, James Frank Institute, University of Chicago; b and c, Charles D. Winters)*

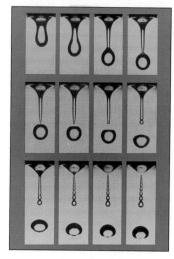

(a) A series of photographs showing the different stages when a water drop falls. The drop was illuminated by a strobe light of 5-ms duration. The total time for this sequence was 0.05 s.

(b) Capillary action. Polar water molecules are attracted to the —OH bonds in paper fibers, and water rises in the paper. If a line of ink is placed in the path of the rising water, the different components of the ink are attracted differently to the water and paper and are separated in a process called chromatography.

(c) Water (top layer) forms a concave meniscus, while mercury (bottom layer) forms a convex meniscus. The different shapes are determined by the adhesive forces of the molecules of the liquid with the walls of the tube and the cohesive forces between molecules of the liquid.

13.6 SOLID-STATE CHEMISTRY: METALS

Many kinds of solids exist in the world around us (Chapter Focus, page 508). Solid-state chemistry is one of the booming areas of science, especially because it relates to the development of interesting new materials. As we describe various kinds of solids, we hope to provide a glimpse of the reason this area is exciting.

Solid-state chemistry can be organized by classifying the common types of solids (Table 13.6). This section will describe the solid-state structures of common metals and the next section will take up ionic solids [CD-ROM, Screens 13.12 and 13.13]. Next we examine molecular and network solids in Section 13.8 [CD-ROM, Screens 13.14–13.16]. Finally, Section 13.9 outlines important properties of solids.

Crystal Lattices and Unit Cells

In both gases and liquids, molecules move continually and randomly, and rotate and vibrate as well. Because of this movement, orderly arrangement of molecules in the gas or liquid state is not possible. In solids, however, the molecules, atoms, or ions cannot move (although they vibrate and occasionally rotate). Thus, a regular, repeating pattern of atoms or molecules within the structure — a long-range order — is a characteristic of the solid state.

The beautiful, external (macroscopic) regularity of a crystal of salt (Figure 13.21 and page 510) suggests that it has an internal symmetry, a symmetry involving the ions that make up the solid. Structures of solids can be described as three-dimensional lattices of atoms, ions, or molecules. For a crystalline solid, the **unit**

Figure 13.21 Crystalline NaCl. The regular internal structure of a salt (NaCl) crystal is reflected by its outward appearance. *(Charles D. Winters)*

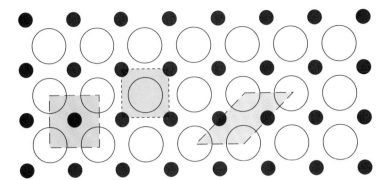

Figure 13.22 Unit cells for a flat, two-dimensional solid made from circular "atoms." Several unit cells are possible, two of the most obvious being squares. The lattice can be built by translating the unit cells throughout the plane of the figure. All unit cells contain a net of one black circle and one white circle.

cell is the smallest repeating unit that has all of the symmetry characteristic of the atomic, ionic, or molecular arrangement.

To understand unit cells, consider first a two-dimensional lattice model, the repeating pattern of circles shown in Figure 13.22. The green square shaded in this figure is a unit cell. The complete lattice could be created from a group of unit cells identical to this by joining the unit cells edge to edge. It is also a requirement that a unit cell reflect the stoichiometry of the solid. Here the square unit cell clearly contains one fourth of each of four white circles and one black circle, giving a total of one white and one black circle per two-dimensional unit cell.

You may recognize that it is possible to draw other unit cells for the two-dimensional lattice. Another is a square (red) that fully encloses a single white circle and parts of black circles that add up to one net black circle. Yet another possible unit cell is a parallelogram (yellow). Other unit cells may be possible, but it is conventional to try to draw unit cells in which atoms or ions are placed at the **lattice points,** that is, at the corners of the cube or other geometric object that constitutes the unit cell.

Solids (three-dimensional lattices) can be built by assembling three-dimensional unit cells much like building blocks, as the artist M. C. Escher illustrated in a well-known drawing (Figure 13.23). The corners (or lattice points) defining each unit cell in simple solids represent identical environments for the ions in an ionic solid, metal atoms in a metallic solid, or molecules in a molecular solid. Such points are equivalent to one another, and collectively they define the **crystal lattice.**

Chapter Goals • Revisited

- Describe intermolecular forces and their effects and the importance of hydrogen bonding.
- Understand the properties of liquids.
- **Understand cubic unit cells.**
- Relate unit cells for ionic compounds to compound formulas.
- Describe the properties of solids.
- Understand the nature of phase diagrams.

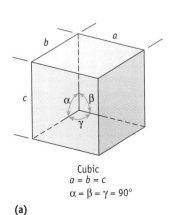

Cubic
$a = b = c$
$\alpha = \beta = \gamma = 90°$

(a)

(b)

Figure 13.23 Cubic unit cells.
(a) The cube, one of the seven basic unit cells that describe crystal systems.
(b) "Cubic Space Division" by M. C. Escher. The artist has depicted three-dimensional space built up by stacking many, many cubes (with each corner of each cube itself a smaller cube). Each cube is a unit cell of the whole, because the whole can be made up from the smaller cubes. *(M. C. Escher's "Cubic Space Division"* © *2002 Cordon Art B.V. — Baarn, The Netherlands. All rights reserved.)*

Table 13.6 • **Structures and Properties of Various Types of Solid Substances**

Type	Examples	Structural Units
Ionic	NaCl, K_2SO_4, $CaCl_2$, $(NH_4)_3PO_4$	Positive and negative ions; no discrete molecules
Metallic	Iron, silver, copper, other metals and alloys	Metal atoms (positive metal ions with delocalized electrons)
Molecular	H_2, O_2, I_2, H_2O, CO_2, CH_4, CH_3OH, CH_3CO_2H	Molecules
Network	Graphite, diamond, quartz, feldspars, mica	Atoms held in an infinite two- or three-dimensional network
Amorphous	Glass, polyethylene, nylon	Covalently bonded networks with no long-range regularity

To construct crystal lattices, nature uses seven, three-dimensional unit cells. These differ from one another in that their sides have different relative lengths and their edges meet at different angles. The simplest of the seven crystal lattices is the **cubic unit cell,** a cell with edges of equal length that meet at 90° angles (Figure 13.23). We shall look in detail only at this structure, not only because cubic unit cells are easily visualized but also because they are common.

Within the cubic class, three cell symmetries occur: **primitive,** or **simple cubic (sc); body-centered cubic (bcc);** and **face-centered cubic (fcc)** (Figure 13.24). All three have eight identical atoms, molecules, or ions at the corners of the cubic unit cell. The bcc and fcc arrangements, however, differ from the primitive cube because they have additional particles, of the same type as those at the corners, at other locations. The bcc structure is called "body-centered" because it has an ad-

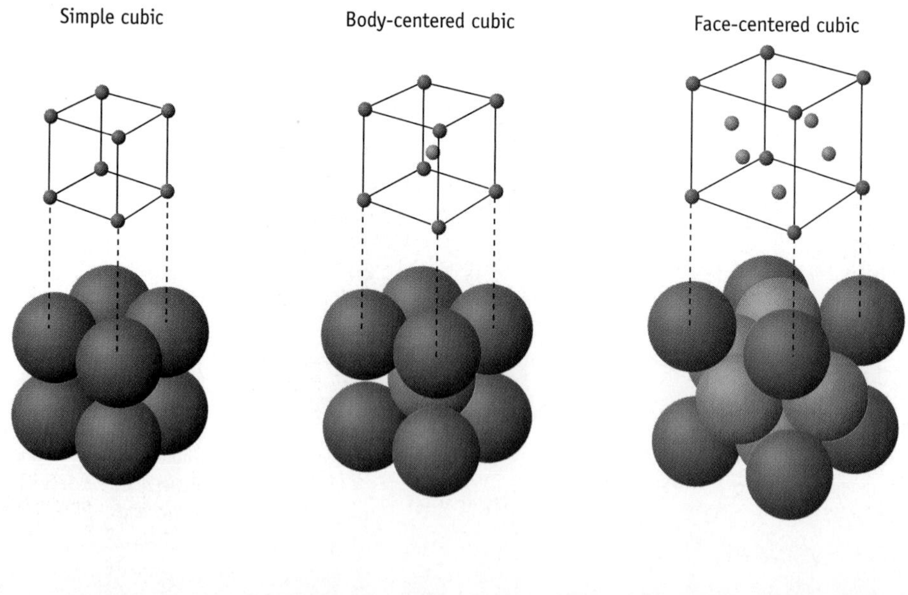

Simple cubic   Body-centered cubic   Face-centered cubic

Figure 13.24   The three cubic unit cells. The top row shows the lattice points of the three cells, and the bottom row shows the same cells using space-filling spheres. The spheres in each figure represent identical atoms or ions centered on the lattice points; different colors were used in the drawing only to call attention to the spheres in the center of the cube or on the faces.

Forces Holding Units Together	Typical Properties
Ionic; attractions among charges on positive and negative ions	Hard; brittle; high melting point; poor electric conductivity as solid, good as liquid; often water-soluble
Metallic; electrostatic attraction among metal ions and electrons	Malleable; ductile; good electric conductivity in solid and liquid; good heat conductivity; wide range of hardness and melting points
Dispersion forces, dipole–dipole forces, hydrogen bonds	Low to moderate melting points and boiling points; soft; poor electric conductivity in solid and liquid
Covalent; directional electron-pair bonds	Wide range of hardnesses and melting points (three-dimensional bonding > two-dimensional bonding); poor electric conductivity, with some exceptions
Covalent; directional electron-pair bonds	Noncrystalline; wide temperature range for melting; poor electric conductivity, with some exceptions

ditional particle at the center or "body" of the cube. The fcc arrangement is called "face-centered" because it has a particle of the same type as the corner atoms in the center of each of the six faces of the cube. Metals may assume any of these structures. The alkali metals, for example, are body-centered cubic, whereas nickel, copper, and aluminum are face-centered cubic (Figure 13.25).

When the cubes pack together to make a three-dimensional crystal of a metal, the atom at each corner is shared among eight cubes (Figure 13.26a). Because of this, only one eighth of each corner atom is actually within a given unit cell. Furthermore, because a cube has eight corners, and because one eighth of the

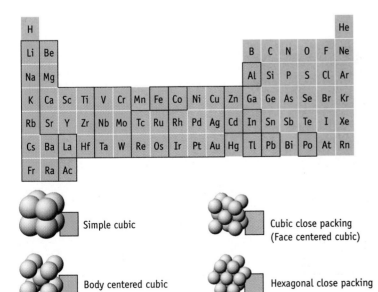

Simple cubic

Cubic close packing
(Face centered cubic)

Body centered cubic

Hexagonal close packing

Figure 13.25   Metals use four different unit cells. Three are based on the cube, and the fourth is a hexagonal unit cell.

536   *Chapter 13   Intermolecular Forces, Liquids, and Solids*

Figure 13.26   Atom sharing at cube corners and faces.
(a) In any cubic lattice, each corner particle is shared equally among eight cubes, so one eighth of the particle is within a particular cubic unit cell.
(b) In a face-centered lattice, each particle on a cube face is shared equally between two unit cells. One half of each particle of this type is within the given unit cell.

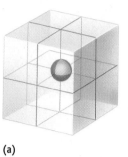
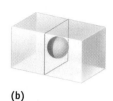

(a)   **(b)**

atom at each corner "belongs to" a particular unit cell, the corner atoms contribute a net of one atom to a given unit cell.

$$\text{(8 corners of a cube)}\left(\frac{1}{8}\text{ of each corner atom within a unit cell}\right) =$$

1 net atom per unit cell

A primitive or simple cubic arrangement has one net atom within the unit cell.

    In contrast to the simple cube, a body-centered cube has an additional atom, and it is wholly within the unit cell at the cube's center. The center atom is present in addition to those at the cube corners, so *a body-centered cubic arrangement has a net of two atoms within the unit cell.*

    In an fcc arrangement, there is an atom in each of the six faces of the cube in addition to those at the cube corners. One half of each atom on a face belongs to the given unit cell (Figure 13.26b). Three net particles are therefore contributed by the particles on the faces of the cube; that is,

$$\text{(6 faces of a cube)}\left(\frac{1}{2}\text{ of an atom within a unit cell}\right) =$$

3 net face-centered atoms within a unit cell

Thus, *the face-centered cubic arrangement has a net of four atoms within the unit cell,* one contributed by the corner atoms and another three contributed by the ones centered in the six faces.

● **Atoms per Unit Cell**

Cell Type	Atoms Per Unit Cell
Simple cubic (sc)	1
Body-centered cubic (bcc)	2
Face-centered cubic (fcc)	4

A piece of aluminum metal and a face-centered cubic unit cell. *(Charles D. Winters)*

A Closer Look

Using X-rays to Determine Crystal Structure

How do chemists determine the distance between the atoms in a metal, the ions in a salt, or the atoms of a molecule? In 1912 Max von Laue (1879–1960), a German physicist, found that crystalline solids diffract x-rays. Somewhat later, the English scientists William (1862–1942) and Lawrence (1890–1971) Bragg (father and son), showed that x-ray diffraction by crystalline solids could be used to determine distances between atoms. X-ray crystallography, the science of determining atomic-scale crystal structures, is now a procedure used extensively by chemists.

To determine distances on an atomic scale we have to use a probe that can locate atoms or ions. The probe must have dimensions not much larger than the atoms, otherwise it would pass over them unperturbed. The probe must therefore be only a few picometers in size. X-rays meet this requirement.

Solids are generally studied because the atoms or molecules are in fixed positions, so they can be located with x-ray photons and because the ions, atoms, or molecules of the crystal are arranged in an orderly manner and give rise to a regular diffraction pattern.

To "see" radiation interact with atoms we have to rely on a change in the radia-

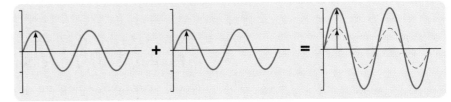

(a) Constructive interference

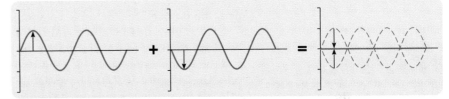

(b) Destructive interference

Figure B **(a)** Constructive interference occurs when two in-phase waves combine to produce a wave of greater amplitude. **(b)** Destructive interference results from the combination of two waves of equal magnitude that are exactly out of phase.

tion as its passes through the crystal. The change is observed as a scattering, or "diffraction," of the photons of the radiation from their original path (Figure A). The diffraction of x-rays produces a regularly arranged series of spots on a fluorescent screen or photographic film.

Structural information from x-ray crystallography comes from the fact that crystals scatter x-rays depending on the locations of the atoms in the crystal. We can understand this by considering the wave properties of photons. You know that waves have peaks and valleys at different points in space. If two waves meet at some point, and the peak of one wave meets the valley of another, the waves cancel each other at that point. If the waves meet peak-to-peak, however, they reinforce each other, and the radiation produces a detectable signal or spots on a photographic film (Figure B). This condition is met when the x-rays are scattered at special values of the angle ϕ, an angle related to the distances between atoms in the solid.

The experiment as it is really done is more complicated than we have described. Nonetheless, with modern instruments and computers, chemists and physicists can usually determine quite readily the location of atoms in a crystal and the distances between them. Indeed, the technique has provided so much structural information in the past 30 years or so that the science of chemistry has itself been revolutionized. Many of the structural models you have seen in this book are based on the results of x-ray crystallography.

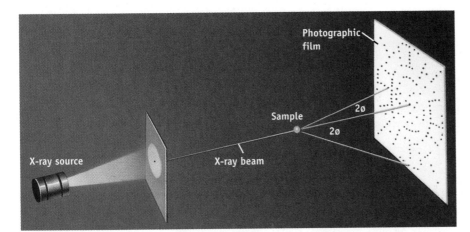

Figure A In the x-ray diffraction experiment, a beam of x-rays is directed at a crystalline solid. The photons of the x-ray beam are scattered by the atoms of the solid. The angle of scattering depends on the locations of the atoms in the crystal. The scattered x-rays are detected by a photographic film or an electronic detector.

An experimental technique, x-ray crystallography, can be used to determine the structure of a crystalline substance (page 537). Once the structure is known, the information can be combined with other experimental information to calculate such useful parameters as the radius of an atom (Study Questions 13.62 and 13.63).

13.7   SOLID-STATE CHEMISTRY: STRUCTURES AND FORMULAS OF IONIC SOLIDS

The lattices of many ionic compounds are built by taking a simple cubic or face-centered cubic lattice of spherical ions of one type and placing ions of opposite charge in the holes within the lattice. This produces a three-dimensional lattice of regularly placed ions. The smallest repeating unit in these structures is, by definition, the unit cell for the ionic compound.

The choice of the lattice and the number and location of the holes that are filled are the keys to understanding the relationship between the lattice structure and the formula of a salt. This is illustrated with the ionic compound cesium chloride, CsCl (Figure 13.27). The structure of CsCl has a primitive cubic unit cell of chloride ions. The cesium ion fits into a hole in the center of the cube. Notice that the Cs^+ ion has eight Cl^- ions as nearest neighbors in the lattice on the right.

Next consider the structure for NaCl. An extended view of the lattice and one unit cell are illustrated in Figure 13.28a and 13.28b. The Cl^- ions are arranged in a face-centered cubic unit cell, and the Na^+ ions are arranged in a regular manner between these ions. Notice that each Na^+ ion is surrounded by six Cl^- ions. An octahedral geometry is assumed by the ions surrounding an Na^+ ion so the Na^+ ions are said to be in **octahedral holes** (Figure 13.28c).

The formula for NaCl can be related to this structure by counting the number of cations and anions contained in one unit cell. A face-centered cubic lattice of Cl^- ions has a net of four Cl^- ions within the unit cell. There is one Na^+ in the center of the unit cell, contained totally within the unit cell. In addition, there are 12 Na^+ ions along the edges of the unit cell. Each of these Na^+ ions is shared among four unit cells, so each contributes one fourth of an Na^+ ion to the unit cell, giving three additional Na^+ ions within the unit cell. This accounts for all the ions contained in the unit cell: 4 Cl^- and 4 Na^+ ions. Thus, a unit cell of NaCl has a 1:1 ratio of Na^+ and Cl^- ions as the formula requires.

Another common unit cell has ions of one type in a face-centered cubic unit cell. Ions of the other type are then located in **tetrahedral holes** (Figure 13.29). In ZnS (zinc blende), the sulfide ions (S^{2-}) compose a face-centered cubic unit cell. The zinc ions (Zn^{2+}) then occupy four of eight possible tetrahedral holes, each of which is surrounded by four S^{2-} ions. The unit cell consists of a net of four S^{2-} ions at the lattice points and four Zn^{2+} ions, which are contained wholly within the unit cell.

• Lattice Ions and Holes
The holes in the lattice are smaller than the ions used to create an ionic lattice. A lattice, therefore, is usually built out of the larger ions, and the smaller ions are placed into the holes in the lattice. For NaCl, for example, an fcc lattice is built out of the Cl^- ions (radius = 181 pm), and the smaller Na^+ cations (radius = 98 pm) are placed in appropriate holes in the lattice.

Chapter Goals • Revisited

- Describe intermolecular forces and their effects and the importance of hydrogen bonding.
- Understand the properties of liquids.
- Understand cubic unit cells.
- **Relate unit cells for ionic compounds to compound formulas.**
- Describe the properties of solids.
- Understand the nature of phase diagrams.

Figure 13.27   Cesium chloride (CsCl) unit cell. The unit cell of CsCl may be viewed in two ways. The only requirement is that the unit cell must have a net of one Cs^+ ion and one Cl^- ion. Either way, it is a simple cubic unit cell of ions of one type (Cl^- on the left or Cs^+ on the right). Generally, ionic lattices are assembled by placing the larger ions (here Cl^-) at the lattice points and placing the smaller ions (here Cs^+) in the lattice holes.

Cl^-, radius = 181 pm Cs^+, radius = 165 pm

Cl^- ions at each cube corner = 1 net Cl^- ion in the unit cell.

One Cs^+ ion at each cube corner gives one net Cs^+ ion in the unit cell.

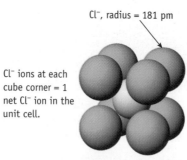

Cl^- lattice and Cs^+ in lattice hole

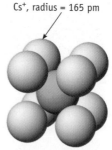

Cs^+ lattice and Cl^- in lattice hole

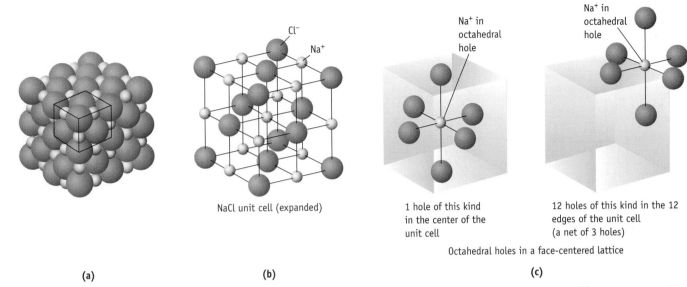

Octahedral holes in a face-centered lattice

(a)   (b)   (c)

Figure 13.28   Sodium chloride. (a) An extended view of a sodium chloride lattice. The smaller Na^+ ions (silver) are packed into a face-centered cubic lattice of larger Cl^- ions (yellow). One unit cell is outlined. **(b)** An expanded view of the NaCl face-centered cubic unit cell. The lines only represent the connections between lattice points. **(c)** A close-up view of the octahedral holes in the lattice.

Example 13.6 **Ionic Structure and Formula**

Problem • One unit cell of the mineral perovskite is illustrated here. The compound is composed of calcium and titanium cations and oxide anions. Based on the unit cell, what is the formula of perovskite?

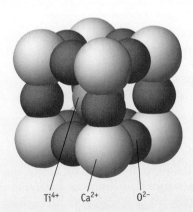

Strategy • Identify the ions involved in the unit cell and their location in the unit cell. Decide on the net number of ions of each kind in the cell.

Solution • The unit cell has Ca^{2+} ions at the corners of the cubic unit cell, a titanium ion (Ti^{4+}) in the center of the cell, and oxide ions (O^{2-}) on the cube edges.

Number of Ca^{2+} ions:

$$(8\ Ca^{2+}\ \text{ions at cube corners})\left(\frac{1}{8}\ \text{of each ion inside unit cell}\right)$$
$$= 1\ \text{net}\ Ca^{2+}\ \text{ion}$$

Number of titanium ions:

One ion is in the cube center = 1 net titanium ion

Number of O^{2-} ions:

$$(12\ O^{2-}\ \text{ions in cube edges})\left(\frac{1}{4}\ \text{of each ion inside cell}\right)$$
$$= 3\ \text{net}\ O^{2-}\ \text{ions}$$

This means the formula of perovskite is $CaTiO_3$.

Comment • This is a reasonable formula. A Ca^{2+} ion and three O^{2-} ions would require a titanium ion with a 4+ charge. Titanium is in group 4B of the periodic table, so Ti^{4+} is a predictable ion.

Exercise 13.8 **Structure and Formula**

If an ionic solid has an fcc lattice of anions (X) and all of the tetrahedral holes are occupied by metal cations (M), is the formula of the compound MX, MX_2, or M_2X?

Figure 13.29   Two views of the ZnS or zinc blende unit cell. This unit cell is an example of a face-centered cubic lattice of ions of one type with ions of the opposite type in tetrahedral holes. [Notice that there are twice as many tetrahedral lattice holes (8) as ions composing the fcc lattice (4).]

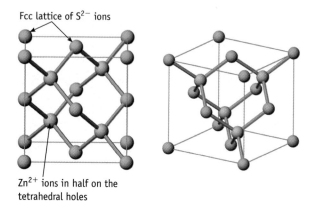

Fcc lattice of S^{2-} ions

Zn^{2+} ions in half on the tetrahedral holes

In summary, compounds with the formula MX commonly form one of three possible crystal structures:

1. M^{n+} ions occupying all the cubic holes of a simple cubic X^{n-} lattice
2. M^{n+} ions in all the octahedral holes in a face-centered cubic X^{n-} lattice
3. M^{n+} ions occupying half of the tetrahedral holes in a face-centered cube lattice of X^{n-} ions

CsCl is a good example of (1) (see Figure 13.27), NaCl is the usual example of (2), and ZnS is an example of (3). Chemists and geologists in particular have observed that the sodium chloride, or "rock salt," structure is adopted by many ionic compounds, most especially by all the alkali metal halides (except CsCl, CsBr, and CsI), all the oxides and sulfides of the alkaline earth metals, and all the oxides of formula MO of the transition metals of the fourth period. Finally, the formulas of compounds in general must be reflected in the structures of their unit cells (see Example 13.6); therefore, the formula can always be derived from the unit cell structure.

13.8   OTHER KINDS OF SOLID MATERIALS

So far we have described the structures of metals and simple ionic solids. Next, we will look briefly at the other categories of solids: molecular solids, network solids, and amorphous solids.

Molecular Solids

Molecules such as H_2O and CO_2 are found in the solid state under appropriate conditions. In these cases, it is molecules, rather than atoms or ions, that pack in a regular fashion in a three-dimensional lattice [CD-ROM, Screen 13.14]. We have commented already on one such structure, the structure of ice (see Section 13.3, Figure 13.10).

The way molecules are arranged in a crystalline lattice depends on the shape of the molecules and the types of intermolecular forces. Molecules tend to pack in the most efficient manner and to align in ways that maximize intermolecular forces of attraction. Thus, the water structure was established to gain the maximum intermolecular attraction through hydrogen bonding. As illustrated in Figure 13.9, organic carboxylic acid molecules often assemble in the solid state as dimers, with two molecules linked by hydrogen bonding.

The greatest interest in structures of molecular solids centers on the structure of the molecules themselves. It is from structural studies on molecular solids that

A Closer Look

Packing Atoms and Ions into Unit Cells

Nature often does things as efficiently as possible. So, if we imagine the atoms or ions of a crystal lattice to be tiny, hard spheres, it is no surprise that they will be packed together to fill space as efficiently as possible. A simple cubic arrangement does not do this well; only 52% of the space is filled. The efficiency of packing in a body-centered unit cell is somewhat better, 68%. A face-centered cubic lattice, however, is the most efficient way to fill space; atoms or ions occupy 74% of the space within the fcc unit cell.

You can appreciate the idea of efficient packing by looking at a two-dimensional analogy — how marbles can be arranged in a layer. In one arrangement (Figure Aa), the marbles are at the corners of a square and each touches four other spheres. In the other (Figure Ab), each marble touches six others at the corners of a hexagon. In both arrangements, space is left between atoms in the layer, but the amount of unfilled space is less in Ab than in Aa. Thus, Ab is the more efficient way of packing spheres in two dimensions.

To fill three-dimensional space, layers of atoms are stacked one on top of another. If you start with the square arrangement (Figure Aa) and stack the next layer directly on top of the first, the arrangement resembles Escher's work (see Figure 13.23); the result is a solid with a simple cubic unit cell.

There are two more efficient ways of packing atoms or ions. They begin with a layer of atoms as in Figure Ab. Succeeding layers of atoms or ions are then stacked one on top of the other in two different ways. Depending on the stacking pattern, you will get either a *cubic close-packed (ccp)* or *hexagonal close-packed (hcp)* arrangement (Figure B). In the hcp arrangement (Figure Ba), additional layers of particles are placed above and below a given layer, fitting into the same depressions on either side of the middle layer. In a three-dimensional crystal, the layers repeat their pattern in the manner ABABAB. . . . Atoms in each A layer are directly above the ones in another A layer; the same holds for the B layers. In the ccp arrangement

 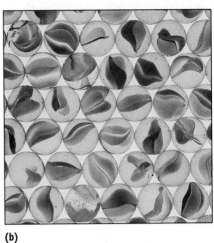

(a)   **(b)**

Figure A   Packing of spheres in one layer. **(a)** Marbles are arranged in a square pattern, and each marble contacts four others. **(b)** Each marble contacts six others at the corners of a hexagon. Pattern (b) is a more efficient packing arrangement than (a). *(Charles D. Winters)*

(Figure Bb), the atoms of the "top" layer rest in depressions in the middle layer, and those of the "bottom" layer are oriented opposite to those in the top layer. In a crystal, the pattern is repeated ABCABCABC. . . . By turning the whole crystal, you can see that the ccp arrangement is the face-centered cubic structure.

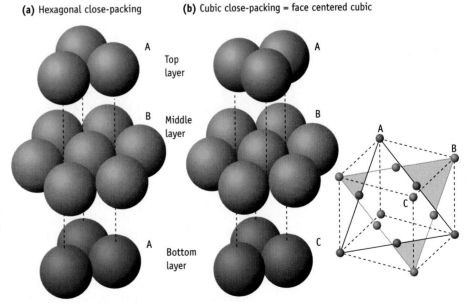

(a) Hexagonal close-packing   **(b)** Cubic close-packing = face centered cubic

Figure B   The most efficient ways to pack atoms or ions in crystalline materials, hexagonal close packing (hcp) and cubic close packing (ccp). The drawing on the right shows that cubic close packing results in the face-centered cubic lattice.

Figure 13.30   Silicates. Naturally oc-curring silicates include clear quartz, sand, sheets of mica, light green talc, and sandstone. All are network solids. *(Charles D. Winters)*

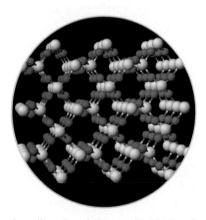

A small portion of the crystalline lattice of quartz, SiO_2. Each Si atom is bound, tetrahedrally, to four O atoms, each of which is connected to another Si atom. *(S. M. Young)*

● **Amorphous Solids**
It is not possible to identify a unit cell, a repeating structural unit that is reproduced in all directions, in the solid phase of an amorphous solid. Such solids do not have a regular, long-range molecular structure.

most of the information on molecular geometries, bond lengths, and bond angles discussed in Chapters 9 through 11 was assembled.

Network Solids

Network solids are substances composed entirely of networks of covalently bonded atoms. Common examples include two allotropes of carbon: graphite and diamond [← FIGURE 2.12]. Elemental silicon is a network solid with a diamond-like structure (← PAGE 408), and silicon dioxide, SiO_2, is also a network solid (Figure 13.30) [🖰 CD-ROM, Screen 13.15].

Diamonds have a low density ($d = 3.51$ g/cm^3), but they are also the hardest material and the best conductor of heat known. They are transparent to visible light, as well as to infrared and ultraviolet radiation. They are electrically insulating but behave as semiconductors with some advantages over silicon. What more could a scientist want in a material — except a cheap, practical way to make it!

In the 1950s, scientists at General Electric in Schenectady, New York, achieved something alchemists had sought for centuries, the synthesis of diamonds from carbon-containing materials, including wood or peanut butter. Their technique is to heat graphite to a temperature of 1500 °C in the presence of a metal, such as nickel or iron, and under a pressure of 50,000–65,000 atm. Under these conditions, the carbon dissolves in the metal and slowly forms diamonds (Figure 13.31). Over $500 million worth of diamonds are made this way annually. Most are used for abrasives and diamond-coated cutting tools.

Silicates, compounds composed of silicon and oxygen, represent an enormous class of chemical compounds [🖰 CD-ROM, Screen 13.16]. You know them in the form of sand, quartz, talc, and mica, or as a major constituent of rocks such as granite (see Figure 13.30). Details of several silicate structures are outlined in Chapter 21.

Most network solids are hard and rigid and are characterized by high melting and boiling points. These characteristics simply reflect the fact that a great deal of energy must be provided to break the covalent bonds in the lattice. The contrasting properties of CO_2 and SiO_2, compounds formed by the lightest members of Group 4A, illustrate this point. Carbon dioxide is a gas at room temperature (although it condenses to solid "dry ice" when cooled to −78 °C), whereas silicon dioxide, a solid, melts above 1600 °C. This difference in properties is explained by their structures. Silicon dioxide consists of tetrahedral silicon atoms linked together through covalent bonds to oxygen in a giant three-dimensional lattice. A high temperature is required to break the covalent bonds between silicon and oxygen and disrupt this stable structure. In contrast, relatively weak dispersion forces hold the molecules of CO_2 together in the solid state, and the molecules separate from one another at relatively low temperatures.

Amorphous Solids

A characteristic property of pure crystalline solids — metals, ionic solids, and molecular solids — is that they melt at a specific temperature. For example, water melts at 0 °C, aspirin at 135 °C, lead at 327.5 °C, and NaCl at 801 °C. Because of their specific and reproducible values, melting points are often used as a means of identifying chemical compounds.

Another property of these crystalline solids is that they form well-defined crystals, with smooth flat faces. When a sharp force is applied to a crystal, it will most often cleave to give smooth flat faces, and the resulting solid particles are smaller versions of the macroscopic crystal (Figure 13.32).

Many common solids, ones that we encounter everyday, do not have these properties, however. Glass is a good example. When glass is heated it softens over a wide temperature range, a property useful for artisans and craftsmen who can create beautiful and functional products for our enjoyment and use. We also know of a property of glass that we would just as soon not have: When glass breaks it leaves randomly shaped pieces. Other materials that behave similarly, with respect to both melting and breaking, are the polymers used extensively in our society. Polyethylene, nylon, and other common plastics have the same basic properties as glass.

The characteristics of these **amorphous solids** relate to their molecular structure. At the particulate level, amorphous solids do not have a regular structure. In many ways such substances look a lot like liquids. Unlike liquids, however, the forces of attraction are strong enough that movement of the molecules is restricted.

13.9   THE PHYSICAL PROPERTIES OF SOLIDS

The outward shape of a crystalline solid is a reflection of its internal structure. But what about the temperatures at which solids melt, their hardness, or their solubility in water? All of these and many other physical properties of solids are of interest to chemists, geologists, engineers, and others.

The **melting point** of a solid is the temperature at which the lattice collapses and the solid is converted to liquid. As in the case of the liquid-to-vapor transformation, melting requires energy, the **enthalpy of fusion** (in kilojoules per mole) [← CHAPTER 6].

Heat energy absorbed on melting = enthalpy of fusion = ΔH_{fusion} (kJ/mol)

Heat energy evolved on freezing = enthalpy of crystallization = $-\Delta H_{fusion}$ (kJ/mol)

Enthalpies of fusion can range from just a few thousand joules per mole to many thousands per mole (Table 13.7). A low melting temperature will certainly

Figure 13.31 A mixture of natural and synthetic industrial diamonds. The colors of diamonds may range from colorless to yellow, brown, or black. Poorer quality diamonds are used extensively in industry, mainly for cutting or grinding tools. Industrial-quality diamonds are produced synthetically at present by heating graphite, along with a metal catalyst, to 1200–1500 °C and a pressure of 65–90 kbars. New techniques are being sought, however, to manufacture them less expensively and to be able to coat surfaces with a diamond film. *(Sinclair Stammers/Science Photo Library/Photo Researchers, Inc.)*

(a) A salt crystal can be cleaved cleanly into smaller and smaller crystals that are duplicates of the larger crystal.

(b) Glass is an amorphous solid composed of linked silicon-oxygen tetrahedra. It has, however, no long-range order as in crystalline quartz.

(c) Glass can be molded and shaped into beautiful forms and, by adding metal oxides, can take on wonderful colors.

Figure 13.32   Crystalline and amorphous solids. *(Photos, Charles D. Winters)*

Table 13.7 • Melting Points and Enthalpies of Fusion of Some Elements and Compounds

Compound	Melting Point (°C)	Enthalpy of Fusion (kJ/mol)	Type of Interparticle Forces
Metals			
Hg	−39	2.29	Metal bonding; see Section 10.4
Na	98	2.60	
Al	660	10.7	
Ti	1668	20.9	
W	3422	35.2	
Molecular Solids: Nonpolar Molecules			
O_2	−219	0.440	Dispersion forces only
F_2	−220	0.510	
Cl_2	−102	6.41	
Br_2	−7.2	10.8	
Molecular Solids: Polar Molecules			
HCl	−114	1.99	All three HX molecules have dipole–dipole forces. Dispersion forces increase with size and molar mass.
HBr	−87	2.41	
HI	−51	2.87	
H_2O	0	6.02	Hydrogen bonding
Ionic Solids			
NaF	996	33.4	All ionic solids have extended ion–ion interactions. Note the general trend is the same as for lattice energies (see Table 9.3, page 332).
NaCl	801	28.2	
NaBr	747	26.1	
NaI	660	23.6	

mean a low value for the enthalpy of fusion, whereas high melting points are associated with high enthalpies of fusion. Figure 13.33 shows the enthalpy of fusion of the metals of the fourth through the sixth period. From this you can conclude (1) that metals that have notably low melting points, such as the alkali metals and mercury (mp = −39 °C), also have low enthalpies of fusion; and (2) that transition metals have high heats of fusion, with those of the third transition series being extraordinarily high. This parallels melting points for these elements. Tungsten, which has the highest melting point of all the known elements except for carbon, also has the highest enthalpy of fusion among the transition metals. These properties relate to the uses of this metal. Tungsten, for example, is used for the filaments in light bulbs; no other material has been found to work better since the invention of the light bulb in 1908.

Chapter Goals • Revisited

- Describe intermolecular forces and their effects and the importance of hydrogen bonding.
- Understand the properties of liquids.
- Understand cubic unit cells.
- Relate unit cells for ionic compounds to compound formulas.
- **Describe the properties of solids.**
- Understand the nature of phase diagrams.

The melting temperature of a solid can convey a great deal of information. Table 13.7 gives you some data for several basic types of substances: metals, polar and nonpolar molecules, and ionic solids. In general, nonpolar substances that form molecular solids have low melting points. Melting points increase in a series of related molecules, however, as the size and molar mass increase. This happens because dispersion forces are larger when the molar mass is larger. Thus, increas-

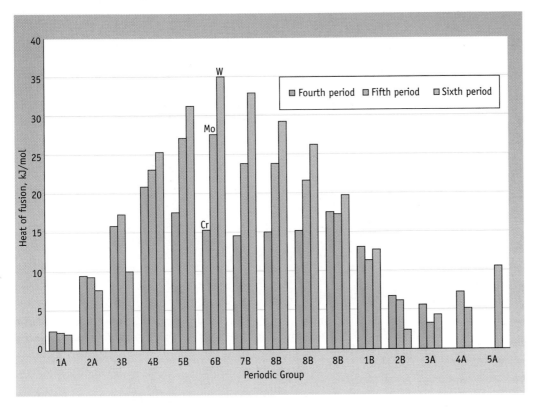

Figure 13.33   Heat of fusion of fourth, fifth, and sixth period metals. Heats of fusion range from 2 to 5 kJ/mol for Group 1A elements to 35.2 kJ/mol for tungsten. Notice that heats of fusion generally increase for B group metals on descending the periodic table.

ing amounts of energy are required to break down the intermolecular forces in the solid, a principle reflected in an increasing enthalpy of fusion.

The ionic compounds in Table 13.7 have higher melting points and higher enthalpies of fusion than molecular solids. This phenomenon is due to strong ion–ion forces in ionic solids, forces that are reflected in high lattice energies (see Section 9.3). Because ion–ion forces depend on ion size (as well as ion charge), there is a good correlation between lattice energy and the position of the metal or halogen in the periodic table [⬅ TABLE 9.3]. That is, as the cation size increases from Li⁺ to Cs⁺, the lattice energy declines for compounds of a given halide ion. A similar decrease occurs in lattice energy as the halide ion size increases from F⁻ to I⁻ when salts of a given cation are considered. As illustrated by the data in Table 13.7, the decline in lattice energy with increasing ion size is accompanied by a decrease in melting point and enthalpy of fusion.

Molecules can escape directly from the solid to the gas phase by **sublimation** (Figure 13.34).

$$\text{Solid} \longrightarrow \text{gas} \qquad \text{heat energy required} = \Delta H_{\text{sublimation}}$$

Sublimation, like fusion and evaporation, is an endothermic process. The heat energy required is called the enthalpy of sublimation. Water, which has a molar enthalpy of sublimation of 51 kJ/mol, can be converted from solid ice to water vapor quite readily. A good example of this is the sublimation of frost from grass and trees as night turns to day on a cold morning in the winter.

In the winter, frost collects on trees and grass during the night. As the sun rises, some frost melts to water, but much of it sublimes, turning directly from solid ice to water vapor. *(Charles D. Winters)*

Figure 13.34 Sublimation. This process is the conversion of a solid directly to its vapor. Here, iodine sublimes when heated. *(Charles D. Winters)*

Iodine sublimes
when heated
→

13.10 PHASE DIAGRAMS

Depending on the conditions of temperature and pressure, a substance exists as a gas, a liquid, or a solid. In addition, under certain specific conditions, two (or even three) states can coexist in equilibrium. It is possible to summarize this information in the form of a graph called a **phase diagram.** Phase diagrams are used to illustrate the relationship between phases of matter and pressure and temperature. A phase diagram for water is shown in Figure 13.35 [🖰 CD-ROM, Screen 13.17].

The lines in a phase diagram identify the conditions under which two phases exist at equilibrium. (Conversely, all points that do not fall on the lines in the figure represent conditions under which only one state exists.) Line *AD* represents

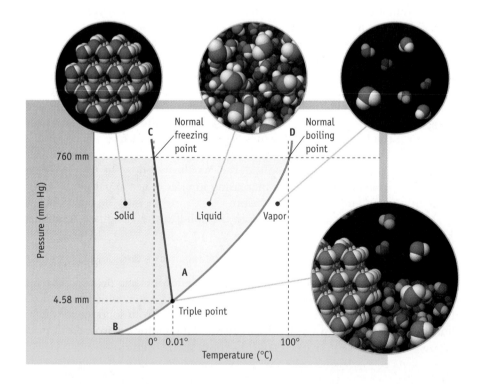

Figure 13.35 Phase diagram for water. The scale is intentionally exaggerated to be able to show the triple point and the negative slope of the line representing the liquid–solid equilibrium.

the combinations of temperature and pressure for which there can be equilibrium between the liquid and vapor phases. Line *AB* represents conditions for solid–vapor equilibrium, and line *AC* for liquid–solid equilibrium. The line from point *A* to point *D*, which represents the equilibrium between liquid and gaseous water, is the same curve plotted for water vapor pressure in Figure 13.16. Recall that the normal boiling point, 100 °C in the case of water, is the temperature at which the equilibrium vapor pressure is 760 mm Hg.

Point *A*, appropriately called the **triple point,** represents conditions under which all three phases can coexist in equilibrium. For water, the triple point is at $P = 4.58$ mm Hg and $T = 0.01$ °C.

The line *AC* shows the conditions of pressure and temperature at which solid–liquid equilibrium exists. (Because no vapor pressure is involved here, the pressure referred to is the external pressure on the liquid.) For water, this line has a negative slope. That is, the higher the external pressure, the lower the melting point (Figure 13.36a). The change for water is approximately 0.01 °C for each atmosphere increase in pressure. The negative slope of the water solid–liquid equilibrium line can be explained from our knowledge of the structure of water and ice. When pressure on an object is increased, common sense tells you the volume of the object will become smaller, giving the substance a higher density. Because ice is less dense than liquid water (due to the open lattice structure of ice), ice and water in equilibrium respond to increased pressure (at constant *T*) by melting ice to form more water because the same mass of water requires less volume.

Ice is slippery stuff, and it had long been assumed you can ski or skate on it because the surface melted slightly from the pressure of a skate blade or ski, or that surface melting occurred because of frictional heating. This has always seemed an unsatisfying explanation, however, because it does not seem possible that just standing or sliding on a piece of ice could produce a pressure or temperature high enough to cause sufficient melting. Recently, though, surface chemists have studied ice surfaces and have come up with a better explanation. They have concluded that water molecules on the surface of ice are vibrating rapidly. In fact, the outermost layer or two of water molecules is almost liquid-like. This makes the surface slippery, and it is for this reason we can ski on snow and skate on ice.

The basic features of the phase diagram for CO_2 (Figure 13.37) are the same as those for water, except that the CO_2 solid–liquid equilibrium line has a positive slope. Solid CO_2 is denser than the liquid, so solid CO_2 sinks to the bottom in a container of liquid CO_2. Notice also that solid CO_2 sublimes directly to CO_2 gas as it warms to room temperature (at 1 atm pressure). This is the reason CO_2 is called *dry ice;* it looks like water ice, but it does not melt.

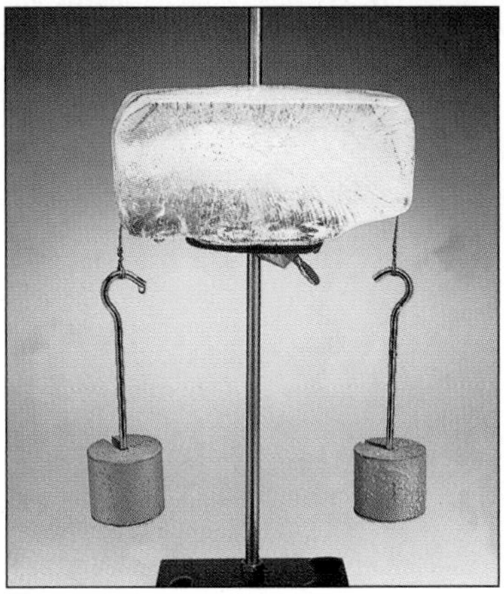

Figure 13.36 Ice. Ice melts under pressure. Here a thin wire is passed over a block of ice, and weights are suspended at either end. With time, the pressure exerted on the ice by the wire causes the ice to melt, and the wire eventually melts through the ice. *(Charles D. Winters)*

• Phase Diagrams

A phase diagram can be defined for any substance. Some substances can exist in many different solid phases.

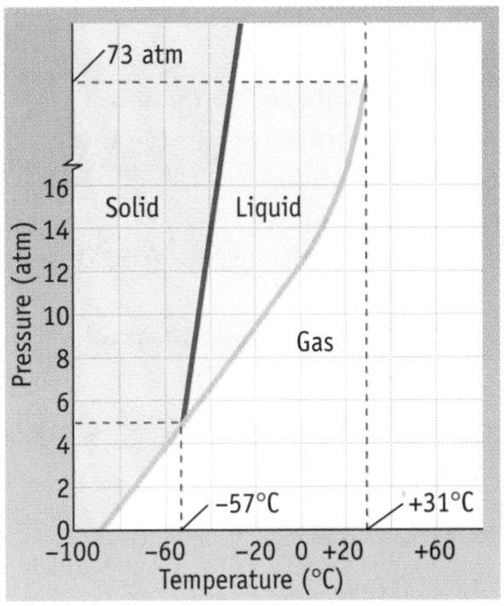

Figure 13.37 The phase diagram of CO_2. Notice in particular the positive slope of the solid–liquid equilibrium line.

In Summary

When you have finished studying this chapter, you should ask if you have met the chapter goals. In particular, you should be able to

- Use the kinetic-molecular theory to define the difference in solids, liquids, and gases (Section 13.1).
- Describe the different intermolecular forces in liquids and solids (Sections 13.2 and 13.4).
- Tell when two molecules can interact through a dipole–dipole attraction or when hydrogen bonding may occur. The latter occurs most strongly when H is attached to O, N, or F (Sections 13.2 and 13.3).

- Identify instances in which molecules interact only by induced dipoles (dispersion forces) (Section 13.2).
- Explain how hydrogen bonding affects the properties of water (Section 13.3).
- Explain the process of evaporation of a liquid or condensation of its vapor, and use the enthalpy of vaporization in calculations (Section 13.5).
- Define and use the concept of the equilibrium vapor pressure of a liquid and its relation to the boiling point of a liquid (Section 13.5).
- Describe the phenomena of the critical temperature, T_c, and critical pressure, P_c, of a substance (Section 13.5).
- Describe how intermolecular interactions affect (a) the cohesive forces between identical liquid molecules, (b) the energy necessary to break through the surface of a liquid (surface tension), and (c) the resistance to flow, or viscosity, of liquids (Section 13.5).
- Characterize different types of solids: metallic (such as copper), ionic (such as NaCl and CaF_2), molecular (for example, water and I_2), network (such as diamond), and amorphous (for example, glass and many synthetic polymers) (Sections 13.6–13.8 and Table 13.6).
- Describe the three types of cubic unit cells: primitive or simple cubic (sc); body-centered cubic (bcc); and face-centered cubic (fcc) (Section 13.6).
- Understand the relation between the unit cell for an ionic compound and its formula (Section 13.7).
- Define and use the enthalpy of fusion (Section 13.9).
- Identify the different points (triple point, normal boiling point, freezing point) and regions (solid, liquid, vapor) of a phase diagram, and use the diagram to evaluate the vapor pressure of a liquid or the relative densities of liquid and solid (Section 13.10).

Key Terms

Section 13.1
intermolecular forces

Section 13.2
dipole–dipole attraction
dipole/induced dipole interaction
polarization
polarizability
induced dipole/induced dipole force
London dispersion force

Section 13.3
hydrogen bond

Section 13.5
evaporation
molar enthalpy of vaporization, ΔH°_{vap}
condensation
dynamic equilibrium

equilibrium vapor pressure
volatility
Clausius-Clapeyron equation
normal boiling point
critical point
critical temperature, T_c
critical pressure, P_c
supercritical fluid
surface tension
capillary action
adhesive forces
cohesive forces

Section 13.6
unit cell
lattice points
crystal lattice
cubic unit cell
primitive or simple cubic (sc)

body-centered cubic (bcc)
face-centered cubic (fcc)

Section 13.7
octahedral holes
tetrahedral holes

Section 13.8
network solid
silicates
amorphous solids

Section 13.9
melting point
enthalpy of fusion
sublimation

Section 13.10
phase diagram
triple point

Key Equation

Equation 13.1 (page 528)

Clausius–Clapeyron equation relates the equilibrium vapor pressures, P_1 and P_2, of a volatile liquid to the molar enthalpy of vaporization (ΔH_{vap}) at a specified temperature, T_1 and T_2. (R is the universal constant, 8.314472 J/K · mol.)

$$\ln \frac{P_2}{P_1} = \frac{\Delta H_{vap}}{R} \left[\frac{1}{T_1} - \frac{1}{T_2} \right]$$

Study Questions

Questions with blue, bold-faced numbers have answers in Appendix O. Use the descriptions, tutorials, and simulations in Chapter 13 of the General Chemistry Interactive CD-ROM, *version 3.0.*

Reviewing Important Concepts

1. Name the types of forces that can be involved between two molecules and between ions and molecules.

2. Explain how a water molecule can interact with a molecule such as CO_2. What intermolecular force is involved?

3. What type of intermolecular force exists between water and iodine molecules?

4. Hydrogen bonding is most important when H is attached to certain very electronegative atoms. Which are those atoms? Why does this lead to strong hydrogen bonding?

5. Explain how hydrogen bonding leads to the decline in density of water from 4 °C to solid ice at 0 °C.

6. Why is the specific heat capacity of water so large compared with many other liquids? How does this affect weather?

7. When you exercise vigorously your body sweats. Why does this help your body cool down?

8. Explain, in terms of intermolecular forces, the trends in boiling points of the molecules EH_3, where E is a Group 5A element (see Figure 13.8).

9. Explain why the viscosity of an oil (a long-chain hydrocarbon) is so much greater than the viscosity of liquid benzene (C_6H_6).

10. What are the three types of cubic unit cells? Explain their similarities and differences.

11. Sketch the body-centered cubic unit cell of potassium metal. What is the net number of potassium atoms within the unit cell?

12. Explain why a cubic salt crystal can be cleaved cleanly into smaller cubes, whereas glass shatters into randomly shaped pieces.

13. What types of solids tend to have very high melting points?

14. Sketch the phase diagram for water. Label the normal boiling point, melting point, and triple point, and show what regions of temperature and pressure are appropriate to solid, liquid, and vapor.

15. Explain why the solid–liquid equilibrium line in the water phase diagram has a negative slope.

Practicing Skills

Intermolecular Forces

(See Examples 13.1–13.3 and CD-ROM Screens 13.3–13.7)

16. What intermolecular force(s) must be overcome to
    (a) Melt ice
    (b) Melt solid I_2
    (c) Convert liquid NH_3 to NH_3 vapor

17. What type of forces must be overcome within solid I_2 when I_2 dissolves in methanol, CH_3OH? What type of forces must be disrupted between CH_3OH molecules when I_2 dissolves? What type of forces exist between I_2 and CH_3OH molecules in solution?

18. What type of intermolecular force must be overcome in converting each of the following from a liquid to a gas?
    (a) Liquid O_2 (c) CH_3I (methyl iodide)
    (b) Mercury   (d) CH_3CH_2OH (ethanol)

19. What type of intermolecular forces must be overcome in converting each of the following from a liquid to a gas?
    (a) CO_2 (c) $CHCl_3$
    (b) NH_3 (d) CCl_4

20. Rank the following atoms or molecules in order of increasing strength of intermolecular forces in the pure substance. Which exist as a gas at 25 °C and 1 atm?
    (a) Ne   (c) CO
    (b) CH_4 (d) CCl_4

21. Rank the following in order of increasing strength of intermolecular forces in the pure substances. Which exist as a gas at 25 °C and 1 atm?
    (a) $CH_3CH_2CH_2CH_3$ (butane)
    (b) CH_3OH (methanol)
    (c) He

22. Which of the following compounds would be expected to form intermolecular hydrogen bonds in the liquid state?

    (a) CH_3OCH_3 (dimethyl ether)

    (b) CH_4

    (c) HF

    (d) CH_3CO_2H (acetic acid)

    (e) Br_2

    (f) CH_3OH (methanol)

23. Which of the following compounds would be expected to form intermolecular hydrogen bonds in the liquid state?

    (a) H_2Se

    (b) HCO_2H (formic acid)

    (c) HI

    (d) Acetone

Liquids

(See Examples 13.4 and 13.5 and CD-ROM Screens 13.8–13.11)

24. Ethanol, CH_3CH_2OH, has a vapor pressure of 59 mm Hg at 25 °C. What quantity of heat energy is required to evaporate 125 mL of the alcohol at 25 °C? The enthalpy of vaporization of the alcohol at 25 °C is 42.32 kJ/mol. The density of the liquid is 0.7849 g/mL.

25. The enthalpy of vaporization of liquid mercury is 59.11 kJ/mol. What quantity of heat is required to vaporize 0.500 mL of mercury at 357 °C, its normal boiling point? The density of mercury is 13.53 g/mL.

26. Answer the following questions using Figure 13.16:

    (a) What is the approximate equilibrium vapor pressure of water at 60 °C? Compare your answer with the data in Appendix G.

    (b) At what temperature does water have an equilibrium vapor pressure of 600 mm Hg?

    (c) Compare the equilibrium vapor pressures of water and ethanol at 70 °C. Which is higher?

27. Answer the following questions using Figure 13.16:

    (a) What is the equilibrium vapor pressure of diethyl ether at room temperature (approximately 20 °C)?

    (b) Place the three compounds in Figure 13.16 in order of increasing intermolecular forces.

    (c) If the pressure in a flask is 400 mm Hg and if the temperature is 40 °C, which of the three compounds (diethyl ether, ethanol, and water) are predominantly liquids and which are gases?

28. Assume you seal 1.0 g of diethyl ether (see Figure 13.16) in an evacuated 100.-mL flask. If the flask is held at 30 °C, what is the approximate gas pressure in the flask? If the flask is placed in an ice bath, does additional liquid ether evaporate or does some ether condense to a liquid?

29. Refer to Figure 13.16 as an aid in answering these questions:

    (a) You put some water at 60 °C in a plastic milk carton and seal the top very tightly so gas cannot enter or leave the carton. What happens when the water cools?

    (b) If you put a few drops of liquid diethyl ether on your hand, does it evaporate completely or remain a liquid?

30. Which member of each of the following pairs of compounds has the higher boiling point?

    (a) O_2 or N_2 (c) HF or HI

    (b) SO_2 or CO_2 (d) SiH_4 or GeH_4

31. Place the following four compounds in order of increasing boiling point:

    (a) SCl_2 (c) CH_4

    (b) NH_3 (d) CO

32. Vapor pressure curves for CS_2 (carbon disulfide) and CH_3NO_2 (nitromethane) are shown here.

    (a) What are the approximate vapor pressures of CS_2 and CH_3NO_2 at 40 °C?

    (b) What type of intermolecular forces exist in the liquid phase of each compound?

    (c) What is the normal boiling point of CS_2? Of CH_3NO_2?

    (d) At what temperature does CS_2 have a vapor pressure of 600 mm Hg?

    (e) At what temperature does CH_3NO_2 have a vapor pressure of 60 mm Hg?

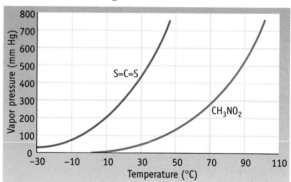

33. Answer each of the following questions with *increases, decreases,* or *does not change.*

    (a) If the intermolecular forces in a liquid increase, the normal boiling point of the liquid _____.

    (b) If the intermolecular forces in a liquid decrease, the vapor pressure of the liquid _____.

    (c) If the surface area of a liquid decreases, the vapor pressure _____.

    (d) If the temperature of a liquid increases, the equilibrium vapor pressure _____.

34. Can carbon monoxide (T_c = 132.9 K; P_c = 34.5 atm) be liquefied at or above room temperature? Explain briefly. (For comparison, critical temperature and pressure of carbon dioxide are 304.2 K and 72.8 atm, respectively.)

35. Methane (CH_4) cannot be liquefied at room temperature, no matter how high the pressure. Propane (C_3H_8), another alkane,

has a critical pressure of 42 atm and a critical temperature of 96.7 °C. Can this compound be liquefied at room temperature?

Metallic and Ionic Solids

(See Example 13.6, Exercise 13.8, and CD-ROM Screens 13.12–13.17)

36. Outline a two-dimensional unit cell for the pattern shown here. If the black squares are labeled A and the white squares are B, what is the simplest formula for a "compound" based on this pattern?

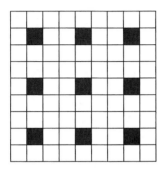

37. Outline a two-dimensional unit cell for the pattern shown here. If the black squares are labeled A and the white squares are B, what is the simplest formula for a "compound" based on this pattern?

38. One way of viewing the unit cell of perovskite was illustrated in Example 13.6. Another way is shown here. Prove that this view also leads to a formula of $CaTiO_3$.

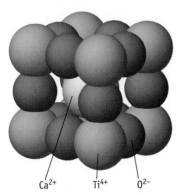

Ca^{2+} Ti^{4+} O^{2-}

39. Rutile, TiO_2, crystallizes in a structure characteristic of many other ionic compounds. How many formula units of TiO_2 are in the unit cell illustrated here? (The oxide ions marked by an x are wholly within the cell; the others are in the cell faces.)

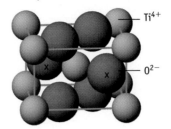

Ti^{4+}

O^{2-}

40. Cuprite is a semiconductor. Oxide ions are at the cube corners and in the cube center. Copper ions are wholly within the unit cell.
    (a) What is the formula of cuprite?
    (b) What is the oxidation number of copper?

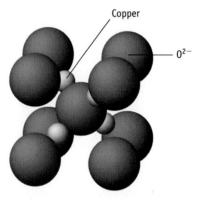

Copper

O^{2-}

41. The mineral fluorite, which is composed of calcium ions and fluoride ions, has the unit cell shown here.
    (a) What type of unit cell is described by the Ca^{2+} ions?
    (b) Where are the F^- ions located, in octahedral holes or tetrahedral holes?
    (c) Based on this unit cell, what is the formula of fluorite?

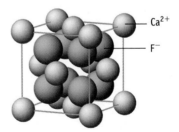

Ca^{2+}

F^-

Other Types of Solids

(See CD-ROM Screens 13.14–13.17)

42. A diamond unit cell is shown here.
    (a) How many carbon atoms are in one unit cell?
    (b) The unit cell can be considered as a cubic unit cell of C atoms with other C atoms in holes in the lattice. What type of unit cell is this (sc, bcc, fcc)? In what holes are other C atoms located, octahedral or tetrahedral holes?

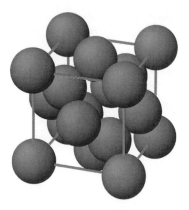

43. The structure of graphite is given in Figure 2.21.

    (a) What type of intermolecular bonding forces exist between the layers of six-member carbon rings?

    (b) Account for the lubricating ability of graphite, that is, why does graphite feel slippery? Why does pencil lead (which is graphite in clay) leave black marks on paper?

Physical Properties of Solids

44. Benzene, C_6H_6, is an organic liquid that freezes at 5.5 °C (see Figure 13.2) to form beautiful, feather-like crystals. How much heat is evolved when 15.5 g of benzene freezes at 5.5 °C? (The heat of fusion of benzene is 9.95 kJ/mol.) If the 15.5-g sample is remelted, again at 5.5 °C, what quantity of heat is required to convert it to a liquid?

45. The specific heat capacity of silver is 0.235 J/g · K. Its melting point is 962 °C, and its heat of fusion is 11.3 kJ/mol. What quantity of heat, in joules, is required to change 5.00 g of silver from solid at 25 °C to liquid at 962 °C?

Phase Diagrams and Phase Changes
(See CD-ROM Screen 13.17)

46. Consider the phase diagram of CO_2 in Figure 13.37.

    (a) Is the density of liquid CO_2 greater or less than that of solid CO_2?

    (b) In what phase do you find CO_2 at 5 atm and 0 °C?

    (c) What is the critical temperature of CO_2?

47. Use the phase diagram of xenon given here to answer the following questions:

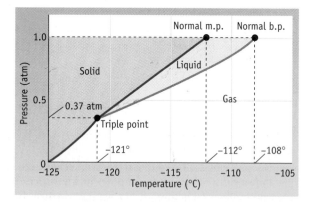

(a) In what phase is xenon found at room temperature and 1.0 atm pressure?

(b) If the pressure exerted on a xenon sample is 0.75 atm, and the temperature is −114 °C, in what phase does the xenon exist?

(c) If you measure the vapor pressure of a liquid xenon sample to be 380 mm Hg, what is the temperature of the liquid phase?

(d) What is the vapor pressure of the solid at −122 °C?

(e) Which is the denser phase — solid or liquid? Explain briefly.

48. Liquid ammonia, $NH_3(\ell)$, was once used in home refrigerators as the heat transfer fluid. The specific heat of the liquid is 4.7 J/g · K and that of the vapor is 2.2 J/g · K. The enthalpy of vaporization is 23.33 kJ/mol at the boiling point. If you heat 12 kg of liquid ammonia from −50.0 °C to its boiling point of −33.3 °C, allow it to evaporate, and then continue warming to 0.0 °C, how much heat energy must you supply?

49. If your air conditioner is more than several years old, it may use the chlorofluorocarbon CCl_2F_2 as the heat transfer fluid. The normal boiling point of CCl_2F_2 is −29.8 °C, and the enthalpy of vaporization is 20.11 kJ/mol. The gas and the liquid have specific heats of 117.2 J/mol · K and 72.3 J/mol · K, respectively. How much heat is evolved when 20.0 g of CCl_2F_2 is cooled from +40 °C to −40 °C?

General Questions

These questions are not designated as to type or location in the chapter. They may combine several concepts. More challenging questions are indicated by an underlined number.

50. Rank the following substances in order of increasing strength of intermolecular forces: (a) Ar, (b) CH_3OH, and (c) CO_2.

51. What types of intermolecular forces are important in the liquid phase of (a) C_2H_6 and (b) $(CH_3)_2CHOH$?

52. Construct a phase diagram for O_2 from the following information: normal boiling point, 90.18 K; normal melting point, 54.8 K; and triple point 54.34 K (at a pressure of 2 mm Hg). Very roughly estimate the vapor pressure of liquid O_2 at −196 °C. Is the density of liquid O_2 greater or less than that of solid O_2?

53. Cooking oil floats on top of water. From this observation, what conclusions can you draw regarding the polarity or hydrogen-bonding ability of molecules found in cooking oil?

54. Acetone, CH_3COCH_3, is a common laboratory solvent. It is usually contaminated with water, however. Why does acetone absorb water so readily? Draw molecular structures showing how water and acetone can interact. What intermolecular force(s) is (are) involved in the interaction?

55. If you place 1.0 L of ethanol (C_2H_5OH) in a room that is 3.0 m long, 2.5 m wide, and 2.5 m high, will all the alcohol evaporate? If some liquid remains, how much will there be? The vapor pressure of ethanol at 25 °C is 59 mm Hg, and the density of the liquid at this temperature is 0.785 g/cm³.

56. Liquid ethylene glycol, $HOCH_2CH_2OH$, is one of the main ingredients in commercial antifreeze. Do you predict its viscosity to be greater or less than that of ethanol, CH_3CH_2OH?

57. Select the substance in each of the following pairs that should have the higher boiling point:

    (a) Br_2 or ICl

    (b) Neon or krypton

    (c) CH_3CH_2OH (ethanol) or C_2H_4O (ethylene oxide, structure shown here)

    $$H_2C \!\!-\!\! CH_2$$
    $$\diagdown \diagup$$
    $$O$$

58. Liquid methanol, CH_3OH, is placed in a glass tube. Is the meniscus of the liquid concave or convex?

59. Use the vapor pressure curves illustrated here to answer the questions that follow.

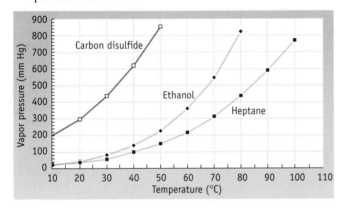

    (a) What is the vapor pressure of ethanol, C_2H_5OH, at 60 °C?

    (b) Considering only carbon disulfide (CS_2) and ethanol, which has the stronger intermolecular forces in the liquid state?

    (c) At what temperature does heptane (C_7H_{16}) have a vapor pressure of 500 mm Hg?

    (d) What are the approximate normal boiling points of each of the three substances?

    (e) At a pressure of 400 mm Hg and a temperature of 70 °C, is each substance a liquid, a gas, or a mixture of liquid and gas?

60. Account for these facts:

    (a) Although ethanol (C_2H_5OH)(bp, 80 °C) has a higher molar mass than water (bp, 100 °C), the alcohol has a lower boiling point.

    (b) Mixing 50 mL of ethanol with 50 mL of water produces a solution with a volume slightly less than 100 mL.

61. Rank the following molecules in order of *increasing* intermolecular forces: CH_3Cl, HCO_2H (formic acid), and CO_2.

62. Silver crystallizes in a face-centered cubic unit cell. Each side of the unit cell has a length of 409 pm. What is the radius of a silver atom? (*Hint:* Assume the atoms just touch each other on the diagonal across the face of the unit cell. That is, each face atom is touching the four corner atoms.)

63. Tungsten crystallizes in the unit cell shown here.

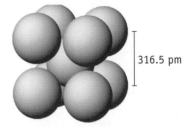

316.5 pm

    (a) What type of cell is this?

    (b) How many tungsten atoms occur per unit cell?

    (c) If the edge of the unit cell is 316.5 pm, what is the radius of a tungsten atom? (*Hint:* The W atoms touch each other along the diagonal line from one corner of the unit cell to the opposite corner of the unit cell.)

64. The noncubic unit cell shown here is for calcium carbide. How many calcium atoms and how many carbon atoms are in each unit cell? What is the formula of calcium carbide?

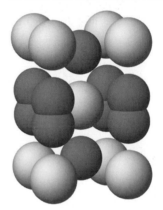

65. Why is it not possible for a salt with the formula M_3X (Na_3PO_4, for example) to have a face-centered cubic lattice of X anions with M cations in the lattice holes?

66. Can $CaCl_2$ have a unit cell like that of sodium chloride? Explain.

67. Rationalize the observation that $CH_3CH_2CH_2OH$, 1-propanol, has a boiling point of 97.2 °C, whereas a compound with the same empirical formula, methyl ethyl ether ($CH_3CH_2OCH_3$) boils at 7.4 °C.

68. Cite two pieces of evidence to support the statement that water molecules in the liquid state exert considerable attractive force on one another.

69. During thunderstorms in the Midwest, very large hailstones can fall from the sky. (Some are the size of golf balls!) To preserve some of these stones, we put them in the freezer compartment of a frost-free refrigerator. Our friend, who is a chemistry student, tells us to use an older model that is not frost-free. Why?

70. The photos illustrate an experiment you can do yourself. Place 10–20 mL of water in an empty soda can and heat the water to boiling. Using tongs or pliers, quickly turn the can over in a pan of cold water, making sure the opening in the

can is below the water level in the pan. Describe what happens and explain it in terms of the subject of this chapter.

(Charles D. Winters)

71. Rank the following compounds in order of increasing molar enthalpy of vaporization: CH_3OH, C_2H_6, HCl.

72. The following figure is a plot of vapor pressure versus temperature for dichlorodifluoromethane, CCl_2F_2. The heat of vaporization of the liquid is 165 kJ/g, and the specific heat of the liquid is about 1.0 J/g · K.

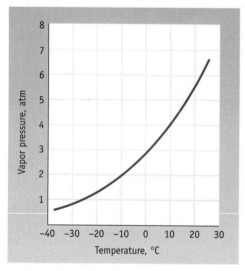

(a) What is the normal boiling point of CCl_2F_2?

(b) A steel cylinder containing 25 kg of CCl_2F_2 in the form of liquid and vapor is set outdoors on a warm day (25 °C). What is the approximate pressure of the vapor in the cylinder?

(c) The cylinder valve is opened and CCl_2F_2 vapor gushes out of the cylinder in a rapid flow. Soon, however, the flow becomes much slower, and the outside of the cylinder is coated with ice frost. When the valve is closed, and the cylinder is reweighed, it is found that 20 kg of CCl_2F_2 is still in the cylinder. Why is the flow fast at first? Why does it slow down long before the cylinder is empty? Why does the outside become icy?

(d) Which of the following procedures would be effective in emptying the cylinder rapidly (and safely)? (i) Turn the cylinder upside down and open the valve. (ii) Cool the cylinder to −78 °C in dry ice and open the valve.

(iii) Knock off the top of the cylinder, valve and all, with a sledge hammer.

73. A fluorocarbon, CF_4, has a critical temperature of −45.7 °C and a critical pressure of 37 atm. Under what conditions (if any) can this compound be a liquid at room temperature? Explain briefly.

74. The following data are the equilibrium vapor pressure of benzene, C_6H_6, at various temperatures.

Temperature (°C)	Vapor Pressure (mm Hg)
7.6	40.
26.1	100.
60.6	400.
80.1	760.

(a) What is the normal boiling point of benzene?

(b) Plot these data so that your plot resembles the one in Figure 13.16. At what temperature does the liquid have an equilibrium vapor pressure of 250 mm Hg? At what temperature is it 650 mm Hg?

(c) Calculate the molar enthalpy of vaporization for benzene using the the Clausius–Clapeyron equation (Equation 13.1, page 528).

75. Vapor pressure data is given here for octane, C_8H_{18}.

Temperature (°C)	Vapor Pressure (mm Hg)
13.6	25.
45.3	50.
127.2	75.
310.8	100.

Use the Clausius–Clapeyron equation (Equation 13.1, page 528) to calculate the molar enthalpy of vaporization of octane and its normal boiling point.

76. The very dense metal iridium has a face-centered cubic unit cell and a density of 22.56 g/cm³. Use this information to calculate the radius of an atom of the element.

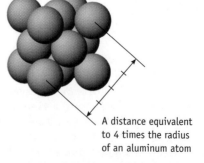

A distance equivalent to 4 times the radius of an aluminum atom

Pictured here is a face-centered cubic lattice of atoms. Measuring the distance shown allows a scientist to estimate the radius of an atom.

77. Calcium metal crystallizes in a face-centered cubic unit cell. The density of the solid is 1.54 g/cm³. What is the radius of a calcium atom? (*Hint:* see Study Question 76.)

78. The density of copper metal is 8.95 g/cm³. If the radius of a copper atom is 127.8 pm, is the copper unit cell simple cubic, body-centered cubic, or face-centered cubic? (*Hint:* see Study Question 76.)

79. Vanadium metal has a density of 6.11 g/cm³. Assuming the vanadium atomic radius is 132 pm, is the vanadium unit cell simple cubic, body-centered cubic, or face-centered cubic? (*Hint:* see Study Question 76.)

80. Iron has a body-centered cubic unit cell with a cell dimension of 286.65 pm. The density of iron is 7.874 g/cm³. Use this information to calculate Avogadro's number.

81. Calcium fluoride is the well-known mineral fluorite. It is known that each unit cell contains four Ca^{2+} ions and eight F^- ions and that the Ca^{2+} ions are arranged in an fcc lattice. The F^- ions fill all the tetrahedral holes. The edge of the CaF_2 unit cell is 5.46295×10^{-8} cm in length. The density of the solid is 3.1805 g/cm³. Use this information to calculate Avogadro's number.

82. Mercury and many of its compounds are dangerous poisons if breathed, swallowed, or even absorbed through the skin. The liquid metal has a vapor pressure of 0.00169 mm Hg at 24 °C. If the air in a small room is saturated with mercury vapor, how many atoms of mercury vapor occur per cubic meter?

83. You can get some idea of how efficiently spherical atoms or ions are packed in a three-dimensional solid by seeing how well circular atoms pack in two dimensions. Using the drawings shown here, prove that B is a more efficient way to pack circular atoms than A. A unit cell of A contains portions of four circles and one hole. In B, packing coverage can be calculated by looking at a triangle that contains portions of three circles and one hole. Show that A fills about 80% of the available space, whereas B fills closer to 90% of the available space.

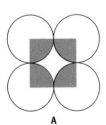
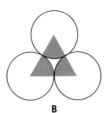

A   B

84. If a simple cubic unit cell is formed so that the spherical atoms or ions just touch one another along the edge, calculate the percentage of empty space within the unit cell. (Recall that the volume of a sphere is $\frac{4}{3}(\pi r^3)$, where r is the radius of the sphere.)

85. Two identical swimming pools are filled with uniform spheres of ice packed as closely as possible. The spheres in the first pool are the size of grains of sand; those in the second pool are the size of oranges. The ice in both pools melts. In which pool, if either, will the water level be higher? (Ignore any differences in filling space at the planes next to the walls and bottom.)

Using Electronic Resources

These questions refer to the General Chemistry Interactive CD-ROM, Version 3.0.

86. *Screen 13.4: Intermolecular Forces (2).* Click on *Dipole–Dipole Forces* and then on the table of *Molar Masses and Boiling Points.*

   (a) What is the connection between the strength of a compound's intermolecular forces and its boiling point?

   (b) What two factors control the strength of intermolecular forces, as reflected in the boiling point table?

87. *Screen 13.5 Intermolecular Forces (3)*

   (a) What does the term "polarizable" mean?

   (b) Rank the halogens F_2, Cl_2, and Br_2 in terms of their expected boiling points. Explain briefly.

88. *Screen 13.12 Solid Structures (Crystalline and Amorphous Solids)*

   (a) What is the difference between crystalline and amorphous solids?

   (b) Watch the simple cubic unit cell animation on this screen. How many unit cells make up the rotating structure?

   (c) Consider the shapes of the following solids and predict if they are crystalline solids or amorphous solids: table salt, ice (think of snow flakes), glass, and wood.

89. *Screen 13.13 Solid Structures (Ionic Solids)*

   (a) Watch the animation of a face-centered cubic (fcc) unit cell on this screen. Why is the sodium ion said to occupy "octahedral" holes?

   (b) Examine the structure of CaO using the model in the *Molecular Models* folder (in the *Solids* folder).

      1. Describe the structure. What type of lattice is described by the Ca^{2+} ions? What type of hole does the O^{2-} occupy in the lattice of Ca^{2+} ions?

      2. How is the formula of CaO related to its unit cell structure? (How many Ca^{2+} ions and how many O^{2-} ions are in one unit cell?)

      3. How is the CaO structure related to the NaCl structure?

   (c) Examine the structure of ZnS using the model in the *Molecular Models* folder (in the *Solids* folder). Describe the structure. What type of lattice is described by the Zn^{2+} ions (silver color)? What type of holes do the S^{2-} ions (yellow color) occupy in the lattice of Zn^{2+} ions? How is the formula of ZnS related to its unit cell structure?

   (d) Lead sulfide, PbS (commonly called galena), has the same formula as ZnS. Does it have the same solid structure? (See the model in the *Molecular Models* folder.) If different, how it is different? How is its unit cell related to its formula?

   (e) Examine the structure of CaF_2 using the model in the *Molecular Models* folder (in the *Solids* folder).

      1. Describe the structure. What type of lattice is described by the Ca^{2+} ions? What type of holes do the F^- ions occupy in the lattice of Ca^{2+} ions?

      2. How is the formula of CaF_2 related to its unit cell structure? (How many Ca^{2+} ions and how many F^- ions are in one unit cell?)

      3. How is the CaF_2 structure related to the ZnS structure?

14

Solutions and Their Behavior

The Killer Lakes of Cameroon

It was evening on Thursday, August 21, 1986. Suddenly people and animals around Lake Nyos in Cameroon, a small nation on the west coast of Africa, collapsed and died. By the next morning nearly 1700 people and hundreds of animals were dead. There was no apparent cause. There was no fire, no earthquake, nor any storm. What had caused this disaster?

Some weeks later the mystery was solved. Lake Nyos and nearby Lake Monoun are crater lakes, formed when cooled volcanic craters filled with water. But these lakes are different from other crater lakes. Lake Nyos was lethal because it contained an enormous amount of dissolved carbon dioxide. On that fateful evening in 1986 carbon dioxide erupted from the lake, shooting up about 260 ft. Because carbon dioxide gas is denser than air, the bubble of carbon dioxide hugged the ground. Moving about 45 miles an hour, it reached villages 12 miles away. The lake released

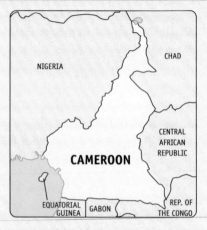

about a cubic kilometer of carbon dioxide. Vital oxygen was displaced, and people and animals were asphyxiated.

The carbon dioxide came from volcanic activity deep in the earth, and the gas dissolved in the water. In most lakes this would not be a problem because lake water "turns over" as the seasons change. In the autumn the top layer of water in a lake cools, the density increases, and the water sinks. This process continues, as warmer water comes to the surface and cooler water sinks. Any dissolved CO_2 would normally be lost in this turnover process.

Geologists found that the lakes in Cameroon are different. The chemocline, the boundary between deep water, rich in gas and minerals, and the fresh upper layer, stays intact. As carbon dioxide gas continues to enter the lake through vents in the bottom of the lake, the water is eventually saturated with gas. A minor disturbance — perhaps a small earthquake, a

▲ **Lake Nyos in Cameroon (western Africa), the site of a natural disaster.** In 1986, a huge bubble of CO_2 escaped from the lake and asphyxiated nearly 1700 people. *(Courtesy of George Kling)*

"This rare natural disaster is complex, little understood, and difficult to study."

G. Kling, University of Michigan

▲ **A soft drink is saturated with CO_2 gas.** When the bottle is opened, the gas erupts from the solution. *(Charles D. Winters)*

States has been working on a solution.

In early 2001 scientists lowered a pipe, about 200 m long, into the lake. Now the pressure of escaping carbon dioxide causes a jet of water to rise as high as 165 ft in the air. Over the course of a year, about 20 million m³ of gas will be released. Although this has been a successful first step, more gas must be removed to make the lake entirely safe, so additional vents are planned.

strong wind, or an underwater landslide — disturbed the delicate equilibrium in August 1986, and the carbon dioxide erupted from the water. This is what happens when you shake a bottle of carbonated beverage and then suddenly open the bottle.

Lake Nyos is still potentially deadly. Geologists estimate the lake contains 10.6–14.1 billion cubic feet (300–400 million m³) of carbon dioxide. This is about 16,000 times the amount in an average lake that size. So, a team of geologists from France and the United

▲ A pipe extending into the depths of Lake Nyos lets gas-rich water spout into the air. *(Courtesy of George Kling)*

Before You Begin

- Review solution concentrations (Section 5.8).
- Review thermochemistry and enthalpy of reactions (Chapter 6).
- Review intermolecular forces (Section 13.2).
- Review properties of liquids (Section 13.5).

557

14

Solutions and Their Properties When a solute dissolves in a solvent, the properties of the solvent are changed. These changes—called colligative properties—depend on the number of molecules or ions of solute relative to the number of molecules solvent.

Effects of dissolving compounds in water

A jar of pure water (left) and a jar containing ethylene glycol (HOC_2H_4OH) and water (right) were placed in a freezer. In a few hours the water froze (and on expansion broke the jar) but the water-glycol mixture did not. What is the explanation for this?

Salt is spread on snow and ice. The snow and ice melt. How can this be explained?

Beer is poured from a bottle into a glass and foam appears. Why does foam appear only when the solution is poured?

Effect of the environment on a biological system

If the egg, with its shell removed, is placed in pure water, the egg swells. Placed in a concentrated sugar solution, the egg shrivels. What is the explanation for these observations? (See Study Question 94, page 597)

(a) A fresh egg is placed in dilute acetic acid. The acid reacts with the $CaCO_3$ of the shell but leaves the egg's membrane intact.

(b) If the egg, with its shell removed, is placed in pure water, the egg swells.

(c) Placed in a concentrated sugar solution, the egg shrivels. What is the explanation for the observations in (b) and (c)?

Charles D. Winters

We come into contact with solutions every day: aqueous solutions of ionic salts, gasoline with additives to improve its properties, and household cleaners such as ammonia in water. We purposely make solutions. Adding sugar, flavoring, and sometimes CO_2 to water produces a palatable soft drink. Athletes drink commercial beverages with dissolved salts to match precisely salt concentrations in body fluids, thus allowing the fluid to be taken into the body more rapidly. In medicine, saline solutions (an aqueous solution with NaCl and other soluble salts) are infused into the body to replace lost fluids.

A **solution** is a homogeneous mixture of two or more substances in a single phase. It is usual to think of the component present in largest amount as the **solvent** and the other component as the **solute.** When you think of solutions, those that occur to you first probably involve a liquid as solvent. Some solutions, however, do not involve a liquid solvent at all; examples include the air you breathe (a solution of nitrogen, oxygen, carbon dioxide, water vapor, and other gases) and solid solutions such as 18 K gold, brass, bronze, and pewter. Although there are many types of solutions, the objective in this chapter is to develop an understanding of gases, liquids, and solids dissolved in liquid solvents.

Experience tells you that adding a solute to a pure liquid will change the properties of the liquid. Indeed, that is the reason some solutions are made. For instance, adding antifreeze to your car's radiator prevents the coolant from boiling in the summer and freezing in the winter. The changes that occur in the freezing and boiling points when a substance is dissolved in a pure liquid are two properties that we will examine in detail. These properties, as well as the osmotic pressure of a solution and changes in vapor pressure, are examples of colligative properties. **Colligative properties** are those that ideally depend only on the number of solute particles per solvent molecule and not on the identity of the solute.

Four major topics are covered in this chapter. First, because colligative properties depend on the relative number of solvent and solute particles in solution, convenient ways of describing solution concentration in these terms are required. Second, we consider how and why solutions form on the molecular and ionic level. This gives us some insight into the third topic, the colligative properties themselves. The chapter concludes with a brief discussion of colloids, mixtures that have properties intermediate between solutions and suspensions and that are important in many biological systems.

Copper chloride is added to water (*top*). Interactions between water molecules and Cu^{2+} and Cl^- ions allow the solid to dissolve. The ions are now sheathed with water molecules (*bottom*). *(Charles D. Winters)*

14.1   UNITS OF CONCENTRATION

To analyze the colligative properties of a solution, we need ways of measuring solute concentrations that reflect the number of molecules or ions of solute per molecule of solvent. Molarity, a concentration unit useful in stoichiometry calculations, does not work when dealing with colligative properties. Recall that molarity (M) is defined as the amount of solute per liter of solution:

$$\text{Molar concentration of solute } A = \frac{\text{amount of } A \text{ (mol)}}{\text{volume of solution (L)}}$$

Molarity does not allow us to identify the exact amount of solvent used to make the solution. This problem is illustrated in Figure 14.1. The flask on the right contains a 0.100 M aqueous solution of potassium chromate. It was made by adding enough water to 0.100 mol of K_2CrO_4 to make 1.000 L of solution. No mention is

---

Chapter Goals • Revisited

- **Learn additional methods of expressing solution concentration.**
- Understand the solution process.
- Understand and use the colligative properties of solutions.
- Describe colloids and their applications.

$V_{soln} > 1.00$ L $V_{soln} = 1.00$ L
V_{H_2O} added = 1.00 L V_{H_2O} added < 1.00 L
0.1 molal solution   0.1 molar solution

Figure 14.1   Preparing 0.100 molal and 0.100 molar solutions. In the flask on the right, 0.100 mol (19.4 g) of K_2CrO_4 was mixed with enough water to make 1.000 L of solution. (The volumetric flask was filled to the mark on its neck, indicating the volume is exactly 1.000 L. Slightly less than 1.000 L of water was added.) Exactly 1.000 kg of water was added to 0.100 mol of K_2CrO_4 in the flask on the left. Notice that the volume of solution is greater than 1.000 L. (The small pile of yellow solid in front of the flask is 0.100 mol of K_2CrO_4.) *(Charles D. Winters)*

● **Symbols, *m*, M, and *M***
The symbol for molality is a small italized *m*. The symbol for molarity is a regular capital M. You will often see an italicized *M*, which stands for molar mass.

● **Mole Fraction**
The mole fraction for components of solutions was first introduced for mixtures of gases, page 488.

made of how much solvent (water) was actually added. If 1.000 L of water had been added to 0.100 mol of K_2CrO_4, as illustrated by the flask on the left in Figure 14.1, the volume of solution would not be 1.000 L. It is slightly greater than a liter, a result that might have been anticipated because the solute molecules are expected to occupy some volume.

Several concentration units reflect the number of molecules or ions of solute per solvent molecule. Four such units are molality, mole fraction, weight percent, and parts per million [CD-ROM, Screen 14.2].

The molality (*m*) of a solution is defined as the amount of solute per kilogram of solvent.

$$\text{Molality of solute } (m) = \frac{\text{amount of solute (mol)}}{\text{kilograms of solvent}} \qquad \textbf{(14.1)}$$

The concentration of the K_2CrO_4 solution in the flask on the left side of Figure 14.1 is 0.100 molal (0.100 *m*). It was prepared from 0.100 mol (19.4 g) of K_2CrO_4 and 1.00 kg (1.00 L × 1.00 kg/L) of water:

$$\text{Molality of } K_2CrO_4 = \frac{0.100 \text{ mol}}{1.00 \text{ kg water}} = 0.100 \ m$$

Notice that different quantities of water were used to make the 0.100 M (0.100 molar) and 0.100 *m* (0.100 molal) solutions of K_2CrO_4. This means the *molarity and molality of a given solution cannot be the same* (although the difference may be negligibly small when the solution is quite dilute).

The **mole fraction,** *X*, of a solution component is defined as the amount of a given component of a mixture divided by the total amount of all of the components of the mixture. Mathematically this is represented as

$$\text{Mole fraction of } A \ (X_A) = \frac{n_A}{n_A + n_B + n_C + \cdots} \qquad \textbf{(14.2)}$$

Consider a solution that contains 1.00 mol (46.1 g) of ethanol, C_2H_5OH, in 9.00 mol (162 g) of water. Here the mole fraction of alcohol is 0.100 and that of water is 0.900:

$$X_{ethanol} = \frac{1.00 \text{ mol ethanol}}{1.00 \text{ mol ethanol} + 9.00 \text{ mol water}} = 0.100$$

$$X_{water} = \frac{9.00 \text{ mol water}}{1.00 \text{ mol ethanol} + 9.00 \text{ mol water}} = 0.900$$

Notice that the sum of the mole fractions of the components in the solution equals 1.000, a fact that is true for the solute and solvent in all solutions:

$$X_{water} + X_{ethanol} = 1.000$$

Weight percent is the mass of one component divided by the total mass of the mixture multiplied by 100%:

$$\text{Weight } \% \ A = \frac{\text{mass of } A}{\text{mass of } A + \text{mass of } B + \text{mass of } C + \cdots} \times 100\% \qquad \textbf{(14.3)}$$

The alcohol–water mixture has 46.1 g of ethanol and 162 g of water, so the total mass of solution is 208 g, and the weight % of alcohol is

$$\text{Weight \% ethanol} = \frac{46.1 \text{ g ethanol}}{46.1 \text{ g ethanol} + 162 \text{ g water}} \times 100\% = 22.2\%$$

Notice that if you know the weight percentage of a solute, you can also determine its mole fraction or molality (or vice versa) because the masses of solute and solvent are known.

Weight percent is a common unit in consumer products (Figure 14.2). Vinegar, for example, is an aqueous solution containing approximately 5% acetic acid and 95% water, and the label on a common household bleach lists its active ingredient as 5.25% sodium hypochlorite (NaOCl) and 94.75% inert ingredients.

Naturally occurring solutions are often very dilute. Environmental chemists, biologists, geologists, oceanographers, and others frequently use **parts per million (ppm)**. The unit ppm refers to relative amounts by mass; 1.0 ppm represents 1.0 g of a substance in a sample with a total mass of 1.0 million grams. Because water at 25 °C has a density of 1.0 g/mL, a concentration of 1.0 mg/L is equivalent to 1.0 mg of solute in 1000 g of water or to 1.0 g of solute in 1,000,000 g of water; that is, units of ppm and mg/L are approximately equivalent.

Figure 14.2   Weight percent. The composition of many common products is given in terms of weight percent. Here the label on household bleach indicates that it contains 5.25% sodium hypochlorite. *(Charles D. Winters)*

Example 14.1   Calculating Mole Fractions, Molality, and Weight Percent

Problem • Assume you add 1.2 kg of ethylene glycol, HOCH$_2$CH$_2$OH, as an antifreeze to 4.0 kg of water in the radiator of your car. What are the mole fraction, molality, and weight percent of the ethylene glycol?

Strategy • Calculate the amounts of glycol and water and then use Equations 14.1–14.3.

Solution • 1.2 kg of ethylene glycol (molar mass = 62.1 g/mol) is equivalent to 19 mol, and 4.0 kg of water represents 220 mol.

Mole fraction:

$$X_{\text{glycol}} = \frac{19 \text{ mol glycol}}{19 \text{ mol glycol} + 220 \text{ mol water}} = 0.080$$

Molality

$$\text{Molality} = \frac{19 \text{ mol glycol}}{4.0 \text{ kg}} = 4.8 \, m$$

Weight percentage:

$$\text{Weight \%} = \frac{1.2 \times 10^3 \text{ g glycol}}{1.2 \times 10^3 \text{ g glycol} + 4.0 \times 10^3 \text{ g water}} \times 100\%$$

$$= 23\%$$

Commercial antifreeze contains ethylene glycol, HOCH$_2$CH$_2$OH, an alcohol that is readily soluble in water. *(Charles D. Winters)*

Example 14.2   Parts per Million

Problem • You dissolve 560 g of NaHSO$_4$ in a swimming pool that contains 4.5×10^5 L of water at 25 °C. What is the sodium ion concentration in parts per million? (Sodium hydrogen sulfate is used to adjust the pH of the pool water because the anion, HSO$_4^-$, can furnish H$^+$(aq) to the solution.)

Strategy • First calculate the quantity of sodium ions (in grams) in 560 g of NaHSO$_4$. Use this mass of sodium and the volume to calculate milligrams per liter, which is equivalent to parts per million.

(Example continues on next page)

Solution •

Mass of Na$^+$

= 560 g NaHSO$_4$ $\left(\dfrac{1 \text{ mol NaHSO}_4}{120. \text{ g NaHSO}_4}\right)\left(\dfrac{1 \text{ mol Na}^+}{1 \text{ mol NaHSO}_4}\right)\left(\dfrac{23.0 \text{ g Na}^+}{1 \text{ mol Na}^+}\right)$

= 110 g Na$^+$

$$\text{Concentration of Na}^+ = \frac{110 \text{ g} \,(1000 \text{ mg/g})}{4.5 \times 10^5 \text{ L}}$$

$$= 0.24 \text{ mg/L or } 0.24 \text{ ppm}$$

Exercise 14.1 **Mole Fraction, Molality, and Weight Percentage**

If you dissolve 10.0 g of sugar (sucrose), C$_{12}$H$_{22}$O$_{11}$ (about one heaping teaspoonful), in a cup of water (250. g), what are the mole fraction, molality, and weight percent of sugar?

Exercise 14.2 **Parts Per Million**

Sea water has a sodium ion concentration of 1.08 × 10^4 ppm. If the sodium is present in the form of dissolved sodium chloride, what mass of NaCl is in each liter of sea water? Sea water is denser than pure water because of dissolved salts. Assume that the density of sea water is 1.05 g/mL.

14.2 SOLUTION PROCESS

If solid CuCl$_2$ is added to a beaker of water, the salt will begin to dissolve [← PAGE 148]. The amount of solid diminishes, and the concentration of Cu^{2+}(aq) and Cl$^-$(aq) in solution increases. If we continue to add CuCl$_2$, however, a point will eventually be reached when no additional CuCl$_2$ seems to dissolve. The concentrations of Cu^{2+}(aq) and Cl$^-$(aq) will not increase further, and any additional solid CuCl$_2$ added after this point will simply remain as a solid at the bottom of the beaker. We say that the solution is **saturated** [🖰 CD-ROM, Screen 14.2].

Although no change is observed on the macroscopic level when the solution is saturated, things are different on the particulate level. The process of dissolving is still going on, with Cu^{2+}(aq) and Cl$^-$(aq) ions leaving the solid state and entering solution. Concurrently, however, a second process is occurring; this is the formation of solid CuCl$_2$(s) from Cu^{2+}(aq) and Cl$^-$(aq). The rates at which CuCl$_2$ is dissolving and reprecipitating are equal in a saturated solution, so that no net change takes place on the macroscopic level.

This is an another example of equilibrium in chemistry, and we can describe the situation in terms of an equation with substances linked by a set of double arrows (⇌).

When CuCl$_2$ is dissolved in water, the Cu^{2+} and Cl$^-$ ions are cloaked in water molecules.

• **Unsaturated**

The term "unsaturated" is used when referring to solutions with concentrations of solute that are less than that of a saturated solution.

$$\text{CuCl}_2(s) \xrightleftharpoons{\text{H}_2\text{O}} \text{Cu}^{2+}(aq) + 2 \text{ Cl}^-(aq)$$

Recall that equilibrium systems were encountered earlier with changes of state [← SECTION 13.5]. In equilibria involving two states of matter, the description was very similar to what we are seeing for a saturated solution: No changes are observable at the macroscopic level, but two opposing processes go on at the particulate level at the same rate.

A saturated solution gives us a way to define precisely the solubility of a solid in a liquid. **Solubility** is the concentration of solute in equilibrium with undissolved solute in a saturated solution [💿 CD-ROM, Screen 14.2]. The solubility of $CuCl_2$, for example, is 70.6 g in 100 mL of water at 0 °C. If we added 100.0 g of $CuCl_2$ to 100 mL of water at 0 °C, we can expect 70.6 g to dissolve, and 29.4 g of solid to remain.

Liquids Dissolving in Liquids

If two liquids mix to an appreciable extent to form a solution, they are said to be **miscible.** In contrast, **immiscible** liquids do not mix to form a solution; they exist in contact with each other as separate layers [⬅ FIGURE 13.5].

The polar compound ethanol (C_2H_5OH) dissolves in water, which is also a polar compound. In fact, ethanol and water are miscible in all proportions. The nonpolar liquids octane (C_8H_{18}) and carbon tetrachloride (CCl_4) are also miscible in all proportions. On the other hand, neither C_8H_{18} nor CCl_4 is miscible with water. Observations like these have led to a familiar rule of thumb described in Section 13.2: *like dissolves like.* That is, two or more nonpolar liquids frequently are miscible, just as are two or more polar liquids.

What is the molecular basis for the "like dissolves like" guideline? In pure water and pure ethanol, the major force between molecules is hydrogen bonding involving —OH groups. When the two liquids are mixed, hydrogen bonding between ethanol and water molecules also occurs and assists in the solution process. Molecules of pure octane or pure CCl_4, both of which are nonpolar, are held together in the liquid phase by dispersion forces [⬅ SECTION 13.2]. When these nonpolar liquids are mixed, the energy associated with these forces of attraction is similar in value to the energy due to the forces of attraction between octane and CCl_4

• **Solubility Data**

Quantitative data on solubility for many compounds are listed in chemical handbooks. (Data are usually given as mass in 100 ml of solvent.) It is from these data that solubility rules such as those given in Figure 5.1 were created.

• **Alcohol–Water Solutions**

Beer, wine, and other alcoholic beverages contain amounts of alcohol ranging from just a few percent to more than 50%. Ethanol commonly used in laboratories has a concentration of 95% ethanol and 5% water.

Chapter Goals • Revisited

- Learn additional methods of expressing solution concentration.
- **Understand the solution process.**
- Understand and use the colligative properties of solutions.
- Describe colloids and their applications.

A Closer Look

Supersaturated Solutions

Although at first glance it may seem a contradiction, it is possible to have a solution in which there is *more* dissolved solute than the amount in a saturated solution. Such solutions are referred to as *supersaturated* solutions. Supersaturated solutions are unstable (like Lake Nyos, page 556), and the excess solid eventually crystallizes from the solution until the equilibrium concentration of the solute is reached.

The solubility of substances often decreases if the temperature is lowered. Supersaturated solutions are usually made by preparing a saturated solution at a given temperature and then carefully cooling it. If the rate of crystallization is slow, the solid may not precipitate when the solubility is exceeded. Going to still lower temperatures results in a solution that has more solute than the amount defined by equilibrium conditions; it is supersaturated.

When disturbed in some manner, a supersaturated solution moves toward equilibrium by precipitating solute. This can occur rapidly and often with the evolution of heat. In fact, supersaturated solutions are used in "heat packs" to apply heat to injured muscles (see page 235). When crystallization of sodium acetate ($NaCH_3CO_2$) from a supersaturated solution in a heat pack is initiated, the temperature of the heat pack rises to close to 50 °C, and crystals of solid sodium acetate are detectable inside the bag.

When a supersaturated solution is disturbed, ▶ the dissolved salt (here sodium acetate, $NaCH_3CO_2$) rapidly crystallizes. See a video of this process on Screen 14.2 of the *General Chemistry Interactive CD-ROM*, Version 3.0. (*Charles D. Winters*)

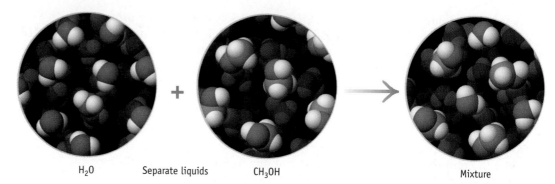

H₂O   Separate liquids   CH₃OH   Mixture

Figure 14.3   Driving the solution process — entropy. When two similar liquids — here water and methanol — are mixed, the molecules are intermingled. The mixture has a less ordered arrangement of molecules than the separate liquids. It is this disordering process that largely drives solution formation. See Chapter 19.

molecules. Thus, little or no energy change occurs when octane–octane and CCl_4–CCl_4 attractive forces are replaced with octane–CCl_4 forces. The solution process is expected to be nearly energy-neutral. So, why do they mix? The answer lies deeper in thermodynamics. As you shall see in Chapter 19, processes that move to less orderly arrangements tend to occur. The tendency is measured by a thermodynamic function called *entropy* (Figure 14.3).

In contrast, polar and nonpolar liquids usually do not mix to an appreciable degree; placed together in a container they separate into two distinct layers (Figure 14.4). The rationale for this behavior is complicated. Experimental data show that the enthalpy of mixing of dissimilar liquids is also near zero, so the energetics of the process is not the primary factor. Apparently interposing nonpolar molecules into a polar solvent such as water causes the internal structure of water at the molecular level to become more ordered. Forming a more ordered system disfavors the process of mixing.

● **Entropy and the Solution Process**
Although the energetics of solution formation are important, it is generally accepted that the more important contributor is the entropy of mixing. As you shall see in Chapter 19, entropy favors the formation of less ordered systems such as solutions. See also T. P. Silverstein: "The real reason why oil and water don't mix." *Journal of Chemical Education*, Vol. 75, pp. 116–118, 1998.

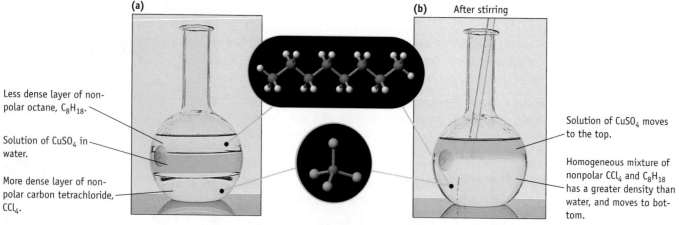

(a)

(b)   After stirring

Less dense layer of non-polar octane, C_8H_{18}.

Solution of $CuSO_4$ in water.

More dense layer of non-polar carbon tetrachloride, CCl_4.

Solution of $CuSO_4$ moves to the top.

Homogeneous mixture of nonpolar CCl_4 and C_8H_{18} has a greater density than water, and moves to bottom.

Figure 14.4   Miscibility. (a) The colorless, denser bottom layer is nonpolar carbon tetrachloride, CCl_4. The blue middle layer is a solution of $CuSO_4$ in water, and the colorless, less dense top layer is nonpolar octane (C_8H_{18}). This mixture was prepared by carefully layering one liquid on top of another, without mixing. **(b)** After stirring the mixture, the two nonpolar liquids form a homogeneous mixture. This layer of mixed liquids is under the water layer because the mixture of CCl_4 and C_8H_{18} has a greater density than water. *(Charles D. Winters)*

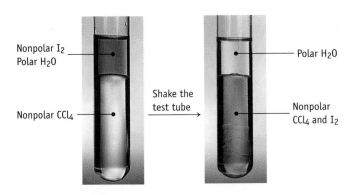

Solids Dissolving in Liquids

The "like dissolves like" guideline also holds for solids dissolving in liquids. For example, I$_2$ is a nonpolar molecular solid in which the I$_2$ molecules are held together by intermolecular dispersion forces. Iodine is expected to dissolve to a greater extent in a nonpolar liquid such as CCl$_4$ than in polar water (Figure 14.5).

Sucrose, a sugar, is also a molecular solid. The molecule is capable of hydrogen bonding, however, because it contains numerous —OH groups. Sucrose is readily soluble in water, a fact that we know well because of its use to sweeten beverages. On the other hand, sugar is not soluble in CCl$_4$ or other nonpolar liquids.

Network solids include graphite, diamond, and quartz sand (SiO$_2$). Experience tells you they do not dissolve readily in water. After all, where would all the beaches be if sand dissolved readily in water? What if the diamond in a ring dissolved when you washed your hands? The normal covalent chemical bonding in network solids is simply too strong to be replaced by weaker hydrogen bonding attraction to water dipoles.

Many ionic solids are soluble in water, largely owing to the attraction between water dipoles and the positive and negative ions of the compound. On entering solution, the ions are surrounded by water molecules, that is, they are **hydrated.**

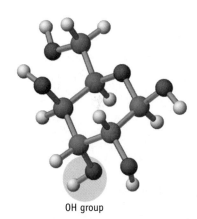

OH group

Sucrose, or cane sugar, has a number of —OH groups on each molecule, groups that allow it to form hydrogen bonds with water molecules.

Heat of Solution

Both ammonium nitrate and sodium hydroxide dissolve readily in water, but the solution becomes colder when NH$_4$NO$_3$ dissolves, and it warms up when NaOH dissolves (Figure 14.6). Why do energy changes occur when substances dissolve? Let us examine the solution process at the particulate level and then ask what thermodynamic information is available [⏣ CD-ROM, Screen 14.4].

The Solution Process at the Particulate Level

To understand the energetics of the solution process let us estimate the energy involved when potassium fluoride, KF, dissolves in water. To do this, we shall focus on the attractive forces between the ions in the crystal lattice and those between the ions and water molecules (Figure 14.7). We can obtain an approximation of the net energy change during the solution process and have some insight into what affects the heat of solution.

For KF, let us imagine the crystal lattice first comes apart into K$^+$ and F$^-$ ions in the gas phase.

$$KF(s) \longrightarrow K^+(g) + F^-(g) \qquad \Delta E = +821 \text{ kJ}$$

The energy involved is the negative of the crystal lattice energy (Table 9.3).

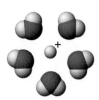

Water surrounding a cation

Water surrounding an anion

A Closer Look

Hydrated Salts

Solid salts with waters of hydration are common. Their formulas are given by appending a specific number of water molecules at the end of the formula as in $BaCl_2 \cdot 2\,H_2O$. Sometimes the water molecules simply fill in empty spaces in a crys-

talline lattice, but often the cation in these salts is directly associated with water molecules. For example, the compound $CoCl_2 \cdot 6\,H_2O$ is better written as $[Co(H_2O)_4Cl_2] \cdot 2\,H_2O$. Four of the six water molecules are associated with the Co^{2+} ion

by ion–dipole attractive forces; the remaining two water molecules are in the lattice. Common examples of hydrated salts are listed in the table.

Compound	Common Name	Uses
$Na_2CO_3 \cdot 10\,H_2O$	Washing soda	Water softener
$Na_2S_2O_3 \cdot 5\,H_2O$	Hypo	Photography
$MgSO_4 \cdot 7\,H_2O$	Epsom salt	Cathartic, dyeing and tanning
$CaSO_4 \cdot 2\,H_2O$	Gypsum	Wallboard
$CuSO_4 \cdot 5\,H_2O$	Blue vitriol	Biocide

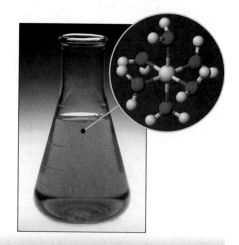

Hydrated cobalt(II) chloride, $CoCl_2 \cdot 6\,H_2O$. In the solid state the compound is best described by ▶ the formula $[Co(H_2O)_4Cl_2] \cdot 2\,H_2O$. The cobalt(II) ion is surrounded by four water molecules and two chloride ions in an octahedral arrangement. In water, the ion is completely hydrated, now being surrounded by six water molecules. Cobalt(II) ions and water molecules interact by ion-dipole forces. This is an example of a coordination compound, a class of compounds discussed in detail in Chapter 22. *(Charles D. Winters)*

Next, imagine that the ions are surrounded by water molecules owing to the attractive forces between the ions and water dipoles.

$$K^+(g) + F^-(g) \longrightarrow K^+(aq) + F^-(aq) \qquad \Delta H_{hydr} = -819\ KJ$$

The enthalpy change involved here is the enthalpy or heat of hydration.

The net energy change for the solution process, the **heat of solution**, ΔH_{soln}, is the sum of the energies of these two steps.

$$KF(s) \longrightarrow K^+(aq) + F^-(aq) \qquad \text{Estimated } \Delta H_{soln} = +2\ kJ$$

Figure 14.6   Heat of solution.
(a) Dissolving NaOH in water is a strongly exothermic process. **(b)** When NH_4NO_3 is dissolved in water, the temperature of the system drops. If a flask containing the mixture is set on a few drops of water on a piece of wood, the water freezes, and the flask is stuck to the board. [Here a mixture of solid $Ba(OH)_2 \cdot 8\,H_2O$ and solid NH_4NO_3 reacted to form $Ba(NO_3)_2$ and NH_3 and release water. When excess NH_4NO_3 dissolved in this small amount of water, the temperature of the system dropped because of the endothermic heat of solution of NH_4NO_3 in water.] *(Charles D. Winters)*

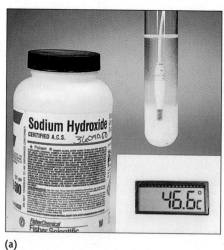

(a)

(b)

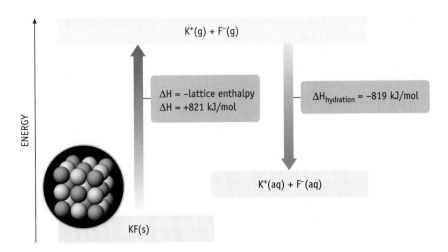

Figure 14.7   A model for estimating energy changes on dissolving. A very rough estimate of the magnitude of the energy change on dissolving an ionic compound in water is achieved by imagining it as occuring in two steps at the particular level. Here KF is first separated into cations and anions in the gas phase with an expenditure of 821 kJ per mol of KF. These ions are then hydrated, with $\Delta H_{hydration}$ estimated to be −819 kJ. Thus, the net estimated energy change is about zero for this compound.

Several aspects of this result are of interest. First, we see that the energy required to break down the KF crystal lattice is essentially returned by the energy of attraction between the ions and the polar water molecules. In some cases, the value of ΔH_{soln} calculated by this method is decidely endothermic, whereas it is exothermic in other cases (Table 14.1).

Another aspect of the solution process that is of interest is that, for simple salts, Table 14.1 shows a correlation between the value of estimated ΔH_{soln} and water solubility. As ΔH_{soln} becomes more negative, the solubility increases. Rubidium fluoride is very soluble in water and has an exothermic enthalpy of solution. Silver chloride is quite insoluble and has a very positive value of ΔH_{soln}. Solubility is apparently favored when the energy required to break down the lattice of the solid (the negative of the lattice energy) is smaller or roughly equal to the energy given off when the ions are hydrated (ΔH_{hydr}).

Recall that the energy of a system with oppositely charged particles — here an ion and a polar molecule — depends on the charge on the particles and the distance between the particles [← SECTION 9.3]. We can assess the heat of hydration of an ion using a similar logic. The amount of energy evolved in hydrating an ion ($\Delta H_{ion\ hydr}$) will depend on

- *The distance between the ion and the dipole:* The closer the ion and dipole, the stronger the attraction.
- *The charge on the ion:* The higher the ion charge, the stronger the attraction.

● **Estimating Heats of Solution**
The particulate level model presented here can only give us very rough estimates of heats of solution. What is important about the model is that it gives us some insight into the factors that control the heat of solution: the forces owing to ion-ion attraction in the crystal lattice, and the ion-dipole forces that come into play in the solution process.

Table 14.1 ● **Estimated Enthalpy of Solution and Water Solubility of Some Ionic Compounds**

Compound	Estimated ΔH_{soln} (kJ/mol)	Solubility in H_2O (g/100 mL)
AgCl	+61	0.000089 (10 °C)
NaCl	+26	35.7 (0 °C)
LiF	+32	0.3 (18 °C)
KF	+2	92.3 (18 °C)
RbF	−3	130.6 (18 °C)

• *The polarity of the solvating molecule:* The greater the magnitude of the dipole, the stronger the attraction.

The enthalpy change associated with the hydration of a cation, $\Delta H_{\text{ion hydr}}$, can be substantial. This quantity cannot be measured directly for an individual ion, but values can be estimated. For the sodium ion, for example, the hydration energy is estimated to be −405 kJ/mol.

$$Na^+(g) + x\, H_2O(\ell) \longrightarrow [Na(H_2O)_x]^+(aq)\ (x\text{ probably} = 6) \qquad \Delta H_{\text{ion hydr}} = -405\text{ kJ}$$

The trend in the enthalpy of hydration of the alkali metal cations illustrates this property.

Cation	Ion Radius (pm)	Enthalpy of Ion Hydration (kJ/mol)
Li^+	78	−515
Na^+	98	−405
K^+	133	−321
Rb^+	149	−296
Cs^+	165	−263

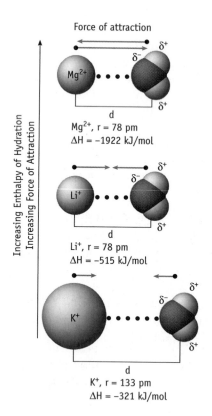

It is interesting to compare these values with the enthalpy of hydration of the H^+ ion, estimated to be −1090 kJ/mol. This extraordinarily large value is due to the tiny size of the H^+ ion.

Example 14.3) Hydration Energy

Problem • Explain why the enthalpy of hydration of Na^+ (-405 kJ/mol) is somewhat more negative than that of Cs^+ (-263 kJ/mol), whereas that of Mg^{2+} is much more negative (-1922 kJ/mol) than that of either Na^+ or Cs^+.

Strategy • The strength of ion–dipole attractions depends directly on the size of the ion charge and the magnitude of the dipole, and inversely on the distance between the ion and the polar molecule. So, in addition to the ion charge, we need ion sizes from Figure 8.13.

Solution • The relevant ion sizes are $Na^+ = 98$ pm, $Cs^+ = 165$ pm, and $Mg^{2+} = 79$ pm. From these values we predict that the

distances between the center of the positive charge on the metal ion and the negative side of the water dipole vary in this order: $Mg^{2+} < Na^+ < Cs^+$. The hydration energy varies in the reverse order (the hydration energy of Mg^{2+} being the most negative value). Notice also that Mg^{2+} has a $+2$ charge, whereas the other ions are $+1$. The greater charge on Mg^{2+} leads to a greater force of ion–dipole attraction than for the other two ions, which only have a $+1$ charge. This means the enthalpy of hydration for Mg^{2+} is much more negative than the hydration enthalpies of the other two ions.

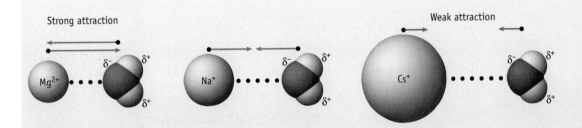

Exercise 14.3) Hydration Energy

Which should have the more negative enthalpy of hydration, F^- or Cl^-? Explain briefly.

Heat of Solution and Thermodynamics

Heats of solution are easily measured using a calorimeter. This is usually done in an open system, such as the coffee cup calorimeter described in Section 6.6, and for an experiment run under standard conditions the resulting measurement produces a value for the enthalpy of solution, $\Delta H°_{soln}$. For solutions, standard conditions refer to a concentration of 1 *m*.

Tables of thermodynamic values often include values for the heats of formation of aqueous solutions of salts. For example, a value of $\Delta H°_f$ for NaCl(aq) of -407.3 kJ/mol is listed in Appendix L. This value refers to the formation of a 1 *m* solution of NaCl from the elements. It may be considered to involve the enthalpies of two steps: (1) the formation of NaCl(s) from the elements Na(s) and Cl_2(g) in their standard states, and (2) the formation of a 1 *m* solution by dissolving solid NaCl in water:

Formation of NaCl(s)	$Na(s) + \frac{1}{2}Cl_2(g) \longrightarrow NaCl(s)$	$\Delta H°_f = -411.2$ kJ/mol
Dissolving NaCl	$NaCl(s) \longrightarrow NaCl(aq, 1\ m)$	$\Delta H°_{soln} = +3.9$ kJ/mol
Net process	$Na(s) + \frac{1}{2}Cl_2(g) \longrightarrow NaCl(aq, 1\ m)$	$\Delta H°_f = -407.3$ kJ/mol

	Table 14.2 • Data for Calculating Enthalpy of Solution	

Compound	$\Delta H_f^\circ(s)$ (kJ/mol)	$\Delta H_f^\circ(aq, 1\ m)$ (kJ/mol)
LiF	−616.9	−611.1
NaF	−573.6	−572.8
KF	−568.6	−585.0
RbF	−557.7	−583.8
LiCl	−408.6	−445.6
NaCl	−411.2	−407.3
KCl	−436.7	−419.5
RbCl	−435.4	−418.3
NaOH	−425.9	−469.2
NH_4NO_3	−365.6	−339.9

Heats of solution, ΔH_{soln}° (Table 14.2), can be calculated using ΔH_f° data. The solution process for NaCl(s), for example, is represented by the equation

$$NaCl(s) \longrightarrow NaCl(aq, 1\ m)$$

The energy of this process is calculated using Equation 6.6:

$$\Delta H_{soln}^\circ = \sum [\Delta H_f^\circ (products)] - \sum [\Delta H_f^\circ (reactants)]$$
$$= \Delta H_f^\circ[NaCl(aq)] - \Delta H_f^\circ[NaCl(s)]$$
$$= -407.3\ kJ/mol - (-411.2\ kJ/mol) = +3.9\ kJ/mol$$

Example 14.4 Calculating an Enthalpy of Solution

Problem • Use the data given in Table 14.2 to determine the heat of solution for NH_4NO_3, the compound used in cold packs (see page 235 and Figure 14.6).

Strategy • Use Equation 6.6 and data from Table 14.2 for reactants and products.

Solution • The solution process for NH_4NO_3 is represented by the equation

$$NH_4NO_3(s) \longrightarrow NH_4NO_3(aq)$$

The energy of this process is calculated using heats of formation given in Table 14.2:

$$\Delta H_{soln}^\circ = \sum [\Delta H_f^\circ(products)] - \sum [\Delta H_f^\circ(reactants)]$$
$$= \Delta H_f^\circ[NH_4NO_3(aq)] - \Delta H_f^\circ[NH_4NO_3(s)]$$
$$= -339.9\ kJ/mol - (-365.6\ kJ/mol) = +25.7\ kJ/mol$$

The process is endothermic as indicated by the fact that ΔH_{soln}° has a positive value.

Exercise 14.4 Calculating an Enthalpy of Solution

Use the data in Table 14.2 to calculate the enthalpy of solution for NaOH.

Product-favored processes are often accompanied by evolution of energy, so we can expect that if the solution process is exothermic a substance is likely to be soluble. This is not the whole story, however. As described earlier (page 564) another factor besides energy influences whether a substance dissolves. This is the driving force to form a mixture from pure substances, or more generally to move from a more ordered to a less ordered system. In Chapter 19 we will describe this when another thermodynamic quantity, called entropy, is introduced.

14.3 FACTORS AFFECTING SOLUBILITY: PRESSURE AND TEMPERATURE

Biochemists and physicians, among others, are interested in the solubility of gases such as CO_2 and O_2 in water or body fluids, and scientists and engineers need to know about the solubility of solids in various solvents. Pressure and temperature are two external factors that influence solubility. Both affect the solubility of gases in liquids, whereas only temperature is an important factor in the solubility of solids in liquids [🖥 CD-ROM, Screens 14.5 and 14.6].

Dissolving Gases in Liquids: Henry's Law

The solubility of a gas in a liquid is directly proportional to the gas pressure. As the gas pressure increases, so does its solubility. This is a statement of **Henry's law,**

$$S_g = k_H P_g \qquad (14.4)$$

where S_g is the gas solubility, P_g is the partial pressure of the gaseous solute, and k_H is Henry's law constant (Table 14.3), a constant characteristic of the solute and solvent.

Carbonated soft drinks illustrate Henry's law. These are packed under pressure in a chamber filled with carbon dioxide gas, some of which dissolves in the drink. When the can or bottle is opened, the partial pressure of CO_2 above the solution drops, which causes the concentration of CO_2 in solution to drop, and gas to bubble out of the solution (Figure 14.8).

Henry's law has important consequences in SCUBA diving (Figure 14.9) [SCUBA stands for self-contained underwater breathing apparatus and was invented by the famous oceanographer Jacques Cousteau (1910–1997)]. When you dive, the pressure of the air you breathe must be balanced against the external pressure of the water. In deeper dives, the pressure of the gases in the SCUBA gear must be several atmospheres and, as a result, more gas dissolves in the blood. This can lead to a problem if you ascend too rapidly. You can experience a painful and potentially lethal condition referred to as "the bends," in which nitrogen gas bubbles form in the blood as the solubility of nitrogen decreases with decreasing pressure. To prevent the bends, divers may use a helium–oxygen mixture (rather than nitrogen–oxygen) because helium is not as soluble in aqueous media as nitrogen.

We can understand the effect of pressure on solubility by examining the system at the particulate level. The solubility of a gas is defined as the concentration of the dissolved gas in equilibrium with the substance in the gaseous state. At equilibrium, the rate at which solute gas molecules escape the solution and enter the gaseous state equals the rate at which gas molecules reenter the solution. An increase in pressure results in more molecules of gas striking the surface of the liquid and entering solution in a given time. The solution eventually reaches equilibrium when the concentration of gas dissolved in the solvent is high enough that the rates of gas molecules escaping and entering the solution are the same.

Table 14.3 • Henry's Law Constants (25 °C)

Gas	k_H (M/mm Hg)
N_2	8.42×10^{-7}
O_2	1.66×10^{-6}
CO_2	4.48×10^{-5}

Source: W. Stumm and J. J. Morgan: *Aquatic Chemistry*, p. 109. New York, Wiley, 1981.

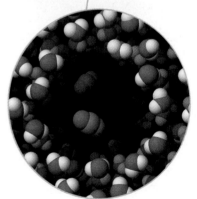

Figure 14.8 Gas solubility and pressure. Carbonated beverages are bottled under CO_2 pressure. When the bottle is opened, the pressure is released and bubbles of CO_2 form within the liquid and rise to the surface. After some time, an equilibrium between dissolved CO_2 and atmospheric CO_2 is reached. Because CO_2 provides some of the taste in the beverage, the beverage tastes flat after most of its dissolved CO_2 is lost. *(Charles D. Winters)*

Figure 14.9 Illustrations of Henry's law. *(a, Tom and Therisa Stack/Tom Stack Associates; b, Peter Arnold Inc.)*

(a) SCUBA divers must pay attention to the solubility of gases in the blood and the fact that solubility increases with pressure.

(b) A hyperbaric chamber. People who have problems breathing can be placed in a hyperbaric chamber where they are exposed to a higher partial pressure of oxygen.

Example 14.5 **Using Henry's Law**

Problem • What is the concentration of O_2 in a fresh water stream in equilibrium with air at 25 °C and 1.0 atm? Express the answer in grams of O_2 per liter of water.

Strategy • To use Henry's law to calculate the molar solubility of oxygen the partial pressure of O_2 in air must first be calculated. Recall that the mole fraction of O_2 in air is 0.21 (see Table 12.2, page 488).

Solution • Taking the mole fraction of O_2 in air as 0.21, and assuming the total pressure is 1.0 atm, the partial pressure of O_2 is 160 mm Hg. Using this pressure for P_g in Henry's law (and k_H from Table 14.3) we have

$$\text{Solubility of } O_2 = \left(\frac{1.66 \times 10^{-6}\,M}{\text{mm Hg}}\right)(160 \text{ mm Hg})$$

$$= 2.66 \times 10^{-4}\,M$$

This concentration can be expressed in grams per liter using the molar mass of O_2:

$$\text{Solubility of } O_2 = \left(\frac{2.66 \times 10^{-4}\text{ mol}}{L}\right)\left(\frac{32.0\text{ g}}{\text{mol}}\right)$$

$$= 0.00850 \text{ g/L}$$

This concentration of O_2 (8.50 mg/L) is quite low, but it is sufficient to provide the oxygen required by aquatic life.

Exercise 14.5 **Using Henry's Law**

Henry's law constant for CO_2 in water at 25 °C is 4.48×10^{-5} M/mm Hg. What is the concentration of CO_2 in water when the partial pressure is 0.33 atm? (Although CO_2 reacts with water to give traces of H_2CO_3, H^+, HCO_3^-, and CO_3^{2-} the reaction occurs to such a small extent that Henry's law is obeyed at low CO_2 partial pressures.)

Temperature Effects on Solubility: Le Chatelier's Principle

The solubility of all gases in water decreases with increasing temperature. You may realize this from everyday observations such as the appearance of bubbles as water is heated to just below the boiling point.

Decreased solubility of gases with increasing temperature has environmental consequences. Fish often seek lower depths in summer because the warmer surface layers of water have lower oxygen concentrations. Thermal pollution, resulting from the use of surface water as a coolant for various industries, can be a special problem for marine life that requires oxygen to survive. Effluent water returned to a natural water source at a warmer temperature will be depleted of oxygen.

To understand the effect of temperature on the solubility of gases, let us re-examine the heat of solution. Gases that dissolve to an appreciable extent in water usually do so in an exothermic process.

$$\text{Gas + liquid solvent} \rightleftharpoons \text{saturated solution + heat}$$
$$\Delta H_{\text{soln}} < 0$$

The reverse process, loss of dissolved gas molecules from a solution, requires heat. These two processes can reach equilibrium, as depicted in an equation where products and reactants are connected by the symbol for an equilibrium process, a double arrow ($\rightleftharpoons$).

To understand how temperature affects solubility, we turn to **Le Chatelier's principle,** which states that a change in any of the factors determining an equilibrium causes the system to adjust to reduce or counteract the effect of the change. If a solution of a gas in a liquid is heated, the equilibrium will shift to absorb some of the added heat energy. That is, the reaction

$$\text{Gas + liquid solvent} \underset{\substack{\text{Exothermic process} \\ \Delta H_{\text{soln}} \text{ is negative.}}}{\rightleftharpoons} \text{Saturated solution + heat}$$

Add heat energy. Equilibrium shifts left.

shifts back to the left if the temperature is raised because heat energy can be consumed in the process that gives free gas molecules and pure solvent. This shift corresponds to less gas dissolved, or a lower solubility, at higher temperature, the observed result.

The solubility of solids in water is also affected by temperature, but unlike the situation involving solutions of gases there is no general pattern of behavior. In Figure 14.10, the solubilities of several salts are plotted versus temperature. The solubility of many salts increases with increasing temperature, but there are notable

● **Solubility and Temperature**
Predicting the solubility of solids in non-aqueous solvents is less complicated. In general, solubility increases with increasing temperature. For more on the subject of solids dissolving in water, see R. S. Treptow: "Le Chatelier's principle applied to the temperature dependence of solubility." *Journal of Chemical Education,* Vol. 61, p. 499, 1984.

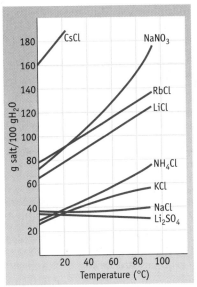

(a) Temperature dependence of the solubility of some ionic compounds.

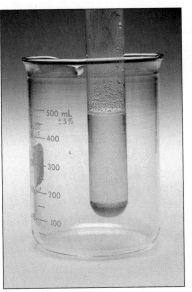

(b) NH$_4$Cl dissolved in water.

(c) NH$_4$Cl precipitates when the solution is cooled in ice.

Figure 14.10   The temperature dependence of the solubility of some ionic compounds in water.
Most compounds decrease in solubility with decreasing temperature. *(Charles D. Winters)*

Figure 14.11   Giant crystals of potassium dihydrogen phosphate. The crystal being measured by this researcher at Lawrence Livermore Laboratory in California weighs 318 kg and measures 66 × 53 × 58 cm. The crystals were grown by suspending a thumbnail-sized seed crystal in a 6-foot tank of supersaturated KH_2PO_4. The temperature of the solution was gradually reduced from 65 °C over a period of about 50 days. The crystals are sliced into thin plates, which are used to convert light from a giant laser from infrared to ultraviolet. *(Charles D. Winters)*

Chapter Goals • Revisited

- Learn additional methods of expressing solution concentration.
- Understand the solution process.
- **Understand and use the colligative properties of solutions.**
- Describe colloids and their applications.

exceptions. Predictions based on whether the heat of solution is positive or negative work most of the time, but exceptions do occur.

The variation of solubility with temperature is used by chemists to purify compounds. An impure sample of a given compound is dissolved in a solvent at high temperature because the solubility is usually increased under these conditions. The solution is cooled to decrease the solubility, and, when the limit of solubility is reached at the lower temperature, crystals of the pure compound form. If the process is done slowly and carefully it is sometimes possible to obtain very large crystals (see Figure 14.11).

14.4   COLLIGATIVE PROPERTIES

When water contains dissolved sodium chloride, the vapor pressure of water over the solution is different from that over pure water, as is the freezing point of the solution, its boiling point, and its osmotic pressure. These colligative properties depend on the relative numbers of solute and solvent particles [CD-ROM, Screens 14.7–14.9].

Changes in Vapor Pressure: Raoult's Law

The equilibrium vapor pressure at a particular temperature is the pressure of the vapor when the rate at which molecules escape the liquid and enter the gaseous state equals the rate at which gaseous molecules condense to re-form the liquid. When the vapor pressure of the solvent over a solution is measured at a given temperature, it is experimentally observed that

- The vapor pressure over the solution is lower than the vapor pressure of pure solvent.
- The vapor pressure of the solvent, P_{solv}, is proportional to the relative number of solvent molecules in the solution; that is, the solvent vapor pressure is proportional to the solvent mole fraction, $P_{solv} \propto X_{solv}$.

Because solvent vapor pressure and the relative number of solvent molecules are proportional, we can write the following equation for the equilibrium vapor pressure of the solvent over a solution at a particular temperature:

$$P_{solv} = X_{solv}\, P^{\circ}_{solv} \qquad\qquad \textbf{(14.5)}$$

This equation, called **Raoult's law,** tells us that the vapor pressure of solvent over a solution (P_{solv}) is some fraction of the pure solvent equilibrium vapor pressure (P°_{solv}) [CD-ROM, Screen 14.7]. For example, if 95% of the molecules in a solution are solvent molecules ($X_{solv} = 0.95$), then the vapor pressure of the solvent (P_{solv}) is 95% of P°_{solv}.

Like the ideal gas law, Raoult's law describes a simplified model of a solution. We say that an **ideal solution** is one that obeys Raoult's law. No solution is ideal, just as no gas is truly ideal, but Raoult's law is a good approximation of solution behavior in many instances, especially at low solute concentration.

When will a solution not be ideal? This brings us to another effect of dissolved solutes, the forces of attraction between solute and solvent molecules. For Raoult's law to hold, the forces between solute and solvent molecules must be the same as those between the solvent molecules in the pure solvent. This is frequently the case when molecules with similar structures are involved. Solutions of one hydrocarbon in another (hexane, C_6H_{14}, dissolved in octane, C_8H_{18}, for example) usually follow Raoult's law quite closely. If solvent–solute interactions are stronger than solvent–solvent interactions, the actual vapor pressure will be lower than calculated by Raoult's law. If the solvent–solute interactions are weaker than solvent–solvent interactions, the vapor pressure will be higher.

Example 14.6 Using Raoult's Law

Problem • Suppose 651 g of ethylene glycol, $HOCH_2CH_2OH$, is dissolved in 1.50 kg of water (a 30.3% solution, a commonly used antifreeze solution for automobiles). What is the vapor pressure of the water over the solution at 90 °C? Assume ideal behavior for the solution.

Strategy • To use Raoult's Law (Equation 14.5), we first must calculate the mole fraction of the solvent (water). We also need the vapor pressure of pure water at 90 °C (= 525.8 mm Hg, Appendix G).

Solution • Let us first calculate the amount of water and ethylene glycol, and, from these, the mole fraction of water.

$$\text{Amount of water} = 1.50 \times 10^3 \text{ g}\left(\frac{1 \text{ mol}}{18.02 \text{ g}}\right)$$

$$= 83.2 \text{ mol water}$$

$$\text{Amount of ethylene glycol} = 651 \text{ g}\left(\frac{1 \text{ mol}}{62.07 \text{ g}}\right)$$

$$= 10.5 \text{ mol glycol}$$

$$X_{\text{water}} = \frac{83.2 \text{ mol water}}{83.2 \text{ mol water} + 10.5 \text{ mol glycol}}$$

$$= 0.888$$

Next apply Raoult's law, calculating the vapor pressure from the mole fraction of water and the vapor pressure of pure water:

$$P_{\text{water}} = X_{\text{water}}P^{\circ}_{\text{water}} = (0.888)(525.8 \text{ mm Hg}) = 467 \text{ mm Hg}$$

The dissolved solute decreases the vapor pressure by 59 mm Hg, or about 11%:

$$\Delta P_{\text{water}} = P_{\text{water}} - P^{\circ}_{\text{water}} = 467 \text{ mm Hg} - 525.8 \text{ mm Hg}$$

$$= -59 \text{ mm Hg}$$

Comment • Ethylene glycol dissolves easily in water, is noncorrosive, and is relatively inexpensive. Because it is nonvolatile it will not boil off. These features make it ideal for use as antifreeze. It is, however, toxic to animals, so it is being replaced by less toxic propylene glycol (page 443).

Exercise 14.6 Using Raoult's Law

Assume you dissolve 10.0 g of sucrose ($C_{12}H_{22}O_{11}$) in 225 mL (225 g) of water and warm the water to 60 °C. What is the vapor pressure of the water over this solution? (Appendix G lists $P^{\circ}(H_2O)$ at various temperatures.)

Adding a nonvolatile solute to a solvent lowers the vapor pressure of the solvent (see Example 14.6), and Raoult's law can be modified to calculate directly the lowering of the vapor pressure, ΔP_{solv}, as a function of the mole fraction of the solute.

$$\Delta P_{\text{solv}} = P_{\text{solv}} - P^{\circ}_{\text{solv}}$$

Substituting Raoult's law for P_{solv}, we have

$$\Delta P_{\text{solv}} = (X_{\text{solv}} P^{\circ}_{\text{solv}}) - P^{\circ}_{\text{solv}} = -(1 - X_{\text{solv}})P^{\circ}_{\text{solv}}$$

In a solution that has only the volatile solvent and one nonvolatile solute, the sum of the mole fraction of solvent and solute must be 1:

$$X_{\text{solv}} + X_{\text{solute}} = 1$$

Therefore, $1 - X_{\text{solv}} = X_{\text{solute}}$, and the equation for ΔP_{solv} can be rewritten as

$$\Delta P_{\text{solv}} = -X_{\text{solute}} P^{\circ}_{\text{solv}} \qquad \textbf{(14.6)}$$

History

François M. Raoult (1830–1901)

Raoult's law is named for Raoult, a professor of chemistry at the University of Grenoble in France, who did the pioneering studies in this area. •

Thus, the change in the vapor pressure of the solvent is *proportional to the mole fraction* (the relative number of particles) *of solute.*

Boiling Point Elevation

Suppose you have a solution of a nonvolatile solute in the volatile solvent benzene. If the solute concentration is 0.200 mol in 100. g of benzene (C_6H_6) (= 2.00 *m*), this means that $X_{benzene} = 0.865$. Using $X_{benzene}$ and applying Raoult's law, we can calculate that the vapor pressure of the solvent at 60 °C will drop from 400. mm Hg for the pure solvent to 346 mm Hg for the solution:

$$X_{benzene} = 0.865$$

$$P_{benzene} = X_{benzene}\, P^{\circ}_{benzene} = (0.865)(400.\ \text{mm Hg}) = 346\ \text{mm Hg}$$

This point is marked on the vapor pressure graph in Figure 14.12. Now, what is the vapor pressure when the temperature of the solution is raised another 10 °C? $P^{\circ}_{benzene}$ becomes larger with increasing temperature, so $P_{benzene}$ for the solution must also become larger. This new point and additional ones calculated in the same way for other temperatures define the vapor pressure curve for the solution (the lower curve in Figure 14.12).

An important observation we can make in Figure 14.12 is that *the vapor pressure lowering caused by the nonvolatile solute leads to an increase in the boiling point* [⊖ CD-ROM, Screen 14.8]. The normal boiling point of a liquid is the temperature at which its vapor pressure is equal to 1 atm or 760 mm Hg [◆ SECTION 13.5]. From Figure 14.12 we see that the normal boiling point of pure benzene (at 760 mm Hg) is about 80 °C. Tracing the vapor pressure curve for the solution, we also see

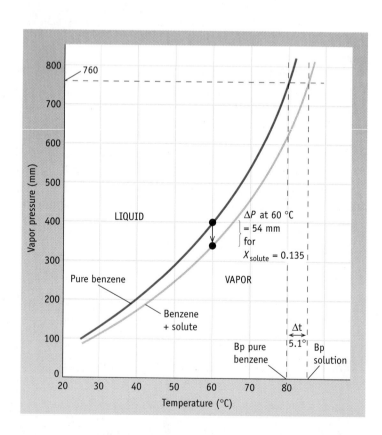

Figure 14.12 Lowering the vapor pressure of benzene by addition of a nonvolatile solute. The curve drawn in red represents the vapor pressure of pure benzene, and the curve below it, in blue, the vapor pressure of a solution containing 0.200 mol of a solute dissolved in 0.100 kg of solvent (2.00 *m*). This graph was created using a series of calculations such as those shown in the text. As an alternative, the graph could be created by measuring the values for the vapor pressure for the solution in a laboratory experiment.

Table 14.4 • Some Boiling Point Elevation and Freezing Point Depression Constants

Solvent	Normal bp (°C) Pure Solvent	K_{bp} (°C/m)	Normal fp (°C) Pure Solvent	K_{fp} (°C/m)
Water	100.00	+0.5121	0.0	−1.86
Benzene	80.10	+2.53	5.50	−5.12
Camphor	207.4	+5.611	179.75	−39.7
Chloroform ($CHCl_3$)	61.70	+3.63	—	—

that the vapor pressure reaches 760 mm Hg at a temperature about 5 °C higher than the boiling point of pure benzene.

The vapor pressure curve and increase in the boiling point shown in Figure 14.12 refer specifically to a 2.00 *m* solution. We might wonder at this point how the boiling point of the solution would vary with solute concentration, and it is possible to reason out the answer. Recall that the change in vapor pressure is directly proportional to the concentration of solute ($\Delta P_{benzene} = -X_{solute} P^{\circ}_{benzene}$, Equation 14.6). Concentrations of solute greater than 2.00 *m* lead to a larger decrease in vapor pressure and consequently to a higher boiling point. Conversely, solute concentrations less than 2.00 *m* show a smaller decrease in vapor pressure and less of an increase in boiling point. In fact, a simple relationship exists between boiling point elevation and molal concentration: the *boiling point elevation, ΔT_{bp}, is directly proportional to the molality of the solute:*

$$\text{Elevation in boiling point} = \Delta T_{bp} = K_{bp} \, m_{solute} \qquad \textbf{(14.7)}$$

In this equation, K_{bp} is a proportionality constant called the **molal boiling point elevation constant.** It has the units of degrees/molal (°C/m). Values for K_{bp} are determined experimentally, and different solvents have different values (Table 14.4). Formally, the value corresponds to the elevation in boiling point for a 1 *m* solution.

Example 14.7 Boiling Point Elevation

Problem • Eugenol, the active ingredient in cloves, has a formula of $C_{10}H_{12}O_2$ (page 103). What is the boiling point of a solution containing 0.144 g of this compound dissolved in 10.0 g of benzene?

Strategy • Equation 14.7 can be used to calculate the change in boiling point. This value must then be added to the boiling point of pure benzene to provide the answer. To use Equation 14.7 you need a value of K_{bp} and the molality of the solution. The value for K_{bp} for benzene is given in Table 14.4, but you need to calculate the molality, *m* first.

Solution •

$$0.144 \text{ g eugenol} \left(\frac{1 \text{ mol eugenol}}{164.2 \text{ g}} \right) = 8.77 \times 10^{-4} \text{ mol eugenol}$$

$$\frac{8.77 \times 10^{-4} \text{ mol eugenol}}{0.0100 \text{ kg benzene}} = 8.77 \times 10^{-2} \text{ m eugenol}$$

Use the value for the molality to calculate the boiling point elevation and then the boiling point:

$$\Delta T_{bp} = (2.53 \text{ °C/m})(0.0877 \text{ m}) = 0.222 \text{ °C}$$

Because the boiling point *rises* relative to that of the pure solvent, the boiling point of the solution is

$$80.10 \text{ °C} + 0.222 \text{ °C} = 80.32 \text{ °C}$$

Exercise 14.7   Boiling Point Elevation

What quantity of ethylene glycol, $HOCH_2CH_2OH$, must be added to 125 g of water to raise the boiling point by 1.0 °C? Express the answer in grams.

The elevation of the boiling point of a solvent on adding a solute has many practical consequences. One of them is the summer protection your car's engine receives from "all-season" antifreeze. The main ingredient of commercial antifreeze is ethylene glycol, $HOCH_2CH_2OH$. The car's radiator and cooling system are sealed to keep the coolant under pressure, so that it will not vaporize at normal engine temperatures. When the air temperature is high in the summer, however, the radiator could "boil over" if it were not protected with "antifreeze." By adding this nonvolatile liquid, the solution in the radiator has a higher boiling point than that of pure water.

Freezing Point Depression

Another consequence of dissolving a solute in a solvent is that the freezing point of the solution is lower than that of the pure solvent (Figure 14.13). For an ideal solution, the depression of the freezing point is given by an equation similar to that for the elevation of the boiling point:

$$\text{Freezing point depression} = \Delta T_{fp} = K_{fp}\, m_{solute} \qquad (14.8)$$

where K_{fp} is the molal **freezing point depression constant** in degrees per molal (°C/m). Values of K_{fp} for a few common solvents are given in Table 14.4. The values are negative quantities, so that the result of the calculation is a negative value for ΔT_{fp}, signifying a decrease in temperature.

The practical aspects of freezing point changes from pure solvent to solution are similar to those for boiling point elevation. The very name of the liquid you add to the radiator in your car, *antifreeze,* indicates its purpose (see Figure 14.13a). The label on the container of antifreeze tells you, for example, to add 6 qt (5.7 L) of antifreeze to a 12-qt (11.4-L) cooling system in order to lower the freezing point to −34 °C and raise the boiling point to +109 °C.

Figure 14.13   Freezing a solution.
(a) Antifreeze. Adding antifreeze to water prevents the water from freezing. Here a jar of pure water (*left*) and a jar of water to which automobile antifreeze had been added (*right*) were kept overnight in the freezing compartment of a home refrigerator. *(Charles D. Winters)*
(b) When a solution freezes, it is pure solvent that solidifies. To take this photo, a purple dye was dissolved in water, and the solution was frozen slowly. Pure ice formed along the walls of the tube, and the dye stayed in solution. The concentration of the solute increased as more and more solvent was frozen out, and the resulting solution had a lower and lower freezing point. At equilibrium, the system contains pure, colorless ice that had formed along the walls of the tube and a concentrated solution of dye in the center of the tube. *(Charles D. Winters)*

(a)

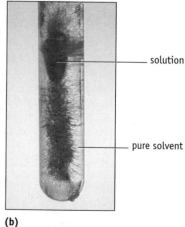

solution

pure solvent

(b)

Example 14.8 Freezing Point Depression

Problem • What mass of ethylene glycol, $HOCH_2CH_2OH$, must be added to 5.50 kg of water to lower the freezing point of the water from 0.0 to -10.0 °C? (This is approximately the situation in your car.)

Strategy • To use Equation 14.8 you need K_{fp} (see Table 14.4). You can then calculate the molality of the solution and, from this, the amount of ethylene glycol required.

Solution • The solute concentration (molality) in a solution with a freezing point depression of -10.0 °C is

$$\text{Solute concentration } (m) = \frac{\Delta T_{fp}}{K_{fp}} = \frac{-10.0 \text{ °C}}{-1.86 \text{ °C}/m}$$

$$= 5.38 \ m$$

Starting with 5.50 kg of water, we need 29.6 mol of ethylene glycol:

$$\left(\frac{5.38 \text{ mol glycol}}{1.00 \text{ kg water}}\right)(5.50 \text{ kg water}) = 29.6 \text{ mol glycol}$$

The molar mass of ethylene glycol is 62.07 g/mol, so the mass required is

$$29.6 \text{ mol glycol}\left(\frac{62.07 \text{ g}}{1 \text{ mol}}\right) = 1840 \text{ g glycol}$$

Comment • The density of ethylene glycol is 1.11 kg/L, so the volume of antifreeze to be added is 1.66 L [1.84 kg (1 L/1.11 kg)].

Exercise 14.8 Freezing Point Depression

In the northern United States, summer cottages are usually closed up for the winter. When doing so the owners "winterize" the plumbing by putting antifreeze in the toilet tanks, for example. Will adding 525 g of $HOCH_2CH_2OH$ to 3.00 kg of water ensure that the water will not freeze at -25 °C?

Colligative Properties and Molar Mass Determination

Early in this book you learned how to calculate a molecular formula from an empirical formula when given the molar mass. But how do you know the molar mass of an unknown compound? An experiment must be done to find this crucial piece of information, and one way to do this is to use a colligative property of a solution of the compound. If the compound is soluble in a solvent of appreciable vapor pressure and a known K_{bp} or K_{fp}, the molar mass can then be determined. All approaches use the same basic logic:

Change in vapor pressure, boiling point elevation, freezing point depression, or osmotic pressure. → Solution conc --Use mass of solvent--> Moles of solute --g solute / mol solute--> Molar mass

Example 14.9 Determining Molar Mass from Boiling Point Elevation

Problem • A solution prepared from 1.25 g of oil of wintergreen (methyl salicylate) in 99.0 g of benzene has a boiling point of 80.31 °C. Determine the molar mass of this compound.

Strategy • Use Equation 14.7 to find the solution molality from this the amount of solute. The ratio of the mass and amount of solute give the molar mass.

Solution • We first use the boiling point elevation to calculate the solution concentration:

$$\text{Boiling point elevation } (\Delta T_{bp}) = 80.31 \text{ °C} - 80.10 \text{ °C} = 0.21 \text{ °C}$$

and then calculate the molality:

$$\text{Molality of solution} = \frac{\Delta T_{bp}}{K_{bp}} = \frac{0.21 \text{ °C}}{2.53 \text{ °C}/m} = 0.083 \ m$$

(Example continues on next page)

The amount of solute in the solution is calculated from the solution concentration.

$$\text{Moles of solute} = \left(\frac{0.083 \text{ mol}}{1.00 \text{ kg}}\right)(0.099 \text{ kg solvent})$$

$$= 0.0082 \text{ mol solute}$$

Now we can combine the amount of solute with its mass:

$$\frac{1.25 \text{ g}}{0.0082 \text{ mol}} = 150 \text{ g/mol}$$

Comment • Methyl salicylate has the formula $C_8H_8O_3$ and a molar mass of 152.14 g/mol.

Exercise 14.9 **Determining Molar Mass by Boiling Point Elevation**

Crystals of the beautiful blue hydrocarbon, azulene (0.640 g), which has an empirical formula of C_5H_4, are dissolved in 99.0 g of benzene. The boiling point of the solution is 80.23 °C. What is the molecular formula of azulene?

Colligative Properties of Solutions Containing Ions

In the northern United States it is common practice to scatter salt on snowy or icy roads or sidewalks. When the sun shines on the snow or patch of ice, a small amount is melted, and the water dissolves some of the salt. As a result of the dissolved solute, the freezing point of the solution is lower than 0 °C. The solution "eats" its way through the ice, breaking it up, and the icy patch is no longer dangerous for drivers or for people walking [CD-ROM, Screen 14.8, Exercise 2].

Salt (NaCl) is the most common substance used on roads because it is inexpensive and dissolves readily in water. Its relatively low molar mass means that the effect per gram is large. In addition, salt is especially effective because it is an electrolyte; it dissolves to give ions in solution:

$$\text{NaCl(s)} \longrightarrow \text{Na}^+\text{(aq)} + \text{Cl}^-\text{(aq)}$$

Putting salt on ice assists in melting the ice. *(Charles D. Winters)*

Remember that colligative properties depend not on what is dissolved but *only on the number of particles of solute per solvent particle.* When 1 mol of NaCl dissolves, 2 mol of ions are formed, which means that the effect on the freezing point of water should be twice as large as that expected for a mole of sugar. This peculiarity was discovered by Raoult in 1884 and studied in detail by Jacobus Henrikus van't Hoff (1852–1911) in 1887. Later in that same year Svante Arrhenius (1859–1927) provided the explanation for the behavior of electrolytes based on ions in solution. A 0.100 *m* solution of NaCl really contains two solutes, 0.100 *m* Na$^+$ and 0.100 *m* Cl$^-$. What we should use to estimate the freezing point depression is the *total* molality of solute particles:

$$m_{\text{total}} = m(\text{Na}^+) + m(\text{Cl}^-) = (0.100 + 0.100)\text{mol/kg} = 0.200 \text{ mol/kg}$$

$$\Delta T_{\text{fp}} = (-1.86 \text{ °C}/m)(0.200 \text{ } m) = -0.372 \text{ °C}$$

To estimate the freezing point depression for an ionic compound, first find the molality of solute from the mass and molar mass of the compound. Then, multiply the number you get by the number of ions in the formula: two for NaCl, three for Na_2SO_4, four for $LaCl_3$, five for $Al_2(SO_4)_3$, and so on.

As it turns out, this model is a reasonable first guess at the effect of ionization of an electrolyte on colligative properties, but it is not exact. Let us look at some

Table 14.5 • Freezing Point Depression of Some Ionic Solutions

Mass %	m (mol/kg)	ΔT_{fp} (measured, °C)	ΔT_{fp} (calculated, °C)	$\dfrac{\Delta T_{fp},\text{ measured}}{\Delta T_{fp},\text{ calculated}}$
NaCl				
0.00700	0.0120	−0.0433	−0.0223	1.94
0.500	0.0860	−0.299	−0.160	1.87
1.00	0.173	−0.593	−0.322	1.84
2.00	0.349	−1.186	−0.649	1.83
Na₂SO₄				
0.00700	0.00493	−0.0257	−0.00917	2.80
0.500	0.0354	−0.165	−0.0658	2.51
1.00	0.0711	−0.320	−0.132	2.42
2.00	0.144	−0.606	−0.268	2.26

experimental data (Table 14.5) for the effect of the dissociation of two ionic compounds, NaCl and Na_2SO_4, on the solution freezing point. The measured freezing point depression is larger than that calculated from Equation 14.8 assuming no ionization. As seen in the last column of the table, however, ΔT_{fp} is not twice the value expected for NaCl, but only about 1.8 times larger. Likewise, ΔT_{fp} for Na_2SO_4 approaches but does not reach a value that is 3 times larger than the value assuming no ionization. The ratio of the experimentally observed value of ΔT_{fp} to the value calculated, assuming no ionization, is called the **van't Hoff factor** and is represented by i.

$$i = \frac{\Delta T_{fp},\text{ measured}}{\Delta T_{fp},\text{ calculated}} = \frac{\Delta T_{fp},\text{ measured}}{K_{fp}\, m}$$

or

$$\Delta T_{fp},\text{ measured} = K_{fp}\, m\, i \qquad\qquad (14.9)$$

The van't Hoff factors in the last column of Table 14.5 can be used in calculations of any colligative property. Vapor pressure lowering, boiling point elevation, freezing point depression, and osmotic pressure are all larger for electrolytes than for nonelectrolytes of the same molality.

The van't Hoff factor approaches a whole number (2, 3, and so on) only in very dilute solutions. In more concentrated solutions, the experimental freezing point depressions tell us that there are fewer ions in solution than expected. This behavior, which is typical of all ionic compounds, is a consequence of the strong attractions between ions. The result is as if some of the positive and negative ions are paired, decreasing the total molality of particles. Indeed, in more concentrated solutions, and especially in solvents less polar than water, ions are extensively associated in ion pairs and in even larger clusters.

Example 14.10 Freezing Point and Ionic Solutions

Problem • A 0.00200 *m* aqueous solution of an ionic compound Co(NH$_3$)$_5$(NO$_2$)Cl freezes at −0.00732 °C. How many moles of ions does 1 mol of the salt produce on being dissolved in water?

Strategy • Calculate ΔT_{fp} of the solution assuming no ions are produced. Compare this value with the actual value of ΔT_{fp}. The ratio will reflect the number of ions produced.

Solution • The freezing point depression expected for a 0.00200 *m* solution assuming that the salt does not dissociate into ions is

$$\Delta T_{fp}, \text{calculated} = K_{fp}\, m = (-1.86\ °C/m)(0.00200\ m)$$
$$= -3.72 \times 10^{-3}\ °C$$

Now compare the calculated freezing point depression with the measured depression. This gives us the van't Hoff factor:

$$i = \frac{\Delta T_{fp}, \text{measured}}{\Delta T_{fp}, \text{calculated}} = \frac{-7.32 \times 10^{-3}\ °C}{-3.72 \times 10^{-3}\ °C} = 1.97 \approx 2$$

It appears that 1 mol of this compound gives 2 mol of ions. In this case, the ions are [Co(NH$_3$)$_5$(NO$_2$)]$^+$ and Cl$^-$.

Exercise 14.10 Freezing Point and Ionic Compounds

Calculate the freezing point of 525 g of water that contains 25.0 g of NaCl. Assume *i*, the van't Hoff factor, is 1.85 for NaCl.

Osmosis

Osmosis is the movement of solvent molecules through a semipermeable membrane from a region of lower to a region of higher solute concentration. This movement can be demonstrated with a simple experiment. The beaker in Figure 14.14 contains pure water, and a concentrated sugar solution is in the bag and tube. The liquids are separated by a semipermeable membrane, a thin sheet of material (such as a vegetable tissue or cellophane) through which only certain types of molecules can pass. Here, water molecules can pass through but larger sugar molecules (or hydrated ions) cannot (Figure 14.15). When the experiment is begun, the liquid levels in the beaker and the tube are the same. Over time, however, the

(a)

(b)

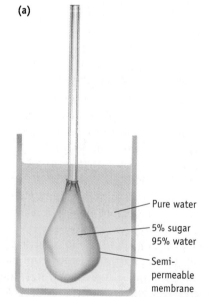
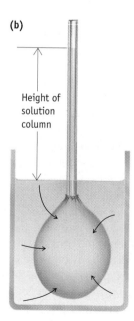

Height of solution column

Pure water

5% sugar 95% water

Semi-permeable membrane

Figure 14.14 The process of osmosis.
(a) The bag attached to the tube contains a solution that is 5% sugar and 95% water. The beaker contains pure water. The bag is made of a material that is semipermeable, meaning that it allows water but not sugar molecules to pass through.
(b) Over time, water flows from the region of low solute concentration (pure water) to the region of higher solute concentration (the sugar solution). Flow continues until the pressure exerted by the column of solution in the tube above the water level in the beaker is great enough to result in equal rates of passage of water molecules in both directions. The height of the column of solution is a measure of the osmotic pressure, Π.

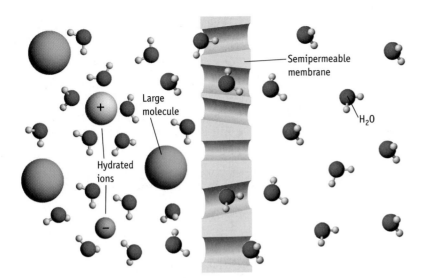

Figure 14.15   Osmosis at the particulate level. Osmotic flow through a membrane that is selectively permeable (semipermeable) to water. Dissolved substances such as hydrated ions or large sugar molecules cannot diffuse through the membrane.

level of the sugar solution inside the tube rises, the level of pure water in the beaker falls, and the sugar solution becomes steadily more dilute. After a while, no further net change occurs; equilibrium is reached [⊕ CD-ROM, Screen 14.9].

From a molecular point of view, the semipermeable membrane does not present a barrier to the movement of water molecules, so they move through the membrane in both directions. Over a given time, more water molecules pass through the membrane from the pure water side to the solution side than in the opposite direction. In effect, water molecules tend to move from regions of low solute concentration to regions of high solute concentration. The same is true for any solvent, as long as the membrane allows solvent molecules but not solute molecules or ions to pass through.

Why does the system eventually reach equilibrium? It is evident that the solution in the tube in Figure 14.14 can never reach zero sugar or salt concentration, which would be required to equalize the number of water molecules moving through the membrane in each direction in a given time. The answer lies in the fact that the solution moves higher and higher in the tube as osmosis continues and water moves into the sugar solution. Eventually the pressure exerted by this column of solution counterbalances the pressure of the water moving through the membrane from the pure water side and no further net movement of water occurs. An equilibrium of forces is achieved. The pressure created by the column of solution for the system at equilibrium is called the **osmotic pressure.** A measure of this pressure is the difference in height between the solution in the tube and the level of pure water in the beaker.

From experimental measurements on dilute solutions, it is known that osmotic pressure (Π) and concentration (c) are related by the equation

$$\Pi = cRT \tag{14.10}$$

In this equation, the quantity c is the molar concentration (in moles per liter), R is the gas constant, and T is the absolute temperature (in kelvins). Using a value for the gas law constant of $0.082057 \ \text{L} \cdot \text{atm}/\text{K} \cdot \text{mol}$ allows calculation of the osmotic pressure Π in atmospheres. This equation is analogous to the ideal gas law ($PV = nRT$), with Π taking the place of P and c equivalent to n/V.

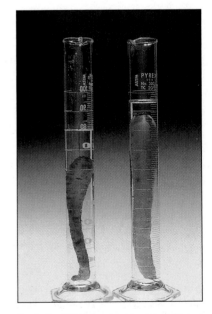

Osmosis. After some hours a carrot in a concentrated NaCl solution shows the effects of osmosis (*left*). Water has flowed out of the cells of the carrot and into the NaCl solution. A carrot in pure water (*right*) is not appreciably affected. (*Charles D. Winters*)

According to the osmotic pressure equation, the pressure exerted by a 0.10 M solution of particles at 25 °C is

$$\Pi = (0.10 \text{ mol/L})(0.0821 \text{ L} \cdot \text{atm/K} \cdot \text{mol})(298 \text{ K}) = 2.4 \text{ atm}$$

Because pressures on the order of 10^{-3} atm are easily measured, concentrations of about 10^{-4} M can be determined through measurements of osmotic pressure. Polymers and biologically important molecules often have a large number of atoms, so osmosis is an ideal method for measuring the molar masses of such molecules.

Example 14.11 Osmotic Pressure and Molar Mass

Problem • Beta-carotene is the most important of the A vitamins. Its molar mass can be determined by measuring the osmotic pressure generated by a given mass of β-carotene dissolved in the solvent chloroform. Calculate the molar mass of β-carotene if 10.0 mL of a solution containing 7.68 mg has an osmotic pressure of 26.57 mm Hg at 25.0 °C.

Strategy • Use Equation 14.10 to calculate the solution concentration from the osmotic pressure. Then, use the volume and concentration of the solution to calculate the amount of solute. Finally, find the molar mass from solute mass and amount.

Solution • The osmotic pressure gives the concentration of β-carotene:

$$\text{Concentration (M)} = \frac{\Pi}{RT}$$

$$= \frac{(26.57 \text{ mm Hg})\left(\dfrac{1 \text{ atm}}{760 \text{ mm Hg}}\right)}{(0.082057 \text{ L} \cdot \text{atm/K} \cdot \text{mol})(298.15 \text{ K})}$$

Concentration $= 1.429 \times 10^{-3}$ M

Now the amount of β-carotene dissolved in 10.0 mL of solvent can be calculated:

$$(1.429 \times 10^{-3} \text{ mol/L})(0.0100 \text{ L}) = 1.429 \times 10^{-5} \text{ mol}$$

This amount of β-carotene (1.429×10^{-5} mol) is equivalent to 7.68 mg (7.68×10^{-3} g). This gives us a way to calculate the molar mass:

$$\frac{7.68 \times 10^{-3} \text{ g}}{1.429 \times 10^{-5} \text{ mol}} = 537 \text{ g/mol}$$

Comment • Beta-carotene is a hydrocarbon with the formula $C_{40}H_{56}$. See page 433.

Exercise 14.11 Osmotic Pressure and Molar Mass

A 1.40-g sample of polyethylene, a common plastic, is dissolved in enough benzene to give exactly 100 mL of solution. The measured osmotic pressure of the solution is 1.86 mm Hg at 25 °C. Calculate the average molar mass of the polymer.

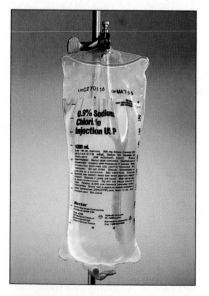

An isotonic saline solution. *(Charles D. Winters)*

Osmosis is of great practical significance for people in the health professions. Patients who become dehydrated through illness often need to be given water and nutrients intravenously. Water cannot simply be dripped into a patient's vein, however. Rather, the intravenous solution must have the same overall solute concentration as the patient's blood: the solution must be isoosmotic or **isotonic.** If pure water was used, the inside of a blood cell would have a higher solute concentration (lower water concentration), and water would flow into the cell. This *hypotonic* situation would cause the red blood cells to burst (lyse) (Figure 14.16). The opposite situation, *hypertonicity,* occurs if the intravenous solution is more concentrated than the contents of the blood cell. In this case the cell would lose water and shrivel up (crenate). To combat this, a dehydrated patient is rehydrated in the hospital with a sterile saline solution that is 0.16 M NaCl, a solution that is isotonic with the fluids in the cells of the body.

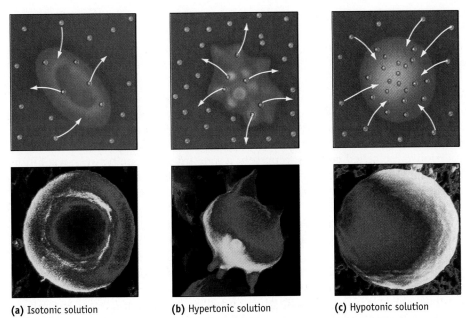

(a) Isotonic solution   **(b)** Hypertonic solution   **(c)** Hypotonic solution

Figure 14.16   Osmosis and living cells. (a) A cell placed in an isotonic solution. The net movement of water into and out of the cell is zero because the concentration of solutes inside and outside the cell is the same. **(b)** In a hypertonic solution, the concentration of solutes outside the cell is greater than that inside. There is a net flow of water out of the cell, causing the cell to dehydrate, shrink (crenate), and perhaps die. **(c)** In a hypotonic solution, the concentration of solutes outside the cell is less than that inside. There is a net flow of water into the cell, causing the cell to swell and perhaps to burst (lyse). *(David Phillips/Science Source/Photo Researchers, Inc.)*

● **Osmosis and Sea Water**
The osmotic pressure of sea water, which contains about 35 g of dissolved salts per kilogram of sea water, is about 27 atm!

A Closer Look

Reverse Osmosis in Tampa Bay

Finding sources of fresh water for human and agricultural use has been a constant battle for centuries. Water has been the cause of conflicts and a target in war. Although the earth has abundant water, 97% of it is too salty to drink or to use for irrigating crops. And a large portion of what is left is in the form of ice in the polar regions.

One of the oldest ways to obtain fresh water from the oceans is evaporation. Heating water, however, requires large quantities of heat, and the salt left behind may not be suitable for consumption.

A more modern way to extract fresh water from the oceans is desalination by reverse osmosis. The best estimate is that about 1% of the world's drinking water is now supplied by about 12,500 desalination plants. The latest, and one of the largest, is planned for Tampa, Florida. This plant will use the cooling water from a conventional power plant to obtain about 25 million gallons (about 95 million liters) per day.

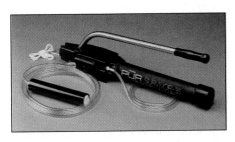

A reverse osmosis device suitable for individual use in a life raft. *(Courtesy of Catadyn North America)*

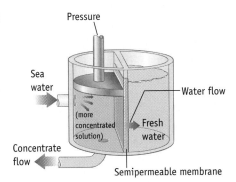

▲ Drinking water can be produced from sea water by reverse osmosis. The osmotic pressure of sea water is approximately 27 atm. Reverse osmosis requires a pressure of about 50 atm. For comparison, bicycle tires are usually pumped up to about 2–3 atm of pressure.

14.5 COLLOIDS

We defined a solution broadly as a homogeneous mixture of two or more substances in a single phase (page 23). To this we should add that, in a true solution, no settling of the solute should be observed and the solute particles should be in the form of ions or relatively small molecules. Thus, NaCl and sugar form true solutions in water. You are also familiar with suspensions, which result, for example, if a handful of fine sand is added to water and shaken vigorously. Sand particles are still visible and gradually settle to the bottom of the beaker or bottle. **Colloidal dispersions,** also called **colloids,** represent a state intermediate between a solution and a suspension. Colloids include many of the foods you eat and the materials around you; among them are Jello, milk, fog, and styrofoam (see Table 14.6) [CD-ROM, Screen 14.10].

Around 1860, the British chemist Thomas Graham (1805–1869) found that substances such as starch, gelatin, glue, and albumin from eggs diffused only very slowly when placed in water, compared with sugar or salt. In addition, the former substances differ significantly in their ability to diffuse through a thin membrane: sugar molecules can diffuse through many membranes, but the very large molecules that make up starch, gelatin, glue, and albumin do not. Moreover, Graham found that he could not crystallize these substances, whereas he could crystallize sugar, salt, and other materials that form true solutions. Graham coined the word "colloid" (from the Greek meaning "glue") to describe this class of substances distinctly different from true solutions and suspensions.

We now know that it is possible to crystallize some colloidal substances, albeit with difficulty, so there really is no sharp dividing line between these classes based on this property. Colloids do, however, have the following distinguishing characteristics: (1) Colloids generally have high molar masses; this is certainly true of substances in human cells such as proteins like hemoglobin that have molar masses in the thousands. (2) The particles of a colloid are relatively large (say 1000 nm in diameter). Because of this, they exhibit the **Tyndall effect;** they scatter visible light when dispersed in a solvent, making the mixture appear cloudy (Figure 14.17). (3) Even though colloidal particles are large, they are not so large that they settle out.

Graham also gave us the words **sol** for a dispersion of a solid substance in a fluid medium, and **gel** for a dispersion that has a structure that prevents it from being mobile. Jello is a sol when the solid is first mixed with boiling water, but it

Table 14.6 • Types of Colloids

Type	Dispersing Medium	Dispersed Phase	Examples
Aerosol	Gas	Liquid	Fog, clouds, aerosol sprays
Aerosol	Gas	Solid	Smoke, airborne viruses, automobile exhaust
Foam	Liquid	Gas	Shaving cream, whipped cream
Foam	Solid	Gas	Styrofoam, marshmallow
Emulsion	Liquid	Liquid	Mayonnaise, milk, face cream
Gel	Solid	Liquid	Jelly, Jello, cheese, butter
Sol	Liquid	Solid	Gold in water, milk of magnesia, mud
Solid sol	Solid	Solid	Milk glass, alloys (for example, steel or brass)

(a)   **(b)**

Figure 14.17   The Tyndall effect. Colloidal dispersions scatter light, a phenomenon known as the Tyndall effect.
(a) Dust in the air scatters the light coming through the trees in a forest along the Oregon coast.
(b) A narrow beam of light from a laser is passed through an NaCl solution (*left*) and then a colloidal mixture of gelatin and water (*right*). *(Charles D. Winters)*

becomes a gel when cooled. Other examples of gels are the gelatinous precipitates of $Al(OH)_3$, $Fe(OH)_3$, and $Cu(OH)_2$ (Figure 14.18).

Colloidal dispersions consist of finely divided particles that, as a result, have a very high surface area. For example, if you have one-millionth of a mole of colloidal particles, each assumed to be a sphere with a diameter of 200 nm, the total surface area of the particles would be on the order of 100 million cm^2, or the size of several football fields. It is not surprising, therefore, that many of the properties of colloids depend on the properties of surfaces.

Types of Colloids

Colloids are classified according to the state of the dispersed phase and the dispersing medium. Table 14.6 lists several types of colloids and gives examples of each.

Colloids with water as the dispersing medium can be classified as **hydrophobic** (from the Greek, meaning "water-fearing") or **hydrophilic** ("water loving"). A hydrophobic colloid is one in which only weak attractive forces exist between the water and the surface of the colloidal particles. Examples are dispersions of metals and of nearly insoluble salts in water. When compounds like AgCl precipitate, the result is often a colloidal dispersion. The precipitation reaction occurs too rapidly for ions to gather from long distances and make large crystals, so the ions aggregate to form small particles that remain suspended.

Why don't the particles come together (coagulate) and form larger particles? The answer seems to be that the colloidal particles carry electric charges. An AgCl particle, for example, will absorb Ag^+ ions if the ions are present in substantial concentration; an attraction occurs between Ag^+ ions in solution and Cl^- ions on the surface of the particle. The colloidal particles thus become positively charged, and attract a secondary layer of anions. The particles, now surrounded by layers of ions, repel one another and are prevented from coming together to form a precipitate (Figure 14.19).

A stable *hydrophobic colloid* can be made to coagulate by introducing ions into the dispersing medium. Milk is a colloidal suspension of hydrophobic particles. When milk ferments, lactose (milk sugar) is converted to lactic acid, which forms lactate ions and hydrogen ions. The protective charge on the surfaces of the colloidal particles is overcome, and the milk coagulates; the milk solids come together in clumps called "curds."

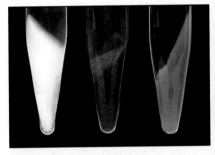

Figure 14.18   Gelatinous precipitates. (*left*) $Al(OH)_3$, (*center*) $Fe(OH)_3$, and (*right*) $Cu(OH)_2$. *(Charles D. Winters)*

Figure 14.19   Hydrophobic colloids. A hydrophobic colloid is stabilized by positive ions absorbed onto each particle and a secondary layer of negative ions. Because the particles bear similar charges they repel one another and precipitation is prevented.

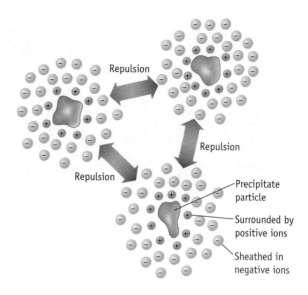

Figure 14.20   Formation of silt. Silt forms at a river delta as colloidal soil particles come in contact with salt water in the ocean. Here the Ashley and Cooper Rivers empty into the Atlantic Ocean at Charleston, SC. The high concentration of ions in salt water causes the colloidal soil particles to coagulate. *(NASA/Peter Arnold, Inc.)*

Soil particles are often carried by water in rivers and streams as hydrophobic colloids. When river water carrying large amounts of colloidal particles meets sea water with its high concentration of salts, the particles coagulate to form the silt seen at the mouth of the river. Municipal water treatment plants often add aluminum salts such as $Al_2(SO_4)_3$ to clarify water. In aqueous solution, aluminum ions exist as $[Al(H_2O)_6]^{3+}$ cations, which neutralize the charge on the hydrophobic colloidal soil particles, causing the soil particles to aggregate and settle out.

Hydrophilic colloids are strongly attracted to water molecules. They often have groups such as —OH and —NH$_2$ on their surfaces. These groups form strong hydrogen bonds to water, thus stabilizing the colloid. Proteins and starch are important examples, and homogenized milk is a familiar example.

Emulsions are colloidal dispersions of one liquid in another, such as oil or fat in water. Familiar examples are salad dressing, mayonnaise, and milk. If vegetable oil and vinegar are mixed to make a salad dressing, the mixture quickly separates into two layers because the nonpolar oil molecules do not interact with polar water and acetic acid (CH_3CO_2H) molecules. So why are milk and mayonnaise apparently homogeneous mixtures that do not separate into layers? The answer is that they contain an **emulsifying agent** such as soap or a protein. Lecithin is a protein found in egg yolks, so mixing egg yolks with oil and vinegar stabilizes the colloidal dispersion known as mayonnaise. To understand the effect of emulsifying agents further, let us look into the functioning of soaps and detergents, substances known as surfactants.

Surfactants

Soaps and detergents are emulsifying agents. Soap is made by heating a fat with sodium or potassium hydroxide (see page 454), which produces the anion of a fatty acid.

$$H_3C(CH_2)_{16} \underbrace{\phantom{H_3C(CH_2)_{16}}}_{\substack{\text{Hydrocarbon tail} \\ \text{Soluble in oil}}} \overset{\displaystyle O}{\underset{\substack{\text{Polar head} \\ \text{Soluble in water}}}{\overset{\displaystyle \|}{C}} - O^- \, Na^+}$$

sodium stearate, a soap

The fatty acid anion has a split personality: it has a nonpolar, hydrophobic hydrocarbon tail that is compatible with other similar hydrocarbons and a polar, hydrophilic head that is compatible with water.

Oil cannot be simply washed away from dishes or clothing with water because oil is nonpolar and thus insoluble in water. Instead, we add soap to the water to clean away the oil. The nonpolar molecules of the oil interact with the nonpolar hydrocarbon tails of the soap molecules, leaving the polar heads of the soap to interact with surrounding water molecules. The oil and water then mix (Figure 14.21). If the oily material on a piece of clothing or a dish also contains some dirt particles, the dirt can now be washed away.

Substances such as soaps that affect the properties of surfaces, and so affect the interaction between two phases, are called surface-active agents, or **surfactants,** for short. A surfactant used for cleaning has come to be called a **detergent.** One function of a surfactant is to lower the surface tension of water, which enhances the cleansing action of the detergent (Figure 14.22) [CD-ROM, Screen 14.11].

Many detergents used in the home and industry are synthetic. One example is sodium lauryl benzenesulfonate, a biodegradable compound.

$$CH_3CH_2CH_2CH_2CH_2CH_2CH_2CH_2CH_2CH_2CH_2CH_2 - \underset{\text{sodium lauryl benzenesulfonate}}{\bigcirc} - SO_3^- \, Na^+$$

In general, synthetic detergents use the sulfonate $-SO_3^-$ group as the polar head instead of the carboxylate group, $-CO_2^-$. The carboxylate anions form an insoluble precipitate with any Ca^{2+} or Mg^{2+} ions present in water. Because hard wa-

● **Soap and Surfactants**
A sodium soap is a solid at room temperature, whereas potassium soaps are usually liquids. About 30 million tons of household and toilet soap, and synthetic and soap-based laundry detergents, is produced annually worldwide.

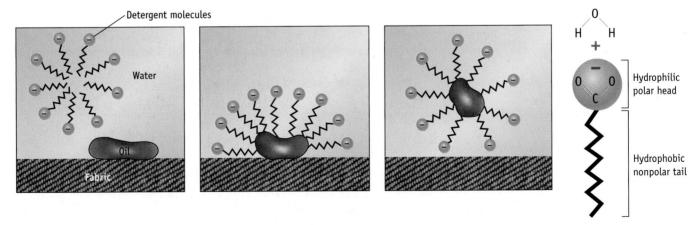

Figure 14.21   The cleaning action of soap. Soap molecules interact with water through the charged, hydrophilic end of the molecule. The long, hydrocarbon end of the molecule is hydrophobic but can bind through dispersion forces with hydrocarbons and other nonpolar substances.

Figure 14.22   Effect of a detergent on the surface tension of water. Sulfur (density = 2.1 g/cm³) is carefully placed on the surface of water (density 1.0 g/cm³) (*left*). The surface tension of the water keeps the denser sulfur afloat. Several drops of detergent are then placed on the surface of the water (*right*). The surface tension of the water is reduced, and the sulfur sinks to the bottom of the beaker. *(Charles D. Winters)*

add surfactant

ter contains high concentrations of these ions, using soaps containing carboxylates produces bathtub rings and graying of white clothing. The synthetic sulfonate detergents have the advantage that they do not form such precipitates because their calcium salts are more soluble in water.

In Summary

When you have finished studying this chapter, you should ask if you have met the chapter goals. In particular, you should be able to

- Define the terms solution, solvent, solute, and colligative properties (Section 14.1).
- Use the following concentration units: molality, mole fraction, weight percent, and parts per million (Section 14.1).
- Describe the process of dissolving a solute in a solvent, including the energy changes that may occur (Section 14.2).
- Understand the distinctions between saturated, unsaturated, and supersaturated solutions (Section 14.2).
- Define and illustrate the terms miscible and immiscible (Section 14.2).
- Relate lattice energy and enthalpy of hydration to the enthalpy of solution for an ionic solute (Section 14.2).
- Describe the effect of pressure and temperature on the solubility of a solute (Sections 14.3).
- Use Henry's law to calculate the solubility of a gas in a solvent (Section 14.3).
- Apply Le Chatelier's principle to the change in solubility of gases with pressure and temperature changes (Section 14.3).
- Calculate the mole fraction of a solute or solvent (X_{solv}) and the effect of a solute on solvent vapor pressure (P_{solv}) using Raoult's law (Section 14.4).
- Calculate the boiling point elevation or freezing point depression caused by a solute in a solvent (Section 14.4).

- Use colligative properties to determine the molar mass of a solute (Section 14.4).
- Characterize the effect of ionic solutes on colligative properties (Section 14.4).
- Use the van't Hoff factor, *i*, in calculations involving colligative properties (Section 14.4).
- Calculate the osmotic pressure (Π) for solutions, and use the equation defining osmotic pressure to determine the molar mass of a solute (Section 14.4).
- Recognize the differences between a homogeneous solution, a suspension, and a colloid (or colloidal dispersion) (Section 14.5).
- Recognize hydrophobic and hydrophilic colloids (Section 14.5).
- Describe the action of a surfactant (Section 14.5).

Key Terms

solution
solvent
solute
colligative properties

Section 14.1
mole fraction
weight percent
parts per million (ppm)

Section 14.2
saturated
solubility
miscible
immiscible
hydrated
heat of solution

Section 14.3
Henry's law
Le Chatelier's principle

Section 14.4
Raoult's law
ideal solution
molal boiling point elevation
    constant
molal freezing point depression
    constant
van't Hoff factor, *i*
osmosis
osmotic pressure
isotonic

Section 14.5
colloidal dispersions
colloids
Tyndall effect
sol
gel
hydrophobic
hydrophilic
emulsions
emulsifying agent
surfactant
detergent

Key Equations

Equation 14.1 (page 560)
Molality is defined as the amount of solute per kilogram of solvent.

$$\text{Molality of solute } (m) = \frac{\text{amount of solute (mol)}}{\text{kilograms of solvent}}$$

Equation 14.2 (page 560)
The mole fraction, *X*, of a solution component is defined as the number of moles of that component of a mixture (*n*, mol) divided by the total number of moles of all of the components of the mixture.

$$\text{Mole fraction of } A \ (X_A) = \frac{n_A}{n_A + n_B + n_C + \cdots}$$

Equation 14.3 (page 560)
Weight percent is the mass of one component divided by the total mass of the mixture (multiplied by 100%):

$$\text{Weight \% } A = \frac{\text{mass of } A}{\text{mass of } A + \text{mass of } B + \text{mass of } C + \cdots} \times 100\%$$

Equation 14.4 (page 571)

Henry's law: the solubility of a gas, S_g, is equal to the product of the partial pressure of the gaseous solute (P_g) and a constant (k_H) characteristic of the solute and solvent.

$$S_g = k_H P_g$$

Equation 14.5 (page 574)

Raoult's law: the equilibrium vapor pressure of a solvent over a solution at a given temperature, P_{solv}, is the product of the mole fraction of the solvent (X_{solv}) and the vapor pressure of the pure solvent (P°_{solv}).

$$P_{solv} = X_{solv} P^\circ_{solv}$$

Equation 14.6 (page 575)

The change in the equilibrium vapor pressure of the solvent at a given temperature, ΔP_{solv}, is the product of the mole fraction of the solute, X_{solute}, and the vapor pressure of the pure solvent.

$$\Delta P_{solv} = -X_{solute} P^\circ_{solv}$$

Equation 14.7 (page 577)

The elevation in boiling point of the solvent in a solution, ΔT_{bp}, is the product of the molality of the solute, m_{solute}, and a constant characteristic of the solvent, K_{bp}.

$$\text{Elevation in boiling point} = \Delta T_{bp} = K_{bp}\, m_{solute}$$

Equation 14.8 (page 578)

The depression of the freezing point of the solvent in a solution, ΔT_{fp}, is the product of the molality of the solute, m_{solute}, and a constant characteristic of the solvent, K_{fp}.

$$\text{Freezing point depression} = \Delta T_{fp} = K_{fp}\, m_{solute}$$

Equation 14.9 (page 581)

This equation takes into account the possible dissociation of the solute. Here i, the van't Hoff factor, is the ratio of the measured freezing point depression and the freezing point depression calculated assuming no solute dissociation. As such it is related to the relative number of particles produced by a solute.

$$\Delta T_{fp},\text{ measured} = K_{fp}\, m\, i$$

Equation 14.10 (page 583)

The osmotic pressure, Π, is the product of the solute concentration, c (mol/L), the universal gas constant ($0.0821\ \text{L} \cdot \text{atom/K} \cdot \text{mol}$), and the temperature in kelvins.

$$\Pi = cRT$$

Study Questions

Questions with blue, bold-faced numbers have answers in Appendix O. See the tutorials, simulations, and descriptions in Chapter 14 of the General Chemistry Interactive CD-ROM, *Version 3.0.*

Reviewing Important Concepts

1. Is the following statement true or false? If false, change it to make it true. "Colligative properties depend on the nature of the solvent and solute and on the concentration of the solute."

2. Name the four colligative properties described in this chapter. Write the mathematical expression that describes each of them.

3. Define molality and tell how it differs from molarity.

4. Name three effects that govern the solubility of a gas in water.

5. If you dissolve equal molar amounts of NaCl and CaCl$_2$ in water, the calcium salt lowers the freezing point of the water almost 1.5 times as much as the NaCl. Why?

6. Explain why a cucumber shrivels up when it is placed in a concentrated solution of salt.

7. Define the term "solubility." Describe an experiment you could use to measure the solubility of NaCl.

8. Use Hess's law (page 226) to explain how the heat of solution of an ionic compound is related to its lattice energy and energies of solvation of the ions.

9. A 100.-g sample of sodium chloride (NaCl) is added to 100 mL of water at 0 °C. After equilibrium is reached, about 64 g of solid remains undissolved. Describe the equilibrium that exists in this system at the particulate level of subdivision.

10. In words, describe an experimental procedure using freezing point depression to determine the molar mass of an unknown compound.

11. The general guideline "like dissolves like" is often cited in chemistry. Explain the meaning of this phrase giving several examples.

12. Which of the following substances is likely to dissolve in water, and which is likely to dissolve in benzene (C_6H_6)?

   (a) $NaNO_3$

   (b) Diethyl ether, $CH_3CH_2OCH_2CH_3$

   (c) Naphthalene ($C_{10}H_8$)

   (d) NH_4Cl

   (e) I_2

13. Explain the differences between a colloidal dispersion, a suspension, and a true solution.

14. Explain the difference between a sol and a gel. Give an example of each.

15. Explain how a surfactant such as a soap or detergent functions.

Practicing Skills

Concentration
(See Examples 14.1 and 14.2 and CD-ROM Screen 14.2)

16. Suppose you dissolve 2.56 g of succinic acid, $C_4H_6O_4$, in half a liter of water (500. mL). Assuming that the density of water is $1.00\ g/cm^3$, calculate the molarity, molality, mole fraction, and weight percentage of acid in the solution.

17. Assume you dissolve 45.0 g of camphor ($C_{10}H_{16}O$) in 425 mL of ethanol, C_2H_5OH. Calculate the molarity, molality, mole fraction, and weight percentage of camphor in this solution. (The density of ethanol is 0.785 g/mL.)

18. Fill in the blanks in the table. Aqueous solutions are assumed.

Compound	Molality	Weight Percentage	Mole Fraction
NaI	0.15	——	——
C_2H_5OH	——	5.0	——
$C_{12}H_{22}O_{11}$	0.15	——	——

19. Fill in the blanks in the table. Aqueous solutions are assumed.

Compound	Molality	Weight Percentage	Mole Fraction
KNO_3	——	10.0	——
CH_3CO_2H	0.0183	——	——
$HOCH_2CH_2OH$	——	18.0	——

20. What mass of Na_2CO_3 must you add to 125 g of water to prepare 0.200 m Na_2CO_3? What is the mole fraction of Na_2CO_3 in the resulting solution?

21. You want to prepare a solution that is 0.0512 m in $NaNO_3$. What mass of $NaNO_3$ must be added to 500. g of water? What is the mole fraction of $NaNO_3$ in the solution?

22. You wish to prepare an aqueous solution of glycerol, $C_3H_5(OH)_3$, in which the mole fraction of the solute is 0.093. What mass of glycerol must you add to 425 g of water to make this solution? What is the molality of the solution?

23. You want to prepare an aqueous solution of ethylene glycol, $HOCH_2CH_2OH$, in which the mole fraction of solute is 0.125. What mass of ethylene glycol, in grams, should you combine with 955 g of water? What is the molality of the solution?

24. Hydrochloric acid is sold as a concentrated aqueous solution. If the molarity of commercial HCl is 12.0 and its density is $1.18\ g/cm^3$, calculate

   (a) The molality of the solution

   (b) The weight percentage of HCl in the solution

25. Concentrated sulfuric acid has a density of $1.84\ g/cm^3$ and is 95.0% by weight H_2SO_4. What is the molality of this acid? What is its molarity?

26. The average lithium ion concentration in sea water is 0.18 ppm. What is the molality of Li^+ in sea water?

27. Silver ion has an average concentration of 28 ppb (parts per billion) in U. S. water supplies.

   (a) What is the molality of the silver ion?

   (b) If you wanted 1.0×10^2 g of silver and could recover it chemically from water supplies, what volume of water (in liters) would you have to treat? (Assume the density of water is $1.0\ g/cm^3$.)

The Solution Process
(See Examples 14.3 and 14.4 and CD-ROM Screens 14.3 and 14.4)

28. Which pairs of liquids will be miscible?

   (a) H_2O and $CH_3CH_2CH_2CH_3$

   (b) C_6H_6 (benzene) and CCl_4

   (c) H_2O and CH_3CO_2H

29. Acetone, CH_3COCH_3, is quite soluble in water. Explain why this should be so.

30. Use the data of Table 14.2 to calculate the enthalpy of solution of LiCl. How does the value differ from that for NaCl?

31. Use the following data to calculate the enthalpy of solution of sodium perchlorate, $NaClO_4$:

$$\Delta H_f^\circ(s) = -382.9 \text{ kJ/mol}$$

$$\Delta H_f^\circ(\text{aq, } 1 \text{ } m) = -369.5 \text{ kJ/mol}$$

32. You make a saturated solution of NaCl at 25 °C. No solid is present in the beaker holding the solution. What can be done to increase the amount of dissolved NaCl in this solution? (See Figure 14.10)

   (a) Add more solid NaCl.

   (b) Raise the temperature of the solution.

   (c) Raise the temperature of the solution and add some NaCl.

   (d) Lower the temperature of the solution and add some NaCl.

33. Some lithium chloride, LiCl, is dissolved in 100 mL of water in one beaker and some Li_2SO_4 is dissolved in 100 mL of water in another beaker. Both are at 10 °C and both are saturated solutions; some solid remains undissolved in each beaker. Describe what you would observe as the temperature is raised. The following data are available to you from a handbook of chemistry:

Solubility (g/100 mL)		
Compound	10 °C	40 °C
Li_2SO_4	35.5	33.7
LiCl	74.5	89.8

34. In each pair of ionic compounds, which is more likely to have the more negative heat of hydration? Briefly explain your reasoning in each case.

   (a) LiCl or CsCl

   (b) $NaNO_3$ or $Mg(NO_3)_2$

   (c) RbCl or $NiCl_2$

35. When salts of Mg^{2+}, Na^+, and Cs^+ are placed in water, the positive ion is hydrated (as is the negative ion). Which of these three cations is most strongly hydrated? Which one is least strongly hydrated?

Henry's Law
(See Example 14.5 and CD-ROM Screen 14.5)

36. The partial pressure of O_2 in your lungs varies from 25 mm Hg to 40 mm Hg. What mass of O_2 can dissolve in 1.0 L of water at 25 °C if the partial pressure of O_2 is 40 mm Hg?

37. Henry's law constant for O_2 in water at 25 °C is 1.66×10^{-6} M/mm Hg. Which of the following is a reasonable constant when the temperature is 50 °C? Explain the reason for your choice.

   (a) 8.80×10^{-7} M/mm Hg

   (b) 3.40×10^{-6} M/mm Hg

   (c) 1.66×10^{-6} M/mm Hg

   (d) 8.40×10^{-5} M/mm Hg

38. An unopened soda can has an aqueous CO_2 concentration of 0.0506 M at 25 °C. What is the pressure of CO_2 gas in the can?

39. Hydrogen gas has a Henry's law constant of 1.07×10^{-6} M/mm Hg at 25 °C when dissolving in water. If the total pressure of gas (H_2 gas plus water vapor) over water is 1.0 atm, what is the concentration of H_2 in the water in grams per milliliter? (See Appendix G for the vapor pressure of water.)

Raoult's Law
(See Example 14.6 and CD-ROM Screen 14.7)

40. A 35.0-g sample of ethylene glycol, $HOCH_2CH_2OH$, is dissolved in 500.0 g of water. The vapor pressure of water at 32 °C is 35.7 mm Hg. What is the vapor pressure of the water–ethylene glycol solution at 32 °C? (Ethylene glycol is nonvolatile.)

41. Urea, $(NH_2)_2CO$, which is widely used in fertilizers and plastics, is quite soluble in water. If you dissolve 9.00 g of urea in 10.0 mL of water, what is the vapor pressure of the solution at 24 °C? Assume the density of water is 1.00 g/mL.

42. Pure ethylene glycol [$HOCH_2CH_2OH$] is added to 2.00 kg of water in the cooling system of a car. The vapor pressure of the water in the system when the temperature is 90 °C is 457 mm Hg. What mass of glycol was added? (Assume that ethylene glycol is not volatile at this temperature. See Appendix G for the vapor pressure of water.)

43. Pure iodine (105 g) is dissolved in 325 g of CCl_4 at 65 °C. Given that the vapor pressure of CCl_4 at this temperature is 531 mm Hg, what is the vapor pressure of the CCl_4–I_2 solution at 65 °C? (Assume that I_2 does not contribute to the vapor pressure.)

Boiling Point Elevation
(See Example 14.7)

44. Verify that 0.200 mol of a nonvolatile solute in 125 g of benzene (C_6H_6) produces a solution whose boiling point is 84.2 °C.

45. What is the boiling point of a solution composed of 15.0 g of urea, $(NH_2)_2CO$, in 0.500 kg of water?

46. What is the boiling point of a solution composed of 15.0 g of $CHCl_3$ and 0.515 g of the nonvolatile solute acenaphthene, $C_{12}H_{10}$, a component of coal tar?

47. What is the boiling point of a solution composed of 0.755 g of caffeine ($C_8H_{10}O_2N_4$) in 95.6 g of benzene, C_6H_6?

48. Phenanthrene, $C_{14}H_{10}$, is an aromatic hydrocarbon. If you dissolve some phenanthrene in 50.0 g of benzene, the boiling point of the solution is 80.51 °C. What mass of the hydrocarbon must have been dissolved?

49. A solution of glycerol, $C_3H_5(OH)_3$, in 735 g of water has a boiling point of 104.4 °C at a pressure of 760 mm Hg. What is the mass of glycerol in the solution? What is the mole fraction of the solute?

Freezing Point Depression

(See Example 14.8 and CD-ROM Screen 14.8)

50. A mixture of ethanol (C_2H_5OH) and water has a freezing point of -16.0 °C.

    (a) What is the molality of the alcohol?

    (b) What is the weight percent of alcohol in the solution?

51. Some ethylene glycol, $HOCH_2CH_2OH$, is added to your car's cooling system along with 5.0 kg of water. If the freezing point of the water–glycol solution is -15.0 °C, what mass of $HOCH_2CH_2OH$ must have been added?

52. You dissolve 15.0 g of sucrose ($C_{12}H_{22}O_{11}$) in a cup of water (225 g). What is the freezing point of the solution?

53. Assume a bottle of wine consists of a 11% by weight solution of ethanol (C_2H_5OH) in water. If the bottle of wine is chilled to -20 °C, will the solution begin to freeze?

Colligative Properties and Molar Mass Determination

(See Example 14.9)

54. You add 0.255 g of an orange, crystalline compound whose empirical formula is $C_{10}H_8Fe$ to 11.12 g of benzene. The boiling point of the benzene rises from 80.10 to 80.26 °C. What is the molar mass and molecular formula of the compound?

55. Butylated hydroxyanisole (BHA) is used as an antioxidant in margarine and other fats and oils; it prevents oxidation and prolongs the shelf life of the food. What is the molar mass of BHA if 0.640 g of the compound, dissolved in 25.0 g of chloroform, produces a solution whose boiling point is 62.22 °C?

56. Benzyl acetate is one of the active components of oil of jasmine. If 0.125 g of the compound is added to 25.0 g of chloroform ($CHCl_3$), the boiling point of the solution is 61.82 °C. What is the molar mass of benzyl acetate?

57. Anthracene, a hydrocarbon obtained from coal, has an empirical formula of C_7H_5. To find its molecular formula you dissolve 0.500 g in 30.0 g of benzene. The boiling point of pure benzene is 80.10 °C, whereas the solution has a boiling point of 80.34 °C. What is the molecular formula of anthracene?

58. An aqueous solution contains 0.180 g of an unknown, non-ionic solute in 50.0 g of water. The solution freezes at -0.040 °C. What is the molar mass of the solute?

59. The organic compound called aluminon is used as a reagent to test for the presence of the aluminum ion in aqueous solution. A solution of 2.50 g of aluminon in 50.0 g of water freezes at -0.197 °C. What is the molar mass of aluminon?

60. The melting point of pure biphenyl ($C_{12}H_{10}$) is found to be 70.03 °C. If 0.100 g of naphthalene is added to 10.0 g of biphenyl, the freezing point of the mixture is 69.40 °C. If K_{fp} for biphenyl is -8.00 °C/m, what is the molar mass of naphthalene?

61. Phenylcarbinol is used in nasal sprays as a preservative. A solution of 0.52 g of the compound in 25.0 g of water has a freezing point of -0.36 °C. What is the molar mass of phenylcarbinol?

Colligative Properties of Ionic Compounds

(See Example 14.10 and CD-Screen 14.8, Exercise)

62. If 52.5 g of LiF is dissolved in 306 g of water, what is the expected freezing point of the solution? (Assume the van't Hoff factor i for LiF is 2.)

63. To make homemade ice cream, you cool the milk and cream by immersing the container in ice and a concentrated solution of rock salt (NaCl) in water. If you want to have a water–salt solution that freezes at $-10.$ °C, what mass of NaCl must you add to 3.0 kg of water? (Assume the van't Hoff factor i for NaCl is 1.85.)

64. List the following aqueous solutions in order of decreasing freezing point. (The last three are all assumed to be 90% dissociated.)

    (a) 0.1 m sugar

    (b) 0.1 m NaCl

    (c) 0.08 m $CaCl_2$

    (d) 0.04 m Na_2SO_4

65. Arrange the following aqueous solutions in order of decreasing freezing point. (The last three are all assumed to be 90% dissociated.)

    (a) 0.20 m ethylene glycol (nonvolatile, nonelectrolyte)

    (b) 0.12 m K_2SO_4

    (c) 0.10 m $MgCl_2$

    (d) 0.12 m KBr

Osmosis

(See Example 14.11 and CD-ROM Screen 14.9)

66. An aqueous solution contains 3.00% phenylalanine ($C_9H_{11}NO_2$) by mass. Assume the phenylalanine is non-ionic and nonvolatile. Find

    (a) The freezing point of the solution

    (b) The boiling point of the solution

    (c) The osmotic pressure of the solution at 25 °C

    In your view, which of these is most easily measured in the laboratory?

67. Estimate the osmotic pressure of human blood at 37 °C. Assume blood is isotonic with a 0.16 M NaCl solution, and assume the van't Hoff factor i is 1.9 for NaCl.

68. An aqueous solution containing 1.00 g of bovine insulin (a protein, not ionized) per liter has an osmotic pressure of 3.1 mm Hg at 25 °C. Calculate the molar mass of bovine insulin.

69. Calculate the osmotic pressure of a 0.0120 M solution of NaCl in water at 0 °C. Assume the van't Hoff factor i is 1.94 for this solution.

Colloids

(See Section 14.5 and CD-ROM Screen 14.10)

70. When solutions of $BaCl_2$ and Na_2SO_4 are mixed, the mixture becomes cloudy. After a few days, a white solid is observed on the bottom of the beaker with a clear liquid above it.

(a) Write a balanced equation for the reaction that occurs.

(b) Why is the solution cloudy at first?

(c) What happens during the few days of waiting?

71. The dispersed phase of a certain colloidal dispersion consists of spheres of diameter 1.0×10^2 nm.

(a) What is the volume $(V = \frac{4}{3}\pi r^3)$ and surface area $(A = 4\pi r^2)$ of each sphere?

(b) How many spheres are required to give a total volume of 1.0 cm^3? What is the total surface area of these spheres in square meters?

General Questions

These questions are not designated as to type or location in the chapter. They may combine several concepts. More challenging questions are indicated by an underlined number.

72. Which of the following salts, Li_2SO_4 or Cs_2SO_4, is expected to have the more exothermic heat of hydration?

73. Solution properties

(a) Which solution is expected to have the higher boiling point: 0.10 m Na_2SO_4 or 0.15 m sugar?

(b) For which aqueous solution is the vapor pressure of water higher: 0.30 m NH_4NO_3 or 0.15 m Na_2SO_4?

74. Arrange the following aqueous solutions in order of (i) increasing vapor pressure of water and (ii) increasing boiling point:

(a) 0.35 m $HOCH_2CH_2OH$ (a nonvolatile solute)

(b) 0.50 m sugar

(c) 0.20 m KBr

(d) 0.20 m Na_2SO_4

75. Making homemade ice cream is one of life's great pleasures. Fresh milk and cream, sugar, and flavorings are churned in a bucket suspended in an ice–water mixture, the freezing point of which has been lowered by adding rock salt. One manufacturer of home ice cream freezers recommends adding 2.50 lb (1130 g) of rock salt (NaCl) to 16.0 lb of ice (7250 g) in a 4-qt freezer. For the solution when this mixture melts calculate

(a) The weight percentage of NaCl

(b) The mole fraction of NaCl

(c) The molality of the solution

76. Dimethylglyoxime [DMG, $(CH_3CNOH)_2$] is used as a reagent to precipitate nickel ion. Assume that 53.0 g of DMG has been dissolved in 525 g of ethanol (C_2H_5OH).

The red, insoluble compound formed between nickel(II) ion and dimethylglyoxime (DMG) is precipitated when DMG is added to a basic solution of Ni^{2+}(aq). *(Charles D. Winters)*

(a) What is the mole fraction of DMG?

(b) What is the molality of the solution?

(c) What is the vapor pressure of the ethanol over the solution at ethanol's normal boiling point of 78.4 °C?

(d) What is the boiling point of the solution? (DMG does not produce ions in solution.) (K_{bp} for ethanol = +1.22 °C/m)

77. A 10.7 m solution of NaOH has a density of 1.33 g/cm^3 at 20 °C. Calculate

(a) The mole fraction of NaOH

(b) The weight percentage of NaOH

(c) The molarity of the solution

78. Concentrated aqueous ammonia has a molarity of 14.8 and a density of 0.90 g/cm^3. What is the molality of the solution? Calculate the mole fraction and weight percentage of NH_3.

79. If you dissolve 2.00 g of $Ca(NO_3)_2$ in 750 g of water, what is the molality of $Ca(NO_3)_2$? What is the total molality of ions in solution? (Assume total dissociation of the ionic solid.)

80. If you want a solution that is 0.100 m in ions, what mass of Na_2SO_4 must you dissolve in 125 g of water? (Assume total dissociation of the ionic solid.)

81. Consider the following aqueous solutions: (i) 0.20 m $HOCH_2CH_2OH$ (nonvolatile, nonelectrolyte); (ii) 0.10 m $CaCl_2$; (iii) 0.12 m KBr; and (iv) 0.12 m Na_2SO_4.

(a) Which solution has the highest boiling point?

(b) Which solution has the lowest freezing point?

(c) Which solution has the highest water vapor pressure?

82. Solution properties

(a) Which solution is expected to have the higher boiling point: 0.20 m KBr or 0.30 m sugar?

(b) Which aqueous solution has the lower freezing point: 0.12 m NH_4NO_3 or 0.10 m Na_2CO_3?

83. The solubility of NaCl in water at 100 °C is 39.1 g/100. g of water. Calculate the boiling point of this solution. (Assume $i = 1.85$ for NaCl.)

84. Instead of using NaCl to melt the ice on your sidewalk, you decide to use $CaCl_2$. If you add 35.0 g of $CaCl_2$ to 150. g of

water, what is the freezing point of the solution? (Assume $i = 2.7$ for $CaCl_2$.)

85. The smell of ripe raspberries is due to *p*-hydroxyphenyl-2-butanone, which has the empirical formula C_5H_6O. To find its molecular formula, you dissolve 0.135 g in 25.0 g of chloroform, $CHCl_3$. The boiling point of the solution is 61.82 °C. What is the molecular formula of the solute?

86. Hexachlorophene has been used in germicidal soap. What is its molar mass if 0.640 g of the compound, dissolved in 25.0 g of chloroform, produces a solution whose boiling point is 61.93 °C?

87. The solubility of ammonium formate, NH_4HCO_2, in 100 g of water is 102 g at 0 °C and 546 g at 80 °C. A solution is prepared by dissolving NH_4HCO_2 in 200 g of water until no more will dissolve at 80 °C. The solution is then cooled to 0 °C. What mass of NH_4HCO_2 precipitates? (Assume no water evaporates and that the solution is not supersaturated.)

88. How much N_2 can dissolve in water at 25 °C if the N_2 partial pressure is 585 mm Hg?

89. Cigars are best stored in a "humidor" at 18 °C and 55% relative humidity. This means the pressure of water vapor should be 55% of the vapor pressure of pure water at the same temperature. The proper humidity can be maintained by placing a solution of glycerol [$C_3H_5(OH)_3$] and water in the humidor. Calculate the percentage by mass of glycerol that will lower the vapor pressure of water to the desired value. (Assume the vapor pressure of glycerol is negligible.)

90. An aqueous solution containing 10.0 g of starch per liter has an osmotic pressure of 3.8 mm Hg at 25 °C.

    (a) What is the molar mass of starch? (Because not all starch molecules are identical, the result will be an average molar mass.)

    (b) What is the freezing point of the solution? Would it be easy to determine the molecular weight of starch by measuring the freezing point depression? (Assume that the molarity and molality are the same for this solution.)

91. Vinegar is a 5% solution (by weight) of acetic acid in water. Determine the mole fraction and molality of acetic acid. What is the concentration of acetic acid in parts per million (ppm)? Explain why it is impossible to calculate the molarity of this solution from the information provided.

92. A solution of 5.00 g of acetic acid in 100. g of benzene freezes at 3.37 °C. A solution of 5.00 g of acetic acid in 100. g of water freezes at −1.49 °C. Find the molar mass of acetic acid from each of these experiments. What can you conclude about the state of the acetic acid molecules dissolved in each of these solvents? Recall the discussion of hydrogen bonding in Section 13.3, and propose a structure for the species in benzene solution.

93. Some ethylene glycol, $HOCH_2CH_2OH$, is added to your car's cooling system along with 5.0 kg of water. If the freez-

ing point of the water–glycol solution is −15.0 °C, what is the boiling point of the solution?

94. If an egg is placed in dilute acetic acid (vinegar), the acid reacts with the calcium carbonate of the shell but the membrane around the egg remains intact (a). If the egg, without its shell, is placed in pure water, the egg swells (b). If, however, the egg is placed in a solution with a high solute concentration (a mixture of equal volumes of water and corn syrup), it shrivels dramatically (c). Explain these observations. (See the Chapter Focus, page 558 and Screen 14.9 of the *General Chemistry Interactive CD-ROM*, Version 3.0.)

95. Account for the fact that alcohols such as methanol (CH_3OH) and ethanol (C_2H_5OH) are miscible with water, whereas an alcohol with a long carbon chain, such as octanol ($C_8H_{17}OH$), is poorly soluble in water.

96. Starch contains C—C, C—H, C—O, and O—H bonds. Hydrocarbons have only C—C and C—H bonds. Both starch and hydrocarbons can form colloidal dispersions in water. Which dispersion is classified as hydrophobic? Which is hydrophilic? Explain briefly.

97. How are the aqueous solubilities of NaCl, KCl, RbCl, and CsCl in Figure 14.10 related to their lattice energies (see Table 9.3)?

98. Calculate the enthalpies of solution for Li_2SO_4 and K_2SO_4. Are the solution processes exothermic or endothermic? Compare them with LiCl and KCl. What similarities or differences do you find?

Compound	ΔH_f° (s) (kJ/mol)	ΔH_f° (aq, 1 *m*) (kJ/mol)
Li_2SO_4	−1436.4	−1464.4
K_2SO_4	−1437.7	−1414.0

99. Water at 25 °C has a density of 0.997 g/cm^3. Calculate the molality and molarity of pure water at this temperature.

100. If a volatile solute is added to a volatile solvent, both substances contribute to the vapor pressure over the solution. Assuming an ideal solution, the vapor pressure of each is given by Raoult's law, and the total vapor pressure is the sum of the vapor pressure of each component. A solution, assumed to be ideal, is made from 1.0 mol of toluene ($C_6H_5CH_3$) and 2.0 mol of benzene (C_6H_6). The vapor pressures of the pure solvents are 22 mm Hg and 75 mm Hg, respectively, at 20 °C. What is the total vapor pressure of the mixture? What is the mole fraction of each component in the liquid and in the vapor?

101. A solution is made by adding 50.0 mL of ethanol (C_2H_5OH) to 50.0 mL of water. What is the total vapor pressure over the solution at 20 °C? (See Study Question 100.) The vapor pressure of ethanol at 20 °C is 43.6 mm Hg.

102. A 2.0% (by mass) aqueous solution of novocainium chloride ($C_{13}H_{21}ClN_2O_2$) freezes at −0.237 °C. Calculate the

van't Hoff factor *i*. How many moles of ions are in solution per mole of compound?

103. A solution is 4.00% (by mass) maltose and 96.00% water. It freezes at −0.229 °C.

(a) Calculate the molar mass of maltose (which is not an ionic compound).

(b) The density of the solution is 1.014 g/mL. Calculate the osmotic pressure of the solution.

104. The following table lists the concentrations of the principal ions in sea water:

Ion	Concentration (ppm)
Cl^-	1.95×10^4
Na^+	1.08×10^4
Mg^{2+}	1.29×10^3
SO_4^{2-}	9.05×10^2
Ca^{2+}	4.12×10^2
K^+	3.80×10^2
Br^-	67

(a) Calculate the freezing point of water.

(b) Calculate the osmotic pressure of sea water at 25 °C. What is the minimum pressure needed to purify sea water by reverse osmosis?

105. A tree is exactly 10 m tall.

(a) What must be the total molarity of solutes if the sap rises to the top of the tree by osmotic pressure at 20 °C? Assume the groundwater outside the tree is pure water and that the density of the sap is 1.0 g/mL. (1 mm Hg = 13.6 mm H_2O)

(b) If the only solute in the sap is sucrose, $C_{12}H_{22}O_{11}$, what is its percentage by mass?

106. A 2.00% solution of H_2SO_4 in water freezes at −0.796 °C.

(a) Calculate the van't Hoff factor *i*.

(b) Which of the following best represents sulfuric acid in a dilute aqueous solution: H_2SO_4, $H^+ + HSO_4^-$, or $2 H^+ + SO_4^{2-}$?

107. A compound is known to be a potassium halide, KX. If 4.00 g of the salt is dissolved in exactly 100 g of water, the solution freezes at −1.28 °C. Identify the halide ion in this formula.

108. A newly synthesized compound containing boron and fluorine is 22.1% boron. Dissolving 0.146 g of the compound in 10.0 g of benzene gives a solution with a vapor pressure of 94.16 mm Hg at 25 °C. (The vapor pressure of pure benzene at this temperature is 95.26 mm Hg.) In a separate experiment, it is found that the compound does not have a dipole moment.

(a) What is the molecular formula for the compound?

(b) Draw a Lewis structure for the molecule, and suggest a possible molecular structure. Give the bond angles in the molecule and the hybridization of the boron atom.

109. In chemical research we often send newly synthesized compounds to commercial laboratories for analysis. These laboratories determine the weight percentage of C and H by burning the compound and collecting the evolved CO_2 and H_2O. They determine the molar mass by measuring the osmotic pressure of a solution of the compound. Calculate the empirical and molecular formulas of a compound, C_xH_yCr, given the following information:

(a) The compound contains 73.94% C and 8.27% H; the remainder is chromium.

(b) At 25 °C, the osmotic pressure of 5.00 mg of the unknown compound dissolved in exactly 100 mL of chloroform solution is 3.17 mm Hg.

Using Electronic Resources

These questions refer to the General Chemistry Interactive CD-ROM, *Version 3.0.*

110. See CD-ROM Screen 14.2: Solubility.

(a) Examine the video that plays when the "Unsaturated" button is clicked. What is the evidence that the final solution of nickel chloride is unsaturated?

(b) How can you test a solution to see if it is supersaturated?

111. See CD-ROM Screen 14.3: The Solution Process. Examine the problem screen associated with this screen.

(a) In which solvent is I_2 more soluble: H_2O or CCl_4? Could you have predicted this?

(b) What do you think will be found if the solubility of hexane (C_6H_{14}) is examined in these same two solvents: water and carbon tetrachloride? In which is hexane more soluble?

112. See CD-ROM Screen 14.5: Henry's Law. Click on the button for the audio explanation of Henry's law. What is meant by a "dynamic equilibrium"?

113. See CD-ROM Screen 14.7: Colligative Properties (1). What substance would have the greater influence on the vapor pressure of water when added to 1000 g of the liquid: 10.0 g of sucrose ($C_{12}H_{22}O_{11}$), or 10.0 g of ethylene glycol [$HOCH_2CH_2OH$]?

114. See CD-ROM Screen 14.8: Colligative Properties (2).

(a) Explain why the boiling point of a liquid is elevated on adding a solute. (Click on the animation for boiling point and do the Exercise on Screen 14.7.)

(b) Which should lower the freezing point to a greater degree: 0.10 *m* NH_4NO_3 or 0.10 *m* $HOCH_2CH_2OH$?

115. See CD-ROM Screen 14.9: Osmosis.

(a) This screen explains the observations first seen on the Chemical Puzzler screen. Osmotic pressure is found to be responsible for the changes in the egg's size. What part of the egg acts as a semipermeable membrane?

(b) If the egg were put in concentrated salt water, what would happen to its size?

(c) Suppose you have two solutions separated by a semi-permeable membrane. One contains 5.85 g of NaCl dissolved in 100 mL of solution and the other 8.88 g of KNO_3 dissolved in 100 mL of solution. In which direction will solvent flow: from the NaCl solution to the KNO_3 solution or from KNO_3 to NaCl? Explain briefly.

116. See CD-ROM Screen 14.11: Surfactants.

(a) Explain how surfactants act to help oil and water form a colloid.

(b) Explain how a fabric softener works.

15 Principles of Reactivity: Chemical Kinetics

Chapter Goals

- Understand rates of reaction and the conditions affecting rates.
- Derive the rate equation, rate constant, and reaction order from experimental data.
- Use integrated rate laws.
- Understand the collision theory of reaction rates and the role of activation energy.
- Relate reaction mechanisms and rate laws.

Faster and Faster

Certain foods such as beans, cabbage, and broccoli contain complex sugars known as oligosaccharides. These are broken down to simple sugars during the digestive process. But some people have

▲ Foods such as beans, cabbage, and broccoli are known to produce flatulence in some people due to incomplete digestion of complex sugars. However, an enzyme added to the food can help break down these complex sugars and avoid "problem gas." *(Charles D. Winters)*

a problem breaking them down completely. This can lead to a condition politely know as flatulence because the undigested material is eventually fermented by anaerobic organisms in the colon to produce gases such as CO_2, H_2, CH_4, and small amounts of smelly compounds. A commercial product, called Beano, is available to help such people. The company's advertising material states that it "is a food enzyme from a natural source that breaks down the complex sugars in gassy foods, making them more digestible."

As you will learn in this chapter, enzymes are biological catalysts. Their role is to speed up chemical reactions. One of the enzymes in Beano, α-galactosidase, speeds up the breakdown of the oligosaccharides in certain foods to the simple sugars galactose and sucrose.

(a)

(b)

(c)

(d)

(e)

▲ **CO_2 in water.** **(a)** A cold solution of CO_2 in water. **(b)** A few drops of a dye (bromthymol blue) are added to a cold solution of CO_2 in water. The yellow color of the dye indicates an acidic solution. **(c)** A less than stoichiometric amount of sodium hydroxide is added, converting the H_2CO_3 to HCO_3^- (and CO_3^{2-}). **(d)** The blue color of the dye indicates a basic solution. **(e)** The blue color begins to fade after some seconds as CO_2 slowly forms more H_2CO_3. The amount of H_2CO_3 formed is finally sufficient to consume the added NaOH and the solution is again acidic. *(Charles D. Winters)*

Chapter Outline

15.1 Rates of Chemical Reactions

15.2 Reaction Conditions and Rate

15.3 Effect of Concentration on Reaction Rate

15.4 Concentration-Time Relationships: Integrated Rate Laws

15.5 Particulate View of Reaction Rates

15.6 Reaction Mechanisms

*"**S**o Toby rose up and called loudly for more bean hash and ale. Then, turning to his fatigued companions said, "Let us give renewed fire to our extincted spirits."*

(W. Shakespeare, *Othello*, 2.1)

$$\text{Oligosaccharides} + H_2O \xrightarrow{\alpha\text{-galactosidase}} \text{galactose} + \text{sucrose}$$

Carbonic anhydrase is one of many enzymes important in biological processes (see page 632). Carbon dioxide dissolves in water to a small extent to produce carbonic acid, which ionizes to give H^+ and HCO_3^- ions.

$$CO_2(g) \rightleftharpoons CO_2(aq) \qquad \textbf{(1)}$$

$$CO_2(aq) + H_2O(\ell) \rightleftharpoons H_2CO_3(aq) \qquad \textbf{(2)}$$

$$H_2CO_3(aq) \rightleftharpoons H^+(aq) + HCO_3^-(aq) \qquad \textbf{(3)}$$

Carbonic anhydrase promotes reactions 1 and 2. Many of the H^+ ions produced by ionization of H_2CO_3 are picked up by hemoglobin in the blood as hemoglobin loses O_2. The resulting HCO_3^- ions are transported back to the lungs. When he-moglobin again takes on O_2 it releases H^+ ions. These ions and HCO_3^- re-form H_2CO_3, from which CO_2 is liberated and exhaled.

You can do an experiment that illustrates the effect of carbonic anhydrase. First, add a small amount of NaOH to a cold, aqueous solution of CO_2 (Figure A). The solution becomes basic immediately because there is not enough H_2CO_3 in the solution to use up the NaOH. After some seconds, however, dissolved CO_2 slowly produces more H_2CO_3, which consumes NaOH, and the solution is again acidic.

Now try the experiment again, this time with a few drops of blood (B). Carbonic anhydrase in blood greatly speeds up reactions 1 and 2, as evidenced by the more rapid reaction under these conditions.

Before You Begin

- Review reaction stoichiometry (Chapters 4 and 5).
- Understand the distribution of molecular energies in a gas (Figure 12.13) and in a liquid (Figure 13.12).

(a)

(b)

(c)

(d)

(e)

▲ **Action of carbonic anhydrase.** **(a)** A few drops of blood are added to a cold solution of CO_2 in water. **(b)** The dye indicates an acidic solution. **(c,d)** A less than stoichiometric amount of sodium hydroxide is added, converting the H_2CO_3 to HCO_3^- (and CO_3^{2-}). The dye's blue color indicates a basic solution. **(e)** The blue color begins to fade after a few seconds as more H_2CO_3 forms and the solution again becomes acidic. The formation of H_2CO_3 is more rapid in the presence an enzyme. *(Charles D. Winters)*

Kinetics Chemical kinetics is concerned with the rate of chemical reactions and external factors that affect reaction rate, including temperature, reactant concentration, and the state of the reactants. Here are some experiments on the "iodine clock reaction" you can do yourself with reagents available in the supermarket. For details see S.W. Wright, "The Vitamin C Clock reaction," *Journal of Chemical Education*, Volume 79, page 41, 2002.

Iodine Clock Reaction

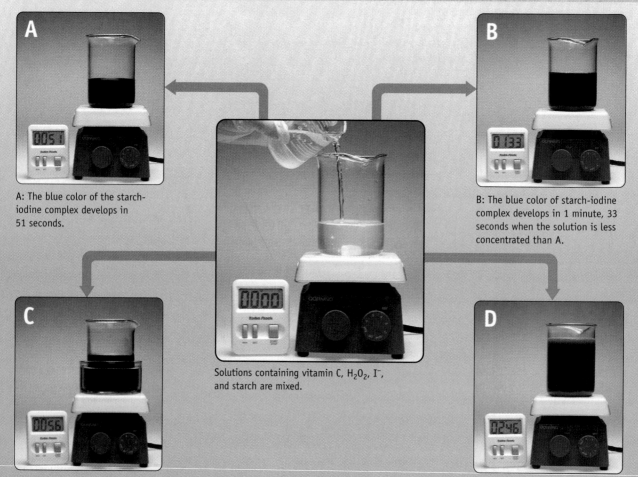

A: The blue color of the starch-iodine complex develops in 51 seconds.

B: The blue color of starch-iodine complex develops in 1 minute, 33 seconds when the solution is less concentrated than A.

Solutions containing vitamin C, H_2O_2, I^-, and starch are mixed.

C: The blue color of the starch-iodine complex develops in 56 seconds when the solution is less concentrated than A but at higher temperature.

D: Here orange juice was used as a source of vitamin C.

The Chemistry of the Iodine Clock Reaction

Hydrogen peroxide slowly oxidizes iodide ion, I^- to I_2.

$$H_2O_2(aq) + 2\ I^-(aq) + 2\ H^+(aq) \rightarrow 2\ H_2O(\ell) + I_2(aq)$$

As soon as I_2 appears in the solution, vitamin C (ascorbic acid), which is also present, rapidly reduces the I_2 back to I^-.

$$I_2(aq) + C_6H_8O_6(aq) \rightarrow C_6H_6O_6(aq) + 2\ H^+(aq) + 2\ I^-(aq)$$

When all of the vitamin C has been consumed, I_2 remains in solution and forms a blue-black complex with starch. The time measured represents how long it has taken for the given amount of vitamin C to react.

As you observe the photos of the iodine clock reaction (and a video in Chapter 15 of the *General Chemistry Interactive CD-ROM*) you might think about the following:

a) How did the amount of water affect time required for the blue-black color to develop?

b) What effect did raising the temperature have?

When carrying out a chemical reaction, chemists are concerned with two issues: the *rate* at which the reaction proceeds and the *extent* to which the reaction is product-favored. Chapter 6 began to address the second question, and Chapters 16 and 20 will develop that topic further. In this chapter we turn to the other part of our question, **chemical kinetics,** a study of the rates of chemical reactions [⊕ CD-ROM, Screen 6.3].

The study of kinetics is divided into two parts. The first part is at the *macroscopic level,* which addresses rates of reactions: what reaction rate means, how to determine a reaction rate experimentally, and how factors such as temperature and the concentrations of reactants influence rates. The second part of this subject considers chemical reactions at the *particulate level.* Here, the concern is with the **reaction mechanism,** the detailed pathway taken by atoms and molecules as a reaction proceeds. The goal is to reconcile data in the macroscopic world of chemistry with an understanding of how and why chemical reactions occur at the particulate level — and then to apply this information to control important reactions.   ●

15.1   RATES OF CHEMICAL REACTIONS

The concept of rate is encountered in many nonchemical circumstances. Common examples are the speed of an automobile in terms of the distance traveled per unit time (for example, kilometers per hour) and the rate of flow of water from a faucet as volume per unit time (liters per minute). In each case a change is measured over an interval of time. The **rate of a chemical reaction** refers to the change in concentration of a substance per unit of time. During a chemical reaction, amounts of reactants decrease with time, and amounts of products increase. It is possible to describe the rate of reaction based on either the increase in concentration of a product or the decrease in concentration of a reactant per unit of time [⊕ CD-ROM, Screen 15.2].

An easy way to gauge the speed of an automobile is to measure how far it travels during a time interval. Two measurements are made: distance traveled and time elapsed. The speed is the distance traveled divided by the time elapsed, or Δ(distance)$/\Delta$(time). If an automobile travels 2.4 miles in 4.5 min (0.075 h), its average speed is (2.4 miles/0.075 h), or 32 mph.

For a rate study, the concentration of a substance undergoing reaction can be determined by a variety of methods. For example, concentrations can be obtained by measuring a property such as the absorbance of light that is related to concentration (Figure 15.1).

Consider the decomposition of N_2O_5 in a solvent, liquid carbon tetrachloride. This reaction occurs according to the following equation:

$$2\,N_2O_5 \longrightarrow 4\,NO_2 + O_2$$

The progress of this reaction can be followed in a number of ways, one of which is by monitoring the increase in O_2 pressure. The number of moles of O_2 formed (calculated from measured values of *P, V,* and *T*), is related to the amount of N_2O_5 that has decomposed: for every 1 mol of O_2 formed, 2 mol of N_2O_5 decomposed. The N_2O_5 concentration in solution at a given time equals the initial concentration of N_2O_5 minus the amount decomposed. Data for a typical experiment done at 30.0 °C are presented as a graph of concentration of N_2O_5 versus time in Figure 15.2.

Chapter Goals • Revisited

- **Understand rates of reaction and the conditions affecting rates.**
- Derive the rate equation, rate constant, and reaction order from experimental data.
- Use integrated rate laws.
- Understand the collision theory of reaction rates and the role of activation energy.
- Relate reaction mechanisms and rate laws.

● **The Concept of Average Rate**
By calculating the speed of an automobile as the distance traveled divided by time we learn the average speed. The speed at any instant may be different.

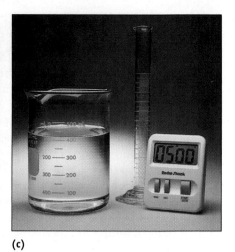

(a)   (b)   (c)

Figure 15.1   An experiment to measure rate of reaction. (a) A few drops of blue food dye were added to water, followed by a solution of bleach. Initially, the concentration of dye was about 3.4×10^{-5} M, and the bleach (NaOCl) concentration was about 0.034 M. (**b** and **c**) The dye faded as it reacted with the bleach. The absorbance of the solution can be measured at various times using a spectrophotometer, and these values can be used to determine the concentration of the dye. *(Charles. D. Winters)*

The rate of this reaction in any interval of time can be expressed as the change in concentration of N_2O_5 divided by the change in time,

● **Representing Concentration**
Recall that square brackets around a formula indicate its concentration in moles per liter (Section 5.8).

$$\text{Rate of reaction} = \frac{\text{change in } [N_2O_5]}{\text{change in time}} = -\frac{\Delta[N_2O_5]}{\Delta t}$$

The minus sign is needed because the concentration of N_2O_5 decreases with time, and the rate is always expressed as a positive quantity.

The rate could also be expressed in terms of the rate of formation of NO_2 or the rate of formation of O_2. Rates expressed in these ways will have a positive sign

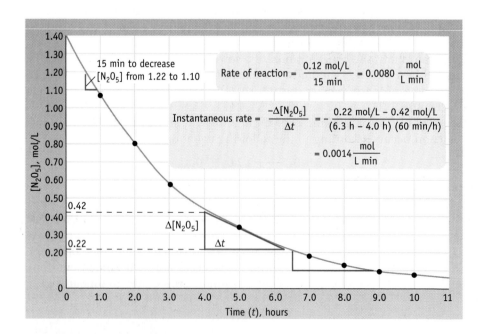

Figure 15.2   A plot of reactant concentration versus time for the decomposition of N_2O_5. The average rate for a 15-min interval from 45 min to 1 h is 0.0080 mol/L · min. The instantaneous rate calculated when $[N_2O_5] = 0.34$ M is 0.0014 mol/L · min.

15 min to decrease $[N_2O_5]$ from 1.22 to 1.10

$$\text{Rate of reaction} = \frac{0.12 \text{ mol/L}}{15 \text{ min}} = 0.0080 \frac{\text{mol}}{\text{L min}}$$

$$\text{Instantaneous rate} = \frac{-\Delta[N_2O_5]}{\Delta t} = -\frac{0.22 \text{ mol/L} - 0.42 \text{ mol/L}}{(6.3 \text{ h} - 4.0 \text{ h}) (60 \text{ min/h})}$$

$$= 0.0014 \frac{\text{mol}}{\text{L min}}$$

because the concentration is increasing. Furthermore, the rate of formation of NO_2 is twice the rate of decomposition of N_2O_5 because the balanced chemical equation tells us that 2 mol of NO_2 is formed from 1 mol of N_2O_5. The rate of formation of O_2 is one half of the rate of decomposition of N_2O_5 because 1 mol of O_2 is formed per 2 mol of N_2O_5 decomposed. For example, the rate of disappearance of N_2O_5 between 40 min and 55 min (see Figure 15.2) is given by

$$\text{Rate} = -\frac{\Delta[N_2O_5]}{\Delta t} = -\frac{(1.10 \text{ mol/L}) - (1.22 \text{ mol/L})}{55 \text{ min} - 40 \text{ min}} = +\frac{0.12 \text{ mol/L}}{15 \text{ min}}$$

$$= 0.0080 \text{ mol/L} \cdot \text{min}$$

• **Calculating Changes**
Recall that when we calculate a change in a quantity, we always do so by subtracting the initial quantity from the final quantity: $\Delta X = X_{final} - X_{initial}$.

Expressing the rate in terms of the rate of appearance of NO_2 produces a rate that is twice the rate of disappearance of N_2O_5.

$$\text{Rate} = \frac{\Delta[NO_2]}{\Delta t} = \left(\frac{0.0080 \text{ mol } N_2O_5 \text{ consumed}}{L \cdot \text{min}}\right)\left(\frac{2 \text{ mol } NO_2 \text{ formed}}{1 \text{ mol } N_2O_5 \text{ consumed}}\right)$$

$$= 0.016 \text{ mol } NO_2 \text{ formed/L} \cdot \text{min}$$

• **Units of Reaction Rates**
Notice that the units used to describe reaction rates are mol/L · time.

In terms of the rate at which O_2 is formed the rate of the reaction is

$$\text{Rate} = \frac{\Delta[O_2]}{\Delta t} = \left(\frac{0.0080 \text{ mol } N_2O_5 \text{ consumed}}{L \cdot \text{min}}\right)\left(\frac{1 \text{ mol } O_2 \text{ formed}}{2 \text{ mol } N_2O_5 \text{ consumed}}\right)$$

$$= 0.0040 \text{ mol } O_2 \text{ formed/L} \cdot \text{min}$$

The graph of concentration versus time in Figure 15.2 is not a straight line because the rate of the reaction changes during the course of the reaction. The concentration of N_2O_5 decreases rapidly at the beginning of the reaction but more slowly near the end. We can verify this by comparing the rate of disappearance of N_2O_5 calculated previously (the concentration decreased by 0.12 mol/L in 15 min) with the rate of reaction calculated for the time interval from 6.5 h to 9.0 h (when the concentration drops by 0.12 mol/L in 2.5 h).

• **Summarizing Rate Expressions**
For the reaction $2 N_2O_5 \longrightarrow 4 NO_2 + O_2$,

$$-\frac{1}{2}\frac{\Delta[N_2O_5]}{\Delta t} = \frac{1}{4}\frac{\Delta[NO_2]}{\Delta t} = \frac{\Delta[O_2]}{\Delta t}$$

To equate rates of disappearance or appearance, we divide $\Delta[\text{reagent}]/\Delta t$ by the stoichiometric coefficient in the balanced equation.

$$-\frac{\Delta[N_2O_5]}{\Delta t} = -\frac{(0.10 \text{ mol/L}) - (0.22 \text{ mol/L})}{540 \text{ min} - 390 \text{ min}} = +\frac{0.12 \text{ mol/L}}{150 \text{ min}}$$

$$= 0.00080 \text{ mol/L} \cdot \text{min}$$

The rate in this later stage of this reaction is only one tenth of the previous value.

The procedure we have used to calculate rate gives us the **average rate** over the chosen time interval. We might also ask what the **instantaneous rate** is at a single point in time. The instantaneous rate is determined by drawing a line tangent to the concentration-time curve at a particular time (see Figure 15.2) and obtaining the rate from the slope of this line. For example, when $[N_2O_5] = 0.34$ mol/L and $t = 5.0$ h, the rate is

• **The Slope of a Line**
The instantaneous rate in Figure 15.2 can be obtained from an analysis of the slope a line. See Appendix A for more on determining the slope of a line.

$$\text{Rate when } [N_2O_5] \text{ is } 0.34 \text{ M} = -\frac{\Delta[N_2O_5]}{\Delta t} = +\frac{0.20 \text{ mol/L}}{140 \text{ min}} = 1.4 \times 10^{-3} \text{ mol/L} \cdot \text{min}$$

This shows that N_2O_5 is being consumed, at that particular moment in time ($t = 5.0$ h), at a rate of 0.0014 mol/L · min.

The difference between an average rate and an instantaneous rate has an analogy in the speed of an automobile. In the previous example, the car traveled 2.4 miles in 4.5 min for an average speed of 32 mph. At any instant in time, however, the car may be moving much slower or much faster. The instantaneous speed at any instant is indicated by the car's speedometer.

Example 15.1   Reaction Rates and Stoichiometry

Problem • Compare the reaction rates for the disappearance of reactants and formation of products for the following reaction:

$$4 \, PH_3(g) \longrightarrow P_4(g) + 6 \, H_2(g)$$

Strategy • In this reaction 4 mol of PH_3 disappears when 1 mol of P_4 and 6 mol of H_2 are formed. To equate rates, we divide $\Delta[\text{reagent}]/\Delta t$ by the stoichiometric coefficient in the balanced equation.

Solution • Because 4 mol of PH_3 disappears for every mole of P_4 formed, the numerical value of the rate of formation of P_4 can only be one fourth of the rate of disappearance of PH_3. Similarly, P_4 is formed at only one sixth of the rate that H_2 is formed.

$$-\frac{1}{4}\left(\frac{\Delta[PH_3]}{\Delta t}\right) = +\frac{\Delta[P_4]}{\Delta t} = +\frac{1}{6}\left(\frac{\Delta[H_2]}{\Delta t}\right)$$

Example 15.2   Rate of Reaction

Problem • Data collected on the concentration of dye as a function of time (see Figure 15.1) is given in the graph. What is the average rate of change of the dye concentration over the first 2 min? What is the average rate of change during the fifth minute (from $t = 4$ to $t = 5$)? Estimate the instantaneous rate at 4 min.

Strategy • To find the average rate calculate the difference in concentration at the beginning and end of a time period ($\Delta c = c_{\text{final}} - c_{\text{initial}}$). Divide by the elapsed time. To find the instantaneous rate at a given time, draw a line tangent to the graph at the given time. The slope of the line (Appendix A) is the instantaneous rate [see 🖱 CD-ROM, Screen 15.2].

Solution • The concentration of dye decreases from 3.4×10^{-5} M at $t = 0$ min, to 1.7×10^{-5} M at $t = 2$ min. The average rate of the reaction in this interval of time is

$$-\frac{\Delta[\text{Dye}]}{\Delta t} = -\frac{(1.7 \times 10^{-5} \, \text{mol/L}) - (3.4 \times 10^{-5} \, \text{mol/L})}{2 \, \text{min}}$$

$$= 8.5 \times 10^{-6} \, \text{mol/L} \cdot \text{min}$$

The concentration of dye decreases from 0.90×10^{-5} M at $t = 4$ min, to 0.60×10^{-5} M at $t = 5$ min. The average rate of the reaction in this interval of time is

$$-\frac{\Delta[\text{Dye}]}{\Delta t} = -\frac{(0.6 \times 10^{-5} \, \text{mol/L}) - (0.9 \times 10^{-5} \, \text{mol/L})}{1 \, \text{min}}$$

$$= 3.0 \times 10^{-6} \, \text{mol/L} \cdot \text{min}$$

Finally, the slope of the line tangent to the curve, the instantaneous rate, at 4 min is 3.5×10^{-6} mol/L · min.

Comment • Notice that the average rate of reaction in the 4–5-min interval is less than half the value in the first minute.

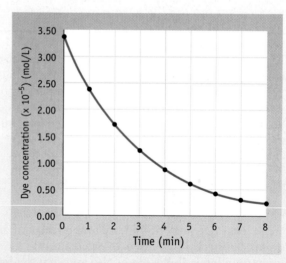

The decrease in concentration of dye as a function of time (see Figure 15.1).

Exercise 15.1   Reaction Rates and Stoichiometry

Compare the rates of appearance or disappearance of each product and reactant, in the decomposition of nitrosyl chloride, NOCl.

$$2 \, NOCl(g) \longrightarrow 2 \, NO(g) + Cl_2(g)$$

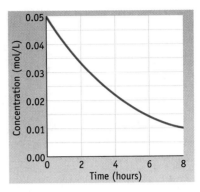

Concentration versus time for the decomposition of sucrose. See Exercise 15.2.

Exercise 15.2 Rate of Reaction

Sucrose decomposes to fructose and glucose in acid solution. A plot of the concentration of sucrose as a function of time is given in the margin. What is the rate of change of the sucrose concentration over the first 2 h? What is the rate of change over the last 2 h? Estimate the instantaneous rate at 4 h.

15.2 REACTION CONDITIONS AND RATE

For a chemical reaction to occur, molecules of the reactants must come together so that atoms can be exchanged or rearranged. Atoms and molecules are mobile in the gas phase or in solution, and so reactions are often carried out using a mixture of gases or using solutions of reactants. Under these circumstances, several factors affect the speed of a reaction.

- *Concentrations of reactants.* Alka-Seltzer contains $NaHCO_3$ and citric acid and when a tablet is placed in water these compounds react to give CO_2 (Figure 15.3). The reaction is faster in pure water than in an ethanol–water mixture in which the concentration of water is low.

- *Temperature.* Cooking involves chemical reactions, and a higher temperature results in foods cooking faster. In the laboratory reaction mixtures are often heated to make reactions occur faster (Figure 15.4).

- *Catalysts.* Catalysts are substances that accelerate chemical reactions but are not themselves transformed. For example, hydrogen peroxide, H_2O_2, decomposes to water and oxygen,

$$2 \, H_2O_2(\ell) \longrightarrow O_2(g) + 2 \, H_2O(\ell)$$

but a solution of H_2O_2 can be stored for many months because the rate of the decomposition reaction is extremely slow. Adding a manganese salt, an iodide-containing salt, or a biological substance called an *enzyme* causes this reaction to occur rapidly, as shown by vigorous bubbling as gaseous oxygen escapes from the solution (Figure 15.5, page 608).

Figure 15.3 Dependence of reaction rate on concentration. The rate of reaction depends on the concentrations of the reactants. Here an Alka-Seltzer tablet is placed in pure water (*right*) or in ethanol to which a trace of water has been added (*left*). Water is a reactant, and the reaction rate is much greater when the concentration of water is high. (*Charles D. Winters*)

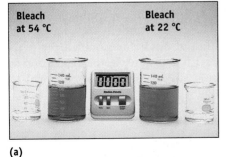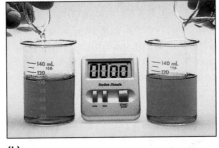

(a)   **(b)**   **(c)**

Figure 15.4 Dependence of reaction rate on temperature. Bleach, at two different temperatures, is poured into water containing blue dye. **(a)** (*far left*) Bleach at 54 °C and dye solution. (*far right*) Bleach at 22 °C and dye solution. **(b)** The bleach is poured into the two solutions. **(c)** The dye treated with warm bleach (*left*) has turned almost colorless after about 4 min, whereas the solution into which the cooler bleach was poured (*right*) is still blue. (*Charles D. Winters*)

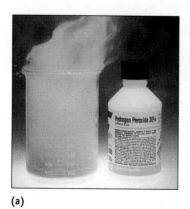
(a)

(b)

(c)

Figure 15.5 Catalyzed decomposition of H₂O₂. (a) The rate of decomposition of hydrogen peroxide is increased by the catalyst MnO₂. Here a 30% solution of H₂O₂, poured onto the black solid MnO₂, rapidly decomposes to O₂ and H₂O. Steam forms because of the high heat of reaction. **(b)** A bombardier beetle uses the catalyzed decomposition of H₂O₂ as a defense mechanism. The heat of the reaction lets the insect eject hot water and other irritating chemicals with explosive force. **(c)** A naturally occurring catalyst, called an enzyme, decomposes hydrogen peroxide. Here the enzyme found in a potato is used to catalyze H₂O₂ decomposition, and bubbles of O₂ gas are seen rising in the solution. *(a and c, Charles D. Winters; b, Thomas Eisner with Daniel Aneshansley, Cornell University)*

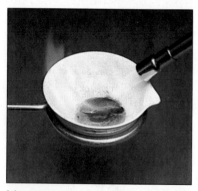
(a)

(b)

Figure 15.6 The combustion of ly-copodium powder. (a) The spores of this common fern burn only with difficulty when piled in a dish. **(b)** If the finely powdered spores are sprayed into a flame, combustion is rapid. *(Charles D. Winters)*

When reactants are in different phases, as when wood burns in air, the area of contact affects the rate. Only molecules at the surface of a solid can come in contact with other reactants. The smaller the particles of a solid, the more molecules there are on the surface. With very small particles the effect of surface area on rate can be quite dramatic (Figure 15.6). Farmers know that dust explosions (in an enclosed silo or at a feed mill) represent a major hazard.

15.3 EFFECT OF CONCENTRATION ON REACTION RATE

One goal in studying kinetics is to determine how concentrations of reactants affect the reaction rate. The effect can be determined by evaluating the rate of a reaction using different concentrations of each reactant (with temperature held constant) [CD-ROM, Screen 15.4]. Consider, for example, the decomposition of N_2O_5 to NO_2 and O_2. Figure 15.2 presented data on the concentration of N_2O_5 as a function of time. We previously calculated that the instantaneous rate of disappearance of N_2O_5 when $[N_2O_5] = 0.34$ mol/L is 0.0014 mol/L · min. An evaluation of the instantaneous rate of the reaction when $[N_2O_5] = 0.68$ mol/L shows a rate of 0.0028 mol/L · min. Doubling the concentration of N_2O_5 has doubled the reaction rate. A similar exercise shows that if $[N_2O_5]$ is 0.17 mol/L, the reaction rate is halved. These results tell us that the reaction rate is directly proportional to reactant concentration for this reaction. That is,

$$\text{Rate of reaction} \propto [N_2O_5]$$

where the symbol $\propto$ means "proportional to."

Different relationships between reaction rate and reactant concentration are encountered in other reactions. For example, the reaction rate could be independent of concentration, or the rate may depend on the concentration of some reactant raised to some power (that is, $[\text{reactant}]^n$). If the reaction involves several reactants, the reaction rate may depend on the concentrations of each of them. Finally, if a catalyst is involved, its concentration may also affect the rate.

Rate Equations

The relationship between reactant concentration and reaction rate is expressed by an equation called a **rate equation,** or **rate law.** For the N_2O_5 decomposition reaction the rate equation is

$$\text{Rate of reaction} = k[N_2O_5]$$

where the proportionality constant, k, is called the **rate constant.** This rate equation tells us that the reaction rate is proportional to the concentration of the reactant.

For a reaction such as

$$a\,A + b\,B \longrightarrow x\,X$$

the rate equation often has the form

$$\text{Rate} = k[A]^m[B]^n$$

The rate equation expresses the fact that the rate of reaction is proportional to the reactant concentrations, each concentration being raised to some power. It is important to recognize that the exponents, m and n are *not necessarily the stoichiometric coefficients* for the balanced chemical equation. *The exponents must be determined by experiment.* They are often positive whole numbers, but they can be negative numbers, fractions, or zero.

If a catalyst is present in solution, its concentration may also be included in the rate equation. Consider, for example, the decomposition of hydrogen peroxide in the presence of a catalyst such as iodide ion.

$$2\,H_2O_2(aq) \xrightarrow{\;I^-(aq)\;} 2\,H_2O(\ell) + O_2(g)$$

Experiments show that the reaction has the following rate equation:

$$\text{Reaction rate} = k[H_2O_2][I^-]$$

Here the exponent on each concentration term is 1, even though the stoichiometric coefficient of H_2O_2 is 2 and I^- does not appear in the balanced equation.

Order of a Reaction

The **order** with respect to a particular reactant is the exponent of its concentration term in the rate expression, and the total reaction order is the sum of the exponents on all concentration terms. The rate equation for the decomposition of N_2O_5

$$2\,N_2O_5 \longrightarrow 4\,NO_2 + O_2$$
$$\text{Rate} = k[N_2O_5]$$

has an exponent of 1 on $[N_2O_5]$, which means the reaction is first order with respect to N_2O_5. If the concentration of N_2O_5 is doubled, the rate of reaction doubles. If the concentration of N_2O_5 decreases by half, the rate is half as fast.

As another example, consider the reaction of NO and Cl_2:

$$2\,NO(g) + Cl_2(g) \longrightarrow 2\,NOCl(g)$$

The rate equation for this reaction is

$$\text{Rate} = k[NO]^2[Cl_2]$$

Chapter Goals • Revisited

- Understand rates of reaction and the conditions affecting rates.
- **Derive the rate equation, rate constant, and reaction order from experimental data.**
- Use the integrated rate laws.
- Understand the collision theory of reaction rates and the role of activation energy.
- Relate reaction mechanisms and rate laws.

● **The Nature of Catalysts**
A catalyst does not appear as a reactant in the balanced, overall equation for the reaction, but it may appear in the rate expression. It is common practice to indicate catalysts by placing them above the reaction arrow, as shown in the examples.

The reaction is second order in NO, first order in Cl_2, and third order overall. We can see how this rate equation is related to experimental data by examining some data for the rate of disappearance of NO.

Experiment	[NO] (mol/L)	[Cl₂] (mol/L)	Rate (mol/L · s)
1	0.250	0.250	1.43×10^{-6}
2	0.500	0.250	5.72×10^{-6}
3	0.250	0.500	2.86×10^{-6}
4	0.500	0.500	$11.4 \ \times 10^{-6}$

- If $[Cl_2]$ is held constant and [NO] is doubled to 0.500 mol/L (compare Experiment 1 and Experiment 2), the reaction rate increases by a factor of 4 (to 5.72×10^{-6} mol/L · s). The ratio of rates is 4/1, and, on dividing the rate equation for Experiment 2 by that for Experiment 1, we also find a 4/1 ratio. Therefore, the rate is proportional to $[NO]^2$, and the reaction is second order in NO.

$$\frac{\text{Rate for Experiment 2}}{\text{Rate for Experiment 1}} = \frac{5.72 \times 10^{-6} \text{ mol/L} \cdot \text{s}}{1.43 \times 10^{-6} \text{ mol/L} \cdot \text{s}} = \frac{4}{1}$$

$$\frac{k[NO]^2[Cl_2] \text{ for Experiment 2}}{k[NO]^2[Cl_2] \text{ for Experiment 1}} = \frac{k[0.500]^2[0.250]}{k[0.250]^2[0.250]} = \frac{4}{1}$$

- If [NO] is held constant and $[Cl_2]$ is doubled to 0.500 mol/L (compare Experiment 1 and Experiment 3), the rate is doubled (to 2.86×10^{-6} mol/L · s). The rate is seen to be proportional to $[Cl_2]$, and the reaction is first order in Cl_2.

The decomposition of ammonia on a platinum surface at 856 °C, like many other surface catalyzed reactions, is zero order:

$$2 \text{ NH}_3(g) \longrightarrow \text{N}_2(g) + 3 \text{ H}_2(g)$$

This means the reaction rate is independent of NH_3 concentration:

$$\text{Rate} = k[NH_3]^0 = k$$

The reaction order is important because it gives some insight into the most interesting question of all—how the reaction occurs. This is described further in Section 15.6.

Rate Constant, *k*

The rate constant, k, is a proportionality constant that relates rate and concentration *at a given temperature*. It is an important quantity because it enables you to find the reaction rate for a new set of concentrations. To see how to use k, consider the substitution of Cl^- ion by water in the cancer chemotherapy agent cisplatin, $Pt(NH_3)_2Cl_2$. The rate expression for this reaction is

$$\text{Rate} = k[Pt(NH_3)_2Cl_2]$$

• Confirming a Rate Equation
Compare Experiments 1 and 4 in the data table for the rate of NO disappearance. Does the increase in rate from 1 to 4 confirm that the rate equation is Rate = $k[NO]^2[Cl_2]$?

$$Pt(NH_3)_2Cl_2(aq) \quad + \quad H_2O(\ell) \quad \longrightarrow \quad [Pt(NH_3)_2(H_2O)Cl]^+(aq) \quad + \quad Cl^-(aq)$$

and the rate constant, k, is 0.090/h. A knowledge of k allows you to calculate the rate at a particular reactant concentration, for example when $[Pt(NH_3)_2Cl_2] =$ 0.018 mol/L:

$$Rate = (0.090/h)(0.018 \text{ mol/L}) = 0.0016 \text{ mol/L} \cdot h$$

Rate constants must have units consistent with the units for the other terms in the rate equation. The units for the reaction rate ($\Delta[R]/\Delta t$) are mol/L · time, where concentrations are given as moles per liter. In a first-order process, such as the $Pt(NH_3)_2Cl_2$ reaction, the units of k are time^{-1}. For a second-order reaction, the units of k are L/mol · time, and for a zero-order reaction the units of k are mol/L · time.

● **Time and Rate Constants**
The time in a rate constant can be seconds, minutes, hours, days, years, or whatever time unit is appropriate.

● **Expressing Time and Rate**
The fraction 1/time can also be written as time^{-1}. For example, 1/years is equivalent to years^{-1}, and 1/s is equivalent to s^{-1}.

Determining a Rate Equation

A rate equation must be determined experimentally [🖱 CD-ROM, Screen 15.5]. One way to do this is the "method of initial rates." The **initial rate** is the instantaneous reaction rate at the start of the reaction (the rate at $t = 0$). An approximate value of the initial rate can be obtained by mixing the reactants and determining $\Delta[\text{product}]/\Delta t$ or $-\Delta[\text{reactant}]/\Delta t$ after 1–2% of the limiting reactant has been consumed. Measuring the rate during the initial stage of a reaction is convenient because initial concentrations can be set to any value, and this method avoids possible complications arising from interference by reaction products or the occurrence of other reactions.

As an example of the determination of a reaction rate by this method, let us look at the reaction of sodium hydroxide with methyl acetate to produce acetate ion and methyl alcohol.

$$CH_3CO_2H \quad + \quad OH^- \quad \longrightarrow \quad CH_3CO_2^- \quad + \quad CH_3OH$$

Data in the table were collected for several experiments at 25 °C:

Experiment	Initial Concentrations		Initial Reaction Rate (mol/L · s) at 25 °C
	[CH₃CO₂CH₃]	**[OH⁻]**	
1	0.050 M ↓no change	0.050 M ↓× 2	0.00034 ↓× 2
2	0.050 M ↓× 2	0.10 M ↓no change	0.00069 ↓× 2
3	0.10 M	0.10 M	0.00137

When the initial concentration of one reactant, either $CH_3CO_2CH_3$ or OH^-, is doubled, and the concentration of the other reactant is held constant, the reaction rate doubles. This rate doubling shows that the rate for the reaction is directly proportional to the concentration of *both* $CH_3CO_2CH_3$ and OH^- so the reaction is first order in each of these reactants. The rate equation that reflects these experimental observations is

$$Rate = k[CH_3CO_2CH_3][OH^-]$$

Using this equation we can predict that doubling *both* concentrations at the same time should cause the rate to go up by a factor of 4. What happens, however, if one concentration is doubled and the other halved? The rate equation tells us the rate should not change, and that is what is observed!

When the rate equation is known, the value for k, the rate constant, can be found by substituting values of rate and concentration into the rate equation. To find k for the methyl acetate/hydroxide ion reaction, for example, data from one of the experiments are substituted into the rate equation. Using the data from the first experiment, we have

$$Rate = 0.00034 \text{ mol/L} \cdot s = k(0.050 \text{ mol/L})(0.050 \text{ mol/L})$$

$$k = \frac{0.00034 \text{ mol/L} \cdot s}{(0.050 \text{ mol/L})(0.050 \text{ mol/L})} = 0.14 \text{ L/mol} \cdot s$$

Example 15.3 Determining a Rate Equation

Problem • The rate of the reaction between CO and NO_2

$$CO(g) + NO_2(g) \longrightarrow CO_2(g) + NO(g)$$

was studied at 540 K starting with various concentrations of CO and NO_2, and the data in the table were collected. Determine the rate equation and the value of the rate constant.

	Initial Concentrations		Initial Rate
Experiment	[CO], mol/L	[NO₂], mol/L	(mol/L · h)
1	5.10×10^{-4}	0.350×10^{-4}	3.4×10^{-8}
2	5.10×10^{-4}	0.700×10^{-4}	6.8×10^{-8}
3	5.10×10^{-4}	0.175×10^{-4}	1.7×10^{-8}
4	1.02×10^{-3}	0.350×10^{-4}	6.8×10^{-8}
5	1.53×10^{-3}	0.350×10^{-4}	10.2×10^{-8}

Strategy • For a reaction involving several reactants, the general approach is to keep the concentration of one reactant constant and then decide how the rate of reaction changes as the concentration of the other reagent is varied. Because the rate is proportional to the concentration of a reactant, say A, raised to some power n (the reaction order)

$$Rate \propto [A]^n$$

we can write

$$\frac{\text{Rate in Experiment 2}}{\text{Rate in Experiment 1}} = \frac{[A_2]^n}{[A_1]^n} = \left(\frac{[A_2]}{[A_1]}\right)^n$$

If [A] doubles ($[A_2] = 2[A_1]$), and the rate doubles from Experiment 1 to 2, then $n = 1$. If [A] doubles, and the rate goes up by 4, then $n = 2$ [⊕ CD-ROM, Screen 15.5].

Solution • In the first three experiments the concentration of CO is constant. In the second experiment the NO_2 concentration has been doubled, leading to an increase in rate of a factor of 2. Thus, $n = 1$ and the reaction is first order in NO_2.

$$\frac{6.8 \times 10^{-8} \text{ mol/L} \cdot h}{3.4 \times 10^{-8} \text{ mol/L} \cdot h} = \left(\frac{0.700 \times 10^{-4}}{0.350 \times 10^{-4}}\right)^n = (2)^1$$

This is confirmed by Experiment 3. Decreasing $[NO_2]$ to half its original value causes the rate to decrease by half.

The data in Experiments 1 and 4 (with constant $[NO_2]$) show that doubling [CO] doubles the rate, whereas data from Experiments 1 and 5 show that tripling the concentration triples the rate. This means the reaction is also first order in [CO], and we now know the rate equation is

$$Rate = k[CO][NO_2]$$

The rate constant k can be found by inserting data for one of the experiments into the rate equation. Using data from Experiment 1, for example, Rate = 3.4×10^{-8} mol/L · h = $k(5.10 \times 10^{-4}$ mol/L$)(0.350 \times 10^{-4}$ mol/L$)$.

$$k = 1.9 \text{ L/mol} \cdot h$$

Example 15.4   **Using a Rate Equation to Determine Rates**

Problem • Using the rate equation and rate constant determined for the reaction of CO and NO_2 at 540 K in Example 15.3, determine the initial rate of the reaction when $[CO] = 3.8 \times 10^{-4}$ and $[NO_2] = 0.650 \times 10^{-4}$.

Strategy • A rate equation consists of three parts: a rate, a rate constant (k), and the concentration terms. If two are known (here k and the concentrations), the third can be calculated.

Solution • Substitute k ($= 1.9$ L/mol · h) and the concentration of each reactant into the rate law determined in Example 15.3.

$$\text{Rate} = k[CO][NO_2]$$
$$= 1.9 \text{ L/mol} \cdot \text{h}(3.8 \times 10^{-4} \text{ mol/L})(0.650 \times 10^{-4} \text{ mol/L})$$
$$\text{Rate} = 4.7 \times 10^{-8} \text{ mol/L} \cdot \text{h}$$

Comment • It is sometimes useful to make an educated guess before carrying out a mathematical solution. In this case, the guess would act as a check on the result that we have just calculated. We know that the reaction is first order in both reactants. Comparing concentration values in this problem with the values of concentration in Experiment 1 in Example 15.3, we notice that [CO] is about three fourths of the value, whereas [NO_2] is almost twice the value. The effects do not precisely offset each other, but we might predict that the difference in rates between this experiment and Experiment 1 will be fairly small, with the rate of this experiment being just a little greater. The calculated value bears this out.

Exercise 15.3   **Determining a Rate Equation**

The initial rate of the reaction of nitrogen monoxide and oxygen

$$2 \text{ NO(g)} + O_2\text{(g)} \longrightarrow 2 \text{ NO}_2\text{(g)}$$

was measured at 25 °C for various initial concentrations of NO and O_2. Data are collected in the table. Determine the rate equation from these data. What is the value of the rate constant and what are the appropriate units of k?

	Initial Concentrations (mol/L)		Initial Rate
Experiment	[NO]	[O_2]	(mol/L ·s)
1	0.020	0.010	0.028
2	0.020	0.020	0.057
3	0.020	0.040	0.114
4	0.040	0.020	0.227
5	0.010	0.020	0.014

Exercise 15.4   **Using Rate Laws**

The rate constant k is 0.090 h^{-1} for the reaction

$$\text{Pt(NH}_3)_2\text{Cl}_2\text{(aq)} + H_2O(\ell) \longrightarrow [\text{Pt(NH}_3)_2(H_2O)\text{Cl}]^+\text{(aq)} + \text{Cl}^-\text{(aq)}$$

and the rate equation is

$$\text{Rate} = k[\text{Pt(NH}_3)_2\text{Cl}_2]$$

Calculate the rate of reaction when the concentration of $\text{Pt(NH}_3)_2\text{Cl}_2$ is 0.020 M. What is the rate of change in the concentration of Cl^- under these conditions?

15.4 CONCENTRATION–TIME RELATIONSHIPS: INTEGRATED RATE LAWS

In many practical applications, it is useful or important to know how long a reaction must proceed to reach a predetermined concentration of some reactant or product, or what the reactant and product concentrations will be after some time has elapsed [⏻ CD-ROM, Screen 15.6]. One way to do this is to collect experimental data and construct graphs such as that shown in Figure 15.2. These methods can be inconvenient and time-consuming, however. It would be easier if we had a mathematical equation relating concentration and time, an equation in which concentration of the reactant and time were the only unknowns. Such an equation could then be used to calculate a concentration at any given time, or the length of time for a given amount of reactant to react.

In effect, we are asking the following question: Can mathematical equations be written that define relationships between the concentration of a reactant and time, such as that shown graphically in Figure 15.2? The answer is yes, so let us proceed to develop and use equations applying to rate laws for zero, first, and second-order reactions.

First-Order Reactions

Suppose the reaction "R $\longrightarrow$ products" is first order. This means the reaction rate is directly proportional to the concentration of R raised to the first power, or, mathematically,

$$-\frac{\Delta[R]}{\Delta t} = k[R]$$

Using the methods of calculus, this relationship for a first-order reaction can be transformed into a very useful equation called an **integrated rate equation** because integral calculus is used in its derivation

$$\ln \frac{[R]_t}{[R]_0} = -kt \qquad \textbf{(15.1)}$$

● **Initial and Final Time, *t***
t = 0 does not need to correspond to the actual beginning of the experiment. It can be the time when instrument readings were started, for example.

Here $[R]_0$ and $[R]_t$ are concentrations of the reactant at time $t = 0$ and at a later time, t. The *ratio* of concentrations, $[R]_t/[R]_0$, is the fraction of reactant that *remains* after a given time has elapsed. In words, the equation says

$$\text{Natural logarithm} \left(\frac{\text{concentration of R after some time}}{\text{concentration of R at start of experiment}} \right)$$

$$= \ln (\text{fraction remaining at time, } t)$$

$$= -(\text{rate constant})(\text{elapsed time})$$

Notice the negative sign in the equation. The ratio $[R]_t/[R]_0$ is less than 1 because $[R]_t$ is always less than $[R]_0$. This means the logarithm of $[R]_t/[R]_0$ is negative, and so the other side of the equation must also bear a negative sign. Equation 15.1 is useful in three ways:

- If $[R]_t/[R]_0$ is measured in the laboratory after some amount of time has elapsed, then k can be calculated.
- If $[R]_0$ and k are known, the concentration of material expected to remain after a given amount of time can be calculated.

• If k is known, Equation 15.1 can be used to calculate the time elapsed until a specific fraction ($[R]_t/[R]_0$) remains.

Finally, notice that k for a first-order reaction is independent of concentration [k has units of time^{-1} (years^{-1} or s^{-1}, for example)]. This means we can choose any convenient unit for $[R]_t$ and $[R]_0$: moles per liter, moles, grams, number of atoms, number of molecules, or pressure.

Example 15.5   The First-Order Rate Equation

Problem • Cyclopropane, C_3H_6, has been used in a mixture with oxygen as an anesthetic. This practice has diminished greatly, however, because the compound is very flammable. When heated, this compound rearranges to propene, in a process that is first-order in cyclopropane.

$$\text{Rate} = k[\text{cyclopropane}] \quad k = 5.4 \times 10^{-2}\,h^{-1}$$

If the initial concentration of cyclopropane is 0.050 mol/L, how much time (hours) must elapse for its concentration to drop to 0.010 mol/L?

Strategy • The reaction is first-order in cyclopropane. You know the rate constant k and the concentrations at $t = 0$ and after some time has elapsed. Use Equation 15.1 to calculate the time (t) elapsed to reach a concentration of 0.010 mol/L.

Solution • The first-order rate equation applied to this reaction is

$$\ln\frac{[\text{cyclopropane}]_t}{[\text{cyclopropane}]_0} = -kt$$

where $[\text{cyclopropane}]_t$, $[\text{cyclopropane}]_0$, and k are given:

$$\ln\frac{[0.010]}{[0.050]} = -(5.4 \times 10^{-2}\,h^{-1})t$$

$$t = \frac{-\ln(0.20)}{5.4 \times 10^{-2}\,h^{-1}}$$

$$t = \frac{-(-1.61)}{5.4 \times 10^{-2}\,h^{-1}}$$

$$t = 30.\,h$$

Comment • Cycloalkanes with fewer than five carbon atoms are strained because the C—C—C bond angles cannot match the preferred 109.5°. As a consequence of this ring strain, the ring opens readily to form propene (see also page 431).

Example 15.6   Using the First-Order Rate Equation

Problem • Hydrogen peroxide decomposes in dilute aqueous sodium hydroxide at 20 °C in a first-order reaction:

$$2\,H_2O_2(aq) \longrightarrow 2\,H_2O(\ell) + O_2(g)$$
$$\text{Rate} = k[H_2O_2] \quad k = 1.06 \times 10^{-3}\,min^{-1}$$

What is the fraction remaining after exactly 100 min if the initial concentration of H_2O_2 is 0.020 mol/L? What is the concentration of the peroxide after exactly 100 min?

Strategy • Because the reaction is first order in H_2O_2, we use Equation 15.1. Here $[H_2O_2]_0$, k, and t are known, and we are asked to find $[H_2O_2]_t$. Recall that

$$\frac{[R]_t}{[R]_0} = \text{fraction remaining}$$

Therefore, once this value is known, and knowing $[H_2O_2]_0$, we can calculate $[H_2O_2]_t$ [CD-ROM, Screen 15.6].

Solution • Substitute the known values into Equation 15.1.

$$\ln\frac{[H_2O_2]_t}{[H_2O_2]_0} = -kt = -(1.06 \times 10^{-3}\,min^{-1})(100.\,min)$$

$$= -0.106$$

Taking the antilogarithm of −0.106 (that is, the inverse of ln (−0.106) or $e^{-0.106}$), we find the fraction remaining to be 0.90.

$$\text{Fraction remaining} = \frac{[H_2O_2]_t}{[H_2O_2]_0} = 0.90$$

Because $[H_2O_2]_0 = 0.020$ mol/L, this gives

$$[H_2O_2]_t = 0.018\,mol/L$$

Exercise 15.5 **Using the First-Order Rate Equation**

Sucrose, a sugar, decomposes in acid solution to glucose and fructose. The reaction is first order in sucrose, and the rate constant at 25 °C is $k = 0.21$ h^{-1}. If the initial concentration of sucrose is 0.010 mol/L, what is its concentration after 5.0 h?

Exercise 15.6 **Using the First-Order Rate Equation**

Gaseous NO_2 decomposes when heated

$$2\ NO_2(g) \longrightarrow 2\ NO(g) + O_2(g)$$

The disappearance of NO_2 is a first-order reaction with $k = 3.6 \times 10^{-3}$ s^{-1} at 300 °C.

(a) A sample of gaseous NO_2 is placed in a flask and heated at 300 °C for 150 s. What fraction of the initial sample remains after this time?

(b) How long must a sample be heated so that 99% of the sample has decomposed?

Second-Order Reactions

Suppose the reaction "R $\longrightarrow$ products" is second order. The rate equation is

$$-\frac{\Delta[R]}{\Delta t} = k[R]^2$$

Using the methods of calculus, this relationship can be transformed into the following equation that relates reactant concentration and time:

$$\frac{1}{[R]_t} - \frac{1}{[R]_0} = kt \qquad (15.2)$$

The same symbolism used with first-order reactions applies: $[R]_0$ is the concentration of reactant at the time $t = 0$, and $[R]_t$ is the concentration at a later time, and k is the second-order rate constant (with units of L/mol · time).

Example 15.7 **Using the Second-Order Integrated Rate Equation**

Problem • The gas phase decomposition of HI

$$HI(g) \longrightarrow \tfrac{1}{2} H_2(g) + \tfrac{1}{2} I_2(g)$$

has the rate equation

$$-\frac{\Delta[HI]}{\Delta t} = k[HI]^2$$

where $k = 30.$ L/mol · min at 443 °C. How much time does it take for the concentration of HI to drop from 0.010 mol/L to 0.0050 mol/L at 443 °C?

Strategy • Substitute the values of $[HI]_0$, $[HI]_t$, and k into Equation 15.2.

Solution • Here $[HI]_0 = 0.010$ mol/L and $[HI]_t = 0.0050$ mol/L. Using Equation 15.2, we have

$$\frac{1}{0.0050\ \text{mol/L}} - \frac{1}{0.010\ \text{mol/L}} = (30.\ \text{L/mol} \cdot \text{min})t$$

$$(2.0 \times 10^2\ \text{L/mol}) - (1.0 \times 10^2\ \text{L/mol}) = (30.\ \text{L/mol} \cdot \text{min})t$$

$$t = 3.3\ \text{min}$$

Exercise 15.7   Using the Second-Order Concentration/Time Equation

Using the rate constant for HI decomposition given in Example 15.7, calculate the concentration of HI after 12 min if $[HI]_0 = 0.010$ mol/L.

Zero-Order Reactions

For a zero-order reaction of the kind "R $\longrightarrow$ products," the rate equation is

$$-\frac{\Delta[R]}{\Delta t} = k[R]^0$$

This equation leads to the integrated rate equation

$$[R]_0 - [R]_t = kt \qquad\qquad \textbf{(15.3)}$$

where the units of k are mol/(L $\cdot$ time).

Graphical Methods for Determining Reaction Order and the Rate Constant

Equations 15.1, 15.2, and 15.3 relating concentration and time for first-, second-, and zero-order reactions suggest a convenient way to determine the order of a reaction and its rate constant [CD-ROM, Screen 15.7]. Rearranged slightly, each of these equations has the form $y = a + bx$. This is the equation for a straight line, where b is the slope of the line and a is the y-intercept (the value of y when x is zero). As illustrated here, $x = t$ in each case.

Zero Order			First Order			Second Order		
$[R]_t = [R]_0 - kt$			$\ln[R]_t = \ln[R]_0 - kt$			$\dfrac{1}{[R]_t} = \dfrac{1}{[R]_0} + kt$		
↓	↓	↓	↓	↓	↓	↓	↓	↓
y	a	bx	y	a	bx	y	a	bx

The decomposition of ammonia on a platinum surface was previously mentioned as a zero-order reaction,

$$2\ NH_3(g) \longrightarrow N_2(g) + 3\ H_2(g) \qquad \text{Rate} = k[NH_3]^0 = k$$

which means the reaction is independent of NH_3 concentration. The straight line, obtained when the concentration at time t, $[R]_t$, is plotted against time (Figure 15.7), is proof that this reaction is zero-order in NH_3 concentration. The rate constant, k, can be determined from the slope of the line. Here the slope $= -k$, so in this case

$$-k = -1.5 \times 10^{-6}\ \text{mol/L} \cdot \text{s}$$
$$k = 1.5 \times 10^{-6}\ \text{mol/L} \cdot \text{s}$$

They y-intercept of the line at $t = 0$ is equal to $[R]_0$.

• **Graphing Software**
A convenient program for graphing and analyzing information (Mr. Plot) is available on the *General Chemistry Interactive CD-ROM*, Version 3.0, that accompanies this book.

• **Finding the Slope of a Line**
See Appendix A.5 for a description of methods for finding the slope of a line. Microsoft Excel or the graphing program Mr. Plot on the *General Chemistry Interactive CD-ROM*, Version 3.0, will also give the slope of a line from experimental data.

Figure 15.7   Plot of a zero-order reaction. A graph of the concentration of ammonia, [NH₃], against time for the decomposition of NH₃

$$2\ NH_3(g) \longrightarrow N_2(g) + 3\ H_2(g)$$

on a metal surface at 856 °C is a straight line, indicating that this is a zero-order reaction. The rate constant k for this reaction is found from the slope of the line; $k = -$slope. (The points chosen to calculate the slope are given in red.)

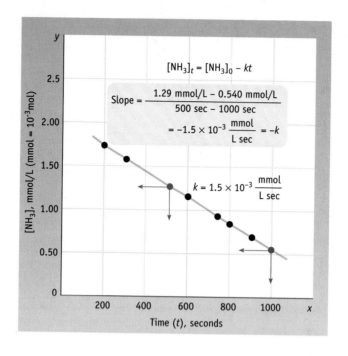

The plot of concentration versus time for a first-order reaction is a curved line (see Figure 15.2). Plotting ln [reactant] versus time, however, produces a straight line with a negative slope when the reaction is first order in that reactant. Consider the decomposition of hydrogen peroxide, a first-order reaction (see Example 15.6).

$$2\ H_2O_2(aq) \longrightarrow 2\ H_2O(\ell) + O_2(g)$$
$$\text{Rate} = k[H_2O_2]$$

Values of the concentration of H_2O_2 ($[H_2O_2]$) as a function of time for a typical experiment are given as the first two columns of numbers in Figure 15.8. The third column contains values of ln $[H_2O_2]$. A graph of ln $[H_2O_2]$ versus time produces

Figure 15.8   The decomposition of H₂O₂. (a) Concentration-versus-time data for the decomposition of hydrogen peroxide

$$2\ H_2O_2(aq) \longrightarrow 2\ H_2O(\ell) + O_2(g)$$

(b) A plot of ln $[H_2O_2]$ against time is a straight line with a negative slope, indicating a first-order reaction. The rate constant $k = -$slope.

Time (min)	$[H_2O_2]$ mol/L	ln $[H_2O_2]$
0	0.0200	−3.912
200	0.0160	−4.135
400	0.0131	−4.335
600	0.0106	−4.547
800	0.0086	−4.76
1000	0.0069	−4.98
1200	0.0056	−5.18
1600	0.0037	−5.60
2000	0.0024	−6.03

(a)

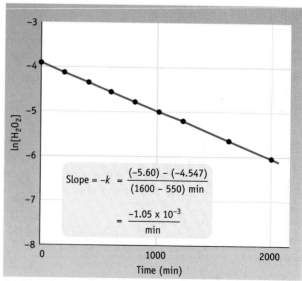

(b)

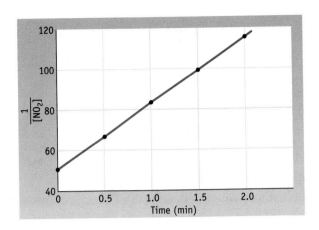

Figure 15.9   A second-order reaction. Concentration-versus-time curve for the decomposition of NO_2 [$2 NO_2(g) \longrightarrow 2 NO(g) + O_2(g)$]. A straight line for the plot of $1/[NO_2]$ versus time confirms that this is a second-order reaction. The slope of the line equals the rate constant for this reaction.

a straight line, showing that the reaction is first order in H_2O_2. The negative of the slope of the line equals the rate constant for the reaction, $1.06 \times 10^{-3}\,min^{-1}$.

The decomposition of NO_2 is a second-order process,

$$NO_2(g) \longrightarrow NO(g) + \tfrac{1}{2} O_2(g)$$

$$Rate = k[NO_2]^2$$

This fact can be verified by showing that a plot of $1/[NO_2]$ versus time is a straight line (Figure 15.9). Here the slope of the line is equal to k.

To determine the reaction order, therefore, a chemist will plot the experimental concentration–time data in different ways until a straight line plot is achieved. The mathematical relationships for zero-, first-, and second-order reactions are summarized in Table 15.1.

Exercise 15.8   **Using Graphical Methods**

Data for the decomposition of N_2O_5 in a solution at 45 °C are as follows:

$[N_2O_5]$ (mol/L)	t (min)
2.08	3.07
1.67	8.77
1.36	14.45
0.72	31.28

Plot $[N_2O_5]$, ln $[N_2O_5]$, and $1/[N_2O_5]$ versus time, t. What is the order of the reaction? What is the rate constant for the reaction?

Table 15.1 • Characteristic Properties of Reactions of the Type R $\longrightarrow$ Products

Order	Rate Equation	Integrated Rate Equation	Straight Line Plot	Slope	k Units
0	$-\Delta[R]/\Delta T = k[R]^0$	$[R]_0 - [R]_t = kt$	$[R]_t$ vs. t	$-k$	mol/L · time
1	$-\Delta[R]/\Delta T = k[R]^1$	$\ln([R]_t/[R]_0) = -kt$	$\ln[R]_t$ vs. t	$-k$	time^{-1}
2	$-\Delta[R]/\Delta T = k[R]^2$	$(1/[R]_t) - (1/[R]_0) = kt$	$1/[R]_t$ vs. t	k	L/mol · time

---

Problem-Solving Tips 15.1

Using Integrated Rate Laws

Integrated rate expressions provide a useful way to determine reaction order. Graphs based on these laws give a straight line when

- [R] is plotted against time for a zero-order process (rate = k).

- ln [R] is plotted against time for a first-order process (rate = $k[R]$).

- 1/[R] is plotted against time for a second-order process with a rate equation of the type rate = $k[R]^2$.

Half-Life and First-Order Reactions

- **Half-Life and Radioactivity**
Half-life is a common term used when dealing with radioactive elements. Radioactive decay is a first-order process, and half-life is commonly used to describe how rapidly a radioactive element decays. See Chapter 23 and Example 15.9.

The **half-life, $t_{1/2}$,** of a reaction is the time required for the concentration of a reactant to decrease to one-half its initial value [CD-ROM, Screen 15.8]. It indicates the rate at which a reactant is consumed in a chemical reaction; the longer the half-life, the slower the reaction. Half-life is used primarily when dealing with first-order processes.

If a reaction is first-order in a reactant R, $t_{1/2}$ is the time when the fraction remaining is $\frac{1}{2}$.

$$t_{1/2} \text{ is when} \qquad \frac{[R]_t}{[R]_0} = \frac{1}{2} \qquad \text{or} \qquad [R]_t = \frac{1}{2}[R]_0$$

Here $[R]_0$ is the initial concentration, and $[R]_t$ is the concentration after the reaction is half completed. To evaluate $t_{1/2}$ we substitute $[R]_t/[R]_0 = \frac{1}{2}$ and $t = t_{1/2}$ into the integrated first-order rate equation (Equation 15.1),

$$\ln\frac{[R]_t}{[R]_0} = -kt$$

$$\ln\left(\frac{1}{2}\right) = -kt_{1/2}$$

Rearranging this equation (and knowing that ln 2 = 0.693), we come to the very useful equation that relates half-life and the first-order rate constant:

$$t_{1/2} = \frac{0.693}{k} \qquad\qquad (15.4)$$

The significant feature of this equation is that $t_{1/2}$ is *independent* of concentration for a first-order reaction.

To illustrate the concept of half-life, consider the first-order decomposition of H_2O_2 (see Figure 15.8):

$$2\ H_2O_2(aq) \longrightarrow 2\ H_2O(\ell) + O_2(g)$$

- **Half-Life Equations for Other Reaction Orders**
For a zero-order reaction,

$$t_{1/2} = \frac{[R]_0}{2k}$$

For a second-order reaction,

$$t_{1/2} = \frac{1}{k[R]_0}$$

The data provided in Figure 15.8 allowed us to determine that the rate constant k for this reaction is 1.06×10^{-3} min^{-1}. Using Equation 15.4, the half-life of H_2O_2 in this reaction can be calculated from the rate constant.

$$t_{1/2} = \frac{0.693}{k} = \frac{0.693}{1.06 \times 10^{-3}\ \text{min}^{-1}} = 654\ \text{min}$$

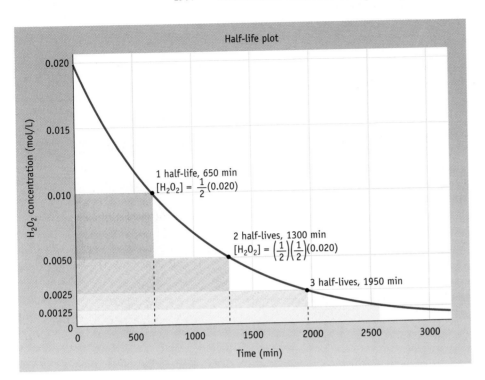

Half-life plot

1 half-life, 650 min
$[H_2O_2] = \frac{1}{2}(0.020)$

2 half-lives, 1300 min
$[H_2O_2] = \left(\frac{1}{2}\right)\left(\frac{1}{2}\right)(0.020)$

3 half-lives, 1950 min

Figure 15.10   Half-life of a first-order reaction. The concentration-versus-time curve for the disappearance of H_2O_2 (where $k = 1.06 \times 10^{-3}$ min^{-1}). The concentration of H_2O_2 is halved every 650 min. (This plot of concentration versus time is similar in shape to those for all other first-order reactions.)

In Figure 15.10, the concentration of H_2O_2 has been plotted as a function of time. This shows that $[H_2O_2]$ decreases by half within each 650-min period. The initial concentration of H_2O_2 is 0.020 M, but this drops to 0.010 M after 650 min. The concentration drops again by half (to 0.0050 M) after another 650 min. That is, after two half-lives (1300 min) the concentration is $(\frac{1}{2})(\frac{1}{2}) = (\frac{1}{2})^2 = \frac{1}{4}$, or 25% of the initial concentration. After three half-lives (1950 min), the concentration has now dropped to $(\frac{1}{2})(\frac{1}{2})(\frac{1}{2}) = (\frac{1}{2})^3 = \frac{1}{8}$, or 12.5% of the initial value; here $[H_2O_2] = 0.0025$ M.

It is hard to visualize whether a reaction is fast or slow from the value of the rate constant. Can you tell from the value of the rate constant, $k = 1.06 \times 10^{-3}$ min, whether the decomposition of H_2O_2 will require seconds, minutes, hours, or days to reach completion? Probably not, but this is easily assessed from the value of half-life for this reaction, 650 min. The half-life is just under 11 h, so you will have to wait several days for most of the H_2O_2 in a sample to decompose.

Example 15.8   Half-Life and a First-Order Process

Problem • Sucrose, $C_{12}H_{22}O_{11}$, decomposes to fructose and glucose in acid solution with the rate law

$$\text{Rate} = k[\text{sucrose}] \quad k = 0.208 \text{ h}^{-1} \text{ at } 25 \text{ °C}$$

What amount of time is required for 87.5% of the initial concentration of sucrose to decompose?

Strategy • After 87.5% of the sucrose has decomposed, 12.5% remains. That is, the fraction remaining is 0.125. To reach this point, three half-lives are required.

Half-life	Fraction Remaining
1	0.5
2	0.25
3	0.125

Therefore, we calculate the half-life from Equation 15.4 and from this the time required.

(Example continues on next page)

Solution • The half-life for the reaction is

$$t_{1/2} = \frac{0.693}{k} = \frac{0.693}{0.208 \text{ h}^{-1}} = 3.33 \text{ h}$$

Three half-lives elapse until the fraction remaining is 0.125, so

Time elapsed = 3 × 3.33 h = 9.99 h

Example 15.9 Half-Life and First-Order Processes

Problem • Radioactive radon-222 gas (^{222}Rn) from natural sources can seep into the basement of a home. The half-life of ^{222}Rn is 3.8 days. If a basement has 4.0×10^{13} atoms of ^{222}Rn per liter of air and the radon gas is trapped in the basement, how many atoms of ^{222}Rn will remain after one month (30 days)?

Strategy • We ultimately need to use Equation 15.1, where we can calculate the number of atoms remaining (= $[R]_t$) knowing the number at the beginning (= $[R]_0$), the elapsed time (30 days), and the rate constant. First, the rate constant, k, must be found from the half-life using Equation 15.4 [CD-ROM, Screen 15.6].

Solution • The rate constant, k, is

$$k = \frac{0.693}{t_{1/2}} = \frac{0.693}{3.8 \text{ d}} = 0.18 \text{ d}^{-1}$$

Now use Equation 15.1 to calculate the number of atoms remaining after 30 days.

$$\ln \frac{[Rn]_t}{4.0 \times 10^{13} \text{ atom/L}} = -(0.18 \text{ d}^{-1})(30 \text{ d}) = -5.4$$

$$\frac{[Rn]_t}{4.0 \times 10^{13} \text{ atom/L}} = e^{-5.4} = 0.0045$$

$$[Rn]_t = 1.8 \times 10^{11} \text{ atom/L}$$

Exercise 15.9 Half-Life of a First-Order Process

Americium is used in smoke detectors and in medicine for the treatment of certain malignancies. One isotope of americium, ^{241}Am, has a rate constant, k, for radioactive decay of 0.0016 year^{-1}. In contrast, radioactive iodine-125, which is used for studies of thyroid functioning, has a rate constant for decay of 0.011 day^{-1}.

(a) What are the half-lives of these isotopes?

(b) Which element decays faster?

(c) If you begin a treatment with iodine-125, and have 1.6×10^{15} atoms, how many remain after 2.0 days?

Chapter Goals • Revisited

- Understand rates of reaction and the conditions affecting rates.
- Derive the rate equation, rate constant, and reaction order from experimental data.
- Use the integrated rate laws.
- **Understand the collision theory of reaction rates and the role of activation energy.**
- Relate reaction mechanisms and rate laws.

15.5 PARTICULATE VIEW OF REACTION RATES

Throughout this book we have turned to the particulate level of chemistry to understand chemical phenomena. Looking at the way reactions occur at the atomic and molecular level provides some insight into the various influences on rates of reactions.

Let us review the macroscopic observations we have made so far concerning reaction rates. We know the wide difference in rates of reaction relates to the specific compounds involved — from very fast reactions such as an explosion that occurs when hydrogen and oxygen are exposed to a spark or flame (see Figure 1.8), to slow reactions like the formation of rust that occur over days, weeks, or years. For a specific reaction, a number of factors influence rate. These include the concentration of reactants, the temperature of the reaction system, and the presence of catalysts. Let us next look at each of these factors in more depth.

Concentration, Reaction Rate, and Collision Theory

Consider the gas phase reaction of nitric oxide and ozone,

$$NO(g) + O_3(g) \longrightarrow NO_2(g) + O_2(g)$$

The rate law for this product-favored reaction is first order in each reactant, that is Rate = $k[NO][O_3]$. How can this reaction have this rate law?

Let us consider the reaction at the particulate level and imagine a flask containing a mixture of NO and O_3 molecules in the gas phase. Both kinds of molecules are in rapid and random motion within the flask. They strike the walls of the vessel and collide with other molecules. For this or any other reaction to occur, the **collision theory** of reaction rates states that three conditions must be met:

1. The reacting molecules must collide with one another.
2. The reacting molecules must collide with sufficient energy.
3. The molecules must collide in an orientation that can lead to rearrangement of the atoms.

We shall discuss each of these within the context of the effects of concentration and temperature on reaction rate [CD-ROM, Screens 15.9–15.11].

To react, molecules must collide with one another. The rate of their reaction is primarily related to the number of collisions and that is related in turn to their concentrations (Figure 15.11). Doubling the concentration of one reagent in the NO + O_3 reaction, say NO, will lead to twice the number of molecular collisions. Figure 15.11a shows a single molecule of one of the reactants (NO) moving randomly among 16 O_3 molecules. In a given time period, it might collide with two O_3 molecules. The number of NO–O_3 collisions will double, however, if the concentration of NO molecules is doubled (to 2, as shown in Figure 15.11b) or if the number of O_3 molecules is doubled (to 32, as in Figure 15.11c). Thus we can explain the dependence of reaction rate on concentration: The number of collisions between the two reactant molecules is directly proportional to the concentrations of each reactant, and the rate of the reaction shows a first-order dependence on each reactant.

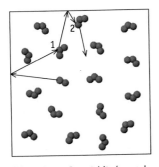 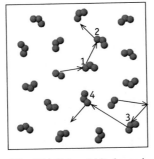 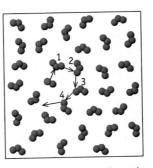

(a) 1 NO: 16 O_3 – 2 hits/second   **(b)** 2 NO: 16 O_3 – 4 hits/second   **(c)** 1 NO: 32 O_3 – 4 hits/second

Figure 15.11   The effect of concentration on the frequency of molecular collisions. (a) A single NO molecule, moving among 16 O_3 molecules, is shown colliding with two of them per second. **(b)** If 2 NO molecules move among 16 O_3 molecules, we would predict that four NO–O_3 collisions would occur per second. **(c)** If the number of O_3 molecules is doubled (to 32), the frequency of NO–O_3 collisions is also doubled, to four per second.

Temperature, Reaction Rate, and Activation Energy

In a laboratory or in the chemical industry, chemical reactions are often carried out at elevated temperature to make the reaction occur more rapidly. Conversely, it is sometimes desirable to lower the temperature to slow down a chemical reaction (to avoid an uncontrollable reaction or an explosion that could be dangerous). Chemists are very aware of the effect of temperature on the rate of a reaction. But how and why does temperature influence reaction rate?

A discussion of the effect of temperature on reaction rate begins with the distribution of energies for molecules in a sample of a gas or liquid. Recall from studying gases and liquids that the molecules in a sample have a wide range of energies, described earlier as a Boltzmann distribution of energies (← FIGURE 12.13 AND FIGURE 13.12). That is, in any sample of a gas or liquid, some molecules have very low energies, others have very high energies, but most have some intermediate energy. As the temperature increases, the average energy of the molecules increases as does the fraction having higher energies (Figure 15.12).

Activation Energy

Molecules require some minimum energy in order to react. Chemists visualize this as an energy barrier that must be surmounted by the reactants for a reaction to occur (Figure 15.13). The energy required to surmount the barrier is called the **activation energy, E_a.** If the barrier is low, the kinetic energy required is low, and a high proportion of the molecules in a sample may have sufficient energy to react. The reaction will be fast. If the barrier is high, the activation energy is high, and only a few reactant molecules in a sample may have sufficient energy. The reaction will be slow.

As an illustration of an activation energy barrier, consider the conversion of *cis*-2-butene to *trans*-2-butene. Heating either butene isomer to about 500 °C leads to a mixture of the two molecules. (There is only a very slight preference, about 4 kJ/mol, favoring the formation of *trans*-2-butene in the product mixture.)

At the molecular level we imagine that interconversion of the two isomers — called a *cis–trans* isomerization — is achieved by twisting one end of the molecule with respect to the other end. This requires considerable energy, however, because for rotation about a double bond to occur, the C=C π-bond must be broken (see p. 395).

● **Butene Isomerization**
See Screen 10.8 of the *General Chemistry Interactive CD-ROM*, Version 3.0, for an animation of the interconversion of butene isomers and the energy barrier to the process.

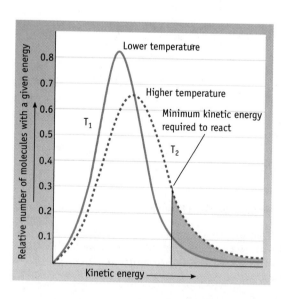

Figure 15.12   Kinetic energy distribution curve. The vertical axis gives the relative number of molecules possessing the energy indicated on the horizontal axis. The graph indicates the minimum energy required for an arbitrary reaction. At a higher temperature, a larger fraction of the molecules have sufficient energy to react. (See Figure 12.13, the Boltzmann distribution function for a collection of gas molecules.)

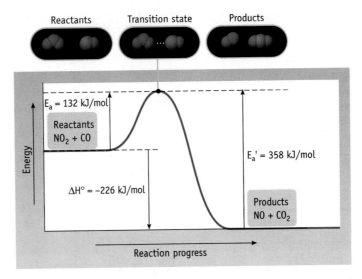

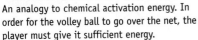

An analogy to chemical activation energy. In order for the volley ball to go over the net, the player must give it sufficient energy.

The reaction of NO_2 and CO (to give NO and CO_2) requires an activation energy of 132 kJ/mol. The reverse reaction (for $NO + CO_2 \longrightarrow NO_2 + CO$) requires 358 kJ/mol. The net enthalpy change for the reaction of NO_2 and CO is –226 kJ/mol.

Figure 15.13 Activation energy. *(Photo: Charles D. Winters)*

We can look at this reaction from the viewpoint of the energy of the system (Figure 15.14). For *cis*-2-butene to undergo a twisting motion the energy of the system must increase, reaching a maximum when the twist is 90°. This maximum in the potential energy diagram represents the activation energy barrier. The activation energy for this reaction (the forward reaction, *cis* $\longrightarrow$ *trans*) is 262 kJ/mol.

As a reaction passes over the activation energy barrier, the **transition state** is the arrangement of reactant molecules and atoms at the maximum point in the *reaction coordinate diagram*. At the transition state, sufficient energy has been concentrated in the appropriate bonds; bonds in the reactants can now break and new bonds can form to give products. The system is poised to proceed to products with a release of energy, or it can return to the reactants. Because the transition state is at a maximum in potential energy, it cannot be isolated nor can its structure be determined experimentally. However, by understanding the structures of reactants and products, we can often infer the structure of the species at the transition state.

Effect of a Temperature Increase

The conversion of *cis*-2-butene to the *trans* isomer at room temperature is slow because only a few of the butene molecules have enough energy to undergo this reaction. The rate can be increased, however, by heating the sample, which has the effect of increasing the fraction of molecules having higher energies. Raising the temperature increases the reaction rate by increasing the fraction of molecules with enough energy to surmount the activation energy barrier.

Effect of Molecular Orientation on Reaction Rate

Not only must the NO and O_3 molecules collide with sufficient energy, but they must come together in the correct orientation. Having a sufficiently high energy is necessary, but this is not sufficient to ensure that reactants will form products. For the reaction of NO and O_3, the N of NO must come together with one of the

A Closer Look

Reaction Coordinate Diagrams

Reaction coordinate diagrams (see Figures 15.13 and 15.14) can convey a great deal of information. For example, the diagram for the butene isomerization shows that, with time, the reaction progresses smoothly in a single step from reactants to products. Another example is the substitution of a halogen atom of CH_3Cl by an ion such as F^-. Here the F^- ion attacks the molecule from the side opposite the Cl substituent. As F^- begins to form a bond to carbon, the C—Cl bond weakens and the CH_3 portion of the molecule changes shape. As time progresses, the products CH_3F and Cl^- are formed. The reaction would have a reaction coordinate diagram like that in Figure 15.14.

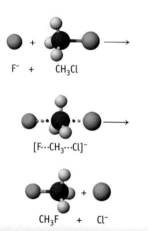

F⁻ + CH₃Cl

[F···CH₃···Cl]⁻

CH₃F + Cl⁻

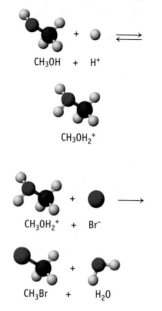

CH_3OH + H^+

$CH_3OH_2^+$

$CH_3OH_2^+$ + Br^-

CH_3Br + H_2O

The diagram in Figure A shows a two-step reaction that involves a reaction intermediate. An example would be the substitution of the —OH group on methanol by a halide ion in the presence of acid.

In the first step, an H^+ ion attaches to the O of the C—O—H group in a rapid, reversible reaction. Activation energy is required to reach this state. The energy of this protonated species, a reaction intermediate, is higher than that of the reactants

and is represented by the dip in the diagram. In the second step, a halide ion, say Br^-, attacks the intermediate in a process that requires further activation energy. The final result is methyl bromide, CH_3Br, and water.

Notice in Figure A, as in Figures 15.14 and 15.15, that the energy of the products is lower than the energy of the reactants. The reactions are exothermic.

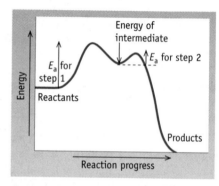

Figure A A reaction coordinate diagram for a two-step reaction, a process involving an intermediate.

O atoms at the end of the O_3 molecule (Figure 15.15). This so-called "steric factor" is important in determining the rate of the reaction and is a factor affecting the value of the rate constant *k*. The lower the probability of achieving the proper alignment, the lower the value of *k*, and the slower the reaction.

Imagine what happens when two or more complicated molecules collide. To have them come together in exactly the correct relationship means only a tiny fraction of the collisions can be effective. No wonder some reactions are so slow. Conversely, it is amazing that so many are so fast!

Arrhenius Equation

The observation that reaction rates depend on the energy and frequency of collisions between reacting molecules, on the temperature, and on whether the collisions have the correct geometry is summarized by the **Arrhenius equation:**

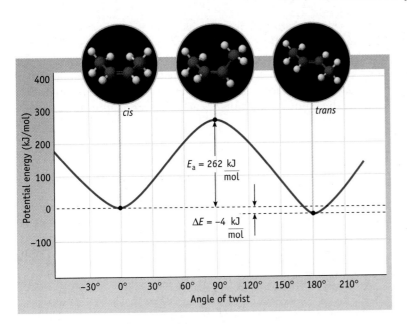

Figure 15.14 **Energy profile for the isomerization of *cis*-2-butene to *trans*-2-butene.** As the molecule twists around the central carbon–carbon bond, the energy of the system increases. The maximum on this curve is 262 kJ/mol higher than the energy of the reactant; this is the activation energy for the reaction of *cis*-2-butene to *trans*-2-butene. Notice that in this case the energy of the product, the *trans* isomer, is lower than that of the reactant, so the reaction is exothermic.

$$k = \text{rate constant} = Ae^{-E_a/RT}$$

Frequency factor · · · · · Fraction of molecules with minimum energy for reaction

(15.5)

where R is the gas constant with a value of 8.314510×10^{-3} kJ/K · mol and T is the temperature in kelvins. The parameter A is called the *frequency factor*. It is related to the number of collisions and to the fraction of collisions that have the correct geometry. The factor $e^{-E_a/RT}$ is interpreted as the *fraction of molecules hav-*

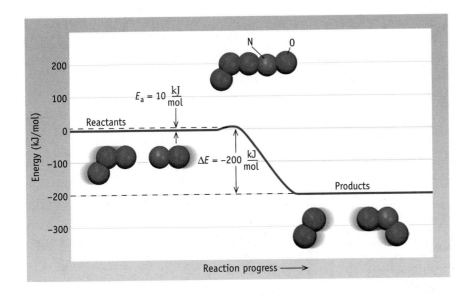

Figure 15.15 **The exothermic reaction of NO and O$_3$.** Reaction is possible only if ozone, O$_3$, and NO approach each other in the proper orientation and with sufficient energy to cross over the activation energy barrier. If O$_3$ approaches the O atom of NO, or if NO approaches the central O atom of O$_3$, for example, reaction should not occur.

ing at least the minimum energy required for reaction; its value is always less than 1. As the table in the margin shows, the factor changes significantly with temperature.

The Arrhenius equation is valuable because it can be used to (1) calculate the value of the activation energy from the temperature dependence of the rate constant and (2) calculate the rate constant for a given temperature if the activation energy and A are known. Taking the natural logarithm of each side of Equation 15.5, we have

$$\ln k = \ln A - \left(\frac{E_a}{RT}\right)$$

and, if we rearrange this slightly, it becomes the equation for a straight line relating $\ln k$ to $(1/T)$:

$$\ln k = \ln A - \frac{E_a}{R}\left(\frac{1}{T}\right) \quad \longleftarrow \text{ Arrhenius equation}$$

$$\downarrow \qquad \downarrow \qquad \downarrow$$

$$y \quad = a \quad + bx \quad \longleftarrow \text{ Equation for straight line}$$

(15.6)

This means that, if the natural logarithm of k ($\ln k$) is plotted versus $1/T$, the result is a downward sloping line with a slope of $(-E_a/R)$. So, now we have a means to calculate E_a from experimental values of k at several temperatures, a calculation illustrated in Example 15.10.

• **Interpreting the Arrhenius Equation**

(a) Fraction of molecules having sufficient energy for reaction as a function of T.

Temperature (K)	Value of $e^{-E_a/RT}$ for $E_a = 40$ kJ/mol
298	9.7×10^{-8}
400	5.9×10^{-6}
600	3.3×10^{-4}

(b) Significance of A. Although a complete understanding of A goes beyond the level of this text, it can be noted that A becomes smaller as the reactants become larger, a reflection of the "steric effect."

Example 15.10   Determination of E_a from the Arrhenius Equation

Problem • Using the experimental data shown in the table, calculate the activation energy E_a for the reaction

$$2\ N_2O(g) \longrightarrow 2\ N_2(g) + O_2(g)$$

Experiment	Temperature (K)	k [L/mol · s]
1	1125	11.59
2	1053	1.67
3	1001	0.380
4	838	0.0011

Strategy • To use the Arrhenius equation, Equation 15.6, we first need to calculate $\ln k$ and $1/T$ for each data point. These data are then plotted, and E_a is calculated from the slope of the resulting straight line (slope $= -E_a/R$).

Solution • The data are expressed as $1/T$ and $\ln k$ [CD-ROM, Screen 15.11].

Experiment	$(1/T)K^{-1}$	$\ln k$
1	8.889×10^{-4}	2.4501
2	9.497×10^{-4}	0.513
3	9.990×10^{-4}	-0.968
4	11.9×10^{-4}	-6.81

Plotting these data gives the graph shown here. Choosing the large blue points on the graph, the slope is found to be

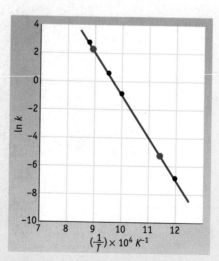

A plot of $\ln k$ versus $1/T$ for the reaction $2\ N_2O(g) \longrightarrow 2\ N_2(g) + O_2(g)$. E_a can be calculated from the slope of the line.

$$\text{Slope} = \frac{\Delta \ln k}{\Delta(1/T)} = \frac{2.0 - (-5.6)}{(9.0 - 11.5) \times 10^{-4}/K}$$

$$= -\frac{7.6}{2.5 \times 10^{-4}} \text{ K}$$

$$= -3.0 \times 10^4 \text{ K}$$

$$-3.0 \times 10^4 \text{ K} = -\frac{E_a}{8.31 \times 10^{-3} \text{ kJ/K} \cdot \text{mol}}$$

$$E_a = 250 \text{ kJ/mol}$$

The activation energy is evaluated from

$$\text{Slope} = -\frac{E_a}{R}$$

In addition to the graphical method for evaluating E_a used in Example 15.10, E_a can also be obtained algebraically. Knowing k at two different temperatures, we can write an equation for each of these conditions:

$$\ln k_1 = \ln A - \left(\frac{E_a}{RT_1}\right) \qquad \text{or} \qquad \ln k_2 = \ln A - \left(\frac{E_a}{RT_2}\right)$$

If one of these equations is subtracted from the other, we have

$$\ln k_2 - \ln k_1 = \ln \frac{k_2}{k_1} = -\frac{E_a}{R}\left[\frac{1}{T_2} - \frac{1}{T_1}\right] \qquad (15.7)$$

• **E_a, Reaction Rates, and Temperature**
A good rule of thumb is that reaction rates double for every 10 °C rise in temperature in the vicinity of room temperature.

an equation used in Example 15.11.

Example 15.11   **Calculating E_a from the Temperature Dependence of k**

Problem • Using values of k determined at two different temperatures, calculate the value of E_a for the decomposition of HI:

$$2 \text{ HI(g)} \longrightarrow \text{H}_2\text{(g)} + \text{I}_2\text{(g)}$$

$k_1 = 2.15 \times 10^{-8}$ L/(mol · s) at $T_1 = 6.50 \times 10^2$ K
$k_2 = 2.39 \times 10^{-7}$ L/(mol · s) at $T_2 = 7.00 \times 10^2$ K

Strategy • Here we are given values of k_1, T_1, k_2, and T_2, and so we use Equation 15.7.

Solution •

$$\ln \frac{2.39 \times 10^{-7} \text{ L/(mol} \cdot \text{s)}}{2.15 \times 10^{-8} \text{ L/(mol} \cdot \text{s)}} =$$

$$-\frac{E_a}{8.315 \times 10^{-3} \text{ kJ/K} \cdot \text{mol}}\left[\frac{1}{7.00 \times 10^2 \text{ K}} - \frac{1}{6.50 \times 10^2 \text{ K}}\right]$$

Solving this equation for E_a, we find

$$E_a = 182 \text{ kJ/mol}$$

Exercise 15.10   **Calculating E_a from the Temperature Dependence of k**

The colorless gas N_2O_4 decomposes to the brown gas NO_2 in a first-order reaction:

$$\text{N}_2\text{O}_4\text{(g)} \longrightarrow 2 \text{ NO}_2\text{(g)}$$

The rate constant $k = 4.5 \times 10^3$ s^{-1} at 274 K and 1.00×10^4 s^{-1} at 283 K. What is the energy of activation, E_a?

Effect of Catalysts on Reaction Rate

Catalysts are substances that speed up the rate of a chemical reaction. Examples in earlier discussions in this chapter are MnO_2 (see Figure 15.5), iodide ion (page 609), an enzyme in a potato (page 608), and hydroxide ion (page 615). In biological systems, catalysts called *enzymes* are used in most reactions (page 632).

Catalysts are not consumed in a chemical reaction [CD-ROM, Screen 15.14]. They are, however, intimately involved in the details of the reaction at the particulate level. Their function is to provide a different pathway with a lower activation energy for the reaction.

To illustrate how a catalyst participates in a reaction, let us again consider the conversion of *cis*-2-butene to *trans*-2-butene.

$$\begin{array}{ccc}
H_3C \qquad CH_3 & & H \qquad CH_3 \\
\quad C = C \quad (g) & \longrightarrow & \quad C = C \quad (g) \\
H \qquad\quad H & & H_3C \qquad H \\
\textit{cis}\text{-2-butene} & & \textit{trans}\text{-2-butene}
\end{array}$$

As noted earlier (page 624), this is a first-order reaction. The rate equation for the reaction is Rate = k[*cis*-2-butene], and the energy profile shows a high activation energy barrier associated with the breaking of the π bond in the molecule (see Figure 15.14). Because of the high activation energy this is a slow reaction, and rather high temperatures are required for it to occur at a reasonable rate.

The butene conversion reaction is greatly speeded up by a trace of iodine, however. The presence of iodine allows the isomerization reaction to be carried out at a temperature several hundred degrees lower than the uncatalyzed reaction. Iodine is not consumed (nor is it a product), and it does not appear in the overall balanced equation. It does appear in the reaction rate law, however; the rate of the reaction depends on the square root of the iodine concentration:

$$\text{Rate} = k\,[\textit{cis}\text{-2-butene}]\,[I_2]^{1/2}$$

The rate of the *cis–trans* conversion changes because the presence of I_2 changes the reaction mechanism (Figure 15.16). The best hypothesis is that iodine molecules first dissociate to form iodine atoms (Step 1). An I atom then adds to one of the C atoms of the C=C double bond (Step 2). This converts the double bond between the carbon atoms to a single bond (the π bond is broken) and allows the

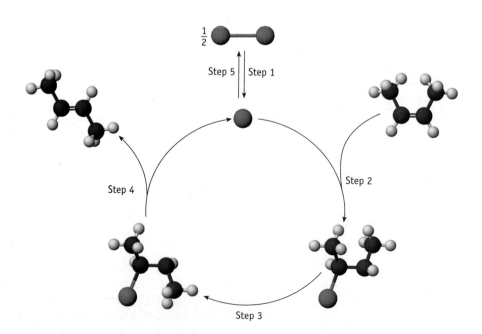

Figure 15.16 The mechanism of the iodine-catalyzed isomerization of *cis*-2-butene. *Cis*-2-butene is converted to *trans*-2-butene in the presence of a catalytic amount of iodine.

ends of the molecule to twist freely relative to each other (Step 3). If the I atom then dissociates from the intermediate, the double bond can re-form in the *trans* configuration (Step 4).

The iodine atom catalyzing the rotation is now free to add to another molecule of *cis*-2-butene. The result is a kind of chain reaction as one molecule of *cis*-2-butene after another is converted to the *trans* isomer. The chain is broken if the iodine atom recombines with another iodine atom to re-form molecular iodine.

An energy profile for the catalyzed reaction (Figure 15.17) contains several interesting features. First, the overall energy barrier has been greatly lowered from the situation in the uncatalyzed reaction. Second, the profile for the reaction includes several steps (five in all), representing each step in the reaction. This diagram includes a series of chemical species called **reaction intermediates,** species formed in one step of the reaction and consumed in a later step. Iodine atoms are intermediates, as are the free radical species formed when an iodine atom adds to *cis*-2-butene.

Five important points are associated with this mechanism.

1. Iodine, I_2, molecules dissociate to atoms and then re-form. On the macroscopic level the concentration of I_2 is unchanged. Iodine does not appear in the balanced, stoichiometric equation even though it appears in the rate equation. This is true of catalysts in general.

2. Both the catalyst I_2 and the reactant *cis*-2-butene are in the gas phase. If a catalyst is present in the same phase as the reacting substance it is called a **homogeneous catalyst.**

3. Iodine atoms and the radical species formed by addition of I to 2-butene are intermediates.

4. The activation energy barrier to reaction is significantly lower because the mechanism changed. (Compare Figures 15.14 and 15.17.) In fact, dropping the activation energy from 262 kJ/mol for the uncatalyzed reaction to 115 kJ/mol for the catalyzed process makes the catalyzed reaction 10^{15} times faster!

5. The diagram of energy-versus-reaction progress has five energy barriers (five humps appear in the curve). This feature in the diagram means that the reaction occurs in a series of five steps.

● **Heterogeneous Catalysts**
Many reactions occurring in solution are catalyzed by solid catalysts. These are referred to as *heterogeneous catalysts.* An example is the catalysis of H_2O_2 decomposition by solid MnO_2 in Figure 15.5a.

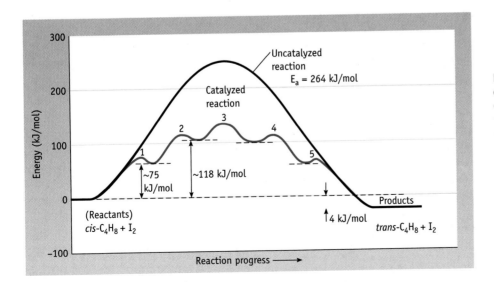

Figure 15.17 Energy profile for the iodine-catalyzed reaction of *cis*-2-butene. A catalyst accelerates a reaction by altering the mechanism so that the activation energy is lowered. With a smaller barrier to overcome, more reacting molecules have sufficient energy to surmount the barrier, and the reaction occurs more readily. The energy profile for the uncatalyzed conversion of *cis*-2-butene to *trans*-2-butene is shown by the black curve, and that for the iodine-catalyzed reaction is represented by the red curve. Notice that the shape of the barrier has changed because the mechanism has changed. (The numbers refer to the steps of the mechanism in Figure 15.16)

A Closer Look

Enzymes, Nature's Catalysts

Enzymes are powerful catalysts, typically producing a reaction rate that is 10^7–10^{14} times faster than the uncatalyzed rate. Many enzymes require metal ions for maximal activity. Carboxypeptidase, for example, contains Zn^{2+} ions at the active site.

In 1913, Leonor Michaelis and Maud L. Menten proposed a general theory of enzyme action based on kinetic observations. They assumed that the substrate, S (the reactant) and the enzyme, E, form a complex, ES. This complex then breaks down, releasing the enzyme and producing product, P.

$$E + S \rightleftharpoons ES \rightleftharpoons E + P$$

The following table lists a few important enzymes. One, carbonic anhydrase, was noted at the beginning of the chapter (page 600), where a simple experiment showed the rate-enhancing ability of this enzyme.

Here is another experiment you can do with carbonic anhydrase. Take a sip of very cold carbonated beverage. The tingling sensation you feel on your tongue and in your mouth is not from the CO_2 bubbles. Rather, it comes from the protons released when carbonic anhydrase accelerates the formation of H^+ ions from dissolved H_2CO_3 (see page 601). Acidification of nerve endings creates the tingling feeling.

The enzymes trypsin, chymotrypsin, and elastase are digestive enzymes, catalyzing the hydrolysis of peptide bonds (◀ PAGE 461). They are synthesized in the pancreas and secreted into the digestive tract.

Acetylcholinesterase is involved in transmission of nerve impulses. Many pesticides interfere with this enzyme, so farm workers are often tested to be sure they have not been overexposed to agricultural toxins.

The liver has the primary role in maintaining blood glucose levels. The liver produces glucose with phosphate groups attached (PO_4^{3-}), so the enzyme glucose phosphatase in the liver has the function of removing the phosphate group before the glucose enters the blood.

The tingling feeling you get when you drink a carbonated beverage comes from the H^+ ions released by H_2CO_3. The acid is formed rapidly from dissolved CO_2 in the presence of the enzyme carbonic anhydrase in your mouth. (*Charles D. Winters*)

Biologically Important Reactions Catalyzed by Enzymes

Enzyme	Enzyme Function or Reaction Catalyzed
Carbonic anhydrase	$CO_2 + H_2O \longrightarrow HCO_3^- + H^+$
Chymotrypsin	Cleavage of peptide linkages in proteins
Urease	$(H_2N)_2CO + 2 H_2O + H^+ \longrightarrow 2 NH_4^+ + HCO_3^-$
Catalase	$2 H_2O_2 \longrightarrow 2 H_2O + O_2$
Acetylcholinesterase	Degrades acetylcholine to acetate and choline after it has transmitted a nerve impulse
Hexokinase and glucokinase	Both enzymes catalyze the formation of a phosphate ester linkage to a —OH group of a sugar. Glucokinase is a liver-specific enzyme, and the liver is the major organ for the storage of excess dietary sugar as glycogen.

What we have described here are *reaction mechanisms*. The uncatalyzed isomerization reaction of *cis*-2-butene is a one-step reaction mechanism, whereas the catalyzed mechanism involves a series of steps. We shall discuss reaction mechanisms in more detail in Section 15.6.

15.6 REACTION MECHANISMS

One of the most important reasons to study the rates of reactions is that rate laws help us to develop *reaction mechanisms*, the sequence of bond-making and bond-breaking steps that occurs during the conversion of reactants to products [CD-

ROM, Screens 15.12 and 15.13]. The study of reaction mechanisms places you squarely in the realm of the particulate level of chemistry. We want to analyze the changes that atoms and molecules undergo when they react. And then, we want to relate this description back to the macroscopic world, to the experimental observations of reaction rates.

You have seen that the rate equation for a reaction is determined by experimentation. Based on the rate equation, and applying chemical intuition, chemists can often make an educated guess about the mechanism for a reaction. In some reactions, the conversion of reactants to products in a single step is envisioned. For example, nitric oxide and ozone react in a single-step reaction, with the reaction occurring as a consequence of a collision between reactant molecules:

$$NO(g) + O_3(g) \longrightarrow NO_2(g) + O_2(g)$$

The uncatalyzed isomerization of *cis*-2-butene to *trans*-2-butene is also best described as a single-step reaction.

Most chemical reactions occur in a sequence of steps, however. We saw an example of this with the iodine-catalyzed 2-butene isomerization reaction. Another example of a reaction that occurs in several steps is the reaction of bromine and NO:

$$Br_2(g) + 2 NO(g) \longrightarrow 2 BrNO(g)$$

A single-step reaction would require that three molecules collide simultaneously in just the right orientation. Clearly, such an event has a very low probability of occurring. Thus, for this reaction it would be reasonable to look for a mechanism that occurs in a series of steps, each step involving only one or two molecules. For example, in one possible mechanism Br_2 and NO combine in an initial step to produce an intermediate species, Br_2NO (Figure 15.18). This intermediate then reacts with another NO to give the reaction products. The equation for the overall reaction is obtained by adding the equations for these two steps:

Step 1: $Br_2(g) + NO(g) \longrightarrow Br_2NO(g)$

Step 2: $\underline{Br_2NO(g) + NO(g) \longrightarrow 2 BrNO(g)}$

Overall Reaction: $Br_2(g) + 2 NO(g) \longrightarrow 2 BrNO(g)$

Each step in a multistep-reaction sequence is an **elementary step,** which is defined as a chemical equation that describes an assumed single molecular event such as the formation or rupture of a chemical bond or the displacement of atoms as a result of a molecular collision. Each step has its own activation energy barrier, E_a, and rate constant, k. The steps must add up to give the balanced equation for the overall reaction, and the time required to complete all of the steps defines the overall reaction rate. A series of steps that satisfactorily explains the kinetic properties of a chemical reaction constitutes a possible reaction mechanism.

Mechanisms of reactions are postulated starting with experimental data. To see how this is done, we first describe three types of elementary steps.

● **Rate Laws and Mechanisms**
Rate laws are macroscopic observations. Mechanisms analyze how reactions occur at the particulate level.

Chapter Goals ● Revisited

● Understand rates of reaction and the conditions affecting rates.

● Derive the rate equation, rate constant, and reaction order from experimental data.

● Use the integrated rate laws.

● Understand the collision theory of reaction rates and the role of activation energy.

● **Relate reaction mechanisms and rate laws.**

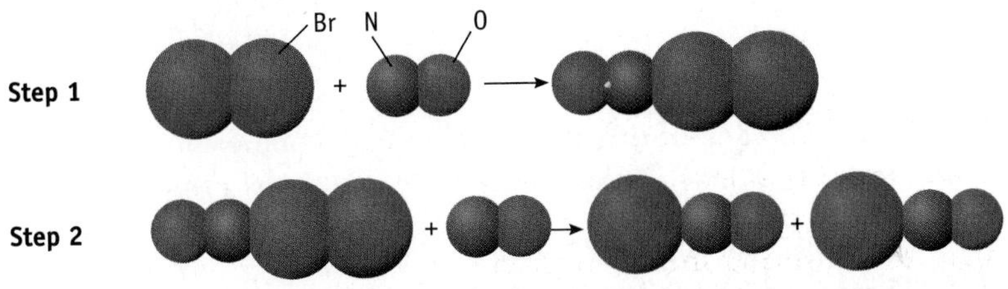

Figure 15.18   A reaction mechanism. A representation of the proposed two-step mechanism by which NO and Br_2 are converted to NOBr.

Molecularity of Elementary Steps

Elementary steps are classified by the number of reactant molecules (or ions, atoms, or free radicals) that come together. This whole, positive number is called the **molecularity** of the elementary step. When one molecule is the only reactant in an elementary step, the reaction is a **unimolecular** process. A **bimolecular** elementary process involves two molecules, which may be identical (A + A $\longrightarrow$ products) or different (A + B $\longrightarrow$ products). For example, a two-step mechanism has been proposed for the decomposition of ozone in the stratosphere:

Step 1:	Unimolecular	$O_3(g) \longrightarrow O_2(g) + O(g)$
Step 2:	Bimolecular	$O_3(g) + O(g) \longrightarrow 2\ O_2(g)$
Overall Reaction:		$2\ O_3(g) \longrightarrow 3\ O_2(g)$

The reaction is suggested to occur by an initial unimolecular step followed by a second bimolecular step.

A **termolecular** elementary step involves three molecules. This could involve three molecules of the same or different types (3 A $\longrightarrow$ products; 2 A + B $\longrightarrow$ products; or A + B + C $\longrightarrow$ products). As might be suspected, the simultaneous collision of three molecules is not likely, unless one of the molecules involved is in high concentration, such as a solvent molecule. In fact, most termolecular processes involve the reaction of two molecules; the function of the third particle is to absorb the excess energy produced when a new chemical bond is formed by the first two molecules in an exothermic step. For example, N_2 is unchanged in a termolecular reaction between oxygen molecules and oxygen atoms that produces ozone in the upper atmosphere:

$$O(g) + O_2(g) + N_2(g) \longrightarrow O_3(g) + \text{energetic } N_2(g)$$

The probability that four or more molecules will simultaneously collide with sufficient kinetic energy and proper orientation to react is so small that molecularities greater than three are never proposed.

Rate Equations for Elementary Steps

As you have already seen, the experimentally determined rate equation for a reaction cannot be predicted from its overall stoichiometry. In contrast, *the rate equation for any elementary step is defined by the stoichiometry of the step*. The rate equation of an elementary step is given by the product of the rate constant and the concentrations of the reactants in that step. This means we can write the rate equation for any elementary step, as shown by examples in the table:

Elementary Step	Molecularity	Rate Equation
A $\longrightarrow$ product	Unimolecular	Rate = $k[A]$
A + B $\longrightarrow$ product	Bimolecular	Rate = $k[A][B]$
A + A $\longrightarrow$ product	Bimolecular	Rate = $k[A]^2$
2 A + B $\longrightarrow$ product	Termolecular	Rate = $k[A]^2[B]$

For example, the rate laws for each of the two steps in the decomposition of ozone are

$$\text{Rate for (unimolecular) Step 1} = k[O_3]$$

$$\text{Rate for (bimolecular) Step 2} = k'\,[O_3][O]$$

When a reaction mechanism consists of two elementary steps, it is likely that the two steps are going to occur at different rates. The two rate constants (k and k' in this example) are not expected to have the same value (nor the same units, if the two steps have different molecularities).

Molecularity and Reaction Order

The molecularity of an elementary step and its order are the same. A unimolecular elementary step must be first order, a bimolecular elementary step must be second order, and a termolecular elementary step must be third order. Such a direct relation between molecularity and order is emphatically *not* true for the *overall* reaction. If you discover experimentally that a reaction is first order, you cannot conclude that it occurs in a single, unimolecular elementary step. Similarly, a second-order rate equation does not imply the reaction occurs in a single, bimolecular elementary step. An example illustrating this is the decomposition of N_2O_5:

$$2\ N_2O_5(g) \longrightarrow 4\ NO_2(g) + O_2(g)$$

Here the rate equation is Rate $= k[N_2O_5]$, but chemists are fairly certain the mechanism involves a series of unimolecular and bimolecular steps.

To see how the experimentally observed rate equation for the *overall reaction* is connected with a possible mechanism or sequence of elementary steps requires some chemical intuition. We will provide only a glimpse of the subject in the next section.

Example 15.12   Elementary Steps

Problem • The hypochlorite ion undergoes self oxidation–reduction to give chlorate, ClO_3^-, and chloride ions:

$$3\ ClO^-(aq) \longrightarrow ClO_3^-(aq) + 2\ Cl^-(aq)$$

It is thought that the reaction occurs in two steps:

Step 1 $ClO^-(aq) + ClO^-(aq) \longrightarrow ClO_2^-(aq) + Cl^-(aq)$

Step 2 $ClO_2^-(aq) + ClO^-(aq) \longrightarrow ClO_3^-(aq) + Cl^-(aq)$

What is the molecularity of each step? Write the rate equation for each reaction step. Show that the sum of these reactions gives the equation for the net reaction.

Strategy • The molecularity is the number of ions or molecules involved in a reaction step. The rate equation involves the concentration of each ion or molecule in an elementary step, raised to the power of its stoichiometric coefficient.

Solution • Because two ions are involved in each elementary step, each step is bimolecular. The rate equation for any elementary step is the product of the concentrations of the reactants. Thus, in this case, the rate equations are

Step 1   Rate $= k[ClO^-]^2$

Step 2   Rate $= k[ClO^-][ClO_2^-]$

Finally, on adding the equations for the two elementary steps, we see that the ClO_2^- ion is a product of the first step and a reactant in the second step. It therefore cancels out, and we are left with the stoichiometric equation for the overall reaction:

Step 1: $ClO^-(aq) + ClO^-(aq) \longrightarrow ClO_2^-(aq) + Cl^-(aq)$

Step 2: $\underline{ClO_2^-(aq) + ClO^-(aq) \longrightarrow ClO_3^-(aq) + Cl^-(aq)}$

Sum of steps: $3\ ClO^-(aq) \longrightarrow ClO_3^-(aq) + 2\ Cl^-(aq)$

Exercise 15.11   Elementary Steps

Nitric oxide is reduced by hydrogen to give nitrogen and water,

$$2\ NO(g) + 2\ H_2(g) \longrightarrow N_2(g) + 2\ H_2O(g)$$

and one possible mechanism to account for this reaction is

$$2\ NO(g) \rightleftharpoons N_2O_2(g)$$

$$N_2O_2(g) + H_2(g) \longrightarrow N_2O(g) + H_2O(g)$$

$$N_2O(g) + H_2(g) \longrightarrow N_2(g) + H_2O(g)$$

What is the molecularity of each of the three steps? What is the rate equation for the third step? Show that the sum of these elementary steps gives the net reaction.

Reaction Mechanisms and Rate Equations

The dependence of rate on concentration is an experimental fact. Mechanisms, by contrast, are constructs of our imagination, intuition, and good "chemical sense." To describe a mechanism we need to make a guess (a *good* guess, we hope) about how the reaction occurs at the particulate level. Several mechanisms can often be proposed that correspond to the observed rate equation, and a postulated mechanism can be wrong. A good mechanism is a worthy goal, however; it allows us to understand the chemistry better. A practical consequence of a good mechanism is that it allows us to predict important things, such as how to control a reaction better and how to design new experiments that expand our knowledge of chemistry.

One of the important aspects of kinetics to understand is that *products of a reaction can never be produced at a rate faster than the rate of the slowest step.* If one step in a multistep reaction is slower than the others, the rate of the overall reaction is limited by the combined rates of all elementary steps up through the slowest step in the mechanism. That is, the overall reaction rate and the rate of the slow step are nearly the same. Then, the slow step determines the rate of the reaction, so it is given the name the **rate-determining step,** or rate-limiting step. You are already familiar with rate-determining steps. No matter how fast you shop in the supermarket, it always seems that the time it takes to finish is determined by the wait in the checkout line.

Imagine a reaction takes place with a mechanism involving two sequential steps and assume we know the rates of both steps. The first step is slow, the second is fast:

Elementary Step 1: $A + B \xrightarrow[\text{Slow, } E_a \text{ large}]{k_1} X + M$

Elementary Step 2: $M + A \xrightarrow[\text{Fast, } E_a \text{ small}]{k_2} Y$

Overall Reaction: $2A + B \longrightarrow X + Y$

In the first step, A and B come together and slowly react to form one of the products (X) plus another reactive species, M. Almost as soon as M is formed, however, it is rapidly consumed by reaction with an additional molecule of A to form the second product Y. The products X and Y are the result of two elementary steps. The rate-determining elementary step is the first step. That is, the rate of the first step is equal to the rate of the overall reaction. This step is bimolecular and so has the rate equation

$$\text{Rate} = k_1[A][B]$$

where k_1 is the rate constant for that step. The overall reaction is expected to follow this same second-order rate equation.

Let us apply these ideas to the mechanism of a real reaction. Experiment shows that the reaction of nitrogen dioxide with fluorine has a second-order rate equation:

Overall Reaction: $2 NO_2(g) + F_2(g) \longrightarrow 2 FNO_2(g)$

$$\text{Rate} = k[NO_2][F_2]$$

The experimental rate equation immediately rules out the possibility that the reaction occurs in a single step. If the reaction was an elementary step, the rate law would have a second-order dependence on $[NO_2]$. Because a single-step reaction is ruled out, it follows that there must be at least two steps in the mechanism. We can also conclude that the rate-determining elementary step must involve NO_2 and

● **Can You Derive a Mechanism?**
At this introductory level you cannot be expected to derive reaction mechanisms. Given a mechanism, however, you can decide if it agrees with experiment.

F_2 in a 1:1 ratio. The simplest possible mechanism is identical with the hypothetical example:

Elementary Step 1 Slow $NO_2(g) + F_2(g) \xrightarrow{k_1} FNO_2(g) + F(g)$

Elementary Step 2 Fast $NO_2(g) + F(g) \xrightarrow{k_2} FNO_2(g)$

Overall Reaction $2\,NO_2(g) + F_2(g) \longrightarrow 2\,FNO_2(g)$

This proposed mechanism suggests that molecules of NO_2 and F_2 first react to produce one molecule of the product (FNO_2) plus an F atom. In a second step, the F atom produced in the first step reacts with additional NO_2 to give a second molecule of product. If we assume that the first, bimolecular step is rate-determining, its rate equation would be Rate $= k_1[NO_2][F_2]$, the rate equation observed experimentally. The experimental rate constant is, therefore, the same as k_1.

The F atom in the first step of the NO_2/F_2 reaction is a reaction **intermediate.** It does not appear in the equation describing the overall reaction. Reaction intermediates usually have a very fleeting existence, but they occasionally have long enough lifetimes to be observed. One of the tests of a proposed mechanism is the detection of an intermediate.

Example 15.13   **Elementary Steps and Reaction Mechanisms**

Problem • Oxygen atom transfer from nitrogen dioxide to carbon monoxide produces nitrogen monoxide and carbon dioxide:

$$NO_2(g) + CO(g) \longrightarrow NO(g) + CO_2(g)$$

It has the following rate equation at temperatures less than 500 K:

$$\text{Rate} = k[NO_2]^2$$

Can this reaction occur in one bimolecular step whose stoichiometry is the same as the overall reaction?

Strategy • Write the rate law based on the equation for the $NO_2 + CO$ reaction occurring as an elementary step. If this rate law corresponds to the observed rate law, then the overall equation reflects the way the reaction occurs [💿 CD-ROM, Screens 15.12 and 15.13].

Solution • If the reaction occurs by the collision of one NO_2 molecule with one CO molecule the rate equation would be

$$\text{Rate} = k[NO_2][CO]$$

This does not agree with experiment, so the mechanism must involve more than a single step. In fact, the reaction is thought to occur in two bimolecular steps:

Elementary Step 1 Slow, rate- $2\,NO_2(g) \longrightarrow NO_3(g) + NO(g)$
determining

Elementary Step 2 Fast $NO_3(g) + CO(g) \longrightarrow NO_2(g) + CO_2(g)$

Overall Reaction $NO_2(g) + CO(g) \longrightarrow NO(g) + CO_2(g)$

The first or rate-determining step indeed has a rate equation that agrees with experiment.

Exercise 15.12   **Elementary Steps and Reaction Mechanism**

The Raschig reaction produces the industrially important reducing agent hydrazine, N_2H_4, from NH_3 and OCl^- in basic, aqueous solution. A proposed mechanism is

Step 1:   Fast $NH_3(aq) + OCl^-(aq) \longrightarrow NH_2Cl(aq) + OH^-(aq)$

Step 2:   Slow $NH_2Cl(aq) + NH_3(aq) \longrightarrow N_2H_5^+(aq) + Cl^-(aq)$

Step 3:   Fast $N_2H_5^+(aq) + OH^-(aq) \longrightarrow N_2H_4(aq) + H_2O(\ell)$

(a) What is the overall stoichiometric equation?

(b) Which step of the three is rate-determining?

(c) Write the rate equation for the rate-determining elementary step.

(d) What reaction intermediates are involved?

Another common two-step reaction mechanism involves an initial fast reaction that produces an intermediate, followed by a slower second step in which the intermediate is converted to the final product. The rate of the reaction is determined by the second step, for which a rate law can be written. The rate of that step, however, depends on the concentration of the intermediate. An important thing to remember, though, is that *the rate law of an elementary step must be written with respect to the reactants only.* An intermediate, whose concentration will probably not be measurable, cannot appear as a term in the rate expression.

The reaction of nitric oxide and oxygen provides an example in this category to discuss in more detail:

$$2\, NO(g) + O_2(g) \longrightarrow 2\, NO_2(g)$$
$$\text{Rate} = k[NO]^2[O_2]$$

The experimentally determined rate law shows second-order dependence on NO_2 and first-order dependence on O_2. Although this rate law would be correct for a termolecular reaction, there is experimental evidence that there is an intermediate in this reaction. A possible two-step mechanism involving an intermediate is

Elementary Step 1:   Fast, Equilibrium $NO(g) + O_2(g) \underset{k_{-1}\ \text{intermediate}}{\overset{k_1}{\rightleftharpoons}} OONO(g)$

Elementary Step 2:   Slow, Rate-determining $\dfrac{NO(g) + OONO(g) \xrightarrow{k_2} 2\, NO_2(g)}{}$

Overall Reaction: $2\, NO(g) + O_2(g) \longrightarrow 2\, NO_2(g)$

The second step of this reaction is the slow step, and the overall rate depends on it. We can write a rate law for the second step:

$$\text{Rate} = k_2[NO][OONO]$$

This rate law cannot be compared directly with the experimental rate law because it contains the concentration of an intermediate, OONO. Recall that the experimental rate law must be written only in terms of compounds appearing in the overall equation. We therefore need to express the postulated rate law in a way that eliminates the intermediate. To do this, we look at the rapid first step in this reaction sequence.

At the beginning of the reaction NO and O_2 react rapidly and produce the intermediate OONO; the rate of formation can be defined by a rate law with a rate constant k_1:

$$\text{Rate of production of OONO} = k_1[NO][O_2]$$

Because the intermediate is consumed only very slowly in the second step, it is possible for the OONO to revert to NO and O_2 before it reacts further:

Rate of reversion of OONO to NO and $O_2 = k_{-1}[OONO]$

As NO and O_2 form OONO, their concentration drops, so the rate of the forward reaction decreases. At the same time, the concentration of OONO builds up, so the rate of the reverse reaction increases. Eventually, the rates of the forward and reverse reactions become the same, and the first elementary step reaches a *state of equilibrium*. The forward and reverse reactions in the first step are so much faster than the second elementary step that equilibrium is established before any significant amount of OONO is consumed by NO to give NO_2. The state of equilibrium for the first step remains throughout the lifetime of the overall reaction.

Because equilibrium is established when the rates of the forward and reverse reactions are the same, this means

Rate of forward reaction = rate of reverse reaction

$$k_1[NO][O_2] = k_{-1}[OONO]$$

Rearranging this equation, we find

$$\frac{k_1}{k_{-1}} = \frac{[OONO]}{[NO][O_2]} = K$$

Both k_1 and k_{-1} are constants (they will change only if the temperature changes). We can define a new constant K, equal to the ratio of these two constants and called an equilibrium constant, which is equal K to the quotient $[OONO]/[NO][O_2]$. From this we can derive an expression for the concentration of OONO:

$$[OONO] = K[NO][O_2]$$

If $K[NO][O_2]$ is substituted for [ONOO] in the rate law for the rate-determining elementary step, we have

$$\text{Rate} = k_2[NO][OONO] = k_2[NO]\{K[NO][O_2]\}$$
$$= k_2 K[NO]^2[O_2]$$

Because both k_2 and K are constants, their product is another constant k', and we have

$$\text{Rate} = k'[NO]^2[O_2]$$

This is exactly the rate law derived from experiment. This means the sequence of reactions on which the rate equation is based may be a reasonable mechanism for this reaction. It is *not* the only possible mechanism, however. This rate equation is consistent with the reaction occurring in a single termolecular step, and one other mechanism is illustrated in Example 15.14.

Example 15.14 **Reaction Mechanism Involving an Equilibrium Step**

Problem • The $NO + O_2$ reaction described in the text could also occur by the following mechanism:

Elementary Step 1: Fast, Equilibrium

$$NO(g) + NO(g) \xrightleftharpoons[k_{-1}]{k_1} N_2O_2(g)$$
$$\text{intermediate}$$

Elementary Step 2: Slow, Rate-determining

$$N_2O_2(g) + O_2(g) \xrightarrow{k_2} 2\ NO_2(g)$$

Overall Reaction: $2\ NO(g) + O_2(g) \longrightarrow 2\ NO_2(g)$

Show that this mechanism leads to the experimental rate law, Rate = $k[NO]^2[O_2]$.

(Example continues on next page)

Strategy • The rate law for the rate determining elementary step is

$$Rate = k_2[N_2O_2][O_2]$$

The compound N_2O_2 is an intermediate, which is consumed in the slow step and cannot appear in the final derived rate law. The postulated rate law must be expressed in a way that eliminates the intermediate. To do this we use the the equilibrium constant expression for the first step.

Solution • $[N_2O_2]$ is related to $[NO]$ by the equilibrium constant.

$$K = \frac{[N_2O_2]}{[NO]^2}$$

If this is solved for $[N_2O_2]$, we have $[N_2O_2] = K[NO]^2$. When this is substituted into the derived rate law

$$Rate = k_2\{K[NO]^2\}[O_2]$$

the resulting equation is seen to be identical with the experimental rate law where $k_2K = k'$.

Comment • The $NO + O_2$ reaction has an experimental rate law for which at least three mechanisms can be proposed. The challenge is to decide which is correct. In this case further experimentation detected the species OONO as a short-lived intermediate, confirming the mechanism involving this intermediate.

Exercise 15.13   Reaction Mechanisms Involving a Fast Initial Step

One possible mechanism for the decomposition of nitryl chloride, NO_2Cl, is

Elementary Step 1:   Fast, Equilibrium $NO_2Cl(g) \underset{k_{-1}}{\overset{k_1}{\rightleftharpoons}} NO_2(g) + Cl(g)$

Elementary Step 2:   Slow $NO_2Cl(g) + Cl(g) \xrightarrow{k_2} NO_2(g) + Cl_2(g)$

What is the overall reaction? What rate law would be derived from this mechanism? What effect does increasing the concentration of the product NO_2 have on the reaction rate?

Problem-Solving Tips 15.2

Relating Rate Equations and Reaction Mechanisms

The connection between an experimental rate equation and the hypothesis of a reaction mechanism is important in chemistry.

1. Experiments must first be performed that define the effect of reactant concentrations on the rate of the reaction. This gives the experimental rate equation.

2. A mechanism for the reaction is proposed on the basis of the experimental rate equation, the principles of stoichiometry and molecular structure and bonding, general chemical experience, and intuition.

3. The *proposed* reaction mechanism is used to *derive* a rate equation. This rate equation must contain only those species present in the overall chemical reaction and not reaction intermediates. If the derived and experimental rate equations are the same, the postulated mechanism *may* be a reasonable hypothesis of the reaction sequence.

4. A rate equation can *only* be derived based on an elementary step in a mechanism. You cannot write a rate equation based on the overall chemical equation.

5. If more than one mechanism can be proposed, and they all predict derived rate equations in agreement with experiment, then more experiments must be done.

In Summary

When you have finished studying this chapter, you should ask if you have met the chapter goals. In particular, you should be able to

• Explain the concept of reaction rate (Section 15.1).

- Derive the average and instantaneous rate of a reaction from experimental information (Section 15.1).
- Describe the various conditions that affect reaction rate (that is, reactant concentrations, temperature, presence of a catalyst, and the state of the reactants) (Section 15.2).
- Define the various parts of a rate equation (the rate constant and order of reaction) and understand their significance(Section 15.3).
- Derive a rate equation from experimental information (Section 15.3).
- Describe and use the relationships between reactant concentration and time for zero-order, first-order, and second-order reactions (Section 15.4 and Table 15.1).
- Apply graphical methods for determining reaction order and the rate constant from experimental data (Section 15.4 and Table 15.1).
- Use the concept of half-life $(t_{1/2})$, especially for first-order reactions (Section 15.4).
- Describe the collision theory of reaction rates (Section 15.5).
- Relate activation energy (E_a) to the rate and thermodynamics of a reaction (Section 15.5) using reaction coordinate diagrams.
- Use collision theory to describe the effect of reactant concentration on reaction rate (Section 15.5).
- Describe the functioning of a catalyst and its effect on the activation energy and mechanism of a reaction (Section 15.5).
- Define homogeneous and heterogeneous catalysts (Section 15.5).
- Understand the effect of molecular orientation on reaction rate (Section 15.5).
- Describe the effect of temperature on reaction rate using the collision theory of reaction rates and the Arrhenius equation (Equation 15.7 and Section 15.5).
- Use Equations 15.5, 15.6, and 15.7 to calculate the activation energy from experimental data (Section 15.5).
- Understand the concept of reaction mechanism (the sequence of bond-making and bond-breaking steps that occurs during the conversion of reactants to products) and the relation of the mechanism to the overall, stoichiometric equation for a reaction (Section 15.6).
- Describe the elementary steps of a mechanism and give their molecularity (Section 15.6).
- Recognize the rate-determining step in a mechanism and identify any reaction intermediates (Section 15.6).

Key Terms

chemical kinetics
reaction mechanisms

Section 15.1
rate of a chemical reaction
average rate
instantaneous rate

Section 15.3
rate equation
rate law
rate constant
order
initial rate

Section 15.4
integrated rate equation
half-life, $t_{1/2}$

Section 15.5
collision theory
activation energy, E_a
Arrhenius equation
reaction intermediate
homogeneous catalyst

Section 15.6
elementary step
molecularity
unimolecular
bimolecular
termolecular
rate-determining step
intermediate

Key Equations

Equation 15.1 (page 614)

Integrated rate equation for a first-order reaction (in which $-\Delta[R]/\Delta t = k[R]$)

$$\ln \frac{[R]_t}{[R]_0} = -kt$$

Here $[R]_0$ and $[R]_t$ are concentrations of the reactant at time $t = 0$ and at a later time, t. The *ratio* of concentrations, $[R]_t/[R]_0$, is the fraction of reactant that *remains* after a given time has elapsed.

Equation 15.2 (page 616)

Integrated rate equation for a second-order reaction (in which $-\Delta[R]/\Delta t = k[R]^2$)

$$\frac{1}{[R]_t} - \frac{1}{[R]_0} = kt$$

The parameters in this equation are the same as those in Equation 15.1.

Equation 15.3 (page 617)

Integrated rate equation for a zero-order reaction (in which $-\Delta[R]/\Delta t = k[R]^0$)

$$[R]_0 - [R]_t = kt$$

The parameters in this equation are the same as those in Equation 15.1.

Equation 15.4 (page 620)

The relation between the half-life $(t_{1/2})$ and the rate constant (k) for a first-order reaction.

$$t_{1/2} = \frac{0.693}{k}$$

Equation 15.5 (page 627)

Arrhenius equation in exponential form

$$k = \text{rate constant} = Ae^{-E_a/RT}$$

Frequency factor   Fraction of molecules with minimum energy for reaction

A is the frequency factor, E_a the activation energy, T the temperature (K), and R is the gas constant ($= 8.314510 \times 10^{-3}$ kJ/mol · mol). The equation shows how the rate constant is related to the activation energy and temperature of reaction.

Equation 15.6 (page 628)

Expanded Arrhenius equation in logarithmic form

$$\ln k = \ln A - \frac{E_a}{R}\left(\frac{1}{T}\right) \longleftarrow \text{Arrhenius equation}$$

$$y = a + bx \longleftarrow \text{Equation for straight line}$$

Equation 15.7 (page 629)

A version of the Arrhenius equation used to calculate the activation energy for a reaction knowing values of the rate constant at two temperatures (in kelvins).

$$\ln k_2 - \ln k_1 = \ln \frac{k_2}{k_1} = -\frac{E_a}{R}\left[\frac{1}{T_2} - \frac{1}{T_1}\right]$$

Study Questions

Questions with blue, bold-faced numbers have answers in Appendix O. See the tutorials, simulations, and descriptions in Chapter 15 of the General Chemistry Interactive CD-ROM, *Version 3.0.*

Reviewing Important Concepts

1. Considering chemical reactions at the particulate level, explain how temperature, reactants concentrations, and catalysts affect reaction rates.

2. Refer to Figure 15.2. After 2.0 h, what is the concentration of NO_2? Of O_2?

3. Using the rate equation Rate = $k[A]^2[B]$, define the order of the reaction with respect to A and B. What is the total order of the reaction?

4. A reaction has the experimental rate equation Rate = $k[A]^2$. How will the rate change if the concentration of A is tripled? If the concentration of A is halved?

5. A reaction has the experimental rate equation Rate = $k[A]^2[B]$. If the concentration of A is doubled, and the concentration of B is halved, what happens to the reaction rate?

6. Write the equation relating concentration of reactant and time for a first-order reaction. Define each term in the equation.

7. After five half-life periods for a first-order reaction, what fraction of reactant remains?

8. Plotting 1/[reactant] versus time for a reaction produces a straight line. What is the order of the reaction? If a straight line of negative slope is observed for a plot of ln[reactant] versus time, what is the order of the reaction?

9. Draw a reaction coordinate diagram for a single-step, exothermic process. Mark the activation energies of the forward and reverse processes. Identify the net energy change for the reaction on this diagram.

10. What experimental information is required to use the Arrhenius equation to calculate the activation energy of a reaction?

12. Define what is meant by the term "mechanism" for a chemical reaction.

13. Define the terms "elementary step" and "rate-determining step" as applied to mechanisms.

14. What is a reaction intermediate? Give an example using one of the reaction mechanisms described in the text.

15. What is a catalyst? What is the effect of a catalyst on the mechanism of a reaction?

Practicing Skills

Reaction Rates
(See Examples 15.1–15.2, Exercises 15.1–15.2, and CD-ROM Screen 15.2)

16. Give the relative rates of disappearance of reactants and formation of products for each of the following reactions:
    (a) $2\,O_3(g) \longrightarrow 3\,O_2(g)$
    (b) $2\,HOF(g) \longrightarrow 2\,HF(g) + O_2(g)$

17. Give the relative rates of disappearance of reactants and formation of products for each of the following reactions:
    (a) $2\,NO(g) + Br_2(g) \longrightarrow 2\,NOBr(g)$
    (b) $N_2(g) + 3\,H_2(g) \longrightarrow 2\,NH_3(g)$

18. In the reaction $2\,O_3(g) \longrightarrow 3\,O_2(g)$ the rate of formation of O_2 is 1.5×10^{-3} mol/L · s. What is the rate of decomposition of O_3?

19. In the synthesis of ammonia, $N_2(g) + 3\,H_2(g) \longrightarrow 2\,NH_3(g)$, if $-\Delta[H_2]/\Delta t = 4.5 \times 10^{-4}$ mol/L min, what is $\Delta[NH_3]/\Delta t$?

20. Experimental data are listed here for the reaction $A \longrightarrow 2\,B$.

Time (s)	[B] (mol/L)
0.00	0.000
10.0	0.326
20.0	0.572
30.0	0.750
40.0	0.890

(a) Prepare a graph from these data, connect the points with a smooth line, and calculate the rate of change of [B] for each 10-s interval from 0.0 to 40.0 s. Does the rate of change decrease from one time interval to the next? Suggest a reason for this result.

(b) How is the rate of change of [A] related to the rate of change of [B] in each time interval? Calculate the rate of change of [A] for the time interval from 10.0 to 20.0 s.

(c) What is the instantaneous rate when [B] = 0.750 mol/L?

21. Phenyl acetate, an ester, reacts with water according to the equation

$$\underset{\text{phenyl acetate}}{CH_3\overset{O}{\overset{\|}{C}}OC_6H_5} + H_2O \longrightarrow \underset{\text{acetic acid}}{CH_3\overset{O}{\overset{\|}{C}}OH} + \underset{\text{phenol}}{C_6H_5OH}$$

The data in the table were collected for this reaction at 5 °C.

Time (s)	[Phenyl acetate] (mol/L)
0.00	0.55
15.0	0.42
30.0	0.31
45.0	0.23
60.0	0.17
75.0	0.12
90.0	0.085

(a) Plot the phenyl acetate concentration versus time, and describe the shape of the curve observed.

(b) Calculate the rate of change of the phenyl acetate concentration during the period 15.0 s to 30.0 s and also during the period 75.0 s to 90.0 s. Compare the values, and suggest a reason why one value is smaller than the other.

(c) What is the rate of change of the phenol concentration during the time period 60.0 s to 75.0 s?

(d) What is the instantaneous rate at 15.0 s?

Concentration and Rate Equations

(See Examples 15.3–15.4, Exercises 15.3–15.4, and CD-ROM Screen 15.4)

22. The reaction between ozone and nitrogen dioxide at 231 K is first order in both $[NO_2]$ and $[O_3]$.

$$2 NO_2(g) + O_3(g) \longrightarrow N_2O_5(s) + O_2(g)$$

(a) Write the rate equation for the reaction.

(b) If the concentration of NO_2 is tripled, what is the change in the reaction rate?

(c) What is the effect on reaction rate if the concentration of O_3 is halved?

23. Nitrosyl bromide, NOBr, is formed from NO and Br_2:

$$2 NO(g) + Br_2(g) \longrightarrow 2 NOBr(g)$$

Experiments show the reaction is second order in NO and first order in Br_2.

(a) Write the rate equation for the reaction.

(b) How does the initial reaction rate change if the concentration of Br_2 is changed from 0.0022 mol/L to 0.0066 mol/L?

(c) What is the change in the initial rate if the concentration of NO is changed from 0.0024 mol/L to 0.0012 mol/L?

24. The data in the table are for the reaction of NO and O_2 at 660 K.

$$2 NO(g) + O_2(g) \longrightarrow 2 NO_2(g)$$

Reactant Concentration (mol/L)		Rate of Appearance of NO_2 (mol/L · s)
[NO]	[H_2]	
0.010	0.010	2.5×10^{-5}
0.020	0.010	1.0×10^{-4}
0.010	0.020	5.0×10^{-5}

(a) Determine the order of the reaction for each reactant.

(b) Write the rate equation for the reaction.

(c) Calculate the rate constant.

(d) Calculate the rate (in mol/L · s) at the instant when $[NO] = 0.015$ mol/L and $[O_2] = 0.0050$ mol/L.

(e) At the instant when NO is reacting at the rate 1.0×10^{-4} mol/L · s, what is the rate at which O_2 is reacting and NO_2 is forming?

25. The reaction

$$2 NO(g) + 2 H_2(g) \longrightarrow N_2(g) + 2 H_2O(g)$$

was studied at 904 °C, and the data in the table were collected.

Reactant Concentration (mol/L)		Rate of Appearance of N_2 (mol/L · s)
[NO]	[H_2]	
0.420	0.122	0.136
0.210	0.122	0.0339
0.210	0.244	0.0678
0.105	0.488	0.0339

(a) Determine the order of the reaction for each reactant.

(b) Write the rate equation for the reaction.

(c) Calculate the rate constant for the reaction.

(d) Find the rate of appearance of N_2 at the instant when $[NO] = 0.350$ mol/L and $[H_2] = 0.205$ mol/L.

26. Data for the reaction $2 NO(g) + O_2(g) \longrightarrow 2 NO_2(g)$ are given in the table.

	Reactant Concentration (mol/L)		Initial rate mol/L · h
	[NO]	[O_2]	
1	3.6×10^{-4}	5.2×10^{-3}	3.4×10^{-8}
2	3.6×10^{-4}	1.04×10^{-2}	6.8×10^{-8}
3	1.8×10^{-4}	1.04×10^{-2}	1.7×10^{-8}
4	1.8×10^{-4}	5.2×10^{-3}	?

(a) What is the rate law for this reaction?

(b) What is the rate constant for the reaction?

(c) What is the initial rate of the reaction in Experiment 4?

27. Data for the reaction $CO(g) + NO_2(g) \longrightarrow CO_2(g) + NO(g)$ are given in the following table.

	Reactant Concentration (mol/L)		Initial rate mol/L · h
	[CO]	[NO_2]	
1	5.0×10^{-4}	0.36×10^{-4}	3.4×10^{-8}
2	5.0×10^{-4}	0.18×10^{-4}	1.7×10^{-8}
3	1.0×10^{-3}	0.36×10^{-4}	6.8×10^{-8}
4	1.5×10^{-3}	0.72×10^{-4}	?

(a) What is the rate law for this reaction?

(b) What is the rate constant for the reaction?

(c) What is the initial rate of the reaction in Experiment 4?

28. Carbon monoxide burns to form carbon dioxide

$$2 CO(g) + O_2(g) \longrightarrow 2 CO_2(g)$$

Information on the reaction is given in the following table:

[CO] (mol/L)	[O₂] (mol/L)	Initial Rate (mol/L · min)
0.02	0.02	3.68×10^{-5}
0.04	0.02	1.47×10^{-4}
0.02	0.04	7.36×10^{-5}

(a) A general expression for the rate law for this reaction is Rate = $k\,[CO]^n[O_2]^m$. Determine the values of n and m using the data in the table.

(b) What is the order of the reaction with respect to CO? What is the order with respect O_2? What is the overall order of the reaction?

(c) What is the value for the rate constant?

29. Data for the reaction

$$H_2PO_4^-(aq) + OH^-(aq) \longrightarrow HPO_4^{2-}(aq) + H_2O(\ell)$$

are provided in the table.

Experiment	[H₂PO₄⁻] (M)	[OH⁻] (M)	Initial Rate (mol/L · min)
1	0.0030	0.00040	0.0020
2	0.0030	0.00080	0.0080
3	0.0090	0.00040	0.0060
4	?	0.00033	0.0020

(a) What is the rate law for this reaction?

(b) What is the value of k?

(c) What is the concentration of $H_2PO_4^-$ in Experiment 4?

Concentration-Time Equations

(See Examples 15.5–15.7, Exercises 15.5–15.7, and CD-ROM Screen 15.6)

30. The rate equation for the hydrolysis of sucrose to fructose and glucose

$$C_{12}H_{22}O_{11}(aq) + H_2O(\ell) \longrightarrow 2\ C_6H_{12}O_6(aq)$$

is $-\Delta[\text{sucrose}]/\Delta t = k[C_{12}H_{22}O_{11}]$. After 2.57 h at 27 °C, the sucrose concentration decreased from 0.0146 M to 0.0132 M. Find the rate constant k.

31. The decomposition of N_2O_5 in CCl_4 is a first-order reaction. If 2.56 mg of N_2O_5 is present initially, and 2.50 mg is present after 4.26 min at 55 °C, what is the value of the rate constant k?

32. The decomposition of SO_2Cl_2 is a first-order reaction:

$$SO_2Cl_2(g) \longrightarrow SO_2(g) + Cl_2(g)$$

The rate constant for the reaction is 2.8×10^{-3} min^{-1} at 600 K. If the initial concentration of SO_2Cl_2 is 1.24×10^{-3} mol/L, how long will it take for the concentration to drop to 0.31×10^{-3} mol/L?

33. The conversion of cyclopropane to propene, described in Example 15.5, occurs with a first-order rate constant of 5.4×10^{-2} h^{-1}. How long will it take for the concentration of cyclopropane to decrease from an initial concentration of 0.080 mol/L to 0.020 mol/L?

34. Ammonium cyanate, NH_4NCO, rearranges in water to give urea, $(NH_2)_2CO$:

$$NH_4NCO(aq) \longrightarrow (NH_2)_2CO(aq)$$

The rate equation for this process is Rate = $k[NH_4NCO]^2$, where $k = 0.0113$ L/mol · min. If the original concentration of NH_4NCO in solution is 0.229 mol/L, how long will it take for the concentration to decrease to 0.180 mol/L?

35. The decomposition of nitrogen dioxide at a high temperature

$$NO_2(g) \longrightarrow NO(g) + \tfrac{1}{2}O_2(g)$$

is second order in this reactant. The rate constant for this reaction is 3.40 L/mol · min. Determine the time needed for the concentration of NO_2 to decrease from 2.00 mol/L to 1.50 mol/L.

36. Hydrogen peroxide, $H_2O_2(aq)$, decomposes to $H_2O(\ell)$ and $O_2(g)$ in a reaction that is first order in H_2O_2 and with a rate constant $k = 1.06 \times 10^{-3}$ min^{-1}.

(a) How long will it take for 15% of a sample of H_2O_2 to decompose?

(b) How long will it take for 85% of the sample to decompose?

37. The thermal decomposition of HCO_2H is a first-order reaction with a rate constant of 2.4×10^{-3} s^{-1} at a given temperature. How long will it take for three-fourths of a sample of HCO_2H to decompose?

Half-Life

(See Example 15.8–15.9, Exercise 15.8, and CD-ROM Screen 15.8)

38. The rate equation for the decomposition of N_2O_5 (giving NO_2 and O_2) is $-\Delta[N_2O_5]/\Delta t = k[N_2O_5]$. For the reaction the value of k is 5.0×10^{-4} s^{-1} at a particular temperature.

(a) Calculate the half-life of N_2O_5.

(b) How long does it take for the N_2O_5 concentration to drop to one tenth of its original value?

39. The decomposition of SO_2Cl_2

$$SO_2Cl_2(g) \longrightarrow SO_2(g) + Cl_2(g)$$

is first order in SO_2Cl_2, and the reaction has a half-life of 245 min at 600 K. If you begin with 3.6×10^{-3} mol of SO_2Cl_2 in a 1.0-L flask, how long will it take for the quantity of SO_2Cl_2 to decrease to 2.00×10^{-4} mol?

40. Gaseous azomethane, $CH_3N{=}NCH_3$, decomposes in a first-order reaction when heated:

$$CH_3N{=}NCH_3(g) \longrightarrow N_2(g) + C_2H_6(g)$$

The rate constant for this reaction at 425 °C is 40.8 min^{-1}.

If the initial quantity of azomethane in the flask is 2.00 g, how much remains after 0.0500 min? What quantity of N_2 is formed in this time?

41. The compound $Xe(CF_3)_2$ is unstable, decomposing to elemental Xe with a half-life of 30. min. If you place 7.50 mg of $Xe(CF_3)_2$ in a flask, how long must you wait until only 0.25 mg of $Xe(CF_3)_2$ remains?

42. The radioactive isotope ^{64}Cu is used in the form of copper(II) acetate to study Wilson's disease. The isotope has a half-life of 12.70 h. What fraction of radioactive copper(II) acetate remains after 64 h?

43. Radioactive gold-198 is used in the diagnosis of liver problems. The half-life of this isotope is 2.7 days. If you begin with a 5.6-mg sample of the isotope, how much of this sample remains after 1.0 day?

44. Formic acid decomposes at 550 °C according to the equation $HCO_2H(g) \longrightarrow CO_2(g) + H_2(g)$. The reaction follows first-order kinetics. In an experiment, it is determined that 75% of a sample of HCO_2H has decomposed in 72 s. Determine $t_{1/2}$ for this reaction.

45. The decomposition of SO_2Cl_2 to SO_2 and Cl_2 at high temperature is a first-order reaction with a half-life of 2.5×10^3 min. What fraction of SO_2Cl_2 will remain after 750 min?

Graphical Analysis of Rate Equations and *k*
(See Exercise 15.8 and CD-ROM Screen 15.7)

46. Common sugar, sucrose, breaks down in dilute acid solution to form glucose and fructose. Both products have the same formula, $C_6H_{12}O_6$:

$$C_{12}H_{22}O_{11}(aq) + H_2O(\ell) \longrightarrow 2\,C_6H_{12}O_6(aq)$$

The rate of this reaction has been studied in acid solution, and the data in the table were obtained.

Time (min)	[$C_{12}H_{22}O_{11}$] (mol/L)
0	0.316
39	0.274
80	0.238
140	0.190
210	0.146

(a) Plot the data in the table as ln[sucrose] versus time and 1/[sucrose] versus time. What is the order of the reaction?

(b) Write the rate equation for the reaction, and calculate the rate constant *k*.

(c) Estimate the concentration of sucrose after 175 min.

47. Data for the reaction of phenyl acetate with water are given in Study Question 21. Plot these data as ln[phenyl acetate] and 1/[phenyl acetate] versus time. Based on the appearance of the two graphs, what can you conclude about the order of the reaction with respect to phenyl acetate? Working from the data and the rate law, determine the rate constant for the reaction.

48. Data for the decomposition of dinitrogen oxide

$$2\,N_2O(g) \longrightarrow 2\,N_2(g) + O_2(g)$$

on a gold surface at 900 °C are given. Verify that the reaction is first order by preparing a graph of $\ln[N_2O]$ versus time. Derive the rate constant from the slope of the line in this graph (see Table 15.1). Using the rate law and value of *k*, determine the decomposition rate at 900 °C when $[N_2O] = 0.035$ mol/L.

Time (min)	[N_2O] (mol/L)
15.0	0.0835
30.0	0.0680
80.0	0.0350
120.0	0.0220

49. Ammonia decomposes when heated according to the equation

$$NH_3(g) \longrightarrow NH_2(g) + H(g)$$

The data in the table for this reaction were collected at 2000 K.

Time (h)	[NH_3] (mol/L)
0	8.00×10^{-7}
25	6.75×10^{-7}
50	5.84×10^{-7}
75	5.15×10^{-7}

Plot $\ln[NH_3]$ versus time and $1/[NH_3]$ versus time. Can you draw a conclusion from these graphs about the order of this reaction with respect to NH_3? Find the rate constant for the reaction from the slope.

50. Gaseous [NO_2] decomposes at 573 K.

$$2\,NO_2(g) \longrightarrow 2\,NO(g) + O_2(g)$$

The concentration of NO_2 was measured as a function of time. A graph of $1/[NO_2]$ versus time gives a straight line with a slope of 1.1 L/mol · s. Based on this information what is the rate law for this reaction? What is the rate constant?

51. The reaction $2\,HOF(g) \longrightarrow 2\,HF(g) + O_2(g)$ occurs at 25 °C. Using data in the following table, determine the rate law and then calculate the rate constant.

[HOF] (mol/L)	Time (min)
0.850	0
0.810	2
0.754	5
0.526	20
0.243	50

52. For the reaction $2\ C_2F_4 \longrightarrow C_4F_8$, a graph of $1/[C_2F_4]$ versus time gives a straight line with a slope of $+0.04$ L/mol · sec. What is the rate law for this reaction?

53. Butadiene, $C_4H_6(g)$, dimerizes when heated, forming 1,5-cyclooctadiene, C_8H_{12}. The data in the table were collected

$$2\ H_2C=CHCH=CH_2 \longrightarrow$$
1,3-butadiene

$$\begin{array}{c} H_2C-CH_2 \\ HC \qquad\quad CH \\ \| \qquad\qquad \| \\ HC \qquad\quad CH \\ H_2C-CH_2 \end{array}$$

1,5-cyclooctadiene

[C₄H₆] (mol/L)	Time (s)
1.0×10^{-2}	0
8.7×10^{-3}	200
7.7×10^{-3}	500
6.9×10^{-3}	800
5.8×10^{-3}	1200

(a) Use a graphical method to verify that this is a second-order reaction.

(b) Calculate the rate constant for this reaction.

Kinetics and Energy
(See Examples 15.10–15.11, Exercise 15.9, and CD-ROM Screens 15.9 and 15.10)

54. Calculate the activation energy, E_a, for the reaction

$$N_2O_5(g) \longrightarrow 2\ NO_2(g) + \tfrac{1}{2}\ O_2(g)$$

from the observed rate constants: k at 25 °C $= 3.46 \times 10^{-5}$ s^{-1} and k at 55 °C $= 1.5 \times 10^{-3}$ s^{-1}.

55. If the rate constant for a reaction triples in value when the temperature rises from 3.00×10^2 K to 3.10×10^2 K, what is the activation energy of the reaction?

56. When heated to a high temperature, cyclobutane, C_4H_8, decomposes to ethylene:

$$C_4H_8(g) \longrightarrow 2\ C_2H_4(g)$$

The activation energy, E_a, for this reaction is 260 kJ/mol. At 800 K, the rate constant $k = 0.0315$ s^{-1}. Determine the value of k at 850 K.

57. When heated, cyclopropane is converted to propene (see Example 15.5). Rate constants for this reaction at 470 °C and 510 °C are $k = 1.10 \times 10^{-4}$ s^{-1} and $k = 1.02 \times 10^{-3}$ s^{-1}, respectively. Determine the activation energy, E_a, from these data.

58. The reaction of H_2 molecules with F atoms

$$H_2(g) + F(g) \longrightarrow HF(g) + H(g)$$

has an activation energy of 8 kJ/mol and an energy change of -133 kJ/mol. Draw a diagram similar to Figure 15.15 for this process. Indicate the activation energy and enthalpy of reaction on this diagram.

59. Answer questions (a) and (b) based on the accompanying reaction coordinate diagram.

(a) Is the reaction exothermic or endothermic?

(b) Does the reaction occur in more than one step? If so, how many?

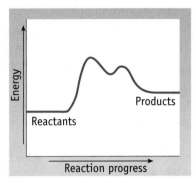

Mechanisms
(See Examples 15.11–15.13, Exercises 15.11–15.14, and CD-ROM Screens 15.12 and 15.13)

60. What is the rate law for each of the following *elementary* reactions?

(a) $NO(g) + NO_3(g) \longrightarrow 2\ NO_2(g)$

(b) $Cl(g) + H_2(g) \longrightarrow HCl(g) + H(g)$

(c) $(CH_3)_3CBr(aq) \longrightarrow (CH_3)_3C^+(aq) + Br^-(aq)$

61. What is the rate law for the following *elementary* reactions?

(a) $Cl(g) + ICl(g) \longrightarrow I(g) + Cl_2(g)$

(b) $O(g) + O_3(g) \longrightarrow 2\ O_2(g)$

(c) $2\ NO_2(g) \longrightarrow N_2O_4(g)$

62. Ozone, O_3, in earth's upper atmosphere decomposes according to the equation $2\ O_3(g) \longrightarrow 3\ O_2(g)$. The mechanism of the reaction is thought to proceed through an initial fast, reversible step followed by a slow second step.

Step 1:	Fast, reversible	$O_3(g) \rightleftharpoons O_2(g) + O(g)$
Step 2:	Slow	$O_3(g) + O(g) \longrightarrow 2\ O_2(g)$

(a) Which of the steps is rate-determining?

(b) Write the rate equation for the rate-determining step.

63. The reaction of $NO_2(g)$ and $CO(g)$ is thought to occur in two steps:

Step 1:	Slow	$NO_2(g) + NO_2(g) \longrightarrow NO(g) + NO_3(g)$
Step 2:	Fast	$NO_3(g) + CO(g) \longrightarrow NO_2(g) + CO_2(g)$

(a) Show that the elementary steps add up to give the overall, stoichiometric equation.

(b) What is the molecularity of each step?

(c) For this mechanism to be consistent with kinetic data, what must be the experimental rate equation?

(d) Identify any intermediates in this reaction.

64. Iodide ion is oxidized in acid solution by hydrogen peroxide. (See the Chapter Focus, page 602).

$$H_2O_2(aq) + 2\,H^+(aq) + 2\,I^-(aq) \longrightarrow I_2(aq) + 2\,H_2O(\ell)$$

A proposed mechanism is

Step 1:   Slow $H_2O_2(aq) + I^-(aq) \longrightarrow$
$$H_2O(\ell) + OI^-(aq)$$

Step 2:   Fast $H^+(aq) + OI^-(aq) \longrightarrow HOI(aq)$

Step 3:   Fast $HOI(aq) + H^+(aq) + I^-(aq) \longrightarrow$
$$I_2(aq) + H_2O(\ell)$$

(a) Show that the three elementary steps add up to give the overall, stoichiometric equation.

(b) What is the molecularity of each step?

(c) For this mechanism to be consistent with kinetic data, what must be the experimental rate equation?

(d) Identify any intermediates in the elementary steps in this reaction.

65. The mechanism for the reaction of CH_3OH and HBr is believed to involve two steps. The overall reaction is exothermic.

Step 1:   Fast, endothermic $CH_3OH + H^+ \rightleftharpoons CH_3OH_2^+$

Step 2:   Slow $CH_3OH_2^+ + Br^- \longrightarrow CH_3Br + H_2O$

(a) Write an equation for the overall reaction.

(b) Draw a reaction coordinate diagram for this reaction.

(c) Show that the rate law for this reaction is

$$\frac{-\Delta[CH_3OH]}{\Delta t} = k[CH_3OH][H^+][Br^-]$$

66. A proposed mechanism for the reaction of NO_2 and CO is

Step 1:   Slow, endothermic $2\,NO_2(g) \longrightarrow$
$$NO(g) + NO_3(g)$$

Step 2:   Fast, exothermic $NO_3(g) + CO(g) \longrightarrow$
$$NO_2(g) + CO_2(g)$$

Overall, exothermic: $NO_2(g) + CO(g) \longrightarrow$
$$NO(g) + CO_2(g)$$

(a) Identify each of the following as reactant, product, or intermediate: $NO_2(g)$, $CO(g)$, $NO_3(g)$, $CO_2(g)$, $NO(g)$.

(b) Draw a reaction coordinate diagram for this reaction. Indicate on this drawing the activation energy for each step and the overall reaction enthalpy.

67. A three-step mechanism for the reaction of $(CH_3)_3CBr$ and H_2O is proposed:

Step 1:   Slow $(CH_3)_3CBr \longrightarrow (CH_3)_3C^+ + Br^-$

Step 2:   Fast $(CH_3)_3C^+ + H_2O \longrightarrow (CH_3)_3COH_2^+$

Step 3:   Fast $(CH_3)_3COH_2^+ + Br^- \longrightarrow$
$$(CH_3)_3COH + HBr$$

(a) Write an equation for the overall reaction.

(b) Which step is rate determining?

(c) What rate law is expected for this reaction?

General Questions

These questions are not designated as to type or location in the chapter. They may combine several concepts. More challenging questions are indicated by an underlined number.

68. To determine the concentration dependence of rate of the reaction

$$H_2PO_3^-(aq) + OH^-(aq) \longrightarrow HPO_3^{2-}(aq) + H_2O(\ell)$$

you might measure $[OH^-]$ as a function of time using a pH meter. (To do this, you would set up conditions under which $[H_2PO_3^-]$ remains constant; this is done by using a large excess of this reactant.) How would you prove a second-order rate dependence for $[OH^-]$?

69. The following statements relate to the reaction $H_2(g) + I_2(g) \longrightarrow 2\,HI(g)$ for which the rate law is Rate $= k[H_2][I_2]$. Determine which of the following statements are true. If a statement is false, indicate why it is incorrect.

(a) The reaction must occur in a single step.

(b) This is a second-order reaction overall.

(c) Raising the temperature will cause the value of k to decrease.

(d) Raising the temperature lowers the activation energy for this reaction.

(e) If the concentrations of both reactants are doubled the rate will double.

(f) Adding a catalyst in the reaction will cause the initial rate to increase.

70. Identify which of the following statements are incorrect. If the statement is incorrect, rewrite it to be correct.

(a) Reactions are faster at a higher temperature because activation energies are lower.

(b) Rates increase with increasing concentration of reactants because there are more collisions between reactant molecules.

(c) At higher temperature a larger fraction of molecules have enough energy to get over the activation energy barrier.

(d) Catalyzed and uncatalyzed reactions have identical mechanisms.

71. The reaction cyclopropane $\longrightarrow$ propene occurs on a platinum metal surface as 200 °C. (The platinum is a catalyst.) The reaction rate is first order in cyclopropane. Indicate how the following quantities change (increase, decrease, or

no change) as this reaction progresses, assuming constant temperature.

(a) [cyclopropane] (d) the rate constant k

(b) [propene] (e) the order of the reaction

(c) [catalyst] (f) the half-life of cyclopropene

72. Gaseous ammonia is made by the reaction

$$N_2(g) + 3 H_2(g) \longrightarrow 2 NH_3(g)$$

Use the information on the formation of NH_3 given in the table to answer questions that follow.

$[N_2]$ (M)	$[H_2]$ (M)	Rate (mol/L · min)
0.03	0.01	4.21×10^{-5}
0.06	0.01	1.68×10^{-4}
0.03	0.02	3.37×10^{-4}

(a) Determine n and m in the rate equation, Rate = $k[N_2]^n[H_2]^m$

(b) Calculate the value of the rate constant.

(c) What is the order of the reaction with respect to $[H_2]$?

(d) What is the overall order of the reaction?

73. The reaction $2 NH_3(g) \longrightarrow N_2(g) + 3 H_2(g)$ is first order with respect to NH_3. (Compare with Study Question 72.)

(a) What is the rate equation for this reaction?

(b) Calculate the rate constant k, given the following data:

$[NH_3]$ (mol/L)	Time (s)
0.67	0
0.26	19

(c) Determine the half-life of NH_3.

74. Data for the reaction $2 NO(g) + Br_2(g) \longrightarrow 2 NOBr(g)$ are given in the following table.

Experiment	$[NO]$ (M)	$[Br_2]$ (M)	Initial rate mol/L · sec
1	1.0×10^{-2}	2.0×10^{-2}	2.4×10^{-2}
2	4.0×10^{-2}	2.0×10^{-2}	0.384
3	1.0×10^{-2}	5.0×10^{-2}	6.0×10^{-2}

(a) What is the order of the reaction with respect to $[NO]$?

(b) What is the order with respect to $[Br_2]$?

(c) What is the overall order of the reaction?

75. The reaction $2 CO_2(g) \longrightarrow 2 CO(g) + O_2(g)$ is first order with respect to the concentration of CO_2. Data on this reaction is provided in the following table.

$[CO_2]$ (mol/L)	Time (s)
0.38	0
0.27	12

(a) Write the rate equation for this reaction.

(b) Use the data to determine the value of k.

(c) What is the half-life of CO_2 under these conditions?

76. The reaction *cis*-2-butene $\longrightarrow$ *trans*-2-butene is first order in reactant and in the absence of a catalyst it is believed to be a single-step reaction (Figure 15.14). The value of the activation energy is 262 kJ/mol; the enthalpy of the reaction is -4 kJ/mol. Draw a reaction coordinate diagram. On your drawing indicate the activation energy and the enthalpy.

77. The isomerization reaction $CH_3NC(g) \longrightarrow CH_3CN(g)$ occurs slowly when CH_3NC is heated. To study the rate of this reaction at 488 K, data on $[CH_3NC]$ were collected at various times. Analysis led to the following graph.

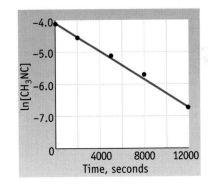

(a) What is the rate law for this reaction?

(b) What is the equation for the straight line in this graph?

(c) What is the rate constant for this reaction? (Include units)

(d) How long does it takes for half of the sample to isomerize?

(e) What is the concentration of CH_3NC after 10,000 s?

78. When heated, tetrafluoroethylene dimerizes to form octafluorocyclobutane.

$$2 C_2F_4(g) \longrightarrow C_4F_8(g)$$

To determine the rate of this reaction at 488 K, the data in the table were collected. Analysis was done graphically, as shown here:

$[C_2F_4]$ (M)	Time (s)
0.100	0
0.080	56
0.060	150
0.040	335
0.030	520

(a) What is the rate law for this reaction?

(b) What is the value of the rate constant?

(c) What is the concentration of C_2F_4 after 600 s?

(d) How long will it take until the reaction is 90% complete?

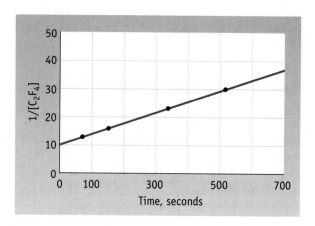

79. Data in the table were collected at 540 K for the following reaction:

$$CO(g) + NO_2(g) \longrightarrow CO_2(g) + NO(g)$$

(a) Derive the rate equation.

(b) Determine the reaction order with respect to each reactant.

(c) Calculate the rate constant, giving the correct units for *k*.

Initial Concentration (mol/L)		Initial Rate
[CO]	**[NO$_2$]**	**(mol/L · h)**
5.1×10^{-4}	0.35×10^{-4}	3.4×10^{-8}
5.1×10^{-4}	0.70×10^{-4}	6.8×10^{-8}
5.1×10^{-4}	0.18×10^{-4}	1.7×10^{-8}
1.0×10^{-3}	0.35×10^{-4}	6.8×10^{-8}
1.5×10^{-3}	0.35×10^{-4}	10.2×10^{-8}

80. Ammonium cyanate, NH_4NCO, rearranges in water to give urea, $(NH_2)_2CO$:

$$NH_4NCO(aq) \longrightarrow (NH_2)_2CO(aq)$$

Time (min)	[NH$_4$NCO] (mol/L)
0	0.458
4.50×10^1	0.370
1.07×10^2	0.292
2.30×10^2	0.212
6.00×10^2	0.114

Using the data in the table

(a) Decide if the reaction is first or second order.

(b) Calculate *k* for this reaction.

(c) Calculate the half-life of ammonium cyanate under these conditions.

(d) Calculate the concentration of NH_4NCO after 12.0 h.

81. Nitrogen oxides, NO_x (a mixture of NO and NO_2 collectively designated as NO_x), play an essential role in the production of pollutants found in photochemical smog. The NO_x in the atmosphere is slowly broken down to N_2 and O_2 in a first-order reaction. The average half-life of NO_x in the smokestack emissions in a large city during daylight is 3.9 h.

(a) Starting with 1.50 mg in an experiment, what quantity of NO_x remains after 5.25 h?

(b) How many hours of daylight must have elapsed to decrease 1.50 mg of NO_x to 2.50×10^{-6} mg?

82. At temperatures below 500 K the reaction between carbon monoxide and nitrogen dioxide

$$NO_2(g) + CO(g) \longrightarrow CO_2(g) + NO(g)$$

has the rate equation Rate $= k[NO_2]^2$. Which of the three mechanisms suggested here best agrees with the experimentally observed rate equation?

Mechanism 1:   Single, elementary step
$$NO_2 + CO \longrightarrow CO_2 + NO$$

Mechanism 2:   Two steps
Slow $NO_2 + NO_2 \longrightarrow NO_3 + NO$
Fast $NO_3 + CO \longrightarrow NO_2 + CO_2$

Mechanism 3:   Two steps
Slow $NO_2 \longrightarrow NO + O$
Fast $CO + O \longrightarrow CO_2$

83. Chlorine atoms contribute to the destruction of the earth's ozone layer by the following sequence of reactions:

$$Cl + O_3 \longrightarrow ClO + O_2$$
$$ClO + O \longrightarrow Cl + O_2$$

where the O atoms in the second step come from the decomposition of ozone by sunlight:

$$O_3(g) \longrightarrow O(g) + O_2(g)$$

What is the net equation on summing these three equations? Why does this lead to ozone loss in the stratosphere? What is the role played by Cl in this sequence of reactions? What name is given to species such as ClO?

84. Nitryl fluoride can be made by treating nitrogen dioxide with fluorine:

$$2 NO_2(g) + F_2(g) \longrightarrow 2 NO_2F(g)$$

Use the rate data in the table to do the following:

(a) Write the rate equation for the reaction.

(b) Indicate the order of reaction with respect to each component of the reaction.

(c) Find the numerical value of the rate constant *k*.

Experiment	Initial Concentrations (mol/L)			Initial Rate (mol/L · s)
	[NO₂]	[F₂]	[NO₂F]	
1	0.001	0.005	0.001	2×10^{-4}
2	0.002	0.005	0.001	4×10^{-4}
3	0.006	0.002	0.001	4.8×10^{-4}
4	0.006	0.004	0.001	9.6×10^{-4}
5	0.001	0.001	0.001	4×10^{-5}
6	0.001	0.001	0.002	4×10^{-5}

85. Describe each of the following statements as true or false. If false, rewrite the sentence to make it correct.

    (a) The rate-determining elementary step in a reaction is the slowest step in a mechanism.

    (b) It is possible to change the rate constant by changing the temperature.

    (c) As a reaction proceeds at constant temperature, the rate remains constant.

    (d) A reaction that is third order overall must involve more than one step.

86. The decomposition of dinitrogen pentaoxide

    $$2\,N_2O_5(g) \longrightarrow 4\,NO_2(g) + O_2(g)$$

    has the rate equation $-\Delta[N_2O_5]/\Delta t = k[N_2O_5]$. It has been found experimentally that the decomposition is 20% complete in 6.0 h at 300 K. Calculate the rate constant and the half-life at 300 K.

87. The data in the table give the temperature dependence of the rate constant for the reaction $N_2O_5(g) \longrightarrow 2\,NO_2(g) + \frac{1}{2}O_2(g)$. Plot these data in the appropriate way to derive the activation energy for the reaction.

T(K)	k(s⁻¹)
338	4.87×10^{-3}
328	1.50×10^{-3}
318	4.98×10^{-4}
308	1.35×10^{-4}
298	3.46×10^{-5}
273	7.87×10^{-7}

88. The decomposition of gaseous dimethyl ether at ordinary pressures is first order. Its half-life is 25.0 min at 500 °C:

    $$CH_3OCH_3(g) \longrightarrow CH_4(g) + CO(g) + H_2(g)$$

    (a) Starting with 8.00 g of dimethyl ether, what mass remains (in grams) after 125 min and after 145 min?

    (b) Calculate the time in minutes required to decrease 7.60 ng (nanograms) to 2.25 ng.

    (c) What fraction of the original dimethyl ether remains after 150 min?

89. The decomposition of phosphine, PH_3, proceeds according to the equation

    $$4\,PH_3(g) \longrightarrow P_4(g) + 6\,H_2(g)$$

    It is found that the reaction has the rate equation Rate = $k[PH_3]$. The half-life of PH_3 is 37.9 s at 120 °C.

    (a) How much time is required for three-fourths of the PH_3 to decompose?

    (b) What fraction of the original sample of PH_3 remains after 1 min?

90. Three mechanisms are proposed for the gas phase reaction of NO with Br_2 to give BrNO:

    Mechanism 1:
    $$NO(g) + NO(g) + Br_2(g) \longrightarrow 2\,BrNO(g)$$

    Mechanism 2:
    Step 1 $\quad NO(g) + Br_2(g) \longrightarrow Br_2NO(g)$
    Step 2 $\quad Br_2NO(g) + NO(g) \longrightarrow 2\,BrNO(g)$

    Mechanism 3:
    Step 1 $\quad NO(g) + NO(g) \longrightarrow N_2O_2(g)$
    Step 2 $\quad N_2O_2(g) + Br_2(g) \longrightarrow 2\,BrNO(g)$

    (a) Write the balanced equation for the net reaction.

    (b) What is the molecularity for each step in each mechanism?

    (c) What are the intermediates formed in Mechanisms 2 and 3?

    (d) Compare the rate laws that are derived from these three mechanisms. How could you differentiate them experimentally?

91. Radioactive iodine-131, which has a half-life of 8.04 days, is used in the form of sodium iodide to treat cancer of the thyroid. If you begin with 25.0 mg of $Na^{131}I$, what quantity of the material remains after 31 days?

92. The ozone in the earth's ozone layer decomposes according to the equation

    $$2\,O_3(g) \longrightarrow 3\,O_2(g)$$

    The mechanism of the reaction is thought to proceed through an initial fast equilibrium and a slow step:

    Step 1: Fast, Reversible $\quad O_3(g) \rightleftharpoons O_2(g) + O(g)$
    Step 2: Slow $\quad O_3(g) + O(g) \longrightarrow 2\,O_2(g)$

    Show that the mechanism agrees with the experimental rate law $-\Delta[O_3]/\Delta t = k[O_3]^2/[O_2]$.

93. Hundreds of different reactions can occur in the stratosphere, among them reactions that destroy earth's ozone layer. The following table lists several (second order) reactions of Cl atoms with ozone and organic compounds; each is given with its rate constant:

Reaction	Rate Constant (238 K, cm³/molecule · s)
(a) $Cl + O_3 \longrightarrow ClO + O_2$	1.2×10^{-11}
(b) $Cl + CH_4 \longrightarrow HCl + CH_3$	1.0×10^{-13}
(c) $Cl + C_3H_8 \longrightarrow HCl + C_3H_7$	1.4×10^{-10}
(d) $Cl + CH_2FCl \longrightarrow HCl + CHFCl$	3.0×10^{-18}

For equal concentrations of Cl and other reactant, which is the slowest reaction? Which is the fastest reaction?

94. Hydrogenation reactions, processes wherein H_2 is added to a molecule, are usually catalyzed. An excellent catalyst is a very finely divided metal suspended in the reaction solvent. Tell why finely divided rhodium, for example, is a much more efficient catalyst than a small block of the metal.

95. It is instructive to use a mathematical model in connection with Study Question 94. Suppose you have 1000 blocks, each of which is 1.0 cm on a side. If all 1000 of these blocks are stacked to give a cube that is 10. cm on a side, what fraction of the 1000 blocks have at least one surface on the outside surface of the cube? Now divide the 1000 blocks into eight equal piles of blocks and form them into eight cubes, 5.0 cm on a side. Now what fraction of the blocks have at least one surface on the outside of the cubes? How does this mathematical model pertain to Study Question 94?

96. Isotopes are often used as "tracers" to follow an atom through a chemical reaction, and the following is an example. Acetic acid reacts with methanol by eliminating a molecule of water and forming methyl acetate (see Chapter 11).

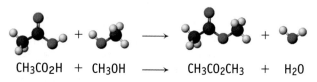

$$CH_3CO_2H + CH_3OH \longrightarrow CH_3CO_2CH_3 + H_2O$$

Explain how you could use the isotope ^{18}O to show whether the oxygen atom in the water comes from the $-OH$ of the acid or the $-OH$ of the alcohol.

97. Examine the reaction coordinate diagram given here.

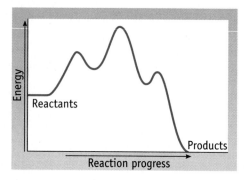

(a) How many steps are in the mechanism for the reaction described by this diagram?

(b) Which is the slowest step in the reaction?

(c) Is the overall reaction exothermic or endothermic?

98. Data for the following reaction

$$[Mn(CO)_5(CH_3CN)]^+ + NC_5H_5 \longrightarrow$$
$$[Mn(CO)_5(NC_5H_5)]^+ + CH_3CN$$

are given in the table. Calculate E_a from a plot of ln k versus $1/T$.

$k(min^{-1})$	$T(K)$
0.0409	298
0.0818	308
0.157	318

99. Draw a reaction coordinate diagram for an exothermic reaction that occurs in a single step. Mark the activation energies, and identify the net energy change for the reaction on this diagram. Draw a second diagram that represents the same reaction in the presence of a catalyst. Identify the activation energy of this reaction and the energy change. Is the activation energy in the two drawings different? Does the energy evolved in the two reactions differ?

100. The gas phase reaction

$$2\,N_2O_5(g) \longrightarrow 4\,NO_2(g) + O_2(g)$$

has an activation energy of 103 kJ, and the rate constant is 0.0900 min^{-1} at 328.0 K. Find the rate constant at 318.0 K.

101. Egg protein albumin is precipitated when an egg is cooked in boiling (100 °C) water. The E_a for this first-order reaction is 52.0 kJ/mol. Estimate the time to prepare a 3-min egg at an altitude at which water boils at 90 °C.

102. Many biochemical reactions are catalyzed by acids. A typical mechanism consistent with the experimental results (in which HA is the acid and X is the reactant) is

Step 1:	Fast, Reversible	$HA \rightleftharpoons H^+ + A^-$
Step 2:	Fast, Reversible	$X + H^+ \rightleftharpoons XH^+$
Step 3:	Slow	$XH^+ \longrightarrow$ products

What rate law is derived from this mechanism? What is the order of the reaction with respect to HA? How would doubling the concentration of acid HA affect the reaction?

103. Hypofluorous acid, HOF, is very unstable, decomposing in a first-order reaction to give HF and O_2, with a half-life of only 30 min at room temperature:

$$HOF(g) \longrightarrow HF(g) + \tfrac{1}{2}O_2(g)$$

If the partial pressure of HOF in a 1.00-L flask is initially 1.00×10^2 mm Hg at 25 °C, what is the total pressure in the flask and the partial pressure of HOF after exactly 30 min? After 45 min?

104. We know that the decomposition of SO_2Cl_2 is first order in SO_2Cl_2,

$$SO_2Cl_2(g) \longrightarrow SO_2(g) + Cl_2(g)$$

with a half-life of 245 min at 600 K. If you begin with a partial pressure of SO_2Cl_2 of 25 mm Hg in a 1.0-L flask, what is the partial pressure of each reactant and product after 245 min? What is the partial pressure of each reactant and product after 12 h?

105. The substitution of CO in $Ni(CO)_4$ by another group L [where L is an electron pair donor such as $P(C_6H_5)_3$] was studied some years ago and led to an understanding of some of the general principles that govern the chemistry of compounds having metal $-CO$ bonds. (See J. P. Day,

F. Basolo, and R. G. Pearson, *Journal of the American Chemical Society,* Vol. 90, p. 6927, 1968.)

A detailed study of the kinetics of the reaction led to the following mechanism [where L = $P(C_6H_5)_3$]:

$$\text{Slow} \qquad Ni(CO)_4 \longrightarrow Ni(CO)_3 + CO$$

$$\text{Fast} \qquad Ni(CO)_3 + L \longrightarrow Ni(CO)_3L$$

(a) What is the molecularity of each of the elementary reactions?

(b) It was found that doubling the concentration of $Ni(CO)_4$ led to an increase in reaction rate by a factor of 2. Doubling the concentration of L had no effect on the reaction rate. Based on this information, write the rate equation for the reaction. Does this agree with the mechanism described?

(c) The experimental rate constant for the reaction is $9.3 \times 10^{-3}\ s^{-1}$ at 20 °C. If the initial concentration of $Ni(CO)_4$ is 0.025 M, what is the concentration of the product after 5.0 min?

Using Electronic Resources

These questions refer to the General Chemistry Interactive CD-ROM, *Version 3.0.*

106. See CD-ROM Screen 15.2: Rates of Chemical Reactions.

(a) What is the difference between an instantaneous rate and an average rate?

(b) Observe the graph of food dye concentration versus time on this screen. (Click on the "tool" icon on this screen.) The plot shows the concentration of dye as the reaction progresses. What does the steepness of the plot at any particular time tell you about the rate of the reaction at that time?

(c) As the reaction progresses, the concentration of dye decreases as it is consumed. What happens to the reaction rate as this occurs? What is the relationship between reaction rate and dye concentration?

107. See CD-ROM Screen 15.4: Control of Reaction Rates (Concentration Dependence).

(a) Watch the video on this screen. How does an increase in HCl concentration affect the rate of the reaction of the acid with magnesium metal?

(b) On the second portion of this screen are data for the rate of decomposition of N_2O_5 (click on "More"). The initial reaction rate is given for three separate experiments, each beginning with a different concentration of N_2O_5. How is the initial reaction rate related to $[N_2O_5]$?

108. See CD-ROM Screen 15.5: Determination of Rate Equation (Method of Initial Rates). This screen describes how to determine experimentally a rate law using the method of initial rates.

(a) Why must the rate of reaction be measured at the very beginning of the process for this method to be valid?

(b) The first experiment shows that the initial rate of NH_4NCO degradation is $2.2 \times 10^{-4}\ mol/L \cdot min$ when $[NH_4NCO] = 0.14$ M. Using the rate law determined on this screen, predict what the rate would be if $[NH_4NCO] = 0.18$ M.

109. See CD-ROM Screen 15.7: Determination of a Rate Equation (Graphical Methods). The rate law for the decomposition of NH_4NCO is Rate = $k[NH_4NCO]$. If you wanted to calculate the concentration of NH_4NCO after some time had elapsed, which equation would you use?

110. See CD-ROM Screen 15.8: Half-Life: First-Order Reactions. Examine the graph and table of concentrations on the second portion of this screen.

(a) What will the concentration of H_2O_2 be after 3270 min? After 3924 min?

(b) What fraction of the original concentration of H_2O_2 remains after each of these times?

111. See CD-ROM Screen 15.9: Microscopic View of Reactions.

(a) According to collision theory, what three conditions must be met for two molecules to react?

(b) Examine the animations that play when numbers 1 and 2 are selected. One of these occurs at a higher temperature than the other. Which one? Explain briefly.

(c) Examine the animations that play when numbers 2 and 3 are selected. Would you expect the reaction of O_3 with N_2

$$O_3(g) + N_2(g) \longrightarrow O_2(g) + ONN(g)$$

to be more or less sensitive to requiring a proper orientation for reaction than the reaction displayed on this screen? Explain briefly.

112. See CD-ROM Screen 15.10: Microscopic View of Reactions. Examine the two animations on the *Reaction Coordinate Diagrams* sidebar to this screen. What is the difference between the ways in which the two reactions occur?

113. See CD-ROM Screen 15.12: Reaction Mechanisms. Examine the two mechanisms on this screen. Why is the second step in each said to be bimolecular?

114. See CD-ROM Screen 15.13: Reaction Mechanisms and Rate Equations.

(a) What is the difference between an overall mechanism and an elementary step?

(b) What is the relationship between the stoichiometric coefficients of the reactants in an elementary step and the rate law for that step?

(c) What is the rate law for Step 2 of Mechanism 2?

(d) Examine the *Isotopic Labeling* sidebar to this screen. If the reaction occurred in a single step transfer of an oxygen atom from NO_2 to CO, would any $N^{16}O^{18}O$ be found if the reaction is started using a mixture of $N^{16}O_2$ and $N^{18}O_2$? Why or why not?

115. See CD-ROM Screen 15.14: Catalysis and Reaction Rate.

(a) Examine the mechanism for the iodide ion-catalyzed decomposition of H_2O_2. Explain how the mechanism shows that I^- is a catalyst.

(b) How does the reaction coordinate diagram show that the catalyzed reaction is expected to be faster than the uncatalyzed reaction?

16 Principles of Reactivity: Chemical Equilibria

Chapter Goals

- Understand the nature and characteristics of chemical equilibria.
- Understand the significance of the equilibrium constant, K.
- Understand how to use the equilibrium constant in quantitative studies of chemical equilibria.

▲ Fritz Haber (1868–1934) developed a method for combining nitrogen from the air with hydrogen to make ammonia, a valuable agricultural chemical. A biographer described him as "verbally and action-oriented rather than contemplative," and a contemporary said that his talent for gaiety and laughter was enormously appealing. *(Oesper Collection in the History of Chemistry/University of Cincinnati)*

Fertilizer and Poison Gas

Nitrogen-containing substances are used around the world to stimulate the growth of field crops. Farmers from Portugal to Tibet have used animal waste for centuries as a natural fertilizer. In the 19th century, industrialized countries imported nitrogen-rich marine bird manure from Peru, Bolivia, and Chile, but the supply was clearly limited. In 1898, William Ramsay (the discoverer of the noble gases) pointed out the depletion of "fixed nitrogen" in the world and predicted world food shortage by the mid-20th century as a result. That this did not occur was due to the work of Fritz Haber.

Fritz Haber was born in 1868 in eastern Germany. His mother died shortly after his birth, and Haber was raised by his father, a trader in dyes and chemicals. When Fritz Haber was a young man, his father insisted that young Fritz come into the business. Haber, however, was restless and spent some years at various universities before earning his Ph.D. in 1891 from the University of Berlin. To prepare to finally join his father's business, young Haber studied chemical technology in distilleries, a soda ash factory, and a salt mine. Unfortunately, these experiences did not help the family business, and young Haber soon lost a large amount of the firm's money.

Fritz Haber truly wanted to become a professor. One problem, however, was that his scientific training was considered unsound. More importantly, he was a Jew, and few Jews were given professorships. Haber solved this by pragmatically converting to Christianity, but his Jewish roots were to follow him the rest of his life.

After leaving the family business, Haber found a place at the famous Karlsruhe Institute of Technology. He read voraciously in physical chemistry and became a largely self-taught expert. His work ranged widely across industrial problems and fundamental studies in electrochemistry. In 1906 he achieved his goal of a professorship, and his interest turned to one of the fundamental chemical problems of the day: how to turn atmospheric nitrogen into a form usable by plants. Ammonia was being produced by German industry in small amounts, but it was not enough for agricultural use, and there was concern that Europe, with its expanding population, would run out of food.

Haber was an ideal person to enter the race to produce ammonia. He had had practical industrial experience and understood the problem theoretically. And he was tireless. He and his assistant, Robert Le Rossignol, discovered that the only way to accomplish the union of nitrogen and hydrogen was at high temperatures (over 200 °C) and high pres-

"By discovering how to convert nitrogen from the air into ammonia for fertilizer, Fritz Haber saved millions of people from starvation."

S. B. McGrayne: *Prometheans in the Lab*, p. 58. New York, McGraw-Hill, 2001.

▲ **Ammonia gas is "drilled" into the soil of a farm field.** Most of the ammonia manufactured in the world is used as a fertilizer because ammonia supplies the nitrogen needed by green plants. Some of the ammonia is also converted to nitric acid, and ammonia and the acid are combined to give ammonium nitrate, another important industrial chemical. *(Arthur C. Smith III, from Grant Heilman Photography)*

sures (200 atm). Although no one had yet developed methods of achieving those conditions in the laboratory, Haber and his coworkers did, and they produced ammonia — but very slowly. To speed up the process, they knew they needed a catalyst. After testing many substances, they found that osmium and uranium metals accelerated the reaction. Haber excitedly told his colleagues, "You have to see how liquid ammonia is pouring out."

German industry was skeptical that an industrial process could run under the severe conditions that Haber prescribed. What is more, a uranium catalyst was out of the question. Nonetheless, the process was so promising that Carl Bosch (1874–1940) and chemists at a large German chemical company conducted over 10,000 experiments and tested 2000

catalysts. A suitable catalyst — based on iron oxide — was finally found, and the process was patented in 1913. Now known as the Haber–Bosch process, the process is still the cheapest way to "fix" atmospheric nitrogen. Not only that but, because Haber got 1 pfennig (German penny) per kilogram of ammonia, he soon became not only famous but rich.

Haber reflected the intense nationalistic spirit in Germany at the time, so, when Germany became embroiled in World War I, he turned his scientific expertise to chemical problems of warfare. He became the director of the German Chemical Warfare Service, whose primary mission was to develop gas warfare. In 1915, Haber supervised the first use of chlorine gas against an opponent at the infamous battle of Ypres in France. Not only was this a tragedy of modern warfare but also for Haber personally. His wife pleaded with him to stop his work on gas warfare, and, when he refused, she committed suicide.

Haber received the Nobel Prize in 1918 for the ammonia synthesis, although this award was widely criticized because of his wartime work on chemical warfare. After the war he did some of his best work, including his studies of thermodynamics (page 331). Because he had a Jewish background, however, he had to leave Germany in 1933. He worked for a short time in England and died in Switzerland in 1934.

Before You Begin

- Review reaction stoichiometry (Chapters 4 and 5).
- Review the concepts of chemical kinetics (Chapter 15).
- Review the behavior of gases (Chapter 12).

The Nature of Chemical Equilibrium These experiments demonstrate the reversibility of chemical reactions. All chemical reactions are in principle reversible and will, given enough time and the proper conditions, achieve a state of dynamic equilibrium.

Forward Reaction

A. Reactants: CaCl₂ on the left, NaHCO₃ on the right

B. The solutions are mixed. Na⁺ and Cl⁻ are spectator ions (not shown)

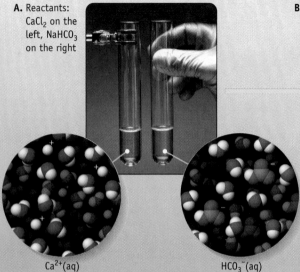

Products: H₂O, a precipitate of CaCO₃ and CO₂ gas

$Ca^{2+}(aq)$ $HCO_3^-(aq)$ $CaCO_3(s)$ $CO_2(g)$

Reaction Equation

$$Ca^{2+}(aq) + 2\ HCO_3^-(aq) \rightleftharpoons CaCO_3(s) + CO_2(g) + H_2O(\ell)$$

Reverse Reaction

D. The CaCO₃ dissolves when the solution has been saturated with CO₂.

C. The reaction can be reversed by bubbling CO₂ gas into the CaCO₃ suspension.

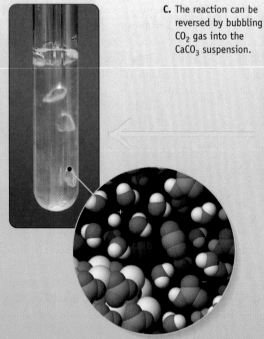

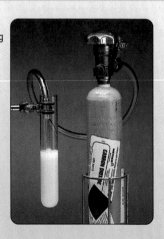

$Ca^{2+}(aq) + 2\ HCO_3^-(aq)$

Because equilibrium is fundamental in chemistry, this and the next two chapters explore this concept as applied to chemical reactions. Our goal is to explore the consequences of the fact that chemical reactions are reversible, that in a closed system a state of equilibrium is eventually achieved between reactants and products, and that outside forces can affect the equilibrium. A major result of this exploration will be an ability to describe "chemical reactivity" in quantitative terms.

Figure 16.1   Cave chemistry. Calcium carbonate stalactites cling to the roof of a cave, and stalagmites grow up from the cave floor. The chemistry producing these formations is a good example of the reversibility of chemical reactions. *(Arthur N. Palmer)*

16.1   NATURE OF THE EQUILIBRIUM STATE

If you ever visited a limestone cave, you were surely impressed with the beautiful stalactites and stalagmites, which are made chiefly of calcium carbonate (Figure 16.1). How did these evolve?

The formation of stalactites and stalagmites depends on the reversibility of chemical reactions. Calcium carbonate is found in underground deposits in the form of limestone, a leftover from ancient oceans. If water seeping through the limestone contains dissolved CO_2, a reaction occurs in which the mineral dissolves, giving an aqueous solution of Ca^{2+} and HCO_3^- ions.

$$CaCO_3(s) + CO_2(aq) + H_2O(\ell) \longrightarrow Ca^{2+}(aq) + 2\ HCO_3^-(aq)$$

When the mineral-laden water reaches a cave, the reverse reaction occurs, with CO_2 being evolved into the cave and solid $CaCO_3$ being deposited.

$$Ca^{2+}(aq) + 2\ HCO_3^-(aq) \longrightarrow CaCO_3(s) + CO_2(g) + H_2O(\ell)$$

Dissolving and reprecipitating limestone can be illustrated by a laboratory experiment with some soluble salts containing the Ca^{2+} and HCO_3^- ions (say $CaCl_2$ and $NaHCO_3$). If you put these salts in an open beaker of water in the laboratory, you will soon see bubbles of CO_2 gas and a precipitate of solid $CaCO_3$ (Chapter Focus). If you bubble CO_2 into the solution, however, the solid $CaCO_3$ redissolves. This experiment illustrates an important feature of chemical reactions: *In principle, all chemical reactions are reversible* [🔘 CD-ROM, Screen 16.3].

It is interesting and informative to ask what happens if the initial solution of Ca^{2+} and HCO_3^- ions is in a closed container (Chapter Focus). As the reaction begins, Ca^{2+} and HCO_3^- react to give the products at some rate [◀ SECTION 15.1]. As the reactants are used up, the rate of this reaction slows. However, the reaction products ($CaCO_3$, CO_2, and H_2O) begin to combine to reform Ca^{2+} and HCO_3^-, at a rate that increases as the amounts of $CaCO_3$ and CO_2 increase. Eventually the rate of the forward reaction, the formation of $CaCO_3$, and the rate of the reverse reaction, the redissolving of $CaCO_3$, become equal. With $CaCO_3$ being formed and redissolving at the same rate, no further *macroscopic* change is observed. The system is at **equilibrium,** a state in which both the forward and reverse reactions continue to occur at equal rates but no *net* change is observed. We depict the situation by writing a balanced equation with reactants and products connected with a double arrow.

$$Ca^{2+}(aq) + 2\ HCO_3^-(aq) \rightleftarrows CaCO_3(s) + CO_2(g) + H_2O(\ell)$$

Chemical reactions are dynamic and and many are reversible. Another example is the ionization of acetic acid, the reaction responsible for the acidity of vinegar.

Chapter Goals • Revisited

- **Understand the nature and characteristics of chemical equilibria.**
  - Understand the significance of the equilibrium constant, *K*.
  - Understand how to use the equilibrium constant in quantitative studies of chemical equilibria.

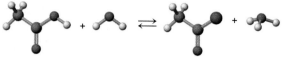

$$CH_3CO_2(aq) \quad + \quad H_2O(\ell) \quad \rightleftharpoons \quad CH_3CO_2^-(aq) \quad + \quad H_3O^+(aq)$$

Acetic acid   Acetate ion   Hydronium ion

Dissolving 1 mol of CH_3CO_2H in enough water to make a 1 M solution will produce a solution having concentrations of 0.0042 M $CH_3CO_2^-$ and 0.0042 M H_3O^+ at equilibrium. An identical set of concentrations can be achieved by dissolving 1 mol of a source of $CH_3CO_2^-$ ions (say $NaCH_3CO_2$) and 1 mol of a source of H_3O^+ ions (say HCl) in the same volume of water.

16.2   THE REACTION QUOTIENT AND EQUILIBRIUM CONSTANT

When a reaction has reached equilibrium the concentrations of reactants and products are related. For the reaction of hydrogen and iodine to produce hydrogen iodide, a very large number of experiments have shown that *at equilibrium* the ratio of the square of the HI concentration to the product of the H_2 and I_2 concentrations is a constant at a given temperature.

$$H_2(g) + I_2(g) \rightleftharpoons 2\,HI(g) \qquad \frac{[HI]^2}{[H_2][I_2]} = \text{constant at equilibrium}$$

The constant is always the same within experimental error for all experiments done at a given temperature. Suppose, for example, the concentrations of H_2 and I_2 in a flask are each initially 0.0175 mol/L at 425 °C; no HI is present. With time, the concentrations of H_2 and I_2 will decline and that of HI will increase; a state of equilibrium will be attained eventually (Figure 16.2). If the gases in the flask are then analyzed, the result would be $[H_2] = [I_2] = 0.0037$ mol/L and $[HI] = 0.0276$ mol/L. The following table — which we call an ICE table — summarizes these results:

Equation	$H_2(g)$	+	$I_2(g)$	$\rightleftharpoons$	2 HI(g)
Initial concentration (M)	0.0175		0.0175		0
Change in concentration as reaction proceeds to equilibrium (M)	−0.0138		−0.0138		+0.0276
Equilibrium concentration (M)	0.0037		0.0037		0.0276

Putting these equilibrium concentration values into the preceding expression

$$\frac{[HI]^2}{[H_2][I_2]} = \frac{(0.0276)^2}{(0.0037)(0.0037)} = 56$$

gives a ratio of about 56 (or 55.64 if the experimental information contains more significant figures). *This quotient is always the same for all experiments at 425 °C, no matter from which direction the equilibrium is approached (mixing H_2 and I_2 or allowing HI to decompose) and no matter what the initial concentrations.*

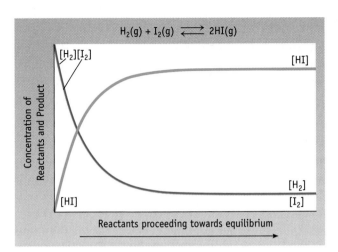

$$H_2(g) + I_2(g) \rightleftharpoons 2HI(g)$$

[H_2][I_2]

[HI]

Concentration of Reactants and Product

[H_2]

[HI]

[I_2]

Reactants proceeding towards equilibrium

Figure 16.2   The reaction of H_2 and I_2 reaches equilibrium. The final concentrations of H_2, I_2, and HI depend on the initial concentrations of H_2 and I_2. If one begins with a different set of initial concentrations, the equilibrium concentrations will be different but the quotient $[HI]^2/[H_2][I_2]$ will always be the same at a given temperature.

See the *General Chemistry Interactive CD-ROM*, Version 3.0, Screen 16.4 where you can observe the effect of changing initial concentrations on the equilibrium concentrations.

The hydrogen–iodine experiment can be generalized. For any chemical reaction under any conditions,

$$a\,A + b\,B \rightleftharpoons c\,C + d\,D$$

the concentrations of reactants and products are always related by a mathematical expression to a value called the **reaction quotient.**

Under Any Reaction Conditions

Product concentrations

$$\text{Reaction quotient} = Q = \frac{[C]^c[D]^d}{[A]^a[B]^b}$$ **(16.1)**

Reactant concentrations

Hundreds of experiments have proved that, *when a reaction has reached equilibrium, the reaction quotient has a constant value.* That constant is called the **equilibrium constant, K** [CD-ROM, Screen 16.4].

When Reaction Is at Equilibrium

$$\text{Reaction quotient} = \text{equilibrium constant} = K = \frac{[C]^c[D]^d}{[A]^a[B]^b}$$ **(16.2)**

We shall henceforth refer to this general mathematical expression (Equation 16.2) by the commonly used term, the **equilibrium constant expression.** In this expression (and in Equation 16.1),

- Product concentrations always appear in the numerator.
- Reactant concentrations always appear in the denominator.
- Each concentration is always raised to the power of its stoichiometric coefficient in the balanced equation.
- When the reaction has reached equilibrium, the value of the constant K depends on the particular reaction and on the temperature. Units are never given with K.

The equilibrium constant for a chemical reaction is very useful because it indicates whether a reaction is product- or reactant-favored, and it can be used to calculate the quantity of reactant or product present at equilibrium (see page 665).

Writing Equilibrium Constant Expressions

Reactions Involving Solids, Water, and Pure Liquids

You should know a few rules about writing equilibrium constant expressions [CD-ROM, Screen 16.6]. For example, the oxidation of solid, yellow sulfur produces colorless sulfur dioxide gas in a product-favored reaction (Figure 16.3).

$$S(s) + O_2(g) \rightleftharpoons SO_2(g)$$

Following the general principle that product concentrations appear in the numerator and reactant concentrations in the denominator, you would initially write

$$K' = \frac{[SO_2]}{[S][O_2]}$$

Because sulfur is a molecular solid and because the concentration of molecules within any solid is fixed by its density, the sulfur concentration is not changed either by reaction or by addition or removal of some solid. Furthermore, it is an experimental fact that the equilibrium concentrations of O_2 and SO_2 are not changed by the amount of sulfur present, as long as some solid sulfur is present at equilibrium. Therefore, *the concentrations of any solid reactants and products are omitted from the equilibrium constant expression*. This means you should write the equilibrium constant expression for the combustion of sulfur as

$$K = \frac{[SO_2]}{[O_2]}$$

There are special considerations for reactions occurring in aqueous solution (or in any solvent). Consider ammonia, which is a weak base owing to its reaction with water (see Figure 5.6).

$$NH_3(aq) + H_2O(\ell) \rightleftharpoons NH_4^+(aq) + OH^-(aq)$$

Because the water concentration is very high in a dilute ammonia solution, the concentration of water is essentially unchanged by the reaction. As for solids, *the molar concentration of water (or of any liquid reactant or product) is omitted from the equilibrium constant expression*. Thus, we write

$$K = \frac{[NH_4^+][OH^-]}{[NH_3]}$$

Reactions Involving Gases

Much of the chemistry with which we shall be concerned involves substances in aqueous solution, so the data used in the equilibrium constant expression are given in moles per liter (M). For reactions of this type the symbol K sometimes has the subscript c for "concentration," as in K_c.

A number of important reactions involve gases, and equilibrium constant expressions for these reactions can also be written in terms of partial pressures of reactants and products. If you rearrange the ideal gas law [$PV = nRT$], and recognize that the "gas concentration," (n/V), is equivalent to P/RT, you see that the partial pressure of a gas is proportional to its concentration [$P = (n/V)RT$]. If reactant and product quantities are given in partial pressures, then K is given the subscript p, as in K_p.

$$H_2(g) + I_2(g) \rightleftharpoons 2\,HI(g) \qquad K_p = \frac{P^2_{HI}}{P_{H_2}P_{I_2}}$$

Figure 16.3 Burning sulfur. Elemental sulfur burns in oxygen with a beautiful blue flame to give SO_2 gas. *(Charles D. Winters)*

● K_c and K_p
The subscript c (K_c) indicates the numerical values of concentrations in the equilibrium constant expression have units of moles per liter. A subscript p (K_p) indicates values in units of pressure. In this chapter we use K for all equilibrium constants except when equilibrium values are in pressure units.

Equilibrium Constant Expressions for Gases — K_c and K_p

Many metal carbonates, such as limestone, decompose on heating to give the metal oxide and CO_2 gas.

$$CaCO_3(s) \rightleftharpoons CaO(s) + CO_2(g)$$

The equilibrium condition for this reaction can be expressed either in terms of the number of moles per liter of CO_2, $K_c = [CO_2]$, or in terms of the pressure of CO_2, $K_p = P_{CO_2}$. From the ideal gas law, you know that

$$P = (n/V)RT$$
$$= \text{(concentration in moles per liter)}RT$$

For this reaction, we can therefore say that $K_p = [CO_2]RT$. Because $K_c = [CO_2]$, the interesting conclusion is that $K_p = K_c(RT)$. That is, the *values* of K_p and K_c are *not* the same; K_p for the decomposition of calcium carbonate is the product of K_c and the factor RT.

The equilibrium constant in terms of partial pressures, K_p, is known for the reaction of N_2 and H_2 to produce ammonia.

$$N_2(g) + 3 H_2(g) \rightleftharpoons 2 NH_3(g)$$
$$K_p = \frac{(P_{NH_3})^2}{(P_{N_2})(P_{H_2})^3} = 5.8 \times 10^5 \quad \text{at 25 °C}$$

Does K_c, the equilibrium constant in terms of concentrations, have the same value or a different value as K_p? We can answer this by substituting for each pressure in K_p the equivalent expression $c(RT)$. That is,

$$K_p = \frac{\{[NH_3](RT)\}^2}{\{[N_2](RT)\}\{[H_2](RT)\}^3}$$
$$= \left\{ \frac{[NH_3]^2}{[N_2][H_2]^3} \right\} \left\{ \frac{1}{(RT)^2} \right\} = \frac{K_c}{(RT)^2}$$

Solving for K_c, we find

$$K_p = 5.8 \times 10^5 = \frac{K_c}{[(0.08206)(298)]^2}$$
$$K_c = 3.5 \times 10^8$$

Once again you see that K_p and K_c are not the same but are related by some function of RT.

Looking carefully at these and other examples, we find that, in general,

$$K_p = K_c(RT)^{\Delta n}$$

where Δn is the change in the number of moles of gas on going from reactants to products.

Δn = total moles of gaseous products − total moles of gaseous reactants

For the decomposition of $CaCO_3$,

$$\Delta n = 1 - 0 = 1$$

whereas the value of Δn for the ammonia synthesis is

$$\Delta n = 2 - 4 = -2$$

In some cases the numerical values of K_c and K_p may be the same, but they are often different. The box, "*A Closer Look: Equilibrium Constant Expressions for Gases — K_c and K_p*" shows in more detail how K_c and K_p are related and how to convert one to the other if necessary.

Example 16.1 **Writing Equilibrium Expressions**

Problem • Write the equilibrium constant expressions in terms of concentrations for

(a) $N_2(g) + 3 H_2(g) \rightleftharpoons 2 NH_3(g)$
(b) $H_2CO_3(aq) + H_2O(\ell) \rightleftharpoons HCO_3^-(aq) + H_3O^+(aq)$

Strategy • Remember that product concentrations always appear in the numerator and reactant concentrations in the denominator. Also, each concentration should be raised to a power equal to the stoichiometric coefficient in the balanced equation.

Solution •

(a) $K = \dfrac{[NH_3]^2}{[N_2][H_2]^3}$

(b) $K = \dfrac{[HCO_3^-][H_3O^+]}{[H_2CO_3]}$

Comment • In part (b) note that water does not appear in the equilibrium expression.

Exercise 16.1 **Writing Equilibrium Constant Expressions**

Write the equilibrium constant expression for each of the following reactions in terms of concentrations:

(a) $PCl_5(g) \rightleftharpoons PCl_3(g) + Cl_2(g)$
(b) $Cu(OH)_2(s) \rightleftharpoons Cu^{2+}(aq) + 2 OH^-(aq)$
(c) $Cu(NH_3)_4^{2+}(aq) \rightleftharpoons Cu^{2+}(aq) + 4 NH_3(aq)$
(d) $CH_3CO_2H(aq) + H_2O(\ell) \rightleftharpoons CH_3CO_2^-(aq) + H_3O^+(aq)$

Meaning of the Equilibrium Constant, K

Table 16.1 illustrates the wide range of equilibrium constants observed for different kinds of reactions. A large value of K means that reactants have been converted largely to products when equilibrium has been achieved. That is, the products are strongly favored over the reactants at equilibrium. An example is the reaction of nitrogen monoxide and ozone.

$$NO(g) + O_3(g) \rightleftharpoons NO_2(g) + O_2(g) \qquad K = \frac{[NO_2][O_2]}{[NO][O_3]} = 6 \times 10^{34} \quad \text{at 25 °C}$$

$$K \gg 1, \text{ so, at equilibrium, } [NO_2][O_2] \gg [NO][O_3]$$

> $K \gg 1$: **Reaction is product-favored.** Equilibrium concentrations of products are greater than equilibrium concentrations of reactants.

The very large value of K indicates that, if stoichiometric amounts of NO and O_3 are mixed and allowed to come to equilibrium, virtually none of the reactants will be found. Essentially all will have been converted to NO_2 and O_2. A chemist would say "the reaction has gone to completion."

Conversely, a small value of K means that very little of the reactants have formed products when equilibrium has been achieved. In other words, the reactants are favored over the products at equilibrium.

$$\tfrac{3}{2} O_2(g) \rightleftharpoons O_3(g) \qquad K = \frac{[O_3]}{[O_2]^{3/2}} = 2.5 \times 10^{-29} \quad \text{at 25 °C}$$

$$K \ll 1, \text{ this means } [O_3] \ll [O_2]^{3/2} \text{ at equilibrium}$$

Table 16.1 • Selected Equilibrium Constant Values

Reaction	Equilibrium Constant K (at 25 °C)	Product- or Reactant-Favored
Combination Reactions of Nonmetals		
$S(s) + O_2(g) \rightleftharpoons SO_2(g)$	4.2×10^{52}	$K > 1$; Product-favored
$2 H_2(g) + O_2(g) \rightleftharpoons 2 H_2O(g)$	3.2×10^{81}	$K > 1$; Product-favored
$N_2(g) + 3 H_2(g) \rightleftharpoons 2 NH_3(g)$	3.5×10^8	$K > 1$; Product-favored
$N_2(g) + O_2(g) \rightleftharpoons 2 NO(g)$	1.7×10^{-3} (at 2300 K)	$K < 1$; Reactant-favored
Ionization of Weak Acids and Bases		
$HCO_2H(aq) + H_2O(\ell) \rightleftharpoons HCO_2^-(aq) + H_3O^+(aq)$ formic acid	1.8×10^{-4}	$K < 1$; Reactant-favored
$CH_3CO_2H(aq) + H_2O(\ell) \rightleftharpoons CH_3CO_2^-(aq) + H_3O^+(aq)$ acetic acid	1.8×10^{-5}	$K < 1$; Reactant-favored
$H_2CO_3(aq) + H_2O(\ell) \rightleftharpoons HCO_3^-(aq) + H_3O^+(aq)$ carbonic acid	4.2×10^{-7}	$K < 1$; Reactant-favored
$NH_3(aq) + H_2O(\ell) \rightleftharpoons NH_4^+(aq) + OH^-(aq)$ ammonia	1.8×10^{-5}	$K < 1$; Reactant-favored
Dissolution of "Insoluble" Solids		
$CaCO_3(s) \rightleftharpoons Ca^{2+}(aq) + CO_3^{2-}(aq)$	3.4×10^{-9}	$K < 1$; Reactant-favored
$AgCl(s) \rightleftharpoons Ag^+(aq) + Cl^-(aq)$	1.8×10^{-10}	$K < 1$; Reactant-favored

$K \ll 1$: **Reaction is reactant-favored.** Equilibrium concentrations of reactants are greater than equilibrium concentrations of products.

The very small value of K indicates that very little O_2 will have been converted to O_3 when equilibrium has been achieved.

When K is close to 1, it may not be immediately clear whether the reactant concentrations are larger than the product concentrations or vice versa. It will depend on the form of K and thus on the reaction stoichiometry. Calculations of the concentrations will have to be done.

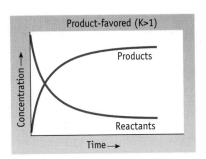

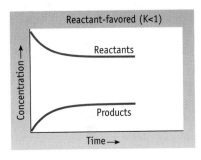

The difference between product- and reactant-favored reactions. When $K \gg 1$, there is much more product than reactant at equilibrium. When $K \ll 1$, there is much more reactant than product at equilibrium.

Exercise 16.2   **The Equilibrium Constant and Extent of Reaction**

(a) Are the following reactions reactant- or product favored?

$$Cu(NH_3)_4^{2+} \rightleftarrows Cu^{2+}(aq) + 4\ NH_3(aq) \qquad\qquad K = 1.5 \times 10^{-13}$$
$$Cd(NH_3)_4^{2+} \rightleftarrows Cd^{2+}(aq) + 4\ NH_3(aq) \qquad\qquad K = 1.0 \times 10^{-7}$$

(b) If each reaction in part (a) has a reactant concentration of 0.10 M, in which solution is the NH_3 concentration greater?

Meaning of the Reaction Quotient, Q

Consider a reaction such as the transformation of butane to isobutane (2-methylpropane) [🖱 CD-ROM, Screen 16.9].

$$\text{Butane} \quad\rightleftarrows\quad \text{Isobutane}$$

$$\qquad\qquad\qquad\qquad\qquad\overset{\displaystyle CH_3}{\underset{\displaystyle |}{}}$$

$$CH_3CH_2CH_2CH_3 \quad\rightleftarrows\quad CH_3CHCH_3$$

$$K_c = \frac{[\text{isobutane}]}{[\text{butane}]} = 2.50 \text{ at } 298 \text{ K}$$

If the concentration of one of the compounds is known, then only one value of the other concentration will satisfy the equation for the equilibrium constant. For example, if [butane] is 1.0 mol/L, then the equilibrium concentration of isobutane, [isobutane], must be 2.5 mol/L. If [butane] is 0.80 M, then [isobutane] at equilibrium must be 2.0 M.

$$[\text{isobutane}] = K\,[\text{butane}] = (2.50)(0.80\text{ M}) = 2.0\text{ M}$$

Choosing a number of possible values for [butane] and solving for the allowed value of [isobutane] would eventually lead to the points on the line in the graph in Figure 16.4. That is, the equilibrium expression $K = [\text{isobutane}]/[\text{butane}]$ is just the equation of a straight line (with a slope equal to K when [isobutane] is plotted against [butane]).

Figure 16.4   Equilibrium in the butane–isobutane system. The straight line represents all the concentrations of butane and isobutane that satisfy the expression $Q = K = $ [isobutane]/[butane] = 2.50. When $Q = K = 2.50$, the system is at equilibrium. When $Q < K$, the system is not at equilibrium and reactants are further converted to products. When $Q > K$, the system is also not at equilibrium, but products must revert to reactants to establish equilibrium. See the text for discussion of the specific points $Q = 6/1$ and $Q = 4/3$.

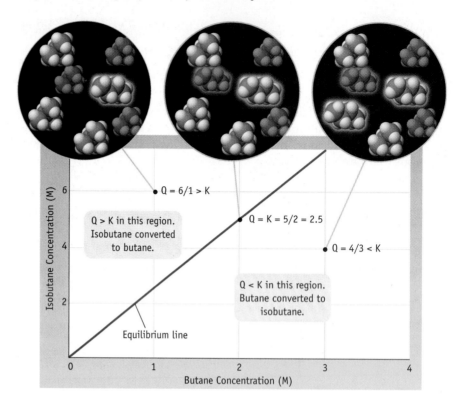

$Q = 6/1 > K$

$Q > K$ in this region. Isobutane converted to butane.

$Q = K = 5/2 = 2.5$

$Q = 4/3 < K$

$Q < K$ in this region. Butane converted to isobutane.

Equilibrium line

Isobutane Concentration (M)

Butane Concentration (M)

The graphical treatment of the simple equilibrium expression $K = $ [isobutane]/[butane] is useful for two reasons.

- The graph in Figure 16.4 illustrates the fact that, for any chemical reaction, there are an infinite number of possible *sets* of equilibrium concentrations of reactants and products. (For the butane–isobutane reaction these sets are represented by the straight line. Other reactions will have more complex graphical representations.)

- Sets of concentrations for butane and isobutane that do not lie along the equilibrium line do not satisfy the equilibrium condition. If a system is not at equilibrium, the concentrations of products and reactants will change until the equilibrium condition is satisfied.

Any mixture of butane and isobutane, whether at equilibrium or not, can be represented by the reaction quotient, $Q (= $ [isobutane]/[butane]) (Equation 16.1). It is important to recognize that Q is equal to the equilibrium constant, K, *only* when the reaction is at equilibrium [see page 659]. Suppose you have a system composed of 3 mol/L of butane and 4 mol/L of isobutane (at 298 K). This means that the ratio of concentrations, Q, is

$$Q = \frac{[\text{isobutane}]}{[\text{butane}]} = \frac{4.0}{3.0} = 1.3$$

a ratio represented by a point to the right of the line in Figure 16.4. This set of concentrations clearly does *not* represent an equilibrium system because $Q < K$. To reach equilibrium, some butane molecules are transformed into molecules of isobutane, thus lowering [butane] and raising [isobutane]. This transformation will continue until the ratio [isobutane]/[butane] = 2.5; that is, until $Q = K$, and the ratio of concentrations in the system is represented by a point on the equilibrium line.

What happens when there is too much isobutane in the system relative to the amount of butane? Suppose [isobutane] = 6.0 M but [butane] is only 1.0 M (see Figure 16.4). Now the reaction quotient Q is greater than K ($Q > K$), and the system is again not at equilibrium. It can proceed to equilibrium by converting isobutane molecules to butane molecules.

The isobutane–butane example illustrates the relationship of K and Q:

> If $Q < K$, the system is not at equilibrium and some reactants will be converted to products.

The ratio of product concentrations to reactant concentrations is too small. More reactants must be converted to products (thus increasing Q) to achieve equilibrium (when $Q = K$).

> If $Q > K$, the system is not at equilibrium and some products will be converted to reactants.

The ratio of product concentrations to reactant concentrations is too large. To reach equilibrium, products must be converted to reactants (thus decreasing the value of Q until $Q = K$ at equilibrium).

> If $Q = K$, the system is at equilibrium.

● Comparing Q and K

Relative Magnitude	Direction of Reaction
$Q < K$	Reactants ⟶ Products
$Q = K$	Reaction at equilibrium
$Q > K$	Reactants ⟵ Products

Example 16.2   The Reaction Quotient

Problem ● The brown gas nitrogen dioxide, NO_2, can form the colorless gas N_2O_4. $K = 171$ at 298 K.

$$2 \, NO_2(g) \rightleftharpoons N_2O_4(g)$$

Suppose the concentration of NO_2 is 0.015 M, and the concentration of N_2O_4 is 0.025 M. Is Q larger than, smaller than, or equal to K? If the system is not at equilibrium, in which direction will the reaction proceed to achieve equilibrium?

Strategy ● Write the expression for Q and substitute the numerical values into the equation. Decide if Q is less than, equal to, or greater than K.

Solution ● When the reactant and product concentrations are substituted into the reaction quotient expression, we have

$$Q = \frac{[N_2O_4]}{[NO_2]^2} = \frac{(0.025)}{(0.015)^2} = 110$$

The value of Q is less than the value of K ($Q < K$), so the reaction is not at equilibrium. It must proceed to equilibrium by converting more NO_2 to N_2O_4, thus increasing $[N_2O_4]$ and decreasing $[NO_2]$ until $Q = K$.

Comment ● When writing Q, make sure you raise each concentration to the power of the stoichiometric coefficient.

Exercise 16.3   The Reaction Quotient

Answer the following questions regarding the butane $\rightleftharpoons$ isobutane equilibrium ($K = 2.50$ at 298 K) and Figure 16.4.

(a) Is the system at equilibrium when [butane] = 0.97 M and [isobutane] = 2.18 M? If it is not at equilibrium, in which direction will the reaction proceed in order to achieve equilibrium?

(b) Is the system at equilibrium when [butane] = 0.75 M and [isobutane] = 2.60 M? If it is not at equilibrium, in which direction will the reaction proceed in order to achieve equilibrium?

Exercise 16.4 **The Reaction Quotient**

At 2000 K the equilibrium constant, K, for the formation of NO(g)

$$N_2(g) + O_2(g) \rightleftharpoons 2\ NO(g)$$

is 4.0×10^{-4}. If, at 2000 K, the concentration of N_2 is 0.50 mol/L, that of O_2 is 0.25 mol/L, and that of NO is 4.2×10^{-3} mol/L, is the system at equilibrium? If not, predict which way the reaction will proceed to achieve equilibrium.

Chapter Goals • Revisited

- Understand the nature and characteristics of chemical equilibria.
- Understand the significance of the equilibrium constant, K.
- **Understand how to use the equilibrium constant in quantitative studies of chemical equilibria.**

16.3 DETERMINING AN EQUILIBRIUM CONSTANT

If the experimental values of the concentrations of all of the reactants and products *at equilibrium* are known, an equilibrium constant can be calculated by substituting the data into the equilibrium constant expression [CD-ROM, Screens 16.4 and 16.8]. Consider this for the oxidation of sulfur dioxide.

$$2\ SO_2(g) + O_2(g) \rightleftharpoons 2\ SO_3(g)$$

In an experiment done at 852 K the equilibrium concentrations are found to be $[SO_2] = 3.61 \times 10^{-3}$ mol/L, $[O_2] = 6.11 \times 10^{-4}$ mol/L, and $[SO_3] = 1.01 \times 10^{-2}$ mol/L. Substituting these data into the equilibrium constant expression we find that K has a large value; the oxidation of sulfur dioxide is clearly product-favored.

$$K = \frac{[SO_3]^2}{[SO_2]^2[O_2]} = \frac{(1.01 \times 10^{-2})^2}{(3.61 \times 10^{-3})^2(6.11 \times 10^{-4})} = 1.28 \times 10^4$$

More commonly, an experiment will provide information on the initial quantities of reactants and the concentration at equilibrium of only one of the reactants or of one of the products. Equilibrium concentrations of the rest of the reactants and products must then be inferred from the balanced chemical equation. As an example, consider again the oxidation of sulfur dioxide to sulfur trioxide. Suppose that 1.00 mol of SO_2 and 1.00 mol of O_2 are placed in a 1.00-L flask, this time at 1000 K. When equilibrium has been achieved, 0.925 mol of SO_3 has been formed. Let us use these data to calculate the equilibrium constant, at 1000 K, for the reaction. After writing the equilibrium constant expression in terms of concentrations, we set up an ICE table showing the initial concentrations, how those concentrations change on proceeding to equilibrium, and the concentrations at equilibrium (see page 658).

Equation	2 SO$_2$	+	O$_2$	$\rightleftharpoons$	2 SO$_3$
Initial (M)	1.00		1.00		0
Change (M)	−0.925		−0.925/2		+0.925
Equilibrium (M)	1.00 − 0.925		$1.00 - \dfrac{0.925}{2}$		0.925
	= 0.075		= 0.537		

The quantities in the ICE table result from the following analysis:

- The amount of SO_2 consumed on proceeding to equilibrium is equal to the amount of SO_3 produced (= 0.925 mol because the stoichiometric factor is [2 mol SO_2 consumed/2 mol SO_3 produced]).
- The amount of O_2 consumed is half of the amount of SO_3 produced (= 0.463 mol because the stoichiometric factor is [1 mol O_2 consumed/2 mol SO_3 produced]).

- As you will observe in other examples, the equilibrium concentration of a reactant is its initial concentration minus the quantity consumed on proceeding to equilibrium.

With the equilibrium concentrations now known, it is possible to calculate K at 1000 K.

$$K = \frac{[SO_3]^2}{[SO_2]^2[O_2]} = \frac{(0.925)^2}{(0.075)^2(0.537)} = 2.8 \times 10^2 \quad \text{at 1000 K}$$

● **K and Temperature**
Notice that K is smaller for the SO_2/O_2 reaction at 1000 K (2.8×10^2) than at 852 K (1.28×10^4). We shall discuss the reasons for the difference in Section 16.6.

Example 16.3   Calculating an Equilibrium Constant

Problem • An aqueous solution of ethanol and acetic acid, each with a concentration of 0.810 M, is heated to 100 °C. At equilibrium, the acetic acid concentration is 0.748 M. Calculate K at 100 °C for the reaction

$$\underset{\text{ethanol}}{C_2H_5OH(aq)} + \underset{\text{acetic acid}}{CH_3CO_2H(aq)} \rightleftharpoons \underset{\text{ethyl acetate}}{CH_3CO_2C_2H_5(aq)} + H_2O(\ell)$$

Strategy • Always focus on defining equilibrium concentrations. The equilibrium concentration of acetic acid is known, so the amount consumed is given by [M initial reactant − M reactant remaining]. Because 1 mol of ethanol reacts per mol of acetic acid, the concentration of ethanol is also known at equilibrium. Finally, the concentration of the product at equilibrium is equivalent to the amount of reactant consumed.

Solution • The amount of acetic acid consumed is 0.810 M − 0.748 M = 0.062 M. This is the same as the amount of ethanol

consumed and as the concentration of ethyl acetate produced. The ICE table for this reaction is therefore

Equation	C_2H_5OH	$+$ CH_3CO_2H	$\rightleftharpoons$ $CH_3CO_2C_2H_5$	$+$ H_2O
Initial (M)	0.810	0.810	0	
Change (M)	−0.062	−0.062	+0.062	
Equilibrium (M)	0.748	0.748	0.062	

The concentration of each substance at equilibrium is now known, and K can be calculated.

$$K = \frac{[CH_3CO_2C_2H_5]}{[C_2H_5OH][CH_3CO_2H]} = \frac{0.062}{(0.748)(0.748)} = 0.11$$

Comment • Notice that a product, liquid water, does not appear in the equilibrium expression.

Example 16.4   Calculating an Equilibrium Constant (K_p) Using Partial Pressures

Problem • Suppose a tank initially contains H_2S with a pressure of 10.00 atm at 800 K. When the reaction

$$2\,H_2S(g) \rightleftharpoons 2\,H_2(g) + S_2(g)$$

reaches equilibrium, the partial pressure of S_2 vapor is 0.020 atm. Calculate K_p.

Strategy • Recall from page 661 that the equilibrium constant expression can be written in terms of gas partial pressures or concentrations. The equilibrium pressure of a gaseous reactant is $P_{initial} - P_{gas\ consumed}$. Here we set up an ICE table to determine the equilibrium partial pressures.

Solution • The equilibrium expression that we want to evaluate is

$$K_p = \frac{P^2_{H_2} P_{S_2}}{P^2_{H_2S}}$$

Let us set up a table that expresses the equilibrium partial pressures of each gas.

Equation	2 H_2S	$\rightleftharpoons$ 2 H_2	$+$ S_2
Initial (atm)	10.00	0	0
Change (atm)	−2(0.020)	+2(0.020)	+0.020
Equilibrium (atm)	10.00 − 2(0.020)	2(0.020)	0.020
	= 9.96	= 0.040	

The balanced equation informs us that for every atmosphere of pressure of S_2 formed, 2 atm of H_2 is also formed, and 2 atm of H_2S is consumed. Because the amount of S_2 present at equilibrium is known from experiment to be 0.020 atm, this means the partial pressure of each gas is known at equilibrium, and K_p can be calculated.

$$K_p = \frac{P^2_{H_2} P_{S_2}}{P^2_{H_2S}} = \frac{(0.040)^2(0.020)}{(9.96)^2} = 3.2 \times 10^{-7}$$

Comment • The value of K_p will be the same as the value of K_c *only* when the number of moles of gaseous reactants is the same as the number of moles of gaseous products. This is not true here, so $K_p \neq K_c$. See "*A Closer Look: Equilibrium Constant Expressions for Gases—K_c and K_p*" (page 661).

Exercise 16.5 Calculating an Equilibrium Constant, K

A solution is prepared by dissolving 0.050 mol of diiodocyclohexane, $C_6H_{10}I_2$, in the solvent CCl_4. The total solution volume is 1.00 L. When the reaction

$$C_6H_{10}I_2 \rightleftharpoons C_6H_{10} + I_2$$

has come to equilibrium at 35 °C, the concentration of I_2 is 0.035 mol/L.

(a) What are the concentrations of $C_6H_{10}I_2$ and C_6H_{10} at equilibrium?

(b) Calculate K_c, the equilibrium constant.

16.4 USING EQUILIBRIUM CONSTANTS IN CALCULATIONS

In many cases the value of K is known as well as the initial amounts of reactants, and you want to know the quantities present at equilibrium [CD-ROM, Screen 16.10]. As we look at several examples of this situation, we shall introduce a further refinement of the ICE table that summarizes initial conditions, final conditions, and changes on proceeding to equilibrium.

Example 16.5 Calculating an Equilibrium Concentration from an Equilibrium Constant

Problem • The equilibrium constant K (= 55.64) for

$$H_2(g) + I_2(g) \rightleftharpoons 2 HI(g)$$

has been determined at 425 °C. If 1.00 mol each of H_2 and I_2 are placed in a 0.500-L flask at 425 °C, what are the concentrations of H_2, I_2, and HI when equilibrium has been achieved?

Strategy • Having written the balanced chemical equation, the next step is to write the equilibrium constant expression.

$$K = \frac{[HI]^2}{[H_2][I_2]} = 55.64$$

Next set up an ICE table to express the concentrations of H_2, I_2, and HI before reaction and on reaching equilibrium. Here, however, we do not know the numerical value of change in H_2 and I_2 concentration on proceeding to equilibrium. We shall therefore express that as the unknown quantity x. That is, x is the quantity of H_2 or of I_2 consumed in the reaction. It follows then that $2x$ is the quantity of HI produced (because the stoichiometric factor is [2 mol HI produced/1 mol H_2 consumed]).

Solution • Set up an ICE table where x is equal to the quantity of H_2 or I_2 consumed in the reaction.

Now the equilibrium concentrations can be substituted into the equilibrium constant expression.

$$55.64 = \frac{(2x)^2}{(2.00 - x)(2.00 - x)} = \frac{(2x)^2}{(2.00 - x)^2}$$

In this case, the unknown quantity x can be found by taking the square root of both sides of the equation,

$$\sqrt{K} = 7.459 = \frac{2x}{2.00 - x}$$

$$7.459(2.00 - x) = 2x$$

$$14.9 - 7.459x = 2x$$

$$14.9 = 9.459x$$

$$x = 1.58$$

With x known, we can now solve for the equilibrium concentrations of the reactants and products.

$$[H_2] = [I_2] = 2.00 - x = 0.42 \text{ M}$$

$$[HI] = 2x = 3.16 \text{ M}$$

Comment • It is always wise to verify the answer by substituting the values back into the equilibrium expression to see if your calculated K agrees with the one given in the problem. In this case $(3.16)^2/(0.42)^2 = 57$. The slight discrepancy with the given value, $K = 55.64$, is because we know $[H_2]$ and $[I_2]$ to only two significant figures.

Equation	H_2	+	I_2	$\rightleftharpoons$	2 HI
Initial (M)	$\dfrac{1.00 \text{ mol}}{0.500 \text{ L}}$ = 2.00 M		$\dfrac{1.00 \text{ mol}}{0.500 \text{ L}}$ = 2.00 M		0 0
Change (M)	$-x$		$-x$		$+2x$
Equilibrium (M)	$2.00 - x$		$2.00 - x$		$2x$

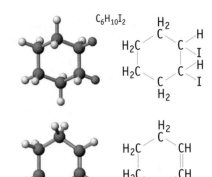

$C_6H_{10}I_2$

C_6H_{10}

Exercise 16.6   Calculating an Equilibrium Concentration Using an Equilibrium Constant

At some temperature, $K = 33$ for the reaction

$$H_2(g) + I_2(g) \rightleftharpoons 2\,HI(g)$$

Assume the initial concentrations of both H_2 and I_2 are 6.00×10^{-3} mol/L. Find the concentration of each reactant and product at equilibrium.

Calculations in Which the Solution Is a Quadratic Expression

When calculating equilibrium concentrations, you may be confronted with an equilibrium constant expression in the form of a quadratic equation: $ax^2 + bx + c = 0$.

Let us say you are studying the decomposition of PCl_5.

$$PCl_5(g) \rightleftharpoons PCl_3(g) + Cl_2(g)$$

You know that $K = 1.20$ at a given temperature. If the initial concentration of PCl_5 is 1.60 M, what will be the equilibrium concentrations of reactant and products?

Reaction	$PCl_5(g)$	$\rightleftharpoons$	$PCl_3(g)$	$+$	$Cl_2(g)$
Initial (M)	1.60		0		0
Change (M)	1.60 − x		+x		+x
Equilibrium (M)	1.60 − x		x		x

Substituting the quantities in the table, we have

$$K = 1.20 = \frac{[PCl_3][Cl_2]}{[PCl_5]} = \frac{(x)(x)}{1.60 - x}$$

Expanding the algebraic expression results in a quadratic equation of the form $ax^2 + b + c = 0$,

$$x^2 + 1.20x - 1.92 = 0$$

Here $a = 1$, $b = 1.20$, and $c = -1.92$. Using the quadratic formula (Appendix A), we find two roots to the equation: $x = 0.91$ and -2.11. Because a negative value of x is not chemically meaningful, the answer is $x = 0.91$. Therefore, we have, at equilibrium,

$$[PCl_5] = 1.60 - 0.91, \text{ or } 0.69 \text{ M}$$
$$[PCl_3] = [Cl_2] = 0.91 \text{ M}$$

Do you always have to solve a quadratic equation when the situation arises or might a mathematically less complicated approximation give you an acceptable answer? To answer this question, consider the meaning of the equilibrium constant. When K is relatively large, the reaction is product-favored; much of the reactants have been consumed and the concentration of products is high. (In our preceding problem, K was larger than 1, and x, which represented the product concentrations, was large.) When K is small, the reaction is reactant-favored, and x in the equilibrium constant expression will be small. In fact, if K is quite small, x can be so small that subtracting it from the original reactant concentration in the denominator of the equilibrium constant expression will not significantly change the

● **Solving Quadratic Equations**
Quadratic equations are usually solved using the quadratic formula (Appendix A). An alternative is the method of successive approximations, also outlined in Appendix A. Most equilibrium expressions can be solved quickly by this method. This will remove the uncertainty of whether K expressions need to be solved exactly.

Problem-Solving Tip 16.1

When Do You Need to Use the Quadratic Equation?

In most equilibrium calculations, the quantity x may be ignored in the denominator of the equation $K = x^2/([A]_0 - x)$ if x is less than 10% of the smallest quantity of reactant initially present. The guideline presented in the text for making the approximation that $[A]_0 - x \approx [A]_0$ when $100K < [A]_0$ reflects this.

In general, when K is about 1 or greater, the approximation *cannot* be made. If K is much less than 1 (you will see many cases in Chapter 17) the approximate expression ($K = x^2/[A]_0$) gives an acceptable answer. If you are not certain, then first make the assumption that the unknown (x) is small and solve the approximate expression. Next compare the "approximate" value of x with $[A]_0$. If x has a value equal to less than 10% of $[A]_0$, then there is no need to solve the full equation using the quadratic formula. For a further discussion of such calculations, see Section 17.7 and Appendix A.

numerical value of the final answer. Knowing this, we can formulate a guideline for deciding when a quadratic expression must be solved and when an approximation can be made.

For the general reaction

$$A \rightleftharpoons B + C$$

the equilibrium constant expression is

$$K = \frac{[B][C]}{[A]} = \frac{(x)(x)}{[A]_0 - x}$$

where K and the initial concentration of A ($= [A]_0$) are known; we wish to find [B] and [C], and we say they are equal to the unknown quantity x. When K is very small, the value of x will be *much* less than $[A]_0$, and $[A]_0 - x \approx [A]_0$. That is, the equilibrium concentration of A is essentially equal to $[A]_0$, and so

$$K = \frac{[B][C]}{[A]} \approx \frac{(x)(x)}{[A]_0}$$

This expression is more readily solved than a quadratic expression, and our guideline for its use is

If $100K < [A]_0$ the approximate expression will give acceptable values of equilibrium concentrations.

This guideline will lead to acceptable values of equilibrium concentrations (to two significant figures). The underlying reasons for the guideline are outlined in Problem-Solving Tip 16.1.

Example 16.6 Calculating an Equilibrium Concentration Using an Equilibrium Constant

Problem • The reaction

$$N_2(g) + O_2(g) \rightleftharpoons 2\ NO(g)$$

contributes to air pollution whenever a fuel is burned in air at a high temperature, as in a gasoline engine. At 1500 K, $K = 1.0 \times 10^{-5}$. Suppose a sample of air has $[N_2] = 0.80$ mol/L and $[O_2] = 0.20$ mol/L before any reaction occurs. Calculate the equilibrium concentrations of reactants and products after the mixture has been heated to 1500 K.

Strategy • Set up an ICE table of equilibrium concentrations and then substitute these concentrations into the equilibrium constant expression. When this is done, you will see that it is a quadratic equation. This can be solved by the methods outlined in Appendix A or we can use the guideline in the text to derive an acceptable, approximate answer.

Solution • As outlined in the Strategy, we first set up an ICE table of equilibrium concentrations.

Equation	N_2	+	O_2	$\rightleftharpoons$	2 NO
Initial (M)	0.80		0.20		0
Change (M)	$-x$		$-x$		$+2x$
Equilibrium (M)	$0.80 - x$		$0.20 - x$		$2x$

Next, the equilibrium concentrations are substituted into the equilibrium constant expression.

$$K = 1.0 \times 10^{-5} = \frac{[NO]^2}{[N_2][O_2]} = \frac{(2x)^2}{(0.80 - x)(0.20 - x)}$$

We refer to our guideline to decide if an approximate solution is possible. Here $100K$ ($= 1.0 \times 10^{-3}$) is smaller than either of the initial reactant concentrations (0.80 and 0.20). This means we can use the approximate expression

$$K = 1.0 \times 10^{-5} = \frac{[NO]^2}{[N_2][O_2]} \approx \frac{(2x)^2}{(0.80)(0.20)}$$

Solving this, we find

$$1.6 \times 10^{-6} = 4x^2$$

$$x = 6.3 \times 10^{-4}$$

This means the reactant and product concentrations at equilibrium are

$$[N_2] = 0.80 - 6.3 \times 10^{-4} \approx 0.80 \text{ M}$$
$$[O_2] = 0.20 - 6.3 \times 10^{-4} \approx 0.20 \text{ M}$$
$$[NO] = 2x = 1.26 \times 10^{-3} \text{ M}$$

Comment • The value of x obtained using the approximation is the same as from the quadratic equation. If the full equilibrium constant expression is expanded, we have

$$(1.0 \times 10^{-5})(0.80 - x)(0.20 - x) = 4x^2$$
$$(1.0 \times 10^{-5})(0.16 - 1.00x + x^2) = 4x^2$$
$$(4 - 1.0 \times 10^{-5})x^2 + (1.0 \times 10^{-5})x + (-0.16 \times 10^{-5}) = 0$$
$$\phantom{(4 - 1.0 \times 10^{-5})}_{ax^2} _{bx} _{c}$$

The two roots to this equation are:

$$x = 6.3 \times 10^{-4} \quad \text{or} \quad x = -6.3 \times 10^{-4}$$

One root is identical to the approximate answer obtained earlier. The approximation is indeed valid in this case.

Exercise 16.7   **Calculating an Equilibrium Partial Pressure Using an Equilibrium Constant**

Graphite and carbon dioxide are kept at 1000 K until the reaction

$$C(graphite) + CO_2(g) \rightleftharpoons 2 CO(g)$$

has come to equilibrium. At this temperature, $K = 0.021$. The initial concentration of CO_2 is 0.012 mol/L. Calculate the equilibrium concentration of CO.

16.5   MORE ABOUT BALANCED EQUATIONS AND EQUILIBRIUM CONSTANTS

Chemical equations can be balanced using different sets of stoichiometric coefficients. For example, the equation for the oxidation of carbon to give carbon monoxide can be written

$$C(s) + \tfrac{1}{2} O_2(g) \rightleftharpoons CO(g)$$

In this case the equilibrium constant expression would be

$$K_1 = \frac{[CO]}{[O_2]^{1/2}} = 4.6 \times 10^{23} \quad \text{at 25 °C}$$

You can write the chemical equation equally well, however, as

$$2 C(s) + O_2(g) \rightleftharpoons 2 CO(g)$$

and the equilibrium constant expression would now be

$$K_2 = \frac{[CO]^2}{[O_2]} = 2.1 \times 10^{47} \quad \text{at 25 °C}$$

When you compare the two equilibrium constant expressions you find that $K_2 = (K_1)^2$; that is,

$$K_2 = \frac{[CO]^2}{[O_2]} = \left\{ \frac{[CO]}{[O_2]^{1/2}} \right\}^2 = K_1{}^2$$

When the stoichiometric coefficients of a balanced equation are multiplied by some factor, the equilibrium constant for the new equation (K_{new}) is the old equilibrium constant (K_{old}) raised to the power of the multiplication factor.

In the case of the oxidation of carbon, the second equation was obtained by multiplying the first equation by two. Therefore, K_2 is the *square* of K_1.

Let us consider what happens if a chemical equation is reversed. Here we will compare the value of *K* for formic acid transferring an H^+ ion to water

$$HCO_2H(aq) + H_2O(\ell) \rightleftharpoons HCO_2{}^-(aq) + H_3O^+(aq)$$

$$K_1 = \frac{[HCO_2{}^-][H_3O^+]}{[HCO_2H]} = 1.8 \times 10^{-4} \quad \text{at 25 °C}$$

with the opposite reaction, the gain of an H^+ ion by the formate ion, $HCO_2{}^-$.

$$HCO_2{}^-(aq) + H_3O^+(aq) \rightleftharpoons HCO_2H(aq) + H_2O(\ell)$$

$$K_2 = \frac{[HCO_2H]}{[HCO_2{}^-][H_3O^+]} = 5.6 \times 10^3 \quad \text{at 25 °C}$$

Here $K_2 = 1/K_1$.

The equilibrium constants for a reaction and its reverse are the reciprocals of each other.

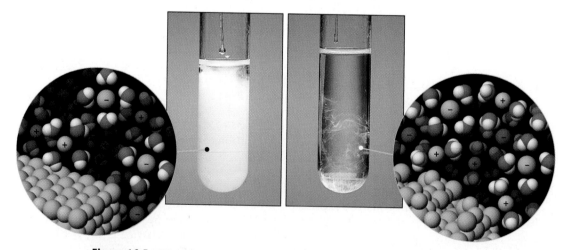

Figure 16.5   Dissolving AgCl in aqueous ammonia. (*left*) A precipitate of AgCl(s) is suspended in water. (*right*) When aqueous ammonia is added, the ammonia reacts with the trace of silver ion in solution, the equilibrium shifts and the silver chloride dissolves. See Figure 5.3 and *General Chemistry Interactive CD-ROM*, Version 3.0, Screen 18.12 for the process of forming a precipitate of AgCl. (*Charles D. Winters*)

It is often useful to add two equations to obtain the equation for a net process. As an example, consider the reactions that take place when silver chloride dissolves in water (to a *very* small extent) and ammonia is added to the solution. The ammonia reacts with the silver ion to form a water-soluble compound, $[Ag(NH_3)_2^+]Cl$ (Figure 16.5). Adding the equation for dissolving solid AgCl to the equation for the reaction of Ag^+ ion with ammonia, gives the equation for net reaction, dissolving solid AgCl in aqueous ammonia. (All equilibrium constants are given at 25 °C.)

$$AgCl(s) \rightleftharpoons Ag^+(aq) + Cl^-(aq) \qquad K_1 = [Ag^+][Cl^-] = 1.8 \times 10^{-10}$$

$$\underline{Ag^+(aq) + 2\,NH_3(aq) \rightleftharpoons Ag(NH_3)_2^+(aq) \qquad K_2 = \frac{[Ag(NH_3)_2^+]}{[Ag^+][NH_3]^2} = 1.6 \times 10^7}$$

Net reaction: $AgCl(s) + 2\,NH_3(aq) \rightleftharpoons Ag(NH_3)_2^+(aq) + Cl^-(aq)$

To obtain the equilibrium constant for the net reaction, K_{net}, we *multiply* the equilibrium constants for the two reactions, K_1 by K_2.

$$K_{net} = K_1 K_2 = [Ag^+][Cl^-]\left\{\frac{[Ag(NH_3)_2^+]}{[Ag^+][NH_3]^2}\right\} = \frac{[Ag(NH_3)_2^+][Cl^-]}{[NH_3]^2}$$

$$K_{net} = K_1 K_2 = 2.9 \times 10^{-3}$$

> *When two or more chemical equations are added to produce a net equation, the equilibrium constant for the net equation is the product of the equilibrium constants for the added equations.*

Example 16.7   **Balanced Equations and Equilibrium Constants**

Problem • A mixture of nitrogen, hydrogen, and ammonia is brought to equilibrium. When the equation is written using whole-number coefficients, as follows, the value of K is 3.5×10^8 at 25 °C.

Equation 1:

$$N_2(g) + 3\,H_2(g) \rightleftharpoons 2\,NH_3(g) \qquad K_1 = 3.5 \times 10^8$$

However, the equation can also be written as

Equation 2:

$$\tfrac{1}{2} N_2(g) + \tfrac{3}{2} H_2(g) \rightleftharpoons NH_3(g) \qquad K_2 = ?$$

What is the value of K_2? What is the value of K_3, the equilibrium constant for the reverse of Equation 1, that is, the decomposition of ammonia to the elements?

Equation 3:

$$2\,NH_3(g) \rightleftharpoons N_2(g) + 3\,H_2(g) \qquad K_3 = ?$$

Strategy • Review what happens to the value of K when the stoichiometric coefficients are changed or the reaction is reversed. See Problem-Solving Tip 16.2.

Solution • To see the relation between K_1 and K_2, first write the equilibrium constant expressions for these two balanced equations.

$$K_1 = \frac{[NH_3]^2}{[N_2][H_2]^3} \qquad \text{and} \qquad K_2 = \frac{[NH_3]}{[N_2]^{1/2}[H_2]^{3/2}}$$

Writing these expressions makes it clear that K_2 is the square root of K_1.

$$K_2 = \sqrt{K_1} = \sqrt{3.5 \times 10^8} = 1.9 \times 10^4$$

Equation 3 is the reverse of Equation 1, and its equilibrium constant expression is

$$K_3 = \frac{[N_2][H_2]^3}{[NH_3]^2}$$

In this case, K_3 is the reciprocal of K_1. That is, $K_3 = 1/K_1$.

$$K_3 = \frac{1}{K_1} = \frac{1}{3.5 \times 10^8} = 2.9 \times 10^{-9}$$

Comment • As a final comment, notice that the production of ammonia from the elements has a large equilibrium constant and is product-favored at 25 °C (see Section 16.7). As expected, the reverse reaction, the decomposition of ammonia to its elements, has a small equilibrium constant and is reactant-favored.

Exercise 16.8 Manipulating Equilibrium Constant Expressions

The conversion of oxygen to ozone has a very small equilibrium constant.

$$\tfrac{3}{2} O_2(g) \rightleftharpoons O_3(g) \qquad\qquad K = 2.5 \times 10^{-29}$$

(a) What is the value of K when the equation is written using whole-number coefficients?

$$3 O_2(g) \rightleftharpoons 2 O_3(g)$$

(b) What is the value of K for the conversion of ozone to oxygen?

$$2 O_3(g) \rightleftharpoons 3 O_2(g)$$

Exercise 16.9 Manipulating Equilibrium Constant Expressions

The following equilibrium constants are given at 500 K:

$$H_2(g) + Br_2(g) \rightleftharpoons 2 HBr(g) \qquad\qquad K_p = 7.9 \times 10^{11}$$
$$H_2(g) \rightleftharpoons 2 H(g) \qquad\qquad K_p = 4.8 \times 10^{-41}$$
$$Br_2(g) \rightleftharpoons 2 Br(g) \qquad\qquad K_p = 2.2 \times 10^{-15}$$

Calculate K_p for the reaction of H and Br atoms to give HBr.

$$H(g) + Br(g) \rightleftharpoons HBr(g) \qquad\qquad K_p = ?$$

16.6 DISTURBING A CHEMICAL EQUILIBRIUM

A chemical reaction at equilibrium may be disturbed in three common ways: (1) by changing the temperature, (2) by changing the concentration of a reactant or product, or (3) by changing the volume (for systems involving gases) (Table 16.2). *A change in any of the factors that determine the equilibrium conditions of a system will cause the system to change in such a manner as to reduce or counteract the effect of the change.* This statement is often referred to as *Le Chatelier's principle* [◀ SECTION 14.2]. It is a shorthand way of describing how a reaction will attempt to adjust the quantities of reactants and products so that equilibrium is restored (that is, so that the reaction quotient is once again equal to the equilibrium constant).

● **Le Chatelier's Principle**
The principles of equilibrium can also be applied to phase changes (Sections 13.5 and 13.10) and the solubility of gases in water (Section 14.3). Both are affected by temperature changes, and you can use Le Chatelier's principle to predict the consequences of these changes.

Problem-Solving Tip 16.2

Balanced Equations and Equilibrium Constants.

You should now know

1. How to write an equilibrium constant expression from the balanced equation, recognizing that the concentrations of pure liquids, solids, and liquids used as solvents do not appear in the expression.

2. That when the stoichiometric coefficients in a balanced equa-

tion are changed by a factor of n, $K_{new} = (K_{old})^n$.

3. That when a balanced equation is reversed, $K_{new} = 1/K_{old}$

4. That when several balanced equations (each with its own equilibrium constant, K_1, K_2, and so on) are added to obtain a net, balanced equation, $K_{net} = K_1 K_2 \cdots$

Table 16.2 • **Effects of Disturbances on Equilibrium Composition**

Disturbance	Change as Mixture Returns to Equilibrium	Effect on Equilibrium	Effect on K
Reactions Involving Solids, Liquids, or Gases			
Rise in temperature	Heat energy is consumed.	Shift in endothermic direction	Change
Drop in temperature	Heat energy is generated.	Shift in exothermic direction	Change
Addition of reactant*	Some of added reactant is consumed.	Product concentration increases.	No change
Addition of product*	Some of added product is consumed.	Reactant concentration increases.	No change
Reactions Involving Gases			
Decrease in volume, increase in pressure	Pressure decreases.	Composition changes to decrease total number of gas molecules.	No change
Increase in volume, decrease in pressure	Pressure increases.	Composition changes to increase total number of gas molecules.	No change

*Does not apply when an insoluble solid or pure liquid reactant or product is added. Recall that their "concentrations" do not appear in the reaction quotient.

Effect of Temperature on Equilibrium Composition

You can make a qualitative prediction about the effect of a temperature change on the equilibrium composition of a chemical reaction if you know whether the reaction is exothermic or endothermic [⌨ CD-ROM, Screens 16.11 and 16.12]. As an example, consider the endothermic reaction of N_2 with O_2 to give NO.

$$N_2(g) + O_2(g) \rightleftharpoons 2\,NO(g) \qquad \Delta H^\circ_{rxn} = +180.5 \text{ kJ}$$

$$K = \frac{[NO]^2}{[N_2][O_2]}$$

Equilibrium Constant	Temperature (K)
4.5×10^{-31}	298
6.7×10^{-10}	900
1.7×10^{-3}	2300

We are surrounded by N_2 and O_2, but you know that they do not react appreciably at room temperature. However, if a mixture of N_2 and O_2 is heated above 700 °C, as in an automobile engine, an equilibrium mixture will contain more NO (see Example 16.6). The experimental equilibrium constants given in the table show that K increases with temperature; the equilibrium concentration of NO becomes larger relative to the concentrations of N_2 and O_2.

To see why the value of K can change with temperature, consider the N_2/O_2 reaction further. The enthalpy change for the reaction is +180.5 kJ, so we might imagine heat as a reactant. If the reaction is at equilibrium, and the temperature increases, the system will attempt to alleviate this "stress." The way to counteract the energy input in this case is to use up some of the added heat by consuming N_2 and O_2 and producing more NO. Because this raises the value of the numerator ($[NO]^2$) and lowers the value of the denominator ($[N_2][O_2]$) in the reaction quotient, K must also increase in value.

As a further example, consider the combination of molecules of the brown gas NO_2 to form colorless N_2O_4. An equilibrium between these compounds is readily achieved in a closed system (Figure 16.6).

$$2\ NO_2(g) \rightleftharpoons N_2O_4(g) \qquad\qquad \Delta H^\circ_{rxn} = -57.2\ kJ$$

$$K = \frac{[N_2O_4]}{[NO_2]^2}$$

Equilibrium Constant, K	Temperature (K)
1300	273
170	298

Here the reaction is exothermic, so we might imagine heat as a reaction product. By lowering the temperature of the reaction, as in Figure 16.6, some heat is removed. The removal of heat can be counteracted if the reaction produces more heat by the combination of NO_2 molecules to give more N_2O_4. Thus, the equilibrium concentration of NO_2 declines, that of N_2O_4 increases, and the values of K in the table become larger as the temperature declines.

In summary,

- When the temperature of a system at equilibrium increases, the equilibrium shifts in the direction that absorbs heat energy (see Table 16.2), that is, in the endothermic direction.

- If the temperature decreases, the equilibrium shifts in the direction that releases heat energy, that is, in the exothermic direction.

- Whether the temperature increases or decreases, the equilibrium composition will change, and the value of K will be different.

Figure 16.6   Effect of temperature on an equilibrium. The tubes in the photograph both contain gaseous NO_2 (brown) and N_2O_4 (colorless). Because K_c for $2\ NO_2(g) \rightleftharpoons N_2O_4(g)$ is larger at the lower temperature, the equilibrium favors colorless N_2O_4 at lower temperatures. This is clearly seen in the tube at the right, where the gas in the ice bath at 0 °C is only slightly brown because there is only a small partial pressure of the brown gas NO_2. At 50 °C (*the tube at the left*), the equilibrium is shifted toward NO_2, as indicated by the darker brown color. *(Photo, Marna G. Clarke)*

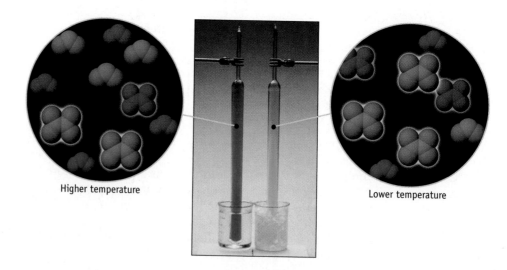

Higher temperature   Lower temperature

Exercise 16.10 Disturbing a Chemical Equilibrium

Consider the effect of temperature changes on the following equilibria.

(a) Does the concentration of NOCl increase or decrease at equilibrium as the temperature of the system is increased?

$$2\ NOCl(g) \rightleftharpoons 2\ NO(g) + Cl_2(g) \qquad \Delta H^\circ_{rxn} = +77.1\ kJ$$

(b) Does the concentration of SO_3 increase or decrease when the temperature increases?

$$2\ SO_2(g) + O_2(g) \rightleftharpoons 2\ SO_3(g) \qquad \Delta H^\circ_{rxn} = -198\ kJ$$

Effect of the Addition or Removal of a Reactant or Product

If the concentration of a reactant or product is changed from its equilibrium value *at a given temperature*, equilibrium will be reestablished eventually [🖱 CD-ROM, Screen 16.12]. The new equilibrium concentrations of reactants and products will be different, but the value of the equilibrium constant expression will still equal K (see Table 16.2). To illustrate this, let us return to the butane–isobutane equilibrium.

$$CH_3CH_2CH_2CH_3 \rightleftharpoons \overset{\overset{\displaystyle CH_3}{\displaystyle |}}{CH_3CHCH_3}$$

butane   isobutane

Suppose the equilibrium mixture consists of two molecules of butane and five of isobutane (Figure 16.7). The ratio of molecules is 5/2 or 2.5/1, the value of the equilibrium constant for the reaction. Now add 7 more molecules of isobutane to the mixture to give a ratio of 12 isobutane molecules to 2 of butane. The ratio or reaction quotient, Q, is 6/1. Q is greater than K, so the system will change to reestablish equilibrium. The only way this can happen is for some molecules of isobutane to be changed into butane molecules, a process that continues until the ratio [isobutane]/[butane] is once again 5/2 or 2.5/1. In this particular case, it means that if

$Q = 5/2 = K$

An equilibrium mixture of 5 isobutane molecules and 2 butane molecules.

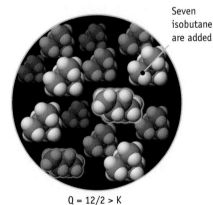

Seven isobutane are added

$Q = 12/2 > K$

Seven isobutane molecules are added, so the system is no longer at equilibrium.

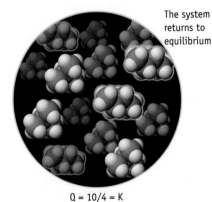

The system returns to equilibrium

$Q = 10/4 = K$

A net of 2 isobutane molecules have changed to butane molecules, to once again give an equilibrium mixture where the ratio of isobutane to butane is 5 to 2 (or 2.5/1).

Figure 16.7   Changing concentrations.

2 of the 12 isobutane molecules change to butane, the *ratio* of isobutane to butane will again be equal to K (= 10/4 = 2.5/1) and equilibrium is reestablished, albeit with different concentrations.

Example 16.8 — The Effect of Concentration Changes on an Equilibrium

Problem • Assume equilibrium has been established in a 1.00-L flask with [butane] = 0.500 mol/L and [isobutane] = 1.25 mol/L.

$$\text{butane} \rightleftharpoons \text{isobutane} \qquad K = 2.5$$

Then 1.50 mol/L of butane is added. What are the new equilibrium concentrations of butane and isobutane?

Strategy • After adding excess butane $Q < K$. To reestablish equilibrium, the concentration of butane (= initial concentration *plus* added butane) must decrease, and that of isobutane must increase. The decrease in butane concentration, and increase in isobutane concentration, is designated as x.

Solution • First organize the information in a modified ICE table.

Equation	Butane	$\rightleftharpoons$	Isobutane
Initial (M)	0.500		1.25
Concentration immediately on adding butane (M)	0.500 + 1.50		1.25
Change in concentration to reestablish equilibrium (M)	−x		+x
Equilibrium (M)	0.500 + 1.50 − x		1.25 + x

The entries in this table were arrived at as follows:

(a) The concentration of butane when equilibrium is reestablished will be the original equilibrium concentration *plus* what was added (1.50 mol/L) *minus* the concentration of butane that must be converted to isobutane in order to reestablish equi-

librium. The quantity of butane converted to isobutane is unknown as yet and so is designated as x.

(b) The concentration of isobutane when equilibrium is reestablished is the concentration that was already present (1.25 mol/L) *plus* the concentration formed (x mol/L) on reestablishing equilibrium.

Having defined [butane] and [isobutane] when equilibrium is reestablished, and remembering that K is a constant (= 2.50), we can write

$$K = 2.50 = \frac{[\text{isobutane}]}{[\text{butane}]}$$

$$2.50 = \frac{1.25 + x}{0.500 + 1.50 - x} = \frac{1.25 + x}{2.00 - x}$$

$$2.50\,(2.00 - x) = 1.25 + x$$

$$5.00 - 2.50x = 1.25 + x$$

$$3.75 = 3.50x$$

$$x = 1.07 \text{ mol/L}$$

We now know the new equilibrium composition is

$$[\text{butane}] = 0.500 + 1.50 - x = 0.93 \text{ mol/L}$$

$$[\text{isobutane}] = 1.25 + x = 2.32 \text{ mol/L}$$

Comment • Be sure to verify that [isobutane]/[butane] = 2.32/0.93 = 2.5. Finally, note that when equilibrium is reestablished, the ratio of isobutane to butane concentration (Q) is again equal to K, but both concentrations are larger than before adding excess butane.

Exercise 16.11 — The Effect of Concentration Changes on an Equilibrium

Equilibrium exists between butane and isobutane when [butane] = 0.20 M and [isobutane] = 0.50 M. What are the concentrations of butane and isobutane, after equilibrium has been attained, if 2.00 mol/L of isobutane is added to the original mixture?

Effect of Volume Changes on Gas-Phase Equilibria

For a reaction that involves gases, what happens to equilibrium concentrations or pressures if the size of the container is changed [CD-ROM, Screen 16.14]? (This occurs, for example, when fuel and air is compressed in an automobile engine.)

To answer this question, recall that concentrations are in moles per liter. If the volume of a gas changes, the concentration therefore must also change, and the equilibrium composition can change. As an example, again consider the equilibrium.

$$2 \, NO_2(g) \;\rightleftharpoons\; N_2O_4(g)$$

brown gas   colorless gas

$$K = \frac{[N_2O_4]}{[NO_2]^2} = 171 \text{ at } 298 \, K$$

What happens to this equilibrium if the volume of the flask holding the gases is suddenly halved? The immediate result is that the concentrations of both gases double. For example, assume equilibrium is established when $[N_2O_4]$ is 0.0280 mol/L and $[NO_2]$ is 0.0128 mol/L. When the volume is halved, $[N_2O_4]$ becomes 0.0560 mol/L and $[NO_2]$ is 0.0256 mol/L. The reaction quotient, Q, under these circumstances is $(0.0560)/(0.0256)^2 = 85.5$, a value clearly less than K. Because Q is less than K, the quantity of product must increase at the expense of the reactants to return to equilibrium, and the new equilibrium composition will have a higher concentration of N_2O_4 than before the volume change.

$$2 \, NO_2(g) \;\rightleftharpoons\; N_2O_4(g)$$

$$\xrightarrow{\text{decrease volume of container}}$$

new equilibrium favors product

This means that 1 molecule of N_2O_4 is formed by consuming two molecules of NO_2. The concentration of NO_2 decreases twice as fast as that of N_2O_4 increases until the reaction quotient, $[N_2O_4]/[NO_2]^2$, is once again equal to K. The conclusions for the NO_2/N_2O_4 equilibrium can be generalized.

- For any reaction involving gases, the stress of a volume decrease (a pressure increase) will be counterbalanced by a change in the equilibrium composition to one having a smaller number of gas molecules.

- For a volume increase (a pressure decrease), the equilibrium composition will shift toward the side of the reaction with the greater number of gas molecules.

- For a reaction in which there is no change in the number of gas molecules, such as

$$H_2(g) + I_2(g) \;\rightleftharpoons\; 2 \, HI(g)$$

a volume change will have no effect.

Exercise 16.12   **Effect of Concentration and Volume Changes on Equilibria**

The formation of ammonia from its elements is an important industrial process.

$$3 \, H_2(g) + N_2(g) \;\rightleftharpoons\; 2 \, NH_3(g)$$

(a) How does the equilibrium composition change when extra H_2 is added? When extra NH_3 is added?

(b) What is the effect on the equilibrium when the volume of the system is increased? Does the equilibrium composition change or is the system unchanged?

16.7 APPLYING THE CONCEPTS OF CHEMICAL EQUILIBRIUM

Haber Process

As described on pages 654–655, one of the greatest advances in agriculture has been the use of manufactured nitrogen-containing fertilizers, largely ammonia and ammonium salts. In 1909, the German chemist Fritz Haber accomplished the feat of making ammonia from its elements, nitrogen and hydrogen.

$$N_2(g) + 3\,H_2(g) \rightleftharpoons 2\,NH_3(g)$$

At 25 °C, K (calculated value) $= 3.5 \times 10^8$ and $\Delta H^\circ_{rxn} = -92.2$ kJ/mol

At 450 °C, K (experimental value) $= 0.16$ and $\Delta H^\circ_{rxn} = -111.3$ kJ/mol

Now ammonia is made for pennies a kilogram and is consistently ranked in the top five chemicals produced in the United States with 15–20 billion kilograms produced annually. Not only is it used as fertilizer but it is also a starting material for making nitric acid and ammonium nitrate, among other things.

The manufacture of ammonia (Figure 16.8) is a good example of the role that kinetics and chemical equilibria play in practical chemistry.

- The $N_2 + H_2$ reaction is exothermic and product-favored ($K > 1$ at 25 °C). The reaction is at 25 °C slow, however, so it is carried out at a higher temperature to increase the reaction rate.

- Although the reaction rate increases with temperature, the equilibrium constant declines, as predicted by Le Chatelier's principle for an exothermic reaction. Thus, the equilibrium concentration of NH_3 is smaller at higher temperatures. In an industrial ammonia plant it is necessary to balance reaction rate (improved at higher temperature) with product yield (K is smaller at higher temperatures).

- To increase the equilibrium concentration of NH_3, the reaction is carried out at a higher pressure. This does not change the value of K, but an increase in pressure can be compensated by converting 4 mol of reactants to 2 mol of product. Higher pressure favors products.

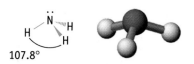

107.8°

Properties of Ammonia

Melting point $= -77$ °C

Boiling point $= -33.4$ °C

Liquid density $= 0.6826$ g/cm^3

Dipole moment $= 1.46$ debyes

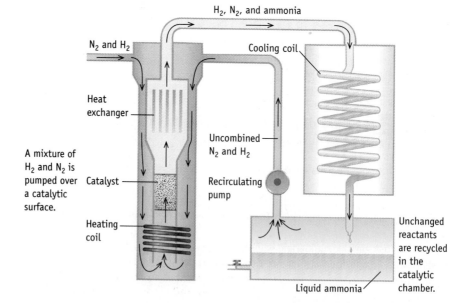

Figure 16.8   The Haber process for ammonia synthesis. A mixture of H_2 and N_2 is pumped over a catalytic surface. The NH_3 is collected as a liquid (at -33 °C), and unchanged reactants are recycled in the catalytic chamber.

H$_2$, N$_2$, and ammonia

N$_2$ and H$_2$

Cooling coil

Heat exchanger

Uncombined N$_2$ and H$_2$

A mixture of H$_2$ and N$_2$ is pumped over a catalytic surface.

Catalyst

Recirculating pump

Heating coil

Unchanged reactants are recycled in the catalytic chamber.

Liquid ammonia

- A catalyst is used to increase the reaction rate further. An effective catalyst for the Haber process is Fe_3O_4 mixed with KOH, SiO_2, and Al_2O_3 (all inexpensive chemicals). Because the catalyst is not effective below 400 °C, however, the optimum temperature is about 450 °C.

In Summary

When you have finished studying this chapter, ask if you have met the chapter goals. In particular, you should be able to

- Understand the nature and characteristics of the state of equilibrium: (a) chemical reactions are reversible and (b) equilibria are dynamic (Section 16.1).
- Write the reaction quotient, Q, for a chemical reaction (Section 16.2). When the system is at equilibrium, the (products/reactants) quotient is called the equilibrium constant expression and has a constant value called the equilibrium constant, which is symbolized by K (Equation 16.2)
- Recognize that the concentrations of solids, pure liquids, and solvents (such as water) are not included in the equilibrium constant expression, Equation 16.2 (Section 16.2).
- Recognize that a large value of K ($K \gg 1$) means the reaction is product-favored, and the product concentrations are greater than the reactant concentrations at equilibrium. A small value of K ($K \ll 1$) indicates a reactant-favored reaction in which the product concentrations are smaller than the reactant concentrations at equilibrium (Section 16.2).
- Appreciate the fact that equilibrium concentrations may be expressed in terms of reactant and product concentrations (expressed in moles per liter), and K is then sometimes designated as K_c. Alternatively, concentrations of gases may be represented by partial pressures, and K for such cases is designated K_p (Section 16.2).
- Use the reaction quotient (Q) to decide if a reaction is at equilibrium ($Q = K$), or if there will be a net conversion of reactants to products ($Q < K$) or products to reactants ($Q > K$) to attain equilibrium (Section 16.2).
- Calculate an equilibrium constant given the reactant and product concentrations at equilibrium (Section 16.3).
- Use equilibrium constants to calculate the concentration of a reactant or product at equilibrium (Section 16.4).
- Know how K changes as different stoichiometric coefficients are used in a balanced equation, or if the equation is reversed, or if several equations are added to give a new, net equation (Section 16.5).
- Know how to predict, using Le Chatelier's Principle, the effect of a disturbance on a chemical equilibrium: a change in temperature, a change in concentration, or a change in volume or pressure for a reaction involving gases (Section 16.6 and Table 16.2).

Key Terms

Section 16.1
  equilibrium

Section 16.2
  reaction quotient, Q
  equilibrium constant, K
  equilibrium constant expression

Key Equations

Equation 16.1 (page 659)

Equation defining the reaction quotient for the general reaction $a\,A + b\,B \rightleftharpoons c\,C + d\,D$. This is the quotient of the product and reactant concentrations, each concentration raised to the power of its stoichiometric coefficient in the balanced equation.

$$\text{Reaction quotient} = Q = \frac{\overset{\text{Product concentrations}}{[C]^c[D]^d}}{\underset{\text{Reactant concentrations}}{[A]^a[B]^b}}$$

Equation 16.2 (page 659)

The equilibrium constant expression. At equilibrium the reaction quotient has a constant value, K (at a particular temperature).

$$K = \frac{[C]^c[D]^d}{[A]^a[B]^b}$$

Study Questions

Questions with blue, bold-faced numbers have answers in Appendix O. Also note that, unless indicated as K_p, all equilibrium constants are for concentrations in moles per liter.

Reviewing Important Concepts

1. Decide if each of the following statements is true or false. If false, change the wording to make it true.

   (a) The magnitude of the equilibrium constant is always independent of temperature.

   (b) When two chemical equations are added to give a net equation, the equilibrium constant for the net equation is the product of the equilibrium constants of the summed equations.

   (c) The equilibrium constant for a reaction has the same value as K for its reverse.

   (d) Only the concentration of CO_2 appears in the equilibrium constant expression for the reaction $CaCO_3(s) \rightleftharpoons CaO(s) + CO_2(g)$.

   (e) For the reaction $CaCO_3(s) \rightleftharpoons CaO(s) + CO_2(g)$, the value of K is numerically the same whether the amount of CO_2 is expressed as moles per liter or as gas pressure.

2. Neither $PbCl_2$ nor PbF_2 is appreciably soluble in water. If solid $PbCl_2$ and solid PbF_2 are placed in equal amounts of water in separate beakers, in which beaker is the concentration of Pb^{2+} greater? Equilibrium constants for these solids dissolving in water are

$$PbCl_2(s) \rightleftharpoons Pb^{2+}(aq) + 2\,Cl^-(aq) \qquad K = 1.7 \times 10^{-5}$$
$$PbF_2(s) \rightleftharpoons Pb^{2+}(aq) + 2\,F^-(aq) \qquad K = 3.7 \times 10^{-8}$$

3. The decomposition of calcium carbonate

$$CaCO_3(s) \rightleftharpoons CaO(s) + CO_2(g)$$

is an endothermic process. Using Le Chatelier's principle, explain how increasing the temperature would affect the equilibrium. If more $CaCO_3$ is added to a flask in which this equilibrium exists, how is the equilibrium affected? What if some additional CO_2 is placed in the flask?

4. The oxidation of NO to NO_2 is exothermic.

$$2\,NO(g) + O_2(g) \rightleftharpoons 2\,NO_2(g)$$

Which way will the equilibrium shift when (a) the temperature is lowered and (b) the volume of the reaction flask is increased?

5. If the reaction quotient is smaller than the equilibrium constant for a reaction such as $A \rightleftharpoons B$, does this mean that reactant A continues to be consumed to form B, or does B form A, as the system moves to equilibrium?

6. Characterize each of the following as product- or reactant-favored.

   (a) $CO(g) + \tfrac{1}{2}O_2(g) \rightleftharpoons CO_2(g)$ $K_p = 1.2 \times 10^{45}$

   (b) $H_2O(g) \rightleftharpoons H_2(g) + \tfrac{1}{2}O_2(g)$ $K_p = 9.1 \times 10^{-41}$

   (c) $CO(g) + Cl_2(g) \rightleftharpoons COCl_2(g)$ $K_p = 6.5 \times 10^{11}$

7. Which of the following is the correct form of the equilibrium constant expression for the reaction of iron(II) ion with tin?

$$Fe^{2+}(aq) + Sn(s) \rightleftharpoons Sn^{2+}(aq) + Fe(s)$$

(a) $K = [Sn^{2+}][Fe]/[Fe^{2+}][Sn]$

(b) $K = [Fe^{2+}][Sn]/[Sn^{2+}][Fe]$

(c) $K = [Fe]/[Sn]$

(d) $K = [Sn^{2+}]/[Fe^{2+}]$

Practicing Skills

Writing Equilibrium Constant Expressions
(See Example 16.1 and CD-ROM Screen 16.6)

8. Write equilibrium constant expressions for the following reactions. For gases use either pressures or concentrations.

(a) $2 H_2O_2(g) \rightleftharpoons 2 H_2O(g) + O_2(g)$

(b) $CO(g) + \frac{1}{2} O_2(g) \rightleftharpoons CO_2(g)$

(c) $C(s) + CO_2(g) \rightleftharpoons 2 CO(g)$

(d) $NiO(s) + CO(g) \rightleftharpoons Ni(s) + CO_2(g)$

9. Write equilibrium constant expressions for the following reactions. For gases use either pressures or concentrations.

(a) $3 O_2(g) \rightleftharpoons 2 O_3(g)$

(b) $Fe(s) + 5 CO(g) \rightleftharpoons Fe(CO)_5(g)$

(c) $(NH_4)_2CO_3(s) \rightleftharpoons 2 NH_3(g) + CO_2(g) + H_2O(g)$

(d) $Ag_2SO_4(s) \rightleftharpoons 2 Ag^+(aq) + SO_4^{2-}(aq)$

The Reaction Quotient
(See Example 16.2 and CD-ROM Screen 16.9)

10. $K = 5.6 \times 10^{-12}$ at 500 K for the dissociation of iodine molecules to iodine atoms.

$$I_2(g) \rightleftharpoons 2 I(g)$$

A mixture has $[I_2] = 0.020$ mol/L and $[I] = 2.0 \times 10^{-8}$ mol/L. Is the reaction at equilibrium (at 500 K)? If not, which way must the reaction proceed to reach equilibrium?

11. The reaction

$$2 NO_2(g) \rightleftharpoons N_2O_4(g)$$

has an equilibrium constant, K, of 171 at 25 °C. If 2.0×10^{-3} mol of NO_2 is present in a 10.-L flask along with 1.5×10^{-3} mol of N_2O_4, is the system at equilibrium? If it is not at equilibrium, does the concentration of NO_2 increase or decrease as the system proceeds to equilibrium?

12. A mixture of SO_2, O_2, and SO_3 at 1000 K contains the gases at the following concentrations: $[SO_2] = 5.0 \times 10^{-3}$ mol/L, $[O_2] = 1.9 \times 10^{-3}$ mol/L, and $[SO_3] = 6.9 \times 10^{-3}$ mol/L. Is the reaction at equilibrium? If not, which way will the reaction proceed to reach equilibrium?

$$2 SO_2(g) + O_2(g) \rightleftharpoons 2 SO_3(g) \qquad K = 279$$

13. The equilibrium constant, K, for the reaction

$$2 NOCl(g) \rightleftharpoons 2 NO(g) + Cl_2(g)$$

is 3.9×10^{-3} at 300 °C. A mixture contains the gases at the following concentrations: $[NOCl] = 5.0 \times 10^{-3}$ mol/L, $[NO] = 2.5 \times 10^{-3}$ mol/L, and $[Cl_2] = 2.0 \times 10^{-3}$ mol/L. Is the reaction at equilibrium at 300 °C? If not, in which direction does the reaction proceed to come to equilibrium?

Calculating an Equilibrium Constant
(See Examples 16.3 and 16.4 and CD-ROM Screens 16.4 and 16.8)

14. The reaction

$$PCl_5(g) \rightleftharpoons PCl_3(g) + Cl_2(g)$$

was examined at 250 °C. At equilibrium $[PCl_5] = 4.2 \times 10^{-5}$ mol/L, $[PCl_3] = 1.3 \times 10^{-2}$ mol/L, and $[Cl_2] = 3.9 \times 10^{-3}$ mol/L. Calculate K for the reaction.

15. An equilibrium mixture of SO_2, O_2, and SO_3 at 1000 K contains the gases at the following concentrations: $[SO_2] = 3.77 \times 10^{-3}$ mol/L, $[O_2] = 4.30 \times 10^{-3}$ mol/L, and $[SO_3] = 4.13 \times 10^{-3}$ mol/L. Calculate the equilibrium constant, K, for the reaction

$$2 SO_2(g) + O_2(g) \rightleftharpoons 2 SO_3(g)$$

16. The reaction

$$C(s) + CO_2(g) \rightleftharpoons 2 CO(g)$$

occurs at high temperatures. At 700 °C, a 2.0-L flask contains 0.10 mol of CO, 0.20 mol of CO_2, and 0.40 mol of C at equilibrium.

(a) Calculate K for the reaction at 700 °C.

(b) Calculate K for the reaction, also at 700 °C, if the amounts at equilibrium in the 2.0-L flask are 0.10 mol of CO, 0.20 mol of CO_2, and 0.80 mol of C.

(c) Compare the results of parts (a) and (b). Does the quantity of carbon affect the value of K? Explain.

17. Hydrogen and carbon dioxide react at a high temperature to give water and carbon monoxide.

$$H_2(g) + CO_2(g) \rightleftharpoons H_2O(g) + CO(g)$$

(a) Laboratory measurements at 986 °C show that there is 0.11 mol each of CO and H_2O vapor and 0.087 mol each of H_2 and CO_2 at equilibrium in a 1.0-L container. Calculate the equilibrium constant for the reaction at 986 °C.

(b) Suppose 0.050 mol each of H_2 and CO_2 are placed in a 2.0-L container. When equilibrium is achieved at 986 °C, what amounts of CO(g) and $H_2O(g)$, in moles, would be present? [Use the value of K calculated in part (a).]

18. A mixture of CO and Cl_2 is placed in a reaction flask: $[CO] = 0.0102$ mol/L and $[Cl_2] = 0.00609$ mol/L. When the reaction

$$CO(g) + Cl_2(g) \rightleftharpoons COCl_2(g)$$

has come to equilibrium at 600 K, $[Cl_2] = 0.00301$ mol/L.

(a) Calculate the concentrations of CO and $COCl_2$ at equilibrium.

(b) Calculate K.

19. You place 3.00 mol of pure SO_3 in an 8.00-L flask at 1150 K. At equilibrium, 0.58 mol of O_2 has been formed. Calculate K for the reaction at 1150 K.

$$2\ SO_3(g) \rightleftharpoons 2\ SO_2(g) + O_2(g)$$

Using Equilibrium Constants

(See Examples 16.5 and 16.6 and CD-ROM Screen 16.10)

20. The value of K for the interconversion of butane and isobutane is 2.5 at 25 °C.

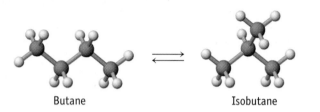

Butane   Isobutane

If you place 0.017 mol of butane in a 0.50-L flask at 25 °C and allow equilibrium to be established, what will be the equilibrium concentrations of the two forms of butane?

21. Cyclohexane, C_6H_{12}, a hydrocarbon, can isomerize into methylcyclopentane, a compound of the same formula ($C_5H_9CH_3$) but with a different molecular structure.

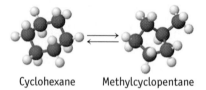

Cyclohexane   Methylcyclopentane

The equilibrium constant has been estimated to be 0.12 at 25 °C. If you originally placed 0.045 mol of cyclohexane in a 2.8-L flask, what are the concentrations of cyclohexane and methylcyclopentane when equilibrium is established?

22. The equilibrium constant for the dissociation of iodine molecules to iodine atoms

$$I_2(g) \rightleftharpoons 2\ I(g)$$

is 3.76×10^{-3} at 1000 K. Suppose 0.105 mol of I_2 is placed in a 12.3-L flask at 1000 K. What are the concentrations of I_2 and I when the system comes to equilibrium?

23. The equilibrium constant for the reaction

$$N_2O_4(g) \rightleftharpoons 2\ NO_2(g)$$

at 25 °C is 5.88×10^{-3}. Suppose 15.6 g of N_2O_4 is placed in a 5.00-L flask at 25 °C. Calculate

(a) The number of moles of NO_2 present at equilibrium

(b) The percentage of the original N_2O_4 that is dissociated

24. Carbonyl bromide decomposes to carbon monoxide and bromine.

$$COBr_2(g) \rightleftharpoons CO(g) + Br_2(g)$$

K is 0.190 at 73 °C. If you place 0.500 mol of $COBr_2$ in a 2.00-L flask and heat it to 73 °C, what are the equilibrium concentrations of $COBr_2$, CO, and Br_2? What percentage of the original $COBr_2$ decomposed at this temperature?

25. Iodine dissolves in water, but its solubility in a nonpolar solvent such as CCl_4 is greater.

Nonpolar I_2
Polar H_2O

Nonpolar CCl_4

Shake the test tube

Polar H_2O

Nonpolar CCl_4 and I_2

Extracting iodine (I_2) from water with the nonpolar solvent CCl_4. *(left)* Iodine is dissolved in water *(upper layer)*. *(right)* I_2 is more soluble in CCl_4 than in water. After shaking a mixture of water and CCl_4, the I_2 has accumulated in the more dense CCl_4 layer. *(Charles D. Winters)*

The equilibrium constant is 85.0 for the reaction

$$I_2(aq) \rightleftharpoons I_2(CCl_4)$$

You place 0.0340 g of I_2 in 100.0 mL of water. After shaking it with 10.0 mL of CCl_4, how much I_2 remains in the water layer?

Balanced Equations and Equilibrium Constants

(See Example 16.7)

26. Which of the following correctly relates the two equilibrium constants for the two reactions shown?

$$A + B \rightleftharpoons 2\ C \qquad\qquad K_1$$
$$2\ A + 2\ B \rightleftharpoons 4\ C \qquad\qquad K_2$$

(a) $K_2 = 2\ K_1$ (c) $K_2 = 1/K_1$

(b) $K_2 = K_1^2$ (d) $K_2 = 1/K_1^2$

27. Which of the following correctly relates the two equilibrium constants for the two reactions shown?

$$A + B \rightleftharpoons 2\ C \qquad\qquad K_1$$
$$C \rightleftharpoons \tfrac{1}{2}A + \tfrac{1}{2}B \qquad\qquad K_2$$

(a) $K_2 = 1/(K_1)^{1/2}$ (c) $K_2 = K_1^2$

(b) $K_2 = 1/K_1$ (d) $K_2 = -K_1^{1/2}$

28. Consider the following equilibria involving $SO_2(g)$ and their corresponding equilibrium constants.

$$SO_2(g) + \tfrac{1}{2}O_2(g) \rightleftharpoons SO_3(g) \qquad\qquad K_1$$
$$2\ SO_3(g) \rightleftharpoons 2\ SO_2(g) + O_2(g) \qquad\qquad K_2$$

Which of the following expressions relates K_1 to K_2?

(a) $K_2 = K_1{}^2$ (d) $K_2 = 1/K_1$
(b) $K_2{}^2 = K_1$ (e) $K_2 = 1/K_1{}^2$
(c) $K_2 = K_1$

29. The equilibrium constant K for the reaction

$$CO_2(g) \rightleftharpoons CO(g) + \tfrac{1}{2} O_2(g)$$

is 6.66×10^{-12} at 1000 K. Calculate K for the reaction

$$2 CO(g) + O_2(g) \rightleftharpoons 2 CO_2(g)$$

30. Calculate K for the reaction

$$SnO_2(s) + 2 CO(g) \rightleftharpoons Sn(s) + 2 CO_2(g)$$

given the following information:

$$SnO_2(s) + 2 H_2(g) \rightleftharpoons Sn(s) + 2 H_2O(g) \qquad K = 8.12$$
$$H_2(g) + CO_2(g) \rightleftharpoons H_2O(g) + CO(g) \qquad K = 0.771$$

31. Calculate K for the reaction

$$Fe(s) + H_2O(g) \rightleftharpoons FeO(s) + H_2(g)$$

given the following information:

$$H_2O(g) + CO(g) \rightleftharpoons H_2(g) + CO_2(g) \qquad K = 1.6$$
$$FeO(s) + CO(g) \rightleftharpoons Fe(s) + CO_2(g) \qquad K = 0.67$$

Disturbing a Chemical Equilibrium

(See Example 16.8 and CD-ROM Screens 16.11–16.14)

32. Dinitrogen trioxide decomposes to NO and NO_2 in an endothermic process ($\Delta H = 40.5$ kJ/mol).

$$N_2O_3(g) \rightleftharpoons NO(g) + NO_2(g)$$

Predict which way the equilibrium will shift (left, right, or no change) when each of the following changes is made:
(a) Adding more $N_2O_3(g)$
(b) Adding more $NO_2(g)$
(c) Increasing the volume of the reaction flask
(d) Lowering the temperature

33. K_p for the following reaction is 0.16 at 25 °C.

$$2 NOBr(g) \rightleftharpoons 2 NO(g) + Br_2(g)$$

The enthalpy change for the reaction at standard conditions is +16.1 kJ. Predict which way the equilibrium will shift (left, right, or no change) when each of the following changes is made:
(a) Adding more $Br_2(g)$
(b) Removing some $NOBr(g)$
(c) Decreasing the temperature
(d) Increasing the container volume

34. Consider the isomerization of butane with an equilibrium constant of $K = 2.5$. (See Study Question 20.) The system is originally at equilibrium with [butane] = 1.0 M and [isobutane] = 2.5 M.
   (a) If 0.50 mol/L of isobutane is suddenly added and the system shifts to a new equilibrium position, what is the equilibrium concentration of each gas?

(b) If 0.50 mol/L of butane is added, and the system shifts to a new equilibrium position, what is the equilibrium concentration of each gas?

35. The decomposition of NH_4HS

$$NH_4HS(s) \rightleftharpoons NH_3(g) + H_2S(g)$$

is an endothermic process. Using Le Chatelier's principle, explain how increasing the temperature would affect the equilibrium. If more NH_4HS is added to a flask in which this equilibrium exists, how is the equilibrium affected? What if some additional NH_3 is placed in the flask? What will happen to the pressure of NH_3 if some H_2S is removed from the flask?

General Questions

More challenging questions are marked with an underlined number.

36. Suppose 0.086 mol of Br_2 is placed in a 1.26-L flask and heated to a high temperature at which the halogen dissociates to atoms

$$Br_2(g) \rightleftharpoons 2 Br(g)$$

If Br_2 is 3.7% dissociated at this temperature, calculate K.

37. The equilibrium constant for the reaction

$$N_2(g) + O_2(g) \rightleftharpoons 2 NO(g)$$

is 1.7×10^{-3} at 2300 K.
   (a) What is K for the reaction when written as

$$\tfrac{1}{2} N_2(g) + \tfrac{1}{2} O_2(g) \rightleftharpoons NO(g)$$

   (b) What is K for the reaction

$$2 NO(g) \rightleftharpoons N_2(g) + O_2(g)$$

38. K_p for the formation of phosgene, $COCl_2$, is 6.5×10^{11} at 25 °C.

$$CO(g) + Cl_2(g) \rightleftharpoons COCl_2(g)$$

What is the value of K_p for the dissociation of phosgene?

$$COCl_2(g) \rightleftharpoons CO(g) + Cl_2(g)$$

39. The equilibrium constant, K_c, for the following reaction is 1.05 at 350 K.

$$2 CH_2Cl_2(g) \rightleftharpoons CH_4(g) + CCl_4(g) \qquad K_c = 1.05$$

If an equilibrium mixture of the three gases at 350 K contains 0.0206 M $CH_2Cl_2(g)$ and 0.0163 M CH_4, what is the equilibrium concentration of CCl_4?

40. Carbon tetrachloride can be produced by the following reaction:

$$CS_2(g) + 3 Cl_2(g) \rightleftharpoons S_2Cl_2(g) + CCl_4(g)$$

Suppose 1.2 mol of CS_2 and 3.6 mol of Cl_2 are placed in a 1.00-L flask. After equilibrium has been achieved, the mixture contains 0.90 mol CCl_4. Calculate K.

41. Equal numbers of moles of H_2 gas and I_2 vapor are mixed in a flask and heated to 700 °C. The initial concentration of each gas is 0.0088 mol/L, and 78.6% of the I_2 is consumed when equilibrium is achieved according to the equation

$$H_2(g) + I_2(g) \rightleftharpoons 2\,HI(g)$$

Calculate K for this reaction.

42. Consider a gas-phase reaction where a colorless compound C produces a blue compound B.

$$2\,C \rightleftharpoons B$$

After reaching equilibrium, the size of the flask is halved.

(a) What color change (if any) is observed immediately on reducing the flask size by half?

(b) What color change (if any) is observed after equilibrium has been reestablished in the flask?

43. The equilibrium constant for the butane $\rightleftharpoons$ isobutane isomerization reaction is 2.5 at 25 °C. (See Study Question 20.) If 1.75 mol of butane and 1.25 mol of isobutane are mixed, is the system at equilibrium? If not, when it proceeds to equilibrium which reagent increases in concentration? Calculate the concentrations of the two compounds when the system reaches equilibrium.

44. At 2300 K the equilibrium constant for the formation of NO(g) is 1.7×10^{-3}.

$$N_2(g) + O_2(g) \rightleftharpoons 2\,NO(g)$$

(a) Analysis shows that the concentrations of N_2 and O_2 are both 0.25 M, and that of NO is 0.0042 M under certain conditions. Is the system at equilibrium?

(b) If the system is not at equilibrium, in which direction does the reaction proceed?

(c) When the system is at equilibrium, what are the equilibrium concentrations?

45. Which of the following correctly relates the equilibrium constants for the two reactions shown?

$$NOCl(g) \rightleftharpoons NO(g) + \tfrac{1}{2}\,Cl_2(g) \qquad K_1$$
$$2\,NO(g) + Cl_2(g) \rightleftharpoons 2\,NOCl(g) \qquad K_2$$

(a) $K_2 = -K_1^2$
(b) $K_2 = 1/(K_1)^{1/2}$
(c) $K_2 = 1/K_1^2$
(d) $K_2 = 2\,K_1$

46. Sulfur dioxide is readily oxidized to sulfur trioxide.

$$2\,SO_2(g) + O_2(g) \rightleftharpoons 2\,SO_3(g) \qquad K = 279$$

If we add 3.00 g of SO_2 and 5.00 g of O_2 to a 1.0-L flask, approximately what quantity of SO_3 will be in the flask once the reactants and product reach equilibrium?

(a) 2.21 g   (c) 3.61 g
(b) 4.56 g   (d) 8.00 g

(*Note:* The full solution to this problem results in a cubic equation. Do not try to solve it exactly. Decide only which of the answers is most reasonable.)

47. Heating a metal carbonate leads to decomposition.

$$BaCO_3(s) \rightleftharpoons BaO(s) + CO_2(g)$$

Predict the effect on the equilibrium of each change listed here. Answer by choosing (i) no change, (ii) shifts left, or (iii) shifts right.

(a) Add $BaCO_3$.
(b) Add CO_2.
(c) Add BaO.
(d) Raise the temperature.
(e) Increase the volume of the flask containing the reaction.

48. Carbonyl bromide decomposes to carbon monoxide and bromine.

$$COBr_2(g) \rightleftharpoons CO(g) + Br_2(g)$$

K is 0.190 at 73 °C. Suppose you placed 0.500 mol of $COBr_2$ in a 2.00-L flask and heated it to 73 °C (see Study Question 24). After equilibrium had been achieved, you added an additional 2.00 mol of CO.

(a) How is the equilibrium mixture affected by adding additional CO?

(b) When equilibrium is reestablished, what are the new equilibrium concentrations of $COBr_2$, CO, and Br_2?

(c) How has the addition of CO affected the percentage of $COBr_2$ that decomposed?

49. The photo shows the result of heating and cooling a solution of Co^{2+} ions in water containing hydrochloric acid. The equation for the equilibrium existing in these solutions is

$$Co(H_2O)_6{}^{2+}(aq) + 4\,Cl^-(aq) \rightleftharpoons CoCl_4{}^{2-}(aq) + 6\,H_2O(\ell)$$

The $Co(H_2O)_6{}^{2+}$ ion is pink, whereas the $CoCl_4{}^{2-}$ ion is blue. Is the transformation of $Co(H_2O)_6{}^{2+}$ to $CoCl_4{}^{2-}$ exothermic or endothermic?

Solution of Co^{2+} ion in water containing HCl(aq). The solution at the left is in a beaker of hot water, whereas the solution at the right is in a beaker of ice water. (*Charles D. Winters*)

50. Phosphorus pentachloride decomposes at higher temperatures.

$$PCl_5(g) \rightleftharpoons PCl_3(g) + Cl_2(g)$$

An equilibrium mixture at some temperature consists of 3.120 g of PCl_5, 3.845 g of PCl_3, and 1.787 g of Cl_2 in a 1.00-L flask. If you add 1.418 g of Cl_2, how will the equilibrium be affected? What will be the concentrations of PCl_5, PCl_3, and Cl_2 when equilibrium is reestablished?

51. The equilibrium constant is 4.5 at 600 °C for the reaction

$$2 SO_2(g) + O_2(g) \rightleftharpoons 2 SO_3(g)$$

Some SO_3 gas was placed in a 1.00-L flask. When the reaction reached equilibrium at 600 °C, the flask contained 2.0 mol of O_2 gas (as well as SO_2 and SO_3). What quantity of SO_3 was originally added to the flask?

52. Ammonium hydrogen sulfide decomposes on heating.

$$NH_4HS(s) \rightleftharpoons NH_3(g) + H_2S(g)$$

If K_p for this reaction is 0.11 at 25 °C (when the partial pressures are measured in atmospheres), what is the total pressure in the flask at equilibrium?

53. Ammonium iodide dissociates reversibly to ammonia and hydrogen iodide if the salt is heated to a sufficiently high temperature.

$$NH_4I(s) \rightleftharpoons NH_3(g) + HI(g)$$

Some ammonium iodide is placed in a flask, which is then heated to 400 °C. If the total pressure in the flask when equilibrium has been achieved is 705 mm Hg, what is the value of K_p (when partial pressures are in atmospheres)?

54. When solid ammonium carbamate sublimes it dissociates completely into ammonia and carbon dioxide according to the equation

$$(NH_4)(H_2NCO_2)(s) \rightleftharpoons 2 NH_3(g) + CO_2(g)$$

At 25 °C, experiment shows that the total pressure of the gases in equilibrium with the solid is 0.116 atm. What is the equilibrium constant, K_p?

55. The equilibrium constant K_p for $N_2O_4(g) \rightleftharpoons 2 NO_2(g)$ is 0.15 at 25 °C. If the pressure of N_2O_4 at equilibrium is 0.85 atm, what is the total pressure of the gas mixture (N_2O_4 + NO_2) at equilibrium?

56. In the gas phase, acetic acid exists as an equilibrium of monomer and dimer molecules. (The dimer consists of two molecules linked through hydrogen bonds.)

The equilibrium constant K_c at 25 °C for the monomer–dimer equilibrium

$$2 CH_3CO_2H(g) \rightleftharpoons (CH_3CO_2H)_2(g)$$

has been determined to be 3.2×10^4. Assume that acetic acid is present initially at a concentration of 5.4×10^{-4} mol/L at 25 °C and that no dimer is present initially.

(a) What percentage of acetic acid is converted to dimer?

(b) As the temperature goes up, in which direction does the equilibrium shift? (Recall that hydrogen-bond formation is an exothermic process.)

57. At 450 °C, 3.60 mol of ammonia is placed in a 2.00-L vessel and allowed to decompose to the elements.

$$2 NH_3(g) \rightleftharpoons N_2(g) + 3 H_2(g)$$

If the experimental value of K is 6.3 for this reaction at this temperature, calculate the equilibrium concentration of each reagent. What is the total pressure in the flask?

58. The total pressure for a mixture of N_2O_4 and NO_2 is 1.5 atm. If $K_p = 6.75$ (at 25 °C), calculate the partial pressure of each gas in the mixture.

$$2 NO_2(g) \rightleftharpoons N_2O_4(g)$$

59. K_c for the decomposition of ammonium hydrogen sulfide is 1.8×10^{-4} at 25 °C.

$$NH_4HS(s) \rightleftharpoons NH_3(g) + H_2S(g)$$

(a) When the pure salt decomposes in a flask, what are the equilibrium concentrations of NH_3 and H_2S?

(b) If NH_4HS is placed in a flask already containing 0.020 mol/L of NH_3 and then the system is allowed to come to equilibrium, what are the equilibrium concentrations of NH_3 and H_2S?

60. A sample of liquid water is sealed in a container. Over time some of the liquid evaporates, but equilibrium is reached eventually. At this point you can measure the equilibrium vapor pressure of the water. Is the process $H_2O(g) \rightleftharpoons H_2O(\ell)$ a dynamic equilibrium? Explain the changes that take place in reaching equilibrium in terms of the rates of the competing processes of evaporation and condensation.

61. An ice cube is placed in a beaker of water at 20 °C. The ice cube partially melts, and the temperature of the water is lowered to 0 °C. At this point, both ice and water are at 0 °C, and no further change is apparent. Is the system at equilibrium? Is this a dynamic equilibrium? That is, are events still occurring at the molecular level? Suggest an experiment to test whether this is so. (*Hint:* Consider using D_2O.)

62. A reaction important in smog formation is

$$O_3(g) + NO(g) \rightleftharpoons O_2(g) + NO_2(g) \qquad K = 6.0 \times 10^{34}$$

(a) If the initial concentrations are $[O_3] = 1.0 \times 10^{-6}$ M, $[NO] = 1.0 \times 10^{-5}$ M, $[NO_2] = 2.5 \times 10^{-4}$ M, and

$[O_2] = 8.2 \times 10^{-3}$ M, is the system at equilibrium? If not, in which direction does the reaction proceed?

(b) If the temperature is increased, as on a very warm day, will the concentrations of the products increase or decrease? (*Hint:* You have to calculate the enthalpy change for the reaction to find out if it is exothermic or endothermic.)

63. The equilibrium reaction $N_2O_4(g) \rightleftharpoons 2\,NO_2(g)$ has been thoroughly studied (see Figures 16.6 and 16.8). If the total pressure in a flask containing NO_2 and N_2O_4 gas at 25 °C is 1.50 atm, and the value of K_p at this temperature is 0.148, what fraction of the N_2O_4 has dissociated to NO_2? What happens to the fraction dissociated if the volume of the container is increased so that the total equilibrium pressure falls to 1.00 atm?

64. Lanthanum oxalate decomposes when heated to lanthanum oxide, CO, and CO_2.

$$La_2(C_2O_4)_3(s) \rightleftharpoons La_2O_3(s) + 3\,CO(g) + 3\,CO_2(g)$$

(a) If, at equilibrium, the total pressure in a 10.0-L flask is 0.200 atm, what is the value of K_p?

(b) Suppose 0.100 mol of $La_2(C_2O_4)_3$ was originally placed in the 10.0-L flask. What amount of $La_2(C_2O_4)_3$ remains unreacted at equilibrium?

65. The ammonia complex of trimethylborane, $(H_3N)B(CH_3)_3$, dissociates at 100 °C to its components with $K_p = 4.62$.

$$(NH_3)B(CH_3)_3 \rightleftharpoons B(CH_3)_3 + NH_3$$

If NH_3 is changed to some other molecule, the equilibrium constant is different.

For $[(CH_3)_3P]B(CH_3)_3$	$K_p = 0.128$
For $[(CH_3)_3N]B(CH_3)_3$	$K_p = 0.472$
For $(H_3N)B(CH_3)_3$	$K_p = 4.62$

(a) If you begin an experiment by placing 0.010 mol of each complex in a flask, which would have the largest partial pressure of $B(CH_3)_3$ at 100 °C?

(b) If 0.73 g (0.010 mol) of $(NH_3)B(CH_3)_3$ is placed in a 100.-mL flask and heated to 100 °C, what is the partial pressure of each gas in the equilibrium mixture and what is the total pressure? What is the percent dissociation of $(NH_3)B(CH_3)_3$?

66. Sulfuryl chloride, SO_2Cl_2, is a compound with very irritating vapors; it is used as a reagent in the synthesis of organic compounds. When heated to a sufficiently high temperature it decomposes to SO_2 and Cl_2.

$$SO_2Cl_2(g) \rightleftharpoons SO_2(g) + Cl_2(g) \qquad K = 0.045 \text{ at } 375\,°C$$

(a) 6.70 g of SO_2Cl_2 is placed in a 1.00-L flask and then heated to 375 °C. What is the concentration of each of the compounds in the system when equilibrium is achieved? What fraction of SO_2Cl_2 has dissociated?

(b) What are the concentrations of SO_2Cl_2, SO_2, and Cl_2 at equilibrium in the 1.00-L flask at 375 °C if you begin with a mixture of SO_2Cl_2 (6.70 g) and Cl_2 (1.00 atm)? What fraction of SO_2Cl_2 has dissociated?

(c) Compare the fraction of SO_2Cl_2 in parts (a) and (b). Do they agree with your expectation based on Le Chatelier's principle?

67. Limestone decomposes at high temperatures.

$$CaCO_3(s) \rightleftharpoons CaO(s) + CO_2(g)$$

At 1000 °C, $K_p = 3.87$. If pure $CaCO_3$ is placed in a 5.00-L flask and heated to 1000 °C, what quantity of $CaCO_3$ must decompose to achieve the equilibrium pressure of CO_2?

68. Hemoglobin (Hb) can form a complex with both O_2 and CO. For the reaction

$$HbO_2(aq) + CO(g) \rightleftharpoons HbCO(aq) + O_2(g)$$

at body temperature, K is about 2.0×10^2. If the ratio $[HbCO]/[HbO_2]$ comes close to 1, death is probable. What partial pressure of CO in the air is likely to be fatal? Assume the partial pressure of O_2 is 0.20 atm.

69. At 1800 K, oxygen dissociates very slightly into its atoms.

$$O_2(g) \rightleftharpoons 2\,O(g) \qquad K_p = 1.2 \times 10^{-10}$$

If you place 1.0 mole of O_2 in a 10.-L vessel and heat it to 1800 K, how many O atoms are present in the flask when equilibrium is achieved?

70. Nitrosyl bromide, NOBr, dissociates readily at room temperature.

$$NOBr(g) \rightleftharpoons NO(g) + \tfrac{1}{2} Br_2(g)$$

Some NOBr is placed in a flask at 25 °C and allowed to dissociate. The total pressure at equilibrium is 190 mm Hg and the compound is found to be 34% dissociated. What is the value of K_p?

Using Electronic Resources

These questions refer to the General Chemistry Interactive CD-ROM, *Version 3.0.*

71. See CD-ROM Screen 16.3: The Principle of Microscopic Reversibility. What is the "principle of microscopic reversibility?" How is this illustrated by the equilibrium

$$PbCl_2(s) \rightleftharpoons Pb^{2+}(aq) + 2\,Cl^-(aq)$$

72. See CD-ROM Screen 16.4: The Equilibrium Constant. The reaction

$$Fe(H_2O)_6^{3+}(aq) + SCN^-(aq) \rightleftharpoons$$
$$Fe(H_2O)_5(SCN)^{2+}(aq) + H_2O(\ell)$$

is illustrated on Screen 16.3. What would you observe if more SCN^- ion is added to the solution at equilibrium? What happens to the concentration of each species in solution?

73. See CD-ROM Screen 16.3: The Equilibrium State. See the Simulation on this screen.

(a) Set the concentration of Fe^{3+} at 0.0050 M and that of SCN^- at 0.0070 M. Press the "React" button. Does the concentration of Fe^{3+} go to zero? When equilibrium is reached what are the concentrations of the reactants and products? What is the equilibrium constant?

(b) Begin with $[Fe^{3+}] = [SCN^-] = 0.0$ M and $[FeSCN^{2+}] = 0.0080$ M. When equilibrium is reached, what ion has the largest concentration in solution?

(c) Begin with $[Fe^{3+}] = 0.0010$ M, $[SCN^-] = 0.0020$ M, and $[FeSCN^{2+}] = 0.0030$ M. Describe the result of allowing this system to come to equilibrium.

74. See CD-ROM Screen 16.6: Writing Equilibrium Expressions. See the Simulation on this screen. Find the value of the equilibrium constant for each of the four hypothetical reactions you can explore on this screen.

(a) $A \rightleftharpoons B$

(b) $B \rightleftharpoons 2\,B$

(c) $HA \rightleftharpoons H^+ + A^-$

(d) $M + 4\,L \rightleftharpoons ML_4$

17

Principles of Reactivity: Chemistry of Acids and Bases

Nature's Acids

When we were small, our mothers and grandmothers told us never to eat the leaves of the rhubarb plant because it would make us very sick! Many people grow rhubarb in gardens because the stalks of the plant, when stewed with sugar, make a wonderful dessert or filling for a pie or tart. But why not eat the leaves?

Rhubarb leaves are a source of oxalic acid, $H_2C_2O_4$, an organic acid.

$$H_2O(\ell) + H_2C_2O_4(aq) \rightleftharpoons$$
$$H_3O^+(aq) + HC_2O_4^-(aq)$$

$$H_2O(\ell) + HC_2O_4^-(aq) \rightleftharpoons$$
$$H_3O^+(aq) + C_2O_4^{2-}(aq)$$

Rhubarb leaves contain a very large amount of this acid. The problem with ingesting a large amount of oxalic acid is that it interferes with essential elements in the body such as iron, magnesium, and, especially, calcium. The Ca^{2+} ion and oxalic acid react to form insoluble calcium oxalate.

$$Ca^{2+}(aq) + H_2C_2O_4(aq) \longrightarrow$$
$$CaC_2O_4(s) + 2\,H^+(aq)$$

Not only does this effectively remove calcium ions from the body but calcium oxalate crystals can grow into painful kidney and bladder stones. Because of this, people susceptible to kidney stones are put on a low-oxalic acid diet. Such people also have to be careful of taking too much vitamin C, a compound that can be turned into oxalic acid in the body. People have also died from accidentally drinking antifreeze because the ethylene glycol

◀ **Rhubarb, a source of many organic acids.** The leaves and stalks of rhubarb are the source of acids, such as oxalic, citric, acetic, and succinic, among others. *(David Turnley/2002 Corbis Images)*

Our mothers said, "Don't eat those rhubarb leaves!"

▲ Calcium oxalate precipitates when solutions of calcium chloride and the organic acid oxalic acid are mixed. *(Charles D. Winters)*

in the antifreeze is turned into oxalic acid in the body. Symptoms of oxalate poisoning include nausea, vomiting, abdominal pain, and hemorrhaging.

Oxalic acid is in fact found in the stems and leaves of many plants, such as cabbage, spinach, and beets. Because it also occurs in other edible substances, including cocoa, peanuts, and tea, the average person consumes about 150 mg of oxalic acid a day. But will it kill you? For a person weighing about 145 lb (65.7 kg), the lethal dose is around 24 g of pure oxalic acid. You would have to eat a field of rhubarb leaves or drink an ocean of tea to come close to a fatal dose of oxalic acid. What would happen first, however, is that you may have severe diarrhea. Your gut knows oxalic acid is a natural toxin and is stimulated to get rid of it.

In spite of the minor health risk from eating too much rhubarb, the plant has been cultivated for thousands of years for its healthful properties. The Chinese particularly have used it in traditional medicine for centuries. Indeed, it was so important that emperors of China in the 18th and 19th centuries forbade its export. It was, however, cultivated in Russia and later in England. It made its appearance in the United States in about 1800.

Before You Begin

- Review the pH scale in Section 5.9.
- Review the outline of acids and bases and their chemistry in Chapter 5 (pages 156–164).
- Review the principles of chemical equilibria in Chapter 16.

Acids and Bases The focus of this chapter is the chemistry of acids and bases. Here we illustrate some experiments with the strong acid HCl, the weak acetic acid, CH_3CO_2H, and the weak base ammonia, NH_3. In this chapter we describe the differences in the chemistry of strong and weak acids and bases, discuss how the outcome of their reactions can be predicted, and examine theories of acid-base behavior.

Strong Acid

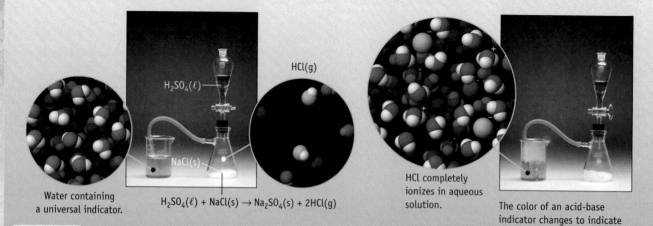

Water containing a universal indicator.

$H_2SO_4(\ell) + NaCl(s) \rightarrow Na_2SO_4(s) + 2HCl(g)$

HCl(g)

HCl completely ionizes in aqueous solution.

The color of an acid-base indicator changes to indicate an acidic solution.

Weak Acid

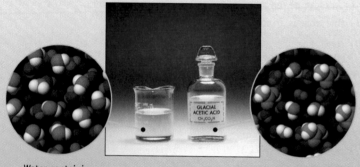

Water containing a universal indicator.

$CH_3CO_2H(aq)$

Acetic acid, CH_3CO_2H, ionizes only slightly in aqueous solution.

The indicator changes color to indicate an acidic solution.

Weak Base

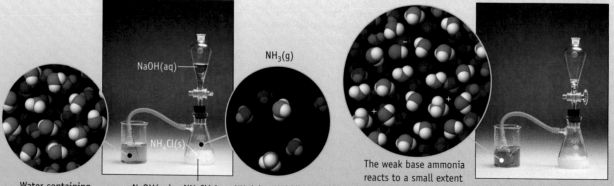

NaOH(aq)

$NH_3(g)$

$NH_4Cl(s)$

Water containing a universal indicator.

$NaOH(aq) + NH_4Cl(s) \rightarrow NH_3(g) + H_2O(\ell) + NaCl(aq)$

The weak base ammonia reacts to a small extent with water to give a weakly basic solution.

The indicator color indicates a basic solution.

Acids and bases are among the most common substances in nature. Amino acids are the building blocks of proteins. The pH of the lakes, rivers, and oceans is affected by dissolved acids and bases, and your bodily functions depend on acids and bases. This chapter and the next take up the chemistry of these substances because they are so important to chemists, biochemists, physicians, geologists, nutritionists, and engineers.   ●

17.1   ACIDS, BASES, AND THE EQUILIBRIUM CONCEPT

An acid was described in Chapter 5 as any substance that, when dissolved in water, can increase the concentration of hydrogen ions, H^+ [◀ SECTION 5.3]. A base was defined as any substance that increases the concentration of hydroxide ions, OH^-, when dissolved in water. Two other features of acids and bases were also introduced.

- Acids and bases can be divided roughly into those that are strong electrolytes (such as HCl, HNO_3, and $NaOH$) and those that are weak electrolytes (such as CH_3CO_2H and NH_3) [◀ TABLE 5.2, COMMON ACIDS AND BASES, PAGE 158].

- An H^+ ion — the nucleus of the hydrogen atom — simply cannot exist separately in water. Therefore, when an acid is dissolved in water, the proton produced by the acid combines with water to produce the hydronium ion, H_3O^+ [◀ "A CLOSER LOOK: H^+ IONS IN WATER," PAGE 160].

$$HCl(aq) \quad + \quad H_2O(\ell) \quad \longrightarrow \quad H_3O^+(aq) \quad + \quad Cl^-(aq)$$

hydrochloric acid   water   hydronium ion   chloride ion
strong electrolyte
= 100% ionized

Now let us take a closer look at what is meant by a strong or weak electrolyte. Hydrochloric acid is a strong acid, so 100% of the acid ionizes to produce hydronium and chloride ions. In contrast, acetic acid is a weak electrolyte. This acid ionizes only to a very small extent in water. That is, the acid, its anion, and the hydronium ion are in equilibrium in solution. From your introduction to the concept of chemical equilibrium in Chapter 16, you recognize that we can describe the extent to which the acid ionizes in terms of the equilibrium constant for the ionization process.

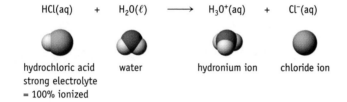

Acetic acid   Water   Acetate ion   Hydronium ion

$$K = \frac{[CH_3CO_2^-][H_3O^+]}{[CH_3CO_2H]} = 1.8 \times 10^{-5}$$

The equilibrium constants for the ionization of many weak acids, often called **ionization constants,** are a measure of the extent to which these substances ionize in water.

The notion that the chemistry of an acid or base is reflected by its equilibrium constant is important. Indeed, useful organizing principles are

- Acids or bases that ionize extensively, with $K > 1$, are referred to as *strong acids* or *bases*.
- Acids or bases that do not ionize extensively, with $K < 1$, are referred to as *weak acids* or *bases*.

This chapter is about the chemistry of acids and bases, especially weak acids and bases, and how we measure and use their equilibrium constants.

Chapter Goals • Revisited

- **Use the Brønsted–Lowry and Lewis concepts of acids and bases.**
- Apply the principles of chemical equilibrium to acids and bases in aqueous solution.
- Predict the outcome of reactions of acids and bases.
- Understand the influence of structure and bonding on acid–base properties.

Brønsted–Lowry Theory
The Brønsted concept of acid–base behavior is often called the Brønsted–Lowry theory. It broadens the definition of acids and bases first given in Chapter 5 (pages 156–158). Note that the theory is not restricted to compounds in water.

17.2   BRØNSTED CONCEPT OF ACIDS AND BASES

In 1923, Johannes N. Brønsted (1879–1947) in Copenhagen (Denmark) and Thomas M. Lowry (1874–1936) in Cambridge (England) independently suggested a new concept of acid and base behavior [CD-ROM, Screen 17.2]. They proposed that an *acid* is any substance that can *donate a proton* to any other substance. Thus, **Brønsted acids** can be molecular compounds such as nitric acid,

$$HNO_3(aq) + H_2O(\ell) \longrightarrow NO_3^-(aq) + H_3O^+(aq)$$
Acid

or they can be cations such as NH_4^+ or hydrated metal ions

$$NH_4^+(aq) + H_2O(\ell) \rightleftharpoons NH_3(aq) + H_3O^+(aq)$$
Acid

$$Fe(H_2O)_6^{3+}(aq) + H_2O(\ell) \rightleftharpoons Fe(H_2O)_5(OH)^+(aq) + H_3O^+(aq)$$
acid

or anions.

$$H_2PO_4^-(aq) + H_2O(\ell) \rightleftharpoons HPO_4^{2-}(aq) + H_3O^+(aq)$$
acid

According to Brønsted and Lowry, a **Brønsted base** is a substance that can *accept a proton* from any other substance. Brønsted bases can also be molecular compounds,

$$NH_3(aq) + H_2O(\ell) \rightleftharpoons NH_4^+(aq) + OH^-(aq)$$
Base

or they can be anions

$$CO_3^{2-}(aq) + H_2O(\ell) \rightleftharpoons HCO_3^-(aq) + OH^-(aq)$$
Base

or cations.

$$Al(H_2O)_5(OH)^{2+}(aq) + H_2O(\ell) \rightleftharpoons Al(H_2O)_6^{3+}(aq) + OH^-(aq)$$
Base

A wide variety of Brønsted acids are known, and you are familiar with many of them [← TABLE 5.2, PAGE 158]. Acids such as HF, HCl, HNO_3, and CH_3CO_2H (acetic acid) are all capable of donating one proton and so are called **monoprotic acids.** Other acids, called **polyprotic acids** (Table 17.1), are capable of donating two or more protons. An example is sulfuric acid.

$$H_2SO_4(aq) + H_2O(\ell) \longrightarrow HSO_4^-(aq) + H_3O^+(aq)$$

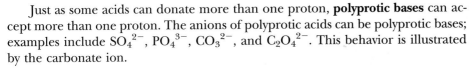

$$HSO_4^-(aq) + H_2O(\ell) \rightleftharpoons SO_4^{2-}(aq) + H_3O^+(aq)$$

Just as some acids can donate more than one proton, **polyprotic bases** can accept more than one proton. The anions of polyprotic acids can be polyprotic bases; examples include SO_4^{2-}, PO_4^{3-}, CO_3^{2-}, and $C_2O_4^{2-}$. This behavior is illustrated by the carbonate ion.

$$CO_3^{2-}(aq) + H_2O(\ell) \rightleftharpoons HCO_3^-(aq) + OH^-(aq)$$
base

$$HCO_3^-(aq) + H_2O(\ell) \rightleftharpoons H_2CO_3(aq) + OH^-(aq)$$
base

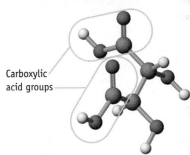

Tartaric acid, $H_2C_4H_4O_6$, is a naturally occurring diprotic acid. Tartaric acid and its potassium salt are found in many fruits. *(Charles D. Winters)*

Table 17.1 • Polyprotic Acids and Bases

Acid Form	Amphiprotic Form	Base Form
H_2S (hydrosulfuric acid or hydrogen sulfide)	HS^- (hydrogen sulfide ion)	S^{2-} (sulfide ion)
H_3PO_4 (phosphoric acid)	$H_2PO_4^-$ (dihydrogen phosphate ion) / HPO_4^{2-} (hydrogen phosphate ion)	PO_4^{3-} (phosphate ion)
H_2CO_3 (carbonic acid)	HCO_3^- (hydrogen carbonate ion or bicarbonate ion)	CO_3^{2-} (carbonate ion)
$H_2C_2O_4$ (oxalic acid)	$HC_2O_4^-$ (hydrogen oxalate ion)	$C_2O_4^{2-}$ (oxalate ion)

Some molecules or ions can behave as either a Brønsted acid or base. These are called **amphiprotic,** and one example is the hydrogen phosphate anion (see Table 17.1). For example,

$$HPO_4^{2-}(aq) + H_2O(\ell) \rightleftharpoons H_3O^+(aq) + PO_4^{3-}(aq)$$
acid

$$HPO_4^{2-}(aq) + H_2O(\ell) \rightleftharpoons H_2PO_4^-(aq) + OH^-(aq)$$
base

A final, important point is illustrated by the preceding chemical equations: *water is amphiprotic.* It can accept a proton to form H_3O^+,

$$H_2O(\ell) + HCl(aq) \rightleftharpoons H_3O^+(aq) + Cl^-(aq)$$
base   acid

or it can donate a proton to form the OH^- ion.

$$H_2O(\ell) + NH_3(aq) \rightleftharpoons NH_4^+(aq) + OH^-(aq)$$
acid   base

Exercise 17.1   Brønsted Acids and Bases

(a) Write a balanced equation for the reaction that occurs when H_3PO_4, phosphoric acid, donates a proton to water to form the dihydrogen phosphate ion. Is the dihydrogen phosphate ion an acid, a base, or amphiprotic?

(b) Write a balanced equation for the reaction that occurs when the cyanide ion, CN^-, accepts a proton from water to form HCN. Is CN^- a Brønsted acid or base?

Conjugate Acid–Base Pairs

In each of the chemical equations written so far in this chapter *a proton has been transferred* to or from water. For example, a reaction important in the control of acidity in biological systems is the reaction of the hydrogen carbonate ion, a Brønsted acid, with water.

The weak acid, HCO_3^- (hydrogen carbonate ion) is found in common baking soda. Its conjugate base, the CO_3^{2-} ion (carbonate ion) is widely distributed in nature as limestone, seashells, many other minerals. *(Charles D. Winters)*

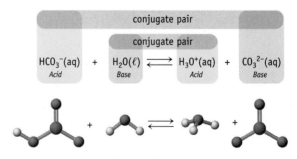

This equation exemplifies a feature of all reactions involving Brønsted acids and bases. The HCO_3^- and CO_3^{2-} ions are related by the loss or gain of H^+, as are H_2O and H_3O^+. A pair of compounds or ions that differ by the presence of one H^+ ion is called a **conjugate acid–base pair.** We say that HCO_3^- is the conjugate acid of the base CO_3^{2-} or that CO_3^{2-} is the conjugate base of the acid HCO_3^-. *Every reaction between a Brønsted acid and Brønsted base involves H^+ transfer and has two conjugate acid–base pairs.* To convince yourself of this, look at the previous reactions and those in Table 17.2.

Table 17.2 • Conjugate Acid–Base Pairs

Name	Acid 1	Base 2		Base 1	Acid 2*
Hydrochloric acid	HCl	$+ H_2O$	$\longrightarrow$	Cl^-	$+ H_3O^+$
Hydrogen carbonate	HCO_3^-	$+ H_2O$	$\rightleftharpoons$	CO_3^{2-}	$+ H_3O^+$
Acetic acid	CH_3CO_2H	$+ H_2O$	$\rightleftharpoons$	$CH_3CO_2^-$	$+ H_3O^+$
Ammonia	H_2O	$+ NH_3$	$\rightleftharpoons$	OH^-	$+ NH_4^+$
Carbonate ion	H_2O	$+ CO_3^{2-}$	$\rightleftharpoons$	OH^-	$+ HCO_3^-$
Water	H_2O	$+ H_2O$	$\rightleftharpoons$	OH^-	$+ H_3O^+$

*Acid 1 and base 1 are a conjugate pair, as are base 2 and acid 2.

Exercise 17.2   Conjugate Acids and Bases

In the following reaction, identify the acid on the left and its conjugate base on the right. Similarly, identify the base on the left and its conjugate acid on the right.

$$HNO_3(aq) + NH_3(aq) \rightleftharpoons NH_4^+(aq) + NO_3^-(aq)$$

17.3   WATER AND THE pH SCALE

The properties of water are a recurring theme of this book [← SECTIONS 5.1, 13.3]. Because we generally use aqueous solutions of acids and bases, and because the acid–base reactions in your body happen in your aqueous interior, we come again to the behavior of water.

Water Autoionization and the Water Ionization Constant, K_w

An acid such as HCl does not need to be present for the hydronium ion to exist in water. In fact, two water molecules interact with each other to produce a hydronium ion and a hydroxide ion by proton transfer from one water molecule to the other.

$$2\ H_2O(\ell) \rightleftharpoons H_3O^+(aq) + OH^-(aq)$$

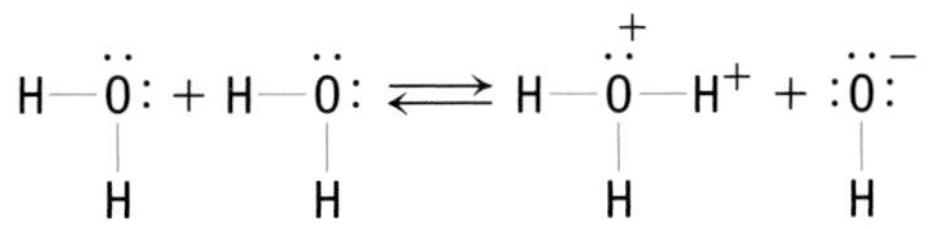

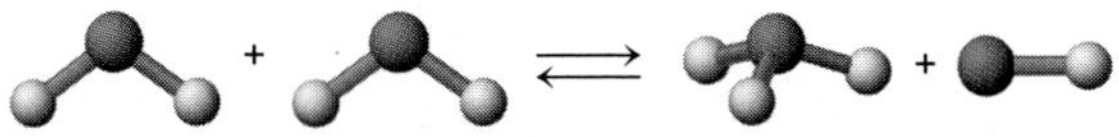

This **autoionization** reaction of water was demonstrated many years ago by Friedrich Kohlrausch (1840–1910). He found that, even after water is painstakingly purified, it still conducts electricity to a very small extent because autoionization produces very low concentrations of H_3O^+ and OH^- ions. Water autoionization is the cornerstone of our concepts of aqueous acid–base behavior [📀 CD-ROM, Screen 17.3].

● *K_w and Temperature*
The equation $K_w = [H_3O^+][OH^-]$ is valid in pure water and in any aqueous solution. K_w is temperature-dependent because the autoionization reaction is endothermic. K_w increases with temperature.

°C	K_w
10	0.29×10^{-14}
15	0.45×10^{-14}
20	0.68×10^{-14}
25	1.01×10^{-14}
30	1.47×10^{-14}
50	5.48×10^{-14}

When water autoionizes, the equilibrium lies far to the left side. In fact, in pure water at 25 °C only about two out of a billion (10^9) water molecules are ionized at any instant. To express this idea more quantitatively, we can write the equilibrium constant expression for autoionization.

$$K = \frac{[H_3O^+][OH^-]}{[H_2O]^2}$$

In pure water or in dilute aqueous solutions, the concentration of water can be considered to be a constant (55.5 M). For this reason $[H_2O]^2$ is included in the constant K and the equilibrium constant expression becomes

$$K[H_2O]^2 = [H_3O^+][OH^-]$$

$$\boxed{K_w = [H_3O^+][OH^-] = 1.0 \times 10^{-14} \text{ at 25 °C}} \qquad (17.1)$$

This equilibrium constant is given a special symbol, **K_w**, and is known as the **ionization constant for water**. In pure water, the transfer of a proton between two water molecules leads to one H_3O^+ and one OH^-. Because this is the only source of these ions in pure water, we know that $[H_3O^+]$ must equal $[OH^-]$. Electrical conductivity measurements of pure water show that $[H_3O^+] = [OH^-] = 1.0 \times 10^{-7}$ M at 25 °C, so K_w has a value of 1.0×10^{-14} at 25 °C.

In pure water the hydronium ion and hydroxide ion concentrations are both 1.0×10^{-7} M at 25 °C, and the water is said to be *neutral*. If some acid or base is added to pure water, however, the equilibrium

$$2 H_2O(\ell) \rightleftharpoons H_3O^+(aq) + OH^-(aq)$$

is disturbed. Adding acid raises the concentration of the H_3O^+ ions, and so the solution is *acidic*. To oppose this increase, Le Chatelier's principle [← SECTION 16.6] predicts that a small fraction of the H_3O^+ ions reacts with OH^- ions from water autoionization to form water. This lowers $[OH^-]$ until the product of $[H_3O^+]$ and $[OH^-]$ is again equal to 1.0×10^{-14} at 25 °C. Similarly, adding a base to pure water gives a *basic* solution because the OH^- ion concentration has increased. Le Chatelier's principle predicts that some of the added OH^- ions will react with H_3O^+ ions present in the solution from water autoionization, thereby lowering $[H_3O^+]$ until the value of the product $[H_3O^+][OH^-]$ equals 1.0×10^{-14} at 25 °C.

Thus, for aqueous solutions at 25 °C, we can say that

- In a *neutral solution* $[H_3O^+] = [OH^-]$
  Both are equal to 1.0×10^{-7} M.

- In a *acidic solution* $[H_3O^+] . [OH^-]$.
  $[H_3O^+] > 1.0 \times 10^{27}$ M and $[OH^-] < 1.0 \times 10^{-7}$ M.

- In a *basic solution* $[H_3O^+] < [OH^-]$.
  $[H_3O^+] < 1.0 \times 10^{-7}$ M and $[OH^-] > 1.0 \times 10^{-7}$ M.

Example 17.1   Ion Concentrations in a Solution of a Strong Base

Problem • What are the hydroxide and hydronium ion concentrations in a 0.0012 M solution of NaOH?

Strategy • NaOH, a strong base, is 100% dissociated into ions in water, so we assume the OH^- ion concentration is the same as the NaOH concentration. The H_3O^+ ion concentration can then be calculated using Equation 17.1.

Solution • The initial concentration of OH^- is 0.0012 M.

0.0012 mol NaOH per liter
$$\longrightarrow 0.0012\ M\ Na^+(aq) + 0.0012\ M\ OH^-(aq)$$

Substituting the OH^- concentration into Equation 17.1, we have

$$K_w = 1.0 \times 10^{-14} = [H_3O^+][OH^-] = [H_3O^+](0.0012)$$

and so

$$[H_3O^+] = \frac{1.0 \times 10^{-14}}{0.0012} = 8.3 \times 10^{-12}\ M$$

Comment • Why didn't we take into account the H_3O^+ and OH^- ions produced by water autoionization? This is expected to

add OH^- ions to the solution (say x mol/L) and H_3O^+ (also x mol/L). When equilibrium is achieved in this case, this means the H_3O^+ concentration is x, and

$$[OH^-] = (0.0012\ M + OH^-\ \text{from water autoionization})$$
$$[OH^-] = (0.0012\ M + x)$$

In pure water, the amount of OH^- ion generated is 1.0×10^{-7} M. Le Chatelier's principle [◀ SECTION 16.6] suggests that the amount should be even smaller when OH^- ions are already present in solution from NaOH; that is, x should be $\ll 1.0 \times 10^{-7}$ M. This means x in the term $(0.0012 + x)$ is insignificant compared with 0.0012. (Following the rules for significant figures, the sum of 0.0012 and a number even smaller than 1.0×10^{-7} is 0.0012.) Thus, the equilibrium concentration of OH^- is just equivalent to the concentration of NaOH added.

What about the Na^+ ion? As described later [see Section 17.4], alkali metal ions have no effect on the acidity or basicity of a solution.

Exercise 17.3   Hydronium Ion Concentration in a Solution of a Strong Acid

A solution of the strong acid HCl has [HCl] = 4.0×10^{-3} M. What are the concentrations of H_3O^+ and OH^- in this solution? (Recall that because HCl is a strong acid, it is 100% ionized in water.)

pH Scale

The **pH** of a solution is defined as the negative of the base-10 logarithm (log) of the hydronium ion concentration [◀ pH, SECTION 5.9, PAGE 181] [💿 CD-ROM, Screen 17.4].

$$pH = -\log[H_3O^+] \qquad \textbf{(5.2 and 17.2)}$$

In a similar way, we can now define the pOH of a solution as the negative of the base-10 logarithm of the hydroxide ion concentration.

$$pOH = -\log[OH^-] \qquad \textbf{(17.3)}$$

• **The pK Scale**
In general, $-\log X = pX$, so $-\log K = pK$.

$-\log[H_3O^+] = pH$
$-\log[OH^-] = pHO$

In pure water, the hydronium and hydroxide ion concentrations are both 1.0×10^{-7} M. Therefore, for pure water at 25 °C

$$pH = -\log (1.0 \times 10^{-7}) = 7.00$$

In the same way, you can show that the pOH of pure water is also 7.00 at 25 °C.

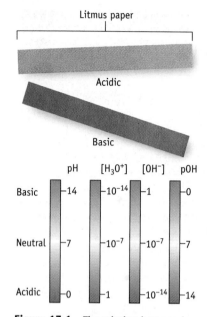

Figure 17.1 **The relation between hydronium ion and hydroxide ion concentrations and pH and pOH.** *(Charles D. Winters)*

If we take the negative logarithms of both sides of the expression $K_w = [H_3O^+][OH^-]$, we obtain another useful equation.

$$K_w = 1.0 \times 10^{-14} = [H_3O^+][OH^-]$$

$$-\log K_w = -\log(1.0 \times 10^{-14}) = -\log([H_3O^+][OH^-])$$

$$pK_w = 14.00 = -\log([H_3O^+]) + (-\log[OH^-])$$

$$\boxed{pK_w = 14.00 = pH + pOH} \tag{17.4}$$

The sum of the pH and pOH of a solution must be equal to 14.00 at 25 °C.

As illustrated in Figures 5.17 and 17.1, solutions with pH less than 7.00 (at 25 °C) are acidic, and solutions with pH greater than 7.00 are basic. Solutions with pH = 7.00 at 25 °C are neutral.

Determining and Calculating pH

The approximate pH of a solution can be determined using an acid–base indicator, a substance that changes color in some known pH range (Figure 17.2). The indicators we use in the laboratory, such as phenolphthalein, are Brønsted acids or bases for which the acid and its conjugate base have different colors (➡ SECTION 18.3).

Although an approximate pH may be obtained by using an appropriate chemical indicator, a modern pH meter is far preferable for accurately determining pH (Figure 5.18).

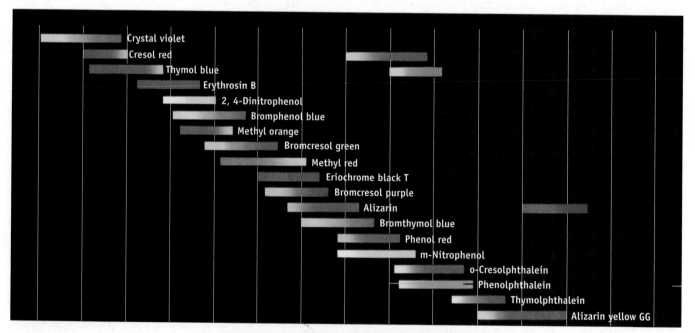

Figure 17.2 **Common acid–base indicators.** The color changes occur over a range of pH values. Notice that a few indicators have color changes over two different pH ranges. (See also Figure 5.18.) *(Hach Company)*

Exercise 17.4 Reviewing pH Calculations

For a review of pH calculations, see Example 5.11.

(a) What is the pH of a 0.0012 M NaOH solution?

(b) The pH of a diet soda is 4.32 at 25 °C. What are the hydronium and hydroxide ion concentrations in the soda?

(c) If the pH of a solution of the strong base $Sr(OH)_2$ is 10.46, what is the concentration of the $Sr(OH)_2$ in mol/L?

17.4 EQUILIBRIUM CONSTANTS FOR ACIDS AND BASES

How can we define quantitatively the extent to which an acid or a base reacts with water? That is, how can we define the relative strengths of acids and bases?

One way to define the relative strengths of a series of acids would be to measure the pH of solutions of acids of equal concentration: the lower the pH, the stronger the acid.

- For a strong acid, $[H_3O^+]$ in solution is equal to the original acid concentration (see Exercise 17.3). Similarly, for a strong base, $[OH^-]$ equals the original base concentration (see Example 17.1) [⊕ CD-ROM, Screen 17.5].

- For a weak acid $[H_3O^+]$ is much less than the original acid concentration. That is, $[H_3O^+]$ is smaller than if the acid were a strong acid of the same concentration. Similarly, a weak base gives a smaller $[OH^-]$ than if the base were a strong base of the same concentration [⊕ CD-ROM, Screen 17.6].

- For a series of weak monoprotic acids (of the type HA) of the same concentration, $[H_3O^+]$ increases (and the pH declines) as the acids become stronger. Similarly, for a series of weak bases, $[OH^-]$ will increase (and the pH increases) as the bases become stronger.

The relative strength of an acid or base can be expressed quantitatively with an equilibrium constant. For the general acid HA, we can write

$$HA(aq) + H_2O(\ell) \rightleftharpoons H_3O^+(aq) + A^-(aq)$$

$$K_a = \frac{[H_3O^+][A^-]}{[HA]} \qquad \textbf{(17.5)}$$

where the equilibrium constant, K, has a subscript "a" to indicate that it is an equilibrium constant for an *acid* in water. For weak acids, the value of K is less than 1 because the product $[H_3O^+][A^-]$ is less than the equilibrium concentration of the weak acid, $[HA]$. For a series of acids of the same concentration, the value of K_a increases as the acid strength increases, that is, as the acids ionize to a greater and greater extent.

Similarly, we can write the equilibrium expression for a *weak base* B in water. Here we label K with a subscript "b." Its value is also less than 1.

$$B(aq) + H_2O(\ell) \rightleftharpoons BH^+(aq) + OH^-(aq)$$

$$K_b = \frac{[BH^+][OH^-]}{[B]} \qquad \textbf{(17.6)}$$

- **pH Calculations**
Because we make pH measurements to determine solution H_3O^+ and OH^- concentrations, it is useful to be able to convert experimental pH readings to concentrations. Review Example 5.11, pH of Solutions, and check yourself with Exercise 17.4.

Chapter Goals • Revisited

- Use the Brønsted–Lowry and Lewis concepts of acids and bases.
- **Apply the principles of chemical equilibrium to acids and bases in aqueous solution.**
- Predict the outcome of reactions of acids and bases.
- Understand the influence of structure and bonding on acid–base properties.

- **Weak Acid or Weak Base**
If an acid or base is weak, a dilute aqueous solution of the acid or base (say 0.1 M) will have pH values in the following ranges.

Weak acid	Small $[H_3O^+]$ $(10^{-2} - 10^{-7})$	pH ≈ 2 – 7
Weak base	Small $[OH^-]$ $(10^{-2} - 10^{-7})$	pH ≈ 12 – 7

Some acids and bases are ordered on the basis of their relative abilities to donate or accept protons in aqueous solution in Table 17.3, where each acid and base is listed with its value of K_a and K_b, respectively. The following are important ideas concerning Table 17.3.

- Acids are listed in Table 17.3 at the left and their conjugate bases are on the right.

Table 17.3 • Ionization Constants for Some Acids and Their Conjugate Bases

Acid Name	Acid	K_a	Base	K_b	Base Name
Perchloric acid	$HClO_4$	large	ClO_4^-	very small	perchlorate ion
Sulfuric acid	H_2SO_4	large	HSO_4^-	very small	hydrogen sulfate ion
Hydrochloric acid	HCl	large	Cl^-	very small	chloride ion
Nitric acid	HNO_3	large	NO_3^-	very small	nitrate ion
Hydronium ion	H_3O^+	1.0	H_2O	1.0×10^{-14}	water
Sulfurous acid	H_2SO_3	1.2×10^{-2}	HSO_3^-	8.3×10^{-13}	hydrogen sulfite ion
Hydrogen sulfate ion	HSO_4^-	1.2×10^{-2}	SO_4^{2-}	8.3×10^{-13}	sulfate ion
Phosphoric acid	H_3PO_4	7.5×10^{-3}	$H_2PO_4^-$	1.3×10^{-12}	dihydrogen phosphate ion
Hexaaquairon(III) ion	$Fe(H_2O)_6^{3+}$	6.3×10^{-3}	$Fe(H_2O)_5OH^{2+}$	1.6×10^{-12}	pentaaquahydroxoiron(III) ion
Hydrofluoric acid	HF	7.2×10^{-4}	F^-	1.4×10^{-11}	fluoride ion
Nitrous acid	HNO_2	4.5×10^{-4}	NO_2^-	2.2×10^{-11}	nitrite ion
Formic acid	HCO_2H	1.8×10^{-4}	HCO_2^-	5.6×10^{-11}	formate ion
Benzoic acid	$C_6H_5CO_2H$	6.3×10^{-5}	$C_6H_5CO_2^-$	1.6×10^{-10}	benzoate ion
Acetic acid	CH_3CO_2H	1.8×10^{-5}	$CH_3CO_2^-$	5.6×10^{-10}	acetate ion
Propanoic acid	$CH_3CH_2CO_2H$	1.3×10^{-5}	$CH_3CH_2CO_2^-$	7.7×10^{-10}	propanoate ion
Hexaaquaaluminum ion	$Al(H_2O)_6^{3+}$	7.9×10^{-6}	$Al(H_2O)_5OH^{2+}$	1.3×10^{-9}	pentaaquahydroxoaluminum ion
Carbonic acid	H_2CO_3	4.2×10^{-7}	HCO_3^-	2.4×10^{-8}	hydrogen carbonate ion
Hexaaquacopper(II) ion	$Cu(H_2O)_6^{2+}$	1.6×10^{-7}	$Cu(H_2O)_5OH^+$	6.25×10^{-8}	pentaaquahydroxocopper(II) ion
Hydrogen sulfide	H_2S	1×10^{-7}	HS^-	1×10^{-7}	hydrogen sulfide ion
Dihydrogen phosphate ion	$H_2PO_4^-$	6.2×10^{-8}	HPO_4^{2-}	1.6×10^{-7}	hydrogen phosphate ion
Hydrogen sulfite ion	HSO_3^-	6.2×10^{-8}	SO_3^{2-}	1.6×10^{-7}	sulfite ion
Hypochlorous acid	$HClO$	3.5×10^{-8}	ClO^-	2.9×10^{-7}	hypochlorite ion
Hexaaqualead(II) ion	$Pb(H_2O)_6^{2+}$	1.5×10^{-8}	$Pb(H_2O)_5OH^+$	6.7×10^{-7}	pentaaquahydroxolead(II) ion
Hexaaquacobalt(II) ion	$Co(H_2O)_6^{2+}$	1.3×10^{-9}	$Co(H_2O)_5OH^+$	7.7×10^{-6}	pentaaquahydroxocobalt(II) ion
Boric acid	$B(OH)_3(H_2O)$	7.3×10^{-10}	$B(OH)_4^-$	1.4×10^{-5}	tetrahydroxoborate ion
Ammonium ion	NH_4^+	5.6×10^{-10}	NH_3	1.8×10^{-5}	ammonia
Hydrocyanic acid	HCN	4.0×10^{-10}	CN^-	2.5×10^{-5}	cyanide ion
Hexaaquairon(II) ion	$Fe(H_2O)_6^{2+}$	3.2×10^{-10}	$Fe(H_2O)_5OH^+$	3.1×10^{-5}	pentaaquahydroxoiron(II) ion
Hydrogen carbonate ion	HCO_3^-	4.8×10^{-11}	CO_3^{2-}	2.1×10^{-4}	carbonate ion
Hexaaquanickel(II) ion	$Ni(H_2O)_6^{2+}$	2.5×10^{-11}	$Ni(H_2O)_5OH^+$	4.0×10^{-4}	pentaaquahydroxonickel(II) ion
Hydrogen phosphate ion	HPO_4^{2-}	3.6×10^{-13}	PO_4^{3-}	2.8×10^{-2}	phosphate ion
Water	H_2O	1.0×10^{-14}	OH^-	1.0	hydroxide ion
Hydrogen sulfide ion*	HS^-	1×10^{-19}	S^{2-}	1×10^5	sulfide ion
Ethanol	C_2H_5OH	very small	$C_2H_5O^-$	large	ethoxide ion
Ammonia	NH_3	very small	NH_2^-	large	amide ion
Hydrogen	H_2	very small	H^-	large	hydride ion

Vertical label left side: Increasing Acid Strength ↑

Vertical label right side: Increasing Base Strength ↓

*The values of K_a for HS^- and K_b for S^{2-} are estimates.

- A large value of *K* indicates ionization products are strongly favored, whereas a small value of *K* indicates reactants are favored.

- The strongest acids are at the upper left. They have the largest K_a values. K_a values become smaller on descending the chart as the acid strength declines.

- The strongest bases are at the lower right. They have the largest K_b values. K_b values become larger on descending the chart as base strength increases.

- The weaker the acid, the stronger its conjugate base. That is, the smaller the value of K_a, the larger the corresponding value of K_b.

- Some acids or bases are listed as having K_a or K_b values that are large or very small. Acids that are stronger than H_3O^+ are completely ionized so their K_a values are "large." Their conjugate bases do not produce meaningful concentrations of OH^- ions, so their K_b values are "very small." Similar arguments follow for strong bases and their conjugate acids.

To illustrate these ideas, let us use some common acids and bases. For example, nitric acid, a strong acid, is *much* stronger than the related weak acid nitrous acid.

HNO₃, $K_a \gg 1$ $\gg$ HNO₂, $K_a = 4.5 \times 10^{-4}$

Their conjugate bases, however, are reversed in their relative strength. Indeed, the NO_3^- ion is such a weak base that it has no effect on solution pH.

The three organic acids pictured in the margin decline in strength as more carbon atoms are added (Table 17.3). The opposite ordering occurs for their conjugate bases, as expected. That is, the propanoate ion, $CH_3CH_2CO_2^-$ ($K_b = 7.7 \times 10^{-10}$) is a stronger base than the formate ion (HCO_2^-, $K_b = 5.6 \times 10^{-11}$).

Nature abounds in weak bases as well as weak acids (Figure 17.3). Ammonia and its conjugate acid, the ammonium ion, are part of the nitrogen cycle in the

- K_a, K_b, [H_3O^+], and pH

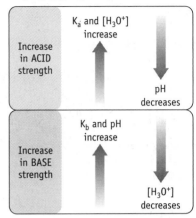

K_a increases; acid strength increases

K_b of conjugate base increases

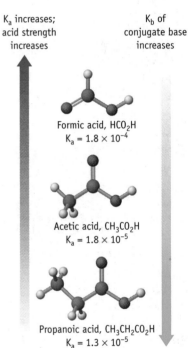

Formic acid, HCO_2H
$K_a = 1.8 \times 10^{-4}$

Acetic acid, CH_3CO_2H
$K_a = 1.8 \times 10^{-5}$

Propanoic acid, $CH_3CH_2CO_2H$
$K_a = 1.3 \times 10^{-5}$

Three weak organic acids. These acids decline in strength as the number of carbon atoms increases.

Problem-Solving Tip 17.1

Strong or Weak?

How can you tell whether an acid or base is weak? The easiest way is to remember those few that are strong (see Table 5.2 and the following short list). All others are probably weak.

The strong acids are:

   Hydrohalic acids: HCl, HBr, and HI
   Nitric acid: HNO₃

Sulfuric acid: H_2SO_4 (for loss of first H^+ only)
Perchloric acid: $HClO_4$

Some common strong bases are:

   All Group 1A hydroxides: LiOH, NaOH, KOH, RbOH, CsOH
   Group 2A hydroxides: Sr(OH)₂ and Ba(OH)₂

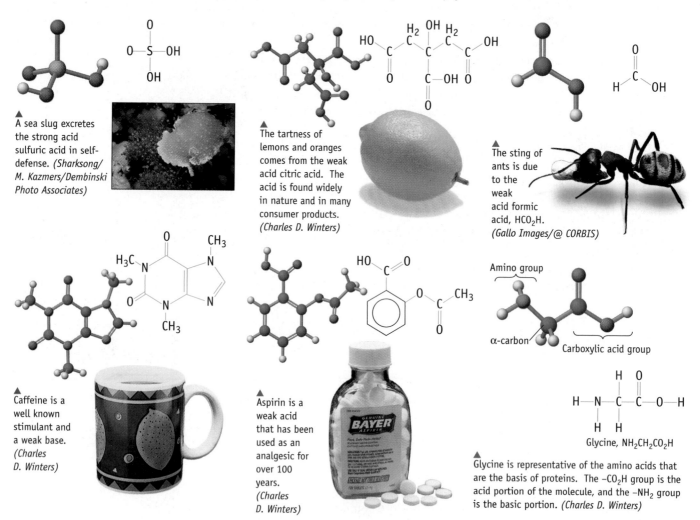

A sea slug excretes the strong acid sulfuric acid in self-defense. (Sharksong/ M. Kazmers/Dembinski Photo Associates)

The tartness of lemons and oranges comes from the weak acid citric acid. The acid is found widely in nature and in many consumer products. (Charles D. Winters)

The sting of ants is due to the weak acid formic acid, HCO_2H. (Gallo Images/@ CORBIS)

Caffeine is a well known stimulant and a weak base. (Charles D. Winters)

Aspirin is a weak acid that has been used as an analgesic for over 100 years. (Charles D. Winters)

Amino group

α-carbon

Carboxylic acid group

Glycine, $NH_2CH_2CO_2H$

Glycine is representative of the amino acids that are the basis of proteins. The $-CO_2H$ group is the acid portion of the molecule, and the $-NH_2$ group is the basic portion. (Charles D. Winters)

Figure 17.3 Acids and bases. Hundreds of acids and bases are found in nature. Our foods contain a wide variety, and biochemically important molecules are acids and bases.

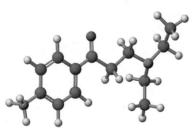

Novocaine

Many substances with $-NH_2$ groups are weak bases. Under natural conditions, however, the group is protonated to give $-NH_3^+$ group. That is, they exist in the form of their conjugate acid.

environment. Biological systems reduce nitrate ion to NH_3 and NH_4^+ and incorporate nitrogen into amino acids and proteins. Many bases can be thought of as derivatives of NH_3 by replacement of the H atoms with organic groups.

Ammonia
$K_b = 1.8 \times 10^{-5}$

Methylamine
$K_b = 5.0 \times 10^{-4}$

Aniline
$K_b = 4.0 \times 10^{-10}$

Ammonia is a weaker base than methylamine (K_b for NH_3 < K_b for CH_3NH_2). Correspondingly, the conjugate acid of ammonia, NH_4^+ ($K_a = 5.6 \times 10^{-10}$) is stronger than the conjugate acid of methylamine ($CH_3NH_3^+$, $K_a = 2.0 \times 10^{-11}$).

Exercise 17.5 Strengths of Acids and Bases

Use Table 17.3 to answer the following questions.

(a) Which is the stronger acid, H_2SO_4 or H_2SO_3?

(b) Is benzoic acid, $C_6H_5CO_2H$, stronger or weaker than acetic acid?

(c) Which has the stronger conjugate base, acetic acid or boric acid?

(d) Which is the stronger base, ammonia or the acetate ion?

(e) Which has the stronger conjugate acid, ammonia or the acetate ion?

Aqueous Solutions of Salts

A number of the acids and bases listed in Table 17.3 are cations or anions [CD-ROM, Screen 17.11]. As described earlier, anions in particular can act as Brønsted bases as they can accept a proton from water to form the conjugate acid of the base and hydroxide ion.

$$CO_3^{2-}(aq) + H_2O(\ell) \rightleftharpoons HCO_3^-(aq) + OH^-(aq) \qquad K_b = 2.1 \times 10^{-4}$$

Although they do not appear at first glance to be acids, be sure to notice that many hydrated metal cations in water are effective Brønsted acids.

$$Al(H_2O)_6^{3+}(aq) + H_2O(\ell) \rightleftharpoons Al(H_2O)_5(OH)^{2+}(aq) + H_3O^+(aq) \qquad K_a = 7.9 \times 10^{-6}$$

Table 17.4 is a summary of the acid–base properties of some of the common cations and anions in Table 17.3. As you look over this table, notice the following points:

- Anions that are conjugate bases of strong acids (for example, Cl^- and NO_3^-) are such weak bases that they *have no effect on solution pH.*
- There are numerous basic anions (such as CO_3^{2-}). All are the conjugate bases of weak acids.
- Acidic anions arise from polyprotic acids (and such anions are amphiprotic as well).

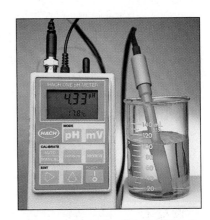

A pH measurement of a dilute solution of a copper(II) sulfate shows that the solution is decidedly acidic. Of the common cations, Al^{3+} and transition metal ions form acidic solutions in water. *(Charles D. Winters)*

Table 17.4 • Acid and Base Properties of Some Ions in Aqueous Solution

	Neutral		Basic			Acidic
Anions	Cl^-	NO_3^-	$CH_3CO_2^-$	CN^-	SO_4^{2-}	HSO_4^-
	Br^-	ClO_4^-	HCO_2^-	PO_4^{3-}	HPO_4^{2-}	$H_2PO_4^-$
	I^-		CO_3^{2-}	HCO_3^-	SO_3^{2-}	HSO_3^-
			S^{2-}	HS^-	OCl^-	
			F^-	NO_2^-		
Cations	Li^+	Mg^{2+}	$Al(H_2O)_5(OH)^{2+}$			$Al(H_2O)_6^{3+}$ and hydrated
	Na^+	Ca^{2+}	and analogous ions			transition metal cations
	K^+	Ba^{2+}				$(Fe(H_2O)_6^{3+})$
						NH_4^+

> ## Problem-Solving Tip 17.2
>
> ### Aqueous Solutions of Salts
>
> Because aqueous solutions of salts are found in our bodies and throughout our economy and environment, it is important to know how to predict their acid and base properties. The necessary information to determine the relative pH of an aqueous solution of a salt is summarized in Table 17.4. Consider also the following examples:
>
> | Cation | Anion | pH of the Solution |
> |--------|-------|---------------------|
> | From strong base (Na^+) | From strong acid (Cl^-) | $= 7$ (neutral) |
> | From strong base (K^+) | From weak acid ($CH_3CO_2^-$) | > 7 (basic) |
> | From weak base (NH_4^+) | From strong acid (Cl^-) | < 7 (acidic) |
> | From any weak base (BH^+) | From any weak acid (A^-) | Depends on relative strengths of acid and base |

- Alkali metal and alkaline earth cations have no measurable effect on solution pH.
- All metal cations are hydrated in water. That is, they form ions such as $M(H_2O)_6^{n+}$. However, only when M is a $+2$ or $+3$ ion, particularly a transition metal ion, does the ion act as an acid.

Example 17.2 Acid–Base Properties of Salts

Problem • Decide whether each of the following will give rise to an acidic, basic, or neutral solution in water.

(a) $NaNO_3$ (d) $NaHCO_3$
(b) K_3PO_4 (e) NH_4F
(c) $FeCl_2$

Strategy • First, decide on the cation and anion in each salt. Next, use Tables 17.3 and 17.4 to describe the properties of each ion.

Solution •

(a) **$NaNO_3$:** This salt gives a neutral, aqueous solution (pH = 7). The Na^+ ion does not react with water to an appreciable extent. Similarly, the nitrate ion, NO_3^-, is the *very* weak conjugate base of a strong acid and so does not affect the solution pH.

(b) **K_3PO_4:** An aqueous solution of K_3PO_4 should be basic (pH > 7) because PO_4^{3-} is the conjugate base of the weak acid HPO_4^{2-}. In contrast, the K^+ ion, like the Na^+ ion, does not react with water appreciably.

(c) **$FeCl_2$:** An aqueous solution of $FeCl_2$ should be weakly acidic (pH < 7). The Fe^{2+} ion in water, $Fe(H_2O)_6^{2+}$, is a Brønsted acid. In contrast, Cl^- is the *very* weak conjugate base of the strong acid HCl and so does not contribute excess OH^- ions to the solution.

(d) **$NaHCO_3$:** Some additional explanation is needed concerning salts of amphiprotic anions such as HCO_3^- and $H_2PO_4^-$.

Because they have an ionizable hydrogen, they can act as acids.

$$HCO_3^-(aq) + H_2O(\ell) \rightleftharpoons CO_3^{2-}(aq) + H_3O^+(aq)$$
$$K_a = 4.8 \times 10^{-11}$$

but they are also the conjugate bases of weak acids.

$$HCO_3^-(aq) + H_2O(\ell) \rightleftharpoons H_2CO_3(aq) + OH^-(aq)$$
$$K_b = 2.4 \times 10^{-8}$$

Whether the solution is acidic or basic will depend on the *relative* magnitude of K_a and K_b. In the case of the hydrogen carbonate anion, K_b is larger than K_a, so $[OH^-]$ is larger than $[H_3O^+]$, and an aqueous solution of $NaHCO_3$ will be slightly basic.

(e) **NH_4F:** What happens if you have a salt based on an acidic cation and a basic anion? One example is ammonium fluoride. Here the ammonium ion would decrease the pH, and the fluoride ion would increase the pH.

$$NH_4^+(aq) + H_2O(\ell) \rightleftharpoons H_3O^+(aq) + NH_3(aq)$$
$$K_a(NH_4^+) = 5.6 \times 10^{-10}$$
$$F^-(aq) + H_2O(\ell) \rightleftharpoons HF(aq) + OH^-(aq)$$
$$K_b(F^-) = 1.4 \times 10^{-11}$$

Because $K_a(NH_4^+) > K_b(F^-)$, the ammonium ion is a stronger acid than fluoride ion is a base. The resulting solution should be slightly acidic.

Comment • There are several important points to notice here:

- Anions such as Cl^- and NO_3^- that are conjugate bases of strong acids have no effect on solution pH.
- Alkali metal cations (Na^+, K^+) have no effect on solution pH.

- In general, *for a salt that has an acidic cation and a basic anion, the pH of the solution will be determined by the ion that is the stronger acid or base of the two.*

Exercise 17.6 Acid–Base Properties of Salts in Aqueous Solution

For each of the following salts in water, predict whether the pH will be greater than, less than, or equal to 7.

(a) KBr   (b) NH_4NO_3 (c) $AlCl_3$ (d) Na_2HPO_4

A Logarithmic Scale of Relative Acid Strength, pK_a

Many chemists and biochemists use a logarithmic scale to report and compare acid strengths.

$$pK_a = -\log K_a \qquad (17.7)$$

The pK_a of an acid is the negative log of the K_a value, (just as pH is the negative log of the hydronium ion concentration). For example, acetic acid has a pK_a value of 4.74.

$$pK_a = -\log (1.8 \times 10^{-5}) = 4.74$$

The pK_a value becomes smaller as the acid strength increases.

Acid strength increases →

Propanoic acid	**Acetic acid**	**Formic acid**
$CH_3CH_2CO_2H$	CH_3CO_2H	HCO_2H
$K_a = 1.3 \times 10^{-5}$	$K_a = 1.8 \times 10^{-5}$	$K_a = 1.8 \times 10^{-4}$
$pK_a = 4.89$	$pK_a = 4.74$	$pK_a = 3.74$

← *pK_a increases*

Exercise 17.7 A Logarithmic Scale for Acid Strengths, pK_a

(a) What is the pK_a value for benzoic acid, $C_6H_5CO_2H$?

(b) Is chloroacetic acid ($ClCH_2CO_2H$), p$K_a = 2.87$, a stronger or weaker acid than benzoic acid?

(c) What is the pK_a for the conjugate acid of ammonia? Is this acid stronger or weaker than acetic acid?

Relating the Ionization Constants for an Acid and Its Conjugate Base

Let us look again at Table 17.3. On descending the table, the strengths of the acids decline (K_a becomes smaller) and the strengths of their conjugate bases increase (the values of K_b increase). Examining a few cases shows that the product of K_a

for an acid and K_b for its conjugate base is equal to a constant, specifically K_w [🖱 CD-ROM, Screen 17.6].

● **A Relation Among pK Values**
A useful relationship derived from Equation 17.8 is

$$pK_w = pK_a + pK_b$$

$$K_aK_b = K_w \tag{17.8}$$

Consider the specific case of the ionization of a weak acid, say HCN, and the interaction of its conjugate base, CN^-, with H_2O.

Weak acid:

$$HCN(aq) + H_2O(\ell) \rightleftarrows H_3O^+(aq) + CN^-(aq) \qquad K_a = 4.0 \times 10^{-10}$$

Conjugate base:

$$\underline{CN^-(aq) + H_2O(\ell) \rightleftarrows HCN(aq) + OH^-(aq)} \qquad K_b = 2.5 \times 10^{-5}$$
$$2\,H_2O(\ell) \rightleftarrows H_3O^+(aq) + OH^-(aq) \qquad K_w = 1.0 \times 10^{-14}$$

Adding the equations gives the chemical equation for the autoionization of water, and the numerical value of $K_a \cdot K_b$ is indeed 1.0×10^{-14}. That is,

$$K_aK_b = \left(\frac{[H_3O^+][CN^-]}{[HCN]}\right)\left(\frac{[HCN][OH^-]}{[CN^-]}\right) = [H_3O^+][OH^-] = K_w$$

Equation 17.8 is useful because K_b can be calculated from K_a. The value of K_b for the cyanide ion, for example, is

$$K_b \text{ for } CN^- = \frac{K_w}{K_a \text{ for HCN}} = \frac{1.0 \times 10^{-14}}{4.0 \times 10^{-10}} = 2.5 \times 10^{-5}$$

Exercise 17.8   **Using the Equation $K_aK_b = K_w$**

K_a for lactic acid, $CH_3CHOHCO_2H$, is 1.4×10^{-4}. What is K_b for the conjugate base of this acid, $CH_3CHOHCO_2^-$? Where does this base fit in Table 17.3?

(**Chapter Goals** ● Revisited)

● Use the Brønsted–Lowry and Lewis concepts of acids and bases.

● Apply the principles of chemical equilibrium to acids and bases in aqueous solution.

● **Predict the outcome of reactions of acids and bases.**

● Understand the influence of structure and bonding on acid–base properties.

17.5   EQUILIBRIUM CONSTANTS AND ACID–BASE REACTIONS

According to the Brønsted concept, all acid–base reactions can be written as equilibria involving the acid and base and their conjugates [🖱 CD-ROM, Screen 17.7].

Acid + Base $\rightleftarrows$ Conjugate base of the acid + Conjugate acid of the base

In the preceding section we used equilibrium constants to provide quantitative information about the relative strengths of acids and bases. Now we want to show how the constants can be used to decide whether a particular acid–base reaction is product- or reactant-favored. If the reaction is product-favored, what is the nature of the solution when reaction is complete?

Predicting the Direction of Acid–Base Reactions

Hydrochloric acid is a strong Brønsted acid. Its equilibrium constant for reaction with water is very large, the equilibrium effectively lying completely to the right.

$$HCl(aq) + H_2O(\ell) \longrightarrow H_3O^+(aq) + Cl^-(aq)$$

Strong acid ($\approx 100\%$ $[H_3O^+] \approx$ initial
ionized), $K \gg 1$ concentration of the acid

For the reaction of all strong acids with water, the acid on the reactant side of the equation is stronger than the acid on the product side (and the base on the reactant side is stronger than the base on the product side).

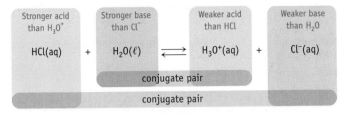

Of the two acids here, HCl is stronger than H_3O^+. Of the two bases, H_2O and Cl^-, water is the stronger base and wins out in the competition for the proton. The equilibrium lies to the side of the chemical equation having the weaker acid and base.

In contrast to HCl and other strong acids, acetic acid ionizes only to a very small extent and so is considered a *weak* Brønsted acid (see Table 17.3)

$$CH_3CO_2H(aq) + H_2O(\ell) \rightleftharpoons H_3O^+(aq) + CH_3CO_2^-(aq)$$

Weak acid ($< 100\%$ ionized), $[H_3O^+] \ll$ initial
$K = 1.8 \times 10^{-5}$ concentration of the acid

Thus, when equilibrium is achieved in a 0.1 M aqueous solution of CH_3CO_2H, the concentration of $H_3O^+(aq)$ and $CH_3CO_2^-(aq)$ are each only about 0.001 M. About 99% of the acetic acid is *not* ionized.

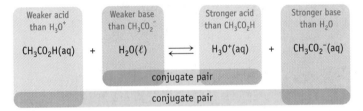

Again, the equilibrium lies toward the side of the reaction having the weaker acid and base.

These two examples of the relative extent of acid–base reactions illustrate another general principle: *all proton transfer reactions proceed from the stronger acid and base to the weaker acid and base.* That is, an equilibrium favors the weaker acid and base. Using this principle and Table 17.3 you can predict which reactions are product-favored and which are reactant-favored. Consider the possible reaction of phosphoric acid and acetate ion to give acetic acid and the dihydrogen phosphate ion. Table 17.3 informs us that H_3PO_4 is a stronger acid ($K_a = 7.5 \times 10^{-3}$) than acetic acid ($K_a = 1.8 \times 10^{-5}$), and the acetate ion ($K_b = 5.6 \times 10^{-10}$) is a stronger base than the dihydrogen phosphate ion ($K_b = 1.3 \times 10^{-12}$).

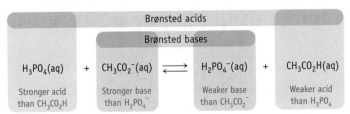

Thus, mixing phosphoric acid with sodium acetate would produce a significant amount of dihydrogen phosphate ion and acetic acid. That is, the equilibrium is predicted to lie to the right because the reaction has proceeded from the stronger acid–base combination to the weaker acid–base combination.

Example 17.3 | Reactions of Acids and Bases

Problem • Write a balanced, net ionic equation for the reaction that occurs between acetic acid and sodium bicarbonate. Decide whether the equilibrium lies predominantly to the left or right.

Strategy • First, identify the products of the acid–base reaction, which arise by H^+ transfer from the acid to the base. Next, identify the two acids (or the two bases) in the reaction. Finally, using Table 17.3 decide which is the weaker of the two acids (or the weaker of the two bases). The reaction will proceed from the stronger acid (or base) to the weaker acid (or base).

Solution • Acetic acid is clearly one acid involved (and its conjugate base is the acetate ion, $CH_3CO_2^-$). The other reactant, $NaHCO_3$, is a water-soluble salt that forms Na^+ and HCO_3^- ions in water. Because acetic acid can only function as an acid, whereas the HCO_3^- ion can be an acid or a base, the HCO_3^- ion in this case must be the Brønsted base. Thus, hydrogen ion transfer from the acid to the base (HCO_3^- ion) could lead to the following net ionic equation.

$$CH_3CO_2H(aq) + HCO_3^-(aq) \rightleftharpoons CH_3CO_2^-(aq) + H_2CO_3(aq)$$

According to Table 17.3, H_2CO_3 is a weaker acid ($K_a = 4.2 \times 10^{-7}$) than CH_3CO_2H ($K_a = 1.8 \times 10^{-5}$), and $CH_3CO_2^-$ is a weaker base ($K_b = 5.6 \times 10^{-10}$) than HCO_3^- ($K_b = 2.4 \times 10^{-8}$). The stronger base, HCO_3^- is expected to win out in the competition for the H^+ ion, and the equilibrium lies to the right.

Comment • See the photograph here of the reaction of acetic acid and $NaHCO_3$. The reaction clearly favors the weaker acid

(H_2CO_3) and base ($CH_3CO_2^-$). However, the equilibrium lies even further to the right because H_2CO_3 dissociates into CO_2 and H_2O, and the CO_2 bubbles out of the solution.

$$H_2CO_3(aq) \rightleftharpoons CO_2(g) + H_2O(\ell)$$

See the discussion of gas-forming reactions in Section 5.5 and of Le Chatelier's principle in Section 16.6.

Predicting the direction of an acid–base reaction. This is the reaction of the weak acid acetic acid and the weak base HCO_3^- from sodium hydrogen carbonate. Based on the values of equilibrium constants the reaction is predicted to proceed to the right. *(Charles D. Winters)*

Exercise 17.9 | Relative Strengths of Acids and Bases: Predicting the Direction of an Acid–Base Reaction

(a) Which is the stronger Brønsted acid, HCO_3^- or NH_4^+? Which has the stronger conjugate base?

(b) Is a reaction between HCO_3^- ions and NH_3 product- or reactant-favored?

$$HCO_3^-(aq) + NH_3(aq) \rightleftharpoons CO_3^{2-}(aq) + NH_4^+(aq)$$

Exercise 17.10 | Reaction of an Acid and a Base

Write the net ionic equation for the possible reaction between acetic acid and sodium hydrogen sulfate, $NaHSO_4$. Does the equilibrium lie to the left or right?

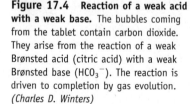

Figure 17.4 Reaction of a weak acid with a weak base. The bubbles coming from the tablet contain carbon dioxide. They arise from the reaction of a weak Brønsted acid (citric acid) with a weak Brønsted base (HCO_3^-). The reaction is driven to completion by gas evolution. *(Charles D. Winters)*

17.6 TYPES OF ACID–BASE REACTIONS

The reaction of hydrochloric acid and sodium hydroxide is the classic example of a strong acid/strong base reaction, whereas the reaction of citric acid and bicarbonate ion represents the reaction of a weak acid and weak base (Figure 17.4). But there are two other types as well (Table 17.5). The reaction of an acid with a base is one of the most important classes of chemical reactions. It is therefore useful for you to know the outcome of the various types of acid–base reactions.

Table 17.5 • Characteristics of Acid–Base Reactions

Type	Example	Net Ionic Equation	Species Present After Equal Molar Amounts Are Mixed; pH
Strong acid + strong base	HCl + NaOH	$H_3O^+(aq) + OH^-(aq) \rightleftharpoons 2\,H_2O(\ell)$	Cl^-, Na^+, pH = 7
Strong acid + weak base	HCl + NH$_3$	$H_3O^+(aq) + NH_3(aq) \rightleftharpoons NH_4^+(aq) + H_2O(\ell)$	Cl^-, NH_4^+, pH < 7
Weak acid + strong base	HCO$_2$H + NaOH	$HCO_2H(aq) + OH^-(aq) \rightleftharpoons HCO_2^-(aq) + H_2O(\ell)$	HCO_2^-, Na^+, pH > 7
Weak acid + weak base	HCO$_2$H + NH$_3$	$HCO_2H(aq) + NH_3(aq) \rightleftharpoons HCO_2^-(aq) + NH_4^+(aq)$	HCO_2^-, NH_4^+, pH dependent on K_a and K_b of conjugate acid and base.

Reaction of a Strong Acid with a Strong Base

Strong acids and bases are effectively 100% ionized in solution. Therefore, the total ionic equation for the reaction of HCl (strong acid) and NaOH (strong base) is

$$H_3O^+(aq) + Cl^-(aq) + Na^+(aq) + OH^-(aq) \rightleftharpoons 2\,H_2O(\ell) + Na^+(aq) + Cl^-(aq)$$

which leads to the following net ionic equation:

$$H_3O^+(aq) + OH^-(aq) \rightleftharpoons 2\,H_2O(\ell) \qquad K = 1/K_w = 1.0 \times 10^{14}$$

The net ionic equation for the reaction of any strong acid with any strong base is always simply the union of hydronium ion and hydroxide ion to give water [← SECTION 5.4]. Because this reaction is the reverse of the autoionization of water, it has an equilibrium constant of $1/K_w$. This very large value of K shows that, for all practical purposes, the reactants are completely consumed to form products. Thus, if equal numbers of moles of NaOH and HCl are mixed, the result is just a solution of NaCl in water. Because the constituents of NaCl, Na^+ and Cl^- ions, arise from a strong base and a strong acid, respectively, they produce a neutral aqueous solution. For this reason reactions of strong acids and bases are often called "neutralizations."

> Mixing equal molar quantities of a strong base with a strong acid produces a neutral solution (pH = 7 at 25 °C).

Reaction of a Weak Acid with a Strong Base

Consider the reaction of the naturally occurring weak acid formic acid (see Chapter Focus, page 692), HCO$_2$H, with sodium hydroxide. The net ionic equation is

$$HCO_2H(aq) + OH^-(aq) \rightleftharpoons H_2O(\ell) + HCO_2^-(aq)$$

In the reaction of formic acid with NaOH, OH^- is a much stronger base than HCO_2^- ($K_b = 5.6 \times 10^{-11}$), and the reaction is predicted to proceed to the right. If equal molar quantities of weak acid and base are mixed, the final solution will contain sodium formate (NaHCO$_2$), a salt that is 100% dissociated in water. The Na^+ ion is the cation of a strong base and so gives a neutral solution. The formate ion, however, is the conjugate base of a weak acid (see Table 17.3), so the solution is basic. This example leads to a useful general conclusion:

● **Formic Acid + NaOH**
The equilibrium constant for the reaction of formic acid and sodium hydroxide is 1.8×10^{10}. Can you confirm this? (See Study Question 17.109.)

> Mixing equal molar quantities of a strong base with a weak acid produces a salt whose anion is the conjugate base of the weak acid. The solution is basic, with the pH depending on K_b for the anion.

Reaction of a Strong Acid with a Weak Base

When HCl and the weak base ammonia are combined we assume that HCl is 100% ionized in solution and that the H_3O^+ ion produced by the acid reacts with the weak base. The net ionic equation for the reaction of aqueous HCl and NH_3 is

$$H_3O^+(aq) + NH_3(aq) \rightleftharpoons H_2O(\ell) + NH_4^+(aq)$$

The hydronium ion, H_3O^+, is a much stronger acid than NH_4^+ ($K_a = 5.6 \times 10^{-10}$), and NH_3 is a stronger base ($K_b = 1.8 \times 10^{-5}$) than H_2O. Therefore, the reaction is predicted to proceed to the right and essentially to completion. Thus, after mixing equal molar quantities of HCl and NH_3, the solution contains the salt ammonium chloride, NH_4Cl. The Cl^- ion has no effect on the solution pH (Tables 17.3 and 17.4). However, the NH_4^+ ion is the conjugate acid of the weak base NH_3, so the solution at the conclusion of the reaction is acidic. In general, we can conclude that

> Mixing equal molar quantities of a strong acid and a weak base produces a salt whose cation is the conjugate acid of the weak base. The solution is acidic, with the pH depending on K_a for the cation.

- **Ammonia + HCl**

The equilibrium constant for the reaction of a strong acid with aqueous ammonia is 1.8×10^9. Can you confirm this? (See Study Question 17.110.)

Reaction of a Weak Acid with a Weak Base

If acetic acid, a weak acid, is mixed with ammonia, a weak base, the following reaction occurs.

$$CH_3CO_2H(aq) + NH_3(aq) \rightleftharpoons NH_4^+(aq) + CH_3CO_2^-(aq)$$

You know that the reaction is product-favored because CH_3CO_2H is a stronger acid than NH_4^+ and NH_3 is a stronger base than $CH_3CO_2^-$ (see Table 17.3). Thus, if equal molar quantities of the acid and base are mixed, the resulting solution contains ammonium acetate, $NH_4CH_3CO_2$. Is this solution acidic or basic? On page 706, you learned that this depends on the relative values of K_a for the conjugate acid (here NH_4^+; $K_a = 5.6 \times 10^{-10}$) and K_b for the conjugate base (here $CH_3CO_2^-$; $K_b = 5.6 \times 10^{-10}$). In this case the values of K_a and K_b are the same, so the solution is predicted to be neutral.

A weak acid reacting with a weak base. Baking powder contains the weak acid calcium dihydrogen phosphate, $Ca(H_2PO_4)_2$. This can react with the basic HCO_3^- ion in baking soda to give HPO_4^{2-}, CO_2 gas, and water. (*Charles D. Winters*)

> Mixing equal molar quantities of a weak acid and a weak base produces a salt whose cation is the conjugate acid of the weak base and whose anion is the conjugate base of the weak acid. The solution pH depends on the relative K_a and K_b values.

- **K for Reaction of Weak Acid and Weak Base**

The equilibrium constant for the reaction between a weak acid and a weak base is $K_{net} = \dfrac{K_w}{K_a K_b}$. Can you confirm this? (See Study Question 17.111.)

Exercise 17.11 **Acid–Base Reactions**

(a) Equal molar quantities of HCl(aq) and NaCN(aq) are mixed. Is the resulting solution acidic, basic, or neutral?

(b) Equal molar quantities of acetic acid and sodium sulfite, Na_2SO_3, are mixed. Is the resulting solution acidic, basic, or neutral?

17.7   CALCULATIONS WITH EQUILIBRIUM CONSTANTS

Determining K from Initial Concentrations and Measured pH

The K_a and K_b values found in Table 17.3 and in the more extensive tables in Appendices H and I were all determined by experiment. Several experimental methods are available, but one approach, illustrated by the following example, is to determine the pH of the solution.

Example 17.4 | **Calculating a K_a Value from a Measured pH**

Problem • Lactic acid is a monoprotic acid that occurs naturally in sour milk and arises from metabolism in the human body (see page 424). A 0.10 M aqueous solution of lactic acid, $CH_3CHOHCO_2H$, has a pH of 2.43. What is the value of K_a for lactic acid?

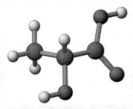

Lactic acid, $CH_3CH(OH)CO_2H$

Strategy • To calculate K_a, we must know the equilibrium concentration of each species. The pH of the solution directly tells us the equilibrium concentration of H_3O^+, and we can derive the other equilibrium concentrations from this.

Solution • The equation for the equilibrium interaction of lactic acid with water is

$$CH_3CHOHCO_2H(aq) + H_2O(\ell) \rightleftharpoons CH_3CHOHCO_2^-(aq) + H_3O^+(aq)$$

    lactic acid   lactate ion

and the equilibrium constant expression is

$$K_a = \frac{[H_3O^+][CH_3CHOHCO_2^-]}{[CH_3CHOHCO_2H]}$$

We begin by converting the pH to $[H_3O^+]$.

$$[H_3O^+] = 10^{-pH} = 10^{-2.43} = 3.7 \times 10^{-3}\ M$$

Next, prepare an ICE table of the concentrations in the solution before equilibrium is established, the change occurring as the re-action proceeds to equilibrium, and the concentrations when equilibrium has been achieved [⬅ EXAMPLES 16.3–16.5].

Equilibrium	$CH_3CHOHCO_2H$ + H_2O	$\rightleftharpoons$	$CH_3CHOHCO_2^-$	+	H_3O^+
Initial (M)	0.10		0		0
Change	$-x$		$+x$		$+x$
Equilibrium (M)	$(0.10 - x)$		x		x

The following points can be made concerning the ICE table.

• The quantity x represents the equilibrium concentrations of hydronium ion and lactate ion. That is, at equilibrium $x = [H_3O^+] \approx [CH_3CHOHCO_2^-] = 3.7 \times 10^{-3}$ M.

• By stoichiometry, x is also the amount of acid that ionized on proceeding to equilibrium. With these points in mind, we can calculate K_a for lactic acid.

$$K_a = \frac{[H_3O^+][CH_3CHOHCO_2^-]}{[CH_3CHOHCO_2H]}$$

$$= \frac{(3.7 \times 10^{-3})(3.7 \times 10^{-3})}{0.10 - 0.0037} = 1.4 \times 10^{-4}$$

Comparing this value of K_a with others in Table 17.3, we see that lactic acid is similar to formic acid in its strength.

Comment • Hydronium ion, H_3O^+, is present in solution from lactic acid ionization *and* from water autoionization. However, because $[H_3O^+]$ from water must be less than 10^{-7} M, the pH is almost completely a reflection of H_3O^+ from lactic acid [⬅ EXAMPLE 17.1, PAGE 699].

Exercise 17.12 | **Calculating a K_a Value from a Measured pH**

A solution prepared from 0.055 mol of butanoic acid dissolved in sufficient water to give 1.0 L of solution has a pH of 2.72. Determine K_a for butanoic acid. The acid ionizes according to the balanced equation

$$CH_3CH_2CH_2CO_2H(aq) + H_2O(\ell) \rightleftharpoons H_3O^+(aq) + CH_3CH_2CH_2CO_2^-(aq)$$

There is an important point to notice in Example 17.4. The lactic acid concentration at equilibrium was given by $(0.10 - x)$, where x was found to be 3.7×10^{-3} M. By the usual rules governing significant figures, $(0.10 - 0.0037)$ is equal to 0.10. The acid is weak, so very little of it ionizes (approximately 4%), and the equilibrium concentration of lactic acid is essentially equal to the initial acid concentration. Ignoring the subtraction of 0.0037 from 0.10 has no effect on the answer.

Like lactic acid, most weak acids (HA) are so weak that the equilibrium concentration of the acid, [HA], is effectively its initial concentration $(=[HA]_0)$. This leads to the useful conclusion that the denominator in the equilibrium constant expression for dilute solutions of most weak acids is simply $[HA]_0$, the original or initial concentration of the weak acid.

$$HA(aq) + H_2O(\ell) \rightleftharpoons H_3O^+(aq) + A^-(aq)$$

$$K_a = \frac{[H_3O^+][A^-]}{[HA]_0 - [H_3O^+]} \approx \frac{[H_3O^+][A^-]}{[HA]_0} \text{ when } [H_3O^+] \ll [HA].$$

Error analysis shows that

> The approximation that $[HA]_{equilibrium}$ is effectively equal to $[HA]_0$
>
> $([HA]_{equilibrium} = [HA]_0 - [H_3O^+] \approx [HA]_0)$
>
> is valid whenever $[HA]_0$ is greater than or equal to $100 K_a$.

This is the same approximation we derived in Chapter 16 when deciding whether we needed to solve quadratic equations exactly (◀ PROBLEM-SOLVING TIPS, PAGE 670).

What Is the pH of an Aqueous Solution of a Weak Acid or Base?

Knowing values of the equilibrium constants for weak acids and bases enables us to calculate the pH of a solution of a weak acid or base.

Example 17.5   Calculating Equilibrium Concentrations and pH from K_a

Problem • Calculate the pH of a 0.020 M solution of benzoic acid ($C_6H_5CO_2H$) if $K_a = 6.3 \times 10^{-5}$ for the acid.

$$C_6H_5CO_2H(aq) + H_2O(\ell) \rightleftharpoons H_3O^+(aq) + C_6H_5CO_2^-(aq)$$

Strategy • This is similar to Examples 16.5 and 16.6 where we wanted to find the concentration of a reaction product. The strategy is the same: designate the quantity of product (here $[H_3O^+]$) by x and derive the other concentrations from that starting point.

Solution • Organize the information in an ICE table.

Equilibrium	$C_6H_5CO_2H + H_2O \rightleftharpoons C_6H_5CO_2^- + H_3O^+$		
Initial (M)	0.020	0	0
Change (M)	$-x$	$+x$	$+x$
Equilibrium (M)	$(0.020 - x)$	x	x

According to reaction stoichiometry,

$$[H_3O^+] = [C_6H_5CO_2^-] = x \text{ at equilibrium}$$

Stoichiometry also tells us that the amount of acid ionized is x. Thus, the benzoic acid concentration at equilibrium is

$$[C_6H_5CO_2H] = \text{initial acid concentration} - \text{amount of acid that ionized}$$

$$[C_6H_5CO_2H] = [C_6H_5CO_2H]_0 - x$$

$$[C_6H_5CO_2H] = 0.020 - x$$

Substituting these equilibrium concentrations into the K_a expression, we have

$$K_a = \frac{[H_3O^+][C_6H_5CO_2^-]}{[C_6H_5CO_2H]}$$

$$= 6.3 \times 10^{-5} = \frac{(x)(x)}{0.020 - x}$$

The value of x is small compared with 0.020 (because $[HA]_0 >$ $100K_a$; here 0.020 M $> 6.3 \times 10^{-3}$). Therefore,

$$K_a = 6.3 \times 10^{-5} = \frac{x^2}{0.020}$$

Solving for x, we have

$$x = \sqrt{K_a \times (0.020)} = 0.0011 \text{ M}$$

and we find that

$$[H_3O^+] \approx [C_6H_5CO_2^-] = 0.0011 \text{ M}$$

and

$$[C_6H_5CO_2H] = (0.020 - x) = 0.019 \text{ M}$$

Finally, the pH of the solution is found to be

$$pH = -\log(1.1 \times 10^{-3}) = 2.96$$

Comment • Let us think again about the result. Because benzoic acid is weak, we made the approximation that $(0.020 - x)$ ≈ 0.020. If we do *not* make the approximation and instead solve the exact expression, $x = [H_3O^+] = 1.1 \times 10^{-3}$. This is the same answer to two significant figures that we obtained from the "approximate" expression.

Finally, notice that we again ignored any H_3O^+ that arises from water ionization (see Example 17.4).

Example 17.6   Calculating Equilibrium Concentrations and pH from K_a and Using the Method of Successive Approximations

Problem • What is the pH of a 0.0010 M solution of formic acid? What is the concentration of formic acid at equilibrium? The acid is moderately weak, with $K_a = 1.8 \times 10^{-4}$.

$$HCO_2H(aq) + H_2O(\ell) \rightleftharpoons HCO_2^-(aq) + H_3O^+(aq)$$

Strategy • This is similar to Example 17.5 except that formic acid is a stronger acid than benzoic acid so an approximate solution will not be possible.

Solution • The ICE table is shown here

Equilibrium	HCO_2H + H_2O $\rightleftharpoons$ HCO_2^- + H_3O^+		
Initial (M)	0.0010	0	0
Change	$-x$	$+x$	$+x$
Equilibrium (M)	$(0.0010 - x)$	x	x

Substituting the values in the table into the K_a expression we have

$$K_a = \frac{[H_3O^+][HCO_2^-]}{[HCO_2H]} = 1.8 \times 10^{-4} = \frac{(x)(x)}{0.0010 - x}$$

Formic acid is a weak acid because it has a value of K_a much less than 1. In this situation, however, $[HA]_0$ (= 0.0010 M) is *not* greater than $100K_a$ (= 1.8×10^{-2}), so the usual approximation is not reasonable. Thus, we have to find the equilibrium concentrations by solving the "exact" expression. This can be solved with the quadratic formula [← EXAMPLE 16.5] or by successive approximations (Appendix A). Let us use the successive approximation method here.

Begin by solving the approximate expression for x.

$$1.8 \times 10^{-4} = \frac{(x)(x)}{0.0010}$$

Solving this, $x = 4.2 \times 10^{-4}$. Put this value into the expression for x in the denominator of the exact expression.

$$1.8 \times 10^{-4} = \frac{(x)(x)}{0.0010 - x} = \frac{(x)(x)}{0.0010 - 4.2 \times 10^{-4}}$$

Solving this equation for x, we find $x = 3.2 \times 10^{-4}$. Again put this value into the denominator and solve for x.

$$1.8 \times 10^{-4} = \frac{(x)(x)}{0.0010 - x} = \frac{(x)(x)}{0.0010 - 3.2 \times 10^{-4}}$$

Continue this procedure until the value of x does not change from one cycle to the next. In this case, two more steps give us the result that

$$x = [H_3O^+] = [HCO_2^-] = 3.4 \times 10^{-4} \text{ M}$$

Thus,

$$[HCO_2H] = 0.0010 - x \approx 0.0007 \text{ M}$$

and the pH of the formic acid solution is

$$pH = -\log(3.4 \times 10^{-4}) = 3.47$$

Comment • If we had used the approximate expression to find the H_3O^+ concentration ($[H_3O^+] = [K_a(0.0010)]^{1/2}$), we would have obtained a value of $[H_3O^+] = 4.2 \times 10^{-4}$ M. A simplifying assumption led to a large error, about 24%. The approximate solution fails in this case because (a) the acid concentration is small and (b) the acid is not all that weak. These conditions made invalid the approximation that $[HA]_{equilibrium} \approx [HA]_{initial}$.

Exercise 17.13 **Calculating Equilibrium Concentrations and pH from K_a**

What are the equilibrium concentrations of acetic acid, acetate ion, and H_3O^+ for a 0.10 M solution of acetic acid ($K_a = 1.8 \times 10^{-5}$)? What is the pH of the solution?

Exercise 17.14 **Calculating Equilibrium Concentrations and pH from K_a**

What are the equilibrium concentrations of HF, fluoride ion, and H_3O^+ when a 0.015 M solution of HF is allowed to come to equilibrium? What is the pH of the solution?

Just as acids can be molecular species or ions, so too can bases be molecular or ionic (Figures 17.3–17.5). Many molecular bases are based on nitrogen, with ammonia being the simplest. Many other nitrogen-containing bases occur naturally; caffeine and nicotine are two that are well known. The anionic conjugate bases of weak acids make up another group of bases. The following example describes the calculation of the pH for a solution of sodium acetate.

Example 17.7 **The pH of a Solution of a Weakly Basic Salt, Sodium Acetate**

Problem • What is the pH of a 0.015 M solution of sodium acetate, $NaCH_3CO_2$?

Strategy • Sodium acetate is basic in water because the acetate ion, the conjugate base of a weak acid, acetic acid, reacts with water to form OH^- (see Tables 17.3 and 17.4). (The sodium ion of sodium acetate does not affect the solution pH; Table 17.4.) We shall calculate the hydroxide ion concentration in a manner similar to the calculation in Example 17.5.

Solution • The value of K_b for the acetate ion is 5.6×10^{-10} (Table 17.3).

$$CH_3CO_2^-(aq) + H_2O(\ell) \rightleftharpoons CH_3CO_2H(aq) + OH^-(aq)$$

Set up an ICE table to summarize the initial and equilibrium concentrations of the species in solution.

Equilibrium	$CH_3CO_2^-$ + H₂O ⇌ CH_3CO_2H + OH^-		
Initial (M)	0.015	0	0
Change	$-x$	$+x$	$+x$
Equilibrium (M)	$(0.015 - x)$	x	x

Next substitute the values in the table into the K_b expression.

$$K_b = 5.6 \times 10^{-10} = \frac{[CH_3CO_2H][OH^-]}{[CH_3CO_2^-]} = \frac{x^2}{0.015 - x}$$

The acetate ion is a very weak base, as reflected by the very small value of K_b. Therefore, we assume that x, the concentration of hydroxide ion generated by reaction of acetate ion with water, is very small, and we use the approximate expression to solve for x.

$$K_b = 5.6 \times 10^{-10} = \frac{x^2}{0.015}$$

$$x = [OH^-] = [CH_3CO_2H] = \sqrt{(5.6 \times 10^{-10})(0.015)}$$

$$= [OH^-] = [CH_3CO_2H] = 2.9 \times 10^{-6}\ M$$

To calculate the pH of the solution, we need the hydronium ion concentration. In aqueous solutions, it is always true that

$$K_w = 1.0 \times 10^{-14} = [H_3O^+][OH^-]$$

Therefore,

$$[H_3O^+] = \frac{K_w}{[OH^-]} = \frac{1.0 \times 10^{-14}}{2.9 \times 10^{-6}} = 3.5 \times 10^{-9}\ M$$

$$pH = -\log(3.5 \times 10^{-9}) = 8.46$$

Comment • The hydroxide ion concentration (x) is indeed quite small relative to the initial base concentration. (We would have predicted this from our "rule of thumb": that $100K_b$ should be less than the initial base concentration if we wish to use the approximate expression.)

Ammonia, NH_3
$K_b = 1.8 \times 10^{-5}$

Caffeine, $C_8H_{10}N_4O_2$
$K_b = 2.5 \times 10^{-4}$

Benzoate ion, $C_6H_5CO_2^-$
$K_a = 1.6 \times 10^{-10}$

Phosphate ion, PO_4^{3-}
$K_b = 2.8 \times 10^{-2}$

Figure 17.5 Types of Bases. Brønsted bases are generally of two types, molecular and anionic. Molecular bases often have one or more N atoms capable of accepting an H^+ ion. Anionic bases are conjugate bases of weak acids. *(Charles D. Winters)*

Exercise 17.15   **The pH of the Solution of the Conjugate Base of a Weak Acid**

Sodium hypochlorite, NaOCl, is used as a disinfectant in swimming pools and water treatment plants. What are the concentrations of HOCl and OH^- and the pH of a 0.015 M solution of NaOCl?

What Is the pH of the Solution after an Acid–Base Reaction?

In Section 17.6 you learned how to predict the relative pH of the solution resulting from an acid–base reaction. Whether a solution will be acidic, basic, or neutral depends on the reactants, and the results are summarized in Table 17.5. Let us turn now to the way in which you can calculate a value for the pH after such a reaction.

Example 17.8   **Calculating the pH After the Reaction of a Base with an Acid**

Problem • What is the pH of the solution that results from mixing 25 mL of 0.016 M NH_3 and 25 mL of 0.016 M HCl?

Strategy • This question involves three problems in one:

(a) Writing Balanced Equations: First, we have to write a balanced equation for the reaction that occurs and then decide whether the reaction products are acidic or basic. Here NH_4^+ is the product of interest, and it is a weak acid.

(b) Stoichiometry Problem: To find the "initial" NH_4^+ concentration is a stoichiometry problem: what amount of NH_4^+ (in

moles) is produced in the HCl + NH_3 reaction, and in what volume of solution is the NH_4^+ ion found?

(c) Equilibrium Problem: Calculating the pH involves solving an equilibrium problem. The crucial piece of information needed here is the "initial" concentration of NH_4^+ from part (b).

Solution • If equal molar quantities of base (NH_3) and acid (HCl) are mixed, the result should be an acidic solution because the significant species remaining in solution upon completion of the reaction is NH_4^+, the conjugate acid of the weak base (see

(Example continues on next page)

Tables 17.3 and 17.5). The chemistry can be summarized by the following net ionic equations.

(a) Writing Balanced Equations:

Reaction of HCl (the supplier of hydronium ion) with NH_3 to give NH_4^+:

$$NH_3(aq) + H_3O^+(aq) \longrightarrow NH_4^+(aq) + H_2O(\ell)$$

The reaction of NH_4^+ with water:

$$NH_4^+(aq) + H_2O(\ell) \rightleftharpoons H_3O^+(aq) + NH_3(aq)$$

(b) Stoichiometry Problem

Amount of HCl and NH_3 consumed:

$$(0.025 \text{ L HCl})(0.016 \text{ mol/L}) = 4.0 \times 10^{-4} \text{ mol HCl}$$
$$(0.025 \text{ L } NH_3)(0.016 \text{ mol/L}) = 4.0 \times 10^{-4} \text{ mol } NH_3$$

Amount of NH_4^+ produced on completion of the reaction:

$$4.0 \times 10^{-4} \text{ mol } NH_3 \left(\frac{1 \text{ mol } NH_4^+}{1 \text{ mol } NH_3} \right) = 4.0 \times 10^{-4} \text{ mol } NH_4^+$$

Concentration of NH_4^+:

Combining 25 mL each of HCl and NH_3 gives a total solution volume of 50 mL. Therefore, the concentration of NH_4^+ is

$$[NH_4^+] = \frac{4.0 \times 10^{-4} \text{ mol}}{0.050 \text{ L}} = 8.0 \times 10^{-3} \text{ M}$$

(c) Acid–Base Equilibrium Problem

With the initial concentration of ammonium ion known, set up an ICE table to find the equilibrium concentration of hydronium ion.

Equilibrium	$NH_4^+ + H_2O \rightleftharpoons$	NH_3	+	H_3O^+
Initial (M)	0.0080	0		0
Change	$-x$	$+x$		$+x$
Equilibrium (M)	$(0.0080 - x)$	x		x

Next, substitute the values in the table into the K_a expression for the ammonium ion. Thus, we have

$$K_a = 5.6 \times 10^{-10} = \frac{[H_3O^+][NH_3]}{[NH_4^+]} = \frac{(x)(x)}{0.0080 - x}$$

The ammonium ion is a very weak acid, as reflected by the very small value of K_a. Therefore, x, the concentration of hydronium ion generated by reaction of ammonium ion with water, is assumed to be very small, and the approximate expression is used to solve for x. (Here $100K_a$ is much less than the original acid concentration.)

$$K_a = 5.6 \times 10^{-10} \approx \frac{x^2}{0.0080}$$
$$x = \sqrt{(5.6 \times 10^{-10})(0.0080)}$$
$$= [H_3O^+] = [NH_3] = 2.1 \times 10^{-6} \text{ M}$$
$$pH = -\log (2.1 \times 10^{-6}) = 5.67$$

Comment • As predicted (Table 17.5), the solution after mixing equal molar quantities of a strong acid and weak base is weakly acidic.

Exercise 17.16 **What Is the pH After the Reaction of a Weak Acid and Strong Base?**

Calculate the pH after mixing 15 mL of 0.12 M acetic acid with 15 mL of 0.12 M NaOH. What are the major species in solution at equilibrium (besides water) and what are their concentrations?

17.8 POLYPROTIC ACIDS AND BASES

Some important acids are capable of donating more than one proton; they are *polyprotic* (see Table 17.1). Indeed, many occur in nature. Examples include oxalic acid in rhubarb (page 690), citric acid in citrus fruit, malic acid in apples, and tartaric acid in grapes (page 695).

Phosphoric acid is commonly used in the food industry and the hydrogen phosphate anions are important biochemically (➜ PAGE 750). The acid ionizes in three steps:

First ionization step:

$$H_3PO_4(aq) + H_2O(\ell) \rightleftharpoons H_3O^+(aq) + H_2PO_4^-(aq) \qquad K_{a1} = 7.5 \times 10^{-3}$$

Problem-Solving Tip 17.3

What Is the pH After Mixing an Acid and a Base?

Table 17.5 summarizes the outcome of mixing various types of acids and bases. But how do you calculate a numerical value for the pH, particularly in the case of mixing a weak acid with a strong base or a weak base with a strong acid? The strategy (Example 17.8) is to recognize that this involves two related calculations: a stoichiometry calculation and an equilibrium calculation. The key to this calculation is that you need to know the concentration of the weak acid or weak base produced when the acid and base are mixed. You should ask yourself the following questions:

(a) What amounts of acid and base are used (in moles)? If they are the same, then proceed to Step (b). [If not, you need

additional techniques as described in Chapter 18.] (This is a stoichiometry problem.)

(b) What is the total volume of the solution after mixing the acid and base solutions?

(c) What is the concentration of the weak acid or base produced on mixing the acid and base solutions?

(d) Using the concentration found in Step (c), what is the pH of the solution? (This is an equilibrium problem.)

Second ionization step:

$$H_2PO_4^-(aq) + H_2O(\ell) \rightleftharpoons H_3O^+(aq) + HPO_4^{2-}(aq) \qquad K_{a2} = 6.2 \times 10^{-8}$$

Third ionization step:

$$HPO_4^{2-}(aq) + H_2O(\ell) \rightleftharpoons H_3O^+(aq) + PO_4^{3-}(aq) \qquad K_{a3} = 3.6 \times 10^{-13}$$

Notice that the K_a values for each successive step are smaller and smaller because it is more difficult to remove H^+ from a negatively charged ion, such as $H_2PO_4^-$, than from a neutral molecule such as H_3PO_4. The larger the negative charge of the anionic acid, the more difficult it is to remove H^+.

For many inorganic acids, such as phosphoric acid, carbonic acid, and hydrogen sulfide, each successive loss of a proton is about 10^4–10^6 more difficult than the previous ionization step. This means that the first ionization step of a polyprotic acid produces up to about a million times more H_3O^+ than the second step. For this reason, *the pH of many inorganic polyprotic acids depends primarily on the hydronium ion generated in the first ionization step; the hydronium ion produced in the second step can be neglected.* The same principles apply to the conjugate bases of polyprotic acids. This is illustrated by the calculation of the pH of a solution of carbonate ion, an important base in our environment (Example 17.9).

Alpha-Hydroxy acid. Malic acid is a diprotic acid occurring in apples. It has an —OH group on the C atom next to the —CO₂H group (in the alpha position). It is one of a larger group of natural acids such as lactic acid, citric acid, and ascorbic acid. Alpha-hydroxy acids have been touted as an ingredient in "anti-aging" skin creams. They work by accelerating the natural process by which the skin replaces the outer layer of cells with new cells. *(Charles D. Winters)*

Example 17.9   **Calculating the pH of the Solution of a Polyprotic Base**

Problem • The carbonate ion, CO_3^{2-}, is a base in water, forming the hydrogen carbonate ion, which in turn can form carbonic acid (see Chapter Focus, page 692).

$$CO_3^{2-}(aq) + H_2O(\ell) \rightleftharpoons HCO_3^-(aq) + OH^-(aq)$$
$$K_{b1} = 2.1 \times 10^{-4}$$

$$HCO_3^-(aq) + H_2O(\ell) \rightleftharpoons H_2CO_3(aq) + OH^-(aq)$$
$$K_{b2} = 2.4 \times 10^{-8}$$

What is the pH of a 0.10 M solution of Na_2CO_3?

Strategy • The second ionization constant, K_{b2}, is much smaller than the first, K_{b1}, so the hydroxide ion concentration in the solution results almost entirely from the first step. Therefore, let us calculate the OH^- concentration produced in the first ionization step, but test the conclusion that OH^- produced in the second step is negligible.

Solution • Set up an ICE table for the reaction of the carbonate ion (Equilibrium Table 1).

(Example continues on next page)

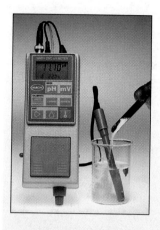

Sodium carbonate. This common substance is a base in aqueous solution. Its primary use is in the glass industry. Although it was once manufactured, it is now mined as the mineral trona, $Na_2CO_3 \cdot NaHCO_3 \cdot 2 H_2O$. *(Charles D. Winters)*

Using this value of $[OH^-]$ to calculate the pH, we find

$$pOH = -\log(4.6 \times 10^{-3}) = 2.34$$
$$pH = 14 - pOH = 11.66$$

Comment • It is instructive to ask what the concentration of H_2CO_3 in the solution might be. If HCO_3^- were to react significantly with water to produce H_2CO_3, the pH of the solution would be affected. Let us set up a second ICE table.

Equilibrium Table 2 • Reaction of HCO_3^- Ion

Equilibrium	$HCO_3^- + H_2O \rightleftharpoons H_2CO_3 + OH^-$		
Initial (M)	4.6×10^{-3}	0	4.6×10^{-3}
Change (M)	$-y$	$+y$	$+y$
Equilibrium (M)	$(4.6 \times 10^{-3} - y)$	y	$(4.6 \times 10^{-3} + y)$

Because K_{b2} is so small, the second step occurs to a *much* smaller extent than the first step. This means the amount of H_2CO_3 and OH^- produced in the second step ($= y$) is *much* smaller than 4.6×10^{-3} M. Therefore, it is reasonable that both $[HCO_3^-]$ and $[OH^-]$ are very close to 4.6×10^{-3} M.

$$K_{b2} = 2.4 \times 10^{-8} = \frac{[H_2CO_3][OH^-]}{[HCO_3^-]} = \frac{(y)(4.6 \times 10^{-3})}{4.6 \times 10^{-3}}$$

Because $[HCO_3^-]$ and $[OH^-]$ have nearly identical values, they cancel from the expression, and we find that $[H_2CO_3]$ is simply equal to K_{b2}.

$$y = [H_2CO_3] = K_{b2} = 2.4 \times 10^{-8} M$$

For the carbonate ion, where K_1 and K_2 differ by about 10^4, the hydroxide ion is essentially all produced in the first equilibrium process.

Equilibrium Table 1 • Reaction of CO_3^{2-} Ion

Equilibrium	$CO_3^{2-} + H_2O \rightleftharpoons HCO_3^- + OH^-$		
Initial (M)	0.10	0	0
Change (M)	$-x$	$+x$	$+x$
Equilibrium (M)	$(0.10 - x)$	x	x

Based on this table, the equilibrium concentration of OH^- ($= x$) can then be calculated.

$$K_{b1} = 2.1 \times 10^{-4} = \frac{[HCO_3^-][OH^-]}{[CO_3^{2-}]} = \frac{x^2}{0.10 - x}$$

Because K_{b1} is relatively small, it is reasonable to make the approximation that $(0.10 - x) \approx 0.10$. Therefore,

$$x = [HCO_3^-] = [OH^-] = \sqrt{(2.1 \times 10^{-4})(0.10)}$$
$$= 4.6 \times 10^{-3} M$$

Exercise 17.17 Calculating the pH of the Solution of a Polyprotic Acid

What is the pH of a 0.10 M solution of oxalic acid, $H_2C_2O_4$? What are the concentrations of H_3O^+, $HC_2O_4^-$, and the oxalate ion, $C_2O_4^{2-}$?

Chapter Goals • Revisited

• **Use the Brønsted–Lowry and Lewis concepts of acids and bases.**

• Apply the principles of chemical equilibrium to acids and bases in aqueous solution.

• Predict the outcome of reactions of acids and bases.

• Understand the influence of structure and bonding on acid–base properties.

17.9 LEWIS CONCEPT OF ACIDS AND BASES

The concept of acid–base behavior advanced by Brønsted and Lowry in the 1920s works well for reactions involving proton transfer. A more general concept, however, was developed by Gilbert N. Lewis in the 1930s [CD-ROM, Screens 17.12–17.14]. This concept is based on the sharing of electron pairs between acid and base. A **Lewis acid** is a substance that can accept a pair of electron from another atom to form a new bond, and a **Lewis base** is a substance that can donate a pair of electrons to another atom to form a new bond. This means that an acid–base reaction in the Lewis sense occurs when a molecule (or ion) donates a pair of electrons to another molecule (or ion).

$$A + B: \longrightarrow B:A$$

acid   base   adduct

The product is often called an **acid–base adduct.** In Section 9.6 this type of chemical bond was called a *coordinate covalent bond.*

Formation of a hydronium ion from H^+ and water is a good example of a Lewis acid–base reaction. The H^+ ion has no electrons in its valence ($1s$) orbital, and the water molecule has two unshared pairs of electrons (located in sp^3 hybrid orbitals). One of the O atom lone pairs can be shared between an H^+ ion and a water molecule, thus forming an $O-H$ bond. A similar interaction occurs between H^+ and the base ammonia to form the ammonium ion.

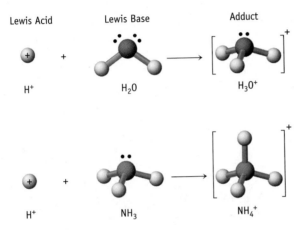

Such reactions are very common. In general, they involve Lewis acids that are cations or neutral molecules with an available, empty valence orbital and bases that are anions or neutral molecules with a lone electron pair.

Cationic Lewis Acids

All metal cations are known to interact with water molecules to form hydrated cations, ions in which the metal ion is surrounded by water molecules (Figure 17.6). In these species, coordinate covalent bonds form between the metal cation and a lone pair of electrons on the O atom of each water. For example, an iron(II) ion, Fe^{2+}, forms six coordinate covalent bonds to water.

$$Fe^{2+}(aq) + 6\ H_2O(\ell) \longrightarrow [Fe(H_2O)_6]^{2+}(aq)$$

Similar structures formed by transition metal cations are generally very colorful (Figure 17.6 and Section 22.3). Chemists call these **complex ions** or, because of the coordinate covalent bond, **coordination complexes.** Several are listed in Table 17.3 as acids, and their behavior is described further in Section 17.10 and Chapter 22.

Like water, ammonia is an excellent Lewis base and combines with metal cations to give adducts (complex ions), which are often very colorful. For example, copper(II) ions, light blue in aqueous solution (see Figure 17.6), react with ammonia to give a deep blue adduct with four ammonia molecules surrounding each Cu^{2+} ion (Figure 17.7).

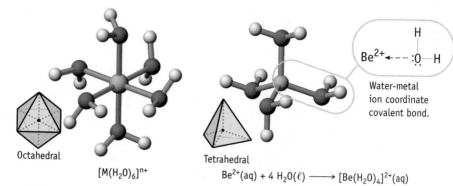

Octahedral

$[M(H_2O)_6]^{n+}$

Tetrahedral

Water-metal ion coordinate covalent bond.

$$Be^{2+}(aq) + 4 H_2O(\ell) \longrightarrow [Be(H_2O)_4]^{2+}(aq)$$

Figure 17.6   Metal cations in water. (a) Solutions of the nitrate salts of iron(III), nickel(II), and copper(II). All have characteristic colors (see Chapter 22). **(b)** Models of complex ions (Lewis acid–base adduct) formed between a metal cation and water molecules. Such complexes generally have six water molecules arranged octahedrally around the metal cation.   *(Charles D. Winters)*

$$Cu^{2+}(aq) + 4 NH_3(aq) \longrightarrow [Cu(NH_3)_4]^{2+}(aq)$$

light blue   deep blue

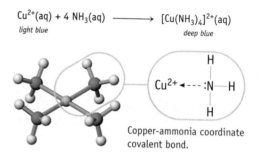

Copper-ammonia coordinate covalent bond.

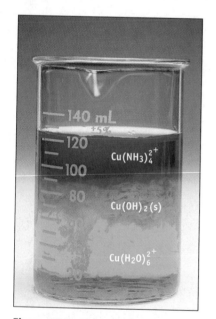

Figure 17.7   The Lewis acid–base complex ion $[Cu(NH_3)_4]^{2+}$. Here aqueous ammonia was added to aqueous $CuSO_4$ (the light blue solution at the bottom of the beaker). The small concentration of OH^- in $NH_3(aq)$ first formed insoluble blue-white $Cu(OH)_2$ (the solid in the middle of the beaker). With additional NH_3, however, the deep blue, soluble complex ion formed (the solution at the top of the beaker). *(Charles D. Winters)*

The hydroxide ion, OH^-, is also an excellent Lewis base and binds readily to metal cations to give metal hydroxides. An important feature of the chemistry of some metal hydroxides is that they are **amphoteric.** An amphoteric metal hydroxide can behave as an acid or a base (Table 17.6). One of the best examples of this behavior is aluminum hydroxide, $Al(OH)_3$ (Figure 17.8). Adding OH^- to a precipitate of $Al(OH)_3$ produces the water-soluble $[Al(OH)_4]^-$ ion.

$$Al(OH)_3(s) + OH^-(aq) \longrightarrow [Al(OH)_4]^-(aq)$$
acid   base

If acid is added to the $Al(OH)_3$ precipitate, it dissolves. This time, however, the aluminum hydroxide is acting as a base.

$$Al(OH)_3(s) + 3 H_3O^+(aq) \longrightarrow Al^{3+}(aq) + 6 H_2O(\ell)$$
base   acid

Molecular Lewis Acids

Lewis's acid–base concept accounts nicely for the fact that oxides of nonmetals behave as acids (◀ SECTION 5.3). Two important examples of acidic oxides are carbon dioxide and sulfur dioxide.

Table 17.6 • Some Common Amphoteric Metal Hydroxides

Hydroxide	Reaction as a Base	Reaction as an Acid
$Al(OH)_3$	$Al(OH)_3(s) + 3\ H_3O^+(aq) \longrightarrow Al^{3+}(aq) + 6\ H_2O(\ell)$	$Al(OH)_3(s) + OH^-(aq) \longrightarrow [Al(OH)_4]^-(aq)$
$Zn(OH)_2$	$Zn(OH)_2(s) + 2\ H_3O^+(aq) \longrightarrow Zn^{2+}(aq) + 4\ H_2O(\ell)$	$Zn(OH)_2(s) + 2\ OH^-(aq) \longrightarrow [Zn(OH)_4]^{2-}(aq)$
$Sn(OH)_4$	$Sn(OH)_4(s) + 4\ H_3O^+(aq) \longrightarrow Sn^{4+}(aq) + 8\ H_2O(\ell)$	$Sn(OH)_4(s) + 2\ OH^-(aq) \longrightarrow [Sn(OH)_6]^{2-}(aq)$
$Cr(OH)_3$	$Cr(OH)_3(s) + 3\ H_3O^+(aq) \longrightarrow Cr^{3+}(aq) + 6\ H_2O(\ell)$	$Cr(OH)_3(s) + OH^-(aq) \longrightarrow [Cr(OH)_4]^-(aq)$

Because oxygen is electronegative, the C—O bonding electrons in CO_2 are polarized away from carbon and toward oxygen. This causes the carbon atom to be slightly positive, and it is this atom that the negatively charged Lewis base OH^- can attack to give, ultimately, the bicarbonate ion.

This reaction is the first step in the precipitation of $CaCO_3$ when CO_2 is bubbled into a solution of $Ca(OH)_2$ (page 692).

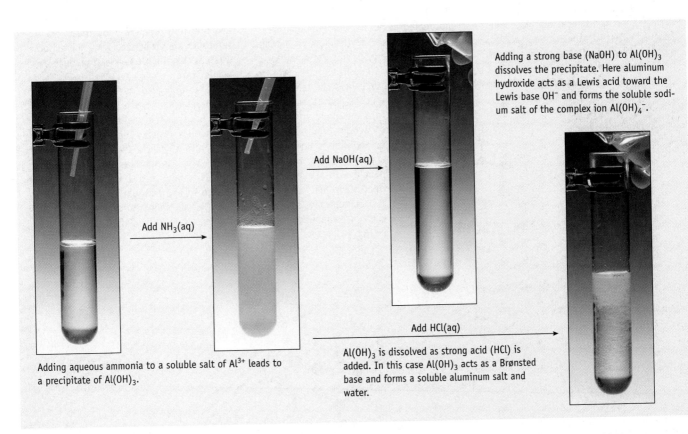

Adding a strong base (NaOH) to $Al(OH)_3$ dissolves the precipitate. Here aluminum hydroxide acts as a Lewis acid toward the Lewis base OH^- and forms the soluble sodium salt of the complex ion $Al(OH)_4^-$.

Add NaOH(aq)

Add NH₃(aq)

Add HCl(aq)

Adding aqueous ammonia to a soluble salt of Al^{3+} leads to a precipitate of $Al(OH)_3$.

$Al(OH)_3$ is dissolved as strong acid (HCl) is added. In this case $Al(OH)_3$ acts as a Brønsted base and forms a soluble aluminum salt and water.

Figure 17.8   The amphoteric nature of Al(OH)₃. *(Charles D. Winters)*

Exercise 17.18   **Lewis Acids and Bases**

Describe each of the following as a Lewis acid or a Lewis base.

(a) PH_3 **(c)** H_2S
(b) BCl_3 **(d)** HS^-

Hint: In each case draw the Lewis electron dot structure of the molecule or ion. Are there lone pairs of electrons on the central atom? If so, it can be a Lewis base. Does the central atom lack an electron pair? If so it can behave as a Lewis acid.

17.10   MOLECULAR STRUCTURE, BONDING, AND ACID–BASE BEHAVIOR

One of the most interesting aspects of chemistry is the correlation between molecular structure, bonding, and observed properties. Here it is useful to analyze the connection between the structure and bonding in some acids and their relative strengths.

Chemical Perspectives

Lewis and Brønsted Bases: Adrenaline and Serotonin

You are going to take a chemistry exam — your heart races and you begin to sweat. These actions of your nervous system are affected by a chemical compound *epinephrine*, which is also called by the alternative name *adrenaline*. The compound has a basic NH_2 group which is protonated at normal pHs. Epinephrine, which is produced in the body in a chain of reactions starting with the amino acid phenylalanine, is known as the "flight-or-fight" hormone. It causes the release of glucose and other nutrients into the blood and also stimulates brain function. Epinephrine is now used as a bronchodilator to treat people with asthma and to treat glaucoma.

Epinephrine is in a class of compounds called neurotransmitters. This class includes serotonin, another Lewis and Brønsted base. Very low levels of serotonin are associated with depression, whereas very high levels can produce a manic state.

Serotonin is derived from the amino acid tryptophan. Some people take tryptophan because they believe it makes them feel good and helps them sleep at night. Milk proteins have a high level of tryptophan, which is the reason you may enjoy a glass of milk or dish of ice cream before you go to bed at night.

(See J. Mann: *Murder, Magic, and Medicine*, New York, Oxford University Press, 1994.)

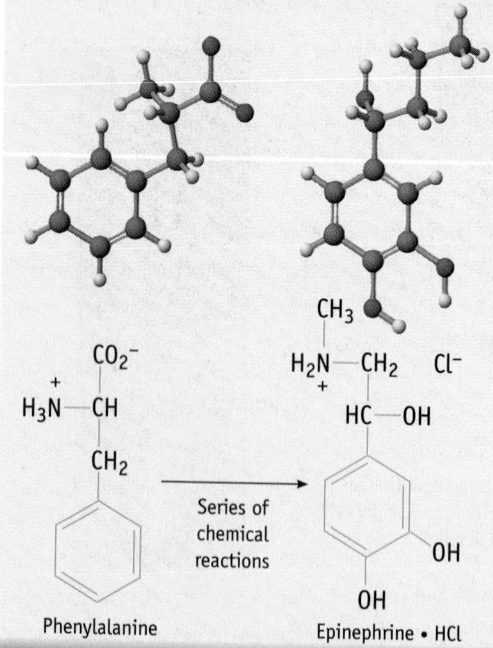

Phenylalanine

Epinephrine • HCl

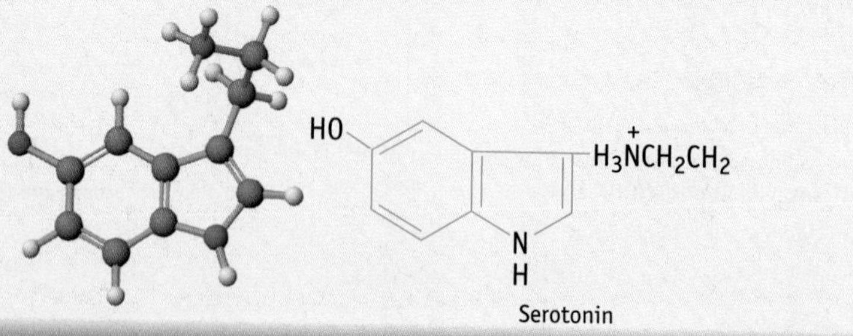

Serotonin

Why Is HF a Weak Acid but HCl is a Strong Acid?

Aqueous HF is a weak Brønsted acid in water, whereas the other hydrohalic acids — aqueous HCl, HBr, and HI — are all strong acids.

$$HX(aq) + H_2O(\ell) \longrightarrow H_3O^+(aq) + X^-(aq)$$

Experiments show that the acid strength increases in the order HF $\ll$ HCl $<$ HBr $<$ HI. The reason the strength of these acids increases on descending Group 7A involves several factors, among them the electron affinity of the halogen and the energy of solvation of the acid and the anion. The most significant factor determining acid strength, however, is the H—X bond energy.

	$\longrightarrow$ Increasing acid strength $\longrightarrow$			
	HF	**HCl**	**HBr**	**HI**
pK_a	+3.14	−7	−9	−10
H—X bond strength (kJ/mol)	565	432	366	299

The weakest acid, HF, has the strongest H—X bond, whereas the strongest acid, HI, has the weakest H—X bond.

Why Is HNO₂ a Weak Acid but HNO₃ Is a Strong Acid?

Nitrous acid (HNO₂) and nitric acid (HNO₃) are representative of several series of **oxoacids.** Oxoacids contain an atom (usually a nonmetal atom) bonded to one or more oxygen atoms, some with hydrogen atoms attached. Besides those based on N, you are familiar with the sulfur- and chlorine-based oxoacids (Table 17.7). In all these series of related compounds, the acid strength increases as the number of oxygen atoms bonded to the central element increases.

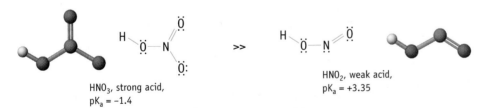

HNO₃, strong acid, pK_a = −1.4   >>   HNO₂, weak acid, pK_a = +3.35

Thus, nitric acid (HNO₃) is a stronger acid than nitrous acid (HNO₂), and the order of acid strength for the chlorine-based oxoacids is HOCl $<$ HOClO $<$ HOClO₂ $<$ HOClO₃ (see Table 17.7).

Acid strength is determined by the extent to which the ionization reaction is favored, that is, the extent to which the reaction of an acid HX to form H⁺ and the conjugate base (X⁻) is product-favored. It is necessary to assume that acid strength will be related to characteristics of *both* the reactant (HX) *and* the products (H⁺ and X⁻).

In an oxoacid molecule, attention focuses on the O—H bond and the influence of other atoms in the molecule on this bond. Because of the difference in electronegativities of O and H, the O—H bond is polar ($O^{\delta-}$—$H^{\delta+}$), but how do the other atoms in the molecule affect the O—H bond polarity? One view is that electrons in the O—H bond are attracted by other electronegative atoms or groups in the molecule and are drawn away from the O—H hydrogen. This in-

Chapter Goals • Revisited

- Use the Brønsted–Lowry and Lewis concepts of acids and bases.
- Apply the principles of chemical equilibrium to acids and bases in aqueous solution.
- Predict the outcome of reactions of acids and bases.
- **Understand the influence of structure and bonding on acid–base properties.**

• **pK_a Values for Hydrogen Halides**
The acids HCl, HBr, and HI have *negative* pK_a values. A negative pK_a indicates a K_a value greater than 1. The more negative the value the larger the value of K_a and the stronger the acid.

Table 17.7 • Oxoacids

Acid	pK_a
Cl-Based Oxoacids	
HOCl	7.46
HOClO (HClO₂)	$\approx$ 2
HOClO₂ (HClO₃)	$\approx$ −1
HOClO₃ (HClO₄)	$\approx$ −10
S-Based Oxoacids	
(HO)₂SO [H₂SO₃]	1.92, 7.21
(HO)₂SO₂ [H₂SO₄]	$\approx$ −3, 1.92

Linus Pauling stated for oxoacids with the general formula (HO)$_n$E(O)$_m$, the value of pK_a is about 8−5m. When $n > 1$, the pK_a increases by about 5 for each successive loss of a proton.

creases the O—H bond polarity and makes hydrogen easier to separate from the molecule as H^+.

The extent to which adjacent atoms or groups of atoms attract electrons from another part of a molecule is called the **inductive effect,** the attraction of electrons from adjacent bonds by more electronegative atoms. Inductive effects are used to explain many properties of molecules. This analysis, related to acid strength, is only one example. For nitric and nitrous acid we are comparing the ability of an NO group (in HONO) and an NO_2 group (in $HONO_2$) to attract electrons and increase the polarity of the O—H bond. Two oxygen atoms are bonded to the central nitrogen atom in the NO_2 group, and only one oxygen is bonded to the nitrogen atom in the NO group. By attaching more electronegative oxygen atoms to nitrogen, the inductive effect is greater, and the O—H bond becomes more polar. Thus, the H atom in the O—H group is more positive in HNO_3 than in HNO_2, and HNO_3 is a stronger acid

Hydrogen bond to H atom leads to O—H bond breaking in HNO_3.

Electrons in H—O bond flow toward electronegative O atoms owing to inductive effect.

Additional oxygen atoms in an oxoacid also have the effect of stabilizing the anion formed by removal of H^+ from the oxoacid. This added stability arises because the negative charge on the anion can be dispersed over more atoms. In nitrate ion, for example, the negative charge is shared equally over the three oxygen atoms. This is represented symbolically in the three resonance structures for this ion.

In nitrite ion, only two atoms share the negative charge. Greater stabilization of the products formed by ionizing the acid contributes to increased acidity.

In summary, a molecule can behave as a Brønsted acid if electronegative atoms increase the polarization of the O—H bond. In addition, the anion created by loss of H^+ should be stable and able to accommodate the negative charge. These conditions are promoted by

- The presence of electronegative atoms attached to the central atom
- The possibility of resonance structures for the anion, which lead to delocalization of the negative charge over the anion and thus to a stable ion

Why Are Carboxylic Acids Brønsted Acids?

Another important question is why carboxylic acids (such as, acetic acid, CH_3CO_2H) are Brønsted acids and what is the source of the ionized H atom in these acids. The arguments used to explain acidity of oxoacids can also be applied to carboxylic

● **Anion Solvation**
The solvation of the anion is a further contributing factor in determining the relative strength of an acid.

acids. The O—H bond in these compounds is polar, a prerequisite for ionization.

C—H bonds
not broken
in water

Polar O—H bond broken by
interaction of positively charged H
atom with hydrogen-bonded H₂O

In addition, carboxylate anions are stabilized by delocalizing the negative charge over the two oxygen atoms.

The simple carboxylic acids, RCO_2H in which R is an alkyl group [◄ SECTION 11.4], do not differ markedly in acid strength (compare acetic acid, $pK_a = 4.74$, and propanoic acid, $pK_a = 4.89$ [see Table 17.3]. The acidity of carboxylic acids is enhanced, however, if electronegative substituents replace the hydrogens in the alkyl group. Compare for example, the pK_a values of a series of acetic acids in which hydrogen is replaced sequentially by the more electronegative element chlorine.

Acid		pK_a Value
CH_3CO_2H	acetic acid	4.74
$ClCH_2CO_2H$	chloroacetic acid	2.85
Cl_2CHCO_2H	dichloroacetic acid	1.49
Cl_3CCO_2H	trichloroacetic acid	0.7

increasing acid strength

We can again rationalize the trend in acidity based on the inductive effect of increasingly substituting electronegative Cl atoms for H atoms.

Finally, why are the C—H hydrogens of carboxylic acids not dissociated as H^+ instead of (or in addition to) the O—H hydrogen atom? Recall that the stability of the product ion is an important part of promoting ionization. In carboxylic acids the C atom is not sufficiently electronegative to accommodate the negative charge left if the bond breaks as $C—H \longrightarrow C:^- + H^+$ (see Figure 17.9).

Why Are Hydrated Metal Cations Brønsted Acids?

When a coordinate covalent bond is formed between a metal cation (a Lewis acid) and a water molecule (a Lewis base), the positive charge of the metal ion and its small size means that the electrons of the $H_2O—M^{n+}$ bond are very strongly attracted to the metal. As a result of this inductive effect, the O—H bonds of the bound water molecules are polarized, just as in oxoacids and carboxylic acids. The net effect is that an H atom of a coordinated water molecule is removed as H^+ more readily than in an uncoordinated water molecule. Thus,

(a) Lewis electron dot structure of acetic acid, CH_3CO_2H

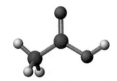

(b) Ball and stick model of acetic acid

C atom with partial positive charge

O atoms with partial negative charge

Strongly polarized O–H bond

Note that the H atoms of the CH_3 group have a very small charge

(c) Computer calculated partial charges on atoms of CH_3CO_2H

Figure 17.9   Acetic acid, a carboxylic acid. The model in part (c) was generated by computer to show the partial + and − charges on the H, C, and O atoms. Bonding electrons in the molecule flow toward the O atoms, away from the H atom, leaving the H atom positively charged and readily removed by interaction with a polar water molecule. In the computer-generated picture in (c), atoms with a positive charge are red and those with a negative charge are yellow. The relative size of the partial charge is reflected by the relative size of the red or yellow sphere.

● **Polarization of O—H Bonds**
Water molecules attached to a metal cation have strongly polarized O—H bonds.

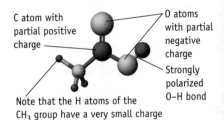

a hydrated metal cation functions as a Brønsted acid or proton donor (see Figure 17.5).

$$[Cu(H_2O)_6]^{2+} + H_2O(\ell) \rightleftharpoons [Cu(H_2O)_5(OH)]^+(aq) + H_3O^+(aq)$$

The inductive effect of a metal ion increases with increasing charge. Consulting Table 17.3, you see that the Brønsted acidity of +3 ions (for example hydrated Al^{3+} and Fe^{3+}) is greater than for +2 cations (hydrated Cu^{2+}, Pb^{2+}, Co^{2+}, Fe^{2+}, Ni^{2+}). Ions with a single positive charge such as Na^+ and K^+ are not acidic.

Why Are Anions Brønsted Bases?

Anions, particularly oxoanions such as PO_4^{3-}, are Brønsted bases. The negatively charged anion interacts with the positively charged H atom of a polar water molecule and a H^+ ion is transferred to the anion.

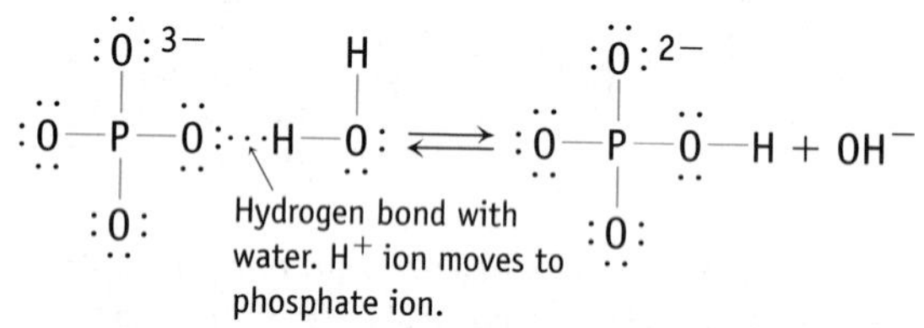

The data in Table 17.8 show that, in a series of related anions, the basicity of an anionic base increases as the negative charge of the anion increases.

Why Are Organic Amines Brønsted and Lewis Bases?

Ammonia is the parent compound of an enormous number of compounds that behave as Brønsted and Lewis bases. These molecules have an N atom surrounded by three other atoms as well as a lone pair of electrons.

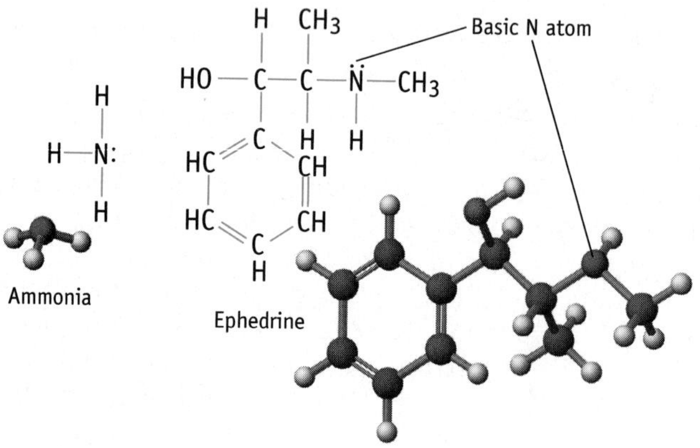

Ammonia

Ephedrine

In each case, the positively charged H atom of a polar water molecule can interact with the lone pair of electrons on the electronegative N atom. An H^+ ion transfers from water to the nitrogen atom, and an OH^- ion enters the solution.

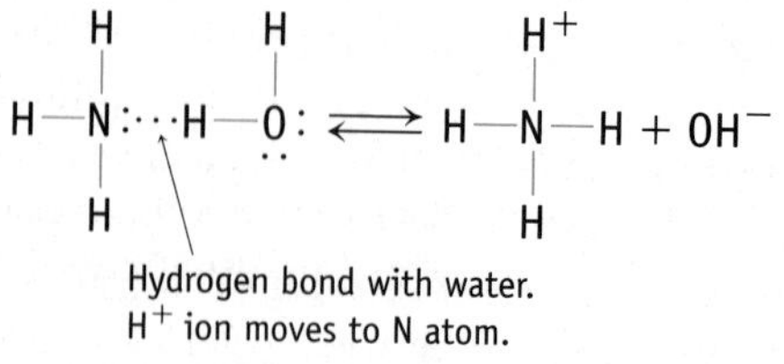

Table 17.8 • Basic Oxoanions

Anion	pK_b
PO_4^{3-}	1.55
HPO_4^{2-}	6.80
$H_2PO_4^-$	11.89
CO_3^{2-}	3.68
HCO_3^-	7.62
SO_3^{2-}	6.80
HSO_3^-	12.08

• **Ephedra**

Ma Huang, an extract from the *Ephedra* species of plants, contains ephedrine. The Chinese have used it for over 5000 years to treat asthma. Recently, however, the substance has been used in diet pills that can be purchased over the counter in herbal medicine shops. As this book was being written *very* serious concerns about these pills have been raised as there have been reports of serious heart problems. In fact, ephedra has been banned by the National Football League in the United States.

Exercise 17.19 Molecular Structure, Acids, and Bases

(a) Which is the stronger acid, H_2TeO_3 or H_2TeO_4?

(b) Which is the stronger acid, $Fe(H_2O)_6^{2+}$ or $Fe(H_2O)_6^{3+}$?

(c) Which is the stronger acid, $HOCl$ or $HOBr$?

(d) The molecule whose structure is illustrated here is amphetamine, a stimulant. Is the compound a Brønsted acid, a Lewis acid, a Brønsted base, a Lewis base, or some combination of these?

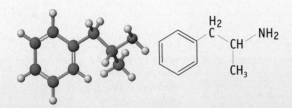

In Summary

When you have finished studying this chapter, ask if you have met the goals of the chapter. In particular, you should be able to

- Define and use the Brønsted concept of acids and bases (Sections 17.1 and 17.2).

- Recognize common monoprotic and polyprotic acids and bases and write balanced equations for their ionization in water (Section 17.2).

- Appreciate when a substance can be amphiprotic (Section 17.2).

- Recognize the Brønsted acid and base in a reaction and identify the conjugate partner of each (Section 17.2).

- Understand the concept of water autoionization and its role in Brønsted acid–base chemistry. Use the water ionization constant, K_w (Section 17.3).

- Use the pH concept (Section 5.9).

- Identify common strong acids and bases (Table 17.3).

- Recognize some common weak acids, including neutral molecules (such as acetic acid), cations (such as NH_4^+ or hydrated metal ions such as $Fe(H_2O)_6^{2+}$) and anions (such as HCO_3^-) (Table 17.3).

- Write equilibrium constant expressions for weak acids and bases (Section 17.4).

- Calculate pK_a from K_a (or K_a from pK_a) and understand how pK_a is correlated with acid strength (Section 17.4).

- Understand the relation of K_a for a weak acid to K_b for its conjugate base (Section 17.4).

- Write acid–base reactions and decide whether they are product- or reactant-favored (Section 17.5 and Table 17.5).

- Recognize the types of acid–base reactions and describe their result (Section 17.6).

- Calculate the equilibrium constant for a weak acid (K_a) or weak base (K_b) from experimental information (such as pH, $[H_3O^+]$ or $[OH^-]$) (Section 17.7 and Example 17.4).

- Use the equilibrium constant and other information to calculate the pH of a solution of a weak acid or weak base (Section 17.7 and Examples 17.5, 17.6, and 17.7).

- Describe the acid–base properties of salts and calculate the pH of a solution of a salt of a weak acid or of a weak base (Section 17.7 and Example 17.7).

- Calculate the pH after an acid–base reaction (Section 17.7 and Example 17.8).
- Calculate the pH of a solution of a polyprotic acid or base (Section 17.8 and Example 17.9).
- Characterize a compound as a Lewis base (an electron pair donor) or Lewis acid (an electron pair acceptor) (Section 17.9).
- Appreciate the connection between the structure of a compound and its acidity or basicity (Section 17.10).

Key Terms

Section 17.1
ionization constant

Section 17.2
Brønsted acid
Brønsted base
monoprotic acid
polyprotic acid
polyprotic base

amphiprotic
conjugate acid–base pair

Section 17.3
autoionization
water ionization constant K_w
pH

Section 17.9
Lewis acid

Lewis base
acid–base adduct
complex ion
coordination complex
amphoteric

Section 17.10
oxoacids
inductive effect

Key Equations

Equation 17.1 (page 698)
Water ionization constant

$$K_w = [H_3O^+][OH^-] = 1.0 \times 10^{-14} \quad \text{at } 25\ °C$$

Equation 17.2 (page 699)
Definition of pH (see also Equation 5.2)

$$pH = -\log[H_3O^+]$$

Equation 17.3 (page 699)
Definition of pOH

$$pOH = -\log[OH^-]$$

Equation 17.4 (page 699)
Relate pK_w to pH and pOH.

$$pK_w = 14.00 = pH + pOH$$

Equation 17.5 (page 701)
Equilibrium expression for a weak acid, HA, in water

$$HA(aq) + H_2O(\ell) \rightleftharpoons H_3O^+(aq) + A^-(aq)$$

$$K_a = \frac{[H_3O^+][A^-]}{[HA]}$$

Equation 17.6 (page 701)
Equilibrium expression for a weak base, B, in water

$$B(aq) + H_2O(\ell) \rightleftharpoons BH^+(aq) + OH^-(aq)$$

$$K_b = \frac{[BH^+][OH^-]}{[B]}$$

Equation 17.7 (page 707)
Definition of pK_a

$$pK_a = -\log K_a$$

Equation 17.8 (page 708)
Relationship of K_a, K_b, and K_w

$$K_a K_b = K_w$$

Study Questions

Questions with blue, bold-faced numbers have answers in Appendix O. See the descriptions, exercises, tutorials, and simulations in Chapter 17 of the General Chemistry Interactive CD-ROM, *Version 3.0.*

Reviewing Important Concepts

1. Why can water be both a Brønsted base and a Lewis base? Can water be a Brønsted acid? A Lewis acid?

2. Define the term "amphoteric." What are the requirements for a substance to be amphoteric?

3. The nickel(II) ion exists as $[Ni(H_2O)_6]^{2+}$ in aqueous solution. Why is such a solution acidic? As part of your an-

swer include a balanced equation to depict what happens with $[Ni(H_2O)_6]^{2+}$ when it interacts with water.

4. Describe an experiment that will allow you to place the following three bases in order of increasing base strength: NaCN, CH_3NH_2, and Na_2CO_3.

5. The data in the table compare the strength of acetic acid with a related series of acids where the H atoms of the CH_3 group in acetic acid are successively replaced by Cl.

Acid	K_a
CH_3CO_2H	1.8×10^{-5}
$ClCH_2CO_2H$	1.4×10^{-3}
Cl_2CHCO_2H	3.3×10^{-2}
Cl_3CCO_2H	2.0×10^{-1}

(a) What trend in acid strength do you observe as H is successively replaced by Cl? Can you suggest a reason for this trend?

(b) Suppose each of the acids in the table was present as a 0.10 M aqueous solution. Which would have the highest pH? The lowest pH?

6. Which should be the stronger acid and why: H_2SeO_4 or H_2SeO_3? Describe an experiment by which you could confirm your prediction.

7. Perchloric acid behaves as an acid even when it is dissolved in concentrated sulfuric acid.

(a) Write a balanced equation showing how perchloric acid can transfer a proton to sulfuric acid.

(b) Draw a Lewis electron dot structure for sulfuric acid. How can sulfuric acid function as a base?

8. You purchase a bottle of water. On checking its pH you find it is not neutral as you might have expected. Instead, it is slightly acidic? Why?

9. You have three solutions labeled A, B, and C. You only know that each contains a different cation, Na^+, NH_4^+, or H^+. Each has an anion that does not contribute to solution pH (for example, Cl^-). You also have two other solutions, Y and Z, each containing a different anion, Cl^- or OH^-, with a cation that does not influence solution pH (such as, K^+). If equal amounts of B and Y are mixed, the result is an acidic solution. Mixing A and Z gives a neutral solution, whereas B and Z give a basic solution. Identify the five unknown solutions. (Adapted from D. H. Barouch: *Voyages in Conceptual Chemistry*, Boston, Jones and Bartlett, 1997.)

	Y	Z
A		neutral
B	acidic	basic
C		

Practicing Skills

The Brønsted Concept
(See Exercises 17.1 and 17.2 and CD-ROM Screen 17.2)

10. Write the formula and give the name of the conjugate base of each of the following acids:

(a) HCN   (b) HSO_4^- (c) HF

11. Write the formula and give the name of the conjugate acid of each of the following bases:

(a) NH_3 (b) HCO_3^- (c) Br^-

12. What are the products of each of the following acid–base reactions? Indicate the acid and its conjugate base and the base and its conjugate acid.

(a) $HNO_3 + H_2O \longrightarrow$

(b) $HSO_4^- + H_2O \longrightarrow$

(c) $H_3O^+ + F^- \longrightarrow$

13. What are the products of each of the following acid–base reactions? Indicate the acid and its conjugate base and the base and its conjugate acid.

(a) $HClO_4 + H_2O \longrightarrow$

(b) $NH_4^+ + H_2O \longrightarrow$

(c) $HCO_3^- + OH^- \longrightarrow$

14. Write balanced equations showing how the hydrogen oxalate ion, $HC_2O_4^-$, can be both a Brønsted acid and a Brønsted base.

15. Write balanced equations showing how the HPO_4^{2-} ion of sodium monohydrogen phosphate, Na_2HPO_4, can be a Brønsted acid or a Brønsted base.

16. In each of the following acid–base reactions, identify the Brønsted acid and base on the left and their conjugate partners on the right.

(a) $HCO_2H(aq) + H_2O(\ell) \rightleftharpoons HCO_2^-(aq) + H_3O^+(aq)$

(b) $NH_3(aq) + H_2S(aq) \rightleftharpoons NH_4^+(aq) + HS^-(aq)$

(c) $HSO_4^-(aq) + OH^-(aq) \rightleftharpoons SO_4^{2-}(aq) + H_2O(\ell)$

17. In each of the following acid–base reactions, identify the Brønsted acid and base on the left and their conjugate partners on the right.

(a) $C_5H_5N(aq) + CH_3CO_2H(aq) \rightleftharpoons$
$C_5H_5NH^+(aq) + CH_3CO_2^-(aq)$

(b) $N_2H_4(aq) + HSO_4^-(aq) \rightleftharpoons N_2H_5^+(aq) + SO_4^{2-}(aq)$

(c) $[Al(H_2O)_6]^{3+}(aq) + OH^-(aq) \rightleftharpoons$
$[Al(H_2O)_5OH]^{2+}(aq) + H_2O(\ell)$

pH Calculations
(See Examples 5.11 and 17.1, Exercise 17.4; CD-ROM Screens 5.17 and 17.3–17.4)

18. An aqueous solution has a pH of 3.75. What is the hydronium ion concentration of the solution? Is it acidic or basic?

19. A saturated solution of milk of magnesia, $Mg(OH)_2$, has a pH of 10.52. What is the hydronium ion concentration of the solution? What is the hydroxide ion concentration? Is the solution acidic or basic?

20. What is the pH of a 0.0075 M solution of HCl? What is the hydroxide ion concentration of the solution?

21. What is the pH of a 1.2×10^{-4} M solution of KOH? What is the hydronium ion concentration of the solution?

22. What is the pH of a 0.0015 M solution of $Ba(OH)_2$?

23. The pH of a solution of $Ba(OH)_2$ is 10.66 at 25 °C. What is the hydroxide ion concentration in the solution? If the solution volume is 125 mL, what mass of $Ba(OH)_2$ must have been dissolved?

Equilibrium Constants for Acids and Bases
(See Exercise 17.5, Example 17.2, and CD-ROM Screen 17.6)

24. Several acids are listed here with their respective equilibrium constants:

$$C_6H_5OH(aq) + H_2O(\ell) \rightleftharpoons H_3O^+(aq) + C_6H_5O^-(aq)$$
$$K_a = 1.3 \times 10^{-10}$$

$$HCO_2H(aq) + H_2O(\ell) \rightleftharpoons H_3O^+(aq) + HCO_2^-(aq)$$
$$K_a = 1.8 \times 10^{-4}$$

$$HC_2O_4^-(aq) + H_2O(\ell) \rightleftharpoons H_3O^+(aq) + C_2O_4^{2-}(aq)$$
$$K_a = 6.4 \times 10^{-5}$$

(a) Which is the strongest acid? Which is the weakest?

(b) Which acid has the weakest conjugate base?

(c) Which acid has the strongest conjugate base?

25. Several acids are listed here with their respective equilibrium constants.

$$HF(aq) + H_2O(\ell) \rightleftharpoons H_3O^+(aq) + F^-(aq)$$
$$K_a = 7.2 \times 10^{-4}$$

$$HPO_4^{2-}(aq) + H_2O(\ell) \rightleftharpoons H_3O^+(aq) + PO_4^{3-}(aq)$$
$$K_a = 3.6 \times 10^{-13}$$

$$CH_3CO_2H(aq) + H_2O(\ell) \rightleftharpoons H_3O^+(aq) + CH_3CO_2^-(aq)$$
$$K_a = 1.8 \times 10^{-5}$$

(a) Which is the strongest acid? Which is the weakest?

(b) What is the conjugate base of the acid HF?

(c) Which acid has the weakest conjugate base?

(d) Which acid has the strongest conjugate base?

26. State which of the following ions or compounds has the strongest conjugate base and briefly explain your choice: (a) HSO_4^-, (b) CH_3CO_2H, and (c) HOCl.

27. Which of the following compounds or ions has the strongest conjugate acid? Briefly explain your choice: (a) CN^-, (b) NH_3, (c) SO_4^{2-}.

28. Dissolving K_2CO_3 in water gives a basic solution. Write a balanced equation showing how the carbonate ion is responsible for this effect.

29. Dissolving ammonium bromide in water gives an acidic solution. Write a balanced equation showing how this can occur.

30. If each of the salts listed here were dissolved in water to give a 0.10 M solution, which solution would have the highest pH? Which would have the lowest pH?

(a) Na_2S (d) NaF

(b) Na_3PO_4 (e) $NaCH_3CO_2$

(c) NaH_2PO_4 (f) $AlCl_3$

31. Which of the following common food additives would give a basic solution when dissolved in water?

(a) $NaNO_3$ (used as a meat preservative)

(b) $NaC_6H_5CO_2$ (sodium benzoate; used as a soft drink preservative)

(c) Na_2HPO_4 (used as an emulsifier in the manufacture of pasteurized Kraft cheese)

pK_a, A Logarithmic Scale of Acid Strength
(See Exercise 17.7 and CD-ROM Screen 17.6)

32. A weak acid has a K_a of 6.5×10^{-5}. What is the value of pK_a for the acid?

33. If K_a for a weak acid is 2.4×10^{-11}, what is the value of pK_a?

34. Epinephrine (page 724) has a pK_a value of 9.53. What is the value of K_a? Where does the acid fit in Table 17.3?

35. An organic acid has $pK_a = 8.95$. What is its K_a value? Where does the acid fit in Table 17.3?

36. Which is the stronger of the two acids below?

(a) Benzoic acid, $C_6H_5CO_2H$, $pK_a = 4.20$

(b) 2-Chlorobenzoic acid, $ClC_6H_4CO_2H$, $pK_a = 2.88$

37. Which is the stronger of the two acids below?

(a) Acetic acid, CH_3CO_2H, $K_a = 1.8 \times 10^{-5}$

(b) Chloroacetic acid, $ClCH_2CO_2H$, $pK_a = 2.87$

Ionization Constants for Weak Acids and Their Conjugate Bases
(See Exercise 17.8 and CD-ROM Screen 17.8)

38. Chloroacetic acid ($ClCH_2CO_2H$) has $K_a = 1.36 \times 10^{-3}$. What is the value of K_b for the chloroacetate ion ($ClCH_2CO_2^-$)?

39. A weak base has $K_b = 1.5 \times 10^{-9}$. What is the value of K_a for the conjugate acid?

40. The trimethylammonium ion, $(CH_3)_3NH^+$, is the conjugate acid of the weak base trimethylamine, $(CH_3)_3N$. A chemical handbook gives 9.80 as the pK_a of $(CH_3)_3NH^+$. What is the value of K_b for $(CH_3)_3N$?

41. The chromium(III) ion in water, $Cr(H_2O)_6^{3+}$, is a weak acid with $pK_a = 3.95$. What is the value of K_b for its conjugate base, $Cr(H_2O)_5OH^{2+}$?

Predicting the Direction of Acid–Base Reactions
(See Example 17.3 and CD-ROM Screen 17.7)

42. Acetic acid and sodium hydrogen carbonate, $NaHCO_3$, are mixed in water. Write a balanced equation for the acid–base reaction that could, in principle, occur. Using Table 17.3, decide if the equilibrium lies predominantly to the right or left.

43. Ammonium chloride and sodium dihydrogen phosphate, NaH_2PO_4, are mixed in water. Write a balanced equation for the acid–base reaction that could, in principle, occur.

Using Table 17.3, decide if the equilibrium lies predominantly to the right or left.

44. For each of the following reactions, predict whether the equilibrium lies predominantly to the left or to the right. Explain your prediction briefly.

    (a) $NH_4^+(aq) + Br^-(aq) \rightleftharpoons NH_3(aq) + HBr(aq)$

    (b) $HPO_4^{2-}(aq) + CH_3CO_2^-(aq) \rightleftharpoons$
        $PO_4^{3-}(aq) + CH_3CO_2H(aq)$

    (c) $Fe(H_2O)_6^{3+}(aq) + HCO_3^-(aq) \rightleftharpoons$
        $Fe(H_2O)_5(OH)^{2+}(aq) + H_2CO_3(aq)$

45. For each of the following reactions, predict whether the equilibrium lies predominantly to the left or to the right. Explain your prediction briefly.

    (a) $H_2S(aq) + CO_3^{2-}(aq) \rightleftharpoons HS^-(aq) + HCO_3^-(aq)$

    (b) $HCN(aq) + SO_4^{2-}(aq) \rightleftharpoons CN^-(aq) + HSO_4^-(aq)$

    (c) $SO_4^{2-}(aq) + CH_3CO_2H(aq) \rightleftharpoons$
        $HSO_4^-(aq) + CH_3CO_2^-(aq)$

Types of Acid–Base Reactions
(See Exercise 17.11 and CD-ROM Screen 17.7)

46. Solutions with equal molar quantities of sodium hydroxide and sodium hydrogen phosphate (Na_2HPO_4) are mixed.

    (a) Write the balanced, net ionic equation for the acid–base reaction that can, in principle, occur.

    (b) Is the resulting solution acidic, basic, or neutral?

47. Solutions with equal molar quantities of hydrochloric acid and sodium hypochlorite (NaOCl) are mixed.

    (a) Write the balanced, net ionic equation for the acid–base reaction that occurs.

    (b) Is the resulting solution acidic, basic, or neutral?

48. Solutions with equal molar quantities of acetic acid and sodium hydrogen phosphate (Na_2HPO_4) are mixed.

    (a) Write a balanced, net ionic equation for an acid–base reaction that occurs.

    (b) Is the resulting solution acidic, basic, or neutral?

49. Solutions with equal molar quantities of ammonia and sodium dihydrogen phosphate (NaH_2PO_4) are mixed.

    (a) Write a balanced, net ionic equation for an acid–base reaction that occurs.

    (b) Is the resulting solution acidic, basic, or neutral?

Using pH to Calculate Ionization Constants
(See Example 17.4 and CD-ROM Screen 17.8)

50. A 0.015 M solution of hydrogen cyanate, HOCN, has a pH of 2.67.

    (a) What is the hydronium ion concentration in the solution?

    (b) What is the ionization constant, K_a, for the acid?

51. A 0.10 M solution of chloroacetic acid, $ClCH_2CO_2H$, has a pH of 1.95. Calculate K_a for the acid.

52. A 0.025 M solution of hydroxylamine has a pH of 9.11. What is the value of K_b for this weak base?

    $H_2NOH(aq) + H_2O(\ell) \rightleftharpoons H_3NOH^+(aq) + OH^-(aq)$

53. Methylamine, CH_3NH_2, is a weak base.

    $CH_3NH_2(aq) + H_2O(\ell) \rightleftharpoons CH_3NH_3^+(aq) + OH^-(aq)$

    If the pH of a 0.065 M solution of the amine is 11.70, what is the value of K_b?

54. A 2.5×10^{-3} M solution of an unknown acid has a pH of 3.80 at 25 °C.

    (a) What is the hydronium ion concentration of the solution?

    (b) Is the acid a strong acid, a moderately weak acid (K_a of about 10^{-5}), or a very weak acid (K_a of about 10^{-10})?

55. A 0.015 M solution of an unknown base has a pH of 10.09.

    (a) What are the hydronium and hydroxide ion concentrations of this solution?

    (b) Is the base a strong base, a moderately weak base (K_b of about 10^{-5}), or a very weak base (K_b of about 10^{-10})?

Using Ionization Constants
(See Examples 17.5–17.7; CD-ROM Screens 17.9–17.11)

56. What are the equilibrium concentrations of hydronium ion, acetate ion, and acetic acid in a 0.20 M aqueous solution of acetic acid?

57. The ionization constant of a very weak acid HA is 4.0×10^{-9}. Calculate the equilibrium concentrations of H_3O^+, A^-, and HA in a 0.040 M solution of the acid.

58. What are the equilibrium concentrations of H_3O^+, CN^-, and HCN in a 0.025 M solution of HCN? What is the pH of the solution?

59. Phenol (C_6H_5OH), commonly called carbolic acid, is a weak organic acid.

    $C_6H_5OH(aq) + H_2O(\ell) \rightleftharpoons C_6H_5O^-(aq) + H_3O^+(aq)$
    $K_a = 1.3 \times 10^{-10}$

    If you dissolve 0.195 g of the acid in enough water to make 125 mL of solution, what is the equilibrium hydronium ion concentration? What is the pH of the solution?

60. What are the equilibrium concentrations of NH_3, NH_4^+, and OH^- in a 0.15 M solution of ammonia? What is the pH of the solution?

61. A weak base has $K_b = 5.0 \times 10^{-4}$. Calculate the equilibrium concentrations of the base, its conjugate acid, and OH^- in a 0.15 M solution of the base.

62. The weak base methylamine, CH_3NH_2, has $K_b = 4.2 \times 10^{-4}$. It reacts with water according to the equation

    $CH_3NH_2(aq) + H_2O(\ell) \rightleftharpoons CH_3NH_3^+(aq) + OH^-(aq)$

    Calculate the equilibrium hydroxide ion concentration in a 0.25 M solution of the base. What are the pH and pOH of the solution?

63. Calculate the pH of a 0.12 M aqueous solution of the base aniline, $C_6H_5NH_2$ ($K_b = 4.0 \times 10^{-10}$).

    $C_6H_5NH_2(aq) + H_2O(\ell) \rightleftharpoons C_6H_5NH_3^+(aq) + OH^-(aq)$

64. Calculate the pH of a 0.0010 M aqueous solution of HF.

65. A solution of hydrofluoric acid, HF, had a pH of 2.30. Calculate the equilibrium concentrations of HF, F^-, and H_3O^+, and calculate the amount of HF originally dissolved per liter.

Acid–Base Properties of Salts
(See Example 17.7 and CD-ROM Screen 17.11)

66. Calculate the hydronium ion concentration and pH in a 0.20 M solution of ammonium chloride, NH_4Cl.

67. Calculate the hydronium ion concentration and pH for a 0.015 M solution of sodium formate, $NaHCO_2$.

68. Sodium cyanide is the salt of the weak acid HCN. Calculate the concentrations of H_3O^+, OH^-, HCN, and Na^+ in a solution prepared by dissolving 10.8 g of NaCN in enough water to make 5.00×10^2 mL of solution at 25 °C.

69. The sodium salt of propanoic acid, $NaCH_3CH_2CO_2$, is used as an antifungal agent by veterinarians. Calculate the equilibrium concentration of H_3O^+ and OH^-, and the pH, for a solution of 0.10 M $NaCH_3CH_2CO_2$.

pH After an Acid–Base Reaction
(See Example 17.8 and CD-ROM Screen 17.7)

70. Calculate the hydronium ion concentration and pH of the solution that results when 22.0 mL of 0.15 M acetic acid, CH_3CO_2H, is mixed with 22.0 mL of 0.15 M NaOH.

71. Calculate the hydronium ion concentration and the pH when 50.0 mL of 0.40 M NH_3 is mixed with 50.0 mL of 0.40 M HCl.

72. For each of the following cases, decide whether the pH is less than, equal to, or greater than 7.
    (a) Equal volumes of 0.10 M acetic acid, CH_3CO_2H, and 0.10 M KOH are mixed
    (b) 25 mL of 0.015 M NH_3 is mixed with 25 mL of 0.015 M HCl
    (c) 150 mL of 0.20 M HNO_3 is mixed with 75 mL of 0.40 M NaOH

73. For each of the following cases, decide whether the pH is less than, equal to, or greater than 7.
    (a) 25 mL of 0.45 M H_2SO_4 is mixed with 25 mL 0.90 M NaOH
    (b) 15 mL of 0.050 M formic acid, HCO_2H, is mixed with 15 mL of 0.050 M NaOH
    (c) 25 mL of 0.15 M $H_2C_2O_4$ (oxalic acid) is mixed with 25 mL of 0.30 M NaOH. (Both H^+ ions of oxalic acid are removed with NaOH.)

Polyprotic Acids and Bases
(See Example 17.9)

74. Sulfurous acid, H_2SO_3, is a weak acid capable of providing two H^+ ions.
    (a) What is the pH of a 0.45 M solution of H_2SO_3?
    (b) What is the equilibrium concentration of the sulfite ion, SO_3^{2-}, in the 0.45 M solution of H_2SO_3?

75. Ascorbic acid (vitamin C, $C_6H_8O_6$) is a diprotic acid ($K_{a1} = 6.8 \times 10^{-5}$ and $K_{a2} = 2.7 \times 10^{-12}$). What is the pH of a

solution that contains 5.0 mg of acid per milliliter of solution?

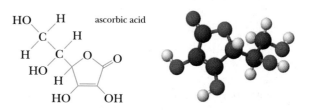

76. Hydrazine, N_2H_4, can interact with water in two steps.

$$N_2H_4(aq) + H_2O(\ell) \rightleftharpoons N_2H_5^+(aq) + OH^-(aq)$$
$$K_{b1} = 8.5 \times 10^{-7}$$
$$N_2H_5^+(aq) + H_2O(\ell) \rightleftharpoons N_2H_6^{2+}(aq) + OH^-(aq)$$
$$K_{b2} = 8.9 \times 10^{-16}$$

    (a) What are the concentrations of OH^-, $N_2H_5^+$, and $N_2H_6^{2+}$ in a 0.010 M aqueous solution of hydrazine?
    (b) What is the pH of the 0.010 M solution of hydrazine?

77. Ethylenediamine, $H_2NCH_2CH_2NH_2$, can interact with water in two steps, forming OH^- in each step (see Appendix I). If you have a 0.15 M aqueous solution of the amine, calculate the concentrations of $[H_3NCH_2CH_2NH_3]^{2+}$ and OH^-.

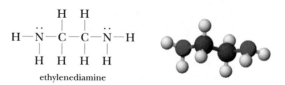

ethylenediamine

Lewis Acid and Bases
(See Exercise 17.18; CD-ROM Screens 17.12–17.14)

78. Decide if each of the following substances should be classified as a Lewis acid or base.
    (a) H_2NOH in the reaction

$$H_2NOH(aq) + HCl(aq) \longrightarrow [H_3NOH]Cl(aq)$$

    (b) Fe^{2+}
    (c) CH_3NH_2

79. Decide if each of the following substances should be classified as a Lewis acid or a Lewis base.
    (a) BCl_3
    (b) H_2NNH_2, hydrazine
    (c) The reactants in the reaction

$$Ag^+(aq) + 2\,NH_3(aq) \rightleftharpoons [Ag(NH_3)_2]^+(aq)$$

80. Carbon monoxide forms complexes with low-valent metals. For example, $Ni(CO)_4$ and $Fe(CO)_5$ are well known. CO also forms complexes with the iron(II) ion in hemoglobin, which prevents the hemoglobin from acting in its normal way. Is CO a Lewis acid or a Lewis base?

81. Trimethylamine, $(CH_3)_3N$, is a common reagent. It interacts readily with diborane gas, B_2H_6. The latter dissociates to BH_3, and this forms a complex with the amine, $(CH_3)_3N \rightarrow BH_3$. Is the BH_3 fragment a Lewis acid or a Lewis base?

Molecular Structure, Bonding, and Acid–Base Behavior
(See Section 17.10 and Exercise 17.19)

82. Which should be the stronger acid, HOCN or HCN? Explain briefly. (In HOCN, the H^+ ion is attached to the O atom of the OCN^- ion.)

83. Which should be the stronger Brønsted acid, $V(H_2O)_6^{2+}$ or $V(H_2O)_6^{3+}$?

84. Explain why benzenesulfonic acid is a Brønsted acid.

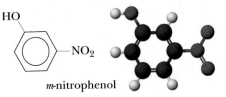

benzenesulfonic acid

85. The structure of ethylenediamine is illustrated in Question 77. Is this compound a Brønsted acid, a Brønsted base, a Lewis acid, or a Lewis base, or some combination of these?

General Questions on Acids and Bases

More challenging questions are marked with an underlined number.

86. About this time, you may be wishing you had an aspirin. Aspirin is an organic acid (page 449) with a K_a of 3.27×10^{-4} for the reaction

    $$HC_9H_7O_4(aq) + H_2O(\ell) \rightleftharpoons C_9H_7O_4^-(aq) + H_3O^+(aq)$$

    If you have two tablets, each having 0.325 g of aspirin (mixed with a neutral "binder" to hold the tablet together) and dissolve them in a glass of water (225 mL), what is the pH of the solution?

87. Consider the following ions: NH_4^+, CO_3^{2-}, Br^-, S^{2-}, and ClO_4^-.
    (a) Which of these ions might lead to an acidic solution and which can lead to a basic solution?
    (b) Which of these anions will have no effect on the pH of an aqueous solution?
    (c) Which ion is the strongest base?
    (d) Write a chemical equation for the reaction of each basic anion with water.

88. You have 0.010 M solutions of benzoic acid, $C_6H_5CO_2H$ (K_a = 6.3×10^{-5}) and 4-chlorobenzoic acid ($ClC_6H_4CO_2H$) ($K_a = 1.0 \times 10^{-4}$). Which solution will have the higher pH?

89. Place the following acids in order of (a) increasing strength and (b) increasing pH. Assume you have a 0.10 M solution of each acid.

    (a) 4-Chlorobenzoic acid, $ClC_6H_4CO_2H$, $K_a = 1.0 \times 10^{-4}$
    (b) Bromoacetic acid, $BrCH_2CO_2H$, $K_a = 1.3 \times 10^{-3}$
    (c) Trimethylammonium ion, $(CH_3)_3NH^+$, $K_a = 1.6 \times 10^{-10}$

90. Hydrogen sulfide, H_2S, and sodium acetate, $NaCH_3CO_2$, are mixed in water. Using Table 17.3, write a balanced equation for the acid–base reaction that could, in principle, occur. Does the equilibrium lie toward products or reactants?

91. For each of the following reactions, predict whether the equilibrium lies predominantly to the left or to the right. Explain your prediction briefly.
    (a) $HCO_3^-(aq) + SO_4^{2-}(aq) \rightleftharpoons CO_3^{2-}(aq) + HSO_4^-(aq)$
    (b) $HSO_4^-(aq) + CH_3CO_2^-(aq) \rightleftharpoons SO_4^{2-}(aq) + CH_3CO_2H(aq)$
    (c) $Co(H_2O)_6^{2+}(aq) + CH_3CO_2^-(aq) \rightleftharpoons Co(H_2O)_5(OH)^+(aq) + CH_3CO_2H(aq)$

92. A monoprotic acid HX has $K_a = 1.3 \times 10^{-3}$. Calculate the equilibrium concentration of HX and H_3O^+ and the pH for a 0.010 M solution of the acid.

93. The base $Ca(OH)_2$ is almost insoluble in water; only 0.50 g can be dissolved in 1.0 L of water at 25 °C. If the dissolved substance is completely dissociated into its constituent ions, what is the pH of a saturated solution?

94. *m*-Nitrophenol, a weak acid, can be used as a pH indicator because it is yellow at a pH above 8.6 and colorless at a pH below 6.8. If the pH of a 0.010 M solution of the compound is 3.44, calculate the pK_a of the compound.

HO

—NO_2

m-nitrophenol

95. The butylammonium ion, $C_4H_9NH_3^+$, has a K_a of 2.3×10^{-11}.

    $$C_4H_9NH_3^+(aq) + H_2O(\ell) \rightleftharpoons H_3O^+(aq) + C_4H_9NH_2(aq)$$

    (a) Calculate K_b for the conjugate base, $C_4H_9NH_2$ (butylamine).
    (b) Place the butylammonium ion and its conjugate base in Table 17.3. Name an acid weaker than $C_4H_9NH_3^+$ and a base stronger than $C_4H_9NH_2$.
    (c) What is the pH of a 0.015 M solution of the butylammonium ion?

96. The local anesthetic Novocain is the hydrogen chloride salt of an organic base, procaine (see page 704).

    $$C_{13}H_{20}N_2O_2(aq) + HCl(aq) \longrightarrow [HC_{13}H_{20}N_2O_2]^+Cl^-(aq)$$
    $$\text{procaine} \qquad\qquad\qquad \text{Novocain}$$

    The pK_a for Novocain is 8.85. What is the pH of a 0.0015 M solution of Novocain?

97. The anilinium ion, $C_6H_5NH_3^+$, is the conjugate acid of the weak base organic base aniline. If the anilinium ion has a pK_a of 4.60, what is the pH of a 0.080 M solution of anilinium hydrochloride, $[C_6H_5NH_3]Cl$?

98. The base ethylamine $(CH_3CH_2NH_2)$ has a K_b of 4.3×10^{-4}. A closely related base, ethanolamine $(HOCH_2CH_2NH_2)$, has a K_b of 3.2×10^{-5}.

    (a) Which of the two bases is stronger?

    (b) Calculate the pH of a 0.10 M solution of the stronger base.

99. Chloroacetic acid, $ClCH_2CO_2H$, is a moderately weak acid $(K_a = 1.40 \times 10^{-3})$. If you dissolve 94.5 mg of the acid in 125 mL of water, what is the pH of the solution?

100. Pyridine is a weak organic base $(K_b = 1.5 \times 10^{-9})$ and readily forms a salt with hydrochloric acid.

$$C_5H_5N(aq) + HCl(aq) \longrightarrow C_5H_5NH^+(aq) + Cl^-(aq)$$

    pyridine   pyridinium ion

What is the pH of a 0.025 M solution of pyridinium hydrochloride, $[C_5H_5NH^+]Cl^-$?

101. Saccharin $(HC_7H_4NO_3S)$ is a weak acid with a $pK_a = 2.32$ at 25 °C. It is used in the form of sodium saccharide, $NaC_7H_4NO_3S$. What is the pH of a 0.10 M solution of sodium saccharide at 25 °C?

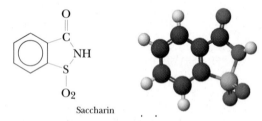

         Saccharin

102. For each of the following salts, predict whether an aqueous solution has a pH less than, equal to, or greater than 7.

    (a) $NaHSO_4$ (f) $NaNO_3$

    (b) NH_4Br (g) Na_2HPO_4

    (c) $KClO_4$ (h) $LiBr$

    (d) Na_2CO_3 (i) $FeCl_3$

    (e) $(NH_4)_2S$

103. Given the following solutions:

    (a) 0.1 M NH_3 (e) 0.1 M NH_4Cl

    (b) 0.1 M Na_2CO_3 (f) 0.1 M $NaCH_3CO_2$

    (c) 0.1 M $NaCl$ (g) 0.1 M $NH_4CH_3CO_2$

    (d) 0.1 M CH_3CO_2H

    (i) Which of the solutions are acidic?

    (ii) Which of the solutions are basic?

    (iii) Which of the solutions is most acidic?

104. Arrange the following 0.1 M solutions in order of increasing pH:

    (a) $NaCl$ (d) $NaCH_3CO_2$

    (b) NH_4Cl (e) KOH

    (c) HCl

105. Nicotinic acid, $C_6H_5NO_2$, is found in minute amounts in all living cells, but appreciable amounts occur in liver, yeast, milk, adrenal glands, white meat, and corn. Whole wheat flour contains about 60. µg/g of flour. One gram (1.00 g) of the acid dissolves in 60. mL of solution and gives a pH of 2.70. What is the approximate value of K_a for the acid?

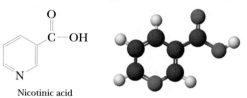

      Nicotinic acid

106. Oxalic acid is a relatively weak acid capable of losing two protons. Calculate the equilibrium constant for the overall reaction from K_{a1} and K_{a2}. (See Appendix H for the required K_a values.)

$$H_2C_2O_4(aq) + 2\,H_2O(\ell) \rightleftharpoons C_2O_4{}^{2-}(aq) + 2\,H_3O^+(aq)$$

107. Nicotine, $C_{10}H_{14}N_2$, has two basic nitrogen atoms (page 704), and both can react with water to give a basic solution.

$$Nic(aq) + H_2O(\ell) \rightleftharpoons NicH^+(aq) + OH^-(aq)$$
$$NicH^+(aq) + H_2O(\ell) \rightleftharpoons NicH_2{}^{2+}(aq) + OH^-$$

K_{b1} is 7.0×10^{-7} and K_{b2} is 1.1×10^{-10}. Calculate the approximate pH of a 0.020 M solution.

108. A hydrogen atom in the organic base pyridine, C_5H_5N, can be substituted by various atoms or groups to give XC_5H_4N, where X is an atom such as Cl or a group such as CH_3. The following table gives K_a values for the conjugate acids of a variety of substituted pyridines.

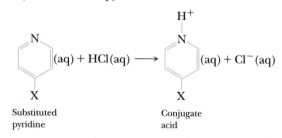

Substituted   Conjugate
pyridine   acid

Atom or Group X	K_a of Conjugate Acid
NO_2	5.9×10^{-2}
Cl	1.5×10^{-4}
H	6.8×10^{-6}
CH_3	1.0×10^{-6}

(a) Suppose each conjugate acid is dissolved in sufficient water to give a 0.050 M solution. Which solution would have the highest pH? The lowest pH?

(b) Which of the substituted pyridines is the strongest Brønsted base? Which one is the weakest Brønsted base?

109. The equilibrium constant for the reaction of formic acid and sodium hydroxide is 1.8×10^{10} (page 711). Confirm this value.

110. The equilibrium constant for the reaction of hydrochloric acid and ammonia is 1.8×10^{9} (page 713). Confirm this value.

111. Consider a salt of a weak base and a weak acid such as ammonium cyanide. Both NH_4^+ and CN^- ion hydrolyze in aqueous solution, but the net reaction can be considered as a proton transfer from NH_4^+ to CN^-.

$$NH_4^+(aq) + CN^-(aq) \rightleftarrows NH_3(aq) + HCN(aq)$$

   (a) Show that the equilibrium constant for this reaction, K_{net}, is

$$K_{net} = \frac{K_w}{K_a K_b}$$

   where K_a is the ionization constant for the weak acid HCN and K_b is the constant for the weak base NH_3.

   (b) Prove that the hydronium ion concentration in this solution must be given by

$$[H_3O^+] = \sqrt{\frac{K_w K_a}{K_b}}$$

   (c) What is the pH of a 0.15 M solution of ammonium cyanide?

112. Listed in the table are values of pK_a for some compounds.

Acid	pK_a
Benzoic acid, $C_6H_5CO_2H$	4.20
Benzylammonium ion, $C_6H_5CH_2NH_3^+$	9.35
Chloroacetic acid, $ClCH_2CO_2H$	2.87
Conjugate acid of cocaine	8.41
Thioacetic acid, $HSCH_2CO_2H$	3.33

   (a) Which is the strongest acid?

   (b) Which acid has the strongest conjugate base?

   (c) List the acids in order of increasing strength.

113. To what volume should 1.00×10^2 mL of any weak acid HA with a concentration 0.20 M be diluted in order to double the percentage ionization?

114. Equilibrium constants can be measured for the dissociation of Lewis acid–base complexes such as the dimethyl ether complex of BF_3, $(CH_3)_2O \rightarrow BF_3$. The value of K (here K_p) for the reaction is 0.17.

$$(CH_3)_2O - BF_3(g) \rightleftarrows BF_3(g) + (CH_3)_2O(g)$$

   (a) Describe each product as a Lewis acid or Lewis base.

   (b) If you place 0.100 g of the complex in a 565-mL flask at 25 °C, what is the total pressure in the flask? What are the partial pressures of the Lewis acid, the Lewis base, and the complex?

115. Iodine, I_2, is much more soluble in a water solution of potassium iodide, KI, than it is in pure water. The anion found in solution is I_3^-.

   (a) Draw an electron dot structure for I_3^-.

   (b) Write an equation for this reaction, indicating the Lewis acid and Lewis base.

116. Sulfanilic acid, which is used in making dyes, is made by reacting aniline with sulfuric acid.

$$H_2SO_4(aq) + C_6H_5NH_2(aq) \longrightarrow$$

aniline

sulfanilic acid

The acid has a pK_a value of 3.23. The sodium salt of the acid, $Na(H_2NC_6H_4SO_3)$, is quite soluble in water. If you dissolve 1.25 g of the salt in 125 mL of solution, what is the pH of the solution?

Using Electronic Resources

These questions refer to the General Chemistry Interactive CD-ROM, *Version 3.0.*

117. See CD-ROM Screen 17.4: The Acid–Base Properties of Water. Explore the pH of water at different temperatures using the Simulation.

   (a) What is the value of K_w at 50 °C?

   (b) What is $[H_3O^+]$ at 50 °C in a neutral solution? What is the pH of the solution?

   (c) Is the value of K_w at 5 °C larger or smaller than at 25 °C?

   (d) A solution has pH = 7.00 at 5 °C. Is the solution acidic, basic, or neutral?

118. See CD-ROM Screen 17.7: Acid–Base Reactions. Explore the reactions of acid and bases using the Simulation.

   (a) Write a balanced net ionic equation for the reaction of HF(aq) with NO_2^-(aq). What is the extent of reaction? Which is the stronger base in this reaction, F^- or NO_2^-?

   (b) Which is the stronger of the two bases NO_2^- and HCO_3^-?

   (c) To what extent will NH_3(aq) react with HF(aq)? To a large extent? To a small extent?

119. See CD-ROM Screen 17.11: The Acid–Base Properties of Salts. Explore the pH of various solutions using the Simulation.

   (a) What is the pH of a 0.50 M solution of NH_4Cl? What happens to the pH of the solution as the NH_4Cl concentration increases? When $[NH_4Cl]$ doubles, does the pH double or halve?

   (b) When the cation is changed from Na^+ to NH_4^+ for a given anion, what happens to the pH? Explain the observation.

   (c) What combination of cation and anion leads to the lowest pH? To the highest pH?

18 Principles of Reactivity: Other Aspects of Aqueous Equilibria

Chapter Goals

- Understand the common ion effect.
- Understand the control of pH in aqueous solutions with buffers.
- Evaluate the pH in the course of acid–base titrations.
- Apply chemical equilibrium concepts to the solubility of ionic compounds.

Roses Are Red, Violets Are Blue, and Hydrangeas Are Red or Blue

Like the colors of some dyes (← PAGE 418), the colors of some flowers change with the pH of their environment. This is certainly true of the beautiful, showy flowers of the hydrangea bush.

We learned from a gardening book that "Hydrangea flowers need aluminum to maintain a good blue. Aluminum is

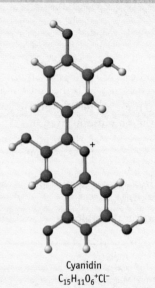

Cyanidin
$C_{15}H_{11}O_6{}^+Cl^-$

▲ Hydrangea blossoms in a Buddhist temple in Kunming, China.
(a, John C. Kotz; b, Terry Donnelly/Dembinsky Photo Associates)

easily absorbed from acidic lower pH soils. . . . The quickest way is to add aluminum sulfate to the soil." What does pH have to do with the color? What is the role of aluminum? And what does this have to do with chemical equilibria?

The pigment in hydrangea blossoms belongs to the class of molecules called anthocyanins, or, more particularly, it is a cyanidin. Cyanidins are responsible for the red color of roses, strawberries, raspberries, apple skins, rhubarb, and cherries and for the purple color of blueberries. What is more, the color of cyanidins depend on the pH. Red cabbage juice is only red in acid; it is purple in a solution closer to a pH of 7 (← FIGURE 5.5). The reason for this shift in color

Why are some flowers red, others blue, and still others yellow?

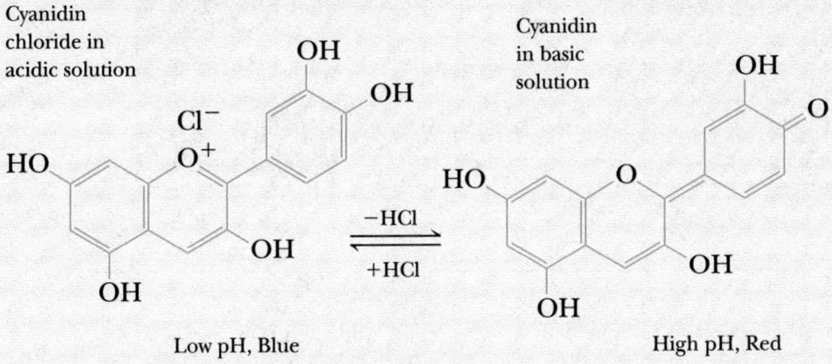

Cyanidin chloride in acidic solution

Cyanidin in basic solution

$-HCl$ ⇌ $+HCl$

Low pH, Blue   High pH, Red

▲ When cyanidin functions as an acid and loses an H^+ ion, the color shifts from blue to red.

with pH is that the pigment is an acid and can donate an H^+ ion. As the pH increases, the conjugate base is formed, and this is accompanied by a significant color change. If the pH could be controlled, then the color could be controlled.

But why does the color of hydrangea blossoms depend on the presence of aluminum ions? One idea is that the cyanidin is stabilized in the blue, acid form by complexation with Al^{3+}, a Lewis acid–base interaction. Aluminum is an avid seeker of oxygen atoms, so it not only interacts with the O atoms of cyanidin but it also forms $Al(H_2O)_6^{3+}$ in acidic solutions and insoluble $Al(OH)_3$ in basic solutions.

$$Al(H_2O)_6^{3+}(aq) + 3\ OH^-(aq) \rightleftharpoons Al(OH)_3(s) + 6\ H_2O(\ell)$$

Thus, soil should be on the acid side if Al^{3+} ions are to be available to bind to the cyanidin pigment and give blue flowers. Indeed, the ideal pH range to promote blue hydrangeas is 5–5.5, whereas a pH of 6–6.5 is better suited to pink blossoms. For this reason, potting soil, which is usually somewhat acidic, will produce blue hydrangeas. In an area with significant rainfall, the result will also be blue blossoms, perhaps owing to the acidity of the rain. On the other hand, in more arid regions, soils are more alkaline, and pink blossoms result.

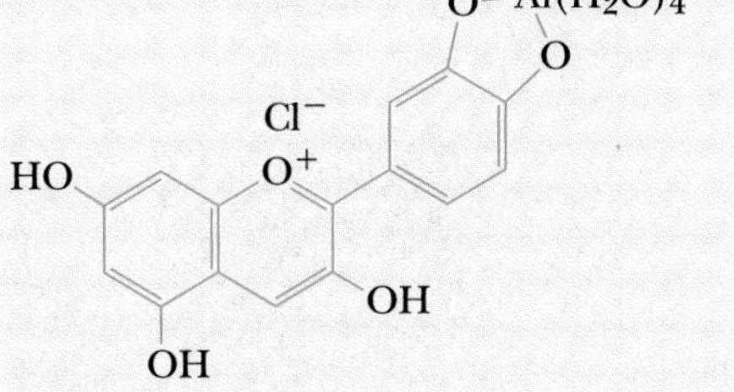

▲ The presence of Al^{3+} ions stabilizes the blue form of cyanidin.

Before You Begin

- Review the principles of acid–base equilibria (Chapter 17).
- Review the solubility rules (Figure 5.1, page 151).
- Review the stoichiometry of acid–base reactions (Section 5.10).

The focus of this chapter is to apply chemical equilibrium concepts to two of the main types of chemical reactions: acid-base reactions and precipitation reactions. This series of photographs illustrates the chemistry of lead.

The chemistry of lead

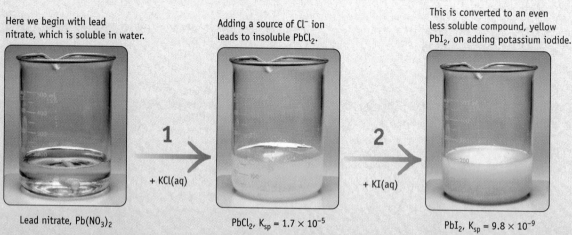

Here we begin with lead nitrate, which is soluble in water.

1 + KCl(aq)

Adding a source of Cl^- ion leads to insoluble $PbCl_2$.

2 + KI(aq)

This is converted to an even less soluble compound, yellow PbI_2, on adding potassium iodide.

Lead nitrate, $Pb(NO_3)_2$ $PbCl_2$, $K_{sp} = 1.7 \times 10^{-5}$ PbI_2, $K_{sp} = 9.8 \times 10^{-9}$

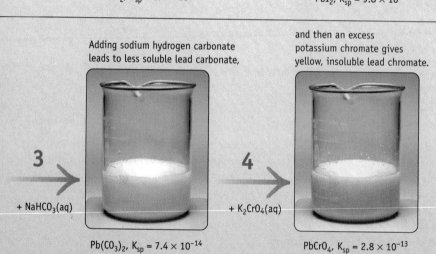

3 + NaHCO$_3$(aq)

Adding sodium hydrogen carbonate leads to less soluble lead carbonate,

4 + K$_2$CrO$_4$(aq)

and then an excess potassium chromate gives yellow, insoluble lead chromate.

$Pb(CO_3)_2$, $K_{sp} = 7.4 \times 10^{-14}$ $PbCrO_4$, $K_{sp} = 2.8 \times 10^{-13}$

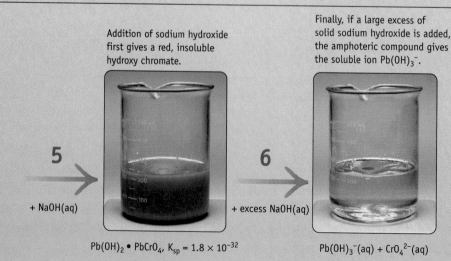

For more information on insoluble compounds and their equilibrium constants (solubility product constant, K_{sp}) see Section 18.4.

5 + NaOH(aq)

Addition of sodium hydroxide first gives a red, insoluble hydroxy chromate.

6 + excess NaOH(aq)

Finally, if a large excess of solid sodium hydroxide is added, the amphoteric compound gives the soluble ion $Pb(OH)_3^-$.

$Pb(OH)_2 \cdot PbCrO_4$, $K_{sp} = 1.8 \times 10^{-32}$ $Pb(OH)_3^-$(aq) + CrO_4^{2-}(aq)

In Chapter 5 we described four fundamental types of chemical reactions: gas-forming reactions, acid–base reactions, precipitation reactions, and oxidation–reduction reactions (← SECTIONS 5.4–5.7). In the present chapter we want to apply the principles of chemical equilibria to an understanding of two of these kinds of reactions: acid–base reactions and precipitation reactions.

We began to study acid–base reactions in Chapter 17 (Sections 17.5 and 17.6). Now we turn to the following questions:

- How can we control the pH in a solution?
- What happens when an acid and base are mixed in any amount?

Precipitation reactions are common (Chapter Focus), and they can also be understood in terms of chemical equilibria. For example, we want to know

- If aqueous solutions of two ionic compounds are mixed, will a precipitation occur?
- To what extent does an apparently insoluble substance actually dissolve?
- What chemical reactions can be used to redissolve a precipitate?

18.1   COMMON ION EFFECT

Lactic acid is a weak acid found in sour milk, apples, beer, wine, and sore muscles.

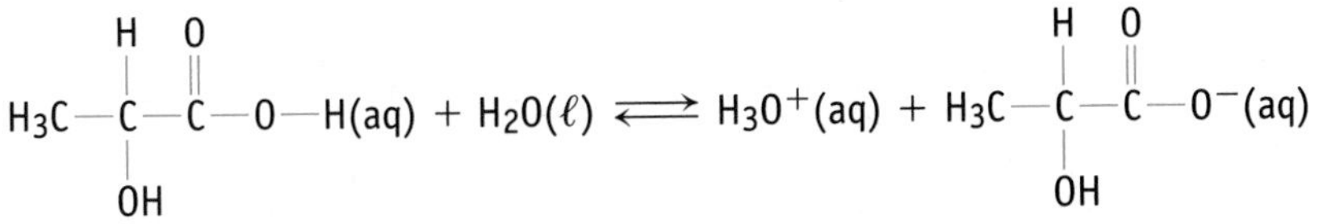

lactic acid ($HC_3H_5O_3$)
$K_a = 1.4 \times 10^{-4}$

lactate ion ($C_3H_5O_3^-$)

Suppose you add 0.025 mol of sodium hydroxide, a strong base, to an aqueous solution containing 0.10 mol of the acid.

$$HC_3H_5O_3(aq) + OH^-(aq) \longrightarrow C_3H_5O_3^-(aq) + H_2O(\ell)$$

lactic acid   lactate ion

You have added only one fourth as much base as is needed to completely consume the acid. Therefore, when reaction is complete, the solution contains both a weak acid (unreacted lactic acid) and a weak base (the product, lactate ion). The extent to which lactic acid can ionize under these conditions — and thus the pH of the solution — is clearly affected by the presence of the lactate ion. Recall Le Chatelier's principle!

The **common ion effect** is the limiting of the ionization of an acid (or base) by the presence of a significant concentration of its conjugate base (or acid) [CD-ROM, Screen 18.2]. In the ionization of lactic acid, the presence of some conjugate base, the lactate ion, in solution limits the extent to which the acid ionizes and thus affects the pH of the solution (Figure 18.1). Work through Example 18.1 to examine the effect of a common ion on the ionization of a weak acid, and then go to Example 18.2 to see the outcome of the lactic acid question.

Chapter Goals • Revisited

- **Understand the common ion effect.**
- Understand the control of pH in aqueous solutions with buffers.
- Evaluate the pH in the course of acid–base titrations.
- Apply chemical equilibrium concepts to the solubility of ionic compounds.

Figure 18.1   The common ion effect.
An acid (0.25 M acetic acid) is mixed with a base (0.10 M sodium acetate). The pH meter shows the resulting solution (*center*) has a lower hydronium ion concentration (a pH of about 5) than the acetic acid solution (pH is about 2.7). (Each solution contains universal indicator. This dye is red in low pH, yellow in slightly acidic media, and green in neutral to weakly basic media.) *(Charles D. Winters)*

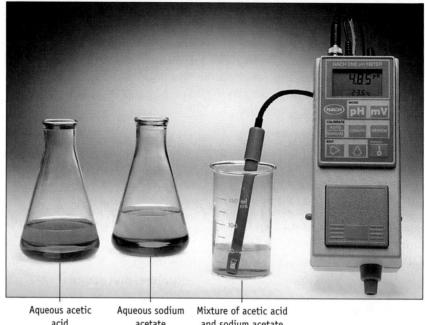

Aqueous acetic acid

Aqueous sodium acetate

Mixture of acetic acid and sodium acetate

Example 18.1   Common Ion Effect

Problem • The pH of 0.25 M acetic acid is 2.67. What is the pH if the solution is 0.25 M in acetic acid and 0.10 M in sodium acetate?

Strategy • Sodium acetate, $NaCH_3CO_2$, is 100% dissociated into its ions, Na^+ and $CH_3CO_2^-$ in water. Sodium ion has no effect on the pH of a solution (← TABLE 17.4, EXAMPLE 17.7). Thus, the important components of the solution are a weak acid (CH_3CO_2H) and its conjugate base ($CH_3CO_2^-$); the latter is an ion "common" to the ionization equilibrium reaction of acetic acid.

$$CH_3CO_2H(aq) + H_2O(\ell) \rightleftharpoons H_3O^+(aq) + CH_3CO_2^-(aq)$$

Assume the acid ionizes to give H_3O^+ and $CH_3CO_2^-$, both in the amount x. This means that, relative to their initial concentrations, CH_3CO_2H decreases in concentration slightly (by an amount x) and $CH_3CO_2^-$ increases slightly (by an amount x). Thus, because we can define the equilibrium concentrations of acid and conjugate base, and we know K_a, the hydronium ion concentration (= x) and the pH can be calculated.

Solution • First set out the concentrations of the various species in an ICE table.

Equation	CH_3CO_2H + H_2O $\rightleftharpoons$	H_3O^+ +	$CH_3CO_2^-$
Initial (M)	0.25	0	0.10
Change	$-x$	$+x$	$+x$
Equilibrium (M)	$0.25 - x$	x	$0.10 + x$

To solve for the hydronium ion concentration, set up the usual equilibrium constant expression.

$$K_a = 1.8 \times 10^{-5} = \frac{[H_3O^+][CH_3CO_2^-]}{[CH_3CO_2H]}$$

$$= \frac{(x)(0.10 + x)}{0.25 - x}$$

Now, because acetic acid is a weak acid, and because it is ionizing in the presence of a significant concentration of its conjugate base, let us assume x is quite small. That is, it is reasonable to assume that $(0.10 + x)M \approx 0.10$ M and that $(0.25 - x)M \approx 0.25$M. This leads to the "approximate" expression.

$$K_a = 1.8 \times 10^{-5} = \frac{[H_3O^+][CH_3CO_2^-]}{[CH_3CO_2H]} \approx \frac{(x)(0.10)}{0.25}$$

Solving this, we find that

$$x = [H_3O^+] = 4.5 \times 10^{-5} M$$
$$pH = 4.35$$

Comment • Without the added salt $NaCH_3CO_2$, which provides the "common ion" $CH_3CO_2^-$, ionization of 0.25 M acid will produce H_3O^+ and $CH_3CO_2^-$ ions in a concentration of 0.0021 M (to give a pH of 2.67). Le Chatelier's principle, however, predicts that the added common ion causes the reaction to proceed less far to the right. Hence, $x = [H_3O^+]$ must be less than 0.0021 M in the presence of added acetate ion, and this is indeed the case. Our approximations to solve this problem were reasonable.

Example 18.2 Reaction of Lactic Acid with a Deficiency of Sodium Hydroxide

Problem • What is the pH of the solution that results from adding 25.0 mL of 0.0500 M NaOH to 25.0 mL of 0.100 M lactic acid? (K_a for lactic acid = 1.4×10^{-4})

$$HC_3H_5O_3(aq) + OH^-(aq) \longrightarrow C_3H_5O_3^-(aq) + H_2O(\ell)$$

lactic acid   lactate ion

Strategy • This problem ___ ___ parts: a stoichiometry problem followed by an eq___ ___m. We first calculate the concentrations of l___ ___ion that are present following the reac___ ___H. Then, with the acid and conjugat___ ___ follow the strategy of Example 18.___

Solution •

Part 1: Stoichiometry
First, consider what sp___ ___ acid–base reaction, and what ___ ___ies are.

(a) Amounts of NaOH and la___
(0.0250 L NaOH)(0.0500 ___
(0.0250 L lactic acid)(0.100___

(b) Amount of lactate ion produced ___
Recognizing that NaOH is the limit___

$$(1.25 \times 10^{-3} \text{ mol NaOH})\left(\frac{1 \text{ mol lactate}}{1 \text{ mol NaOH}}\right)$$
$$= 1.25 \times 10^{-3} \text{ mol la___}$$

(c) Amount of lactic acid consumed

$$(1.25 \times 10^{-3} \text{ mol NaOH})\left(\frac{1 \text{ mol lactic acid}}{1 \text{ mol NaOH}}\right)$$
$$= 1.25 \times 10^{-3} \text{ mol lactic acid co___}$$

(d) Amount of lactic acid remaining when reaction is complete___
2.50×10^{-3} mol lactic acid available -1.25×10^{-3} mol lactic acid consumed = 1.25×10^{-3} mol lactic acid remaining

(e) Concentrations of lactic acid and lactate ion after reaction. Note that the total solution volume after reaction is 0.050 L

$$[\text{Lactic acid}] = \frac{1.25 \times 10^{-3} \text{ mol lactic acid}}{0.050 \text{ L}}$$
$$= 2.50 \times 10^{-2} \text{ M}$$

Because the amount of lactic acid remaining is the same as the amount of lactate ion produced, we have

$$[\text{Lactic acid}] = [\text{Lactate ion}] = 2.5 \times 10^{-2} \text{ M}$$

Part 2: Equilibrium Calculation
With the "initial" concentrations known, construct a table summarizing equilibrium concentrations.

Equation	$HC_3H_5O_3 + H_2O \rightleftharpoons$	$C_3H_5O_3^-$ +	H_3O^+
Initial (M)	0.0250	0.0250	0
Change	$-x$	$+x$	$+x$
Equilibrium (M)	$(0.0250 - x)$	$(0.0250 + x)$	x

Substituting the concentrations into the equilibrium expression, we have

$$K_a(\text{lactic acid}) = 1.4 \times 10^{-4} = \frac{[H_3O^+][C_3H_5O_3^-]}{[HC_3H_5O_3]}$$
$$= \frac{(x)(0.0250 + x)}{0.0250 - x}$$

___ssumption that x is small with respect to 0.0250 M ___ 18.1), we see that

$$= [H_3O^+] = K_a = 1.4 \times 10^{-4} \text{ M}$$

___.85.

___two final points to be made:

___ $\ll$ 0.0250 is valid.

___ining only 0.100 M lactic acid ___ of the added conjugate base ___ to increase the pH.

Exercise 18.1 Common Ion Effect

Assume you have a 0.30 M solution of formic acid (HCO_2H) and have added enough sodium formate ($NaHCO_2$) to make the solution 0.10 M in the salt. Calculate the pH of the formic acid solution before and after adding sodium formate.

Exercise 18.2 Mixing an Acid and a Base

What is the pH of the solution that results from adding 30.0 mL of 0.100 M NaOH to 45.0 mL of 0.100 M acetic acid?

18.2   CONTROLLING pH: BUFFER SOLUTIONS

The normal pH of human blood is 7.4. However, the addition of a small amount of strong acid or base to blood, say 0.01 mol, to a liter of blood, leads to a change in pH of only about 0.1 pH units. In comparison, if you add 0.01 mol of HCl to 1.0 L of pure water, the pH drops from 7 to 2. Addition of 0.01 mol of NaOH increases the pH from 7 to 12. Blood, and many other body fluids, are said to be buffered. A **buffer** causes solutions to be resistant to a change in pH on addition of a strong acid or base (Figure 18.2) [CD-ROM, Screens 18.3 and 18.4].

There are two requirements for a buffer:

- Two substances are needed: an acid capable of reacting with added OH^- ions and a base that can consume added H_3O^+ ions.
- The acid and base must not react with each another.

These requirements mean a buffer is usually prepared from roughly equal quantities of a conjugate acid–base pair: (a) a weak acid and its conjugate base (acetic acid and acetate ion, for example), or (b) a weak base and its conjugate acid (ammonia and ammonium ion, for example). Thus, *the action of a buffer is simply a special case of the common ion effect.* Some buffers commonly used in the laboratory are given in Table 18.1.

To see how a buffer works, let us consider an acetic acid/acetate ion buffer. Acetic acid, the weak acid, is needed to consume any added hydroxide ion.

$$CH_3CO_2H(aq) + OH^-(aq) \longrightarrow CH_3CO_2^-(aq) + H_2O(\ell) \qquad K = 1.8 \times 10^9$$

The equilibrium constant for the reaction is very large because OH^- is a much stronger base than acetate ion, $CH_3CO_2^-$ (SECTION 17.6, TABLE 17.3). This means that any OH^- entering the solution from an outside source is consumed completely. In a similar way, any hydronium ion added to the solution reacts with the acetate ion present in the buffer.

$$H_3O^+(aq) + CH_3CO_2^-(aq) \longrightarrow H_2O(\ell) + CH_3CO_2H(aq) \qquad K = 5.6 \times 10^4$$

Before

After adding 0.10 M HCl

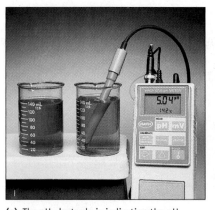
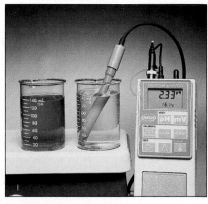

(a) The pH electrode is indicating the pH of water that contains a trace of acid (and bromphenol blue indicator; see Figure 17.2). The solution at the left is a buffer solution with a pH of about 7. (It also contains bromphenol blue dye.)

(b) When 5 mL of 0.10 M HCl is added to each solution, the pH of the water drops several units, whereas the pH of the buffer stays constant as implied by the fact that the indicator color did not change.

Figure 18.2   Buffer solutions. *(Charles D. Winters)*

Table 18.1 • **Some Commonly Used Buffer Systems**

Table 18.1 • **Some Commonly Used Buffer Systems**

Weak Acid	Conjugate Base	Acid K_a (pK_a)	Useful pH Range
Phthalic acid, $C_6H_4(CO_2H)_2$	Hydrogen phthalate ion $C_6H_4(CO_2H)(CO_2)^-$	1.3×10^{-3} (2.89)	1.9–3.9
Acetic acid, CH_3CO_2H	Acetate ion, $CH_3CO_2^-$	1.8×10^{-5} (4.74)	3.7–5.7
Dihydrogen phosphate ion, $H_2PO_4^-$	Hydrogen phosphate ion, HPO_4^{2-}	6.2×10^{-8} (7.21)	6.2–8.2
Hydrogen phosphate ion, HPO_4^{2-}	Phosphate ion, PO_4^{3-}	3.6×10^{-13} (12.44)	11.3–13.3

The equilibrium constant for this reaction is also quite large because H_3O^+ is a much stronger acid than CH_3CO_2H.

Now that we have established that a buffer solution will effectively consume small amounts of added acid or base, work through the next several examples. First, you see how to calculate the pH of a buffer solution and then how to prepare a buffer solution. Finally, you learn more quantitatively how the pH of a solution can be maintained.

Example 18.3 **pH of a Buffer Solution**

Problem • What is the pH of an acetic acid/sodium acetate buffer with $[CH_3CO_2H] = 0.700$ M and $[CH_3CO_2^-] = 0.600$ M?

Strategy • This is similar to Example 18.1. Knowing the concentrations of weak acid, its conjugate base, and K_a, we can calculate the hydronium ion concentration [CD-ROM, Screen 18.4].

Solution • Write a balanced equation for the ionization of acetic acid and set up an ICE table.

Equation	$CH_3CO_2H + H_2O \rightleftharpoons H_3O^+ + CH_3CO_2^-$		
Initial (M)	0.700	0	0.600
Change	$-x$	$+x$	$+x$
Equilibrium (M)	$0.700 - x$	x	$0.600 + x$

The appropriate equilibrium constant expression is

$$K_a = 1.8 \times 10^{-5} = \frac{[H_3O^+][CH_3CO_2^-]}{[CH_3CO_2H]}$$

$$= \frac{(x)(0.600 + x)}{0.700 - x}$$

As explained in Example 18.1, the value of x will be very small with respect to 0.700 or 0.600, so we can use the "approximate expression" to find x, the hydronium ion concentration.

$$[H_3O^+] = x = \frac{[CH_3CO_2H]}{[CH_3CO_2^-]} \cdot K_a$$

$$= \frac{(0.700)}{0.600}(1.8 \times 10^{-5}) = 2.1 \times 10^{-5} \text{ M}$$

$$pH = -\log(2.1 \times 10^{-5}) = 4.68$$

Exercise 18.3 **pH of a Buffer Solution**

What is the pH of a buffer solution composed of 0.50 M formic acid (HCO_2H) and 0.70 M sodium formate ($NaHCO_2$)?

General Expressions for Buffer Solutions

In Example 18.3 we solved for the hydronium ion concentration of the acetic acid/acetate ion buffer solution by rearranging the K_a expression to give

$$[H_3O^+] = \frac{[CH_3CO_2H]}{[CH_3CO_2^-]} \cdot K_a$$

● **Buffer Solutions**

You will find it generally useful to consider all buffer solutions as composed of a weak acid and its conjugate base (and we shall do so in this book). Suppose, for example, a buffer is composed of the weak base ammonia and its conjugate acid ammonium ion. The hydronium ion concentration can be found from Equation 18.1 by assuming the buffer is composed of the weak acid NH_4^+ and its conjugate base, NH_3.

This result can be generalized for any buffer. For a buffer solution based on *a weak acid and its conjugate base*, the hydronium ion concentration is given by

$$[H_3O^+] = \frac{[acid]}{[conjugate\ base]} \cdot K_a \qquad (18.1)$$

It is often convenient to use Equation 18.1 in a different form. If we take the negative logarithm of each side of the equation we have

$$-\log[H_3O^+] = \left[-\log\frac{[acid]}{[conjugate\ base]}\right] + (-\log K_a)$$

You know that $-\log[H_3O^+]$ is defined as pH, and $-\log K_a$ is equivalent to pK_a (← SECTIONS 17.3 AND 17.4). Furthermore, because

$$-\log\frac{[acid]}{[conjugate\ base]} = +\log\frac{[conjugate\ base]}{[acid]}$$

the preceding equation can be rewritten as

● **Henderson–Hasselbalch Equation**

The Henderson–Hasselbalch equation is generally valid when the ratio of [conjugate base]/[acid] is no larger than 10 and no smaller than 0.1. Many handbooks of chemistry list acid ionization constants in terms of pK_a values, so the approximate pH values of possible buffer solutions are readily apparent.

$$pH = pK_a + \log\frac{[conjugate\ base]}{[acid]} \qquad (18.2)$$

This equation, known as the **Henderson–Hasselbalch equation,** shows that the pH of the solution of a weak acid and its conjugate base is controlled primarily by the strength of the acid (as expressed by pK_a). The "fine" control of the pH is then given by the relative amounts of conjugate base and acid. When the concentrations of conjugate base and acid are the same in a solution, the ratio [conjugate base]/[acid] is 1. The log of 1 is zero, so $pH = pK_a$ under these circumstances. If there is more of the conjugate base in the solution than acid, for example, then $pH > pK_a$.

Example 18.4   Using the Henderson–Hasselbalch Equation

Problem ● Benzoic acid ($C_6H_5CO_2H$, 2.00 g) and sodium benzoate ($NaC_6H_5CO_2$, 2.00 g) are dissolved in enough water to make 1.00 L of solution. Calculate the pH of the solution using the Henderson–Hasselbalch equation.

Strategy ● The Henderson–Hasselbalch equation requires the pK_a of the acid, and this is obtained from the K_a for the acid (← TABLE 17.3 OR APPENDIX H). You will also need the acid and conjugate base concentrations.

Solution ● K_a for benzoic acid as 6.3×10^{-5}. Therefore,

$$pK_a = -\log(6.3 \times 10^{-5}) = 4.20$$

Next, we need the concentrations of the acid (benzoic acid) and its conjugate base (benzoate ion).

$$2.00\ g\ benzoic\ acid\left(\frac{1\ mol}{122.1\ g}\right) = 0.0164\ mol\ benzoic\ acid$$

$$2.00\ g\ sodium\ benzoate\left(\frac{1\ mol}{144.1\ g}\right) = 0.0139\ mol\ sodium\ benzoate$$

Because the solution volume is 1.00 L, the concentrations are [benzoic acid] = 0.0164 M and [sodium benzoate] = 0.0139 M. Therefore, using Equation 18.2, we have

$$pH = 4.20 + \log\frac{0.0139}{0.0164}$$
$$= 4.20 + \log(0.848)$$
$$= 4.13$$

Comment ● Notice that the pH is less than the pK_a because the acid predominates in solution (the ratio of conjugate base to acid concentration is less than 1).

Exercise 18.4   **Using the Henderson–Hasselbalch Equation**

Suppose you dissolve 15.0 g of $NaHCO_3$ and 18.0 g of Na_2CO_3 in enough water to make 1.00 L of solution. Use the Henderson–Hasselbalch equation to calculate the pH of the solution. (Consider this buffer as a solution of the weak acid HCO_3^- with CO_3^{2-} as its conjugate base.)

Preparing Buffer Solutions

A buffer solution must meet two requirements [CD-ROM, Screen 18.5].

- *pH Control:* It should control the pH at the desired value. The equation for the hydronium ion concentration of a buffer shows us how this can be done.

$$[H_3O^+] = \frac{[acid]}{[conjugate\ base]} \cdot K_a$$

First, we want $[acid] \approx [conjugate\ base]$ so that the solution can buffer equal amounts of added acid or base. Next, we want to choose an acid whose K_a is near the intended value of $[H_3O^+]$. The exact value of $[H_3O^+]$ is then achieved by adjusting the acid/conjugate base ratio. Example 18.5 illustrates this approach.

- *Buffer capacity:* The buffer should have the capacity to control the pH after the addition of reasonable amounts of acid and base. That is, the concentration of acetic acid in an acetic acid/acetate ion buffer, for example, must be sufficient to consume all the hydroxide ion that may be added (see Example 18.6). Buffers are usually prepared as 0.10–1.0 M solutions of reagents. However, it is possible to exceed the buffer capacity if too much strong acid or base is added.

Example 18.5   **Preparing a Buffer Solution**

Problem • You wish to prepare 1.0 L of a buffer solution with a pH of 4.30. A list of possible acids (and their conjugate bases) is

Acid	Conjugate Base	K_a	pK_a
HSO_4^-	SO_4^{2-}	1.2×10^{-2}	1.92
CH_3CO_2H	$CH_3CO_2^-$	1.8×10^{-5}	4.74
HCO_3^-	CO_3^{2-}	4.8×10^{-11}	10.32

Which combination should be selected and what should the ratio of acid to conjugate base be?

Strategy • Use either the general equation for a buffer (Equation 18.1) or the Henderson–Hasselbalch equation (18.2). Equation 18.1 informs you that $[H_3O^+]$ should be close to the acid K_a value, or Equation 18.2 tells you that pH should be close to the acid pK_a value. This will establish which acid you will use.

Having decided which acid to use, convert pH to $[H_3O^+]$ to use Equation 18.1. If you use Equation 18.2, use the pK_a value in the table. Finally, calculate the ratio of acid to conjugate base.

Solution • The hydronium ion concentration for the buffer is found from the desired pH.

$$pH = 4.30, so [H_3O^+] = 10^{-pH}$$
$$= 10^{-4.30} = 5.0 \times 10^{-5}\ M$$

Of the acids given, only acetic acid (CH_3CO_2H) has a K_a value close to that of the desired $[H_3O^+]$ (or a pK_a close to a pH of 4.30). Therefore, you need only to adjust the ratio $[CH_3CO_2H]/[CH_3CO_2^-]$ to achieve the desired hydronium ion concentration.

$$[H_3O^+] = 5.0 \times 10^{-5} = \frac{[CH_3CO_2H]}{[CH_3CO_2^-]}(1.8 \times 10^{-5})$$

(Example continues on next page)

To satisfy the previous expression, find the ratio $[CH_3CO_2H]/[CH_3CO_2^-]$.

$$\frac{[CH_3CO_2H]}{[CH_3CO_2^-]} = \frac{5.0 \times 10^{-5}}{1.8 \times 10^{-5}} = \frac{2.8 \text{ mol/L}}{1.0 \text{ mol/L}}$$

Therefore, if you add 0.28 mol of acetic acid and 0.10 mol of sodium acetate (or any other pair of molar quantities in the ratio 2.8/1) to enough water to make 1.0 L of solution, the solution constitutes a buffer with a pH at 4.30.

Comment • If you prefer to use the Henderson–Hasselbalch equation, you would have

$$pH = 4.30 = 4.74 + \log \frac{[CH_3CO_2^-]}{[CH_3CO_2H]}$$

$$\log \frac{[CH_3CO_2^-]}{[CH_3CO_2H]} = 4.30 - 4.74 = -0.44$$

$$\frac{[CH_3CO_2^-]}{[CH_3CO_2H]} = 10^{-0.44} = 0.36$$

The ratio of conjugate base to acid, $[CH_3CO_2^-]/[CH_3CO_2H]$, is 0.36. The reciprocal of this ratio $\{= [CH_3CO_2H]/[CH_3CO_2^-] = 1/0.35)\}$ is 2.8. This is the same result obtained above using Equation 18.1.

Exercise 18.5 Preparing a Buffer Solution

Using an acetic acid/sodium acetate buffer solution, what ratio of acid to conjugate base will you need to maintain the pH at 5.00? Explain how you would make up such a solution.

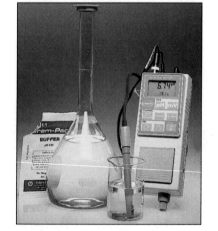

Figure 18.3 A commercial buffer solution. The solid acid and conjugate base in the packet are mixed with water to give a solution with the indicated pH. The quantity of water used does not matter because the ratio [acid]/[conjugate base] does not depend on the solution volume. *(Charles D. Winters)*

Example 18.5 illustrates a very important point concerning buffer solutions. The hydronium ion concentration depends not only on the K_a value of the acid but also on the ratio of acid and conjugate base concentrations. However, even though we write these ratios in terms of reagent concentrations, it is *the relative number of moles of acid and conjugate base that is important in determining the pH of a buffer solution.* Because both reagents are dissolved in the same solution, their concentrations depend on the same solution volume. In Example 18.5, the ratio 2.8/1 for acetic acid and sodium acetate implied that 2.8 mol of the acid and 1.0 mol of sodium acetate were dissolved per liter.

$$\frac{[CH_3CO_2H]}{[CH_3CO_2^-]} = \frac{2.8 \text{ mol } CH_3CO_2H/L}{1.0 \text{ mol } CH_3CO_2^-/L} = \frac{2.8 \text{ mol } CH_3CO_2H}{1.0 \text{ mol } CH_3CO_2^-}$$

On dividing one concentration by the other, the volumes "cancel." This means that we only need to ensure that the ratio of moles of acid to moles of conjugate base is 2.8 to 1 in this example. The acid and its conjugate base could have been dissolved in any reasonable amount of water. The mole ratio of acid to its conjugate base is maintained, no matter what the solution volume may be, so *diluting a buffer solution will not change its pH.* Commercially available buffer solutions are often sold as premixed, dry ingredients. To use them, you only need to mix the ingredients in some volume of pure water (Figure 18.3).

How Does a Buffer Maintain pH?

Now let us explore quantitatively how a given buffer solution can maintain the pH of a solution on adding a small amount of strong acid [CD-ROM, Screen 18.6].

Example 18.6 How Does a Buffer Maintain a Constant pH?

Problem • What is the change in pH when 1.00 mL of 1.00 M HCl is added to (1) 1.000 L of pure water and to (2) 1.000 L of acetic acid/sodium acetate buffer with $[CH_3CO_2H] = 0.700$ M and

$[CH_3CO_2^-] = 0.600$ M. (See Example 18.3; pH of this acetic acid/acetate ion buffer was found to be 4.68.)

Strategy • HCl is a strong acid, so we are only concerned that it ionizes completely to supply H_3O^+ ions. Part 1 involves two steps: (a) Finding the H_3O^+ concentration when diluting 1.0 mL of acid to 1.0 L. (b) Converting the value of $[H_3O^+]$ for the dilute solution to pH.

Part 2 involves three steps: (a) A stoichiometry calculation to find how the concentrations of acid and conjugate base change on adding H_3O^+ in the form of HCl. (b) An equilibrium calculation to find $[H_3O^+]$ for a buffer solution in which the concentrations of CH_3CO_2H and $CH_3CO_2^-$ are slightly altered owing to the reaction of $CH_3CO_2^-$ with added H_3O^+. (c) Finding pH from $[H_3O^+]$.

Solution •
Part 1. Adding Acid to Pure Water.
1.00 mL of 1.00 M HCl represents 0.00100 mol of acid. If this is added to 1.000 L of pure water, the H_3O^+ concentration of the water changes from 10^{-7} to 10^{-3},

$$c_1V_1 = c_2V_2$$
$$(1.00\ M)(0.00100\ L) = c_2(1.00\ L)$$
$$c_2 = [H_3O^+] \text{ in diluted solution} = 1.00 \times 10^{-3}\ M$$

and so the pH falls from 7 to 3.

Part 2. Adding Acid to an Acetic Acid/Acetate Buffer Solution.
HCl is a strong acid that is 100% ionized in water and supplies H_3O^+, which reacts completely with the base (acetate ion) in the solution according to the following equation:

$$H_3O^+(aq) + CH_3CO_2^-(aq) \longrightarrow H_2O(\ell) + CH_3CO_2H(aq)$$

	H_3O^+ from added HCl	$CH_3CO_2^-$ from buffer	CH_3CO_2H from buffer
Initial amount of acid or base ($mol = cV$)	0.00100	0.600	0.700
Change (mol)	−0.00100	−0.00100	+0.00100
After reaction (mol)	0	0.599	0.701
Concentrations after reaction ($c = mol/V$)	0	0.598	0.701

Because the added HCl reacts *completely* with acetate ion to produce acetic acid, the solution after this reaction (with $V = 1.001$ L) is once again a buffer containing only the weak acid and its salt. Now we only need to use Equation 18.1 (or the Henderson–Hasselbalch equation) to find $[H_3O^+]$ and the pH in the buffer solution as in Examples 18.3 and 18.4.

Equation	$CH_3CO_2H + H_2O \rightleftharpoons H_3O^+ + CH_3CO_2^-$		
Initial (M)	0.701	0	0.599
Change	$-x$	$+x$	$+x$
Equilibrium (M)	$0.701 - x$	x	$0.599 + x$

As usual, we make the approximation that x, the amount of H_3O^+ formed by ionizing acetic acid in the presence of acetate ion, is very small compared with 0.701 M or 0.599 M. Therefore, because $x = [H_3O^+]$, we can use Equation 18.1.

$$[H_3O^+] = x = \frac{[CH_3CO_2H]}{[CH_3CO_2^-]} \cdot K_a$$
$$= \left(\frac{0.701\ mol}{0.598\ mol}\right)(1.8 \times 10^{-5}) = 2.1 \times 10^{-5}\ M$$
$$pH = 4.68$$

Comment • Within the number of significant figures allowed, the *pH does not change* in the buffer solution after adding HCl, even though it changed by 4 units when 1 mL of 1.0 M HCl was added to 1.0 L of pure water. Buffer solutions do indeed "buffer." In this case the acetate ion consumed the added hydronium ion, and the solution resisted a change in pH.

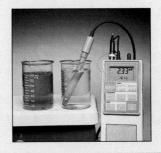

Buffering. Acid was added to two beakers, one with a buffer solution (*left*) and one having only pure water (*right*). (Both contain the indicator methyl red). Adding 1.0 mL of 1.0 M HCl to each gives no change in pH in the buffer solution (the indicator color does not change), but pH drops to about 2.3 in the beaker with no buffer. (*Charles D. Winters*)

Exercise 18.6 Buffer Solutions

Calculate the pH of 0.500 L of a buffer solution composed of 0.50 M formic acid (HCO_2H) and 0.70 M sodium formate ($NaHCO_2$) before and after adding 10.0 mL of 1.0 M HCl. (See Exercise 18.3)

18.3 ACID–BASE TITRATIONS

A *titration* is one of the most important ways of determining accurately the quantity of an acid, a base, or some other substance in a mixture or of ascertaining the purity of a substance. You learned how to perform the stoichiometry calculations

Chemical Perspectives

Buffers in Biochemistry

Maintenance of pH is vital to the cells of all living organisms because enzyme activity is influenced by pH. The primary protection against harmful pH changes in cells is provided by buffer systems. The intracellular pH of most cells is maintained at about 7.4, and two important biological buffers systems can control pH in this range: the phosphate system ($HPO_4^{2-}/H_2PO_4^-$) and the bicarbonate/carbonic acid system (HCO_3^-/H_2CO_3).

Phosphate ions are abundant in cells, both as the ions themselves and as important substituents on organic molecules. Most importantly, the pK_a for the $H_2PO_4^-$ ion is 7.20, which is very close to the low end of the normal pH range.

$$H_2PO_4^-(aq) + H_2O(\ell) \rightleftharpoons$$
$$H_3O^+(aq) + HPO_4^{2-}(aq)$$

This means that if the buffer is to control the pH at about 7.4, the ratio of HPO_4^{2-} to $H_2PO_4^-$ must be 1.58.

$$pH = pK_a + \log\frac{[HPO_4^{2-}]}{[H_2PO_4^-]}$$

$$7.40 = 7.20 + \log\frac{[HPO_4^{2-}]}{[H_2PO_4^-]}$$

$$\frac{[HPO_4^{2-}]}{[H_2PO_4^-]} = 1.58$$

If the total phosphate concentration in a cell, $[HPO_4^{2-}] + [H_2PO_4^-]$, is 2.0×10^{-2} M, a typical concentration, you can

calculate that $[HPO_4^{2-}]$ should be about 1.2×10^{-2} M and $[H_2PO_4^-]$ should be about 7.6×10^{-3} M.

The bicarbonate/carbonic acid buffer is important in blood plasma. Three equilibria are important here.

$$CO_2(g) \rightleftharpoons CO_2(\text{dissolved})$$
$$CO_2(\text{dissolved}) + H_2O(\ell) \rightleftharpoons H_2CO_3(aq)$$
$$H_2CO_3(aq) + H_2O(\ell) \rightleftharpoons$$
$$H_3O^+(aq) + HCO_3^-(aq)$$

Alkalosis. If blood pH is too high, alkalosis results. This can be reversed by breathing into a bag, an action that recycles exhaled CO_2. This affects the carbonic acid buffer system in the body, raising the blood hydronium ion concentration. The blood pH drops back to a more normal level of 7.4. *(Charles D. Winters)*

The overall equilibrium constant for the second and third steps is $pK_{overall} = 6.3$ at 37 °C, the temperature of the human body. Thus,

$$7.4 = 6.3 + \log\frac{[HCO_3^-]}{[CO_2(\text{dissolved})]}$$

Although the value of $pK_{overall}$ is about 1 pH unit away from the blood pH, the natural partial pressure of CO_2 in the alveoli of the lungs (about 40 mm Hg) is sufficient to keep $[CO_2(\text{dissolved})]$ at about 1.2×10^{-3} M and $[HCO_3^-]$ at about 1.5×10^{-2} M.

If blood pH rises above about 7.45, you can suffer from a condition called *alkalosis*. This can arise from hyperventilation, from severe anxiety, or from an oxygen deficiency at high altitude. It can ultimately lead to overexcitability of the central nervous system, muscle spasms, convulsions, and death. One way to treat acute alkalosis is to breath into a paper bag. The CO_2 you exhale is recycled. This raises the blood CO_2 level and causes the equilibria above to shift to the right, thus raising the hydronium ion concentration.

Acidosis is the opposite of alkalosis. This can arise from inadequate exhalation of CO_2. Acidosis can be reversed by breathing rapidly and deeply. Doubling the breathing rate increases the blood pH by about 0.23 units.

involved in titrations in Chapter 5 (◀ SECTION 5.10). In Chapter 17, you were introduced to more information about acid–base reactions (◀ SECTION 17.6):

- The pH at the equivalence point of a strong acid/strong base titration is 7. The solution at the equivalence point is truly "neutral" *only* when a strong acid is titrated with a strong base and vice versa.

- If the substance being titrated is a weak acid or base, then the pH at the equivalence point is not 7 (◀ TABLE 17.5 AND PAGE 711).

  (a) Weak acid titrated with strong base $\longrightarrow$ pH > 7 at equivalence point due to the conjugate base of the weak acid.

  (b) Weak base titrated [🖱 CD-ROM, Screen 9.4] with strong acid $\longrightarrow$ pH < 7 at equivalence point due to the conjugate acid of the weak base.

● **Weak Acid/Weak Base Titrations**
Titrations combining a weak acid and weak base are generally not done because the equivalence point cannot be accurately judged.

Problem-Solving Tip 18.1

Buffer Solutions

The following is a summary of important aspects of buffer solutions.

- A buffer resists changes in pH on adding small quantities of strong acid or base.
- A buffer contains a weak acid and its conjugate base.
- The hydronium ion concentration of a buffer solution can be calculated from Equation 18.1,

$$[H_3O^+] = \frac{[acid]}{[conjugate\ base]} \cdot K_a$$

or the pH can be calculated from the Henderson–Hasselbalch equation (Equation 18.2).

$$pH = pK_a + \log \frac{[conjugate\ base]}{[acid]}$$

- The pH depends primarily on the K_a of the weak acid and secondarily on the relative amounts of acid and conjugate base.
- The weak acid of a buffer consumes added base, and the conjugate base consumes added acid. Such reactions affect the relative quantities of weak acid and its conjugate base. Because this ratio has only a secondary effect on the pH, the pH can be maintained relatively constant.
- The buffer must have sufficient capacity to react with reasonable quantities of added acid or base.

Now that you know about buffers, we want to describe how the pH changes in the course of a titration [CD-ROM, Screen 18.7]. This will give you more insight into the nature of acids, bases, and their conjugates, and of buffer solutions.

Titration of a Strong Acid with a Strong Base

Figure 18.4 illustrates what happens to the pH as 0.100 M NaOH is slowly added to 50.0 mL of 0.100 M HCl.

$$HCl(aq) + NaOH(aq) \longrightarrow NaCl(aq) + H_2O(\ell)$$

Net ionic equation: $H_3O^+(aq) + OH^-(aq) \longrightarrow 2\ H_2O(\ell)$

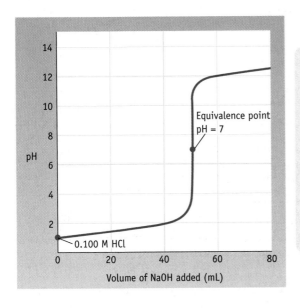

Amount of base added	pH
100.0	12.52
80.0	12.36
60.0	11.96
55.0	11.68
51.0	11.00
50.0	7.00
49.0	3.00
48.0	2.69
45.0	2.28
40.0	1.95
20.0	1.37
10.0	1.18
0.0	1.00
very large amount	13.00 (maximum)

50.0 mL of 0.100 M HCl titrated with 0.100 M NaOH

Figure 18.4 The change in pH as a strong acid is titrated with a strong base. In this case 50.0 mL of 0.100 M HCl is titrated with 0.100 M NaOH. The pH at the equivalence point is 7.0 for the reaction of a strong acid with a strong base.

Let us focus on four regions on this plot.

• pH of the initial solution
• pH as NaOH is added to the HCl solution before the equivalence point
• pH at the equivalence point
• pH after the equivalence point

Before the titration begins, the solution is 0.100 M in HCl, so the pH is 1.00. As NaOH is added to the acid solution, the amount of HCl declines, and the acid remaining is dissolved in an ever-increasing volume of solution. Thus, $[H_3O^+]$ decreases, and the pH slowly increases. As an example, let us find the pH of the solution after 10.0 ml of 0.100 M NaOH has been added to 50.0 mL of 0.100 M HCl. Here we will set up a table to list the amounts of acid and base before reaction, the changes in those amounts, and the amounts remaining after reaction. Be sure to notice that the volume of the solution after reaction is the sum of the combined volumes of NaOH and HCl (60.0 mL or 0.0600 L in this case).

	H_3O^+ (aq)	+	OH^- (aq) $\longrightarrow$ 2 $H_2O(\ell)$
Initial amount (mol = cV)	0.00500		0.00100
Change (mol)	−0.00100		−0.00100
After reaction (mol)	0.00400		0
After reaction (c = mol/V)	0.00400 mol/0.0600 L = 0.0667 M		0

The final solution has a hydronium ion concentration of 0.0667 M, and so the pH is

$$pH = -\log[H_3O^+] = -\log(0.0667) = 1.176$$

After 49.5 mL of base have been added, that is, just before the equivalence point, you can use the same approach to show that

$$pH = -\log[H_3O^+] = -\log(5.03 \times 10^{-4}) = 3.298$$

The solution being titrated is still quite acidic even very close to the equivalence point.

The equivalence point in any acid–base titration is identified as the midpoint in the vertical portion of the pH versus volume of titrant curve. (The **titrant** is the substance being added during the titration.) In the HCl/NaOH titration illustrated in Figure 18.4 you see that the pH increases very rapidly near the equivalence point. In fact, in this case the pH rises 7 units (the H_3O^+ concentration increases by a factor of 10 million!) when only a drop or two of the NaOH solution is added, and the midpoint of the vertical portion of the curve is a pH of 7.

> The pH of the solution at the equivalence point in a strong acid/strong base reaction is always 7 (at 25 °C).

After all of the HCl has been consumed and the slightest excess of NaOH has been added, the solution will be basic and the pH will continue to increase slightly as more NaOH is added (and the solution volume increases). For example, if we calculate the pH of the solution after 55.0 mL of 0.100 M NaOH has been added to 50.0 mL of 0.100 M HCl, we find

	$H_3O^+(aq)$	+	$OH^-(aq)$ ⟶ 2 $H_2O(\ell)$
Initial amount (mol = cV)	0.00500		0.00550
Change (mol)	−0.00500		−0.00500
After reaction (mol)	0		0.00050
After reaction (c = mol/V)	0		0.00050 mol/0.1050 L = 0.0048 M

The final solution has a hydroxide ion concentration of 0.0048 M, and so the pH is

$$pOH = -\log[OH^-] = 2.32$$
$$pH = 14 - pOH = 11.68$$

Exercise 18.7   **Titration of a Strong Acid with a Strong Base**

What is the pH after 25.0 mL of 0.100 M NaOH has been added to 50.0 mL of 0.100 M HCl?
What is the pH after 50.50 mL of NaOH has been added?

Titration of a Weak Acid with a Strong Base

The titration of a weak acid with a strong base is somewhat different from the strong acid/strong base titration. Look carefully at the titration curve for 100.0 mL of 0.100 M acetic acid with 0.100 M NaOH (Figure 18.5).

$$CH_3CO_2H(aq) + NaOH(aq) \longrightarrow NaCH_3CO_2(aq) + H_2O(\ell)$$

Let us focus on three important points on this curve:

- *The pH before titration begins.* The pH before any base is added is found from the weak acid K_a value and the acid concentration (⬅ EXAMPLE 17.5).

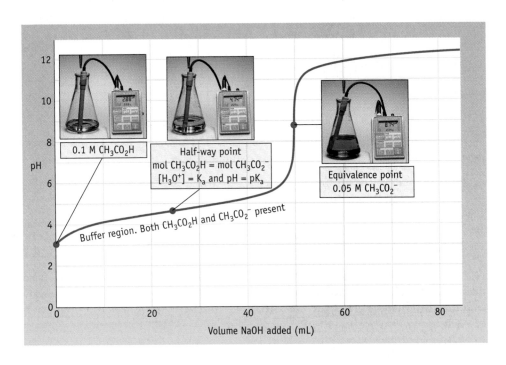

Figure 18.5   The change in pH during the titration of a weak acid with a strong base. Here 100.0 mL of 0.100 M acetic acid is titrated with 0.100 M NaOH. Note especially that **(a)** Acetic acid is a weak acid so the pH of the original solution is 2.87. **(b)** The pH at the point at which half the acid has reacted with base is equal to the pK_a for the acid (pH = pK_a = 4.74). **(c)** At the equivalence point the solution consists of acetate ion, a weak base. Therefore, the solution is basic with a pH of 8.72. *(Charles D. Winters)*

- *The pH at the equivalence point.* At the equivalence point the solution contains only sodium acetate, the CH_3CO_2H and NaOH having been consumed. The pH is controlled by the acetate ion, the conjugate base of acetic acid (see Example 18.2)(◀ TABLE 17.3, PAGE 702).

- *The pH at the halfway point (half-equivalence point) of the titration.* This is an important point, because here the pH is equal to the pK_a of the weak acid, a conclusion that needs more discussion.

The halfway point in a titration is the point at which half of the acid has been consumed. As NaOH is added to acetic acid, for example, the base is consumed and sodium acetate is produced. Thus, at every point between the beginning of the titration (when only acetic acid is present) and the equivalence point (when only sodium acetate is present) the solution contains both acetic acid and its salt, sodium acetate. These are the components of a *buffer solution,* and the hydronium ion concentration can be found from Equation 18.1.

$$[H_3O^+] \text{ at any point in the titration} = \frac{[\text{weak acid remaining}]}{[\text{conjugate base produced}]} \cdot K_a \qquad (18.3)$$

or

● **Finding K_a and pK_a Values**
One way to determine values of K_a (and pK_a) is to observe the pH at the halfway point in a titration.

$$pH \text{ at any point in the titration} = pK_a + \log\frac{[\text{conjugate base produced}]}{[\text{weak acid remaining}]} \qquad (18.4)$$

The presence of a weak acid and its conjugate base at any point between the beginning of the titration and the equivalence point means that the solution is buffered. This is the reason that the pH of the solution only rises slowly after a few milliliters titrant has been added.

What happens when *exactly* half of the acid has been consumed by base? Half of the acid (say CH_3CO_2H) has been converted to the conjugate base ($CH_3CO_2^-$), and half remains. Therefore, the concentration of weak acid remaining is equal to the concentration of conjugate base produced {$[CH_3CO_2H] = [CH_3CO_2^-]$}, and so

> At halfway point in the titration of an acid
> $$[H_3O^+] = K_a \quad \text{and} \quad pH = pK_a \qquad (18.5)$$

In the particular case of the titration of acetic acid with strong base, $[H_3O^+] = 1.8 \times 10^{-5}$ M at the midpoint, and so the pH is 4.74. This result is of general importance because we can say to a good approximation that

- At the halfway point in any weak acid + strong base titration, $[H_3O^+] = K_a$ (or $pH = pK_a$) of the weak acid.

- At the halfway point in any weak base + strong acid titration, $[OH^-] = K_b$ of the weak base (see Example 18.8). (Alternatively, the $pH = pK_a$ of the conjugate acid of the weak base.)

Equation 18.4, an adaptation of the Henderson–Hasselbalch equation, also makes it clear that at the halfway point, where [weak acid remaining] = [conjugate base produced], $pH = pK_a$. Before that point, $pH < pK_a$, and after that point $pH > pK_a$.

Example 18.7 **Titration of Acetic Acid with Sodium Hydroxide**

Problem • What is the pH of the solution when 90.0 mL of 0.100 M NaOH has been added to 100.0 mL of 0.100 M acetic acid (see Figure 18.5)?

Strategy • This problem is a combination of Examples 18.2 and 18.3. It involves two major steps: (a) A stoichiometry calculation to find the amount of acid remaining and the amount of conjugate base formed after adding NaOH. (b) An equilibrium calculation to find $[H_3O^+]$ for a buffer solution where the amounts of CH_3CO_2H and $CH_3CO_2^-$ are known from the first part of the calculation.

Solution • The pH is desired at a point after the halfway point of the titration (pH = 4.74; see Equation 18.5) but before the equivalence point (pH = 8.72) (◀ EXAMPLE 17.8 AND EXERCISE 17.16, PAGE 717). Let us first calculate the amounts of reactants before reaction (= concentration × volume) and then use the principles of stoichiometry to calculate the amounts of reactants and products after reaction.

Equation	CH_3CO_2H +	OH^-	$\longrightarrow$ $CH_3CO_2^-$	+ H_2O
Initial (mol)	0.0100	0.00900	0	
Change (mol)	−0.00900	−0.00900	+0.00900	
After reaction (mol)	0.0010	0	0.00900	

On page 748 you learned that the ratio of amounts (moles) of acid and conjugate base is the same as the ratio of their concentrations. Therefore, we can use the amounts of weak acid remaining and conjugate base formed to find the pH from Equation 18.3 (where we use amounts and not concentrations).

$$[H_3O^+] = \frac{mol\ CH_3CO_2H}{mol\ CH_3CO_2^-} \cdot K_a$$

$$= \left(\frac{0.0010\ mol}{0.00900\ mol}\right)(1.8 \times 10^{-5}) = 2.0 \times 10^{-6}\ M$$

$$pH = -\log(2.0 \times 10^{-6}) = 5.70$$

The pH is 5.70, in agreement with Figure 18.5.

Comment • If you use the Henderson–Hasselbalch equation (18.2), the solution is

$$pH = pK_a + \log\frac{mol\ CH_3CO_2^-}{mol\ CH_3CO_2H} = 4.74 + \log\frac{0.00900}{0.0010}$$

$$= 4.74 + (0.95) = 5.69$$

The titration of 0.100 M acetic acid with 0.100 M NaOH is described in the text. What is the pH of the solution when 35.0 mL of the base has been added to 100.0 mL of 0.100 M acetic acid?

Titration of Weak Polyprotic Acids

The titrations illustrated thus far have been for the reaction of a monoprotic acid (HA) with a base such as NaOH. It is possible to extend the discussion of titrations to polyprotic acids such as oxalic acid, $H_2C_2O_4$.

$$H_2C_2O_4(aq) + H_2O(\ell) \rightleftharpoons HC_2O_4^-(aq) + H_3O^+(aq) \qquad K_{a1} = 5.9 \times 10^{-2}$$
$$HC_2O_4^-(aq) + H_2O(\ell) \rightleftharpoons C_2O_4^{2-}(aq) + H_3O^+(aq) \qquad K_{a2} = 6.4 \times 10^{-5}$$

Figure 18.6 illustrates the curve for the titration of 100 mL of 0.100 M oxalic acid with 0.100 M NaOH. The first significant rise in pH is experienced after 100 mL of base has been added, indicating that the first proton of the acid has been titrated.

$$H_2C_2O_4(aq) + OH^-(aq) \longrightarrow HC_2O_4^-(aq) + H_2O(aq)$$

When the second proton of oxalic acid is titrated the pH again rises significantly.

$$HC_2O_4^-(aq) + OH^-(aq) \longrightarrow C_2O_4^{2-}(aq) + H_2O(\ell)$$

Figure 18.6   Titration curve for a diprotic acid. The curve for the titration of 100. mL of 0.100 M oxalic acid ($H_2C_2O_4$, a weak diprotic acid) with 0.100 M NaOH. The first equivalence point (at 100 mL) occurs when the first H ion of $H_2C_2O_4$ is titrated, and the second (at 200 mL) occurs at the completion of the reaction. The curve for pH versus volume of NaOH added shows an initial rise at the first equivalence point and then another rise at the second equivalence point.

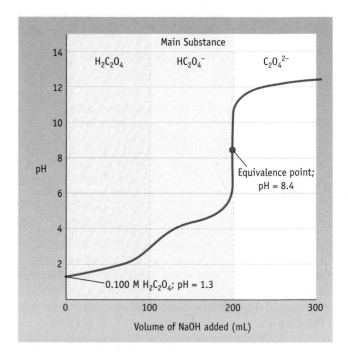

The pH at the second equivalence point is controlled by the conjugate base of the hydrogen oxalate ion, $C_2O_4^{2-}$.

$$C_2O_4^{2-}(aq) + H_2O(\ell) \rightleftharpoons HC_2O_4^-(aq) + OH^-(aq) \qquad K_b = \frac{K_w}{K_{a2}} = 1.6 \times 10^{-10}$$

Calculation of the pH at the equivalence point indicates that it should be about 8.5, as observed.

Titration of a Weak Base with a Strong Acid

Finally, it is useful to consider the titration of a weak base with a strong acid. Figure 18.7 illustrates the pH curve for the titration of 100.0 mL of 0.100 M NH_3 with 0.100 M HCl.

$$NH_3(aq) + HCl(aq) \longrightarrow NH_4^+(aq) + Cl^-(aq)$$

The initial pH for a 0.100 M NH_3 solution is 11.12, as expected for this weak base. At the halfway point of the titration, half of the original ammonia has been converted to ammonium chloride, and half of the ammonia remains. Thus, the important species in solution are the weak acid NH_4^+ and its conjugate base NH_3. This means we could base our calculations on the chemical equation

$$NH_4^+(aq) + H_2O(\ell) \rightleftharpoons NH_3(aq) + H_3O^+(aq) \qquad K_a = 5.6 \times 10^{-10}$$

$$[H_3O^+] = \frac{[NH_4^+]}{[NH_3]} \cdot K_a = \left(\frac{[NH_4^+]}{[NH_3]}\right)(5.6 \times 10^{-10})$$

At the halfway point, the concentrations of weak acid (NH_4^+) and conjugate base (NH_3) are the same, so

$$[H_3O^+] = K_a$$

$$pH = pK_a = -\log(5.6 \times 10^{-10}) = 9.25$$

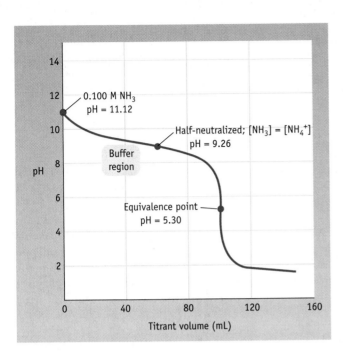

Figure 18.7 Titration of a weak base with a strong acid. The change in pH during the titration of a weak base (100.0 mL of 0.100 M NH_3) with a strong acid (0.100 M HCl). The pH at the half-neutralization point is equal to the pK_a for the conjugate acid (NH_4^+) of the weak base (NH_3) (pH = pK_a = 9.25). At the equivalence point the solution contains the NH_4^+ ion, a weak acid, so the pH is about 5.

As the addition of HCl to NH_3 continues, the pH declines slowly because of the buffering action of the NH_3/NH_4^+ combination. Near the equivalence point, however, the pH drops rapidly. At the equivalence point, the solution contains only ammonium chloride, a weak Brønsted acid, and the solution is weakly acidic.

Example 18.8   Titration of Ammonia with HCl

Problem • What is the pH of the solution at the equivalence point in the titration of 100.0 mL of 0.100 M ammonia with 0.100 M HCl (see Figure 18.7)?

Strategy • This problem closely resembles Example 17.8: (a) A stoichiometry calculation to find the concentration of NH_4^+ at the equivalence point. (b) An equilibrium calculation to find $[H_3O^+]$ for a solution of the weak acid NH_4^+.

Solution •

Part 1: Stoichiometry Problem
Here we are titrating 0.0100 mol of NH_3 (= cV), so 0.0100 mol of HCl is required. Thus, 100.0 mL of 0.100 M HCl (= 0.0100 mol HCl) must be used in the titration.

Equation	NH_3	+ H_3O^+	$\rightleftharpoons$ NH_4^+ + H_2O
Initial (mol = cV)	0.0100	0	0
Amount HCl added (mol)	——	0.0100	——
Change on reaction (mol)	−0.0100	−0.0100	+0.0100
After reaction (mol)	0	0	0.0100
Concentration (M)	0	0	$\dfrac{0.0100 \text{ mol}}{0.200 \text{ L}}$ = 0.0500 M

Part 2: Equilibrium Problem
When the equivalence point is reached, the solution consists of 0.0500 M NH_4^+. The pH is determined by the ionization of this acid

Equation	NH_4^+ + H_2O	$\rightleftharpoons$	NH_3	+	H_3O^+
Initial (M)	0.0500		0		0
Change (M)	−x		+x		+x
Equilibrium (M)	0.0500 − x		x		x

Using K_a for the ionization of the weak acid NH_4^+ (and assuming x ≪ 0.0500), we have

$$K_a = 5.6 \times 10^{-10} = \frac{[NH_3][H_3O^+]}{[NH_4^+]}$$

$$[H_3O^+] = \sqrt{(5.6 \times 10^{-10})(0.0500)} = 5.3 \times 10^{-6} \text{ M}$$

$$pH = 5.28$$

The pH at the equivalence point in this weak base/strong acid titration is indeed slightly acidic as expected.

> **Exercise 18.9**   Titration of a Weak Base with a Strong Acid
>
> Calculate the pH after 75.0 mL of 0.100 M HCl has been added to 100.0 mL of 0.100 M NH_3 with 0.100 M HCl. Compare your answer with Figure 18.7.

pH Indicators

● **Mauve, A Synthetic Dye**
Mauve, the synthetic dye discovered by Perkin, is a weak acid. See Chapter 11, page 418.

The blossoms of hydrangeas are red or blue, depending on the pH of the medium. Indeed, many organic compounds, both natural and synthetic, have a color that changes with pH (← FIGURE 17.2). Not only does this add beauty and variety to our world, but it is a useful property in chemistry.

You have likely carried out an acid–base titration in the laboratory, and, before starting the titration, you would have added an indicator. The **acid–base indicator** is usually an organic compound that is itself a weak acid or weak base, and the anthocyanin dyes of flowers (page 738) are typical. In aqueous solution the acid form is in equilibrium with its conjugate base. Abbreviating the indicator's acid formula as HInd and the formula of its conjugate base as Ind^-, we can write the equilibrium equation

$$HInd(aq) + H_2O(\ell) \rightleftharpoons H_3O^+(aq) + Ind^-(aq)$$

Acid–base indicators are compounds whose color is sensitive to pH. The acid form of the compound (HInd) has one color, and the conjugate base (Ind^-) has another (Figure 18.8). To see how such compounds can be used as equivalence point indicators, let us write the usual equilibrium constant expressions for the dependence of hydronium ion concentration or pH on the indicator's ionization constant (K_a) and on the relative quantities of the acid and conjugate base.

$$[H_3O^+] = \frac{[HInd]}{[Ind^-]} \cdot K_a \qquad \text{or} \qquad pH = pK_a + \log \frac{[Ind^-]}{[HInd]} \qquad \text{(18.6)}$$

> ### Problem-Solving Tip 18.2
>
> #### Calculating the pH at Various Stages of an Acid–Base Reaction
>
> Finding the pH at or before the equivalence point for an acid–base reaction always involves several calculation steps. There are no shortcuts. Consider the *titration of a weak base, B, with a strong acid* as in Example 18.8. (The same principles apply to other acid–base reactions.)
>
> $$H_3O^+(aq) + B(aq) \longrightarrow BH^+(aq) + H_2O(\ell)$$
>
> **Step 1.** *Stoichiometry problem.* Up to the equivalence point, acid is consumed completely to leave a solution containing some base (B) and its conjugate acid (BH^+). Use the principles of stoichiometry to calculate (a) the amount of acid added, (b) the amount of base consumed, (c) the amount of conjugate acid (BH^+) formed and (d) the amount of base remaining.
>
> **Step 2.** *Calculate the concentrations of base, [B], and conjugate acid, $[BH^+]$* by dividing amounts by total solution volume.
>
> Recognize that the volume of the solution at any point is the sum of the original volume of base solution plus the volume of acid solution added.
>
> **Step 3.** *Calculate pH before the equivalence point.* At any point before the equivalence point, the solution is a buffer solution because both the base and its conjugate acid are present. Calculate $[H_3O^+]$ using the concentrations (or amounts) of Step 2 and the value of K_a for the conjugate acid of the weak base.
>
> **Step 4.** *pH at the equivalence point.* Calculate the concentration of the conjugate acid using the procedure of Steps 1 and 2. Use the value of K_a for the conjugate acid of the weak base and the procedure outlined in Example 18.8.

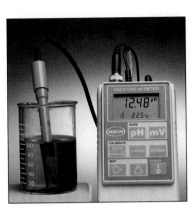

Figure 18.8 **Phenolphthalein, a common acid–base indicator.** Phenolphthalein, a weak acid, is colorless. As the pH increases, the pink conjugate base form predominates, and the color of the solution changes. The change in color is most noticeable around pH 9 (see Figure 18.4). For other suitable indicator dyes, see Figure 17.2. *(Charles D. Winters)*

Phenolphthalein, Brønsted acid, colorless

Conjugate base of phenolphthalein, Brønsted base, pink

$(aq) + 2 H_2O(\ell) \rightleftharpoons 2 H_3O^+(aq) +$

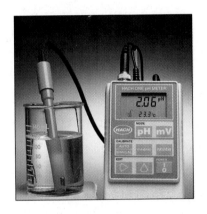

These equations inform us that

- When the hydronium ion concentration is equivalent to the value of K_a (or when pH = pK_a), then [HInd] = [Ind$^-$].
- When [H$_3$O$^+$] > K_a (or pH < pK_a), then [HInd] > [Ind$^-$].
- When [H$_3$O$^+$] < K_a (or pH > pK_a), then [HInd] < [Ind$^-$].

Now let us apply these conclusions to, for example, the titration of an acid with a base using an indicator whose pK_a value is close to the numerical value of the pH at the equivalence point. At the beginning of the titration, the pH is low and [H$_3$O$^+$] is high; the acid form of the indicator (HInd) predominates. Its color is the one observed. As the titration progresses and the pH increases ([H$_3$O$^+$] decreases), less of the acid HInd and more of its conjugate base exist in solution. Finally, by the time we reach the equivalence point, [Ind$^-$] is much larger than HInd, and the color of Ind$^-$ is observed.

Several obvious questions remain to be answered. If you are trying to analyze for an acid and add an indicator that is a weak acid, won't this affect the analysis? Recall that you used only a tiny amount of an indicator in a titration. Although the indicator molecules also react with the base as the titration progresses, so little indicator is present that the analysis is not in error.

Another question is whether you could accurately determine the pH by observing the color change of an indicator. In practice, your eyes are not quite that good. Usually, you see the color of HInd when [HInd]/[Ind$^-$] is about 10/1, and the color of Ind$^-$ when [HInd]/[Ind$^-$] is about 1/10. This means the color change is observed over a hydronium ion concentration interval of about 2 pH units. However, as you can see in Figures 18.4–18.7, on passing through the equivalence point of these titrations the pH changes by as many as 7 units.

As Figure 17.2 shows, a variety of indicators is available, each changing color in a different pH range. If you are analyzing a weak acid or base by titration, you must choose an indicator that changes color in a range that includes the pH to be observed at the equivalence point. This means that an indicator that changes color in the pH range 7 ± 2 should be used for a strong acid/strong base titration. On the other hand, the pH at the equivalence point in the titration of a weak acid with a strong base is greater than 7, and you should choose an indicator that changes color at a pH near the anticipated equivalence point.

Exercise 18.10   **Indicators**

Use Figure 17.2 to decide which indicator is best to use in the titration of NH_3 with HCl shown in Figure 18.7.

18.4 SOLUBILITY OF SALTS

Precipitation reactions (◄ SECTION 5.2) are exchange reactions in which one of the products is a water-insoluble compound such as $CaCO_3$,

$$CaCl_2(aq) + Na_2CO_3(aq) \longrightarrow CaCO_3(s) + 2\ NaCl(aq)$$

that is, a compound having a water solubility of less than about 0.01 mol of dissolved material per liter of solution (Figure 18.9).

How do you know when to predict an insoluble compound as the product of a reaction? In Chapter 5 we listed some guidelines for predicting solubility (Figure 5.1) and mentioned a few important minerals that are insoluble in water. Now we want to make our estimates of solubility more quantitative and to explore conditions under which some compounds precipitate and others do not [🔵 CD-ROM, Screen 18.8].

(a) Metal sulfides in a black smoker (page 119).

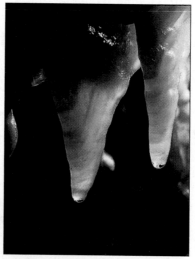

(b) $CaCO_3$ stalactites.

(c) Eggshells are $CaCO_3$.

Figure 18.9   **Insoluble substances in nature.** *(a, National Oceanic and Atmospheric Administration; b, Arthur Palmer; c, Charles D. Winters)*

Solubility Product Constant, K_{sp}

Silver bromide, AgBr, is used in photographic film (Figure 18.10). If some AgBr is placed in pure water, a tiny amount of the compound dissolves, and an equilibrium is established.

$$AgBr(s) \rightleftharpoons Ag^+(aq, 7.35 \times 10^{-7} \text{ M}) + Br^-(aq, 7.35 \times 10^{-7} \text{ M})$$

When the AgBr has dissolved to the greatest extent possible, the solution is said to be saturated (⬅ SECTION 14.2), and experiments show that the concentrations of the silver and bromide ions in the solution are each about 7.35×10^{-7} M at 25 °C. The extent to which an insoluble salt dissolves can be expressed in terms of the equilibrium constant for the dissolving process. In this case the appropriate expression is

$$K = [Ag^+][Br^-]$$

Geochemists and geologists are often interested in knowing the water solubility of a compound. This can be estimated by determining the concentration of the cation or anion when the compound dissolves. For example, if we find that AgBr dissolves to give a silver ion concentration of 7.35×10^{-7} mol/L, we know that 7.35×10^{-7} mol of AgBr must have dissolved per liter of solution (and that the bromide ion concentration also equals 7.35×10^{-7} M). Therefore, the calculated value of the equilibrium constant for the dissolving process is

$$K = [Ag^+(aq)][Br^-(aq)] = (7.35 \times 10^{-7})(7.35 \times 10^{-7}) = 5.40 \times 10^{-13} \text{ (at 25 °C)}$$

Equilibrium constants for the dissolving of other insoluble salts can be calculated in the same manner.

The equilibrium constant that reflects the solubility of a compound is referred to as its **solubility product constant.** We use the notation K_{sp} for such constants, the subscript "sp" denoting a "solubility product." For AgBr, $K_{sp} = 5.4 \times 10^{-13}$ [💿 CD-ROM, Screen 18.9].

The solubility product constant, K_{sp}, for any salt always has the form

$$A_xB_y(s) \rightleftharpoons x A^{y+}(aq) + y B^{x-}(aq) \qquad K_{sp} = [A^{y+}]^x[B^{x-}]^y$$

This means that you would write K_{sp} expressions for the dissolving of other salts as follows

$$CaF_2(s) \rightleftharpoons Ca^{2+}(aq) + 2\,F^-(aq) \qquad K_{sp} = [Ca^{2+}][F^-]^2 = 5.3 \times 10^{-11}$$
$$Ag_2SO_4(s) \rightleftharpoons 2\,Ag^+(aq) + SO_4^{2-}(aq) \qquad K_{sp} = [Ag^+]^2[SO_4^{2-}] = 1.2 \times 10^{-5}$$

The numerical values of K_{sp} for a few salts are given in Table 18.2, and more values are collected in Appendix J.

Finally, do not confuse the solubility of a compound with its solubility product constant. The *solubility* of a salt is the amount present in some volume of a saturated solution, expressed in moles per liter, grams per 100 mL, or other units. The *solubility product constant* is an equilibrium constant. Nonetheless, there is a connection between them: if one is known, the other can be calculated.

● **Writing K Expressions for Solubility**
Recall that the "concentration" of solids does not appear in an equilibrium constant expression (⬅ PAGE 660).

Figure 18.10 Photographic film is coated with water-insoluble silver bromide, AgBr. *(Charles D. Winters)*

Exercise 18.11   Writing K_{sp} Expressions

Write K_{sp} expressions for the following insoluble salts and look up numerical values for the constant in Appendix J.

(a) AgI   **(b)** BaF₂   **(c)** Ag₂CO₃

Table 18.2 • Some Common, Slightly Soluble Compounds and Their K_{sp} Values*

Formula	Name	K_{sp} (25 °C)	Common Names/Uses
$CaCO_3$	Calcium carbonate	3.4×10^{-9}	Calcite, Iceland spar
$MnCO_3$	Manganese(II) carbonate	2.3×10^{-11}	Rhodochrosite (forms rose-colored crystals)
$FeCO_3$	Iron(II) carbonate	3.1×10^{-11}	Siderite
CaF_2	Calcium fluoride	5.3×10^{-11}	Fluorite (source of HF and other inorganic fluorides)
$AgCl$	Silver chloride	1.8×10^{-10}	Chlorargyrite
$AgBr$	Silver bromide	5.4×10^{-13}	Used in photographic film
$CaSO_4$	Calcium sulfate	4.9×10^{-5}	Hydrated form is commonly called gypsum
$BaSO_4$	Barium sulfate	1.1×10^{-10}	Barite (used in "drilling mud" and as a component of paints)
$SrSO_4$	Strontium sulfate	3.4×10^{-7}	Celestite
$Ca(OH)_2$	Calcium hydroxide	5.5×10^{-5}	Slaked lime

*The values reported in this table were taken from *Lange's Handbook of Chemistry,* 15th Edition, McGraw Hill Publishers, New York, NY (1999). Additional K_{sp} values are given in Appendix J.

Relating Solubility and K_{sp}

Solubility product constants are determined by careful laboratory measurements. The examples that follow illustrate how to calculate K_{sp} from experimental information [⌘ CD-ROM, Screen 18.10].

Example 18.9 K_{sp} from Solubility Measurements

Problem • The solubility of silver iodide, AgI, is 9.2×10^{-9} mol/L at 25 °C. Calculate K_{sp} for AgI.

Strategy • We first write the K_{sp} expression for AgI and then substitute the numerical values for the equilibrium concentrations of the ions.

Solution • When silver iodide dissolves in solution, the following balanced chemical equation and K_{sp} expression are appropriate.

$$AgI(s) \rightleftharpoons Ag^+(aq) + I^-(aq) \qquad K_{sp} = [Ag^+][I^-]$$

We know from experiments that the solubility of AgI is 9.2×10^{-9} mol/L at 25 °C. Because each mole of AgI that dissolves provides 1 mol of Ag^+ ions and 1 mol of I^- ions, the concentration of each ion is 9.2×10^{-9} M. Therefore,

$$K_{sp} = [Ag^+][I^-]$$
$$= (9.2 \times 10^{-9})(9.2 \times 10^{-9}) = 8.5 \times 10^{-17}$$

Example 18.10 K_{sp} from Solubility Measurements

Problem • Calcium fluoride dissolves to a slight extent in water.

$$CaF_2(s) \rightleftharpoons Ca^{2+}(aq) + 2 F^-(aq) \qquad K_{sp} = [Ca^{2+}][F^-]^2$$

Calculate the K_{sp} value for CaF_2 if the calcium ion concentration has been found to be 2.4×10^{-4} mol/L.

Strategy • We first write the K_{sp} expression for CaF_2 and then substitute the numerical values for the equilibrium concentrations of the ions.

Solution • When CaF_2 dissolves to a small extent in water, the balanced equation shows that the concentration of F^- ion must be twice the Ca^{2+} ion concentration.

If $[Ca^{2+}] = 2.4 \times 10^{-4}$ M,

then $[F^-] = 2 \times [Ca^{2+}] = 4.8 \times 10^{-4}$ M

This means the solubility product constant is

$$K_{sp} = [Ca^{2+}][F^-]^2 = (2.4 \times 10^{-4})(4.8 \times 10^{-4})^2$$
$$= 5.5 \times 10^{-11}$$

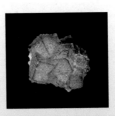

◄ **Fluorite.** The mineral fluorite is composed of CaF_2, a compound with a very low water-solubility. The mineral can vary widely in color from purple, to green, to colorless.

Exercise 18.12 K_{sp} from Solubility Measurements

The barium ion concentration, $[Ba^{2+}]$ in a saturated solution of barium fluoride is 3.6×10^{-3} M. Calculate the value of the K_{sp} for BaF_2.

$$BaF_2(s) \rightleftharpoons Ba^{2+}(aq) + 2 F^-(aq)$$

K_{sp} values for insoluble salts are an invaluable aid because they can be used to estimate the solubility of a solid salt or to determine if a solid will precipitate when solutions of its anion and cation are mixed. Let us first look at an example of the estimation of the solubility of a salt from its K_{sp} value. Later we shall see how to use these predictions to plan the separation of ions that are mixed in solution (page 776) [⏺ CD-ROM, Screen 18.11].

Example 18.11 Solubility from K_{sp}

Problem • The K_{sp} for $BaSO_4$ (as the mineral barite) is 1.1×10^{-10} at 25 °C. Calculate the solubility of barium sulfate in pure water in (a) moles per liter and (b) grams per liter.

Strategy • When 1 mol of $BaSO_4$ dissolves, 1 mol of Ba^{2+} ions and 1 mol of SO_4^{2-} ions are produced. Thus, the solubility of $BaSO_4$ can be estimated by calculating the concentration of either Ba^{2+} or SO_4^{2-} from the solubility product constant.

Solution • The equation for the solubility of $BaSO_4$ is

$$BaSO_4(s) \rightleftharpoons Ba^{2+}(aq) + SO_4^{2-}(aq)$$
$$K_{sp} = [Ba^{2+}][SO_4^{2-}] = 1.1 \times 10^{-10}$$

Let us denote the solubility of $BaSO_4$ (in moles per liter) by x; that is, x moles of $BaSO_4$ dissolves per liter. Therefore, both $[Ba^{2+}]$ and $[SO_4^{2-}]$ must also equal x at equilibrium.

Equation	$BaSO_4(s) \rightleftharpoons$	Ba^{2+} +	$SO_4^{2-}(aq)$
Initial (M)		0	0
Change (M)		$+x$	$+x$
Equilibrium (M)		x	x

◄ **Barium sulfate.** Barium sulfate, a white solid, is quite insoluble in water ($K_{sp} = 1.1 \times 10^{-10}$). **(a)** A sample of the mineral barite, which is mostly barium sulfate. **(b)** Barium sulfate is opaque to x-rays. Therefore, if you drink a "cocktail" containing $BaSO_4$ it does not dissolve in your stomach or intestines, and its progress through your digestive organs can be followed by x-ray analysis. This photo is an x-ray film of a gastrointestinal tract after a person ingested barium sulfate. It is fortunate that $BaSO_4$ is so insoluble, because water- and acid-soluble barium salts are toxic. *(a, Charles D. Winters; b, Susan Leavines/Science Source/Photo Researchers, Inc.)*

(a)　　　　(b)

(Example continues on next page)

Because K_{sp} is the product of the barium and sulfate ion concentrations, K_{sp} is the square of the solubility, x,

$$K_{sp} = [Ba^{2+}][SO_4^{2-}] = 1.1 \times 10^{-10} = (x)(x) = x^2$$

and so the value of x is

$$x = [Ba^{2+}] = [SO_4^{2-}] = \sqrt{1.1 \times 10^{-10}} = 1.0 \times 10^{-5} \text{ M}$$

The solubility of $BaSO_4$ in pure water is 1.0×10^{-5} mol/L. To find its solubility in grams per liter, we need only multiply by the molar mass of $BaSO_4$.

Solubility in grams per liter = $(1.0 \times 10^{-5}$ mol/L$)(233$ g/mol$)$
$$= 0.0024 \text{ g/L}$$

Comment • It will always be the case that the solubility of a salt having one cation and one anion (MX) can be estimated from

Solubility of MX (mol/L) = $\sqrt{K_{sp}}$

Example 18.12 Solubility from K_{sp}

Problem • Knowing that the K_{sp} value for MgF_2 is 5.2×10^{-11}, calculate the solubility of the salt in (a) moles per liter and (b) grams per liter.

Strategy • The problem is to define the salt solubility in terms that will allow us to solve the K_{sp} expression for this value. We know that, if 1 mol of MgF_2 dissolves, 1 mol of Mg^{2+} and 2 mol of F^- appear in the solution. This means the MgF_2 solubility (in moles dissolved per liter) is equivalent to the concentration of Mg^{2+} ion in the solution. Thus, if the solubility of MgF_2 is x mol/L, then $[Mg^{2+}] = x$ and $[F^-] = 2x$.

Solution • As usual we begin by writing the equilibrium equation and the K_{sp} expression.

$$MgF_2(s) \rightleftharpoons Mg^{2+}(aq) + 2 F^-(aq)$$
$$K_{sp} = [Mg^{2+}][F^-]^2 = 5.2 \times 10^{-11}$$

The MgF_2 solubility in moles per liter is equivalent to the concentration of Mg^{2+} that appears in the solution. If the solubility of MgF_2 is given by the unknown quantity x, then $[Mg^{2+}] = x$ and $[F^-] = 2x$.

Equation	$MgF_2(s) \rightleftharpoons$	$Mg^{2+}(aq)$	+	$2 F^-(aq)$
Initial (M)		0		0
Change		$+x$		$+2x$
Equilibrium (M)		x		$2x$

Substituting these values into the K_{sp} expression, we find

$$K_{sp} = [Mg^{2+}][F^-]^2 = (x)(2x)^2 = 4x^3$$

Solving the equation for x,

$$x = \sqrt[3]{\frac{K_{sp}}{4}} = \sqrt[3]{\frac{5.2 \times 10^{-11}}{4}} = 2.4 \times 10^{-4}$$

we find that 2.4×10^{-4} mol of MgF_2 dissolves per liter to give

$$[Mg^{2+}] = x = 2.4 \times 10^{-4} \text{ M}$$
$$[F^-] = 2x = 2 \times (2.4 \times 10^{-4}) = 4.8 \times 10^{-4} \text{ M}$$

Finally, the solubility of MgF_2 in grams per liter is

$$(2.4 \times 10^{-4} \text{ mol/L})(62.3 \text{ g/mol}) = 0.015 \text{ g } MgF_2/\text{L}$$

Comment • Problems like this one often provoke our students to ask such questions as "Aren't you counting things twice when you multiply x by 2 and then square it as well?" in the expression $K_{sp} = (x)(2x)^2$. The answer is no. The quantities that belong in the equation are the concentration of Mg^{2+} and the concentration of F^-.

$$K_{sp} = (\text{concentration of } Mg^{2+})(\text{concentration of } F^-)^2$$

From the formula of the compound we know that the concentration of F^- must be twice that of Mg^{2+}: if $[Mg^{2+}] = x$ then $[F^-]$ must equal $2x$. Having defined the ion concentration, we then square the concentration of F^- as the equation demands.

Exercise 18.13 Salt Solubility from K_{sp}

Using the value of K_{sp} in Appendix J, calculate the solubility of $Ca(OH)_2$ in moles per liter and grams per liter.

The *relative* solubilities of salts can often be deduced by comparing values of solubility product constants, but you must be careful! For example, the K_{sp} for silver chloride is

$$AgCl(s) \rightleftharpoons Ag^+(aq) + Cl^-(aq) \qquad K_{sp} = 1.8 \times 10^{-10}$$

whereas that for silver carbonate is

$$Ag_2CO_3(s) \rightleftharpoons 2\ Ag^+(aq) + CO_3{}^{2-}(aq) \qquad K_{sp} = 8.5 \times 10^{-12}$$

In spite of the fact that Ag_2CO_3 has a numerically smaller K_{sp} value than AgCl, the carbonate salt is about ten times *more* soluble than the chloride salt. If you solve for their solubilities as in the previous examples, you would find the solubility of AgCl is 1.3×10^{-5} mol/L, whereas that of Ag_2CO_3 is 1.3×10^{-4} mol/L. From this example and countless others, we conclude that

> Direct comparisons of the solubility of two salts on the basis of their K_{sp} values can only be made for salts having the same ion ratio.

This means, for example, that you can directly compare solubilities of 1:1 salts such as the silver halides by comparing their K_{sp} values.

$$AgI\ (K_{sp} = 8.5 \times 10^{-17}) < AgBr\ (K_{sp} = 5.4 \times 10^{-13}) < AgCl\ (K_{sp} = 1.8 \times 10^{-10})$$

──────increasing K_{sp} and increasing solubility──────→

Similarly, you can compare 1:2 salts such as the lead halides.

$$PbI_2\ (K_{sp} = 9.8 \times 10^{-9}) < PbBr_2\ (K_{sp} = 6.6 \times 10^{-6}) < PbCl_2\ (K_{sp} = 1.7 \times 10^{-5})$$

──────increasing K_{sp} and increasing solubility──────→

but you cannot directly compare a 1:1 salt (AgCl) with a 2:1 salt (Ag_2CrO_4).

Exercise 18.14 **Comparing Salt Solubilities**

Using K_{sp} values, predict which salt in each pair is more soluble in water.

(a) AgCl or AgCN   **(b)** $Mg(OH)_2$ or $Ca(OH)_2$ **(c)** $Ca(OH)_2$ or $CaSO_4$

Solubility and the Common Ion Effect

The test tube on the left in Figure 18.11 contains a precipitate of silver acetate, $AgCH_3CO_2$, in water. The solution is saturated, and silver ions and acetate ions in solution are in equilibrium with solid silver acetate.

$$AgCH_3CO_2(s) \rightleftharpoons Ag^+(aq) + CH_3CO_2{}^-(aq)$$

● = Ag^+
△ = $CH_3CO_2{}^-$
■ = $NO_3{}^-$

Figure 18.11   Common ion effect. The tube at the left contains a saturated solution of silver acetate, $AgCH_3CO_2$. When 1 M $AgNO_3$ is added to the tube (*right*), more solid silver acetate forms. (*Charles D. Winters*)

A Closer Look

Solubility Calculations

The K_{sp} value reported for lead(II) chloride, $PbCl_2$, is 1.7×10^{-5}. If we assume the appropriate equilibrium in solution is

$$PbCl_2(s) \rightleftharpoons Pb^{2+}(aq) + 2\,Cl^-(aq)$$

the calculated solubility of $PbCl_2$ is 0.016 M. The experimental value for the solubility of the salt, however, is 0.036 M. The experimental value is more than twice the calculated value! What is the problem? There are several, as summarized by the diagram below.

The main problem in the lead(II) chloride case, and in many others, is that the compound dissolves but is not 100% dissociated into its constituent ions. Instead, it dissolves as the undissociated salt or forms ion pairs.

Other problems that lead to discrepancies between calculated and experimental solubilities are the reactions of ions with water (particularly anions) and complex ion formation. An example of the former effect is the reaction of sulfide ion with water, that is, hydrolysis.

$$S^{2-}(aq) + H_2O(\ell) \rightleftharpoons HS^-(aq) + OH^-(aq)$$

Sulfide ion is a strong base ($K_b = 1 \times 10^5$), and its hydrolysis reaction is strongly product-favored. It is this hydrolysis of a basic anion that drives the equilibrium reaction for dissolving a metal sulfide to the right. That is to say, the solubility of a metal sulfide is not accurately expressed by the chemical equation for solubility product constants.

$$MS(s) \rightleftharpoons M^{2+}(aq) + S^{2-}(aq)$$
$$K_{sp} = [M^{2+}][S^{2-}]$$

The solubility of the sulfide is greater than expressed by the K_{sp} value.

The true solubility of a metal sulfide is better represented by a modified solubility product constant, K_{spa}, which is defined as follows:

$$MS(s) \rightleftharpoons M^{2+}(aq) + S^{2-}(aq)$$
$$K_{sp} = [M^{2+}][S^{2-}]$$
$$S^{2-}(aq) + H_2O(\ell) \rightleftharpoons HS^-(aq) + OH^-(aq)$$
$$K_b = [HS^-][OH^-]/[S^{2-}]$$

Net:
$$MS(s) + H_2O(\ell) \rightleftharpoons$$
$$HS^-(aq) + OH^-(aq)M^{2+}(aq)$$
$$K_{spa} = [M^{2+}][HS^-][OH^-] = K_{sp} \times K_b$$

Values for K_{spa} for several metal sulfides are included in Appendix J. A similar analysis applies to the water solubility of any insoluble salt containing a basic anion (such as CO_3^{2-} and PO_4^{3-}).

Complex ion formation is illustrated by the fact that lead chloride is more soluble in the presence of *excess* chloride ion, owing to the formation of the complex ion $PbCl_4^{2-}$.

$$PbCl_2(s) + 2\,Cl^-(aq) \rightleftharpoons PbCl_4^{2-}(aq)$$

Both hydrolysis and complex ion formation are discussed further on pages 768–770 and 773–775.

For further information on these issues see:

(a) L. Meites, J. S. F. Pode, and H. C. Thomas: *Journal of Chemical Education*, Vol. 43, pp. 667–672, 1966.

(b) S. J. Hawkes: *Journal of Chemical Education*, Vol. 75, pp. 1179–1181, 1998.

(c) R. W. Clark and J. M. Bonicamp: *Journal of Chemical Education*, Vol. 75, pp. 1182–1185, 1998.

(d) R. J. Myers: *Journal of Chemical Education*, Vol. 63, pp. 687–690, 1986.

$PbCl_2(aq)$ $\xrightarrow{\;\; K = 0.63 \;\;}$ $PbCl^+(aq) + Cl^-(aq)$

Undissociated salt dissolved in water   Ion pairs

$K = 0.0011 \updownarrow$ $\updownarrow K = 0.026$

$PbCl_2(s)$ $\xrightarrow{\;\; K = 1.7 \times 10^{-5} \;\;}$ $Pb^{2+}(aq) + 2\,Cl^-(aq)$

Slightly soluble salt   100% dissociated into ions

But what would happen if the silver ion concentration is increased, say by adding silver nitrate? Le Chatelier's principle ($\Leftarrow$ SECTION 16.6) suggests — and we observe — that more silver acetate precipitate should form because a product ion has been added, causing the equilibrium to shift back toward silver acetate formation.

The ionization of weak acids and bases is affected by the presence of an ion common to the equilibrium process (see Section 18.1), and the effect of adding silver ion to a saturated silver acetate solution is another example of the common ion effect. Adding a common ion will lower the salt solubility [CD-ROM, Screen 18.13].

Example 18.13 **Common Ion Effect and Salt Solubility**

Problem • If solid AgCl is placed in 1.00 L of 0.55 M NaCl, what mass of AgCl will dissolve?

Strategy • The presence of an ion common to the equilibrium always suppresses the solubility of a salt. As a measure of the solubility of the salt under these circumstances, calculate the concentration of the ion other than the common ion (Ag^+ ion in this case).

Solution • In pure water, the solubility of AgCl is equal to *either* $[Ag^+]$ *or* $[Cl^-]$.

$$AgCl(s) \rightleftharpoons Ag^+(aq) + Cl^-(aq) \qquad K_{sp} = 1.8 \times 10^{-10}$$

Solubility of AgCl in pure water = $[Ag^+]$ or $[Cl^-]$

$$= \sqrt{K_{sp}}$$

$$= 1.3 \times 10^{-5} \text{ mol/L or } 0.0019 \text{ g/L}$$

However, in water already containing a common ion, here the Cl^- ion, Le Chatelier's principle predicts that the solubility is less than 1.3×10^{-5} mol/L. In this case the solubility of AgCl is equivalent to the concentration of Ag^+ ion in solution, so we set up an ICE table to show the concentrations of Ag^+ and Cl^- when equilibrium is attained.

Equation	AgCl (s)	$\rightleftharpoons$	Ag^+(aq)	+	Cl^-(aq)
Initial (M)			0		0.55
Change (M)			+x		+x
Equilibrium (M)			x		0.55 + x

Some AgCl dissolves in the presence of chloride ion and produces Ag^+ and Cl^- ion concentrations of x mol/L. Some chloride ion was already present, however, so the total chloride ion concentration is what was already there (0.55 M) *plus* the amount supplied

by AgCl dissociation (equals x).

The equilibrium concentrations from the table are substituted into the K_{sp} expression,

$$K_{sp} = 1.8 \times 10^{-10} = [Ag^+][Cl^-] = (x)(0.55 + x)$$

and rearranged to

$$x^2 + 0.55\,x - K_{sp} = 0$$

This is a quadratic equation and can be solved by the methods in Appendix A. An easier approach, however, is to make the approximation that x is *very* small with respect to 0.55 [and so (0.55 + x) ≈ 0.55)]. This is a reasonable assumption because we know that the solubility equals 1.3×10^{-5} M *without* the common ion Cl^-, and that it will be even smaller in the presence of added Cl^-. Therefore,

$$K_{sp} = 1.8 \times 10^{-10} \approx (x)(0.55)$$

$$x = [Ag^+] \approx 3.3 \times 10^{-10} \text{ M} \qquad \text{or} \qquad 4.7 \times 10^{-8} \text{ g/L}$$

As predicted by Le Chatelier's principle, the solubility of AgCl in the presence of added Cl^- is less (3.3×10^{-10} M) than in pure water (1.3×10^{-5} M).

Comment • As a final step, check the approximation by substituting the calculated value of x into the exact expression $K_{sp} = (x)(0.55)$. If the product $(x)(0.55)$ is the same as the given value of K_{sp}, the approximation is valid.

$$K_{sp} = (x)(0.55 + x)$$

$$= (3.3 \times 10^{-10})(0.55)$$

$$\approx 1.8 \times 10^{-10}$$

The approximation we made here is similar to the approximations we make in acid–base equilibrium problems (⬅ SEE PAGE 714 AND EXAMPLE 17.5).

Exercise 18.15 **The Common Ion Effect and Salt Solubility**

Calculate the solubility of $BaSO_4$ (a) in pure water and (b) in the presence of 0.010 M $Ba(NO_3)_2$, K_{sp} for $BaSO_4$ is 1.1×10^{-10}.

Example 18.13 conveys two important general ideas:

• The solubility of a salt will be reduced by the presence of a common ion, in accordance with Le Chatelier's principle.

• We made the approximation that the amount of common ion added to the solution was very large in comparison with the amount of that ion coming from the insoluble salt, and this allowed us to simplify our calculations. This is almost always the case, but you should check to be sure.

Example 18.14   Common Ion Effect and Salt Solubility

Problem • Calculate the solubility of silver chromate, Ag_2CrO_4, at 25 °C in the presence of 0.0050 M K_2CrO_4 solution.

$$Ag_2CrO_4(s) \rightleftharpoons 2\ Ag^+(aq) + CrO_4^{2-}(aq)$$

$$K_{sp} = [Ag^+]^2[CrO_4^{2-}] = 1.1 \times 10^{-12}$$

For comparison, the solubility of the salt in pure water is calculated to be 6.5×10^{-5} mol/L.

Strategy • In the presence of chromate ion from the water-soluble salt K_2CrO_4, the concentration of Ag^+ ions produced by Ag_2CrO_4 will be less than in pure water. Assume the solubility of Ag_2CrO_4 is x mol/L. This means the concentration of Ag^+ ions will be $2x$ mol/L, whereas the concentration of CrO_4^{2-} ions will be x mol/L *plus* the amount of CrO_4^{2-} already in the solution.

Solution •

Equation	$Ag_2CrO_4(s) \rightleftharpoons 2\ Ag^+(aq)\ +\ CrO_4^{2-}(aq)$	
Initial (M)	0	0.0050
Change (M)	$+2x$	$+x$
Equilibrium (M)	$2x$	$0.0050 + x$

Substituting the equilibrium amounts into the K_{sp} expression, we have

$$K_{sp} = 1.1 \times 10^{-12} = [Ag^+]^2[CrO_4^{2-}]$$
$$= (2x)^2(0.0050 + x)$$

Again make the approximation that x is very small with respect to 0.0050, and so $(0.0050 + x) \approx 0.0050$. (This is reasonable because $[CrO_4^{2-}]$ is 6.5×10^{-5} M without added chromate ion, and it is certain that x is even smaller in the presence of extra chromate ion.) Therefore, the approximate expression is

$$K_{sp} = 1.1 \times 10^{-12} = [Ag^+]^2[CrO_4^{2-}] = (2x)^2(0.0050)$$

and so

$$x = [CrO_4^{2-}] \text{ from } Ag_2CrO_4 = 7.4 \times 10^{-6} \text{ M}$$

The solubility of Ag_2CrO_4 with added CrO_4^{2-} ion is 7.4×10^{-6} M. This also means the silver ion concentration in the presence of the common ion is

$$[Ag^+] = 2x = 1.5 \times 10^{-5} \text{ M}$$

This silver ion concentration is less than its value in pure water (1.3×10^{-4} M), and you again see the result of adding an ion "common" to the equilibrium.

Exercise 18.16   Common Ion Effect and Salt Solubility

Calculate the solubility of $Zn(CN)_2$ at 25 °C (a) in pure water and (b) in the presence of 0.10 M $Zn(NO_3)_2$. (K_{sp} for $Zn(CN)_2$ is 8.0×10^{-12}.)

Effect of Basic Anions on Salt Solubility

The next time you are tempted to wash a supposedly insoluble salt down the kitchen or laboratory drain, stop and consider the consequences. Many metal ions such as lead, chromium, and mercury are toxic in the environment. Even if a so-called insoluble salt of one of these cations does not appear to dissolve, its solubility in water may be greater than you think, in part owing to the possibility that the anion of the salt is a weak base or the cation is a weak acid [🖰 CD-ROM, Screen 18.16].

Lead sulfide, PbS, which is found in nature as the mineral galena (Figure 18.12), provides an example of the effect of the acid–base properties of an ion on salt solubility. When placed in water, a trace amount dissolves,

$$PbS(s) \rightleftharpoons Pb^{2+}(aq) + S^{2-}(aq)$$

and one product of the reaction is the sulfide ion. This anion is a strong base,

$$S^{2-}(aq) + H_2O(\ell) \rightleftharpoons HS^-(aq) + OH^-(aq) \qquad K_b = 1 \times 10^5$$

and it undergoes extensive hydrolysis (reaction with water) (← TABLE 17.3). The hydrolysis reaction lowers the S^{2-} concentration, and additional PbS dissolves in an attempt to re-establish equilibrium. The result is that the equilibrium process for dissolving PbS shifts to the right, and the lead ion concentration in solution is greater than expected from the simple ionization of the salt.

The lead sulfide example leads to the following general observation:

> Any salt containing an anion that is the conjugate base of a weak acid will dissolve in water to a greater extent than given by K_{sp}.

This means that salts of phosphate, acetate, carbonate, and cyanide, as well as sulfide, can be affected, because all of these anions undergo the general hydrolysis reaction

$$X^-(aq) + H_2O(\ell) \rightleftharpoons HX(aq) + OH^-(aq)$$

in aqueous solution.

The observation that ions from insoluble salts can undergo hydrolysis leads to yet another useful, general conclusion:

> Insoluble salts in which the anion is the conjugate base of a weak acid will dissolve in strong acids.

Insoluble salts containing such anions as acetate, carbonate, hydroxide, phosphate, and sulfide dissolve in strong acids. For example, you know that if a strong acid is added to a water-insoluble metal carbonate such as $CaCO_3$, the salt dissolves (← SECTION 5.5).

$$CaCO_3(s) + 2\,H_3O^+(aq) \longrightarrow Ca^{2+}(aq) + 3\,H_2O(\ell) + CO_2(g)$$

You can think of this as the net result of a series of reactions.

$$CaCO_3(s) \rightleftharpoons Ca^{2+}(aq) + CO_3^{2-}(aq) \quad K = K_{sp} = 3.4 \times 10^{-9}$$

$$CO_3^{2-}(aq) + H_2O(\ell) \rightleftharpoons HCO_3^-(aq) + OH^-(aq) \quad K = K_{b1} = 2.1 \times 10^{-4}$$

$$HCO_3^-(aq) + H_2O(\ell) \rightleftharpoons H_2CO_3(aq) + OH^-(aq) \quad K = K_{b2} = 2.4 \times 10^{-8}$$

$$2[OH^-(aq) + H_3O^+(aq) \rightleftharpoons 2\,H_2O(\ell)] \qquad K = \left(\frac{1}{K_w}\right)^2 = \left[\frac{1}{(1 \times 10^{-14})}\right]^2$$

Overall: $\quad CaCO_3(s) + 2\,H_3O^+(aq) \rightleftharpoons Ca^{2+}(aq) + 2\,H_2O(\ell) + H_2CO_3(aq)$

$$K_{net} = (K_{sp})(K_{b1})(K_{b2})\left(\frac{1}{K_w}\right)^2 = 1.7 \times 10^8$$

Carbonic acid, a product of this reaction, is not stable,

$$H_2CO_3(aq) \rightleftharpoons CO_2(g) + H_2O(\ell) \qquad\qquad K \approx 10^5$$

and you see CO_2 bubbling out of the solution, a process that moves the $CaCO_3 + H_3O^+$ equilibrium even further to the right. Calcium carbonate dissolves completely in strong acid!

In addition to metal carbonates, many metal sulfides are soluble in strong acids

$$FeS(s) + 2\,H_3O^+(aq) \rightleftharpoons Fe^{2+}(aq) + H_2S(aq) + 2\,H_2O(\ell)$$

Figure 18.12 Lead sulfide. This and other metal sulfides dissolve in water to a greater extent than expected, in part because the sulfide ion reacts with water to form HS^- and OH^-.

$$PbS(s) + H_2O(\ell) \rightleftharpoons$$
$$Pb^{2+}(aq) + HS^-(aq) + OH^-(aq)$$

The model of PbS shows that the unit cell is cubic, a feature reflected by the cubic crystals of the mineral galena. *(Charles D. Winters)*

Figure 18.13   Effect of the anion on salt solubility in acid. (*left*) A precipitate of AgCl (white) and Ag_3PO_4 (yellow). (*right*) Adding a strong acid dissolves Ag_3PO_4 (and leaves insoluble AgCl). The basic anion PO_4^{3-} reacts with acid to give H_3PO_4, whereas Cl^- is too weakly basic to form HCl. (*Charles D. Winters*)

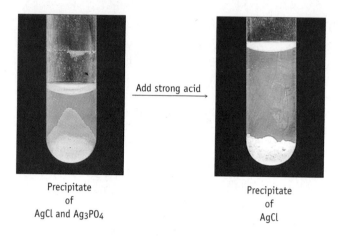

Add strong acid

Precipitate
of
AgCl and Ag3PO4

Precipitate
of
AgCl

as are metal phosphates (Figure 18.13),

$$Ag_3PO_4(s) + 3\,H_3O^+(aq) \rightleftharpoons 3\,Ag^+(aq) + H_3PO_4(aq) + 3\,H_2O(\ell)$$

and metal hydroxides.

$$Mg(OH)_2(s) + 2\,H_3O^+(aq) \longrightarrow Mg^{2+}(aq) + 4\,H_2O(\ell)$$

In general, the solubility of a salt containing the conjugate base of a weak acid is increased by addition of a stronger acid to the solution. In contrast, salts are not soluble in strong acid if the anion is the conjugate base of a strong acid. For example, AgCl is not soluble in strong acid

$$AgCl(s) \rightleftharpoons Ag^+(aq) + Cl^-(aq) \qquad K_{sp} = 1.8 \times 10^{-10}$$

$$H_3O^+(aq) + Cl^-(aq) \rightleftharpoons HCl(aq) + H_2O(\ell) \qquad\qquad K \ll 1$$

because Cl^- is a very weak base (see Table 17.3) and so is not removed by the strong acid H_3O^+ (Figure 18.13). You can see that this same conclusion would apply to insoluble salts of Br^- and I^-.

18.5   PRECIPITATION REACTIONS

Metal-bearing ores contain the metal in the form of an insoluble salt (Figure 18.14), and, to complicate matters, ores often contain several such metal salts. Many industrial methods for separating metals from their ores involve dissolving metal salts to obtain the metal ion or ions in solution. The solution is then concentrated in some manner, and a precipitating agent is added to precipitate selectively only one type of metal ion as an insoluble salt. In the case of iron, for example, the ion can be precipitated as insoluble iron(II) sulfide (Appendix J) or iron(II) carbonate.

$$Fe^{2+}(aq) + HS^-(aq) + H_2O(\ell) \rightleftharpoons FeS(s) + H_3O^+(aq) \qquad K = \frac{1}{K_{spa}} = 1.7 \times 10^{18}$$

$$Fe^{2+}(aq) + CO_3^{2-}(aq) \rightleftharpoons FeCO_3(s) \qquad K = \frac{1}{K_{sp}} = 3.2 \times 10^{10}$$

The final step in obtaining the metal itself is to reduce the metal cation to the metal either chemically or electrochemically (Chapter 20).

Our immediate goal is to work out methods to determine if a precipitate will form under a given set of conditions. For example, if Ag^+ and Cl^- are present at some given concentrations, will AgCl precipitate from the solution [⊕ CD-ROM, Screen 18.12]?

Figure 18.14   Minerals. Minerals are often insoluble salts. Besides iron pyrite (see page 12) and galena (PbS, Figure 18.12) we have such minerals as light purple fluorite (calcium fluoride), black hematite [iron(III) oxide], and rust brown goethite, a mixture of iron(III) oxide and iron(III) hydroxide. (*Charles D. Winters*)

K_{sp} and the Reaction Quotient, Q

Silver chloride, like silver bromide, is used in photographic films. It dissolves to a very small extent in water and has a correspondingly small value of K_{sp}.

$$AgCl(s) \rightleftharpoons Ag^+(aq) + Cl^-(aq) \qquad K_{sp} = [Ag^+][Cl^-] = 1.8 \times 10^{-10}$$

But let us look at the problem from the other direction: if a solution contains Ag^+ and Cl^- ions at some concentration, will AgCl precipitate from solution? This is the same question we asked in Section 16.2 when we wanted to know if a given mixture of reactants and products was an equilibrium mixture, if the reactants continued to form products, or if products would revert to reactants. The solution there was to calculate the reaction quotient, Q.

For silver chloride, the expression for the reaction quotient, Q, is

$$Q = [Ag^+][Cl^-]$$

Recall that *the difference between Q and K is that the concentrations used in the reaction quotient expression may or may not be those at equilibrium.* For the case of a slightly soluble salt such as AgCl, we can reach the following conclusions (← SECTION 16.2).

1. If $Q = K_{sp}$ the solution is saturated.

   When the product of the ion concentrations is equal to K_{sp}, the ion concentrations have reached their equilibrium value.

2. If $Q < K_{sp}$ the solution is not saturated.

   This means one of two things: (i) If solid AgCl is present, more will dissolve until equilibrium is achieved (when $Q = K_{sp}$). (ii) If solid AgCl is not already present, more $Ag^+(aq)$ or more $Cl^-(aq)$ (or both) can be added to the solution until precipitation of solid AgCl begins (when $Q > K_{sp}$).

3. If $Q > K_{sp}$, the system is not at equilibrium; the solution is supersaturated.

   The concentrations of Ag^+ and Cl^- in solution are too high, and AgCl will precipitate (until $Q = K_{sp}$).

Example 18.15   **Solubility and the Reaction Quotient**

Problem • Some solid AgCl has been placed in a beaker of water. After some time, experiment shows that the concentrations of Ag^+ and Cl^- are each 1.2×10^{-5} mol/L. Has the system yet reached equilibrium? That is, is the solution saturated? If not, will more AgCl dissolve?

Strategy • Use the experimental ion concentrations to calculate the reaction quotient Q. Compare Q and K_{sp} to decide if the system is at equilibrium (when $Q = K_{sp}$).

Solution • For this AgCl case,

$$Q = [Ag^+][Cl^-] = (1.2 \times 10^{-5})(1.2 \times 10^{-5}) = 1.4 \times 10^{-10}$$

Here Q is less than K_{sp} (1.8×10^{-10}). The solution is not yet saturated, and AgCl will continue to dissolve until $Q = K_{sp}$, at which point $[Ag^+] = [Cl^-] = 1.3 \times 10^{-5}$ M. That is, an additional 0.1×10^{-5} mol of AgCl (about 0.14 mg) will dissolve per liter.

Exercise 18.17   **Solubility and the Reaction Quotient**

Solid PbI_2 ($K_{sp} = 9.8 \times 10^{-9}$) is placed in a beaker of water. After a period of time, the lead(II) concentration is measured and found to be 1.1×10^{-3} M. Has the system yet reached equilibrium? That is, is the solution saturated? If not, will more PbI_2 dissolve?

K_{sp}, the Reaction Quotient, and Precipitation Reactions

Knowing the reaction quotient, we can decide (1) if a precipitate will form when the ion concentrations are known or (2) what concentrations of ions are required to begin the precipitation of an insoluble salt.

Example 18.16   Deciding Whether a Precipitate Will Form

Problem • Suppose the concentration of aqueous magnesium(II) ion in a solution is 1.5×10^{-6} M. If enough NaOH is added to make the solution 1.0×10^{-4} M in hydroxide ion, OH^-, will precipitation of $Mg(OH)_2$ occur ($K_{sp} = 5.6 \times 10^{-12}$)? If not, will it occur if the concentration of OH^- is increased to 1.0×10^{-2} M?

Strategy • As in Example 18.15, use the ion concentrations to calculate the value of Q. Compare Q with K_{sp} to decide if the system is at equilibrium (when $Q = K_{sp}$).

Solution • The chemical equation for the dissolving of insoluble $Mg(OH)_2$ is

$$Mg(OH)_2(s) \rightleftharpoons Mg^{2+}(aq) + 2\,OH^-(aq)$$

When the concentrations of magnesium(II) and hydroxide ion are those stated earlier,

$$Q = [Mg^{2+}][OH^-]^2$$
$$= (1.5 \times 10^{-6})(1.0 \times 10^{-4})^2 = 1.5 \times 10^{-14}$$
$$Q\,(1.5 \times 10^{-14}) < K_{sp}\,(5.6 \times 10^{-12})$$

The reaction quotient is less than the value of K_{sp}. Therefore, the solution is not yet saturated, and precipitation does not occur.

If $[OH^-]$ is increased to 1.0×10^{-2} M, Q is 1.5×10^{-10},

$$Q = (1.5 \times 10^{-6})(1.0 \times 10^{-2})^2$$
$$= 1.5 \times 10^{-10}$$

a value *larger* than K_{sp}. Precipitation of $Mg(OH)_2$ occurs and will continue until the Mg^{2+} and OH^- concentrations have declined to the point where their product is equal to K_{sp}.

Exercise 18.18   Deciding Whether a Precipitate Will Form

If the concentration of strontium ion, Sr^{2+}, is 2.5×10^{-4} M, will precipitation of $SrSO_4$ occur when enough of the soluble salt Na_2SO_4 is added to make the solution 2.5×10^{-4} M in SO_4^{2-}? (K_{sp} for $SrSO_4$ is 3.4×10^{-7}).

Now that we know how to decide if a precipitate will form when the concentration of each ion is known, let us turn to the problem of deciding how much of the precipitating agent is required to begin the precipitation of an ion at a given concentration level.

Example 18.17   Ion Concentrations Required to Begin Precipitation

Problem • The concentration of barium ion, Ba^{2+}, in a solution is 0.010 M.

(a) What concentration of sulfate ion, SO_4^{2-}, is required to just begin precipitating $BaSO_4$?

(b) When the concentration of sulfate ion in the solution reaches 0.015 M, what concentration of barium ion will remain in solution?

Strategy • The K_{sp} is the product of the cation and anion concentrations. That is, there are three variables in the K_{sp} expres-

sion. Here we know K_{sp} (1.1×10^{-10}) and one of the ion concentrations. We can then calculate the third value.

Solution • Let us begin by writing the balanced equation for the equilibrium that will exist when $BaSO_4$ has been precipitated.

$$BaSO_4(s) \rightleftharpoons Ba^{2+}(aq) + SO_4^{2-}(aq)$$
$$K_{sp} = [Ba^{2+}][SO_4^{2-}] = 1.1 \times 10^{-10}$$

(a) From the K_{sp} expression, we know that, when the product of the ion concentrations exceeds 1.1×10^{-10}, that is, when

$Q > K_{sp}$, precipitation will occur. The Ba^{2+} ion concentration is known (0.010 M), so the $SO_4{}^{2-}$ ion concentration necessary for precipitation can be calculated.

$$K_{sp} = 1.1 \times 10^{-10} = [0.010][SO_4{}^{2-}]$$

$$[SO_4{}^{2-}] = \frac{K_{sp}}{[Ba^{2+}]} = \frac{1.1 \times 10^{-10}}{0.010} = 1.1 \times 10^{-8} \text{ M}$$

The result tells us that if the sulfate ion is just slightly greater than 1.1×10^{-8} M, $BaSO_4$ will begin to precipitate; $Q = [Ba^{2+}][SO_4{}^{2-}]$ would then be greater than K_{sp}.

(b) If the sulfate ion concentration is increased to 0.015 M, the maximum concentration of Ba^{2+} ion that can exist in solution (in equilibrium with $BaSO_4$) is

$$[Ba^{2+}] = \frac{K_{sp}}{[SO_4{}^{2-}]} = \frac{1.1 \times 10^{-10}}{0.015} = 7.3 \times 10^{-9} \text{ M}$$

Comment • The fact that the barium ion concentration is so small when $[SO_4{}^{2-}] = 0.015$ M means that the Ba^{2+} ion has been essentially completely removed from solution. (It began at 0.010 M and has dropped by a factor of about 10 million.) You would say that Ba^{2+} ion precipitation is, for all practical purposes, complete.

Example 18.18 K_{sp} **and Precipitations**

Problem • Suppose you mix 100. mL of 0.0200 M $BaCl_2$ with 50.0 mL of 0.0300 M Na_2SO_4. Will $BaSO_4$ ($K_{sp} = 1.1 \times 10^{-10}$) precipitate?

Strategy • Here we mix two solutions, one containing Ba^{2+} ions and the other $SO_4{}^{2-}$ ions. First, find the concentration of each of these ions *after* mixing. Then, knowing the ion concentrations in the diluted solution, calculate Q and compare it with the K_{sp} value for $BaSO_4$.

Solution • Are the concentrations of the Ba^{2+} and $SO_4{}^{2-}$ ions *after the solutions are mixed* large enough to lead to $BaSO_4$ precipitation? First use the equation $c_1V_1 = c_2V_2$ (◄ SECTION 5.8) to

calculate c_2, the concentration of the Ba^{2+} or $SO_4{}^{2-}$ ions after mixing to give a new solution with a volume of 150. mL ($= V_2$).

$$[Ba^{2+}] \text{ after mixing} = \frac{(0.0200 \text{ mol/L})(0.100 \text{ L})}{0.150 \text{ L}} = 0.0133 \text{ M}$$

$$[SO_4{}^{2-}] \text{ after mixing} = \frac{(0.0300 \text{ mol/L})(0.0500 \text{ L})}{0.150 \text{ L}} = 0.0100 \text{ M}$$

Now the reaction quotient can be calculated.

$$Q = [Ba^{2+}][SO_4{}^{2-}] = (0.0133)(0.0100) = 1.33 \times 10^{-4}$$

Q is larger than K_{sp} so $BaSO_4$ precipitates.

Exercise 18.19 **Ion Concentrations Required to Begin Precipitation**

What is the minimum concentration of I^- that can cause precipitation of PbI_2 from a 0.050 M solution of $Pb(NO_3)_2$? K_{sp} for PbI_2 is 9.8×10^{-9}. What concentration of Pb^{2+} ions remains in solution when the concentration of I^- is 0.0015 M?

Exercise 18.20 K_{sp} **and Precipitations**

You have 100.0 mL of 0.0010 M silver nitrate. Will AgCl precipitate if you add 5.0 mL of 0.025 M HCl?

18.6 SOLUBILITY AND COMPLEX IONS

Calcium carbonate does not dissolve in water, but it does dissolve in strong acids. In contrast, silver chloride dissolves neither in water nor in strong acid — but it does dissolve in ammonia. How do we explain this observation and how can we use it in a practical way [CD-ROM, Screen 18.17]?

Metal ions exist in aqueous solution as complex ions (◄ SECTION 17.9) (Figure 18.15). Complex ions consist of the metal ion and other molecules or ions, bound

Dimethylglyoxime complex of Ni^{2+} ion

— Ni(NH$_3$)$_6$$^{2+}$

— Ni(H$_2$O)$_6$$^{2+}$

Figure 18.15 Complex ions. The green solution contains soluble Ni(H$_2$O)$_6$$^{2+}$ ions in which water molecules are bound to Ni^{2+} ions by ion–dipole forces. This ion, which gives the solution its green color, is referred to as a "complex ion" because it is formed by a combination of a metal ion and molecules of water. The blue color is due to a complex ion of Ni^{2+} and NH$_3$. The red, insoluble solid is the dimethylglyoximate complex of the Ni^{2+} ion [Ni(C$_4$H$_7$O$_2$N$_2$)$_2$] *(model at top right)*. Formation of this beautiful red insoluble compound is the classical test for the presence of the aqueous Ni^{2+} ion. *(Charles D. Winters)*

● **Successive Equilibria**
Dissolving AgCl in NH$_3$ is an example of what chemists often refer to as "successive equilibria," in which one product-favored reaction affects an overall equilibrium. Another example is the increase in solubility of an insoluble salt owing to the reaction of its anion with water (page 766) and yet another is dissolving CaCO$_3$ in strong acid (page 769).

into a single entity. In water, metal ions are always surrounded by water molecules, the negative end of the polar water molecule, the oxygen atom, attracted to the positive metal ion. Indeed, any negative ion or polar molecule, such as NH$_3$, can be attracted to a metal ion.

As you shall see in Chapter 22, complex ions are prevalent in chemistry and are the basis of such biologically important substances as hemoglobin and vitamin B$_{12}$. For our present purposes, they are important because a water-soluble complex ion can often be formed in preference to an insoluble salt. For example, adding sufficient aqueous ammonia to solid AgCl will cause the insoluble salt to dissolve owing to the formation of the complex ion Ag(NH$_3$)$_2$$^+$ (Figure 18.16)

$$AgCl(s) + 2\ NH_3(aq) \rightleftharpoons Ag(NH_3)_2{}^+(aq) + Cl^-(aq)$$

We can view dissolving AgCl(s) in this way as a two step process. First, AgCl dissolves minimally in water giving Ag$^+$(aq) and Cl$^-$(aq) ion. Then, the Ag$^+$(aq) ion combines with NH$_3$ to give the ammonia complex. Lowering the Ag$^+$(aq) concentration through complexation with NH$_3$ shifts the solubility equilibrium to the right, and more solid AgCl dissolves.

$$AgCl(s) \rightleftharpoons Ag^+(aq) + Cl^-(aq) \qquad K_{sp} = 1.8 \times 10^{-10}$$
$$Ag^+(aq) + 2\ NH_3(aq) \rightleftharpoons Ag(NH_3)_2{}^+(aq) \qquad K = 1.6 \times 10^7$$

This is an example of combining two (or more) equilibria, one of which is a product-favored reaction and the other is reactant-favored.

The equilibrium constant for the formation of a complex ion such as Ag(NH$_3$)$_2$$^+$ is called a **formation constant**, $K_{\mathbf{form}}$.

$$Ag^+(aq) + 2\ NH_3(aq) \rightleftharpoons Ag(NH_3)_2{}^+(aq) \qquad K_{form} = 1.6 \times 10^7$$

The large value of this equilibrium constant means that the equilibrium lies well to the right and provides the driving force for dissolving AgCl in the presence of NH$_3$. If we combine K_{form} with K_{sp} we obtain the net equilibrium constant for dissolving AgCl.

$$K_{net} = K_{sp}\ K_{form} = (1.8 \times 10^{-10})(1.6 \times 10^7) = 2.9 \times 10^{-3} = \frac{[Ag(NH_3)_2^+][Cl^-]}{[NH_3]^2}$$

Even though the value of K_{net} seems small, we use a large concentration of NH$_3$, so the value of [Ag(NH$_3$)$_2$$^+$] in solution can be high. Silver chloride is thus much more soluble in the presence of ammonia than in pure water.

Formation constants have been measured for many complex ions. This makes it possible to compare the stabilities of various complex ions by comparing values of their formation constants. For the silver(I) ion, a few other values are

Formation Equilibrium	$K_{\mathbf{form}}$
Ag$^+$(aq) + 2 Cl$^-$(aq) $\rightleftharpoons$ AgCl$_2$$^-$(aq)	2.5×10^5
Ag$^+$(aq) + 2 S$_2$O$_3$$^{2-}$(aq) $\rightleftharpoons$ Ag(S$_2$O$_3$)$_2$$^{3-}$	2.0×10^{13}
Ag$^+$(aq) + 2 CN$^-$(aq) $\rightleftharpoons$ Ag(CN)$_2$$^-$(aq)	5.6×10^{18}

● **Formation Constants**
Values of K_{form} for other complex ions are given in Appendix K.

The formation of all three silver complexes is strongly product-favored. The cyanide complex ion [Ag(CN)$_2$$^-$] is the most stable of the three.

Figure 18.16 shows what happens as complex ions are formed. Beginning with a precipitate of AgCl, adding aqueous ammonia dissolves the precipitate to give

AgCl(s),
$K_{sp} = 1.8 \times 10^{-10}$

(a) AgCl is precipitated by adding NaCl(aq) to AgNO$_3$(aq).

NH$_3$(aq) →

[Ag(NH$_3$)$_2$]$^+$(aq)

(b) The precipitate of AgCl dissolves on adding aqueous NH$_3$ to give Ag(NH$_3$)$_2^+$ (see Figure 5.3).

NaBr(aq) →

AgBr(s),
$K_{sp} = 5.4 \times 10^{-13}$

(c) The silver-ammonia complex ion is changed to insoluble AgBr on adding NaBr(aq).

Na$_2$S$_2$O$_3$(aq) →

[Ag(S$_2$O$_3$)$_2$]$^{3-}$(aq)

(d) The precipitate of AgBr is dissolved on adding Na$_2$S$_2$O$_3$(aq). The product is the complex ion Ag(S$_2$O$_3$)$_2^{3-}$.

Figure 18.16   Forming and dissolving precipitates. This is also illustrated by the chemistry of lead(II) ions on page 740. *(Charles D. Winters)*

the complex ion Ag(NH$_3$)$_2^+$. Silver bromide is even more stable than Ag(NH$_3$)$_2^+$, so AgBr forms in preference to the complex ion on adding bromide ion. If thiosulfate ion, S$_2$O$_3^{2-}$, is then added, AgBr dissolves due to the formation of Ag(S$_2$O$_3$)$_2^{3-}$, a complex ion with a large formation constant.

Example 18.19   Complex Ions and Solubility

Problem • What is the value of the equilibrium constant, K_{net}, for dissolving AgBr in a solution saturated with the thiosulfate ion, S$_2$O$_3^{2-}$ (see Figure 18.16)? Explain why AgBr can dissolve readily on adding aqueous sodium thiosulfate to the solid.

Strategy • Summing several equilibrium processes gives the net chemical equation. K_{net} is the product of the values of K of the summed chemical equations (⬅ SECTION 16.5)

Solution • The overall reaction for dissolving AgBr is the sum of two equilibrium processes.

$$AgBr(s) \rightleftharpoons Ag^+(aq) + Br^-(aq)$$
$$K_{sp} = 5.4 \times 10^{-13}$$

$$Ag^+(aq) + 2\ S_2O_3^{2-}(aq) \rightleftharpoons Ag(S_2O_3)_2^{3-}(aq)$$
$$K_{form} = 2.0 \times 10^{13}$$

Net chemical equation:

$$AgBr(s) + 2\ S_2O_3^{2-}(aq) \rightleftharpoons Ag(S_2O_3)_2^{3-}(aq) + Br^-(aq)$$
$$K_{net} = K_{sp}K_{form} = 1.1 \times 10^1$$

The value of K_{net} is greater than 1, indicating a product-favored reaction. AgBr is predicted to dissolve readily in aqueous Na$_2$S$_2$O$_3$, as observed (see Figure 18.16).

Exercise 18.21   Complex Ions and Solubility

What is the value of the equilibrium constant, K_{net}, for dissolving Cu(OH)$_2$ in aqueous ammonia (to form the complex ion Cu(NH$_3$)$_4^{2+}$)? (⬅ FIGURE 17.7)

18.7 SOLUBILITY, ION SEPARATIONS, AND QUALITATIVE ANALYSIS

In many introductory chemistry courses a portion of the laboratory work is devoted to the qualitative analysis of aqueous solutions, the identification of anions and metal cations. The purpose of such laboratory work is (1) to introduce some basic chemistry of various ions and (2) to illustrate how the principles of chemical equilibria can be applied. This section of the chapter gives examples of some of those ideas [📀 CD-ROM, Screen 18.14].

Assume you have an aqueous solution that contains the metal ions Ag^+, Pb^{2+}, and Cu^{2+}. Your objective is to separate the ions from one another so that each type of ion ends up in a separate test tube; the presence or absence of the ion can then be confirmed. As a first step in this process, you want to find one reagent that will form a precipitate with one or more of the cations and leave the others in solution. This is done by comparing K_{sp} values for salts of cations with various anions (say S^{2-}, OH^-, Cl^-, or SO_4^{2-}), looking for an anion that gives insoluble salts for some cations but not others.

Looking over the list of solubility products in Appendix J, you notice that the ions in our example solution all form very insoluble sulfides (Ag_2S, PbS, and CuS). However, only two of them form insoluble chlorides, $AgCl$ and $PbCl_2$. Thus, your "magic reagent" for partial cation separation could be aqueous HCl, which will form precipitates with Ag^+ and Pb^{2+} ($AgCl$ and $PbCl_2$) while Cu^{2+} ions remain in solution (Figure 18.17).

Now we have another problem to solve. How can we separate $AgCl$ and $PbCl_2$, which are now present in the same test tube [part (c) of Figure 18.17] into separate test tubes? This turns out to be relatively easy: $PbCl_2$ (with a value of K_{sp} of 1.7×10^{-5}) is much more soluble than $AgCl$, and the lead salt dissolves readily in hot water, whereas $AgCl$ remains insoluble.

We now have three test tubes: one containing $PbCl_2$, another with $AgCl$, and a third with Cu^{2+} ions. How can we verify the presence of $PbCl_2$ in one of the test

Some Insoluble Sulfides and Chlorides

Compound*	K_{spa} at 25 °C
Ag_2S	6×10^{-51}
PbS	3×10^{-28}
CuS	6×10^{-37}
$AgCl$	1.8×10^{-10}
$PbCl_2$	1.7×10^{-5}

*See Appendix J for the K_{spa} values of metal sulfides.

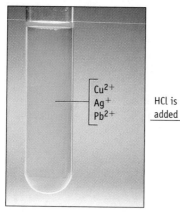

(a) The solution contains nitrate salts of Ag^+, Pb^{2+}, and Cu^{2+}. (The Cu^{2+} ion in water is light blue; the others are colorless.)

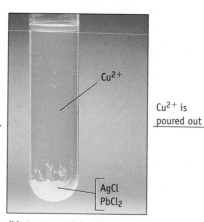

(b) Aqueous HCl is added in an amount sufficient to precipitate completely the white solids $AgCl$ and $PbCl_2$.

(c) The blue solution containing the Cu^{2+} ion is poured carefully into another test tube, leaving the white precipitates in the first test tube.

Figure 18.17 Ion separations by solubility difference. *(Charles D. Winters)*

tubes? The answer is to convert it to an even more insoluble salt with a distinctive appearance. Consider two common lead compounds: white lead(II) chloride and bright yellow lead(II) chromate.

$$PbCl_2, \ K_{sp} = 1.7 \times 10^{-5} \quad \text{and} \quad PbCrO_4, \ K_{sp} = 2.8 \times 10^{-13}$$

If you add a few drops of K_2CrO_4 to a small amount of a white precipitate of $PbCl_2$ and shake the mixture, the solid will be changed to yellow $PbCrO_4$ (Figure 18.18). That this is possible is evident from the equilibrium constant, K_{net}, for the process

$$PbCl_2(s) \rightleftharpoons Pb^{2+}(s) + 2 \ Cl^-(aq) \qquad K_1 = K_{sp} = 1.7 \times 10^{-5}$$

$$Pb^{2+}(aq) + CrO_4{}^{2-}(aq) \rightleftharpoons PbCrO_4(s) \qquad K_2 = \frac{1}{K_{sp}} = \frac{1}{2.8 \times 10^{-13}}$$

Net chemical equation:

$$PbCl_2(s) + CrO_4{}^{2-}(aq) \rightleftharpoons PbCrO_4(s) + 2 \ Cl^-(aq) \qquad K_{net} = 6.1 \times 10^7$$

The value of $K_{net} \ (= K_1 K_2)$ is very large, indicating that the reaction should proceed from left to right as observed (Figure 18.18)

How can we verify the presence of AgCl in one of the test tubes? One way is to attempt to dissolve the solid AgCl in aqueous ammonia (see Figure 18.16), a characteristic reaction of AgCl. In contrast, $PbCl_2$ does not react with aqueous ammonia.

Finally, how might we verify that the blue solution in the third test tube really contains aqueous Cu^{2+} ions? The classical test is to add aqueous ammonia. This first precipitates blue-white copper(II) hydroxide (owing to the presence of OH^- ion in solution of aqueous ammonia).

$$Cu^{2+}(aq) + 2 \ NH_3(aq) + 2 \ H_2O(\ell) \rightleftharpoons Cu(OH)_2(s) + 2 \ NH_4{}^+(aq)$$
$$K_{sp} \text{ for } Cu(OH)_2 = 2.2 \times 10^{-20}$$

As more ammonia is added, however, copper(II) hydroxide is converted to the very stable complex ion $Cu(NH_3)_4{}^{2+}$, which has a very distinctive deep blue color (see Figure 17.7).

$$Cu(OH)_2(s) + 4 \ NH_3(aq) \rightleftharpoons Cu(NH_3)_4{}^{2+}(aq) + 2 \ OH^-(aq)$$
$$K_{form} \text{ for } Cu(NH_3)_4{}^{2+} = 6.8 \times 10^{12}$$

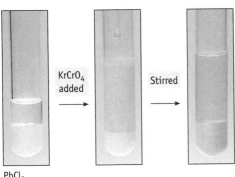

KrCrO$_4$ added

Stirred

PbCl$_2$ precipitate

Figure 18.18   Lead(II) chloride and lead(II) chromate. The test tube at the left contains a precipitate of white $PbCl_2$. The test tube at the right shows what happens when the $PbCl_2$ precipitate is stirred in the presence of K_2CrO_4. The white solid $PbCl_2$ has been transformed into the less soluble yellow $PbCrO_4$. *(Charles D. Winters)*

Exercise 18.22   Ion Separation

The cations of each pair given here appear together in one solution.

(a) Ag^+ and Bi^{3+} **(b)** Fe^{2+} and K^+

You may add only one reagent to precipitate one cation and not the other. Consult the solubility product table in Appendix J (and Figure 5.1) and tell whether you would use Cl^-, S^{2-}, or OH^- as the precipitating ion in each case. (The precipitating ions are introduced in the form of HCl, $(NH_4)_2S$, or NaOH, for example.)

In Summary

When you have finished studying this chapter, ask if you have met the chapter goals. In particular, you should be able to

- Predict the effect of the addition of a common ion on the pH of the solution of a weak acid or base (Section 18.1 and Examples 18.1 and 18.2).
- Describe the functioning of buffer solutions (Section 18.2 and Example 18.3).
- Use the Henderson–Hasselbalch equation (Equation 18.3) to calculate the pH of a buffer solution of given composition (Section 18.2 and Example 18.4).
- Describe how a buffer solution of a given pH can be prepared (Section 18.2 and Example 18.5).
- Calculate the pH of a buffer solution before and after adding excess acid or base (Section 18.2 and Example 18.6).
- Predict the pH of an acid–base reaction at its equivalence point (Section 18.3; see also Section 17.5).

Acid	Base	pH at Equivalence Point
Strong	Strong	$= 7$ (neutral)
Strong	Weak	< 7 (acidic)
Weak	Strong	> 7 (basic)
Weak	Weak	Depends on K values of conjugate base and acid

- Calculate the pH at the equivalence point in the reaction of a strong acid with a strong base or weak base, or in the reaction of a strong base with a weak acid (see Examples 18.7 and 18.8).
- Understand the differences between the titration curves for a strong acid/strong base titration versus cases in which one of the substances is weak (Section 18.3).
- Describe how an indicator functions in an acid–base titration (Section 18.3).
- Write the equilibrium constant expression — the solubility product constant, K_{sp} — for any insoluble salt (Section 18.4).
- Calculate K_{sp} values from experimental data (Section 18.4 and Examples 18.9 and 18.10).
- Estimate the solubility of a salt from the value of K_{sp} (Section 18.4 and Examples 18.11 and 18.12).
- Calculate the solubility of a salt in the presence of a common ion (Section 18.4 and Example 18.13 and 18.14).
- Understand the effect of basic anions on the solubility of a salt (Section 18.4).

- Decide if a precipitate will form when the ion concentrations are known (Section 18.5 and Example 18.15).
- Calculate the ion concentrations that are required to begin the precipitation of an insoluble salt (Section 18.5 and Examples 18.16–18.18).
- Understand that the formation of a complex ion can increase the solubility of an insoluble salt (Section 18.6 and Example 18.19).
- Use K_{sp} values to devise a method of separating ions in solution from one another (Section 18.7).

Key Terms

Section 18.1
common ion effect

Section 18.2
buffer
Henderson–Hasselbalch equation

Section 18.3
titrant
acid–base indicator

Section 18.4
solubility product constant, K_{sp}

Section 18.6
formation constant, K_{form}

Key Equations

Equation 18.1 (page 746)
Hydronium ion concentration in a buffer solution composed of a weak acid and its conjugate base.

$$[H_3O^+] = \frac{[acid]}{[conjugate\ base]} \cdot K_a$$

Equation 18.2 (page 746)
Henderson–Hasselbalch equation. To calculate the pH of a buffer solution composed of a weak acid and its conjugate base.

$$pH = pK_a + \log\frac{[conjugate\ base]}{[acid]}$$

Equation 18.5 (page 754)
The relationship between the pH of the solution and the pK_a of the weak acid at the halfway point in the titration of a weak acid with a strong base (or of a weak base with a strong acid).

$$pH = pK_a$$

Study Questions

Questions with blue, bold-faced numbers have answers in Appendix O. Use the descriptions, simulations, and tutorials in Chapter 18 of the General Chemistry Interactive CD-ROM, *Version 3.0.*

Reviewing Important Concepts

1. What experiment can you do to decide when a sample of acid has been completely consumed by a base?

2. Decide if the resulting pH is equal to, less than, or greater than 7 after combining equal molar amounts of: (a) a weak base with a strong acid, (b) a strong base with a strong acid, and (c) a strong base with a weak acid.

3. Briefly describe how a buffer solution can control the pH of a solution when a strong acid is added. Use the NH_3/NH_4Cl buffer as an example.

4. You can use a dye (an indicator) to follow an acid–base reaction. When the dye changes color, is this necessarily the point at which the acid has been consumed *exactly* by the base (the equivalence point)?

5. Sketch the general shape of the titration curve when (a) a strong base is titrated with a strong acid and (b) a weak base is titrated with a strong acid. In each case indicate if the pH at the equivalence point is less than, equal to, or greater than 7.

6. Briefly describe how a buffer solution can control the pH of a solution when a strong base is added. Use a solution of acetic acid and sodium acetate as an example.

7. Use the Henderson–Hasselbalch equation to decide how the pH of a buffer solution based on a weak acid and its conjugate base changes when (a) the ionization constant of the weak acid increases and (b) when the acid concentration is decreased relative to the concentration of its conjugate base.

8. Explain why the concentration of $CaCO_3$ does not appear in the K_{sp} expression for

$$CaCO_3(s) \rightleftharpoons Ca^{2+}(aq) + CO_3{}^{2-}(aq)$$

9. What is a "reaction quotient" and how does its value differ from an equilibrium constant? Use the following equilibrium in your discussion.

$$AgCl(s) \rightleftharpoons Ag^+(aq) + Cl^-(aq)$$

10. Find two salts in Appendix J that have K_{sp} values less than 1×10^{-6}.

    (a) Write balanced equations to show the equilibria existing when the compounds dissolve in water.

    (b) Write the K_{sp} expression for each compound.

11. Explain the terms "saturated," "not saturated" (or unsaturated), and "supersaturated." Use the following equilibrium in your discussion.

$$CaF_2(s) \rightleftharpoons Ca^{2+}(aq) + 2\ F^-(aq)$$

12. What is the common ion effect? Use the following equilibrium in your discussion.

$$Fe(OH)_2(s) \rightleftharpoons Fe^{2+}(aq) + 2\ OH^-(aq)$$

13. Explain why the solubility of Ag_3PO_4 can be greater in water than is calculated from the K_{sp} of the salt.

14. Silver chloride is insoluble in water. If aqueous ammonia is added, however, the precipitate of AgCl dissolves. Explain.

15. The CO_2/H_2CO_3 buffer is one mechanism by which your blood pH is controlled. Explain how breathing into a paper bag can lower the pH of your blood.

16. A solution contains $NaNO_3$ and $Fe(NO_3)_2$. Describe how you could separate the cations, so that Na^+ ions appear in one test tube and Fe^{2+} ions appear in another.

17. In each of the following cases, decide if a precipitate will form when mixing the indicated reagents, and write a balanced equation for the reaction.

    (a) $NaBr(aq) + AgNO_3(aq)$

    (b) $KCl(aq) + Pb(NO_3)_2(aq)$

    (c) $Na_2SO_4(aq) + Mg(NO_3)_2(aq)$

Practicing Skills

Common Ion Effect
(See Examples 18.1 and 18.2 and CD-ROM Screen 18.2)

18. Does the pH of the solution increase, decrease, or stay the same when you

    (a) Add solid ammonium chloride to a dilute aqueous solution of NH_3?

    (b) Add solid sodium acetate to a dilute aqueous solution of acetic acid?

    (c) Add solid NaCl to a dilute aqueous solution of NaOH?

19. Does the pH of the solution increase, decrease, or stay the same when you

    (a) Add solid sodium oxalate, $Na_2C_2O_4$, to 50.0 mL of 0.015 M oxalic acid, $H_2C_2O_4$?

    (b) Add solid ammonium chloride to 75 mL of 0.016 M HCl?

    (c) Add 20.0 g of NaCl to 1.0 L of 0.10 M sodium acetate, $NaCH_3CO_2$?

20. What is the pH of a solution that is 0.20 M with respect to ammonia, NH_3, and 0.20 M with respect to ammonium chloride, NH_4Cl?

21. What is the pH of a solution if 100.0 mL of the solution is 0.15 M with respect to acetic acid and contains 1.56 g of sodium acetate, $NaCH_3CO_2$?

22. What is the pH of the solution that results from adding 30.0 mL of 0.015 M KOH to 50.0 mL of 0.015 M benzoic acid?

23. What is the pH of the solution that results from adding 25.0 mL of 0.12 M HCl to 25.0 mL of 0.43 M NH_3?

Buffer Solutions
(See Example 18.3, CD-ROM Screens 18.3 and 18.4)

24. What is the pH of the buffer solution that results when 2.2 g of NH_4Cl is added to 250 mL of 0.12 M NH_3? Is the final pH lower or higher than the pH of the original ammonia solution?

25. Lactic acid ($CH_3CHOHCO_2H$) is found in sour milk, in sauerkraut, and in muscles after activity (see page 741). (K_a for lactic acid $= 1.4 \times 10^{-4}$)

    (a) If 2.75 g of $NaCH_3CHOHCO_2$, sodium lactate, is added to 5.00×10^2 mL of 0.100 M lactic acid, what is the pH of the resulting buffer solution?

    (b) Is the final pH lower or higher than the pH of the lactic acid solution?

26. What mass of sodium acetate, $NaCH_3CO_2$, must be added to 1.00 L of 0.10 M acetic acid to give a solution with a pH of 4.50?

27. What mass of ammonium chloride, NH_4Cl, would have to be added to exactly 5.00×10^2 mL of 0.10 M NH_3 solution to give a solution with a pH of 9.00?

Using the Henderson–Hasselbalch Equation
(See Example 18.4)

28. Calculate the pH of a solution that has an acetic acid concentration of 0.050 M and a sodium acetate concentration of 0.075 M.

29. Calculate the pH of a solution that has an ammonium chloride concentration of 0.050 M and an ammonia concentration of 0.045 M.

30. A buffer is composed of formic acid and its conjugate base, the formate ion.

    (a) What is the pH of a solution that has a formic acid concentration of 0.050 M and a sodium formate concentration of 0.035 M?

    (b) What must the ratio be of acid to conjugate base in order to increase the pH by 0.5?

31. A buffer solution is composed of 1.360 g of KH_2PO_4 and 5.677 g of Na_2HPO_4.

    (a) What is the pH of the buffer solution?

    (b) What mass of KH_2PO_4 must be added to decrease the buffer solution pH by 0.5?

Preparing a Buffer Solution
(See Example 18.5, CD-ROM Screen 18.5)

32. Which of the following combinations would be the best to buffer the pH at approximately 9?

    (a) HCl and NaCl

    (b) NH_3 and NH_4Cl

    (c) CH_3CO_2H and $NaCH_3CO_2$

33. Which of the following combinations would be the best choice to buffer the pH of a solution at approximately 7?

    (a) Na_3PO_4 and NaH_2PO_4

    (b) NaH_2PO_4 and Na_2HPO_4

    (c) Na_2HPO_4 and Na_3PO_4

Adding an Acid or Base to a Buffer Solution
(See Example 18.6, CD-ROM Screen 18.6)

34. A buffer solution was prepared by adding 4.95 g of sodium acetate, $NaCH_3CO_2$, to 2.50×10^2 mL of 0.150 M acetic acid, CH_3CO_2H.

    (a) What is the pH of the buffer?

    (b) What is the pH of 1.00×10^2 mL of the buffer solution if you add 82 mg of NaOH to the solution?

35. You dissolve 0.425 g of NaOH in 2.00-L of a solution that has $[H_2PO_4^-] = [HPO_4^{2-}] = 0.132$ M. What was the pH of the solution before adding NaOH? After adding NaOH?

36. A buffer solution is prepared by adding 0.125 mol of ammonium chloride to 5.00×10^2 mL of 0.500 M solution of ammonia.

    (a) What is the pH of the buffer?

    (b) If 0.0100 mol of HCl gas is bubbled into 5.00×10^2 mL of the buffer, what is the new pH of the solution?

37. What will be the pH change when 20.0 mL of 0.100 M NaOH is added to 80.0 mL of a buffer solution consisting of 0.169 M NH_3 and 0.183 M NH_4Cl?

More About Acid-Base Reactions: Titrations
(See Examples 18.7 and 18.8, CD-ROM Screens 17.7 and 18.7)

38. Phenol, C_6H_5OH, is a weak organic acid. Suppose 0.515 g of the compound is dissolved in exactly 125 mL of water. The resulting solution is titrated with 0.123 M NaOH.

    $$C_6H_5OH(aq) + OH^-(aq) \longrightarrow C_6H_5O^-(aq) + H_2O(\ell)$$

    (a) What is the pH of the original solution of phenol?

    (b) What are the concentrations of all of the following ions at the equivalence point: Na^+, H_3O^+, OH^-, and $C_6H_5O^-$?

    (c) What is the pH of the solution at the equivalence point?

39. Assume you dissolve 0.235 g of the weak acid benzoic acid, $C_6H_5CO_2H$, in enough water to make 1.00×10^2 mL of solution and then titrate the solution with 0.108 M NaOH.

    $$C_6H_5CO_2H(aq) + OH^-(aq) \longrightarrow C_6H_5CO_2^-(aq) + H_2O(\ell)$$

    (a) What is the pH of the original benzoic acid solution?

    (b) What are the concentrations of all of the following ions at the equivalence point: Na^+, H_3O^+, OH^-, and $C_6H_5CO_2^-$?

    (c) What is the pH of the solution at the equivalence point?

40. You require 36.78 mL of 0.0105 M HCl to reach the equivalence point in the titration of 25.0 mL of aqueous ammonia.

    (a) What was the concentration of NH_3 in the original ammonia solution?

    (b) What are the concentrations of H_3O^+, OH^-, and NH_4^+ at the equivalence point?

    (c) What is the pH of the solution at the equivalence point?

41. A solution of the weak base aniline, $C_6H_5NH_2$, in 25.0 mL of water requires 25.67 mL of 0.175 M HCl to reach the equivalence point.

    $$C_6H_5NH_2(aq) + H_3O^+(aq) \longrightarrow C_6H_5NH_3^+(aq) + H_2O(\ell)$$

    (a) What was the concentration of aniline in the original solution?

    (b) What are the concentrations of H_3O^+, OH^-, and $C_6H_5NH_3^+$ at the equivalence point?

    (c) What is the pH of the solution at the equivalence point?

Titration Curves and Indicators
(See Figures 18.4–18.7, CD-ROM Screen 18.7)

42. Without doing detailed calculations, sketch the curve for the titration of 30.0 mL of 0.10 M NaOH with 0.10 M HCl. Indicate the approximate pH at the beginning of the titration and at the equivalence point. What is the total solution volume at the equivalence point?

43. Without doing detailed calculations, sketch the curve for the titration of 50 mL of 0.050 M pyridine, C_5H_5N (a weak base), with 0.10 M HCl. Indicate the approximate pH at the beginning of the titration and at the equivalence point. What is the total solution volume at the equivalence point?

44. You titrate 25.0 mL of 0.10 M NH_3 with 0.10 M HCl.
    (a) What is the pH of the NH_3 solution before the titration begins?
    (b) What is the pH at the equivalence point?
    (c) What is the pH at the midpoint of the titration?
    (d) What indicator in Figure 17.2 could be used to detect the equivalence point?
    (e) Calculate the pH of the solution after adding 5.00, 15.0, 20.0, 22.0, and 30.0 mL of the acid. Combine this information with that from parts (a)–(d) and plot the titration curve.

45. Construct a rough plot of pH versus volume of base for the titration of 25.0 mL of 0.050 M HCN with 0.075 M NaOH.
    (a) What is the pH before any NaOH is added?
    (b) What is the pH at the half-neutralization point?
    (c) What is the pH when 95% of the required NaOH has been added?
    (d) What volume of base, in milliliters, is required to reach the equivalence point?
    (e) What is the pH at the equivalence point?
    (f) What indicator would be most suitable? (See Figure 17.2.)
    (g) What is the pH when 105% of the required base has been added?

46. Using Figure 17.2, suggest an indicator to use in each of the following titrations:
    (a) The weak base pyridine is titrated with HCl.
    (b) Formic acid is titrated with NaOH.
    (c) Hydrazine, a weak diprotic base, is titrated with HCl.

47. Using Figure 17.2, suggest an indicator to use in each of the following titrations:
    (a) $NaHCO_3$ is titrated to CO_3^- with NaOH.
    (b) Hypochlorous acid is titrated with NaOH.
    (c) Trimethylamine is titrated with HCl.

Solubility Guidelines
(Review Section 5.1, Table 5.1, and Example 5.1; CD-ROM Screens 5.4 and 18.8)

48. Name two insoluble salts of each of the following ions:
    (a) Cl^-
    (b) Zn^{2+}
    (c) Fe^{2+}

49. Name two insoluble salts of the following ions:
    (a) SO_4^{2-}
    (b) Ni^{2+}
    (c) Br^-

50. Using the table of solubility guidelines (Figure 5.1), predict whether each of the following is insoluble or soluble in water.
    (a) $(NH_4)_2CO_3$
    (b) $ZnSO_4$
    (c) NiS
    (d) $BaSO_4$

51. Predict whether each of the following is insoluble or soluble in water.
    (a) $Pb(NO_3)_2$
    (b) $Fe(OH)_3$
    (c) $ZnCl_2$
    (d) CuS

Writing Solubility Product Constant Expressions
(See Exercise 18.11 and CD-ROM Screen 18.9)

52. For each of the following insoluble salts, write a balanced equation showing the equilibrium occurring when the salt is added to water and write the K_{sp} expression.
    (a) AgCN
    (b) $NiCO_3$
    (c) $AuBr_3$

53. For each of the following insoluble salts, write a balanced equation showing the equilibrium occurring when the salt is added to water and write the K_{sp} expression.
    (a) $PbSO_4$
    (b) BaF_2
    (c) Ag_3PO_4

Calculating K_{sp}
(See Examples 18.9 and 18.10 and CD-ROM Screen 18.10)

54. When 1.55 g of solid thallium(I) bromide is added to 1.00 L of water, the salt dissolves to a small extent.

$$TlBr(s) \rightleftharpoons Tl^+(aq) + Br^-(aq)$$

The thallium(I) and bromide ions in equilibrium with TlBr each have a concentration of 1.9×10^{-3} M. What is the value of K_{sp} for TlBr?

55. At 20 °C, a saturated aqueous solution of silver acetate, $AgCH_3CO_2$, contains 1.0 g of the silver compound dissolved in 100.0 mL of solution. Calculate K_{sp} for silver acetate.

$$AgCH_3CO_2(s) \rightleftharpoons Ag^+(aq) + CH_3CO_2^-(aq)$$

56. When 250 mg of SrF_2, strontium fluoride, is added to 1.00 L of water, the salt dissolves to a very small extent.

$$SrF_2(s) \rightleftharpoons Sr^{2+}(aq) + 2\,F^-(aq)$$

At equilibrium, the concentration of Sr^{2+} is found to be 1.03×10^{-3} M. What is the value of K_{sp} for SrF_2?

57. Calcium hydroxide, $Ca(OH)_2$, dissolves in water to the extent of 1.04 g/L. What is the value of K_{sp} for $Ca(OH)_2$?

$$Ca(OH)_2(s) \rightleftharpoons Ca^{2+}(aq) + 2\,OH^-(aq)$$

58. You add 0.979 g of $Pb(OH)_2$ to 1.00 L of pure water at 25 °C. The pH is 9.15. Estimate the value of K_{sp} for $Pb(OH)_2$.

59. You place 1.234 g of solid $Ca(OH)_2$ in 1.00 L of pure water at 25 °C. The pH of the solution is found to be 12.68. Estimate the value of K_{sp} for $Ca(OH)_2$.

Estimating Salt Solubility from K_{sp}
(See Examples 18.11 and 18.12, Exercise 8.14, and CD-ROM Screen 18.11)

60. Estimate the solubility of silver iodide in pure water at 25 °C in (a) moles per liter and (b) grams per liter.

$$AgI(s) \rightleftharpoons Ag^+(aq) + I^-(aq)$$

61. What is the molar concentration of $Au^+(aq)$ in a saturated solution of AuCl in pure water at 25 °C?

$$AuCl(s) \rightleftharpoons Au^+(aq) + Cl^-(aq)$$

62. Estimate the solubility of magnesium fluoride, MgF_2, in (a) moles per liter and (b) grams per liter of pure water.

$$MgF_2(s) \rightleftharpoons Mg^{2+}(aq) + 2\,F^-(aq)$$

63. Estimate the solubility of lead(II) bromide in (a) moles per liter and (b) grams per liter of pure water.

64. The K_{sp} value for radium sulfate, $RaSO_4$, is 3.7×10^{-11}. If 25 mg of radium sulfate is placed in 1.00×10^2 mL of water, does all of it dissolve? If not, how much dissolves?

65. If 55 mg of lead(II) sulfate is placed in 250 mL of pure water, does all of it dissolve? If not, how much dissolves?

66. Use K_{sp} values to decide which compound in each of the following pairs is the more soluble.
    (a) $PbCl_2$ or $PbBr_2$
    (b) HgS or FeS
    (c) $Fe(OH)_2$ or $Zn(OH)_2$

67. Use K_{sp} values to decide which compound in each of the following pairs is the more soluble.
    (a) AgBr or AgSCN
    (b) $SrCO_3$ or $SrSO_4$
    (c) AgI or PbI_2
    (d) MgF_2 or CaF_2

Common Ion Effect and Salt Solubility
(See Examples 18.13 and 18.14 and CD-ROM Screen 18.13)

68. Calculate the molar solubility of silver thiocyanate, AgSCN, in pure water and in water containing 0.010 M NaSCN.

69. Calculate the solubility of silver bromide, AgBr, in moles per liter, in pure water. Compare this with the molar solubility of AgBr in 225 mL of water to which 0.15 g of NaBr has been added.

70. Compare the solubility, in milligrams per milliliter, of silver iodide, AgI, in (a) pure water and (b) in water that is 0.020 M in $AgNO_3$.

71. What is the solubility, in milligrams per milliliter, of BaF_2, (a) in pure water and (b) in water containing 5.0 mg/mL of KF?

Effect of Basic Anions on Salt Solubility
(See pages 768–770; CD-ROM Screen 18.15)

72. Which insoluble compound in each pair should be more soluble in nitric acid than in pure water?
    (a) $PbCl_2$ or PbS
    (b) Ag_2CO_3 or AgI
    (c) $Al(OH)_3$ or AgCl

73. Which compound in each pair is more soluble in water than is predicted by a calculation from K_{sp}?
    (a) AgI or Ag_2CO_3
    (b) $PbCO_3$ or $PbCl_2$
    (c) AgCl or AgCN

Precipitations Reactions
(See Examples 18.15–18.18 and CD-ROM Screen 18.12)

74. You have a solution that has a lead(II) concentration of 0.0012 M.

$$PbCl_2(s) \rightleftharpoons Pb^{2+}(aq) + 2\,Cl^-(aq)$$

If enough soluble chloride-containing salt is added so that the Cl^- concentration is 0.010 M, will $PbCl_2$ precipitate?

75. Sodium carbonate is added to a solution in which the concentration of Ni^{2+} ion is 0.0024 M.

$$NiCO_3(s) \rightleftharpoons Ni^{2+}(aq) + CO_3^{2-}(aq)$$

Will precipitation of $NiCO_3$ occur (a) when the concentration of the carbonate ion is 1.0×10^{-6} M or (b) when it is 100 times greater (or 1.0×10^{-4} M)?

76. If the concentration of Zn^{2+} in exactly 10 mL of water is 1.6×10^{-4} M, will zinc hydroxide, $Zn(OH)_2$, precipitate when 4.0 mg of NaOH is added?

77. You have 95 mL of a solution that has a lead(II) concentration of 0.0012 M. Will $PbCl_2$ precipitate when 1.20 g of solid NaCl is added?

78. If the concentration of Mg^{2+} ion in sea water is 1350 mg/L, what OH^- concentration is required to precipitate $Mg(OH)_2$?

79. Will a precipitate of $Mg(OH)_2$ form when 25.0 mL of 0.010 M NaOH is combined with 75.0 mL of a 0.10 M solution of magnesium chloride?

Solubility and Complex Ions

(See Example 18.19 and CD-ROM Screen 18.16)

80. Solid gold(I) chloride, AuCl, dissolves when excess cyanide ion, CN^-, is added to give a water-soluble complex ion.

$$AuCl(s) + 2\ CN^-(aq) \rightleftharpoons Au(CN)_2^-(aq) + Cl^-(aq)$$

Show that this equation is the sum of two other equations, one for dissolving AuCl to give its ions and the other for the formation of the $Au(CN)_2^-$ ion from Au^+ and CN^-. Calculate K_{net} for the overall reaction.

81. Solid silver iodide, AgI, can be dissolved by adding aqueous sodium cyanide.

$$AgI(s) + 2\ CN^-(aq) \rightleftharpoons [Ag(CN)_2]^-(aq) + I^-(aq)$$

Show that this equation is the sum of two other equations, one for dissolving AgI to give its ions and the other for the formation of the $[Ag(CN)_2]^-$ ion from Ag^+ and CN^-. Calculate K_{net} for the overall reaction.

Separations

(See Exercise 18.21)

82. Each of the following pairs of ions is found together in aqueous solution. Using the table of solubility product constants in Appendix J, devise a way to separate these ions by precipitating one of them as an insoluble salt and leaving the other in solution.
    (a) Ba^{2+} and Na^+
    (b) Ni^{2+} and Pb^{2+}

83. Each of the following pairs of ions is found together in aqueous solution. Using the table of solubility product constants in Appendix J, devise a way to separate these ions by adding one reagent to precipitate one of them as an insoluble salt and leave the other in solution.
    (a) Cu^{2+} and Ag^+
    (b) Al^{3+} and Fe^{3+}

General Questions

These questions concern basic concepts of chemical equilibria but may also call on concepts from previous chapters. More challenging questions are marked with an underlined number.

84. If you mix 48 mL of 0.0012 M $BaCl_2$ with 24 mL of 1.0×10^{-6} M H_2SO_4 will a precipitate of $BaSO_4$ form? The equilibrium process involved is

$$BaSO_4(s) \rightleftharpoons Ba^{2+}(aq) + SO_4^{2-}(aq)$$

85. Calculate the hydronium ion concentration and pH of the solution that results when 20.0 mL of 0.15 M acetic acid, CH_3CO_2H, is mixed with 10.0 mL of 0.15 M NaOH.

86. Calculate the hydronium ion concentration and the pH when 50.0 mL of 0.40 M NH_3 is mixed with 50.0 mL of 0.40 M HCl.

87. For each of the following cases, decide whether the pH is less than, equal to, or greater than 7.
    (a) Equal volumes of 0.10 M acetic acid, CH_3CO_2H, and 0.10 M KOH are mixed.
    (b) 25 mL of 0.015 M NH_3 is mixed with 12 mL of 0.015 M HCl.
    (c) 150 mL of 0.20 M HNO_3 is mixed with 75 mL of 0.40 M NaOH.
    (d) 25 mL of 0.45 M H_2SO_4 is mixed with 25 mL of 0.90 M NaOH.

88. Rank the following compounds in order of increasing solubility in water: Na_2CO_3, $BaCO_3$, and Ag_2CO_3.

89. A sample of hard water contains about 2.0×10^{-3} M Ca^{2+}. A soluble fluoride-containing salt such as NaF is added to "fluoridate" the water (to aid in the prevention of dental caries). What is the maximum concentration of F^- that can be present without precipitating CaF_2?

Dietary sources of fluoride ion. Adding fluoride ion to drinking water (or toothpaste) prevents the formation of dental caries. *(Charles D. Winters)*

90. If a buffer solution is prepared from 5.15 g of NH_4NO_3 and 0.10 L of 0.15 M NH_3, what is the pH of the solution? What is the new pH if the solution is diluted with pure water to a volume of 5.00×10^2 mL?

91. The weak base ethanolamine, $HOCH_2CH_2NH_2$, can be titrated with HCl.

$$HOCH_2CH_2NH_2(aq) + H_3O^+(aq) \longrightarrow$$
$$HOCH_2CH_2NH_3^+(aq) + H_2O(\ell)$$

Assume you have 25.0 mL of a 0.010 M solution of ethanolamine and titrate it with 0.0095 M HCl. (K_b for ethanolamine is 3.2×10^{-5})

(a) What is the pH of the ethanolamine solution before the titration begins?

(b) What is the pH at the equivalence point?

(c) What is the pH at the midpoint of the titration?

(d) What indicator in Figure 17.2 would be best to detect the equivalence point?

(e) Calculate the pH of the solution after adding 5.00, 10.0, 20.0, and 30.0 mL of the acid.

(f) Combine the information in parts (a), (b), and (e) and plot an approximate titration curve.

92. Aniline hydrochloride, $[C_6H_5NH_3]Cl$, is a weak acid. (Its conjugate base is the weak base aniline, $C_6H_5NH_2$.) The acid can be titrated with a strong base such as NaOH.

$$C_6H_5NH_3^+(aq) + OH^-(aq) \longrightarrow C_6H_5NH_2(aq) + H_2O(\ell)$$

Assume 50.0 mL of 0.100 M aniline hydrochloride is titrated with 0.185 M NaOH. (K_a for aniline hydrochloride is 2.4×10^{-5}.)

(a) What is the pH of the $[C_6H_5NH_3]Cl$ solution before the titration begins?

(b) What is the pH at the equivalence point?

(c) What is the pH at the midpoint of the titration?

(d) What indicator in Figure 17.2 could be used to detect the equivalence point?

(e) Calculate the pH of the solution after adding 10.0, 20.0, and 30.0 mL of the base.

(f) Combine the information in parts (a), (b), and (e) and plot an approximate titration curve.

93. If you place 5.0 mg of $SrCO_3$ in 1.0 L of pure water, will all of the salt dissolve before equilibrium is established, or will some salt remain undissolved?

94. To have a buffer with a pH of 2.50, what volume of 0.150 M NaOH must be added to 100. mL of 0.230 M H_3PO_4?

95. What mass of Na_3PO_4 must be added to 80.0 mL of 0.200 M HCl to obtain a buffer with a pH of 7.75?

96. For the titration of 50.0 mL of 0.150 M ethylamine, $C_2H_5NH_2$, with 0.100 M HCl, find the pH at each of the following points and then use that information to sketch the titration curve and decide on an appropriate indicator.

(a) At the beginning, before HCl is added

(b) At the halfway point in the titration

(c) When 75% of the required acid has been added

(d) At the equivalence point

(e) When 10.0 mL more HCl has been added than is required

(f) Sketch the titration curve.

(g) Suggest an appropriate indicator for this titration.

97. What volume of 0.120 M NaOH must be added to 100. mL of 0.100 M $NaHC_2O_4$ (sodium hydrogen oxalate) to reach a pH of 4.70?

98. Describe the effect on the pH of

(a) Adding sodium acetate, $NaCH_3CO_2$ to 0.100 M CH_3CO_2H

(b) Adding $NaNO_3$ to 0.100 M HNO_3

(c) Why are the effects not the same?

99. A buffer solution is prepared by dissolving 1.50 g each of benzoic acid, $C_6H_5CO_2H$, and sodium benzoate, $NaC_6H_5CO_2$, in 150.0 mL of solution.

(a) What is the pH of this buffer solution?

(b) Which buffer component must be added and what quantity is needed to change the pH to 4.00?

(c) What volume of 2.0 M NaOH or 2.0 M HCl must be added to the original buffer solution to change the pH to 4.00?

100. A buffer solution with a pH of 12.00 consists of Na_3PO_4 and Na_2HPO_4. The volume of solution is 200.0 mL.

(a) Which component of the buffer is present in a larger amount?

(b) If the concentration of Na_3PO_4 is 0.400 M, what mass of Na_2HPO_4 is present?

(c) Which component of the buffer must be added to change the pH to 12.25? What mass of that component is required?

101. What volume of 0.200 M HCl must be added to 500.0 mL of 0.250 M NH_3 to have a buffer with a pH of 9.00?

102. The cations Ba^{2+} and Sr^{2+} can be precipitated as very insoluble sulfates. If you add sodium sulfate to a solution containing these metal cations, each with a concentration of 0.1 M, which is precipitated first, $BaSO_4$ or $SrSO_4$?

103. You often work with salts of Fe^{2+}, Pb^{2+}, and Al^{3+} in the laboratory. (All are found in nature, and all are important economically.) If you have a solution containing these three ions, each at a concentration of 0.1 M, what is the order in which their hydroxides precipitate as aqueous NaOH is slowly added? [K_{sp} for $Al(OH)_3$ is 1.3×10^{-33}.]

104. What is the equilibrium constant for the following reaction?

$$AgCl(s) + I^-(aq) \rightleftharpoons AgI(s) + Cl^-(aq)$$

Does the equilibrium lie predominantly to the left or right? Will AgI form if iodide ion, I^-, is added to a saturated solution of AgCl?

105. Calculate the equilibrium constant for the following reaction.

$$Zn(OH)_2(s) + 2 CN^-(aq) \rightleftharpoons Zn(CN)_2(s) + 2 OH^-(aq)$$

Does the equilibrium lie predominantly to the left or right? Can zinc hydroxide be transformed into zinc cyanide by adding a soluble salt of the cyanide ion?

106. In principle the ions Ba^{2+} and Ca^{2+} can be separated by the difference in solubility of their fluorides, BaF_2 and CaF_2.

If you have a solution that is 0.10 M in both Ba^{2+} and Ca^{2+}, CaF_2 will begin to precipitate first as fluoride ion is added slowly to the solution.

   (a) What concentration of fluoride ion will precipitate the maximum amount of Ca^{2+} ion without precipitating BaF_2?

   (b) What concentration of Ca^{2+} remains in solution when BaF_2 just begins to precipitate?

107. A solution contains 0.10 M iodide ion, I^-, and 0.10 M carbonate ion, CO_3^{2-}.

   (a) If solid $Pb(NO_3)_2$ is slowly added to the solution, which salt will precipitate first, PbI_2 or $PbCO_3$?

   (b) What will be the concentration of the first ion that precipitates (CO_3^{2-} or I^-) when the second or more soluble salt begins to precipitate?

108. A solution contains Ca^{2+} and Pb^{2+} ions, both at a concentration of 0.010 M. You wish to separate the two ions from each other as completely as possible by precipitating one but not the other using aqueous Na_2SO_4 as the precipitating agent.

   (a) Which will precipitate first as sodium sulfate is added, $CaSO_4$ or $PbSO_4$?

   (b) What will be the concentration of the first ion that precipitates (Ca^{2+} or Pb^{2+}) when the second or more soluble salt begins to precipitate?

109. Which of the following barium salts should be soluble in a strong acid such as HCl: $Ba(OH)_2$, $BaSO_4$, or $BaCO_3$?

110. Suggest a method for separating a precipitate consisting of a mixture of CuS and $Cu(OH)_2$.

Using Electronic Resources

These questions refer to the General Chemistry Interactive CD-ROM, *Version 3.0.*

111. See CD-ROM Screen 18.2: The Common Ion Effect.

   (a) What fraction of the acid molecules in a 0.10 M solution of $HClO_2$ ionize?

   (b) What happens to the fraction of acid molecules ionized when 0.010 M ClO_2^- ions are added?

112. See CD-ROM Screen 18.4: Buffer Solutions. Choose the *Simulation.* Set the concentrations of NH_4Cl and NH_3 in a buffer each to 0.1 M. What occurs when you raise the concentration of NH_4Cl? What happens as you raise the concentration of NH_3?

113. See CD-ROM Screen 18.5: Buffer Solutions. Choose the *Simulation.* What concentrations of $NaHCO_3$ and Na_2CO_3 would you need to have a buffer solution with a pH of exactly 10?

114. See CD-ROM Screen 18.7: Titration Curves. Choose the *Simulation.* Titrate 25.0 mL of 0.30 M acetic acid with 0.50 M NaOH.

   (a) What volume of NaOH is required?

   (b) What is the pH at the equivalence point? Explain why the pH has this value.

   (c) Which of the three indicators available are best used in this titration?

115. See CD-ROM Screen 18.10: Determining K_{sp}.

   (a) In the example on this screen, we determine the K_{sp} value for $PbCl_2$ by measuring the concentration of Pb^{2+} ions in solution. Why do we not need to measure the concentration of Cl^- ions in this case?

   (b) Watch the video on this screen describing the use of an atomic absorption spectrometer. Explain how the spectrometer works.

116. See CD-ROM Screen 18.11: Estimating Salt Solubility. This screen describes how the solubility of $BaSO_4$ is estimated using its K_{sp} value. The calculation assumes that the ions produced, Ba^{2+} and SO_4^{2-}, do not react further in solution. What would happen to the solubility of $BaSO_4$ if some of the sulfate ion reacted with acid in solution?

$$SO_4^{2-}(aq) + H_3O^+(aq) \rightleftharpoons HSO_4^-(aq) + H_2O(\ell)$$

117. See CD-ROM Screen 18.12: Can a Precipitation Reaction Occur?

   (a) Choose the *Simulation.* When you have an Fe^{2+} concentration of 1.0×10^{-4} M in a solution, what concentration of OH^- is required to just begin precipitation of $Fe(OH)_2$?

   (b) Examine the $Q > K_{sp}$ video and animation. Describe the silver chloride precipitation at the particulate level. (See also Figure 5.3.)

118. See CD-ROM Screen 18.13: The Common Ion Effect. The animation on the *Description* screen illustrates the common ion effect for the case of adding extra chloride ion to an equilibrium system containing $PbCl_2(s)$, $Pb^{2+}(aq)$, and $Cl^-(aq)$. Explain the changes you see in terms of the solubility product constant expression for this system.

119. See CD-ROM Screen 18.14: Using Solubility. Examine the flowchart for the separation of the ions Ag^+, Pb^{2+}, and Cu^{2+} in aqueous solution.

   (a) What is the initial step in the separation, and why does it work?

   (b) In what way does the solubility of $PbCl_2$ differ from that of $AgCl$?

   (c) How do we know that Pb^{2+} ions are in solution?

120. See CD-ROM Screen 18.15: Simultaneous Equilibria.

   (a) The video on this screen shows that the white solid $PbCl_2$ can be converted into the yellow solid $PbCrO_4$. It might appear that the reaction occurs in the solid state, even though it actually occurs in aqueous solution. Write chemical equations to show how it occurs in solution.

   (b) Why is the experiment on this screen good evidence that chemical equilibria are dynamic as opposed to static?

121. See CD-ROM Screen 18.16: Solubility and pH.
   (a) Explain how the example on this screen, the solubility of $Co(OH)_2(s)$, illustrates Le Chatelier's principle.
   (b) Notice that the solubility of the compound increases as the pH decreases. What effect do you think a pH decrease would have on the solubility of $MgCO_3$, considering the fact that carbonate ion, CO_3^{2-}, is a weak base?

   (c) Examine the pH and Solubility Table. Explain why the solubility of $Co(OH)_2$ increases by 100 for each 1.0 decrease in pH.

122. See CD-ROM Screen 18.17: Complex Ion Formation and Solubility. Examine the sidebar to this screen. How does the chemistry of floor wax support the idea that reactions can be reversible?

19 Principles of Reactivity: Entropy and Free Energy

Chapter Goals

- Understand the concept of entropy and how it relates to spontaneity.
- Predict whether a process is product- or reactant-favored.
- Use tables of data in thermodynamic calculations.
- Define and use a new thermodynamic function, free energy.

Perpetual Motion Machines

Economists have often said, "There is no free lunch," meaning that every desirable thing requires the expenditure of some effort, money, or energy. But people have certainly tried.

Wouldn't it be wonderful if someone could design a perpetual motion ma-chine, one that produces energy but re-quires no energy itself? Over the centuries many people have tried to do this. But as you will see in this chapter, such a machine violates fundamental laws of thermodynamics.

Among the best known early machines was one proposed by Robert Fludd in the 1600s. Fludd was a well known scientist, who also proposed that the sun and not the earth was the center of our universe, and that our blood carries the gases important to life throughout the body. But he also believed that lightning was simply an act of God.

In 1812 Charles Redheffer of Philadelphia set up a machine that, he claimed, required no source of energy to run. He applied to the city government for the funds to build an even larger version, but did not allow the city commissioners to get too close to the machine. The commissioners were suspicious, however, and asked a local engineer, Isaiah Lukens, to build a machine that worked on the same principle as Redheffer's

◀ The fanciful *Waterfall* by M. C. Escher, 1961. This is reminiscent of Fludd's perpetual motion machine. (M. C. Escher's "Waterfall" ©Cardon Art B. V.—Baarn, Holland. All rights reserved.)

"You can't get something for nothing."

▲ A late 17th-century version of Fludd's proposed "perpetual motion machine." Water flows from a reservoir at the right, turning a water wheel. Work could be extracted from this motion, and it was also used to turn an Archimedes screw, the long rod to the left, that transported the water back to the top where it again could flow downward onto the water wheel. *(Image courtesy of University of Kentucky Library/Special Collections)*

machine. When Redheffer saw Lukens's replica, Redheffer decided it would be best to leave town. He moved to New York City, where he once again tried to interest investors in his machine.

Redheffer was again exposed as a charlatan, this time by the inventor Robert Fulton. When Fulton was invited to view the machine, he noticed it was operating in a wobbly manner. Fulton challenged Redheffer that he could expose the secret source of energy. Fulton also offered to compensate Redheffer for damages if Redheffer was falsely accused. Redheffer should never have agreed. When Fulton removed some boards in a wall near the machine, a thin cord was discovered to lead to yet another room. There Fulton found an elderly gentleman, eating a crust of bread with one hand and turning a crank with the other hand — not so smoothly — to run Redheffer's machine! Redheffer disappeared.

The vast majority of the so-called perpetual motion machines violate the first law of thermodynamics. Falling water in Fludd's machine can indeed turn a crank to perform useful work. But then there is insufficient energy available to lift the water to the reservoir to begin the cycle again. Besides, any machine loses energy due to friction.

In the 1880s John Gamgee invented an "ammonia engine" and tried to persuade the U.S. Navy to use it for ship propulsion. Even President Garfield took time to inspect Gamgee's engine. The engine worked by using the heat of ocean water to evaporate liquid ammonia. The expanding ammonia vapor was supposed to drive a piston, in the same way that the combustion of gasoline in an automobile's engine drives the pistons of the engine. In the ammonia engine, the ammonia vapor cools when it expands, which would lead to the condensation of the ammonia. The reservoir of liquid ammonia is thereby replenished, and the cycle begins anew.

Gamgee's engine sounded good — but, like the others, it violates the laws of thermodynamics, the subject of this chapter.

Before You Begin

- Review enthalpy and the first law of thermodynamics (Chapter 6).
- Review the concepts of chemical equilibrium (Chapter 16).

Spontaneous and product- or reactant-favored processes These photographs illustrate the principal theme of this chapter: When will a process be spontaneous? What drives a chemical or physical change to be product-favored and how can we predict whether a process is product- or reactant-favored?

Heat transfer

Heat transfers spontaneously from a hotter object to a colder object.

Product-favored exothermic reaction

Reaction of sodium with chlorine.

Reactant-favored endothermic reaction

The decomposition of water using electrical energy.

Product-favored endothermic process

Gallium melts from the heat of your hand.

Energy neutral

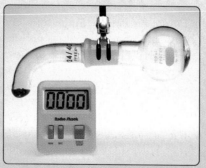

Here a gas diffuses spontaneously from the flask in which it was generated to another, attached flask.

Charles D. Winters

Chapter 6 (thermochemistry) and Chapter 19 (entropy and free energy) together provide an introduction to the subject of thermodynamics. Chapter 6 focused on heat and heat transfer with discussion guided by the first law of thermodynamics, the law of conservation of energy. In this chapter we encounter two more laws of thermodynamics. These laws determine the spontaneity of chemical and physical processes, explain the driving forces that compel the changes that a system undergoes to achieve equilibrium, and guide us to understand when reactions are product–favored or reactant–favored.   ●

19.1   SPONTANEOUS CHANGE AND EQUILIBRIUM

Change is central to the study of chemistry, so it is important to understand the factors that determine if a change will occur. In chemistry, we encounter many examples of both chemical changes (chemical reactions) and physical changes (the formation of mixtures, expansion of gases, and changes of state to name a few). When describing changes, chemists use the term **spontaneous.** A spontaneous change is one that occurs without outside intervention. It does not say anything about the rate of the change, only that a spontaneous change is one that is naturally occurring and unaided. Furthermore, a spontaneous change leads inexorably to equilibrium.

If a piece of hot metal is placed in a beaker of cold water, heat transfer from the hot metal to the cooler water occurs (Chapter Focus). The process of heat transfer continues until the two objects are at the same temperature and thermal equilibrium is attained. Similarly, chemical reactions proceed until equilibrium is reached. Some chemical reactions greatly favor products at equilibrium, as in the reaction of sodium and chlorine (page 13). In Chapters 16–18 we describe them as *product-favored* reactions (page 662). In other instances, the position of the equilibrium favors the reactants. One example of such *reactant-favored* processes would be the dissolution of an "insoluble salt" like $PbCl_2$. If you place a handful of $PbCl_2$ in a small amount of water, the sample will dissolve spontaneously until equilibrium is reached with most of the salt remaining undissolved.

When considering both physical and chemical processes, the focus is always on the changes that must occur to achieve the condition of equilibrium. The factors

---

Problem-Solving Tip 19.1

A Review of Concepts of Thermodynamics

To understand the thermodynamic concepts introduced in this chapter, be sure to review the ideas of Chapter 6.

System: The part of the universe under study.

Surroundings: The rest of the universe exclusive of the system.

Exothermic: Thermal energy transfer occurs from the system to the surroundings.

Endothermic: Thermal energy transfer occurs from the surroundings to the system.

First law of thermodynamics: The law of the conservation of energy, $\Delta E = q + w$. The change in internal energy of a

system is the sum of heat transferred to or from the system and the work done on or by the system.

Enthalpy change: The thermal energy transferred at constant pressure.

State function: A quantity whose value is determined only by the initial and final states of a system.

Standard conditions: Pressure of 1 bar (1 bar = 0.98692 atm) and solution concentration of 1 m.

Standard enthalpy of formation, ΔH_f°: The enthalpy change occurring when a compound is formed from its elements in their standard states.

that determine the directionality of a process are the topic of this chapter. Given two objects in contact and at the same temperature, it will never happen that one heats up while the other gets colder. Gas molecules will never spontaneously congregate at one end of a flask. A chemical system at equilibrium will not spontaneously change in a way that results in the system no longer being at equilibrium. Neither will a chemical system not at equilibrium change in the direction that takes it further from the equilibrium condition.

19.2   HEAT AND SPONTANEITY

In previous chapters, we have encountered many product-favored chemical reactions: Hydrogen and oxygen combine, forming water; methane burns to give CO_2 and H_2O; Na and Cl_2 react to form NaCl; HCl(aq) and NaOH(aq) react to form H_2O and NaCl(aq). These reactions proceed spontaneously from reactants to products and go substantially to completion when equilibrium is reached. These chemical reactions and many others have a common feature: They are exothermic processes. It might be tempting to conclude that evolution of heat is the criterion that determines whether a reaction is spontaneous. Further inspection will reveal flaws in this reasoning. This is especially evident with the inclusion of some common physical changes that are endothermic or energy-neutral:

- *Expansion of a gas into a vacuum* (see Chapter Focus). If a flask containing a gas is connected through a valve to an evacuated flask and the valve is opened, the gas will flow from one vessel to the other until the pressure is the same throughout. The expansion of an ideal gas is energy-neutral, heat being neither evolved nor required.

- *Phase changes.* Melting of ice is an endothermic process; to melt one mole of H_2O requires about 6 kJ. At temperatures above 0 °C, the melting of ice is spontaneous. Below 0 °C, ice does not melt, however; melting is not a spontaneous process. At 0 °C, the system is at equilibrium and there is no net change. This example illustrates that temperature has a role in determining whether a process is spontaneous. We will return to this important issue later in this chapter (Section 19.5).

- *Heat transfer.* The temperature of a cold soft drink will rise until the beverage reaches ambient temperature. The heat required for this process comes from the surroundings. The endothermic process illustrates that heat transfer from a hotter object (the surroundings) to a cooler object (the soft drink) is spontaneous.

- *Dissolving NH_4Cl.* The ionic compound NH_4Cl dissolves spontaneously in water in an endothermic process with $\Delta H° = +14.89$ kJ/mol.

We can have further insight into spontaneity if we think about a chemical system, $H_2(g) + I_2(g) \longrightarrow 2\,HI(g)$. Equilibrium in this system can be approached from either direction [← CHAPTER 16, PAGE 658]. The reaction of H_2 and I_2 is endothermic, but the reaction occurs spontaneously until an equilibrium mixture of H_2, I_2, and HI is formed. The reverse reaction, the decomposition of HI to form H_2 and I_2 until equilibrium is achieved, is also spontaneous, but in this case the process evolves heat. Equilibrium is approached spontaneously from either direction.

By these examples, and many others, we must conclude that evolution of heat is not a sufficient criterion to determine whether a process is spontaneous. In ret-

● **Spontaneity and Product-favored Reactions**

A spontaneous reaction proceeds to equilibrium without outside intervention. Such a reaction may or may not be product-favored, a reaction in which products predominate over reactants at equilibrium.

rospect, this makes sense. The first law of thermodynamics tells us that in any process energy must be conserved. If energy is evolved by a system, then the same amount of energy must be absorbed by the surroundings. The exothermicity of the system must be balanced by the endothermicity of the surroundings in order that the energy content of the universe remains unchanged. If energy evolution were the only criterion for spontaneity, then for every spontaneous exothermic change in a system there would be a corresponding non-spontaneous change in the surroundings. We must search further than heat evolution and the first law to determine whether a process is spontaneous.

19.3   ENTROPY AND THE SECOND LAW

A search for a better way to predict if a process is spontaneous leads to a new thermodynamic function, **entropy, S,** a property associated with disorder in the system. It also leads to the **second law of thermodynamics,** which states that in *spontaneous process the entropy of the universe increases.* Ultimately this law will allow us to predict the conditions at equilibrium as well as the direction of spontaneous change toward equilibrium. That is, in the end we will have a relation between how far a system is from equilibrium and how strong the tendency is to get there.

The concept of entropy is built around the idea that *spontaneous processes result in dispersal of matter and energy.* The logic underlying this statement is statistical. Let us look at several examples to illustrate the statistical nature of entropy [CD-ROM, Screen 19.3-19.6].

Dispersal of Matter

Matter dispersal is illustrated by the expansion of a gas into a vacuum. Consider a flask containing a gas connected by a valve to an evacuated flask having the same volume (Figure 19.1a). If the valve is opened, it is highly probable that gas will flow from one flask into the other until the pressures are equal. The opposite process, in which all the gas molecules in the apparatus congregate in one of the two flasks, is highly improbable. (Similarly, it is highly probable that, when a water-soluble compound is placed in water at room temperature (Figure 19.1b), the molecules or ions of the compound will ultimately be distributed throughout the solution.)

To understand how probability drives gas expansion, consider Figure 19.1. Assume that, when the valve connecting the flasks is opened, the two molecules originally in flask A can move randomly throughout flasks A and B. At any given instant, the molecules will be in one of four possible configurations: two molecules in flask A, two molecules in flask B, or one molecule in each flask. The probability of having one molecule in each flask is 50%. There is a 25% probability that the two molecules will be simultaneously in flask A and a 25% probability that they will be in flask B.

If we consider a system consisting of three molecules originally in flask A, we would find there is only a one-in-eight chance that the three molecules will remain in the original flask. With ten molecules there is only one chance in 1024 to find all the molecules in flask A. The probability of n molecules remaining in the initial flask in this two flask system is $(1/2)^n$, where n is the number of molecules. If flask A contained 1 mol of a gas, the probability of all the molecules being found in that flask when the connecting valve is opened is $(1/2)^N$ (N being Avogadro's number), a probability almost too small to comprehend! The most likely scenario is that there will be equal numbers of molecules in each flask.

Chapter Goals • Revisited

- **Understand the concept of entropy and how it relates to spontaneity.**
- Predict whether a process is product- or reactant-favored.
- Use tables of data in thermodynamic calculations.
- Define and use a new thermodynamic function, free energy.

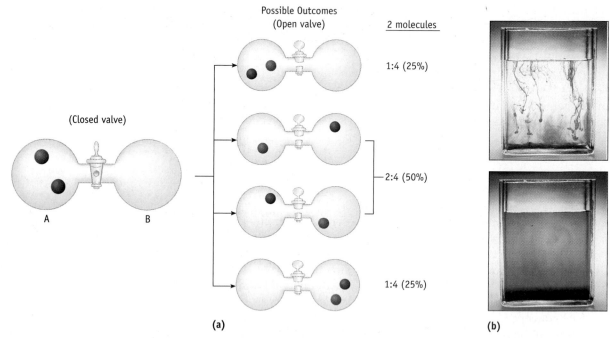

Possible Outcomes
(Open valve)

2 molecules

(Closed valve)

1:4 (25%)

2:4 (50%)

1:4 (25%)

A   B

(a)   (b)

Figure 19.1   Dispersal of matter. (a) The expansion of a gas into a vacuum. There are four possible arrangements for two molecules in this apparatus. There is a 50% probability that there will be one molecule in each flask at any instant of time. **(b)** A small quantity of purple $KMnO_4$ is added to water (*top*). With time, the solid dissolves and the highly colored MnO_4^- ions (and the K^+ ions) are dispersed throughout the solution. *(Charles D. Winters)*

The logic applied to the expansion of a gas into a vacuum can be used to rationalize the mixing of two gases. If flasks containing O_2 and N_2 are connected (in an experimental set-up like that in Figure 19.1a), the two gases will diffuse together, eventually leading to a mixture in which molecules of O_2 and N_2 are evenly distributed throughout the total volume. A mixture of N_2 and O_2 will never separate into samples of each component of its own accord. The important point is that mixing N_2 and O_2 will always lead to a less ordered system.

The tendency to move from order to disorder is also the reason that the $KMnO_4$ sample in Figure 19.1b eventually disperses over the entire container. The process leads to a mixture, a less ordered system.

It is also evident from these examples of matter dispersal that if we wanted to put all of the gas molecules in Figure 19.1a back into one flask, or recover the $KMnO_4$ crystals, we would have to intervene in some way. We could use a pump to force all of the gas molecules from one side to the other or we could lower the temperature drastically (Figure 19.2). In the latter case the molecules move "downhill" energetically to a place of very low kinetic energy.

Does the formation of a mixture always lead to greater disorder? It does when gases are mixed. With liquids and solids this is usually the case too, but, not always. Exceptions can be found, especially when considering aqueous solutions. For example, Li^+ and OH^- ions are highly ordered in the solid lattice (◄ SECTION 13.6), and they become more disordered when they enter the solution. However, the solution process is accompanied by solvation, a process in which water molecules become tightly bound to the ions. Because of this, the water molecules are constrained

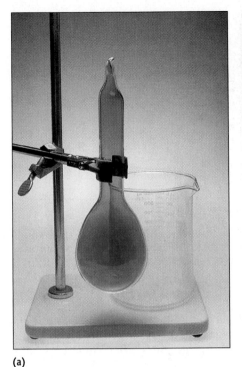

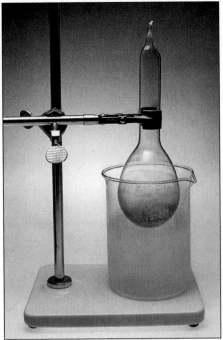

(a)   (b)

Figure 19.2 **Reversing the process of matter dispersal.** (a) Brown NO_2 gas is dispersed evenly throughout the flask. **(b)** If the flask is immersed in liquid nitrogen at -196 °C, the kinetic energy of the NO_2 molecules is reduced to the point that they become a solid and collect on the cold walls of the flask. *(Charles D. Winters)*

to a more ordered arrangement than in pure water. Thus, there are two opposing effects here: a decrease in order for the Li^+ and OH^- ions when dispersed in the water but an increased order for water (Figure 19.3). The higher degree of ordering due to solvation dominates, and the result is a higher degree of order overall in the system.

Dispersal of Energy

The dispersal of energy over as many atoms or molecules as possible is also a favored process. Let us explore the situation in which heat flows from a hot object to a cold object until both have the same temperature (Figure 19.4). A model involving gaseous atoms again provides a basis for this analysis. Start by placing a sample of hot atoms in contact with a sample of cold atoms. The atoms move randomly, and hot atoms will sometimes collide with colder atoms. Energy can be transferred between the atoms when they collide. The collisions obey the first law; energy lost by one atom is gained by the second. Eventually the system stabilizes at an average temperature so that any sample of the gas has the same distribution of energies (◀ SECTION 12.6).

Let us again use a statistical explanation to show how energy is dispersed in a system. Start with two hot atoms, A1* and A2* and two cold atoms, A3 and A4. The asterisks on atoms A1 and A2 represent packets of energy. Collisions between atoms allow energy to be transferred so that all distributions of the two packets of energy over four atoms are eventually achieved. There are ten different ways to distribute the energy over the four atoms (Figure 19.5). In only three of the ten cases, the packets of energy remain with atoms A1 and A2. In this four-atom sample, there is a 70% probability that some or all of energy will be transferred from A1 and A2 to A3 or A4.

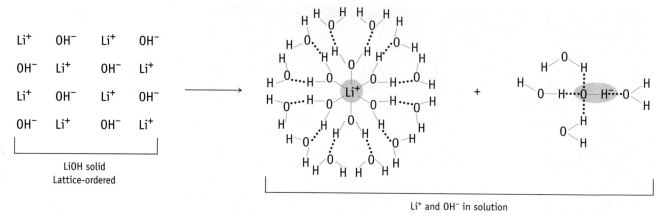

Figure 19.3   **Order and disorder in solid and solution.** Lithium hydroxide is a crystalline solid. When placed in water, it dissolves, and water molecules interact with the Li$^+$ cations and OH$^-$ anions. This leads to a more orderly arrangement for the aqueous solution than in pure water.

With a larger collection of atoms and many energy packets, the likelihood of energy dispersal over all the atoms in the system becomes larger and larger. In a mole of atoms, the most probable situation disperses energy over the largest number of atoms.

Summary: Matter and Energy Dispersal

In summary, the final state of a system can be more probable than the initial state in either or both of two ways: (1) the atoms and molecules can be more disordered and (2) energy can be dispersed over a greater number of atoms and molecules.

* If energy and matter are both dispersed, a process is definitely spontaneous because both the products and the distribution of energy are more probable.
* If only energy or matter is dispersed, then quantitative information is needed to decide which effect is greater.

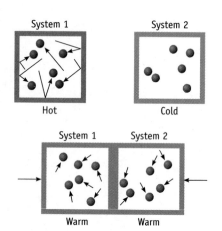

Figure 19.4   **Energy transfer between molecules in the gas phase.**

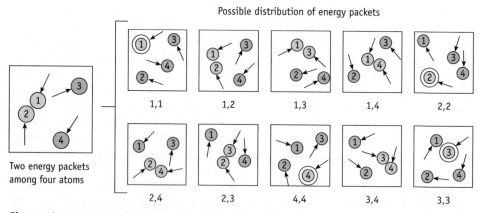

Figure 19.5   **Energy dispersal.** Possible ways of distributing two packets of energy among four atoms.

- If neither matter nor energy is more spread out after a process occurs, then that process will never be spontaneous.

19.4 ENTROPY AND THE THIRD LAW OF THERMODYNAMICS

Entropy is used to quantify the extent of disorder resulting from dispersal of energy and matter. For any substance under a given set of conditions a numerical value for entropy can be determined. The greater the disorder in a system, the greater the entropy and the larger the value of S [CD-ROM, Screens 19.4–19.6].

Like internal energy (E) and enthalpy (H), entropy is a state function. This means that the *change in entropy* for any process depends only on the initial and final states of the system and not on the pathway by which the process occurs (SECTION 6.4). This is an important point because it bears on how the value of S is determined, as we will see later.

The point of reference for entropy values is established by the **third law of thermodynamics.** Defined by Ludwig Boltzmann, the third law states that there is no disorder in a perfect crystal at 0 K, that is, $S = 0$. The entropy of an element or compound under any set of conditions is the entropy gained by converting the substance from 0 K to the defined conditions. The entropy (disorder) of a substance at any temperature can be obtained by measuring the heat added to raise the temperature from 0 K, but with a specific provision that the conversion is carried out by a **reversible process.** Adding heat in very small increments approximates a reversible process. The entropy added by each incremental change is calculated using Equation 19.1.

History

Ludwig Boltzmann (1844–1906)
Engraved on his tombstone in Vienna, Austria is his equation relating entropy and chaos, $S = k \log W$. The symbol k is a constant now known as the Boltzmann constant, and W stands for the number of ways that atoms or molecules can be arranged in a given state. *(Oesper Collection in the History of Chemistry/University of Cincinnati.)*

$$\Delta S = \frac{q_{rev}}{T} \qquad (19.1)$$

In this equation, q_{rev} is the heat absorbed and T is the Kelvin temperature at which the change occurs. Adding the entropy changes for the incremental steps gives the total entropy of a substance. (Because it is necessary to add heat to raise the temperature, all substances have positive entropy values at temperatures above 0 K. Based on the third law of thermodynamics, negative values of entropy cannot occur.)

The **standard entropy, $S°$,** of a substance is the entropy gained by converting it from a perfect crystal at 0 K to standard state conditions (1 bar, 1 m solution). The units for entropy are J/K · mol. A few values of $S°$ are given in Table 19.1. Generally, values of $S°$ found in tables of data refer to a temperature of 298 K.

Some interesting and useful generalizations can be drawn from the data given in Table 19.1 (and Appendix L).

- *When comparing the same or similar substances, entropies of gases are much larger than those for liquids, and entropies of liquids are larger than those for solids.* In a solid the particles have fixed positions in the solid lattice. When a solid melts, its particles have more freedom to assume different positions, resulting in an increase in disorder (Figure 19.6). When a liquid evaporates, restrictions due to forces between the particles nearly disappear, and another large increase in entropy takes place. For example, the entropies of $I_2(s)$, $Br_2(\ell)$, and $Cl_2(g)$ are 116.1, 152.2, and 223.1, respectively, and the entropies of C(s, graphite) and C(g) are 5.6 and 158.1, respectively.

- As a general rule, *larger molecules have larger entropies than smaller molecules, and molecules with more complex structures have larger entropies than simpler molecules.*

● **Entropy Values**
There is a longer list of standard entropy values in Appendix L and extensive lists of $S°$ values can be found in standard chemical reference sources (such as the NIST tables and at http://webbook.nist.gov).

Table 19.1 ● Some Standard Molar Entropy Values at 298 K

Element	Entropy, $S°$ (J/K · mol)	Compound	Entropy, $S°$ (J/K · mol)
C(graphite)	5.6	$CH_4(g)$	186.3
C(diamond)	2.377	$C_2H_6(g)$	229.2
C(vapor)	158.1	$C_3H_8(g)$	270.3
Ca(s)	41.59	$CH_3OH(\ell)$	127.2
Ar(g)	154.9	CO(g)	197.7
$H_2(g)$	130.7	$CO_2(g)$	213.7
$O_2(g)$	205.1	$H_2O(g)$	188.84
$N_2(g)$	191.6	$H_2O(\ell)$	69.95
$F_2(g)$	202.8	HCl(g)	186.2
$Cl_2(g)$	223.1	NaCl(s)	72.11
$Br_2(\ell)$	152.2	MgO(s)	26.85
$I_2(s)$	116.1	$CaCO_3(s)$	91.7

$S°(J · K \cdot mol)$

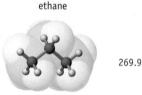

186.3

methane

229.6

ethane

269.9

propane

These generalizations work best in series of related compounds. In a more complicated molecule there are more ways for the molecule to rotate, twist, and vibrate in space. Entropies (in J/K · mol) for methane (CH_4), ethane (C_2H_6), and propane (C_3H_8) are 186.3, 229.2, and 270.3, respectively. The effect of molecular structure can be seen with atoms or molecules of similar molar mass: Ar, CO_2, and C_3H_8 have entropies of 154.9, 213.7, and 270.3 (in J/K · mol), respectively.

● For a given substance, entropy increases as the temperature is raised. Large increases in entropy accompany changes of state (Figure 19.7).

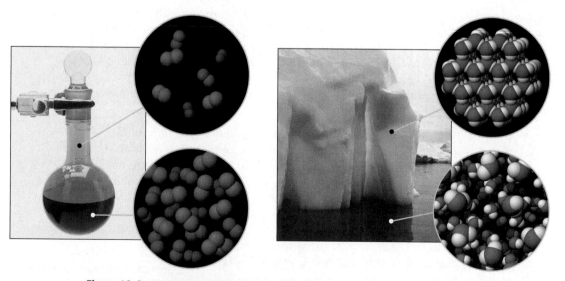

Figure 19.6   Entropy and states of matter. (a) The standard entropy of liquid bromine, $Br_2(\ell)$, is 152.2 J/K · mol and that for bromine vapor is 175.0 J/K · mol. **(b)** The standard entropy of ice, which has a highly ordered molecular arrangement, is smaller than that for disordered liquid water. *(Charles D. Winters)*

Reversible and Irreversible Processes

To determine the entropy change experimentally, heat transfer must be measured for a reversible process. What is a reversible process and why is this important in this discussion?

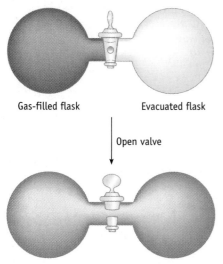

Gas-filled flask   Evacuated flask

Open valve

When the valve is opened the gas expands irreversibly to fill both flasks.

◄ Gas expansion into a vacuum is irreversible. To return to the original pressure and volume work must be done on the system.

The melting of ice/freezing of water at 0 °C is an example of a reversible process. Given a mixture of ice and water at equilibrium, adding heat in small increments will convert ice to water; removing heat in small increments will convert water back to ice. The test for reversibility is that after carrying out a change along a given path (in this instance, heat added), it is possible to return to the starting point by the same path (heat taken away) without altering the surroundings.

Reversibility is closely associated with equilibrium. Assume that we have a system at equilibrium. Changes can then be made by slightly perturbing the equilibrium and letting the system readjust. Melting or freezing water is carried out by adding or removing heat in small increments.

Spontaneous processes are not reversible: the process occurs in one direction. Suppose a gas is allowed to expand into a vacuum, a spontaneous process. No work is done in this process because no force is resisting this expansion. To return the system to its original state, it will be necessary to compress the gas, but doing this means doing work on the system. In doing so, the energy content of the surroundings will decrease by the amount of work expended by the surroundings. The system can be restored to its original state, but the surroundings will be altered in the process.

In summary, two important points must be made concerning reversibility:

- At every step along a reversible pathway between two states the system remains at equilibrium.
- Spontaneous processes follow irreversible pathways and involve nonequilibrium conditions.

To determine the entropy change for a process, it is necessary to identify a reversible pathway. Only then can an entropy change for the process be calculated from the measured heat change for the process, q_{rev}, and the temperature at which it occurs.

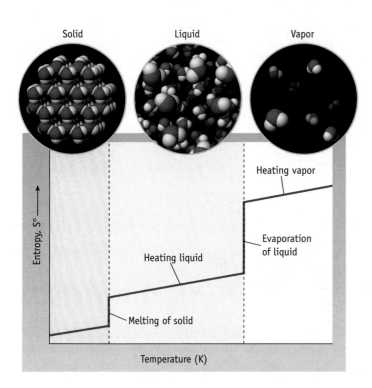

Solid   Liquid   Vapor

Heating vapor

Evaporation of liquid

Heating liquid

Melting of solid

Entropy, $S°$

Temperature (K)

Figure 19.7 Entropy and temperature. For each of the three states of matter, entropy increases with increasing temperature. An especially large increase in entropy accompanies a phase change from a solid to a liquid to a gas.

Example 19.1 Entropy Comparisons

Problem • Which substance has the higher entropy under standard conditions? Explain your reasoning. Check your answer against data in Appendix L.

(a) $NO_2(g)$ or $N_2O_4(g)$
(b) $I_2(g)$ or $I_2(s)$

Strategy • Use the general guidelines on entropy listed in the text: that entropy decreases in the order gas > liquid > solid; that larger molecules have greater entropy than smaller molecules.

Solution •
(a) Both NO_2 and N_2O_4 are gases. Dinitrogen tetraoxide, N_2O_4,

the larger molecule, is expected to have the higher standard entropy. $S°$ values (Appendix L) confirm this prediction: $S°$ for $NO_2(g)$ is 240.04 J/K · mol, $S°$ for $N_2O_4(g)$ is 304.38 J/K · mol.

(b) Gases have higher entropies than solids. $S°$ for $I_2(g)$ is 260.69 J/K · mol, $S°$ for $I_2(s)$ is 116.135 J/K · mol.

Comment • A prediction is a useful check on your numerical result. If an error is inadvertently made, then the prediction will alert you to reconsider your work.

Exercise 19.1 Entropy Comparisons

Predict which substance has the higher entropy and explain your reasoning.

(a) $O_2(g)$ or $O_3(g)$
(b) $SnCl_4(\ell)$ or $SnCl_4(g)$

Entropy Changes in Physical and Chemical Processes

The entropy change ($\Delta S°$) for chemical and physical changes for a system at standard conditions can be calculated from values of $S°$. The procedure used to calculate $\Delta S°$ is similar to that used to obtain values of $\Delta H°$ (◀ EQUATION 6.6). The entropy change is the sum of the entropies of the products minus the sum of the entropies of reactants (Equation 19.2).

$$\Delta S°_{sys} = \sum S°(\text{products}) - \sum S°(\text{reactants}) \tag{19.2}$$

To illustrate, let us calculate $\Delta S°_{rxn}$ ($= \Delta S°_{sys}$) for the oxidation of NO with O_2.

$$2\,NO(g) + O_2(g) \longrightarrow 2\,NO_2(g)$$

Here we subtract the entropies of the reactants (2 mol NO and 1 mol O_2) from the entropy of the products (2 mol NO_2).

$$\begin{aligned}\Delta S°_{rxn} &= (2\text{ mol }NO_2)(240.0\text{ J/K · mol})\\ &\quad - [(2\text{ mol }NO)(210.8\text{ J/K · mol}) + (1\text{ mol }O_2)(205.1\text{ J/K · mol})]\\ &= -146.7\text{ J/K}\end{aligned}$$

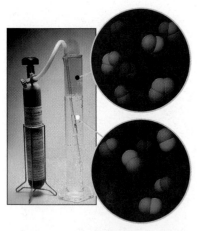

The reaction of NO with O_2. The entropy of the system decreases because two molecules of gas are produced from three molecules of gas. *(Charles D. Winters)*

or −73.35 J/K for 1 mol of NO_2 formed. Notice that the entropy of the system decreases, as predicted for a reaction that converts three molecules of gas into two molecules of another gas.

Problem-Solving Tip 19.2

Summary of Common Entropy-Favored Processes

The discussion to this point, along with the Examples and Exercises, allows the listing of several general principles involving entropy changes.

- A substance becomes increasingly disordered going from a solid to a liquid to a gas. Significant increases in entropy correspond to these phase changes.
- Entropy of any substance increases with temperature (see Figure 19.7). Heat must be added to a system to increase

its temperature (that is, $q > 0$) so q_{rev}/T is necessarily positive.

- Entropy of a gas increases with an increase in volume. A larger volume provides a larger number of places for molecules to "reside."
- Reactions that increase the number of moles of gases in a system are accompanied by an increase in entropy.

Example 19.2 Predicting and Calculating $\Delta S°$

Problem • Calculate the standard entropy changes for the following processes. Do the calculations match predictions?

(a) Evaporation of 1.00 mol of liquid ethanol to ethanol vapor.

$$C_2H_5OH(\ell) \longrightarrow C_2H_5OH(g)$$

(b) The oxidation of one mole of ethanol vapor.

$$C_2H_5OH(g) + 3\,O_2(g) \longrightarrow 2\,CO_2(g) + 3\,H_2O(g)$$

Strategy • Standard entropy changes are calculated from values of standard entropies (Appendix L) using Equation 19.2. Predictions are made using the guidelines given in the text: entropy increases going from solid to liquid to gas (see Figure 19.7), and entropy increases in a chemical reaction if the number of moles of gases in the system increases.

Solution •
(a) Evaporating ethanol.

$$\Delta S° = \sum S°(\text{products}) - \sum S°(\text{reactants})$$

$$= S°[C_2H_5OH(g)] - S°[C_2H_5OH(\ell)]$$

$$= 1 \text{ mol } (282.70 \text{ J/K} \cdot \text{mol}) - 1 \text{ mol } (160.7 \text{ J/K} \cdot \text{mol})$$

$$= +122.0 \text{ J/K}$$

A large positive value for the entropy change can be predicted because the process converts ethanol from a more ordered condition (liquid) to a less ordered condition (vapor).

(b) Oxidation of ethanol vapor

$$\Delta S° = 2\,S°[CO_2(g)] + 3\,S°[H_2O(g)]$$
$$\quad - \{S°[C_2H_5OH(g)] + 3\,S°[O_2(g)]\}$$

$$= 2 \text{ mol } (213.74 \text{ J/K} \cdot \text{mol}) + 3 \text{ mol } (188.84 \text{ J/K} \cdot \text{mol})$$
$$\quad - [1 \text{ mol } (282.70 \text{ J/K} \cdot \text{mol}) + 3 \text{ mol } (205.07 \text{ J/K} \cdot \text{mol})]$$

$$= +96.09 \text{ J/K}$$

An increase in entropy can be predicted for this reaction because the number of moles of gases increases from 4 to 5.

Comment • Values of entropies in tables are based on 1 mol of the compound. In part (b), the number of moles of reactants and products is defined by the stoichiometric coefficients in the balanced chemical equation.

Exercise 19.2 Calculating $\Delta S°$

Calculate the standard entropy changes for the following processes using entropy values in Appendix L. Do the calculated values of $\Delta S°$ match predictions?

(a) Dissolving 1.0 mol of $NH_4Cl(s)$ in water: $NH_4Cl(s) \longrightarrow NH_4Cl(aq)$.

(b) The formation of 2.0 mol of $NH_3(g)$ from $N_2(g)$ and $H_2(g)$:

$$N_2(g) + 3\,H_2(g) \longrightarrow 2\,NH_3(g)$$

19.5 ENTROPY CHANGES

What processes are spontaneous and how is this predicted? Entropy is the basis for this determination. But, how can this be useful when, as you have seen earlier, the entropy of a system may either decrease (oxidation of NO) or increase (evaporation and oxidation of ethanol). The second law of thermodynamics provides an answer. This law states that *a spontaneous process is always accompanied by an increase in the entropy in the universe*. This criterion requires assessing entropy changes in both the system under study and in the surroundings [☉ CD-ROM, Screens 19.5 AND 19.6].

At first glance the statement of the second law may seem curious. Ordinarily our thinking is not so expansive as to consider the whole universe; we think mainly about a given system. Recall that the term "system" is defined as "that part of the universe being studied." However, it will become apparent that this view is not so complicated as it might first sound.

The "universe" (= univ) has two parts: the system (= sys) and the surroundings (= surr) (◀ SECTION 6.1). The entropy change for the universe is the sum of the entropy changes for the system and its surroundings (Equation 19.3).

$$\Delta S_{univ} = \Delta S_{sys} + \Delta S_{surr} \tag{19.3}$$

The second law states that $\Delta S_{univ} > 0$ for a spontaneous process. A negative value of ΔS_{univ} means the process cannot be spontaneous as written. If $\Delta S_{univ} = 0$, the system is at equilibrium.

A similar equation can be written for the entropy change for a process under standard conditions.

$$\Delta S°_{univ} = \Delta S°_{surr} + \Delta S°_{sys} \tag{19.4}$$

The value $\Delta S°_{univ}$ represents the entropy change for a process in which all reactants and products are in their standard states.

The standard entropy change for a reaction can be calculated from values found in tables such as Appendix L. The value $\Delta S°_{univ}$ calculated in this way is the entropy change when reactants are completely converted to products with all species at standard conditions. The sign of $\Delta S°_{univ}$ carries with it the same connotation concerning spontaneity as ΔS_{univ}.

As an example of the calculation of $\Delta S°_{univ}$, consider the reaction currently used to manufacture methanol, (CH_3OH).

$$CO(g) + 2 H_2(g) \longrightarrow CH_3OH(\ell)$$

If the change in entropy for the universe is positive, the conversion of 1 mol of CO and 2 mol of H_2 to 1 mol of CH_3OH will be spontaneous under standard conditions.

Calculating $\Delta S°_{sys}$, the Entropy Change for the System

To calculate $\Delta S°_{sys}$ start by defining the system to include the reactants and products. This means that dispersal of matter in this process occurs entirely within the system. Calculation of the entropy change for the synthesis of methanol follows the procedure given in Example 19.2 [☉ CD-ROM, Screen 19.5].

$$
\begin{aligned}
\Delta S°_{sys} &= \sum S°(\text{products}) - \sum S°(\text{reactants}) \\
&= S°[CH_3OH(\ell)] - \{S°\,[CO(g)] + 2\,S°\,[H_2(g)]\} \\
&= 1\ \text{mol}\ (127.2\ \text{J/K} \cdot \text{mol}) - \{1\ \text{mol}\ (197.7\ \text{J/K} \cdot \text{mol}) \\
&\quad + 2\ \text{mol}\ (130.7\ \text{J/K} \cdot \text{mol})\} \\
&= -331.9\ \text{J/K}
\end{aligned}
$$

• Spontaneity and the Second Law
Spontaneous change is always accompanied by an increase in entropy. This is in contrast to enthalpy and internal energy. According to the first law, the energy contained in the universe is constant.

• Using $\Delta S°_{univ}$
For a spontaneous process:

$$\Delta S°_{univ} > 0$$

For a system at equilibrium:

$$\Delta S°_{univ} = 0$$

For a non-spontaneous process:

$$\Delta S°_{univ} < 0$$

Chapter Goals • Revisited

- **Understand the concept of entropy and how it relates to spontaneity.**
- Predict whether a process is product- or reactant-favored.
- **Use tables of data in thermodynamic calculations.**
- Define and use a new thermodynamic function, free energy.

A decrease in entropy for the system is expected because 3 mol of gaseous reactants are converted to 1 mol of a liquid product.

Calculating $\Delta S°_{surr}$, the Entropy Change for the Surroundings

The entropy change resulting from the dispersal of energy produced in this exothermic chemical reaction is evaluated from the enthalpy change for the reaction. There is no dispersal of heat within the system, which stays at a constant temperature, 298 K. The heat evolved in this reaction is transferred to the surroundings, so $q_{surr} = -\Delta H°_{sys}$, and

$$\Delta S°_{surr} = \frac{q_{surr}}{T} = -\frac{\Delta H°_{sys}}{T}$$

An exothermic reaction ($\Delta H°_{sys} < 0$) is therefore accompanied by an increase in the entropy of the surroundings.

The enthalpy change for the synthesis of methanol is

$$\Delta H°_{sys} = \sum \Delta H°_f(\text{products}) - \sum \Delta H°_f(\text{reactants})$$

$$= \Delta H°_f[CH_3OH(\ell)] - \{\Delta H°_f[CO(g)] + 2\ \Delta H°_f[H_2(g)]\}$$

$$= (1\ \text{mol})(-238.4\ \text{kJ/mol}) - [(1\ \text{mol})(-110.5\ \text{kJ/mol}) + (2\ \text{mol})(0)]$$

$$= -127.9\ \text{kJ}$$

Therefore, assuming the process is reversible and occurs at a constant temperature, the entropy change for the surroundings in the methanol synthesis can then be calculated.

$$\Delta S°_{surr} = -\frac{\Delta H°_{sys}}{T} = -\frac{(-127.9\ \text{kJ})}{298\ \text{K}}(1000\ \text{J/kJ}) = +429.2\ \text{J/K}$$

• **Kelvin Temperature**
The temperature in kelvins (K) must be used in these calculations.

Calculating $\Delta S°_{univ}$, the Total Entropy Change for System and Surroundings

The pieces are now in place to calculate the entropy change in the universe. For the formation of $CH_3OH(\ell)$ from $CO(g)$ and $H_2O(g)$, the $\Delta S°_{univ}$ is

$$\Delta S°_{univ} = \Delta S°_{sys} + \Delta S°_{surr}$$

$$= -331.9\ \text{J/K} + 429.2\ \text{J/K}$$

$$= +97.3\ \text{J/K}$$

The positive value indicates an increase in the entropy of the universe. It follows from the second law of thermodynamics that this reaction is spontaneous.

Example 19.3 | **Determining if a Process Is Spontaneous**

Problem • We know that NaCl is a soluble salt. Show that $\Delta S°_{univ} > 0$ for this process.

Strategy • The problem is divided into two parts: determining $\Delta S°_{sys}$ and determining $\Delta S°_{surr}$ (calculated from $\Delta H°$ for the process). The sum of these two entropy changes is $\Delta S°_{univ}$. Values of $\Delta H°$ and $S°$ for NaCl(s) and NaCl(aq) are obtained in Appendix L.

Solution • The process occurring is $NaCl(s) \longrightarrow NaCl(aq)$ and its entropy change, $\Delta S°_{sys}$, can be calculated from values of $S°$ for the two species using Equation 19.2. The calculation is based on 1 mol of NaCl.

$$\Delta S°_{sys} = S°[NaCl(aq)] - S°[NaCl(s)]$$

$$= (1\ \text{mol})(115.5\ \text{J/K} \cdot \text{mol}) - (1\ \text{mol})(72.11\ \text{J/K} \cdot \text{mol})$$

$$= +43.4\ \text{J/K}$$

(Example continues on next page)

The heat of solution is determined from values of ΔH_f° for solid and aqueous sodium chloride. The solution process is slightly endothermic, indicating heat transfer from the surroundings to the system:

$$\Delta H_{sys}^\circ = \Delta H_f^\circ[NaCl(aq)] - \Delta H_f^\circ[NaCl(s)]$$

$$= (1 \text{ mol})(-407.27 \text{ kJ/mol}) - (1 \text{ mol})(-411.12 \text{ kJ/mol})$$

$$= +3.85 \text{ kJ}$$

The heat transferred to the surroundings has the same numerical value but is opposite in sign: $q_{surr} = -\Delta H_{sys}^\circ = -3.85$ kJ.

The entropy change of the surroundings is determined by dividing q_{surr} by the Kelvin temperature.

$$\Delta S_{surr}^\circ = \frac{q_{surr}}{T} =$$

$$= \frac{(-3.85 \text{ kJ})}{298 \text{ K}}(1000 \text{ J/kJ}) = -12.9 \text{ J/K}$$

We see that ΔS_{surr}° has a negative sign. Heat was transferred from the surroundings to the system in this endothermic process.

The overall entropy change, the change of entropy in the universe, is the sum of the values for the system and the surroundings.

$$\Delta S_{univ}^\circ = \Delta S_{sys}^\circ + \Delta S_{surr}^\circ$$

$$= (+43.4 \text{ J/K}) + (-12.9 \text{ J/K})$$

$$= 30.5 \text{ J/K}$$

Comment • The sum of the two entropy quantities is positive, indicating that, overall, entropy in the universe increases. However, notice that the process is favored based on dispersal of matter ($\Delta S_{sys}^\circ > 0$) and disfavored by dispersal of energy ($\Delta S_{surr}^\circ < 0$).

Summary: Spontaneous or Not?

In the preceding examples, predictions were made using values of ΔS_{sys}° and ΔH_{sys}° calculated from tables of thermodynamic data. It will be useful to look at the possibilities that result from the interplay of these two quantities. Four outcomes are possible when these two quantities are matched (Table 19.2). In two, ΔH_{sys}° and ΔS_{sys}° work in concert (Types 1 and 4 in Table 19.2). In the other two, the two quantities are opposed (Types 2 and 3).

Processes favored by both energy and matter dispersal (Type 1) are always spontaneous. Processes disfavored by both energy and matter dispersal (Type 4) can never be spontaneous. Let us consider examples that illustrate each situation.

Combustion reactions are always exothermic and often produce a larger number of product molecules from a few reactant molecules. The equation for the combustion of butane is

$$2 \text{ C}_4\text{H}_{10}(g) + 13 \text{ O}_2(g) \longrightarrow 8 \text{ CO}_2(g) + 10 \text{ H}_2\text{O}(g)$$

For this reaction $\Delta H^\circ = -5315.1$ kJ and $\Delta S^\circ = +310.8$ J/K. Both values indicate that this reaction, like all combustion reactions, must be spontaneous.

Table 19.2 • **Predicting if a Process Is Spontaneous**

Type	ΔH_{sys}°	ΔS_{sys}°	Spontaneous Process?
1	Exothermic process $\Delta H_{sys}^\circ < 0$	Less order $\Delta S_{sys}^\circ > 0$	Spontaneous under all conditions $\Delta S_{univ}^\circ > 0$
2	Exothermic process $\Delta H_{sys}^\circ < 0$	More order $\Delta S_{sys}^\circ < 0$	Depends on relative magnitudes of ΔH and ΔS More favorable at *lower* temperatures
3	Endothermic process $\Delta H_{sys}^\circ > 0$	Less order $\Delta S_{sys}^\circ > 0$	Depends on relative magnitudes of ΔH and ΔS More favorable at *higher* temperatures
4	Endothermic process $\Delta H_{sys}^\circ > 0$	More order $\Delta S_{sys}^\circ < 0$	Not spontaneous under all conditions $\Delta S_{univ}^\circ < 0$

Hydrazine, N_2H_4, is used as a high-energy rocket fuel. Synthesis of N_2H_4 from gaseous N_2 and H_2 would be attractive because these reactants are inexpensive.

$$N_2(g) + 2\ H_2(g) \longrightarrow N_2H_4(\ell)$$

However, neither matter nor energy dispersal are favorable (Type 4). The reaction is endothermic ($\Delta H° = +50.63$ kJ $\cdot$ mol^{-1}), and the entropy change $\Delta S°$ is negative, $\Delta S° = -331.4$ J/K (1 mol of liquid is produced from 3 mol of gases).

In the two other possible outcomes, entropy changes from energy and matter dispersal oppose each other. A process could be favored by energy dispersal but disfavored by matter dispersal (Type 2), or vice versa (Type 3). In either instance, whether a process is spontaneous depends on which factor is more important.

Temperature also influences the value of $\Delta S°_{univ}$ because $\Delta S°_{surr}$ (the dispersal of energy) varies with temperature. Because the enthalpy change in the surroundings is divided by temperature to obtain $\Delta S°_{surr}$, the numerical value of $\Delta S°_{surr}$ is smaller (either less positive or less negative) at higher temperatures. In contrast, $\Delta S°_{sys}$ does not depend on the temperature. Thus, the effect of $\Delta S°_{surr}$ relative to $\Delta S°_{sys}$ diminishes at higher temperature. Stated another way, at higher temperatures energy dispersal becomes less of a factor relative to matter dispersal. Consider the two cases where $\Delta H°_{sys}$ and $\Delta S°_{sys}$ are in opposition (see Table 19.2):

- **Type 2:** *Exothermic processes (favored by heat dispersal) but entropy-disfavored (disfavored by matter dispersal).* Such processes become less favorable with an increase in temperature.
- **Type 3:** *Endothermic processes (disfavored by energy dispersal) but entropy-favored (favored by matter dispersal).* These become more favorable as the temperature increases.

The effect of temperature is illustrated by two examples. The first is the reaction of N_2 and H_2 to form NH_3, one of the most important industrial chemical processes. The reaction is exothermic, that is, favored by energy dispersal. The entropy change for the system is unfavorable, however, because the reaction, $N_2(g) + 3\ H_2(g) \longrightarrow 2\ NH_3(g)$, converts 4 mol of gaseous reactants to 2 mol of gaseous products. To maximize the yield of ammonia, the lowest possible temperature should be used.

The second example considers the thermal decomposition of NH_4Cl (Figure 19.8). At room temperature, NH_4Cl is a stable, white crystalline salt. When heated strongly, it decomposes to $NH_3(g)$ and $HCl(g)$. The reaction is endothermic (enthalpy-disfavored) but entropy-favored because of the formation of 2 mol of gas from 1 mol of a solid reactant.

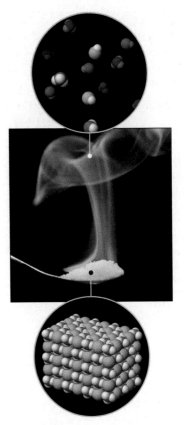

Figure 19.8   Thermal decomposition of $NH_4Cl(s)$. White, solid ammonium chloride is being heated in a spoon. At high temperatures, $NH_4Cl(s)$ decomposes to form $NH_3(g)$ and $HCl(g)$ in a spontaneous reaction. At lower temperatures the reverse reaction forming $NH_4Cl(s)$ is spontaneous. As $HCl(g)$ and $NH_3(g)$ above the heated solid cool, they recombine to form solid NH_4Cl, the white "smoke" seen in this photo. *(Charles D. Winters)*

Exercise 19.3   Is a Reaction Spontaneous?

Classify the following reactions as one of the four types of reactions summarized in Table 19.2:

Reaction	$\Delta H°_{rxn}$ (298 K) (kJ)	$\Delta S°_{sys}$ (298 K) (J/K)
(a) $CH_4(g) + 2\ O_2(g) \longrightarrow 2\ H_2O(\ell) + CO_2(g)$	-890.6	-242.8
(b) $2\ Fe_2O_3(s) + 3\ C(graphite) \longrightarrow 4\ Fe(s) + 3\ CO_2(g)$	$+467.9$	$+560.7$
(c) $C(graphite) + O_2(g) \longrightarrow CO_2(g)$	-393.5	$+3.1$
(d) $N_2(g) + 3\ F_2(g) \longrightarrow 2\ NF_3(g)$	-264.2	-277.8

A reaction requiring a higher temperature so that ΔS°_{univ} is positive. Iron is obtained in a blast furnace by heating iron ore and coke (carbon). See Exercise 19.5. *(Courtesy Bethlehem Steel)*

Exercise 19.4 Is a Reaction Spontaneous?

Is the reaction of hydrogen and chlorine to give hydrogen chloride gas predicted to be spontaneous?

$$H_2(g) + Cl_2(g) \longrightarrow 2\ HCl(g)$$

Answer the question by calculating the values for ΔS°_{sys} and ΔS°_{surr} (at 298 K) and then summing them to determine ΔS°_{univ}.

Exercise 19.5 Effect of Temperature on Spontaneity

Iron is produced in a blast furnace by reducing iron oxide using carbon. For the reaction $2\ Fe_2O_3(s) + 3\ C(graphite) \longrightarrow 4\ Fe(s) + 3\ CO_2(g)$ the following parameters are determined: $\Delta H^\circ_{rxn} = +467.9$ kJ and $\Delta S^\circ_{rxn} = +560.7$ J/K. Show that it is necessary that this reaction be carried out at a high temperature.

19.6 Gibbs Free Energy

The method used so far to determine whether a process is spontaneous required evaluation of two quantities, ΔS°_{sys} and ΔS°_{surr}. Wouldn't it be convenient to have a single thermodynamic function to serve the same purpose? A function associated with a system only — one that does not require assessment of the surroundings — would be even better. In fact there is such a function. It is called Gibbs free energy, the name honoring J. Willard Gibbs. **Gibbs free energy, G,** often referred to simply as "free energy," is defined mathematically as

$$G = H - TS$$

in which H is enthalpy, T is the Kelvin temperature and S is entropy. In this equation, G, H, and S all refer to a system. Because enthalpy and entropy are state functions (◀ SECTION 6.4), free energy is also a state function [🖱 CD-ROM, Screen 19.7].

Every substance possesses a specific quantity of free energy. The actual quantity of free energy is seldom known, however, or even of interest. Instead, we are concerned with *changes* in free energy, ΔG, in the course of a chemical or physical process. We do not need to know the free energy of a substance to determine ΔG. In this sense, free energy and enthalpy are similar. A substance possesses some amount of enthalpy, but we do not have to know what the actual value of H is to obtain or use ΔH.

Let us see how to use free energy as a way to determine if a reaction is spontaneous and then ask further about the meaning of the term "free" energy and its use in deciding if a reaction is product-favored.

ΔG° and Spontaneity

Recall the equation defining the entropy change for the universe (Equation 19.3).

$$\Delta S_{univ} = \Delta S_{surr} + \Delta S_{sys}$$

On page 803 we noted that the entropy change of the surroundings equals the negative of the change in enthalpy of the system divided by T. Thus

$$\Delta S_{univ} = (-\Delta H_{sys}/T) + \Delta S_{sys}$$

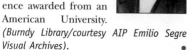

Chapter Goals • Revisited

- Understand the concept of entropy and how it relates to spontaneity.
- Predict whether a process is product- or reactant-favored.
- Use tables of data in thermodynamic calculations.
- **Define and use a new thermodynamic function, free energy.**

Multiplying through this equation by $-T$, we have

$$-T\Delta S_{univ} = \Delta H_{sys} - T\Delta S_{sys}$$

Gibbs defined the free energy function so that

$$\Delta G_{sys} = -T\Delta S_{univ}$$

Therefore, we have

$$\Delta G_{sys} = \Delta H_{sys} - T\Delta S_{sys}$$

The connection between $\Delta G_{rxn}(= \Delta G_{sys})$ and spontaneity is the following:

- If the value of ΔG_{rxn} is negative, a reaction is spontaneous.
- If $\Delta G_{rxn} = 0$, the reaction is at equilibrium.
- If the value of ΔG_{rxn} is positive, the reaction is not spontaneous.

The free energy change can also be defined under standard conditions.

$$\Delta G^{\circ}_{sys} = \Delta H^{\circ}_{sys} - T\Delta S^{\circ}_{sys} \qquad (19.5)$$

ΔG°_{sys} is also used as a criterion of reaction spontaneity, and as you shall see, it *is directly related to the value of the equilibrium constant and hence to product favorability.*

What Is "Free" Energy?

The term "free energy" was not arbitrarily chosen. In any given process, the free energy represents the maximum energy available to do useful work (or, mathematically, $\Delta G = w_{max}$). In this context, the word "free" means "available."

To illustrate the reasoning behind this relationship, consider a reaction in which heat is evolved ($\Delta H_{rxn} < 0$) and entropy decreases ($\Delta S_{rxn} < 0$).

$$C(graphite) + 2\ H_2(g) \longrightarrow CH_4(g)$$
$$\Delta H^{\circ}_{rxn} = -74.9\ kJ \quad \text{and} \quad \Delta S^{\circ}_{rxn} = -80.7\ J/K$$
$$\Delta G^{\circ}_{rxn} = -50.9\ kJ$$

The enthalpy of a reaction, ΔH°_{rxn}, is the heat produced by the reaction. At first thought it might seem that all this energy could be transferred to the surroundings and would thus be available to do work, but this is not true. Because the entropy term is unfavorable, some portion of this energy will be used to create a more ordered system and will not be available to do work. What is left is "free," or available, to do work. Here that free energy amounts to 50.8 kJ, as shown in Example 19.4.

The logic applies to any combination of ΔH° and ΔS°. The available "free" energy is the sum of the energies available from dispersal of energy (the enthalpy term) and dispersal of matter (the entropy term). In a process that is favored by both the enthalpy and entropy changes, the free energy available exceeds what is available based on enthalpy alone.

Calculating ΔG°_{rxn}, the Free Energy Change for a Reaction

Enthalpy and entropy changes can be calculated for chemical reactions using values of ΔH°_f and S° for substances in the reaction. Then, ΔG°_{rxn} ($= \Delta G^{\circ}_{sys}$) can be found from the resulting values of ΔH°_{rxn} and ΔS°_{rxn} using Equation 19.5, as illustrated in the following example and exercise.

Example 19.4   Calculating ΔG°_{rxn} from ΔH°_{rxn} and ΔS°_{rxn}

Problem • Calculate the standard free energy change, ΔG°, for the formation of methane at 298 K:

$$C(graphite) + 2\ H_2(g) \longrightarrow CH_4(g)$$

Strategy • Values for ΔH°_f and S° are provided in Appendix L. These are first combined to find ΔH°_{rxn} and ΔS°_{rxn}. With these known, ΔG°_{rxn} can be calculated using Equation 19.5. When doing so, recall that S° values are in units of J/K · mol whereas ΔH° values are in units of kJ/mol.

Solution •

	C(graphite)	H_2(g)	CH_4(g)
ΔH°_f (kJ/mol)	0	0	−74.9
S° (J/K · mol)	5.6	130.7	186.3

From these values, we can find both ΔH° and ΔS° for the reaction:

$$\Delta H^\circ_{rxn} = \Delta H^\circ_f[CH_4(g)] - \{\Delta H^\circ_f[C(graphite)] + 2\ \Delta H^\circ_f[H_2(g)]\}$$
$$= -74.9\ kJ - (0 + 0)$$
$$= -74.9\ kJ$$

$$\Delta S^\circ_{rxn} = S^\circ[CH_4(g)] - \{S^\circ[C\ (graphite)] + 2\ S^\circ[H_2(g)]\}$$

$$= 186.3\ J/K \cdot mol$$
$$- [1\ mol\ (5.6\ J/K \cdot mol) + 2\ mol\ (130.7\ J/K \cdot mol)]$$
$$= -80.7\ J/K$$

Both the enthalpy change and the entropy change for this reaction are negative. This is a case when the reaction is predicted to be spontaneous at "low temperature" (see Table 19.2). These values alone do not tell us if the temperature is low enough, however. By combining them in the Gibbs equation, and calculating ΔG°_{rxn} for a temperature of 298 K, we can predict the outcome of the reaction:

$$\Delta G^\circ_{rxn} = \Delta H^\circ_{rxn} - T\ \Delta S^\circ_{rxn}$$
$$= -74.9\ kJ - (298\ K)(-80.7\ J/K)(1\ kJ/1000\ J)$$
$$= -74.9\ kJ - (-24.0\ kJ)$$
$$= -50.8\ kJ$$

ΔG°_{rxn} is negative at 298 K, so the reaction is predicted to be spontaneous.

Comment • In this example $T\Delta S^\circ$ is negative and smaller than ΔH°_{rxn} because the entropy change is relatively small. Chemists call this an "enthalpy-driven reaction" because the exothermic nature of the reaction overcomes the decrease in entropy of the system.

Exercise 19.6   Calculating ΔG°_{rxn} from ΔH°_{rxn} and ΔS°_{rxn}

Using values of ΔH°_f and S° to find ΔH°_{rxn} and ΔS°_{rxn}, respectively, calculate the free energy change, ΔG°, for the formation of 2 mol of NH_3(g) from the elements at standard conditions (and 25 °C): $N_2(g) + 3\ H_2(g) \longrightarrow 2\ NH_3(g)$.

Standard Free Energy of Formation

The **standard free energy of formation** of a compound, ΔG°_f, is the free energy change when forming one mole of the compound from the component elements, with products and reactants in their standard states. By defining ΔG°_f in this way, *the free energy of formation of an element in its standard states is zero* (Table 19.3).

Just as the standard enthalpy change for a reaction can be calculated using values of ΔH°_f, the standard free energy change for a reaction can also be calculated from values of ΔG°_f (Equation 19.6).

$$\Delta G^\circ_{rxn} = \sum \Delta G^\circ_f(\text{products}) - \sum \Delta G^\circ_f(\text{reactants}) \qquad (19.6)$$

Table 19.3 • Standard Molar Free Energies of Formation of Some Substances at 298 K

Element/Compound	$\Delta G_f^\circ (kJ \cdot mol^{-1})$	Element/Compound	$\Delta G_f^\circ (kJ \cdot mol^{-1})$
$H_2(g)$	0	$CO_2(g)$	-394.4
$O_2(g)$	0	$CH_4(g)$	-50.87
$N_2(g)$	0	$H_2O(g)$	-228.6
C(graphite)	0	$H_2O(\ell)$	-237.2
C(diamond)	2.900	$NH_3(g)$	-16.4
CO(g)	-137.2	$Fe_2O_3(s)$	-742.2

Example 19.5 Calculating ΔG_{rxn}° from ΔG_f°

Problem • Calculate the standard free energy change for the combustion of 1.0 mol of methane from the standard free energies of formation of the products and reactants.

Strategy • Use Equation 19.6 with values obtained from Table 19.3 or Appendix L.

Solution • First, write a balanced equation for the reaction and then find values of ΔG_f° for each reactant and product.

$$CH_4(g) + 2 O_2(g) \longrightarrow 2 H_2O(g) + CO_2(g)$$

ΔG_f° -50.9 0 -228.6 -394.4
(kJ/mol)

Because ΔG_f° values are given for 1 mol of each substance (the units are kJ/mol) each value of ΔG_f° must be multiplied by the number of moles defined by the stoichiometric coefficient in the balanced chemical equation.

$$\Delta G_{rxn}^\circ = 2 \, \Delta G_f^\circ[H_2O(g)] + \Delta G_f^\circ[CO_2(g)]$$
$$- \{\Delta G_f^\circ[CH_4(g)] + 2 \, \Delta G_f^\circ[O_2(g)]\}$$
$$= 2 \, mol(-228.6 \, kJ/mol) + 1 \, mol \, (-394.4 \, kJ/mol)$$
$$- [1 \, mol \, (-50.9 \, kJ/mol) + 2 \, mol \, (0 \, kJ/mol)]$$
$$= -800.7 \, kJ$$

The large, negative value of ΔG_{rxn}° indicates that the reaction is spontaneous under standard conditions.

Comment • The most common errors made by students in this calculation are (1) ignoring the stoichiometric coefficients in the equation, and (2) confusion with signs for each term when using Equation 19.6.

Exercise 19.7 Calculating ΔG_{rxn}° from ΔG_f°

Calculate the standard free energy change for the oxidation of 1.00 mol of $SO_2(g)$ to form $SO_3(g)$.

Free Energy and Temperature

The definition for free energy, $G = H - TS$, shows that free energy is a function of temperature, and ΔG will change as the temperature changes [CD-ROM, Screen 19.8].

A consequence of this dependence on temperature is that, in certain instances, reactions can be spontaneous at one temperature and not spontaneous at another. Those instances arise when the ΔH° and $T\Delta S^\circ$ terms work in opposite directions.

- Processes that are entropy-favored ($\Delta S^\circ > 0$) and enthalpy-disfavored ($\Delta H^\circ > 0$).
- Processes that are enthalpy-favored ($\Delta H^\circ < 0$) and entropy-disfavored ($\Delta S^\circ < 0$).

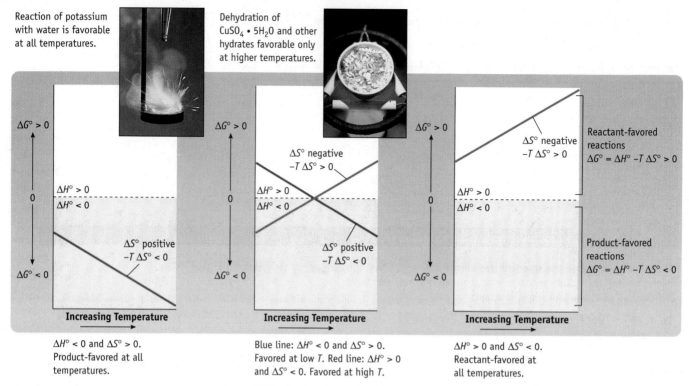

Reaction of potassium with water is favorable at all temperatures.

Dehydration of $CuSO_4 \cdot 5H_2O$ and other hydrates favorable only at higher temperatures.

$\Delta H^\circ < 0$ and $\Delta S^\circ > 0$. Product-favored at all temperatures.

Blue line: $\Delta H^\circ < 0$ and $\Delta S^\circ > 0$. Favored at low T. Red line: $\Delta H^\circ > 0$ and $\Delta S^\circ < 0$. Favored at high T.

$\Delta H^\circ > 0$ and $\Delta S^\circ < 0$. Reactant-favored at all temperatures.

Figure 19.9   Product- or reactant-favored? Changes in ΔG° with temperature. *(Charles D. Winters)*

Let us explore the relationship of ΔG° and T further and illustrate how it can be used to advantage.

Calcium carbonate is a common substance. Among other things, it is the primary component of limestone, marble, and seashells. Heating $CaCO_3$ produces lime, CaO, along with gaseous CO_2. The data below are from Appendix L.

	$CaCO_3(s)$ $\longrightarrow$	$CaO(s)$	+	$CO_2(g)$
ΔG_f° (kJ $\cdot$ mol^{-1})	−1129.2	−603.4		−394.36
ΔH_f° (kJ $\cdot$ mol^{-1})	−1207.6	−635.0		−393.51
S° (J $\cdot$ K^{-1} $\cdot$ mol^{-1})	91.7	38.2		213.74

For the conversion of limestone, $CaCO_3(s)$ to lime, $CaO(s)$, $\Delta G_{rxn}^\circ = +131.4$ kJ, $\Delta H_{rxn}^\circ = +179.1$ kJ, and $\Delta S_{rxn}^\circ = 160.2$ J/K. Although the reaction is entropy-favored, the large positive and unfavorable enthalpy change dominates at this temperature. Thus, the free energy change is positive at 298 K and 1 bar, indicating that overall the reaction is not spontaneous under these conditions.

Although the formation of CaO from $CaCO_3$ is unfavorable at 298 K, the temperature dependence of ΔG° provides a means to make this a spontaneous reaction. Notice that the entropy change in the reaction is positive, indicating that mass dispersal is favored. (The formation of a gas, CO_2, is the reason.) Thus, raising the temperature results in the value of $T\Delta S^\circ$ becoming increasingly large, and at a high enough temperature the effect of $T\Delta S^\circ$ will outweigh ΔH° and the process becomes favorable.

How high must the temperature be for this reaction to become spontaneous? An estimate of the temperature can be obtained using Equation 19.5. To do this, let us calculate the temperature at which $\Delta G° = 0$. Above that temperature, $\Delta G°$ will have a negative value. (Note that in this calculation the enthalpy must be in joules, not kilojoules, to match the units used in the entropy term. Keeping track of the units will show that the temperature will have the units of kelvins.)

$$\Delta G° = 0 = \Delta H° - T\Delta S°$$

$$\Delta H° = T\Delta S°$$

$$179.1 \text{ kJ } (1000 \text{ J/kJ}) = T(160.2 \text{ J/K})$$

$$T = 1118 \text{ K (or 845 °C)}$$

How accurate is this result? We can only obtain an approximate answer from this calculation. The biggest source of error is the assumption that $\Delta H°$ and $\Delta S°$ do not vary with temperature, which is not strictly true. There is usually a small variation in these values when the temperature changes, not too important if the temperature range is narrow but potentially a problem over wider temperature ranges. As an estimate, however, a temperature in the 800 °C range for this reaction is reasonable. In fact, the pressure of CO_2 in an equilibrium system $[CaCO_3(s) \rightleftharpoons CaO(s) + CO_2(g)]$ is 1 bar at about 880 °C, very close to our estimated temperature.

Example 19.6 **Effect of Temperature on $\Delta G°$**

Problem • Use thermodynamic parameters to estimate the boiling point of methanol.

Strategy • At the boiling point, liquid and gas exist at equilibrium, and $\Delta G° = 0$ at the boiling point. Values of $\Delta H_f°$ and $S°$ from Appendix L for the process $CH_3OH(\ell) \longrightarrow CH_3OH(g)$, are used to calculate $\Delta H°$ and $\Delta S°$. T is the unknown in Equation 19.5.

Solution • Values for $\Delta G_f°$, $\Delta H_f°$, and $S°$ are obtained from Appendix L for CH_3OH liquid and vapor.

	CH₃OH(ℓ)	CH₃OH(g)
$\Delta G_f°$ (kJ/mol)	−166.14	−162.3
$\Delta H_f°$ (kJ/mol)	−238.4	−201.0
$S°$ (J/K · mol)	127.19	239.9

For a process in which 1 mol of liquid is converted to 1 mol of gas: $\Delta G°_{rxn} = +3.8$ kJ, $\Delta H°_{rxn} = +37.4$ kJ, and $\Delta S°_{rxn} = 112.7$ J/K. The process is endothermic (as expected, heat being required to convert a liquid to a gas), and entropy favors the process (the vapor has a higher degree of disorder than the liquid). These two quantities oppose one another. The free energy change per mole for this process under standard conditions is +3.8 kJ. The positive value indicates that this process is not spontaneous under standard conditions (298 K and 1 bar).

If we use the values of $\Delta H°_{rxn}$ and $\Delta S°_{rxn}$, along with the criterion that $\Delta G° = 0$ at the boiling point, we have

$$\Delta G° = \Delta H° - T\Delta S°$$

$$0 = +37400 \text{ J} - [T (+112.7 \text{ J/K})]$$

$$T = 332 \text{ K (or 59 °C)}$$

Comment • The calculated boiling temperature is close to the observed value of 65.0 °C.

Exercise 19.8 **Effect of Temperature on $\Delta G°$**

Oxygen was first prepared by Joseph Priestley (1733–1804) by heating HgO. Use the thermodynamic data in Appendix L to estimate the temperature required to decompose HgO(s) into Hg(ℓ) and $O_2(g)$.

19.7 $\Delta G°$, K, and Product-Favorability

The terms *product-fovored* and *reactant-favored* were introduced in Chapter 4, and in Chapter 16 we pointed out how these terms were related to the values of equilibrium constants. *Reactions for which K is large are product-favored and those with small K values are reactant-favored.* We now return to this important topic, to relate the value of the equilibrium constant K_e, and thus product- and reactant-favorability, to the standard free energy change, $\Delta G°$, for a chemical reaction.

The standard free energy change for a reaction, $\Delta G°$, is the increase or decrease in free energy as the reactants in their standard states are converted *completely* to the products in their standard states. But complete conversion is not often observed in practice. A product-favored reaction proceeds largely to products, but some reactants may remain when equilibrium is achieved. A reactant-favored reaction proceeds only partially to products before achieving equilibrium. To discover the relationship of $\Delta G°$ and the equilibrium constant K, let us use Figure 19.10. The free energy of the pure reactants in their standard states is indicated at the left, and the free energy of the pure products in their standard states is at the right. The difference in these values is $\Delta G°$. In this example $\Delta G° = G°_{\text{prod}} - G°_{\text{react}}$ has a negative value and the reaction is product-favored.

When the reactants are mixed in a chemical system, the system will proceed spontaneously to a position of lower free energy, and the system will eventually achieve equilibrium. At any point along the way from the pure reactants to equilibrium, the reactants are not at standard conditions. The change in free energy under these nonstandard conditions, ΔG, is related to $\Delta G°$ by the equation

$$\Delta G = \Delta G° + RT \ln Q \tag{19.7}$$

where R is the universal gas constant, T is the temperature in kelvins, and Q is the reaction quotient (◄ SECTION 16.2). Recall that, for the general reaction of A and B giving products C and D

$$a\,A + b\,B \longrightarrow c\,C + d\,D$$

the reaction quotient, Q, is

$$Q = \frac{[C]^c[D]^d}{[A]^a[B]^b}$$

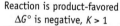

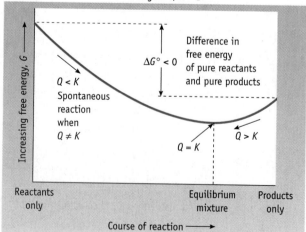

Figure 19.10 Free energy changes as a reaction approaches equilibrium. Equilibrium represents a minimum in free energy for a system. The reaction portrayed here has $\Delta G° < 0$, $K > 1$, and is product-favored.

Equation 19.7 informs us that, at a given temperature, the difference in free energy between that of the pure reactants and a mixture of reactants and products is determined by the values of $\Delta G°$ and Q. Further, as long as ΔG is negative — the reaction is "descending" from the free energy of the pure reactants to the equilibrium position — the reaction is spontaneous.

Eventually, however, the system reaches equilibrium. Because no further change in concentration of reactants and products is seen at this point, ΔG must be zero, that is, there is no further change in free energy in the system. Substituting $G = 0$ into equation 19.7 gives

$$0 = \Delta G° + RT \ln K \quad \text{(at equilibrium)}$$

Rearranging this equation leads to a useful relationship between the the standard free energy change for a reaction and the equilibrium constant, K.

$$\Delta G°_{rxn} = -RT \ln K \qquad \textbf{(19.8)}$$

From Equation 19.8 we learn that, *when $\Delta G°_{rxn}$ is negative, indicating a product-favored reaction, K must be greater than 1.* The more negative the value of $\Delta G°$, the larger the equilibrium constant. This makes sense because you learned in Chapter 16 that large equilibrium constants are associated with product-favored reactions. *For reactant-favored reactions, $\Delta G°$ is positive and K is less than 1.* If K happens to be 1, then $\Delta G° = 0$. (This is a rare situation, requiring that $[A]^a[B]^b/[C]^c[D]^d = 1$ for the reaction $a\,A + b\,B \longrightarrow c\,C + d\,D$.)

Free Energy, the Reaction Quotient, and the Equilibrium Constant

Let us summarize the relationship among $\Delta G°$, ΔG, Q, and K [CD-ROM, Screen 19.9].

- The solid line in Figure 19.10 shows how free energy decreases to a minimum as a system approaches equilibrium. *The free energy at equilibrium is lower than either the free energy of the pure reactants or the free energy of the pure products.*

- $\Delta G°_{rxn}$ gives the position of equilibrium and may be calculated from

$$\Delta G°_{rxn} = \sum \Delta G°_f(\text{products}) - \sum \Delta G°_f(\text{reactants})$$
$$= \Delta H°_{rxn} - T\Delta S°_{rxn}$$
$$= -RT \ln K$$

- ΔG_{rxn} predicts the direction in which a reaction proceeds to reach equilibrium and may be calculated from

$$\Delta G_{rxn} = \Delta G°_{rxn} + RT \ln Q$$

When $\Delta G_{rxn} < 0$, $Q < K$, and the reaction proceeds spontaneously to convert reactants to products until equilibrium is attained.

When $\Delta G_{rxn} > 0$, $Q > K$, and the reaction proceeds spontaneously to convert products to reactants until equilibrium is attained.

When $\Delta G_{rxn} = 0$, $Q = K$, and the reaction is at equilibrium.

Using the Relationship of $\Delta G°_{rxn}$ and *K*

Equation 19.8 provides a direct route to determine the standard free energy change from experimentally determined equilibrium constants. Alternatively, it allows calculation of an equilibrium constant from thermochemical data contained in tables or obtained from an experiment.

Spontaneous but not product-favored. If a sample of yellow lead iodide is placed in pure water, the compound will begin to dissolve. The dissolving process will be spontaneous ($\Delta G < 0$) until equilibrium is reached. However, because PbI_2 is insoluble with $K_{sp} = 9.8 \times 10^{-9}$, the process of dissolving the compound is strongly reactant-favored, and value of $\Delta G°$ is positive. *(Charles D. Winters)*

Example 19.7) Calculating K_p from ΔG°_{rxn}

Problem • Determine the standard free energy change, ΔG°_{rxn}, for the formation of 1.00 mol of ammonia from nitrogen and hydrogen, and use this value to calculate the equilibrium constant for this reaction at 25 °C.

Strategy • The free energy of formation of ammonia represents the free energy change to form 1.0 mol of $NH_3(g)$ from the elements. The equilibrium constant for this reaction is calculated from ΔG° using Equation 19.8. Because the reactants and products are gases, the calculated value will be K_p.

Solution • Begin by specifying a balanced equation for the chemical reaction under investigation.

$$\tfrac{1}{2} N_2(g) + \tfrac{3}{2} H_2(g) \longrightarrow NH_3(g)$$

The free energy change for this reaction is −16.37 kJ/mol. ($\Delta G^\circ_{rxn} = \Delta G^\circ_f$ for $NH_3(g)$; Appendix L). In a calculation of K_p

using Equation 19.8, we will need consistent units. The gas constant, R, is 8.3145 J/K · mol, so the value of ΔG° must also be in joules (not kilojoules). The temperature is 298 K.

$$\Delta G^\circ = -RT \ln K$$
$$-16{,}370 \text{ J/mol} = -(8.3145 \text{ J/K} \cdot \text{mol})(298.15 \text{ K}) \ln K_p$$
$$\ln K_p = 6.604$$
$$K_p = 7.38 \times 10^2$$

Comment • As illustrated by this example, it is possible to calculate equilibrium constants from thermodynamic data. This gives us another use for these valuable tables. For the reaction of N_2 and H_2 to form NH_3, at 298 K (25 °C), the value of the equilibrium constant is quite large, indicating that the reaction is product-favored.

Example 19.8) Calculating ΔG°_{rxn} from K_c

Problem • The value of K_{sp} for AgCl(s) at 25 °C is 1.8×10^{-10}. Use this value in Equation 19.8 to determine ΔG° for the process $Ag^+(aq) + Cl^-(aq) \longrightarrow AgCl(s)$ at 25 °C.

Strategy • The chemical equation given is the opposite of the equation used to define K_{sp}; therefore its equilibrium constant is $1/K_{sp}$. This value is used to calculate ΔG°.

Solution • For $Ag^+(aq) + Cl^-(aq) \longrightarrow AgCl(s)$,

$$K = \frac{1}{K_{sp}} = \frac{1}{1.8 \times 10^{-10}} = 5.6 \times 10^9$$

$$\Delta G^\circ = -RT \ln K$$
$$= -(8.3145 \text{ J/K} \cdot \text{mol})(298.15 \text{ K}) \ln (5.6 \times 10^9)$$
$$= -56 \text{ kJ/mol (to two significant figures)}$$

Comment • The negative value of ΔG° indicates that precipitation of AgCl from $Ag^+(aq)$ and $Cl^-(aq)$ is a product-favored process.

Exercise 19.9) Calculating K_p from ΔG°_{rxn}

Determine the value of ΔG°_{rxn} for the reaction $C(s) + CO_2(g) \longrightarrow 2 CO(g)$ from thermodynamic data in Appendix L. Use this result to calculate the equilibrium constant.

Exercise 19.10) Calculating ΔG°_{rxn} from K_c

The formation constant for $Ag(NH_3)_2^+$ is 1.6×10^7. Use this value to calculate the ΔG° for the reaction

$$Ag^+(aq) + 2 NH_3(aq) \longrightarrow [Ag(NH_3)_2]^+(aq)$$

19.8   THERMODYNAMICS, TIME, AND LIFE

Chapters 6 and 19 have brought together the three laws of thermodynamics.

First law:	The total energy of the universe is a constant.
Second law:	The total entropy of the universe is always increasing.
Third law:	The entropy of a pure, perfectly formed crystalline substance at 0 K is zero.

Some cynic long ago paraphrased the first two laws. The first law was transmuted into a statement, "You can't win!," referring to the fact that the energy will always be conserved, and a process in which you get back more energy than you put in is impossible. The paraphrase of the second law? "You can't break even!" The Gibbs free energy provides a rationale for this interpretation. Only part of the energy from a chemical reaction can be converted to useful work, the rest will be committed to the redistribution of matter or energy.

The second law tells us the entropy of the universe is continually increasing. A snowflake will melt in a warm room, but you won't see a glassful of water molecules reassemble into snowflakes at any temperature above 0 °C. Molecules of perfume will diffuse throughout a room, but they won't collect again on your body. All spontaneous processes result in a more disordered universe. This is what scientists mean when they say that the second law is an expression of time in a physical — as opposed to psychological — form. In fact, entropy has been called "time's arrow."

Neither the first nor the second law of thermodynamics has ever been proven. It is just that there never has been a single example showing otherwise. No less a scientist than Albert Einstein once remarked that thermodynamic theory ". . . is the only physical theory of the universe content [which], within the framework of applicability of its basic concepts, will never be overthrown." Einstein's statement does not mean that people have not tried (and are continuing to try) to disprove the laws of thermodynamics. Claims to have invented machines that perform useful work without expending energy — perpetual motion machines — are frequently made. However, no perpetual motion machine has ever been shown to work. We can feel safe with the assumption that such a machine never will be built.

Roald Hoffmann, a chemist who shared the 1981 Nobel Prize in chemistry, has said that "One amusing way to describe synthetic chemistry, the making of molecules that is at the intellectual and economic center of chemistry, is that it is the local defeat of entropy" [*American Scientist*, Nov-Dec. 1987, pages 619–621]. "Local defeat" in this context refers to an unfavorable entropy change in a system because, of course, the entropy of the universe must increase if this process is to occur. An important point to make is that chemical syntheses are often entropy-disfavored. Chemists find ways to accomplish them balancing the unfavorable changes in the system with favorable changes in surroundings to make these reactions occur.

Finally, thermodynamics speaks to one of the great mysteries: the origin of life. Life as we know it requires the creation of extremely complex molecules such as proteins and nucleic acids. Their formation must have occurred from atoms and small molecules. Assembling thousands of atoms into a highly ordered state in biochemical compounds clearly requires a local decrease in entropy. Some have said that life is a violation of the second law of thermodynamics. A more logical view, however, is that the local decrease is offset by an increase of entropy in the rest of the universe. Here again, thermodynamics is unchallenged.

● **Time's Arrow**
If you are interested in the theories of the origin of the universe, and in "time's arrow," read *A Brief History of Time, From the Big Bang to Black Holes* by Stephen W. Hawking, Bantam Books, New York, NY, 1998.

● **Entropy and Time**
The second law requires that disorder increase with time. Because all natural processes that occur do so as time progresses and result in increased disorder, it is evident that increasing entropy and time "point" in the same direction.

Chemical Perspectives

Thermodynamics and Speculation on the Origin of Life

Early earth was very different from the planet we inhabit. The atmosphere was made up of simple molecular substances such as N_2, H_2, CO_2, CO, NH_3, CH_4, H_2S, and H_2O. Elemental oxygen was not origi-

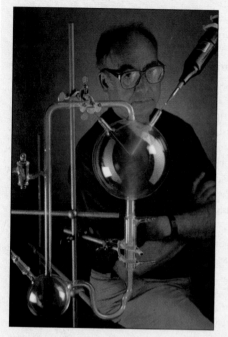

(James A. Sugar/©2002 CORBIS)

nally present; this atmospheric gas, now necessary for most life, would eventually be produced by plants via photosynthesis more than a billion years after earth was formed.

The crust of the planet was rocky, consisting mostly of silicon and aluminum oxides. What was available was an abundance of energy, from solar radiation (without O_2 and O_3, a high level of ultraviolet radiation reached the surface), from lightning, and from the heat in earth's core. How did the molecules found in living organisms form under these conditions?

A classic experiment to probe the question of how more complex molecules arose was performed in the late 1950s. In the Miller–Urey experiment, as it is now identified, a mixture of these gases was subjected to an electric discharge, the spark simulating lightning. After a few days HCN and formaldehyde (CH_2O), along

◀ **The Miller-Urey experiment.** An electric discharge through a mixture of gases leads to formation of organic molecules including formaldehyde and amino acids.

with amino acids and other organic molecules, had accumulated in this system. Although scientists no longer believe molecules necessary for life were formed this way, it is nonetheless a starting point for new experiments in this direction.

Other theories on the formation of organic molecules have their supporters. One is that organic molecules formed in aqueous solution on the surface of clay particles. Another is that important molecules formed near cracks in the earth's crust on the floor of the oceans. Such processes have also been shown to occur in labs.

It is, of course, a long way from small organic molecules to larger molecules, and the conversion of inanimate molecules to living beings is an even greater leap. Creating larger and larger molecules from small molecules requires energy and means that entropy must decrease in the "system." For the process to occur, however, the entropy of the universe must increase. So, the creation of living systems means a corresponding increase must take place in the entropy of the "surroundings" elsewhere in the universe. One thing remains clear: the laws of thermodynamics must be obeyed.

In Summary

When you have finished studying this chapter, ask if you have met the chapter goals. In particular, you should be able to

- Understand that entropy is a measure of matter dispersal or disorder (Section 19.3).

- Recognize that entropy can be determined experimentally as the heat change of a reversible process ("A Closer Look: Reversible and Irreversible Processes," Section 19.4).

- Know how to calculate entropy changes (Section 19.4).

- Identify common processes that are entropy-favored (Section 19.4).

- Use entropy and enthalpy changes to predict whether a reaction is spontaneous. (Section 19.5 and Table 19.2).

- Recognize how temperature influences whether a reaction is spontaneous (Section 19.5).

- Understand the connection between enthalpy and entropy changes and the Gibbs free energy change for a process (Section 19.6).
- Calculate the change in free energy at standard conditions for a reaction from the enthalpy and entropy changes or from the standard free energy of formation of reactants and products (ΔG_f°) (Section 19.6).
- Know how free energy changes with temperature (Section 19.6).
- Describe the relationship between the free energy change, product favorability and equilibrium constants. Calculate K from ΔG_{rxn}° (Section 19.7).

Key Terms

Section 19.1
  spontaneous

Section 19.3
  entropy, S

reversible processes
second law of thermodynamics

Section 19.4
  third law of thermodynamics
  standard entropy, S°

Section 19.6
  Gibbs free energy, G
  standard free energy of formation, ΔG_f°

Key Equations

Equation 19.1 (page 797)
Calculate entropy change from the heat absorbed in a reversible process (q_{rev}) process and the temperature at which it occurs.

$$\Delta S = q_{rev}/T$$

Equation 19.2 (page 800)
Calculate the entropy change for a process from the tabulated entropies of the products and reactants.

$$\Delta S_{sys}^\circ = \sum S^\circ(\text{products}) - \sum S^\circ(\text{reactants})$$

Equation 19.4 (page 802)
Calculate the total entropy change for the universe from the entropy changes for a system and its surroundings.

$$\Delta S_{univ}^\circ = \Delta S_{sys}^\circ + \Delta S_{surr}^\circ$$

Equation 19.5 (page 807)
Calculate the free energy change for a process from the enthalpy and entropy change for the process.

$$\Delta G_{sys}^\circ = \Delta H_{sys}^\circ - T\,\Delta S_{sys}^\circ$$

Equation 19.6 (page 807)
Calculate the free energy change for a reaction using tabulated values of ΔG_f°, the standard free energy of formation.

$$\Delta G_{rxn}^\circ = \sum \Delta G_f^\circ(\text{products}) - \sum \Delta G_f^\circ(\text{reactants})$$

Equation 19.8 (page 813)
The relationship between the standard free energy change for a reaction and its equilibrium constant.

$$\Delta G_{rxn}^\circ = -RT \ln K$$

Study Questions

Questions with blue, bold-faced numbers have answers in Appendix O. Many of these questions require thermodynamic data. If required data are not given in the question, consult the tables in this chapter or Appendix L.

Reviewing Important Concepts

1. What are the three laws of thermodynamics? How would you explain each law to a nonscientist?

2. Chemical reactions and physical processes are classified as spontaneous and non-spontaneous. What is meant by these terms? Give an example of each kind of chemical reaction and each type of physical process.

3. Define the following terms: system, surroundings, universe, standard conditions, state function, spontaneous.

4. Based on your experience and common sense, which of the following processes would you describe as product-favored and which are reactant-favored under standard conditions?

   (a) $Hg(\ell) \longrightarrow Hg(s)$

   (b) $H_2O(g) \longrightarrow H_2O(\ell)$

   (c) $2\ HgO(s) \longrightarrow 2\ Hg(\ell) + O_2(g)$

   (d) $C(s) + O_2(g) \longrightarrow CO_2(g)$

   (e) $NaCl(s) \longrightarrow NaCl(aq)$

   (f) $CaCO_3(s) \longrightarrow Ca^{2+}(aq) + CO_3{}^{2-}(aq)$

5. Explain why each of the following statements is incorrect.

   (a) The entropy decreases in all spontaneous reactions.

   (b) Reactions with a negative standard free energy change ($\Delta G°_{rxn} < 0$) are product-favored and occur with rapid transformation of reactants to products.

   (c) All spontaneous processes are exothermic.

   (d) Endothermic processes are never product-favored.

6. Decide if each of the following statements is true or false. If false, rewrite it to make it true.

   (a) The entropy of a substance increases on going from the liquid to the vapor state at any temperature.

   (b) An exothermic reaction will always be product-favored.

   (c) Reactions with a positive $\Delta H°_{rxn}$ and a positive $\Delta S°_{rxn}$ can never be product-favored.

   (d) If $\Delta G°_{rxn}$ for a reaction is negative, the reaction will have an equilibrium constant greater than 1.

7. Under what conditions is the entropy of a pure substance $0\ J/K \cdot mol$? Could a substance at standard conditions have a value of $0\ J/K \cdot mol$? A negative value of entropy value? Under which conditions (if any) will a substance have negative entropy? Explain your answer.

8. Write a chemical equation for the oxidation of $C_2H_6(g)$ by $O_2(g)$ to form $CO_2(g)$ and $H_2O(g)$. Defining this as the system, predict whether the signs of $\Delta S°_{sys}$, $\Delta S°_{surr}$, and $\Delta S°_{univ}$ are greater than, equal to, or less than zero. Explain your prediction.

9. For the system in Question 8, predict the signs of $\Delta H°$ and $\Delta G°$. Explain how you made this prediction.

10. Consider the system in Question 8. Will the value of K_p be very large, very small, or near 1? Will the equilibrium constant K_p for this system be larger or smaller at temperatures greater than 298 K? Explain how you made this prediction.

11. Explain why the entropy of the system increases on dissolving solid NaCl in water $\{S°[NaCl(s)] = 72.1\ J/K \cdot mol$ and $S°[NaCl(aq)] = 115.5\ J/K \cdot mol\}$.

Practicing Skills

Entropy Comparisons
(See Example 19.1 and CD-ROM Screen 19.4)

12. Which substance has the higher entropy in each of the following pairs?

    (a) Dry ice (solid CO_2) at $-78\ °C$ or $CO_2(g)$ at $0\ °C$

    (b) Liquid water at $25\ °C$ or liquid water at $50\ °C$

    (c) Pure alumina, $Al_2O_3(s)$, or ruby. (Ruby is Al_2O_3 in which some of the Al^{3+} ions in the crystalline lattice are replaced with Cr^{3+} ions)

    (d) One mole of $N_2(g)$ at 1 bar pressure or 1 mol of $N_2(g)$ at 10 bar pressure (both at 298 K)

13. Which substance has the higher entropy in each of the following pairs?

    (a) A sample of pure silicon (to be used in a computer chip) or a piece of silicon containing a trace of some other atoms such as B or P.

    (b) $O_2(g)$ at $0\ °C$ or $O_2(g)$ at $-50\ °C$

    (c) $I_2(s)$ or $I_2(g)$, both at room temperature

    (d) One mole of $O_2(g)$ at 1 bar pressure or 1 mol of $O_2(g)$ at 0.01 bar pressure (both at 298 K)

14. By comparing the formulas for each set of compounds, decide which is expected to have the higher standard entropy. Check your answers using data in Appendix L.

    (a) $O_2(g)$ or $CH_3OH(g)$ (two substances with the same molar mass)

    (b) $HF(g)$, $HCl(g)$, or $HBr(g)$

    (c) $NH_4Cl(s)$ or $NH_4Cl(aq)$

    (d) $HNO_3(g)$, $HNO_3(\ell)$, or $HNO_3(aq)$

15. By comparing the formulas for each set of compounds, decide which is expected to have the higher entropy. Assume both are at the same temperature. Check your answers using data in Appendix L.

    (a) $NaCl(s)$, $NaCl(g)$, or $NaCl(aq)$

    (b) $H_2O(g)$ or $H_2S(g)$

    (c) $C_2H_4(g)$ or $N_2(g)$ (two substances with the same molar mass)

    (d) $H_2SO_4(\ell)$ or $H_2SO_4(aq)$

Predicting and Calculating Entropy Changes

(See Example 19.2 and CD-ROM Screen 19.5)

16. Use $S°$ values to calculate the entropy change, $\Delta S°$, for each of the following processes and comment on the sign of the change.
    (a) $KOH(s) \longrightarrow KOH(aq)$
    (b) $Na(g) \longrightarrow Na(s)$
    (c) $Br_2(\ell) \longrightarrow Br_2(g)$
    (d) $HCl(g) \longrightarrow HCl(aq)$

17. Use $S°$ values to calculate the entropy change, $\Delta S°$, for each of the following changes and comment on the sign of the change.
    (a) $NH_4Cl(s) \longrightarrow NH_4Cl(aq)$
    (b) $C_2H_5OH(\ell) \longrightarrow C_2H_5OH(g)$
    (c) $CCl_4(g) \longrightarrow CCl_4(\ell)$
    (d) $NaCl(s) \longrightarrow NaCl(g)$

18. Calculate the standard entropy change for the formation of 1 mol of gaseous ethane (C_2H_6) at 25 °C.
$$2\ C(graphite) + 3\ H_2(g) \longrightarrow C_2H_6(g)$$

19. Using standard entropy values, calculate the standard entropy change for a reaction forming 1.00 mol of $NH_3(g)$ from $N_2(g)$ and $H_2(g)$.
$$\tfrac{1}{2}\ N_2(g) + \tfrac{3}{2}\ H_2(g) \longrightarrow NH_3(g)$$

20. Calculate the standard molar entropy change for the formation of each of the following compounds from the elements at 25 °C.
    (a) $HCl(g)$ (b) $Ca(OH)_2(s)$

21. Calculate the standard molar entropy change for the formation of each of the following compounds from the elements at 25 °C.
    (a) $H_2S(g)$ (b) $MgCO_3(s)$

22. Calculate the standard molar entropy change for each of the following reactions at 25 °C. Comment on the sign of $\Delta S°$.
    (a) $2\ Al(s) + 3\ Cl_2(g) \longrightarrow 2\ AlCl_3(s)$
    (b) $2\ CH_3OH(\ell) + 3\ O_2(g) \longrightarrow 2\ CO_2(g) + 4\ H_2O(g)$

23. Calculate the standard molar entropy change for each of the following reactions at 25 °C. Comment on the sign of $\Delta S°$.
    (a) $2\ Na(s) + 2\ H_2O(\ell) \longrightarrow 2\ NaOH(aq) + H_2(g)$
    (b) $Na_2CO_3(s) + 2\ HCl(aq) \longrightarrow$
$$2\ NaCl(aq) + H_2O(\ell) + CO_2(g)$$

$\Delta S°_{univ}$ and Spontaneity

(See Example 19.3 and CD-ROM Screen 19.6)

24. Is the reaction
$$Si(s) + 2\ Cl_2(g) \longrightarrow SiCl_4(g)$$
spontaneous? Answer this question by calculating $\Delta S°_{sys}$, $\Delta S°_{surr}$, and $\Delta S°_{univ}$. (The reactants and products are defined as the system.)

25. Is the reaction
$$Si(s) + 2\ H_2(g) \longrightarrow SiH_4(g)$$

spontaneous? Answer this question by calculating $\Delta S°_{sys}$, $\Delta S°_{surr}$, and $\Delta S°_{univ}$. (The reactants and products are defined as the system.)

26. Calculate the standard enthalpy and entropy changes for the decomposition of liquid water to form gaseous hydrogen and oxygen. Is this reaction spontaneous? Explain your answer briefly.

27. Calculate the standard enthalpy and entropy changes for the formation of $HCl(g)$ from gaseous hydrogen and chlorine. Is this reaction spontaneous? Explain your answer briefly.

28. Classify each of the reactions according to one of the four reaction types summarized in Table 19.2.
    (a) $Fe_2O_3(s) + 2\ Al(s) \longrightarrow 2\ Fe(s) + Al_2O_3(s)$
$$\Delta H° = -851.5\ kJ \qquad \Delta S° = -375.2\ J/K$$
    (b) $N_2(g) + 2\ O_2(g) \longrightarrow 2\ NO_2(g)$
$$\Delta H° = 66.2\ kJ \qquad \Delta S° = -121.6\ J/K$$

29. Classify each of the reactions according to one of the four reaction types summarized in Table 19.2.
    (a) $C_6H_{12}O_6(s) + 6\ O_2(g) \longrightarrow 6\ CO_2(g) + 6\ H_2O(\ell)$
$$\Delta H° = -673\ kJ \qquad \Delta S° = 60.4\ J/K$$
    (b) $MgO(s) + C(graphite) \longrightarrow Mg(s) + CO(g)$
$$\Delta H° = 490.7\ kJ \qquad \Delta S° = 197.9\ J/K$$

Effect of Temperature on Reactions

(See Example 19.4 and CD-ROM Screen 19.8)

30. Heating some metal carbonates, among them magnesium carbonate, leads to their decomposition.
$$MgCO_3(s) \longrightarrow MgO(s) + CO_2(g)$$
    (a) Calculate $\Delta H°$ and $\Delta S°$ for the reaction.
    (b) Is the reaction spontaneous at 298 K?
    (c) Is the reaction predicted to be spontaneous at higher temperatures?

31. Calculate $\Delta H°$ and $\Delta S°$ for the reaction of tin(IV) oxide with carbon.
$$SnO_2(s) + C(s) \longrightarrow Sn(s) + CO_2(g)$$
    (a) Is the reaction spontaneous at 298 K?
    (b) Is the reaction predicted to be spontaneous at higher temperatures?

Changes in Free Energy

(See Example 19.5; use $\Delta G° = \Delta H° - T\,\Delta S°$; see CD-ROM Screen 19.7)

32. Using values of $\Delta H°_f$ and $S°$, calculate $\Delta G°_{rxn}$ for each of the following reactions.
    (a) $2\ Pb(s) + O_2(g) \longrightarrow 2\ PbO(s)$
    (b) $NH_3(g) + HNO_3(aq) \longrightarrow NH_4NO_3(aq)$

    Which of these reactions is (are) predicted to be product-favored? Are the reactions enthalpy- or entropy-driven?

33. Using values of $\Delta H°_f$ and $S°$, calculate $\Delta G°_{rxn}$ for each of the following reactions.

(a) $Ca(s) + 2 H_2O(\ell) \longrightarrow Ca(OH)_2(aq) + H_2(g)$

(b) $6 C(graphite) + 3 H_2(g) \longrightarrow C_6H_6(\ell)$

Which of these reactions is (are) predicted to be product-favored? Are the reactions enthalpy- or entropy-driven?

34. Using values of ΔH_f° and S°, calculate the standard molar free energy of formation, ΔG_f°, for each of the following compounds:

    (a) $CS_2(g)$ (b) $NaOH(s)$ (c) $ICl(g)$

    Compare your calculated values of ΔG_f° with those listed in Appendix L.

35. Using values of ΔH_f° and S°, calculate the standard molar free energy of formation, ΔG_f°, for each of the following compounds:

    (a) $Ca(OH)_2(s)$ (b) $Cl(g)$ (c) $Na_2CO_3(s)$

    Compare your calculated values of ΔG_f° with those listed in Appendix L.

Free Energy of Formation

(See Example 19.6; use $\Delta G_{rxn}^\circ = \Sigma \Delta G_f^\circ$ (products) $- \Sigma \Delta G_f^\circ$ (reactions); see CD-ROM Screen 19.7)

36. Using values of ΔG_f°, calculate ΔG_{rxn}° for each of the following reactions. Which are product-favored?

    (a) $2 K(s) + Cl_2(g) \longrightarrow 2 KCl(s)$

    (b) $2 CuO(s) \longrightarrow 2 Cu(s) + O_2(g)$

    (c) $4 NH_3(g) + 7 O_2(g) \longrightarrow 4 NO_2(g) + 6 H_2O(g)$

37. Using values of ΔG_f°, calculate ΔG_{rxn}° for each of the following reactions. Which are product-favored?

    (a) $HgS(s) + O_2(g) \longrightarrow Hg(\ell) + SO_2(g)$

    (b) $2 H_2S(g) + 3 O_2(g) \longrightarrow 2 H_2O(g) + 2 SO_2(g)$

    (c) $SiCl_4(g) + 2 Mg(s) \longrightarrow 2 MgCl_2(s) + Si(s)$

38. For the reaction $BaCO_3(s) \longrightarrow BaO(s) + CO_2(g)$, $\Delta G_{rxn}^\circ = +219.7$ kJ. Using this value and other data available in Appendix L, calculate the value of ΔG_f° for $BaCO_3(s)$.

39. For the reaction $TiCl_2(s) + Cl_2(g) \longrightarrow TiCl_4(\ell)$, $\Delta G_{rxn}^\circ = -272.8$ kJ. Using this value and other data available in Appendix L, calculate the value of ΔG_f° for $TiCl_2(s)$.

Effect of Temperature on ΔG

(See Example 19.8 and CD-ROM Screen 19.8)

40. Determine whether the reactions listed here are entropy-favored or -disfavored under standard conditions. Predict how an increase in temperature will affect the value of ΔG_{rxn}°.

(a) $N_2(g) + 2 O_2(g) \longrightarrow 2 NO_2(g)$

(b) $2 C(s) + O_2(g) \longrightarrow 2 CO(g)$

(c) $CaO(s) + CO_2(g) \longrightarrow CaCO_3(s)$

(d) $2 NaCl(s) \longrightarrow 2 Na(s) + Cl_2(g)$

41. Determine whether the reactions listed here are entropy-favored or -disfavored under standard conditions. Predict how an increase in temperature will affect the value of ΔG_{rxn}°.

    (a) $I_2(g) \longrightarrow 2 I(g)$

    (b) $2 SO_2(g) + O_2(g) \longrightarrow 2 SO_3(g)$

    (c) $SiCl_4(\ell) + 2 H_2O(\ell) \longrightarrow SiO_2(s) + 4 HCl(g)$

    (d) $P_4(s, white) + 6 H_2(g) \longrightarrow 4 PH_3(g)$

42. Estimate the temperature required to decompose $HgS(s)$ into $Hg(\ell)$ and $S(g)$.

43. Estimate the temperature required to decompose $CaSO_4(s)$ into $CaO(s)$ and $SO_3(g)$.

Free Energy and Equilibrium Constants

(See Example 19.8; use $\Delta G^\circ = -RT \ln K$; see CD-ROM Screen 19.9)

44. The formation of $NO(g)$ from its elements

$$\tfrac{1}{2} N_2(g) + \tfrac{1}{2} O_2(g) \longrightarrow NO(g)$$

has a standard free energy change, ΔG_f°, of $+86.58$ kJ/mol at 25 °C. Calculate K_p at this temperature. Comment on the connection between the sign of ΔG° and the magnitude of K_p.

45. The formation of $O_3(g)$ from $O_2(g)$ has a standard free energy change, ΔG_f°, of $+163.2$ kJ/mol at 25 °C. Calculate K_p at this temperature. Comment on the connection between the sign of ΔG° and the magnitude of K_p.

46. Calculate ΔG° and K_p at 25 °C for the reaction

$$C_2H_4(g) + H_2(g) \longrightarrow C_2H_6(g)$$

Comment on the sign of ΔG° and the magnitude of K_p. Is the reaction product-favored?

47. Calculate ΔG° and K_p at 25 °C for the reaction

$$2 HBr(g) + Cl_2(g) \longrightarrow 2 HCl(g) + Br_2(\ell)$$

Is the reaction predicted to be product-favored under standard conditions? Comment on the sign of ΔG° and the magnitude of K_p.

General Questions on Thermodynamics

More challenging questions are marked with an underlined number.

48. Calculate the standard molar entropy change, ΔS°, for each of the following reactions:

    1. $C(s) + 2 H_2(g) \longrightarrow CH_4(g)$

    2. $CH_4(g) + \tfrac{1}{2} O_2(g) \longrightarrow CH_3OH(\ell)$

    3. $C(s) + 2 H_2(g) + \tfrac{1}{2} O_2(g) \longrightarrow CH_3OH(\ell)$

    Verify that these values are related by the equation $\Delta S_1^\circ + \Delta S_2^\circ = \Delta S_3^\circ$ What general principle is illustrated here?

49. Hydrogenation, the addition of hydrogen to an organic compound, is a reaction of considerable industrial importance. Calculate ΔH°, ΔS°, and ΔG° at 25 °C for the hydrogenation of octene, C_8H_{16}, to give octane, C_8H_{18}. Is the reaction product- or reactant-favored under standard conditions?

$$C_8H_{16}(g) + H_2(g) \longrightarrow C_8H_{18}(g)$$

The following information is required, in addition to data in Appendix L.

Compound	ΔH°_f (kJ/mol)	S° (J/K · mol)
Octene	−82.93	462.8
Octane	−208.45	463.6

50. Is the combustion of ethane, C_2H_6, a spontaneous reaction?

$$C_2H_6(g) + \tfrac{7}{2} O_2(g) \longrightarrow 2\ CO_2(g) + 3\ H_2O(g)$$

Answer the question by calculating the value of ΔS°_{univ}, using values of ΔH°_f and S° in Appendix L. Does your calculated answer agree with your preconceived idea of this reaction?

51. "Heater Meals" are food packages that contain their own heat source. Just pour water into the heater unit, wait a few minutes, and you have a hot meal.

(Charles D. Winters)

The heat for the heater unit is developed by the reaction of magnesium with water.

$$Mg(s) + 2\ H_2O(\ell) \longrightarrow Mg(OH)_2(s) + H_2(g)$$

Confirm that this is a spontaneous reaction.

52. Write a balanced equation that depicts the formation of 1 mol of $Fe_2O_3(s)$ from its elements. What is the standard free energy of formation of 1.00 mol of $Fe_2O_3(s)$? What is the value of ΔG°_{rxn} when 454 g (1 lb) of $Fe_2O_3(s)$ are formed from the elements?

53. The reaction used by Joseph Priestley to prepare oxygen is

$$2\ HgO(s) \longrightarrow 2\ Hg(\ell) + O_2(g)$$

Defining the reaction as the system under study,

(a) Predict whether the signs of ΔS°_{sys}, ΔS°_{surr}, ΔS°_{univ}, ΔH°, and ΔG° are greater than zero, equal to zero, or less than zero and explain your prediction. Using data in Appendix L, calculate the values of each of these quantities to verify your predictions.

(b) Calculate a value of K_p for this reaction at 298 K. Is this reaction product-favored?

54. When vapors from hydrochloric acid and aqueous ammonia come in contact the following reaction takes place, producing a white "cloud" of solid NH_4Cl.

$$HCl(g) + NH_3(g) \longrightarrow NH_4Cl(s)$$

Defining this as the system under study:

(a) Predict whether the signs of ΔS°_{sys}, ΔS°_{surr}, ΔS°_{univ}, ΔH°, and ΔG° are greater than, equal to, or less than zero, and explain your prediction. Verify your predictions by calculating values for each of these quantities.

(b) Calculate a value of K_p for this reaction at 298 K.

55. The formation of diamond from graphite is a process of considerable importance.

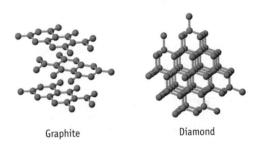

Graphite   Diamond

(a) Using data in Appendix L, calculate ΔS°_{univ}, ΔH°, and ΔG° for this process.

(b) The calculations will suggest that this process is impossible under any conditions. However, the synthesis of diamonds by this reaction is a commercial process. How can this contradiction be rationalized? (Note: In the synthesis, high pressure and temperature are used.)

56. Calculate ΔS°_{sys}, ΔS°_{surr}, ΔS°_{univ} for each of the following processes:

(a) $NaCl(s) \longrightarrow NaCl(aq)$

(b) $NaOH(s) \longrightarrow NaOH(aq)$

Comment on how these systems differ.

57. For the process $LiOH(s) \longrightarrow LiOH(aq)$ calculate ΔS° and ΔH°. Use these values to calculate ΔG°. Compare this value with the one calculated from values of ΔG°_f.

58. Methanol is now widely used as a fuel in race cars such as those that compete in the Indianapolis 500 race (see page 439). Consider the following reaction as a possible synthetic route to methanol.

$$C(graphite) + \tfrac{1}{2} O_2(g) + 2\ H_2(g) \longrightarrow CH_3OH(\ell)$$

Calculate K_p for the formation of methanol at 25 °C using this reaction. Would a different temperature be better for this reaction? Comment on the connection between the sign of ΔG° and the magnitude of K_p.

59. The enthalpy of vaporization of liquid diethyl ether, $(C_2H_5)_2O$, is 26.0 kJ/mol at the boiling point of 35.0 °C. Calculate ΔS° for (a) liquid to vapor and (b) vapor to liquid at 35.0 °C.

60. Calculate the entropy change, ΔS°, for the vaporization of ethanol, C_2H_5OH, at its normal boiling point, 78.0 °C. The enthalpy of vaporization of the ethanol is 39.3 kJ/mol.

61. If gaseous hydrogen can be produced cheaply, it could be burned directly as a fuel or converted to another fuel, methane (CH_4), for example.

$$3 H_2(g) + CO(g) \longrightarrow CH_4(g) + H_2O(g)$$

Calculate $\Delta G°$, $\Delta S°$, and $\Delta H°$ at 25 °C for the reaction. Is it predicted to be product- or reactant-favored under standard conditions?

62. Using thermodynamic data, estimate the normal boiling point of ethanol. (Recall that liquid and vapor are in equilibrium at 1.0 atm pressure at the normal boiling point.) The actual normal boiling point is 78 °C. How well does your calculated result agree with the actual value?

63. Estimate the vapor pressure of ethanol at 37 °C using thermodynamic data. Express the result in millimeters of mercury.

64. The following reaction is not product-favored at room temperature.

$$COCl_2(g) \longrightarrow CO(g) + Cl_2(g)$$

Would you have to raise or lower the temperature to make it product-favored?

65. When calcium carbonate is heated strongly, CO_2 gas is evolved. The equilibrium pressure of the gas is 1.00 atm at 897 °C, and $\Delta H°_{rxn}$ at 298 K = 179.1 kJ.

$$CaCO_3(s) \longrightarrow CaO(s) + CO_2(g)$$

Estimate the value of $\Delta S°$ at 298 K for the reaction.

66. The normal melting point of benzene, C_6H_6, is 5.5 °C. For the process of melting, what is the sign of
   (a) $\Delta H°$ (d) $\Delta G°$ at 0.0 °C
   (b) $\Delta S°$ (e) $\Delta G°$ at 25.0 °C
   (c) $\Delta G°$ at 5.5 °C

67. Estimate the boiling point of water in Denver, Colorado (where the altitude is 1.60 km, and the atmospheric pressure is 630 mm Hg).

68. Sodium reacts violently with water according to the equation)

$$Na(s) + H_2O(\ell) \longrightarrow NaOH(aq) + \tfrac{1}{2} H_2(g)$$

Without doing calculations, predict the signs of $\Delta H°$ and $\Delta S°$ for the reaction. Next, verify your prediction with a calculation.

69. For each of the following processes, give the algebraic sign of $\Delta H°$, $\Delta S°$, and $\Delta G°$. No calculations are necessary; use your common sense.
   (a) The decomposition of liquid water to give gaseous oxygen and hydrogen, a process that requires a considerable amount of energy.
   (b) Dynamite is a mixture of nitroglycerin, $C_3H_5N_3O_9$, and diatomaceous earth. The explosive decomposition of nitroglycerin gives gaseous products such as water, CO_2, and others; much heat is evolved.

(c) The combustion of gasoline in the engine of your car, as exemplified by the combustion of octane.

$$2 C_8H_{18}(g) + 25 O_2(g) \longrightarrow 16 CO_2(g) + 18 H_2O(g)$$

70. Yeast can produce ethanol by the fermentation of glucose ($C_6H_{12}O_6$); this is the basis for the production of most alcoholic beverages.

$$C_6H_{12}O_6(aq) \longrightarrow 2 C_2H_5OH(\ell) + 2 CO_2(g)$$

Calculate $\Delta H°$, $\Delta S°$, and $\Delta G°$ for the reaction. Is the reaction spontaneous as written? (In addition to the thermodynamic values in Appendix L, you will need the following for $C_6H_{12}O_6(aq)$: $\Delta H_f° = -1260.0$ kJ/mol; $S° = 289$ J/K · mol; and $\Delta G_f° = -918.8$ kJ/mol.)

71. Elemental boron, in the form of thin fibers, can be made by reducing a boron halide with H_2.

$$BCl_3(g) + \tfrac{3}{2} H_2(g) \longrightarrow B(s) + 3 HCl(g)$$

Calculate $\Delta H°$, $\Delta S°$, and $\Delta G°$ at 25 °C for this reaction. Is the reaction predicted to be product-favored under standard conditions? If product-favored, is it enthalpy-driven or entropy-driven? [$S°$, for B(s) is 5.86 J/K · mol.]

72. The equilibrium constant, K_p, for $N_2O_4(g) \rightleftharpoons 2 NO_2(g)$ is 0.14 at 25 °C. Calculate $\Delta G°$ from this constant, and compare your calculated value with that determined from the $\Delta G_f°$ values in Appendix L.

73. Most metal oxides can be reduced with hydrogen to the pure metal. (Although such reactions work well, this is an expensive method and not used often for large-scale preparations.) The equilibrium constant for the reduction of iron(II) oxide is 0.422 at 700 °C. Estimate $\Delta G°_{rxn}$.

$$FeO(s) + H_2(g) \longrightarrow Fe(s) + H_2O(g)$$

74. The equilibrium constant for the butane $\rightleftharpoons$ isobutane equilibrium at 25 °C is 2.5.

Butane $\rightleftharpoons$ Isobutane

$$CH_3CH_2CH_2CH_3 \rightleftharpoons \begin{array}{c} CH_3 \\ | \\ CH_3CHCH_3 \end{array}$$

$$K_c = \frac{[\text{isobutane}]}{[\text{butane}]} = 2.50 \text{ at } 298 \text{ K}$$

Calculate $\Delta G°_{rxn}$ at this temperature in units of kJ/mol.

75. Almost 5 billion kilograms of benzene, C_6H_6, are made each year. It is used as a starting material for many other com-

pounds and as a solvent (although it is also a carcinogen, and its use is restricted). One compound that can be made from benzene is cyclohexane, C_6H_{12}.

$$C_6H_6(\ell) + 3\,H_2(g) \longrightarrow C_6H_{12}(\ell)$$
$$\Delta H°_{rxn} = -206.7\text{ kJ} \qquad \Delta S°_{rxn} = -361.5\text{ J/K}$$

Is this reaction predicted to be spontaneous under standard conditions at 25 °C? If spontaneous, is the reaction enthalpy or entropy driven?

76. Iodine, I_2, dissolves readily in carbon tetrachloride. For this process, $\Delta H° = 0$ kJ/mol.

$$I_2(s) \longrightarrow I_2 \text{ (in CCl}_4 \text{ solution)}$$

What is the sign of $\Delta G°_{rxn}$? Is the dissolving process entropy-driven or enthalpy-driven? Explain briefly.

77. A crucial reaction for the production of synthetic fuels is the conversion of steam to H_2 with coal. The chemical reaction is

$$C(s) + H_2O(g) \longrightarrow CO(g) + H_2(g)$$

   (a) Calculate $\Delta G°_{rxn}$ for this reaction at 25 °C assuming $C(s)$ is graphite.
   (b) Calculate K_p for the reaction at 25 °C.
   (c) Is the reaction predicted to be product-favored under standard conditions? If not, at what temperature will it become so?

78. Calculate $\Delta G°_{rxn}$ for the decomposition of sulfur trioxide to sulfur dioxide and oxygen.

$$2\,SO_3(g) \longrightarrow 2\,SO_2(g) + O_2(g)$$

   (a) Is the reaction product-favored under standard conditions, at 25 °C?
   (b) If the reaction is not product-favored at 25 °C, is there a temperature at which it will become so? Estimate this temperature.
   (c) What is the equilibrium constant for the reaction at 1500 °C?

79. Methanol is relatively inexpensive to produce. Much consideration has been given to using it as a precursor to other fuels such as methane, which could be obtained by the decomposition of the alcohol.

$$CH_3OH(\ell) \longrightarrow CH_4(g) + \tfrac{1}{2}\,O_2(g)$$

   (a) What are the sign and magnitude of the entropy change for the reaction? Does the sign of $\Delta S°$ agree with your expectation? Explain briefly.
   (b) Is the reaction spontaneous under standard conditions at 25 °C? Use thermodynamic values to prove your answer.
   (c) If not spontaneous at 25 °C, at what temperature does the reaction become spontaneous?

80. A cave in Mexico was recently discovered to have some in-

teresting chemistry (see Chapter 21). Hydrogen sulfide, H_2S, reacts with oxygen in the cave to give sulfuric acid, which drips from the ceiling in droplets with a pH less than 1. If the reaction occurring is

$$H_2S(g) + 2\,O_2(g) \longrightarrow H_2SO_4(\ell)$$

calculate $\Delta H°$, $\Delta S°$, and $\Delta G°$ at 25 °C. Is the reaction product-favored? Is it enthalpy- or entropy-driven?

81. Wet limestone is used to scrub SO_2 gas from the exhaust gases of power plants. One possible reaction gives hydrated calcium sulfite,

$$CaCO_3(s) + SO_2(g) + \tfrac{1}{2}\,H_2O(\ell) \rightleftharpoons$$
$$CaSO_3 \cdot \tfrac{1}{2}\,H_2O(s) + CO_2(g)$$

and another reaction gives hydrated calcium sulfate:

$$CaCO_3(s) + SO_2(g) + \tfrac{1}{2}\,H_2O(\ell) + \tfrac{1}{2}\,O_2(g) \rightleftharpoons$$
$$CaSO_4 \cdot \tfrac{1}{2}\,H_2O(s) + CO_2(g)$$

Which is the more product-favored reaction? Use the data in the following table below and any other needed in Appendix L.

	$CaSO_3 \cdot \tfrac{1}{2}\,H_2O(s)$	$CaSO_4 \cdot \tfrac{1}{2}\,H_2O(s)$
$\Delta H°_f$ (kJ/mol)	-1311.7	-1574.65
$S°$ (J/K · mol)	121.3	134.8

82. Some metal oxides can be decomposed to the metal and oxygen under reasonable conditions. Is the decomposition of silver(I) oxide product-favored at 25 °C?

$$2\,Ag_2O(s) \longrightarrow 4\,Ag(s) + O_2(g)$$

If not, can it become so if the temperature is raised? At what temperature is the reaction product-favored?

83. Calculate the entropy change for dissolving HCl gas in water. Is the sign of $\Delta S°$ what you expected? Why or why not?

84. Sulfur undergoes a phase transition between 80 and 100 °C.

$$S_8(\text{rhombic}) \longrightarrow S_8(\text{monoclinic})$$
$$\Delta H°_{rxn} = 3.213\text{ kJ/mol} \qquad \Delta S°_{rxn} = 8.7\text{ J/K · mol}$$

   (a) Estimate $\Delta G°$ for the transition at 80.0 °C and 110.0 °C. What do these results tell you about the stability of the two forms of sulfur at each of these temperatures?
   (b) Calculate the temperature at which $\Delta G° = 0$. What is the significance of this temperature?

85. Copper(II) oxide, CuO, can be reduced to copper metal with hydrogen at higher temperatures.

$$CuO(s) + H_2(g) \longrightarrow Cu(s) + H_2O(g)$$

Is this reaction product- or reactant-favored under standard conditions at 298 K?

If copper metal is heated in air, a black film of CuO forms on the surface. In this photo the heated bar, with the black CuO film, has been bathed in hydrogen gas. Black, solid CuO is reduced rapidly to copper at higher temperatures. *(Charles D. Winters)*

86. The *"Desert Refrigerator"* was described on page 200. Explain how the second law applies to this simple but useful device.

87. Oxygen dissolved in water can cause corrosion in hot water heating systems. To remove oxygen, hydrazine (N_2H_4) is often added. Hydrazine reacts with dissolved O_2 to form water and N_2.

    (a) Write a balanced chemical equation for the reaction of hydrazine and oxygen. Identify oxidizing and reducing agent in this redox reaction.

    (b) Calculate $\Delta H°$, $\Delta S°$, and $\Delta G°$ for this reaction.

88. Consider the formation of $NO(g)$ from its elements.

$$N_2(g) + O_2(g) \longrightarrow 2\,NO(g)$$

    (a) Calculate K_p at 25 °C. Is the reaction product-favored at this temperature?

    (b) Assuming $\Delta H°_{rxn}$ and $\Delta S°_{rxn}$ are nearly constant with temperature, calculate $\Delta G°_{rxn}$ at 700 °C. Estimate K_p from the new value of $\Delta G°_{rxn}$ at 700 °C. Is the reaction product-favored at 700 °C?

    (c) Using K_p at 700 °C, calculate the equilibrium partial pressures of the three gases if you mix 1.00 bar each of N_2 and O_2.

89. Calculate $\Delta G°_f$ for HI(g) at 350 °C, given the following equilibrium partial pressures: $P(H_2) = 0.132$ bar, $P(I_2) = 0.295$ bar, and $P(HI) = 1.61$ bar. At 350 °C and 1 bar, I_2 is a gas.

$$\tfrac{1}{2}\,H_2(g) + \tfrac{1}{2}\,I_2(g) \rightleftharpoons HI(g)$$

90. Silver(I) oxide can be formed by the reaction of silver metal and oxygen.

$$4\,Ag(s) + O_2(g) \longrightarrow 2\,Ag_2O(s)$$

    (a) Calculate $\Delta H°_{rxn}$, $\Delta S°_{rxn}$, and $\Delta G°_{rxn}$ for the reaction.

    (b) What is the pressure of O_2 in equilibrium with Ag and Ag_2O at 25 °C?

    (c) At what temperature would the pressure of O_2 in equilibrium with Ag and Ag_2O become equal to 1.00 bar?

91. Draw a diagram like that in Figure 19.10 for a reactant-favored process.

92. In Chapter 14 you learned that entropy, as well as enthalpy, plays a role in the solution process. If $\Delta H°$ for the solution process is zero, explain how the process can be driven by entropy.

93. Mercury vapor is dangerous because it can be breathed into the lungs. We wish to estimate the vapor pressure of mercury at two different temperatures from the following data:

	$\Delta H°_f$ (kJ/mol)	$S°$ (J/K · mol)	$\Delta G°_f$ (kJ/mol)
Hg(ℓ)	0	76.02	0
Hg(g)	61.38	174.97	31.88

Estimate the temperature at which K_p for the process Hg(ℓ) $\longrightarrow$ Hg(g) is equal to (a) 1.00 atm. and (b) 1 mm Hg. What is the vapor pressure at each of these temperatures? (Experimental vapor pressures are 1 mm Hg at 126.2 °C and 1 atm. at 356.6 °C.) (*Note:* The temperature at which $P = 1.00$ bar can be calculated from thermodynamic data. To find the other temperature, you will need to use the temperature for $P = 1.00$ atm and the Clausius–Clapeyron equation on page 528.)

Using Electronic Resources

These questions refer to the General Chemistry Interactive CD-ROM, *Version 3.0.*

94. CD-ROM Screen 19.2: Reaction Favorability. If gaseous H_2 and O_2 are carefully mixed and left alone, they can remain intact for millions of years. Is this "stability" a function of thermodynamics or of kinetics? (You may wish to review Screen 6.3.)

95. CD-ROM Screen 19.3: Directionality of Reactions.

    (a) Which of the dispersal mechanisms involves an enthalpy change?

    (b) Is energy dispersed during the process described in the Chemical Puzzler (Screen 19.1)?

$$NH_4NO_3(s) \longrightarrow NH_4NO_3(aq)$$

96. Screen 19.4: Entropy.

    (a) Which of the dispersal mechanisms described on this screen involve changes in a system's entropy?

    (b) Why does the entropy of a substance increase with temperature?

97. Screen 19.6: The Second Law of Thermodynamics. How is the second law of thermodynamics illustrated by the reaction of H_2 and O_2?

98. Screen 19.6: The Second Law of Thermodynamics— Simulation. Select the reaction of NO and Cl_2 to produce NOCl.

(a) What is ΔS°_{sys} at 400 K for this reaction?

(b) Does ΔS°_{sys} change with temperature?

(c) Does ΔS°_{surr} change with temperature?

(d) Does ΔS°_{surr} always change with an increase in temperature?

(e) Do exothermic reactions always lead to positive values of ΔS°_{sys}?

(f) Is the NO + Cl_2 reaction spontaneous at 400 K? At 700 K?

(g) Does the spontaneity of the decomposition of CH_3OH change as the temperature increases? Is there a temperature between 400 K and 1000 K at which the decomposition is spontaneous?

99. Screen 19.8: Free Energy and Temperature—Simulation

(a) Consider the reaction of Fe_2O_3 and C. How does ΔG°_{rxn} vary with temperature? Is there a temperature at which the reaction is spontaneous?

(b) Consider the reaction of HCl and Na_2CO_3. Is there a temperature at which the reaction is no longer spontaneous?

(c) How does the spontaneity of a reaction depend on temperature?

100. Screen 19.9: Thermodynamics and the Equilibrium Constant —Simulation. This screen considers the outcome of the reaction sequence

$$A \rightleftharpoons B \text{ and } B \rightleftharpoons C$$

(a) Set up the activation energy diagram with $\Delta E_a(1) = 18$ kJ, $\Delta G^\circ(1) = 7$ kJ, $\Delta E_a(2) = 13$ kJ, and $\Delta G^\circ(2) = -22$ kJ. (Energies need only be approximate.) When the system reaches equilibrium, will there be an appreciable concentration of B? Which species predominates? Explain the results.

(b) Leave the activation energy values the same as above but make $\Delta G^\circ(1) = 10$ kJ and $\Delta G^\circ(2) = -10$ kJ. When the equilibrium is attained, what are the relative amounts of A, B, and C? Explain the results.

20 Principles of Reactivity: Electron Transfer Reactions

Chapter Goals

- Balance net ionic equations for oxidation–reduction reactions.
- Understand the principles of voltaic cells.
- Understand how to use electrochemical potentials.
- Explore electrolysis, the use of electrical energy to produce chemical change.

Blood Gases

Today, in hundreds of hospitals and clinics around the country, doctors will order a BGA, a blood gas analysis. A complete analysis will give the physician information on pH and the partial pressures of O_2 and CO_2 in arterial blood. The pH is a measure of the body's acid–base equilibrium, arterial O_2 pressure measures oxygenation in the blood, and arterial CO_2 pressure is a measure of the body's ability to excrete CO_2. An elevated CO_2 pressure may indicate a problem with lung function, whereas a low CO_2 pressure may suggest a metabolic problem.

Because the solubility of O_2 in water is low, the blood must contain a substance that transports the gas to tissues. That role is filled by hemoglobin in blood and myoglobin in muscles. Both are proteins that have at their center a heme group, and it is an iron at the center of that group that binds O_2.

Myoglobin and hemoglobin have different affinities for O_2, reflecting their different physiological functions. Myoglobin, the oxygen delivery system in muscles, has a higher affinity for O_2 than hemoglobin at all O_2 pressures. Hemoglobin, the oxygen carrier in blood, becomes saturated with O_2 in the lungs, where the O_2 partial pressure is about 100 mm Hg. In tissue capillaries, however, the O_2 partial pressure drops, and hemoglobin releases O_2. Some is stored in myoglobin for use in times of immediate need, such as during strenuous exercise.

Monitoring blood O_2 levels is important, especially in a newborn. Oxygen in high concentrations can damage lungs and eyes, so blood O_2 levels are kept as low as possible.

Because the level of O_2 in blood is so important, a number of devices have been developed for its measurement. Among them is a device that is a tiny electrochemical cell. One half of this cell is a reference electrode; the second half is an electrode that detects and measures oxygen. The sensing electrode is placed in a sample of arterial blood, and a small voltage applied to the cell

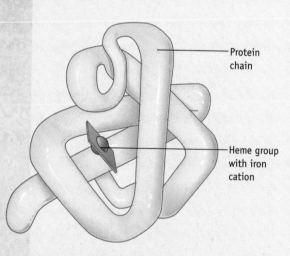

Myoglobin (Mb)

▲ **Myoglobin, the protein that binds O_2 in muscles.** At the center of the molecule is a heme group. The oxygen binds to the iron in the center of this group. The red color of muscle comes from the oxygenated protein.

Protein chain

Heme group with iron cation

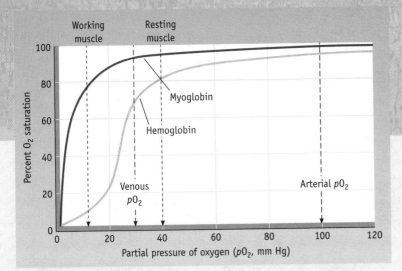

▲ O₂ binding by myoglobin and hemoglobin at a pH of 7.4 in a normal adult.
Interestingly, fetal blood has a low pressure of O₂, about 30 mm Hg, equivalent to living at an altitude of 26,000 feet.

causes the reduction of dissolved oxygen in the sample.

$$O_2(aq) + 2 H_2O(\ell) + 4 e^- \longrightarrow 4 OH^-(aq)$$

The device "counts" the number of electrons moving through the circuit and from that determines the amount of dissolved oxygen reduced.

This device has been miniaturized to fit into the tip of a narrow catheter so that it can be inserted into an artery, allowing a direct measure of oxygen concentration within the body.

Measurement of oxygen concentration using electrochemistry has been a well-established laboratory procedure for some time, but applying this technology to a mixture as complicated as blood did not work well because many other substances in blood interfered with the measurement. The key development that led to this device was incredibly simple. Biochemist L. C. Clark, Jr., modified the laboratory device, covering the electrode with a polyethylene membrane. The membrane screens most of the components of blood from the electrode, but it is permeable to oxygen which can pass through to reach the electrode where it is detected. Clark obtained a patent in 1956, and the Clark electrode remains the basis for this important medical diagnostic procedure to this day.

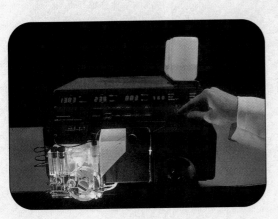

◄ pH/Blood gas analyzer. (Hank Morgan/Photo Researchers, Inc.)

Before You Begin

- Review definitions for oxidation, reduction, reducing agent, oxidizing agent, and oxidation number (Section 5.7).
- Review common redox reactions and common oxidizing and reducing agents (Section 5.7).
- Review the concepts of thermodynamics (Chapters 6 and 19).

Electron Transfer Reactions The transfer of electrons can occur in a product-favored, oxidation-reduction reaction. Such reactions are the basis of batteries. Using an outside source of energy, electron transfer can also occur in an electrolysis reaction.

The oxidation of copper metal by silver ions

A clean piece of copper wire will be placed in a solution of silver nitrate, AgNO₃.

With time, the copper reduces Ag⁺ ions to silver metal crystals, and the copper metal is oxidized to copper ions, Cu²⁺.

The blue color of the solution is due to the presence of aqueous copper(II) ions. (C. D. Winters)

Add Ag⁺(aq)

After several days

Silver ions in solution

Surface of copper wire

Cu²⁺

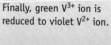

Reduction of an ion containing vanadium(V) with zinc metal

The yellow color of the VO₂⁺ ion in acid solution.

Zn added. With time the yellow VO₂⁺ ion is reduced to blue VO²⁺ ion.

With time the blue VO²⁺ ion is further reduced to green V³⁺ ion.

Finally, green V³⁺ ion is reduced to violet V²⁺ ion.

Add Zn

VO₂⁺

VO²⁺

V³⁺

V²⁺

Photo: Charles D. Winters

Let us introduce you to electrochemistry and electron transfer reactions with a simple chemistry experiment. Place a piece of copper in an aqueous solution of silver nitrate. After a short time, metallic silver deposits on the copper and the solution takes on the blue color typical of aqueous Cu^{2+} ions (Chapter Focus). The following oxidation–reduction (redox) reaction has occurred.

$$Cu(s) + 2\,Ag^+(aq) \longrightarrow Cu^{2+}(aq) + 2\,Ag(s)$$

At the microscopic level, Ag^+ ions in solution come into direct contact with the copper surface where the transfer of electrons occurs. Two electrons are transferred from a Cu atom to two Ag^+ ions. Copper ions, Cu^{2+}, enter the solution and silver atoms deposit on the copper surface. This product-favored reaction continues until one or both of the reactants are consumed.

The reaction between copper metal and silver ions can be used to generate an electric current. The experiment must be carried out in a different way, however, because the reactants, Cu(s) and Ag^+(aq), must not be in direct contact. If they are in contact, electrons will be transferred directly from copper atoms to silver ions, and the energy produced will be in the form of heat rather than electricity. Instead, the reaction is carried out in an apparatus that allows the electrons to be transferred from one reactant to the other through an electrical circuit. The movement of electrons through the circuit is the electric current that can be used to light a light bulb or run a motor (see Chapter Focus).

Devices that use chemical reactions to produce an electric current are called **voltaic cells** or **galvanic cells,** names honoring Count Alessandro Volta (1745–1827) and Luigi Galvani (1737–1798). All voltaic cells work in the same general way. They use product-favored redox reactions. The cell is constructed so that electrons produced when one reactant is oxidized are transferred through an electric circuit to the second reactant, which is reduced.

Chemical energy is converted to electrical energy in a voltaic cell. The opposite process, the use of electric energy to effect a chemical change, occurs in a process called **electrolysis.** Electrolysis of water is an example. Electrical energy splits water into its component elements, hydrogen and oxygen. Electrolysis is also used to electroplate one metal onto another, to obtain aluminum from its common ore (bauxite, Al_2O_3), and to prepare important chemicals such as chlorine [CHAPTER 21 ➡].

Electrochemistry is the field of chemistry that considers chemical reactions that produce or are caused by electrical energy. Because all electrochemical reactions are oxidation–reduction (redox) reactions, we begin our exploration of this subject by first describing how to balance equations for redox reactions.   ●

20.1   OXIDATION–REDUCTION REACTIONS

In a redox reaction, electron transfer occurs between a reducing agent and an oxidizing agent [⬅ SECTION 5.7] [🕮 CD-ROM, Screen 20.2]. The essential features of all electron transfer reactions are

- One reactant is oxidized and one is reduced.
- The extent of oxidation and reduction must balance.
- The oxidizing agent (the chemical species causing oxidation) is reduced.
- The reducing agent (the chemical species causing reduction) is oxidized.
- Oxidation numbers [⬅ SECTION 5.7, PAGE 171] can be used to determine whether

● Two Types of Electrochemical Processes
- Chemical change can produce an electric current in a voltaic cell.
- Electric energy can cause chemical change in the process of electrolysis.

⟮ *Chapter Goals* • Revisited ⟯

- **Balance net ionic equations for oxidation-reduction reactions.**
- Understand the principles of voltaic cells.
- Understand how to use electrochemical potentials.
- Explore electrolysis, the use of electrical energy to produce chemical change.

a substance is oxidized or reduced. An element is oxidized if its oxidation number increases. In reduction, the oxidation number decreases.

These aspects of a redox reaction are illustrated here for the reaction of copper metal and silver ion (see Chapter Focus).

Cu oxidized, oxidation number increases;
Cu is the reducing agent

$$Cu(s) + 2\ Ag^+(aq) \longrightarrow Cu^{2+}(aq) + 2\ Ag(s)$$

Ag^+ reduced, oxidation number decreases;
Ag^+ is the oxidizing agent

Balancing Oxidation–Reduction Equations

All redox equations must be balanced for both mass and charge. The same number of atoms appear in the products and reactants of an equation, and the sum of electric charges of all the species on each side of the equation arrow must be the same. Charge balance guarantees that the number of electrons produced in oxidation equals the number of electrons consumed in reduction [🕮 CD-ROM, Screen 20.3].

Balancing redox equations can be complicated, but fortunately there are procedures that can be used in these cases. Here, we describe the half-reaction method, a procedure that involves writing separate, balanced equations for the oxidation and reduction processes called **half-reactions.** One half-reaction describes the oxidation part of the reaction, and a second half-reaction describes the reduction part. The equation for the overall reaction is the sum of the two half-reactions, after adjustments have been made (if necessary) in one or both half-reaction equations so that the number of electrons transferred from reducing agent to oxidizing agent balances. For example, the half-reactions for the reaction of copper metal with silver ions are

Reduction half-reaction: $Ag^+(aq) + e^- \longrightarrow Ag(s)$
Oxidation half-reaction: $Cu(s) \longrightarrow Cu^{2+}(aq) + 2\ e^-$

Notice that the equations for each half-reaction are themselves balanced for mass and charge. In the copper half-reaction there is one Cu atom on each side of the equation (*mass balance*). The electric charge on the right side of the equation is 0 (the sum of +2 for the ion and −2 for two electrons), as it is on the left side (*charge balance*).

To produce the net chemical equation, we add the two half-reactions. First, however, we must multiply the silver half-reaction by 2.

$$2\ Ag^+(aq) + 2\ e^- \longrightarrow 2\ Ag(s)$$

Each mole of copper atoms produces two moles of electrons, and two moles of Ag^+ ions are required to consume those electrons.

Finally, adding the two half-reactions, and canceling electrons from both sides, leads to the net ionic equation for the reaction.

Reduction half-reaction: $2\ Ag^+(aq) + 2\ \cancel{e^-} \longrightarrow 2\ Ag(s)$
Oxidation half-reaction: $Cu(s) \longrightarrow Cu^{2+}(aq) + 2\ \cancel{e^-}$
Balanced net ionic equation: $Cu(s) + 2\ Ag^+(aq) \longrightarrow Cu^{2+}(aq) + 2\ Ag(s)$

The net ionic equation is likewise balanced for mass and charge.

Example 20.1 Balancing Oxidation–Reduction Equations

Problem • Balance the equation

$$Al(s) + Cu^{2+}(aq) \longrightarrow Al^{3+}(aq) + Cu(s)$$

Identify the oxidizing agent, the reducing agent, the substance oxidized, and the substance reduced. Write balanced half-reactions and the balanced net ionic equation.

Strategy • First, make sure the reaction is an oxidation–reduction reaction by checking each element to see if the oxidation numbers change. Next, separate the equation into half-reactions, identifying what has been reduced (oxidizing agent) and oxidized (reducing agent). Then balance the half-reactions, first for mass and then for charge. Finally, add the two half-reactions, after ensuring that the reducing agent half-reaction provides the same number of moles of electrons as the oxidizing agent half-reaction requires.

(a)   (b)

Reduction of Cu^{2+} by Al. **(a)** A ball of aluminum foil is added to a solution of $Cu(NO_3)_2$ and NaCl. **(b)** A coating of copper is soon seen on the surface of the aluminum, and the reaction generates sufficient heat to boil the water. (A coating of Al_2O_3 on the surface of aluminum protects the metal from reaction. However, in the presence of Cl^- ion, the coating is breached and reaction occurs.) *(Charles D. Winters)*

Solution •
Step 1. *Recognize the reaction as an oxidation–reduction reaction.* Here the oxidation number for aluminum changes from 0 to +3 and the oxidation number of copper changes from +2 to 0. Aluminum is oxidized and serves as the reducing agent. Copper(II) ions are reduced, and Cu^{2+} is the oxidizing agent.

Step 2. *Separate the process into half-reactions.*

Reduction: $Cu^{2+}(aq) \longrightarrow Cu(s)$
            (Oxidation number of Cu decreases.)

Oxidation: $Al(s) \longrightarrow Al^{3+}(aq)$
            (Oxidation number of Al increases.)

Step 3. *Balance each half-reaction for mass.* Both half-reactions are already balanced for mass.

Step 4. *Balance each half-reaction for charge.* The equations are balanced for charge by adding electrons to the more positive side of each half-reaction.

Reduction: $2 e^- + Cu^{2+}(aq) \longrightarrow Cu(s)$
            Each Cu^{2+} ion adds two electrons.

Oxidation: $Al(s) \longrightarrow Al^{3+}(aq) + 3 e^-$
            Each Al atom releases three electrons.

Step 5. *Multiply each half-reaction by an appropriate factor.* The reducing agent must donate as many electrons as the oxidizing agent acquires. Three Cu^{2+} ions are required to take on the 6 electrons produced by two Al atoms. Thus, we multiply the Cu^{2+}/Cu half-reaction by 3 and the Al/Al^{3+} half-reaction by 2.

Reduction: $3[2 e^- + Cu^{2+}(aq) \longrightarrow Cu(s)]$
Oxidation: $2[Al(s) \longrightarrow Al^{3+}(aq) + 3 e^-]$

Step 6. *Add the half-reactions to produce the overall balanced equation.*

Reduction: $6\,e^- + 3\,Cu^{2+}(aq) \longrightarrow 3\,Cu(s)$
Oxidation: $2\,Al(s) \longrightarrow 2\,Al^{3+}(aq) + 6\,e^-$
_____
Net ionic equation: $3\,Cu^{2+}(aq) + 2\,Al(s) \longrightarrow$
                                $3\,Cu(s) + 2\,Al^{3+}(aq)$

Step 7. *Simplify by eliminating reactants and products that appear on both sides.* This is not required here.

Comment • You should always check the overall equation to ensure there is a mass and charge balance. In this case there are three Cu atoms and two Al atoms on each side. The net electric charge on each side is +6. The equation is balanced.

Exercise 20.1 Balancing Oxidation–Reduction Equations

Aluminum reacts with nonoxidizing acids to give $Al^{3+}(aq)$ and $H_2(g)$. The (unbalanced) equation is

$$Al(s) + H^+(aq) \longrightarrow Al^{3+}(aq) + H_2(g)$$

Write balanced half-reactions and the balanced net ionic equation. Identify the oxidizing agent, the reducing agent, the substance oxidized, and the substance reduced.

When balancing equations for redox reactions in aqueous solution, it is sometimes necessary to add water molecules (H_2O) and either $H^+(aq)$ in acidic solution or $OH^-(aq)$ in basic solution to the equation. Equations that include oxoanions such as SO_4^{2-}, NO_3^-, ClO^-, CrO_4^{2-}, and MnO_4^- fall into this category. The process is outlined in Example 20.2 for the reduction of an oxocation in acidic solution and in Example 20.3 for a reaction in a basic solution.

Example 20.2   Balancing Equations for Oxidation–Reduction Reactions in Acid Solution

Problem • Balance the net ionic equation for the reaction of the dioxovanadium(V) ion, VO_2^+, with zinc in acid solution to form VO^{2+} (see Chapter Focus).

$$VO_2^+(aq) + Zn(s) \longrightarrow VO^{2+}(aq) + Zn^{2+}(aq)$$

Strategy • Follow the strategy outlined in Example 20.1. One exception is that water and H^+ ions will appear as product and reactant, respectively, in the half-reaction for the reduction of VO_2^+ ion.

Solution •

Step 1. *Recognize the reaction as an oxidation–reduction reaction.* The oxidation number of V changes from +5 in VO_2^+ to +4 in VO^{2+}. The oxidation number of Zn changes from 0 in the metal to +2 in Zn^{2+}.

Step 2. *Separate the process into half-reactions.*

Oxidation: $Zn(aq) \longrightarrow Zn^{2+}(aq)$
          Zn(s) is oxidized and is the reducing agent.

Reduction: $VO_2^+(aq) \longrightarrow VO^{2+}(aq)$
          VO_2^+ (aq) is reduced and is the oxidizing agent.

Step 3. *Balance half-reactions for mass.* Begin by balancing all atoms except H and O. (These atoms are always the last to be balanced because they often appear in more than one reactant or product.)

Zinc half-reaction: $Zn(aq) \longrightarrow Zn^{2+}(aq)$

This half-reaction is already balanced for mass.

Vanadium half-reaction: $VO_2^+(aq) \longrightarrow VO^{2+}(aq)$

The V atoms in this half-reaction are already balanced. An oxygen-containing species must be added to the right side of the equation to achieve an O atom balance, however.

$$VO_2^+(aq) \longrightarrow VO^{2+}(aq) + (\text{need 1 O atom})$$

In acid solution, add H_2O to the side requiring O atoms, one H_2O molecule for each O atom required.

$$VO_2^+(aq) \longrightarrow VO^{2+}(aq) + H_2O(\ell)$$

This means there are now two unbalanced H atoms on the right. Because the reaction occurs in an acidic solution, H^+ ions are

present. Therefore, in acidic solution a mass balance for H can be achieved by adding H^+ to the side of the equation deficient in H atoms. Here two H^+ ions are added to the left side of equation.

$$2\,H^+(aq) + VO_2^+(aq) \longrightarrow VO^{2+}(aq) + H_2O(\ell)$$

Step 4. *Balance the half-reactions for charge.*

Zinc half-reaction: $Zn(s) \longrightarrow Zn^{2+}(aq) + 2\,e^-$

The mass-balanced VO_2^+ equation has a net charge of +3 on the left side and +2 on the right. Therefore, 1 e^- is added to the more positive left side.

Vanadium half-reaction:
$$1\,e^- + 2\,H^+(aq) + VO_2^+(aq) \longrightarrow VO^{2+}(aq) + H_2O(\ell)$$

Step 5. *Multiply the half-reactions by appropriate factors so that the reducing agent donates as many electrons as the oxidizing agent consumes.* Here the reducing agent half-reaction supplies 2 mol of electrons per mol of Zn and the oxidizing agent half-reaction consumes 1 mol of electrons per mol of VO_2^+. Therefore, the oxidizing half-reaction must be multiplied by 2. Now 2 mol of the oxidizing agent (VO_2^+) consumes the 2 mol of electrons provided per mole of the reducing agent (Zn).

$$Zn(s) \longrightarrow Zn^{2+}(aq) + 2\,e^-$$
$$2[1\,e^- + 2\,H^+(aq) + VO_2^+(aq) \longrightarrow VO^{2+}(aq) + H_2O(\ell)]$$

Step 6. *Add the half-reactions to give the balanced, overall equation.*

Reduction: $2\,e^- + 4\,H^+(aq) + 2\,VO_2^+(aq) \longrightarrow$
                              $2\,VO^{2+}(aq) + 2\,H_2O(\ell)$

Oxidation: $\underline{Zn(s) \longrightarrow Zn^{2+}(aq) + 2\,e^-}$

Net ionic equation: $Zn(s) + 4\,H^+(aq) + 2\,VO_2^+(aq) \longrightarrow$
                              $Zn^{2+}(aq) + 2\,VO^{2+}(aq) + 2\,H_2O(\ell)$

Step 7. *Simplify by eliminating reactants and products that appear on both sides.* This is not required here.

Comment • Check the overall equation to ensure that there is a mass and charge balance.

Mass balance: 1 Zn, 2 V, 4 H, and 4 O

Charge balance: Each side has a net charge of +6.

Exercise 20.2 Balancing Equations for Oxidation–Reduction Reactions in Acid Solution

The yellow dioxovanadium (V) ion, VO_2^+ (aq) is reduced by zinc metal in three steps. The first step reduces it to blue VO^{2+}(aq) (Example 20.2) and it is further reduced to green V^{3+}(aq) in the second step. In a third step, V^{3+} can be reduced to violet V^{2+}(aq). In each step zinc is oxidized to Zn^{2+}(aq). Write balanced net ionic equations for steps two and three. (This reduction sequence is shown in the Chapter Focus.)

Exercise 20.3 Balancing Equations for Oxidation–Reduction Reactions in Acid Solution

The permanganate ion, MnO_4^-, is an oxidizing agent [← SECTION 5.7 AND TABLE 5.4]. A common laboratory analysis for iron is to titrate aqueous iron(II) ion with a solution of potassium permanganate of precisely known concentration (Figure 20.1) [← FIGURE 5.15, PAGE 177]. Use the half-reaction method to write the balanced net ionic equation for the reaction in acidic solution.

$$MnO_4^-(aq) + Fe^{2+}(aq) \longrightarrow Mn^{2+}(aq) + Fe^{3+}(aq)$$

Figure 20.1 **The reaction of the purple permanganate ion (MnO_4^-) with iron(II) ion in aqueous solution.** The reaction gives the nearly colorless ions Mn^{2+} and Fe^{3+}. See Exercise 20.3. *(Charles D. Winters)*

Example 20.2 and Exercises 20.2 and 20.3 illustrate the technique of balancing equations for redox reactions involving oxocations and oxoanions that occur in acidic solution. Under these conditions, H^+ ion or the H^+/H_2O pair can be used to achieve a balanced equation if required. Conversely, in basic solution, only OH^- ion or the OH^-/H_2O pair can be used.

Problem-Solving Tip 20.1

Balancing Oxidation–Reduction Equations — A Summary

- Hydrogen balance can only be achieved with H^+/H_2O (in acid) or OH^-/H_2O (in base). Never add H or H_2 to balance hydrogen.

- Use H_2O or OH^- as appropriate to balance oxygen. Never add O atoms, O^{2-} ions, or O_2 for O balance.

- Never include H^+(aq) and OH^-(aq) in the same equation. A solution can be either acidic or basic, never both.

- The number of electrons in a half-reaction corresponds to the change in oxidation state of the element being oxidized or reduced.

- Electrons are always a component of half-reactions but should never appear in the overall equation.

- Remember to include charges in the formulas for ions. Omitting the charge, or writing the charge incorrectly, is one of the most common errors seen on student papers.

- The best way to become competent in balancing redox equations is to practice, practice, practice.

Example 20.3 Balancing Equations for Oxidation–Reduction Reactions in Basic Solution

Problem • Aluminum metal is oxidized in aqueous base with water serving as the oxidizing agent. The products of the reaction are $Al(OH)_4^-$(aq) and H_2(g). Write a balanced net ionic equation for this reaction.

Strategy • First identify the oxidizing and reducing half-reactions and then balance them for mass and charge. Finally, add the balanced half-reactions to obtain the balanced net ionic equation for the reaction.

(Example continues on next page)

Solution •

Step 1. *Recognize the reaction as an oxidation–reduction reaction.* The unbalanced equation is

$$Al(s) + H_2O(\ell) \longrightarrow Al(OH)_4^-(aq) + H_2(g)$$

Here aluminum is oxidized, its oxidation number changing from 0 to +3. Hydrogen is reduced, with its oxidation number decreasing from +1 to zero.

Step 2. *Separate the process into half-reactions.*

Oxidation half-reaction: $Al(s) \longrightarrow Al(OH)_4^-(aq)$
                    (Al oxidation number increases from 0 to +3.)

Reduction half-reaction: $H_2O(\ell) \longrightarrow H_2(g)$
                    (H oxidation number decreases from +1 to 0.)

Step 3. *Balance half-reactions for mass.* Addition of OH^- and/or H_2O is required for mass balance in both half-reactions. In the case of the aluminum half-reaction we simply add OH^- ions to the left side.

Oxidation half-reaction: $Al(s) + 4 OH^-(aq) \longrightarrow Al(OH)_4^-(aq)$

Reduction half-reaction: To balance the half-reaction for water reduction, notice that an O-containing species must appear on the right side of the equation. Because H_2O is a reactant, we use OH^-, which is present in this basic solution, as the other product.

$$2 H_2O(\ell) \longrightarrow H_2(g) + 2 OH^-(aq)$$

Step 4. *Balance half-reactions for charge.* Electrons are added to balance charge.

Oxidation half-reaction:

$$Al(s) + 4 OH^-(aq) \longrightarrow Al(OH)_4^-(aq) + 3 e^-$$

Reduction half-reaction:

$$2 H_2O(\ell) + 2 e^- \longrightarrow H_2(g) + 2 OH^-(aq)$$

Step 5. *Balance oxidation and reduction.* Here electron balance is achieved by using 2 mol of Al to provide 6 mol of e^-, which are then acquired by 6 mol of H_2O.

Oxidation half-reaction:

$$2[Al(s) + 4 OH^-(aq) \longrightarrow Al(OH)_4^-(aq) + 3 e^-]$$

Reduction half-reaction:

$$3[2 H_2O(\ell) + 2 e^- \longrightarrow H_2(g) + 2 OH^-(aq)]$$

Step 6. *Add half-reactions.*

$$2 Al(s) + 8 OH^-(aq) \longrightarrow 2 Al(OH)_4^-(aq) + \cancel{6 e^-}$$
$$\underline{6 H_2O(\ell) + \cancel{6 e^-} \longrightarrow 3 H_2(g) + 6 OH^-(aq)}$$

Net ionic equation:
$$2 Al(s) + 8 OH^-(aq) + 6 H_2O(\ell) \longrightarrow$$
$$2 Al(OH)_4^-(aq) + 3 H_2(g) + 6 OH^-(aq)$$

Reaction of aluminum metal with water in basic solution. Aluminum metal is converted to the $Al(OH)_4^-$ ion by oxidation with water in a basic solution. Hydrogen gas is the product of water reduction.

Step 7. *Simplify by eliminating reactants and products that appear on both sides.* Six OH^- ions can be canceled from the two sides of the equation. This gives

$$2 Al(s) + 2 OH^-(aq) + 6 H_2O(\ell) \longrightarrow 2 Al(OH)_4^-(aq) + 3 H_2(g)$$

Comment • The final equation is balanced for mass and charge.

Mass balance: 2 Al, 14 H, and 8 O

Charge balance: There is a net −2 charge on each side.

Exercise 20.4   Balancing Equations for Oxidation–Reduction Reactions in Basic Solution

Voltaic cells based on the oxidation of sulfur are under development. One such cell involves the reaction of sulfur with aluminum under basic conditions.

$$Al(s) + S(s) \longrightarrow Al(OH)_3(s) + HS^-(aq)$$

(a) Balance this equation showing each balanced half-reaction.

(b) Identify the oxidizing and reducing agents, the substance oxidized, and the substance reduced.

Problem-Solving Tip 20.2

An Alternative Method for Balancing Equations in Basic Solution

Another way to balance equations for reactions in basic solution is to first do so in acidic solution and then add enough OH^- ions to both sides of the equation so that any H^+ ions are converted to water. Taking the half-reaction for the reduction of ClO^- ion to Cl_2, we have

(a) Balance in acid.

$$4\,H^+(aq) + 2\,ClO^-(aq) + 2\,e^- \longrightarrow Cl_2(g) + 2\,H_2O(\ell)$$

(b) Add 4 OH^- ions to both sides.

$$4\,OH^-(aq) + 4\,H^+(aq) + 2\,ClO^-(aq) + 2\,e^- \longrightarrow$$
$$Cl_2(g) + 2\,H_2O(\ell) + 4\,OH^-(aq)$$

(c) Combine OH^- and H^+ to form water where appropriate.

$$4\,H_2O(\ell) + 2\,ClO^-(aq) + 2\,e^- \longrightarrow$$
$$Cl_2(g) + 2\,H_2O(\ell) + 4\,OH^-(aq)$$

(d) Simplify.

$$2\,H_2O(\ell) + 2\,ClO^-(aq) + 2\,e^- \longrightarrow Cl_2(g) + 4\,OH^-(aq)$$

20.2   SIMPLE VOLTAIC CELLS

Let us use the reaction of copper metal and silver ions (see Chapter Focus) as the basis of a voltaic cell. To do this, we place the components of each half-cell in separate compartments (Figure 20.2). This prevents the copper metal from transferring electrons directly to silver ions. Instead, electrons are transferred through an external circuit, and useful work can potentially be done [🖳 CD-ROM, Screen 20.4].

The copper half-cell (on the left in Figure 20.2) contains copper metal that serves as one electrode and a solution containing copper(II) ions. The half-cell on the right uses a silver electrode and a solution containing silver(I) ions. Important features of this simple cell are

- *The two half-cells are connected with a **salt bridge** that allows cations and anions to move between the two half-cells.* The electrolyte chosen for the salt bridge

Chapter Goals • Revisited

- Balance net ionic equations for oxidation–reduction reactions.
- **Understand the principles of voltaic cells.**
- Understand how to use electrochemical potentials.
- Explore electrolysis, the use of electrical energy to produce chemical change.

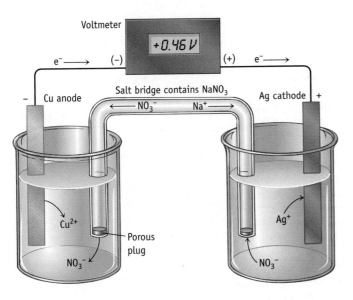

Net reaction: Cu(s) + 2 Ag$^+$(aq) ⟶ Cu^{2+}(aq) + 2 Ag(s)

Figure 20.2   A voltaic cell using Cu(s)|Cu²⁺(aq) and Ag(s)|Ag⁺(aq) half-cells. Electrons flow through the external circuit from the anode (the copper electrode) to the cathode (silver electrode). In the salt bridge, which contains aqueous NaNO₃, negative NO₃⁻(aq) ions migrate toward the copper half-cell, and positive Na⁺(aq) ions migrate toward the silver half-cell. Using 1.0 M Cu²⁺(aq) and 1.0 M Ag⁺(aq) solutions, this cell will generate 0.46 V.

should contain ions that will not react with chemical reagents in the cell. In the example in Figure 20.2, $NaNO_3$ is used.

- *In all electrochemical cells the **anode** is the electrode at which oxidation occurs. The electrode at which reduction occurs is always the **cathode**.* (In Figure 20.2, the copper electrode is the anode and the silver electrode is the cathode.)
- *A minus sign can be assigned to the anode in a voltaic cell, and the cathode is marked with a positive sign.* The chemical oxidation occurring at the anode, which produces electrons, gives it a negative charge. Electric current in the external circuit of a voltaic cell consists of electrons moving from the negative to the positive electrode, that is, from anode to cathode.
- *In the salt bridge, cations move from anode to cathode, and anions move from cathode to anode.*

The chemistry occurring in the cell pictured in Figure 20.2 is summarized by the following half-reactions and net ionic equation:

Cathode, reduction:	$2\ Ag^+(aq) + 2\ e^- \longrightarrow 2\ Ag(s)$
Anode, oxidation:	$\underline{Cu\ (s) \longrightarrow Cu^{2+}(aq) + 2\ e^-}$
Net ionic equation:	$Cu(s) + 2\ Ag^+(aq) \longrightarrow Cu^{2+}(aq) + 2\ Ag(s)$

The salt bridge is required in a voltaic cell for the reaction to proceed. In the Cu/Ag^+ voltaic cell, anions move in the salt bridge toward the copper half-cell and cations move toward the silver half-cell (Figure 20.2). As $Cu^{2+}(aq)$ ions are formed in the copper half-cell by oxidation, negative ions enter that cell from the salt bridge (and positive ions leave the cell) so that the numbers of positive and negative charges in the half-cell compartment remain in balance. Likewise, in the silver half-cell, negative ions move out of the half-cell into the salt bridge and positive ions

• Salt Bridges
A simple salt bridge can be made using a solution of an electrolyte and adding gelatin. Gelatin makes the contents semirigid so the salt bridge is easier to handle. Porous disks and permeable membranes are alternatives to a salt bridge. These devices allow ions to traverse from one half-cell to the other while keeping the two solutions from mixing.

• Electron and Ion Flow
It is helpful to notice that in an electrochemical cell the negative electrons and negatively charged anions make a "circle." That is, electrons move from anode to cathode in the external circuit, and negative anions move from the cathode compartment, through the salt bridge, to the anode compartment.

A Closer Look

Redox Reactions in Acidic and Basic Solution

Balancing equations for the redox reactions of oxoanions can be complicated. There is a way to think about this, however, that can simplify the task.

Metal oxoanions such as VO_3^-, MnO_4^-, and $Cr_2O_7^{2-}$ are excellent oxidizing agents in acidic solution. (See Chapter Focus and Figure 20.2). Let us *imagine* that these ions shed oxide ions as the metal ion is reduced.

$$MnO_4^-(aq) \longrightarrow$$
$$Mn^{2+}(aq) + [4\ O^{2-}] + 5\ e^-$$

The oxide ion is a very strong base, so it reacts with H^+ to form water.

$$[O^{2-}] + 2\ H^+(aq) \longrightarrow H_2O(\ell)$$

This tells us that for each O^{2-} ion "lost" by the oxidizing agent, two H^+ ions and one H_2O molecule are required to balance the equation.

$$MnO_4^-(aq) + 8\ H^+(aq) \longrightarrow$$
$$Mn^{2+}(aq) + 4\ H_2O(\ell) + 5\ e^-$$

When metals or metal ions are used as reducing agents, their oxidation number increases. When transition metals are involved, these reactions often occur in basic solution. Transition metal ions with higher oxidation numbers are more stable in the form of oxoanions or oxocations, so some reagent that can supply the required oxygen atoms must be present. For example, vanadium(II) ion can be a reducing agent, but it must acquire oxide ions to stabilize it as a vanadium(IV) or (V) ion.

$$V^{2+}(aq) + 2\ [O^{2-}] \longrightarrow VO_2^+(aq) + 3\ e^-$$

Oxide ions do not exist in water, but hydroxide ions are an oxygen-rich species in basic solution. For purposes of balancing the half-reaction, we *imagine* that two OH^- ions can act as the source of each oxygen required to stabilize the oxocation or oxoanion.

$$2\ OH^-(aq) \longrightarrow H_2O(\ell) + [O^{2-}]$$

One H_2O molecule is the byproduct for each oxygen supplied.

$$V^{2+}(aq) + 4\ OH^-(aq) \longrightarrow VO_2^+(aq) + 2\ H_2O(\ell) + 3\ e^-$$

A final thought: This approach is useful when trying to decide where to place H^+ or OH^- ions and H_2O molecules when balancing half-reactions.

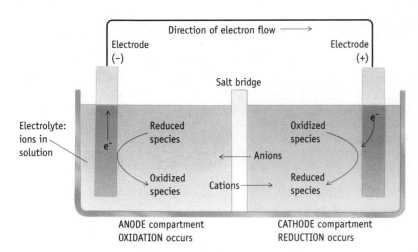

Figure 20.3   Summary of terms used in a voltaic cell. Electrons move from the anode, the site of oxidation, through the external circuit to the cathode, the site of reduction. Charge balance in each half-cell is achieved by migration of ions through the salt bridge. Negative ions move from the reduction half-cell to the oxidation half-cell and positive ions move in the opposite direction.

move into the cell as $Ag^+(aq)$ ions are reduced to silver metal. A complete circuit is required for current to flow. If the salt bridge is removed, the circuit is not complete and reactions at the electrodes will cease.

In Figure 20.2, the electrodes are connected by wires to a voltmeter. The connections might instead be to a light bulb or some other device that uses electricity. Electrons are produced by oxidation of copper, and $Cu^{2+}(aq)$ ions enter the solution. The electrons traverse the external circuit to the silver electrode where they reduce $Ag^+(aq)$ ions to silver metal. To balance the extent of oxidation and reduction, two $Ag^+(aq)$ ions are reduced for every $Cu^{2+}(aq)$ ion formed. The main features of this and of all other voltaic cells are summarized in Figure 20.3.

Example 20.4 **Electrochemical Cells**

Problem • Describe how to set up a voltaic cell to generate an electric current using the reaction

$$Fe(s) + Cu^{2+}(aq) \longrightarrow Cu(s) + Fe^{2+}(aq)$$

Which electrode is the anode and which is the cathode? In what direction do electrons flow in the external circuit? In what direction do the positive and negative ions flow in the salt bridge? Write equations for the half-reactions that occur at each electrode.

Strategy • The first step is to identify the two different half-cells that make up the cell. Next, decide in which half-cell oxidation occurs and in which one reduction occurs.

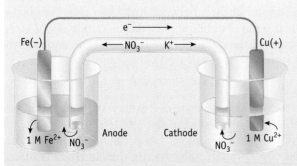

Solution • This voltaic cell is similar to the one diagrammed in Figure 20.2. One half-cell contains an iron electrode and a solution of an iron(II) salt such as $Fe(NO_3)_2$. The other half-cell contains a copper electrode and a soluble copper(II) salt such as $Cu(NO_3)_2$. The two half-cells are linked with a salt bridge containing an electrolyte such as $NaNO_3$. Iron is oxidized, and the iron electrode is therefore the anode:

Oxidation, anode: $Fe(s) \longrightarrow Fe^{2+}(aq) + 2\,e^-$

Because copper(II) ions are reduced, the copper electrode is the cathode. The cathodic half-reaction is

Reduction, cathode: $Cu^{2+}(aq) + 2\,e^- \longrightarrow Cu(s)$

In the external circuit, electrons flow from the iron electrode (anode) to the copper electrode (cathode). In the salt bridge, negative ions flow toward the $Fe|Fe^{2+}(aq)$ half-cell and positive ions flow in the opposite direction.

Comment • Recall that oxidation always occurs at the anode and reduction always occurs at the cathode and that electrons always flow from the anode to the cathode. See Figure 20.3.

Exercise 20.5 **Electrochemical Cells**

Describe how to set up a voltaic cell using the following half-reactions:

Reduction half-reaction: $Ag^+(aq) + e^- \longrightarrow Ag(s)$

Oxidation half-reaction: $Ni(s) \longrightarrow Ni^{2+}(aq) + 2\,e^-$

Which is the anode and which is the cathode? What is the overall cell reaction? What is the direction of electron flow in an external wire connecting the two electrodes? Describe the ion flow in a salt bridge connecting the cell compartments.

Voltaic Cells with Inert Electrodes

In the half-cells described so far, the metal used as an electrode is also a reactant or a product in the redox reaction. Not all half-reactions involve a metal as a reactant or product, however. With the exception of carbon as graphite, most non-metals are unsuitable as electrode materials because they do not conduct electricity. It is not possible to make an electrode from a gas, liquid, or a solution. Ionic solids do not make satisfactory electrodes because the ions are locked tightly in a crystal lattice, and these materials do not conduct electricity.

In situations where reactants and products cannot serve as the electrode material, inert electrodes must be used. Such electrodes are made of materials that conduct an electric current but that are neither oxidized nor reduced in the cell.

Consider constructing a voltaic cell to accommodate the following product-favored reaction:

$$2\,Fe^{3+}(aq) + H_2(g) \longrightarrow 2\,Fe^{2+}(aq) + 2\,H^+(aq)$$

Reduction half-reaction: $Fe^{3+}(aq) + e^- \longrightarrow Fe^{2+}(aq)$

Oxidation half-reaction: $H_2(g) \longrightarrow 2\,H^+(aq) + 2\,e^-$

Neither the reactants nor the products can be used as an electrode material. Therefore, the two half-cells are set up so that the reactants and products come in contact with an electrode such as graphite where they can accept or give up electrons. Graphite is a commonly used electrode material: It is a conductor of electricity, inexpensive (an essential feature in commercial cells), and not readily oxidized under the conditions encountered in most cells. Platinum and gold are also commonly used in laboratory experiments because both are chemically inert under most circumstances. They are generally too costly for commercial cells, however.

The *hydrogen electrode* (Figure 20.4) is particularly important in the field of electrochemistry because it is used as a reference in assigning cell voltages (see Section 20.4) The electrode itself is platinum, chosen because hydrogen adsorbs on the metal's surface. In operation, hydrogen is bubbled over the electrode and a large surface area maximizes the contact of the gas and the electrode. The aqueous solution contains $H^+(aq)$. The half-reactions involving $H^+(aq)$ and $H_2(g)$

$$2\,H^+(aq) + 2\,e^- \longrightarrow H_2(g) \qquad \text{or} \qquad H_2(g) \longrightarrow 2\,H^+(aq) + 2\,e^-$$

take place at the platinum electrode, and the electrons involved in the reaction are conducted to or from the reaction site by the electrode.

A half-cell using the reduction of $Fe^{3+}(aq)$ to $Fe^{2+}(aq)$ can also be set up with a platinum electrode. In this case the solution surrounding the electrode contains

● **Electrodes**
Electrodes are made of materials that conduct electricity. If a reactant is a metal, that metal can be used as an electrode material. Graphite, also a good conductor, is another common material used for electrodes.

Figure 20.4   Hydrogen electrode.
Hydrogen gas is bubbled over a platinum electrode in a solution containing H^+ ions. Such electrodes function best if they have a large surface area. Often platinum wires are woven into a gauze or the metal surface is roughened either by abrasion or by chemical treatment to increase the surface area. *(Charles D. Winters)*

Serendipity and Inert Electrodes

In the 1960s Barnett Rosenberg set out to study a problem that had interested him for some time — the effect of electrical fields on living cells — but the results of the experiment were quite different from his expectations. He and his students had placed an aqueous suspension of live *Escherichia coli* bacteria in an electric field between supposedly inert platinum plates. Much to their surprise they found that cell growth was affected; cell division had stopped! After careful experimentation, the effect on cell division was traced to tiny amounts of a compound now called cisplatin [Pt(NH$_3$)$_2$Cl$_2$], which was produced by reaction of the platinum with electrically charged chemical species in the water.

Much to the benefit of cancer chemotherapy, Rosenberg recognized that these laboratory results had wider implications, and subsequent experiments led to compounds now used to treat cancer patients. He recently said that the use of cisplatin has meant that "Testicular cancer went from a disease that normally killed about 80% of the patients, to one which is close to 95% curable. This is probably the most exciting development in the treatment of cancers that we have had in the past 20 years. It is now the treatment of first choice in ovarian, bladder, and osteogenic sarcoma [bone] cancers as well."

Dr. Barnett Rosenberg, Head of the Barros Research Institute in East Lansing, Michigan. Rosenberg discovered the anticancer activity of the platinum compounds cisplatin and carboplatin — some of the most widely used chemotherapy drugs. *(Doug Elbinger)*

iron ions in two different oxidation states. Transfer of electrons to or from the reactant occurs at the electrode surface.

A voltaic cell involving the reduction of Fe^{3+}(aq, 1.0 M) to Fe^{2+}(aq, 1.0 M) with H$_2$ gas is illustrated in Figure 20.5. In this cell, the hydrogen electrode is the anode (H$_2$ is oxidized to H$^+$), and the iron-containing compartment is the cathode (Fe^{3+} is reduced to Fe^{2+}). The cell produces 0.77 V.

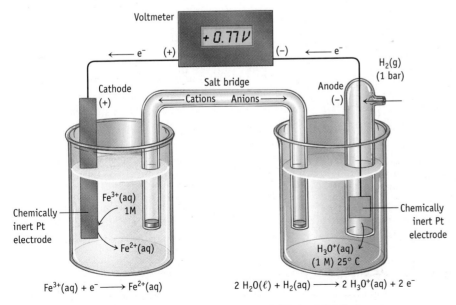

Figure 20.5 A voltaic cell with a hydrogen electrode. This cell has Fe^{2+}(aq, 1.0 M) and Fe^{3+}(aq, 1.0 M) in the cathode compartment and H$_2$(g) and H$^+$(aq, 1.0 M) in the anode compartment. At 25 °C, the cell generates 0.77 V.

Electrochemical Cell Conventions

Chemists often use a shorthand notation to simplify cell descriptions. For example the cell involving the reduction of silver ion with copper metal is written as

$$Ag^+(aq, 1.0\ M) + Cu(s) \longrightarrow Ag(s) + Cu^{2+}(aq, 1.0\ M)$$

$$Cu(s)|Cu^{2+}(aq, 1.0\ M)\|Ag^+(aq, 1.0\ M)|Ag(s)$$

Anode information   Cathode information

and the cell using H_2 gas to reduce Fe^{3+} ions is

$$2\ Fe^{3+}(aq) + H_2(g) \longrightarrow 2\ Fe^{2+}(aq) + 2\ H^+(aq)$$

$$Pt|H_2(P = 1\ atm)|H^+(aq, 1.0\ M)\|Fe^{2+}(aq, 1.0\ M),\ Fe^{3+}(aq, 1.0\ M)|Pt$$

Anode information   Cathode information

By convention, the anode and information with respect to the solution with which it is in contact are always written on the left. A single vertical line (|) indicates a phase boundary, and double vertical lines (‖) indicate a salt bridge.

20.3   COMMERCIAL VOLTAIC CELLS

The cells described so far illustrate the principles by which voltaic cells produce an electric current. Cells like these are unlikely to have practical use, however. They are neither compact nor robust, high priorities for most uses. In most situations, it is also important that the cell produce a constant voltage, but a problem with the cells described so far is that the voltage produced varies as the concentrations of ions in solution change (see Section 20.5). Also, the rate of current production is low. Attempting to draw a large current results in a drop in voltage because the current depends on how fast ions in solution migrate to the electrode. Ion concentrations near the electrode become depleted if current is drawn rapidly, resulting in a decline in voltage.

The amount of current that can be drawn from a voltaic cell depends on the quantity of reagents consumed. A voltaic cell must have a large mass of reactants to produce current over a prolonged period. In addition, a voltaic cell that can be recharged is attractive. Recharging a cell means returning the reagents to their original sites in the cell. In the cells described so far, the movement of ions in the cell mixes the reagents, and they cannot be unmixed after the cell has been running.

Commercial voltaic cells, usually called *batteries*, are designed to overcome the limitation of simpler voltaic cells. Batteries can be classified as primary and secondary [CD-ROM, Screen 20.9]. **Primary batteries** use redox reactions that cannot be returned to their original state by recharging, so when the reactants are consumed, the battery is "dead" and must be discarded. **Secondary batteries** are often called **storage batteries,** or **rechargeable batteries.** The reactions in these batteries can be reversed; the battery can be recharged.

Years of development have led to many different commercial voltaic cells to meet specific needs (Figure 20.6). Several common ones are described below. All adhere to the principles that have been developed in earlier discussions.

Primary Batteries: Dry Cells and Mercury Batteries

If you buy an inexpensive flashlight battery or **dry cell battery,** it will probably be a modern version of a voltaic cell invented by George LeClanché in 1866 (Figure 20.7). A zinc can serves as the anode, and the cathode is a graphite rod down the

An electrochemical cell can be made by inserting copper and zinc electrodes into almost any conductive material. *(Charles D. Winters)*

● **Batteries**
The word "battery" has become part of our common language, designating any self-contained device that generates an electric current. The term battery has a more precise scientific meaning, however; it refers to a collection of two or more voltaic cells. For example, the 12-V battery used in automobiles is made up of six voltaic cells. Each voltaic cell develops a voltage of 2 V. Six cells connected in series produce 12 V.

Galvani, Volta, Batteries, and Frogs

Voltaic cells are also called galvanic cells after an Italian physician Luigi Galvani (1737–1798), who carried out early studies of what he called "animal electricity," studies that brought new words into our language — among them "galvanic" and "galvanize."

Around 1780 Galvani observed that the electric current from a static electricity generator caused the contraction of the muscles in a frog's leg. Following Ben Franklin's suggestion that lightning was an electric spark, Galvani found that the

◄ Luigi Galvani experimenting on the effect of electricity on muscles (1792). *(Archivo Iconografico, S. A./CORBIS)*

frog's leg contracted when he attached it with a brass hook to a metal railing during a thunderstorm. Investigating further, he found he could induce contraction when the muscle was contacted with two different metals. In this case, no external source of electricity was applied to the muscles, so Galvani concluded the frog's muscles were generating electricity by themselves. This was evidence, he believed, of a kind of "vital energy" or "animal electricity," which was related to but different from "natural electricity" generated by machines or lightning.

Alessandro Volta repeated Galvani's experiments, with the same results, but he came to different conclusions. Volta proposed that an electric current was generated by the contact between two different metals — an explanation we now

know to be correct — and that the muscle was a detector of the small current generated. To prove his hypothesis, Volta built the first "electric pile" in 1800. This was a series of metal disks of two kinds (silver and zinc), separated by paper disks soaked in acid or salt solutions. Soon after Volta announced his discovery, Carlisle and Nicholson in England used the electricity from a "pile" to decompose water into hydrogen and oxygen. Within a few years, the great English chemist Humphry Davy used a more powerful voltaic pile to isolate potassium and sodium metals by electrolysis.

In 1797 Alexander von Humboldt found that Galvani had in fact discovered "bioelectricity," that a muscle of one frog would contract when touched by the nerve of another. Indeed, scientists now know that there are tiny differences in electric potential across cell membranes, an effect arising from the unequal distribution of ions such as Na^+, K^+, and Mg^{2+} inside and outside cells.

Figure 20.6 Some commercial voltaic cells. Commercial voltaic cells provide energy for a wide range of devices, come in a myriad of sizes and shapes, and produce different voltages. Some are rechargeable, and others are thrown away after use. One might think that there is nothing further to learn about batteries, yet research on these devices is one of the more active areas of research in the chemical community. *(Charles D. Winters)*

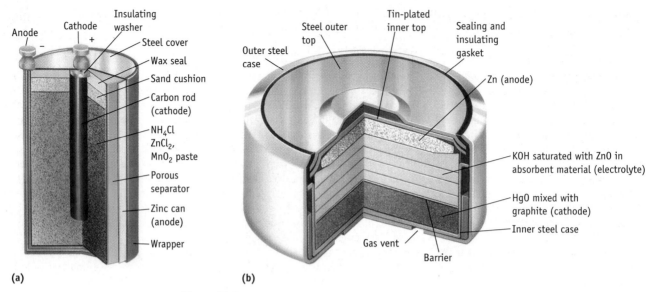

Figure 20.7 Two common commercial voltaic cells. (a) A LeClanché dry cell. **(b)** A mercury "button" battery. Zinc is the reducing agent in both (and is the anode).

center. These cells are often called "dry cells" because there is no visible liquid phase. However, the cell contains a moist paste of NH_4Cl, $ZnCl_2$, and MnO_2. The moisture is necessary because the ions must be in a medium in which they can migrate from one electrode to the other. The cell generates a potential of 1.5 V using the following half-reactions:

Cathode, reduction: $\quad 2\,NH_4^+(aq) + 2\,e^- \longrightarrow 2\,NH_3(g) + H_2(g)$

Anode, oxidation: $\quad\quad Zn(s) \longrightarrow Zn^{2+}(aq) + 2\,e^-$

The two gases formed at the cathode will build up pressure and could cause the cell to rupture. This is avoided, however, by two other reactions in the cell. Ammonia molecules bind to Zn^{2+} ions, and hydrogen gas is oxidized by MnO_2.

$$Zn^{2+}(aq) + 2\,NH_3(g) + 2\,Cl^-(aq) \longrightarrow Zn(NH_3)_2Cl_2(s)$$
$$2\,MnO_2(s) + H_2(g) \longrightarrow Mn_2O_3(s) + H_2O(\ell)$$

LeClanché cells are widely used because of their low cost, but they have several disadvantages. If current is drawn from the battery rapidly, the gaseous products cannot be consumed rapidly enough, the cell resistance rises, and the voltage drops. In addition, the zinc electrode and ammonium ion are in contact in the cell, and these chemicals react slowly. Recall that zinc reacts with acid to form hydrogen. The ammonium ion, $NH_4^+(aq)$, is a weak Brønsted acid [← TABLE 17.4] and it reacts slowly with zinc. Because of this reaction these voltaic cells cannot be stored indefinitely, a fact you may have learned from experience. When the zinc outer shell deteriorates, the battery can leak acid and perhaps damage the flashlight or other appliance in which it is contained.

Not satisfied with a simple flashlight battery? You can spend a little more money to buy **alkaline cells** for your flashlight. They generate current up to 50% longer than the same size dry cell battery. The chemistry of alkaline cells is quite similar to that in a LeClanché cell except that the material inside the cell is basic (alkaline). Alkaline cells use the oxidation of zinc and the reduction of MnO_2 to generate a current, but NaOH or KOH is used in the cell instead of the acidic salt NH_4Cl.

| *Cathode, reduction:* | $2\ MnO_2(s) + H_2O(\ell) + 2\ e^- \longrightarrow Mn_2O_3(s) + 2\ OH^-(aq)$ |
| *Anode, oxidation:* | $Zn(s) + 2\ OH^-(aq) \longrightarrow ZnO(s) + H_2O(\ell) + 2\ e^-$ |

Alkaline cells, which produce 1.54 V (approximately the same voltage as the LeClanché cell), have the further advantage that the cell potential does not decline under high current loads because no gases are formed.

Mercury cells (Figure 20.7) are also alkaline cells. They are typically used in calculators, cameras, watches, heart pacemakers, and other devices. The anode in a mercury cell is again metallic zinc, but the cathode is mercury(II) oxide.

| *Cathode, reduction:* | $HgO(s) + H_2O(\ell) + 2\ e^- \longrightarrow Hg(\ell) + 2\ OH^-(aq)$ |
| *Anode, oxidation:* | $Zn(s) + 2\ OH^-(aq) \longrightarrow ZnO(s) + H_2O(\ell) + 2\ e^-$ |

Zinc and ZnO are separated from a moist paste of HgO and either NaOH or KOH by a moistened paper that also serves as the salt bridge. The cell produces 1.35 V. These cells are widely used, but their disposal is an environmental problem, and there are ongoing efforts to phase them out. Mercury and its compounds are poisonous, so mercury batteries should be reprocessed to recover the metal at the end of their useful life.

Secondary or Rechargeable Batteries

When a LeClanché cell or an alkaline cell ceases to produce a usable electric current, it is discarded. In contrast, some types of cells can be recharged, often hundreds of times. Recharging requires applying an external electric current to restore the cell to its original state [CD-ROM, Screen 20.13].

An automobile battery — the **lead storage battery** — is probably the best-known rechargeable battery (Figure 20.8). The 12-V version of this battery contains six voltaic cells, each generating 2.04 V. The lead storage battery can produce a large initial current, a feature essential to starting an automobile engine.

The anode of a lead storage battery is metallic lead. The cathode is also made of lead, but it is covered with a layer of compressed, insoluble lead(IV) oxide, PbO_2.

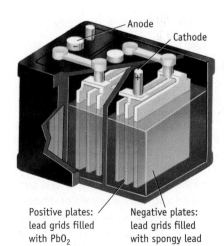

Positive plates: lead grids filled with PbO_2

Negative plates: lead grids filled with spongy lead

(a) Lead storage battery. The negative plates (anode) are lead grids filled with spongy lead. The positive plates (cathode) are lead grids filled with lead(IV) oxide, PbO_2. Each cell of the battery generates 2 V.

(b) Nickel-cadmium batteries in a recharger. Each is a 1.5 V battery.

Figure 20.8   Lead storage battery and NiCd batteries, secondary or rechargeable batteries.
(b, Charles D. Winters)

The electrodes, arranged alternately in a stack and separated by thin fiberglass sheets, are immersed in aqueous sulfuric acid. When the cell supplies electrical energy, the lead anode is oxidized to lead(II) sulfate, an insoluble substance that adheres to the electrode surface. The two electrons produced per lead atom move through the external circuit to the cathode where PbO_2 is reduced to Pb^{2+} that, in the presence of H_2SO_4, also forms lead(II) sulfate.

Cathode, reduction:
$$PbO_2(s) + 4\,H^+(aq) + SO_4^{2-}(aq) + 2\,e^- \longrightarrow$$
$$PbSO_4(s) + 2\,H_2O(\ell)$$

Anode, oxidation:
$$\underline{Pb(s) + SO_4^{2-}(aq) \longrightarrow PbSO_4(s) + 2\,e^-}$$

Net ionic equation:
$$Pb(s) + PbO_2(s) + 2\,H_2SO_4(aq) \longrightarrow 2\,PbSO_4(s) + 2\,H_2O(\ell)$$

When current is generated, sulfuric acid is consumed and water is formed. Because water is less dense than sulfuric acid, the density of the solution decreases during this process. One way to determine if a lead storage battery needs to be recharged is to measure the density of the solution.

A lead storage battery is recharged by supplying electrical energy. The $PbSO_4$ coating the surfaces of the electrodes is converted back to metallic lead and PbO_2, and sulfuric acid is regenerated. Recharging this battery is possible because the reactants and products remain attached to the electrode surface. The lifetime of a lead storage battery is limited because the coatings of PbO_2 and $PbSO_4$ flake off the surface and fall to the bottom of the battery case.

Scientists and engineers would like to find an alternative to lead storage batteries, especially for use in electric cars. Lead storage batteries have the disadvantage of being large and heavy. In addition, lead and its compounds are toxic and their disposal adds a further complication. At this time, however, the advantages of lead storage batteries outweigh their disadvantages.

Nickel–cadmium ("ni-cad") batteries, used in a variety of cordless appliances, such as telephones, video camcorders, and cordless power tools, are lightweight and rechargeable. The chemistry of the cell utilizes the oxidation of cadmium and the reduction of nickel(III) oxide under basic conditions. As with the lead storage battery, the reactants and products formed when producing a current are solids that adhere to the electrodes.

Cathode, reduction:
$$NiO(OH)(s) + H_2O(\ell) + e^- \longrightarrow Ni(OH)_2(s) + OH^-(aq)$$

Anode, oxidation:
$$Cd(s) + 2\,OH^-(aq) \longrightarrow Cd(OH)_2(s) + 2\,e^-$$

Ni-cad batteries produce a nearly constant voltage. However, their cost is relatively high and there are restrictions on their disposal because cadmium compounds are toxic and present an environmental hazard.

Fuel Cells

History

William Grove (1811–1896) demonstrated a fuel cell in 1839 at the Royal Institute in London. At the time, Michael Faraday directed the institute. •

An advantage of voltaic cells is that they are small and portable, but their size is also a limitation. The amount of electric current produced is limited by the quantity of reagents contained in the cell. When one of the reactants is completely consumed, the cell will no longer generate a current. **Fuel cells** avoid this limitation. Fuel cells are electrochemical cells in which the reactants (fuel and oxidant) are supplied continuously from an external reservoir.

Although the first fuel cells were constructed more than 150 years ago, little was done to develop this technology until the space program rekindled interest in these devices. Hydrogen–oxygen fuel cells have been used in NASA's Gemini, Apollo, and Space Shuttle programs. Not only are they lightweight and efficient, but they have the added benefit that they generate drinking water for the ship's crew. The fuel cells on board the Space Shuttle deliver the same power as batteries weighing ten times as much. On a typical seven-day mission, the fuel cells consume 680 kg of hydrogen and produce 720 L of water.

In a hydrogen–oxygen fuel cell (Figure 20.9), hydrogen is pumped onto the anode of the cell, and O_2 (or air) is directed to the cathode where the following reactions occur:

Cathode, reduction: $O_2(g) + 2\,H_2O(\ell) + 4\,e^- \longrightarrow 4\,OH^-(aq)$

Anode, oxidation: $H_2(g) \longrightarrow 2\,H^+(aq) + 2\,e^-$

The two halves of the cell are separated by a special material called a proton exchange membrane (PEM). Protons, $H^+(aq)$, formed at the anode traverse the PEM and react with the hydroxide ions produced at the cathode, forming water. The net reaction in the cell is thus the formation of water from H_2 and O_2. Cells currently in use run at a temperature of 70–140 °C and produce a voltage of about 0.9 V.

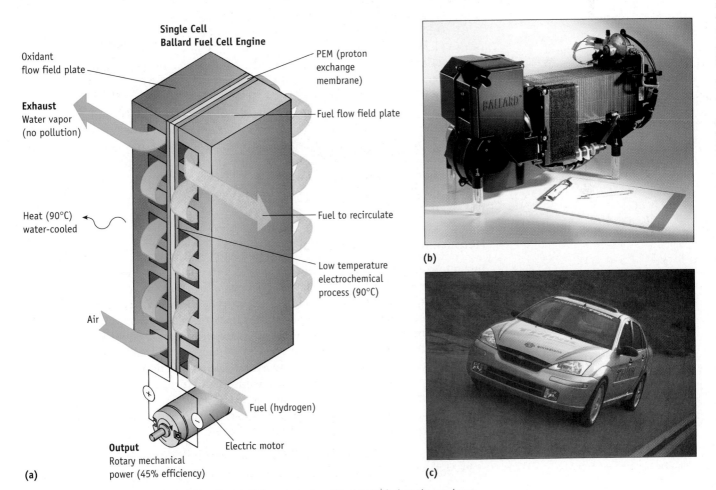

Figure 20.9   A hydrogen–oxygen fuel cell. (a) Hydrogen gas is oxidized to $H^+(aq)$ at the anode surface. On the other side of the proton exchange membrane (PEM), oxygen gas is reduced to $OH^-(aq)$. The $H^+(aq)$ ions travel through the PEM and combine with $OH^-(aq)$, forming water. **(b)** A fuel cell "stack." Individual fuel cells are stacked together and connected in series to obtain a higher voltage. **(c)** A car using a hydrogen–oxygen fuel cell. *(Courtesy Ballard Power Systems, Inc.)*

● **Electrochemical Units**
The coulomb (abbreviated C) is the standard (SI) unit of electrical charge (Appendix C.3)
- 1 Joule = 1 volt · 1 coulomb
- 1 coulomb = 1 amp · 1 sec

20.4   STANDARD ELECTROCHEMICAL POTENTIALS

Different electrochemical cells produce different voltages: 1.5 V for the LeClanché and alkaline cells, 1.35 V for a mercury cell, and about 2.0 V for the individual cells in a lead storage battery. In this section, we want to identify factors affecting cell voltages and develop procedures to calculate the voltage of a cell based on the chemistry in the cell and the conditions used [🖱 CD-ROM, Screen 20.5].

Electromotive Force (EMF)

Electrons generated at the anode of an electrochemical cell move through the external circuit toward the cathode, and the force needed to move the electrons arises from a difference in potential energy of electrons at the two electrodes. This difference in potential energy per electrical charge is called the **electromotive force** or **emf,** for which the literal meaning is "force causing electrons to move." Emf has units of volts (V); one volt is the potential difference needed to impart one joule of energy to an electric charge of one coulomb (1 Joule = 1 volt × 1 coulomb). *One coulomb is the quantity of charge that passes a point in an electric circuit when a current of one ampere flows for one second (1 coulomb = 1 amp × 1 sec).*

It would be possible to calculate cell voltages if a value for the potential energy of the electron for each half-cell were known. The problem is that the potential energy of the electron for a single half-cell (not connected to anything else) is not easily determined. Instead, the emf of a voltaic cell is equated with the cell voltage, E_{cell}, which can be measured.

You may recognize a similarity between emf, enthalpy (H), and free energy (G). Changes in enthalpy and free energy (ΔH and ΔG) can be measured but the value of H or G for a specific substance is generally not known. Values for ΔH_f° and ΔG_f° (Appendix L) were established by choosing a reference point, the elements in their standard states. As you will see next, data on cell potentials are all referenced to the standard hydrogen electrode. All other voltages are measured against this reference cell.

Measuring Standard Potentials

Imagine you were going to study cell voltages in a laboratory with two objectives: (a) to understand the factors that affect these values and (b) to be able to predict the potential of a voltaic cell. You might construct a number of different half-cells, link them together in various combinations to form voltaic cells (as in Figure 20.5), and measure the cell potentials. After a few experiments, it would become apparent that cell voltages depend on a number of factors: the half-cells used (that is, the reactions in each half cell and the overall or net reaction in the cell), the concentrations of reactants and products in solution, the pressure of gaseous reactants, and the temperature [🖱 CD-ROM, Screen 20.6].

So that the potential of one half-cell can be compared with another, all cell voltages are measured under **standard conditions:**

- Reactants and products are present as pure liquids or solids.
- Solutes in aqueous solution have a concentration of 1.0 M.
- Gaseous reactants or products have a pressure of 1.0 atm.

A cell potential measured under these conditions is called the **standard potential** and is given the symbol E°_{cell}. Unless otherwise specified, all values of E°_{cell} refer to measurements at 298 K (25 °C).

Your Next Car?

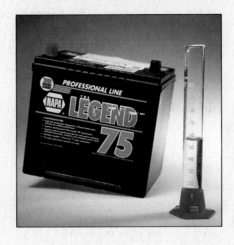

As a response to federally mandated clean air standards, several major car manufacturers have designed prototype electric cars that use various types of batteries to provide the power to drive the car. The most commonly used battery for automotive use is the lead storage battery, but they are problematic owing to their mass. To produce a mole of electrons requires 321 g of reactants in a lead storage battery. As a result, these batteries rank very low among various options in power per kilogram of battery weight. In fact, the power available from any type of battery is much less than is available from an equivalent mass of gasoline. The best values achieved by several high-tech batteries currently being developed are still almost 100 times less than what is available from gasoline.

A 15-kg battery has the same amount of stored energy as 59 mL (47 g) of gasoline. *(Charles D. Winters)*

Chemical System	Watt-hour/kg (1 W · h = 3600 J)
Lead–acid battery	18–56
Nickel–cadmium battery	33–70
Sodium-sulfur battery	80–140
Lithium polymer battery	150
Gasoline–air combustion engine	12,200

Fuel cells improve the situation. Hydrogen–oxygen fuel cells operate at 50–65% efficiency, and they meet most of the requirements for use in automobiles. They operate at room temperature or slightly above, start rapidly, and develop a high current density. Cost is a serious problem, however, and it appears that a substantial move away from the internal combustion engine is still a long way off. The cost–performance of the internal combustion engine cannot be matched. A further problem with the H_2/O_2 fuel cell is the transport and storage of hydrogen gas. This problem would vanish, however, if fuels cells could be built to use other fuels such as gasoline or methanol or ethanol and air.

Design for a car using a fuel cell.
The hydrogen for the fuel cell is produced from gasoline in the car.

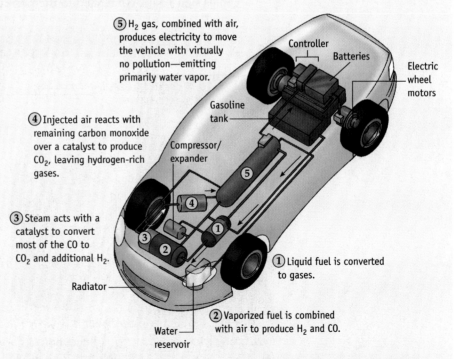

⑤ H_2 gas, combined with air, produces electricity to move the vehicle with virtually no pollution—emitting primarily water vapor.

④ Injected air reacts with remaining carbon monoxide over a catalyst to produce CO_2, leaving hydrogen-rich gases.

③ Steam acts with a catalyst to convert most of the CO to CO_2 and additional H_2.

Controller

Batteries

Electric wheel motors

Gasoline tank

Compressor/ expander

Radiator

Water reservoir

① Liquid fuel is converted to gases.

② Vaporized fuel is combined with air to produce H_2 and CO.

Design for a car using a fuel cell. The hydrogen for the fuel cell is produced from gasoline in this car. (1) By applying heat, liquid fuel is converted to a gas. (2) Vaporized fuel is combined with air over a catalyst to produce H_2 and CO. (3) Most of the CO reacts with steam to form CO_2 and H_2. (4) In the "preferential oxidation stage," injected air reacts with the remaining CO over a catalyst to produce CO_2, leaving hydrogen-rich gases. (5) Hydrogen gas, with air, produces electricity to move the vehicle. There is virtually no pollution; the emission consists of water vapor and CO_2.

Suppose you set up a number of standard half-cells and connect each in turn to a **standard hydrogen electrode (SHE).** Your apparatus would look like the voltaic cell in Figure 20.10. There are three important things you can learn:

1. *The reaction that occurs.* The reaction occurring in the cell pictured in Figure 20.10 could be *either* the reduction of Zn^{2+} ions with H_2 gas

$$Zn^{2+}(aq) + H_2(g) \longrightarrow Zn(s) + 2\,H^+(aq)$$

$Zn^{2+}(aq)$ is the oxidizing agent and H_2 is the reducing agent.
Standard hydrogen electrode would be the anode, or negative electrode.

or the reduction of $H^+(aq)$ ions by $Zn(s)$.

$$Zn(s) + 2\,H^+(aq) \longrightarrow Zn^{2+}(aq) + H_2(g)$$

Zn is the reducing agent and $H^+(aq)$ is the oxidizing agent.
Standard hydrogen electrode would be the cathode, or positive electrode.

All of the substances named in these equations are present in the cell. The reaction that actually occurs is the product-favored reaction. That is, we predict that *the reaction occurring is the one in which the reactants are stronger reducing and oxidizing agents than the products.*

2. *Direction of electron flow in the external circuit.* In a voltaic cell, electrons always flow from the anode (negative electrode) to the cathode (positive electrode). That is, *electrons move from the electrode of higher potential energy to the one of lower potential energy.* We can tell the direction of electron movement by placing a voltmeter in the circuit. A positive potential is observed if the voltmeter terminal with a plus sign (+) is connected to the positive electrode (and the terminal with the

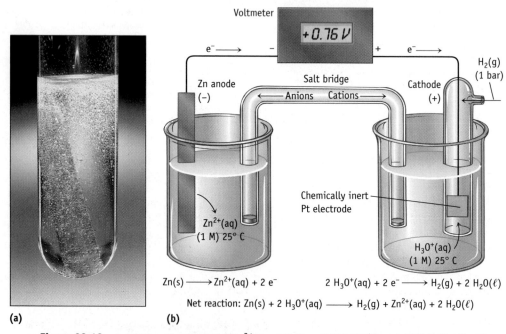

(a)   **(b)**

Figure 20.10   A voltaic cell using $Zn(s)|Zn^{2+}$ (aq, 1.0 M) and $H_2(g)|H^+$ (aq, 1.0 M) half-cells. The reaction of zinc and hydrogen ion is product-favored **(a).** This is reflected by the potential of 0.76 V generated by this cell **(b).** The electrode in the $H_2(g)|H^+$ (aq, 1.0 M) half-cell is the cathode and the Zn electrode is the anode. Electrons flow in the external circuit to the hydrogen half-cell from the zinc half cell. The positive sign of the measured voltage indicates that the hydrogen electrode is the cathode or positive electrode. *(a, Charles D. Winters)*

minus sign ($-$) is connected to the negative electrode). Connected in the opposite way (plus to minus and minus to plus) will give a negative value on the digital readout.

3. *Cell potential.* In Figure 20.10 the voltmeter is hooked up with its positive terminal connected to the hydrogen half-cell, and +0.76 V is observed. The hydrogen electrode is thus the positive electrode or cathode, and the reactions occurring in this cell must be

Reduction, cathode:	$2 H^+(aq) + 2 e^- \longrightarrow H_2(g)$
Oxidation, anode:	$Zn(s) \longrightarrow Zn^{2+}(aq) + 2 e^-$
Net cell reaction:	$Zn(s) + 2 H^+(aq) \longrightarrow Zn^{2+}(aq) + H_2(g)$

This result confirms that, of the two oxidizing agents present in the cell, $H^+(aq)$ is better than $Zn^{2+}(aq)$, and Zn metal is a better reducing agent than H_2 gas.

The potential of +0.76 V measured for the oxidation of zinc with hydrogen ion represents the difference in potential energy of an electron at each electrode. From the direction of flow of electrons in the external circuit (Zn electrode $\longrightarrow$ H_2 electrode) we conclude that the potential energy of an electron at the zinc electrode is 0.76 V higher than the potential energy of the electron at the hydrogen electrode.

Hundreds of electrochemical cells such as that in Figure 20.10 can be set up and allow us to determine the relative oxidizing or reducing ability of various chemical species and to determine the potential generated by the reaction under standard conditions. A few results are given in Figure 20.11, where half-reactions are listed as reductions. That is, the chemical species on the left is acting as an oxidizing agent. In Figure 20.11, we list them in *descending* ability to act as oxidizing agents.

Standard Reduction Potentials

By doing experiments such as that as illustrated by Figure 20.10, not only can we determine the *relative* oxidizing and reducing ability of various chemical species, but we can also rank them quantitatively.

If E°_{cell} is a measure of the standard potential for the cell, then $E^\circ_{cathode}$ and E°_{anode} can be taken as measures of electrode potential energy. Because E°_{cell} reflects

---

A Closer Look

EMF, Cell Potential, and Voltage

The terms emf and cell potential (E_{cell}) are usually used synonymously but they are subtly different. The distinction is that E_{cell} is a measured quantity and its value is affected by how the measurement is made. To understand this point, consider as an analogy water in a pipe under pressure. Water pressure can be viewed as analogous to emf; it represents a force that will cause

water in the pipe to move. If we open a faucet, water will flow. Opening the faucet will, however, decrease the pressure in the system.

Emf is the potential difference when no current flows. To measure E_{cell}, a voltmeter is placed in the external circuit. Although voltmeters have high internal resistance to minimize current flow, a small

current flows nonetheless. As a result, the value of E_{cell} will be slightly different from the emf.

Finally, there is a difference between a potential and a voltage. The voltage of a cell has a magnitude but no sign. In contrast, the potential of a half-reaction or a cell has a sign (+ or $-$) and a magnitude.

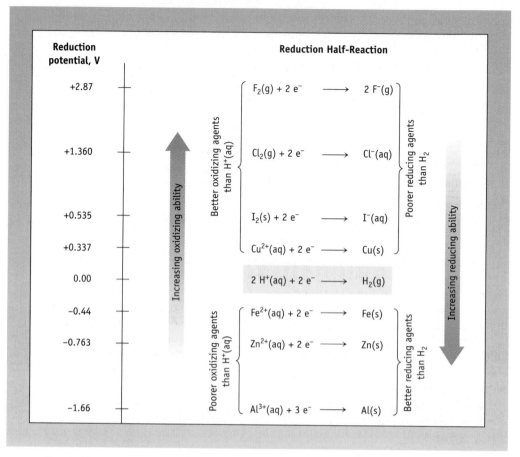

Figure 20.11   A potential ladder for reduction half-reactions. The relative position of a half-reaction on this potential ladder reflects the relative ability of the species at the left to act as an oxidizing agent. The higher the compound or ion is in the list, the better it is as an oxidizing agent. Conversely, the atoms or ions on the right are reducing agents. The lower they are in the list, the better they are as a reducing agent. The potential for each half-reaction is given with its reduction potential, $E°_{cathode}$. (For more information, see J. R. Runo and D. G. Peters, *Journal of Chemical Education*, Vol. 70, page 708, 1993.)

the *difference* in electrode potential energies, this means $E°_{cell}$ is the difference between $E°_{cathode}$ and $E°_{anode}$.

$$E°_{cell} = E°_{cathode} - E°_{anode} \qquad\qquad (20.1)$$

● Equation 20.1
Equation 20.1 is another example of calculating an energy change from $E_{final} - E_{initial}$. Electrons move to the cathode (the "final" state) from the anode (the "initial" state). Thus, Equation 20.1 resembles equations you have seen previously in this book (such as Equations 6.6 and 7.7).

Here, $E°_{cathode}$ and $E°_{anode}$ are the *standard reduction potentials* for the half-cell reactions that occur at the cathode and anode, respectively. Equation 20.1 is important for three reasons:

- If we have values of $E°_{cathode}$ and $E°_{anode}$ we can calculate the standard potential, $E°_{cell}$, for a voltaic cell.
- *When the calculated value of $E°_{cell}$ has a positive value, the reaction is predicted to be product-favored as written.* Conversely, if the calculated value of $E°_{cell}$ is negative, the reaction is predicted to be reactant-favored. The reaction will be product-favored in a direction opposite to that in which it is written.

- If we measure E°_{cell}, and know either $E^\circ_{cathode}$ or E°_{anode}, we can calculate the other value. This value would tell us how one half-cell reaction compares with others in terms of relative oxidizing or reducing ability.

But here is a dilemma. One cannot measure individual half-cell potentials. Instead, what scientists have done is assign a potential of 0.00 V to the half-reaction that occurs at a standard hydrogen electrode (SHE).

$$2\,H^+(aq,\,1\,M) + 2\,e^- \longrightarrow H_2(g,\,1\,bar) \qquad E^\circ = 0.00\,V$$

With this standard, we can now set up experiments such as that in Figures 20.5 or 20.10 to determine E° for half-cells by measuring E°_{cell} where one of the electrodes is the standard hydrogen electrode. With this we can quantify the information in Figure 20.11 and are then use these values to make predictions about E°_{cell} for new voltaic cells.

Tables of Standard Reduction Potentials

The experimental approach just described leads to tables of E° values such as Figure 20.11 and Table 20.1. Some important points concerning this table are listed here and are illustrated in the discussion and examples that follow [CD-ROM, Screen 20.6].

1. As in Figure 20.11 reactions are written as "oxidized form + electrons $\longrightarrow$ reduced form." The species on the left side of the reaction is an oxidizing agent, and the species on the right is a reducing agent. Therefore, *all potentials are for reduction reactions.*

2. The more positive the value of E° for the reactions in Table 20.1, the better the oxidizing ability of the ion or compound on the left side of the reaction. This means $F_2(g)$ *is the best oxidizing agent in the table.*

$$F_2(g,\,1\,atm) + 2\,e^- \longrightarrow 2\,F^-(aq,\,1\,M) \qquad E^\circ = +2.87\,V$$

Lithium ion at the bottom left corner of the table is the poorest oxidizing agent because its E° value is the most negative.

3. The oxidizing agents in the table (ions, elements, and compounds at the left) increase in strength from the bottom to the top of the table.

4. The more negative the value of the reduction potential, E°, the less likely the reaction will occur as a reduction and the more likely the reverse reaction will occur (as an oxidation). Thus, Li(s) is the strongest reducing agent in the table, and F^- is the weakest reducing agent. The reducing agents in the table (the ions, elements, or compounds at the right) increase in strength from the top to the bottom.

5. When a half-reaction is reversed to give ["reduced form $\longrightarrow$ oxidized form + electrons"], the sign of E° is reversed but the value of E° is unaffected.

$$Fe^{3+}(aq,\,1\,M) + e^- \longrightarrow Fe^{2+}(aq,\,1\,M) \qquad E^\circ = +0.771\,V$$
$$Fe^{2+}(aq,\,1\,M) \longrightarrow Fe^{3+}(aq,\,1\,M) + e^- \qquad E^\circ = -0.771\,V$$

6. *The reaction between any substance on the left in this table (an oxidizing agent) with any substance lower than it on the right (a reducing agent) is product-favored under standard conditions.* This has been called the *northwest–southeast rule:* Product-favored reactions will always involve a reducing agent that is "southeast" of the proposed oxidizing agent.

● *E°* **Values**

An extensive listing of E° values is found in Appendix M, and still larger tables of data can be found in chemistry reference books. A common convention, used in Appendix M, lists standard reduction potentials in two groups, one for acid and neutral solutions and the other for basic solutions.

Table 20.1 • **Standard Reduction Potentials in Aqueous Solution at 25 °C***

Reduction Half-Reaction		$E°$ (V)
$F_2(g) + 2 e^-$	$\longrightarrow 2 F^-(aq)$	+2.87
$H_2O_2(aq) + 2 H^+(aq) + 2 e^-$	$\longrightarrow 2 H_2O(\ell)$	+1.77
$PbO_2(s) + SO_4^{2-}(aq) + 4 H^+(aq) + 2 e^-$	$\longrightarrow PbSO_4(s) + 2 H_2O(\ell)$	+1.685
$MnO_4^-(aq) + 8 H^+(aq) + 5 e^-$	$\longrightarrow Mn^{2+}(aq) + 4 H_2O(\ell)$	+1.52
$Au^{3+}(aq) + 3 e^-$	$\longrightarrow Au(s)$	+1.50
$Cl_2(g) + 2 e^-$	$\longrightarrow 2 Cl^-(aq)$	+1.360
$Cr_2O_7^{2-}(aq) + 14 H^+(aq) + 6 e^-$	$\longrightarrow 2 Cr^{3+}(aq) + 7 H_2O(\ell)$	+1.33
$O_2(g) + 4 H^+(aq) + 4 e^-$	$\longrightarrow 2 H_2O(\ell)$	+1.229
$Br_2(\ell) + 2 e^-$	$\longrightarrow 2 Br^-(aq)$	+1.08
$NO_3^-(aq) + 4 H^+(aq) + 3 e^-$	$\longrightarrow NO(g) + 2 H_2O(\ell)$	+0.96
$OCl^-(aq) + H_2O(\ell) + 2 e^-$	$\longrightarrow Cl^-(aq) + 2 OH^-(aq)$	+0.89
$Hg^{2+}(aq) + 2 e^-$	$\longrightarrow Hg(\ell)$	+0.855
$Ag^+(aq) + e^-$	$\longrightarrow Ag(s)$	+0.80
$Hg_2^{2+}(aq) + 2 e^-$	$\longrightarrow 2 Hg(\ell)$	+0.789
$Fe^{3+}(aq) + e^-$	$\longrightarrow Fe^{2+}(aq)$	+0.771
$I_2(s) + 2 e^-$	$\longrightarrow 2 I^-(aq)$	+0.535
$O_2(g) + 2 H_2O(\ell) + 4 e^-$	$\longrightarrow 4 OH^-(aq)$	+0.40
$Cu^{2+}(aq) + 2 e^-$	$\longrightarrow Cu(s)$	+0.337
$Sn^{4+}(aq) + 2 e^-$	$\longrightarrow Sn^{2+}(aq)$	+0.15
$2 H^+(aq) + 2 e^-$	$\longrightarrow H_2(g)$	0.00
$Sn^{2+}(aq) + 2 e^-$	$\longrightarrow Sn(s)$	−0.14
$Ni^{2+}(aq) + 2 e^-$	$\longrightarrow Ni(s)$	−0.25
$V^{3+}(aq) + e^-$	$\longrightarrow V^{2+}(aq)$	−0.255
$PbSO_4(s) + 2 e^-$	$\longrightarrow Pb(s) + SO_4^{2-}(aq)$	−0.356
$Cd^{2+}(aq) + 2 e^-$	$\longrightarrow Cd(s)$	−0.40
$Fe^{2+}(aq) + 2 e^-$	$\longrightarrow Fe(s)$	−0.44
$Zn^{2+}(aq) + 2 e^-$	$\longrightarrow Zn(s)$	−0.763
$2 H_2O(\ell) + 2 e^-$	$\longrightarrow H_2(g) + 2 OH^-(aq)$	−0.8277
$Al^{3+}(aq) + 3 e^-$	$\longrightarrow Al(s)$	−1.66
$Mg^{2+}(aq) + 2 e^-$	$\longrightarrow Mg(s)$	−2.37
$Na^+(aq) + e^-$	$\longrightarrow Na(s)$	−2.714
$K^+(aq) + e^-$	$\longrightarrow K(s)$	−2.925
$Li^+(aq) + e^-$	$\longrightarrow Li(s)$	−3.045

Increasing strength of oxidizing agents (left margin, pointing up)

Increasing strength of reducing agents (right margin, pointing down)

*In volts (V) versus the standard hydrogen electrode.

For example, Zn can reduce Fe^{2+}, H^+, Cu^{2+}, and I_2, but Cu can only reduce I_2.

Reduction Half-Reaction

$$I_2(s) + 2 e^- \longrightarrow 2 I^-(aq)$$
$$Cu^{2+}(aq) + 2 e^- \longrightarrow Cu(s)$$
$$2 H^+(aq) + 2 e^- \longrightarrow H_2(g)$$
$$Fe^{2+}(aq) + 2 e^- \longrightarrow Fe(s)$$
$$Zn^{2+}(aq) + 2 e^- \longrightarrow Zn(s)$$

The northwest-southeast rule: The reducing agent always lies to the southeast of the oxidizing agent in a product-favored reaction.

● **Northwest–Southeast Rule**
This guideline is a reflection of the idea of moving down a potential energy "ladder" in a product-favored reaction.

7. The algebraic sign of the half-reaction potential is the sign of the electrode when it is attached to the H_2/H^+ standard cell (see Figures 20.5 and 20.10).

8. Electrochemical potentials depend on the nature of the reactants and products and their concentrations, not on the quantities of material used. Therefore, changing the stoichiometric coefficients for a half-reaction does not change the value of $E°$. For example, the reduction of Fe^{3+} has an $E°$ of +0.771 V whether the reaction is written as

$$Fe^{3+}(aq, 1\ M) + e^- \longrightarrow Fe^{2+}(aq, 1\ M) \qquad E° = +0.771\ V$$

or as

$$2\ Fe^{3+}(aq, 1\ M) + 2\ e^- \longrightarrow 2\ Fe^{2+}(aq, 1\ M) \qquad E° = +0.771\ V$$

● **Changing Stoichiometric Coefficients**
The volt is defined as "energy/charge" (V = J/C). Multiplying a reaction by some number causes both the energy and the charge to be multiplied by that number. Thus, the ratio "energy/charge = volt" does not change.

Using Tables of Standard Reduction Potentials

Tables or "ladders" of standard reduction potentials are immensely useful. They allow you to predict the potential of a new voltaic cell, provide information that can be used to balance redox equations, and help predict which redox reactions are product-favored. Let us expand on each of these ideas [🖰 CD-ROM, Screen 20.6, *Simulation*].

Calculating Cell Potentials, $E°_{cell}$

Standard reduction potentials were obtained by measuring cell potentials. It makes sense, therefore, that these values can be combined to give the potential of some new cell.

The net reaction occurring in a voltaic cell using silver and copper half-cells is

$$2\ Ag^+(aq) + Cu(s) \longrightarrow 2\ Ag(s) + Cu^{2+}(aq)$$

The silver electrode is the cathode and the copper electrode the anode. We know this because silver ion is reduced (to silver metal) and copper metal is oxidized (to Cu^{2+} ions). (Recall that oxidations always occur at the anode and reductions at the cathode.) Also notice that the $Cu^{2+}|Cu$ half-reaction is "southeast" of the $Ag^+|Ag$ half-reaction in the potential ladder (Table 20.1) (see point 6 on page 851).

$$E°_{cathode} = +0.80\ V \qquad Ag^+(aq) + e^- \longrightarrow Ag(s)$$

"Distance" from $E°_{cathode}$ to $E°_{anode}$ is 0.80 V − 0.337 V = 0.46 V.

Cu is "southeast" of Ag^+

$$E°_{anode} = +0.337\ V \qquad Cu^{2+}(aq) + 2e^- \longrightarrow Cu(s)$$

The equations for the half-reactions at each electrode and the standard reduction potentials (Table 20.1) are

Reduction, cathode: $2\ Ag^+(aq) + 2\ e^- \longrightarrow 2\ Ag(s)$

     Standard reduction for $Ag^+|Ag$: $E° = +0.80\ V$

Oxidation, anode: $Cu(s) \longrightarrow Cu^{2+}(aq) + 2\ e^-$

     Standard reduction for $Cu^{2+}|Cu$: $E° = +0.337\ V$

The potential for the voltaic cell is the difference between the standard reduction potentials.

$$E°_{cell} = E°_{cathode} - E°_{anode}$$
$$E°_{cell} = (+0.80\ V) - (+0.337\ V)$$
$$E°_{cell} = +0.46\ V$$

Chemical Perspectives

An Electrochemical Toothache!

It was recently reported that a 66-year-old woman had intense pain that was traced to her dental work [*N. Engl. J. Med.*, Vol. 342, 2003 (2000)]. A root canal job had moved a mercury amalgam filling on one tooth slightly closer to a gold alloy crown on an adjacent tooth. Eating acidic foods caused her intense pain. When dental amalgams of dissimilar metals come in contact with saliva, a voltaic cell is formed that generates potentials up to several hundred millivolts — and you feel it! You can do it, too, if you chew a foil gum wrapper with teeth that have been filled with a dental amalgam. Ouch!

Notice that the value of $E°_{cell}$ is the "distance" between the cathode and anode reactions on the potential ladder. The products have a lower potential energy than the reactants (we have moved down the potential ladder) and the cell potential, $E°_{cell}$, has a positive value.

A positive potential calculated for the $Ag^+|Ag$ and $Cu^{2+}|Cu$ cell ($E°_{cell} = +0.46$ V) confirms that the reduction of silver ions in water with copper metal is product-favored (see Chapter Focus). We might ask, however, what would be the value of $E°_{cell}$ if a reactant-favored equation had been selected? For example, what is $E°_{cell}$ for the reduction of copper ions with silver metal?

Cathode, reduction: $Cu^{2+}(aq) + 2\,e^- \longrightarrow Cu(s)$

Anode, oxidation: $2\,Ag(s) \longrightarrow 2\,Ag^+(aq) + 2\,e^-$

Net ionic equation: $2\,Ag(s) + Cu^{2+}(aq) \longrightarrow 2\,Ag^+(aq) + Cu(s)$

Cell Voltage Calculation

$$E°_{cathode} = +0.34 \text{ V} \quad \text{and} \quad E°_{anode} = +0.80 \text{ V}$$

$$E°_{cell} = E°_{cathode} - E°_{anode} = (+0.34 \text{ V}) - (0.80 \text{ V})$$

$$E°_{cell} = -0.46 \text{ V}$$

The negative sign for $E°_{cell}$ indicates that the reaction as written is reactant-favored. The products of the reaction (Ag^+ and Cu) have a higher potential energy than the reactants (Ag and Cu^{2+}). We have moved *up* the potential ladder. For the indicated reaction to occur, a potential of at least 0.46 V would have to be imposed on the system by an external source of electricity (see Section 20.7).

Exercise 20.6 Calculating Standard Cell Potentials

The net reaction that occurs in a voltaic cell is

$$Zn(s) + 2\,Ag^+(aq) \longrightarrow Zn^{2+}(aq) + 2\,Ag$$

Assuming standard conditions, identify the half-reactions that occur at the anode and the cathode and calculate a potential for the cell.

Relative Strengths of Oxidizing and Reducing Agents

Five half-reactions, selected from Table 20.1, are arranged from the half-reaction with the highest (most positive) $E°$ value to the one with the lowest (most negative) value.

E°, V		Reduction Half-Reaction
+1.36		$Cl_2(g) + 2\ e^- \longrightarrow 2\ Cl^-(aq)$
+0.80		$Ag^+(aq) + e^- \longrightarrow Ag(s)$
0.00		$2\ H^+(aq) + 2\ e^- \longrightarrow H_2(g)$
−0.25		$Ni^{2+}(aq) + 2\ e^- \longrightarrow Ni(s)$
−0.76		$Zn^{2+}(aq) + 2\ e^- \longrightarrow Zn(s)$

Increasing strength as oxidizing agents ↑

Look first at the chemical species identified in these half-reactions. The species on the left side of the arrow in these equations are oxidizing agents (Cl_2, Ag^+, H^+, Ni^{2+}, and Zn^{2+}); that is, they can be reduced. On the right are substances (Cl^-, Ag, H_2, Ni, Zn) that are reducing agents; they can be oxidized. Listing half-reactions in this order matches important trends in chemical behavior (see point 3 on page 851).

- The list on the left is headed by Cl_2, an element that is a strong oxidizing agent and is thus easily reduced. At the bottom of the list is Zn^{2+}(aq), an ion not easily reduced and thus a poor oxidizing agent.

- On the right, the list is headed by Cl^-(aq), an ion that can be oxidized to Cl_2 only with difficulty. It is a very poor reducing agent. At the bottom of this list is zinc metal, which is, in contrast, a metal that is quite easy to oxidize and a good reducing agent.

By arranging these half-reactions based on $E°$ values, we have also arranged the chemical species on the two sides of the equation in order of their strengths as oxidizing or reducing agents. In this list, from strongest to weakest, the order is

Oxidizing agents: $Cl_2 > Ag^+ > H^+ > Ni^{2+} > Zn^{2+}$

strong ⟶ *weak*

Reducing agents: $Cl^- < Ag < H_2 < Ni < Zn$

weak ⟶ *strong*

This example illustrates the use of the table of standard reduction potentials to provide information on relative strengths of oxidizing and reducing agents.

Finally, notice that the value of $E°_{cell}$ is greater the further apart the oxidizing and reducing agents are on the potential ladder. For example,

$$Zn(s) + Cl_2(g) \longrightarrow Zn^{2+}(aq) + 2\ Cl^-(aq) \qquad E°_{cell} = +2.123\ V$$

is more strongly product-favored than the reduction of hydrogen ions with nickel metal.

$$Ni(s) + 2\ H^+(aq) \longrightarrow Ni^{2+}(aq) + H_2(g) \qquad E°_{cell} = +0.25\ V$$

Example 20.5 Ranking Oxidizing and Reducing Agents

Problem • Use the table of standard reduction potentials (Table 20.1)

(a) to rank the halogens in order of their strength as oxidizing agents.

(b) to decide if hydrogen peroxide (H_2O_2) in acidic solution is a stronger oxidizing agent than Cl_2.

(c) to decide which of the halogens is capable of oxidizing gold metal to Au^{3+}(aq).

Strategy • The ability of a species on the left side of Table 20.1 to function as an oxidizing agent declines on descending the list (see points 2–4, page 851).

(Example continues on next page)

Solution •

(a) *Ranking halogens according to oxidizing ability.* The halogens (F_2, Cl_2, Br_2, and I_2) appear in the top left portion of the table, with F_2 highest followed in order by the other three species. Their strengths as oxidizing agents are $F_2 > Cl_2 > Br_2 > I_2$.

(b) *Comparing hydrogen peroxide and chlorine.* H_2O_2 lies just below F_2 but well above Cl_2 in the potential ladder (see Table 20.1). Thus, H_2O_2 is a weaker oxidizing agent than F_2 but a stronger one than Cl_2. (Note that the $E°$ value for H_2O_2 refers to an acidic solution and standard conditions.)

(c) *Which halogen will oxidize gold metal to gold(III) ions?* The $Au^{3+}|Au$ half-reaction is listed below the $F_2|F^-$ half-reaction and just above the $Cl_2|Cl^-$ half-reaction. This tells us that, among the halogens, only F_2 is capable of oxidizing Au to Au^{3+} under standard conditions. That is, in the reaction of Au and F_2,

Oxidation, anode: $2[Au(s) \longrightarrow Au^{3+}(aq) + 3\,e^-]$

Reduction, cathode: $3[F_2(aq)\,2\,e^- + \longrightarrow 2\,F^-(aq)]$

Net ionic equation: $3\,F_2(aq) + 2\,Au(s) \longrightarrow$
$$6\,F^-(aq) + 2\,Au^{3+}(aq)$$

$$E°_{cell} = E°_{cathode} - E°_{anode} = +1.37\ V$$

F_2 is a stronger oxidizing agent than Au^{3+}, so the reaction proceeds from left to right as written. (This is confirmed by a positive value of $E°_{cell}$) For the reaction of Cl_2 and Au, Table 20.1 shows us that Cl_2 is a *weaker* oxidizing agent than Au^{3+}, so the reaction would be expected to proceed in the opposite direction.

Oxidation, anode: $2[Au(s) \longrightarrow Au^{3+}(aq) + 3\,e^-]$

Reduction, cathode: $3[Cl_2(aq) + 2\,e^- \longrightarrow 2\,Cl^-(aq)]$

Net ionic equation: $3\,Cl_2(aq) + 2\,Au(s) \longrightarrow$
$$6\,Cl^-(aq) + 2\,Au^{3+}(aq)$$

$$E°_{cell} = E°_{cathode} - E°_{anode} = -0.14\ V$$

This is confirmed by a negative value of $E°_{cell}$.

Comment • In part (c) we calculated $E°_{cell}$ for two reactions. To achieve a balanced net ionic equation we added the half-reactions, but only after multiplying the gold half-reaction by 2 and the halogen half-reaction by 2. (This means 6 mol of electrons were transferred from 2 mol of Au to 3 mol of Cl_2.) Notice that this does not change the value of $E°$ for the half-reactions. As described in point 7 on page 853, cell potentials do not depend on the quantity of material.

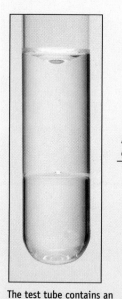

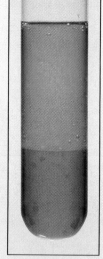

The test tube contains an aqueous solution of KI (top layer) and immiscible CCl_4 (bottom layer).

Add Br_2 to solution of KI and shake.

After adding a few drops of Br_2 in water the I_2 produced collects in the bottom CCl_4 layer and gives it a purple color. (The top layer contains excess Br_2 in water.)

The reaction of bromine and iodide ion. This experiment proves that Br_2 is a better oxidizing agent than I_2. *(Charles D. Winters)*

Exercise 20.7   **Relative Oxidizing and Reducing Ability**

Which metal in the following list is easiest to oxidize? Which metal is the most difficult to oxidize: Fe, Ag, Zn, Mg, Au?

Balanced Equations for Oxidation–Reduction Reactions

A table of standard reduction potentials (such as Table 20.1 or Appendix M) is a valuable resource for writing redox equations (see Section 20.1).

The mechanics of writing a redox equation from half-reactions is illustrated by the oxidation of tin(II) ions by permanganate ions in an acidic solution.

$$Sn^{2+}(aq) + MnO_4^-(aq) \longrightarrow Sn^{4+}(aq) + Mn^{2+}(aq) \qquad \textit{(not balanced)}$$

1. Select two half reactions from the table that include the reactants and products shown in the unbalanced equation. (Here the $MnO_4^-|Mn^{2+}$ half-reaction should be taken from the list of reactions occurring in acid.) Use one as it appears in the table (a reduction reaction) and write the second in the reverse direction (an oxidation process).

   Reduction, cathode: $MnO_4^-(aq) + 8 H^+(aq) + 5 e^- \longrightarrow Mn^{2+}(aq) + 4 H_2O(\ell)$

   Oxidation, anode: $Sn^{2+}(aq) \longrightarrow Sn^{4+}(aq) + 2 e^-$

2. Multiply one or both equations by an integer so that when the two equations are added together the electrons will cancel out. Here the reduction half-reaction is multiplied by 2 and the oxidation half-reaction by 5. This means that 10 mol of electrons are transferred to 2 mol of MnO_4^- ions from 5 mol of Sn^{2+} ions.

$$2\left[MnO_4^-(aq) + 8 H^+(aq) + 5 e^- \longrightarrow Mn^{2+}(aq) + 4 H_2O(\ell)\right]$$
$$5\left[Sn^{2+}(aq) \longrightarrow Sn^{4+}(aq) + 2 e^-\right]$$

3. Add the two half-reactions together. Simplify, if necessary, and check the result for mass and charge balance.

$$2 MnO_4^-(aq) + 16 H^+(aq) + 5 Sn^{2+}(aq) \longrightarrow 2 Mn^{2+}(aq) + 8 H_2O(\ell) + 5 Sn^{4+}(aq)$$

A redox equation constructed from two randomly chosen half-reactions could be either product- or reactant-favored. We can determine which in several ways. One way utilizes Equation 20.1 to calculate a cell voltage from the standard reduction potentials of the two half-reactions. For the Sn^{2+}/MnO_4^- reaction,

$$E^\circ_{cell} = E^\circ_{cathode} - E^\circ_{anode} = (+1.52 \text{ V}) - (+0.15 \text{ V}) = +1.37 \text{ V}$$

The positive value indicates that the reaction is product-favored.

Selecting the half-reaction higher on the potential ladder as the oxidizing agent in the overall reaction (the cathode reaction) will automatically ensure a product-favored process. To illustrate, $MnO_4^-(aq)$, a strong oxidizing agent, is high on the list of oxidizing agents. It is capable of oxidizing any species below the $MnO_4^-|Mn^{2+}$ half-reaction and on the right side of the table. Thus, permanganate is capable of oxidizing such species as Cl^-, Br^-, Sn^{2+}, Sn, Fe, Zn, Al, and Li. In contrast, a reaction between $MnO_4^-(aq)$ and F^- (a species on the right, *higher* up on the potential ladder) is reactant-favored. The following example further illustrates this point.

Example 20.6   **Using a Table of Standard Reduction Potentials to Predict Chemical Reactions**

Problem •

(a) Which of the following metals will react with $H^+(aq)$ to produce H_2 in a product-favored reaction: Cu, Al, Ag, Fe, Zn?

(b) Select from Table 20.1 three oxidizing agents that are capable of oxidizing $Cl^-(aq)$ to Cl_2.

Strategy • Recall that the reaction between any substance on the left in Table 20.1 (an oxidizing agent) with any substance lower than it on the right (a reducing agent) is product-favored under standard conditions (see point 5 on page 851).

Solution •

(a) Hydrogen ion is an oxidizing agent, so product-favored reactions will occur between H^+ and reducing agents found on the right side of the table and below the $H^+(aq)|H_2(g)$ half-reaction. Of the metals on our list, three meet this criterion: Al, Fe, and Zn (see figure). In contrast, Cu and Ag are located above the $H^+(aq)|H_2$ half-reaction in the table. Reactions of these metals and $H^+(aq)$ are not product-favored.

To confirm these conclusions, E°_{cell} has been calculated for each reaction.

(Example continues on next page)

The reaction of zinc metal with acid is product-favored. (Charles D. Winters)

Reaction	$E°_{cell}$ V
Product-favored	
$2 Al(s) + 6 H^+(aq) \longrightarrow 2 Al^{3+}(aq) + 3 H_2(g)$	+1.66
$Zn(s) + 2 H^+(aq) \longrightarrow Zn^{2+}(aq) + H_2(g)$	+0.763
$Fe(s) + 2 H^+(aq) \longrightarrow Fe^{2+}(aq) + H_2(g)$	+0.44

Reaction	$E°_{cell}$ V
Reactant-favored	
$Cu(s) + 2 H^+(aq) \longrightarrow Cu^{2+}(aq) + H_2(g)$	−0.337
$2 Ag(s) + 2 H^+(aq) \longrightarrow 2 Ag^+(aq) + H_2(g)$	−0.80

(b) Locate the $Cl_2(g)|Cl^-(aq)$ half-reaction in Table 20.1. Chloride ion, $Cl^-(aq)$, is quite high on the list of species that can be oxidized, so only strong oxidizing agents are capable of oxidizing this ion to Cl_2. (Conversely, Cl_2 is a strong oxidizing agent, so an oxidant even more powerful than Cl_2 is required to convert Cl^- to Cl_2.) Five substances in Table 20.1 can oxidize $Cl^-(aq)$ to Cl_2: F_2, H_2O_2, $PbO_2(s)$, $MnO_4^-(aq)$, and $Au^{3+}(aq)$. (Notice that reactions with H_2O_2, $PbO_2(s)$, or $MnO_4^-(aq)$ require acid conditions.)

$$H_2O_2(aq) + 2 H^+(aq) + 2 Cl^-(aq) \longrightarrow 2 H_2O(\ell) + Cl_2(g)$$

Comment • We checked our predictions by calculating $E°_{cell}$. (The reaction of Cl^- with H_2O_2, for example, has a positive potential ($E°_{cell} = +0.41$ V) as expected for a product-favored reaction.) A calculation is not necessary, however, as the position of half-reactions in the table is sufficient to decide if a reaction is product- or reactant-favored.

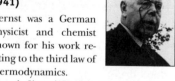
Exercise 20.8   **Using a Table of Standard Reduction Potentials to Predict Chemical Reactions**

Determine whether the following redox equations are product-favored. Assume standard conditions.

(a) $Ni^{2+}(aq) + H_2(g) \longrightarrow Ni(s) + 2 H^+(aq)$

(b) $2 Fe^{3+}(aq) + 2 I^-(aq) \longrightarrow 2 Fe^{2+}(aq) + I_2(s)$

(c) $Br_2(\ell) + 2 Cl^-(aq) \longrightarrow 2 Br^-(aq) + Cl_2(g)$

(d) $Cr_2O_7^{2-}(aq) + 6 Fe^{2+}(aq) + 14 H^+(aq) \longrightarrow 2 Cr^{3+}(aq) + 6 Fe^{3+}(aq) + 7 H_2O(\ell)$

20.5   ELECTROCHEMICAL CELLS UNDER NONSTANDARD CONDITIONS

Electrochemical cells seldom operate under standard conditions in the real world. Even if the cell is constructed with all dissolved species at 1 M, reactant concentrations decrease and product concentrations increase in the course of the reaction. Changing concentrations of reactants and products will affect the cell voltage. Thus, we need to ask what happens to cell potentials under nonstandard conditions [CD-ROM, Screen 20.7].

The Nernst Equation

Based on both theory and experimental results, it has been determined that cell potentials are related to concentrations of reactants and products, and to temperature, by Equation 20.2:

$$E = E° - (RT/nF) \ln Q \qquad \text{(20.2)}$$

In this equation, known as the **Nernst equation,** R is the gas constant (8.314510 J/K · mol), T is the temperature (K), and n is the number of moles of electrons transferred between oxidizing and reducing agents (as determined by the balanced equation for the reaction). The symbol F represents the **Faraday constant** (9.6485309×10^4 J/V · mol). *One Faraday is the quantity of electric charge carried by one mole of electrons.* The term Q is the reaction quotient, an expression relating the concentrations of the products and reactants raised to an appropriate power as defined by the stoichiometric coefficients in the balanced, net equation [← EQUATION 16.1, SECTION 16.2]. Substituting values for the constants in Equation 20.2, and using 298 K as the temperature, gives Equation 20.3:

$$E = E° - \frac{0.0257 \text{ V}}{n} \ln Q \qquad \text{at } 25°C \qquad \text{(20.3)}$$

In essence, the Nernst equation "corrects" the standard potential $E°$ for nonstandard conditions of concentrations at 298 K.

Example 20.7 **Using the Nernst Equation**

Problem • A voltaic cell is set up at 25 °C with the half-cells $Al^{3+}(0.0010 \text{ M})|Al$ and $Ni^{2+}(0.50 \text{ M})|Ni$. Write an equation for the reaction that occurs when the cell generates an electric current and determine the cell potential.

Strategy • The first step is to determine which substance is oxidized (Al or Ni) by looking at the appropriate half-reactions in Table 20.1 and deciding which is the better reducing agent (see Examples 20.5 and 20.6). Next, add the half-reactions to determine the net ionic equation and calculate $E°$. Finally, use the Nernst equation to calculate E, the nonstandard potential.

Solution • Aluminum metal is a stronger reducing agent than Ni metal. (Conversely, Ni^{2+} is a better oxidizing agent than Al^{3+}.) Therefore, Al is oxidized and the $Al^{3+}|Al$ compartment is the anode.

Cathode, reduction: $3[Ni^{2+}(aq) + 2 e^- \longrightarrow Ni(s)]$

Anode, oxidation: $2[Al(s) \longrightarrow Al^{3+}(aq) + 3 e^-]$

Net ionic equation: $2 \text{ Al}(s) + 3 \text{ Ni}^{2+}(aq) \longrightarrow$
$$2 \text{ Al}^{3+}(aq) + 3 \text{ Ni}(s)$$

$$E°_{cell} = E°_{cathode} - E°_{anode}$$
$$E°_{cell} = (-0.25 \text{ V}) - (-1.66 \text{ V}) = 1.41 \text{ V}$$

The expression for Q is written based on the cell reaction. In the net reaction, $Al^{3+}(aq)$ has a coefficient of 2, so this concentration is squared. Similarly, $[Ni^{2+}(aq)]$ is cubed. Solids are not included in the expression for Q [← SECTION 16.2].

$$Q = \frac{[Al^{3+}]^2}{[Ni^{2+}]^3}$$

The net equation requires transfer of 6 electrons from two Al atoms to three Ni^{2+} ions, so $n = 6$. Substituting for $E°$, n, and Q in the Nernst equation gives

$$E_{cell} = E°_{cell} - \frac{0.0257}{n} \ln \frac{[Al^{3+}]^2}{[Ni^{2+}]^3}$$

$$E_{cell} = +1.41 \text{ V} - \frac{0.0257}{6} \ln \frac{[0.0010]^2}{[0.50]^3}$$

$$E_{cell} = +1.41 \text{ V} - 0.00428 \ln (8.0 \times 10^{-6})$$

$$E_{cell} = +1.41 \text{ V} - 0.00428 \, (-1.17)$$

$$E_{cell} = +1.41 \text{ V} + 0.050 \text{ V}$$

$$E_{cell} = +1.46 \text{ V}$$

Comment • Notice that E_{cell} is larger than $E°_{cell}$ because the product concentration, $[Al^{3+}]$, is much smaller than 1.0 M. Generally, when product concentrations are smaller initially than the reactant concentrations in a product-favored reaction, the cell potential is *more* positive than $E°$.

Exercise 20.9 **Using the Nernst Equation**

The half-cells $Fe^{2+}(aq, 0.024 \text{ M})|Fe(s)$ and $H^+(aq, 0.056 \text{ M})|H_2$ (g, 1.0 bar) are linked by a salt bridge to create a voltaic cell. Determine the cell potential (E_{cell}) at 298 K.

Example 20.7 demonstrates the calculation of a cell potential if concentrations are known. It is also useful to apply the Nernst equation in the opposite sense, using a measured cell potential to determine an unknown concentration. A device that does just this is the pH meter (Figure 20.12). In an electrochemical cell in which $H^+(aq)$ is a reactant or product, the cell voltage will vary predictably with the hydrogen ion concentration. The cell voltage is measured and the value used to calculate pH. Example 20.8 illustrates how E_{cell} varies with the hydrogen ion concentration in a simple cell.

Example 20.8 Variation of E_{cell} with Concentration-Measuring pH

Problem • A voltaic cell is set up with copper and hydrogen half-cells. Standard conditions are employed in the copper half-cell, Cu^{2+}(aq, 1.0 M)|Cu(s). The hydrogen gas pressure is 1.00 bar, whereas [H^+(aq)] in the hydrogen half-cell is the unknown. A value of 0.49 V is recorded for E_{cell} at 298 K. Determine the pH of the solution.

Strategy • We first decide which is the better oxidizing and reducing agent in order to decide on the net reaction occurring in the cell. With this known, E°_{cell} can be calculated. The only unknown quantity in the Nernst equation is the concentration of hydrogen ion, from which we can calculate the solution pH.

Solution • Hydrogen is a better reducing agent than copper metal, so Cu(s)|Cu^{2+}(aq, 1.0 M) is the cathode, and H_2(g, 1.00 bar)|H^+(aq, ?M) is the anode.

Cathode, reduction: $Cu^{2+}(aq) + 2\,e^- \longrightarrow Cu(s)$

Anode, oxidation: $H_2(g) \longrightarrow 2\,H^+(aq) + 2\,e^-$

Net ionic equation: $H_2(g) + Cu^{2+}(aq) \longrightarrow Cu(s) + 2\,H^+(aq)$

$$E^\circ_{cell} = E^\circ_{cathode} - E^\circ_{anode}$$

$$E^\circ_{cell} = (+0.34\ V) - (0.00\ V) = +0.34\ V$$

The reaction quotient Q is derived from the balanced net ionic equation.

$$Q = \frac{[H^+]^2}{[Cu^{2+}]P_{H_2}}$$

The net equation requires transfer of two electrons so $n = 2$. The value of [Cu^{2+}] is 1.00 M (standard conditions), but [H^+] is unknown. Substitute this information into the Nernst equation (and don't overlook the fact that [H^+] is squared in the expression for Q).

$$E_{cell} = E^\circ_{cell} - \frac{0.0257}{n} \ln \frac{[H^+]^2}{[Cu^{2+}]P_{H_2}}$$

$$0.49\ V = 0.34\ V - \frac{0.0257}{2} \ln \frac{[H^+]^2}{(1.0)(1.0)}$$

$$-12 = \ln [H^+]^2$$

$$[H^+] = 2.5 \times 10^{-3}\ M$$

$$pH = 2.61$$

Exercise 20.10 Variation of E_{cell} with Concentration

A voltaic cell is set up with an aluminum electrode in a 0.025 M Al(NO$_3$)$_3$(aq) solution and an iron electrode in a 0.50 M Fe(NO$_3$)$_2$(aq) solution. Calculate the voltage produced by this cell.

In practice, using a hydrogen electrode in a pH meter is not practical. The apparatus is clumsy, anything but robust, and platinum (for the electrode) is costly. Common pH meters now use a glass electrode, so called because it contains a thin glass membrane separating the cell from the solution whose pH is to be measured (Figure 20.12). Inside the glass electrode is a silver wire coated with AgCl and a solution of HCl; outside is the solution of unknown pH to be evaluated. A calomel electrode, a common reference electrode using a mercury–mercury(I) redox couple (Hg$_2$Cl$_2$|Hg), is the second electrode of the cell. The potential across the glass membrane depends on [H^+]. Common pH meters give a direct readout of pH.

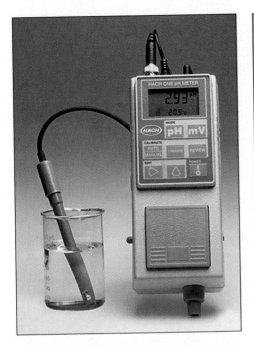

Figure 20.12 A pH meter and pH electrode. *(Charles D. Winters)*

Concentrations of ions other than $H^+(aq)$ can be measured using electrochemistry. Collectively, electrodes used to measure ion concentrations are known as **ion-selective electrodes.** In many areas of the United States, houses are equipped with water softeners. Their function is to remove ions such as Ca^{2+} and Mg^{2+} from household water and replace the alkaline earth cations with Na^+ ions. This is done using a material called an "ion-exchange resin." To test if the resin is functioning, the water is periodically sampled for Ca^{2+} ions, and one type of water softener has a built-in ion-selective electrode to detect the concentration of this ion. When the electrode indicates that the Ca^{2+} concentration has reached a designated level, it sends a signal to begin regenerating the ion-exchange resin.

20.6 ELECTROCHEMISTRY AND THERMODYNAMICS

Work and Free Energy

The first law of thermodynamics [◄ SECTION 6.4] states that the internal energy change in a system (ΔE) is related to two quantities, heat (q) and work (w): $\Delta E = q + w$. This equation also applies to chemical changes that occur in a voltaic cell. As current flows, energy is transferred from the system (the voltaic cell) to the surroundings [CD-ROM, Screen 20.8].

In a voltaic cell, the decrease in internal energy in the system will manifest itself ideally as electric work done on the surroundings by the system. The maximum work done by an electrochemical system (ideally, assuming no heat is generated) is proportional to the potential difference (volts) and the quantity of charge (coulombs).

$$w_{max} = nFE$$

E in this equation is the cell voltage and the quantity nF is the quantity of electric charge (in units of coulombs) transferred from anode to cathode.

History

Michael Faraday (1791–1867)

The terms "anion," "cation," "electrode," and "electrolyte" originated with Michael Faraday, one of the most influential men in the history of chemistry. He made fundamental discoveries in

- Electromagnetic induction, which led to the invention of the first transformer and electric motor
- The laws of electrolysis
- The discovery of the magnetic properties of matter
- The discovery of benzene and other organic chemicals
- The introduction of the concept of electric and magnetic fields

(Oesper Collection in the History of Chemistry/ University of Cincinnati)

The free energy change for a process is, by definition, the maximum amount of work that can be obtained [← SECTION 19.6]. Because the maximum work and the cell potential are related, $E°$ and $\Delta G°$ can be related mathematically (taking care to assign signs correctly). The maximum work done on the surroundings when electricity is produced by a voltaic cell is $+nFE$, the positive sign denoting an increase in energy in the surroundings. The energy content of the cell decreases by this amount. Thus, ΔG for the voltaic cell has the opposite sign in the following equation.

$$\Delta G = -nFE$$

Under standard conditions, the appropriate equation is

$$\Delta G° = -nFE° \tag{20.4}$$

• **Units in Equation 20.4**
n has units of mol e⁻ and F has units of (coulombs/mol e⁻). Therefore, nF has units of coulombs (C). Because 1 J = 1 C · V, the product nFE will have units of energy (J).

This shows that the more positive the value of $E°_{cell}$, the more negative the value of $\Delta G°$ for the reaction becomes. That is, the further apart the half-reactions are on the potential ladder, the more strongly product-favored the reaction.

Example 20.9 | **The Relation Between $E°$ and $\Delta G°$**

Problem • The standard cell potential $E°_{cell}$ for the reduction of silver ions with copper metal (Figure 20.2) is $+0.46$ V at 25 °C. Calculate $\Delta G°$ for this reaction.

Strategy • We use Equation 20.4, where F is a constant, and $E°_{cell}$ is given in the problem. We need to know the value of n, the number of moles of electrons transferred between copper metal and silver ions in the balanced equation.

Solution • In this cell, copper is the anode and silver the cathode. The overall cell reaction is

$$Cu(s) + 2\,Ag^+(aq) \longrightarrow Cu^{2+}(aq) + 2\,Ag(s)$$

which means that each mole of copper transfers two moles of electrons to two moles of Ag^+ ions. That is, $n = 2$. Now use Equation 20.4.

$$\Delta G° = -nFE°$$
$$= -(2\text{ mol e}^-)(96{,}500\text{ C/mol e}^-)(0.46\text{ V})$$
$$= -89{,}000\text{ C} \cdot V$$

Because 1 C · V = 1 J, we have

$$\Delta G° = -89{,}000\text{ J or } -89\text{ kJ}$$

Comment • This represents a very effective method of obtaining thermodynamic values from relatively simple electrochemical experiments.

Exercise 20.11 | **The Relation Between $E°$ and $\Delta G°_{rxn}$**

The following reaction has an $E°_{cell}$ value of -0.76 V.

$$H_2(g) + Zn^{2+}(aq) \longrightarrow Zn(s) + 2\,H^+(aq)$$

Calculate $\Delta G°$ for this reaction. Is the reaction product- or reactant-favored?

$E°$ and the Equilibrium Constant

When a voltaic cell produces an electric current, the reactant concentrations decrease and the product concentrations increase. The cell voltage also changes; as reactants are converted to products, the value of E_{cell} decreases. Eventually the cell potential reaches zero, no further reaction occurs, and equilibrium has been reached.

This situation can be analyzed using the Nernst equation. When $E_{cell} = 0$, the reactants and products are at equilibrium and the reaction quotient Q is equal to the equilibrium constant, K. Substituting the appropriate symbols and values into the Nernst equation

● **K and E°**
The further apart half-reactions are on the potential ladder, the larger the value of K.

$$E = 0 = E° - \frac{0.0257 \text{ V}}{n} \ln K$$

and collecting terms gives an equation that relates the cell potential and equilibrium constant (Equation 20.5).

$$\ln K = \frac{nE°}{0.0257 \text{ V}} \quad \text{at 25 °C} \qquad \textbf{(20.5)}$$

This equation can be used to determine values for equilibrium constants, as illustrated in the Example and Exercise below [⟳ CD-ROM, Screen 20.8, *Tutorial 2*].

Example 20.10 E° and Equilibrium Constants

Problem • Calculate the equilibrium constant for the reaction

$$Fe(s) + Cd^{2+}(aq) \rightleftharpoons Fe^{2+}(aq) + Cd(s)$$

Strategy • First determine $E°_{cell}$ from $E°$ values for the two half-reactions (see Examples 20.5 and 20.6). This also gives us the value of n, the other parameter required in Equation 20.5.

Solution • The half-reactions and $E°$ values are

Cathode, reduction: $Cd^{2+}(aq) + 2\,e^- \longrightarrow Cd(s)$

Anode, oxidation: $Fe(s) \longrightarrow Fe^{2+}(aq) + 2\,e^-$

Net ionic equation: $Fe(s) + Cd^{2+}(aq) \rightleftharpoons Fe^{2+}(aq) + Cd(s)$

$E°_{cell} = E°_{cathode} - E°_{anode}$

$E°_{cell} = (-0.40 \text{ V}) - (-0.44 \text{ V}) = +0.04 \text{ V}$

Now substitute $n = 2$ and $E°_{cell}$ into Equation 20.5.

$$\ln K = \frac{nE°}{0.0257 \text{ V}} = \frac{(2)(0.04 \text{ V})}{0.0257 \text{ V}} = 3$$

$$K = 20$$

Comment • The relatively small positive voltage (0.040 V) for the standard cell indicates that the cell reaction is only mildly product-favored. A value of 20 for the equilibrium constant is in accord with this observation.

Exercise 20.12 E° and Equilibrium Constants

Calculate the equilibrium constant for the reaction

$$2\,Ag^+(aq) + Hg(\ell) \rightleftharpoons 2\,Ag(s) + Hg^{2+}(aq)$$

The relationship between $E°$ and K can be used to obtain equilibrium constants for many different chemical systems. For example, let us construct an electrode in which an insoluble ionic compound is a component of the half-cell. For this purpose a silver electrode with a surface layer of AgCl can be prepared. The reaction occurring at this electrode is then

$$AgCl(s) + e^- \longrightarrow Ag(s) + Cl^-(aq)$$

The standard reduction potential for this half-cell (Appendix M) is +0.222 V. When this half-reaction is paired with a standard silver electrode in an electrochemical cell, the cell reactions are

Cathode, reduction: $AgCl(s) + e^- \longrightarrow Ag(s) + Cl^-(aq)$

Anode, oxidation: $\underline{Ag(s) \longrightarrow Ag^+(aq) + e^-}$

Net ionic equation: $AgCl(s) \longrightarrow Ag^+(aq) + Cl^-(aq)$

$$E^\circ_{cell} = E^\circ_{cathode} - E^\circ_{anode} = (+0.22\text{ V}) - (+0.80\text{ V}) = -0.58\text{ V}$$

The equation for the net reaction represents the equilibrium of solid AgCl and its ions. The cell potential is negative, indicating a reactant-favored process, as expected based on the low solubility of AgCl. Using Equation 20.5, the value of the equilibrium constant [$\leftarrow K_{sp}$, SECTION 19.7] can be obtained from E°_{cell}.

$$\ln K = \frac{nE^\circ}{0.0257\text{ V}} = \frac{(1)(-0.58\text{ V})}{0.0257\text{ V}} = -23$$

$$K_{sp} = e^{-23} = 1.0 \times 10^{-10}$$

Exercise 20.13   **Determining an Equilibrium Constant**

In Appendix M the following standard reduction potential is reported:

$$[Zn(CN)_4]^{2-}(aq) + 2\,e^- \longrightarrow Zn(s) + 4\,CN^-(aq) \qquad E^\circ = -1.26\text{ V}$$

Use this information along with the data on the $Zn^{2+}(aq)|Zn(s)$ half-cell to calculate the equilibrium constant for the reaction

$$Zn^{2+}(aq) + 4\,CN^-(aq) \rightleftharpoons [Zn(CN)_4]^{2-}(aq)$$

The value calculated is the formation constant for this complex ion.

Chapter Goals • Revisited

- Balance net ionic equations for oxidation–reduction reactions.
- Understand the principles of voltaic cells.
- Understand how to use electrochemical potentials.
- **Explore electrolysis, the use of electrical energy to produce chemical change.**

20.7   ELECTROLYSIS: CHEMICAL CHANGE USING ELECTRICAL ENERGY

Electrolysis of water (Figure 20.13a) is a classic chemistry experiment. An electric current is passed through water containing a small amount of an electrolyte, and gaseous hydrogen and oxygen are formed at the electrodes. This experiment is used to illustrate stoichiometry and gas laws and to show how an energetically disfavored reaction can be carried out [CD-ROM, Screen 20.11].

Electroplating (Figure 20.13b) is another example of electrolysis. Here, an electric current is passed through a solution containing a salt of the metal to be plated. The object to be plated is the cathode, and when metal ions in solution are reduced, the metal deposits on its surface.

Electrolysis is also widely used in the refining of metals such as aluminum and in the production of important commodity chemicals such as chlorine. These important topics are visited in Chapter 21.

Electrolysis of Molten Salts

In an electrolysis experiment, the material to be electrolyzed, either a molten salt or a solution, is contained in an electrolysis cell. As was the case with voltaic cells, ions must be present in the liquid or solution for a current to flow. The movement

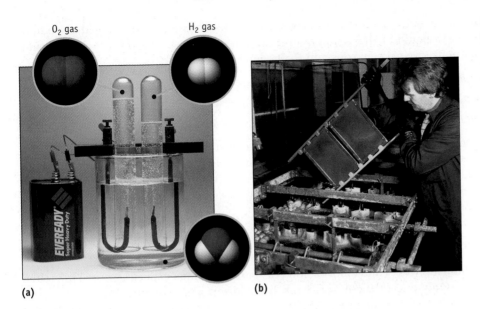

(a)   (b)

Figure 20.13   Electrolysis. (a) Electrolysis of water produces hydrogen and oxygen gas. **(b)** Electroplating adds a layer of metal to the surface of an object, either to protect it from corrosion or to improve the physical appearance. The procedure uses an electrolysis cell set up with the object to be plated as the cathode and a solution containing a salt of the metal to be plated. *(a, Charles D. Winters; b, Tom Hollyman, Photo Researchers, Inc.)*

of ions constitutes the electric current within the cell. The cell has two electrodes that are connected to a source of DC (direct-current) voltage. If a high enough voltage is applied, chemical reactions occur at the two electrodes. Reduction occurs at the negatively charged cathode, with electrons being transferred from that electrode to a chemical species in the cell. Oxidation occurs at the positive anode with electrons from a chemical species being transferred to that electrode.

Let us first focus our attention on the chemical reactions that occur at each electrode in the electrolysis of a molten salt. In molten NaCl (Figure 20.14) sodium ions (Na^+) and chloride ions (Cl^-) ion are freed from their rigid arrangement in the crystalline lattice above 800 °C. If a potential is applied to the cell, sodium ions are attracted to the negative electrode and chloride ions are attracted to the positive electrode. Applying a potential results in chemical reactions at each electrode.

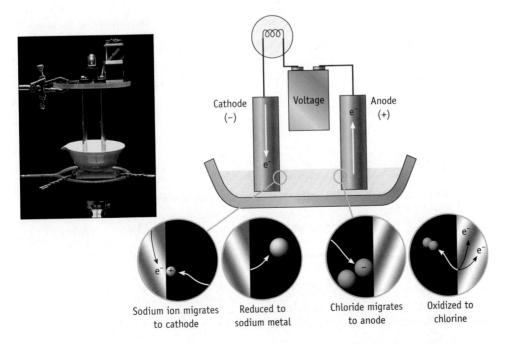

Cathode (−)   Voltage   Anode (+)

Sodium ion migrates to cathode   Reduced to sodium metal   Chloride migrates to anode   Oxidized to chlorine

Figure 20.14   The preparation of sodium and chlorine by the electrolysis of molten NaCl. In the molten state, sodium ions migrate to the negative cathode where they are reduced to sodium metal. Chloride ions migrate to the positive anode where they are oxidized to elemental chlorine. *(Charles D. Winters)*

At the negative cathode, Na^+ ions accept electrons and are reduced to sodium metal (a liquid at this temperature). Simultaneously, at the positive anode, chloride ions give up electrons and form elemental chlorine.

Cathode (−), reduction:	$2\,Na^+ + 2\,e^- \longrightarrow 2\,Na(\ell)$
Anode (+), oxidation:	$2\,Cl^- \longrightarrow Cl_2(g) + 2\,e^-$
Net reaction:	$2\,Na^+ + 2\,Cl^- \longrightarrow 2\,Na(\ell) + Cl_2(g)$

Electrons move through the external circuit under the force exerted by the applied potential, and the movement of positive and negative ions in the molten salt constitutes the current within the cell. Finally, it is important to recognize that the reaction is *not* product-favored. The energy required for this reaction to occur has been provided by the electric current.

Electrolysis of Aqueous Solutions

Sodium ion (Na^+) and chloride ion (Cl^-) are the primary species present in molten NaCl. Only one of these (Cl^-) can be oxidized and only one (Na^+) can be reduced. Electrolyses of aqueous solutions are more complicated than electrolyses of molten salts, however, because water is now present. Water is an *electroactive* substance; it can be oxidized or reduced in a electrochemical process.

    Consider the electrolysis of aqueous sodium iodide (Figure 20.15). In this experiment the electrolysis cell contains $Na^+(aq)$, $I^-(aq)$, and H_2O molecules. Table 20.1 shows that possible *reduction reactions* at the *cathode* include

$$Na^+(aq) + e^- \longrightarrow Na(s)$$
$$2\,H_2O(\ell) + 2\,e^- \longrightarrow H_2(g) + 2\,OH^-(aq)$$

Possible *oxidation reactions* at the *anode* are

$$2\,I^-(s) \longrightarrow I_2(aq) + 2\,e^-$$
$$2\,H_2O(\ell) \longrightarrow O_2(g) + 4\,H^+(aq) + 4\,e^-$$

In the electrolysis of aqueous NaI, experiment shows (Figure 20.15) that $H_2(g)$ and $OH^-(aq)$ are formed by reduction of water at the cathode, and iodine is formed at the anode. Thus, the overall cell process can be summarized by the following equations:

Cathode(−), reduction:	$2\,H_2O(\ell) + 2\,e^- \longrightarrow H_2(g) + 2\,OH^-(aq)$
Anode(+), oxidation:	$2\,I^-(s) \longrightarrow I_2(aq) + 2\,e^-$
Net ionic equation:	$2\,H_2O(\ell) + 2\,I^-(s) \longrightarrow H_2(g) + 2\,OH^-(aq) + I_2(aq)$

where E°_{cell} has a negative value.

Problem-Solving Tip 20.3

Electrochemical Conventions: Voltaic Cells and Electrolysis Cells

Whether you are describing a voltaic cell or electrolysis cell, the terms "anode" and "cathode" always refer to the electrodes at which oxidation and reduction occur, respectively. The polarity of the electrodes is reversed, however.

Type of Cell	Electrode	Function	Polarity
Voltaic	Anode	Oxidation	−
	Cathode	Reduction	+
Electrolysis	Anode	Oxidation	+
	Cathode	Reduction	−

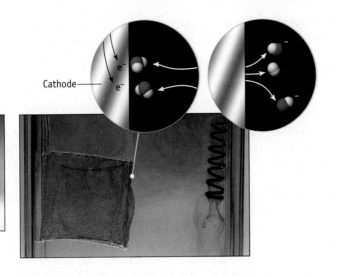

Figure 20.15 **Electrolysis of aqueous NaI.** A solution of NaI(aq) is electrolyzed by a potential applied by an external source of electricity. A drop of an acid-base indicator (phenolphthalein) has been added in this experiment so that the formation of $OH^-(aq)$ can be detected. Iodine forms at the anode, and H_2 and OH^- form at the cathode. (The hydroxide ion formed in the solution is detected by the red color of the indicator.) *(Charles D. Winters)*

Cathode

$$E^\circ_{cell} = E^\circ_{cathode} - E^\circ_{anode} = (-0.828\ V) - (+0.535\ V) = -1.363\ V$$

This is a reactant-favored process, which means a potential of *at least* 1.363 V must be *applied* to the cell in order for these reactions to occur. If the process had involved the oxidation of water instead of iodide ion at the anode, the required potential would be -2.06 V [$E^\circ_{cathode} - E^\circ_{anode} = (-0.83\ V) - (+1.23\ V)$]. The reaction occurring is the one requiring the smaller applied potential, so the net cell reaction in the electrolysis of NaI(aq) is the oxidation of iodide and reduction of water.

What happens if an aqueous solution of some other metal halide such as $SnCl_2$ is electrolyzed? Consult Table 20.1, and, considering all possible half-reactions, find the net reaction that requires the smallest (least negative) applied potential. In this case aqueous Sn^{2+} ion is much more easily reduced ($E^\circ = -0.14$ V) than water ($E^\circ = -0.83$ V), so tin metal is produced at the cathode. At the anode, there are two possible oxidations, $Cl^-(aq)$ to $Cl_2(g)$ or $H_2O(\ell)$ to $O_2(g)$. Experiments show that chloride ion is generally oxidized in preference to water, so the reactions occurring on electrolysis of aqueous tin(II) chloride are (Figure 20.16)

Cathode, reduction: $Sn^{2+}(aq) + 2\ e^- \longrightarrow Sn(s)$

Anode, oxidation: $\underline{2\ Cl^-(aq) \longrightarrow Cl_2(g) + 2\ e^-}$

Net reaction: $Sn^{2+}(aq) + 2\ Cl^-(aq) \longrightarrow Sn(s) + Cl_2(g)$

$$E^\circ_{cell} = E^\circ_{cathode} - E^\circ_{anode} = (-0.14\ V) - (+1.36\ V) = -1.50\ V$$

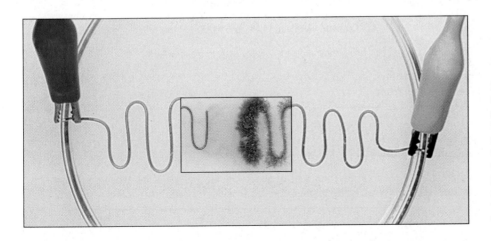

Figure 20.16 **Electrolysis of aqueous tin(II) chloride.** Tin metal collects at the negative cathode. Chlorine gas collects at the positive anode. Elemental chlorine is formed in the cell in spite of the fact that the potential for the oxidation of Cl^- is more negative than that for oxidation of water. (That is, chlorine should be less readily oxidized than water.) This is the result of relative reaction rates and illustrates the complexity of some aqueous electrochemistry. *(Charles D. Winters)*

Formation of Cl_2 at the anode in the electrolysis of $SnCl_2(aq)$ is contrary to a prediction based on $E°$ values. If the electrode reactions were

Cathode, reduction: $Sn^{2+}(aq) + 2 e^- \longrightarrow Sn(s)$

Anode, oxidation: $2 H_2O(\ell) \longrightarrow O_2(g) + 4 H^+(aq) + 4 e^-$

$$E°_{cell} = (-0.14 \text{ V}) - (+1.23 \text{ V}) = -1.37 \text{ V}$$

a smaller applied potential is seemingly required. To explain the formation of chlorine instead of oxygen, we must take into account rates of reaction. In the commercially important electrolysis of aqueous NaCl, a voltage high enough to oxidize both Cl^- and H_2O will be used. However, chloride ion is oxidized much faster than H_2O, and Cl_2, the less favored product based on reduction potentials, is the major product in this electrolysis.

Another instance in which rates are important concerns electrode materials. Graphite, commonly used to make inert electrodes, can be oxidized. For the half-reaction $CO_2(g) + 4 H^+(aq) + 4 e^- \longrightarrow C(s) + 2 H_2O(\ell)$, $E°$ is $+0.20$ V; this indicates that carbon is slightly easier to oxidize than copper ($E° = +0.34$ V), for example. Based on this value, oxidation of a graphite electrode might reasonably be expected to occur during an electrolysis. And indeed it does, albeit slowly; graphite electrodes used in electrolysis cells slowly deteriorate and eventually have to be replaced.

One other factor, the concentration of electroactive species in solution, must be taken into account when discussing electrolyses. As shown in Section 20.6, the potential at which a species in solution is oxidized or reduced depends on concentration. Unless standard conditions are used, predictions based on $E°$ values are only qualitative.

● **Overvoltage**
Voltages higher than the minimum are typically used to speed up reactions that would otherwise be slow. The term "overvoltage" is often used when describing experiments; this refers to the voltage needed to make a reaction occur at a reasonable rate.

Example 20.11   **Electrolysis of Aqueous Solutions**

Problem • Predict how products of electrolysis of aqueous solutions of NaF, NaCl, NaBr, and NaI are likely to be different. (The electrolysis of NaI is illustrated in Figure 20.15.)

Strategy • The main criterion used to predict the chemistry in an electrolytic cell should be the ease of oxidation and reduction, an assessment based on values of $E°$.

Solution • The cathode reaction in all four examples presents no problem — water is reduced to hydroxide ion in preference to reduction of $Na^+(aq)$ (as in the electrolysis of aqueous NaI).

Thus, the primary cathode reaction in all cases is:

$$2 H_2O(\ell) + 2 e^- \longrightarrow H_2(g) + 2 OH^-(aq)$$
$$E°_{cathode} = -0.83 \text{ V}$$

At the anode, we need to assess the ease of oxidation of halide ions relative to water. Based on $E°$ values, this is $I^-(aq) > Br^-(aq) > Cl^-(aq) \gg F^-(aq)$. Fluoride ion is much more difficult

to oxidize than water, and electrolysis of an aqueous solution containing this ion results exclusively in O_2 formation.

The primary anode reaction in NaF(aq) is:

$$2 H_2O(\ell) \longrightarrow O_2(g) + 4 H^+(aq) + 4 e^-$$

In this case,

$$E°_{cell} = (-0.83 \text{ V}) - (+1.23 \text{ V}) = -2.06 \text{ V}$$

Bromide and iodide ion are considerably easier to oxidize than chloride ion. Recalling that chlorine is the primary product in the electrolysis of chloride salts (Figure 21.16), the formation of Br_2 and I_2 may also be expected as primary products in the electrolysis of aqueous NaBr and NaI. For NaBr(aq), the primary anode reaction is

$$2 Br^-(aq) \longrightarrow Br_2(\ell) + 2 e^-$$

and so $E°_{cell}$ is

$$E°_{cell} = (-0.83 \text{ V}) - (+1.08 \text{ V}) = -1.91 \text{ V}$$

Exercise 20.14   **Electrolysis of Aqueous Solutions**

Predict the main chemical reaction that occurs in the electrolysis of an aqueous sodium hydroxide solution. What is the minimum voltage needed to cause this reaction to occur?

20.8   COUNTING ELECTRONS

Metallic silver is produced at the cathode in the electrolysis of aqueous $AgNO_3$, the reaction being $Ag^+(aq) + e^- \longrightarrow Ag(s)$. One mole of electrons is required to produce one mole of silver. In contrast, two moles of electrons are required to produce one mole of tin (Figure 20.16).

$$Sn^{2+}(aq) + 2\,e^- \longrightarrow Sn(s)$$

It follows that if the number of moles of electrons flowing through the electrolysis cell could be measured, the number of moles of silver or tin produced could be calculated. Conversely, if the amount of silver or tin produced is known, then the number of moles of electrons moving through the circuit could be measured [CD-ROM, Screen 20.12].

The number of moles of electrons consumed or produced in an electron transfer reaction is obtained by measuring the current flowing in the external electric circuit in a given time. The *current* flowing in an electrical circuit is the amount of charge (in units of coulombs, C) per unit time, and the usual unit for current is the ampere (A). (One ampere equals the passage of one coulomb of charge per second.)

$$\text{Current (amperes, A)} = \frac{\text{electric charge (coulombs, C)}}{\text{time (seconds, s)}} \qquad \textbf{(20.6)}$$

The current passing through an electrochemical cell, and the time during which the current flows, are easily measured quantities. Therefore, the charge (in coulombs) that passes through a cell can be obtained by multiplying the current (in amperes) by the time (in seconds). Knowing the charge, and using the Faraday constant as a conversion factor, the number of moles of electrons that pass through an electrochemical cell can be calculated. In turn, this quantity can then be used to calculate the quantity of reactants and products.

Example 20.12   Using the Faraday Constant

Problem • A current of 2.40 A is passed through a solution containing $Cu^{2+}(aq)$ for 30.0 min. What mass of copper, in grams, is deposited at the cathode? The reaction at the cathode is $Cu^{2+}(aq) + 2\,e^- \longrightarrow Cu(s)$.

Strategy • A roadmap for this calculation is given to the right.

Solution •

1. Calculate the charge (number of coulombs) passing through the cell in 30.0 min.

$$\begin{aligned}
\text{Charge (coulombs, C)} &= \text{current (A)} \times \text{time (s)} \\
&= (2.40\ \text{A})(30.0\ \text{min})(60.0\ \text{s/min}) \\
&= 4.32 \times 10^3\ \text{C}
\end{aligned}$$

2. Calculate the number of moles of electrons (i.e., the number of Faradays of electricity)

$$\begin{aligned}
\text{mol e}^- &= (4.32 \times 10^3\ \text{C})\left(\frac{1\ \text{mol e}^-}{96{,}500\ \text{C}}\right) \\
&= 4.48 \times 10^{-2}\ \text{mol e}^-
\end{aligned}$$

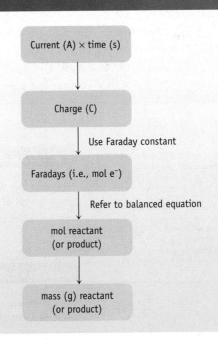

Current (A) × time (s)

↓

Charge (C)

↓ Use Faraday constant

Faradays (i.e., mol e⁻)

↓ Refer to balanced equation

mol reactant (or product)

↓

mass (g) reactant (or product)

(Example continues on next page)

3. Calculate the number of moles of copper and, from this, the mass of copper.

Mass of copper (g) =

$$(4.48 \times 10^{-2} \text{ mol e}^-)\left(\frac{1 \text{ mol Cu}}{2 \text{ mol e}^-}\right)\left(\frac{63.55 \text{ g}}{1 \text{ mol Cu}}\right)$$

Mass of copper = 1.42 g

Comment • The key relation in this calculation is Current = charge/time. All situations will involve knowing two of these three quantities from experiment and calculating the third.

Example 20.13 Using the Faraday Constant

Problem • How long must a current of 0.800 A flow in order to form 2.50 g of silver metal in an electroplating experiment? The cathode reaction is $Ag^+(aq) + e^- \longrightarrow Ag(s)$.

Strategy • This problem is the reverse of the problem in the previous example. We shall use the roadmap to the right. Here, we calculate the amount of Ag (mol Ag), then the number of electrons (mol e$^-$), then the charge (C), and finally the time.

Solution •

1. Calculate moles of electrons required.

$$\text{mol e}^- = (2.50 \text{ g Ag})\left(\frac{1 \text{ mol Ag}}{107.9 \text{ g Ag}}\right)\left(\frac{1 \text{ mol e}^-}{1 \text{ mol Ag}}\right)$$

$$= 2.32 \times 10^{-2} \text{ mol e}^-$$

2. Calculate the quantity of charge (coulombs).

$$\text{Charge (C)} = (2.32 \times 10^{-2} \text{ mol e}^-)\left(\frac{96,500 \text{ C}}{1 \text{ mol e}^-}\right)$$

$$= 2240 \text{ C}$$

3. Use charge and the current (amperes) to calculate the time required.

$$\text{Charge (coulombs)} = \text{current (A)} \times \text{time (s)}$$

$$2240 \text{ C} = (0.800 \text{ A})(\text{time, s})$$

$$\text{Time} = 2.80 \times 10^3 \text{ s (or 46.6 min)}$$

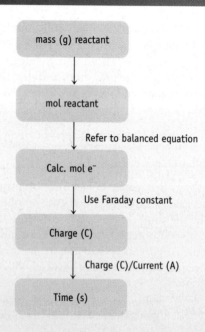

mass (g) reactant

↓

mol reactant

Refer to balanced equation

↓

Calc. mol e$^-$

Use Faraday constant

↓

Charge (C)

Charge (C)/Current (A)

↓

Time (s)

Exercise 20.15 Using the Faraday Constant

Calculate the mass of O_2 produced in the electrolysis of water, using a current of 0.445 A for a period of 45 minutes.

Exercise 20.16 Using the Faraday Constant

In the commercial production of sodium by electrolysis, the cell operates at 7.0 V and a current of 25×10^3 amps. What mass of sodium can be produced in one hour?

In Summary

When you have finished studying this chapter, you should ask if you have met the chapter goals. In particular, you should be able to

- Balance equations for oxidation–reduction reactions in acidic or basic solutions using the half-reaction approach (Section 20.1).
- In a voltaic cell identify the half-reactions occurring at the anode and cathode, the polarity of the electrodes, the direction of electron flow in the external connection, and the direction of ion flow in the salt bridge (Section 20.2).
- Know the chemistry and recognize the advantages and disadvantages of dry cells, alkaline batteries, mercury batteries, lead storage batteries, and Ni-Cad batteries (Section 20.3).
- Understand how fuel cells work (Section 20.3).
- Understand the process by which standard reduction potentials are determined and identify standard conditions as applied to electrochemistry (Section 20.4).
- Describe the standard hydrogen electrode ($E° = 0.00$ V) and explain how it is used to determine standard potentials of half-reactions (Section 20.4).
- Know how to use standard reduction potentials to determine cell voltages for cells under standard conditions (Equation 20.1).
- Know how to use a Table of Standard Reduction Potentials (Table 20.1 and Appendix M) to rank the strengths of oxidizing and reducing agents, to predict which substances can reduce or oxidize another species, and to predict whether redox reactions are product-favored or reactant-favored (Section 20.4).
- Use the Nernst equation (Equations 20.2 and Equation 20.3) to calculate the cell potential under nonstandard conditions (Section 20.5).
- Explain how cell voltage relates to the concentration of ions and explain how this relationship allows the determination of pH (Section 20.6) and other ion concentrations.
- Use the relationships between cell voltage ($E°_{cell}$) and free energy ($\Delta G°$) (Equation 20.4) and between $E°_{cell}$ and an equilibrium constant for the cell reaction (Equation 20.5) (Section 20.6).
- Describe the chemical processes occurring in an electrolysis. Recognize the factors that determine which substances are oxidized and reduced at the electrodes (Section 20.7).
- Relate the amount of a substance oxidized or reduced to the amount of current and the time the current flows (Section 20.8).

Key Terms

voltaic cells
galvanic cells
electrolysis

Section 20.1
half-reactions

Section 20.2
salt bridge
anode
cathode

Section 20.3
primary batteries
secondary batteries
storage batteries
rechargeable batteries
dry cell battery
alkaline cells
mercury cells
lead storage battery
nickel–cadmium batteries
fuel cells

Section 20.4
electromotive force, emf
standard conditions
standard potential
standard hydrogen electrode (SHE)

Section 20.5
Nernst equation
Faraday constant
ion-selective electrodes

Key Equations

Equation 20.1 (page 850)
Calculating a standard cell potential, E°_{cell}, from standard half-cell potentials.

$$E^\circ_{cell} = E^\circ_{cathode} - E^\circ_{anode}$$

Equation 20.3 (page 859)
Nernst equation.

$$E = E^\circ - \frac{0.0257 \text{ V}}{n} \ln Q \qquad \text{at 25 °C}$$

Here E is the cell potential under nonstandard conditions, n is the number of electrons transferred from reducing agent to oxidizing agent (according to the balanced equation), and Q is the reaction quotient.

Equation 20.4 (page 862)
Free energy and standard cell potential.

$$\Delta G^\circ = -nFE^\circ$$

The constant F is the Faraday constant, 96,485 C/mol e^-.

Equation 20.5 (page 863)
Relation between the equilibrium constant and the standard cell potential for a reaction.

$$\ln K = \frac{nE^\circ}{0.0257 \text{ V}} \qquad \text{at 25 °C}$$

Equation 20.6 (page 869)
Relation between current, electric charge, and time.

$$\text{Current (amperes, A)} = \frac{\text{electric charge (coulombs, C)}}{\text{time (seconds, s)}}$$

Study Questions

Questions with blue, bold-faced numbers have answers in Appendix O. Many of these questions require values of E°, which may be found in Table 20.1 or Appendix M.

Reviewing Important Concepts

1. Which of the following are redox reactions? For redox reactions, determine which element is oxidized and which is reduced, and identify the oxidizing agent and the reducing agent.

   (a) $Zn(s) + 2 HNO_3(aq) \longrightarrow Zn(NO_3)_2(aq) + H_2(g)$

   (b) $P_4O_{10}(s) + 6 H_2O(\ell) \longrightarrow 4 H_3PO_4(aq)$

   (c) $3 Cu(s) + 2 NO_3^-(aq) + 8 H^+(aq) \longrightarrow$
   $\qquad\qquad 2 NO(g) + 3 Cu^{2+}(aq) + 4 H_2O(\ell)$

   (d) $HOCl(aq) + H^+(aq) + Cl^-(aq) \longrightarrow Cl_2(g) + H_2O(\ell)$

2. Magnesium metal is oxidized and silver ions are reduced in a voltaic cell using $Mg^{2+}(aq, 1 M)|Mg(s)$ and $Ag^+(aq, 1 M)|Ag(s)$ half-cells.

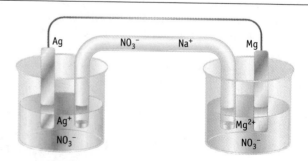

   (a) Label each part of the cell.

   (b) Write equations for the half-reactions occurring at the anode and cathode, and write an equation for the net reaction in the cell.

3. For the cell in Question 2, trace the movement of electrons in the external circuit. Assuming the salt bridge contains $NaNO_3$, trace the movement of the Na^+ and NO_3^- ions in the salt bridge that occurs when the voltaic cell produces current. Why is a salt bridge required in the cell?

4. Identify the anode and cathode in a lead storage battery and indicate the half-reactions occurring at each electrode. What key feature makes it possible to recharge this battery?

5. Describe the experiment used to determine the standard reduction potential of zinc.

6. What factors affect the cell voltage in a voltaic cell?

7. Use the table of standard reduction potentials (Table 20.1, page 852) for this question. Standard conditions are assumed.

   (a) Arrange the following chemical species: $Zn(s)$, $Cl^-(aq)$, $Cu(s)$, $K(s)$, and $H_2(g)$, in order of their ease of oxidation, from easiest to oxidize to hardest to oxidize.

   (b) Arrange the following chemical species: $H^+(aq)$, $I_2(s)$, $Na^+(aq)$, $Ag^+(aq)$, and $H_2O(\ell)$ in order of their ease of reduction, from easiest to reduce to hardest to reduce.

8. Which of the following reactions is/are product-favored?

   (a) $Zn(s) + I_2(s) \longrightarrow Zn^{2+}(aq) + 2\,I^-(aq)$

   (b) $2\,Cl^-(aq) + I_2(s) \longrightarrow Cl_2(g) + 2\,I^-(aq)$

   (c) $2\,Na^+(aq) + 2\,Cl^-(aq) \longrightarrow 2\,Na(s) + Cl_2(g)$

   (d) $2\,K(s) + 2\,H_2O(\ell) \longrightarrow$
   $$2\,K^+(aq) + H_2(g) + 2\,OH^-(aq)$$

9. What are the signs of $E°$ and $\Delta G°$ for a product-favored oxidation–reduction reaction?

10. Draw a picture of the apparatus used for electrolysis of aqueous $NaCl$. Identify the cathode and the anode. Write equations for reaction(s) that take place at each electrode. Trace the movement of electrons in the external circuit and ions in the cell.

11. What mathematical relationships relate ampere-seconds, coulombs, and the Faraday? How many seconds would be required to electroplate 1.00 mol of Ag from a solution of $Ag^+(aq)$ using a current of 1.00 amp?

Practicing Skills

Balancing Equations for Oxidation–Reduction Reactions
(*See Examples 20.1–20.3 and CD-ROM Screen 20.3*)
When balancing the following redox equations, it may be necessary to add $H^+(aq)$ or $H^+(aq)$ and H_2O for reactions in acid, and $OH^-(aq)$ or $OH^-(aq)$ and H_2O for reactions in base.

12. Write balanced equations for the following half-reactions. Specify whether each is an oxidation or reduction.

    (a) $Cr(s) \longrightarrow Cr^{3+}(aq)$ (in acid)

    (b) $AsH_3(g) \longrightarrow As(s)$ (in acid)

    (c) $VO_3^-(aq) \longrightarrow V^{2+}(aq)$ (in acid)

    (d) $Ag(s) \longrightarrow Ag_2O(s)$ (in base)

13. Write balanced equations for the following half-reactions. Specify whether each is an oxidation or reduction.

    (a) $H_2O_2(aq) \longrightarrow O_2(g)$ (in acid)

    (b) $H_2C_2O_4(aq) \longrightarrow CO_2(g)$ (in acid)

    (c) $NO_3^-(aq) \longrightarrow NO(g)$ (in acid)

    (d) $MnO_4^-(aq) \longrightarrow MnO_2(s)$ (in base)

14. Balance the following redox equations. All occur in acidic solution.

    (a) $Ag(s) + NO_3^-(aq) \longrightarrow NO_2(g) + Ag^+(aq)$

    (b) $MnO_4^-(aq) + HSO_3^-(aq) \longrightarrow Mn^{2+}(aq) + SO_4^{2-}(aq)$

    (c) $Zn(s) + NO_3^-(aq) \longrightarrow Zn^{2+}(aq) + N_2O(g)$

    (d) $Cr(s) + NO_3^-(aq) \longrightarrow Cr^{3+}(aq) + NO(g)$

15. Balance the following redox equations. All occur in acidic solution.

    (a) $Sn(s) + H^+(aq) \longrightarrow Sn^{2+}(aq) + H_2(g)$

    (b) $Cr_2O_7^{2-}(aq) + Fe^{2+}(aq) \longrightarrow Cr^{3+}(aq) + Fe^{3+}(aq)$

    (c) $MnO_2(s) + Cl^-(aq) \longrightarrow Mn^{2+}(aq) + Cl_2(g)$

    (d) $CH_2O(aq) + Ag^+(aq) \longrightarrow HCO_2H(aq) + Ag(s)$

16. Balance the following redox equations. All occur in basic solution.

    (a) $Al(s) + OH^-(aq) \longrightarrow Al(OH)_4^-(aq) + H_2(g)$

    (b) $CrO_4^{2-}(aq) + SO_3^{2-}(aq) \longrightarrow Cr(OH)_3(s) + SO_4^{2-}(aq)$

    (c) $Zn(s) + Cu(OH)_2(s) \longrightarrow Zn(OH)_4^{2-}(aq) + Cu(s)$

    (d) $HS^-(aq) + ClO_3^-(aq) \longrightarrow S(s) + Cl^-(aq)$

17. Balance the following redox equations. All occur in basic solution.

    (a) $Fe(OH)_3(s) + Cr(s) \longrightarrow Cr(OH)_3(s) + Fe(OH)_2(s)$

    (b) $NiO_2(s) + Zn(s) \longrightarrow Ni(OH)_2(s) + Zn(OH)_2(s)$

    (c) $Fe(OH)_2(s) + CrO_4^{2-}(aq) \longrightarrow$
    $$Fe(OH)_3(s) + Cr(OH)_4^-(aq)$$

    (d) $N_2H_4(aq) + Ag_2O(s) \longrightarrow N_2(g) + Ag(s)$

Constructing Voltaic Cells
(*See Example 20.4 and CD-ROM Screen 20.4*)

18. A voltaic cell is constructed using the reaction of chromium metal and iron(II) ion.

$$2\,Cr(s) + 3\,Fe^{2+}(aq) \longrightarrow 2\,Cr^{3+}(aq) + 3\,Fe(s)$$

Complete the following sentences: Electrons in the external circuit flow from the _____ electrode to the _____ electrode. Negative ions move in the salt bridge from the _____ half-cell to the _____ half-cell. The half-reaction at the anode is _____ and that at the cathode is _____.

19. A voltaic cell is constructed using the reaction

$$Mg(s) + 2\,H^+(aq) \longrightarrow Mg^{2+}(aq) + H_2(g)$$

   (a) Write equations for the oxidation and reduction half-reactions.

   (b) Which half-reaction occurs in the anode compartment and which occurs in the cathode compartment?

   (c) Complete the following sentences: Electrons in the external circuit flow from the _____ electrode to the

_____ electrode. Negative ions move in the salt bridge from the _____ half-cell to the _____ half-cell. The half-reaction at the anode is _____ and that at the cathode is _____.

20. The half-cells $Fe^{2+}(aq)|Fe(s)$ and $O_2(g)|H_2O(\ell)$ (in acidic solution) are linked to create a voltaic cell.

    (a) Write equations for the oxidation and reduction half-reactions and for the overall (cell) reaction.

    (b) Which half-reaction occurs in the anode compartment and which occurs in the cathode compartment?

    (c) Complete the following sentences: Electrons in the external circuit flow from the _____ electrode to the _____ electrode. Negative ions move in the salt bridge from the _____ half-cell to the _____ half-cell.

21. The half-cells $Ag^+(aq)|Ag(s)$ and $Cl_2(g)|Cl^-(aq)$ are linked to create a voltaic cell.

    (a) Write equations for the oxidation and reduction half-reactions and for the overall (cell) reaction.

    (b) Which half-reaction occurs in the anode compartment and which occurs in the cathode compartment?

    (c) Complete the following sentences: Electrons in the external circuit flow from the _____ electrode to the _____ electrode. Negative ions move in the salt bridge from the _____ half-cell to the _____ half-cell.

Commercial Cells

22. What are the similarities and differences between dry cells, alkaline batteries, and mercury batteries?

23. What reactions occur when a lead storage battery is recharged?

Standard Electrochemical Potentials

(See Examples 20.5–20.7 and CD-ROM Screens 20.5 and 20.6.

24. Calculate the value of $E°$ for each of the following reactions. Decide if each is product-favored in the direction written.

    (a) $2 I^-(aq) + Zn^{2+}(aq) \longrightarrow I_2(g) + Zn(s)$

    (b) $Zn^{2+}(aq) + Ni(s) \longrightarrow Zn(s) + Ni^{2+}(aq)$

    (c) $2 Cl^-(aq) + Cu^{2+}(aq) \longrightarrow Cu(s) + Cl_2(g)$

    (d) $Fe^{2+}(aq) + Ag^+(aq) \longrightarrow Fe^{3+}(aq) + Ag(s)$

25. Calculate the value of $E°$ for each of the following reactions. Decide if each is product-favored in the direction written. [Reaction (d) occurs in basic solution.]

    (a) $Br_2(\ell) + Mg(s) \longrightarrow Mg^{2+}(aq) + 2 Br^-(aq)$

    (b) $Zn^{2+}(aq) + Mg(s) \longrightarrow Zn(s) + Mg^{2+}(aq)$

    (c) $Sn^{2+}(aq) + 2 Ag^+(aq) \longrightarrow Sn^{4+}(aq) + 2 Ag(s)$

    (d) $2 Zn(s) + O_2(g) + 2 H_2O(\ell) + 4 OH^-(aq) \longrightarrow$ $2 Zn(OH)_4{}^{2-}(aq)$

26. Balance each of the following unbalanced equations, then calculate the standard potential, $E°$, and decide whether each is product-favored as written. (All are in acidic solution.)

    (a) $Sn^{2+}(aq) + Ag(s) \longrightarrow Sn(s) + Ag^+(aq)$

    (b) $Al(s) + Sn^{4+}(aq) \longrightarrow Sn^{2+}(aq) + Al^{3+}(aq)$

    (c) $ClO_3{}^-(aq) + Ce^{3+}(aq) \longrightarrow Cl^-(aq) + Ce^{4+}(aq)$

    (d) $Cu(s) + NO_3{}^-(aq) \longrightarrow Cu^{2+}(aq) + NO(g)$

27. Balance each of the following unbalanced equations, then calculate the standard potential, $E°$, and decide whether each is product-favored as written. (All occur in acidic solution.)

    (a) $I_2(s) + Br^-(aq) \longrightarrow I^-(aq) + Br_2(\ell)$

    (b) $Fe^{2+}(aq) + Cu^{2+}(aq) \longrightarrow Cu(s) + Fe^{3+}(aq)$

    (c) $Fe^{2+}(aq) + Cr_2O_7{}^{2-}(aq) \longrightarrow Fe^{3+}(aq) + Cr^{3+}(aq)$

    (d) $MnO_4{}^-(aq) + HNO_2(aq) \longrightarrow Mn^{2+}(aq) + NO_3{}^-(aq)$

28. Consider the following half-reactions:

Half-Reaction	$E°(V)$
$Cu^{2+}(aq) + 2 e^- \longrightarrow Cu(s)$	+0.34
$Sn^{2+}(aq) + 2 e^- \longrightarrow Sn(s)$	−0.14
$Fe^{2+}(aq) + 2 e^- \longrightarrow Fe(s)$	−0.44
$Zn^{2+}(aq) + 2 e^- \longrightarrow Zn(s)$	−0.76
$Al^{3+}(aq) + 3 e^- \longrightarrow Al(s)$	−1.66

    (a) Based on $E°$ values, which metal is the most easily oxidized?

    (b) Which metals on this list are capable of reducing $Fe^{2+}(aq)$ to Fe?

    (c) Write a balanced chemical equation for the reaction of $Fe^{2+}(aq)$ with $Sn(s)$. Is this reaction product-favored or reactant-favored?

    (d) Write a balanced chemical equation for the reaction of $Zn^{2+}(aq)$ with $Sn(s)$. Is this reaction product-favored or reactant-favored?

29. Consider the following half-reactions:

Half-Reaction	$E°(V)$
$MnO_4{}^-(aq) + 8 H^+(aq) + 5 e^- \longrightarrow$ $Mn^{2+}(aq) + 4 H_2O(\ell)$	+1.51
$BrO_3{}^-(aq) + 6 H^+(aq) + 6 e^- \longrightarrow$ $Br^-(aq) + 3 H_2O(\ell)$	+1.47
$Cr_2O_7{}^{2-}(aq) + 14 H^+(aq) + 6 e^- \longrightarrow$ $2 Cr^{3+}(aq) + 7 H_2O(\ell)$	+1.33
$NO_3{}^-(aq) + 4 H^+(aq) + 3 e^- \longrightarrow$ $NO(g) + 2 H_2O(\ell)$	+0.96
$SO_4{}^{2-}(aq) + 4 H^+(aq) + 2 e^- \longrightarrow$ $SO_2(g) + 2 H_2O(\ell)$	+0.20

    (a) Choosing from among the reactants in these half-reactions, identify the strongest and weakest oxidizing agent.

    (b) Which of the oxidizing agents listed is (are) capable of oxidizing $Br^-(aq)$ to $BrO_3{}^-(aq)$ (in acidic solution)?

    (c) Write a balanced chemical equation for the reaction of $Cr_2O_7{}^{2-}(aq)$ with $SO_2(g)$ in acidic solution. Is this reaction product-favored or reactant-favored?

(d) Write a balanced chemical equation for the reaction of $Cr_2O_7{}^{2-}(aq)$ with $Mn^{2+}(aq)$. Is this reaction product-favored or reactant-favored?

Ranking Oxidizing and Reducing Agents
(See Examples 20.6 and 20.7 and CD-ROM Screen 20.6.)
Use a table of standard reduction potentials (Table 20.1 or Appendix M) to answer Questions 30–35.

30. Which element among those listed below is the best reducing agent?
   (a) Cu   (d) Ag
   (b) Zn   (e) Cr
   (c) Fe

31. From the list below, identify those elements that are easier to oxidize than $H_2(g)$.
   (a) Cu   (d) Ag
   (b) Zn   (e) Cr
   (c) Fe

32. Which ion among those listed below is most easily reduced?
   (a) $Cu^{2+}(aq)$ (d) $Ag^+(aq)$
   (b) $Zn^{2+}(aq)$ (e) $Al^{3+}(aq)$
   (c) $Fe^{2+}(aq)$

33. From the list below, identify the ions that are more easily reduced than $H^+(aq)$.
   (a) $Cu^{2+}(aq)$ (d) $Ag^+(aq)$
   (b) $Zn^{2+}(aq)$ (e) $Al^{3+}(aq)$
   (c) $Fe^{2+}(aq)$

34. Halogen chemistry:
   (a) Which halogen is most easily reduced: F_2, Cl_2, Br_2, or I_2?
   (b) Identify the halogens that are better oxidizing agents than $MnO_2(s)$.

35. Halide ion chemistry:
   (a) Which ion among those listed below is most easily oxidized to the elemental halogen: F^-, Cl^-, Br^-, I^-?
   (b) Identify the halide ions that are more easily oxidized than $H_2O(\ell)$.

Electrochemical Cells Under Nonstandard Conditions
(See Examples 20.8 and 20.9 and CD-ROM Screen 20.7)

36. Calculate the voltage delivered by a voltaic cell using the following reaction if all dissolved species are 2.5×10^{-2} M.
$$Zn(s) + 2\,H_2O(\ell) + 2\,OH^-(aq) \longrightarrow$$
$$Zn(OH)_4{}^{2-}(aq) + H_2(g)$$

37. Calculate the voltage delivered by a voltaic cell using the following reaction if all dissolved species are 0.015 M.
$$2\,Fe^{2+}(aq) + H_2O_2(aq) + 2\,H^+(aq) \longrightarrow$$
$$2\,Fe^{3+}(aq) + 2\,H_2O(\ell)$$

38. One half-cell in a voltaic cell is constructed with a silver wire dipped into a 0.25 M solution of $AgNO_3$. The other half-cell consists of a zinc electrode in a 0.010 M solution of $Zn(NO_3)_2$. Calculate the cell potential.

39. One half-cell in a voltaic cell is constructed with a copper wire dipped into a 4.8×10^{-3} M solution of $Cu(NO_3)_2$. The other half-cell consists of a zinc electrode in a 0.40 M solution of $Zn(NO_3)_2$. Calculate the cell potential.

40. One half-cell in a voltaic cell is constructed with a silver wire dipped into a $AgNO_3$ solution of unknown concentration. The other half-cell consists of a zinc electrode in a 1.0 M solution of $Zn(NO_3)_2$. A voltage of 1.48 V is measured for this cell. Use this information to calculate the concentration of $Ag^+(aq)$.

41. One half-cell in a voltaic cell is constructed with an iron wire dipped into a $Fe(NO_3)_2$ solution of unknown concentration. The other half-cell is a standard hydrogen electrode. A voltage of 0.49 V is measured for this cell. Use this information to calculate the concentration of $Fe^{2+}(aq)$.

Electrochemistry, Thermodynamics, and Equilibrium
(See Examples 20.10 and 20.11 and CD-ROM Screen 20.8)

42. Calculate $\Delta G°$ and the equilibrium constant for the following reactions:
   (a) $2\,Fe^{3+}(aq) + 2\,I^-(aq) \longrightarrow 2\,Fe^{2+}(aq) + I_2(s)$
   (b) $I_2(s) + 2\,Br^-(aq) \longrightarrow 2\,I^-(aq) + Br_2(\ell)$

43. Calculate $\Delta G°$ and the equilibrium constant for the following reactions:
   (a) $Zn^{2+}(aq) + Ni(s) \longrightarrow Zn(s) + Ni^{2+}(aq)$
   (b) $Cu(s) + 2\,Ag^+(aq) \longrightarrow Cu^{2+}(aq) + 2\,Ag(s)$

44. Use standard reduction potentials (Appendix M) for the half-reactions $AgBr(s) + e^- \longrightarrow Ag(s) + Br^-(aq)$ and $Ag^+(aq) + e^- \longrightarrow Ag(s)$ to calculate the value of K_{sp} for AgBr.

45. Use the standard reduction potentials (Appendix M) for the half-reactions $Hg_2Cl_2(s) + 2\,e^- \longrightarrow 2\,Hg(\ell) + 2\,Cl^-(aq)$ and $Hg_2{}^{2+}(aq) + 2\,e^- \longrightarrow 2\,Hg(\ell)$ to calculate the value of K_{sp} for Hg_2Cl_2.

46. Use the standard reduction potentials (Appendix M) for the half-reactions $AuCl_4{}^-(aq) + 3\,e^- \longrightarrow Au(s) + 4\,Cl^-(aq)$ and $Au^{3+}(aq) + 3\,e^- \longrightarrow Au(s)$ to calculate the value of $K_{formation}$ for the complex ion $AuCl_4{}^-(aq)$.

47. Use the standard reduction potentials (Appendix M) for the half-reactions $[Zn(OH)_4]^{2-}(aq) + 2\,e^- \longrightarrow Zn(s) + 4\,OH^-(aq)$ and $Zn^{2+}(aq) + 2\,e^- \longrightarrow Zn(s)$ to calculate the value of $K_{formation}$ for the complex ion $[Zn(OH)_4]^{2-}$.

48. Iron(II) ion undergoes a disproportionation reaction to give Fe(s) and the iron(III) ion. That is, iron(II) ion is both oxidized and reduced within the same reaction.
$$3\,Fe^{2+}(aq) \rightleftharpoons Fe(s) + 2\,Fe^{3+}(aq)$$
   (a) What two half-reactions make up the disproportionation reaction?
   (b) Use the values of the standard reduction potentials for the two half-reactions in part (a) to determine if this disproportionation reaction is product-favored.
   (c) What is the equilibrium constant for this reaction?

49. Copper(I) ion disproportionates to copper metal and copper(II) ion.

$$2\,Cu^+(aq) \rightleftharpoons Cu(s) + Cu^{2+}(aq)$$

(a) What two half-reactions make up the disproportionation reaction?

(b) Use values of the standard reduction potentials for the two half-reactions in part (a) to determine if this disproportionation reaction is product-favored.

(c) What is the equilibrium constant for this reaction?

Electrolysis

(See Section 20.7 and Example 20.11 and CD-ROM Screen 20.11)

50. Diagram the apparatus used to electrolyze molten NaCl. Identify the anode and cathode. Trace the movement of electrons through the external circuit and the movement of ions in the electrolysis cell.

51. Diagram the apparatus used to electrolyze aqueous $CuCl_2$. Identify the anode and cathode. Trace the movement of electrons through the external circuit and the movement of ions in the electrolysis cell.

52. Which product, O_2 or F_2, is more likely to be formed at the anode in the electrolysis of an aqueous solution of KF? Explain your reasoning.

53. Which product, Ca or H_2, is more likely to be formed at the cathode in the electrolysis of $CaCl_2$? Explain your reasoning.

54. An aqueous solution of KBr is placed in a beaker with two inert platinum electrodes. When the cell is attached to an external source of electrical energy, electrolysis occurs.

(a) Hydrogen gas and hydroxide ion are formed at the cathode. Write an equation for the half-reaction that occurs at this electrode.

(b) Bromine is the primary product at the anode. Write an equation for its formation.

55. An aqueous solution of Na_2S is placed in a beaker with two inert platinum electrodes. When the cell is attached to an external battery, electrolysis occurs.

(a) Hydrogen gas and hydroxide ion are formed at the cathode. Write an equation for the half-reaction that occurs at this electrode.

(b) Sulfur is the primary product at the anode. Write an equation for its formation.

Counting Electrons

(See Examples 20.12 and 20.13 and CD-ROM Screen 20.12)

56. In the electrolysis of a solution containing $Ni^{2+}(aq)$, metallic Ni(s) deposits on the cathode. Using a current of 0.150 A for 12.2 min, what mass of nickel will form?

57. In the electrolysis of a solution containing $Ag^+(aq)$, metallic Ag(s) deposits on the cathode. Using a current of 1.12 A for 2.40 h, what mass of silver forms?

58. Electrolysis of a solution of $CuSO_4(aq)$ to give copper metal is carried out using a current of 0.66 A. How long should

electrolysis be carried out to produce 0.50 g of copper?

59. Electrolysis of a solution of $Zn(NO_3)_2(aq)$ to give zinc metal is carried out using a current of 2.12 A. How long should electrolysis be carried out to prepare 2.5 g of zinc?

60. A voltaic cell can be built using the reaction between Al metal and O_2 from the air. If the Al anode of this cell consists of 84 g of aluminum, how many hours can the battery produce 1.0 amp of electricity, assuming an unlimited supply of O_2?

61. Assume the specifications of a Ni-Cd voltaic cell include delivery of 0.25 A of current for 1.00 h. What is the minimum mass of cadmium that must be used to make the anode in this cell?

General Questions

More challenging questions are marked with an underlined number.

62. Write balanced equations for the following half-reactions.

(a) $UO_2^+(aq) \longrightarrow U^{4+}(aq)$ (acid solution)

(b) $ClO_3^-(aq) \longrightarrow Cl^-(aq)$ (acid solution)

(c) $N_2H_4(aq) \longrightarrow N_2(g)$ (basic solution)

(d) $OCl^-(aq) \longrightarrow Cl^-(aq)$ (basic solution)

63. Balance the following equations.

(a) $Zn(s) + VO^{2+}(aq) \longrightarrow$
$$Zn^{2+}(aq) + V^{3+}(aq) \quad \text{(acid solution)}$$

(b) $Zn(s) + VO_3^-(aq) \longrightarrow$
$$V^{2+}(aq) + Zn^{2+}(aq) \quad \text{(acid solution)}$$

(c) $Zn(s) + ClO^-(aq) \longrightarrow$
$$Zn(OH)_2(s) + Cl^-(aq) \quad \text{(basic solution)}$$

(d) $ClO^-(aq) + Cr(OH)_4^-(aq) \longrightarrow$
$$Cl^-(aq) + CrO_4^{2-}(aq) \quad \text{(basic solution)}$$

64. You want to set up a series of voltaic cells with specific cell voltages. A Zn^{2+}(aq, 1 M)|Zn(s) half-cell is in one compartment. Identify several half-cells that you could use so that the cell voltage is close to the following values (a) 1.1 V, (b) 0.5 V. Consider cells in which zinc can be either the cathode or the anode.

65. You want to set up a series of voltaic cells with specific cell voltages. The Ag^+(aq, 1 M)|Ag(s) half-cell is one of the compartments. Identify several half-cells that you could use so that the cell voltage will be close to the following values: (a) 1.7 V, (b) 0.5 V. Be sure to consider cells in which silver can be either the cathode or the anode.

66. In the table of standard reduction potentials, locate the half-reactions for the reductions of the following metal ions to the metal: $Sn^{2+}(aq)$, $Au^+(aq)$, $Zn^{2+}(aq)$, $Co^{2+}(aq)$, $Ag^+(aq)$, $Cu^{2+}(aq)$.

Among the metal ions and metals that make up these half-reactions

(a) Which metal ion is the weakest oxidizing agent?

(b) Which metal ion is the strongest oxidizing agent?

(c) Which metal is the strongest reducing agent?

(d) Which metal is the weakest reducing agent?

(e) Will Sn(s) reduce $Cu^{2+}(aq)$ to Cu(s)?

(f) Will $Ag(s)$ reduce $Co^{2+}(aq)$ to $Co(s)$?

(g) Which metal ions on the list can be reduced by $Sn(s)$?

(h) What metals can be oxidized by $Ag^+(aq)$?

67. In the table of standard reduction potentials, locate the half-reactions for the reductions of the following nonmetals: F_2, Cl_2, Br_2, I_2 (reduction to halide ions); and O_2, S, Se (reduction to H_2X in aqueous acid). Among the elements and the ions or compounds that make up these half-reactions

(a) Which element is the weakest oxidizing agent?

(b) Which element is the weakest reducing agent?

(c) Which of the elements listed is (are) capable of oxidizing H_2O to O_2?

(d) Which of these elements listed is (are) capable of oxidizing H_2S to S?

(e) Is O_2 capable of oxidizing I^- to I_2, in acidic solution?

(f) Is S capable of oxidizing I^- to I_2?

(g) Is the reaction $H_2S(aq) + Se(s) \longrightarrow H_2Se(aq) + S(s)$ product-favored?

(h) Is the following reaction product-favored?
$H_2S(aq) + I_2(s) \longrightarrow 2 H^+(aq) + 2 I^- (aq) + S(s)$

68. Four voltaic cells are set up. In each, one half-cell contains a standard hydrogen electrode. The second half-cell is one of the following: $Cr^{3+}(aq, 1.0\ M)|Cr(s)$, $Fe^{2+}(aq, 1.0\ M)|Fe(s)$, $Cu^{2+}(aq, 1.0\ M)|Cu(s)$, and $Mg^{2+}(aq, 1.0\ M)|Mg(s)$.

(a) In which of the voltaic cells is the hydrogen electrode the cathode?

(b) Which voltaic cell produces the highest voltage? Which produces the lowest voltage?

69. The following half-cells are available: $Ag^+(aq, 1.0\ M)|Ag(s)$, $Zn^{2+}(aq, 1.0\ M)|Zn(s)$, $Cu^{2+}(aq, 1.0\ M)|Cu(s)$, and $Co^{2+}(aq, 1.0\ M)|Co(s)$. Given four different half-cells, six voltaic cells are possible. These are labeled, for simplicity, Ag-Zn, Ag-Cu, Ag-Co, Zn-Cu, Zn-Co, and Cu-Co.

(a) In which of the voltaic cells is the copper electrode the cathode? In which of the voltaic cells is the cobalt electrode the anode?

(b) Which combination of half-cells generates the highest voltage? Which combination of half-cells generates the lowest voltage?

70. Consider an electrochemical cell based on the half-reactions $Ni^{2+}(aq) + 2 e^- \longrightarrow Ni(s)$ and $Cd^{2+}(aq) + 2 e^- \longrightarrow Cd(s)$.

(a) Diagram the cell and label each of the components (including the anode, cathode, and salt bridge).

(b) Use the equations for the half-reactions to write a balanced, net ionic equation for the overall cell reaction.

(c) What is the polarity (charge) of each electrode?

(d) What is the value of E°_{cell}?

(e) In which direction do electrons flow in the external circuit?

(f) Assume that a salt bridge containing $NaNO_3$ connects the two half-cells. In which direction do the $Na^+(aq)$ ions move? In what direction do the $NO_3^-(aq)$ ions move?

(g) Calculate the equilibrium constant for the reaction.

(h) If the concentration of Cd^{2+} is reduced to 0.010 M, and $[Ni^{2+}] = 1.0$ M, what is the value of E_{cell}? Is the net reaction still the reaction given in part (b)?

(i) If 0.050 A is drawn from the battery, how long can it last if you begin with 1.0 L of each of the solutions, and each was initially 1.0 M in dissolved species? The electrodes each weigh 50.0 g in the beginning.

71. The reaction occurring in the cell in which Al_2O_3 and aluminum salts are electrolyzed is $Al^{3+}(aq) + 3 e^- \longrightarrow Al(s)$. If the electrolysis cell operates at 5.0 V and 1.0×10^5 amps, what mass of aluminum metal can be produced in a 24-h day?

72. A potential of -0.146 V is calculated (under standard conditions) for a voltaic cell constructed using the following half-reactions:

Anode: $Ag(s) \longrightarrow Ag^+(aq) + e^-$

Cathode: $Ag_2SO_4(s) + 2 e^- \longrightarrow 2 Ag(s) + SO_4^{2-}(aq)$

(a) What is the standard reduction potential for the cathode reaction?

(b) Calculate the solubility product, K_{sp}, for Ag_2SO_4.

73. A potential of -0.142 V is calculated (under standard conditions) for a voltaic cell constructed using the following half-reactions:

Anode: $PbCl_2(s) + 2 e^- \longrightarrow Pb(s) + 2 Cl^-(aq)$

Cathode: $Pb(s) \longrightarrow Pb^{2+}(aq) + 2 e^-$

Net: $PbCl_2(s) \longrightarrow Pb^{2+}(aq) + 2 Cl^-(aq)$

(a) What is the standard reduction potential for the anode reaction?

(b) Estimate the solubility product, K_{sp}, for $PbCl_2$.

74. The standard voltage, E°, for the reaction of $Zn(s)$ and $Cl_2(g)$ is $+2.12$ V. What is the standard free energy change, ΔG°, for the reaction?

75. The standard potential for the reaction of $Mg(s)$ with $I_2(s)$ is $+2.91$ V. What is the standard free energy change, ΔG°, for the reaction?

76. An electrolysis cell for aluminum production operates at 5.0 V and a current of 1.0×10^5 amps. Calculate the number of kilowatt-hours of energy required to produce 1 metric ton (1.0×10^3 kg) of aluminum. (1 kwh = 1 kilowatt-hour = 3.6×10^6 J and 1 J = 1 coulomb · volt)

77. Electrolysis of molten NaCl is done in cells operating at 7.0 V and 4.0×10^4 amps. What mass of $Na(s)$ and $Cl_2(g)$ can be produced in one day in such a cell? What is the energy consumption in kilowatt-hours? (1 kwh = 1 kilowatt-hour = 3.6×10^6 J and 1 J = 1 coulomb · volt)

78. A current of 0.0100 amp is passed through a solution of rhodium sulfate, causing reduction of the metal ion to the metal. After 3.00 h, 0.038 g of Rh has been deposited. What is the charge on the rhodium ion, Rh^{n+}? What is the formula for rhodium sulfate?

79. A current of 0.44 amp is passed through a solution of ruthenium nitrate causing reduction of the metal ion to the

metal. After 25.0 min, 0.345 g of Ru has been deposited. What is the charge on the ruthenium ion, Ru^{n+}? What is the formula for ruthenium nitrate?

80. The total charge that can be delivered by a large dry cell battery before its voltage drops too low is usually about 35 amp-hours. (One amp-hour is the charge that passes through a circuit when 1 amp flows for 1 h.) What mass of Zn is consumed when 35 amp-hours are drawn from the cell?

81. Chlorine gas is obtained commercially by electrolysis of brine (a concentrated aqueous solution of NaCl). If the electrolysis cells operate at 4.6 V and 3.0×10^5 amps, what mass of chlorine can be produced in a 24-h day?

82. An old method of measuring the current flowing in a circuit was to use a "silver coulometer." The current passed first through a solution of $Ag^+(aq)$ and then into another solution containing an electroactive species. The amount of silver metal deposited at the cathode was weighed. From the mass of silver, the number of atoms of silver is calculated, and since the reduction of a silver ion requires one electron, this equals the number of electrons passing through the circuit. If the time was noted, the average current could be calculated. If, in such an experiment, 0.052 g of Ag is deposited during 450 s, how much current flowed in the circuit?

83. A "silver coulometer" (Question 82) was used in the past to measure the current flowing in an electrochemical cell. Suppose you found that the current flowing through an electrolysis cell deposited 0.089 g of Ag metal at the cathode after exactly 10 min. If this same current then passed through a cell containing gold(III) ion in the form of $[AuCl_4]^-$, how much gold was deposited at the cathode in that electrolysis cell?

84. Write balanced equations for the following reduction half-reactions involving organic compounds:

(a) $HCO_2H \longrightarrow HCHO$ (acid solution)

(b) $C_6H_5CO_2H \longrightarrow C_6H_5CH_3$ (acid solution)

(c) $CH_3CH_2CHO \longrightarrow CH_3CH_2CH_2OH$ (acid solution)

(d) $CH_3OH \longrightarrow CH_4$ (acid solution)

85. Balance the following equations involving organic compounds:

(a) $Ag^+(aq) + C_6H_5CHO(aq) \longrightarrow$
           $Ag(s) + C_6H_5CO_2H(aq)$ (acid solution)

(b) $CH_3CH_2OH(aq) + Cr_2O_7^{2-}(aq) \longrightarrow$
           $CH_3CO_2H(aq) + Cr^{3+}(aq)$ (acid solution)

86. A voltaic cell is constructed in which one half-cell uses a silver wire in an aqueous solution of $AgNO_3$. The other half-cell consists of an inert platinum wire in an aqueous solution containing $Fe^{2+}(aq)$ and $Fe^{3+}(aq)$.

(a) Calculate the voltage of the cell, assuming standard conditions.

(b) Write the net ionic equation for the reaction occurring in the cell.

(c) In this voltaic cell, which electrode is the anode and which is the cathode?

(d) If $[Ag^+]$ is 0.10 M, and $[Fe^{2+}]$ and $[Fe^{3+}]$ are both 1.0 M, what is the cell voltage? Is the net cell reaction still that in part (b) above? If not, what is the net reaction under the new conditions?

87. An expensive but lighter alternative to the lead storage battery is the silver–zinc battery. Zinc is the reducing agent and silver oxide is the oxidizing agent.

$$Ag_2O(s) + Zn(s) + H_2O(\ell) \longrightarrow Zn(OH)_2(s) + 2 Ag(s)$$

The electrolyte is 40% KOH, and silver/silver oxide electrodes are separated from zinc/zinc hydroxide electrodes by a plastic sheet permeable to hydroxide ion. Under normal operating conditions, the battery has a potential of 1.59 V. (R. C. Plumb, *Journal of Chemical Education*, Vol. 50, page 857, 1973.)

(a) How much energy can be produced per gram of reactants in the silver–zinc battery? Assume the battery produces a current of 0.10 amp.

(b) How much energy can be produced per gram of reactants in the standard lead storage battery? Assume the battery produces a current of 0.10 amp at 2.0 V.

(c) Which battery produces the greater energy per gram of reactants?

88. Is it easier to reduce water in acid or base? To evaluate this, consider the half-reaction

$$2 H_2O(\ell) + 2 e^- \longrightarrow 2 OH^-(aq) + H_2(g)$$
$$E° = -0.83 V$$

What is the reduction potential for water for solutions at pH = 7 (neutral) and pH = 0 (acid)? Comment on the value of $E°$ at pH = 1. (Assume $P(H_2) = 1$ atm.)

89. A solution of KI is added dropwise to a pale blue solution of $Cu(NO_3)_2$. The solution changes to a brown color and a precipitate forms. In contrast, no change is observed if solutions of KCl and KBr are added to aqueous $Cu(NO_3)_2$. Consult the table of standard reduction potentials to explain the dissimilar results seen with the different halides. Write an equation for the redox reaction that occurs when solutions of KI and $Cu(NO_3)_2$ are mixed.

90. Four metals, A, B, C, and D, exhibit the following properties:

(a) Only A and C react with 1.0 M hydrochloric acid to give $H_2(g)$.

(b) When C is added to solutions of the ions of the other metals, metallic B, D, and A are formed.

(c) Metal D reduces B^{n+} to give metallic B and D^{n+}. Based on the information above, arrange the four metals in order of increasing ability to act as reducing agents.

91. The specifications for a lead storage battery include delivery of a steady 1.5 A of current for 15 h.

(a) What is the minimum mass of lead that will be used in the anode?

(b) What mass of PbO_2 must be used in the cathode?

(c) Assume that the volume of the battery is 0.50 L. What is the minimum concentration of H_2SO_4 necessary?

92. Fluorinated organic compounds are important commercially, as they are used as herbicides, flame retardants, and fire extinguishing agents, among other things. A reaction such as

$$CH_3SO_2F + 3\,HF \longrightarrow CF_3SO_2F + 3\,H_2$$

is carried out electrochemically in liquid HF as the solvent.

   (a) If you electrolyze 150 g of CH_3SO_2F, what mass of HF is required and what mass of each product can be isolated?

   (b) Is H_2 produced at the anode or cathode of the electrolysis cell?

   (c) A typical electrolysis cell operates at 8.0 V and 250 amps. How many kilowatt-hours of energy does one such cell consume in 24 h?

93. The free energy change for a reaction, $\Delta G°_{rxn}$, is the maximum energy that can be extracted from the process, whereas $\Delta H°_{rxn}$ is the total chemical potential energy change. The efficiency of a fuel cell is the ratio of these two quantities.

$$\text{Efficiency} = \frac{\Delta G°_{rxn}}{\Delta H°_{rxn}} \cdot 100\%$$

Consider the hydrogen–oxygen fuel cell where the net reaction is

$$H_2(g) + \tfrac{1}{2} O_2(g) \longrightarrow H_2O(\ell)$$

   (a) Calculate the efficiency of the fuel cell under standard conditions.

   (b) Calculate the efficiency of the fuel cell if the product is water vapor instead of liquid water.

   (c) Does the efficiency depend on the product being a liquid or gas? Explain why or why not.

94. A hydrogen–oxygen fuel cell operates on the simple reaction

$$H_2(g) + \tfrac{1}{2} O_2(g) \longrightarrow H_2O(\ell)$$

If the cell is designed to produce 1.5 A of current, and if the hydrogen is contained in a 1.0-L tank at 200. atm pressure at 25 °C, how long can the fuel cell operate before the hydrogen runs out? (Assume there is an unlimited supply of O_2.)

95. Living organisms derive energy from the oxidation of food, typified by glucose.

$$C_6H_{12}O_6(aq) + 6\,O_2(g) \longrightarrow 6\,CO_2(g) + 6\,H_2O(\ell)$$

Electrons in this redox process are transferred from glucose to oxygen in a series of at least 25 steps. It is instructive to calculate the total daily current flow in a typical organism and the rate of energy expenditure (power). (See T. P. Chirpich, *Journal of Chemical Education*, Vol. 52, page 99, 1975.)

   (a) The molar enthalpy of combustion of glucose is -2800 kJ. If you are on a typical daily diet of 2400 Cal (kilocalories), what amount of glucose (in mol) must be consumed in a day if glucose is assumed to be the only source of energy? What quantity of O_2 must be consumed in the oxidation process?

   (b) How many moles of electrons must be supplied to reduce the amount of O_2 calculated in part (a)?

   (c) Based on the answer in part (b), calculate the current flowing, per second, in your body from the combustion of glucose.

   (d) If the average standard potential in the electron transport chain is 1.0 V, what is the rate of energy expenditure in watts (1 watt = 1 J/s)?

Using Electronic Resources

These questions refer to the General Chemistry Interactive CD-ROM, *Version 3.0.*

97. CD-ROM Screen 20.2: Redox Reactions. What is the difference between a direct redox reaction and an indirect redox reaction?

98. CD-ROM Screen 20.5: Electrochemical Cells and Potentials

   (a) When the contents of the half-cells in the photos are changed, the standard potential for the cells also change. What does this imply about the tendency of different metals to either hold onto or release electrons?

   (b) Rank the three metals used on this screen, Ag, Cu, and Zn, in increasing tendency to release electrons.

   (c) Rank the three metal ions used in the reactions on this screen, Ag^+, Cu^{2+}, and Zn^{2+}, in increasing tendency to accept electrons.

   (d) What is the connection between the two rankings you have just made?

99. CD-ROM Screen 20.6: Electrochemical Cells and Potentials —Simulation

   (a) Using the reduction potential table here, what species can be reduced by Cu(s)?

   (b) What metals can be oxidized by Cd^{2+}?

100. CD-ROM Screen 20.8: Batteries

   (a) What common feature is shared by the dry cell battery, the alkaline battery, and the mercury battery?

   (b) What is the anode material in a ni-cad battery?

101. CD-ROM Screen 20.9: Corrosion

   (a) What are the half-reactions in the corrosion of iron?

   (b) What is responsible for the fact that aluminum does not corrode, thermodynamics or kinetics?

21

The Chemistry of the Main Group Elements

Chapter Goals

- Relate the formulas and properties of compounds to the periodic table.
- Describe the chemistry of the main group or A-group elements, particularly H; Na and K; Mg and Ca; Al; Si; N and P; O and S; and Cl.

Sulfur Chemistry and Life on the Edge

Whether life exists elsewhere in our solar system is one of the unanswered questions of science. Extremely harsh conditions on planets other than Earth almost surely make human life impossible, but scientists hold out hope that simpler life forms may exist. One reason for this belief is that life on Earth has been found to exist under unimaginably severe conditions: extreme heat and cold, high pressure, and highly acidic, highly basic, and highly saline conditions.

How do organisms — often called extremophiles — survive under severe conditions? How and when did they originate and where do they fit into current ecosystems? Chemistry has a major role in answering these questions, and the nonmetal sulfur is a key player in the existence of some extremophiles.

Exposure of sulfur-containing minerals such as pyrite (FeS_2) to air and water leads to the formation of hydronium and sulfate ions. The sulfide ion in minerals is oxidized by air or iron(III) ion, Fe^{3+}. Using pyrite, FeS_2, as an example, one possible reaction is

$$FeS_2 + 14\ Fe^{3+} + 8\ H_2O \longrightarrow$$
$$15\ Fe^{2+} + 2\ SO_4^{2-} + 16\ H^+$$

Significantly, several species of bacteria (such as *Thiobacillus ferrooxidans*) thrive

▲ **Acid mine drainage.** The water flowing from mines is often acidic owing to the production of sulfuric acid from sulfur-bearing minerals. *(Simon Fraser/Science Photo Library/Photo Researchers, Inc.)*

in highly acidic environments and greatly speed up this mineral degradation. The net result is that water in contact with the waste products of mines is often highly acidic, a serious pollution problem around mines that contain sulfide ores. In addition to harming plants and animals, the acid can extract arsenic and other toxic elements from minerals that are otherwise locked up tight in rocks. The importance of this process is measured by the fact that about half of the sulfate ion that enters the oceans is produced in this manner.

Sulfur chemistry can be important in cave formation, and there is a spectacular example in the jungles of southern Mexico. Toxic hydrogen sulfide gas spews from the Cueva de Villa Luz along with water that is milky white with suspended sulfur particles. The cave can be followed downward to a large underground stream and a maze of actively enlarging cave passages. Water rises into the cave from underlying sulfur-bearing strata, releasing hydrogen sulfide at concentrations up to 150 ppm. Yellow sulfur crystallizes on the cave walls around the inlets. The sulfur and sulfuric acid are produced by the following reactions:

$$2\ H_2S(g) + O_2(g) \longrightarrow 2\ S(s) + 2\ H_2O(\ell)$$
$$2\ S(s) + 2\ H_2O(\ell) + 3\ O_2(g) \longrightarrow$$
$$2\ H_2SO_4(aq)$$

The cave atmosphere is poisonous to humans, so gas masks are essential. But surprisingly, the cave is teeming with life. Once again, sulfur-oxidizing bacteria speed the chemical reactions and thrive

> "*The cave is a spectacular example of an ecosystem driven not by light but by inorganic chemistry.*"
>
> From C. Petit, *U.S. News & World Report*, February 9, 1998, page 59.

◀ **Jillian Banfield, geochemist and MacArthur Fellow.** Professor Jillian Banfield and her students at the University of California in Berkeley have been at the forefront of studying extremophiles. Findings from these studies are linked to environmental issues, particularly to the environmental damage caused by drainage from mines. Her work brought her to the attention of the MacArthur Foundation, which awarded her a fellowship, a so-called "genius award." *(Courtesy of Jillian Banfield)*

on the large amounts of energy released even in the absence of all other food sources. They use the chemical energy to obtain carbon for their bodies from calcium carbonate and carbon dioxide, both of which are abundant in the cave. Bacterial filaments hang from the walls and ceilings in bundles. (Because the filaments look like something coming from a runny nose, cave explorers refer to

them as "snot-tites.") Other microbes feed on the bacteria, and so on up the food chain — which includes spiders, gnats, and pygmy snails — all the way to sardine-like fish that swim in the cave stream. This entire ecosystem is supported by reactions involving sulfur within the cave.

But this is no ordinary cave environment. Droplets of water seeping in from the surface absorb both hydrogen sulfide and oxygen from the cave air. As illustrated by the reactions shown previously, these dissolved gases react to produce sulfuric acid. This depletes the concentration of both gases in the droplets, allowing more to be absorbed from the air. Meanwhile the droplets grow more acidic. The longer the droplets remain on the ceiling, the lower their pH becomes. The droplets clinging to the bacterial filaments in the photo had average pH values of 1.4, with some as low as zero! Drops that landed on explorers in the cave burned their skin and disintegrated their clothing.

◀ **Snot-tites.** Filaments of sulfur-oxidizing bacteria (dubbed "snot-tites") hang from the ceiling of a Mexican cave containing an atmosphere rich in hydrogen sulfide. The bacteria thrive on the energy released by oxidation of the hydrogen sulfide, forming the base of a complex food chain. Droplets of sulfuric acid on the filaments have the pH of battery acid. *(Arthur N. Palmer)*

Before You Begin

- Review electron configurations and periodic trends (Sections 8.4 and 8.5).
- Review the structure and bonding of main group elements (Sections 9.3 and 9.4).
- Review types of reactions (Chapter 5).

The Main Group Elements The main group elements are those in the A-groups of the periodic table. They range in properties from nonmetals to metalloids and to metals. They also include some of the most abundant elements in the solar system (H and He) and some of the least abundant (Li, Be, B). The plot below shows the abundances of elements 1–18 in the solar system. Abundances are given as the number of atoms per 10^{12} H atoms.

Some main group elements and their abundances

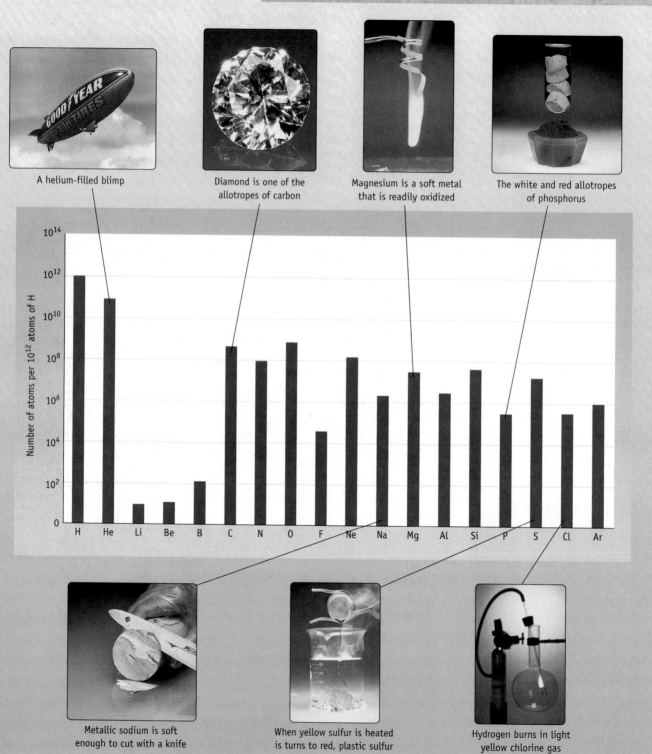

A helium-filled blimp

Diamond is one of the allotropes of carbon

Magnesium is a soft metal that is readily oxidized

The white and red allotropes of phosphorus

Metallic sodium is soft enough to cut with a knife

When yellow sulfur is heated is turns to red, plastic sulfur

Hydrogen burns in light yellow chlorine gas

The main group or Group A elements occupy an important place in the world of chemistry. Eight of the ten most abundant elements on Earth are in this group. Of the top ten chemicals produced by the U. S. chemical industry, all are main group elements or their compounds.

Because main group elements and their compounds are of economic importance — and because they also have an interesting chemistry — we devote this chapter to a brief survey of this group of elements.   •

Abundance of Elements in Earth's Crust

O	49.5%	Na	2.6%
Si	25.7%	K	2.4%
Al	7.4%	Mg	1.9%
Fe	4.7%	H	0.9%
Ca	3.4%	Ti	0.6%

21.1   THE PERIODIC TABLE—A GUIDE TO THE ELEMENTS

As you learned in Chapter 2, similarities in the properties of the elements guided creation of the first periodic table. Mendeleev placed elements in groups based partly on the composition of their common compounds with oxygen and hydrogen, as illustrated in Table 21.1. We now understand that the elements can be grouped according to the arrangements of their valence electrons.

Recall that the metallic character of the elements declines on moving from left to right in the periodic table. The first group on the left side of the table — Group 1A, the alkali metals — encompasses the most metallic elements in the periodic table. Elements on the far right are nonmetals, and in between are the metalloids. Metallic character increases from the top of a group to the bottom, as illustrated by Group 4A. Carbon, at the top of the group, is a nonmetal, silicon and germanium are metalloids, and tin and lead are metals (Figure 21.1).

Valence Electrons

The ns and np electrons are the valence electrons for main group elements [◀ SECTION 8.4], and their chemistry is determined by these electrons.

A useful reference point is the noble gases (Group 8A), elements having filled electron subshells. Helium has an electron configuration $1s^2$; the others have ns^2np^6 valence electron configurations. The dominant characteristic of the noble gases is their lack of reactivity. Indeed, the first three elements in the group are not known to form any isolable compounds. The other three elements are now known to have limited chemistry, however, and the discovery of xenon compounds in the 1960s ranks as one of the most interesting developments in modern chemistry.

• **Pictures of Elements**
See the periodic table tool on the CD-ROM for a photo of each element and a listing of its properties.

Figure 21.1   Group 4A elements. A nonmetal, carbon (graphite crucible); a metalloid, silicon (the round lustrous bar); and metals tin (chips of metal) and lead (a bullet, a toy, and a sphere). *(Charles D. Winters)*

Table 21.1 • Similarities within Periodic Groups*

Group	1A	2A	3A	4A	5A	6A	7A
Common oxide	M_2O	MO	M_2O_3	EO_2	E_4O_{10}	EO_3	E_2O_7
Common hydride	MH	MH_2	MH_3	EH_4	EH_3	EH_2	EH
Highest oxidation state	$+1$	$+2$	$+3$	$+4$	$+5$	$+6$	$+7$
Common oxoanion			BO_3^{3-}	CO_3^{2-}	NO_3^-	SO_4^{2-}	ClO_4^-
				SiO_4^{4-}	PO_4^{3-}		

*M denotes a metal and E denotes a nonmetal.

Chapter Goals • Revisited

• **Relate the formulas and properties of compounds to the periodic table.**

• Describe the chemistry of the main group or A-group elements, particularly H; Na and K; Mg and Ca; Al; Si; N and P; O and S; and Cl.

Ionic Compounds of Main Group Elements

Ions with filled s and p subshells are very common — justifying the often-seen reference that elements react in ways that achieve a "noble gas configuration." The elements in Groups 1A and 2A form $+1$ and $+2$ ions with electron configurations the same as those for the previous noble gases. All common compounds of these elements (e.g., NaCl, $CaCO_3$) are ionic (Table 21.2). As expected for ionic compounds, these are crystalline solids with high melting points that conduct electricity in the molten state. Many compounds of Group 3A elements (except the metalloid boron) contain $+3$ ions. For example, most compounds of aluminum contain the Al^{3+} ion.

Elements of Groups 6A and 7A can reach a noble gas configuration by adding electrons. Thus, in many reactions, the Group 7A elements (halogens) form anions with a -1 charge (the halide ions, F^-, Cl^-, Br^-, I^-), and the Group 6A elements form anions with a -2 charge (O^{2-}, S^{2-}, Se^{2-}, Te^{2-}). In Group 5A chemistry, -3 ions with a noble gas configuration (such as the nitride ion, N^{3-}) are occasionally encountered. The energy required to form highly charged anions is large, however; this means that other types of chemical behavior will often take precedence.

Table 21.2 • Some Reactions of Group 1A, 2A, and 3A Metals

Metal	+ Nonmetal	⟶ Product
K(s), Group 1A	$Br_2(\ell)$, Group 7A	KBr(s), ionic
Ba(s), Group 2A	$Cl_2(g)$, Group 7A	$BaCl_2(s)$, ionic
Al(s), Group 3A	$F_2(g)$, Group 7A	$AlF_3(s)$, ionic
Na(s), Group 1A	$S_8(s)$, Group 6A	$Na_2S(s)$, ionic
Mg(s), Group 2A	$O_2(g)$, Group 6A	MgO(s), ionic

Example 21.1 Reactions of Group 1A–3A Elements

Problem • Give the formula and name for the product in each of the following reactions. Write a balanced chemical equation for the reaction.

(a) $Ca(s) + S_8(s)$

(b) $Rb(s) + I_2(s)$

(c) Lithium and chlorine

(d) Aluminum and oxygen

Strategy • Predictions are based on the assumption that ions are formed with electron configurations of the nearest noble gas. Group 1A elements form $+1$ ions, Group 2A elements form $+2$ ions, and metals in Group 3A generally form to $+3$ ions. In their

reactions with metals, halogen atoms typically add a single electron to give anions with a -1 charge and Group 6A elements add two electrons to form anions with a -2 charge. For names of products, refer to the nomenclature discussion on page 92.

Solution •

Balanced Equation	Product Name
(a) $8 Ca(s) + S_8(s) \longrightarrow 8 CaS(s)$	calcium sulfide
(b) $2 Rb(s) + I_2(s) \longrightarrow 2 RbI(s)$	rubidium iodide
(c) $2 Li(s) + Cl_2(g) \longrightarrow 2 LiCl(s)$	lithium chloride
(d) $4 Al(s) + 3 O_2(g) \longrightarrow 2 Al_2O_3(s)$	aluminum oxide

Exercise 21.1 Main Group Element Chemistry

Write a balanced chemical equation for a reaction forming the following compounds from the elements.

(a) NaBr (b) CaSe (c) K_2O (d) $AlCl_3$

Molecular Compounds of Main Group Elements

Unlike the metals of Groups 1A–3A, many avenues of reactivity are open to the nonmetallic elements of Groups 3A–7A. Specifically, molecular compounds are formed in reactions between two or more nonmetals, whereas combining a metal and a nonmetal gives an ionic compound.

Molecular compounds are encountered with some Group 3A elements, especially boron (Figure 21.2), and the chemistry of carbon is dominated by molecular compounds with covalent bonds [← CHAPTER 11]. Similarly, nitrogen chemistry is dominated by molecular compounds. Consider ammonia, NH_3; the ammonium ion, NH_4^+; the various nitrogen oxides; nitric acid, HNO_3; and the nitrate ion, NO_3^-. In each of these species, nitrogen bonds covalently to another nonmetallic element. Also in Group 5A, phosphorus reacts with chlorine to produce the molecular compounds PCl_3 and PCl_5 (see Chapter Focus, page 120).

The valence electron configuration of elements determines the composition of molecular compounds. Involving all the valence electrons in the formation of a compound is a frequent occurrence in main group element chemistry. We should not be surprised to discover halogen compounds in which the central element has the highest possible oxidation number (such as P in PF_5). The highest oxidation number is readily determined: It has the value of the group number. Thus the highest (and only) oxidation number of sodium in its compounds is +1, the highest oxidation number of C is +4, and the highest oxidation number of phosphorus is +5 (Tables 21.1 and 21.3).

With this overview, we can now predict the results of reactions between elements of Groups 1A, 2A, and 3A and elements of Group 6A or 7A. In general, we expect the metal to be oxidized, the nonmetal to be reduced, and ionic compounds to be formed.

Figure 21.2   Boron halides. Liquid BBr_3 (*left*) and solid BI_3 (*right*). Formed from a metalloid and a nonmetal, both are molecular compounds. (*Charles D. Winters*)

Table 21.3 • Fluorine Compounds Formed by Main Group Elements

Group	Compound	Bonding
1A	NaF	Ionic
2A	MgF_2	Ionic
3A	AlF_3	Ionic
4A	SiF_4	Covalent
5A	PF_5	Covalent
6A	SF_6	Covalent
7A	IF_7	Covalent
8A	XeF_4	Covalent

Example 21.2   Predicting Formulas for Compounds of Main Group Elements

Problem • Predict formulas for each of the following:

(a) The product of a reaction between germanium and excess oxygen.

(b) The product of the reaction of arsenic and fluorine.

(c) A compound formed from phosphorus and excess chlorine.

(d) An anion of selenic acid.

Strategy • We will predict that in each reaction the element other than hydrogen or oxygen has its highest oxidation number.

Solution •

(a) The Group 4A element germanium should have a maximum oxidation number of +4. Thus, its oxide has the formula GeO_2.

(b) Arsenic, in Group 5A, reacts vigorously with fluorine to form AsF_5, in which arsenic has an oxidation number of +5.

(c) PCl_5 is the product formed when the Group 5A element phosphorus reacts with chlorine.

(d) The chemistry of S and Se is similar. Sulfur, in Group 6A, has a maximum oxidation number of +6, and forms SO_3 and sulfuric acid, H_2SO_4. Selenium, also in Group 6A, has analogous chemistry, forming SeO_3 and selenic acid, H_2SeO_4. The anion of this acid is the selenate ion, SeO_4^{2-}.

Exercise 21.2   Predicting Formulas for Main Group Compounds

Write formulas for:

(a) Hydrogen telluride

(b) Sodium arsenate

(c) Selenium hexachloride

(d) Perbromic acid

You should expect many similarities among elements in the same periodic group. This means you can use compounds of more common elements as examples when compounds of less common elements are encountered. For example, water, H_2O, is the simplest hydrogen compound of oxygen. Therefore, you can reasonably expect the hydrogen compounds of other Group 6A elements to be H_2S, H_2Se, and H_2Te; all are well known.

Example 21.3   Predicting Formulas

Problem • Predict formulas for

(a) A compound of hydrogen and phosphorus.

(b) The hypobromite ion.

(c) Germane (the simplest hydrogen compound of germanium).

(d) Two oxides of tellurium.

Strategy • Recall as examples some of the compounds of lighter elements in a group and then assume other elements in that group will form analogous compounds.

Solution •

(a) Phosphine, PH_3, has a composition analogous to ammonia, NH_3.

(b) Hypobromite ion, BrO^-, is similar to the hypochlorite ion, ClO^-, the anion of hypochlorous acid (HClO).

(c) GeH_4 is analogous to CH_4 and SiH_4, other Group 4A compounds.

(d) Te and S are in Group 6A. TeO_2 and TeO_3 are analogs of the oxides of sulfur, SO_2 and SO_3.

Example 21.4   Recognizing Incorrect Formulas

Problem • One formula is incorrect in each of the following groups. Pick out the incorrect formula and indicate why it is incorrect.

(a) $CsSO_4$, KCl, $NaNO_3$, Li_2O

(b) MgO, CaI_2, Ba_2SO_4, $CaCO_3$

(c) CO, CO_2, CO_3

(d) PF_5, PF_4^+, PF_2, PF_6^-

Strategy • Look for errors such as incorrect charges on ions or an oxidation number exceeding the maximum possible for the periodic group.

Solution •

(a) $CsSO_4$. Sulfate ion has a -2 charge, so this formula would require a Cs^{2+} ion. Cesium, in Group 1A, forms only $+1$ ions.

(b) Ba_2SO_4. This formula requires a Ba^+ ion (because sulfate is SO_4^{2-}). The ion charge does not equal the group number.

(c) CO_3. Given that O has an oxidation number of -2, carbon would have an oxidation number of $+6$. Carbon is in Group 4A, however, and can have a maximum oxidation number of $+4$.

(d) PF_2. This species has an odd number of electrons. There are very few isolable odd-electron molecules. PF_2 is unlikely to exist.

Comment • To chemists, this exercise is second nature. Incorrect formulas stand out. You will find that your ability to write and recognize correct formulas will grow as you learn more chemistry.

Exercise 21.3   Predicting Formulas

Identify a compound or ion based on a second-period element that has a formula and Lewis structure analogous to each of the following:

(a) PH_4^+ (c) P_2H_4

(b) S_2^{2-} (d) PF_3

Exercise 21.4   Recognizing Incorrect Formulas

Explain why compounds with the following formulas would not be expected to exist: ClO, Na_2Cl, $CaCH_3CO_2$, C_3H_7.

21.2   HYDROGEN

Chemical and Physical Properties of Hydrogen

Hydrogen has three isotopes, two of them stable (protium and deuterium) and one radioactive (tritium).

Hydrogen Isotope Mass (amu)	Symbol	Name
1.0078	^{1}H (H)	Hydrogen (protium)
2.0141	^{2}H (D)	Deuterium
3.0160	^{3}H (T)	Tritium

Of the three isotopes, only H and D are found in nature in measurable quantities. In contrast, tritium, which is produced by cosmic ray bombardment of water in the atmosphere, is found to the extent of 1 atom per 10^{18} atoms of ordinary hydrogen. Tritium is radioactive, half of a sample of the element disappearing in 12.26 years.

Under standard conditions, hydrogen is a colorless gas. Its very low boiling point, 20.7 K, reflects its nonpolar character and low molar mass. It is the least dense gas known and so is ideal for filling lighter-than-air craft.

Deuterium compounds have been thoroughly studied. One important observation is that, since D has twice the mass of H, reactions involving D atom transfer are slightly slower than those involving H atoms. This leads to a way to produce D_2O, or "heavy water." Hydrogen can be produced, albeit expensively, by electrolysis (Figure 21.3).

$$H_2O(\ell) + \text{electrical energy} \longrightarrow H_2(g) + \tfrac{1}{2}O_2(g)$$

Figure 21.3   Electrolysis. Electrolysis of a dilute aqueous solution of H_2SO_4, to give H_2 (*left*) and O_2 (*right*). (*Charles D. Winters*)

(**A Closer Look**)

Hydrogen, Helium, and Balloons

In 1783 Jacques Charles first used hydrogen to fill a balloon large enough to float above the French countryside (page 470), a method used in World War I to float observation balloons. The *Graf Zeppelin*, a passenger-carrying dirigible built in Germany in 1928, was also filled with hydrogen. She carried over 13,000 people between Germany and the United States until 1937, when she was replaced by the *Hindenburg*. The *Hindenburg* was designed to be filled with helium. However, World War II was approaching, and the United States, which has much of the world's supply of helium, would not sell the gas to Germany. The *Hindenburg* had to use hydrogen. Unfortunately, the Hindenburg exploded and burned when landing in Lakehurst, New Jersey in May, 1939. Of the 62 people on board, only about half escaped uninjured. The result is that hydrogen has acquired a reputation as a very dangerous substance. Actually, it is as safe to handle as any other fuel, as evidenced by the large quantities used in rockets today.

The hydrogen-filled dirigible *Hindenburg* crashed ▶ in Lakehurst, New Jersey, in May, 1939. There is some speculation that the aluminum paint coating the skin of the dirigible was more responsible for the fire than the hydrogen. (*© Mary Evans Picture Library/Photo Researchers, Inc.*)

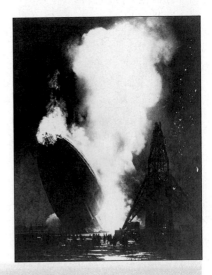

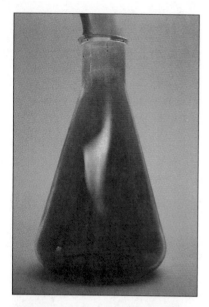

Figure 21.4 The reaction of H₂ and Br₂. Hydrogen gas burns in an atmosphere of bromine vapor to give hydrogen bromide. *(Charles D. Winters)*

In any sample of natural water, there is always a tiny concentration of D_2O. When water is electrolyzed, H_2 is liberated about six times more rapidly than D_2. Thus, as the electrolysis proceeds the liquid remaining is increasingly D_2O. Heavy water is valuable as a moderator of some nuclear reactions used in power generation.

Hydrogen combines chemically with virtually every other element except the noble gases. Although hydrogen-containing compounds seem to come in a bewildering variety, there are only three different types of binary hydrogen-containing compounds.

Ionic metal hydrides are formed in the reaction of H_2 with a Group 1A, 2A, or 3A metal:

$$2\,Na(s) + H_2(g) \longrightarrow 2\,NaH(s)$$
$$Ca(s) + H_2(g) \longrightarrow CaH_2(s)$$

These compounds contain the hydride ion, H^-, in which hydrogen has a -1 oxidation number.

Molecular compounds (such as H_2O, HF, and NH_3) are generally formed by direct combination of the elements (Figure 21.4). In such compounds, covalent bonds to hydrogen are the rule. The oxidation number of the hydrogen atom in these compounds is $+1$.

$$N_2(g) + 3\,H_2(g) \longrightarrow 2\,NH_3(g)$$
$$F_2(g) + H_2(g) \longrightarrow 2\,HF(g)$$

Hydrogen is absorbed by many metals to form *interstitial hydrides,* the third general class of hydrogen compounds. This name refers to the fact that hydrogen atoms reside in the spaces between metal atoms (called interstices) in the crystal lattice. For example, when a piece of palladium metal is used as an electrode for the electrolysis of water, the metal can soak up a thousand times its volume of hydrogen (at STP). Most interstitial hydrides are nonstoichiometric, that is, the ratio of metal and hydrogen is not a whole number. When interstitial hydrides are heated, H_2 is driven out. This allows these materials to store H_2, just as a sponge can store water, and provides one way to store hydrogen for use as a fuel in automobiles [⬅ PAGE 238].

Preparation of Hydrogen

Production of water gas. Water gas, also called synthesis gas, is a mixture of CO and H₂. It is produced by treating coal, coke, or some other hydrocarbon with steam at high temperatures in plants such as is pictured here. Methane has the advantage that it gives more total H₂ per gram than other hydrocarbons (such as coal) and the ratio of the byproduct CO₂ to H₂ is lower. No matter what method is chosen, there are environmental problems: CO₂ or CaCO₃ is produced in large quantities and must be disposed of. *(© Roger Ressmeyer/CORBIS)*

About 300 billion L (STP) of hydrogen gas is produced annually worldwide, and virtually all is used immediately in the manufacture of ammonia [⬅ SECTION 16.7], methanol, or other chemicals.

Some hydrogen is made from coal in the *water gas reaction*, a reaction that has been used for more than 100 years.

$$C(s) + H_2O(g) \longrightarrow \underbrace{H_2(g) + CO(g)}_{\text{water gas or synthesis gas}}$$

The reaction is carried out by injecting water into a bed of red hot coke, and the mixture of gases produced is called "water gas." This mixture was used until about 1950 as a fuel for cooking, heating, and lighting. However, it has serious drawbacks. It produces only about half the heat produced by an equal amount of methane, and the flame is almost invisible. Moreover, as it contains carbon monoxide, water gas is toxic.

The largest quantity of hydrogen is now produced by the *catalytic steam reformation* of hydrocarbons such as methane in natural gas. Methane reacts with steam at high temperature to give H_2 and CO.

$$CH_4(g) + H_2O(g) \longrightarrow 3\,H_2(g) + CO(g) \qquad \Delta H^\circ_{rxn} = +206\ kJ$$

The reaction is rapid at 900–1000 °C and goes nearly to completion. More hydrogen can be obtained in a second step in which the CO that is formed reacts with more water. This so-called *water gas shift reaction* is run at 400–500 °C and is slightly exothermic:

● **Water Gas Reaction in a Car**
The car on page 847 uses the water gas reaction to produce H_2 for a fuel cell.

$$H_2O(g) + CO(g) \longrightarrow H_2(g) + CO_2(g) \qquad \Delta H^\circ_{rxn} = -41 \text{ kJ}$$

The CO_2 formed in the process is removed by reaction with CaO (to give $CaCO_3$), leaving fairly pure hydrogen.

Perhaps the cleanest way to make hydrogen on a relatively large scale is the electrolysis of water (Figure 21.3). This not only provides hydrogen gas but high-purity O_2 as well. Because electricity is quite expensive, however, this method is not generally used commercially.

Table 21.4 and Figure 21.5 provide examples of reactions used to produce H_2 gas in the laboratory to form hydrogen. The most common is the reaction of a metal with an acid [← FIGURE 5.5, PAGE 157]. Alternatively, the reaction of aluminum with NaOH (Figure 21.5b) also generates hydrogen as one product. During World War II, this reaction was used to obtain hydrogen to inflate small balloons for weather observation and to raise radio antennas. Metallic aluminum was plentiful because it came from damaged aircraft.

Table 21.4 ● Methods for Preparing H_2 in the Laboratory

1. Metal + acid $\longrightarrow$ metal salt + H_2

   $Mg(s) + 2 HCl(aq) \longrightarrow MgCl_2(aq) + H_2(g)$

2. Metal + H_2O or base $\longrightarrow$ metal hydroxide or oxide + H_2

   $2 Na(s) + 2 H_2O(\ell) \longrightarrow 2 NaOH(aq) + H_2(g)$

   $2 Fe(s) + 3 H_2O(\ell) \longrightarrow Fe_2O_3(s) + 3 H_2(g)$

   $2 Al(s) + 2 KOH(aq) + 6 H_2O(\ell) \longrightarrow 2 KAl(OH)_4(aq) + 3 H_2(g)$

3. Metal hydride + H_2O $\longrightarrow$ metal hydroxide + H_2

   $CaH_2(s) + 2 H_2O(\ell) \longrightarrow Ca(OH)_2(s) + 2 H_2(g)$

(a) The reaction of magnesium and acid. The products are hydrogen gas and a magnesium salt.

(b) The reaction of aluminum and NaOH. The products of this reaction are hydrogen gas and a solution of $Na[Al(OH)_4]$.

(c) The reaction of CaH_2 and water. The products are hydrogen gas and $Ca(OH)_2$.

Figure 21.5 **Producing hydrogen gas.** *(Charles D. Winters)*

Alkali Metals and Water. Sodium and the other alkali metals react vigorously with water (see page 288). *(Charles D. Winters)*

The combination of a metal hydride and water (Table 21.4 and Figure 21.5c) is an efficient but expensive way to synthesize H_2 in the laboratory. The reaction is more commonly used to dry solvents because the metal hydride reacts with traces of water present in the solvent.

Exercise 21.5 Hydrogen Chemistry

Use bond energies [← PAGE 356] to calculate the enthalpy change for the reaction of methane and water to give hydrogen and carbon monoxide (with all compounds in the gas phase). Considering the bond energies of reactants and products, suggest a reason the reaction is endothermic.

21.3 SODIUM AND POTASSIUM

Sodium and potassium are the sixth and seventh most abundant elements in Earth's crust, 2.6% and 2.4%, respectively, by mass. Both metals, as well as the other Group 1A elements, are highly reactive with oxygen, water, and other oxidizing agents [← PAGE 65]. In all cases, compounds of the Group 1A metals contain the element as a +1 ion. None is ever found in nature as the uncombined element.

Most sodium and potassium compounds are water-soluble [← SOLUBILITY RULES, FIGURE 5.1], so it is not surprising that sodium and potassium compounds are found either in the oceans or in underground deposits that are the residue of ancient seas. To a much smaller extent, these elements are also found in minerals, like Chilean saltpeter ($NaNO_3$) and borax ($Na_2B_4O_7 \cdot 10\ H_2O$) [FIGURE 2.11, PAGE 67].

Some NaCl is essential in the diet of humans and animals because many biological functions are controlled by the concentrations of Na^+ and Cl^- ions (Figure 21.6). The fact that salt has been important for a long time is evident in surprising ways. We are paid a "salary" for work done. The word is derived from the Latin word *salarium,* which meant "salt money" because Roman soldiers were paid in salt.

Preparation and Properties of Sodium and Potassium

Sodium is produced by electrolysis of molten NaCl (Figure 21.7) [← SECTION 20.7]. Electrolysis is necessary because common chemical reducing agents are not powerful enough to convert sodium ions to sodium metal. Potassium can also be made by electrolysis, but molten potassium is soluble in molten KCl, making separation of the metal difficult. The preferred preparation of potassium uses the reaction of sodium vapor with molten KCl, with potassium being continually removed from the equilibrium mixture.

$$Na(g) + KCl(\ell) \rightleftharpoons K(g) + NaCl(\ell)$$

Sodium and potassium are silvery metals that are soft and easily cut with a knife (see page 13). Their densities are just a bit less than the density of water, and their melting points are quite low (93.5 °C for sodium and 65.65 °C for potassium).

All the alkali metals are highly reactive. When exposed to moist air, the metal surface is quickly coated with film of oxide or hydroxide. Consequently, the metals must be stored in a way that avoids contact with air; this is often done by placing them in kerosene or mineral oil.

The great reactivity of Group 1A metals is exemplified by their reaction with water, which generates an aqueous solution of the metal hydroxide and hydrogen (page 288),

$$2\ Na(s) + 2\ H_2O(\ell) \longrightarrow 2\ Na^+(aq) + 2\ OH^-(aq) + H_2(g)$$

Figure 21.6 The importance of salt. All animals, including humans, need a certain amount of salt in their diet. Sodium ion is important in maintaining electrolyte balance and in regulating osmotic pressure. *(Charles D. Winters)*

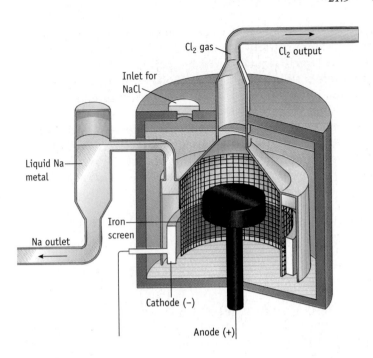

Cl₂ gas

Cl₂ output

Inlet for
NaCl

Liquid Na
metal

Iron
screen

Na outlet

Cathode (−)

Anode (+)

Figure 21.7   A Downs Cell for preparing sodium. A circular iron cathode is separated from the graphite anode by an iron screen. At the temperature of the electrolysis, about 600 °C, sodium is a liquid. It floats to the top and is drawn off periodically. Chlorine gas is produced at the anode and collected inside the inverted cone in the center of the cell.

and their reaction with any of the halogens to yield a metal halide [⬅ PAGE 13]:

$$2 \ Na(s) + Cl_2(g) \longrightarrow 2 \ NaCl(s)$$

$$2 \ K(s) + Br_2(\ell) \longrightarrow 2 \ KBr(s)$$

Chemistry sometimes produces surprises. Group 1A metal oxides, M_2O, are known, but they are not the principal products of reactions between the Group 1A elements and oxygen. The primary product of the reaction of sodium and oxygen is sodium *peroxide*, Na_2O_2, whereas the principal product from the reaction of potassium and oxygen is KO_2, potassium *superoxide*.

$$2 \ Na(s) + O_2(g) \longrightarrow Na_2O_2(s)$$

$$K(s) + O_2(g) \longrightarrow KO_2(s)$$

Both Na_2O_2 and KO_2 are ionic compounds. The Group 1A cation is paired with either the peroxide ion (O_2^{2-}) or the superoxide ion, O_2^-. These compounds are not just laboratory curiosities. They are used in oxygen generation devices in places where people are confined, such as submarines, aircraft, and spacecraft, or when an emergency supply is needed. When a person breathes, 0.82 L of CO_2 is exhaled for every liter of O_2 inhaled. Thus, a requirement of an O_2 generation system is that it should produce a larger volume of O_2 than the volume of CO_2 taken in. This requirement is met with peroxides and superoxides. With KO_2 the reaction is

$$4 \ KO_2(s) + 2 \ CO_2(g) \longrightarrow 2 \ K_2CO_3(s) + 3 \ O_2(g)$$

Important Sodium Compounds

Electrolysis of aqueous sodium chloride (*brine*) is the basis of one of the largest chemical industries in the United States.

$$2 \ NaCl(aq) + 2 \ H_2O(\ell) \longrightarrow Cl_2(g) + 2 \ NaOH(aq) + H_2(g)$$

The products from this process are chlorine, sodium hydroxide, and hydrogen, and give the industry its name: the chlor-alkali industry. Over 10 billion kilograms of NaOH and Cl_2 are produced annually in the United States.

A closed-circuit breathing apparatus that generates its own oxygen. One source of oxygen is potassium superoxide (KO_2). Both carbon dioxide and moisture exhaled by the wearer into the breathing tube react with the KO_2 to generate oxygen. Because the rate of the chemical reaction is determined by the quantity of moisture and carbon dioxide exhaled, the production of oxygen is regulated automatically. With each exhalation, more oxygen is produced by volume than is required by the user. (*Courtesy of the Mine Safety Appliances Company.*)

Sodium carbonate, Na_2CO_3, is another commercially important compound of sodium. This compound is also known by two common names, soda ash and washing soda. In the past it was largely manufactured by combining NaCl and CO_2 in the Solvay process (and this is still the method of choice in many countries). In the United States, however, it is obtained from naturally occurring deposits of the mineral *trona*, $Na_2CO_3 \cdot NaHCO_3 \cdot 2\,H_2O$ (Figure 21.8).

Owing to the environmental problems associated with chlorine and its byproducts, there is considerable interest in manufacturing sodium hydroxide by methods other than brine electrolysis. This has led to a revival of the old "soda-lime process," which produces NaOH from inexpensive lime (CaO) and soda (Na_2CO_3):

$$Na_2CO_3(aq) + CaO(s) + H_2O(\ell) \longrightarrow 2\,NaOH(aq) + CaCO_3(s)$$

The insoluble calcium carbonate byproduct is filtered off and is recycled into the process by heating it (calcining) to recover lime:

$$CaCO_3(s) \longrightarrow CaO(s) + CO_2(g)$$

Sodium bicarbonate, $NaHCO_3$, also known as *baking soda,* is another common compound of sodium. Not only is $NaHCO_3$ used in cooking but it is also added in small amounts to table salt. This is because NaCl is often contaminated with small amounts of $MgCl_2$. The magnesium salt is hygroscopic; that is, it picks water up from the air and, in doing so, causes the NaCl to clump. Adding $NaHCO_3$ converts $MgCl_2$ to magnesium carbonate, a non-hygroscopic salt:

$$MgCl_2(s) + 2\,NaHCO_3(s) \longrightarrow MgCO_3(s) + 2\,NaCl(s) + H_2O(\ell) + CO_2(g)$$

(a) (Above) A mine in California. The mineral trona is taken from a mine 1600 feet deep. *(Jack Dermid/Photo Researchers, Inc.)*

(b) (Right) Blocks of trona are cut from the face of the mine. *(Courtesy of FMC Wyoming Corp.)*

Figure 21.8   Producing soda ash. Trona mined in California is processed into soda ash and other sodium chemicals. Soda ash is the ninth most widely used chemical in the United States. Domestically, about half of the soda ash is used in making glass. The remainder is used to make chemicals including sodium silicate, sodium phosphate, sodium cyanide, and detergents. Some is also used in the pulp and paper industry and in water treatment.

Exercise 21.6 Sodium Production

What current must be used in a Downs cell operating at 7.0 V to produce 1.00 metric ton (exactly 1000 kg) of sodium per day? Assume 100% efficiency.

21.4   TWO ALKALINE EARTH ELEMENTS: CALCIUM AND MAGNESIUM

The "earth" part of the name *alkaline earth* is from the days of medieval alchemy. To alchemists, any solid that did not melt and was not changed by fire into another substance was called an "earth." Compounds of the 2A elements, such CaO, were alkaline according to experimental tests of the alchemists: they had a bitter taste and neutralized acids. With very high melting points, these compounds were unaffected by fire.

Calcium and magnesium rank fifth and seventh in abundance on Earth. Both elements form many commercially important compounds, and we shall focus our attention on this chemistry.

Like the Group 1A elements, the Group 2A elements are very reactive, so they are only found in nature combined with other elements. Unlike Group 1A metals, however, many compounds of the Group 2A elements have low water solubility, which explains their occurrence as various minerals. Common calcium minerals include limestone ($CaCO_3$), gypsum ($CaSO_4 \cdot 2\,H_2O$), and fluorite (CaF_2) (Figure 21.9). Magnesite ($MgCO_3$), talc or soapstone ($3\,MgO \cdot 4\,SiO_2 \cdot H_2O$), and asbestos ($3\,MgO \cdot 4\,SiO_2 \cdot 2\,H_2O$) are common magnesium-containing minerals. The mineral dolomite, $MgCO_3 \cdot CaCO_3$ contains both magnesium and calcium.

Limestone, a sedimentary rock, is found widely on Earth's surface (Figure 21.9). These deposits are the fossilized remains of marine life. Other forms of calcium carbonate include marble and Icelandic spar, which forms large, clear crystals (Figure 21.9).

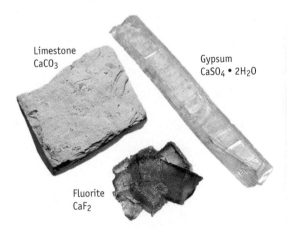

Limestone
$CaCO_3$

Gypsum
$CaSO_4 \cdot 2H_2O$

Fluorite
CaF_2

Common minerals of Group 2A elements.

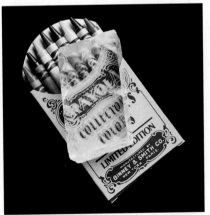

Icelandic spar. This mineral, one of a number of crystalline forms of $CaCO_3$, displays birefringence, a property in which a double image is formed when light passes through the crystal.

The walls of the Grand Canyon in Arizona are largely limestone or dolomite.

Figure 21.9   Various minerals containing calcium and magnesium. *(a and b, Charles D. Winters; c, James Cowlin/Image Enterprises/Phoenix, AZ)*

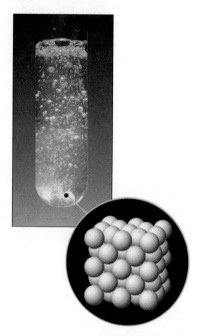

Figure 21.10 The reaction of calcium and warm water. Hydrogen bubbles are seen rising from the metal surface. The inset is a model of hexagonal closed packed calcium metal. *(Charles D. Winters)*

Calcium and magnesium are fairly high-melting, silvery metals. Their chemical properties present few surprises. They are oxidized by a wide range of oxidizing agents to form ionic compounds that contain the M^{2+} ion. For example, these elements combine with halogens to form MX_2, with oxygen or sulfur to form MO or MS — the reaction of Mg and O_2 [← FIGURE 4.1], and with water to form hydrogen and the metal hydroxide, $M(OH)_2$ (Figure 21.10). With acids, hydrogen is evolved (see Figure 21.5) and a salt of the metal and the anion of the acid results.

Metallurgy of Magnesium

Several hundred thousand tons of magnesium are produced annually, largely for use in lightweight alloys. (Magnesium has a very low density, 1.74 g/cm^3.) In fact, most aluminum used today contains about 5% magnesium to improve its mechanical properties and to make it more resistant to corrosion. Other alloys having more magnesium than aluminum are used when a high strength-to-weight ratio is needed and when corrosion resistance is important, such as in aircraft and automotive parts and in lightweight tools.

Interestingly, magnesium-containing minerals are not the source of this element. Most magnesium is obtained from sea water, in which Mg^{2+} ion is present in a concentration of about 0.05 M (Figure 21.11). To obtain the element, Mg^{2+} is first precipitated from sea water as the relatively insoluble hydroxide (K_{sp} for $Mg(OH)_2 = 5.6 \times 10^{-12}$). The source of OH^- in this reaction, $Ca(OH)_2$, is prepared in a sequence of reactions beginning with $CaCO_3$, which may be in the form of seashells. Heating $CaCO_3$ gives CO_2 and CaO, and addition of water to CaO gives calcium hydroxide. When $Ca(OH)_2$ is added to sea water, $Mg(OH)_2$ precipitates:

$$Mg^{2+}(aq) + Ca(OH)_2(aq) \longrightarrow Mg(OH)_2(s) + Ca^{2+}(aq)$$

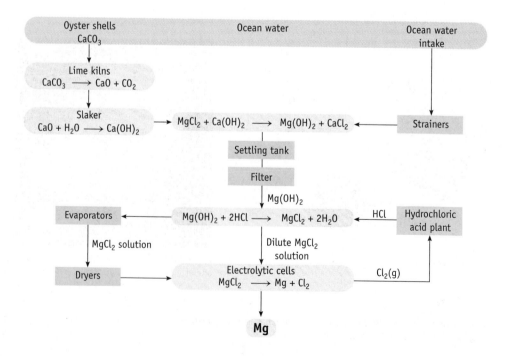

Figure 21.11 Diagram of the process used to produce magnesium metal from the magnesium in sea water.

Magnesium hydroxide is isolated by filtration and then neutralized with hydrochloric acid:

$$Mg(OH)_2(s) + 2\ HCl(aq) \longrightarrow MgCl_2(aq) + 2\ H_2O(\ell)$$

After evaporating the water, anhydrous magnesium chloride remains. Solid $MgCl_2$ melts at 708 °C and the molten salt is electrolyzed to give the metal and chlorine:

$$MgCl_2(\ell) \longrightarrow Mg(s) + Cl_2(g)$$

Calcium Minerals and Their Applications

The most common calcium minerals are the fluoride, phosphate, and carbonate salts of the element. Fluorite, CaF_2, and fluorapatite $[CaF_2 \cdot 3\ Ca_3(PO_4)_2]$ are important as commercial sources of fluorine. Almost half of the CaF_2 mined is used in the steel industry, where it is added to the mixture of materials that is melted to make crude iron. The CaF_2 acts to remove some impurities and improves the separation of molten metal from silicate impurities and other byproducts that come from reducing iron ore to the metal [CHAPTER 22 ➡]. A second major application of fluorite is in the manufacture of hydrofluoric acid by a reaction of the mineral with concentrated sulfuric acid:

$$CaF_2(s) + H_2SO_4(\ell) \longrightarrow 2\ HF(g) + CaSO_4(s)$$

Apatite is a mineral with the general formula of $CaX_2 \cdot 3\ Ca_3(PO_4)_2$ (X = F, Cl, OH). (The apatite is the elongated crystal in the center of a matrix of other rock.) *(Charles D. Winters)*

Chemical Perspectives

Of Romans, Limestone, and Champagne

The stones of the Appian Way in Italy, a road conceived by the Roman senate in about 310 BC, are cemented with mortar made from limestone. The purpose of the Appian Way was to serve as a military road linking Rome to seaports from which soldiers could embark to Greece and other Mediterranean ports. The road stretches

A limestone cave in the champagne region ▶ of France. The cave is used for storing and aging champagne. *(Roy/Explorer/Photo Researchers, Inc.)*

560 kilometers (350 miles) from Rome to Brindisi on the Adriatic Sea at the heel of the Italian boot. It took almost 200 years to construct. The road had a standard width of 14 Roman feet, approximately 20 feet, large enough to allow two chariots, one per lane, and featured two sidewalks of 4 feet each. Every 10 miles or so there were horse-changing stations with taverns, shops and latrinæ, the famous Roman restrooms.

All over the Roman Empire buildings, temples, and aqueducts were constructed of limestone and marble. Mortar was made by "burning" chips from the stonecutting.

◀ The Appian Way in Italy. *(Louis Goldman/Photo Researchers, Inc.)*

Where there was no limestone or marble, as in central France, the Romans dug chalk (also $CaCO_3$) from the ground for cementing sandstone blocks. This created huge caves that remain to this day and are used for aging and storing champagne.

Hydrofluoric acid is used to make cryolite, Na_3AlF_6, a material needed in aluminum production (see Section 21.5) and in the manufacture of fluorocarbons such as tetrafluoroethylene, the precursor to Teflon [◀ TABLE 11.12].

Apatites have the general formula $CaX_2 \cdot 3\ Ca_3(PO_4)_2$ (X = F, Cl, OH). Over 100 million tons of apatite are mined annually; Florida alone accounts for about one third of the world's output. Most is converted to phosphoric acid by reaction with sulfuric acid. (Phosphoric acid is needed in the manufacture of a multitude of products including fertilizer and detergents, baking powder, and various food products) (see Section 21.7).

$$CaF_2 \cdot 3\ Ca_3(PO_4)_2(s) + 10\ H_2SO_4(aq) \longrightarrow 10\ CaSO_4(s) + 6\ H_3PO_4(aq) + 2\ HF(g)$$
fluorapatite

Calcium carbonate and calcium oxide (lime) are of special interest. The thermal decomposition of $CaCO_3$ to lime is one of the oldest chemical reactions known. Lime is in the top ten industrial chemicals produced today, with about 20 billion kilograms produced annually.

Limestone, which consists mostly of calcium carbonate, has been used in agriculture for centuries. It is spread on fields to neutralize acidic compounds in the soil and to supply the essential nutrient Ca^{2+}. Because magnesium carbonate is often present in limestone, "liming" a field also supplies Mg^{2+}, another important plant nutrient.

For several thousand years, lime has been used in mortar (a lime, sand, and water paste) to secure stones to one another in building houses, walls, and roads. The Chinese used it to set stones in the Great Wall. The Romans perfected its use, and the fact that many of their constructions still stand is testament both to their skill and the usefulness of lime. The famous Appian Way used lime mortar between several layers of its stones.

The utility of mortar depends on some simple chemistry. Mortar consists of one part lime to three parts sand, with water added to make a thick paste. The first reaction that occurs, referred to as *slaking*, produces calcium hydroxide, which is known as *slaked lime*. When the mortar is placed between bricks or stone blocks, it slowly absorbs CO_2 from the air and the slaked lime reverts to calcium carbonate.

$$Ca(OH)_2(s) + CO_2(g) \longrightarrow CaCO_3(s) + H_2O(\ell)$$

The sand grains are bound together by the particles of calcium carbonate.

"Hard water" contains dissolved metal ions, chiefly Ca^{2+} and Mg^{2+}. Water containing dissolved CO_2 reacts with limestone [◀ PAGE 656]:

$$CaCO_3(s) + H_2O(\ell) + CO_2(g) \rightleftharpoons Ca^{2+}(aq) + 2\ HCO_3^-(aq)$$

Figure 21.12 Limestone from hard water. The inside of the pipe is almost blocked with $CaCO_3$ deposited by hard water flowing through it. *(Sheila Terry/Science Photo Library/Photo Researchers, Inc.)*

Alkaline Earths and Biology

Plants and animals derive energy from the oxidation of a sugar, glucose, with oxygen. Plants are unique, however, in being able to synthesize glucose from CO_2 and H_2O using sunlight as an energy source. The process is initiated by chlorophyll, a very large, magnesium-based molecule.

In your body the metal ions Na^+, K^+, Mg^{2+}, and Ca^{2+} serve regulatory functions. Although the two alkaline earth metal ions are required by living systems, the other Group 2A elements are toxic. Beryllium compounds are carcinogenic, and soluble barium salts are poisons. You may be concerned if your physician asks you to drink a "barium cocktail" to check the condition of your digestive tract. There is no problem, though, because the "cocktail" contains very insoluble $BaSO_4$ ($K_{sp} = 1.1 \times 10^{-10}$). Barium sulfate is opaque to x-rays, so its path through your organs appears on an x-ray photograph.

The calcium-containing compound hydroxyapatite is the main component of tooth enamel. Cavities in your teeth form when acids decompose the weakly basic apatite coating.

$$Ca_5(PO_4)_3(OH)(s) + 4\,H^+(aq) \longrightarrow$$
$$5\,Ca^{2+}(aq) + 3\,HPO_4^{2-}(aq) + H_2O(\ell)$$

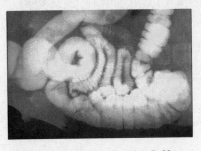

X-ray of an intestinal tract using $BaSO_4$ to make the organs visible. *(Susan Leavines/Science Source/Photo Researchers, Inc.)*

This can be prevented, however, by converting hydroxyapatite to a more acid-resistant coating of fluoroapatite.

$$Ca_5(PO_4)_3(OH)(s) + F^-(aq) \longrightarrow$$
$$Ca_5(PO_4)_3(F)(s) + OH^-(aq)$$

The source of fluoride ion can be sodium fluoride or sodium monofluorophosphate (Na_2FPO_3, commonly known as MFP) in your toothpaste.

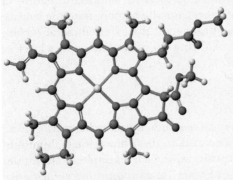

A molecule of chlorophyll, which has magnesium as the central element.

This reaction can be reversed. When hard water is heated, the solubility of CO_2 drops, and the equilibrium shifts to the left. If this happens in a heating system or steam-generating plant, the walls of the hot water pipes can become coated or even blocked with solid $CaCO_3$ (Figure 21.12). In your home, you may notice a coating of calcium carbonate on the inside of cooking pots.

The previous equation also describes the chemistry occurring inside limestone caves (page 656). The acidic oxide CO_2 reacts with $Ca(OH)_2$ to produce white, solid $CaCO_3$. When further CO_2 is available, however, the $CaCO_3$ dissolves due to the formation of aqueous Ca^{2+} and HCO_3^- ions.

Exercise 21.7

Beryllium, the lightest element in Group 2A, has some important industrial uses, but exposure (by breathing) some of its compounds can cause berylliosis. Search the World Wide Web for the uses of the element and the causes and symptoms of berylliosis.

21.5 ALUMINUM

Aluminum is the third most abundant element in Earth's crust (7.4%). Its low cost and the excellent characteristics of its alloys with other metals (low density, strength, ease of handling in fabrication, and inertness toward corrosion, among others)

A large passenger plane may use more than 50 tons of aluminum alloy. *(Allen Green/Photo Researchers, Inc.)*

have led to wide use. You know it best in the form of aluminum foil, aluminum cans, and parts of aircraft.

Pure aluminum is soft and weak; moreover, it loses strength rapidly above 300 °C. What we call aluminum is actually aluminum alloyed with small amounts of other elements to strengthen the metal and improve its properties. A typical alloy may contain about 4% copper with smaller amounts of silicon, magnesium, and manganese. Softer, more corrosion-resistant alloys for window frames, furniture, highway signs, and cooking utensils may include only manganese.

The standard reduction potential of aluminum ($Al^{3+}(aq) + 3\ e^- \longrightarrow Al(s)$; $E° = -1.66$ V) tells us that aluminum is easily oxidized. From this, we might expect aluminum to be highly susceptible to corrosion but in fact it is quite resistant. Aluminum's corrosion resistance is due to the formation of a thin, tough, and transparent skin of Al_2O_3 that adheres to the metal surface. An important feature of the protective oxide layer is that it rapidly self-repairs. If you penetrate the surface coating by scratching it or using some chemical agent, the exposed metal surface immediately reacts with oxygen (or other oxidizing agent) to form a new layer of oxide over the damaged area (Figure 21.13).

Metallurgy of Aluminum

The first preparations of aluminum were carried out by reducing $AlCl_3$ using sodium or potassium. This was a costly process and, in the 19th century, aluminum was a precious metal. At the 1855 Paris Exposition, in fact, a sample of aluminum was exhibited along with the crown jewels of France. In an interesting coincidence two men, Frenchman Paul Heroult (1863–1914) and American Charles Hall, in 1886, simultaneously and independently conceived of the electrochemical method used today. The Hall–Heroult method bears the names of the two discoverers.

Aluminum is found in nature as aluminosilicates, minerals such as clay that are based on aluminum, silicon, and oxygen. As these minerals weather, they break down to various forms of hydrated aluminum oxide, $Al_2O_3 \cdot n\,H_2O$, called *bauxite*. Mined in huge quantities, bauxite is the raw material from which aluminum is obtained. The first step is to purify the ore, separating Al_2O_3 from iron and silicon

(a)   (b)

Figure 21.13 Corrosion of aluminum. (a) A ball of aluminum foil is added to a solution of copper(II) nitrate and sodium chloride. Normally, the coating of chemically inert Al_2O_3 on the surface of aluminum protects the metal from further oxidation. **(b)** In the presence of the Cl^- ion, however, the coating of Al_2O_3 is breached, and aluminum reduces copper(II) ions to copper metal. The reaction is rapid and is so exothermic that the water can boil on the surface of the foil. (Notice that the blue color of copper(II) ions has faded, as these ions are consumed in the reaction.) *(Charles D. Winters)*

oxides. This is done by the *Bayer process*, which relies on the amphoteric, basic, or acidic nature of the various oxides. Silica, SiO_2, is an acidic oxide, Al_2O_3 is amphoteric, and Fe_2O_3 is a basic oxide. Silica and Al_2O_3 dissolve in a hot concentrated solution of caustic soda (NaOH), leaving insoluble Fe_2O_3 to be filtered out:

$$Al_2O_3(s) + 2\ NaOH(aq) + 3\ H_2O(\ell) \longrightarrow 2\ Na[Al(OH)_4](aq)$$

$$SiO_2(s) + 2\ NaOH(aq) + 2\ H_2O(\ell) \longrightarrow Na_2[Si(OH)_6](aq)$$

By treating the solution containing aluminate and silicate anions with CO_2, Al_2O_3 precipitates and the silicate ion remains in solution. Recall that CO_2 is an acidic oxide that forms the weak acid H_2CO_3 in water, so the Al_2O_3 precipitation in this step is an acid–base reaction

$$H_2CO_3(aq) + 2\ Na[Al(OH)_4](aq) \longrightarrow Na_2CO_3(aq) + Al_2O_3(s) + 5\ H_2O(\ell)$$

Metallic aluminum is obtained from the purified bauxite by electrolysis (Figure 21.14). High-melting bauxite is first mixed with cryolite, Na_3AlF_6 to give a lower melting mixture (melting temperature = 980 °C) that is electrolyzed in a cell with graphite electrodes. The cell operates at a relatively low voltage (4.0–5.5 V) but with an extremely high current (50,000–150,000 amps). Aluminum is produced at the cathode and oxygen at the anode. To produce a kilogram of aluminum requires 13 to 16 kilowatt-hours of energy plus the energy required to maintain the high temperature.

Aluminum Compounds

Aluminum dissolves in hydrochloric acid but not in nitric acid. The latter rapidly oxidizes the surface of aluminum, and the film of Al_2O_3 protects the metal from further attack. This protection allows nitric acid to be shipped in aluminum tanks.

Aluminum does not react with nitric acid. Nitric acid, a strongly oxidizing acid, reacts vigorously with copper (*left*) but aluminum (*right*) is untouched. *(Charles D. Winters)*

● **Recycling Aluminum**
The energy costs in the metallurgy of aluminum are high. Recycling of aluminum cans and other aluminum scrap is a viable proposition because these costs are largely avoided.

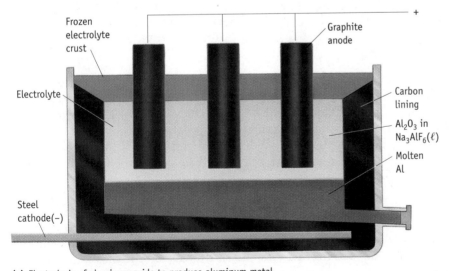

(a) Electrolysis of aluminum oxide to produce aluminum metal.

(b) Molten aluminum from recycled metal.

Figure 21.14   Industrial production of aluminum by electrolysis. (a) Purified aluminum-containing ore (bauxite), essentially Al_2O_3, and cryolite (Na_3AlF_6) gives a mixture that melts at a lower temperature than Al_2O_3 alone. The aluminum-containing substances are reduced at the cathode to give molten aluminum. Oxygen is produced at the carbon anode, and the gas reacts slowly with the carbon to give CO_2, leading to eventual destruction of the electrode. **(b)** Molten aluminum alloy, produced from recycled metal, at 760 °C, in 1.6×10^4-kg-capacity crucibles. *(b, Courtesy, Allied Metal Company, Chicago, IL)*

Figure 21.15 Sapphires and rubies.
Both are minerals based on Al_2O_3 in which a few Al^{3+} ions have been replaced by ions such as Cr^{3+}, Fe^{2+}, or Ti^{4+}. (*top*) The Star of Asia sapphire. (*middle*) Various sapphires. (*bottom*) Uncut corundum. (*Chip Clark/Smithsonian Museum of Natural History*)

Various salts of aluminum dissolve in water, giving the hydrated $Al^{3+}(aq)$ ion. These solutions are acidic [TABLE 17.3, PAGE 702], owing to the following equilibrium:

$$[Al(H_2O)_6]^{3+}(aq) + H_2O(\ell) \rightleftharpoons [Al(H_2O)_5(OH)]^{2+}(aq) + H_3O^+(aq)$$

Adding acid shifts the equilibrium to the left, whereas base causes the equilibrium to shift to the right. Addition of sufficient hydroxide ion results in precipitation of the hydrated oxide $Al_2O_3 \cdot 3\ H_2O$.

Aluminum oxide, Al_2O_3, formed by dehydrating the hydrated oxide, is quite insoluble in water and generally resistant to chemical attack. In the crystalline form, aluminum oxide is known as *corundum*. This material is extraordinarily hard, a property that leads to its use as the abrasive in grinding wheels, "sandpaper," and toothpaste.

Some gems are impure aluminum oxide. Rubies, beautiful red crystals prized for jewelry and used in some lasers, are Al_2O_3 contaminated with a small amount of Cr^{3+} (Figure 21.15). The Cr^{3+} ions replace some of the Al^{3+} ions in the crystal lattice. Blue sapphires consist of Al_2O_3 with Fe^{2+} and Ti^{4+} impurities in place of Al^{3+} ions. Synthetic rubies were first made in 1902, and the worldwide capacity is now about 200,000 kg/year; much of this production is used for jewel bearings in watches and instruments.

Exercise 21.8   Aluminum, Sewage Treatment, and Paper

Aluminum sulfate, $Al_2(SO_4)_3$, is used to treat sewage and in making paper. Both uses are related to the acid–base properties of the aluminum ion. When dissolved in water, is aluminum ion an acid or a base? Write a balanced equation to show how it behaves.

21.6   SILICON

Silicon is second after oxygen in abundance in the Earth's crust, so it is not surprising that we are surrounded by silicon-containing materials: bricks, pottery, porcelain, lubricants, sealants, computer chips, and solar cells. The computer revolution is based on the semiconducting properties of silicon.

Reasonably pure silicon can be made in large quantities by heating pure silica sand with purified coke to approximately 3000 °C in an electric furnace:

$$SiO_2(s) + 2\ C(s) \longrightarrow Si(\ell) + 2\ CO(g)$$

The molten silicon is drawn off the bottom of the furnace and allowed to cool to a shiny blue-gray solid. Because extremely high purity silicon is needed for the electronics industry, purifying raw silicon requires several steps. First, the silicon in the impure sample is allowed to react with chlorine to convert the silicon to liquid silicon tetrachloride.

$$Si(s) + 2\ Cl_2(g) \longrightarrow SiCl_4(\ell)$$

Silicon tetrachloride (boiling point of 57.6 °C) is carefully purified by distillation and then reduced to silicon using magnesium:

$$SiCl_4(g) + 2\ Mg(s) \longrightarrow 2\ MgCl_2(s) + Si(s)$$

The magnesium chloride is washed out with water, and the silicon is remelted and cast into bars. A final purification is carried out by *zone refining*, a process in which a special heating device is used to melt a narrow segment of the silicon rod. The heater is moved slowly down the rod. Impurities contained in the silicon tend to remain in the liquid phase because the melting point of a mixture is lower than that of the pure element (Chapter 14). The silicon that crystallizes above the heated zone is therefore of a higher purity (Figure 21.16).

Silicon Dioxide

The simplest oxide of silicon is SiO_2, commonly called *silica*, a constituent of many rocks such as granite and sandstone. Quartz is a pure crystalline form of silica. Impurities in quartz produce gemstones such as amethyst (Figure 21.17).

Silica and CO_2 are oxides of two elements in the same chemical group, so similarities might be expected. In fact, SiO_2 is a high-melting solid (quartz melts at 1610 °C), whereas CO_2 is a gas at room temperature and 1 bar. This great disparity arises from the different structures of these two oxides. Carbon dioxide is a molecular compound, with the carbon atom linked to each oxygen by a double bond. In contrast, SiO_2 is a network solid [CD-ROM, Models Folder], which is the preferred structure because the energy of two $Si = O$ double bonds is much less than the energy of four $Si - O$ single bonds. The contrast between SiO_2 and CO_2 exemplifies a more general phenomenon. Multiple bonds, often encountered between second-period elements, are rare among elements in the lower periods. There are many compounds with multiple bonds to carbon, but *very* few compounds with multiple bonds to silicon.

Quartz crystals are used to control the frequency of almost all radio and television transmissions. Because these and related applications use so much quartz, there is not enough natural quartz to fulfill demand, and so quartz is synthesized. Noncrystalline, or vitreous, quartz, made by melting pure silica sand, is placed in a steel "bomb" and dilute aqueous NaOH is added. A "seed" crystal is placed in the mixture, just as you might use a seed crystal in a hot sugar solution to grow rock candy. When the mixture is heated above the critical temperature of water (above 400 °C and 1700 atm) over a period of days, pure quartz crystallizes.

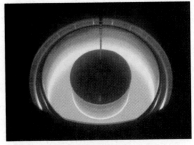

Figure 21.16   Making pure silicon. A rod of pure silicon is being withdrawn from a vat of molten silicon. To form the rod, a seed crystal was mechanically rotated in molten silicon. Pure silicon formed on the seed crystal as the rod was slowly withdrawn from the vat. *(Photo compliments of Bullen Ultrasonics, Inc. Eaton, OH))*

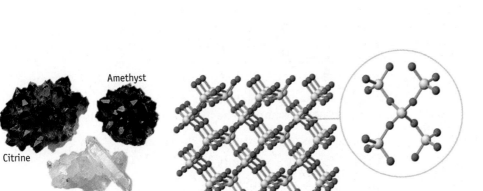

(a) Pure quartz is colorless, but the presence of small amounts of impurities adds color. Purple amethyst and brown citrine crystals are quartz with iron impurities.

(b) Quartz is a network solid in which each Si atom is bound tetrahedrally to four O atoms, each O atom linked to another Si atom. The basic structure consists of interlocking chains of Si and O atoms. See the Molecular Models folder on the *Interactive General Chemistry CD-ROM*.

Figure 21.17   Various Forms of Quartz. *(a, Charles D. Winters)*

Packages of silica gel (solid, noncrystalline SiO₂) are often used to keep electronic equipment dry when stored. The gel is also used to clarify beer; passing beer through a bed of silica gel removes minute particles that make the brew cloudy. Yet another use is as kitty litter. *(Charles D. Winters)*

Silica is resistant to attack by all acids except HF, with which it reacts to give SiF_4 and H_2O. It also dissolves slowly in hot, molten NaOH or Na_2CO_3 to give Na_4SiO_4:

$$SiO_2(s) + 4\ HF(\ell) \longrightarrow SiF_4(g) + 2\ H_2O(\ell)$$

$$SiO_2(s) + 2\ Na_2CO_3(\ell) \longrightarrow Na_4SiO_4(s) + 2\ CO_2(g)$$

After the molten mixture has cooled, hot water under pressure is added. This partially dissolves the material to give a solution of sodium silicate. After filtering off insoluble sand or glass, the solvent is evaporated to leave sodium silicate, called *water glass*. The biggest single use of this material is in household and industrial detergents because a sodium silicate solution maintains pH by its buffering ability. Additionally, sodium silicate is used in various adhesives and binders, especially for gluing corrugated cardboard boxes.

If sodium silicate is treated with acid, a gelatinous precipitate of SiO_2 called "silica gel" is obtained. Washed and dried, silica gel is a highly porous material with dozens of uses. It is a drying agent, readily absorbing up to 40% of its own weight of water. Small packets of silica gel are often placed in packing boxes of merchandise during storage. The material is frequently stained with $(NH_4)_2CoCl_4$, a humidity detector that is pink when hydrated and blue when dry.

Silicate Minerals

Silicate minerals are a world in themselves. All silicates are built from tetrahedral SiO_4 units, but they have different properties owing to the way these tetrahedral SiO_4 units link together.

The simplest silicates, *orthosilicates*, contain SiO_4^{4-} anions. The -4 charge of the anion is balanced by four M^+ ions, two M^{2+} ions, or a combination of ions. Calcium orthosilicate, Ca_2SiO_4, is a component of Portland cement. Olivine, an important mineral in Earth's mantle, contains Mg^{2+}, Fe^{2+}, and Mn^{2+}, the Fe^{2+} ion giving the mineral its characteristic olive color.

A group of minerals called *pyroxenes* have as their basic structural unit a chain of SiO_4 tetrahedra.

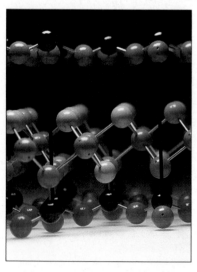

If two such chains link by sharing oxygen atoms, the result is an *amphibole*, of which the asbestos minerals are one example. The molecular chain results in asbestos being a fibrous material.

Linking many silicate chains together produces a sheet of SiO_4 tetrahedra (Figure 21.18). This is the basic structural feature of some of Earth's most important minerals, particularly the clay minerals (such as china clay), the micas, and the chrysotile form of asbestos.

The sheet structure leads to the characteristic feature of mica, which is often found as "books" of thin silicate sheets. Mica is used in furnace windows and as insulation, and flecks of mica give the glitter to "metallic" paints.

Mica, a large family of clays, and asbestos are actually *aluminosilicates*, substances containing both aluminum and silicon. In kaolinite clay, for example, the sheet of SiO_4 tetrahedra is bonded to a sheet of AlO_6 octahedra. In addition, some Si atoms can be replaced by Al atoms. Because Si^{4+} ions are being replaced by Al^{3+} ions with

The basic structural feature of many clays, kaolinite in particular, is a sheet of SiO_4 tetrahedra (see Figure 21.17) bonded to a sheet of AlO_6 octahedra. *(Charles D. Winters)*

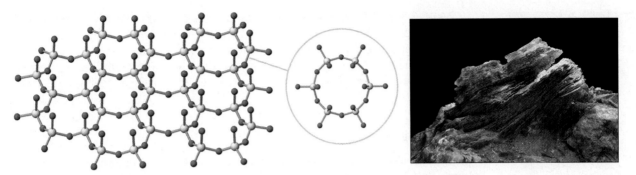

Figure 21.18   Mica, a sheet silicate. The sheet-like structure of mica explains its physical appearance. As in the pyroxenes, each silicon is bonded to four oxygen atoms, but the Si and O atoms form a sheet of six-member rings of Si atoms with O atoms in each edge. The ratio of Si to O is 1:2.5. *(Charles D. Winters)*

a smaller charge, nature adds positive ions such as Na$^+$ and Mg^{2+} for charge balance in the lattice. This feature leads to some interesting uses of clays, one being in medicine (Figure 21.19). In certain cultures, clay is eaten for medicinal purposes. Several remedies for the relief of upset stomach contain highly purified clays that absorb excess stomach acid as well as potentially harmful bacteria and their toxins by exchanging the intersheet cations in the clays for the toxins, which are often organic cations.

Other aluminosilicates are the feldspars, common minerals that make up about 60% of Earth's crust, and zeolites (Figure 21.19). Both are composed of SiO$_4$ tetrahedra with some Si atoms replaced by Al atoms, along with alkali and alkaline earth ions for charge balance. The main feature of zeolites is their regularly shaped tunnels and cavities. Hole diameters are between 300 and 1000 pm, and small molecules such as water can fit into the cavities. As a result, zeolites can be used as drying agents to selectively absorb water from air or a solvent. Small amounts of zeolites are often sealed into multipane windows to keep the air dry between the panes.

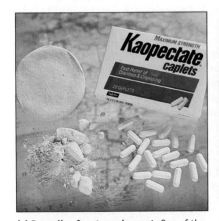

(a) Remedies for stomach upset. One of the ingredients in Kaopectate is kaolin, one form of clay. The off-white objects are pieces of clay purchased in a market in Ghana, West Africa. This clay was made to be eaten as a remedy for stomach ailments. Eating clay is widespread among the world's different cultures.

(b) The stucture of a zeolite. Zeolites, which have Si, Al, and O linked in a polyhedral framework, are often portrayed in drawings like this. Each edge consists of a Si-O-Si, Al-O-Si, or Al-O-Al bond. The channels in the framework can selectively capture small molecules or ions or act a catalytic sites.

(c) Apophyllite, a crystalline zeolite.

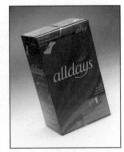

(d) Consumer products that remove odor-causing molecules from the air often contain zeolites.

Figure 21.19   Aluminosilicates. **(a)** Clay. **(b, c, d)** Zeolites. (In the zeolite model, Na$^+$ ions are yellow, O atoms red, Si and Al atoms are white.) *(a, c, and d, Charles D. Winters; b, Alfred Pasieka/Science Photo Library/Photo Researchers, Inc.)*

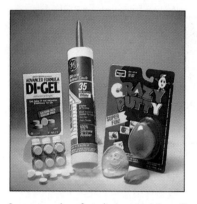

Some examples of products containing silicones, polymers with repeating
— R_2Si — O — units. *(Charles D. Winters)*

Zeolites are also used as catalysts. Mobil Oil Corporation, for example, has patented a process in which methanol, CH_3OH, is converted to gasoline in the presence of specially tailored zeolites. In addition, zeolites are added to detergents, where they function as water-softening agents because the sodium ions of the zeolite can be exchanged for Ca^{2+} ions in hard water, effectively removing Ca^{2+} from the water.

Silicone Polymers

Silicon and methyl chloride (CH_3Cl) react at 300 °C in the presence of a catalyst, Cu powder. The primary product of this reaction is $(CH_3)_2SiCl_2$.

$$Si(s) + 2\ CH_3Cl(g) \longrightarrow (CH_3)_2SiCl_2(\ell)$$

Chemical Perspectives

Lead Pollution, Old and New

Anchoring the bottom of Group 4A is the element lead. One of a handful of elements known since ancient times, it is also a metal with modern uses. It is an essential commodity in the modern industrial world, ranking fifth in consumption behind iron, copper, aluminum, and zinc. The major uses of the metal and its compounds are in storage batteries (page 843), pigments, ammunition, solders, plumbing, and bearings. Its chief source is lead sulfide, PbS, known commonly as galena (page 87).

Lead and its compounds are cumulative poisons, particularly in children. At a blood level as low as about 50 ppb (parts per billion), blood pressure is elevated; intelligence is affected at about 100 ppb; and blood levels higher than about 800 ppb can lead to coma and possible death. Health experts believe that more than 200,000 children become ill from lead poisoning annually. This is caused chiefly by eating paint containing lead pigments. Older homes often contain lead-based paint because white lead [2 $PbCO_3 \cdot Pb(OH)_2$] was the pigment used in white paint until about 50 years ago, when it was replaced by TiO_2. Lead salts have a sweet taste, which may contribute to the tendency of children to chew on painted objects.

Until just a few years ago a major use of lead was as tetraethyllead, $Pb(C_2H_5)_4$. This was added to gasoline to improve the burning properties of fuel. The compound has been phased out, however, in part because of the hazards from tons of lead compounds being spewed into the environment and in part because lead poisons the catalyst in catalytic exhaust systems. All gasoline sold in the United States now bears the label "unleaded" or "no lead."

Lead poisoning is often implicated as one cause of the collapse of the Roman Empire. At the height of the Roman

Uses of lead. The primary use of lead is in storage batteries (where the metal is used as the anode, and PbO_2 is used as the cathode.) Other uses include as plumbers' lead and as a component of solder and pewter. Lead is also used to frame the glass pieces in stained glass windows. *(Charles D. Winters)*

Empire (about AD 500) the production of lead was as much as 80,000 tons per year because lead had so many uses: for roofing, water pipes, kitchenware, and coffins, for lining the hulls of ships, and to make bronze. The Romans used lead to make pipes to carry water, and they cooked in lead vessels. Because many foods are acidic, and lead reacts slowly with acid, lead could be incorporated into food.

Scientists have recently obtained evidence that extensive lead mining and smelting operations during the Roman Empire contributed to atmospheric pollution on a global scale. Drilling in the Greenland ice cap produced an ice core, a cylinder of ice 9938 feet long. The core contains trace remnants of the atmosphere for the last 9000 years, and it is possible to obtain very accurate dates on the various segments of this core.

Among the trace elements in the polar ice core was lead, in concentrations in the parts per trillion (ppt) range. Samples of lead have isotope ratios (such as $^{208}Pb/^{206}Pb$) that depend on the source of the element. (Analysis of isotope ratios is carried out using sensitive mass spectrometric techniques; see Figure 2.5.) In the segment of the ice core between 300 BC and AD 600, samples had lead isotope ratios identical to the lead from Roman mines in southwestern Spain! Lead polluted the earth's atmosphere 2000 years ago.

Halides of Group 4A elements other than carbon hydrolyze readily. The reaction of $(CH_3)_2SiCl_2$ with water, for example, initially produces $(CH_3)_2Si(OH)_2$. On standing, these molecules rapidly condense together to form a polymer, eliminating water. The polymer, called polydimethylsiloxane, a member of the *silicone* family of polymers.

$$(CH_3)_2SiCl_2 + 2\ H_2O \longrightarrow (CH_3)_2Si(OH)_2 + 2\ HCl$$

$$n\ (CH_3)_2Si(OH)_2 \longrightarrow \pm(CH_3)_2SiO\mp_n + n\ H_2O$$

Silicone polymers are nontoxic and have good stability to heat, light, and oxygen; they are chemically inert and have valuable antistick and antifoam properties. They can be oils, greases, and resins. Some have rubber-like properties; "Silly Putty," for example, is a silicone polymer. Approximately 3×10^6 tons of silicone polymers are made worldwide annually. These materials are used in a wide variety of products: lubricants, peel-off labels, lipstick, suntan lotion, car polish, and building caulk.

Exercise 21.9 **Silicon Chemistry**

Silicon–oxygen rings are a common structural feature in silicate chemistry. Draw the structure for the anion $Si_3O_9^{6-}$, which is found in minerals such as benitoite. The ring has three Si and three O atoms, and there are two other O atoms on each Si atom.

21.7 NITROGEN AND PHOSPHORUS

Nitrogen is found primarily as N_2 in the atmosphere, where it constitutes 78.1% by volume (75.5% by weight). In contrast, phosphorus occurs in Earth's crust in solids. More than 200 different phosphorus-containing minerals are known; all contain the tetrahedral phosphate ion, PO_4^{3-}, or a derivative of this ion. By far the most abundant minerals are apatites [← PAGE 895].

Nitrogen and its compounds play a key role in our economy, with ammonia being particularly notable. Phosphoric acid is an important commodity chemical, its major use being in fertilizers.

Both phosphorus and nitrogen are part of every living organism. Phosphorus is contained in biochemicals called nucleic acids and phospholipids, and nitrogen occurs in proteins and nucleic acids. Indeed, the element phosphorus was first derived from human waste (see *A Closer Look: Making Phosphorus.*)

Mining phosphate rock in Morocco. In the United States, most phosphate comes from Florida, which also supplies about 25% of the world's needs annually. Phosphates are used for fertilizer and to prepare phosphoric acid. *(William Felger/Grant Heilman Photography)*

Properties of Nitrogen and Phosphorus

Nitrogen (N_2) is a colorless gas that liquefies at 77 K (-196 °C). Its most notable feature is its reluctance to react with other elements or compounds. This is because the $N{\equiv}N$ triple bond has a large dissociation energy (945 kJ/mol) and because the molecule is nonpolar. However, nitrogen does react with hydrogen to give ammonia in the presence of a catalyst (page 680) and with a few metals to give metal nitrides, compounds containing the N^{3-} ion.

$$3\ Mg(s) + N_2(g) \longrightarrow Mg_3N_2(s)$$
magnesium nitride

Elemental nitrogen is a very useful material. Because of its lack of reactivity, it is used to provide a nonoxidizing atmosphere for packaged foods and wine and to

A Closer Look

Making Phosphorus

He stoked his small furnace with more charcoal and pumped the bellows until his retort glowed red hot. Suddenly something strange began to happen. Glowing fumes filled the vessel and from the end of the retort dripped a shining liquid that burst into flames.

J. Emsley, *The 13th Element*, John Wiley, New York, 2000, page 5.

John Emsley begins his story of phosphorus, its discovery and its uses by imagining what the German alchemist Hennig Brandt must have seen in his laboratory that day in 1669. Brandt was in search of the philosopher's stone, the magic elixir that would turn the crudest substance into gold.

Brandt was experimenting with urine, which had been the source of useful chemicals since Roman times, and it is not surprising that phosphorus could be extracted from this source. Humans consume much more phosphorus, in the form of phosphate, than they require, and the excess phospho-

rus (about 1.4 g per day) is passed in the urine. It is nonetheless extraordinary that Brandt was able to isolate the element. According to an 18th century chemistry book, about 30 g of phosphorus could be obtained from 60 gallons of urine. And the process was not simple. Another 18th century recipe states that "50 or 60 pails full" of urine was to be used. "Let it lie steeping . . . till it putrefy and breed worms." Then the chemist was to reduce the whole to a paste and finally heat the paste very strongly in a retort. After some days phosphorus distilled from the mixture and was collected in water. (We know now that carbon from the organic compounds in the urine would have served to reduce the

phosphate to phosphorus.) Phosphorus was made in this manner for well over 100 years.

The Alchymist in Search of the Philosopher's ▶ Stone, Discovers Phosphorus. Painted by J. Wright of Derby (1734–1797). *(Derby Museum and Art Gallery, Derbyshire, UK/Bridgeman Art Library)*

Biological samples — such as embryos or semen from animals and humans — can be stored in liquid nitrogen for long periods of time. *(Simon Fraser/MRC Unit, Newcastle General Hospital/Science Photo Library)*

pressurize electric cables and telephone wires. Liquid nitrogen is valuable as a coolant in freezing biological samples such as blood and semen, in freeze-drying food, and for other applications that require extremely low temperatures.

Elemental phosphorus is produced by the reduction of phosphate minerals:

$$2\ Ca_3(PO_4)_2(s) + 10\ C(s) + 6\ SiO_2(s) \longrightarrow P_4(g) + 6\ CaSiO_3(s) + 10\ CO(g)$$

The standard state of phosphorus is white phosphorus, but red phosphorus is the most stable allotrope. Rather than occurring as a diatomic molecule with a triple bond, like its second-period relative nitrogen (N_2), phosphorus is made up of tetrahedral P_4 molecules, with each P atom joined to three others via single bonds.

Nitrogen Compounds

A feature of the chemistry of nitrogen is the wide diversity of compounds of this element. Compounds with nitrogen in all oxidation numbers between -3 and $+5$ are known (Figure 21.20).

Hydrogen Compounds of Nitrogen: Ammonia and Hydrazine

Ammonia is a gas at room temperature and pressure. It has a very penetrating odor and condenses to a liquid at $-33\ °C$ under 1 bar pressure. Solutions in water, often referred to as ammonium hydroxide, are basic due to the reaction of ammonia with water [◀ SECTION 17.5 AND FIGURE 5.6].

$$NH_3(aq) + H_2O(\ell) \rightleftharpoons NH_4^+(aq) + OH^-(aq) \qquad K_b = 1.8 \times 10^{-5} \text{ at } 25\ °C$$

Ammonia is a major industrial chemical. It is prepared by the *Haber process* [← PAGE 650], largely for use as a fertilizer.

Hydrazine, N_2H_4, is a colorless fuming liquid with an ammonia-like odor (mp, 2.0 °C; bp, 113.5 °C). Almost a million kilograms of hydrazine are produced annually by the *Raschig process*—the oxidation of ammonia with alkaline sodium hypochlorite in the presence of gelatin (which is added to suppress metal-catalyzed side reactions that lower the yield of hydrazine):

$$2\ NH_3(aq) + NaClO(aq) \longrightarrow N_2H_4(aq) + NaCl(aq) + H_2O(\ell)$$

Hydrazine, like ammonia, is a Brønsted and Lewis base:

$$N_2H_4(aq) + H_2O(\ell) \rightleftharpoons N_2H_5^+(aq) + OH^-(aq) \qquad K_b = 8.5 \times 10^{-7}$$

It is also a strong reducing agent, as reflected in the $E°$ value in basic solution:

$$N_2(g) + 4\ H_2O(\ell) + 4\ e^- \longrightarrow N_2H_4(aq) + 4\ OH^-(aq) \qquad E° = -1.16\ V$$

Hydrazine's reducing ability is exploited in its use in waste water treatment for chemical plants. It removes ions such as CrO_4^{2-} by reducing them and thus preventing chromium from entering the environment. A related use is the treatment of water boilers in large electric-generating plants. Oxygen dissolved in the water is a serious problem in these plants because the dissolved gas can oxidize the metal of the boiler and pipes and lead to corrosion. Hydrazine reduces the dissolved oxygen to water:

$$N_2H_4(aq) + O_2(g) \longrightarrow N_2(g) + 2\ H_2O(\ell)$$

Oxides of Nitrogen

Nitrogen is unique among elements in the number of binary oxides it forms (Table 21.5). All are thermodynamically unstable with respect to decomposition to N_2 and O_2; that is, all have positive $\Delta G_f°$ values. Most are slow to decompose, however, and so are described as kinetically stable. [⊙ CD-ROM, Screen 6.3]

Dinitrogen oxide (nitrous oxide), N_2O, is a nontoxic, odorless, and tasteless gas having nitrogen with the lowest oxidation number (+1) among nitrogen oxides. It can be made by the careful decomposition of ammonium nitrate at 250 °C:

$$NH_4NO_3(s) \longrightarrow N_2O(g) + 2\ H_2O(g)$$

Dinitrogen oxide is used as an anesthetic in minor surgery. It has been called "laughing gas" because of its effects. Because it is soluble in vegetable fats, the largest commercial use of N_2O is as a propellant and aerating agent in cans of whipped cream.

Nitrogen monoxide (nitric oxide), NO, is an odd-electron molecule. It has 11 valence electrons, giving it one unpaired electron and making it a free radical [← PAGE 345]. The compound has recently been the subject of intense research because it has been found to be important in a number of biochemical processes.

Nitrogen dioxide, NO_2, is the brown gas you see when a bottle of nitric acid is allowed to stand in the sunlight:

$$2\ HNO_3(aq) \longrightarrow 2\ NO_2(g) + H_2O(\ell) + \tfrac{1}{2}\ O_2(g)$$

Nitrogen dioxide is also a culprit in air pollution. Nitrogen monoxide is often present in urban polluted air. Its source is the combination of atmospheric nitrogen and oxygen when they are heated in internal combustion engines. In excess oxygen, NO rapidly forms NO_2, which, if not removed by a catalytic exhaust system, enters the atmosphere:

$$2\ NO(g) + O_2(g) \longrightarrow 2\ NO_2(g)$$

Compound & Oxidation Number of N

 Ammonia, –3

 Hydrazine, –2

 Dinitrogen, 0

 Dinitrogen oxide, +1

 Nitrogen monoxide, +2

 Nitrogen dioxide, +4

 Nitric acid, +5

Figure 21.20 Compounds of nitrogen. In its compounds, the N atom can have oxidation states ranging from –3 to +5.

Nitrous oxide, N_2O, dissolves in fats. The gas is added, under pressure, to cans of cream. When the valve is opened, the gas expands, thus "expanding" or whipping the cream. Like other fat-soluble molecules, N_2O is also an anesthetic and is considered safe when used for medical purposes. However, there are significant dangers to using it as a recreational drug. Long-term use can result in a condition called neuropathy, in which nerve fibers are permanently damaged, causing such problems as weakness, tingling, and a loss of feeling. *(Charles D. Winters)*

Table 21.5 • Some Oxides of Nitrogen

Formula	Name	Structure	Nitrogen Oxidation Number	Description
N_2O	Dinitrogen monoxide (nitrous oxide)	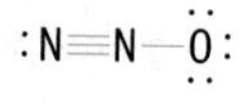 :N≡N—Ö: \n linear	+1	Colorless gas (laughing gas)
NO	Nitrogen monoxide (nitric oxide)	*	+2	Colorless gas, odd-electron molecule (paramagnetic)
N_2O_3	Dinitrogen trioxide	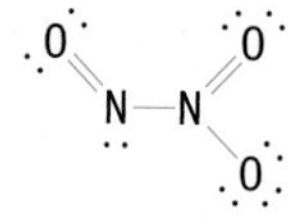 (structure shown) \n planar	+3	Blue solid (mp, −100.7 °C), reversibly dissociates to NO and NO_2
NO_2	Nitrogen dioxide	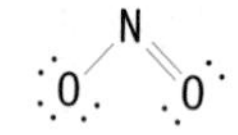 (structure shown)	+4	Brown, paramagnetic gas
N_2O_4	Dinitrogen tetraoxide	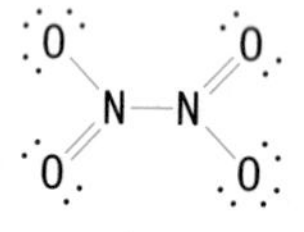 (structure shown) \n planar	+4	Colorless liquid, dissociates to NO_2 (see Figure 16.6)
N_2O_5	Dinitrogen pentaoxide	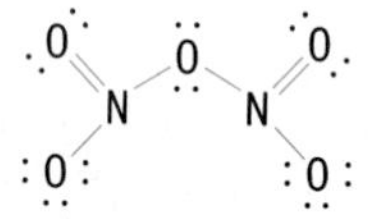 (structure shown)	+5	Colorless solid

*It is not possible to draw a Lewis structure that accurately represents the electronic structure of NO. See Chapter 10.

Nitrogen dioxide has 21 valence electrons and so is also an odd-electron molecule. Because the odd electron largely resides on the N atom, two NO_2 molecules combine, forming an N—N bond and producing N_2O_4, *dinitrogen tetraoxide.*

$$2\ NO_2(g) \rightleftharpoons N_2O_4(g)$$

deep brown gas   colorless
(mp, −11.2 °C)

When N_2O_4 is frozen (mp, −11.2 °C), the colorless solid consists entirely of N_2O_4 molecules. As the solid melts and the temperature increases to the boiling point, colorless N_2O_4 dissociates to form brown NO_2. At the boiling point (21.5 °C) and 1 bar pressure, the distinctly brown gas phase consists of 15.9% NO_2 and 84.1% N_2O_4 [◀ PAGE 676].

When NO_2 is bubbled into water, *nitric acid, HNO₃,* forms.

$$2\ NO_2(g) + H_2O(\ell) \longrightarrow HNO_3(aq) + HNO_2(aq)$$

nitric acid   nitrous acid

Nitric acid has been known for centuries and has become an important chemical in our economy. The oldest way to make the acid is to treat $NaNO_3$ with sulfuric acid (Figure 21.21):

$$2\ NaNO_3(s) + H_2SO_4(aq) \longrightarrow 2\ HNO_3(aq) + Na_2SO_4(s)$$

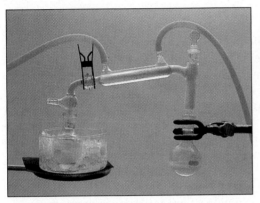

(a) Preparation of nitric acid.

(b) Reaction of HNO_3 with copper.

Figure 21.21 The preparation and properties of nitric acid. (a) Nitric acid is prepared by the reaction of sulfuric acid and sodium nitrate. Pure HNO_3 is colorless, but some acid decomposes to give brown NO_2, and it is this gas that fills the apparatus and colors the liquid in the distillation flask. (b) When concentrated nitric acid reacts with copper, the metal is oxidized to copper(II) ions, and NO_2 gas is a reaction product. *(Charles D. Winters)*

Enormous quantities of nitric acid are now produced industrially by the oxidation of ammonia in the multistep *Ostwald process*. The acid has many uses, but by far the greatest amount is turned into ammonium nitrate (for use as a fertilizer) by the reaction of nitric acid and ammonia.

Nitric acid is a powerful oxidizing agent, as the large, positive $E°$ values for the following half-reactions illustrate:

$$NO_3^-(aq) + 4 H_3O^+(aq) + 3 e^- \longrightarrow NO(g) + 6 H_2O(\ell) \qquad E° = +0.96 \text{ V}$$
$$NO_3^-(aq) + 2 H_3O^+(aq) + e^- \longrightarrow NO_2(g) + 3 H_2O(\ell) \qquad E° = +0.80 \text{ V}$$

Concentrated nitric acid attacks and oxidizes most metals. In this process, the nitrate ion is reduced to one of the nitrogen oxides. Which oxide is formed depends on the metal and on reaction conditions. In the case of copper, for example, either NO or NO_2 is produced, depending on the concentration of the acid:

In dilute acid:

$$3 Cu(s) + 8 H_3O^+(aq) + 2 NO_3^-(aq) \longrightarrow 3 Cu^{2+}(aq) + 12 H_2O(\ell) + 2 NO(g)$$

In concentrated acid:

$$Cu(s) + 4 H_3O^+(aq) + 2 NO_3^-(aq) \longrightarrow Cu^{2+}(aq) + 6 H_2O(\ell) + 2 NO_2(g)$$

Four metals (Au, Pt, Rh, and Ir) that are not attacked by nitric acid are often described as the "noble metals." The alchemists of the 14th century, however, knew that if they mixed HNO_3 with HCl in a ratio of about 1:3, this *aqua regia,* or "kingly water," would attack even gold, the noblest of metals.

$$10 Au(s) + 6 NO_3^-(aq) + 40 Cl^-(aq) + 36 H^+(aq) \longrightarrow$$
$$10 [AuCl_4]^-(aq) + 3 N_2(g) + 18 H_2O(\ell)$$

Exercise 21.10 **Nitrogen Oxide Chemistry**

Dinitrogen oxide can be made by the decomposition of NH_4NO_3.

(a) A Lewis electron dot structure of N_2O is given in Table 21.5. Is this the only possible structure? If other structures are possible, is the one in Table 21.5 the most important?

(b) Is the decomposition of $NH_4NO_3(s)$ to give $N_2O(g)$ and $H_2O(g)$ endo- or exothermic?

Figure 21.22   Sulfur spewing from a volcano in Indonesia. *(© Roger Ressmeyer/CORBIS)*

21.8   OXYGEN AND SULFUR

Oxygen is by far the most abundant element in Earth's crust, representing just under 50% by weight. It is present as elemental oxygen in the atmosphere, and is combined with other elements in water and in many minerals. Scientists believe that elemental oxygen did not appear on this planet until about 2 billion years ago, when it was formed by plants in the process of photosynthesis [← PAGE 489].

Sulfur, 15th in abundance in Earth's crust, is also found in its elemental form in nature, but only in certain concentrated deposits. Sulfur-containing compounds occur in natural gas and oil. In minerals, sulfur occurs as the sulfide ion (for example, in cinnabar, HgS, and galena, PbS), as the disulfide ion (in iron pyrite, FeS_2, or "fool's gold"), and as sulfate ion (for example, in gypsum, $CaSO_4 \cdot 2\ H_2O$). Sulfur oxides (SO_2 and SO_3) also occur in nature, primarily as products of volcanic activity (Figure 21.22). Sulfur chemistry supports some interesting life, as described in the story at the beginning of this chapter (page 880).

In the United States, most sulfur, about 10 million tons per year, is obtained from deposits of the element along the Gulf of Mexico. These occur typically at a depth of 150 to 750 m below the surface in layers about 30 m thick. These deposits are thought to have been formed by anaerobic ("without air") bacteria acting on sedimentary sulfate deposits such as gypsum.

Pyrite, FeS_2

Stibnite, Sb_2S_3

Orpiment, As_2S_3

Galena, PbS

Sulfide-containing minerals. *(Charles D. Winters)*

Preparation and Properties of the Elements

Pure oxygen is obtained by fractionation of air and is third among industrial chemicals produced in the United States. Oxygen can be made in the laboratory by electrolysis of water and by the catalyzed decomposition of metal chlorates such as $KClO_3$:

$$2\ KClO_3(s) \xrightarrow{\text{catalyst}} 2\ KCl(s) + 3\ O_2(g)$$

At room temperature and pressure oxygen is a colorless gas, but it is pale blue when condensed to the liquid at $-183\ °C$ [← FIGURE 10.16]. As described in Section 10.3, diatomic oxygen is paramagnetic because it has two unpaired electrons.

An allotrope of oxygen, ozone (O_3), is a blue, diamagnetic gas with an odor so strong that it can be detected in concentrations as low as 0.05 ppm. Ozone is synthesized by passing O_2 through an electric discharge, or by irradiation of O_2 with ultraviolet light. Ozone is in the news regularly because Earth's protective layer of ozone in the stratosphere is being disrupted by chlorofluorocarbons and other chemicals.

Sulfur has numerous allotropes. The most common and most stable allotrope is the yellow, orthorhombic form, which consists of S_8 molecules with the sulfur atoms arranged in a crown-shaped ring (Figure 21.23). Less stable allotropes are known that have rings of 6 to 20 sulfur atoms. Another form of sulfur called plastic sulfur has a molecular structure with chains of sulfur atoms (Figure 21.23).

Sulfur is obtained from underground deposits by a process developed by Herman Frasch (1851–1914) about 1900. Superheated water (at 165 °C) and then air are forced into the deposit. The sulfur melts (mp, 113 °C) and is forced to the surface as a frothy, yellow stream where it solidifies.

The largest use of sulfur is the production of sulfuric acid, H_2SO_4, the compound produced in largest quantity by the chemical industry [← PAGE 159]. In the United States, roughly 70% of the sulfuric acid is used to manufacture fertilizer.

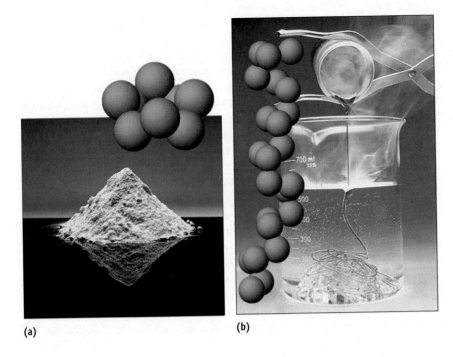

(a)   (b)

Smaller amounts are used in the conversion of ilmenite, a titanium-bearing ore, to TiO_2, which is then used as a white pigment in paint, plastics, and paper. The acid is also used to make iron and steel, petroleum products, synthetic polymers, and paper.

Sulfur Compounds

Hydrogen sulfide, H_2S, like the water molecule has a bent molecular geometry. Unlike water, however, H_2S is a gas under standard conditions (mp, $-85.6\ °C$; bp, $-60.3\ °C$) because the intermolecular forces are weak compared with the strong hydrogen bonding in water [◀ FIGURE 13.8]. Hydrogen sulfide is poisonous, comparable in toxicity to hydrogen cyanide, but it fortunately has a terrible odor and is detected in concentrations as low as 0.02 ppm. You must be careful with H_2S, however, because it has an anesthetic effect; your nose rapidly loses its ability to detect it. Death occurs at H_2S concentrations of 100 ppm.

Sulfur is often found as the sulfide ion in conjunction with metals because all metal sulfides (except those based on Group 1A metals) are insoluble in water. The recovery of metals from their sulfide ores usually begins by heating the ore in air. This converts the metal sulfide to either a metal oxide or the metal itself, the sulfur appearing as SO_2, as in the reactions of zinc or lead sulfide with oxygen (see *A Closer Look: Lead Pollution, Past and Present*, page 904).

Some common household products containing sulfur or sulfur-based compounds. *(Charles D. Winters)*

$$2\ ZnS(s) + 3\ O_2(g) \longrightarrow 2\ ZnO(s) + 2\ SO_2(g)$$
$$2\ PbO(s) + PbS(s) \longrightarrow 3\ Pb(s) + SO_2(g)$$

  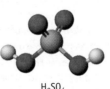

H_2S SO_2 SO_3 H_2SO_4

Models of some common sulfur-containing molecules: H_2S, SO_2, SO_3, and H_2SO_4.

Sulfur dioxide (SO_2) is produced on an enormous scale by the combustion of sulfur and by roasting sulfide ores in air. The combustion of sulfur in sulfur-containing coal and fuel oil creates particularly large environmental problems. It has been estimated that about 2.0×10^8 tons of sulfur oxides are released into the atmosphere each year by human activities, primarily in the form of SO_2; this is more than half of the total emitted by all natural sources of sulfur in the environment.

Sulfur dioxide is a colorless, toxic gas with a sharp odor. It readily dissolves in water. The most important reaction of this gas is oxidation to SO_3:

$$SO_2(g) + \tfrac{1}{2} O_2(g) \longrightarrow SO_3(g) \qquad \Delta H^\circ = -98.9 \text{ kJ/mol}$$

Sulfur trioxide is almost never isolated but is converted directly to sulfuric acid by reaction with water [← PAGE 159].

Exercise 21.11 Sulfur Chemistry

Metal sulfides roasted in air produce metal oxides.

$$2 \, ZnS(s) + 3 \, O_2(g) \longrightarrow 2 \, ZnO(s) + 2 \, SO_2(g)$$

Use thermodynamics [← PAGE 809] to estimate the temperature at which this reaction becomes product-favored. Is the temperature low enough that it is reasonable to expect that this process can be done on an industrial scale?

21.9   CHLORINE

Chlorine is the most abundant halogen. In nature, it occurs as chloride ion in sea water and in brine wells. Chlorine, the element, is made from these sources in enormous quantities so that the element ranks in the top ten industrial chemicals. It is used for the production of organic chemicals, as a bleach, and as a disinfectant.

Preparation of Chlorine

Elemental chlorine was first made by the Swedish chemist Karl Wilhelm Scheele (1742–1786) in 1774 by combining sodium chloride with an oxidizing agent in an acidic solution (Figure 21.24).

Industrially, chlorine is made by electrolysis of brine (concentrated aqueous NaCl). The other product of the electrolysis, NaOH, is also a valuable industrial chemical. About 80% of chlorine produced is made using an electrochemical cell similar to the one in Figure 21.25. Oxidation of chloride ion to Cl_2 gas occurs at the anode and reduction of water occurs at the cathode:

Anode reaction, oxidation: $2 \, Cl^-(aq) \longrightarrow Cl_2(g) + 2 \, e^-$

Cathode reaction, reduction: $2 \, H_2O(\ell) + 2 \, e^- \longrightarrow H_2(g) + 2 \, OH^-(aq)$

Activated titanium is used for the anode, and stainless steel or nickel is preferred for the cathode. The membrane separating the anode and cathode is not permeable to water, but it does allow Na^+ ions to pass; that is, the membrane is just a bridge between anode and cathode compartments. To maintain charge balance in the cell, sodium ions pass through the membrane. The energy consumption of these cells is in the range of 2000–2500 kwh/ton of NaOH produced.

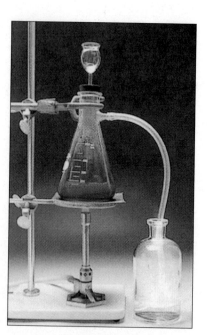

Figure 21.24 Chlorine preparation. Chlorine is prepared by oxidation of chloride ion using a strong oxidizing agent. Here, oxidation of NaCl is accomplished using $K_2Cr_2O_7$ in H_2SO_4. *(Charles D. Winters)*

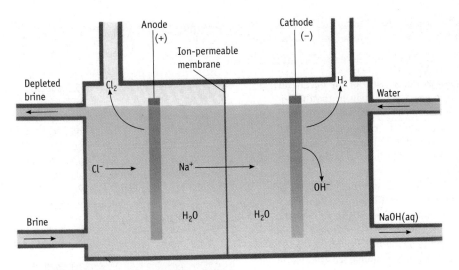

Figure 21.25   A membrane cell for the production of NaOH and Cl₂ gas from a saturated, aqueous solution of NaCl (brine). Here the anode and cathode compartments are separated by a water-impermeable but ion-conducting membrane. A widely used membrane is made of Nafion, a fluorine-containing polymer that is a relative of polytetrafluoroethylene (Teflon). Brine is fed into the anode compartment and dilute sodium hydroxide or water into the cathode compartment. Overflow pipes carry the evolved gases and NaOH away from the chambers of the electrolysis cell.

Chlorine Compounds

Hydrogen Chloride

Hydrochloric acid, an aqueous solution of hydrogen chloride, is a valuable industrial chemical. Hydrogen chloride gas can be prepared by the reaction of hydrogen and chlorine, but the rapid, exothermic reaction is difficult to control. The classical method of making HCl in the laboratory uses the reaction of NaCl and sulfuric acid, a procedure that takes advantage of the fact that HCl is a gas and that H_2SO_4 will not oxidize the chloride ion.

$$2\ NaCl(s) + H_2SO_4(\ell) \longrightarrow Na_2SO_4(s) + 2\ HCl(g)$$

Hydrogen chloride gas has a sharp, irritating odor. Both gaseous and aqueous HCl react with metals and metal oxides to give metal chlorides and, depending on the reactant, hydrogen or water:

$$Mg(s) + 2\ HCl(g) \longrightarrow MgCl_2(s) + H_2(g)$$
$$ZnO(s) + 2\ HCl(g) \longrightarrow ZnCl_2(s) + H_2O(g)$$

Oxoacids of Chlorine

Oxoacids of chlorine range from HClO, in which chlorine has an oxidation number of +1, to HClO₄, in which the oxidation number is equal to the group number, +7. All are strong oxidizing agents.

Acid	Name	Anion	Name
HClO	Hypochlorous	ClO^-	Hypochlorite
$HClO_2$	Chlorous	ClO_2^-	Chlorite
$HClO_3$	Chloric	ClO_3^-	Chlorate
$HClO_4$	Perchloric	ClO_4^-	Perchlorate

Hypochlorous acid, HClO, forms when chlorine dissolves in water. In this reaction, half of the chlorine is oxidized to hypochlorite ion and half is reduced to

Chemical Perspectives

Chlorine and Green Chemistry

(John C. Kotz)

Elemental chlorine is the starting point for the production of at least 10,000 compounds that contain chlorine and that are currently important in our economy. The major uses of chlorine and chlorine-containing compounds are in water purification, as solvents for the preparation of pharmaceuticals, in making plastics, and in pulp and paper manufacturing. Chlorine-containing compounds are also used as pesticides. Dichlorodiphenyl-trichloroethane, commonly called DDT, was until recently widely used to eradicate mosquitoes that carried malaria. It was so effective that the World Health Organization once regarded shortages of the chemical as a threat to public health.

Most chlorine-containing compounds are very stable. Although this property makes them valuable in some applications, it also means that they can persist in the environment for a considerable time. Chlorine-containing compounds accumulate in animals, reaching higher and higher concentrations the farther up the food chain you go. For example, biologists found that accumulation of DDT in bald

eagles was a cause of the decline in the bald eagle population in the United States in the 1960s. The pesticide apparently interferes with the calcium carbonate structure of the eggs, creating thin shells and causing the eagles' reproductive rate to drop severely.

Other environmental problems arise from the formation of toxic chlorine-containing compounds in the environment. For example, the largest use of chlorine gas is in bleaching paper. The problem is that unreacted chlorine in the effluent from a paper mill can react with naturally occurring compounds in water and turn them into toxins.

Recognition of the toxicity of chlorine-containing compounds has increasingly led chemists to find new ways of producing useful products with minimal environmental impact. This exciting new field, called "green chemistry," is causing chemists to think of ways to prevent waste from chemical processes, to develop better and more efficient syntheses, and to design safer chemicals. As "green chemist" Terry Collins has said, "If young chemists were rigorously taught about the toxicity of the substances they synthesize and study, their efforts would naturally be directed towards avoiding persistent toxic

substances (see *Nature,* Vol. 406, page 18, July 2000).

For a well-researched and reasonably balanced view of the chlorine industry, see J. Thornton, *Pandora's Poison: Chlorine, Health, and a New Environmental Strategy,* MIT Press, 2000. See also M. C. Cann and M. E. Connelly, *Real-World Cases in Green Chemistry,* American Chemical Society, 2001.

All of these products contain organochlorine compounds. Such compounds are the subject of a fierce debate because of their environmental impact or the impact of the methods of their production. A new emphasis in chemistry, "green chemistry," has arisen to address these problems. *(Charles D. Winters)*

● **Disproportionation**
A reaction in which an element or compound is simultaneously oxidized and reduced is called a *disproportionation reaction*. Here Cl_2 is oxidized to ClO^- and reduced to Cl^-.

chloride ion in a *disproportionation* reaction (a reaction in which the same element is both oxidized and reduced).

$$Cl_2(g) + 2\ H_2O(\ell) \rightleftharpoons H_3O^+(aq) + HClO(aq) + Cl^-(aq)$$

Chlorine, chloride ion, and hypochlorous acid exist in equilibrium. A low pH favors the reactants, whereas a high pH favors the products.

If, instead of dissolving Cl_2 in pure water, the element is added to cold aqueous NaOH, hypochlorite ion and chloride ion form:

$$Cl_2(g) + 2\ OH^-(aq) \rightleftharpoons ClO^-(aq) + Cl^-(aq) + H_2O(\ell)$$

The resulting alkaline solution is the "liquid bleach" used in home laundries. Note that the reaction equation bears a close resemblance to the equation for the reaction of Cl_2 with water. Under basic conditions the equilibrium shifts far to the right. The bleaching action of this solution is a result of the oxidizing ability of ClO^-. Most dyes are colored organic compounds, and hypochlorite ion oxidizes dyes to colorless products [◆ PAGE 604].

When calcium hydroxide is combined with Cl_2 in place of sodium hydroxide, solid $Ca(ClO)_2$ is the product. This compound is easily handled and is the "chlorine" that is sold for swimming pool disinfection.

When a basic solution of hypochlorite ion is heated, disproportionation again occurs to form chlorate ion and chloride ion:

$$3\ ClO^-(aq) \longrightarrow ClO_3^-(aq) + 2\ Cl^-(aq)$$

Sodium and potassium chlorates are made in large quantities by this reaction. The sodium salt can be reduced to ClO_2, a compound used for bleaching paper pulp. Some $NaClO_3$ is also converted to potassium chlorate, $KClO_3$, the preferred oxidizing agent in fireworks and a component of safety matches.

Perchlorates, salts containing ClO_4^-, are powerful oxidants. Pure perchloric acid, $HClO_4$, is a colorless liquid that explodes if shocked. It explosively oxidizes organic materials and rapidly oxidizes silver and gold. Dilute aqueous solutions of the acid are safe to handle, however.

Perchlorate salts of most metals are usually relatively stable but they are unpredictable. Great care should be used when handling any perchlorate salt. Ammonium perchlorate, for example, bursts into flame if heated above 200 °C.

$$2\ NH_4ClO_4(s) \longrightarrow N_2(g) + Cl_2(g) + 2\ O_2(g) + 4\ H_2O(g)$$

This property of the ammonium salt accounts for its use as the oxidizer in the solid booster rockets for the Space Shuttle. The solid propellant in these rockets is largely NH_4ClO_4, the remainder being the reducing agent, powdered aluminum. Each Shuttle launch requires about 750 tons of ammonium perchlorate, and more than half of the sodium perchlorate currently manufactured is converted to the ammonium salt for this purpose. The process for doing this is an exchange reaction that takes advantage of the fact that ammonium perchlorate is less soluble in water than sodium perchlorate:

$$NaClO_4(aq) + NH_4Cl(aq) \longrightarrow NaCl(aq) + NH_4ClO_4(s)$$

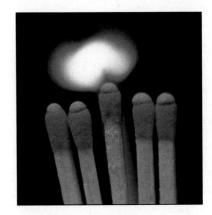

The head of a "strike anywhere" match contains, among other things, P_4S_3, and an oxidizing agent, potassium chlorate. Safety matches have sulfur (3–5%) and $KClO_3$ (45–55%) in the head and red phosphorus (page 882) in the striking strip. *(Charles D. Winters)*

The solid fuel booster rockets of the Space Shuttle are fueled by a mixture of NH_4ClO_4 (oxidizing agent) and Al powder (reducing agent). *(NASA)*

Exercise 21.12 Reactions of the Halogen Acids

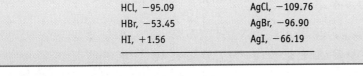

Metals generally react with hydrogen halides such as HCl to give the metal halide and hydrogen. The reaction is product-favored if ΔG°_{rxn} is negative. Is this true for reactions of silver with all of the hydrogen halides?

$$Ag(s) + HCl(g) \longrightarrow AgCl(s) + \tfrac{1}{2} H_2(g)$$

The required free energies of formation are (in kJ/mol):

Hydrogen Halide	Metal Halide
HF, −273.2	AgF, −193.8
HCl, −95.09	AgCl, −109.76
HBr, −53.45	AgBr, −96.90
HI, +1.56	AgI, −66.19

In Summary

When you have finished studying this chapter, you should ask if you have met the chapter goals. In particular, you should be able to

- Predict several chemical reactions of the Group A elements (Section 21.1).

- Predict similarities and differences among the elements in a given group, based on periodic properties (Section 21.1).
- Know which reactions produce ionic compounds, and predict formulas for common ions and common ionic compounds based on electron configurations (Section 21.1).
- Recognize when a formula is incorrectly written, based on general principles governing electron configurations (Section 21.1).
- Know the most abundant elements, know how they are obtained, and list some of their common chemical and physical properties (Sections 21.2–21.9).
- Be able to summarize briefly a series of facts about the most common compounds of main group elements (ionic or covalent structure, color, solubility, simple reaction chemistry) (Sections 21.2–21.9).
- Identify uses of common elements and compounds and understand the chemistry that relates to their usage (Sections 21.2–21.9).

Study Questions

Questions with blue, bold-faced numbers have answers in Appendix O.

Reviewing Important Concepts

1. The periodic table is one of the most useful sources of information available to a chemist. List at least three types of information that can be obtained from the periodic table.

2. Give formulas for the following common acids: nitric acid, sulfuric acid, hydrobromic acid, perchloric acid, carbonic acid. What is the oxidation number of the central atom in each of these compounds? How are these oxidation numbers related to the periodic group number of the element?

3. Define the term "amphoteric." Write chemical equations to illustrate the amphoteric character of $Al(OH)_3$.

4. Give examples of two basic oxides. Write equations illustrating the formation of each oxide from its component elements. Write another chemical equation that illustrates the basic character of each oxide.

5. Give examples of two acidic oxides. Write equations illustrating the formation of each oxide from its component elements. Write another chemical equation that illustrates the acidic character of each oxide.

6. How many valence electrons are present on each second-period atom?

7. What is the general electron configuration of the elements in Group 4A in the periodic table? Relate this to the formulas of some compounds formed by these elements.

8. Give the name and symbol of each element having the outer shell configuration ns^2np^1.

9. Give symbols and names for four monatomic ions that have the same electron configuration as argon.

10. Describe the structure of NaCl (page 330). What forces hold these particles together?

11. Select one of the alkali metals and write a balanced chemical equation for its reaction with chlorine. Is the reaction likely to be exothermic or endothermic? Is the product ionic or molecular?

12. Select one of the alkaline earth metals and write a balanced chemical equation for its reaction with oxygen. Is the reaction likely to be exothermic or endothermic? Is the product ionic or molecular?

13. For the product of the reaction you wrote for Question 11, predict the following physical properties: color, state of matter (s, ℓ, or g), solubility in water.

14. For the product of the reaction you wrote for Question 12, predict the following physical properties: color, state of matter (s, ℓ, or g), solubility in water.

15. List, in order, the ten most abundant elements in Earth's crust (see page 883). Which of these elements are main group elements? Identify one or more chemical species that occur in Earth's crust containing these main group elements.

16. Would you expect to find calcium occurring naturally in Earth's crust as a free element? Why or why not?

17. Which of the first ten elements in the periodic table are found as the free element in Earth's crust? Which elements in this group occur in Earth's crust only as part of a chemical compound?

18. Place the following oxides in order of increasing basicity: CO_2, SiO_2, and SnO_2.

19. Place the following oxides in order of increasing basicity: Na_2O, Al_2O_3, SiO_2, and SO_3.

20. Complete and balance equations for the following reactions. [Assume an excess of oxygen for (d).]
    (a) $Na(s) + Br_2(\ell)$
    (b) $Mg(s) + O_2(g)$
    (c) $Al(s) + F_2(g)$
    (d) $C(s) + O_2(g)$

21. Complete and balance equations for the following reactions:
    (a) $K(s) + I_2(s)$
    (b) $Ba(s) + O_2(g)$
    (c) $Al(s) + S_8(s)$
    (d) $Si(s) + Cl_2(g)$

Practicing Skills

Hydrogen

22. Write balanced chemical equations for the reaction of hydrogen gas with oxygen, chlorine, and nitrogen.

23. Write an equation for the reaction of potassium and hydrogen. Name the product. Is it ionic or covalent? Predict one physical property and one chemical property of this compound.

24. A procedure recently suggested for the preparation of hydrogen (and oxygen) from water proceeds as follows: (a) Sulfuric acid and hydrogen iodide are formed from sulfur dioxide, water, and iodine. (b) The sulfuric acid from the first step is decomposed by heat to water, sulfur dioxide, and oxygen. (c) The hydrogen iodide from the first step is decomposed with heat to hydrogen and iodine. Write a balanced equation for each of these steps and show that their sum is the decomposition of water to form hydrogen and oxygen.

25. Write a balanced chemical equation for the preparation of H_2 and CO_2 by the reaction of CH_4 and water. Using data in Appendix L, calculate $\Delta H°$, $\Delta S°$, and $\Delta G°$ for this reaction.

26. Using values in Appendix L, calculate $\Delta H°$, $\Delta S°$, and $\Delta G°$ for the reaction of carbon and water to give CO and H_2.

27. You are given a stoppered flask that contains one of the following: hydrogen, nitrogen, or oxygen. Suggest an experiment to identify the gas.

Alkali Metals

28. Write equations for the reaction of sodium with each of the halogens. Predict at least two physical properties that are common to all of the alkali metal halides.

29. Sodium peroxide is the primary product when sodium metal is burned in oxygen. Write a balanced equation for this reaction. What ions make up this compound?

30. Write balanced equations for the reaction of lithium, sodium, and potassium with O_2. Specify which metal forms an oxide, which one forms a peroxide, and which one forms a superoxide.

31. A piece of sodium catches on fire in the laboratory! Why shouldn't you try to extinguish the fire using water?

32. The electrolysis of aqueous NaCl gives NaOH, Cl_2, and H_2.
    (a) Write a balanced equation for the process.
    (b) In 1995 in the United States, 1.19×10^{10} kg of NaOH and 1.14×10^{10} kg of Cl_2 were produced. Does the ratio of masses of NaOH and Cl_2 produced agree with the ratio of masses expected from the balanced equation? If not, what does this tell you about the way in which NaOH or Cl_2 is actually produced? Is the electrolysis of aqueous NaCl the only source of these chemicals?

33. Write equations for the half-reactions that occur at the cathode and anode when an aqueous solution of KCl is electrolyzed. What chemical species is oxidized, and what chemical species is reduced in this reaction?

Alkaline Earth Elements

34. When magnesium burns in air, it forms both an oxide and a nitride. Write balanced equations for the formation of both compounds.

35. Calcium reacts with hydrogen gas at elevated temperatures to form a hydride. This compound reacts readily with water, so it is an excellent drying agent for organic solvents.
    (a) Write a balanced equation showing the formation of calcium hydride from Ca and H_2.
    (b) Write a balanced equation for the reaction of calcium hydride with water (see Figure 21.5).

36. Name three uses of limestone. Write a balanced equation for the reaction of limestone with CO_2 in water.

37. Explain what is meant by "hard water." What causes hard water, and what problems are associated with hard water?

38. Calcium oxide, CaO, is used to remove SO_2 from power plant exhaust. These two compounds react to give solid $CaSO_3$. What mass of SO_2 can be removed using 1.2×10^3 kg of CaO?

39. $Ca(OH)_2$ has a K_{sp} of 5.5×10^{-5}, whereas that for $Mg(OH)_2$ is 5.6×10^{-12}. Calculate the equilibrium constant for the reaction

$$Ca(OH)_2(s) + Mg^{2+}(aq) \rightleftharpoons Ca^{2+}(aq) + Mg(OH)_2(s)$$

Explain why this reaction can be used in the commercial isolation of magnesium from sea water.

Aluminum

40. Write balanced equations for the reactions of aluminum with $HCl(aq)$, Cl_2, and O_2.

41. Write an equation for the reaction of Al and $H_2O(\ell)$ to produce H_2 and Al_2O_3. Using thermodynamic data in Appendix L, calculate $\Delta H°$, $\Delta S°$, and $\Delta G°$ for this reaction. Do these data indicate that the reaction should favor the products? Why is aluminum metal unaffected by water?

42. Aluminum dissolves readily in hot aqueous NaOH to give the aluminate ion, $Al(OH)_4^-$, and H_2. Write a balanced equation for this reaction. If you begin with 13.2 g of Al, what volume (mL) of H_2 gas is produced when the gas is measured at 735 mm Hg and 22.5 °C?

43. Alumina, Al_2O_3, is amphoteric. Among examples of its amphoteric character are the reactions that occur when Al_2O_3 is heated strongly or "fused" with acidic oxides and basic oxides.
    (a) Write a balanced equation for the reaction of alumina with silica, an acidic oxide, to give aluminum metasilicate, $Al_2(SiO_3)_3$.

(b) Write a balanced equation for the reaction of alumina with the basic oxide CaO to give calcium aluminate, $Ca(AlO_2)_2$.

44. Aluminum sulfate (1995 worldwide production of about 3×10^9 kg) is the most commercially important aluminum compound after aluminum oxide and aluminum hydroxide. Write a balanced equation for the reaction of aluminum oxide with sulfuric acid to give aluminum sulfate. To manufacture 1.00 kg of aluminum sulfate, what mass (kg) of aluminum oxide and sulfuric acid must be used?

45. Gallium hydroxide, like aluminum hydroxide, is amphoteric. Write balanced equations for the reaction of solid $Ga(OH)_3$ with solutions of HCl and NaOH. What volume of 0.0112 M HCl is needed to react completely with 1.25 g of $Ga(OH)_3$?

46. Halides of the Group 3A elements are excellent Lewis acids. When a Lewis base such as Cl^- interacts with $AlCl_3$, the ion $AlCl_4^-$ is formed. Write a Lewis electron dot structure for this anion. What structure is predicted for $AlCl_4^-$? What hybridization is assigned to the aluminum atom in $AlCl_4^-$?

47. "Aerated" concrete bricks are widely used building materials. They are obtained by mixing gas-forming additives with a moist mixture of lime, sand, and possibly cement. Industrially, the following reaction is important:

$$2\,Al(s) + 3\,Ca(OH)_2(s) + 6\,H_2O(\ell) \longrightarrow$$
$$[3\,CaO \cdot Al_2O_3 \cdot 6\,H_2O](s) + 3\,H_2(g)$$

Assume that the mixture of reactants contains 0.56 g of Al for each brick. What volume of hydrogen gas do you expect at 26 °C and atmospheric pressure (745 mm Hg)?

Silicon

48. Describe the structures of SiO_2 and CO_2. Explain why SiO_2 has a very high melting point, whereas CO_2 is a gas.

49. Describe how ultrapure silicon can be produced from sand.

50. One material needed to make silicones is dichlorodimethylsilane, $(CH_3)_2SiCl_2$. It is made by treating silicon powder at about 300 °C with CH_3Cl in the presence of a copper-containing catalyst.

    (a) Write a balanced equation for the reaction.

    (b) Assume you carry out the reaction on a small scale with 2.65 g of silicon. To measure the CH_3Cl gas, you fill a 5.60-L flask at 24.5 °C. What pressure of CH_3Cl gas must you have in the flask in order to have the stoichiometrically correct amount of the compound?

    (c) What mass of $(CH_3)_2SiCl_2$ is produced? (Assume 100% yield.)

51. Describe the structure of pyroxenes (see page 901). What is the ratio of silicon to oxygen in this type of compound?

Nitrogen and Phosphorus

52. Consult the data in Appendix L. Are any of the nitrogen oxides listed there stable with respect to decomposition to N_2 and O_2?

53. Use data in Appendix L to calculate the enthalpy change for the reaction

$$2\,NO_2(g) \rightleftharpoons N_2O_4(g)$$

Is this reaction exothermic or endothermic?

54. Use data in Appendix L to calculate the enthalpy change for the reaction

$$2\,NO(g) + O_2(g) \longrightarrow 2\,NO_2(g)$$

Is this reaction exothermic or endothermic?

55. The overall reaction involved in the industrial synthesis of nitric acid is

$$NH_3(g) + 2\,O_2(g) \rightleftharpoons HNO_3(aq) + H_2O(\ell)$$

Calculate $\Delta G°$ for the reaction and then the equilibrium constant, at 25 °C.

56. A major use of hydrazine, N_2H_4, is in steam boilers in power plants.

    (a) The reaction of hydrazine with O_2 dissolved in water gives N_2 and water. Write a balanced equation for this reaction.

    (b) O_2 dissolves in water to the extent of 3.08 mL (gas at STP) in 100. mL of water at 20 °C. To consume all of the dissolved O_2 in 3.00×10^4 L of water (enough to fill a small swimming pool), what mass of N_2H_4 is needed?

57. Before hydrazine came into use to remove dissolved oxygen in the water in steam boilers, Na_2SO_3 was commonly used:

$$2\,Na_2SO_3(aq) + O_2(aq) \longrightarrow 2\,Na_2SO_4(aq)$$

What mass of Na_2SO_3 is required to remove O_2 from 3.00×10^4 L of water as outlined in Question 56?

58. A common analytical method for hydrazine involves its oxidation with iodate ion, IO_3^- in acid solution. In the process, hydrazine is a four-electron reducing agent:

$$N_2(g) + 5\,H_3O^+(aq) + 4\,e^- \longrightarrow N_2H_5^+(aq) + 5\,H_2O(\ell)$$
$$E° = -0.23\,V$$

Write the balanced equation for the reaction of hydrazine in acid solution ($N_2H_5^+$) with $IO_3^-(aq)$ to give N_2 and I_2. Calculate $E°$ for this reaction.

59. The steering rockets in the Space Shuttle use N_2O_4 and a derivative of hydrazine, 1,1-dimethylhydrazine (page 247). This mixture is called a hypergolic fuel because it ignites when the reactants come into contact:

$$H_2NN(CH_3)_2(\ell) + 2\,N_2O_4(\ell) \longrightarrow$$
$$3\,N_2(g) + 4\,H_2O(g) + 2\,CO_2(g)$$

    (a) Identify the oxidizing agent and the reducing agent in this reaction.

    (b) The same propulsion system was used by the Lunar Lander on moon missions in the 1970s. If the Lander used 4100 kg of $H_2NN(CH_3)_2$, what mass (kg) of N_2O_4 was required to react with it? What mass (kg) of each of the reaction products was generated?

60. Unlike carbon, which can form extended chains of atoms, nitrogen can form chains of very limited length. Draw the Lewis electron dot structure of the azide ion, N_3^-.

61. $CaHPO_4$ is used as an abrasive in toothpaste. Write a balanced equation showing a possible preparation for this compound from readily accessible compounds.

Oxygen and Sulfur

62. In the "contact process" for making sulfuric acid, sulfur is first burned to SO_2. Environmental restrictions allow no more than 0.30% of this SO_2 to be vented to the atmosphere.

    (a) If enough sulfur is burned in a plant to produce 1.80×10^6 kg of pure, anhydrous H_2SO_4 per day, what is the maximum amount of SO_2 that is allowed to be exhausted to the atmosphere?

    (b) One way to prevent any SO_2 from reaching the atmosphere is to "scrub" the exhaust gases with slaked lime, $Ca(OH)_2$:

    $$Ca(OH)_2(s) + SO_2(g) \longrightarrow CaSO_3(s) + H_2O(\ell)$$
    $$2\,CaSO_3(s) + O_2(g) \longrightarrow 2\,CaSO_4(s)$$

    What mass of $Ca(OH)_2$ (in kilograms) is needed to remove the SO_2 calculated in part (a)?

63. A sulfuric acid plant produces an enormous amount of heat. To keep costs as low as possible, much of this heat is used to make steam to generate electricity. Some of the electricity is used to run the plant, and the excess is sold to the local electrical utility. Three reactions are important in sulfuric acid production: (1) burning S to SO_2; (2) oxidation of SO_2 to SO_3; and (3) reaction of SO_3 with H_2O:

    $$SO_3(g) + H_2O(\text{in 98\% } H_2SO_4) \longrightarrow H_2SO_4(\ell)$$

    The enthalpy change of the third reaction is -130 kJ/mol. Estimate the total heat produced per mole of H_2SO_4 produced. How much heat is produced per ton (907 kg) of H_2SO_4?

64. In addition to the sulfide ion, S^{2-}, there are polysulfides, S_n^{2-}. (Such ions are chains of sulfur atoms, not rings.) Draw a Lewis electron dot structure for the S_2^{2-} ion. The S_2^{2-} ion is the disulfide ion, an analogue of the peroxide ion. It occurs in iron pyrites, FeS_2.

65. Sulfur forms a range of compounds with fluorine. Draw Lewis electron dot structures for S_2F_2 (connectivity is FSSF), SF_2, SF_4, SF_6, and S_2F_{10}. What is the formal oxidation number of sulfur in each of these compounds?

Chlorine

66. The halogen oxides and oxoanions are good oxidizing agents. For example, the reduction of bromate ion has an $E°$ value of 1.44 V in acid solution:

    $$2\,BrO_3^-(aq) + 12\,H^+(aq) + 10\,e^- \longrightarrow$$
    $$Br_2(aq) + 6\,H_2O(\ell)$$

    Is it possible to oxidize aqueous 1.0 M Mn^{2+} to aqueous MnO_4^- with 1.0 M bromate ion?

67. The hypohalite ions, XO^-, are the anions of weak acids. Calculate the pH of a 0.10 M solution of NaClO. What is the concentration of HClO in this solution?

68. Bromine is obtained from sea water. The process involves treating water containing bromide ion with Cl_2 and extracting the Br_2 from the solution using an organic solvent. Write a balanced equation for the reaction of Cl_2 and Br^-. What are the oxidizing and reducing agents in this reaction? Using the table of standard reduction potentials (Appendix M), verify that this is a product-favored reaction.

69. To prepare chlorine from chloride ion, a strong oxidizing agent is required. The dichromate ion, $Cr_2O_7^{2-}$, is one example (see Figure 21.24). Consult the table of standard reduction potentials (Appendix M) and identify several other oxidizing agents that may be suitable. Write balanced equations for the reactions of these substances with chloride ion.

General Questions

More challenging questions are indicated by an underlined number.

70. For each of the third-period elements (Na through Ar), identify the following:

    (a) Whether the element is a metal, nonmetal, or metalloid

    (b) The color and appearance of the element

    (c) The state of the element (s, ℓ, or g) under standard conditions

    For help in this question, consult Figure 2.7 or use the periodic table "tool" on the *General Chemistry Interactive CD-ROM*, Version 3.0. The latter provides a picture of each element and a listing of its properties.

71. For each of the second-period elements (Li through Ne), identify the following:

    (a) Whether the element is a metal, nonmetal, or metalloid

    (b) The color and appearance of the element

    (c) The state of the element (s, ℓ, or g) under standard conditions

    Consult Figure 2.7 or use the periodic table "tool" on the *General Chemistry Interactive CD-ROM*, Version 3.0.

72. Let us consider the chemistry of the elements sodium, magnesium, aluminum, silicon, phosphorus, and sulfur. (See Figure 1.2 and Chapter Focus, page 882.)

    (a) Write a balanced chemical equation depicting the reaction of each of the elements with elemental chlorine.

    (b) Describe the bonding in each of the products of the reactions with chlorine as ionic or covalent.

    (c) Draw Lewis electron dot structures for the products of the reactions of silicon and phosphorus with chlorine. What are their electron-pair and molecular geometries?

73. Let us consider the chemistry of the elements of Group 4A.

    (a) Write a balanced chemical equation to depict the reaction of each of the elements with elemental chlorine.

    (b) Describe the bonding in each of the products of the reactions with chlorine as ionic or covalent.

(You have not seen reactions of some of these elements in the text, but you have been given enough information to be able to predict the reactions that can occur.)

74. One of the pieces of evidence for the hydride ion in metal hydrides comes from electrochemistry. Predict the reactions that occur at each electrode when molten LiH is electrolyzed.

75. To store 2.88 kg of gasoline with an energy equivalence of 1.43×10^8 J requires a volume of 4.1 L. In comparison, 1.0 kg of H_2 has the same energy equivalence. What volume is required if this quantity of H_2 is to be stored at 25 °C and 1.0 atm of pressure?

76. Using data in Appendix L, calculate $\Delta G°$ values for the decomposition of MCO_3 to MO and CO_2 where M = Mg, Ca, and Ba. What is the relative tendency of these carbonates to decompose?

77. Ammonium perchlorate is used as the oxidizer in the solid fuel booster rockets of the Space Shuttle. If one launch requires 700 tons (6.35×10^5 kg) of the salt, and the salt decomposes according to the equation on page 915, what mass of water is produced? What mass of O_2 is produced? If the O_2 produced is assumed to react with the powdered aluminum present in the rocket engine, what mass of aluminum is required to use up all of the O_2, and how much Al_2O_3 is produced?

78. Metals react with hydrogen halides (such as HCl) to give the metal halide and hydrogen:

$$M(s) + n\,HX(g) \longrightarrow MX_n(s) + \tfrac{1}{2}\,n\,H_2(g)$$

The free energy change for the reaction is $\Delta G°_{rxn} = \Delta G°_f (MX_n) - n\,\Delta G°_f\,[HX(g)]$

(a) $\Delta G°_f$ for HCl(g) is -95.1 kJ/mol. What must be the value for $\Delta G°_f$ for MX_n in order for the reaction to be product-favored?

(b) Which of the following metals is (are) predicted to react with HCl(g): Ba, Pb, Hg, or Ti?

79. The boron atom in boric acid, $B(OH)_3$, is bonded to three $-OH$ groups. (In the solid state, the $-OH$ groups are in turn hydrogen-bonded to $-OH$ groups in neighboring molecules.)

(a) Draw the Lewis structure for boric acid.

(b) What is the hybridization of the boron in the acid?

(c) Sketch a picture showing how hydrogen bonding can occur between neighboring molecules.

(d) Boric acid is an acid because it reacts with water to give the borate and hydronium ions:

$$B(OH)_3(aq) + 2\,H_2O(\ell) \longrightarrow B(OH)_4{}^-(aq) + H_3O^+(aq)$$

Given the bonding in boric acid, suggest why this reaction is possible. Is boric acid acting as a Lewis acid or Lewis base in this reaction?

80. The structure of nitric acid is illustrated on pages 338 and 725.

(a) Why are the $N-O$ bonds the same length, and why are both shorter than the $N-OH$ bond length?

(b) Rationalize the bond angles in the molecule.

(c) What is the hybridization of the central N atom? What orbitals overlap to form the $N-O$ π bond?

81. Like many metals, aluminum reacts readily with halogens (page 120). Aluminum bromide has the structure illustrated here in the solid and gaseous phases. That is, it is a dimer of the monomer units $AlBr_3$ with a Br atom bridging from one Al atom to another. (This form of halogen bridging is fairly common in chemistry.) The $Br-Al-Br$ angle is 115° and the $Al-Br-Al$ angle is 87°.

Model of aluminum bromide dimer.

(a) What is the hybridization of the Al atoms?

(b) Draw the Lewis structure for an $AlBr_3$ monomer.

(c) How might a Br atom form a bridge from one Al atom to another? Describe the bonding in the Al_2Br_6 molecule.

82. When 1.00 g of a white solid, alkaline earth compound, A, is strongly heated, you obtain another white solid, B, and a gas. An experiment is carried out on the gas, showing that it exerts a pressure of 209 mm Hg in a 450-mL flask at 25 °C. Bubbling the gas into a solution of $Ca(OH)_2$ gives another white solid, C. If the white solid B is added to water, the resulting solution turns red litmus paper blue. Addition of 40.6 mL of 0.125 M HCl to the solution of B and evaporation of the resulting solution to dryness yields 1.055 g of a white solid D. When D is placed in a Bunsen burner flame, it colors the flame green. Finally, if the aqueous solution of B is treated with sulfuric acid, a white precipitate, E, forms. Identify the lettered compounds in the reaction scheme.

83. You have a 1.0-L flask that contains a mixture of argon and hydrogen. The pressure inside the flask is 745 mm Hg and the temperature is 22 °C. Describe an experiment that you could use to determine the percentage of hydrogen in this mixture.

84. Use $\Delta H°_f$ data in Appendix L to calculate the enthalpy change of the reaction

$$2\,N_2(g) + 5\,O_2(g) + 2\,H_2O(\ell) \longrightarrow 4\,HNO_3(aq)$$

Speculate on whether such a reaction could be used to "fix" nitrogen. Would research to find ways to accomplish this reaction be a good endeavor?

85. You are given air and water for starting materials, along with whatever laboratory equipment you need. Describe how you could synthesize ammonium nitrate from these reagents.

86. Magnesium chemistry:

(a) Magnesium is obtained from sea water. If the concentration of Mg^{2+} in sea water is 0.050 M, what volume of sea water (in liters) must be treated to obtain 1.00 kg of magnesium metal? What mass of lime (CaO, in kilograms) must be used to precipitate the magnesium in this volume of sea water?

(b) When 1.2×10^3 kg of molten $MgCl_2$ is electrolyzed to produce magnesium, what mass of metal (kg) is produced at the cathode? What is produced at the anode? What mass of the other product (kg) is produced? What is the total number of Faradays of electricity used in the process?

(c) One industrial process has an energy consumption of 18.5 kwh/kg of Mg. How many joules are therefore required per mole (1 kwh = 1 kilowatt-hour = 3.6×10^6 J)? How does this energy compare with the energy of the following process?

$$MgCl_2(s) \longrightarrow Mg(s) + Cl_2(g)$$

87. Assume an electrolysis cell that produces chlorine from aqueous sodium chloride (called "brine") operates at 4.6 V (with a current of 3.0×10^5 amps). Calculate the number of kilowatt-hours of energy required to produce 1.00 kg of chlorine (1 kwh = 1 kilowatt-hour = 3.6×10^6 J).

88. Sodium metal is produced by electrolysis from molten sodium chloride. The cell operates at 7.0 V with a current of 25×10^3 amps.

(a) What mass of sodium can be produced in one hour?

(b) How many kilowatt-hours of electricity are used to produce 1.00 kg of sodium metal (1 kwh = 1 kilowatt-hour = 3.6×10^6 J)?

22

The Chemistry of the Transition Elements

Chapter Goals

- Identify and explain chemical and physical properties of the transition elements.
- Understand the composition, structure, and bonding in coordination compounds.
- Show how bonding is used to explain the magnetism and spectra of coordination compounds.

Memory Metal

In the early 1960s, metallurgical engineer William J. Buehler, a researcher at the Naval Ordinance Laboratory in White Oak, Maryland, was experimenting with alloys made up of two metals.

▲ These sunglasses frames are made of nitinol, so they snap back to the proper fit even after being twisted like a pretzel. The nitinol used in these frames has a critical temperature below room temperature, so the metal readily returns to its "memorized" shape. Similar alloys are used for wires in dental braces and surgical anchors. *(NASA/Science Source/Photo Researchers, Inc.)*

He was looking for a material that was impact- and heat-resistant because it was intended for use in the nose cone of a Navy missile. It was also important that the material be fatigue-resistant; that is, it should not lose its desirable properties when bent and shaped. An alloy of two transition metals, nickel and titanium, appeared to have the desirable properties. Buehler prepared long thin strips of

this alloy to demonstrate that it could be folded and unfolded many times without breaking. At a meeting to discuss this material, one of his associates decided to see what happened when the strip was heated. He held a cigarette lighter to a folded-up piece of metal and was amazed to observe that the metal strip immediately unfolded to its original shape. Thus, memory metal was discovered. This unusual alloy is now called nitinol, a name constructed out of Nickel, Titanium, Naval Ordinance Lab.

Memory metal is an alloy with about the same number of Ni and Ti atoms. The alloy remembers its shape because of the arrangement of these atoms in the solid phase. When the atoms are arranged in the highly symmetric phase (austenite), the alloy is relatively rigid. The metal can be trained to remember its shape by twisting or bending it to the desired shape when in this phase. When the alloy is cooled below a temperature called its "phase transition temperature,"

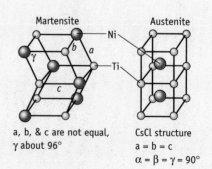

a, b, & c are not equal, γ about 96°

CsCl structure
a = b = c
α = β = γ = 90°

▲ Two phases of nitinol. *(Mike Condren/UW/MRSEC)*

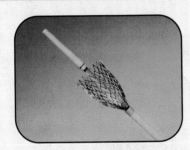

▲ A stent for reinforcing blood vessels, woven from nitinol wire. *(Courtesy Nitinol Devices & Components, a Johnson and Johnson Company)*

the most interesting are in the medical area. One application is the fabrication of vascular stents to reinforce blood vessels. The stent is crushed and inserted into a blood vessel through a very fine needle. When the nitinol stent warms to body temperature, it returns to its memorized shape and reinforces the walls of the blood vessel.

Nitinol can also be used in orthodontics; braces of nitinol remember their shape and apply a steady, constant pressure to move teeth into position.

it enters a less symmetric but flexible phase (martensite). Below its transition temperature, the metal is fairly soft and may be bent and twisted out of shape. When warmed, the metal returns to its original shape. The temperature at which the change in shape occurs varies with small differences in the nickel-to-titanium ratio. Depending on composition, materials that change shape at temperatures ranging from −125 °C to about 70 °C are possible, greatly increasing the number of possible uses for this intriguing material.

Memory metal never made it into missile nose cones, but it has found a wide variety of other applications. Some of

▲ Orthodontic devices can also be made of nitinol. *(© Françoise Gervais/CORBIS)*

Before You Begin

- Recall the position of transition elements in the periodic table (Section 2.6).
- Know the formulas and names for transition metals, their ions, and their compounds (Section 3.3).
- Be able to balance chemical equations (Sections 4.1 and 20.1).
- Review electronic structure, spectra, and magnetism (Chapters 8 and 9).

The chemistry of nickel, a transition metal The transition metals and their ions form a wide range of compounds, many with beautiful colors and interesting shapes. The purpose of this chapter is to explore some commonly observed shapes and to explain how compounds of the transition metals can be so colorful.

Nickel and its coordination compounds

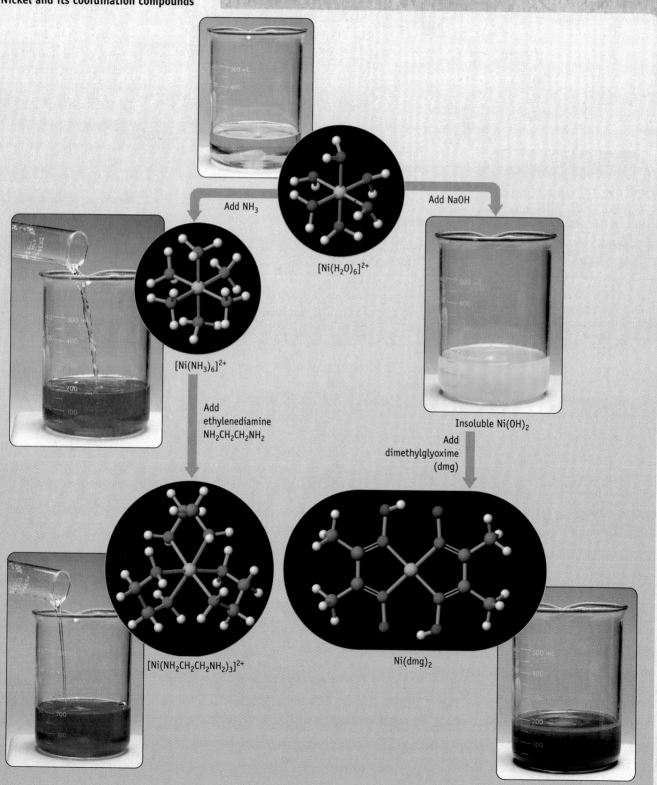

$[Ni(H_2O)_6]^{2+}$

Add NH$_3$

Add NaOH

$[Ni(NH_3)_6]^{2+}$

Insoluble Ni(OH)$_2$

Add ethylenediamine
NH$_2$CH$_2$CH$_2$NH$_2$

Add dimethylglyoxime (dmg)

$[Ni(NH_2CH_2CH_2NH_2)_3]^{2+}$

Ni(dmg)$_2$

Charles D. Winters

The transition elements are the large block of elements in the central portion of the periodic table. All are metals and bridge the *s*-block elements at the left and the *p*-block elements on the right (Figure 22.1). The transition elements are divided into two types of elements, depending on the valence electrons involved. First are the **d-block elements,** because their occurrence in the periodic table coincides with the filling of the *d* orbitals. The second type are the **f-block elements,** characterized by filling *f* orbitals. Contained within this group of elements are two subgroups: the *lanthanides,* elements that occur between La and Hf, and the *actinides,* elements that occur between Ac and Rf.

In this chapter, the primary focus is on the *d*-block elements, and within this group we concentrate mainly on the elements in the fourth period, that is, the elements of the first transition series, scandium to zinc.

22.1   PROPERTIES OF THE TRANSITION ELEMENTS

The *d*-block metals include elements with a wide range of properties. They encompass the most common metal used in construction and manufacturing (iron), metals that are valued for their beauty (gold, silver, and platinum), and metals used in coins (nickel, copper, and zinc). There are metals used in modern technology (titanium) and metals known and used in early civilizations (copper, silver, gold, and iron). The *d*-block contains the densest elements (osmium, $d = 22.49$ g/cm^3, and iridium, $d = 22.41$ g/cm^3), the metals with the highest and lowest melting points (tungsten, mp = 3410 °C, and mercury, mp = -38.9 °C), and one of two radioactive elements with atomic numbers less than 83 [technetium (Tc), atomic number 43; the other is promethium (Pm), atomic number 61, in the *f*-block].

With the exception of mercury, transition elements are solids, often with high melting and boiling points. They have a metallic sheen and are conductors of electricity and heat. They react with various oxidizing agents to give ionic compounds. There is considerable variation in such reactions among the elements, however. Most react readily, but silver, gold, and platinum resist oxidation and are resistant to becoming tarnished, they are used for jewelry and decorative items.

Certain *d*-block elements are particularly important in living organisms. Cobalt is the crucial element in vitamin B$_{12}$, which is part of a catalyst essential for several biochemical reactions. Hemoglobin and myglobin, oxygen-carrying and storage proteins, contain iron (see page 939). Molybdenum and iron, together with sulfur, form the reactive portion of nitrogenase, a biological catalyst used by nitrogen-fixing organisms to convert atmospheric nitrogen into ammonia.

Chapter Goals • Revisited

- **Identify and explain chemical and physical properties of the transition elements.**

- Understand the composition, structure, and bonding in coordination compounds.

- Show how bonding is used to explain the magnetism and spectra of coordination compounds.

Gold is quite resistant to oxidation and so retains its luster even under harsh conditions. Consequently, it has been used for decorative items for centuries. (*© Reuters NewMedia Inc./CORBIS*)

Figure 22.1   The transition metals. The *d*-block elements are highlighted in a darker shade of purple. (*Charles D. Winters*)

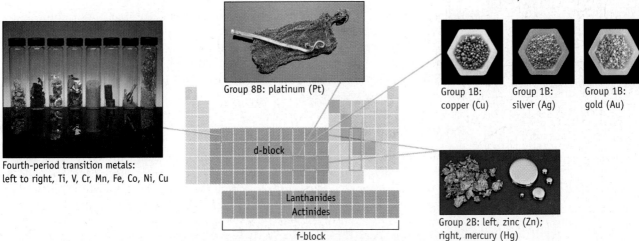

Fourth-period transition metals: left to right, Ti, V, Cr, Mn, Fe, Co, Ni, Cu

Group 8B: platinum (Pt)

d-block

Lanthanides
Actinides

f-block

Group 1B: copper (Cu)

Group 1B: silver (Ag)

Group 1B: gold (Au)

Group 2B: left, zinc (Zn); right, mercury (Hg)

(a) Paint pigments: yellow, Cds; green, Cr_2O_3; white, TiO_2 and ZnO; purple, $Mn_3(PO_4)_2$; blue, Co_2O_3 and Al_2O_3; ochre, Fe_2O_3.

(b) Small amounts of transition metal compounds are used to color glass: blue, Co_2O_3; green, copper or chromium oxides; purple, nickel or cobalt oxides; red, copper oxide; iridescent green, uranium oxide.

(c) Traces of transition metal ions are responsible for the colors in purple amethyst (iron), green jade (iron), red corundum (chromium), and blue lapis lazuli and turquoise (copper).

Figure 22.2   Colorful chemistry. Transition metal compounds are often colored, a property that leads to specific uses. *(Charles D. Winters)*

Many transition metal compounds are highly colored, which makes them useful as pigments in paints and dyes (Figure 22.2). Prussian blue, $Fe_4[Fe(CN)_6]_3$, is a "bluing agent" used in engineering blueprints and in the laundry to brighten white cloth. A common pigment (artist's cadmium-yellow) contains cadmium sulfide (CdS), and the white in most white paints is titanium(IV) oxide, TiO_2.

The presence of transition metal ions in crystalline silicates or alumina transforms these common materials into gemstones. Iron(II) causes the yellow color in citrine, and chromium(III) causes the red color of a ruby. Transition metal complexes in small quantities add color to glass. Blue glass contains a small amount of cobalt(III) oxide, and addition of chromium oxide to glass gives a green color. Old window panes sometimes take on a purple color over time as a consequence of oxidation of traces of manganese(II) ion present in the glass to permanganate ion (MnO_4^-).

In the next few pages we will examine the properties of the transition elements, concentrating on the underlying principles that govern these properties.

Electron Configurations

Because chemical behavior is related to electron structure, it is important to know the electron configurations of the *d*-block elements (Table 22.1) and their common ions [← SECTIONS 8.4 AND 8.5]. Recall that the configuration of these metals has the general form [noble gas core]$ns^a(n-1)d^b$; that is, valence electrons for the transition elements reside in the *ns* and $(n-1)d$ subshells [← TABLES 8.3 AND 8.4] [CD-ROM, Screen 8.7, Simulation].

Oxidation and Reduction

A characteristic chemical property of all metals is that they undergo oxidation by a wide range of oxidizing agents—oxygen, halogens, and aqueous acids being only a few of many examples. Standard reduction potentials for the elements of the first transition series can be used to predict which elements will be oxidized by a given

Table 22.1 • Electron Configurations of the Fourth-Period Transition Elements

	spdf Configuration	Box Notation		
			3d	4s
Sc	[Ar]$3d^14s^2$		↑	↑↓
Ti	[Ar]$3d^24s^2$		↑ ↑	↑↓
V	[Ar]$3d^34s^2$		↑ ↑ ↑	↑↓
Cr	[Ar]$3d^54s^1$		↑ ↑ ↑ ↑ ↑	↑
Mn	[Ar]$3d^54s^2$		↑ ↑ ↑ ↑ ↑	↑↓
Fe	[Ar]$3d^64s^2$		↑↓ ↑ ↑ ↑ ↑	↑↓
Co	[Ar]$3d^74s^2$		↑↓ ↑↓ ↑ ↑ ↑	↑↓
Ni	[Ar]$3d^84s^2$		↑↓ ↑↓ ↑↓ ↑ ↑	↑↓
Cu	[Ar]$3d^{10}4s^1$		↑↓ ↑↓ ↑↓ ↑↓ ↑↓	↑
Zn	[Ar]$3d^{10}4s^2$		↑↓ ↑↓ ↑↓ ↑↓ ↑↓	↑↓

Chemical Perspectives

Corrosion of Iron

It is hard not to be aware of corrosion. Those of us who live in the northern part of the United States are well aware of the problems of rust on our automobiles. It is estimated that 20% of iron production each year is solely to replace iron that has rusted away.

The corrosion or rusting of iron results in major economic loss. *(Charles D. Winters)*

Qualitatively, we describe corrosion as the deterioration of metals by a product-favored oxidation reaction. The corrosion of iron, for example, converts iron metal to red-brown rust, which is hydrated iron(III) oxide, $Fe_2O_3 \cdot H_2O$. The process requires both air and water, and it is enhanced if the water contains dissolved ions and if the metal is stressed (dents, cuts, and scrapes on the surface).

The corrosion process occurs in what is essentially a small electrochemical cell. There is an anode and cathode, an electrical connection between the two (the metal itself), and an electrolyte with which both anode and cathode are in contact. When a metal corrodes, there are anodic areas of the metal surface where the metal is oxidized:

Anode, oxidation: $M(s) \longrightarrow M^{n+} + n\,e^-$

and cathodic areas where the electrons are consumed by several possible half-reactions:

Cathode, reduction:
$H^+(aq) + 2\,e^- \longrightarrow H_2(g)$

$2\,H_2O(\ell) + 2\,e^- \longrightarrow H_2(g) + 2\,OH^-(aq)$

$O_2(g) + 2\,H_2O(\ell) + 4\,e^- \longrightarrow 4\,OH^-(aq)$

With iron, the rate of the corrosion process is controlled by the rate of the cathodic process. Of the three possible cathodic reactions, the one that is fastest is determined by acidity and the amount of oxygen present. If little or no oxygen is present—as when a piece of iron is buried in soil such as moist clay—hydrogen ion or water is reduced and $H_2(g)$ and hydroxide ions are the products. Iron(II) hydroxide is relatively insoluble, and it will precipitate on the metal surface, further inhibiting the formation of Fe^{2+}.

Anode: $Fe(s) \longrightarrow Fe^{2+}(aq) + 2\,e^-$
Cathode:
$2\,H_2O(\ell) + 2\,e^- \longrightarrow H_2(g) + 2\,OH^-(aq)$

Precipitation:
$Fe^{2+}(aq) + 2\,OH^-(aq) \longrightarrow Fe(OH)_2(s)$

Net reaction:
$Fe(s) + 2\,H_2O(\ell) \longrightarrow H_2(g) + Fe(OH)_2(s)$

If both water and O_2 are present, the corrosion reaction is about 100 times faster than without oxygen.

Anode: $2\,Fe(s) \longrightarrow 2\,Fe^{2+}(aq) + 4\,e^-$
Cathode:
$O_2(g) + 2\,H_2O(\ell) + 4\,e^- \longrightarrow 4\,OH^-(aq)$
Precipitation:
$2\,Fe^{2+}(aq) + 4\,OH^-(aq) \longrightarrow 2\,Fe(OH)_2(s)$

Net reaction:
$2\,Fe(s) + 2\,H_2O(\ell) + O_2(g) \longrightarrow$
$\qquad\qquad\qquad\qquad 2\,Fe(OH)_2(s)$

If oxygen is not freely available, further oxidation of the iron(II) hydroxide leads to the formation of magnetic iron oxide (which can be thought of as a mixed oxide of Fe_2O_3 and FeO).

$6\,Fe(OH)_2(s) + O_2(g) \longrightarrow$
$\qquad 2\,Fe_3O_4 \cdot H_2O(s) + 4\,H_2O(\ell)$
green hydrated magnetite

$Fe_3O_4 \cdot H_2O(s) \longrightarrow H_2O(\ell) + Fe_3O_4(s)$
black magnetite

It is the black magnetite you find coating an iron object that has corroded by resting in moist soil.

If the iron object has free access to oxygen and water, as in the open or in flowing water, red-brown iron(III) oxide will form.

$4\,Fe(OH)_2(s) + O_2(g) \longrightarrow$
$\qquad 2\,Fe_2O_3 \cdot H_2O(s) + 2\,H_2O(\ell)$
red-brown

This is the familiar rust you see on cars and buildings, and the substance that colors the water red in some mountain streams or in your home.

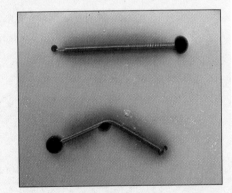

Anode and cathode reactions in iron corrosion. Two iron nails were placed in an agar gel that contains phenolphthalein and $Na_3[Fe(CN)_6]$. Iron(II) ion, formed at the tip and where the nail is bent, reacts with $[Fe(CN)_6]^{3-}$ to form blue-green $Fe_3[Fe(CN_6)]_2$ (Prussian blue). Hydrogen (H_2) and $OH^-(aq)$ are formed at the other parts of the surface of the nail, the latter being detected by the red color of the acid–base indicator. In this electrochemical cell, the ends and the bent region of the nail are anodes, and the remainder of the surface is the cathode. *(Charles D. Winters)*

oxidizing agent. For example, all of these metals except vanadium and copper are oxidized by aqueous HCl (Table 22.2). The products of oxidation are generally ionic compounds containing a metal cation. This feature, which dominates the chemistry of these elements, is sometimes highly undesirable (see *Chemical Perspectives: Corrosion of Iron* and *CD-ROM, Screen 20.10*).

When a transition metal is oxidized, the outermost s electrons are removed, followed by one or more d electrons. With a few exceptions, transition metal ions have the electron configuration [noble gas core]$(n - 1)d^x$. In contrast to ions formed by main group elements, these ions do not have noble gas configurations and their compounds often possess unpaired electrons, resulting in paramagnetism [← PAGE 290]. They are frequently colored as well, due to the absorption of light in the visible region of the spectrum. Color and magnetism figure prominently in a discussion of the properties and bonding of these elements, as you shall see shortly.

In the first transition series, the most commonly encountered metal ions have oxidation numbers of +2 and +3. (Table 22.2). With iron, for example, oxidation usually converts Fe ([Ar]$3d^6 4s^2$) to Fe^{2+} ([Ar]$3d^6$) or to Fe^{3+} ([Ar]$3d^5$). Iron reacts with chlorine to give $FeCl_3$, and it reacts with aqueous acids to produce Fe^{2+}(aq) and H_2 (Figure 22.3). Despite the preponderance of +2 and +3 ions in compounds of the first transition metal series, the range of possible oxidation states is broad (Figure 22.4). In examples earlier in this text, we encountered chromium with a +6 oxidation number (CrO_4^{2-}, $Cr_2O_7^{2-}$), manganese with an oxidation number of +7 (MnO_4^-), and silver and copper as +1 ions. The most common oxidation number of titanium is +4, found, for example, in the common minerals of this element: TiO_2 (rutile) and ilmenite ($FeTiO_3$).

Higher oxidation numbers are more common in compounds of the elements in the second and third transition series. For example, the naturally occurring sources of molybdenum and tungsten are the ores molybdenite (MoS_2) and wolframite (WO_3). This general trend is carried over in the f-block. The lanthanides form +3 ions, whereas actinide elements usually also have higher oxidation numbers, such as +4 and even +6 in typical compounds. For example, UO_3 is a common oxide of uranium, and UF_6 is a compound important in processing uranium fuel for nuclear reactors [→ SECTION 23.6].

Table 22.2 • Products from Reactions of the Elements in the First Transition Series with O_2, Cl_2, or Aqueous HCl

Element	Reaction with O_2	Reaction with Cl_2	Reaction with Aqueous HCl
Scandium	Sc_2O_3	$ScCl_3$	Sc^{3+}(aq)
Titanium	TiO_2	$TiCl_4$	Ti^{3+}(aq)
Vanadium	V_2O_5	VCl_4	NR*
Chromium	Cr_2O_3	$CrCl_3$	Cr^{2+}(aq)
Manganese	Mn_3O_4	$MnCl_2$	Mn^{2+}(aq)
Iron	Fe_2O_3	$FeCl_3$	Fe^{2+}(aq)
Cobalt	Co_3O_4	$CoCl_2$	Co^{2+}(aq)
Nickel	NiO	$NiCl_2$	Ni^{2+}(aq)
Copper	CuO	$CuCl_2$	NR*
Zinc	ZnO	$ZnCl_2$	Zn^{2+}(aq)

*NR = no reaction

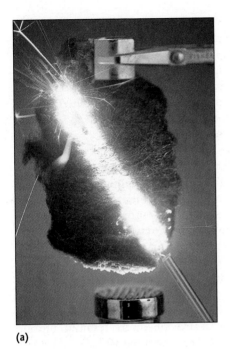

(a)

(b)

(c)

Figure 22.3 Typical reactions of transition metals. These metals react with oxygen, with halogens, and with acids under appropriate conditions. **(a)** Here steel wool reacts with O_2 and **(b)** with chlorine gas, Cl_2, and **(c)** iron filings react with aqueous HCl. *(Charles D. Winters)*

Periodic Trends in the *d*-Block: Size, Density, Melting Point

The periodic table is the most useful single reference source for a chemist. Not only does it provide data that have everyday use, but it organizes the elements with respect to chemical and physical properties. Let us look at three physical properties of the transition elements that vary periodically: atomic radii, density and melting point.

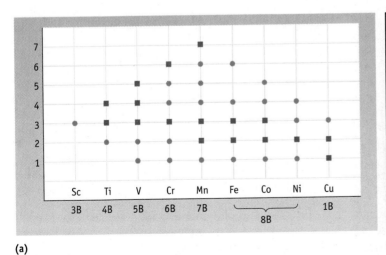

(a)

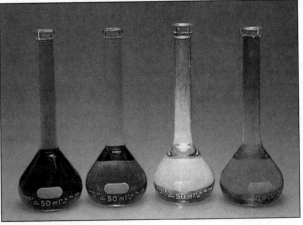

(b)

Figure 22.4 Oxidation numbers of the transition elements in the first transition series. (a) The most common oxidation numbers are indicated with red squares; less common oxidation numbers are indicated with blue dots. **(b)** Aqueous solutions of chromium compounds with two different oxidation numbers: +3 in $Cr(NO_3)_3$ (violet) and $CrCl_3$ (green), and +6 in K_2CrO_4 (yellow) and $K_2Cr_2O_7$ (orange). *(Charles D. Winters)*

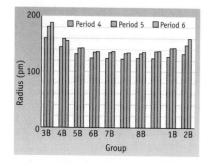

Trends in the radii of the transition elements in the fourth, fifth, and sixth periods. For a large version, see Figure 8.10 on page 307.

Metal Atom Radii

The variation in atomic radii for the transition elements in the fourth, fifth, and sixth periods is illustrated in the margin and in Figure 8.10. The radii of the transition elements vary over a fairly narrow range, with a small decrease to a minimum around the middle of this group of elements. This variation can be understood based on electron configurations. Atomic size is determined by the radius of the outermost occupied orbital, which for these elements is the ns orbital ($n = 4, 5$, or 6). Progressing from left to right in the periodic table, the size decline expected from increasing the number of protons in the nucleus is mostly canceled out by an opposing effect, repulsion from additional electrons in the $(n - 1)d$ orbitals.

The radii of the d-block elements in the fifth and sixth periods in each group are almost identical. The reason is that the lanthanide elements immediately precede the third series of d-block elements. The filling of $4f$ orbitals is accompanied by a steady contraction in size (not unlike the overall size decrease across each series of transition elements). At the point where the $5d$ orbitals begin to fill again, the radii have decreased to a size similar to that of elements in the previous period. The decrease in size that results from the filling of the $4f$ orbitals is given a specific name, **lanthanide contraction.**

The similarity in size between the second and third rows of d-block elements has significant consequences for their chemistry. For example, the "platinum group metals" (Ru, Os, Rh, Ir, Pd, and Pt) form similar compounds. Thus, it is not surprising that minerals containing these metals are found in the same geological zones on Earth. Nor is it surprising that it is difficult to separate these elements from one another.

Density

The variation in metal radii causes the densities of the transition elements to first increase and then decrease across a period (Figure 22.5a). Although the overall change in radii among these elements is small, the effect is magnified because the atomic volume is actually changing with the cube of the radius ($V = 4/3\pi r^3$).

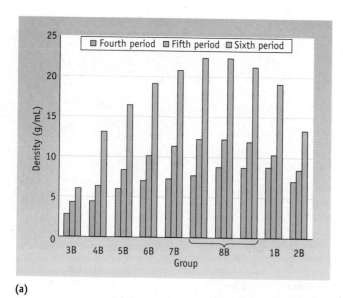

(a)

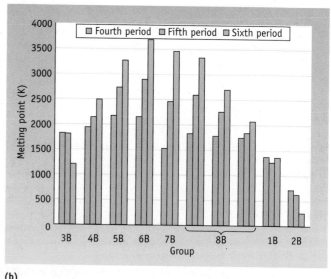

(b)

Figure 22.5 Periodic properties in the transition series. Density **(a)** and melting point **(b)** of the d-block elements.

The lanthanide contraction is the reason that elements in the sixth period have the highest density. The relatively small radii of sixth-period transition metals, combined with the fact that their atomic masses are considerably larger than their counterparts in the fifth period, cause sixth-period metal densities to be very large.

Melting Point

The metals with the highest melting points occur in the middle of each series (Figure 22.5b). The melting point of any substance reflects the forces of attraction between the atoms, molecules, or ions that compose the solid. With transition elements, the melting points rise to a maximum around the middle of the series, then descend. Again, electron configurations provide us with an explanation. The variation in melting point indicates the strongest metallic bonds occur when the *d* subshell is about half-filled. This is also the point at which the largest number of unpaired electrons occurs in the isolated atoms. We can reasonably conclude that the *d* electrons are playing an important role in metallic bonding.

22.2   METALLURGY

A few metals occur in nature as the free elements. These include copper (Figure 22.6), silver, and gold. Most metals, however, are found as oxides, sulfides, halides, carbonates, or other ionic compounds (Figure 22.7). Some metal-containing mineral deposits are of little economic value, either because the concentration of the metal is too low or because the metal is difficult to separate from impurities. The relatively few minerals from which elements can be obtained profitably are called **ores** (Figure 22.7). **Metallurgy** is the general name given to the process of obtaining metals from their ores.

Very few ores are chemically pure substances. The desired mineral is usually mixed with large quantities of impurities such as sand and clay, called **gangue** (pronounced "gang"). Generally, the first step in a metallurgical process is to separate the mineral from the gangue. Then the ore is converted to the metal, a reduction process. Pyrometallurgy and hydrometallurgy are two methods of recovering metals from their ores. As the names imply, **pyrometallurgy** involves high temperatures and **hydrometallurgy** uses aqueous solutions (and thus is limited to the relatively low temperatures at which water is a liquid). Iron and copper metallurgy illustrate these two methods of metal production.

Figure 22.6   Naturally occurring copper. Copper occurs as the metal (native copper) and as minerals such as blue azurite [2 $CuCO_3 \cdot Cu(OH)_2$] and green malachite [$CuCO_3 \cdot Cu(OH)_2$]. *(Charles D. Winters)*

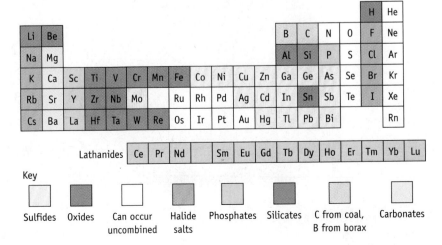

Figure 22.7   Sources of the elements. A few transition metals such as copper and silver occur naturally as the metal. Most of these elements are found naturally as oxides or sulfides, however. (Boxes with no symbol represent radioactive elements.)

Pyrometallurgy: Iron Production

The production of iron from its ores is carried out in a blast furnace (Figure 22.8). The furnace is charged with a mixture of ore (usually hematite, Fe_2O_3), coke (which is primarily carbon), and limestone ($CaCO_3$). A blast of hot air forced in at the bottom causes the coke to burn with such an intense heat that the temperature at the bottom is about 1500 °C. The quantity of air is controlled so that carbon monoxide is the primary product. Both carbon and carbon monoxide participate in the reduction of iron(III) oxide to give impure metal:

$$Fe_2O_3(s) + 3\ C(s) \longrightarrow 2\ Fe(\ell) + 3\ CO(g)$$
$$Fe_2O_3(s) + 3\ CO(g) \longrightarrow 2\ Fe(\ell) + 3\ CO_2(g)$$

Much of the carbon dioxide formed in the reduction process (and from heating the limestone) is reduced on contact with unburned coke and produces more reducing agent:

$$CO_2(g) + C(s) \longrightarrow 2\ CO(g)$$

The molten iron flows down through the furnace and collects at the bottom where it is tapped off through an opening in the side. This impure iron is called *cast iron* or *pig iron*. Usually, the impure metal is either brittle or soft (undesirable properties for most uses), due to the presence of small amounts of impurities such as elemental carbon, phosphorus, and sulfur.

● **Coke, a Reducing Agent**
Coke is made by heating coal in a tall, narrow oven that is sealed to keep out oxygen. Heating drives off volatile compounds including benzene and ammonia. What remains is nearly pure carbon.

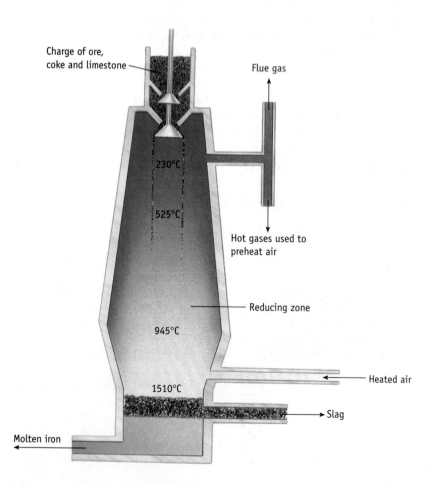

Figure 22.8 A blast furnace. The largest modern furnaces have hearths 14 m in diameter. They can produce up to 10,000 tons of iron per day.

Iron ores generally contain silicate minerals and silicon dioxide. Lime (CaO), formed when limestone is heated, reacts with these materials to give calcium silicate.

$$SiO_2(s) + CaO(s) \longrightarrow CaSiO_3(\ell)$$

This is an acid–base reaction because CaO is a basic oxide and SiO_2 is an acidic oxide. The calcium silicate, molten at the temperature of the blast furnace and less dense than molten iron, floats on the iron. Other nonmetal oxides dissolve in this layer and the mixture, called *slag*, is easily removed.

Pig iron from the blast furnace may contain up to 4.5% carbon, 0.3% phosphorus, 0.04% sulfur, and as much as 15% silicon, and sometimes other elements. The impure iron must be purified to remove these nonmetal impurities. Several processes are available, but the most important uses the *basic oxygen furnace* (Figure 22.9). The process in the furnace removes much of the carbon and all of the phosphorus, sulfur, and silicon. Pure oxygen is blown into the molten pig iron and oxidizes phosphorus to P_4O_{10}, sulfur to SO_2, and carbon to CO_2. These nonmetal oxides either escape as gases or react with basic oxides such as CaO that are added or are used to line the furnace. For example,

$$P_4O_{10}(g) + 6\ CaO(s) \longrightarrow 2\ Ca_3(PO_4)_2(\ell)$$

The result is ordinary carbon steel. Almost any degree of flexibility, hardness, strength, and malleability can be achieved in carbon steel by reheating and cooling in a process called *tempering*. The resulting material can then be put to a wide variety of uses. The major disadvantages with carbon steel are that it corrodes easily and loses its properties when heated strongly.

Other transition metals, such as chromium, manganese, and nickel, can be added during the steel-making process, giving *alloys* (solid solutions of two or more metals) that have specific physical, chemical, and mechanical properties. A well-known alloy is stainless steel; this contains 18–20% Cr and 8–12% Ni. Stainless steel is much more resistant to corrosion than is carbon steel. Another alloy of iron is Alnico V. This alloy, used in loudspeaker magnets because of its permanent magnetism, contains five elements: Al (8%), Ni (14%), Co (24%), Cu (3%), and Fe (51%).

Figure 22.9 **Molten iron being poured from a basic oxygen furnace.** *(Bethlehem Steel Corp.)*

Hydrometallurgy: Copper Production

In contrast to iron ores, which are mostly oxides, most copper ores are sulfides. Copper-bearing minerals include chalcopyrite ($CuFeS_2$), chalcocite (Cu_2S), and covellite (CuS) (Figure 22.6). Because ores containing these minerals generally have a very low percentage of copper, enrichment is necessary, and this is carried out by a process known as *flotation*. First, the ore is finely powdered. Oil is then added and the mixture agitated with soapy water in a large tank (Figure 22.10). At the same time, compressed air is forced through the mixture, and the lightweight, oil-covered copper sulfide particles are carried to the top as a frothy mixture. The heavier gangue settles to the bottom of the tank, and the copper-laden froth is skimmed off.

Hydrometallurgy can be used to obtain copper from an enriched ore. In one method, enriched chalcopyrite ore is treated with a solution of copper(II) chloride. A reaction ensues that leaves copper in the form of solid insoluble CuCl, which is easily separated from the iron that remains in solution as aqueous $FeCl_2$:

$$CuFeS_2(s) + 3\ CuCl_2(aq) \longrightarrow 4\ CuCl(s) + FeCl_2(aq) + 2\ S(s)$$

Figure 22.10 **Enriching copper ore by the flotation process.** The less dense particles of Cu_2S are trapped in the soap bubbles and float. The denser gangue settles to the bottom.

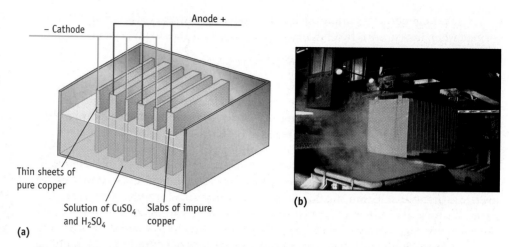

(a)

(b)

Figure 22.11 Electrolytic refining of copper. (a) Slabs of impure copper, called "blister copper," form the anode and pure copper is deposited at the cathode. **(b)** The electrolysis cells at a copper refinery. *(b, Simon Fraser/Northumbria Circuits/Science Photo Library/Photo Researchers, Inc.)*

Aqueous NaCl is then added and CuCl dissolves because of the formation of the complex ion $[CuCl_2]^-$:

$$CuCl(s) + Cl^-(aq) \longrightarrow [CuCl_2]^-(aq)$$

Copper(I) compounds are unstable with respect to Cu(0) and Cu(II). Thus, $[CuCl_2]^-$ disproportionates to the metal and $CuCl_2$, and the latter is used to treat more ore.

$$2\,[CuCl_2]^-(aq) \longrightarrow Cu(s) + CuCl_2(aq) + 2\,Cl^-(aq)$$

Approximately 10% of the copper produced in the United States is obtained with the aid of bacteria. Acidified water is sprayed onto copper-mining wastes that contain low levels of copper. As the water trickles down through the crushed rock, the bacterium *Thiobacillus ferrooxidans* [◄ PAGE 880] breaks down the iron sulfides in the rock and oxidizes iron(II) to iron(III). Iron(III) ion in turn oxidizes the sulfide ion of copper sulfide to sulfate, leaving copper(II) ion in solution. Then the copper(II) ion is reduced to metallic copper by reaction with iron:

$$Cu^{2+}(aq) + Fe(s) \longrightarrow Cu(s) + Fe^{2+}(aq)$$

The purity of copper at this stage is about 99%, but this is not acceptable because even traces of impurities greatly diminish the electrical conductivity of the metal. Consequently, a further purification step is needed, and this is done using electrolysis (Figure 22.11). Thin sheets of pure copper metal and slabs of impure copper are immersed in a solution containing $CuSO_4$ and H_2SO_4. The pure copper sheets are the cathode of an electrolysis cell, and the impure slabs are the anode. Copper in the impure sample is oxidized to copper(II) ions at the anode; copper(II) ions in solution are reduced to pure copper at the cathode.

22.3 COORDINATION COMPOUNDS

When metal salts dissolve, water molecules cluster around the ions (◄ Chapter Focus, page 924). The negative end of each polar water molecule is attracted to the positively charged metal ion, and the positive end of the water molecule is attracted to the anion. As noted earlier [← SECTION 13.2], the energy of the ion–solvent interaction (solvation energy) is an important aspect of the solution process. But, there is much more to this story.

Complexes and Ligands

A green solution formed by dissolving nickel(II) chloride in water contains Ni^{2+} (aq) and Cl^- (aq) ions (see Chapter Focus). If the water is removed, a green crystalline solid is obtained. The formula of the solid is often written as $NiCl_2 \cdot 6\,H_2O$, and the compound is called hydrated nickel(II) chloride. Addition of ammonia to the aqueous nickel(II) chloride solution gives a lilac-colored solution from which another compound, $NiCl_2 \cdot 6\,NH_3$, can be isolated. This formula looks very similar to the formula for the hydrate, with ammonia substituted for water.

What are these two nickel species? The formulas identify composition but fail to give information about structure. Because properties of compounds derive from their structures, we need to evaluate their structures in more detail. Typically, metal compounds are ionic, and solid ionic compounds have structures with cations and anions arranged in a regular array. The structure of hydrated nickel chloride contains cations with the formula $[Ni(H_2O)_6]^{2+}$ and chloride anions. The structure of the ammonia-containing compound is similar to the hydrate; it is made up of $[Ni(NH_3)_6]^{2+}$ cations and chloride anions (see page 924).

Ions such as $[Ni(H_2O)_6]^{2+}$ and $[Ni(NH_3)_6]^{2+}$, in which a metal ion and either water or ammonia molecules compose a single structural unit, are examples of **coordination complexes,** also known as **complex ions** (Figure 22.12). Compounds containing a coordination complex as part of the structure are called **coordination compounds** and their chemistry is known as coordination chemistry. Although the older "hydrate" formulas are still used, the preferred method of writing the formula for coordination compounds places the metal atom or ion and the molecules or anions directly bonded to it within brackets to show that this is a single structural entity. Thus, the formula for the nickel(II)–ammonia compound is better written as $[Ni(NH_3)_6]Cl_2$.

All coordination complexes contain a metal atom or ion as the central part of the structure. Bonded to the metal are molecules or ions called **ligands** (from the Latin verb ligare, "to bind"). In the above examples, water and ammonia are the ligands. The number of ligands attached to the metal defines the **coordination number** of the metal. The geometry described by the attached ligands is called the **coordination geometry.** In the nickel complex ion $[Ni(NH_3)_6]^{2+}$ (Figure 22.12), the six ligands are arranged in a regular octahedral geometry around the central metal ion.

Ligands can be either neutral molecules or anions (and in rare instances, cations). The characteristic feature of a ligand is that it contains a lone pair of electrons. In the classical description of bonding in a coordination complex, the lone pair of electrons on a ligand is shared with the metal ion. The attachment is a coordinate covalent bond [← SECTION 9.6], because the electron pair being shared was originally on the ligand. The name "coordination complex" derives from the name given to this kind of bonding.

If the ligands are ions, the net charge on a coordination complex is the sum of the charges on the metal and its attached ligands. Complexes can be cations (as in the two nickel complexes used as examples here) or anions, or they can be uncharged.

Ligands such as H_2O and NH_3, which coordinate to the metal via a single Lewis base atom, are termed **monodentate.** The word "dentate" comes from the Latin

Chapter Goals • Revisited

- Identify and explain chemical and physical properties of the transition elements.
- **Understand the composition, structure, and bonding in coordination compounds.**
- Show how bonding is used to explain the magnetism and spectra of coordination compounds.

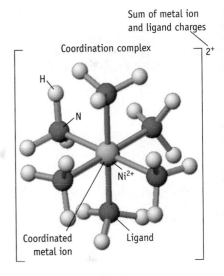

$[Ni(NH_3)_6]^{2+}$

Figure 22.12 A coordination complex. In this complex, $[Ni(NH_3)_6]^{2+}$, the ligands are NH_3 molecules. They are bound to the Ni^{2+} ion by coordinate covalent bonds. Because the metal has a +2 charge, and the ligands have no charge, the charge on the complex ion is +2.

• **Ligands Are Lewis Bases**
Ligands are Lewis bases, because they furnish one or more electron pairs, and the metal ion is a Lewis acid because it accepts electron pairs [← SECTION 17.9]. Thus, the bond between metal and ligand is formally a Lewis acid–Lewis base interaction.

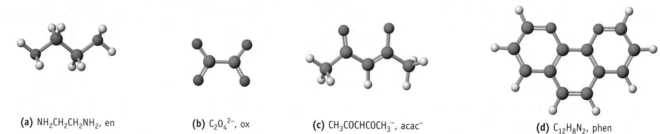

(a) $NH_2CH_2CH_2NH_2$, en   **(b)** $C_2O_4^{2-}$, ox   **(c)** $CH_3COCHCOCH_3^-$, acac$^-$ **(d)** $C_{12}H_8N_2$, phen

Figure 22.13   Common bidentate ligands. (a) Ethylenediamine, $H_2NCH_2CH_2NH_2$; **(b)** oxalate ion, $C_2O_4^{2-}$; **(c)** acetylacetonate ion, $CH_3COCHCOCH_3$; **(d)** phenanthroline, $C_{12}H_8N_2$. Coordination as bidentate ligands results in five-member rings and no ring strain.

● **Complex Ions**
Ions such as $Ag(NH_3)_2^+$, $Ag(CN)_2^-$, $Cu(NH_3)_4^{2+}$ [◀ SECTION 18.6] and the $CuCl_2^-$ ion, mentioned in the discussion of copper metallurgy (page 934), are examples of coordination complexes.

● **Bidentate Ligands**
All common bidentate ligands bind to *adjacent* sites on the metal.

word *dentis* meaning "tooth," so NH_3 is a "one-toothed" ligand. Some ligands attach to the metal with more than one donor atom. These ligands are called **polydentate.** Ethylenediamine (1,2-diaminoethane), $H_2NCH_2CH_2NH_2$, often abbreviated as en, oxalate ion, $C_2O_4^{2-}$ (ox^{2-}), and phenanthroline (phen) are examples of a wide variety of *bidentate ligands* (Figure 22.13). Structures and samples of some complex ions with bidentates ligand are illustrated in Figure 22.14.

Polydentate ligands are also called **chelating ligands,** or just chelates (pronounced "keylates"). The name derives from the Greek word *chele* meaning "claw." Because two bonds must be broken to separate the ligand from the metal, complexes with chelated ligands have greater stability than those with monodentate ligands. Chelated complexes are important in everyday life. One way to clean the rust out of water-cooled automobile engines and steam boilers is to add a solution of oxalic acid. Iron oxide dissolves in the presence of oxalic acid to give a water-soluble iron oxalate complex ion:

$$Fe_2O_3(s) + 6\ H_2C_2O_4(aq) \longrightarrow 2\ [Fe(C_2O_4)_3]^{3-}(aq) + 6\ H^+(aq) + 3\ H_2O(\ell)$$

Ethylenediaminetetraacetate ion (EDTA^{4-}), a *hexadentate* ligand, is an excellent chelating ligand (Figure 22.15). It can wrap around a metal ion, encapsulat-

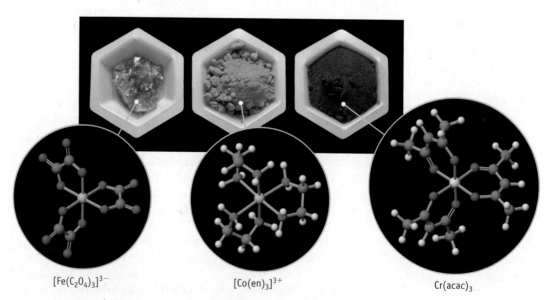

$[Fe(C_2O_4)_3]^{3-}$ $[Co(en)_3]^{3+}$ $Cr(acac)_3$

Figure 22.14   Complex ions with bidentate ligands. See Figure 22.13 for abbreviations. *(Photos, Charles D. Winters)*

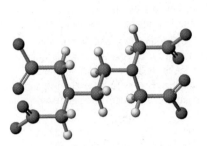

(a) Ethylenediaminetetraacetate, EDTA⁴⁻

(b) [Co(EDTA)]⁻

Figure 22.15   EDTA⁴⁻, a hexadentate ligand.
(a) Ethylenediaminetetraacetate, EDTA⁴⁻. **(b)** [Co(EDTA)]⁻.
Notice the five- and six-member rings created when this ligand
bonds to the metal.

ing it. Salts of this anion are often added to commercial salad dressing to remove traces of free metal ions from solution because these metal ions can otherwise act as catalysts for the oxidation of the oils in the dressing. Without $EDTA^{4-}$, the dressing would quickly become rancid. Another use is in bathroom cleansers. The $EDTA^{4-}$ ion removes deposits of $CaCO_3$ and $MgCO_3$ left by hard water by coordinating to Ca^{2+} or Mg^{2+} to create a soluble complex ion.

Complexes with polydentate ligands have a particularly important role in biochemistry, as described in *A Closer Look: Hemoglobin* (page 939).

Formulas of Coordination Compounds

It is useful to be able to predict the formula of a coordination complex, given the metal ion and ligands, and to derive the oxidation number of the coordinated metal ion, given the formula in a coordination compound. The following examples explore these issues.

Example 22.1   Formulas of Coordination Compounds

Problem • Give the formula of the coordination complex in which

(a) A Ni^{2+} ion is bound to two water molecules and two bidentate oxalate ions.

(b) A Co^{3+} ion is bound to one Cl^- ion, one ammonia molecule, and two bidentate ethylenediamine (en) molecules.

Strategy • The problem requires determining the net charge, which equals the sum of the charges of each of the component parts of the complex ion. With that information, the metal and ligands are assembled in the formula, which is placed in brackets, and the net charge is attached.

Solution •

(a) This complex ion is constructed from two neutral H_2O molecules, two $C_2O_4^{2-}$ ions, and a Ni^{2+} ion, so the net charge on the complex is −2. The formula for the complex ion is $[Ni(C_2O_4)_2(H_2O)_2]^{2-}$.

(b) This cobalt(III) complex combines two en molecules and one NH_3 molecule, both having no charge, one Cl^- ion and a Co^{3+} ion. The net charge is +2. The formula for this complex (writing out the entire formula for ethylenediamine) is $[Co(H_2NCH_2CH_2NH_2)_2(NH_3)Cl]^{2+}$.

Example 22.2   Coordination Compounds

Problem • In each of the following complexes, determine the metal's oxidation number and coordination number.

(a) $[Co(en)_2(NO_2)_2]Cl$

(b) $Pt(NH_3)_2(C_2O_4)$

(c) $Pt(NH_3)_2Cl_4$

(d) $[Co(NH_3)_5Cl]SO_4$

Strategy • Each formula consists of a complex ion or molecule made up of the metal ion, neutral and/or anionic ligands (the

(Example continues on next page)

part inside the square brackets), and, if required, a counterion (outside the brackets). The oxidation number of the metal is the charge necessary to balance the sum of the negative charges associated with anionic ligands and counterion. The coordination number is the number of donor atoms in the ligands that are bonded to the metal. Be sure to remember that the bidentate ligands in these examples (en, oxalate ion) attach to the metal at two sites and that the counterion is not part of the complex ion; that is, it is not a ligand.

Solution •

(a) The chloride ion with a -1 charge, outside the brackets, shows that the charge on the complex ion must be $+1$. There are two nitrite ions (NO_2^-) and two neutral bidentate ethylenediamine ligands in the complex. To give a $+1$ charge on the complex ion, the cobalt ion must have an oxidation number of $+3$; that is, the sum of -2 (two nitrites), 0 (two en ligands), and $+3$ (the cobalt ion) $= +1$. Each en ligand fills

two coordination positions, and the two nitrite ions fill two more positions. The coordination number of the metal is 6.

(b) There is an oxalate ion ($C_2O_4^{2-}$) and two neutral ammonia ligands. To balance that charge on the oxalate ion, platinum must have a $+2$ charge; that is, it has an oxidation number of $+2$. The coordination number is 4, with an oxalate ligand filling two coordination positions and each ammonia molecule filling one.

(c) There are four chloride ions (Cl^-) and two neutral ammonia ligands. This is a complex in which the oxidation number of the metal is $+4$ and the coordination number is 6.

(d) There is one chloride ion (Cl^-) and five neutral ammonia ligands. The counterion is sulfate with a -2 charge, so the overall charge on the complex is $+2$. The oxidation number of the metal is $+3$ and the coordination number is 6 (sulfate is not coordinated to the metal).

Exercise 22.1   **Formulas of Coordination Compounds**

(a) What is the formula of a complex ion composed of one Co^{3+} ion, three ammonia molecules, and three Cl^- ions?

(b) Determine the oxidation number and coordination number of the metal ion in (i) $K_3[Co(NO_2)_6]$ and in (ii) $Mn(NH_3)_4Cl_2$.

Naming Coordination Compounds

Just as there are rules for naming inorganic and organic compounds, coordination compounds are named according to an established system. The three compounds just below are named according to the rules that follow.

Compound	Systematic Name
$[Ni(H_2O)_6]SO_4$	Hexaaquanickel(II) sulfate
$[Cr(en)_2(CN)_2]Cl$	Dicyanobis(ethylenediamine)chromium(III) chloride
$K[Pt(NH_3)Cl_3]$	Potassium amminetrichloroplatinate(II)

1. In naming a coordination compound that is a salt, name the cation first and then the anion. (This is how salts are commonly named.)

2. When giving the name of the complex ion or molecule, name the ligands first, in alphabetical order, followed by the name of the metal. (When determining alphabetical order, the prefix is not considered part of the name.)

3. Ligands and their names:

   (a) If a ligand is an anion whose name ends in *-ite* or *-ate*, the final *e* is changed to *o* (sulfate $\longrightarrow$ sulfato or nitrite $\longrightarrow$ nitrito).

   (b) If the ligand is an anion whose name ends in *-ide*, the ending is changed to *o* (chloride $\longrightarrow$ chloro, cyanide $\longrightarrow$ cyano).

Hemoglobin

Metal-containing coordination compounds figure prominently in many biochemical reactions. Perhaps the best-known example is hemoglobin, the chemical in the blood responsible for O_2 transport. It is also one of the most completely studied bioinorganic compounds.

Hemoglobin (Hb) is a large iron-containing protein, made up of over 10,000 atoms and having a molar mass of around 68,000. Despite its very large size, the intimate details of its structure are known. There are four polypeptide segments to this structure. Two identical segments contain 141 amino acids; the other two, also identical, contain 146 amino acids. Each segment contains an iron(II) ion locked inside a heme unit and coordinated to a nitrogen atom from another part of the protein. A sixth site is available to attach to oxygen.

Porphine ring. The tetradentate ligand surrounding iron(II) ion in hemoglobin is a dianion of a substituted molecule called a porphine. Because of the double bonds in this structure, all the carbon and nitrogen atoms in the dianion of the porphine lie in a plane. In addition, the nitrogen lone pairs are directed toward the center of the molecule and the molecular dimensions are such that a metal ion may fit nicely into the cavity.

In contrast with the hemoglobin molecule, myoglobin (the oxygen storage protein in muscle) has only one polypeptide chain with an enclosed heme group. Here you see that the iron-containing heme group is encased in a polypeptide chain. The iron ion in the porphine ring is shown. The first and sixth coordination positions are taken up by N atoms from amino acids of the polypeptide chain.

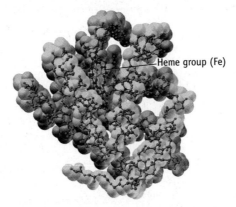

Heme group (Fe)

The heme group in myoglobin, a close relative of hemoglobin, is encased in a polypeptide chain. The heme group with its iron ion is shown. *(Source: Adapted from original illustration of Myoglobin by Irving Geis)*

Hemoglobin functions by reversibly adding oxygen to the sixth coordination position of each iron, giving a complex called oxyhemoglobin.

Because there are four iron centers in hemoglobin, up to four molecules of oxygen

Oxygen binds to the iron of the heme group in hemoglobin and in myoglobin. Interestingly, the Fe—O—O angle is bent.

will bind to the molecule. The binding to oxygen is cooperative; that is, binding one molecule enhances the tendency to bind a second, third, and fourth. Formation of the oxygenated complex is favored, but not too highly, because oxygen must also be released by the molecule to body tissues. Interestingly, an increase in acidity leads to a decrease in the stability of the oxygenated complex. This is known as the *Bohr effect,* named for Christian Bohr, the father of Niels Bohr (page 264). Release of oxygen in tissues is facilitated by a rise in acidity that results from the presence of CO_2 formed by metabolism.

Among the properties of hemoglobin is its ability to form a complex with carbon monoxide. This complex is very stable, the equilibrium constant for the following reaction being about 200 (where Hb is hemoglobin).

$$HbO_2(aq) + CO(g) \rightleftharpoons HbCO(aq) + O_2(g)$$

When CO forms a complex with iron, the oxygen-carrying capacity of hemoglobin is lost. Consequently, CO is highly toxic to humans. Exposure to even small amounts of CO greatly reduces the capacity of the blood to transport oxygen.

Hemoglobin abnormalities are well known. One of the most common abnormalities in the structure of hemoglobin causes sickle cell anemia, a disease that takes its name from the odd shape of red blood cells. The structural difference between normal hemoglobin and sickle cell hemoglobin is minute — a single substitution of one amino acid (valine) for another (glutamic acid) at position 6 in the two longer polypeptide chains. However, the physiological effect is striking. Having a nonpolar side chain at position 6 [in valine, $R = CH(CH_3)_3$] instead of the polar side chain of glutamic acid ($R = -CH_2CH_2CO_2H$) results in a tendency of the red blood cells to aggregate, thus restricting the flow of blood within the capillaries and decreasing the amount of oxygen that blood carries.

For more on hemoglobin see the earlier discussion of "Blood Gas" on page 826.

Porphine

$\downarrow$ −2 H⁺

Porphine²⁻

• Naming Conventions

Coordination complexes can be considerably more complicated than those described in this chapter; then, even more rules of nomenclature must be applied. The rules outlined here can be used to assign names to most complexes, however.

(c) If the ligand is a neutral molecule, its common name is usually used, with several important exceptions: water as a ligand is referred to as *aqua*, ammonia is called *ammine*, and CO is called *carbonyl*.

(d) When there is more than one of a particular monodentate ligand with a simple name, the number of ligands is designated by the appropriate prefix: *di, tri, tetra, penta,* or *hexa*. If the ligand name is complicated (whether monodentate or bidentate), the prefix changes to *bis, tris, tetrakis, pentakis,* or *hexakis,* followed by the ligand name in parentheses.

4. If the complex ion is an anion, the suffix *-ate* is added to the metal name.

5. Following the name of the metal, the oxidation number of the metal is given in Roman numerals.

Example 22.3 Naming Coordination Compounds

Problem • Name the following compounds:

(a) $[Cu(NH_3)_4]SO_4$
(b) $K_2[CoCl_4]$
(c) $Co(phen)_2Cl_2$
(d) $[Co(en)_2(H_2O)Cl]Cl_2$

Strategy • Apply the rules for nomenclature given above.

Solution •

(a) The complex ion (in square brackets) is composed of four NH_3 molecules (named *ammine* in a complex) and the copper ion. To balance the -2 charge on the sulfate counterion, copper must have a $+2$ charge. The compound's name is

> tetraamminecopper(II) sulfate.

(b) The complex ion $CoCl_4^{2-}$ has a -2 charge. With four Cl^- ligands, the cobalt ion must have a $+2$ charge so the sum of

charges is -2. The name of the compound is

> potassium tetrachlorocobaltate(II).

(c) This is a neutral coordination compound. The ligands include two Cl^- ions and two neutral bidentate *phen* (phenanthroline) ligands. The metal ion must have a $+2$ charge (Co^{2+}). The name, listing ligands in alphabetical order, is

> dichlorobis(phenanthroline)cobalt(II)

Note the use of *bis* with phenanthroline.

(d) The complex ion has a $+2$ charge because it is paired with two uncoordinated Cl^- ions. The cobalt ion is Co^{3+} because it is bonded to two neutral en ligands, one neutral water, and one Cl^-. The name is

> aquachlorobis(ethylenediamine)cobalt(III) chloride.

Exercise 22.2 Naming Coordination Compounds

Name the following coordination compounds.

(a) $[Ni(H_2O)_6]SO_4$ (c) $K[Pt(NH_3)Cl_3]$
(b) $[Cr(en)_2(CN)_2]Cl$ (d) $K[CuCl_2]$

22.4 STRUCTURES OF COORDINATION COMPOUNDS

Common Coordination Geometries

The geometry of a coordination complex is defined by the arrangement of donor atoms of the ligands around the central metal ion. Metal ions in coordination compounds can have a coordination number from 2 to 12. Complexes with coordination numbers of 2, 4, and 6 are common, however, so we will concentrate on species such as $[ML_2]^{n\pm}$, $[ML_4]^{n\pm}$, and $[ML_6]^{n\pm}$, where M is the metal ion, and L is a monodentate ligand. Within these stoichiometries, the following geometries are encountered:

- All $[ML_2]^{n\pm}$ complexes are linear. The two ligands are on opposite sides of the metal, and the L—M—L bond angle is 180°. Common examples include $[Ag(NH_3)_2]^+$ and $[CuCl_2]^-$.

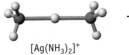
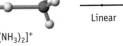

$[Ag(NH_3)_2]^+$ Linear

- Tetrahedral geometry occurs in many $[ML_4]^{n\pm}$ complexes. Examples include $TiCl_4$, $[CoCl_4]^{2-}$, $[NiCl_4]^{2-}$, and $[Zn(NH_3)_4]^{2+}$.

- The $[ML_4]^{n\pm}$ complexes can also have square planar geometry. This geometry is most often seen with metal ions that have eight d electrons. Examples include $Pt(NH_3)_2Cl_2$, $[Ni(CN)_4]^{2-}$, and the square planar nickel complex with dimethylglyoxime shown in the Chapter Focus (page 924).

$Pt(NH_3)_2Cl_2$ Square-planar

- Octahedral geometry is found in complexes with the stoichiometry $[ML_6]^{n\pm}$ (see Chapter Focus).

$[NiCl_4]^{2-}$ Tetrahedral

Isomerism

Isomerism is one of the most interesting aspects of molecular structure. Recall that the chemistry of organic compounds is greatly enlivened by the multitude of isomeric compounds that are known.

Isomers are classified as follows:

- *Structural isomers* have the same molecular formula but different bonding arrangements of atoms.

- *Stereoisomers* have the same atom-to-atom bonding sequence, but the atoms differ in their arrangement in space. There are two types of stereoisomers: geometric isomers (such as *cis* and *trans* alkenes) [← PAGE 423] and optical isomers (non-superimposable mirror images that have the unique property of rotating plane polarized light) [← PAGE 423].

$[M(H_2O)_6]^{n+}$ Octahedral

All three types of isomerism—structural, geometric, and optical—are encountered in coordination chemistry.

Structural Isomerism

The two most important types of structural isomerism in coordination chemistry are coordination isomerism and linkage isomerism. **Coordination isomerism** occurs if it is possible to exchange a coordinated ligand and a noncoordinated ligand. For example, dark violet $[Co(NH_3)_5Br]SO_4$ and red $[Co(NH_3)_5SO_4]Br$ are coordination isomers. In the first compound, bromide ion is a ligand and sulfate is a counterion; in the second, sulfate is a ligand and bromide is the counterion. A diagnostic test for this kind of isomer is often made based on chemical reactions. For example, these two compounds can be distinguished by precipitation reactions. Addition of $Ag^+(aq)$ to a solution of $[Co(NH_3)_5SO_4]Br$ gives a precipitate of AgBr (indicating the presence of bromide ion in solution), whereas there is no reaction if $Ag^+(aq)$ is added to a solution of $[Co(NH_3)_5Br]SO_4$. In the latter complex, bromide ion is attached to Co^{3+} and is not a free ion in solution.

$$[Co(NH_3)_5Br]SO_4(aq) + Ag^+(aq) \longrightarrow \text{no reaction}$$
$$[Co(NH_3)_5SO_4]Br(aq) + Ag^+(aq) \longrightarrow AgBr(s) + [Co(NH_3)_5SO_4]^+(aq)$$

Linkage isomers occur if it is possible to attach a ligand to the metal through different atoms. The two most common ligands for which linkage isomerism occur are thiocyanate, SCN^-, and nitrite, NO_2^-. The Lewis structure of the thiocyanate ion shows that there are lone pairs of electrons on sulfur and nitrogen. The ligand can attach to a metal either through sulfur (called *S*-bonded thiocyanate) or

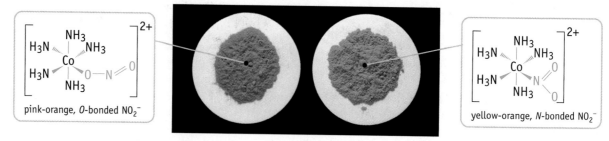

Figure 22.16   Linkage isomers, $[Co(NH_3)_5(ONO)]^{2+}$ and $[Co(NH_3)_5NO_2]^{2+}$. These complexes, whose systematic names are pentaammine(O-nitrito)cobalt(III) and pentaammine(N-nitrito)cobalt(III), were the first known example of this type of isomerism. *(Charles D. Winters)*

through nitrogen (called *N*-bonded thiocyanate). Nitrite ion can attach either at oxygen or at nitrogen. The former are called "O-nitrito" complexes, the latter, "N-nitrito" complexes (Figure 22.16).

Ligands forming linkage isomers

Geometric Isomerism

Geometric isomerism results when the atoms bonded directly to the metal have a different spatial arrangement. The simplest example of geometric isomerism in coordination chemistry is *cis–trans* isomerism, which occurs in both square-planar and octahedral complexes. An example of *cis–trans* isomerism is seen in the square-planar complex $Pt(NH_3)_2Cl_2$ (Figure 22.17a). In this complex, the two Cl^- ions can be either adjacent to each other (*cis*) or on opposite sides of the metal (*trans*). The *cis* isomer is effective in the treatment of testicular, ovarian, bladder, and osteogenic sarcoma cancers, but the *trans* isomer has no effect on these diseases [◄ SECTION 20.2, PAGE 839].

Cis–trans isomerism is illustrated in $[Co(H_2NCH_2\ CH_2NH_2)_2Cl_2]^+$, an octahedral complex with two bidentate ethylenediamine ligands and two chloride ligands.

● **Cis–Trans Isomerism**
Cis–trans isomerism is not possible for tetrahedral complexes. All L—M—L angles in tetrahedron geometry are 109.5°, and all positions are equivalent in this three-dimensional structure.

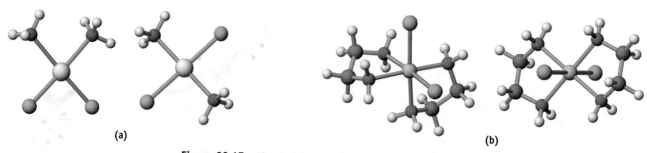

(a)   **(b)**

Figure 22.17   Cis–trans isomers. (a) The complex $Pt(NH_3)_2Cl_2$ can exist in two geometries, *cis* and *trans*. **(b)** Similarly, *cis* and *trans* isomers are possible for $[Co(en)_2Cl_2]^+$.

Figure 22.18   *Fac* and *mer* **isomers of Cr(NH₃)₃Cl₃.** In the *fac* isomer, the three chloride ligands (and the three ammonia ligands) are arranged at the corners of a triangular face of the octahedron. In the *mer* isomer, the three similar ligands follow a meridian.

fac isomer   *mer* isomer

In this complex, the two Cl⁻ ions occupy positions that are either adjacent (the purple *cis* isomer) or opposite (the green *trans* isomer) (Figure 22.17b).

Another common type of geometrical isomerism occurs for octahedral complexes with the general formula MX₃Y₃. A *fac* isomer has three similar ligands lying at the corners of a triangular face of an octahedron defined by the ligands (*fac* = facial), whereas in the *mer* isomer, the ligands follow a meridian (*mer* = meridianal). *Fac* and *mer* isomers of Cr(NH₃)₃Cl₃ are shown in Figure 22.18.

Optical Isomerism

Optical isomerism (chirality) occurs for some octahedral coordination compounds. Such isomers are possible in octahedral complexes when the metal ion coordinates to three bidentate ligands or when the metal ion coordinates to two bidentate ligands and two monodentate ligands in *cis* positions. The complexes [Co(en)₃]³⁺ and *cis*-[Co(en)₂Cl₂]⁺, illustrated in Figure 22.19, are examples of chiral complexes. The diagnostic test for chirality is met with both species: Mirror images of these molecules are not superimposable. Solutions of the optical isomers rotate plane-polarized light in opposite directions.

● **Optical Isomerism**
Square-planar complexes are incapable of optical isomerism based at the metal center; mirror images are always superimposable. Chiral tetrahedral complexes are possible, but examples of stable complexes with a metal bonded tetrahedrally to four different monodentate ligands are rare.

Example 22.4   Isomerism in Coordination Chemistry

Problem • For which of the following compounds do isomers exist? If isomers are possible, identify the type of isomerism (structural or geometric). Determine whether the coordination compound is capable of exhibiting optical isomerism.

(a) [Co(NH₃)₄Cl₂]⁺
(b) Pt(NH₃)₂(CN)₂ (square planar)

(c) Co(NH₃)₃Cl₃
(d) Zn(NH₃)₂Cl₂ (tetrahedral)
(e) K₃[Fe(C₂O₄)₃]
(f) [Co(NH₃)₅SCN]²⁺

Strategy • Determine the number of ligands attached to the metal and whether the ligands are monodentate or bidentate.

(Example continues on next page)

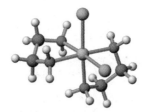

[Co(en)₃]³⁺   [Co(en)₃]³⁺ mirror image   *cis*-[Co(en)₂Cl₂]⁺   *cis*-[Co(en)₂Cl₂]⁺ mirror image

Figure 22.19   Chiral metal complexes. Both [Co(en)₃]³⁺ and *cis*-[Co(en)₂Cl₂]⁺ are chiral. Notice that mirror images of the two compounds are not superimposable.

Knowing how many donor atoms are coordinated to the metal (the coordination number) will allow you to establish the metal geometry. The only isomerism possible for square-planar complexes is geometric (*cis* and *trans*). Tetrahedral complexes do not have isomers. Six-coordinate metals with formulas MA_4B_2 can be either *cis* or *trans*. *Mer* and *fac* isomers are possible for a stoichiometry MA_3B_3. Optical activity arises for metal complexes of the formula *cis*-M(bidentate)$_2$X$_2$ and M(bidentate)$_3$. Drawing pictures of the molecules is a useful exercise, and it will help you to visualize isomers.

Solution •

(a) Two geometric isomers can be drawn for octahedral complexes with a formula MA_4B_2, such as this one. One isomer has two Cl^- ions in *cis* positions (adjacent positions, at a 90° angle), and the other isomer has the Cl^- ligands in *trans* positions (with a 180° angle between the ligands). Optical isomers are not possible.

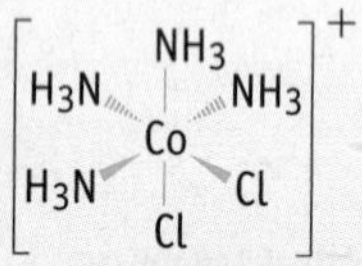

cis isomer 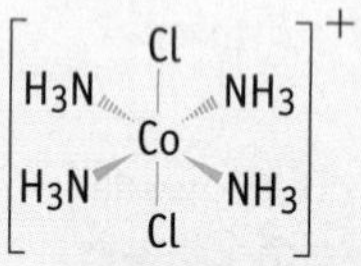   *trans* isomer

(b) In this square-planar complex, the two NH_3 ligands (and the two CN^- ligands) can be either *cis* or *trans*. These are geometric isomers. Optical isomers are not possible.

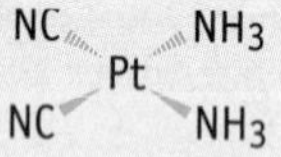

cis isomer 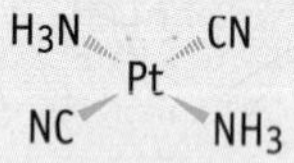   *trans* isomer

(c) Two geometric isomers of this octahedral complex, with chloride ligands either *fac* or *mer*, are possible. In the *fac* isomer the three Cl^- ligands are all at 90°, in the *mer* isomer, two Cl^- ligands are at 180°, the third 90° from the other two. Optical isomers are not possible.

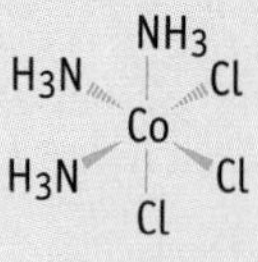

fac isomer 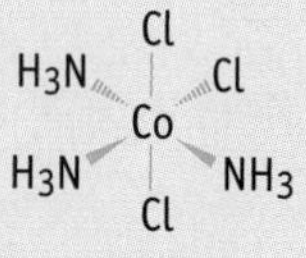   *mer* isomer

(d) Only a single structure is possible for tetrahedral complexes such as $Zn(NH_3)_2Cl_2$.

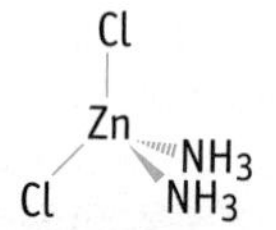

(e) Ignore the counterions, K^+. The anion is an octahedral complex — remember that the bidentate oxalate ion occupies two coordination sites of the metal and that, with three oxalate ligands, the metal has a coordination number of six. Mirror images of complexes of the stoichiometry M(bidentate)$_3$ do not superimpose; therefore, two optical isomers are possible. (Here the ligands, $C_2O_4^{2-}$, are drawn abbreviated as 0 — 0.)

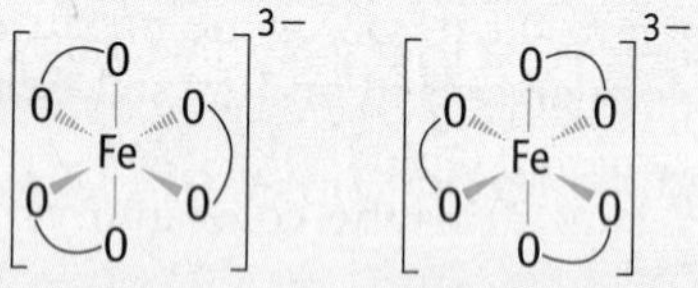

Nonsuperimposable mirror images of [Fe(ox)$_3$]$^{3-}$

(f) $[Co(NH_3)_5SCN]^{2+}$. Only linkage isomerism (structural isomerism) is possible for this octahedral cobalt complex. Either the sulfur or the nitrogen of the SCN^- anion can be attached to the cobalt(III) ion in this complex.

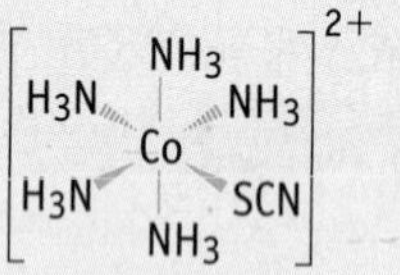

S-bonded SCN^- 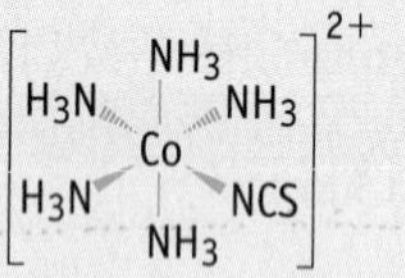   N-bonded SCN^-

---

Exercise 22.3   **Isomers in Coordination Complexes**

What type of isomers, if any, are possible for the following complexes?

(a) $K[Co(NH_3)_2Cl_4]$

(b) $Pt(en)Cl_2$ (square-planar)

(c) $[Co(NH_3)_5Cl]^{2+}$

(d) $[Ru(phen)_3]Cl_3$

(e) $Na_2[MnCl_4]$ (tetrahedral)

(f) $[Co(NH_3)_5NO_2]^{2+}$

22.5   BONDING IN COORDINATION COMPOUNDS

Chapter Goals • Revisited

Metal–ligand bonding in a coordination complex was described earlier in this chapter as being covalent, resulting from the sharing of an electron pair between the metal and the ligand donor atom. Although frequently used, this simple model is not capable of explaining the color and magnetic behavior of complexes. As a consequence, the covalent bonding picture has now largely been superseded by two other bonding models, molecular orbital theory and crystal field theory.

The bonding model using molecular orbital theory assumes that metal and ligand bond through the molecular orbitals formed by atomic orbital overlap between metal and ligand. The **crystal field model,** in contrast, focuses on repulsion (and destabilization) of electrons in the metal coordination sphere. The crystal field model also assumes that the positive metal ion and negative ligand lone pair are attracted electrostatically; that is, the bond arises due to a positively charged metal ion attracting a negative ion or the negative end of a polar molecule. For the most part, the molecular orbital and crystal field models predict similar results regarding color and magnetic behavior. We will focus on the crystal field approach and illustrate how it explains the color and magnetism of transition metal complexes.

- Identify and explain chemical and physical properties of the transition elements.
- Understand the composition, structure, and bonding in coordination compounds.
- **Show how bonding is used to explain the magnetism and spectra of coordination compounds.**

The *d* Orbitals: Crystal Field Theory

To understand crystal field theory, it is necessary to look again at the *d* orbitals, particularly with regard to their orientation relative to the positions of ligands in a metal complex.

We look first at octahedral complexes. Assume the ligands in an octahedral complex lie along the *x*-, *y*-, and *z*-axes. This results in the five *d* orbitals (Figure 22.20) being subdivided into two sets: the $d_{x^2-y^2}$ and d_{z^2} orbitals in one set and the d_{xy}, d_{xz}, and d_{yz} orbitals in the second. The $d_{x^2-y^2}$ and d_{z^2} orbitals are directed along the *x*-, *y*-, and *z*-axes, whereas the orbitals of the second group are aligned between these axes.

In an isolated atom or ion, the five *d* orbitals have the same energy. For a metal atom or ion in a coordination complex, however, the *d* orbitals have different energies. According to the crystal field model, repulsion between *d* electrons on the metal and the electron pairs of the ligands destabilizes electrons that reside in the *d* orbitals; that is, it causes their energy to become higher. Electrons in the various *d* orbitals are not affected equally, however, because of their orientation in space relative to the position of the ligand lone pairs (Figure 22.21). Electrons in the $d_{x^2-y^2}$ and d_{z^2} orbitals experience a larger repulsion because these orbitals point

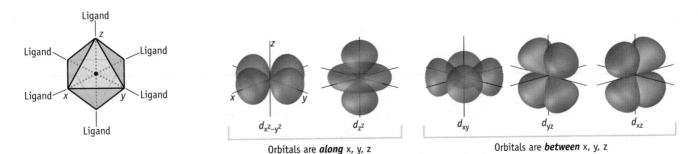

Figure 22.20   The *d* orbitals. The five *d* orbitals and their spatial relation to the ligands on the *x*-, *y*-, and *z*-axes. (*Orbital models: Patrick A. Harman and Charles D. Hamper*)

Figure 22.21   Crystal field splitting for an octahedral complex. The d orbital energies increase as the ligands approach the metal along the x-, y-, and z-axes. The d_{xy}, d_{xz} and d_{yz} orbitals, not pointed toward the ligands, are less destabilized than the $d_{x^2-y^2}$ and d_{z^2} orbitals. Thus, the d_{xy}, d_{xz} and d_{yz} orbitals are at lower energy.

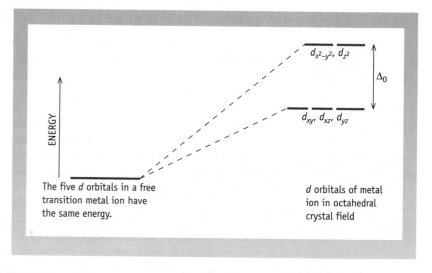

• **Crystal Field Splitting**
The crystal field or electrostatic field of the ligands splits the d orbitals into two sets.

directly at the ligand electron pairs. A smaller repulsion is experienced by electrons in the d_{xy}, d_{xz}, and d_{yz} orbitals. The difference in repulsion means an energy difference exists between the two sets of orbitals in a coordination complex. This difference, called the **crystal field splitting** and given the symbol Δ_o (for an octahedral complex), is a function of the metal and the ligands and varies predictably from one complex to another.

A different splitting pattern is encountered with square-planar complexes (Figure 22.22a). Assume that the four ligands are along the x- and y-axes. The $d_{x^2-y^2}$ orbital also points along these axes, and thus this orbital has the highest energy. The d_{z^2} orbital is next highest in energy followed by the d_{xy} orbital (which also lies in the xy-plane, but is not pointing at the ligands). The d_{xz} and d_{yz} orbitals, both of which are partially pointing in the z-direction, have the lowest energy.

The d orbital splitting pattern for a tetrahedral complex is the reverse of the pattern for octahedral complexes. Three orbitals (d_{xz}, d_{xy}, d_{yz}) are higher in energy, whereas the $d_{x^2-y^2}$ and d_{z^2} orbitals are below them in energy (Figure 22.22b).

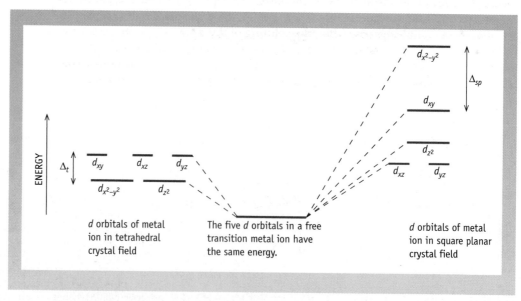

Figure 22.22   Splitting of the d orbitals in (a) tetrahedral (b) square-planar geometries.

Electron Configurations and Magnetic Properties

The *d*-orbital splitting in coordination complexes provides the means to explain both the magnetic behavior and color of these complexes. To understand this explanation, however, we must first understand how to assign electrons to the various orbitals in the most common geometries, octahedral and square-planar.

A gaseous Cr^{2+} ion has the electron configuration $[Ar]3d^4$. The term "gaseous" in this context is used to denote a single, isolated atom or ion with all other particles an infinite distance away. In this situation, the five $3d$ orbitals have the same energy. The four electrons reside singly in different *d* orbitals, according to Hund's rule, and the Cr^{2+} ion has four unpaired electrons.

Cr(II) electron configuration

Cr^{2+} $[Ar]3d^44s^0$ ↑|↑|↑|↑| | ☐

                                           3*d*   4*s*

When the Cr^{2+} ion is part of an octahedral complex, however, the five *d* orbitals do not have identical energies. As illustrated in Figure 22.21, these orbitals divide into two sets, with the d_{xy}, d_{xz}, and d_{yz} orbitals at a lower energy than the $d_{x^2-y^2}$ and d_{z^2} orbitals. Having two sets of orbitals means that two different electron configurations are possible (Figure 22.23). Three of the four *d* electrons in Cr^{2+} are assigned to the lower energy d_{xy}, d_{xz}, and d_{yz} orbitals. The fourth electron can either be assigned to an orbital in the higher energy $d_{x^2-y^2}$ and d_{z^2} set or pair up with an electron already in the lower energy set. The first arrangement is called a **high-spin complex** because it has the maximum number of unpaired electrons, four in the case of Cr^{2+}. In contrast, the second arrangement is called a **low-spin complex** because it has the minimum number of unpaired electrons possible.

At first glance, a high-spin configuration appears to contradict conventional thinking. It seems logical that the most stable situation would occur when electrons occupy the lowest energy orbitals. A second factor intervenes, however. Two electrons paired in a single orbital are constrained to the same region of space. Because electrons are negatively charged, repulsion increases when they are assigned to the same orbital. This is a destabilizing effect and bears the name **pairing energy.** The preference for an electron to be in the lowest energy orbital and the pairing energy have opposing effects (Figure 22.23).

Low-spin complexes arise when the splitting of the *d* orbitals by the crystal field is large; that is, there is a large value of Δ_0. In this situation, the energy gained by putting all the electrons in the lowest energy level is the dominant effect. High-spin complexes occur if the value of Δ_0 is small, and pairing energy dominates.

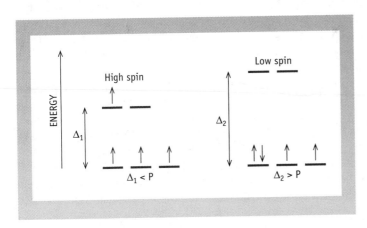

Figure 22.23   High- and low-spin cases for an octahedral chromium(II) complex. If the crystal field splitting (Δ_1) is smaller than the pairing energy (P), the electrons are placed in different orbitals and the complex has four unpaired electrons. If the splitting (Δ_2) is larger than the pairing energy, all four electrons will be in the lower-energy orbital set. This requires pairing two electrons in one of the orbitals and the complex will have only two unpaired electrons.

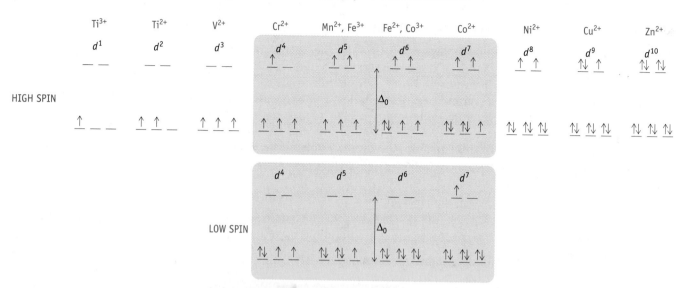

Figure 22.24   High- and low-spin complexes. *d*-Orbital occupancy for octahedral complexes of metal ions. Only the d^4 through d^7 cases have both high- and low-spin configurations.

For octahedral complexes, high- and low-spin complexes can occur only for configurations d^4 through d^7 (Figure 22.24). Complexes of the d^6 metal ion, Fe^{2+}, for example, can have either high spin or low spin configurations. The complex formed when the Fe^{2+} ion is placed in water, $[Fe(H_2O)_6]^{2+}$, is high-spin, whereas $[Fe(CN)_6]^{4-}$ is low-spin.

It is possible to tell the difference between low- and high-spin complexes by determining the magnetic behavior of the substance. The high-spin complex $[Fe(H_2O)_6]^{2+}$ has four unpaired electrons and is *paramagnetic* (attracted by a magnet), whereas the low-spin $[Fe(CN)_6]^{4-}$ complex has no unpaired electrons and is *diamagnetic* (repelled by a magnet) [← SECTION 8.1].

Most complexes of Pd^{2+} and Pt^{2+} ions are square-planar, the electron configuration of these metals being [noble gas]$(n-1)d^8$. In a square-planar complex, there are four sets of orbitals (Figure 22.22). All but the highest-energy orbital is filled, and all electrons are paired, resulting in diamagnetic (low-spin) complexes.

Nickel, above palladium in the periodic table, forms both square-planar and tetrahedral complexes. The complex $[Ni(CN)_4]^{2-}$ is square planar, whereas $[NiCl_4]^{2-}$ is tetrahedral. Magnetism allows us to differentiate between these two geometries. Based on the crystal field splitting pattern, the cyanide complex is expected to be diamagnetic, whereas the chloride complex is paramagnetic with two unpaired electrons.

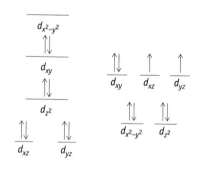

The anion $[NiCl_4]^{2-}$ is a paramagnetic tetrahedral complex ion. In contrast, $[Ni(CN)_4]^{2-}$ is a diamagnetic square-planar complex.

Problem-Solving Tip 22.1

Crystal Field Theory

A summary of the concepts of crystal field theory may help you to keep the broader picture in mind.

- Ligand–metal bonding is coordinate covalent, it results from the attraction of a metal cation and an anion or polar molecule.

- The ligands define a coordination geometry. Common geometries are linear (coordination number = 2), tetrahedral and square-planar (coordination number = 4), and octahedral (coordination number = 6).

- The placement of the ligands around the metal causes the d orbitals on the metal to have different energies. In an octahedral complex, for example, the d orbitals divide into two groups, a higher-energy group ($d_{x^2-y^2}$ and d_{z^2}) and a lower-energy group (d_{xy}, d_{xz}, and d_{yz}).

- Electrons are placed in the metal d orbitals in a manner that leads to the lowest energy. Two competing features determine the placement: the relative energy of the sets of orbitals and the electron-pairing energy.

- For the electron configurations d^4, d^5, d^6, and d^7, two electron configurations are possible: high-spin, occurring if the orbital splitting is small, and low-spin, occurring with a large orbital splitting. To determine whether a complex is high-spin or low-spin, measure its magnetism to determine the number of unpaired electrons.

- The d orbital splitting (the energy difference between the metal d orbital energies) often corresponds to the energy associated with visible light. As a consequence, many metal complexes absorb visible light and thus are colored.

Example 22.5 High- and Low-Spin Complexes and Magnetism

Problem • Depict the electron configurations for each of the following complexes. How many unpaired electrons are present in each complex? Are the complexes paramagnetic or diamagnetic?

(a) Low-spin $[Co(NH_3)_6]^{3+}$

(b) High-spin $[CoF_6]^{3-}$

Strategy • These are complexes of Co^{3+}, which has a d^6 valence electron configuration. Set up an energy level diagram for an octahedral complex. In low-spin complexes, the electrons are added preferentially to the lower set of orbitals. In high-spin complexes, the first five electrons are added singly to each of the five orbitals, then additional electrons are paired with electrons in orbitals in the lower set.

Solution •

(a) The six electrons fill the lower energy set of orbitals entirely. This d^6 complex ion has no unpaired electrons and is diamagnetic.

(b) To obtain the electron configuration in high-spin $[CoF_6]^{3-}$, place one electron in each of the five d orbitals, and then place the sixth electron in one of the lower-energy orbitals. The complex has four unpaired electrons and is paramagnetic.

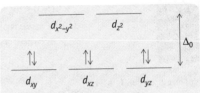

Electron configuration of low-spin, octahedral $[Co(NH_3)_6]^{3+}$

(a)

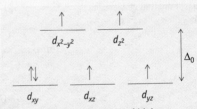

Electron configuration of high-spin, octahedral $[CoF_6]^{3-}$

(b)

Exercise 22.4 High- and Low-Spin Configurations and Magnetism

For each of the following complex ions, give the oxidation number of the metal ion, depict the low- and high-spin configurations, give the number of unpaired electrons in each state, and tell whether each is paramagnetic or diamagnetic.

(a) $[Ru(H_2O)_6]^{2+}$ **(b)** $[Ni(NH_3)_6]^{2+}$

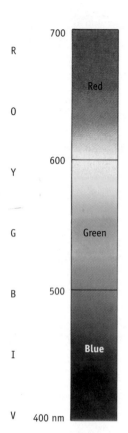

The ROY G BIV spectrum of colors of visible light. The colors used in printing this book are cyan, magenta, yellow, and black. The blue in ROY G BIV is actually cyan, according to color industry standards. Magenta doesn't have its own wavelength region. Rather, it is a mixture of blue and red, and orange is yellow with a little red mixed in.

22.6 COLORS OF COORDINATION COMPOUNDS

One of the most interesting features of compounds of the transition elements is their colors (Figure 22.25). In contrast, compounds of main group metals are usually colorless. The color in these species is a result of *d*-orbital splitting. Before discussing how *d*-orbital splitting is involved, however, we need to look more closely at what we mean by color.

Color

Visible light, radiation with wavelengths from 400 nm to 700 nm [◀ SECTION 7.1], represents a very small portion of the electromagnetic spectrum. Within this region are all the colors you see when white light is passed through a prism: red, orange, yellow, green, blue, indigo, and violet (ROY G BIV). Each color is identified with a portion of the wavelength range.

Isaac Newton did experiments with light and established that the mind's perception of color requires only three colors! When we see white light, we are seeing a mixture of all of the colors, that is, the superposition of red, green, and blue. If one or more of these colors is absent, the light of the other colors that reaches your eyes is interpreted by your mind as color.

Figure 22.26 will help you in analyzing perceived colors. This color wheel shows the three primary colors — red, green, and blue — as overlapping discs arranged in a triangle. The secondary colors — cyan, magenta, and yellow — appear where two discs overlap. The overlap of all three discs in the center produces white light.

The colors we perceive are determined as follows:

- Light of a single primary color is perceived as that color: Red light is perceived as red, green light as green, blue light as blue.
- Light made up of two primary colors is perceived as the color shown where the discs in Figure 22.26 overlap: Red and green light together appears yellow, green and blue light together are perceived as cyan, and red and blue light are perceived as magenta.
- Light made up of the three primary colors is white (colorless).

When discussing the color of a substance such as a coordination complex *in solution,* we have to remember that the color of the solution is the color of the light *not absorbed* by the solution. It is the color of the *transmitted light.*

Figure 22.25 Aqueous solutions of some transition metal ions. Compounds of transition metal elements are often colored, whereas those of main group metals are usually colorless. Pictured here, from left to right, are solutions of the nitrate salts of Fe^{3+}, Co^{2+}, Ni^{2+}, Cu^{2+}, and Zn^{2+}. *(Charles D. Winters)*

Figure 22.26   Using color discs to analyze colors. The three primary colors are red, green, and blue. Adding light of two primary colors gives the secondary colors yellow (= red + green), cyan (= green + blue), and magenta (= red + blue). Adding all three primary colors results in white light.

Yellow = Red + Green

Green

Cyan = Green + Blue

Red

Blue

Magenta = Red + Blue

- Red light is the result of the absence of green and blue light from white light. That is, if a solution appears red, it is because red light is transmitted; green and blue light must have been absorbed.
- A green solution results if red and blue light have been absorbed.
- A blue solution results if red and green have been absorbed.

A Closer Look

A Spectrophotometer

The qualitative conclusions that we have drawn concerning absorption of light are determined more precisely in the laboratory using an instrument called a *spectrophotometer*. A schematic drawing of a spectrophotometer measuring absorption of visible light is shown in the figure below.

White light from a glowing filament is first passed through a prism or diffraction grating, devices that divide the light into its separate frequencies. The instrument selects a specific frequency to pass through a solution of the compound to be studied. After passing through the solution, the intensity of the light is measured by a detector. If light of a given frequency is not absorbed, its intensity is unchanged when it emerges from the sample. If light is absorbed, the light emerging from the sample has a lower intensity.

An absorption spectrum is a graph of the frequency or wavelength of the light *versus* the intensity of light absorbed at that frequency or wavelength. When light in a certain frequency range is absorbed, the plot shows an absorption band. For example, $[Co(NH_3)_6]^{3+}$ absorbs in the blue region, leaving its complementary color yellow (R + G = Y) to pass through to the eye.

Scientists are not limited to the visible region of the spectrum when making measurements on absorption of electromagnetic radiation. Ultraviolet and infrared spectrometers are common pieces of scientific apparatus in a chemistry laboratory. They differ from a visible spectrometer in that the light source must generate radiation in the correct region of the electromagnetic spectrum, and the detector must be able to detect this radiation.

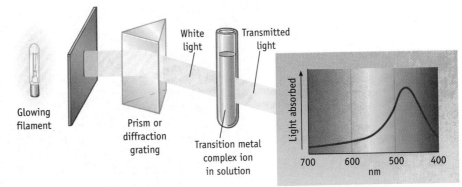

An absorption spectrometer. A beam of white light is passed through a prism or grating that splits the light into its component wavelengths. After passing through a sample, the light reaches a detector. The spectrometer "scans" all wavelengths of light and determines whether or not light is absorbed. The output of a spectrometer is a spectrum, a graph plotting absorption as a function of wavelength (or frequency).

The secondary colors are rationalized similarly. Absorption of blue light will give yellow (the color across from blue in Figure 22.26), absorption of red light results in cyan, and absorption of green light results in magenta.

Now we can apply these ideas to explain colors in transition metal complexes. Focus on what light is *absorbed*. A solution of $[Ni(H_2O)_6]^{2+}$ is green. Green light is the result of removing red and blue light from white light. As white light passes through an aqueous solution of Ni^{2+}, red and blue light are absorbed and green light is allowed to pass (Figure 22.27). Similarly, the $[Co(NH_3)_6]^{3+}$ ion is yellow because blue light has been absorbed and red and green light have been permitted to pass.

The Spectrochemical Series

Recall that atomic spectra are obtained when electrons are excited from one energy level to another [◀ SECTION 7.3]. The energy of the light absorbed or emitted is related to the energy levels of the atom or ion under study. The concept that light is absorbed when electrons move from lower to higher energy levels applies to all substances, not just atoms. This is the basic premise for the absorption of light by transition metal coordination complexes.

In coordination complexes, the splitting between *d* orbitals often corresponds to the energy of visible light, so light in the visible region of the spectrum is absorbed when electrons move from a lower-energy *d* orbital to a higher-energy *d* orbital. This change, as an electron moves between two orbitals having different energies in a complex, is called a **d-to-d transition**. Qualitatively, such a transition for $[Co(NH_3)_6]^{3+}$ might be represented using an energy level diagram such as that shown here.

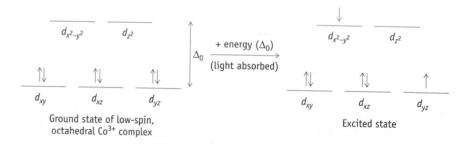

Figure 22.27 **Light absorption and color.** The color of a solution is due to the color of the light not absorbed by the solution. Here a solution of Ni^{2+} ion in water absorbs red and blue light and so appears green. *(Charles D. Winters)*

Experiments with coordination complexes have revealed that, for a given metal ion, some ligands cause a small energy separation of the d orbitals, whereas others cause a large separation. In other words, some ligands create a small *crystal field,* and others create a large one. An example of this is seen in the spectroscopic data for several cobalt(III) complexes presented in Table 22.3.

- Both $[Co(NH_3)_6]^{3+}$ and $[Co(en)_3]^{3+}$ are yellow-orange, because they absorb light in the blue portion of the visible spectrum. These compounds have very similar spectra, to be expected because both have six amine-type donor atoms (NH_3 or RNH_2).
- Although $[Co(CN)_6]^{3-}$ does not have an absorption band in the visible region, it is pale yellow. Light absorption is in the ultraviolet region, but the absorption is broad and extends minimally into the visible region.
- $[Co(C_2O_4)_3]^{3-}$ and $[Co(H_2O)_6]^{3+}$ have similar absorptions, in the yellow and violet regions. Their colors are shades of green with a small difference due to the relative amount of light of each color being absorbed.

The absorption maxima among the listed complexes ranges from 700 nm for $[CoF_6]^{3-}$ to 310 nm for $[Co(CN)_6]^{3-}$. From this information we can conclude that the energy of the light absorbed by the complex is related to different crystal field splittings, Δ_0, caused by the different ligands. Fluoride ion causes the smallest splitting of the d orbitals among the complexes listed in Table 22.3, whereas cyanide causes the largest splitting.

Spectra of complexes of other metals provide similar results. Based on this information, ligands can be listed in order of their ability to split the d orbitals. This list is called the **spectrochemical series** because it was determined by spectroscopy. A short list, with some of the more common ligands, is given below.

Spectrochemical Series

Halides $< C_2O_4^{2-} < H_2O < NH_3 = $ en $<$ phen $< CN^-$

Small orbital splitting   Large orbital splitting
Small Δ_0 Large Δ_0

The spectrochemical series is applicable to a wide range of metal complexes. The ability of crystal field theory to explain differences in the color of transition metal complexes is one of the strengths of this theory.

Table 22.3 • The Colors of Some Co³⁺ Complexes*

Complex Ion	Wavelength of Light Absorbed (nm)	Color of Light Absorbed	Color of Complex
$[CoF_6]^{3-}$	700	Red	Green
$[Co(C_2O_4)_3]^{3-}$	600, 420	Yellow, violet	Dark green
$[Co(H_2O)_6]^{3+}$	600, 400	Yellow, violet	Blue-green
$[Co(NH_3)_6]^{3+}$	475, 340	Blue, ultraviolet	Yellow-orange
$[Co(en)_3]^{3+}$	470, 340	Blue, ultraviolet	Yellow-orange
$[Co(CN)_6]^{3-}$	310	Ultraviolet	Pale yellow

*The complex with fluoride ion, $[CoF_6]^{3-}$, is high-spin and has one absorption band. The other complexes are low-spin and have two absorption bands. In all but one case, one of these absorptions is in the visible region of the spectrum.

From the relative position of a ligand in the series, predictions can be made about a compound's magnetic behavior. Recall that d^4, d^5, d^6, and d^7 complexes can be high- or low-spin, depending on the crystal field splitting Δ_0. Complexes formed with ligands near the left end of the spectrochemical series are expected to have small Δ_0 values and thus are likely to be high-spin. In contrast, complexes with ligands near the right end are expected to have large Δ_0 values and have low-spin configurations. The complex $[CoF_6]^{3-}$ is high-spin, whereas $[Co(NH_3)_6]^{3+}$ and the others shown in Table 22.3 are low-spin.

Example 22.6   Spectrochemical Series

Problem • An aqueous solution of $[Fe(H_2O)_6]^{2+}$ is light blue-green. Do you expect the d^6 Fe^{2+} ion in this complex to have a high- or low-spin configuration? How would you determine this by an experiment?

Strategy • Use the color wheel in Figure 22.26. The color of the complex, blue-green, tells us what light is transmitted (blue and green); from this we learn what light has been absorbed (red). Red light is at the low energy end of the visible spectrum. From this, we can predict that d-orbital splitting must be small. Our answer to the question derives from that conclusion.

Solution • The low energy of the light absorbed suggests that $[Fe(H_2O)_6]^{2+}$ has a small d-orbital splitting and so is likely to be a high-spin complex.

If the complex is high-spin, it will have four unpaired electrons and will be paramagnetic, whereas if it is low-spin it will have no unpaired electrons and will be diamagnetic. Determining the presence of four unpaired electrons by a measurement of the compound's magnetism can be used to verify the high-spin configuration.

In Summary

When you have studied this chapter, you should ask if you have met the chapter goals. In particular, you should be able to

- Identify the general classes of elements: transition elements, lanthanides, and actinides (Section 22.1).
- Identify the more common transition metals from their symbols and positions in the periodic table, and recall some of their physical and chemical properties (Section 22.1).
- Understand the electrochemical nature of corrosion (page 927).
- Know the general features of pyrometallurgy and hydrometallurgy, and describe the metallurgy of iron and copper (Section 22.2).
- Given the formula for a coordination complex, be able to identify the metal and its oxidation state, the ligands, the coordination number and coordination geometry, and the overall charge on the complex (Section 22.3).
- Provide systematic names for complexes from their formulas, and write formulas if the name is given (Section 22.3).
- Given the formula for a complex, be able to recognize whether isomers will exist and draw their structures (Section 22.4).
- Describe the bonding in coordination complexes (Section 22.5).
- Use crystal field theory to determine the number of unpaired electrons in a complex (Section 22.5).
- Understand why substances are colored and be able to predict colors knowing the color of light absorbed (Section 22.6).
- Understand the relationship between the crystal field splitting, magnetism, and color of metal complexes (Section 22.6).

Key Terms

d-block elements
f-block elements

Section 22.1
lanthanide contraction

Section 22.2
ore
metallurgy
gangue
pyrometallurgy
hydrometallurgy

Section 22.3
coordination complex, complex ion
coordination compound
ligand
coordination number
coordination geometry
monodentate
polydentate
chelating ligand

Section 22.4
coordination isomerism

linkage isomers

Section 22.5
crystal field model
crystal field splitting (Δ_0)
high-spin complex
low-spin complex
pairing energy (*P*)

Section 22.6
d-to-*d* transition
spectrochemical series

Study Questions

Questions with blue, bold-faced numbers have answers in Appendix O.

Reviewing Important Concepts

1. Explain the meaning of the terms *s*-block, *p*-block, *d*-block, and *f*-block. How many elements in the sixth period in the periodic table (Cs through Rn) are in each block?

2. Write electron configurations using the *spdf* notation for V, Mn, Ni, and Cu.

3. Identify the common ions for Cr, Fe, Co, and Zn (*hint:* see Figure 3.7) and give their electron configuration.

4. Identify two physical properties and two chemical properties of iron, nickel, copper, and zinc.

5. Identify a common use for each of the following elements: Ti, Cr, Ni, and Cu.

6. Describe the general features of the two major types of processes for the recovery of metals from their ores.

7. Define the terms "coordinate covalent bond" and "Lewis acid–Lewis base reaction" and explain how these concepts relate to coordination chemistry.

8. Define coordination number and coordination geometry, as pertaining to coordination complexes.

9. What structural feature is required for a molecule or ion to be a ligand? What feature(s) would be necessary for a molecule or ion to act as a bidentate ligand?

10. Define (a) chelate and (b) bidentate. Use a specific chemical example to illustrate each answer.

11. Define the terms "diamagnetic" and "paramagnetic." What feature of electronic structure is associated with these properties? What experiment would you do to determine whether a complex is diamagnetic or paramagnetic?

12. Give the formula of a cobalt(III) complex that displays (a) geometric isomerism and (b) optical isomerism, and name each example.

13. According to the crystal field model, what causes the splitting of metal *d* orbitals into two sets in an octahedral complex?

14. What factors determine whether a complex will be high- or low-spin?

15. A compound in solution absorbs green light. What is the color of the solution?

Practicing Skills

Properties of Transition Elements
(See Section 22.1 and Example 8.3)

16. Give the electron configuration for each of the following ions, and tell whether each is paramagnetic or diamagnetic.
    (a) Cr^{3+} (c) Ni^{2+}
    (b) V^{2+} (d) Cu^+

17. Identify two transition metal ions with the following electron configurations:
    (a) $[Ar]3d^6$ (c) $[Ar]3d^5$
    (b) $[Ar]3d^{10}$ (d) $[Ar]3d^8$

18. Identify an ion of a first series transition metal that is isoelectronic with each of the following:
    (a) Fe^{3+} (c) Fe^{2+}
    (b) Zn^{2+} (d) Cr^{3+}

19. Match up the isoelectronic ions on the following list.

    Cu^+ Mn^{2+} Fe^{2+} Co^{3+} Fe^{3+} Zn^{2+} Ti^{2+} V^{3+}

20. The following equations represent various ways of obtaining transition metals from their compounds. Balance each equation.
    (a) $Cr_2O_3(s) + Al(s) \longrightarrow Al_2O_3(s) + Cr(s)$
    (b) $TiCl_4(\ell) + Mg(s) \longrightarrow Ti(s) + MgCl_2(s)$
    (c) $[Ag(CN)_2]^-(aq) + Zn(s) \longrightarrow Ag(s) + [Zn(CN)_4]^{2-}(aq)$
    (d) $Mn_3O_4(s) + Al(s) \longrightarrow Mn(s) + Al_2O_3(s)$

21. Identify the products of each reaction and balance the equation.

(a) $CuSO_4(aq) + Zn(s) \longrightarrow$

(b) $Zn(s) + HCl(aq) \longrightarrow$

(c) $Fe(s) + Cl_2(g) \longrightarrow$

(d) $V(s) + O_2(g) \longrightarrow$

Formulas of Coordination Compounds
(See Examples 22.1 and 22.2)

22. Which of the following ligands is expected to be monodentate and which might be polydentate?

(a) CH_3NH_2 (d) en

(b) CH_3CN (e) Br^-

(c) N_3^- (f) phen

23. One of the following nitrogen compounds or ions is not capable of serving as a ligand: NH_4^+, NH_3, NH_2^-. Identify this species and explain your answer.

24. Give the oxidation number of the metal ion in each of the following compounds:

(a) $[Mn(NH_3)_6]SO_4$ (c) $[Co(NH_3)_4Cl_2]Cl$

(b) $K_3[Co(CN)_6]$ (d) $Cr(en)_2Cl_2$

25. Give the oxidation number of the metal ion in each of the following complexes:

(a) $[Fe(NH_3)_6]^{2+}$ (c) $[Co(NH_3)_5(NO_2)]^+$

(b) $[Zn(CN)_4]^{2-}$ (d) $[Cu(en)_2]^{2+}$

26. Give the formula of a complex constructed from one Ni^{2+} ion, one ethylenediamine ligand, three ammonia molecules, and one water molecule. Is the complex neutral or is it charged? If charged, give the charge.

27. Give the formula of a complex constructed from one Cr^{3+} ion, two ethylenediamine ligands, and two ammonia molecules. Is the complex neutral or is it charged? If charged, give the charge.

Naming Coordination Compounds
(See Example 22.3)

28. Write formulas for the following ions or compounds:

(a) Dichlorobis(ethylenediamine)nickel(II)

(b) Potassium tetrachloroplatinate(II)

(c) Potassium dicyanocuprate(I)

(d) Tetraamminediaquairon(II)

29. Write formulas for the following ions or compounds:

(a) Diamminetriaquahydroxochromium(II) nitrate

(b) Hexaammineiron(III) nitrate

(c) Pentacarbonyliron(0) (where the ligand is CO)

(d) Ammonium tetrachlorocuprate(II)

30. Name the following ions or compounds:

(a) $[Ni(C_2O_4)_2(H_2O)_2]^{2-}$ (c) $[Co(en)_2(NH_3)Cl]^{2+}$

(b) $[Co(en)_2Br_2]^+$ (d) $Pt(NH_3)_2(C_2O_4)$

31. Name the following ions or compounds:

(a) $[Co(H_2O)_4Cl_2]^+$ (c) $[Pt(NH_3)Br_3]^-$

(b) $Co(H_2O)_3F_3$ (d) $[Co(en)(NH_3)_3Cl]^{2+}$

32. Give the name or formula for each ion or compound, as appropriate:

(a) pentaaquahydroxoiron(III) ion

(b) $K_2[Ni(CN)_4]$

(c) $K[Cr(C_2O_4)_2(H_2O)_2]$

(d) Ammonium tetrachloroplatinate(II)

33. Give the name or formula for each ion or compound, as appropriate:

(a) Dichlorotetraaquachromium(III) chloride

(b) $[Cr(NH_3)_5SO_4]Cl$

(c) Sodium tetrachlorocobaltate(II)

(d) $[Fe(C_2O_4)_3]^{3-}$

Isomerism
(See Example 22.4)

34. Draw all possible geometric isomers of

(a) $Fe(NH_3)_4Cl_2$

(b) $Pt(NH_3)_2(SCN)(Br)$
(SCN^- is bonded to Pt^{2+} through S)

(c) $Co(NH_3)_3(NO_2)_3$
(NO_2^- is bonded to Co^{3+} through N)

(d) $[Co(en)Cl_4]^-$

35. In which of the following complexes are geometric isomers possible? If isomers are possible, draw their structures and label them as *cis* or *trans*, or as *fac* or *mer*.

(a) $[Co(H_2O)_4Cl_2]^+$ (c) $[Pt(NH_3)Br_3]^-$

(b) $Co(NH_3)_3F_3$ (d) $[Co(en)_2(NH_3)Cl]^{2+}$

36. Determine whether the following complexes have a chiral metal center:

(a) $[Fe(en)_3]^{2+}$

(b) *trans*-$[Co(en)_2Br_2]^+$

(c) *fac*-$[Co(en)(H_2O)Cl_3]$

(d) square-planar $Pt(NH_3)(H_2O)(Cl)(NO_2)$

37. Four geometric isomers are possible for the ion $[Co(en)(NH_3)_2(H_2O)Cl]^+$. Draw the structures of all four of these isomers. (Two of the four are chiral and so each has a nonsuperimposable mirror image.)

Magnetic Properties of Complexes
(See Example 22.5)

38. The following are low-spin complexes. Use the crystal field model to find the electron configuration of each ion. Determine which are diamagnetic. Give the number of unpaired electrons for the paramagnetic complexes.

(a) $[Mn(CN)_6]^{4-}$ (c) $[Fe(H_2O)_6]^{3+}$

(b) $[Co(NH_3)_6]Cl_3$ (d) $[Cr(en)_3]SO_4$

39. The following are high-spin complexes. Use the crystal field model to find the electron configuration of each ion and determine the number of unpaired electrons in each.

(a) $K_4[FeF_6]$ (c) $[Cr(H_2O)_6]^{2+}$

(b) $[MnF_6]^{4-}$ (d) $(NH_4)_3[FeF_6]$

40. Determine the number of unpaired electrons in the following tetrahedral complexes. All tetrahedral complexes are high-spin.

(a) $[FeCl_4]^{2-}$ (c) $[MnCl_4]^{2-}$

(b) $Na_2[CoCl_4]$ (d) $(NH_4)_2[ZnCl_4]$

41. Determine the number of unpaired electrons in the following tetrahedral complexes. All tetrahedral complexes are high-spin.

(a) $[Zn(H_2O)_4]^{2+}$ (c) $Mn(NH_3)_2Cl_2$

(b) $VOCl_3$ (d) $[Cu(en)_2]^{2+}$

42. For the high-spin complex $[Fe(H_2O)_6]SO_4$ identify

(a) The coordination number of iron

(b) The coordination geometry for iron

(c) The oxidation number of iron

(d) The number of unpaired electrons

(e) Whether the complex is diamagnetic or paramagnetic

43. For the low-spin complex $[Co(en)(NH_3)_2Cl_2]ClO_4$ identify

(a) The coordination number of cobalt

(b) The coordination geometry for cobalt

(c) The oxidation number of cobalt

(d) The number of unpaired electrons

(e) Whether the complex is diamagnetic or paramagnetic.

44. The coordination compound formed in an aqueous solution of cobalt(III) sulfate is diamagnetic. If NaF is added, the complex in solution becomes paramagnetic. Why does the magnetism change?

45. An aqueous solution of iron(II) sulfate is paramagnetic. If NH_3 is added, the solution becomes diamagnetic. Why does the magnetism change?

Spectroscopy of Complexes

(See Example 22.6)

46. In water, the titanium(III) ion, $[Ti(H_2O)_6]^{3+}$, has a broad absorption band at about 500 nm. What color light is absorbed by the ion? What is the color of the solution?

47. In water, the chromium(II) ion, $[Cr(H_2O)_6]^{2+}$, absorbs light with a wavelength of about 700 nm. What color is the solution?

General Questions

More challenging questions have an underlined number.

48. Describe an experiment that would determine whether nickel in $K_2[NiCl_4]$ is square-planar or tetrahedral.

49. Which of the following low-spin complexes has the highest number of unpaired electrons?

(a) $[Cr(H_2O)_6]^{3+}$ (c) $[Fe(H_2O)_6]^{2+}$

(b) $[Mn(H_2O)_6]^{2+}$ (d) $[Ni(H_2O)_6]^{2+}$

50. How many unpaired electrons are expected for high-spin and low-spin octahedral complexes of Fe^{2+}?

51. Excess silver nitrate is added to a solution containing 1.0 mol of $[Co(NH_3)_4Cl_2]Cl$. How many moles of AgCl will precipitate?

52. Which one of the following complexes is (are) square planar?

(a) $[Ti(CN)_4]^{2-}$ (c) $[Zn(CN)_4]^{2-}$

(b) $[Ni(CN)_4]^{2-}$ (d) $[Mn(CN)_4]^{2-}$

53. Which of the following complexes containing the oxalate ion is (are) chiral?

(a) $[Fe(C_2O_4)Cl_4]^{2-}$

(b) *cis*-$[Fe(C_2O_4)_2Cl_2]^{2-}$

(c) *trans*-$[Fe(C_2O_4)_2Cl_2]^{2-}$

54. How many geometric isomers are possible for the square planar complex $[Pt(NH_3)(CN)Cl_2]^-$?

55. For a tetrahedral complex of a metal in the first transition series, which of the following statements concerning energies of the 3d orbitals is correct?

(a) The five d orbitals have the same energy.

(b) The $d_{x^2-y^2}$ and d_{z^2} orbitals are higher in energy than the d_{xz}, d_{yz}, and d_{xy} orbitals.

(c) The d_{xz}, d_{yz}, and d_{xy} orbitals are higher in energy than the $d_{x^2-y^2}$ and d_{z^2} orbitals.

(d) The d orbitals all have different energies.

56. A transition metal complex absorbs 425-nm light. What is its color?

(a) red   (c) yellow

(b) green   (d) blue

57. For the low-spin complex $[Fe(en)_2Cl_2]Cl$ identify:

(a) The oxidation number of iron

(b) The coordination number for iron

(c) The coordination geometry for iron

(d) The number of unpaired electrons per metal atom

(e) Whether the complex is diamagnetic or paramagnetic

(f) The number of possible geometric isomers with this formula.

58. For the high-spin complex $Mn(NH_3)_4Cl_2$ identify

(a) The oxidation number of manganese

(b) The coordination number for manganese

(c) The coordination geometry for manganese

(d) The number of unpaired electrons per metal atom

(e) Whether the complex is diamagnetic or paramagnetic

(f) The number of possible geometric isomers

59. A platinum-containing compound, known as Magnus's green salt, has the formula $[Pt(NH_3)_4][PtCl_4]$ (in which both platinum ions are Pt^{2+}). Name the cation and the anion.

60. Early in the 20th century, complexes sometimes were named based on their colors. Two compounds with the formula $CoCl_3 \cdot 4\ NH_3$ were named praseo-cobalt chloride (praseo = green) and violio-cobalt chloride (violet color). We now know that these compounds are octahedral cobalt complexes and that they are *cis* and *trans* isomers. Draw structures of these two compounds and name them using systematic nomenclature.

61. Give the formula and name of a square-planar complex of Pt^{2+} with one nitrite ion (NO_2^-, which binds to Pt^{2+} through N), one chloride ion, and two ammonia molecules as

ligands. Are isomers possible? If so, draw the structure of each isomer, and tell what type of isomerism is observed.

62. Give the formula of a complex formed from one Co^{3+} ion, two ethylenediamine molecules, one water molecule, and one chloride ion. Is the complex neutral or charged? If charged, give the net charge on the ion.

63. How many geometric isomers of the complex $[Cr(dmen)_3]^{3+}$ can exist? (dmen is the bidentate ligand 1,1-dimethylethylenediamine.)

$$(CH_3)_2\overset{..}{N}CH_2CH_2\overset{..}{N}H_2$$

1,1-dimethylethylenediamine, dmen

64. Diethylenetriamine (dien), is capable of serving as a tridentate ligand.

$$H_2\overset{..}{N}CH_2CH_2\overset{..}{-}\overset{..}{N}\overset{|}{-}CH_2CH_2\overset{..}{N}H_2$$
$$\overset{|}{H}$$

diethylenetriamine, dien

(a) Draw the structure of the compounds *fac*-Cr(dien)Cl$_3$ and *mer*-Cr(dien)Cl$_3$.

(b) Two different geometric isomers of *mer*-Cr(dien)Cl$_2$Br are possible. Draw the structure for each.

(c) Three different geometric isomers are possible for $[Cr(dien)_2]^{3+}$. Two have the dien ligand in a *fac* configuration, and one has the ligand in the *mer* orientation. Draw the structure of each isomer.

65. From experiment we know that $[CoF_6]^{3-}$ is paramagnetic and $[Co(NH_3)_6]^{3+}$ is diamagnetic. Using the crystal field model, depict the electron configuration for each ion and use this model to explain the magnetic property. What can you conclude about the effect of the ligand on the magnitude of Δ_0?

66. $[Co(en)(NH_3)_2(H_2O)_2]^{3+}$ has three geometric isomers. One of the three is chiral; that is, it has a nonsuperimposable mirror image. Draw the structures of the three isomers. Which one is chiral?

67. The square-planar complex Pt(en)Cl$_2$ has chloride ligands in a *cis* configuration. No *trans* isomer is known. Based on the bond lengths and bond angles of carbon and nitrogen in the ethylenediamine ligand, explain why the *trans* compound is not possible.

68. The complex $[Mn(H_2O)_6]^{2+}$ has five unpaired electrons, whereas $[Mn(CN)_6]^{4-}$ has only one. Using the crystal field model, depict the electron configuration for each ion. What can you conclude about the effect of the different ligands on the magnitude of Δ_0?

69. Experiments show that $K_4[Cr(CN)_6]$ is paramagnetic and has two unpaired electrons. The related complex $K_4[Cr(SCN)_6]$ is paramagnetic with four unpaired electrons. Account for the magnetism of each compound using the crystal field model. Predict where the SCN$^-$ ion occurs in the spectrochemical series relative to CN$^-$.

70. Two different coordination compounds containing a cobalt(III) ion, five ammonia molecules, a bromide ion, and a sulfate ion exist. The dark violet form (A) gives a precipitate upon addition of aqueous BaCl$_2$. No reaction is seen upon addition of aqueous BaCl$_2$ to the violet red form (B). Suggest structures for these two compounds and write a chemical equation for the reaction of (A) with aqueous BaCl$_2$.

71. The complex ion $[Co(CO_3)_3]^{3-}$, an octahedral complex with bidentate carbonate ions as ligands, has one absorption in the visible region of the spectrum at 640 nm. From this information

(a) Predict the color of this complex, and explain your reasoning.

(b) Place the carbonate ion in the proper place in the spectrochemical series.

(c) Predict whether $[Co(CO_3)_3]^{3-}$ will be paramagnetic or diamagnetic.

72. The stability of analogous complexes $[ML_6]^{n+}$ (relative to ligand dissociation) is in the general order Mn^{2+}, Fe^{2+}, Co^{2+}, Ni^{2+}, Cu^{2+}, Zn^{2+}. This order of ions is called the Irving–Williams series. Look up the values of formation constants for ammonia complexes Co^{2+}, Ni^{2+}, Cu^{2+}, and Zn^{2+} in Appendix K and verify this statement.

73. The glycinate ion, $H_2NCH_2CO_2^-$, formed by deprotonation of the amino acid glycine, can function as a bidentate ligand, coordinating to a metal ion through the nitrogen of the amino group and one of the oxygen atoms.

Glycinate ion, a bidentate ligand

$$H_3C-\overset{\overset{\displaystyle H}{|}}{C}-\overset{\overset{\displaystyle :O:}{\|}}{C}-\overset{..}{\underset{..}{O}}:$$
$$\underset{\displaystyle H_2N:}{|}\qquad\qquad\overset{|}{}$$

Site of bonding to transition metal ion

An octahedral copper complex of this ligand has the formula Cu(H$_2$NCH$_2$CO$_2$)$_2$(H$_2$O)$_2$. For this complex, determine

(a) The oxidation state of copper

(b) The coordination number of copper

(c) The number of unpaired electrons

(d) Whether the complex is diamagnetic or paramagnetic

74. Draw structures for the five possible geometric isomers of Cu(H$_2$NCH$_2$CO$_2$)$_2$(H$_2$O)$_2$. Are any of these species chiral? (See the structure of the ligand in Study Question 73.)

75. Three different compounds of chromium(III) with water and chloride ion have the same composition: 19.51% Cr, 39.92% Cl, and 40.57% H$_2$O. One of the compounds is violet and dissolves in water to give a complex ion with a +3 charge and three chloride ions. All three chloride ions precipitate immediately as AgCl on adding AgNO$_3$. Draw the structure of the complex ion and name the compound. Write

a net ionic equation for the reaction of this compound with silver nitrate.

76. In this question, we explore the differences between metal coordination by monodentate and bidentate ligands. Formation constants, K_f, for $[Ni(NH_3)_6]^{2+}(aq)$ and $[Ni(en)_3]^{2+}(aq)$ are:

$$Ni^{2+}(aq) + 6 NH_3(aq) \rightleftharpoons [Ni(NH_3)_6]^{2+}(aq)$$
$$K_f = 10^8$$
$$Ni^{2+}(aq) + 3 en(aq) \rightleftharpoons [Ni(en)_3]^{2+}(aq)$$
$$K_f = 10^{18}$$

The values for K_f indicate a higher stability of the chelated complex, caused by the *chelate effect*. Recall that K is related to the standard free energy of the reaction by the equation $\Delta G° = -RT \ln K$ and $\Delta G° = \Delta H° - T \Delta S°$. We know from experiment that $\Delta H°$ for the NH_3 reaction is -109 kJ/mol, and $\Delta H°$ for the ethylenediamine reaction is -117 kJ/mol. Is the difference in $\Delta H°$ sufficient to account for the 10^{10} difference in K_f? Comment on the role of entropy in the second reaction.

77. Titanium is strong, lightweight, and resistant to corrosion; these properties lead to its use in aircraft engines. To obtain metallic titanium, ilmenite ($FeTiO_3$), an ore of titanium, is first treated with sulfuric acid to form $FeSO_4$ and $Ti(SO_4)_2$. After separating these compounds, the latter substance is converted to TiO_2 in basic solution:

$$FeTiO_3(s) + 3 H_2SO_4(aq) \longrightarrow$$
$$FeSO_4(aq) + Ti(SO_4)_2(aq) + 3 H_2O(\ell)$$
$$Ti^{4+}(aq) + 4 OH^-(aq) \longrightarrow TiO_2(s) + 2 H_2O(\ell)$$

What volume of 18.0 M H_2SO_4 is required to react completely with 1.00 kg of ilmenite? What mass of TiO_2 can theoretically be produced by this sequence of reactions?

78. In the process described in Study Question 77, ilmenite ore is leached with sulfuric acid. This leads to the significant environmental problem of disposal of the iron(II) sulfate (which, in its hydrated form, is commonly called "copperas"). To avoid this, it has been suggested that HCl be used to leach ilmenite so that the iron-containing product is $FeCl_2$. This can be treated with water and air to give commercially useful iron(III) oxide and regenerate HCl by the reaction

$$2 FeCl_2(aq) + 2 H_2O(\ell) + \tfrac{1}{2} O_2(g) \longrightarrow$$
$$Fe_2O_3(s) + 4 HCl(aq)$$

(a) Write a balanced equation for the treatment of ilmenite with aqueous HCl to give iron(II) chloride, titanium(IV) oxide, and water.

(b) If the equation written in part (a) is combined with the equation for oxidation of $FeCl_2$ to Fe_2O_3, is the HCl used in the first step recovered in the second step?

(c) What mass of iron(III) oxide can be obtained from one ton (908 kg) of ilmenite in this process?

23

Nuclear Chemistry

Chapter Goals

- Identify radioactive elements and describe natural and artificial nuclear reactions.
- Calculate and understand nuclear binding energies.
- Understand rates of radioactive decay and the stability of nuclei.
- Understand issues of health and safety with respect to radioactivity.
- Be aware of some uses of radioactive isotopes in science and medicine.

Nuclear Medicine

The primary function of the thyroid gland, located in your neck, is the production of the thyroxine (3,5,3′,5′-tetraiodothyronine) and 3,5,3′-triiodothyronine. These chemical compounds are hormones that help to regulate the rate of *metabolism*, a term that refers to all of the chemical reactions that take place in the body. In particular, the thyroid hormones have an important role in the processes that convert food to energy.

Abnormally low levels of thyroxin result in a condition known as *hypothyroidism*. Symptoms include lethargy and feeling cold much of the time. The remedy for this condition is medication, thyroxin pills. The opposite condition, *hyperthyroidism*, also occurs in some people. Here, too much of the hormone is produced by the body. The condition is diagnosed by symptoms such as nervousness, heat intolerance, increased appetite, and muscle weakness and fatigue when blood sugar is too rapidly depleted. The standard remedy for hyperthyroidism is to destroy part of the thyroid gland, and one way to do this uses radioactive iodine-131.

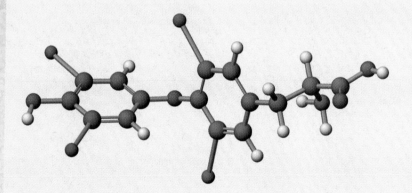

3,5,3′,5′-tetraiodothyronine (thyroxine)

▲ The hormone 3,5,3′,5′-tetraiodothyronine (thyroxin) exerts a stimulating effect on metabolism.

used as one of the raw materials in making thyroxine.

The fact that iodine concentrates in the thyroid tissue is essential to the procedure for using radioiodine therapy as a treatment for hyperthyroidism. Typically this radioisotope is administered orally as an aqueous NaI solution in which a small fraction of iodide consists of the radioactive isotopes iodine-131 or iodine-123, and the rest is nonradioactive iodine-127. The radioactivity destroys thyroid tissue, resulting in a decrease in thyroid activity.

▲ Most salt contains a small amount of NaI; this supplies the body with sufficient iodine for its needs. *(Charles D. Winters)*

To understand this procedure, you need to know something about iodine in the body. Iodine is an essential element. Some diets provide iodine naturally (seaweed, for example, is a good source of iodine) but in the Western world most iodine taken up by the body comes from iodized salt, NaCl containing a few percent of NaI. An adult man or woman of average size should take in about 150 μg (micrograms) of iodine (1 μg = 10^{-6} g) in the daily diet. In the body, iodide ion is transported to the thyroid, where it is

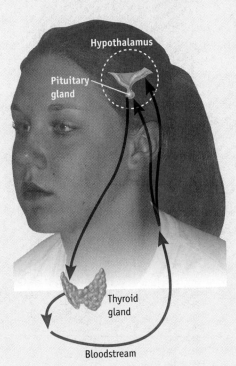

Hypothalamus

Pituitary gland

Thyroid gland

Bloodstream

▲ The thyroid gland, located in the neck, is the source of the hormone thyroxine.

Nuclear Chemistry The atomic age began with the explosion of the first atomic bomb in New Mexico on July 16, 1945. The energy locked in the atomic nucleus has since been harnessed for many uses in medicine, for food preservation, and for the generation of power.

The beginning of the atomic age

The first atomic bomb was detonated at White Sands, New Mexico at 5:30 AM on July 16, 1945. Two weeks later the United States used this weapon against Japan, bringing about the end of World War II. (Los Alamos National Laboratory)

For many years the respected *Bulletin of Atomic Scientists* has used the symbol of a clock with its hands near midnight, illustrating the danger the world faces from atomic weapons. In 1947, when the clock was first used, it was set at 7 minutes to midnight. It moved as close as 2 minutes to midnight in 1953, and was 17 minutes to midnight in 1991 when the U.S. and the Soviet Union signed a Strategic Arms Reduction Treaty. Now, with the rejection of arms control treaties by the U.S., and the threat of terrorism, the hands have been reset to 7 minutes to midnight. (The Bulletin of Atomic Scientists)

The uses of atomic energy

A patient inhales radioactive xenon, which is taken up and carried by the bloodstream throughout the body. The helmet on the patient's head detects gamma rays from the decay, providing a visualization of blood flow in the brain. (Will and Deni McIntyre/Photo Researchers, Inc.)

Atomic power plants supply about 20% of the electricity generated in the United States. (Joe Azzara/Getty/The Image Bank)

Treating foods with radiation kills pathogens and makes food safer. (MDS Nordian)

A household smoke detector uses radioactive americium-241. This alpha emitter has a half-life of 470 years. In a smoke detector the emission ionizes smoke particles to activate the alarm. (Charles D. Winters)

History of science scholars cite three pillars of modern chemistry: technology, medicine, and alchemy. The third, alchemy, was pursued in many cultures on three continents for well over 1000 years. Simply stated, the goal of the ancient alchemists was to turn less valuable materials into gold. We now recognize the futility of these efforts because this goal is not reachable by chemical processes. However, we now also know that it is possible to transmute one element into another. This happens naturally in the decomposition of uranium and other radioactive elements, and scientists can intentionally carry out such reactions in the laboratory. The goal is no longer to make gold, however. There are far more important and valuable products of nuclear reactions.

Nuclear chemistry encompasses a wide range of topics that have one thing in common: they involve changes in the nucleus of an atom (Chapter Focus). While "chemistry" is a major focus in nuclear chemistry, the subject cuts across many areas of science. Radioactive isotopes are used in medicine. Nuclear power provides a sizable fraction of energy for modern society. And, then there are nuclear weapons. . . .

23.1 NATURAL RADIOACTIVITY

In the late 19th century, while studying radiation emanating from uranium and thorium, Ernest Rutherford (1871–1937) [⬅ PAGE 50] stated that "There are present at least two distinct types of radiation — one that is readily absorbed, which will be termed for convenience **α (alpha) radiation,** and the other of a more penetrative character, which will be termed **β (beta) radiation.**" Subsequently, charge-to-mass ratio measurements showed that α radiation is composed of helium nuclei (He^{2+}) and β radiation is composed of electrons (e^-) (Table 23.1).

Rutherford hedged his bet when he said there were at least two types of radiation. A third type was later discovered by the French scientist Paul Villard (1860–1934); he named it **γ (gamma) radiation,** using the third letter in the Greek alphabet in keeping with Rutherford's scheme. Unlike α and β radiation, γ radiation is not affected by electric and magnetic fields. It is a form of electromagnetic radiation like x-rays but even more energetic.

Early studies measured the penetrating power of the three types of radiation (Figure 23.1). Alpha radiation is the least penetrating; it can be stopped by several sheets of ordinary paper or clothing. Aluminum that is at least 0.3-cm thick is needed to stop β particles, and they can penetrate several millimeters of living bone or tissue. Gamma radiation is the most penetrating. Thick layers of lead or concrete are required to shield the body from this radiation, and γ-rays can pass completely through the human body.

Alpha and β particles typically possess high kinetic energies. The energy of γ radiation is similarly very high. The energy associated with this radiation is transferred to any material used to stop the particle or absorb the radiation. This is

Chapter Goals • Revisited

- **Identify radioactive elements and describe natural and artificial nuclear reactions.**
- Calculate and understand nuclear binding energies.
- Understand rates of radioactive decay and the stability of nuclei.
- Understand issues of health and safety with respect to radioactivity.
- Be aware of some uses of radioactive isotopes in science and medicine.

● **Discovery of Radioactivity**
The discovery of radioactivity by Henri Becquerel and the isolation of radium and polonium from pitchblende, a uranium ore, by Marie Curie were described in Chapter 2 (page 47).

● **Common Symbols: α and β**
Symbols used to represent alpha and beta particles do not include a superscript to show that they have a charge.

Table 23.1 • Characteristics of α, β, and γ Radiation

Name	Symbols	Charge	Mass (g/particle)
Alpha	^4_2He, $^4_2\alpha$	+2	6.65×10^{-24}
Beta	$^0_{-1}e$, $^0_{-1}\beta$	−1	9.11×10^{-28}
Gamma	γ	0	0

Figure 23.1 The relative penetrating ability of α, β, and γ radiation. Highly charged α particles interact strongly with matter and are stopped by a piece of paper. Beta particles, with less mass and a lower charge, interact to a lesser extent with matter and thus can penetrate farther. Gamma radiation is the most penetrating.

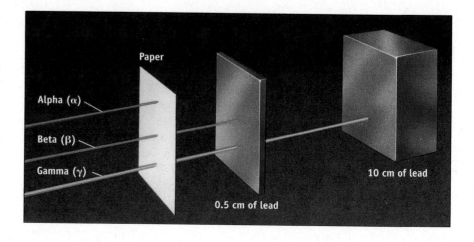

important because the damage caused by radiation is related to the energy absorbed (see Section 23.8).

23.2   NUCLEAR REACTIONS AND RADIOACTIVE DECAY

Equations for Nuclear Reactions

In 1903, Rutherford and Frederick Soddy (1877–1956) proposed that radioactivity is the result of a natural change of an isotope of one element into an isotope of a different element. Such processes are called **nuclear reactions.** In a nuclear reaction, there is a change in the atomic number of an atom and often a change in mass number as well.

Consider a reaction in which radium-226 (the isotope of radium with mass number 226) emits an α particle to form radon-222. The equation for this reaction is

$$^{226}_{88}\text{Ra} \longrightarrow \, ^{4}_{2}\alpha + \, ^{222}_{86}\text{Rn}$$

In a nuclear reaction the sum of the mass numbers of reacting particles must equal the sum of the mass numbers of products. Furthermore, to maintain charge balance, the sum of the atomic numbers of the products must equal the sum of the atomic numbers of the reactants. These principles are illustrated using the preceding nuclear equation:

	$^{226}_{88}\text{Ra}$	$\longrightarrow$	$^{4}_{2}\alpha$	$+$	$^{222}_{86}\text{Rn}$
	radium-226	$\longrightarrow$	α particle	$+$	radon-222
Mass number: (protons + neutrons)	226	=	4	+	222
Atomic number: (protons)	88	=	2	+	86

Alpha particle emission causes a decrease of two units in atomic number and four units in the mass number.

Similarly, mass and charge balance accompany β particle emission by uranium-239:

	$^{239}_{92}\text{U}$	$\longrightarrow$	$^{0}_{-1}\beta$	$+$	$^{239}_{93}\text{Np}$
	uranium-239	$\longrightarrow$	β particle	$+$	neptunium-239
Mass number: (protons + neutrons)	239	=	0	+	239
Atomic number: (protons)	92	=	−1	+	93

The β particle has a charge of -1. Charge balance requires that the atomic number of the product be one unit greater than the atomic number of the reacting nucleus. The mass number does not change in this process.

How does a nucleus, composed of protons and neutrons, eject an electron? This is a complex process, but the net result is the conversion within the nucleus of a neutron to a proton and an electron.

$$\underset{\text{neutron}}{{}^{1}_{0}\text{n}} \longrightarrow \underset{\text{electron}}{{}^{0}_{-1}\text{e}} + \underset{\text{proton}}{{}^{1}_{1}\text{p}}$$

Notice that mass and charge numbers balance in this equation.

Exercise 23.1   Mass and Charge Balance in Nuclear Reactions

Write equations for the following nuclear reactions and confirm that they are balanced with respect to mass and charge.

(a) The emission of an α particle by radon-222 to form polonium-218.
(b) The emission of a β particle by polonium-218 to form astatine-218.

What is the origin of the gamma radiation that accompanies most nuclear reactions? Recall that a photon of visible light is emitted when an atom undergoes a transition from an excited electronic state to a lower energy state [⬅ SECTION 7.3]. Gamma radiation originates in much the same way except that transitions between nuclear energy levels occur. Nuclear reactions often result in the formation of a product nucleus in an excited nuclear state. One option is to return to the ground state by emitting the excess energy as a photon. The high energy of γ radiation is a measure of the large energy difference between the energy levels in the nucleus.

Exercise 23.2   Gamma Ray Energies

Calculate the energy per photon, and the energy per mole of photons, for γ radiation with a wavelength of 2.0×10^{-12} m. (*Hint:* Review similar calculations on the energy of photons of visible light, Section 7.2.)

● Energy Units
Gamma ray energies are often reported with the unit *MeV*, which stands for millions of electron volts. One electron volt (1 eV) is the energy of an electron that has been accelerated by a potential of one volt. The conversion factor between eV and Joules is $1\ \text{eV} = 1.60218 \times 10^{-19}$ J.

Radioactive Decay Series

Several naturally occurring radioactive isotopes are found to decay to form a product that is also radioactive. When this happens, the initial nuclear reaction is followed by a second nuclear reaction, and if the situation is repeated, a third and a fourth, and so on. Eventually, a nonradioactive isotope is formed to end the series. Such a sequence of nuclear reactions is called a **radioactive decay series.** In this nuclear reaction sequence, the reactant nucleus is often called the *parent* and the product called the *daughter.*

Uranium-238, the most abundant of three naturally occurring uranium isotopes, heads one of several radioactive decay series. This series begins with the loss of an α particle from ${}^{238}_{92}\text{U}$ to form radioactive ${}^{234}_{90}\text{Th}$. Thorium-234 then decomposes by β emission to ${}^{234}_{91}\text{Pa}$, which emits a β particle to give ${}^{234}_{92}\text{U}$. Uranium-234 is an α emitter, forming ${}^{230}_{90}\text{Th}$. Further α and β emissions follow until the series ends with formation of the nonradioactive isotope, $^{206}_{82}$Pb. In all, this radioactive decay series converting $^{238}_{92}$U to $^{206}_{82}$Pb is made up of 14 reactions, with eight α and six β particles being emitted. The series is portrayed graphically by plotting atomic number and mass number (Figure 23.2). An equation can be written for each step in the sequence. Equations for the first four steps in the uranium-238 radioactive decay series are

Step 1. $\quad ^{238}_{92}\text{U} \longrightarrow ^{234}_{90}\text{Th} + ^{4}_{2}\alpha$

Step 2. $\quad ^{234}_{90}\text{Th} \longrightarrow ^{234}_{91}\text{Pa} + ^{0}_{-1}\beta$

Step 3. $\quad ^{234}_{91}\text{Pa} \longrightarrow ^{234}_{92}\text{U} + ^{0}_{-1}\beta$

Step 4. $\quad ^{234}_{92}\text{U} \longrightarrow ^{230}_{90}\text{Th} + ^{4}_{2}\alpha$

Uranium ore contains trace quantities of the radioactive elements formed in the radioactive decay series. A significant development in nuclear chemistry was the discovery in 1898 of radium and polonium by Marie Curie as trace components of pitchblende, a uranium ore. The amount of each of these elements is small because the isotopes of these elements have short half-lives. It is reported that she isolated only a single gram of radium from 7 tons of ore. It is a credit to Marie Curie's skills as a chemist that she extracted sufficient amounts of radium and polonium from uranium ore to identify these elements.

The uranium-238 radioactive decay series is also the source of the environmental hazard radon. Trace quantities of uranium are often present naturally in

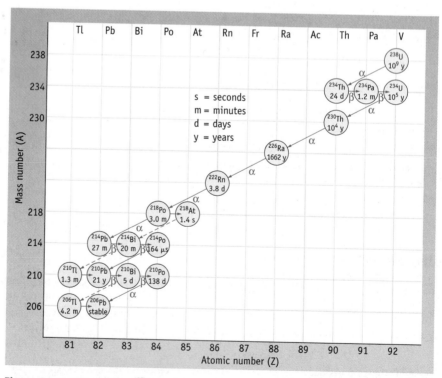

Figure 23.2 **The uranium-238 radioactive decay series.** The steps in this radioactive decay series are shown graphically in this plot of mass number versus atomic number. Each α decay step lowers the atomic number by two units and the mass number by four units. Beta particle emission does not change the mass but raises the atomic number by one unit. Half-lives of the isotopes are included on the chart.

the soil and rocks in which radon-222 is being continuously formed. Because radon is chemically inert, it is not trapped by chemical processes occurring in soil or water and is free to seep into mines or into homes through pores in cement block walls, cracks in the basement floor or walls, or around pipes. More dense than air, radon tends to collect in low spots, so its concentration can build up in a basement if steps are not taken to remove it.

The major health hazard from radon, when it is inhaled by humans, arises not from radon itself but from its decomposition product, polonium.

$$^{222}_{86}\text{Rn} \longrightarrow {}^{218}_{84}\text{Po} + {}^{4}_{2}\alpha \qquad t_{1/2} = 3.82 \text{ days}$$

$$^{218}_{84}\text{Po} \longrightarrow {}^{214}_{82}\text{Pb} + {}^{4}_{2}\alpha \qquad t_{1/2} = 3.11 \text{ minutes}$$

Radon does not undergo chemical reactions or form compounds that can be taken up in the body. Polonium, however, is not chemically inert. Polonium-218 can lodge in body tissues where it undergoes α decay to give lead-214, another radioactive isotope. The range of an α particle in body tissue is quite small, perhaps 0.7 mm. This is approximately the thickness of the epithelial cells of the lungs, however, so α particle radiation can cause serious damage to lung tissues.

Virtually every home in the United States has some level of radon, and kits can be purchased to test for the presence of this gas. If radon gas is detected in your home, you should take corrective actions such as sealing cracks around the foundation and in the basement. It may be reassuring that the health risks associated with radon are low. The likelihood of getting lung cancer from exposure to radon is about the same as the likelihood of dying in an accident in your home.

Radon detector. This kit is for use in the home to detect radon gas. The small device is placed in the home's basement for a given time period and is then sent to a laboratory to measure the amount of radon that might be present. *(Charles D. Winters)*

Example 23.1 ▶ Radioactive Decay Series

Problem • A second radioactive decay series begins with $^{235}_{92}\text{U}$ and ends with $^{207}_{82}\text{Pb}$.

(a) How many α and β particles are emitted in this series?

(b) The first three steps of this series are (in order) α, β, and α emission. Write an equation for each of these steps.

Strategy •

First, find the total change in atomic number and mass number. A combination of α and β particles is required that will decrease the total mass by 28 (235–207) and at the same time decrease the atomic number by 10 (92–82).

Each equation must give symbols for the parent and daughter nuclei and the emitted particle. In the equations, the sums of the atomic numbers and mass numbers for reactants and products must be equal.

Solution •

(a) Mass declines by 28 mass units (235–207). Because a decrease of 4 mass units occurs with each α emission, seven α particles must be emitted. Also, for each α emission, the atomic number decreases by 2. Emission of seven α particles would cause the atomic number to decrease by 14. The actual

decrease in atomic number is 10 (92–82). This means that four β particles must also have been emitted because each β emission *increases* the atomic number of the product by one unit. Thus, the radioactive decay sequence involves emission of seven α and four β particles.

(b) Step 1. $^{235}_{92}\text{U} \longrightarrow {}^{231}_{90}\text{Th} + {}^{4}_{2}\alpha$

   Step 2. $^{231}_{90}\text{Th} \longrightarrow {}^{231}_{91}\text{Pa} + {}^{0}_{-1}\beta$

   Step 3. $^{231}_{91}\text{Pa} \longrightarrow {}^{227}_{89}\text{Ac} + {}^{4}_{2}\alpha$

Comment • Notice in Figure 23.2 that all daughter nuclei for the series beginning with $^{238}_{92}\text{U}$ have mass numbers differing by four units 238, 234, 230, . . . , 206. This series is sometimes called the $4n + 2$ *series* because each mass number (M) fits the equation $4n + 2 = M$ where n is an integer. For the series headed by $^{235}_{92}\text{U}$, the mass numbers are 235, 231, 227, . . . , 207; this is the $4n + 3$ *series*.

Two other decay series are possible. One, called the $4n$ series and beginning with ^{232}Th, is also found in nature, but the other, the $4n + 1$ series, is not. No member of this series has a very long half-life. During the 5 billion years since this planet was formed, all members of this series have completely decayed.

Exercise 23.3 **Radioactive Decay Series**

(a) Six α and five β particles are emitted in the thorium-232 radioactive decay series before a stable isotope is reached. What is the final product in this series?

(b) The first three steps in the thorium-232 decay series (in order) are α, β, and β emission. Write a nuclear equation for each step.

Other Types of Radioactive Decay

• Positrons
Positrons were discovered by Carl Anderson (1905–1971) in 1932. The positron is one of a group of particles that are known as *antimatter*. If matter and antimatter particles collide, mutual annihilation occurs, with energy being emitted.

• Neutrinos and Antineutrinos
Beta particles having a wide range of energies are emitted. To balance energy associated with β decay, it is necessary to postulate the concurrent emission of another particle, the *antineutrino*. Similarly, neutrino emission accompanies positron emission. Much study has gone into detecting neutrinos and antineutrinos. These massless, chargeless particles are not usually included in nuclear equations.

Most naturally occurring radioactive elements decay by emission of α, β, and γ radiation. Other nuclear decay processes became known, however, when new radioactive elements were synthesized by artificial means. Other modes of decomposition are **positron ($_{+1}^{0}\beta$) emission** and **electron capture.**

Positrons ($_{+1}^{0}\beta$) and electrons have the same mass but opposite charge. The positron is the antimatter analogue to an electron. Positron emission by polonium-207 results in the formation of bismuth-207, for example.

$$_{84}^{207}\text{Po} \longrightarrow _{+1}^{0}\beta + _{83}^{207}\text{Bi}$$

polonium-207 $\longrightarrow$ positron + bismuth-207

Mass number: (protons + neutrons)	7	=	0	+ 7
Atomic number: (protons)	84	=	+1	+ 83

To retain charge balance, positron decay results in a decrease in the atomic number.

In *electron capture*, an extranuclear electron is captured by the nucleus. The mass number is unchanged and the atomic number is reduced by 1. (In an old nomenclature the innermost electron shell was called the K shell, and electron capture was called *K capture*.)

$$_{4}^{7}\text{Be} + _{-1}^{0}\text{e} \longrightarrow _{3}^{7}\text{Li}$$

beryllium-7 + electron $\longrightarrow$ lithium-7

Mass number: (protons + neutrons)	7	+ 0	=	7
Atomic number: (protons)	4	+ −1	=	3

In summary, most unstable nuclei decay by one of four paths: α and β decay, positron emission, and electron capture. Gamma radiation often accompanies these processes. We will introduce a fifth way that nuclei decompose, *fission*, in Section 23.6.

Example 23.2 **Nuclear Reactions**

Problem • Complete the following equations. Give the symbol, mass number, and atomic number of the product species.

(a) $_{18}^{37}\text{Ar} + _{-1}^{0}\text{e} \longrightarrow$?   **(c)** $_{16}^{35}\text{S} \longrightarrow _{17}^{35}\text{Cl} + $?

(b) $_{6}^{11}\text{C} \longrightarrow _{5}^{11}\text{B} + $?   **(d)** $_{15}^{30}\text{P} \longrightarrow _{+1}^{0}\beta + $?

Strategy • The missing product in each reaction can be determined by recognizing that sums of mass numbers and atomic numbers for products and reactants must be equal. When you know the mass and charge of the product, you can identify it with its appropriate symbol.

Solution •
(a) This is an electron capture reaction. The product has a mass

number of $37 + 0 = 37$ and an atomic number of $18 - 1 = 17$. Therefore, the symbol for the product is $_{17}^{37}\text{Cl}$.

(b) This missing particle has a mass of zero and a charge of +1; these are the characteristics of a positron, $_{+1}^{0}\beta$. If this particle is included in the equation, the sums of the atomic numbers ($6 = 5 + 1$) and the mass numbers (11) on either side of the equation are equal.

(c) A beta particle, $_{-1}^{0}\beta$, is required to balance the mass numbers (35) and atomic numbers ($16 = 17 - 1$) in the equation.

(d) The product nucleus has mass number 30 and atomic number 14. This identifies the unknown as $_{14}^{30}\text{Si}$.

Exercise 23.4   Nuclear Reactions

Indicate the symbol, the mass number, and the atomic number of missing product in each of the following nuclear equations.

(a) $^{13}_{7}\text{N} \longrightarrow \, ^{13}_{6}\text{C} + \, ?$ **(c)** $^{90}_{38}\text{Sr} \longrightarrow \, ^{90}_{39}\text{Y} + \, ?$

(b) $^{41}_{20}\text{Ca} + \, ^{0}_{-1}\text{e} \longrightarrow \, ?$ **(d)** $^{22}_{11}\text{Na} \longrightarrow \, ? + \, ^{0}_{-1}\beta$

23.3   STABILITY OF ATOMIC NUCLEI

We can learn something about nuclear stability from Figure 23.3. In this plot, the horizontal axis represents the number of protons and the vertical axis is the number of neutrons for known isotopes. Each circle represents an isotope identified by the number of neutrons and protons contained in its nucleus. The black circles represent stable (nonradioactive) isotopes, some 300 in number, and the red circles represent some of the known radioactive isotopes. For example, the three isotopes of hydrogen are $^{1}_{1}\text{H}$ and $^{2}_{1}\text{H}$ (neither is radioactive) and $^{3}_{1}\text{H}$ (tritium, radioactive). For lithium, the third element, isotopes with mass numbers 6, 7, 8, and 9 are known. The first two of these isotopes with masses of 6 and 7 (shown in black) are not radioactive, whereas the third and fourth (in red) are radioactive.

Figure 23.3 contains the following information about nuclear stability:

- Stable isotopes fall in a very narrow range called the **band of stability.** It is remarkable how few isotopes are stable.

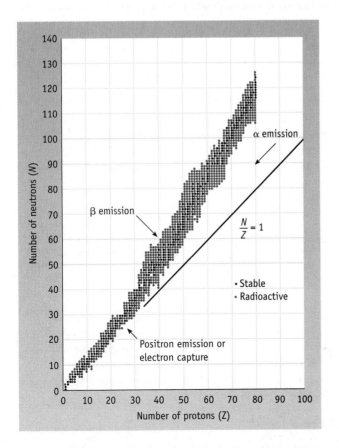

Figure 23.3   Stable and unstable isotopes. A graph of the number of neutrons (N) versus the number of protons (Z) for stable (black circles) and radioactive (red circles) isotopes from hydrogen to bismuth. This graph is used to assess criteria for nuclear stability and to predict modes of decay for unstable nuclei.

- Only two stable isotopes ($^1_1$H and $^3_2$He) have more protons than neutrons.

- Up to Ca ($Z = 20$), stable isotopes often have equal numbers of protons and neutrons or only one or two more neutrons than protons.

- Beyond calcium, the neutron–proton ratio is always greater than 1. As the mass increases, the band of stable isotopes deviates more and more from a line in which $N = Z$.

- Beyond bismuth (83 protons and 126 neutrons) all isotopes are unstable and radioactive. There is apparently no nuclear "superglue" strong enough to hold heavy nuclei together.

- If we were to look up half-lives, we would also learn that the lifetimes of unstable nuclei are shorter for the heaviest nuclei. For example, half of a sample of $^{238}_{92}$U disintegrates in 4.5 billion years, whereas half of a sample of $^{257}_{103}$Lr is gone in only 0.65 s. We would also learn that isotopes that fall farther from the band of stability tend to have shorter half-lives than do unstable isotopes nearer to the band of stability.

- Elements of even atomic number have more stable isotopes than do those of odd atomic number. More stable isotopes have an even number of neutrons than have an odd number. Roughly 200 isotopes have an even number of neutrons and an even number of protons. Only about 120 isotopes have an odd number of either protons or neutrons. Only five stable isotopes ($^1_1$H, $^6_3$Li, $^{10}_5$B, $^{14}_7$N, and $^{180}_{73}$Ta) have odd numbers of both protons and neutrons.

The Band of Stability and Radioactive Decay

Besides being a criterion for stability, the neutron–proton ratio can assist in predicting what type of radioactive decay will be observed. Unstable nuclei decay in a manner that brings them toward a stable neutron–proton ratio, that is, toward the band of stability [CD-ROM, Screen 2.17].

- All elements beyond bismuth ($Z = 83$) are unstable. To reach the band of stability starting with these elements, a process that decreases the atomic number is needed. Alpha emission is an effective way to lower Z, the atomic number, because each emission decreases the atomic number by 2. For example, americium, the radioactive element used in smoke detectors (Chapter Focus), decays by α emission:

$$^{243}_{95}\text{Am} \longrightarrow {}^4_2\alpha + {}^{239}_{93}\text{Np}$$

- Beta emission occurs for isotopes that have a high neutron–proton ratio, that is, isotopes above the band of stability. During β decay the atomic number increases by 1 and the mass number remains constant, resulting in a lower n/p ratio.

$$^{60}_{27}\text{Co} \longrightarrow {}^0_{-1}\beta + {}^{60}_{28}\text{Ni}$$

- Isotopes with a low neutron–proton ratio, below the band of stability, decay by positron emission or by electron capture. Both processes lead to product nuclei with a lower atomic number and the same mass number:

$$^{13}_7\text{N} \longrightarrow {}^0_{+1}\beta + {}^{13}_6\text{C}$$
$$^{41}_{20}\text{Ca} + {}^0_{-1}\text{e} \longrightarrow {}^{41}_{19}\text{K}$$

Example 23.3 Predicting Modes of Radioactive Decay

Problem • Identify probable mode(s) of decay for each isotope and write an equation for the decay process.

(a) Oxygen-15, $^{15}_{8}O$

(b) Uranium-234, $^{234}_{92}U$

(c) Fluorine-20, $^{20}_{9}F$

(d) Manganese-56, $^{56}_{25}Mn$

Strategy • In parts (a), (c), and (d), determine whether there are too many or too few neutrons in the nucleus. To do this, compare the mass number with the atomic weight of the element. If the mass number of the isotope is higher than the atomic weight, then there are too many neutrons and β emission is likely. If the mass number is lower than the atomic weight, then there are too few neutrons, and positron emission or electron capture is the more likely process. It is not possible to choose between the latter two modes of decay without further information. Part (b) is addressed by noting that isotopes with atomic number greater than 83 are likely to be α emitters.

Solution •

(a) Oxygen-15 has 7 neutrons and 8 protons so the n/p ratio is less than 1 — too low for ^{15}O to be stable. Nuclei with too few neutrons are expected to decay either by positron emission or by electron capture. In this instance, the process is $^{0}_{+1}\beta$ emission and the equation is $^{15}_{8}O \longrightarrow {}^{0}_{+1}\beta + {}^{15}_{7}N$

(b) Alpha emission is a common mode of decay for isotopes of elements with atomic numbers over 83. The decay of uranium-234 is one example: $^{234}_{92}U \longrightarrow {}^{230}_{90}Th + {}^{4}_{2}\alpha$

(c) Fluorine-20 has 11 neutrons and 9 protons, a high n/p ratio. The ratio is lowered by β emission: $^{20}_{9}F \longrightarrow {}^{0}_{-1}\beta + {}^{20}_{10}Ne$

(d) The atomic weight of manganese is 54.85. The higher mass number, 56, suggests that this radioactive isotope has an excess of neutrons, in which case it would be expected to decay by β emission: $^{56}_{25}Mn \longrightarrow {}^{0}_{-1}\beta + {}^{56}_{26}Fe$

Comment • Remember that predictions made in this manner will be right much of the time, but exceptions will sometimes be found.

Exercise 23.5 Predicting Modes of Radioactive Decay

Write an equation for the probable mode of decay for each of the following unstable isotopes and write an equation for that nuclear reaction.

(a) Silicon-32, $^{32}_{14}Si$

(b) Titanium-45, $^{45}_{22}Ti$

(c) Plutonium-239, $^{239}_{94}Pu$

(d) Potassium-42, $^{42}_{19}K$

Nuclear Binding Energy

An atomic nucleus can contain as many as 83 protons and still be stable. For stability, nuclear binding (attractive) forces must be greater than the electrostatic repulsive forces between the closely packed protons in the nucleus. **Nuclear binding energy,** E_b, is defined as the energy required to separate the nucleus of an atom into protons and neutrons [⊙ CD-ROM, Screen 2.18]. For example, the nuclear binding energy for deuterium is the energy required to convert a mole of deuterium ($^{2}_{1}H$) nuclei into a mole of protons and a mole of neutrons.

$$^{2}_{1}H \longrightarrow {}^{1}_{1}p + {}^{1}_{0}n \qquad E_b = 2.15 \times 10^8 \text{ kJ/mol}$$

The positive sign for E_b indicates that energy is required for this process. A deuterium nucleus is more stable than an isolated proton and an isolated neutron, just as the H_2 molecule is more stable than two isolated H atoms. Recall, however, that the H—H bond energy is only 436 kJ/mol. The energy holding a proton and a neutron together in a deuterium nucleus, 2.15×10^8 kJ/mol, is about 500,000 times larger than the typical covalent bond energies.

To further understand nuclear binding energy, we turn to an experimental observation and a theory. The experimental observation is that the mass of a nucleus is always less than the sum of the masses of its constituent protons and neutrons. The theory is that the "missing mass," called the **mass defect** [⬅ A CLOSER LOOK, PAGE 56] is equated with energy that holds the nuclear particles together.

Chapter Goals • Revisited

• Identify radioactive elements and describe natural and artificial nuclear reactions.

• **Calculate and understand nuclear binding energies.**

• Understand rates of radioactive decay and the stability of nuclei.

• Understand issues of health and safety with respect to radioactivity.

• Be aware of some uses of radioactive isotopes in science and medicine.

The mass defect for deuterium is the difference between the mass of a deuterium nucleus and the sum of the masses of a proton and a neutron. Mass spectrometric measurements [⬅ PAGE 56] give masses of each of these particles to a high level of accuracy, providing the numbers needed to carry out calculations of mass defects.

Masses of atomic nuclei are not generally listed in reference tables, but masses of atoms are. Calculation of the mass defect can be carried out using masses of atoms instead of masses of nuclei. By using atomic masses, we are including in this calculation the masses of extranuclear electrons in the reactant and product. Since the same number of extranuclear electrons appears in products and reactants, this does not affect the result. For one mole of deuterium nuclei, the mass defect is found as follows:

$$\underset{\text{2.01410 g/mol}}{{}^2_1\text{H}} \longrightarrow \underset{\text{1.007825 g/mol}}{{}^1_1\text{H}} + \underset{\text{1.008665 g/mol}}{{}^1_0\text{n}}$$

$$\text{Mass defect} = \Delta m = \text{mass of products} - \text{mass of reactants}$$
$$= [1.007825 \text{ g/mol} + 1.008665 \text{ g/mol}] - 2.01410 \text{ g/mol}$$
$$= 0.00239 \text{ g/mol}$$

The relationship between mass and energy is contained in Albert Einstein's 1905 theory of special relativity, which holds that mass and energy are different manifestations of the same quantity. Einstein defined the energy–mass relationship: Energy is equivalent to mass times the square of the speed of light; that is, $E = mc^2$. In the case of atomic nuclei, it is assumed that the missing mass (that is, the mass defect, Δm) is equated with the binding energy holding the nucleus together.

$$E_b = (\Delta m)c^2 \qquad\qquad \textbf{(23.1)}$$

If Δm is given in kilograms and the speed of light is in meters per second, E_b will have the units joules (because $1 \text{ J} = 1 \text{ kg} \cdot \text{m}^2/\text{s}^2$). For the decomposition of one mole of deuterium nuclei from one mole of protons and one mole of neutrons, we have

$$E_b = (2.39 \times 10^{-6} \text{ kg/mol})(3.00 \times 10^8 \text{ m/s})^2$$
$$= 2.15 \times 10^{11} \text{ J/mol of } {}^2_1\text{H nuclei} (= 2.15 \times 10^8 \text{ kJ/mol of } {}^2_1\text{H nuclei})$$

Nuclear stabilities of different elements are compared using the **binding energy per mole of nucleons.** (**Nucleon** is the general name given to nuclear particles, that is, protons and neutrons.) A deuterium nucleus contains two nucleons, so the binding energy per mole of nucleons, E_b/n, is 2.15×10^8 kJ/mol divided by 2 or 1.08×10^8 kJ/mol nucleon.

$$E_b/n = \left(\frac{2.15 \times 10^8 \text{ kJ}}{\text{mol }{}^2\text{H nuclei}}\right)\left(\frac{1 \text{ mol }{}^2\text{H nuclei}}{2 \text{ mol nucleons}}\right)$$
$$E_b/n = 1.08 \times 10^8 \text{ kJ/mol nucleons}$$

The binding energy per nucleon can be calculated for any atom whose mass is known. Then, to compare nuclear stabilities, binding energies per nucleon are plotted as a function of mass number (Figure 23.4). The greater the binding energy per nucleon, the greater is the stability of the nucleus. From the graph in Figure 23.4, the point of maximum nuclear stability occurs at a mass of 56 (that is, at iron in the periodic table).

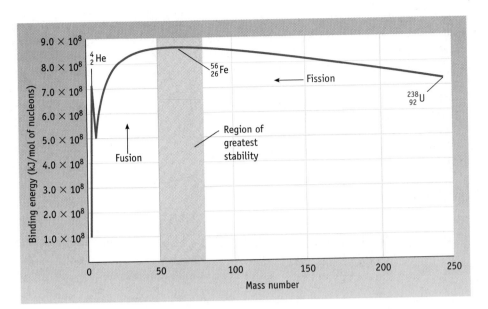

Figure 23.4   Relative stability of nuclei. Binding energy per nucleon for the most stable isotope of elements between hydrogen and uranium is plotted as a function of mass number.

Example 23.4   Nuclear Binding Energy

Problem • Calculate the binding energy, E_b (in kJ/mol), and the binding energy per nucleon, E_b/n (in kJ/mol nucleon), for carbon-12.

Strategy • First determine the mass defect, then use Equation 23.1 to determine the binding energy. There are 12 nuclear particles in carbon-12, so dividing the nuclear binding energy by 12 will give the binding energy per nucleon.

Solution • The mass of ^1_1H is 1.00783 g/mol, and the mass of ^1_0n is 1.00867 g/mol. Carbon-12, $^{12}_6C$, is the standard for the atomic masses in the periodic table, and its mass is defined as exactly 12 g/mol (12.0000 g/mol)

$$\Delta m = [(6 \times \text{mass } ^1_1H) + (6 \times \text{mass } ^1_0n)] - \text{mass } ^{12}_6C$$
$$= [(6 \times 1.00783) + (6 \times 1.00867)] - 12.0000$$
$$= 9.9000 \times 10^{-2} \text{ g/mol nuclei}$$

The binding energy is calculated using Equation 23.1. Using the mass in kilograms and the speed of light in meters per second gives the binding energy in joules:

$$E_b = (\Delta m)c^2$$
$$= (9.9000 \times 10^{-5} \text{ kg/mol})(3.00 \times 10^8 \text{ m/s})^2$$
$$= 8.91 \times 10^{12} \text{ J/mol nuclei} (= 8.91 \times 10^9 \text{ kJ/mol nuclei})$$

The binding energy per nucleon, E_b/n, is determined by dividing the binding energy by 12 (the number of nucleons)

$$\frac{E_b}{n} = \frac{8.91 \times 10^9 \text{ kJ/mol nuclei}}{12 \text{ mol nucleons/mol nuclei}}$$
$$= 7.43 \times 10^8 \text{ kJ/mol nucleons}$$

Exercise 23.6   Nuclear Binding Energy

Calculate the binding energy per nucleon, in kilojoules per mole, for the formation of lithium-6. The molar mass of ^6_3Li is 6.015125 g/mol.

23.4   RATES OF NUCLEAR DECAY

Half-Life

When a radioactive isotope is identified, its *half-life* is usually mentioned. Half-life ($t_{1/2}$) is used in nuclear chemistry in the same way it is used when discussing the

Figure 23.5   Decay of 20.0 mg of oxygen-15. After each half-life period of 2.0 min, the amount of oxygen-15 decreases by one half.

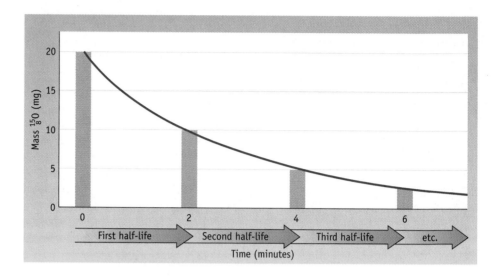

kinetics of first-order chemical reactions [← SECTION 15.4]: It is the time required for half of a sample to decay to products (Figure 23.5). Recall that for first-order kinetics the half-life is independent of the amount of sample [⊕ CD-ROM, Screens 2.19 and 15.8].

Half-lives for radioactive isotopes cover a wide range of values. Uranium-238 has one of the longer half-lives, 4.47×10^9 years, a length of time close to the age of the earth (estimated to be $4.5–4.6 \times 10^9$ years). Roughly half of the uranium-238 present when the planet was formed is still around. At the other end of the range of half-lives are isotopes like element 112, whose 277 isotope has a half-life of 240 microseconds (μs, $1 \ \mu s = 1 \times 10^{-6}$ s).

Half-life provides an easy way to estimate the time required before a radioactive element is no longer a health hazard. Strontium-90, for example, is a β emitter with a half-life of 29.1 years. Significant quantities of strontium-90 were dispersed into the environment in nuclear bomb atmospheric tests in the 1960s and 70s and from the half-life we know that about half is still around. The health problems associated with strontium-90 arise because calcium and strontium have similar chemical properties. Strontium-90 is taken into the body and deposited in bone, taking the place of calcium. Radiation damage by strontium-90 in bone is directly linked to cancer.

Chapter Goals • Revisited

- Identify radioactive elements and describe natural and artificial nuclear reactions.
- Calculate and understand nuclear binding energies.
- **Understand rates of radioactive decay and the stability of nuclei.**
- Understand issues of health and safety with respect to radioactivity.
- Be aware of some uses of radioactive isotopes in science and medicine.

● **Half-life and Temperature**
Unlike what is observed in chemical kinetics, temperature does not affect the rate of nuclear decay.

Example 23.5 Using Half-Life

Problem • Radioactive iodine-131, used to treat hyperthyroidism, has a half-life of 8.04 days.

(a) If you have 8.2 μg (micrograms) of this isotope, what mass remains after 32.2 days?

(b) How long will it take for a sample of iodine-131 to decay to one eighth of its activity?

(c) Estimate the length of time necessary for the sample to decay to 10% of its original activity.

Strategy • This problem asks you to use half-life to qualitatively assess the rate of decay. After one half-life, half of the sample remains. After another half-life, the amount of sample is again decreased by half to one fourth of its original value. (This is illustrated in Figure 23.5.) To answer these questions, we will assess the number of half-lives that have elapsed and use this information to determine the amount of sample remaining.

Solution •

(a) The time elapsed, 32.2 days, is 4 half-lives (32.2/8.04 = 4). The amount of iodine-131 has decreased to 1/16 of the original amount [1/2 × 1/2 × 1/2 × 1/2 = $(1/2)^4$ = 1/16]. The amount of iodine remaining is 8.8 μg × $(1/2)^4$ or 0.55 μg.

(b) After 3 half-lives, the amount of iodine-131 remaining is 1/8 (= $1/2)^3$ of the original amount. The amount remaining is 8.8 μg × $(1/2)^3$ = 1.1 μg.

(c) After 3 half-lives, 1/8 (12.5%) of the sample remains and after 4 half-lives, 1/16 (6.25%) remains. It will take between 3 and 4 half-lives, between 24.15 and 32.2 days, to decrease the amount of sample to 10% of its original value.

Comment • You will find it useful to make approximations as we have done in (c).

Exercise 23.7   Using Half-Life

Tritium ($^3_1$H), a radioactive isotope of hydrogen [◀ *CHEMICAL PERSPECTIVES*, PAGE 54], has a half-life of 12.3 years.

(a) Starting with 1.5 mg of this isotope, how many milligrams remain after 49.2 years?

(b) How long will it take for a sample of tritium to decay to one eighth of its activity?

(c) Estimate the length of time necessary for the sample to decay to 1% of its original activity.

Kinetics of Nuclear Decay

The rate of nuclear decay is determined from measurements of the **activity** (A) of a sample. Activity refers to the number of disintegrations observed per unit time, a quantity that can be measured readily with devices such as a Geiger–Muller counter (Figure 23.6). *Activity is proportional to the number of radioactive atoms present (N),* as shown in Equation 23.2; k in this equation is the proportionality constant.

$$A = kN \qquad \textbf{(23.2)}$$

A (disintegrations/time) = k [disintegrations/(number of atoms · time)] × N (number of atoms)

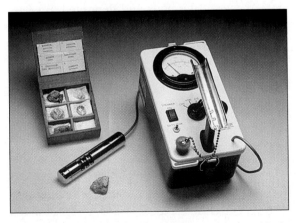
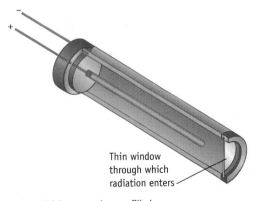

Thin window through which radiation enters

Figure 23.6   A Geiger–Muller counter. A charged particle (an α or β particle) enters the gas-filled tube (diagram at the right) and ionizes the gas. The gaseous ions migrate to electrically charged plates and are recorded as a pulse of electric current. The current is amplified and used to operate a counter. A sample of carnotite, a mineral containing uranium oxide, is also shown in the photograph. *(Charles D. Winters)*

● **Ways of Expressing Activity**
Common units for activity are dps (decompositions per second) and dpm (decompositions per minute).

If the number of radioactive nuclei N is reduced to half, the activity of the sample will be half as large. Doubling N will double the activity. This is evidence that the rate of decomposition is first-order with respect to N. Consequently, the equations describing rates of radioactive decay are the same as those used to describe first-order chemical reactions; the change in the number of radioactive atoms N per unit of time is proportional to N:

$$\frac{\Delta N}{\Delta t} = -kN \tag{23.3}$$

The integrated rate equation can be written in two ways depending on the data used:

$$\ln\left(\frac{N}{N_0}\right) = -kt \tag{23.4}$$

or

$$\ln\left(\frac{A}{A_0}\right) = -kt \tag{23.5}$$

Here, N_0 and A_0 are the number of atoms and the activity of the sample initially, and N and A are the number of atoms and the activity of the sample after time t. Thus, N/N_0 is the fraction of atoms remaining after a given time (t) and A/A_0 is the fraction of the activity remaining after the same period. In these equations k is the rate constant (decay constant) for the isotope in question. The relationship between half-life and the first-order rate constant is the same as seen with chemical kinetics [EQUATION 15.4, PAGE 620]:

$$t_{1/2} = 0.693/k \tag{23.6}$$

Equations 23.3–23.6 are useful in several ways:

- If the activity (A) or the number of radioactive nuclei (N) is measured in the laboratory over some period t, then k can be calculated. The decay constant k can then be used to determine the half-life of the sample.
- If k is known, the fraction of a radioactive sample (N/N_0) still present after some time t has elapsed can be calculated.
- If k is known, the time required for that isotope to decay to a fraction of the original activity (A/A_0) can be calculated.

Example 23.6 Determination of Half-Life

Problem • A sample of radon-222 has an initial α particle activity (A_0) of 7.0×10^4 dps (disintegrations per second). After 6.6 days, its activity (A) is 2.1×10^4 dps. What is the half-life of radon-222?

Strategy • Values for A, A_0, and t are given. The problem can be solved using Equation 23.5 with k as the unknown. Once k is found, the half-life can be calculated using Equation 23.6.

Solution •

$$\ln (2.1 \times 10^4 \text{ dps}/7.0 \times 10^4 \text{ dps}) = -k\,(6.6 \text{ day})$$

$$\ln (0.30) = -k(6.6 \text{ day})$$
$$k = 0.18 \text{ days}^{-1}$$

From k we obtain $t_{1/2}$:

$$t_{1/2} = 0.693/0.18 \text{ days}^{-1} = 3.8 \text{ days}$$

Comments • Notice that the activity decreased to between one half and one fourth of its original value. The 6.6 days elapsed time represents one full half-life and part of another half-life.

Example 23.7   Time Required for a Radioactive Sample to Partially Decay

Problem • Gallium citrate, containing the radioactive isotope gallium-67, is used medically as a tumor-seeking agent. It has a half-life of 78.2 h. How long will it take for a sample of gallium citrate to decay to 10.0% of its original activity?

Strategy • Equation 23.5 is used to solve this problem. In this case, the unknown is the time t. The rate constant k is calculated from the half-life using Equation 23.6. Although we do not have specific values of activity, the value of A/A_0 is known. Because A is 10% of A_0, the value of A/A_0 is 0.10.

Solution •
First determine k: $k = 0.693/t_{1/2} = 0.693/78.2$ h

$$k = 8.86 \times 10^{-3} \text{ h}^{-1}$$

Then substitute the given values of A/A_0 and k into Equation 23.5:

$$\ln (A/A_0) = -kt$$
$$\ln (0.100) = -(8.86 \times 10^{-3} \text{ h}^{-1}) t$$
$$t = 2.60 \times 10^2 \text{ h}$$

Comments • The time required is between three half-lives (3×77.9 h = 234 h) and four half-lives (4×77.9 h = 312 h).

Exercise 23.8   Determination of Half-Life

A sample of $Ca_3(PO_4)_2$ containing phosphorus-32 has an activity of 3.35×10^3 dpm. Two days later, the activity is 3.18×10^3 dpm. Calculate the half-life of phosphorus-32.

Exercise 23.9   Time Required for a Radioactive Sample to Partially Decay

A highly radioactive sample of nuclear waste products with a half-life $t_{1/2}$ of 200. years is stored in an underground tank. How long will it take for the activity to diminish from an initial activity of 6.50×10^{12} disintegrations per minute (dpm) to a fairly harmless activity of 3.00×10^3 dpm?

Radiocarbon Dating

In certain situations, the age of a material can be determined based on the rate of decay of a radioactive isotope. The best-known example of this procedure is the use of carbon-14 to date historical artifacts.

Carbon is primarily carbon-12 and carbon-13 with isotopic abundances of 98.9% and 1.1%, respectively. In addition, traces of a third isotope, carbon-14, are also present to the extent of about 1 in 10^{12} atoms in atmospheric CO_2 and in living materials. Carbon-14 is a β emitter with a half-life of 5730 years. A one-gram sample of carbon from living material will show about 14 decompositions per minute, not a lot of radioactivity but detectable by modern methods.

Carbon-14 is formed in the upper atmosphere by nuclear reactions initiated by neutrons in cosmic radiation:

$$^{14}_{7}\text{N} + ^{1}_{0}\text{n} \longrightarrow ^{14}_{6}\text{C} + ^{1}_{1}\text{H}$$

Once formed, carbon-14 is oxidized to $^{14}\text{CO}_2$. This enters the carbon cycle, circulating through the atmosphere, oceans, and biosphere.

The usefulness of carbon-14 for dating comes about in the following way. Plants absorb CO_2 and convert it to organic compounds, thereby incorporating the carbon-14 into living tissue. As long as a plant is alive, this process will continue and the percent of carbon that is carbon-14 in the plant will equal the percent in

the atmosphere. When the plant dies, however, carbon-14 will no longer be taken up. Radioactive decay continues, however, with the carbon-14 activity decreasing over time. In 5730 years, the activity will be 7 dpm/g, in 11,460 years it will be 3.5 dpm/g, and so on. By measuring the activity of a sample, and knowing the half-life of carbon-14, it is possible to calculate when a plant (or an animal that was eating plants) died.

As with all experimental procedures, carbon-14 dating has limitations. The procedure assumes that the amount of carbon-14 in the atmosphere hundreds or thousands of years ago is the same as it is now. We know that this isn't exactly true; the percentage has varied by as much as 10% (Figure 23.7). Furthermore, it is not possible to use carbon-14 to date an object that is less than about 100 years old; the radiation level from carbon-14 will not change enough in this short time period to accurately detect a difference from the initial value. In most instances, the accuracy of the measurement is, in fact, only about ±100 years. Finally, it is not possible to determine ages of objects much older than about 40,000 years. By then, after nearly 7 half-lives, the radioactivity will have decreased virtually to zero. But for the span of time between 100 and 40,000 years, this technique has provided important information (Figure 23.8).

Example 23.8   Radiochemical Dating

Problem • To test the concept of carbon-14 dating, J. R. Arnold and W. F Libby applied this technique to analyze samples of acacia and Cyprus wood whose ages were already known. (Acacia wood, supplied by the Metropolitan Museum of Art in New York, was from the tomb of Zoser, the first Egyptian Pharaoh to be entombed in a pyramid. The cyprus wood was from the tomb of Sneferu.) The average activity based on five determinations was 7.04 dpm per gram of carbon. Assume (as Arnold and Libby did) that the activity of carbon-14, A_0, is 12.6 dpm per gram of carbon, and calculate the approximate age of the sample.

Strategy • First, determine the rate constant for the decay of carbon-14 from its half-life ($t_{1/2}$ for ^{14}C is 5.73×10^3 years). Then, use Equation 23.5.

Solution •

$$k = 0.693/t_{1/2} = 0.693/5730 \text{ yr}$$
$$= 1.21 \times 10^{-4} \text{ yr}^{-1}$$
$$\ln(A/A_0) = -kt$$
$$\ln\left(\frac{7.04 \text{ dpm/g}}{12.6 \text{ dpm/g}}\right) = (-1.21 \times 10^{-4} \text{ yr}^{-1})\,t$$
$$t = 4.7 \times 10^3 \text{ yr}$$

The wood is about 4700 years old.

Comment • This problem uses real data from an early research paper in which the carbon-14 dating method was being tested. The age of the wood was known to be 4750 ±' 250 years. [See J. R. Arnold and W. F. Libby, *Science*, Vol. 110, page 678 (1949).]

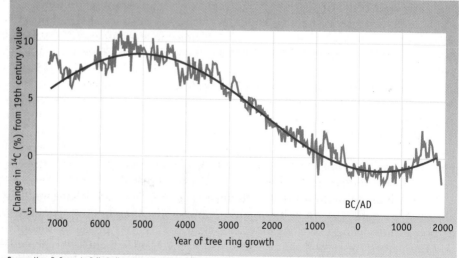

Figure 23.7   Variation of atmospheric carbon-14 activity. The amount of carbon-14 has varied with variation in cosmic ray activity. In modern times, the burning of fossil fuels has also had a significant effect on carbon-14 in the atmosphere. To obtain the data for the pre-1990 part of the curve shown in this graph, scientists carried out carbon-14 dating of artifacts for which the age was accurately known (often through written records). Similar results can be obtained using carbon-14 dating of tree rings.

Source: Hans E. Suess, La Jolla Radiocarbon Laboratory.

A sample of the inner part of a redwood tree felled in 1874 was shown to have ^{14}C activity of 9.32 dpm/g. Calculate the approximate age of the tree when it was felled. Compare this age with that obtained from tree ring data, which estimated that the tree began to grow in 979 ± 52 BC. Use 13.4 dpm/g for the value of A_0.

Figure 23.8 The Ice Man. The world's oldest preserved human remains were discovered in the ice of a glacier high in the Alps. Carbon-14 dating techniques allowed scientists to determine that he lived about 5300 years ago. *(Reuters NewMedia Inc./©Corbis)*

23.5 ARTIFICIAL NUCLEAR REACTIONS

How many different isotopes are found on Earth? All of the stable isotopes occur naturally. A few unstable (radioactive) isotopes that have long half-lives are found in nature in substantial amounts; the best-known examples are uranium 235, uranium-238, and thorium-232. Trace quantities of other radioactive isotopes with short half-lives are present because they are being formed continuously by nuclear reactions. These isotopes include radium, polonium and radon, along with other elements produced in the radioactive decay series, and carbon-14, formed in a nuclear reaction initiated by cosmic radiation.

Naturally occurring isotopes account for only a very small fraction of the radioactive isotopes that are currently known, however. The rest, several thousand, have been synthesized by artificial nuclear reactions.

The first artificial nuclear reaction was identified by Rutherford about 80 years ago. Recall the classic experiment that led to the nuclear model of the atom [◄ FIGURE 2.3] in which gold foil was bombarded with α particles. In the years that followed that experiment, Rutherford and his coworkers bombarded many other elements with α particles. In 1919, one of these experiments led to an unexpected result: When nitrogen atoms were bombarded with α particles, protons were detected among the products. Rutherford correctly concluded that a nuclear reaction had occurred. Nitrogen had undergone a transmutation to oxygen:

$$^{4}_{2}He + {}^{14}_{7}N \longrightarrow {}^{17}_{8}O + {}^{1}_{1}H$$

During the next decade, other nuclear reactions were discovered by bombarding other elements with α particles. Progress was slow, however, because in most cases α particles are simply scattered by target nuclei. The bombarding particles cannot get close enough to the nucleus to react because of strong repulsive forces between the positively charged α particle and the positively charged atomic nucleus.

In 1932, however, two advances were made that greatly extended nuclear reaction chemistry. The first was the use of particle accelerators to create high-energy particles as projectiles. The second was the use of neutrons as the bombarding particles.

Alpha particles used in the early studies on nuclear reactions were from naturally radioactive materials such as uranium and they had relatively low energies, at least by today's standards. Particles with higher energy were needed and J. D. Cockcroft (1897–1967) and E. T. S. Walton (1903–1995), in Rutherford's laboratory in Cambridge, England, turned to protons. Protons are formed when hydrogen atoms ionize in a cathode-ray tube, and it was known that they could be accelerated to higher energy by applying a high voltage. Cockcroft and Walton found that when energetic protons struck a lithium target, the following reaction occurs:

$$^{7}_{3}Li + {}^{1}_{1}p \longrightarrow 2\,{}^{4}_{2}He$$

This was the first example of reaction initiated by a particle that had been artificially accelerated to high energy. Since that time, the technique has been developed

Chapter Goals • Revisited

- **Identify radioactive elements and describe natural and artificial nuclear reactions.**
- Calculate and understand nuclear binding energies.
- Understand rates of radioactive decay and the stability of nuclei.
- Understand issues of health and safety with respect to radioactivity.
- Be aware of some uses of radioactive isotopes in science and medicine.

A Closer Look

The Search for New Elements

By 1936, guided first by Mendeleev's predictions and later by atomic theory, chemists had identified all but two of the elements with atomic numbers between 1 and 92. From this point onward, all new elements to be discovered came from artificial nuclear reactions. Two gaps in the periodic table were filled when radioactive technetium and promethium, the last two elements with atomic numbers less than 92, were identified in 1937 and 1942, respectively. The first success in the search for elements with atomic numbers above 92 was the discovery in 1940 of neptunium and plutonium, elements with atomic numbers higher than uranium.

Since 1950, laboratories in the United States (the Lawrence Berkeley National Laboratory), Russia (the Joint Institute for Nuclear Research at Dubna, near Moscow), and Europe (the Institute for Heavy Ion

Research at Darmstadt, Germany) have competed to make new elements. Syntheses of new transuranium elements use a standard methodology. An element of fairly high atomic number is bombarded by a beam of high-energy particles. Initially, neutrons were used; later, helium nuclei were used, then larger nuclei such as ^{11}B and ^{12}C; and, more recently, highly charged ions of elements like calcium, chromium, cobalt, and zinc were chosen. The bombarding particle fuses with the nucleus of a target atom, forming a new nucleus that lasts for a short time before decomposing. New elements are detected by their decomposition products, a signature of particles with specific energies.

By using bigger particles and higher energies, the list of known elements reached 106 by the end of the 1970s. To further extend the search, Russian scientists employed a new idea, matching precisely

the energy of the bombarding particle with the energy required to fuse the nuclei. This enabled the synthesis of elements 107, 108, and 109 in Darmstadt in the early 1980s, and the synthesis of elements 110, 111, and 112 in the following decade. Lifetimes of these elements were in the millisecond range; the 277 isotope of element 112, for example, had a half-life of 240 μs.

Yet another breakthrough was needed to extend the list further. Scientists have long known that isotopes with specific *magic numbers* of neutrons and protons are more stable. Elements with 2, 8, 20, 50, and 82 protons are in this category, as are elements with 126 neutrons. The magic numbers correspond to filled shells in the nucleus. Their significance is analogous to the significance of filled shells for electronic structure. Theory had predicted that the next magic numbers are 114 protons and 184 neutrons. Using this information, researchers discovered element 114 in early 1999. The Dubna group reporting this discovery found that the 289 isotope had an exceptionally long half-life, about 20 s.

At the time this book was written, 113 elements were known. Will research yield further new elements? It would be hard to say no, given past successes in this area of research, but the going is ever more difficult as scientists now work at the very limits of nuclear stability.

◀ **Fermilab.** The tunnel housing the four-mile-long particle accelerator at Batavia, IL. *(Fermilab Visual Media Services, Batavia, IL)*

much further, and the use of particle accelerators in nuclear chemistry is now commonplace. Particle accelerators operate on the principle that a charged particle placed between charged plates will be accelerated to a high speed and energy. Modern examples of this process are seen in the synthesis of the transuranium elements, several of which are described in more detail in *A Closer Look: The Search for New Elements.*

Experiments using neutrons as bombarding particles were first carried out in both the United States and Great Britain in 1932. Nitrogen, oxygen, fluorine, and neon were bombarded with energetic neutrons, and α particles were detected among the products. Using neutrons made sense; because neutrons have no charge,

● **Discovery of Neutrons**
Neutrons had been predicted to exist for over a decade before they were identified in 1932 by James Chadwick (1891–1974). Chadwick produced neutrons in a nuclear reaction between α particles and beryllium: $^{4}_{2}\alpha + ^{9}_{4}Be \longrightarrow ^{12}_{6}C + ^{1}_{0}n$.

it was reasoned that these particles would not be repelled by the nucleus as were the positively charged particles. Thus, neutrons did not need high energies to react.

In 1934, Enrico Fermi (1901–1954) and his coworkers showed that nuclear reactions using neutrons are more favorable if the neutrons had low energy. A low-energy neutron was simply captured by the nucleus, giving a product in which the mass number was increased by one unit. Because of the low energy of the bombarding particle, the product nucleus does not have sufficient energy to fragment in these reactions. The new nucleus was produced in an excited state, and when the nucleus returned to the ground state, a γ-ray was emitted. Reactions in which a neutron is captured and a γ-ray emitted are called **(n, γ) reactions.**

The (n, γ) reactions are the source of many of the radioisotopes used in medicine and chemistry. An example is radioactive phosphorus, $^{32}_{15}P$, which is used in chemical studies tracing the uptake of phosphorus in the body.

$$^{31}_{15}P + ^{1}_{0}n \longrightarrow ^{32}_{15}P + \gamma$$

Transuranium elements, elements with an atomic number greater than 92, were first made in a nuclear reaction sequence beginning with an n,γ reaction. Scientists at the University of California at Berkeley bombarded uranium-238 with neutrons. Among the products identified were neptunium-239 and plutonium-239. These new elements were formed when ^{239}U decayed by β radiation.

$$^{238}_{92}U + ^{1}_{0}n \longrightarrow ^{239}_{92}U$$
$$^{239}_{92}U \longrightarrow ^{239}_{93}Np + ^{0}_{-1}\beta$$
$$^{239}_{93}Np \longrightarrow ^{239}_{94}Pu + ^{0}_{-1}\beta$$

Four years later a similar reaction sequence was used to make americium-241. Plutonium-239 was found to add two neutrons to form plutonium-241, which decays by β emission to give americium-241.

History

Glenn T. Seaborg (1912–1999)
Figured out that thorium and the elements that followed it fit under the lanthanides in the periodic table. For this insight, he and Edwin McMillan shared the 1951 Nobel prize in chemistry. Over a 21-year period, Seaborg and his scientific colleagues synthesized 10 new transuranium elements (Pu through Lr). In honoring Seaborg's contributions, the name seaborgium was assigned to element 106. This was the first time an element was named for a living person. *(Lawrence Berkeley Laboratory)* •

• **Transuranium Elements in Nature**
Neptunium, plutonium, and americium were unknown prior to their preparation via these nuclear reactions. Later these elements were found to be present in trace quantities in uranium ores.

Example 23.9 **Nuclear Reactions**

Problem • Write equations for the nuclear reactions described below.

(a) Fluorine-19 undergoes an (n, γ) reaction to give a radioactive product that decays by $^{0}_{-1}\beta$ emission. (Write equations for both nuclear reactions.)

(b) A common neutron source is a plutonium–beryllium alloy. Plutonium-239 is an α emitter. When beryllium-9 (the only stable isotope of beryllium) reacts with α particles emitted by plutonium, neutrons are ejected. (Write equations for both reactions.)

Strategy • The equations are written so that both mass and charge are balanced.

Solution •
(a) $^{19}_{9}F + ^{1}_{0}n \longrightarrow ^{20}_{9}F$
$^{20}_{9}F \longrightarrow ^{20}_{10}Ne + ^{0}_{-1}\beta$

(b) $^{239}_{94}Pu \longrightarrow ^{235}_{92}U + ^{4}_{2}\alpha$
$^{4}_{2}\alpha + ^{9}_{4}Be \longrightarrow ^{12}_{6}C + ^{1}_{0}n$

23.6   NUCLEAR FISSION

In 1938, two chemists, Otto Hahn (1879–1968) and Fritz Strassman (1902–1980), isolated and identified barium in a sample of uranium that had been bombarded with neutrons. How was barium formed? The answer to that question explained one of the most significant scientific discoveries of the 20th century. The uranium nucleus had split into smaller pieces in the process we now call **nuclear fission.**

● Fission Reactions

In the fission of uranium-236, a large number of different fission products (different elements) are formed. Barium was the element first identified, and its identification provided the key that led to recognition that fission had occurred.

● The Atomic Bomb

In an atomic bomb, each nuclear fission step produces 3 neutrons, which leads to about 3 more fissions and 9 more neutrons, which leads to 9 fission steps and 27 neutrons, and so on. The rate depends on the number of neutrons, so the nuclear reaction occurs faster and faster as more and more neutrons are formed, leading to an enormous output of energy in a short time span.

The details of nuclear fission were unraveled through the work of a number of scientists. It was determined that a uranium-235 nucleus initially captured a neutron to form uranium-236. This isotope underwent nuclear fission to produce two new nuclei, one with a mass number around 140 and the other with a mass around 90, along with several neutrons (Figure 23.9). The nuclear reactions that led to formation of barium when a sample of ^{235}U was bombarded with neutrons are

$$^{235}_{92}U + ^{1}_{0}n \longrightarrow ^{236}_{92}U$$
$$^{236}_{92}U \longrightarrow ^{141}_{56}Ba + ^{92}_{36}Kr + 3\,^{1}_{0}n$$

An important aspect of fission reactions is that they produce more neutrons than are used to initiate the process. Under the right circumstances, these neutrons then serve to continue the reaction. If one or more of these neutrons is captured by another ^{235}U nucleus, then a further reaction can occur releasing still more neutrons. This sequence repeats over and over. A mechanism like this, in which each step generates a reactant to continue the reaction, is called a **chain reaction.**

A nuclear fission chain reaction has three general steps.

- *Initiation:* The reaction of a single atom is needed to start the chain. Fission of ^{235}U is initiated by the absorption of a neutron.
- *Propagation:* This part of the process repeats itself over and over, each step producing more product. The fission of ^{236}U releases neutrons that initiate the fission of other uranium atoms.
- *Termination:* Eventually the chain will end. This could come about if the reactant (^{235}U) is used up, or if the neutrons that continue the chain escape from the sample without being captured by ^{235}U.

To harness the energy produced in a nuclear reaction, it is necessary to control the rate at which a fission reaction occurs. This is managed by balancing the propagation and termination steps by limiting the number of neutrons available. In a nuclear reactor this balance is accomplished using cadmium rods to absorb neutrons. By withdrawing or inserting the rods, the number of neutrons available to propagate the chain can be varied, and the rate of the fission reaction (and the rate of energy production) can be increased or decreased.

Uranium-235 and plutonium-239 are the common fissionable isotopes used in power reactors. Natural uranium contains only 0.72% of ^{235}U; more than 99% of

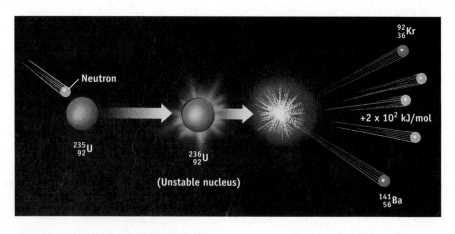

Figure 23.9   Nuclear fission. Neutron capture by $^{235}_{92}$U produces $^{236}_{92}$U. This isotope undergoes fission into several fragments along with several neutrons. These neutrons initiate further nuclear reactions by adding to other $^{235}_{92}$U nuclei. The process is highly exothermic, producing about 2×10^{10} kJ/mol.

the natural element is uranium-238. The percentage of uranium-235 in natural uranium is too small to sustain a chain reaction, however, and uranium used for nuclear fuel must be enriched in this isotope. One way to do this is by gaseous diffusion [← SECTION 12.7]. Plutonium, which occurs naturally in only trace quantities, must be made via a nuclear reaction. The raw material for this nuclear synthesis is the more abundant uranium isotope, ^{238}U. Addition of a neutron to ^{238}U gives ^{239}U, which, as noted above, undergoes two β emissions to form plutonium-239.

Currently there are 103 operating nuclear power plants in the United States and 435 worldwide. About 20% of this nation's electricity (and 17% of the world's energy) comes from nuclear power (Table 23.2). Although one might imagine that nuclear energy would be called upon to meet the ever increasing needs of society, no new nuclear power plants are now under construction in the United States. Among other things, the disasters at Chernobyl (in the former Soviet Union) and Three Mile Island (in Pennsylvania) have sensitized the public to the issue of safety. The cost to construct a nuclear power plant (dollars per kilowatt-hour of power) is considerably more than the cost for a natural gas–powered facility, and the regulatory restrictions for nuclear power are burdensome. Disposal of highly radioactive nuclear waste is a further problem, with 20 metric tons of waste being generated per year at each reactor.

Table 23.2 • Percent of Electricity Produced Using Nuclear Power Plants

Country (rank)	Total power from nuclear energy (%)
1. France	75.0
2. Lithuania	73.1
3. Belgium	57.7
4. Bulgaria	47.1
5. Slovak Republic	47.0
6. Sweden	46.8
. . .	
19. United States	19.9
20. Russia	14.4
21. Canada	12.7

Source: *Chemical and Engineering News*, Oct. 2, 2000, page 42.

23.7   NUCLEAR FUSION

In a **nuclear fusion** reaction, several small nuclei react to form a larger nucleus. Tremendous amounts of energy can be generated by such reactions. An example is the fusion of deuterium and tritium nuclei to form $^4_2$He and a neutron:

$$^3_1H + {}^2_1H \longrightarrow {}^4_2He + {}^1_0n \qquad \Delta E = -1.7 \times 10^9 \text{ kJ/mol}$$

Fusion reactions are the source of the energy of our sun and other stars. It has been the dream of scientists to harness fusion to provide power. To do this, a temperature of 10^6 to 10^7 K, like that in the interior of the sun, is required to bring the positively charged nuclei together with enough energy to overcome nuclear repulsions. At the very high temperatures needed for a fusion reaction, matter does not exist as atoms or molecules; instead, matter is in the form of a *plasma* made up of unbound nuclei and electrons.

Three critical requirements must be met before nuclear fusion is a viable energy source. First, the temperature must be high enough for fusion to occur. The fusion of deuterium and tritium, for example, requires a temperature of 10^8 K or more. Second, the plasma must be confined long enough to release a net output of energy. Third, the energy must be recovered in some usable form.

Harnessing a nuclear fusion reaction for a peaceful use has not yet been achieved. However, many attractive features encourage continuing research in this field. The hydrogen used as "fuel" is cheap and available in almost unlimited amounts. As a further benefit, most radioisotopes produced by fusion have short half-lives and so they are a radiation hazard for only a short time.

23.8   RADIATION HEALTH AND SAFETY

Units for Measuring Radiation

Several units of measurement are used to describe levels and doses of radioactivity. As is the case in everyday life, units used in the United States are not the same as the SI units of measurement.

• **The Roentgen**
The roentgen (R) is an older unit of radiation exposure. It is defined as the amount of x-rays or γ radiation that will produce 2.08×10^9 ions in 1 cm^3 of dry air. The roentgen and the rad are similar in size.

In the United States, the degree of radioactivity is often measured in **curies** (Ci). Less commonly used in the United States is the SI unit, the **becquerel** (Bq). Both units measure the number of decompositions per second; 1 Ci is 3.7×10^{10} dps (decomposition per second), while 1 Bq represents 1 dps. The curie and the becquerel are used to report the amount of radioactivity when multiple kinds of unstable nuclei are decaying and to report amounts necessary for medical purposes.

By itself, the degree of radioactivity does not provide a good measure of the amount of energy in the radiation or the amount of damage that the radiation can cause to living tissue. Two additional kinds of information are necessary. One is the amount of energy absorbed, the second is the effectiveness of the particular kind of radiation in causing tissue damage. The amount of energy absorbed by living tissue is measured in **rads**. *Rad* is an acronym for "radiation absorbed dose." One rad represents 0.01 J of energy absorbed per kilogram of tissue. Its SI equivalent is the **gray** (Gy); 1 Gy denotes the absorption of 1 J per kilogram of tissue.

Different forms of radiation cause different amounts of biological damage. The amount of damage depends on how strongly a form of radiation interacts with matter. Alpha particles cannot penetrate the body any farther than the outer layer of skin, but if α particles are emitted within the body they will do between 10 and 20 times the amount of damage of γ-rays, which can go entirely through a human body without being stopped. In determining the amount of biological damage to living tissue, differences in damaging power are accounted for using a "*quality factor*." This quality factor has been set at 1 for β and γ radiation, 5 for low-energy protons and neutrons, and 20 for α particles or high-energy protons and neutrons.

Biological damage is quantified in a unit called the **rem** (an abbreviation for "roentgen equivalent man"). A dose of radiation in rem is determined by multiplying the energy absorbed in rads by the quality factor for that kind of radiation. The rad and the rem are very large in comparison to normal exposures to radiation, so it is more common to express exposures in millirem (mrem). The SI equivalent of the rem is the **sievert (Sv)**, determined by multiplying the dose in grays by the quality factor.

Radiation: Doses and Effects

Exposure to a small amount of radiation is unavoidable. Earth is constantly being bombarded with radioactive particles from outer space. There is also some exposure to radioactive elements that occur naturally on Earth including ^{14}C, ^{40}K (a radioactive isotope that occurs naturally in 0.0117% abundance), ^{238}U, and ^{232}Th. Radioactive elements in the environment that were created artificially (in the fallout from nuclear bomb tests, for example) contribute further. For some people, medical procedures using radioisotopes are a major contributor.

The average dose of background radioactivity in the United States is about 200 mrem per year (Table 23.3). Well over half comes from natural sources over which we have no control. Of the 60–70 mrem per year exposure that comes from artificial sources, nearly 90% is delivered in medical procedures such as x-ray examinations and radiation therapy. Considering the controversy surrounding nuclear power, it is interesting to know that less than 0.5% of the total annual background dose of radiation that one receives can be attributed to the nuclear power industry.

Describing biological effects of a dose of radiation precisely is not a simple matter. The amount of damage done depends not only on the kind of radiation and the amount of energy absorbed, but also on the particular tissues exposed and the rate at which the dose builds up. A great deal has been learned about the effects

Table 23.3 • **Radiation Exposure of an Individual for One Year from Natural and Artificial Sources**

	Millirem/yr	Percentage
Natural Sources		
Cosmic radiation	50.0	25.8
The earth	47.0	24.2
Building materials	3.0	1.5
Inhaled from the air	5.0	2.6
Elements found naturally in human tissue	21.0	10.8
Subtotal	126.0	64.9
Medical Sources		
Diagnostic x-rays	50.0	25.8
Radiotherapy	10.0	5.2
Internal diagnosis	1.0	0.5
Subtotal	61.0	31.5
Other Artificial Sources		
Nuclear power industry	0.85	0.4
Luminous watch dials, TV tubes	2.0	1.0
Fallout from nuclear tests	4.0	2.1
Subtotal	6.9	3.5
Total	193.9	99.9

of radiation on the human body by studying the survivors of the bombs dropped over Japan in World War II and the workers exposed to radiation from the reactor disaster at Chernobyl. From studies of the health of these survivors, it has been learned that the results of radiation are not generally observable below a single dose of 25 rem. At the other extreme, a single dose of 450 rem is fatal to about half the population (Table 23.4).

Our information is more accurate when dealing with single, large doses than it is for the effects of chronic, smaller doses of radiation. One current issue of debate in the scientific community is how to judge the effects of multiple smaller doses or long-term exposure (see *A Closer Look: What is Safe Exposure?*).

Table 23.4 • **Effects of a Single Dose of Radiation**

Dose (rem)	Effect
0–25	No effect observed
26–50	Small decrease in white blood cell count
51–100	Significant decrease in white blood cell count, lesions
101–200	Loss of hair, nausea
201–500	Hemorrhaging, ulcers, death in 50% of population
>500	Death

What Is a Safe Exposure?

Is the exposure to natural background radiation totally without effect? Can you equate the effect of a single dose and the effect of cumulative doses that are small and spread out over a long period of time? The assumption generally made is that there is no "safe maximum dose", or a level below which absolutely no damage will occur. The accuracy of this assumption has been questioned. These issues are not testable with human subjects, and tests based on animal studies are not completely reliable because of the uncertainty of species-to-species variations.

The model used by government regulators to set exposure limits assumes that the relationship between exposure to radiation and incidence of radiation-induced problems, such as cancer, anemia, and immune system problems, is linear. Under this assumption, if a dose of 2X rem causes damage in 20% of the population, then a dose of X rem will cause damage in 10% of the

population. But is this true? It is known that cells have mechanisms for repairing damage. Many scientists believe that this self-repair mechanism makes the human body less susceptible to damage from smaller doses of radiation, because the damage they cause will be repaired as part of the normal course of events. They argue that, at extremely low doses of radiation, the self-repair response results in less damage.

The bottom line is that there is still much to be learned in this area. And the stakes are significant.

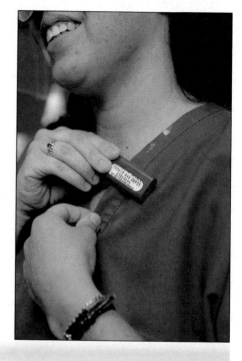

A film badge. These badges, worn by scientists using radioactive materials, are used to monitor cumulative exposure to radiation. *(Cliff Moore/Photo Researchers, Inc.)*

- Identify radioactive elements and describe natural and artificial nuclear reactions.
- Calculate and understand nuclear binding energies.
- Understand rates of radioactive decay and the stability of nuclei.
- Understand issues of health and safety with respect to radioactivity.
- **Be aware of some uses of radioactive isotopes in science and medicine.**

23.9 APPLICATIONS OF NUCLEAR CHEMISTRY

We tend to think about nuclear chemistry in terms of power plants and bombs. The truth is, radioactive elements are now used in all areas of science and medicine, and they are of ever increasing importance to our lives. To describe all of the uses would take several books, so we have selected only a few examples to illustrate the diversity of applications of radioactivity.

Nuclear Medicine: Medical Imaging

Diagnostic procedures using nuclear chemistry are essential in medical imaging, the creation of images of specific parts of the body. There are three principal components to constructing a radioisotope-based image:

- A radioactive isotope of an element, administered on its own or incorporated into a compound, that concentrates the radioactive isotope in the tissue to be imaged
- A method of detecting radiation
- A computer to assemble the information from the detector into a meaningful image

The choice of a radioisotope and how it is administered is determined by the tissue in question. A compound containing the isotope must be absorbed more by the diseased tissue than by the rest of the body. Table 23.5 lists radioisotopes com-

Table 23.5 • Radioisotopes Used in Medical Diagnostic Procedures

Radioisotope	Half-life (h)	Imaging
^{99m}Tc	6.0	Thyroid, brain, kidneys
^{201}Tl	73.0	Heart
^{123}I	13.3	Thyroid
^{67}Ga	78.2	Various tumors and abscesses
^{18}F	1.8	Brain, sites of metabolic activity

monly used in nuclear imaging processes, their half-lives, and the tissues they are used to image. All the isotopes in this table are γ emitters; γ radiation is preferred for imaging because it is less damaging to the body in small doses than is either α or β radiation.

Technetium-99m is used in more than 85% of the diagnostic scans done in hospitals each year (see *A Closer Look: Technetium-99m*). The m stands for *metastable,* a term used to identify an unstable state that exists for a finite period of time. Recall that atoms in excited electronic states emit visible, infrared, and ultraviolet radiation [← CHAPTER 7]. Similarly, a nucleus in an excited state gives up its excess energy, but in this case a much higher energy is involved and the emission occurs as γ radiation. The γ-rays given off by ^{99m}Tc are detected to produce the image (Figure 23.10).

Another medical imaging technique based on nuclear chemistry is positron emission tomography (PET). In PET, an isotope that decays by positron emission is incorporated into a carrier compound and given to the patient. When emitted, the positron travels no more than a few millimeters before undergoing matter–antimatter annihilation.

$$_{+1}^{0}\beta + {}_{-1}^{0}\beta \longrightarrow 2\,\gamma$$

The two emitted γ-rays travel in opposite directions. By determining where high numbers of γ-rays are being emitted, it is possible to construct a map showing where the positron emitter is located in the body.

An isotope often used in PET is ^{15}O. A patient is given gaseous O_2 that contains ^{15}O. This travels throughout the body in the bloodstream. Images of the brain and bloodstream (Figure 23.11) can then be obtained. Because positron emitters are typically very short-lived, PET facilities must be located near a cyclotron where the radioactive nuclei are prepared and then immediately incorporated into a carrier compound.

Nuclear Medicine: Radiation Therapy

To use radiation to treat most cancers, it is necessary to use radiation that can penetrate the body to the location of the tumor. Gamma radiation from a cobalt-60 source is commonly used. Unfortunately, the penetrating ability of γ-rays makes it virtually impossible to destroy diseased tissue without also damaging healthy tissue in the process. Still, this is a regularly sanctioned procedure and its successes are well known.

To avoid the side effects associated with more traditional forms of radiation therapy, a new form of treatment has been explored in the last 10 to 15 years. It is called *boron neutron capture therapy* (BNCT). BNCT is unusual in that boron-10,

Figure 23.10   Bone imaging. If technetium is administered to the body as technetium pyrophosphate. $Tc_2P_2O_7$, the radioactive isotope concentrates in the skeleton. This is a false-color image of a healthy knee joint. The *femur* (thigh bone) is positioned horizontally and connects at the knee joint to the *tibia* (shin bone) (*lower center*). The articular surfaces of these bones meet in cartilage (red), and the joint is united by ligaments (yellow). The image shows the distribution of ^{99m}Tc, which has concentrated in the bone. (*Jean-Perrin/CNRI/Science Photo Library/ Photo Researchers, Inc.*)

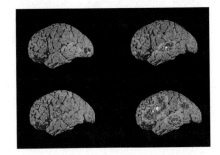

Figure 23.11   PET scans of the brain. These scans show the left side of the brain; red indicates an area of highest activity. At the upper left, *sight* activates the visual area in the occipital cortex at the back of the brain. At the upper right, *hearing* activates the auditory area in the superior temporal cortex of the brain. At the lower left, speaking activates the speech centers in the insula and motor cortex. At the lower right, thinking about verbs, and speaking them, generates high activity, including in the hearing, speaking, temporal, and parietal areas. (*Wellcome Department of Neurology/Science Photo Library/Photo Researchers, Inc.*)

A Closer Look

Technetium-99m

Technetium was the first new element to be made artificially. One might think that this element would be a chemical rarity but this is not so: Its importance in medical imaging has given it a great deal of attention. Although all the technetium in the world must be synthesized by nuclear reactions, the element is readily available and even

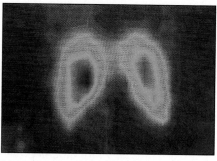

(a) Healthy human thyroid gland.

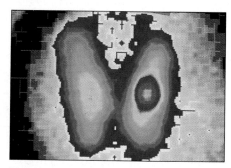

(b) Thyroid gland showing effect of hyperthyroidism.

Thyroid imaging. Technetium-99 concentrates in sites of high activity. Images of this gland, at the base of the neck, were obtained by recording γ-ray emission after the patient was given radioactive technetium-99. Current technology creates a computer color-enhanced scan. *(CNRI/Science Photo Library/Photo Researchers, Inc.)*

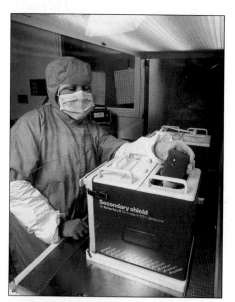

A technetium-99m generator. A technician is loading a sample containing the MoO_4^{2-} ion into the device that will convert molybdate ion to technetium-99–labeled TcO_4^-. *(David Parker/Science Photo Library/Photo Researchers, Inc.)*

inexpensive; its 1999 price was $60 per gram, only about four times the price of gold.

Technetium-99m is formed when molybdenum-99 decays by β emission. Technetium-99m decays to ^{99}Tc, the ground state with a 6.02-h half-life, giving off a 140-KeV γ-ray in the process. (Technetium-99 is itself unstable, decaying to stable ^{99}Ru with a half-life of 2.1×10^5 years.)

Technetium-99m is produced in hospitals using a molybdenum–technetium generator. Sheathed in lead shielding, the generator contains the artificially synthesized isotope ^{99}Mo in the form of the molybdate ion, MoO_4^{2-} adsorbed on a column of alumina (Al_2O_3). In the nuclear reaction MoO_4^{2-} is converted into the pertechnate

ion, $^{99m}TcO_4^-$. The $^{99m}TcO_4^-$ is washed from the column using a saline solution. Technetium-99m may be used as the pertechnate ion or converted into other compounds. The pertechnate ion or radio-pharmaceuticals made from it are administered intravenously to the patient. Such small quantities are needed that 1 μg (microgram) of technetium-99m is sufficient for a hospital's daily imaging needs.

One of the uses of ^{99m}Tc is for imaging the thyroid gland. Because I^-(aq) and TcO_4^-(aq) ions have very similar sizes, the thyroid will (mistakenly) take up TcO_4^-(aq) along with iodide ion. This concentrates ^{99m}Tc in the thyroid and allows a physician to obtain images such as the one shown in the figure.

the particular isotope of boron used as part of the treatment, is not radioactive. BNCT is possible because boron-10 is effective in capturing neutrons, 2500 times better than boron-11, and eight times better than uranium-235. When the nucleus of a boron-10 atom captures a neutron, the resulting boron-11 nucleus has so much energy that it fragments to an α particle and a lithium-7 atom. The α particles do a great deal of damage, but because their penetrating power is so low, the damage is confined to an area not much larger than one or two cells in diameter.

In a typical BNCT treatment, a solution of a boron compound is injected into the tumor. After a few hours, the tumor is bombarded with neutrons. The α particles are produced only at the site of the tumor, and the production stops when the neutron bombardment ends.

Analytical Methods: The Use of Radioactive Isotopes as Tracers

Radioactive isotopes can be used to help determine the fate of compounds in a body or the environment. These studies begin with a compound that contains a radioactive isotope of one of its component elements. In biology, for example, scientists can use radioactive isotopes to measure the uptake of nutrients. Plants take up phosphorus-containing compounds from the soil through their roots. By adding a small amount of radioactive ^{32}P, a β emitter with a half-life of 14.3 days, to fertilizer and then measuring the rate at which the radioactivity appears in the leaves, plant biologists can determine how fast phosphorus is taken up. The outcome can assist scientists studying hybrid strains of plants that can absorb phosphorus quickly, resulting in faster maturing crops, better yields per acre, and more food or fiber at less expense.

To measure pesticide levels, a pesticide can be tagged with a radioisotope and then applied to a test field. By counting the disintegrations of the radioactive tracer, information can be obtained on the amount of pesticide that accumulates in the soil, that is taken up by the plant, and that is carried off in runoff surface water. After these tests are completed, the radioactive isotope decays to harmless levels in a few days or a few weeks because of the short half-lives of the isotopes used.

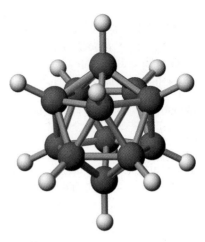

One of the compounds used in BNCT is $Na_2[B_{12}H_{12}]$. The structure of the $B_{12}H_{12}^{2-}$ anion is a regular polyhedron with 20 sides called a icosadecahedron.

Analytical Methods: Isotope Dilution

Imagine, for the moment, that you wanted to estimate the volume of blood in an animal subject. How might you do this? Obviously, draining the blood and measuring it in volumetric glassware is not a desirable option.

One way uses a method called isotope dilution. In this process, a small amount of radioactive isotope is injected into the bloodstream. After a period of time to allow the isotope to distribute throughout the body, a blood sample is taken and its radioactivity measured. The calculation used to determine the total blood volume is illustrated in the next example.

Example 23.10) Analysis Using Isotope Dilution

Problem • A 1.00-mL solution containing 0.240 μCi of tritium is injected into a dog's bloodstream. After a period of time to allow the sample to be dispersed, a 1.00-mL sample of blood was drawn. The radioactivity of this sample was found to be 4.3 $\times$ 10^{-4} μCi/mL. What is the blood volume?

Strategy • For any solution, concentration equals the amount of solute divided by volume of solution. In this problem, we relate the activity of the sample (in Ci) to concentration. The total amount of solute is 0.240 μCi, and the concentration (measured on the small sample of blood) is 4.3 $\times$ 10^{-4} μCi/mL. The unknown is the volume.

Solution • The blood contains a total of 0.240 μCi of the radioactive material. We can represent its concentration as 0.240 μCi/x where x is the total blood volume. After dilution in the bloodstream, 1 mL of blood is found to have an activity of 4.3 $\times$ 10^{-4} μCi/mL.

$$0.240 \ \mu\text{Ci}/x \ \text{mL} = 4.3 \times 10^{-4} \ \mu\text{Ci}/1 \ \text{mL}$$

$$x = 560 \ \text{mL}$$

Exercise 23.11) Analysis by Isotope Dilution

Suppose you hydrolyze a 10.00-g sample of a protein. Add to this a 3.00 mg (milligram) sample of ^{14}C-labeled amino acid threonine with a specific activity of 3000 dpm. After mixing, part of the threonine (60.0 mg) is separated and isolated from the mixture. The activity of the isolated sample is 1200 dpm. How much threonine was present in the original sample?

Table 23.6 • Rare Earth Analysis of Moon Rock Sample 10022 (A fine-grain igneous rock)

Element	Concentration (ppm)
La	26.4
Ce	68
Nd	66
Sm	21.2
Eu	2.04
Gd	25
Tb	4.7
Dy	31.2
Ho	5.5
Er	16
Yb	17.7
Lu	2.55

Source: Haskin, L. A, Helmke, P. A., and Allen, R. O. *Science,* 167, 487, 1970. The concentrations of rare earths were quite similar to values in terrestrial rocks except that the europium concentration is much depleted. The January 30, 1970, issue of *Science* was devoted to analysis of moon rocks.

Space Science: Neutron Activation Analysis and the Moon Rocks

The first manned space mission to the moon brought back a number of samples of soil and rock, a treasure trove for scientists. One of the first goals was to analyze these samples to determine their identity and composition. Most analytical methods require chemical reactions using at least a small amount of material; however, this was not a desirable option, considering that the moon rocks were at the time the most valuable rocks in the world.

A few lucky scientists got a chance to work on this unique project, and one of the analytical tools used was **neutron activation analysis.** This is a nondestructive process in which a sample is irradiated with neutrons. Most isotopes add a neutron to form a new isotope one mass unit higher in an excited nuclear state. When the nucleus decays to its ground state, a γ-ray is emitted. The energy of the γ-ray identifies the element and the number of γ-rays can be counted to determine the amount of the element in the sample. Using neutron activation analysis, it is possible to analyze for a number of elements in a single experiment (Table 23.6).

Neutron activation analysis has many other uses. This analytical procedure produces a kind of fingerprint that can be used to identify a substance. For example, this technique has been applied in determining whether an art work is real or fraudulent. Analysis of pigments in paints on a painting can be carried out without damaging the painting to determine whether the composition resembles modern paints or paints used hundreds of years ago.

Food Science: Food Irradiation

Refrigeration, canning, and chemical additives provide significant protection, but in some parts of the world these procedures are unavailable and stored-food spoilage may claim up to 50% of the food crop. Irradiation with γ-rays from sources such as ^{60}Co and ^{137}Cs is an option for prolonging the shelf life of foods. Relatively low levels of radiation retard the growth of organisms, such as bacteria, molds, and yeasts that cause food spoilage. After irradiation, milk in a sealed container has a minimum shelf life of 3 months without refrigeration. Chicken normally has a three-day refrigerated shelf life; after irradiation, it may have a three-week refrigerated shelf life.

Higher levels of radiation, in the 1- to 5-Mrad (1 Mrad = 1×10^6 rad) range, kill every living organism. Foods irradiated at these levels keep indefinitely when sealed in plastic or aluminum-foil packages. Ham, beef, turkey, and corned beef sterilized by radiation have been used on many Space Shuttle flights. An astronaut said "The beautiful thing was that it didn't disturb the taste, which made the meals much better than the freeze-dried and other types of foods we had."

These procedures are not without their opponents, and the public has not fully accepted irradiation of foods yet. An interesting argument favoring irradiated foods is that radiation is less harmful than other methodologies for food preservation. This type of sterilization offers greater worker safety because it lessens chances of exposure to harmful chemicals, and it protects the environment by avoiding contamination of water supplies with toxic chemicals.

Food irradiation is commonly used in European countries, Canada, and Mexico. Its use in the United States is currently regulated by the the United States Food and Drug Administration (FDA) and the Department of Agriculture (USDA). In 2000, the USDA approved the irradiation of refrigerated and frozen uncooked meat to control pathogens and extend shelf life and the FDA approved the irradiation of eggs to control *Salmonella* infection (Table 23.7).

Table 23.7 • Approvals of Food Irradiation

1963	FDA approves irradiation to control insects in wheat and flour.
1964	FDA approves irradiation to inhibit sprouting of white potatoes.
1985	FDA approves irradiation at specific doses to control *Trichinella spiralis* infection of pork.
1986	FDA approves irradiation at specific doses to delay maturation, inhibit growth, and disinfect foods including vegetables and spices.
1992	USDA approves irradiation of poultry to control infection by *Salmonella* and bacteria.
1997	FDA permits use of ionizing radiation as a source of irradiation to treat refrigerated and frozen meat and other meat products.
2000	USDA approves irradiation of eggs to control *Salmonella* infection.

Source: www.foodsafety.gov/~fsg/irradiat.html

In Summary

When you have finished studying this chapter, you should ask if you have met the chapter goals. In particular, you should be able to

- Identify α, β, and γ radiation, the three major types of radiation from radioactive elements in nature (Section 23.1).
- Write balanced equations for nuclear reactions (Section 23.2).
- Predict whether a radioactive isotope will decay by α or β emission, or by positron emission or electron capture (Sections 23.2).
- Understand the nature and origin of γ radiation (Section 23.2).
- Calculate the binding energy and binding energy per nucleon for a particular isotope (Section 23.3).
- Recognize the significance of a graph of binding energy per nucleon versus mass number and understand how it relates to nuclear stability (Section 23.3).
- Understand and use mathematical equations that characterize rates of radioactive decay (Section 23.4).
- Use the half-life to estimate the time required for an isotope to decay to a particular activity (Section 23.4).
- Describe how artificial nuclear reactions are carried out (Section 23.5).
- Describe nuclear chain reactions, nuclear fission, and nuclear fusion (Sections 23.6 and 23.7).
- Describe the units used to measure intensity and understand how they pertain to health issues (Section 23.8).
- Describe some uses of radioisotopes (Section 23.9).

Key Terms

Section 23.1
α (alpha) radiation
β (beta) radiation
γ (gamma) radiation

Section 23.2
nuclear reaction
radioactive decay series
positron emission
electron capture

Section 23.3
band of stabililty
nuclear binding energy

mass defect
nucleon
binding energy per nucleon

Section 23.4
activity

Section 23.5
(n, γ) reactions
transuranium elements

Section 23.6
nuclear fission
chain reaction

Section 23.7
nuclear fusion

Section 23.8
curie
becquerel
rad
gray
rem
sievert

Section 23.9
neutron activation analysis

Key Equations

Equation 23.1 (page 972)
The equation relating interconversion of mass (m) and energy (E).

$$E = (\Delta m)c^2$$

Equation 23.2 (page 976)
The activity of a radioactive sample (A) is proportional to the number of atoms (N).

$$A = kN$$

Equation 23.4 (page 976)
The rate law for nuclear decay based on number of atoms initially present (N_0) and the number N after time t.

$$\ln(N/N_0) = -kt$$

Equation 23.5 (page 976)
The rate law for nuclear decay based on the measured activity of a sample (A).

$$\ln(A/A_0) = -kt$$

Equation 23.6 (page 976)
The relationship of the half-life and the rate constant for a nuclear decay process.

$$t_{1/2} = 0.693/k$$

Study Questions

Questions with blue, bold-faced numbers have answers in Appendix O.

Reviewing Important Concepts

1. Some important discoveries in scientific history that contributed to the development of nuclear chemistry are listed. Briefly, describe each of these discoveries, identify prominent scientists who contributed to them, and comment on the significance of each to the development of this field.

   (a) 1896, the discovery of radioactivity
   (b) 1898, the identification of radium and polonium
   (c) 1918, the first artificial nuclear reaction
   (d) 1932, (n, γ) reactions
   (e) 1939, fission reactions

2. In Chapter 2, the law of conservation of mass was introduced as an important principle in chemistry. The discovery of nuclear reactions forced scientists to modify this law. Explain why this is so, and give an example illustrating that mass is not conserved in a nuclear reaction.

3. A graph of binding energy per nucleon is shown in the text as Figure 23.4. Explain how data used to construct this graph were obtained.

4. How is Figure 23.3 used to predict the type of decomposition for unstable (radioactive) isotopes?

5. Outline how nuclear reactions are carried out in the laboratory. Describe the artificial nuclear reactions used to make an element with an atomic number greater than 92.

6. What mathematical equations define the rates of decay for radioactive elements?

7. Explain how carbon-14 is used to obtain ages of archeological artifacts. What are the limitations for use of this technique?

8. Describe how the concept of half-life for nuclear decay is used.

9. What is a radioactive decay series? Explain why radium and polonium are found in uranium ores.

10. There are both positive and negative consequences to the interaction of radiation with matter. Discuss briefly the hazards of radiation and the way that radiation can be used in medicine.

11. Define the terms curie, rad, and rem.

Practicing Skills

Nuclear Reactions
(See Examples 23.1 and 23.2)

12. Complete the following nuclear equations. Write the mass number and atomic number for the remaining particle, as well as its symbol.

(a) $^{54}_{26}\text{Fe} + ^{4}_{2}\text{He} \longrightarrow 2\,^{1}_{0}\text{n} + ?$

(b) $^{27}_{13}\text{Al} + ^{4}_{2}\text{He} \longrightarrow ^{30}_{15}\text{P} + ?$

(c) $^{32}_{16}\text{S} + ^{1}_{0}\text{n} \longrightarrow ^{1}_{1}\text{H} + ?$

(d) $^{96}_{42}\text{Mo} + ^{2}_{1}\text{H} \longrightarrow ^{1}_{0}\text{n} + ?$

(e) $^{98}_{42}\text{Mo} + ^{1}_{0}\text{n} \longrightarrow ^{99}_{43}\text{Tc} + ?$

(f) $^{18}_{9}\text{F} \longrightarrow ^{18}_{8}\text{O} + ?$

13. Complete the following nuclear equations. Write the mass number, atomic number, and symbol for the remaining particle.

(a) $^{9}_{4}\text{Be} + ? \longrightarrow ^{6}_{3}\text{Li} + ^{4}_{2}\text{He}$

(b) $? + ^{1}_{0}\text{n} \longrightarrow ^{24}_{11}\text{Na} + ^{4}_{2}\text{He}$

(c) $^{40}_{20}\text{Ca} + ? \longrightarrow ^{40}_{19}\text{K} + ^{1}_{1}\text{H}$

(d) $^{241}_{95}\text{Am} + ^{4}_{2}\text{He} \longrightarrow ^{243}_{97}\text{Bk} + ?$

(e) $^{246}_{96}\text{Cm} + ^{12}_{6}\text{C} \longrightarrow 4\,^{1}_{0}\text{n} + ?$

(f) $^{238}_{92}\text{U} + ? \longrightarrow ^{249}_{100}\text{Fm} + 5\,^{1}_{0}\text{n}$

14. Complete the following nuclear equations. Write the mass number, atomic number, and symbol for the remaining particle.

(a) $^{111}_{47}\text{Ag} \longrightarrow ^{111}_{48}\text{Cd} + ?$

(b) $^{87}_{36}\text{Kr} \longrightarrow ^{0}_{-1}\beta + ?$

(c) $^{231}_{91}\text{Pa} \longrightarrow ^{227}_{89}\text{Ac} + ?$

(d) $^{230}_{90}\text{Th} \longrightarrow ^{4}_{2}\text{He} + ?$

(e) $^{82}_{35}\text{Br} \longrightarrow ^{82}_{36}\text{Kr} + ?$

(f) $? \longrightarrow ^{24}_{12}\text{Mg} + ^{0}_{-1}\beta$

15. Complete the following nuclear equations. Write the mass number, atomic number, and symbol for the remaining particle.

(a) $^{19}_{10}\text{Ne} \longrightarrow ^{0}_{+1}\beta + ?$

(b) $^{59}_{26}\text{Fe} \longrightarrow ^{0}_{-1}\beta + ?$

(c) $^{40}_{19}\text{K} \longrightarrow ^{0}_{-1}\beta + ?$

(d) $^{37}_{18}\text{Ar} + ^{0}_{-1}\text{e}$ (electron capture) $\longrightarrow ?$

(e) $^{55}_{26}\text{Fe} + ^{0}_{-1}\text{e}$ (electron capture) $\longrightarrow ?$

(f) $^{26}_{13}\text{Al} \longrightarrow ^{25}_{12}\text{Mg} + ?$

16. The uranium-235 radioactive decay series, beginning with $^{235}_{92}\text{U}$ and ending with $^{207}_{82}\text{Pb}$, occurs in the following sequence: $\alpha, \beta, \alpha, \beta, \alpha, \alpha, \alpha, \alpha, \beta, \beta, \alpha$. Write an equation for each step in this series.

17. The thorium-232 radioactive decay series, beginning with $^{232}_{90}\text{Th}$ and ending with $^{208}_{82}\text{Pb}$, occurs in the following sequence: $\alpha, \beta, \beta, \alpha, \alpha, \alpha, \alpha, \beta, \beta, \alpha$. Write an equation for each step in this series.

Nuclear Stability and Nuclear Decay
(See Examples 23.3 and 23.4)

18. What particle is emitted in the following nuclear reactions? Write an equation for each reaction.

(a) Gold-198 decays to mercury-198.

(b) Radon-222 decays to polonium-218.

(c) Cesium-137 decays to barium-137.

(d) Indium-110 decays to cadmium-110.

19. What is the product of the following nuclear decay processes? Write an equation for each process.

(a) Gallium-67 decays by electron capture.

(b) Potassium-38 decays with positron emission.

(c) Technetium-99m decays with γ emission.

(d) Manganese-56 decays by β emission.

20. Predict the probable mode of decay for each of the following radioactive isotopes and write an equation to show the products of decay:

(a) Bromine-80m   (c) Cobalt-61

(b) Californium-240   (d) Carbon-11

21. Predict the probable mode of decay for each of the following radioactive isotopes and write an equation to show the products of decay:

(a) Manganese-54   (c) Silver-110

(b) Americium-241   (d) Mercury-197m

22. (a) Which of the following nuclei decay by $^{0}_{-1}\beta$ decay?

$$^{3}\text{H} \quad ^{16}\text{O} \quad ^{20}\text{F} \quad ^{13}\text{N}$$

(b) Which of the following nuclei decays by $^{0}_{+1}\beta$ decay?

$$^{238}\text{U} \quad ^{19}\text{F} \quad ^{22}\text{Na} \quad ^{24}\text{Na}$$

23. (a) Which of the following nuclei decay by $_{-1}^{0}\beta$ decay?

$$^{1}\text{H} \quad ^{23}\text{Mg} \quad ^{32}\text{P} \quad ^{20}\text{Ne}$$

(b) Which of the following nuclei decays by $_{+1}^{0}\beta$ decay?

$$^{235}\text{U} \quad ^{35}\text{Cl} \quad ^{38}\text{K} \quad ^{24}\text{Na}$$

24. Boron has two stable isotopes, ^{10}B and ^{11}B. Calculate the binding energies per nucleon of these two nuclei. The required masses (in grams per mole) are $_{1}^{1}\text{H} = 1.00783$; $_{0}^{1}\text{n} = 1.00867$; $_{5}^{10}\text{B} = 10.01294$; and $_{5}^{11}\text{B} = 11.00931$.

25. Calculate the binding energy in kilojoules per mole of nucleons of P for the formation of ^{30}P and ^{31}P. The required masses (in grams per mole) are $_{1}^{1}\text{H} = 1.00783$; $_{0}^{1}\text{n} = 1.00867$; $_{15}^{30}\text{P} = 29.97832$; and $_{15}^{31}\text{P} = 30.97376$.

26. Calculate the binding energy per nucleon for calcium-40, and compare your result with the value for calcium-40 in Figure 23.4. Masses needed for this calculation are $_{1}^{1}\text{H} = 1.00783$; $_{0}^{1}\text{n} = 1.00867$; and $_{20}^{40}\text{Ca} = 39.96259$.

27. Calculate the binding energy per nucleon for iron-56. Masses needed for this calculation are $_{1}^{1}\text{H} = 1.00783$; $_{0}^{1}\text{n} = 1.00867$; $_{26}^{56}\text{Fe} = 55.9349$. Compare the result of your calculation to the value for iron-56 in the graph in Figure 23.4, page 973.

28. Calculate the binding energy per mole of nucleons for ^{16}O. Masses needed for this calculations are $_{1}^{1}\text{H} = 1.00783$; $_{0}^{1}\text{n} = 1.00867$; $_{8}^{16}\text{O} = 15.99492$.

29. Calculate the binding energy per nucleon for nitrogen-14. The mass of nitrogen-14 is 14.003074.

Rates of Radioactive Decay

(See Examples 23.5 and 23.6)

30. Copper acetate containing ^{64}Cu is used to study brain tumors. This isotope has a half-life of 12.7 h. If you begin with 25.0 μg of ^{64}Cu, what mass in micrograms remains after 64 h?

31. Gold-198 is used as the metal in the diagnosis of liver problems. The half-life of ^{198}Au is 2.69 days. If you begin with 2.8 μg of this gold isotope, what mass remains after 10.8 days?

32. Iodine-131 is used to treat thyroid cancer.

(a) The isotope decays by β particle emission. Write a balanced equation for this process.

(b) Iodine-131 has a half-life of 8.04 days. If you begin with 2.4 μg of radioactive ^{131}I, what mass remains after 40.2 days?

33. Phosphorus-32 is used in the form of Na_2HPO_4 in the treatment of chronic myeloid leukemia, among other things.

(a) The isotope decays by β particle emission. Write a balanced equation for this process.

(b) The half-life of ^{32}P is 14.3 days. If you begin with 4.8 μg of radioactive ^{32}P in the form of Na_2HPO_4, what mass remains after 28.6 days (about one month)?

34. Gallium-67 ($t_{1/2} = 78.25$ h) is used in the medical diagnosis of certain kinds of tumors. If you ingest a compound containing 0.015 mg of this isotope, what mass in milligrams remains in your body after 13 days? (Assume none is excreted.)

35. Iodine-131 ($t_{1/2} = 8.04$ days), a β emitter, is used to treat thyroid cancer.

(a) Write an equation for the decomposition of ^{131}I.

(b) If you ingest a sample of NaI containing ^{131}I, how much time is required for the activity to decrease to 35.0% of its original value?

36. Radon has been the focus of much attention recently because it is often found in homes. Radon-222 emits α particles and has a half-life of 3.82 days.

(a) Write a balanced equation to show this process.

(b) How long does it take for a sample of ^{222}Rn to decrease to 20.0% of its original activity?

37. A sample of wood from a Thracian chariot found in an excavation in Bulgaria has a ^{14}C activity of 11.2 dpm/g. Estimate the age of the chariot and the year it was made. ($t_{1/2}$ for ^{14}C is 5.73×10^3 years, and the activity of ^{14}C in living material is 14.0 dpm/g.)

38. A piece of charred bone found in the ruins of an American Indian village has a ^{14}C-to-^{12}C ratio that is 72% of that found in living organisms. Calculate the age of the bone fragment.

39. Strontium-90 is a hazardous radioactive isotope that resulted from atmospheric nuclear testing. A sample of strontium carbonate containing ^{90}Sr is found to have an activity of 1.0×10^3 dpm. One year later the activity of this sample is 975 dpm.

(a) Calculate the half-life of strontium-90 from this information.

(b) How long will it take for the activity of this sample to drop to 1.0% of the initial value?

40. Radioactive cobalt-60 is used extensively in nuclear medicine as a γ-ray source. It is made by a neutron capture reaction from cobalt-59, and it is a β emitter; β emission is accompanied by strong γ radiation. The half-life of cobalt-60 is 5.27 years.

(a) How long will it take for a cobalt-60 source to decrease to one eighth of its original activity?

(b) What fraction of the activity of a cobalt-60 source remains after 1.0 year?

41. Scandium occurs in nature as a single isotope, scandium-45. Neutron irradiation produces scandium-46, a β emitter, with a half-life of 83.8 days. Draw a graph showing disintegrations per minute as a function of time during a period of one year.

42. Phosphorus occurs in nature as a single isotope, phosphorus-31. Neutron irradiation of phosphorus-31 produces phosphorus-32, a β emitter, with a half-life of 14.28 days. Assume you have a sample containing phosphorus-32 that has a rate of decay of 3.2×10^6 dpm. Draw a graph showing disintegrations per minute as a function of time during a period of one year.

43. Sodium-23 (in a sample of NaCl) is subjected to neutron bombardment in a nuclear reactor to produce ^{24}Na. When removed from the reactor, the sample is radioactive with β

activity of 2.54×10^4 dpm. The decrease in radioactivity over time was studied, producing the following data:

Activity (dpm)	Time (h)
2.54×10^4	0
2.42×10^4	1
2.31×10^4	2
2.00×10^4	5
1.60×10^4	10
1.01×10^4	20

(a) Write equations for the neutron capture reaction and for the reaction in which the product of this reaction decays by β emission.

(b) Determine the half-life of sodium-24.

44. The isotope of polonium most likely isolated by Madame Curie in her pioneering studies is polonium-210. A sample of this element was prepared in a nuclear reaction. Initially its activity (α emission) was 7840 dpm. Measuring radioactivity over time produced the following data:

Activity (dpm)	Time (days)
7840	0
7570	7
7300	14
5920	56
5470	72

Determine the half-life of polonium-210.

Nuclear Reactions

(See Example 23.9)

45. There are two isotopes of americium, both with half-lives sufficiently long to allow the handling of large quantities. Americium-241, with a half-life of 432 years, is an α emitter. It is used in smoke detectors. The isotope is formed from ^{239}Pu by absorption of two neutrons followed by emission of a β particle. Write a balanced equation for this process.

46. Americium-240 is made by bombarding plutonium-239 with α particles. In addition to ^{240}Am, the products are a proton and two neutrons. Write a balanced equation for this process.

47. To synthesize the heavier transuranium elements, a nucleus must be bombarded with a relatively large particle. If you know the products are californium-246 and four neutrons, with what particle would you bombard uranium-238 atoms?

48. Element $^{287}114$ was made by firing a beam of ^{48}Ca ions at ^{242}Pu. Three neutrons were ejected in the reaction. Write a balanced nuclear equation for the synthesis of this superheavy element.

49. Element $^{287}114$ decayed by α emission with a half-life of about 5 s. Write an equation for this process.

50. Deuterium nuclei ($^2_1$H) are particularly effective as bombarding particles to carry out nuclear reactions. Complete the following equations:

(a) $^{114}_{48}\text{Cd} + ^2_1\text{H} \longrightarrow ? + ^1_1\text{H}$

(b) $^6_3\text{Li} + ^2_1\text{H} \longrightarrow ? + ^1_0\text{n}$

(c) $^{40}_{20}\text{Ca} + ^2_1\text{H} \longrightarrow ^{38}_{19}\text{K} + ?$

(d) $? + ^2_1\text{H} \longrightarrow ^{65}_{30}\text{Zn} + \gamma$

51. Some of the reactions explored by Rutherford and others are listed below. Identify the unknown species in each reaction:

(a) $^{14}_7\text{N} + ^4_2\text{He} \longrightarrow ^{17}_8\text{O} + ?$

(b) $^9_4\text{Be} + ^4_2\text{He} \longrightarrow ? + ^1_0\text{n}$

(c) $? + ^4_2\text{He} \longrightarrow ^{30}_{15}\text{P} + ^1_0\text{n}$

(d) $^{239}_{94}\text{Pu} + ^4_2\text{He} \longrightarrow ? + ^1_0\text{n}$

52. Boron is an effective absorber of neutrons. When boron-10 adds a neutron, an α particle is emitted. Write an equation for this nuclear reaction.

53. Tritium, $^3_1$H, is one of the nuclei used in fusion reactions. This isotope is radioactive with a half-life of 12.3 years. Like carbon-14, tritium is formed in the upper atmosphere from cosmic radiation, and it is found in trace amounts on Earth. In amounts required for a fusion reaction, however, it must be made via a nuclear reaction. The reaction of $^6_3$Li with a neutron produces tritium and an α particle. Write an equation for this nuclear reaction.

General Questions

More challenging questions are marked with an underlined number.

54. A technique to date geological samples uses rubidium-87, a long-lived radioactive isotope of rubidium ($t_{1/2} = 4.8 \times 10^{10}$ years). Rubidium-87 decays by β emission to strontium-87. If the rubidium-87 is part of a rock or mineral, then strontium-87 will remain trapped within the crystalline structure of the rock. The age of the rock dates back to the time when the rock solidified. Chemical analysis of the rock gives the amounts of ^{87}Rb and ^{87}Sr, and from this the fraction of ^{87}Rb that remains can be calculated.

Analysis of a stony meteorite determined that 1.8 mmol of ^{87}Rb and 1.6 mmol of ^{87}Sr were present. Estimate the age of the meteorite. (*Hint:* The amount of ^{87}Rb at t_0 is moles ^{87}Rb + moles ^{87}Sr.)

55. The oldest known fossil found in South Africa has been dated based on the decay of Rb-87.

$$^{87}\text{Rb} \longrightarrow ^{87}\text{Sr} + ^{\ 0}_{-1}\beta \qquad t_{1/2} = 4.8 \times 10^{10} \text{ years}$$

If the ratio of the present quantity of ^{87}Rb to the original quantity is 0.951, calculate the age of the fossil.

56. The age of minerals can sometimes be determined by measuring the amounts of ^{206}Pb and ^{238}U in a sample. This determination assumes that all of the ^{206}Pb in the sample comes from the decay of ^{238}U. The date obtained relates to the time when the rock solidified. Assume that the ratio of ^{206}Pb to

^{238}U in an igneous rock sample is 0.33. Calculate the age of the rock. ($t_{1/2}$ for ^{238}U is 4.5×10^9 years.)

57. In June, 1972, natural fission reactors, which operated billions of years ago, were discovered in Oklo, Gabon. At present, natural uranium contains 0.72% ^{235}U. How many years ago did natural uranium contain 3.0% ^{235}U, the amount needed to sustain a natural reactor? ($t_{1/2}$ for ^{235}U is 7.04×10^8 years.)

58. If there were a shortage in worldwide supplies of fissionable uranium, it is possible to use other fissionable nuclei. Plutonium, one such fuel, can be made in "breeder" reactors that manufacture more fuel than they consume. The sequence of reactions by which plutonium is made is

    (a) A ^{238}U nucleus undergoes an (n, γ) to produce ^{239}U.

    (b) ^{239}U decays by β emission ($t_{1/2} = 23.5$ min) to give an isotope of neptunium.

    (c) This neptunium isotope decays by β emission to give a plutonium isotope.

    (d) The plutonium isotope is fissionable. On collision of one of these plutonium isotopes with a neutron, fission occurs, with at least two neutrons and two other nuclei as products.

    Write an equation for each of the nuclear reactions.

59. When a neutron is captured by an atomic nucleus, energy is released as γ radiation. This energy can be calculated based on the change in mass in converting reactants to products. For the nuclear reaction $^6_3\text{Li} + ^1_0\text{n} \longrightarrow ^7_3\text{Li} + \gamma$

    (a) Calculate the energy evolved in this reaction (per atom). Masses needed for this calculation are $^6_3\text{Li} = 6.01512$, $^1_0\text{n} = 1.00867$, and $^7_3\text{Li} = 7.01600$.

    (b) Use the answer in part (a) to calculate the wavelength of the γ-rays emitted in the reaction.

60. The average energy output of a good grade of coal is 2.6×10^7 kJ/ton. Fission of 1 mol of ^{235}U releases 2.1×10^{10} kJ. Find the number of tons of coal needed to produce the same energy as 1 lb of ^{235}U. (See Appendix C for conversion factors.)

61. Collision of an electron and a positron results in formation of two γ-rays. In the process, their masses are converted completely into energy.

    (a) Calculate the energy evolved from the annihilation of an electron and a positron, in kJ.

    (b) Using Planck's equation (Equation 7.2) determine the frequency of the γ-rays emitted in this process.

62. To measure the volume of the blood system of an animal, the following experiment was done. A 1.0-mL sample of an aqueous solution containing tritium with an activity of 2.0×10^6 dps was injected into the bloodstream. After time was allowed for complete circulatory mixing, a 1.0-mL blood sample was withdrawn and found to have an activity of 1.5×10^4 dps. What was the volume of the circulatory system? (The half-life of tritium is 12.3 years, so this experiment assumes that only a negligible amount of tritium has decayed in the time of the experiment.)

63. Suppose that you hydrolyze 10.00 g of a protein to form a mixture of different amino acids. To this is added a 3.00-mg sample of ^{14}C-labeled threonine (one of the amino acids present). The activity of this small sample is 3000 dpm. A chromatographic separation of the amino acids is carried out, and a small sample of pure threonine is separated. This sample has an activity of 1200 dpm. What fraction of the threonine present was separated? What is the total amount of threonine in the sample?

64. The principle in the isotope dilution method can be applied to many kinds of problems. Let us imagine that you, a marine biologist, want to estimate the number of fish in a lake. You release 1000 tagged fish, and after allowing an adequate amount of time for the fish to disperse evenly in the lake, you catch 5250 fish and find that 27 of them have tags. How many fish are in the lake?

65. Radioactive isotopes are often used as "tracers" to follow an atom through a chemical reaction, and the following is an example. Acetic acid reacts with methanol, CH_3OH, by eliminating a molecule of H_2O to form methyl acetate, $CH_3CO_2CH_3$. Explain how you would use the radioactive isotope ^{15}O to show whether the oxygen atom in the water product comes from the —OH of the acid or the —OH of the alcohol.

66. Radioactive decay series begin with a very long-lived isotope. For example, the half-life of ^{238}U is 4.5×10^9 years. Each series is identified by the name of the long-lived parent isotope of highest mass.

    (a) The uranium-238 radioactive decay series is sometimes referred to as the $4n + 2$ series because masses of all 13 members of this series can be expressed by the equation $m = 4n + 2$, where m is the mass number and n is an integer. Explain why the masses correlate in this way.

    (b) Two other radioactive decay series identified in minerals in Earth's crust are the thorium-232 series and the uranium-235 series. Do the masses of the isotopes in these series conform to a simple mathematical equation? If so, identify the equation.

    (c) Identify the radioactive decay series to which each of the following isotopes belong: $^{226}_{88}\text{Ra}$, $^{215}_{86}\text{At}$, $^{228}_{90}\text{Th}$, $^{210}_{83}\text{Bi}$.

    (d) Evaluation reveals one series of elements, the $4n + 1$ series, is not present in Earth's crust. Speculate why this is so.

67. You might wonder how it is possible to determine the half-life of long-lived radioactive isotopes like ^{238}U. With a half-life of more than 10^9 years, the radioactivity of a sample of uranium will not measurably change in your lifetime. The answer is that you can calculate the half-life using the mathematics governing first-order reactions.

    It can be shown that a 1.0-mg sample of ^{238}U decays at the rate of 12 α emissions per second. Set up a mathematical equation for the rate of decay, $\Delta N/\Delta t = -kN$, where N is the number of nuclei in the 1.0-mg sample and $\Delta N/\Delta t$ is 12 dps. Next, solve this equation for the rate constant for this process and then relate the rate constant to the half-life of the reaction. Carry out this calculation, and compare your result with the literature value, 4.5×10^9 years.

68. The last unknown element between bismuth and uranium was discovered by Lise Meitner (1878–1968) and Otto Hahn (1879–1968) in 1918. They obtained ^{231}Pa by chemical extraction of pitchblende, in which its concentration is about 1 ppm. This isotope, an α emitter, has a half-life of 3.27×10^4 years.

    (a) What radioactive decay series (the uranium-235, uranium-238, or thorium-232 series) contains ^{231}Pa as a member?

    (b) Suggest a possible sequence of nuclear reactions starting with the long-lived isotope that forms this isotope?

    (c) What quantity of ore would be required to isolate 1.0 g of ^{231}Pa, assuming 100% yield?

    (d) Write an equation for the radioactive decay process for ^{231}Pa.

List of Appendices

Some Mathematical Operations

The mathematical skills required in this introductory course are basic skills in algebra and a knowledge of (1) exponential (or scientific) notation, (2) logarithms, and (3) quadratic equations. This appendix reviews each of the final three topics.

A.1 ELECTRONIC CALCULATORS

● The directions for calculator use in this section are given for calculators using "algebraic" logic. Such calculators are the most common type used by students in introductory courses. The procedures differ slightly for calculators using RPN logic (such as those made by Hewlett-Packard).

The advent of inexpensive electronic calculators a few years ago has made calculations in introductory chemistry much more straightforward. You are well advised to purchase a calculator that has the capability of performing calculations in scientific notation, has both base-10 and natural logarithms, and is capable of raising any number to any power and of finding any root of any number. In the following discussion, we point out how these functions of your calculator can be used.

Although electronic calculators have greatly simplified calculations, they have also forced us to focus again on significant figures. A calculator easily handles eight or more significant figures, but real laboratory data are never known to this accuracy. You are therefore urged to review Section 1.7 on handling numbers.

A.2 EXPONENTIAL (SCIENTIFIC) NOTATION

In exponential, or scientific, notation, a number is expressed as a product of two numbers: $N \times 10^n$. The first number, N, is the so-called *digit term* and is a number between 1 and 10. The second number, 10^n, the *exponential term*, is some integer power of 10. For example, 1234 is written in scientific notation as 1.234×10^3, or 1.234 multiplied by 10 three times:

$$1234 = 1.234 \times 10^1 \times 10^1 \times 10^1 = 1.234 \times 10^3$$

Conversely, a number less than 1, such as 0.01234, is written as 1.234×10^{-2}. This notation tells us that 1.234 should be divided twice by 10 to obtain 0.01234:

$$0.01234 = \frac{1.234}{10^1 \times 10^1} = 1.234 \times 10^{-1} \times 10^{-1} = 1.234 \times 10^{-2}$$

Some other examples of scientific notation are

$$10000 = 1 \times 10^4 \qquad\qquad 12345 = 1.2345 \times 10^4$$
$$1000 = 1 \times 10^3 \qquad\qquad 1234 = 1.234 \times 10^3$$
$$100 = 1 \times 10^2 \qquad\qquad 123 = 1.23 \times 10^2$$
$$10 = 1 \times 10^1 \qquad\qquad 12 = 1.2 \times 10^1$$
$$1 = 1 \times 10^0 \qquad (\text{any number to the zero power} = 1)$$
$$1/10 = 1 \times 10^{-1} \qquad\qquad 0.12 = 1.2 \times 10^{-1}$$
$$1/100 = 1 \times 10^{-2} \qquad\qquad 0.012 = 1.2 \times 10^{-2}$$
$$1/1000 = 1 \times 10^{-3} \qquad\qquad 0.0012 = 1.2 \times 10^{-3}$$
$$1/10000 = 1 \times 10^{-4} \qquad\qquad 0.00012 = 1.2 \times 10^{-4}$$

When converting a number to scientific notation, notice that the exponent n is positive if the number is greater than 1 and negative if the number is less than 1. The value of n is the number of places by which the decimal is shifted to obtain the number in scientific notation:

$$1\;\;2\;\;3\;\;4\;\;5. = 1.2345 \times 10^4$$

Decimal shifted 4 places to the left. Therefore, n is positive and equal to 4.

$$0.0\;\;0\;\;1\;\;2 = 1.2 \times 10^{-3}$$

Decimal shifted 3 places to the right. Therefore, n is negative and equal to 3.

If you wish to convert a number in scientific notation to the usual form, the procedure is simply reversed:

$$6\;.\;2\;\;7\;\;3 \times 10^2 = 627.3$$

Decimal point moved 2 places to the right, since n is positive and equal to 2.

$$0\;\;0\;\;6.273 \times 10^{-3} = 0.006273$$

Decimal point shifted 3 places to the left, since n is negative and equal to 3.

Two final points must be made concerning scientific notation. First, if you are used to working on a computer, you may be in the habit of writing a number such as 1.23×10^3 as 1.23E3 or 6.45×10^{-5} as 6.45E-5. Second, some electronic calculators allow you to convert numbers readily to the scientific notation. If you have such a calculator, you can change a number shown in the usual form to scientific notation simply by pressing the EE or EXP key and then the "=" key.

1. Adding and Subtracting Numbers

When adding or subtracting two numbers, first convert them to the same powers of 10. The digit terms are then added or subtracted as appropriate:

$$(1.234 \times 10^{-3}) + (5.623 \times 10^{-2}) = (0.1234 \times 10^{-2}) + (5.623 \times 10^{-2})$$
$$= 5.746 \times 10^{-2}$$

$$(6.52 \times 10^2) - (1.56 \times 10^3) = (6.52 \times 10^2) - (15.6 \times 10^2)$$
$$= -9.1 \times 10^2$$

2. Multiplication

The digit terms are multiplied in the usual manner, and the exponents are added algebraically. The result is expressed with a digit term with only one nonzero digit to the left of the decimal:

$$(1.23 \times 10^3)(7.60 \times 10^2) = (1.23)(7.60) \times 10^{3+2}$$
$$= 9.35 \times 10^5$$

$$(6.02 \times 10^{23})(2.32 \times 10^{-2}) = (6.02)(2.32) \times 10^{23-2}$$
$$= 13.966 \times 10^{21}$$
$$= 1.40 \times 10^{22} \text{ (answer in three significant figures)}$$

3. Division

The digit terms are divided in the usual manner, and the exponents are subtracted algebraically. The quotient is written with one nonzero digit to the left of the decimal in the digit term:

$$\frac{7.60 \times 10^3}{1.23 \times 10^2} = \frac{7.60}{1.23} \times 10^{3-2} = 6.18 \times 10^1$$

$$\frac{6.02 \times 10^{23}}{9.10 \times 10^{-2}} = \frac{6.02}{9.10} \times 10^{(23)-(-2)} = 0.662 \times 10^{25} = 6.62 \times 10^{24}$$

4. Powers of Exponentials

When raising a number in exponential notation to a power, treat the digit term in the usual manner. The exponent is then multiplied by the number indicating the power:

$$(1.25 \times 10^3)^2 = (1.25)^2 \times 10^{3 \times 2}$$
$$= 1.5625 \times 10^6 = 1.56 \times 10^6$$

$$(5.6 \times 10^{-10})^3 = (5.6)^3 \times 10^{(-10) \times 3}$$
$$= 175.6 \times 10^{-30} = 1.8 \times 10^{-28}$$

Electronic calculators usually have two methods of raising a number to a power. To square a number, enter the number and then press the "x^2" key. To raise a number to any power, use the "y^x" key. For example, to raise 1.42×10^2 to the fourth power,

1. Enter 1.42×10^2.
2. Press "y^x".
3. Enter 4 (this should appear on the display).
4. Press "=" and 4.0659×10^8 appears on the display.

As a final step, express the number in the correct number of significant figures (4.07×10^8) in this case.

5. Roots of Exponentials

Unless you use an electronic calculator, the number must first be put into a form in which the exponential is exactly divisible by the root. The root of the digit term is found in the usual way, and the exponent is divided by the desired root:

$$\sqrt{3.6 \times 10^7} = \sqrt{36 \times 10^6} = \sqrt{36} \times \sqrt{10^6} = 6.0 \times 10^3$$

$$\sqrt[3]{2.1 \times 10^{-7}} = \sqrt[3]{210 \times 10^{-9}} = \sqrt[3]{210} \times \sqrt[3]{10^{-9}} = 5.9 \times 10^{-3}$$

To take a square root on an electronic calculator, enter the number and then press the "$\sqrt{x}$" key. To find a higher root of a number, such as the fourth root of 5.6×10^{-10},

1. Enter the number.
2. Press the "$\sqrt[x]{y}$" key. (On most calculators, the sequence you actually use is to press "2ndF" and then "$\sqrt[x]{y}$." Alternatively, you press "INV" and then "y^x.")
3. Enter the desired root, 4 in this case.
4. Press " = ". The answer here is 4.8646×10^{-3}, or 4.9×10^{-3}.

A general procedure for finding any root is to use the "y^x" key. For a square root, x is 0.5 (or $\frac{1}{2}$), whereas it is 0.33 (or $\frac{1}{3}$) for a cube root, 0.25 (or $\frac{1}{4}$) for a fourth root, and so on.

A.3 LOGARITHMS

Two types of logarithms are used in this text: (1) common logarithms (abbreviated log) whose base is 10 and (2) natural logarithms (abbreviated ln) whose base is e (= 2.71828):

$$\log x = n, \text{ where } x = 10^n$$
$$\ln x = m, \text{ where } x = e^m$$

Most equations in chemistry and physics were developed in natural, or base e, logarithms, and we follow this practice in this text. The relation between log and ln is

$$\ln x = 2.303 \log x$$

Despite the different bases of the two logarithms, they are used in the same manner. What follows is largely a description of the use of common logarithms.

A common logarithm is the power to which you must raise 10 to obtain the number. For example, the log of 100 is 2, since you must raise 10 to the second power to obtain 100. Other examples are

$$\log 1000 = \log (10^3) = 3$$
$$\log 10 = \log (10^1) = 1$$
$$\log 1 = \log (10^0) = 0$$
$$\log 0.1 = \log (10^{-1}) = -1$$
$$\log 0.0001 = \log (10^{-4}) = -4$$

To obtain the common logarithm of a number other than a simple power of 10, you must resort to a log table or an electronic calculator. For example,

$$\log 2.10 = 0.3222, \text{ which means that } 10^{0.3222} = 2.10$$
$$\log 5.16 = 0.7126, \text{ which means that } 10^{0.7126} = 5.16$$
$$\log 3.125 = 0.49485, \text{ which means that } 10^{0.49485} = 3.125$$

To check this on your calculator, enter the number, and then press the "log" key. When using a log table, the logs of the first two numbers can be read directly from the table. The log of the third number (3.125), however, must be interpolated. That is, 3.125 is midway between 3.12 and 3.13, so the log is midway between 0.4942 and 0.4955.

To obtain the natural logarithm ln of the numbers shown here, use a calculator having this function. Enter each number and press "ln:"

$$\ln 2.10 = 0.7419, \text{ which means that } e^{0.7419} = 2.10$$
$$\ln 5.16 = 1.6409, \text{ which means that } e^{1.6409} = 5.16$$

To find the common logarithm of a number greater than 10 or less than 1 with a log table, first express the number in scientific notation. Then find the log of each part of the number and add the logs. For example,

$$\log 241 = \log (2.41 \times 10^2) = \log 2.41 + \log 10^2$$
$$= 0.382 + 2 = 2.382$$

$$\log 0.00573 = \log (5.73 \times 10^{-3}) = \log 5.73 + \log 10^{-3}$$
$$= 0.758 + (-3) = -2.242$$

Significant Figures and Logarithms

Notice that the mantissa has as many significant figures as the number whose log was found. (So that you could more clearly see the result obtained with a calculator or a table, this rule was not strictly followed until the last two examples.)

● **Logarithms and Nomenclature** The number to the left of the decimal in a logarithm is called the **characteristic,** and the number to the right of the decimal is the **mantissa.**

Obtaining Antilogarithms

If you are given the logarithm of a number, and find the number from it, you have obtained the "antilogarithm," or "antilog," of the number. Two common procedures used by electronic calculators to do this are:

Procedure A	Procedure B
1. Enter the log or ln.	1. Enter the log or ln.
2. Press 2ndF.	2. Press INV.
3. Press 10^x or e^x.	3. Press log or ln x.

Test one or the other of these procedures with the following examples:

1. Find the number whose log is 5.234:

   Recall that $\log x = n$, where $x = 10^n$. In this case $n = 5.234$. Enter that number in your calculator, and find the value of 10^n, the antilog. In this case,

   $$10^{5.234} = 10^{0.234} \times 10^5 = 1.71 \times 10^5$$

   Notice that the characteristic (5) sets the decimal point; it is the power of 10 in the exponential form. The mantissa (0.234) gives the value of the number x. Thus, if you use a log table to find x, you need only look up 0.234 in the table and see that it corresponds to 1.71.

2. Find the number whose log is -3.456:

   $$10^{-3.456} = 10^{0.544} \times 10^{-4} = 3.50 \times 10^{-4}$$

   Notice here that -3.456 must be expressed as the sum of -4 and $+0.544$.

Mathematical Operations Using Logarithms

Because logarithms are exponents, operations involving them follow the same rules used for exponents. Thus, multiplying two numbers can be done by adding logarithms:

$$\log xy = \log x + \log y$$

For example, we multiply 563 by 125 by adding their logarithms and finding the antilogarithm of the result:

$$\log 563 = 2.751$$
$$\log 125 = \underline{2.097}$$
$$\log xy = 4.848$$
$$xy = 10^{4.848} = 10^4 \times 10^{0.848} = 7.05 \times 10^4$$

One number (x) can be divided by another (y) by subtraction of their logarithms:

$$\log \frac{x}{y} = \log x - \log y$$

For example, to divide 125 by 742,

$$\log 125 = 2.097$$
$$-\log 742 = \underline{2.870}$$
$$\log \frac{x}{y} = -0.773$$

$$\frac{x}{y} = 10^{-0.773} = 10^{0.227} \times 10^{-1} = 1.68 \times 10^{-1}$$

Similarly, powers and roots of numbers can be found using logarithms.

$$\log x^y = y(\log x)$$
$$\log \sqrt[y]{x} = \log x^{1/y} = \frac{1}{y} \log x$$

As an example, find the fourth power of 5.23. We first find the log of 5.23 and then multiply it by 4. The result, 2.874, is the log of the answer. Therefore, we find the antilog of 2.874:

$$(5.23)^4 = ?$$
$$\log (5.23)^4 = 4 \log 5.23 = 4(0.719) = 2.874$$
$$(5.23)^4 = 10^{2.874} = 748$$

As another example, find the fifth root of 1.89×10^{-9}:

$$\sqrt[5]{1.89 \times 10^{-9}} = (1.89 \times 10^{-9})^{1/5} = ?$$
$$\log (1.89 \times 10^{-9})^{1/5} = \frac{1}{5} \log (1.89 \times 10^{-9}) = \frac{1}{5}(-8.724) = -1.745$$

The answer is the antilog of -1.745:

$$(1.89 \times 10^{-9})^{1/5} = 10^{-1.745} = 1.80 \times 10^{-2}$$

A.4   QUADRATIC EQUATIONS

Algebraic equations of the form $ax^2 + bx + c = 0$ are called **quadratic equations.** The coefficients a, b, and c may be either positive or negative. The two roots of the equation may be found using the *quadratic formula:*

$$x = \frac{-b \pm \sqrt{b^2 - 4ac}}{2a}$$

As an example, solve the equation $5x^2 - 3x - 2 = 0$. Here $a = 5$, $b = -3$, and $c = -2$. Therefore,

$$x = \frac{3 \pm \sqrt{(-3)^2 - 4(5)(-2)}}{2(5)}$$

$$= \frac{3 \pm [2(5)/\sqrt{9 - (-40)}]}{10} = \frac{3 \pm \sqrt{49}}{10} = \frac{3 \pm 7}{10}$$

$$= 1 \text{ and } -0.4$$

How do you know which of the two roots is the correct answer? You have to decide in each case which root has physical significance. It is *usually* true in this course, however, that negative values are not significant.

When you have solved a quadratic expression, you should always check your values by substitution into the original equation. In the previous example, we find that $5(1)^2 - 3(1) - 2 = 0$ and that $5(-0.4)^2 - 3(-0.4) - 2 = 0$.

The most likely place you will encounter quadratic equations is in the chapters on chemical equilibria, particularly in Chapters 16 through 18. Here you will often be faced with solving an equation such as

$$1.8 \times 10^{-4} = \frac{x^2}{0.0010 - x}$$

This equation can certainly be solved using the quadratic equation (to give $x = 3.4 \times 10^{-4}$). You may find the *method of successive approximations* to be especially convenient, however. Here we begin by making a reasonable approximation of x. This approximate value is substituted into the original equation, and this is solved to give what is hoped to be a more correct value of x. This process is repeated until the answer converges on a particular value of x, that is, until the value of x derived from two successive approximations is the same.

Step 1: First assume that x is so small that $(0.0010 - x) \approx 0.0010$. This means that

$$x^2 = 1.8 \times 10^{-4}(0.0010)$$
$$x = 4.2 \times 10^{-4} \text{ (to 2 significant figures)}$$

Step 2: Substitute the value of x from Step 1 into the denominator of the original equation, and again solve for x:

$$x^2 = 1.8 \times 10^{-4}(0.0010 - 0.00042)$$
$$x = 3.2 \times 10^{-4}$$

Step 3: Repeat Step 2 using the value of x found in that step:

$$x = \sqrt{1.8 \times 10^{-4}(0.0010 - 0.00032)} = 3.5 \times 10^{-4}$$

Step 4: Continue repeating the calculation, using the value of x found in the previous step:

$$x = \sqrt{1.8 \times 10^{-4}(0.0010 - 0.00035)} = 3.4 \times 10^{-4}$$

Step 5: $\quad x = \sqrt{1.8 \times 10^{-4}(0.0010 - 0.00034)} = 3.4 \times 10^{-4}$

Here we find that iterations after the fourth step give the same value for x, indicating that we have arrived at a valid answer (and the same one obtained from the quadratic formula).

Here are several final thoughts on using the method of successive approximations. First, in some cases the method does not work. Successive steps may give answers that are random or that diverge from the correct value. In Chapters 16 through 18, you confront quadratic equations of the form $K = x^2/(C - x)$. The method of approximations works as long as $K < 4C$ (assuming one begins with $x = 0$ as the first guess, that is, $K \approx x^2/C$). This is always going to be true for weak acids and bases (the topic of Chapters 17 and 18), but it may *not* be the case for problems involving gas phase equilibria (Chapter 16), where K can be quite large.

Second, values of K in the equation $K = x^2/(C - x)$ are usually known only to two significant figures. We are therefore justified in carrying out successive steps until two answers are the same to two significant figures.

Finally, we highly recommend this method of solving quadratic equations, especially those in Chapters 17 and 18. If your calculator has a memory function, successive approximations can be carried out easily and rapidly.

A.5   GRAPHING

In a number of instances in this text, graphs are used when analyzing experimental data with a goal in obtaining a mathematical equation. The procedure used will plot data points in a way that gives a straight line. The equation that corresponds to a straight line on a graph is

$$y = a + bx$$

In this equation, y is usually referred to as the *dependent variable;* its value is determined from (that is, dependent on) the values of x, a, and b. In this equation x is called *the independent variable,* b is the *slope* of the line, and a is the *y-intercept.* Let us use an example to investigate two things: (a) how to construct a graph from a set of data points, and (b) how to derive an equation for the line generated on the graph by this data.

A set of data points to be graphed is presented in the table below. We must first plot these points on a graph. This will require establishing values along x and y axis of a graph and then identifying each data point by a mark such as a dot. Be sure that the values along each axis are evenly spaced. After plotting the points on the graph, a straight line is drawn that best matches the points. Because there is always some inaccuracy in experimental data, the line may not pass exactly through every point (See Figure).

The fact that these data plotted as a straight line indicates that the equation for the line will have the form $y = a + bx$. To identify the specific equation corresponding to these data, we must determine the y-intercept and slope. The y-intercept is the point at which $x = 0$, that is, 3.38. The slope is determined by selecting two points on the line and calculating the difference in values of y (Δy) and x (Δx). In the example, the slope of the line, $(\Delta y)/(\Delta x)$, is negative, -1.82. With slope and intercept now calculated, we can write the equation for the line

$$y = 3.38 - 1.82x$$

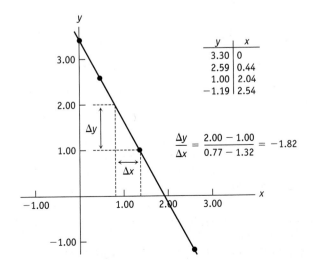

y	x
3.30	0
2.59	0.44
1.00	2.04
−1.19	2.54

$$\frac{\Delta y}{\Delta x} = \frac{2.00 - 1.00}{0.77 - 1.32} = -1.82$$

Some Important Physical Concepts

B.1 MATTER

The tendency to maintain a constant velocity is called inertia. Thus, unless acted on by an unbalanced force, a body at rest remains at rest, and a body in motion remains in motion with uniform velocity. Matter is anything that exhibits inertia; the quantity of matter is its mass.

B.2 MOTION

Motion is the change of position or location in space. Objects can have the following classes of motion:

- Translation occurs when the center of mass of an object changes its location. Example: a car moving on the highway.
- Rotation occurs when each point of a moving object moves in a circle about an axis through the center of mass. Examples: a spinning top, a rotating molecule.
- Vibration is a periodic distortion of and then recovery of original shape. Examples: a struck tuning fork, a vibrating molecule.

B.3 FORCE AND WEIGHT

Force is that which changes the velocity of a body; it is defined as

$$\text{Force} = \text{mass} \times \text{acceleration}$$

The SI unit of force is the **newton,** N, whose dimensions are kilograms times meter per second squared ($kg \cdot m/s^2$). A newton is therefore the force needed to change the velocity of a mass of 1 kilogram by 1 meter per second in a time of 1 second.

Because the earth's gravity is not the same everywhere, the weight corresponding to a given mass is not a constant. At any given spot on earth gravity is constant, however, and therefore weight is proportional to mass. When a balance tells

---

*Adapted from F. Brescia, J. Arents, H. Meislich, et al.: *General Chemistry,* 5th ed. Philadelphia, Harcourt Brace, 1988.

us that a given sample (the "unknown") has the same weight as another sample (the "weights," as given by a scale reading or by a total of counterweights), it also tells us that the two masses are equal. The balance is therefore a valid instrument for measuring the mass of an object independently of slight variations in the force of gravity.

B.4   PRESSURE*

Pressure is force per unit area. The SI unit, called the pascal, Pa, is

$$1 \text{ pascal} = \frac{1 \text{ newton}}{m^2} = \frac{1 \text{ kg} \cdot m/s^2}{m^2} = \frac{1 \text{ kg}}{m \cdot s^2}$$

The International System of Units also recognizes the bar, which is 10^5 Pa and which is close to standard atmospheric pressure (Table 1).

Table 1 • Pressure Conversions

From	To	Multiply By
Atmosphere	mm Hg	760 mm Hg/atm (exactly)
Atmosphere	lb/in.^{2}	14.6960 lb/(in.^{2} atm)
Atmosphere	kPa	101.325 kPa/atm
Bar	Pa	10^5 Pa/bar (exactly)
Bar	lb/in.^{2}	14.5038 lb/(in.^{2} bar)
mm Hg	torr	1 torr/mm Hg (exactly)

Chemists also express pressure in terms of the heights of liquid columns, especially water and mercury. This usage is not completely satisfactory, because the pressure exerted by a given column of a given liquid is not a constant but depends on the temperature (which influences the density of the liquid) and the location (which influences gravity). Such units are therefore not part of the SI, and their use is now discouraged. The older units are still used in books and journals, however, and chemists must still be familiar with them.

The pressure of a liquid or a gas depends only on the depth (or height) and is exerted equally in all directions. At sea level, the pressure exerted by the earth's atmosphere supports a column of mercury about 0.76 m (76 cm, or 760 mm) high.

One **standard atmosphere** (atm) is the pressure exerted by exactly 76 cm of mercury at 0 °C (density, 13.5951 g/cm^3) and at standard gravity, 9.80665 m/s^2. The **bar** is equivalent to 1.01325 atm. One **torr** is the pressure exerted by exactly 1 mm of mercury at 0 °C and standard gravity.

B.5   ENERGY AND POWER

The SI unit of energy is the product of the units of force and distance, or kilograms times meter per second squared (kg · m/s^2) times meters ($\times$ m), which is kg · m^2/s^2; this unit is called the **joule, J**. The joule is thus the work done when a force of 1 newton acts through a distance of 1 meter.

---

*See Section 12.1

Work may also be done by moving an electric charge in an electric field. When the charge being moved is 1 coulomb (C), and the potential difference between its initial and final positions is 1 volt (V), the work is 1 joule. Thus,

$$1 \text{ joule} = 1 \text{ coulomb volt (CV)}$$

Another unit of electric work that is not part of the International System of Units but is still in use is the **electron volt,** eV, which is the work required to move an electron against a potential difference of 1 volt. (It is also the kinetic energy acquired by an electron when it is accelerated by a potential difference of 1 volt.) Because the charge on an electron is 1.602×10^{-19} C, we have

$$1 \text{ eV} = 1.602 \times 10^{-19} \text{ CV} \cdot \frac{1 \text{ J}}{1 \text{ CV}} = 1.602 \times 10^{-19} \text{ J}$$

If this value is multiplied by Avogadro's number, we obtain the energy involved in moving 1 mole of electron charges (1 faraday) in a field produced by a potential difference of 1 volt:

$$1 \frac{\text{eV}}{\text{particle}} = \frac{1.602 \times 10^{-19} \text{ J}}{\text{particle}} \cdot \frac{6.022 \times 10^{23} \text{ particles}}{\text{mol}} \cdot \frac{1 \text{ kJ}}{1000 \text{ J}} = 96.49 \text{ kJ/mol}$$

Power is the amount of energy delivered per unit time. The SI unit is the watt, W, which is a joule per second. One kilowatt, kW, is 1000 W. Watt hours and kilowatt hours are therefore units of energy (Table 2). For example, 1000 watts, or 1 kilowatt, is

$$1.0 \times 10^3 \text{ W} \cdot \frac{1 \text{ J}}{1 \text{ W} \cdot \text{s}} \cdot \frac{3.6 \times 10^3 \text{ s}}{1 \text{ h}} = 3.6 \times 10^6 \text{ J}$$

Table 2 • Energy Conversions

From	To	Multiply By
Calorie (cal)	joule	4.184 J/cal (exactly)
Kilocalorie (kcal)	cal	10^3 cal/kcal (exactly)
Kilocalorie	joule	4.184×10^3 J/kcal (exactly)
Liter atmosphere (L · atm)	joule	101.325 J/L · atm
Electron volt (eV)	joule	1.60218×10^{-19} J/eV
Electron volt per particle	kilojoules per mole	96.485 kJ · particle/eV · mol
Coulomb volt (CV)	joule	1 CV/J (exactly)
Kilowatt hour (kWh)	kcal	860.4 kcal/kWh
Kilowatt hour	joule	3.6×10^6 J/kWh (exactly)
British thermal unit (Btu)	calorie	252 cal/Btu

Abbreviations and Useful Conversion Factors

Table 3 • Some Common Abbreviations and Standard Symbols

Term	Abbreviation	Term	Abbreviation
Activation energy	E_a	Face-centered cubic	fcc
Ampere	A	Faraday constant	F
Aqueous solution	aq	Gas constant	R
Atmosphere, unit of pressure	atm	Gibbs free energy	G
Atomic mass unit	amu	Standard free energy	$G°$
Avogadro constant	N_A	Standard free energy of formation	$\Delta G_f°$
Bar, unit of pressure	bar	Free energy change for reaction	$\Delta G_{rxn}°$
Body-centered cubic	bcc	Half-life	$t_{1/2}$
Bohr radius	a_0	Heat	q
Boiling point	bp	Hertz	Hz
Celsius temperature, °C	T	Hour	h
Charge number of an ion	z	Joule	J
Coulomb, electric charge	C	Kelvin	K
Curie, radioactivity	Ci	Kilocalorie	kcal
Cycles per second, hertz	Hz	Liquid	ℓ
Debye, unit of electric dipole	D	Logarithm, base 10	log
Electron	e^-	Logarithm, base e	ln
Electron volt	eV	Minute	min
Electronegativity	χ	Molar	M
Energy	E	Molar mass	M
Enthalpy	H	Mole	mol
Standard enthalpy	$H°$	Osmotic pressure	Π
Standard enthalpy of formation	$\Delta H_f°$	Planck's constant	h
Standard enthalpy of reaction	$\Delta H_{rxn}°$	Pound	lb
Entropy	S	Pressure	
Standard entropy	$S°$	Pascal, unit of pressure	Pa
Entropy change for reaction	$\Delta S_{rxn}°$	In atmospheres	atm
Equilibrium constant	K	In millimeters of mercury	mm Hg
Concentration basis	K_c	Proton number	Z
Pressure basis	K_p	Rate constant	k
Ionization weak acid	K_a	Simple cubic (unit cell)	sc
Ionization weak base	K_b	Standard temperature and pressure	STP
Solubility product	K_{sp}	Volt	V
Formation constant	K_{form}	Watt	W
Ethylenediamine	en	Wavelength	λ

C.1   FUNDAMENTAL UNITS OF THE SI SYSTEM

The metric system was begun by the French National Assembly in 1790 and has undergone many modifications. The International System of Units or *Système International* (SI), which represents an extension of the metric system, was adopted by the 11th General Conference of Weights and Measures in 1960. It is constructed from seven base units, each of which represents a particular physical quantity (Table 4).

Table 4 • SI Fundamental Units

Physical Quantity	Name of Unit	Symbol
Length	meter	m
Mass	kilogram	kg
Time	second	s
Temperature	kelvin	K
Amount of substance	mole	mol
Electric current	ampere	A
Luminous intensity	candela	cd

The first five units listed in Table 4 are particularly useful in general chemistry and are defined as follows:

1. The *meter* was redefined in 1960 to be equal to 1,650,763.73 wavelengths of a certain line in the emission spectrum of krypton-86.
2. The *kilogram* represents the mass of a platinum–iridium block kept at the International Bureau of Weights and Measures at Sèvres, France.
3. The *second* was redefined in 1967 as the duration of 9,192,631,770 periods of a certain line in the microwave spectrum of cesium-133.
4. The *kelvin* is 1/273.15 of the temperature interval between absolute zero and the triple point of water.
5. The *mole* is the amount of substance that contains as many entities as there are atoms in exactly 0.012 kg of carbon-12 (12 g of ^{12}C atoms).

C.2   PREFIXES USED WITH TRADITIONAL METRIC UNITS AND SI UNITS

Decimal fractions and multiples of metric and SI units are designated by using the prefixes listed in Table 5. Those most commonly used in general chemistry appear in italics.

C.3   DERIVED SI UNITS

In the International System of Units, all physical quantities are represented by appropriate combinations of the base units listed in Table 4. A list of the derived units frequently used in general chemistry is given in Table 6.

Table 5 • Traditional Metric and SI Prefixes

Factor	Prefix	Symbol	Factor	Prefix	Symbol
10^{12}	tera	T	10^{-1}	*deci*	d
10^{9}	giga	G	10^{-2}	*centi*	c
10^{6}	mega	M	10^{-3}	*milli*	m
10^{3}	*kilo*	k	10^{-6}	micro	μ
10^{2}	hecto	h	10^{-9}	*nano*	n
10^{1}	deka	da	10^{-12}	*pico*	p
			10^{-15}	femto	f
			10^{-18}	atto	a

Table 6 • Derived SI Units

Physical Quantity	Name of Unit	Symbol	Definition
Area	square meter	m^2	
Volume	cubic meter	m^3	
Density	kilogram per cubic meter	kg/m^3	
Force	newton	N	$kg \cdot m/s^2$
Pressure	pascal	Pa	N/m^2
Energy	joule	J	$kg \cdot m^2/s^2$
Electric charge	coulomb	C	$A \cdot s$
Electric potential difference	volt	V	$J/(A \cdot s)$

Table 7 • Common Units of Mass and Weight

1 Pound = 453.39 Grams

1 kilogram = 1000 grams = 2.205 pounds

1 gram = 10 decigrams = 100 centigrams = 1000 milligrams

1 gram = 6.022×10^{23} atomic mass units

1 atomic mass unit = 1.6605×10^{-24} gram

1 short ton = 2000 pounds = 907.2 kilograms

1 long ton = 2240 pounds

1 metric tonne = 1000 kilograms = 2205 pounds

Table 8 • Common Units of Length

1 inch = 2.54 centimeters (Exactly)

1 mile = 5280 feet = 1.609 kilometers
1 yard = 36 inches = 0.9144 meter
1 meter = 100 centimeters = 39.37 inches = 3.281 feet = 1.094 yards
1 kilometer = 1000 meters = 1094 yards = 0.6215 mile
1 Ångstrom = 1.0×10^{-8} centimeter = 0.10 nanometer = 100 picometers
    = 1.0×10^{-10} meter = 3.937×10^{-9} inch

Table 9 • Common Units of Volume

1 quart = 0.9463 liter
1 liter = 1.0567 quarts

1 liter = 1 cubic decimeter = 1000 cubic centimeters = 0.001 cubic meter
1 milliliter = 1 cubic centimeter = 0.001 liter = 1.056×10^{-3} quart
1 cubic foot = 28.316 liters = 29.924 quarts = 7.481 gallons

Physical Constants

Table 10

Quantity	Symbol	Traditional Units	SI Units
Acceleration of gravity	g	980.6 cm/s	9.806 m/s
Atomic mass unit (1/12 the mass of ^{12}C atom)	amu or u	1.6605×10^{-24} g	1.6605×10^{-27} kg
Avogadro's number	N	$6.02214199 \times 10^{23}$ particles/mol	$6.02214199 \times 10^{23}$ particles/mol
Bohr radius	a_0	0.52918 Å 5.2918×10^{-9} cm	5.2918×10^{-11} m
Boltzmann constant	k	1.3807×10^{-16} erg/K	1.3807×10^{-23} J/K
Charge-to-mass ratio of electron	e/m	1.7588×10^{8} C/g	1.7588×10^{11} C/kg
Electronic charge	e	1.6022×10^{-19} C 4.8033×10^{-10} esu	1.6022×10^{-19} C
Electron rest mass	m_e	9.1094×10^{-28} g 0.00054858 amu	9.1094×10^{-31} kg
Faraday constant	F	96,485 C/mol e^- 23.06 kcal/V · mol e^-	96,485 C/mol e^- 96,485 J/V · mol e^-
Gas constant	R	$0.08206 \dfrac{L \cdot atm}{mol \cdot K}$ $1.987 \dfrac{cal}{mol \cdot K}$	$8.3145 \dfrac{Pa \cdot dm^3}{mol \cdot K}$ 8.3145 J/mol · K
Molar volume (STP)	V_m	22.414 L/mol	22.414×10^{-3} m^3/mol 22.414 dm^3/mol
Neutron rest mass	m_n	1.67493×10^{-24} g 1.008665 amu	1.67493×10^{-27} kg
Planck's constant	h	6.6261×10^{-27} erg · s	$6.62606876 \times 10^{-34}$ J · s
Proton rest mass	m_p	1.6726×10^{-24} g 1.007276 amu	1.6726×10^{-27} kg
Rydberg constant	R_α	3.289×10^{15} cycles/s 2.1799×10^{-11} erg	1.0974×10^{7} m^{-1} 2.1799×10^{-18} J
Velocity of light (in a vacuum)	c	2.9979×10^{10} cm/s (186,282 miles/s)	2.9979×10^{8} m/s

$\pi = 3.1416$

$e = 2.7183$

$\ln X = 2.303 \log X$

Table 11 • Specific Heats and Heat Capacities for Some Common Substances at 25 °C

Substance	Specific Heat J/g · K	Molar Heat Capacity J/mol · K
Al(s)	0.897	24.2
Ca(s)	0.646	25.9
Cu(s)	0.385	24.5
Fe(s)	0.449	25.1
Hg(ℓ)	0.140	28.0
H_2O(s), ice	2.06	37.1
H_2O(ℓ), water	4.184	75.4
H_2O(g), steam	1.86	33.6
C_6H_6(ℓ), benzene	1.74	136
C_6H_6(g), benzene	1.06	82.4
C_2H_5OH(ℓ), ethanol	2.44	112.3
C_2H_5OH(g), ethanol	1.41	65.4
$(C_2H_5)_2O$(ℓ), diethyl ether	2.33	172.6
$(C_2H_5)_2O$(g), diethyl ether	1.61	119.5

Table 12 • Heats of Transformation and Transformation Temperatures of Several Substances

Substance	MP (°C)	Heat of Fusion		BP (°C)	Heat of Vaporization	
		J/g	kJ/mol		J/g	kJ/mol
*Elements**						
Al	660	395	10.7	2518	12083	294
Ca	842	212	8.5	1484	3767	155
Cu	1085	209	13.3	2567	4720	300
Fe	1535	267	13.8	2861	6088	340
Hg	−38.8	11	2.29	357	295	59.1
Compounds						
H_2O	0.00	338	6.09	100.0	2260	40.7
CH_4	−182.5	58.6	0.94	−161.5	511	8.2
C_2H_5OH	−114	109	5.02	78.3	838	38.6
C_6H_6	5.48	127.4	9.95	80.0	393	30.7
$(C_2H_5)_2O$	−116.3	98.1	7.27	34.6	357	26.5

*Data for the elements are taken from J. A. Dean, *Lange's Handbook of Chemistry*, 15th Edition (1999).

Naming Organic Compounds

It seems a daunting task — to devise a systematic procedure that gives each organic compound a unique name — but that is what has been done. A set of rules was developed to name organic compounds by the International Union of Pure and Applied Chemistry, IUPAC. The IUPAC nomenclature allows chemists to write a name for any compound based on its structure or to identify the formula and structure for a compound from its name. In this book, we have generally used the IUPAC nomenclature scheme when naming compounds.

In addition to the systematic names, many compounds also have common names. The common names came into existence before the nomenclature rules were developed, and they have continued in use. For some compounds, these names are so well entrenched that they are used most of the time. One such compound is acetic acid, which is almost always referred to by that name and not by its systematic name, ethanoic acid.

The general procedure for systematic naming of organic compounds begins with nomenclature for hydrocarbons. Other organic compounds are then named as derivatives of hydrocarbons. Nomenclature rules for simple organic compounds are given in the following section.

E.1 HYDROCARBONS

Alkanes

The names of alkanes end in "-ane." When naming a specific alkane, the root of the name identifies the longest carbon chain in a compound. Specific substituent groups attached to this carbon chain are identified by name and position.

Alkanes with chains of from one to ten carbon atoms are given in Table 11.3. After the first four compounds, the names derive from Latin numbers — pentane, hexane, heptane, octane, nonane, decane — and this regular naming continues for higher alkanes. For substituted alkanes, the substituent groups on a hydrocarbon chain must be identified both by a name and by the position of substitution; this information precedes the root of the name. The position is indicated by a number that refers to the carbon atom to which the substituent is attached. (Numbering of the carbon atoms in a chain should begin at the end of the carbon chain that allows the substituent groups to have the lowest numbers.)

Names of hydrocarbon substituents are derived from the name of the hydrocarbon. The group —CH_3, derived by taking a hydrogen from methane, is called the methyl group; the C_2H_5 group is the ethyl group. The nomenclature scheme is easily extended to derivatives of hydrocarbons with other substituent groups like

-Cl (chloro), -NO$_2$ (nitro), -CN (cyano), -D (deuterio), and so on (Table 13). If two or more of the same substituent groups occur, the prefixes di-, tri-, and tetra- are added. When different substituent groups are present, they are generally listed in alphabetical order.

Table 13 • Names of Common Substituent Groups

Formula	Name	Formula	Name
—CH$_3$	methyl	—D	deuterio
—C$_2$H$_5$	ethyl	—Cl	chloro
—CH$_2$CH$_2$CH$_3$	1-propyl (*n*-propyl)	—Br	bromo
—CH(CH$_3$)$_2$	2-propyl (isopropyl)	—F	fluoro
—CH=CH$_2$	ethenyl (vinyl)	—CN	cyano
—C$_6$H$_5$	phenyl	—NO$_2$	nitro
—OH	hydroxo		
—NH$_2$	amino		

Example:

$$\underset{\text{CH}_3\text{CH}_2\overset{\displaystyle |}{\text{C}}\text{HCH}_2\overset{\displaystyle |}{\text{C}}\text{HCH}_2\text{CH}_3}{\overset{\text{CH}_3 \qquad \text{C}_2\text{H}_5}{}}$$

Step	Information to include	Contribution to name
1.	An alkane	name will end in "-ane"
2.	Longest chain is 7 carbons	name as a *heptane*
3.	—CH$_3$ group at carbon 3	3-*methyl*
4.	—C$_2$H$_5$ group at carbon 5	5-*ethyl*

Name: 5-ethyl-3-methylheptane

Cycloalkanes are named based on the ring size adding the prefix "cyclo"; for example, the cycloalkane with a six-member ring of carbons is called cyclohexane.

Alkenes

Alkenes have names ending in "-ene." The name of an alkene must specify the length of the carbon chain, the position of the double bond (and when appropriate, the configuration, either *cis* or *trans*). As with alkanes, both identity and position of substituent groups must be given. The carbon chain is numbered from the end that gives the double bond the lowest number.

Compounds with two double bonds are called dienes and they are named similarly—specifying the positions of the double bonds and the name and position of any substituent groups.

For example, the compound H$_2$C=C(CH$_3$)CH(CH$_3$)CH$_2$CH$_3$ has a five-carbon chain with a double bond between carbon atoms 1 and 2 and methyl groups on carbon atoms 2 and 3. Its name using IUPAC nomenclature is **2,3-dimethyl-1-pentene.** The compound CH$_3$CH=CHCCl$_3$ with a *cis* configuration around the double bond is named **1,1,1-trichloro-*cis*-2-butene.** The compound $H_2C=C(Cl)CH=CH_2$ is **2-chloro-1,3-butadiene.**

Alkynes

The naming of alkynes is similar to the naming of alkenes, except that *cis–trans* isomerism isn't a factor. The ending "-yne" on a name identifies a compound as an alkyne.

Benzene Derivatives

The carbon atoms in the six-member ring are numbered 1 through 6, and the name and position of substituent groups are given. The two examples shown here are **1-ethyl-3-methylbenzene** and **1,4-diaminobenzene.**

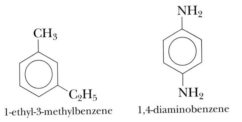

1-ethyl-3-methylbenzene   1,4-diaminobenzene

ε.2   DERIVATIVES OF HYDROCARBONS

The names for alcohols, aldehydes, ketones, and acids are based on the name of the hydrocarbon with an appropriate suffix to denote the class of compound, as follows:

- **Alcohols:** Substitute "-ol" for the final "-e" in the name of the hydrocarbon, and designate the position of the —OH group by the number of the carbon atom. For example, $CH_3CH_2CHOHCH_3$ is named as a derivative of the 4-carbon hydrocarbon butane. The —OH group is attached to the second carbon, so the name is 2-butanol.
- **Aldehydes:** Substitute "-al" for the final "-e" in the name of the hydrocarbon. The carbon atom of an aldehyde is, by definition, carbon-1 in the hydrocarbon chain. For example, the compound $CH_3CH(CH_3)CH_2CH_2CHO$ contains a 5-carbon chain with the aldehyde functional group being carbon 1 and the —CH_3 group at position 4; thus the name is **4-methylpentanal.**
- **Ketones:** Substitute "-one" for the final "-e" in the name of the hydrocarbon. The position of the ketone functional group (the carbonyl group) is indicated by the number of the carbon atom. For example, the compound $CH_3COCH_2CH(C_2H_5)CH_2CH_3$ has the carbonyl group at the 2 position and an ethyl group at the 4 position of a 6-carbon chain; its name is **4-ethyl-2-hexanone.**
- **Carboxylic acids (organic acids):** Substitute "-oic" for the final "-e" in the name of the hydrocarbon. The carbon atoms in the longest chain are counted beginning with the carboxylic carbon atom. For example, *trans*-$CH_3CH=CHCH_2CO_2H$ is named as a derivative of *trans*-3-pentene, that is, ***trans*-3-pentenoic acid.**

An **ester** is named as a derivative of the alcohol and acid from which it is made. The name of an ester is obtained by splitting the formula RCO_2R' into two parts, the RCO_2— portion and the —R′ portion. The —R′ portion comes from the alcohol and is identified by the hydrocarbon group name; derivatives of ethanol, for example, are called *ethyl* esters. The acid part of the compound is named by dropping the "-oic" ending for the acid and replacing it by "-oate." The compound $CH_3CH_2CO_2CH_3$ is named **methyl propanoate.**

Notice that an anion derived from a carboxylic acid by loss of the acidic proton is named the same way. Thus, $CH_3CH_2CO_2^-$ is the **propanoate anion,** and the sodium salt of this anion, $Na(CH_3CH_2CO_2)$, is **sodium propanoate.**

Values for the Ionization Energies and Electron Affinities of the Elements

1A (1)	2A (2)	3B (3)	4B (4)	5B (5)	6B (6)	7B (7)	8B (8,9,10)			1B (11)	2B (12)	3A (13)	4A (14)	5A (15)	6A (16)	7A (17)	8 (18)
H 1312																	He 2371
Li 520	Be 899											B 801	C 1086	N 1402	O 1314	F 1681	Ne 2081
Na 496	Mg 738											Al 578	Si 786	P 1012	S 1000	Cl 1251	Ar 1521
K 419	Ca 599	Sc 631	Ti 658	V 650	Cr 652	Mn 717	Fe 759	Co 758	Ni 757	Cu 745	Zn 906	Ga 579	Ge 762	As 947	Se 941	Br 1140	Kr 1351
Rb 403	Sr 550	Y 617	Zr 661	Nb 664	Mo 685	Tc 702	Ru 711	Rh 720	Pd 804	Ag 731	Cd 868	In 558	Sn 709	Sb 834	Te 869	I 1008	Xe 1170
Cs 377	Ba 503	La 538	Hf 681	Ta 761	W 770	Re 760	Os 840	Ir 880	Pt 870	Au 890	Hg 1007	Tl 589	Pb 715	Bi 703	Po 812	At 890	Rn 1037

Table 14 • Electron Affinity Values for Some Elements (kJ/mol)*

H −72.77							
Li −59.63	Be 0[†]	B −26.7	C −121.85	N 0	O −140.98	F −328.0	
Na −52.87	Mg 0	Al −42.6	Si −133.6	P −72.07	S −200.41	Cl −349.0	
K −48.39	Ca 0	Ga −30	Ge −120	As −78	Se −194.97	Br −324.7	
Rb −46.89	Sr 0	In −30	Sn −120	Sb −103	Te −190.16	I −295.16	
Cs −45.51	Ba 0	Tl −20	Pb −35.1	Bi −91.3	Po −180	At −270	

*Data taken from H. Hotop and W. C. Lineberger: *Journal of Physical Chemistry, Reference Data,* Vol. 14, p. 731, 1985. (This paper also includes data for the transition metals.) Some values are known to more than two decimal places.

[†] Elements with an electron affinity of zero indicate that a stable anion A⁻ of the element does not exist in the gas phase.

Vapor Pressure of Water at Various Temperatures

Table 15 • Vapor Pressure of Water at Various Temperatures

Temperature °C	Vapor Pressure torr	Temperature °C	Vapor Pressure torr	Temperature °C	Vapor Pressure torr	Temperature °C	Vapor Pressure torr
−10	2.1	21	18.7	51	97.2	81	369.7
−9	2.3	22	19.8	52	102.1	82	384.9
−8	2.5	23	21.1	53	107.2	83	400.6
−7	2.7	24	22.4	54	112.5	84	416.8
−6	2.9	25	23.8	55	118.0	85	433.6
−5	3.2	26	25.2	56	123.8	86	450.9
−4	3.4	27	26.7	57	129.8	87	468.7
−3	3.7	28	28.3	58	136.1	88	487.1
−2	4.0	29	30.0	59	142.6	89	506.1
−1	4.3	30	31.8	60	149.4	90	525.8
0	4.6	31	33.7	61	156.4	91	546.1
1	4.9	32	35.7	62	163.8	92	567.0
2	5.3	33	37.7	63	171.4	93	588.6
3	5.7	34	39.9	64	179.3	94	610.9
4	6.1	35	42.2	65	187.5	95	633.9
5	6.5	36	44.6	66	196.1	96	657.6
6	7.0	37	47.1	67	205.0	97	682.1
7	7.5	38	49.7	68	214.2	98	707.3
8	8.0	39	52.4	69	223.7	99	733.2
9	8.6	40	55.3	70	233.7	100	760.0
10	9.2	41	58.3	71	243.9	101	787.6
11	9.8	42	61.5	72	254.6	102	815.9
12	10.5	43	64.8	73	265.7	103	845.1
13	11.2	44	68.3	74	277.2	104	875.1
14	12.0	45	71.9	75	289.1	105	906.1
15	12.8	46	75.7	76	301.4	106	937.9
16	13.6	47	79.6	77	314.1	107	970.6
17	14.5	48	83.7	78	327.3	108	1004.4
18	15.5	49	88.0	79	341.0	109	1038.9
19	16.5	50	92.5	80	355.1	110	1074.6
20	17.5						

Ionization Constants for Weak Acids at 25 °C

Table 16 • Ionization Constants for Weak Acids at 25 °C

Acid	Formula and Ionization Equation	K_a
Acetic	$CH_3CO_2H \rightleftharpoons H^+ + CH_3CO_2^-$	1.8×10^{-5}
Arsenic	$H_3AsO_4 \rightleftharpoons H^+ + H_2AsO_4^-$	$K_1 = 2.5 \times 10^{-4}$
	$H_2AsO_4^- \rightleftharpoons H^+ + HAsO_4^{2-}$	$K_2 = 5.6 \times 10^{-8}$
	$HAsO_4^{2-} \rightleftharpoons H^+ + AsO_4^{3-}$	$K_3 = 3.0 \times 10^{-13}$
Arsenous	$H_3AsO_3 \rightleftharpoons H^+ + H_2AsO_3^-$	$K_1 = 6.0 \times 10^{-10}$
	$H_2AsO_3^- \rightleftharpoons H^+ + HAsO_3^{2-}$	$K_2 = 3.0 \times 10^{-14}$
Benzoic	$C_6H_5CO_2H \rightleftharpoons H^+ + C_6H_5CO_2^-$	6.3×10^{-5}
Boric	$H_3BO_3 \rightleftharpoons H^+ + H_2BO_3^-$	$K_1 = 7.3 \times 10^{-10}$
	$H_2BO_3^- \rightleftharpoons H^+ + HBO_3^{2-}$	$K_2 = 1.8 \times 10^{-13}$
	$HBO_3^{2-} \rightleftharpoons H^+ + BO_3^{3-}$	$K_3 = 1.6 \times 10^{-14}$
Carbonic	$H_2CO_3 \rightleftharpoons H^+ + HCO_3^-$	$K_1 = 4.2 \times 10^{-7}$
	$HCO_3^- \rightleftharpoons H^+ + CO_3^{2-}$	$K_2 = 4.8 \times 10^{-11}$
Citric	$H_3C_6H_5O_7 \rightleftharpoons H^+ + H_2C_6H_5O_7^-$	$K_1 = 7.4 \times 10^{-3}$
	$H_2C_6H_5O_7^- \rightleftharpoons H^+ + HC_6H_5O_7^{2-}$	$K_2 = 1.7 \times 10^{-5}$
	$HC_6H_5O_7^{2-} \rightleftharpoons H^+ + C_6H_5O_7^{3-}$	$K_3 = 4.0 \times 10^{-7}$
Cyanic	$HOCN \rightleftharpoons H^+ + OCN^-$	3.5×10^{-4}
Formic	$HCO_2H \rightleftharpoons H^+ + HCO_2^-$	1.8×10^{-4}
Hydrazoic	$HN_3 \rightleftharpoons H^+ + N_3^-$	1.9×10^{-5}
Hydrocyanic	$HCN \rightleftharpoons H^+ + CN^-$	4.0×10^{-10}
Hydrofluoric	$HF \rightleftharpoons H^+ + F^-$	7.2×10^{-4}
Hydrogen peroxide	$H_2O_2 \rightleftharpoons H^+ + HO_2^-$	2.4×10^{-12}
Hydrosulfuric	$H_2S \rightleftharpoons H^+ + HS^-$	$K_1 = 1 \times 10^{-7}$
	$HS^- \rightleftharpoons H^+ + S^{2-}$	$K_2 = 1 \times 10^{-19}$
Hypobromous	$HOBr \rightleftharpoons H^+ + OBr^-$	2.5×10^{-9}
Hypochlorous	$HOCl \rightleftharpoons H^+ + OCl^-$	3.5×10^{-8}
Nitrous	$HNO_2 \rightleftharpoons H^+ + NO_2^-$	4.5×10^{-4}
Oxalic	$H_2C_2O_4 \rightleftharpoons H^+ + HC_2O_4^-$	$K_1 = 5.9 \times 10^{-2}$
	$HC_2O_4^- \rightleftharpoons H^+ + C_2O_4^{2-}$	$K_2 = 6.4 \times 10^{-5}$
Phenol	$C_6H_5OH \rightleftharpoons H^+ + C_6H_5O^-$	1.3×10^{-10}

(continued)

Table 16 • continued

Acid	Formula and Ionization Equation	K_a
Phosphoric	$H_3PO_4 \rightleftharpoons H^+ + H_2PO_4^-$	$K_1 = 7.5 \times 10^{-3}$
	$H_2PO_4^- \rightleftharpoons H^+ + HPO_4^{2-}$	$K_2 = 6.2 \times 10^{-8}$
	$HPO_4^{2-} \rightleftharpoons H^+ + PO_4^{3-}$	$K_3 = 3.6 \times 10^{-13}$
Phosphorus	$H_3PO_3 \rightleftharpoons H^+ + H_2PO_3^-$	$K_1 = 1.6 \times 10^{-2}$
	$H_2PO_3 \rightleftharpoons H^+ + HPO_3^{2-}$	$K_2 = 7.0 \times 10^{-7}$
Selenic	$H_2SeO_4 \rightleftharpoons H^+ + HSeO_4^-$	$K_1 =$ very large
	$HSeO_4^- \rightleftharpoons H^+ + SeO_4^{2-}$	$K_2 = 1.2 \times 10^{-2}$
Selenous	$H_2SeO_3 \rightleftharpoons H^+ + HSeO_3^-$	$K_1 = 2.7 \times 10^{-3}$
	$HSeO_3^- \rightleftharpoons H^+ + SeO_3^{2-}$	$K_2 = 2.5 \times 10^{-7}$
Sulfuric	$H_2SO_4 \rightleftharpoons H^+ + HSO_4^-$	$K_1 =$ very large
	$HSO_4^- \rightleftharpoons H^+ + SO_4^{2-}$	$K_2 = 1.2 \times 10^{-2}$
Sulfurous	$H_2SO_3 \rightleftharpoons H^+ + HSO_3^-$	$K_1 = 1.7 \times 10^{-2}$
	$HSO_3^- \rightleftharpoons H^+ + SO_3^{2-}$	$K_2 = 6.4 \times 10^{-8}$
Tellurous	$H_2TeO_3 \rightleftharpoons H^+ + HTeO_3^-$	$K_1 = 2 \times 10^{-3}$
	$HTeO_3^- \rightleftharpoons H^+ + TeO_3^{2-}$	$K_2 = 1 \times 10^{-8}$

Appendix I

Ionization Constants for Weak Bases at 25 °C

Table 17 • Ionization Constants for Weak Bases at 25 °C

Base	Formula and Ionization Equation	K_b
Ammonia	$NH_3 + H_2O \rightleftharpoons NH_4{}^+ + OH^-$	1.8×10^{-5}
Aniline	$C_6H_5NH_2 + H_2O \rightleftharpoons C_6H_5NH_3{}^+ + OH^-$	4.0×10^{-10}
Dimethylamine	$(CH_3)_2NH + H_2O \rightleftharpoons (CH_3)_2NH_2{}^+ + OH^-$	7.4×10^{-4}
Ethylenediamine	$H_2NCH_2CH_2NH_2 + H_2O \rightleftharpoons H_2NCH_2CH_2NH_3{}^+ + OH^-$	$K_1 = 8.5 \times 10^{-5}$
	$H_2NCH_2CH_2NH_3{}^+ + H_2O \rightleftharpoons H_3NCH_2CH_2NH_3{}^{2+} + OH^-$	$K_2 = 2.7 \times 10^{-8}$
Hydrazine	$N_2H_4 + H_2O \rightleftharpoons N_2H_5{}^+ + OH^-$	$K_1 = 8.5 \times 10^{-7}$
	$N_2H_5{}^+ + H_2O \rightleftharpoons N_2H_6{}^{2+} + OH^-$	$K_2 = 8.9 \times 10^{-16}$
Hydroxylamine	$NH_2OH + H_2O \rightleftharpoons NH_3OH^+ + OH^-$	6.6×10^{-9}
Methylamine	$CH_3NH_2 + H_2O \rightleftharpoons CH_3NH_3{}^+ + OH^-$	5.0×10^{-4}
Pyridine	$C_5H_5N + H_2O \rightleftharpoons C_5H_5NH^+ + OH^-$	1.5×10^{-9}
Trimethylamine	$(CH_3)_3N + H_2O \rightleftharpoons (CH_3)_3NH^+ + OH^-$	7.4×10^{-5}
Ethylamine	$C_2H_5NH_2 + H_2O \rightleftharpoons C_2H_5NH_3{}^+ + OH^-$	4.3×10^{-4}

Solubility Product Constants for Some Inorganic Compounds at 25 °C

Table 18A • Solubility Product Constants (25 °C)

Cation	Compound	K_{sp}	Cation	Compound	K_{sp}
Ba^{2+}	*$BaCrO_4$	1.2×10^{-10}	Mg^{2+}	$MgCO_3$	6.8×10^{-6}
	$BaCO_3$	2.6×10^{-9}		MgF_2	5.2×10^{-11}
	BaF_2	1.8×10^{-7}		$Mg(OH)_2$	5.6×10^{-12}
	*$BaSO_4$	1.1×10^{-10}	Mn^{2+}	$MnCO_3$	2.3×10^{-11}
Ca^{2+}	$CaCO_3$ (calcite)	3.4×10^{-9}		*$Mn(OH)_2$	1.9×10^{-13}
	*CaF_2	5.3×10^{-11}	Hg_2^{2+}	*Hg_2Br_2	6.4×10^{-23}
	*$Ca(OH)_2$	5.5×10^{-5}		Hg_2Cl_2	1.4×10^{-18}
	$CaSO_4$	4.9×10^{-5}		*Hg_2I_2	2.9×10^{-29}
$Cu^{+,2+}$	$CuBr$	6.3×10^{-9}		Hg_2SO_4	6.5×10^{-7}
	CuI	1.3×10^{-12}	Ni^{2+}	$NiCO_3$	1.4×10^{-7}
	$Cu(OH)_2$	2.2×10^{-20}		$Ni(OH)_2$	5.5×10^{-16}
	$CuSCN$	1.8×10^{-13}	Ag^+	*$AgBr$	5.4×10^{-13}
Au^+	$AuCl$	2.0×10^{-13}		*$AgBrO_3$	5.4×10^{-5}
Fe^{2+}	$FeCO_3$	3.1×10^{-11}		$AgCH_3CO_2$	1.9×10^{-3}
	$Fe(OH)_2$	4.9×10^{-17}		$AgCN$	6.0×10^{-17}
Pb^{2+}	$PbBr_2$	6.6×10^{-6}		Ag_2CO_3	8.5×10^{-12}
	$PbCO_3$	7.4×10^{-14}		*$Ag_2C_2O_4$	5.4×10^{-12}
	$PbCl_2$	1.7×10^{-5}		*$AgCl$	1.8×10^{-10}
	$PbCrO_4$	2.8×10^{-13}		Ag_2CrO_4	1.1×10^{-12}
	PbF_2	3.3×10^{-8}		*AgI	8.5×10^{-17}
	PbI_2	9.8×10^{-9}		$AgSCN$	1.0×10^{-12}
	$Pb(OH)_2$	1.4×10^{-15}		*Ag_2SO_4	1.2×10^{-5}
	$PbSO_4$	2.5×10^{-8}			

(continued)

Table 18A • continued

Cation	Compound	K_{sp}	Cation	Compound	K_{sp}
Sr^{2+}	$SrCO_3$	5.6×10^{-10}	Zn^{2+}	$Zn(OH)_2$	3×10^{-17}
	SrF_2	4.3×10^{-9}		$Zn(CN)_2$	8.0×10^{-12}
	$SrSO_4$	3.4×10^{-7}			
Tl^{+}	$TlBr$	3.7×10^{-6}			
	$TlCl$	1.9×10^{-4}			
	TlI	5.5×10^{-8}			

The values reported in this table were taken from *Lange's Handbook of Chemistry,* 15th Edition, McGraw Hill Publishers, New York, NY (1999). Values have been rounded off to two significant figures.

*Calculated solubility from these K_{sp} values will match experimental solubility for this compound within a factor of 2. Experimental values for solubilities are given in R. W. Clark and J. M. Bonicamp, *J. Chem. Ed.,* **75,** 1182 (1998).

Table 18B • K_{spa} Values* for Some Metal Sulfides (25 °C)

Substance	K_{spa}
HgS (red)	4×10^{-54}
HgS (black)	2×10^{-53}
Ag_2S	6×10^{-51}
CuS	6×10^{-37}
PbS	3×10^{-28}
CdS	8×10^{-28}
SnS	1×10^{-26}
FeS	6×10^{-19}

*The equilibrium constant value K_{spa} for metal sulfides refers to the equilibrium $MS(s) + H_2O(\ell) \rightleftharpoons M^{2+}(aq) + OH^-(aq) + HS^-(aq)$; see R. J. Myers, *J. Chem. Ed.,* **63,** 687 (1986).

Formation Constants for Some Complex Ions in Aqueous Solution

Table 19 • Formation Constants for Some Complex Ions in Aqueous Solution

Formation Equilibrium	K
$Ag^+ + 2\ Br^- \rightleftharpoons [AgBr_2]^-$	1.3×10^7
$Ag^+ + 2\ Cl^- \rightleftharpoons [AgCl_2]^-$	2.5×10^5
$Ag^+ + 2\ CN^- \rightleftharpoons [Ag(CN)_2]^-$	5.6×10^{18}
$Ag^+ + 2\ S_2O_3^{2-} \rightleftharpoons [Ag(S_2O_3)_2]^{3-}$	2.0×10^{13}
$Ag^+ + 2\ NH_3 \rightleftharpoons [Ag(NH_3)_2]^+$	1.6×10^7
$Al^{3+} + 6\ F^- \rightleftharpoons [AlF_6]^{3-}$	5.0×10^{23}
$Al^{3+} + 4\ OH^- \rightleftharpoons [Al(OH)_4]^-$	7.7×10^{33}
$Au^+ + 2\ CN^- \rightleftharpoons [Au(CN)_2]^-$	2.0×10^{38}
$Cd^{2+} + 4\ CN^- \rightleftharpoons [Cd(CN)_4]^{2-}$	1.3×10^{17}
$Cd^{2+} + 4\ Cl^- \rightleftharpoons [CdCl_4]^{2-}$	1.0×10^4
$Cd^{2+} + 4\ NH_3 \rightleftharpoons [Cd(NH_3)_4]^{2+}$	1.0×10^7
$Co^{2+} + 6\ NH_3 \rightleftharpoons [Co(NH_3)_6]^{2+}$	7.7×10^4
$Cu^+ + 2\ CN^- \rightleftharpoons [Cu(CN)_2]^-$	1.0×10^{16}
$Cu^+ + 2\ Cl^- \rightleftharpoons [CuCl_2]^-$	1.0×10^5
$Cu^{2+} + 4\ NH_3 \rightleftharpoons [Cu(NH_3)_4]^{2+}$	6.8×10^{12}
$Fe^{2+} + 6\ CN^- \rightleftharpoons [Fe(CN)_6]^{4-}$	7.7×10^{36}
$Hg^{2+} + 4\ Cl^- \rightleftharpoons [HgCl_4]^{2-}$	1.2×10^{15}
$Ni^{2+} + 4\ CN^- \rightleftharpoons [Ni(CN)_4]^{2-}$	1.0×10^{31}
$Ni^{2+} + 6\ NH_3 \rightleftharpoons [Ni(NH_3)_6]^{2+}$	5.6×10^8
$Zn^{2+} + 4\ OH^- \rightleftharpoons [Zn(OH)_4]^{2-}$	2.9×10^{15}
$Zn^{2+} + 4\ NH_3 \rightleftharpoons [Zn(NH_3)_4]^{2+}$	2.9×10^9

Selected Thermodynamic Values*

Table 20 • Selected Thermodynamic Values*

Species	ΔH_f° (298.15 K) kJ/mol	S° (298.15 K) J/K · mol	ΔG_f° (298.15 K) kJ/mol
Aluminum			
Al(s)	0	28.3	0
AlCl$_3$(s)	−705.63	109.29	−630.0
Al$_2$O$_3$(s)	−1675.7	50.92	−1582.3
Barium			
BaCl$_2$(s)	−858.6	123.68	−810.4
BaCO$_3$(s)	−1213	112.1	−1134.41
BaO(s)	−548.1	72.05	−520.38
BaSO$_4$(s)	−1473.2	132.2	−1362.2
Boron			
BCl$_3$(g)	−402.96	290.17	−387.95
Beryllium			
Be(s)	0	9.5	0
Be(OH)$_2$(s)	−902.5	51.9	−815.0
Bromine			
Br(g)	111.884	175.022	82.396
Br$_2$(ℓ)	0	152.2	0
Br$_2$(g)	30.91	245.42	3.12
BrF$_3$(g)	−255.60	292.53	−229.43
HBr(g)	−36.29	198.70	−53.45
Calcium			
Ca(s)	0	41.59	0
Ca(g)	178.2	158.884	144.3
Ca^{2+}(g)	1925.90	—	—
CaC$_2$(s)	−59.8	70.	−64.93
CaCO$_3$(s, calcite)	−1207.6	91.7	−1129.16

(continued)

---

*Most thermodynamic data are taken from the NIST Webbook at http://webbook.nist.gov.

Table 20 • *continued*

Species	ΔH_f° (298.15 K) kJ/mol	S° (298.15 K) J/K · mol	ΔG_f° (298.15 K) kJ/mol
$CaCl_2(s)$	−795.8	104.6	−748.1
$CaF_2(s)$	−1219.6	68.87	−1167.3
$CaH_2(s)$	−186.2	42	−147.2
$CaO(s)$	−635.09	38.2	−603.42
$CaS(s)$	−482.4	56.5	−477.4
$Ca(OH)_2(s)$	−986.09	83.39	−898.43
$Ca(OH)_2(aq)$	−1002.82	−74.5	−868.07
$CaSO_4(s)$	−1434.52	106.5	−1322.02
Carbon			
$C(s, graphite)$	0	5.6	0
$C(s, diamond)$	1.8	2.377	2.900
$C(g)$	716.67	158.1	671.2
$CCl_4(\ell)$	−128.4	214.39	−57.63
$CCl_4(g)$	−95.98	309.65	−53.61
$CHCl_3(\ell)$	−134.47	201.7	−73.66
$CHCl_3(g)$	−103.18	295.61	−70.4
$CH_4(g, methane)$	−74.87	186.26	−50.8
$C_2H_2(g, ethyne)$	226.73	200.94	209.20
$C_2H_4(g, ethene)$	52.47	219.36	68.35
$C_2H_6(g, ethane)$	−83.85	229.2	−31.89
$C_3H_8(g, propane)$	−104.7	270.3	−24.4
$C_6H_6(\ell, benzene)$	48.95	173.26	124.21
$CH_3OH(\ell, methanol)$	−238.4	127.19	−166.14
$CH_3OH(g, methanol)$	−201	126.8	−161.96
$C_2H_5OH(\ell, ethanol)$	−277.0	160.7	−174.7
$C_2H_5OH(g, ethanol)$	−235.3	282.70	−168.49
$CO(g)$	−110.525	197.674	−137.168
$CO_2(g)$	−393.509	213.74	−394.359
$CS_2(\ell)$	89.41	151	65.2
$CS_2(g)$	116.7	237.8	66.61
$COCl_2(g)$	−218.8	283.53	−204.6
Cesium			
$Cs(s)$	0	85.23	0
$Cs^+(g)$	457.964	—	—
$CsCl(s)$	−443.04	101.17	−414.53
Chlorine			
$Cl(g)$	121.3	165.19	105.3
$Cl^-(g)$	−233.13	—	—
$Cl_2(g)$	0	223.08	0
$HCl(g)$	−92.31	186.2	−95.09
$HCl(aq)$	−167.159	56.5	−131.26

Table 20 • *continued*

Species	ΔH_f° (298.15 K) kJ/mol	S° (298.15 K) J/K · mol	ΔG_f° (298.15 K) kJ/mol
Chromium			
Cr(s)	0	23.62	0
Cr_2O_3(s)	−1134.7	80.65	−1052.95
$CrCl_3$(s)	−556.5	123.0	−486.1
Copper			
Cu(s)	0	33.17	0
CuO(s)	−156.06	42.59	−128.3
$CuCl_2$(s)	−220.1	108.07	−175.7
$CuSO_4$(s)	−769.98	109.05	−660.75
Fluorine			
F_2(g)	0	202.8	0
F(g)	78.99	158.754	61.91
F^-(g)	−255.39	—	—
F^-(aq)	−332.63	−13.8	−278.79
HF(g)	−271.1	173.779	−273.2
HF(aq)	−332.63	88.7	−278.79
Hydrogen			
H_2(g)	0	130.7	0
H(g)	217.965	114.713	203.247
H^+(g)	1536.202	—	—
$H_2O(\ell)$	−285.83	69.95	−237.15
H_2O(g)	−241.83	188.84	−228.59
$H_2O_2(\ell)$	−187.78	109.6	−120.35
Iodine			
I_2(s)	0	116.135	0
I_2(g)	62.438	260.69	19.327
I(g)	106.838	180.791	70.250
I^-(g)	−197	—	—
ICl(g)	17.51	247.56	−5.73
Iron			
Fe(s)	0	27.78	0
FeO(s)	−272	—	—
Fe_2O_3(s, hematite)	−824.2	87.40	−742.2
Fe_3O_4(s, magnetite)	−1118.4	146.4	−1015.4
$FeCl_2$(s)	−341.79	117.95	−302.30
$FeCl_3$(s)	−399.49	142.3	−344.00
FeS_2(s, pyrite)	−178.2	52.93	−166.9
$Fe(CO)_5(\ell)$	−774.0	338.1	−705.3
Lead			
Pb(s)	0	64.81	0
$PbCl_2$(s)	−359.41	136.0	−314.10

(continued)

Table 20 • continued

Species	ΔH_f° (298.15 K) kJ/mol	S° (298.15 K) J/K · mol	ΔG_f° (298.15 K) kJ/mol
PbO(s, yellow)	−219	66.5	−196
PbO$_2$(s)	−277.4	68.6	−217.39
PbS(s)	−100.4	91.2	−98.7
Lithium			
Li(s)	0	29.12	0
Li$^+$(g)	685.783	—	—
LiOH(s)	−484.93	42.81	−438.96
LiOH(aq)	−508.48	2.80	−450.58
LiCl(s)	−408.701	59.33	−384.37
Magnesium			
Mg(s)	0	32.67	0
MgCl$_2$(s)	−641.62	89.62	−592.09
MgCO$_3$(s)	−1111.69	65.84	−1028.2
MgO(s)	−601.24	26.85	−568.93
Mg(OH)$_2$(s)	−924.54	63.18	−833.51
MgS(s)	−346.0	50.33	−341.8
Mercury			
Hg(ℓ)	0	76.02	0
HgCl$_2$(s)	−224.3	146.0	−178.6
HgO(s, red)	−90.83	70.29	−58.539
HgS(s, red)	−58.2	82.4	−50.6
Nickel			
Ni(s)	0	29.87	0
NiO(s)	−239.7	37.99	−211.7
NiCl$_2$(s)	−305.332	97.65	−259.032
Nitrogen			
N$_2$(g)	0	191.56	0
N(g)	472.704	153.298	455.563
NH$_3$(g)	−45.90	192.77	−16.37
N$_2$H$_4$(ℓ)	50.63	121.52	149.45
NH$_4$Cl(s)	−314.55	94.85	−203.08
NH$_4$Cl(aq)	−299.66	169.9	−210.57
NH$_4$NO$_3$(s)	−365.56	151.08	−183.84
NH$_4$NO$_3$(aq)	−339.87	259.8	−190.57
NO(g)	90.29	210.76	86.58
NO$_2$(g)	33.1	240.04	51.23
N$_2$O(g)	82.05	219.85	104.20
N$_2$O$_4$(g)	9.08	304.38	97.73
NOCl(g)	51.71	261.8	66.08
HNO$_3$(ℓ)	−174.10	155.60	−80.71
HNO$_3$(g)	−135.06	266.38	−74.72

Table 20 • continued

Species	ΔH_f° (298.15 K) kJ/mol	S° (298.15 K) J/K · mol	ΔG_f° (298.15 K) kJ/mol
$HNO_3(aq)$	−207.36	146.4	−111.25
Oxygen			
$O_2(g)$	0	205.07	0
$O(g)$	249.170	161.055	231.731
$O_3(g)$	142.67	238.92	163.2
Phosphorus			
$P_4(s, white)$	0	41.1	0
$P_4(s, red)$	−17.6	22.80	−12.1
$P(g)$	314.64	163.193	278.25
$PH_3(g)$	22.89	210.24	30.91
$PCl_3(g)$	−287.0	311.78	−267.8
$P_4O_{10}(s)$	−2984.0	228.86	−2697.7
$H_3PO_4(\ell)$	−1279.0	110.5	−1119.1
Potassium			
$K(s)$	0	64.63	0
$KCl(s)$	−436.68	82.56	−408.77
$KClO_3(s)$	−397.73	143.1	−296.25
$KI(s)$	−327.90	106.32	−324.892
$KOH(s)$	−424.72	78.9	−378.92
$KOH(aq)$	−482.37	91.6	−440.50
Silicon			
$Si(s)$	0	18.82	0
$SiBr_4(\ell)$	−457.3	277.8	−443.9
$SiC(s)$	−65.3	16.61	−62.8
$SiCl_4(g)$	−662.75	330.86	−622.76
$SiH_4(g)$	34.31	204.65	56.84
$SiF_4(g)$	−1614.94	282.49	−1572.65
$SiO_2(s, quartz)$	−910.86	41.46	−856.97
Silver			
$Ag(s)$	0	42.55	0
$Ag_2O(s)$	−31.1	121.3	−11.32
$AgCl(s)$	−127.01	96.25	−109.76
$AgNO_3(s)$	−124.39	140.92	−33.41
Sodium			
$Na(s)$	0	51.21	0
$Na(g)$	107.3	153.765	76.83
$Na^+(g)$	609.358	—	—
$NaBr(s)$	−361.02	86.82	−348.983
$NaCl(s)$	−411.12	72.11	−384.04
$NaCl(g)$	−181.42	229.79	−201.33
$NaCl(aq)$	−407.27	115.5	−393.133

(continued)

Table 20 • continued

Species	ΔH_f° (298.15 K) kJ/mol	S° (298.15 K) J/K · mol	ΔG_f° (298.15 K) kJ/mol
NaOH(s)	−425.93	64.46	−379.75
NaOH(aq)	−469.15	48.1	−418.09
Na_2CO_3(s)	−1130.77	134.79	−1048.08
Sulfur			
S(s, rhombic)	0	32.1	0
S(g)	278.98	167.83	236.51
S_2Cl_2(g)	−18.4	331.5	−31.8
SF_6(g)	−1209	291.82	−1105.3
H_2S(g)	−20.63	205.79	−33.56
SO_2(g)	−296.84	248.21	−300.13
SO_3(g)	−395.77	256.77	−371.04
$SOCl_2$(g)	−212.5	309.77	−198.3
$H_2SO_4(\ell)$	−814	156.9	−689.96
H_2SO_4(aq)	−909.27	20.1	−744.53
Tin			
Sn(s, white)	0	51.08	0
Sn(s, gray)	−2.09	44.14	0.13
$SnCl_4(\ell)$	−511.3	258.6	−440.15
$SnCl_4$(g)	−471.5	365.8	−432.31
SnO_2(s)	−577.63	49.04	−515.88
Titanium			
Ti(s)	0	30.72	0
$TiCl_4(\ell)$	−804.2	252.34	−737.2
$TiCl_4$(g)	−763.16	354.84	−726.7
TiO_2(s)	−939.7	49.92	−884.5
Zinc			
Zn(s)	0	41.63	0
$ZnCl_2$(s)	−415.05	111.46	−369.398
ZnO(s)	−348.28	43.64	−318.30
ZnS(s, sphalerite)	−205.98	57.7	−201.29

Standard Reduction Potentials in Aqueous Solution at 25 °C

Table 21 • Standard Reduction Potentials in Aqueous Solution at 25 °C

Acidic Solution	Standard Reduction Potential, $E°$ (volts)
F_2 (g) + 2 e^- ⟶ 2 F^- (aq)	2.87
Co^{3+} (aq) + e^- ⟶ Co^{2+} (aq)	1.82
Pb^{4+} (aq) + 2 e^- ⟶ Pb^{2+} (aq)	1.8
H_2O_2 (aq) + 2 H^+ (aq) + 2 e^- ⟶ 2 H_2O	1.77
NiO_2 (s) + 4 H^+ (aq) + 2 e^- ⟶ Ni^{2+} (aq) + 2 H_2O	1.7
PbO_2 (s) + SO_4^{2-} (aq) + 4 H^+ (aq) + 2 e^- ⟶ $PbSO_4$ (s) + 2 H_2O	1.685
Au^+ (aq) + e^- ⟶ Au (s)	1.68
2 HClO (aq) + 2 H^+ (aq) + 2 e^- ⟶ Cl_2 (g) + 2 H_2O	1.63
Ce^{4+} (aq) + e^- ⟶ Ce^{3+} (aq)	1.61
$NaBiO_3$ (s) + 6 H^+ (aq) + 2 e^- ⟶ Bi^{3+} (aq) + Na^+ (aq) + 3 H_2O	≈ 1.6
MnO_4^- (aq) + 8 H^+ (aq) + 5 e^- ⟶ Mn^{2+} (aq) + 4 H_2O	1.51
Au^{3+} (aq) + 3 e^- ⟶ Au (s)	1.50
ClO_3^- (aq) + 6 H^+ (aq) + 5 e^- ⟶ $\frac{1}{2}$ Cl_2 (g) + 3 H_2O	1.47
BrO_3^- (aq) + 6 H^+ (aq) + 6 e^- ⟶ Br^- (aq) + 3 H_2O	1.44
Cl_2 (g) + 2 e^- ⟶ 2 Cl^- (aq)	1.36
$Cr_2O_7^{2-}$ (aq) + 14 H^+ (aq) + 6 e^- ⟶ 2 Cr^{3+} (aq) + 7 H_2O	1.33
$N_2H_5^+$ (aq) + 3 H^+ (aq) + 2 e^- ⟶ 2 NH_4^+ (aq)	1.24
MnO_2 (s) + 4 H^+ (aq) + 2 e^- ⟶ Mn^{2+} (aq) + 2 H_2O	1.23
O_2 (g) + 4 H^+ (aq) + 4 e^- ⟶ 2 H_2O	1.229
Pt^{2+} (aq) + 2 e^- ⟶ Pt (s)	1.2
IO_3^- (aq) + 6 H^+ (aq) + 5 e^- ⟶ $\frac{1}{2}$ I_2 (aq) + 3 H_2O	1.195
ClO_4^- (aq) + 2 H^+ (aq) + 2 e^- ⟶ ClO_3^- (aq) + H_2O	1.19
Br_2 (ℓ) + 2 e^- ⟶ 2 Br^- (aq)	1.08
$AuCl_4^-$ (aq) + 3 e^- ⟶ Au (s) + 4 Cl^- (aq)	1.00
Pd^{2+} (aq) + 2 e^- ⟶ Pd (s)	0.987
NO_3^- (aq) + 4 H^+ (aq) + 3 e^- ⟶ NO (g) + 2 H_2O	0.96

(continued)

Table 21 • *continued*

Acidic Solution	Standard Reduction Potential, $E°$ (volts)
NO_3^- (aq) + 3 H^+ (aq) + 2 e^- ⟶ HNO_2 (aq) + H_2O	0.94
2 Hg^+ (aq) + 2 e^- ⟶ Hg_2^{2+} (aq)	0.920
Hg^{2+} (aq) + 2 e^- ⟶ Hg (ℓ)	0.855
Ag^+ (aq) + e^- ⟶ Ag (s)	0.7994
Hg_2^{2+} (aq) + 2 e^- ⟶ 2 Hg (ℓ)	0.789
Fe^{3+} (aq) + e^- ⟶ Fe^{2+} (aq)	0.771
$SbCl_6^-$ (aq) + 2 e^- ⟶ $SbCl_4^-$ (aq) + 2 Cl^- (aq)	0.75
$[PtCl_4]^{2+}$ (aq) + 2 e^- ⟶ Pt (s) + 4 Cl^- (aq)	0.73
O_2 (g) + 2 H^+ (aq) + 2 e^- ⟶ H_2O_2 (aq)	0.682
$[PtCl_6]^{2-}$ (aq) + 2 e^- ⟶ $[PtCl_4]^{2-}$ (aq) + 2 Cl^- (aq)	0.68
H_3AsO_4 (aq) + 2 H^+ (aq) + 2 e^- ⟶ H_3AsO_3 (aq) + H_2O	0.58
I_2 (s) + 2 e^- ⟶ 2 I^- (aq)	0.535
TeO_2 (s) + 4 H^+ (aq) + 4 e^- ⟶ Te (s) + 2 H_2O	0.529
Cu^+ (aq) + e^- ⟶ Cu (s)	0.521
$[RhCl_6]^{3-}$ (aq) + 3 e^- ⟶ Rh (s) + 6 Cl^- (aq)	0.44
Cu^{2+} (aq) + 2 e^- ⟶ Cu (s)	0.337
Hg_2Cl_2 (s) + 2 e^- ⟶ 2 Hg (ℓ) + 2 Cl^- (aq)	0.27
AgCl (s) + e^- ⟶ Ag (s) + Cl^- (aq)	0.222
SO_4^{2-} (aq) + 4 H^+ (aq) + 2 e^- ⟶ SO_2 (g) + 2 H_2O	0.20
SO_4^{2-} (aq) + 4 H^+ (aq) + 2 e^- ⟶ H_2SO_3 (aq) + H_2O	0.17
Cu^{2+} (aq) + e^- ⟶ Cu^+ (aq)	0.153
Sn^{4+} (aq) + 2 e^- ⟶ Sn^{2+} (aq)	0.15
S (s) + 2 H^+ + 2 e^- ⟶ H_2S (aq)	0.14
AgBr (s) + e^- ⟶ Ag (s) + Br^- (aq)	0.0713
2 H^+ (aq) + 2 e^- ⟶ H_2 (g) (reference electrode)	0.0000
N_2O (g) + 6 H^+ (aq) + H_2O + 4 e^- ⟶ 2 NH_3OH^+ (aq)	−0.05
Pb^{2+} (aq) + 2 e^- ⟶ Pb (s)	−0.126
Sn^{2+} (aq) + 2 e^- ⟶ Sn (s)	−0.14
AgI (s) + e^- ⟶ Ag (s) + I^- (aq)	−0.15
$[SnF_6]^{2-}$ (aq) + 4 e^- ⟶ Sn (s) + 6 F^- (aq)	−0.25
Ni^{2+} (aq) + 2 e^- ⟶ Ni (s)	−0.25
Co^{2+} (aq) + 2 e^- ⟶ Co (s)	−0.28
Tl^+ (aq) + e^- ⟶ Tl (s)	−0.34
$PbSO_4$ (s) + 2 e^- ⟶ Pb (s) + SO_4^{2-} (aq)	−0.356
Se (s) + 2 H^+ (aq) + 2 e^- ⟶ H_2Se (aq)	−0.40
Cd^{2+} (aq) + 2 e^- ⟶ Cd (s)	−0.403
Cr^{3+} (aq) + e^- ⟶ Cr^{2+} (aq)	−0.41
Fe^{2+} (aq) + 2 e^- ⟶ Fe (s)	−0.44
2 CO_2 (g) + 2 H^+ (aq) + 2 e^- ⟶ $H_2C_2O_4$ (aq)	−0.49
Ga^{3+} (aq) + 3 e^- ⟶ Ga (s)	−0.53
HgS (s) + 2 H^+ (aq) + 2 e^- ⟶ Hg (ℓ) + H_2S (g)	−0.72
Cr^{3+} (aq) + 3 e^- ⟶ Cr (s)	−0.74
Zn^{2+} (aq) + 2 e^- ⟶ Zn (s)	−0.763
Cr^{2+} (aq) + 2 e^- ⟶ Cr (s)	−0.91
FeS (s) + 2 e^- ⟶ Fe (s) + S^{2-} (aq)	−1.01
Mn^{2+} (aq) + 2 e^- ⟶ Mn (s)	−1.18
V^{2+} (aq) + 2 e^- ⟶ V (s)	−1.18

Table 21 • *continued*

	Standard Reduction Potential, $E°$ (volts)
Acidic Solution	
$CdS\ (s) + 2\ e^- \longrightarrow Cd\ (s) + S^{2-}\ (aq)$	-1.21
$ZnS\ (s) + 2\ e^- \longrightarrow Zn\ (s) + S^{2-}\ (aq)$	-1.44
$Zr^{4+}\ (aq) + 4\ e^- \longrightarrow Zr\ (s)$	-1.53
$Al^{3+}\ (aq) + 3\ e^- \longrightarrow Al\ (s)$	-1.66
$Mg^{2+}\ (aq) + 2\ e^- \longrightarrow Mg\ (s)$	-2.37
$Na^+\ (aq) + e^- \longrightarrow Na\ (s)$	-2.714
$Ca^{2+}\ (aq) + 2\ e^- \longrightarrow Ca\ (s)$	-2.87
$Sr^{2+}\ (aq) + 2\ e^- \longrightarrow Sr\ (s)$	-2.89
$Ba^{2+}\ (aq) + 2\ e^- \longrightarrow Ba\ (s)$	-2.90
$Rb^+\ (aq) + e^- \longrightarrow Rb\ (s)$	-2.925
$K^+\ (aq) + e^- \longrightarrow K\ (s)$	-2.925
$Li^+\ (aq) + e^- \longrightarrow Li\ (s)$	-3.045
Basic Solution	
$ClO^-\ (aq) + H_2O + 2\ e^- \longrightarrow Cl^-\ (aq) + 2\ OH^-\ (aq)$	0.89
$OOH^-\ (aq) + H_2O + 2\ e^- \longrightarrow 3\ OH^-\ (aq)$	0.88
$2\ NH_2OH\ (aq) + 2\ e^- \longrightarrow N_2H_4\ (aq) + 2\ OH^-\ (aq)$	0.74
$ClO_3^-\ (aq) + 3\ H_2O + 6\ e^- \longrightarrow Cl^-\ (aq) + 6\ OH^-\ (aq)$	0.62
$MnO_4^-\ (aq) + 2\ H_2O + 3\ e^- \longrightarrow MnO_2\ (s) + 4\ OH^-\ (aq)$	0.588
$MnO_4^-\ (aq) + e^- \longrightarrow MnO_4^{2-}\ (aq)$	0.564
$NiO_2\ (s) + 2\ H_2O + 2\ e^- \longrightarrow Ni(OH)_2\ (s) + 2\ OH^-\ (aq)$	0.49
$Ag_2CrO_4\ (s) + 2\ e^- \longrightarrow 2\ Ag\ (s) + CrO_4^{2-}\ (aq)$	0.446
$O_2\ (g) + 2\ H_2O + 4\ e^- \longrightarrow 4\ OH^-\ (aq)$	0.40
$ClO_4^-\ (aq) + H_2O + 2\ e^- \longrightarrow ClO_3^-\ (aq) + 2\ OH^-\ (aq)$	0.36
$Ag_2O\ (s) + H_2O + 2\ e^- \longrightarrow 2\ Ag\ (s) + 2\ OH^-\ (aq)$	0.34
$2\ NO_2^-\ (aq) + 3\ H_2O + 4\ e^- \longrightarrow N_2O\ (g) + 6\ OH^-\ (aq)$	0.15
$N_2H_4\ (aq) + 2\ H_2O + 2\ e^- \longrightarrow 2\ NH_3\ (aq) + 2\ OH^-\ (aq)$	0.10
$[Co(NH_3)_6]^{3+}\ (aq) + e^- \longrightarrow [Co(NH_3)_6]^{2+}\ (aq)$	0.10
$HgO\ (s) + H_2O + 2\ e^- \longrightarrow Hg\ (\ell) + 2\ OH^-\ (aq)$	0.0984
$O_2\ (g) + H_2O + 2\ e^- \longrightarrow OOH^-\ (aq) + OH^-\ (aq)$	0.076
$NO_3^-\ (aq) + H_2O + 2\ e^- \longrightarrow NO_2^-\ (aq) + 2\ OH^-\ (aq)$	0.01
$MnO_2\ (s) + 2\ H_2O + 2\ e^- \longrightarrow Mn(OH)_2\ (s) + 2\ OH^-\ (aq)$	-0.05
$CrO_4^{2-}\ (aq) + 4\ H_2O + 3\ e^- \longrightarrow Cr(OH)_3\ (s) + 5\ OH^-\ (aq)$	-0.12
$Cu(OH)_2\ (s) + 2\ e^- \longrightarrow Cu\ (s) + 2\ OH^-\ (aq)$	-0.36
$S\ (s) + 2\ e^- \longrightarrow S^{2-}\ (aq)$	-0.48
$Fe(OH)_3\ (s) + e^- \longrightarrow Fe(OH)_2\ (s) + OH^-\ (aq)$	-0.56
$2\ H_2O + 2\ e^- \longrightarrow H_2\ (g) + 2\ OH^-\ (aq)$	-0.8277
$2\ NO_3^-\ (aq) + 2\ H_2O + 2\ e^- \longrightarrow N_2O_4\ (g) + 4\ OH^-\ (aq)$	-0.85
$Fe(OH)_2\ (s) + 2\ e^- \longrightarrow Fe\ (s) + 2\ OH^-\ (aq)$	-0.877
$SO_4^{2-}\ (aq) + H_2O + 2\ e^- \longrightarrow SO_3^{2-}\ (aq) + 2\ OH^-\ (aq)$	-0.93
$N_2\ (g) + 4\ H_2O + 4\ e^- \longrightarrow N_2H_4\ (aq) + 4\ OH^-\ (aq)$	-1.15
$[Zn(OH)_4]^{2-}\ (aq) + 2\ e^- \longrightarrow Zn\ (s) + 4\ OH^-\ (aq)$	-1.22
$Zn(OH)_2\ (s) + 2\ e^- \longrightarrow Zn\ (s) + 2\ OH^-\ (aq)$	-1.245
$[Zn(CN)_4]^{2-}\ (aq) + 2\ e^- \longrightarrow Zn\ (s) + 4\ CN^-\ (aq)$	-1.26
$Cr(OH)_3\ (s) + 3\ e^- \longrightarrow Cr\ (s) + 3\ OH^-\ (aq)$	-1.30
$SiO_3^{2-}\ (aq) + 3\ H_2O + 4\ e^- \longrightarrow Si\ (s) + 6\ OH^-\ (aq)$	-1.70

Answers to Exercises

Chapter 1

1.1 (a) Na = sodium; Cl = chlorine; Cr = chromium

    (b) Zinc = Zn; nickel = Ni; potassium = K

1.2 (a) Iron: lustrous solid, metallic, good conductor of heat and electricity, malleable, ductile, attracted to a magnet.

    (b) Water: Colorless liquid (at room temperature), melting point is 0 °C and boiling point is 100 °C.

    (c) Table salt: Solid, white crystals, soluble in water

    (d) Oxygen: Colorless gas (at room temperature).

1.3 $15.5 \text{ g} \cdot (1 \text{ cm}^3/1.18 \times 10^{-3} \text{ g}) = 1.31 \times 10^4 \text{ cm}^3$

1.4 $77 \text{ K} - 273.15 \text{ K} (1 \text{ °C/K}) = -196 \text{ °C}$

1.5 (a) $(1.293 \times 10^{-3} \text{ g/cm}^3)(1000 \text{ cm}^3/\text{L}) = 1.293 \text{ g/L}$

    (b) Density decreases by about 0.025 g/cm^3 for each 10 °C increase in temperature. The density at 30 °C is expected to be about $13.546 \text{ g/cm}^3 - 0.025$ $\text{g/cm}^3 = 13.521 \text{ g/cm}^3$.

1.6 Chemical changes: the fuel in the campfire burns in air (combustion). Physical changes: water boils. Energy evolved in combustion is transferred to the water, to the water container, and to the surrounding air.

1.7 Based on appearance, the beaker on the left appears to contain a homogeneous solution. The beaker on the right appears to contain a heterogeneous mixture of at least two solid substances.

1.8 Length: 25.3 cm (1 m/100 cm) = 0.253 m;
        25.3 cm (10 mm/1 cm) = 253 mm

Width: 21.6 (1 m/100 cm) = 0.216 m;
        21.6 cm (10 mm/1 cm) = 216 mm

Area = length · width = (25.3 cm)(21.6 cm)
    = 546 cm^2; 546 cm^2(1 m/100 cm)^2 = 0.0546 m^2

1.9 Area of sheet = $(2.50 \text{ cm})^2 = 6.25 \text{ cm}^2$

Volume = 1.656 g(1 cm^3/21.45 g) = 0.07220 cm^3

Thickness = volume/area = (0.07220 cm^3/6.25 cm^2)
    = 0.0124 cm

0.0124 cm(10 mm /1 cm) = 0.124 mm

1.10 (a) 750 mL(1 L/1000 mL) = 0.75 L
       0.75 L(10 dL/L) = 7.5 dL

    (b) 2.0 qt = 0.50 gal
       0.50 gal(3.786 L/gal) = 1.9 L
       1.9 L (1 dm^3/1 L) = 1.9 dm^3

1.11 (a) mass in grams = 325 mg(1 g/1000 mg) = 0.325 g
       Mass in kilograms = 0.325 g(1 kg/1000 g)
             $= 3.25 \times 10^{-4}$ kg

    (b) $(19{,}320 \text{ kg/m}^3)(1000 \text{ g/1 kg})(1 \text{ m}^3/1 \times 10^{-6} \text{ cm}^3)$
       = 19.32 g/cm^3

    (c) Diameter = 5.0 mm (1 cm/10 mm)
            = 0.50 cm; radius = 0.25 cm
      Volume = π (radius)^2(length)
           = (3.1416)(0.25 cm)^2(3.0 cm) = 0.59 cm^3
      Mass = (0.59 cm^3)(21.450 g/1 cm^3) = 13 g

1.12 Student A: average = −0.1 °C; average deviation = 0.2 °C; error = −0.1 °C. Student B: average = 273.16 K; average deviation = 0.02 K; error = +0.01 K. Student B's values are more accurate and more precise.

1.13 (a) The product of 10.26 and 0.063 is 0.65, a number with two significant figures. (10.26 has four significant figures whereas 0.063 has two.)
The sum of 10.26 and 0.063 is 10.32. The number 10.26 has only two numbers to the right of the decimal, so the sum must also.

    (b) $x = 3.9 \times 10^6$. The difference between 110.7 and 64 is 47. Dividing 47 by 0.056 and 0.00216 gives an answer with two significant figures.

1.14 Change dimensions to centimeters: 7.6 m = 760 cm; 2.74 m = 274 cm; and 0.13 mm = 0.013 cm.

Volume of paint = (760 cm)(274 cm)(0.013 cm)
            $= 2.7 \times 10^3$ cm^3

Volume (L) = $(2.7 \times 10^3 \text{ cm}^3)(1 \text{ L}/10^3 \text{ cm}^3)$ = 2.7 L

Mass = $(2.7 \times 10^3 \text{ cm}^3)(0.914 \text{ g/cm}^3) = 2.5 \times 10^3$ g

Chapter 2

2.1 The ratio of the atom radius to the radius of the nucleus is 1×10^5 to 1. If the radius of the atom were 100 m, then the radius of the nucleus is 0.0010 m or 1.0 mm and the diameter is 2.0 mm. The head of an ordinary pin has a diameter of about 1.0 mm.

2.2 (a) Mass number with 26 protons and 30 neutrons is 56 amu

    (b) 58.93079 amu$(1.661 \times 10^{-24}$ g/ amu$) = 9.955 \times 10^{-23}$ g

    (c) ^{64}Zn has 30 protons, 30 electrons, and $(64 - 30) = 34$ neutrons.

2.3 (a) Argon has an atomic number of 18. ^{36}Ar, ^{38}Ar, ^{40}Ar

    (b) ^{69}Ga: 31 protons, 38 neutrons; ^{71}Ga: 31 protons, 40 neutrons; % abundance of ^{71}Ga = 39.9 %

2.4 $(0.7577)(34.96885$ amu$) + (0.2423)(36.96590$ amu$) = 35.45$ amu

2.5 (a) 1.5 mol Si$(28.1$ g/mol$) = 42.$ g Si

    (b) 454 g S (1 mol S/32.07 g) = 14.2 mol S

    14.2 mol S $(6.022 \times 10^{23}$ atoms/mol$) = 8.53 \times 10^{24}$ atoms S

    (c) $(32.07$ g S/1 mol S$)(1$ mol S/6.022×10^{23} atoms$) = 5.325 \times 10^{-23}$ g / atom

2.6 2.6×10^{24} atoms$(1$ mol/6.022×10^{23} atoms$)$ $(197.0$ g Au/1 mol$) = 850$ g Au

    Volume = 850 g Au $(1$ cm$^3/19.32$ g$) = 44$ cm^3

    Volume = 44 cm^3 = (thickness)(area)

              = (0.10 cm)(area)

    Area = 440 cm^2

    Length = width = $(440$ cm$^2)^{1/2} = 21$ cm

2.7 There are 8 elements in the third period. Sodium (Na), magnesium (Mg), and aluminum (Al) are metals. Silicon (Si) is a metalloid. Phosphorus (P), sulfur (S), chlorine (Cl), and argon (Ar) are nonmetals.

CHAPTER 3

3.1 The molecular formula for styrene C_8H_8; the condensed formula, $C_6H_5CH{=}CH_2$ is more descriptive of its structure.

3.2 Cysteine's molecular formula is $C_3H_7NO_2S$. You will often see its formula written as $HSCH_2CH(N^+H_3)CO_2{}^-$ to emphasize the molecule's structure.

3.3 (a) K^+ is formed if K loses one electron. K^+ has the same number of electrons as Ar.

    (b) Se^{2-} is formed by adding two electrons to an atom of Se. It has the same number of electrons as Kr.

    (c) Ba^{2+} is formed if Ba loses two electrons; Ba^{2+} has same number of electrons as Xe.

    (d) Cs^+ is formed if Cs loses one electron. It has same number of electrons as Xe.

3.4 (a) (1) NaF: 1 Na^+ and 1 F^- ion. (2) $Cu(NO_3)_2$: 1 Cu^{2+} and 2 $NO_3{}^-$ ions. (3) $NaCH_3CO_2$: (4) 1 Na^+ and 1 $CH_3CO_2{}^-$ ion.

    (b) $FeCl_2$ and $FeCl_3$

    (c) Na_2S, Na_3PO_4, BaS, $Ba_3(PO_4)_2$

3.5 1. (a) NH_4NO_3 (b) $CoSO_4$ (c) $Ni(CN)_2$ (d) V_2O_3 (e) $Ba(CH_3CO_2)_2$ (f) $Ca(ClO)_2$

    2. (a) magnesium bromide (b) lithium carbonate (c) potassium hydrogen sulfite (d) potassium permanganate (e) ammonium sulfide (f) copper(I) chloride and copper(II) chloride

3.6 The force of attraction between ions is proportional to the product of the ion charges (Coulomb's law). The force of attraction between Mg^{2+} and O^{2-} ions in MgO is approximately four times greater than the force of attraction between Na^+ and Cl^- ions in NaCl so a much higher temperature is required to disrupt the orderly array of ions in crystalline MgO.

3.7 1. (a) CO_2 (b) PI_3 (c) SCl_2 (d) BF_3 (e) O_2F_2 (f) XeO_3

    2. (a) dinitrogen tetrafluoride (b) hydrogen bromide (c) sulfur tetrafluoride (d) boron trichloride (e) tetraphosphorus decaoxide (f) chlorine trifluoride

3.8 (a) Citric acid: 192.1 g/mol. Magnesium carbonate: 84.3 g/mol.

    (b) 454 g citric acid$(1$ mol/192.1 g$) = 2.36$ mol citric acid

    (c) 0.125 mol $MgCO_3$ $(84.3$ g/mol$) = 10.5$ g $MgCO_3$

3.9 (a) 1.00 mol NaCl (molar mass 58.45 g/mol) contains 23.0 g of Na (39.3 %) and 35.5 g of Cl (60.7 %).

    (b) 1.00 mol C_8H_{18} (molar mass 114.2 g/mol) contains 96.08 g of C (84.13 %) and 18.14 g of H (15.88 %).

    (c) 1.00 mol $(NH_4)_2CO_3$ (molar mass 96.07 g/mol) has 28.01 g of N (29.16 %), 8.06 g of H (8.39%), 12.01 g of C (12.50 %), and 48.0 g of O (49.96 %).

3.10 454 g C_8H_{18} $(1$ mol C_8H_{18}/114.2 g$)$ $(8$ mol C/1 mole $C_8H_{18})$ $(12.01$ g C/1 mol C$) = 382$ g C

3.11 (a) C_5H_4 (b) $C_2H_4O_2$

3.12 88.17 g C$(1$ mole C/12.011 g C$) = 7.341$ mol C

    11.83 g H$(1$ mol H/1.008 g H$) = 11.74$ mol H

    11.74 mol H/7.341 mol C = 1.6 mol H/1 mole C = (8/5)mol H/1 mol C = 8 mol H/5 mol C. Thus, the empirical formula is C_5H_8. The molar mass, 68.11 g/mol, matches this formula so C_5H_8 is also the molecular formula.

3.13 78.90 g C$(1$ mole C/12.011 g C$) = 6.569$ mol C

    10.59 g H$(1$ mol H/1.008 g H$) = 10.51$ mol H

    10.51 g O$(1$ mol O/16.00 g O$) = 0.6569$ mol O

    10.51 mol H/0.6569 mol O = 16 mol H/1 mol O

    6.569 mol C/0.6569 mol O = 10 mol C/1 mol O

    Thus, the empirical formula is $C_{10}H_{16}O$.

3.14 0.586 g K (1 mol K/39.10 g K) = 0.0150 mol K

    0.480 g O (1 mol O/16.00 g O) = 0.0300 mol O

    The ratio of moles K to moles O atoms is 1 to 2 and the empirical formula is KO_2

3.15 mass of water lost on heating is 0.235 g − 0.128 g = 0.107 g

    0.107 g H_2O(1 mol H_2O/18.016 g H_2O) = 0.00594 mol H_2O

0.128 g $NiCl_2$(1 mol $NiCl_2$/129.6 g $NiCl_2$) = 0.000988 mol $NiCl_2$

mole ratio = 0.00594 mol H_2O/0.000988 mol $NiCl_2$ = 6.01

The formula for the hydrate is $NiCl_2 \cdot 6\ H_2O$

CHAPTER 4

4.1 (a) Reactants: Fe: iron, a solid, and Cl_2, chlorine, a gas. Product: $FeCl_3$, iron(III) chloride, a solid

(b) Stoichiometric coefficients: 2 for Fe, 3 for Cl_2, and 2 for $FeCl_3$

(c) 8000 atoms of Fe require (3/2)8000 = 12,000 molecules of Cl_2

4.2 (a) $2\ C_4H_{10}(g) + 13\ O_2(g) \longrightarrow 8\ CO_2(g) + 10\ H_2O(g)$

(b) $2\ Pb(C_2H_5)_4(\ell) + 27\ O_2\ (g) \longrightarrow$
$$2\ PbO(s) + 16\ CO_2(g) + 20\ H_2O(\ell)$$

4.3 454 g C_3H_8(1 mol C_3H_8/44.09 g C_3H_8) = 10.3 mol C_3H_8

10.3 mol C_3H_8(5 mol O_2/1 mol C_3H_8)(32.00 g O_2/mol O_2) = 1650 g O_2

10.3 mol C_3H_8(3 mol CO_2/1 mol C_3H_8)(44.01 g CO_2/mol CO_2) = 1360 g CO_2

10.3 mol C_3H_8(4 mol H_2O/1 mol C_3H_8)(18.02 g H_2O/mol H_2O) = 742 g H_2O

4.4 Amount C = 125 g C(1 mol C/12.01 g C) = 10.41 mol C

Amount Cl_2 = 125 g Cl_2(1 mole Cl_2/70.90 g Cl_2) = 1.763 mol Cl_2

Mol C/mol Cl_2 = 10.41/1.763 = 5.90. This is more than the 1/2 ratio required so the limiting reactant is Cl_2.

Mass $TiCl_4$ = 1.763 mol Cl_2(1 mol $TiCl_4$/2 mol Cl_2) (189.7 g $TiCl_4$/1 mol $TiCl_4$) = 167 g $TiCl_4$

4.5 Amount Mg = 225 g Mg(1 mol Mg/24.31 g Mg) = 9.255 mol Mg

Amount $SiCl_4$ = 225 g $SiCl_4$(1 mole $SiCl_4$/169.9 g $SiCl_4$) = 1.324 mol $SiCl_4$

Mol Mg/mol $SiCl_4$ = 9.255/1.324 = 6.99. This is more than the 2/1 ratio required so the limiting reactant is $SiCl_4$.

Mass Si = 1.324 mol $SiCl_4$(1 mol Si/1 mol $SiCl_4$)(28.09 g Si/1 mol Si) = 37.2 g Si

4.6 theoretical yield = 125 g CH_3OH (1 mol CH_3OH/32.04 g CH_3OH)(2 mol H_2/1 mol CH_3OH)(2.016 g H_2/1 mol H_2) = 15.7 g H_2

Percent yield = (13.6 g/15.7 g)(100%) = 86.6%

4.7 0.143 g O_2(1 mol O_2/32.00 g O_2)(3 mol TiO_2/3 mol O_2) (79.88 g TiO_2/1 mol TiO_2) = 0.357 g TiO_2

Percent TiO_2 in sample = (0.357 g/2.367 g)(100%) = 15.1%

4.8 1.612 g CO_2(1 mol CO_2/44.01 g CO_2) (1 mol C/1 mol CO_2) = 0.03663 mol C

0.7425 g H_2O(1 mol H_2O/18.01 g H_2O)(2 mol H/1 mol H_2O) = 0.08245 mol H

0.08245 mol H/0.03663 mol = 2.25 H/1 C = 9 H/4 C.

Thus the empirical formula is C_4H_9, which has a molar mass of 57 g/mol. This is one-half of the molar mass, so the molecular formula is $(C_4H_9)_2$ or C_8H_{18}.

CHAPTER 5

5.1 Epsom salt is an electrolyte and methanol is a non-electrolyte.

5.2 (a) $LiNO_3$ is soluble and gives Li^+ and NO_3^- ions.

(b) $CaCl_2$ is soluble and gives Ca^{2+} and Cl^- ions.

(c) CuO is not water-soluble.

(d) $NaCH_3CO_2$ is soluble and gives Na^+ and $CH_3CO_2^-$ ions.

5.3 (a) $Na_2CO_3(aq) + CuCl_2(aq) \longrightarrow$
$$2\ NaCl(aq) + CuCO_3(s)$$

(b) No reaction; no insoluble compound is produced.

(c) $NiCl_2(aq) + 2\ KOH(aq) \longrightarrow$
$$Ni(OH)_2(s) + 2\ KCl(aq)$$

5.4 (a) $AlCl_3(aq) + Na_3PO_4(aq) \longrightarrow$
$$AlPO_4(s) + 3\ NaCl(aq)$$
$Al^{3+}(aq) + PO_4^{3-}(aq) \longrightarrow AlPO_4(s)$

(b) $FeCl_3(aq) + 3\ KOH(aq) \longrightarrow$
$$Fe(OH)_3(s) + 3\ KCl(aq)$$
$Fe^{3+}(aq) + 3\ OH^-(aq) \longrightarrow Fe(OH)_3(s)$

(c) $Pb(NO_3)_2(aq) + 2\ KCl(aq) \longrightarrow$
$$PbCl_2(s) + 2\ KNO_3(aq)$$
$Pb^{2+}(aq) + 2\ Cl^-(aq) \longrightarrow PbCl_2(s)$

5.5 (a) $H^+(aq)$ and $NO_3^-(aq)$

(b) $Ba^{2+}(aq)$ and $2\ OH^-(aq)$

5.6 Metals form basic oxides, non-metals form acidic oxides. Thus: (a) SeO_2 is an acidic oxide; (b) MgO is a basic oxide; and (c) P_4O_{10} is an acidic oxide.

5.7 $Mg(OH)_2(s) + 2\ HCl(aq) \longrightarrow MgCl_2(aq) + 2\ H_2O(\ell)$

Net ionic equation: $Mg(OH)_2(s) + 2\ H^+(aq) \longrightarrow$
$$Mg^{2+}(aq) + 2\ H_2O(\ell)$$

5.8 (a) $BaCO_3(s) + 2\ HNO_3(aq) \longrightarrow$
$$Ba(NO_3)_2(aq) + CO_2(g) + H_2O(\ell)$$

Barium carbonate and nitric acid produce barium nitrate, carbon dioxide, and water.

(b) $(NH_4)_2SO_4(aq) + 2\ NaOH(aq) \longrightarrow$
$$2\ NH_3(g) + Na_2SO_4(aq) + 2\ H_2O(\ell)$$

5.9 (a) Gas-forming reaction: $CuCO_3(s) + H_2SO_4(aq) \longrightarrow$
$$CuSO_4(aq) + H_2O(\ell) + CO_2(g)$$

Net ionic equation: $CuCO_3(s) + 2\ H^+(aq) \longrightarrow$
$$Cu^{2+}(aq) + H_2O(\ell) + CO_2(g)$$

(b) Acid-base reaction: $Ba(OH)_2(s) + 2\ HNO_3(aq) \longrightarrow$
$$Ba(NO_3)_2(aq) + 2\ H_2O(\ell)$$

Net ionic equation: $Ba(OH)_2(s) + 2\ H^+(aq) \longrightarrow$
$$Ba^{2+}(aq) + 2\ H_2O(\ell)$$

(c) Precipitation reaction: $CuCl_2(aq) + (NH_4)_2S(aq) \longrightarrow CuS(s) + 2\ NH_4Cl(aq)$

Net ionic equation: $Cu^{2+}(aq) + S^{2-}(aq) \longrightarrow CuS(s)$

5.10 (a) Fe in Fe_2O_3, +3; (b) S in H_2SO_4, +6; (c) C in CO_3^{2-}, +4; (d) N in NO_2^+, +5

5.11 (a) An acid-base reaction ($H^+ + OH^- \longrightarrow H_2O$).

(b) An oxidation-reduction reaction; Cu is oxidized (oxidation number changes from 0 to +2 in $CuCl_2$), and Cl_2 is reduced (oxidation number for each Cl changes from 0 to -1). Cu is the reducing agent and Cl_2 is the oxidizing agent.

(c) A gas-forming reaction (gaseous CO_2 is evolved).

(d) An oxidation-reduction reaction. S in $S_2O_3^{2-}$ is oxidized (oxidation number changes from +2 to +2.5 in $S_4O_6^{2-}$), and I_2 is reduced (oxidation number changes from 0 to -1). $S_2O_3^{2-}$ is the reducing agent and I_2 is the oxidizing agent.

5.12 Dichromate ion is the oxidizing agent and is reduced. (Cr with a +6 oxidation number is reduced to Cr^{3+} with a +3 oxidation number). Ethanol is the reducing agent and is oxidized. (The C atoms in ethanol have an oxidation number of -2 whereas the oxidation number of carbon is 0 in acetic acid.)

5.13 26.3 g(1 mol $NaHCO_3$/84.01 g $NaHCO_3$) = 0.313 mol $NaHCO_3$

0.313 mol $NaHCO_3$/0.200 L = 1.57 M

The concentration of the two ions formed, $[Na^+]$ and $[HCO_3^-]$, is 1.57 M.

5.14 First, determine the mass of $AgNO_3$ required.

Amount of $AgNO_3$ required = (0.0200 M)(0.250 L) = 5.00×10^{-3} mol

Mass of $AgNO_3$ = 5.00×10^{-3} mol(169.9 g/mol) = 0.850 g $AgNO_3$

Weigh out 0.850 g $AgNO_3$; then, dissolve it in a small amount of water in the volumetric flask. After the solid is dissolved, fill the flask to the mark.

5.15 (0.15 M)(0.0060 L) = (0.0100 L)c_{dilute}

c_{dilute} = 0.090 M

5.16 (2.00 M)V_{conc} = (1.00 M)(0.250 L); so V_{conc} = 0.125 L

To prepare solution: measure accurately 125 mL of 2.00 M NaOH into a 250 mL volumetric flask, and add water to give a total volume of 250 mL.

5.17 (a) pH = $-\log(2.6 \times 10^{-2})$ = 1.6

(b) $-\log[H^+]$ = 12.45; $[H^+]$ = 3.5×10^{-13} M

5.18 HCl is the limiting reagent.

(0.350 mol HCl/1 L)(0.075 L)(1 mol CO_2/2 mol HCl)(44.01 g CO_2/1 mol CO_2) = 0.58 g CO_2

5.19 (0.953 mol NaOH/1 L)(0.02833 L NaOH) = 0.0270 mol NaOH

(0.0270 mol NaOH)(1 mol CH_3CO_2H /1 mol NaOH) = 0.0270 mol CH_3CO_2H

0.0270 mol CH_3CO_2H (60.05 g/mol) = 1.62 g CH_3CO_2H

0.0270 mol CH_3CO_2H/0.0250 L = 1.08 M

5.20 (0.100 mol HCl/1 L)(0.02967 L) = 0.00297 mol HCl

0.00297 mol HCl(1 mol NaOH/1 mol HCl) = 0.00297 mol NaOH

0.00297 mol NaOH/0.0250 L = 0.119 M NaOH

5.21 (0.196 mol $Na_2S_2O_3$/1 L)(0.02030 L) = 0.00398 mol $Na_2S_2O_3$

0.00398 mol $Na_2S_2O_3$(1 mol I_2/2 mol $Na_2S_2O_3$) = 0.00199 mol I_2

0.00199 mol I_2 was excess, not used in reaction with ascorbic acid.

I_2 originally added = (0.0520 mol I_2/1 L)(0.05000 L) = 0.00260 mol I_2

I_2 used in reaction with ascorbic acid = 0.00260 mol $-$ 0.00199 mol = 6.1×10^{-4} mol I_2

$(6.1 \times 10^{-4}$ mol $I_2)$(1 mol $C_6H_8O_6$ / 1 mol I_2)(176.1 g/mol) = 0.11 g $C_6H_8O_6$

CHAPTER 6

6.1 The chemical potential energy of a battery can be converted to work (to run a motor), heat (an electric space heater), and light (a light bulb).

6.2 (a) (3800 calories)(4.184 J/calorie) = 1.6×10^4 J

(b) (250 Calories)(1000 calories/Calorie)(4.184 J/calorie)(1 kJ/1000 J) = 1050 kJ

6.3 C = 59.8 J/[(25.0 g)(1 K)] = 2.39 J/g $\cdot$ K

6.4 (15.5 g)(C_{metal})(18.9 °C $-$ 100.0 °C) + (55.5 g)(4.184 J/g $\cdot$ K)(18.9 °C $-$ 16.5 °C) = 0

C_{metal} = 0.44 J/g $\cdot$ K

6.5 (400. g iron)(0.451 J/g $\cdot$ K)(32.8 °C $-$ $T_{initial}$) + (1000. g)(4.184 J/g $\cdot$ K)(32.8 °C $-$ 20.0 °C) = 0

$T_{initial}$ = 330. °C

6.6 To heat methanol:

(25.0 g CH_3OH)(2.53 J/g $\cdot$ K)(64.6 °C $-$ 25.0 °C) = 2.50×10^3 J

To evaporate methanol:

(25.0 g CH_3OH)(2.00×10^3 J/g) = 5.00×10^4 J

Total heat energy:

0.250×10^4 J + 5.00×10^4 J = 5.25×10^4 J

6.7 Heat transferred from tea + heat to melt ice = 0

(250 g)(4.2 J/g $\cdot$ K)(291.4 K $-$ 273.2 K) + x g (333 J/g) = 0

x = 57.4 g

Thus, 57 g of ice melts with heat supplied by cooling 250 g of tea from 18.2 °C (291.4 K) to 0 °C (273.2 K)

Mass of ice remaining = Mass of ice initially $-$ mass of ice melted

Mass of ice remaining = 75 g $-$ 57 g = 18 g

6.8 (a) (12.6 g H_2O)(1 mol /18.02 g)(285.8 kJ/mol) = 2.00×10^2 kJ

(b) (15.0 g C_2H_6)(1 mol/30.2 g) = 0.50 mol C_2H_6

ΔH = 15.0 g(1 mol C_2H_6/30.0 g)(-2857.3 kJ/2 mol C_2H_6) = -710 kJ

6.9 Mass of final solution = 400. g; ($T = 27.78\ °C - 25.10$ °C = 2.68 °C = 2.68 K

Amount of HCl used = amount of NaOH used = $c \cdot V = 0.0800$ mol

Heat transferred by acid-base reaction + heat absorbed to warm solution = 0

$q_{rxn} + (4.20\ J/g \cdot K)(400.\ g)(2.68\ K) = 0$

$q_{rxn} = -4.50 \times 10^3\ J$ (This represents heat evolved in reaction of 0.0800 mol HCl.)

Heat per mole = $\Delta H_{rxn} = -4.50\ kJ/0.0800$ mol HCl = $-56.3\ kJ/mol$ of HCl

6.10 (a) Heat evolved in reaction + heat absorbed by H_2O + heat absorbed by bomb = 0

$q_{rxn} + (1.50 \times 10^3\ g)(4.20\ J/g \cdot K)(27.32\ K - 25.00\ K) + (837\ J/K)(27.32\ K - 25.00\ K) = 0$

$q_{rxn} = -16,500\ J$ (heat evolved burning 1.0 g sucrose)

(b) Heat per mole = $-16.5\ kJ/g$ (342.2 g sucrose/1 mol sucrose) = $-5,650\ kJ/mol$

6.11 $2\ CO_2(g) + 3\ H_2O(\ell) \longrightarrow C_2H_6(g) + 7/2\ O_2(g)$
$$\Delta H_1 = +1559.7\ kJ$$

$2\ C(s) + 2\ O_2(g) \longrightarrow 2\ CO_2(g) \quad \Delta H_2 = 2(-393.5\ kJ)$

$3\ H_2(g) + 3/2\ O_2(g) \longrightarrow 3\ H_2O(\ell)$
$$\Delta H_3 = 3(-285.8\ kJ)$$

NET $2\ C(s) + 3\ H_2(g) \longrightarrow C_2H_6(g)$
$$\Delta H_{net} = \Delta H_1 + \Delta H_2 + \Delta H_3 = -84.7\ kJ$$

6.12 (a) $C(graphite) + O_2(g) \longrightarrow CO_2(g)$
$$\Delta H_1 = -393.5\ kJ$$

$CO_2(g) \longrightarrow C(diamond) + O_2(g)$
$$\Delta H_2 = +395.4\ kJ$$

NET $C(graphite) \longrightarrow C(diamond)$
$$\Delta H_{net} = \Delta H_1 + \Delta H_2 = +1.9\ kJ$$

(b)
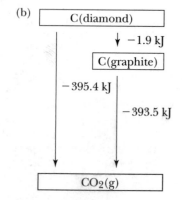

6.13 $C(s) + O_2(g) \longrightarrow CO_2(g) \qquad \Delta H_1 = -393.5\ kJ$

$2[S(s) + O_2(g) \longrightarrow SO_2(g)] \quad \Delta H_2 = 2(-148.4\ kJ)$

$CO_2(g) + 2\ SO_2(g) \longrightarrow CS_2(g) + 3\ O_2(g)$
$$\Delta H_3 = +1103.9\ kJ$$

NET $C(s) + 2\ S(s) \longrightarrow CS_2(g)$
$$\Delta H_{net} = \Delta H_1 + \Delta H_2 + \Delta H_3 = +116.8\ kJ$$

6.14 Standard states: $Br_2(l)$, $Hg(l)$, $Na_2SO_4(s)$, $C_2H_5OH(l)$

6.15 $Fe(s) + 3/2\ Cl_2(g) \longrightarrow FeCl_3(s)$

$12\ C(s, graphite) + 11\ H_2(g) + 11/2\ O_2(g) \longrightarrow$
$$C_{12}H_{22}O_{11}(s)$$

6.16 $\Delta H°_{rxn} = 6\ \Delta H°_f\ [CO_2(g)] + 3\ \Delta H°_f\ [H_2O(\ell)]$
$$- \{\Delta H°_f\ [C_6H_6(\ell)] + 15/2\ \Delta H°_f\ [O_2(g)]\}$$
$$= (6\ mol)(-393.5\ kJ/mol) + (3\ mol)$$
$$(-285.8\ kJ/mol) - (1\ mol)(+49.0\ kJ/mol) - 0$$
$$= -3,267.4\ kJ$$

6.17 (a) $\Delta H°_{rxn} = -2\ \Delta H°_f\ [HBr(g)] = -2\ mol(36.99\ kJ/mol)$
$$= +72.58\ kJ$$

The reaction is reactant-favored.

(b) $\Delta H°_{rxn} = \Delta H°_f\ [C(graphite)] - \Delta H°_f\ [C(diamond)]$
$$= 0 - 1.895\ kJ = -1.895\ kJ$$

The reaction is product-favored.

6.18 Volume of water (in m^3) = $(2.6 \times 10^6\ m^2)(12\ m)$
$$= 3.1 \times 10^7\ m^3$$

Volume of water (in cm^3) = $(3.1 \times 10^7\ m^3)(1 \times 10^6\ cm^3/ 1\ m^3) = 3.1 \times 10^{13}\ cm^3$

Mass of lake water = $3.1 \times 10^{13}\ g$

$q_{lake} = (4.184\ J/g \cdot K)(3.1 \times 10^{13}\ g)(-1.0\ K)(1\ kJ/1000\ J)$
$$= 1.3 \times 10^{11}\ kJ$$

CHAPTER 7

7.1 (a) 10 cm

(b) 5 cm

(c) 4 waves. There are 9 nodes, 7 in the middle and 2 at the ends.

7.2 (a) Highest frequency, violet; lowest frequency, red

(b) The FM radio frequency, 91.7 MHz, is lower than the frequency of a microwave oven, 2.45 GHz.

(c) The wavelength of x-rays is longer than the wavelength of ultraviolet light.

7.3 Blue light: $4.00 \times 10^2\ nm = 4.00 \times 10^{-7}\ m$

$\nu = (2.998 \times 10^8\ m/s)/4.00 \times 10^{-7}\ m = 7.50 \times 10^{14}\ s^{-1}$

$E = (6.626 \times 10^{-34}\ J \cdot s/photon)(7.50 \times 10^{14}\ s^{-1})$
$$(6.022 \times 10^{23}\ photons/mol)$$
$$= 2.99 \times 10^5\ J/mol$$

Microwave: $E = (6.626 \times 10^{-34}\ J \cdot s/photon)(2.45 \times 10^9\ s^{-1})(6.022 \times 10^{23}\ photons/mol) = 0.978\ J/mol$

E(blue light)/E(microwave) = 3.1×10^5; blue (400 nm) light is 310,000 times more energetic than 2.45 GHz microwaves.

7.4 (a) E(per atom) = $-Rhc/n^2$
$$= (-2.179 \times 10^{-18}/3^2)\ J/atom$$
$$= -2.421 \times 10^{-19}\ J/atom$$

(b) E(per mole) = $(-2.421 \times 10^{-19}\ J/atom)$
$$(6.022 \times 10^{23}\ atoms/mol)(1\ kJ/10^3\ J)$$
$$= -145.8\ kJ/mol$$

7.5 The least energetic line is from the electron transition from n = 2 to n = 1

$$\Delta E = -Rhc[(1/1^2) - (1/2^2)])$$
$$= -(2.421 \times 10^{-19} \text{ J/atom})(3/4)$$
$$= -1.634 \times 10^{-18} \text{ J/atom}$$

$$\nu = \Delta E/h = -1.634 \times 10^{-18} \text{ J/atom}/(6.626 \times 10^{-34} \text{ J} \cdot \text{s}) = 2.466 \times 10^{15} \text{ s}^{-1}$$

$$\lambda = c/\nu \ (2.998 \times 10^8 \text{ m/s})/(2.466 \times 10^{15} \text{ s}^{-1})$$
$$= 1.216 \times 10^{-7} \text{ m (or 121.6 nm)}$$

7.6 First, calculate the velocity of the neutron:

$$\nu = [2E/m]^{1/2} = [2(6.21 \times 10^{-21} \text{ kg m}^2/\text{s}^2)/(1.675 \times 10^{-27} \text{ kg})]^{1/2} = 2,720 \text{ m/s}$$

Next, use this value in the de Broglie equation

$$\lambda = hc/nu = (6.626 \times 10^{-34} \text{ kg m}^2/\text{s}^2)/(1.675 \times 10^{-31} \text{ kg})(2,720 \text{ m/s}) = 1.45 \times 10^{-10} \text{ m}$$

7.7 (a) $\ell = 0$ or 1; (b) $m_l = -1$, 0, or +1; (c) d subshell; (d) $\ell = 0$ and $m_l = 0$; (e) 3 orbitals in the p subshell; (f) 7 values of m_l and 7 orbitals.

7.8 (a)

orbital	n	ℓ
6s	6	0
4p	4	1
5d	5	2
4f	4	3

(b) A 4p orbital has one nodal plane and a 6d orbital has 2 nodal planes.

CHAPTER 8

8.1 (a) 4s ($n + \ell = 4$) filled before 4p ($n + \ell = 5$)

(b) 6s ($n + \ell = 6$) filled before 5d ($n + \ell = 7$)

(c) 5s ($n + \ell = 5$) filled before 4f ($n + \ell = 7$)

8.2 (a) chlorine (Cl)

(b) $1s^2 2s^2 2p^6 3s^2 3p^3$

$$\text{[Ne]} \begin{array}{cc} 3s & 3p \\ \boxed{\uparrow\downarrow} & \boxed{\uparrow\ \uparrow\ \uparrow} \end{array}$$

(c) Calcium has two valence electrons in the 4s subshell. Quantum numbers for these two electrons are $n = 4$, $\ell = 0$, $m_l = 0$, and $m_s = \pm 1/2$

8.3 Obtain answers from Table 8.3

8.4

V^{2+} [Ar] $\boxed{\uparrow\ \uparrow\ \uparrow\ \ \ }$ (3d) $\boxed{\ }$ (4s)

V^{3+} [Ar] $\boxed{\uparrow\ \uparrow\ \ \ \ }$ (3d) $\boxed{\ }$ (4s)

Co^{3+} [Ar] $\boxed{\uparrow\downarrow\ \uparrow\ \uparrow\ \uparrow\ \uparrow}$ (3d) $\boxed{\ }$ (4s)

All three ions are paramagnetic with 3, 2, and 4 unpaired electrons, respectively.

8.5 Increasing atomic radius: C < Si < Al

8.6 (a) H—O distance: 37 pm + 66 pm = 103 pm

(b) H—S distance: 37 pm + 104 pm = 141 pm

8.7 (a) Increasing atomic radius: C < B < Al

(b) Increasing ionization energy: Al < B < C

(c) Carbon is predicted to have the most negative electron affinity

8.8 Decreasing ionic radius: $N^{3-} > O^{2-} > F^-$. In this series of isoelectronic ions, the size decreases with increased nuclear charge.

8.9 MgCl$_3$, if it existed, would contain a Mg^{3+} ion (and three Cl$^-$ ions). The formation of Mg^{3+} is energetically unfavorable, a huge input of energy being required to remove the third electron (a core electron). Formation of MgCl$_3$ is energetically unfavorable.

CHAPTER 9

9.1 ·Ba· Ba, Group 2A, has 2 valence electrons.

·Äs· As, Group 5A, has 5 valence electrons.

·B̈r: Br, Group 7A, has 7 valence electrons.

9.2 ΔH_f° for Na(g) = +107.3 kJ/mol

ΔH_f° for I(g) = +106.8 kJ/mol

ΔH°[for Na(g) ⟶ Na$^+$(g)] = +502.0 kJ/mol

ΔH°[for I(g) ⟶ I$^-$(g)] = −295.2 kJ/mol

ΔH°[for Na$^+$(g) + I$^-$(g)] = −702 kJ/mol

The sum of these values = ΔH_f° [NaI(s)] = −281 kJ/mol; the literature value (from calorimetry) is −287.8 kJ/mol.

9.3

$$\left[\begin{array}{c} H \\ | \\ H-N-H \\ | \\ H \end{array}\right]^+ \quad :C\equiv O: \quad :N\equiv O:^+ \quad \left[\begin{array}{c} :\ddot{O}: \\ | \\ :\ddot{O}-S-\ddot{O}: \\ | \\ :\ddot{O}: \end{array}\right]^{2-}$$

9.4

$$\begin{array}{c} H \\ | \\ H-C-\ddot{O}-H \\ | \\ H \end{array} \qquad H-\ddot{N}-\ddot{O}-H$$

methanol   hydroxylamine

9.5

$$\left[\begin{array}{c} :\ddot{O}: \\ | \\ H-\ddot{O}-P-\ddot{O}-H \\ | \\ :\ddot{O}: \end{array}\right]^-$$

9.6 (a) The acetylide ion, C$_2^{2-}$, and the N$_2$ molecule have the same number of valence electrons (10) and identical electronic structures; that is, they are isoelectronic.

(b) Ozone, O$_3$, is isoelectronic with NO$_2^-$; hydroxide ion, OH$^-$, is isoelectronic with HF.

9.7 Resonance structures for the nitrate ion:

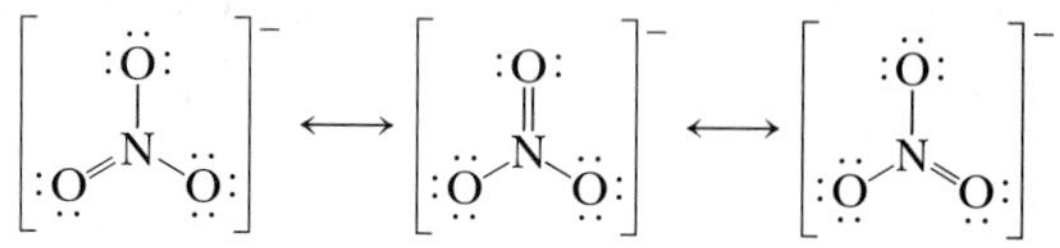

Lewis electron dot structure for nitric acid, HNO_3:

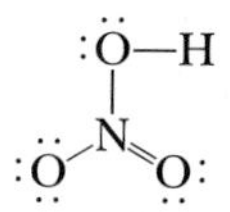

9.8 $\left[:\overset{..}{F}-\overset{..}{\underset{..}{Cl}}-\overset{..}{F}:\right]^{+}$ ClF_2^{+}, 2 bond pairs and 2 lone pairs

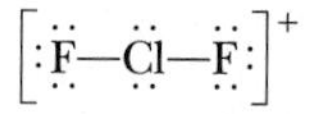

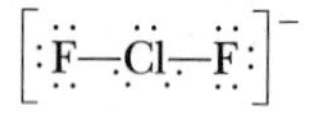

$\left[:\overset{..}{F}-\overset{..}{\underset{..}{Cl}}-\overset{..}{F}:\right]^{-}$ ClF_2^{-}, 2 bond pairs and 3 lone pairs

9.9 (a) CN^-: Formal charge on C is -1, formal charge on N is 0.

(b) SO_3: Formal charge on S is $+2$, formal charge on each O is $-2/3$.

9.10 (a) The H atom is more positive in each case. H—F ($\Delta\chi = 1.9$) is more polar than H—I ($\Delta\chi = 0.4$).

(b) B—F ($\Delta\chi = 2.0$) is more polar than B—C ($\Delta\chi = 0.5$). In B—F, F is the negative pole and B the positive pole. In B—C, C is the negative pole and B the positive pole.

(c) C—Si ($\Delta\chi = 0.7$) is more polar than C—S ($\Delta\chi = 0$). In C—Si, C is the negative pole and Si the positive pole.

9.11

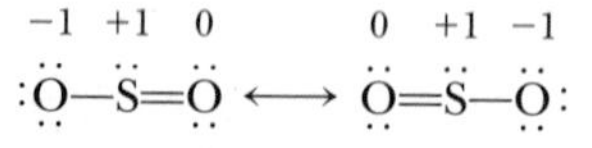

The S—O bonds are polar, with the negative end being the O atom. (The O atom is more electronegative than the S atom.) Formal charges show that these bonds are in fact polar, with the O atom being the more negative atom.

9.12 (a) C—N: bond order 1; C=N: bond order 2; C≡N: bond order 3. Trend in bond length: C—N > C=N > C≡N

(b)

The bond order in NO_2^- is 1.5. Therefore, the NO bond length (124 pm) should be between the length of a N—O single bond (136 pm) and a N=O double bond (115 pm)

9.13 $CH_4(g) + 2\ O_2(g) \longrightarrow CO_2(g) + 2\ H_2O(g)$

Break 4 C—H bonds and 2 O=O bonds: (4 mol)(413 kJ/mol) + (2 mol)(498 kJ/mol) = 2,648 kJ

Make 2 C=O bonds and 4 H—O bonds: (2 mol)(732 kJ/mol) + (4 mol)(463 kJ/mol) = 3,316 kJ

ΔH°_{rxn} = 2,648 kJ $-$3,316 kJ = $-$668 kJ

9.14 Tetrahedral geometry around carbon. The Cl—C—Cl bond angle will be close to 109.5°

9.15 For each species, the electron pair geometry and molecular shape is the same: BF_3: trigonal planar; BF_4^-: tetrahedral. Adding F^- to BF_3 adds an electron pair to the central atom and changes the shape.

9.16 The electron-pair geometry around the I atom is trigonal-bipyramidal. The molecular geometry of the ion is linear.

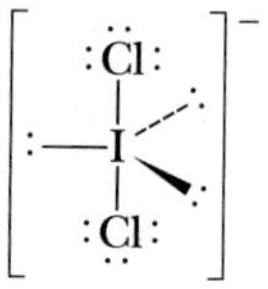

9.17 (a) In PO_4^{3-} there is tetrahedral electron pair geometry. The molecular geometry is tetrahedral.

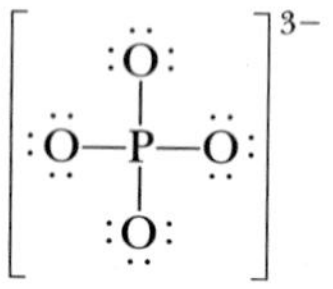

(b) In SO_3^{2-} there is tetrahedral electron pair geometry. The molecular geometry is trigonal pyramidal.

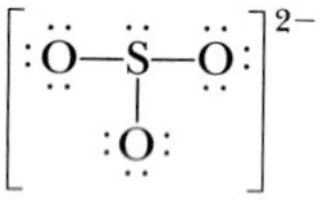

(c) In IF_5 there is octahedral electron pair geometry. The molecular geometry is square pyramidal.

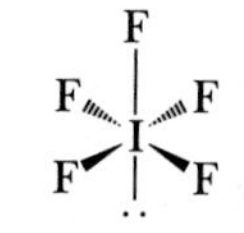

9.18 (a) $BFCl_2$, polar, negative side is the F atom because F is the most electronegative atom in the molecule.

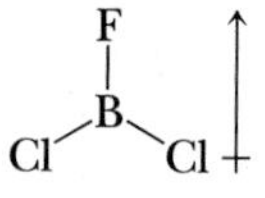

(b) NH_2Cl, polar, negative side is the Cl atom.

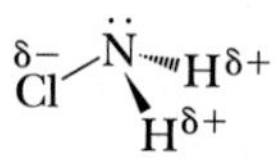

(c) SCl_2, polar, Cl atoms are on the negative side.

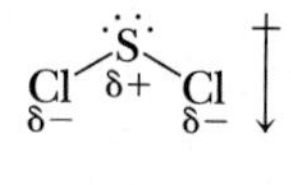

CHAPTER 10

10.1 The oxygen atom in H_3O^+, is hybridized sp^3. The three O—H bonds are formed by overlap of oxygen sp^3 and hydrogen $1s$ orbitals. the fourth sp^3 orbital contains a lone pair of electrons.

Carbon and nitrogen atoms in CH_3NH_2 are hybridized sp^3. The C—H bonds arise from overlap of carbon sp^3 orbitals and hydrogen $1s$ orbitals. The bond between C and N is formed by overlap of sp^3 orbitals from these atoms. Overlap of nitrogen sp^3 and hydrogen $1s$ orbitals gives the two N—H bonds, and there is a lone pair in the remaining sp^3 orbital on nitrogen.

10.2 The Lewis structure and the electron pair and molecular geometries are shown below. The Xe atom is hybridized sp^3d^2. Lone pairs of electrons reside in two of these orbitals; the four others overlap with $2p$ orbitals on fluorine forming sigma bonds.

Electron dot structure   Molecular geometry

10.3 (a) BH_4^-, tetrahedral electron pair geometry, sp^3

(b) SF_5^-, octahedral electron pair geometry, sp^3d^2

(c) SOF_4, trigonal bipyramidal electron pair geometry, sp^3d

(d) ClF_3, trigonal bipyramidal electron pair geometry, sp^3d

(e) BCl_3, trigonal planar electron pair geometry, sp^2

(f) XeO_6^{4-}, octahedral electron pair geometry, sp^3d^2

10.4 The two CH_3 carbon atoms are sp^3 hybridized and the center carbon is sp^2 hybridized. For each of the carbon atoms in the methyl groups, three orbitals are used to form C—H bonds and the fourth is used to bond to the central carbon atom. Overlap of carbon and oxygen sp^2 orbitals gives the sigma bond between these elements; the pi bond arises by overlap of p orbitals on these elements.

10.5 A triple bond links the two nitrogen atoms, each of which also has one lone pair. Each nitrogen is hybridized sp. One sp orbital contains the lone pair, the other is used to form the sigma bond between the two atoms. Two pi bonds arise by overlap of p orbitals on the two atoms.

10.6 Bond angles: H—C—H = 109.5°, H—C—C = 180°, C—C—N = 180°.

10.7 H_2^+: $(\sigma_{1s})^1$ The ion has a bond order of 1/2 and is expected to exist. Bond orders of 1/2 are predicted for He_2^+ and H_2^- both of which are predicted to have electron configurations $(\sigma_{1s})^2 (\sigma^*_{1s})^1$.

10.8 Li_2^- is predicted to have an electron configuration $(\sigma_{1s})^2 (\sigma^*_{1s})^2 (\sigma_{2s})^2 (\sigma^*_{2s})^1$ and a bond order of 1/2, implying that the ion might exist.

10.9 O_2^+: [core electrons] $(\sigma_{2s})^2 (\sigma^*_{2s})^2 (\pi_{2p})^4 (\sigma_{2p})^2 (\pi^*_{2p})^1$. The bond order is 2.5. The ion is paramagnetic with one unpaired electron.

CHAPTER 11

11.1 (a) Isomers of C_7H_{16}

$CH_3CH_2CH_2CH_2CH_2CH_3$ heptane

$CH_3CH_2CH_2CH_2CHCH_3$ 2-methylhexane

$CH_3CH_2CH_2CHCH_2CH_3$ 3-methylhexane

$CH_3CH_2CHCHCH_3$ 2,3-dimethylpentane

$CH_3CH_2CH_2CCH_3$ 2,2-dimethylpentane

$CH_3CH_2CCH_2CH_3$ 3,3-dimethylpentane

$CH_3CHCH_2CHCH_3$ 2,4-dimethylpentane

2-ethylpentane is pictured on page 490

CH_3C—$CHCH_3$ 2,2,3-trimethylbutane

(b) One isomer, 3-methylhexane, is chiral.

11.2 The names accompany the structures in the answer to Q11.1.

11.3 Some isomers of C_6H_{12} (in which the longest chain has six C atoms):

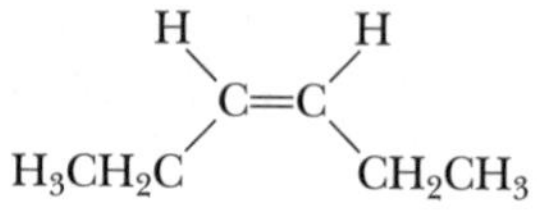

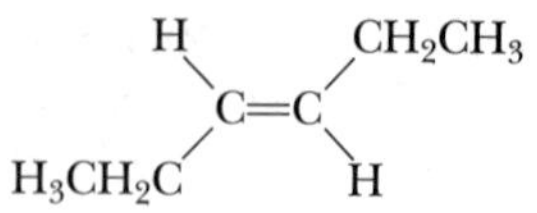

Names: 1-hexene, *cis*-2-hexene, *trans*-2-hexene, cis-3-hexene, *trans*-3-hexene. None of these isomers are chiral.

11.4   Reactions of alkanes:

(a)

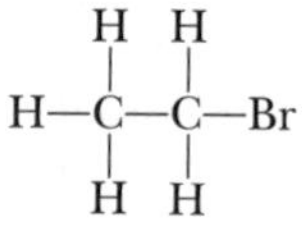

bromoethane

(b)

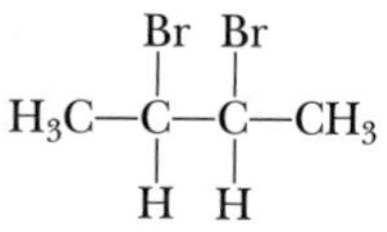

2,3-dibromobutane

11.5   1,4-diaminobenzene

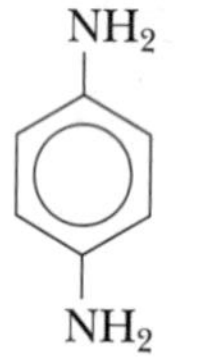

11.6 $CH_3CH_2CH_2CH_2OH$ 1-butanol

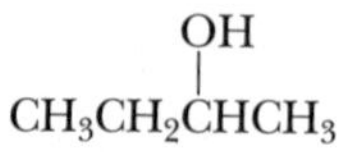

2-butanol

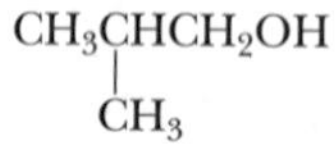

2-methyl-1-propanol

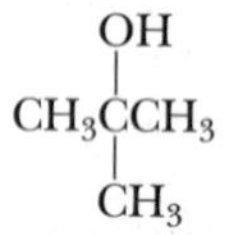

2-methyl-2-propanol

11.7   (a)

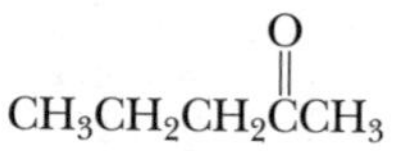

2-pentanone

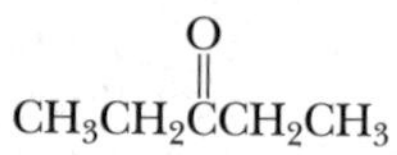

3-pentanone

pentanal

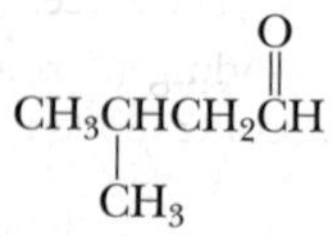

3-methylbutanal

(b)

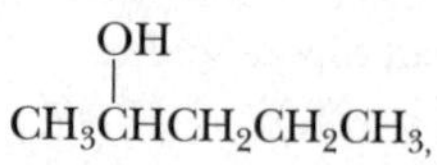

2-pentanol

11.8   (a) 1-butanol gives butanal

(b) 2-butanol gives butanone

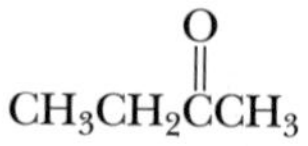

(c) 2-methyl-1-propanol gives 2-methylpropanal

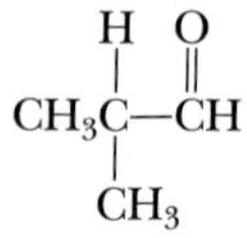

The three alcohols are structural isomers.

11.9   (a) 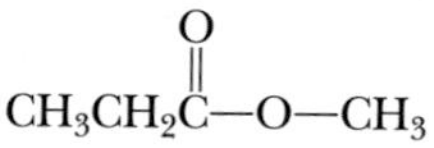   methyl propanoate

(b)

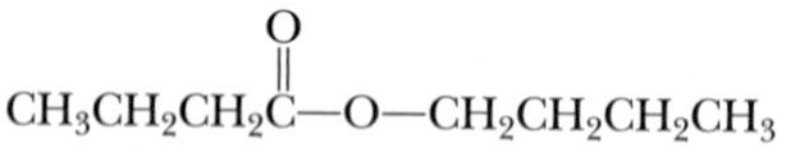

butyl butanoate

(c)

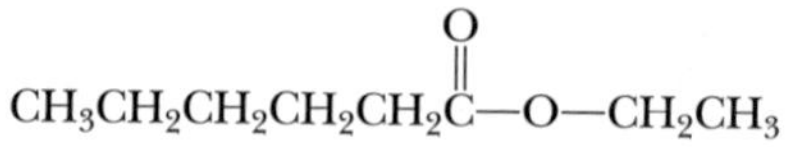

ethyl hexanoate

11.10   (a) Propyl acetate from acetic acid and propanol

(b) 3-Methylpentyl benzoate from benzoic acid and 3-methylpentanol

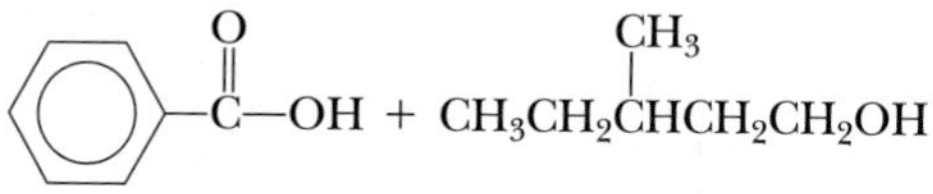

(c) Ethyl salicylate from salicylic acid and ethanol

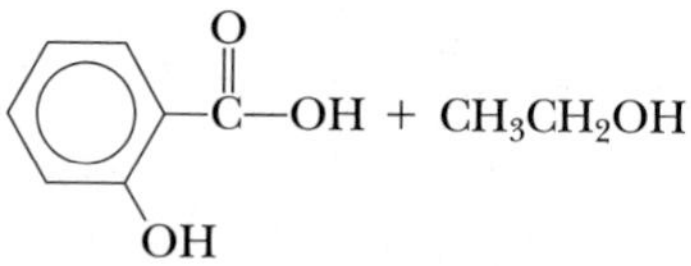

11.11   (a) $CH_3CH_2CH_2OH$: 1-propanol, has an alcohol (OH) group

CH_3CO_2H: ethanoic acid (acetic acid), has a carboxylic acid ($-CO_2H$) group

$CH_3CH_2NH_2$: ethylamine, has an amino ($-NH_2$) group

(b) 1-propyl ethanoate (propyl acetate)

(c) Oxidation of this primary alcohol gives, first, an aldehyde (propanal, CH_3CH_2CHO). Further oxidation gives a carboxylic acid (propanoic acid, $CH_3CH_2CO_2H$)

(d) N-ethylacetamide, $CH_3CONHCH_2CH_3$

(e) The amine will be protonated by hydrochloric acid, forming ethylammonium chloride, $[CH_3CH_2NH_3]Cl$, (an ionic compound, with $CH_3CH_2NH_3^+$ and Cl^- ions)

11.12 The polymer is a polyester.

CHAPTER 12

12.1 0.83 bar (0.82 atm) > 75 kPa (0.74 atm) > 0.63 atm > 250. mm Hg (0.329 atm)

12.2 $P_1 = 55$ mm Hg and $V_1 = 125$ mL; $P_2 = 78$ mm Hg and $V_2 = ?$
$V_2 = V_1(P_1/P_2) = (125$ mL$)(55$ mm Hg $/78$ mm Hg$)$
$= 88$ mL

12.3 $V_1 = 45$ L and $T_1 = 298$ K; $V_2 = ?$ and $T_2 = 263$ K
$V_2 = V_1(T_2/T_1) = (45$ L$)(263$ K$/298$ K$) = 40.$ L

12.4 $V_2 = V_1(P_1/P_2)(T_2/T_1)$
$= (22$ L$)(150$ atm$/0.993$ atm$)(295$ K$/304$ K$)$
$= 3200$ L

At 5.0 L per balloon, there is sufficient He to fill 640 balloons.

12.5 44.8 L of O_2 are required; 44.8 L of $H_2O(g)$ and 22.4 L $CO_2(g)$ are produced.

12.6 $PV = nRT$
$(750/760$ atm$)(V) = 1300$ mol$(0.082057$ L $\cdot$ atm/mol $\cdot$ K$)(296$ K$)$
$V = 3.2 \times 10^4$ L

12.7 $d = PM/RT = (1.00$ atm$)(28.96$ g/mol$)/(0.082057$ L $\cdot$ atm/mol $\cdot$ K$)(288$ K$)$

12.8 $PV = nRT$; $(0.738$ atm$)(0.125$ L$) = n(0.082057$ L $\cdot$ atm/mol $\cdot$ K$)(296.2$ K$)$
$n = 3.80 \times 10^{-3}$ mol
molar mass $= (0.105$ g$)/(3.80 \times 10^{-3}$ mol$)$
$= 27.7$ g/mol

12.9 $n(H_2) = PV/RT = (542/760$ atm$)(355$ L$)/(0.082057$ L $\cdot$ atm/mol $\cdot$ K$)(298$ K$)$
$n(H_2) = 10.4$ mol
$n(NH_3) = 10.4$ mol H_2 $(2$ mol $NH_3/3$ mol $H_2)$
$= 6.90$ mol NH_3
$PV = nRT$; $P(125$ L$) =$
6.90 mol $(0.082057$ L $\cdot$ atm/mol $\cdot$ K$)$ $(298$ K$)$
$P(NH_3) = 1.35$ atm

12.10 180 g N_2H_4 (1 mol $N_2H_4/32$ g N_2H_4)(1 mol $O_2/1$ mol N_2H_4) $= 5.60$ mol O_2
$PV = nRT$; $(750/760$ atm$)(V) =$
5.60 mol $(0.082057$ L $\cdot$ atm/mol $\cdot$ K$)$ $(294$ K$)$
Volume $O_2 = 140$ L

12.11 $P_{halothane}(5.00$ L$) =$
$(0.0760$ mol$)(0.082057$ L $\cdot$ atm/mol $\cdot$ K$)$ $(298.2$ K$)$
$P_{halothane} = 0.372$ atm (or 283 mm Hg)
$P_{oxygen}(5.00$ L$) =$
$(0.734$ mol$)$ $(0.082057$ L $\cdot$ atm/mol $\cdot$ K$)$ $(298.2$ K$)$
$P_{oxygen} = 3.59$ atm (or 2730 mm Hg)
$P_{total} = P_{halothane} + P_{oxygen}$
$= 283$ mm Hg $+ 2730$ mm Hg $= 3010$ mm Hg

12.12 Use equation 12.9, with M $= 4.00 \times 10^{-3}$ kg/mol, $T = 298$ K, and R $= 8.314$ J/mol $\cdot$ K to calculate the rms speed of 1360 m/s for He. A similar calculation for N_2, with M $= 28.01 \times 10^{-3}$ kg/mol, gives a rms speed of 515 m/s.

12.13 The molar mass of CH_4 is 16.0 g/mol. Therefore,
$$\frac{\text{Rate for } CH_4}{\text{Rate for unk}} = \frac{n \text{ molecules}/1.50 \text{ min}}{n \text{ molecules}/4.73 \text{ min}} = \sqrt{\frac{M_{unk}}{16.0}}$$
$M_{unk} = 159$ g/mol

12.14 Pressure calculated by the ideal gas law, use $PV = nRT$
$P(1.00$ L$) =$
10.0 mol $(0.082057$ L $\cdot$ atm/mol $\cdot$ K$)$ $(298$ K$)$
$P = 245$ atm
Pressure calculated by the van der Waals equation is 320 atm.

CHAPTER 13

13.1 Water is a polar solvent while hexane and CCl_4 are nonpolar. London dispersions forces of attraction occur between all pairs of dissimilar solvents. For mixtures of water with the other solvents, dipole-induced dipole forces will also be present. When mixed, the three liquids will form two separate layers, the first water and the second a mixture of the two non-polar liquids.

13.2

Hydrogen bonding in methanol is the attraction of the hydrogen atom bearing a partial positive charge $(\delta+)$ on one molecule to the oxygen bearing a partial negative charge $(\delta-)$ on a second molecule. The strong attractive force of hydrogen bonding will cause the boiling point and the heat of vaporization of methanol to be quite high.

13.3 (a) O_2: induced dipole-induced dipole forces only.

(b) CH_3OH: Strong hydrogen bonding (dipole-dipole forces) as well as induced dipole-induced dipole forces.

(c) O_2 in H_2O: dipole-induced dipole forces as well as induced dipole-induced dipole forces.

The relative strength is a < c < b

13.4 $(1.00 \times 10^3 \text{ g})(1 \text{ mol}/32.04 \text{ g})(35.2 \text{ kJ/mol} = 1.10 \times 10^3 \text{ kJ}$

13.5   (a) At 40 °C, the vapor pressure of ethanol is about 120 mm Hg.

     (b) The equilibrium vapor pressure of ethanol at 60 °C is about 370 mm Hg. At 60 °C and 600 mm Hg, ethanol is a liquid. If vapor is present, it will condense to a liquid.

13.6 $PV = nRT$; $P = 0.50 \text{ g } (1 \text{ mol}/18.02 \text{ g})$
$(0.0821 \text{ L} \cdot \text{atm/mol} \cdot \text{K})(333 \text{ K})/5.0 \text{ L}$

     $P = 0.15$ atm (= 120 mm Hg). The vapor pressure of water at 60 °C is 149 mm Hg (Appendix G). The calculated pressure is lower than this so all the water (0.50 g) evaporates. If 2.0 g of water is used, however, the calculated pressure, 460 mm Hg, is greater than the vapor pressure. In this case, only part of the water will evaporate.

13.7   Glycerol is predicted to have a higher viscosity. Glycerol is a larger molecule than ethanol, and there are higher forces of attraction between molecules because each molecule has 3 OH groups that hydrogen bond to other molecules.

13.8 M_2X; In a fcc unit cell, there are 4 anions (X^-) and 8 tetrahedral holes in which to place metal ions. All the tetrahedral holes are inside the unit cell, so the ratio of atoms in the unit cell is 2:1, 8M to 4 X.

CHAPTER 14

14.1   10.0 g sucrose = 0.0292 moles; 250 g H_2O = 13.9 mol
$X = (0.0292 \text{ mol})/(0.0292 \text{ mol} + 13.9 \text{ mol}) = 0.00210$
$(0.0292 \text{ mol sucrose})/(0.250 \text{ kg solvent}) = 0.117 \text{ m}$
% sucrose = $(10.0 \text{ g sucrose})(/260. \text{ g soln})(\times 100) = 3.85$ %

14.2 1.08×10^4 ppm $\equiv 1.08 \times 10^4$ mg NaCl per 1000 g soln
$(1.08 \times 10^4 \text{ mg Na}/1000 \text{ g soln})(1050 \text{ g soln}/1 \text{ L}) = 1.13 \times 10^4 \text{ mg Na/L} =$
11.3 g Na/L (11.3 g Na/L)(58.45 g NaCl/23.0 g Na) = 28.7 g NaCl/L

14.3   Because F^- is the smaller ion, water molecules can approach more closely and interact more strongly. Thus, F^- should have the most negative heat of hydration.

14.4 $\Delta H^\circ_{\text{soln}} = \Delta H^\circ_f [\text{NaOH(aq)}] - \Delta H^\circ_f [\text{NaOH(s)}]$
$= -470.1 \text{ kJ/mol} - (-425.6 \text{ kJ/mol})$
$= -44.5 \text{ kJ/mol}$

14.5   Solubility(CO_2) =
$(4.48 \times 10^{-5} \text{ M/mm Hg})(251 \text{ mm Hg}) = 1.12 \times 10^{-2} \text{ M}$

14.6   The solution contains sucrose (0.0292 mol) in water (12.5 mole)
$X_{\text{water}} = (12.5 \text{ mol } H_2O)/(12.5 \text{ mol} + 0.0292 \text{ mol})$
$= 0.998$
$P_{\text{water}} = 0.998 (149.4 \text{ mm Hg}) = 149 \text{ mm Hg}$

14.7 $m = \Delta T_{\text{bp}}/K_{\text{bp}} = 1.0 \text{ °C}/(0.512 \text{ °C}/m) = 2.0 \text{ m}$
$(2.0 \text{ mol/kg})(0.125 \text{ kg})(62.02 \text{ g/mol}) = 16 \text{ g glycol}$

14.8   concentration = 525 g(1 mol/62.07 g)/(3.00 kg)
$= 2.82 \text{ m}$
$\Delta T_{\text{fp}} = K_{\text{fp}} \cdot m = (-1.86 \text{ °C}/m)(2.82 \text{ } m) = -5.24 \text{ °C}$
You will be protected only to about −5 °C and not to −25 °C.

14.9 $\Delta T_{\text{bp}} = 80.23 \text{ °C} - 80.10 \text{ °C} = 0.13 \text{ °C}$
$m = \Delta T_{\text{bp}}/K_{\text{bp}} = 0.13 \text{ °C}/(2.53 \text{ °C}/m) = 0.051 \text{ } m$
$(0.051 \text{ mol/kg})(0.099 \text{ kg}) = 0.0051 \text{ mol}$
molar mass = 0.640 g/0.0051 mol = 130 g/mol. The formula $C_{10}H_8$ (molar mass = 128.2 g/mol) is the closest match to this value.

14.10   concentration = (25.0 g NaCl)(1 mol/58.45 g)/
(0.525 kg) = 0.815 m
$\Delta T_{\text{fp}} = K_{\text{fp}} \cdot m \cdot i = (-1.86 \text{ °C}/m)(0.815 \text{ } m)(1.85)$
$= -2.80 \text{ °C}$

14.11 $M = \Pi/RT = [(1.86 \text{ mm Hg})(1 \text{ atm}/760 \text{ mm Hg})]/(0.0821 \text{ L} \cdot \text{atm/mol} \cdot \text{K})(298 \text{ K})$
$= 1.00 \times 10^{-4} \text{ M}$
$1.00 \times 10^{-4} \text{ mol/L})(0.100 \text{ L}) = 1.0 \times 10^{-3} \text{ mol}$
molar mass = $1.40 \text{ g}/1.00 \times 10^{-3} \text{ mol} = 1.4 \times 10^3$ g/mol. Assuming the polymer is composed of CH_2 units, the polymer is about 100 units long.

CHAPTER 15

15.1 $-1/2 (\Delta[\text{NOCl}]/\Delta t) = 1/2 (\Delta[\text{NO}]/\Delta t) = (\Delta[\text{Cl}_2]/\Delta t)$

15.2   for the first 2 hours:
$\Delta[\text{sucrose}]/\Delta t = [(0.34 - 0.50)\text{mol/L}]/(2 \text{ h })$
$= -0.0080 \text{ mol/L} \cdot \text{h}$
For the last two hours:
$\Delta[\text{sucrose}]/\Delta t = [(0.010 - 0.0150) \text{ mol/L}]/(2 \text{ h})$
$= -0.0025 \text{ mol/L} \cdot \text{h}$
Instantaneous rate at 4 h = −0.0045 mol/L · h

15.3   Compare experiments 1 and 2: doubling $[O_2]$ causes the rate to double, so the rate is first order in $[O_2]$. Comparing experiments 2 and 4: doubling [NO] causes the rate to increase by a factor of 4, so the rate is second order in [NO]. Thus, the rate law is
Rate = $k[\text{NO}]^2[O_2]$
Using the data in experiment 1 to determine k:
$0.028 \text{ mol/L} \cdot \text{s} = k[0.020 \text{ mol/L}]^2[0.010 \text{ mol/L}]$
$k = 7.0 \times 10^3 \text{ L}^2/\text{mol}^2 \cdot \text{s}$

15.4   Rate = $k[\text{Pt}(NH_3)_2Cl_2] = (0.090 \text{ h}^{-1})(0.020 \text{ mol/L})$
$= 0.0018 \text{ mol/L} \cdot \text{h}$
The rate of formation of Cl^- is the same value, 0.0018 mol/L · h

15.5 $\ln([\text{sucrose}]/[\text{sucrose}]_o) = -kt$
$\ln([\text{sucrose}]/[0.010]) = -(0.21 \text{ h}^{-1})(5.0 \text{ h})$
[sucrose] = 0.0035 mol/L

15.6   (a) The fraction remaining is $[\text{NO}_2]/[\text{NO}_2]_0$
$\ln\{[\text{NO}_2]/[\text{NO}_2]_o\} = -(3.6 \times 10^{-3} \text{ s}^{-1})(150 \text{ s})$
$[\text{NO}_2]/[\text{NO}_2]_o = 0.58$

(b) The fraction remaining after the reaction is 99 % complete is 0.010

$\ln(0.010) = -(3.6 \times 10^{-3} \text{ s}^{-1})(t)$

$t = 1300$ s

15.7 $1/[HI] - 1/[HI]_o = kt$

$1/[HI] - 1/[0.010 \text{ M}] = (30. \text{ L/mol} \cdot \text{min})(12 \text{ min})$

$[HI] = 0.0022$ M

15.8

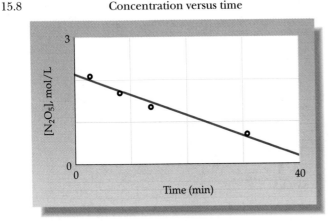

Concentration versus time

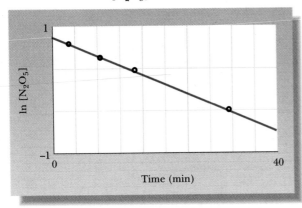

ln [N₂O₅] versus time

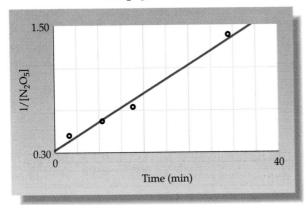

1/[N₂O₅] versus time

The plot of $\ln[N_2O_5]$ vs. time is linear, indicating that this is a first order reaction. The rate constant is determined from the slope: $k = -\text{slope} = 0.038 \text{ min}^{-1}$

15.9 (a) For ^{241}Am, $t_{1/2} = 0.693/k = 0.693/(0.0016 \text{ y}^{-1})$
$= 430$ y

For ^{125}I, $t_{1/2} = 0.693/(0.011 \text{ d}^{-1}) = 63$ d

(b) ^{125}I decays much faster

(c) $\ln[(n)/(1.6 \times 10^{15} \text{ atoms})] =$
$-(0.011 \text{ day}^{-1})(2.0 \text{ day})$

$n/1.6 \times 10^{15}$ atoms $= 0.978$; $n = 1.57 \times 10^{-15}$ atoms
(Since the answer should have 2 significant figures, we should round off the answer the answer to 1.6×10^{15} atoms. The approximately 2% that has decayed is not noticeable within the limits of accuracy of the data presented.)

15.10 $\ln(k_2/k_1) = (-E_a/R)(1/T_2 - 1/T_1)$

$\ln[(100 \times 10^4)/(4.5 \times 10^3) = -(E_a/8.31 \times 10^{-3}$
kJ/mol $\cdot$ K$)(1/283 \text{ K} - 1/274 \text{ K})$

$E_a = 57$ kJ/mol

15.11 All three steps are bimolecular

For step 3: Rate $= k[N_2O_2][H_2]$

When the 3 equations are added, N_2O_2 (a product in the first step and a reactant in the second) and N_2O (a product of the second step and a reactant in the third) cancel, leaving the net equation, 2 NO(g) + 2 H₂(g) $\longrightarrow$ N₂(g) + 2 H₂O(g)

15.12 (a) 2 NH₃(aq) + OCl⁻(aq) $\longrightarrow$
N₂H₄(aq) + Cl⁻(aq) + H₂O(ℓ)

(b) The second step is the rate-determining step.

(c) Rate $= k[NH_2Cl][NH_3]$

(d) NH_2Cl, $N_2H_5^+$, and OH^- are intermediates

15.13 Overall reaction: 2 NO₂Cl(g) $\longrightarrow$ 2 NO₂(g) + Cl₂(g)

Rate $= k[NO_2Cl]^2/[NO_2]$

The presence of NO₂ causes the reaction rate to decrease.

CHAPTER 16

16.1 (a) $K = [PCl_3][][Cl_2]/[PCl_5]$

(b) $K = [Cu^{2+}][OH^-]^2$

(c) $K = [Cu^{2+}][NH_3]^4/[Cu(NH_3)_4^{2+}]$

(d) $K = [H_3O^+][CH_3CO_2^-]/[CH_3CO_2H]$

16.2 (a) Both reactions are reactant favored ($K << 1$)

(b) $[NH_3]$ in the second solution is greater. K for this reaction is larger so the reactant, $Cd(NH_3)_4^{2+}$, dissociates to a greater extent.

16.3 (a) $Q = [2.18]/[0.97] = 2.25$. The system is not at equilibrium; $Q < K$; to reach equilibrium, [isobutane] will increase and [butane] will decrease.

(b) $Q = [2.60]/[0.75] = 3.5$. The system is not at equilibrium; $Q > K$; to reach equilibrium, [butane] will increase and [isobutane] will decrease.

16.4 $Q = [NO]^2/[N_2][O_2] = [4.2 \times 10^{-3}]^2/[0.50][0.25]$
     $= 1.4 \times 10^{-4}$

$Q < K$ so the reaction is not at equilibrium. To reach equilibrium, [NO] will increase and $[N_2]$ and $[O_2]$ will decrease.

16.5   (a)

Equation	$C_6H_{10}I_2$	$\rightleftharpoons$	C_6H_{10}	+	I_2
Initial (M)	0.050		0		0
Change (M)	−0.035		+0.035		+0.035
Equilibrium (M)	0.015		0.035		0.035

(b) $K = (0.035)(0.035)/(0.015) = (0.082)$

16.6

Equation	H_2	+	I_2	$\rightleftharpoons$	2 HI
Initial (M)	6.00×10^{-3}		6.00×10^{-3}		0
Change (M)	$-x$		$-x$		$+2x$
Equilibrium (M)	$0.00600 - x$		$0.00600 - x$		$+2x$

$$K_c = 33 = \frac{(2x)^2}{(0.00600 - x)^2}$$

$x = 0.0045$ M, so $[H_2] = [I_2] = 0.0015$ M and [HI] $= 0.0090$ M

16.7

Equation	C(s)	+	$CO_2(g)$	$\rightleftharpoons$	2 CO(g)
Initial (M)			0.012		0
Change (M)			$-x$		$+2x$
Equilibrium (M)			$0.012 - x$		$2x$

$$K_c = 0.021 = \frac{(2x)^2}{(0.012 - x)}$$

$x = [CO_2] = 0.0057$ M and $2x = [CO] = 0.011$ M

16.8   (a) $K' = K^2 = (2.5 \times 10^{-29})^2 = 6.3 \times 10^{-58}$
       (b) $K'' = 1/K^2 = 1/(6.3 \times 10^{-58}) = 1.6 \times 10^{57}$

16.9   Manipulate the equations and equilibrium constants as follows:

$1/2\ H_2(g) + 1/2\ Br_2(g) \rightleftharpoons HBr(g)$
                 $K_1' = (K_1)^{1/2} = 8.9 \times 10^5$
$H(g) \rightleftharpoons 1/2\ H_2(g)$ $K_2' = 1/(K_2)^{1/2} = 1.4 \times 10^{20}$
$Br(g) \rightleftharpoons 1/2\ Br_2(g)$ $K_2' = 1/(K_3)^{1/2} = 2.1 \times 10^7$
Net: $H(g) + Br(g) \rightleftharpoons HBr(g)$
                 $K_{net} = K_1'K_2'K_3' = 2.6 \times 10^{33}$

16.10 (a)   [NOCl] decreases with an increase in temperature
        (b) $[SO_3]$ decreases with an increase in temperature

16.11

Equation	butane	$\rightleftharpoons$	isobutane
Initial (M)	0.20		0.50
After adding 2.0 M more isobutane	0.20		2.0 + 0.50
Change (M)	$+x$		$-x$
Equilibrium (M)	$0.20 + x$		$2.50 - x$

$$K = \frac{[\text{isobutane}]}{[\text{butane}]} = \frac{(2.50 - x)}{(0.20 + x)}.$$

Solving for x gives $x = 0.57$ M. Therefore, [isobutane] = 1.93 M and [butane] = 0.77 M.

16.12 (a)   Adding H_2 shifts the equilibrium to the right, increasing $[NH_3]$; adding NH_3 shifts the equilibrium to the left, increasing $[N_2]$ and $[H_2]$.

        (b)   An increase in volume shifts the equilibrium to the left.

CHAPTER 17

17.1   (a) $H_3PO_4(aq) + H_2O(l) \rightleftharpoons$
                        $H_3O^+(aq) + H_2PO_4^-(aq)$
           $H_2PO_4^-$ is amphiprotic
        (b) $CN^-(aq) + H_2O(l) \rightleftharpoons HCN(aq) + OH^-(aq)$
           CN^- is a Bronsted base

17.2   (a) NO_3^- is the conjugate base of the acid HNO_3; NH_4^+ is the conjugate acid of the base NH_3
        (b)   The conjugate base of H_2S is HS^-; the conjugate base of NH_4^+ is NH_3
        (c)   The conjugate acid of Br^- is HBr; the conjugate acid of HPO_4^{2-} is $H_2PO_4^-$.

17.3 $[H_3O^+] = 4.0 \times 10^{-3}$ M; $[OH^-] = K_w/[H_3O^+]$
                       $= 2.5 \times 10^{-12}$ M

17.4   (a)   pOH $= -\log[0.0012] = 2.92$; pH $= 14 -$ pOH
                        $= 11.08$
        (b) $[H^+] = 4.8 \times 10^{-5}$ M; $[OH^-] = 2.1 \times 10^{-10}$
        (c)   pOH $= 14 - 10.46 = 3.54$; $[OH^-] = 2.9 \times 10^{-4}$ M. The solubility of $Sr(OH)_2$ is half of this value, 1.5×10^{-4} M because dissolving 1 mole of $Sr(OH)_2$ will give 2 moles of OH^- in solution.

17.5   Answer this question by comparing values of K_a and K_b from Table 17.3
        (a) H_2SO_4 is stronger than H_2SO_3
        (b) $C_6H_5CO_2H$ is a stronger acid than CH_3CO_2H
        (c)   The conjugate base of boric acid, $B(OH)_4^-$, is a stronger base than the conjugate base of acetic acid, $CH_3CO_2^-$.
        (d)   Ammonia is a stronger base than acetate ion
        (e)   The conjugate acid of acetate ion (CH_3CO_2H) is a stronger acid than the conjugate acid of ammonia (NH_4^+)

17.6 (a) pH = 7

(b) pH < 7 (NH_4^+ is an acid)

(c) pH < 7 [$Al(H_2O)_6^{3+}$ is an acid]

(d) pH > 7 (HPO_4^{2-} is a base)

17.7 (a) $pK_a = -\log[6.3 \times 10^{-5}] = 4.20$

(b) $ClCH_2CO_2H$ is stronger (pK_a of 2.87 is less than pK_a of 4.20)

(c) pK_a for NH_4^+, the conjugate acid of NH_3, is $-\log[5.6 \times 10^{-10}] = 9.25$. It is a weaker acid than acetic acid whose $K_a = 1.8 \times 10^{-5}$.

17.8 K_b for the lactate ion = $K_w/K_a = 7.1 \times 10^{-11}$. It is a slightly stronger base than the formate, nitrite, and fluoride ions.

17.9 (a) NH_4^+ is a stronger acid than HCO_3^-. Thus, CO_3^{2-}, the conjugate base of HCO_3^-, is a stronger base than NH_3, the conjugate base of NH_4^+.

(b) Reactant-favored, the reactants are the weaker acid and base.

(c) Reactant-favored, the reactants are the weaker acid and base.

17.10 $CH_3CO_2H(aq) + HSO_4^-(aq) \rightleftarrows CH_3CO_2^-(aq) + H_2SO_4(aq)$. The equilibrium favors the weaker acid and base, which in this example are the reactants.

17.11 (a) The two compounds react and form a solution containing HCN and NaCl. The solution is acidic (HCN is an acid).

(b) $CH_3CO_2H(aq) + SO_3^{2-}(aq) \rightleftarrows$
$HSO_3^-(aq) + CH_3CO_2^-(aq)$

The solution is acidic, because HSO_3^- is a stronger acid than $CH_3CO_2^-$ is a base.

17.12 From the pH we can calculate $[H_3O^+] = 1.9 \times 10^{-3}$ M. Also, $[butanoate^-] = [H_3O^+] = 1.9 \times 10^{-3}$ M. Use these values along with [butanoic acid] to calculate K_a.
$K_a = [1.9 \times 10^{-3}][1.9 \times 10^{-3}]/(0.055 - 1.9 \times 10^{-3})$
$= 6.8 \times 10^{-5}$

17.13 $K_a = 1.8 \times 10^{-5} = [x][x]/(0.10 - x)$
$x = [H_3O^+] = [CH_3CO_2^-] = 1.3 \times 10^{-3}$ M; pH = 2.89

17.14 $K_a = 7.2 \times 10^{-4} = [x][x]/(0.015 - x)$
x in the denominator cannot be dropped. This equation must be solved with the quadratic formula or by successive approximations.
$x = [H_3O^+] = [F^-] = 2.9 \times 10^{-3}$ M; [HF] = $0.015 - 2.9 \times 10^{-3} = 0.012$ M; pH = 2.53

17.15 $OCl^-(aq) + H_2O(l) \rightleftarrows HOCl(aq) + OH^-(aq)$
$K_b = 2.9 \times 10^{-7} = [x][x]/(0.015 - x)$
$x = [OH^-] = [HOCl] = 6.6 \times 10^{-5}$ M; pOH = 4.18; pH = 9.82

17.16 Equivalent amounts of acid and base were used. The solution will contain $CH_3CO_2^-$ and Na^+. Acetate ion hydrolyzes to a small extent, giving CH_3CO_2H and OH^-.

We need to determine $[CH_3CO_2^-]$ and then solve a weak base equilibrium problem to determine $[OH^-]$
amount $CH_3CO_2^-$ = moles base = 0.12 M · 0.015 L
$= 1.8 \times 10^{-3}$ mol.

total volume = 0.030 L; so $[CH_3CO_2^-] = 0.060$ M
$CH_3CO_2^-(aq) + H_2O(\ell) \rightleftarrows$
$CH_3CO_2H(aq) + OH^-(aq)$
$K_b = 5.6 \times 10^{-10} = [x][x]/(0.060 - x)$
$x = [OH^-] = [CH_3CO_2H] = 5.8 \times 10^{-6}$ M; pOH = 5.24; pH = 8.76

17.17 $H_2C_2O_4(aq) + H_2O(\ell) \rightleftarrows H_3O^+(aq) + HC_2O_4^-(aq)$
$K_{a1} = 5.9 \times 10^{-2} = [x][x]/(0.10 - x)$
x in the denominator cannot be dropped. This equation must be solved with the quadratic formula or by successive approximations.
$x = [H_3O^+] = [HC_2O_4^-] = 5.3 \times 10^{-2}$ M; and pH = 1.28
$[C_2O_4^{2-}] = K_{a2} = 6.4 \times 10^{-5}$ M

17.18 (a) Lewis base (electron pair donor); (b) Lewis acid (electron pair acceptor); (c) Lewis base (electron pair donor); (d) Lewis base (electron pair donor)

17.19 (a) H_2TeO_4

(b) $Fe(H_2O)_6^{3+}$

(c) HOCl

(d) amphetamine is a primary amine and a (weak) base.

CHAPTER 18

18.1 pH of 0.30 M HCO_2H:
$K_a = [H_3O^+][HCO_2^-]/[HCO_2H]$
$1.8 \times 10^{-4} = [x][x]/[0.30 - x]$; $x = 7.3 \times 10^{-3}$ M and pH = 2.14

pH of 0.30 M formic acid + 0.10 M $NaHCO_2$
$K_a = [H_3O^+][HCO_2^-]/[HCO_2H]$
$1.8 \times 10^{-4} = [x][0.10 + x]/(0.30 - x)$
$x = 5.4 \times 10^{-4}$ M; pH = 3.27

18.2 NaOH: (0.100 mol/L)(0.0300 L) = 3.00×10^{-3} mol
CH_3CO_2H: (0.100 mol/L)(0.0450 L) = 4.50×10^{-3} mol

3.00×10^{-3} mol NaOH reacts with 3.00×10^{-3} mol CH_3CO_2H, forming 3.00×10^{-3} mol $CH_3CO_2^-$; 1.5×10^{-3} mol unreacted CH_3CO_2H remains in solution. The total volume is 75.0 mL. Use these values to calculate $[CH_3CO_2H]$ and $[CH_3CO_2^-]$ and use the concentrations in a weak acid equilibrium calculation to obtain $[H_3O^+]$ and pH
$[CH_3CO_2H] = 1.5 \times 10^{-3}$ mol/0.075 L = 0.200 M
$[CH_3CO_2^-] = 3.0 \times 10^{-3}$ mol/0.075 L = 0.400 M
$K_a = [H_3O^+][CH_3CO_2^-]/[CH_3CO_2H]$
$1.8 \times 10^{-5} = [x][0.400 + x]/(0.200 - x)$
$x = [H_3O^+] = 9.0 \times 10^{-6}$ M; pH = 5.05

18.3   pH = pK_a + log{[base]/[acid]}

pH = $-\log(1.8 \times 10^{-4})$ + log{[0.70]/[0.50]}

pH = 3.74 + 0.15 = 3.89

18.4   15.0 g $NaHCO_3$ (1 mol/84.01 g) = 0.179 mol $NaHCO_3$, and 18.0 g Na_2CO_3(1 mol/106.01 g) = 0.170 mole Na_2CO_3

pH = pK_a + log{[base]/[acid]}

pH = $-\log(4.8 \times 10^{-11})$ + log{[0.170]/[0.179]}

pH = 10.32 − 0.02 = 10.30

18.5   pH = pK_a + log{[base]/[acid]}

5.00 = $-\log(1.8 \times 10^{-5})$ + log{[base]/[acid]}

5.00 = 4.74 + log{[base]/[acid]}

[base]/[acid] = 1.8

18.6   Initial pH (before adding acid)

pH = pK_a + log{[base]/[acid]}

= $-\log(1.8 \times 10^{-4})$ + log{[0.70]/[0.50]}

= 3.74 + 0.15 = 3.89

After adding acid:

The added HCl will react with the weak base (formate ion) and form more formic acid. The net effect is to change the ratio of [base]/[acid] in the buffer solution

Initial amount HCO_2H = 0.50 M · 0.500 L = 0.250 mol

Initial amount HCO_2^- = 0.70 M · 0.50 L = 0.350 mol

Amount HCl added = 1.0 M · 0.010 L = 0.010 mol

Amount HCO_2H after HCl addition = 0.250 mol + 0.010 = 0.026 mol

Initial amount HCO_2^- after HCl addition = 0.350 mol − 0.010 = 0.034 mol

pH = pK_a + {log[moles base]/[moles acid]}

pH = $-\log(1.8 \times 10^{-4})$ + log{[0.340]/[0.260]}

pH = 3.74 + 0.12 = 3.86

18.7   After addition of 25 mL base, half of the acid has been neutralized.

Initial amount HCl = 0.10 M · 0.050 L = 0.005 mol

Amount NaOH added = 0.10 M · 0.025 L = 0.0025 mol

Amount HCl after reaction: 0.0050 − 0.0025 = 0.0025 mol HCl

[HCl] after reaction = 0.0025 mol/0.075 L = 0.0333 M.

This is a strong acid and completely ionized so $[H_3O^+]$ = 0.0333 M and pH = 1.48

After 50.5 mL base added, a small excess of base is present in the 100.5 mL of solution. (Volume of excess base added is 0.50 mL = 5.0×10^{-4} L)

amount excess base = 0.100 M · 5.0×10^{-4} L = 5.0×10^{-5} mol

$[OH^-]$ = 5.0×10^{-5} mol/0.105 L) = 4.8×10^{-4} M

So, pOH = $-\log(4.8 \times 10^{-4})$ = 3.32, pH = 14 − pOH = 10.68

18.8   35 mL base will partially neutralize the acid.

Initial amount CH_3CO_2H = 0.100 M · 0.1000 L = 0.0100 mol

Amount NaOH added = 0.10 M · 0.035 L = 0.0035 mol

Amount CH_3CO_2H after reaction = 0.0100 − 0.0035 = 0.0065 mol

Amount $CH_3CO_2^-$ after reaction = 0.0035 mol

$[CH_3CO_2H]$ after reaction = 0.00650 mol/0.135 L = 0.0481 M

$[CH_3CO_2^-]$ after reaction = 0.00350 mol/0.135 L = 0.0259 M

$K_a = [H_3O^+][CH_3CO_2^-]/[CH_3CO_2H]$

1.8×10^{-5} = [x][0.0259 + x]/[0.0481 − x]

x = $[H_3O^+]$ = 3.34×10^{-5} M; pH = 4.476

18.9   75 mL acid will partially neutralize the base.

Initial amount NH_3 = 0.100 M · 0.1000 L = 0.0100 mol

Amount HCl added = 0.10 M · 0.0750 L = 0.00750 mol

Amount NH_3 after reaction = 0.0100 − 0.00750 = 0.00250 mol

Amount NH_4^+ after reaction = 0.0075 mol

Solve using the Henderson-Hasselbach equation; use K_a for the weak acid NH_4^+

pH = pK_a + {log[moles base]/[moles acid]}

pH = $-\log(5.6 \times 10^{-10})$ + log{[0.00250]/[0.00750]}

pH = 9.252 − 0.477 = 8.775

18.10 An indicator that changes color near the pH at the equivalence point is required. Possible indicators include methyl red and bromcresol green and Eriochrome black T; all change color in the pH range of 5–6.

18.11 (a) AgI $\rightleftharpoons$ Ag^+ + I^-

$K_{sp} = [Ag^+][I^-]$; $K_{sp} = 8.5 \times 10^{-17}$

(b) BaF_2 $\rightleftharpoons$ Ba^{2+} + 2 F^-

$K_{sp} = [Ba^{2+}][F^-]^2$; $K_{sp} = 1.8 \times 10^{-7}$

(c) Ag_2CO_3 $\rightleftharpoons$ 2 Ag^+ + CO_3^{2-}

$K_{sp} = [Ag^+]^2[CO_3^{2-}]$; $K_{sp} = 8.5 \times 10^{-12}$

18.12 $[Ba^{2+}]$ = 3.6×10^{-3} M; $[F^-]$ = 7.2×10^{-3} M

$K_{sp} = [Ba^{2+}][F^-]^2$

$K_{sp} = [3.6 \times 10^{-3}][7.2 \times 10^{-3}]^2 = 1.9 \times 10^{-7}$

18.13 $Ca(OH)_2$ $\rightleftharpoons$ Ca^{2+} + 2 OH^-

$K_{sp} = [Ca^{2+}][OH^-]^2$; $K_{sp} = 5.5 \times 10^{-5}$

5.5×10^{-5} = [x][2x]^2 (where x = solubility in mol/L)

x = 2.4×10^{-2} mol/L

18.14 (a) AgCN

(b) $Ca(OH)_2$

(c) Because these compound have different stoichiometries, the most soluble cannot be identified without doing a calculation. The solubility of $Ca(OH)_2$ is 2.4×10^{-2} M, (from Exercise 18.13) and it is more soluble than $CaSO_4$ whose solubility is 7.0×10^{-3} M {$K_{sp} = [Ca^{2+}][SO_4^{2-}]$; 4.9×10^{-5} = [x][x]; x = 7.0×10^{-3} M}

18.15 (a) In pure water:

$K_{sp} = [Ba^{2+}][SO_4^{2-}]$; $1.1 \times 10^{-10} = [x][x]$;

$x = 1.0 \times 10^{-5}$ M

(b) In 0.010 M $Ba(NO_3)_2$, which furnishes 0.010 M Ba^{2+} in solution:

$K_{sp} = [Ba^{2+}][SO_4^{2-}]$; $1.1 \times 10^{-10} = [0.010 + x][x]$;

$x = 1.1 \times 10^{-8}$ M

18.16 (a) In pure water:

$K_{sp} = [Zn^{2+}][CN^-]^2$; $8.0 \times 10^{-12} = [x][2x]^2 = 4x^3$

solubility $= x = 1.3 \times 10^{-4}$ M

(b) In 0.10 M $Zn(NO_3)_2$, which furnishes 0.10 M Zn^{2+} to solution:

$K_{sp} = [Zn^{2+}][CN^-]^2$; $8.0 \times 10^{-12} = [0.10 + x][2x]^2$

solubility $= x = 4.5 \times 10^{-7}$ M

18.17 When $[Pb^{2+}] = 1.1 \times 10^{-3}$ M, $[I^-] = 2.2 \times 10^{-3}$ M

$Q = [Pb^{2+}][I^-]^2 = [1.1 \times 10^{-3}][2.2 \times 10^{-3}]^2 = 5.3 \times 10^{-9}$. This value is less than the K_{sp} which means the system has not yet reached equilibrium and more PbI_2 will dissolve.

18.18 $Q = [Sr^{2+}][SO_4^{2-}] = [2.5 \times 10^{-4}][2.5 \times 10^{-4}] = 6.3 \times 10^{-8}$. This value is less than K_{sp} which means the system has not yet reached equilibrium. Precipitation will not occur.

18.19 $K_{sp} = [Pb^{2+}][I^-]^2$ Let x be the concentration of I^- required at equilibrium; if an amount greater than x is used precipitation will occur.

$9.8 \times 10^{-9} = [0.050][x]^2$

$x = [I^-] = 4.4 \times 10^{-5}$ M

Let x be the concentration of Pb^{2+} in solution, in equilibrium with 0.0015 M I^-

$9.8 \times 10^{-9} = [x][1.5 \times 10^{-3}]^2$

$x = [Pb^{2+}] = 4.4 \times 10^{-3}$ M

18.20 First determine the concentrations of Ag^+ and Cl^-, then calculate Q and see whether it is greater than or less than K_{sp}. Concentrations are calculated using the final volume, 105 mL using the equation $c_{dil} \cdot V_{dil} = M_{conc} \cdot V_{conc}$

$[Ag^+] \cdot 0.105$ L $= 0.0010$ mol/L $\cdot 0.100$ L,

$[Ag^+] = 9.5 \times 10^{-4}$ M

$[Cl^-] \cdot 0.105$ L $= 0.025$ M $\cdot 0.005$ L,

$[Cl^-] = 1.2 \times 10^{-3}$ M

$Q = [Ag^+][Cl^-] = [9.5 \times 10^{-4}][1.2 \times 10^{-3}]$

$= 1.1 \times 10^{-6}$

$Q > K_{sp}$, precipitation occurs.

18.21 $Cu(OH)_2 \rightleftharpoons Cu^{2+} + 2\,OH^-$ $K_{sp} = [Cu^{2+}][OH^-]^2$

$Cu^{2+} + 4\,NH_3 \rightleftharpoons Cu(NH_3)_4^{2+}$

$K_{form} = [Cu(NH_3)_4^{2+}]/[Cu^{2+}][NH_3]^4$

Net: $Cu(OH)_2 + 4\,NH_3 \rightleftharpoons Cu(NH_3)_4^{2+} + 2\,OH^-$

$K_{net} = K_{sp} \cdot K_{form} = (2.2 \times 10^{-20})(6.8 \times 10^{12})$

$= 1.5 \times 10^{-7}$

18.22 (a) Add NaCl(aq). AgCl will precipitate; Bi^3Cl_3 is soluble

(b) Add Na_2S(aq) or NaOH(aq) to precipitate FeS or $Fe(OH)_2$; K_2S and KOH are soluble.

CHAPTER 19

19.1 (a) O_3; larger molecules generally have higher entropies than smaller molecules

(b) $SnCl_4$(g); gases have a higher entropy than liquids.

19.2 (a) $\Delta S° = \Sigma S°(\text{products}) - \Sigma S°(\text{reactants})$

$\Delta S° = S°[NH_4Cl(aq)]) - S°[NH_4Cl(s)]$

$\Delta S° = (1\text{ mol})(169.9 \text{ J/mol}\cdot\text{K}) - (1\text{ mol})(94.85 \text{ J/mol}\cdot\text{K}) = 77.1$ J/K

A gain in entropy for the formation of a mixture (solution) is expected.

(b) $\Delta S° = 2\,S°(NH_3) - [S°(N_2) + 3\,S°(H_2)]$

$\Delta S° = (2\text{ mol})(192.77 \text{ J/mol}\cdot\text{K}) - [(1\text{ mol})(191.56 \text{ J/mol}\cdot\text{K}) + (3\text{ mol})(130.7 \text{ J/mol}\cdot\text{K})]$

$\Delta S° = -198.1$ J/K

A decrease in entropy is expected because there is a decrease in the number of moles of gases.

19.3 (a) Type 2 (b) Type 3 (c) Type 1 (d) Type 2

19.4 $\Delta S° = 2\,S°(HCl) - [S°(H_2) + S°(Cl_2)]$

$\Delta S°_{sys} = (2\text{ mol})(186.2 \text{ J/mol}\cdot\text{K}) - [(1\text{ mol})(130.7 \text{ J/mol}\cdot\text{K}) + (1\text{ mol})(223.08 \text{ J/mol}\cdot\text{K})]$

$= 18.6$ J/K

$\Delta S°_{surr} = -\Delta H°_{sys}/T = -(-184,620 \text{ J}/298 \text{ K})$

$= 616.5$ J/K

$\Delta S°_{univ} = \Delta S°_{sys} + \Delta S°_{surr} = 18.6 \text{ J/K} + 616.5 \text{ J/K}$

$= 635.1$ J/K

19.5 At 298 K, $\Delta S°_{surr} = -(467,900 \text{ J/K})/298 \text{ K} = -1570$ J/K

$\Delta S°_{univ} = \Delta S°_{sys} + \Delta S°_{surr} = 560.7 \text{ J/K} - 1570 \text{ J/K} = -1010$ J/K. The negative sign indicates the process is not spontaneous. At higher temperature, the value of $-\Delta H°_{sys}/T$ will be smaller. At a high enough temperature, matter dispersal will outweigh the energy dispersal in the system and the reaction will be spontaneous.

19.6 For the reaction: $N_2(g) + 3\,H_2(g) \longrightarrow 2\,NH_3(g)$

$\Delta H°_{rxn} = 2\,\Delta H°_f$ for NH_3(g) $= (2\text{ mol})(-45.90 \text{ kJ/mol})$

$= -91.80$ kJ

$\Delta S° = 2\,S°(NH_3) - [S°(N_2) + 3\,S°(H_2)]$

$\Delta S° = (2\text{ mol})(192.77 \text{ J/mol}\cdot\text{K}) - [(1\text{ mol})(191.56 \text{ J/mol}\cdot\text{K}) + (3\text{ mol})(130.7 \text{ J/mol}\cdot\text{K})]$

$\Delta S° = -198.1$ J/K

$\Delta G°_f = \Delta H°_f - T\Delta S° = -91.80 \text{ kJ} - (298 \text{ K})(-0.198 \text{ kJ/K})$

$\Delta G°_f = -32.80$ kJ

19.7 $SO_2(g) + 1/2\,O_2(g) \longrightarrow SO_3(g)$

$\Delta G° = \Sigma\,\Delta G°_f(\text{products}) - \Sigma\,\Delta G°_f(\text{reactants})$

$\Delta G° = \Delta G°_f[SO_3(g)] - \{\Delta G°_f[SO_2(g)] + 1/2\,\Delta G°_f[O_2(g)]\}$

$\Delta G° = -371.04 \text{ kJ} - \{300.13 \text{ kJ} + 0\} = -70.94$ kJ

19.8 $HgO(s) \longrightarrow Hg(\ell) + 1/2\ O_2(g)$; Determine the temperature at which $\Delta G_f^\circ = 0$ in which case $\Delta H_f^\circ - T\Delta S^\circ = 0$. T is the unknown in this problem.

$\Delta H_f^\circ = -90.83$ kJ

$\Delta S^\circ = S^\circ[Hg(\ell)] + 1/2\ S^\circ(O_2) - S^\circ(HgO)(s)$

$\Delta S^\circ = (1\ mol)(76.02\ J/mol\cdot K) + (0.5\ mol)(205.07\ J/mol\cdot K) - (1\ mol)(70.29\ J/mol\cdot K)$

$\quad = 108.26$ J/K

$\Delta H_f^\circ - T\Delta S^\circ = -90,830\ J - T(108.26\ J/K) = 0$

$T = 839$ K (566 °C)

19.9 $C(s) + CO_2(g) \rightleftharpoons 2\ CO(g)$

$\Delta G_{rxn}^\circ = 2\ \Delta G_f^\circ\ (CO) - \Delta G_f^\circ\ CO_2)$

$\Delta G_{rxn}^\circ = (2\ mol)(-137.17\ kJ/mol) - (1\ mol)(-394.36\ kJ/mol)$

$\Delta G_{rxn}^\circ = 120.02$ kJ

$\Delta G_{rxn}^\circ = -RT\ln K$

$120,020 = -(8.31\ J/mol\cdot K)(298\ K)\ln K$

$K = 8.94 \times 10^{-22}$

19.10 $\Delta G_{rxn}^\circ = -RT\ln K$

$\quad = -(8.31\ J/mol\cdot K)(298\ K)\ln (1.6 \times 10^7)$

$\Delta G_{rxn}^\circ = -41,100$ J/mol

CHAPTER 20

20.1 Oxidation half-reaction: $Al(s) \longrightarrow Al^{3+}(aq) + 3\ e^-$

Reduction half-reaction: $2\ H^+(aq) + 2\ e^- \longrightarrow H_2(g)$

Overall reaction: $2\ Al(s) + 6\ H^+(aq) \longrightarrow 2\ Al^{3+}(aq) + 3\ H_2(g)$

Al is the reducing agent and is oxidized, $H^+(aq)$ is the oxidizing agent and is reduced.

20.2 $2\ VO^{2+}(aq) + Zn(s) + 4\ H^+(aq) \longrightarrow Zn^{2+}(aq) + 2\ V^{3+}(aq) + 2\ H_2O(\ell)$

$2\ V^{3+}(aq) + Zn(s) \longrightarrow 2\ V^{2+}(aq) + Zn^{2+}(aq)$

20.3 Oxidation half-reaction: $Fe^{2+}(aq) \longrightarrow Fe^{3+}(aq) + e^-$

Reduction half-reaction: $MnO_4^-(aq) + 8\ H^+(aq) + 5\ e^- \longrightarrow Mn^{2+}(aq) + 4\ H_2O(\ell)$

Overall reaction: $MnO_4^-(aq) + 8\ H^+(aq) + 5\ Fe^{2+}(aq) \longrightarrow Mn^{2+}(aq) + 5\ Fe^{3+}(aq) + 4\ H_2O(\ell)$

20.4 (a) Oxidation half-reaction: $Al(s) + 3\ OH^-(aq) \longrightarrow Al(OH)_3(s) + 3\ e^-$

Reduction half-reaction: $S(s) + H_2O(\ell) + 2\ e^- \longrightarrow HS^-(aq) + OH^-(aq)$

Overall reaction: $2\ Al(s) + 3\ S(s) + 3\ H_2O(\ell) + 3\ OH^-(aq) \longrightarrow 2\ Al(OH)_3(s) + 3\ HS^-(aq)$

(b) Aluminum is the reducing agent and is oxidized, sulfur is the oxidizing agent and is reduced.

20.5 Construct two half cells, the first with a silver electrode and a solution containing $Ag^+(aq)$ and the second with a nickel electrode and a solution containing $Ni^{2+}(aq)$. Connect the two half cells with a salt bridge. When the electrodes are connected through an external circuit, elec-trons will flow from the anode (the nickel electrode) to the cathode (the silver electrode). The overall cell reaction is $Ni(s) + 2\ Ag^+(aq) \longrightarrow Ni^{2+}(aq) + 2\ Ag(s)$. To maintain electrical neutrality in the two half cells, negative ions will flow from the $Ag|Ag^+$ half-cell to the $Ni|Ni^{2+}$ half cell, and positive ions will flow in the opposite direction.

20.6 Anode reaction: $Zn(s) \longrightarrow Zn^{2+}(aq) + 2\ e^-$

Cathode reaction: $2\ Ag^+(aq) + 2\ e^- \longrightarrow 2\ Ag(s)$

$E_{cell}^\circ = E_{cathode}^\circ - E_{anode}^\circ = 0.80\ V - (-0.76\ V)$

$\quad = 1.56$ V

20.7 Mg is easiest to oxidize, and Au is the most difficult. (See Table 20.1)

20.8 Use the "northwest-southeast rule" or calculate the cell voltage to determine if a reaction is product-favored. Reactions (a) and (c) are reactant-favored; reactions (b) and (d) are product-favored.

20.9 Overall Reaction: $Fe(s) + 2\ H^+(aq) \longrightarrow Fe^{2+}(aq) + H_2(g)$

$(\Delta E_{cell}^\circ = 0.44\ V, n = 2)$

$E_{cell} = E_{cell}^\circ - (0.0257/n)\ \ln\{[Fe^{2+}]/[H^+]^2\}$

$\quad = 0.44 - (0.0257/2)\ \ln\{[0.024]/[0.056]^2\}$

$\quad = 0.44\ V - 0.026\ V = 0.41$ V

20.10 Overall Reaction: $2\ Al(s) + 3\ Fe^{2+}(aq) \longrightarrow 2\ Al^{3+}(aq) + 3\ Fe(s)$

$(E_{cell}^\circ = 1.22\ V, n = 6)$

$E_{cell} = E_{cell}^\circ - (0.0257/n)\ \ln\{[Al^{3+}]^2/[Fe^{2+}]^3\}$

$\quad = 1.22 - (0.0257/6)\ \ln\{[0.025]^2/[0.50]^3\}$

$\quad = 1.22\ V - (-0.023)\ V = 1.24$ V

20.11 $\Delta G^\circ = -nFE^\circ = -(2\ mol\ e^-)(96500\ coul/mol\ e^-)$ $(-0.76\ V)(1\ J/1\cdot coul\ V) = 146,680\ J\ (= 147\ kJ)$ The negative value of E° and the positive value of ΔG° both indicate a reactant-favored reaction.

20.12 $E_{cell}^\circ = E_{cathode}^\circ - E_{anode}^\circ = 0.80\ V - 0.855\ V = -0.055\ V$, and n = 2

$E^\circ = (0.0257/n)\ln K$

$-0.055 = (0.0257/2)\ln K; K = 1.4 \times 10^{-2}$

20.13 cathode: $Zn^{2+}(aq) + 2\ e^- \longrightarrow Zn(s)$

$\qquad\qquad\qquad\qquad\qquad E_{cathode}^\circ = -0.76$ V

anode: $Zn(s) + 4\ CN^-(aq) \longrightarrow [Zn(CN)_4^{2-}] + 2\ e^-$

$\qquad\qquad\qquad\qquad\qquad E_{anode}^\circ = -1.26$ V

Overall: $Zn^{2+}(aq) + 4\ CN^-(aq) \longrightarrow [Zn(CN)_4^{2-}]$

$\qquad\qquad\qquad\qquad\qquad E_{cell}^\circ = 0.50$ V

$E^\circ = 0.0257/n\ \ln K$

$0.50 = (0.0257/2)\ln K; K = 7.9 \times 10^{16}$

20.14 Cathode: $2\ H_2O(\ell) + 2\ e^- \longrightarrow 2\ OH^-(aq) + H_2(g)$

$\qquad\qquad\qquad\qquad\qquad E_{cathode}^\circ = -0.83$ V

Anode: $4\ OH^-(aq) \longrightarrow O_2(g) + 2\ H_2O(\ell) + 4\ e^-$

$\qquad\qquad\qquad\qquad\qquad E_{anode}^\circ = 0.40$ V

Overall: $2\ H_2O(\ell) \longrightarrow 2\ H_2(g) + O_2(g)$

$E_{cell}^\circ = E_{cathode}^\circ - E_{anode}^\circ = -0.83\ V - 0.40\ V = -1.23$ V. This is the minimum voltage needed to cause this reaction to occur.

20.15 O_2 is formed at the anode, by the reaction $H_2O(\ell) \longrightarrow$
$4H^+(aq) + O_2(g) + 4\,e^-$

(0.445 amp)(45 min)(60 sec/min)(1 coul/
1 amp · sec)(1 mol e^-/96,500 coul)(1 mol O_2/
4 mol e^-)(32 g O_2/mol O_2) = 0.10 g O_2

20.16 Cathode reaction (electrolysis of molten NaCl) is $Na^+ +$
$e^- \longrightarrow Na$

$(25 \times 10^3$ amp)(60 min)(60 sec/min)(1 coul/1 amp ·
sec)(1 mol e^-/96,500 coul)(1 mol Na/mol e^-)
(23 g Na/mol Na) = 21,450 g Na = 21 kg Na

CHAPTER 21

21.1 (a) $2\,Na(s) + Br_2(\ell) \longrightarrow 2\,NaBr(s)$

(b) $Ca(s) + Se(s) \longrightarrow CaSe(s)$

(c) $4\,K(s) + O_2(g) \longrightarrow 2\,K_2O(s)$ (K_2O is one of the
possible products of this reaction. The primary
product from the reaction of potassium and oxygen
is KO_2, potassium superoxide.

(d) $2\,Al(s) + 3\,Cl_2(g) \longrightarrow 2\,AlCl_3(s)$

21.2 (a) H_2Te

(b) Na_3AsO_4

(c) $SeCl_6$

(d) $HBrO_4$

21.3 (a) NH_4^+ (ammonium ion)

(b) O_2^{2-} (peroxide ion)

(c) N_2H_4 (hydrazine)

(d) NF_3 (nitrogen trifluoride)

21.4 (a) ClO is an odd-electron molecule with Cl having the
unlikely oxidation number of +2.

(b) In Na_2Cl, chlorine would have the unlikely charge
of -2 (to balance the two positive charges of two
Na^+ ions.) The ionization energy of Cl (Chapter 8)
is high, so loss of two electrons is unlikely.

(c) This compound would require either calcium ion to
have the formula Ca^+ or acetate ion to be
$CH_3CO_2^{2-}$. In all of its compounds, calcium occurs
as the Ca^{2+} ion. The acetate ion, formed from
acetic acid by loss of H^+, has a -1 charge.

(d) No octet structure for C_3H_7 can be drawn. This
species has an odd number of electrons.

21.5 $CH_4(g) + H_2O(g) \longrightarrow 3\,H_2(g) + CO(g)$

Bonds broken: 4 C—H and 2 O—H (Sum = 2578 kJ)

4 D(C—H) = 4(413 kJ) = 1652 kJ

2 D(O—H) = 4(463 kJ) = 926 kJ

Bonds formed: 3 H—H and 1 C≡O (Sum = 2354 kJ)

3 D(H—H) = 3(436 kJ) = 1308 kJ

D(CO) = 1046 kJ

Estimated Energy of Reaction = 2578 kJ − 2354 kJ
    = +224 kJ

The triple bond energy is low, relative to three single
bonds. Six bonds were broken and made, but the
incremental energy to form the second and third bond
between C and O is not sufficient to overcome the energy required to break two single bonds.

21.6 Cathode reaction: $Na^+ + e^- \longrightarrow Na(\ell)$; 1 F, or 96,485
coulombs, is required to form 1 mole of Na. There are
24 h(60min/h)(60 s/min) = 86,400 s in a day. 1000. kg
$= 1.00 \times 10^6$ g.

1.000×10^6 g Na(1 mol Na/23.00 g Na)
(95,485 coul/mol Na)(1 amp · sec/1 coul)(1/86400 s) =
4.855×10^4 amp

21.7 Some interesting topics: gemstones of the mineral beryl,
uses of Be in the aerospace industry and in nuclear reactors, beryllium-copper alloy, severe health hazards
when beryllium or its compounds get into the lungs.

21.8 $Al_2(SO_4)_3$ is weakly acidic. The formula pairs a weak
acid, Al^{3+} ($K_a = 7.9 \times 10^{-6}$), and a very weak base
SO_4^{2-} (aq) ($K_b = 8.3 \times 10^{-13}$), and the acid strength is
greater than the base strength. The equation for the
acid dissociation is

$[Al(H_2O)_6]^{3+} + H_2O(\ell) \rightleftharpoons$
$\qquad\qquad\qquad H_3O^+(aq) + [Al(H_2O)_5OH]^{2+}$

21.9

21.10 (a) :N≡N—Ö: ⟷ :N̈=N=Ö:

The first resonance structure places the a negative
charge on oxygen, the second places it on the terminal
nitrogen. Because oxygen is more electronegative and
better able to accommodate the negative charge, the
first structure is favored.

(b) $NH_4NO_3(s) \longrightarrow N_2O(g) + 2\,H_2O(g)$
$\Delta H°$(reaction) = $\Delta H_f°$(products) − $\Delta H_f°$(reactants)
$\Delta H°$(reaction) = $\Delta H_f°(N_2O) + 2\,\Delta H_f°(H_2O) -$
$\Delta H_f°(NH_4NO_3)$ = 82.05 kJ + 2(−241.83 kJ)
$\qquad\qquad\qquad$ − (−365.56 kJ)
$\qquad\qquad$ = −36.32 kJ; the reaction is
$\qquad\qquad\qquad$ exothermic

21.11 First, calculate $\Delta G°$, $\Delta H°$ and $\Delta S°$ for this reaction, using
data from Appendix L.

$\Delta G° = \Delta G_f°$(products) − $\Delta G_f°$(reactants)

$\Delta G° = 2\,\Delta G_f°(ZnO) + 2\,\Delta G_f°(SO_2)$
$\qquad -2\,\Delta G_f°(ZnS) - 3\,\Delta G_f°(O_2)$
$\qquad = 2(-318.30$ kJ) + 2(−300.13 kJ)
$\qquad\quad - 2(-201.29$ kJ) − 0

$\Delta G° = -834.28$ kJ

$\Delta H° = 2\,\Delta H_f°(ZnO) + 2\,\Delta H_f°(SO_2)$
$\qquad - 2\,\Delta H_f°(ZnS) - 3\,\Delta H_f°(O_2)$
$\qquad = 2(-348.28$ kJ) + 2(−296.84 kJ)
$\qquad\quad - [2(-205.98$ kJ) + 0] = −878.28 kJ

$\Delta S° = 2\,S°(ZnO) + 2\,S°(SO_2) - 2\,S°(ZnS) - 3\,S°(O_2)$
$= 2(43.64\ J/K) + 2(248.21\ J/K) - 2(57.7\ J/K)$
$+ 3(205.07\ J/K)] = -46.9\ J/K$

$\Delta G° = \Delta H° - T(\Delta S°)$

$-834{,}280\ J = -878{,}280\ J - T(46.9\ J/K)$

$T = 938$ K, this reaction is spontaneous and product favored below about 900 K.

21.12 For the reaction: $HX(g) + Ag(s) \longrightarrow AgX(s) + 1/2\ H_2(g)$

$\Delta G° = \Delta G_f°(products) - \Delta G_f°(reactants)$

$\Delta G° = \Delta G_f°(AgX) - \Delta G_f°(HX)$

For HF: $\Delta G° = +79.4$ kJ Reactant-favored

For HCl: $\Delta G° = -14.67$ kJ Product-favored

For HBr: $\Delta G° = -43.45$ kJ Product-favored

CHAPTER 22

22.1 (a) $Co(NH_3)_3Cl_3$

(b) (i) $K_3[Co(NO_2)_6]$: a complex of cobalt(III) with a co-ordination number of 6
(ii) $Mn(NH_3)_2Cl_2$: a complex of manganese(II) with a coordination number of 6

22.2 (a) hexaaquanickel(II) sulfate

(b) dicyanobis(ethylenediamine)chromium(III) chloride

(c) potassium amminetrichloroplatinate(II)

(d) potassium dichlorocuprate(I)

22.3 (a) geometric isomers (with Cl in *cis* and *trans* positions)

(b) Only a single structure is possible.

(c) Only a single structure is possible.

(d) This compound is chiral; there are two optical isomers.

(e) Only a single structure is possible.

(f) Two structural isomers are possible based on coordination of the NO_2^- ligand through oxygen or nitrogen.

22.4 (a) $[Ru(H_2O)_6]^{2+}$: A complex with ruthenium(II) (d^6). Low spin complexes have no unpaired electrons and are diamagnetic. High spin complexes have 4 unpaired electrons and are paramagnetic.

high-spin Ru^{2+} low-spin Ru^{2+}

(b) $[Ni(NH_3)_6]^{2+}$: A complex with nickel(II) (d^8). Only one electron configuration is possible, this has 2 unpaired electrons and is paramagnetic.

Ni^{2+} ion (d^8)

CHAPTER 23

23.1 (a) $^{222}_{86}Rn \longrightarrow ^{218}_{84}Po + ^{4}_{2}\alpha$

(b) $^{218}_{84}Po \longrightarrow ^{218}_{85}At + ^{0}_{-1}\beta$

23.2 (a) E per photon = $h\nu = hc/\lambda$
$E = [(6.626 \times 10^{-34}\ J \cdot s/photon)(3.00 \times 10^8$
$m/s)]/(2.0 \times 10^{-12}\ m)$
$E = 9.94 \times 10^{-14}\ J/photon$
$E\ (per\ mol) = (9.94 \times 10^{-14}\ J/photon)(6.022$
$\times 10^{23}\ photons/mol)$
$E\ (per\ mol) = 5.99 \times 10^{10}\ J/mol$

23.3 (a) Emission of 6 α will lead to a decrease in the mass number of 24 and a decrease of 12 in atomic number. Emission of 5 β will increase the atomic number by 5, and not affect the mass. The final product of this series will have a mass number of 232 − 24 = 208 and an atomic number of 90 − 12 + 5 = 83. This identifies it as $^{208}_{83}Bi$.

(b) Step 1: $^{232}_{90}Th \longrightarrow ^{228}_{88}Ra + ^{4}_{2}\alpha$
Step 2: $^{228}_{88}Ra \longrightarrow ^{228}_{89}Ac + ^{0}_{-1}\beta$
Step 3: $^{228}_{89}Ac \longrightarrow ^{224}_{87}Fr + ^{4}_{2}\alpha$

23.4 (a) $^{0}_{1}\beta^+$ (b) $^{41}_{19}K$ (c) $^{0}_{-1}\beta$ (d) $^{22}_{12}Mg$

23.5 (a) $^{32}_{14}Si \longrightarrow ^{32}_{15}P + ^{0}_{-1}\beta$

(b) $^{45}_{22}Ti \longrightarrow ^{45}_{21}Sc + ^{0}_{1}\beta^+$ or $^{45}_{22}Ti + ^{0}_{-1}e \longrightarrow ^{45}_{21}Sc$

(c) $^{239}_{94}Pu \longrightarrow ^{235}_{92}U + ^{4}_{2}\alpha$

(d) $^{42}_{19}K \longrightarrow ^{42}_{20}Ca + ^{0}_{-1}\beta$

23.6 $\Delta m = 0.03428$ g/mol
$E_b = (3.438 \times 10^{-5}\ kg/mol)(2.998 \times 10^8)^2$
$= 3.090 \times 10^{12}\ J/mol\ (= 3.090 \times 10^9\ kJ/mol)$
$E_b/n = 5.150 \times 10^8\ kJ/mol$ nucleons

23.7 (a) 49.2 years is exactly 4 half-lives; quantity remaining
$= 1.5\ mg(1/2)^4 = 0.094$ mg

(b) 3 half lives, 36.9 years

(c) 1% is between 6 half-lives, 73.8 years ($1/64^{th}$ remains) and 7 half-lives, 86.1 years ($1/128^{th}$ remains)

23.8 $\ln([A]/[A_0]) = -kt$
$\ln([3.18 \times 10^3]/[3.35 \times 10^3]) = -k(2.00\ d)$
$k = 0.0260\ d^{-1}$
$t_{1/2} = 0.693/k = 0.693/(0.0260\ d^{-1}) = 26.7$ d

23.9 $k = 0.693/t_{1/2} = 0.693/200.\ y = 3.47 \times 10^{-3}\ y^{-1}$
$\ln([A]/[A_0]) = -kt$
$\ln(3.00 \times 10^3]/[6.50 \times 10^{12}]) = -(3.47 \times 10^{-3}\ y^{-1})t$
$\ln(4.61 \times 10^{-10}) = -(3.47 \times 10^{-3}\ y^{-1})t$
$t = 5860$ y

23.10 $\ln([A]/[A_0]) = -kt$
$\ln(10.99]/[12.56]) = -(1.21 \times 10^{-4}\ y^{-1})t$
$t = 1100$ y

23.11 3000 dpm/x = 1200 dpm/60 mg
$x = 150$ mg

Answers to Study Questions

CHAPTER 1

1.1 The crystals are cubic, implying that the Ca and F are arranged in a cubic manner.

1.3 The mixture is heterogeneous. Iron is magnetic, so a small magnet will attract the iron chips and remove them from the non-magnetic sand.

1.5 (a) Physical property

(b) Chemical property

(c) Chemical property

(d) Physical property

(e) Physical property

(f) Physical property

1.7 Mercury and water are liquids, and copper is a solid. The most dense substance is mercury, and the least dense is water.

1.8 (a) 5

(b) 3

(c) 3

(d) 4

1.9 (a) Al, Si, O

(b) All are solids. Aluminum metal (Al) and silicon (Si) chips are shiny, whereas the aquamarine crystal is a blue-gray color.

1.11 Particulate: particles of sodium (gray) and chlorine (Cl, yellow) are arranged in a cubic arrangement.
Macroscopic: the salt is composed of a transparent, crystalline material. See also Figure 3.11, page 94.

1.12 (a) C, carbon

(b) K, potassium

(c) Cl, chlorine

(d) P, phosphorus

(e) Mg, magnesium

(f) Ni, nickel

1.14 (a) Ba, barium

(b) Ti, titanium

(c) Cr, chromium

(d) Pb, lead

(e) As, arsenic

(f) Zn, zinc

1.16 Na (element) and NaCl (compound)

(b) Sugar (compound) and carbon (element)

(c) Gold (element) and gold chloride (compound)

1.18 (a) Physical (colorless) and chemical (burns in air)

(b) Physical (shiny metal, orange liquid) and chemical (reacts with bromine).

1.20 557 g

1.22 2.79 cm^3 or 2.79 mL

1.24 Al, aluminum

1.26 298 K

1.28 (a) 289 K

(b) 97 °C

(c) 310 K

1.30 4.00×10^4 m; 24.9 miles

1.32 5.3 cm^2; 5.3×10^{-4} m^2

1.34 250 cm^3; 0.25 L, 2.5×10^{-4} m^3; 0.25 dm^3

1.36 2.52×10^3 g

1.38 (a) Method A with all data included: average = 2.4 g/cm^3 and average deviation = 0.2 g/cm^3.

Method B with all data included: average = 3.480 g/cm^3 and average deviation = 1.166 g/cm^3.

For B the reading of 5.811 can be excluded because it is more than twice as large as all other readings. Using only the first three readings, you find average = 2.703 g/cm^3 and average deviation = 0.001 g/cm^3

(b) Method A: error = 0.3 g/cm^3 or about 10%

Method B: error = 0.001 g/cm^3

(c) Before excluding a data point for B, method A gives more accurate and more precise answer. After excluding data for B, this method gives a more accurate and more precise result.

1.40 (a) 3

(b) 3

(c) 5

(d) 4

1.42 0.122; three significant figures

1.44 0.0854 cm³

1.46 22 g

1.48 20,100 L

1.50 Normal body temperature is 37 °C. Therefore, Ga should melt if held in your hand.

1.52 0.197 nm; 197 pm

1.54 The volume increases from 250. mL to 272 mL on freezing. The ice will likely expand out of the can. (See page 20.)

1.56 Given equal masses, the less dense substance will occupy a larger volume, so water has the larger volume.

1.58 The lab partner is on the right track in estimating the density to lie between that at 15 °C and that at 25 °C. If one assumes is it halfway between those densities, a better estimate would be 0.99810 g/cm³. A more accurate analysis, however, shows that the decrease in density with increasing temperature is not linear. A more careful analysis would lead to a value of perhaps 0.99820 g/cm³ (not far from the actual value of 0.99840 g/cm³).

1.60 The density of the plastic is less than that of CCl₄, so the plastic will float on the liquid CCl₄. On the other hand, the aluminum is more dense than CCl₄, so aluminum will sink when placed in CCl₄.

1.62 Mass of lead is 170 g

1.64

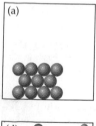
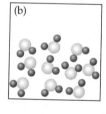
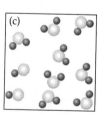
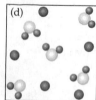
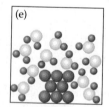
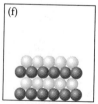

1.66 One could check for an odor, check the boiling or freezing point, or determine the density. If the density is approximately 1 g/cm³ at room temperature, the liquid could be water. If it boils at about 100 °C and freezes about 0 °C, that would be consistent with water. To check for the presence of salt, boil the liquid away. If a substance remains, it could be salt, but further testing would be required.

1.68

1.70 A copper-colored metal could be copper, but it may also be an alloy of copper, for example, brass or bronze. Testing the material's density and melting temperature would be one way to find out if it is copper.

1.72 (a) Solid potassium metal reacts with liquid water to produce gaseous hydrogen and a homogeneous mixture (solution) of potassium hydroxide in liquid water.

(b) The reaction is a chemical change.

(c) The reactants are potassium and water. Products are hydrogen gas and a water (aqueous) solution of potassium hydroxide. Heat and light are also evolved.

(d) Among the qualitative observations are (i) the reaction is violent; and (ii) heat and light (a purple flame) are produced.

1.74 0.018 mm

1.76 Thickness = 2×10^{-7} cm. This is likely related to the "length" of oil molecules.

CHAPTER 2

2.6 Radius is about 600,000 cm or about 6 km.

2.8

Element	Titanium	Thallium
Symbol	Ti	Tl
Atomic Number	22	81
Atomic Weight	47.87	204.38
Group	4B	3A
Period	4	6

Both elements are metals.

2.10 (a) 0.5 mol Si (14 g) > 0.5 mol Na (11 g)

(b) 0.5 mol Na (11 g) > 9.0 g Na

(c) Molar mass of Fe is greater than molar mass of K. Therefore, 10 atoms of Fe have more mass than 10 atoms of K.

2.12 A group or family are the elements in a vertical column of the periodic table. The elements in a horizontal row of the table constitute a period.

2.14 Metals are elements given in a purple color in the periodic table in the front of the book. Nonmetals are in yellow, whereas metalloids are in green. For example iron (Fe) is a metal in Group 8B and in period 4. Silicon (Si) is a metalloid in Group 4A and period 3. Br is a nonmetal in Group 7A and period 4.

2.16 Nickel (Ni) is a transition element (as are all those in a darker purple on the periodic table inside the front cover). Halogens are those elements in Group 7A (17) (e.g., F). The noble gases are in Group 8A (18) (e.g., Ne). The alkali metals are in Group 1A (1) (e.g., Na).

2.18 Madame Curie discovered radium (Ra, element 88) and Polonium (Po, element 84). The former is named because the element emits radiation, and the latter is named for her homeland, Poland. On the World Wide Web go to the website for the Nobel Foundation (www.nobel.se).

2.20 (a) ^{27}Mg, mass number = 27

(b) ^{48}Ti, mass number = 48

(c) ^{62}Zn, mass number = 62

2.22 (a) $^{39}_{19}$K

(b) $^{84}_{36}$Kr

(c) $^{60}_{27}$Co

2.24
Element	^{24}Mg	^{119}Sn	^{232}Th
Electrons	12	50	90
Protons	12	50	90
Neutrons	12	69	142

2.26 ^{99}Tc has 43 protons, 43 electrons, and 56 neutrons

2.28 $^{57}_{27}$Co, $^{58}_{27}$Co, $^{60}_{27}$Co

2.30 ^{205}Tl is more abundant (70.5%) than ^{203}Tl (29.5%). Its atomic weight is closer to 205 than to 203.

2.32 $(0.0750)(6.015121) + (0.9250)(7.016003) = 6.94$

2.34 ^{69}Ga, 60.12%; ^{71}Ga, 39.88%

2.36 (a) 68 g Al

(b) 0.0698 g Fe

(c) 0.60 g Ca

(d) 1.32×10^4 g Ne

2.38 (a) 1.9998 mol Cu

(b) 0.0017 mol Li

(c) 2.1×10^{-5} mol Am

(d) 0.250 mol Al

2.40 1.0552×10^{-22} g

2.42 Five elements. Nitrogen (N) and phosphorus (P) are nonmetals. Arsenic (As) and antimony (Sb) are metalloids, and bismuth (Bi) is a metal.

2.44 Eight elements: periods 2 and 3. Eighteen elements: periods 4 and 5. Thirty two elements: period 6.

2.46
Symbol	^{58}Ni	^{33}S	^{20}Ne	^{55}Mn
Protons	28	16	10	25
Neutrons	30	17	10	30
Electrons	28	16	10	25
Name	Nickel	Sulfur	Neon	Manganese

2.48 Potassium has an atomic weight of 39.0983. This mass is close to the mass of the ^{39}K isotope. Therefore, the abundance of ^{41}K is low (6.73%).

2.50 (a) Mg

(b) H

(c) Si

(d) Fe

(e) F and Cl, and Br. Chlorine in more abundant.

2.52 a, b, and c are all possible. d is impossible because one atom of S has a mass of 5.325×10^{-23} g. Therefore, one mole of molecules consisting of 8 S atoms cannot be less than the mass of one atom.

2.54 (a) Three elements — Co, Ni, and Cu — have densities of about 9 g/cm^3.

(b) Boron in the 2nd period and aluminum in the 3rd period have the largest densities. Both are in Group 3A.

(c) Elements that have very low densities are all gases. These include hydrogen, helium, nitrogen, oxygen, fluorine, neon, chlorine, argon, and krypton.

2.56 (a) Beryllium, magnesium, calcium, strontium, barium, radium

(b) Sodium, magnesium, aluminum, silicon, phosphorus, sulfur, chlorine, argon

(c) carbon

(d) sulfur

(e) iodine

(f) magnesium

(g) krypton

(h) sulfur

(i) germanium or arsenic

2.58 9.42×10^{-5} mol Kr; 5.67×10^{19} atoms Kr

2.60 1.44 cm

2.62 1.9×10^{22} atoms Cr

2.64 Ratio of masses = 1.9375 g P/1.0000 g O (or 1.94 g P/1.00 g O to three significant figures). If mass of O is assumed to be 16.000, then mass of P is $(1.9375)(16.000) = 31.0000$ (or 31.0 to three significant figures).

2.66 1.0028×10^{23} atoms C. If the accuracy is ± 0.0001 g, the maximum mass could be 2.0001 g, which also represents 1.0028×10^{23} atoms C.

2.68 6.0×10^{29} atoms/ton-mol

2.70 The simulation indicates that Ca has 5 stable isotopes 20, 22, 23, 24, and 26 neutrons. ($^{48}_{20}$Ca, with a very low abundance, is also possible.) The ratio of neutrons to protons rapidly becomes greater than 1 as the number of protons increases. An isotope such as $^{108}_{54}$Xe is not possible, because the neutron/proton ratio is only 1.

CHAPTER 3

3.1 One molecule has 2 N atoms and 6 H atoms. There are 6 mol of H atoms (3.6×10^{24} H atoms) in 1 mol of cisplatin. The molar mass of cisplatin is 300.1 g/mol.

3.3 Sr atom has 38 electrons. Atom loses 2 electrons to form Sr^{2+}. The ion has 36 electrons, the same as a Kr atom.

3.5 Adenine, $C_5H_5N_5$, molar mass = 135.1 g/mol

(a) 40 g represents 0.30 mol, whereas 3.0×10^{23} molecules represent about 0.5 mol.

(b) All of the bases have at least a six-member ring consisting of C and N atoms.

3.7   BaCl$_2$ molar mass = 208.2 g/mol and SiCl$_4$ molar mass = 169.9 g/mol. Therefore, 0.5 mol of BaCl$_2$ has more mass than 0.5 mol of SiCl$_4$.

3.9   H$_2$O has 88.7% O whereas CH$_3$OH has 49.9% O.

3.10 (a) C$_7$H$_{16}$O

      (b) C$_6$H$_8$O$_6$

      (c) C$_{14}$H$_{18}$N$_2$O$_5$

3.12 (a) 1 Ca atom, 2 C atoms, and 4 O atoms

      (b) 7 C atoms, 6 H atoms, and 1 O atom

      (c) 1 Co atom, 6 N atoms, 15 H atoms, 2 O atoms, and 2 Cl atoms

      (d) 4 K atoms, 1 Fe atom, and 6 C atoms, and 6 N atoms

3.14 Sulfuric acid, H$_2$SO$_4$. The structure is not flat. Chemists describe the structure as a tetrahedron of O atoms around the S atom.

3.16 (a) Mg^{2+}

      (b) Zn^{2+}

      (c) Ni^{2+}

      (d) Ga^{3+}

3.18 (a) Ba^{2+}

      (b) Ti^{4+}

      (c) PO$_4^{3-}$

      (d) HCO$_3^-$

      (e) S^{2-}

      (f) ClO$_4^-$

      (g) Co^{2+}

      (h) SO$_4^{2-}$

3.20 K loses 1 electron per atom to form a K$^+$ ion. It has the same number of electrons as an Ar atom.

3.22 Ba^{2+} and Br$^-$ ions. Compound formula is BaBr$_2$.

3.24 (a) Two K$^+$ ions and one S^{2-} ion

      (b) One Co^{2+} ion and one SO$_4^{2-}$ ion

      (c) One K$^+$ ion and one MnO$_4^-$ ion

      (d) Three NH$_4^+$ ions and one PO$_4^{3-}$ ion

      (e) One Ca^{2+} ion and two ClO$^-$ ions

3.26 Co^{2+} gives CoO and Co^{3+} gives Co$_2$O$_3$

3.28 (a) AlCl$_2$ should be AlCl$_3$ (based on a Al^{3+} ion and three Cl$^-$ ions)

      (b) KF$_2$ should be KF, which is based on a K$^+$ ion and a F$^-$ ion.

      (c) Ga$_2$O$_3$ is correct

      (d) MgS is correct

3.30 (a) Mg^{2+} forms MgO with O^{2-} and Mg$_3$(PO$_4$)$_2$ with Mg^{2+} and PO$_4^{3-}$.

      (b) Al^{3+} forms Al$_2$O$_3$ and AlPO$_4$.

3.32 (a) potassium sulfide

      (b) cobalt(II) sulfate

(c) ammonium phosphate

(d) calcium hypochlorite

3.34 (a) (NH$_4$)$_2$CO$_3$

      (b) CaI$_2$

      (c) CuBr$_2$

      (d) AlPO$_4$

      (e) AgCH$_3$CO$_2$

3.36 Compounds with Na$^+$: Na$_2$CO$_3$ and NaI

      Compounds with Ba^{2+}: BaCO$_3$ and BaI$_2$

3.38 The force of attraction is stronger in NaF than in NaI because the distance between ion centers is smaller in NaF (235 pm) than in NaI (322 pm).

3.40 (a) nitrogen trifluoride

      (b) hydrogen iodide

      (c) boron triiodide

      (d) phosphorus pentafluoride

3.42 (a) SCl$_2$

      (b) N$_2$O$_5$

      (c) SiCl$_4$

      (d) B$_2$O$_3$

3.44 (a) 159.7 g/mol

      (b) 117.2 g/mol

      (c) 176.1 g/mol

3.46 (a) 290.8 g/mol

      (b) 249.7 g/mol

3.48 (a) 0.0166 mol

      (b) 0.00555 mol

      (c) 0.00555 mol

3.50 60.9 mol CH$_3$CN

3.52 Amount of SO$_2$ = 12.5 mol

      Number of molecules = 7.52 × 10^{24} molecules

      Number of S atoms = 7.52 × 10^{24} atoms

      Number of O atoms = 2.26 × 10^{25} atoms

3.54 (a) 86.60% Pb and 13.40% S

      (b) 81.71% C and 18.29% H

      (c) 79.96% C, 9.394% H, and 10.65% O

3.56 8.66 g Pb

3.58 15.0 g CuS

3.60 C$_4$H$_6$O$_4$

3.62 (a) CH, 26.0 g/mol; C$_2$H$_2$

      (b) CHO, 116.1 g/mol; C$_4$H$_4$O$_4$

      (c) CH$_2$, 112.2 g/mol, C$_8$H$_{16}$

3.64 Empirical formula, CH; molecular formula C$_2$H$_2$

3.66 Empirical formula, C$_3$H$_4$; molecular formula C$_9$H$_{12}$

3.68 Empirical and molecular formulas are both C$_8$H$_8$O$_3$

3.70 Formula is MgSO$_4$ · 7H$_2$O

3.72 XeF$_2$

3.74 ZnI$_2$

3.76 2 × 10^{21} molecules of water

3.78 (a) 25.86% Cu, 5.742% H, 22.80% N, 32.55% O, 13.05% S

    (b) 2.72 g Cu and 0.770 g H_2O in 10.5 g of compound (molar mass = 245.72 g/mol)

3.80 1200 kg

3.82 $C_4H_6O_5$

3.84 FeC_2O_4

3.86 (a) $C_{10}H_{15}NO$, molar mass = 165.23 g/mol

    (b) 72.69% C

    (c) 7.57×10^{-4} mol

    (d) 4.56×10^{20} molecules and 4.56×10^{21} C atoms

3.88 Ionic compounds

    (c) LiS, lithium sulfide

    (d) In_2O_3, indium oxide

    (g) CaF_2, calcium fluoride

3.90 (a) NaClO, ionic

    (b) BI_3

    (c) $Al(ClO_4)_3$, ionic

    (d) $Ca(CH_3CO_2)_2$, ionic

    (e) $KMnO_4$, ionic

    (f) $(NH_4)_2SO_3$, ionic

    (g) KH_2PO_4, ionic

    (h) S_2Cl_2

    (i) ClF_3

    (j) PF_3

3.92 Empirical formula = C_5H_4; molecular formula = $C_{10}H_8$

3.94 Empirical formula and molecular formula = $C_5H_{14}N_2$

3.96 $C_9H_7MnO_3$

3.98 Empirical formula = ICl_3; molecular formula = I_2Cl_6

3.100 7.35 kg of iron

3.102 Statements (a)–(c) are true. Statement (d) is not true; 57.1 g of octane contains 9.08 g H.

3.104 0.346 g Bi

3.106 The unknown element is carbon.

3.108 Water is most strongly attracted to Al^{3+} ion. The ion has the smallest size and the largest charge.

3.110 Students did not find the ratio of mol H_2O/mol $CaCl_2$ to be a small, whole number. They should heat again and reweigh.

3.112 (a) 0.766 g Ni or 0.0130 mol

    (b) NiF_2

    (c) nickel(II) fluoride

3.114 When words are written with the red, hydrated compound, the words are not visible. However, when heated, the hydrated salt loses water to form anhydrous $CoCl_2$, which is deep blue, and the words are visible.

3.116 According to Coulomb's law, the force of attraction between oppositely charged ions increases with the ion charges and with decreasing ion-ion separation.

CHAPTER 4

4.2 $N_2(g) + 3 H_2(g) \longrightarrow 2 NH_3(g)$

4.4 Stoichiometric factor = (2 mol NH_3/1 mol N_2)

4.6 Ratio of CO to Fe_2O_3 should be (3 mol CO/1 mol Fe_2O_3). Here we have (65 mol CO/25 mol Fe_2O_3) = 2.6 mol CO/1 mol Fe_2O_3. There is not enough CO to consume the Fe_2O_3, so CO is the limiting reactant. A related way to view this is as follows: 25 mol of Fe_2O_3 requires 3 × 25 mol of CO or 75 mol. Only 65 mol of CO is available, so CO is the limiting reactant.

4.8 (a) $4 Cr(s) + 3 O_2(g) \longrightarrow 2 Cr_2O_3(s)$

    (b) $Cu_2S(s) + O_2(g) \longrightarrow 2 Cu(s) + SO_2(g)$

    (c) $C_6H_5CH_3(\ell) + 9 O_2(g) \longrightarrow 4 H_2O(\ell) + 7 CO_2(g)$

4.10 (a) $Fe_2O_3(s) + 3 Mg(s) \longrightarrow 3 MgO(s) + Fe(s)$

    Reactants = iron(III) oxide, magnesium

    Products = magnesium oxide, iron

    (b) $AlCl_3(s) + 3 H_2O(\ell) \longrightarrow Al(OH)_3(s) + 3 HCl(aq)$

    Reactants = aluminum chloride, water

    Products = aluminum hydroxide, hydrochloric acid

    (c) $2 NaNO_3(s) + H_2SO_4(\ell) \longrightarrow Na_2SO_4(s) + 2 HNO_3(\ell)$

    Reactants = sodium nitrate, sulfuric acid

    Products = sodium sulfate, nitric acid

    (d) $NiCO_3(s) + 2 HNO_3(aq) \longrightarrow Ni(NO_3)_2(aq) + CO_2(g) + H_2O(\ell)$

    Reactants = nickel(II) carbonate, nitric acid

    Products = nickel(II) nitrate, water, carbon dioxide

4.12 4.5 mol O_2; 310 g Al_2O_3

4.14 22.7 g Br_2; 25.3 g Al_2Br_6

4.16 (a) $4 Fe(s) + 3 O_2(g) \longrightarrow 2 Fe_2O_3(s)$

    (b) 3.83 g Fe_2O_3

    (c) 1.15 g O_2

4.18 (a) 242 g $CaCO_3$; (b) 329 g $CaSO_4$

4.20 F_2 is the limiting reactant

4.22 (a) CH_4 is the limiting reactant

    (b) 375 g H_2

    (c) Excess H_2O = 1390 g

4.24 (a) 68.0 g NH_3; (b) 10. g NH_4Cl remains

4.26 (332 g/407 g)100% = 81.6%

4.28 (a) 14.3 g $Cu(NH_3)_4SO_4$

    (b) 88.3% yield

4.30 91.9% hydrate

4.32 84.3% $CaCO_3$

4.34 1.467% Tl_2SO_4

4.36 Empirical formula = CH

4.38 Empirical formula = CH_2; molecular formula = C_5H_{10}

4.40 Empirical formula = KO

4.42 Empirical formula = $Ni(CO)_4$

4.44 (a) $CO_2(g) + 2 NH_3(g) \longrightarrow NH_2CONH_2(s) + H_2O(\ell)$

(b) $UO_2(s) + 4\,HF(aq) \longrightarrow UF_4(s) + 2\,H_2O(\ell)$

$UF_4(s) + F_2(g) \longrightarrow UF_6(s)$

(c) $TiO_2(s) + 2\,Cl_2(g) + 2\,C(s) \longrightarrow$
$$TiCl_4(\ell) + 2\,CO(g)$$
$TiCl_4(\ell) + 2\,Mg(s) \longrightarrow Ti(s) + 2\,MgCl_2(s)$

4.46 (a) Products = $CO_2(g)$ and $H_2O(g)$

(b) $CH_4(g) + O_2(g) \longrightarrow CO_2(g) + H_2O(g)$

(c) 63.99 g O_2

(d) 80.03 g products

4.48 71.1 mg

4.50 (a) $2\,Fe(s) + 3\,Cl_2(g) \longrightarrow 2\,FeCl_3(s)$

(b) 19.0 g Cl_2 required; 29.0 g $FeCl_3$ produced.

(c) 63.7% yield

4.52 (a) 507 g H_2O produced

(b) 1760 g O_2 required

4.54 The H/B ratio is 1.4/1.0. The empirical formula is B_5H_7.

4.56 Empirical formula = $C_{10}H_{20}O$

4.58 The metal is most likely copper, Cu.

4.60 Ti_2O_3

4.62 19.9 mg

4.64 (a) In the reactions represented by the sloping portion of the graph, Fe is the limiting reactant. At the point at which the yield of product begins to be constant (at 2.0 g Fe), the reactants are present in stoichiometric amounts. That is, 10.6 g of product contains 2.0 g Fe and 8.6 g Br_2.

(b) 2.0 g Fe = 0.036 mol Fe; 8.6 g Br_2 = 0.054 mol Br_2. The mol ratio is 1.5 Br_2 to 1.0 mol Fe.

(c) The mol ratio of 1.5 mol Br_2/1.0 mol Fe = 3 Br/1 Fe. The empirical formula is $FeBr_3$.

(d) $2\,Fe(s) + 3\,Br_2(\ell) \longrightarrow 2\,FeBr_3(s)$

(e) iron(III) bromide

(f) Statement (i) is correct.

CHAPTER 5

5.5 Water-soluble: $Cu(NO_3)_2$, $CuCl_2$

5.7 NO_3^- is the only spectator ion
$2\,H^+(aq) + Mg(OH)_2(s) \longrightarrow 2\,H_2O(\ell) + Mg^{2+}(aq)$

5.9 Br^- has been oxidized to Br_2 and Cl_2 has been reduced to Cl^-. Cl_2 is the oxidizing agent and Br^- is the reducing agent.

5.11 HNO_3, Cl_2, O_2, and $KMnO_4$ are oxidizing agents. Sodium (Na) is a reducing agent.

5.13 0.100 mol NaCl (58.4 g/mol)= 5.84 g and 0.0600 mol Na_2CO_3 (106 g/mol) = 6.36 g

5.15 (a) Reactant-favored. This is the reverse of an acid-base reaction.

(b) Product-favored. A precipitation occurs.

5.17 250. mL of 0.500 M KCl requires 0.125 mol of KCl or 9.32 g. Weigh out this mass of KCl, place it in a 250.-mL

volumetric flask. Add water to the mark on the flask (and mix thoroughly).

5.19 The equivalence point is that point at which the amount of hydroxide ion added, for example, is exactly equivalent to the amount of hydrogen ions available from an acid in solution.

5.20 (a) $CuCl_2$

(b) $AgNO_3$

(c) All are water-soluble

5.22 (a) K^+ and OH^-

(b) $2\,K^+$ and SO_4^{2-}

(c) Li^+ and NO_3^-

(d) $2\,NH_4^+$ and SO_4^{2-}

5.24 (a) Water-soluble; $2\,Na^+$ and CO_3^{2-}

(b) Water-soluble; Cu^{2+} and SO_4^{2-}

(c) Not water-soluble

(d) Water-soluble; Ba^{2+} and $2\,Br^-$

5.26 $CdCl_2(aq) + 2\,NaOH(aq) \longrightarrow$
$$Cd(OH)_2(s) + 2\,NaCl(aq)$$
$Cd^{2+}(aq) + 2\,OH^-(aq) \longrightarrow Cd(OH)_2(s)$

5.28 (a) $NiCl_2(aq) + (NH_4)_2S(aq) \longrightarrow$
$$NiS(s) + 2\,NH_4Cl(aq)$$

(b) $3\,Mn(NO_3)_2(aq) + 2\,Na_3PO_4(aq) \longrightarrow$
$$Mn_3(PO_4)_2(s) + 6\,NaNO_3(aq)$$

5.30 (a) $(NH_4)_2CO_3(aq) + Cu(NO_3)_2(aq) \longrightarrow$
$$CuCO_3(s) + 2\,NH_4NO_3(aq)$$
$CO_3^{2-}(aq) + Cu^{2+}(aq) \longrightarrow CuCO_3(s)$

(b) $Pb(OH)_2(s) + 2\,HCl(aq) \longrightarrow PbCl_2(s) + 2\,H_2O(\ell)$
$Pb(OH)_2(s) + 2\,H^+(aq) + 2\,Cl^-(aq) \longrightarrow$
$$PbCl_2(s) + 2\,H_2O(\ell)$$

(c) $BaCO_3(s) + 2\,HCl(aq) \longrightarrow$
$$BaCl_2(aq) + H_2O(\ell) + CO_2(g)$$
$BaCO_3(s) + 2\,H^+(aq) \longrightarrow$
$$Ba^{2+}(aq) + H_2O(\ell) + CO_2(g)$$

5.32 (a) $AgNO_3(aq) + KI(aq) \longrightarrow AgI(s) + KNO_3(aq)$
$Ag^+(aq) + I^-(aq) \longrightarrow AgI(s)$

(b) $Ba(OH)_2(aq) + 2\,HNO_3(aq) \longrightarrow$
$$Ba(NO_3)_2(aq) + 2\,H_2O(\ell)$$
$OH^-(aq) + H^+(aq) \longrightarrow H_2O(\ell)$

(c) $Na_3PO_4(aq) + Ni(NO_3)_2(aq) \longrightarrow$
$$Ni_3(PO_4)_2(s) + NaNO_3(aq)$$
$2PO_4^{3-}(aq) + 3Ni^{2+}(aq) \longrightarrow Ni_3(PO_4)_2(s)$

5.34 $HNO_3(aq) \longrightarrow H^+(aq) + NO_3^-(aq)$

5.36 $H_2C_2O_4(aq) \longrightarrow H^+(aq) + HC_2O_4^-(aq)$
$HC_2O_4^-(aq) \longrightarrow H^+(aq) + C_2O_4^{2-}(aq)$

5.38 $MgO(s) + H_2O(\ell) \longrightarrow Mg(OH)_2(s)$

5.40 (a) Acetic acid reacts with magnesium hydroxide to give magnesium acetate and water.
$2\,CH_3CO_2H(aq) + Mg(OH)_2(s) \longrightarrow$
$$Mg(CH_3CO_2)_2(aq) + 2\,H_2O(\ell)$$

(b) Perchloric acid reacts with ammonia to give ammonium perchlorate

$$HClO_4(aq) + NH_3(aq) \longrightarrow NH_4ClO_4(aq)$$

5.42 $Ba(OH)_2(s) + 2\ HNO_3(aq) \longrightarrow$
$$Ba(NO_3)_2(aq) + 2\ H_2O(\ell)$$

5.44 $FeCO_3(s) + 2\ HNO_3(aq) \longrightarrow$
$$Fe(NO_3)_2(aq) + CO_2(g) + H_2O(\ell)$$

Iron(II) carbonate reacts with nitric acid to give iron(II) nitrate, carbon dioxide, and water.

5.46 (a) Acid-base

$$Ba(OH)_2(s) + 2\ HCl(aq) \longrightarrow BaCl_2(aq) + H_2O(\ell)$$

(b) Gas-forming

$$2\ HNO_3(aq) + CoCO_3(s) \longrightarrow$$
$$Co(NO_3)_2(aq) + H_2O(\ell) + CO_2(g)$$

(c) Precipitation

$$2\ Na_3PO_4(aq) + 3\ Cu(NO_3)_2(aq) \longrightarrow$$
$$Cu_3(PO_4)_2(s) + 6\ NaNO_3(aq)$$

5.48 (a) Precipitation

$$MnCl_2(aq) + Na_2S(aq) \longrightarrow MnS(s) + 2\ NaCl(aq)$$

(b) Precipitation

$$K_2CO_3(aq) + ZnCl_2(aq) \longrightarrow ZnCO_3(s) + 2\ KCl(aq)$$

5.50 (a) Precipitation of CuS

(b) Formation of water in an acid-base reaction

5.52 (a) Br = +5 and O = −2

(b) C = +3 each and O = −2

(c) F = −1

(d) Ca = +2 and H = −1

(e) H = +1, Si = +4, and O = −2

(f) H = +1, S = +6, and O = −2

5.54 (a) Oxidation-reduction

Zn is oxidized from 0 to +2, and N in NO_3^- is reduced from +5 to +4 in NO_2.

(b) Acid-base reaction

(c) Oxidation-reduction

Calcium is oxidized from 0 to +2 in $Ca(OH)_2$, and H is reduced from +1 in H_2O to 0 in H_2.

5.56 (a) O_2 is the oxidizing agent (as it is always is) and so C_2H_4 is the reducing agent. As is always the case in a combustion, O_2 oxidizes the other reactant (a C-containing compound), which is reduced.

(b) Si is oxidized from 0 in Si to +4 in $SiCl_4 \cdot Cl_2$ is reduced from 0 in Cl_2 to −1 in Cl^-.

5.58 $[Na_2CO_3] = 0.254$ M; $[Na^+] = 0.508$ M; $[CO_3^{2-}] = 0.254$ M

5.60 0.494 g $KMnO_4$

5.62 5.08×10^3 mL

5.64 (a) 0.50 M NH_4^+ and 0.25 M SO_4^{2-}

(b) 0.246 M Na^+ and 0.123 M CO_3^{2-}

(c) 0.056 M H^+ and 0.056 M NO_3^-

5.66 A mass of 1.06 g of Na_2CO_3 is required. After weighing out this quantity of Na_2CO_3, transfer it to a 500.-mL volumetric flask. Rinse any solid from the neck of the flask while filling the flask with distilled water. Add water until the bottom of the meniscus of the water is at the top of the scribed mark on the neck of the flask.

5.68 0.0750 M

5.70 Method (a) is correct. Method (b) gives an acid concentration of 0.15 M.

5.72 $[H^+] = 10^{-pH} = 4.0 \times 10^{-4}$ M

5.74 HNO_3 is a strong acid, so $[H^+] = 0.0013$ M. pH = 2.89.

5.76

	pH	$[H^+]$	Acid/Base
(a)	1.00	0.10 M	Acidic
(b)	10.50	3.2×10^{-11} M	Basic
(c)	4.89	1.3×10^{-5} M	Acidic
(d)	7.64	2.3×10^{-8} M	Basic

5.78 268 mL

5.80 210 g NaOH and 190 g Cl_2

5.82 174 mL

5.84 1500 mL

5.86 44.6 mL

5.88 1.052 M

5.90 104 g/mol

5.92 12.8% Fe

5.94 260 mL

5.96 (a) $Mg(s) + 4\ HNO_3(aq) \longrightarrow$
$$Mg(NO_3)_2(aq) + 2\ NO_2(g) + 2\ H_2O(\ell)$$

(b) magnesium, nitric acid, magnesium nitrate, nitrogen dioxide, water

(c) $Mg(s) + 4\ H^+(aq) + 2\ NO_3^-(aq) \longrightarrow$
$$Mg^{2+}(aq) + 2\ NO_2(g) + 2\ H_2O(\ell)$$

(d) Mg is the reducing agent and HNO_3 is the oxidizing agent

5.98 (a) H_2O, NH_3, NH_4^+, and OH^- (and a trace of H^+)

(b) H_2O, CH_3CO_2H, $CH_3CO_2^-$, and H^+ (and a trace of OH^-)

(c) H_2O, Na^+, and OH^- (and a trace of H^+)

(d) H_2O, H^+, and Br^- (and a trace of OH^-)

5.100 15.0 g of $NaHCO_3$ require 1190 mL of 0.15 M acetic acid. Therefore, acetic acid is the limiting reactant. (Conversely, 125 mL of 0.15 M acetic acid requires only 1.58 g of $NaHCO_3$.)

5.102 16.1% oxalic acid

5.104 (a) Water soluble: $Cu(NO_3)_2$ [copper(II) nitrate] or $CuCl_2$ [copper(II) chloride].

Water-insoluble: CuS [copper(II) sulfide] or $CuCO_3$ [copper(II) carbonate]

(b) Water-soluble: $BaCl_2$ (barium chloride) or $Ba(NO_3)_2$ (barium nitrate)

Water-insoluble: $BaSO_4$ (barium sulfate) or barium phosphate [$Ba_3(PO_4)_2$].

5.106 (a) Gas-forming reaction

$$K_2CO_3(aq) + 2\ HClO_4(aq) \longrightarrow$$
$$2\ KClO_4(aq) + CO_2(g) + H_2O(\ell)$$
$$CO_3^{2-}(aq) + 2\ H^+(aq) \longrightarrow CO_2(g) + H_2O(\ell)$$

(b) Precipitation

$$FeCl_2(aq) + (NH_4)_2S(aq) \longrightarrow$$
$$FeS(s) + 2\ NH_4Cl(aq)$$
$$Fe^{2+}(aq) + S^{2-}(aq) \longrightarrow FeS(s)$$

(c) Precipitation

$$Fe(NO_3)_2(aq) + Na_2CO_3(aq) \longrightarrow$$
$$FeCO_3(s) + 2\ NaNO_3(aq)$$
$$Fe^{2+}(aq) + CO_3^{2-}(aq) \longrightarrow FeCO_3(s)$$

5.108 (a) pH = 0.979

(b) For pH = 2.56, $[H^+] = 2.8 \times 10^{-3}$ M; acidic

(c) For pH = 9.67, $[H^+] = 2.1 \times 10^{-10}$ M; basic

5.110 HCl concentration = 2.92 M and pH = −0.465

5.112 mol H^+ in original solution = 3.01×10^{-3} mol

mol NaOH added = 2.63×10^{-3} mol

H^+ in excess after adding NaOH = 3.81×10^{-4} mol

$[H^+] = 7.61 \times 10^{-4}$ M, so pH = 3.13

5.114 (a) Reactants: Na(+1), I(−1), H(+1), S(+6), O(−2), Mn(+4)

Products: Na(+1), I(0), H(+1), S(+6), O(−2), Mn(+2)

(b) I^- is the reducing agent. It was oxidized from −1 to 0 in I_2. MnO_2 is the oxidizing agent. (The Mn was reduced from Mn^{4+} to Mn^{2+}.)

(c) NaI is the limiting reactant. 16.9 g of I_2 is produced.

5.116 Precipitation reaction: $BaCl_2(aq) + Na_2SO_4(aq) \longrightarrow$
$$BaSO_4(s) + 2\ NaCl$$

Gas-forming reaction: $BaCO_3(s) + H_2SO_4(aq) \longrightarrow$
$$BaSO_4(s) + CO_2(g) + H_2O(\ell)$$

5.118 (a) Acid-base reaction

$$Zn(OH)_2(s) + 2\ HCl(aq) \longrightarrow$$
$$ZnCl_2(aq) + 2\ H_2O(\ell)$$

When the reaction is complete, evaporate the water to obtain the $ZnCl_2$.

(b) Gas-forming reaction

$$ZnCO_3(s) + 2\ HCl(aq) \longrightarrow$$
$$ZnCl_2(aq) + CO_2(g) + H_2O(\ell)$$

Evaporate to dryness to obtain the $ZnCl_2$.

(c) Oxidation-reduction reaction

$$Zn(s) + 2\ HCl(aq) \longrightarrow ZnCl_2(aq) + H_2(g)$$

5.120 They will both calculate the same concentration of HCl. Diluting samples with pure water does not change the amount (mol) of HCl present, so each student should use the same volume of NaOH in the titration.

5.122 (a) Au is the reducing agent (and is oxidized from 0 to +1 in $NaAu(CN)_2$). O_2 is the oxidizing agent (and is reduced from 0 to −2 in water.)

(b) 26 L of NaCN solution

5.124 67.3% copper

5.126 (a) $(NH_4)_2PtCl_4(aq) + 2\ NH_3(aq) \longrightarrow$
$$Pt(NH_3)_2Cl_2(aq) + 2\ NH_4Cl(aq)$$

(b) 15.54 g $(NH_4)_2PtCl_4$ and 667 mL NH_3

(c) $x = 2$; $Pt(NH_3)_2Cl_2(C_5H_5N)_2$

5.128 100 mL of 0.10 M HCl contains 0.010 mol of HCl. This requires 0.0050 mol of Zn or 3.27 g for complete reaction. Thus, in flask 2 the reaction just uses all of the Zn and produces 0.0050 mol H_2 gas. In flask 1, containing 7.00 g of Zn, some Zn remains after the HCl has been consumed; 0.005 mol H_2 gas is produced. In flask 3, there is insufficient Zn, so less hydrogen is produced.

CHAPTER 6

6.1 (a) exothermic

(b) exothermic (all combustion reaction are exothermic)

(c) exothermic

(d) endothermic

6.4 State functions: balloon volume, temperature of water, and potential energy.

6.10 Product-favored reactions: the amount of products predominates over the amount of reactants. When the reaction is complete, virtually all reactants have been converted to products. It is usually the case that product-favored reactions are exothermic.

6.12 5.0×10^6 J

6.14 0.140 J/g · K

6.16 2.44 kJ

6.18 32.8 °C

6.20 20.7 °C

6.22 47.8 °C

6.24 0.40 J/g · K

6.26 330 kJ

6.28 49.3 kJ

6.30 273 J

6.32 9.97×10^5 J

6.34 Reaction is exothermic because ΔH°_{rxn} is negative. The heat evolved is 2.38 kJ.

6.36 3.3×10^4 kJ

6.38 $\Delta H_{rxn} = -56$ kJ/mol

6.40 0.52 J/g · K

6.42 $\Delta H = +23$ kJ/mol

6.44 297 kJ/mol SO_2

6.46 3.09×10^3 kJ/mol

6.48 0.236 J/g · K

6.50 (a) $\Delta H^\circ_{rxn} = -126$ kJ

(b)

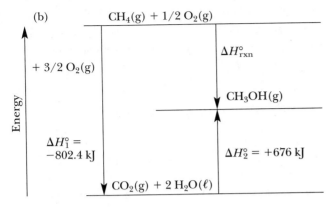

6.52 $\Delta H^\circ_{rxn} = +90.3$ kJ

6.54 $\Delta H^\circ_f = -238.4$ kJ/mol

6.56   (a) $4\ Cr(s) + 3\ O_2(g) \longrightarrow 2\ Cr_2O_3(s)$
$\Delta H^\circ_f = -1134.7$ kJ/mol

   (b)   2.4 g is equivalent to 0.046 mol. This will produce 52 kJ of heat energy.

6.58   (a) $\Delta H^\circ_{rxn} = -24$ kJ for 1.0 g of phosphorus

   (b) $\Delta H^\circ_{rxn} = -18$ kJ for 0.2 mol NO.

   (c) $\Delta H^\circ_{rxn} = -16.9$ kJ for the formation of 2.40 g of NaCl(s)

   (d) $\Delta H^\circ_{rxn} = -1.8 \times 10^3$ kJ for the oxidation of 250 g of iron

6.60   (a) $\Delta H^\circ_{rxn} = -906.2$ kJ (or -226.5 kJ for 1.00 mol of NH_3)

   (b)   For oxidation of 10.0 g of NH_3 the heat evolved is 133 kJ.

6.62   (a) $\Delta H^\circ_{rxn} = +80.8$ kJ

   (b)

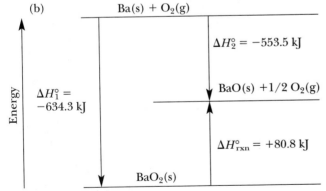

6.64 $\Delta H^\circ_f = +77.7$ kJ/mol for naphthalene

6.66   (a) $\Delta H^\circ_{rxn} = -705.63$ kJ; reaction is expected to be product-favored.

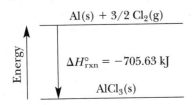

6.66   (b)   Decomposition of HgO(s) has $\Delta H^\circ_{rxn} = +90.83$ kJ. The reaction is expected to be reactant-favored.

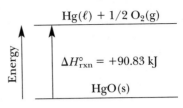

6.68   Specific heat capacity of lead is 0.121 J/g · K

6.70   Final temperature = 6.2 °C

6.72   Total heat required = 180 kJ

6.74   Mass of melted ice = 126 g. Mass of floating ice = 9 g.

6.76   To chill the cola to 0 °C requires removing 1.05×10^4 J of heat energy. If all of the ice melted, it would absorb 1.50×10^4 J. Therefore, the answer is (a), the temperature is 0 °C and some ice (13 g) remains.

6.78   The amount of AgCl produced in the reaction is 0.045 mol. The heat evolved is $q = -2.8 \times 10^3$ J. Therefore, the enthalpy change is -61 kJ per mol of AgCl.

6.80   The total heat evolved by 7.647 g of NH_4NO_3 is 3440 J. Because 7.647 g is equivalent to 0.09554 mol, the heat evolved is 36.0 kJ/mol.

6.82   (a)   When summed, the following equations give the balanced equation for the formation of $B_2H_6(g)$.

$2\ B(s) + 3/2\ O_2(g) \longrightarrow B_2O_3(s)$
$\Delta H^\circ_{rxn} = -1271.9$ kJ

$B_2O_3(s) + 3\ H_2O(g) \longrightarrow B_2H_6(g) + 3\ O_2(g)$
$\Delta H^\circ_{rxn} = +2032.9$ kJ

$3\ H_2(g) + 3/2\ O_2(g) \longrightarrow 3\ H_2O(g)$
$\Delta H^\circ_{rxn} = -725.4$ kJ

   (b)   Summing the values of ΔH°_{rxn} gives +35.6 kJ/mol for the enthalpy of formation of $B_2H_6(g)$.

   (c)

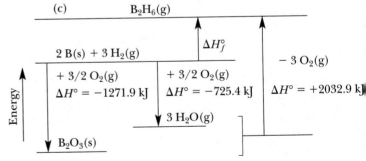

   (d)   The formation of $B_2H_6(g)$ from the elements is reactant-favored.

6.84   Standard enthalpy change, ΔH°_{rxn}, is -352.88 kJ. The quantity of magnesium needed is 4.3 g.

6.86   (a) $\Delta H^\circ_{rxn} = +131.31$ kJ

   (b)   Reactant-favored

   (c) 1.0932×10^7 kJ

6.88 Assuming $CO_2(g)$ and $H_2O(\ell)$ are the products of combustion:

ΔH°_{rxn} for isooctane is -5461.3 kJ/mol or -47.81 kJ per gram

ΔH°_{rxn} for liquid methanol is -726.77 kJ/mol or -22.682 kJ per gram

6.90 The molar heat capacities of metals are quite similar:

Al: $(0.897$ J/g $\cdot$ K$)(26.98$ g/mol$) = 24.2$ J/mol $\cdot$ K

Fe: 25.1 J/mol $\cdot$ K

Au: 25.4 J/mol $\cdot$ K

Cu: 24.5 J/mol $\cdot$ K

The average value if 24.8 J/mol $\cdot$ K. Using this to calculate silver's specific heat capacity, we find C(Ag) = $(24.8$ J/mol $\cdot$ K$)(1$mol Ag/107.9 g$) = 0.230$ J/g $\cdot$ K. This agrees well with the observed value of 0.236 J/g $\cdot$ K.

6.92 The condenser and motor evolve more heat to cool the refrigerator's interior than the amount of cooling available when the door is first opened.

6.94 (a) Your body expends about 47 kJ to warm the soda.

(b) 47 kJ is equivalent to 11.2 Cal. Therefore, you consume more energy warming the soda (11.2 Cal) than you gain (1 Cal).

6.96 120 g CH_4 required

6.98 Butene energy level diagram.

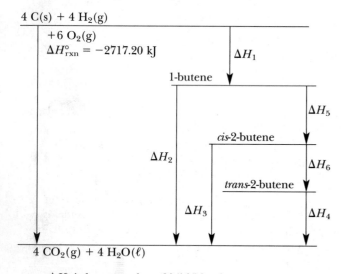

4 C(s) + 4 H_2(g)

+6 O_2(g)
$\Delta H^\circ_{rxn} = -2717.20$ kJ

ΔH_1

1-butene

ΔH_5

cis-2-butene

ΔH_2

ΔH_6

trans-2-butene

ΔH_3

ΔH_4

4 CO_2(g) + 4 $H_2O(\ell)$

ΔH_1 is known to be -20.5 kJ/mol.

ΔH_2 = enthalpy of combustion of 1-butene = -2696.7 kJ

ΔH_3 = enthalpy of combustion of *cis*-2-butene = -2687.5 kJ

ΔH_4 = enthalpy of combustion of *trans*-2-butene = -2684.2 kJ

ΔH_5 = difference in enthalpies of combustion of 1-butene and *cis*-2-butene = $\Delta H_2 - \Delta H_3 = -9.2$ kJ

$\Delta H_6 = \Delta H$ for *cis*-2-butene to *trans*-2-butene = -3.3 kJ

ΔH°_f of *cis*-2-butene = $\Delta H_1 + \Delta H_5 = -29.7$ kJ

ΔH°_f of *trans*-2-butene = $\Delta H_1 + \Delta H_5 + \Delta H_6 = -33.0$ kJ

6.100 $\Delta H^\circ_{rxn} = -305.3$ kJ

6.102 Assuming the lot receives 2.6×10^7 J of energy per square meter per day, the whole parking lot receives 4.2×10^{11} J per day.

Chapter 7

7.15 Nodal surfaces: s, 0; p, 1; d, 2; and f, 3.

7.17 $\ell = 3$, f orbital

$\ell = 0$, s orbital

$\ell = 1$, p orbital

$\ell = 2$, d orbital

7.19 s subshell has 1 orbital in subshell; 0 nodal surfaces

p subshell has 3 orbitals in subshell; 1 nodal surface

d subshell has 5 orbitals in subshell; 2 nodal surfaces

f subshell has 7 orbitals in subshell; 3 nodal surfaces

7.20 (a) microwaves

(b) red light

(c) infrared light

7.22 (a) Green has a higher frequency than amber light.

(b) 5.04×10^{14} s^{-1}

7.24 Frequency = 6.0×10^{14} s^{-1}; Energy per photon = 4.0×10^{-19} J; energy per mol of photos = 2.4×10^5 J.

7.26 302 kJ/mol of photons

7.28 In order of increasing energy: FM station < Microwaves < Yellow light < X-rays

7.30 Light with a wavelength as long as 600. nm would be sufficient. This is in the visible region.

7.32 (a) Light of the shortest wavelength (253.652 nm)

(b) Frequency = 1.18190×10^{15} s^{-1}. Energy = 7.83139×10^{-19} J/photon.

(c) The lines at 404 and 436 nm are in the violet region of the visible spectrum.

7.34 The color is violet. $n_{initial} = 6$ and $n_{final} = 2$.

7.36 (a) $n = 3$ to $n = 2$

(b) $n = 4$ to $n = 1$

7.38 (a) 6 lines are possible

(b) From $n = 4$ to $n = 3$

(c) Shortest wavelength emission has the greatest energy. This involves $n = 4$ to $n = 1$.

7.40 Wavelength = 102.6 nm and frequency = 2.923×10^{15} s^{-1}. Light with these properties is in the ultraviolet region.

7.42 Wavelength is 0.29 nm

7.44 The wavelength is 2.2×10^{-34} m. (Calculated from $\lambda = h/m \cdot v$, where m is the ball's mass in kg and v is the velocity). To have a wavelength of 5.6×10^{-3} nm, the ball would have to travel at 1.2×10^{-21} m/s.

7.46 (a) $n = 4$, $\ell = 0, 1, 2, 3$

(b) When $\ell = 2$, $m_\ell = -2, -1, 0, 1, 2$

(c) For a 4s orbital, $n = 4$, $\ell = 0$, and $m_\ell = 0$

(d) For a 4f orbital, $n = 4$, $\ell = 3$, and $m_\ell = -3, -2, -1, 0, 1, 2, 3$

7.48   Set 1: $n = 4$, $\ell = 1$, and $m_\ell = -1$

      Set 2: $n = 4$, $\ell = 1$, and $m_\ell = 0$

      Set 3: $n = 4$, $\ell = 1$, and $m_\ell = +1$

7.50   4 subshells. (The number of subshells in a shell is always equal to n.)

7.52   (a) ℓ must have a value no greater than $n - 1$.

      (b) m_ℓ can only equal 0 in this case

      (c) m_ℓ can only equal zero in this case

7.54   (a) None. The quantum number set is not possible. Here m_ℓ can only be equal to zero.

      (b) 3 orbitals

      (c) 11 orbitals

      (d) 1 orbital

7.56 $2d$ and $3f$ cannot exist. The $n = 2$ shell consists only of s and p subshells. The $n = 3$ shell consists only of s, p, and d subshells.

7.58   (a) For $2p$: $n = 2$, $\ell = 1$, and $m_\ell = -1, 0, +1$

      (b) For $3d$: $n = 3$, $\ell = 2$, and $m_\ell = -2, -1, 0, +1, +2$

      (c) For $4f$: $n = 4$, $\ell = 3$, and $m_\ell = -3, -2, -1, 0, +1, +2, +3$

7.60 $4d$

7.62   (a) $2s$ has 0 nodal surfaces

      (b) $5d$ has 2 nodal surfaces

      (c) $5f$ has three nodal surfaces

7.64   The energy absorbed = 2.093×10^{-18} J

7.66   (a) Green light

      (b) Red light has a wavelength of 680 nm, and green light has a wavelength of 500 nm.

      (c) Green light has a higher frequency than red light.

7.68   (a) Wavelength = 0.35 m

      (b) Energy = 0.34 J/mol

      (c) Blue light (with $\lambda = 420$ nm) has an energy of 290 kJ/mol.

      (d) Therefore, blue light has an energy (per mol of photons) that is 840,000 times greater than a mole of photons from a cell phone.

7.70   The ionization energy for He^+ is 5248 kJ/mol. This is four times the ionization energy for the H atom.

7.72   (i)   Smallest energy: $n = 7$ to $n = 6$

      (ii)   Highest frequency: $n = 7$ to $n = 1$

      (iii)   Shortest wavelength: $n = 7$ to $n = 1$

7.74   (a) $3p$, 3 orbitals

      (b) $4p$, 3 orbitals

      (c) $4p_x$, 1 orbital

      (d) $6d$, 5 orbitals

      (e) $5d$, 5 orbitals

      (f) $5f$, 7 orbitals

      (g) $n = 5$, 25 orbitals ($= n^2$)

      (h) $7s$, 1 orbital

7.76   In Bohr's model, we would know the precise distance of the electron from the nucleus and its energy at any given instant. According to the uncertainty principle, however, we can only know one of these two parameters precisely. Thus, Bohr's concept of fixed orbits violates the uncertainty principle.

7.78   b, e, f, g, j. Note, though, that atoms and molecules can be observed with scanning tunneling microscopes (STM).

7.80   3 orbitals in each of 3 shells for a total of 9 orbitals

7.82   Frequency = 2.836×10^{20} s^{-1}; wavelength = 1.057×10^{-12} m

7.84   260 s or 4.3 min

7.86   (a) size

      (b) ℓ

      (c) more

      (d) 7 (when $\ell = 3$ these are f orbitals)

      (e) one orbital

      (f) (left to right) d, s, and p

      (g) $\ell = 0, 1, 2, 3, 4$ (or 5 subshells)

      (h) 16 orbitals (1 s; 3 p; 5 d; and 7 f) ($= n^2$)

7.88   (a) Group 7B and period 5

      (b) $n = 5$, $\ell = 0$, and $m_\ell = 0$

      (c) Frequency = 3.41×10^{19} s^{-1} and wavelength = 8.79×10^{-12} m

      (d) $HTcO_4(aq) + NaOH(aq) \longrightarrow NaTcO_4(aq) + H_2O\ (\ell)$

      0.0085 g of $NaTcO_4$ can be made using 0.0018 g NaOH.

CHAPTER 8

8.3   *spdf* notation for Li: $1s^2 2s^1$

Box notation: [↑↓]   [↑]

            $1s$ $2s$

8.8   Ge, germanium. This element should have a core of 18 electrons (equivalent to Ar), 2 electrons in the 4s orbital, 10 electrons in 3d orbitals, and 2 electrons in 4p orbitals.

8.10   (a) Phosphorus: $1s^2 2s^2 2p^6 3s^2 3p^3$

      [↑↓] [↑↓] [↑↓][↑↓][↑↓] [↑↓] [↑][↑][↑]

      $1s$ $2s$ $2p$ $3s$ $3p$

      The element is in the 3rd period in Group 5A. Therefore, it has 5 electrons in the third shell.

      (b) Chlorine: $1s^2 2s^2 2p^6 3s^2 3p^5$

      [↑↓] [↑↓] [↑↓][↑↓][↑↓] [↑↓] [↑↓][↑↓][↑]

      $1s$ $2s$ $2p$ $3s$ $3p$

      The element is in the 3rd period and in Group 7A. Therefore, it has 7 electrons in the third shell.

8.12 (a) Chromium: $1s^22s^22p^63s^23p^63d^54s^1$

(b) Iron: $1s^22s^22p^63s^23p^63d^64s^2$

8.14 (a) Arsenic: $1s^22s^22p^63s^23p^63d^{10}4s^24p^3$

(b) Krypton: $1s^22s^22p^63s^23p^63d^{10}4s^24p^6$

8.16 (a) Tantalum: This is the third element in the transition series in the 6th period. Therefore, it has a core equivalent to Xe plus 2 6s electrons, 14 4f electrons, and 3 electrons in 5d: $[Xe]4f^{14}5d^36s^2$

(b) Platinum: This is the eighth element in the transition series in the 6th period. Therefore, it has a core equivalent to Xe plus 2 6s electrons, 14 4f electrons, and 8 electrons in 5d: $[Xe]4f^{14}5d^86s^2$. The actual configuration (Table 8.3) is $[Xe]4f^{14}5d^96s^1$.

8.18 Americium: $[Rn]5f^77s^2$ (see Table 8.3)

8.20 (a) Mg^{2+} ion

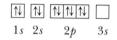
1s 2s   2p   3s

(b) K^+ ion

1s 2s   2p   3s   3p

(c) Cl^- (Note that Cl^- and K^+ have the same electron configuration; both are equivalent to Ar.)

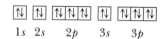
1s 2s   2p   3s   3p

(d) O^{2-} ion

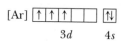
1s 2s   2p

8.22 (a) V (paramagnetic; three unpaired electrons)

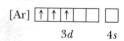
        3d   4s

(b) V^{2+} ion (paramagnetic, three unpaired electrons)

[Ar] ↑ ↑ ↑ ☐ ☐   ☐
        3d   4s

(c) V^{5+} The ion has an electron configuration equivalent to argon. It is diamagnetic with no unpaired electrons.

8.24 (a) Manganese

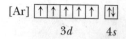
        3d   4s

(b) Mn^{2+} ion:

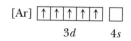
        3d   4s

(c) The +2 ion is paramagnetic to the extent of 5 unpaired electrons.

(d) 5

8.26 (a) The spin quantum number cannot be 0. The set is correct is $m_s = \pm 1/2$.

(b) m_ℓ cannot be larger than ℓ. The set is correct if $m_\ell = -1$, 0, or +1.

(c) ℓ can be no larger than $n - 1$. The set is correct if $\ell = 1$ or 2.

8.28 (a) 14

(b) 2

(c) 0 (because ℓ cannot be equal to n)

8.30 Magnesium: $1s^22s^22p^63s^2$

[Ar] ↑↓ ↑↓ ↑↓↑↓↑↓ ↑↓
1s 2s   2p   3s

Quantum numbers for the two electrons in the 3s orbital:
$n = 3$, $\ell = 0$, $m_\ell = 0$, and $m_s = +1/2$
$n = 3$, $\ell = 0$, $m_\ell = 0$, and $m_s = -1/2$

8.32 Gallium: $1s^22s^22p^63s^23p^63d^{10}4s^24p^1$

[Ar] ↑↓↑↓↑↓↑↓↑↓ ↑↓ ↑ ☐ ☐
          3d   4s   4p

Quantum number for the 4p electron:
$n = 4$, $\ell = 1$, $m_\ell = +1$, and $m_s = +1/2$

8.34 Increasing size: C < B < Al < Na < K

8.36 (a) Cl^-

(b) Al

(c) In

8.38 (c)

8.40 (a) Largest radius, Na

(b) Most negative electron affinity: O

(c) Ionization energy: Na < Mg < P < O

8.42 (a) Increasing ionization energy: S < O < F
S is less than O because the IE decreases down a group. F is greater than O because IE generally increases across a period.

(b) Largest IE: O. See part (a).

(c) Most negative electron affinity: Cl
Electron affinity becomes more and more negative across the periodic table and on ascending a group.

(d) Largest radius: O^{2-}. See page 313.

8.44 Rutherfordium, element 104:
$[Rn]5f^{14}6d^27s^2$

8.46 (a) Cerium: $[Xe]4f^15d^16s^2$

(b) Holmium: $[Xe]4f^{11}6s^2$ (see Table 8.3)

Cerium(III) ion: $[Xe]4f^1$

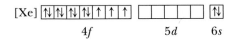

Holmium(III) ion: $[Xe]4f^{10}$

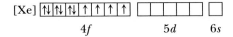

8.48 Meitnerium, element 109: $[Rn]5f^{14}6d^77s^2$

8.50 There are six possible sets:

$n = 4$, $\ell = 1$, $m_\ell = +1$, and $m_s = \pm 1/2$

$n = 4$, $\ell = 1$, $m_\ell = 0$, and $m_s = \pm 1/2$

$n = 4$, $\ell = 1$, $m_\ell = -1$, and $m_s = \pm 1/2$

8.52 (a) Phosphorus

(b) Beryllium

(c) Nitrogen

(d) Tc, technetium

(e) Chlorine (see Figure 8.12)

(f) Zinc

8.54 $Cl^- < Cl < Ca^{2+}$

Ionizing a negative ion should occur more readily than the related atom. A cation is the hardest to ionize.

8.56 (a) A is a metal

(b) B is a nonmetal

(c) B should have the largest IE

(d) B has the smaller radius

8.58 Decreasing size: $S^{2-} > Cl^- > K^+ > Ca^{2+}$

8.60 (a) Increasing size: $Ca^{2+} < K^+ < Cl^-$

(b) Increasing IE: $Cl^- < K^+ < Ca^{2+}$

(c) Increasing EA: $Cl^- < K^+ < Ca^{2+}$

The affinity for an electron increases in the order above. The value of EA will become larger and more negative.

8.62 Tc and Rh

8.64 (a) Cobalt

(b) Paramagnetic (has three unpaired electrons)

(c) The +3 ion will have four unpaired electrons.

8.66 (a) The decrease in size is attributed to the increasing effective nuclear charge on moving across a period.

(b) See the last paragraph on page 306.

8.68 Li. The first electron is removed from the 2s orbital, and the second electron comes from a shell of lower n. Removing an electron from an inner shell — in the atom's core — always requires much greater energy.

8.70 To make K^{2+} or Al^{4+} would require an electron to be removed from the atom's core. To form F^{2-} the second electron would have to be added in a shell beyond the 2p shell.

8.72 The first ionization energy of Ca is greater than that of K because the effective nuclear charge is larger in Ca than in K. However, to remove the second electron of K requires removal of a core electron (from 3p), whereas to form Ca^{2+} both electrons removed are from a 4s orbital.

8.74 CaF_3 requires that Ca be a +3 ion. This would require removing an electron from the core of the Ca atom.

8.76 The size declines across this series of elements while their mass increases. Thus, the mass per volume, the density, increases.

8.78 The ionization energy of B is less than expected. On going from Be to B the next electron is placed in a 2p orbital, a higher energy orbital than the 2s orbital. Thus, although Z* is greater for B than Be, the higher energy orbital from which the electron is removed (to form B^+) leads to a lower ionization energy.

8.80 On, 90 pm; M, 120 pm; E, 140 pm; and Ch, 180 pm.

8.82 (a) The reducing agent is Na. The low ionization energy of sodium plays a major role in making it a good reducing agent.

(b) The oxidizing agent is Cl_2. Among other properties, the element has a high electron affinity.

(c) Na_2Cl would have a Cl^{2-} ion. Adding a second electron to Cl^- means placing an electron in a higher energy electron shell. Conversely, $NaCl_2$ would have a Na^{2+} ion. Here one would have to remove the second electron from the atom's core. See page 314.

8.84 (a) Z* is a modified nuclear charge, modified by the screening of inner electrons. Only in the case of H would Z and Z* be the same.

(b) Effective nuclear charge increases across the periodic table.

(c) The effective nuclear charge drops on going from Ne to Na. It is difficult to ionize Ne, whereas Na readily forms a +1 ion.

8.86 (a) Mg has a higher ionization energy than Al for the same reason IE(Be) is greater than IE(B). See Question 8.78.

(b) The orbital energies drop, thus raising the element ionization energies.

CHAPTER 9

9.4 KI and MgS are ionic and CS_2 and P_4O_{10} are covalent

9.6 $CaCl_4$ is not likely to exist. This formula implies the calcium is Ca^{4+}. It is energetically too expensive to create Ca^{4+}.

9.10 NO$_2$ and O$_2^-$ do not obey the octet rule

9.12 The C—H bonds in C$_2$H$_2$ have a bond order of 1, whereas the carbon-carbon bond has an order of 3. In phosgene the C—Cl bonds are single bonds, whereas the C=O bond is a double bond of order 2.

9.16 The bond dissociation energy is the energy required to break a bond in a molecule with the reactants and products in the gas phase. The bond dissociation energy is always a positive number because energy is required for bond breaking.

9.18 To estimate the enthalpy change we need energies for the following bonds: O=O, H—H, and H—O.

Energy to break bonds = 498 kJ (for O=O) + 2 × 436 kJ (for H—H) = +1370 kJ

Energy evolved when bonds are made = 4 × 463 kJ (for O—H) = −1852 kJ

Total energy = −482 kJ

9.20 The electron affinity of an atom is a measure of its ability to attract an electron in the gas phase. Electronegativity measures the electron attracting ability in a molecule.

9.24 In water there are four electron pairs around the O atom. The electron-pair geometry is the geometry adopted by these four pairs. The molecular geometry is the geometry described by the atoms of the molecule. In water the electron-pair geometry is tetrahedral, whereas the molecular geometry is bent.

9.26 A molecule with four electron pairs can be pyramidal if one of the pairs is a lone pair and the other three are bond pairs (as in NH$_3$).

9.28 (a) Group 6A, 6 valence electrons
(b) Group 3A, 3 valence electrons
(c) Group 1A, 1 valence electrons
(d) Group 2A, 2 valence electrons
(e) Group 7A, 7 valence electrons
(f) Group 6A, 6 valence electrons

9.30 See page 336

Group 3A, 3 bonds

Group 4A, 4 bonds

Group 5A, three bonds (for a neutral compound)

Group 6A, 2 bonds (for a neutral compound)

Group 7A, 1 (for a neutral compound)

9.32 Most negative, MgS. Least negative, NaCl.

9.34 Increasing lattice energy: RbI < LiI < LiF < CaO

9.36 As the ion-ion distance decreases, the force of attraction between the ions increases. This should make the lattice more stable, and more energy should be required to melt the compound.

9.38 (a) NF$_3$, 26 valence electrons   (b) ClO$_3^-$, 26 valence electrons

$$:\!\ddot{F}\!-\!\ddot{N}\!-\!\ddot{F}\!:$$
$$\overset{|}{:\!\ddot{F}\!:}$$

$$\left[:\!\ddot{O}\!-\!\ddot{Cl}\!-\!\ddot{O}\!:\right]^-$$
$$\overset{|}{:\!\ddot{O}\!:}$$

(c) HOBr, 14 valence electrons   (d) SO$_3^{2-}$, 26 valence electrons

$$H\!-\!\ddot{O}\!-\!\ddot{Br}\!:$$

$$\left[:\!\ddot{O}\!-\!\overset{|}{S}\!-\!\ddot{O}\!:\right]^{2-}$$
$$\overset{|}{:\!\ddot{O}\!:}$$

9.40 (a) CHClF$_2$, 26 valence electrons   (b) CH$_3$CO$_2$H, 24 valence electrons

$$\overset{\displaystyle H}{\underset{\displaystyle :\!\ddot{F}\!:}{:\!\ddot{Cl}\!-\!\overset{|}{\underset{|}{C}}\!-\!\ddot{F}\!:}}$$

$$H\!-\!\overset{\displaystyle H}{\underset{\displaystyle H}{\overset{|}{\underset{|}{C}}}}\!-\!\overset{\displaystyle :\ddot{O}:}{C}\!-\!\ddot{O}\!-\!H$$

(c) CH$_3$CN, 16 valence electrons   (d) CH$_3$OH, 14 valence electrons

$$H\!-\!\overset{\displaystyle H}{\underset{\displaystyle H}{\overset{|}{\underset{|}{C}}}}\!-\!C\!\equiv\!N\!:$$

$$H\!-\!\overset{\displaystyle H}{\underset{\displaystyle H}{\overset{|}{\underset{|}{C}}}}\!-\!\ddot{O}\!-\!H$$

9.42 (a) SO$_2$, 18 valence electrons

$$:\!\ddot{O}\!-\!S\!=\!\ddot{O} \longleftrightarrow \ddot{O}\!=\!S\!-\!\ddot{O}\!:$$

(b) NO$_2^-$, 18 valence electrons

$$\left[:\!\ddot{O}\!-\!\ddot{N}\!=\!\ddot{O}\right]^- \longleftrightarrow \left[\ddot{O}\!=\!\ddot{N}\!-\!\ddot{O}\!:\right]^-$$

(c) SCN$^-$, 16 valence electrons

$$\left[\ddot{S}\!=\!C\!=\!\ddot{N}\right]^- \longleftrightarrow \left[:\!S\!\equiv\!C\!-\!\ddot{N}\!:\right]^- \longleftrightarrow \left[:\!\ddot{S}\!-\!C\!\equiv\!N\!:\right]^-$$

9.44 (a) BrF$_3$, 28 valence electrons

(b) I$_3^-$, 22 valence electrons

(c) XeO$_2$F$_2$, 34 valence electrons

$$\overset{\displaystyle :\ddot{F}:}{\underset{\displaystyle :\ddot{F}:}{:\!\ddot{O}\!-\!\overset{|}{\underset{|}{Xe}}\!-\!\ddot{O}\!:}}$$

(d) XeF_3^+, 28 valence electrons

$$\left[\begin{array}{c} :\ddot{F}: \\ | \\ :Xe-\ddot{F}: \\ | \\ :\ddot{F}: \end{array}\right]^+$$

9.46 (a) N = 0; H = 0

(b) P = +1; O = −1

(c) B = −1; H = 0

(d) All are zero

9.48 (a) N = +1; O = 0

(b) The central N is 0. The singly bonded O atom is −1, and the doubly bonded O atom is 0.

$$\left[:\ddot{O}-\ddot{N}=\ddot{O}\right]^- \longleftrightarrow \left[\ddot{O}=\ddot{N}-\ddot{O}:\right]^-$$

(c) N = 0; F = 0

(d) The central N is +1, one of the O atoms is −1, and the other two O atoms are both 0.

$$\overset{0}{H}-\overset{+1}{\ddot{O}}-\overset{0}{N}=\ddot{O}$$
$$\underset{-1}{\overset{|}{:\ddot{O}:}}$$

9.50 (a) $\overset{\longrightarrow}{\underset{+\delta\ -\delta}{C-O}}$ $\overset{\longrightarrow}{\underset{+\delta\ -\delta}{C-N}}$
CO is more polar

(b) $\overset{\longrightarrow}{\underset{+\delta\ -\delta}{P-Cl}}$ $\overset{\longrightarrow}{\underset{+\delta\ -\delta}{P-Br}}$
PCl is more polar

(c) $\overset{\longrightarrow}{\underset{+\delta\ -\delta}{B-O}}$ $\overset{\longrightarrow}{\underset{+\delta\ -\delta}{B-S}}$
BO is more polar

(d) $\overset{\longrightarrow}{\underset{+\delta\ -\delta}{B-F}}$ $\overset{\longrightarrow}{\underset{+\delta\ -\delta}{B-I}}$
BF is more polar

9.52 (a) CH and CO bonds are polar.

(b) The CO bond is the most polar, and O is the most negative atom.

9.54 (a) Even though the formal charge on B is −1 and on F is 0, F is much more electronegative. The four F atoms therefore likely bear the −1 charge of the ion, and the bonds are polar with the F atom the negative end.

(b) Even though the formal charge on B is −1 and on H is 0, H is slightly more electronegative. The four H atoms therefore likely bear the −1 charge of the ion. The BH bonds are polar with the H atom the negative end.

(c) The formal charge on O is −1 and on H it is 0. This conforms with the relative electronegativities. The bond is polar with O the negative end.

(d) The CH and CO bonds are polar. The negative charge of the CO bonds lies on the O atoms.

9.56 $\underset{A}{:\overset{-2}{\ddot{N}}-\overset{+1}{N}\equiv\overset{+1}{O}:} \longleftrightarrow \underset{B}{\overset{-1}{\ddot{N}}=\overset{+1}{N}=\overset{0}{\ddot{O}}} \longleftrightarrow \underset{C}{:\overset{0}{N}\equiv\overset{+1}{N}-\overset{-1}{\ddot{O}}:}$

Structure C is the most reasonable: The charges are as small as possible and the negative charge resides on the more electronegative atom.

9.58 $\left[:\ddot{O}-\ddot{N}=\ddot{O}\right]^- \longleftrightarrow \left[\ddot{O}=\ddot{N}-\ddot{O}:\right]^-$

If an H^+ ion were to attack, it would attach to an O atom.

9.60 (a) 2 C—H bonds, bond order 1; 1 C=O bond, bond order 2

(b) 3 S—O single bonds, bond order 1

(c) 2 nitrogen–oxygen double bonds, bond order 2

(d) 1 N=O bond, bond order 2; 1 N—Cl bond, bond order 1

9.62 (a) B—Cl

(b) C—O

(c) P—O

(d) C=O

9.64 NO bond orders: 2 in NO_2^+; 1.5 in NO_2^-; 1.33 in NO_3^-. The NO bond is longest in NO_3^- and shortest in NO_2^+.

9.66 The CO bond in carbon monoxide is a triple bond, so it is both shorter and stronger than the CO double bond in H_2CO.

9.68 $\Delta H^\circ_{rxn} = -126$ kJ

9.70 O—F bond dissociation energy = 192 kJ/mol

9.72 (a) $:\ddot{Cl}-\underset{\underset{H}{|}}{N}-H$ — Electron-pair geometry around N is tetrahedral. Molecular geometry is pyramidal.

(b) $:\ddot{Cl}-\ddot{O}-\ddot{Cl}:$ — Electron-pair geometry around O is tetrahedral. Molecular geometry is bent.

(c) $\left[\ddot{S}=C=\ddot{N}\right]^-$ — Electron-pair geometry around C is linear. Molecular geometry is linear.

(d) $H-\ddot{O}-\ddot{F}:$ — Electron-pair geometry around O is tetrahedral. The molecular geometry is bent.

9.74 (a) $\ddot{O}=C=\ddot{O}$ — Electron-pair geometry around C is linear. Molecular geometry is linear.

(b) $\left[:\ddot{O}-N=\ddot{O}\right]^{-}$ Electron-pair geometry around O is trigonal-planar. Molecular geometry is bent.

(c) $\ddot{O}=\ddot{O}-\ddot{O}:$ Electron-pair geometry around O is trigonal-planar. Molecular geometry is bent.

(d) $\left[:\ddot{O}-\ddot{C}l-\ddot{O}:\right]^{-}$ Electron-pair geometry around O is tetrahedral. Molecular geometry is bent.

All have two atoms attached to the central atom. As the bond and lone pairs vary, the molecular geometries vary from linear to bent.

9.76 (a) $\left[:\ddot{F}-\ddot{C}l-\ddot{F}:\right]^{-}$ Electron-pair geometry is trigonal-bipyramidal. Molecular geometry is linear.

(b) $:\ddot{F}-\ddot{C}l-\ddot{F}:$ with $:\ddot{F}:$ Electron-pair geometry is trigonal-bipyramidal. Molecular geometry is T-shaped.

(c) $\left[\begin{array}{c}:\ddot{F}: \\ :\ddot{F}-\ddot{C}l-\ddot{F}: \\ :\ddot{F}:\end{array}\right]^{-}$ Electron pair geometry is octahedral. Molecular geometry is square planar

(d) $\ddot{F}$ ... $\ddot{F}$ Cl $\ddot{F}$ ... $\ddot{F}$ Electron pair geometry is octahedral. Molecular geometry is square pyramid

9.78 (a) Ideal O—S—O angle = 120°

(b) 120°

(c) 120°

(d) H—C—H = 109° and C—C—N angle = 180°

9.80 1 = 120°; 2 = 109°; 3 = 120°; 4 = 109°; 5 = 109°

The electron-pair geometry around the middle C atom is tetrahedral, so the chain of atoms cannot be linear.

9.82 (i) Most polar bonds are in H_2O (O and H have the largest difference in electronegativity).

(ii) Not polar: CO_2, CCl_4

(iii) F

9.84 (a) Not polar; linear molecule

(b) HBF_2, polar, trigonal-planar molecule with F atoms negative end of dipole and H atom the positive end

(c) CH_3Cl, polar, Cl atom is the negative end and the 3 H atoms are on the positive side of the molecule.

(d) SO_3, not polar, trigonal-planar molecule

9.86 All the molecules in the series have 16 valence electrons and all are linear.

(a) $\ddot{O}=C=\ddot{O} \longleftrightarrow :\ddot{O}-C\equiv O: \longleftrightarrow :O\equiv C-\ddot{O}:$

(b) $\left[\ddot{N}=N=\ddot{N}\right]^{-} \longleftrightarrow \left[:\ddot{N}-N\equiv N:\right]^{-} \longleftrightarrow \left[:N\equiv N-\ddot{N}:\right]^{-}$

(c) $\left[\ddot{O}=C=\ddot{N}:\right]^{-} \longleftrightarrow \left[:\ddot{O}-C\equiv N:\right]^{-} \longleftrightarrow \left[:O\equiv C-\ddot{N}:\right]^{-}$

9.88 NO_2^{-} has a smaller bond angle (about 120°) than NO_2^{+} (180°). The former has a trigonal-planar electron-pair geometry, whereas the latter is linear.

9.90 The negative (formal) charge resides on C, so the H^{+} should attack that atom and form H—CN.

$$\left[:C\equiv N:\right]^{-}$$

9.92 Assume the structure of N_2O is as drawn below, ΔH_{rxn} = −96 kJ

$$:N\equiv N-\ddot{O}:$$

9.94 The F^{-} ion has a much smaller radius than the other halide ions (see Table 8.13). Thus, the ion-ion distance in NaF is smaller than in the other sodium halides. Coulomb's law (page 93) indicates that the force of attraction increases as distance decreases, so the ion-ion attractions in NaF are greater than in the other sodium halides.

9.96 (a) 1 = 120°, 2 = 180°, 3 = 120°

(b) The C=C double bond is shorter than the C—C single bond.

(c) The C=C double bond is stronger than the C—C single bond.

(d) The CN triple bond is slightly more polar than the CH bonds.

9.98 (a) The three lone pairs of XeF_2 occupy the equatorial positions. There the angles between lone pairs are 120°, so there is less lone pair/lone pair repulsion with this arrangement.

(b) The two lone pairs on the Cl atom occupy the equatorial positions. The reasoning is the same as for XeF_2.

9.100 (a) Angle 1 = 109°
Angle 2 = 120°
Angle 3 = 109°
Angle 4 = 109°
Angle 5 = 109°

(b) The O—H bonds are the most polar.

9.102 ΔH_{rxn} = +146 kJ = 2 (D_{C-N}) + $D_{C=O}$ − $[D_{N-N}$ + $D_{C=O}]$

9.104 (a) Both the S and O atoms have a formal charge of 0.

(b) Angle 1 = 109°; angle 2 = 109°; and angle 3 = 120°

(c) The shorter carbon-carbon bonds are the double bonds.

(d) The most polar bonds are the CO bonds.

(e) The molecule is polar.

(f) The four C atoms are planar and trigonal, so the ring as a whole is planar.

9.106 One of the resonance structures has a double bond to the O atom bearing the acidic H atom. This O atom, as a consequence, has a formal charge of +1. Because O is quite electronegative, it is highly unlikely that the atom would bear such a charge.

$$0 \quad +1 \quad 0 \qquad\qquad 0 \quad +1 \quad -1 \qquad\qquad +1 \quad +1 \quad -1$$
$$H-\ddot{O}-N{=}\ddot{O} \longleftrightarrow H-\ddot{O}-N-\ddot{O}: \longleftrightarrow H-\ddot{O}{=}N-\ddot{O}:$$
$$\qquad\quad \underset{-1}{\overset{|}{:\ddot{O}:}} \qquad\qquad\quad \underset{0}{\overset{|}{:\ddot{O}:}} \qquad\qquad\quad \underset{-1}{\overset{|}{:\ddot{O}:}}$$

9.108 (a) The C=C bond is stronger than the C—C bond.

(b) The C—C single bond is longer than the C=C double bond.

(c) Ethylene is not polar, but acrolein is polar.

(d) Exothermic. The heat of the reaction is −45 kJ.

9.110 (a) Angle 1 = 109°
Angle 2 = 120°
Angle 3 = 120°
Angle 4 = 109°
Angle 5 = 109°

(b) The OH bonds are the most polar bonds in the molecule.

9.111 The data collected from the models on the CD-ROM, show that carbon-carbon bonds are generally shorter as the bond order increases.

Name	Bond Distance (Å)	Bond Order
Ethane	1.540	1
Butane	1.540	1
Ethylene	1.352	2
Acetylene	1.226	3
Benzene	1.397	1.5

CHAPTER 10

10.2 A C atom may form 2, 3, or 4 hybrid orbitals.

10.4 If the atom is sp hybridized, two unhybridized p orbitals remain, so two π bonds can be formed.

10.6 (a) CF_4

(b) SiF_4 has 32 valence electrons, whereas SF_4 has 34 valence electrons. They are not isoelectronic.

(c) BF_4^-, sp^3; SiF_4, sp^3; SF_4, sp^3d

10.8 A third period element may form up to six hybrid orbitals. (S can be d^2sp^3 hybridized.)

10.12 Bond length decreases as bond order increases. See the answer to SQ 9.111.

10.16 An insulator has a large band gap between valence and conduction bands, whereas the bands in a metal have no separation.

10.18 The electron-pair geometry of NF_3 is tetrahedral, whereas the molecular geometry is pyramidal. An sp^3 hybrid orbital on the N atom overlaps a p orbital on the F atom to form a sigma bond.

$$:\ddot{F}-\overset{\displaystyle N}{|}-\ddot{F}:$$
$$\overset{|}{:\ddot{F}:}$$

10.20 The electron-pair and molecular geometry of $CHCl_3$ are both tetrahedral. An sp^3 hybrid orbital on the C atom overlaps a p orbital on a Cl atom to form a sigma bond. A C—H sigma bond is formed by a C atom orbital overlapping an H atom 1s orbital.

$$:\ddot{C}l:$$
$$H-\overset{\displaystyle |}{C}-\ddot{C}l:$$
$$\overset{|}{:\ddot{C}l:}$$

10.22 (a) sp^2 (b) sp

(c) sp^3 (d) sp^2

10.24 (a) C, sp^3; O, sp^3

(b) CH_3, sp^3; middle C, sp^2; CH_2, sp^2

(c) CH_2, sp^3; CO_2H, sp^2; N, sp^3

10.26 (a)

Electron-pair geometry is octahedral. Molecular geometry is octahedral. S: sp^3d^2

(b)

Electron-pair geometry is trigonal-bipyramidal. Molecular geometry is seesaw. Se: sp^3d

(c)

Electron-pair geometry is trigonal-bipyramidal. Molecular geometry is linear. I: sp^3d

(d)

Electron-pair geometry is octahedral. Molecular geometry is square-planar. Xe: sp^3d^2

10.28 There are 32 valence electrons in both HPO_2F_2 and its anion. Both have a tetrahedral molecular geometry and so the P atom in both is sp^3 hybridized.

10.30 The C atom is sp^2-hybridized. Two of the sp^2 hybrid orbitals are used to form C—Cl s bonds. The third is used to form the C—O s bond. The p orbital not used in the C atom hybrid orbitals is used to form the CO p bond.

10.32

cis isomer   *trans* isomer

10.34 H_2^+ ion: $(\sigma_{1s})^1$. Bond order is 0.5. The bond in H_2^+ is weaker (bond order = 0.5) than in H_2 (bond order = 1).

10.36 MO diagram for C_2^{2-} ion.

The ion has 10 valence electrons (isoelectronic with N_2). There are 1 net σ bond and 2 net π bonds, for a bond order of 3. The bond order increases by 1 on going from C_2 to C_2^{2-}. The ion is not paramagnetic.

10.38 (a) CO has 10 valence electrons:
$[core](\sigma_{2s})^2(\sigma_{2s}^*)^2(\pi_{2p})^4(\sigma_{2p})^2$

(b) HOMO, σ_{2p}

(c) Diamagnetic

(d) σ bonds, 1; π bonds, 2; bond order, 3

10.40 100 Mg atoms, each with a 2s and three 2p valence orbitals, can form 400 molecular orbitals. Because 100 Mg atoms have $2 \times 100 = 200$ electrons, only-fourth of the molecular orbitals are occupied by pairs of electrons.

10.42

Molecule/Ion	O—S—O Angle	Hybrid Oritals
SO_2	120°	sp^2
SO_3	120°	sp^2
SO_3^{2-}	109°	sp^3
SO_4^{2-}	109°	sp^3

10.44 $\left[:\ddot{O}—\ddot{N}=\ddot{O} \right]^- \longleftrightarrow \left[\ddot{O}=\ddot{N}—\ddot{O}: \right]^-$

The electron-pair geometry is trigonal and planar. The molecular geometry is bent or angular. The O—N—O angle will be about 120°, the average N—O bond order is 3/2, and the N atom is sp^2 hybridized.

10.46 The resonance structures, with formal charges, are shown here.

A   B   C

The central N atom is sp hybridized in all structures.

10.48 (a) Both C atoms are sp^2 hybridized.

(b) All three angles are about 120°.

(c) *cis–trans* isomerism is not possible because C atom 1 has two of the same kind of atoms attached to it.

10.50 (a) All three have the formula C_2H_4O. They are usually referred to as structural isomers.

(b) *Ethylene oxide:* both C atoms are sp^3 hybridized, and the bond angles in the ring are only 60° (which leads to a relatively unstable molecule).
Acetaldehyde: The CH_3 carbon atom has sp^3 hybridization (bond angles of 109°), and the other C atom is sp^2 hybridized (bond angles of 120°.
Vinyl alcohol: Both C atoms are sp^2 hybridized, and all bond angles are 120°.

(c) Vinyl alcohol has the strongest carbon-carbon bond, and acetaldehyde has the strongest CO bond.

(d) All are polar.

(e) Acetaldehyde has the strong CO bond, and vinyl alcohol has the strongest C—C bond.

10.52 (a) 1 π bond and 11 σ bonds

(b) $C(1) = sp^3$; $C(2) = sp^2$; $C(3) = sp^3$

(c) The C=O double bond is the strongest and shortest carbon-oxygen bond in the molecule.

10.54 (a) The geometry about the B atom is trigonal planar in BF_3 but tetrahedral in $H_3N \longrightarrow BF_3$.

(b) Boron is sp^2 hybridized in BF_3, but it is sp^3 hybridized in $H_3N \longrightarrow BF_3$

(c) Yes.

10.56 (a) The Sb in SbF_5 is sp^3d hybridized, whereas it is sp^3d^2 hybridized in SbF_6^-.

(b) The molecular geometry of the H_2F^+ ion is bent or angular, and the F atom is sp^3 hybridized.

10.58 (a) C=O bond is most polar.

(b) 18 σ and 5 π bonds

(c) *cis-trans* isomers

trans isomer   *cis* isomer

(d) All C atoms are sp^2 hybridized

(e) All three bond angles are 120°.

10.60 (a) The peroxide ion has a bond order of 1.

$$\left[:\ddot{O}-\ddot{O}: \right]^{2-}$$

(b) [core electrons] $(\sigma_{2s})^2(\sigma_{2s}^*)^2(\pi_{2p})^4(\sigma_{2p})^2(\pi_{2p}^*)^4$
This configuration also leads to a bond order of 1.

(c) Both theories lead to a diamagnetic ion with a bond order of 1.

10.62 See Table 10.1 on page 403.

(a) Paramagnetic diatomic molecules: B_2, O_2

(b) Bond order of 1: Li_2, B_2, F_2

(c) Bond order of 2: C_2, O_2

(d) Highest bond order is N_2

10.64 CN, 9 valence electrons:
[core electrons] $(\sigma_{2s})^2(\sigma_{2s}^*)^2(\pi_{2p})^4(\sigma_{2p})^1$

(a) HOMO, σ_{2p}

(b, c) Bond order = 2.5 (0.5 σ bond and 2 π bonds)

(d) Paramagnetic

10.66 (a) All C atoms are sp^3 hybridized.

(b) About 109°

(c) Polar

(d) The six-membered ring cannot be planar owing to the tetrahedral C atoms of the ring. The bond angles are all 109°.

10.68 In a conductor, the valence band is only partially filled, allowing for relatively easy promotion to slightly higher energy levels. In an insulator, however, the valence band is completely filled, and the band gap between the valence band and conduction band is relatively large.

10.70 Germanium should be a better conductor of electricity than diamond because of the smaller band gap in Ge. However, Ge should be a much poorer conductor than Li.

10.72 (a) C, sp^2; N, sp^3

(b) The amide link has two resonance structures (shown here with formal charges on the O and N atoms). Structure B is less favorable owing to the separation of charges.

$$\underset{A}{\overset{:O:^0}{\underset{|}{\overset{\|}{-C-\overset{..}{N}^0-}}}\ \overset{}{\underset{H}{|}}} \longleftrightarrow \underset{B}{\overset{:\ddot{O}:^{-1}}{\underset{|}{\overset{|}{-C=N^{+1}}}}\ \overset{}{\underset{H}{|}}}$$

(c) The fact that the amide link is planar indicates that structure B has some importance.

10.74 (a) 60°

(b) sp^3

(c) Based on hybridization, the angles would be expected to be about 109°. Because the angles are significantly less than this, the ring structure is strained and relatively easy to break. (The molecule is reactive.)

10.76 (a) The number of hybrid orbitals is always equal to the number of atomic orbitals used.

(b) No. All hybrid orbital sets involve an s orbital.

(c) The energy of the hybrid orbital set is the weighted average of the energy of the combining atomic orbitals.

(d) The shapes are identical. They are oriented in different directions in space.

(e) As in (d), these hybrid orbitals have the same shape, but they are oriented in different planes.

10.78 The molecule with the double bond requires a great deal more energy because the π bond must be broken in order for the ends of the molecule to rotate relative to each other.

(b) C—CH_3

(c) No. Both CC bonds are double bonds.

CHAPTER 11

11.4 The chiral C atom is the atom to which the OH group is attached. Mirror images of the molecule are not be superimposable.

$$CH_3CH_2\underset{CH_3}{\overset{H}{\underset{|}{\overset{|}{C}}}}\!\!\!^{\backprime\backprime\backprime\backprime}OH \qquad HO\cdots\underset{H_3C}{\overset{H}{\underset{|}{\overset{|}{C}}}}CH_2CH_3$$

Mirror images of 2-butanol

11.6 Butene isomers.

$$CH_3CH_2CH{=}CH_2 \qquad \underset{H_3C}{\overset{H}{\diagdown}}C{=}C\underset{CH_3}{\overset{H}{\diagup}}$$

1-butene   *cis*-2-butene

$$\underset{CH_3C{=}CH_2}{\overset{CH_3}{\underset{|}{}}} \qquad \underset{H}{\overset{H_3C}{\diagup}}C{=}C\underset{CH_3}{\overset{H}{\diagdown}}$$

2-methylpropene   *trans*-2-butene

11.8 (a)
$$CH_3CH_2CH_2\overset{O}{\overset{\|}{C}}OH \quad \text{butanoic acid}$$

(b)
$$CH_3CH_2\overset{O}{\overset{\|}{C}}CH_3 \quad \text{2-butanone}$$

(c) Cannot be oxidized

(d)
$$H_3C-\underset{\underset{CH_3}{|}}{\overset{\overset{H}{|}}{C}}-\overset{\overset{O}{\|}}{C}-OH \quad \text{2-methylpropanoic acid}$$

11.10 (a)

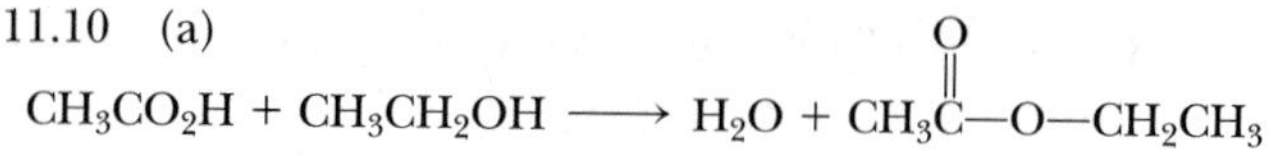

$$CH_3CO_2H + CH_3CH_2OH \longrightarrow H_2O + CH_3\overset{\displaystyle O}{\overset{\|}{C}}-O-CH_2CH_3$$

(b)

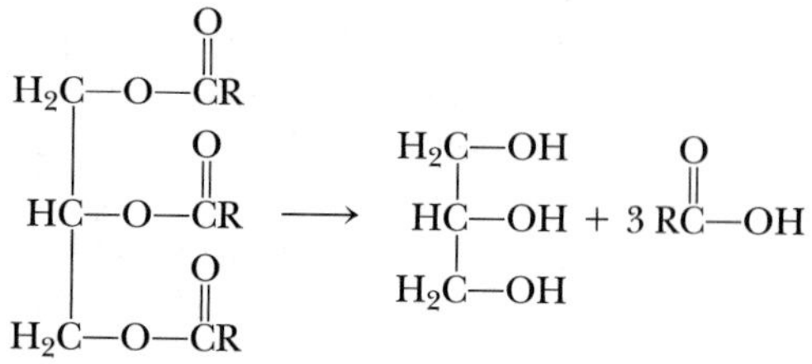

R = (CH_2)_{16}CH_3

11.12 (a)

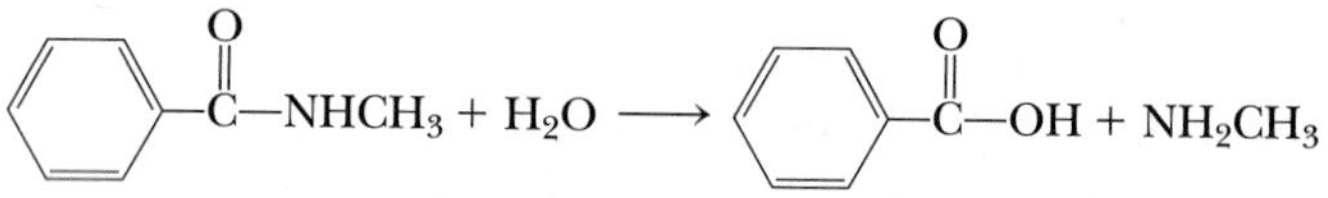

(b)

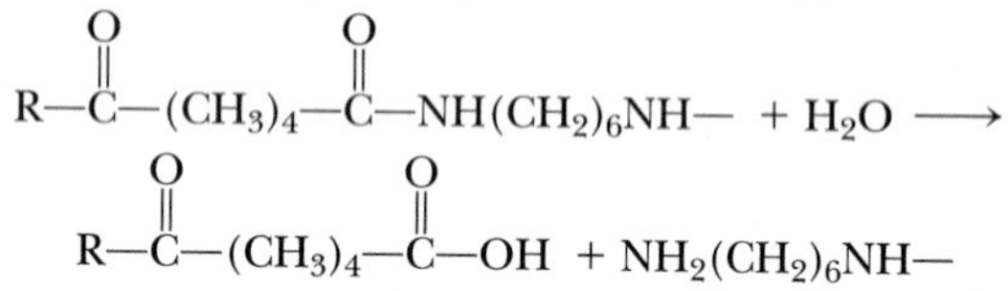

11.14 (a) Cross-linking polyethylene causes it to be more rigid and inflexible. See page 457.

    (b) The OH group of PVA allow it to be somewhat water-soluble.

    (c) Hydrogen bonding allows the chains of a polypeptide to interact (as in Figure 11.19, page 460).

11.16 heptane

11.18 (c) $C_{14}H_{30}$ is an alkane

    (d) C_5H_{10} could be a cycloalkane

11.20 2,3-dimethylbutane

11.22 (a) 2,3-dimethylhexane

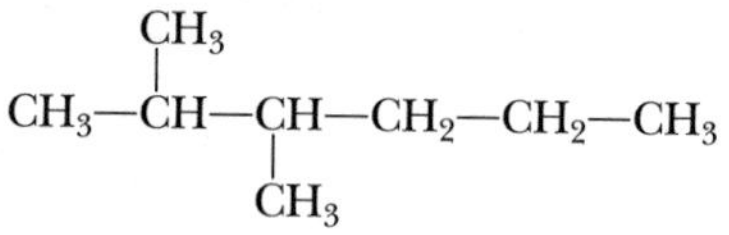

$$CH_3-\underset{\underset{\displaystyle CH_3}{|}}{CH}-\underset{\underset{\displaystyle CH_3}{|}}{CH}-CH_2-CH_2-CH_3$$

    (b) 2,3-dimethoctane

$$CH_3-\underset{\underset{\displaystyle CH_3}{|}}{CH}-\underset{\underset{\displaystyle CH_3}{|}}{CH}-CH_2-CH_2-CH_2-CH_2-CH_3$$

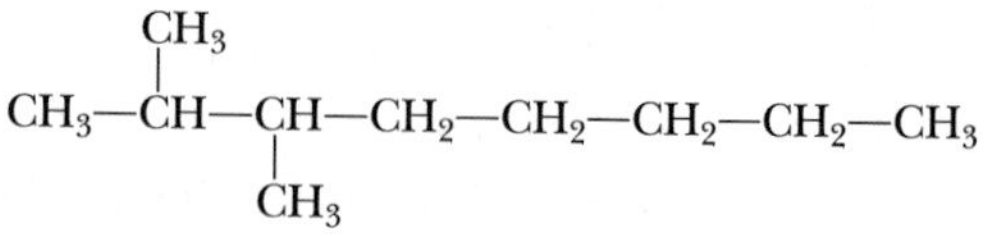

    (c) 3-ethylheptane

$$CH_3-CH_2-\underset{\underset{\displaystyle CH_2CH_3}{|}}{CH}-CH_2-CH_2-CH_2-CH_3$$

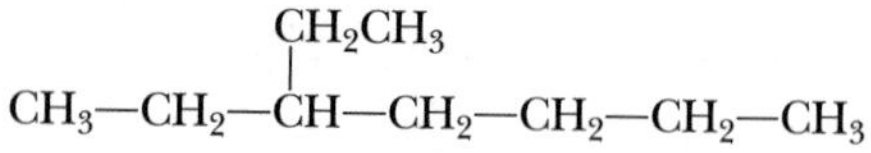

    (d) 2-methyl-3-ethylhexane

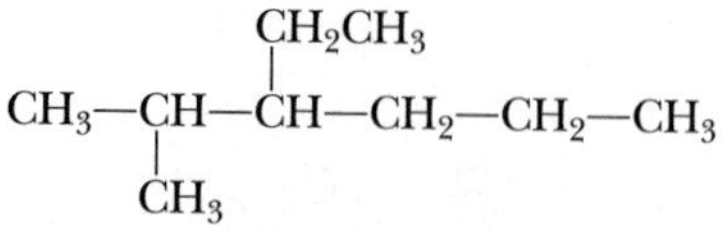

$$CH_3-\underset{\underset{\displaystyle CH_3}{|}}{CH}-\underset{\overset{\displaystyle CH_2CH_3}{|}}{CH}-CH_2-CH_2-CH_3$$

11.24

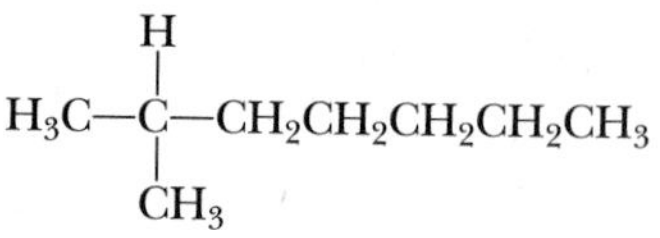

2-methylheptane

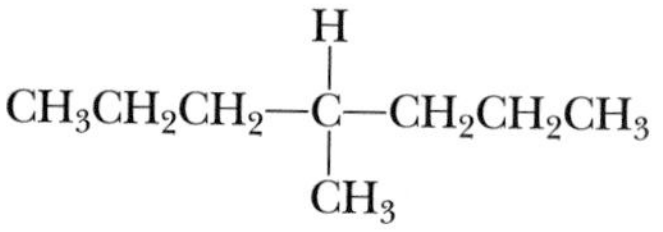

4-methylheptane

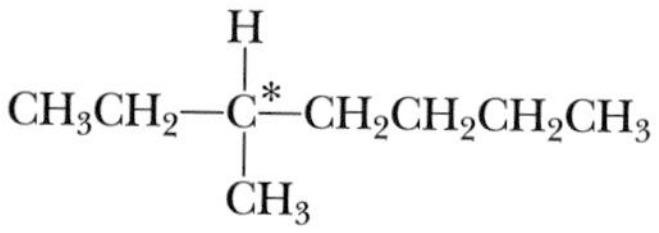

3-methylheptane. The C atom with an asterisk is chiral.

11.26

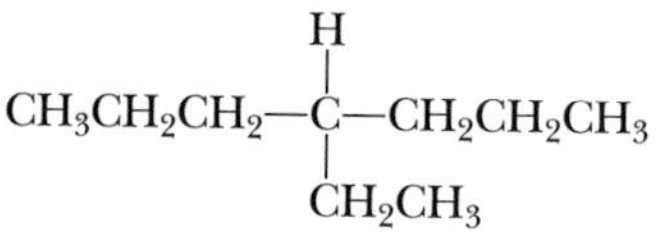

4-ethylheptane. The compound is not chiral.

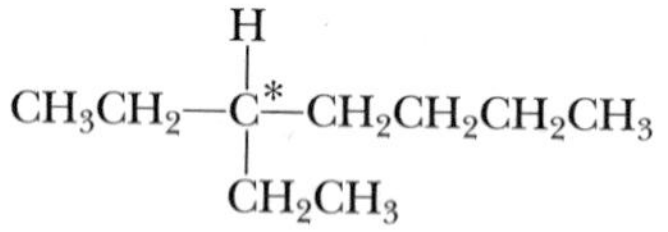

3-ethylheptane. The C atom with an asterisk is chiral.

11.28 C_4H_{10}, butane: A low molecular weight fuel gas at room temperature and pressure. Not soluble in water. $C_{12}H_{26}$, dodecane: A colorless liquid at room temperature. Expected to be insoluble in water but quite soluble in nonpolar solvents.

11.30

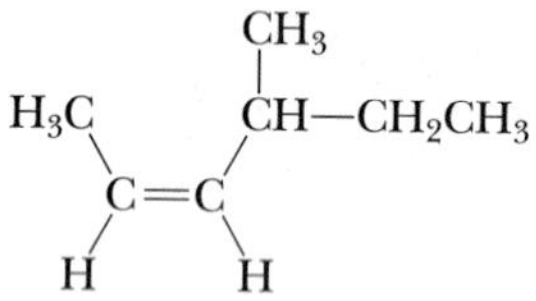

cis-4-methyl-2-hexene

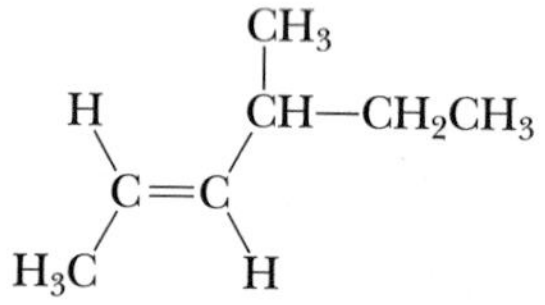

trans-4-methyl-2-hexene

11.32 (a)

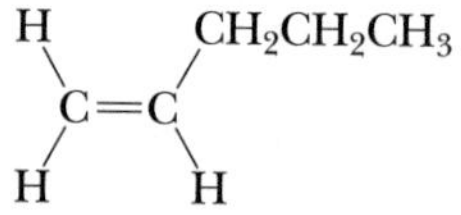

1-pentene

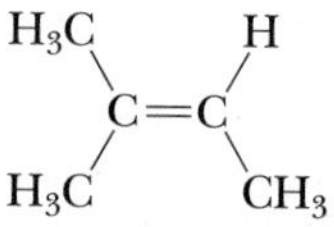

2-methyl-2-butene

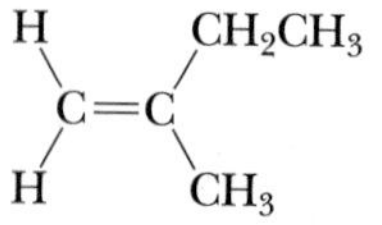

2-methyl-1-butene

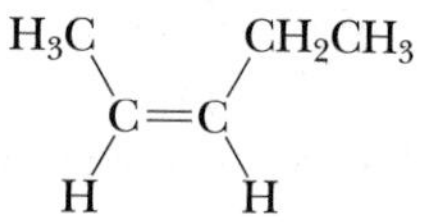

cis-2-pentene

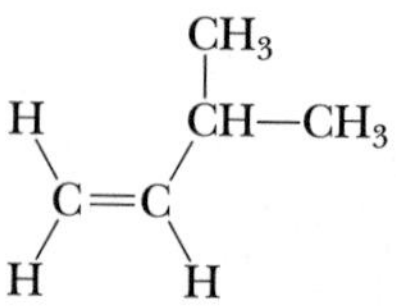

3-methyl-1-butene

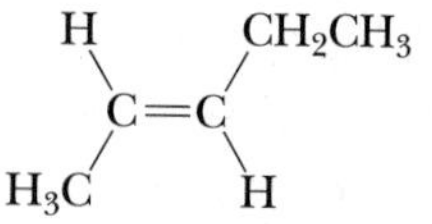

trans-2-pentene

(b)

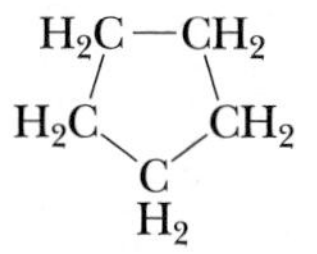

cyclopentane

11.34 (a) 1,2-dibromopropane, $CH_3CHBrCH_2Br$

       (b) pentane, $CH_3CH_2CH_2CH_2CH_3$

11.36 1-butene, $CH_3CH_2CH=CH_2$

11.38 Four isomers are possible.

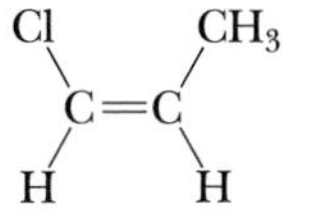

cis-1-chloropropene 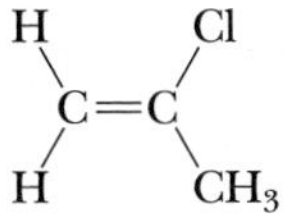   2-chloropropene

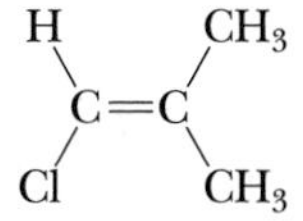

trans-1-chloropropene 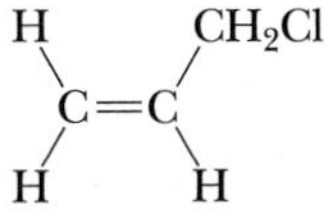   3-chloro-1-propene

11.40

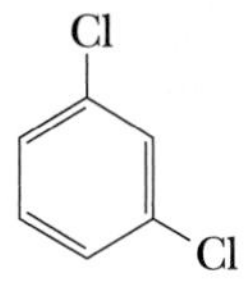

m-dichlorobenzene    *o*-bromotoluene

11.42

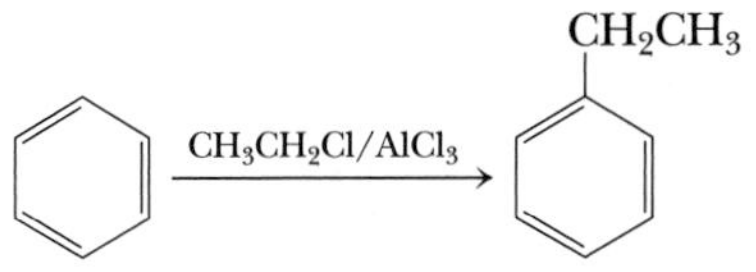

ethylbenzene

11.44

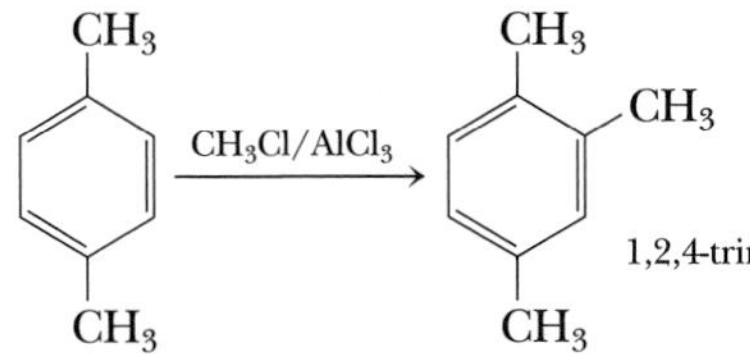

1,2,4-trimethylbenzene

11.46 (a) 1-propanol, primary

       (b) 1-butanol, primary

       (c) 2-methyl-2-propanol, tertiary

       (d) 2-methyl-2-butanol, tertiary

11.48 (a) ethylamine, $CH_3CH_2NH_2$

       (b) dipropylamine, $(CH_3CH_2CH_2)_2NH$

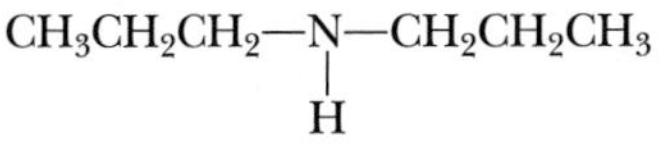

       (c) Butyldimethylamine

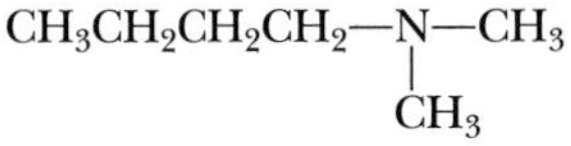

(d) Triethylamine

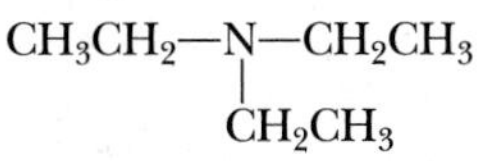

11.50 (a) 1-butanol, $CH_3CH_2CH_2CH_2OH$

       (b) 2-butanol

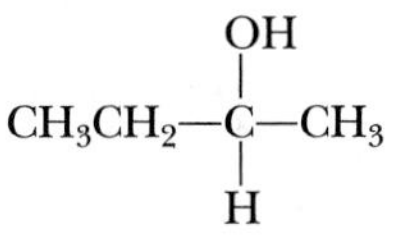

       (c) 2-methyl-1-propanol

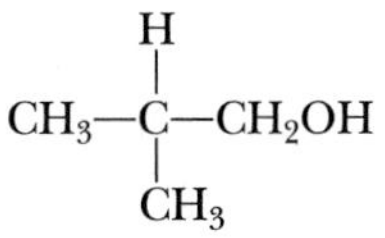

       (d) 2-methyl-2-propanol

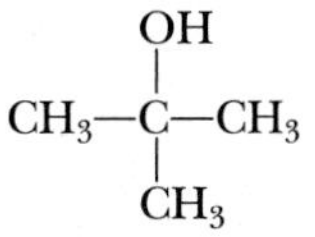

11.52 (a) $C_6H_5NH_2(aq) + HCl(aq) \longrightarrow (C_6H_5NH_3)Cl$

       (b) $(CH_3)_3N(aq) + H_2SO_4(aq) \longrightarrow$
            $[(CH_3)_3NH]HSO_4$

11.54 (a) 2-pentanone

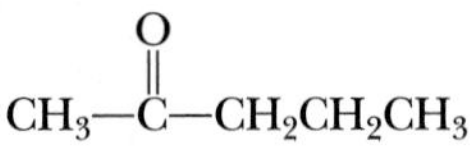

       (b) hexanal

       (c) pentanoic acid

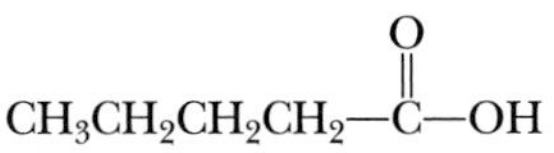

11.56 (a) acid, 3-methylpentanoic acid

       (b) ester, methyl propanoate

       (c) ester, butyl acetate (or butyl ethanoate)

       (d) acid, *p*-bromobenzoic acid

11.58 (a) pentanoic acid (see Question 11.54c)

       (b) 1-pentanol, $CH_3CH_2CH_2CH_2CH_2OH$

       (c) 2-octanol

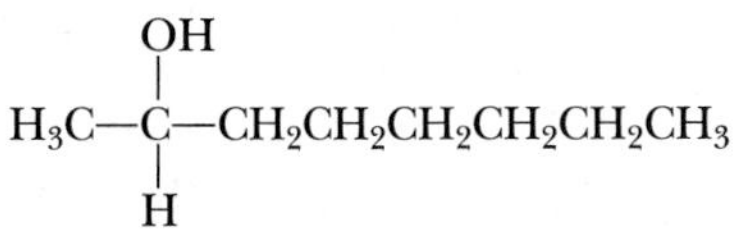

       (d) No reaction. A ketone is not oxidized by $KMnO_4$.

11.60   Step 1: oxidize 1-propanol to propanoic acid

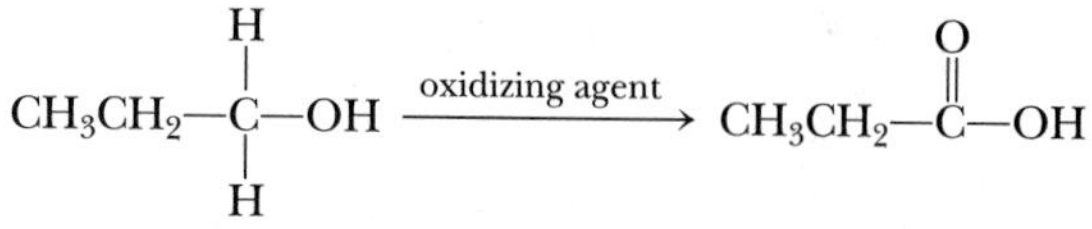

$$CH_3CH_2-\underset{\underset{H}{|}}{\overset{\overset{H}{|}}{C}}-OH \xrightarrow{\text{oxidizing agent}} CH_3CH_2-\overset{\overset{O}{\|}}{C}-OH$$

Step 2: combine propanoic acid and 1-propanol

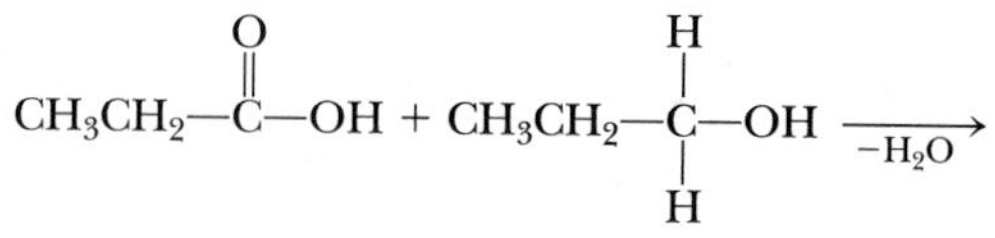

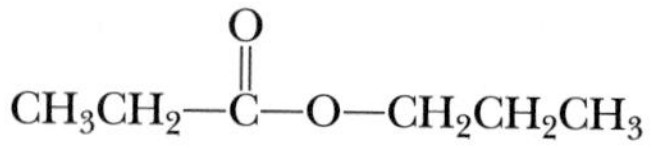

$$CH_3CH_2-\overset{\overset{O}{\|}}{C}-OH + CH_3CH_2-\underset{\underset{H}{|}}{\overset{\overset{H}{|}}{C}}-OH \xrightarrow[-H_2O]{}$$

$$CH_3CH_2-\overset{\overset{O}{\|}}{C}-O-CH_2CH_2CH_3$$

11.62   Sodium acetate ($NaCH_3CO_2$) and 1-butanol, $CH_3CH_2CH_2CH_2OH$.

11.64   (a)   trigonal planar

(b)   120°

(c)   The molecule is chiral. The are four different groups around the carbon marked 2.

(d)   The acidic H atom is the H attached to the CO_2H (carboxyl) group.

11.66   (a)   alcohol   (c)   acid

(b)   amide   (d)   ester

11.68   (a)   Prepare polyvinylacetate (PVA) from vinylacetate

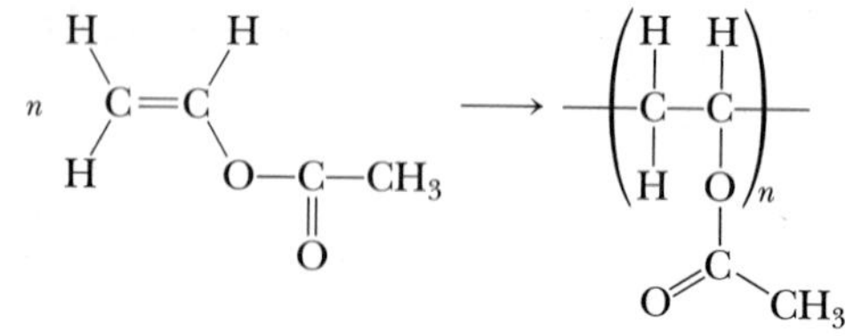

(b)   Three units of PVA

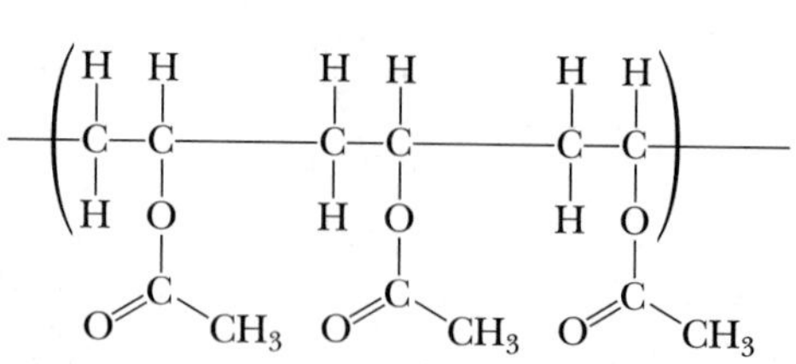

11.70   Illustrated here is a segment of a copolymer composed of two units of 1,1-dichloroethylene and two of chloroethylene.

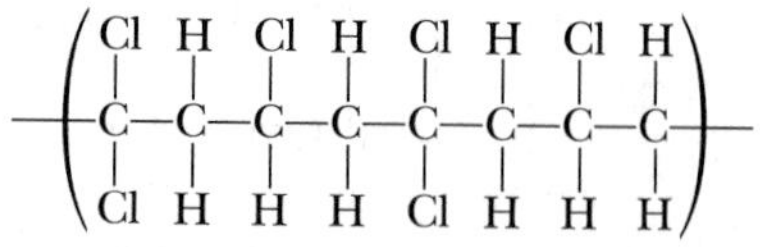

11.72   One of the 18 possible tripeptides that can from from a mixture of glycine, alanine, and histidine.

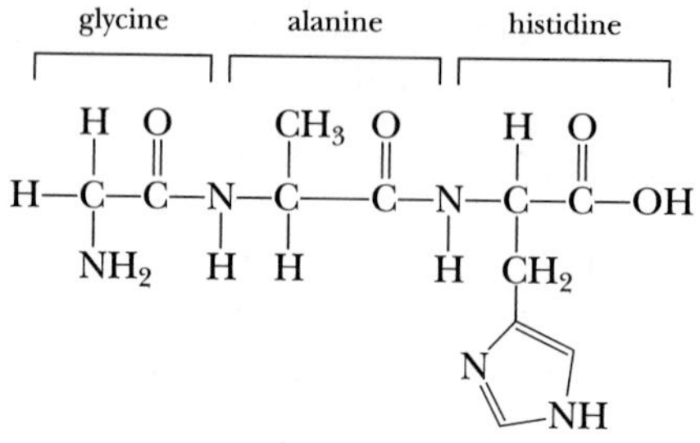

11.74   (a)   2,2-dimethylpentane

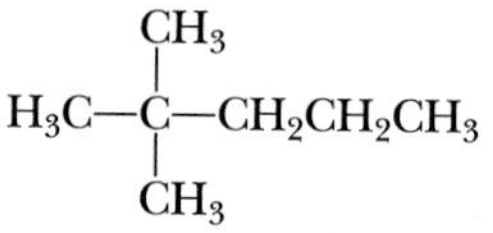

$$H_3C-\underset{\underset{CH_3}{|}}{\overset{\overset{CH_3}{|}}{C}}-CH_2CH_2CH_3$$

(b)   3,3-diethylpentane

$$CH_3CH_2-\underset{\underset{CH_2CH_3}{|}}{\overset{\overset{CH_2CH_3}{|}}{C}}-CH_2CH_3$$

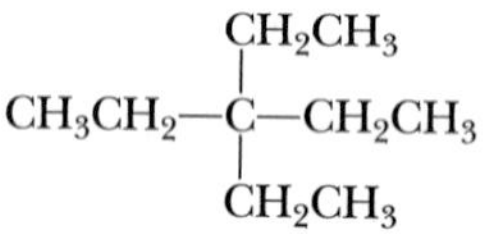

(c)   2-methyl-3-ethylpentane

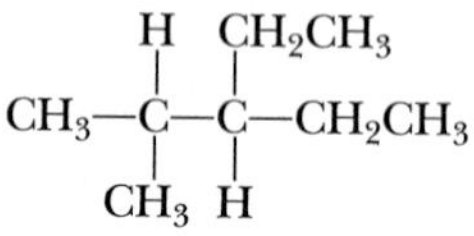

$$CH_3-\underset{\underset{CH_3}{|}}{\overset{\overset{H}{|}}{C}}-\underset{\underset{H}{|}}{\overset{\overset{CH_2CH_3}{|}}{C}}-CH_2CH_3$$

(d)   3-ethylhexane

$$CH_3CH_2-\underset{\underset{H}{|}}{\overset{\overset{CH_2CH_3}{|}}{C}}-CH_2CH_2CH_3$$

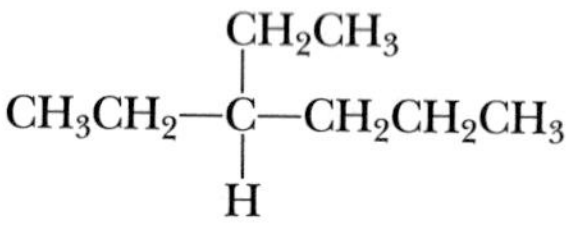

11.76

1,1-dichloropropane $H-\underset{\underset{Cl}{|}}{\overset{\overset{Cl}{|}}{C}}-CH_2CH_3$

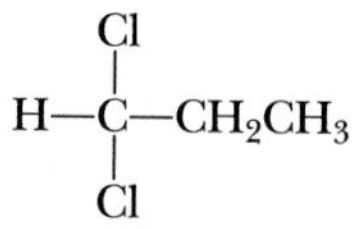

1,2-dichloropropane $H-\underset{\underset{H}{|}}{\overset{\overset{Cl}{|}}{C}}-\underset{\underset{H}{|}}{\overset{\overset{Cl}{|}}{C}}-CH_3$

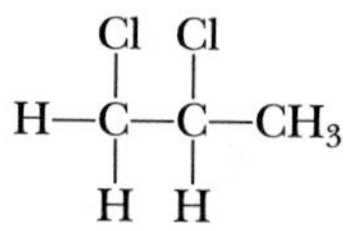

1,3-dichloropropane $H-\underset{\underset{H}{|}}{\overset{\overset{Cl}{|}}{C}}-\underset{\underset{H}{|}}{\overset{\overset{H}{|}}{C}}-\underset{\underset{H}{|}}{\overset{\overset{Cl}{|}}{C}}-H$

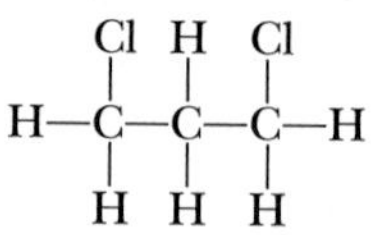

2,2-dichloropropane $H-\underset{\underset{H}{|}}{\overset{\overset{H}{|}}{C}}-\underset{\underset{Cl}{|}}{\overset{\overset{Cl}{|}}{C}}-\underset{\underset{H}{|}}{\overset{\overset{H}{|}}{C}}-H$

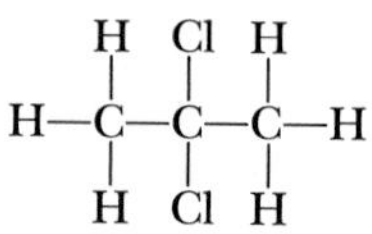

11.78

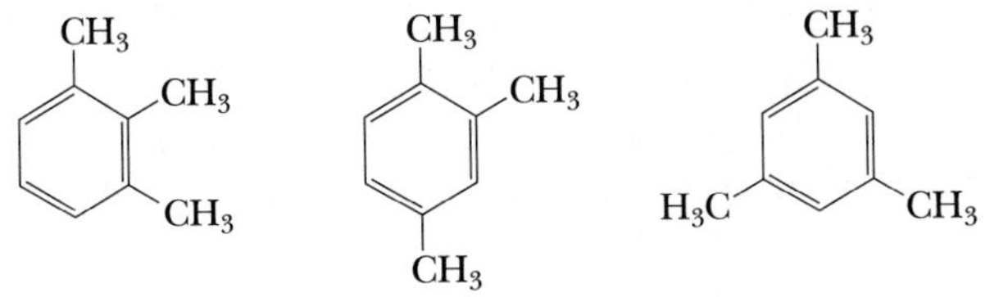

1,2,3-trimethylbenzene 1,2,4-trimethylbenzene 1,3,5-trimethylbenzene

11.80 Replace the carboxylic acid group with an H atom.

11.82

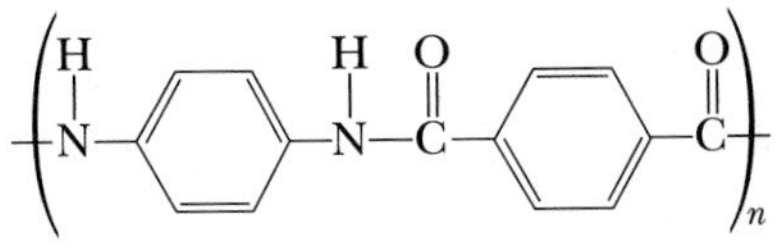

11.84

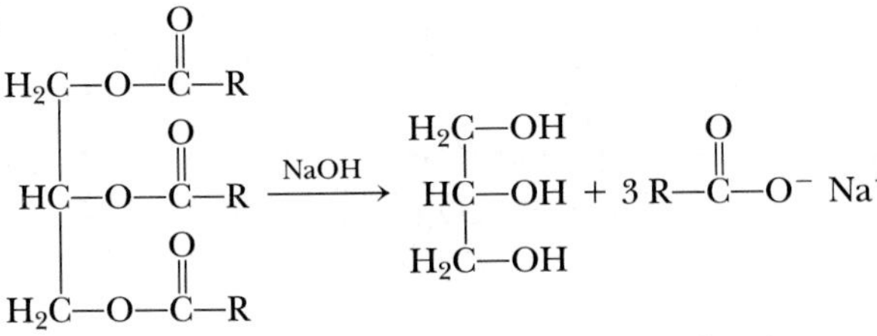

glyceryl trilaurate   glycerol   sodium laurate

11.86 Cyclohexene, a cyclic alkene will add Br_2 readily (to give $C_6H_{12}Br_2$). Benzene, however, needs much more stringent conditions to react with bromine and then Br_2 will substitute for H atoms on benzene and not add to the ring.

11.88 (a) The compound is either propane, a ketone, or propanal, an aldehyde.

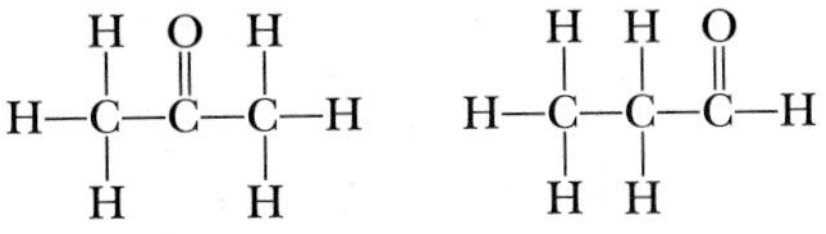

propanone   propanal
(a ketone)   (an aldehyde)

(b) The ketone will not oxidize, but the aldehyde will be oxidized to the acid, $CH_3CH_2CO_2H$. Thus, the unknown is likely propanal.

(c) Propanoic acid

11.90

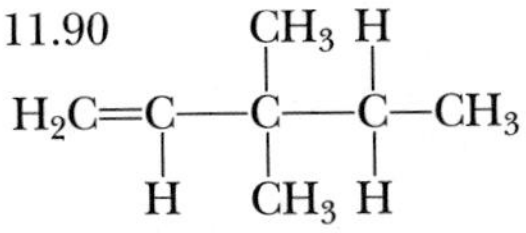

X = 3,3-dimethyl-2-pentanol

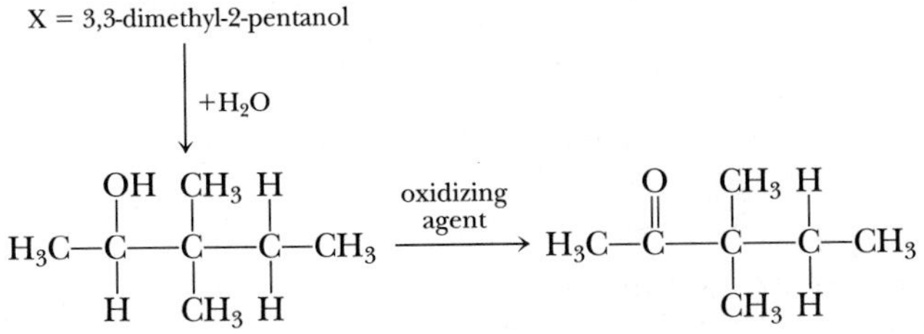

$+H_2O$

Y = 3,3-dimethyl-2-pentanol   3,3-dimethyl-2-pentanone

11.92

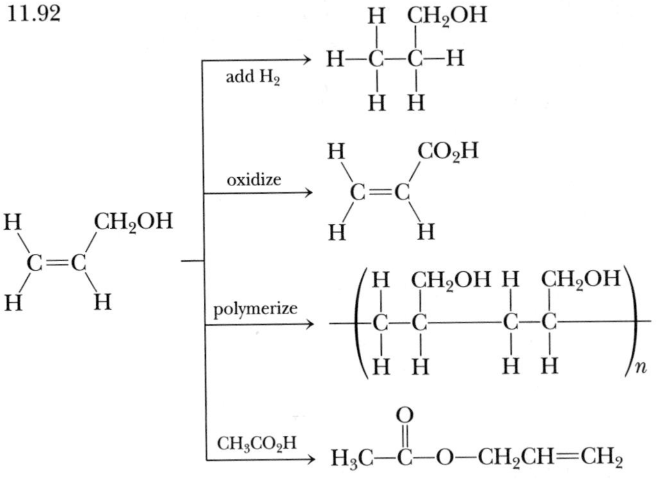

11.94 Pyridine is isoelectronic with benzene. A CH in benzene has been replaced by an N atom in pyridine. Both benzene and pyridine have two resonance structures.

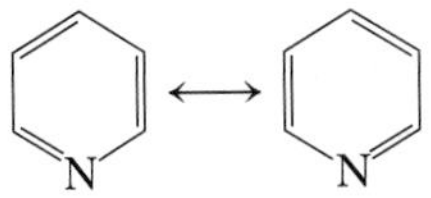

11.96 The alkene undergoes an addition reaction with bromine, whereas cyclopentane does not. When a solution of bromine in carbon tetrachloride is added to the alkene, the color of bromine rapidly disappears. (See the reaction of bacon with bromine on page 434.)

11.98

Symbol	Plastic	Use
1 PETE	polyethyl-eneterephthalate	soda bottles
2 HDPE	high density polyethylene	milk containers
3 V	polyvinyl chloride	shampoo bottles
4 LDPE	low density polyethylene	cosmetic containers
5 PP	polypropylene	syrup containers

The PET plastics should sink in water, whereas HDPE and PP will float.

11.100 (a) The C atoms of benzene are sp^2 hybridized (see page 396.) The C atoms of cyclohexane, C_6H_{12}, are sp^3 hybridized (page 431).

(b) π electron delocalization can occur in benzene because each C atom has an unhybridized p orbital perpendicular to the ring. Overlap of these orbitals leads to alternating π bonds.

(c) Cyclohexane cannot be planar because the geometry around each C atom is tetrahedral with bond angles of 109°.

11.102 (a) sp^3

(b) sp^2

(c) sp^2

(d) sp^3

(e) sp^3

11.104 (a) addition

(b) Oils generally have C=C double bonds in the long carbon chains of the fatty acids. Fats tend to have only C—C single bonds in these carbon chains (and the acid residues in fats are called saturated fatty acids). See page 454.

(c) As noted in (b), oils have unsaturated fatty acids, which makes the fatty acid chains less flexible.

11.106 (a) Double bonds. See also page 455.

(b) Termination occurred when the chain reached 14 atoms. It could have been terminated earlier than this, or the chain could have continued to grow.

(c) The termination step.

(d) Addition

CHAPTER 12

12.4 Avogadro's hypothesis states that equal volumes of gases (at the same T and P) have equal numbers of molecules. Thus, we can represent amounts of compounds by volumes in calculations. In the reaction 2 $H_2(g)$ + $O_2(g) \longrightarrow$ 2 $H_2O(g)$, if 3 mol of H_2 is used (67.2 L at STP), this requires 33.6 L of O_2 and produces 67.2 L of H_2O vapor.

12.10 (a) 0.58 atm

(b) 0.59 bar

(c) 59 kPa

12.12 (a) 0.754 bar

(b) 534 mm Hg

(c) 934 kPa

12.14 2.70×10^2 mm Hg

12.16 3.7 L

12.18 1.8 L

12.20 3.2×10^2 mm Hg

12.22 9.72 atm

12.24 (a) 75 mL O_2

(b) 150 mL NO_2

12.26 0.919 atm

12.28 V = 2.9 L

12.30 1.9×10^6 g He

12.32 3.7×10^{-4} g/L

12.34 34.0 g/mol

12.36 57.5 g/mol

12.38 Molar mass = 74.9 g/mol; B_6H_{10}

12.40 0.0394 mol H_2; 0.096 atm; 73 mm Hg

12.42 170 g NaN_3

12.44 1.7 atm O_2

12.46 4.1 atm H_2; 1.6 atm Ar; Total pressure = 5.7 atm

12.48 (a) 0.30 mol halothane/1 mol O_2

(b) 290 g halothane

12.50 (a) They all have the same kinetic energy.

(b) Average speed of H_2 molecules is greater than the average speed of the CO_2 molecules.

(c) The number of CO_2 molecules is greater than the number of H_2 molecules [$n(CO_2) = 1.8n(H_2)$]

(d) The mass of CO_2 is greater than the mass of H_2.

12.52 Average speed of CO_2 molecule = 3.65×10^4 cm/s

12.54 Average speed (and molar mass) increases in the order $CH_2F_2 <$ Ar $< N_2 < CH_4$.

12.56 (a) F_2 (38 g/mol) effuses faster than CO_2 (44 g/mol)

(b) N_2 (28 g/mol) effuses faster than O_2 (32 g/mol)

(c) C_2H_4 (28.1 g/mol) effuses faster than C_2H_6 (30.1 g/mol)

(d) $CFCl_3$ (137 g/mol) effuses faster than $C_2Cl_2F_4$ (171 g/mol)

12.58 36 g/mol

12.60 P from the van der Waals equation = 29.5 atm

P from the ideal gas law = 49.3 atm

12.62 (a) Standard atmosphere: 1 atm; 760 mm Hg; 101.325 kPa; 1.013 bar.

(b) N_2 partial pressure: 0.780 atm; 593 mm Hg; 79.1 kPa; 0.791 bar

(c) H_2 pressure: 131 atm; 9.98×10^4 mm Hg; 1.33×10^4 kPa; 133 bar

(d) Air: 0.333 atm; 253 mm Hg; 33.7 kPa; 0.337 bar

12.64 C_2H_7N

12.66 T = 290. K or 17 °C

12.68 (a) There are more molecules of H_2 than atoms of He.

(b) The mass of He is greater than the mass of H_2.

12.70 4 mol

12.72 Ni is the limiting reactant; 1.31 g $Ni(CO)_4$

12.74 204 g/mol; empirical formula = CCl_2F; molecular formula = $C_2Cl_4F_2$

12.76 P(total) = 0.031 atm; P(N_2) = 0.0061 atm; P(H_2O) = 0.024 atm;

12.78 64.2% $NaNO_2$

12.80 P(He) = 87 mm Hg; P(Ar) = 140 mm Hg; P(total) = 230 mm Hg

12.82 Empirical formula = XeF_4

12.84 0.0127 mol He

12.86 7.71×10^{17} molecules/cm^3

12.88 163 mm Hg

12.90   (a)   10.0 g of O_2 represents more molecules than 10.0 g of CO_2

     (b)   Average speed of O_2 molecules is greater than the average of speed of CO_2 molecules.

     (c)   The gases are at the same temperature and so have the same average kinetic energy.

12.92   (a) $P(C_2H_2) > P(CO)$

     (b)   There are more molecules in the C_2H_2 container than in the CO container.

12.94   (a)   Not a gas. A gas would expand to an infinite volume.

     (b)   Not a gas. A density of 8.2 g/mL is typical of a solid.

     (c)   Insufficient information.

     (d)   Gas

12.96   (a)   Because *P*, *V*, and *T* are the same for each tire, each contains the same number of gas molecules

     (b)   The unknown gas molecules are 10.0 times heavier than He atoms.

     (c)   The kinetic energies of the gases are the same, but He has the largest average speed.

12.98   Rate = 10. L/min

12.100   Amount of CO_2 = amount of MCO_3 = 0.00107 mol

Molar mass of MCO_3 = 148 g/mol, so the molar mass of the metal M is 88 g/mol. A likely choice is Sr.

12.102   (a)   3.8 g NiO

     (b) $P(O_2)$ = 190 mm Hg; $P(Cl_2)$ = 120 mm Hg; P(total) = 310 mm Hg.

12.104   (a)   46.0 g NaN_3

     (b)

$$\left[\ddot{\text{N}}=\text{N}=\ddot{\text{N}}\right]^{-} \longleftrightarrow \left[:\ddot{\text{N}}-\text{N}\equiv\text{N}:\right]^{-} \longleftrightarrow \left[:\text{N}\equiv\text{N}-\ddot{\text{N}}:\right]^{-}$$

     (c)   The N_3^{-} ion is linear.

12.106   (a)   A gas will likely become a solid long before reaching the vicinity of absolute zero.

     (b)   The N_2 and H_2 are in the correct stoichiometric ratio and will produce 8 molecules of NH_3. The volume of NH_3 will be twice that of the volume of N_2 and 2/3 that of the H_2 (at the same T and P).

12.108   (a)   All of these foods have distinctive odors, which reach our noses by diffusion.

     (b)   Speed of gas molecules is related to the square root of the absolute temperature, so a doubling of the temperature will lead to an increase of about $(2)^{1/2}$ or 1.4.

     (c)   No.

CHAPTER 13

13.16   (a)   Dipole–dipole interactions (and hydrogen bonds)

     (b)   Induced-dipole/induced-dipole forces

     (c)   Dipole–dipole interactions (and hydrogen bonds)

13.18   (a)   Induced-dipole/induced-dipole forces

     (b)   Induced-dipole/induced-dipole forces

     (c)   Dipole–dipole forces

     (d)   Dipole–dipole forces (and hydrogen bonding)

13.20   The predicted order of increasing strength is: Ne < CH_4 < CO < CCl_4. In this case, however, prediction does not quite agree with reality. The boiling points are Ne (−246 °C) < CO (−192 °C) < CH_4 (−162 °C) < CCl_4 (77 °C)

13.22   (c) HF; (d) acetic acid; (f) CH_3OH

13.24   *q* = +90.1 kJ

13.26   (a)   Water vapor pressure is about 150 mm Hg at 60 °C. (Appendix G gives a value of 149.4 mm Hg at 60 °C.)

     (b)   600 mm Hg at about 93 °C

     (c)   At 70 °C, ethanol has a vapor pressure of about 520 mm Hg, whereas that of water is about 225 mm Hg.

13.28   At 30 °C the vapor pressure of ether is about 590 mm Hg. (This pressure requires 0.23 g of ether in the vapor phase at the given conditions, so there is sufficient ether in the flask.) At 0 °C the vapor pressure is about 160 mm Hg, so some ether condenses when the temperature declines.

13.30   (a) O_2 (−183 °C) (Bp of N_2 = −196 °C)

     (b) SO_2 (−10 °C) (CO_2 sublimes at −78 °C)

     (c)   HF (+19.7 °C) (HI, −35.6 °C)

     (d) GeH_4 (−90.0 °C) (SiH_4, −111.8 °C)

13.32   (a) CS_2, about 620 mm Hg; CH_3NO_2, about 80 mm Hg

     (b) CS_2, induced-dipole/induced-dipole forces; CH_3NO_2, dipole–dipole forces

     (c) CS_2, about 46 °C; CH_3NO_2, about 100 °C

     (d)   About 39 °C

     (e)   About 34 °C

13.34   Because the critical pressure of CO is −140.3 °C, no amount of pressure applied at or above room temperature will liquefy CO.

13.36   Two possible unit cells are illustrated here. The simplest formula is AB_8.

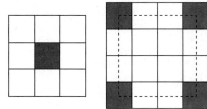

13.38 Ti^{4+} at 8 corners = 1 net Ti^{4+} ion

O^{2-} on 12 edges = 3 net O^{2-} ions

Ca^{2+} in center of unit cell = 1 net Ca^{2+}

Formula = $CaTiO_3$

13.40 (a) There are 8 O^{2-} ions at the corners and one in the center for a net of 2 O^{2-} ions per unit cell. There are four Cu ions in the interior in tetrahedral holes. The ratio of ions is Cu_2O.

(b) The oxidation number of copper must be +1.

13.42 (a) 8 C atoms per unit cell. (There are 8 corners (= 1 net C atom), 6 faces (= 3 net C atoms), and 4 internal C atoms.)

(b) fcc with C atoms in the tetrahedral holes.

13.44 q (for fusion) = -1.97 kJ; q (for melting) = $+1.97$ kJ

13.46 (a) Density of liquid CO_2 is less than that of solid CO_2.

(b) CO_2 is a gas at 5 atm and 0 °C.

(c) Critical temperature = 31 °C

13.48 q (to heat liquid) = 9.4×10^2 kJ

q (to vaporize NH_3) = 1.6×10^4 kJ

q (to heat the vapor) = 8.8×10^2 kJ

q_{total} = 1.8×10^4 kJ

13.50 $Ar < CO_2 < CH_3OH$

13.52 O_2 phase diagram. (i) Note the slight positive slope of the solid–liquid equilibrium line. This indicates the density of solid O_2 is greater than that of liquid O_2. (ii) Using the diagram here, the vapor pressure of O_2 at 74 K is between 150 mm Hg and 200 mm Hg.

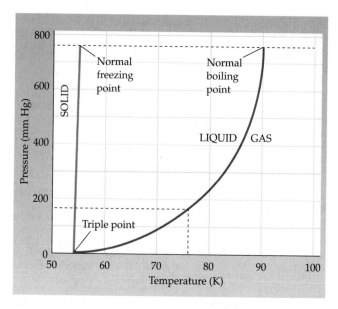

13.54 Acetone and water can interact by hydrogen bonding.

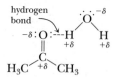

13.56 Glycol's viscosity will be greater than ethanol's owing to the greater hydrogen bonding capacity of glycol.

13.58 The meniscus should be concave because there are adhesive forces between the methanol and the silicate of glass.

13.60 (a) Water has two OH bonds and two lone pairs, whereas the O atom of ethanol has only one OH bond (and two lone pairs). More extensive H-bonding is likely for water.

(b) Water and ethanol interact extensively through H-bonding, so the volume is expected to be slightly smaller than the sum of the two volumes.

13.62 Radius of silver = 145 pm

13.64 Ca^{2+}: there are 8 corner Ca^{2+} ions and 1 internal Ca^{2+} ion or a total of 2 Ca^{2+} ions.

C atoms: there are 8 C atoms on edges. At 1/4 per atom, there are 2 within the unit cell. There are 2 more C atoms internal to the cell. Thus, there is a total of 4 C atoms per unit cell.

The formula is CaC_2.

13.66 No. NaCl has a 1:1 ratio of cations and anions in the unit cell, whereas the unit cell of $CaCl_2$ must have a 1:2 ratio of cations to anions.

13.68 Two pieces of evidence for $H_2O(\ell)$ having considerable intermolecular attractive forces:

(a) Based on the boiling points of the Group 6A hydrides (Figure 13.8), the boiling point of water should be approximately -80 °C. The actual boiling point of 100 °C reflects the significant H-bonding that occurs.

(b) Liquid water has a specific heat capacity that is higher than almost any other liquid. This reflects the fact that a relatively larger amount of energy is necessary to overcome intermolecular forces and raise the temperature of the liquid.

13.70 When the can is inverted in cold water, the water vapor pressure in the can, that was approximately 760 mm Hg, drops rapidly, say to 9 mm Hg at 10 °C. This creates a partial vacuum in the can, and the can is crushed because of the difference in pressure inside the can and the pressure of the atmosphere pressing down on the outside of the can.

13.72 (a) About -27 °C

(b) Pressure is about 6.5 atm.

(c) As the more energetic molecules leave the liquid phase, enter the gas phase, and escape from the tank, only lower energy molecules remain. These have a lower temperature. The tank is thereby cooled, and water vapor can condense on the tank's surface. Further, as the temperature drops, the vapor pressure of the remaining liquid drops and the flow of gas out of the tank slows.

(d) Cool the tank in dry ice (to -78 °C). The vapor pressure of the liquid is less than one atmosphere, so the tank can be opened safely and the liquid poured out.

13.74 (a) 80.1 °C

(b) At about 47 °C the liquid has a vapor pressure of 250 mm Hg. The vapor pressure is 600 mm Hg at 77 °C.

(c) 31 kJ/mol

13.76 1.356×10^{-8} cm (literature value is 1.357×10^{-8} cm)

13.78 Copper is face-centered cubic. (One approach to solving the problem is to use the copper radius to calculate the metal density, assuming each different cell type. Only when fcc is assumed do the calculated and experimental densities match.)

13.80 An approach to calculating Avogadro's number is as follows:

Cell side = 2.8665×10^{-8} cm

Cell volume = 2.3554×10^{-23} cm^3/unit cell

Molar volume = 0.1410 mol/cm^3

Using the information that there are 2 atoms per unit cell, the cell volume, and the molar volume, Avogadro's number is calculated.

13.82 5.49×10^{19} mercury atoms per cubic meter

13.84 Empty space = 47.6%

13.86 See Screen 13.4 for more detailed answers:

(a) In a related series of compounds, the boiling point generally increases with increasing intermolecular forces.

(b) Polarity and mass.

13.88 See Screen 13.12 for more detailed answers:

(a) Crystalline solids have long-range order.

(b) 12 unit cells

(c) Salt and ice, crystalline; glass and wood, amorphous.

CHAPTER 14

14.16 (a) Concentration (M) = 0.0434 mol/L. (This assumes the volume of the solution is 500. mL)

(b) Concentration (m) = 0.0434 m

(c) Mole fraction = 0.000781

(d) Weight percent = 0.509%

14.18 NaI: 0.15 m; 2.2%; X = 2.7×10^{-3}

CH$_3$CH$_2$OH: 1.1 m; 5.0%; X = 0.020

C$_{12}$H$_{22}$O$_{11}$: 0.15 m; 4.9%; X = 2.7×10^{-3}

14.20 2.65 g Na$_2$CO$_3$; X(Na$_2$CO$_3$) = 3.59×10^{-3}

14.22 220 g glycol; 5.7 m

14.24 16.2 m; 37.1%

14.26 Molality = 2.6×10^{-5} m (assuming that 1 kg of sea water is equivalent to 1 kg of solvent)

14.28 (b) and (c)

14.30 $\Delta H^{\circ}_{\text{solution}}$ for LiCl = -37.0 kJ/mol. This is an exothermic heat of solution, as compared with the very slightly endothermic value for NaCl.

14.32 Above about 40 °C the solubility increases with temperature; therefore, add more NaCl and raise the temperature.

14.34 See the discussion and data on page 568.

(a) The heat of hydration of LiCl is more negative than that for CsCl because the Li$^+$ ion is much smaller than the Cs$^+$ ion.

(b) The heat of hydration for Mg(NO$_3$)$_2$ is larger than for NaNO$_3$ owing to the +2 charge on the Mg^{2+} ion.

(c) The heat of hydration is greater for NiCl$_2$ than for RbCl because Ni^{2+} has a larger charge than Rb$^+$, and the Ni^{2+} ion is smaller than the Rb$^+$ ion.

14.36 2×10^{-3} g O$_2$

14.38 1130 mm Hg or 1.49 atm

14.40 35.0 mm Hg

14.42 X(H$_2$O) = 0.869; 16.7 mol glycol; 1040 g glycol

14.44 Calculated boiling point = 84.2 °C

14.46 ΔT_{bp} = 0.808 °C; Solution boiling point = 62.51 °C

14.48 Molality = 0.162 m; 0.00810 mol solute; 1.4 g solute

14.50 Molality = 8.60 m; 28.4%

14.52 Molality = 0.195 m; ΔT_{fp} = -0.362 °C

14.54 Molar mass = 360 g/mol; C$_{20}$H$_{16}$Fe$_2$

14.56 Molar mass = 150 g/mol

14.58 Molar mass = 170 g/mol

14.60 Molar mass = 130 g/mol

14.62 Freezing point = -24.6 °C

14.64 0.080 m CaCl$_2$ < 0.10 m NaCl < 0.040 m Na$_2$SO$_4$ < 0.10 sugar

14.66 (a) ΔT_{fp} = -0.348 °C; fp = -0.348 °C

(b) ΔT_{bp} = $+0.0959$ °C; bp = 100.0959 °C

(c) Π = 4.58 atm

The osmotic pressure is large and can be measured with a small experimental error.

14.68 Molar mass = 6.0×10^3 g/mol

14.70 (a) BaCl$_2$(aq) + Na$_2$SO$_4$(aq) $\longrightarrow$ BaSO$_4$(s) + 2 NaCl(aq)

(b) Initially the BaSO$_4$ particles form a colloidal suspension.

(c) Over time the particles of BaSO$_4$(s) grow and precipitate.

14.72 Li$_2$SO$_4$ should have the more negative heat of hydration than Cs$_2$SO$_4$ owing to the fact that the Li$^+$ ion is smaller than the Cs$^+$ ion.

14.74 (a) Increasing vapor pressure of water

0.20 m Na$_2$SO$_4$ < 0.50 m sugar < 0.20 m KBr < 0.35 m ethylene glycol

(b) increase in boiling point

0.35 m ethylene glycol < 0.20 m KBr < 0.50 m sugar < 0.20 m Na$_2$SO$_4$

14.76   (a) 0.456 mol DMG and 11.4 mol ethanol; X(DMG) = 0.0385

      (b) 0.869 *m*

      (c) VP ethanol over the solution at 78.4 °C = 730.7 mm Hg

      (d) bp = 79.5 °C

14.78   For ammonia: 23 *m*; 28%; X(NH$_3$) = 0.29

14.80   0.592 g Na$_2$SO$_4$

14.82   (a) 0.20 *m* KBr

      (b) 0.10 *m* Na$_2$CO$_3$

14.84   Freezing point = −11 °C

14.86 4.0×10^2 g/mol

14.88 4.93×10^{-4} mol/L

14.90   (a) Molar mass = 4.9×10^4 g/mol

      (b) $\Delta T_{fp} = -3.8 \times 10^{-4}$ °C

14.92   Molar mass in benzene = 1.20×10^2 g/mol; molar mass in water = 62.4 g/mol. The actual molar mass of acetic acid is 60.1 g/mol. In benzene the molecules of acetic acid form "dimers." That is, two molecules form a single unit through hydrogen bonding. See Figure 13.9 on page 518.

14.94   When the egg is placed in water, it swells because the concentration of solute is higher inside the egg than outside. There is therefore a net flow of water into the egg. The situation is opposite when the egg is placed in corn syrup. Water passes out of the egg from a solution of "low" solute concentration to one of relatively higher concentration.

14.96   The C—O and O—H bonds in hydrocarbons being nonpolar would tend to make such dispersions hydrophobic (water-hating). The C—O and C—H bonds in starch present opportunities for hydrogen bonding with water. Hence, starch is expected to be more hydrophilic.

14.98 $\Delta H^{\circ}_{\text{solution}}$ [Li$_2$SO$_4$] = −28.0 kJ/mol

      $\Delta H^{\circ}_{\text{solution}}$ [LiCl] = −37.0 kJ/mol

      $\Delta H^{\circ}_{\text{solution}}$ [K$_2$SO$_4$] = +23.7 kJ/mol

      $\Delta H^{\circ}_{\text{solution}}$ [KCl] = +17.2 kJ/mol

      Both lithium compounds have exothermic heats of solution, whereas both potassium compounds have endothermic values. Consistent with this is the fact that lithium salts (LiCl) are often more water-soluble than potassium salts (KCl) (see Figure 14.10).

14.100 X(benzene in solution) = 0.67 and X(toluene in solution) = 0.33

$$P_{\text{total}} = P_{\text{toluene}} + P_{\text{benzene}} = 7.3 \text{ mm Hg} + 50. \text{ mm Hg}$$
$$= 57 \text{ mm Hg}$$

$$\text{X(toluene in vapor)} = \frac{7.3 \text{ mm Hg}}{57 \text{ mm Hg}} = 0.13$$

$$\text{X(benzene in vapor)} = \frac{50. \text{ mm Hg}}{57 \text{ mm Hg}} = 0.87$$

14.102 *i* = 1.7. That is, there is 1.7 mol of ions in solution per mol of compound.

14.104 (a) Calculate the number of moles of ions in 10^6 g H$_2$O: 550. mol Cl$^-$; 470. mol Na$^+$; 53.1 mol Mg^{2+}; 9.42 mol SO$_4^{2-}$; 10.3 mol Ca^{2+}; 9.72 mol K$^+$; 0.83 mol Br$^-$. Total moles of ions = 1.103×10^3 per 10^6 g water. This gives ΔT_{fp} of −2.05 °C.

      (b) Π = 27.0 atm. This means that a minimum pressure of 27 atm would have to be used in a reverse osmosis device.

14.106 (a) *i* = 2.06

      (b) There are approximately 2 particles in solution, so H$^+$ + HSO$_4^-$ best represents H$_2$SO$_4$ in aqueous solution.

14.108 (a) Molar mass = 98.9 g/mol. Empirical formula = BF$_2$, and molecular formula = B$_2$F$_4$.

      (b)

14.110 (a) Additional NiCl$_2$ can be dissolved.

      (b) Disturbing the solution in some way will lead to precipitation.

14.112 The rate at which molecules of gas enter the solution equals the rate at which they leave the solution to return to the gas phase.

14.114 (a) Boiling point elevation is the consequence of the decrease in solvent vapor pressure, owing to the dissolution of solute.

      (b) The ionic solute 0.10 m NH$_4$NO$_3$

14.116 See the sidebar on CD-ROM Screen 14.11.

CHAPTER 15

15.2   After 2.0 hours the N$_2$O$_5$ concentration is about 0.80 M. If N$_2$O$_5$ began at 1.40 M, this means that 0.60 M N$_2$O$_5$ has been consumed. If 0.60 M N$_2$O$_5$ was consumed, then 1.2 M NO$_2$ and 0.30 M O$_2$ were formed.

15.4   If [A] is tripled the rate is increased by 9. If [A] is halved the rate is decreased to 1/4 of the original rate.

15.8   See Table 15.1 on page 619.

15.16   (a) $-\dfrac{1}{2}\dfrac{\Delta[\text{O}_3]}{\Delta t} = \dfrac{1}{3}\dfrac{\Delta[\text{O}_2]}{\Delta t}$

      (b) $-\dfrac{1}{2}\dfrac{\Delta[\text{HOF}]}{\Delta t} = \dfrac{1}{2}\dfrac{\Delta[\text{HF}]}{\Delta t} = \dfrac{\Delta[\text{O}_2]}{\Delta t}$

15.18 $\dfrac{1}{3}\dfrac{\Delta[\text{O}_2]}{\Delta t} = -\dfrac{1}{2}\dfrac{\Delta[\text{O}_3]}{\Delta t}$ or $\dfrac{\Delta[\text{O}_2]}{\Delta t} = -\dfrac{2}{3}\dfrac{\Delta[\text{O}_2]}{\Delta t}$

      so $\Delta[\text{O}_3]/\Delta t = -1.0 \times 10^{-3}$ mol/L · s

15.20   (a) The graph of [B] (product concentration) versus time shows [B] increasing from zero. The line is curved, indicating the rate changes with time; this means that the rate depends on concentration. Rates for the four 10-s intervals are: from 0–10 s, 0.0326 mol/L s; from 10–20 s, 0.0246 mol/L s; from 20–30 s, 0.0178 mol/L s; from 30–40 s, 0.0140 mol/L s.

(b) $-\dfrac{\Delta[A]}{\Delta t} = \dfrac{1}{2}\dfrac{\Delta[B]}{\Delta t}$ throughout the reaction

In the interval from 10–20 s, $\dfrac{\Delta[A]}{\Delta t} =$

$-0.0123 \dfrac{mol}{L \cdot s}$

(c) Instantaneous rate when $[B] = 0.750$ mol/L

$= \dfrac{\Delta[B]}{\Delta t} = 0.0163 \dfrac{mol}{L \cdot s}$

15.22 (a) Rate $= k[NO_2][O_3]$

(b) If $[NO_2]$ is tripled, the rate triples.

(c) If $[O_3]$ is halved, the rate is halved.

15.24 (a) The reaction is second order in [NO] and first order in $[O_2]$.

(b) $\dfrac{\Delta[NO]}{\Delta t} = -k[NO]^2[O_2]$

(c) $k = 12$ L^2/mol$^2 \cdot$ s

(d) $\dfrac{\Delta[NO]}{\Delta t} = -1.4 \times 10^{-5}$ mol/L $\cdot$ s

(e) When $\Delta[NO]/\Delta t = -1.0 \times 10^{-4}$ mol/L $\cdot$ s, $\Delta[O_2]/\Delta t = 5.0 \times 10^{-5}$ mol/L $\cdot$ s and $\Delta[NO_2]/\Delta t = 1.0 \times 10^{-4}$ mol/L $\cdot$ s.

15.26 (a) Rate $= k[NO]^2[O_2]$

(b) $k = 50.$ L^2/mol$^2 \cdot$ h

(c) Rate $= 8.4 \times 10^{-9}$ mol/L $\cdot$ h

15.28 (a) $n = 2$ and $m = 1$

(b) third order overall

(c) $k = 5$ L^2/mol$^2 \cdot$ min

15.30 $k = 0.0392$ h^{-1}

15.32 5.0×10^2 min

15.34 105 min

15.36 (a) 153 min

(b) 1790 min

15.38 (a) $t_{1/2} = 1400$ s

(b) 4600 s

15.40 4.48×10^{-3} mol (0.260 g) of azomethane remain; 0.0300 mol N_2 formed.

15.42 Fraction of ^{64}Cu remaining $= 0.030.$

15.44 72 s represents two half-lives, so $t_{1/2} = 36$ s.

15.46 (a) A graph of ln[sucrose] versus time produces a straight line, indicating that the reaction is first order in [sucrose]

(b) Δ[sucrose]$/\Delta t = -k$[sucrose]; $k = 3.7 \times 10^{-3}$ min^{-1}

(c) At 175 min, [sucrose] $= 0.167$ M

15.48 The straight line obtained in a graph of ln[N_2O] versus time indicates a first-order reaction.

$k = (-slope) = 0.0127$ min^{-1}

The rate when [N_2O] $= 0.035$ mol/L is 4.4×10^{-4} mol/L min.

15.50 The graph of $1/[NO_2]$ versus time gives a straight line, indicating the reaction is second order with respect to $[NO_2]$ (see Table 15.1 on page 619). The slope of the line is k, so $k = 1.1$ L/mol $\cdot$ s.

15.52 $-\Delta[C_2F_4]/\Delta t = 0.04$ L/mol $\cdot$ s$[C_2F_4]^2$

15.54 102 kJ/mol

15.56 0.3 s^{-1}

15.58

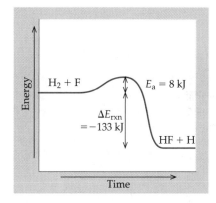

15.60 (a) Rate $= k[NO_3][NO]$

(b) Rate $= k[Cl][H_2]$

(c) Rate $= k[(CH_3)_3CBr]$

15.62 (a) The second step

(b) Rate $= k[O_3][O]$

15.64 (a) The substances OI^- and HOI cancel out to give the equation for the overall reaction.

(b) Steps 1 and 2 are bimolecular, whereas step 3 is termolecular.

(c) Rate $= k[H_2O_2][I^-]$

(d) OI^- and HOI are intermediates. They are produced and consumed during the reaction and do not appear in the equation for the overall reaction.

15.66 NO_2 is a reactant in the first step and a product in the second step. CO is a reactant in the second step. NO_3 is an intermediate, and CO_2 and NO are products.

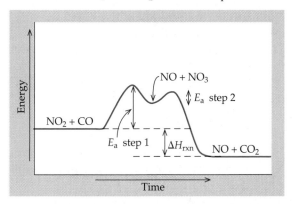

15.68 After measuring pH as a function of time, one could then calculate pOH and then $[OH^-]$. Finally, a plot of $1/[OH^-]$ versus time would give a straight line with a slope equal to k.

15.70 (a) Incorrect. Reactions are faster at a higher temperature because more reactants have an energy greater than the energy of activation.

(b) Correct

(c) Correct

(d) Incorrect. Catalysts provide a different pathway than uncatalyzed reactions; the catalyzed pathway has a lower activation energy than the uncatalyzed pathway.

15.72 (a) $n = 2$ and $m = 3$

(b) $k = 5 \times 10^4 \ L^4/mol^4 \cdot min$

(c) Third order with respect to $[H_2]$

(d) Fifth order overall.

15.74 (a) The reaction is second order with respect to NO.

(b) The reaction is first order with respect to Br_2.

(c) Third order overall.

15.76

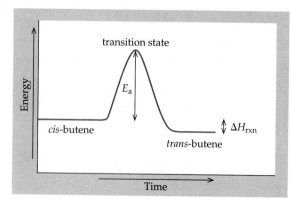

15.78 (a) A plot of $1/[C_2F_4]$ versus time indicates the reaction is second order with respect to $[C_2F_4]$. The rate law is Rate $= k[C_2F_4]^2$.

(b) The rate constant ($=$ slope of the line) is about $0.045 \ L/mol \cdot s$. (The graph does not allow a very accurate calculation.)

(c) Using $k = 0.045 \ L/mol \cdot s$, the concentration after 600 s is 0.03 M (to 1 significant figure).

(d) time $= 2000$ s (using k from part a).

15.80 (a) A plot of $1/[NH_4NCO]$ versus time is linear, so the reaction is second order with respect to NH_4NCO.

(b) Slope $= k = 0.0109 \ L/mol \cdot min$.

(c) $t_{1/2} = 200. \ min$

(d) $[NH_4NCO] = 0.0997 \ mol/L$

15.82 Mechanism 2 gives a rate equation of Rate $= k[NO_2]^2$

15.84 (a) Rate $= k[F_2][NO_2]$

(b) The reaction is first order with respect to each of the reactants, NO_2 and F_2.

(c) $k = 40 \ L/mol \cdot s$

15.86 $k = 0.037 \ h^{-1}$ and $t_{1/2} = 19$ h

15.88 (a) After 125 min, 0.251 g remain. After 145, 0.144 g remain.

(b) Time $= 43.9$ min

(c) Fraction remaining $= 0.016$

15.90 (a) $2 \ NO(g) + Br_2(g) \longrightarrow 2 \ BrNO(g)$

(b) Mechanism 1 is termolecular

Mechanism 2 has two bimolecular steps.

Mechanism 3 has two bimolecular steps.

(c) Br_2NO is the intermediate in Mechanism 2 and N_2O_2 is the intermediate in Mechanism 3.

(d) Assuming step 1 in each mechanism is the slow step, the rate equations will all differ. Mechanism 1 would be second order in NO and first order in Br_2. Mechanism 2 would be first order in both NO and Br_2. Finally, Mechanism 3 would be second order in NO and zero order in Br_2.

15.92 The rate equation for the slow step is Rate $= k[O_3][O]$. The equilibrium constant, K, for step 1 is $K = [O_2][O]/[O_3]$. Solving this for $[O]$, we have $[O] = K[O_3]/[O_2]$. Finally, substituting the expression for $[O]$ into the rate equation we find

Rate $= k[O_3]\{K[O_3]/[O_2]\} = kK[O_3]^2/[O_2]$

15.94 The finely divided rhodium metal will have a significantly greater surface area than the small block of metal. This leads to a large increase in the number of reaction sites, and vastly increases the reaction rate.

15.96 Carry out the reaction with ^{18}O-labeled methanol. If the product, methyl acetate, contains ^{18}O, this will prove that the oxygen in the ester linkage comes from the alcohol. The water would therefore come from the —OH group of the acid.

15.98 The slope of the ln k versus $1/T$ plot is -6114. From slope $= -E_a/R$, we derive $E_a = 51 \ kJ/mol$.

15.100 $k = 0.0274 \ min^{-1}$

15.102 Rate $= kK_1K_2[X][HA]/[A^-]$ where K_1 and K_2 are the equilibrium constants for steps 1 and 2, respectively.

15.104 Starting with an initial pressure of 25.0 mm Hg, the total pressure after one half-life (245 min) is 37.5 mm Hg. From the half-life we find $k = 2.83 \times 10^{-3} \ min^{-1}$. Using the rate constant, we find the fraction remaining of SO_2Cl_2 after 12 h (720 min) is 0.13. Thus, the pressure of SO_2Cl_2 after this time is 3.3 mm Hg. The pressure of each product is 21.7 mm Hg, and the total pressure is 46.7 mm Hg.

15.106 (a) The average rate is calculated over a period of time, whereas the instantaneous rate is the rate of reaction at some instant in time.

(b) The reaction rate decreases with time as the dye concentration decreases.

(c) See (b) above.

15.108 (a) Chemists use initial rates because the presence of products can affect the mechanism of the reaction.

(b) $3.6 \times 10^{-4} \ mol/L \cdot min$

15.110 3270 min represents 5 half-lives, so 1/32 is the fraction remaining. The concentration will be 6.3×10^{-4} M. After 3924 min, the fraction remaining is 1/64 because 6 half-lives have elapsed. The concentration at this point is 3.1×10^{-4} M.

15.112 One mechanism is a bimolecular substitution reaction. The reaction passes through the transition state as an incoming ion displaces an atom attached to the C atom. In the other mechanism a high energy intermediate is formed, and the reaction has two transition states.

15.114 (a) An overall reaction can be the result of a sequence of two or more elementary reactions or steps. The net result of these elementary steps is the overall reaction.

(b) The stoichiometric coefficient of a reactant in an elementary step is the order of the reaction for that reactant in that step.

(c) Rate = $k[NO_3][CO]$

(d) No. Only in the two-step mechanism is there isotopic scrambling in O atom exchange between NO_2 molecules. If CO reacts directly with NO_2, the CO_2 can have either ^{16}O or ^{18}O, and the product NO molecules will be either $N^{16}O$ or $N^{18}O$.

CHAPTER 16

16.2 The Pb^{2+} concentration is greater in the $PbCl_2$ beaker.

16.3 Increasing the temperature will shift the equilibrium to the right. This would be observed as an increase in CO_2 partial pressure. If more $CaCO_3$ is added, there is no effect on the equilibrium as solid $CaCO_3$ does not appear in the equilibrium constant expression. If additional CO_2 is added, the equilibrium will shift to the left.

16.4 (a) shifts to the right

(b) shifts to the left

16.6 (a) product-favored

(b) reactant-favored

(c) product-favored

16.8 (a) $K = \dfrac{[H_2O]^2[O_2]}{[H_2O_2]^2}$

(b) $K = \dfrac{[CO_2]}{[CO][O_2]^{1/2}}$

(c) $K = \dfrac{[CO]^2}{[CO_2]}$

(d) $K = \dfrac{[CO_2]}{[CO]}$

16.10 $Q = (2.0 \times 10^{-8})^2/(0.020) = 2.0 \times 10^{-14}$

$Q < K$ so reaction proceeds right

16.12 $Q = 1.0 \times 10^3$, so $Q > K$ and the reaction is not at equilibrium. It proceeds to the left to convert products to reactants.

16.14 $K = 1.2$

16.16 (a) $K = 0.025$

(b) $K = 0.025$

(c) The amount of solid does not affect the equilibrium.

16.18 (a) $[COCl_2] = 0.00308$ M; $[CO] = 0.00712$ M

(b) $K = 144$

16.20 [isobutane] = 0.024 M; [butane] = 0.010 M

16.22 $[I_2] = 6.14 \times 10^{-3}$ M; $[I] = 4.79 \times 10^{-3}$ M

16.24 $[COBr_2] = 0.107$ M; $[CO] = [Br_2] = 0.143$ M

57.1% of the $COBr_2$ has decomposed

16.26 (b)

16.28 Answer e, $K_2 = 1/(K_1)^2$

16.30 $K = 13.7$

16.32 (a) Equilibrium shifts to the right.

(b) Equilibrium shifts to the left.

(c) Equilibrium shifts to the right.

(d) Equilibrium shifts to the left.

16.34 Equilibrium concentrations are the same under both circumstances: [butane] = 1.1 M and [isobutane] = 2.9 M

16.36 $K = 3.9 \times 10^{-4}$

16.38 For decomposition of $COCl_2$, K = 1/(K for $COCl_2$ formation) = $1/(6.5 \times 10^{11}) = 1.5 \times 10^{-12}$

16.40 $K = 4$

16.42 (a) [B] will immediately increase, and the blue color will become darker.

(b) The reaction equilibrium will shift to the side with fewer molecules (because the flask size has been halved). Thus, it will shift toward the blue molecules, and the blue color will become even more pronounced.

16.44 (a) No, Q is not equal to K

(b) Q is less than K, so the equilibrium will shift to the right.

(c) $[N_2] = [O_2] = 0.25$ M and [NO] = 0.0102 M

16.46 The value of K is relatively large, so let us make the assumption that the reaction proceeds completely to the right. We would then calculate that 3.75 g of SO_3 could be produced (where SO_2 is the limiting reactant). Because the value of K is not infinite, however, we know the reaction does not proceed completely to the right, so 3.61 g is the only reasonable answer.

16.48 (a) The equilibrium will shift to the left.

(b) $[COBr_2] = 0.211$ M, $[CO] = 1.04$ M, and $[Br_2] = 0.039$ M

(c) Addition of CO has decreased the $COBr_2$ decomposition from 57.1% to 16%.

16.50 (a) The equilibrium will shift to the left on adding more Cl_2.

(b) Here K is calculated (from the quantities of reactants and products at equilibrium) to be 0.0470. After Cl_2 is added, you find that the concentrations are: $[PCl_5] = 0.0199$ M, $[PCl_3] = 0.0231$ M, and $[Cl_2] = 0.0403$ M.

16.52 $P(NH_3) = P(H_2S) = 0.33$ atm so $P(total) = 0.66$ atm

16.54 $K_p = 2.31 \times 10^{-4}$

16.56 (a) 84%

(b) The equilibrium shifts to the left.

16.58 $P(NO_2) = 0.40$ atm; $P(N_2O_4) = 1.1$ atm

16.60 This is a dynamic equilibrium. Initially, the rate of evaporation is greater than the rate of condensation. At equilibrium, the two rates are equal.

16.62 (a) The system is not at equilibrium. It must shift right to reach equilibrium.

(b) $\Delta H°_{rxn} = -199.9$ kJ. Raising the temperature shifts the equilibrium to the left, increasing the concentrations of the reactants.

16.64 (a) Because $P(total) = P(CO) + P(CO_2) = 0.200$ atm, this means that $P(CO_2) = P(CO) = 0.100$ atm. Therefore, $K_p = 1.00 \times 10^{-6}$.

(b) At 100 °C, $n(CO) = 0.0327$ mol. This means 0.0109 mol of $La_2(C_2O_4)_3$ decomposed, so 0.089 mol remains.

16.66 (a) $[SO_2Cl_2] = 0.020$ M and $[SO_2] = [Cl_2] = 0.030$ M The fraction of SO_2Cl_2 decomposed is 0.60

(b) $[SO_2Cl_2] = [SO_2] = 0.025$ M and $[Cl_2] = 0.044$ M The fraction of SO_2Cl_2 decomposed is 0.50

(c) Le Chatelier's principle predicts that the addition of Cl_2 should shift the equilibrium to the left, thus lowering the extent of SO_2Cl_2 decomposition. This is confirmed by these calculations.

16.68 $P(CO) = 1.0 \times 10^{-3}$ atm

16.70 $K_p = 0.098$

16.72 Adding SCN^- ion shifts the equilibrium to the right, increasing the concentration of the red $Fe(H_2O)_5(SCN)^{2+}$ ion. The color of the solution should become more red, as observed.

16.74 (a) $K = 4.5$

(b) $K = 2.3$

(c) $K = 6.0$

(d) $K = 5.6$

CHAPTER 17

17.1 Water is a proton donor (a Brønsted acid) and an electron pair donor (a Lewis base). It is also a proton acceptor (a Brønsted base). However, the molecule is not an electron-pair acceptor, a Lewis acid.

17.5 (a) The increasing acidity as Cl atoms replace H atoms is due to the inductive effect of the Cl atoms. See page 727.

(b) The strongest acid (Cl_3CCO_2H) will have the lowest pH, and the weakest acid (CH_3CO_2H) will have the highest pH.

17.6 The greater the number of O atoms attached to the central atom, the stronger the acid. Thus, H_2SeO_4 is a stronger acid than H_2SeO_3. See Table 17.7.

17.9 A = HCl, B = NH_4Cl, C = NaCl, Y = KCl, and Z = KOH. The following combinations occur:

A + Y give an acidic solution

C + Y give a neutral solution

C + Z give a basic solution

17.10 (a) CN^-, cyanide ion

(b) SO_4^{2-}, sulfate ion

(c) F^-, fluoride ion

17.12 (a) $H_3O^+(aq) + NO_3^-(aq)$; $H_3O^+(aq)$ is the conjugate acid of H_2O, and $NO_3^-(aq)$ is the conjugate base of HNO_3.

(b) $H_3O^+(aq) + SO_4^{2-}(aq)$; $H_3O^+(aq)$ is the conjugate acid of H_2O, and $SO_4^{2-}(aq)$ is the conjugate base of HSO_4^-.

(c) $H_2O + HF$; H_2O is the conjugate base of H_3O^+, and HF is the conjugate acid of F^-.

17.14 Brønsted acid: $HC_2O_4^-(aq) + H_2O(\ell) \rightleftharpoons H_3O^+(aq) + C_2O_4^{2-}(aq)$

Brønsted base: $HC_2O_4^-(aq) + H_2O(\ell) \rightleftharpoons H_2C_2O_4(aq) + OH^-(aq)$

17.16

Acid (A)	Base (B)	Conjugate base of A	Conjugate acid of B
(a) HCO_2H	H_2O	HCO_2^-	H_3O^+
(b) H_2S	NH_3	HS^-	NH_4^+
(c) HSO_4^-	OH^-	SO_4^{2-}	H_2O

17.18 $[H_3O^+] = 1.8 \times 10^{-4}$ M; acidic

17.20 HCl is a strong acid, so $[H_3O^+]$ = concentration of the acid. $[H_3O^+] = 0.0075$ M and $[OH^-] = 1.3 \times 10^{-12}$ M; pH = 2.12.

17.22 $Ba(OH)_2$ is a strong base, so $[OH^-] = 2 \times$ concentration of the base.

$[OH^-] = 3.0 \times 10^{-3}$ M, pOH = 2.52 and pH = 11.48

17.24 (a) The strongest acid is HCO_2H (largest K_a) and the weakest acid is C_6H_5OH (smallest K_a)

(b) Strongest acid (HCO_2H) has the weakest conjugate base

(c) The weakest acid has the strongest conjugate base.

17.26 Answer is (c). HOCl, the weakest acid in this list (Table 17.3), has the strongest conjugate base.

17.28 $CO_3^{2-}(aq) + H_2O(\ell) \rightleftharpoons HCO_3^-(aq) + OH^-(aq)$

17.30 Highest pH, Na_2S; lowest pH, $AlCl_3$ (which gives the weak acid $[Al(H_2O)_6]^{3+}$ in solution).

17.32 $pK_a = 4.19$

17.34 $K_a = 3.0 \times 10^{-10}$

17.36 2-chlorobenzoic acid has the smaller pK_a value

17.38 $K_b = 7.4 \times 10^{-12}$

17.40 $K_b = 6.3 \times 10^{-5}$

17.42 $CH_3CO_2H(aq) + HCO_3^-(aq) \rightleftharpoons CH_3CO_2^-(aq) + H_2CO_3(aq)$

Equilibrium lies predominantly to the right because CH_3CO_2H is a stronger acid than H_2CO_3.

17.44   (a) Left; NH_3 and HBr are the stronger base and acid, respectively.

(b) Left; PO_4^{3-} and CH_3CO_2H are the stronger base and acid.

(c) Right; $Fe(H_2O)_6^{3+}$ and HCO_3^- are the stronger acid and base.

17.46 $OH^-(aq) + HPO_4^{2-}(aq) \rightleftharpoons H_2O(\ell) + PO_4^{3-}(aq)$

OH^- is a stronger base than PO_4^{3-}, but the equilibrium does not lie strongly to the right. Therefore, the solution is likely to be basic.

17.48 $CH_3CO_2H(aq) + HPO_4^{2-}(aq) \rightleftharpoons$
$CH_3CO_2^-(aq) + H_2PO_4^-(aq)$

CH_3CO_2H is a stronger acid than $H_2PO_4^-$, so the equilibrium will lie to the right. The solution is likely to be weakly acidic.

NOTE: neither of the answers to 46 and 48 is very satisfactory.

17.50   (a) 2.1×10^{-3} M; (b) $K_a = 3.5 \times 10^{-4}$

17.52 $K_b = 6.6 \times 10^{-9}$

17.54   (a) $[H_3O^+] = 1.6 \times 10^{-4}$ M

(b) Moderately weak; $K_a = 1.1 \times 10^{-5}$

17.56 $[CH_3CO_2^-] = [H_3O^+] = 1.9 \times 10^{-3}$ M and $[CH_3CO_2H] = 0.20$ M

17.58 $[H_3O^+] = [CN^-] = 3.2 \times 10^{-6}$ M; $[HCN] = 0.025$ M; pH = 5.50

17.60 $[NH_4^+] = [OH^-] = 1.64 \times 10^{-3}$ M; $[NH_3] = 0.15$ M; pH = 11.22

17.62 $[OH^-] = 0.0102$ M; pH = 12.01; pOH = 1.99

17.64   pH = 3.25

17.66 $[H_3O^+] = 1.1 \times 10^{-5}$ M; pH = 4.98

17.68 $[HCN] = [OH^-] = 3.3 \times 10^{-3}$ M; $[H_3O^+] = 3.0 \times 10^{-12}$ M; $[Na^+] = 0.441$ M

17.70 $[H_3O^+] = 1.5 \times 10^{-9}$ M; pH = 8.81

17.72   (a) Reaction produces acetate ion, the conjugate base of acetic acid; solution is weakly basic. pH is greater than 7.

(b) Reaction produces NH_4^+, the conjugate acid of NH_3. Solution is weakly acidic. pH is less than 7.

(c) Reaction mixes equal molar amounts of strong base and strong acid. Solution will be neutral. pH will be 7.

17.74   (a) pH = 1.17; (b) $[SO_3^{2-}] = 6.2 \times 10^{-8}$ M

17.76   (a) $[OH^-] = [N_2H_5^+] = 9.2 \times 10^{-5}$ M; $[N_2H_6^{2+}] = 8.9 \times 10^{-16}$ M

(b) pH = 9.96

17.78   (a) Lewis base

(b) Lewis acid

(c) Lewis base (owing to lone pair of electrons on the N atom)

17.80   CO is a Lewis base in its reactions with transition metal atoms. The CO donates a lone pair of electrons on the C atom.

17.82   HOCN should be a stronger acid than HCN because the H atom in HOCN is attached to a highly electronegative O atom. This induces a positive charge on the H atom, making it more readily removed by an interaction with water.

17.84   The S atom is surrounded by four highly electronegative O atoms. The inductive effect of these atoms induces a positive charge on the H atom, making it susceptible to removal by water.

17.86   pH = 2.671

17.88   The weaker acid (smaller K_a) will have the higher pH in solution. Thus, the pH of a benzoic acid solution is higher than that of 4-chlorobenzoic acid.

17.90 $H_2S(aq) + CH_3CO_2^-(aq) \rightleftharpoons$
$CH_3CO_2H(aq) + HS^-(aq)$

The equilibrium lies to the left and favors the reactants.

17.92 $[X^-] = [H_3O^+] = 3.0 \times 10^{-3}$ M; $[HX] = 0.007$ M; pH = 2.52

17.94 $K_a = 1.4 \times 10^{-5}$; $pK_a = 4.86$

17.96   pH = 5.84

17.98   (a) Ethylamine is a stronger base than ethanolamine.

(b) For ethylamine, the pH of the solution is 11.82.

17.100   The K_b for pyridine (Appendix I) is 1.5×10^{-9}. Therefore, K_a for the conjugate acid, the pyridinium ion, is $K_a = K_w/K_b = 6.7 \times 10^{-6}$. The pH of the pyridinium hydrochloride solution is 3.39.

17.102   Acidic: $NaHSO_4$, NH_4Br, and $FeCl_3$

Neutral: $KClO_4$, $NaNO_3$, and LiBr

Basic: Na_2CO_3, $(NH_4)_2S$, Na_2HPO_4

17.104   HCl < NH_4Cl < NaCl < $NaCH_3CO_2$ < KOH

17.106 $K_{net} = K_{a1} \cdot K_{a2} = 3.8 \times 10^{-6}$

17.108   (a) Highest pH, $CH_3C_5H_4NH^+$; lowest pH, $O_2NC_5H_4NH^+$

(b) Strongest base, $CH_3C_5H_4N$; weakest base, $O_2NC_5H_4N$.

17.110   Reaction of ammonia with acid is
$H_3O^+(aq) + NH_3(aq) \rightleftharpoons NH_4^+(aq) + H_2O(\ell)$

This is the reverse of the reaction for ammonium ion in water. K for the HCl/NH_3 reaction is the reciprocal of K_a for NH_4^+ or $1/(5.6 \times 10^{-10}) = 1.8 \times 10^9$.

17.112   (a) Chloroacetic acid

(b) Benzylammonium ion

(c) In order of increasing acid strength: benzylammonium ion < conjugate acid of cocaine < benzoic acid < thioacetic acid < chloroacetic acid.

17.114   (a) BF_3 is a Lewis acid. (The empty p orbital on the B atom makes the molecule an electron pair acceptor). Dimethyl ether is a Lewis base. A Lewis structure shows two lone electron pairs on the O atom.

(b) For 0.100 g of $(CH_3)_2OBF_3$, P of BF_3 and $(CH_3)_2O$ = 0.032 atm and P of $(CH_3)_2OBF_3$ = 0.006 atm. P(total) = 0.0700 atm.

17.116 pH = 7.97

CHAPTER 18

18.2   See Table 17.5 on page 711.

(a) Weak base + strong acid gives pH < 7

(b) Strong acid + strong base give pH = 7

(c) Strong base + weak acid gives pH > 7.

18.4   The pH changes over a large value at the equivalence point. Therefore, we only need a dye that changes color somewhere in the pH range of the equivalence point.

18.7   (a) If K_a increases, pK_a decreases. Therefore, the pH should drop.

(b) The ratio [conjugate base]/[acid] will increase. The log term will increase, so the pH should increase. (In other words, as the conjugate base concentration increases, the solution will become more basic, and the pH increases.)

18.13   When Ag_3PO_4 dissolves slightly, it produces a small concentration of PO_4^{3-} ion. The phosphate ion is a strong base and readily hydrolyzes to HPO_4^{2-}. As this removes PO_4^{3-} from the equilibrium with Ag_3PO_4, the equilibrium shifts to the right,

$$Ag_3PO_4(s) \rightleftharpoons 3\ Ag^+(aq) + PO_4^{3-}(aq)$$

and Ag_3PO_4 dissolves to a greater extent than expected from the K_{sp} value.

18.17   (a) yes, AgBr is insoluble

(b) yes, $PbCl_2$ is insoluble

(c) no, $MgSO_4$ and $NaNO_3$ are both water-soluble

18.18   (a) Decrease pH; (b) Increase pH; (c) No change in pH

18.20   pH = 9.25

18.22   pH = 4.38

18.24   pH = 9.11; pH of buffer is lower than the pH of the original solution of NH_3.

18.26   4.7 g

18.28   pH = 4.92

18.30   (a) pH = 3.59; (b) $[HCO_2H]/[HCO_2^-]$ = 0.4

18.32   (b) $NH_3 + NH_4Cl$

18.34   (a) pH = 4.95; (b) pH = 5.05

18.36   (a) pH = 9.55; (b) pH = 9.50

18.38   (a) Original pH = 5.62

(b) $[Na^+]$ = 0.032 M, $[OH^-]$ = 1.5 × 10^{-3} M, $[H_3O^+]$ = 6.5 × 10^{-12} M, and $[C_6H_5O^-]$ = 0.031 M

(c) pH = 11.19

18.40   (a) Original NH_3 concentration = 0.0154 M

(b) At the equivalence point $[H_3O^+]$ = 1.9 × 10^{-6} M, $[OH^-]$ = 5.3 × 10^{-9} M, and $[NH_4^+]$ = 6.25 × 10^{-3} M

(c) pH at equivalence point = 5.73

18.42   The titration curve begins at pH = 13.00 and drops slowly as HCl is added. Just before the equivalence point (when 30.0 mL of acid has been added), the curve falls steeply. The pH at the equivalence point is exactly 7. Just after the equivalence point, the curve flattens again and begins to approach the final pH of just over 1.0. The total volume at the equivalence point is 60.0 mL.

18.44   (a) Starting pH = 11.12

(b) pH at equivalence point = 5.28

(c) pH at midpoint (half-neutralization point) = 9.25

(d) Methyl red, bromcresol green

(e)

Acid Added (mL)	pH
5.00	9.85
15.0	9.08
20.0	8.65
22.0	8.39
30.0	2.04

18.46   See Figure 17.2 on page 700.

(a) thymol blue or bromphenol blue

(b) phenolphthalein

(c) bromcresol green

18.48   (a) Silver chloride, AgCl, and lead chloride, $PbCl_2$

(b) Zinc carbonate, $ZnCO_3$, zinc sulfide, ZnS

(c) Iron(II) carbonate, $FeCO_3$, iron (II) oxalate, FeC_2O_4

18.50   (a) and (b) are soluble, (c) and (d) are insoluble.

18.52   (a) $AgCN(s) \rightleftharpoons Ag^+(aq) + CN^-(aq)$, $K_{sp} = [Ag^+][CN^-]$

(b) $NiCO_3(s) \rightleftharpoons Ni^{2+}(aq) + CO_3^{2-}(aq)$, $K_{sp} = [Ni^{2+}][CO_3^{2-}]$

(c) $AuBr_3(s) \rightleftharpoons Au^{3+}(aq) + 3\ Br^-(aq)$, $K_{sp} = [AU^{3+}][Br^-]^3$

18.54 $K_{sp} = (1.9 × 10^{-3})^2 = 3.6 × 10^{-6}$

18.56 $K_{sp} = 4.37 × 10^{-9}$

18.58 $K_{sp} = 1.4 × 10^{-15}$

18.60   (a) 9.2 × 10^{-9} M; (b) 2.2 × 10^{-6} g/L

18.62   (a) 2.4 × 10^{-4} M; (b) 0.015 g/L

18.64   Only 2.0 × 10^{-4} g dissolves

18.66   (a) $PbCl_2$; (b) FeS; (c) $Fe(OH)_2$

18.68   Solubility in pure water = 1.0 × 10^{-6} mol/L; solubility in 0.010 M SCN^- = 1.0 × 10^{-10} mol/L

18.70   (a) Solubility in pure water = 2.2 × 10^{-6} mg/mL

(b) Solubility in 0.020 M $AgNO_3$ = 1.0 × 10^{-12} mg/mL

18.72 (a) PbS doe to HS$^-$ formation (as well as oxidation of the S^{2-} ion to S)

(b) Ag$_2$CO$_3$ due to the formation of CO$_2$ and H$_2$O

(c) Al(OH)$_3$ due to the formation of water

18.74 Q < K$_{sp}$, so no precipitate forms.

18.76 Q > K$_{sp}$; Zn(OH)$_2$ will precipitate

18.78 [OH$^-$] must exceed 1.0×10^{-5} M

18.80 AuCl(s) $\rightleftharpoons$ Au$^+$(aq) + Cl$^-$(aq)

Au$^+$(aq) + 2 CN$^-$(aq) $\rightleftharpoons$ Ag(CN)$_2{}^-$(aq)

Net: AuCl(s) + 2 CN$^-$(aq) $\rightleftharpoons$ Au(CN)$_2{}^-$(aq) + Cl$^-$(aq)

K$_{net}$ = K$_{sp} \cdot$ K$_f$ = 4.0×10^{25}

18.82 (a) Add H$_2$SO$_4$, precipitating BaSO$_4$ and leaving Na$^+$(aq) in solution.

(b) Add HCl or another source of chloride ion. PbCl$_2$ will precipitate, but NiCl$_2$ is water-soluble.

18.84 Q > K$_{sp}$ so BaSO$_4$ precipitates

18.86 [H$_3$O$^+$] = 1.1×10^{-5} M; pH = 4.98

18.88 BaCO$_3$ < Ag$_2$CO$_3$ < Na$_2$CO$_3$

18.90 Original pH = 8.62; dilution will not affect the pH.

18.92 (a) pH = 2.81

(b) pH at equivalence point = 8.72

(c) pH at the midpoint = pKa = 4.62

(d) phenolphthalein

(e) After 10.0 mL, pH = 4.39

After 20.0 mL, pH = 5.07

After 30.0 mL, pH = 11.84

(f) A plot of pH versus volume of NaOH added would begin at a pH of 2.81, rise slightly to the midpoint at pH = 4.62 and then begin to rise more steeply as the equivalence point is approached (when the volume of NaOH added is 27.0 mL). The pH rises vertically through the equivalence point, and then begins to level off above a pH of about 11.0

18.94 110 mL NaOH

18.96 The K$_b$ value for ethylamine (4.27×10^{-4}) is found in Appendix I.

(a) pH = 11.89

(b) Midpoint pH = 10.63

(c) pH = 10.15

(d) pH = 5.93 at the equivalence point

(e) pH = 2.13

(g) Alizarin or bromcresol purple (see Figure 17.2 on page 700).

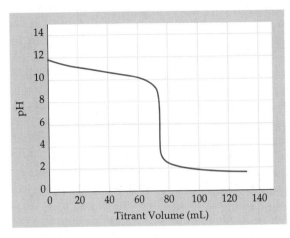

18.98 (a) 0.100 M acetic acid has a pH of 2.87. Adding sodium acetate slowly raises the pH.

(b) Adding NaNO$_3$ to 0.100 M HNO$_3$ has no effect on the pH.

(c) In (a) adding the conjugate acid of a weak acid creates a buffer solution. In (b), HNO$_3$ is a strong acid, and its conjugate base (NO$_3{}^-$) is so weak that the base has no effect on the complete ionization of the acid.

18.100 (a) HPO$_4{}^{2-}$

(b) 26 g of Na$_2$HPO$_4$

(c) Add 10. g of the base, Na$_3$PO$_4$, to raise the pH.

18.102 BaSO$_4$ will precipitate first.

18.104 K = 2.1×10^6; yes, AgI forms

18.106 (a) [F$^-$] = 1.3×10^{-3} M; (b) [Ca^{2+}] = 2.9×10^{-5} M

18.108 (a) PbSO$_4$ will precipitate first.

(b) [Pb^{2+}] = 5.1×10^{-6} M

18.110 CuS has K$_{spa}$ (Appendix J) of 6×10^{-37}, and Cu(OH)$_2$ has K$_{sp}$ = 2.2×10^{-22}. Both are quite insoluble. However, Cu(OH)$_2$ should much more readily dissolve in a strong acid like HCl.

CHAPTER 19

19.4 (a) Reactant-favored (mercury is a liquid under standard conditions)

(b) Product-favored (water vapor will condense to liquid) at 25 °C

(c) Reactant-favored (a continuous supply of energy is required)

(d) product-favored (carbon will burn)

(e) product-favored (salt will dissolve in water)

(f) Reactant-favored (calcium carbonate is insoluble)

19.12 (a) CO$_2$(g) has more disorder than the solid

(b) Liquid water at 50 °C

(c) Ruby

(d) One mol of N$_2$ at 1 bar

19.14 (a) $CH_3OH(g)$ (c) $NH_4Cl(aq)$

   (b) HBr (d) $HNO_3(g)$

19.16 (a) $\Delta S° = +12.7$ J/K. Solution is more disordered than the solid.

   (b) $\Delta S° = -102.55$ J/K. Significant ordering when going from gas to solid.

   (c) $\Delta S° = +93.2$ J/K. Greater disorder in the vapor.

   (d) $\Delta S° = -129.7$ J/K. The solution is more ordered (with H^+ forming H_3O^+ and hydrogen bonding occurring) than HCl in the gaseous state.

19.18 $\Delta S° = -174.1$ J/K

19.20 (a) $\Delta S° = +9.3$ J/K; (b) $\Delta S° = -293.97$ J/K

19.22 (a) $\Delta S° = -507.3$ J/K; (b) $\Delta S° = +313.25$ J/K

19.24 $\Delta S°_{sys} = -134.18$ J/K; $\Delta H°_{sys} = -662.75$ kJ and so $\Delta S°_{surr} = +2222.9$ J/K; $\Delta S°_{univ} = +2088.7$ J/K

19.26 $\Delta S°_{sys} = +163.3$ J/K; $\Delta H°_{sys} = +285.83$ kJ and so $\Delta S°_{surr} = -958.68$ J/K; $\Delta S°_{univ} = -795.4$ J/K

   The reaction is not spontaneous as the overall entropy change in the universe is negative. The reaction is disfavored by energy dispersal.

19.28 (a) *Type 2*. Reaction is enthalpy-favored but entropy disfavored. More favorable at low temperatures.

   (b) *Type 4*. This endothermic reaction is not favored by the enthalpy change nor is it favored by entropy change. It is not spontaneous under any conditions.

19.30 (a) $\Delta S°_{sys} = +174.75$ J/K; $\Delta H°_{surr} = +116.94$ kJ and so $\Delta S°_{surr} = -392.4$ J/K

   (b) $\Delta S°_{univ} = -217.67$ J/K. Reaction is not spontaneous at 298 K.

   (c) As the temperature increases, $\Delta S°_{surr}$ becomes less important, so $\Delta S°_{univ}$ will become positive at a sufficiently high temperature.

19.32 (a) $\Delta H°_{rxn} = -438$ kJ; $\Delta S°_{rxn} = -201.7$ J/K; $\Delta G°_{rxn} = -378$ kJ.

   Reaction is product-favored and is enthalpy-driven.

   (b) $\Delta H°_{rxn} = -86.61$ kJ; $\Delta S°_{rxn} = -79.4$ J/K; $\Delta G°_{rxn} = -62.95$ kJ.

   Reaction is product-favored. The enthalpy change favors the reaction but not the entropy change.

19.34 (a) $\Delta H°_{rxn} = +116.7$ kJ; $\Delta S°_{rxn} = +168.0$ J/K; $\Delta G°_f = +66.6$ kJ/mol

   (b) $\Delta H°_{rxn} = -425.93$ kJ; $\Delta S°_{rxn} = -154.6$ J/K; $\Delta G°_f = -379.82$ kJ/mol

   (c) $\Delta H°_{rxn} = +17.51$ kJ; $\Delta S°_{rxn} = +77.95$ J/K; $\Delta G°_f = -5.73$ kJ/mol

19.36 (a) $\Delta G°_{rxn} = -817.54$ kJ. Product-favored

   (b) $\Delta G°_{rxn} = +256.6$ kJ. Not product-favored

   (c) $\Delta G°_{rxn} = -1101.14$ kJ. Product-favored

19.38 $\Delta G°_f [BaCO_3(s)] = -1134.4$ kJ/mol

19.40 (a) $\Delta H°_{rxn} = +66.2$ kJ; $\Delta S°_{rxn} = -121.62$ J/K; $\Delta G°_{rxn} = +102.5$ kJ

   Reaction is reactant-favored by both the enthalpy and entropy changes. There is not temperature at which it will become product-favored. This is a case like that in the right panel in Figure 19.9 and is a Type 4 reaction (Table 19.2).

   (b) $\Delta H°_{rxn} = -221.05$ kJ; $\Delta S°_{rxn} = +179.1$ J/K; $\Delta G°_{rxn} = -274.45$ kJ

   Reaction is favored by both enthalpy and entropy and is product-favored at all temperatures. This is a case like that in the left panel in Figure 19.9 and is a Type 1 reaction.

   (c) $\Delta H°_{rxn} = -179.00$ kJ; $\Delta S°_{rxn} = -160.2$ J/K; $\Delta G°_{rxn} = -131.4$ kJ

   The reaction is favored by the enthalpy change but disfavored by the entropy change. The reaction becomes less product-favored as the temperature increases; it is a case like the upper line in the middle panel of Figure 19.9.

   (d) $\Delta H°_{rxn} = +822.2$ kJ; $\Delta S°_{rxn} = +181.28$ J/K; $\Delta G°_{rxn} = +768.1$ kJ

   The reaction is not favored by the enthalpy change but favored by the entropy change. The reaction becomes more product-favored as the temperature increases; it is a case like the lower line in the middle panel of Figure 19.9.

19.42 $\Delta H°_{rxn} = +337.2$ kJ and $\Delta S°_{rxn} = +161.5$ J/K. When $\Delta G°_{rxn} = 0$, T = 2088 K.

19.44 $K = 6.8 \times 10^{-16}$. Note that K is very small and that $\Delta G°$ is positive. Both indicate a reactant-favored process.

19.46 $\Delta G°_{rxn} = -100.24$ kJ and $K_p = 3.64 \times 10^{17}$. Both the free energy change and K indicate a product-favored process.

19.48 Reaction 1: $\Delta S°_1 = -80.7$ J/K
   Reaction 2: $\Delta S°_2 = -161.60$ J/K
   Reaction 3: $\Delta S°_3 = -242.3$ J/K
   $\Delta S°_1 + \Delta S°_2 = \Delta S°_3$

19.50 $\Delta H°_{rxn} = -1428.66$ kJ; $\Delta S°_{rxn} = +47.1$ J/K; $\Delta S°_{univ} = +4838.8$ J.

   Combustion reaction are spontaneous, and this is confirmed by the sign of $\Delta S°_{univ}$.

19.52 $2 Fe(s) + 3/2 O_2(g) \longrightarrow Fe_2O_3(s)$
   $\Delta G°_{rxn}$ for formation of 1 mol $= \Delta G°_f = -742.2$ kJ
   $\Delta G°_{rxn}$ for 454 g $Fe_2O_3 = -2110$ kJ

19.54 The reaction occurs spontaneously and is product-favored. Therefore, $\Delta S°_{univ}$ is positive and $\Delta G°_{rxn}$ is negative. The reaction is likely to be exothermic, so $\Delta H°_{rxn}$ is negative, and $\Delta S°_{surr}$ is positive. $\Delta S°_{sys}$ is expected to be negative because two moles of gas form a mol of solid. The calculated values are as follows:

   $\Delta S°_{sys} = -284.2$ J/K

$\Delta H^{\circ}_{rxn} = -176.34$ kJ and so $\Delta S^{\circ}_{surr} = +591.45$ J/K

$\Delta S^{\circ}_{univ} = +307.3$ J/K

$\Delta G^{\circ}_{rxn} = -91.64$ kJ and $K_p = 1.13 \times 10^{16}$

19.56 (a) $\Delta S^{\circ}_{sys} = +43.4$ J/K

$\Delta H^{\circ}_{rxn} = +3.85$ kJ and so $\Delta S^{\circ}_{surr} = -12.9$ J/K

$\Delta S^{\circ}_{univ} = +30.5$ J/K

Dissolving NaCl is a spontaneous process, so the sign of ΔS°_{univ} should be positive. The process is slightly endothermic, so the reaction is driven by material dispersal ($\Delta S^{\circ}_{sys} > 0$).

(b) $\Delta S^{\circ}_{sys} = -16.4$ J/K

$\Delta H^{\circ}_{rxn} = -43.22$ kJ and so $\Delta S^{\circ}_{surr} = +145.0$ J/K

$\Delta S^{\circ}_{univ} = +128.6$ J/K

Dissolving NaOH is a spontaneous process, and this is reflected by a positive value of ΔS°_{univ}. The process is driven by energy dispersal ($\Delta S^{\circ}_{surr} > 0$) because the process is quite exothermic. (The decline in entropy on dissolving is seen frequently with substances that can form hydrogen bonds with water.)

19.58 $K_p = 1.3 \times 10^{29}$ at 298 K ($\Delta G^{\circ} = -166.1$ kJ). The reaction is already extremely product-favored at 298 K. A higher temperature, however, would make the reaction less product-favored because ΔS°_{rxn} has a negative value (-242.3 J/K).

19.60 At the boiling point $\Delta G^{\circ} = 0 = \Delta H^{\circ} - T\Delta S^{\circ}$.

Here $\Delta S^{\circ} = \Delta H^{\circ}/T = 112$ J/K · mol at 351.15 K.

19.62 For $C_2H_5OH(\ell) \longrightarrow C_2H_5OH(g)$, $\Delta S^{\circ} = +122.0$ J/K and $\Delta H^{\circ} = +41.7$ kJ.

$T = \Delta H^{\circ}/\Delta S^{\circ} = 341.8$ K or 68.7 °C.

19.64 ΔS°_{rxn} is $+137.2$ J/K. A positive entropy change means that raising the temperature will increase the product-favorability of the reaction (because $T\Delta S^{\circ}$ will become more negative).

19.66 (a) ΔH° is positive (heat is required to melt benzene)

(b) ΔS° is positive (substance changes from solid to liquid)

(c) ΔG° is 0 at the melting point

(d) ΔG° is positive at 0 °C, a temperature lower than the melting point.

(e) ΔG° is negative at 25.0 °C, a temperature higher than the melting point

19.68 The reaction is exothermic, so ΔH°_{rxn} should be negative. Also a gas and an aqueous solution are formed, so ΔS°_{rxn} should be positive. The calculated values are:

$\Delta H^{\circ}_{rxn} = -183.32$ kJ (with a negative sign as expected)

$\Delta S^{\circ}_{rxn} = -7.7$ J/K.

The entropy change is slightly negative, not positive as predicted. The reason for this is the negative entropy change for dissolving NaOH (see Question 19.56b).

19.70 $\Delta S^{\circ}_{rxn} = +460.$ J/K; $\Delta H^{\circ}_{rxn} = -81.0$ kJ; $\Delta G^{\circ}_{rxn} = -218.1$ kJ

19.72 ΔG° at 298 from K_p is 4.87 kJ

ΔG° at 298 from free energies of formation is 4.73 kJ

19.74 $\Delta G^{\circ}_{rxn} = -2.3$ kJ

19.76 Iodine dissolves readily so the process is favorable and ΔG° must be less than zero. Because $\Delta H^{\circ} = 0$, the process is entropy-driven.

19.78 (a) $\Delta G^{\circ}_{rxn} = +141.82$ kJ, so the reaction is reactant-favored

(b) $\Delta H^{\circ}_{rxn} = +197.86$ kJ and $\Delta S^{\circ}_{rxn} = +187.95$ J/K

$T = \Delta H^{\circ}_{rxn}/\Delta S^{\circ}_{rxn} = 1052.7$ K or 779.6 °C

(c) ΔG°_{rxn} at 1500 °C (1773 K) $= -135.4$ kJ

K_p at 1500 °C $= 1 \times 10^4$

19.80 $\Delta S^{\circ}_{rxn} = -459.0$ J/K; $\Delta H^{\circ}_{rxn} = -793$ kJ; $\Delta G^{\circ}_{rxn} = -657$ kJ

The reaction is product-favored and enthalpy-driven.

19.82 $\Delta H^{\circ}_{rxn} = +62.2$ kJ and $\Delta S^{\circ}_{rxn} = +132.7$ J/K

$T = \Delta H^{\circ}_{rxn}/\Delta S^{\circ}_{rxn} = 469$ K or 196 °C

ΔG°_{rxn} at 25 °C $= +22.6$ kJ

The reaction is reactant-favored at 298 K, but it becomes product-favored at 469 K ($= \Delta H^{\circ}/\Delta S^{\circ}$).

19.84 (a) ΔG° at 80.0 °C $= +0.14$ kJ

ΔG° at 110.0 °C $= -0.12$ kJ

Rhombic sulfur is more stable than monoclinic sulfur at 80 °C, but the reverse is true at 110 °C.

(b) $T = 370$ K or about 96 °C

19.86 Abba's refrigerator has liquid water on the cloth covering the pots and moistened sand in between the pots. The evaporation process is spontaneous at "room temperature," so $\Delta G < 0$. But we also know that $\Delta H^{\circ} > 0$ (evaporation absorbs heat), and $\Delta S^{\circ} > 0$ (evaporation increases disorder). This means at "room temperature" the process is entropy driven. Eventually, however, as the system approaches equilibrium ($\Delta G = 0$), the temperature must drop in order that $\Delta H^{\circ} = T\Delta S^{\circ}$.

19.88 (a) $\Delta S^{\circ}_{rxn} = +24.89$ J/K; $\Delta H^{\circ}_{rxn} = +180.58$ kJ; $\Delta G^{\circ}_{rxn} = +173.16$ kJ

$K_p = 4.62 \times 10^{-31}$, the reaction is reactant-favored at 298 K

(b) at 700 °C (973 K), $\Delta G^{\circ}_{rxn} = +156.6$ kJ

$K_p = 4 \times 10^{-9}$, the reaction is still reactant-favored at 700 °C, but less so than at 298 K.

(c) $P(NO) = 6 \times 10^{-5}$; $P(O_2) = P(N_2) = 1$ bar

19.90 (a) $\Delta S^{\circ}_{rxn} = -132.7$ J/K; $\Delta H^{\circ}_{rxn} = -62.2$ kJ; $\Delta G^{\circ}_{rxn} = -22.6$ kJ

(See Question 19.82, which is the opposite reaction.)

(b) From ΔG°_{rxn} we find that $\ln K_p = 9.11$.

$K_p = 1/P(O_2)$, so $P(O_2) = 1.1 \times 10^{-4}$ bar

(c) When $K_p = 1/P(O_2) = 1$, $\Delta G^{\circ} = 0$, and

$$T = \Delta H^{\circ}_{rxn}/\Delta S^{\circ}_{rxn} = 469 \text{ K.}$$

19.92 Dissolving a solid such as NaCl in water is a spontaneous process. Thus, $\Delta G^{\circ} < 0$. If ΔH° is zero, then the only way the free energy change can be negative is if $\Delta S°$ is positive. Generally the entropy change is the driving force in forming a solution.

CHAPTER 20

20.2 (a) The electrode at the right is a magnesium anode; magnesium metal supplies electrons and is oxidized to Mg^{2+} ions. Electrons pass through the wire to the silver cathode where Ag^+ ions are reduced to silver. Nitrate ions in the salt bridge move from the $AgNO_3$ solution to the $Mg(NO_3)_2$ solution (and Na^+ ions move in the opposite direction).

    (b) Anode: $Mg(s) \longrightarrow Mg^{2+}(aq) + 2\ e^-$

           Cathode: $Ag^+(aq) + e^- \longrightarrow Ag(s)$

           Net: reaction: $Mg(s) + 2\ Ag^+(aq) \longrightarrow$
$$Mg^{2+}(aq) + 2\ Ag(s)$$

20.3 See part (a) of Question 2.

20.4 Anode: lead(IV) oxide is reduced to lead metal
$$PbO_2(s) + 4\ H^+(aq) + SO_4^{2-}(aq) + 2\ e^- \longrightarrow$$
$$PbSO_4(s) + 2\ H_2O(\ell)$$

     Cathode: lead is oxidized to lead(II) ions
$$Pb(s) + SO_4^{2-}(aq) \longrightarrow PbSO_4(s) + 2\ e^-$$

     The battery can be recharged because the reaction products are attached to the electrode surface. See page 844.

20.7 (a) Ease of oxidation (equivalent to declining ability to act as a reducing agent): $K > Zn > H_2 > Cu > Cl^-$

    (b) Ease of reduction: $Ag^+ > I_2 > H^+ > H_2O > Na^+$

20.9 A product-favored reaction has $E°$ positive and $\Delta G°$ negative.

20.11 Current (A) = Charge (C)/time (s). Each mole of electrons carries a charge of 96, 485 C. To produce 1.00 mol of Ag requires 1.00 mol of electrons or 96,500 C. If the current is 1.00 A, the time is 96,500 s.

20.12 (a) $Cr(s) \longrightarrow Cr^{3+}(aq) + 3\ e^-$

     Cr is a reducing agent; this is an oxidation reaction.

    (b) $AsH_3(g) + 3\ H_2O(\ell) \longrightarrow$
$$As(s) + 3\ H_3O^+(aq) + 3\ e^-$$

     AsH_3 is a reducing agent; this is an oxidation reaction.

    (c) $VO_3^-(aq) + 6\ H_3O^+(aq) + 3\ e^- \longrightarrow$
$$V^{2+}(aq) + 9\ H_2O(\ell)$$

     $VO_3^-(aq)$ is an oxidizing agent; this is a reduction reaction.

    (d) Silver is a reducing agent; this is an oxidation reaction.
$$2\ Ag(s) + 2\ OH^-(aq) \longrightarrow Ag_2O(s) + H_2O(\ell) + 2e^-$$

20.14 (a) $Ag(s) \longrightarrow Ag^+(aq) + e^-$
$$e^- + NO_3^-(aq) + 2\ H^+(aq) \longrightarrow NO_2(g) + H_2O(\ell)$$

$\overline{Ag(s) + NO_3^-(aq) + 2\ H^+(aq) \longrightarrow}$
$$Ag^+(aq) + NO_2(g) + H_2O(\ell)$$

    (b) $2[MnO_4^-(aq) + 8\ H^+(aq) + 5\ e^- \longrightarrow$
$$Mn^{2+}(aq) + 4\ H_2O(\ell)]$$
$5[HSO_3^-(aq) + H_2O(\ell) \longrightarrow SO_4^{2-}(aq) + 3\ H^+(aq) + 2\ e^-]$

$\overline{2\ MnO_4^-(aq) + H^+(aq) + 5\ HSO_3^-(aq) \longrightarrow}$
$$2\ Mn^{2+}(aq) + 3\ H_2O(\ell) + 5\ SO_4^{2-}(aq)$$

    (c) $4[Zn(s) \longrightarrow Zn^{2+}(aq) + 2\ e^-]$
$2\ NO_3^-(aq) + 10\ H^+(aq) + 8\ e^- \longrightarrow N_2O(g) + 5\ H_2O(\ell)$

$\overline{4\ Zn(s) + 2\ NO_3^-(aq) + 10\ H^+(aq) \longrightarrow}$
$$4\ Zn^{2+}(aq) + N_2O(g) + 5\ H_2O(\ell)$$

    (d) $Cr(s) \longrightarrow Cr^{3+}(aq) + 3\ e^-$
$3\ e^- + NO_3^-(aq) + 4\ H^+(aq) \longrightarrow NO(g) + 2\ H_2O(\ell)$

$\overline{Cr(s) + NO_3^-(aq) + 4\ H^+(aq) \longrightarrow}$
$$Cr^{3+}(aq) + NO(g) + 2\ H_2O(\ell)$$

20.16 (a) $2[Al(s) + 4\ OH^-(aq) \longrightarrow Al(OH)_4^-(aq) + 3\ e^-]$
$3[2\ H_2O(\ell) + 2\ e^- \longrightarrow H_2(g) + 2\ OH^-(aq)]$

$\overline{2\ Al(s) + 2\ OH^-(aq) + 6\ H_2O(\ell) \longrightarrow}$
$$2\ Al(OH)_4^-(aq) + 3\ H_2(g)$$

    (b) $2[CrO_4^{2-}(aq) + 4\ H_2O(\ell) + 3\ e^- \longrightarrow$
$$Cr(OH)_3(s) + 5\ OH^-(aq)]$$
$3[SO_3^{2-}(aq) + 2\ OH^-(aq) \longrightarrow$
$$SO_4^{2-}(aq) + H_2O(\ell) + 2\ e^-]$$

$\overline{2\ CrO_4^{2-}(aq) + 3\ SO_3^{2-}(aq) + 5\ H_2O(\ell) \longrightarrow}$
$$2\ Cr(OH)_3(s) + 3\ SO_4^{2-}(aq) + 4\ OH^-(aq)$$

    (c) $Zn(s) + 4\ OH^-(aq) \longrightarrow Zn(OH)_4^{2-}(aq) + 2\ e^-$
$Cu(OH)_2(s) + 2\ e^- \longrightarrow Cu(s) + 2\ OH^-(aq)$

$\overline{Zn(s) + 2\ OH^-(aq) + Cu(OH)_2(s) \longrightarrow}$
$$Zn(OH)_4^{2-}(aq) + Cu(s)$$

    (d) $3[HS^-(aq) + OH^-(aq) \longrightarrow S(s) + H_2O(\ell) + 2\ e^-]$
$ClO_3^-(aq) + 3\ H_2O(\ell) + 6\ e^- \longrightarrow Cl^-(aq) + 6\ OH^-(aq)$

$\overline{3\ HS^-(aq) + ClO_3^-(aq) \longrightarrow}$
$$3\ S(s) + Cl^-(aq) + 3\ OH^-(aq)$$

20.18 Electrons flow from the *Cr* electrode *Fe* electrode. Negative ions move in the salt bridge from the Fe/Fe^{2+} half-cell to the Cr/Cr^{3+} half-cell.

     Anode (oxidation): $Cr(s) \longrightarrow Cr^{3+}(aq) + 3\ e^-$

     Cathode (reduction): $Fe^{2+}(aq) + 2\ e^- \longrightarrow Fe(s)$

20.20 (a) Oxidation: $Fe(s) \longrightarrow Fe^{2+}(aq) + 2\ e^-$

         Reduction: $O_2(g) + 4\ H^+(aq) + 4\ e^- \longrightarrow 2\ H_2O(\ell)$

         Overall: $2\ Fe(s) + O_2(g) + 4\ H^+(aq) \longrightarrow$
$$2\ Fe^{2+}(aq) + 2\ H_2O(\ell)$$

         Anode, oxidation: $Fe(s) \longrightarrow Fe^{2+}(aq) + 2\ e^-$

         Cathode, reduction: $O_2(g) + 4\ H^+(aq) + 4\ e^- \longrightarrow$
$$2\ H_2O(\ell)$$

    (c) Electrons flow from the negative anode (Fe) to the positive cathode (site of the O_2 half-reaction). Negative ions move from the cathode compartment in which the O_2 reduction occurs to the anode compartment in which Fe oxidation occurs.

20.22 See pages 840–843.

    (a) All are primary batteries, not rechargeable.

    (b) All have Zn as the anode.

(c) Dry cells have an acidic environment, whereas the environment is alkaline for alkaline and mercury cells.

20.24 (a) $E°_{cell} = -1.298$ V, not product-favored

(b) $E°_{cell} = -0.51$ V, not product-favored

(c) $E°_{cell} = -1.023$ V, not product-favored

(d) $E°_{cell} = +0.029$ V, product-favored

20.26 (a) $Sn^{2+}(aq) + 2 Ag(s) \longrightarrow Sn(s) + 2 Ag^+(aq)$

$E°_{cell} = -0.94$ V, not product-favored

(b) $3 Sn^{4+}(aq) + 2 Al(s) \longrightarrow 3 Sn^{2+}(aq) + 2 Al^{3+}(aq)$

$E°_{cell} = +1.81$ V, product-favored

(c) $ClO_3^-(aq) + 6 Ce^{3+}(aq) + 6 H^+(aq) \longrightarrow Cl^-(aq) + 6 Ce^{4+}(aq) + 3 H_2O(\ell)$

$E°_{cell} = -0.99$ V, not product-favored

(d) $3 Cu(s) + 2 NO_3^-(aq) + 8 H^+(aq) \longrightarrow 3 Cu^{2+}(aq) + 2 NO(g) + 4 H_2O(\ell)$

$E°_{cell} = +0.62$ V, product-favored

20.28 (a) Al

(b) Zn and Al

(c) $Fe^{2+}(aq) + Sn(s) \longrightarrow Fe(s) + Sn^{2+}(aq)$, reactant-favored

(d) $Zn^{2+}(aq) + Sn(s) \longrightarrow Zn(s) + Sn^{2+}(aq)$, not product-favored

20.30 Best reducing agent, Zn(s)

20.32 Ag^+

20.34 See Example 20.5, page 855.

(a) F_2, most readily reduced

(b) F_2 and Cl_2

20.36 $E°_{cell} = +0.3923$ V. When $[Zn(OH)_4^{2-}] = [OH^-] = 0.025$ M and $P(H_2) = 1.0$ atm, $E_{cell} = 0.345$ V

20.38 $E°_{cell} = +1.563$ V and $E_{cell} = +1.58$ V

20.40 When $E°_{cell} = +1.563$ V, $E_{cell} = 1.48$ V, $n = 2$, and $[Zn^{2+}] = 1.0$ M, the concentration of $Ag^+ = 0.040$ M

20.42 (a) $\Delta G° = -45.5$ kJ; $K = 9 \times 10^7$

(b) $\Delta G° = +110$ kJ; $K = 4 \times 10^{-19}$

20.44 See page 864.

$E°_{cell}$ for $AgBr(s) \rightleftharpoons Ag^+(aq) + Br^-(aq)$ is -0.7281.

$K_{sp} = 4.9 \times 10^{-13}$

20.46 $K_{formation} = 2 \times 10^{25}$

20.48 (a) $Fe^{2+}(aq) + 2 e^- \longrightarrow Fe(s)$

$\dfrac{2[Fe^{2+}(aq) \longrightarrow Fe^{3+}(aq) + e^-]}{3 Fe^{2+}(aq) \longrightarrow Fe(s) + 2 Fe^{3+}(aq)}$

(b) $E°_{cell} = -1.21$ V; reaction is not product-favored

(c) $K = 1 \times 10^{-41}$

20.50 See Figure 20.14.

20.52 O_2 from the oxidation of water is more likely than F_2. See Example 20.11.

20.54 See Example 20.11.

(a) Cathode: $2 H_2O(\ell) + 2 e^- \longrightarrow H_2(g) + 2 OH^-(aq)$

(b) Anode: $2 Br^-(aq) \longrightarrow Br_2(\ell) + 2 e^-$

20.56 Mass of Ni = 0.0334 g

20.58 Time = 2300 s or 38 min

20.60 Time = 250 h

20.62 (a) $UO_2^+(aq) + 4 H^+(aq) + e^- \longrightarrow U^{4+}(aq) + 2 H_2O(\ell)$

(b) $ClO_3^-(aq) + 6 H^+(aq) + 6 e^- \longrightarrow Cl^-(aq) + 3 H_2O(\ell)$

(c) $N_2H_4(aq) + 4 OH^-(aq) \longrightarrow N_2(g) + 4 H_2O(\ell) + 4 e^-$

(d) $OCl^-(aq) + H_2O(\ell) + 2 e^- \longrightarrow Cl^-(aq) + 2 OH^-(aq)$

20.64 (a) For 1.1 V could use as the anode $Al^{3+}(aq) + 3 e^- \longrightarrow Al(s)$, $E° = -1.66$ V

(b) For 0.5 V could use as the cathode $Ni^{2+}(aq) + 2 e^- \longrightarrow Ni(s)$, $E° = -0.25$ V

20.66 (a) $Zn^{2+}(aq)$; (b) $Au^+(aq)$; (c) Zn(s); (d) Au(s)

(e) Yes, Sn(s) will reduce Cu^{2+} (as well as Ag^+ and Au^+)

(f) No, Ag(s) can only reduce $Au^+(aq)$.

(g) See part e

(h) $Ag^+(aq)$ can oxidize Cu, Sn, Co, and Zn.

20.68 (a) The cathode is the site of reduction, so the half-reaction must be $2 H^+(aq) + 2 e^- \longrightarrow H_2(g)$. This is the case with the following half-reactions: $Cr^{3+}(aq)|Cr(s)$, $Fe^{2+}(aq)|Fe(s)$, and $Mg^{2+}(aq)|Mg(s)$.

(b) Choosing from the half-cells above, the reaction of Mg(s) and $H^+(aq)$ would produce the most positive potential (2.37 V), and the reaction with Cr(s) would produce the least positive potential (+0.41 V).

20.70 (a) Cell diagram

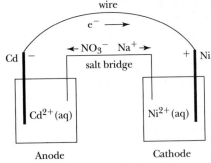

(b) Anode: $Cd(s) \longrightarrow Cd^{2+}(aq) + 2 e^-$

Cathode: $Ni^{2+}(aq) + 2 e^- \longrightarrow Ni(s)$

Net: $Cd(s) + Ni^{2+}(aq) \longrightarrow Cd^{2+}(aq) + Ni(s)$

(c) The anode is negative and the cathode is positive.

(d) $E°_{cell} = E°_{cathode} - E°_{anode} = (-0.25$ V$) - (-0.40$ V$) = +0.15$ V

(e) Electrons flow from anode (Cd) to cathode (Ni).

(f) Na^+ ions move from the cathode compartment to the anode compartment. Anions move in the opposite direction.

(g) $K = 1 \times 10^5$

(h) $E_{cell} = 0.21$ V

(i) 480 h

20.72 (a) $E^°_{cathode} = +0.65$ V

(b) $K_{sp} = 1.2 \times 10^{-5}$

20.74 $\Delta G^° = -409$ kJ

20.76 1.5×10^4 kwh

20.78 Rhodium is present in solution as Rh^{3+}

20.80 43 g Zn

20.82 0.10 A

20.84 (a) $HCO_2H(aq) + 2 H^+(aq) + 2 e^- \longrightarrow$
$\qquad\qquad HCHO(aq) + H_2O(\ell)$

(b) $C_6H_5CO_2H(aq) + 6 H^+(aq) + 6 e^- \longrightarrow$
$\qquad\qquad C_6H_5CH_3(\ell) + 2 H_2O(\ell)$

(c) $CH_3CH_2CHO(aq) + 2 H^+(aq) + 2 e^- \longrightarrow$
$\qquad\qquad CH_3CH_2CH_2OH(aq)$

(d) $CH_3OH(aq) + 2 H^+(aq) + 2 e^- \longrightarrow$
$\qquad\qquad CH_4(g) + H_2O(\ell)$

20.86 (a) $E^°_{cell} = E^°_{cathode} - E^°_{anode} = (+0.80$ V$) - (+0.771$ V$)$
$\qquad = +0.029$ V

(b) $Fe^{2+}(aq) + Ag^+(aq) \longrightarrow Ag(s) + Fe^{3+}(aq)$

(c) Anode: $Fe^{2+}(aq) \longrightarrow Fe^{3+}(aq) + e^-$
Cathode: $Ag^+(aq) + e^- \longrightarrow Ag(s)$

(d) $E_{cell} = -0.030$ V. The negative sign indicates the opposite reaction occurs under these nonstandard conditions. The equation for the reaction now occurring is:
$Ag(s) + Fe^{3+}(aq) \longrightarrow Fe^{2+}(aq) + Ag^+(aq)$

20.88 At pH = 7, $[OH^-] = 1 \times 10^{-7}$ M. Under these conditions $E_{cell} = -0.42$ V
At pH = 1, the pOH = 13, and $[OH^-] = 1 \times 10^{-13}$ M. Under these conditions, $E_{cell} = -0.06$ V. It is much more advantageous to reduce water in an acidic solution.

20.90 Conclusions: (a) A and C are stronger reducing agents than H_2; B, and D are weaker. (b) C is the strongest reducing agent. (c) D is a stronger reducing agent than B. Ordered relative to strength as a reducing agent: B < D < A < C.

20.92 (a) 92 g HF

(b) 230 g CF_3SO_2F

(c) 9.3 g H_2

20.94 290 h

CHAPTER 21

21.18 Increasing basicity: $CO_2 < SiO_2 < SnO_2$

21.20 (a) $2 Na(s) + Br_2(\ell) \longrightarrow 2 NaBr(s)$

(b) $2 Mg(s) + O_2(g) \longrightarrow 2 MgO(s)$

(c) $2 Al(s) + 3 F_2(g) \longrightarrow 2 AlF_3(s)$

(d) $C(s) + O_2(g) \longrightarrow CO_2(g)$

21.22 $2 H_2(g) + O_2(g) \longrightarrow 2 H_2O(g)$
$H_2(g) + Cl_2(g) \longrightarrow 2 HCl(g)$
$3 H_2(g) + N_2(g) \longrightarrow 2 NH_3(g)$

21.24 *Step 1:* $2 SO_2(g) + 4 H_2O(\ell) c + 2 I_2(s) \longrightarrow$
$\qquad\qquad\qquad 2 H_2SO_4(\ell) + 4 HI(g)$

Step 2: $2 H_2SO_4(\ell) \longrightarrow 2 H_2O(\ell) + 2 SO_2(g) + O_2(g)$

Step 3: $4 HI(g) \longrightarrow 2 H_2(g) + 2 I_2(g)$

Net: $2 H_2O(\ell) \longrightarrow 2 H_2(g) + O_2(g)$

21.26 $\Delta H^° = +131.31$ kJ; $\Delta S^° = +133.9$ J/k; and $\Delta G^° = +91.42$ kJ

21.28 $2 Na(s) + F_2(g) \longrightarrow 2 NaF(s)$
$2 Na(s) + Cl_2(g) \longrightarrow 2 NaCl(s)$
$2 Na(s) + Br_2(\ell) \longrightarrow 2 NaBr(s)$
$2 Na(s) + I_2(s) \longrightarrow 2 NaI(s)$
The alkali metal halides are white, crystalline solids. They have high melting and boiling points, and are soluble in water.

21.30 $4 Li(s) + O_2(g) \longrightarrow 2 Li_2O(s)$
$\qquad\qquad$ lithium oxide
$2 Na(s) + O_2(g) \longrightarrow Na_2O_2(s)$
$\qquad\qquad$ sodium peroxide
$K(s) + O_2(g) \longrightarrow KO_2(s)$
$\qquad\qquad$ potassium superoxide

21.32 $2 Cl^-(aq) + 2 H_2O(\ell) \longrightarrow$
$\qquad\qquad Cl_2(g) + H_2(g) + 2 OH^-(aq)$

If this were the only process used to produce chlorine, the mass of Cl_2 reported for industrial production would be 0.88 times the mass of NaOH produced. (2 mol NaCl, 117 g, would yield 2 mol NaOH, 80 g, and 1 mol of Cl_2, 70 g). The amounts quoted indicate a Cl_2 to NaOH mass ratio of 0.96. So, chlorine is presumably also prepared by other routes than this.

21.34 $2 Mg(s) + O_2(g) \longrightarrow 2 MgO(s)$
$3 Mg(s) + N_2(g) \longrightarrow Mg_3N_2(s)$

21.36 $CaCO_3$ is used in agriculture to neutralize acidic soil, to prepare CaO for use in mortar, and in steel production.
$CaCO_3(s) + H_2O(\ell) + CO_2(g) \longrightarrow$
$\qquad\qquad Ca^{2+}(aq) + 2 HCO_3^-(aq)$

21.38 1.4×10^6 g

21.40 $2 Al(s) + 6 HCl(aq) \longrightarrow$
$\qquad\qquad 2 Al^{3+}(aq) + 6 Cl^-(aq) + 3 H_2(g)$
$2 Al(s) + 3 Cl_2(g) \longrightarrow 2 AlCl_3(s)$
$4 Al(s) + 3 O_2(g) \longrightarrow 2 Al_2O_3(s)$

21.42 $2 Al(s) + 2 OH^-(aq) + 6 H_2O(\ell) \longrightarrow$
$\qquad\qquad 2 Al(OH)_4^-(aq) + 3 H_2(g)$
Volume of H_2 obtained = 18.4 L

21.44 $Al_2O_3(s) + 3 H_2SO_4(aq) \longrightarrow Al_2(SO_4)_3(s) + 3 H_2O(\ell)$
Mass of H_2SO_4 required = 0.860 kg and mass of Al_2O_3 required = 0.298 g

21.46

$$\left[\begin{array}{c} :\ddot{\text{Cl}}: \\ | \\ :\ddot{\text{Cl}}-\text{Al}-\ddot{\text{Cl}}: \\ | \\ :\ddot{\text{Cl}}: \end{array}\right]^{-}$$

The ion has tetrahedral geometry. Aluminum is sp^3-hybridized.

21.48 SiO_2 is a network solid, with tetrahedral silicon atoms covalently bonded to four oxygens in an infinite array; CO_2 consists of individual molecules, with oxygen atoms double bonded to carbon. Melting SiO_2 requires breaking very stable Si—O bonds. Weak intermolecular forces of attraction between CO_2 molecules result in this substance being a gas at ambient conditions.

21.50 (a) $2\ CH_3Cl(g) + Si(s) \longrightarrow (CH_3)_2SiCl_2(\ell)$

(b) 0.823 atm

(c) 12.2 g

21.52 Consider the general decomposition reaction:

$$N_xO_y \longrightarrow {}^x\!/_2\ N_2 + {}^y\!/_2\ O_2$$

The value of $\Delta G°$ can be obtained for all N_xO_y molecules because $\Delta G°_{rxn} = -\Delta G°_f$. These data show that the decomposition reaction is product-favored for all of the nitrogen oxides. All are unstable with respect to decomposition to the elements.

Compound	$-\Delta G°_f$ (kJ/mol)
NO(g)	-86.58
NO_2	-51.23
N_2O	-104.20
N_2O_4	-97.73

21.54 $\Delta H°_{rxn} = -114.4$ kJ; exothermic

21.56 (a) $N_2H_4(aq) + O_2(g) \longrightarrow N_2(g) + 2\ H_2O(\ell)$

(b) 1.32×10^3 g

21.58 $5\ N_2H_5{}^+(aq) + 4\ IO_3{}^-(aq) \longrightarrow$
$5\ N_2(g) + 2\ I_2(aq) + H^+(aq) + 12\ H_2O(\ell)$

$E°_{net} = 1.43$ V

21.60

$$\left[\ddot{\text{N}}=\text{N}=\ddot{\text{N}}\right]^{-} \longleftrightarrow \left[:\ddot{\text{N}}-\text{N}\equiv\text{N}:\right]^{-} \longleftrightarrow \left[:\text{N}\equiv\text{N}-\ddot{\text{N}}:\right]^{-}$$

21.62 (a) 3.5×10^6 g; (b) 4.1×10^6 g

21.64 The Lewis structure of $S_2{}^{2-}$ is

$$\left[:\ddot{\text{S}}-\ddot{\text{S}}:\right]^{2-}$$

21.66 $E°_{cell} = E°_{cathode} - E°_{anode} = +1.44\ V - (+1.52\ V) = -0.08$ V. The reaction is not product-favored under standard conditions.

21.68 $Cl_2(aq) + 2\ Br^-(aq) \longrightarrow 2\ Cl^-(aq) + Br_2(aq)$

Cl_2 is the oxidizing agent, Br^- is the reducing agent; $E°_{cell} = 0.28$ V

21.70

Element	Appearance	State
Na, Mg, Al	silvery metals	solids
Si	black, shiny metalloid	solid
P	white, red, and black allotropes; nonmetal	solid
S	yellow nonmetal	solid
Cl	pale green, nonmetal	gas
Ar	colorless, nonmetal	gas

21.72 (a) $2\ Na(s) + Cl_2(g) \longrightarrow 2\ NaCl(s)$

$Mg(s) + Cl_2(g) \longrightarrow MgCl_2(s)$

$2\ Al(s) + 3\ Cl_2(g) \longrightarrow 2\ AlCl_3(s)$

$Si(s) + 2\ Cl_2(g) \longrightarrow SiCl_4(\ell)$

$P_4(s) + 10\ Cl_2(g) \longrightarrow 4\ PCl_5(s)$ (excess Cl_2)

$S_8(s) + 8\ Cl_2(g) \longrightarrow 8\ SCl_2(s)$

(b) NaCl and $MgCl_2$ are ionic, the other products are covalent.

(c) $SiCl_4$ is tetrahedral, PCl_5 is trigonal-bipyramidal.

21.74 *Cathode:* $Li^+(\ell) + e^- \longrightarrow Li(\ell)$

Anode: $2\ H^-(\ell) \longrightarrow H_2(g) + 2\ e^-$

Formation of H_2 at the anode is evidence for the presence of H^-.

21.76 Mg: $\Delta G°_{rxn} = +64.9$ kJ

Ca: $\Delta G°_{rxn} = +131.4$ kJ

Ba: $\Delta G°_{rxn} = +219.7$ kJ

Relative tendency to decompose: $MgCO_3 > CaCO_3 > BaCO_3$

21.78 (a) $\Delta G°_f$ should be more negative than $(-95.1\ kJ) \times n$

(b) Ba, Pb, and Ti

21.80 (a) We rationalize the equivalence of the two N—O bond lengths by writing two resonance structures for this molecule.

(b) The central atom, nitrogen, has three sets of bonding electrons in its valence shell; VSEPR predicts that this atom is trigonal-planar. Oxygen, in the —OH group, has four electron pairs in its valence shell, arranged tetrahedrally. Two are bonding pairs, defining the bent molecular geometry.

(c) sp^2. There is an empty p orbital on N that is perpendicular to the plane of the molecule; this can overlap with p orbitals on the two terminal oxygens to form a delocalized π bond.

21.82 A through E, in order: $BaCO_3$; BaO; $CaCO_3$; $BaCl_2$; $BaSO_4$

21.84 $\Delta H^\circ_{rxn} = -257.78$ kJ. This reaction is entropy-disfavored, however, with $\Delta S^\circ_{rxn} = -963$ J/K because of the decrease in the number of moles of gases. Combining these values gives $\Delta G^\circ_{rxn} = +29.19$ kJ, indicating that under standard conditions at 298 K the reaction is not favorable. (The reaction has a favorable ΔG°_{rxn} below 268 K, indicating that further research on this system might be worthwhile. Note, however, that at that temperature water is a solid.)

21.86 (a) 820 L seawater; 2.31 kg CaO

(b) 310 kg Mg, 890 kg Cl_2, 2.5×10^4 Faradays

(c) Energy required $= 1.6 \times 10^6$ J/mol. From thermodynamics, we find $\Delta H^\circ = +641.62$ kJ.

21.88 (a) Mass produced in one hour = 21 kg

(b) Power used = 8.2 kwh

Chapter 22

22.16 (a) Cr^{3+}: $[Ar]3d^3$, paramagnetic

(b) V^{2+}: $[Ar]3d^3$, paramagnetic

(c) Ni^{2+}: $[Ar]3d^8$, paramagnetic

(d) Cu^+: $[Ar]3d^{10}$, diamagnetic

22.18 (a) Fe^{3+} ($[Ar]3d^5$) isoelectronic with Mn^{2+}

(b) Zn^{2+} ($[Ar]3d^{10}$) isoelectronic with Cu^+

(c) Fe^{2+} ($[Ar]3d^6$) isoelectronic with Co^{3+}

(d) Cr^{3+} ($[Ar]3d^3$) isoelectronic with V^{2+}

22.20 (a) $Cr_2O_3(s) + 2\,Al(s) \longrightarrow Al_2O_3(s) + 2\,Cr(s)$

(b) $TiCl_4(\ell) + 2\,Mg(s) \longrightarrow Ti(s) + 2\,MgCl_2(s)$

(c) $2\,[Ag(CN)_2]^-(aq) + Zn(s) \longrightarrow$
$2\,Ag(s) + [Zn(CN)_4]^{2-}(aq)$

(d) $3\,Mn_3O_4(s) + 8\,Al(s) \longrightarrow 4\,Al_2O_3(s) + 9\,Mn(s)$

22.22 Monodentate: CH_3NH_2, CH_3CN, N_3^-, Br^-

Bidentate: en, phen (see Figure 22.13)

22.24 (a) Mn^{2+} (c) Co^{3+}

(b) Co^{3+} (d) Cr^{2+}

22.26 $[Ni(en)(NH_3)_3(H_2O)]^{2+}$

22.28 (a) $Ni(en)_2Cl_2$ (en = $H_2NCH_2CH_2NH_2$)

(b) $K_2[PtCl_4]$

(c) $K[Cu(CN)_2]$

(d) $[Fe(NH_3)_4(H_2O)_2]^{2+}$

22.30 (a) Diaquabis(oxalato)nickelate(II) ion

(b) Dibromobis(ethylenediamine)cobalt(III) ion

(c) Amminechlorobis(ethylenediamine)cobalt(III) ion

(d) Diammineoxalatoplatinum(II)

22.32 (a) $[Fe(H_2O)_5OH]^{2+}$

(b) Potassium tetracyanonickelate(II)

(c) Potassium diaquabis(oxalato)chromate(III)

(d) $(NH_4)_2[PtCl_4]$

22.34 (a)

cis   *trans*

(b)

cis   *trans*

(c)

fac   *mer*

(d)

Only one structure possible. (N—N is the bidentate ethylenediamine ligand.)

22.36 For a discussion of chirality, see Chapter 11, pages 423–425).

(a) Fe^{2+} is a chiral center

(b) Co^{3+} is not a chiral center

(c) Co^{3+} is a chiral center

(d) No. Square planar complexes are never chiral.

22.38 (a) $[Mn(CN)_6]^{4-}$: d^5, low spin complex is paramagnetic.

(b) $[Co(NH_3)_6]^{3+}$: d^6, low spin complex is diamagnetic.

(c) $[Fe(H_2O)_6]^{3+}$: d^5, low spin complex is paramagnetic (1 unpaired electron).

(d) $[Cr(en)_3]^{3+}$: d^3, complex is paramagnetic (3 unpaired electrons).

(b) d^6 low spin configuration (diamagnetic)

$$\underline{\uparrow\downarrow} \quad \underline{\uparrow\downarrow} \quad \underline{\uparrow\downarrow}$$

(a and c) d^5 low spin configuration (paramagnetic)

$$\underline{\uparrow\downarrow} \quad \underline{\uparrow\downarrow} \quad \underline{\uparrow}$$

(d) d^3 configuration (paramagnetic)

$$\underline{\uparrow} \quad \underline{\uparrow} \quad \underline{\uparrow}$$

22.40 (a) Fe^{2+}, d^6, paramagnetic, 4 unpaired electrons

(b) Co^{2+}, d^7, paramagnetic, 3 unpaired electrons

(c) Mn^{2+}, d^5, paramagnetic, 5 unpaired electrons

(d) Zn^{2+}, d^{10}, diamagnetic, 0 unpaired electrons

22.42 (a) 6; (b) octahedral; (c) +2; (d) 4 unpaired electrons (high spin); (e) paramagnetic

22.44 When $Co_2(SO_4)_3$ dissolves in water it forms $[Co(H_2O)_6]^{3+}$; addition of fluoride converts this to

$[CoF_6]^{3-}$. The hexaaquo complex is low spin (diamagnetic, no unpaired electrons), and the fluoride complex is high spin (paramagnetic, 4 unpaired electrons.) Notice that fluoride is a weaker field ligand than water.

22.46 The light absorbed is in the blue region of the spectrum (page 905). Therefore, the light transmitted — which is the color of the solution — is yellow.

22.48 Determine the magnetic properties of the complex. Square planar Ni^{2+} (d^8) complexes are diamagnetic, whereas tetrahedral complexes are paramagnetic.

22.50 Fe^{2+} has a d^6 configuration. Low spin octahedral complexes are diamagnetic, whereas high spin complexes have 4 unpaired electrons and are paramagnetic.

22.52 Square planar complexes most often arise from d^8 transition metal ions. Therefore, it is likely that $[Ni(CN)_4]^{2-}$ (Ni^{2+}, d^8) is square planar.

22.54 Two geometric isomers are possible. See Question 22.34b.

22.56 Absorbing at 425 nm means the complex is absorbing light in the blue-violet end of the spectrum. Therefore, red and green light are transmitted, and the complex appears yellow (see Figure 22.26).

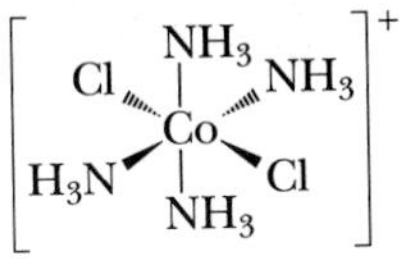

22.58 (a) Mn^{2+}; (b) 6; (c) octahedral; (d) 5; (e) paramagnetic; (f) *cis* and *trans* isomers exist.

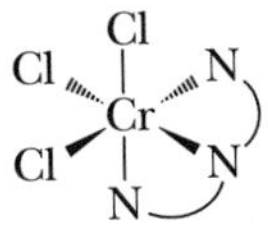

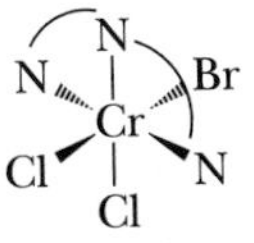

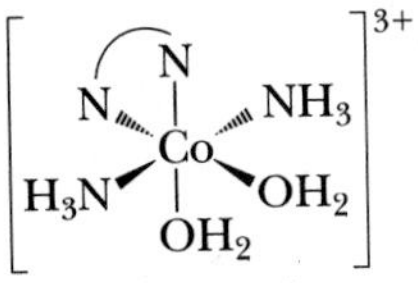

22.60 Name: tetraamminedichlorocobalt(III)

cis

trans

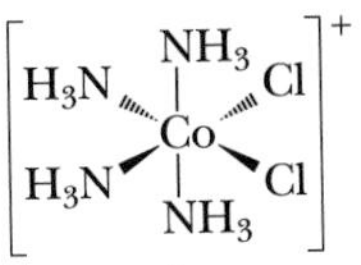

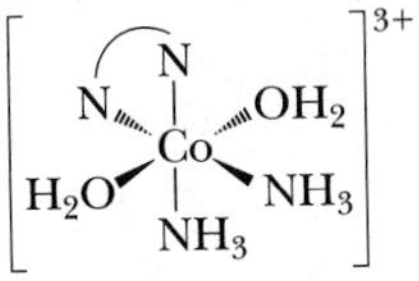

22.62 $[Co(en)_2(H_2O)Cl]^{2+}$

22.64. (a)

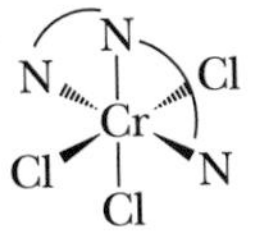

mer

fac

(b)

trans chlorides

cis chlorides

(c)

22.66

H_2O and NH_3 *cis*, chiral

H_2O *cis* and NH_3 *trans*, not chiral

H_2O *trans* and NH_3 *cis*, not chiral

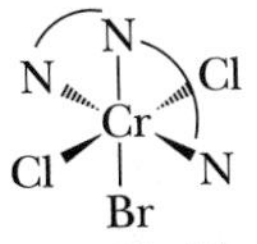

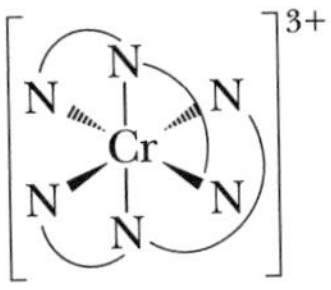

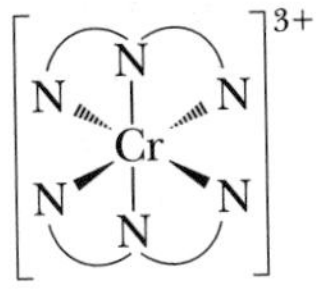

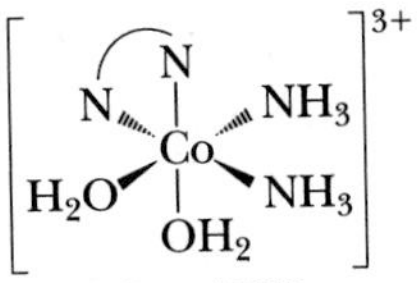

22.68 In $[Mn(H_2O)_6]^{2+}$ and $[Mn(CN)_6]^{4-}$, Mn has an oxidation number of +2 (Mn is a d^5 ion).

$[Mn(H_2O)_6Br]^{2+}$ paramagnetic, 5 unpaired e^-

$[Mn(CN)_6Br]^{4-}$ paramagnetic, 1 unpaired e^-

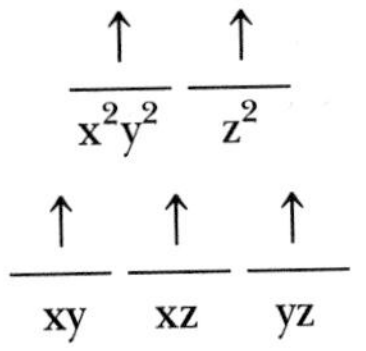

This shows that Δ_o for CN^- is greater than for H_2O.

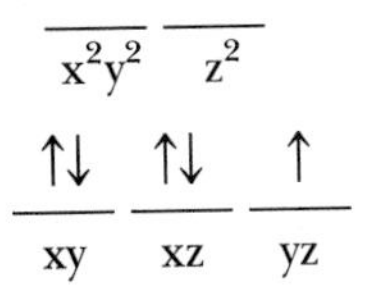

22.70 A, dark violet isomer: $[Co(NH_3)_5Br]SO_4$

B, violet-red isomer: $[Co(NH_3)_5(SO_4)]Br$

$$[Co(NH_3)_5Br]SO_4(aq) + BaCl_2(aq) \longrightarrow [Co(NH_3)_5Br]Cl_2(aq) + BaSO_4(s)$$

22.72

Ion	$K_{formation}$ (ammine complexes)
Co^{2+}	7.7×10^4
Ni^{2+}	5.6×10^8
Cu^{2+}	6.8×10^{12}
Zn^{2+}	2.9×10^9

22.74 N⌒O = $H_2N—CH_2—CO_2^-$

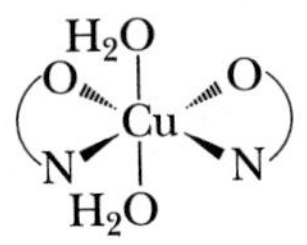

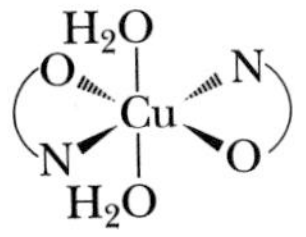

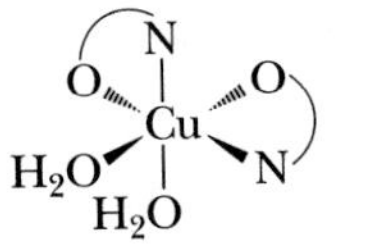

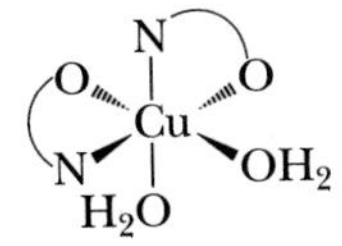

enantiomeric pair

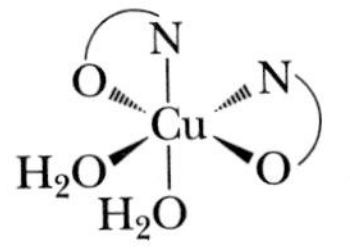

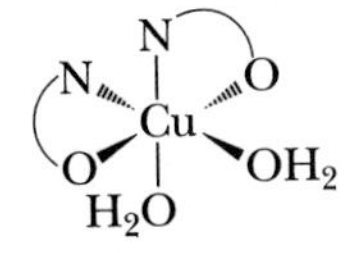

enantiomeric pair

enantiomeric pair

22.76 Substituting 10^8 and 10^{18} into the expression $(-RT\ln K)$ produces values of ΔG of -45.6 kJ (ammine) and -102.7 kJ (en). Because differences in ΔH values are much less than this (about 8 kJ), entropy must play a role. While there are fewer molecules in the second reaction, the change in entropy (as the much larger en ligands form the complex) is greater.

22.78 (a) $FeTiO_3(s) + 2\ HCl(aq) \longrightarrow$
$$FeCl_2(aq) + TiO_2(s) + H_2O(\ell)$$
    (b) Yes

    (c) 4.78×10^5 g Fe_2O_3

CHAPTER 23

23.12 (a) $^{56}_{28}Ni$; (b) $^{1}_{0}n$; (c) $^{32}_{15}P$; (d) $^{97}_{43}Tc$; (e) $^{0}_{-1}\beta$;
    (f) $^{0}_{1}e$ (positron)

23.14 (a) $^{0}_{-1}\beta$; (b) $^{87}_{37}Rb$; (c) $^{4}_{2}\alpha$; (d) $^{226}_{88}Ra$; (e) $^{0}_{-1}\beta$; (f) $^{24}_{11}Na$

23.16 $^{235}_{92}U \longrightarrow ^{231}_{90}Th + ^{4}_{2}\alpha$
$^{231}_{90}Th \longrightarrow ^{231}_{91}Pa + ^{0}_{-1}\beta$
$^{231}_{91}Pa \longrightarrow ^{227}_{89}Ac + ^{4}_{2}\alpha$
$^{227}_{89}Ac \longrightarrow ^{227}_{90}Th + ^{0}_{-1}\beta$
$^{227}_{90}Th \longrightarrow ^{223}_{88}Ra + ^{4}_{2}\alpha$
$^{223}_{88}Ra \longrightarrow ^{219}_{86}Rn + ^{4}_{2}\alpha$
$^{219}_{86}Rn \longrightarrow ^{215}_{84}Po + ^{4}_{2}\alpha$
$^{215}_{84}Po \longrightarrow ^{211}_{82}Pb + ^{4}_{2}\alpha$
$^{211}_{82}Pb \longrightarrow ^{211}_{83}Bi + ^{0}_{-1}\beta$
$^{211}_{83}Bi \longrightarrow ^{211}_{84}Po + ^{0}_{-1}\beta$
$^{211}_{84}Po \longrightarrow ^{207}_{82}Pb + ^{4}_{2}\alpha$

23.18 (a) $^{198}_{79}Au \longrightarrow ^{198}_{80}Hg + ^{0}_{-1}\beta$
    (b) $^{222}_{86}Rn \longrightarrow ^{218}_{84}Po + ^{4}_{2}\alpha$
    (c) $^{137}_{55}Cs \longrightarrow ^{137}_{56}Ba + ^{0}_{-1}\beta$
    (d) $^{110}_{49}In \longrightarrow ^{110}_{48}Cd + ^{0}_{1}e$

23.20 (a) $^{80}_{35}Br$ has a high neutron/proton ratio of $45/35$. Beta decay will allow the ratio to decrease: $^{80}_{35}Br \longrightarrow ^{80}_{36}Kr + ^{0}_{-1}\beta$

    (b) Alpha decay is likely: $^{240}_{98}Cf \longrightarrow ^{236}_{96}Cm + ^{4}_{2}\alpha$

    (c) Cobalt-61 has a high n/p ratio so beta decay is likely:
$$^{61}_{27}Co \longrightarrow ^{61}_{28}Kr + ^{0}_{-1}\beta$$

    (d) Carbon-11 has only 5 neutrons so K-capture or positron emission may occur:
$$^{11}_{6}C + ^{0}_{-1}e \longrightarrow ^{11}_{5}B$$
$$^{11}_{6}C \longrightarrow ^{11}_{5}B + ^{0}_{1}e$$

23.22 Generally beta decay will occur when the n/p ratio is high, whereas positron emission will occur when the n/p ratio is low.

    (a) Beta decay: $^{20}_{9}F \longrightarrow ^{20}_{10}Ne + ^{0}_{-1}\beta$
$$^{3}_{1}H \longrightarrow ^{3}_{2}He + ^{0}_{-1}\beta$$
    (b) Positron emission
$$^{22}_{11}Na \longrightarrow ^{22}_{10}Ne + ^{0}_{1}\beta$$

23.24 Binding energy per nucleon for $^{11}B = 6.70 \times 10^8$ kJ
Binding energy per nucleon for $^{10}B = 6.26 \times 10^8$ kJ

23.26 8.256×10^8 kJ/nucleon

23.28 7.700×10^8 kJ/nucleon

23.30 0.781 micrograms

23.32 (a) $^{131}_{53}I \longrightarrow ^{131}_{54}Xe + ^{0}_{-1}\beta$

    (b) 0.075 micrograms

23.34 9.5×10^{-4} mg

23.36 (a) $^{222}_{86}Rn \longrightarrow ^{218}_{84}Po + ^{4}_{2}\alpha$

    (b) Time $= 8.87$ d

23.38 About 2700 years old

23.40 (a) 15.8 y; (b) 88%

23.42 If $t_{1/2} = 14.28$ d, then $k = 4.854 \times 10^{-2}$ d^{-1}. If the original disintegration rate is 3.2×10^6 dpm, then (from the integrated first order rate equation), the rate after 365 d is 0.065 dpm. The plot will resemble Figure 23.5.

23.44 Plot $\ln$(activity) versus time. The slope of the plot is $-k$, the rate constant for decay. Here $k = 0.0050$ d^{-1}, so $t_{1/2} = 140$ d.

23.46 $^{239}_{94}Pu + ^{4}_{2}\alpha \longrightarrow ^{240}_{95}Am + ^{1}_{1}H + 2\ ^{1}_{0}n$

23.48 $^{48}_{20}Ca + ^{242}_{94}Pu \longrightarrow ^{287}_{114}Uuq + 3\ ^{1}_{0}n$

23.50 (a) $^{115}_{48}Cd$ (b) $^{7}_{4}Be$ (c) $^{4}_{2}\alpha$ (d) $^{63}_{29}Cu$

23.52 $^{10}_{5}B + ^{1}_{0}n \longrightarrow ^{7}_{3}Li + ^{4}_{2}\alpha$

23.54 Time $= 2.0 \times 10^{10}$ y

23.56 Time $= 1.9 \times 10^9$ y

23.58 (a) $^{238}_{92}U + ^{1}_{0}n \longrightarrow ^{239}_{92}U$

    (b) $^{239}_{92}U \longrightarrow ^{239}_{93}Np + ^{0}_{-1}\beta$

    (c) $^{239}_{93}Np \longrightarrow ^{239}_{94}Pu + ^{0}_{-1}\beta$

    (d) $^{239}_{94}Pu + ^{1}_{0}n \longrightarrow 2\ ^{1}_{0}n + $ energy $+$ other nuclei

23.60 Energy obtained from 1.000 lb (453.6 g) of $^{235}U = 4.05 \times 10^{-10}$ kJ
Mass of coal required $= 1.6 \times 10^3$ ton (or about 3 million pounds of coal)

23.62 130 mL

23.64 27 fish tagged fish out of 5250 fish caught represents 0.51% of the fish in the lake. Therefore, 1000 fish put into the lake represent 0.51% of the fish in the lake, or 0.51% of 190,000 fish.

23.66 (a) The mass decreases by 4 units (with an $^{4}_{2}\alpha$ emission) or is unchanged (with a $^{0}_{-1}\beta$ emission) so the only masses possible are 4 units apart.

    (b) ^{232}Th series, $m = 4n$; ^{235}U series $m = 4n + 3$

    (c) ^{226}Ra and ^{210}Bi, $4n + 2$ series; ^{215}At, $4n + 3$ series; ^{228}Th, $4n$ series)

    (d) Each series is headed by a long-lived isotope (in the order of 10^9 years, the age of the earth.) The $4n +$

1 series is missing because there is no long-lived isotope in this series. Over geologic time, all the members of this series have decayed completely.

23.68 (a) ^{231}Pa isotope belongs to the ^{235}U decay series (see Question 23.66b).

   (b) $^{235}_{92}\text{U} \longrightarrow \, ^{231}_{90}\text{Th} + \, ^{4}_{2}\alpha$

       $^{231}_{90}\text{Th} \longrightarrow \, ^{231}_{91}\text{Pa} + \, ^{0}_{-1}\beta$

(c) Pa-231 is present to the extent of 1 part per million. Therefore, 1 million grams of pitchblende need to be used to obtain 1 g of Pa-231.

(d) $^{231}_{91}\text{Pa} \longrightarrow \, ^{227}_{89}\text{Ac} + \, ^{4}_{2}\alpha$

Index/Glossary

Italicized page numbers indicate pages containing illustrations, and those followed by "t" indicate tables. Glossary terms, printed in boldface, are defined here as well as in the text.

Abba, Mohammed Bah, 200

abbreviations, A-15

absolute temperature scale. *See* Kelvin temperature scale.

absolute zero The lowest possible temperature, equivalent to −273.15 °C, used as the zero point of the Kelvin scale, 16, 477, *478*

zero entropy at, 797

absorption spectrum A plot of the intensity of light absorbed by a sample as a function of the wavelength of the light, 951

abundance(s),

of elements in Earth's crust, 67t, 70t

of isotopes, 53–55

acceptor level, in semiconductor, 409

accuracy The agreement between the measured quantity and the accepted value, 31

acetaldehyde, structure of, 413

acetaminophen, structure of, 452

acetate ion, buffer solution of, 745t

acetic acid,

as weak electrolyte, 150, 693

buffer solution of, 745t

formation of, 242

hydrogen bonding in, 518

ionization equilibrium of, 658

orbital hybridization in, 393

production of, 445, 447

quantitative analysis of, 134

reaction with calcium carbonate, 163

reaction with ethanol, 667

reaction with sodium bicarbonate, 709

reaction with sodium hydroxide, 753

structure of, 421

structure of, 444

acetic anhydride, structure of, 132

acetone,

hydrogenation of, 357

structure of, *143*, 444

acetonitrile,

formula of, 113

structure of, 374, *395*

acetylacetone,

enol and keto forms, 416

structure of, 375

acetylcholinesterase, 632t

acetylene,

orbital hybridization in, 393–394

production of, 505

structure of, 422

acetylsalicylic acid. *See* aspirin.

acid ionization constant (K_a) The equilibrium constant for the ionization of an acid in aqueous solution, 701, 702t

determining, 754

relation to conjugate base ionization constant, 708

acid(s) A substance that, when dissolved in pure water, increases the concentration of hydrogen ions, 156. *See also* **Brønsted acid(s), Lewis acid(s)**

bases and, 690–737. *See also* acid–base reaction(s).

Brønsted definition, 694

carboxylic. *See* carboxylic acid(s).

common, 158t

Lewis definition of, 720–724

molecular structure of, 724–729

neutral molecules as, 722

properties of, *157*

reaction with bases, 162–164

strengths of, 701, 703

direction of reaction and, 708–710

strong. *See* strong acid.

weak. *See* weak acid.

acid–base pairs, conjugate, 696, 697t

acid–base reaction(s) An exchange reaction between an acid and a base producing a salt and water, 162–166

equivalence point of, 185, 750

titration using, 185–187, 749–760

acidic oxide(s) An oxide of a nonmetal that acts as an acid, 160

as Lewis acids, 722

acidic solution A solution in which the concentration of hydronium ions is greater than the concentration of

hydroxide ion, 698

acidosis, 750

acrolein,

formation of, 377

structure of, 373, 413

acrylonitrile, structure of, 82, 376

actinide(s) The series of elements between actinium and rutherfordium in the periodic table, 71, 302, 925

activation energy (E_a) The minimum amount of energy that must be absorbed by a system to cause it to react, 624

experimental determination, 628–629

activity (A) A measure of the rate of nuclear decay, the number of disintegrations observed in a sample per unit time, 975

actual yield The measured amount of product obtained from a chemical reaction, 132

addition polymer(s) A synthetic organic polymer formed by directly joining monomer units, 455–458

production from ethylene derivatives, 456t

addition reaction(s), of alkenes and alkynes, 434

adduct, acid–base, 721

adenine,

hydrogen bonding to thymine, 520

structure of, 111

adhesive force A force of attraction between molecules of two different substances, 531

adipoyl chloride, 460

adrenaline, 724

aerosol, 586t

air bags, 472

air,

components of, 488t

density of, 15, 484

oxygen obtained from, 910

alanine, 461

albite, dissolved by rain water, 146

alchemy, 963

dimerization of, 647
butane,
  combustion of, balanced equation for, 124
  conversion to isobutane, 663, 677
  structural isomers of, 426
butanedione, structure of, 420
1-butene,
  hydrogenation of, 374
  structure of, 432
2-butene,
  cis-trans isomers of, 423
  iodine-catalyzed isomerization, 630
  isomers of, 432
  rate of conversion, 624
butylated hydroxyanisole (BHA), 595

cacodyl, 116
cadaverine, 116, 444, 467
cadmium, in nuclear reactor, 982
cadmium sulfide, as pigment, 926
caffeine, extraction with supercritical carbon
    dioxide, 530
calcium,
  chemistry of, 893–897
  reaction with oxygen, 327
  reaction with water, 894
calcium carbide, 242
  reaction with water, 505
  unit cell of, 553
calcium carbonate,
  decomposition of, 232
    temperature and spontaneity, 810–811
  in limestone, 657, 893
  reaction with acetic acid, 163
  reaction with sulfur dioxide, 141
  solubility, 760
calcium fluoride,
  in fluorite, 893
  solubility of, 762
calcium hypochlorite, 915
calcium orthosilicate, 902
calcium oxide, 161, 163
calcium silicate, in blast furnace, 933
calcium sulfate, in gypsum, 107, 893
calculator,
  exponential notation on, 30, A-3
  pH and, 182
  significant figures and, 35, A-3
calomel electrode, 860
calorie (cal) The quantity of energy re-
    quired to raise the temperature of
    1.00 g of pure liquid water from
    14.5 °C to 15.5 °C, 208, A-13
calorimetry The experimental determina-
    tion of the enthalpy changes of reac-
    tions, 222–225
camphor, boiling point elevation and
    freezing point depression constants
    for, 578t
canal rays, *46*
capacity, of buffer solution, 747
capillary action, 531, *532*

capsaicin, formula of, 114
carbohydrates,
  molecules in space, 323
  structure of, 448
carbon,
  allotropes of, 67–68, 421
  as reducing agent, 173t
  nanotubes, 26
  organic compounds of, 418–469
  radioactive isotopes of, 977
carbon dioxide,
  as Lewis acid, 722
  bond order in, 352
  Henry's law constant, 571t
  in Lake Nyos, 556
  Lewis structure of, 333
  molecular geometry, 363
  molecular polarity of, 365
  phase diagram of, 547
  reaction with potassium superoxide, 502
  resonance structures, 351
  standard enthalpy of formation of, 229
  sublimation of, 216, 547
  supercritical, 529–530
carbon disulfide,
  enthalpy of formation, 247
  vapor pressure of, 550, 553
carbon monoxide,
  bond order in, 352
  enthalpy of formation, 226–227
  in water gas, 888
  oxidation of, 128
  reaction with hemoglobin, 939
  reaction with iron(III) oxide, 140, 168
  reaction with methanol, 447
  reaction with nitrogen dioxide, 637
carbon steel, 933
carbon tetrachloride,
  density, 42
  structure of, 337
carbonate ion,
  as polyprotic base, 719
  bond order in, 354
  formal charges in, 348
  molecular geometry, 363
  resonance structures, 340
  solubility in strong acids, 769
carbonic acid, 164
  as polyprotic acid, 695
carbonic anhydrase, 601, 632
carbonyl bromide, 684, 686
carbonyl group The functional group that
    characterizes aldehydes and ketones,
    consisting of a carbon atom doubly
    bonded to an oxygen atom, 444
carboxyl group The functional group that
    consists of a carbonyl group bonded
    to a hydroxyl group, 445
carboxylic acid(s) Any of a class of organic
    compounds characterized by the
    presence of a carboxyl group, 445,
    447

acid strengths of, 726–727
  naming of, A-23
carnauba wax, 509
β-carotene, 380, 433, 584
Carothers, Wallace, 459
carvone, 114, 420
cast iron, 932
catalase, 632t
catalyst(s) A substance that increases the
    rate of a reaction while not being con-
    sumed in the reaction, 435, 607
  effect on reaction rates, 629–632
  heterogeneous versus homogeneous, 631
  in rate equation, 609, 630
  zeolites as, 904
catalytic steam reformation, hydrogen pro-
    duction by, 888
cathode The electrode of an electrochemical
    cell at which reduction occurs, 836
  in corrosion, 927
cathode ray, 48
cation(s) An ion with a positive electrical
    charge, 86
  as Lewis acids, 721
  in living cells, 149t
  naming, 90
  sizes of, 312
Cavendish, Henry, 470
cell(s),
  electrochemical, 835–840
  galvanic, 829
  unit, 532
  voltaic, 829, 835–840
Celsius, Anders, 16
Celsius temperature scale A scale
    defined by the freezing and boiling
    points of pure water, defined as 0 °C
    and 100 °C, 16
cerrusite, 135
cesium chloride, structure of, 538
Chadwick, James, 50, 980
chain reaction, 982
chair conformation, 431
chalcocite, 933
chalcogens, 69
chalcopyrite, 933
characteristic, of logarithm, A-7
charge,
  balanced in chemical equation, 155
  conservation of, 830
charge distribution, in covalent compounds,
    345–352
Charles, Jacques, 470, 477, 887
Charles's law, 477
  kinetic-molecular theory and, 493
chelating ligand A ligand that forms more
    than one coordinate covalent bond
    with the central metal ion in a com-
    plex, 936
chemical analysis The determination of the
    amounts or identities of the compo-
    nents of a mixture, 133–138

hydrazine,
   as fuel, 247
   formula of, 101
   production of, 907
   reaction with oxygen, 502
hydrides,
   reaction with water, 889
   types of, 888
hydrocarbon(s),
   catalytic steam reformation of, 888
   combustion analysis of, 137–139
   derivatives of, naming, A-23
   immiscibility in water, 563
   Lewis structures of, 336
   molecules in space, 323
   strained, 431
   types of, 426–438
hydrochloric acid, 913. *See also* hydrogen
      chloride.
   reaction with iron, 502
hydrofluoric acid, production of, 895
hydrogen,
   as fuel, 238
   as reducing agent, 173t
   balloons and, 470–471
   binary compounds of, 95
   chemistry of, 887–890
   compounds of, Lewis structures of,
      335–337
   electrode, 838, 848
   fusion of, 44, 56, 983
   Henry's law constant, 594
   in fuel cell, 845
   ionization energy of, 268
   line emission spectrum, *263*
      explanation of, 264–268
   molecular orbital energy level diagram, 398
   orbital box diagram, 292
   potential energy during bond forma-
      tion, 382
   reaction with iodine, 658, 668
   reaction with oxygen, 19
   hydrogen bonding Attraction between a hy-
      drogen atom and a very electronega-
      tive atom to produce an unusually
      strong dipole–dipole attraction, 442,
      517–521
      in polyamides, 460
   hydrogen carbonate ion, resonance struc-
      tures, 340
   hydrogen chloride,
      emitted by volcanoes, 147
      ionization in water, 157
      reaction with ammonia, 164, 491, 756
      reaction with sodium hydroxide, 162, 751
   hydrogen economy, 238
   hydrogen electrode, standard, 848
   hydrogen fluoride,
      reaction with silica, 902
      sigma bond in, 383
   hydrogen halides, standard enthalpies of for-
      mation of, 231t

hydrogen iodide,
   decomposition of, 658
   equilibrium with hydrogen and iodine, 792
   second-order decomposition of, 616
hydrogen ion. *See* hydronium ion.
hydrogen peroxide,
   catalyzed decomposition of, 607
   first-order decomposition of, 615
hydrogen phosphate ion, buffer solution
      of, 745t
hydrogen sulfide, as energy source for
      life, 119
   as polyprotic acid, 695
   dissociation of, 667
   properties of, 911
   reaction with chlorine, 173
   sulfur-oxidizing bacteria and, 880
hydrogenation An addition reaction in
      which the reagent is molecular hydro-
      gen, 355, 435
   of oils in foods, 454
hydrohalic acids, acidity and structure of,
      725
hydrolysis reaction A reaction with water in
      which a bond to oxygen is broken, 450
   of anions of insoluble salt, 769
   of fats, 454
   of salts, 769
hydrometallurgy Recovery of metals from
      their ores by reactions in aqueous so-
      lution, 933
hydronium ion (H_3O^+(aq)), 160, 693
   as Lewis adduct, 721
hydrophilic and hydrophobic colloids, 587
hydroxide(s), precipitation of, 154
hydroxide ion (OH^-(aq)), 157
   formal charges in, 347
   solubility in strong acids, 770
hydroxyapatite, 897
p-hydroxyphenyl-2-butanone, 446
hydroxyproline, structure of, 376
hygroscopic salt, 892
hyperbaric chamber, *572*
hypergolic fuel, 918
hyperthyroidism, treatment of, 960
hypertonic solution, 584
hypochlorite ion, 915
   Lewis structure of, 335
   self oxidation-reduction, 635
hypochlorous acid, 913
hypothesis A tentative explanation of or
      prediction derived from experimental
      observations, 5
hypotonic solution, 584

ice,
   hydrogen bonding in, 519
   slipperiness of, 547
   structure of, 83, 519
ice calorimeter, 243
Ice Man, radiochemical dating of, *979*
Icelandic spar, 893

ideal gas, 481
ideal gas law A law that relates pressure, vol-
      ume, number of moles, and tempera-
      ture for an ideal gas, 481–485
   departures from, 496–498
   osmotic pressure equation and, 583
   stoichiometry and, 485–487
ideal solution A solution that obeys Raoult's
      law, 574
ilmenite, 911
   formula of, 114
imaging, medical, 986
indicator(s) A substance used to signal the
      equivalence point of a titration by a
      change in some physical property such
      as color, 185
   acid–base, 182, 185, 700, 758–760
induced dipole(s) Separation of charge in
      a normally nonpolar molecule,
      caused by the approach of a polar
      molecule, 514
induced dipole/induced dipole attraction
      The electrostatic force between two
      neutral molecules, both having in-
      duced dipoles, 515–516
inductive effect The attraction of electrons
      from adjacent bonds by an electroneg-
      ative atom, 725
inert gas(es). *See* noble gas(es).
infrared (IR) radiation, 256, *257*
initial rate The instantaneous reaction
      rate at the start of the reaction, 611
insoluble compound, 760
insulator, electrical, 407
integrated rate equation, 614–622
integrity, in science, 6
intensive properties Physical properties that
      do not depend on the amount of mat-
      ter present, 18
intercept, of straight line, A-10
interference, constructive and destruc-
      tive, 537
intermediate. *See* **reaction intermediate.**
intermolecular forces Interactions between
      molecules, between ions, or between
      molecules and ions, 497, 508–555
   determining types of, 521–522
   energies of, 512
internal energy The sum of the potential
      and kinetic energies of the particles
      in the system, 217
   relation to enthalpy change, 219
interstitial hydrides, 888
intravenous solution(s), tonicity of, 584
intrinsic semiconductor, 408
iodine,
   as catalyst, 630
   in thyroid gland, 960
   preparation of, 197
   reaction with hydrogen, 658, 668
   solubility in carbon tetrachloride, 684
   solubility in liquids, 515–516

Physical and Chemical Constants

Avogadro's number	$N = 6.02214199 \times 10^{23}/mol$
Electronic charge	$e = 1.60217646 \times 10^{-19}$ C
Faraday's constant	$F = 9.6485342 \times 10^4$ C/mol electrons
Gas constant	$R = 8.314472$ J/K · mol
	$= 0.082057$ L · atm/K · mol

π	$\pi = 3.1415926536$
Planck's constant	$h = 6.62606876 \times 10^{-34}$ J · sec
Speed of light (in a vacuum)	$c = 2.99792458 \times 10^8$ m/sec

Useful Conversion Factors and Relationships

Length
SI unit: Meter (m)
1 kilometer = 1000 meters
$\quad$ = 0.62137 mile
1 meter = 100 centimeters
1 centimeter = 10 millimeters
1 nanometer = 1.00×10^{-9} meter
1 picometer = 1.00×10^{-12} meter
$\quad$ 1 inch = 2.54 centimeter (exactly)

Mass
SI unit: Kilogram (kg)
1 kilogram = 1000 grams
$\quad$ 1 gram = 1000 milligrams
$\quad$ 1 pound = 453.59237 grams = 16 ounces
$\quad$ 1 ton = 2000 pounds

Volume
SI unit: Cubic meter (m^3)
1 liter (L) = 1.00×10^{-3} m^3
$\quad$ = 1000 cm^3
$\quad$ = 1.056710 quarts
1 gallon = 4.00 quarts

Energy
SI unit: Joule (J)
1 joule = 1 kg · m^2/s^2
$\quad$ = 0.23901 calorie
$\quad$ = 1 C $\times$ 1 V
1 calorie = 4.184 joules

Pressure
SI unit: Pascal (Pa)
1 pascal = 1 N/m^2
$\quad$ = 1 kg/m · s^2
1 atmosphere = 101.325 kilopascals
$\quad$ = 760 mm Hg = 760 torr
$\quad$ = 14.70 lb/in^2
$\quad$ = 1.01325 bar
1 bar = 10^5 Pa (exactly)

Temperature
SI unit: kelvin (K)
0 K = -273.15 °C
K = °C + 273.15°C
? °C = (5 °C/9 °F)(°F $-$ 32 °F)
? °F = (9 °F/5 °C)°C + 32 °F

Location of Useful Tables and Figures

Atomic and Molecular Properties

Atomic radii	Figures 8.9, 8.10
Bond energies	Table 9.9
Bond lengths	Table 9.8
Electron configurations	Table 8.3
Electronegativity	Figure 9.9
Ionic radii	Figure 8.13
Ionization energies	Figure 8.11, Appendix F

Thermodynamic Properties

Enthalpy, free energy, entropy	Appendix L
Enthalpies of formation (selected values)	Table 6.2
Specific heats	Table 6.1
Lattice energies	Table 9.3

Acids, Bases and Salts

Common acids and bases	Table 5.2
Names and composition of polyatomic ions	Table 3.1
Solubility guidelines	Figure 5.1
Ionization constants for weak acids and bases	Table 17.3

Miscellaneous

Selected alkanes	Table 11.2
Oxidizing and reducing agents	Table 5.4
Charges on common monatomic cations and anions	Figure 3.7
Hybrid orbitals	Figure 10.5
Common polymers	Table 11.12